COMPREHENSIVE MEDICINAL CHEMISTRY II

COMPREHENSIVE MEDICINAL CHEMISTRY II

Editors-in-Chief

Dr John B Taylor

Former Senior Vice-President for Drug Discovery, Rhône-Poulenc Rorer, Worldwide, UK

Professor David J Triggle

State University of New York, Buffalo, NY, USA

Volume 5

ADME-Tox APPROACHES

Volume Editors

Professor Bernard Testa

University Hospital Centre, Lausanne, Switzerland

Dr Han van de Waterbeemd

AstraZeneca, Macclesfield, UK

ELSEVIER

AMSTERDAM BOSTON HEIDELBERG LONDON NEW YORK OXFORD
PARIS SAN DIEGO SAN FRANCISCO SINGAPORE SYDNEY TOKYO

20167325

Elsevier Ltd.
The Boulevard, Langford Lane, Kidlington, Oxford OX5 1GB, UK

First edition 2007

British Library Cataloguing in Publication Data
A catalogue record for this book is available from the British Library

Library of Congress Catalog Number: 2006936669

ISBN-13: 978-0-08-044513-7
ISBN-10: 0-08-044513-6

For information on all Elsevier publications
visit our website at books.elsevier.com

Printed and bound in Spain

06 07 08 09 10 10 9 8 7 6 5 4 3 2 1

Working together to grow
libraries in developing countries

www.elsevier.com | www.bookaid.org | www.sabre.org

ELSEVIER   BOOK AID International   Sabre Foundation

Disclaimer

Both the Publisher and the Editors wish to make it clear that the views and opinions expressed in this book are strictly those of the Authors. To the extent permissible under applicable laws, neither the Publisher nor the Editors assume any responsibility for any loss or injury and/or damage to persons or property as a result of any actual or alleged libellous statements, infringement of intellectual property or privacy rights, whether resulting from negligence or otherwise.

Knowledge and best practice in this field are constantly changing. As new research and experience broaden our knowledge, changes in practice, treatment and drug therapy may become necessary or appropriate. Readers are advised to check the most current information provided (i) on procedures featured or (ii) by the manufacturer of each product to be administered, to verify the recommended dose or formula, the method and duration of administration, and contraindications. It is the responsibility of the practitioner, relying on their own experience and knowledge of the patient, to make diagnoses, to determine dosages and the best treatment for each individual patient, and to take all appropriate safety precautions. To the fullest extent of the law, neither the Publisher, nor Editors, nor Authors assume any liability for any injury and/or damage to persons or property arising out or related to any use of the material contained in this book.

Contents

In Silico Tools in Absorption, Distribution, Metabolism, Excretion, and Toxicity

Contents of all Volumes

Preface

The first edition of *Comprehensive Medicinal Chemistry* was published in 1990 and was intended to present an integrated and comprehensive overview of the then rapidly developing science of medicinal chemistry from its origins in organic chemistry. In the last two decades, the field has grown to embrace not only all the sophisticated synthetic and technological advances in organic chemistry but also major advances in the biological sciences. The mapping of the human genome has resulted in the provision of a multitude of new biological targets for the medicinal chemist with the prospect of more rational drug design (CADD). In addition, the development of sophisticated in silico technologies for structure–property relationships (ADMET) enables a much better understanding of the fate of potential new drugs in the body with the subsequent development of better new medicines.

It was our ambitious aim for this second edition, published 16 years after the first edition, to provide both scientists and research managers in all relevant fields with a comprehensive treatise covering all aspects of current medicinal chemistry, a science that has been transformed in the twenty-first century. The second edition is a complete reference source, published in eight volumes, encompassing all aspects of modern drug discovery from its mechanistic basis, through the underlying general principles and exemplified with comprehensive therapeutic applications. The broad scope and coverage of *Comprehensive Medicinal Chemistry II* would not have been possible without our panel of authoritative Volume Editors whose international recognition in their respective fields has been of paramount importance in the enlistment of the world-class scientists who have provided their individual 'state of the science' contributions. Their collective contributions have been invaluable.

Volume 1 (edited by Peter D Kennewell) overviews the general socioeconomic and political factors influencing modern R&D in both the developed and developing worlds. Volume 2 (edited by Walter H Moos) addresses the various strategic and organizational aspects of modern R&D. Volume 3 (edited by Hugo Kubinyi) critically reviews the multitude of modern technologies that underpin current discovery and development activities. Volume 4 (edited by Jonathan S Mason) highlights the historical progress, current status, and future potential in the field of computer-assisted drug design (CADD). Volume 5 (edited by Bernard Testa and Han van de Waterbeemd) reviews the fate of drugs in the body (ADMET), including the most recent progress in the application of 'in silico' tools. Volume 6 (edited by Michael Williams) and Volume 7 (edited by Jacob J Plattner and Manoj C Desai) cover the pivotal roles undertaken by the medicinal chemist and pharmacologist in integrating all the preceding scientific input into the design and synthesis of viable new medicines. Volume 8 (edited by John B Taylor and David J Triggle) illustrates the evolution of modern medicinal chemistry with a selection of personal accounts by eminent scientists describing their lifetime experiences in the field, together with some illustrative case histories of successful drug discovery and development.

We believe that this major work will serve as the single most authoritative reference source for all aspects of medicinal chemistry for the next decade and it is intended to maintain its ongoing value by systematic electronic upgrades. We hope that the material provided here will serve to fulfill the words of Antoine de Saint-Exupery (1900–44) and allow future generations of medicinal chemists to discover the future.

'As for the future, your task is not to foresee it but to enable it'
Citadelle (1948)

John B Taylor and David J Triggle

Preface to Volume 5

To be effective, a drug must be absorbed (when administered orally) and then distributed to reach its sites of action. Other pharmacokinetic processes that influence the nature and intensity of a drug's therapeutic activity are its metabolism and excretion. In today's pharmacological parlance, the processes of absorption, distribution, metabolism, and excretion translate into the ADME acronym, which is usually completed with the Tox or T abbreviation (toxicity) to become ADME-Tox or ADMET. Both of these are in common use, and their success is due to the fact that they encapsulate in a few letters some of the main causes of drug attrition in drug discovery and development.

This volume addresses major ADMET issues and the tools to assess them. To this end, we designed a logical sequence that begins with the ultimate objective, namely, the desirable clinical pharmacokinetic criteria, continues with in vivo, in vitro, physicochemical and in silico tools, and concludes with enabling strategies and technologies.

In concrete terms, the volume is subdivided into five parts:

- Part I entitled 'Biological and in vivo aspects of ADMET' includes chapters presenting the biology of transporters, the biochemistry of drug metabolism, and the toxification/detoxification argument.
- Part II pertains to 'Biological in vitro tools in ADMET' and covers the various tissular, cellular, and subcellular models used to assess ADME. Emphasis is given to the in vivo relevance of such tools.
- Part III is dedicated to 'Physicochemical tools in ADMET' and includes the many technologies in current use to measure physicochemical properties of pharmacokinetic relevance. Here, there is emphasis on in vitro and in vivo relevance.
- Part IV is the largest and indeed the hottest one, since it covers 'In silico tools in ADMET.' This part deals with the following themes: theoretical backgrounds, methods and techniques, and physicochemical and biological relevance.
- Part V serves as a conclusion by presenting a few multidisciplinary chapters dedicated to 'Enabling ADMET Strategies and Technologies in Early Development.' These include the biopharmaceutics classification system, metabonomics, and prodrugs.

Our job as Editors has been a challenging and exciting endeavor. This was made all the more easy since we are colleagues and friends for over 25 years. Also, our views and perceptions of drug research and development nicely complement each other. We hope our friendship and complementarity are apparent in this volume and will help you enjoy studying it as much as we enjoyed editing it.

Bernard Testa and Han van de Waterbeemd

Editors-in-Chief

John B Taylor, DSc, was formerly Senior Vice President for Drug Discovery at Rhône-Poulenc Rorer. He obtained his BSc in chemistry from the University of Nottingham in 1956 and his PhD in organic chemistry at the Imperial College of Science and Technology with Nobel Laureate Professor Sir Derek Barton in 1962. He subsequently undertook postdoctoral research fellowships at the Research Institute for Medicine and Chemistry in Cambridge (US) with Sir Derek and at the University of Liverpool (UK), before entering the pharmaceutical industry.

During his career in the pharmaceutical industry Dr Taylor spent more than 30 years covering all aspects of research and development in an international environment. From 1970 to 1985 he held a number of positions in the Hoechst Roussel organization, ultimately as research director for Roussel Uclaf (France). In 1985 he joined Rhône-Poulenc Rorer holding various management positions in the research groups worldwide before becoming Senior Vice President for Drug Discovery in Rhône-Poulenc Rorer.

Dr Taylor is the co-author of two books on medicinal chemistry and has more than 50 publications and patents in medicinal chemistry. He was joint executive editor for the first edition of Comprehensive Medicinal Chemistry, a visiting professor for medicinal chemistry at the City University (London) from 1974 to 1984 and was awarded a DSc in medicinal chemistry from the University of London in 1991.

David J Triggle, PhD, is the University Professor and a Distinguished Professor in the School of Pharmacy and Pharmaceutical Sciences at the State University of New York at Buffalo. Professor Triggle received his education in the UK with a BSc degree in chemistry at the University of Southampton and a PhD degree in chemistry at the University of Hull working with Professor Norman Chapman. Following postdoctoral fellowships at the University of Ottawa (Canada) with Bernard Belleau and the University of London (UK) with Peter de la Mare he assumed a position in the School of Pharmacy at the University at Buffalo. He served as Chairman of the Department of Biochemical Pharmacology from 1971 to 1985 and as Dean of the School of Pharmacy from 1985 to 1995. From 1996 to 2001 he served as Dean of the Graduate School and from 1999 to 2001 was also the University Provost. He is currently the University Professor, in which capacity he teaches bioethics and science policy, and is President of the Center for Inquiry Institute, a secular think tank located in Amherst, New York.

Professor Triggle is the author of three books dealing with the autonomic nervous system and drug–receptor interactions, the editor of a further dozen books, some 280 papers, some 150 chapters and reviews, and has presented over 1000 invited lectures worldwide. The Institute for Scientific Information lists him as one of the 100 most highly cited scientists in the field of pharmacology. His principal research interests have been in the areas of drug–receptor interactions, the chemical pharmacology of drugs active at ion channels, and issues of graduate education and scientific research policy.

Editors of Volume 5

Bernard Testa studied pharmacy because he was unable to choose between medicine and chemistry. Because he was incapable of working in a community pharmacy, he undertook a PhD thesis on the physicochemistry of drug–macromolecule interactions. Since he felt himself ungifted for the pharmaceutical industry, he went for two years to Chelsea College, University of London, for postdoctoral research under the supervision of Prof Arnold H Beckett. And because these were easy times, he was called as assistant professor to the University of Lausanne, Switzerland, to become full professor and Head of Medicinal Chemistry in 1978. Since then, he has tried to repay his debts by fulfilling a number of local and international commitments (e.g., Dean of the Faculty of Sciences (1984–86)), Director of the Geneva-Lausanne School of Pharmacy (1994–96 and 1999–2001), and President of the University Senate (1998–2000). He has written 4 books and edited 29 others, and co-authored 450 research and review articles in the fields of drug design and drug metabolism. During the years 1994–98, he was the Editor-Europe of Pharmaceutical Research, the flagship journal of the American Association of Pharmaceutical Scientists (AAPS), and he is now the co-editor of the new journal Chemistry and Biodiversity. He is also a member of the Editorial Board of several leading journals (e.g., Biochemical Pharmacology, Chirality, Drug Metabolism Reviews, Helvetica Chimica Acta, Journal of Pharmacy and Pharmacology, Medicinal Research Reviews, Pharmaceutical Research, Xenobiotica). He holds Honorary Doctorates from the universities of Milan, Montpellier and Parma, and was the 2002 recipient of the Nauta Award on Pharmacochemistry given by the European Federation for Medicinal Chemistry. He was elected as a Fellow of the AAPS, and is a member of a number of scientific societies such as the French Academy of Pharmacy, the Royal Academy of Medicine of Belgium, and the American Chemical Society. His recently granted Emeritus status has freed him from administrative duties and gives him more time for writing, editing and collaborating in research projects. His hobbies, interests, and passions include jogging, science fiction, epistemology, teaching, and scientific exploration.

Han van de Waterbeemd studied physical organic chemistry at the Technical University of Eindhoven, The Netherlands, and did a PhD in medicinal chemistry at the University of Leiden, The Netherlands. After a postdoc with Bernard Testa at the School of Pharmacy of the University of Lausanne, Switzerland, he held a five-year faculty position at the same institution. He has also been teaching medicinal chemistry to pharmacy students at the universities of Berne and Basel in Switzerland from 1987 to 1997. In 1988, he joined F. Hoffmann-La Roche Ltd in Basel, as head of the Molecular Properties Group. He then moved to Pfizer Central Research, UK, in 1997, later onto Pfizer Global Research and Development, and held various positions in the Department of Drug Metabolism, later called PDM (Pharmacokinetics, Dynamics and Metabolism), including Head of Discovery, and Head of Automation and In Silico ADME Technologies. In 2005, he moved to AstraZeneca to become the global project leader of their Molecular Properties & In Silico ADMET Modelling system.

He published more than 135 peer-reviewed papers and book chapters, and coedited 11 books. His research interests include physicochemical and structural molecular properties and their role in drug disposition, as well as the in silico modeling of ADMET properties. Han was secretary of the QSAR and Modelling Society from 1995 to 2005.

Contributors to Volume 5

N J Abbott
King's College London, London, UK

M H Abraham
University College London, London, UK

B Abrahamsson
AstraZeneca, Mölndal, Sweden

S Agatonovic-Kustrin
The University of Western Australia, Perth, WA, Australia

T B Andersson
Institute of Environmental Medicine, Karolinska Institutet, Stockholm, Sweden

A M Aronov
Vertex Pharmaceuticals Inc., Cambridge, MA, USA

P Artursson
Uppsala University, Uppsala, Sweden

R P Austin
AstraZeneca R&D Charnwood, Loughborough, UK

A Avdeef
pION Inc., Woburn, MA, USA

P Barton
AstraZeneca R&D Charnwood, Loughborough, UK

M Bertrand
Technologie Servier, Orléans, France

M-J Bossant
Technologie Servier, Orléans, France

S Cagnani
University of Parma, Parma, Italy

G Caron
University of Torino, Torino, Italy

G Colmenarejo
GlaxoSmithKline, Madrid, Spain

G Colombo
University of Ferrara, Ferrara, Italy

P Colombo
University of Parma, Parma, Italy

J E A Comer
Sirius Analytical Instruments Ltd, Forest Row, UK

M T D Cronin
Liverpool John Moores University, Liverpool, UK

L Cucurull-Sanchez
Pfizer Global Research and Development, Sandwich, UK

M Danhof
Leiden University, Leiden, The Netherlands

M J De Groot
Pfizer Global Research and Development, Sandwich, UK

L Di
Wyeth Research, Princeton, NJ, USA

M Dickins
Pfizer Global Research and Development, Sandwich, UK

G Ermondi
University of Torino, Torino, Italy

C Esser
Institut für Umweltmedizinische Forschung, Heinrich-Heine University, Düsseldorf, Germany

B Faller
Novartis Institutes for Biomedical Research, Basel, Switzerland

R E Fessey
AstraZeneca R&D Charnwood, Loughborough, UK

A Foreman
pION Inc., Woburn, MA, USA

R Fraczkiewicz
Simulations Plus, Inc., Lancaster, CA, USA

C Funk
F. Hoffmann-La Roche Ltd, Basel, Switzerland

A Galetin
University of Manchester, Manchester, UK

D Giron
Novartis Pharma AG, Basel, Switzerland

N Greene
Pfizer Global Research and Development, Groton, CT, USA

L H Hall
Eastern Nazarene College, Quincy, MA, USA and Hall Associates Consulting, Quincy, MA, USA

L M Hall
Hall Associates Consulting, Quincy, MA, USA

D R Hawkins
Huntingdon Life Sciences, Alconbury, UK

A Hersey
GSK Medicines Research Centre, Stevenage, UK

M Hewitt
Liverpool John Moores University, Liverpool, UK

H Jones
F. Hoffmann-La Roche Ltd, Basel, Switzerland

D M Jonker
Novo Nordisk A/S, Bagsværd, Denmark

E H Kerns
Wyeth Research, Princeton, NJ, USA

L B Kier
Virginia Commonwealth University, Richmond, VA, USA.

T Lavé
F. Hoffmann-La Roche Ltd, Basel, Switzerland

H Lennernäs
Uppsala University, Uppsala, Sweden

D F V Lewis
University of Surrey, Guildford, UK

D J Livingstone
ChemQuest, Sandown, Isle of Wight, UK

P Matsson
Uppsala University, Uppsala, Sweden

S Modi
GlaxoSmithKline Research and Development, Stevenage, UK

S Neuhoff
Uppsala University, Uppsala, Sweden

J K Nicholson
Imperial College, London, UK

U Norinder
AstraZeneca R&D, Södertälje, Sweden

B Oesch-Bartlomowicz
University of Mainz, Mainz, Germany

F Oesch
University of Mainz, Mainz, Germany

I K Pajeva
Bulgarian Academy of Sciences, Sofia, Bulgaria

Y Parmentier
Technologie Servier, Orléans, France

N Parrott
F. Hoffmann-La Roche Ltd, Basel, Switzerland

G Pasut
University of Padua, Padua, Italy

A Pedretti
Institute of Medicinal Chemistry, University of Milan, Milan, Italy

N Proctor
Simcyp, Sheffield, UK

A E Rettie
University of Washington, Seattle, WA, USA

M Ridderström
AstraZeneca R&D Mölndal, Mölndal, Sweden

P Russo
University of Salerno, Fisciano, Italy

P Santi
University of Parma, Parma, Italy

R A Scherrer
BIOpK, White Bear Lake, MN, USA

J-M Scherrmann
University of Paris, Paris, France

D A Smith
Pfizer Global Research and Development, Sandwich, UK

F Sonvico
University of Parma, Parma, Italy

K Sugano
Pfizer Inc., Aichi, Japan

J Taskinen
University of Helsinki, Helsinki, Finland

S Tavelin
Umeå University, Umeå, Sweden

B Testa
University Hospital Centre, Lausanne, Switzerland

I V Tetko
Institute of Bioorganic and Petrochemistry, Kiev, Ukraine

J-P Tillement
Faculty of Medicine of Paris XII, Paris, France

R A Totah
University of Washington, Seattle, WA, USA

W F Trager
University of Washington, Seattle, WA, USA

D Tremblay
AFSSAPS Consultant (French Agency for Drugs), France

J V Turner
The University of Queensland, Brisbane, Qld, Australia

A-L Ungell
AstraZeneca R&D Mölndal, Mölndal, Sweden

P H Van der Graaf
Pfizer Global Research and Development, Sandwich, UK

H van de Waterbeemd
AstraZeneca, Macclesfield, UK

F M Veronese
University of Padua, Padua, Italy

S A G Visser
AstraZeneca R&D, Södertälje, Sweden

G Vistoli
Institute of Medicinal Chemistry, University of Milan, Milan, Italy

D Voloboy
pION Inc., Woburn, MA, USA

B Walther
Technologie Servier, Orléans, France

J Wang
Novartis Institutes for Biomedical Research, Cambridge, MA, USA

M Wiese
University of Bonn, Bonn, Germany

I D Wilson
AstraZeneca, Macclesfield, UK

S Winiwarter
AstraZeneca R&D Mölndal, Mölndal, Sweden

I Zamora
Universitat Pompeu Fabra, Barcelona, Spain

K P Zuideveld
F. Hoffmann-La Roche Ltd, Basel, Switzerland

5.01 The Why and How of Absorption, Distribution, Metabolism, Excretion, and Toxicity Research

H Van de Waterbeemd, AstraZeneca, Macclesfield, UK
B Testa, University Hospital Centre, Lausanne, Switzerland

5.01.1 Evolving Paradigm in Drug R&D

Not so very long ago, pharmacokinetics, drug metabolism, and toxicology of selected clinical candidates were studied mainly during preclinical and clinical development. In those days the mission of medicinal chemistry was to discover and supply very potent compounds, with less interest being given to their behavior in the body. However, the R&D paradigm in the pharmaceutical industry has undergone dramatic changes since the 1970s and particularly since the mid-1990s. High-throughput biological assays were developed that have enabled large series of compounds to be screened. This was driven by the increasing size of proprietary depositories and the availability of new reagents and detection technologies.

Simultaneously, medicinal chemists have developed new synthetic strategies such as combinatorial chemistry and parallel synthesis. The number of compounds synthesized increased dramatically. In addition, specialized biotech companies as well as universities began offering compound collections and focused libraries. As a result, much attention is currently being paid to the design and/or purchase criteria of lead- and drug-like compounds.[1–3] Increasingly, this includes considerations on ADME-related physicochemical properties as well as ADME properties themselves. The concept of property-based design,[4] in addition to structure-based design where target structures are available, is now commonly used to address ADME issues as early as possible. Thus, the former traditional in vivo animal ADME evaluation could no longer cope with the demand and in vitro ADME screens became widely used. Despite rapid advances in the use of automation and robotics to increase the throughput of the in vitro ADME assays,[5] screening all available compounds was not necessarily the most efficient and cost-effective strategy. It thus became reasonable and even essential to develop in silico tools to predict and simulate various physicochemical and ADME properties and to balance these in decision-making processes together with combined in vivo and in vitro approaches (in combo).

Rigorous analyses of the root causes of attrition during development revealed that lack of efficacy, toxicity, as well as inappropriate absorption, distribution, metabolism, and excretion (ADME) are among the major determinants of the failure of candidates.[6,7] Lack of efficacy, in addition to insufficient response of the target, may of course be caused by poor absorption, inadequate distribution, and/or rapid metabolism, leading to drug concentrations at the target site that are too low.[8] And since toxicity is also a major factor of attrition, the development of compounds is often halted even before a detailed human pharmacokinetics or efficacy study can be performed. Hence, the real impact of absorption, distribution, metabolism, excretion, and toxicity (ADMET) processes remains somewhat hidden in (incomplete) attrition data. During the 1990s, it became good practice to collect ADME and toxicity data during the drug discovery stage in order to use them in decision making to select the best clinical candidates.[9] Today, the drug discovery process has also become strongly dependent on departments providing data, guidance, and insight on issues of drug metabolism and pharmacokinetics (DMPK), toxicology, and safety.

Yet, the success of this move to early involvement of DMPK has been questioned. According to some practitioners the main contribution of a discovery (research) department of drug metabolism and pharmacokinetics has been to enable the design of pharmacokinetically adequate rather than optimal compounds and thus to make it possible to work on difficult targets.[10] However, despite all preclinical efforts there will always remain an essential need for extensive clinical pharmacokinetics to lay the ground for safe prescription once the drug is on the market.[10]

This bird's-eye view of ADMET research can be summarized in two statements. First, and in a more industrial perspective, it is now entirely clear that ADMET profiling must be initiated as early as possible in the discovery process, using high-throughput and in silico methods characterized by the best possible balance between good relevance to clinical properties on the one hand, and high speed, efficiency, and capacity on the other. Second, and in a more fundamental perspective, it took decades for pharmacologists and biologists to realize that there is an unseverable relation between pharmacodynamic effects (what the drug does to the organism) and pharmacokinetic effects (what the organism does to the drug) (**Figure 1**).[11–13] [A word of caution is necessary here about the meaning of the noun 'pharmacokinetics' and the adjective 'pharmacokinetic.' In its first meaning, the adjective refers to everything 'the organism does to the drug,' in other words to its disposition (absorption, distribution, metabolism, and excretion = ADME) in the body. This meaning is seen in the expression 'pharmacokinetic effects.' However, such effects can be investigated and described at the qualitative level (e.g., drug D yields metabolite M but not metabolite N), at the quantitative level (e.g., the oral availability of drug D in humans is in the range 70–90%), and at the kinetic (quantitative versus time) level (e.g., the half-life of drug D in humans is in the range 3–4 h). In its second meaning, the adjective refers strictly to the kinetic levels (e.g., when one speaks of 'pharmacokinetic parameters'). In contrast, the noun 'pharmacokinetics' refers exclusively to the kinetic aspects of drug disposition, at least in the English language. Unfortunately, this unambiguous and restrictive meaning is not found in all languages; witness the French language where 'la pharmacocinétique' can mean either disposition or pharmacokinetics.] For well over a century, these two components of the interaction between drug and organism were investigated separately and in complete ignorance of any influence the other component might have. The importance now given to early ADME screening is a belated recognition of the interdependence of pharmacodynamic and pharmacokinetic effects. Indeed, the influence of pharmacokinetic effects on a drug's actions is common knowledge, be it in the duration and intensity of these actions, or even in their nature when active metabolites are produced. As for the changes in its disposition that result directly from a drug's pharmacodynamic effects, these may be due to modifications in blood flow, gastrointestinal transit time, or enzyme responses, to name a few.

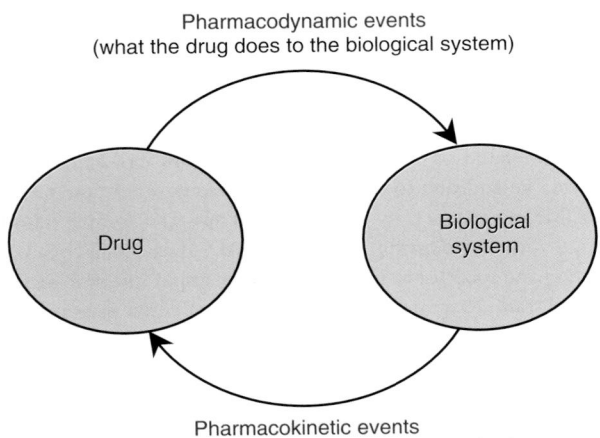

Figure 1 The two basic modes of interaction between bioactive agents and biological systems, namely pharmacodynamic events (activity and toxicity) and pharmacokinetic events (ADME). (Reproduced with modifications from Testa, B.; Vistoli, G.; Pedretti, A. *Chem. Biodiv.* **2005**, *2*, 1411–1427 with the kind permission of the copyright owner, Verlag Helvetica Chimica Acta in Zurich.)

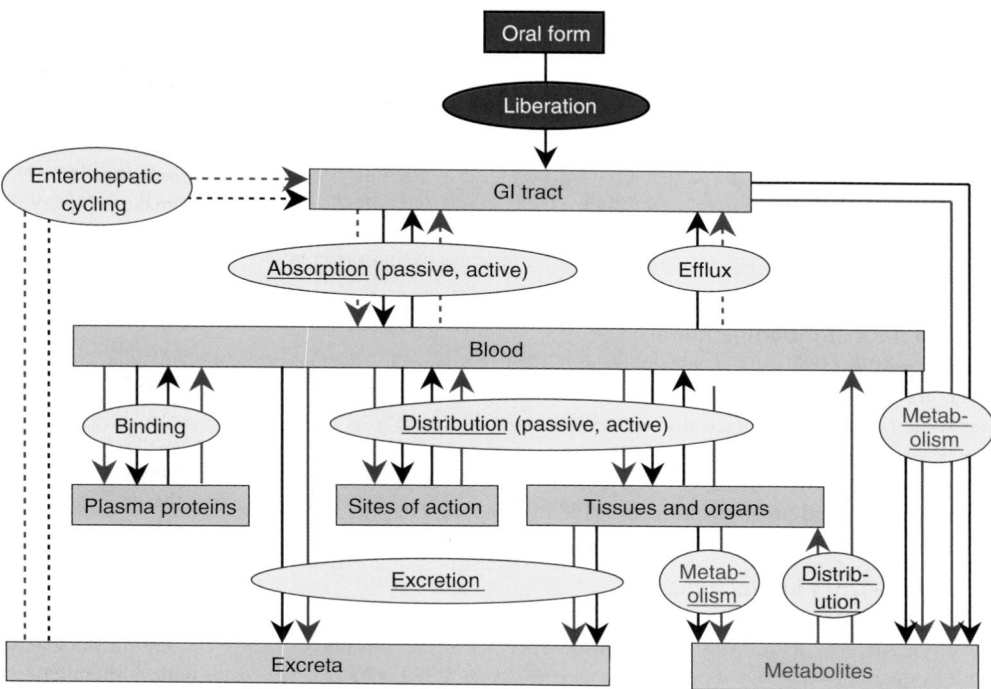

Figure 2 A schematic description of the major processes of drug disposition showing absorption (passive and active), distribution (passive including binding, and active including efflux), metabolism (= biotransformation), and excretion (passive and active including efflux). Elimination is not indicated explicitly, since metabolism is chemical elimination and excretion is physical elimination.

5.01.2   The Increasing Role of Absorption, Distribution, Metabolism, Excretion, and Toxicity (ADME-Tox) in Pharmaceutical R&D

ADME studies aim at obtaining an early estimate of human pharmacokinetic and metabolic profiles.[14] However, drug behavior in the body is a highly complex process involving numerous components, as presented in very simplified form in **Figure 2**.[15] This diversity and complexity is reflected in the ADME studies themselves, which include absorption, bioavailability, clearance and its mechanism, volume of distribution, plasma half-life, involvement of major metabolizing enzymes, nature and level of metabolites, dose estimates, dose intervals, potential for drug–drug interactions, etc.

Early toxicology and safety studies should weed out compounds before they enter lengthy and costly clinical trials. As a result of the recent withdrawal of a number of marketed drugs, more pressure is now being put on pharmaceutical companies by regulatory agencies such as the Food and Drug Administration (FDA) and the European Agency for the Evaluation of Medicinal Products (EMEA) with regard to safety evaluations including pharmacological and toxicological safety. Investigation of the potential to cause QT prolongation is now routine. However, the interpretation of data is not straightforward, since many marketed drugs can prolong the QT interval.[16]

The increased role of early and preclinical ADME and safety/tox studies has led to an important growth of the supporting departments and a considerable development of various technologies to address the key issues. A short overview of the hot issues is given below[17]; the important challenges in investigating the disposition of new chemical entities (NCEs) and candidates are highlighted first, followed by some technological issues.

5.01.2.1   Issues in Absorption, Distribution, Metabolism, Excretion, and Toxicity

5.01.2.1.1   Transporters

Transporter proteins constitute a significant fraction of membrane-bound proteins. They are typically expressed in all organs involved in the uptake, distribution, and elimination of drugs, including the gastrointestinal tract, the blood–brain barrier, the liver, and the kidneys. The sequence of many transporters is known (see 5.04 The Biology and Function of

Transporters; 5.32 In Silico Models for Interactions with Transporters). There is hope that in the near future the experimental 3D structures of the key transporters will be elucidated. Interaction of drugs with transporters can alter their behavior in membrane transport, which may result in, for example, active uptake, efflux, and rapid elimination. In other words, the pharmacokinetics of a drug may be influenced by transporters. Apart from a metabolic component (*see* Section 5.01.2.1.3), drug–drug and drug–nutrient interactions may involve transporters.[18,19] However, while the basic knowledge on transporters is rapidly growing, their real clinical significance remains open to debate despite some convincing theories.[20] A number of P-glycoprotein (P-gp) assays have been developed, as well as some double- and triple-transfected assays. More transporter assays will be available in the near future. The challenge will be to translate the flood of experimental data into relevant information for drug design projects.

5.01.2.1.2   Metabolite identification

With increasing resolution of mass spectrometry and nuclear magnetic resonance (NMR), it is now feasible to detect minute amounts of metabolites. Debate is ongoing as to how to define major versus minor metabolites. Some metabolites might be pharmacologically active and contribute to the overall pharmacokinetics (PK). Reactive metabolites might bind to proteins and cause idiosyncratic reactions.[21] The challenge remains to detect these metabolites as early as possible. It has been suggested that time-dependent inhibition should become routine in vitro screening protocols.[22] Good progress has been made in computational (in silico) metabolite prediction.[23–25]

5.01.2.1.3   Drug–Drug interactions

Regulatory authorities require information to be submitted on the potential for interactions to cause adverse effects. Thanks to the availability of in vitro systems this aspect is often considered during the early stages of discovery, including hit evaluation. Oxidative metabolism by cytochromes P450 (CYPs) is the major route of elimination of most drugs. Since CYPs are also able to metabolize multiple substrates, their inhibition is the major focus of drug–drug interaction (DDI) studies. Thus, CYP3A4 is not only the most abundant hepatic CYP, but is also present in the gut wall and is responsible for the metabolism of 50–60% of all drugs. Therefore, this enzyme is highly susceptible to both reversible and irreversible (mechanism-based) inhibition.[26] Most CYP3A4 substrates or inhibitors are also P-gp substrates or inhibitors. It is believed that CYP3A4 and P-gp in the gastrointestinal tract work in concert to limit uptake of xenobiotics including drugs.[27] Current inhibition studies are based on K_i or IC_{50} values,[28] but more kinetic approaches would be of benefit.[29] Great progress has been made in the reliable simulation of DDIs, even taking into account variability in the population.[30] A further question is how metabolites contribute to DDIs; a better understanding of the allosteric kinetics of CYPs is also needed.[31]

Although clinically somewhat less important than enzyme inhibition, enzyme induction is also an inescapable issue and adequate protocols are being developed for its characterization. It remains to be agreed when to carry out such assays or screens during the discovery process.[22]

5.01.2.1.4   Toxicology and safety prediction

Drug safety is of great concern to patients, medical professionals, and regulatory bodies. As a result, early toxicity predictions and safety estimates are receiving ever-increasing attention in all drug discovery programs. Simple in vitro screening assays, e.g., for hERG and other cardiac ion channels, genetic toxicology, and cytotoxicity, are now routinely added to the growing battery of biology, ADME, and tox/safety screens. Despite encouraging progress,[32] in silico predictive toxicology is still in its infancy.[33] A promising tool is the integration of ADME-Tox, and pharmacology data to predict side effects, as in the BioPrint approach.[34]

5.01.2.2   Technological Issues

5.01.2.2.1   In vitro screening

Based on the experience gained with pharmacodynamic high-throughput screening, many in vitro ADME screens can now run in medium or high-throughput modes using automation, robotics, and miniaturization.[35–37] Physicochemical properties are now recognized to play a key role in modulating DMPK properties,[4,38–40] and their assessment and understanding are therefore receiving greater attention. Owing to the nature of many high-throughput physicochemistry and ADME assays, the typical analytical endpoint is often liquid chromatography–mass spectrometry (LC/MS). Cell-based assays such as the Caco-2 screen for permeability/absorption are quite expensive due to considerable reagent costs, particularly when run in screening mode with many compounds. The trend is to invest in either cheaper in vitro alternatives such as the PAMPA (parallel artificial membrane permeability assay) method or in silico approaches. A proper synergistic hybrid combination of in vitro and in silico methods,[41] which has been called the 'in combo' approach,[42] may be the most cost-effective approach to ADME screening in drug discovery.

5.01.2.2.2 In silico absorption, distribution, metabolism, and excretion

Prediction and simulation of various ADME properties is considerably cheaper than in vitro screening. Therefore, great efforts have been made to turn all available data into predictive computational models using quantitative structure–activity relationship (QSAR) methods and molecular modeling.[42,43] In vitro data are now generated for many ADME and physicochemical endpoints and can be used to build more robust models. Model updating will need to be automated and fitted in the data generation cycle. There is a need for both local (project-specific) as well as global (general, encompassing a wide range of chemotypes) models. Unfortunately, there is still a paucity of human in vivo data, and thus models based upon these will need to be handled with care. Of course interindividual variability within the population is another key factor to take into account for human predictions. Predictions will be ranges rather than hard numbers. Many drug companies have web-based cheminformatics and ADME predictions deployed via their intranet giving the medicinal chemist easy access to 'web screening.'[44] For the development of potential drugs, predictions may also contribute to high-throughput pharmaceutics and rational drug delivery.[45]

5.01.2.2.3 Simulations: physiologically-based pharmacokinetic (PBPK) and pharmacokinetic/ pharmacodynamic (PK/PD)

Physiologically-based pharmacokinetic (PBPK) models rely on principles well known in chemical engineering and describe the human or animal body as a series of pipes and tanks.[46,47] Convenient software packages are now available and make PBPK modeling more accessible to drug discovery and development to predict various PK parameters and concentration–time profiles in the body. Population variability[48] and specific groups such as children and the elderly can also be taken into account. The more experimental data that are available, the better the simulations, making these approaches of interest in drug discovery and clinical development. Some commercial programs have put considerable effort into simulating the absorption process, which is of interest in optimizing pharmaceutical formulations.

Pharmacokinetic/pharmacodynamic (PK/PD) modeling links dose–concentration relationships (PK) to concentration–effect relationships (PD). This approach helps to simulate the time course of drug effects depending on the dose regimen.[49] Specialized software and a better understanding of the applications may help to speed up clinical development. As a whole the pharmaceutical industry lags behind engineering-based industries such as car and airplane manufacturers in using prediction, modeling, and simulation. Often, resources remain allocated to producing experimental data rather than developing the information technology (IT) tools to move toward in silico pharma; in other words, toward web screening rather than wet screening and to a 'will do' rather than a 'can do' attitude.

5.01.3 Concept of the Volume

ADME-Tox issues were far from the forefront in the first edition of *Comprehensive Medicinal Chemistry*, being implicit rather than explicit in a number of chapters on 'drug design.' Numerous and fast developments since the early 1990s now fully justify a complete volume being devoted to the strategies and technologies currently aimed at evaluating and assessing biotransformation, pharmacokinetics, toxicology, and safety of new chemical entities in the preclinical phases. A schematic diagram of drug discovery, development, and clinical assessment is shown in **Figure 3a**. Pharmacodynamics (i.e., activity) is obviously the first object of study, but the new paradigm of drug R&D now dictates that ADMET screening must be initiated rapidly. Activity (PD) and ADMET (PK) screening and evaluation thus run in parallel throughout the preclinical phases, and this is when medicinal chemists find themselves in close collaboration with pharmacologists, biologists, biochemists, bioanalysts, physicochemists, computer scientists, and other experts. Assessment of efficacy and tolerance, to merge into utility assessment, then become the objectives of clinical trials.

The present volume is organized mainly according to the methodologies and technologies available to researchers, with due consideration being given to the biochemical/biological background underlying them. However, such a concept taken per se offers a necessary but not sufficient rationale for a logical and mission-oriented organization of this volume. The second concept underlying the organization of the volume is the overwhelming fact that the ultimate 'target' of medicinal chemistry and drug discovery is the human patient, and that all methodologies, technologies, and tools used during the preclinical phases to screen, assess, and understand the biological behavior of hits, leads, and candidates must be relevant to this ultimate 'target.' In other words, each and every biological model, physicochemical method, and computer software package used in drug discovery and development must demonstrate a useful degree of predictivity to behavior in the human patient.

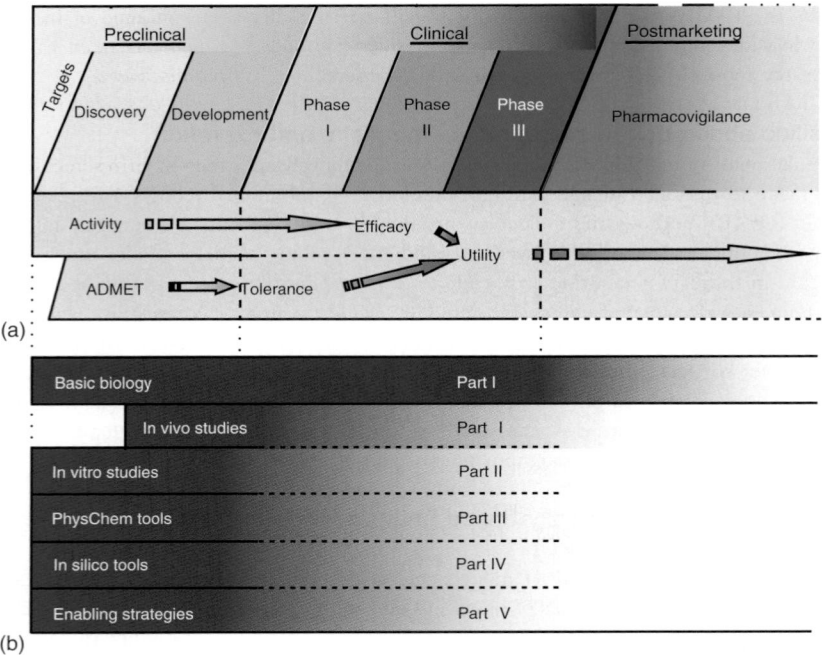

Figure 3 A schematic representation of modern drug research and development (a), presented in parallel with the layout of this volume (b). (a) The major steps in drug R&D, showing target validation, discovery, and development, followed by the clinical phases and the postmarketing phase. Note the overlaps, which reflect the fact that many knowledge-expanding investigations are carried out after the next phases have begun, e.g., in vivo research in animals continues during the clinical phases, and clinical studies continue postmarketing for confirmatory and indication-enlarging purposes. Preclinical R&D has two objectives: activity and ADMET. These translate as efficacy and tolerance in the early clinical phases, to merge subsequently into the global objective of utility. (b) This part of the figure shows the content of this volume and how its parts correspond to the ADMET methodologies and technologies used in drug discovery and development. (The upper part is reproduced from Testa, B.; Krämer, S. D. *Chem. Biodiv.* (in press) with the kind permission of the copyright owner, Verlag Helvetica Chimica Acta in Zurich.)

As a result of this dual rationale, this volume has been organized to begin with the human patient; concretely with the clinical pharmacokinetic criteria good candidates are expected to fulfill. The first chapter (5.02 Clinical Pharmacokinetic Criteria for Drug Research), written by two renowned clinical pharmacologists, is thus the platform on which the volume is built. Part I (see **Figure 3b**) then examines in vivo animal models used in discovery and development (*see* 5.03 In Vivo Absorption, Distribution, Metabolism, and Excretion Studies in Discovery and Development). This is followed by a number of chapters presenting the biochemical bases of drug metabolism and toxification/detoxification.

Part 2 is dedicated to in vitro biological tools used to investigate the absorption, distribution, and metabolism of lead candidates, leads, and clinical candidates. As stressed above, relevance to in vivo and clinical behavior is a major concern. Part 3 covers the major physicochemical tools used in discovery and mainly in development to characterize ionization profile, solubility and dissolution, lipophilicity and permeation, and stability of leads and clinical candidates. Relevance to biological and clinical behavior is considered.

Part 4 is dedicated to in silico tools. The term 'in silico' was coined by the molecular biologist Antoine Danchin, whose quotation below offers a concise and cogent justification to our decision to dedicate the largest part of the volume to this methodology:

[I]nformatics is a real aid to discovery when analyzing biological functions…. [I] was convinced of the potential of the computational approach, which I called *in silico*, to underline its importance as a complement to in vivo and in vitro experimentation.[50]

Indeed, in silico tools have gained an irreplaceable significance in virtual experimentation by allowing medicinal chemists to predict physicochemical and ADMET properties of projected and existing molecules. But here again, relevance to 'wet' physicochemistry, biological properties, and clinical behavior is of paramount importance.

The volume ends with the short Part 5 where global tools (the biopharmaceutics classification system and metabonomics) and enabling strategies are presented.

With 45 chapters, this volume aims at informing by providing data and examples, and instructing by presenting a conceptual and logical framework. As such, careful students of this volume may find in it a didactic value even greater than that of the sum of its individual chapters. The hope is that this volume will retain its usefulness for many years to come, and we thank our contributors for their dedication and enthusiasm.

References

1. Lipinski, C. A.; Lombardo, F.; Dominy, B. W.; Feeney, P. *J. Adv. Drug Deliv. Rev.* **1997**, *23*, 3–25.
2. Lipinski, C. A. *Ann. Rep. Comput. Chem.* **2005**, *1*, 155–168.
3. Leeson, P. D.; Davis, A. M.; Steele, J. *DDT Technol.* **2004**, *1*, 189–195.
4. Van de Waterbeemd, H.; Smith, D. A.; Beaumont, K.; Walker, D. K. *J. Med. Chem.* **2001**, *44*, 1313–1333.
5. Van de Waterbeemd, H. *Curr. Opin. Drug Disc. Dev.* **2002**, *5*, 33–43.
6. Kennedy, T. *Drug Disc. Today* **1997**, *2*, 436–444.
7. Kola, I.; Landis, J. *Nature Rev. Drug Disc.* **2004**, *3*, 711–713.
8. Kubinyi, H. *Nature Rev. Drug Disc.* **2003**, *2*, 665–668.
9. Eddershaw, P.; Beresford, A. P.; Bayliss, M. K. *Drug Disc. Today* **2000**, *5*, 409–414.
10. Smith, D.; Schmid, E.; Jones, B. *Clin. Pharmacokinet.* **2002**, *41*, 1005–1019.
11. Testa, B.; Vistoli, G.; Pedretti, A. *Chem. Biodiv.* **2005**, *2*, 1411–1427.
12. Testa, B. *Trends Pharmacol. Sci.* **1987**, *8*, 381–383.
13. Testa, B. *Pharm. News* **1996**, *3*, 10–12.
14. Palmer, A. M. *Drug News Perspect* **2003**, *16*, 57–62.
15. Testa, B.; Krämer, S. D. *Chem. Biodiv.*, in press.
16. Abriel, H.; Schlaepfer, J.; Keller, D. I.; Gavillet, B.; Buclin, T.; Biollaz, F.; Stoller, R.; Kappenberger, L. *Swiss Med. Wkly* **2004**, *134*, 685–694.
17. Van de Waterbeemd, H. *Expert Opin. Drug Metab. Toxicol.* **2005**, *1*, 1–4.
18. Tucker, G.; Houston, J. B.; Huang, S. M. *Eur. J. Pharm. Sci.* **2001**, *13*, 417–428.
19. Dantzig, A. H.; Hillgren, K. M.; De Alwis, D. P. *Ann. Rep. Med. Chem.* **2004**, *39*, 279–291.
20. Balayssac, D.; Authier, N.; Coudore, F. *Lett. Pharmacol.* **2004**, *18*, 76–80.
21. Kalgutkar, A. S.; Gardner, I.; Obach, R. S.; Shaffer, C. L.; Callegari, E.; Henne, K. R.; Mutlib, A. E.; Dalvie, D. K.; Lee, J. S.; Nakai, Y. et al. *Curr. Drug Metab.* **2005**, *6*, 161–225.
22. Ayrton, A.; Morgan, P. *Xenobiotica* **2001**, *31*, 469–497.
23. Testa, B.; Balmat, A. L.; Long, A. *Pure Appl. Chem.* **2004**, *76*, 907–914.
24. Testa, B.; Balmat, A. L.; Long, A.; Judson, P. *Chem. Biodiv.* **2005**, *2*, 872–885.
25. Ekins, S.; Andreyev, S.; Ryabov, A.; Kirillov, E.; Rakhmatulin, E. A.; Bugrim, A.; Nikolskaya, T. *Exp. Opin. Drug Metab. Toxicol.* **2005**, *1*, 303–324.
26. Zhou, S.; Chan, E.; Lim, L. Y.; Boelsterli, U. A.; Li, S. C.; Wang, J.; Zhang, Q.; Huang, M.; Xu, A. *Curr. Drug Metab.* **2004**, *5*, 415–442.
27. Cummins, C. L.; Jaconsen, W.; Benet, L. Z. *J. Pharmacol. Exp. Ther.* **2002**, *300*, 1036–1045.
28. Shou, M. *Curr. Opin. Drug Disc. Dev.* **2005**, *8*, 66–77.
29. Blanchard, N.; Richert, L.; Coassolo, P.; Lavé, T. *Curr. Drug Metab.* **2004**, *5*, 147–156.
30. Rostami-Hodjegan, A.; Tucker, G. *DDT Technol.* **2004**, *1*, 441–448.
31. Atkins, W. M. *Drug Disc. Today* **2004**, *9*, 478–484.
32. Helma, C. *Curr. Opin. Drug Disc. Dev.* **2005**, *8*, 27–31.
33. Egan, W. J.; Zlokarnik, G.; Grootenhuis, P. D. J. *DDT Technol.* **2004**, *1*, 381–387.
34. Krejsa, C. M.; Horvath, D.; Rogalski, S. L.; Penzotti, J. E.; Mao, B.; Barbosa, F.; Migeon, J. C. *Curr. Opin. Drug Disc. Dev.* **2003**, *6*, 470–480.
35. Saunders, K. C. *DDT Technol.* **2004**, *1*, 373–380.
36. Jenkins, K. M.; Angeles, R.; Quintos, M. T.; Xu, R.; Kassel, D. B.; Rourick, R. A. *J. Pharm. Biomed. Anal.* **2004**, *34*, 989–1004.
37. Wang, J.; Urban, L. *Drug Disc. World* **2004** (Fall), 73–86.
38. Smith, D. A.; Jones, B. C.; Walker, D. K. *Med. Res. Rev.* **1996**, *16*, 243–266.
39. Testa, B.; Van de Waterbeemd, H.; Folkers, G.; Guy, R. *Pharmacokinetic Optimization in Drug Research*; VHCA: Zurich, Switzerland and Wiley-VCH: Weinheim, Germany, 2001.
40. Smith, D. A.; Van de Waterbeemd, H.; Walker, D. K. *Pharmacokinetics and Metabolism in Drug Design*; Wiley-VCH: Weinheim, Germany, 2001 and 2006.
41. Yu, H.; Adedoyin, A. *Drug Disc. Today* **2003**, *8*, 852–860.
42. Dickins, M.; Van de Waterbeemd, H. *DDT Biosilico* **2004**, *2*, 38–45.
43. Van de Waterbeemd, H.; Gifford, E. *Nat. Rev. Drug Disc.* **2003**, *2*, 192–204.
44. Ertl, P.; Selzer, P.; Muhlbacher, J. *DDT Biosilico* **2004**, *2*, 201–207.
45. Varma, M. V. S.; Khandavillis, S.; Ashokraj, Y.; Jain, A.; Dhanikula, A.; Sood, A.; Thomas, N. S.; Pillai, O.; Sharma, P.; Gandhi, R. et al. *Curr. Drug Metab.* **2004**, *5*, 375–388.
46. Leahy, D. E. *Curr. Topics Med. Chem.* **2003**, *3*, 1257–1268.
47. Poggesi, I. *Curr. Opin. Drug Disc. Dev.* **2004**, *7*, 100–111.
48. Yang, J.; Rostami-Hodjegan, A.; Tucker, G. *J. Clin. Pharmacol.* **2001**, *52*, 465–486.
49. Zuideveld, K. P.; Van der Graaf, P. H.; Newgreen, D.; Thurlow, R.; Petty, N.; Jordan, P.; Peletier, L. A.; Danhof, M. *J. Pharmacol. Exp. Ther.* **2004**, *308*, 1012–1020.
50. Danchin, A. *The Delphic Boat – What Genomes Tell Us*; Harvard University Press: Cambridge, MA, 2002, p 30 (translated by A. Quayle).

Biographies

Han Van de Waterbeemd studied physical organic chemistry at the Technical University of Eindhoven, the Netherlands, and then took a PhD in medicinal chemistry at the University of Leiden, the Netherlands. After carrying out postdoctoral research with Bernard Testa at the School of Pharmacy of the University of Lausanne, Switzerland, he took up a position at the same institution for 5 years. He also taught medicinal chemistry to pharmacy students at the universities of Berne and Basel in Switzerland from 1987 to 1997. In 1988 Han joined F Hoffmann-La Roche Ltd in Basel as head of the Molecular Properties Group. In 1997 he moved to Pfizer Central Research UK (now Pfizer Global Research and Development), and held various positions in the Department of Drug Metabolism, later called PDM (Pharmacokinetics, Dynamics, and Metabolism), including Head of Discovery and Head of Automation and In Silico ADME Technologies. In 2005 he moved to AstraZeneca to become global project leader of their Molecular Properties and In Silico ADMET Modeling system.

He has published more than 135 peer-reviewed papers and book chapters, and has (co)-edited 11 books. His research interests include physicochemical and structural molecular properties and their role in drug disposition, as well as the in silico modeling of ADMET properties. Han was secretary of The QSAR and Modeling Society between 1995 and 2005.

Bernard Testa studied pharmacy because he was unable to choose between medicine and chemistry. Because he was incapable of working in a community pharmacy, he undertook a PhD thesis on the physicochemistry of drug–macromolecule interactions. Because he felt himself to be not gifted enough for the pharmaceutical industry, he applied for a postdoctoral research position at Chelsea College, University of London, where he worked for 2 years under the supervision of Prof Arnold H Beckett. And because these were easy times, he was called upon to become assistant professor at the University of Lausanne, Switzerland, becoming full professor and Head of Medicinal Chemistry in 1978. Since then, he has tried to repay his debts by fulfilling a number of local and international commitments, e.g., Dean of the Faculty of Sciences (1984–86), Director of the Geneva-Lausanne School of Pharmacy (1994–96 and 1999–2001), and President of the University Senate (1998–2000). He has written four books and edited 29 others, and (co)-authored 450 research and review articles in the fields of drug design and drug metabolism. During the years 1994–98, he was the Editor-Europe of Pharmaceutical Research, the flagship journal of the American

Association of Pharmaceutical Scientists (AAPS), and he is now the co-editor of the new journal *Chemistry and Biodiversity*. He is also a member of the Editorial Board of several leading journals (e.g., *Biochemical Pharmacology*, *Chirality*, *Drug Metabolism Reviews*, *Helvetica Chimica Acta*, *Journal of Pharmacy and Pharmacology*, *Medicinal Research Reviews*, *Pharmaceutical Research*, and *Xenobiotica*). He holds Honorary Doctorates from the Universities of Milan, Montpellier, and Parma, and was the 2002 recipient of the Nauta Award on Pharmacochemistry given by the European Federation for Medicinal Chemistry. He was elected a Fellow of the AAPS, and is a member of a number of scientific societies such as the French Academy of Pharmacy, the Royal Academy of Medicine of Belgium, and the American Chemical Society. His recently granted Emeritus status has freed him from administrative duties and gives him more time for writing, editing and collaborating in research projects. His hobbies, interests, and passions include jogging, science fiction, epistemology, teaching, and scientific exploration.

Comprehensive Medicinal Chemistry II
ISBN (set): 0-08-044513-6

ISBN (Volume 5) 0-08-044518-7; pp. 1–9

5.02 Clinical Pharmacokinetic Criteria for Drug Research

J-P Tillement, Faculty of Medicine of Paris XII, Paris, France
D Tremblay, AFSSAPS Consultant (French Agency for Drugs), France

5.02.1 Introduction

Combinatorial chemistry and in silico approaches have created an avalanche of new chemical entities (NCEs), which will compete to become drug candidates. Most of these NCEs will be rejected following a thorough selection process, which must be carried out as early as possible for obvious reasons of cost and time.[1] The criteria of selection of a drug candidate involve many diverse aspects such as required pharmacological activity, an absence of toxicity at least within a large range of doses including the active ones, and pharmacokinetic characteristics that enable it to develop its activity in human disease.[2] These criteria should be weighted simultaneously and balanced. Indeed, the selected agent is not obviously the most powerful one, but the one that satisfies efficacy, safety, and appropriate kinetics. A highly powerful agent that is eliminated from the body too quickly would not be of therapeutic interest. Similarly, a molecule that exhibits a very long half-life ($t_{1/2}$) may be difficult to monitor and may lead to accumulation in the body. As a result, the selected drug candidate is often a compromise between the needed activity and the toxicological and pharmacokinetic characteristics. The latter allow the fulfillment of this activity and they facilitate prescription, safe use, comedication, and monitoring.

This chapter describes the pharmacokinetic criteria a drug must meet, the principles of their evaluation, and the information needed for rational and safe use in humans, and highlights some new trends in pharmacokinetics.

5.02.2   Pharmacokinetic Requirements for Drug Development

The pharmacological characteristics of a drug involve its pharmacodynamic (PD) and pharmacokinetic (PK) properties. The former include its effects, their mechanisms, and intensity, while the latter include the conditions of the drug's effectiveness, namely its ability to reach and bind to its receptors at pharmacological concentrations over a period of time sufficient for its effects to develop.

Pharmacokinetic studies observe and measure the variations of drug levels in the body as a function of time. They enable the corresponding parameters to be calculated in order to draw models of the fate of the drug allowing for prediction under various conditions. Pharmacokinetics involves studying the concentrations in various parts of the body, namely organs, tissues, blood, and excreta (urine, bile, feces, sweat, saliva, etc.). It involves not only the parent drug (the administered agent) but also its metabolites, be they active, partly active, or inactive.

Drug disposition consists of four phases (**Figure 1**):

- absorption;
- distribution;
- metabolism (i.e., chemical elimination); and
- excretion (i.e., physical elimination).

These four phases are not successive but can occur more or less simultaneously in vivo. However, they can be investigated separately in in vitro assays.

Appropriate pharmacokinetic characteristics are needed to fulfill the following therapeutic objectives:

1. To reach the selected targets in amounts high enough to produce pharmacological effects. This objective supposes a body distribution selective enough to avoid high concentrations in tissues or organs where the drug does not act (no receptors) or may lead to toxic effects.
2. To maintain the drug concentrations at sufficient levels over a period of time for the effect to develop.
3. To excrete the parent drug and its metabolites according to stable and predictable processes, avoiding as much as possible interindividual variations due to genetic factors and/or pathophysiological conditions such as diseases that alter drug distribution or excretion. Both in vitro and in vivo assays are needed during drug research and development in order to select a pharmacologically effective agent that will exhibit the best profile for human use.

A number of properties must be assessed during preclinical studies, for example, stability at physiological pH, ability to cross cell membranes by passive diffusion or active transfer, blood distribution, and metabolic stability and profiles. These studies also allow selection of the compulsory second animal species for toxicological testing (the first one often being the rat), since its metabolic profile needs to be close to the human one. This can be done in vitro before long-term toxicological studies begin and the first administration to humans. Metabolic profiles are checked using

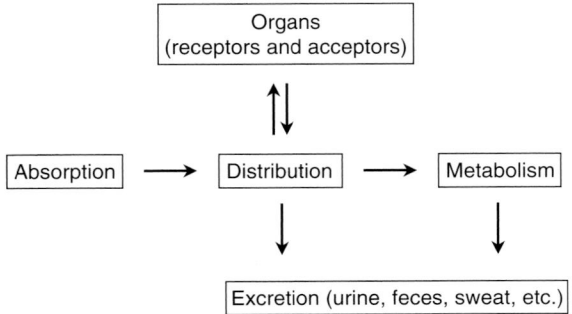

Figure 1   The simplest model of drug disposition inside the body.

microsomes, S9 fractions, animal and human hepatocytes, etc. These assays may also be used to check the possible formation of toxic reactive metabolites and the conditions of their production, for instance, when biotransformation processes are saturated or genetically limited.

Each pharmacodynamic effect will then be correlated with the concentrations in the corresponding organ and blood, including those of the active metabolites. This step initiates the fundamental pharmacokinetic–pharmacodynamic (PK/PD) evaluation of the drug and an assessment of its dose–concentration–effect relationships.[3] Toxic effects are also correlated with the relevant concentrations measured during toxicological studies (toxicokinetics).[4] In humans, PK/PD evaluations will be made using the corresponding data, quantified effects, and relevant drug levels. They will be used to define the administered dose, the number of daily administrations, drug dosage regimens, duration of treatment, etc.

Variations in drug concentrations, in parallel with a quantification of effects, may also bring to light drug–drug interactions, potentiation or antagonism, metabolic induction or inhibition, and the effects of pathological states, as discussed throughout this volume.

5.02.3 The Fate of Drugs in the Body: The Absorption, Distribution, Metabolism, and Excretion (ADME) Paradigm

5.02.3.1 An Overview of Kinetics

The fate of a drug in the body is related to the efficacy of clearing organs and to the fact that most of its interactions in the body (effects, storage, and metabolism) involve reversible binding to specific proteins, receptors, acceptors, and enzymes. Covalent bonds may sometimes be involved, which prolong effects or cause toxicity.

The fate of a drug in the body depends on a series of transfers through and between membranes and into, out of, or between cells during which it may be transformed. Kinetics of transfer and transformation result from either passive or active processes. Passive transfers are described according to Fick's law:

$$\frac{dM}{dt} = k_f \frac{S}{d} \frac{\Delta C}{\sqrt{MW}} \qquad [1]$$

where dM/dt is the amount of drug transferred during time dt, ΔC is the gradient of concentration, MW is the molecular weight of the transferred solute, and k_f is a constant related to its lipophilicity (solubility in membranes) and hydrophilicity (solubility in water). As a general rule, drugs transferred by this process have a MW below 500. S is the surface available for transfer and d is the thickness of this surface or membrane.

This equation is valid for linear kinetics where rate and concentration are directly related. The membrane transfer depends also on other parameters (see 5.12 Biological In Vitro Models for Absorption by Nonoral Routes; 5.13 In Vitro Models for Examining and Predicting Brain Uptake of Drugs; 5.19 Artificial Membrane Technologies to Assess Transfer and Permeation of Drugs in Drug Discovery; 5.28 In Silico Models to Predict Oral Absorption; 5.29 In Silico Prediction of Oral Bioavailability; 5.30 In Silico Models to Predict Passage through the Skin and Other Barriers; 5.37 Physiologically-Based Models to Predict Human Pharmacokinetic Parameters; 5.42 The Biopharmaceutics Classification System).

Active transfers are energy consuming as defined in Michaelis kinetics by the general equation:

$$\frac{dM}{dt} = \frac{V_{max} \times C}{K_m + C} \qquad [2]$$

where dM/dt is the amount of drug transferred or metabolized during time dt, C is the drug concentration, V_{max} is the maximum rate of kinetics, and K_m is the Michaelis constant. Since drug concentrations are generally very low, C is often negligible compared to K_m and the equation becomes:

$$\frac{dM}{dt} \approx \frac{V_{max}}{K_m} C \qquad [3]$$

This approximation yields linear kinetics.

5.02.3.2 Absorption

Absorption is the early phase of distribution when the drug enters the bloodstream from its site of administration. Two main parameters describe absorption: the ratio of absorbed dose over given dose and the rate of absorption. Both parameters influence the initial concentrations of the drug in blood, hence its concentration gradient and intensity of effects (when concentration dependent).

The main sites of absorption are the gut (oral route), connective tissues (subcutaneous route), muscles (intramuscular route), skin, eyes, nose, and various mucosa. These routes of administration are extravascular. The intravascular routes (intraarterial or intravenous) bypass the absorption phase as the drug is administered directly into the blood.

5.02.3.3   Distribution

Distribution is the next phase, when the drug is transferred from the blood into tissues.[5] To do so, it exits the vessels and enters the extravascular circulation, first into interstitial fluids, then into cells. Distribution may be subdivided into two successive phases. The first phase involves dilution of the absorbed dose in the whole blood as a free fraction, and binding to plasma proteins and circulating cells (e.g., erythrocytes). In the second phase, the drug distributes to organs. When the drug is injected intravenously as a bolus, a high local concentration occurs in blood before dilution, thus creating a high transitory gradient that reaches organs immediately downstream. This high gradient is sometimes required in order to obtain a quick and intense effect in well-irrigated organs (e.g., in the brain for general anesthetics). In many other cases, it may be dangerous as it promotes over dosage and possible toxic effects in heart and lungs.

In the blood, drugs are bound to plasma proteins according to the law of mass action (reversible equilibrium). The binding to a plasma protein may have two different meanings, either pharmacokinetic or pharmacodynamic. In the first case, the amount of bound drug and the characteristics of binding can alter the overall drug distribution. In the second, the drug is bound to its circulating receptor.

Many plasma proteins can bind drugs according to eqns [4] and [5]:

$$[\text{Drug}] + [\text{Protein}] \rightleftarrows [\text{Drug} - \text{Protein}] \tag{4}$$

$$K_a = [C_b]/[C_u \cdot (C_P - C_b)] \tag{5}$$

which gives:

$$f_u = 1/[1 + K_a \cdot (C_P - C_b)] \tag{6}$$

where K_a is the association constant, C_P is the protein concentration, C_b is the concentration of protein-bound drug, C_u is the free (unbound) drug concentration (with $C_b + C_u = C$, the total drug concentration in plasma), f_u is the free fraction of drug in plasma (with $f_u = C_u/C$), and $100(1 - f_u)$ is the percentage of bound drug.

According to eqn [6], f_u is not constant but increases when C_b increases. But when C_b is negligibly small compared to C_P ($C_b \ll C_P$) f_u will be constant if the protein concentration is constant. Then:

$$f_u = 1/[1 + K_a \cdot C_P] \tag{7}$$

Human serum albumin (HSA) and α_1-acid glycoprotein (AGP) bind significant amounts of drugs as a consequence of their concentration and affinities. HSA concentration is high (ca. $630\,\mu M$), implying that $C_b \ll C_P$ is generally fulfilled, and HSA binding capacity for acidic drugs is generally high. In contrast, AGP concentrations are low (ca. $10\,\mu M$, except in some genetic or pathological states),[6] and its affinity is mainly toward basic drugs, with the result that condition $C_b \ll C_P$ is not always fulfilled, binding is quickly saturated, and f_u increases when drug concentration increases and binding percentage decreases.

Significant amounts of a drug may also be taken up by red cell membranes, or in the case of basic drugs in red cell cytosol at pH 7.20. Generally, uptake and binding to red cells are nonsaturable processes. When they are saturable, saturating concentrations will induce nonlinear kinetics.

Lipophilic drugs may also interact with lipoproteins,[7] their dissolution in the lipid core of lipoproteins being roughly proportional to their lipid content (VLDL>LDL>HDL). Binding to the protein core has also been described but its pharmacokinetic significance is not yet established. The design of these various binding modes to plasma proteins has been described elsewhere.[8] Other plasma proteins of pharmacokinetic interest are fibrinogen, α_2-glycoprotein, and gamma globulins.

From the blood, drugs will distribute into organs at different rates corresponding to different membrane structures and different local blood flow rates. Schematically, drug distribution inside an organ is superimposable to the rate of blood flow. This observation explains why the blood cannot selectively distribute a drug into a given organ.

In every organ and every cell, a drug is partly free and partly bound to functional proteins, both forms being in reversible equilibrium. According to Brodie,[9] it is assumed that at any time and in every tissue, an instantaneous equilibrium exists between blood and organ concentrations. This equilibrium is characterized by the same free drug

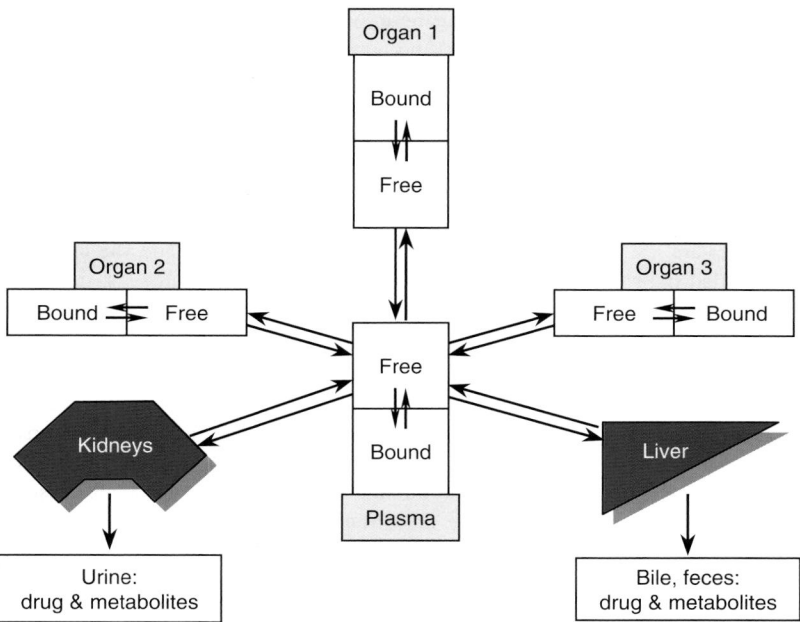

Figure 2 Equilibria between blood and various organs, with liver and kidneys as excretion pathways. It is assumed that each binding is governed by a reversible equilibrium. At any one time, the free drug concentrations are assumed to be rapidly exchangeable. The liver (by biotransformation and bile excretion) and kidneys (by urine excretion) continuously disrupt the general equilibrium by removing some of the free drug. At steady-state, the free drug concentration is assumed to be identical in plasma and organs.

concentration (C_u) in all organs, which is considered to be instantaneously exchangeable with the plasma. However, this equilibrium is disrupted since clearing organs excrete a fraction of the free drug. The decrease in C_u is quickly compensated, and a new equilibrium is reached with a decrease in overall drug concentration (**Figure 2**). At the end of this cascade of equilibria, the entire dose is cleared from the body.

5.02.3.4   Metabolism

Drug metabolism is the phase of biochemical transformation of the drug. It is highly variable among drugs and depends on biological conditions. The metabolism phase is absent for the few drugs that are not transformed. As explained in great detail in other chapters (*see* 5.05 Principles of Drug Metabolism 1: Redox Reactions; 5.06 Principles of Drug Metabolism 2: Hydrolysis and Conjugation Reactions; 5.07 Principles of Drug Metabolism 3: Enzymes and Tissues; 5.08 Mechanisms of Toxification and Detoxification which Challenge Drug Candidates and Drugs; 5.09 Immuno-toxicology; 5.10 In Vitro Studies of Drug Metabolism; 5.33 Comprehensive Expert Systems to Predict Drug Metabolism; 5.43 Metabonomics), biotransformations may involve one or more successive reactions:

- Phase 1 transformations (reactions of functionalization) involve the creation of a functional group or the modification of an existing one by oxidation, reduction, or hydrolysis.
- Phase 2 transformations (reactions of conjugation) couple a drug or a metabolite to an endogenous conjugating molecule such as glucuronic acid, sulfuric acid, acetic acid, glutathione, etc.

From a physicochemical point of view, drug metabolism is expected to yield metabolites of lower lipophilicity relative to the parent drug, e.g., by adding an ionizable group. As a result, metabolites are often excreted faster than the parent drug, but there are exceptions. From a pharmacological point of view, it is essential to check the pharmacodynamic consequences of these metabolic reactions. Often but far from always, biotransformation involves inactivation or detoxification. Activation concerns prodrugs, but also active compounds (drugs) giving rise to active metabolites. The latter may exhibit a PK profile different from that of the parent drug, and/or a qualitatively different activity. Prodrugs receive specific treatment in Chapters 5.44 and 5.45.

Some enzymes involved in metabolism present a genetic polymorphism, which separates populations of patients according to their phenotypes (i.e., very fast, 'normal,' and poor metabolizers). This is the field of pharmacogenetics. Independently of any pathological state, individuals who are very fast or poor metabolizers need to be identified and have their dosages adjusted.[10] Specific monitoring must also be applied for drugs with a low therapeutic index resulting in a low safety margin due to relatively vicinal effective and toxic doses.

5.02.3.5   Excretion

The kidneys and the liver are the main organs that clear drugs from the body, and the only ones that elicit active excretion processes provided the compounds are ionized. Rates of excretion are always faster for hydrophilic than for lipophilic compounds, however, metabolites cannot be excreted more rapidly than the parent drug, as their excretion is limited by their formation from the parent drug.

5.02.3.5.1   Renal excretion

Three separate processes are involved in renal excretion, namely, glomerular filtration, active tubular excretion, and passive tubular reabsorption.

Glomerular filtration is a simple filtration of plasma through the pores of the glomeruli, whose size allows only the plasmatic free drug and metabolites to be cleared. The amount of compound eliminated is restricted by the glomerular filtration rate and as a result by the renal blood flow. Thus, it may be modified by comedications that increase the rate of elimination of other drugs (e.g., diuretics), and obviously by various pathophysiological conditions.

Active tubular excretion is the process by which tubules excrete ionized compounds, a process requiring energy supply and specific transporters. P-glycoprotein and multidrug resistance-associated proteins (type 2) transfer amphiphilic anions and most of the conjugates. Other transporters transfer organic cations. Most of the renal transporters work from plasma to urine and to a lesser extent back from urine to plasma. The fact that this process involves only ionized compounds underlines the prominent role of plasma pH as a factor influencing the degree of ionization.

Passive tubular reabsorption takes place at proximal and distal tubules. Following filtration, nonionized compounds may return to capillaries provided that a favorable gradient exists. This process involves mostly the more lipophilic parent compounds. It is directed by the pH of elementary urine as a determinant of percentage of the reabsorbed nonionized form.

The total elimination of a parent compound and metabolites requires the integrity of these three separate processes, which themselves depend on plasma pH, urine pH, and renal blood flow. All active processes may be restricted by the amount of available transporters and energy supply, which may lead to competitive drug–drug interactions.

5.02.3.5.2   Hepatic elimination

Three main processes may be involved in the hepatic elimination of a drug and its metabolites, namely, uptake, metabolism, and biliary secretion. These processes can be minor or even lacking depending on the physicochemical properties of the drug (e.g., its hydrophilicity). As in the kidney, hepatocytes excrete unchanged drugs and metabolites by active processes involving transporters. As a result, competition for biotransformation and excretion may also take place in the liver. Following bile secretion, the excreted compounds reach the duodenum where they may be either excreted from the body via the feces, partly reabsorbed in enterohepatic recycling, or partly metabolized by the intestinal flora (as an example, conjugated morphine is partly deconjugated). Metabolites may also be excreted in plasma from which they are cleared by kidneys.

In conclusion, prescribing of drugs in a rational way requires a knowledge of what causes pharmacological activity to stop, namely:

- in which organs (liver, kidneys, or others);
- by which processes (biotransformation to inactive metabolites, biliary secretion without reabsorption, or urinary excretion); and
- at which rates (clearance and half-life).

An alteration in any of these processes may necessitate a dosage adjustment or even a drug change.

5.02.3.5.3   Other routes of excretion

Fecal excretion of drugs can result from incomplete intestinal absorption following oral administration, biliary elimination, or direct excretion by enterocytes. Pulmonary excretion is found for volatile agents, even lipophilic ones, but in situ transformation may also occur. Excretion in saliva, sweat, and tears results from passive diffusion processes. Assaying concentrations in saliva may be useful to monitor some drugs or toxins as they reflect free plasma concentrations.

Special attention should be given to the secretion of drugs in breast milk, which, being more acidic than blood plasma, can concentrate basic compounds. Milk also contains a high proportion of lipids, a factor favoring the secretion of lipophilic drugs. The amount of drug transferred by this route from mother to baby is usually low, but it can be sufficient to induce side effects in the newborn, especially in the CNS since the blood–brain barrier is lacking at birth.

Drug disposition in hair has no pharmacokinetic relevance but may indicate drug intake.

5.02.4   Pharmacokinetic Parameters

5.02.4.1   Introduction

Pharmacokinetic parameters are assessed by monitoring variations in concentration of the drug and/or its metabolites in physiological fluids that are easy to access (i.e., plasma and urine). Plasma concentrations are usually checked, and in addition biopsies can be taken from animals and sometimes from humans. Pharmacokinetic parameters give an overall indication of the behavior of the drug in the body; the basic parameters are listed in **Table 1**.[11–13] (**Figures 3** and **4**)

Since the kinetics of absorption, distribution, metabolism, and excretion of a drug are usually linear, a polyexponential function may be adjusted to the plasma concentrations allowing the calculation of pharmacokinetic parameters. However, adjustment is not always feasible and is not mandatory. It is sufficient to consider the terminal part of the concentrations-versus-time curve, assuming that absorption and distribution, being almost complete, would have no significant influence on both metabolism and excretion. A monoexponential function can then be fitted, according to eqn [8]:

$$C = C_0 \times e - \lambda_1 t \qquad [8]$$

The logarithmic transformation gives eqn [9]:

$$\ln C = \ln C_0 - \lambda_1 t \qquad [9]$$

which is the equation of a straight line. C and λ_1 can be calculated by linear regression, C being the concentration at time t, C_0 the concentration extrapolated to the origin, and λ_1 the apparent elimination constant.

5.02.4.2   Calculation of Individual Parameters

The lag time (t_l) is the time between administration and the first measurable concentration. In the case of oral administration, for instance, it is not nil but corresponds to the time needed for the tablet to disintegrate in the gastrointestinal tract, the drug to dissolve, and absorption to occur. C_{max} is deduced from the observation of time-related plasma concentrations. After intravascular bolus administration, C_{max} is obviously the first concentration measured, however rapidly the first sampling was taken. After an infusion, C_{max} is the concentration measured at the end of the infusion. The time that corresponds to C_{max} is t_{max}.

Table 1 Basic pharmacokinetic parameters

Parameters obtained from plasma concentrations
  After an intravascular administration (**Figure 3**)
    AUC = area under the curve of a plasma concentration versus time profile
    CL = total plasma clearance
    V_D = volume of distribution
    $t_{1/2}$ = elimination half-life
  After an extravascular administration (**Figure 4**)
    t_l = lag time
    C_{max} = maximum concentration
    t_{max} = time to reach C_{max}
    AUC = area under the curve of a plasma concentration versus time profile
    $t_{1/2}$ = elimination half-life
Parameters obtained from urine concentrations (whatever the route of administration)
    M_U = amount excreted in urine
    CL_R = renal clearance when plasma concentrations are known

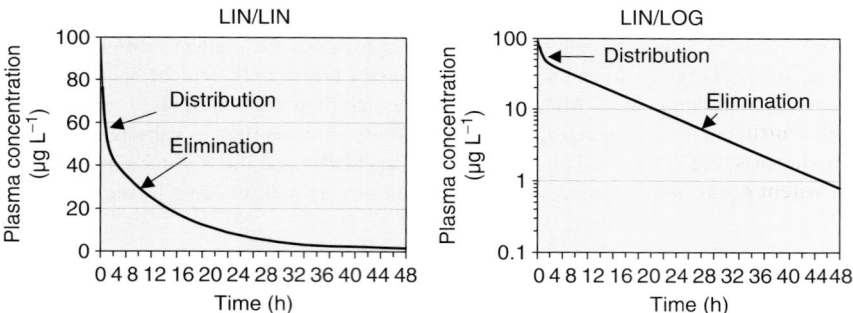

Figure 3 Time course of plasma concentration following administration by the intravascular (iv bolus) route. The same data are displayed as linear (LIN/LIN) and semilogarithmic (LIN/LOG) plots.

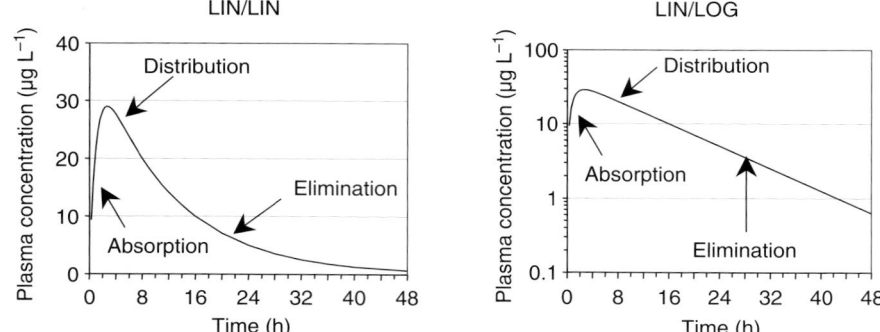

Figure 4 Time course of the plasma concentration of a drug administered by the extravascular route. The same data are displayed as linear (LIN/LIN) and semilogarithmic (LIN/LOG) plots. Dose: 5 mg, C_{max}: 30 µg L^{-1}, AUC: 420 h µg L^{-1}, and $t_{1/2}$ 8 h. No lag time is observed here.

The AUC (area under the curve of a plasma concentration versus time profile) from time 0 to the last sampling is calculated directly from concentrations C and their corresponding t by using eqn [10] (trapezoidal rule):

$$AUC_{0-t} = \frac{1}{2} \sum (C_n + C_{n-1})(t_n - t_{n-1})$$ [10]

The extrapolation to infinite may be done as per eqn [11]:

$$AUC_{0-\infty} = AUC_{0-t} + AUC_{t-\infty} = AUC_{0-t} + \frac{C_{last}}{\lambda_1}$$ [11]

AUC is expressed as time multiplied by concentration. When the kinetics is linear, AUC is proportional either to the amount of drug that reaches the body without any transformation, or to the amount of metabolite when its concentrations have been measured.

The total plasma clearance indicates the elimination of the drug from the body whatever the process involved (metabolism, urinary excretion, bile secretion, pulmonary elimination).[14] It is the part of the distribution volume (V_D) cleared by time unit (eqns [12–15]).

$$CL = V_D \frac{\text{amount of drug metabolized and/or excreted per unit of time}}{\text{amount of drug in the body}}$$ [12]

$$CL = V_D \frac{dM/dt}{V_D C} = \frac{dM/dt}{C}$$ [13]

$$\frac{dM}{dt} = CL \times C$$ [14]

$$dM = CL \cdot C \cdot dt$$ [15]

Making the assumption that CL is constant, the integration as a function of time gives eqn [16]:

$$M = \mathrm{CL} \cdot \mathrm{AUC} \tag{16}$$

And making the further assumption that: (1) the drug is administered by the intravascular route; (2) the drug is eliminated from the plasma without any reintroduction (no enterohepatic cycling, no reversible biotransformation regenerating the parent drug from metabolites); and (3) the total amount M of excreted drug is equal to the dose, then:

$$\mathrm{CL} = \frac{\mathrm{Dose}}{\mathrm{AUC}} \tag{17}$$

When the drug is administered by an extravascular route, the total amount reaching the body is unknown and CL cannot be calculated.

The product of urinary concentration and urinary volume gives the amount eliminated in urine (M_U). When the compound measured in plasma and urine is the same (parent or metabolite), eqns [14] and [15] apply and become eqns [18] and [19]:

$$\frac{\mathrm{d}M_U}{\mathrm{d}t} = \mathrm{CL_R} \times C \tag{18}$$

$$\mathrm{d}M_U = \mathrm{CL_R} \cdot C \cdot \mathrm{d}t \tag{19}$$

where $\mathrm{CL_R}$ is the renal clearance, which can be calculated from urine fractions collected during Δt (eqns [20] and [21]):

$$\Delta M_U = \mathrm{CL_R} \cdot C \cdot \Delta t \tag{20}$$

$$\mathrm{CL_R} = \frac{\Delta M_U}{\Delta \mathrm{AUC}} \tag{21}$$

ΔM_U is the amount eliminated in urine during Δt and $\Delta \mathrm{AUC}$ is the area corresponding to Δt. The calculation can be made using the global kinetics; therefore, eqn [17] becomes eqn [22]:

$$\mathrm{CL_R} = \frac{M_U}{\mathrm{AUC}} \tag{22}$$

The only assumption here is that $\mathrm{CL_R}$ remains constant.

Hepatic and renal clearances are additive; lung clearance is multiplicative since it acts upstream of liver and kidneys. When neither the lungs nor any other organ plays a significant role in the overall clearing process, the total plasma clearance is the sum of hepatic and renal clearances:

$$\mathrm{CL} = \mathrm{CL_R} + \mathrm{CL_H} \tag{23}$$

and $\mathrm{CL_H}$ can be obtained by difference.

Clearance of drugs occurs by blood perfusing the organ of extraction. Extraction refers to the proportion of the drug arriving at the organ and removed irreversibly by excretion or metabolism. Hepatic intrinsic clearance ($\mathrm{CL_{int}}$) refers to the global capacity of the hepatocytes to contribute to the processes of the hepatic elimination. $\mathrm{CL_{int}}$ can be estimated from in vitro studies using hepatocytes and the drug as substrate and scaling up to whole liver dimensions:

$$\mathrm{CL_{int}} = \frac{V_{\max}}{K_m} \times \frac{\text{number of hepatocytes in the liver}}{\text{number of cultured hepatocytes}} \tag{24}$$

For drugs highly extracted by the liver (high $\mathrm{CL_{int}}$), clearance depends on the blood flow perfusing the liver, and causes a large hepatic first-pass effect. Conversely, for a drug with a low $\mathrm{CL_{int}}$, clearance depends on C_u, the free concentration of the drugs, which is only available for extraction, and on enzymatic activity. Equation [14] then becomes eqn [25]:

$$\frac{\mathrm{d}M}{\mathrm{d}t} = \mathrm{CL_H} \times C = \mathrm{CL_{H,u}} \times C_u \tag{25}$$

in which $\mathrm{CL_{H,u}}$ is the clearance based on unbound concentration. When $\mathrm{CL_{int}}$ is low, $\mathrm{CL_{int}} \approx \mathrm{CL_{H,u}}$.

From eqn [25], one obtains eqn [26]:

$$CL_H = CL_{H,u} \frac{C}{C_u} = CL_{H,u} \times f_u \qquad [26]$$

Changes in f_u (e.g., due to saturation or displacement) and/or in $CL_{H,u}$ (e.g., by enzymatic inhibition or induction) will induce changes in CL_H.

Protein binding is one of the factors that influences CL_R. Filtration and secretion depend on f_u and change in the same way.[15]

At steady-state, increase of f_u will decrease C, but C_u will be the same, as CL_u should not be affected and the activity of the drug should not change, despite the decrease of C.

The distribution volume V_D represents the theoretical body volume inside which the drug would be distributed if it had the same concentration everywhere in the body and in plasma. V_D equals the amount of drug in the body divided by the plasma concentration:

$$V_D = \frac{M}{C} \qquad [27]$$

This definition is simple to use but not sufficient. Indeed, according to this equation V_D increases between the moment the drug reaches the body and the moment its distribution is complete. A condition must be added, that the drug be in equilibrium in all parts of the distribution volume, a variation in one part being immediately reflected in all other parts. This condition is fulfilled during the elimination phase when plasma and tissue concentrations decrease in parallel, which means that the ratio of tissue concentration over plasma concentration is constant.

With this condition, the amount of drug remaining in the body (M_t) at time t after an intravascular administration is equal to the initial dose minus the amount eliminated:

$$M_t = \text{Dose} - CL \cdot (AUC_{0-\infty} - AUC_{t-\infty}) \qquad [28]$$

$$M_t = \text{Dose} - CL \cdot AUC_{0-\infty} + CL \cdot AUC_{t-\infty} = \text{Dose} - \text{Dose} + CL\frac{C_t}{\lambda_1} = CL\frac{C_t}{\lambda_1} \qquad [29]$$

Coming back to the definition of V_D according to the eqn [27]:

$$V_D = \frac{M_t}{C_t} = \frac{CL\dfrac{C_t}{\lambda_1}}{C_t} \qquad [30]$$

$$V_D = \frac{CL}{\lambda_1} \qquad [31]$$

After an extravascular administration, the amount of drug that reaches the body can be different from the administrated dose, thus in that case the calculation of V_D cannot be done.

The distribution volume of a drug depends on: (1) its respective affinities for plasma proteins and blood circulating cells that retain it in plasma; (2) its ability to leave the vessels; and (3) its affinities for the tissues that can attract it. Consider the following mass balance relationship[12]:

$$V_D \cdot C = V_P \cdot C + V_T \cdot C_T \qquad [32]$$

where V_P is the plasma volume and V_T and C_T are the tissue volume and the tissue concentration, respectively. Knowing that $f_u = C_u/C$ and similarly for tissues $f_{u,T} = C_u/C_T$, C_u at steady-state being the same in all parts of the body, eqn [32] becomes:

$$V_D \times \frac{C_u}{f_u} = V_P \times \frac{C_u}{f_u} + V_T \times \frac{C_u}{f_{u,T}} \qquad [33]$$

Dividing by C_u/f_u yields eqn [34]:

$$V_D = V_P + V_T \times \frac{f_u}{f_{u,T}} \qquad [34]$$

This approach shows that V_D depends on f_u and $f_{u,T}$. The distribution volume does not correspond to a physiological one. However, it is interesting to compare it to the volume of exchangeable water as related to body weight, $V_D = 0.6\,L\,kg^{-1}$. A smaller V_D indicates that the distribution of the drug is quantitatively limited inside the body due to several possible causes (f_u is small and/or $f_{u,T}$ is large). Either the drug is not transferable because it is not lipophilic enough (e.g., many antibiotics) or its molecular weight is too high. The drug may also be able to diffuse through membranes but is retained in the blood flow because of a high affinity to plasma proteins (restricted distribution).[16]

A larger V_D ($>1\,\mathrm{L\,kg^{-1}}$) indicates an important distribution linked to a high tissue affinity ($f_{u,T}$ is small). When V_D is large, the corresponding C_Ts of the drug are large. If we consider now C_u, eqn [27] and eqn [33] give:

$$V_{D,u} = \frac{M}{C_u} = \frac{M}{f_u \times C} = \frac{V_D}{f_u} = \frac{V_P}{f_u} + \frac{V_T}{f_{u,T}} \qquad [35]$$

The distribution volume $V_{D,u}$ represents the theoretical body volume inside which the unbound drug would be distributed. The smaller the free fractions f_u and $f_{u,T}$, the larger $V_{D,u}$ will be.

The half-life ($t_{1/2}$) is the time needed for a plasma concentration to decrease by one-half. During elimination, when the decrease in plasma concentrations follows a nonexponential process, $t_{1/2}$ is a constant, which can be calculated as follows from the apparent elimination constant:

$$t_{1/2} = (\ln 2)/\lambda_1 = (\ln 2) \cdot V_D/\mathrm{CL} \qquad [36]$$

The half-life $t_{1/2}$ may be deduced directly from the graph of concentrations-versus-time, whereas V_D and CL are calculated. However, $t_{1/2}$ also depends on V_D and CL as shown above.

5.02.4.3   Clinical Considerations

Most drugs are administered chronically at a constant dose and at regular time intervals. When the next dose is given before complete elimination of the previous one, the drug accumulates until elimination compensates absorption. When kinetics is linear, CL, V_D, and thus $t_{1/2}$ are constant. Hence, a steady state is reached after repeated administration and can be predicted from a unique administration. The time needed to reach the steady state usually represents 4–6 half-lives (**Figure 5**).

In the simplest case (intravascular bolus injection), distribution and elimination follow a nonexponential process, and C_{max} at steady state is:

$$C_{max,ss} = \frac{\text{Dose}}{V_D(1 - e^{-\lambda_1 \tau})} \qquad [37]$$

and C_{min} at steady state is:

$$C_{min,ss} = C_{max,ss} \cdot e^{-\lambda_1 \tau} \qquad [38]$$

where τ is the time between two doses.

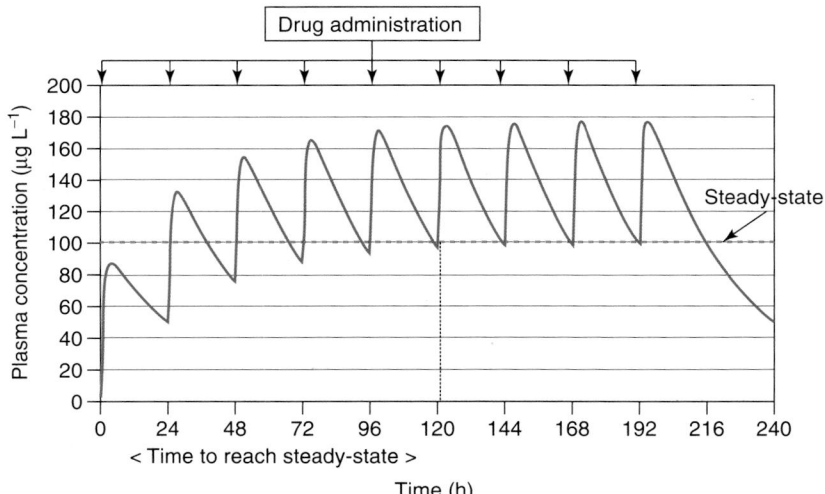

Figure 5 A typical curve of the variation of plasma concentration following repeated administrations by the extravascular route ($t_{1/2} = 24\,\mathrm{h}$).

If accumulation of the drug induces a saturation or an inhibition of its own elimination process, i.e., if its CL decreases with the amount of drug in the body, the steady-state concentrations may be higher than expected. In the worst-case scenario, accumulation proceeds without reaching a steady state and elicits toxic effects. More information about the determination of pharmacokinetic parameters can be found elsewhere.[17] In contrast, if the drug induces an increase of its own elimination process, the steady-state levels will be lower than predicted.

Inhibition and induction can be caused by a coadministered drug. It is thus necessary to verify whether a steady state is reached after repeated administration and to compare the relevant calculated and measured concentrations. If they are not in accordance, the discrepancy must be explained and the drugs changed or dosed differently.

5.02.5   Bioavailability and Bioequivalence

The concept of bioavailability (biological availability) has been created to evaluate the pharmacological effects of a drug in humans when these effects are delayed (e.g., with antidepressants, digitalis) or not easily measured (e.g., for hormones, glucocorticoids). Instead of measuring more or less precisely the corresponding clinical effects, it was proposed: (1) to evaluate the concentration of drug in the receptor compartment (making the assumption of concentration-related effects); (2) to measure the maximal plasma concentration and the time needed to reach it; and (3) to develop a pharmacokinetic approach to predict the intensity of effects.

5.02.5.1   Bioavailability

Bioavailability means the extent and rate to which a drug becomes available in the general circulation. The extent is measured by the fraction (F) of the administered dose which reaches intact the general circulation. When the drug is administered by the intravascular route, $F = 1$ and the relevant rate is that of the injection. When extravascular routes are used, for instance, the oral route, $F \leq 1$ and depends on the absorbed fraction of the ingested dose and on the lost part of the dose metabolized in the gut during absorption or in the liver before reaching the general blood flow. These latter possibilities are named first-pass effects (intestinal and/or hepatic).

The amount of drug M that reaches the general circulation is given by eqn [16]. Thus:

$$F = CL \cdot AUC/Dose \qquad [39]$$

The rate of absorption is estimated by two parameters, C_{max} and t_{max}, with C_{max} being related to: (1) total plasma clearance; (2) the fraction of dose that reaches the general circulation without being metabolized; (3) the rate of absorption; and (4) the rates of distribution and elimination. As for t_{max}, it depends on: (1) the rate of absorption; and (2) the rates of distribution and elimination.

After oral administration, bioavailability may be estimated by comparison with either the intravascular route, yielding the absolute bioavailability, or with another pharmaceutical form of the drug by the same route, yielding the relative bioavailability. Absolute bioavailability can be measured after intravascular administration:

$$Dose = CL \cdot AUC_{iv} \qquad [40]$$

or after oral administration:

$$F \cdot Dose = CL \cdot AUC_{po} \qquad [41]$$

If the doses used for both routes are identical, assuming a linear and identical kinetics (i.e., a constant CL), then:

$$F = \frac{AUC_{po}}{AUC_{iv}} \qquad [42]$$

This test is of interest for several reasons. When a drug has not been previously investigated by the i.v. route, the i.v. results allow the calculation of CL and V_D and can lead to a basic understanding of the drug's kinetics. Moreover, they yield the characteristics of oral absorption, which are needed when oral administration follows i.v. injection. A typical example of such a test is given in **Figure 6**.

The relative bioavailability (F_{rel}) is used to compare two different pharmaceutical forms, one serving as a reference:

$$F_{rel} = \frac{AUC_{test}}{AUC_{ref}} \qquad [43]$$

This equation may be used if the respective doses of the two pharmaceutical forms are the same and if the clearance is constant. It also allows a comparison of C_{max} and t_{max}.

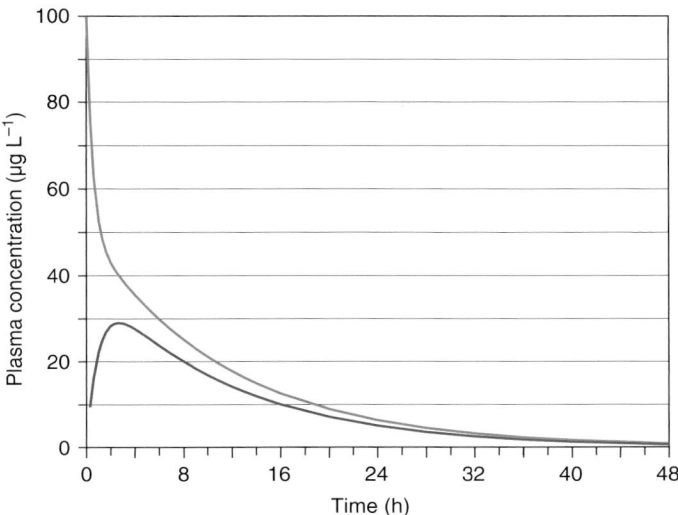

Figure 6 Absolute bioavailability of a drug administered at the same dose by the intravascular (red) and extravascular (blue) route. Assuming CL to be constant and identical for the two administrations, the ratio of the AUC is 0.7, meaning that 70% of the administered dose reaches the systemic circulation intact.

5.02.5.2 Bioequivalence

The concept of bioequivalence was born from the following observation. Digoxin, administered orally at the same dose of 1 mg but as a tablet made by two different pharmaceutical companies using different components, induced quantitatively different pharmacological effects. The two tablets were considered as being nonbioequivalent, their nonbioequivalence being a consequence of the difference in formulation between the two tablets, which released digoxin at different rates and in different amounts. Thus, one tablet could not replace the other.

The many pharmaceutical forms of the same drug that are now available, either during its development or as generics, need to be checked to see whether they will produce the same quantitative effects; in other words, that one may be substituted for another without any change in therapeutic effects. Making the likely assumption that the same plasma concentrations develop the same clinical effects, bioequivalence may be assessed by the equivalence of the relative bioavailabilities of the two pharmaceutical forms of a drug. Such equivalence is checked by comparing AUC, C_{max}, and t_{max}, assuming that pharmacokinetic parameters remain constant during the comparison.[13] These tests are carried out in healthy volunteers, using a randomized two-period, two-sequence crossover design. The statistical method for testing bioequivalence is based upon the 90% confidence interval for the ratio of the means for the parameters under consideration (test/reference). Concerning C_{max} and AUC, the confidence interval is calculated using the residual of the analysis of variance (ANOVA), the data should be transformed prior to analysis using a logarithmic transformation. The analysis for t_{max} should be nonparametric and should be applied to untransformed data. Bioequivalence is concluded when confidence intervals are within 0.8 and 1.25 (**Figure 7**).

This procedure is required for any generic of a drug that has already been marketed in its original pharmaceutical form. More information about the requirements on bioavailability and bioequivalence can be found in the guidances of the European Community,[18] FDA,[19] and ICH.[20]

5.02.6 Which Pharmacokinetic Profiles for Drugs?

The overall objective of a pharmacokinetic design is to harmonize the effects of the active compounds (parent and/or active metabolites). It has to allow the drug to develop its effects under the best conditions of efficacy, intensity and duration, interindividual predictability, safety, and comfort to the patient. These requirements are obviously difficult to fulfill simultaneously, although many drugs come very close.

Specific designs are needed according to the medical use of the drug. For instance, some antibiotics are designed to target intracellular germs and must therefore be extensively distributed to tissues, whereas others are targeted to cure vascular infections. V_D in the former case must be larger than in the latter. Thus, the design may be done in accordance with the specifications of the treatment. Only some general proposals in the design of pharmacokinetics are presented here based on observations of the evolution of the characteristics of recent drugs.

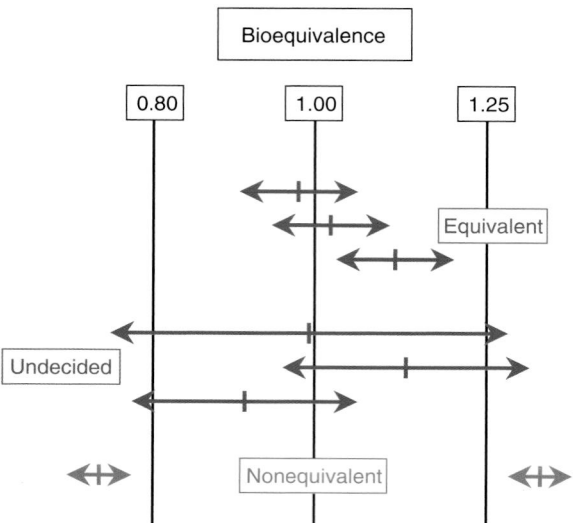

Figure 7 Acceptance or rejection of bioequivalence according to the confidence intervals of ratio of transformed C_{max} and AUC of test and reference medicine calculated from residual of variance analysis.

5.02.6.1   A Limited Distribution in the Body

This simple idea is based on the fact that drugs can distribute throughout the whole body whereas their targets are limited to some organs. As an example, the liver receives almost 100% of the amount of the drug absorbed after oral administration and about 50% after the drug has reached the general circulation. Thus, a significant amount of drug is lost as a fraction distributed in organs where it is useless at best and toxic at worst. Considering that most effects, either pharmacological or toxic, are concentration dependent, the first option is to restrict as much as possible the distribution of the drug to the nontargeted organs. This can be achieved by decreasing its V_D to the lowest value compatible with activity. A limited lipophilicity may also fulfill this goal by decreasing transfer through membranes.

Another possibility is to design chemical structures ionized at physiological pH, i.e., poorly transferable by passive diffusion, but such compounds should retain good oral absorption. An elegant solution is obtained with zwitterionic molecules, which are mostly ionized at plasma pH yet well absorbed from the gut. This is possible when the isoelectric pH of the compound falls within the range of pH variations inside the gut (e.g., cetirizine).[21,22]

Another way to avoid an extensive distribution is to design agents that are bound to plasma proteins (mostly to HSA or to AGP) with an association constant high enough to retain a large amount of drug inside the compartment of distribution of the protein. This design creates the conditions for a restricted distribution (the drug is characterized by a low V_D) avoiding an extensive and ubiquitous distribution of the drug in the body but delivering it only to sites of high affinity.

Such a design redirects the quantitative distribution to selective targets provided that they have a higher retention power for the drug. More precisely, the leading parameter is the binding capacity (C_a) of the protein:

$$C_a = N \cdot K_a \qquad [44]$$

where

$$N = n \cdot C_P \qquad [45]$$

n being the number of available binding sites per mole of protein.

The right solution is to design a compound whose plasma C_a is high enough to prevent the drug from an extensive distribution, but lower than the C_a of the targeted sites. Another example adds to this design the transfer of the plasma-binding carrier and its bound drug to the pathological sites. This is illustrated by the kinetics of many nonsteroidal anti-inflammatory drugs (e.g., profens, mefenamic acid, and oxicams), all of which are acidic drugs ionized at plasma pH and extensively bound to HSA, their corresponding V_D ranging between 0.1 and 0.4 L kg^{-1}. The early inflammation process induces a vasodilatation and an increase in the capillary permeability followed by a plasma transudation including HSA and its bound drug. As a result, the distribution of the drug is directed to its target by the pathological process. When inflammation stops, the impermeability of the capillaries is restored and no further drug is transferred to the previously inflamed tissue.

The best solution to limiting drug distribution in the body remains the discovery of very effective and selective agents. This allows the use of small doses, and limits their amounts in nontarget tissues. Unfortunately, agents with optimal in vitro behavior rarely have the kinetic characteristics necessary to become a drug. Favorable cases are those encountered when the pathological process induces specific enzymes not found in healthy tissues, e.g., the pathological increases in dihydrofolate reductases (DHFRs) and tyrosine kinases (TKs). DHFR inhibitors include trimethoprim (an antibacterial agent), chloroguanide (an antimalarial agent), and methotrexate (an antitumoral agent) while TK inhibitors include the antitumoral agents imatinib, erlotinib, and geftinib. Other specific inhibitors of cell enzymes are the statins and the coxibs.

Specific markers of pathological states, especially of cancers, are generally used as tests for the diagnosis of the disease and the prognosis of its evolution. As they are secreted by the tumor, they may be considered as drug targets. Antibodies are generated against these markers; in some cases they are effective by themselves, and in other cases active drugs are linked to them and thus targeted to the tumor.

A limited distribution may be obtained when the selected targets of a drug and its potential toxic sites are located in separated organs. This is a likely possibility when the target cells are more easily reached from the blood flow than the toxic sites. This is seen with the H1-antihistamines whose therapeutic targets are always on the external sides of intravascular cells, on circulating eosinophils and neutrophils, or on the internal wall of veins and smaller vessels, or in perivascular tissues (connective tissue and basophils).[23] Thus, the kinetics of cetirizine and that of its eutomer levocetirizine[24] show that one can avoid most intracellular side effects of H1-antihistamines, namely inside the heart, brain and liver, by limiting their distribution to their therapeutic target sites. The V_D of cetirizine, for example, is $0.4\,L\,kg^{-1}$, i.e., less than the exchangeable water volume. In other words, the idea is to avoid organ toxicity by limiting the access of the drugs to these organs.[25] Other examples include α_1-antagonists (antihypertensive agents acting on vascular targets), inhibitors of angiotensin-converting enzyme, and angiotensin II receptor antagonists. In these cases, an intravascular distribution is enough to reach their targets, whereas their distribution to tissues serves no purpose.

To summarize, the kinetics of drugs can be improved by developing molecules whose affinity for and activity toward their targets are as high as possible, with a significant but smaller affinity for plasma proteins and the lowest possible tissue affinity.

5.02.6.2 Stable and Predictable Elimination Processes

The effects of a drug are limited in time by the efficacy of physiological clearing functions, namely:

- metabolism when the metabolites are inactive;
- bile secretion of active compounds (parent drug and/or metabolites) when they are excreted fecally without enterohepatic cycling; and
- urinary excretion of active compounds (parent drug and active metabolites).

The liver, kidneys, and lungs (mainly for volatile agents) are the major organs of elimination whose overall clearing activities are expressed by clearance. The respective roles of these organs are investigated by means of two parameters: the first is the fraction of administered dose that they inactive or excrete; and the second is the rate of these processes. Such data are needed for rational prescription of drugs and should be carefully observed by those who prescribe.

For many drugs, target and clearing organs are different. With the exception of some drugs such as diuretics and urinary antibiotics, most are cleared by the liver and/or the kidneys where they have no role to play. The liver and kidneys fulfill their elimination function by very different mechanisms. Metabolic transformations are mainly hepatic and are carried out by enzymes whose activity varies among humans (interindividual variability) and as a function of time for the same individual (intraindividual variability). The fact that these enzymes act on chemically very diverse drugs explains the occurrence of drug interactions when two drugs compete for the same enzymes. Frequently, a competitive interaction ensues where the metabolism of one is decreased by the other (potentiation). On the other hand, enzymatic induction or inhibition may also occur following administration of some drugs or intake of some foods (e.g., grapefruit). All these potential causes of variation explain why the duration of action of drugs when controlled by hepatic enzymes may differ among patients. Such a route of elimination thus brings a certain degree of uncertainty in the duration of action.

When a drug is excreted mainly unchanged in urine, the margin of uncertainty is lessened if renal functions are normal and the pH of urine is controlled. The most likely explanation of the differences between hepatic degradation and kidney excretion is linked to identical renal functions among individuals (creatinine clearance is a stable parameter, $120\,mL\,min^{-1}$), whereas the activities of normal hepatic enzymes differ to a large extent. This observation remains true when renal functions are impaired. If the pathological process is stable with a decreased but stable creatinine clearance, it is possible to adjust drug dosage knowing the relationship between the clearance of the drug and creatinine clearance. This relationship is determined by specific pharmacokinetic studies including normal volunteers and patients with various degrees of renal insufficiency. This is done for instance for digoxin and aminoside antibiotics.

When a choice is possible, physicians are thus likely to prefer a drug excreted mainly unchanged in urine. To avoid hepatic metabolism as much as possible, one needs a low lipophilicity and sterically protected target groups in the molecule, inasmuch as these features are compatible with activity. This objective is valid for all drug classes for which no benefit can be expected from hepatic extraction. This is the case for drugs acting in peripheral organs, for example anti-inflammatory drugs, most antibiotics, antihypertensive agents, and H1-antihistamines. However, it appears that this proposal cannot be applied to drugs acting in the CNS, which must be able to cross the blood–brain barrier and require a sufficient lipophilicity also synonymous with liver uptake (e.g., psychotropic drugs).

5.02.6.3   Linear Pharmacokinetic Parameters

The main interest of pharmacokinetic parameters (CL, V_D, $t_{1/2}$, t_{max}) is that they are constants applicable to all doses. This argument, however, is restricted to drugs whose kinetics is linear. With nonlinear pharmacokinetics, the corresponding parameters vary according to dose. Nonlinearity may be the consequence of two main processes, namely, plasma binding or liver transformation. When the overall plasma concentrations of a drug increase beyond the upper limit of protein binding capacity, the excess of drug cannot be bound. Being free and completely distributed (saturation of the restricted distribution process), it induces an increase of f_u which leads to a modification of pharmacokinetic parameters: V_D increases (see eqn [34]) and $V_{D,u}$ decreases (see eqn [35]). Moreover, CL_R increases and when CL_{int} is low, CL_H increases (see eqn [26]) but $CL_{H,u}$ is not affected. Another cause is a quickly saturated reaction of transformation, which leads to a decrease of CL_H because $CL_{H,u}$ decreases, f_u being constant (see eqn [26]) and either results in accumulation or is replaced by an untoward transformation, for instance, into a toxic reactive metabolite (e.g., acetaminophen).

Nonlinear pharmacokinetics is a handicap for drug prescription. Schematically, infra- and suprasaturating doses exhibit different pharmacokinetic parameters due to different kinetics at enzymes or transporters. Changing doses from infra to supra will induce an increased effect, the intensity of which may be higher than expected.

In summary, the therapeutic use of drugs exhibiting nonlinear pharmacokinetics is not easy but cannot be rejected a priori. Indeed, the threshold concentration beyond which nonlinear processes begin varies among individuals according to their plasma binding capacity and liver enzymatic activities. A simple solution is to use only low, infrasaturating doses, a drug with linear kinetics remaining the best option when a choice exists.

5.02.6.4   One Oral Dose Once a Day

The oral route is obviously the best route of administration especially when chronic treatment is needed. The rule 'one tablet once a day' is very well known, but it requires some specifications. Thus, absorption has to be almost complete, unmodified by food intake, stable according to time, and produce an effect of constant intensity for 24 h meaning that $t_{1/2}$ should be adequate, close to 24 h. Specific formulations may fulfill this objective.

A pharmacokinetic solution to prolonging effects is to design a drug with a restricted distribution that is bound to HSA, and whose release from this protein is delayed. This is achieved with a low dissociation rate of the drug–protein complex, as successfully shown by some sulfonamides, which are active for 2 weeks after a single administration.

5.02.7   Which Pharmacokinetics for Populations at Risk?

Populations at risk involve apparently healthy people as well as those with health problems. The latter group includes the extremes of age, namely embryos, fetuses, newborns, children, and elderly persons. The second group includes patients presenting with either a disease that affects drug elimination or rapidly evolving multipathological processes. It is not possible here to propose specific pharmacokinetic profiles according to the encountered risk, but guidelines can be given to obtain the best benefit/risk ratio, as exemplified below.

During pregnancy, the amount of drug transferred through the placenta to the fetus is limited to the free drug plasma concentration of the mother. The rate of transfer follows Fick's law (eqn [1]). In this specific case, S is the placenta surface available for drug transfer, d is the thickness of this surface, and ΔC represents the gradient of plasma free drug concentrations between the mother and the fetus. During pregnancy, S increases and d decreases, and both modifications favor the transfer of drugs to the fetus. The consequence is that more drug is transferred to the fetus at the end of the pregnancy with a maximum at the time of delivery, S being then maximal and d minimal. From a pharmacological point of view, where extensive drug diffusion to the fetus needs to be avoided, one can reduce its transfer by decreasing the maternal–fetal gradient. This can be achieved by oral administration in small repeated doses distributed over the course of a day. In some specific and rare circumstances, the intravenous route may be used to reach a high gradient, i.e., when the drug must be administered to the fetus via the mother. But in most cases transfer of drug to the fetus is not wanted and can be limited by binding to maternal plasmatic proteins (restricted distribution). Use of such binding can also limit drug transfer to breast milk after delivery. Moreover, as the newborn will not have a functional blood–brain barrier (BBB) until 3 weeks after birth, a strong binding to plasma proteins can be useful to limit CNS distribution when not required therapeutically.

At birth, renal excretion and liver metabolism are not immediately functional but they improve progressively. Unfortunately, many drugs are not adapted to this age group, and clearly the physicians must favor drugs known to be quickly eliminated in adults. Special efforts to design adapted drugs should be encouraged.[26]

In the elderly, several physiological alterations can be observed.[27] The adipose mass is increased even in lean people and liquid volumes are decreased. As a result, hydrophilic drugs will reach higher plasma concentrations and doses must be decreased accordingly. In contrast, lipophilic drugs are stored for longer periods inducing a longer duration of effects in the elderly as compared to adults, also increasing the risk of drug interactions. Moreover, the liver mass is reduced in elderly people and the number of functional hepatocytes is decreased accordingly; hence, metabolic processes are usually slower and hepatic clearance decreased especially when cytochrome P450 dependent since these enzymes are very sensitive to decreased protein synthesis. Conjugation reactions that occur in the cytosol appear to be less modified. This is why, for instance, hydroxylated benzodiazepines are preferred for elderly patients over those that are oxidized in vivo. More generally, drugs having a short $t_{1/2}$ are preferable.

Elderly people frequently develop a chronic renal insufficiency (decreased renal clearance), a situation which must be taken into account when setting the doses of drugs excreted renally.

Hepatocellular insufficiency[28] and chronic renal disease[29] are the two main pathological states that induce a significant alteration of the kinetics of drugs (mainly decreased total plasma clearance). The knowledge of PK parameters in healthy subjects, and more precisely the respective roles of liver and kidneys in overall elimination processes, allows it to be anticipated whether drug elimination will be normal or decreased in a given pathological state. Thus, it is usual to prefer extensively metabolized drugs in cases of renal insufficiency and urine-excreted, unmetabolized drugs in those of hepatic diseases. This procedure is now questioned, as it appears that it is easier to adjust the dosage of a drug excreted in urine during renal insufficiency than to use a drug inactivated by liver enzymes. The best example of this is shown by digoxin and its alternative digitoxin. Digoxin is excreted mainly in urine without transformation (85%) and secondarily as hepatic metabolites (15%). Digitoxin is excreted mainly as inactive metabolites (85%) and partly in urine as the parent drug (15%). However, their $t_{1/2}$ are 1.6 ± 0.5 and 6.7 ± 1.7 days for digoxin and digitoxin, respectively. Both have a low therapeutic index (2 or so) and a dose-dependent toxicity. Digoxin is selected despite renal insufficiency, as the risk of accumulation is lower with this drug than with digitoxin. Here again, renal excretion appears to be more easily controlled than liver metabolism.

It is not possible to define drug kinetics for emergency situations as the encountered pathological states differ vastly and evolve quickly. The selection of the intravenous route bypasses the absorption phase, which is a pharmacological lag phase. Emergency wards tend to use drugs whose effects are immediate and brief. These drugs are mainly hydrophilic, a requisite to intravenous injections, and are excreted by the kidneys.

5.02.8   Conclusion

Besides efficacy and safety, there is now a growing interest in other qualities expected from a new drug. Regularity of effects and facility of use are important factors that improve the quality of life. Adequate pharmacokinetic parameters help to achieve this. Considering the relevance of current pharmacokinetic tools, it is obvious that methodological progress is needed in the screening and evaluation of new chemical entities. One advance is a more precise evaluation of drug distribution in and between organs; V_D affords an idea of drug concentrations in blood and organs, but it cannot establish in which of them the drug is predominantly located. Improvements will probably be afforded by new techniques such as imaging, which allows the respective locations of drugs and their specific targets to be observed using labeled markers and positron emission tomography (PET) scans. The location of a drug inside cells should also be studied, and the following need to be considered: binding to nuclei, which are the starting point of many pharmacological effects; binding to the endoplasmic reticulum as a regulator of Ca^{2+} homeostasis; and binding to mitochondria as the main source of energy transmission (ATP). Subcellular pharmacokinetics should also be developed.

A more ambitious yet realistic research trend involves avoiding the phase of vascular transfer by going directly to the site of action. One solution may be the direct application of drugs inside the targeted organs, permitting the use of high doses and limiting blood transfer and nonspecific distribution. Such projects need new technologies such as nanoparticles and probably new modes of administration adapted to each patient. These are some of the challenges for the near future.

References

1. Sinko, P. J. *Curr. Opin. Drug Disc. Dev.* **1999**, *2*, 42–48.
2. Boobis, A.; Gundert-Remy, U.; Kremers, P.; Macheras, P.; Pelkonen, O. *Eur. J. Pharm. Sci.* **2002**, *17*, 183–193.
3. ICH Topic S3B, Step 5 Note for Guidance on Pharmacokinetics: Guidance for Repeated Dose Tissue Distribution Studies (CPMP/ICH385/95 – adopted November 1994). http://www.ich.org (accessed April 2006).
4. ICH Topic S3A, Step 5 Note for Guidance on Toxicokinetics: A Guidance for Assessing Systemic Exposure in Toxicology Studies (CPMP/ICH384/95 – adopted November 1994). http://www.ich.org (accessed April 2006).
5. Barré, J.; Urien, S. Distribution of Drugs. In *Clinical Pharmacology*; Sirtori, C., Kulhman, J., Tillement, J. P., Vrhovac, B., Reidenberg, M., Eds.; Clinical Medicine Series McGraw-Hill: London, UK, 2000, pp 37–43.
6. Hervé, F.; Gomas, E.; Duché, J. C.; Tillement, J. P. *Br. J. Clin. Pharmacol.* **1993**, *36*, 241–249.
7. Tillement, J. P.; Houin, G.; Zini, R.; Urien, S.; Albengres, E.; Barré, J.; Lecomte, M.; d'Athis, P.; Sébille, B. *Adv. Drug Res.* **1984**, *13*, 60–90.
8. Urien, S.; Tillement, J. P.; Barré, J. In *Pharmacokinetic Optimization in Drug Research*; Testa, B., van de Waterbeemd, H., Folkers, G., Guy, R., Eds.; VHCA: Zurich, Switzerland, 2001, pp 189–215.
9. Brodie, B. B.; Kurtz, H.; Schanker, L. J. *J. Pharmacol. Exp. Ther.* **1980**, *130*, 20–25.
10. Bechtel, P.; Bechtel, Y. In *Clinical Pharmacology*; Sirtori, C., Kuhlmann, J., Tillement, J. P., Vrhovac, B., Reidenberg, M., Eds.; Clinical Medicine Series; McGraw-Hill: London, UK, 2000, pp 25–36.
11. Gibaldi, M.; Perrier, D. *Pharmacokinetics,* 2nd ed.; Dekker: New York, 1982.
12. Rowland, M.; Tozer, T. N. *Clinical Pharmacokinetics – Concepts and Applications*, 3rd ed.; Williams & Wilkins: Baltimore, MD, 1995.
13. Shargel, L.; Yu, A. B. C. *Applied Biopharmaceutics et Pharmacokinetics*, 4th ed.; Appleton Century Crofts: New York, 1984.
14. Wilkinson, G. R. *Pharmacol. Rev.* **1987**, *39*, 1–47.
15. Smith, D. A.; van de Waterbeemd, H.; Walker, D. K. *Pharmacokinetics and Metabolism in Drug Design*; Wiley-VCH: Weinheim, Germany, 2001, pp 19–37.
16. Tillement, J. P.; Albengres, E.; Barré, J.; Rihoux, J. P. *Dermatol. Ther.* **2000**, *13*, 337–343.
17. Balant, L. P.; Gex-Fabry, M.; Balant-Gorgia, E. A. In *Clinical Pharmacology*; Sirtori, C., Kuhlmann, J., Tillement, J. P., Vrhovac, B., Reidenberg, M., Eds.; Clinical Medicine Series; McGraw-Hill: London, UK, 2000, pp 9–24.
18. C.P.M.P. Committee For Proprietary Medicinal Products. Note for Guidance on the Investigation of Bioavailability and Bioequivalence. C.P.M.P/EWP/QWP/1401/98 (adapted July 2001); The European Agency for the Evaluation of Medicinal Products: London, 2001. http://www.emea.eu.int/index/indexh1.htm (accessed April 2006).
19. Regulatory and Scientific Guidances. http://www.fda.gov/cder/regulatory/default.htm (accessed July 2006).
20. http://www.ich.org (accessed July 2006).
21. Pagliara, A.; Testa, B.; Carrupt, P. A.; Jolliet, P.; Morin, D.; Morin, C.; Urien, S.; Tillement, J. P.; Rihoux, J. P. *J. Med. Chem.* **1998**, *41*, 853–863.
22. Baltes, E.; Coupez, R.; Giezeck, H.; Voss, G.; Meherhoff, C.; Strohlin-Benedetti, M. *Fund. Clin. Pharmacol.* **2001**, *15*, 269–277.
23. Tillement, J. P.; Albengres, E. *Eur. Ann. Allergy Clin. Immunol.* **1996**, *28*, 1–4.
24. Tillement, J. P.; Testa, B.; Brée, F. *Biochem. Pharmacol.* **2003**, *66*, 1123–1126.
25. Tillement, J. P. *Allergy* **1995**, *50*, 12–16.
26. Milne, P. C. The Pediatric Studies Initiative. In *New Drug Development: A Regulatory Overview*, 6th ed; Mathieu, M., Ed.; Parexel: Waltham, MA, 2002, pp 303–327.
27. Beers, M. H.; Ouslander, J. G. *Drugs* **1989**, *37*, 105–112.
28. C.P.M.P/EWP/2339/02 Guideline on Evaluation of the Pharmacokinetics of Medicinal Products in Patients with Impaired Hepatic Function (CHMP adopted February 2005). http://www.emea.eu.int/index/indexh1.htm (accessed April 2006).
29. C.P.M.P/EWP/225/02 Note for Guidance on the Evaluation of the Pharmacokinetics of Medicinal Products in Patients with Impaired Renal Function (CHMP adopted June 2004). http://www.emea.eu.int/index/indexh1.htm (accessed April 2006).

Biographies

Jean-Paul Tillement is MD, PharmD MSc, Professor of Pharmacology. From 1985 to 2003 he held the position of Full Professor of Fundamental and Clinical Pharmacology and Head of Department at the Faculty of Medicine Paris-XII. His responsibilities during this time included teaching and research at the Faculty, as well as activities in clinical pharmacology as part of drug treatments of patients at the University Hospital as a qualified biologist Head of the Hospital Department of Pharmacology. He held various national appointments as an expert at the French Ministry of Health (still ongoing) and public research structures (INSERM, INRA), and at the French Ministry of Justice (still ongoing). Prof Tillement was also involved as an expert in the COST project, Brussels, Belgium (Criteria for the choice and definition of healthy volunteers and or patients for phase I and II studies in drug development, 1995). He has organized international scientific meetings especially on drug–protein interactions, his first recognized field of expertise, which he has in parallel broadened to include the pharmacological preservation of mitochondrial functions. He is past-president of the Société Française de Pharmacologie Clinique et Thérapeutique; past member of the jury of Prix Galien (1988–2005); current member of 10 medical and scientific societies; of the board of 9 international journals; current referee of 7 international journals; full member of the French National Academy of Pharmacy and corresponding member of the French National Academy of Medicine; Adjunct Professor at the Georgetown University Medical Center (Washington DC, USA) since 2002. He serves as a consultant in pharmacology to pharmaceutical companies. He has published more than 560 scientific works (December 2005) consisting in Research & Educational papers, Reviews and Books, Communications and Posters, and conducted 35 thesis.

Dominique Tremblay, PhD (Pharmacy), has obtained certificates in General and Human Biochemistry, and in Statistics Applied to Medical Biology, Clinical Pharmacology and Pharmacokinetics. From 1991 to 1998, he held the position of director of Preclinical Development at Hoescht-Marion-Roussel in Romainville (France). In this position he was in charge of the departments of Toxicology, Pharmacokinetics (preclinical and clinical), Quality, Pharmaceutics, Quality Assurance and Scientific Coordination and Documentation. He actively participated in the development and introduction of various drugs (antibiotics, hormones and antihormones, nonsteroidal antiinflammatory drugs, and cardiovascular drugs) into the international market. After he left Hoescht-Marion-Roussel, he became an expert in Pharmacokinetics, Toxicokinetics and Toxicology at the French medical agency (Agence Française de Sécurité Sanitaire des Produits de Santé, AFSSaPS). At present, he is a member of the Marketing Committee Authorization and chairman of the Preclinical Working Party at the AFSSaPS, and a member of the Safety Working Party at the European Agency for the Evaluation of Medicinal Products (EMEA). He gives lectures at the University of Paris on toxicokinetics and pharmacokinetics. He participates as a tutor in the Pharmacokinetic Workshop under the guidance of Professors Malcolm Rowland and Thomas Tozer. Dominique Tremblay's interests are in experimental and applied pharmacokinetics. He has published about 50 scientific publications, abstracts and posters in this area. His main field of interest is how pharmacokinetics and toxicokinetics can be best used to predict and/or understand the effects of drugs on both man and animal in terms of efficacy and safety.

Comprehensive Medicinal Chemistry II
ISBN (set): 0-08-044513-6

ISBN (Volume 5) 0-08-044518-7; pp. 11–30

5.03 In Vivo Absorption, Distribution, Metabolism, and Excretion Studies in Discovery and Development

T Lavé and C Funk, F. Hoffmann-La Roche Ltd, Basel, Switzerland

© 2007 Elsevier Ltd. All Rights Reserved.

5.03.1   Objectives of Absorption, Distribution, Metabolism, and Excretion (ADME) Studies

In recent years, there has been an increased awareness about the importance of pharmacokinetics and metabolism data in all stages of the drug discovery and development process. In order to understand better the impact of in vivo drug metabolism and pharmacokinetic studies, it is important to first outline the major objectives of drug metabolism and pharmacokinetics (DMPK) when applied at the discovery and nonclinical development stages.

5.03.1.1   Support to Drug Discovery

DMPK is now a routine part of the lead identification, lead optimization, and clinical candidate selection processes. At these different stages an appraisal of DMPK issues alongside other 'developability' factors is performed. Thus, the deployment of available in vivo and in vitro methods is based on a sound understanding of the relevant DMPK issues and not simply on the capacity to process large numbers of compounds.

5.03.1.2   Support to Drug Metabolism

The main goal of drug metabolism or biotransformation for the body is to eliminate potentially harmful xenobiotics via urine and/or bile. This is typically realized in a stepwise process: the often lipophilic drug molecules are metabolized (biotransformation) to typically inactive, nontoxic, and more hydrophilic products, which can then be readily excreted in urine or bile.[1] In some cases, however, drug metabolites might be toxic or might represent activated products (e.g., acylglucuronides), which can potentially lead to organ toxicity or immune-mediated toxicity. Therefore, an extensive knowledge of drug metabolism is required for an overall understanding of the pharmacological and safety properties of new drug molecules.[2]

5.03.1.3   Support to Pharmacodynamics

The monitoring of pharmacodynamic studies for exposure and the rigorous design of pharmacokinetic/pharmacodynamic (PK/PD) studies in animals are key to a number of questions that need to be addressed during preclinical development. These questions include: (1) identification of potential pharmacodynamics endpoints for efficacy in animal models; (2) development of mechanism-based models for efficacy; (3) determination of in vivo potency and intrinsic activity and prediction in humans; (4) dosage form and dosage regimen optimization; and (5) supporting dose selection for phase 1 studies.

Thus, the use of PK/PD modeling in early preclinical development to define the dose-concentration-pharmacological effects and dose–concentration–toxicity relationships, as well as the extrapolation of these results to humans using a combination of in vitro and in vivo data, can be particularly helpful in determining the appropriate dosing regimen for phase 1 studies. Preclinical PK/PD studies may also prompt a series of important mechanistic studies to explore any dissociation between plasma concentration and duration of pharmacological effect (i.e., active metabolites or long half-life in the effect compartment, etc.). For PK/PD models and their use in drug discovery (*see* 5.38 Mechanism-Based Pharmacokinetic–Pharmacodynamic Modeling for the Prediction of In Vivo Drug Concentration–Effect Relationships – Application in Drug Candidate Selection and Lead Optimization).

5.03.1.4   Support to Toxicology

An important function of animal DMPK studies is in support of preclinical safety evaluation. Thus, determination of pharmacokinetics and exposures during the course of a toxicological experiment is key to interpreting toxicological findings. For example, it is essential to compare exposures in toxicological studies with the exposures expected or achieved in humans to ensure sufficient safety margin. Furthermore, metabolite profiles obtained from the species used in toxicology studies are compared to those in humans. In this case, emphasis is placed on qualitative similarity in metabolite profiles between humans and species used in toxicology studies in order to ensure that both are exposed to the parent drug as well as the same metabolites, any of which may contribute to toxicity.

5.03.1.5   Support to Formulation Testing

During the discovery and development process, the formulation scientist develops formulations to ensure appropriate dosing of the test compound during early DMPK, safety, and pharmacodynamics studies. Various formulations also should be investigated when the in vivo plasma concentration versus time profiles need, for example, to be extended with controlled release dosage forms and when oral bioavailability needs to be improved. In this context, formulations with a desirable pattern in vitro are submitted to in vivo oral absorption studies in animals. Once a favorable plasma concentration versus time profile in animals is achieved, the formulations are developed for testing in humans.

The judicious selection of animal species to support formulation testing is of key importance at this stage. To this end it is important to understand the similarities and differences of the physiology of the gastrointestinal (GI) tract between animals and humans. These issues are discussed in detail later in this chapter. However, it is a fact that the dog has become the most commonly used species for bioavailability studies in the context of formulation testing. Because of some species differences between humans and dogs, the findings in dogs and in other animal species have to be interpreted cautiously for their relevance to humans.

5.03.2 In Vivo Drug Metabolism and Pharmacokinetics Screening Studies

In recent years, several pharmaceutical companies have published reports on the simultaneous administration of several compounds to a single animal (cassette dosing or 'N-in-One' dosing)[3–7] as a means to rapidly rank order compounds on the basis of their in vivo pharmacokinetic properties. Compared with conventional pharmacokinetic studies, this method has the advantage of speed, because the slow steps of animal dosing, blood collection, and sample analysis are minimized. Another advantage is that animal usage is greatly reduced. The enabling technology for cassette dosing is liquid chromatography coupled to tandem mass spectrometry (LC/MS), which allows many compounds to be analyzed simultaneously.

Although cassette dosing has been reported to yield useful results when used as a screen, especially to rank order drug candidates, the potential for large errors was shown recently both theoretically and experimentally.[4] Consequently, it is recommended that the pharmacokinetic parameters derived from cassette dosing are interpreted very cautiously.

5.03.3 Pharmacokinetic Studies

A pharmacokinetic study involves dosing and sampling of animals or subjects, bioanalysis of the biological samples (e.g., blood, plasma, tissues) and analysis of the resulting blood, plasma, or serum concentration versus time data using for example noncompartmental or compartmental pharmacokinetic methods. Low aqueous solubility of molecules often necessitates significant formulation work prior to dosing. The bioanalytical or assay phase typically involves sample extraction with organic solvents and LC/MS or LC/MS/MS separation and detection of analytes. During drug discovery, pharmacokinetic studies are most commonly conducted in rodents and/or the species used for the assessment of in vivo efficacy. Subsequently, experiments in larger animal species such as dog or monkey are performed to further characterize the compound of interest, to support toxicology studies, and to generate data useful in predicting human pharmacokinetics (see discussion on scaling below). The most common routes of compound administration are oral and intravenous.

5.03.4 Metabolism Studies

Two aspects are of importance in the context of drug metabolism studies. The first aspect is the chemical nature of the metabolites formed in respect to their safety profile. All metabolites that are formed and are systemically available in man should ideally also be formed and reach at least similar systemic exposure in one animal species used in the preclinical toxicity program.[8] Second, the enzymes mainly responsible for the primary metabolic clearance steps might exhibit species-specific expression patterns, inducibility, or inhibition potential. Many of the clinically relevant drug–drug interactions are based on an induction or inhibition of enzymes or other active processes involved in drug metabolism and elimination (e.g., transporters).

While numerous in vitro tools are available to study the main processes involved in drug metabolism and the metabolites formed, quantitative correlations between in vitro and in vivo metabolism are often difficult to make.[9] In the early lead optimization process, the metabolic stability, biotransformations, and metabolite structures of new chemical entities (NCEs) are typically studied first using appropriate in vitro tools such as expressed human enzymes or cellular systems (e.g., hepatocytes).[9] Often a ranking of compounds in terms of metabolic stability and metabolites formed is sufficient as output from these in vitro studies.[9] Later on during drug development, differences in the systemic exposure to metabolites and the overall excretion pathways have to be addressed. As compounds are increasingly optimized for metabolic stability, extrahepatic metabolic routes and direct excretion of unchanged drug molecules are often seen, requiring in vivo studies in animals and man. The completeness of excretion, the excretory pathways, and the metabolite patterns in plasma and excreta are studied sequentially in the animal species involved in toxicity studies and finally in man using radiolabeled drugs. This chapter will focus on in vivo metabolism studies; in vitro studies are discussed in detail in Chapter 5.10.

5.03.4.1   The Use of Radiolabeled Compounds

Radiolabeled drug molecules are useful for most in vitro and in vivo drug metabolism studies. The total sum of all drug-related molecules can be easily quantified in different biological matrices and the recoveries of sample work-up procedures can be determined. Also, chromatographic separation of the individual metabolites with easy quantification and detection of unknown metabolites for which no chemical standard is available is greatly facilitated by radiolabeled isotopes.[10] Therefore, radiolabeled drugs are used in most in vivo ADME studies; ^{14}C-labeled compounds are preferred for theses studies due to the higher metabolic stability of this isotope in the molecule as compared to the ^{3}H label. The isotope is typically placed on the metabolically stable core group of the molecules. However, for more specific questions or depending on the metabolic steps involved, the label might be placed both on stable and labile moieties. Furthermore, double-labeled compounds with different isotopes (^{13}C/^{14}C or ^{3}H/^{14}C) might be synthesized to aid in metabolite identification and quantification of the individual moieties.[10]

5.03.4.2   Chromatographic Separation and Quantification of Drug Metabolites

High-performance liquid chromatography (HPLC) is the method of choice to analyze and quantify the parent drug and any metabolites, both from in vitro drug metabolism assays or in vivo animal studies using radiolabeled compounds. A new technique with a much higher sensitivity for the detection of ^{3}H and ^{14}C isotopes is accelerated mass spectroscopy (AMS), which has potential especially in human ADME studies.[10–12] ADME studies in man can be performed with radioactive doses of very low specific activity, the total amount of radioactivity being in the range of nanocuries. Information on the completeness of excretion (mass balance), the excretion pathways, and also limited information on metabolites formed can be obtained very early on in drug development.

5.03.4.3   In Vivo Animal Models Used for Absorption, Distribution, Metabolism, and Excretion Studies

A number of different animal models are used for in vivo drug metabolism studies. Nonoperated, naive animals are used for excretion balance studies and to assess the metabolite profiles in plasma, urine, and feces. For compounds with a high proportion of biliary elimination of total drug-related material, bile duct-cannulated animals are often used in addition. The total gastrointestinal drug absorption, the importance of first pass metabolism, and also secretion of drug and metabolites into the gastrointestinal tract (after intravenous administration) can be evaluated in this animal model.[10] However, operated rats often show inflammatory reactions leading to induced or repressed phase I and phase II enzyme systems (cytochrome P450 and glucuronyltransferases) thereby significantly altering metabolic capacity. For specific questions, knock-out animals or strains with deficiencies in certain metabolic enzymes or transporters can be used to complement in vitro findings. P-glycoprotein knock-out mice are used to study the contribution of this transporter for limitations in brain uptake and to a minor degree also drug absorption in the GI tract. Humanized mice can be used to study the impact of the respective cytochrome P450 isoenzymes on the metabolism and the drug–drug interaction potential in an in vivo environment.[13]

5.03.4.4   Mass Balance and Route of Excretion Studies

In animals the full urinary and fecal balance is typically assessed at different developmental stages in the main species involved in safety studies, e.g., rats and dog or monkey.[8] The nonrodent species used for teratology studies, typically the rabbit, and the second species used for carcinogenicity studies, typically the mouse, are not routinely investigated to this extent.[14] The results are used for the preparation of the ADME study using a radiolabeled dose in man, which is typically initiated during clinical development phase I.

The mass balance study in rats is most often performed prior to initiation of clinical phase I studies[14] in order to evaluate the kinetics and completeness of excretion and the excretion routes in this species. Both oral and intravenous administration are studied for most compounds and three to five animals are typically used per administration mode. The doses selected are within the pharmacologically relevant range and typically about 20 μCi of radiolabel are used per animal enabling enough sensitivity for quantification. A 1:1 mixture of cold and radiolabeled drug is ideal for later metabolite identification by LC/MS/MS. If the excretion is incomplete within 5 to 7 days (<90–95% recovery), the remaining carcasses are analyzed after dissolution for any remaining radiolabel. For the second species used (dogs or monkeys) the excretion balance study is often performed later, in parallel to the phase I studies, in preparation for the human mass balance study. Urine and feces samples from both rat and nonrodent studies can be further used for the analysis of metabolite patterns. Plasma samples are typically not taken from rats, so as not to interfere with the main objective of the study, the completeness of mass balance; for larger animals, however, samples can be taken for additional analyses.

5.03.4.5   Metabolite Profiling in Animal Species and Man

One of the main goals of in vivo ADME studies is the evaluation of metabolite profiles in plasma and excreta and to investigate any species differences.

5.03.4.5.1   Systemically available metabolites in plasma

The plasma exposure of parent drug and major metabolites along with any related interspecies differences are of special interest for interpretation of toxicology or carcinogenicity studies. However, it is only data from the radiolabeled human ADME study that allows safety margins for all the relevant metabolites to be finally established, as only then the systemic availability of the metabolites formed in man is known. It is possible to detect known or postulated metabolites early on in plasma samples from the first clinical studies in man with newer, more sensitive, and selective technologies such as LC/MS/MS. This technology, however, does not allow the quantification of all those metabolites for which no reference material is available. Furthermore, it is extremely difficult to detect all unknown metabolites.

Special attention should be drawn to major human metabolites, which account for a considerable amount of the AUC (area under the plasma concentration–time curve) relative to the parent drug, and metabolites, that are only seen in man.[15] For both types of metabolites, a quantitative analytical assay should be set up in order to establish their kinetics and exposure in animals and man. For the human-specific metabolites and metabolites that do not reach comparable systemic exposure in at least one animal species used in the different toxicity studies, separate toxicity studies should be considered.[15]

Major metabolites in plasma are not necessarily major metabolites of the overall metabolism, as both the rates of formation and elimination as well as other kinetic parameters are important factors. An example of species-specific differences in the exposure of a major metabolite is given in **Figure 1**. Most of the radiolabeled test drug was eliminated in the feces in both man and cynomolgus monkey. The oxidative metabolite D was formed in addition to other oxidative metabolites in both the monkey and man to a similar extent (~15% of the dose) based on analysis of pooled feces samples. However, this metabolite was only a minor peak in cynomolgus monkey plasma, while in man it reached systemic plasma exposures comparable to, or even exceeding, that of the parent drug. Species-specific differences in the rates of elimination of this metabolite might be one reason for this observation, which was only apparent once human plasma samples were analyzed quantitatively for these metabolites.

5.03.4.5.2   Metabolites found in excreta

The main aim of analyzing metabolites in excreta is to quantitatively evaluate the contribution of the different excretion routes and metabolic pathways to the overall elimination and metabolism of a test drug. This information is used to estimate the fraction metabolized by various pathways. Furthermore, metabolites can be purified from urine for elucidation of their structures. Purification from bile and feces, although more difficult, can also be successful.

For mostly renally cleared drug-related material, typically smaller molecular weight entities or some phase II metabolites, analysis and quantification is in most cases easy to accomplish. For compounds that are mainly eliminated via bile and feces, the metabolite patterns should be studied in both bile and feces for an overall evaluation of the nature and route of drug metabolites excreted. A feasible approach in the rat is to analyze urine and feces samples from nonoperated animals. This can easily be done with samples from the excretion balance study, allowing an overall quantitative assessment of the individual metabolites in excreta. In parallel studies bile and feces samples from bile duct-cannulated animals after intravenous and oral drug administration should be analyzed. This helps to understand the nature of biliary eliminated metabolites, their stability in feces, the fraction absorbed, the effect of first pass metabolism and bioavailability, and the potential for direct secretion of drug/metabolites into feces.[10] A similar approach can be taken for dogs and monkeys, while for man the analysis of excreta is typically limited to urine and feces. It must be emphasized, however, that bile duct-cannulated animals often show altered metabolic activities or incomplete excretion within the study period. Therefore, it is very important to compare these observations with those in naive animals.

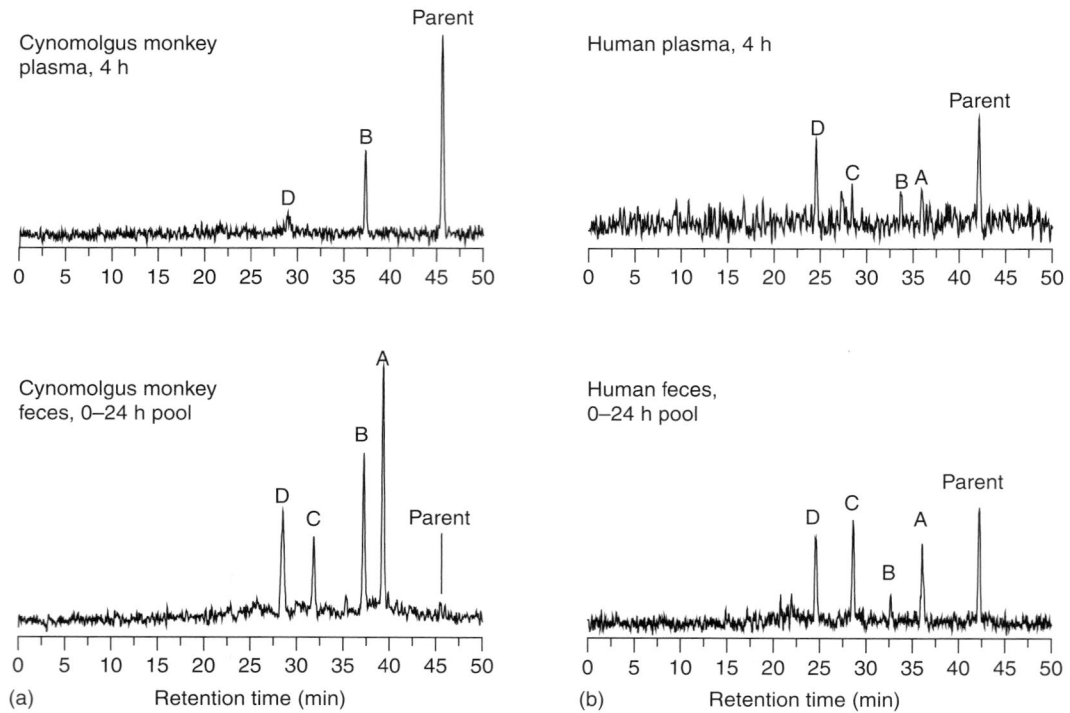

Figure 1   Metabolite patterns in plasma and feces collected from cynomolgus monkey (a) and human (b) excretion balance studies. The identity of parent drug and metabolites A, B, C, and D were confirmed by LC/MS/MS analysis.

5.03.4.5.3   Limitations of animal models for drug metabolism

Species differences in drug metabolism limit the value of preclinical in vivo studies. Such differences can be either of a quantitative (same metabolic route although at different relative rates) or qualitative (different metabolic routes, different metabolites formed) nature.[16] No single in vivo animal model can be said to be most suitable for man. Species differences are discussed in more detail at the end of this chapter.

A different type of limitation of animal models is caused by inflammatory reactions often observed in operated animals. Upon cannulation acute phase response markers, such as α_1-acid glycoprotein and tumor necrosis factor (TNF), can be observed.[17,18] As a consequence of this different groups of drug-metabolizing enzymes are either induced or repressed, resulting in significantly altered drug kinetics and metabolic pathways (see for example **Figure 2**).

In nonoperated rats, the parent drug was metabolized by cytochromes P450 (CYPs) to M1 and both parent drug and M1 were conjugated with glucuronic acid. M1 and M1-glucuronide were the main peaks observed in a liver homogenate prepared from naive rats 5 min after dosing. However, in bile duct-cannulated rats, 4 days after the operation, metabolization to M1 was hardly observed, while the main product of metabolism was parent drug conjugated with glucuronic acid. The significantly reduced activity of the CYP enzyme involved could be demonstrated in vitro using liver microsomes prepared from bile duct-cannulated rats (**Figure 3**).

5.03.4.6   Covalent Binding Studies: Reactive Metabolites

Although most drug metabolites formed are stable and can be readily detected, some metabolites are short lived and are reactive intermediate products. For these metabolites other detection methods have to be used, such as trapping with glutathione or measuring stable degradation products.[19] In cells these reactive products are typically trapped by glutathione. These very reactive metabolites can readily react with proteins resulting in covalent drug-protein adducts that can potentially trigger immune-mediated idiosyncratic toxicity.[20] These reactions might also occur in situations where glutathione is depleted. Such underlying processes are typically addressed early on in the drug discovery and

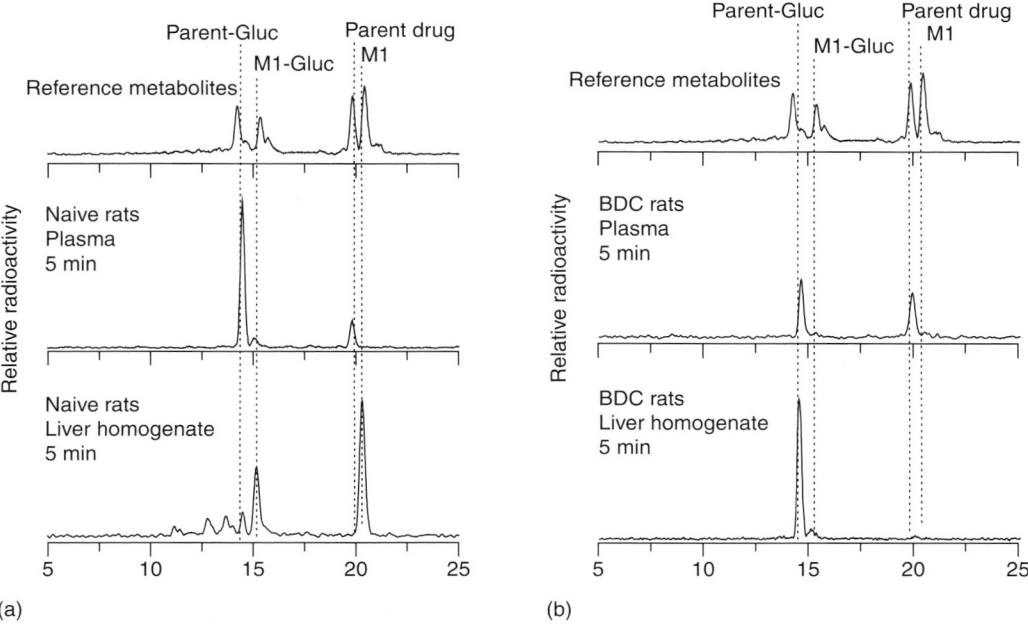

Figure 2 Metabolite patterns from plasma and liver homogenate of naive (a) and bile duct-cannulated (b) rats, compared to in vitro metabolite patterns.

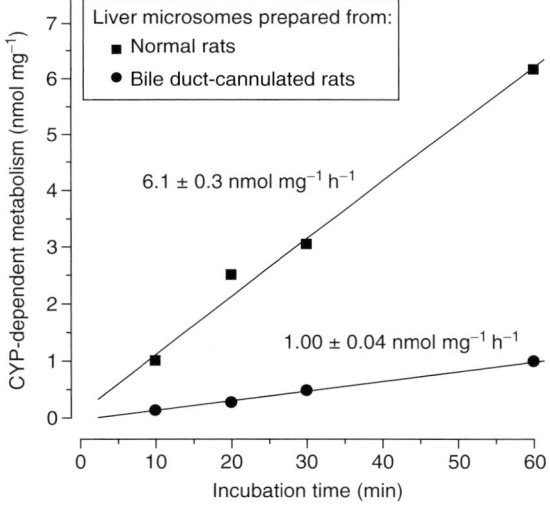

Figure 3 Cytochrome P450-dependent in vitro metabolism using liver microsomes prepared from nonoperated rats and animals 5 days after bile duct cannulation.

development process using appropriate in vitro assays.[19] In vivo studies can help to assess the relevance of these findings for the in vivo situation. The radiolabeled drug, covalently bound to plasma, liver, or microsomal proteins can be quantified and the binding can be compared relative to the dose and in relation to the binding of positive test compounds under the same conditions.[10]

5.03.4.7   Enzyme Induction and Inhibition Studies

Studies addressing the enzyme induction and inhibition potential for new test drugs are typically performed in specific in vitro studies, since studies with animals are of limited predictive value for man for these two mechanisms.

The inhibition of major drug-metabolizing enzymes of the cytochrome P450 family by a test drug is studied using human liver microsomal preparations or the recombinantly expressed human enzymes. Based on the result, in vivo studies are then performed to study the relevance of the in vitro findings. The affected enzymes can be studied using selective marker substrates for the inhibited isoenzyme, or by using a 'cocktail' addressing all major cytochrome P450 enzymes in one study.[21,22] Animal experiments are generally not considered as an alternative to human studies due to significant differences in substrate specificities.[21,23] Simulation of drug–drug interactions are reviewed elsewhere in this book (see 5.35 Modeling and Simulation of Pharmacokinetic Aspects of Cytochrome P450-Based Metabolic Drug–Drug Interactions).

Enzyme induction by a new test drug might have already been observed in multiple dose toxicity studies by decreasing drug exposure over time. However, the regulation of drug-metabolizing enzymes in animals and man is different, making animals a poor model for enzyme induction studies. In addition induction of a number of isoenzymes might be overlooked if the test compound is not significantly metabolized by these isoenzymes. Therefore, the enzyme-inducing properties of a new test drug are typically studied using human hepatocytes.[23] Such information is usually required early on in drug development to exclude potentially harmful comedications from the first phase of clinical studies.

5.03.5   Mechanistic Pharmacokinetic Studies

5.03.5.1   Intestinal Absorption Studies

Absorption studies can be carried out using a variety of dosing routes. In this chapter we will focus on intestinal absorption since oral dosing is the most common route of administration for drugs. Models for absorption by nonoral routes are in Chapter 5.12.

There are a number of different approaches to estimate absorption. Artificial membranes such as parallel artificial membrane permeability assay (PAMPA) or cell-based assays using Caco-2 and Madin–Darby canine kidney (MDCK) cell lines are commonly utilized for assessing the intestinal permeability of a drug. These methods are well suited to high- or medium-throughput screening and have been shown to produce permeability values that correlate with intestinal absorption in man.[24–27] However, these cell lines are mostly used to predict passive absorption, and the results obtained are greatly affected by experimental parameters such as pH and co-solvents. In silico and in vitro transport models for oral absorption are discussed in detail elsewhere in this book (see 5.28 In Silico Models to Predict Oral Absorption; 5.11 Passive Permeability and Active Transport Models for the Prediction of Oral Absorption). The in situ perfusion approaches described in this section provide experimental conditions closest to those encountered following oral administration, with a lower sensitivity to pH variations due to a preserved microclimate above the epithelial cells.[28] In this model, the animal is anesthetized and a segment of gut is cannulated and perfused with a solution containing the test drug. The concentrations measured in the perfusate before and after perfusion are used to quantify the absorption of the compound. These techniques maintain an intact blood supply to the intestine, and can be used to estimate the impact of clearance pathways, such as enzymes and transporters, that are present in the gut. Moreover, drug permeability[29] and expression of drug-metabolizing enzymes and transporters have been shown to vary along the intestinal tract, which can be investigated using intestinal perfusion of the various regions. In addition, intestinal permeability measured in rats with this perfusion technique was recently reported[28] to correlate very closely with that obtained in man (**Figure 4**).

5.03.5.2   Liver and Intestinal First Pass Studies

The use of hepatic portal vein-cannulated animals can be helpful in determining specific causes of poor bioavailability. Numerous factors contribute to the low and erratic oral bioavailability of drugs such as low aqueous solubility, poor dissolution properties, and poor apparent permeability due to intrinsically low absorptive membrane permeability. Extensive CYP3A-mediated metabolism in the liver and, to some extent, in intestinal tissues is also considered a significant factor. More recently, secretory membrane transporters (e.g., P-glycoprotein (P-gp) and MRP2) have also been implicated in controlling the oral bioavailability and variability of drug absorption. The oral bioavailability (F) can be described as:

$$F = f_a \cdot (1 - f_g) \cdot (1 - f_h) \qquad [1]$$

where f_a is the fraction of dose absorbed from the gastrointestinal lumen and f_g and f_h are the fractions of drug metabolized by the gut wall and liver, respectively, during the first pass. In order to gain insights into the absorption

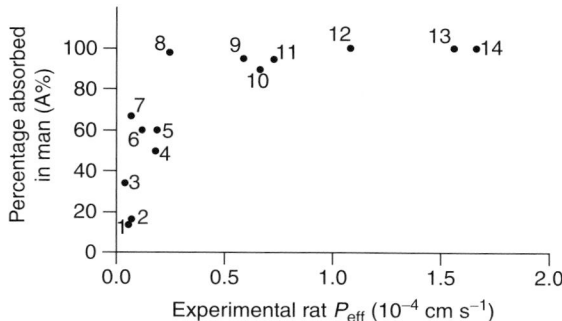

Figure 4 Percentage of drug dose absorbed in man (A%) versus rat P_{eff} (effective intestinal permeability) value. 1, Sulfasalazine; 2, aciclovir; 3, nadolol; 4, atenolol; 5, furosemide; 6, terbutaline; 7, hydrochlorothiazide; 8, cefalexin; 9, metoprolol; 10, propranolol; 11, antipyrine; 12, ketoprofen; 13, dexamethasone; 14, naproxen.

processes and the main factors limiting the oral bioavailability it is possible to perform dosing and/or sampling of the hepatic portal vein in animals in addition to the traditional methods of oral and intravenous dosing coupled with intravenous sampling. Thus, for example, f_g and f_h can be easily derived from these experiments and the respective contribution of gut and liver first pass effects can be easily quantified. Studies to investigate the relative contributions of the gut and liver to the first-pass loss of a number of compounds have been performed using in vivo intestinal-vascular access port models in various species including rat, dog, and rabbit. In order to assess and differentiate the roles of the intestine and liver with respect to metabolism and secretion, test compounds needed to be coadministered with specific P-gp and CYP3A inhibitors.

5.03.5.3   Studies to Assess Blood–Brain Barrier Permeation

Drugs that are effective against diseases in the central nervous system and reach the brain via the blood compartment must pass the blood–brain barrier (BBB). This is considered to be the most important barrier for drug transport to the brain because its surface area is 5000 times larger than the blood-cerebrospinal fluid barrier located at the choroid plexus.[30] Compounds can cross the BBB by passive processes including paracellular transport for small hydrophilic compounds and transcellular transport for lipophilic compounds. Other possibilities include transcytosis with involvement of transporters. A number of transporters have been studied. Among these, the discovery of the presence of P-gp has greatly contributed to our understanding of the transport of compounds into the brain.[30]

Various methods are available to study the transport of compounds into the brain including in silico (*see* 5.31 In Silico Models to Predict Brain Uptake), in vitro (*see* 5.13 In Vitro Models for Examining and Predicting Brain Uptake of Drugs), and in vivo methods. The in vivo methods available to study brain transport include the single carotid injection technique, the internal carotid artery perfusion technique, the intravenous injection technique with brain tissue of cerebrospinal fluid (CSF) sampling, and in vivo brain microdialysis. These methods have been reviewed recently[31] and are summarized in this section.

5.03.5.3.1   Carotid artery injection technique

With this technique the test compound is rapidly injected or perfused into the carotid artery[32] This technique has been applied to various animal species including rats, mice, and guinea pigs. The brain is collected 5–15 s after administration of the compound to assess brain permeability. This technique has been widely used to study BBB permeability and the role of protein binding and transport in penetration to the brain.

5.03.5.3.2   Intravenous injection technique with brain collection

In this case, the compound is usually administered in the femoral vein through bolus injection or through infusion. Then, animals are sacrificed at different time points to allow brain and blood collection. The concentrations measured in brain and plasma are then used to assess brain permeability or determine the brain:plasma ratio at steady-state.

5.03.5.3.3   Cerebrospinal fluid sampling technique

The determination of CSF concentration has been attractive because of its low protein content and analysis of CSF samples is simple. Furthermore, there have been a number of examples indicating that CSF exposure might be related to exposure at the site of action. However, these observations cannot be generalized as a number of examples have also shown that the permeability at the blood-CSF barrier (the choroid plexus) could also be different from the BBB permeability.

For drugs that are passively transported, the CSF concentration may approximate unbound concentration in plasma. Jezequel[33] compared the CSF to plasma concentration ratio with the free fraction in plasma for 50 drugs. About 50% of the compounds had a CSF to plasma concentration ratio similar to free fraction in plasma. For the other compounds, the unbound concentration in CSF was usually lower than the corresponding unbound concentration in plasma, most likely as a result of the involvement of active transport processes at the blood-CSF barrier.

5.03.5.3.4 Intracerebral microdialysis

These techniques have been reviewed recently.[30] Intracerebral microdialysis involves the stereotactic implantation of a microdialysis probe in the brain. The probe comprises a semipermeable membrane partly covered with impermeable coating. The probe can be positioned at a specific site in the brain and may be used to sample extracellular fluid but also to deliver compound to the brain. One important issue of microdialysis experiments is the careful determination of in vivo concentration recovery for each experiment to characterize the relation between brain extracellular concentration and dialyzate concentration.

Intracerebral microdialysis is particularly suitable to estimate extracellular unbound drug concentration as a function of time. In addition unbound concentrations can be measured at the same time in blood and compared to unbound brain concentration to characterize drug transport across the BBB. The microdialysis technique offers the possibility to take multiple samples from individual animals and to sample from different brain regions including diseased brain sites for example.

A number of examples illustrate the applicability of microdialysis to study drug transport to the brain.[34] Thus, the extracellular concentrations of atenolol and acetaminophen were shown to follow very closely the corresponding plasma concentrations. As a result of higher lipophilicity leading to larger lipophilic diffusion across the BBB, the extracellular fluid concentrations of acetaminophen were much larger than those of atenolol.[34]

5.03.6   Interspecies Differences in Pharmacokinetics

5.03.6.1   Species Differences in Absorption

Drug absorption is influenced by a variety of physiological and physicochemical factors. The physiological factors are species dependent and include gastric and intestinal transit time, blood flow rate, gastrointestinal pH and first-pass metabolism, while the physicochemical factors correspond to drug-specific and species-independent properties, such as pK_a, molecular size, solubility, and lipophilicity. The oral bioavailability of a drug is defined as the fraction of an oral dose that reaches the systemic circulation unchanged. The oral bioavailability (F) can be described by eqn [1]. The f_a is determined mainly by solubility and stability of the drug in the gastrointestinal tract and its permeability across the intestinal membrane. Models for human bioavailability are given in Chapter 5.29.

In comparing oral bioavailability across animal species, it is not unusual to observe marked interspecies differences (**Figure 5**).

For many compounds, these differences reflect presystemic (intestinal and/or hepatic) drug metabolism. By contrast to presystemic metabolism and oral bioavailability, some similarities across species[35–38] have been reported for fraction absorbed (f_a) as described below in more detail.

5.03.6.1.1 Physiology of the gastrointestinal tract

A recent review article has highlighted a number of similarities and differences in anatomy and physiology of the GI tract in rats and humans, which need to be taken into consideration in the interpretation of species differences in oral absorption[39] For example, the human GI tract is capable of absorbing materials faster and to a greater extent than that of the rat. Such differences are likely to influence the extent to which drugs are absorbed. Overall, the rat's gastrointestinal tract is organized in the same way as the human's, but with a few important differences. For example, the relative lengths of the small intestine in the rat differ from those in man, in that the jejunum makes up nearly the entire small intestine in the rat. Another important difference is that the human intestinal tract is only about 5.5 times the length of that in the rat, despite man's much larger body size (70 kg) compared to the rat (0.25 kg). The absolute surface area of the human intestinal tract is about 200 times that of the rat, which when normalized on the basis of body surface area amounts to a factor of 4 times. The physiological consequences of these anatomical differences are twofold.

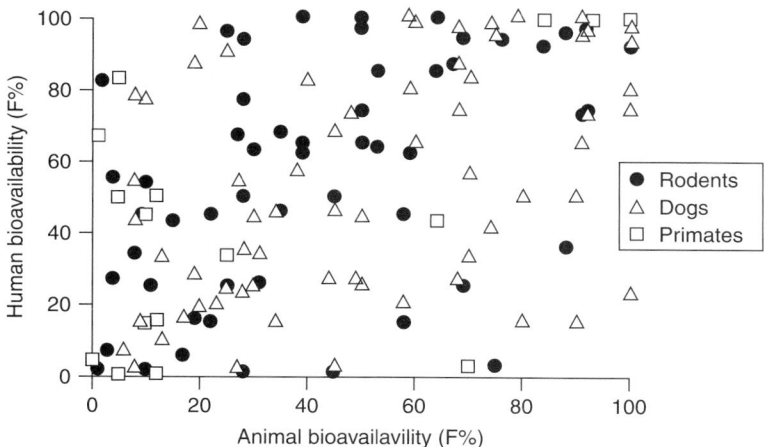

Figure 5 Species differences in oral bioavailability. (Reproduced from Grass, G. M.; Sinko, P. J. *Adv. Drug Deliv. Rev.* **2002**, *54*, 433–451.[80])

First, well-absorbed substances are likely to be absorbed more quickly in humans (i.e., man will exhibit a higher absorption rate). Second, substances that are poorly or incompletely absorbed by both species are likely to be absorbed to a greater extent in man.[39]

With regard to interspecies differences in secretions into the intestinal tract, an important anatomical difference between rats and humans is the absence of a gall bladder in rats. This means that in the rat, bile enters the duodenum continuously as it is made. By contrast the bile in humans is released only when chyme is present. These anatomical differences can be responsible for pronounced species differences in the concentration versus time profile of drugs that undergo enterohepatic recycling.

There are also some important similarities and differences in gastrointestinal physiology[40–42] between dogs and humans, which are discussed below. The physiology of the stomach is relatively similar under fasting conditions in both dog and man. The higher intestinal pH observed in the dog is the main difference between the species under fasting conditions. This could lead to species differences in the absorption of compounds for which solubility is a function of pH in the region of 5–7 and for enteric-coated formulations with dissolution in this pH range. Dogs also exhibit short intestinal transit time, which could lead to lower absorption for compounds slowly absorbed in the colon and for compounds in slow-release dosage forms. Postprandially, differences occur in both the gastric and intestinal physiology. For example, meal emptying rates are much slower in the dog. Furthermore, gastric and intestinal pH appears to be more acidic in dogs than in humans. These lower pH values may result in different absorption rates between dog and man for compounds whose intestinal permeability or solubility are affected by fraction ionized. These species differences in physiology between fasted and fed state might also result in different food effects on pharmacokinetics between human and dog.

Solubilization of lipophilic drugs is a critical step in ensuring its bioavailability. This solubilization process is highly dependent upon the presence and nature of the bile salts contained within the intestinal fluids. However, the marked interspecies differences in bile flow and composition can significantly affect the nature of this solubilization process.

The solubility of poorly soluble compounds was recently compared in dog and humans. For danazol, felodipine, and griseofulvin, the solubility in dog intestinal fluid was significantly higher than in human intestinal fluid, indicating that dog might not be a suitable species to predict human absorption for these poorly soluble compounds.[43] This higher solubility in dog intestinal fluid reflects the influence of higher bile salt concentration in dog as compared to man.

5.03.6.1.2 Interspecies comparison of fraction absorbed

Despite the differences in physiology, nearly identical oral absorptions in humans and rats have been reported for 64 drugs with markedly varying physicochemical properties and a wide range of intestinal permeabilities (**Figure 6**).[38]

In addition, similar oral bioavailabilities in rats and humans have been reported for 16 PEG compounds with molecular weights ranging from 280 to 950.[44]

The monkey may be another excellent animal model for predicting oral absorption in humans.[35] Chiou and Buehler found that absorption values in monkeys were similar or identical to those in humans (**Figure 7**). In contrast to rat and monkey, the dog appears to be less predictive for man. In general there is a tendency for dogs to absorb compounds

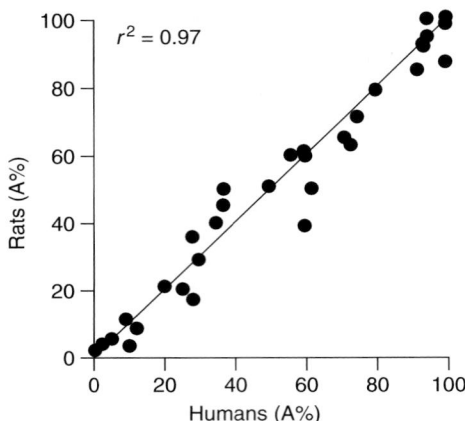

Figure 6 Oral absorption in rat versus human of 64 drugs with markedly different physicochemical properties. (Reproduced from Chiou, W. L.; Barve, A. *Pharm. Res.* **1998**, *15*, 1792–1795.[38])

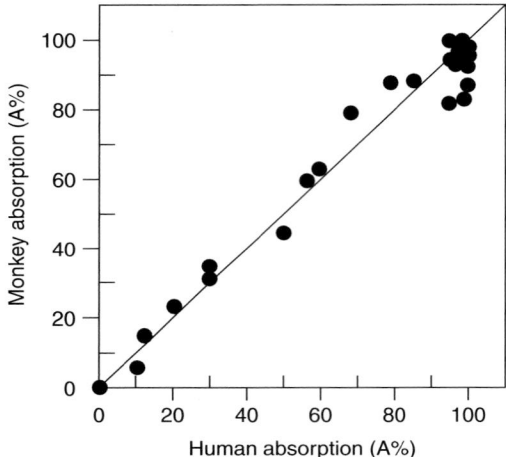

Figure 7 Oral absorption in cynomolgus monkey versus human of 43 drugs with markedly different physicochemical properties. (Reproduced from Chiou, W. L.; Buehler, P. W. *Pharm. Res.* **2002**, *19*, 868–874.[35])

better than man. For the 16 PEG compounds the absorption in dogs was better than in humans.[44] Also for 43 drugs, the correlation coefficient between the fraction absorbed in humans and dogs ($r^2 = 0.51$) was much lower than that ($r^2 = 0.97$) reported between humans and rats[36,38] (**Figure 8**).

For acyclovir and nadolol, the fraction absorbed in dogs was about 100% while humans only absorbed about 20% of the dose. Great differences were also found for atenolol (50% in man versus 100% in dogs), methyldopa (43% in man versus 100% in dogs), ranitidine (63% in man versus 100% in dogs), sumatriptan (60% in man versus 97% in dogs), and xamoterol (8.6% in man versus 36% in dogs).

A number of hypotheses have been proposed to explain the higher absorption in dog as compared to man. For example, the dog has longer villi, which may compensate for a shorter intestinal transit time. Furthermore, the higher bile salt secretion rate in the dog could increase the permeability of the intestinal membrane and might, through a solubilizing effect, also facilitate the absorption of poorly water-soluble drugs.[36,38] In addition the size and frequency of the tight junction for paracellular transport may be greater in dogs than in humans, which might explain the greater extent of absorption of small hydrophilic compounds, such as polyethylene glycol oligomers.[44]

5.03.6.2   Interspecies Differences in Distribution

The PK parameter that is most commonly used to characterize the distribution of a drug is its volume of distribution at steady state ($V_{D,ss}$). $V_{D,ss}$ represents the equivalent volume into which a given dose of drug is apparently distributed

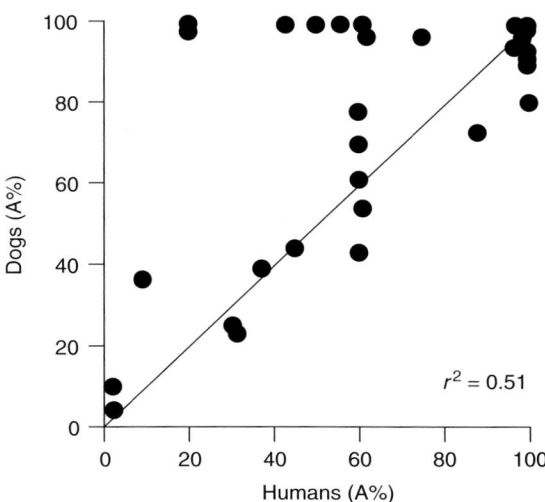

Figure 8 Oral absorption in dog versus human of 43 drugs with markedly different physicochemical properties. (Reproduced from Chiou, W. L.; Jeong, H. Y.; Chung, S. M.; Wu, T. C. *Pharm. Res.* **2000**, *17*, 135–140.[36])

within the body and, as such, it includes the extent to which the drug is bound to tissue and/or plasma proteins. The numerical value of $V_{D,ss}$ that is recorded can be as small as the blood volume or, for compounds that are extensively bound to tissues, can exceed the volume of total body water. The following equation has been proposed to relate $V_{D,ss}$ to plasma and tissue binding:

$$V_{D,ss} = V_P + f_{u,p} \cdot V_E + V_P \cdot R_{e/i} \cdot (1 - f_u) + V_R \cdot f_{u,p}/f_{u,t} \qquad [2]$$

where V_P, V_E, and $R_{e/i}$ are the plasma and extracellular volumes and the ratio of extravascular to intravascular proteins, respectively. V_R is the volume in which the drug distributes minus the extracellular space. $f_{u,p}$ and $f_{u,t}$ are the fractions of unbound drug in plasma and tissue.[45]

$V_{D,ss}$ can also be represented in terms of tissue/plasma partition coefficients (K_p) and plasma volume (V_P):

$$V_{D,ss} = V_P + V_{ery} \cdot ery/plasma + \sum V_T \cdot K_p \qquad [3]$$

Models for plasma protein binding and tissue storage are given in Chapter 5.14. Plasma protein binding of drugs often differs considerably across species, and more so than tissue binding. This might be because plasma binding frequently involves the interaction of a drug with a specific binding site on a specific protein, which differs structurally from one species to another, whereas tissue binding involves interaction with many constituents, thereby reflecting an average value.[46,47] This is supported by the observation that using the same K_{pu} ($K_p/f_{u,p}$) value for a tissue across species and correcting for differences in plasma protein binding tends to produce better prediction of tissue distribution than when no correction is made.[48] This is illustrated with the example of propranolol for which the large species differences in free fraction are responsible for species differences in volume of distribution derived from total plasma concentrations. When the volume is derived from unbound plasma concentrations, the volume is very similar across species (**Figure 9**).

Along these lines, in a study with a series of 40 drugs with widely differing physicochemical characteristics, Fichtl and coworkers demonstrated close interspecies correlations in binding to muscle tissue[49] (**Figure 10**).

However, some differences in tissue binding have also been reported. Thus, for lipophilic compounds that distribute into and bind more extensively to tissues, interspecies differences in K_{pu} values have been reported.[50] These may be due to differences between rodents and humans in the lipid content and composition of certain tissues, such as muscle[50,51] (**Table 1**) where the fraction of neutral lipids is about 10-fold higher in humans compared to rats.

Some studies demonstrated interspecies differences of up to threefold in K_{pu} values of heart muscle and lung determined in rat and rabbit.[50,51] For drugs whose distribution is diffusion rate limited, species differences in membrane permeability could also influence the K_{pu} values. In addition, a different degree of active transport processes

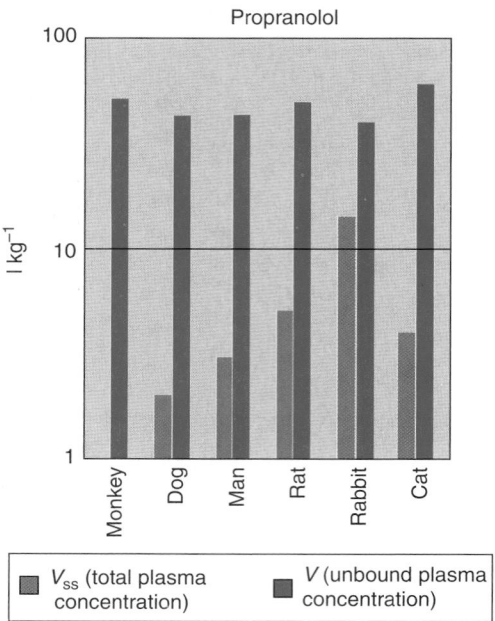

Figure 9 Interspecies differences in volume of distribution and protein binding for propranolol.

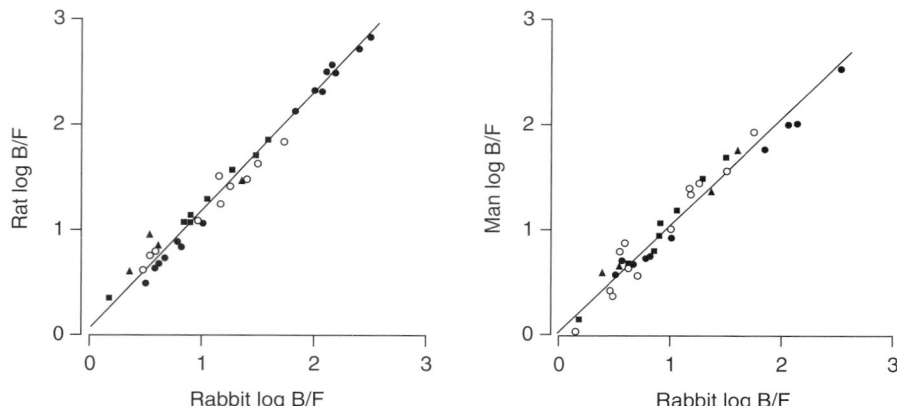

Figure 10 Correlation among different species in binding of drugs to muscle tissue in vitro. B/F represents the ratio between bound (B) and free (F) concentrations. (Reproduced from Fichtl, B.; Nieciecki, A.; Walter, K. In *Advances in Drug Research*; Mordenti, C. A., Ed.; Academic Press: London, 1991, pp 117–177.[49])

(e.g., Na$^+$/K$^+$ ATPase) in animals and humans can lead to significant interspecies differences in $V_{D,ss-u}$. For example, $V_{D,ss-u}$ (volume of distribution of the unbound concentration) of digitoxin is 2.1 L kg^{-1} in rat but 9.1 L kg^{-1} in man.[51]

Overall, however, species differences in tissue binding seem to be less pronounced than the corresponding differences in plasma binding.

5.03.6.3   Species Differences in Elimination

5.03.6.3.1   Interspecies differences in metabolic clearance

Hepatic clearance can be described by theoretical models that relate the intrinsic clearance (CL$_{int}$) to the other physiological parameters that influence clearance, namely hepatic blood flow and blood binding. Hepatic clearance in the well-stirred model is described by:

$$CL_H = [Q_H \cdot f_{u,b} \cdot CL_{int}] / [Q_H + f_{u,b} \cdot CL_{int}] \qquad [4]$$

Table 1 Physiological parameters for tissue composition in the adult male human and in rat

Tissue	Neutral lipids (V_n) (fraction of wet tissue weight)		Phospholipids (V_{ph}) (fraction of wet tissue weight)		Water (V_w) (fraction of wet tissue weight)	
	Human	Rat	Human	Rat	Human	Rat
Adipose	0.79	0.853	0.002	0.002	0.18	0.12
Bone	0.074	0.0273	0.0011	0.0027	0.439	0.446
Brain	0.051	0.0392	0.0565	0.0533	0.77	0.788
Gut	0.0487	0.0292	0.0163	0.0138	0.718	0.749
Heart	0.0115	0.014	0.0166	0.0118	0.758	0.779
Kidney	0.0207	0.0123	0.0162	0.0284	0.783	0.771
Liver	0.0348	0.0138	0.0252	0.0303	0.751	0.705
Lung	0.003	0.0219	0.009	0.014	0.811	0.79
Muscle	0.0238	0.00147	0.0072	0.00083	0.76	0.96
Plasma	0.0035	0.01	0.00225	0.009	0.945	0.756
Skin	0.0284	0.0239	0.0111	0.018	0.718	0.651
Spleen	0.0201	0.0077	0.0198	0.0136	0.788	0.771

where Q_H is the liver blood flow, $f_{u,b}$ is the fraction unbound in the blood, and CL_{int} (the intrinsic clearance) is a measure of the maximal ability of the liver to metabolize a drug in the absence of protein binding or blood flow limitations.

For drugs that are rapidly metabolized (i.e., very high CL_{int} leading to $f_{u,b} \cdot CL_{int} \gg Q_H$) the value of CL approaches Q_H and hepatic clearance is limited by the liver blood flow. For these compounds, interspecies differences in clearances will reflect interspecies differences in liver blood flow. One such example is propranolol. This drug is mainly eliminated by the liver, and the intrinsic clearance is so rapid that CL is limited by hepatic blood flow. Its CL is 90 mL min^{-1} kg^{-1} in the rat, 34 mL min^{-1} kg^{-1} in the dog, 18 mL min^{-1} kg^{-1} in the monkey, and 15 mL min^{-1} kg^{-1} in man.[52] The CL value in each species approximates their hepatic blood flow rate.

For drugs with low extraction ratios ($f_{u,b} \cdot CL_{int} \ll Q_H$) the value of CL is directly related to $f_{u,b}$ and CL_{int} ($CL = f_{u,b} \cdot CL_{int}$). Thus, the clearance of such compounds is mainly determined by their plasma protein binding and intrinsic clearance, which can both be subject to large species differences. Species differences in rate of metabolism (CL_{int}) have been shown with many different drugs undergoing both phase I and phase II metabolism. In general, drug metabolism in mammalian species is lower than in nonmammalian species[53] and decreases with increasing body weight. These species differences are the result of differences in the amino acid sequence, substrate specificity, and levels of the enzymes (e.g., cytochrome P450). For example, the hepatic levels of CYP1A, CYP2C and CYP3A in rats are approximately 28, 638, and 165 pmol mg^{-1} microsomal protein, respectively,[54] and the corresponding values for humans are 37, 55, and 87 pmol mg^{-1} microsomal protein.[55]

In general, deficiencies are well known for some animal species often used in experimental studies. Such deficiencies are for example a lack of acetylation of some compounds in dogs and guinea pigs, glucuronidation deficiency in cats, and some deficiencies in sulfotransferase activities in pigs.[16] In addition to these species differences in phase II conjugation reactions, differences are well known in phase I cytochrome P450 isoenzyme activities. However, for this group of enzymes it is more the altered substrate specificities and quantitative differences that lead to differences in metabolite patterns. The relevance of animal species depends on the cytochrome P450 isoenzyme involved. While dog seems to be a good species for CYP 2D-dependent reactions, CYP2E1 and CYP1A1/2 appear more appropriately represented in rats, (mini)pigs, and rhesus monkey.[56] In addition to species differences significant gender differences are apparent, especially in rats. In general, male rats metabolize many drugs faster than female rats, while the opposite is reported for mice. In rats many of these differences are due to the male-specific cytochrome 2C11,[57] which is often involved in the metabolism of compounds in man via CYP3A . New, humanized mouse models might overcome some of these limitations.[13] Mice expressing human CYP2D6 and CYP3A4 have been successfully generated; however, many other components of human drug metabolism and transport might be important for consideration in such a holistic model to adequately represent the interplay of drug metabolism in man.

Large species differences in hepatic metabolic clearance have been reported for many drugs and have made extrapolation of metabolic clearance from animal to man highly questionable. Bosentan is one example of a drug characterized by large interspecies differences in clearance.[58] Bosentan, exhibits a high blood clearance in rabbits, which approximates the liver blood flow. Intermediate to high clearance values were observed in mice, marmosets, and rats, while dogs exhibited a low blood clearance, which represented less than 5% of the corresponding liver blood flow. From these examples, it appears that it is not recommended to predict human metabolic clearance based on animal data. Fortunately, the availability of in vitro models in humans allows predictions of clearances to be performed based on physiologically-based scaling methods with satisfactory prediction accuracy.[59,60]

Comprehensive expert systems to predict drug metabolism are discussed in 5.33 Comprehensive Expert Systems to Predict Drug Metabolism.

5.03.6.3.2   Interspecies differences in renal clearance

Many drugs and drug metabolites are excreted by the kidneys. Renal clearance is the net result of three interrelated processes: glomerular filtration, tubular secretion, and tubular reabsorption. Both glomerular filtration and active tubular excretion eliminate a drug from the blood, while tubular reabsorption allows drug that has entered the renal tubules to re-enter the circulation.

A mechanistic relationship, in which the contributions of tubular secretion and glomerular filtration are additive, has been proposed by Levy.[61] This describes renal clearance (CL_R) in the following terms:

$$CL_R = (1 - f_r) \cdot f_u \cdot GFR + (1 - f_r) \cdot Q_R [f_u \cdot CL_{int,us}]/[Q_R + f_u \cdot CL_{int,us}] \qquad [5]$$

where CL_R represents renal clearance, f_r is the fraction of drug reabsorbed in the tubules, f_u is the free fraction in plasma, GFR the glomerular filtration rate, Q_R the renal blood flow, and $CL_{int,us}$ is the intrinsic tubular secretion rate. This model assumes that tubular secretion and glomerular filtration are fully independent processes.

If tubular secretion is absent then CL_R is directly proportional to GFR and the free fraction in plasma. For compounds with these characteristics, the unbound renal clearance in different species will reflect species differences in GFR. Similarly, if the secretory pathway is very efficient at removing drug then CL_R approximates Q_R, and in this case species differences in the CL_R will reflect species differences in renal blood flow. *para*-Aminohippuric acid is one example of a compound whose renal clearance is limited by blood flow.

Lin[62,63] recently provided a number of examples where mechanisms of renal excretion were compared across species. Based on these examples, depending on whether renal clearance is less than, equal to, or greater than the GFR, it may be possible to predict the mechanisms for renal elimination in man. Thus, famotidine has a renal clearance in rats of $1.68 \, L \, h^{-1} \, kg^{-1}$ which exceeds its GFR ($0.522 \, L \, h^{-1} \, kg^{-1}$), indicating that a net tubular secretion is occurring. In humans, the renal clearance of famotidine is $0.266 \, L \, h^{-1} \, kg^{-1}$, which is also greater its GFR ($0.108 \, L \, h^{-1} \, kg^{-1}$). The ratio of renal clearance to GFR is 2.5 and 3.2 in humans and rats, respectively. Therefore, for famotidine, there is reasonably good agreement between renal clearance in humans and that expected from animal models. Extending this approach to the data for the six compounds published by Sawada[64] tends to support this observation. Comparing the unbound renal clearance to GFR in the rat and human gave comparable ratios for cefoperazone (0.5 and 0.9), moxalactam (1.7 and 1.8), and cefazolin (7.7 and 3.6). For cefmetazole, the mechanism of renal excretion was similar in rat and human but the net secretion was much more pronounced in human (ratios of 1.6 in rat and 6 in human). For cefpiramide the ratios in rat (1.1) and human (0.7) were similar but the values indicated different excretion processes, namely a net secretion in rat and a net reabsorption in human.

Nevertheless, important species differences might arise when active transport is involved in the renal disposition of compounds. For example, Giacomini and colleagues[65] have shown that the human organic cation transporter (hOCT1) system is functionally different from the OCT1 homolog in other mammals, and it is reasonable to assume that these differences will modify their interactions with drugs. In this context, Dresser and co-workers[66] determined the kinetics and substrate selectivity of the OCT1 homologs from mouse, rat, rabbit, and human, and found that the human homolog is functionally different from that of the rodent and rabbit.

Species differences in renal clearance might also occur through differences in urinary pH.[67] Such effects might influence not only the extent of diffusion and reabsorption across the renal tubule, but could also result in artificial differences due to changes in the stability at different urinary pHs.[67]

5.03.6.3.3   Interspecies differences in biliary clearance

In general, mice, rats, and dogs are good biliary excretors, while guinea pigs, monkeys, and humans are relatively poor biliary excretors. The species differences in biliary excretion become less marked when the molecular size of the drug being excreted exceeds 700 Da.[53] Species variation in the efficiency of the transport systems is the more probable cause. However, as discussed earlier, differences in biliary anatomy (i.e., presence or absence of a gallbladder) can give rise to species differences in pharmacokinetics.[53]

The role of active transport processes and their impact on species differences in drug disposition have been reviewed recently.[68] Species differences in active transport may result from differences in the nature, levels, and/or regulation of the various transport proteins, as well as from their sequence substrate/inhibitor specificities. Where data are available on rat and human orthologs, sequence identity is reasonably high (70–80%) for the OATP,[69] OCT,[70,71] and MDR[72] transporter families. There is some evidence to suggest that the rate of transport in rat is greater than in humans. For example, some basic compounds were shown to be taken up $\sim$10-fold faster in rat than human hepatocytes[73] Also K_m and K_i for a range of bases were significantly lower for rat OCT1 than human OCT1,[70,71] suggesting a greater substrate affinity in rat. Species differences in the transport across the bile canalicular membrane of certain organic anions (temocaprilat, 2,4-dinitrophenyl-S-gluthatione and taurocholate, which are specific substrates of MRP2 and BSEP transporters) were shown to be due mainly to differences in V_{max} rather than K_m.[74]

5.03.7 Prediction to Humans

Preclinical in vivo studies are helpful to guide the design of the first clinical studies. A number of approaches including empirical methods and physiologically-based models are being used to predict human pharmacokinetics based on preclinical data (*see* 5.37 Physiologically-Based Models to Predict Human Pharmacokinetic Parameters). Among the empirical methods, allometric scaling explores the mathematical relationships between pharmacokinetic parameters from various animal species in order to make predictions in humans. Such methods are relatively easy to apply but place demands on resources for the collection of in vivo data in animals. Nevertheless, their application has led to useful predictions of individual pharmacokinetic parameters (e.g., clearance, fraction absorbed, volume of distribution). In general, however, such methods can only take account of 'passive' transport differences between species.

Allometric scaling is based on a power function of the form:

$$y = a \cdot W^x \qquad [6]$$

where y is the dependent variable (e.g., clearance, volume of distribution, half-life), W (body weight) is the independent variable, and a and x are the allometric coefficient and exponent, respectively. The values of a and x can be estimated by linear least squares regression of the log transformed allometric equations, i.e.,

$$\log(\text{clearance, volume of distribution, half-life, etc.}) = \log(a) + x \cdot \log(W) \qquad [7]$$

The sign and magnitude of the exponent indicate how the physiological or pharmacokinetic variable is changing as a function of W[75-77]: for $b < 0$, y decreases as W increases; for $0 < b < 1$, y increases as the species get larger, but does not increase as rapidly as W; when $b = 1$, y increases in direct proportion to the increases in W; and when $b > 1$, y increases faster than W.

The allometric equations for the pharmacokinetic parameters tend to be of the same order of magnitude as those for the corresponding physiologic variables, i.e.:

1. half-life (min) $\sim aW^{0.2}$ to $aW^{0.4}$;
2. clearance (mL min^{-1}) $\sim aW^{0.6}$ to $aW^{0.8}$; and
3. volume of distribution (mL) $\sim aW^{0.8}$ to $aW^{1.0}$.

The allometric approach can be particularly useful for compounds that are primarily eliminated by physical transport processes and when distribution occurs through passive processes. In such cases, when their values depend upon underlying physical or physiological processes that scale according to body weight, parameters such as clearance and volume of distribution do lend themselves well to allometric scaling. However, it may be less reliable for compounds that exhibit marked species differences in distribution, metabolism, or excretion mechanisms and a number of authors have cautioned against its use for selecting doses for the first phase of human studies.[78,79]

Physiologically-based models (PBPK) can be used to explore, and help to explain, the mechanisms that lie behind species differences in pharmacokinetics and drug metabolism. Physiologically-based models are discussed in more details in Chapter 5.37. Such models can provide a rational basis for interspecies scaling of individual parameters, which can then be integrated to provide quantitative and time-dependent estimates of both the plasma and tissue concentrations in humans. Furthermore, being mechanistically based they can be used diagnostically to generate information on new compounds and to understand the sensitivity of the in vivo profile to compound properties. Recent developments of in silico and in vitro models, which can provide estimates of the input parameters for PBPK models, have dramatically reduced the amount of experimental work required to make use of them.

References

1. Lin, J.; Sahakian, D. C.; de Morais, S. M.; Xu, J. J.; Polzer, R. J.; Winter, S. M. *Curr. Top. Med. Chem.* **2003**, *3*, 1125–1154.
2. Garattini, S. *Exp. Toxic. Pathol.* **1996**, *48*, 142–151.
3. Christ, D. *Drug Metab. Dispos.* **2001**, *29*, 935.
4. White, R. E.; Manitpisitkul, P. *Drug Metab. Dispos.* **2001**, *29*, 391–400.
5. Allen, M. C.; Shah, T. S.; Day, W. W. *Pharm. Res.* **1998**, *15*, 93–97.
6. Bayliss, M. K.; Frick, L. W. *Curr. Opin. Drug Disc. Devel.* **1999**, *2*, 20–25.
7. Frick, L. W.; Adkison, K. L.; Wells-Knecht, K. J.; Woolard, P.; Higton, D. M. *Pharmaceut. Sci. Technol. Today* **1998**, *1*, 12–18.
8. Miwa, G. T. *Toxicol. Pathol.* **1995**, *23*, 131–135.
9. Eddershaw, P. J.; Beresford, A. P.; Bayliss, M. K. *Drug Disc. Today* **2000**, *5*, 409–414.
10. Marathe, P. H.; Shyu, W. C.; Humphreys, W. G. *Curr. Pharm. Des.* **2004**, *10*, 2991–3008.
11. Garner, R. C. *Curr. Drug Metab.* **2000**, *1*, 205–213.
12. Combes, R. D.; Berridge, T.; Connelly, J.; Eve, M. D.; Garner, R. C.; Toon, S.; Wilcox, P. *Eur. J. Pharm. Sci.* **2003**, *19*, 1–11.
13. Gonzalez, F. J. *Hum. Genomics* **2004**, *1*, 300–306.
14. Campbell, D. B. *Eur. J. Drug Metab. Pharmacokinet.* **1994**, *19*, 283–293.
15. Baillie, T. A.; Cayen, M. N.; Fouda, H.; Gerson, R. J.; Green, J. D.; Grossman, S. J.; Klunk, L. J.; LeBlanc, B.; Perkins, D. G.; Shipley, L. A. *Toxicol. Appl. Pharmacol.* **2002**, *182*, 188–196.
16. Caldwell, J.; Gardner, I.; Swales, N. *Toxicol. Pathol.* **1995**, *23*, 102–114.
17. de Jong, W. H.; Timmerman, A.; van Raaij, M. T. *Lab. Anim.* **2001**, *35*, 243–248.
18. Martin, L. F.; Vary, T. C.; Davis, P. K.; Munger, B. L.; Lynch, J. C.; Spangler, S.; Remick, D. G. *Arch. Surg.* **1991**, *126*, 1087–1093.
19. Evans, D. C.; Watt, A. P.; Nicoll-Griffith, D. A.; Baillie, T. A. *Chem. Res. Toxicol.* **2004**, *17*, 3–16.
20. Uetrecht, J. P. *Chem. Res. Toxicol.* **1999**, *12*, 387–395.
21. Tucker, G. T.; Houston, J. B.; Huang, S. M. *Eur. J. Pharm. Sci.* **2001**, *13*, 417–428.
22. Zhou, H.; Tong, Z.; McLeod, J. F. *J. Clin. Pharmacol.* **2004**, *44*, 120–134.
23. Bjornsson, T. D.; Callaghan, J. T.; Einolf, H. J.; Fischer, V.; Gan, L.; Grimm, S.; Kao, J.; King, S. P.; Miwa, G.; Ni, L. et al. *J. Clin. Pharmacol.* **2003**, *43*, 443–469.
24. Rubas, W.; Cromwell, M. E.; Shahrokh, Z.; Villagran, J.; Nguyen, T. N.; Wellton, M.; Nguyen, T. H.; Mrsny, R. J. *J. Pharm. Sci.* **1996**, *85*, 165–169.
25. Artursson, P.; Artursson, P. *Crit. Rev. Ther. Drug Carrier Syst.* **1991**, *8*, 305–330.
26. Artursson, P.; Karlsson, J. *Biochem. Biophys. Res. Commun.* **1991**, *175*, 880–885.
27. Yamashita, S.; Tanaka, Y.; Endoh, Y.; Taki, Y.; Sakane, T.; Nadai, T.; Sezaki, H. *Pharm. Res.* **1997**, *14*, 486–491.
28. Salphati, L.; Childers, K.; Pan, L.; Tsutsui, K.; Takahashi, L. *J. Pharm. Pharmacol* **2001**, *53*, 1007–1013.
29. Ungell, A. L.; Nylander, S.; Bergstrand, S.; Sjöberg, A.; Lennernäs, H. *J. Pharm. Sci.* **1998**, *87*, 360–366.
30. De Boer, A. G.; de Lange, E. C. M.; van der Sandt, I. C. J.; Breimer, D. In *The Blood–Brain Barrier and Drug Delivery to the CNS*; Begley, D. J., Bradbury, M. W., Kreuter, J., Eds.; Culinary and Hospitality Industry Publications Services: London, UK, 2000, pp 77–92.
31. Feng, M. R. *Curr. Drug Metab.* **2002**, *3*, 647–657.
32. Oldendorf, W. H. *Brain. Res.* **1970**, *24*, 372–378.
33. Jezequel, S. G. In *Progress in Drug Metabolism*; Gibson, G. G., Ed.; Taylor & Francis: London, UK, 1992; Vol. 13, pp 141–178.
34. de Lange, E. C. M. *Brain Res.* **1994**, *666*, 1–8.
35. Chiou, W. L.; Buehler, P. W. *Pharm. Res.* **2002**, *19*, 868–874.
36. Chiou, W. L.; Jeong, H. Y.; Chung, S. M.; Wu, T. C. *Pharm. Res.* **2000**, *17*, 135–140.
37. Chiou, W. L.; Ma, C.; Chung, S. M.; Wu, T. C.; Jeong, H. Y. *Int. J. Clin. Pharmacol. Ther.* **2000**, *38*, 532–539.
38. Chiou, W. L.; Barve, A. *Pharm. Res.* **1998**, *15*, 1792–1795.
39. DeSesso, J. M.; Jacobson, C. F. *Food Chem. Toxicol.* **2001**, *39*, 209–228.
40. Kararli, T. T. *Biopharm. Drug Dispos.* **1995**, *16*, 351–380.
41. Martinez, M.; Amidon, G.; Clarke, L.; Jones, W. W.; Mitra, A.; Riviere, J. *Adv. Drug Deliv. Rev.* **2002**, *54*, 825–850.
42. Dressman, J. B. *Pharm. Res.* **1986**, *3*, 123–131.
43. Carlsson, A.; Kostewicz, E.; Hanisch, G.; Krumk, K.; Nilsson, R.; Abrahamsson, B. Proceedings of the AAPS Pharmaceutica, Toronto, Canada, 2002.
44. He, Y. L.; Murby, S.; Warhurst, G.; Gifford, L.; Walker, D.; Ayrton, J.; Eastmond, R.; Rowland, M. *J. Pharm. Sci.* **1998**, *87*, 626–633.
45. Rowland, M.; Tozer, T. N. *Clinical Pharmacokinetics: Concept and Applications*, 3rd ed.; Williams & Wilkins Media, London, 1995.
46. Rodgers, T.; Leahy, D.; Rowland, M. *J. Pharm. Sci.* **2005**, *94*, 1237–1248.
47. Rodgers, T.; Leahy, D.; Rowland, M. *J. Pharm. Sci.* **2005**, *94*, 1259–1276.
48. Sawada, Y.; Hanano, M.; Sugiyama, Y.; Harashima, H.; Iga, T. *J. Pharmacokinet. Biopharm.* **1984**, *12*, 587–596.
49. Fichtl, B.; Nieciecki, A.; Walter, K. In *Advances in Drug Research*; Mordenti, C. A., Ed. Academic Press: London, 1991, pp 117–177.
50. Poulin, P.; Theil, F. P. *J. Pharm. Sci.* **2000**, *89*, 16–35.
51. Theil, F. P.; Guentert, T. W.; Haddad, S.; Poulin, P. *Toxicol. Lett* **2003**, *138*, 29–49.
52. Lave, T.; Dupin, S.; Schmitt, C.; Chou, R. C.; Jaeck, D.; Coassolo, P. *J. Pharm. Sci.* **1997**, *86*, 584–590.
53. McNamara, P. J. In ; Welling, P. G., Tse, F. L. S., Dighe, S. V., Eds.; Marcel Dekker: New York, 1991; Vol. 48, pp 267–300.
54. de Waziers, I.; Bouguet, J.; Beaune, P. H.; Gonzalez, F. J.; Ketterer, B.; Barouki, R. *Pharmacogenetics* **1992**, *2*, 12–18.
55. Guengerich, F. P. *Environ. Health Perspect.* **1995**, *103*, 25–28.
56. Anzenbacher, P.; Anzenbacherova, E. *Nova Acta Leopoldina* **2003**, *329*, 11–20.
57. Funae, Y.; Imaoka, S. In *Cytochrome P450*; Greim, H., Ed.; Springer Verlag: Berlin, Germany, 1993; Vol. 105, pp 221–238.

58. Ubeaud, G.; Schmitt, C.; Jaeck, D.; Lave, T.; Coassolo, P. *Xenobiotica* **1995**, *25*, 1381–1390.
59. Zuegge, J.; Schneider, G.; Coassolo, P.; Lavé, T. *Clin. Pharmacokinet.* **2001**, *40*, 553–563.
60. Ito, K.; Houston, J. B. *Pharm. Res.* **2005**, *22*, 103–112.
61. Levy, G. *J. Pharm. Sci.* **1980**, *69*, 482–483.
62. Lin, J. H. *Ernst Schering Res. Found Workshop* **2002**, *37*, 33–47.
63. Lin, J. H. *Drug Metab. Dispos.* **1995**, *23*, 1008–1021.
64. Sawada, Y.; Hanano, M.; Sugiyama, Y.; Iga, T. *J. Pharmacokinet. Biopharm.* **1984**, *12*, 241–261.
65. Giacomini, K. M. *J. Pharmacokinet. Biopharm.* **1997**, *25*, 731–741.
66. Dresser, M. J.; Gray, A. T.; Giacomini, K. M. *J. Pharmacol. Exp. Ther.* **2000**, *292*, 1146–1152.
67. Bonate, P. L.; Reith, K.; Weir, S. *Clin. Pharmacokinet.* **1998**, *34*, 375–404.
68. Ayrton, A.; Morgan, P. *Xenobiotica* **2001**, *31*, 469–497.
69. Hsiang, B.; Zhu, Y.; Wang, Z.; Wu, Y.; Sasseville, V.; Yang, W. P.; Kirchgessner, T. G. *J. Biol. Chem.* **1999**, *274*, 37161–37168.
70. Zhang, L.; Brett, C. M.; Giacomini, K. M. *Annu. Rev. Pharmacol. Toxicol.* **1998**, *38*, 431–460.
71. Zhang, L.; Schaner, M. E.; Giacomini, K. M. *J. Pharmacol. Exp. Ther.* **1998**, *286*, 354–361.
72. Klein, I.; Sarkadi, B.; Varadi, A. *Biochim. Biophys. Acta* **1999**, *1461*, 237–262.
73. Sandker, G. W.; Weert, B.; Olinga, P.; Wolters, H.; Slooff, M. J.; Meijer, D. K.; Groothuis, G. M. *Biochem. Pharmacol.* **1994**, *47*, 2193–2200.
74. Ishizuka, H.; Konno, K.; Shiina, T.; Naganuma, H.; Nishimura, K.; Ito, K.; Suzuki, H.; Sugiyama, Y. *J. Pharmacol. Exp. Ther.* **1999**, *290*, 1324–1330.
75. Mordenti, J. *Antimicrob. Agents Chemother.* **1985**, *27*, 887–891.
76. Mordenti, J. *J. Pharm. Sci.* **1986**, *75*, 1028–1040.
77. Mordenti, J.; Chen, S. A.; Moore, J. A.; Ferraiolo, B. L.; Green, J. D. *Pharm. Res.* **1991**, *8*, 1351–1359.
78. Bonate, P. L.; Howard, D. *J. Clin. Pharm.* **2000**, *40*, 665–670.
79. Boxenbaum, H.; Dilea, C. *J. Clin. Pharmacol.* **1995**, *35*, 957–966.
80. Grass, G. M.; Sinko, P. J. *Adv. Drug Deliv. Rev.* **2002**, *54*, 433–451.

Biographies

Thierry Lavé received his PharmD and PhD in Pharmaceutical Sciences from the University of Strasbourg in 1992 and 1993. He also received a statistical degree from the University of Paris in 1992 and did an internship in Hospital Pharmacy from 1988 to 1992 in Strasbourg. He joined Roche's Department of Preclinical Pharmacokinetics and Drug Metabolism in 1992 where he served for 2 years as a postdoctoral fellow. Thierry Lavé is currently a Scientific Expert at F Hoffmann-La Roche and took responsibility for the global modeling and simulation group in preclinical research in 2002. His main area of research is in preclinical modeling of safety, pharmacokinetics and formulation, physiologically based pharmacokinetics, interspecies scaling, drugdrug interactions, and all aspects related to early predictions of pharmacokinetics/pharmacodynamics in man during drug discovery and early drug development. He has published a number of papers in these areas and also teaches pharmacokinetics at the Universities of Basel, Strasbourg, Helsinki, and Besançon.

Christoph Funk studied Biology at the Swiss Federal Institute of Technology (ETH) in Zürich, Switzerland in the early 1980s. Following graduation in 1985, he then completed his PhD thesis at the same University on the metabolism of several natural products in plants and plant tissue cultures. Dr Funk continued these studies between 1990 and 1993 as a postdoctoral fellow at the Institute of Biological Chemistry at Washington State University in Pullman, USA). His research focused on the metabolism of several mono- and diterpenoids in plants, including the characterization of several cytochrome P450 monooxygenases and other metabolic enzymes involved in secondary product formation.

Dr Funk returned to Switzerland where he took up a post-doctoral post at the Botanical Institute of the University of Basel. In 1994 he became Laboratory Head of Drug Metabolism in the Nonclinical Drug Development unit of F Hoffmann-La Roche in Basel. Besides the metabolic characterization of many new drug molecules, the research focused on the setup of in vitro drug transport assays and their characterization. The goal is to understand the relevant enzymatic steps involved in drug distribution and elimination. Since 2000, Dr Funk has been responsible for the drug metabolism and pharmacokinetics (DMPK) technologies group, bringing together the disciplines of drug metabolism, drug transport, kinetics, and mechanistic toxicology, with the aim of better characterization of the absorption, distribution, metabolism, excretion, and toxicity (ADMET) properties of potential new drugs.

Comprehensive Medicinal Chemistry II
ISBN (set): 0-08-044513-6

ISBN (Volume 5) 0-08-044518-7; pp. 31–50

5.04 The Biology and Function of Transporters

J-M Scherrmann, University of Paris, Paris, France

5.04.1 Introduction

Pharmacokinetics has been defined over the past several decades as the study of drug absorption, distribution, metabolism, and excretion (ADME) when the drug is introduced into a biological system such as the human body. Most of the molecular processes responsible for the fate of a drug in the body were attributed to the passive crossing of the bilayer cell membranes. These explained a large part of how a drug was believed to be absorbed, distributed, and eliminated. Several drug descriptors, such as molecular weight, degree of ionization, and solubility, were used to explain how a drug crossed biological membranes. Other pharmacokinetic effectors were considered, such as the metabolizing enzymes, that played an important role in drug biotransformation and the circulating plasma proteins defining the so-called 'fraction of free drug' available for tissue distribution. Elimination was also defined as the endpoint of all pharmacokinetic events, by which the more hydrophilic drugs or their metabolites were excreted in urine or bile.

Pharmacokineticists learned quite early that the overall renal clearance of a drug results from a combination of three processes: glomerular filtration, tubular secretion, and reabsorption. These last two processes have long been known to be mediated by carrier systems. In these systems, membrane proteins transport solutes in oriented directions. It was gradually realized that drugs could cross biological membranes by these carrier-mediated processes as well as by passive diffusion. This view of pharmacokinetics is now challenged by the growing importance of transporters, a relatively new and potentially major contributor. The impact of drug transporters on pharmacokinetics has suffered from the fact that less is known about them than about drug-metabolizing enzymes. Schwenk[1] published a review in 1987 that stated that "drug transport in the intestine occurs mainly by diffusion and the contribution of intestinal drug carriers seems to be of minor importance." He also pointed that "the liver is the organ where carrier-mediated drugs predominate" and that there may be four hypothetical carrier systems, called "carrier 1, 2, 3 and 4," for drug uptake by hepatic cells. At the same time, the main enzymes implicated in drug metabolism were identified and the factors causing variations, such as the regulation and genetic polymorphism of these enzymes, were extensively studied. We are now in a position to complete the list of drug transporters, the tissues in which they occur and function, how they are regulated, and the clinical relevance of their presence in normal and diseased tissues. Nevertheless, despite the tremendous efforts that have been made in recent years, we still lack a significant amount of information about the fundamental contributions that transporters make to the pharmacokinetic processes that regulate the disposition of drugs in the body.

5.04.2   Membrane Transport Proteins in Biological Systems

5.04.2.1   Passive and Active Permeation

The exchange of solutes between body compartments depends to a considerable extent on the properties of the body that allow easy communication between tissues and compartments via pores and fenestra on the walls of the blood vessel or gap junctions between cells. These features allow free solutes to move in both directions through biological membranes by the so-called 'paracellular pathway.' But the organs of the body and pharmacological targets are not readily accessible to exogenous molecules because of the integrity of the lipid bilayer membranes that protect the interiors of cells. Some physiological barriers like the blood–brain barrier (BBB), the blood–placenta barrier, and the blood–testis barrier are so impermeable that solutes can only cross the lipid bilayer by a transcellular pathway. It has been established for some time that with advantageous gradient conditions only small nonpolar, uncharged molecules, such as oxygen, carbon dioxide, water, and ethanol, can diffuse easily through the membranes, whereas charged small ions, such as sodium and potassium, and large molecules, such as glucose (180 Da), are considerably less able to permeate than water (**Figure 1**). As the delivery of many polar molecules, such as anions and cations, vitamins, sugars, nucleosides, amino acids, peptides, bile acids, and porphyrins, to cells is essential for life, essential transporter proteins anchored in the lipid bilayer have evolved to permit their exchange between cells and their environment.

There are two principal classes of transport proteins: the proteins that form channels or pores, and the transporters. Channels are responsible for the facilitated diffusion of ions and small nutrient molecules down their electrochemical gradients. The flow of ions or solutes through channels is controlled by the opening of the channels via gating mechanisms and the single-channel conductance (number of charges per second at a given voltage). Some channels are more selective and have specific binding sites; they undergo changes of conformation and these changes regulate their opening to solute traffic.[2] The second class of transporters or carriers undergoes a conformational change during the translocation of a solute across the membrane. Transporters that facilitate the movement of solutes into the cell are called influx (import) transporters, while those that remove substances from the cytosol of the cell are efflux (export) transporters.

5.04.2.2   Transport Modes

Influx and efflux transporters can be further classified on the basis of their energy requirements.[3] Like protein channels, passive transporters, or facilitated transporters or uniporters, allow net solute flow down an electrochemical concentration gradient.

In contrast, active transport leads to the movement of solutes against an electrochemical gradient across the membrane. Active transport always occurs in a specific direction because it is coupled to an energy source. Several types of transporters have been identified; they differ by their energy source and the direction of transport (**Figure 2**). The

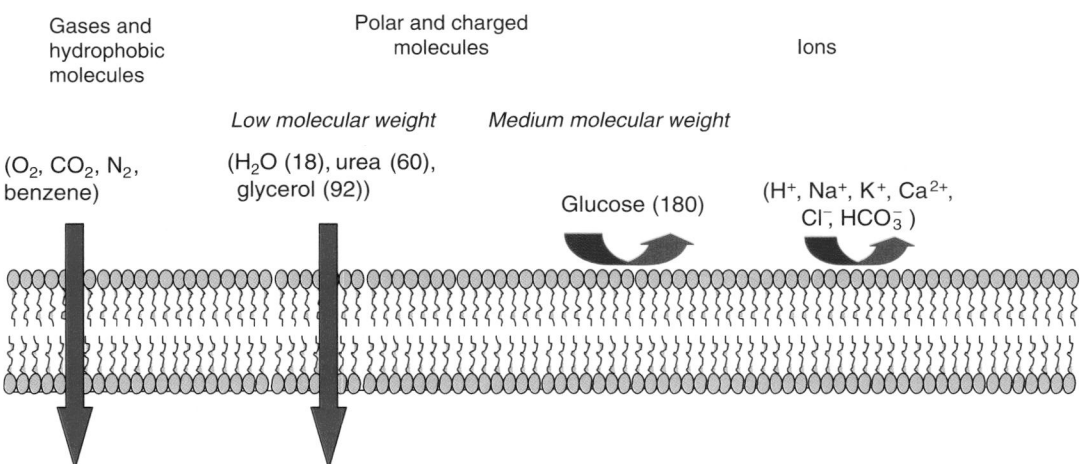

Figure 1 Permeation of ions and molecules through the membrane lipid bilayer. Small, lipophilic molecules with no charge are more likely to cross the membrane by passive diffusion. Ions and high molecular weight hydrophilic compounds cannot cross the lipid bilayer in the absence of pores, channels, or specific transporters.

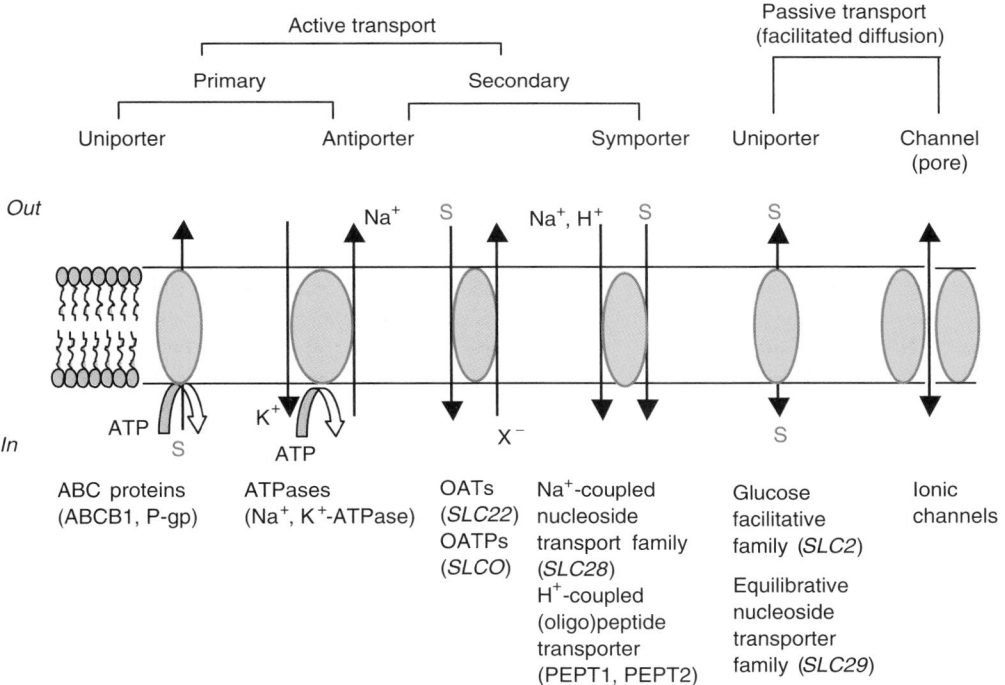

Figure 2 The classification of modes of solute (S) transport translocation between the extracellular (*out*) and intracellular (*in*) compartments. A few examples of transporter families are included.

primary active transport systems are coupled to an energy source that results in solar, electrical or chemical reactions, such as the hydrolysis of ATP by ion pumps (ATPases) and the ATP-binding cassette (ABC) transporters that transport the substrate in one defined direction. They are called primary transporters because no additional biochemical step is needed for solute transport. The combination of an energy-yielding reaction and transport occurs within a single catalytic cycle of the protein. The second group of transporters are the co-transporters; these use a voltage or/and ion gradient to transport both ions and solutes together. They are uniporters when only one species is transported, and symporters when both species are transported in the same direction, whereas the antiporters transport solutes and ions in opposite directions. The H^+ ion is the most common form of energy in prokaryotes, while Na^+ is more frequently

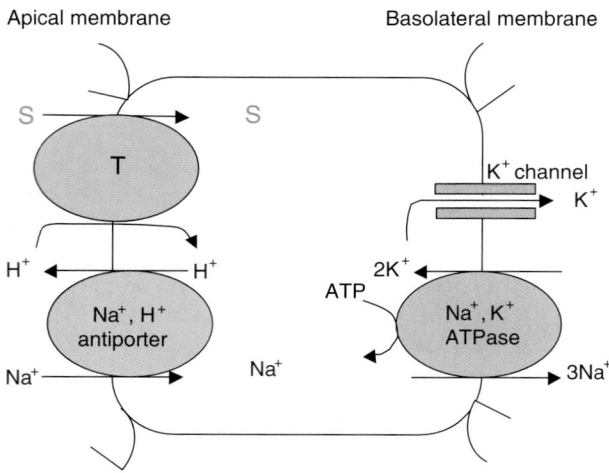

Apical membrane   Basolateral membrane

Figure 3 This model shows how solutes (S) are transported in a mammalian epithelial cell having K^+ channels and three types of transporters at the apical or basolateral membranes. The Na^+,K^+-ATPase (the primary active transporter) generates a Na^+ gradient, which provides the driving force for activating Na^+ and H^+ fluxes via the Na^+,H^+ antiporter (the secondary active transporter). T is the tertiary active transporter, which transports S molecules down the H^+ gradient in symporter mode.

encountered in eukaryotic cells.[3] For example, several secondary transporters are involved in the entrance and exit of Na^+ at both apical and basolateral membranes of intestinal and renal epithelial cells. Thus the rate of transport of solutes like sugars or amino acids depends on the extracellular concentration of Na^+. The main role of the Na^+, K^+-ATPase system is to activate the cascades of multiple co-transport processes. These co-transporters are also called secondary or even tertiary transporters because the machinery of ion transport must be activated by one or two pumps before solute transport occurs (**Figure 3**). Thus the energy component is clearly critical and the consequence of a lack of ATP in cells, as occurs in ischemia, leads to a loss of all transport by the ABC transporters. Other sources of energy involve glutathione (GSH), HCO_3^-, and dicarboxylate α-ketoglutarate (α-KG).

The biological significance of these proteins is highlighted by their universal distributions in both prokaryotes and eukaryotes. Transporters are involved in the biology of yeasts, plants, bacteria, and parasites, so that it is possible that these molecules in bacteria and parasites may raise resistance to drug therapy. Many of the major body functions in mammals, like gut absorption, glucose uptake by tissues, bile acid secretion into the gut from the liver, and the renal secretion or reabsorption of nutrients, are linked to the presence of protein transporters on both the plasma membranes of cells and in intracellular organelles like the endoplasmic reticulum, mitochondria, and endosomes. Therefore, the extracellular and intracellular trafficking of any solute, including exogenous compounds like drugs, depends directly on this network of protein transporters.[4]

5.04.3   Biology of Membrane Transport Proteins

5.04.3.1   Genes and Classification

Following the determination of the prokaryotic and eukaryotic genomes, Paulsen et al.[5] determined the distribution of membrane transport proteins for all those organisms whose genomes had been completely sequenced. Paulsen predicted that 15% of the 23 000 genes in the human genome code for transport proteins; this would result in nearly 3500 transporters. Ward et al.[6] assumed that 3–12% of the genes in bacterial genomes encode membrane transport proteins that can be vital for the efflux of antibiotics, the secretion of proteins or toxins, and other functions. But only about 10% of these proteins with known physiological function have been identified in mammalian cells, and only the members of two superfamilies are presently known to affect drug transport. They are the ABCs and solute carriers (SLCs). ABC proteins are widespread in all organisms, from bacteria to mammals, with about 600 referenced transporters, but only 48 genes have been identified in humans and no more than around nine ABCs have been shown to affect drug pharmacokinetics and pharmacodynamics.[7] The SLC family, which may contain about 2000 members, is presently known to have 46 families, including 362 transporter genes with documented transport functions.[8]

The current status of transporter nomenclature and classification is in the same disarray as was that of the metabolizing enzymes in the early 1980s. However, the Human Genome Organization (HUGO) Nomenclature

Committee Database has provided a list of the ABC and SLC genes and defined families of these transporters.[8a] There are other classification systems, such as the Transport Classification Data Base (TCDB). This is analogous to the Enzyme Commission system for the classification of enzymes but also incorporates phylogenetic information.[8b] This review uses the HUGO as the primary reference for identifying genes and proteins. Human proteins (*genes*) are shown in capitals (e.g., ABCB1 (*MDR1*)), while rat and mouse proteins (*genes*) are indicated by an initial capital followed by small letters (proteins) and small letters (genes) (e.g., Abcc1 (*mdr1*)). Some transporter databases provide only certain properties of a specific class or group of membrane transporter proteins. But, altogether, they contain information on cDNA and amino sequences, gene family, putative membrane topology, driving force, transport direction, lists of substrates, inhibitors, inducers, transport kinetic data, tissue distribution in both humans and mice and rats, drug–drug interactions involving transporters, and altered functions caused by mutations or polymorphisms and the influence of associated diseases.[9]

5.04.3.2 Basic Structure

The primary and secondary structures of many transporters are known. The structures of many ABCs and SLCs remain to be determined due to the inherent difficulties in crystallizing transporters for the x-ray crystallographic analyses needed to prepare high-resolution three-dimensional (3D) structures. This needs to be addressed. Biochemical analyses of transporter proteins are more challenging than studies on soluble proteins because they are hydrophobic, high molecular weight proteins and represent a small percentage of the total cell proteins. They are also rather inaccessible because they are tightly inserted into the lipid bilayer. Nevertheless, considerable progress has been made in the past 10 years by combining several biophysical approaches to determine the structure and topology of membrane transporters in the absence of a crystal structure. For example, several homology models predicting the pharmacophore pattern for the transporter binding site(s) are now available tools for 3D structure–activity relationship (3D-QSAR) analyses (*see* 5.32 In Silico Models for Interactions with Transporters).

Drug transporters are integral membrane proteins that typically have 12 transmembrane domains (TMDs), although there are exceptions with 6, 8, 10, 11, 13, or even 17 TMDs. The TMDs are folded in α-helical structures within the membrane and linked at both sides by amino acid sequences floating in the internal or external cell environment. The amino acids in the external loop domains are frequently *N*-glycosylated, while those of the intracellular loops of both SLC and ABC proteins bear phosphorylation sites and one or two ATP binding domains in the ABCs. When observed in 3D structure, TMDs form a crown and look like a channel allowing communication between the two fluid spaces separated by the lipid bilayer. Many SLC transporters have 300–800 amino acid residues and a molecular mass of 40–90 kDa, while the ABC transporters are larger, with 1200–1500 residues and a mass of 140–190 kDa.[10,11] Knowledge of the amino acid sequence can be used to classify transporters by family and subfamily according to the degree of amino acid homology. For example, a transporter protein is assigned to a specific SLC family if it has an amino acid sequence that is at least 20–25% identical to those of other members of that family. A new nomenclature system was recently proposed that is based on the classification of drug-metabolizing enzymes; the transport proteins of a superfamily are arranged in clusters of families ($\geqslant$40% identity) and subfamilies ($\geqslant$60% identity).[12] Knowledge of the amino acid sequences is also extremely helpful for assessing the effect of a single mutation within the sequence that can induce a change in the conformation of the transport protein and alter its transport functions.

5.04.3.3 Transport Kinetics

Transport proteins may have one or more binding sites formed by amino acid sequences of the TDMs or the extracellular loops. Environmental factors, such as the extracellular or intracellular pH or the lipid bilayer fluidity, may influence the transport process. For example, the probability of a solute interacting with a membrane transport protein can depend on the permeability coefficient (P_e), which determines how fast a solute diffuses passively across the lipid bilayer. This condition is particularly relevant when solute molecules interact with the transport binding sites of the TDM inserted in the plasma membrane. The higher the P_e ($< 10^{-6}\,\mathrm{cm\,s^{-1}}$) the lower the probability of the binding site interacting with the substrate because the residence time of the substrate molecules within the membrane space is too short. In contrast, a long substrate residence time within the membrane domain may facilitate the interaction between the transporter and the substrate.

Like drug-metabolizing enzymes, transporters are substrate specific and the kinetics of transport obey the capacity limit rules described by the Michaelis–Menten equation. The influx or efflux flux rate (J_{in} or J_{ex}) for a single permeation process mediated by one transporter depends on the affinity of the transporter for its substrate (K_m), which is the concentration at the half-maximal transport rate (V_{max}):

$$J_{in} \ \text{or} \ J_{ex} = V_{max}C/(K_m + C) \qquad [1]$$

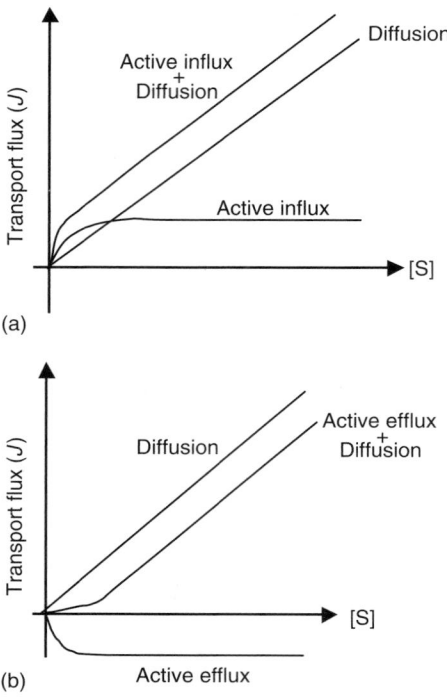

Figure 4 Transport kinetics: relationships between the transport flux (J) and the concentration of substrate [S] in (a) influx and (b) efflux modes. Multiple components (diffusion and one or more forms of active transport) participate in the overall transport process. Kinetic parameters such as K_m and V_{max} (J_{max}) can be derived from these plots.

The active process is more often coupled to a passive diffusion component, which is directly, proportional to the substrate concentration by a rate constant (k_d) leading to an additional term in eqn [1]:

$$J_{in} \ \text{or} \ J_{ex} = [V_{max}C/(K_m + C)] + k_dC \qquad [2]$$

The permeation of a solute through biological membranes may be more complex than that described by the above equations, and results from the combination of several components associating more than one active process with the diffusion component. In this case, the net flux of the transport (J_{net}) is calculated by adding all influx components and subtracting the efflux ones:

$$J_{net} = J_{diff} + J_{in} - J_{ex} + \cdots \qquad [3]$$

This last feature is not simply hypothetical; more and more examples of multicomponent permeation are now known. The impact of all these kinetic processes of varying complexity is illustrated in **Figure 4**. It can be seen that the superimposing passive diffusion and active influx and/or efflux processes leads to nonlinear relationships between J_{net} and the substrate concentration. These relationships can be dissected into their diffusion and active component processes. This can be used to calculate the kinetic parameters of active transport (K_m and V_{max}) and to anticipate the risks of permeation saturation in situations such as exposure to a high dose of a drug or following drug–drug interactions.

5.04.4   Transporters in Drug Disposition

This section covers only those transporters that influence the A, D, or E of drugs and xenobiotics. They belong to either the ABC or the SLC superfamilies and are often classified according to the chemical nature of their substrates. Hence, they translocate organic anions or cations, peptides, or nucleosides. Most of them were first named according to their specific chemical substrate, such as the organic cation transporters (OCTs), or the organic anion transporters (OATs), before they were named using the HUGO nomenclature rules. They all have some common general properties: a broad specificity with frequent overlaps of substrate recognition, making them 'polyspecific transporters,' and they are present in several body tissues and organs. They frequently differ in the type of energy mechanism that catalyzes the transport reaction.

5.04.4.1   Adenosine Triphosphate-Binding Cassette Transporter Superfamily

The transporters of the ABC superfamily have one of three types of structure, but they all have one or two nucleotide-binding domains (NBDs) (**Figure 5**). Two sequence motifs located 200 amino acids apart in each NBD, designated Walker A and Walker B, are conserved among all ABC transporter superfamily members. The lysine residue in the Walker A motif is involved in the binding of the β-phosphate of ATP while the aspartic acid residue in the Walker B motif interacts with Mg^{2+} An additional element, the signature (C) motif with a highly conserved amino acid sequence, lies just upstream of the Walker B site. The precise function of this sequence has not yet been determined although it is directly implicated in the recognition, binding, and hydrolysis of ATP.[13] The two NBDs are located in the cytoplasm and transfer the energy from ATP hydrolysis to transport the substrate. The main differences between the ABC transporters is in the number of TMDs in the hydrophobic membrane-spanning domain (MSD). The MSD itself consists of several transmembrane α-helices. The ABC transporters ABCB1 (MDR1) and ABCC4, 5, 8, and 9 (multidrug resistance associated proteins (MRPs)) have a classical 12-TMD body. The structures of ABCC1, 2, 3, 6, and 7 are similar in that they possess two ATP binding sites, but they also contain an additional MSD with five TMDs at the amino-terminal end, giving them a total of 17 TMDs (see **Figure 5**). At the other extreme, ABBG2, also called a 'half-transporter', contains six TMDs and one ATP binding region on the amino-terminal side, but this half-transporter is thought to homodimerize to form a functional transporter. Most of the ABC transporters in eukaryotes move compounds from the cytoplasm to the outside of the cell or into intracellular organelles, such as the endoplasmic

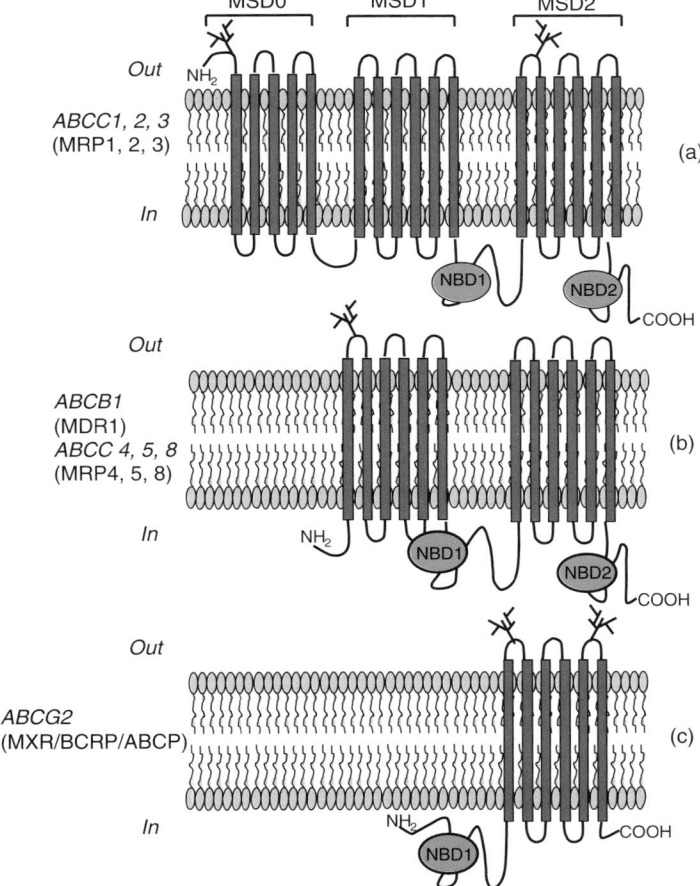

Figure 5   Topological models of the best characterized ABC transporters that confer drug resistance and transport drugs and toxicants. (a) The model of MRP1 resembles MRP2, 3; (b) the model of MDR1 (P-gp) resembles MRP4, 5, 8; and (c) the model of BCRP. MSD, membrane-spanning domain; NBD, nucleotide binding domain. Transmembrane α-helical domains (TMDs) are represented by red rectangular bars; *N*-glycosylation sites (indicated by branches) are present on extracellular protein loops. Cytoplasmic (in) and extracellular (out) orientations are indicated.

reticulum, mitochondria, and peroxisomes. This means that they are frequently called efflux pumps.[14,15] Phylogenetic analysis has grouped the eukaryotic ABC genes into seven subfamilies (A–G).[7] Only three of these subfamilies, B, C, and G, contain transporters that influence drug disposition. **Table 1** shows the main properties of the nine single ABC transporters belonging to these three subfamilies.

5.04.4.1.1 ABCB subfamily

These full transporters include ABCB1 (MDR1), ABCB4 (MDR3), and ABCB11 (sister P-gp or BSEP). ABCB1 is also called P-glycoprotein (P-gp); it is a 170-kDa protein which was the first human ABC transporter cloned and has been shown to be responsible for the multidrug resistance (MDR) phenomenom that occurs with such anticancer agents as the anthracyclines, vinca alkaloids, and taxanes.[16,17] P-gp is the product of two MDR genes in humans, *MDR1* and *MDR2* (also called *MDR3*), and only the MDR1 protein is involved in the MDR phenotype; the human *MDR2* gene functions as a phospholipid translocase at the canalicular membrane of hepatocytes where it is important for the secretion of phosphatidyl choline into the bile. Two genes, *mdr1a* and *1b*, result in a similar MDR phenotype in rodents.

P-gp is present mainly on the apical membrane of many secretory cells, including those of the intestine, liver, kidney, choroid plexus, and adrenal gland. In the placenta, P-gp is found on the apical surface of syncytiotrophoblasts, where it can protect the fetus from toxic xenobiotics. P-gp is also abundant on hematopoietic stem cells, where it may protect the cells from toxins, and on the luminal surface of endothelial cells forming physiological barriers such as the blood–testis barrier and the BBB.[18]

P-gp transports not only antineoplastic agents, but also a wide variety of substrates that are not structurally closely related: they are mostly hydrophobic compounds that are either neutral or positively charged, and are probably presented to the transporter binding sites directly from the lipid bilayer. The range of substrates that P-gp recognizes overlaps with those of the main drug-metabolizing enzymes CYP3A4/5. These enzymes are known to metabolize about 50% of the drugs currently on the market. Immunosuppressive agents (cyclosporin A and its analog PSC833), cardiac glycosides (digoxin), protease inhibitors (saquinavir and indinavir), antibiotics (rifampicin), calcium channel blockers (verapamil), and quinoline (quinidine) have been found to interact with P-gp as both substrates and inhibitors.[19] It is now well accepted that P-gp has several drug-binding domains lying within the TMDs (*see* 5.32 In Silico Models for Interactions with Transporters). This helps to clarify why so many substrates are transported by P-gp. A complete catalytic cycle includes the hydrolysis of two ATP molecules, but the two events do not occur simultaneously, although both are needed to transport one substrate molecule. The substrate becomes bound to P-gp and the first ATP hydrolysis causes a change in the conformation of P-gp that releases the substrate to either the outer leaflet of membrane or the extracellular space. The second ATP hydrolysis allows the 're-set' of P-gp so that it can bind another molecule of substrate.[20]

The use of P-gp inhibitors to improve bioavailability of P-gp substrate drugs taken orally and their brain penetration has many clinical interests and is not limited to anticancer therapy. Extensive pharmacotherapy programs have tested three generations of chemical inhibitors. The first-generation inhibitors of P-gp were drugs like verapamil and cyclosporin A that act as competitive inhibitors. They failed in clinical application because they were bound poorly by P-gp; this meant that high doses had to be used, and that led to the concomitant inhibition of the main drug-metabolizing enzyme, CYP3A4, and hence pharmacokinetic interactions. The second-generation P-gp inhibitors, like the cyclosporin A analogue valspodar (PSC 833), are more potent but they also inhibit CYP3A4, again limiting their clinical application. The third-generation inhibitors, such as elacridar (GF120918), zosuquidar (LY335979), and tariquidar (XR9576), are more clinically useful as they are potent inhibitors of P-gp and do not interact with CYP3A4.[21]

The combination of P-gp on the apical or luminal membranes of epithelial or endothelial cells with its capacity to extrude compounds from the cell cytosol or to impede their entry into cells makes this transporter a strategic pharmacokinetic effector that limits absorption of a drug by the intestine, its distribution to the brain, testis, placenta, and stem cells, and enhances its elimination in the bile and urine. The pharmacological and toxicological effects of active drug transport by P-gp have been demonstrated in knockout mice that lack one or both of the murine homologs of the genes encoding human MDR1, Mdr1a, and Mdr1b.[22] The two murine isoforms have approximately 30% amino acid identity with MDR1. Murine P-gps confer resistance to a similar spectrum of anticancer agents as does MDR1. Nevertheless, the two rodent isoforms differ in their tissue distribution. Mdr1a is relatively abundant in the BBB, placenta, liver, and kidney, while Mdr1b is found in the adrenal gland, uterus, ovary, placenta, liver, and kidney. Studies using the Mdr1a or 1b or Mdr1a/1b knockout-mice, which are viable and fertile but more susceptible to developing colitis than wild-type mice, have made a marked contribution to elucidation of the function of P-gp in pharmacology and toxicology.[21]

The discovery in recent years of *MDR1* polymorphisms has amplified the question of the importance of P-gp in the response to drugs and their toxicity in humans. A total of 29 single nucleotide polymorphisms (SNPs) have been reported in the *MDR1* gene.[23] SNPs in exons 26 (C3435 T), 21 (G2677T/A), and 12 (C1236T) are the most frequent genotypic combinations and the allele frequencies of the three variant sites differ markedly between populations of

Table 1 Current human ABC protein, *gene* nomenclature and properties

Human gene name	Protein name	Common name (aliases)	Predominant substrates	Transport mode energy source	Tissue distribution (cellular expression)	Predominant transport impact
ABCB1 (MDR1)	ABCB1 (MDR1)	PGP (P-gp)	Neutral and basic amphiphatic xenobiotics	Primary, ATP	Intestine, liver, kidney, BBB, placenta, plasma membrane (apical in epithelial cells)	Efflux
ABCB11	ABCB11	BSEP (SPGP)	Bile salts, paclitaxel, vinblastine	Primary, ATP	Liver (bile canicular membrane)	Efflux
ABCC1	ABCC1	MRP1	Organic anions; GSH and GSH-, Gluc-, Sulf- conjugates; Leukotriene C4 (LTC4); Anticancer drugs and neutral xenobiotics	Primary, ATP, GSH	All tissues (basolateral in epithelial cells)	Efflux
ABCC2	ABCC2	MRP2 (cMOAT)	Organic anions, GSH and other conjugates similar to MRP1	Primary, ATP, GSH	Liver, intestine, kidney, BBB, placenta (apical in epithelial cells)	Efflux
ABCC3	ABCC3	MRP3 (MOAT-D)	GSH and other conjugates similar to MRP1 bile acids	Primary, ATP, GSH	Liver, intestine, kidney, pancreas (basolateral in epithelial cells)	Efflux
ABCC4	ABCC4	MRP4 (MOAT-B)	cAMP, cGMP, nucleotide analogues and organic anions	Primary, ATP	Lung, kidney, liver, intestine, prostate, testis, ovary, BBB (apical in epithelial cells)	Efflux
ABCC5	ABCC5	MRP5 (MOAT-C)	cAMP, cGMP, nucleotide analogues and organic anions	Primary, ATP	All tissues (basolateral in epithelial cells)	Efflux
ABCC8	ABCC8	MRP8	Cyclic nucleotides, nucleoside analogues, and anionic conjugates	Primary, ATP	Liver, kidney	Efflux
ABCG2	ABCG2	BCRP MXR ABCP	Mitoxantrone, prazosin, topotecan, irinotecan, gleevec, flavopiridol	Primary, ATP	Liver, intestine, placenta, BBB, stem cells, mammary gland (apical in epithelial cells)	Efflux

different ethnic origin. The effect of *MDR1* polymorphisms on the pharmacokinetics of drugs such as digoxin, fexofenadine, cyclosporin A, and tacrolimus remains very controversial, but it seems to involve mainly their absorption by the gut.[23] The C3435T SNP is most frequently associated with nortriptyline-induced postural hypotension in patients with major depression who are being treated with nortriptyline and fluoxetine.[24] However, other studies have shown no association between antidepressant-induced respiratory depression and the *MDR1* C3435T variation.[25] These studies are considerably hampered by the lack of *MDR1* haplotype analysis. A recent study by Yi *et al.*[26] examined exons 12, 21, and 26 of the three most frequent SNPs and showed that the distribution of fexofenadine varied considerably among healthy Korean male volunteers. These subjects were allocated to one of six groups, depending on their drug pharmacokinetics. *MDR1* polymorphism may also influence the susceptibility of an individual to diseases. The exon 26 allele is significantly more frequent in subjects with drug-resistant epilepsy, and the same exon seems to protect patients with parkinsonism, especially those with a history of exposure to pesticides (which are frequently substrates of P-gp).[27,28] Additional studies that focus on haplotypes, environmental factors, and patient size and selection are needed to demonstrate just how *MDR1* polymorphisms influence drug pharmacokinetics.

MDR1 activity can also be altered by a modulation of gene transcription. Transcription of the *MDR1* gene can be modulated by exposure to xenobiotics or mediators of inflammation, as well as by cell stress. The pregnane X receptor (PXR) controls the transcription of *MDR1* and is activated by several steroids, dietary compounds, toxicants, and a number of currently used drugs. PXR regulates several genes encoding phase I and II metabolism enzymes and transporters, like the ABC proteins MRP2 and MRP3 in the liver and intestine.[29] This raises the question of whether there is a coordinated defense system. The body seems to have at least three lines of defense that can be activated in response to xenobiotics. P-gp could be the first reducing the uptake of substrates; phase I and II enzymes producing metabolites could be the second; while the third line of defense could be other transporters, like the MRPs, that excrete the metabolites.

ABCB11 is a second member of this family that can affect drug disposition. This transporter was first reported as the sister of P-gp and called SPGP, or BSEP (bile salt export pump). It is found mainly in the bile canicular membranes and participates in the secretion of bile salts, such as taurocholate, but it can confer resistance to paclitaxel and vinblastine.[30] BSEP appears to be a key target of drug-induced cholestasis, which results in the intracellular accumulation of bile salts, whose detergent actions promote hepatocellular damage by interfering with mitochondrial functions. Cyclosporin A, rifampin, glibenclamide, and bosentan have been shown to inhibit rat BSEP-mediated transport of bile salts.[31] This pump also influences drug absorption by modulating the production of bile salts. These combine with biliary cholesterol and phospholipids to form the micelles that facilitate the solubilization of drugs and their absorption by the intestine.

5.04.4.1.2   ABCC subfamily

Although P-gp is considered to be the major transporter responsible for drug export at the plasma membrane of many cells, MDR phenotypes that are not P-gp-mediated have been linked to several MRPs or ABCCs. The MRP subfamily of proteins was first described in 1992 when Cole and Deeley cloned ABCC1.[32] Other MRPs followed in 1996, and several isoforms of the 12 members of the ABCC subfamily have now been found at various sites in the body. At least four of them, MRP2, 3, 4, 5, and recently MRP8, are likely to be involved in mediating drug resistance and to affect drug pharmacokinetics. Although several molecules are substrates of P-gp when the unconjugated cationic (vincristine and doxorubicin) and neutral (etoposide) compounds are transported, these MRPs preferentially transport anions (like many phase II metabolites of drugs) conjugated to GSH, glucuronate, or sulfate. The MRP4 and MRP5 proteins mainly confer resistance to cyclic nucleosides and purine analogs. They transport substrates by a different mechanism from P-gp; there may even be multiple mechanisms that include co-transport with GSH. All these isoforms are concentrated on specific areas of polarized cells, like the epithelial cells of the gut and kidney, and probably also in the brain microvessel endothelial cells. MRP2 and MRP4 are, like P-gp, found in the apical (luminal) membrane, while MRP1, MRP3, and MRP5 are found in the basolateral (abluminal) membrane.[33,34]

5.04.4.1.2.1   ABCC1 (MPR1)

The 190-kDa MRP1 is the founding member of this subfamily. Despite its modest degree of sequence similarity with P-gp (approximately 15%), MRP1 confers similar degrees of resistance to anthracyclines, vinca alkaloids, camptothecins, and methotrexate, but not to taxanes, which are an important component of the P-gp profile. The substrate specificities of the two pumps differ markedly; MRP1 can transport lipophilic anions, including structurally diverse conjugates of GSH, glucuronate, and sulfate, such as the cysteinyl leukotriene LTC4, an important mediator of inflammatory responses, the estradiol 17β-D-glucuronide (E₂17βG), and sulfated bile acids. LTC4 is the substrate for which MRP1 has the highest affinity, with a K_m of about 100 nM. The ability of MRP1 to transport GSH conjugates indicates that it is a ubiquitous GS-X pump.[35] The precise mechanism by which GSH participates in MRP1-mediated efflux is still unsettled. At least four different mechanisms have been proposed.

The predominant idea is that GSH is a direct, low-affinity substrate for MRP1, or a co-substrate that allows the co-transport of substrates like vincristine, daunorubicin, and aflatoxin B_1. Alternatively, GSH may stimulate the transport of certain compounds by MRP1, but is itself not translocated across the membrane. Conversely, GSH transport can be enhanced by certain substances, like verapamil, that are not themselves substrates of MRP1.[35] Thus, the exact mechanism by which GSH is transported and interacts with MRP1 is quite complex. Although there are several potent inhibitors of P-gp, there are only a few inhibitors of MRP1. One of the more potent is the leukotriene antagonist MK571.[36] Indomethacin and the isoflavonoid genestein have been reported to inhibit the transport function of MRP1 in vitro.[37] Unfortunately, we have no specific inhibitors for each MRP homolog, so that it is quite possible to misinterpret results obtained using inhibitors in whole-cell systems that contain multiple isoforms of MRP.

The murine ortholog of MRP1, Mrp1, is 88% amino acid identical to human MRP1, but shows some differences in substrate specificity despite this great identity. Mouse and rat Mrp1 confer negligible resistance to anthracyclines compared with the human MRP1.[38] Mrp1 ($-/-$) knockout mice were developed and found to be viable, healthy, and fertile, but they have altered immune responses that are attributed to decreased LTC4 secretion.[39,40] Studies on Mrp1 ($-/-$) mice indicate that Mrp1 is involved in the defense against xenobiotics and the regulation of GSH. The GSH concentration is elevated in the tissues of Mrp1 ($-/-$) mice that normally contain high concentrations of Mrp1, whereas the GSH concentrations in tissues normally containing little Mrp1 are the same as in those of wild-type mice. The capacity of Mrp1 to function as an in vivo resistance factor is supported by the finding that Mrp1-deficient mice are hypersensitive to etoposide and the penetration of etoposide into the cerebrospinal fluid (CSF) is increased 10-fold resulting from the lack of Mrp1 at the choroid plexus.[41]

5.04.4.1.2.2 ABCC2 (MRP2, cMOAT)

Prior to the identification of the MRP2 molecule, this protein was known as the canalicular multispecific organic anion transporter (cMOAT). MRP2 has 48% amino acid identity with MRP1 and lies at the apical membranes of epithelial cells. Its function was deduced from investigations in humans and rats that were genetically deficient in the pump.[42] Dubin–Johnson syndrome, the Sprague–Dawley Eisai hyperbilirubinuric rat (EHBR), the Wistar transport deficient (TR$^-$) rat, and Groningen yellow (GY) rats have all been used to assess the role of MRP2 in the excretion of conjugates in the bile.[35] These rat strains have reduced biliary excretion of methotrexate and irinotecan.[43] The drug resistance profile of MRP2 is similar to that of MRP1, but it is less potent. MRP2 also transports uncharged and cationic substrates such as saquinavir, rifampicin, sulfinpyrazone, and ceftriaxone, the flavonoids epicatechin (found in tea) and chrysin, and the meat-derived heterocyclic amine PhIP, a carcinogen with genotoxic properties.[44] The MRP1 inhibitor MK-571 also blocks transport by MRP2. GSH plays a critical role in the MRP2-mediated transport of etoposide, vincristine, and vinblastine, as it does in MRP1, and GSH itself is actively transported into the bile by MRP2. Like MRP1, the transport protein has at least two binding sites that give rise to cooperative binding of substrates or modulators. Thus, the transport of $E_2 17\beta G$ is stimulated 30-fold by adding sulfamitran and sixfold by adding indomethacin.[45]

High concentrations of MRP2 mRNA are present in the human liver, and lower ones in the duodenum, brain, placenta, and kidney.[46] These findings indicate that MRP2 plays a major role in the hepatobiliary excretion of drugs and conjugates. MRP2 is also an important component of the detoxification system of hepatocytes. Recent immunohistochemical and confocal microscopy studies showed that Mrp2 lies in the luminal membranes of the endothelial cells isolated from the brain capillaries of fish, rats, and pigs.[47] This group also demonstrated that the luminal accumulation of the substrate sulforhodamine 101 was inhibited by other MRP substrates, implicating Mrp2 in the transport of drugs from the brain to blood. MRP2 has also been found in the luminal side of the human BBB and is overproduced in patients with resistant epilepsy.[48] Potschka et al.[49] showed that the concentrations of phenytoin and its anticonvulsant activity were significantly higher in the brains of Mrp2-deficient TR$^-$ rats than in normal rats. The authors suggest that MRP2 is involved, like P-gp, in the efflux of a wide spectrum of drugs, sometimes overlapping with that of P-gp. Thus MRP2 might play an important role in the detoxification and protection of the brain, similar to its action in hepatocytes. MRP2 is most abundant in the duodenum of rats and humans, and its concentration decreases toward the terminal ileum and colon, where there is very little. The production of MRP2 in the duodenum is inducible; it responds to signals from several nuclear receptors, including the retinoid X receptor (RXR), the retinoic acids receptor (RAR), and the farnesoid X receptor (FXR). FXR is also involved in bile acid homeostasis in the liver and gut.[45]

5.04.4.1.2.3   ABCC3 (MRP3, MOAT-D)

MRP3 is the human MRP that is most like MRP1, with 58% of amino acid identity. The concentration of *MRP3* mRNA is high in the liver, duodenum, colon, and adrenal glands, and lower in the lung, kidney, bladder, spleen, and ovary.[46,50] Despite the high concentration of MRP3 transcripts in the liver, the protein concentration is relatively low and *MRP3* immunostaining is found in the bile duct epithelial cells and in the basolateral membrane of hepatocytes.[46] The basolateral membranes of hepatocytes of cholestatic patients and those with Dubin–Johnson syndrome contain high concentrations of MRP3, suggesting that there is a compensatory induction of MRP3 in hepatocytes. MRP3 is mainly a GSH S-conjugate transporter that probably does not transport GSH itself. Its presence in hepatocytes and enterocytes indicates that MRP3 is involved in the enterohepatic circulation of nonsulfated and sulfated bile salts such as glycocholates and taurocholates. Thus MRP3 protects hepatocytes and enterocytes against endogenous bile salts and provides an efflux pump that can compensate for a deficient MRP2. The MRP3 substrate profile overlaps with that of MRP1 and MRP2, but the affinity of MRP3 and its capacity to produce drug resistance are not as great as those of either MRP1 or MRP2. MRP3 has a high affinity for glucuronidated compounds like etoposide-glucuronide, $E_2$17βG, and the glucuronides of morphine. A recent study demonstrated that Mrp3 $(-/-)$ mice are unable to excrete morphine-3-glucuronide (M3G) from the liver into the bloodstream.[51] This results in increased concentrations of M3G in the liver and bile and a 50-fold reduction in its plasma concentration. These data, plus the fact that the expression of MRP3 varies greatly from one human to another because of genetic variations and its induction by other compounds, raise questions about its function, particularly its influence on the differences in the pharmacokinetics and pharmacodynamics of morphine seen among individual patients.[52,53]

5.04.4.1.2.4   ABCC4 (MRP4, MOAT-B), ABCC5 (MRP5, MOAT-C), and ABCC8 (MRP8)

MRP1, 2, and 3 all have a third MSD, but MRP4, 5, and 8 do not (**Figure 5**). Thus their functions might be distinct from those of the previous members of the MRP family.

MRP4 and MRP5 do not confer resistance to anthracyclines or vinca alkaloids, but transport organic anions like typical MRP1 substrates ($E_2$17βG, methotrexate and reduced folates). Both pumps also selectively transport the nucleotides cAMP and cGMP, but the extent to which they influence cyclic nucleotide homeostasis seems to be limited by the highly efficient phosphodiesterase system, and they are probably more effective at mediating the extrusion of cAMP and cGMP.[54] Certain nucleoside analogues, such as PMEA (9-(2-phosphonylmethoxyethyl adenine)), employed in the treatment of hepatitis B, and anticancer derivatives like 6-mercaptopurine (6MP) and 6-thioguanine, are also transported by both pumps. Neither transporter translocates uncharged purine and pyrimidine base analogues, but they do carry their metabolites. For example, 6MP is metabolized to its monophosphate intracellular cytotoxic metabolite, which is effluxed by MRP4 and MRP5. It remains to be determined whether these pumps are induced in treated patients, and if they contribute to the resistance of cells to the anticancer agents used in the maintenance therapy of childhood acute lymphoblastic leukemia or to antivirals such as PMEA and zalcitabine (ddc). GSH appears to be a substrate for both MRP4 and MRP5 and the co-transport of bile salts with GSH has been reported.[55] However, the co-transport mechanism is not necessary for all substrates of MRP4.

MRP4 is found in the lung, kidney, bladder, liver, prostate, testis, ovary, and brain. Although the concentration in the liver appears to be low in most species, a recent study found MRP4/Mrp4 at the basolateral membrane in human, mouse and rat hepatocytes.[56] Unlike MRP4, MRP5 is ubiquitous, with high concentrations of its mRNA in the brain, skeletal muscle, lung, and heart.[46,50]

The recent development of *mrp4*-deficient mice has shown that the anticancer drug topotecan, an Mrp4 substrate, accumulates in the brain parenchyma and CSF of these mice.[57] Immunocytochemical analyses showed that Mrp4 is concentrated at the luminal membrane of the brain capillary endothelial cells and the basolateral membrane of the choroid plexus. This dual distribution of Mrp4 suggests that Mrp4 protects the brain from cytotoxins at both BBBs.

The human *MRP8* (*ABCC8*) gene is very similar to *MRP5* and is expressed in various tissues, including the liver and kidney. Like MRP4 and MRP5, MRP8 enhances the extrusion of cyclic nucleotides from cells and confers resistance to nucleoside analogues. LCT4, dehydroepiandrosterone sulfate (DHEAS), $E_2$17βG, monoanionic bile salts, and methotrexate were shown recently to be transported by MRP8.[35]

5.04.4.1.3   **ABCG subfamily**

The members of the ABCG subfamily differ from the more common ABC transporters, like P-gp and MRPs, in that they are half-transporters that are composed of a single NBD followed by one MSD with six transmembrane α-helices. The domains of P-gp and MRP1 are arranged in an opposite fashion, with the MSD followed by the NBD (**Figure 5**).

The second unique feature of the ABCG proteins is that they may work as homodimers or homo-oligomer or hetero-oligomer held together by disulfide bonds.

There are presently four known human members in the G subfamily: ABCG1, ABCG2, ABCG5, and ABCG8. They are all implicated in lipid transport, except ABCG2, which is important for drug resistance and drug disposition. This transporter was cloned independently by three different groups and called breast cancer resistance protein (BCRP), mitoxantrone-resistance protein (MXR), and placenta-specific ABC protein (ABCP), before it was designated ABCG2.[58–60] This second member of the G subfamily confers resistance to anticancer agents by ensuring their energy-dependent efflux. The drug substrates include anthracyclines, mitoxantrone, the camptothecins, topotecan and irinotecan, and flavopiridol, a flavonoid-like cell cycle inhibitor, but not paclitaxel, cisplatin, or vinca alkaloids. BCRP also actively transports structurally diverse organic molecules, both conjugated and unconjugated, such as SN38, the metabolite of irinotecan and its glucuronide conjugate SN38-G, estrone-3-sulfate, 17β-E$_2$G, DHEAS and organic anions like methotrexate.

Hence, BCRP seems to transport sulfated conjugates of steroids and xenobiotics rather than GSH and glucuronide metabolites. Other BCRP substrates include tyrosine kinase inhibitors like imatinib mesylate (Gleevec), which has also been proposed as a potent inhibitor of BCRP, nucleotide reverse transcriptase inhibitors, zidovudine (AZT), its active metabolite AZT5′-monophosphate, lamivudine (3TC), and the proton pump inhibitor pantoprazole.[61] BCRP can transport chemical toxins, in addition to chemotherapeutic agents and conjugated or unconjugated organic anions. These chemicals include pheophorbide a, a breakdown product of dietary chlorophyll that is phototoxic, and the small heterocyclic amine carcinogen PhIP, which induces mammary and prostate cancers. Bcrp1-knockout mice suffer from diet-induced ear phototoxicity that results from accumulation of pheophorbide a in the brain. This occurs because these mice lack the Bcrp1 needed to export pheophorbide a at both the intestine and BBB.[62]

The third-generation P-gp inhibitor, elacridar (GF120918), was recently found to be an efficient inhibitor of human BCRP and mouse Bcrp1. More specific BCRP inhibitors include the fungal toxin derivative fumitremorgin C (FTC); some analogs like KO143 are less toxic than FTC in vivo, but are more potent inhibitors than FTC. Several dietary flavonoids are also potent inhibitors of BCRP. Both 6-prenylchrysin and tectochrysin are more potent inhibitors than elacridar, and are promising new specific inhibitors for the reversal of ABCG2-mediated drug transport.[63] BCRP is found in many hematological malignancies and solid tumors, where it confers resistance to chemotherapeutic agents. BCRP lies primarily in the plasma membrane and at the apical membrane of polarized epithelia, as does P-gp. High concentrations of BCRP are found in the placental syncytiotrophoblasts, the apical membrane of the epithelium of the small intestine, the liver canicular membrane, and at the luminal surface of the endothelial cells of the brain microvessels that form the BBB. Thus ABCG2 is found mainly in organs that are important for absorption (proximal part of the small intestine), distribution (placenta and the BBB), and elimination (liver and small intestine). Its tissue distribution overlaps considerably that of P-gp.[21]

Orally administered topotecan is much more bioavailable when it is given together with elacridar to either mice or humans. This is due to its more efficient intestinal uptake when BCRP is inhibited by elacridar, and its decreased hepatobiliary excretion since canalicular BCRP is also responsible for excretion into the bile.[64] These data suggest that BCRP plays a major role in drug absorption and elimination.[21] BCRP inhibitors might also be used to improve the penetration of BCRP anticancer drugs into the central nervous system (CNS). BCRP might be an important component of the efflux activity of the BBB. Recent studies of the P-gp and BCRP substrate imatinib in Bcrp1 knockout mice and Mdr1a and Mdr1b double-knockout mice show that BCRP, together with P-gp, is involved in the efflux of imatinib at the BBB.[65] The chemotherapy treatment of primary or metastatic brain tumors could be improved by increasing the distribution of anticancer agents into the CNS by adding P-gp and BCRP inhibitors to the chemotherapy cocktail.[21]

BCRP has been found in stem cells and progenitor cells from the interstitial spaces of mammalian skeletal muscle, human pancreas islets, the human liver and heart, and the hematopoietic compartment.[66] The BCRP in stem cells seems to protect them from cytotoxic substrates; this is consistent with the physiological role attributed to BCRP at other sites in the body. BCRP was also recently shown to secrete drugs or toxins into milk.[67] BCRP lies in the apical membrane of the mammary gland alveolar epithelial cells, at the main site of milk production. It is present in the mammary glands of mice, cows, and humans and its concentration increases greatly during late pregnancy, and particularly during lactation in mice. The milk-to-plasma ratios of several drugs, such as aciclovir, cimetidine, and nitrofurantoin, were found to be high even before they were known to be BCRP substrates.[68] Experimental studies with wild-type and Bcrp1 knockout mice have demonstrated that topotecan, the PhIP carcinogen and the hepatocarcinogen aflatoxin B1 become concentrated in milk.[66] As the substrates transported by mouse and human BCRP are very similar, various types of compounds, such as drugs, pesticides, and carcinogens, are very likely to be concentrated in milk, placing infants and consumers of dairy products at risk of dangerously increased exposure. The secretion of xenobiotics into milk by BCRP is puzzling because this function exposes the suckling infant to a range of drugs and toxins. BCRP appears to protect against xenobiotics everywhere else in the body. Its physiological function in the mammary gland remains intriguing.

BCRP seems to have a large number of polymorphic variants, unlike other ABC proteins, and these may yield important functional differences. The first cloned BCRP cDNA was later found to encode a mutant BCRP that deviated from the 'wild type' BCRP at Arg 482 (R482); it was replaced by either threonine (R482T) or glycine (R482G). Mutations in BCRP alter its capacity to transport doxorubicin, rhodamine 123, topotecan, and methotrexate, but not the transport of mitoxantrone.[69] However, these mutants have only been detected in cell lines overexpressing BCRP, and never in any normal human population or cancer specimens. Since the oral bioavaibility and clearance of drugs that are BCRP substrates, such as topotecan, varies greatly from one individual to another, analysis of the BCRP gene for SNPs can provide interesting data. Several mutations have been identified in the coding region of BCRP from healthy individuals and patients.[69] Two of them, V12M and Q141 K, are polymorphic in several populations, with significant differences in allele frequencies between different populations around the world. The allele Q141 K is most frequent in Asian populations (approximately 30–60%) and relatively rare in Caucasians and African-Americans (approximately 5–10%). This variant of BCRP is less abundant at the plasma membrane of transfected cells, and therefore confers lower resistance to drugs than the wild-type protein.[69,70] A recent clinical study correlated the Q141 K polymorphism with changes in the pharmacokinetic properties of diflomotecan. The plasma concentrations of diflomotecan in five heterozygote patients were much higher (299%) than those of 15 patients with wild-type alleles, while 11 known variants in the ABCB1, ABCC2, CYP3A4, and CYP3A5 genes did not affect diflomotecan disposition.[71]

Thus, BCRP can transport a structurally and functionally wide range of organic substrates, and this range overlaps extensively with those of P-gp, MRP1, and MRP2. The tissue distribution of BCRP varies greatly, but overlaps that of P-gp. We infer from this that BCRP has a very similar role to P-gp in the pharmacology and toxicology of substrates.

5.04.4.2   Solute Carrier Transporter Superfamily

While the transporters of the ABC superfamily are mainly concerned with the efflux of substrates from cells, the transporters of the second superfamily, the SLCs, are influx transporters. Unlike the ABC transporters, the physiological roles of which are seldom clearly understood, the SLC proteins are indispensable for the uptake of many essential nutrients by cells. They also play a critical role in the absorption, excretion and toxicity of many xenobiotics. Several inherited diseases, like the glucose-galactose and bile acid malabsorption syndromes, are attributed to mutations of SLC transporters.[72,73]

A total of 362 mammalian genes have, to date, been assigned to the SLC superfamily, divided into 46 families.[8a] The system for naming members of the SLC superfamily differs somewhat from the ABC nomenclature. The genes are usually named using the root symbol SLC, followed by a number corresponding to the family (e.g., SLC22, solute carrier family 22), the letter A, and finally the number of the individual transporter (e.g., SLC22A2). But there may be differences between families. The SLC21 family encoding the organic anion transporting proteins (OATPs) has been reclassified as a superfamily with families and subfamilies much like the classification of drug-metabolizing enzymes. The gene symbol then becomes SLCO (i.e., the '21' and the 'A' have been replaced by the letter 'O' for organic transporter) and the 'OATP' symbol has been kept for protein nomenclature (e.g., SLCO1A2 for the gene and OATP1A2 for the protein).

Tables 2–6 list the main characteristics of many of the SLC proteins implicated in pharmacology and toxicology. The SLC21 and SLC22 families contain several members that transport a variety of structurally diverse organic anions, cations, and uncharged compounds. Most of the members of the SLC21 family of OATPs transport anionic amphipathic compounds. The SLC22 family includes OCTs, the carnitine OCTs novel type (OCTNs), the urate anion-exchanger (URAT1), and several OATs. Other families that may be involved in pharmacology and toxicology include transporters with more restricted substrate specificities. These include the hexose transporters (SLC2 and SLC5), the amino acid (SLC7) transporters, which may be used for drug delivery, and the SLC16 that transport monocarboxylic acids. The main transporters include the SLC15 proteins that carry di- and tripeptides, and the SLC28 and SLC29 proteins that transport nucleoside analogues.

5.04.4.2.1   OATP (SLC21/SLCO) transporters

The nomenclature of the OATP is the most confusing of all the drug transporter families because different names have been coined for individual proteins and no orthologous human OATP genes have been identified for many rodent Oatp genes, indicating no one-to-one relationship between rodent Oatps and human OATPs. All Oatps/OATPs contain 12 TMDs; their structural features include a large extracellular domain between TMDs 9 and 10, N-glycosylation sites in

Table 2 Current human OATP protein, SLCO gene nomenclature and properties

Human gene name (old gene symbol)	Protein name	Common name (aliases)	Predominant substrates	Transport mode and energy source	Tissue distribution (cellular expression)	Predominant transport impact
SLCO1A2 (SLC21A3)	OATP1A2	OATP-A	BSP, bile salts, organic anions and cations (amphiphatic xenobiotics)	Exchanger HCO_3^-, GSH	Liver BBB Kidney (apical in proximal tubule), lung, testis	Uptake Uptake/efflux Reabsorption
SLCO1B1 (SLC21A6)	OATP1B1	OATP-C, LST-1,[a] OATP2	BSP, bile salts, organic anions	Exchanger HCO_3^-, GSH	Liver	Uptake
SLCO1B3 (SLC21A8)	OATP1B3	OATP8	BSP, bile salts, organic anions, digoxin	Exchanger HCO_3^-, GSH	Liver, eye (ciliary body)	Uptake
SLCO1C1 (SLC21A14)	OATP1C1	OATP-F	BSP, thyroid hormones	Exchanger HCO_3^-, GSH	Brain, testis	
SLCO2A1 (SLC21A2)	OATP2A1	PGT[b]	Eicosanoids	Exchanger HCO_3^-, GSH	Ubiquitous	
SLCO2B1 (SLC21A9)	OATP2B1	OATP-B	BSP, DHEAS, E-3-S	Exchanger HCO_3^-, GSH	Liver, placenta, lung, kidney, small intestine, brain	Uptake
SLCO3A1 (SLC21A11)	OATP3A1	OATP-D	E-3-S, prostaglandins	Exchanger HCO_3^-, GSH	Ubiquitous	
SLCO4A1 (SLC21A12)	OATP4A1	OATP-E	Taurocholate, T3, prostaglandins	Exchanger HCO_3^-, GSH	Ubiquitous (liver, heart, placenta, pancreas)	Uptake
SLCO4C1 (SLC21A20)	OATP4C1	OATP-H		Exchanger HCO_3^-, GSH	Kidney (basolateral in proximal tubule)	
SLCO5A1 (SLC21A15)	OATP5A1	OATP-J		Exchanger HCO_3^-, GSH		
SLCO6A1 (SLC21A19)	OATP6A1	OATP-I		Exchanger HCO_3^-, GSH	Testis	

[a] LST, liver-specific transporter.
[b] PGT, prostaglandin transporter.

Table 3 Current human OCT protein, *SLC22* gene nomenclature and properties

Human gene name	Protein name	Common name (aliases)	Predominant substrates	Transport mode and energy source	Tissue distribution (cellular expression)	Predominant transport impact
SLC22A1	OCT1	–	Organic cations, polyspecific	Facilitator	Liver, kidney, intestine, heart, placenta (basolateral in epithelial cells)	Uptake Efflux
SLC22A2	OCT2	–	Organic cations, polyspecific	Facilitator	Kidney (basolateral in proximal tubule), brain, neurons, placenta	Secretion
SLC22A3	OCT3	EMT	Organic cations, polyspecific	Facilitator	Liver, kidney, skeletal muscle, heart, lung, brain, placenta (apical and basolateral in epithelial cells)	Efflux
SLC22A4	OCTN1	–	Organic cations, polyspecific	Exchanger H^+/OC	Kidney (apical in proximal tubule), small intestine, placenta	Secretion Reabsorption Efflux
SLC22A5	OCTN2	CT1	L-Carnitine, organic cations, polyspecific	Exchanger Na^+/carnitine facilitator	Kidney (apical in proximal tubule), heart, skeletal muscle, placenta	Secretion Reabsorption
SLC22A16	OCTN3	CT2	L-Carnitine	Facilitator	Testis	Uptake

CT, carnitine transporter; EMT, extraneuronal monoamine transporter.

the extracellular loops 2 and 5, and the OATP 'superfamily signature' at the border between extracellular loop 3 and TMD6 (**Figure 6**); 13 rat, 11 mouse, and 11 human Oatps/OATPs have been identified to date. The amino acid contents of the mammalian proteins are 31–82% identical, and these similarities have been used to arrange individual proteins into families and subfamilies. The amino acid profiles of the rat, mouse, and human OATP-3 proteins (r Oatp11, m Oatp11, and OATP3A1) are almost identical (97% and 98%).[12]

The OATPs were originally characterized as uptake transporters, although some may function primarily for efflux.[74] The driving force for OATP-mediated transport is still not clear, but it is independent of ATP or sodium gradients. There is experimental evidence that bidirectional transmembrane transport can be mediated by anion (HCO_3^-) or GSH exchange. The GSH gradient may be a powerful driving force. It is due to the high intracellular GSH concentration (approximately 10 mM), the low extracellular concentration (approximately 0.01 mM), the negative charge on GSH at physiological pH, and the negative intracellular potential (-30 to -60 mV).[75] This is well documented for rat Oatp 1 and Oatp 2, but many other OATPs remain to be checked. Oatp 1 (Oatp1a1) was first cloned as a bromosulfophthalein (BSP) and taurocholate uptake system of the rat liver. It has now been shown that many OATPs are polyspecific OATs with partially overlapping substrate specificities for a wide range of solutes, including bile salts, organic dyes (BSP), steroid conjugates (DHEAS, 17βE$_2$G estrone-3-sulfate (E-3-S)), thyroid hormones, neuroactive peptides ((D-penicillamine-2,5) enkephalin (DPDPE), Leu-enkephalin and deltorphin II), and numerous drugs and toxins such as the cardiotonic digoxin, the angiotensin-converting enzyme inhibitors enalapril and temocaprilat, and the 3-hydroxy-3-methylglutaryl coenzyme A (HMG-CoA) reductase inhibitor pravastatin (**Table 2**). OATP substrates are mainly high molecular weight (>450 Da) amphiphatic molecules, mostly bound to albumin, that have a steroid nucleus or linear and cyclic peptides.[76] Most OATPs, mainly those of the OATP1 family, are found in many tissues and though to be part of the overall body detoxification system, helping to remove xenobiotics from the systemic circulation (e.g., drug uptake into hepatocytes). Others, those of families 2–4, may act more specifically in selected organs, such as transporting thyroid hormones or steroids. The rat Oatp1a1, Oatp1a4, and Oatp1b2, and human OATP1B3, OATP1B1, and OATP2B1 are found in the sinusoid membrane of hepatocytes, where they are responsible for the uptake of xenobiotics for hepatic clearance. The hepatic OATPs may have a strategic role in

Table 4 Current human OAT protein, *SLC22* gene nomenclature and properties

Human gene name	Protein name	Common name (aliases)	Predominant substrates (inhibitors)	Transport mode and energy source	Tissue distribution (cellular expression)	Predominant transport impact
SLC22A6	OAT1	PAH	PAH, fluorescein, dicarboxylates, antivirals, antibiotics, diurectics, NSAIDs, heavy metal chelators, thiol conjugates of heavy metals, ochratoxin A (probenecid)	Exchanger Na^+, α-KG	Kidney (basolateral in proximal tubule)	Secretion
SLC22A7	OAT2	NLT	PAH, dicarboxylates, salicylate, (probenecid)	Facilitator	Kidney (basolateral in proximal tubule), liver	Secretion
SLC22A8	OAT3	–	PAH, sulfate and glucuronide steroid conjugates (E-3-S, $E_2 17\beta G$, DHEAS) antivirals, NSAIDs, ochratoxin A (probenecid)	Exchanger, Na^+, α-KG	Kidney (basolateral in proximal tubule), liver	Secretion
SLC22A11	OAT4	–	E-3-S, DHEAS, prostaglandins, ochratoxin A	Facilitator	Kidney (apical in proximal tubule), placenta, liver	Reabsorption Secretion
SLC22A12	URAT1	RST	Urate (monocarboxylic acid, NSAIDs, probenecid, diuretics)	Exchanger, urate, OAs	Kidney (apical in proximal tubule)	Urate reabsorption Secretion

NLT, novel liver-specific transporter; PAH, *p*-aminohippurate transporter; RST, renal-specific transporter.

drug–drug interactions and hepatotoxicity. For example, rifampicin is a potent inhibitor of both OATP transporters and CYP3A4. Thus giving it together with OATP substrates may reduce hepatic first-pass clearance and increase the bioavailability and decrease the efficacy of an intrahepatically active drug like pravastatin.[77] On the other hand, induction of *OATP* gene expression could increase the hepatic uptake and the total body clearance of the substrate. Expression of the rat *Slco1a4* gene is induced by phenobarbital and pregnenolone-16α-carbonitrile (PCN), a well-known inducer of CYP3A4, via the PXR nuclear receptor pathway.[72]

OATPs have also been found at the two brain–blood interfaces. Oatp1a4 and OATP1A2 were found at the luminal and abluminal membranes of the BBB, where they could mediate the efflux of conjugated metabolites and the brain uptake of drugs like digoxin and DPDPE.[72] Oatp1a1 and Oatp1a4 are present in the epithelium of the rat choroid plexus, Oatp1a1 at the apical membrane and Oatp1a4 at the basolateral membrane, where they can account for the secretion of GSH and the removal of LTC4 from the CSF and the uptake of thyroid hormones into the brain.

Only a few OATPs (rat Oatp1a1 and OAT-K1/2, human OATP1A2 and OATP4C1) have been found in the kidney, where they could be responsible for the reabsorption of organic anions.[76,78] The human proteins OATP2B1, OATP3A1, and OATP4A1 and the rat transporters Oatp2b1 and Oatp4a1 have been found at the trophoblast epithelium between the fetal and maternal circulations. The OATP system may facilitate the transport of bile acids and steroid hormones, which are extensively synthesized by the fetal liver in utero, across the trophoblast epithelium from the fetus to the maternal circulation.[79]

No Oatp knockout mice have yet been generated, but studies on several disease models, such as cholestatic liver diseases or cholate feeding, have shown downregulation of hepatocellular *Oatps*. Hence, the loss of basolateral hepatocellular Oatps during cholestasis and massive liver regeneration can explain the impaired transport under these pathological conditions.

Table 5   Current human oligopeptide transporter, *SLC15* gene nomenclature and properties

Human gene name	Protein name	Common name (aliases)	Predominant substrates	Transport mode and energy source	Tissue distribution (cellular expression)	Predominant transport impact
SLC15A1	PEPT1	Oligopeptide transporter 1 H$^+$/peptide transporter 1	Di- and tripeptides, cephalosporins, ACE inhibitors, penicillins, ester pro-drugs (valacyclovir)	Exchanger, H$^+$/S	Small intestine, kidney (apical in epithelial cells)	Uptake Reabsorption
SLC15A2	PEPT2	Oligopeptide transporter 2 H$^+$/peptide transporter 2	Di- and tripeptides and drugs similar to PEPT1	Exchanger, H$^+$/S	Kidney (apical in proximal tubule), brain, lung, mammary gland	Reabsorption
SLC15A3	PHT2, PTR3	Peptide/histidine transporter 2 Human peptide transporter 3	Histidine, oligopeptides	Exchanger, H$^+$/S	Lung, spleen, thymus	–
SLC15A4	PHT1, PTR4	Peptide/histidine transporter 1 Human peptide transporter 4	Histidine, oligopeptides	Exchanger, H$^+$/S	Brain, retina, placenta	–

5.04.4.2.2   OCT (*SLC22*) transporters

The OCTs of the human *SLC22* family include three potential sensitive proteins (OCT1, OCT2, and OCT3) and three H$^+$-driven transporters of carnitine and/or cations (OCTN1, OCTN2, and CT2, also known as *SLC22A4*, *SLC22A5*, and either *FLIPT2* or *SLC22A16*, respectively). Like most members of the *SLC22* family they have 12 TMDs. The amino and carboxyl termini are cytoplasmic and the proteins have glycosylated extracellular loops and intracellular loops that contain phosphorylation sites (**Figure 6**).

OCT1 and OCT2 were originally cloned from the rat kidney[80,81]; the human homologs were then isolated and OCT3 cloned.[82] Both OCT1 and OCT2 lie in the basolateral membranes of epithelial cells, but there is no conclusive evidence as to whether OCT3 is located in the basolateral or apical membrane of polarized cells. The substrate and inhibitor specificities of OCT1, OCT2, and OCT3 overlap extensively, but the OCT subtypes differ significantly in their affinity and maximal transport rates. All three OCTs recognize a variety of organic cations (OCs), including endogenous bioactive amines such as acetylcholine, choline, epinephrine, norepinephrine, dopamine, and serotonin, and drugs like cimetidine, quinine, quinidine, prazosin, desipramine, verapamil, and morphine. Some uncharged compounds and anions are also transported, such as some anionic prostaglandins.[83] The nitrogen moiety of the weak bases bears a net positive charge at physiological pH, allowing electrostatic interaction with the binding sites of the OCTs. The 'type 1' and 'type 2' classifications of OCs were developed to study their uptake by the liver.[84] Type 1 OCs are small (60–350 Da), monovalent, hydrophilic compounds such as tetraethylammonium (TEA) and the parkinsonian neurotoxin 1-methyl-4-phenyl pyridinium (MPP+). In contrast, type 2 OCs are usually bulkier (>500 Da; e.g., anthracyclines) and polyvalent (e.g., d-tubocurarine). This classification helps to differentiate the mechanisms by which they are transported across polarized cells. Type 2 OCs are believed to diffuse across the basolateral membrane and to be exported across the apical membrane by MDR1. In contrast, the basolateral entry of type 1 OCs involves one or more transporters, including OCT1, 2, and 3, and their efflux at the apex may be mediated by OCTN 1 and 2.[85] The OCTs generally mediate bidirectional transport of substrate molecules, and this depends mainly on the membrane potential and not directly on the transmembrane gradients of Na$^+$ or H$^+$. There is thus an electrogenic facilitated diffusion of the monovalent cations in a direction that is defined by the prevailing electrical and chemical (i.e., substrate concentration) gradients across the membrane.[86]

Both OCT1 and OCT2 are found primarily in the major excretory organs (the kidney and the liver) and to a lesser extent in the intestine and the brain. OCT1 is highly species-specific. Human OCT1 is most abundant in the liver, where it mediates the uptake of type1 OCs across the sinusoidal membrane of hepatocytes; it is less abundant in the

Table 6 Current human concentrative and equilibrative nucleoside transporters, *SLC28* and *SLC29* gene nomenclature and properties

Human gene name	Protein name	Common name (aliases)	Predominant substrates	Transport mode and energy source	Tissue distribution (cellular expression)	Predominant transport impact
SLC28A1	CNT1	–	Pyrimidine nucleosides, adenosine AZT, 3TC, ddC, Arac, dFdc	Exchanger, Na$^+$/S	Liver, small intestine, kidney (apical in proximal tubule)	–
SLC28A2	CNT2	SPNT	Purine nucleosides, uridine, ddI, ribavirin	Exchanger, Na$^+$/S	Kidney liver, heart, brain, placenta (apical in epithelial cells)	–
SLC28A3	CNT3	–	Purine and pyrimidine nucleosides, anticancer nucleoside analogs	Exchanger, Na$^+$/S	Pancreas, bone marrow, mammary gland	–
SLC29A1	ENT1	–	Purine and pyrimidine nucleosides	Facilitator	Ubiquitous (basolateral in epithelial cells)	Uptake and Efflux
SLC29A2	ENT2	–	Purine and pyrimidine nucleosides, nucleobases (AZT, ddC, ddI)	Facilitator	Ubiquitous (basolateral in epithelial cells)	Uptake and Efflux
	ENT3	–	Purine and pyrimidine nucleosides	ND	Ubiquitous (intracellular?)	–
	ENT4	–	Adenosine	ND	Ubiquitous	–

ND, not determined; SPNT, sodium-dependent purine nucleoside transporter.

intestine and kidney, where it is less than 1% of that of OCT2.[85] There is enough OCT1 in the kidneys of rats, mice, and rabbits for it to play a significant role in substrate transport. OCT2 is mainly found in the kidneys of rats and humans. The amount of OCT2 in the kidney depends on gender and age. The kidneys of males contain more *oct2* mRNA and protein than do those of females, whereas the concentrations of OCT2 in young male and female rats are similar.[87] Although the physiological relevance of this gender dimorphism is unknown, it might be responsible for differences in the renal elimination of OCs and responses and sensitivity to drugs. OCT2 is also present at the choroid plexus and in various dopamine-rich regions of the brain (substantia nigra, nucleus accumbens, and striatum), where it might act as a 'background' transporter in the removal of monoamine neurotransmitters that have escaped reuptake by high-affinity monoamine transporters.[83]

OCT3, unlike OCT1 and OCT2, is much more widely distributed. *hOCT3* mRNA has been detected in the aorta, skeletal muscle, prostate, adrenal gland, and placenta.[88] OCT3 is also abundant in the mammalian brain, where it has been identified as an extraneuronal monoamine transporter like OCT2.[89] OCT3 also plays a very limited role in the transport of type 1 OCs (TEA) in the human and rodent kidney. Nevertheless, OCT3 seems to play a significant role in the handling of amantadine, a putative dopaminergic compound, by the rat and rabbit kidneys.[86]

The quantitative significance of the OCTs in drug pharmacokinetics was clearly established in studies with mice lacking the three Octs. All three knockout strains are viable, fertile, and display no physiological abnormalities.[83] However, the knockout of Octs markedly affects the pharmacokinetics of OC substrates. For example, the knockout of both Oct1 and Oct2 transporters eliminates the renal tubular secretion of TEA and reduces the hepatic uptake and clearance of TEA by about 80% in the Oct1 knockout mice due to the lack of Oct1 at the sinusoidal membrane of the hepatocytes.[89] The intestinal secretion of TEA is reduced by about 50%, indicating that Oct1 also mediates the basolateral uptake of TEA into enterocytes. Similar observations were made with the liver uptake of several biguanides (metformin and buformin); these antidiabetes drugs are sometimes responsible for lactic acidosis. The accumulation of metformin in the liver was

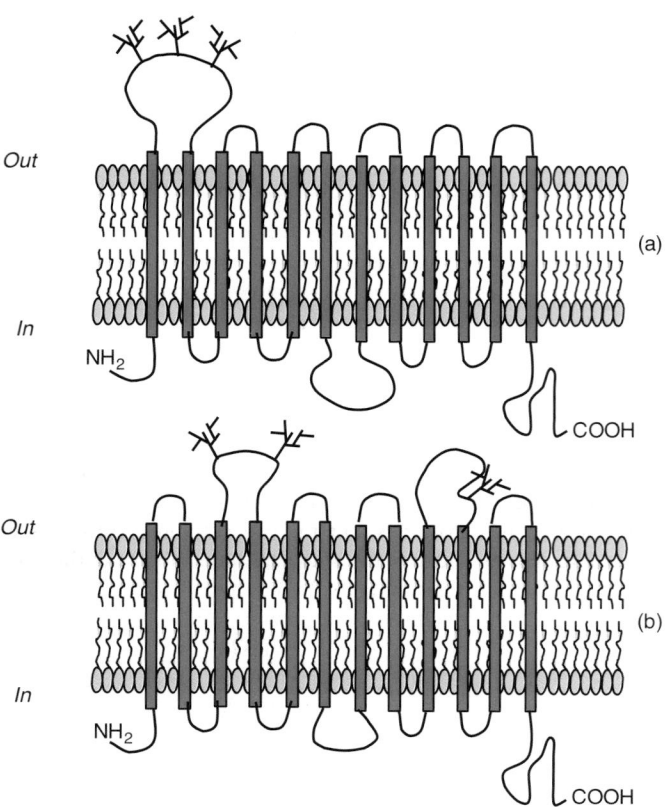

Figure 6 Topological models of the human *SLC22*; (a) organic cation transporters (OCT), organic anion transporters (OAT), and *SLCO*; (b) the organic anion polypeptide transporter (OATP). Features common to all members of the OCT, OAT, OATP transporter family include 12 transmembrane-spanning domains (TMDs) with intracellular amino and carboxyl termini. Transmembrane α-helical domains (TMD) indicated by red rectangular bars; *N*-glycosylation sites (indicated by branches) are present on extracellular protein loops. Cytoplasmic (*in*) and extracellular (*out*) orientations are indicated.

more than 30-fold lower than in wild-type mice. The metformin-induced blood lactate concentration is also decreased in the knockout mice, suggesting that Oct1 is involved in the hepatotoxicity of biguanides.[90]

The OCTNs differ markedly in their mode of action, unlike the OCTs, which have a common energy-supply mechanism (**Table 3**). OCTN1 supports electroneutral OC/H^+ exchange, OCTN2 supports both Na^+-dependent co-transport (e.g., carnitine) and electrogenic-facilitated diffusion (e.g., TEA and type 1 OCs), and OCTN3 mediates the electrogenic transport of carnitine. OCTN3 and CT2 are present only in the testes of mice and humans,[91] where transported carnitine improves sperm quality and fertility. OCTN1 and OCTN2 were originally cloned from human fetal liver and placenta.[92,93] OCTN1 is most abundant in the kidney (at the apical membrane of the tubular cells), small intestine, bone marrow, and fetal liver (but not in the adult liver).[92] OCTN2 is mainly found in the heart, placenta, skeletal muscle, kidney, and pancreas.[93] Both OCTN1 and OCTN2 have a low affinity for MPP^+, cimetidine, and TEA, and OCTN2 plays a major role in carnitine homeostasis.[85,93]

5.04.4.2.3 OAT (*SLC22*) transporters

As their name implies, small organic anions (300–500 Da) possess a net negative charge at physiological pH and their transepithelial transport into the negatively charged environment of the cell requires energy. This is largely to the OATs (*SLC22* family) that are found mainly in cells playing a critical role in the excretion and detoxification of xenobiotics. The OAT family contains six members (OAT1, OAT2, OAT3, OAT4, OAT5, and URAT1), present mainly in the liver, kidney, placenta, brain capillaries, and choroid plexus (**Table 4**). Their topological structures are very similar to those of the OCTs (**Figure 6**), but they have more complex energy requirements.[94] The members of the OAT1, OAT2, and OAT3 group form a tertiary active system, with the first driving element being a Na^+ gradient furnished by the Na^+,

K^+-ATPase pump. The inward movement of Na^+ drives the uptake of dicarboxylate α-ketoglutarate (α-KG) by a second transport protein, the Na^+/dicarboxylate co-transporter 3 (NaDC3). This, in conjunction with mitochondrial α-KG production, maintains an outward α-KG gradient.[95] Finally, the third transporter in the chain is one of the three OATs; it functions like an α-KG/organic anion exchanger for translocation of organic anion substrates into the cell. This differs slightly from the mechanism depicted in **Figure 3**. As a result, a negatively charged molecule can enter the cell against its chemical concentration gradient and the negative electrical potential of the cell. While these three OATs are present at the basolateral membrane and function mainly as influx systems, OAT4 and URAT1 are found at the cell apex and may function as either Na^+-independent organic anion exchangers or membrane potential-facilitators allowing the bidirectional transport of organic anions.[96]

The first OAT, rat and mouse Oat1, was cloned by several groups in 1997, and its orthologs were subsequently isolated from the kidneys of rabbits, humans, and pigs.[97] OAT1 is mainly a kidney-specific transporter, found at the basolateral membrane of the renal proximal tubular cells. OAT1 was earlier known as the *p*-aminohippurate (PAH) transporter. As PAH is almost completely extracted during a single pass through the kidney, it has been used as the prototypical substrate for the renal organic anion transport system and as a biological marker of renal plasma flow, which is equal to the renal clearance of PAH.[98]

OAT2 was originally isolated from the rat liver as a novel liver-specific transporter (NLT) labeling the sinusoidal membrane of rat hepatocytes; it was later found in the kidney and choroid plexus.[95] The concentration of Oat2 is higher in the kidney than in the liver of female rats, but the concentrations are reversed in male rats. *OAT3* gene expression has been detected in mammalian kidneys and in the apical membrane of the choroid plexus and basolateral membrane of the endothelial cells at the BBB from rodents. No orthologs of human OAT4 have been identified. OAT4 is present in the kidney, liver, and placenta.[96] Oat5 is the first rodent Oat to be detected in the kidney, but not in the choroid plexus, in marked contrast to Oat1, Oat2, and Oat3. URAT1 is expressed in the luminal membrane of the kidney proximal tubules, and was originally identified in the mouse as a renal specific transporter (RST). URAT1 is involved in renal reabsorbtion of urate and helps to maintain blood levels of uric acid. It is the in vivo target for the uricosuric acid drugs probenicid and benzbromarone and the antiuricosuric acid drug pyrazinamide.[96]

The OAT proteins play a critical role in the excretion and detoxification of a wide variety of drugs, toxins, hormones, and neurotransmitter metabolites. Uremic toxins, which accumulate in the blood during chronic renal failure, inhibit renal transport, hepatic drug metabolism and serum protein binding. Several uremic toxins (indoxyl sulfate, indoleacetic acid, hippuric acid, etc.) are substrates or inhibitors of both rat and human OAT1 and OAT3; this could explain changes in the pharmacokinetics of other OAT substrates, including drugs.[96] Several acidic metabolites of neurotransmitters are excreted into the CSF via OAT transporters, which is consistent with the presence of OATs in the choroid plexus. A number of common nonsteroidal anti-inflammatory drugs (NSAIDs), including acetyl salicylate and salicylate, acetaminophen, diclofenac, ibuprofen, ketoprofen, indomethacin, and naproxen, are substrates of one or more OAT isoforms, so that there can be significant interactions between NSAIDs and other drugs. The β-lactam antibiotics (penicillins, cephalosporins, and penems) and the antiviral nucleosides adefovir, cidofovir, aciclovir, and AZT are also substrates of one or more OAT isoforms and are actively excreted in the urine.[98] Toxins like chlorinated phenoxyacetic acid herbicides, mercuric conjugates, cadmium, and ochratoxin A are also transported either into renal tubular or hepatocyte cells via the OAT network, and this predisposes these tissues to nephrotoxicity or hepatotoxicity.[95]

5.04.4.2.4 PEPT1 (*SLC15A1*) and PEPT2 (*SLC15A2*) transporters

Peptide transporters 1 and 2 (PEPT1 and PEPT2, *SLC15A*1 and *SLC15A2*, respectively) are H^+-coupled oligopeptide symporters (**Figure 3**) whose predicted membrane topology anticipates the 12 TMDs of the SLC family (**Figure 6**). More recently, two new type 1 (PHT1, *SLC15A4*) and type 2 (PHT2, *SLC15A3*) peptide/histidine transporters have been identified in mammals. PEPT1 and PEPT2 translocate dipeptides and tripeptides produced by protein catabolism. Considering all possible combinations, they can transport 400 dipeptides and 8000 tripeptides derived from the 20 L-α amino acids present in proteins.[99] Their pharmacological importance results from their ability to transport a wide variety of peptide-mimetic drugs, such as β-lactam antibiotics of the cephalosporin and penicillin classes and drugs like captopril, enalapril, and fosinopril. Other drugs include the dopamine D2 receptor antagonist sulpiride and the peptidase inhibitor bestatin.[99] PHT1 and PHT2 transport histidine and certain di- and tripeptides, but their location on the cell or lysosomal membranes remains as questionable as their implication in pharmacotherapy.[100] PEPT1 and PEPT2 differ in their transport properties and tissue distributions (**Table 5**). PEPT1 is the low-affinity (millimolar range), high-capacity transporter that is mainly found in the apical membranes of enterocytes in the small intestine, in renal proximal tubular cells of the S1 segment, and in bile duct epithelial cells. In contrast, PEPT2 is a high-affinity (micromolar range), low-capacity transporter that is more widely distributed in the apical membranes of kidney tubular cells of the S2 and S3 segments, brain astrocytes, and epithelial cells of the choroid plexus. They are involved in the uptake of their substrates, leaving a basal transporter(s) to account for the exit. This basal transporter could be PHT1 and/or PHT2, or the amino acid transporters of the *SLC1* and *SLC7* families.[99] Both PEPT1 and PEPT2 can mediate the renal reabsorption of the filtered compounds in kidney tubules, whereas PEPT2 may be responsible for the removal of brain-derived peptide substrates from the CSF via the choroid plexus.[101] Recent studies on *pept2*-deficient mice (which are viable and have no kidney or brain abnormalities) have confirmed that Pept2 is involved in the uptake role of peptides into the cells of the choroid plexus and the proximal tubule.[102] The pharmaceutical relevance of these peptide transporters is closely linked to the design of drug delivery strategies mediated by the intestinal PEPT1. One successful approach has been to produce peptide derivatives of parent compounds as substrates for PEPT1. The pharmacophoric pattern for the transporter includes the rules that the peptide bond is not a prerequisite for a substrate and that 5'amino acid esterification, mostly using L-valine or L-alanine, markedly improves recognition by PEPT1.[103] This prodrug strategy was used to improve the bioavaibility of oral enalapril from 3–12% to 60–70% for the ester enalaprilat, which resembles the structure of the tripeptide Phe-Ala-Pro. The oral bioavaibility of the nucleoside antiviral aciclovir (22%) was similarly improved by adding a valine residue to give valaciclovir (70%).[103] Current studies on the regulation of PEPT1 and PEPT2 synthesis in inflammatory intestinal diseases may provide helpful information on the variations in bioavaibility of oral PEPT1 drug substrates.[99]

5.04.4.2.5 CNT (*SLC28*) and ENT (*SLC29*) transporters

The members of the human *SLC28* and *29* families catalyze mainly the transport of purine and pyrimidine nucleosides.[104] The hydrophilic nucleosides, like the purine adenosine, are important signaling molecules that control both neurotransmission and cardiovascular activity. They are also precursors of nucleotides, the constitutive elements of DNA and RNA, and are the basic elements of a variety of antineoplastic and antiviral drugs. The *SLC28* proteins in the apical membranes of polarized cells work in tandem with the *SLC29* proteins found in the basolateral membrane. Thus, the members of these two families are critical for the transepithelial transport of nucleosides – they form a coordinated transport system.

The *SLC28* family consists of three dependent concentrative nucleoside transporters, CNT1, CNT2, and CNT3 (*SLC28A1*, *SLC28A2*, and *SLC28A3*, respectively). They co-transport Na^+ and the substrate in a symporter mode and are considered to be a high-affinity, low-capacity transporter.[105] The members of the *SLC28* family have 13 TMDs, unlike the 12 TMDs of most of the SLC, with the Na^+ and substrate recognition sites in the carboxyl half of the protein. The three subtypes differ in their substrate specificities. CNT1 transports naturally occurring pyrimidine nucleosides plus the purine adenosine. Several antiviral analogues, like AZT, lamivudine (3TC), and ddc, are substrates of CNT1. The cytotoxic cytidine analogues, cytarabine (AraC) and gemcitabine (dFdc) are also transported by CNT1.[106] CNT1 is primarily found at the apical membrane in epithelial tissues including the small intestine, kidney, and liver (**Table 6**).

Human CNT2 is widely distributed in the kidney, liver, heart, brain, intestine, skeletal muscle, pancreas, and placenta. CNT2 transports purine nucleosides and uridine. Pharmaceutical substrates include the antiviral didanosine (ddI) and ribavirin.[107]

Human CNT3 has, like CNT2, a wide tissue distribution with high concentrations in the pancreas, bone marrow, and mammary gland. CNT3 is broadly selective and transports both purine and pyrimidine nucleosides in a 2:1 Na^+/nucleoside coupling ratio, in contrast to the 1:1 ratio employed by CNT1 and CNT2. CNT3 transports several anticancer nucleoside analogues, including cladrabine, dFdc, fludarabine, and zebularine.[108]

A number of coding region SNPs have been reported for human CNT1 and CNT2,[108a] and the determination of their function is an important goal for future investigations.[105]

The human *SLC29* family contains four members. These equilibrative nucleoside transporters (ENTs) include the well-characterized, low-affinity facilitators ENT1 (*SLC29A1*) and ENT2 (*SLC29A2*), but the cation-dependence, and the tissue and subcellular distribution of ENT3 have not yet been fully evaluated. The ENT proteins have 11 TMDs, in which the amino-terminus is in the cytoplasm and the carboxyl-terminus is extracellular. ENT1 and ENT2 can both transport adenosine, but differ in their abilities to transport other nucleosides and nucleobases.[109] ENT1 is almost ubiquitously distributed in human and rodent tissues and transports purine and pyrimidine nucleosides with K_m values of 50 μM (adenoside) to 680 μM (cytidine). The antiviral drugs ddC and ddI are also poorly transported.[110] ENT2 is present in a wide range of tissues, including the brain, heart, pancreas, prostate, and kidney, and is particularly abundant in skeletal muscle. ENT2 differs from ENT1 in that it can also transport nucleobases like hypoxanthine and AZT.[111] ENTs also mediate the uptake and efflux of several nucleoside drugs because of their bidirectional transport property. Selective inhibition of ENTs may be a strategy for improving therapy with nucleoside drugs. For example, the vasodilator draflazine increases and prolongs the cardiovascular effects of adenosine by inhibiting nucleoside uptake into endothelial cells. Dipyridamole, dilazep, and imatinib all selectively inhibit ENTs and modulate the toxic or therapeutic effects of the endogenous nucleosides.[104]

5.04.5 Expression and Function Properties of Transporters in Tissues

About 40 transporters belonging to the ABC and SLC superfamilies are presently known to influence the pharmacokinetics, pharmacodynamics, and toxicity of drugs and xenobiotics. While there are still many gaps in our knowledge of the function of transporters, it is possible to outline their main properties.

5.04.5.1 Transporters at the Plasma Membrane and Within Cells

There are about 200 different types of cells in human tissues, and all their plasma membranes and the membranes of their organelles contain transporters. As outlined above, they can import or export endogenous and exogenous compounds, thereby regulating intracellular ions and essential nutrients and controlling the traffic of exogenous compounds into, from, and within cells. They are thus the fundamental effectors of physiological and pharmacological processes. The drug transporters at the organelles may well become most important in future studies. This was recently documented during a dramatic phase II trial in which the nucleoside antiviral fialuridine (FIAU) caused the death of subjects as a result of severe toxicity, including hepatotoxicity, pancreatis, neuropathy, or myopathy.[112] These toxic events were clearly linked to mitochondrial damage due to the inactivation of the mitochodrial DNA polymerase γ by FIAU or its triphosphorylated metabolite. Before this can happen, FIAU must be transported into the cytosol, and then it, or its phosphorylated metabolite, must be transported into the mitochondria. Molecular and imaging techniques were used to show that FIAU is transported by ENT1, which is located in both the plasma and mitochondria membranes, but not in the nuclear envelope, endosomes, lysosomes or Golgi complex. But the mitochondrial membranes of rodents do not contain Ent1, and they are not affected by this toxic action. This variation in FIAU toxicity and ENT1 expression between species dramatically highlights the inability of preclinical studies on rodents to predict the toxicity of nucleoside analogs for human mitochondria. It also shows how the toxicity of antiviral drugs like AZT, stavudine, and ddI for mitochondria is linked to the role of a transporter.[112]

5.04.5.2 Tissue Specificity

The distribution of certain transporters may be restricted to a limited range of cells or tissues. Thus, MRP2, and MRP3, OCT1, and OATPs are mainly found in the liver, while OAT1 is restricted to the kidney. In contrast others, like MRP1, the CNTs, and ENTs, are ubiquitous, being present throughout the human body. This indicates that the transporters with a limited distribution are more specialized. For example, MRP2 and MRP3 are concerned with the export of conjugated metabolites from hepatocytes, while the ubiquitous transporters are implicated in essential physiological functions such as the export of LTC4 and GSH (MRP1) and uptake of nucleosides (CNTs and ENTs), and their influence on pharmacokinetics may be less organ-specific.[35,104]

5.04.5.3 Polarized Cellular Distribution

The cellular location of transporters at the plasma membrane is a critical issue because most of the cells involved in the A, D, and E pharmacokinetic processes are polarized. Hence their apical (luminal) and basolateral (abluminal) membranes do not have the same populations of transporters. The same transporter is found at both the apical and basolateral membranes in only a few cases. Good examples are the glucose transporter GLUT1 (*SLC2*) and the monocarboxylic transporter MCT1 (*SLC7*), and they ensure the efficient transepithelial transport of their substrates.[8] However, most of the ABC and SLC transporters are found at either the apical or the basolateral epithelial membranes, and their location helps define the direction of substrate transport and the resulting pharmacokinetic event. Efflux transporters are usually found in the apical membrane of epithelial cells, where their main function is to reduce the uptake of substrates. Most of the ABC efflux transporters (P-gp, BCRP, MRP2, and MRP4) lie in the apical membrane, although MRP1, MRP3, and MRP5 are more usually found in basolateral membranes.[55] In contrast, influx transporters are more equally distributed on both apical and basolateral membranes. Although the great majority of transporters are specific to the apical or basolateral sides, there are exceptions. One is the rat Oatp1a1, which is found on the basolateral membrane of hepatocytes, where it takes up substrates from the blood, and on the apical membrane of renal tubular cells, where it reabsorbs substrates from the urine.[113] Another is MRP4, which is localized in the basolateral membrane in prostatic glandular cells and in apical membrane in rat kidney tubule cells.[114] The basolateral segment of the epithelial cells of the small intestine and the kidney renal tubules face the blood, while the apical sides are in contact with the gut lumen or urine. However, the blood (luminal) side of the BBB is in contact with the apical side of the brain endothelial cells, while their basolateral sides face the brain. Thus, each tissue must be carefully examined to determine the apical/basolateral distributions of transporters and the correlation between this and the cell's environment (blood, CSF, extracellular fluid, urine, etc.). The mode of transport, such as bidirectional facilitated diffusion or one-directional active transport, must also be considered before assigning a direction to the transport of the substrate.

5.04.5.4   Coordinated and Vectorial Transports

The net pharmacokinetic effect of active transport processes is frequently due to the involvement of several transporters that may not always belong to the same superfamily or subfamily. For example, the OATPs (*SLCO*) on the sinusoidal (basolateral) membrane of hepatocytes take up organic anions, while the MRP2 (*ABC*) on the apical bile canicular membrane excrete them. The combined activities of these two transporters thus results in the vectorial transport of drugs from the blood to the bile.[113] Similarly, the basolateral transporters OCT1, OCT2, and OCT3 of the kidney tubular cells act in a coordinated, vectorial manner with the apical transporters OCTN1 and OCTN2 to secrete OCs from the blood to the urine, whereas the basolateral OAT1, OAT2, and OAT3 are coupled to the apical OAT4, and URAT1 for transporting organic anions in the same tissue.[85,95] Unfortunately, the apical or basolateral partners of several coordinated systems have not yet been identified. Two examples are the apical peptide transporters PEPT1 and PEPT2; their basolateral partners have not yet been found.[99]

5.04.5.5   Coordination between Transporters and Enzymes

Drug metabolism was considered to be one of the main processes of removing xenobiotics prior to the emergence of transporters. The CYP isoenzymes catalyze the first step of biotransformation and this function was called phase I metabolism. The subsequent conjugation step was called phase II metabolism. We now know that these two phases occur in specialized cells like the hepatocytes and enterocytes, and that they are preceded and followed by two other phases controlled by transporters. Efflux or influx transporters reduce or increase the uptake of substrates, and these actions help regulate the amounts of xenobiotics reaching the enzyme binding sites or the rate at which the metabolites produced are eliminated. The first step has been called 'phase 0' and the second 'phase III,' indicating a close relationship between transporters and enzymes. They provide the cell with a suite of events and operate both in parallel and in series. This integrated biological function of combined transport and metabolic processes is strongly supported by the presence of common regulation pathways that act via similar nuclear receptors, such as PXR, RXR, and others, to induce or repress the genes encoding enzymes and transporters.[45,115]

5.04.5.6   Polyspecific Transport and Inhibition

The substrate specificities of transporters are often very broad, as indicated by the many overlaps of substrates and inhibitors, much like the specificity of the drug metabolism enzymes. Thus, probenecid was initially known to produce many drug interactions by blocking the secretion by the kidney of many drugs, including the penicillins and the antiviral Tamiflu.[116] Probenicid is today known to be a polyspecific inhibitor of several MRPs, OATs, OATPs, and even OCTs.[117] Similarly, MRP1, MRP2, and MRP3 have broad, overlapping substrate specificities, while OCT1, OCT2, and OCT3 transport a wide range of similar OCs. P-gp interacts with a multitude of xenobiotics, many of which are metabolized by CYP3A4/5, and some of them are also substrates of MRP1 and BCRP.[21] The tyrosine kinase inhibitor imatinib is effluxed by both P-gp and BCRP, and imported by OCT1.[118] Thus all ionized chemicals, peptides, and nucleosides that cannot diffuse freely across membranes are very likely to interact with one or more transporters. A review of the entry and export of 14 antiretroviral nucleoside agents into and from cells found that 14 distinct transporters of the *ABC* and *SLC* superfamilies were involved in translocating these drugs, illustrating this unique property of polyspecificity.[104]

5.04.5.7   Variations in Transport Kinetics

As each transporter has a limited capacity, it can be saturated by substrate concentrations greater than its K_m. The K_m of transporters can vary from nanomolar to millimolar values and the risk of saturating transport will depend on the amount of substrate in the transporter environment. It was mentioned above that the efflux of P-gp at the BBB can be saturated with a taxane derivative, leading to its nonlinear penetration into the brain parenchyma.[119]

Transport can also be inhibited in a competitive or noncompetitive manner, in the same way as the drug-metabolizing enzymes, so that transporters can promote drug–drug interactions that were initially thought to be due to the drug-metabolizing enzymes alone. The calcium channel blocker mibefradil, which was first believed to be only a CYP3A4 inhibitor, is now known to inhibit P-gp.[120] Similarly, the effect of grapefruit juice on the disposition of fexofenadine, which was attributed to the inhibition of CYP3A4, is now known to result from its ability to inhibit the OATP-mediated uptake of fexofenadine.[121] In vitro transporter assays are increasingly being used to assess the potential risks of drug–drug interactions mediated by transporters. The in vitro inhibition constant (K_i) can be measured and used to predict changes in the clearance or systemic exposure by measuring the area under the curve (AUC).[122] Transport kinetics may also depend on the amount of transporter, which will depend on the actions of drugs, nutrients, and disease states on the nuclear receptor pathways mentioned above. The most recent area of variation concerns the presence of genetic polymorphisms; these have been described for MDR1, BCRP, and nucleoside transporters. However, studies on the pharmacogenetics of most drug transporters have only recently begun. SNPs have been identified in drug-uptake transporters such as OCT1 and OATP1B1, and those in OATP1B1 may have an impact on the therapeutic efficacy and toxicity of HMG-CoA reductase inhibitors such as pravastatin.[123]

5.04.6 Roles of Transporters in Pharmacokinetics

Transporters are now recognized to be as important as the metabolizing enzymes in the modulation of the main steps controlling the fate and action of xenobiotics in the body. They affect all the main pharmacokinetic events, such as the oral bioavailability, distribution, and clearance of any substrates. As most drugs, toxicants, and carcinogens are taken orally, we focus here on their impact on the enterocytes, as these are the main intestinal cells involved in the passage of compounds from the intestine to the blood. The absorbed compounds are then cleared by the liver, where the hepatocytes may produce, like the enterocytes, coordinated transporter–enzyme systems that influence the overall availability of compounds taken via the mouth before they can be distributed throughout the body. Tissue distribution depends on many factors, such as the degree to which a drug is bound to plasma proteins, the organ perfusion rate, and the rate at which the drug permeates biological membranes. Here too, transporter-mediated effects are particularly important at the so-called barriers between the blood and specific organs, like the BBB, the blood–CSF barrier at the choroid plexus, and the blood–testis, blood–prostate, and blood–placenta barriers. All these barriers protect tissues from external agents, and transporters are now considered to be the predominant actors in this defense system. They are frequently described as the gatekeepers of these physiological barriers between the exterior and interior environments. Finally, the last step concerns the elimination of parent compounds and their metabolites. The two most frequent routes for the removal of xenobiotics that do not require their metabolism are excretion in the bile or urine. **Figure 7** illustrates all these steps from the absorption to the elimination of a xenobiotic, including the key organs in drug disposition and the networks of transporters at the apical or basolateral poles of their cells.

5.04.6.1 Intestinal Absorption

Drugs can be absorbed from the intestine by passive transcellular, paracellular, and carrier-mediated transport. The development of various in vitro experimental techniques in the 1980s, such as the human colon adenocarcinoma cell line Caco-2, which forms confluent monolayers of well-differentiated enterocyte-like cells, facilitated the characterization of intestinal transepithelial transport. Numerous SLC transporters translocating a wide range of natural and essential nutrients were found on both the intestinal brush border and the basolateral membrane.[124] Some of them, like the amino acid (*SLC7*) and monocarboxylic acid transporters (MCT, *SLC16*), may also transport drugs and xenobiotics.[125] Both influx and efflux transporters modulating drug absorption are present in the epithelium of the various segments of the intestine.[126] PEPT1, OATP2B1, OATP3A1, and OATP4A1 are all on the apical membrane and mostly import substrates from the lumen to the circulation (**Figure 8**). PEPT1 is the best-characterized drug transporter in the small intestine of mammals, and is widely used to improve the absorption of poorly absorbed oral drugs using a prodrug strategy (*see* Section 5.04.4.2.4). ABC transporters, including MDR1, MRP2, and BCRP, are also found on the apical membrane, where they either limit the intestinal uptake of their substrates or contribute to the active secretion of drugs from the blood to the intestinal lumen.[127] For example, the antineoplastic agent, paclitaxel, a P-gp substrate, is poorly absorbed when taken orally in humans (only 5% is bioavailable), but when it is administered with the P-gp competitor cyclosporin A, its bioavailability is increased to 50%.[127] The roles of basolateral transporters are much less well known. Oct1 is present on the basolateral sides of cells, and studies using Oct1 knockout mice indicate that Oct1 is important for the secretion of OCs into the lumen of the small intestine. The intestinal distribution of human OCT1 and OCT3 is still poorly documented, whereas the functional activity of OCT1, 2, and 3 has been demonstrated using Caco-2 cells.[128] The network of OCTs in the small intestine remains to be identified.

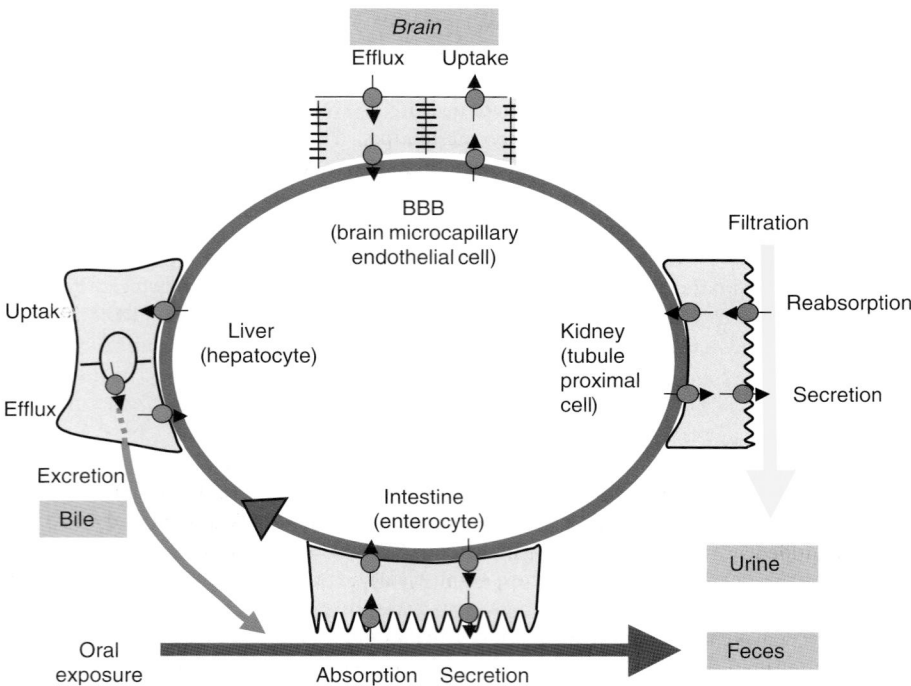

Figure 7   Drug absorption, distribution, and elimination: the ABC and/or SLC transporter network mediate vectorial transport in each of the key body organs, with transporters at the basolateral or apical membranes of the organ-specific cells indicated in brackets.

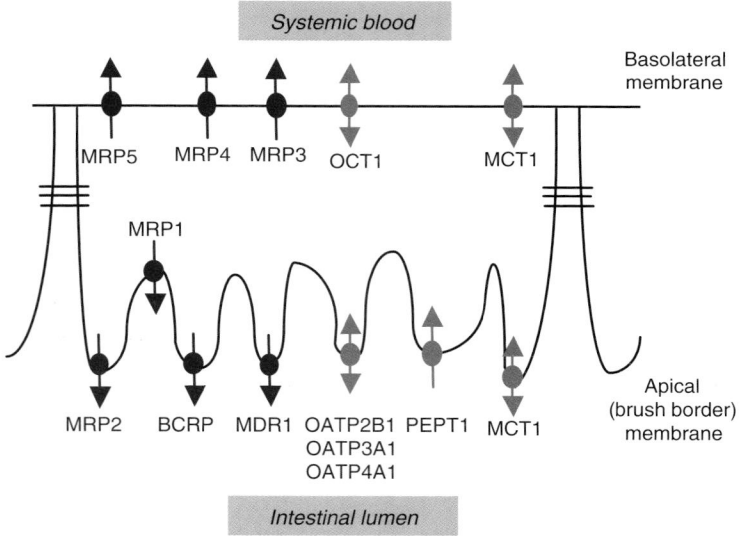

Figure 8   Distribution of the main drug ABC (blue) and SLC (red) transporters on the apical and basolateral membranes of intestinal enterocytes. All apical transporters (except MRP1) lie at the top of the villi.

The ABC transporter MRP3 is concentrated in the basolateral membranes, where it mediates the transfer of bile acids to the blood; its role in the intestinal absorption of drugs needs further clarification. The mRNAs of *MRP4* and *MRP5* were recently detected in the human intestine, suggesting that they are involved in the transport of nucleosides.[127] The intestinal transporters are not identically distributed along the crypt–villus axis. Many of those implicated in the absorption of drugs, like PEPT1, MDR1, BCRP, MRP2, and MRP3, are villus-specific.[113] This restriction of transporters to the villus is also correlated with the presence of CYP3A in intestinal cells, suggesting coordinated phase 0 and 1 activities of MDR1 and CYP3A4 in the so-called intestinal first-pass effect.[115] By contrast, MRP1 is predominantly found on the membranes of crypt cells in the small intestine, the site of enterocyte renewal, where it can protect these cells from toxins. The function and interplay of all these intestinal transporters must be very carefully considered when looking at their roles in drug absorption. A major concern is the way in which their densities vary along the gastrointestinal tract. For example, MRP3 is the most abundant ABC protein throughout the human intestine, except for the terminal ileum where MDR1 is most abundant. Similarly, the concentration of MDR1 increases from the duodenum to the colon, whereas BCRP is found throughout the small intestine and colon, and MRP2 is most prevalent in the duodenum and becomes undetectable toward the terminal ileum and colon.[45,113] MCT1 has also been detected throughout the intestinal tract from the stomach to the large intestine, but it is most abundant in more proximal regions of the duodenum–ileum.[129] These diverse densities of the intestinal transporters may have dramatic pharmaceutical consequences. The pharmaceutical form of an oral drug can vary from a simple solution to a solid, controlled-release complex, and this can influence the gastrointestinal site (stomach, duodenum, jejunum, ileum, or colon) at which the active compound is released. Such differences may also influence the efficacy of the carrier-mediated transporters, as these may vary from one region of the intestine to another.

This raises the question of how useful in vitro Caco-2 cells are. They are representative only of the colon, whereas the majority of drugs are absorbed more proximally. Caco-2 cells can also bear some OCTs, as indicated above, although these have not yet been identified in vivo, but they lack BCRP and CYP3A4, which are now known to be important for effluxing drugs. The great risk of saturating active transport can also affect the kinetics of drug absorption. This can occur when a large amount of drug is rapidly dissolved in the intestinal lumen and ready to be absorbed by a relatively small area of intestine. Active transport can be saturated by a relatively high concentration of substrate (**Figure 4a** and **b**), so shifting absorption toward diffusion. Here, too, the properties of the oral preparation, like its rate of dissolution, may influence the contribution of active transport to drug absorption.

In summary, the intestinal drug transporters play two major roles. First, they take part in drug influx – the absorption of drugs such as PEPT1 may be used to develop drug-delivery strategies for poorly absorbed drugs. Second, they are very important for drug efflux, either limiting the intestinal uptake of xenobiotics or mediating the secretion into the intestine of substrates circulating in the blood.

5.04.6.2   Liver and Hepatic Clearance

The liver is the most important drug-metabolizing organ in the body and acts an apparently 'homogeneous' pool of enzymatic activity. Its parallel sinusoids lined with fenestrated endothelial cells form a freely accessible extracellular space surrounded by plates of hepatocytes. The hepatocyte plasma membrane is the only barrier to the entry of drugs into the hepatocytes. Hepatocytes are polarized cells with basolateral (sinusoidal and lateral) and apical (canilicular) membranes (**Figure 9**). Molecules may be excreted from the hepatocytes across their canalicular membrane into the

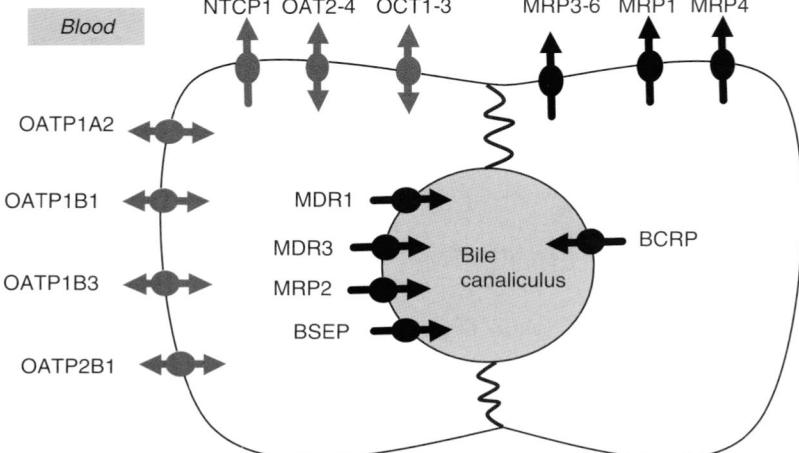

Figure 9   Distribution of the main drug ABC (blue) and SLC (red) transporters on the basolateral (sinusoïd) and apical (bile canaliculus) membranes of hepatocytes.

bile, or across the basolateral membrane into sinusoidal blood, from which they are subsequently removed by other organs (e.g., the kidney). Hepatic clearance is a combination of metabolic (phase I and II) and biliary clearance. As previously indicated, hepatocytes can take up drugs by diffusion or active transport (phase 0).

The basolateral membrane OATs include OAT2, OAT4, OATP1A2, -1B1, -1B3, and -2B1, the organic cation transporter OCT1 and the Na-taurocholate co-transporting polypeptide NTCP (*SLC10A1*).[130] They are responsible for the uptake of a wide variety of drugs by the liver because of their broad, overlapping substrate specificities. Phase III, which follows phases 0, I, and II, results in the elimination of the intact drug and/or metabolite(s) via efflux transporters on the apical and basolateral membranes. The hepatobiliary transporters include several ABC proteins (MDR1, MDR3, MRP2, BSEP, and BCRP) that are the main mediators of the excretion of numerous endogenous conjugated and unconjugated bile salts and drugs via the bile. Phase III also includes the efflux of compounds from hepatocytes back into the systemic circulation via basolateral membrane efflux transporters. Some of the OATPs, OATs, and OCT1 are bidirectional and may facilitate efflux, but the main exporters are the ABC proteins, which transport a wide range of glucuronides, and sulfated and GSH conjugates. The main ones are MRP1 and MRP3, the synthesis of which is readily induced, and the cyclic nucleoside transporters MRP4 and MRP5.[113,130]

This huge network of hepatobiliary transporters can give rise to variations in drug disposition between individuals by modulating the uptake or the exit of drugs and their metabolites from hepatocytes. A change in hepatic uptake may have clinical consequences. It may modulate the pharmacological activity of drugs that act via the intrahepatocellular transduction pathways, it may cause hepatotoxicity, or give rise to drug–drug interactions. The concentration of the cholesterol-lowering HMGCoA inhibitors in hepatocytes must be adequate for their pharmacological activity, and most of the statins (e.g., pravastatin, simvastatin, lovastatin, cerivastatin, and pitavastatin) enter hepatocytes via OATP1B1, and to a lesser degree via OAT1B3.[131] Recently identified genetic polymorphisms like the *SLCO1B3* haplotype *17 are associated with reduced statin clearance by the liver and lower concentrations in hepatocytes; they thus have less effect on cholesterol synthesis.[132] Large-scale clinical studies are needed to confirm the impact of OATP1B1 polymorphisms on the considerable variation between individuals to therapy with hypolipidemic agents. The clinical efficacy and adverse effects of oral antidiabetic drugs similarly vary greatly between individuals. Polymorphisms of CYP2C8/2C9, the main enzymes catalyzing the transformation of sulfonylureas, and the meglitinide-class drugs, such as repaglinide and nateglinide, have been advanced to explain these variations, but one *SLCO1B1* genotype was recently shown to affect markedly the pharmacokinetics of repaglinide and its effect on blood glucose.[133] The clinical efficacy of the antidiabetic drug metformin, which is not significantly metabolized, mainly depends on its hepatic uptake by OCT1 (*see* Section 5.04.4.2.2), which can also be affected by several SNPs.[134]

Transporters can also mediate hepatotoxicity. For example, the sulfate conjugate of the antidiabetic troglitazone can cause troglitazone hepatotoxicity by inhibiting OATP1B1 and OATP1B3.[135] Phalloidin, the major toxin of the mushroom *Amanita phalloides*, enters hepatocytes via OATP1B1 and -1B3, and cyclosporin A is reported to be the most potent competitive inhibitor of OATP1B1-mediated phalloidin transport in the liver.[136] Several other drugs also inhibit the basolateral OATPs. The fibrate gemfibrozil interacts with statins by inhibiting OATP1B1.[137] Thus drug–drug interactions do not concern only the inhibition of metabolic enzyme, but may involve the first line of hepatocyte transporters. These hepatic impacts of the basolateral transporters have their counterpart at the apical pole. The multiple ABC transporters may also be responsible for variable drug disposition. For example, giving patients receiving digoxin the P-gp inhibitor verapamil decreases the biliary clearance of digoxin by 43% and increases its plasma concentration by 44%.[138] Furthermore, each of the apical ABC proteins contains several genetic polymorphisms and is very sensitive to liver diseases.[139] Both the influx and efflux hepatic transporters are thus critical influences on drug efficacy and toxicity. They are also very important for selecting the appropriate in vitro system for evaluating hepatic clearance and predicting hepatic clearance in vivo. The absence of transporters from the hepatic microsomal assays widely used to assess the metabolism of xenobiotics means that both suspensions and sandwich-cultures of primary hepatocytes are presently the most relevant in vitro models for studying hepatic uptake, biliary excretion, and drug–drug interactions, because they integrate the entire hepatobiliary transporter network.[130]

5.04.6.3   Blood Barriers and Tissue Distribution

The tissue distribution of a drug can be affected by transporters because they lie on either the luminal or abluminal membranes of the endothelial cells of the tissue blood vessels, or on the membranes of the specific cells of the underlying organ. The transporters on the membranes of the blood vessels may be several key physiological components of the blood barriers throughout the human body if tight junctions seal adjacent cells and prevent the paracellular exchange of solutes. In contrast, solutes can freely communicate between extracellular spaces when blood vessels are fenestrated, as in the liver sinusoids, and transporters on the plasma membranes of the tissue cells (e.g., the

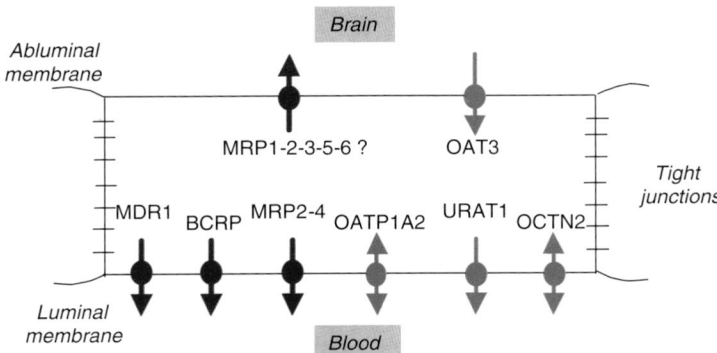

Figure 10 Distribution of the main drug ABC (blue) and SLC (red) transporters on the abluminal (facing brain extracellular fluid) and luminal membranes of the brain microcapillary endothelial cells constituting the BBB.

hepatocyte membranes) become the first barrier regulating the import and export of solutes. Several organs, including the brain, nose, eyes, testes, prostate, and placenta, are protected by endothelial barriers that contain extensive networks of transporters.[79,140–143]

Figure 10 illustrates the luminal and abluminal distributions of several transporters at the BBB. The two ABC proteins MDR1 and BCRP are most abundant on the luminal side of the endothelial cells and are most important for protecting the brain from numerous xenobiotics (*see* Sections 5.04.4.1.1 and 5.04.4.1.3).[144] Most of the MRPs have been also identified in the brain microvessel endothelial cells, but their luminal or abluminal location remains questionable and no important functional effects on drug transport across the BBB have been documented, except in a few cases for MRP2 and MRP4 (*see* Section 5.04.4.1.2). Few SLCs have been characterized in the rat BBB, except for the important network of SLC transporters that allows the blood–brain exchange of amino acids and sugars. Rat Oatp1a4 is found on both the luminal and abluminal membranes of the brain capillaries.[145] The human isoform OATP1A2 is also present, but its membrane location has not been determined. Both OATPs can mediate uptake or efflux transport because of their bidirectional transport characteristic (*see* Section 5.04.4.2.1). The members of the *SLC22*, OAT3, OCTN2, and URAT1, have been found in the BBB.[146] OAT3 is abluminal and effluxes benzylpenicillin, cimetidine, PAH, and several acidic metabolite of neurotransmitters from the brain to the blood. The luminal position of URAT1 enables this vectorial transport of the OAT3 substrates. OCTN2, which is believed to be luminal, can simultaneously transport carnitine into the brain and efflux OCs from the brain to the blood.

The other BBB, the blood–CSF barrier (BCSFB), which regulates the solute exchanges between the blood and the CSF, is a monolayer of epithelial cells at the choroid plexus floating in the brain ventricles. The apical and basolateral locations of ABC and SLC transporters makes this second barrier an important interface for brain homeostasis and drug disposition.[147] Here, too, drug transporters on the membranes of physiological barriers or on specific membranes of the tissue cells can affect drug distribution and, consequently, the fraction of the drug available for binding to intracellular receptors or other biological targets.

5.04.6.4   Kidney and Renal Clearance

The presence of a drug in the urine is the net result of filtration, secretion, and reabsorption. Filtration occurs by passive glomerular filtration of unbound plasma solutes, whereas secretion and reabsorption are generally carrier mediated. They can occur in the proximal tubule, which has three segments (S1, S2, and S3): the loop of Henle, the distal tubule, and the collecting tubule. These specific anatomical and functional regions of the kidney must be carefully considered, just like the regions of the intestine, because region-specific distributions of transporters define their action in renal clearance. Most renal transporters lie on the apical and basolateral membranes of the proximal tubule cells, with fewer on the epithelial membrane of the other components of the nephron. The resulting vectorial transport from the peritubular capillaries to the tubule lumen, or vice versa, can produce either secretion or reabsorption. **Figure 11** shows the locations of the major drug transporters in the human proximal renal tubule cells. Multiple SLC transporters at their basolateral membrane (close to peritubular capillaries) mediate drug uptake into the tubule cells. Although by nature bidirectional, the direction of the transmembrane driving gradients favors tubular uptake rather than the efflux of organic anions and cations. Organic anions enter these cells via OAT1, OAT2, and OAT3 – and probably via OATP1A2 and OATP4C1, which was recently identified, and transports digoxin and

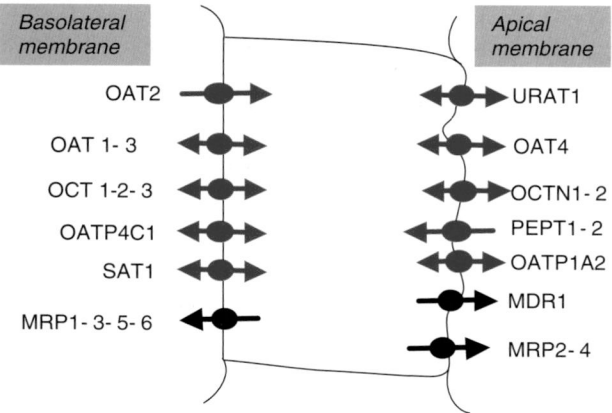

Figure 11   Distribution of the main drug ABC (blue) and SLC (red) transporters on the basolateral (peritubular fluid) and apical (glomerular filtrate) membranes of the kidney proximal tubule cells.

methotrexate.[148] OCs are similarly transported by OCT 1, OCT2, and OCT3; the efflux transporters MRP1, MRP3, MRP5, and MRP6 mediate their efflux back into the systemic circulation.[114,134] The sulfate-anion antiporter 1 (SAT-1; *SLC26A1*) is responsible for the sodium-dependent movement of sulfate across the basolateral membrane of the proximal tubule in exchange for HCO_3^-, but organic anions can also be substrates or inhibitors of SAT-1.[98] At the apical membrane, OAT4 and URAT1, OCTN1, and OCTN2 can mediate drug transport with bidirectional properties, either secretion or reabsorption. For example, OCTN2 secretes OCs and reabsorbs zwitterions. OATP1A2, PEPT1, and PEPT2 mediate the reabsorption of their substrates from the tubule lumen (*see* Section 5.04.4.2.4).

The ABC transporters MDR1, MRP2, and MRP4 are also present on the apical membrane and efflux compounds by secretion. As indicated above, transporters are not evenly distributed along the nephron: MDR1, MRP2, MRP4, and MRP6 are found mainly within the three segments of the proximal tubule; MRP3 lies in the distal convoluted tubule; and MRP1 is found in the epithelial cells of the loop of Henle and the distal and collecting duct tubule cells, but not in proximal tubule cells.[114] The regional distributions of the SLC transporters are also specific. OAT1 is found only on the basolateral membrane of the S2 segment cells of the proximal tubule, whereas OAT3 is present on the cells of the S1, S2, and S3 segments. Once again, this raises the question of the most appropriate in vitro kidney models containing the whole transport network. As with Caco-2 intestinal cells, the Madin–Darby canine kidney (MDCK) cell lines are widely used in wild-type or transfected cell systems for examining particular transport processes.[149,150] They are useful in vitro tools for assessing the risks of drug–drug interactions, nephrotoxicity, and drug efficacy mediated by the reabsorptive and secretory capacities of the kidney.

If the renal clearance of a drug is equal to or more than the overall body clearance, renal transporters can be important in clinical efficacy or toxicity. For example, the cephalosporin antibiotics are primary eliminated via the kidney. Creatinine clearance is normally $100–140\,mL\,min^{-1}$, but the renal clearance of cephalosporins is $16.8–469\,mL\,min^{-1}$, suggesting that some of them, like cefotaxine and cefadroxil, are excreted into the urine by tubular secretion, whereas others, like ceftriaxone and cefazodin, the renal clearance of which is less than that of creatinine, are reabsorbed.[151]

OAT1, OAT2, and OAT3 are located on the basolateral side of the proximal tubule and mediate the uptake of most of the cephalosporins into the proximal tubule from the peritubular capillary. The apical OAT4 mediates both the uptake (reabsorption from the tubular lumen) and the efflux (secretion) of these anionic antibiotics.[95] Like the basolateral transporters of hepatocytes, which can modulate the pharmacological activity of drugs acting via intrahepatocyte targets or induce hepatotoxicity, the basolateral OATs can make some cephalosporins cause nephrotoxicity, which may lead to acute proximal tubular necrosis. This toxicity is mainly due to the accumulation of cephalosporin in the renal cortex because of the lack of efficient vectorial transtubular transport.[95,98,152] This transport-mediated nephrotoxicity also results in the adverse effect of cisplastin drugs via their basolateral uptake in the proximal tubule by OCT2 and the toxic effects depend on the platinum complex used, as does the structure-dependent nephrotoxicity of cephalosporins.[153] Nephrotoxicity also limits the use of the nucleoside phosphonates, adefovir and cidofovir, in the treatment of human immunodeficiency virus. The toxicity of these drugs appears to be a function of both OAT1-mediated proximal tubular accumulation and decreased efflux at the luminal membrane by MRP2. A small dose of the OAT1 inhibitor, probenecid, may reduce the nephrotoxicity of cidofovir.[154]

The use of transporter inhibitors to reduce nephrotoxicity suggests that drug–drug interactions affecting anionic and cationic drugs can be mediated via competition at the basolateral and luminal tubular transporters. Multiple drug–drug interactions have been reported with probenecid and cimetidine, and there have been fatal cases with methotrexate and NSAIDs following the inhibition of the basolateral OAT1 and OAT3.[155]

Finally, renal transporters can be critical for the action of diuretics. Tubular secretion is the main route by which diuretics act in the kidney and are excreted. The diuretic drugs, such as the thiazides, the loop diuretics bumetanide and furosemide, and the carbonic anhydrase inhibitors, are all competitive inhibitors of the renal OATs, although their affinities and specificities vary.[156]

5.04.6.5 Transporters and the Biopharmaceutics Classification System

Amidon et al.[157] have devised a biopharmaceutics classification system (BCS) that divides drugs into four classes according to their solubility and permeability (see 5.42 The Biopharmaceutics Classification System). The BCS has produced rates and extents by which oral drugs are absorbed. Extended versions of the BCS have been proposed by associating the four classes with the main routes of elimination and the effects of efflux and influx transporters. Their objectives were to predict the pharmacokinetic performance of a drug product in vivo from measurements of permeability and solubility. Figure 12 combines the first elements of the original BCS and additional information on the main pathways of drug elimination and the impact of drug transporters.[158] Most class 1 and class 2 compounds, which are highly permeable, are eliminated via metabolism, whereas class 3 and 4 compounds, which are poorly permeable, are primarily eliminated unchanged into the urine and/or bile. The effect of intestinal transporters is expected to be minimal for class 1 compounds, even if they interact with transporters, as their great permeability and solubility allows high concentrations in the gut lumen to saturate any transporter. In contrast, many large ($> 500\,\mathrm{Da}$), lipophilic and poorly water-soluble new BCS class 2 molecules that deviate the crude Lipinski filter rule of 5 are often the target of gut efflux transporters such as MDR1 and BCRP.[158] Sufficient amounts of class 3 drugs will be available for gut absorption because of their great solubility, but their poor permeability requires an absorptive transporter. The poor oral bioavailability and transporter effects are likely to determine the oral absorption of class 4 compounds.

The postabsorptive pharmacokinetics can also be anticipated with the BCS. Most of the drugs with high hepatic extraction ratios and clearances approaching the hepatic blood flow are class 1 compounds, and the target mediating drug–drug interactions is primarily metabolic. In contrast, both uptake and efflux transporters are critical determinants of the disposition of class 2, 3, and 4 compounds, the biliary secretion and renal secretion or reabsorption-mediated transport of which can dramatically modulate the systemic concentrations of these drugs. Thus, transporters may be the primary mediators of drug–drug interactions.

	High solubility	Low solubility
High permeability	Class 1 (high solubility, high permeability) *Metabolism* 'High hepatic extraction' Transporter effects minimal	Class 2 (low solubility, high permeability) *Metabolism* Efflux transporter effects predominate
Low permeability	Class 3 (high solubility, low permeability) *Renal and/or biliary elimination* Absorptive transporter effects predominate	Class 4 (low solubility, low permeability) *Renal and/or biliary elimination* Absorptive and efflux transporter effects could be important

Figure 12 The biopharmaceutics classification system (BCS), as defined by Amidon et al.[157] and extended by Wu and Benet,[158] showing the predominant routes of drug elimination and the transporter effects on drug disposition.

There was very little interest in the importance of drug transporters in drug bioavailability when the BCS was first developed. But the involvement of drug transport at all steps of the pharmacokinetic processes has provided a clearer correlation between the solubility and permeability of a drug and its fate within the human body. Hence the action of transporters is now included in ADMET predictive models.

5.04.7   Conclusion

The recent expansion of our knowledge of the involvement of drug transporters in pharmacokinetics has added a new layer of complexity to our understanding of the mechanisms underlying the absorption, distribution, and elimination of drugs. New transporters undoubtedly remain to be identified, and the functions of some identified transporters remain poorly understood. Nevertheless, it is clear that the drug transporters in organs such as the intestine, liver, brain, and kidney are significant determinants of variations in drug responsiveness between individuals. Together with the drug-metabolizing enzymes, they determine drug–drug interactions, drug-induced organ toxicities, and diseases. Detailed knowledge of genetic polymorphisms in transporters and how they affect transporter function will help to optimize drug therapies and identify unknown, residual factors that influence intersubject variations.

Transporters are now an integral part of the drug discovery and development processes. They are attractive markers in the creation of drugs that are readily absorbed and accurately targeted. The incorporation of transport properties into structure–activity models should help medicinal chemists design more efficient and safer new medicines.

A few years ago, the body factors governing the main phases of drug pharmacokinetics were believed to be completely understood. The transporter adventure has now opened up a new horizon for pharmacokinetics that focuses on the cells and molecules that govern the main events in the body, so promoting pharmacokinetics to a discipline that integrates multiple interacting biological systems.

References

1. Schwenk, M. *Arch. Toxicol.* **1987**, *60*, 37–42.
2. Dunbar, L. A.; Caplan, M. J. *Eur. J. Cell Biol.* **2000**, *79*, 557–563.
3. Kaplan, J. H. *Cell* **1993**, *72*, 13–18.
4. Sadee, W.; Drubbisch, V.; Amidon, G. L. *Pharm. Res.* **1995**, *12*, 1823–1837.
5. Paulsen, I. T.; Sliwinski, M. K.; Nelissen, B.; Goffeau, A.; Saier, M. H., Jr. *FEBS Lett.* **1998**, *430*, 116–125.
6. Ward, A.; Hoyle, C.; Palmer, S.; O'Reilly, J.; Griffith, J.; Pos, M.; Morrison, S.; Poolman, B.; Gwynne, M.; Henderson, P. *J. Mol. Microbiol. Biotechnol.* **2001**, *3*, 193–200.
7. Dassa, E.; Bouige, P. *Res. Microbiol.* **2001**, *152*, 211–229.
8. Hediger, M. A.; Romero, M. F.; Peng, J. B.; Rolfs, A.; Takanaga, H.; Bruford, E. A. *Pflugers Arch.* **2004**, *447*, 465–468.
8a. HUGO Gene Nomenclature Committee. http://www.gene.ucl.ac.uk/nomenclature (accessed June 2006).
8b. Transport Classification Database. http://www.tcdb.org (accessed June 2006).
9. Ozawa, N.; Shimizu, T.; Morita, R.; Yokono, Y.; Ochiai, T.; Munesada, K.; Ohashi, A.; Aida, Y.; Hama, Y.; Taki, K.; Maeda, K.; Kusuhara, H.; Sugiyama, Y. *Pharm. Res.* **2004**, *21*, 2133–2134.
10. Tusnady, G. E.; Sarkadi, B.; Simon, I.; Varadi, A. *FEBS Lett.* **2006**, *580*, 1017–1022.
11. Zhang, E. Y.; Knipp, G. T.; Ekins, S.; Swaan, P. W. *Drug Metab. Rev.* **2002**, *34*, 709–750.
12. Hagenbuch, B.; Meier, P. J. *Biochim. Biophys. Acta.* **2003**, *160*, 91–118.
13. Walker, J. E.; Saraste, M.; Runswick, M. J.; Gay, N. J. *EMBO J.* **1982**, *1*, 945–951.
14. Gottesman, M. M.; Fojo, T.; Bates, S. E. *Nat. Rev. Cancer* **2002**, *2*, 48–58.
15. Leslie, E. M.; Deeley, R. G.; Cole, S. P. *Toxicol. Appl. Pharmacol.* **2005**, *204*, 216–237.
16. Biedler, J. L.; Riehm, H. *Cancer Res.* **1970**, *30*, 1174–1184.
17. Juliano, R. L.; Ling, V. *Biochim. Biophys. Acta* **1976**, *455*, 152–162.
18. Schinkel, A. H. *Semin. Cancer Biol.* **1997**, *8*, 161–170.
19. Schinkel, A. H.; Jonker, J. W. *Adv. Drug Deliv. Rev.* **2003**, *55*, 3–29.
20. Litman, T.; Skovsgaard, T.; Stein, W. D. *J. Pharmacol. Exp. Ther.* **2003**, *307*, 846–853.
21. Breedveld, P.; Beijnen, J. H.; Schellens, J. H. *Trends Pharmacol. Sci.* **2006**, *27*, 17–24.
22. Schinkel, A. H. *Adv. Drug Deliv. Rev.* **1999**, *36*, 179–194.
23. Marzolini, C.; Paus, E.; Buclin, T.; Kim, R. B. *Clin. Pharmacol. Ther.* **2004**, *75*, 13–33.
24. Roberts, R. L.; Joyce, P. R.; Mulder, R. T.; Begg, E. J.; Kennedy, M. A. *Pharmacogenom. J.* **2002**, *21*, 91–196.
25. De Luca, V.; Mundo, E.; Trakalo, J.; Wong, G. W.; Kennedy, J. L. *Pharmacogenom. J.* **2003**, *3*, 297–299.
26. Yi, S. Y.; Hong, K. S.; Lim, H. S.; Chung, J. Y.; Oh, D. S.; Kim, J. R.; Jung, H. R.; Cho, J. Y.; Yu, K. S.; Jang, I. J.; Shin, S. G. *Clin. Pharmacol. Ther.* **2004**, *76*, 418–427.
27. Drozdzik, M.; Bialecka, M.; Mysliwiec, K.; Honczarenko, K.; Stankiewicz, J.; Sych, Z. *Pharmacogenetics* **2003**, *13*, 259–263.
28. Siddiqui, A.; Kerb, R.; Weale, M. E.; Brinkmann, U.; Smith, A.; Goldstein, D. B.; Wood, N. W.; Sisodiya, S. M. N. *Engl. J. Med.* **2003**, *348*, 1442–1448.
29. Fardel, O.; Payen, L.; Courtois, A.; Vernhet, L.; Lecureur, V. *Toxicology* **2001**, *167*, 37–46.
30. Childs, S.; Yeh, R. L.; Hui, D.; Ling, V. *Cancer Res.* **1998**, *58*, 4160–4167.

31. Fattinger, K.; Funk, C.; Pantze, M.; Weber, C.; Reichen, J.; Stieger, B.; Meier, P. J. *Clin. Pharmacol. Ther.* **2001**, *69*, 223–231.
32. Cole, S. P.; Bhardwaj, G.; Gerlach, J. H.; Mackie, J. E.; Grant, C. E.; Almquist, K. C.; Stewart, A. J.; Kurz, E. U.; Duncan, A. M.; Deeley, R. G. *Science* **1992**, *258*, 1650–1654.
33. Borst, P.; Evers, R.; Kool, M.; Wijnholds, J. *J. Natl. Cancer Inst.* **2000**, *92*, 1295–1302.
34. Kruh, G. D.; Zeng, H.; Rea, P. A.; Liu, G.; Chen, Z. S.; Lee, K.; Belinsky, M. G. *J. Bioenerg. Biomembr.* **2001**, *33*, 493–501.
35. Kruh, G. D.; Belinsky, M. G. *Oncogene* **2003**, *22*, 7537–7552.
36. Leier, I.; Jedlitschky, G.; Buchholz, U.; Cole, S. P.; Deeley, R. G.; Keppler, D. *J. Biol. Chem.* **1994**, *269*, 27807–27810.
37. Leslie, E. M.; Mao, Q.; Oleschuk, C. J.; Deeley, R. G.; Cole, S. P. *Mol. Pharmacol.* **2001**, *59*, 1171–1180.
38. Godinot, N.; Iversen, P. W.; Tabas, L.; Xia, X.; Williams, D. C.; Dantzig, A. H.; Perry, W. L., Jr. *Mol. Cancer Ther.* **2003**, *2*, 307–316.
39. Lorico, A.; Rappa, G.; Finch, R. A.; Yang, D.; Flavell, R. A.; Sartorelli, A. C. *Cancer Res.* **1997**, *57*, 5238–5242.
40. Wijnholds, J.; Evers, R.; van Leusden, M. R.; Mol, C. A.; Zaman, G. J.; Mayer, U.; Beijnen, J. H.; van der Valk, M.; Krimpenfort, P.; Borst, P. *Nat. Med.* **1997**, *3*, 1275–1279.
41. Wijnholds, J.; deLange, E. C.; Scheffer, G. L.; van den Berg, D. J.; Mol, C. A.; van der Valk, M.; Schinkel, A. H.; Scheper, R. J.; Breimer, D. D.; Borst, P. *J. Clin. Invest.* **2000**, *105*, 279–285.
42. Paulusma, C. C.; Bosma, P. J.; Zaman, G. J.; Bakker, C. T.; Otter, M.; Scheffer, G. L.; Scheper, R. J.; Borst, P.; Oude Elferink, R. P. *Science* **1996**, *271*, 1126–1128.
43. Kitamura, T.; Jansen, P.; Hardenbrook, C.; Kamimoto, Y.; Gatmaitan, Z.; Arias, I. M. *Proc. Natl. Acad. Sci. USA* **1990**, *87*, 3557–3561.
44. Dietrich, C. G.; de Waart, D. R.; Ottenhoff, R.; Bootsma, A. H.; van Gennip, A. H.; Elferink, R. P. *Carcinogenesis* **2001**, *22*, 805–811.
45. Dietrich, C. G.; Geier, A.; Oude Elferink, R. P. *Gut* **2003**, *52*, 1788–1795.
46. Kool, M.; de Haas, M.; Scheffer, G. L.; Scheper, R. J.; van Eijk, M. J.; Juijn, J. A.; Baas, F.; Borst, P. *Cancer Res.* **1997**, *57*, 3537–3547.
47. Miller, D. S.; Nobmann, S. N.; Gutmann, H.; Toeroek, M.; Drewe, J.; Fricker, G. *Mol. Pharmacol.* **2000**, *58*, 1357–1367.
48. Dombrowski, S. M.; Desai, S. Y.; Marroni, M.; Cucullo, L.; Goodrich, K.; Bingaman, W.; Mayberg, M. R.; Bengez, L.; Janigro, D. *Epilepsia* **2001**, *42*, 1501–1506.
49. Potschka, H.; Fedrowitz, M.; Loscher, W. *J. Pharmacol. Exp. Ther.* **2003**, *306*, 124–131.
50. Belinsky, M. G.; Bain, L. J.; Balsara, B. B.; Testa, J. R.; Kruh, G. D. *J. Natl. Cancer Inst.* **1998**, *90*, 1735–1741.
51. Zelcer, N.; van de, W. K.; Hillebrand, M.; Sarton, E.; Kuil, A.; Wielinga, P. R.; Tephly, T.; Dahan, A.; Beijnen, J. H.; Borst, P. *Proc. Natl. Acad. Sci. USA* **2005**, *102*, 7274–7279.
52. Cherrington, N. J.; Slitt, A. L.; Maher, J. M.; Zhang, X. X.; Zhang, J.; Huang, W.; Wan, Y. J.; Moore, D. D.; Klaassen, C. D. *Drug Metab. Dispos.* **2003**, *31*, 1315–1319.
53. Lang, T.; Hitzl, M.; Burk, O.; Mornhinweg, E.; Keil, A.; Kerb, R.; Klein, K.; Zanger, U. M.; Eichelbaum, M.; Fromm, M. F. *Pharmacogenetics* **2004**, *14*, 155–164.
54. Jedlitschky, G.; Burchell, B.; Keppler, D. *J. Biol. Chem.* **2000**, *275*, 30069–30074.
55. Borst, P.; Balzarini, J.; Ono, N.; Reid, G.; de Vries, H.; Wielinga, P.; Wijnholds, J.; Zelcer, N. *Antiviral Res.* **2004**, *62*, 1–7.
56. Rius, M.; Nies, A. T.; Hummel-Eisenbeiss, J.; Jedlitschky, G.; Keppler, D. *Hepatology* **2003**, *38*, 374–384.
57. Leggas, M.; Adachi, M.; Scheffer, G. L.; Sun, D.; Wielinga, P.; Du, G.; Mercer, K. E.; Zhuang, Y.; Panetta, J. C.; Johnston, B.; Scheper, R. J.; Stewart, C. F.; Schuetz, J. D. *Mol. Cell Biol.* **2004**, *24*, 7612–7621.
58. Doyle, L. A.; Yang, W.; Abruzzo, L. V.; Krogmann, T.; Gao, Y.; Rishi, A. K.; Ross, D. D. *Proc. Natl. Acad. Sci. USA* **1998**, *95*, 15665–15670.
59. Miyake, K.; Mickley, L.; Litman, T.; Zhan, Z.; Robey, R.; Cristensen, B.; Brangi, M.; Greenberger, L.; Dean, M.; Fojo, T.; Bates, S. E. *Cancer Res.* **1999**, *59*, 8–13.
60. Allikmets, R.; Schriml, L. M.; Hutchinson, A.; Romano-Spica, V.; Dean, M. *Cancer Res.* **1998**, *58*, 5337–5339.
61. Mao, Q.; Unadkat, J. D. *AAPS J.* **2005**, *7*, E118–E133.
62. Jonker, J. W.; Buitelaar, M.; Wagenaar, E.; Van Der Valk, M. A.; Scheffer, G. L.; Scheper, R. J.; Plosch, T.; Kuipers, F.; Elferink, R. P.; Rosing, H.; Beijnen, J. H.; Schinkel, A. H. *Proc. Natl. Acad. Sci. USA* **2002**, *99*, 15649–15654.
63. Ahmed-Belkacem, A.; Pozza, A.; Munoz-Martinez, F.; Bates, S. E.; Castanys, S.; Gamarro, F.; Di Pietro, A.; Perez-Victoria, J. M. *Cancer Res.* **2005**, *65*, 4852–4860.
64. Jonker, J. W.; Smit, J. W.; Brinkhuis, R. F.; Maliepaard, M.; Beijnen, J. H.; Schellens, J. H.; Schinkel, A. H. *J. Natl. Cancer Inst.* **2000**, *92*, 1651–1656.
65. Breedveld, P.; Zelcer, N.; Pluim, D.; Sonmezer, O.; Tibben, M. M.; Beijnen, J. H.; Schinkel, A. H.; van Tellingen, O.; Borst, P.; Schellens, J. H. *Cancer Res.* **2004**, *64*, 5804–5811.
66. van Herwaarden, A. E.; Schinkel, A. H. *Trends Pharmacol. Sci.* **2006**, *27*, 10–16.
67. Jonker, J. W.; Merino, G.; Musters, S.; van Herwaarden, A. E.; Bolscher, E.; Wagenaar, E.; Mesman, E.; Dale, T. C.; Schinkel, A. H. *Nat. Med.* **2005**, *11*, 127–129.
68. Merino, G.; Jonker, J. W.; Wagenaar, E.; van Herwaarden, A. E.; Schinkel, A. H. *Mol. Pharmacol.* **2005**, *67*, 1758–1764.
69. Cervenak, J.; Andrikovics, H.; Ozvegy-Laczka, C.; Tordai, A.; Nemet, K.; Varadi, A.; Sarkadi, B. *Cancer Lett.* **2006**, *234*, 62–72.
70. Zamber, C. P.; Lamba, J. K.; Yasuda, K.; Farnum, J.; Thummel, K.; Schuetz, J. D.; Schuetz, E. G. *Pharmacogenetics* **2003**, *13*, 19–28.
71. Sparreboom, A.; Gelderblom, H.; Marsh, S.; Ahluwalia, R.; Obach, R.; Principe, P.; Twelves, C.; Verweij, J.; McLeod, H. L. *Clin. Pharmacol. Ther.* **2004**, *76*, 38–44.
72. Wright, E. M.; Turk, E.; Zabel, B.; Mundlos, S.; Dyer, J. *J. Clin. Invest.* **1991**, *88*, 1435–1440.
73. Wong, M. H.; Oelkers, P.; Dawson, P. A. *J. Biol. Chem.* **1995**, *270*, 27228–27234.
74. Hagenbuch, B.; Gao, B.; Meier, P. J. *News Physiol. Sci.* **2002**, *17*, 231–234.
75. Ballatori, N.; Hammond, C. L.; Cunningham, J. B.; Krance, S. M.; Marchan, R. *Toxicol. Appl. Pharmacol.* **2005**, *204*, 238–255.
76. Hagenbuch, B.; Meier, P. J. *Pflugers Arch.* **2004**, *447*, 653–665.
77. Shitara, Y.; Sato, H.; Sugiyama, Y. *Ann. Rev. Pharmacol. Toxicol.* **2005**, *45*, 689–723.
78. Masuda, S. *Drug Metab Pharmacokinet.* **2003**, *18*, 91–103.
79. Unadkat, J. D.; Dahlin, A.; Vijay, S. *Curr. Drug Metab.* **2004**, *5*, 125–131.
80. Grundemann, D.; Gorboulev, V.; Gambaryan, S.; Veyhl, M.; Koepsell, H. *Nature* **1994**, *372*, 549–552.
81. Gorboulev, V.; Ulzheimer, J. C.; Akhoundova, A.; Ulzheimer-Teuber, I.; Karbach, U.; Quester, S.; Baumann, C.; Lang, F.; Busch, A. E.; Koepsell, H. *DNA Cell Biol.* **1997**, *16*, 871–881.
82. Wu, X.; Kekuda, R.; Huang, W.; Fei, Y. J.; Leibach, F. H.; Chen, J.; Conway, S. J.; Ganapathy, V. *J. Biol. Chem.* **1998**, *273*, 32776–32786.
83. Jonker, J. W.; Schinkel, A. H. *J. Pharmacol. Exp. Ther.* **2004**, *308*, 2–9.

84. Meijer, D. K.; Mol, W. E.; Muller, M.; Kurz, G. *J. Pharmacokinet. Biopharm.* **1990**, *18*, 35–70.
85. Wright, S. H. *Toxicol. Appl. Pharmacol.* **2005**, *204*, 309–319.
86. Koepsell, H. *Trends Pharmacol. Sci.* **2004**, *25*, 375–381.
87. Urakami, Y.; Nakamura, N.; Takahashi, K.; Okuda, M.; Saito, H.; Hashimoto, Y.; Inui, K. *FEBS Lett.* **1999**, *461*, 339–342.
88. Verhaagh, S.; Schweifer, N.; Barlow, D. P.; Zwart, R. *Genomics* **1999**, *55*, 209–218.
89. Jonker, J. W.; Wagenaar, E.; Van Eijl, S.; Schinkel, A. H. *Mol. Cell Biol.* **2003**, *23*, 7902–7908.
90. Wang, D. S.; Kusuhara, H.; Kato, Y.; Jonker, J. W.; Schinkel, A. H.; Sugiyama, Y. *Mol. Pharmacol.* **2003**, *63*, 844–848.
91. Enomoto, A.; Wempe, M. F.; Tsuchida, H.; Shin, H. J.; Cha, S. H.; Anzai, N.; Goto, A.; Sakamoto, A.; Niwa, T.; Kanai, Y.; Anders, M. W.; Endou, H. *J. Biol. Chem.* **2002**, *277*, 36262–36271.
92. Tamai, I.; Yabuuchi, H.; Nezu, J.; Sai, Y.; Oku, A.; Shimane, M.; Tsuji, A. *FEBS Lett.* **1997**, *419*, 107–111.
93. Wu, X.; Huang, W.; Prasad, P. D.; Seth, P.; Rajan, D. P.; Leibach, F. H.; Chen, J.; Conway, S. J.; Ganapathy, V. *J. Pharmacol. Exp. Ther.* **1999**, *290*, 1482–1492.
94. Burckhardt, G.; Wolff, N. A. *Am. J. Physiol. Renal Physiol.* **2000**, *278*, F853–F866.
95. Sweet, D. H. *Toxicol. Appl. Pharmacol.* **2005**, *204*, 198–215.
96. Enomoto, A.; Endou, H. *Clin. Exp. Nephrol.* **2005**, *9*, 195–205.
97. Sekine, T.; Watanabe, N.; Hosoyamada, M.; Kanai, Y.; Endou, H. *J. Biol. Chem.* **1997**, *272*, 18526–18529.
98. Robertson, E. E.; Rankin, G. O. *Pharmacol. Ther.* **2006**, *109*, 399–412.
99. Daniel, H.; Rubio-Aliaga, I. *Am. J. Physiol. Renal Physiol.* **2003**, *284*, F885–F892.
100. Sakata, K.; Yamashita, T.; Maeda, M.; Moriyama, Y.; Shimada, S.; Tohyama, M. *Biochem. J.* **2001**, *356*, 53–60.
101. Shu, C.; Shen, H.; Teuscher, N. S.; Lorenzi, P. J.; Keep, R. F.; Smith, D. E. *J. Pharmacol. Exp. Ther.* **2002**, *301*, 820–829.
102. Rubio-Aliaga, I.; Frey, I.; Boll, M.; Groneberg, D. A.; Eichinger, H. M.; Balling, R.; Daniel, H. *Mol. Cell Biol.* **2003**, *23*, 3247–3252.
103. Yang, C. Y.; Dantzig, A. H.; Pidgeon, C. *Pharm. Res.* **1999**, *16*, 1331–1343.
104. Pastor-Anglada, M.; Cano-Soldado, P.; Molina-Arcas, M.; Lostao, M. P.; Larrayoz, I.; Martinez-Picado, J.; Casado, F. J. *Virus Res.* **2005**, *107*, 151–164.
105. Gray, J. H.; Owen, R. P.; Giacomini, K. M. *Pflugers Arch.* **2004**, *447*, 728–734.
106. Ritzel, M. W.; Yao, S. Y.; Huang, M. Y.; Elliott, J. F.; Cass, C. E.; Young, J. D. *Am. J. Physiol.* **1997**, *272*, C707–C714.
107. Wang, J.; Su, S. F.; Dresser, M. J.; Schaner, M. E.; Washington, C. B.; Giacomini, K. M. *Am. J. Physiol.* **1997**, *273*, F1058–F1065.
108. Ritzel, M. W.; Ng, A. M.; Yao, S. Y.; Graham, K.; Loewen, S. K.; Smith, K. M.; Ritzel, R. G.; Mowles, D. A.; Carpenter, P.; Chen, X. Z.; Karpinski, E.; Hyde, R. J.; Baldwin, S. A.; Cass, C. E.; Young, J. D. *J. Biol. Chem.* **2001**, *276*, 2914–2927.
108a. http://www.pharmgkb.org/ (accessed June 2006) and http://www.pharmacogenetics.ucsf.edu (accessed June 2006).
109. Baldwin, S. A.; Beal, P. R.; Yao, S. Y.; King, A. E.; Cass, C. E.; Young, J. D. *Pflugers Arch.* **2004**, *447*, 735–743.
110. Yao, S. Y.; Ng, A. M.; Sundaram, M.; Cass, C. E.; Baldwin, S. A.; Young, J. D. *Mol. Membr. Biol.* **2001**, *18*, 161–167.
111. Yao, S. Y.; Ng, A. M.; Vickers, M. F.; Sundaram, M.; Cass, C. E.; Baldwin, S. A.; Young, J. D. *J. Biol. Chem.* **2002**, *277*, 24938–24948.
112. Lai, Y.; Tse, C. M.; Unadkat, J. D. *J. Biol. Chem.* **2004**, *279*, 4490–4497.
113. Ito, K.; Suzuki, H.; Horie, T.; Sugiyama, Y. *Pharm. Res.* **2005**, *22*, 1559–1577.
114. van de Water, F. M.; Masereeuw, R.; Russel, F. G. *Drug Metab. Rev.* **2005**, *37*, 443–471.
115. Cummins, C. L.; Jacobsen, W.; Benet, L. Z. *J. Pharmacol. Exp. Ther.* **2002**, *300*, 1036–1045.
116. Hill, G.; Cihlar, T.; Oo, C.; Ho, E. S.; Prior, K.; Wiltshire, H.; Barrett, J.; Liu, B.; Ward, P. *Drug Metab. Dispos.* **2002**, *30*, 13–19.
117. Izzedine, H.; Launay-Vacher, V.; Deray, G. *AIDS* **2005**, *19*, 455–462.
118. Thomas, J.; Wang, L.; Clark, R. E.; Pirmohamed, M. *Blood* **2004**, *104*, 3739–3745.
119. Cisternino, S.; Bourasset, F.; Archimbaud, Y.; Semiond, D.; Sanderink, G.; Scherrmann, J. M. *Br. J. Pharmacol.* **2003**, *138*, 1367–1375.
120. Wandel, C.; Kim, R. B.; Guengerich, F. P.; Wood, A. J. *Drug Metab. Dispos.* **2000**, *28*, 895–898.
121. Dresser, G. K.; Bailey, D. G. *Eur. J. Clin. Invest.* **2003**, *33*, 10–16.
122. Endres, C. J.; Hsiao, P.; Chung, F. S.; Unadkat, J. D. *Eur. J. Pharm. Sci.* **2006**, *27*, 501–517.
123. Tachibana-Iimori, R.; Tabara, Y.; Kusuhara, H.; Kohara, K.; Kawamoto, R.; Nakura, J.; Tokunaga, K.; Kondo, I.; Sugiyama, Y.; Miki, T. *Drug Metab. Pharmacokinet.* **2004**, *19*, 375–380.
124. Tsuji, A.; Tamai, I. *Pharm. Res.* **1996**, *13*, 963–977.
125. Halestrap, A. P.; Meredith, D. *Pflugers Arch.* **2004**, *447*, 619–628.
126. Anderle, P.; Sengstag, T.; Mutch, D. M.; Rumbo, M.; Praz, V.; Mansourian, R.; Delorenzi, M.; Williamson, G.; Roberts, M. A. *BMC Genomics* **2005**, *6*, 69.
127. Takano, M.; Yumoto, R.; Murakami, T. *Pharmacol. Ther.* **2006**, *109*, 137–161.
128. Katsura, T.; Inui, K. *Drug Metab. Pharmacokinet.* **2003**, *18*, 1–15.
129. Sai, Y. *Drug Metab. Pharmacokinet.* **2005**, *20*, 91–99.
130. Chandra, P.; Brouwer, K. L. *Pharm. Res.* **2004**, *21*, 719–735.
131. Hirano, M.; Maeda, K.; Shitara, Y.; Sugiyama, Y. *J. Pharmacol. Exp. Ther.* **2004**, *311*, 139–146.
132. Niemi, M.; Neuvonen, P. J.; Hofmann, U.; Backman, J. T.; Schwab, M.; Lutjohann, D.; von Bergmann, K.; Eichelbaum, M.; Kivisto, K. T. *Pharmacogenet. Genomics* **2005**, *15*, 303–309.
133. Niemi, M.; Backman, J. T.; Kajosaari, L. I.; Leathart, J. B.; Neuvonen, M.; Daly, A. K.; Eichelbaum, M.; Kivisto, K. T.; Neuvonen, P. J. *Clin. Pharmacol. Ther.* **2005**, *77*, 468–478.
134. Koepsell, H.; Endou, H. *Pflugers Arch* **2004**, *447*, 666–676.
135. Nozawa, T.; Sugiura, S.; Nakajima, M.; Goto, A.; Yokoi, T.; Nezu, J.; Tsuji, A.; Tamai, I. *Drug Metab. Dispos.* **2004**, *32*, 291–294.
136. Fehrenbach, T.; Cui, Y.; Faulstich, H.; Keppler, D. *Naunyn Schmiedebergs Arch. Pharmacol.* **2003**, *368*, 415–420.
137. Shitara, Y.; Hirano, M.; Sato, H.; Sugiyama, Y. *J. Pharmacol. Exp. Ther.* **2004**, *311*, 228–236.
138. Hedman, A.; Angelin, B.; Arvidsson, A.; Beck, O.; Dahlqvist, R.; Nilsson, B.; Olsson, M.; Schenck-Gustafsson, K. *Clin. Pharmacol. Ther.* **1991**, *49*, 256–262.
139. Ho, R. H.; Kim, R. B. *Clin. Pharmacol. Ther.* **2005**, *78*, 260–277.
140. Graff, C. L.; Pollack, G. M. *Pharm. Res.* **2005**, *22*, 86–93.
141. Hosoya, K.; Tomi, M. *Biol. Pharm. Bull.* **2005**, *28*, 1–8.
142. Obligacion, R.; Murray, M.; Ramzan, I. *J. Androl.* **2006**, *27*, 138–150.
143. Augustine, L. M.; Markelewicz, R. J., Jr.; Boekelheide, K.; Cherrington, N. J. *Drug Metab. Dispos.* **2005**, *33*, 182–189.

144. Cisternino, S.; Mercier, C.; Bourasset, F.; Roux, F.; Scherrmann, J. M. *Cancer Res.* **2004**, *64*, 3296–3301.
145. Kusuhara, H.; Sugiyama, Y. *NeuroRx* **2005**, *2*, 73–85.
146. Ohtsuki, S. *Biol. Pharm. Bull.* **2004**, *27*, 1489–1496.
147. Graff, C. L.; Pollack, G. M. *Curr. Drug Metab.* **2004**, *5*, 95–108.
148. Mikkaichi, T.; Suzuki, T.; Onogawa, T.; Tanemoto, M.; Mizutamari, H.; Okada, M.; Chaki, T.; Masuda, S.; Tokui, T.; Eto, N.; Abe, M.; Satoh, F.; Unno, M.; Hishinuma, T.; Inui, K.; Ito, S.; Goto, J.; Abe, T. *Proc. Natl. Acad. Sci. USA* **2004**, *101*, 3569–3574.
149. Sasaki, M.; Suzuki, H.; Ito, K.; Abe, T.; Sugiyama, Y. *J. Biol. Chem.* **2002**, *277*, 6497–6503.
150. Pritchard, J. B.; Miller, D. S. *Toxicol. Appl. Pharmacol.* **2005**, *204*, 256–262.
151. Khamdang, S.; Takeda, M.; Babu, E.; Noshiro, R.; Onozato, M. L.; Tojo, A.; Enomoto, A.; Huang, X. L.; Narikawa, S.; Anzai, N.; Piyachaturawat, P.; Endou, H. *Eur. J. Pharmacol.* **2003**, *465*, 1–7.
152. Jung, K. Y.; Takeda, M.; Shimoda, M.; Narikawa, S.; Tojo, A.; Kim, D. K.; Chairoungdua, A.; Choi, B. K.; Kusuhara, H.; Sugiyama, Y.; Sekine, T.; Endou, H. *Life Sci.* **2002**, *70*, 1861–1874.
153. Ciarimboli, G.; Ludwig, T.; Lang, D.; Pavenstadt, H.; Koepsell, H.; Piechota, H. J.; Haier, J.; Jaehde, U.; Zisowsky, J.; Schlatter, E. *Am. J. Pathol.* **2005**, *167*, 1477–1484.
154. Izzedine, H.; Launay-Vacher, V.; Deray, G. *Am. J. Kidney Dis.* **2005**, *45*, 804–817.
155. Miller, D. S. *J. Pharmacol. Exp. Ther.* **2001**, *299*, 567–574.
156. Hasannejad, H.; Takeda, M.; Taki, K.; Shin, H. J.; Babu, E.; Jutabha, P.; Khamdang, S.; Aleboyeh, M.; Onozato, M. L.; Tojo, A.; Enomoto, A.; Anzai, N.; Narikawa, S.; Huang, X. L.; Niwa, T.; Endou, H. *J. Pharmacol. Exp. Ther.* **2004**, *308*, 1021–1029.
157. Amidon, G. L.; Lennernas, H.; Shah, V. P.; Crison, J. R. *Pharm. Res.* **1995**, *12*, 413–420.
158. Wu, C. Y.; Benet, L. Z. *Pharm. Res.* **2005**, *22*, 11–23.

Biography

Jean-Michel Scherrmann, PharmD, PhD, is Professor and Chair at the Department of Clinical Pharmacy and Pharmacokinetics, Faculty of Pharmacy, the University of Paris 5. He received his PharmD and PhD in analytical radiochemistry from the University René Descartes at Paris. He currently leads the Neuropsychopharmacology Unit at the French Institute of Health and Medical Research (INSERM) and the National Center of Scientific Research (CNRS). Professor Scherrmann has made major contributions to the development of drug radioimmunoassay, drug detoxification by immunotherapy, and drug redistribution concepts in pharmacokinetics. He pioneered the first clinical application of colchicine immunotherapy in acute colchicine overdose, and is now focusing his research on the role of drug transporters in neuropharmacokinetics and drug delivery strategies to the brain. His work has resulted in two patents, 290 scientific articles, 45 book chapters, and more than 300 presentations. He has served on the editorial boards of several journals in the analytical and pharmaceutical sciences. He has mentored over 30 postgraduate and 44 doctoral students. Professor Scherrmann has been a recipient of the 1992 American Academy of Clinical Toxicology Award and the 1999 French National Academy of Medicine Achievement Award. He is a Fellow of the American Association of Pharmaceutical Scientists and a member of the French National Academy of Pharmacy.

Comprehensive Medicinal Chemistry II
ISBN (set): 0-08-044513-6

ISBN (Volume 5) 0-08-044518-7; pp. 51–85

5.05 Principles of Drug Metabolism 1: Redox Reactions

W F Trager, University of Washington, Seattle, WA, USA

5.05.1 Introduction

Environmental exposure to foreign compounds that might be harmful to normal biological function has necessitated the development of a generalized defense system in virtually all higher life forms. To cope with the challenges that exposure brings requires that the system be sophisticated enough to effectively operate on chemical entities to which the living organism has never been exposed. This has been partially achieved by the evolution of the group of enzyme systems outlined in **Table 1**. Virtually any organic molecule can be transformed by one or more of the enzymes in this group. Most are oxidoreductases, and thus have dual reaction modes. They can activate and utilize molecular oxygen to oxidize xenobiotics or utilize electrons supplied by cofactors to reduce xenobiotics. For a few, reduction is the preferred course of reaction. The products invariably have altered biological properties since they are structurally distinct from the parents. Their increased polarity results in their increased water solubility and more rapid elimination from the body. An excellent discussion of the role these enzymes play in drug metabolism can be found in the book by Testa.[1]

Table 1   General properties of the major oxidoreductases that catalyze the metabolism of xenobiotics in the human

Enzyme system	Size (kDa)	Location	Cofactors	Primary substrates	Site of attack
CYP[a] (multiple isoforms in three families)	~50	Endoplasmic reticulum; primarily liver	NADPH[b], heme, CYP reductase	Lipoidal organics	Primarily C also S and N
FMO[c] (six isoforms)	~60	Endoplasmic reticulum; primarily liver	FAD[d], NADPH	Heteroatom-containing lipoidal organics	Nucleophilic heteroatoms N, S, P, Se, etc.
MAO[e] (two forms; A and B)	~60	Outer mitochondrial membrane; liver, brain, gut, kidney	FAD	Arylalkylamine neurotransmitters	N
XOR[f] (two interconvertible forms)	~300 (homodimeric)	Cytosol primarily primarily liver and intestine	Molybdopterin, two iron–sulfur centers, FAD	Purines, N-containing heterocycles	Ring C
Aldehyde oxidase	~300 (homodimeric)	Cytosol	Molybdopterin, two iron–sulfur centers, FAD	Aldehydes purines, N heterocycles	Carbonyl C, ring C
Aldehyde dehydrogenase (many isoforms)	~50–60	Cytosol, microsomes, mitochondria	NAD, NADP	Aldehydes	Carbonyl C
Alcohol dehydrogenase (superfamily)	Variable	Cytosol	Zn, NAD	Alcohols	C bound to O
AKR[g] (superfamily)	~35–40	Cytosol primarily liver	NADPH	Aldehydes, ketones	Carbonyl C
Carbonyl reductase (superfamily)	~30	Cytosol	NADPH, NADH	Aldehydes, ketones	Carbonyl C

[a] Cytochrome P450.
[b] Reduced form of nicotinamide adenine dinucleotide phosphate.
[c] Flavin-containing monooxygenase.
[d] Flavin adenine dinucleotide.
[e] Monoamine oxidase.
[f] Xanthine oxidoreductase (xanthine oxidase and xanthine dehydrogenase).
[g] Aldo-keto reductase.

The organism's defense mechanism is completed by two additional classes of enzymes. The conjugating enzymes catalyze covalent bond formation between highly polar species (e.g., glucuronic acid) and a suitable functional group (usually hydroxyl) of the xenobiotic. The hydrolytic enzymes operate primarily on esters and amides, generating alcohols and acids or amines and acids, respectively. Both classes serve as mechanisms to enhance the elimination of highly lipoidal target molecules by increasing their water solubility through chemical modification.

The content of this chapter is focused on the metabolic reactions catalyzed by the enzyme systems listed in **Table 1**. Conjugation and hydrolytic reactions are the subjects of Chapter 5.06.

5.05.2   Oxidations

5.05.2.1   Cytochromes P450

5.05.2.1.1   Occurrence, multiplicity, catalytic cycle, oxygen activation, and selectivity

As components of a biological defense system the cytochrome P450 enzymes[2] are by far the most dominant and important (**Table 1**). They embody a superfamily of monooxygenases that are found in living organisms, ranging from bacteria to humans.[3] In humans and other mammals the cytochromes P450 are found in highest concentration in the endoplasmic reticulum of the liver, and in lower concentration in other tissues (e.g., brain, lung, kidney, and intestine). In addition to defense, some cytochromes P450 play a significant role in normal physiology. In the human, for example, specialized cytochromes P450 found in the adrenal cortex are involved in the syntheses of steroid hormones, while others are involved in the oxidative transformation of fatty acids and the production of prostaglandins. However, the focus of this chapter will be on the cytochromes P450 involved in drug metabolism. These specific cytochromes P450 are found primarily in three families,[4] CYP1, CYP2, and CYP3 (*see* 5.06 Principles of Drug Metabolism 2: Hydrolysis and Conjugation Reactions; 5.07 Principles of Drug Metabolism 3: Enzymes and Tissues, for an extensive discussion of the biological and structural properties of the cytochromes P450). Of the cytochromes P450, CYP3A4 is, arguably, the most important, as it has been found to contribute significantly to the metabolism of approximately half of all drugs that have been administered to humans or are in current use.[5]

All the cytochromes P450 work by activating O_2. In the cycle[6,7] that describes the catalytic role, O_2 is split into two oxygen atoms; one atom is reduced to water, while the second is transferred to the substrate (**Figure 1**).

In the resting state the charge of heme Fe^{3+} is counterbalanced by two negatively charged nitrogen atoms from the porphyrin ring plus a negatively charged cysteine thiolate group that anchors heme to the protein. Heme Fe^{3+} is primarily in the low-spin form characterized by hexacoordinated heme Fe^{3+} in which Fe^{3+} lies in the plane of the porphyrin ring. Four of the ligand sites are occupied by the four imidazole nitrogen atoms, and the fifth by cysteine thiolate. The sixth is presumed to be occupied by a molecule of water.

- Step 1 – the substrate, RH, associates with the active site of the enzyme, and perturbs the spin state equilibrium. Water is ejected from the active site, and the high-spin pentacoordinated substrate-bound form becomes dominant. Fe^{3+} is puckered out and above the plane in the direction of the sixth ligand site. The change in spin state alters the redox potential of the system. Substrate bound enzyme is now more easily reduced.
- Step 2 – NADPH-dependent cytochrome P450 reductase transfers an electron to heme Fe^{3+}.

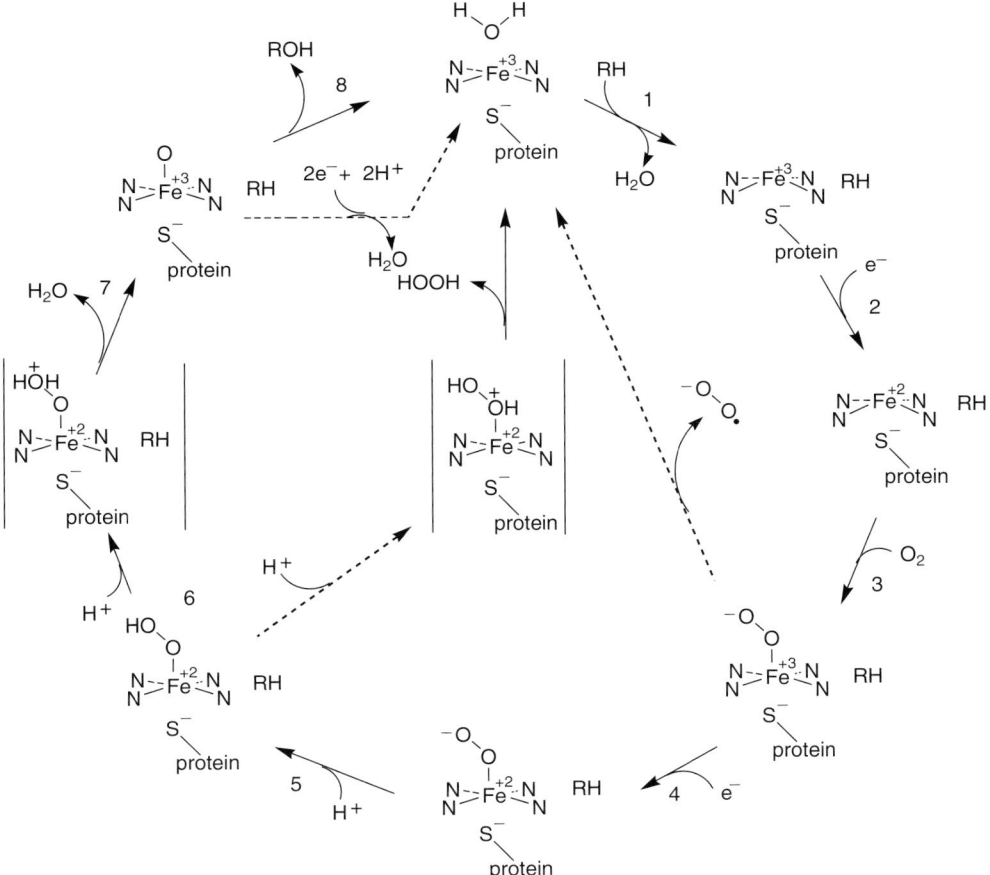

Figure 1   The cytochrome P450 catalytic cycle.

- Step 3 – O_2 binds, but can also dissociate, causing the enzyme to revert to the heme Fe^{3+} resting state while generating the superoxide radical anion.
- Step 4 – a second electron, via cytochrome P450 reductase or in some instances cytochrome $b5$, is added to the system, generating a heme-bound peroxide dianion formally equivalent to FeO_2^{1+}.
- Step 5 – H^+ adds to the system, generating a heme-bound hydroperoxide anion complex formally equivalent to heme FeO_2H^{2+}.
- Step 6 – a second H^+ is added. If H^+ adds to the inner oxygen of heme FeO_2H^{2+}, decoupling occurs, H_2O_2 is released, and the enzyme reverts to the heme Fe^{3+} form.
- Step 7 – the second H^+ adds to the outer oxygen atom of heme FeO_2H^{2+}; water is formed and released. Residual heme FeO^{3+} bears an oxygen atom (oxene) complexed to heme Fe^{3+}, a species considered to be analogous to compound 1 (heme in which $Fe^{IV}O$ is complexed to a porphyrin π radical cation), the active oxidant of the peroxidases. Decoupling can again occur via a two-electron reduction of FeO^{3+} plus the addition of two protons. This generates a molecule of water and the heme Fe^{3+} resting state of the enzyme. The degree to which this process occurs depends on the relative rates of heme FeO^{3+} reduction versus oxygen atom transfer to the substrate, as outlined in the next step.[8,9]
- Step 8 – an oxygen atom is transferred from heme FeO^{3+} to the substrate, forming the oxidized product, which is then released. The enzyme reverts to its heme Fe^{3+} resting state.

5.05.2.1.2   Oxidation of sp³ carbon–hydrogen bonds
5.05.2.1.2.1   Mechanism

Unlike other enzymes, which are generally highly specific, the cytochromes P450 are of necessity general oxidants that must be able to oxidize an unlimited array of organic compounds of divergent structure. This is achieved by their ability to generate a species, an oxygen atom, that is reactive enough to oxidize virtually any organic molecule that it comes in contact with.[10,11] This high reactivity suggests that the importance of protein composition and active site architecture to catalysis is fundamentally different for the cytochromes P450 than it is for most other enzymes. Cytochromes P450 have the ability to control access to the active site oxidant. In most other enzymes the active site architecture promotes a specific active site binding orientation that approaches the transition state geometry for the reaction and thereby lowers the necessary energy of activation.[10,11] Since all cytochromes P450 have the same heme cofactor and the same active oxidant, it is the spectrum of specific substrates that are able to access the active site of a given cytochrome P450 that distinguishes one cytochrome P450 from another.

The consensus mechanism for the cytochrome P450-catalyzed oxidation of covalent C–H bonds of simple normal hydrocarbons has been the two-step hydrogen atom abstraction–oxygen rebound mechanism where heme FeO^{3+} abstracts a hydrogen atom from the substrate to form heme $FeOH^{3+}$ and a carbon-based substrate radical in the first step. In the second step, a hydroxy radical from heme $FeOH^{3+}$ combines with the substrate radical to form the hydroxylated product and regenerate the resting state of the enzyme.[12–16]

While the evidence for a radical rebound mechanism is strong, it is not without problems,[14,15,17,18] and has been challenged by Newcomb and co-workers,[18–25] who suggest a single-step direct insertion mechanism.

A potential resolution to this dilemma has been offered by Shaik and co-workers,[26–32] based on theoretical calculations. These investigators describe a two-state reactivity paradigm for heme FeO^{3+} that rationalizes the properties of a two-step radical recombination mechanism[12,13] and those of a single-step direct insertion mechanism.[20,22] An excellent review that focuses on the mechanisms of O–O bond activation and assesses the potential role of multiple active intermediates has recently appeared in the literature.[33]

5.05.2.1.2.2   Primary, secondary, and tertiary carbon atoms

Despite the complexity of the mechanism for cytochrome P450-catalyzed oxidation of sp³ C–H bonds, the electrophilic character of a carbon-based radical or radical-like species generated by the two-state reactivity paradigm state suggests that the ease of formation of either of these species by the two states should in general mirror the relative ease of formation of carbonium ions, that is, hydroxylated product formation should follow the order tertiary (3°) > secondary (2°) > primary (1°). Experimentally, this is indeed what is found when simple normal- or branched-chain hydrocarbons are exposed to cytochrome P450. Even though they are the most sterically hindered, 3° carbons tend to be preferentially hydroxylated as are ω-1 2° carbons (i.e., 2° carbons immediately adjacent to terminal methyl groups).

It is important to note that while 3°, 2°, and 1° is the order usually followed by members of the CYP1, CYP2, and CYP3 families, the CYP4 family contains specialized cytochromes P450 only marginally involved in drug metabolism that are selective for ω-hydroxylation, particularly in relation to the metabolism of fatty acids. For example,

Powell *et al.*[34] have identified CYP4A11 as the major lauric acid ω-hydroxylase found in human liver, while Alexander *et al.*[35] report that rat brain tissue has high fatty acid ω-hydroxylase activity, leading to the formation of dicarboxylic acids. Sawamura *et al.*[36] found that CYP4A7 hydroxylates the prostaglandins PGA_1 (**1**) and PGA_2 (**3**) exclusively in the ω-position (eqns [1] and [2]).

These results not only emphasize the capacity of heme FeO^{3+} to selectively oxidize a C–H bond as unreactive as a methyl group attached to a saturated aliphatic chain but they also highlight the importance of active site architecture in dictating what part of the substrate molecule is exposed to the oxygen atom of heme FeO^{3+}. Members of CYP families 1, 2, and 3, as major drug-metabolizing enzymes, would be expected to have relatively open active and less constraining active sites to accommodate a greater variety of molecules. Conversely, members of CYP families, such as CYP4, that are involved in the metabolism of endogenous bioactive molecules, such as the prostaglandins or the steroids, would be expected to have active sites that confine the oxidation of specific molecules to specific sites within the molecule. This indeed appears to be the case.

That nature has demonstrated the possibility of constructing cytochromes P450 that can selectively oxidize sites within complex molecules that are neither energetically nor statistically favored suggests that it may be possible to bioengineer cytochromes P450 to obtain catalysts for the production of important intermediates or end products that can only be obtained with great difficulty or at great expense. Catalysts could also be engineered for processing recalcitrant environmental pollutants such as polychlorinated aromatic compounds (e.g., dioxin). Such possibilities have provoked a great deal of interest within the scientific community, and their exploration is beginning to meet with some success. While a thorough discussion of the area is beyond the scope of this chapter, a few examples are worth noting.

Fisher *et al.*[37] covalently linked rat cytochrome P450 reductase to rat CYP4A1, a cytochrome P450 responsible for the ω-hydroxylation of lauric acid (**5**) (eqn [3]). The new fusion protein is extremely active, and catalyzes this reaction at the remarkable turnover rate of 300 nmol (nmol P450)$^{-1}$ min^{-1}.[38] The exceptional rate is presumably achieved by compelling the reductase and cytochrome P450 to maintain close proximity by covalently linking the two proteins. Thus, the price of entropy is paid as the reductase is always in position to supply reducing equivalents to the cytochrome P450 and fuel the reaction.

Purging the environment of polychlorinated aromatics has proven to be a major problem because of their toxicity, widespread presence, chemical inertness, and resistance to bacterial degradation. To address the problem, Jones *et al.*[39] genetically altered the bacterial enzyme cytochrome P450$_{cam}$ (CYP101), one of the most well-characterized and

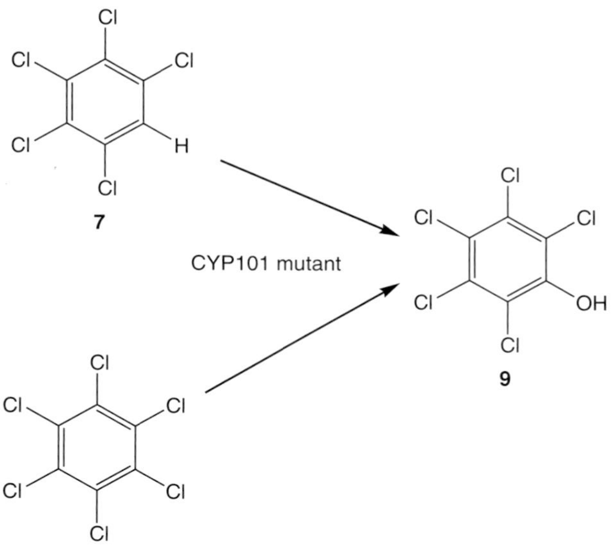

Scheme 1

well-studied of all known cytochromes P450. While wild-type CYP101 can oxidize dichloro- and trichlorobenzenes to chlorophenols, it is not able to metabolize the more highly substituted pentachlorobenzene (**7**) and hexachlorobenzene (**8**). Polychlorinated phenols do not present a significant hazard, as they are degraded by microorganisms in the environment. Thus, finding or constructing an enzyme that would metabolize highly chlorinated aromatics such as **7** and **8** would represent a significant advance toward a solution. Once constructed, the genes encoding the CYP101 system could be genetically incorporated into chlorophenol-degrading microorganisms to convert chlorobenzenes to chlorophenols that would then be further degraded by the host.[39] The authors' challenge then was to bioengineer CYP101 so it would be able to hydroxylate **7** and **8** to form **9** (**Scheme 1**). Reasoning that changes that increased hydrophobicity and decreased the size of the active site would be beneficial to that end, three specific mutations were introduced. The mutant enzyme was not only able to metabolize **7** and **8**, but exceeded the hydroxylase activity of wild type by three orders of magnitude.

5.05.2.1.2.3   Benzylic and allylic carbon atoms

If cytochrome P450-catalyzed hydroxylation of benzylic or allylic C–H bonds proceeds by the same mechanism operative for the simple saturated systems discussed above, they would be expected to be favored processes. Resonance stabilization of a radical intermediate would lower the activation energy for both benzylic and allylic hydroxylation. This is generally what is found in drug metabolism studies.

White *et al.*[40] reinvestigated the stereochemistry of hydroxylation of the prochiral benzylic carbon atom of phenylethane (**10**), using enantiomerically pure (*R*)- and (*S*)-phenylethane-1-d as substrates (**Scheme 2**). General findings of particular note were (1) hydroxylation occurs almost exclusively, greater than 99%, at the benzylic position to give the isomeric α-methylbenzyl alcohols (**11**), a result consistent with the benzylic position being a favored site of attack, and (2) the percentage yield of the minor metabolites, 2-phenylethanol (**12**) and 4-ethylphenol, (**13**), more than triples when phenylethane-1-d$_2$ is used as the substrate. The isotopically driven switching to other sites of metabolism not only indicates the operation of a significant isotope effect but the tripling of such metabolites, particularly 4-ethylphenol, also indicates that the substrate has considerable freedom of motion within the active site and the potential to form multiple catalytically productive active site binding orientations.[8,41,42] It is not uncommon for a single cytochrome P450 to catalyze the formation of multiple regioisomeric products from the same substrate,[43,44] where the primary metabolite is often formed by oxidation at the energetically most favored position.[45]

An early example of allylic hydroxylation was provided by Licht and Coscia,[46] who reported that the CYP2B 4-catalyzed hydroxylation of the terpenes geraniol (**14**) and nerol (**16**) occurred almost exclusively at the C-10 (*E*)-methyl group of both compounds (eqns [4] and [5]). What is informative about this example is that the adjacency of a double bond converts the methyl group to a major site of oxidative attack.

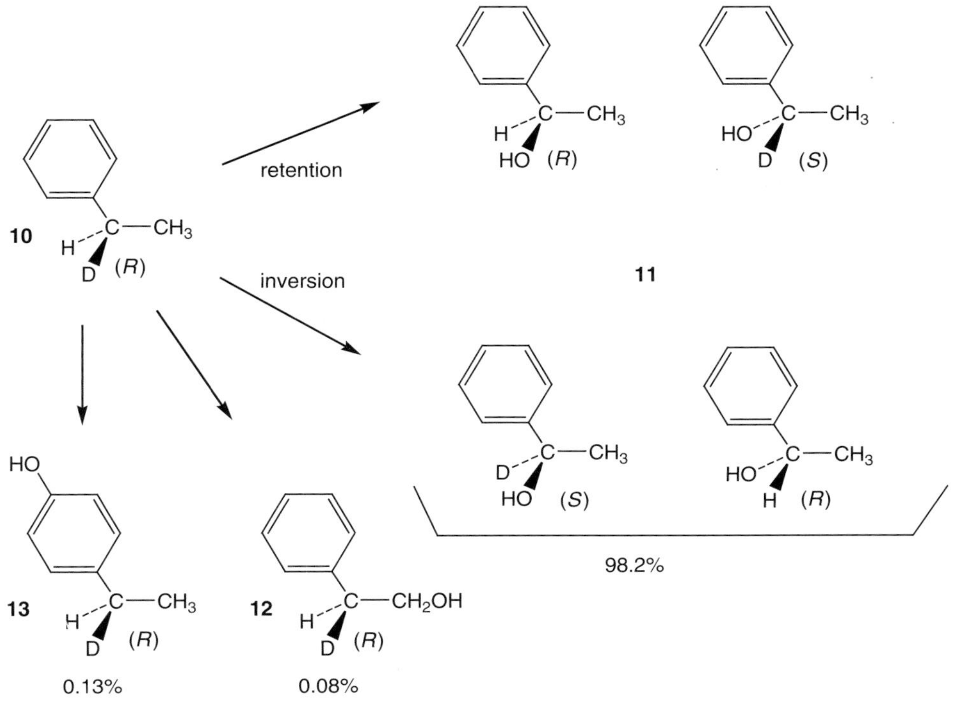

retention

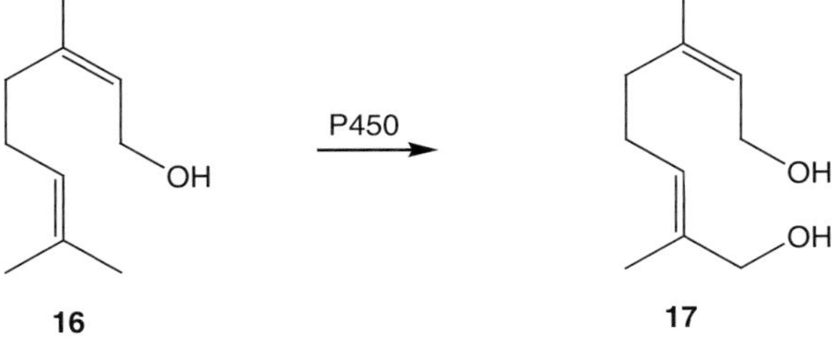

inversion

11

98.2%

13   0.13%

12   0.08%

Scheme 2

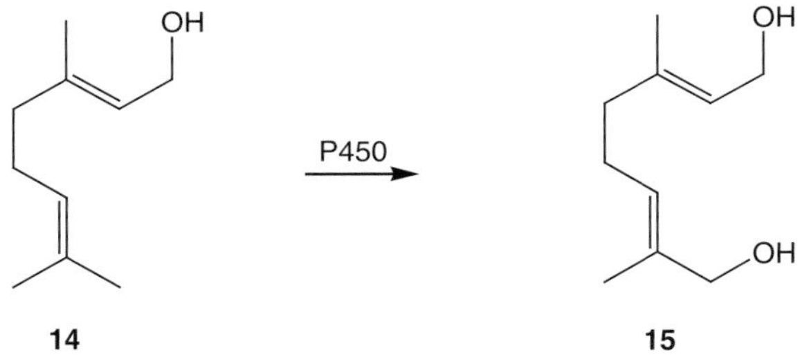

P450   [4]

14   15

P450   [5]

16   17

   The mechanism of cytochrome P450-catalyzed allylic hydroxylation was found to be consistent with a hydrogen abstraction–oxygen rebound mechanism by Groves and Subramanian.[47] A consequence of generating a radical intermediate, as the mechanism requires, is the possible production of a rearranged product. This is seen in the metabolism of (R)-pulegone (18), the major constituent of pennyroyal oil, a volatile plant oil that has been used as an abortifacient and causes major toxicity at high doses. Menthofuran (19), previously identified as a metabolite of pulegone, appeared to arise from initial cytochrome P450-catalyzed oxidation of one of the allylic methyl groups.[48] To probe the mechanism of the reaction, the (E)-methyl-d$_3$ analog of pulegone (18) was synthesized and incubated with

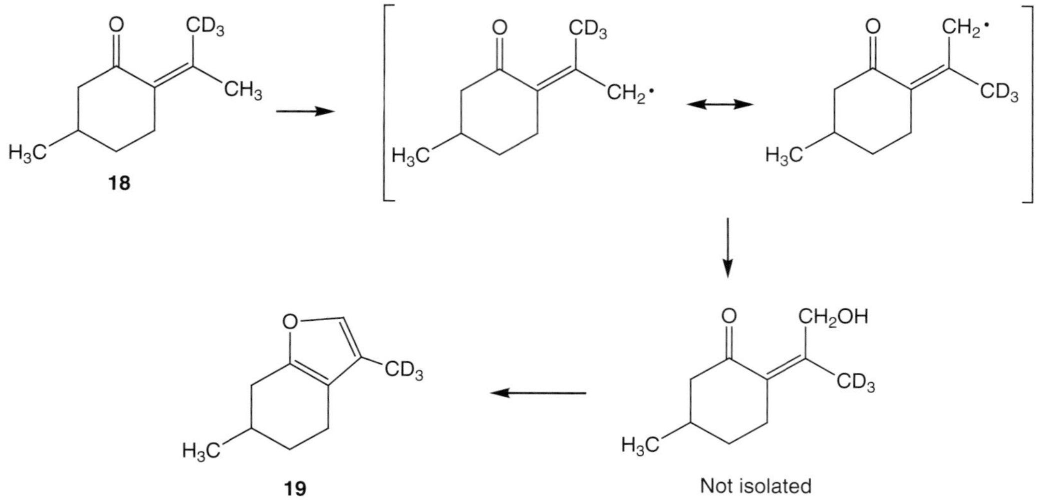

Scheme 3

microsomal cytochrome P450, to form **19** (**Scheme 3**).[49] Isolated **19** contained a furano-trideuteromethyl group, indicating that a labile intermediate must have been formed during the course of the reaction to allow interchange of the positions of the two allylic methyl groups prior to hydroxylation, ring closure, and aromatization.

5.05.2.1.2.4   Carbon atoms α to a heteroatom (N, O, S, and X)

N-Dealkylation is a frequently encountered metabolic reaction. It is often responsible for the production of the major metabolite obtained from an *N*-alkyl containing drug. Its prominence does not simply derive from the commonality of an alkyl-substituted amino group as an important part of the structural motif of many drugs. It arises primarily as a consequence of *N*-dealkylation being energetically favored.[10] Typical examples include the *N*-demethylation of methamphetamine (**20**)[50] (eqn [6]), the *N*-deethylation of lidocaine (**22**)[51] (eqn [7]), and the *N*-deisopropylation of propranolol (**24**)[52] (eqn [8]), to form the secondary amines **21**, **23**, and **25**, respectively.

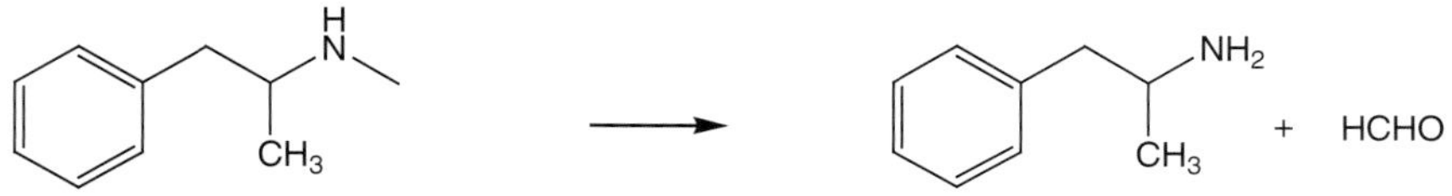

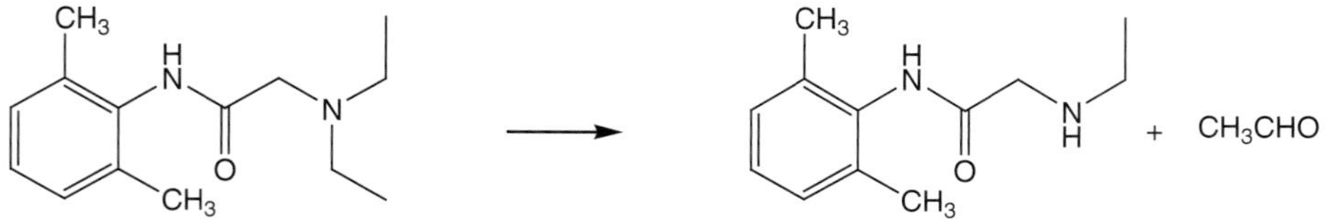

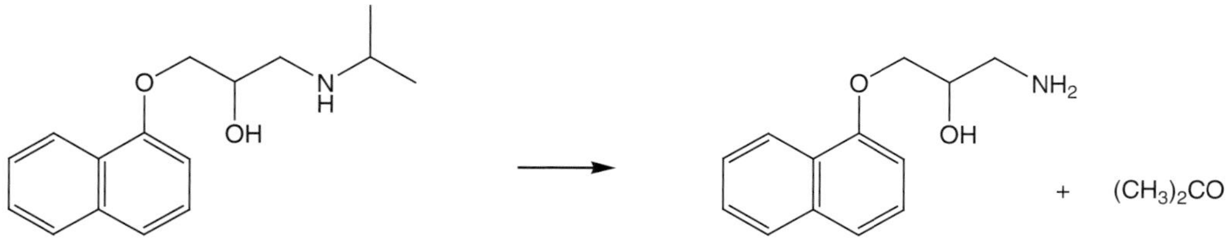

There are two competing mechanisms for oxidative N-dealkylation: the single-electron transfer (SET) mechanism, championed by Guengerich and McDonald,[53–59] and the hydrogen atom transfer (HAT) mechanism, advocated by Dinnocenzo and Jones,[60–64] Both mechanisms postulate the intermediacy of a carbon-based radical, but differ in the mechanistic events leading to its formation (**Figure 2**). The HAT pathway postulates direct formation of the radical by transfer of a hydrogen atom from the α-carbon atom to heme FeO^{3+}, to form heme $FeOH^{3+}$. In contrast, the SET pathway requires two steps, initiated by single-electron transfer from the nitrogen lone pair to heme FeO^{3+}, to form heme FeO^{2+}, followed by transfer of H^+ from the α-carbon atom to heme FeO^{2+}, forming heme $FeOH^{3+}$ and the α-carbon radical. Oxygen rebound by either mechanism then forms the carbinolamine. Carbinolamines are chemically unstable, and dissociate to generate the secondary amine and aldehydes as products, or eliminate water to generate the iminium ion. The iminium ion if formed can reversibly add water to reform the carbinolamine, or add a different nucleophile if present (**Figure 2**). If the nucleophile is within the same molecule and four or five atoms removed from the iminium carbon atom, cyclization can occur and form a stable five- or six-membered ring system. For

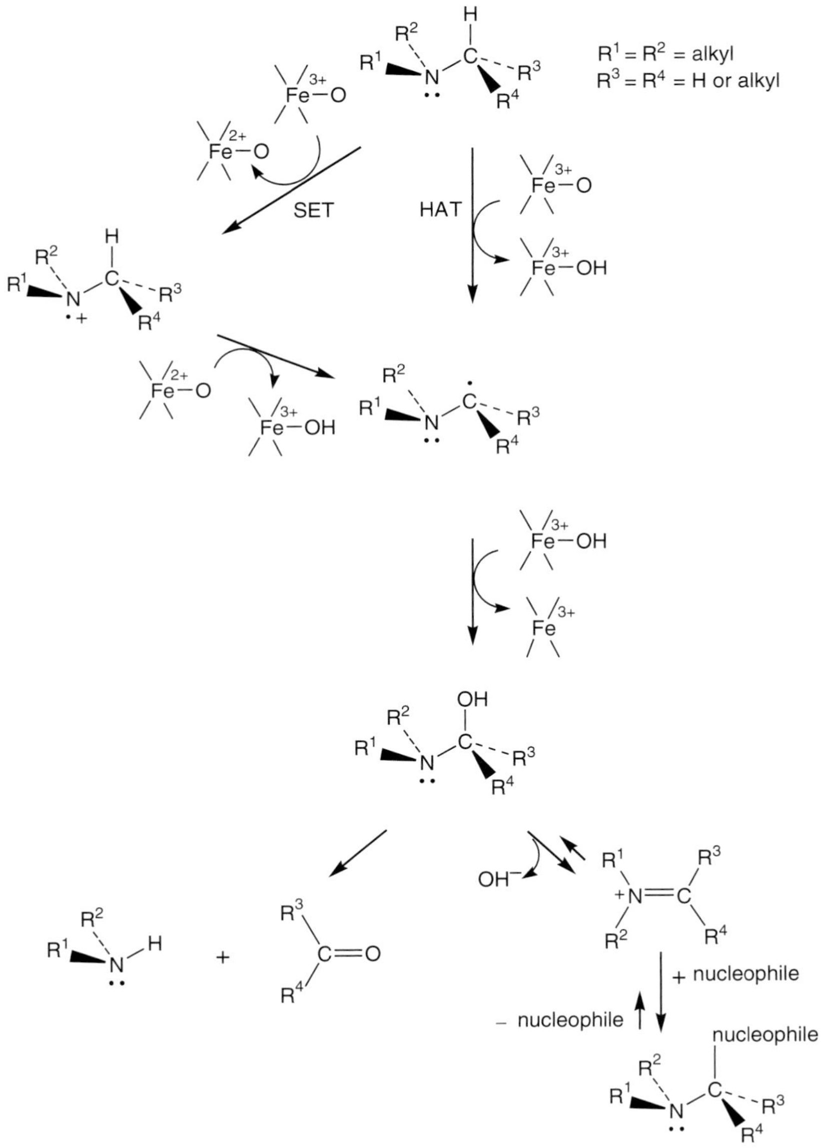

Figure 2 Mechanisms for oxidative N-dealkylation: SET and HAT.

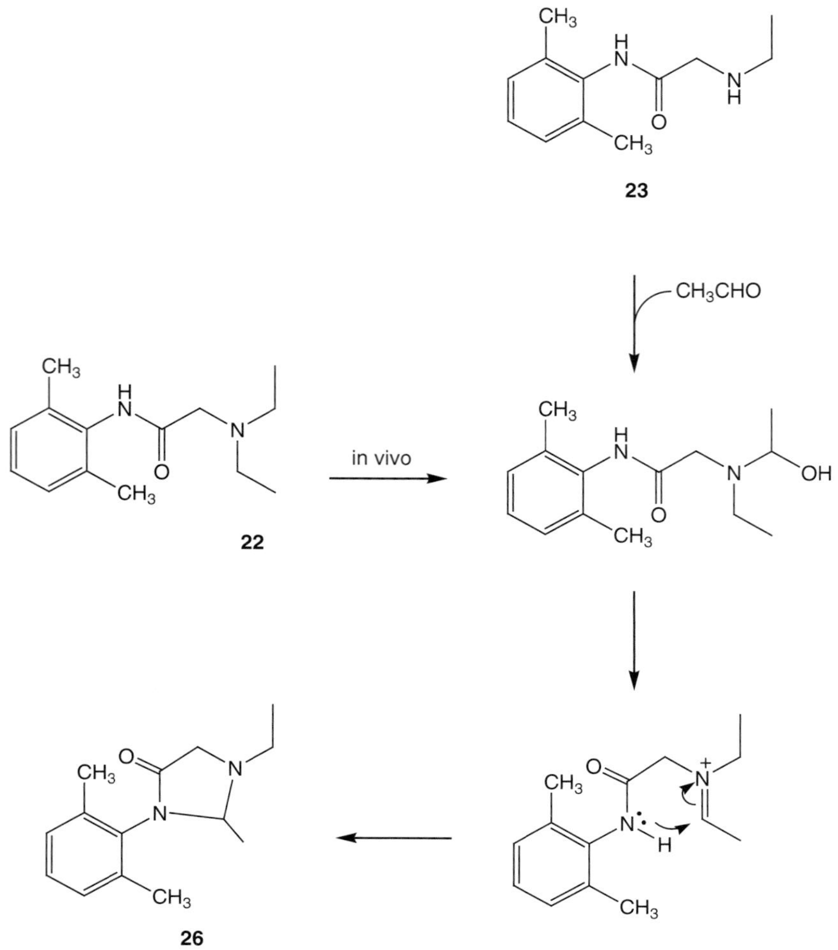

Scheme 4

example, the 4-imidazolidinone **26** is a major in vivo metabolite of lidocaine (**22**) (**Scheme 4**). It can also be formed upon isolation of the N-deethyl metabolite of lidocaine (**23**), if a trace of acetaldehyde happens to be present in the extraction solvent[51] (**Scheme 4**). A related example[65] is the formation of the stable 3,3-diphenylpyrrolidine **29** that is generated by the intramolecular cyclization of N-desmethylmethadone (**28**), the major metabolite of methadone (**27**) (**Scheme 5**).

Since iminium ions are reactive electrophiles, it is not surprising that they have been associated with toxicity. An interesting example is provided by the psychotomimetic agent phencylclidine (**30**), whose use can lead to long-term psychoses. When phencyclidine is incubated with rabbit liver microsomes, a reactive intermediate is formed that can be trapped by the addition of cyanide ions (**Scheme 6**). The structure of the cyanide adduct (**32**) is consistent with being formed from reaction of cyanide with the phencyclidine iminium ion (**31**).[66,67] Further NADPH-dependent metabolism of the iminium ion leads to the production of the conjugated pyridone **33**, a reactive electrophile that is the likely species responsible for stable covalent binding with critical bio-macromolecules and toxicity (**Scheme 6**).[68]

Oxidative attack on a C–H bond of an alkyl group α to a nitrogen atom is not restricted to saturated aliphatic amines. In fact, X in an X–N–CH– structural subunit can be virtually any common atomic grouping that can be found in stable organic molecules. For example, α-carbon hydrogen atoms of N-alkyl-substituted aromatic cyclic amines (**34**),[69] aryl amines (**36**),[60] amides (**38**),[70,71] amidines (**41**),[72] and N-nitrosoalkylamines (**43**)[73] (eqns [9]–[13]), are all subject to oxidative attack, carbinolamine formation, and, in most cases, release of an aldehyde or ketone, depending on the substitution pattern (1° or 2°) (eqns [11]–[13]) In some cases (e.g., N-alkyl aromatic cyclic amines), carbinolamines are stable enough to be isolated (e.g., **35**) (eqn [9]).

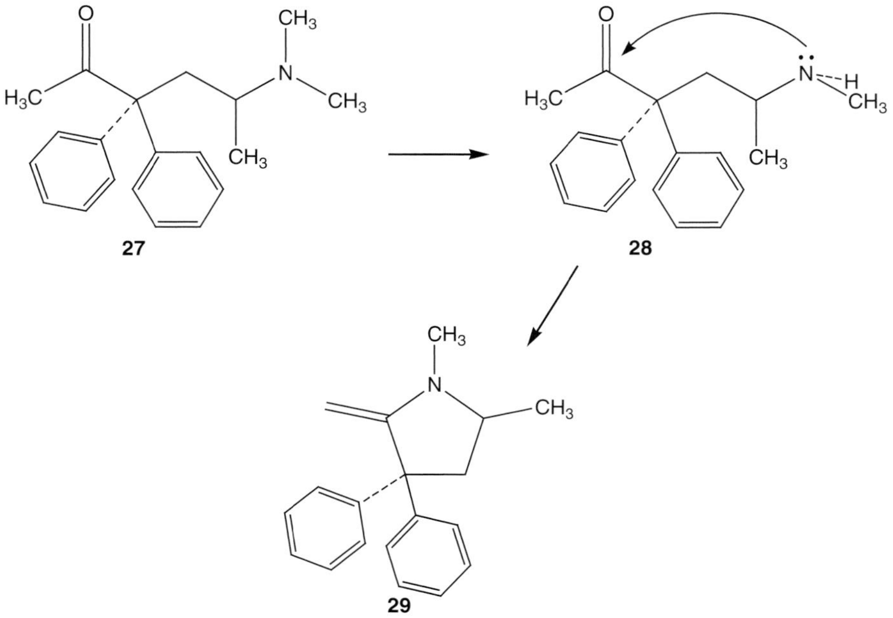

Scheme 5

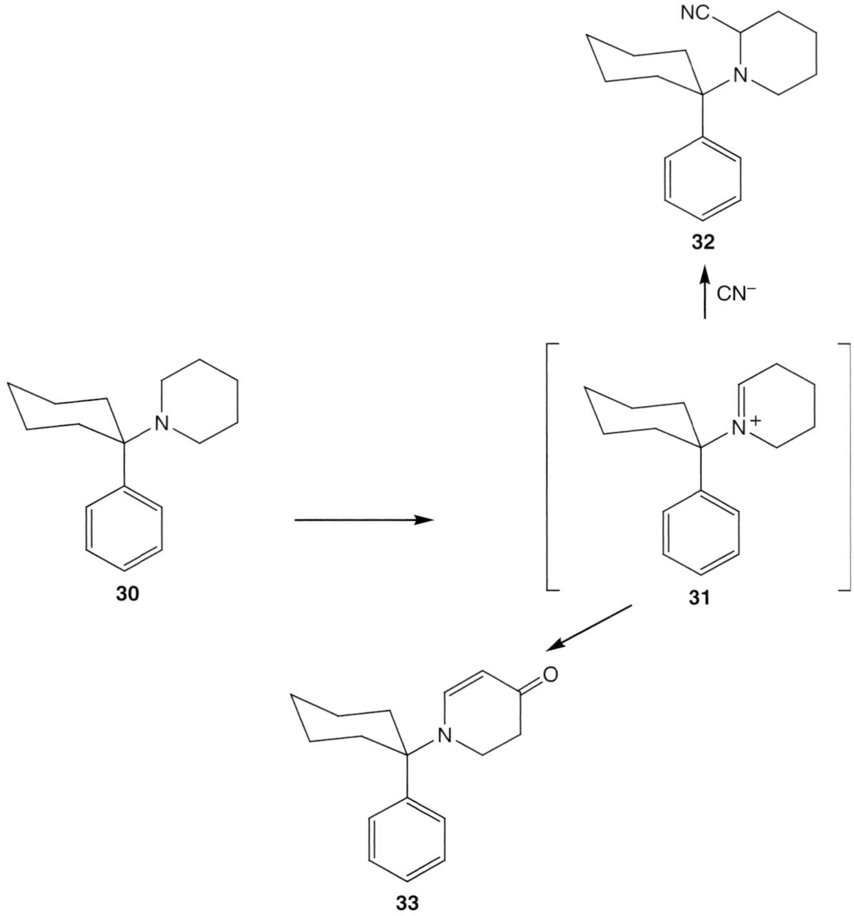

Scheme 6

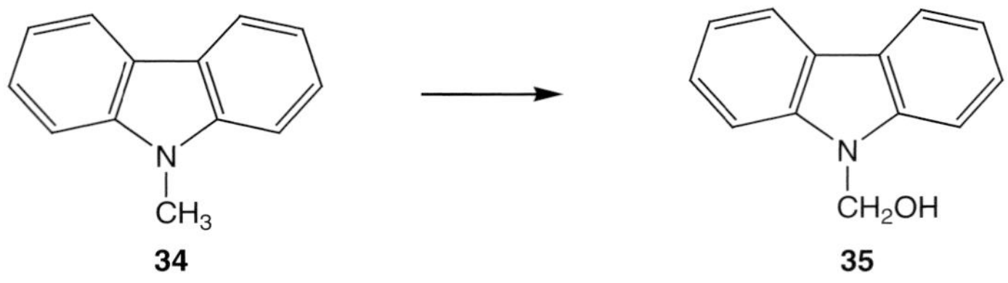

[9]

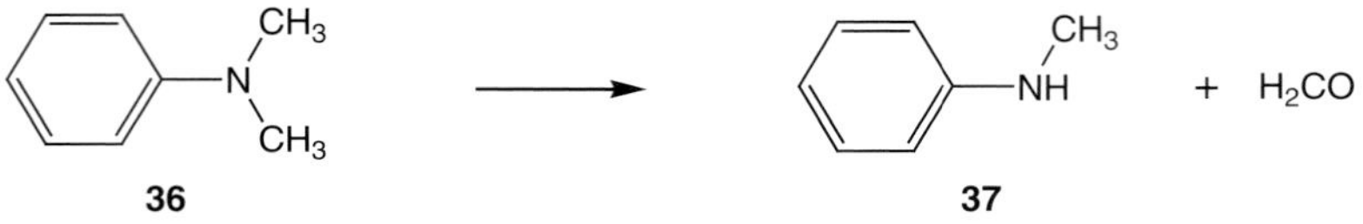

[10]

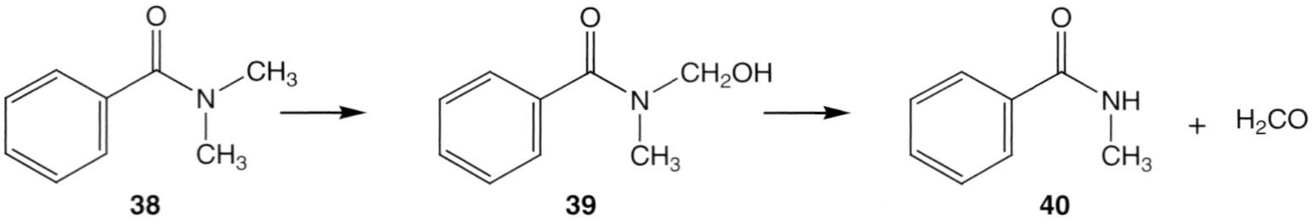

[11]

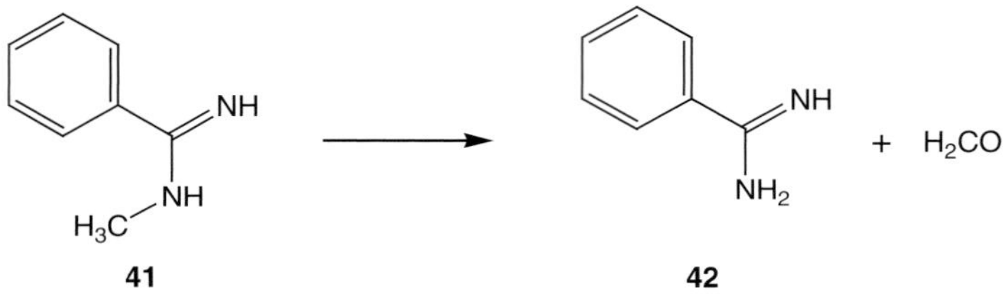

[12]

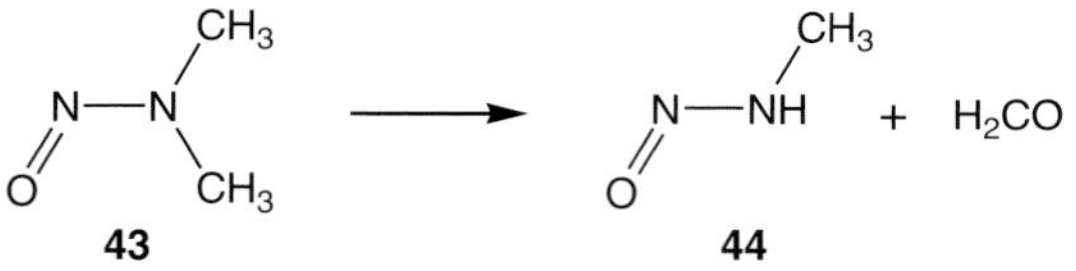

[13]

Cytochrome P450 can also catalyze hydroxylation of a C–H bond α to the oxygen atom in both alcohols and ethers. However, in contrast to ethers, the primary oxidants of alcohols appear to be the dehydrogenase enzymes and not the cytochromes P450, as will be discussed later.

The *O*-dealkylation of ethers, while not encountered as frequently as *N*-dealkylation, is still a common metabolic pathway. Mechanistically, it is less controversial than *N*-dealkylation. It is generally believed to proceed by the HAT pathway (**Figure 3**). The product of the reaction is unstable, being a hemiacetal or hemiketal, which dissociates to generate an alcohol and an aldehyde or ketone.

Energetically, *O*-dealkylation is less favored than *N*-dealkylation.[44] This is not surprising, as the greater electronegativity of oxygen relative to nitrogen would make abstraction of an α-hydrogen atom more difficult. Examples of drugs in which *O*-dealkylation plays a significant role are phenacetin (**45**),[74] dextromethorphan (**47**),[75] codeine (**49**),[76] and metoprolol (**51**)[77] (eqns [14]–[17]).

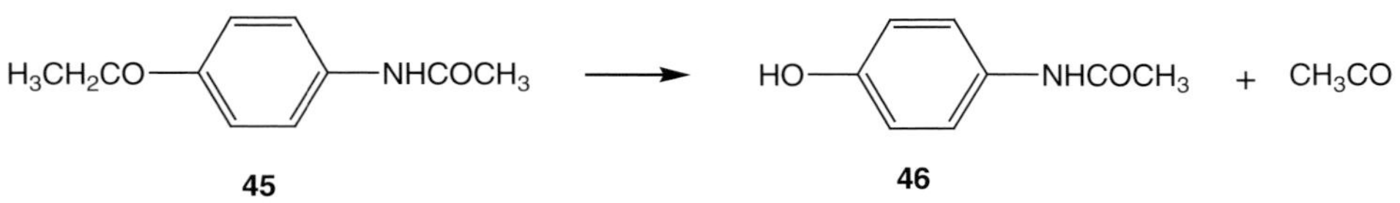

[14]

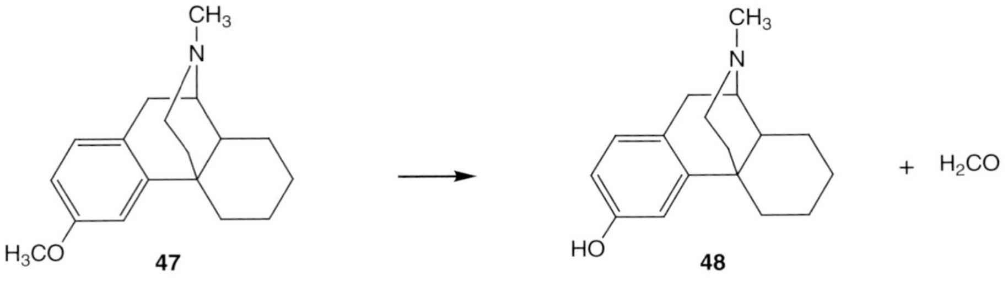

[15]

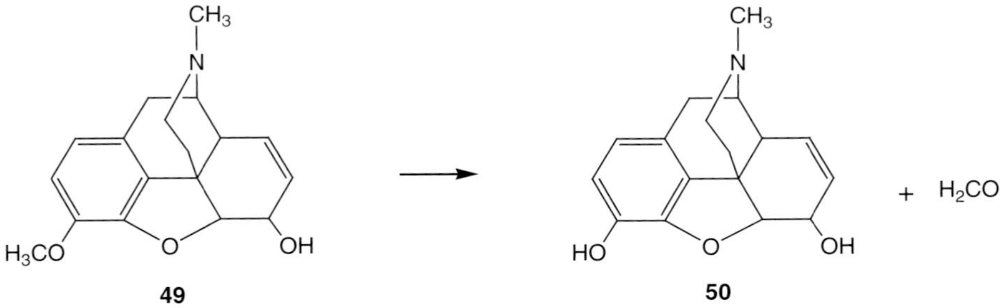

[16]

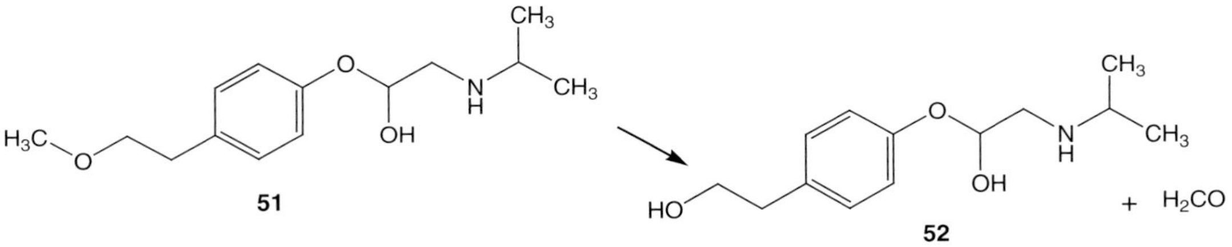

[17]

A methylene dioxy group in aromatic compounds is subject to *O*-dealkylation (e.g., 3,4-methylenedioxyampheta-mine (**53**)[78]) (eqn [18]). The process generates formic acid and the catechol metabolite **54** as final products. However, in the course of the reaction a portion of the enzyme can be inactivated by formation of what has been termed a metabolic intermediate (MI) complex[79,80] characterized by an absorption peak maximum at 455 nm in the difference spectrum of reduced cytochrome P450. The complexing species is believed to be a carbene that associates with heme Fe^{2+} (**55**), in much the same way that carbon monoxide[81,82] does. The complex is of moderate stability, and thus a quasi-irreversible inhibitor of the enzyme. It ultimately dissociates to generate ferric cytochrome P450, carbon monoxide, and the catechol **54**.[83]

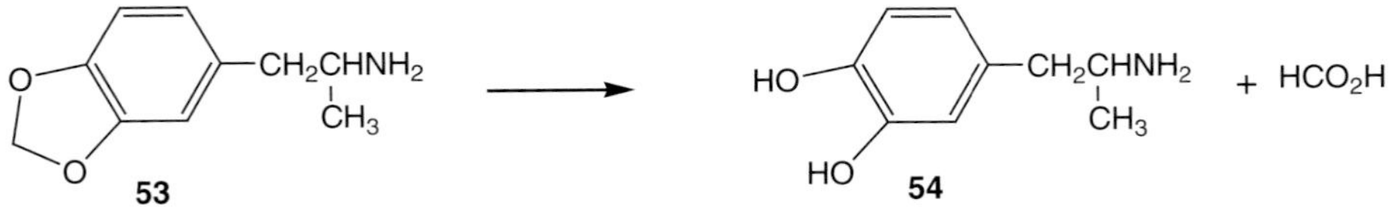

[18]

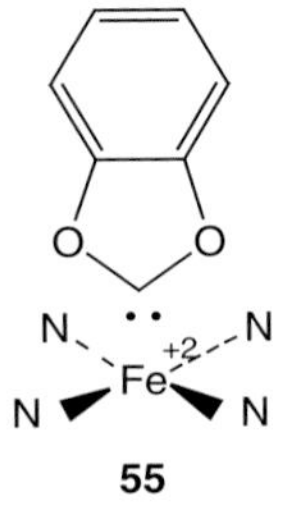

55

S-Dealkylation, unlike *N*- or *O*-dealkylation, is uncommon, generally not a major metabolic pathway, and in some cases does not even contribute to the metabolic profile of a sulfide-containing drug. This is probably due to two factors: (1) sulfide-containing drugs represent a small percentage of available drugs and (2) the sulfur atom itself is more

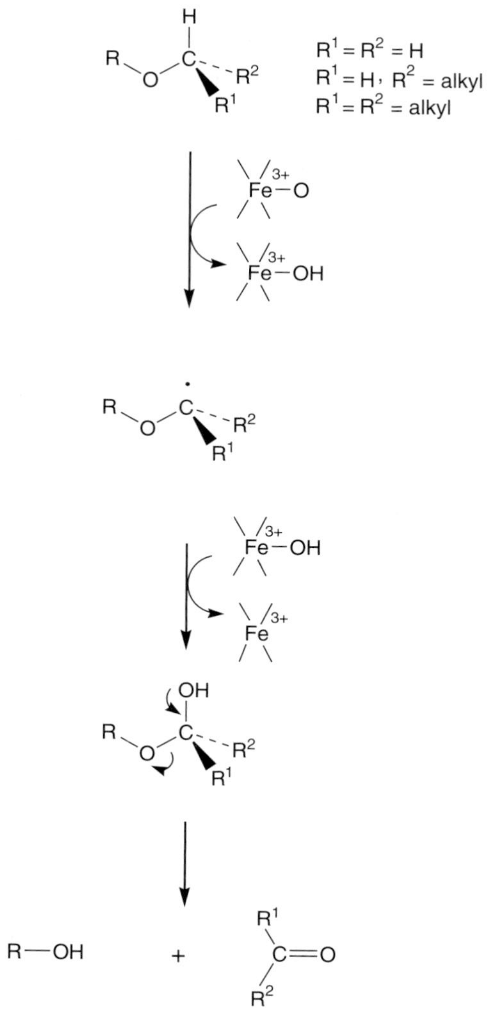

Figure 3 Mechanism for the *O*-dealkylation of ethers.

susceptible to oxidation than is the adjacent α-C–H bond. Nevertheless, *S*-dealkylation does occur, but whether it is driven by cytochrome P450 or some other enzyme system is not clear.

Halogen dealkylation mirrors *O*-dealkylation both in terms of mechanism and the commonality of the process, that is, while an aliphatic halogen substituent is not a common structural component, virtually any drug that contains a C–H bond adjacent to a halogen atom will be subject to cytochrome P450-catalyzed oxidative dehalogenation. Halogen atoms can also be removed either reductively, as will be discussed later, or by glutathione displacement, and as such represent a chemical group that is fairly labile in a biological environment.

Aliphatic halogen atoms are present in a number of common solvents and industrial chemicals to which people are exposed. For example, the fuel additive and suspected human carcinogen 1,2-dibromoethane (**56**) is oxidatively transformed to bromoacetaldehyde (**57**) by CYP2E (eqn [19]).[84] Aliphatic halogen atoms are also a major structural component of most inhalation anesthetics.

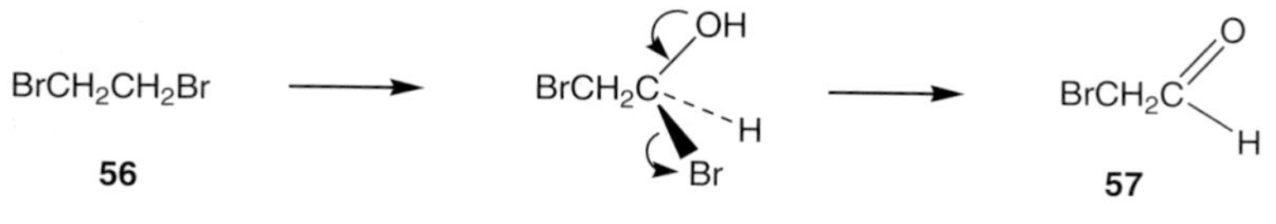

$$\text{BrCH}_2\text{CH}_2\text{Br} \longrightarrow \text{BrCH}_2\text{C} \longrightarrow \text{BrCH}_2\text{C(=O)H} \qquad [19]$$

$$\mathbf{56} \qquad\qquad\qquad\qquad \mathbf{57}$$

Since many of these compounds contain two halogen atoms on a terminal methyl group, their metabolism can lead to serious toxicity because of the generation of reactive intermediates such as acyl halides.[85] For example, chloroform (**58**)-induced nephrotoxicity in mice has been correlated with the ability of the animals to metabolically convert chloroform to the highly reactive acid chloride phosgene (59)[86–88] (eqn [20]). Similarly, the inhalation anesthetic halothane (60) is converted to trifluoroacetyl chloride (61) (eqn [21]), which in turn can covalently bind to protein, generating liver protein neo-antigens.[89,90] In susceptible individuals these neo-antigens stimulate the production of anti-trifluoroacetyl-protein antibodies, which can cause fatal halothane hepatitis upon re-exposure to the anesthetic. The related anesthetics enflurane (62) and isoflurane (63) are also subject to acyl halide formation and an ensuing hepatic dysfunction similar to that caused by halothane.[91–93]

$$CHCl_3 \longrightarrow \quad \longrightarrow \qquad\qquad [20]$$

58   **59**

$$CF_3CHBrCl \longrightarrow \quad \longrightarrow \qquad\qquad [21]$$

60   **61**

$$CHF_2OCF_2CHFCl \qquad\qquad\qquad CHF_2OCHClCF_3$$

62   **63**

5.05.2.1.3   Oxidation of sp^2 carbon atoms

5.05.2.1.3.1   Isolated or *exo*-ring double bonds

The cytochrome P450-catalyzed oxidation of a $C=C$ bond results in the formation of an epoxide. For example, the major metabolite of the anticonvulsant carbamazepine (64) is carbamazepine-10,11-epoxide (65)[94] (eqn [22]), and the potential carcinogen butadiene (66) is converted to the mono-epoxide 67, by human liver microsomes (eqn [23]).[95]

64   **65**   [22]

66   **67**   [23]

Mechanistically, the reaction is bounded by two extremes. At one extreme, FeO^{3+} adds to the double bond in a single step, while at the other a two-step reaction involving the generation of an intermediate is operative. Evidence exists for both pathways, but again the two-state reactivity paradigm of Shaik and co-workers appears to resolve the dilemma.[96,97]

5.05.2.1.3.2   Aromatic rings

The frequency of aromatic hydroxylation as a metabolic event is undoubtedly a reflection of the commonality of an aromatic ring as a structural component of many drugs. While a phenol is usually not the major metabolite of such a drug, it often is found as a significant contributor to the metabolism of that drug. In general, phenol formation follows the rules of electrophilic aromatic hydroxylation established by the linear free energy relationships of physical organic chemistry (i.e., *para* > *ortho* > *meta*). This order prevails unless the system is deactivated by a substituent that on balance withdraws electron density from the ring (e.g., nitro group). In such a case, *meta* substitution dominates, since it is the site that is the least deactivated toward electrophilic attack. In the case of cytochrome P450 catalysis, an exception would occur if the stearic demands of the active site architecture of the enzyme for a specific substrate favored *meta* hydroxylation.

These general observations suggest that if the enzyme has a sterically permissive active site that is not overly restrictive to substrate motion, the electronic properties of the substrate should determine regioselectivity of hydroxylation. Thus, the development of computational models for predicting aliphatic hydroxylation, aromatic hydroxylation, or a combination of the two, pioneered by Jones, Korzekwa, and colleagues,[10,45,98,99] is not only promising but has already met with considerable success.

Establishing the exact mechanism for aromatic hydroxylation has proved to be difficult. If deuterium is present in the substrate at the site of hydroxylation, a fractional amount of deuterium will almost always be retained in the product owing to migration to the adjacent carbon atom during the process of phenol formation. This is the well-known NIH shift.[100] It was believed to occur upon ring opening of an initially formed epoxide (arene oxide) (pathway 1 in **Figure 4**). Mechanistically, epoxide formation can occur either by a concerted addition of oxygen to form the epoxide in a single step or by a stepwise process. The stepwise process would involve: (1) the initial addition of oxygen to a

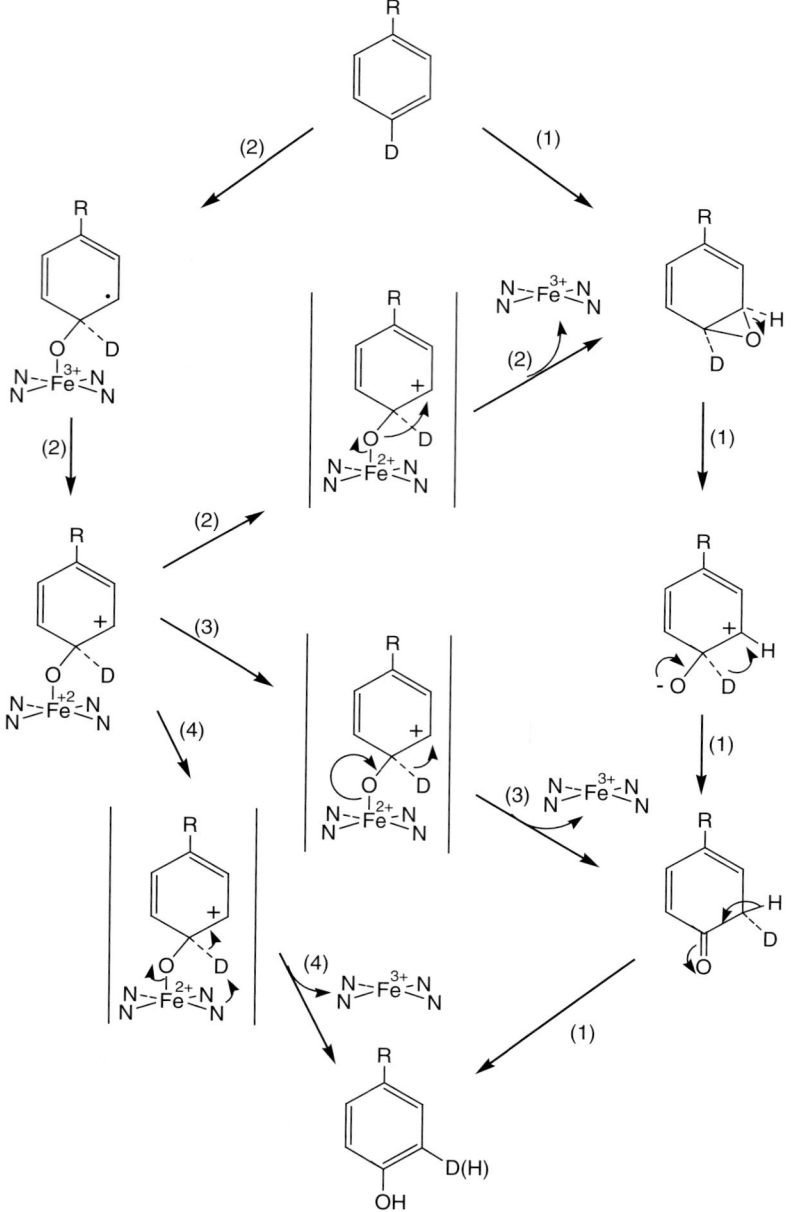

Figure 4   Mechanism for aromatic hydroxylation.

specific carbon atom, to form a tetrahedral intermediate; (2) electron transfer from the aryl group to heme, to form a carbonium ion adjacent to the oxygen adduct; followed by (3) ring closure, to form the epoxide (pathway 2, **Figure 4**). From this point on to the formation of the final product phenol, both mechanisms are identical: the ring opens, generating an adjacent carbonium ion, and the hydride moiety shifts to satisfy the adjacent carbonium ion, as a pair of electrons from the oxygen atom moves in to satisfy the positive charge being developed as the hydride leaves; this overall process leads to the formation of a ketone, which then tautomerizes to generate the phenol. The stepwise mechanism does not necessarily have to close to epoxide. At step 2 it can bypass epoxide formation, to form the ketone directly, then the phenol (pathway 3, **Figure 4**), or it can form the phenol directly by proton transfer to a pyrrole nitrogen atom (pathway 4, **Figure 4**).

A concerted mechanism would require obligatory epoxide formation on the path to phenol, whereas a stepwise mechanism would not necessarily have to pass through the epoxide. Since epoxides can be toxic or even carcinogenic (e.g., epoxides of some polycyclic aromatic hydrocarbons), the question of whether their formation is obligatory is significant for drug design and development.

A theoretical study[101] and several experimental studies[102–105] have provided strong evidence for the stepwise mechanism, and, most recently, de Visser and Shaik[106] examined the problem using density functional calculations, and came to the same basic conclusion.

5.05.2.1.3.3   Carbonyl carbon atoms

The conversion of acetaldehyde to acetic acid by cytochrome P450 has been clearly established.[107] But of the aldehydes that have been shown to be metabolized by cytochrome P450, allylic or aryl aldehydes appear to be the best substrates.[108] For example, retinal (**68**)[109] and the 11-oxo metabolite of tetrahydrocannabinol (**70**)[110] are oxidized to carboxylic acids (eqns [24] and [25]).

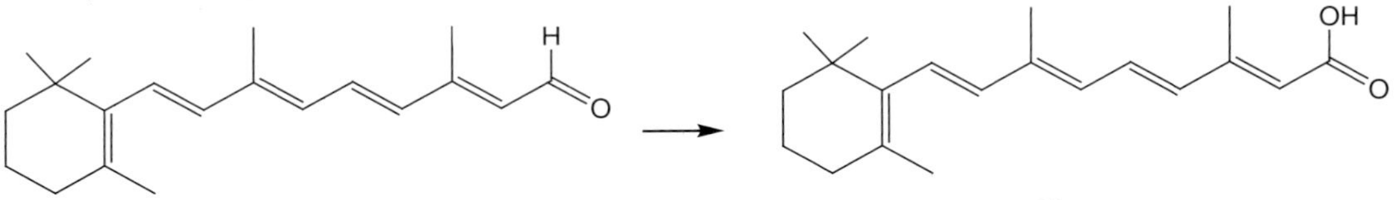

[24]

68   **69**

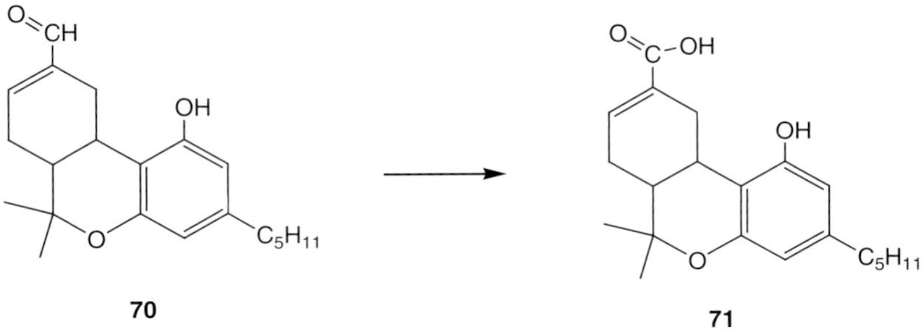

[25]

70   **71**

Selective cleavage of an unactivated C–C bond is chemically difficult. However, the highly specialized steroidogenic cytochromes P450 aromatase and lanosterol-14α-demethylase accomplish this difficult transformation with relative ease. In the case of aromatase, androstenedione (**72**) is converted to estrone (**73**) by three sequential oxidations of the C-19 methyl group, culminating in the elimination of formic acid and the aromatization of the A ring (eqn [26]). While the first two steps (sequential hydroxylations of C-19 to form a hydrated aldehyde) are well understood, the final oxidation leading to loss of formic acid and aromatization is not. An attractive possibility suggested by Akhtar et al.[111] is that an O_2-derived peroxide might be involved in the final C–C bond cleavage step.

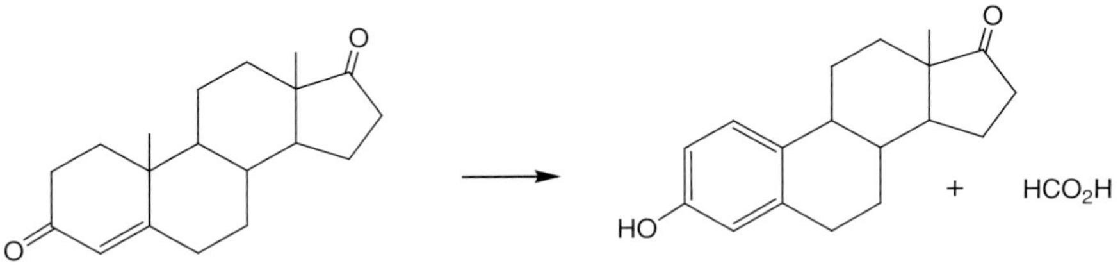

[26]

72   **73**

To investigate this possibility, Vaz *et al.*[112] used cyclohexanecarboxaldehyde (**74**) as a simple model of the C-19 aldehyde of androstendione. Upon treatment with CYP2B4, NADPH, and cytochrome P450 reductase, the aldehyde–cyclohexyl ring C–C bond of **74** was cleaved, forming cyclohexene (**75**) and formic acid (eqn [27]). The reaction was supported if hydrogen peroxide replaced NADPH and cytochrome P450 reductase, but was not supported if other oxidants such as iodosobenzene, *m*-perchlorobenzene, and cumyl hydroperoxide were used. The authors propose that an O_2-derived heme-iron-bound peroxide attacks the carbonyl carbon, to form an enzyme-bound peroxyhemiacetal-like intermediate (**76**). The intermediate rearranges either by a concerted or sequential mechanism, to yield the observed products.

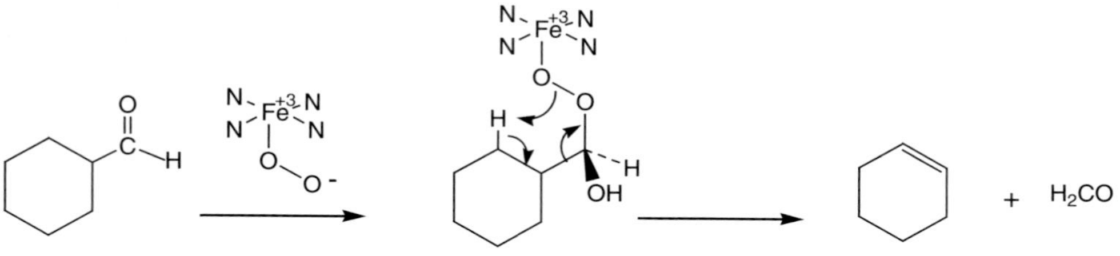

[27]

In a subsequent paper, Roberts *et al.*[113] found that CYP2B4 would selectively deformylate a number of simple α- or β-branched-chain, but not normal-chain, aldehydes, to generate alkenes. Still later, Vaz and co-workers[114–117] provided strong evidence that cytochrome P450 heme FeO_2^{1+}, or FeO_2H^{2+}, rather than FeO^{3+}, is the active oxidant in deformylation reactions.

5.05.2.1.4 Oxidation of sp carbon atoms

The oxidation of terminal aryl alkynes by cytochrome P450 forms the corresponding substituted aryl acetic acid. Early on it was shown[118] that if the alkyne hydrogen atom of 4-ethynylbiphenyl was replaced with deuterium (**77**), it would be quantitatively retained on the α-carbon of the acid metabolite (**79**) (**Scheme 7**). The reaction proceeds by heme FeO^{3+} addition to the terminal acetylenic carbon, with concerted migration of the hydrogen atom to the adjacent carbon atom, leading to the formation of a substituted ketene (**78**). The reactive ketene is hydrolyzed, to generate the final product (**79**). If heme FeO^{3+} adds to the inner carbon of the acetylene group, the reaction takes an entirely different course. This latter reaction pathway leads to alkylation of a heme nitrogen, and destruction of the enzyme.[119,120]

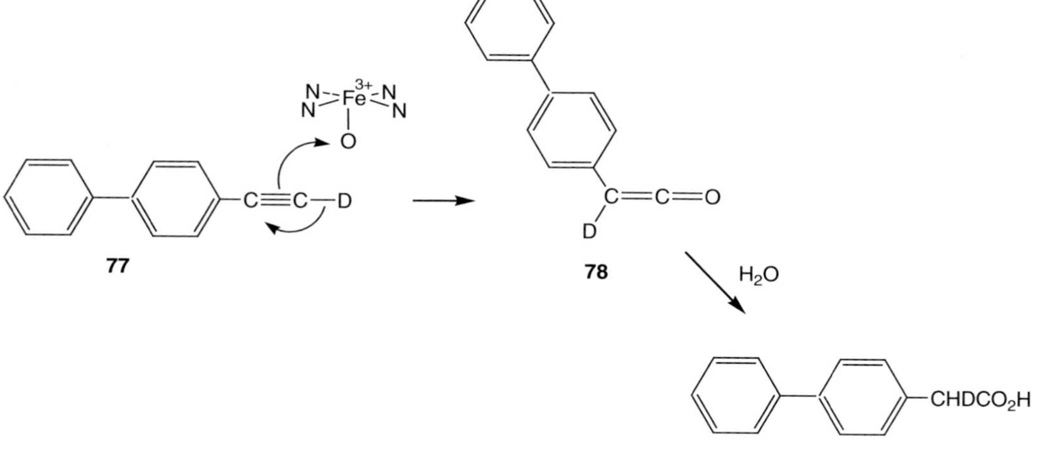

Scheme 7

5.05.2.1.5   Oxidation at a heteroatom

5.05.2.1.5.1   Mechanism for oxidation at a nitrogen atom

It is clear from the literature that while *N*-oxides and *N*-hydroxylated compounds are observable metabolites, they are minor relative to the products of *N*-dealkylation. They appear to only become more significant when *N*-dealkylation is not an option.[57] A caveat, however, is that *N*-hydroxy compounds are not all that stable, particularly in vivo, where they can be reduced back to the amine or further oxidized to even less stable compounds. The degree to which they contribute to nitrogen metabolism may be underestimated because of their relative instability.

5.05.2.1.5.2   *N*-Hydroxylation

N-Hydroxylation is most frequently found occurring with primary and, to a lesser extent, secondary alkyl and aryl amines or with aryl amides. For example, the β-phenethylamine phentermine (**80**),[121] the arylamine dapsone (**82**),[122] and the aryl amide 2-acetoaminofluorene (**84**)[123] are all oxidized, to yield significant amounts of the corresponding *N*-hydroxy metabolite (eqns [28]–[30]).

N-Hydroxylamines as a class of compounds are problematic. The alkyl-*N*-hydroxylamines tend to be toxic, and many of the hydroxylamines of arylamines and arylamides are carcinogenic.[123–125] Alkyl- and aryl-*N*-hydroxylamines are readily oxidized to chemically reactive nitroso compounds that are at least as toxic as the parent. They form quasi-irreversible inhibitors of the enzyme that produces them,[126] through generation of tightly bound heme Fe^{2+} complexes characterized by an ultraviolet spectrum with an absorbance maximum at around 455 nm, analogous to the MI complexes generated from methylenedioxy substrates. The phenomenon is general, and encompasses many widely used drugs,[127] including major antibiotics such as troleandomycin and erythromycin,[128–130] thereby raising its potential as a source of drug–drug interactions.

N-Hydroxylation is not restricted to amines or amides. Other nitrogen-based functional groups that have at least one N–H bond are susceptible to cytochrome P450-catalyzed *N*-hydroxylation. Examples include the amidino group of the antiprotozoal drug pentamidine (**86**)[131] (eqn [31]), the guanidino group of the antihypertensive debrisoquine (**88**)[132] (eqn [32]), or the iminoguanidino group of the antihypertensive gaunabenz (**90**)[133] (eqn [33]).

$$[32]$$

$$[33]$$

The oxidation of alkyl and aromatic tertiary nitrogen-containing substrates by flavin-containing monooxygenase (FMO) and, to a lesser extent, by cytochrome P450, generates N-oxide metabolites. Since FMO is the primary driver of N-oxidation, specific examples will be presented in the section on FMO-catalyzed reactions that follows.

5.05.2.1.5.3   Mechanism for oxidation at a sulfur atom

Where N-hydroxylation and N-oxide formation are minor pathways relative to N-dealkylation, the exact opposite is true of sulfur oxidation. S-Dealkylation is a minor pathway of metabolism, while direct oxidation of sulfur to a sulfoxide and/ or a sulfone is a major pathway. Both FMO and cytochrome P450 can catalyze sulfur oxidation, but cytochrome P450 is often the major contributor. If the reaction is initiated by electron abstraction to generate a sulfur radical cation (the SET mechanism), it could serve as a common intermediate for both sulfoxidation and S-dealkylation. Product formation would reflect competition between the two pathways. However, independent mechanisms, a SET mechanism for sulfoxidation versus a HAT mechanism for S-dealkylation, could account for the data, since the energy required for ionization is much lower for sulfur than it is for nitrogen.

5.05.2.1.5.4   *S*-Sulfoxidation and sulfone formation

Several examples of drugs that are subject to S-oxidation follow. Tazofelone (**92**) is a new drug that has been shown to be highly effective in the treatment of inflammatory bowel disease in animal models but has poor bioavailability because of rapid turnover. Determination of the metabolic profile of **92** revealed that a mixture of tazofelone sulfoxide diastereoisomers (**93**) was a major metabolite of the drug, and that CYP3A4 was primarily responsible for its formation (eqn [34]).[134]

$$[34]$$

The major human metabolites[135] of SNI-2011 (**94**), a new agent being developed to treat a chronic autoimmune disorder known as Sjögren's syndrome, are a mixture of sulfoxide diastereoisomers (**95**) and, to a lesser extent, the N-oxide (**96**) (**Scheme 8**). Interestingly, while **95** arises from the action of cytochrome P450, **96** is formed by FMO in the kidney.

A major human metabolite of sulfinpyrazone (**97**), a sulfoxide-containing uricosuric agent, is the sulfone[136] **98** (eqn [35]). Similarly, a sulfone is a significant human metabolite of the proton pump inhibitor omeprazole[137] (**99**) (eqn [36]).

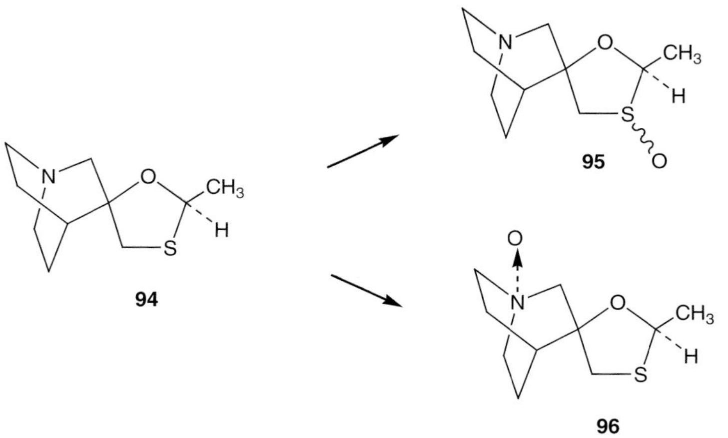

Scheme 8

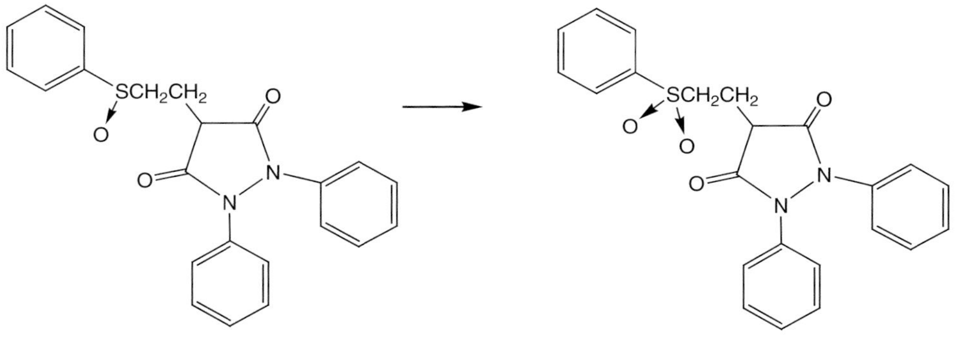

[35]

[36]

5.05.2.1.6 Other cytochrome P450 oxidative reactions

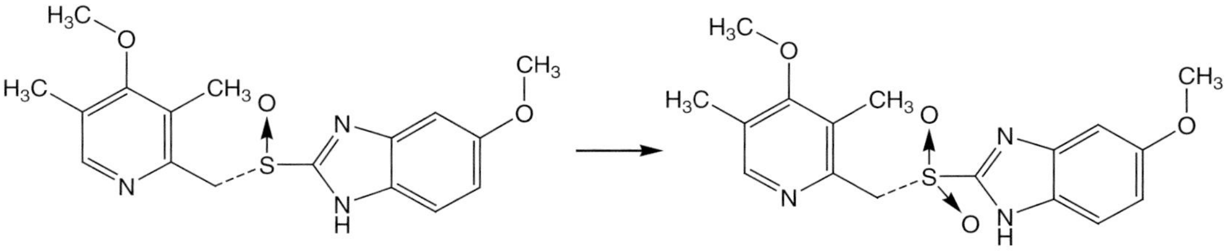

5.05.2.1.6.1 Oxidative dehydrogenation

The anticonvulsant valproic acid (**101**) undergoes normal cytochrome P450-catalyzed transformation, to generate the ω and ω-1 hydroxylated metabolites, (**102** and **103**, respectively; **Scheme 9**). However, the unsaturated substrate, 4-ene valproic acid (**104**), is also produced, suggesting product partitioning between the alcohol and alkene. While **104** is a minor metabolic product, its formation is important both from a mechanistic perspective and from the fact that it is toxic. Using selectively deuterated valproic acid analogs, Rettie *et al.*[138,139] established that **104** forms in competition with **103** after initial abstraction of a hydrogen atom from the ω-1 carbon, to form the radical **105**. Abstraction of a ω-hydrogen atom to form the radical **106** only leads to **102**.

Similarly, testosterone (**107**) is oxidized by CYP2A1 to four metabolites,[140,141] 7α-hydroxy- and 6α-hydroxytestosterone (**108** and **109**, respectively), Δ⁶-testosterone (**110**), and Δ⁶-testosterone epoxide (**111**) (**Scheme 10**), the epoxide being formed from further oxidation of **110**. Again using selectively deuterated analogs, Korzekwa *et al.*[141] established that **110** was formed in competition with **109** after initial hydrogen atom abstraction from C-6 to form the radical **112**. Little if any **108** formed upon initial hydrogen abstraction from C-7.

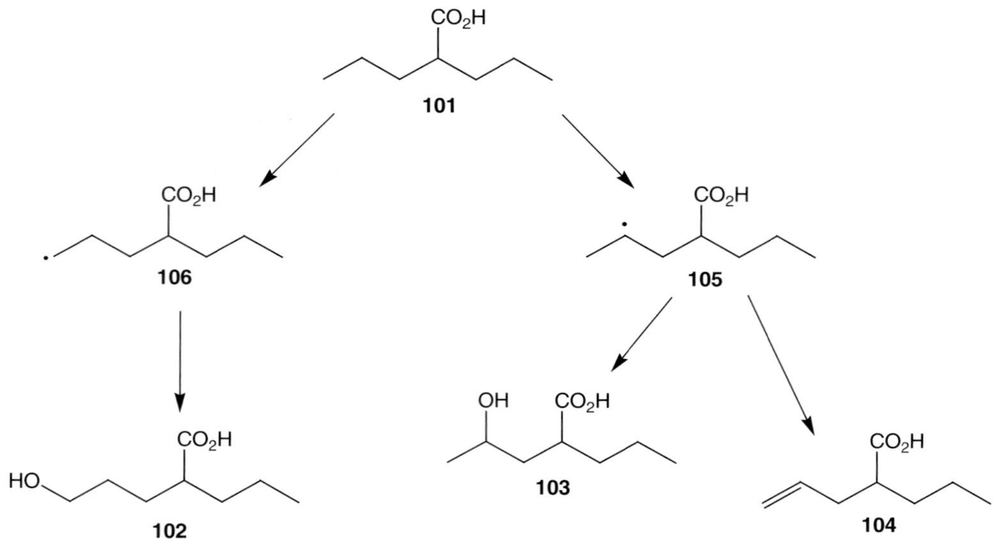

Scheme 9

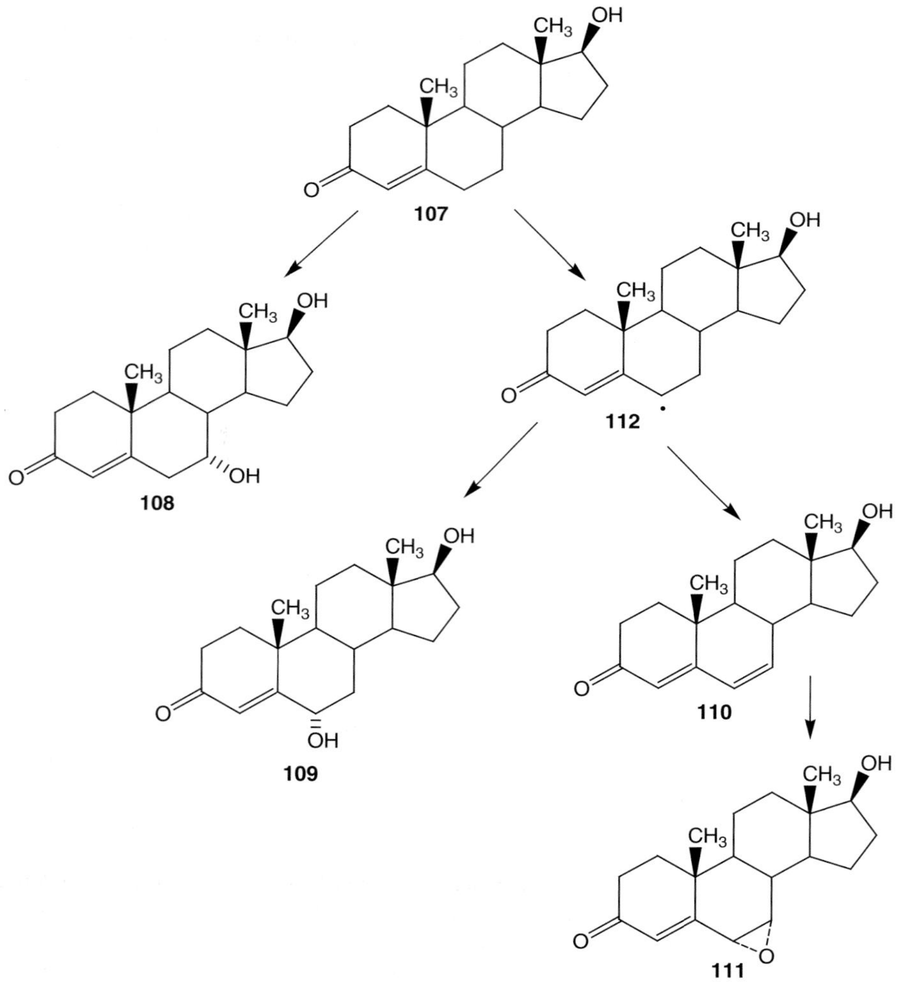

Scheme 10

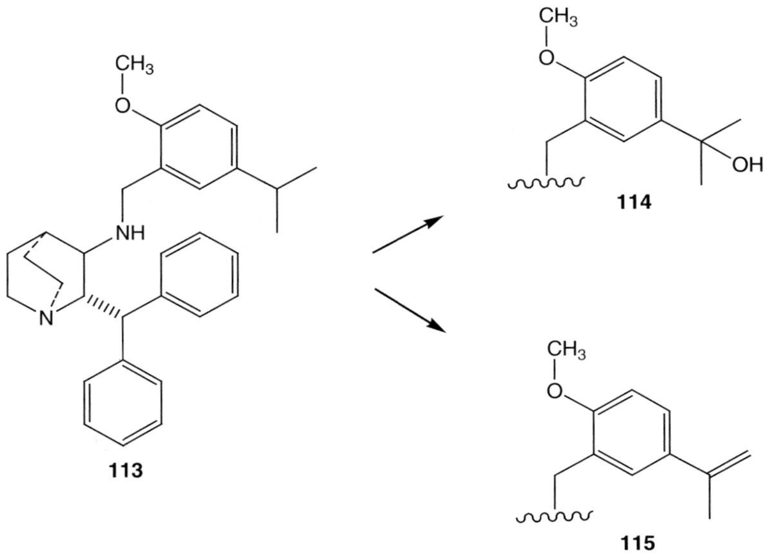

Scheme 11

A third example can be found with ezlopitant (**113**), a drug being developed as a potential substance P receptor antagonist. Metabolism of **113** yields two major metabolites, the benzyl alcohol **114** and the corresponding alkene[142] **115** (**Scheme 11**).

5.05.2.1.6.2   Oxidative cleavage of esters and amides

If esters and amides are also susceptible to oxidative attack, one would expect that the site of attack to be the carbon atom α to an oxygen atom (esters) or α to a nitrogen atom (amides) by analogy to O- and N-dealkylations. This is indeed the case. Guengerich *et al.*[143] demonstrated that the pyridine diester **116** was oxidatively cleaved, to yield the mono-acid **117** (eqn [37]). Subsequently, Peng *et al.*[144] established that oxidative cleavage is a general cytochrome P450 reaction for commonly used esters and amides. The reaction has largely gone unrecognized, probably because of being obscured by hydrolysis, the expected mode of reaction.

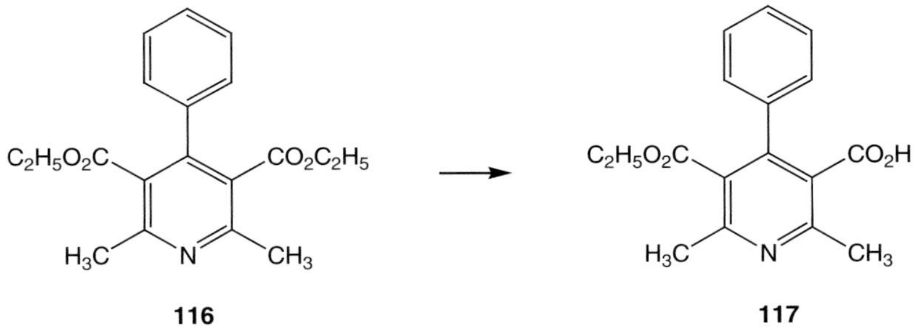

[37]

5.05.2.1.6.3   Oxidative transformation of nitriles

Cytochrome P450-catalyzed C–C bond cleavage is a relatively rare event. The conversion of aldehydes to alkenes[112,113] was presented earlier. A second example is the conversion of a nitrile to a cyanide ion and an aldehyde or ketone. Grogan *et al.*[99] modeled the reaction using a set of 26 structurally diverse nitriles. They found that acute toxicity in the mouse correlated with the ease of hydrogen abstraction α to the nitrile (cyanide release) relative to oxidative attack at other intramolecular sites (no cyanide release).

Cytochrome P450 also appears to generate nitriles from alkyl and arylaldoximes by catalyzing their dehydration.[145,146] The reaction was first reported for the conversion of n-butyraldoxime (**118**) to butyronitrile (**119**), and found to require reduced cytochrome P450 (heme Fe^{2+}) and be inhibited by excess O_2 (eqn [38]).[145] Subsequently, the reaction was established for both alkyl- and arylaldoximes[146] (e.g., benzaldoxime (**120**) (eqn [39]), and phenylacetaldoxime (**122**) (eqn [40])), suggesting that while it is apparently restricted to the (Z) isomer, it might be general.

Nitriles can also be oxidized to amides. Zhang *et al.*[147] found that CYP3A4 would convert the cyano group of the potassium channel-opening agent pinacidil (**124**) to the corresponding amide (**125**) (eqn [41]).

5.05.2.1.6.4   Oxidative transformation of a tertiary carbon atom adjacent to a nitrogen atom

The CYP3A4-catalyzed oxidation of the synthetic opioid alfentanil (**126**) follows two major pathways[148]: *N*-dealkylation, to form noralfentanil (**127**), and cleavage of the spiro center, to generate *N*-phenylpropionamide (**128**) (**Scheme 12**). Moreover, **128** is found to come directly from alfentanil and not **127**. The mechanism of how the C–N bond of the spiro center is cleaved is unknown.

The treatment of 1-phenylcyclobutylamine (**129**) with cytochrome P450 results in ring expansion and the production of 2-pheny-1-pyrroline (**130**). Presumably, the mechanism involves sequential formation of the aminium radical cation, homolytic ring opening, ring closure, then, finally, a second one-electron oxidation to form the pyrroline (**Scheme 13**).[149]

5.05.2.1.6.5   Ipso substitution

A relatively new aspect of cytochrome P450-catalyzed aromatic hydroxylation is the replacement of the substituent of a *para*-substituted phenol with a hydroxy group; a phenomenon termed ipso substitution. In 1994, Ohe *et al.*[150] reported that cytochrome P450 catalyzed the partial conversion of both *p*-methoxyphenol (**131**) and *p*-phenoxyphenol (**132**) to hydroquinone (**133**) (**Scheme 14**). In a subsequent paper, the scope of the reaction was investigated, and found to require the phenolic group.[151] Nine *p*-substituted phenols (F, Cl, Br, NO_2, CN, CH_2OH, $COCH_3$, COPh, CO_2H) of diverse structure were incubated with rat liver microsomes, and the amount of **133** formed from each substrate was determined. All substrates gave measurable or significant levels of **133**, except perhaps for *p*-tolylphenol (**134**). In the case of **134**, the reaction stopped at the formation of the ipso adduct, *p*-toluquinol (**135**) (eqn [42]).

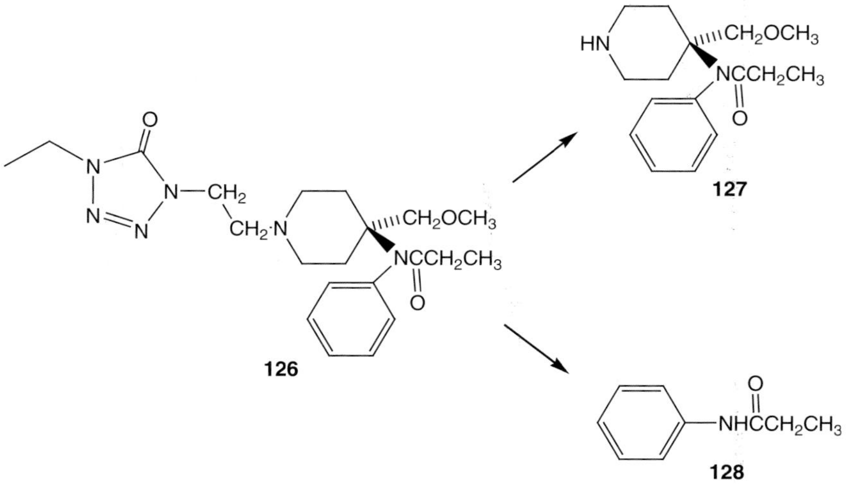

Scheme 12

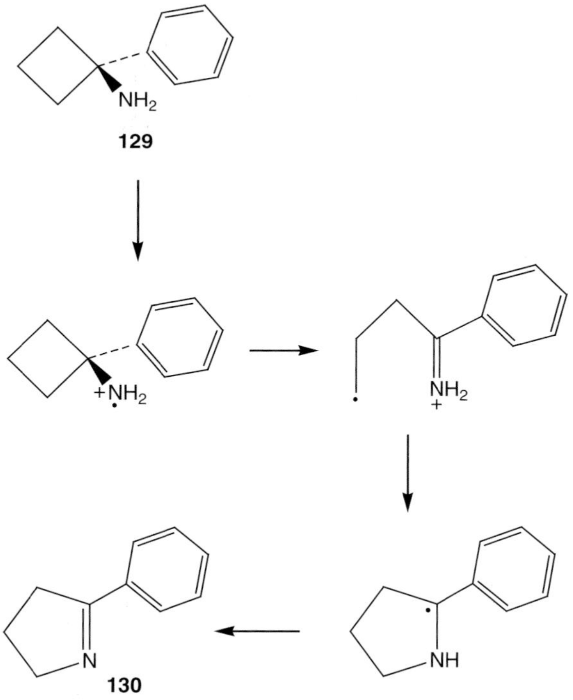

Scheme 13

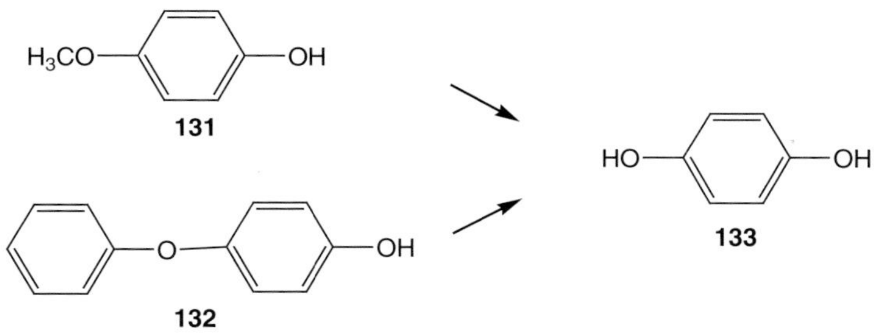

Scheme 14

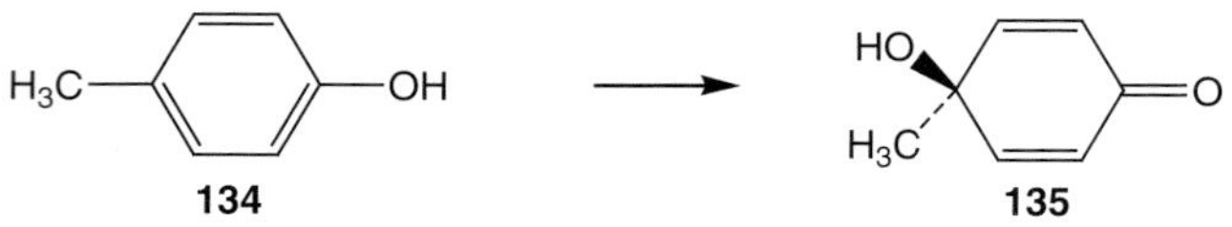

[42]

5.05.2.2   Flavin Monooxygenases

5.05.2.2.1   Occurrence, multiplicity, catalytic cycle, oxygen activation, and selectivity

The oxidative activity of the NADPH and O_2-dependent FMOs are largely complementary to that of the cytochromes P450.[152] Like the cytochromes P450, the FMOs are widely distributed in nature, have multiple family members, with six (FMO1–FMO6) having been characterized.[153] In humans they are found in close association with cytochrome P450 in the endoplasmic reticulum of most tissues, with the greatest concentration being in liver (FMO3)[154,155] FMO1 is the major form in human fetal liver,[156] but it localizes in the kidney in the adult.[157] The FMOs display broad substrate selectivity, and like cytochrome P450 the primary criterion for reaction is substrate access to the active oxidizing species.[152,158]

Unlike the cytochromes P450 the FMOs appear not to be inducible, and operate by the two-electron oxidation characteristic of peroxides. Nature seems to have designed the FMOs to be selective for a class of compounds that the cytochromes P450 are less efficient at oxidizing: xenobiotics that contain electron-rich polarizable nucleophilic groups. These are largely compounds containing the elements sulfur, nitrogen, selenium, and phosphorus (thiols, sulfides, disulfides, amines (1°, 2°, and 3°), imines, hydrazines, hydroxylamines, selenols, selenides, phosphines, etc.).[152]

The active oxygen species utilized by the FMOs is the C(4a)-hydroperoxyflavin **136**, generated from O_2 addition to $FADH_2$, formed upon reduction of flavin adenine dinucleotide (FAD) by enzyme-bound NADPH (**Figure 5**). The substrate is oxidized, water eliminated from the residual C(4a)-hydroxyflavin, and FAD regenerated to complete the catalytic cycle.[159–161]

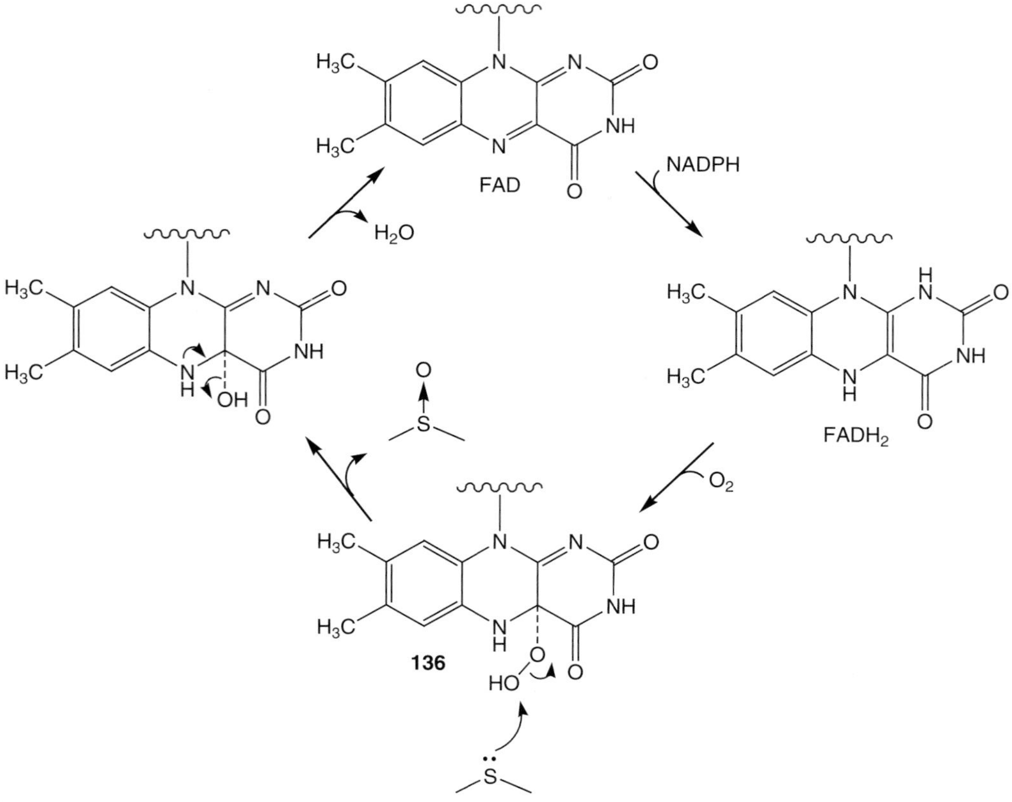

Figure 5   The FMO catalytic cycle.

Since the FMOs and cytochromes P450 are closely associated and can catalyze some of the same reactions, it is important to be able to determine which enzyme is primarily responsible for the formation of a given metabolite. This is of particular importance for in vitro–in vivo correlation and drug interaction studies. Experimentally, the enzymes can be distinguished[136] by heat and by running the reactions at higher pH. If the microsomal preparation is heated (45 °C for 5 min) before NADPH is added to initiate the reaction, cytochrome P450 maintains its activity while that of FMO is lost. If in a separate experiment the pH of the incubation mixture is raised to 9, FMO maintains activity but cytochrome P450 activity is virtually abolished, particularly in the presence of a detergent.

5.05.2.2.2 Oxidation at a nitrogen atom

FMO N-hydroxylates both amphetamine (**21**) and methamphetamine (**20**), to generate the hydroxylamines **137** and **138**, respectively (**Scheme 15**).[162] It then catalyzes a second N-hydroxylation in both metabolites. The two N,N-dihydroxy intermediates eliminate water, to generate the oxime in the case of **139** and the nitrone in the case of **140**.

Both FMO3 and extra-hepatic FMO1 have been shown to be effective in oxidizing the nonsteroidal anti-inflammatory agent benzydamine (**141**)[163] to benzydamine N-oxide (**142**), with only a minor contribution from cytochrome P450 (eqn [43]). FMO is also the major catalyst for the conversion of the antipsychotic clozapine (**143**)[164] and the anticancer agent tamoxifen (**145**),[165] and for the cerebral metabolism of the psychoactive drug imipramine (**147**),[166] to their corresponding N-oxides (**144**, **146**, and **148**, respectively) (eqns [44]–[46]).

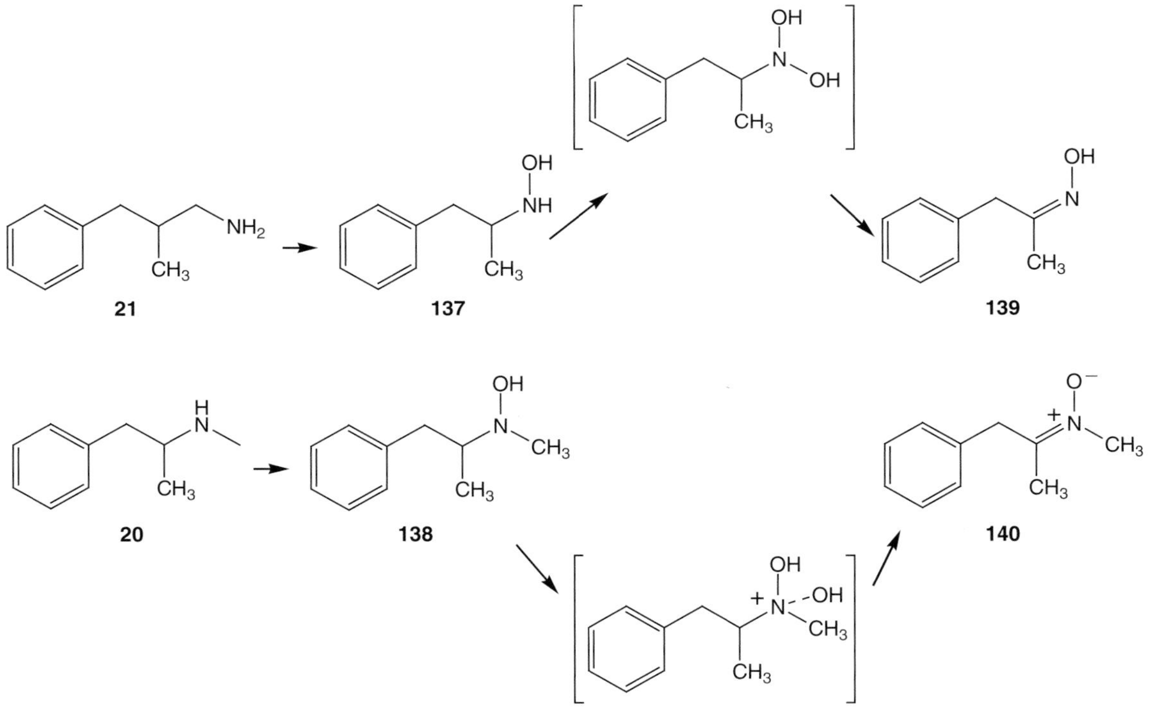

Scheme 15

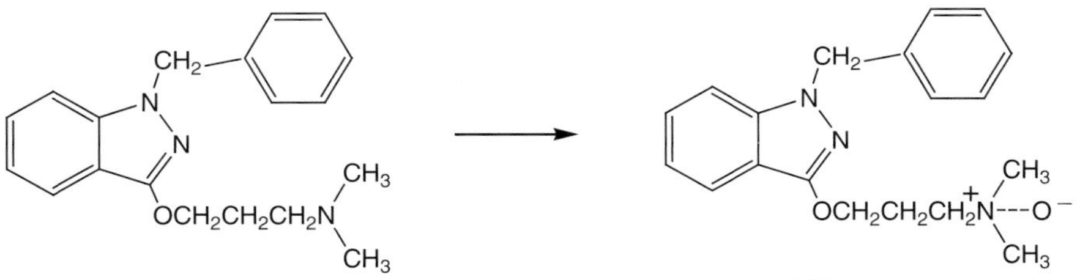

[43]

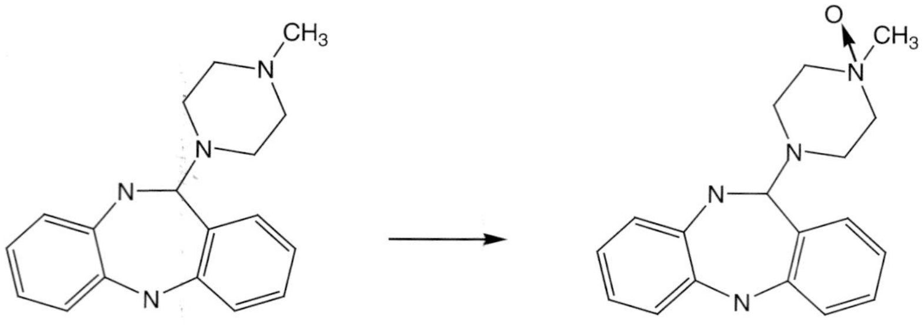

[44]

143   **144**

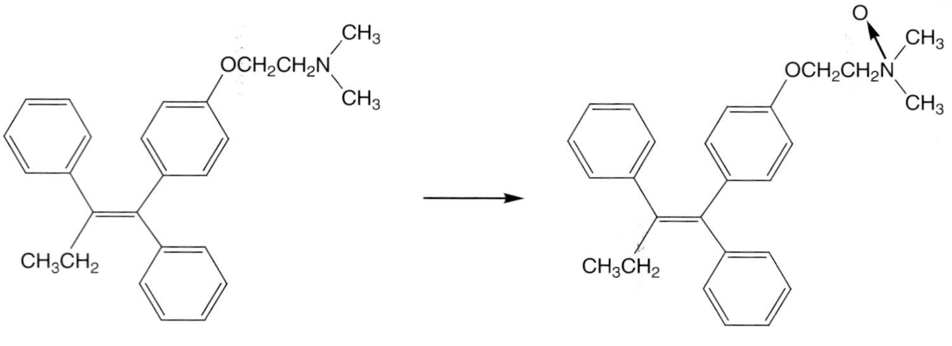

[45]

145   **146**

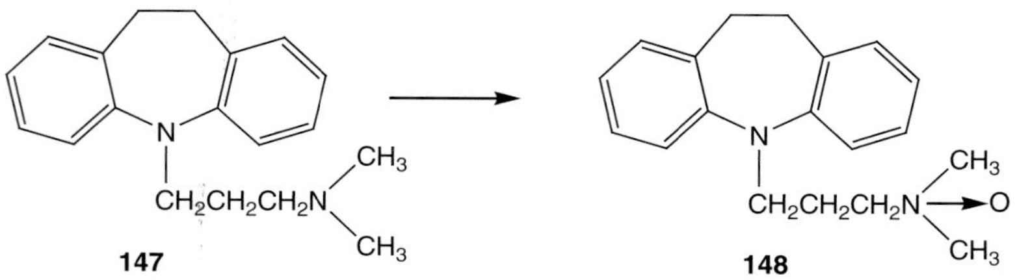

[46]

147   **148**

5.05.2.2.3 Oxidation at a sulfur atom

Sulindac sulfide (**149**), a metabolite of the nonsteroidal anti-inflammatory agent sulindac, is re-oxidized by FMO[167] with a high degree of stereoselectivity toward the (*R*) enantiomer back to enantiomerically enriched sulindac (**150**) (eqn [47]).

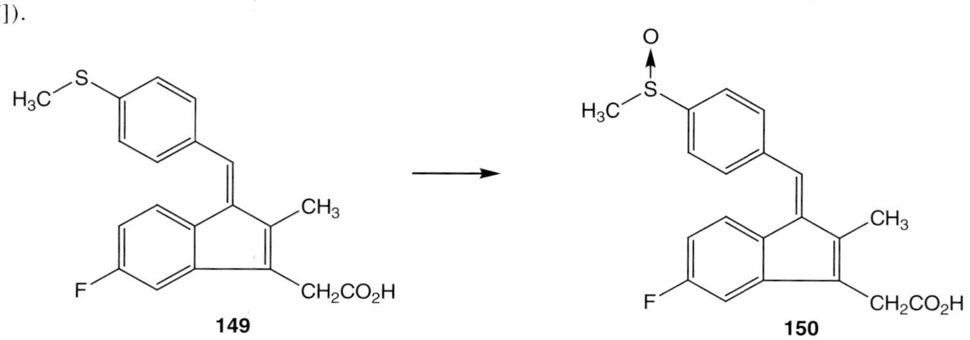

[47]

149   **150**

Metabolic activation of the antitubercular pro-drug ethionamide (**151**) occurs within the organism *Mycobacterium tuberculosis*. FMO[168] has been identified as the enzyme that activates **151** by catalyzing two sequential oxidations. The

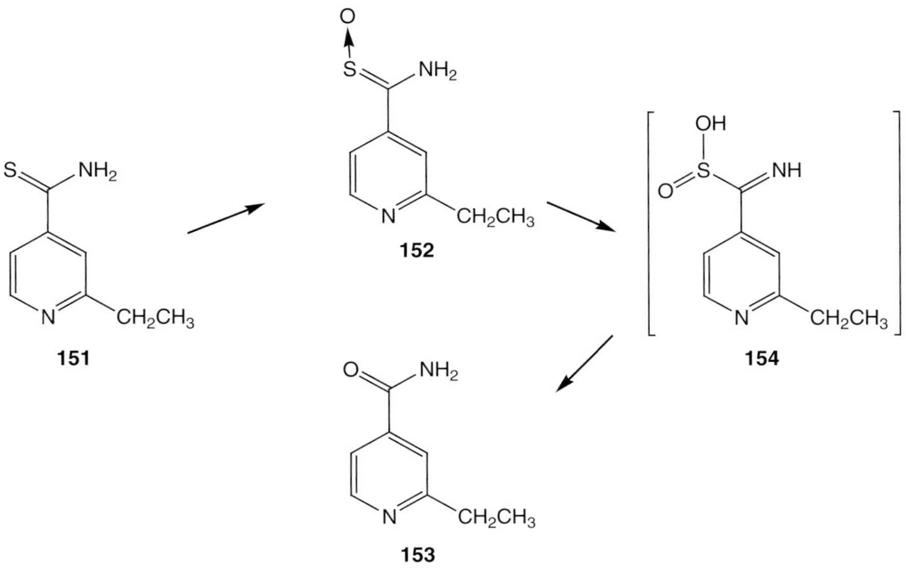

Scheme 16

first is to the inactive *S*-oxide **152**, while the second oxidation generates the active agent, but a stable, noncytotoxic amide **153** is the only product isolated (**Scheme 16**). These results imply that the cytotoxic agent is a reactive intermediate that can break down to **153**. It has been postulated[168] that the reactive intermediate is a sulfinic acid **154**, which upon hydrolysis yields **153**.

5.05.2.3   Monoamine Oxidase (MAO)

5.05.2.3.1   Occurrence, multiplicity, distribution, oxygen activation, and selectivity

MAO, like FMO, is an enzyme that relies on the redox properties of FAD for its oxidative machinery. The enzyme exists as two isoforms, MAO-A and MAO-B, that share a sequence homology of approximately 70%.[169] They are found in the outer mitochondrial membrane, and are widely distributed in mammalian tissues. The isoforms differ in substrate selectivity[170] and tissue distribution.[171] MAO-A is located primarily in the placenta, gut, and liver, while MAO-B is predominant in the brain, liver, and platelets. In humans, MAO-B constitutes about 80% of the enzyme found in the liver.[172] The primary function of both isoforms appears to be to catalyze the oxidative deamination of the arylalkylamino neurotransmitters, to form the corresponding aldehydes. MAO-A is selective for serotonin and norepinephrine, and is selectively inhibited[173] by the mechanism-based inhibitor clorgyline (**155**). MAO-B is selective for β-phenethylamine and tryptamine, and is selectively inhibited[173] by the mechanism-based inhibitors deprenyl (**156**) and pargyline (**157**). Recently, both MAO-A[174] and MAO-B[175] were structurally characterized by x-ray crystallography.

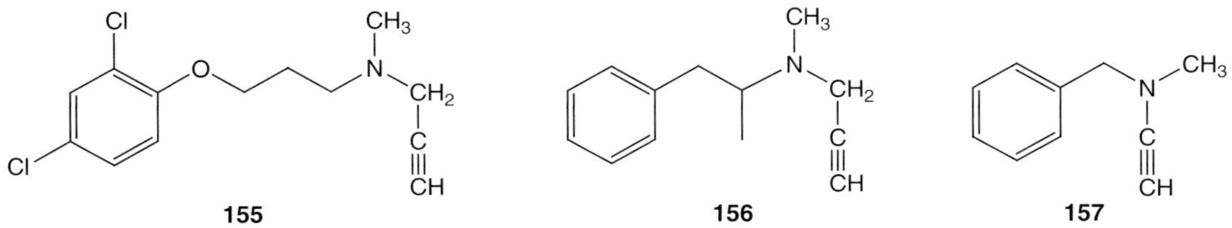

The role of MAO-A and MAO-B in regulating the concentration of the amine neurotransmitters offered a strategy for treating pathologies that involve these critical biomolecules. Thus, drugs have been developed that are used to treat the symptoms of depression by inhibiting MAO-A[176] or neurodegenerative diseases such as parkinsonism by inhibiting MAO-B.[177]

5.05.2.3.2   Mechanism of monoamine oxidase oxygen activation and oxidation of substrate

Overall, the catalytic activity of MAO can be characterized as two half-reactions.[171] In the first the amine substrate is oxidized, and the FAD cofactor is reduced to $FADH_2$, while in the second the imine product is released, and the $FADH_2$ cofactor is re-oxidized to FAD, generating peroxide in the process. The released imine chemically hydrolyzes to the corresponding aldehyde.

A mechanism involving a SET has generally been accepted[178] (**Figure 6a**), although evidence for a HAT mechanism[179,180] exists (**Figure 6b**), and a polar nucleophilic mechanism has also been advanced[181,182] (**Figure 6c**).

5.05.2.3.3   Oxidation of xenobiotic amines

The most well-documented case of a MAO-B-catalyzed oxidation of a nonbiogenic amine[183] is the conversion of *N*-methyl-4-phenyl-1,2,3,6-tetrahydropyridine (MPTP) (**158**) to *N*-methyl-4-phenyl-2,3-dihydropyridine (MPDP$^+$)

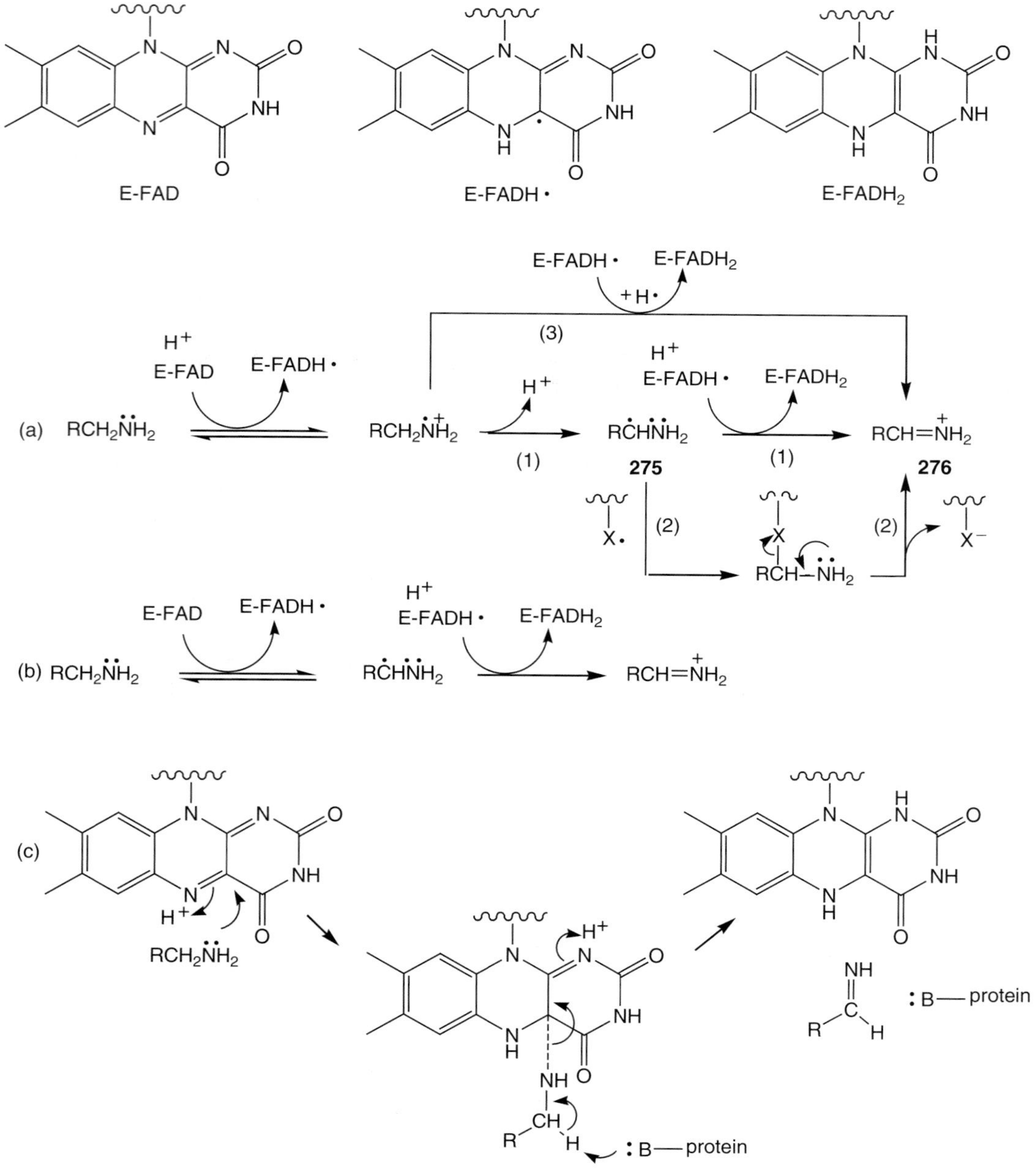

Figure 6 Mechanisms for the catalytic activity of MAO: (a) SET, (b) HAT, and (c) via a polar nucleophile.

(159), on the path to formation of *N*-methyl-4-phenylpyridine (MPP$^+$) (160), the neurotoxin that causes a parkinsonism-like syndrome. The reaction is not restricted to MPTP alone but will oxidize a number of MPTP analogs in which the *N*-methyl group[184,185] and/or the 4-aryl substituent[172,185] has been altered. A structure–activity relationship study[185] suggested that the maximum size of MPTP analogs acceptable as substrates by MAO-B is 12 Å, as measured along the N-1–C-4 axis.

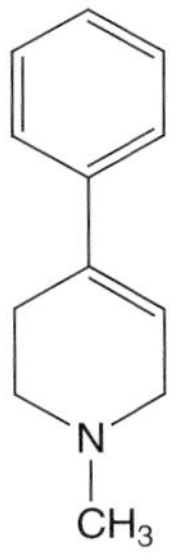

158 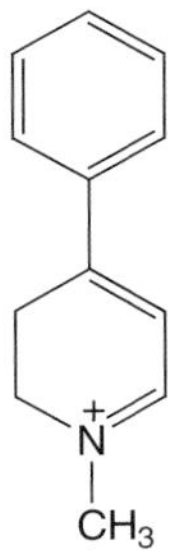   159 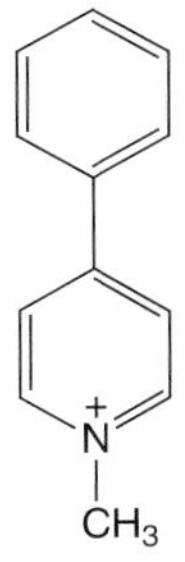   160

While MAO does not appear to be a major catalyst in the metabolism of amine-containing drugs, it can contribute. The extent to which it does remains to be determined.

5.05.2.4   Xanthine Oxidoreductase

Xanthine oxidoreductase is a complex homodimeric 300 kDa cytosolic enzyme. Each subunit contains a molybdopterin cofactor, FAD, and two nonidentical iron–sulfur centers.[186–188] It is widely distributed, and exists in two interconvertible forms in mammals: xanthine dehydrogenase and xanthine oxidase. Xanthine dehydrogenase predominates in vivo, while xanthine oxidase is the form that is generally isolated.[186–188] Maximum concentrations of xanthine oxidoreductase have been found in liver, intestine, and lactating mammary gland.[189] In patients with liver disease, xanthine oxidoreductase activity has been found to be 10–20-fold higher than that found in healthy liver tissue.[190]

Its primary role appears to be in the metabolism of purines (e.g., it catalyzes the sequence of oxidations that convert hypoxanthine (161) to xanthine (162), then to uric acid (163) (Scheme 17)). Excess uric acid production can lead to flare-ups of symptomatic gout. One of the effective treatments for gout is the administration of allopurinol (164). Allopurinol is both a competitive inhibitor of xanthine oxidoreductase and a substrate, as xanthine oxidoreductase slowly oxidizes 164 to alloxanthine (165) (eqn [48]). Since 165 is also an inhibitor of xanthine oxidoreductase, the therapeutic effectiveness of 164 is not significantly compromised by its conversion to 165. However, if 164 and the anticancer agent 6-mercaptopurine (166) are coadministered, inhibition of xanthine oxidoreductase can be problematic. If the extensive first-pass metabolism of 166 to 6-mercaptouric acid (167), catalyzed by xanthine oxidoreductase (eqn [49]), is inhibited by 164, it can result in potentially toxic plasma concentrations of 166 fivefold higher than normal.[191]

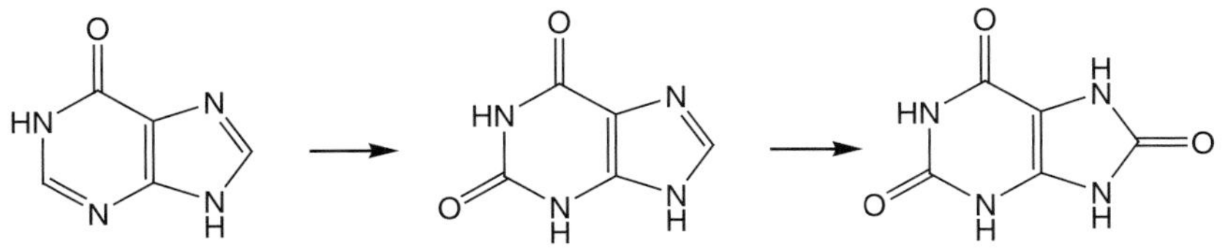

161   162   163

Scheme 17

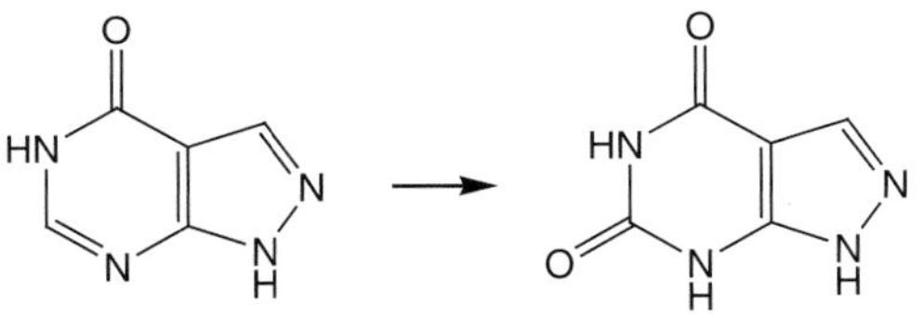

164   165

[48]

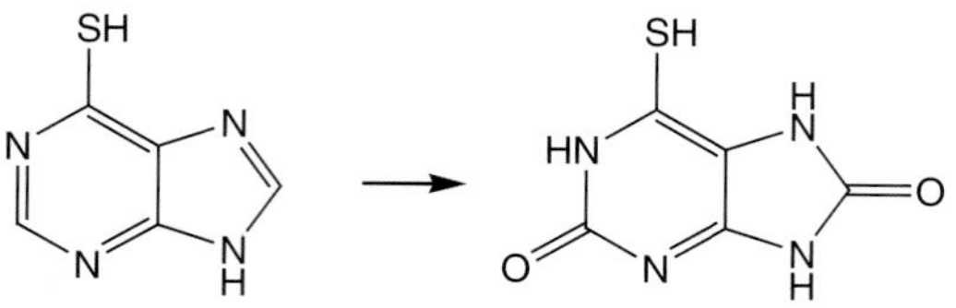

[49]

Xanthine oxidoreductase is an unusual oxidative enzyme in that the source of the oxygen atom that is transferred to the substrate, X–H, originates in water rather than molecular oxygen.[192] The electrons gained through oxidation of a water molecule by the molybdenum cofactor to form the active oxidizing species are ultimately transferred to molecular oxygen via the FAD and iron–sulfur active site components[193,194] (**Figure 7**).

Since xanthine oxidoreductase is a ready source of electrons that can be transferred to molecular oxygen to form reactive oxygen species such as superoxides and peroxides, it is thought to be involved in free radical-generated tissue injury. It has been implicated in the pathogenesis of ischemia–reperfusion damage and, more recently, in the production of peroxynitrite (**168**)[187] and the carbonate radical anion **169**,[195] both potent biological oxidants. Since xanthine oxidoreductase is present in high concentrations in the liver, it may be released into the circulation if the liver is injured, and bind to vascular endothelium, causing vascular dysfunction.[196]

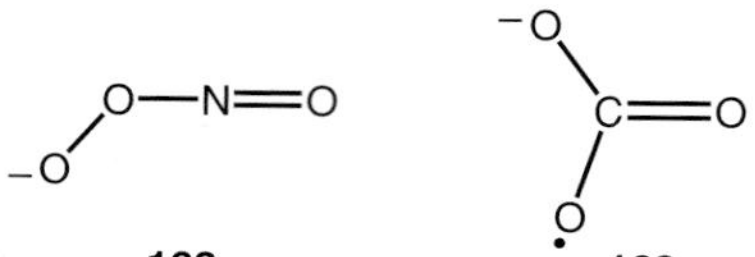

While much has been learned about xanthine oxidoreductase, much remains to be uncovered. Understanding its exact role in lipid peroxidation, inflammation, and infection is particularly important. Its significance to drug metabolism remains to be determined.

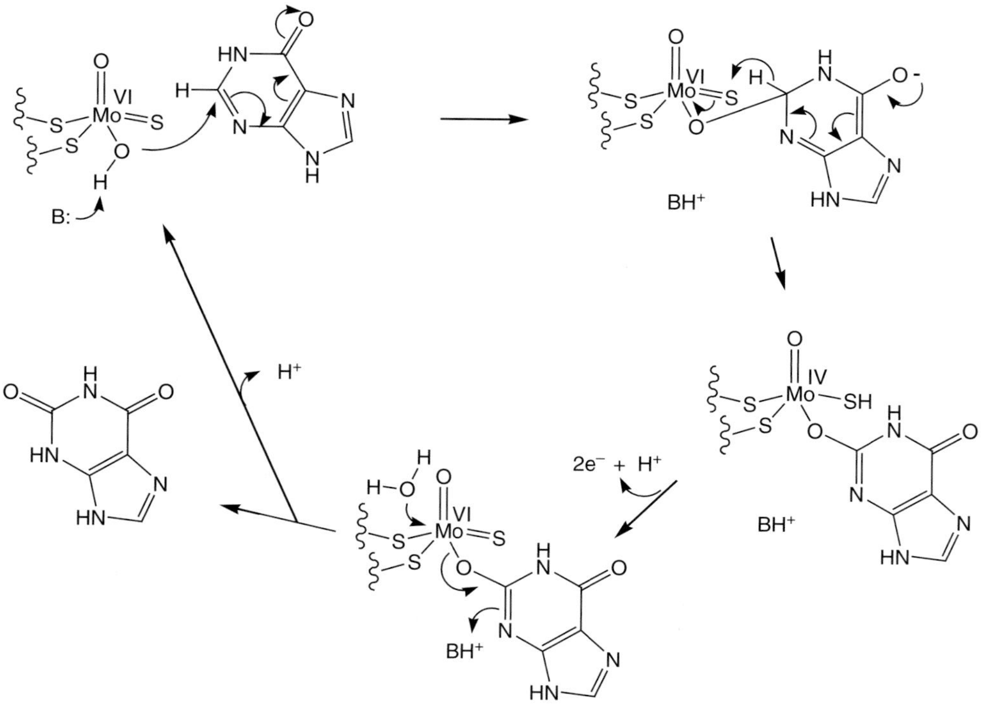

Figure 7   Mechanism of action for xanthine oxidoreductase.

5.05.2.5   Aldehyde Oxidase (AOX)

AOX is closely related to xanthine oxidoreductase, and like xanthine oxidoreductase is a member of the structurally related molybdoflavoenzymes. AOX and xanthine oxidoreductase have overlapping substrate selectivities, and operate by the same chemical mechanism.[196] As the name implies, one of the primary reactions that AOX catalyzes is the oxidation of aldehydes to carboxylic acids. For example, it contributes to the oxidation of acetaldehyde resulting from alcohol ingestion. Unfortunately, it also seems to be implicated in alcohol-induced liver injury because of the free radicals it generates in the process of oxidizing acetaldehyde.[197]

In oxidizing aldehydes, AOX appears to operate by the same mechanism it utilizes to oxidize purines. This is illustrated for the conversion of 2,5-dihydroxyacetaldehyde (170) to 2,5-dihydroxybenzoic acid[193] (171) (Figure 8).

AOX is also effective in metabolizing a wide range of nitrogen-containing heterocycles, such as purines, pyrimidines, pteridines, quinolines, and diazanaphthalenes.[198,199] For example, phthalazine (172) is rapidly converted to 1-phthalazinone (173) (eqn [50]),[200] the prodrug 5-ethynyl-2-(1*H*)-pyrimidone (174) is oxidized to the dihydropyrimidine dehydrogenase mechanism-based inhibitor 5-ethynyluracil (175) (eqn [51]),[201] and the prodrug famciclovir (176) is first hydrolyzed to 6-desoxypanciclovir (177), before being oxidized to the active antiviral agent panciclovir (178) by AOX (Scheme 18).[202]

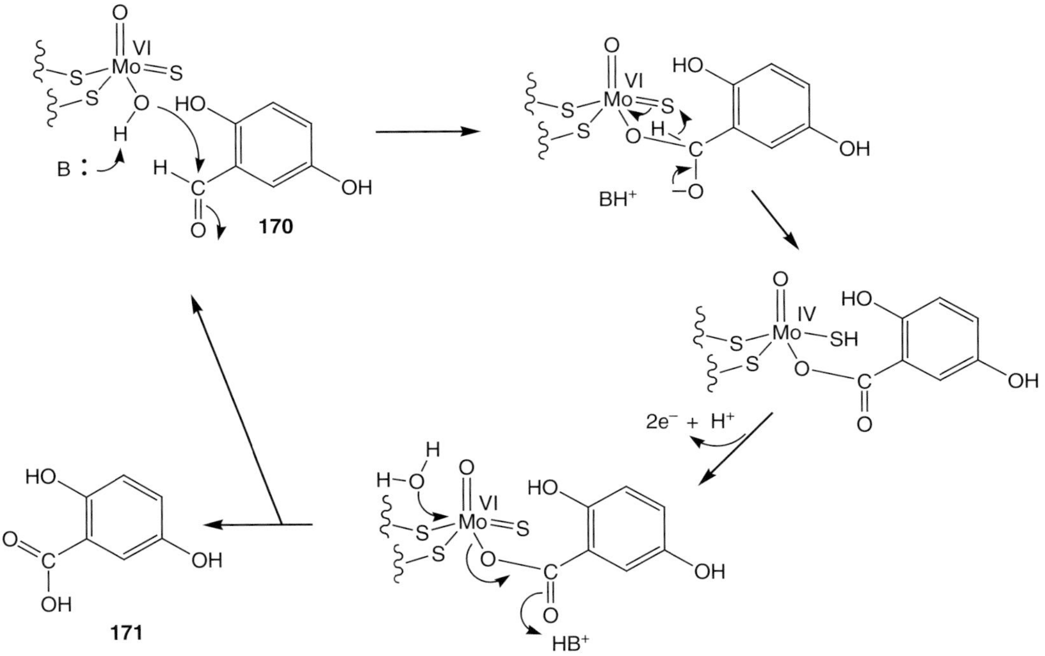

Figure 8 Mechanism for the oxidation of aldehydes by xanthine oxidoreductase.

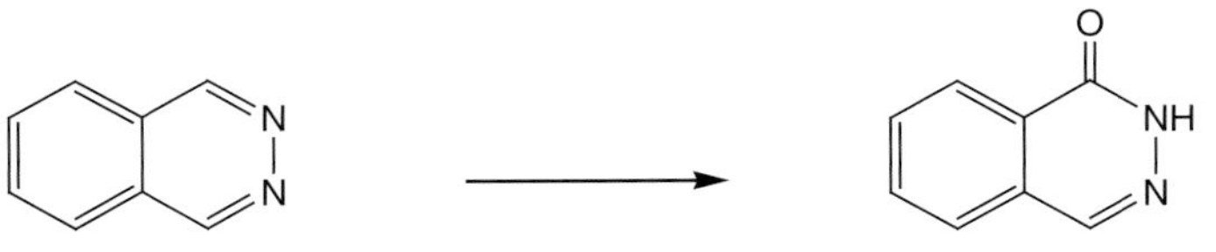

[50]

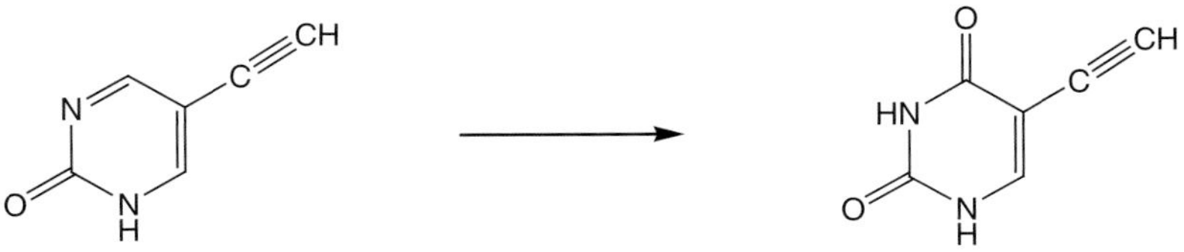

[51]

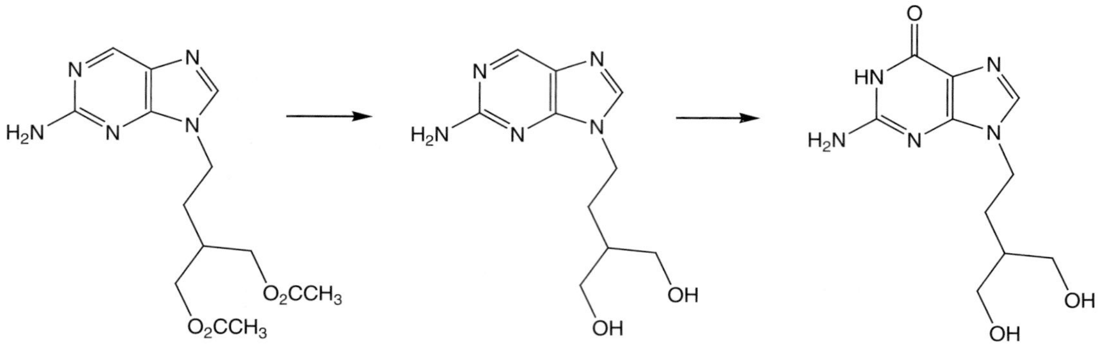

Scheme 18

As more information becomes available, it is becoming increasingly probable that the contributions of AOX and xanthine oxidoreductase to the metabolism of drugs that incorporate aryl nitrogen-containing heterocycles have been largely unrecognized and underestimated.

5.05.2.6   Aldehyde Dehydrogenases

The aldehyde dehydrogenases are members of a superfamily of pyridine nucleotide $(NAD(P)^+)$-dependent oxido-reductases that catalyze the oxidation of aldehydes to carboxylic acids, are widely distributed in mammals, and are found in cytosol, mitochondria, and microsomes.[203–205] Seventeen genes have been identified in the human genome that code for aldehyde dehydrogenases, attesting to the importance of these enzymes to normal physiological function.

Aldehyde dehydrogenases can be placed in two broad categories: (1) those that are highly substrate-selective and are critical for normal development, and (2) those that are less substrate-selective and protect the organism from potentially toxic aldehydes contained in food or generated from xenobiotics.[204] An example of aldehyde dehydrogenases that fall into the first category are those that oxidize retinal (68) to retinoic acid (69) (eqn [24]), a molecule important for growth and development.[206]

Aldehyde dehydrogenases that fall into the second category are almost invariably detoxifying. The most common example is the oxidation of alcohol-derived acetaldehyde to acetic acid. However, sometimes it is advantageous to inhibit their detoxifying effects. For example, inhibition of acetaldehyde oxidation by the administration of disulfiram (179) is a common treatment for alcohol abuse.[207] In vivo, 179 is converted to diethylthiomethylcarbamate (180), the active inhibitor (eqn [52]). Inhibition results in potentially toxic levels of acetaldehyde upon ingestion of alcohol by the abuser.

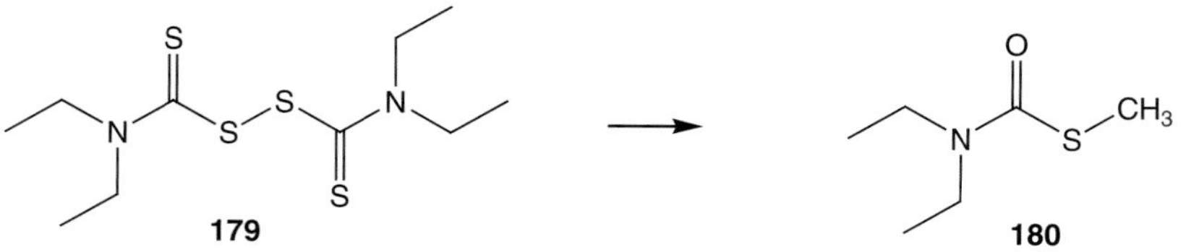

[52]

A second example is provided by the anticancer agent cyclophosphamide (181).[208] Upon administration of 181, cytochrome P450 catalyzes hydroxylation at the 4-position, to produce 182, leading to the formation of the ring-opened aldehyde 183, or further oxidation to the cyclic amide 184 (Scheme 19). In the absence of aldehyde dehydrogenase, the aldehyde spontaneously eliminates to generate acrolein (185) and the phosphoramide mustard 186, both of which are cytotoxic species responsible for the anticancer effects. In the presence of aldehyde dehydrogenase, the ring-opened aldehyde is converted to the acid carboxyphosphamide (187), a species that no longer eliminates 185 and has no anticancer affect.

The major set of substrates for aldehyde dehydrogenase include the aldehydes generated in MAO-catalyzed deamination reactions, cytochrome P450-catalyzed N- and, O-dealkylation, oxidative dehalogenation, and oxidation of aryl and alkyl methyl groups.

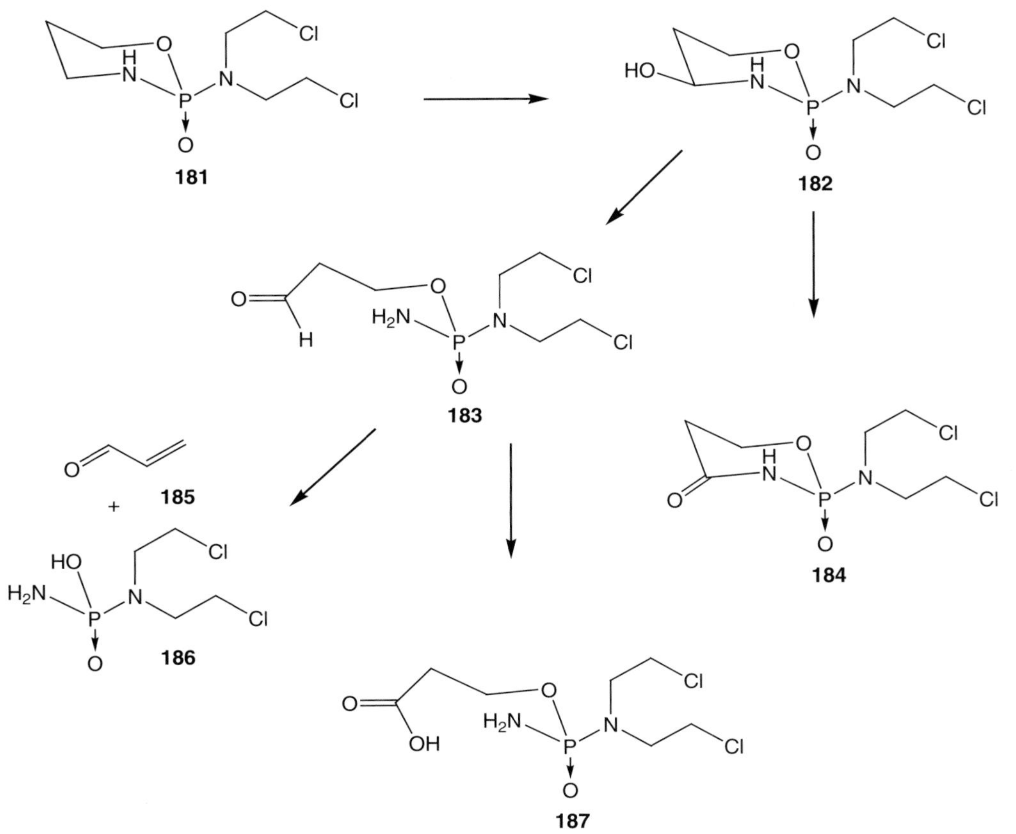

Scheme 19

5.05.2.7   Alcohol Dehydrogenases (ADHs)

ADHs, a subset of the medium-chain dehydrogenases/reductases,[209] constitute a highly complex superfamily of widely distributed zinc containing cytosolic NAD^+ dependant enzymes that are presently divided into seven classes[210] and numerous isozymes and allelic variants.[211] While they can reversibly catalyze the oxidation of the alcohol/aldehyde redox pair in a variety of substrates, including ethanol,[212] retinol,[206] and other aliphatic alcohols,[212] they function primarily as oxidative catalysts.

5.05.3   Reductions

5.05.3.1   Cytochrome P450

5.05.3.1.1   Reductive dehalogenation

Under certain conditions, particularly anaerobic conditions, cytochrome P450 can function as a reductase. The most well-recognized reaction in this regard is probably reductive dehalogenation. In a series of papers in the early 1980s, Ullrich and co-workers[213–215] established that cytochrome P450 could catalyze the reductive removal of halogens from polyhalogenated alkanes such as hexachloroethane (**188**), to yield the corresponding carbon-based radical (**Scheme 20**). The radical would then undergo a second one-electron addition, to yield the carbanion. Elimination of a chloride ion or the addition of a proton yielded the observed products, tetrachloroethylene (**189**) and pentachloroethane (**190**), respectively.

Carbon tetrachloride (**191**) is a solvent that is chemically inert and highly resistant to oxidation, but biologically toxic. Despite its chemical stability, cytochrome P450 is able to convert **191** to a reactive species (**Figure 9**). Reduced cytochrome P450 transfers an electron to **191**, to form a chloride anion and the reactive trichloromethyl radical **192**,[216] in accordance with the mechanism proposed by Ullrich.[213–215] The trichloromethyl radical has a number of

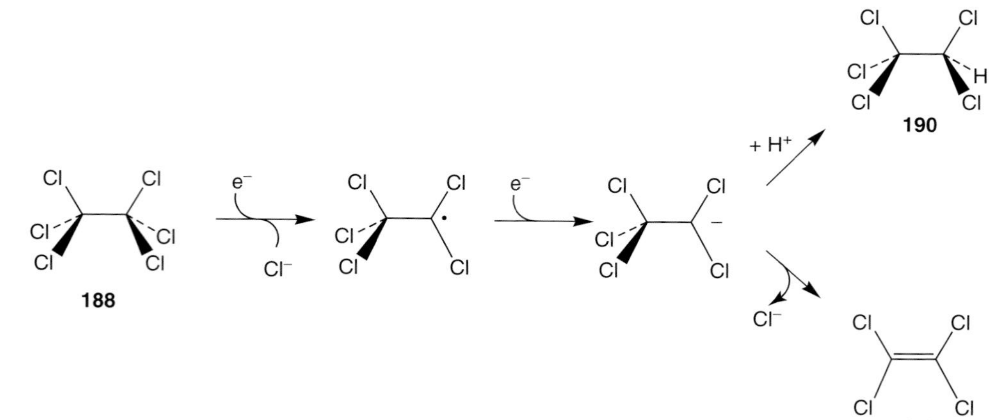

Scheme 20

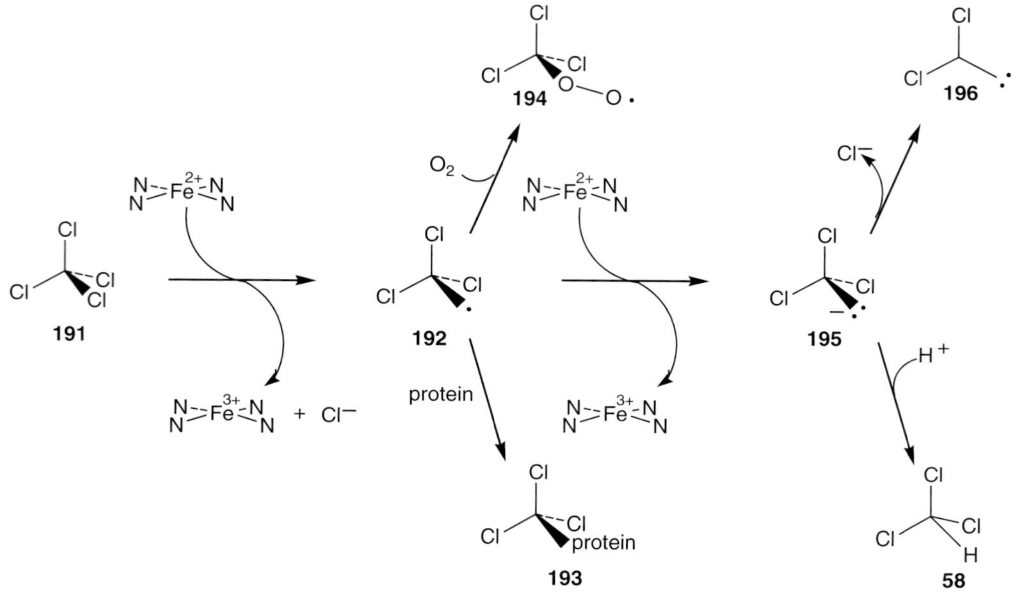

Figure 9 Mechanism for reductive dehalogenation.

pathways open to further reaction. It can adduct to the protein,[217] to give **193**, react with molecular oxygen, to form the trichloromethylperoxy radical (**194**),[216] or undergo a second one-electron reduction, to generate the trichloromethyl anion (**195**), which protonates to yield chloroform (**58**), or eliminates a second chloride anion, to generate dichlorocarbene (**196**).[218]

Both oxidative and reductive metabolism of the general anesthetic halothane (**60**) have been associated with hepatotoxicity (**Scheme 21**). Both processes lead to covalent adduction of a reactive metabolite to various proteins. In the case of oxidative metabolism, it is the trifluoroacetylated protein **197**, formed by reaction with the trifluoroacetyl chloride metabolite of halothane, as discussed earlier, while in the case of reductive metabolism it is the 2,2,2-trifluoro-1-chloroethylated protein **198**, formed from reaction with the 2,2,2-trifluoro-1-chloroethyl radical, which in turn is generated from the reductive elimination of a bromide anion from halothane.[214,219]

5.05.3.1.2   Reduction of nitrogen-containing functional groups

Cytochrome P450 is also capable of reducing nitrogen-containing functional groups of various oxidation states back to the corresponding saturated nitrogen-containing functional group (amine, hydrazine, etc.). So, N-oxides, imines,

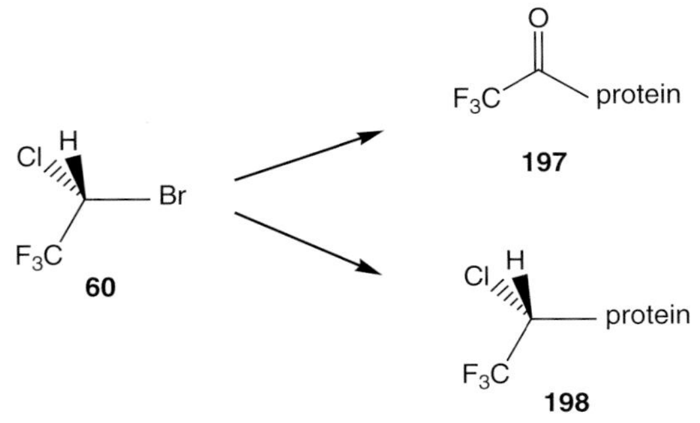

Scheme 21

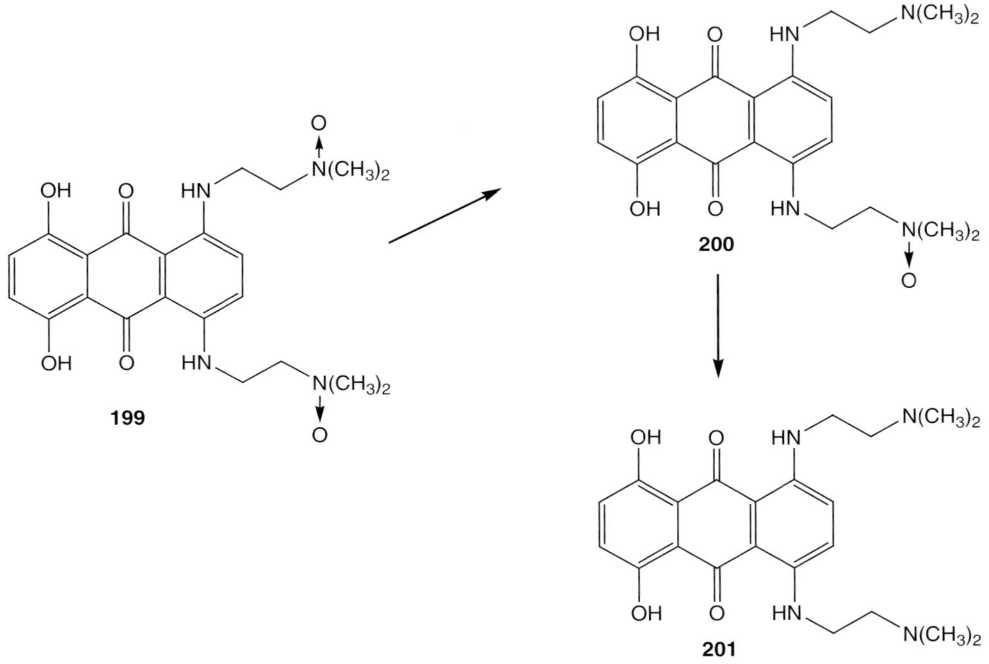

Scheme 22

hydroxylamines, nitroso groups, nitro groups, and azo dyes can all be reduced by cytochrome P450, especially under anaerobic conditions.

N-Oxide metabolites are generally nontoxic or less toxic than the parent drug. Their lack of toxicity coupled to the likely regeneration of the parent in a hypoxic bio-environment suggested a prodrug strategy for treating solid tumors (for an extensive discussion of prodrugs, *see* 5.44 Prodrug Objectives and Design). Administration of an *N*-oxide-containing anticancer agent should reduce systemic toxicity but maintain the effectiveness of the drug, since reduction of the *N*-oxide back to the parent compound at the tumor site would be highly probable. This strategy has been successfully employed[220,221] with the anthraquinone derivative AQ4N (**199**), which is currently in clinical trials (**Scheme 22**). After reaching the hypoxic regions of the solid tumor, the drug undergoes two sequential two-electron reductions. The reductions (CYP3A4) generate AQ4M (**200**), the mono-*N*-oxide, followed by AQ4 (**201**), the di-tertiary amine, and active anticancer agent. The cytotoxic potency of **201** is of the order of 1000-fold greater than that of **199**.[220]

The reduction of an aromatic nitro group to an amine, a process that is frequently associated with toxicity, proceeds through the nitroso compound and the hydroxylamine. Early on it was established[222] that nitrobenzene (**202**), nitrosobenzene (**203**), and phenylhydroxylamine (**204**) could all be reduced to aniline (**205**) by cytochrome P450 (**Scheme 23**). Cytochrome P450 under anaerobic conditions also appears to play a central role in the reductive

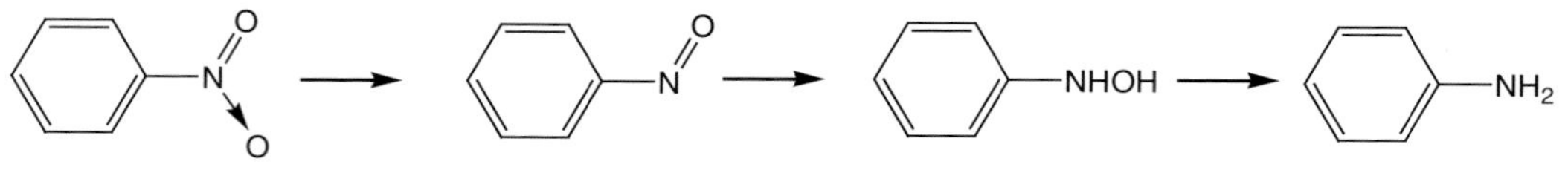

202   **203**   **204**   **205**

Scheme 23

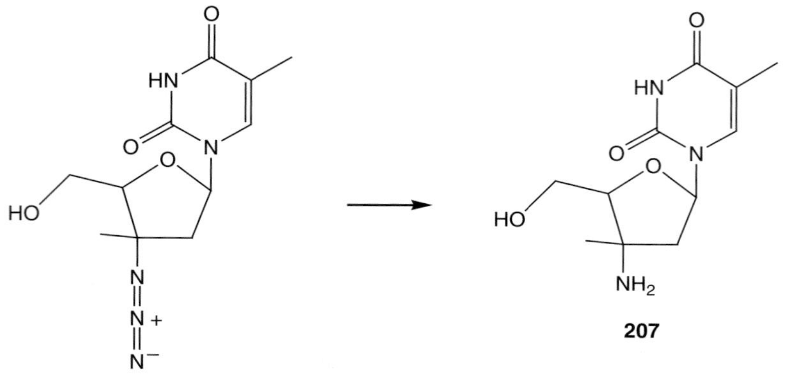

206   **207**

Scheme 24

metabolism of the carcinogens 1-nitropyrene,[223] 2-nitropyrene, 4-nitrobiphenyl, and 1-nitronaphthalene[224] to the corresponding hydroxylamines and amines. In a somewhat similar vein, the 3'-azido group of the anti-HIV drug zidovudine (**206**) has been shown to be susceptible to reduction to its toxic 3'-amino metabolite **207** under anaerobic conditions both by cytochrome P450 and by cytochrome $b5$ and its reductase (**Scheme 24**).[225]

5.05.3.1.3   Benzamidoxime reductase

A cytochrome P450-dependent, oxygen-insensitive reducing system that catalyzes the efficient reduction of strongly basic primary N-hydroxylated functional groups such as amidines, guanidines, and amidinohydrazones was isolated from pig liver and characterized by Clement et al.[226] The reconstituted system, termed benzamidoxime reductase, is composed of NADH, cytochrome $b5$, its reductase, phosphatidylcholine, and a cytochrome P450 from the 2D family.

In subsequent investigations this enzyme system was found to be an efficient reductant of aliphatic hydroxylamines (e.g., amphetamine hydroxylamine (**208**)), in liver microsomal preparations,[227] mitochondria, and other organs and organelles.[228] Its reducing efficiency in conjunction with the wider bio-distribution of the activity suggested its potential usefulness in a pro-drug strategy for administering strongly basic drugs. Such drugs are generally poorly adsorbed because they are virtually 100% protonated throughout the gastrointestinal tract. Administration of the less basic hydroxylamine analogs should increase adsorption, allowing the pro-drug to reach the systemic circulation, where it could be reduced to generate the parent drug.[228] This approach was tried with ximelagatran (**209**), a pro-drug of the thrombin inhibitor melagatran (**210**), and was found to increase the oral bioavailability of **210** from 3–7% to 18–24% (**Scheme 25**).[229] While the hydroxylamine was efficiently and rapidly reduced in humans, the specific cytochrome P450 that catalyzed the reaction could not be identified.[230]

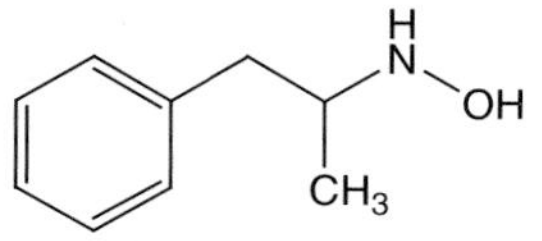

208

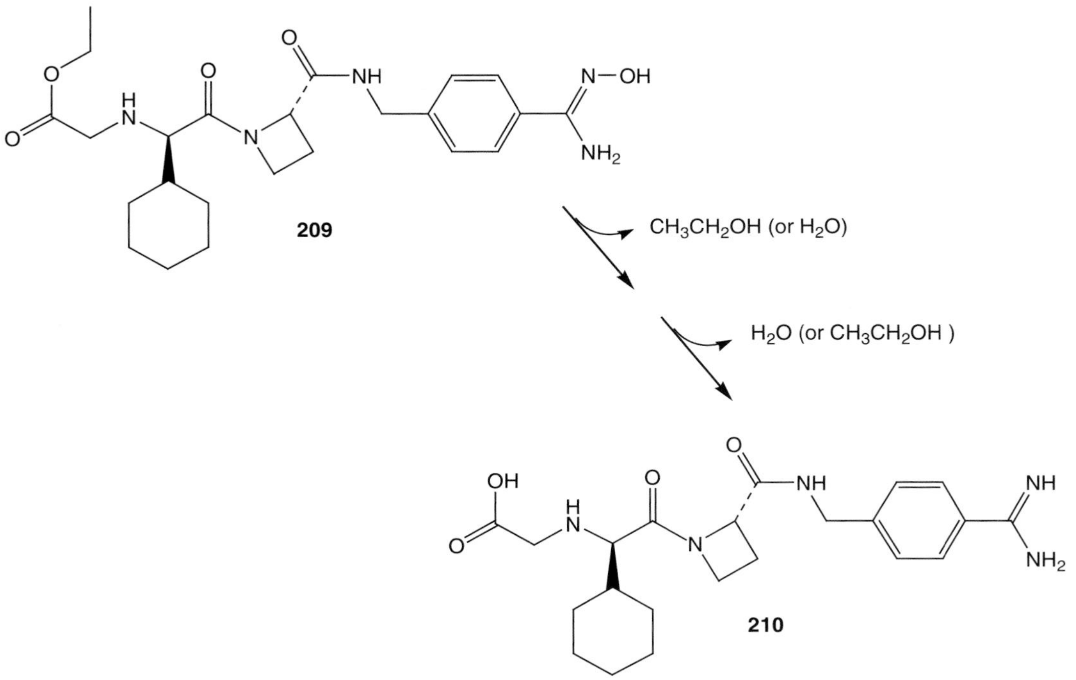

Scheme 25

The benzamidoxime system has recently[231] been found capable of reducing the *N*-hydroxylamines of sulfamethoxazole (**211**) and dapsone (**82**), without cytochrome P450. The notion that the reductase system between the human and pig might differ with regard to the relevance of cytochrome P450 has been proposed by Andersson *et al.*[232] They concluded that neither cytochrome P450 nor FMO is an essential element of the reductase system in the human, based on selective inhibitors, while cytochrome *b*5 reductase and nonheme iron probably are.

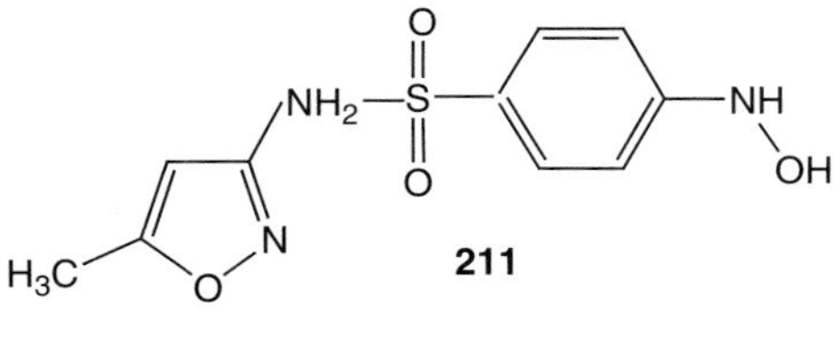

5.05.3.2   Xanthine Oxidoreductase and Aldehyde Oxidase

Cytochrome P450-containing enzyme systems are not the only ones capable of reducing nitrogen-containing functional groups. In fact, the cytosolic molybdenum-containing enzymes xanthine oxidoreductase and, in particular, AOX are often the major contributors. For example, they are major contributors in the reduction of aromatic nitro compounds,[222–224] and perhaps *N*-oxides.[233]

5.05.3.3   Aldo-Keto Reductases (AKRs)

The AKRs are a superfamily of dehydrogenases/reductases.[234] Unlike the ADHs, they require $NADP^+(H)$ as a cofactor, and generally catalyze reductive[235] rather than oxidative reactions. While they operate on the same kinds of endogenous substrates as the ADHs (e.g., ketosteroids,[236] retinal,[237] and lipid peroxidation products),[238] they also tend to operate on xenobiotics.

At least three AKR isoforms (1C1, 1C2, and 1C4) have been isolated from human liver cytosol that contribute to the reduction of the tobacco-specific carcinogen 4-methynitrosoamino-1-(3-pyridyl)-1-butanone (**212**) to the less toxic 4-methynitrosoamino-1-(3-pyridyl)-1-butanonol (**213**), which is susceptible to conjugation and elimination.[239]

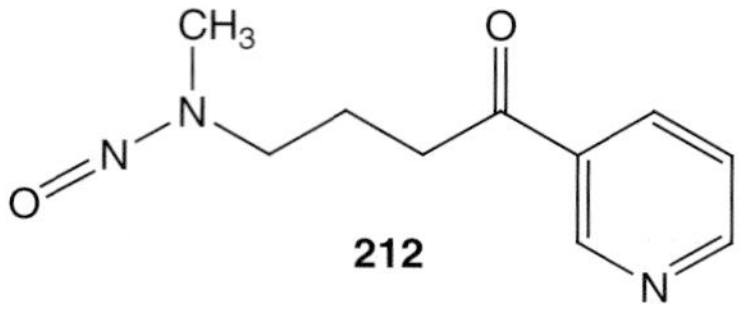

212

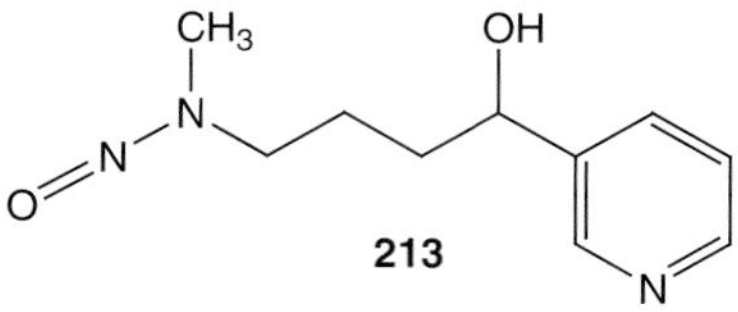

213

Yanmano *et al.*[240] have isolated and purified two AKRs of the 1C subfamily from rabbit liver cytosol that will reduce the opioid receptor antagonist naloxone (**214**) to the α- and β-diastereomeric alcohols **215** and **216**, respectively (**Scheme 26**). Interestingly, one AKR, termed NR1, stereospecifically produces **215**, while the second, termed NR2, stereospecifically produces **216**. Naltrexone (**217**) and dihydromorphinone (**218**) were good substrates for NR2 but not NR1. Both enzymes reduced aliphatic and aromatic aldehydes, cyclic and aromatic ketones, and quinones. They also catalyzed the dehydrogenation of 17β-hydroxysteroids with low K_m values. The main human AKRs have been cloned and expressed, and their substrate reactivity profiles and tissue and organ distribution determined.[241]

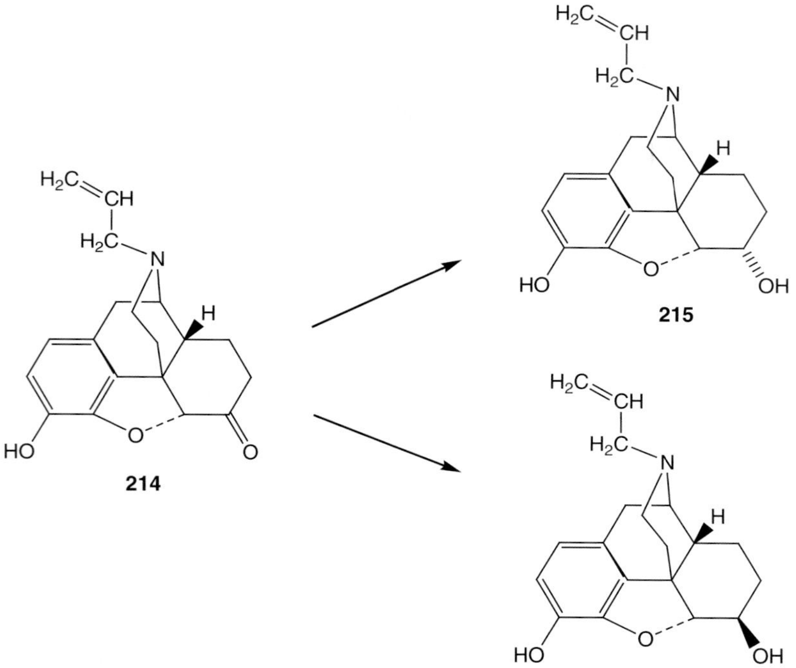

214 215 216

Scheme 26

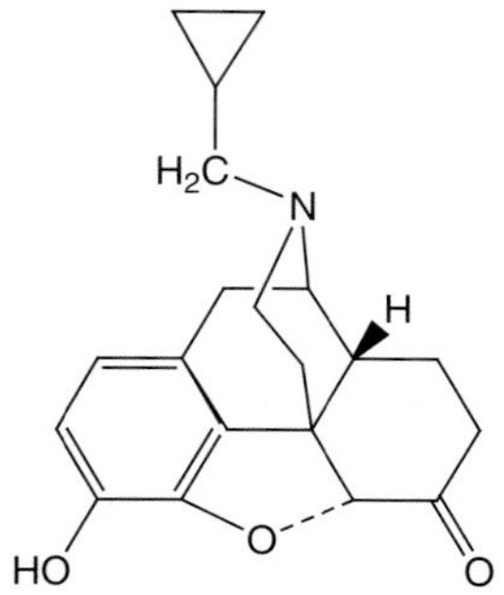

217

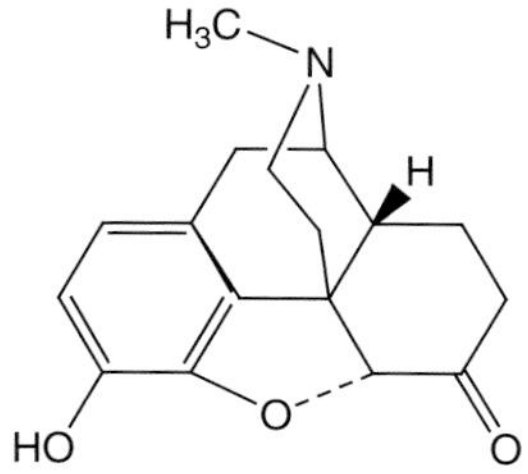

218

5.05.3.4   Carbonyl Reductases (CBRs)

CBRs are a subset of short-chain dehydrogenase/reductases[242,243] that belong to a superfamily of oxidoreductases.[244] They are found in cytosol, have broad substrate selectivity, and encompass both endobiotic and xenobiotic carbonyl compounds, including prostaglandins, steroids, and quinones, plus a wide array of aromatic and aliphatic aldehydes and ketones. For recent reviews, see Oppermann and Maser[235] and Rosemond and Walsh.[245] Typical examples of CBR-catalyzed reduction of quinone carbonyls or keto groups contained in a xenobiotic include the reduction of the anthracycline antibiotic doxorubicin (**219**),[246] halopyridol (**220**),[247] metyrapone (**221**),[247] acetohexamide (**222**),[247] and warfarin (**223**).[248]

5.05.4   Conclusion

The explosion of metabolic information since the 1990s has served to accentuate the importance of drug metabolism to clinical pharmacology and therapeutics. This new knowledge emphasizes the need for a common understanding among investigators of the different disciplines that impact drug development. To help meet that need, my goal in writing this chapter was to provide a comprehensive overview of the scope of the chemistry and major enzyme systems involved in drug metabolism. Many of the topics could not receive the detailed coverage they merit because of the constraints of space. However, it is hoped that nothing of importance has been missed, and that the references provided serve as a key to deeper exploration by the interested reader.

References

1. Testa, B. *The Metabolism of Drugs and Other Xenobiotics – Biochemistry of RedoxReactions*; Academic Press: London, UK, 1995.
2. Nelson, D. R.; Strobel, H. W. *Mol. Biol. Evol.* **1987**, *4*, 572–593.
3. Nelson, D. R.; Koymans, L.; Kamataki, T.; Stegeman, J. J.; Feyereisen, R.; Waxman, D. J.; Waterman, M. R.; Gotoh, O.; Coon, M. J.; Estabrook, R. W. et al. *Pharmacogenetics* **1996**, *6*, 1–42.
4. Nebert, D. W.; Adesnik, M.; Coon, M. J.; Estabrook, R. W.; Gonzalez, F. J.; Guengerich, F. P.; Gunsalus, I. C.; Johnson, E. F.; Kemper, B.; Levin, W. et al. *DNA* **1987**, *6*, 1–11.

5. Benet, L. Z. Parmacokinetic. In *Goodman and Gilman's The Pharmacological Basis of Therapeutics*, 9th ed.; Hardman, J. G., Limbird, L., Molinoff, P. B., Ruddon, R. W., Gilman, A. G., Eds.; McGraw Hill: New York, NY, 1996, p 14.
6. White, R. E.; Coon, M. J. *Annu. Rev. Biochem.* **1980**, *49*, 315–356.
7. Schenkman, J. B.; Gibson, G. G. *Trends Pharmacol. Sci.* **1981**, *2*, 150–152.
8. Atkins, W. M.; Sligar, S. G. *J. Am. Chem. Soc.* **1987**, *109*, 3754–3760.
9. Gorsky, L. D.; Koop, D. R.; Coon, M. J. *J. Biol. Chem.* **1984**, *259*, 6812–6817.
10. Korzekwa, K. R.; Jones, J. P.; Gillette, J. R. *J. Am. Chem. Soc.* **1990**, *112*, 7042–7046.
11. Korzekwa, K. R.; Jones, J. P. *Pharmacogenetics* **1993**, *3*, 1–18.
12. Groves, J. T.; McClusky, G. A. *Biochem. Biophys. Res. Commun.* **1978**, *81*, 154–160.
13. Groves, J. T. *J. Chem. Educ.* **1985**, *62*, 928–931.
14. Ortiz de Montellano, P. R.; Stearns, R. A. *J. Am. Chem. Soc.* **1987**, *109*, 3415–3420.
15. Bowry, B. W.; Ingold, K. U. *J. Am. Chem. Soc.* **1991**, *113*, 5699–5707.
16. Manchester, J. I.; Dinnocenzo, J. P.; Higgins, L. A.; Jones, J. P. *J. Am. Chem. Soc.* **1997**, *119*, 5069–5070.
17. Atkinson, J. K.; Ingold, K. U. *Biochemistry* **1993**, *32*, 9209–9214.
18. Atkinson, J. K.; Hollenberg, P. F.; Ingold, K. U.; Johnson, C. C.; Le Tadic, M. H.; Newcomb, M.; Putt, D. A. *Biochemistry* **1994**, *33*, 10630–10637.
19. Newcomb, M.; Le Tadic, M.-H.; Putt, D. A.; Hollenberg, P. F. *J. Am. Chem. Soc.* **1995**, *117*, 3312–3313.
20. Newcomb, M.; Le Tadic, M.-H.; Chestney, D. L.; Roberts, E. S.; Hollenberg, P. F. *J. Am. Chem. Soc.* **1995**, *117*, 12085–12091.
21. Toy, P. H.; Newcomb, M.; Hollenberg, P. F. *J. Am. Chem. Soc.* **1998**, *120*, 7719–7729.
22. Newcomb, M.; Shen, R.; Choi, S.-Y.; Toy, P. H.; Hollenberg, P. F.; Vaz, A. D. N.; Coon, M. J. *J. Am. Chem. Soc.* **2000**, *122*, 2677–2686.
23. Newcomb, M.; Hollenberg, P. F.; Coon, M. J. *Arch. Biochem. Biophys.* **2003**, *409*, 72–79.
24. Newcomb, M.; Aebisher, D.; Shen, R.; Chandrasena, R. E.; Hollenberg, P. F.; Coon, M. J. *J. Am. Chem. Soc.* **2003**, *125*, 6064–6065.
25. Chandrasena, R. E.; Vatsis, K. P.; Coon, M. J.; Hollenberg, P. F.; Newcomb, M. *J. Am. Chem. Soc.* **2004**, *126*, 115–126.
26. Shaik, S.; Filatov, M.; Schröder, D.; Schwarz, H. *Chem. Eur. J.* **1998**, *4*, 193–199.
27. Harris, N.; Cohen, S.; Filatov, M.; Ogliaro, F.; Shaik, S. *Angew. Chem. Int. Ed.* **2000**, *39*, 2003–2007.
28. Ogliaro, F.; Harris, N.; Cohen, S.; Filatov, M.; de Visser, S. P.; Shaik, S. *J. Am. Chem. Soc.* **2000**, *122*, 8977–8989.
29. Schoneboom, J. C.; Lin, H.; Reuter, N.; Thiel, W.; Cohen, S.; Ogliaro, F.; Shaik, S. *J. Am. Chem. Soc.* **2002**, *124*, 8142–8151.
30. Kumar, D.; de Visser, S. P.; Sharma, P. K.; Cohen, S.; Shaik, S. *J. Am. Chem. Soc.* **2004**, *126*, 1907–1920.
31. Schoneboom, J. C.; Cohen, S.; Lin, H.; Shaik, S.; Thiel, W. *J. Am. Chem. Soc.* **2004**, *126*, 4017–4034.
32. Meunier, B.; de Visser, S. P.; Shaik, S. *Chem. Rev.* **2004**, *104*, 3947–3980.
33. Hlavica, P. *Eur. J. Biochem.* **2004**, *271*, 4335–4360.
34. Powell, P. K.; Wolf, I.; Lasker, J. M. *Arch. Biochem. Biophys.* **1996**, *335*, 219–226.
35. Alexander, J. J.; Snyder, A.; Tonsgard, J. H. *Neurochem. Res.* **1998**, *23*, 227–233.
36. Sawamura, A.; Kusunose, E.; Satouchi, K.; Kusunose, M. *Biochim. Biphys. Acta* **1993**, *1168*, 30–36.
37. Fisher, C. W.; Shet, M. S.; Caudle, D. L.; Martin-Wixtrom, C. A.; Estabrook, R. W. *Proc. Natl. Acad. Sci. USA* **1992**, *89*, 10817–10821.
38. Shet, M.; Fisher, C. W.; Holmans, P. L.; Estabrook, R. W. *Arch. Biochem. Biophys.* **1996**, *330*, 199–208.
39. Jones, J. P.; O'Hare, E. J.; Wong, L. L. *Eur. J. Biochem.* **2001**, *268*, 1460–14677.
40. White, R. E.; Miller, J. P.; Favreau, L. V.; Bhattacharyya, A. *J. Am. Chem. Soc.* **1986**, *108*, 6024–6031.
41. Jones, J. P.; Trager, W. F. *J. Am. Chem. Soc.* **1986**, *108*, 7074–7078 (*J. Am. Chem. Soc.* **1988**, *110*, 2018).
42. Korzekwa, K. R.; Trager, W. F.; Gillette, J. R. *Biochemistry* **1989**, *28*, 9012–9018.
43. White, R. E.; McCarthy, M. B.; Egeberg, K. D.; Sligar, S. G. *Arch. Biochem. Biophys.* **1984**, *228*, 493–502.
44. Higgins, L.; Korzekwa, K. R.; Rao, S.; Shou, M.; Jones, J. P. *Arch. Biochem. Biophys.* **2001**, *385*, 220–230.
45. Jones, J. P.; Mysinger, M.; Korzekwa, K. R. *Drug Metab. Dispos.* **2002**, *30*, 7–12.
46. Licht, H. J.; Coscia, C. J. *Biochemistry* **1978**, *17*, 5638–5646.
47. Groves, J. T.; Subramanian, D. V. *J. Am. Chem. Soc.* **1986**, *106*, 2177–2181.
48. Gordon, W. P.; Huitric, A. C.; Seth, C. L.; McClanahan, R. H.; Nelson, S. D. *Drug Metab. Dispos.* **1987**, *15*, 589–594.
49. McClanahan, R. H.; Huitric, A. H.; Pearson, P. G.; Desper, J. C.; Nelson, S. D. *J. Am. Chem. Soc.* **1988**, *110*, 1979–1980.
50. Baba, T.; Yamada, H.; Oguri, K.; Yoshimura, H. *Xenobiotica* **1988**, *18*, 474–484.
51. Nelson, S. D.; Breck, G. D.; Trager, W. F. *J. Med. Chem.* **1973**, *16*, 1106–1112.
52. Bargetzi, M. J.; Aoyama, T.; Gonzalez, F. J.; Meyer, U. A. *Clin. Pharmacol. Ther.* **1989**, *46*, 521–527.
53. Shea, J. P.; Nelson, S. D.; Ford, J. P. *J. Am. Chem. Soc.* **1983**, *105*, 5451–5454.
54. Burka, L. T.; Guengerich, F. P.; Willard, R. J.; Macdonald, T. L. *J. Am. Chem. Soc.* **1985**, *107*, 2549–2551.
55. Lindsay Smith, J. R.; Mortimer, D. N. *J. Chem. Soc. Perkin Trans.* **1986**, *2*, 1743–1749.
56. Okazaki, O.; Guengerich, F. P. *J. Biol. Chem.* **1993**, *268*, 1546–1552.
57. Seto, Y.; Guengerich, F. P. *J. Biol. Chem.* **1993**, *268*, 9986–9997.
58. Guengerich, F. P.; Okazaki, O.; Seto, Y.; Macdonald, T. L. *Xenobiotica* **1995**, *25*, 689–709.
59. Guengerich, F. P.; Yun, C.-H.; Macdonald, T. L. *J. Biol. Chem.* **1996**, *271*, 27321–27329.
60. Dinnocenzo, J. P.; Karki, S. B.; Jones, J. P. *J. Am. Chem. Soc.* **1993**, *115*, 7111–71116.
61. Carlson, T. J.; Jones, J. P.; Peterson, L.; Castagnoli, N., Jr.,; Iyer, K. R.; Trager, W. F. *Drug Metab. Dispos.* **1995**, *23*, 749–756.
62. Karki, S. B.; Dinnocenzo, J. P.; Jones, J. P.; Korzekwa, K. R. *J. Am. Chem. Soc.* **1995**, *117*, 3657–3664.
63. Karki, S. B.; Dinnocenzo, J. P. *Xenobiotica* **1995**, *25*, 711–724.
64. Shaffer, C. L.; Harriman, S.; Koen, Y. M.; Hanzlik, R. P. *J. Am. Chem. Soc.* **2002**, *124*, 8268–8274.
65. Ferrari, A.; Coccia, C. P.; Bertolini, A.; Sternieri, E. *Pharmacol. Res.* **2004**, *50*, 551–559.
66. Hoag, M. K.; Trevor, A. J.; Asscher, Y.; Weissman, J.; Castagnoli, N., Jr. *Drug. Metab. Dispos.* **1984**, *12*, 371–375.
67. Hoag, M. K.; Trevor, A. J.; Kalir, A.; Castagnoli, N., Jr. *Drug Metab. Dispos.* **1987**, *15*, 485–490.
68. Hoag, M. K.; Schmidt-Peetz, M.; Lampen, P.; Trevor, A.; Castagnoli, N., Jr. *Chem. Res. Toxicol.* **1988**, *1*, 128–131.
69. Shen, T.; Hollenberg, P. F. *Chem. Res. Toxicol.* **1994**, *7*, 231–238.
70. Hall, L. R.; Hanzlik, R. P. *J. Biol. Chem.* **1990**, *265*, 12349–12355.
71. Hall, L. R.; Hanzlik, R. P. *Xenobiotica* **1991**, *21*, 1127–1138.
72. Clement, B.; Zimmermann, M. *Biochem. Pharmacol.* **1987**, *36*, 3127–3133.

73. Keefer, L. K.; Anjo, T.; Wade, D.; Wang, T.; Yang, C. S. *Cancer Res.* **1987**, *47*, 447–452.
74. Garland, W. A.; Nelson, S. D.; Sasame, H. A. *Biochem. Biophys. Res. Commun.* **1976**, *72*, 539–545.
75. Schmid, B.; Bircher, J.; Preisig, R.; Kupfer, A. *Clin. Pharmacol. Ther.* **1985**, *38*, 618–624.
76. Duquette, P. H.; Peterson, F. J.; Crankshaw, D. L.; Lindemann, N. J.; Holtzman, J. L. *Drug Metab. Dispos.* **1983**, *11*, 477–480.
77. Murthy, S. S.; Shetty, H. U.; Nelson, W. L.; Jackson, P. R.; Lennard, M. S. *Biochem. Pharmacol.* **1990**, *40*, 1637–1644.
78. Kumagi, Y.; Lin, L. Y.; Philpot, R. M.; Yamada, H.; Oguri, K.; Yoshimura, H.; Cho, A. K. *Mol. Pharmacol.* **1992**, *42*, 695–702.
79. Franklin, M. R. *Xenobiotica* **1971**, *1*, 581–591.
80. Philpot, R. M.; Hodgson, E. *Life Sci. II* **1971**, *10*, 503–512.
81. Mansuy, D.; Battioni, J. P.; Chottard, J. C.; Ullrich, V. *J. Am. Chem. Soc.* **1979**, *101*, 3971–3973.
82. Ortiz de Montellano, P. R.; Reich, N. O. Inhibition of Cytochrome P450 Enzymes. In *Cytochrome P450, Structure, Mechanism and Biochemistry*, 1st ed.; Ortiz de Montellano, P. R., Ed.; Plenum Press: New York, NY, 1986; Chapter 8, pp 280–283.
83. Wormhoudt, L. W.; Ploemen, J. H.; de Waziers, I.; Commandeur, J. N.; Beaune, P. H.; van Bladeren, P. J.; Vermeulen, N. P. *Chem. Biol. Interact.* **1996**, *101*, 175–192.
84. Osawa, Y.; Highet, R. J.; Pohl, L. R. *Xenobiotica* **1992**, *22*, 1147–1156.
85. Pohl, L. R.; Bhooshan, B.; Whittaker, N. F.; Krishna, G. *Biochem. Biophys. Res. Commun.* **1977**, *79*, 684–691.
86. Pohl, L. R.; Martin, J. L.; George, J. W. *Biochem. Pharmacol.* **1980**, *29*, 3271–3276.
87. Pohl, L. R.; George, J. W.; Satoh, H. *Drug Metab. Dispos.* **1984**, *12*, 304–308.
88. Gandolfi, A. J.; White, R. D.; Sipes, I. G.; Pohl, L. R. *J. Pharmacol. Exp. Ther.* **1980**, *214*, 721–725.
89. Pohl, L. R.; Kenna, J. G.; Satoh, H.; Christ, D.; Martin, J. L. *Drug Metab. Rev.* **1989**, *20*, 203–217.
90. Spracklin, D. K.; Hankins, D. C.; Fisher, J. M.; Thummel, K. E.; Kharasch, E. D. *J. Pharmacol. Exp. Ther.* **1997**, *281*, 400–411.
91. Christ, D. D.; Kenna, J. G.; Kammerer, W.; Satoh, H.; Pohl, L. R. *Anesthesiology* **1988**, *69*, 833–838.
92. Martin, J. L.; Keegan, M. T.; Vasdev, G. M.; Nyberg, S. L.; Bourdi, M.; Pohl, L. R.; Plevak, D. J. *Anesthesiology* **2001**, *95*, 551–553.
93. Njoku, D. B.; Shrestha, S.; Soloway, R.; Duray, P. R.; Tsokos, M.; Abu-Asab, M. S.; Pohl, L. R.; West, A. B. *Anesthesiology* **2002**, *96*, 757–761.
94. Bertilsson, L. *Clin. Pharmacokinet.* **1978**, *3*, 128–143.
95. Duescher, R. J.; Elafarra, A. A. *Arch. Biochem. Biophys.* **1994**, *311*, 342–349.
96. Shaik, S.; de Visser, S. P.; Ogliaro, F.; Schwarz, H.; Schroder, D. *Curr. Opin. Chem. Biol.* **2002**, *6*, 556–567.
97. de Visser, S. P.; Ogliaro, F.; Sharma, P. K.; Shaik, S. *J. Am. Chem. Soc.* **2002**, *124*, 11809–11826.
98. Yin, H.; Anders, M. W.; Korzekwa, K. R.; Higgins, L.; Thummel, K. E.; Kharasch, E. D.; Jones, J. P. *Proc. Natl. Acad. Sci. USA* **1995**, *92*, 11076–11080.
99. Grogan, J.; De Vito, S. C.; Pearlman, R. S.; Korzekwa, K. R. *Chem. Res. Toxicol.* **1992**, *5*, 548–552.
100. Jerina, D. M.; Daly, J. W. *Science* **1974**, *185*, 573–582.
101. Korzekwa, K.; Trager, W.; Gouterman, M.; Spangler, D.; Loew, G. H. *J. Am. Chem. Soc.* **1985**, *107*, 4273–4279.
102. Hanzlik, R. P.; Hogberg, K.; Judson, C. M. *Biochemistry* **1984**, *23*, 3048–3055.
103. Bush, E. D.; Trager, W. F. *J. Med. Chem.* **1985**, *28*, 992–996.
104. Korzekwa, K. R.; Swinney, D. C.; Trager, W. F. *Biochemistry* **1989**, *28*, 9019–9027.
105. Darbyshire, J. F.; Iyer, K. R.; Grogan, J.; Korzekwa, K. R.; Trager, W. F. *Drug Metab. Dispos.* **1996**, *24*, 1038–1045.
106. de Visser, S. P.; Shaik, S. *J. Am. Chem. Soc.* **2003**, *125*, 7413–7424.
107. Terelius, Y.; Norsten-Hoog, C.; Cronholm, T.; Ingelman-Sundberg, M. *Biochem. Biophys. Res. Commun.* **1991**, *179*, 689–694.
108. Watanabe, K.; Matsunaga, T.; Narimatsu, S.; Yamamoto, I.; Yoshimura, H. *Biochem. Biophys. Res. Commun.* **1992**, *188*, 114–119.
109. Roberts, E. S.; Vaz, A. D.; Coon, M. J. *Mol. Pharmacol.* **1992**, *41*, 427–433.
110. Watanabe, K.; Matsunaga, T.; Kimura, T.; Funahashi, T.; Funae, Y.; Ohshima, T.; Yamamoto, I. *Drug Metab. Pharmacokinet.* **2002**, *17*, 516–521.
111. Akhtar, M.; Calder, M. R.; Corina, D. L.; Wright, J. N. *Biochem. J.* **1982**, *201*, 569–580.
112. Vaz, A. D. N.; Roberts, E. S.; Coon, M. J. *J. Am. Chem. Soc.* **1991**, *113*, 5886–5887.
113. Roberts, E. S.; Vaz, A. D. N.; Coon, M. J. *Proc. Natl. Acad. Sci. USA* **1991**, *88*, 8963–8966.
114. Vaz, A. D. N.; Pernecky, S. J.; Raner, G. M.; Coon, M. J. *Proc. Natl. Acad. Sci. USA* **1996**, *93*, 8963–8966.
115. Raner, G. M.; Chiang, E. W.; Vaz, A. D. N.; Coon, M. *Biochemistry* **1997**, *36*, 4895–4902.
116. Vaz, A. D. N.; McGinnity, D. F.; Coon, M. J. *Proc. Natl. Acad. Sci. USA* **1998**, *95*, 3555–3560.
117. Coon, M. J.; Vaz, A. D. N.; McGinnity, D. F.; Peng, H.-W. *Drug Metab. Dispos.* **1998**, *26*, 1190–1193.
118. Ortiz de Montellano, P. R; De Voss, J. J. Substrate Oxidation by Cytochrome P450 Enzymes. In *Cytochrome P450, Structure, Mechanism and Biochemistry*, 3rd ed.; Ortiz de Montellano, P. R., Ed.; Kluwer Academic/Plenum Press: New York, NY, 2005; Chapter 6, pp 198–200.
119. Ortiz de Montellano, P. R.; Komives, E. A. *J. Biol. Chem.* **1985**, *260*, 3330–3336.
120. Komives, E. A.; Ortiz de Montellano, P. R. *J. Biol. Chem.* **1987**, *262*, 9793–9802.
121. Cho, A. K.; Lindeke, B.; Hodshon, B. J. *Res. Commun. Chem. Pathol. Pharmacol.* **1972**, *4*, 519–528.
122. Fleming, C. M.; Branch, R. A.; Wilkinson, G. R.; Guengerich, F. P. *Mol. Pharmacol.* **1992**, *41*, 975–980.
123. Johnson, E. F.; Levitt, D. S.; Muller-Eberhard, U; Thorgeirsson, S. S. *Cancer Res.* **1980**, *40*, 4456–4459.
124. Poupko, J. M.; Radomski, T.; Santella, R. M.; Radomski, J. L. *J. Natl. Cancer Inst.* **1983**, *70*, 1077–1080.
125. Yun, C.-H.; Shimada, T.; Guengerich, F. P. *Carcinogenesis*. **1992**, *13*, 217–222.
126. Correia, M. A.; Ortiz de Montellano, P. R. Inhibition of Cytochrome P450 Enzymes. In *Cytochrome P450, Structure, Mechanism and Biochemistry*, 3rd ed.; Ortiz de Montellano, P. R., Ed.; Kluwer Academic/Plenum Press: New York, NY, 2005; Chapter 7, pp 265–267.
127. Bensoussan, C.; Delaforge, M.; Mansuy, D. *Biochem. Pharmacol.* **1995**, *49*, 602–691.
128. Pershing, L. K.; Franklin, M. R. *Xenobiotica* **1982**, *12*, 687–699.
129. Larrey, D.; Funk Bretano, C. F.; Breil, P.; Vitaux, J.; Theodore, C.; Babany, G.; Pessayre, D. *Biochem. Pharmacol.* **1983**, *32*, 1063–1068.
130. Delaforge, M.; Ladam, P.; Bouillé, G.; Gharbi Benarous, J.; Jaouen, M.; Girault, J. P. *Chem.-Biol. Interactions* **1992**, *85*, 215–227.
131. Clement, B.; Jung, F. *Drug Metab. Dispos.* **1994**, *22*, 486–497.
132. Clement, B.; Schultze-Mosgau, M. H.; Wohlers, H. *Biochem. Pharmacol.* **1993**, *46*, 2249–2267.
133. Clement, B.; Demesmaeker, M. *Drug Metab. Dispos.* **1997**, *25*, 1266–1271.
134. Surapaneni, S. S.; Clay, M. P.; Spangle, L. A.; Pascal, J. W.; Lindstrom, T. D. *Drug Metab. Dispos.* **1997**, *25*, 1383–1388.
135. Washio, T.; Arisawa, H.; Kohsaka, K.; Yasuda, H. *Biol. Pharm. Bull.* **2001**, *24*, 1263–1266.

136. He, M.; Rettie, A. E.; Neal, J.; Trager, W. F. *Drug Metab. Dispos.* **2001**, *29*, 701–711.
137. Äbelö, A.; Andersson, T. B.; Antonsson, M.; Naudot, A. K.; Skånberg, I.; Weidolf, L. *Drug Metab. Dispos.* **2000**, *28*, 966–972.
138. Rettie, A. E.; Rettenmeier, A. W.; Howald, W. N.; Baillie, T. A. *Science* **1987**, *235*, 890–893.
139. Rettie, A. E.; Boberg, M.; Rettenmeier, A. W.; Baillie, T. A. *J. Biol. Chem.* **1988**, *263*, 13733–13738.
140. Nagata, K.; Liberato, D. J.; Gillette, J. R.; Sasame, H. A. *Drug Metab. Dispos.* **1986**, *14*, 559–565.
141. Korzekwa, K. R.; Trager, W. F.; Nagata, K.; Parkinson, A.; Gillette, J. R. *Drug Metab. Dispos.* **1990**, *18*, 974–979.
142. Obach, S. *Drug Metab. Dispos.* **2001**, *29*, 1599–1607.
143. Guengerich, F. P.; Peterson, L. A.; Bocker, R. H. *J. Biol. Chem.* **1988**, *263*, 8176–8183.
144. Peng, H. M.; Raner, G. M.; Vaz, A. D.; Coon, M. J. *Arch. Biochem. Biophys.* **1995**, *318*, 333–339.
145. DeMaster, E. G.; Shirota, F. N.; Nagasawa, H. T. *J. Org. Chem.* **1992**, *57*, 5074–5075.
146. Bucher, J.-L.; Delaforge, M.; Mansuy, D. *Biochemistry* **1994**, *33*, 7811–7818.
147. Zhang, Z.; Li, Y.; Stearns, R. A.; Ortiz de Montellano, P. R.; Baillie, T. A.; Tang, W. *Biochemistry* **2002**, *41*, 2712–2718.
148. Labroo, R. B.; Thummel, K. E.; Kunze, K. L.; Podoll, T.; Trager, W. F.; Kharasch, E. D. *Drug Metab. Dispos.* **1995**, *23*, 490–496.
149. Bondon, A.; Macdonald, T. L.; Harris, T.; Guengerich, F. P. *J. Biol. Chem.* **1989**, *264*, 1988–1997.
150. Ohe, T.; Mashino, T.; Hirobe, M. *Arch. Biochim. Biophys.* **1994**, *310*, 402–409.
151. Ohe, T.; Mashino, T.; Hirobe, M. *Drug Metab. Dispos.* **1997**, *25*, 116–122.
152. Ziegler, D. M. *Annu. Rev. Pharmacol. Toxicol.* **1993**, *33*, 179–199.
153. Cashman, J. R.; Zhang, J. *Drug Metab. Dispos.* **2002**, *30*, 1043–1052.
154. Lomri, R.; Gu, Q.; Cashman, J. R. *Proc. Natl. Acad. Sci. USA* **1992**, *89*, 1685–1689.
155. Lomri, R.; Yang, Z. C.; Cashman, J. R. *Chem. Res. Toxicol.* **1993**, *6*, 800–807.
156. Koukouritaki, S. B.; Simpson, P.; Yeung, C. K.; Rettie, A. E.; Hines, R. N. *Pediatr. Res.* **2002**, *51*, 236–243.
157. Yeung, C. K.; Lang, D. H.; Thummel, K. E.; Rettie, A. E. *Drug Metab. Dispos.* **2000**, *28*, 1107–1111.
158. Ziegler, D. M. *Drug Metab. Rev.* **2002**, *34*, 503–511.
159. Poulsen, L. L.; Ziegler, D. M. *J. Biol. Chem.* **1979**, *254*, 6449–6455.
160. Beaty, N. B.; Ballou, D. P. *J. Biol. Chem.* **1981**, *256*, 4611–4618.
161. Beaty, N. B.; Ballou, D. P. *J. Biol. Chem.* **1981**, *256*, 4619–4625.
162. Cashman, J. R.; Xiong, Y. N.; Xu, L.; Janowsky, A. *J. Pharmacol. Exp. Therap.* **1999**, *288*, 1251–1260.
163. Lang, D. H.; Rettie, A. E. *Br. J. Clin. Pharmacol.* **2000**, *50*, 311–314.
164. Tugnait, M.; Hawes, E. M.; McKay, G.; Rettie, A. E.; Haining, R. L.; Midha, K. K. *Drug Metab. Dispos.* **1997**, *25*, 524–527.
165. Mani, C.; Hodgson, E.; Kupfer, D. *Drug Metab. Dispos.* **1993**, *21*, 657–661.
166. Bhamre, S.; Bhagwat, S. V.; Shankar, S. K.; Boyd, M. R.; Ravindranath, V. *Brain Res.* **1995**, *672*, 276–280.
167. Hamman, M. A.; Haehner-Daniels, B. D.; Wrighton, S. A.; Rettie, A. E.; Hall, S. D. *Biochem. Pharmacol.* **2000**, *60*, 7–17.
168. Vannelli, T. A.; Dykman, A.; Ortis de Montellano, P. R. *J. Biol. Chem.* **2002**, *277*, 12824–12829.
169. Bach, A. W.; Lan, N. C.; Johnson, D. L.; Abell, C. W.; Bembeneck, M. E.; Kwan, S. W.; Seeberg, P. H.; Shih, J. C. *Proc. Natl. Acad. Sci. USA* **1988**, *85*, 4934–4938.
170. Lyles, G. A.; Shaffer, C. *Biochem. Pharmacol.* **1979**, *28*, 1099–1106.
171. Weyler, W.; Hsu, Y. P.; Breakefield, X. O. *Pharmcol. Ther.* **1990**, *47*, 391–417.
172. Inoue, H.; Castagnoli, K.; Van Der Schyf, C.; Mabic, S.; Igarashi, N.; Castagnoli, N., Jr. *J. Pharmacol. Exptl. Therap.* **1999**, *291*, 856–864.
173. Fowler, C. J.; Mantle, T. J.; Tipton, K. F. *Biochem. Pharmacol.* **1982**, *31*, 3555–3561.
174. Ma, J.; Kubota, F.; Yoshimura, M.; Yamashita, E.; Nakagawa, A.; Ito, A.; Tsukihara, T. *Acta Cryst.* **2004**, *D60*, 317–319.
175. Binda, C.; Newton-Vinson, P.; Hubalek, F.; Edmondson, D. E.; Mattevi, A. *Nat. Struct. Biol.* **2002**, *9*, 22–26.
176. Knoll, J. *Med. Res. Rev.* **1992**, *12*, 505–524.
177. Tetrud, V. W.; Langston, J. W. *Science.* **1989**, *245*, 519–522.
178. Silverman, R. B. *Acc. Chem. Res.* **1995**, *26*, 335–342.
179. Walker, M. C.; Edmondson, D. E. *Biochemistry.* **1994**, *33*, 7088–7098.
180. Miller, J. R.; Edmondson, D. E.; Grissom, C. B. *J. Am. Chem. Soc.* **1995**, *117*, 7830–7831.
181. Kim, J. M.; Hoegy, S. E.; Mariano, P. S. *J. Am. Chem. Soc.* **1995**, *117*, 100–105.
182. Edmondson, D. E.; Mattevi, A.; Binda, C.; Li, M.; Hubalek, F. *Curr. Med. Chem.* **2004**, *11*, 1983–1993.
183. Castagnoli, N., Jr.; Trevor, A.; Singer, T.; Sparatore, A.; Leung, L.; Shinka, T.; Wu, E.; Booth, R. Central Aspects. In *Progress in Catecholamine Research, Part B: Central Aspects*; Sandler, M., Dahlstrom, A., Eds.; Alan R. Liss: New York, NY, 1988, pp 93–100.
184. Kuttab, S.; Shang, J.; Castagnoli, N., Jr. *Bioorg. Med. Chem.* **2001**, *9*, 1685–1689.
185. Mabic, S.; Castagnoli, N., Jr. *J. Med. Chem.* **1996**, *39*, 3694–3700.
186. Massey, V.; Harris, C. M. *Biochem. Soc. Trans.* **1997**, *25*, 750–757.
187. Harrison, R. *Free Radic. Biol. Med.* **2002**, *33*, 774–797.
188. Garattini, E.; Mendel, R.; Romao, M. J.; Wright, R.; Terao, M. *Biochem. J.* **2003**, *372* (Part 1), 15–32.
189. Linder, N.; Rapola, J.; Raivio, K. O. *Lab. Invest.* **1999**, *79*, 967–974.
190. Martin, H. M.; Moore, K. P.; Bosmans, S.; Davies, S.; Burroughs, A. K.; Dhillon, A. P.; Tosh, D.; Harrison, R. *Free Radical Biol. Med.* **2004**, *37*, 1214–1223.
191. van Meerten, E.; Verweij, J.; Schellens, J. H. *Drug Safety* **1995**, *12*, 168–182.
192. Hille, R. *Chem. Rev.* **1996**, *96*, 2757–2816.
193. Xia, M.; Dempski, R.; Hille, R. *J. Biol. Chem.* **1999**, *274*, 3323–3330.
194. Doonan, C. J.; Stockert, A.; Hille, R.; George, G. N. *J. Am. Chem. Soc.* **2005**, *127*, 4518–4522.
195. Bonini, M. G.; Miyamoto, S.; Di Mascio, P.; Augusto, O. *J. Biol. Chem.* **2004**, *279*, 51836–51843.
196. Huber, R.; Hof, P.; Duarte, R. O.; Moura, J. J. G.; Moura, I.; Liu, M.-Y.; LeGall, J.; Hille, R.; Archer, M.; Romão, M. J. *Proc. Natl. Acad. Sci. USA* **1996**, *93*, 8846–8851.
197. Shaw, S.; Jayatilleke, E. *Biochem. J.* **1990**, *268*, 579–583.
198. Beedham, C. *Drug Metab. Rev.* **1985**, *16*, 119–156.
199. Beedham, C.; Critchley, D. J.; Rance, D. J. *Arch. Biochem. Biophys.* **1995**, *319*, 481–490.
200. Panoutsopoulos, G. I.; Beedham, C. *Acta Biochim. Pol.* **2004**, *51*, 943–951.
201. Rooseboom, M.; Commandeur, J. N.; Vermeulen, N. P. *Pharmacol. Rev.* **2004**, *56*, 53–102.

202. Rashidi, M. R.; Smith, J. A.; Clarke, S. E.; Beedham, C. *Drug Metab. Dispos.* **1997**, *25*, 805–813.
203. Vasiliou, V.; Pappa, A.; Petersen, D. R. *Chem. Biol. Interact.* **2000**, *129*, 1–19.
204. Sladek, N. E.; Tri, N. *J. Biochem. Mol. Toxicol.* **2003**, *17*, 7–23.
205. Vasiliou, V.; Pappa, A.; Estey, T. *Drug Metab. Rev.* **2004**, *36*, 279–299.
206. Duester, G. *Eur. J. Biochem.* **2000**, *267*, 4315–4324.
207. Petersen, E. N. *Acta Psychiatr. Scand. Suppl.* **1992**, *369*, 7–13.
208. Bunting, K. D.; Lindahl, R.; Townsend, A. J. *J. Biol. Chem.* **1994**, *269*, 23197–23203.
209. Nordling, E.; Jönvall, H.; Persson, B. *Eur. J. Biochem.* **2002**, *269*, 4267–4276.
210. Duester, G.; Farres, J.; Felder, M. R.; Holmes, R. S.; Hoog, J. O.; Pares, X.; Plapp, B. V.; Yin, S. J.; Jörnvall, H. *Biochem. Pharmacol.* **1999**, *58*, 389–395.
211. Jörnvall, H.; Hoog, J. O.; Persson, B.; Pares, X. *Pharmacology* **2000**, *61*, 184–191.
212. Crosas, B.; Allali-Hassani, A.; Martinez, S. E.; Martras, S.; Persson, B.; Jörnvall, H.; Pares, X.; Farrés, J. *J. Biol. Chem.* **2000**, *275*, 25180–25187.
213. Nastainczyk, W.; Ahr, H.; Ulrich, V. *Adv. Exp. Med. Biol. A* **1981**, *136*, 799–808.
214. Ahr, H. J.; King, L. J.; Nastainczyk, W.; Ullrich, V. *Biochem. Pharmacol.* **1982**, *31*, 383–390.
215. Nastainczyk, W.; Ahr, H. J.; Ullrich, V. *Biochem. Pharmacol.* **1982**, *31*, 391–396.
216. Mico, B. A.; Pohl, L. R. *Arch. Biochem. Biophys.* **1983**, *225*, 596–609.
217. Manno, M.; De Matteis, F.; King, L. J. *Biochem. Pharmacol.* **1988**, *37*, 1981–1990.
218. Pohl, L. R.; George, J. W. *Biochem. Biophys. Res. Commun.* **1983**, *117*, 367–371.
219. Trudell, J. R.; Bösterling, B.; Trevor, A. J. *Mol. Pharmacol.* **1982**, *21*, 710–717.
220. Patterson, L. H. *Cancer Metastasis Rev.* **1993**, *12*, 119–134.
221. Patterson, L. H. *Drug Metab. Rev.* **2002**, *34*, 581–592.
222. Harada, N.; Omura, T. *J Biochem. (Tokyo)* **1980**, *87*, 1539–1554.
223. Saito, K.; Kamataki, T.; Kato, R. *Cancer Res.* **1984**, *44*, 3169–3173.
224. Tatsumi, K.; Kitamura, S.; Narai, N. *Cancer Res.* **1986**, *46*, 1089–1093.
225. Pan-Zhou, X. R.; Cretton-Scott, E.; Zhou, X. J.; Yang, M. X.; Lasker, J. M.; Sommadossi, J. P. *Biochem. Pharmacol.* **1998**, *55*, 757–766.
226. Clement, B.; Lomb, R.; Moller, W. *J. Biol. Chem.* **1997**, *272*, 19615–19620.
227. Clement, B.; Behrens, D.; Moller, W.; Cashman, J. R. *Chem. Res. Toxicol.* **2000**, *13*, 1037–1045.
228. Clement, B. *Drug Metab. Rev.* **2002**, *34*, 565–579.
229. Gustafsson, D.; Nystrom, J.; Carlsson, S.; Bredberg, U.; Eriksson, U.; Gyzander, E.; Elg, M.; Antonsson, T.; Hoffmann, K.; Ungell, A. et al. *Thromb. Res.* **2001**, *101*, 171–181.
230. Clement, B.; Lopian, K. *Drug Metab. Dispos.* **2003**, *31*, 645–651.
231. Clement, B.; Behrens, D.; Amschler, J.; Matschke, K.; Wolf, S.; Havemeyer, A. *Life Sci.* **2005**, *77*, 205–219.
232. Andersson, S.; Hofmann, Y.; Nordling, A.; Li, X.-Q.; Nevelius, S.; Andersson, T. B.; Ingelman-Sundberg, M.; Johansson, I. *Drug Metab. Dispos.* **2005**, *33*, 570–578.
233. Kitamura, S.; Tatsumi, K. *Biochem. Biophys. Res. Commun.* **1984**, *121*, 749–754.
234. Jez, J. M.; Penning, T. M. *Chem. Biol. Interact.* **2001**, *130–132*, 499–525.
235. Oppermann, U. C.; Maser, E. *Toxicology* **2000**, *144*, 71–81.
236. Penning, T. M. *Hum. Reprod. Update* **2003**, *9*, 193–205.
237. Crosas, B.; Cederlund, E.; Torres, D.; Jörnvall, H.; Farrés, J.; Pares, X. *J. Biol. Chem.* **2000**, *275*, 25180–25187.
238. Burczynski, M. E.; Sridhar, G. R.; Palackal, N. T.; Penning, T. M. *J. Biol. Chem.* **2001**, *276*, 2890–2897.
239. Atalla, A.; Breyer-Pfaff, U.; Maser, E. *Xenobiotica* **2000**, *30*, 755–769.
240. Yamano, S.; Ichinose, F.; Todaka, T.; Toki, S. *Biol. Pharm. Bull.* **1999**, *22*, 1038–1046.
241. O'Connor, T.; Ireland, L. S.; Harrison, D. J.; Hayes, J. D. *Biochem. J.* **1999**, *343* (Part 2), 487–504.
242. Kaliberg, Y.; Oppermann, U.; Jörnvall, H.; Persson, B. *Eur. J. Biochem.* **2002**, *269*, 4409–4417.
243. Filling, C.; Berndt, K. D.; Benach, J.; Knapp, S.; Prozorovski, T.; Nordling, E.; Ladenstein, R.; Jörnvall, H.; Oppermann, U. *J. Biol. Chem.* **2002**, *277*, 25677–25684.
244. Forrest, G. L.; Gonzalez, B. *Chem. Biol. Interact.* **2000**, *129*, 21–40.
245. Rosemond, M. J.; Walsh, J. S. *Drug Metab. Rev.* **2004**, *36*, 335–3361.
246. Merk, H. F.; Jugert, F. K. *Skin Pharmacol.* **1991**, *4*, 95–100.
247. Ohara, H.; Miyabe, Y.; Deyashiki, Y.; Matsuura, K.; Hara, A. *Biochem. Pharmacol.* **1995**, *50*, 221–227.
248. Kaminsky, L. S.; Zhang, Z. Y. *Pharmacol. Ther.* **1997**, *73*, 67–74..

Biography

William F Trager is currently Emeritus Professor of Medicinal Chemistry of the School of Pharmacy, University of Washington, Seattle, Washington. He is a naturalized citizen from Winnipeg, Canada. In 1960, he received his BS in Chemistry from the University of San Francisco, San Francisco, California and in 1965 his PhD in Medicinal Chemistry from the University of Washington under the direction of Professor Alain C Huitric working in the area of conformational analysis using NMR. After receiving his PhD, he spent 2 years as a Postdoctoral Fellow with Professor Arnold H Beckett at the Chelsea School of Science and Technology in London, England. His postdoctoral studies involved the structural analysis of mitragyna alkaloids of unknown structure by spectral means (NMR, CD, IR, and UV).

In 1967 Dr Trager joined the faculty of the School of Pharmacy, University of California at San Francisco as an Assistant Professor of Medicinal Chemistry. While at the University of California he was responsible for the High Resolution Mass Spectral Facility and developed his on-going interest in drug metabolism. In 1972 he returned to the University of Washington as an Associate Professor of Medicinal Chemistry. In 1977 he was promoted to full Professor and received a Career Development Award from the National Institutes of Health. In 1978 spent 3 months at the NIH working in the Laboratory of Chemical Pharmacology with Dr James Gillette. In 1980 he was appointed as Science Advisor to the FDA, Seattle District, Chairman of the Department of Medicinal Chemistry in the School of Pharmacy and Adjunct Professor of Chemistry in the Department of Chemistry. In 1984 he left the Chair and returned to his academic positions. From 1983 to 1987, Dr Trager served as a committee member of the Pharmacological Sciences Review Committee of the National Institutes of General Medical Sciences and in subsequent years he frequently served as an ad hoc member of various study sections and site visit teams.

Dr Trager's primary research interests are in drug metabolism and he was the P.I. of an NIH Program Project Grant for the past 20 years investigating the molecular basis of the phenomena of drug interactions. More specific interests include quantitative mass spectrometry using stable isotopes, stereoselective metabolism, and the use of deuterium isotope effects to study mechanisms of cytochrome P450-catalyzed reactions and active site structure. In 2004 Dr Trager retired from the University and was converted to Emeritus status.

5.06 Principles of Drug Metabolism 2: Hydrolysis and Conjugation Reactions

B Testa, University Hospital Centre, Lausanne, Switzerland

5.06.1 Introduction

Reactions of xenobiotic metabolism are traditionally categorized into phase I and phase II reactions. Such a nomenclature is based on the fact that phase I reactions generally precede phase II, but it conveys no biochemical information whatsoever, and it neglects the very many examples of phase II reactions occurring without prior phase I reactions and even preceding them. This had led a number of authors to use the less euphonic but more informative terms of functionalization and conjugation reactions.[1,2]

Functionalization (phase I) reactions involve the creation of a functional group in the substrate molecule or the transformation of an existing functional group. The majority of such metabolic reactions are oxidations, but numerous

Table 1 Classification of reactions of xenobiotic metabolism according to the chemical entity being transferred to or from the substrate

Functionalizations (phase I)		*Conjugations (phase II)[a]*	
Redox reactions	*Hydrolyses*		
O	H_2O	Methyl group	
O_2	HO^-	Sulfate and phosphate moieties	
e^-		Glucuronic acid and some sugars	
$2 e^-$		Acetyl and other acyl groups	
H^- (hydride)		Following conjugation with coenzyme A	Glycine and other amino acids
			Diglycerides
			Cholesterol and other sterols
			Chiral inversion[b]
			β-Oxidation[b]
			Chain elongation by two-carbon units[b,c]
		Glutathione	
		Acetaldehyde, pyruvic acid, and other carbonyl compounds[c]	
		CO_2[c]	

[a] Except for boxes with the superscript *b*, the right column lists endogenous conjugating moieties (endocons).
[b] Conjugation is the necessary initial step in these metabolic reactions, which cannot be considered as conjugations from the point of view of the final metabolite.
[c] Not discussed in the text.

reactions of reduction are also known. In the simplest and broadest perspective, metabolic redox reactions involve the transfer of one oxygen atom (monooxygenations), two oxygen atoms (dioxygenations), one or two electrons, or a hydride anion (**Table 1**).[3,4] This vast topic of metabolic redox reactions receives comprehensive coverage in Chapter 5.05.

There is, however, an additional class of functionalizations, namely, the reactions of hydrolysis.[4,5] Here, the degree of oxidation of the substrate remain unchanged, hence the term 'nonredox' reactions, which is sometimes found in the literature. A few authors like to consider hydrolyses as conjugation reactions, an unfortunate and confusing viewpoint given that are they fulfil none of the criteria by which conjugations are recognized (see later).

Reactions of conjugations are by far the most varied when considering the chemical entities being transferred to the substrate, or better here, coupled to the substrate. For example, the endogenous conjugating moiety (the 'endocon,' as it has been termed) can be a small and hydrophobic fragment (the methyl group), a large and hydrophobic molecule (cholesterol or a diglyceride), or a large and hydrophilic one (glucuronic acid or glutathione). **Table 1** is offered here to help the reader acquire a synthetic and schematic understanding of the reactions of drug metabolism.

5.06.2   Reactions of Hydrolysis and Hydration

The two terms of hydrolysis and hydration both imply bond breakage with addition of a molecule of H_2O. In this text, we prefer to apply the term 'hydrolysis' to the cleavage of esters (carboxylesters, lactones, inorganic esters), amides (carboxamides, sulfamates, phosphoamides, lactams, etc.), and glycosides. In contrast, the term 'hydration' will be restricted to epoxides, although the enzymes catalyzing this reaction are also classified as hydrolases (see below). Readers in search of a much more extensive treatment of hydrolases and their metabolic reactions of hydrolysis and hydration might want to consult a recent book coauthored by this writer.[5]

5.06.2.1   Hydrolases

The classification adopted by the Nomenclature Committee (NC) of the International Union of Biochemistry and Molecular Biology (IUBMB) classifies all hydrolases as 'EC 3.' and divides them into classes and subclasses according to the group they hydrolyze and the nature of their substrates. A further criterion in the classification of peptidases is based on the nature of their catalytic site. A constantly updated version with supplements is available online,[6] as are all available PDB (Protein Data Bank) entries classified as recommended by the NC-IUBMB.[7]

A small and somewhat subjective selection of hydrolases of known or potential activity toward drugs and other xenobiotics is presented in **Table 2**. The objective of this table is first and foremost to testify to the impressive variety of functional groups that hydrolases can cleave. Also, the table illustrates the huge variety of enzymes that have evolved to cleave chemical bonds by comparatively simple nucleophilic mechanisms necessitating no high-energy electron transfer. But only a limited number of entries are of foremost significance in drug metabolism, namely:

- carboxylesterases (EC 3.1.1.1), whose sequence is encoded by a superfamily of genes (e.g., *CES1A1, CES1A2, CES1A3, CES1B, CES1C, CES2, CES3, CES4*)[5,8,9];
- cholinesterases (EC 3.1.1.8, also known as butyrylcholinesterase);
- paraoxonases (EC 3.1.8.1, also known as aryldialkylphosphatase), whose sequence is encoded by the *PON1, PON2*, and *PON3* genes[10,11];
- epoxide hydrolases (EC 3.3.2.3), to be discussed separately in Section 5.06.2.3; and
- amidases and peptidases in general, although their individual roles in xenobiotic metabolism remain very poorly understood.

The enzymatic hydrolysis of carboxylic derivatives is far more effective than chemical hydrolysis. For example, subtilisin (EU 3.4.21.62) accelerates the hydrolysis of amide bonds at least 10^9- to 10^{10}-fold. All hydrolases include the following three catalytic features in their active site, which enormously accelerate rates of hydrolysis (**Figure 1**). First, they contain an electrophilic component, which increases the polarization of the carbonyl group in the substrate (Z^+ in **Figure 1**). Second, they use a nucleophile (Y: in **Figure 1**) to attack the carbonyl carbon, leading to the formation of a tetrahedral intermediate. And finally, they use a proton donor (H–B in **Figure 1**) to transform the –OR′ or –NR′R′ moiety into a better leaving group.[5,12]

5.06.2.2   Substrates of Hydrolases

5.06.2.2.1   Esters

Several examples of hydrolysis of esters, lactones, and amides are discussed elsewhere in this Volume (*see* 5.44 Prodrug Objectives and Design). To complement this, we present here a few examples of (active) drugs whose metabolism is catalyzed to a large extent by hydrolases and which serve to illustrate some important concepts.

(−)-Cocaine (**1, Figure 2**) has led to unexpected insights into relations between metabolism and molecular toxicology (reviewed in[5]). This xenobiotic is metabolized by various routes including *N*-demethylation (to form norcocaine), *N*-oxygenation, aromatic hydroxylation, and hydrolysis of its two ester groups. The latter route is particularly important since it is globally a route of detoxification that accounts for as much as 90% of the dose in humans, who excrete high urinary concentrations of benzoylecgonine (**2**), ecgonine methyl ester (**3**), and ecgonine (**4**).

Studies using human serum and liver showed enzymatic hydrolysis of the benzoyl ester bridge, whereas the hydrolysis of the methyl ester group occurs both nonenzymatically and enzymatically. Furthermore, marked tissue differences in the formation of benzoylecgonine from cocaine were seen in the rat, with more than one enzyme activity being involved. Three human enzymes are now known to be involved in the hydrolysis of cocaine. One is a liver carboxylesterase (designated hCE-1), which catalyzes the hydrolysis of the methyl ester group. As for the benzoyl ester goup, it is hydrolyzed by the liver carboxylesterase hCE-2 and serum cholinesterase. Natural (−)-(2R;3S)-cocaine (**1**) is a relatively poor substrate of hepatic carboxylesterases and plasma cholinesterase (EC 3.1.1.8), and also a potent competitive inhibitor of the latter enzyme. In contrast, its unnatural enantiomer (+)-(2S;3R)-cocaine is a good substrate of carboxylesterases and cholinesterase. Because hydrolysis is a route of detoxification for cocaine and its stereoisomers, such metabolic differences have a major impact on their monooxygenase-catalyzed toxification, a reaction of particular effectiveness for (−)-cocaine.

There is an additional factor that contributes to the toxicity of cocaine, namely its interaction with ethanol.[13,14] Many cocaine abusers simultaneously ingest ethanol probably to experience a potentiation of effects and a decrease in headaches. It is now known that ethanol interferes in two ways with the metabolism of cocaine, first by inhibiting its hydrolysis and second by allowing a transesterification to form benzoylecgonine ethyl ester (**5, Figure 2**). This reaction of transesterification is also catalyzed by hCE-1. The benzoylecgonine ethyl ester is commonly known as cocaethylene, and it is of clear significance since it retains the pharmacological and toxicological properties of cocaine, in particular its CNS and hepatic effects.

Table 2   A selection of hydrolases (EC 3) with known or potential activity toward xenobiotic substrates[6]

EC 3.1 Acting on ester bonds

EC 3.1.1 Carboxylic ester hydrolases

    EC 3.1.1.1 Carboxylesterase
    EC 3.1.1.2 Arylesterase
    EC 3.1.1.6 Acetylesterase
    EC 3.1.1.7 Acetylcholinesterase
    EC 3.1.1.8 Cholinesterase
    EC 3.1.1.10 Tropinesterase
    EC 3.1.1.13 Sterol esterase
    EC 3.1.1.55 Acetylsalicylate deacetylase
    EC 3.1.1.60 Bis(2-ethylhexyl)phthalate esterase
    EC 3.1.1.67 Fatty-acyl-ethyl-ester synthase

EC 3.1.2 Thiolester hydrolases

    EC 3.1.2.1 Acetyl-CoA hydrolase
    EC 3.1.2.2 Palmitoyl-CoA hydrolase
    EC 3.1.2.3 Succinyl-CoA hydrolase
    EC 3.1.2.7 Glutathione thiolesterase

EC 3.1.3 Phosphoric monoester hydrolases

    EC 3.1.3.1 Alkaline phosphatase
    EC 3.1.3.2 Acid phosphatase

EC 3.1.4 Phosphoric diester hydrolases
EC 3.1.6 Sulfuric ester hydrolases

    EC 3.1.4.1 Phosphodiesterase I
    EC 3.1.6.1 Arylsulfatase
    EC 3.1.6.2 Steryl-sulfatase

EC 3.1.7 Diphosphoric monoester hydrolases

    EC 3.1.7.1 Prenyl-diphosphatase
    EC 3.1.7.3 Monoterpenyl-diphosphatase

EC 3.1.8 Phosphoric triester hydrolases

    EC 3.1.8.1 Paraoxonase
    EC 3.1.8.2 Diisopropyl-fluorophosphatase

EC 3.2 Glycosylases
EC 3.2.1 Glycosidases, i.e., enzymes hydrolyzing O- and S-glycosyl compounds

    EC 3.2.1.20 α-Glucosidase
    EC 3.2.1.21 β-Glucosidase
    EC 3.2.1.31 β-Glucuronidase

EC 3.2.2 Hydrolyzing N-glycosyl compounds

    EC 3.2.2.1 Purine nucleosidase

EC 3.3 Acting on ether bonds
EC 3.3.2 Ether hydrolases

    EC 3.3.2.3 Epoxide hydrolase

EC 3.4 Acting on peptide bonds (peptidases)
EC 3.4.11 Aminopeptidases

    EC 3.4.11.1 Leucyl aminopeptidase
    EC 3.4.11.2 Membrane alanyl aminopeptidase
    EC 3.4.11.3 Cystinyl aminopeptidase
    EC 3.4.11.5 Prolyl aminopeptidase
    EC 3.4.11.6 Arginyl aminopeptidase
    EC 3.4.11.7 Glutamyl aminopeptidase
    EC 3.4.11.15 Lysyl aminopeptidase
    EC 3.4.11.17 Tryptophanyl aminopeptidase
    EC 3.4.11.18 Methionyl aminopeptidase
    EC 3.4.11.19 D-Stereospecific aminopeptidase
    EC 3.4.11.21 Aspartyl aminopeptidase
    EC 3.4.11.22 Aminopeptidase I

EC 3.4.13 Dipeptidases

    EC 3.4.13.18 Cytosol nonspecific dipeptidase
    EC 3.4.13.19 Membrane dipeptidase
    EC 3.4.13.21 Dipeptidase E

EC 3.4.14 Dipeptidyl-peptidases and tripeptidyl-peptidases

    EC 3.4.14.1 Dipeptidyl-peptidase I
    EC 3.4.14.2 Dipeptidyl-peptidase II
    EC 3.4.14.4 Dipeptidyl-peptidase III
    EC 3.4.14.5 Dipeptidyl-peptidase IV

EC 3.4.16 Serine-type carboxypeptidases

    EC 3.4.16.2 Lysosomal Pro-Xaa carboxypeptidase
    EC 3.4.16.5 Carboxypeptidase C
    EC 3.4.16.6 Carboxypeptidase D

EC 3.4.17 Metallocarboxypeptidases

    EC 3.4.17.1 Carboxypeptidase A
    EC 3.4.17.2 Carboxypeptidase B

Table 2 Continued

	EC 3.4.17.10 Carboxypeptidase E
	EC 3.4.17.12 Carboxypeptidase M
	EC 3.4.17.15 Carboxypeptidase A2
	EC 3.4.17.20 Carboxypeptidase U
EC 3.4.18 Cysteine-type carboxypeptidases	EC 3.4.18.1 Cathepsin X
<u>EC 3.4.21 Serine endopeptidases</u>	EC 3.4.21.1 Chymotrypsin
	EC 3.4.21.2 Chymotrypsin C
	EC 3.4.21.4 Trypsin
	EC 3.4.21.7 Plasmin
	EC 3.4.21.35 Tissue kallikrein
	EC 3.4.21.36 Pancreatic elastase
	EC 3.4.21.64 Endopeptidase K
EC 3.4.22 Cysteine endopeptidases	EC 3.4.22.1 Cathepsin B
	EC 3.4.22.16 Cathepsin H
	EC 3.4.22.27 Cathepsin S
	EC 3.4.22.41 Cathepsin F
	EC 3.4.22.42 Cathepsin O
	EC 3.4.22.52 Calpain-1
	EC 3.4.22.53 Calpain-2
EC 3.4.23 Aspartic endopeptidases	EC 3.4.23.1 Pepsin A
	EC 3.4.23.4 Chymosin
	EC 3.4.23.5 Cathepsin D
	EC 3.4.23.34 Cathepsin E
EC 3.4.24 Metalloendopeptidases	EC 3.4.24.11 Neprilysin
	EC 3.4.24.15 Thimet oligopeptidase
	EC 3.4.24.18 Meprin A
	EC 3.4.24.59 Mitochondrial intermediate peptidase
EC 3.4.25 Threonine endopeptidases	EC 3.4.25.1 Proteasome endopeptidase complex

EC 3.5 Acting on carbon–nitrogen bonds, other than peptide bonds

<u>EC 3.5.1 In linear amides</u>	EC 3.5.1.1 Asparaginase
	EC 3.5.1.2 Glutaminase
	EC 3.5.1.3 ω-Amidase
	EC 3.5.1.4 Amidase
	EC 3.5.1.11 Penicillin amidase
	EC 3.5.1.13 Aryl-acylamidase
	EC 3.5.1.14 Aminoacylase
	EC 3.5.1.19 Nicotinamidase
	EC 3.5.1.32 Hippurate hydrolase
	EC 3.5.1.39 Alkylamidase
	EC 3.5.1.49 Formamidase
	EC 3.5.1.50 Pentanamidase
<u>EC 3.5.2 In cyclic amides</u>	EC 3.5.2.1 Barbiturase
	EC 3.5.2.2 Dihydropyrimidinase
	EC 3.5.2.4 Carboxymethylhydantoinase
	EC 3.5.2.6 β-Lactamase
EC 3.5.3 In linear amidines	EC 3.5.3.1 Arginase
EC 3.5.5 In nitriles	EC 3.5.5.1 Nitrilase
	EC 3.5.5.5 Arylacetonitrilase
	EC 3.5.5.7 Aliphatic nitrilase

EC 3.8 Acting on halide bonds

EC 3.8.1 In C-halide compounds	EC 3.8.1.3 Haloacetate dehalogenase
	EC 3.8.1.5 Haloalkane dehalogenase

EC 3.10 Acting on sulfur–nitrogen bonds

	EC 3.10.1.2 Cyclamate sulfohydrolase

Underlined entries are of particular interest in drug metabolism.

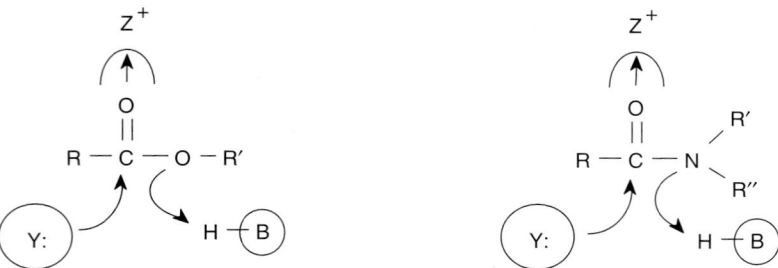

Figure 1 Common catalytic groups in hydrolases involved in ester and amide bond hydrolysis: Z^+ = electrophilic component polarizing the carbonyl group; Y: = nucleophilic group attacking the carbonyl carbon; H–P = proton donor transforming the –OR′ or –NR′R′ moiety into a better leaving group. (Reproduced from Testa, B.; Mayer, J. M. *Hydrolysis in Drug and Prodrug Metabolism – Chemistry, Biochemistry and Enzymology*; Wiley-Verlag Helvetica Chimica Acta: Zurich, Switzerland, 2003, with the kind permission of the copyright owner, Verlag Helvetica Chimica Acta in Zurich.)

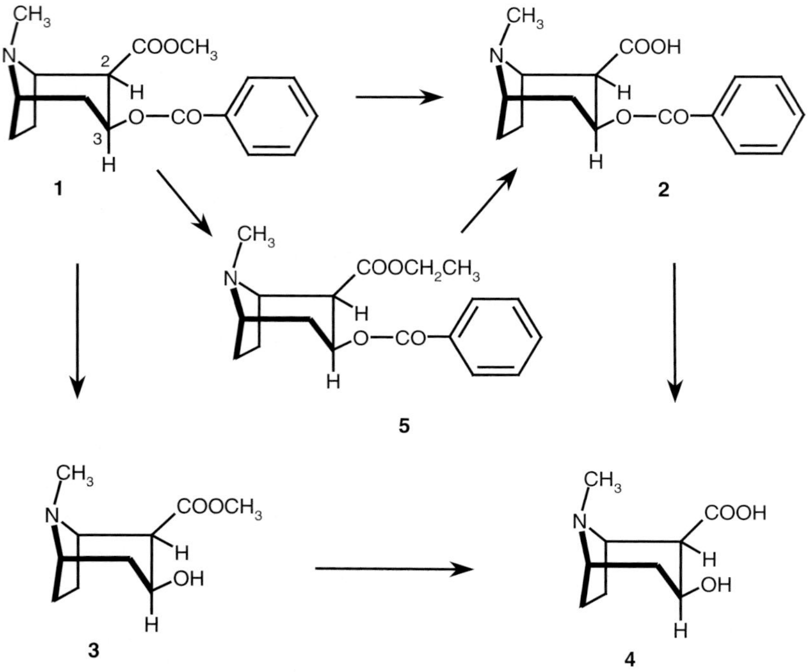

Figure 2 The hydrolytic metabolism of cocaine (**1**) to form benzoylecgonine (**2**), ecgonine methyl ester (**3**), and ecgonine (**4**). In the presence of ethanol, benzoylecgonine ethyl ester (**5**, cocaethylene) is also formed enzymatically as discussed in the text. (Reproduced from Testa, B.; Mayer, J. M. *Hydrolysis in Drug and Prodrug Metabolism – Chemistry, Biochemistry and Enzymology*; Wiley-Verlag Helvetica Chimica Acta: Zurich, Switzerland, 2003, with the kind permission of the copyright owner, Verlag Helvetica Chimica Acta in Zurich.)

5.06.2.2.2 Amides

Lidocaine (**6, Figure 3**) is a local anesthetic and antiarrhythmic drug in wide use since over five decades. The compound is very resistant to chemical hydrolysis even in strongly acidic or basic media. Yet, despite this high chemical stability, metabolic hydrolysis to 2,6-xylidine (**7, Figure 3**) represents a major pathway in mammals, with considerable interspecies differences in its extent. This may be due in part to the fact that *N*-dealkylated and *para*-hydroxylated metabolites of lidocaine also undergo enzymatic hydrolysis, and do so with differing substrate selectivities. In humans, hydrolysis of the amide bond represents the major metabolic pathway, accounting for about 75% of the amount excreted in urine.[5] Investigations using human liver slices have shown that 2,6-xylidine is produced by direct hydrolysis of lidocaine.[15] Furthermore, a liver microsomal carboxylesterase very effectively hydrolyzed lidocaine and its mono-*N*-deethylated metabolite, but not its di-*N*-deethylated metabolite.[16] These findings prove the different substrate selectivities of the oxidized metabolites of lidocaine toward hydrolases. More importantly, the demonstration that carboxylesterases are able to

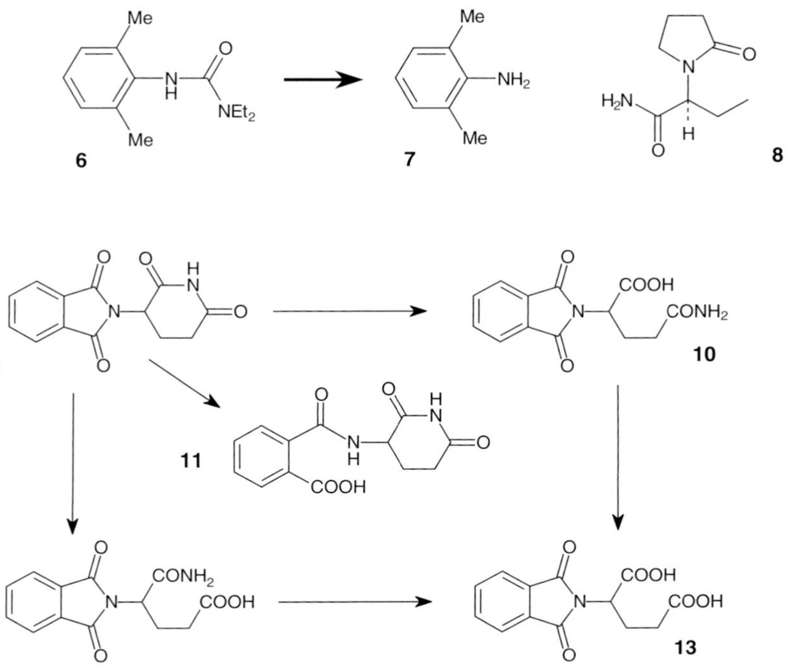

Figure 3 The structure of lidocaine (**6**), 2,6-xylidine (**7**), and levetiracetam (**8**). The figure also shows thalidomide (**9**) and some of its products of spontaneous hydrolysis observed in vivo and in vitro.

efficiently hydrolyze some amides points to the near impossibility of correctly assessing their full range of potential substrates.

Another interesting amide is levetiracetam (**8**, **Figure 3**), a recent antiepileptic agent whose metabolism in humans is almost exclusively CYP-independent (oxidative metabolites represent ⩽ 2.5% of a dose).[17] Indeed, this hydrophilic compound was excreted renally unchanged for two-thirds of the dose, whereas a quarter of the dose was accounted for by its acidic metabolite resulting from hydrolysis of the primary carboxamide group. In vitro investigations have confirmed that the reaction was catalyzed by a serine hydrolase, most probably a carboxylesterase found in human liver, red cells, and most likely other tissues.

Thalidomide (**9**, **Figure 3**) is well known for its high teratogenicity and for causing the greatest drug-related tragedy in history. Recently, new clinical uses have been discovered that render thalidomide useful in alleviating symptoms in lepra and even HIV infections. Although the underlying mechanisms are only poorly understood, some of the activities of thalidomide may be related to its capacity to inhibit the production of tumor necrosis factor alpha (TNF-α).[5,18]

Parallel to rapid inversion of configuration and very low rates of hydroxylations, thalidomide is rapidly hydrolyzed to ring-opened products.[18–21] All four amide bonds of the molecule are susceptible to hydrolytic cleavage at pH > 6, and the reactions are nonenzymatic and base-catalyzed. The main urinary metabolites in humans were shown to be 2-phthalimidoglutaramic acid (**10**, about 50% of a dose) and α-(o-carboxybenzamido)-glutarimide (**11**, about 30% of a dose). Metabolites **12** and **13** were minor. Metabolite **11** was the main product in rats and dogs. Toxicological investigations revealed that of the 12 hydrolysis products of thalidomide only the three containing the intact phthalimido moiety retained teratogenic activity, namely metabolites **10**, **12**, and **13**.

5.06.2.2.3 Peptides

A huge amount of information is known about peptidases,[22] but much remains to be understood regarding their interactions with xenobiotic substrates. This is rendered all the more imperative by an ever increasing interest in therapeutic peptides and peptidomimetics. A number of such peptidases (EC 3.4) are listed in **Table 2**, but this list is based on indirect evidence and hypotheses much more than on solid data. As a reminder, we show in **Figure 4** the classification of peptidases based on the bonds they cleave along the polypeptide chain. Two examples have been selected for presentation here: one (oxytocin) to illustrate the multiplicity of enzymes acting on clinically used natural peptides, and the other (cetrorelix) to hint at the impossibility of predicting the metabolites of a synthetic peptide containing nonproteinogenic acids.

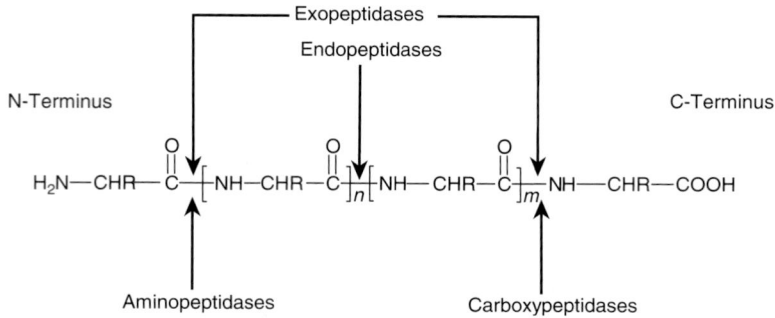

Figure 4 Classification of peptidases according to the bonds they cleave along the polypeptide chain. (Reproduced from Testa, B.; Mayer, J. M. *Hydrolysis in Drug and Prodrug Metabolism – Chemistry, Biochemistry and Enzymology*; Wiley-Verlag Helvetica Chimica Acta: Zurich, Switzerland, 2003, with the kind permission of the copyright owner, Verlag Helvetica Chimica Acta in Zurich.)

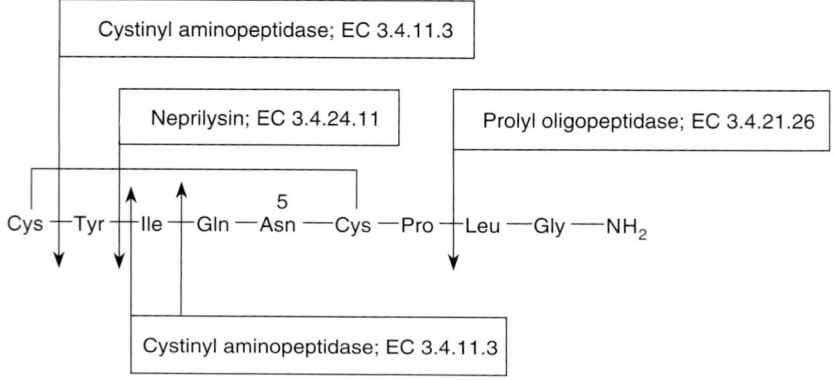

Figure 5 Major peptidases acting on oxytocin. Descending arrows represent primary cleavage reactions, whereas ascending arrows indicate secondary reactions, i.e., cleavage sites in shorter peptides. (Reproduced from Testa, B.; Mayer, J. M. *Hydrolysis in Drug and Prodrug Metabolism – Chemistry, Biochemistry and Enzymology*; Wiley-Verlag Helvetica Chimica Acta: Zurich, Switzerland, 2003, with the kind permission of the copyright owner, Verlag Helvetica Chimica Acta in Zurich.)

Oxytocin (**Figure 5**) is a potent and specific stimulant of myometrial contractions commonly used to induce labor. This peptide is of interest here due to it being a natural peptide and being degraded by a variety of peptidases. Thus, cleavage of the N-terminal cysteine is catalyzed by an aminopeptidase now known as cystinyl aminopeptidase (EC 3.4.11.3; oxytocinase), an enzyme found in the placenta and in the serum of pregnant women.[23] The enzyme acts efficiently to hydrolyze the Cys1-Tyr2 bond, thus opening the ring structure of oxytocin, and then cleaves successive residues from the N-terminal end. At the C-terminus, prolyl oligopeptidase (EC 3.4.21.26; postproline endopeptidase) cleaves the C-terminal dipeptide.[24] The resulting oxytocin-(1–7) is also a substrate for aminopeptidase activity. Furthermore, neprilysin (EC 3.4.24.11) can also play a role in oxytocin degradation, although it seems to act with less efficiency than the two other enzymes.

Our last example here is cetrorelix (**Figure 6**), a potent antagonist of LHRH receptors. The compound is an N- and C-protected decapeptide containing D-Nal, D-(*p*-chloro)Phe, D-(3-pyridyl)Ala, D-Orn and D-Ala as D- or artificial residues. Following subcutaneous administration, the plasma half-life was 35–40 h in rats and 100–130 h in dogs, indicating a remarkable metabolic stability.[25] In both species, only unchanged compound was found in urine, whereas the bile contained up to four metabolites, namely the (1–9)-nonapeptide, the (1–7)-heptapeptide, the (1–6)-hexapeptide, and the (1–4)-tetrapeptide. This indicates that peptidase-related products were the only products identified. Interestingly, the relative abundance of the metabolites changed between bile and feces, indicating additional intestinal breakdown by enteral peptidases or bacteria.

5.06.2.3   Epoxide Hydrolases and Their Reactions

5.06.2.3.1   Epoxide hydrolases

Epoxide hydrolases (see **Table 2**) belong to the ether hydrolases and are a group of related enzymes sharing a similar catalytic mechanism but differing in substrate specificity and biochemical characteristics.[5,26–28] These enzymes are

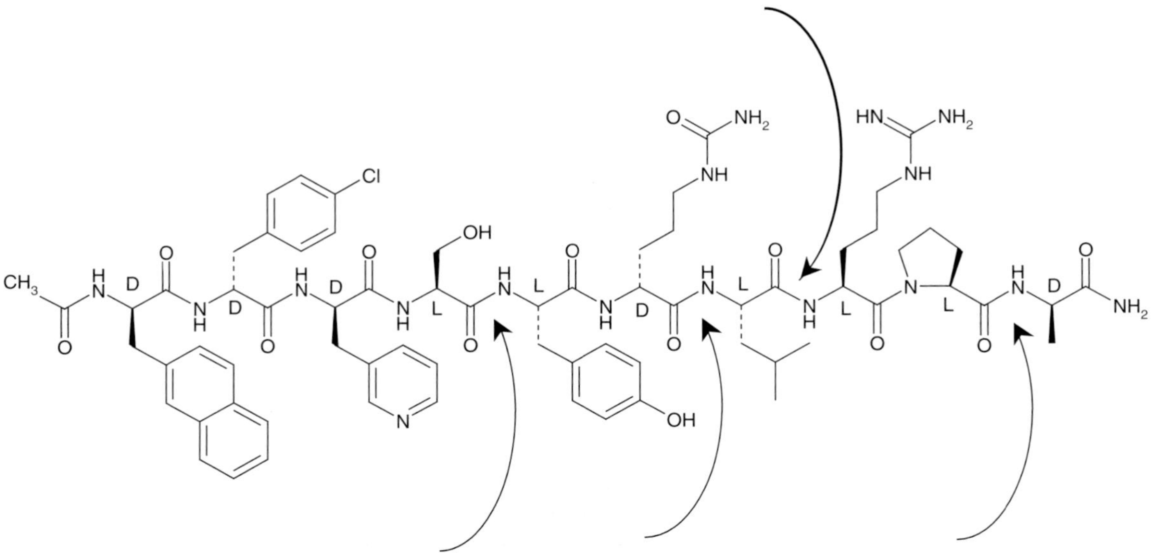

Figure 6 The structure of cetrorelix, an N- and C-protected decapeptide containing D- and artificial residues in addition to natural amino acids. The arrows indicate the sites of attack by peptidases leading to the four identified metabolites.[25]

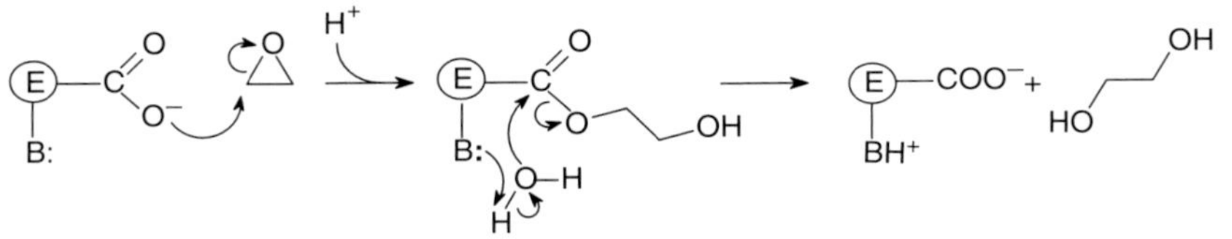

Figure 7 A simplified model showing that the nucleophilic attack of the substrate is mediated by a carboxylate group in the catalytic site to form an ester intermediate. Only in the second step is the intermediate hydrolyzed by an activated water molecule, leading to enzyme reactivation and product liberation. (Reproduced from Testa, B.; Mayer, J. M. *Hydrolysis in Drug and Prodrug Metabolism – Chemistry, Biochemistry and Enzymology*; Wiley-Verlag Helvetica Chimica Acta: Zurich, Switzerland, 2003, with the kind permission of the copyright owner, Verlag Helvetica Chimica Acta in Zurich.)

located in many organs and tissues. They play essential physiological roles, e.g., vitamin K1 oxide reductase, and are also major modifiers of biological activity in the metabolism of xenobiotics. In mammals, epoxide hydrolases (EH) of broad and complementary substrate specificity are the microsomal EH and the soluble EH.

The microsomal epoxide hydrolases (mEH) are predominantly found in the endoplasmic reticulum. They catalyze the hydration of both alkene and arene oxides, including oxides of polycyclic aromatic hydrocarbons, and do so with regio- and stereoselectivity. The human mEH contains 455 amino acids (52.5 kDa) and is the product of the *EPHX1* gene. The human soluble epoxide hydrolase (sEH, also known as cytosolic EH, cEH) has 554 amino acids (62.3 kDa) and is the product of the *EPHX2* gene. Its specific substrate is *trans*-stilbene oxide, and it appears unable to hydrate epoxides of bulky steroids or polycyclic aromatic hydrocarbons.[5]

The overall reaction catalyzed by epoxide hydrolases is the addition of a water molecule to an epoxide. Alkene oxides thus yield diols whereas arene oxides yield dihydrodiols. As shown in **Figure 7**, the nucleophilic attack of the substrate is recognized to be mediated by a catalytic carboxylate group to form an ester intermediate. In a second step, this intermediate is hydrolyzed by an activated water molecule, leading to enzyme reactivation and product liberation. This mechanism involves a catalytic triad consisting of a nucleophile, a general base, and a charge relay acid, in close analogy with many other hydrolases (*see* Section 5.06.2.1).

5.06.2.3.2 Arene oxide substrates

Together with glutathione conjugation, hydration is a major pathway in the inactivation and detoxification of arene oxides. As a rule, these are good substrates of microsomal epoxide hydrolase. But when reading the literature, the variety of intertwined metabolic pathways centered around arene oxides is not always easy to grasp. The general scheme presented in **Figure 8** should contribute to clarify the metabolic context of arene oxides. These arise by CYP-catalyzed oxidation of arenes (*see* 5.05 Principles of Drug Metabolism 1: Redox Reactions) and are a crossroads to a number of metabolic routes, namely: (1) conjugations, (2) adduct formation, (3) proton-catalyzed isomerization to phenols followed by oxidation to diphenols and even to quinones, and (4) EH-catalyzed hydration to *trans*-dihydrodiols. The latter reaction is the topic of this section.

In phenyl and naphthyl rings, the proton-catalyzed isomerization of epoxides to phenols is an extremely fast reaction that markedly reduces the likelihood of the epoxide being hydrated by epoxide hydrolase. This chemical instability decreases for chemicals with three or more fused rings, but such compounds are no longer of medicinal interest and will not be discussed here. Yet, despite the high reactivity of benzene epoxides, the characterization of a dihydrodiol metabolite has been achieved for a limited number of phenyl-containing drugs, and particularly for neurodepressant drugs such as hypnotics (e.g., glutethimide) and antiepileptics (e.g., ethotoin and phenytoin).

For example, phenytoin (diphenylhydantoin) is a good substrate of cytochrome P450. Incubations with rat liver 9000 g supernatant produced the *para*-phenol (4′-hydroxy-phenytoin) as the major metabolite, and the dihydrodiol in smaller proportions. Interestingly, aromatic oxidation of phenytoin in humans and other mammals except the dog occurs almost exclusively at the pro-*S* phenyl ring, the *para*-phenol having the (*S*)-configuration at C5, i.e., (5*S*)-5-(4-hydroxyphenyl)-5-phenylhydantoin. The dihydrodiol similarly has the (*S*)-configuration at its C5, while its C3′ and C4′ atoms in the oxidized phenyl ring have the *trans*-(*R;R*)-configuration, i.e., (5*S*)-5-[(*3R;4R*)-3,4-dihydroxy-1, 5-cyclohexadien-1-yl]-5-phenylhydantoin (**14, Figure 9**) (for a review see[5]).

A more recent example is that of rofecoxib, a potent and selective cyclooxygenase-2 (COX-2) inhibitor. In rats and dogs, phenyl oxidation produced 4′-hydroxy-rofecoxib and rofecoxib-3′,4′-dihydrodiol (**15, Figure 9**) as urinary metabolites of intermediate quantitative importance.[29]

5.06.2.3.3 Alkene oxide substrates

Alkene oxides are generally quite stable chemically, indicating a much reduced chemical reactivity compared to arene oxides. Under physiologically relevant conditions, they have little capacity to undergo rearrangement reactions and are resistant to uncatalyzed hydration. As a result, many alkene oxides formed as metabolites are stable enough to allow their isolation in the absence of degrading enzymes.

Some drugs contain an unconjugated alkene group undergoing CYP-catalyzed epoxidation followed by EH-catalyzed hydration. Thus, the anti-inflammatory agent alclophenac contains an *O*-allyl group. Its epoxide (**16, Figure 9**) was found as a stable metabolite in the urine of mice and humans, and so was the diol, proving the involvement of the epoxide-diol pathway in the metabolism of this drug. The epoxide proved mutagenic, but only in the absence of a rat liver S-9 suspension (which contains EH).[30] A few old barbiturates also contain an allylic group in the 5-position, e.g., allobarbital (**17, Figure 9**; R = allyl), and secobarbital (**17, Figure 9**; R = 1-methylbutyl). These compounds were substrates of the epoxide-diol pathway in rats and guinea pigs. The relative importance of this metabolic route was species dependent, but was also markedly influenced by the nature of the other C5-substituent.[31]

Cycloalkenes are found particularly among natural products (i.e., terpenoids) and as such are of medicinal interest. Limonene (**18, Figure 9**) undergoes epoxidation to the C1-C2 and C8-C9 double bonds. The two epoxide groups are then hydrated by epoxide hydrolase, but at a different rate. Indeed, incubations in rat liver microsomes showed that the hydrolysis of limonene 1,2-epoxide was 70 times slower than that of the 8,9-epoxide, a result explained by steric hindrance on the basis of these and other findings.[32]

A number of neurotropic agents contain a conjugated alkene group incorporated in an iminostilbene (**19, Figure 9**; X = >NR) or dibenzosuberene (**19, Figure 9**; X = >CHR or >C=CHR) ring system. Examples include the anticonvulsant carbamazepine and the antidepressants protriptyline and cyclobenzaprine. As a rule, these drugs are oxidized by cytochrome P450 to the corresponding epoxide (**20, Figure 9**), but hydration to the dihydrodiol (**21**) is usually low for reasons of unfavorable positioning in the catalytic site (for a review see[5]).

As a result, related tricyclic drugs yield modest or very low proportions of dihydrodiols, despite the 10,11-oxides being consistently formed. For example, both the epoxide and the dihydrodiol were characterized in the urine of rats given protriptyline (**19**, in **Figure 9**; X = >CHCH₂CH₂CH₂NHMe), whereas cyclobenzaprine (**19, Figure 9**; X = >C=CHCH₂CH₂NMe₂) did not yield the dihydrodiol despite the epoxide and other oxygenated metabolites being formed in vivo and in vitro.[5]

Carbamazepine (**19, Figure 9**; X = >NCONH₂) is a major antiepileptic drug whose metabolism has interested biochemists and clinical pharmacologists for many years. Well over 30 metabolites of this drug have been characterized

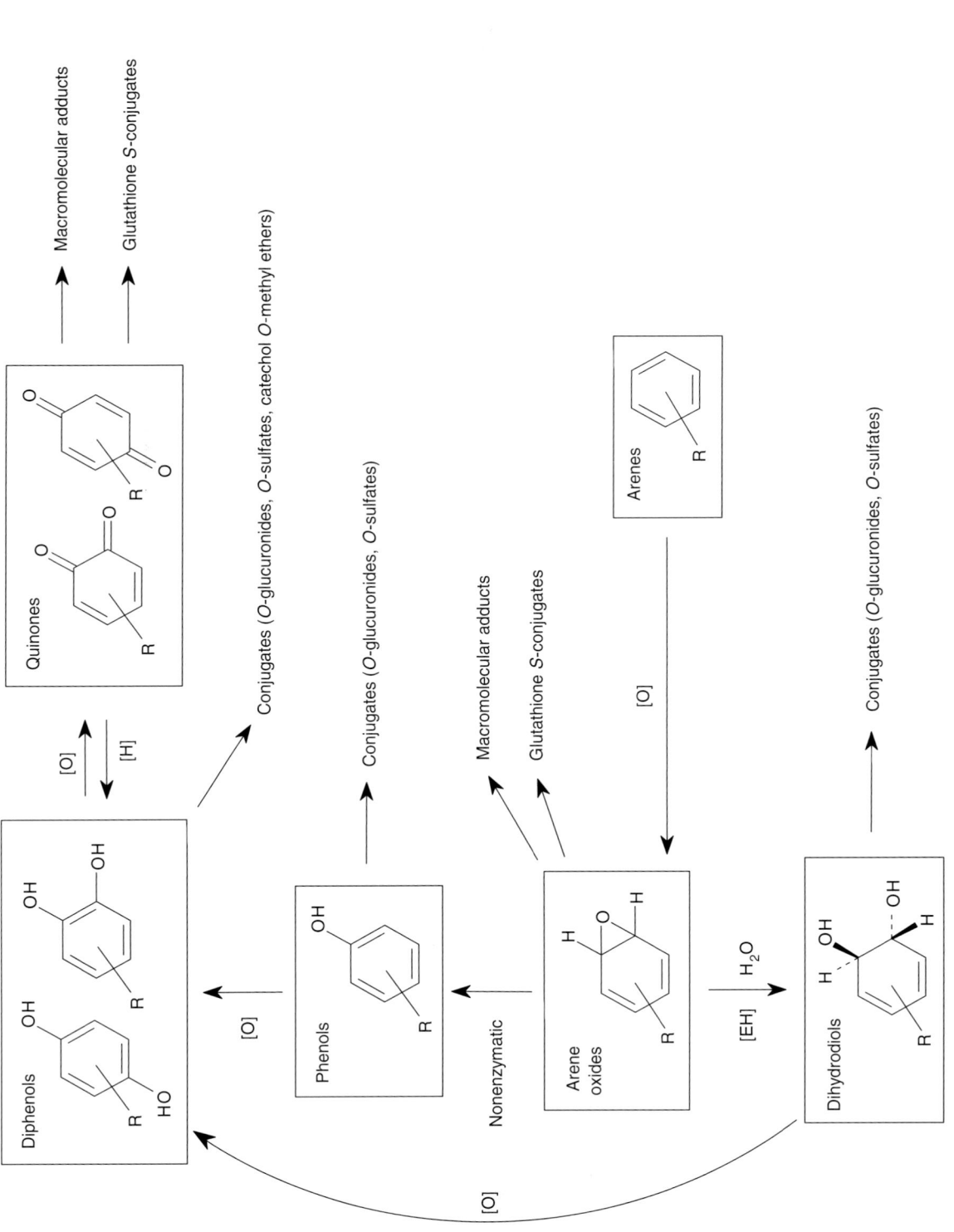

Figure 8 Metabolic reactions centered around arene oxides. (Reproduced from Testa, B.; Mayer, J. M. *Hydrolysis in Drug and Prodrug Metabolism – Chemistry, Biochemistry and Enzymology*; Wiley-Verlag Helvetica Chimica Acta: Zurich, Switzerland, 2003, with the kind permission of the copyright owner, Verlag Helvetica Chimica Acta in Zurich.)

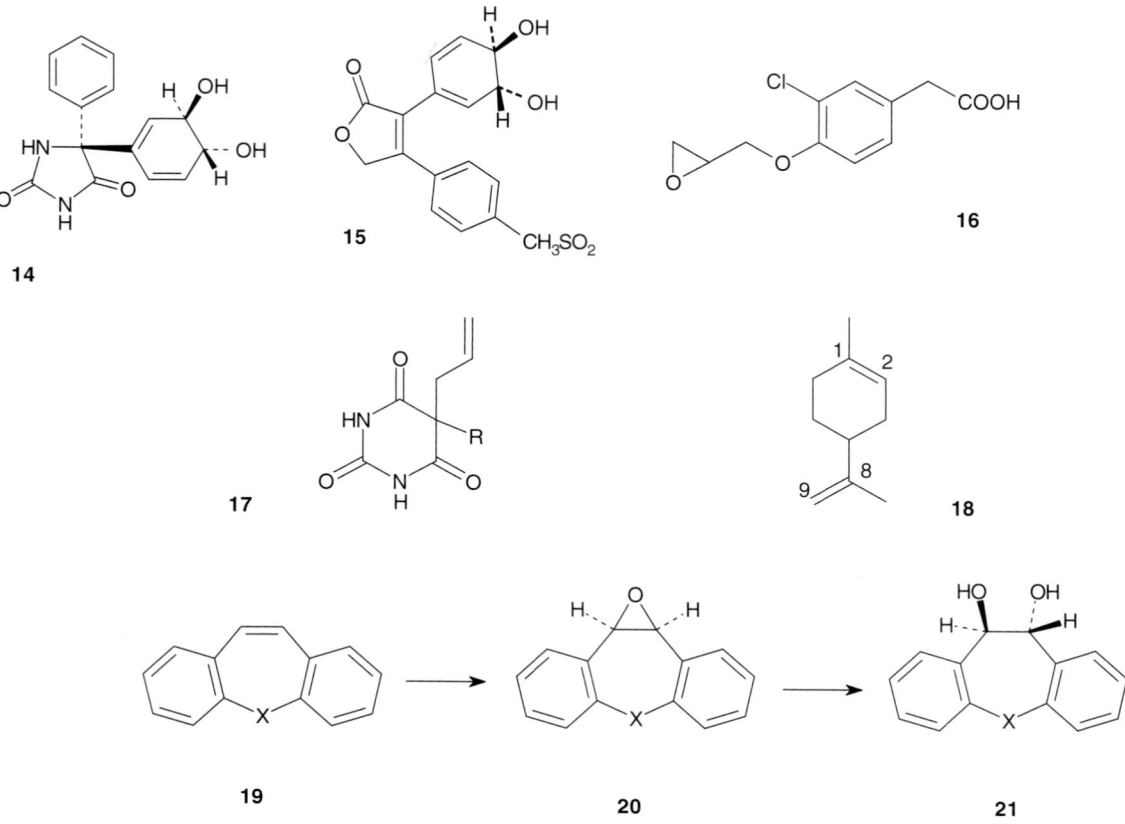

Figure 9 Alkenes, epoxides, and dihydrodiols of medicinal interest discussed in the text.

in vivo and/or in vitro. Thus, the 10,11-epoxide and the 10,11-dihydrodiol are urinary metabolites in humans and rats given the drug. In epileptic patients, the range of plasma concentrations of the epoxide and the diol was approximately 0.8–17 μM and 0.8–36 μM, respectively, i.e., a predominance of the latter.[33] A number of studies have also addressed the origin of toxic reactions seen in some patients, e.g., CNS symptoms, gastrointestinal and hepatic disturbances, and hypersensitivity. Whereas no single factor seems to account for such toxic effects, the pharmacologically active 10,11-epoxide appears to contribute to clinical toxicity.[34] In this perspective, the EH-catalyzed hydrolysis of the epoxide appears as a reaction of detoxification.

Interestingly, there is a marked species difference in the in vitro hydrolysis of carbamazepine 10,11-epoxide, such that the reaction was observable only in human liver microsomes but not in liver microsomal or cytosolic preparations from dogs, rabbits, hamsters, rats, or mice.[35] Thus, carbamazepine appears to be a very poor substrate of epoxide hydrolases, in analogy with its simpler analogs (**19**, **Figure 9**; X = >NR, >CHR or >C=CHR). The human enzyme is exceptional in this respect.

5.06.3   Reactions of Conjugation

The reactions of conjugation listed in **Table 1** are of different quantitative importance in drug metabolism, some involving a very large number of xenobiotic substrates (e.g., glucuronidation) and others a comparatively modest number (e.g., methylation). But their toxicological significance does not necessarily parallel the number of their known substrates, some reactions being particularly effective in quenching reactive intermediates whereas others may lead to toxic conjugates. Before examining in turn major and lesser reactions of conjugation, it appears useful to discuss the criteria that allow conjugation reactions to be recognized as such.[4,36–40]

5.06.3.1   Criteria of Conjugation

Conjugation reactions are characterized by a number of criteria, namely:

1. The substrate is coupled to an endogenous conjugating molecule or moiety (sometimes called an 'endocon'), producing a metabolite known as a conjugate.
2. This endogenous molecule or moiety is highly polar (hydrophilic) and has a molecular weight (MW) in the 100–300 Da range.
3. Conjugation reactions are catalyzed by enzymes known as transferases.
4. They involve a cofactor that binds to the enzyme in close proximity to the substrate and carries the endogenous molecule or moiety to be transferred.

The first criterion is an essential one, and it is the criterion of decision. The other criteria are neither sufficient nor necessary to define conjugation reactions. They are not sufficient, since for example in hydrogenation reactions (i.e., typical reactions of functionalization), the hydride is also transferred from a cofactor (NADPH or NADH). And they are not necessary, since they show some important exceptions, which will become apparent below.

5.06.3.2   Reactions of Methylation

5.06.3.2.1   Mechanism and multiplicity of methyltransferases

Reactions of methylation imply the transfer of a methyl group from the cofactor S-adenosyl-L-methionine (**22, Figure 10**; SAM) to the substrate It is important to note that the methyl group in SAM is bound to a sulfonium center, giving it a marked electrophilic character and explaining its reactivity. Furthermore, it becomes pharmacokinetically relevant to distinguish methylated metabolites in which the positive charge has been retained or lost.

A number of methyltransferases are able to methylate small molecules.[41–43] Thus, reactions of methylation fulfil only three of the four criteria defined above, since the methyl group is small compared to the substrate.

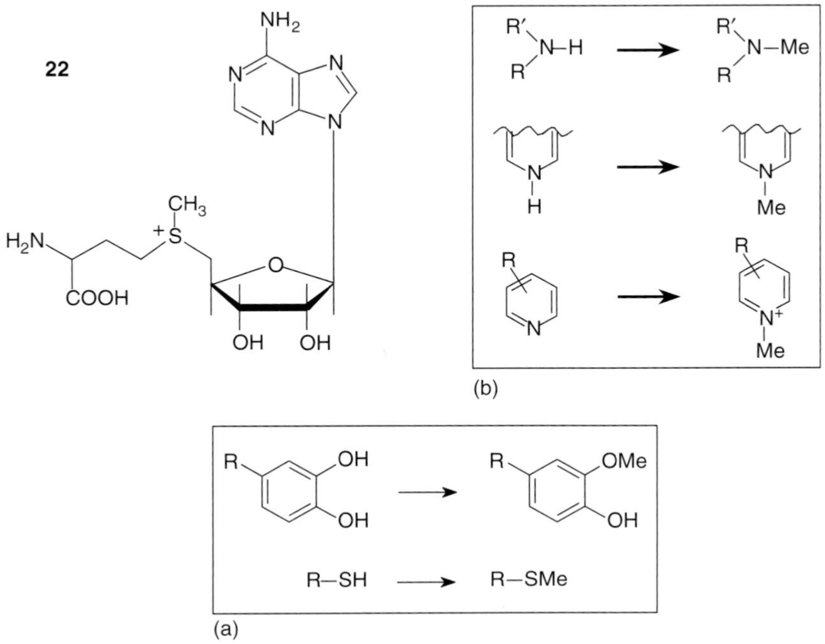

Figure 10 Structure of the cofactor S-adenosyl-L-methionine (**22**; SAM); generic reactions of O- and S-methylation (a) and of N-methylation (b).

Methyltransferases involved in xenobiotic metabolism include:

- the main enzyme responsible for *O*-methylation, namely catechol *O*-methyltransferase (EC 2.1.1.6; COMT), which is mainly cytosolic but also exists in membrane-bound form;
- various enzymes involved in *N*-methylation and characterized by different substrate specificities, e.g., nicotinamide *N*-methyltransferase (EC 2.1.1.1), histamine *N*-methyltransferase (EC 2.1.1.8), phenylethanolamine *N*-methyltransferase (EC 2.1.1.28), and a nonspecific arylamine *N*-methyltransferase (EC 2.1.1.49); and
- enzymes catalyzing *S*-methylation, namely the membrane-bound thiol methyltransferase (EC 2.1.1.9) and the cytosolic thiopurine methyltransferase (EC 2.1.1.67).

This classification makes explicit the three types of functionalities undergoing biomethylation, namely hydroxy (phenolic), amino, and thiol groups.

5.06.3.2.2 Substrates of methyltransferases

Figure 10 also shows the main methylation reactions seen in drug metabolism. *O*-Methylations (**Figure 10a**) are common reactions of compounds containing a catechol moiety, with a usual regioselectivity for the *meta* position. The substrates can be xenobiotics and particularly drugs, L-DOPA being a classic example. More frequently, however, *O*-methylation occurs as a late event in the metabolism of aryl groups, after they have been oxidized to catechols (see **Figure 8**). This sequence was seen, for example, in the metabolism of the anti-inflammatory drug diclofenac, which in humans yielded 3'-hydroxy-4'-methoxy-diclofenac (**23**, **Figure 11**) as a major metabolite with a very long plasma half-life.[44] The rates of *O*-methylation of about 50 substrates in recombinant human soluble COMT has been published and analyzed by partial least squares (PLS) QSAR and 3D-QSAR.[45,46] The compounds examined were natural products. The results showed that increased acidity of the catechol group and a larger size of the adjacent substituents strongly decreased the rate of methylation.

S-Methylation of thiol groups (**Figure 10a**) is documented for such drugs as 6-mercaptopurine (**24**, **Figure 11**) and captopril. More recently, it has been shown to be one of the major routes of human metabolism of the vasopeptidase inhibitor omapatrilat (**25**, **Figure 11**).[47] Other substrates are metabolites (mainly thiophenols) resulting from the S-C cleavage of (aromatic) glutathione and cysteine conjugates (see later). Once formed, such methylthio metabolites can be further processed to sulfoxides and sulfones before excretion (*see* 5.05 Principles of Drug Metabolism 1: Redox Reactions).

Three basic types of *N*-methylation reaction have been recognized (**Figure 10b**). A number of primary amines (e.g., amphetamine) have been shown to be in vitro substrates of amine *N*-methyltransferase. The same is true of some

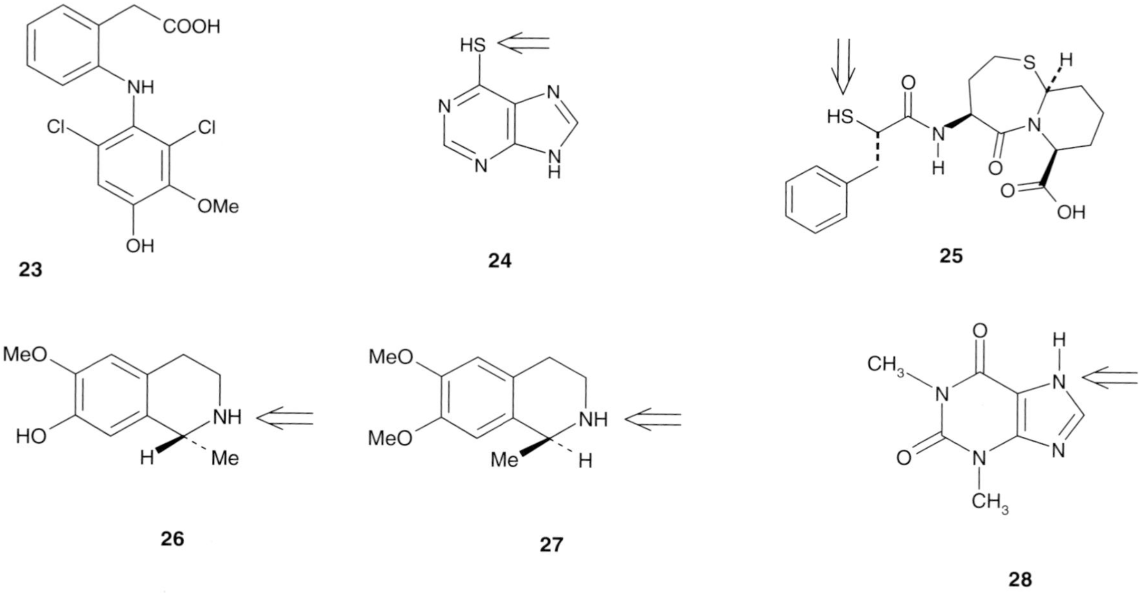

Figure 11 Compounds discussed in the text in connection with reactions of O-, S-, and N-methylation.

secondary amines such as tetrahydroisoquinolines produced in plants and animals. Examples of tetrahydroisoquinolines are isosalsoline and salsolidine; their N-methylation by bovine liver amine N-methyltransferase has been shown to be substrate enantioselective, the preferred substrates being $(+)$-(R)-isosalsoline and $(-)$-(S)-salsolidine (**26** and **27**, respectively, in **Figure 11**).[48] Some phenylethanolamines and analogs are also N-methylated, in this case by phenylethanolamine N-methyltransferase. However, such reactions are seldom of quantitative significance in vivo, presumably due to effective oxidative N-demethylation. A comparable situation involves the N-H group in an imidazole ring, as exemplified by histamine.[49] A therapeutically relevant example is that of theophylline (**28**, **Figure 11**) whose reaction of $N(9)$-methylation to caffeine is very effective in newborn humans with the danger of caffeine intoxication, whereas in adults it is masked by a comparatively fast N-demethylation (futile cycling).

N-Methylation of pyridine-type nitrogen atoms appears to be of greater in vivo pharmacological significance than the two other types of N-methylation, and this is for two reasons. First, the resulting metabolites, being quaternary amines, are more stable than tertiary or secondary amines toward N-demethylation. Second, these metabolites are also more polar than the parent compounds, in contrast to all other products of methylation. Good substrates are pyridine itself, whose N-methylation is a genuine reaction of detoxification, nicotinamide, and a number of monocyclic and bicyclic derivatives.[49]

From **Figure 10**, methylation reactions can clearly be subdivided into two classes:

1. those where the substrate and the product have the same electrical state, a proton in the substrate having been exchanged for a positively charged methyl group; and
2. those where the product has acquired a positive charge, namely has become a pyridine-type quaternary ammonium.

Looking back at **Table 1** and at the criteria discussed above for conjugation reactions, it is apparent that the methyl group being transferred is the smallest of all endocons. Furthermore, and with the exception of pyridine-type substrates, the metabolites formed by methylation are less hydrophilic than the corresponding substrates.

5.06.3.3   Formation of Sulfate and Phosphate Conjugates

5.06.3.3.1   Mechanism and multiplicity of sulfotransferases

Sulfation reactions consist in a sulfate being transferred from the cofactor 3′-phosphoadenosine 5′-phosphosulfate (**29**, **Figure 12**; PAPS) to the substrate under catalysis by a sulfotransferase. All criteria of conjugation are met in these reactions, since they are enzymatic and the moiety transferred (sulfuric acid or rather sulfate) is of medium MW, ionized and highly hydrophilic, and is carried by a coenzyme.

In PAPS, the sulfate and phosphate moieties are linked by an anhydride bond whose cleavage is exothermic and supplies enthalpy to the catalytic reaction. The nucleophilic hydroxy group (phenols, alcohols, and hydroxylamines, **Figure 12a**) or primary or secondary amino groups (**Figure 12b**) in the substrate will react with the leaving SO_3^- moiety, forming an ester sulfate or a sulfamate (**Figure 12**).[50] Some of these conjugates are unstable under biological conditions and may undergo heterolytic cleavage to form electrophilic intermediates of considerable toxicological significance, as discussed later.

Sulfotransferases involved in the metabolism of small endogenous and exogenous molecules are soluble (cytosolic) enzymes. Following major advances in molecular biology, they are now recognized as being encoded by a gene superfamily of which about 50 mammalian genes are known and whose products are classified into families ($>45\%$ residue identity) and subfamilies ($>60\%$ residue identity) according to their degree of homology. Thus, human sulfotransferases include:

- the SULT1A subfamily, which contains the enzymes 1A1, 1A2, and 1A3 (phenol[= aryl] sulfotransferases, with some correspondence with EC 2.8.2.1);
- the subfamily SULT1B with the enzyme 1B1 (thyroid hormone sulfotransferase);
- the subfamily SULT1C with the enzymes 1C1 and 1C2;
- the subfamily SULT1E with 1E1 (estrogen sulfotransferase, EC 2.8.2.4);
- the subfamily SULT2A with 2A1 (alcohol[= hydroxysteroid] sulfotransferase, EC 2.8.2.2);
- the subfamily SULT2B with the two transcript variants 2B1a and 2B1b; and
- the family SULF4 with 4A1.[51–60]

A number of other sulfotransferase activities are recognized by the NC-IUBMB, including steroid sulfotransferase (EC 2.8.2.15) and cortisol sulfotransferase (glucocorticosteroid sulfotransferase; EC 2.8.2.18). An important family is

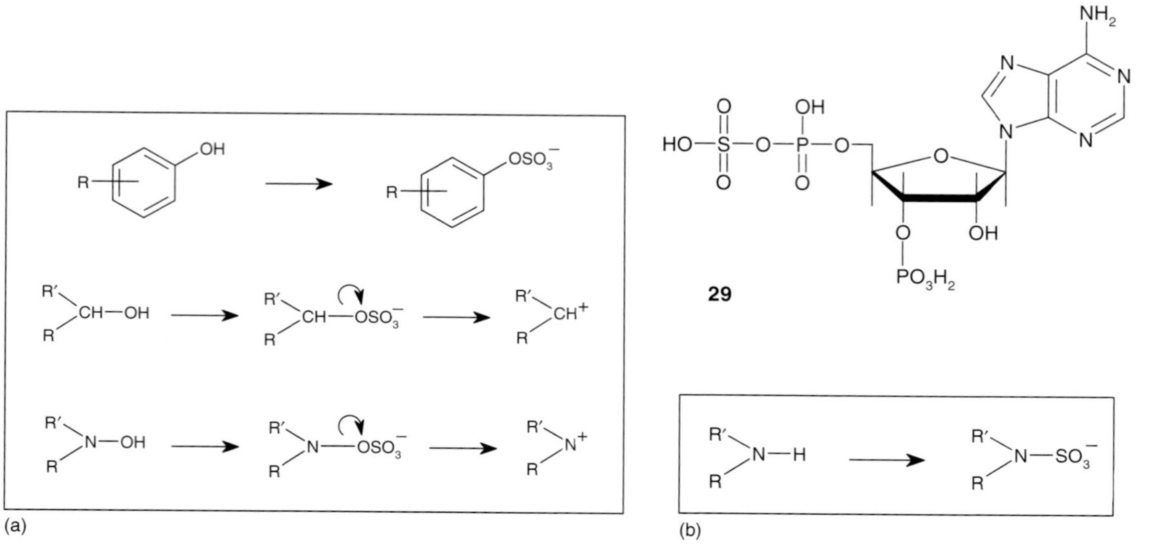

Figure 12 Structure of the cofactor 3′-phosphoadenosine 5′-phosphosulfate (**29**; PAPS); generic reactions of *N*-sulfation (b) and of *O*-sulfation (a).

SULT3, which is involved in amine sulfation and should correspond to amine sulfotransferase (EC 2.8.2.3). However, there are marked and condition-dependent overlaps in the substrate specificity of all these sulfotransferases, a fact that prevents any strong and one-to-one correspondence between the IUBMB and the homology-based nomenclatures. Nevertheless, it is well established that the enzymes of greatest significance in the sulfation of xenobiotics are the aryl sulfotransferases, the alcohol sulfotransferases, and the amine sulfotransferases.

5.06.3.3.2 Substrates of sulfotransferases

Sulfotransferases act on a very large variety of compounds bearing a phenolic, alcoholic, hydroxylamino, or amino function. These substrates include endogenous compounds (e.g., catecholamines, steroids, and bile acids), dietary constituents (e.g., flavonoids), procarcinogens (e.g., benzylic alcohols and heterocyclic aromatic amines, see below), and drugs.

Phenols form stable aryl sulfate esters (**Figure 12a**). The reaction is usually of high affinity (i.e., rapid), but the limited availability of PAPS restricts the amounts of conjugate being produced. Given that phenols are also substrates of glucuronyltransferases (*see* Section 5.06.3.4), a competition for the two reactions is frequently seen, with the high-affinity but low-capacity sulfation predominating at lower doses, and the lower affinity but higher capacity glucuronidation predominating at higher doses. This typical situation is aptly exemplified by contrasting the relative sulfation and glucuronidation of the high-dose drug paracetamol (**30**, **Figure 13**), whose major metabolite is an aryl glucuronide, with those of the low-dose drug troglitazone (**31**), whose major metabolite in human plasma is the aryl sulfate rather than the aryl glucuronide.[61] Sulfation, like glucuronidation (discussed in Section 5.06.3.4), is also a route of significance in the conjugation of phenols generated as phase I metabolites of aromatic drugs, e.g., 4-hydroxypropranol.[62]

Quantitative structure–activity relationship (QSAR) investigations based on sulfation rates obtained with recombinant human SULT enzymes have uncovered some valuable trends for SULT1A3.[46] Factors favoring the rate of sulfation were, for example, lipophilicity and a basic amino group or a zwitterionic nature of the compound; unfavorable factors included a carboxy group and the number of H-bond acceptor groups.

The sulfoconjugation of alcohols (**Figure 12a**) leads to metabolites of different stabilities. Endogenous hydroxysteroids (i.e., cyclic secondary alcohols) form relatively stable sulfates, while some secondary alcohol metabolites of allylbenzenes (e.g., safrole and estragole) form highly genotoxic carbocations.[63] Primary alcohols, e.g., methanol and ethanol, can also form sulfates whose alkylating capacity is well known and results from the heterolytic cleavage of the C–O bond. Similarly, polycyclic hydroxymethylarenes yield reactive sulfates believed to account for their mutagenicity and carcinogenicity.[52] The reaction is not without medicinal relevance, since, for example, the sulfate ester (**32**, **Figure 13**) of α-hydroxytamoxifen, a well-known metabolite of the antiestrogen tamoxifen, has been

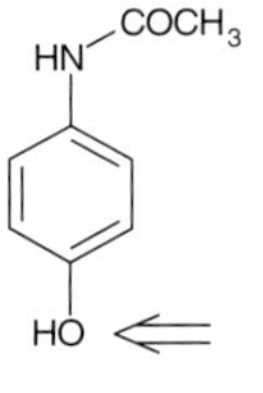

Figure 13 Compounds discussed in the text in connection with reactions of *O*- and *N*-sulfation.

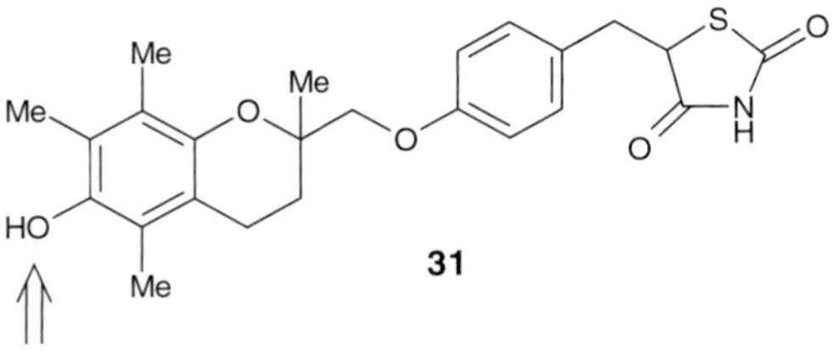

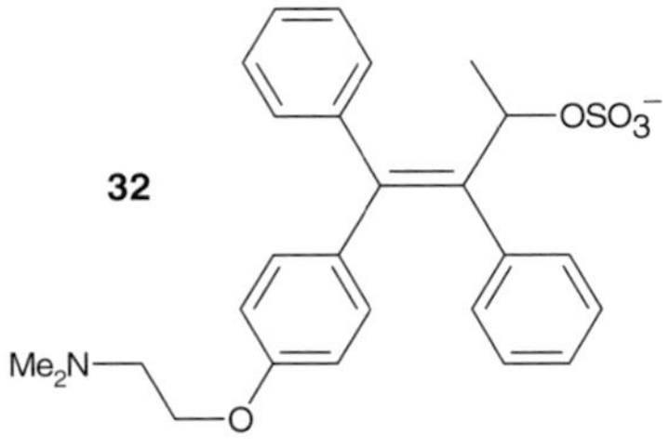

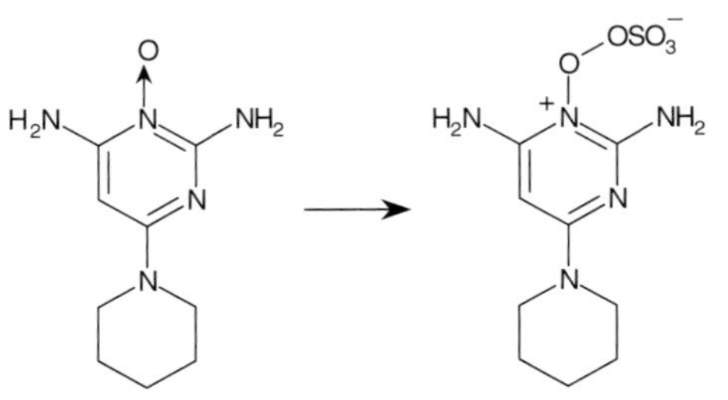

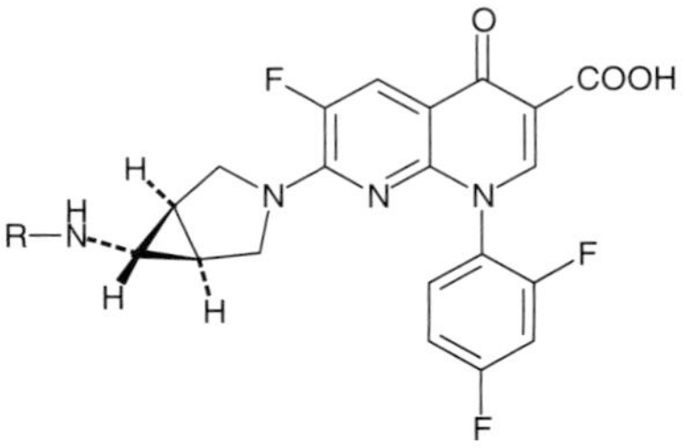

shown to be highly reactive toward 2'-deoxyguanosine.[64] While this study only afforded indirect evidence, its results become meaningful in view of the propensity of tamoxifen to induce endometrial cancers in tumor patients.

Aromatic hydroxylamines and hydroxylamides are good substrates for some sulfotransferases and yield unstable sulfate esters (**Figure 12a**). Indeed, heterolytic N–O cleavage produces a highly electrophilic nitrenium ion. This is a mechanism believed to account for part or all of the cytotoxicity of arylamines and arylamides (e.g., phenacetin). An intriguing and very rare reaction of conjugation occurs for minoxidil (**33**, **Figure 13**), a hypotensive agent also producing hair growth. This drug is an *N*-oxide, and the actual active form responsible for the different therapeutic effects is the stable *N-O*-sulfate ester (**34**).[65]

In contrast to the unstable hydroxylamine sulfates, significantly more stable products are obtained upon *N*-sulfoconjugation of amines (**Figure 12b**). Some alicyclic amines, and primary and secondary alkyl- and arylamines can all yield sulfamates in the presence of human SULT2A1.[66] However, the clinical significance of these reactions in humans remains poorly understood.

One medicinally relevant example is that of trovafloxacin (**35**, **Figure 13**; R = H), a quinolone antibacterial agent. Human volunteers who had been given the drug excreted it partly unchanged and partly as three major conjugates (*see* Section 5.06.3.5).[67] One of these was the sulfamate (**35**, **Figure 13**; R = SO₃H), which accounted for about 10% of the dose and was excreted fecally, indicating its stability against biodegradation by the gut microflora.

5.06.3.3.3 Formation of phosphate conjugates

Phosphate conjugates are rare compared to sulfates, yet they are of primary significance in the metabolism of anticancer and antiviral agents impacting on endogenous nucleotides. Indeed, phosphorylation is an essential metabolic step in the bioactivity of these agents, and numerous in vitro and in vivo studies document their stepwise phosphorylation to mono-, di-, and triphosphates. Such reactions are sometimes, and correctly, labeled as anabolic (i.e., biosynthetic) ones.[68,69] They are known or postulated to be catalyzed by some among the very many phosphotransferases (EC 2.7), for example adenosine kinase (EC 2.7.1.20), thymidine kinase (2.7.1.21), uridine kinase (2.7.1.48), deoxycytidine kinase (EC 2.7.1.74), deoxyadenosine kinase (EC 2.7.1.76), nucleoside phosphotransferase (EC 2.7.1.77), creatine kinase (EC 2.7.3.2), adenylate kinase (EC 2.7.4.3), nucleoside-phosphate kinase (EC 2.7.4.4), guanylate kinase (EC 2.7.4.8), and (deoxy)nucleoside-phosphate kinase (EC 2.7.4.13).

A therapeutically relevant example is afforded by the well-known anti-HIV agent zidovudine (AZT) (**36, Figure 14**). The concentrations of its phosphate anabolites were measured in the peripheral blood mononuclear cells of AIDS patients treated with the drug.[70] The monophosphate was the predominant compound, and the diphosphate and triphosphate were present in comparable amounts.

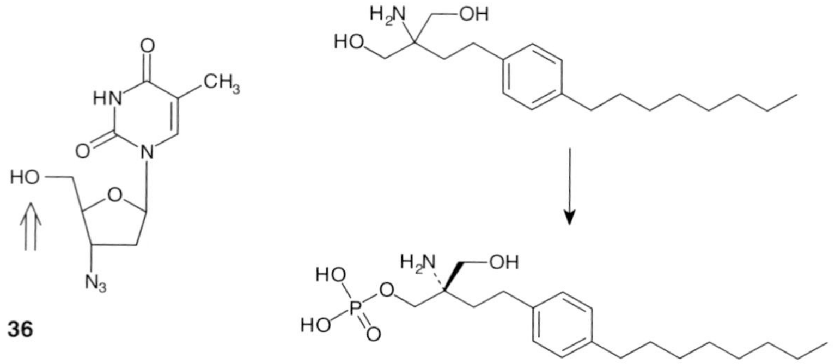

Figure 14 Compounds discussed in the text in connection with reactions of phosphorylation.

The unexpected (and mostly unexplored) activity of phosphotransferases toward xenobiotic substrates is forcefully illustrated by the recently reported activation of FTY720 (**37, Figure 14**), a novel immunomodulator showing great promise in transplantations and to treat autoimmunity.[71] The agent itself appears inactive, being phosphorylated in rats and humans to the active monophosphate. Recent studies have implicated spingosine kinases as the catalysts and have shown that the reaction is highly product enantioselective. Indeed, FTY720 itself is prochiral (it bears two enantiotopic –CH₂OH groups), and the enzymatic reaction results exclusively in the phosphorylated enantiomer of (S)-configuration (**38**), which is also the only active one.

5.06.3.4   Conjugations with Glucuronic Acid

Glucuronidation is considered by many to be the most important reaction of conjugation; qualitatively for the diversity of functional groups to which glucuronic acid can be coupled, and quantitatively for the vast number and diversity of its substrates.

5.06.3.4.1   Mechanism and multiplicity of UDP-glucuronyltransferases

Glucuronidation consists in a molecule of glucuronic acid being transferred from the cofactor uridine-5′-diphospho-α-D-glucuronic acid (**39, Figure 15**; UDPGA) to the substrate. This cofactor is produced endogenously by C6 oxidation of UDP-α-D-glucose, and it is recognized that about 5 g are synthesized daily in the adult human body, hence the high capacity of this metabolic route.

Glucuronic acid exists in UDPGA in the 1α-configuration, but the products of conjugation are β-glucuronides (**40, Figure 15**). This is due to the mechanism of the reaction being a nucleophilic substitution with inversion of configuration. Indeed, all functional groups able to undergo glucuronidation are nucleophiles, a common characteristic they share despite their great chemical variety. As a consequence of this diversity, the products of glucuronidation are classified as O-, N-, S-, and C-glucuronides. Practically all functional groups that are glucuronidated are shown in **Figure 15**. The hydroxy groups (phenols, alcohols, carboxylic acids, carbamic acids, hydroxylamines, and hydroxylamides) form O-glucuronides and are grouped in **Figure 15a** together with strongly acidic enolic acids, a very few of which are known to form C-glucuronides. The N-glucuronides are grouped in **Figure 15b** and are generated from amides, sulfonamides, aromatic amines, heterocyclic amines, and aliphatic amines; also shown in **Figure 15b** are the thiols, which lead to S-glucuronides.

The enzymes catalyzing these highly diverse reactions are known as UDP-glucuronyltransferases (UDP-glucuronosyltransferases; EC 2.4.1.17; UDPGT) and consist in a number of proteins coded by genes of the *UGT* superfamily (**Table 3**). The human UDPGT known to metabolize xenobiotics are the products of two gene families, *UGT1* and *UGT2*. Their major substrates classes are also summarized in **Table 3**.[72–78]

5.06.3.4.2   Substrates of UDP-glucuronyltransferases

Phenols are important substrates of UDP-glucuronyltransferases, be they drugs, drug metabolites, natural products, industrial chemicals, or metabolites thereof. An example of an extensively glucuronidated drug is paracetamol (**30 in Figure 13**; *see* Section 5.06.3.3.2); the major catalysts of the reaction are UGT1A6 and mainly 1A7.[79] Interesting QSARs have been published for the glucuronidation of a variety of natural, medicinal, and model phenols catalyzed by UGT1A1, 1A6, 1A9, 2B7, and 2B15.[46]

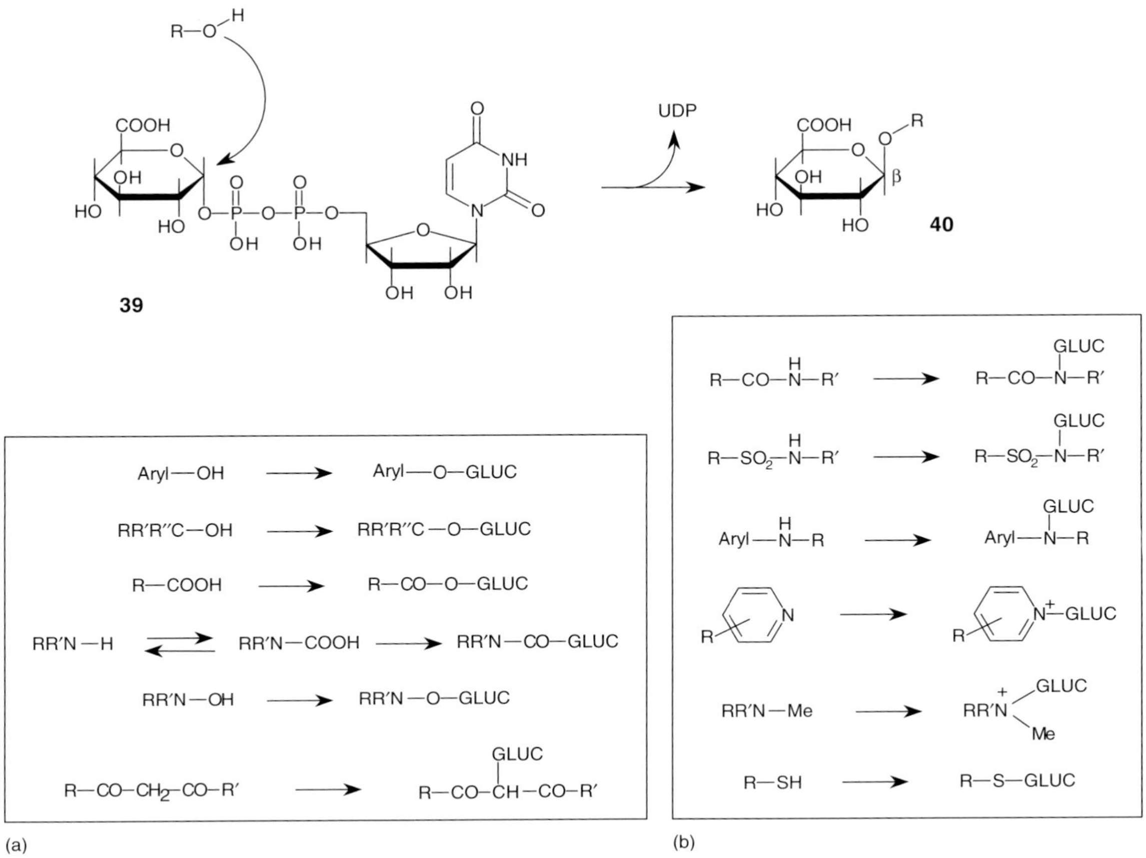

Figure 15 Structure of the cofactor uridine-5'-diphospho-α-D-glucuronic acid (**39**; UDPGA); generic reactions of O- and C-glucuronidation (a) and of N- and S-glucuronidation (b).

Another major group of substrates are alcohols, be they primary, secondary, or tertiary (**Figure 15a**). Medicinal examples include oxazepam (UGT1A9, 2B7, and 2B15) and zidovudine (**36, Figure 14**; UGT2B7).[80,81] Another important example is that of morphine (**41, Figure 16**), which is conjugated on its phenolic and secondary alcohol groups to form the 3-O-glucuronide (a weak opiate antagonist) and the 6-O-glucuronide (a strong opiate agonist), respectively.[82]

Hydroxylamines and hydroxylamides may also form O-glucuronides (**Figure 15a**). Thus, a few drugs and a number of aromatic amines are known to be N-hydroxylated and then O-glucuronidated. A recent example has been found in the metabolism of a new oral hypoglycemic agent 9-[(1S;2R)-2-fluoro-1-methylpropyl]-2-methoxy-6-(1-piperazinyl) purine.[83] When administered to monkeys, more than half of a dose was recovered as a compound found to be the O-glucuronide of the N-hydroxylated metabolite (**42, Figure 16**).

An important pathway of O-glucuronidation is the formation of acyl-glucuronides (**Figure 15a**). Substrates are numerous nonsteroidal anti-inflammatory arylacetic and 2-arylpropionic acids (e.g., ketoprofen, **43** in **Figure 16**) and aliphatic acids (e.g., valproic acid, **44** in **Figure 16**). More recent drug classes such as statins and endothelin receptor antagonists may also yield acyl-glucuronides. Aromatic acids do not appear to be good substrates, but there are exceptions. The significance of acyl-glucuronides has long been underestimated. Indeed, these metabolites are quite reactive, rearranging to positional isomers and binding covalently to plasma and seemingly also tissue proteins.[84,85] Thus, acyl-glucuronide formation cannot be viewed solely as a reaction of inactivation and detoxification.

Second in importance to O-glucuronides are the N-glucuronides (**Figure 15b**) formed from carboxamides and sulfonamides, aromatic and pyridine-type amines, and basic amines.[86] The glucuronidation of primary and secondary amides and amines yields the 'plain' N-glucuronides; these will be discussed first, followed by the quaternary N-glucuronides derived from tertiary amines.

The N-glucuronidation of carboxamides is exemplified by the primary amide carbamazepine (**19, Figure 9**; X = >NCONH$_2$) and by the secondary amide phenytoin (**45, Figure 16**). The reaction has beneficial significance for

Table 3 Human UDP-glucuronyltransferases and their preferred substrates[72–78]

Families and subfamilies	Enzymes	Preferred substrates
Family UGT1		
UGT1A	1A1	Bilirubin, estrogens, natural phenols,[a] simple and complex phenols
	1A3	Estrogens, natural phenols,[a] simple and complex phenols; carboxylic acids; amines
	1A4	Steroids; primary amines, secondary amines, tertiary amines
	1A5[b]	
	1A6	Natural phenols,[a] simple and complex phenols; primary amines
	1A7	Simple and complex phenols
	1A8	Estrogens, natural phenols,[a] simple and complex phenols
	1A9	Aliphatic alcohols, estrogens, natural phenols,[a] simple and complex phenols, thyroid hormone; carboxylic acids; primary amines
	1A10	Natural phenols,[a] other phenols; heterocyclic amines
Family UGT2		
UGT2A	2A1	Aliphatic alcohols, natural phenols,[a] simple and complex phenols, steroids; carboxylic acids
	2A2[b]	
	2A3[b]	
UGT2B	2B4	Catechol estrogens, natural phenols,[a] simple and complex phenols, steroids
	2B7	Aliphatic alcohols, estrogens, opioids; carboxylic acids
	2B10[b]	
	2B15	Natural phenols,[a] other phenols, steroids
	2B17	Androgens
	2B28	Bile acids, natural phenols,[a] steroids
Family UGT3		
UGT3A	3A1[b]	
	3A2[b]	
Family UGT8		
UGT8A	8A1[b]	

[a] Flavonoids, coumarins, etc.
[b] No data or very weak activities.

antibacterial sulfonamides such as sulfadimethoxine (**46, Figure 16**) since it produces highly water-soluble metabolites that show no risk of crystallizing in the kidneys.

N-Glucuronidation of aniline-type aromatic amines has been observed in a few cases only (e.g., benzidine). Similarly, there are a few observations that primary and secondary basic amines can be *N*-glucuronidated, for example, imidazole derivatives such as atipamezole (**47, Figure 16**), a substrate of UGT1A4.[87] *N*-Glucuronidations of particular significance in humans are seen in the formation of quaternary glucuronides, also known as N⁺-glucuronides.[86–89] More and more lipophilic, basic tertiary amines containing one or two methyl groups (e.g., antihistamines and neuroleptics)

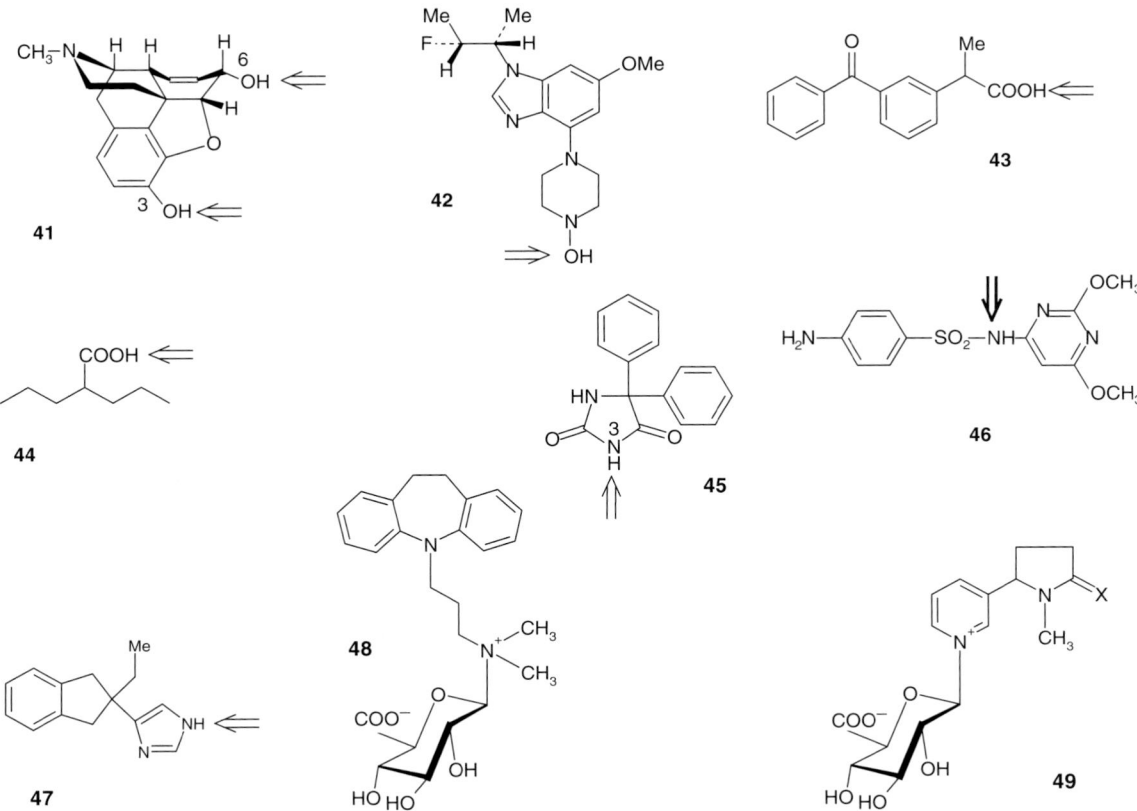

Figure 16 Compounds discussed in the text in connection with reactions of hydroxy, carboxy, carboxamido, sulfamido, and amino glucuronidation.

are found to undergo this reaction to a marked extent in humans, as exemplified by the N^+-glucuronide of imipramine (**48**, **Figure 16**). The same is true for pyridine-type tertiary amines, as exemplified by two important conjugates of nicotine in humans, namely the N^+-glucuronide of nicotine (**49**; X = 2 H) and cotinine (**49**; X = O).

UGT1A4 plays a particularly important role in *N*-glucuronidation, but there are extensive overlaps in the specificity of the various UGT enzymes toward the various classes of substrates discussed above (**Table 3**). QSAR, 3D-QSAR, and pharmacophore modeling are beginning to clarify these specificities and offer preliminary predictive models.[46,90,91]

5.06.3.5   Reactions of Acetylation

The factor involved in the acetylation reaction is acetyl-coenzyme A (**50**, **Figure 17**; R = COCH₃; acetyl-CoA). The acetyl moiety is in activated form, being bound to CoA by a thioester bridge of increased reactivity compared to an oxoester. This endocon is transferred to a nucleophilic function in the substrate, namely an amino, hydroxy, or thiol group. As discussed below, weak amino groups are by far the best targets of acetylation in xenobiotic substrates.

Various enzymes are known to catalyze acetylation reactions:

- arylamine *N*-acetyltransferase (arylamine acetylase; EC 2.3.1.5; NAT) is the major enzyme system involved in xenobiotic acetylation; two enzymes have been characterized in humans, namely NAT1 and the highly variable NAT2[92–94];
- aromatic-hydroxylamine *O*-acetyltransferase (EC 2.3.1.56) and *N*-hydroxyarylamine *O*-acetyltransferase (EC 2.3.1.118) are also involved in the acetylation of aromatic amines and hydroxylamines (see below); and
- other acetyltransferases include diamine *N*-acetyltransferase (putrescine acetyltransferase; EC 2.3.1.57) and aralkylamine *N*-acetyltransferase (serotonin acetyltransferase; EC 2.3.1.87), but their involvement in xenobiotic metabolism does not appear to be sufficiently characterized.

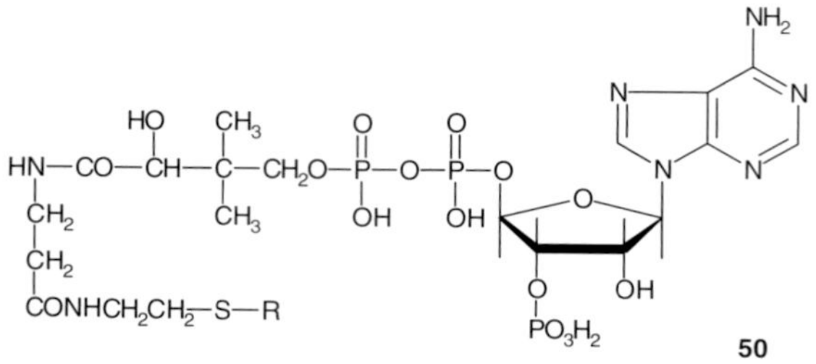

50

Aryl—NH_2 ⟶ Aryl—NH—$COCH_3$

RNH—NH_2 ⟶ RNH—NH—$COCH_3$

RNH_2 ⟶ RNH—$COCH_3$

Aryl—NHOH ⟶ Aryl—NH—O—$COCH_3$

Figure 17 Structure of the cofactor acetyl coenzyme A (**50**; acetyl-CoA; R = acetyl), and generic reactions of *N*- and *O*-acetylation shown by drugs and other xenobiotics (box).

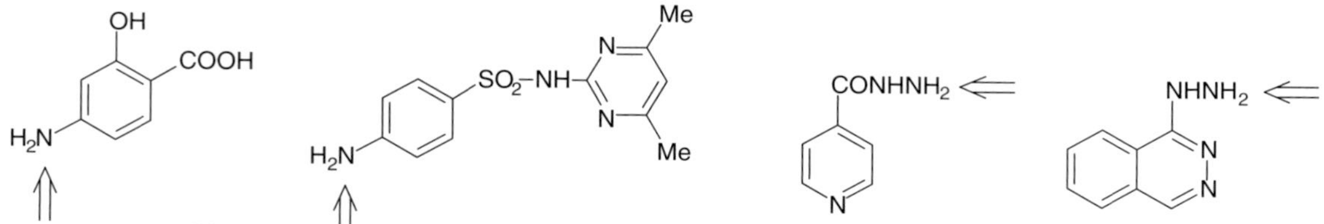

51 **52** **53** **54**

Figure 18 Medicinal substrates discussed in the text in connection with reactions of *N*- and *O*-acetylation.

The xenobiotic substrates of acetylation (box in **Figure 17**) are mainly primary amines of medium basicity, namely arylamines, hydrazines (R = H or aryl), and hydrazides (R = COR′). Medicinal examples include *para*-aminosalicylic acid (**51**, **Figure 18**), sulfamethazine (**52**) and first-generation sulfonamides, isoniazid (**53**) and hydralazine (**54**). Xenobiotics of toxicological interest include hydrazine itself, and many carcinogenic arylamines such as benzidine. Few basic primary amines of medicinal interest have been reported to form *N*-acetylated metabolites, but a noteworthy recent example is that of trovafloxacin (**35**, **Figure 13**; R = H). Human volunteers who had been given the drug excreted it partly unchanged and partly as the sulfamate (*see* Section 5.06.3.3.2), the acyl-glucuronide (the major circulating metabolite), and the *N*-acetyl conjugate (**35**, **Figure 13**; R = $COCH_3$), which accounted for about 6% of the dose.[67]

Xenobiotic arylhydroxylamines can also be acetylated, but the reaction is one of *O*-acetylation (box in **Figure 17**). This is the reaction formally catalyzed by EC 2.3.1.118 with acetyl-CoA acting as the acetyl donor, the *N*-hydroxy metabolites of a number of arylamines being known substrates. The same conjugates can be formed by intramolecular *N*,*O*-acetyl transfer, when an arylhydroxamic acid is *N*-acetylated.[95]

A very different type of reaction is represented by the conjugation of xenobiotic alcohols with fatty acids, yielding highly lipophilic metabolites accumulating in tissues. Thus, ethanol and haloethanols form esters with, e.g., palmitic acid, oleic acid, linoleic acid, and linolenic acid; enzymes catalyzing such reactions are cholesteryl ester synthase (EC 3.1.1.13) and fatty-acyl-ethyl-ester synthase (EC 3.1.1.67).[96] Larger xenobiotics such as tetrahydrocannabinols and codeine are also acylated with fatty acids, possibly by sterol *O*-acyltransferase (EC 2.3.1.26).

5.06.3.6   Conjugations with Coenzyme A and Subsequent Reactions

5.06.3.6.1   Xenobiotic acyl-CoA conjugates as metabolic crossroads

For at least four reasons, the conjugation of xenobiotic acids (R-COOH) with coenzyme A (CoA-SH) is an exception among phase II reactions. First, the reaction begins with the activation of the substrate to a transient enzyme-bound acyl-adenylate intermediate. This intermediate is then reacted with CoA-S- to form an acyl-CoA conjugate (**50**, **Figure 17**; R = xenobiotic acyl group).[97] Second, these R-CO-SCoA conjugates are never excreted and can only be observed transiently in vitro and be inferred indirectly. Third, and most importantly, the R-CO-SCoA conjugates are a crossroads for a number of anabolic or catabolic reactions, which are discussed later. Fourth, and as a consequence of this, it is in the formation and processing of CoA-conjugates that drug metabolism mostly interferes with endogenous metabolic pathways, specifically lipid biochemistry. Such interactions may be of pharmacological or toxicological significance,[97,100] as insufficiently appreciated.

A number of enzymes involved in lipid biochemistry can catalyze the conjugation of xenobiotic acids to coenzyme A. They belong to the ligases forming carbon-sulfur bonds (EC 6.2), and more precisely to the acid-thiol ligases (EC 6.2.1)[6]:

- short-chain fatty acyl-CoA synthetase (EC 6.2.1.1; acetyl-CoA ligase);
- propionate-CoA synthetase (EC 6.2.1.17);
- medium-chain acyl-CoA synthetase (EC 6.2.1.2; butyrate-CoA ligase), which acts on C4 to C11 acids;
- long-chain acyl-CoA synthetase (EC 6.2.1.3), which acts on C6 to C20 and sometimes even C24 acids;
- cholate-CoA ligase (EC 6.2.1.7; cholate thiokinase);
- benzoyl-coenzyme A synthetase (EC 6.2.1.25; benzoate-CoA ligase), which acts on various benzoic acids; and
- phenacyl-CoA synthetase (EC 6.2.1.30; phenylacetyl-CoA ligase), which acts on a variety of arylalkanoic acids.

This list is based for the most part on the recognition of endogenous substrates. There has been a growing interest in xenobiotic-CoA ligases for a number of years, with numerous enzymes of overlapping substrate specificity being discovered.[97–99]

Given that xenobiotic-CoA conjugates are metabolic intermediates characterizable in in vitro studies only, it is of great importance to be aware of the secondary metabolites likely to be seen in in vivo and clinical investigations. As shown in **Table 1**, xenobiotic-CoA conjugates may undergo a variety of secondary reactions resulting in the formation of stable metabolites, some of which are conjugates (**Figure 19**) and others not. The secondary routes open to xenobiotic-CoA conjugates depend primarily on their molecular structure, and in part also on biological conditions. Below, we focus on amino acid conjugation and formation of hybrid triglycerides, and briefly discuss cholesteryl ester formation, chiral inversion, and β-oxidation.

5.06.3.6.2   Amino acid conjugation

Amino acid conjugation is a significant metabolic route for a number of carboxylic acids and involves the formation of an amide bond between the xenobiotic acyl-CoA and the amino acid. Glycine is the amino acid most frequently used for conjugation of small aromatic acids (reaction A in **Figure 19**), while a few glutamine conjugates (reaction B in **Figure 19**) have been characterized in humans. In addition, taurine (reaction C in **Figure 19**) and perhaps a few other amino acids and dipeptides can be used for conjugation in various animal species.[37] The enzymes catalyzing these transfer reactions are known or believed to be various *N*-acyltransferases, for example:

- glycine *N*-acyltransferase (EC 2.3.1.13);
- glutamine *N*-phenylacetyltransferase (EC 2.3.1.14);
- glutamine *N*-acyltransferase (EC 2.3.1.68);
- glycine *N*-benzoyltransferase (EC 2.3.1.71); and
- bile acid-CoA:amino acid *N*-acyltransferase (EC 2.3.1.65; glycine-taurine *N*-acyltransferase).

The xenobiotic acids undergoing glycine conjugation are mainly substituted benzoic acids, e.g., benzoic acid itself (**55**, **Figure 20**); the discovery of hippuric acid, its glycine conjugate, in the urine of horses over 150 years ago marks the birth of the science of drug metabolism. Salicylic acid (**56**) similarly yields salicyluric acid, which accounts for about 3/4 of a dose of aspirin in humans. Similarly, *m*-trifluoromethylbenzoic acid (**57**), a major metabolite of fenfluramine, is excreted as the glycine conjugate. A number of studies have been published on the structure–metabolism relationships of benzoic acid derivatives and analogs in glycine conjugation.[98,101]

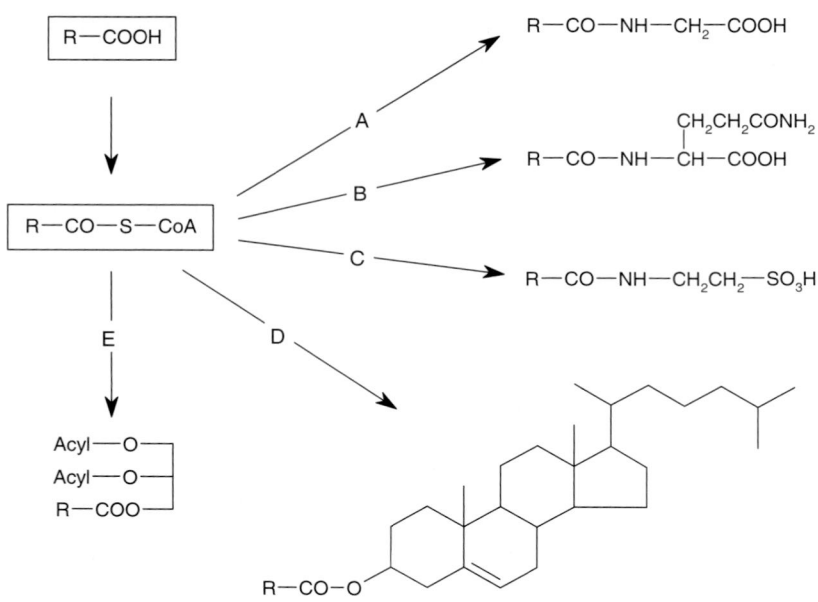

Figure 19 Metabolic conjugations involving acyl-CoA intermediates of xenobiotic acids, namely conjugation to glycine (reaction A), glutamine (reaction B), taurine (reaction C), and cholesterol (reaction D), and formation of hybrid triglycerides (reaction E).

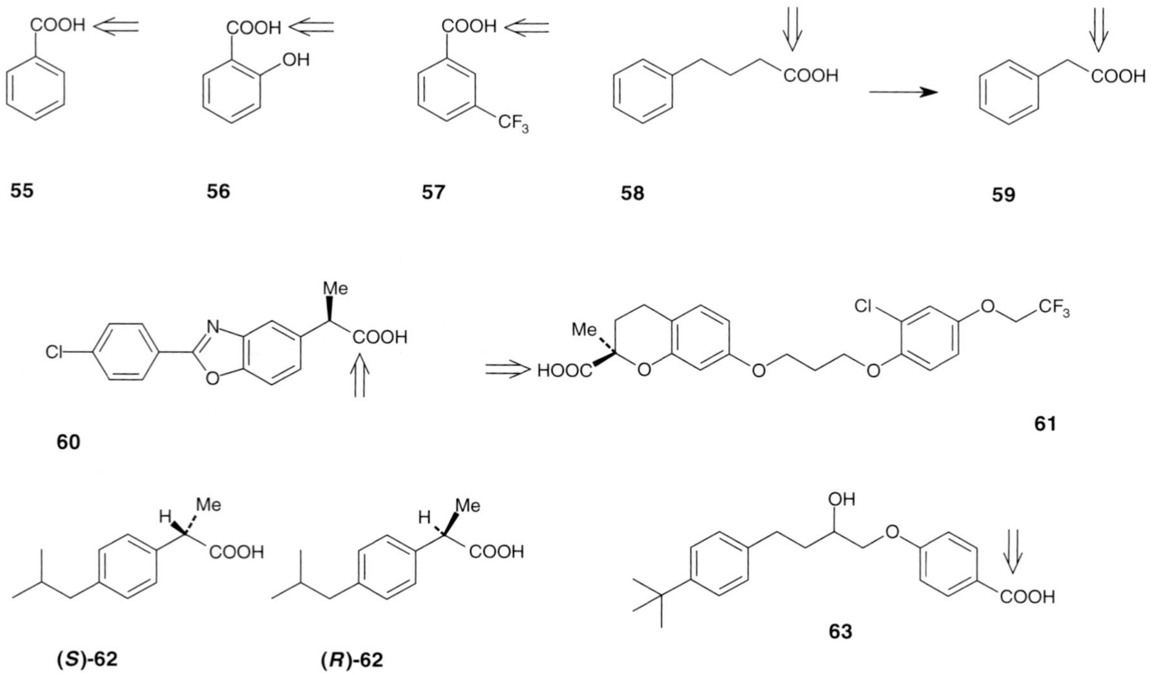

55   **56**   **57**   **58**   **59**

60   **61**

(S)-62   **(R)-62**   **63**

Figure 20 Medicinal substrates discussed in the text in connection with conjugation to amino acids and the unidirectional chiral inversion of profens.

Some phenylalkanoic acids can yield glycine and glutamine conjugates (reactions A and B in **Figure 19**), as illustrated by 4-phenylbutyric acid (**58, Figure 20**), a drug used to treat patients with inborn errors of ureagenesis.[102] In humans and rats, this drug undergoes β-oxidation (*see* Section 5.06.3.6.4) to phenylacetic acid (**59**) via a number of identified oxygenated metabolites. 4-Phenylbutyric acid and phenylacetic acid both yielded the corresponding acyl-glucuronide. Of relevance to the present context is the fact that phenylacetylglycine was formed in the rat, whereas both phenylacetylglutamine and 4-phenylbutyrylglutamine are excreted as the major metabolites in the urine of normal subjects treated with 4-phenylbutyric acid.

Valproic acid (**44, Figures 16** and **21**) is of major interest in the context of lipid biochemistry. Coupling of this antiepileptic drug to coenzyme A opens the door to β-oxidation, as discussed later, and to amino acid conjugation. Indeed, epileptic patients treated with valproic acid produced and excreted trace amounts of valproyl glycine, and more significant amounts of valproyl glutamine and the unexpected glutamic acid conjugate, i.e., valproyl glutamate.[103]

There have been a number of reports in recent years of medicinal carboxylic acids yielding taurine conjugates (reaction C in **Figure 19**). This is exemplified by the withdrawn nonsteroidal anti-inflammatory drug (NSAID) benoxaprofen, which in bile duct-cannulated rats gave both the acyl-glucuronide and taurine conjugate.[104] Only the (*R*)-enantiomer (**60, Figure 20**) formed the acyl-CoA intermediate leading to the taurine conjugate, but racemization of this acyl-CoA conjugate (*see* Section 5.06.3.6.4) ultimately led to minor amounts of the (*S*)-benoxaprofen taurine conjugate also being excreted in the bile.

Another interesting example is that of MRL-II (**61, Figure 20**), an agonist of the α-peroxisome proliferator-activated receptor (αPPAR). When administered to dogs, dose-dependent amounts of the acyl-glucuronide and taurine conjugates were excreted in the bile.[105] Low doses favored the taurine conjugate, whose proportion decreased at higher doses. It is interesting to note that MRL-I, a closely related compound, yielded exclusively the acyl-glucuronide and no taurine conjugate. This fact, together with literature data showing the taurine conjugation of some retinoic acids and synthetic prostaglandins, suggests the involvement of a physiological enzyme such as bile acid-CoA:amino acid *N*-acyltransferase (EC 2.3.1.65).

5.06.3.6.3 Formation of hybrid triacylglycerides and cholesteryl esters

Incorporation of xenobiotic acids into lipids forms highly lipophilic metabolites that may burden the body as long-retained residues. In the majority of cases, cholesterol esters or triacylglycerol analogs are formed. Candidate enzymes for such reactions are respectively:

- sterol *O*-acyltransferase (EC 2.3.1.26; acyl coenzyme A-cholesterol-*O*-acyltransferase, ACAT);
- diacylglycerol *O*-acyltransferase (EC 2.3.1.20); and
- 2-acylglycerol *O*-acyltransferase (EC 2.3.1.22) and other *O*-acyltransferases.

Some phospholipid analogs, as well as some esters to the 3-hydroxy group of biliary acids, have also been characterized.[4,106]

The number of drugs and other xenobiotics known to form glyceryl or cholesteryl esters is currently limited. To the best of our knowledge, only a few nonmedicinal xenobiotics but no drugs have been reported to form cholesteryl esters (reaction D in **Figure 19**). In contrast, results are accumulating on many acidic drugs and other xenobiotics forming triacylglycerol analogs, also called hybrid triglycerides (reaction E in **Figure 19**).[107] One telling example is that of ibuprofen (**62, Figure 20**; *see* Section 5.06.3.6.4), a much used NSAID whose (*R*)-enantiomer forms hybrid triglycerides detectable in rat liver and in human and rat adipose tissues.[100] A similar formation of hybrid triglycerides has been reported for, e.g., the NSAID ketoprofen (**43, Figure 16**) and the antihyperlipidemic agent lifibrol (**63, Figure 20**).[108] The covalent incorporation of valproic acid (**44** in **Figure 16** and **Figure 21**) into phospholipids in neurons is also of note.[109]

5.06.3.6.4 Chiral inversion and β-oxidation

Ibuprofen (**62, Figure 20**) and other arylpropionic acids (i.e., profens) are chiral drugs existing as the (+)-(*S*)-eutomer and the (−)-(*R*)-distomer. Some of these compounds undergo an intriguing metabolic reaction such that the inactive (*R*)-enantiomer is converted to the active (*S*)-enantiomer, while the reverse reaction is negligible. This unidirectional chiral inversion is thus a reaction of bioactivation, and its mechanism is now reasonably well understood.[100] The initial step is the formation of an acyl-CoA conjugate, a reaction that is highly substrate stereoselective since only the (*R*)-form but not the (*S*)-form produces the acyl-CoA conjugate, indicating a high enantiospecificity of the ligase. The (*R*)-acyl-CoA conjugate then undergoes a reaction of epimerization believed to be catalyzed by 2-methylacyl-CoA 2-epimerase (EC 5.1.99.4), resulting in a mixture of the (*R*)-profenoyl-CoA and (*S*)-profenoyl-CoA conjugates. Both epimers can then be hydrolyzed or undergo other reactions such as hybrid triglyceride formation (*see* Section 5.06.3.6.3). This reaction of chiral inversion is not limited to profens, and a few other structurally related xenobiotics are known to undergo this metabolic route.

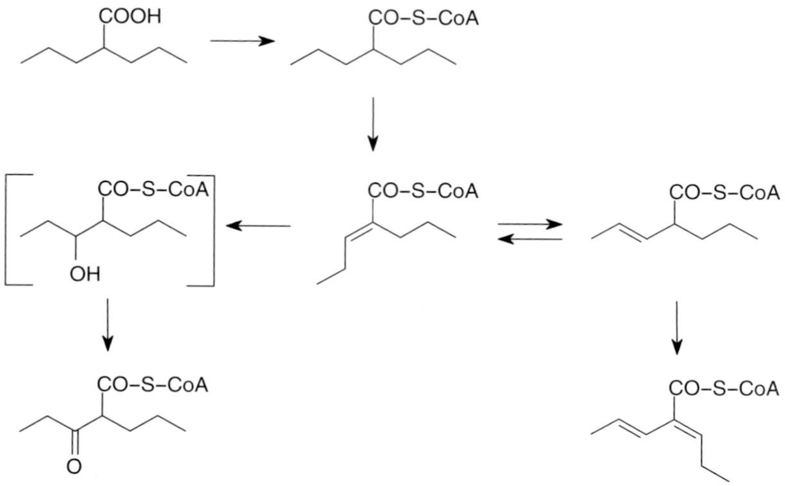

Figure 21   Mitochondrial β-oxidation of valproic acid (**44**).

In some cases, acyl-CoA conjugates formed from xenobiotic acids can also enter the physiological pathways of fatty acids catabolism or anabolism. A few examples are known of xenobiotic alkanoic and arylalkanoic acids undergoing two-carbon chain elongation. Other cases involve the opposite, namely two-, four-, or even six-carbon chain shortening by β-oxidation. An example of two-carbon chain shortening is mentioned in Section 5.06.3.6.2 for 4-phenylbutyric acid (**58, Figure 20**).

In addition, intermediate metabolites of β-oxidation may also be seen, as illustrated by valproic acid (**44, Figure 21**). Approximately 50 metabolites of this drug have been characterized; they are formed by cytochrome P450-catalyzed hydroxylations, dehydrogenations, β-oxidation, glucuronidation, and other routes of conjugation (*see* Sections 5.06.3.4.2, 5.06.3.6.2, and 5.06.3.6.3). **Figure 21** shows the β-oxidation of valproic acid seen in mitochondrial preparations.[110,111] The resulting metabolites have also been found in unconjugated form in the urine of humans or animals dosed with the drug.

5.06.3.7   Conjugations with Glutathione

5.06.3.7.1   Glutathione and glutathione transferases

Glutathione (**64, Figure 22**; GSH) is a thiol-containing tripeptide of major significance in the detoxification of drugs and other xenobiotics. In the body, it exists in a redox equilibrium between the reduced form (GSH) and an oxidized form (GSSG). The metabolism of glutathione (i.e., its synthesis, redox equilibrium, and degradation) is quite complex and involves a number of enzymes.[112–115]

Glutathione reacts in a variety of ways, one of which is its redox capacity. Indeed, GSH can reduce peroxides (a reaction catalyzed by glutathione peroxidase) and organic nitrates; in its GSSG form, glutathione can oxidize the superoxide anion radical. Of major significance in detoxification reactions is the capacity of GSH (and other endogenous thiols including albumin) to scavenge free radicals, in particular radical oxygen species (e.g., R$^\bullet$, HO$^\bullet$, HOO$^\bullet$, ROO$^\bullet$). As such, glutathione and other thiols have a critical role to play in cellular protection.[116] The reactions involved are highly complex and can be simplified as follows:

- GSH + X$^\bullet$ → GS$^\bullet$ + XH
- GS$^\bullet$ + GS$^\bullet$ → GSSG
- GS$^\bullet$ + O$_2$ → GS-OO$^\bullet$
- GS-OO$^\bullet$ + GSH → GS-OOH + GS$^\bullet$

In this chapter, we focus on the conjugation reactions of glutathione as catalyzed by glutathione transferases. The glutathione transferases (EC 2.5.1.18; GSTs) are multifunctional proteins coded by two multigene superfamilies (**Table 4**).[117–123] They can act as enzymes as well as binding proteins. The microsomal superfamily contains the homotrimer GST enzymes, whereas the cytosolic GST enzymes are homodimers and heterodimers. Seven classes are now known in humans, e.g., the alpha class with the following dimers: A1–1, A1–2, A2–2, A3–3, A4–4, and A5–5. The

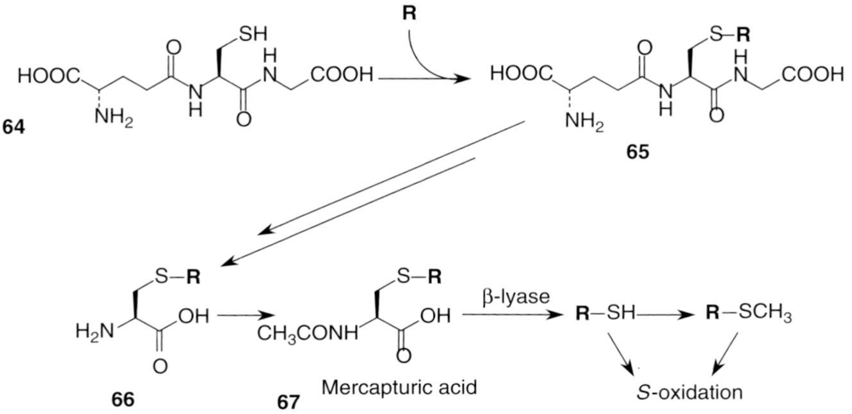

Figure 22 The structure of glutathione (**64**; GSH), its conjugation with substrate R to yield a glutathione conjugate (**65**), and the processing of such conjugates to cysteine conjugates (**66**), mercapturic acids (**67**), and further metabolites discussed in the text.

Table 4 The glutathione transferase gene superfamilies and the corresponding human glutathione transferase enzymes[117–123]

GST genes	Human enzymes (EC 2.5.1.18)
Microsomal GST superfamily (⇒ homotrimers)	
MGST1, MGST2, MGST3	GST1, GST2, GST3
Cytoplasmic GST superfamily (⇒ homodimers, and a few heterodimers)	
GST A1, A2, A3, A4, A5	Alpha class: GST A1-1, A1-2, A2-2, A3-3, A4-4, A5-5
GST K1	Kappa class: GST K1-1 (peroxisomal)
GST M1, M2, M3, M4, M5	Mu class: GST M1-1, M2-2, M3-3, M4-4, M5-5
GST O1, O2	Omega class: GST O1-1, O2
GST P1	Pi class: GST P1-1
GST T1	Theta class: GST T1-1, T2
GST Z1	Zeta class: GST Z1

GST A1–2, A2–2, and P1–1 display selenium-independent glutathione peroxidase activity, a property also characterizing the selenium-containing enzyme glutathione peroxidase (EC 1.11.1.9). The GST A1–1 and A1–2 are also known as ligandin when they act as binding or carrier proteins, a property also displayed by mu class GSTs. In the latter function, these enzymes bind and transport a number of active endogenous compounds (e.g., bilirubin, cholic acid, steroid and thyroid hormones, and hematin), as well as some exogenous dyes and carcinogens.

The conjugating reactivity of glutathione is due to its thiol group (pKa 9.0), which makes it a highly effective nucleophile. The nucleophilic character of the thiol group is greatly enhanced by deprotonation to a thiolate. In fact, an essential component of the catalytic mechanism of glutathione transferases is the marked increase in acidity (pKa decreased by 2–3 units) experienced by the thiol group upon binding of glutathione to the active site of the enzyme.[118] As a result, GSTs transfer glutathione to a very large variety of electrophilic groups; depending on the nature of the substrate, the reactions can be categorized as nucleophilic substitutions or nucleophilic additions. And with compounds of sufficient reactivity, these reactions can also occur nonenzymatically.[118,124,125]

Once formed, glutathione conjugates (**65, Figure 22**; R-SG) are seldom excreted as such (they are best characterized in vitro or in the bile of laboratory animals), but usually undergo further biotransformation prior to urinary or fecal excretion. Cleavage of the glutamyl moiety by glutamyl transpeptidase (EC 2.3.2.2) and of the glycyl moiety by cysteinylglycine dipeptidase (EC 3.4.11.2; aminopeptidase M) leaves a cysteine conjugate (**66**; R-S-Cys), which is further N-acetylated by cysteine-S-conjugate N-acetyltransferase (EC 2.3.1.80) to yield an N-acetylcysteine conjugate (**67**; R-S-CysAc). The latter type of conjugates are known as mercapturic acids, a name which clearly indicates that they were first characterized in urine. This, however, does not imply that the degradation of unexcreted glutathione conjugates must stop at this stage, since cysteine conjugates can be substrates of cysteine-S-conjugate β-lyase (EC 4.4.1.13) to yield thiols (R-SH). These in turn can rearrange, be oxidized, or be S-methylated and then S-oxygenated to yield thiomethyl conjugates (R-S-Me), sulfoxides (R-SO-Me), and sulfones (R-SO$_2$-Me).

5.06.3.7.2   Substrates of glutathione transferases

Most known cases of glutathione conjugation are nucleophilic attacks of electron-deficient carbon atoms, but attack of a nitrogen atom (e.g., in an aromatic nitroso group) or a sulfur atom (in thiols) is also documented. Very often, the target carbon atom is sp^3- or sp^2-hybridized, but a few examples document the reactivity of sp-hybridized carbons. From a mechanistic viewpoint, it may be convenient to subdivide glutathione conjugations into nucleophilic additions and nucleophilic substitutions (i.e., addition-eliminations).[4,37,40]

Nucleophilic additions can involve metabolites arising from oxidation reactions, but they can also occur as primary metabolic reactions. Frequent cases of GSH addition are to α,β-unsaturated carbonyls, a typical xenobiotic substrate being the toxin acrolein ($CH_2=CH-CHO$). Attack occurs at the activated CH_2 group. The mechanism of GSH addition to α,β-unsaturated carbonyls is summarized in **Figure 23** with the diuretic drug ethacrynic acid (**68**), whose glutathione conjugate (**69**) has been known for many years. As shown, the exomethylene is electron deficient and is thus the obvious target of glutathione.

Quinones (*ortho*- and *para*-) and quinoneimines are structurally very similar to α,β-unsaturated carbonyls. They react with glutathione by two distinct and competitive routes, one of which is a reduction to the hydroquinone or aminophenol, where GSH does not react covalently with the substrate but emerges in the oxidized form (GSSG). The other route is relevant to the present context, being a nucleophilic addition to form a conjugate. The reaction has physiological significance since endogenous metabolites such as quinone metabolites of estrogens are conjugated to glutathione. A medicinal example is provided by the toxic quinoneimine metabolite (**71**, Figure 23; *see* 5.05 Principles of Drug Metabolism 1: Redox Reactions; 5.08 Mechanisms of Toxification and Detoxification which Challenge Drug Candidates and Drugs) of paracetamol (**30**). Its glutathione conjugate (**71**) is not excreted as such in humans dosed with the drug, but as the mercapturic acid (**67**, **Figure 22**). The reaction is one of major detoxification, the quinoneimine being extremely hepatotoxic and resulting in liver necrosis, liver failure, and even death when produced at levels and rates that oversaturate the GSH conjugation pathway. Nevertheless, the GSH conjugation of quinones and quinonimines is not always a reaction of detoxification, as some of these conjugates are known to undergo further transformations leading to reactive products.[126]

The addition of glutathione to isocyanates and isothiocyanates (**72**, **Figure 23**; X=O and S, respectively) has received some attention due in particular to its reversible character.[127] Substrates of the reaction are xenobiotics such as the infamous toxin methyl isocyanate, whose glutathione conjugate (**73**) behaves as a transport form able to carbamoylate various macromolecules, enzymes, and membrane components. The reaction is also of interest from a

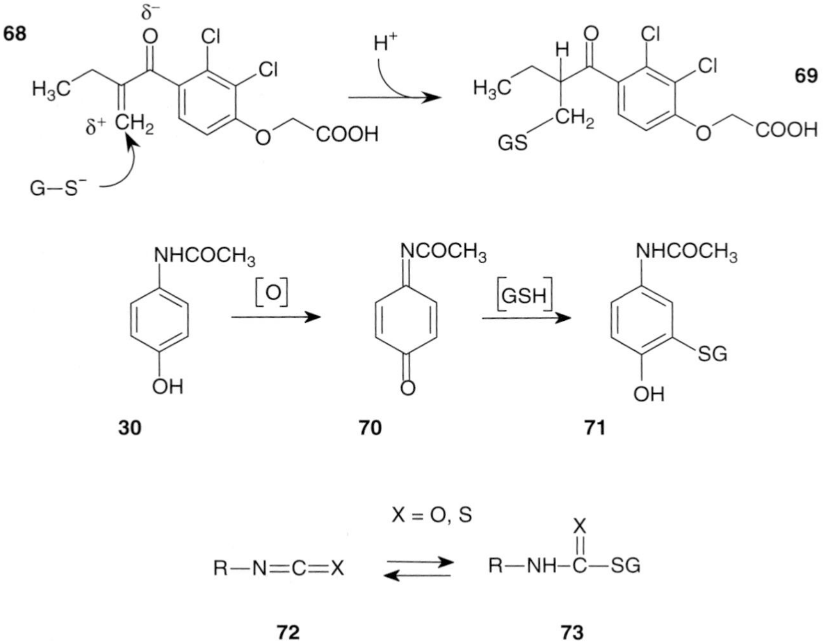

Figure 23 Examples of glutathione conjugation by nucleophilic addition involving ethacrynic acid (**68**), the quinoneimine (**70**) reactive metabolite of paracetamol (**30**), and toxic isocyanates (**72**; X=O) and isothiocyanates (**72**; X=S).

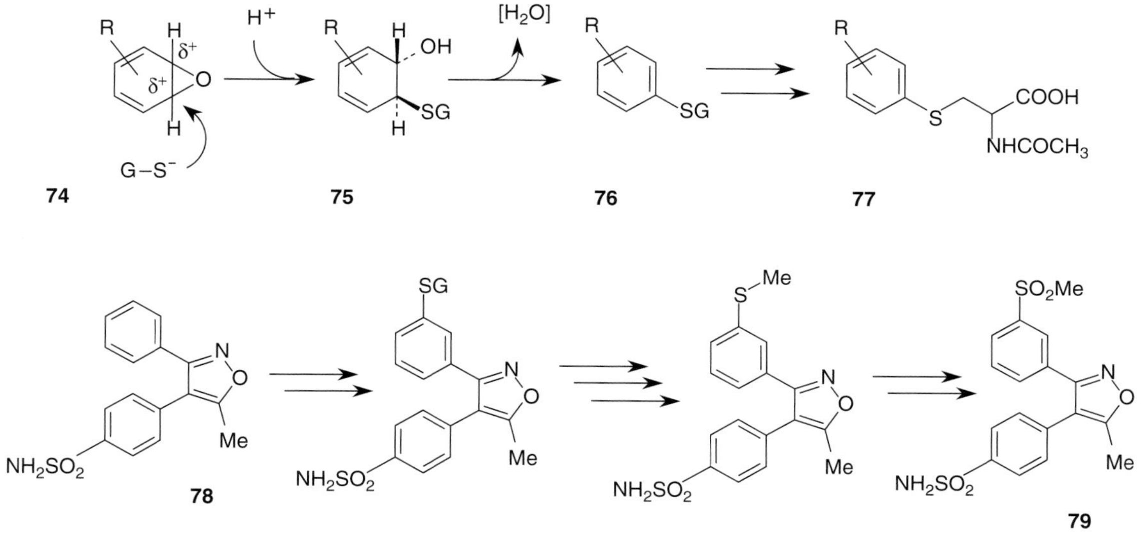

Figure 24 The metabolic pathway from arene oxides (**74**) to aromatic mercapturic acids (**77**) is presented here; the first product of addition is **75**, which dehydrates and aromatizes to **76** before being cleaved and N-acetylated. Also shown here is the logical pathway leading from valdecoxib (**78**) to its methyl sulfone conjugate (**79**).

medicinal viewpoint since anticancer agents such as methylformamide appear to work by undergoing activation to isocyanates and then to the glutathione conjugate.

An important role of GSH is in the conjugation of arene oxides, particularly those with slower rearrangement to the phenol and which are poor substrates of epoxide hydrolase (*see* Section 5.06.2.3.2 and **Figure 8**). The first reaction is again a nucleophilic addition to the epoxide (**74**, **Figure 24**), the target carbon atom here being sp^3-hybridized state. The resulting nonaromatic conjugate (**75**) then dehydrates to an aromatic GSH conjugate (**76**), followed by a cascade leading to the mercapturic acid (**77**) as also shown in **Figure 22**. This is a common reaction of metabolically produced arene oxides, as documented for naphthalene and numerous drugs and xenobiotics containing a phenyl moiety. Note that the same reaction can also occur readily for epoxides of olefins.

An interesting example of GSH addition to an arene oxide is found in the metabolism of the COX-2 inhibitor valdecoxib (**78**, **Figure 24**).[128] When administered to mice, the compound was almost completely metabolized by phase I and phase II reactions, with 16 metabolites being identified in blood, urine, and feces. One of these was the methyl sulfone derivative **79**, whose formation can only be understood by epoxidation, GSH conjugation, mercapturic acid formation, β-lyase C-S cleavage to form the free thiol, *S*-methylation, and finally *S*-oxygenation (**Figure 22**).

Glutathione conjugation by a mechanism of *nucleophilic substitution* (addition-elimination) appears more frequent with industrial xenobiotics than with drugs. Thus, compounds having an activated alkyl moiety of general structure **80** (**Figure 25**) become electrophilic when X is an electron-withdrawing leaving group such as a halogen atom or a sulfate ester group of metabolic origin. Such a reaction occurs for example at the –CHCl$_2$ group of chloramphenicol. At first sight, the reaction of glutathione with alkylating cytostatic drugs having a nitrogen mustard group (e.g., melphalan **81**, **Figure 25**) to yield a conjugate such as **83** occurs by a substitution reaction. However, this is true only if one neglects the intermediate aziridinium ion (**82**), which is formed by chloride elimination independently of the presence of GSH. In other words, the real reaction is again one of nucleophilic addition.[129]

With good leaving groups (e.g., halogen, nitro, sulfoxide, sulfone), nucleophilic aromatic substitution reactions also occur at aromatic rings containing additional electron-withdrawing substituents and/or heteroatoms (i.e., compounds **84**, **Figure 25**).

Acyl halides (**85**; X = F, Cl, Br) are highly reactive industrial compounds, but some can also arise as toxic metabolic intermediate. Their detoxification by glutathione to yield a thioester type of conjugate is a highly effective mechanism of protection. A good example is provided by phosgene (O = CCl$_2$), an extremely toxic metabolite of chloroform which is inactivated to the diglutathionyl conjugate O = C(SG)$_2$.

While not nearly as reactive as acyl halides, acyl-glucuronides (**85**, **Figure 25**; X = O-Gluc) can nevertheless cause damage by forming adducts with proteins (*see* Section 5.06.3.4.2). Interestingly, they can also be conjugated with and detoxified by GSH. This has been demonstrated, for example, with zomepirac, an NSAID of the arylacetic acid class.

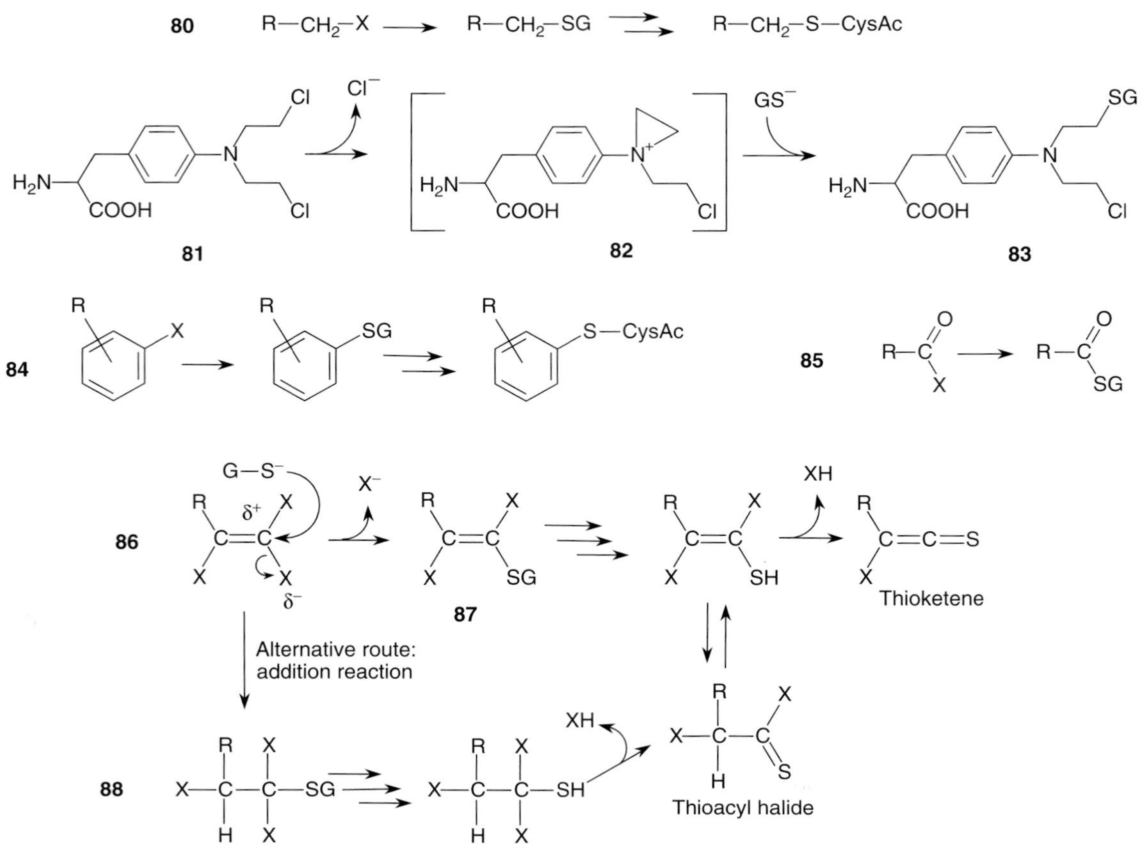

Figure 25 Examples of glutathione conjugation by nucleophilic substitution involving activated alkyl moieties (**80**; X = halogen, -OSO$_3$H), melphalan (**81**) and other alkylating cytostatic drugs (for which chloride elimination does not depend on the presence of GSH), adequately substituted aromatic rings (**84**; X = halogen, nitro, sulfoxide, sulfone), acyl halides (**85**; X = F, Cl, Br) and acyl-glucuronides (**85**; X = O-Gluc), and haloalkenes (**86**). Note that two mechanisms of GSH conjugation exist for haloalkenes, namely substitution (producing conjugates **87**) and addition (producing conjugates **88**); both routes then lead to highly reactive metabolites such as thioacyl halides and thioketenes.

When incubated with rat hepatocytes, zomepirac yielded its *S*-acylglutathione thioester within minutes, but this conjugate then underwent rapid enzymatic hydrolysis.[130] Rats dosed with the drug excreted zomepirac as the *S*-acylglutathione conjugate in their bile. All evidence pointed to the acyl-glucuronide as the reactive intermediate formed in vitro and in vivo.

Haloalkenes (**86**, **Figure 25**) are a special group of substrates of GS-transferases since they may react with GSH either by substitution to form an alkene conjugate (**87**) or by addition to form an alkane conjugate (**88**). Formation of mercapturic acids occurs as for other glutathione conjugates, but in both routes S-C cleavage of the *S*-cysteinyl or *N*-acetyl-*S*-cysteinyl conjugates by renal β-lyase yields thiols of significant toxicity. Indeed, these thiols rearrange by hydrohalide expulsion to form highly reactive thioketenes and/or thioacyl halides.[131,132]

5.06.4   Conclusion

This chapter was written with a number of objectives in mind. As a priority, it was felt necessary to demonstrate to medicinal chemists the fact that metabolic reactions of hydrolysis and conjugation are not second in importance to the redox reactions comprehensively surveyed in 5.05 Principles of drug metabolism 1: Redox reactions The variety of substrates and products presented here is just as ample as that discussed in 5.05 Principles of drug metabolism 1: Redox reactions, and the toxicological consequences (detoxification and toxification) of hydrolyses and conjugations are as important as those of redox reactions.

But differences do exist, and they may explain why hydrolyses and conjugations do not always receive the attention they deserve in drug discovery and development. First, numerous sites of potential enzymatic attack (think of Csp^3 and Csp^2 atoms) exist in substrates of oxidoreductases, most particularly in substrates of cytochrome P450s. As a result, about 90% or more of candidates, drugs, and other xenobiotics are substrates of CYPs. In contrast, hydrolyses and conjugations need specific target groups whose occurrence is less frequent than that of the target groups of CYPs. As a result, each of the metabolic reactions discussed in this chapter knows fewer substrates than CYP-catalyzed monooxygenations.

But this comparatively limited number of xenobiotic substrates per metabolic reaction is compensated by two important facts. The first is the impressive variety of existing transferases and the variety of endocons they transfer. In other words, the chemical nature of the functional groups introduced into substrates by conjugation reactions is markedly more diverse than that resulting from redox reactions. And the second compensating factor is the fact that while CYPs do indeed recognize more substrates than transferases during phase I metabolism, a more balanced situation prevails when considering the further processing of metabolites, in other words secondary metabolic reactions. Indeed, CYP-catalyzed monooxygenations do create the target groups for glucuronidation and other conjugations, meaning that there is markedly more scope for conjugation during phase II metabolism, particularly in in vivo investigations.

In closing, it appears from the above that the many distinct functionalities that differentiate substrates from their products of hydrolysis or conjugation offer a constant challenge to the expertise and intuition of pharmacochemists, bioanalysts, pharmacokineticists, pharmacologists, and toxicologists. The ultimate objective of this chapter is to appeal by providing enough factual evidence and educate by presenting a structured conceptual scaffold.

References

1. Testa, B.; Jenner, P. *Drug Metab. Rev.* **1978**, 7, 325–369.
2. Jenner, P.; Testa, B. *Xenobiotica* **1978**, 8, 1–25.
3. Testa, B. *The Metabolism of Drugs and Other Xenobiotics-Biochemistry of Redox Reactions*; Academic Press: London, UK, 1995.
4. Testa, B.; Soine, W. In *Burger's Medicinal Chemistry and Drug Discovery*, 6th ed.; Abraham, D. J., Ed.; Wiley-Interscience: Hoboken, NJ, 2003; Vol. 2, pp 431–498.
5. Testa, B.; Mayer, J. M. *Hydrolysis in Drug and Prodrug Metabolism – Chemistry, Biochemistry and Enzymology*; Wiley-Verlag Helvetica Chimica Acta: Zurich, Switzerland, 2003.
6. www.chem.qmul.ac.uk/iubmb/enzyme/ (accessed May 2006).
7. www.biochem.ucl.ac.uk/bsm/enzymes/ (accessed May 2006).
8. Satoh, T.; Hosokawa, M. *Annu. Rev. Pharmacol. Toxicol.* **1998**, 38, 257–288.
9. Satoh, T.; Taylor, P.; Bosron, W. F.; Sanghani, S. P.; Hosokawa, M.; La Du, B. N. *Drug Metab. Dispos.* **2002**, 30, 488–493.
10. Teiber, J. F.; Draganov, D. I.; La Du, B. N. *Biochem. Pharmacol.* **2003**, 66, 887–896.
11. Costa, L. G.; Vitalone, A.; Cole, T. B.; Furlong, C. E. *Biochem. Pharmacol.* **2005**, 69, 541–550.
12. Dodson, G.; Wlodawer, A. *Trends Biol. Sci.* **1998**, 23, 347–352.
13. Roberts, S. M.; Phillips, D. L.; Tebbett, I. R. *Drug Metab. Dispos.* **1995**, 23, 664–666.
14. Laizure, S. C.; Mandrell, T.; Gades, N. M.; Parker, R. B. *Drug Metab. Dispos.* **2003**, 31, 16–20.
15. Parker, R. J.; Collins, J. M.; Strong, J. M. *Drug Metab. Dispos.* **1996**, 24, 1167–1173.
16. Alexson, S. E. H.; Diczfalusy, M.; Halldin, M.; Swedmark, S. *Drug Metab. Dispos.* **2002**, 30, 643–647.
17. Strolin-Benedetti, M.; Whomsley, R.; Nicolas, J. M.; Young, C.; Baltes, E. *Eur. J. Clin. Pharmacol.* **2003**, 59, 621–630.
18. Reist, M.; Carrupt, P. A.; Francotte, E.; Testa, B. *Chem. Res. Toxicol.* **1998**, 11, 1521–1528.
19. Schumacher, H.; Smith, R. L.; Williams, R. T. *Br. J. Pharmacol.* **1965**, 25, 338–351.
20. Chen, T. L.; Vogelsang, G. B.; Petty, B. G.; Brundrett, R. B.; Noe, D. A.; Santos, G. W.; Colvin, O. M. *Drug Metab. Dispos.* **1989**, 17, 402–405.
21. Lu, J.; Helsby, N.; Palmer, B. D.; Tingle, M.; Baguley, B. C.; Kestell, P.; Ching, L. M. *J. Pharmacol. Exp. Therap.* **2004**, 310, 571–577.
22. Barrett, A. J.; Rawlings, N. D.; Woessner, J. F.; Eds.; *Handbook of Proteolytic Enzymes*, 2nd ed.; Elsevier: Oxford, UK, 2004, 2 volumes.
23. Naruki, M.; Mizutani, S.; Goto, K.; Tsujimoto, M.; Nakazato, H.; Itakura, A.; Mizuno, M.; Kurauchi, O.; Kikkawa, F.; Tomoda, Y. *Peptides* **1996**, 17, 257–261.
24. Mitchell, B. F.; Wong, S. *J. Clin. Endocrinal. Metab.* **1995**, 80, 2729–2733.
25. Schwahn, M.; Schupke, H.; Gasparic, A.; Krone, D.; Peter, G.; Hempel, R.; Kronbach, T.; Locher, M.; Jahn, W.; Engel, J. *Drug Metab. Dispos.* **2000**, 28, 10–20.
26. Armstrong, R. N. *Drug Metab. Rev.* **1999**, 31, 71–86.
27. Armstrong, R. N.; Cassidy, C. S. *Drug Metab. Rev.* **2000**, 32, 327–338.
28. Seidegard, J.; DePierre, J. W. *Biochim. Biophys. Acta* **1983**, 695, 251–270.
29. Halpin, R. A.; Geer, L. A.; Zhang, K. E.; Marks, T. M.; Dean, D. C.; Jones, A. N.; Melillo, D.; Doss, G.; Vyas, K. P. *Drug Metab. Dispos.* **2000**, 28, 1244–1254.
30. Slack, J. A.; Ford-Hutchinson, A. W.; Richold, M.; Choi, B. C. K. *Chem.-Biol. Interact.* **1981**, 34, 95–107.
31. Harvey, D. J.; Glazener, L.; Johnson, D. B.; Butler, C. M.; Horning, M. G. *Drug Metab. Dispos.* **1977**, 5, 527–546.
32. Watabe, T.; Hiratsuka, A.; Ozawa, N.; Isobe, M. *Xenobiotica* **1981**, 11, 333–344.
33. Svinarov, D. A.; Pippenger, C. E. *Ther. Drug Monit.* **1996**, 18, 660–665.
34. Green, V. J.; Pirmohamed, M.; Kitteringham, N. R.; Gaedigk, A.; Grant, D. M.; Boxer, M.; Burchell, B.; Park, B. K. *Biochem. Pharmacol.* **1995**, 50, 1353–1359.

35. Kitteringham, N. R.; Davis, C.; Howard, N.; Pirmohamed, M.; Park, B. K. *J. Pharmacol. Exp. Ther.* **1996**, *278*, 1018–1027.
36. Testa, B.; Jenner, P. *Drug Metabolism. Chemical and Biochemical Aspects*; Marcel Dekker: New York, 1976.
37. Mulder, G. J., Ed. *Conjugation Reactions in Drug Metabolism*; Taylor & Francis: London, 1990.
38. Kauffman, F. C., Ed. *Conjugation-Deconjugation Reactions in Drug Metabolism and Toxicity*; Springer Verlag: Berlin, Germany, 1994.
39. Woolf, T. F., Ed. *Handbook of Drug Metabolism*; Marcel Dekker: New York, 1999.
40. Ioannides, C., Ed. *Enzyme Systems that Metabolise Drugs and Other Xenobiotics*; John Wiley & Sons: Chichester, UK, 2002.
41. Fujioka, M. *Int. J. Biochem.* **1992**, *24*, 1917–1924.
42. Weinshilboum, R. M.; Otterness, D. M.; Szumlanski, C. L. *Annu. Rev. Pharmacol. Toxicol.* **1999**, *39*, 19–52.
43. Weinshilboum, R. M. *Pharm. News* **2000**, 7, 19–25.
44. Faigle, J. W.; Böttcher, I.; Godbillon, J.; Kriemler, H. P.; Schlumpf, E.; Schneider, W.; Schweizer, A.; Stierlin, H.; Winkler, T. *Xenobiotica* **1988**, *18*, 1191–1197.
45. Lautala, P.; Ulmanen, I.; Taskinen, J. *Mol. Pharmacol.* **2001**, *59*, 393–402.
46. Taskinen, J.; Ethell, B. T.; Puhlaristo, P.; Hood, A. M.; Burchell, B.; Coughtrie, M. W. H. *Drug Metab. Dispos.* **2003**, *31*, 1187–1197.
47. Iyer, R. A.; Malhotra, B.; Khan, S.; Mitroka, J.; Bonacorsi, S., Jr.; Waller, S. C.; Rinehart, J. K.; Kripalani, K. *Drug Metab. Dispos.* **2003**, *31*, 67–75.
48. Bahnmaier, A. H.; Woesle, B.; Thomas, H. *Chirality* **1999**, *11*, 160–165.
49. Crooks, P. A.; Godin, C. S.; Damani, L. A.; Ansher, S. S.; Jakoby, W. B. *Biochem. Pharmacol.* **1988**, *37*, 1673–1677.
50. Armstrong, J. I.; Bertozzi, C. R. *Curr. Opin. Drug Disc. Dev.* **2000**, *3*, 502–515.
51. Raftogianis, R. B.; Wood, T. C.; Weinshilboum, R. M. *Biochem. Pharmacol.* **1999**, *58*, 605–616.
52. Banoglu, E. *Curr. Drug Metab.* **2000**, *1*, 1–30.
53. Nagata, K.; Yamazoe, Y. *Annu. Rev. Pharmacol. Toxicol.* **2000**, *40*, 159–176.
54. Coughtrie, M. W. H.; Johnston, L. E. *Drug Metab. Dispos.* **2001**, *29*, 522–528.
55. Glatt, H.; Boeing, H.; Engelke, C. E. H.; Ma, L.; Kuhlow, A.; Pabel, U.; Pomplun, D.; Teubner, W.; Meinl, W. *Mutat. Res.* **2001**, *482*, 27–40.
56. Tabrett, C. A.; Coughtrie, M. W. H. *Biochem. Pharmacol.* **2003**, *66*, 2089–2097.
57. Kauffman, F., C. *Drug Metab. Rev.* **2004**, *36*, 823–843.
58. Chen, G. *Biochem. Pharmacol.* **2004**, *67*, 1355–1361.
59. Tsoi, C.; Swedmark, S. *Curr. Drug Metab.* **2005**, *6*, 275–285.
60. Runge-Morris, M.; Kocarek, T. A. *Curr. Drug Metab.* **2005**, *6*, 299–307.
61. Yamazaki, H.; Shibata, A.; Suzuki, M.; Nakajima, M.; Shimada, N.; Guengerich, F. P.; Yokoi, T. *Drug Metab. Dispos.* **1999**, *27*, 1260–1266.
62. Miyano, J.; Yamamoto, S.; Hanioka, N.; Narimatsu, S.; Ishikawa, T.; Ogura, K.; Watabe, T.; Nishimura, M.; Ueda, N.; Naito, S. *Biochem. Pharmacol.* **2005**, *69*, 941–950.
63. Tsai, R. S.; Carrupt, P. A.; Testa, B.; Caldwell, J. *Chem. Res. Toxicol.* **1994**, 7, 73–76.
64. Dasaradhi, L.; Shibutani, S. *Chem. Res. Toxicol.* **1997**, *10*, 189–196.
65. Meisheri, K. D.; Johnson, G. A.; Puddington, L. *Biochem. Pharmacol.* **1993**, *45*, 271–279.
66. Shiraga, T.; Hata, T.; Yamazoe, Y.; Ohno, Y.; Iwasaki, K. *Xenobiotica* **1999**, *29*, 341–347.
67. Dalvie, D. K.; Khosla, N.; Vincent, J. *Drug Metab. Dispos.* **1997**, *25*, 423–427.
68. Zhu, C.; Johansson, M.; Permert, J.; Karlsson, A. *Biochem. Pharmacol.* **1998**, *56*, 1035–1040.
69. Feng, J. Y.; Parker, W. B.; Krajewski, M. L.; Deville-Bonne, D.; Veron, M.; Krishnan, P.; Cheng, Y. C.; Borroto-Esoda, K. *Biochem. Pharmacol.* **2004**, *68*, 1879–1888.
70. Peters, K.; Gambertoglio, J. G. *Clin. Pharmacol. Therap.* **1996**, *60*, 168–176.
71. Albert, R.; Hinterding, K.; Brinkmann, V.; Guerini, D.; Müller-Hartwieg, C.; Knecht, H.; Simeon, C.; Streiff, M.; Wagner, T.; Welzenbach, K. et al. *J. Med. Chem.* **2005**, *48*, 5373–5377.
72. som.flinders.edu.au/FUSA/ClinPharm/UGT (accessed May 2006).
73. Radominska-Pandya, A.; Czernik, P. J.; Little, J. M.; Battaglia, E.; Mackenzie, P. I. *Drug Metab. Rev.* **1999**, *31*, 817–899.
74. Tukey, R. H.; Strassburg, C. P. *Annu. Rev. Pharmacol. Toxicol.* **2000**, *40*, 581–616.
75. Fisher, M. B.; Paine, M. F.; Strelevitz, T. J.; Wrighton, S. A. *Drug Metab. Rev.* **2001**, *33*, 273–297.
76. Bock, K. W. *Biochem. Pharmacol.* **2003**, *66*, 691–696.
77. Wells, P.; Mackenzie, P. I.; Roy, J.; Guillemette, C.; Gregory, P. A.; Ishii, Y.; Hansen, A. J.; Kessler, F. K.; Kim, P. M.; Chowdhury, N. R.; Ritter, J. K. *Drug Metab. Dispos.* **2004**, *32*, 281–290.
78. Maruo, Y.; Iwai, M.; Mori, A.; Sato, H.; Takeuchi, Y. *Curr. Drug Metab.* **2005**, *6*, 91–99.
79. Kessler, F. K.; Kessler, M. R.; Auyeung, D. J.; Ritter, J. K. *Drug Metab. Dispos.* **2002**, *30*, 324–330.
80. Court, M. H.; Duan, S. X.; Guillemette, C.; Journault, K.; Krishnaswamy, S.; von Moltke, L. L.; Greenblatt, D. J. *Drug Metab. Dispos.* **2002**, *30*, 1257–1265.
81. Barbier, O.; Turgeon, D.; Girard, C.; Green, M. D.; Tephly, T. R.; Hum, D. W.; Bélanger, A. *Drug Metab. Dispos.* **2000**, *28*, 497–502.
82. Milne, R. W.; Nation, R. L.; Somogyi, A. A. *Drug Metab. Rev.* **1996**, *28*, 345–472.
83. Miller, R. R.; Doss, G. A.; Stearns, R. A. *Drug Metab. Dispos.* **2004**, *32*, 178–185.
84. Spahn-Langguth, H.; Benet, L. Z. *Drug Metab. Rev.* **1992**, *24*, 5–48.
85. Skordi, E.; Wilson, I. D.; Lindon, J. C.; Nicholson, J. K. *Xenobiotica* **2005**, *35*, 715–725.
86. Green, M. D.; Tephly, T. R. *Drug Metab. Dispos.* **1998**, *26*, 860–867.
87. Kaivosaari, S.; Salonen, J. S.; Taskinen, J. *Drug Metab. Dispos.* **2002**, *30*, 295–300.
88. Lee Chiu, S. H.; Huskey, S. W. *Drug Metab. Dispos.* **1998**, *26*, 838–847.
89. Hawes, E. M. *Drug Metab. Dispos.* **1998**, *26*, 830–837.
90. Smith, P. A.; Sorich, M. J.; McKinnon, R. A.; Miners, J. O. *J. Med. Chem.* **2003**, *46*, 1617–1626.
91. Sorich, M. J.; Miners, J. O.; McKinnon, R. A.; Smith, P. A. *Mol. Pharmacol.* **2004**, *65*, 301–308.
92. Grant, D. M.; Blum, M.; Meyer, U. A. *Xenobiotica* **1992**, *22*, 1073–1081.
93. Agundez, J. A. G.; Olivera, M.; Martinez, C.; Ladero, J. M.; Benitez, J. *Pharmacogenetics* **1996**, *6*, 423–428.
94. Kawamura, A.; Graham, J.; Mushtaq, A.; Tsiftsoglou, S. A.; Vath, G. M.; Hanna, P. E.; Wagner, C. R.; Sim, E. *Biochem. Pharmacol.* **2005**, *69*, 347–359.
95. Mattano, S. S.; Land, S.; King, C. M.; Weber, W. W. *Mol. Pharmacol.* **1989**, *35*, 599–609.
96. Bhat, H. K.; Ansari, G. A. S. *Chem. Res. Toxicol.* **1990**, *3*, 311–317.
97. Knights, K. M.; Drogemuller, C. J. *Curr. Drug Metab.* **2000**, *1*, 49–66.

98. Kasuya, F.; Igarashi, K.; Fukui, M. *Biochem. Pharmacol.* **1996**, *51*, 805–809.
99. Brugger, R.; Reichel, C.; Garcia Alia, B.; Brune, K.; Yamamoto, T.; Tegeder, I.; Geissinger, G. *Biochem. Pharmacol.* **2001**, *61*, 651–656.
100. Mayer, J. M.; Roy-de Vos, M.; Audergon, C.; Testa, B. *Arch. Toxicol.* **1995**, *S17*, 499–513.
101. Kanazu, T.; Yamaguchi, T. *Drug Metab. Dispos.* **1997**, *25*, 149–153.
102. Kasumov, T.; Brunengraber, L. L.; Comze, C.; Puchowicz, M. A.; Jobbins, K.; Thomas, K.; David, F.; Kinman, R.; Wehrli, S.; Dahms, W. et al. *Drug Metab. Dispos.* **2004**, *32*, 10–19.
103. Gopaul, V. S.; Tang, W.; Farrell, K.; Abbott, F. S. *Drug Metab. Dispos.* **2003**, *31*, 114–121.
104. Mohri, K.; Okada, K.; Benet, L. Z. *Pharm. Res.* **2005**, *22*, 79–85.
105. Kim, M. S.; Shen, Z.; Kochansky, C.; Lynn, K.; Wang, S.; Wang, Z.; Hora, D.; Brunner, J.; Franklin, R . B.; Vincent, S. H. *Xenobiotica* **2004**, *34*, 665–674.
106. Dodds, P. F. *Life Sci.* **1991**, *49*, 629–649.
107. Vickery, S.; Dodds, P. F. *Xenobiotica* **2004**, *34*, 1025–1042.
108. Sun, E. L.; Feenstra, K. L.; Bell, F. P.; Sanders, P. E.; Slatter, J. G.; Ulrich, R. G. *Drug Metab. Dispos.* **1996**, *24*, 221–231.
109. Siafaka-Kapadai, A.; Patiris, M.; Bowden, C.; Javors, M. *Biochem. Pharmacol.* **1998**, *56*, 207–212.
110. Bjorge, S. M.; Baillie, T. A. *Drug Metab. Dispos* **1991**, *19*, 823–829.
111. Hulsman, J. *Pharm. Weekbl. Sci. Ed.* **1992**, *14*, 98–100.
112. Reed, D. J. *Annu. Rev. Pharmacol. Toxicol.* **1990**, *30*, 603–631.
113. Sies, H. *Free Rad. Biol. Med.* **1999**, *27*, 916–921.
114. Dickinson, D. A.; Forman, H. J. *Biochem. Pharmacol.* **2002**, *64*, 1019–1026.
115. Pompella, A.; Visvikis, A.; Paolicchi, A.; De Tala, V.; Casini, A. F. *Biochem. Pharmacol.* **2003**, *66*, 1499–1503.
116. Ross, D. *Pharmacol. Therap.* **1988**, *37*, 231–249.
117. Hayes, J. D.; Pullford, D. J. *Crit. Rev. Biochem. Mol. Biol.* **1995**, *30*, 445–600.
118. Armstrong, R. N. *Chem. Res. Toxicol.* **1997**, *10*, 2–18.
119. Sheenan, D.; Meade, G.; Foley, V. M.; Dowd, C. *Biochem. J.* **2001**, *360*, 1–16.
120. Jowsey, I. R.; Thomson, A. M.; Flanagan, J. U.; Murdock, P. R.; Moore, G. B. T.; Meyer, D. J.; Murphy, G. J.; Smith, S. A.; Hayes, J. D. *Biochem. J.* **2001**, *359*, 507–516.
121. Schnekenburger, M.; Morceau, F.; Duvoix, A.; Delhalle, S.; Trentesaux, C.; Dicato, M.; Diederich, M. *Biochem. Pharmacol.* **2004**, *68*, 1269–1277.
122. Winayanuwattikun, P.; Ketterman, A. J. *Biochem. J.* **2004**, *382*, 751–757.
123. Robinson, A.; Huttley, G. A.; Booth, H. S.; Board, P. G. *Biochem. J.* **2004**, *379*, 541–552.
124. Ketterer, B. *Drug Metab. Rev.* **1982**, *13*, 161–187.
125. Testa, B. *Drug Metab. Rev.* **1982**, *13*, 25–50.
126. Monks, T. J.; Lau, S. S. *Chem. Res. Toxicol.* **1997**, *10*, 1296–1313.
127. Baillie, T. A.; Slatter, J. G. *Acc. Chem. Res.* **1991**, *24*, 264–270.
128. Zjang, J. Y.; Yuan, J. J.; Wang, Y. F.; Bible, R. H., Jr.; Breau, A. P. *Drug Metab. Dispos.* **2003**, *31*, 491–501.
129. Dirven, H. A. A. M.; van Ommen, B.; van Bladeren, P. J. *Chem. Res. Toxicol.* **1996**, *9*, 351–360.
130. Grillo, M. P.; Hua, F. *Drug Metab. Dispos.* **2003**, *31*, 1429–1436.
131. Anders, M. W.; Dekant, W. *Annu. Rev. Pharmacol. Toxicol.* **1998**, *38*, 501–537.
132. Anders, M. W. *Drug Metab. Rev.* **2004**, *36*, 583–594.

Biography

Bernard Testa studied pharmacy because he was unable to choose between medicine and chemistry. Because he was incapable of working in a community pharmacy, he undertook a PhD thesis on the physicochemistry of drug–macromolecule interactions. Because he felt himself ungifted for the pharmaceutical industry, he went for 2 years to Chelsea College, University of London, for postdoctoral research under the supervision of Prof Arnold H Beckett. And because these were easy times, he was called as assistant professor to the University of Lausanne, Switzerland, to become full professor and Head of Medicinal Chemistry in 1978. Since then, he has tried to repay his debts by fulfilling a number of local and international commitments, e.g., Dean of the Faculty of Sciences (1984–86), Director of the Geneva-Lausanne School of Pharmacy (1994–96 and 1999–2001), and President of the University Senate (1998–2000). He has written four books and edited 29 others, and (co)-authored 450 research and review articles in the fields of drug design and drug metabolism. During the years 1994–98, he was the Editor-Europe of Pharmaceutical Research, the flagship journal of the American Association of Pharmaceutical Scientists (AAPS), and he is now the co-editor of the new journal Chemistry and Biodiversity. He is also a member of the Editorial Board of several leading journals (e.g., Biochemical Pharmacology, Chirality, Drug Metabolism Reviews, Helvetica Chimica Acta, Journal of Pharmacy and Pharmacology, Medicinal Research Reviews, Pharmaceutical Research, Xenobiotica). He holds honorary doctorates from the universities of Milan, Montpellier, and Parma, and was the 2002 recipient of the Nauta Award on Pharmacochemistry given by the European Federation for Medicinal Chemistry. He was elected a Fellow of the AAPS, and is a member of a number of scientific societies such as the French Academy of Pharmacy, the Royal Academy of Medicine of Belgium, and the American Chemical Society. His recently granted Emeritus status has freed him from administrative duties and gives him more time for writing, editing, and collaborating in research projects. His hobbies, interests, and passions include jogging, science-fiction, epistemology, teaching, and scientific exploration.

5.07 Principles of Drug Metabolism 3: Enzymes and Tissues

R A Totah and A E Rettie, University of Washington, Seattle, WA, USA

5.07.1   Overview

The enzymes of drug metabolism mainly act to protect the cell from insult by foreign agents, such as drugs, industrial chemicals, and environmental pollutants. However, these versatile proteins also overlap with normal cellular function by participating in the anabolism and catabolism of a variety of endogenous compounds, including steroids, bile salts, vitamins, and products of the arachidonic acid cascade, to name but a few.[1]

In terms of their xenobiotic function, the drug-metabolizing enzymes can create or unmask hydrophilic moieties in such xenobiotics in order to facilitate their renal clearance (e.g., cytochrome P450 (P450), flavin-containing monooxygenase (FMO), esterases, glucuronyltransferases, sulfotransferases), or they can detoxify electrophilic species that may be damaging to the cell (e.g., epoxide hydrolase, carbonyl reductases and glutathione (GSH) transferases). This enzymatic division is not a strict one, as certain oxidative P450 reactions and conjugation reactions, for example, can be viewed as bioactivation processes that convert a relatively benign compound into a more reactive species.[2] Regardless, these diverse proteins can be divided operationally into three main categories: (1) redox enzymes, that catalyze oxidation and reduction reactions; (2) hydrolases, that catalyze reaction of water with esters, amides, and epoxides; and (3) transferases, that conjugate xenobiotics with relatively small polar molecules.

With regard to the metabolism of therapeutic agents, a survey of the top 200 prescribed drugs from 2002 indicated that more than two-thirds were cleared by metabolic processes, and that of this number about 75% owed their clearance mainly to metabolism by redox enzymes (mostly P450s), up to 15% were cleared by transferases (mostly UDP-glucuronyltransferases (UGTs)), and some 10% were cleared by hydrolase enzymes (mostly esterases).[3] Essential characteristics of each of these important enzyme groups are discussed below that highlight their diversity, structure–function relationships, tissue localization, and development. For additional information, *see* 5.08 Mechanisms of Toxification and Detoxification which Challenge Drug Candidates and Drugs; 5.10 In Vitro Studies of Drug Metabolism.

5.07.2   Oxidases and Reductases

5.07.2.1   Cytochromes P450

The cytochromes P450 (EC 1.14.14.1) are a superfamily of heme-containing monooxygenases that catalyze a myriad of oxidative metabolic reactions upon generation of the high-valent iron-oxo species formed during the enzyme's catalytic cycle. Key characteristics of the P450s are presented in **Table 1** and compared with the other drug-metabolizing enzymes under discussion.

P450s are represented in every living organism examined to date, and as of January 2005, over 4500 unique P450 sequences had been identified from animals, bacteria, fungi, and lower eukaryotes.[134] The plant kingdom is particularly rich in P450 genes where they can comprise up to 1% of the plant genome.[4] Mammalian species exhibit less diversity, with humans possessing 57 functional P450s which are grouped into 18 structurally related families, termed CYP1–5, CYP7–8, CYP11, CYP17, CYP19–21, CYP24, CYP26–27, CYP39, CYP46, and CYP51.[5] These enzyme families are primarily associated with the metabolism of drugs and other xenobiotics or endogenous compounds such as sterols, bile acids, vitamins, and fatty acids. Several members of the CYP1, CYP2, and CYP3 families, present at relatively high concentrations in the endoplasmic reticulum of liver cells, exhibit the broadest substrate selectivity and collectively represent the 'drug-metabolizing' P450s (**Table 2**). *See* 5.05 Principles of Drug Metabolism 1: Redox Reactions for further details on redox reactions in drug metabolism.

Table 1 Human drug-metabolizing enzymes

Enzyme family	Enzymes	Subunit molecular weight (kDa)	Prosthetic group(s)/ external cofactor	Chromosomal location	Enzymes of major importance to drug metabolism/toxicity
Cytochromes P450	57	55–60	Heme/NADPH	Various (7q21, 22q13, 10q24)	CYP3A4, CYP2D6, CYP2C9
Flavin monooxygenase	5	55–60	FAD/NADPH	1q23-25	FMO3
Monoamine oxidases	2	60	FAD	Xp11	MAO-A, MAO-B
Molybdo-flavozymes	5	300	FAD, Mo-pterin	2q33(AOX), 2p22-23 (XOR)	AOX, XOR
Aldehyde dehydrogenases	19	55 (tetramer)	NADPH	Various	ALDH1A1, ALDH1A3, ALDH7
Carbonyl reductases	2	30–35	NADPH	21q21-22	CBR1
Aldo-keto reductases	?	30–40	NADPH	10 (HSD cluster)	AKR1, AKR7
Quinone reductases	2	55 (dimers)	FAD/NAD(P)H	16q22 (NQO1), 6p25 (NQO2)	NQO1
Carboxylesterases	?	55–60	–	–	CES1A1, CES2
Epoxide hydrolases	5	50	–	1q (EPHX1), 8p (EPHX2)	EPHX1 (mEH)
Glucuronyltransferases	17	50–60	UDPGA	2q37 (UGT1), 4q13 (UGT2)	UGT1A1, UGT2B7
N-Acetyltransferases	2	35	Acetyl CoA	8p23.1-p21.3	NAT1, NAT2
Sulfotransferases	10	35	PAPS	Various 16p, 4q, 2q,19q)	SULT1A1
Glutathione S-transferases	16	25 (dimers)	GSH	Various (1p, 6p, 10q, 11q, 14q, 22q)	GSTA1, GSTM1, GSTP1

Table 2 Major human P450 drug-metabolizing enzymes with their most common diagnostic substrates and inhibitors

Isoform (pmol [mg prot]$^{-1}$)	Typical substrate	Typical inhibitor
1A2 (10–50)	Theophylline	Furafylline[2]
2A6 (0–20)	Coumarin	(R)-Tranylcypromine
2B6 (0–50)	Bupropion	Thiotepa, clopidogrel
2C8 (10–25)	Taxol, amodiaquine	Montelukast
2C9 (40–80)	(S)-Warfarin	Sulfaphenazole, benzbromarone
2C19 (0–30)	(S)-Mephenytoin	(R)-Benzylphenobarbital[1]
2D6 (0–15)	Dextromethorphan	Quinidine
2E1 (10–50)	Chlorzoxazone	Disulfiram (diethyldithiocarbamate)
3A4 (50–350)	Midazolam	Ketoconazole, troleandomycin

Parentheses indicate typical hepatic microsomal concentration ranges.

5.07.2.1.1 Nomenclature

The current nomenclature for the P450s is the product of numerous iterations over the past 20 years that currently provides a structured and uniform mechanism for identifying, classifying, and discussing P450 enzymes, regardless of their source.[6] A recent update also makes recommendations for nomenclature of P450 allelic variants and pseudogenes.[7] Key aspects of P450 taxonomy are summarized below:

- When describing a P450 gene, *CYP1A2* for example, *CYP* is italicized and designates the gene as a segment coding for cytochrome **P450**. The first arabic numeral designates the P450 family. This is followed by a capital letter designating the subfamily, and another arabic numeral to distinguish members within a subfamily. By 2005, the enormous biodiversity of P450s required consumption of P450 family 'space' out to *CYP748*!
- When describing a P450 enzyme, italics are not used and either CYP or P450 can be used in front of the family designation; for example, P450 1A2 or CYP1A2.
- P450 isoforms are assigned to specific families on the basis of amino acid sequence identity. If a novel form of the enzyme was discovered and its amino acid sequence found to be less than 40% similar to any of the known families, it would constitute the first member of a new gene family.
- If two P450 sequences have more than 55% homology, they would be placed in the same subfamily.
- The 40% and 55% sequence homology cut-offs for family and subfamily membership are useful guidelines that have been widely applied to the taxonomy of other drug-metabolizing enzyme superfamilies (see below).
- When considering genetic variants of a P450 gene, an asterisk is placed after the arabic numeral for subfamily designation, and each allelic form is assigned an arabic number, e.g., *CYP2C9*2* represents the first allelic form of this gene discovered (relative to the reference sequence, which usually has the *1* designation). Again, when referring to the gene, italics should be used. When referring to the protein encoded by an allelic form of the enzyme, it is recommended that a period replace the asterisk (e.g., CYP2C9.2). However, this has not been widely adopted and most investigators simply use the asterisk designation without italics to denote a polymorphic form of the protein (e.g., CYP2C9*2).

5.07.2.1.2 Multiplicity and ligand selectivity

Within the human CYP1–3 families, some 23 functional forms of P450 can be identified.[5] However, arguably only CYP1A2, CYP2C8, CYP2C9, CYP2C19, CYP2D6, CYP2E1, CYP3A4, and CYP3A5 are critical to drug metabolism, as these enzymes constitute the bulk of the hepatic P450 complement and are known to have wide substrate selectivities. Other extensively studied human P450s, such as CYP1A1 and CYP1B1, have predominantly an extrahepatic location and are not significantly involved in the metabolic clearance of drugs, although they may be important for localized bioactivation reactions that can lead to xenobiotic toxicity.[8] Some newer members of the human P450 family, identified through large-scale sequencing projects, e.g., CYP2S1, CYP2U1, and CYP2W1, have unknown functions and have been referred to as 'orphan' P450s.

Diagnostic substrates and inhibitors for the major hepatic forms of human P450 are listed in **Table 2**. These ligands, together with commercially available recombinant forms of all the major human P450s, are invaluable tools for dissecting the participation of specific human microsomal P450 isoforms that contribute to a given metabolic pathway.[9] Because highly specific substrates and inhibitors can be identified for certain P450s, it could be expected that each isoform would possess a unique active site architecture complementary to key hydrophobic, electrostatic, and steric features of their preferred ligands. Recent crystallization of mammalian forms of P450 is beginning to shed light on these interactions.

5.07.2.1.3 Structure

In general, P450s are 450–500-amino-acid (55–60 kDa) proteins with a completely conserved C-terminal cysteine residue that provides the thiolate coordinate bond to the iron atom which tethers the heme prosthetic group. Intracellularly, P450s reside in the soluble fraction of the cell (bacterial enzymes), the inner membrane of the mitochondrion (several steroid-metabolizing P450s), or the endoplasmic reticulum (drug-metabolizing P450s); targeting is dependent on the nature of the N-terminal leader sequence.[10] Drug-metabolizing P450s contain a hydrophobic string of 20–30 amino acids that is inserted into the endoplasmic reticulum and anchors the enzyme to the membrane, with the bulk of the protein oriented towards the cytosol in a monotopic fashion.

Until 2000, P450 tertiary structure information was limited to the soluble bacterial forms of the enzyme (e.g., CYP101/P450cam and CYP102/P450 BM3). The availability of the first mammalian P450s structure, an N-terminally truncated form of rabbit CYP2C5 also engineered internally to limit additional membrane or oligomer interactions, confirmed that the global features of this enzyme evident for the early bacterial structures were conserved, viz., a heart-shaped protein with an alpha-helix-rich C-terminus and a beta-sheet-rich N-terminus.[11] At the time of writing, crystal structures for human CYP2C9, CYP2C8, CYP3A4, and CYP2A6 have been solved.[12–15] Notably, the structure for

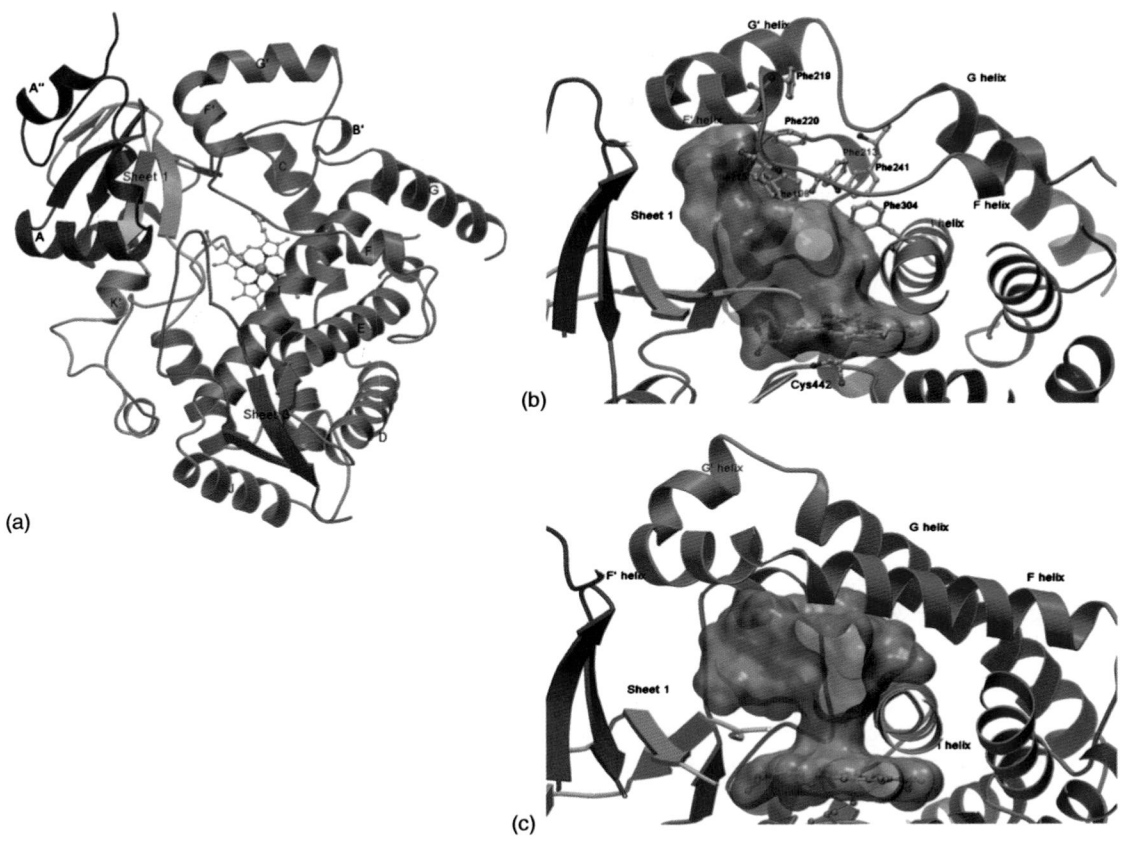

Figure 1 Crystal structures of CYP3A4 and CYP2C9. (a) The overall structural fold of CYP3A4 and location of the heme prosthetic group (ball and stick). Secondary structure elements are represented as coils for helices and arrows for strands. (b) and (c) contrast the solvent accessible molecular surfaces of the active sites of CYP3A4 (green) and CYP2C9 (magenta). These structures highlight differences in volume and shape of P450 active sites. (Reproduced with permission from Williams, P. A.; Cosme, J.; Vinkovic, D. M.; Ward, A.; Angove, H. C.; Day, P. J.; Vonrhein, C.; Tickle, I. J.; Jhoti, H. *Science* **2004**, *305*, 683–686. Copyright (2004) AAAS.)

CYP2C9 with the flurbiprofen bound shows an important salt bridge interaction between an active site Arg108 residue and the charged carboxylate of this nonsteroidal anti-inflammatory drug, which helps explain this isoform's relative selectivity for acidic substrates.[16] A comparison of the active site of CYP3A4 and CYP2C9 is depicted in **Figure 1**. In some cases, ligands have been crystallized within the P450 active site in multiple conformations, highlighting the plasticity of mammalian P450 active sites. This is in accord with the long-held observations of the broad substrate specificity of these enzymes.[11]

5.07.2.1.4 Tissue distribution

Early studies on P450 enzymes focused on the liver as the main drug-detoxifying organ and the adrenal gland for the metabolism of endogenous substrates, but it was soon recognized that other organs, notably those that serve as portals of entry to the body, such as the respiratory and the gastrointestinal tracts, also express P450s and other biotransformation enzymes. Each organ or tissue has its own profile of P450 enzymes that will determine the sensitivity of that tissue to a xenobiotic. It should be noted that most extrahepatic P450 enzymes are also present in the liver, often at a higher concentration, but can be regulated differently in different tissues, leading to altered xenobiotic exposures. **Table 3** summarizes the tissue distribution of P450 enzymes that are commonly detected at either the mRNA (or, less commonly, the protein) level in various human organs.

Although P450 enzymes have been found in all tissues except skeletal muscles (**Table 3**) and red blood cells,[17] the liver and intestine are organs of special significance with regard to drug metabolism. The liver is an impressive drug-metabolizing organ containing an abundance of P450s, UGTs, sulfotransferases (SULTs), and other drug-metabolizing enzymes. With regard to first-pass metabolism, P450s represent the most important hepatic enzymes.

Table 3 Tissue expression of cytochrome P450 enzymes involved in human drug metabolism

Tissue	Cytochrome P450 expression
Liver	1A1, 1A2, 2A6, 2A13, 2B6, 2C8, 2C9, 2C18, 2C19, 2D6, 2E1, 2J2, 2S1, 3A4, 3A5, 3A7 (fetal), 3A43, 4A11, 4F2, 4F11
Small intestine	1A1, 1B1, 2C8, 2C9, 2E1, 2J2, 3A4, 3A5, 4F12
Kidney	1B1, 2C8, 2J2, 3A5, 3A7 (fetal), 3A43, 4A11, 4B1
Heart	1B1, 2J2, 4F12
Lung	1A1, 1B1, 2A6, 2A13, 2B6, 2C18, 2D6, 2E1, 2F1, 2S1, 3A4, 4B1
Brain	2A13, 2B6, 2C8, 2D6, 2E1
Skin	2C18
Placenta	1A1
Ovary	1B1, 2C8
Prostate	1B1, 2A13
Spleen	1B1
Thymus	1B1
Nasal mucosa	2A6, 2A13, 2E1

Table is not exhaustive and reflects only P450 enzymes commonly identified in these organs.

Drug-metabolizing P450s are concentrated in the endoplasmic reticulum of the liver, which represents approximately 5% of adult body weight ($\sim 1500\,g$) and contains $\sim 25\,000$ nmol of total P450. Amongst the P450s, CYP3A4 is the most abundant liver isoform ($\sim 200\,pmol\,mg^{-1}$ $\sim 40\%$ of total hepatic P450), but CYP1A2, CYP2C9, CYP2C19, and CYP2D6 also contribute significantly to hepatic drug metabolism (**Table 3**). CYP3A5, which possesses 83% sequence homology to CYP3A4, is expressed at much lower levels than CYP3A4 in the liver, but is the main CYP3A isoform expressed in the kidney.[18]

The gastrointestinal tract is an absorptive organ involved in the uptake of orally administered drugs, and also an important site for biotransformation. The average total P450 content of human intestine, about $20\,pmol\,mg^{-1}$ microsomal protein, is substantially lower than liver,[19] but nonetheless the intestine contributes significantly to the first-pass effect for some orally administered drugs.[20,21] This reflects both the large surface area of the organ and the fact that CYP3A4 accounts for the majority of total microsomal P450 found in the mucosal epithelium.[22] Other P450 isozymes, including CYP2C9, CYP2J2, and CYP4F12, are also expressed in the small intestine, but at much lower levels than CYP3A4 (**Table 3**). In contrast to the liver, in which the distribution of P450 enzymes is relatively homogeneous, the distribution of P450 enzymes is not uniform along the length of the small intestine, or along the villi within a cross-section of mucosa. CYP3A4 expression decreases down the intestinal tract.[23] The localization and distribution of CYP3A4 along the villi in human small intestine are also variable, with the columnar absorptive epithelial cells of the villi exhibiting the strongest immunoreactivity, whereas no protein is detectable in the goblet cells or the epithelial cells in the crypts.[22] Examples of drugs that undergo significant biotransformation by the gut wall include cyclosporine, nifedipine, midazolam, saquinavir, and verapamil, all of which are CYP3A4 substrates. In the case of midazolam, first extraction by the gut wall seems to be comparable to hepatic extraction.

5.07.2.1.5 Ontogenic development

Organ maturation during fetal life and adulthood has a profound effect on the ability of that tissue to metabolize drugs.[24,25] Maturation of drug-metabolizing enzymes is perhaps the major factor accounting for age-associated changes in the nonrenal clearance of drugs. Individual variability in drug metabolism during different stages of infancy, childhood, and adolescence can partially be explained by the gradual developement of drug-metabolizing enzymes, and failure to recognize this effect can have important consequences in therapeutics. For example, lack of awareness of the ontogenic regulation of UGT activity led to the therapeutic disaster in pediatrics known as 'gray-baby' syndrome caused by administration of chloramphenicol, a UGT substrate, to neonates and young infants with an underdeveloped UGT system.[26] In these young subjects, chloramphenicol accumulated to toxic concentrations and precipitated cardiovascular collapse. **Table 4** contains a list of important hepatic drug-metabolizing enzymes known to be subject to developmental regulation.

In general, P450 expression is low, but detectable, in the developing human fetal liver, increases after birth, and reaches adult levels during the first 10 years of life.[27] CYP1A2 is typical in this regard, with fetal hepatic expression of CYP1A2 protein or mRNA essentially absent in samples from individuals in early gestational stages,[28,29] but steadily increasing in infants to reach 50% of levels seen in adults at 1 year of age.[30] However, temporal expression patterns of individual human liver P450s can vary widely, often among closely related isoforms with the same subfamily. Here the CYP3A family in humans is instructive. Whereas CYP3A4 is the predominant liver isoform in adults, CYP3A7 is the major P450 detected in human embryonic, fetal, and newborn liver. CYP3A7 activity is abundant during embryonic and fetal life and rapidly decreases during the first week of life. In contrast, CYP3A4 levels are very low before birth, but increase steadily thereafter and reach 50% of adulthood levels by 12 months.[31] Conversely, CYP3A5 activity and expression levels exhibit large interindividual variability throughout all stages of development and display no obvious developmental pattern.

In humans, the CYP2 subfamily is the most diverse P450 gene family, with the CYP2C and CYP2D subfamilies contributing substantially to xenobiotic metabolism. However, only the ontogenic development of CYP2C9 and CYP2C19 has been investigated in detail.[32] CYP2C9 protein levels and catalytic activity in first-trimester human fetal liver reflected only 1–2% of adult values, but increased gradually in later fetal life to reach approximately 30% of adult values. In contrast, CYP2C19 development was much slower, implying that different developmental regulatory mechanisms exist for these two closely related P450 isozymes.

The developmental regulation of P450 enzymes associated with the metabolism of endogenous compounds as well as xenobiotic bioactivation is also of interest. CYP1A1 has been closely associated with the bioactivation of polycyclic aromatic hydrocarbons and other chemical carcinogens. CYP1A1 mRNA levels are detectable in fetal liver, adrenal, and lung, but not the kidney, between 6 and 12 weeks of gestation.[33] In all cases mRNA levels decreased with increasing fetal age and were barely detectable in the adult liver, indicating that suppression of CYP1A1 activity occurs some time in the prenatal period, either at birth or during early childhood development.[34] CYP2A enzymes have been associated with the metabolism and bioactivation of tobacco-related xenobiotics. None of the human CYP2A enzymes (CYP2A6, CYP2A13, or CYP2A7) were detected in human fetal liver,[35] but by 1 year of age, hepatic CYP2A6 concentration was 50% of adult expression levels. In the fetal nasal mucosa, CYP2A6 and CYP2A13, but not CYP2A7, were identified (1–5% of adult expression levels) in fetal tissue samples.[36] CYP2J2 has been associated with the epoxidation of endogenous arachidonic acid to biologically active eicosanoids and is abundantly expressed in the adult heart.[37,38] This P450 enzyme appears to have a distinctive developmental pattern because CYP2J2 protein levels in fetal liver and olfactory mucosa samples from fetal tissues were comparable to adult levels.[36] The impact of ontogeny on P450 enzymes should be taken into account when prescribing drugs, whose clearance highly depends on this pathway, to neonates.

5.07.2.2   Flavin-Containing Monooxygenases

Like P450s, flavin monooxygenases (FMO, EC 1.14.13.8) are NADPH-dependent enzymes that couple a two-electron reduction of oxygen to water with two-electron oxidation of its substrates. With FMO, this chemistry occurs at the noncovalently bound flavin center of the molecule. During catalysis, the FAD prosthetic group is activated to a C4a-hydroperoxo species that carries out oxidative reactions on nucleophilic centers. FMOs have a restricted portfolio of reactions that they can catalyze because the activated FAD is a weaker oxidant than the activated perferryl species of P450. Consequently, their role in oxidative drug metabolism is much less pronounced than that of the P450s. FMOs typically form mostly N-oxide and S-oxide metabolites[39] from xenobiotics and also serve an important physiological function in converting trimethylamine to trimethylamine N-oxide.[40]

5.07.2.2.1   Multiplicity and ligand selectivity

Historically, the FMO enzyme system was first characterized with regard to its substrate specificity and catalytic mechanism using the enzyme isolated in the 1970s from hog liver by Ziegler and associates.[41] Molecular biology techniques have now revealed the existence of up to 13 *FMO* genes, only five of which are functional and expressed in all mammalian species examined to date.[42] Compared to the P450s, FMO taxonomy is greatly simplified by the absence of gene family expansion, so, for example, there is only a single FMO1, as no subfamilies have been identified.[43] The five active forms of the enzyme are termed FMO1–FMO5. These enzymes exhibit only 50–55% sequence homology, whereas species orthologs (i.e., rat FMO, human FMO1, dog FMO1, etc.) share 80–90% homology. Human FMO6 is inactive due to alternative splicing that results in a nonfunctional protein.[44] Although five functional mammalian forms of the enzyme are recognized, in practice only two – FMO3 (major) and FMO1 (minor) – appear to be relevant to human drug metabolism.

Table 4 Ontogeny of major hepatic drug-metabolizing enzymes

Enzyme	Prenatal trimester			Neonate (<4 weeks)	<1 year	1–10 years	Adult
	1	2	3				
Oxidases							
CYP1A1[a]	+	+	?	−	−	−	−
CYP1B1	?	?	?	?	−	−	−
CYP1A2[b]	−	−	−	−	+	+	+ +
CYP2A[b]	−	−	−	?	+	+	+ +
CYP2B6[b]	−	−	−	−	?	+	+ +
CYP2C	−	+	+	+	+	+	+ +
CYP2D6	−	−/+	−/+	+	+	+	+ +
CYP2E1	?	+	+	+	+	+	+ +
CYP2J	?	+	?	?	?	+	+ +
CYP3A7[a]	+	+	+	+	−	−	−
CYP3A4/5[b]	−	−	−	+	+	+	+ +
FMO1[a]	+	+	+	−	−	−	−
FMO3[b]	−/+	−	−	−/+	+	+	+ +
MAO-A	+	+	+	+	+	+	+ +
MAO-B	+	+	+	+	+	+	+ +
Hydrolases							
EPHX1	+	+	+	+	+	+	+ +
EPHX2	?	+	+	+	+	+	+ +
Transferases							
UGT1A1[b]	−	−	−	+	+	+	+ +
UGT1A3	?	+	+	+	+	+	+ +
UGT1A6[b]	−	−	−	+	+	+	+ +
UGT2B7	?	+	+	+	+	+	+ +
UGT2B17	?	+	+	+	+	+	+ +
SULT1A1	?	+	+	+	+	+	+ +
SULT1A3	?	+	+	+	+	+	+ +
SULT2A1	−	−	+	+	+	+	+ +
GSTA1/A2	+	+	+	+	+	+	+ +
GSTM	+	+	+	+	+	+	+ +
GSTP1[a]	+	+	+	+	−	−	−

+, protein or activity detected, −, protein or activity not detected, ?, not determined, −/+ protein or activity detected in a fraction of the samples.
Adapted and modified from[24,25].
[a] Fetal-specific enzymes.
[b] Adult-specific enzymes.

FMO1 possesses a very broad substrate specificity and can metabolize most soft nucleophiles it encounters, but its extrahepatic location (see below) minimizes its impact on drug clearance. FMO3 is the major form of the enzyme present in human liver. While it too has a wide substrate specificity in vitro, the enzyme participates significantly in the metabolic clearance of only a few drugs and xenobiotics, e.g., itopride, ranitidine, moclobemide, nicotine.[40] However, FMO3 plays a major role in the conversion of the dietary breakdown product, trimethylamine, to trimethylamine N-oxide, and genetic defects in *FMO3* are responsible for the rare metabolic disorder, trimethylaminuria.[45] Genetic polymorphisms also render FMO2 largely nonfunctional in most human ethnic groups. Very little is known about the substrate specificity of FMO4 and FMO5.

5.07.2.2.2 Structure

Human FMOs are FAD-containing flavoproteins comprising 532–558 amino acids (i.e., 60–65 kDa proteins). Two glycine-rich motifs forming part of the Rossman folds for NADPH and FAD binding are found in the N-terminal half of the enzyme. The C-terminal portion is pronouncedly hydrophobic and may represent a membrane anchor domain. However, deletion of this hydrophobic region has generally been accompanied by loss of enzyme activity and crystallization of the protein has yet to be accomplished. Homology models based on soluble GSH reductase and/or phenylacetone monooxygenase may hold promise in the interim for visualizing some aspects of FMO structure.[39]

5.07.2.2.3 Tissue expression and ontogenic development

As with P450, FMO isoforms are ubiquitously distributed in the animal kingdom and in the body where the highest concentrations are generally found in the liver, although certain species have high enzyme concentrations in the lung (rabbit) and kidney (mouse). FMO1 is the major liver form in most experimental animals, but is not detectable in adult human liver, where FMO3 and FMO5 are the major isoforms.[46] FMO1 is relatively abundant in human kidney and present at much lower levels in the intestine and the brain.[47] FMO5 and FMO2 are the major isoforms expressed, at the mRNA level, in human small intestine and lung respectively,[48] although the functional significance of these two enzymes in these organs is unclear.

Hepatic FMO1 is restricted to the fetus, peaking in the early embryo, and decreasing steadily within 3 days postpartum (**Table 4**).[49] Conversely, hepatic FMO3 protein is not detectable in the fetal state, and its onset of expression after birth can be slow. As noted above, FMO3 is necessary for the conversion of trimethylamine to its nonodoriferous N-oxide metabolite, so delayed onset of expression of this enzyme may contribute to cases of transient childhood trimethylaminuria.[50] The temporal switch for these FMO isoforms contrasts with that observed for CYP3A7 and CYP3A4/5, in which the suppression of CYP3A7 expression is accompanied by a simultaneous increase in CYP3A4/5, and the net hepatic CYP3A content remains constant.[51] The relatively high expression levels of FMO1 during prenatal development and the embryonic periods, in particular, might argue for a role for FMO1 in the metabolism of endogenous substrates that are important for development, but this remains to be determined.

5.07.2.3 Monoamine Oxidases

Monoamine oxidases (MAO, EC 1.4.3.4) are FAD-containing flavoproteins, recognized primarily for their physiological role in metabolically regulating levels of neurotransmitters in the brain. Unlike FMO, the FAD group functions only as an electron acceptor (FAD→FADH$_2$) during oxidative deamination of these biogenic amines. In addition to its physiological role, MAO can participate in the metabolism of certain xenobiotics, especially drug molecules that are structurally related to the endogenous substrates.[52]

5.07.2.3.1 Multiplicity and ligand specificity

Two forms of the enzyme exist, MAO-A and MAO-B, that are encoded by separate genes.[53] These enzymes are ~80% similar, and possess overlapping, albeit sometimes distinctive, substrate specificities. Serotonin, norepinephrine, and epinephrine are the major endogenous substrates for MAO-A. The indoleamine nucleus of serotonin appears in a few drug classes, notably the triptan class of antimigraine drugs. These drugs are metabolized by MAO-A via pathways that ultimately generate carboxylic acid metabolites.[40] The acetylenic compound, clorgyline, is a selective mechanism-based inhibitor of MAO-A, and reversible inhibitors of the enzyme are under development as antidepressant drugs. Amongst the neurotransmitters, dopamine is selectively metabolized by MAO-B. Another acetylenic compound, deprenyl, is a selective mechanism-based inhibitor of MAO-B and the levo enantiomer, selegiline, is marketed as Eldepryl and used as an adjunct to L-DOPA in Parkinsonism. The rationale here is to minimize metabolism of dopamine and reduce the dose of L-DOPA required therapeutically. MAO-B is also implicated in the bioactivation of the environmental neurotoxin 1-methyl-4-phenyl-1,2,3,6-tetrahydropyridine (MPTP).[54]

5.07.2.3.2   Structure

Human MAO-A and MAO-B are FAD-containing flavoproteins containing 527 and 520 amino acids ($\sim$63 kDa), respectively. Models for the membrane topography of MAO suggests that it is monotopically localized in, and anchored to, the mitochondrial membrane by a hydrophobic C-terminus. In contrast to the FMOs, MAOs have their prosthetic group linked covalently to a cysteine residue located in the C-terminal half of the protein. Both human forms of the enzyme have been crystallized, revealing differences in the volume of the active site activity, 550 and 700 $\overset{\circ}{A}^3$, for human MAO-A and MAO-B, respectively.[55] The ligand specificity of the two enzymes is controlled, in part, by a tyrosine (MAO-B) to isoleucine (MAO-A) change in the C-terminal half of the protein.[56]

5.07.2.3.3   Tissue distribution and ontogenic development

MAO is mainly located in the outer membrane of mitochondria of presynaptic nerve terminals, where oxidative deamination of norepinephrine, epinephrine, and serotonin by MAO inactivates these neurotransmitters and abrogates the neural stimulus. Both MAO-A and MAO-B are expressed in several human tissues, with the highest concentrations evident in liver, followed by myocardium, renal cortex, and intestine.[57] MAO-A is selectively expressed in the placenta while MAO-B is selectively expressed in blood platelets.[58] Intestinal MAO is implicated in the breakdown of dietary amines, notably the indirectly acting sympathomimetic tyramine, which is present in high concentrations in aged cheeses and red wine. Normally, tyramine is metabolized by intestinal MAO before it enters systemic circulation. When MAOIs, like isoniazid and tranylcypromine and foods high in tyramine are taken concurrently, large amounts of dietary tyramine can reach the systemic circulation and precipitate a hypertensive crisis.[59]

Limited studies on the ontogenic development of MAO have been performed and most available data deal with developmental aspects of MAO in the human brain.[58] In the fetal brain, lung, aorta, and digestive tract, MAO-A activity is detected earlier than MAO-B, and in the fetal brain the levels of MAO-A are significantly higher than MAO-B (**Table 4**). Interestingly, during aging, in which a general decrease in most enzymatic activity is observed, MAO-B content and activity appear to increase, due perhaps to astroglial proliferation and the associated increased need to metabolize the resulting biogenic amines.[58]

5.07.2.4   Aldehyde Oxidases and Xanthine Oxidases

Aldehyde oxidases (AOX, EC 1.2.3.1) and xanthine oxidases (XOR, EC 1.17.3.2) are molybdo-flavoproteins, with complex structural and mechanistic features.[60] The enzymes principally carry out oxidation reactions on nitrogen heterocycles, ultimately inserting one atom of oxygen from water without the need for an external source of reducing equivalents. In the main, AOX and XOR target sp^2-hybridized carbon atoms rendered electron-deficient by a nitrogen atom to which they are linked by a double bond ($CH=N$), with the resulting formation of lactam metabolites.[61] Historically, the majority of studies with these proteins have been carried out with the rabbit liver form of AOX, and XOR isolated from cow milk, because these are rich and readily available sources of the two enzymes, and because of difficulties in expressing recombinant forms of these complex enzymes in high yield.[62]

5.07.2.4.1   Multiplicity and ligand specificity

Enzymes utilizing a molybdnemum cofactor are known throughout evolution. In mammals, five distinct molybdo-flavoenzymes have been identified: XOR; AOX; and aldehyde oxidase homologs 1–3 (AOH1, AOH2, and AOH3). XOR and AOX have been characterized in a variety of species, including humans, whereas AOH1, AOH2, and AOH3 appear to be important only in rodent and bird species. These five enzymes are the products of distinct genes.[63] The substrate specificities of only AOX and XOR have been studied in detail and much less is known about the catalytic functions of AOH1–3. AOX and XOR exhibit $\sim$50% amino acid homology and overlapping substrate specificities.

XOR catalyzes the oxidation of hypoxanthine to xanthine and xanthine to uric acid and represents a key enzyme in the metabolism of purines. In most tissues, XOR exists in its dehydrogenase form, where the physiological electron acceptor is NAD^+. A number of factors, including ischemia and the purification process itself, promote facile conversion to the oxidase form where the electron acceptor is oxygen. Consequently, this enzyme is associated strongly with the production of reactive oxygen species (ROS), such as hydroxyl radical and superoxide. The enzyme can also produce nitric oxide by reduction of nitrites, which has further stimulated interest in the enzyme from a toxicological viewpoint.[64] Allopurinol is a diagnostic inhibitor of the enzyme. This drug is used clinically to treat hyperuricemia, following its metabolic conversion to oxypurinol, a tight binding inhibitor of XOR.[61]

Abundant levels of AOX are present in the liver where it metabolizes numerous drugs, including zaleplon and 6-mercaptopurine. Guanine derivatives based on aciclovir are useful antiviral agents, but are poorly absorbed after oral administration. Attempts to improve bioavailability have centered around 6-deoxy prodrugs, e.g., 6-deoxy penciclovir, which are bioactivated by AOX (or XOR). AOX also plays an important role in the detoxification of potentially reactive iminium ions that can be generated by P450 or MAO, often from cyclic tertiary amines, e.g., nicotine.[40] Menadione, isovanillin, and raloxifene are useful diagnostic inhibitors of AOX.[65]

5.07.2.4.2 Structure

AOR and XOR are large (~ 300 kDa) multimeric enzymes composed of two subunits. Each subunit contains one atom of molybdenum (Mo) in a pyranopterin complex, one FAD, and two Fe-S centers. All four centers can function as redox groups in the intramolecular transfer of electrons from a reducing substrate (electron donor) to an oxidizing substrate (electron acceptor). Several bacterial members of this enzyme family have been crystallized, indirectly lending insight into the structure and function of XOR and AOR.[60]

5.07.2.4.3 Tissue distribution and ontogenic development

Limited information is available on the ontogenic development of XOR and AOX, except that both enzymes are expressed in fetal tissues at levels similar to those found in adults. Regarding tissue distribution, most studies on XOR have been conducted in experimental animals rather than humans. In rodents the highest expression levels of mRNA and protein of XOR are observed in the liver, lung, kidney, and in the epithelial lining of the duodenum on the proximal side of the small intestine, decreasing steadily towards the distal portion of the small intestine. Similar expression patterns in the intestine have also been reported in humans. The mouse lung also expresses XOR, in contrast to human lung, where the enzyme is barely detectable. Other studies have also reported expression of XOR in the human brain and heart. Intracellularly, XOR is almost always localized exclusively to the cytoplasm. Interestingly, intact XOR is found in milk and other body fluids, including blood.

Tissue expression of AOX is highly variable across species, with high hepatic concentrations present in rabbit and baboon, and low concentrations found in rat and dog liver. In humans, high concentrations are found in the liver and the enzyme has also been detected immunohistochemically in respiratory, urogenital, endocrine, and digestive tissues.[66] In these latter studies AOX was most abundant in the epithelial cells of the trachea and bronchium in addition to the alveolar cells of the respiratory system. In the digestive system, AOX is found in the surface epithelia of the small and large intestines. In the kidney, the proximal, distal, and collecting tubes but not the glomerulus expressed AOR. In mice, AOR exhibits gender differences where it is expressed at higher levels in males.[60] It has been proposed that AOX is induced by high levels of testosterone but this induction is tissue-specific as no gender differences in the levels of XOR were observed in mice lung. No gender differences in expression have been demonstrated in humans.[60]

5.07.2.5   Miscellaneous Dehydrogenases

5.07.2.5.1 Aldehyde dehydrogenases

Aldehyde dehydrogenase (ALDH, EC 1.2.1.3) enzymes catalyze the $NAD(P)^+$-dependent oxidation of aldehydes to carboxylic acids. Nineteen ALDH genes coding for enzymes with broad substrate specificities have been identified in the human genome. These enzymes cluster into nine families, ALDH1–9, with members of the ALDH1 and ALDH3 families most strongly asssociated with xenobiotic metabolism.[67] The nomenclature for this enzyme system follows closely that used for the P450s. ALDH1A1, 1A2, and 1A3 are cytosolic and mitochondrial enzymes that are involved in the biosynthesis of retinoic acid from retinal in humans. Retinoic acid is important for modulating cell differentiation, particularly during embryogenesis. ALDH1 forms are also implicated in the metabolic pathway leading to inactivation of antineoplastic agent, cyclophosphamide. ALDH2 is a mitochondrial enzyme that is mainly responsible for the metabolism of ethanol-derived acetaldehyde. An E487 K mutation (*ALDH2*2*) results in a large increase in the K_m for NAD^+, which almost inactivates the enzyme.[68] This polymorphism is responsible for alcohol flushing in up to 40% of Asians. Members of the ALDH3A subfamily are dioxin-inducible and are overexpressed in some tumors. These enzymes are involved in the metabolism of medium- to long-chain aliphatic and aromatic aldehydes. Individual members of gene families *ALDH4–9* metabolize a variety of usually endogenous aldehydes, although some may also have activity towards some xenobiotic substrates. Structurally, these enzymes are homotetramers, with subunit molecular weights of ~ 55 kDa.[69]

5.07.2.5.2 Alcohol dehydrogenases

Alcohol dehydrogenase (ADH, EC 1.1.1.1) enzymes are zinc-containing dimeric enzymes, found in the cytosolic fraction of the cell, that catalyze the NAD(H)-dependent 'reversible' oxidation of low-molecular-weight alcohols to aldehydes. ADH enzymes perform numerous important metabolic functions, especially with endogenous substrates, but the prototypic reaction is the oxidation of ethanol to acetaldehyde, during which they transfer the pro-R hydrogen from NADH to substrate. Globally, these enzymes constitute part of the medium-chain dehydrogenase/reductase superfamily.[70] In humans, five classes (classes I–V) of ADHs are known: members of different classes share less than 70% amino-acid sequence identity within a species.[71] Three genes found within the human class 1 enzymes – *ADH1*, *ADH2*, and *ADH3* are expressed in most adult tissue, with the exception of brain, kidney, and placenta. The absence of ADH1 enzymes, the most efficient ethanol-metabolizing enzymes among the ADH family, in the placenta and brain argues against a substantial contribution of this enzyme in local ethanol developmental central nervous system toxicity. On the other hand, polymorphisms in the human *ADH2* gene have been associated with alcoholism.[72]

5.07.2.6   Reductases

The reductive enzymology of drug metabolism is the most poorly characterized, but perhaps most complex, of the common drug metabolism processes. This is due, in part, to the host of enzyme systems that can catalyze this reaction, including, but not limited to, ADH, aldo-keto reductases (AKR), carbonyl reductases, quinone reductases, P450 reductase, and even P450 itself where conditions of low oxygen tension prevail. Moreover, reductase enzyme activity can be found throughout nature and in a variety of subcellular fractions, although NADPH-dependent cytosolic enzymes generally predominate. Another facet of the complexity of reductase enzymology is the reversible nature of many of the reactions, made possible by the redox behavior of the common nicotinamide cofactors, NADH and NADPH. For example, although ADHs can catalyze the (reductive) bioactivation of chloral hydrate to trichloroethanol by transferring a hydride from NADH to the substrate, the enzyme is more commonly associated with its important dehydrogenase (oxidative) function in the NAD^+-dependent metabolism of ethanol to acetaldehyde. A common feature of carbonyl (and quinone) reductases is their two-electron reduction mechanism. In contrast, P450 and P450 reductase catalyze one electron reduction processes often associated with xenobiotic toxicity, but also productively in the bioactivation of some anticancer drugs.

5.07.2.6.1   Carbonyl (and quinone) reductases

The carbonyl moiety of aldehydes, ketones, and quinones, often encountered in xenobiotic molecules, is the most common target of reductase enzymes, resulting in the formation of hydrophilic alcohols and hydroquinones that can readily undergo conjugation prior to excretion. This is an important biological function in that it provides a protective mechanism against a host of potentially reactive compounds that are formed intracellularly. Carbonyl-reducing enzymes belong to four main categories: (1) the short-chain dehydrogenase/reductase (SDR, EC 1.1.1) superfamily; (2) the AKR (EC 1.1.1.184) superfamily; (3) the ADH (EC 1.1.1.1) family; or (4) the nicotinamide quinone oxidoreductases (NQO1 and NQO2, EC 1.6.5.2).[73] These enzymes are all principally cytosolic and NADPH-dependent. Few recombinant forms of these enzymes are widely available, greatly complicating efforts to distinguish between the host of carbonyl-reducing enzymes in the cell for metabolism of a given xenobiotic. However, consideration of enzyme localization, cofactor preference, optimal reaction pH, and chemical inhibitor sensitivity, as outlined in a recent review,[73] provide a basis for distinguishing between the major enzyme classes that contribute to carbonyl reduction of xenobiotics.

5.07.2.6.1.1   Short-chain dehydrogenase/reductase superfamily

SDRs are enzymes of great functional diversity found throughout nature. In humans, cytosolic carbonyl reductase (CBR1) is a major member of the SDR superfamily that metabolizes a wide variety of xenobiotics, including the anticoagulant warfarin, anthracycline derivatives like daunorubicin, and aldehyde and ketone products of lipid peroxidation.[74] A recent mouse knockout study demonstrated a critical role for CBR1 in the doxorubicin cardiotoxicity that is attributed to the reduced metabolite, doxorubicinol.[75] CBR3 is a second member of the cytosolic human carbonyl reductases, but its substrate specificity is not well documented. The major human microsomal carbonyl reductase, 11β-hydroxysteroid dehydrogenase, also belongs to the SDR family. Each enzyme demonstrates a cofactor preference for NADPH, transferring the *pro*-S hydrogen to the substrate, i.e., the opposite of ADH.

5.07.2.6.1.2   Aldo-keto reductase superfamily

The AKRs perform oxidoreduction on a wide variety of natural and foreign substrates. A systematic nomenclature for the AKR superfamily similar to the P450 nomenclature has been adopted.[76,135] The superfamily contains several hundred proteins expressed in prokaryotes and eukaryotes that are distributed over 14 families (AKR1–AKR14). The AKR1 family contains aldehyde reductases (AKR1A), aldose reductases (AKR1B), dihydrodiol dehydrogenases, and 3α, 17β and 20α hydroxysteroid dehydrogenases (AKR1C) and keto steroid 5β-reductases (AKR1D). Another family of drug metabolism interest is AKR7, the aflatoxin aldehyde reductases.[77] Crystal structures of many AKRs and their complexes with ligands are available.[78] Each structure has the characteristic (α/β) 8-barrel motif of the superfamily, a conserved cofactor binding site, and a catalytic tetrad comprised of conserved Tyr, Asp, His, and Lys residues.

5.07.2.6.1.3   Quinone oxidoreductases

Two forms of cystosolic quinone oxidoreductase have been described. NQO1 (also known as DT-diaphorase) and NQO2 are FAD-containing enzymes that utilize NAD(P)H and dihydronicotinamide riboside, respectively, as electron donors.[79] From a functional standpoint, NQO1 has been more extensively studied, and is known to act as a chemoprotective enzyme cellular defenses against the electrophilic and oxidizing metabolites of a wide variety of xenobiotic quinones.[80] Obligatory two-electron reduction by the enzyme bypasses the formation of semiquinone radicals, which is important because the semiquinone radical can be reoxidized by molecular oxygen 'futile cycling' with concomitant production of ROS that can lead to cellular damage. NQO1 also participates in reduction of endogenous quinones, such as vitamin E quinone and ubiquinone, generating antioxidant forms of these molecules. Because NQO1 is overexpressed in some tumour types, its enzyme activity has been exploited in the design of anticancer drugs that require reductive bioactivation.[81]

5.07.2.6.1.4   Miscellaneous reductases

Intestinal microflora present in the largely anaerobic environment of the lower gastrointestinal tract are well known to exhibit nitroreductase and azoreductase activities. In terms of specific reductase enzymes, P450 reductase (CPR, EC 1.6.2.4) is a 76 kDa membrane-bound flavoprotein, best recognized in the drug metabolism arena for its role as a coenzyme in P450-dependent oxidative reactions. There the enzyme's role is to transfer electrons, one at a time via its FAD and flavin mononucleotide (FMN) cofactors, from NADPH to P450. However, CPR can also directly reduce a variety of xenobiotics, notably quinones, by one-electron reduction. Under conditions of low oxygen tension, and depending on the redox potential of the ligands, substrates other than molecular oxygen can compete for ferrous cytochrome P450. Notable substrates for the reductive activity of P450 are halogenated hydrocarbons which can undergo reductive dehalogenation. P450-dependent reductive dehalogenation of carbon tetrachloride yielding a trichloroacyl radical that can react with molecular oxygen and initiate lipid peroxidation may be involved in the hepatoxicity of this solvent.

5.07.2.6.2   **Tissue distribution and ontogenic development**

As described above, reductase activity is distributed amongst a host of different enzymes, and systematic studies on many aspects of this diverse enzymology, including tissue distribution and developmental studies of the human enzymes, are lacking. However, carbonyl-reducing enzyme activity is generally present in most human tissues, including the liver, lung, brain, heart, kidney, spleen, testis, and blood, and intracellularly, with a few exceptions, most of these enzymes are cytosolic in origin.[73] Interestingly, NQO1, which has been associated with protection against carcinogenesis and mutagenesis, is expressed in many tissues, including epithelium of the thyroid, breast, colon, and eye, but is expressed at very low levels in the human liver. This contrasts with rat liver, which contains high levels of NQO1 and has been the enzyme's major source for purification.[82]

5.07.3   **Hydrolases**

Hydrolytic reactions in drug metabolism are commonly catalyzed by esterases, amidases, phosphatases, and epoxide hydrolases. While these reactions appear initially quite diverse, in fact each of these enzymes is linked structurally, by membership in the α/β hydrolase-fold family of proteins, and mechanistically, by their reliance on a 'catalytic triad' of acidic and basic residues to effect hydrolysis.[83] Peptidases are not addressed in this chapter but are discussed in depth in 5.06 Principles of Drug Metabolism 2: Hydrolysis and Conjugation Reactions and a recent book.[84]

5.07.3.1   **Epoxide Hydrolases**

Epoxide-containing compounds are cyclic ethers that can constitute a toxic hazard to the cell due to the reactivity of the strained three-membered ring. For example, some reactive polycyclic aromatic hydrocarbon epoxides bind to DNA, leading to mutagenic and carcinogenic effects.[85] In most cases, however, epoxides are of lower reactivity and often constitute intermediates in biological pathways, notably those involving arachidonic acid epoxides.[86] Regardless of the pathway involved, epoxide hydrolases (EC 3.3.2.3) catalyze the nucleophilic addition of water to the substrate.[87]

In most cases, reaction is initiated by attack of an enzyme nucleophile (usually Asp) on the epoxide to generate an acyl intermediate that is subsequently hydrolyzed by a water molecule that has been activated by the other members of the catalytic triad. This results in the formation of vicinal diols, except in the case of leukotriene A4 hydrolases, which catalyze formation of a 5,12-diol product due to the conjugated nature of the substrate.

5.07.3.1.1   Multiplicity and ligand selectivity

Five types of mammalian epoxide hydrolase have been described: (1) microsomal epoxide hydrolase; (2) soluble epoxide hydrolase; (3) cholesterol epoxide hydrolase; (4) leukotriene A4 hydrolase; and (5) hepoxilin A3 hydrolase. The latter three enzymes have tight substrate specificities and do not appear to function in the metabolism of drugs and xenobiotics. The discussion below is, therefore, limited to microsomal epoxide hydrolase, also termed EPHX1, and soluble epoxide hydrolase, or EPHX2.

EPHX1 is a microsomal enzyme that exhibits about 90% sequence homology across species. This enzyme hydrates a variety of alkene and arene oxides, in most cases acting as a detoxifying enzyme by converting electrophilic epoxides to vicinal *trans*-dihydrodiol metabolites which are much less reactive.[88] The enzyme exhibits a broad substrate specificity with a preference for hydrophobic epoxides without extensive substitution. Styrene oxide and benzo(*a*)pyrene 4,5-oxide are common substrate probes for the in vitro determination of EPHX1 activity. Drug substrates are uncommon, but include carbamazepine 10,11-epoxide, an active metabolite of the anticonvulsant drug, carbamazepine, that is inactivated by microsomal epoxide hydrolase. Trichloropropene oxide and cyclohexene oxide are often used as diagnostic inhibitors of the enzyme, exhibiting IC_{50}s around $10\,\mu M$. Valproic acid and valpromide are inhibitors of human EPHX1 at therapeutically relevant concentrations.

EPHX2 is found in high levels in the mouse cytosol, from which the enzyme was originally purified. The enzyme contains an imperfect peroxisome-targeting sequence at the C-terminus and so activity has also been detected in peroxisomes. Like EPHX1, EPHX2 is encoded by a single gene, but there is very little sequence similarity between the two enzymes. EPHX2 is a known hydrate range of aliphatic and extensively substituted, hindered arene oxides. However, current focus is on its physiological role, especially the metabolism of arachidonic, linoleic, and other fatty acid epoxides. These endogenous chemical mediators play an important role in blood pressure regulation and inflammation. It has been proposed that inhibitors of EPHX2 might be useful in the treatment of hypertension or as antiinflammatory agents and several potent and selective amide, carbamate, and urea-based inhibitors of EPHX2 have been developed.[87]

5.07.3.1.2   Structure

Human EPHX1 and EPHX2 contain 455 (55 kDa) and 554 amino acids (65 kDa), respectively. Murine and human EPHX2 have been crystallized as a dimer, both with and without ligands bound.[89] The catalytic domain is located in the C-terminal region of the enzyme. The catalytic triad in EPHX2 is represented by Asp333, His523, and Asp495. In addition, from mutagenesis studies, Tyr381and Tyr465 in the active site have been identified as essential and have been assigned as general acid catalysts that donate a hydrogen bond to the oxygen.

5.07.3.1.3   Tissue distribution and ontogenic development

EPHX1 protein levels have been detected in a wide variety of tissues, including the liver, small intestine, kidney, lung, and urinary bladder. However, significant catalytic activities for styrene oxide hydrolysis were only demonstrated in the liver, lungs, kidneys, and gut.[90] EPHX2 has been detected in all tissues studied except the bile ducts, glomeruli, and the thyroid, and appears to be most abundantly expressed in the hepatocytes, endocrine, proximal tubules in the kidney, and the lymph nodes.[37] Expression of EPHX2 in the pituitary gland could be significant for drug metabolism because the pituitary lacks a blood–brain barrier but contains a high concentration of drug-metabolizing enzymes compared to other regions of the brain.

In the liver and adrenal gland, the activity of EPHX1 was three times higher than that observed in the kidney and lung and could be observed as early as 8 weeks' gestation (**Table 4**). At 22 weeks, EPHX1 activity had increased to about half that observed in adult liver.[25] EPHX2 is also expressed in the fetus, although fewer ontogenic studies have focused on this isoform. EPHX2 activity was demonstrated as early as 14 weeks in fetal liver with no change in activity occurring with time.[25] Adult hepatic activity of EPHX2 was fivefold higher than that observed in the fetus. However, additional studies are needed to examine the change in activity between 30 weeks' gestation and 30 years of age. EPHX2 is also observed in extrahepatic fetal tissue, including the kidney, adrenal glands, intestine, and lungs, but developmental expression in these tissues was less evident.[90]

5.07.3.2   Esterases

These proteins constitute a large and diverse group of enzymes that can hydrolyze peptides, amides, and halides, in addition to carboxylesters (EC 3.1.1), thioesters (EC 3.1.2), and phosphate esters (EC 3.1.3). Carboxylesterases (CESs) are important from a clinical viewpoint because ester derivatives of therapeutic agents are widely used as prodrugs to improve solubility, taste, absorption, bioavailability, and stability and to prolong duration of action.[91] The physiological function of the CESs is obscure, although recent work has shown that they may play a role in cholesterol and fatty acid homeostasis and affect the trafficking of other endoplasmic reticulum proteins such as C-reactive protein.[92] Certain esterases are also of toxicological interest, especially those involved in the metabolism of neurotoxic organophosphates (OPs).[93]

5.07.3.2.1   Nomenclature

The broad early division into A and B esterase classes, proposed by Aldrich in 1953, occurred as a consequence of their differential interaction with OPs such as the insecticide, paraoxon. Whereas A-esterases hydrolyze (inactivate) OP substrates, OPs are mechanism-based inhibitors of B-esterases.[94] OP compounds are widely used as pesticides (mostly insecticides) and a few are nerve gases. The pesticides are usually applied as the relatively nontoxic sulfur derivatives which are bioactivated by microsomal monooxygenases to the oxon forms which are then substrates for the A-esterases. Nerve gases cause neurotoxicity by inhibiting acetylcholinesterase (a B-type esterase). Paraoxonase is an A-type serum esterase. The hydrolysis of neurotoxic OPs is a beneficial, fortuitous activity of serum paraoxonase. The greater toxicity of OP compounds towards birds compared to mammals is due to the absence of serum paraoxonase activity in avian species. B-esterases include the CES which catalyze the hydrolysis of drugs and many other xenobiotics,[92] and which are the focus of the sections below.

5.07.3.2.2   Multiplicity and ligand selectivity

CESs are largely microsomal in origin, with molecular weights of 55–60 kDa. At least four families (CES1–CES4, EC 3.1.1.1) exist, based on sequence similarity, and in humans, the liver (CES1A1, hCE1) and intestinal forms (CES2, hCE2) appear to play the most important roles in detoxication/bioactivation of xenobiotics.[95] hCE1 metabolizes heroin and cocaine and is relatively selective for several of the angiotensin-converting enzyme inhibitors, such as delapril and imidapril, whereas hCE2 is more selective for irinotecan and oxybutynin. Therefore, hCE1 often appears to be associated with the removal of small (methyl, ethyl) groups, whereas hCE2 seems to prefer larger moieties, although this is far from a strict rule.[92]

Some broad-spectrum esterase inhibitors have been in use for many years. Liver and intestinal microsomal esterases can be inhibited by compounds such as bis-nitrophenyl phosphate (BNPP), which phosphonylates an active-site serine residue in the B-esterases. A-esterases can be inhibited by p-chloromercurobenzoate. More recently, aromatic diones, such as benzil, have been identified as general inhibitors of the CESs, and a series of benzene sulfonamides have been described that act as selective potent inhibitors of hCE2.[96]

5.07.3.2.3   Structure

Microsomal CESs are typically about 60 kDa in size and glycosylated, with the carbohydrate modification seemingly necessary for activity. The crystal structure of hCE1 in complex with various ligands has been solved, revealing the serine hydrolase fold common to the esterase family (see **Figure 2** for an example). The enzyme exhibits a large, hydrophobic active site cavity some 15 Å deep, suitable for the binding of a wide variety of substrate molecules, as has been observed experimentally.[97]

5.07.3.2.4   Tissue distribution and ontogenic expression

CES expression is low in the plasma, in contrast to the highly abundant cholinesterase, but high in liver, intestine, and kidney. The cellular localization of CES is predominantly microsomal, although lower levels have been reported in mitochondria, lysosomes, and cytosol (possibly artifactual). Few studies have been performed as yet on the ontogenic regulation of the individual CES isoforms.

5.07.4   Transferases

5.07.4.1   UDP-Glucuronyltransferases

The UDP-Glucuronyltransferases (UGT, EC 2.4.1.17) superfamily is comprised of the *UGT1*, *UGT2*, *UGT3*, and *UGT8* gene families which code for enzymes that attach glycosyl groups to lipophilic substrates.[136] Xenobiotic glucuronidation

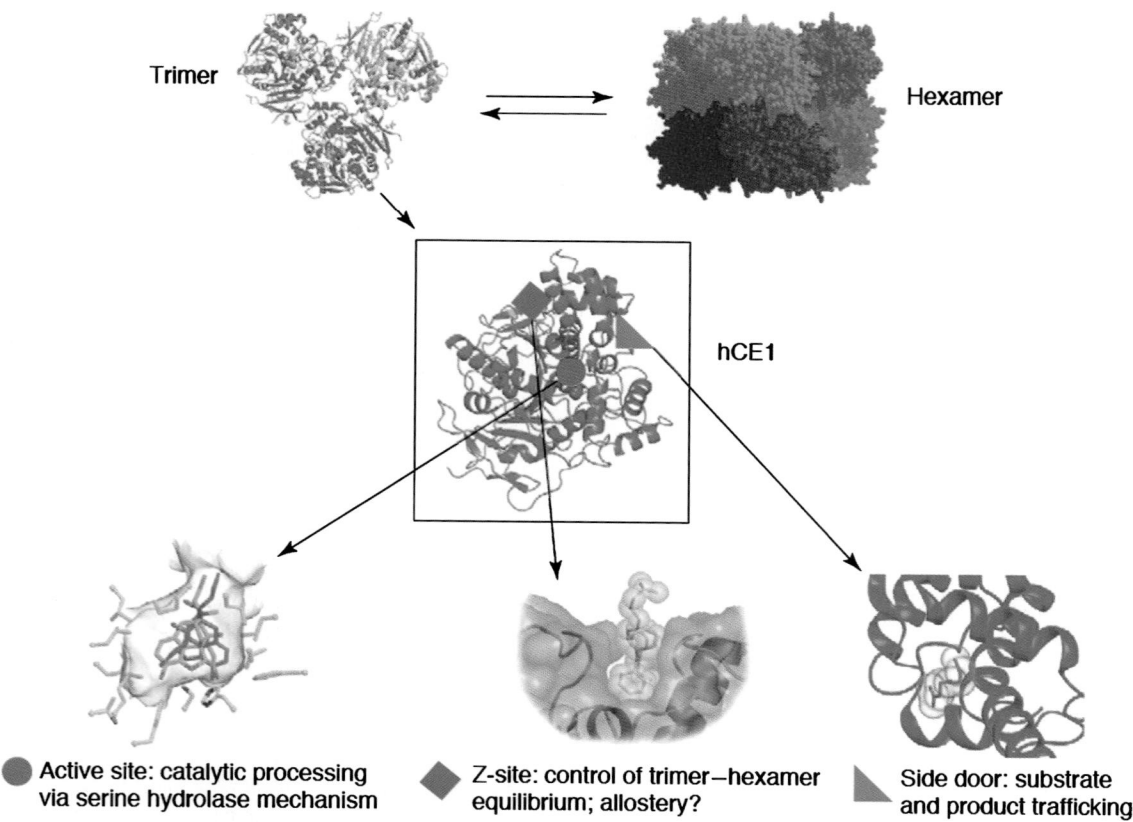

Figure 2 Structural features of human CES 1. The enzyme exists in a trimer–hexamer equilibrium, and each protein monomer contains three ligand-binding sites. The active site (red) incorporates the catalytic triad of the enzyme and facilitates the docking of structurally distinct substrates; in some cases, substrates dock in more than one orientation simultaneously, as illustrated for the heroin analog naloxone (green and gold). The Z-site (blue) is a surface ligand-binding groove that might be allosteric and controls the trimer–hexamer equilibrium of the enzyme. It is relatively nonspecific, and is shown here with the anticancer drug tamoxifen (yellow) bound. The side-door site (magenta) is a secondary pore to the active site that could facilitate the release of product (as indicated here for a fatty acid (cyan)) or the entrance of substrate. (Reproduced from Redinbo, M.; Potter, P. *Drug Disc. Today* **2004**, *10*, 313–325, copyright (2004), with permission from Elsevier.[132])

is catalyzed by glucuronyltransferase enzymes that specifically utilize the cofactor UDPGA).[98] UGTs are ~60 kDa, microsomal (and nuclear) enzymes which, in contrast to the P450s, are localized to the luminal side of the endoplasmic reticulum. In addition to drugs and xenobiotics, there also many 'endogenous' glucuronide acceptors, e.g., bile acids and steroids. Regardless, a common endpoint is enhanced polarity of the original aglycone and facilitated renal excretion. In some cases, glucuronide conjugates may themselves be pharmacologically active. For example, morphine forms both 3-OH and 6-OH glucuronides, with the latter exhibiting more activity than parent drug, at least with some classes of opiate receptors. In addition, acyl glucuronides can have intrinsic chemical reactivity and have been associated with immune toxicities that can occur with some nonsteroidal antiinflammatory drugs.[99]

5.07.4.1.1 Nomenclature

Human UGTs are composed of two gene families, *UGT1* and *UGT2*, and three subfamilies that are designated *UGT1A*, *UGT2A*, and *UGT2B*. Within each family, individual enzymes, e.g., UGT2A1, UGT2B4, share at least 45% sequence homology and subfamily members, e.g., UGT1A1 and UGT1A3, share at least 60% homology. The human *UGT1* gene locus is located on the long arm of chromosome 2, and the *UGT2* genes are found on the long arm of chromosome 4. UGT2 enzymes are each conventionally encoded by separate genes, but, unusually, UGT1 enzymes share a common mRNA transcript that is differentially spliced. Consequently, all UGT1A isoforms share four common C-terminal exons and are distinguished by a variable exon 1 sequence. UGT2 enzymes also share more sequence homology in their

C-termini, and so substrate selectivity is believed to reside largely in the hypervariable N-terminal regions of UGT enzymes.[98] The nomenclature for individual UGT isoforms follows the system established for the P450s, i.e., UGT1A1, UGT2B7, etc.

5.07.4.1.2 Multiplicity and ligand selectivity

At present, 17 functional human UGTs have been identified: UGT1A1, 1A3, 1A4, 1A5, 1A6, 1A7, 1A8, 1A9, 1A10, 2A1, 2B4, 2B7, 2B8, 2B10, 2B11, 2B15, and 2B17. Numerous heterologous expression systems have been developed, and membrane preparations containing most of the recombinant human isoforms are commercially available.[100] Our knowledge of their substrate specificity comes largely from in vitro studies with the recombinant proteins, but also in some cases from clinical and pharamcogenetic studies performed in vivo.[99]

UGT1A1 is probably the most widely studied isoform because it is the most abundant UGT in liver, and it has long been recognized to be the primary enzyme responsible for the glucuronidation of bilirubin. A common genetic polymorphism in the promoter region of the gene (*UGT1A1*28*) underlies the relatively benign hyperbilirubinemia or Gilbert's disease, whereas rare polymorphisms in the coding region of the *UGT1A* gene lead to the much more serious Crigler–Najjar syndromes.[101] In addition to bilirubin, an important substrate for UGT1A1 is SN-38, the active metabolite of the anticancer drug irinotecan. Accumulation of SN-38, prominent in carriers of *UGT1A1*28*, causes dose-limiting toxicity. Human UGT1A4 and UGT2B7 have also attracted considerable attention because of their roles in amine glucuronidation and opiate metabolism, respectively. UGT1A4 is of particular interest because many drugs contain imidazoles and tetrazoles, and because aryl amine glucuronidation may be a modulating factor in colon and bladder cancer.[102]

Selective inhibitors of several human UGT isoforms are known, but these are generally low-potency compounds that are also used as selective substrates for the enzyme. Examples include bilirubin for UGT1A1, propofol for UGT1A9, and 3′-azido-3′-deoxythymidine for UGT2B7.[103]

5.07.4.1.3 Structure

No crystal or solution structures have yet been solved for the UGTs. Evidence does exist that the recombinant proteins can be *N*-glycosylated and phosphorylated[104] but the relevance of these posttranslational modifications to enzyme function is not well understood. The UDPGA cofactor likely binds mainly to the conserved C-terminus of the protein, whereas binding determinants for the aglycone probably reside in the more variable N-terminus. An important step in the catalytic mechanism is activation of the xenobiotic substrate to an optimized nucleophile prior to displacement of glucuronic acid from UDPGA. Site-directed mutagenesis experiments implicate histidine residues in enzyme function,[105] but whether such basic residues are specifically involved in this substrate activation step remains to be determined.

5.07.4.1.4 Tissue distribution

Most of the drug-conjugating UGT isoforms are expressed in the liver, but are also present at variable levels in extrahepatic tissues, particularly tissues that are responsible for the absorption or excretion of drugs, such as intestine, lung, and kidney. Human hepatic UGT enzymes include UGT1A1, UGT1A3, UGT1A4, UGT1A6, UGT1A9, UGT2B4, UGT2B7, UGT2B11, and UGT2B17.[102] The gastrointestinal tract contains UGT1A1, UGT1A8 (restricted to the colon) and UGT1A10, whereas UGT1A7 seems to be localized to in the esophagus, stomach, and lung. Human kidney UGT enzymes include UGT1A8, UGT1A9, UGT1A10, and UGT2B7. In fact, UGT2B7 levels in the kidney are similar to those found in the liver.[106] UGT2B transcripts are found in steroid-sensitive tissues, including the prostate (UGT2B17) and mammary tissues (UGT2B11 and UGT2B28).[99] Glucuronidation reactions are also readily detectable in the brain and placenta.[102]

Intracellularly, UGT enzymes are embedded in the internal membrane and face the lumenal side of the endoplasmic reticulum which provides these enzymes ready access to the metabolic products of drug oxidation. However, this subcellular localization limits the access of substrates, cofactors, and glucuronidated products to and from the active sites of UGT enzymes. It is also responsible for the phenomenon known as 'latency' and contributes to difficulties in extrapolating in vivo effects of UGT enzymes from in vitro experiments using isolated tissue microsomes. To facilitate substrate access for in vitro studies of UGT activity, microsomes are often treated with low concentrations of detergent or a pore-former like allomethicin.

5.07.4.1.5 Ontogenic development

Developmental changes in UGT activity have been extensively studied both in vivo and in vitro using morphine as a substrate. Morphine glucuronidation to both morphine-6-glucuronide and morphine-3-glucuronide is mediated mainly by UGT2B7.[107] Morphine was found to undergo significant glucuronidation by the fetus liver. In vitro studies in hepatic microsomes obtained from fetuses (15–27 weeks) indicated that the glucuronidation rates were 10–20% of that observed in adult microsomes.[108] In addition, the mean rate of morphine glucuronidation in fetal livers obtained after hysterectomy was twofold higher than that obtained from induced abortion livers, suggesting a possible regulatory mechanism for UGT activity related to the birth process (Table 4).[25] The glucuronidation of morphine in vivo has also been demonstrated in premature neonates as young as 24 weeks of gestation. Studies with other substrates which are mainly or partially glucuronidated by UGT2B7, such as naloxone, an opiate agonist, benzodiazepines, and nonsteroidal antiinflammatory drugs are all suggestive of a reduced glucuronidation ability in neonates compared to adults.[109]

The developmental regulation of UGT enzymes was also demonstrated with acetaminophen, which is glucuronidated mainly by UGT1A6 and to some extent by UGT1A9.[110] Acetaminophen glucuronide formation is undetectable in the fetus; low levels are seen after birth and only reach adult levels of glucuronidation after 10 years of age.[111] This lack of UGT activity is compensated, however, by the high activity of sulfotransferases in infants and young children.[112]

5.07.4.2   *N*-Acetyltransferases

Arylamine *N*-acetyltransferases (NAT, EC 2.1.1) transfer an acetyl group from the cofactor acetyl-CoA to lipophilic molecules that contain a primary amino group, notably hydrazines and arylamines.[113] Oftentimes, an arylamine function is first unmasked by N-dealkylation or nitro, azo, or amide hydrolysis prior to acetylation. All NATs contain a conserved active-site cysteine residue that is critical to catalysis, by activating the acetyl CoA cofactor. This cysteine forms part of an active-site catalytic triad composed of Cys-His-Asp that is conserved in all prokaryotic and eukaryotic NAT homologs.[114]

NAT activity has been found in most organisms and mammals, and NAT proteins have been isolated from bacteria, rats, hamsters, and humans. Human NATs are encoded at three separate loci on chromosome 8. One of the loci contains a nonexpressed pseudogene (NAT3). The other two loci contain genes for NAT1 and NAT2 which encode 35 kDa, cytosolic enzymes that share 87% nucleotide and 81% amino acid sequence identities. NAT1 is ubiquitously expressed and catalyzes the acetylation of so-called 'monomorphic' substrates such as *p*-aminosalisylic acid and *p*-aminobenzoic acid. NAT2 is expressed primarily in the liver and intestinal mucosa and catalyzes the acetylation of what have been termed 'polymorphic' substrates, including sulfamethazine, isoniazid, dapsone, sulfamethoxazole, procainamide, hydralazine, and caffeine. NAT polymorphisms underlie polymorphic N-acetylation in drug metabolism. For example, the antituberculosis drug isoniazid displays a bimodal distribution of elimination kinetics within several populations, due to 'fast' versus 'slow' acetylation NAT polymorphisms. This was one of the first documented examples of a polymorphic drug response.[115] Several adverse drug reactions have been linked to altered acetylator status, often related to redox cycling of hydroxylamine metabolites formed by an alternative P450 pathway. Functional polymorphism of NAT2 has also been associated with individual differences in susceptibility to occupational and smoking-related bladder cancer. Many epidemiological studies into NAT-related cancer susceptibility have been carried out and analysis of these combined studies has shown that the role of NAT2 in predisposition to cancer depends on the type of cancer, the ethnicity of the population, and probably also on the level of exposure to carcinogens.[116] Limited information is available on developmental aspects of the NATs.

5.07.4.3   Methyltransferases

The *O*-, *S*-, and *N*-methyltransferases (EC 2.1.2) discussed below all catalyze the transfer of a methyl group from the cofactor *S*-adenosylmethionine to both endogenous and xenobiotic substrates.[117]

5.07.4.3.1   Catechol *O*-methyltransferase

Catechol *O*-methyltransferase (COMT) typically metabolizes catecholamines and estrogens, forming mixtures of *ortho* and *para* methoxy metabolites[118] Drug substrates include L-dopa, 2-hydroxy ethinylestradiol and isoproteranol and COMT inhibitors have therapeutic value in the treatment of Parkinsonism. The enzyme is the product of a single gene located on chromosome 22q11.2, but two forms of the protein exist: soluble COMT (S-COMT, ~ 25 kDa) and membrane-bound COMT (M-COMT, ~ 30 kDa), whose additional 50 amino acids provide the hydrophobic anchor for membrane localization. S-COMT predominates in peripheral tissues and M-COMT is the main form of the enzyme in brain tissue.[119] COMT activity is inherited in an autosomal recessive manner. Individuals with low activity inherit a form of the enzyme that is thermolabile. A single G to A transition at codon 108/158 of the cytosolic/membrane gene results in a Val to Met substitution which forms the molecular basis for the well-recognized interindividual variation in the metabolism and response to the antihypertensive drug, α-methyldopa.[117]

5.07.4.3.2 *S*-Methyltransferases

There are relatively few drugs that undergo *S*-methylation but the enzymes are important for the detoxication of xenobiotic thiol compounds which tend to be toxic, and clinically significant because of polymorphisms that occur in thiopurine methyltransferase (TPMT). The gene for human TPMT, located on chromosome 6p22, codes for a 28 kDa cytosolic enzyme that catalyzes the methylation of aromatic and heteroaromatic thiols.[117] Substrate groups include thiophenols and the oncolytic and immunosuppressive agents 6-mercaptopurine (6-MP) and azathioprine. Polymorphic disposition of these two latter drugs is important in cancer therapy, because a poor metabolizer phenotype is a major risk factor for severe hemotoxicity resulting from an exaggerated pharmacological effect of thioguanine nucleotides in nontarget hemopoietic cells. The distribution of TPMT activity is trimodal: approximately 1 in 300 individuals are poor metabolizers requiring substantial dose reduction of 6-MP and azathioprine. The TPMT polymorphism is as a paradigm for personalized medicine in the field of pharmacogenomics.[120]

5.07.4.3.3 *N*-Methyltransferases

N-Methyltransferases are ~30 kDa cytosolic proteins that are localized in a number of tissues, including the gastrointestinal tract, bronchioles, kidney, and brain. There are only a few examples of drugs which undergo *N*-methylation, but two important endogenous substrates are histamine and nicotinamide. For more details on methyltransferases and *N*-acetyltransferases, *see* 5.06 Principles of Drug Metabolism 2: Hydrolysis and Conjugation Reactions.

5.07.4.4 Glutathione *S*-Transferases

Glutathione *S*-transferase (GST, EC 2.5.1.18) enzymes catalyze conjugation of xenobiotics and endogenous compounds with reduced GSH. Two superfamilies of GST proteins are recognized. Membrane-associated proteins involved in eicosanoid and GSH metabolism (MAPEG) family members are involved in several important physiological processes, including the synthesis of peptido-leukotrienes and the 5-lipoxygenase pathway, but do not seem to have a role in xenobiotic metabolism. Conversely, the cytosolic GSTs (cGSTs) metabolize a wide range of environmental agents and numerous drugs.[121] cGSTs have a particularly important toxicological role in scavenging reactive electrophilic species generated within the cell by various processes, and a large body of literature exists on the role of cGSTs in cancer.[122]

5.07.4.4.1 Multiplicity and ligand specificity

cGST isoforms, like the other drug-metabolizing superfamilies we have already considered, are classified on the basis of sequence homology. However, cGST nomenclature is quite distinct because each of the six gene subfamilies is represented by a letter; i.e. A- (alpha-), M- (mu-), O- (omega), P- (pi), T- (theta), and Z- (zeta).[123] Within these six subfamilies there exist a total of 16 functional mammalian *cGST* genes, i.e., *GSTA1–GSTA5, GSTM1–GSTM5, GSTO1–GSTO2, GSTP1, GST1–GST2*, and *GSTZ1*.[123] Because cGSTs exist as dimers, a second number is added to the specific subfamily member to designate the dimer subunit composition, e.g., GSTA1-1 Finally, cGSTs usually exist as homodimers, but heterodimers are also known.

cGSTs are versatile catalysts that typically carry out nucleophilic addition and nucleophilic replacement reactions with GSH. However, these enzymes can also catalyze isomerization of double bonds without net consumption of GSH. 1-Chloro-2,4-dinitrobenzene (CDNB) is considered a 'universal' substrate for the cGSTs, although some T-isoforms have no activity. Class- and isoform-dependent substrate selectivity is broad and overlapping, as with the other detoxification enzymes. A few generalizations are: M-class GSTs have high activity toward planar aromatic hydrocarbon epoxides. T-class have high affinity for aryl sulfates and catalyze desulfation. A-class GSTs have relatively high activity toward organic peroxides. GSTA4-4 is selective for lipid hydroxy-enals (4-hydroxy-nonenal, HNE). GSTP1-1 may have a 'unique' function related to cancer cell response in that it inhibits c-Jun kinase, thereby regulating signal transduction pathways involved in apoptotic/proliferative responses.[124]

Given their functional versatility and anticipated role in a variety of biological processes, GST inhibitors could have the potential to impact a variety of therapeutic categories.[125] Historically, GST inhibitors have been *S*-linked conjugates of GSH with additional hydrophobic groups. In essence, these inhibitors are products of the reaction, and the fact that they are useful in vitro inhibitors reflects the product inhibition kinetics often observed for these enzymes. Such inhibitors tend not to be isoform-selective, which is not surprising because all isoforms have a GSH-binding site (G-site). Some progress has been made towards the development of isoform-selective inhibitors of GSTP1-1, wherein substitution of the GSH backbone is accommodated by this isoform, but not by others.[126]

5.07.4.4.2   Structure

X-ray structures are available for human A-, P-, M-, T-, and S-class cGSTs.[124] The overall subunit structure is very similar across these classes and contains two distinct domains. The N-terminal domain possesses the G-site that binds nucleophilic GS-, which is stabilized, for example in GSTA, by hydrogen bonding to a conserved Arg residue. The much larger C-terminal domain contributes to the xenobiotic binding site (H-site) that lies between the domains. In some cases, the hydrophobic H-site includes an appropriate residue that can aid in the catalysis of specific substrates. For example, in GSTM an H-site Tyr provides a general base for the 'leaving' oxygen of epoxide substrates.

5.07.4.4.3   Tissue distribution and ontogenic development

Human cGST enzymes are expressed in the cytosol of all organ tissues examined, although their involvement in xenobiotic metabolism is due principally to enzymes expressed in the liver, intestine, and kidney.[127]

Tissue-specific ontogenic expression of GST has been demonstrated for hepatic GSTA1 and GSTA2, both of which were detected as early as 10 weeks' gestation age, increasing to adult levels within the first 2 years of life.[25] GSTM(1–5) was also detected in fetal liver samples, albeit at lower levels, and increased to adult levels following birth. In contrast, GSTP1 expression has been detected in fetal liver samples (10–22 weeks' gestation age), but declined in the second and third trimester and was virtually undetectable in adult samples. In the fetal lung, GSTP1 is the major GST isozyme, detectable in lung tissue at less than 20 weeks' gestation, and decreased continuously in the developing fetus and infant after birth. In fetal kidney at less than 20 weeks' gestation, both GSTA1 and GSTA2 were detected and increased to adult levels within the first 2 years of life. GSTM levels, however, decreased between the fetal and postnatal samples, while GSTP1 levels remained constant during kidney development.

5.07.4.5   Sulfotransferases

Sulfation of drugs and xenobiotics is catalyzed by cytosolic sulfotransferases (SULTs, EC 2.8.1) that utilize the cofactor 3′-phosphoadenosine 5′-phosphosulfate (PAPS).[128] Membrane-bound forms of the enzyme are known which sulfate heparins, tyrosines in proteins, and proteoglycans, but these enzymes have little or no activity toward drugs. The reactivity of cytosolic SULTs overlaps with the UGTs, but the substrate range is not as diverse. SULTs are mostly active with phenols, alcohols, and arylamines, but many endogenous substrates are also recognized, e.g., steroids, bile acids, neurotransmitters, and carbohydrates. As with the glucuronidation, sulfation reactions usually result in more polar species that are readily excreted in the urine. However, SULTs are also implicated in toxicological reactions if nonenzymatic loss of SO_4^{2-} occurs to generate a relatively stable carbocation, e.g., from N-arylamines and benzylic alcohols, that can react with proteins and DNA.[129]

5.07.4.5.1   Multiplicity and ligand specificity

At least 10 human cytosolic SULT isoforms belonging to three SULT families are known. The human SULT1 family is the most diverse, being comprised of SULT1A1, 1A2, 1A3, 1B1, 1C2, 1C4, and 1E1. In addition, SULT2A1, 2B1, and 4A1 have been described in humans.[130] As with the UGTs, family members share at least 45% sequence homology, and subfamily members share at least 60% homology. This nomenclature system has replaced earlier, more confusing designations that usually reflected their substrate specificity, e.g., estradiol sulfotransferase, phenol sulfotransferase (PST), or hydroxysteroid sulfotransferase (HST). In terms of their substrate specificities, most information is available for the SULT1A and 1E isoforms which are recognized to metabolize preferentially a variety of simple phenols, estrogens, and catechols. Selective inhibitors for the various SULT isoforms are not available at present, although some general competitive inhibitors for the PAPS-binding site have been described.[131]

5.07.4.5.2   Structure

Crystal structures for several SULTs are available. These enzymes are globular proteins with α-helices on both sides of a characteristic 5 stranded β-sheet. Despite their discrete susbtrate specifcities, crystallographic evidence indicates that membrane-bound and cytosolic SULTs share the same overall structural fold.[128] The PAPS-binding site is highly conserved between cytosolic and membrane SULTS, but the sulfate-acceptor binding site is very different. **Figure 3** depicts the crystal structure of SULT1A1 complexed with an inhibitor as well as PAPS.

Structural and site-directed mutagenesis data support the involvement of His107 as the active site base that activates the incipent nucleophilic substrate, with additional critical roles for Ser137 and Arg47 in the catalytic mechanism of the enzyme.

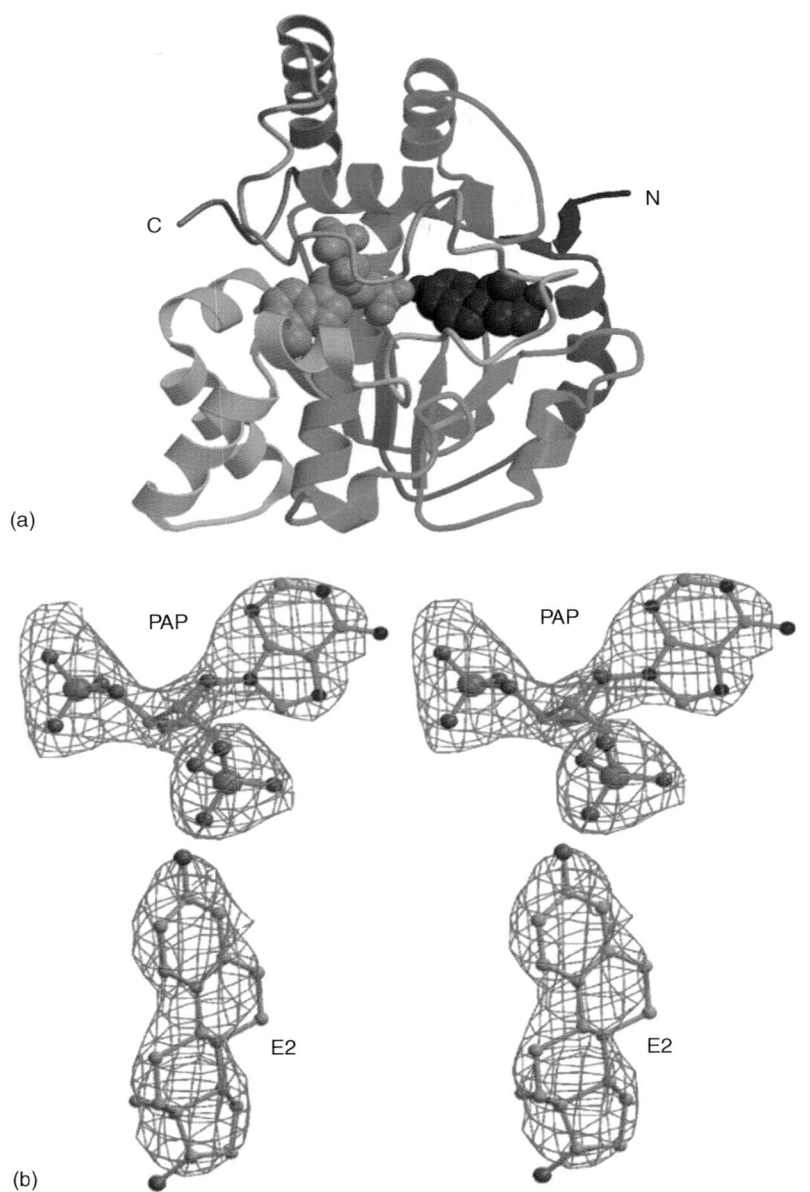

(a)

(b)

Figure 3   The crystal structure of human SULT1A1 complexed with estradiol (E$_2$) and the cofactor (a PAP analog). The Cα trace is colored blue (N-terminus) to red (C-terminus). Secondary structure elements are represented as coils for helices and arrows for strands. The bound ligands are shown as space-filling models: PAP, pink; E$_2$, dark gray. (Reproduced from Gamage, N. U.; Tsvetanov, S.; Duggleby, R. G.; McManus, M. E.; Martin, J. L. *J. Biol. Chem.* **2005**, *280*, 41482–41486, copyright (2005) with permission from the American Society for Biochemistry & Molecular Biology.[133])

5.07.4.5.3   Tissue distribution and ontogenic development

The SULT1 family, which comprises at least 11 isoforms, is highly expressed in the liver, but is also present in other extrahepatic tissues.[130] SULT1A1 is mainly a liver form of the enzyme but is also detectable in the brain, breast, intestine, endometrium, kidney, lung, and platelets. Transcripts for the *SULT1B* subfamily, which is mainly involved in the sulfation of thyroid hormones, have been detected in the liver, colon, small intestine, and blood leukocytes, whereas human *SULT1C2* transcript is detected primarily in the thyroid, stomach, and kidney. *SULT1C4* transcripts are found predominantly in fetal kidney and lung and have been detected in the adult brain and ovary. Human SULT1E1 and 1E2 are found in steroid hormone-responsive tissues including the endometrium, testis, breast, adrenal gland, and placenta, in addition to the liver and intestine. The SULT2 family, including transcripts for *SULT2A1* and *SULT2A2*, have been found in adrenal cortex, liver, brain, and intestine.

The ontogeny of SULTs was studied recently by Richard *et al.*[112] in the developing liver, lung, and brain. SULT1A1 enzymatic activity was higher in fetal tissues than in postnatal liver. Also SULT1A3 was expressed at higher levels in early stages of development, decreased gradually in the late fetal/early neonatal liver and was not observed in adult liver. In the lung, high SULT1A3 activity was observed in the fetus compared to neonatal levels. Therefore, the developing fetus clearly possesses significant sulfation capabilities. This could be a consequence of the important role sulfation plays in the homeostasis of hormones and other endogenous compounds that are important for development or because of the need for xenobiotic detoxification in the fetus where other conjugating enzymes, notably the UGTs, are not expressed at high levels until the neonatal stage.

5.07.5   Closing Remarks

The enzymology of drug metabolism has advanced considerably in the last 20 years, with the advent of molecular biology techniques that have provided detailed sequence information upon which to base rational nomenclature systems, and recombinant enzymes essential to the detailed study of enzyme–ligand interactions. Indeed, contemporary drug discovery and development rely heavily on this advancing lexicon of information to ensure that preclinical drug candidates are chosen with appropriate metabolic stabilities and interaction liability characteristics. Central to these activities is the identification of the specific enzymes involved in the metabolic clearance of a new chemical entity, a process which reaches its zenith for the P450s, due to the wealth of specific substrates, inhibitors, and recombinant enzymes that exist. The increasing availability of crystal structures for human P450s is also propelling efforts to understand and predict P450–ligand interactions in silico. However, impediments to the full implementation of such P450 modeling approaches include the relative dearth of structures with isoform-specific ligands bound and the demonstrated elasticity of mammalian P450 active sites, and research to address these issues will likely be priority items for the future. Much larger knowledge gaps exist for some of the other drug-metabolizing enzymes, notably the reductases, CESs, and glucuronyltransferases, where few isoform-specific substrates and inhibitors are available and recombinant human enzymes are only widely available for the human UGTs. In the case of the reductases, a systematic understanding of the basic enzymology of xenobiotic reduction is still in its infancy. These are important considerations because non-P450 clearance pathways are increasingly emphasized in drug development in order to avoid or minimize drug–drug and drug–gene interactions. Even with a more complete toolbox for the non-P450 enzymes, a much better appreciation of individual tissue isoform concentrations (as opposed to mRNA) will be needed to progress towards realistic scaling of in vitro data to the in vivo situation. While incomplete, experience and advances within the P450 field can serve as a useful template for developing experimental strategies to integrate better non-P450 enzymology into drug development.

References

1. Guengerich, F. P. *Mol. Interv.* **2003**, *3*, 194–204.
2. Ioannides, C.; Lewis, D. F. *Curr. Top. Med. Chem.* **2004**, *4*, 1767–1788.
3. Williams, J. A.; Hyland, R.; Jones, B. C.; Smith, D. A.; Hurst, S.; Goosen, T. C.; Peterkin, V.; Koup, J. R.; Ball, S. E. *Drug Metab. Dispos.* **2004**, *32*, 1201–1208.
4. Nelson, D. R.; Schuler, M. A.; Paquette, S. M.; Werck-Reichhart, D.; Bak, S. *Plant Physiol.* **2004**, *135*, 756–772.
5. Lewis, D. F. *Pharmacogenomics* **2004**, *5*, 305–318.
6. Nelson, D. R. *Methods Mol. Biol.* **1998**, *107*, 15–24.
7. Nelson, D. R.; Zeldin, D. C.; Hoffman, S. M.; Maltais, L. J.; Wain, H. M.; Nebert, D. W. *Pharmacogenetics* **2004**, *14*, 1–18.
8. Shimada, T.; Fujii-Kuriyama, Y. *Cancer Sci.* **2004**, *95*, 1–6.
9. Rodrigues, A. D. *Biochem. Pharmacol.* **1999**, *57*, 465–480.
10. Werck-Reichhart, D.; Feyereisen, R. *Genome Biol.* **2000**, *1*, REVIEWS3003.1–3003.9.
11. Johnson, E. F. *Drug Metab. Dispos.* **2003**, *31*, 1532–1540.
12. Schoch, G. A.; Yano, J. K.; Wester, M. R.; Griffin, K. J.; Stout, C. D.; Johnson, E. F. *J. Biol. Chem.* **2004**, *279*, 9497–9503.
13. Williams, P. A.; Cosme, J.; Vinkovic, D. M.; Ward, A.; Angove, H. C.; Day, P. J.; Vonrhein, C.; Tickle, I. J.; Jhoti, H. *Science* **2004**, *305*, 683–686.
14. Williams, P. A.; Cosme, J.; Ward, A.; Angove, H. C.; Matak Vinkovic, D.; Jhoti, H. *Nature* **2003**, *424*, 464–468.
15. Yano, J. K.; Hsu, M. H.; Griffin, K. J.; Stout, C. D.; Johnson, E. F. *Nat. Struct. Mol. Biol.* **2005**, *12*, 822–823.
16. Wester, M. R.; Yano, J. K.; Schoch, G. A.; Yang, C.; Griffin, K. J.; Stout, C. D.; Johnson, E. F. *J. Biol. Chem.* **2004**, *279*, 35630–35637.
17. Guengerich, F. P. Human Cytochrome P-450 Enzymes. In *Cytochrome P-450 Structure, Mechanism, and Biochemistry*, 3rd ed.; Montellano, O. d., Ed.; Kluwer Academic/Plenum Publishers: New York, 2005, pp 377–463.
18. de Wildt, S. N.; Kearns, G. L.; Leeder, J. S.; van den Anker, J. N. *Clin. Pharmacokinet.* **1999**, *37*, 485–505.
19. Lin, J.; Chiba, M.; Baillie, T. *Pharm. Rev.* **1999**, *51*, 135–157.
20. Ilett, K. F.; Tee, L. B.; Reeves, P. T.; Minchin, R. F. *Pharmacol. Ther.* **1990**, *46*, 67–93.
21. Krishna, D. R.; Klotz, U. *Clin. Pharmacokinet.* **1994**, *26*, 144–160.

22. Ding, X.; Kaminsky, L. S. *Pharmacol. Toxicol. Annu. Rev. Pharmacol. Toxicol.* **2003**, *43*, 149–173.
23. Thummel, K.; Kunze, K.; Shen, D. *Adv. Drug Del. Rev* **1997**, *27*, 99–127.
24. Hines, R. N.; McCarver, D. G. *J. Pharmacol. Exp. Ther.* **2002**, *300*, 355–360.
25. McCarver, D. G.; Hines, R. N. *J. Pharmacol. Exp. Ther.* **2002**, *300*, 361–366.
26. Kearns, G. L. *Curr. Opin. Pediatr.* **1995**, *7*, 220–233.
27. Shimada, T.; Yamazaki, H.; Mimura, M.; Wakamiya, N.; Ueng, Y. F.; Guengerich, F. P.; Inui, Y. *Drug Metab. Dispos.* **1996**, *24*, 515–522.
28. Hakkola, J.; Tanaka, E.; Pelkonen, O. *Pharmacol. Toxicol.* **1998**, *82*, 209–217.
29. Yang, H.-Y.; Namkung, M.; Juchau, M. *Biochem. Pharmacol.* **1995**, *49*, 717–726.
30. Sonnier, M.; Cresteil, T. *Eur. J. Biochem.* **1998**, *251*, 893–898.
31. Stevens, J. C.; Hines, R. N.; Gu, C.; Koukouritaki, S. B.; Manro, J. R.; Tandler, P. J.; Zaya, M. J. *J. Pharmacol. Exp. Ther.* **2003**, *307*, 573–582.
32. Koukouritaki, S. B.; Manro, J. R.; Marsh, S. A.; Stevens, J. C.; Rettie, A. E.; McCarver, D. G.; Hines, R. N. *J. Pharmacol. Exp. Ther.* **2004**, *308*, 965–974.
33. Pasanen, M.; Pelkonen, O.; Kauppila, A.; Park, S.; Friedman, F.; Gelboin, H. *Pharmacol. Ther. Dev. Pharmacol. Ther.* **1987**, *10*, 125–132.
34. Omiecinski, C. J.; Redlich, C. A.; Costa, P. *Cancer Res.* **1990**, *50*, 4315–4321.
35. Tateishi, T.; Nakura, H.; Asoh, M.; Watanabe, M.; Tanaka, M.; Kumai, T.; Takashima, S.; Imaoka, S.; Funae, Y.; Yabusaki, Y. et al. *Life Sci.* **1997**, *61*, 2567–2574.
36. Gu, J.; Su, T.; Chen, Y.; Zhang, Q. Y.; Ding, X. *Toxicol. Appl. Pharmacol.* **2000**, *165*, 158–162.
37. Enayetallah, A. E.; French, R. A.; Thibodeau, M. S.; Grant, D. F. *J. Histochem. Cytochem.* **2004**, *52*, 447–454.
38. Wu, S.; Moomaw, C. R.; Tomer, K. B.; Falck, J. R.; Zeldin, D. C. *J. Biol. Chem.* **1996**, *271*, 3460–3468.
39. Krueger, S. K.; Williams, D. E. *Pharmacol. Ther.* **2005**, *106*, 357–387.
40. Lang, D.; Kalgutkar, A. S. Non-P450 Mediated Oxidative Metabolism of Xenobiotics. In *Drug Metabolizing Enzymes*; Lee, J., Obach, R. S., Fisher, M. B., Eds.; Marcel Dekker: New York, 2003, pp 483–540.
41. Ziegler, D. M. *Drug Metab. Rev.* **2002**, *34*, 503–511.
42. Hernandez, D.; Janmohamed, A.; Chandan, P.; Phillips, I. R.; Shephard, E. A. *Pharmacogenetics* **2004**, *14*, 117–130.
43. Lawton, M. P.; Cashman, J. R.; Cresteil, T.; Dolphin, C. T.; Elfarra, A. A.; Hines, R. N.; Hodgson, E.; Kimura, T.; Ozols, J.; Phillips, I. R. et al. *Arch. Biochem. Biophys.* **1994**, *308*, 254–257.
44. Hines, R. N.; Hopp, K. A.; Franco, J.; Saeian, K.; Begun, F. P. *Mol. Pharmacol.* **2002**, *62*, 320–325.
45. Dolphin, C. T.; Janmohamed, A.; Smith, R. L.; Shephard, E. A.; Phillips, I. R. *Nat. Genet.* **1997**, *17*, 491–494.
46. Overby, L. H.; Carver, G. C.; Philpot, R. M. *Chem. Biol. Interact.* **1997**, *106*, 29–45.
47. Yeung, C. K.; Lang, D. H.; Thummel, K. E.; Rettie, A. E. *Drug Metab. Dispos.* **2000**, *28*, 1107–1111.
48. Zhang, J.; Cashman, J. R. *Drug Metab. Dispos.* **2006**, *34*, 19–26.
49. Koukouritaki, S. B.; Simpson, P.; Yeung, C. K.; Rettie, A. E.; Hines, R. N. *Pediatr. Res.* **2002**, *51*, 236–243.
50. Mayatepek, E.; Kohlmuller, D. *Acta Paediatr.* **1998**, *87*, 1205–1207.
51. Lacroix, D.; Sonnier, M.; Moncion, A.; Cheron, G.; Cresteil, T. *Eur. J. Biochem.* **1997**, *247*, 625–634.
52. Edmondson, D. E.; Mattevi, A.; Binda, C.; Li, M.; Hubalek, F. *Curr. Med. Chem.* **2004**, *11*, 1983–1993.
53. Weyler, W.; Hsu, Y. P.; Breakefield, X. O. *Pharmacol. Ther.* **1990**, *47*, 391–417.
54. Kalgutkar, A. S.; Dalvie, D. K.; Castagnoli, N., Jr.; Taylor, T. J. *Chem. Res. Toxicol.* **2001**, *14*, 1139–1162.
55. De Colibus, L.; Li, M.; Binda, C.; Lustig, A.; Edmondson, D. E.; Mattevi, A. *Proc. Natl. Acad. Sci. USA* **2005**, *102*, 12684–12689.
56. Geha, R. M.; Rebrin, I.; Chen, K.; Shih, J. C. *J. Biol. Chem.* **2001**, *276*, 9877–9882.
57. Saura, J.; Nadal, E.; van den Berg, B.; Vila, M.; Bombi, J. A.; Mahy, N. *Life Sci.* **1996**, *59*, 1341–1349.
58. Nicotra, A.; Pierucci, F.; Parvez, H.; Senatori, O. *Neurotoxicology* **2004**, *25*, 155–165.
59. Shulman, K. I.; Walker, S. E. *J. Clin. Psychiatry* **1999**, *60*, 191–193.
60. Hille, R. *Arch. Biochem. Biophys.* **2005**, *433*, 107–116.
61. Beedham, C. *Pharm. World Sci.* **1997**, *19*, 255–263.
62. Garattini, E.; Mendel, R.; Romao, M. J.; Wright, R.; Terao, M. *Biochem. J.* **2003**, *372*, 15–32.
63. Kurosaki, M.; Terao, M.; Barzago, M. M.; Bastone, A.; Bernardinello, D.; Salmona, M.; Garattini, E. *J. Biol. Chem.* **2004**, *279*, 50482–50498.
64. Harrison, R. *Free Radic. Biol. Med.* **2002**, *33*, 774–797.
65. Obach, R. S.; Huynh, P.; Allen, M. C.; Beedham, C. *J. Clin. Pharmacol.* **2004**, *44*, 7–19.
66. Moriwaki, Y.; Yamamoto, T.; Takahashi, S.; Tsutsumi, Z.; Hada, T. *Histol. Histopathol.* **2001**, *16*, 745–753.
67. Vasiliou, V.; Nebert, D. W. *Hum. Genomics* **2005**, *2*, 138–143.
68. Larson, H. N.; Weiner, H.; Hurley, T. D. *J. Biol. Chem.* **2005**, *280*, 30550–30556.
69. Hurley, T. D.; Steinmetz, C. G.; Xie, P.; Yang, Z. N. *Adv. Exp. Med. Biol.* **1997**, *414*, 291–302.
70. Nordling, E.; Persson, B.; Jornvall, H. *Cell Mol. Life Sci.* **2002**, *59*, 1070–1075.
71. Jornvall, H.; Hoog, J. O. *Alcohol Alcohol* **1995**, *30*, 153–161.
72. Higuchi, S.; Matsushita, S.; Masaki, T.; Yokoyama, A.; Kimura, M.; Suzuki, G.; Mochizuki, H. *Ann. N. Y. Acad. Sci.* **2004**, *1025*, 472–480.
73. Rosemond, M. J.; Walsh, J. S. *Drug Metab. Rev.* **2004**, *36*, 335–361.
74. Forrest, G. L.; Gonzalez, B. *Chem. Biol. Interact.* **2000**, *129*, 21–40.
75. Olson, L. E.; Bedja, D.; Alvey, S. J.; Cardounel, A. J.; Gabrielson, K. L.; Reeves, R. H. *Cancer Res.* **2003**, *63*, 6602–6606.
76. Hyndman, D.; Bauman, D. R.; Heredia, V. V.; Penning, T. M. *Chem. Biol. Interact.* **2003**, *143*, 621–631.
77. O'Connor, T.; Ireland, L. S.; Harrison, D. J.; Hayes, J. D. *Biochem. J.* **1999**, *343*, 487–504.
78. Kozma, E.; Brown, E.; Ellis, E. M.; Lapthorn, A. J. *J. Biol. Chem.* **2002**, *277*, 16285–16293.
79. Bianchet, M. A.; Faig, M.; Amzel, L. M. *Methods Enzymol.* **2004**, *382*, 144–174.
80. Ross, D.; Siegel, D. *Methods Enzymol.* **2004**, *382*, 115–144.
81. Beall, H. D.; Winski, S. I. *Front Biosci.* **2000**, *5*, D639–D648.
82. Ross, D. *Drug Metab. Rev.* **2004**, *36*, 639–654.
83. Holmquist, M. *Curr. Protein Pept. Sci.* **2000**, *1*, 209–235.
84. Testa, B.; Mayer, J. M. *Hydrolysis in Drug and Prodrug Metabolism: Chemistry, Biochemistry and Enzymology*; Wiley-VCH: Weinheim, Germany, 2003.
85. Guengerich, F. P. *Arch. Biochem. Biophys.* **2003**, *409*, 59–71.
86. Spector, A. A.; Fang, X.; Snyder, G. D.; Weintraub, N. L. *Prog. Lipid Res.* **2004**, *43*, 55–90.
87. Morisseau, C.; Hammock, B. D. *Pharmacol. Toxicol. Annu. Rev. Pharmacol. Toxicol.* **2005**, *45*, 311–333.

88. Fretland, A. J.; Omiecinski, C. J. *Chem. Biol. Interact.* **2000**, *129*, 41–59.
89. Gomez, G. A.; Morisseau, C.; Hammock, B. D.; Christianson, D. W. *Biochemistry* **2004**, *43*, 4716–4723.
90. Pacifici, G. M.; Franchi, M.; Bencini, C.; Repetti, F.; Di Lascio, N.; Muraro, G. B. *Xenobiotica* **1988**, *18*, 849–856.
91. Williams, F. M. *Clin. Pharmacokinet.* **1985**, *10*, 392–403.
92. Redinbo, M. R.; Potter, P. M. *Drug Disc. Today* **2005**, *10*, 313–325.
93. Walker, C. H.; Mackness, M. I. *Arch. Toxicol.* **1987**, *60*, 30–33.
94. Nigg, H. N.; Knaak, J. B. *Rev. Environ. Contam. Toxicol.* **2000**, *163*, 29–111.
95. Satoh, T. *Toxicol. Appl. Pharmacol.* **2005**, *207*, 11–18.
96. Wadkins, R. M.; Hyatt, J. L.; Yoon, K. J.; Morton, C. L.; Lee, R. E.; Damodaran, K.; Beroza, P.; Danks, M. K.; Potter, P. M. *Mol. Pharmacol.* **2004**, *65*, 1336–1343.
97. Bencharit, S.; Morton, C. L.; Xue, Y.; Potter, P. M.; Redinbo, M. R. *Nat. Struct. Biol.* **2003**, *10*, 349–356.
98. Tukey, R. H.; Strassburg, C. P. *Pharmacol. Toxicol. Annu. Rev. Pharmacol. Toxicol.* **2000**, *40*, 581–616.
99. Wells, P. G.; Mackenzie, P. I.; Chowdhury, J. R.; Guillemette, C.; Gregory, P. A.; Ishii, Y.; Hansen, A. J.; Kessler, F. K.; Kim, P. M.; Chowdhury, N. R. et al. *Drug Metab. Dispos.* **2004**, *32*, 281–290.
100. Radominska-Pandya, A.; Bratton, S.; Little, J. M. *Curr. Drug Metab.* **2005**, *6*, 141–160.
101. Burchell, B. *Am. J. Pharmacogenomics* **2003**, *3*, 37–52.
102. Kiang, T. K.; Ensom, M. H.; Chang, T. K. *Pharmacol. Ther.* **2005**, *106*, 97–132.
103. Foti, R. S.; Fisher, M. B. *Forens. Sci. Int.* **2005**, *153*, 109–116.
104. Basu, N. K.; Kovarova, M.; Garza, A.; Kubota, S.; Saha, T.; Mitra, P. S.; Banerjee, R.; Rivera, J.; Owens, I. S. *Proc. Natl. Acad. Sci. USA* **2005**, *102*, 6285–6290.
105. Battaglia, E.; Radominska-Pandya, A. *Biochemistry* **1998**, *37*, 258–263.
106. Fisher, M. B.; Paine, M. F.; Strelevitz, T. J.; Wrighton, S. A. *Drug Metab. Rev.* **2001**, *33*, 273–297.
107. Coffman, B. L.; Rios, G. R.; King, C. D.; Tephly, T. R. *Drug Metab. Dispos.* **1997**, *25*, 1–4.
108. Pacifici, G. M.; Sawe, J.; Kager, L.; Rane, A. *Eur. J. Clin. Pharmacol.* **1982**, *22*, 553–558.
109. de Wildt, S. N.; Kearns, G. L.; Leeder, J. S.; van den Anker, J. N. *Clin. Pharmacokinet.* **1999**, *36*, 439–452.
110. Bock, K. W.; Forster, A.; Gschaidmeier, H.; Bruck, M.; Munzel, P.; Schareck, W.; Fournel-Gigleux, S.; Burchell, B. *Biochem. Pharmacol.* **1993**, *45*, 1809–1814.
111. Alam, S. N.; Roberts, R. J.; Fischer, L. J. *J. Pediatr.* **1977**, *90*, 130–135.
112. Richard, K.; Hume, R.; Kaptein, E.; Stanley, E. L.; Visser, T. J.; Coughtrie, M. W. *J. Clin. Endocrinol. Metab.* **2001**, *86*, 2734–2742.
113. Dupret, J. M.; Rodrigues-Lima, F. *Curr. Med. Chem.* **2005**, *12*, 311–318.
114. Sinclair, J. C.; Sandy, J.; Delgoda, R.; Sim, E.; Noble, M. E. *Nat. Struct. Biol.* **2000**, *7*, 560–564.
115. Grant, D. M.; Goodfellow, G. H.; Sugamori, K.; Durette, K. *Pharmacology* **2000**, *61*, 204–211.
116. Hein, D. *Mutat. Res.* **2002**, *506–507*, 65–77.
117. Weinshilboum, R. M.; Otterness, D. M.; Szumlanski, C. L. *Pharmacol. Toxicol. Annu. Rev. Pharmacol. Toxicol.* **1999**, *39*, 19–52.
118. Mannisto, P. T.; Kaakkola, S. *Pharmacol. Rev.* **1999**, *51*, 593–628.
119. Creveling, C. R. *Cell Mol. Neurobiol.* **2003**, *23*, 289–291.
120. Evans, W. E.; Relling, M. V. *Nature* **2004**, *429*, 464–468.
121. Hayes, J. D.; Flanagan, J. U.; Jowsey, I. R. *Pharmacol. Toxicol. Annu. Rev. Pharmacol. Toxicol.* **2005**, *45*, 51–88.
122. Townsend, D.; Tew, K. *Am. J. Pharmacogenomics* **2003**, *3*, 157–172.
123. Nebert, D. W.; Vasiliou, V. *Hum. Genomics* **2004**, *1*, 460–464.
124. Sheehan, D.; Meade, G.; Foley, V. M.; Dowd, C. A. *Biochem. J.* **2001**, *360*, 1–16.
125. Mahajan, S.; Atkins, W. M. *Cell Mol. Life Sci.* **2005**, *62*, 1221–1233.
126. Flatgaard, J. E.; Bauer, K. E.; Kauvar, L. M. *Cancer Chemother. Pharmacol.* **1993**, *33*, 63–70.
127. Rinaldi, R.; Eliasson, E.; Swedmark, S.; Morgenstern, R. *Drug Metab. Dispos.* **2002**, *30*, 1053–1058.
128. Chapman, E.; Best, M. D.; Hanson, S. R.; Wong, C. H. *Angew. Chem. Int. Ed. Engl.* **2004**, *43*, 3526–3548.
129. Kauffman, F. C. *Drug Metab. Rev.* **2004**, *36*, 823–843.
130. Blanchard, R. L.; Freimuth, R. R.; Buck, J.; Weinshilboum, R. M.; Coughtrie, M. W. *Pharmacogenetics* **2004**, *14*, 199–211.
131. Rath, V. L.; Verdugo, D.; Hemmerich, S. *Drug Disc. Today* **2004**, *9*, 1003–1011.
132. Redinbo, M.; Potter, P. *Drug Disc. Today* **2004**, *10*, 313–325.
133. Gamage, N. U.; Tsvetanov, S.; Duggleby, R. G.; McManus, M. E.; Martin, J. L. *J. Biol. Chem.* **2005**, *280*, 41482–41486.
134. Nelson Energy Seal. http://drnelson.com (accessed June 2006).
135. AKR Super family. http://www.med.upenn.edu/akr (accessed June 2006).
136. Mackenzie, P. I.; Walter Bock, K.; Burchell, B.; Guillemette, C.; Ikushiro, S. I.; Iyanagi, T.; Miners, J. O.; Owens, I. S.; Nebert, D. W. *Pharmacogenet. Genomics* **2005**, *15*, 677–685.

Biographies

Rheem A Totah obtained a PhD in Medicinal Chemistry in 2002 from the University of Kansas. Later, she spent 18 months as an Elmer M and Joy B Plein Fellow for Excellence in Pharmacy Education at the University of Washington. Currently Dr Totah is an Acting Assistant Professor in the Medicinal Chemistry Department at the University of Washington. Dr Totah's research interests are focused on studying the mechanism and structure–activity relationship of cytochrome P450 enzymes especially those expressed mainly in extrahepatic tissues.

Allan E Rettie obtained a PhD in Pharmaceutical Sciences following studies at the Wolfson Unit at the University of Newcastle-upon-Tyne, England under the direction of Prof Sir Michael Rawlins. After postdoctoral research training with Drs Mont Juchau and William Trager at the University of Washington, Seattle, Dr Rettie joined the faculty of the School of Pharmacy at the same institution, where he is now Professor and Chair of Medicinal Chemistry. Dr Rettie's research interests revolve around structure–function relationships for the cytochrome P450 enzymes, particularly those in the CYP2 and CYP4 families, and the pharmacogenomics of oral anticoagulants.

Comprehensive Medicinal Chemistry II
ISBN (set): 0-08-044513-6

ISBN (Volume 5) 0-08-044518-7; pp. 167–191

5.08 Mechanisms of Toxification and Detoxification which Challenge Drug Candidates and Drugs

B Oesch-Bartlomowicz and F Oesch, University of Mainz, Mainz, Germany

5.08.1 Introduction

5.08.1.1 Scope of Drug Metabolism toward Xenobiotic Detoxification

Over the course of evolution organisms were exposed to environmental compounds of negligible nutritive value. These have variably been called 'foreign compounds' or 'xenobiotics.' Also the term 'drug' is used for them in contrast to the narrower term 'therapeutic drug' to specifically designate those xenobiotics that are intended for clinical use. Organisms that developed systems enabling them to get rid of these xenobiotics before they could accumulate to toxic levels had an evolutionary advantage. Such systems therefore had an obvious scope toward detoxification.[1]

The major excretory systems of mammals – the renal and the biliary excretion system – require a minimal hydrophilicity of chemical compounds to be excretable. This generated the necessity for lipophilic xenobiotics to be converted to hydrophilic metabolites in order to become excretable.[1–3] This necessity represented one driving force for the metabolism of xenobiotics to develop during evolution.

A further need was the ability to detoxify 'future unknowns.' Therefore, the enzymes exclusively or predominantly catalyzing the metabolism of xenobiotics usually: (1) have a very broad substrate specificity; and (2) exist in large families or superfamilies of enzymes each one of them possessing a broad but distinct substrate specificity, which in part overlaps with the specificities of the other isoenzymes.[1] The catalytically active site of a typical drug-metabolizing enzyme must be able to accommodate a large number of substrates such that the fit of an individual substrate to the active site cannot be very tight. An important consequence of this is that the geometry of the substrate/catalytic site is not optimized for maximal speed of the reaction but rather for substantial flexibility. Therefore, drug-metabolizing enzymes are relatively slow enzymes.[1] Glutathione S-transferases are among the fastest of them (approximately 10^4 mole substrate turned over per minute per mole of enzyme). Even this is very slow compared with fast enzymes such as carboanhydrase (turnover number about $4 \times 10^7\,min^{-1}$).

5.08.1.2   Toxification versus Detoxification

The key point is that a given species was not able to 'predict' what kind of xenobiotics it would be exposed in future since the organisms and their constituents in its environment obviously also underwent evolutionary changes. Thus, excretory systems of low specificity and hence high flexibility provided an evolutionary advantage. However, there was a price to pay for this flexibility. The low substrate specificity of the enzymes catalyzing the reactions allowing for a vast array of 'unforeseen' lipophilic xenobiotics to ultimately be converted to metabolites sufficiently hydrophilic for excretion quite frequently led to toxic intermediary metabolites.[1] Therefore, drug metabolism leads to both detoxification and toxification. Indeed, practically all drug-metabolizing enzymes that have been investigated so far play dual roles: detoxification and toxification. However, for individual structural elements of a given drug it becomes increasingly possible to predict whether a given enzyme will lead to toxification or to detoxification, an important basis for extrapolations from an experimentally used animal species to human when the enzyme in question is expressed to vastly different levels between these species, as is frequently the case.[1]

5.08.2   The Phase Concept of Drug Metabolism and Its Importance for Toxification versus Detoxification

The phase concept of drug metabolism is described elsewhere in this book (*see* 5.05 Principles of Drug Metabolism 1: Redox Reactions; 5.06 Principles of Drug Metabolism 2: Hydrolysis and Conjugation Reactions). With respect to toxification versus detoxification it is important that depending on their chemical nature the functional groups introduced in the phase I of drug metabolism can be classified as electrophilic or nucleophilic (**Figure 1**). Functional groups with an electrophilic carbon are, for example, epoxides or α,β-unsaturated carbonyl groups. Nucleophilic substituents are, for example, hydroxyl, amino, sulfhydryl, or carboxylic groups. Owing to their ability to react with electron-rich substituents in important intracellular steering molecules such as proteins and nucleic acids, electrophilic metabolites can – depending on their individual chemical reactivity – be highly cytotoxic and/or mutagenic. Nucleophilic metabolites, however, usually do not covalently modify endogenous macromolecules and are therefore generally less toxic. On the other hand, they are often determinants of receptor interactions and thus of the biological activity of a xenobiotic compound. Therefore, acutely toxic effects of xenobiotics or their metabolites caused by receptor activation at the wrong time or wrong intensity can depend on specific nucleophilic groups in the compound.[1] Electrophilic substrates (e.g., epoxides, α,β-unsaturated carbonyls) are conjugated by the glutathione S-transferases. Nucleophilic substrates (i.e., those with hydroxyl-, sulfhydryl-, amino- or carboxyl groups) are conjugated by UDP-glucuronosyltransferases (UGTs), sulfotransferases (SULTs), acetyltransferases (ATs), and methyltransferases (MTs). Conjugation reactions usually terminate the potential of electrophiles to react with proteins and DNA, or the ability of nucleophiles to interact with a receptor. Conjugation to both types of substituents usually increases the water solubility of the respective compound greatly.[1–3] Therefore, conjugation reactions usually are the major detoxification step in drug metabolism. However, there are exceptions that are important for predictions of toxification versus detoxification: glutathione conjugation strongly increases the electrophilic reactivity of vicinal dihaloalkanes and enzymatic sulfate ester formation with hydroxymethylated aromatic compounds and with aromatic hydroxylamines (metabolites of aromatic amines) leads to markedly and strongly mutagenic species, respectively.[1]

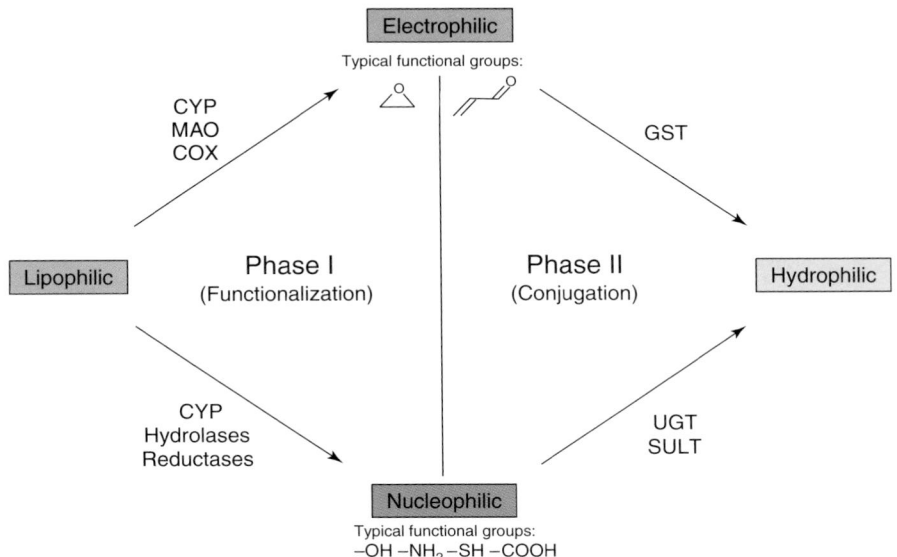

Figure 1 The phases of drug metabolism. Metabolism of lipophilic foreign compounds typically proceeds in sequential steps. In phase I, the compounds are functionalized via introduction or liberation of nucleophilic or electrophilic anchor groups by oxidoreductases or hydrolases, which allow conjugation with strongly polar endogenous building blocks, such as glucuronic acid or glutathione, in phase II of drug metabolism. The resulting metabolites are usually readily water soluble and therefore easily excretable in the urine or bile. The conjugation usually terminates the biological activity of compounds, such as the genotoxic effects of certain electrophiles or the pharmacological effects of therapeutic drugs. (Reproduced from Oesch, F.; Arand, M. Xenobiotic Metabolism. In *Toxicology*; Marquardt, H., Schäfer, S. G., McClellan, R., Welsch, F., Eds.; Academic Press: New York, 1999, pp 83–109, with permission from Elsevier.)

5.08.3   Toxicologically Relevant Properties of Individual Drug-Metabolizing Enzymes

5.08.3.1   Phase I Enzymes

5.08.3.1.1   Oxidoreductases
5.08.3.1.1.1   Cytochromes P450 (CYP; E.C. 1.14.14.1)

5.08.3.1.1.1.1 Multiplicity of CYPs[2,4–8]   The nomenclature of the multiple CYPs is described elsewhere in this book (*see* 5.05 Principles of Drug Metabolism 1: Redox Reactions). When attempting to extrapolate toxicological results obtained in an experimental animal species to human it is often helpful to discriminate whether orthologous CYPs exist in the test species and in man since in many cases, apart from the conservation of a high percentage of the overall primary structure, much of the individual function including substrate specificity is conserved among species in orthologous enzymes. It is possible to define orthologous enzymes in different organisms if there is higher sequence similarity between these forms from different species than there is between the closely related isoenzymes of the same subfamily within the same species. Such orthologous CYPs from different species receive the same designation. This is the case for the individual isoenzymes of CYP family 1. If the assignment of orthologous CYPs is not possible the CYPs within the subfamily are numbered consecutively (many CYPs from families 2, 3, and 4).[1]

5.08.3.1.1.1.2 Enzymatic mechanism of the reactions catalyzed by CYP and toxicological consequences   The CYP-catalyzed oxygenation proceeds by a number of sequential steps forming the CYP reaction cycle (**Figure 2**; *see* 5.05 Principles of Drug Metabolism 1: Redox Reactions).[2,5,6] The prosthetic heme of CYP has its central iron in the ferric state [Fe^{3+}] and, in most CYPs, is predominantly low spin in its d^5-orbitals. Binding of a substrate in the catalytic pocket of CYP in most cases leads to the conversion of the iron from the low- to the high-spin configuration, which favors the reduction of the ferric [Fe^{3+}] to the ferrous [Fe^{2+}] iron of CYP by the NADPH-CYP-reductase by a one-electron transfer. In this state, the iron can bind molecular oxygen to form a $Fe^{2+}–O_2$ complex. The generation of the ultimately activated form of oxygen is now initiated by a further reduction, again by a one-electron transfer. This is

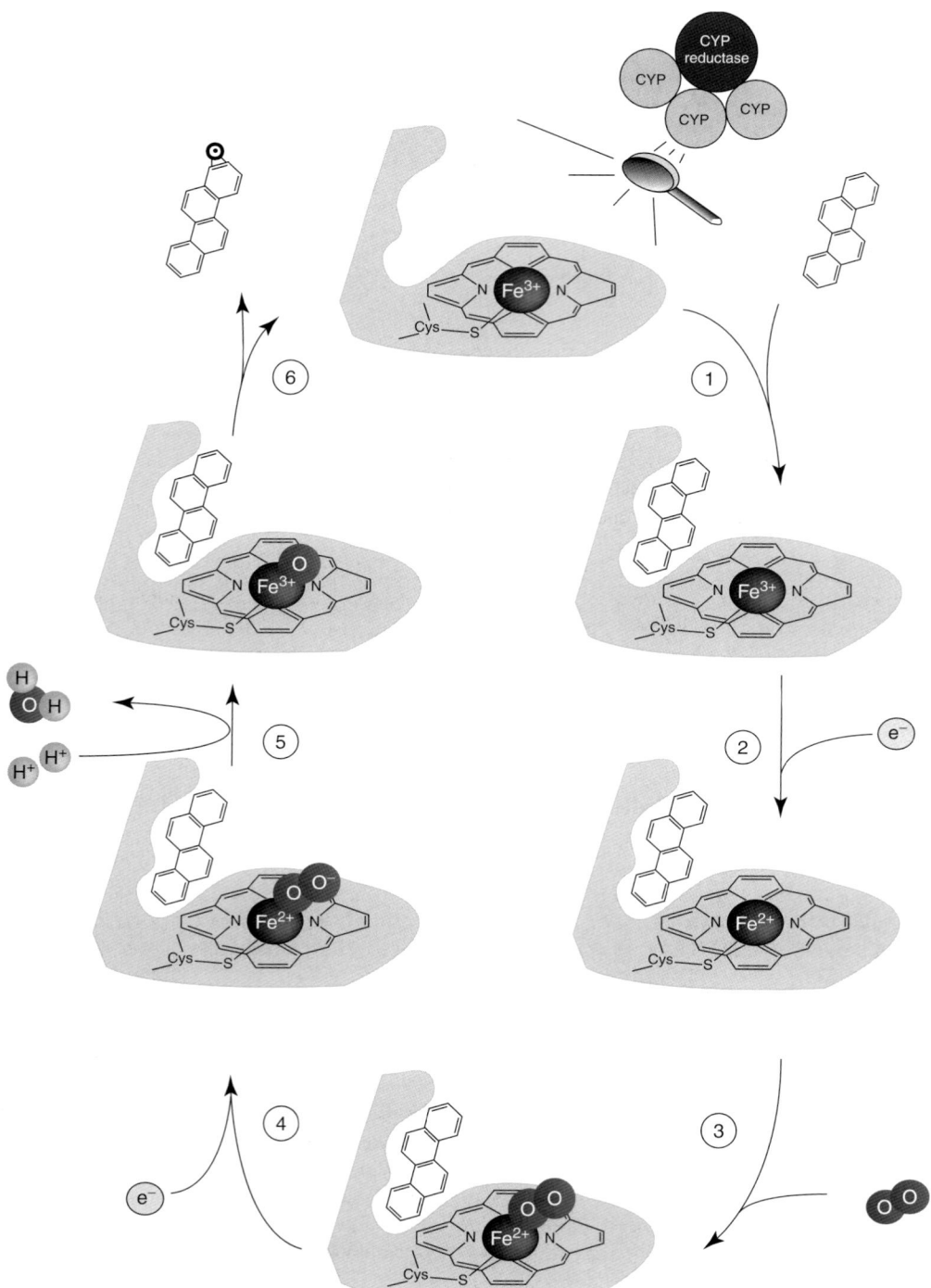

Figure 2 Cytochrome P450 (CYP)-catalyzed reactions: (1) a substrate molecule (depicted example: chrysene) binds in the active site of CYP thereby facilitating the conversion of the heme iron from the low-spin to the high-spin configuration; (2) the ferric Fe^{3+} of CYP is reduced to the ferrous Fe^{2+} by NADPH-CYP-reductase by a one-electron transfer; (3) the iron binds molecular oxygen; (4) a second one-electron reduction either by NADPH-CYP-reductase or by cytochrome b_5 leads to the highly activated $[Fe^{2+}\text{-}O_2^-]$; (5) heterolytic scission of the O–O bond and binding of two protons lead to the release of water; and (6) the reactive $(FeO)^{3+}$ intermediate oxygenates the substrate, which is then released. (Reproduced from Oesch, F.; Arand, M. Xenobiotic Metabolism. In *Toxicology*; Marquardt, H., Schäfer, S. G., McClellan, R., Welsch, F., Eds.; Academic Press: New York, 1999, pp 83–109, with permission from Elsevier.)

catalyzed by the NADPH-CYP-reductase or sometimes (depending on the individual CYP and on the substrate) by cytochrome b_5 (activated by the NADH-dependent cytochrome b_5 reductase).[1,2,5,6] It is of importance toxicologically that the complex is now in a highly activated form (mesomerically between $Fe^{2+}-O_2^-$ and $Fe^{3+}-O_2^{2-}$) that can release active oxygen species, for instance, hydrogen peroxide[9,10] and thereby start cytotoxic processes such as lipid peroxidation. Although the analysis of the unstable intermediates is difficult the available evidence indicates that heterolytic scission of the O–O bond and binding of two protons leads to the formation of a water molecule and a $(FeO)^{3+}$ intermediate with a high oxygen transfer and high oxidation potential, which now oxygenates or oxidizes the substrate in many cases leading to toxic metabolites.[1,2,5,6]

It is of toxicological importance that:

- Many, but not all, CYP bind and activate oxygen only after substrate binding. This partially protects from the potentially destructive generation of reactive oxygen species by CYP;
- Because of their intrinsic ability to activate oxygen, CYP can per se lead to reactions toxic for the organism, especially in the presence of competitive inhibitors that are not (or poor) substrates binding to the active site without themselves being oxidized thereby activating oxygen for a long time.[1]

5.08.3.1.1.1.3 Individual CYPs that are important for xenobiotic toxification and detoxification   CYP-mediated drug metabolism in mammalian species is mainly due to CYPs of families 1–3[11] (**Table 1**). The CYP1 family[8,12–14] has only three isoenzymes CYP1A1, CYP1A2, and CYP1B1. In all mammalian species investigated these isoenzymes are orthologous, that is, the % amino acid sequence identity of the individual isoenzymes between mammalian species is higher than between these isoenzymes within a given species. This allows for some toxicologically important extrapolations between mammalian species although the respective functions of the individual isoenzymes are, of course, not strictly identical in different species.[1]

CYP1A1[8,12,14] metabolizes planar and highly lipophilic compounds such as polycyclic aromatic hydrocarbons (PAH), including many carcinogenic environmental chemicals. CYP1A1 metabolizes these by hydroxylation to phenols, which are conjugated and excreted (detoxification). At the same time CYP1A1 metabolizes them to electrophilically reactive epoxides.[1] In a multistep reaction the very highly reactive dihydrodiol bay region epoxides are formed, which are responsible for much of the genotoxic (DNA damaging, mutagenic) and carcinogenic activity of PAHs.[15,16] **Figure 3** shows the steps leading from benzo[a]pyrene to the ultimate carcinogenic metabolite benzo[a]pyrene-7,8-dihydrodiol-9, 10-epoxide. Both CYP-mediated reactions of this pathway can be catalyzed by CYP1A1. Many PAHs greatly enhance their own metabolism by CYP1A1. The mechanism underlying this induction of the members of the CYP1 family is a transcriptional activation mediated by the aryl hydrocarbon receptor (Ah receptor), which itself is activated by binding of these inducers.[1,8,14,17,18] Just as many xenobiotic metabolizing enzymes have endogenous functions as well, the Ah receptor is also activated in the absence of xenobiotic ligands.[19] Constitutively, i.e., in the absence of exogenous inducers, CYP1A1 expression is virtually absent in human liver but is present in several other tissues such as lung, lymphocytes, and placenta.

CYP1A2, in contrast, is almost exclusively expressed in the liver in humans. It is also inducible by many planar PAHs, markedly but to a much more modest extent than CYP1A1. On average, about 10% of the total liver CYP is CYP1A2, but the level of expression differs considerably among individuals. Many important procarcinogens are metabolically toxified by CYP1A2. This includes the human carcinogenic aromatic amines as well as hetero-cyclic amines such as the protein pyrolysis products 2-amino-3-methylimidazo[4,5-f]quinoline (IQ) and 2-amino-1-methyl-6-phenylimidazo[4,5-b]pyridine (PhIP); these occur in human food after cooking and are mutagenic and carcinogenic, at least in rodents. Also, the reactive epoxide derived from the mycotoxin aflatoxin B_1, which is strongly hepatocarcinogenic for man, is metabolically produced by CYP1A2. In many cases the substrate specificities of CYP1A1 and CYP1A2 overlap: polycyclic aromatic hydrocarbons are preferentially metabolized by CYP1A1 but also by CYP1A2; and aromatic amines are preferentially metabolized by CYP1A2 but also by CYP1A1.[1]

CYP1B1 has been discovered more recently. In several respects it is similar to members of the CYP1A family:

1. It metabolizes some polycyclic aromatic hydrocarbons albeit with a preference for the second step in the toxification to the ultimate carcinogens dihydrodiol bay region epoxides, i.e., the pre-bay dihydrodiols, themselves metabolites of angular PAHs, are good substrates for CYP1B1.
2. It is inducible by many planar PAHs.

CYP1B1 is the most important enzyme for the toxification to fjord-region dihydrodiol epoxides, the PAH metabolites with the highest carcinogenic potency known to date.

Table 1 Toxicologically important human drug-metabolizing cytochromes P450 (CYP)

Iso-enzyme (selection)	Organ[a]	Toxicologically relevant substrates (selection)	Relation to animal CYPs
CYP1A1	Multiple, constitutively very low in liver	Polycyclic aromatic hydrocarbons (PAH)	Orthologous forms with similar biochemical characteristics in different vertebrates
CYP1A2	Liver	Aromatic amines, PAH, aflatoxin B_1, phenacetin, acetaminophen, caffeine, warfarin, imipramine	Orthologous forms with similar biochemical characteristics in different vertebrates
CYP2A6	Liver	6-Aminochrysene, aflatoxin B_1, nitrosamines, coumarin	No clear relationship to a specific animal CYP
CYP2B6	Liver	Cyclophosphamide, nicotine	In contrast to the rat CYP2B1 that is strongly inducible by, e.g., phenobarbital and thereby becoming a major hepatic CYP in this species, CYP2B6 has been found only in rather small amounts in human liver and does not seem to play an important role in the metabolism of most drugs
CYP2C8	Liver	Taxol, retinoids	No clear relationship to a specific animal CYP
CYP2C9	Liver, kidney	Diclofenac, ibuprofen, phenytoin, tolbutamide	No clear relationship to a specific animal CYP
CYP2C19	Liver	(S)-Mephenytoin, omeprazole	No clear relationship to a specific animal CYP
CYP2D6	Liver	Debrisoquin, propranolol, dextromethorphan	No clear relationship to a specific animal CYP
CYP2E1	Liver, gastrointestinal tract	Ethanol, benzene, dimethylnitrosamine, chlorzoxazone	Orthologous enzymes in different species. Only one enzyme in the subfamily in humans and in many animal species
CYP3A4	Liver, small intestine	Aflatoxin B_1, acetaminophen, benzphetamine, nifedipine, steroid hormones, erythromycin	Several very closely related isoenzymes in man (CYP3A3, CYP3A5, and CYP3A7) make it impossible to assign orthologous forms in other species; otherwise, the biochemical characteristics resemble those of animal CYP3A isoenzymes

Reproduced from Oesch, F.; Arand, M. Xenobiotic Metabolism. In *Toxicology*; Marquardt, H., Schäfer, S. G., McClellan, R., Welsch, F., Eds.; Academic Press: New York, 1999, pp 83–109, with permission from Elsevier.
[a] Primarily identified in the indicated organ, but possibly also expressed in other locations.

In the CYP2 family the isoenzymes of major toxicological interest in human are CYP2A6, CYP2C8, CYP2C9, CYP2C19, CYP2D6, and CYP2E1. The expression of CYP2A6[20] in human liver is modest (less than 5% of total CYP). This CYP can toxify the important human hepatocarcinogen aflatoxin B_1 (with high affinity but low capacity), as well as some tobacco-specific nitrosamines and some carcinogenic aromatic amines such as 6-aminochrysene. Coumarin can serve as an in vivo probe substrate. Its 7-hydroxylation is selectively catalyzed by CYP2A6.[20]

CYP2B family members[11,21] are important in many experimental animal species such as the rat and rabbit. However, the only human CYP2B member known, CYP2B6, is expressed at a very low level in the liver and appears to be of limited importance in drug metabolism. One important substrate is the cytostatic drug cyclophosphamide, which is toxified by the human CYP2B6.[22] Inactivation of CYP2B by cAMP-mediated phosphorylation substantially changes the metabolism of cyclophosphamide and its consequent toxicity.[23]

Conversely, the CYP2C family members[24] are highly important in human drug metabolism and apparently much more so than in rodents. They metabolize a large number of drugs (see 5.05 Principles of Drug Metabolism 1: Redox Reactions)

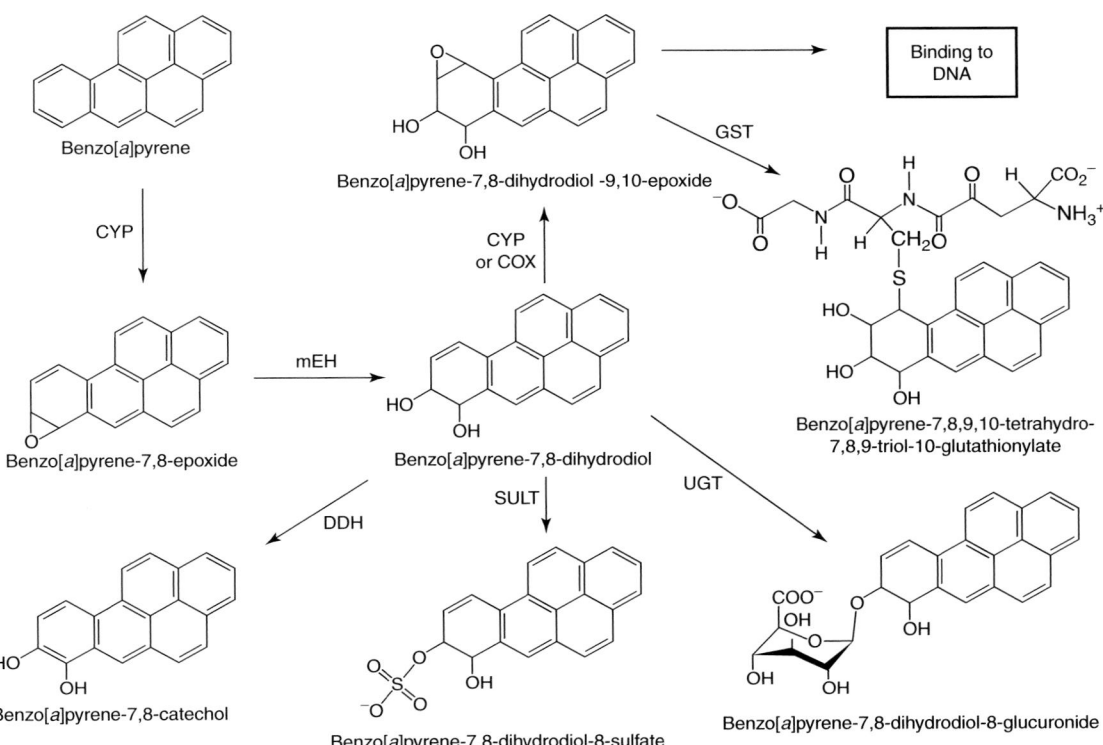

Figure 3 Toxicologicaly important pathways of benzo[a]pyrene (BP) metabolism. BP is metabolized to many primary, secondary, and tertiary metabolites. Depicted is the toxification to the ultimate carcinogen BP-7,8-dihydrodiol-9,10-epoxide. First CYP-mediated oxidation leads to the 7,8-epoxide, which is then hydrolyzed by microsomal epoxide hydrolase to the 7,8-dihydrodiol. Now conjugation to glucuronic acid or sulfate, dehydrogenation to the catechol, and isomerization to the phenol (not shown) compete with the final toxification to the 7,8-dihydrodiol 9,10-epoxide catalyzed by CYP or prostaglandin synthase (COX). The diol epoxide is a highly mutagenic compound that is not a substrate for epoxide hydrolase. Conjugation to glutathione leads to detoxification, but is not fast enough to abolish the genotoxicity. (Reproduced from Oesch, F.; Arand, M. Xenobiotic Metabolism. In *Toxicology*; Marquardt, H., Schäfer, S. G., McClellan, R., Welsch, F., Eds.; Academic Press: New York, 1999, pp 83–109, with permission from Elsevier.)

CYP2D6[25,26] is especially important in humans because of its highly relevant polymorphism, which is due to: (1) inactive and less active alleles resulting in poor metabolizers; and (2) gene amplification of active alleles resulting in ultrarapid metabolizers. CYP2D6 metabolizes many drugs that have a basic nitrogen atom at a distance of 5–7 Å from an easily oxidizable atom. Lack of CYP2D6 activity can lead to accumulation of highly toxic levels of these drugs. About 7% of the Caucasian population have a defective CYP2D6 gene and must be aware of potentially toxic effects caused by standard doses of drugs that are predominantly metabolized by CYP2D6 and have a narrow therapeutic window. There are correlations between CYP2D6 genotype and the incidence of certain forms of the parkinsonian syndrome, which can result from drug-induced toxicity to dopaminergic neurons in the central nervous system.

CYP2E1[27,28] metabolizes a wide array of substrates all of which have a relatively small molecular size. This includes therapeutic drugs such as acetaminophen, environmental carcinogens such as benzene, and several alkylnitrosamines, as well as the compound from which the letter of the family (E) is derived: ethanol. In the presence of inducer, most notably relatively high concentrations of substrate, the half-life of CYP2E1 is increased leading to an accumulation of enzyme protein. In several cases a concomitant stimulation of the transcription occurs. In some cases a stimulation of translation has also been reported. The heme iron of CYP2E1 is predominantly in a high spin state even in the absence of substrate, which is relatively unusual for CYPs. The consequence is that oxygen can bind to CYP2E1 also in the absence of substrate, which leads to the production of activated oxygen that is not used for substrate oxygenation. Thus, induced levels of liver CYP2E1 may lead to hepatotoxicity. Ethanol-induced liver injury may in considerable part be due to oxidative damage caused by increased levels of CYP2E1 as a consequence of the induction by ethanol. Conversely, phosphorylation of CYP2E1 mediated by cAMP and hence by drugs that lead to an increase in cAMP leads to CYP2E1 inactivation.[29]

The CYP3 family[6,30] is the most important CYP family for the metabolism of therapeutic drugs. The most important member of the family is CYP3A4. It is usually the most abundant CYP in human liver (about 30% of the total CYPs). CYP3A4 has extraordinarily broad substrate specificity. This includes a great number of therapeutic drugs (*see* 5.05 Principles of Drug Metabolism 1: Redox Reactions), but also several carcinogens such as the important human hepatocarcinogen aflatoxin B_1 (with high capacity but low affinity) as well as the metabolically formed key intermediary PAH-derived dihydrodiols, which it toxifies to the ultimate carcinogens the dihydrodiolepoxides. CYP3A4 is induced by rifampicin, dexamethasone, pregnenolone carbonitrile, and phenobarbital. The strong induction by rifampicin led to one of the most spectacular cases of drug interactions. The contraceptive steroids were metabolically inactivated as a consequence of CYP3A4 induction by the coincident treatment with rifampicin in tuberculosis patients to such an extent that a great number of unwanted pregnancies resulted.

5.08.3.1.1.2   Flavin-containing monooxygenases (FMOs; E.C. 1.14.13.8)

The flavin-containing monooxygenases (FMOs)[31] possess a broad substrate specificity; however, this is limited to the oxygenation of soft nucleophiles such as the heteroatoms nitrogen and sulfur. **Table 2** shows the substrates of pig liver FMO-1, which has been especially well investigated in this respect. N-oxide and S-oxide formation represent the major FMO-mediated reactions In contrast to most CYP isoenzymes, the FMOs bind and activate molecular oxygen before the substrate binds. First the prosthetic group of FMO, the flavin adenine dinucleotide (FAD), is reduced by NADPH to form $FADH_2$. Then O_2 is incorporated into the FAD, resulting in the formation of the hydroperoxide FADH-4α-OOH. This reactive form is capable to transfer one oxygen atom to the substrate. The toxicologically important question is how does the organism protect itself from the activated oxygen that forms on FMO before the oxygen-accepting substrate is present in the enzyme? The protective mechanism is twofold. First, the oxygen transfer potential of FMO is much lower than that of CYP, restricting exogenous acceptor substrates as well as potential endogenous targets for oxygen transfer-mediated toxicity to soft nucleophiles, usually molecules containing sulfur or nitrogen. Second, charged molecules that carry a single charged group are usually poor FMO substrates, if the charged group is not itself the acceptor site for the oxygen. This substrate specificity protects many important endogenous soft nucleophiles such as glutathione, amino acids, or polyamines from metabolism by FMO.[1]

FMO genetic deficiency leads to the 'fish odor' syndrome, which is caused by the inability to metabolize trimethylamine to its N-oxide, a reaction that is carried out exclusively by FMOs. As a result the unmetabolized amine, which is highly volatile and smells like rotten fish, is exhaled and excreted, which often results in severe social problems.[1]

5.08.3.1.1.3   Monoamine oxidases (MAOs; E.C. 1.4.3.4)

Monoamine oxidases (MAOs) are described elsewhere in this book (*see* 5.05 Principles of Drug Metabolism 1: Redox Reactions). The major function of MAOs consists of the inactivation by oxidative deamination of monoamine

Table 2   Substrate specificity of pig liver flavin-containing monooxygenase (FMO)

Organic nitrogen compounds
  Secondary and tertiary cyclic and acyclic amines
  N-Alkyl- and N,N-dialkyl arylamines
  Hydrazines
Organic sulfur-containing compounds
  Mercaptopurines, -pyrimidines, and -imidazoles
  Cyclic and acyclic sulfides and disulfides
  Thioamides, thiocarbamides, dithiocarbamides, dithioacids
Other organic substrates
  Boronic acid
  Phosphines
  Selenides
  Selenocarbamides
Inorganic compounds
  Sulfur, sulfides, thiocyanates
  Iodine, iodides, hypoiodites

Reproduced from Oesch, F.; Arand, M. Xenobiotic Metabolism. In *Toxicology*; Marquardt, H., Schäfer, S. G., McClellan, R., Welsch, F., Eds.; Academic Press: New York, 1999, pp 83–109, with permission from Elsevier.

neurotransmitters such as serotonin and dopamine. MAO also metabolizes many xenobiotic primary, secondary, or tertiary amines thereby generating an amine (or ammonia), an aldehyde, and hydrogen peroxide:

$$R^1CH_2NR^2R^3 + O_2 + H_2O \Rightarrow R^1CHO + HNR^2R^3 + H_2O_2$$

The hydrogen peroxide formed can reach toxicologically relevant levels.

1-Methyl-4-phenyl-1,2,3,6-tetrahydropyridine (MPTP) is an interesting example of metabolism-related selective toxicity. It is toxified via the intermediate 1-methyl-4-phenyl-2,3-dihydropyridinium salt ($MPDP^+$) to the 1-methyl-4-phenyl pyridinium salt (MPP^+). MPP^+ is taken up by a high-affinity reuptake system specifically localized in the nigrostriatal dopaminergic neurons and blocks their mitochondrial energy metabolism. This leads to death of these neurons, which causes Parkinson's disease. Within the neurons only MAO can metabolize MPTP, while in other tissues CYP- and FMO-catalyzed detoxification reactions compete with MAO for the substrate. This combination of selective uptake into cells possessing a selective pattern of drug-metabolizing enzymes causes the selective neurotoxicity. MPTP is an experimental chemical. Related compounds such as beta-carboline or tetrahydroisoquinoline are present at low concentrations in food. Whether they behave in a similar way to MPTP and therefore are neurotoxicologically important by contributing to the etiology of idiopathic Parkinson's disease is unclear.[1]

5.08.3.1.1.4   Cyclooxygenases (COX; E.C. 1.14.99.1)

The most important function of cyclooxygenases (also called prostaglandin synthases) is endogenous in that they metabolize arachidonic acid to prostaglandins (**Figure 4**). First, the incorporation of two molecules of oxygen into arachidonic acid generates a cyclic peroxide (endoperoxide) and a hydroperoxide substituent. The resulting intermediate is prostaglandin-G_2. Then, the hydroperoxide substituent is reduced to a hydroxyl group. The resulting compound is prostaglandin-H_2. Many sufficiently lipophilic drugs that have a low redox potential act as an oxygen acceptor or electron donor in this second reaction and several of them are toxified in this way. This includes many

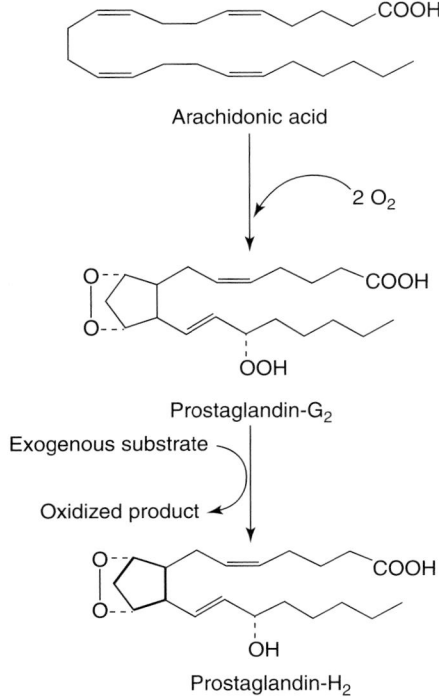

Figure 4 Cyclooxygenase-mediated cooxidation. In the first step of prostaglandin-H_2 synthesis, two molecules of O_2 are incorporated into arachidonic acid, forming the cyclic peroxide (endoperoxide) prostaglandin-G_2, which in addition contains a hydroperoxide substituent. This hydroperoxide substituent is then converted to a hydroxy group, resulting in prostaglandin-H_2. During this reaction, the oxygen can be transferred to an exogenous acceptor substrate, which includes many toxification reactions. (Reproduced from Oesch, F.; Arand, M. Xenobiotic Metabolism. In *Toxicology*; Marquardt, H., Schäfer, S. G., McClellan, R., Welsch, F., Eds.; Academic Press: New York, 1999, pp 83–109, with permission from Elsevier.)

phenolic compounds and aromatic amines. Thus, many drugs are substrates for COX. The ease with which a compound may be co-oxidized by COX can be estimated from the influence of its substituents at the aromatic ring on its redox potential, e.g., electron-withdrawing substituents on the aromatic ring of N-methyl aniline derivatives and N-acetylation of aromatic amines reduce the rate of COX-mediated metabolism.

Some substrates are toxified to radicals during this COX-mediated co-oxidation. The unpaired electron radicals can react with other molecules to yield new radicals thereby, for example, initiating lipid peroxidation or can damage cellular proteins or DNA by covalent modifications. In addition prostaglandin-G_2 can form a peroxyl radical over a number of intermediate steps, which can lead to the formation of highly genotoxic epoxides derived from several compounds including environmental carcinogenic polycyclic aromatic hydrocarbons and the important human hepatocarcinogen aflatoxin B_1.

Two COX isoenzymes are known, COX-1 and COX-2. COX-1 is constitutively expressed and COX-2 is an inducible enzyme, which is increased during inflammation. Many extrahepatic tissues are rich in COX but low in CYP. This makes COX important for the carcinogenic effect of aromatic amines in several extrahepatic organs such as lung, gastrointestinal tract, kidney, urinary bladder, and brain.[1]

5.08.3.1.1.5   Dehydrogenases and reductases

Alcohol dehydrogenase (E.C. 1.1.1) (ADH)[32] toxifies ethanol to acetaldehyde, which is then (predominantly) detoxified by an aldehyde dehydrogenase (E.C. 1.2.1) to acetic acid. The second step, the aldehyde dehydrogenase-mediated oxidation to acetic acid, is inhibited by disulfiram (Antabus), which is used in the treatment of alcohol addiction. After alcohol consumption disulfiram leads to the accumulation of the toxic acetaldehyde. The resulting toxicity provokes headache and nausea, which is intended to keep the alcoholic from further alcohol consumption. Many other aldehydes, such as the α,β-unsaturated aldehydes (lipid peroxidation products), are also markedly toxic. Thus, aldehyde dehydrogenase predominantly leads to detoxification. However, as is the case with all adequately investigated drug-metabolizing enzymes, aldehyde dehydrogenase plays a dual role with respect to toxification/detoxification, the nature of which depends on the substrate in question. Methanol is metabolized via formaldehyde to formic acid. Although formaldehyde is also considerably toxic, the decisive toxic metabolite is formic acid, leading to edema of the retina, blindness, and death. Remarkably, ethanol is an effective antidote, since it has a much higher affinity for the ADH compared with methanol. A relatively high concentration of ethanol inhibits the dehydrogenation of methanol allowing for the excretion of unmetabolized methanol thereby preventing the toxification to formic acid.[1]

ADH is a member of the family of medium-chain alcohol dehydrogenases.[32] The major function of most short-chain alcohol dehydrogenases[32] in mammals is steroid metabolism. Nevertheless, many of them play important roles in drug toxification and detoxification. The 3α-hydroxysteroid dehydrogenase (3α-HSD) of rat and man oxidizes vicinal dihydrodiols of polycyclic aromatic hydrocarbons to catechols and these to quinones.[33] Hence, 3α-HSD is also called dihydrodiol dehydrogenase. This reaction sequesters the pre-bay dihydrodiols of polycyclic aromatic hydrocarbons away from their critical toxification pathway to the ultimate carcinogenic dihydrodiol bay-region epoxides to produce instead the much less toxic catechols[33] (**Figure 3**). In addition, it inactivates the highly mutagenic and carcinogenic bay region diol epoxides.[34] Both of these reactions are protective, but the formation of catechols, which are further oxidized to quinones, is on the other hand also a potential toxification, since quinones frequently undergo redox cycling producing toxic reactive oxygen species. Thus, as usual for drug-metabolizing enzymes, with respect to toxification/detoxification the dihydrodiol dehydrogenase plays a dual role, the nature of which depends on the substrate. Toxification by dihydrodiol dehydrogenase occurs by catechol formation from naphthalene 1,2-dihydrodiol leading to naphthalene-induced cataract formation. In rabbit liver, at least eight enzymes possess dihydrodiol dehydrogenase activity. In rat liver only the 3α-HSD possesses dihydrodiol dehydrogenase activity.[35] This is an interesting example of enzymes of endogenous function in some but not all species having acquired the ability to toxify/detoxify xenobiotics.

The 11β-hydroxysteroid dehydrogenase (11β-HSD) can also metabolize drugs, acting as a carbonyl reductase.[36] The toxification of 4-(methylnitrosamino)-1-(3-pyridyl)-butan-1-one (NNK), a tobacco-specific nitrosamine, to the ultimate mutagenic metabolite can be prevented by 11β-HSD-catalyzed reduction of the keto function followed by conjugation to glucuronic acid.[36]

Similarly, the dimeric flavoprotein NAD(P)H quinone oxidoreductase (NQOR; E.C. 1.6.99.2) (also called DT-diaphorase) protects against the toxicity of quinones by their reduction to catechols followed by their conjugation and excretion. Two NQORs are known. NQOR-1 is induced by 2,3,7,8-tetrachlorodibenzo-p-dioxin (TCDD; also called 'dioxin'). The NQOR works by an interesting 'ordered ping-pong' mechanism. First, an enzyme–cofactor complex is formed, leading to the reduction of the prosthetic group. NADH and NADPH can serve as reducing cofactors. After dissociation of the enzyme–cofactor complex the substrate binds, is reduced, and the product is released. Therefore, the enzymatic reaction can be inhibited by high concentrations of the substrate or the cofactor. Coumarin derivatives, such as warfarin, are naturally occurring inhibitors.

There are important species differences in the relevance of NQOR for drug metabolism and protection against their toxicities. In the rat NQOR is by far the most important enzyme for the reduction of many quinones. In man, however, a less specific carbonyl reductase is quantitatively of much higher importance.[37] This represents an important difference for extrapolations between these species.

NQOR acts on its substrate by a two-electron transfer mechanism. It thereby circumvents the formation of the highly reactive semiquinone radicals from quinones. In this respect NQOR is protective against drug toxicity. On the other hand, NQOR reduces aromatic nitro compounds to aromatic hydroxyl amines, precursors of highly genotoxic reactive esters. Again, with respect to toxification/detoxification NQOR plays a dual role, the nature of which depends on the substrate and can easily be predicted, which is helpful for extrapolations between species.[1]

5.08.3.1.2   Hydrolases
5.08.3.1.2.1   Carboxylic ester hydrolases (E.C. 3.1.1) and amidases (E.C. 3.5.1–3.5.4)
Esterase and amidase reactions are catalyzed by the same enzymes, despite the assignment of different enzyme class numbers for the two different activities. The esterases are grouped according to their interaction with the toxicologically important organophosphates into A-, B-, and C-esterases. Organophosphates are substrates of and are therefore detoxified by A-esterases whereas organophosphates are strong inhibitors of B-esterases, which is the basis of their (intended selective insecticidal as well as unintended overdose mammalian) toxicity. C-esterases do not interact with organophosphates.

Although esters derived from aromatic compounds are preferred substrates of A-esterases (also called aryl esterases), the toxicologically most important substrates are the organophosphates. The selective toxicity toward insects is due to the lack of organophosphate detoxifying A-esterases in nonvertebrates, while mammals are largely protected by their high A-esterase activities.

The prominent member of the B-esterases is the acetyl choline esterase responsible for the inactivation of the neurotransmitter acetyl choline. Organophosphates are mechanism-based inhibitors of B-esterases. Their toxic effects are produced by an accumulation of acetyl choline and therefore by an exaggerated cholinergic activity. The inhibitory effect of the organophosphates is due to the fact that the enzyme phosphoserine ester formed between B-esterases and organophosphates hydrolyzes very slowly, if at all, and thus there is virtually no regeneration of the enzyme. The inhibition of another B-esterase, the neuropathy target esterase (NTE) localized in the central nervous system, leads to neurotoxicity called organophosphorus-induced delayed polyneuropathy (OPIDP). The neurotoxicity observed in some gulf war veterans is suspected to have been caused by a combination of the acetylcholine esterase inhibitor pyridostigmine (used as an antidote against anticholinergic chemical weapons) with organophosphate insecticides. The 'unspecific carboxyl esterase' predominantly expressed in the liver is also a B-esterase. It metabolizes many xenobiotic esters and amides, including several antiarrhythmic and anesthetic drugs. The 'unspecific carboxylesterase' is polymorphic in humans. Individuals lacking this esterase activity are prone to overdose toxicity of those drugs that are preferentially hydrolyzed by this enzyme. In individuals deficient in the unspecific caboxylesterase the neuromuscular-blocking agent suxamethonium leads to overdose toxicity (diaphragm muscle relaxation, cessation of breathing, and death) if the usual artificial respiration is not used or the dose is not adjusted.

C-esterases preferentially hydrolyze acetyl esters. Therefore, they are also called acetyl esterases. Typical substrates are 4-nitrophenyl acetate, propyl chloroacetate, and fluorescein diacetate.[1,3]

5.08.3.1.2.2   Epoxide hydrolases (E.C. 3.3.2.3)
Epoxide hydrolases (EHs)[38] catalyze the hydrolytic cleavage of oxirane rings. Because of the higher electron attracting force of the oxygen atom compared with the two carbon atoms of oxirane rings, epoxides possess two electrophilic carbon atoms, the electrophilic reactivity of which is enhanced by the tension of the three-membered oxirane ring. Depending on the influences of the rest of the molecule this can make epoxides highly cytotoxic, genotoxic, and carcinogenic. Therefore, the EH-catalyzed hydrolytic opening of oxirane rings is normally a detoxification reaction. Some important, and predictable, exceptions are discussed below.

Several mammalian EHs are specialized for endogenous substrates, while two mammalian EHs preferentially metabolize xenobiotic epoxides.[38] These are the microsomal epoxide hydrolase (mEH)[39,40] and the soluble epoxide hydrolase (sEH).[41,42] The two enzymes are α/β hydrolase fold proteins, thus, in spite of a very low similarity of their amino acid sequences (<15%), they share a common three-dimensional structure[43,44] that displays a catalytic triad in its catalytic center: an aspartate as the catalytic nucleophile and a histidine/glutamate pair (mEH) or a histidine/aspartate pair (sEH) as the water-activating charge relay system.[45] The oxirane ring of the substrate is attacked by the aspartate to form an enzyme–substrate ester that is hydrolyzed by a water molecule activated by proton abstraction through the histidine/acidic amino acid pair. This releases the product, a vicinal (*trans*-, if applicable) diol, and restores free enzyme. While the second step of detoxification (hydrolysis of the enzyme–substrate ester to give the diol product) is slow, it is of prime importance that the first step (covalent binding of the epoxide to the enzyme) is very fast. This leads to a practically instantaneous removal of the toxic epoxide from the system as long as the enzyme is present in excess (i.e., as long as the slow regeneration of the free enzyme is unnecessary for the removal of further epoxide).[46] mEH, which is a major constituent of the endoplasmic reticulum, is present in large amounts. This generates for epoxides that are mEH substrates a practical threshold concentration below which genotoxic damage is negligible,[47] an unusual and remarkable situation for genotoxic carcinogens.

mEH[38] is highly expressed in the liver and several steroidogenic organs and at more moderate levels in many tissues of man and other mammals. mEH is induced by phenobarbital, *trans*-stilbene oxide, Aroclor 1254, and by several other xenobiotics including many antioxidants such as ethoxyquin. mEH metabolizes many structurally diverse epoxides. Structural requirements for a mEH substrate are sufficient lipophilicity and lack of *trans*-substitution at the oxirane ring.[39] mEH substrates include epoxides metabolically formed from drugs, such as carbamazepine and phenytoin, from occupational compounds such as styrene, and from environmental compounds such as polycyclic aromatic hydrocarbons.[38] With respect to toxification/detoxification it is important to note that epoxide hydrolysis – normally a detoxification reaction – can be involved in overall toxification pathways in some predictable cases. The mEH-catalyzed hydrolysis of pre-bay epoxides metabolically formed from angular PAHs (e.g., the benzo[*a*]pyrene-7,8-epoxide) prevents its isomerization to the corresponding phenols, products of comparatively low toxicity that are easily conjugated and then excreted. Instead, the pre-bay dihydrodiol (e.g., the benzo[*a*]pyrene-7,8-dihydrodiol) is formed, which can be further metabolized by a variety of enzymes to the highly electrophilically reactive, mutagenic and carcinogenic dihydrodiol bay region epoxides (e.g., benzo[*a*]pyrene-7,8-dihydrodiol-9,10-epoxide), which are not (or very poor) substrates of EHs (see **Figure 3**). Thus, depending on which CYPs are predominantly present (and consequently which metabolic pathways of the large PAHs are predominantly operative) higher amounts of mEH can substantially increase the mutagenicity of angular PAHs such as benzo[*a*]pyrene, i.e., mEHs can paradoxically but predictably contribute to toxification.[15,16]

sEH[42] is predominantly expressed in liver, kidney, heart, brain, and in low amounts in several other organs. It has an unusual and toxicologically important dual localization in the cytosol and in the matrix of peroxisomes, the latter performing several metabolic functions including long chain fatty acid degradation. Several peroxisomal enzymes produce hydrogen peroxide. The fact that sEH is present in peroxisomes and is induced by peroxisome proliferators (which induce peroxisomal beta-oxidation and thereby hydrogen peroxide generation), suggests a protective role of sEH against hydrogen peroxide-induced oxidative damage. Fatty acid epoxides, products of lipid peroxidation, are in fact good substrates of sEH. Some fatty acid epoxides derived from arachidonic acid or linolenic acid are involved in signal transduction. Thus, sEH is likely to have an important regulatory function and sEH inducers or inhibitors are likely to render sEH an important toxification enzyme by disturbance of such regulatory functions. In addition, the sEH catalyzed metabolism of leukotoxin leads to the considerably more toxic 9,10-dihydroxyoctadec-12-enoic acid. This toxification suggests a central role of sEH in the pathogenesis of multiple organ failure, in particular the adult respiratory distress syndrome (ARDS), which develops as a consequence of a leukotoxin overproduction by leukocytes.[42]

The substrate specificity of sEH is complementary to that of mEH. It metabolizes many *trans*-substituted epoxides, such as *trans*-stilbene oxide and *trans*-ethyl styrene oxide, which are not hydrolyzed by mEH. Conversely, in contrast to mEH, sEH does not hydrolyze or poorly hydrolyzes epoxides derived from most polycyclic aromatic hydrocarbons.[38]

For extrapolations of metabolism-related toxicities between species it is important to note that sEH expression displays enormous interspecies differences. The rat has a particularly low level of sEH in the liver (only 0.01% of the total soluble hepatic protein). In contrast, the enzyme is highly abundant in mouse liver (0.3% of the soluble hepatic protein), and is about three times more active than rat sEH. In humans, sEH represents about 0.1% of the soluble liver proteins and possesses a specific activity similar to that of the rat.[38]

5.08.3.1.2.3   Other hydrolases

Glycosidases and sulfatases are further hydrolases of importance for drug metabolism. They primarily metabolize endogenous substrates including glycosaminoglycans and steroids, but they also accept some xenobiotic substrates; this is particularly true for the beta-glucuronidases. The gut flora can deconjugate compounds that are excreted via the bile as glucuronides or sulfates. The deconjugation products are often reabsorbed from the gut leading to enterohepatic circulation. Frequently, the toxicity and mutagenicity of natural products is masked by sugar moieties in glycosides and deglycosylation leads to their toxic effects. Thus, glycosidases are toxifying in many cases. The procarcinogenic glycoside cycasin of the cycad plant, for example, needs glycosidase-mediated toxification. After oral treatment with cycasin, rats with a normal gut flora develop tumors in the liver, kidney, and intestine as a consequence of the bacterial glycosidase-mediated liberation of the genotoxic aglycon methyl azoxymethanol. However, germ-free rats do not suffer from any long-term effects of cycasin.[1]

5.08.3.2 Phase II Enzymes: Transferases

5.08.3.2.1 Glutathione *S*-transferases (E.C. 2.5.1.18)

Glutathione (GSH) and glutathione *S*-transferases (GSTs)[48] represent the prime defense system against electrophile-mediated drug toxicity in mammals However, as is the case with all adequately investigated drug-metabolizing enzymes, GSTs in some predictable cases act as toxifying enzymes and glutathione is itself mutagenic when activated by rat kidney homogenates (S9).[49] The 'normal' detoxifying role of GSTs is due to the fact that they detoxify lipophilic and structurally very diverse electrophiles by conjugating them with the endogenous hydrophilic nucleophile glutathione. This neutralizes the electrophilic reactivity and renders the compound easily excretable.[1]

The structure and catalytic mechanism of GST are described elsewhere in this book (5.06 Principles of Drug Metabolism 2: Hydrolysis and Conjugation Reactions). Almost all GSTs are soluble, dimeric drug-metabolizing enzymes. Subunit dimerization is necessary for their catalytic activity. The substrate specificity is dictated by the subunit and the substrate specificity of the dimer is purely additive of the specificities of the two individual subunits. In addition, there are at least two membrane-bound glutathione-dependent enzymes, one of which is called the microsomal GST, also a drug-metabolizing enzyme.

Glutathione conjugates are released from the cells by an active transport system that belongs to the multidrug resistance (MDR) protein complex. Since the GSTs are often strongly product-inhibited this active removal of the products is important for the efficiency of GST-mediated detoxification.[50]

Glutathione conjugates, predominantly formed in the liver, are usually further processed in the kidney. The glutamate moiety is cleaved off by the γ-glutamyltranspeptidase. Then glycine is released by aminopeptidase M. The resulting cysteinyl conjugate is usually *N*-acetylated to form a mercapturic acid. In competition with the *N*-acetyltransferase, beta-lyase can convert some cysteinyl conjugates to the sulfhydryl metabolite. This reaction can lead to highly toxic metabolites and hence represents a toxification. In these cases the GST-mediated reaction has ultimately led to a highly toxic metabolite.[1]

The superfamily of cytosolic GSTs (*see* 5.06 Principles of Drug Metabolism 2: Hydrolysis and Conjugation Reactions) composed of families (also called classes): the α (alpha), μ (mu), π (pi), θ (theta), σ (sigma), and κ (kappa) families (**Table 3**). The cytosolic GSTs are homodimers or heterodimers of subunits belonging to the same family. The recommendation for a nomenclature system proposes to use 'GST' for the enzyme, preceded by a lower case letter for the species (h for human, r for rat, etc.) followed by a capital letter for the family (A for α, M for μ, P for π, T for θ, K for κ, S for σ) followed by arabic numbers for the subunits (i.e., mGSTA1-1 is the homodimer of the mouse α family subunit 1). For extrapolations of metabolism-controlled toxicities between species it is important to note that, in contrast to the system used for the CYPs, the use of the same number of GST subunits of different animal species does not (necessarily) indicate that they are orthologous; they are usually used in sequence of their discovery.[1]

A number of reactions are catalyzed by GSTs:

- Nucleophilic substitution of electron-withdrawing substituents by GSH such as the formation of *S*-(2,4-dinitrophenyl)glutathione from the broad-spectrum substrate 1-chloro-2,4-dinitrobenzene (CDNB), which is conjugated by almost all GSTs (the exception being the θ family).
- Nucleophilic addition of GSH to epoxides by opening of the oxirane ring and to α,β-unsaturated carbonyl compounds by Michael addition, both detoxification reactions. Many epoxides are good substrates for several GSTs. GSTP1 from most species has been shown to possess significant activity to detoxify the ultimate carcinogen benzo[*a*]pyrene-7,8-diol-9,10-epoxide (**Figure 3**). Also aflatoxin-8,9-epoxide is detoxified by GST. The glutathione conjugate of the endogenous epoxide leukotriene A$_4$ is leukotriene C$_4$ formed by GST and further metabolized to leukotriene D$_4$ both of them being active components of the slow reacting substance of anaphylaxis (SRS-A), which plays an important role in the genesis of anaphylaxis and bronchial asthma. rGSTA4-4 (in older literature 'rat GST 8-8' or 'Y$_k$-Y$_k$') has a particularly high Michael addition rate with the toxic lipid peroxidation product 4-hydroxynon-2-enal, implying an important role for GST in the protection against oxidative stress.

Table 3 Rat cytosolic glutathione S-transferases and typical substrates

Glutathione S-transferases	Selective substrate
Alpha family	
GSTA1-1	Δ5-Androstene-3,17-dione
GSTA2-2	Δ5-Androstene-3,17-dione
GSTA3-3	
GSTA4-4	4-Hydroxynon-2-enal
GSTA5-5	Aflatoxin B_1 exo-8,9-epoxide
Mu family	
GSTM1-1	1,2-Dichloro-4-nitrobenzene
GSTM2-2	trans-Stilbene oxide
GSTM3-3	1,2-Dichloro-4-nitrobenzene
GSTM4-4	
GSTM5-5	
GSTM6-6	
Pi family	
GSTP1-1	(+)-anti-Benzo[a]pyrene-7,8-diol-9,10-epoxide
Theta family	
GSTT1-1	1,2-Epoxy-3-(4-nitrophenoxy)propane
GSTT2-2	Menaphthyl sulfate
Sigma family	
GSTS1-1	Prostaglandin H (isomerization)
Kappa family	
GSTK1-1	Ethacrynic acid

Reproduced from Oesch, F.; Arand, M. Xenobiotic Metabolism. In *Toxicology*; Marquardt, H., Schäfer, S. G., McClellan, R., Welsch, F., Eds.; Academic Press: New York, 1999, pp 83–109, with permission from Elsevier.
This list includes only homodimers. In general, the properties of heterodimers are additive with respect to their subunits, probably because the substrate binding site (H-site) is formed by a single subunit. A broad-spectrum substrate for all GST except GSTTs is 1-chloro-2,4-dinitrobenzene (CDNB). .

- Reduction of organic hydroperoxides, converting them to the corresponding alcohols. GSTs are unable to reduce hydrogen peroxide (in contrast to glutathione peroxidase, which reduces hydrogen peroxide and organic hydroperoxides).
- Isomerizations, e.g. formation of Δ4-androstenedione from Δ5-androstenedione.[1]

In most cases GST-mediated reactions are detoxifications. However, there are at least two pathways by which GST toxifies halogenated hydrocarbons. Vicinal dihaloalkanes (e.g., 1,2-dibromoethane) can be conjugated to GSH, preferentially by GSTT1. The conjugate can undergo an intramolecular nucleophilic substitution reaction releasing the second halogen atom as an anion and forming a highly reactive episulfonium (thiiranium) ion, the thio-analogon of an epoxide with an electrophilic reactivity enhanced by the positive charge of the sulfur atom. This GST-catalyzed reaction, therefore, is a toxification yielding highly cytotoxic and genotoxic metabolites.

The second GST-mediated toxification pathway is particularly relevant for the organic solvent perchloroethylene. A highly cytotoxic and carcinogenic thioketene is formed by the following sequence: formation of the glutathione conjugate, degradation to the cysteinyl conjugate, which is then metabolized by the beta-lyase to the highly toxic thioketene.[50]

5.08.3.2.1.1 Polymorphisms of GSTs

A homozygous null allele, i.e., the absence of a functional gene, of hGSTM1 and hGSTT1 occurs in about 50% and 15% of the Caucasian population, respectively. Lack of a functional hGSTM1 is associated with an increased risk for bladder (possibly also lung) cancer, and lack of a functionally active hGSTT1 is associated with a myelodysplastic syndrome.[51]

5.08.3.2.2 UDP-glucuronosyltransferases (E.C. 2.4.1.17)

Conjugation with glucuronic acid is the most abundant phase-II reaction (see 5.06 Principles of Drug Metabolism 2: Hydrolysis and Conjugation Reactions). UDP-glucuronosyltransferases (UGTs)[52] catalyze the formation of beta-D-glucuronides from a large variety of xenobiotics by their reaction with UDP-glucuronic acid (UDPGA). Hydroxyl-, thiol-, amino-, hydroxylamino- and carboxyl-substituents serve as the anchor to which glucuronic acid can be conjugated (also C-glucuronides can be formed if the hydrogen of the respective C–H bond is sufficiently mobile such as a carbon atom between two carbonyl functions). The xenobiotic substituents engaged in the glucuronic acid conjugation are nucleophilic. Accordingly, the majority of the UGT substrates are not genotoxic. The anchor groups listed above are often determinants or co-determinants of the biological activity of drugs and toxins, since they are frequently involved in the interaction with receptors or enzymes. Therefore, conjugation of these frequently terminates therapeutic or also toxic activity. Additionally, in most cases, glucuronic acid conjugation substantially increases the hydrophilicity of a drug enhancing its excretion. Hence, UGTs are predominantly detoxifying enzymes.[1]

In a few cases, however, UGTs enhance the toxicity of their substrates. This is the case with some nonsteroidal anti-inflammatory drugs (NSAIDs). The resulting ester glucuronides can undergo acyl migration, i.e., the intramolecular transesterification from the C1 hydroxy group of the glucuronic acid to the C2 hydroxy group and further to the C3 and C4 hydroxy groups. This can lead to the formation of a free aldehyde group at C1, which can react with primary amino groups in proteins generating Schiff's bases. Amadori rearrangement can then lead to a stable protein adduct, which may give rise to allergic reactions, a well-known drug toxicity of some NSAIDs. Glucuronic acid conjugation can also result in enhanced genotoxicity. Aromatic amines, including important human carcinogens, are metabolized by CYPs (preferentially CYP1A2) to aromatic hydroxylamines. Glucuronidation of these leads to the formation of a (moderately good) leaving group, which after being cleaved off leaves behind a strongly electrophilic and genotoxic aryl nitrenium ion. Mildly acidic conditions, which are usually present in the urine, but not neutral pH, suffice to produce this situation. The hydroxylamines are converted into glucuronides as transport forms, which survive their journey in the blood from the liver where they are produced to the bladder where they dissociate and lead to bladder cancer (alternative conjugation reactions, sulfation and acetylation, of aromatic hydroxylamines form better leaving groups, which already dissociate at neutral pH).[1,53]

The families and subfamilies of UGTs[52] are described elsewhere in this book (see 5.06 Principles of Drug Metabolism 2: Hydrolysis and Conjugation Reactions). The most important family toxicologically is UGT1.[54] A single gene codes for at least 10 different proteins of this family. Differential splicing results in mRNAs that differ in the sequence of their first exon and are identical in the sequences of exons 2–5. The N-terminal portions of the resulting proteins, which correspond to the variable exon 1, contain the substrate binding sites. Therefore, the corresponding UGTs differ in their substrate specificities. These range from the endogenous bilirubin to pharmacologically active drugs such as morphine, to carcinogen precursors, such as benzo[a]pyrene-7,8-dihydrodiol. Therefore, a defect in this gene can have dramatic consequences. Heritable human diseases that have been attributed to such defects are the Crigler–Najjar syndrome and the Gilbert syndrome. The hyperbilirubinemia that results from the reduced or lost capacity to detoxify bilirubin can lead to an early death (Crigler–Najjar type 1). An experimental model for this disease is the Gunn rat. Owing to a mutation in exon 2 it lacks all UGT1 proteins. Rats are less sensitive than humans to the toxic action of bilirubin. Gunn rats can survive the lack of these UGTs, which makes them useful as a tool for studying the toxicological consequences of this defect.[1]

5.08.3.2.3 Sulfotransferases (E.C. 2.8.2)

Sulfotransferases (SULTs)[55] catalyze the transfer of a sulfonyl group from 3'-phosphoadenosine-5'-phosphosulfate (PAPS) to nucleophilic substituents of their substrates, analogous to the UGTs. Accordingly, SULTs are often detoxifying enzymes for the same reasons discussed above for the UGTs. On the other hand, sulfoconjugation is in a predictable fashion also an important toxification mechanism for several types of precarcinogens.[55,56] Aromatic hydroxylamines, metabolically produced from aromatic amines by CYP, are metabolized by SULT to aromatic N-O-sulfate esters, which heterolytically decompose to the sulfate anion and to the genotoxic nitrenium ions. Sulfation of benzylic alcohols leads to the formation of reactive carbenium ions after cleaving off the sulfate group. Thus, mice with an inherited deficiency in the synthesis of the cofactor required for SULT activity, PAPS, are less prone to develop liver cancer when treated with the hepatocarcinogen 1'-hydroxysafrol.[1]

SULTs have a substrate spectrum similar to that of the UGTs, except they do not conjugate carboxylic acids. Generally, SULTs have a lower K_m than UGTs. Hence, an important detoxification function of SULT is the efficient conjugation of low amounts of xenobiotics. They have a high affinity but only limited capacity for their substrates. The concentrations of the respective cosubstrates are in line with this: the PAPS concentration in the human liver is about 50 μM and UDPGA is present at an approximately 10-fold higher concentration. Thus, in general, drugs are preferentially sulfated at low concentrations and preferentially glucuronidated at high concentrations. When interpreting results from toxicity testing it is important to note that if for a given drug one of these pathways represents a detoxification and the other a toxification, low environmental or moderate therapeutic doses can lead to the opposite effect to the high concentrations used in toxicity tests.[1]

The SULT families[55] are described elsewhere in this book (*see* 5.06 Principles of Drug Metabolism 2: Hydrolysis and Conjugation Reactions).

5.08.3.2.4   Acetyltransferases (included in E.C. 2.3.1)

Two acetyltransferases (ATs) have major implications in drug metabolism: *N*-acetyltransferases 1 and 2 (NAT-1 and NAT-2)[57] (*see* 5.06 Principles of Drug Metabolism 2: Hydrolysis and Conjugation Reactions). Both enzymes catalyze the acetyl transfer from acetyl-CoA to amines or hydroxylamines.

Since, in contrast to most phase II reactions of drug metabolism, the metabolites of NAT reactions (typically amides) are usually less hydrophilic than their parent compounds (typically amines), the function of NAT appears to be primarily the inactivation of biologically active, potentially toxic compounds, rather than a conversion to more hydrophilic metabolites.[1] The role of NATs in the detoxification of the procarcinogenic aryl amines is ambivalent but dependent on the nature of the substrate: *N*-acetylation of the aromatic amine is a detoxification reaction, since it competes with the formation of the hydroxylamine, i.e., with the initiation of the toxification pathway. However, if *N*-oxidation takes place first, the resulting hydroxylamine is also a substrate for NAT leading to the *N*-*O*-acetate ester (an acetoxyamino-group) from which acetate is easily cleaved off, which results in the generation of a reactive nitrenium ion. Thus, the balance between toxification and detoxification is in part determined by the velocity of the oxidative pathway relative to the acetylation pathway, which in turn is in part dependent on the substrate in question. In man, this leads to a high interindividual variability in risk since the expression of CYP1A2, the major contributor to the hydroxylation of aromatic amines, differs greatly among individuals, and NAT-2, which is often quantitatively very important in the acetylation of aryl amines and aryl hydroxylamines, is polymorphic. Slow acetylators are very slow in the acetylation of aromatic amines and their metabolites. The acetylator phenotype appears to be decisive for the organ selectivity of the tumor formation. Rapid acetylators are less susceptible to bladder cancer development caused by (occupational or therapeutic) aromatic amines since rapid acetylation effectively competes against the first step (hydroxylamine formation) in the major toxification pathway. Conversely, rapid acetylators have a higher risk of developing colorectal cancer. The causative agents for these colorectal cancers may be heterocyclic aryl amines arising from amino acids during high-temperature food preparation, such as PhIP. These are poor substrates for both NATs, hence oxidative formation of the corresponding hydroxylamines is favored. This makes NATs into predominately toxifying enzymes in this case. Hence, rapid acetylators are at greater risk for developing colorectal cancer from the food-derived heterocyclic amines in the gut.[58]

The NAT-2 polymorphism also has consequences for drug toxicities. Isoniazide, for example, has a very prolonged half-life in slow acetylators, which can lead to accumulation and toxicity. The acetylator phenotype can be assayed with caffeine as an in vivo probe.[59] One strong cup of coffee is sufficient for this purpose. A ratio between the metabolites 5-acetylamino-6-amino-3-methyl uracil (AAMU) and 1-methylxanthine (1X) greater than two indicates a rapid acetylator.

5.08.3.2.5   Methyltransferases (E.C. 2.1.1)

Methyltransferases are important in drug metabolism related control of toxicities in only a few cases. Beta-lyase products of the glutathione conjugation pathway can be detoxified by methylation of the free SH group competing against the formation of highly reactive and highly toxic thioketenes derived from perchloroethylene (*see* Section 5.08.3.2.1). Some catechol-moieties-containing drugs can be *O*-methylated by the catechol *O*-methyltransferase preventing the formation of toxic quinones. The enzymes catalyzing these reactions are mainly specialized to endogenous substrates. The cosubstrate for these reactions is usually the *S*-adenosyl methionine.[1]

5.08.4   Factors Influencing Drug Metabolism and Consequent Drug Toxicities

5.08.4.1   Modulation of Expression and Activity of Drug-Metabolizing Enzymes and Consequences for Drug Toxicities

Drug-metabolizing enzymes may be modulated by:

- increase of the amount of enzyme protein ('induction' in the broad sense of the term, usually but not always due to transcriptional activation);
- decrease of the amount of enzyme protein ('repression' in the broad sense of the term, sometimes but not always due to transcriptional repression);
- increase of the enzyme's specific activity, i.e., higher activity of the same amount of enzyme protein (activation); and
- decrease of the enzyme's specific activity (inhibition).

Induction by substrate is a strategy to economically meet the demands of drug metabolism without excessive waste of resources. In the strict sense induction is defined as an enhanced rate of transcription. In drug metabolism the term is used to describe an increased amount of enzyme protein regardless of the underlying mechanism. An enhanced stability of the protein and/or mRNA are included in this loose definition. Information on in vitro induction assays can be found elsewhere in this book (see 5.10 In Vitro Studies of Drug Metabolism).

The substrate-induced induction of drug-metabolizing enzymes leads to pharmacokinetic tolerance, an important mechanism of drug interactions and consequent toxicities. As an example treatment with narrow-therapeutic-window anticoagulants (e.g., warfarin), which are substrates of phenobarbital-inducible CYP2Bs, CYP2Cs, and/or CYP3As, simultaneously with inducers of these CYPs, such as the hypnotic drug phenobarbital, lead to a strong increase in the metabolism of these anticoagulants reducing their plasma concentrations. In order to maintain an effective therapy this necessitates a corresponding increase in the dose of the anticoagulant. When the concomitant treatment with the inducer is discontinued this can result in drastic drug toxicity. The decrease in the activity of the induced enzymes leads to accumulation of the anticoagulant, which may result in massive internal bleeding, and eventually death.[1]

Repression of drug-metabolizing enzyme expression in some instances occurs when induction of one set of drug-metabolizing enzymes goes at the expense of the expression of another set of drug-metabolizing enzymes. Some peroxisome proliferators that induce CYP4A, UGT1A1, and sEH decrease the expression of several other CYPs and also that of some GSTs. Cytokines such as interleukin 1β, interleukin 6, and interferon γ repress the expression of several CYPs. Therefore, inflammatory processes may substantially alter the rate of drug metabolism and the metabolite pattern.[60]

Activation of drug-metabolizing enzymes has been reported in relatively few cases. Alpha-naphthoflavone has complex effects including activation of some CYPs. Clotrimazole and isoquinoline enhance the in vitro activity of mEH toward styrene oxide considerably (fivefold). Diethyl ketone and structurally related solvents lead to a strong increase of UGT2B3 activity toward 2-aminophenol, resulting in a marked change in the substrate preferences.[1]

Inhibition of enzyme activities can occur as a result of interaction of the inhibitor with the enzyme resulting in isoenzyme-specific inhibition, or by cofactor depletion, the latter affecting all enzymes that depend on the respective cofactor. CYP1A2 significantly contributes to the metabolic inactivation of caffeine. Drugs/drug candidates that at the intended therapeutic concentrations inhibit CYP1A2 (e.g., fluvoxamine, furafyline) can strongly reduce the caffeine tolerance of individuals with an initially high CYP1A2 activity. This can lead to intoxication by regular coffee intake. Naringenin and related flavone glucosides are effective inhibitors of CYP3A4 at dosages that can be reached by drinking a glass of grapefruit juice. This may reduce the susceptibility to chemical carcinogens that are activated by CYP3A4, but increase unwanted drug effects of CYP3A4 substrates. Since 3A4 is the CYP that metabolizes the greatest number of therapeutic drugs, CYP3A4 inhibition is of great practical importance for drug toxicity.[1]

5.08.4.2   Significance of Interindividual Differences for Drug Metabolism-Associated Toxicities

Genetic polymorphisms are a major cause of interindividual differences in drug metabolism (**Table 4**)[61–67] (see 5.10 In Vitro Studies of Drug Metabolism). Several of them have an important impact on drug metabolism and must be taken into consideration in drug therapy and to avoid drug toxicities. Also, the susceptibility toward chemical carcinogenesis is profoundly influenced by polymorphisms of drug-metabolizing enzymes. The consequences of individual polymorphisms are discussed above where the individual drug-metabolizing enzymes are described (Sections 5.08.3.1 and 5.08.3.2).

Table 4 Polymorphic drug-metabolizing enzymes

Established	*Suspected*
CYP1A1	FMO
CYP2A6	UGT2B7
CYP2C9	SULT1A3
CYP2C18	
CYP2C19	
CYP2D6	
CYP2E1	
mEH	
NAT1	
NAT2	
GSTM1	
GSTP1	
GSTT1	
UGT1	

Reproduced from Oesch, F.; Arand, M. Xenobiotic Metabolism. In *Toxicology*; Marquardt, H., Schäfer, S. G., McClellan, R., Welsch, F., Eds.; Academic Press: New York, 1999, pp 83–109, with permission from Elsevier.
Bold print: especially important.

During ontogenetic development individual drug-metabolizing enzymes develop differently with pronounced consequences for drug toxicities. Human CYPs are already active in the fetus at about the fourth month of gestation, but appear in rodents only 2–3 days before birth. Also, sulfation and acetylation develop early in humans while the glucuronidation of bilirubin is still underdeveloped in the neonate This causes severe problems in neonatal erythroblastosis leading to a toxic accumulation of bilirubin resulting in jaundice. In some CYP families the expression of individual enzymes changes from fetus to adult. Thus, in humans the fetal CYP3A is CYP3A7, which disappears during development and is replaced by the adult CYP3A4.

Sex differences in the expression of drug-metabolizing enzymes are frequent in several experimental animal species, especially the rat, but are usually not pronounced in humans. In the rat there are marked sex-specific differences in the CYP2A, CYP2C, and CYP3A subfamilies. In humans, some drugs that are preferentially metabolized by CYP3A4 are cleared at rates that differ by twofold between the genders. Women have a lower alcohol clearance than men and therefore are more susceptible to alcohol intoxication.[1]

5.08.4.3   Species Differences

During its evolution, every animal species was under an individual selection pressure with regard to drug metabolism, in major part dictated by its preferred diet. Thus, some major species differences in drug metabolism[68,69] may tentatively be rationalized by species-specific requirements. Cats are exclusively carnivores. They almost completely lack the ability to glucuronidate xenobiotics, presumably because they do not have a need to render the easily glucuronidatable plant flavonoids or alkaloids excretable. For less obvious reasons, pigs are slow in drug sulfation and dogs are poor acetylators. The rat has extremely low levels of sEH in the liver.

It is important to choose the appropriate animal species for the analysis of the metabolism of a drug, in order to make predictions for metabolism-dependent toxicities in man. An impressive example of wrong predictions concerning toxification versus detoxification was the use of cynomolgus monkeys as an animal species closely related to man in the assessment of the carcinogenicity of heterocyclic amines present in cooked meat. Cynomolgus monkeys do not possess the heterocyclic amine toxifying CYP1A2. Thus, the study gave false-negative results. In contrast, these heterocyclic amines were potent carcinogens in the marmoset, a primate with a significant level of CYP1A2.

Toxicologically important species differences also occur in the regulation of enzyme expression. Peroxisome proliferators lead to an increase in the number and size of peroxisomes. Peroxisomal beta-oxidation, which produces hydrogen peroxide, is induced together with several drug-metabolizing enzymes by peroxisome proliferators. Treatment of rodents with peroxisome proliferators led to hepatocellular carcinomas. Peroxisome proliferation, the putative reason for the increased liver cancer incidence, did not occur to a significant extent in humans. Therefore, the peroxisome-proliferating hypolipidemic fibrates, which induce liver cancer in rodents, are rightfully still used in human therapy.

Since drug metabolism takes place predominantly in the liver, primary hepatocytes from experimental animal species and from man are often good predictors of drug metabolism-dependent drug toxicities and hence also of the best choice of an animal model for metabolism-dependent drug toxicities.[70,71] An example is the stomach parietal cell proton pump inhibitor pantoprazole that caused death in dogs by formation of the toxic benzimidazourea, which was formed by dog but not by human hepatocytes.[68] An additional excellent model is that of 'humanized' mice in which genes coding for individual xenobiotic metabolizing enzymes have been knocked out and replaced by related human genes.[72]

5.08.4.4 Consequences of Stereoselectivity in Drug Metabolism for Metabolic Toxification

Enzymes are chiral and interact with their substrates stereoselectively and enantioselectively. Enantioselective toxification has been shown for nitrogen mustard-derived cytostatic drugs, such as cyclophosphamide and ifosfamide. Also, metabolic toxification of procarcinogens often proceeds with high stereoselectivity. A striking example is the toxification pathway of benzo[a]pyrene, which ultimately leads to the potent genotoxic metabolite 7,8-dihydrodiol-9,10-epoxide (see **Figure 3**). Four chiral centers are present in this reactive metabolite (carbon atoms 7, 8, 9, and 10). The conformation at position 7 relative to position 8 is usually *trans* and that at position 9 relative to position 10 must be *cis*. This reduces the number of possible diastereomers from 16 to 4. Of these, the (+)-*anti* enantiomer (absolute configuration R,S,S,R) is the most abundant dihydrodiol epoxide formed during the mammalian metabolism of benzo[a]pyrene. Unfortunately, in addition it has the highest tumorigenic potential of the 7,8-dihydrodiol-9,10-epoxides derived from benzo[a]pyrene.[73]

5.08.5 Conclusion

Over the course of evolution organisms were faced with the problem of how to counteract internal accumulation of xenobiotics to toxic levels. These xenobiotics were ever changing in their chemical structures since all other organisms also underwent evolution. Hence, those organisms that developed xenobiotic metabolizing systems that could deal with compounds of highly differing structures were in an advantageous situation for better survival. The problem was solved by (1) developing enzymes catalyzing reactions that work on many different structural elements; and (2) development of many enzymes and isoenzymes with broad and overlapping substrate specificities. For the former, the price that has to be paid is that these reactions because of their lack of specificity convert some substrates into harmless metabolites and others into highly toxic metabolites. Drug metabolism studies have made and will continue to make great progress in elucidating the molecular details of these reactions, which enables us to recognize which combination of structural elements of chemical compounds will be converted by which enzyme to highly toxic or harmless metabolites. This will help us to understand and predict species differences in drug metabolism, to define the risk for humans, and to recognize whether for a given chemical compound more susceptible developmental stages, physiological/pathophysiological conditions or, due to genetic polymorphisms in defined drug-metabolizing enzymes, more susceptible entire populations exist. All this will bring us a great step closer to the development of safer chemicals and drugs and to the safer use of them.

References

1. Oesch, F.; Arand, M. Xenobiotic Metabolism. In *Toxicology*; Marquardt, H., Schäfer, S. G., McClellan, R., Welsch, F., Eds.; Academic Press: New York, 1999, pp 83–109.
2. Testa, B. *The Metabolism of Drugs and Other Xenobiotics – Biochemistry of Redox Reactions*; Academic Press: London, UK, 1995.
3. Testa, B.; Mayer, J. M. *Hydrolysis in Drug and Prodrug Metabolism. Chemistry, Biochemistry and Enzymology*; Verlag Helvetica Chimica Acta: Zurich, Switzerland, 2003.
4. Nelson, D. R.; Koymans, L.; Kamataki, T.; Stegeman, J. J.; Feyereisen, R.; Waxman, D. J.; Waterman, M. R.; Gotoh, O.; Coon, M. J.; Estabrook, R. W. et al. *Pharmacogenetics* **1996**, *6*, 1–42.
5. Estabrook, R. W. *FASEB J.* **1996**, *10*, 202–204.
6. Guengerich, F. P. *Chem. Res. Toxicol.* **2001**, *14*, 611–650.
7. Gonzalez, F. J.; Lee, Y. H. *FASEB J.* **1996**, *10*, 1112–1117.

8. Nebert, D. W.; McKinnon, R. A. *Prog. Liver Dis.* **1994**, *12*, 63–97.
9. Coon, M. J.; Vaz, A. D. N.; Bestervelt, L. L. *FASEB J.* **1996**, *10*, 428–434.
10. Zangar, R. C.; Davydov, D. R.; Verma, S. *Toxicol. Appl. Pharmacol.* **2004**, *199*, 316–331.
11. Ioannides, C.; Lewis, D. F. *Curr. Top. Med. Chem.* **2004**, *4*, 1767–1788.
12. Gonzalez, F. J.; Gelboin, H. V. *Drug Metab. Rev.* **1994**, *26*, 165–183.
13. Kawajiri, K.; Hayashi, S. I. The CYP1 Family. In *Cytochromes P450-Metabolic and Toxicological Aspects*; Ioannides, C., Ed.; CRC Press: Boca Raton, USA, 1996, pp 77–97.
14. Fernandez-Salguero, P. M.; Gonzalez, F. J. Targeted Disruption of Specific Cytochromes P450 and Xenobiotic Receptor Genes. In *Cytochrome P450*; Johnson, E. F., Waterman, M. R., Eds.; Academic Press: San Diego, CA, 1996; Vol. 272, pp 412–430.
15. Conney, A. H.; Chang, R. L.; Cui, X. X.; Schiltz, M.; Yagi, H.; Jerina, D. M.; Wie, S. J. *Adv. Exp. Med. Biol.* **2001**, *500*, 697–707.
16. Oesch, F. *Arch. Toxicol.* **1979**, Suppl. *2*, 215–227.
17. Sogawa, K.; Matsushita, N.; Ema, M.; Fujii-Kuriyama, Y. DNA-Binding Regulatory Factors and Inducible Expression of the P4501A1 Gene. In *Cytochrome P450*; Lechner, M. C., Ed.; John Libbey Eurotext Ltd: Montrouge, France, 1994, pp 75–80.
18. Whitlock, J. P.; Okino, S. T.; Dong, L. Q.; Ko, H. S. P.; Clarke-Katzenberg, R.; Qiang, M.; Li, H. *FASEB J.* **1996**, *10*, 809–818.
19. Oesch-Bartlomowicz, B.; Huelster, A.; Wiss, O.; Antoniou-Lipfert, P.; Dietrich, C.; Arand, M.; Weiss, C.; Bockamp, E.; Oesch, F. *Proc. Natl. Acad. Sci. USA* **2005**, *102*, 9218–9223.
20. Raunio, H.; Rautio, A.; Pelkonen, O. *IARC Sci. Publ.* **1999**, *148*, 197–207.
21. Gervot, L.; Rochat, B.; Gautier, J. C.; Bohnenstengel, F.; Kroemer, H.; de Berardinis, V.; Martin, H.; Beaune, P.; de Waziers, I. *Pharmacogenetics* **1999**, *9*, 295–306.
22. Huang, Z.; Waxman, D. J. *Cancer Gene Ther.* **2001**, *8*, 450–458.
23. Oesch-Bartlomowicz, B.; Richter, B.; Becker, R.; Vogel, S.; Padma, P. R.; Hengstler, J. G.; Oesch, F. *Int. J. Cancer* **2001**, *94*, 733–742.
24. Goldstein, J. A.; de Morais, S. M. F. *Pharmacogenetics* **1994**, *4*, 285–299.
25. Eichelbaum, M.; Ingelman-Sundberg, M.; Evans, W. E. *Annu. Rev. Med.* **2006**, *57*, 119–137.
26. Yu, A. M.; Idle, J. R.; Gonzalez, F. J. *Drug Metab. Rev.* **2004**, *36*, 243–277.
27. Kessova, I.; Cederbaum, A. I. *Curr. Mol. Med.* **2003**, *3*, 509–518.
28. Ronis, M. J. J.; Lindros, K. O.; Ingelman-Sundberg, M. The CYP2E Family. In *Cytochromes P450 – Metabolic and Toxicological Aspects*; Ioannides, C., Ed.; CRC Press: Boca Raton, FL, 1996, pp 211–239.
29. Oesch-Bartlomowicz, B.; Padma, P. R.; Becker, R.; Richter, B.; Hengstler, J. G.; Freeman, J. E.; Wolf, C. R.; Oesch, F. *Exp. Cell Res.* **1998**, *242*, 294–302.
30. Maurel, P. The CYP3 Family. In *Cytochromes P450 – Metabolic and Toxicological Aspects*; Ioannides, C., Ed.; CRC Press: Boca Raton, FL, 1996, pp 241–270.
31. Cashman, J. R. *Chem. Res. Toxicol.* **1995**, *8*, 165–181.
32. Jörnvall, H.; Höög, J. *Alcohol Alcoholism* **1995**, *30*, 153–161.
33. Glatt, H. R.; Vogel, K.; Bentley, P.; Oesch, F. *Nature* **1979**, *277*, 319–320.
34. Glatt, H. R.; Cooper, C. S.; Grover, P. L.; Sims, P.; Bentley, P.; Merdes, M.; Waechter, F.; Vogel, K.; Guenthner, T. M.; Oesch, F. *Science* **1982**, *215*, 1507–1509.
35. Wörner, W.; Oesch, F. *FEBS Lett.* **1984**, *170*, 263–267.
36. Maser, E.; Richter, E.; Friebertshauser, J. *Eur. J. Biochem.* **1996**, *238*, 484–489.
37. Wermuth, B.; Platt, K. L.; Seidel, A.; Oesch, F. *Biochem. Pharmacol.* **1986**, *35*, 1277–1282.
38. Arand, M.; Oesch, F. Mammalian Xenobiotic Epoxide Hydrolases. In *Handbook of Enzyme Systems that Metabolize Drugs and Other Xenobiotics*; Ioannides, C., Ed.; John Wiley: Chichester, UK, 2002, pp 459–483.
39. Oesch, F. *Biochem. J.* **1974**, *139*, 77–88.
40. Oesch, F.; Bentley, P. *Nature* **1976**, *259*, 53–55.
41. Guenthner, T. M.; Hammock, B. D.; Vogel, U.; Oesch, F. *J. Biol. Chem.* **1981**, *256*, 3163–3166.
42. Morisseau, C.; Hammock, B. D. *Annu. Rev. Pharmacol. Toxicol.* **2005**, *45*, 311–333.
43. Arand, M.; Grant, D. F.; Beetham, J. K.; Friedberg, T.; Oesch, F.; Hammock, B. D. *FEBS Lett.* **1994**, *338*, 251–256.
44. Armstrong, R. N. *CRC Crit. Rev. Biochem.* **1987**, *22*, 39–88.
45. Arand, M.; Wagner, H.; Oesch, F. *J. Biol. Chem.* **1996**, *271*, 4223–4229.
46. Oesch, F.; Herrero, M. E.; Hengstler, J. G.; Lohmann, M.; Arand, M. *Toxicol. Pathol.* **2000**, *28*, 382–387.
47. Herrero, M. E.; Arand, M.; Hengstler, J. G.; Oesch, F. *Environ. Mol. Mutagen* **1997**, *30*, 429–439.
48. Armstrong, R. N. *Chem. Res. Toxicol.* **1991**, *4*, 131–140.
49. Glatt, H.; Protic-Sabljic, M.; Oesch, F. *Science* **1983**, *220*, 961–963.
50. Dekant, W.; Vamvakas, S. *Crit. Rev. Toxicol.* **1996**, *26*, 309–334.
51. Chen, H. W.; Sandler, D. P.; Taylor, J. A.; Shore, D. L.; Liu, E.; Bloomfield, C. D.; Bell, D. A. *Lancet* **1996**, *347*, 295–297.
52. Burchell, B.; McGurk, K.; Brierley, C. H.; Clarke, D. J. UDP-Glucuronosyltransferases. In *Comprehensive Toxicology*; Guengerich, F. P., Ed.; Elsevier: Amsterdam, Netherlands, 1997; Vol. 3, pp 401–435.
53. Bock, K. W. *Crit. Rev. Biochem. Mol. Biol.* **1991**, *26*, 129–150.
54. Owens, I. S.; Ritter, J. K. Gene Structure at the Human UGT1 Locus Creates Diversity in Isozyme Structure, Substrate Specificity and Regulation. In *Progress in Nucleic Acid Research and Molecular Biology*; Cohn, W. E., Moldave, K., Eds.; Academic Press: San Diego, CA, 1995; Vol. 51, pp 305–338.
55. Glatt, H. R. Sulphotransferases. In *Handbook of Enzyme Systems that Metabolise Drugs and Other Xenobiotics*; Ioannides, C., Ed.; John Wiley: Chichester, UK, 2002, pp 353–439.
56. Coughtrie, M. W. H. *Hum. Exp. Toxicol.* **1996**, *15*, 547–555.
57. Goodfellow, G. H.; Dupret, J. M.; Grant, D. M. *Biochem J* **2000**, *348*, 159–166.
58. Badawi, A. F.; Stern, S. J.; Lang, N. P.; Kadlubar, F. F. *Prog. Clin. Biol. Res.* **1996**, *395*, 109–140.
59. Vistisen, K.; Poulsen, H. E.; Loft, S. *Carcinogenesis* **1992**, *13*, 1561–1568.
60. Morgan, E. T.; Thomas, K. B.; Swanson, R.; Vales, T.; Hwang, J.; Wright, K. *Biochim. Biophys. Acta* **1994**, *1219*, 475–483.
61. Nebert, D. W.; McKinnon, R. A.; Puga, A. *DNA Cell Biol.* **1996**, *15*, 273–280.
62. Smith, C. A. D.; Smith, G.; Wolf, C. R. *Eur. J. Cancer* **1994**, *30A*, 1921–1935.
63. Meyer, U. A.; Zanger, U. M. *Annu. Rev. Pharmacol. Toxicol.* **1997**, *37*, 269–296.

64. Hengstler, J. G.; Arand, M.; Herrero, M. E.; Oesch, F. *Recent Results Cancer Res.* **1998**, *154*, 47–85.
65. Ingelman-Sundberg, M. *Toxicology* **2002**, *181–182*, 447–452.
66. Brockmöller, J.; Cascorbi, I.; Kerb, R.; Roots, I. *Cancer Res.* **1996**, *56*, 3915–3925.
67. Wormhoudt, L. W.; Commandeur, J. N.; Vermeulen, N. P. *Crit. Rev. Toxicol.* **1999**, *29*, 59–124.
68. Hengstler, J. G.; Oesch, F. Interspecies Differences in Xenobiotic Metabolizing Enzymes and their Importance for Interspecies Extrapolation of Toxicity. In *General and Applied Toxicology*; Ballantyne, B., Marrs, T. C., Syversen, T., Eds.; Macmillan: London, 1999, pp 271–290.
69. Turesky, R. J. *Mol. Nutr. Food Res.* **2005**, *49*, 101–117.
70. Utesch, D.; Glatt, H.; Oesch, F. *Cancer Res.* **1987**, *47*, 1509–1515.
71. Hengstler, J. G.; Utesch, D.; Steinberg, P.; Platt, K. L.; Diener, M.; Ringel, M.; Swales, N.; Fischer, T.; Biefang, K.; Gerl, M. et al. *Drug Metab. Rev.* **2000**, *32*, 81–118.
72. Gonzalez, F. J.; Yu, A. M. *Annu. Rev. Pharmacol. Toxicol.* **2006**, *46*, 41–64.
73. Buening, M. K.; Wislocki, P. G.; Levin, W.; Yagi, H.; Thakker, D. R.; Akagi, H.; Koreeda, M.; Jerina, D. M.; Conney, A. H. *Proc. Natl. Acad. Sci. USA* **1978**, *75*, 5358–5361.

Biographies

Barbara Oesch-Bartlomowicz passed her Medical State Examination at the Pommeranian Medical Academy, Szczecin, Poland in 1978, and was awarded her License to Practice Medicine in 1979. In 1981 she was a Visiting Scientist at the Institute for Biocatalysis, Academy of Sciences, Berlin-Buch, Germany. She was awarded her MD degree in 1983 by the Pommeranian Medical Academy, Szczecin, Poland where she obtained her License in 1984 as a specialist in Pharmacology and Toxicology and as a specialist in Dermatology. In 1987 she was a Visiting Scientist at the Institute of Toxicology, University of Mainz, Germany. Between 1988 and 1989 she was a Scholar of the European Science Foundation. In 1989 she became a Research Associate at the Institute of Toxicology, University of Mainz, Germany, becoming a University Assistant in 1994. In 1999 she was nominated Professor of Pharmacology and Toxicology. In 1996 she was a visiting scientist at the Department of Environmental Toxicology, University of California, Davis, USA. Professor Oesch-Bartlomowicz is currently University Docent, Institute of Toxicology, University of Mainz.

Franz Oesch obtained a PhD from the University of Fribourg, Switzerland in 1969. In 1973 he became Privatdozent at the University of Basel, Switzerland. Between 1974 and 1981 he was Professor (C3) of Pharmacology and Toxicology and Head of Section for Molecular Pharmacology at the University of Mainz, Germany. From 1981 on he was Professor

(C4) of Pharmacology and Toxicology initially (1981–1983) as Chairman of the Department of Molecular Pharmacology, then (from 1983 on) as Director of the Institute of Toxicology, University of Mainz.

Professor Oesch has been awarded the following prizes:

Professor Max Cloetta Prize 1977; American Society of Pharmacology and Experimental Therapeutics Award for the best publication of 1980 in Drug Metabolism and Disposition; Robert Koch Prize 1982; GUM Prize of the Environmental Mutagen Society 1985; Prize of the Rheinisch-Westfälische Academy of Sciences 1989; German Cancer Price 1990; Hoechst-Marion-Roussel-Price 1998; and Prize for Scientific Innovation of the State Rheinland-Pfalz 1999. Other honors include: Most Quoted Scientists: belonging to the top 0.1% (Current Contents No. 41, October 12, 1981); Keynote Lecture, International ISSX Meeting, Amsterdam, 1991; Otto Warburg Memorial Lecture, Berlin, 1992; Election to the Academia Europea, 1992; Keynote Review, International EAPCCT Congress, Zürich, 1998; Keynote Lecture, European Workshop on Drug Metabolism, Kopenhagen, 1998; The Gerhard Zbinden Memorial Lecture Award, EUROTOX, London, 2000; and Werner Heisenberg-Medaille, 2000.

5.09 Immunotoxicology

C Esser, Institut für Umweltmedizinische Forschung, Heinrich-Heine University, Düsseldorf, Germany

5.09.1 Principles of the Immune System

The immune system is the organ that fights infection by pathogenic organisms such as bacteria, viruses, protozoa, or even worms. It also protects against toxins, for instance those from bacteria. The defense system of invertebrates against pathogens is simple, relying mostly on macrophages and bactericidal substances. In vertebrates, however, highly specialized cells have evolved, capable of producing cells with receptors of high binding specifity against literally billions of molecules, be they peptides, lipids, sugars, metal salts, or other chemical classes. The vertebrate immune system is a unique organ in that it is composed of a multitude of cells, molecules, and organized tissue structures, found distributed over the entire body as its field of action (**Table 1**). Immune cells are mobile and capable of communicating directly with each other by cell surface structures, or over considerable distances via lymphokines and chemokines. Close cell–cell contact is often necessary to orchestrate an effective immune response. Lymphoid organs are found at many sites in the body and can provide the relevant spatial structures for this contact. The number of lymphocytes (T and B cells) in the human body is approx. 10^{12}, i.e., about the same as the number of cells of the brain or the liver.

Several features of the immune system must be considered when looking at the possible impact of chemical substances on the immune system: (1) the continuous differentiation and de novo generation of immune cells, (2) the capacity of the specific immune system to distinguish between structures of the own body and foreign structures (called 'self' and 'nonself') and between harmless and harmful molecules, (3) the high mobility of immune cells, (4) the sequestration of certain cell types within the body, (5) immunologic memory, and (6) communication with the environment via specialized signal transduction and cell surface receptors (**Table 2**).

All cells of the immune system, lymphoid and myeloid, are generated throughout life from the common hematopoietic stem cell, and have individual lifespans ranging from a few days (e.g., granulocytes) to many years. For instance, memory T and B cells can live several years.[1–3] Differentiating immune cells pass many checkpoints, which

Table 1 Some components making up the immune system[a]

Component		Target for[b]
Organs		
Primary lymphoid organs	Bone marrow	Benzene
	Thymus	
Secondary lymphoid organs	Lymph nodes	
	Gut-associated lymphoid tissue (GALT)	
	Bronchus-associated lymphoid tissue (BALT)	
Lymph vessels	Thoracic duct	
Cells		
Innate immune response	Macrophages, monocytes	Dioxin
	Granulocytes	
	Natural killer (NK) cells	
Antigen presentation specialists	Dendritic cells (myeloid, lymphoid, plasmacytoid)	Penicillin, procainamid, urishiol, dioxin
	Langerhans cells	
	Kupffer cells	
Adaptive immune response	B cells (CD5$^+$ B cells, other B cells)	Nickel
	T cells (T$_{H1}$, T$_{H2}$, T$_C$, T$_{Reg}$)	
	NK T cells	
Migration	High endothelial venules	
Molecules		
Destruction of bacteria	Complement	
Cell migration	Chemokines, chemokine receptors, leukotrienes	
Lymphocyte function, differentiation, and communication	Lymphokines/interleukins, lymphokine receptors	
Adhesion and trafficking	ICAM-1, integrins, selectins, CD44	Dioxin
Cell destruction and apoptosis	Perforin	
	Fas, FasL	
Recognition of antigen	T cell receptors, B cell receptors, Toll-like receptors, NK receptors	
Direct and indirect neutralization of antigen	Antibodies	
Antigen presentation	MHC class I, MHC class II, CD1	
Second signal in antigen presentation	CD28, ICOS, CD80, CD86, CTLA-4	Nickel
Communication and opsonization	Fc receptors	Organotin
Intracellular signaling and special features of immune cells	NFκB, RAG-1, RAG-2, STAT molecules, suppressors of cytokine signaling (SOCS)	Arsenic
Effector molecules	Histamine, myeloperoxidase, reactive oxygen species	

[a] Note that the list is not exhaustive, but highlights only the most important categories of immune system components.
[b] Some examples of substances, which are known to adversely affect the indicated component of the immune system.

Table 2 Important features of the immune system which can be targeted by toxic substances

Feature	Immunotoxic effect
Differentiation, signaling, and cell interactions	
Continuous de novo cell generation, differentiation, and cell death	Immunosuppression
Cell mobility within the body	
Forming and dissolving functional cell contacts	
Response to exogenous stimuli by special signaling cascasdes	
Antigen presentation and recognition	
Specific recognition via T cell or B cell receptors of billions of antigens	Autoimmunity, allergy, immunosuppression
Tolerance toward self antigens	
Tolerance against harmless nonself antigens	
Defense against harmful nonself antigens	
Sequestration of specific cells	
Eye chamber, brain	Systemic versus local effects
Bone marrow	
Gut-associated lymphoid tissue	
Skin mucosal surfaces	

are either intrinsically programmed or triggered exogenously, e.g., by antigen. Most checkpoints are at the same time 'points of no return,' and there appears to be very limited transdifferentiation. Immunotoxic substances can interfere with cell differentiation, cell homeostasis, or immune functions of cells. Various differentiation stages and cell lineages can become targets of chemical substances, e.g., by affecting intrinsic cell programming or interfering with the quality of the necessary exogenous stimuli such as antigen presentation (**Figure 1**). Thus, it is important to think of the immune system as a continuously dynamic spatial and temporal organ in order to understand any toxic interference by chemicals and to evaluate risks.

The immune system can be divided into the nonadaptive (or innate) part and the adaptive immune system. The adaptive immune system has 'memory,' i.e., it can react faster and better on a renewed contact with the chemical, even after years of not having had any exposure. The cells of the innate immune system use surface receptors to recognize bacterial structures, and initiate defense reactions against them directly. Moreover, the innate defense reaction may include recruitment and instruction of cells of the adaptive immune response, for instance via secretion of cytokines or chemokines. Adaptive immune responses by T cells and B cells can in principle respond to any given structure, i.e., any protein (T cells and B cells), lipid (some specialized T cells, B cells), sugar chain, or chemical substance (B cells only), and mount a humoral or cellular response. Antibody producing B cells, cytokine producing helper T cells, and killer T cells capable of directly killing infected cells recognize pathogenic insult in a uniquely specific way, by virtue of their receptors. These receptors are generated by a genetic process involving stochastic rearrangement of gene segments to finally give rise to coding genes for the millions of different B cell or T cell receptors. B cell receptors in soluble form are also known as antibodies. Every single B cell or T cell has a receptor specificity different from any other B cell or T cell. Because the process of genetic rearrangement is stochastic it will also generate receptors with undesired specificities, i.e., with specificities against molecules of one's own body. Destruction of self – autoimmunity – can be the consequence. The status of not mounting an immune response against self antigens is called tolerance. Not being tolerant against self would pose a strong risk for the body, thus many mechanisms exist to ensure tolerance. The most prominent mechanism is the elimination of autoreactive T cells, which react with good affinity against self proteins; this happens directly after T cells are generated in the thymus.[4] The process has its limits where self proteins are not present in the thymus to begin with,[5] or where self proteins are modified outside of the thymus, e.g., by xenobiotic substances, and thus 'look' foreign. The latter phenomenon is important in immunotoxic effects of many chemical molecules, as will be discussed in detail below.

The immune system also must ignore or not recognize, let alone fight, harmless structures such as food constituents, inhaled pollen, proteins from animal hair, or many xenobiotic chemicals of the environment whether inhaled, ingested, or

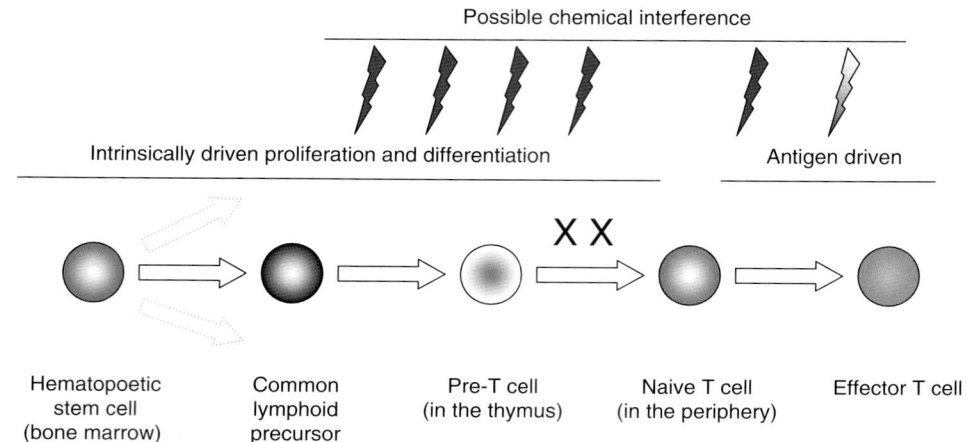

Possible chemical interference

Intrinsically driven proliferation and differentiation   Antigen driven

X X

| Hematopoetic stem cell (bone marrow) | Common lymphoid precursor | Pre-T cell (in the thymus) | Naive T cell (in the periphery) | Effector T cell |

Figure 1 Schematic sequence of immune cell differentiation using T cells as example. Starting with the hematopoietic stem cell in the bone marrow, cells proliferate and differentiate in a well-defined sequence. Transcription factor cascades and regulated expression of surface receptors (to make use of external factors such as interleukin-7 (IL7) produced by the stroma of the bone marrow or the thymus) ensure that the differentiation steps are followed through in a tightly controlled fashion. For T and B cells, selection processes at defined differentiation stages serve to eliminate autoreactive cells (indicated by **X X**). The dotted arrows indicate that from the stem cells other lineages derive as well; eventually T and B cells, granulocytes, makrophages, erythrocytes, platelets, etc. are produced. The late phases of T and B cell differentiation are antigen driven, i.e., the signal to move on into effector cells is antigen contact. For T cells this contact is provided through binding of the T cell receptor with antigenic peptides in association with surface major histocompatibility complex (MHC) molecules on specialized antigen-presenting cells. B cells can be activated by soluble antigen. Black flashes: possibilities for chemicals to act adversely on general immune cell functions. Grey flash: possibility for chemicals to interfere with the specific immune response to antigen (see text).

contacted on the skin. Similar to the tolerance against self, the body is under normal circumstances tolerant against such molecules.[6,7] If this is not the case, we speak of allergy. Allergies and xenobiotic induced autoimmunity are called adverse immune reactions. Adverse immune reactions are common, range from mild to severe, and may even cause death.[8]

5.09.2   The Two Basic Immunotoxic Action of Chemicals: Unwanted Activation and Unwanted Shutdown

The World Health Organization has estimated that at least 100 000 different chemicals are on the market, most of them with unknown immunotoxic potential. The complex nature of the immune system as sketched above allows for two very different mechanisms how low-molecular-weight chemicals such as drugs, food additives, metal salts, naturally occurring chemicals, or any of the many chemicals used in industry, agriculture, or household products can become immunotoxic (**Figure 2**). Chemicals or their metabolites can exert toxic effects on the immune system as the target organ. Any cell type can be affected, granulocytes, stem cells, natural killer cells, lymphocytes, etc. From the point of view of an immunologist, this type of immunotoxicity is nonspecific because it is not mediated by the specific antigen receptors on B or T lymphocytes. This type of immunotoxic effect will not lead to memory to the chemical in an immunological sense. Immunotoxic effects of this type usually cause immunosuppression, whose severeness may vary considerably, depending on the cell(s) affected. However, also examples of unspecific immunostimulation are known, for instance silicosis.[9,10]

Secondly, chemicals or their metabolites can act by the immune system. This means they interfere with the response mediated by the specific antigen receptors of B and/or T cells. Chemicals may be recognized as antigen and elicit immune responses, or they may directly or indirectly change self antigens and thus break tolerance. It is generally accepted by now that T cells are the major players in xenobiotic-induced autoimmunity and allergy. Chemicals acting by the immune system lead to sensitization and a memory response. Memory lymphocytes can expand and mediate a stronger immune response on second contact with the chemical, even if it occurs long after the first contact. Adverse immune reactions of this type may lead to allergy or autoimmunity.

5.09.2.1   Immunological Tolerance

As pointed out above, tolerance is a necessary feature of the immune system and usually exists toward self proteins and against harmless antigens, such as food. Immune tolerance is an active process at both the B cell and

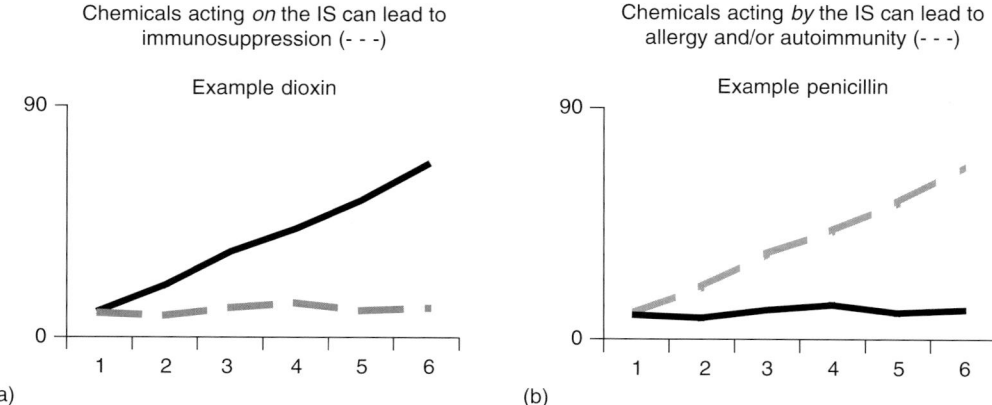

Figure 2 The two principal ways in which chemicals can change normal immune response. (a) They can lead to general dysfunction of the immune system, by destroying cells or disturbing their function. Immunosuppression might lead to more infections, or more severe illness. This type of adverse immune reaction is nonspecific in an immunological sense, i.e., it is not antigen-related. Typical example for this type is the immunotoxicity of dioxins. (b) If chemicals change antigen presentation, the resulting response is antigen-specific, will involve T cells and B cells, will and lead to memory and amplified secondary immune responses. These adverse immune reactions lead to immune responses against self (autoimmunity) or against normally harmless substances (allergy). A typical example is the allergic potential of penicillin. Black line: course of a normal immune response; dotted line: course of an immunotoxic reaction.

T cell level. Potentially harmful autoreactive B cells or T cells are eliminated in the bone marrow or the thymus. In this process, termed negative selection or central tolerance, future T cells die by apoptosis in the thymus, if their T cell receptor can bind with high affinity to self peptides presented on major histocompatibility complex (MHC) molecules.[4] B cells get negatively selected against in the bone marrow.[11] Outside of thymus and bone marrow, i.e., in the periphery (bloodstream, tissues, lymphoid organs), other mechanisms exist.[12,13] T cells need two signals to become activated: one is the antigen contact, the other is provided by surface molecules of the antigen-presenting cell. In general this second signal is not present on cells of uninjured and uninflamed tissue. Moreover, T cells do not enter healthy tissues, so self proteins are sequestered and do not come in contact with potential autoreactive T cells. Some self antigens, which need extra protection, are physically sequestered and not accessible to T cells at all, e.g., proteins in the chamber of the eye, or the brain. A second important mechanisms ensuring tolerance is the existence of regulatory T cells, which have received particular attention in recent years.[14,15] Regulatory T cells, also known as $CD4^+CD25^+$ T cells because they express these two surface molecules, suppress the proliferative response and the production of inflammatory cytokines. Dendritic cells also play a pivotal role in balancing peripheral immune responses and maintaining tolerance. The means how such tolerogenic dendritic cells work are not entirely clear but include inducing T cells to no longer proliferate or produce cytokines (so called anergic T cells), as well as regulatory T cells, and suppressive cytokine secretion. Understanding tolerance mechanisms and manipulating tolerance is currently a highly dynamic research area.

5.09.2.2 Antigen Presentation

In general, T cells only respond to peptide antigens. The peptides of 8–10 or 13–17 amino acids length must be placed in the groove of MHC molecules, which are membrane proteins present on all nucleated cells of the body (MHC class I) or upregulated on lymphoid cells (MHC class II). MHC class II can be significantly induced by γ-interferon.[16] The T cell receptor never binds to ('recognizes') the antigenic peptide alone, but to the peptide–MHC complex. Antigen recognition is the immunologists' expression for an affinity achieved through noncovalent interactions between the T cell receptor and the peptide–MHC proteins, such as hydrophobic, electrophilic or van der Waals forces. If the affinity passes a given threshold, a signal is relayed into the T cell interior,[17,18] triggering the cellular response, i.e., the antigen is 'recognized.'

Antigen presentation on MHC molecules is an immunologic feature, which involves enzymatic digestion of proteins, and several metabolic processes. Briefly, protein antigens are degraded in the cell into peptides, transported onto MHC molecules, which are then brought to the cell surface for possible recognition by T cells.[19] Intracellular self proteins or proteins from pathogens living in the cytoplasm (e.g., viruses) are degraded by proteasomes and transported into the endoplasmatic reticulum for loading onto MHC class I molecules.[20] In contrast, extracellular proteins are taken up by phagosomes, degraded, and loaded onto MHC class II molecules in the lysosomal compartments.[21] Accordingly, xenobiotic substances can have several routes of attack.

5.09.3   Unwanted Activation: Breaking Tolerance by Low-Molecular-Weight Chemicals

5.09.3.1   The Hapten Concept

Low-molecular-weight chemicals (commonly with a molecular mass of less than 1000 Da) are not recognizable by T cells. However, if they are reactive and capable of binding to proteins they may become part of presented peptides (**Figure 3**) as so-called haptens. In particular electrophilic properties of a chemical will enable it to react with nucleophilic groups of proteins such as the thiol group in cysteins (–SH), amino group of lysine (–NH$_2$) or the hydroxy (–OH) group of tyrosine.[22] Known reactive chemicals are isocyanates, quinones, aldehydes, epoxides, beta lactams, and certain nitroaromatics. If a chemical is very reactive, the immune reactions will take place at the site of first contact, e.g., the skin or the lung. Formation of novel antigens recognizable by T cells ('neoantigens') has been shown using the classical hapten trinitrophenol,[23] the sensitizing compound of poison ivy, 3-pentadecyl-catechol (urushiol),[24] or penicillin.[25] Penicillin-induced allergies have been intensively studied. There are immediate type and delayed type forms of penicillin allergy, clinically evident by, e.g., exanthema, urticaria, or specific immunoglobulin E (IgE) formation.[26] The contact sites of T cells with the protein penicilloyl adduct have been mapped. **Table 3** lists examples of known chemicals that form protein adducts (hapten–carrier conjugates in immunological terms) and can lead to adverse immune reactions. Frequently, protein reactive haptens lead to sensitization after dermal contact or inhalation.

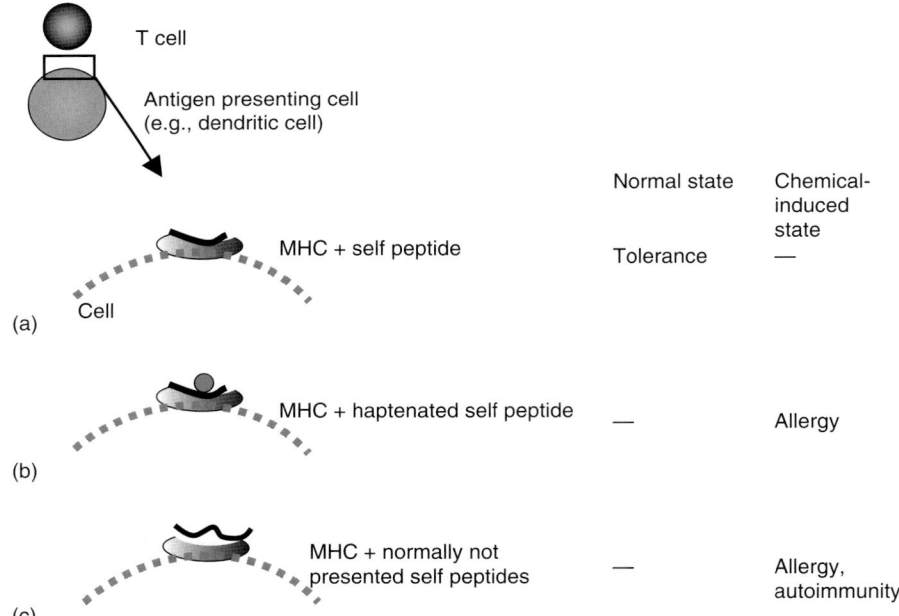

Figure 3 Possible outcomes of the formation of protein adducts with low-molecular-weight compounds. T cells recognize peptides lying in a groove of the cell surface molecule MHC. T cells with receptors that would recognize self peptides are eliminated in the thymus. (a) Those which escape the thymus elimination processes are kept in an inactive state in the healthy situation. (b) Low-molecular-weight chemicals which can bind to presented peptides, covalently or not, change the form of the peptide, which no longer 'looks' like a self antigen. T cells can react. Binding to the peptide can occur inside or outside of the cells, depending on the chemical. (c) Low-molecular-weight chemicals can change either the protein degradation or the loading of peptides in the cell and lead to a changed pattern of presentation and presentation of normally not presented peptides. Thus, even if presented peptides are 'self,' T cells will exist with receptors specific for them, as the selection processes in the thymus will have used the normal set of self peptides, not the cryptic ones. Note that binding of the T cell receptor to the peptide inside the MHC groove is necessary but not sufficient for T cell activation. A second signal, i.e., yet another contact of surface molecules by T cells and antigen-presenting cell, must be provided. This is discussed in the context of the danger hypothesis in the text.

Metals and their salts can also interact with proteins, either by their oxidizing properties, or as haptens, as they can form highly stable coordination bonds with certain amino acids in peptides.[27,28] Nickel salts are a well-known example of this type.

5.09.3.2 The Prohapten Concept

Many low-molecular-weight chemicals are not protein reactive themselves. However, they can become subject to cell metabolism by the xenobiotic metabolizing enzymes (cytochrome P450, glutathione S-transferases, N-acetyltransferases, quinone reductases, etc.). Haptenic reactive intermediates can form and cause adverse immune reactions (Table 4). The chemicals are often called prohaptens and the process bioactivation. While the liver is the major detoxifying organ and expresses the necessary enzymes for bioactivation, the extrahepatic metabolism appears to be more important for the conversion of low-molecular-weight chemicals into haptens of immunotoxic potential. Extrahepatic metabolism is found in macrophages,[29] white blood mononuclear cell, or the skin.[30] Dermal Langerhans cells, or at least Langerhans-like cell lines, can metabolize xenobiotics as shown for dimethyl-benzanthracene and urushiol.[31] Surprisingly, keratinocytes also have bioactivating capacity.[32] For many enzymes of the xenobiotic

Table 3 Examples of some chemicals known to cause adverse immune reactions

Compound	Adverse immune reaction	Hapten–carrier conjugate shown	Specific T cells demonstrated
Carbamazepine	Skin reactions, eosinphilia, systemic symptoms	No	Yes
Ceftriaxone, amoxillin	Drug-induced exanthem	No	Yes
Ciprofloxacin	Hypersensitivity	No	Yes
Lamotrigene	Skin hypersensitivity	No	Yes
Lidocaine	Drug allergy	No	Yes
Penicillin	Anaphylactic shock, urticaria, contact dermatitis, hemolytic anemia	Yes	Yes
Phenobarbital		No	Yes
Trinitrophenol	Contact dermatitis	Yes	Yes

Table 4 Examples of chemicals whose metabolites are associated with adverse immune reactions

Parent compound	Metabolite	Adverse immune reaction[a]
Dihydralazine	Hydralazine radical	Autoimmune hepatitis, drug-induced lupus
Gold(I) antirheumatics	Gold(III)	Dermatitis, glomerulonephritis
Halothane	Trifluoroacetylchloride	Autoimmune hepatitis
p-Phenylenediamine	Bandrowski's base	Contact dermatitis
Practolol	Practolol epoxide	Oculomucocutaneous syndrome
Procaineamide	N-Hydroxyprocainamide	Drug-induced lupus
Propylthiouracil	Propyluracilsulfonic acid	Vasculitis, drug-induced lupus
Sulfamethoxazole	Nitrososulfamethoxazole	Skin reactions
Tienilic acid	Tiophene sulfoxide	Autoimmune hepatitis
Urushiol	3-Pentadecyl-o-quinone	Contact dermatitis

[a] Clinically observable in humans.

metabolizing system, isoforms with differing activity are known. Genetic polymorphisms of these enzymes may influence the generation of reactive metabolites among humans and hence, may be connected with differing susceptibilities to drug-induced adverse immune reaction. For the enzyme N-acetyltransferase clear associations with the slow acetylator genotype and procainamide have been shown.[33] Of patients developing severe erythema multiforme variants following sulfonamide treatment the slow acetylator phenotype is more common than in controls.[34] Other substrates (e.g., dihydrazalazine, dapsone, sulfasalazine, isoniazid) of N-acetyltransferase, which also cause drug-induced adverse immune reactions, have not been tested for associations. Detecting an association of an autoimmune or allergic disease with genetic polymorphism of xenobiotic-metabolizing enzymes indirectly points to a substrate of those enzymes as etiological agents. For psoriasis a protective action associated with the rarer variant alleles of *CYP1A1* was shown, although a possible chemical involved remains enigmatic.[35]

5.09.3.3   The Danger Hypothesis: Low-Molecular-Weight Chemicals as Adjuvants

T cells get activated only if they receive two signals, one via antigen recognition by the T cell receptor, the other from an independent receptor–ligand cell surface interaction, also called the costimulatory signal. Additionally, cytokines produced from cells of the innate immune system appear to be necessary.[16,36] The theory of a danger signal[37] claims that T cells will react only if antigen is presented in the context of danger. Known endogenous danger signals are uric acid, heat shock proteins, nucleotides, reactive oxygen intermediates, extracellular matrix breakdown products, and cytokines like the interferons (IFNs).[37–43] The danger signal can upregulate costimulatory signals on the antigen-presenting cells, notably the dendritic cells, thus providing the second signal necessary for the T cell to react.[40] In adverse immune reactions the danger signal might come either from the chemical or from an independent source, for instance provided by an unrelated ongoing infection, where viral or bacterial products alert the dendritic cells. In agreement with this, drug hypersensitivity reactions are more common in patients with certain concomitant viral infections.[44] Alternatively, the chemical can provide the danger signal itself, either through causing cell damage, or by direct upregulation of costimulatory molecules or cytokines.[45] For instance, 2,4,6-trinitrochlorobenzene (TNCB), 2,4-dinitrofluorobenzene (DNFB), 2,4-dinitrochlorobenzene (DNCB), and nickel sulfate induce IL1β in Langerhans cells,[46–48] or mercuric chloride aids in secretion of IL1 by macrophages.[49,50] Haptens, such as DNCB, TNCB, DNFB, $NiCl_2$, $MnCl_2$, $CoCl_2$, $SnCl_2$, and $CdSO_4$, have been shown to change the expression of the costimulatory molecule CD86 by dendritic cells. In most cases the mechanism of induction of the costimulatory signal is not known. As diverse as the haptens are the patterns of their effects.

One additional level of complexity must be considered. As described above, haptens can provide both signal 1 and signal 2 to T cells, although not all of them are bifunctional in this way. Moreover, haptens can contribute to the pattern of cytokines which are secreted by dendritic cells when they mature.[51–54] As a consequence, T cells differentiate either into T helper 1 or T helper 2 cells, which differ in their cytokine secretion patterns. Thus, the chemicals contribute to what type of an adverse immune response (type 1 or type 2) is triggered.

5.09.3.4   Direct Presentation via Cluster of Differentiations 1 (CD1) Molecules

For decades, the paradigm that T cells recognize only peptide antigens has held. Recently, it was discovered that T cells exist that recognize lipid moieties; recognition is not on MHC molecules, but on evolutionarily related cell surface proteins, the CD1 molecules.[55] Lipids from intracellular pathogens or self lipids can be presented.[56] The CD1 system also allows for polycyclic compounds to be presented, although alkyl chains are preferred.[57] Sulfisomidine, sulfadiazin, sulfasalazine, and celecoxib induce strong hypersensitivity reactions, and are possibly presented by CD1 molecules.[58] Thus, drugs may not only alter peptides and be recognized after covalent binding as haptens, they may simply be presented themselves in a noncovalent way. Apparently, noncovalent binding and associated presentation can also occur via the MHC system.[59] No drug metabolism is required in this case, and the binding is labile. Nonetheless, the immunological consequences may be as dangerous as those of covalent binding.[60] Well-studied examples are sulfamethoxazole, lidocaine, and carbamazepine.[61–63]

5.09.4   Unwanted Shutdown

5.09.4.1   Basic Considerations on Immunosuppression

Proliferation and de novo protein synthesis are necessary for an immune response. Chemicals and drugs can interfere with these physiological functions independent of any activation of antigen-specific cells (as described in the previous sections). Immunotoxic reactions of this type lead either to immunosuppression or to immunostimulation.

Immunosuppression by chemicals can be caused by (1) killing of cells of the immune system, (2) changes in cell differentiation leading to fewer or incapacitated immune cells, and (3) changes in typical cell function such as cytokine secretion or expression of costimulatory molecules. For many drugs and chemicals the exact mechanism by which they change immune functions has not been assessed. Rather, the literature brims over with publications titled "Effects of x on parameter(s) yz." Although often not satisfying for lack of knowledge on the relevant mechanisms, at least these studies show the broad scope of interference. In some cases the unknown mechanism will turn out to be adverse immune reactions as described above, but often action is nonspecific in the immunological sense. Interference may, however, be highly specific in terms of protein chemistry, e.g., interaction with intracellular enzymes, signaling molecules, transcription factors, or other proteins. The action of dioxins, furans, glucocorticoids are very good examples of this (*see* Section 5.09.4.2.1).

Immunosuppression is an operational term describing an immune system operating with less efficiency than normal. Immune responses may start later, or they might be weaker. Immunosuppression might be general, or restricted to certain pathogens. Whether their immunosuppression is caused intentionally by pharmacotherapy, or unintentionally by environmental chemicals, immunosuppressed individuals are more susceptible to infections and the development of spontaneous cancers. Epidemiologically, immunosuppression can be measured by comparing the average healthy population with a particular subgroup, e.g., workers exposed to a toxic substance. However, no easily accessible and universally accepted markers for immunosuppression have been identified in vivo, and the considerable functional reserve of the immune system must be exceeded before immunosuppression becomes clinically relevant.[64,65] Loss of immune cells (unless very large scale as in acquired immunodeficiency syndrome (AIDS)), or a shift in proportion of cells in the blood or in lymphoid organs is of limited diagnostic value in humans. Standard immunotoxicity tests to detect immunosuppression or potentiation have been developed and validated.[66,67] It is worthwhile to keep in mind, however, that both acute clinical illness and small shifts in the susceptibility to normal infections are of medical and economic relevance.

In the following an attempt will be made to look at some examples of actions on the immune system where interference with cellular functions are known in more detail.

5.09.4.2   Signal Transduction and Immunotoxicity

5.09.4.2.1   Single-pathway multieffect chemicals

Perhaps the most important, certainly among the best-studied immunosuppressive environmental chemicals, are the polycyclic halogenated aromatic hydrocarbons (PHAHs), in particular the prototypical substance 2,3,7,8-tetrachloro-dibenzo-p-dioxin (TCDD).[68] Other substances with known immunosuppressive effects are pesticides, organotin compounds, lead, cadmium, and mercury molecules.[69,70] Immunosuppressive PHAHs are polychlorinated biphenyls, dioxins, furans, and others. PHAHs are usually found in mixtures in the environment, and their relative and additive risks can be calculated using the toxic equivalence factor,[71] which describes toxicity in relation to TCDD.

The immune system is a highly sensitive target of TCDD, and thymus atrophy and immunosuppression are hallmarks of even low doses of dioxin exposure in all laboratory species.[68] A dose of only 10 ng TCDD kg^{-1} body weight suffices to induce thymus atrophy in mice.[72,73] The common mechanism of action underlying PHAH toxicity is the activation of an endogenous transcription factor, the arylhydrocarbon receptor (AHR). The exact biochemistry of activation of the AHR has been elucidated in detail. Briefly, lipophilic PHAHs diffuse into the cells, bind with high affinity to the AHR, which then translocates to the nucleus, heterodimerizes with its partner molecule, the arylhydrocarbon receptor nuclear translocator (ARNT), binds to short promoter sequences, and induces gene transcription.[74] Many genes can be targeted by the activated AHR, depending on cell type and cell stage.[75–77] With respect to the immune system, TCDD can cause apoptosis, inhibit cell proliferation of T cell precursors, downregulate IgM production by B cells, upregulate costimulatory molecules on antigen-presenting cells, induce cytokine production, etc.[78,79] While dioxins are immunosuppressive, they also have other effects, such as teratogenicity, tumor promotion, dermatological problems, changes in lipid metabolism, or oxidative stress. The pleiotropic action of the toxic substance (here PHAHs) is not based on a multitude of interactions with various cell proteins, but due to the characteristics of the targeted tissue.

TCDD action differs depending on cell type and cell stage. In immune cells, many TCDD-inducible genes have now been identified using various methods.[75,80] Genes responsive to TCDD include for instance cytokine genes (IL1β, IL2, tumor necrosis factor alpha (TNF-α)), apoptosis genes (FasL, TRAIL), differentiation markers (Notch-1, CD44, CD69), costimulatory signals (CD40, CD80), depending on the cell type and cell differentiation stage. Gene expression profiling of isolated cell types (hepatocytes, thymocytes) has been performed with TCDD-treated cells versus controls. The results of these experiments illustrate several important points: many more genes are modulated after exposure than was originally thought; signaling can be direct or indirect (via AHR-triggered secondary or tertiary events); and exposure regimen is relevant. Moreover, interaction of TCDD with the AHR to achieve transcriptional gene activation is controlled at many levels, e.g., species specificities, cell stage specificities, abundance of the AHR in the cytoplasm, or competition for additional DNA-binding factors. The detailed analysis of TCDD-induced (immuno)toxicity has thus led many scientists to acknowledge a single-pathway–multieffects concept, i.e., the realization that despite a single underlying biochemical pathway, the outcome on the cellular, organ, or systemic level can be very diverse. This has to be considered in any risk assessment, predictive assays using model cell types, or interpretation of molecular and cellular events by TCDD, or any other substance suspected to be immunotoxic.

5.09.4.2.2 Multi-pathway multieffect chemicals

Cells follow their intrinsic programs, but more often they communicate intensely with their environment and adapt to it. Physiological functions such as protein expression or proliferation can be triggered and controlled by chemical or physical exogenous signals, which are relayed into the cells by a limited number of signal transduction pathways. Major pathways are direct ligand activation of latent transcription factors, G protein-coupled receptors, the mitogen activated protein (MAP) kinases, and the Janus kinase (JAK)–STAT pathway.

Tests for immunotoxicity can use typical responses as readouts such as induction or inhibition of lymphocyte proliferation. Indirectly such effects then point to an interference of the chemical in question with one of the signal transduction pathways. Many chemicals have been tested for their ability to inhibit proliferation. However, only for a few chemicals have the underlying mechanisms been addressed. Knowledge of the pathways is necessary for a more rational risk assessment, or for any need to manipulate the effects, e.g., in drug usage. As activated immune cells (in particular T cells and B cells) proliferate extensively, these cells are particularly sensitive to any chemicals that interfere with proliferation.

Two examples shall serve to detail how low-molecular-weight chemicals can interact with signaling pathways, and eventually lead to a shutdown or decrease of immune capacity.

5.09.4.2.2.1   Arsenite

Various forms of inorganic arsenic have been used in agriculture and forestry as components of pesticides and insecticides. Arsenic has become an environmental pollutant of great concern. Worldwide millions of people are exposed to doses believed to be immunosuppressive.[81] Chronic exposure to arsenic is associated with the suppression of hematopoiesis, damage to humoral and cell-mediated immunity,[81,82] and suppression of cytokine production.[83] For arsenic it could be shown that the JAK–STAT pathway is directly inhibited, abolishing STAT activity-dependent expression of suppressors of cytokine signaling (SOCS), and eventually affecting IL6 expression.[84] The JAK–STAT signaling pathway mediates the immune response of various cytokines and growth factors, and thus participates in inflammation. In contrast to TCDD, arsenic is not a single-pathway molecule as, e.g., interaction with the signaling pathway of c-Jun–N-terminal kinase was shown as a causal factor of arsenite mediated apoptosis in tumor cells.[85] Upstream events of c-Jun kinase inhibition were studied in T lymphocytes and membrane structures identified as the first signal affected by arsenite, probably needing interaction between arsenite and protein sulfhydryl groups.[86]

5.09.4.2.2.2   Organotin compounds

Organotin compounds are chemicals with at least one covalent Sn–C bond. They are widely used as polyvinyl chloride (PVC) stabilizers, biocides, or antifouling paints and have given rise to ubiquitous environmental contamination.[87] They are known to be immunotoxic, to induce thymus atrophy in experimental animals, and to decrease proliferation of human lymphocytes.[88] Apoptosis by organotin compounds is apparently mediated by caspase activation triggered via the mitochondrial pathway and the death receptor pathway. Like the pleiotropic effects of PHAHs the effects of organotin differ depending on the cell type and on the exposure regimen.[89,90] In natural killer cells, for instance, the MAP kinases p38 and p44/42 are activated by the organotin tributyltin.[91] Consideration of the dose effect is highlighted in organotins, as different doses have strikingly different outcomes, ranging from necrosis and apoptosis to subtle changes in cytokine secretion by T helper cells.[92]

5.09.4.2.3 Immunosuppressive drugs

Immunosuppressive drugs are of enormous clinical relevance. Their mechanisms of action are known in better detail than for environmental chemicals. The search for new and better immunosuppressive drugs continues and will draw on information about the immune system on one hand, and on the action of chemicals on immune cells and intracellular signaling on the other hand.

In general, the action of immunosuppressive drugs is due to a direct interaction of the drug with immunocompetent cells,[93] in particular with signal transduction. Cyclosporine A and FK506 (tacrolimus) bind to cyclophilin or FKBP12, respectively, and inhibit calcineurin.[94–96] As a consequence the signaling pathway normally triggered by T cell receptor/antigen activation gets blocked and cytokines are not produced as necessary.[97] Another macrolide, rapamycin, also binds to FKBP12, but instead of targeting calcineurin, the complex binds to 'target of rapamycin' (TOR), a serine-threonine kinase with important functions in cell growth and proliferation. TOR inhibition by rapamycin prevents phosphorylation of proteins important for translation.[98] The rapamycin-FKBP12 complex blocks signaling pathways of nuclear factor kappa B (NFκB) as well, and it inhibits cyclin C kinases.[99] The action of rapamycin extends to suppression of actin synthesis by activated T cells, with consequences for cell motility and the formation of the immunological synapse.[100] Other macrolides can block actin polymerization as well.[101]

Another class of immunosuppressants, the glucocorticoids, induce genes directly via the glucocorticoid receptor (a transcription factor) and the respective promoter elements, the glucocorticoid receptor elements (GREs), of the targeted genes. Also, glucocorticoids weaken the activation of T cells by interfering with the immunological synapse.[102] In the thymus, where development of T cells takes place, this leads to a change of the affinity sensitivity window within which T cells are activated or deleted when their T cell receptor binds to antigen. While thymocytes get killed in the thymus upon glucocorticoid treatment, in the periphery the same treatment changes the sensitivity of T cells toward antigen activation. The immunosuppressive action of glucocorticoids thus is not only restricted to gene modulation, but targets the very beginning of T cell activation, namely the integration and relay of the signals from the immunological synapse. The action on the immunological synapse can affect both sides, the T cells and the antigen-presenting cell. However, makrolides and glucocorticoides appear to have contrasting impacts in this regard.[93]

Histone deacetylase inhibitors (HDACIs) are a new emerging class of immunosuppressive drugs with great potential. They are also of interest in cancer treatment. Histone deacetylase inhibitors induce hyperacetylation of core histones modulating chromatin structure and affect gene expression. Treatment with HDACIs eventually can also cause apoptosis. As expected, the mechanisms of HDACI are pleiotropic because of their capacity to modulate gene expression. Downregulation of IL2, changes in CD154 expression after antigen contact, and other changes contribute to their immune cell specific action.[103] In **Table 5** some more examples of immunosuppressive drugs are listed. It must be emphasized that a common pattern of effects does not equate to a common molecular mechanism.

5.09.5 Toxicogenomics

All immune responses in all immune cells pass at some point through the executive step of up- or downregulation of genes. De novo synthesis of cytokine genes, surface molecules, cytoskeleton structures, and so on, are part of the response against antigens, and require controlled gene expression. Thus, the effects of chemicals on the immune response will become evident to some extent by changes in transcription in the individual participating immune cells. This is the basis of toxicogenomics, which has an increasing impact on drug discovery, safety evaluation, elucidation of pathways of toxicity, and risk assessment. The method of gene expression profiles is as follows: short or long DNA sequences of between several hundred genes (usual for complementary DNA (cDNA) microarrays) or several tens of thousands of genes (Affymetrix microarrays covering almost all genes in a genome) are spotted onto a tiny membrane, silicone chip, or glass chip. The RNA of the cells or an organ to be analyzed is isolated and hybridized to the genes on the chip. Either radioactive-based or fluorescent-based methods allow the detection of the formed cDNA hybrids (presence, absence, and relative abundance). Sophisticated computer-assisted evaluation then allows the experimenter to get lists of differentially expressed genes, cluster them into functional groups, and eventually pinpoint detailed expression patterns, and even signaling networks. Gene expression profiles become more and more important in the assessment of risk for new chemicals. Chemicals are usually screened in a tiered approach.[66,67] Tier I screening assays are more coarse and look at general pathology. Tier II assays look closer at immune responses. Smaller-scale microarray chips can be used to screen for immunotoxic effects such as cytokine modulation. The question of which part of immunotoxic risk testing (tier I or tier II) gene expression profiling would be best integrated into is still open. 'Immunotox chips' on which genes for cytokines, chemokines, adhesion molecules, or costimulatory molecules are spotted are already on the market. Global gene expression profiles in addition help in hypothesis generation when the molecular immunotoxic action of a chemical is unknown. New genes regulated under the influence of the chemical can be identified, and further analyzed by classical molecular and cellular approaches.

Table 5 Examples of immunosuppressive drugs[a]

Glucocorticoids
    Dexamethasone
    Methylprednisolone

Macrolides (from soil fungi)
    Everolimus (derivative of rapamycin)
    FK506 (tacrolimus)
    Pimecrolimus
    Rapamycin (sirolimus)

Histone deacetylases
    Apicidin [cyclo(*N-O*-methyl-L-tryptophanyl-L-isoleucinyl-D-pipecolinyl-L-2-amino-8-oxodecanoyl])
    FR23522 (from *Acremonium* sp.)
    FR901228
    Oxamflatin
    Phenylbutyrate
    Pyroxamide
    Sodium butyrate
    Suberoylanilide hydroxamic acid
    Trapoxin
    Trichostatin A

Other
    Cerivastatin (3-hydroxy-3-methylglutaryl coenzyme reductase inhibitor)
    Cyclosporin A
    Leflunomide
    Prodigiosin 25 C (from *Serratia marcescens*)
    Microcolin A (lipopeptide from *Lyngbya majuscula*)
    2,2'-[3-Methoxy-1'-amyl-5'-methyl-4-(1-pyrryl)]dipyrrylmethene (from *Micrococcus* sp.)

[a] Mechanisms of action vary. Note that not all mentioned substances are in clinical use; for some, only animal data are available.

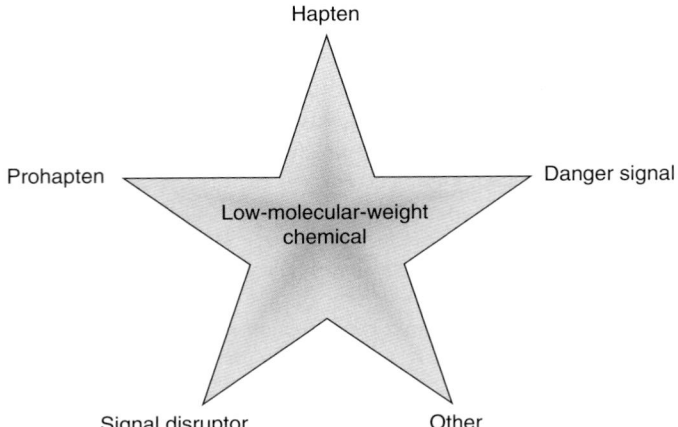

Figure 4 Summary of possible immunotoxic effects of low-molecular-weight chemicals. For details see text.

As the immunotoxic effects are unknown for the vast majority of the chemicals on the market, let alone their mixtures, it is urgent to take toxicology (i.e., immunotoxicology) from a predominantly observational science at the level of disease-specific models to a predominantly predictive science focused upon a broad inclusion of target-specific, mechanism-based, biological observations. At the same time, toxicogenomics might provide routes to valid and robust biomarkers of exposure.

5.09.6 Conclusions

The immune system is an organ composed of a large number of very different cell types with very different functions. Cell–cell interactions, mobility, responsiveness to external stimuli, memory, and continuous cell renewal are typical characteristics. Low-molecular-weight chemicals can biochemically interact with proteins produced by immune cells. Covalent or noncovalent binding with proteins can lead to adverse immune reactions affecting the specific, anamnestic responses of B or T cells and lead, depending on the circumstances, to allergy or autoimmunity. On the other hand, low-molecular-weight chemicals can interact with typical immune cell-signaling pathways, resulting in enhanced or repressed immune responses (**Figure 4**). The underlying mechanisms of both unwanted activation of the adaptive immune response and unwanted shutdown of the immune system in a general way are as diverse as the chemicals and the responsive cells and their targeted proteins. However, increasing knowledge of single substances and their mechanisms of inference leads to the detection of common patterns of reactions. Immunotoxicology and immunopharmacology are on the way from being a descriptive science to a mechanism- and chemical structure-based predictive science.

References

1. Manz, R. A.; Thiel, A.; Radbruch, A. *Nature* **1997**, *388*, 133–134.
2. Sprent, J.; Surh, C. D.; Tough, D. *Res. Immunol.* **1994**, *145*, 328–331.
3. Schittek, B.; Rajewsky, K. *Nature* **1990**, *346*, 749–751.
4. von Boehmer, H.; Teh, H. S.; Kisielow, P. *Immunol. Today* **1989**, *10*, 57–61.
5. Gotter, J.; Kyewski, B. *Curr. Opin. Immunol.* **2004**, *16*, 741–745.
6. de Heer, H. J.; Hammad, H.; Kool, M.; Lambrecht, B. N. *Semin. Immunol.* **2005**, *17*, 295–303.
7. Dubois, B.; Goubier, A.; Joubert, G.; Kaiserlian, D. *Cell Mol. Life Sci.* **2005**, *62*, 1322–1332.
8. Park, B. K.; Kitteringham, N. R.; Powell, H.; Pirmohamed, M. *Toxicology* **2000**, *153*, 39–60.
9. Piguet, P. F.; Collart, M. A.; Grau, G. E.; Sappino, A. P.; Vassalli, P. *Nature* **1990**, *344*, 245–247.
10. Mohr, C.; Gemsa, D.; Graebner, C.; Hemenway, D. R.; Leslie, K. O.; Absher, P. M.; Davis, G. S. *Am. J. Respir. Cell Mol. Biol.* **1991**, *5*, 395–402.
11. Nemazee, D. A.; Bürki, K. *Nature* **1989**, *337*, 562–566.
12. Russell, D. M.; Dembic, Z.; Morahan, G.; Miller, J. F.; Burki, K.; Nemazee, D. *Nature* **1991**, *354*, 308–311.
13. Abbas, A. K. *Cell* **1996**, *84*, 655–657.
14. Sakaguchi, S. *Nat. Immunol.* **2005**, *6*, 345–352.
15. Schwartz, R. H. *Nat. Immunol.* **2005**, *6*, 327–330.
16. Boehm, U.; Klamp, T.; Groot, M.; Howard, J. C. *Annu. Rev. Immunol.* **1997**, *15*, 749–795.
17. Tseng, S. Y.; Dustin, M. L. *Curr. Opin. Cell Biol.* **2002**, *14*, 575–580.
18. Krogsgaard, M.; Davis, M. M. *Nat. Immunol.* **2005**, *6*, 239–245.
19. Germain, R. N. *Cell* **1994**, *76*, 287–299.
20. Shastri, N.; Schwab, S.; Serwold, T. *Annu. Rev. Immunol.* **2002**, *20*, 463–493.
21. Bryant, P. W.; Lennon-Dumenil, A. M.; Fiebiger, E.; Lagaudriere-Gesbert, C.; Ploegh, H. L. *Adv. Immunol.* **2002**, *80*, 71–114.
22. Smith, C. K.; Hotchkiss, S. A. M. *Allergic Contact Dermatitis: Chemical and Metabolic Requirements*; Taylor & Francis: London, 2001.
23. Kohler, J.; Martin, S.; Pflugfelder, U.; Ruh, H.; Vollmer, J.; Weltzien, H. U. *Eur. J. Immunol.* **1995**, *25*, 92–101.
24. Gelber, C.; Gemmell, L.; McAteer, D.; Homola, M.; Swain, P.; Liu, A.; Wilson, K. J.; Gefter, M. *J. Immunol.* **1997**, *158*, 2425–2434.
25. Padovan, E.; Bauer, T.; Tongio, M. M.; Kalbacher, H.; Weltzien, H. U. *Eur. J. Immunol.* **1997**, *27*, 1303–1307.
26. Weltzien, H. U.; Padovan, E. *J. Invest. Dermatol.* **1998**, *110*, 203–206.
27. Sinigaglia, F. *J. Invest. Dermatol.* **1994**, *102*, 398–401.
28. Griem, P.; Takahashi, K.; Kalbacher, H.; Gleichmann, E. *J. Immunol.* **1995**, *155*, 1575–1587.
29. Merk, H. F.; Baron, J.; Kawakubo, Y.; Hertl, M.; Jugert, F. *Clin. Exp. Allergy* **1998**, *28*, 21–24.
30. Baron, J. M.; Merk, H. F. *Curr. Opin. Allergy Clin. Immunol.* **2001**, *1*, 287–291.
31. Anderson, C.; Hehr, A.; Robbins, R.; Hasan, R.; Athar, M.; Mukhtar, H.; Elmets, C. A. *J. Immunol.* **1995**, *155*, 3530–3537.
32. Pirmohamed, M.; Kitteringham, N. R.; Park, B. K. *Drug Saf.* **1994**, *11*, 114–144.
33. von Schmiedeberg, S.; Fritsche, E.; Rönnau, A. C.; Specker, C.; Golka, K.; Richter-Hintz, D.; Schuppe, H. C.; Lehmann, P.; Ruzicka, T.; Esser, C. et al. *Adv. Exp. Med. Biol.* **1999**, *455*, 147–152.
34. Wolkenstein, P.; Carrière, V.; Charue, D.; Bastuji-Garin, S.; Revuz, J.; Roujeau, J. C.; Beaune, P.; Bagot, M. *Pharmacogenetics* **1995**, *5*, 255–258.
35. Richter-Hintz, D.; Thier, R.; Steinwachs, S.; Kronenberg, S.; Fritsche, E.; Sachs, B.; Wulferink, M.; Tönn, T.; Esser, C. *J. Invest. Dermatol.* **2003**, *120*, 765–770.
36. Santana, M. A.; Rosenstein, Y. *J. Cell Physiol.* **2003**, *195*, 392–401.

37. Matzinger, P. *Annu. Rev. Immunol.* **1994**, *12*, 991–1045.
38. Breloer, M.; Dorner, B.; More, S. H.; Roderian, T.; Fleischer, B.; von Bonin, A. *Eur. J. Immunol.* **2001**, *31*, 2051–2059.
39. Gallucci, S.; Matzinger, P. *Curr. Opin. Immunol.* **2001**, *13*, 114–119.
40. Aliberti, J.; Viola, J. P.; Vieira-de-Abreu, A.; Bozza, P. T.; Sher, A.; Scharfstein, J. *J. Immunol.* **2003**, *170*, 5349–5353.
41. Shi, Y.; Evans, J. E.; Rock, K. L. *Nature* **2003**, *425*, 516–521.
42. Powell, J. D.; Horton, M. R. *Immunol. Res.* **2005**, *31*, 207–218.
43. DeMarco, R. A.; Fink, M. P.; Lotze, M. T. *Mol. Immunol.* **2005**, *42*, 433–444.
44. Sullivan, J. R.; Shear, N. H. *Arch. Dermatol.* **2001**, *137*, 357–364.
45. Pirmohamed, M.; Naisbitt, D. J.; Gordon, F.; Park, B. K. *Toxicology* **2002**, *181–182*, 55–63.
46. Enk, A. H.; Katz, S. I. *Proc. Natl. Acad. Sci. USA* **1992**, *89*, 1398–1402.
47. Enk, A. H.; Katz, S. I. *J. Invest. Dermatol.* **1995**, *105*, 80S–83S.
48. Rambukkana, A.; Pistoor, F. H.; Bos, J. D.; Kapsenberg, M. L.; Das, P. K. *Lab. Invest.* **1996**, *74*, 422–436.
49. Zdolsek, J. M.; Soder, O.; Hultman, P. *Immunopharmacology* **1994**, *28*, 201–208.
50. Artik, S.; von Vultee, C.; Gleichmann, E.; Schwarz, T.; Griem, P. *J. Immunol.* **1999**, *163*, 1143–1152.
51. Aiba, S.; Terunuma, A.; Manome, H.; Tagami, H. *Eur. J. Immunol.* **1997**, *27*, 3031–3038.
52. Aiba, S.; Tagami, H. *J. Invest. Dermatol. Symp. Proc.* **1999**, *4*, 158–163.
53. Aeby, P.; Wyss, C.; Beck, H.; Griem, P.; Scheffler, H.; Goebel, C. *J. Invest. Dermatol.* **2004**, *122*, 1154–1164.
54. Tuschl, H.; Kovac, R.; Weber, E. *Toxicol. In Vitro* **2000**, *14*, 541–549.
55. Porcelli, S.; Morita, C. T.; Brenner, M. B. *Nature* **1992**, *360*, 593–597.
56. Beckman, E. M.; Porcelli, S. A.; Morita, C. T.; Behar, S. M.; Furlong, S. T.; Brenner, M. B. *Nature* **1994**, *372*, 691–694.
57. Van, R. I.; Zajonc, D. M.; Wilson, I. A.; Moody, D. B. *Curr. Opin. Immunol.* **2005**, *17*, 222–229.
58. Pichler, W. J.; Zanni, M.; von Greyerz, S.; Schnyder, B.; Mauri-Hellweg, D.; Wendland, T. *Int. Arch. Allergy Immunol.* **1997**, *113*, 177–180.
59. Marin-Esteban, V.; Falk, K.; Rotzschke, O. *J. Biol. Chem.* **2004**, *279*, 50684–50690.
60. Nassif, A.; Bensussan, A.; Dorothee, G.; Mami-Chouaib, F.; Bachot, N.; Bagot, M.; Boumsell, L.; Roujeau, J. C. *J. Invest. Dermatol.* **2002**, *118*, 728–733.
61. Zanni, M. P.; Mauri-Hellweg, D.; Brander, C.; Wendland, T.; Schnyder, B.; Frei, E.; von Greyerz, S.; Bircher, A.; Pichler, W. J. *J. Immunol.* **1997**, *158*, 1139–1148.
62. Zanni, M. P.; von Greyerz, S.; Schnyder, B.; Wendland, T.; Pichler, W. J. *Int. Immunol.* **1998**, *10*, 507–515.
63. Burkhart, C.; Britschgi, M.; Strasser, I.; Depta, J. P.; von Greyerz, S.; Barnaba, V.; Pichler, W. J. *Clin. Exp. Allergy* **2002**, *32*, 1635–1643.
64. Putman, E.; van der Laan, J. W.; van Loveren, H. *Fund. Clin. Pharmacol.* **2003**, *17*, 615–626.
65. Descotes, J. *Drug Saf.* **2005**, *28*, 127–136.
66. Luster, M. I.; Munson, A. E.; Thomas, P. T.; Holsapple, M. P.; Fenters, J. D.; White, K. L., Jr.; Lauer, L. D.; Germolec, D. R.; Rosenthal, G. J.; Dean, J. H. *Fund. Appl. Toxicol.* **1988**, *10*, 2–19.
67. Luster, M. I.; Portier, C.; Pait, D. G.; Rosenthal, G. J.; Germolec, D. R.; Corsini, E.; Blaylock, B. L.; Pollock, P.; Kouchi, Y.; Craig, W. *Fund. Appl. Toxicol.* **1993**, *21*, 71–82.
68. Holsapple, M. P.; Snyder, N. K.; Wood, S. C.; Morris, D. L. *Toxicology* **1991**, *69*, 219–255.
69. Snoeij, N. J.; Penninks, A. H.; Seinen, W. *Environ. Res.* **1987**, *44*, 335–353.
70. Schuppe, H. C.; Rönnau, A. C.; von Schmiedeberg, S.; Ruzicka, T.; Gleichmann, E.; Griem, P. *Clin. Dermatol.* **1998**, *16*, 149–157.
71. van Leeuwen, F. X.; Feeley, M.; Schrenk, D.; Larsen, J. C.; Farland, W.; Younes, M. *Chemosphere* **2000**, *40*, 1095–1101.
72. Esser, C. *Int. Arch. Allergy Immunol.* **1994**, *104*, 126–130.
73. Vogel, C.; Donat, S.; Döhr, O.; Kremer, J.; Esser, C.; Roller, M.; Abel, J. *Arch. Toxicol.* **1997**, *71*, 372–382.
74. Schmidt, J. V.; Bradfield, C. A. *Annu. Rev. Cell Dev. Biol.* **1996**, *12*, 55–89.
75. Majora, M.; Frericks, M.; Temchura, V.; Reichmann, G.; Esser, C. *Int. Immunopharmacol.* **2005**, *5*, 1659–1674.
76. Zeytun, A.; McKallip, R. J.; Fisher, M.; Camacho, I.; Nagarkatti, M.; Nagarkatti, P. S. *Toxicology* **2002**, *178*, 241–260.
77. Puga, A.; Maier, A.; Medvedovic, M. *Biochem. Pharmacol.* **2000**, *60*, 1129–1142.
78. Esser, C. *Recent Res. Dev. Mol. Pharmacol.* **2002**, *1*, 141–155.
79. Kerkvliet, N. I. *Environ. Health Perspect.* **1995**, *103*, 47–53.
80. Donat, S.; Abel, J. *Chemosphere* **1998**, *37*, 1867–1872.
81. Burns, L. A.; Sikorski, E. E.; Saady, J. J.; Munson, A. E. *Toxicol. Appl. Pharmacol.* **1991**, *110*, 157–169.
82. Galicia, G.; Leyva, R.; Tenorio, E. P.; Ostrosky-Wegman, P.; Saavedra, R. *Int. Immunopharmacol.* **2003**, *3*, 671–682.
83. Yen, H. T.; Chiang, L. C.; Wen, K. H.; Chang, S. F.; Tsai, C. C.; Yu, C. L.; Yu, H. S. *Arch. Dermatol. Res.* **1996**, *288*, 716–717.
84. Cheng, H. Y.; Li, P.; David, M.; Smithgall, T. E.; Feng, L.; Lieberman, M. W. *Oncogene* **2004**, *23*, 3603–3612.
85. Huang, C.; Ma, W. Y.; Li, J.; Dong, Z. *Cancer Res.* **1999**, *59*, 3053–3058.
86. Hossain, K.; Akhand, A. A.; Kato, M.; Du, J.; Takeda, K.; Wu, J.; Takeuchi, K.; Liu, W.; Suzuki, H.; Nakashima, I. *J. Immunol.* **2000**, *165*, 4290–4297.
87. Benya, T. J. *Drug Metab. Rev.* **1997**, *29*, 1189–1284.
88. De Santiago, A.; Aguilar-Santelises, M. *Hum. Exp. Toxicol.* **1999**, *18*, 619–624.
89. Stridh, H.; Gigliotti, D.; Orrenius, S.; Cotgreave, I. *Biochem. Biophys. Res. Commun.* **1999**, *266*, 460–465.
90. Stridh, H.; Cotgreave, I.; Muller, M.; Orrenius, S.; Gigliotti, D. *Chem. Res. Toxicol.* **2001**, *14*, 791–798.
91. Aluoch, A.; Whalen, M. *Toxicology* **2005**, *209*, 263–277.
92. Kato, T.; Uchikawa, R.; Yamada, M.; Arizono, N.; Oikawa, S.; Kawanishi, S.; Nishio, A.; Nakase, H.; Kuribayashi, K. *Eur. J. Immunol.* **2004**, *34*, 1312–1321.
93. Matsue, H.; Yang, C.; Matsue, K.; Edelbaum, D.; Mummert, M.; Takashima, A. *J. Immunol.* **2002**, *169*, 3555–3564.
94. Bram, R. J.; Hung, D. T.; Martin, P. K.; Schreiber, S. L.; Crabtree, G. R. *Mol. Cell Biol.* **1993**, *13*, 4760–4769.
95. Griffith, J. P.; Kim, J. L.; Kim, E. E.; Sintchak, M. D.; Thomson, J. A.; Fitzgibbon, M. J.; Fleming, M. A.; Caron, P. R.; Hsiao, K.; Navia, M. A. *Cell* **1995**, *82*, 507–522.
96. Wiederrecht, G.; Hung, S.; Chan, H. K.; Marcy, A.; Martin, M.; Calaycay, J.; Boulton, D.; Sigal, N.; Kincaid, R. L.; Siekierka, J. J. *J. Biol. Chem.* **1992**, *267*, 21753–21760.
97. Kiani, A.; Rao, A.; Aramburu, J. *Immunity* **2000**, *12*, 359–372.
98. Hay, N.; Sonenberg, N. *Genes Dev.* **2004**, *18*, 1926–1945.

99. Gingras, A. C.; Kennedy, S. G.; O'Leary, M. A.; Sonenberg, N.; Hay, N. *Genes Dev.* **1998**, *12*, 502–513.
100. Miyamoto, S.; Safer, B. *Biochem. J.* **1999**, *344*, 803–812.
101. Hackstein, H.; Taner, T.; Zahorchak, A. F.; Morelli, A. E.; Logar, A. J.; Gessner, A.; Thomson, A. W. *Blood* **2003**, *101*, 4457–4463.
102. Van Laethem, F.; Baus, E.; Smyth, L. A.; Andris, F.; Bex, F.; Urbain, J.; Kioussis, D.; Leo, O. *J. Exp. Med.* **2001**, *193*, 803–814.
103. Skov, S.; Rieneck, K.; Bovin, L. F.; Skak, K.; Tomra, S.; Michelsen, B. K.; Odum, N. *Blood* **2003**, *101*, 1430–1438.

Biography

Charlotte Esser studied biology at the universities of Cologne and Tübingen, and at Duke University, North Carolina. She received her PhD in immunology at the University of Cologne, Germany, in 1990. Her research focus at the Institute for Environmental Medical Research has been the immunotoxicology of xenobiotic substances and their interference with signal transduction pathways and T cell differentiation programs. Other topics of her research interest include protective strategies against UVB-induced immunosuppression, and genetic susceptibility loci to autoimmune diseases. She is Professor of Immunology at the University of Düsseldorf.

Comprehensive Medicinal Chemistry II
ISBN (set): 0-08-044513-6

ISBN (Volume 5) 0-08-044518-7; pp. 215–229

5.10 In Vitro Studies of Drug Metabolism

Y Parmentier, M-J Bossant, M Bertrand, and B Walther, Technologie Servier, Orléans, France

© 2007 Elsevier Ltd. All Rights Reserved.

5.10.1 Introduction

During the screening and development of a new chemical entity (NCE) many studies are set up in order to better understand its pharmacokinetic and metabolic characteristics.[1] This knowledge increases during the lifetime of a drug.

At the registration level however, safety and efficacy are the prime concerns of any medical authority and studying metabolism can bring key elements to better understand these issues.

For metabolism studies, two important questions have to be answered: how does the metabolism in animals used for the toxicological/pharmacological evaluations compare and is it as representative as possible of the human metabolism? Is there any risk of drug interactions in the clinic? These two aspects should be the leitmotiv of any screening and development program in metabolism. In this chapter, we discuss how early in vitro metabolism studies can bring important decisional elements always keeping in mind the safety and efficacy profile of the NCE of interest.

Metabolism (see 5.07 Principles of Drug Metabolism 3: Enzymes and Tissues) has been defined as the chemistry of enzymatic and nonenzymatic processes. It covers the identification of the metabolites during early in vitro research programs and in classical in vivo metabolism studies in animals used in the safety program and in humans. It also implies to determine the major characteristics of the enzymes (mainly cytochrome P450 (CYP)) involved in these metabolic reactions. These data will allow one to predict, understand and control potential interactions with co-administered drugs in the clinic, and are of help in explaining metabolic related variability in the activity of a drug.

5.10.2   In Vitro Tools

In vitro tools (Table 1) can be divided into four groups: organs, cells (primary cultures and cell lines), subcellular fractions (S9, cytosol, and microsomes), and isolated enzymes (purified and recombinant enzymes).

5.10.2.1   Organ-Derived Tools

The perfused organ represents the in vitro tool closest to the in vivo situation. The technical difficulties of maintaining the viability of the organs, even with complex perfusion media, together with ethical difficulties (availability of organs), have limited these methods mainly to mechanistic studies in animals.

As an alternative, liver slices, which retain the organ architecture, have been developed with more or less success depending on the incubation conditions. Organ slices contain the complete metabolism machinery with all phase I and phase II drug metabolizing enzymes so that a drug can undergo all possible metabolic reactions. They also contain the different cell subtypes that contribute to the global environment, i.e., interaction between Kupfer or biliary cells with hepatocytes leading to complex regulations in response to drug exposure. Once the problems of oxygen supply and nutrients diffusion up to the inner part of the slices are overcome, the well-preserved cell–cell communications make liver slices a unique model for the differentiated phenotype of hepatocytes, and thus can be applied to metabolism or toxicological studies.

To obtain reproducible, reliable, and predictive results in drug metabolic profiling, preparation of liver slices requires high-precision tissue slicers (e.g., Krumdieck or Brendel-Vitron) producing uniform and thin (less than 250 μm thickness) slices.[2]

However, liver slices are not very often used in drug metabolism studies and even less often in the earliest stage of drug development because of major drawbacks such as the labor-intensive nature of the preparation, which requiries

Table 1 In vitro tools[a] for drug metabolism studies and their applications

	Metabolic profiling	Metabolic clearance	Inhibitory potential	Phenotyping	Induction
Liver slices	+	+	−	+/−	+
Intestine slices	+	+	−	−	−
Hepatocytes	+ +	+ +	+	+/−	+ +
Enterocytes	+	+	−	−	−
S9	+	−	+/−	+/−	−
Cytosol	+	−	+/−	+/−	−
Microsomes (liver, intestine, etc.)	+ +	+ +	+ +	+ +	−
Recombinants	+	+/−	+	+ +	−

[a] + +, model used in first intention; +, model used in second intention; +/−, model only used to address a specific question; −, not applicable.

fresh material. Also, there is no possible automation of the incubation process, no optimal cryopreservation conditions, a short storage time of the preparation (at 4 °C only), a short viability period (5 days), and the problem of damaged cells on the outer edge of the slice with impaired biotransformation.[2]

5.10.2.2   Primary Cultures and Cell Lines

5.10.2.2.1   Hepatocytes

Primary hepatocytes are isolated from the whole liver (for animals[3]) or from liver biopsies (for humans[4]) by collagenase perfusion techniques adapted for human or animal species.[2] After isolation, the cells can be stored at 4 °C for up to 48 h in a solution (University of Wisconsin solution) without relevant loss of viability.[5]

Once isolated, Hepatocytes can also be kept in suspension, in which case they remain viable only for a few hours or can be maintained in monolayer cell culture for at least 4 weeks.[2] Details about the procedures to obtain and use this model have been reviewed by Gebhardt et al.[6]

Cell culture conditions have been optimized to overcome the dedifferentiation process leading to a decrease in enzymatic activity with time (especially for CYP2E1 and 3A4 enzymes[7]) enabling therefore long-term cultures of hepatocyte.[8–10] These culture conditions include the use of specific media (culture media, serum, additional components such as hormones and inducers), different matrix compositions (simple matrices such as rat tail collagen, or complex matrices such as fibronectins or, preferably, Matrigel) or cocultures with epithelial cells derived from primitive biliary cells.[10]

The collagen sandwich model[11,12] has been considered by some authors as the most reliable for studying drug biotransformation, induction, and also transporter-mediated biliary excretion,[2] since it gets closer to reality in terms of cell polarity and of structural and functional normal bile canalicular network or cellular integrity.

Cryopreserved hepatocytes have been shown to retain the activity of most of phase I and phase II enzymes.[2] Fairly successful predictions of in vivo hepatic clearance with cryopreserved hepatocytes in suspension and good correlation when compared to fresh cell cultures have been obtained,[13–15] making them suitable for early metabolism studies.

Moreover these cryopreserved hepatocytes can overcome the problems of interindividual variability and limited time of use after preparation that are encountered with fresh human hepatocytes, since several batches can be pooled (as with liver microsomes).

5.10.2.2.2   Combined culture systems

To study enterohepatic metabolism, a new coculture system combining hepatocytes and intestinal bacteria has been developed. A two-chamber system under aerobic and anaerobic conditions allows analysis of the sequential metabolism of chemicals by liver and microflora in vitro. This model may help to obtain data on hepatic and microbial metabolism during the early phase of drug development.[6]

A new coculture system has also been developed to obtain a full picture of the role of presystemic compartment (including absorption and metabolism) and consists in combining Caco-2 cells and hepatocytes into two linked chambers.

5.10.2.2.3   Hepatoma cell lines

The poor availability of human hepatocytes and their relatively short lifespan in standard cultures have encouraged scientists to develop hepatocyte cell lines. However, even though several sources of cell lines exist, they are still less popular than other in vitro models. This is mainly due to their dedifferentiated cellular characteristics and incomplete expression of the major families of metabolic enzymes.[2]

HepG2 has been the most frequently used and is the best-characterized human hepatoma cell line. It displays different patterns of metabolizing enzyme expressions according to the source and culture conditions limiting its interest as a real alternative for hepatocytes.[16]

Other cell lines such as HLE,[17] THLE,[18] BC2,[19] or Fa2N-4[20] express different metabolizing enzymes but not the full enzymatic profile.

HepaRG, a new hepatoma cell line, developed by Guillouzo's group,[21] has shown high degrees of similarity with fresh hepatocytes in term of morphology, express specific hepatic functions, in particular genotypic profiles for metabolizing enzymes (phase I and II), and transporters, and also in nuclear receptors (AhR, PXR, CAR, PPAR-alpha).[22,23] HepaRG represents a real and reliable alternative to human hepatocytes for drug metabolism and toxicity studies.[22]

In contrast with fresh or cryopreserved hepatocytes, these cell lines can be transfected with vectors that can mediate DNA or RNA constructs to either express, activate, or knockout a specific gene (BC2 cell line). Small interfering RNA (siRNA) technology allows a specific gene to be quenched quite easily and sometimes permanently. The knockout of genes encoding for an enzyme, a transporter, or a nuclear receptor in these cell lines would be of great interest for enzyme or transporter phenotyping or induction mechanisms.

In addition to HepaRG, new avenues of research have been opened, in particular with human medullar cells that might be transformed into human hepatocytes under certain culture conditions.[24,25] Further studies should however be performed to confirmed the full similarity of these cells both in terms of metabolic capacity and inducibility.

5.10.2.3   Subcellular Fractions

Subcellular fractions are obtained by successive centrifugations of tissue homogenates. While the S9 fraction is obtained after the first centrifugation step at about 9000 g, the microsomal fraction requires an ultracentrifugation at 100 000 g to separate the cytosolic soluble fraction from the microsomes.

These systems offer many advantages. They are commercially available from individual livers (inter-individual variation can be studied) or a pool of livers (representative of the average enzyme activities of the population). They can be stored and remain stable at $-80\,^\circ$C for many years and can be thawed easily. They are very easy to handle for several types of incubations and they are suitable for full automation. All of these advantages make them very useful for metabolism studies in the earliest stage of drug development including the screening program.

Subcellular fractions can be obtained from any tissue (liver, intestine, kidney, brain, etc.) with slight differences in the preparation procedure. We will focus on the liver subcellular fractions.

5.10.2.3.1   Human liver S9 fraction

The liver S9 fractions contain both microsomal and cytosolic fractions. They regroup therefore the same metabolic enzyme profile (phase I and phase II enzymes including their cofactors) as hepatocytes with the exception of the membrane barrier and its transporters. Hence, S9 fractions offer a more complete representation of the metabolic profile compared with microsomes and cytosolic fractions.[2] They are therefore of great potential interest for in vitro studies in drug metabolism and drug interaction. However, their low enzyme activities compared to microsomes limit their use for drug metabolism studies.

5.10.2.3.2   Human liver cytosolic fraction

The liver cytosolic fraction contains the soluble phase I enzymes such as esterases, amidases, or epoxyde hydrolases and the soluble phase II enzymes such as most of the sulfotransferases (ST), glutathione s-transferases (GST), and N-acetyltransferases (NAT). For the catalytic activity of the phase II enzymes, some exogenous cofactors such as adenosine 3′-phosphate-5′-phosphosulfate (PAPS) for sulfotransferase activity[2] can be added especially in concentrated (ultrafiltration) cytosolic fractions.

Although this cytosolic fraction cannot be used as a complete metabolic system, it can help to address some issues in metabolic profiling for drugs that are metabolized by soluble enzymes. Coincubation with microsomes is also possible to obtain a more complete system.

5.10.2.3.3   Human liver microsomes

Liver microsomes are by far the most widely used in vitro model, providing an affordable way to give a good indication of the CYP and UDP-glucuronyltransferases (UGT) involved in the metabolism of a drug.[2]

Liver microsomes, as vesicles of the hepatocyte endoplasmic reticulum, contain membrane phase I enzymes namely CYPs, flavine-containing monooxygenases (FMO), esterases, amidases, and epoxide hydrolases, and also the phase II enzymes such as UGTs. For the catalytic activity of both phase I and II enzymes, addition of exogenous cofactors such as NADPH for CYPs and FMO, and UDPGA/alamethicin for UGTs is necessary.

Liver microsomes are the model of choice (with human hepatocytes) for drug metabolism studies (metabolic profile and prediction of hepatic clearance) and drug interaction studies (phenotyping and inhibitory potential studies).

5.10.2.4   Isolated Enzymes

Microsomes containing specific individual human CYPs and UGTs have become extremely important methodological options for in vitro drug metabolism studies, in particular to investigate the contribution of enzymes involved in the biotransformation of an NCE.[2,26]

A variety of recombinant expression systems has been developed including insect, bacterial, yeast, and mammalian models. They have been reviewed in detail by Friedberg et al.[27]

The obvious advantage is that the activity of one specific human CYP or UGT can be studied in isolation. However, one has to keep in mind that such isolated systems have de facto no interaction with the other enzymes and factors present in intact cells. The lipids, the apoprotein nature, and the level of CYP coenzymes can all be different depending on the cell type, leading to potential over- or underestimation of the metabolic reactions.[28]

5.10.3   Rate of Metabolism

Besides the identifications of metabolites, which represented the major part of metabolism work in the past, determination of the rate of in vitro metabolism across species, organs, or enzymes has become more and more important in recent years.

5.10.3.1   Michaelis–Menten Kinetics

The standard approach assumes that the drug obeys to Michaelis–Menten kinetics (**Table 2**):

$$v = \frac{V_{max} \times S}{K_m + S} \qquad [1]$$

where V_{max} is the maximum rate of the reaction, and K_m is the Michaelis–Menten constant (concentration of the substrate S at which v is half maximal). The intrinsic clearance (CL_{int}) can be obtained from

$$CL_{int} = \frac{V_{max}}{K_m + S} \qquad [2]$$

When the substrate concentration is lower than the K_m (negligible as compared to the K_m), eqn [2] can be simplified as follow:

$$CL_{int} = \frac{V_{max}}{K_m} \qquad [3]$$

CL_{int} then represents the slope of the initial linear part of the reaction velocity versus substrate concentration curve.

Proper determination of the enzyme kinetics (K_m, V_{max}, CL_{int}) of a compound requires to measure the rates of metabolite formation at several substrate concentrations and that the data fit to the simple Michaelis–Menten equation in case of a monophasic reaction (substrate metabolized by one enzyme) or to eqns [4] or [5] in case of multiphasic reactions.[26]

$$v = \frac{V_{max(1)} \times S}{K_{m(1)} + S} + \frac{V_{max(2)} \times S}{K_{m(2)} + S} \qquad [4]$$

$$v = \frac{V_{max(1)} \times S}{K_{m(1)} + S} + CL_{int(2)} \times S \qquad [5]$$

Table 2 In vitro enzyme kinetics

Type of inhibition	Apparent K_m	Apparent V_{max}	Michaelis–Menten equation	AUC_i/AUC^a prediction
No inhibition	$= K_m$	$= V_{max}$	$v = \dfrac{V_{max} \times S}{K_m + S}$	
Competitive	$= K_m \times (1 + I/K_{ic})$	$= V_{max}$	$v = \dfrac{V_{max} \times S}{K_m \times (1 + I/K_{ic}) + S}$	$1 + I/K_{ic}$
Uncompetitive	$= \dfrac{K_m}{(1 + I/K_i)}$	$= \dfrac{V_{max}}{(1 + I/K_i)}$	$v = \dfrac{V_{max} \times S}{K_m + S \times (1 + I/K_i)}$	$1 + \left(\dfrac{I}{K_i}\right) \times \left(\dfrac{S}{S + K_m}\right)$
Mixed	$= \dfrac{K_m \times (1 + I/K_{ic})}{(1 + I/K_{iu})}$	$= \dfrac{V_{max}}{(1 + I/K_{iu})}$	$v = \dfrac{V_{max} \times S}{K_m \times (1 + I/K_{ic}) + S \times (1 + I/K_{iu})}$	$1 + I/K_{ic}$
Noncompetitive	$= K_m$	$= \dfrac{V_{max}}{(1 + I/K_i)}$	$v = \dfrac{V_{max} \times S}{K_m \times (1 + I/K_i) + S \times (1 + I/K_i)}$	$1 + I/K_{ic}$
Mechanism-based	$= K_m$	$= V_{max} \times \dfrac{k_{degrad}}{\left(k_{degrad} + \dfrac{I \times k_{inact}}{K_i + I}\right)}$	$\lambda = \dfrac{k_{inact} \times I}{K_i + I}$	$1 + \dfrac{V_{max}}{V_{max}(\text{inactivated})}$

[a] In vitro to in vivo extrapolation.

This implies that the metabolites are available as authentic standards or that the radiolabeled drug has been synthesized. In early stages of drug development, such conditions are not fulfilled and an alternative approach must be considered.

A particularly adapted method for screening programs consists in measuring the intrinsic clearance of a drug using a substrate depletion approach called in vitro $t_{1/2}$ approach described by Obach et al.[29,30] and using eqn [6]. A first-order depletion rate constant for a substrate is measured in an incubation at a single low concentration.

$$CL_{int} = \frac{\ln 2}{t_{1/2} \times [M]}$$
[6]

where $[M]$ represents the concentration of microsomal protein.

This approach implies that the drug concentration used is below the K_m (10^{-7} M is a good compromise for screening purposes) and that the enzymes remain fully active throughout the incubation period. Any autoinactivation or product inhibition can lead to an underestimation of CL_{int} using this approach (see below).[26]

However, many of the enzymes implicated in the metabolism of xenobiotics are highly saturable so that the single substrate concentrations may not be low enough[31] to properly estimate the CL_{int}. To overcome this problem, two substrate depletion approaches have been developed.

The first method, proposed by Obach et al.,[32] consists in monitoring substrate concentration loss (in vitro $t_{1/2}$ approach) at multiple substrate concentrations. The substrate depletion rate constant is plotted against the substrate concentrations and fitted to:

$$k_{dep} = k_{dep(S=0)} \times \left(1 - \frac{S}{S + K_m}\right)$$
[7]

The inflection point of the curve corresponds to the K_m and the maximum depletion rate constant is represented by $k_{dep(S=0)}$. This approach gives a better understanding of the enzyme kinetics of a compound (K_m available) than using the simple in vitro $t_{1/2}$ approach.

The second method, developed in our laboratory,[31,33] and named substrate disappearance kinetic (SDK) method, provides apparent K_m and V_{max} for the overall metabolism of a drug, based on computer fitting of two in vitro disappearance curves using the following Michaelis–Menten integrated equation:

$$-\frac{dS}{dt} = \frac{t_{prot} \times V_{max} e^{-k_d \times t} \times S}{K_m + S}$$
[8]

Here t_{prot} represents the concentration of microsomal protein and k_d the enzyme activity degradation constant (min^{-1}). This model offers the advantage to use only two substrate concentrations (surrounding the K_m) and integrates the loss of enzyme activity with time (denaturation of the enzyme) conversely to the previous approach. This is particularly important for long incubation periods (low clearance drugs). This approach is particularly adapted for studies in early drug development.

5.10.3.2   Non-Michaelis–Menten Kinetics

The observation of atypical enzyme kinetics, particularly involving CYP, has become relatively common. Since predictions of in vivo clearances are based on kinetic parameters observed during in vitro experiments, misidentification of the drug kinetic profile can lead to inaccurate predictions of intrinsic clearance.[34]

The atypical kinetic profiles include autoactivation (sigmoidal kinetics or homotropic cooperativity), substrate inhibition, heteroactivation (heterotropic cooperativity), partial inhibition, and substrate-dependent inhibition (see Hutzler and Tracy[35] for a review of enzymes exhibiting atypical kinetics).

Among these atypical kinetic profiles, autoactivation kinetics have been more and more observed over the past few years, particularly for CYP3A and 2C9. In this case, the v versus S plot yields a sigmoidal curve, attributed to more than one substrate molecule simultaneously binding to the enzyme and stimulating the activity. To prevent the underestimation of the CL_{int} in case of autoactivation, Houston et al.[36] suggested the use of the CL_{max}, describing the steepest part of the v versus S curve and being visualized as the maximum point on a plot of v/S versus S. However this type of model[36,37] remains by far too complex to be integrated in early drug discovery.

5.10.3.3   Metabolic Stability

The prediction of in vivo hepatic metabolic clearance of a drug from its apparent intrinsic clearance requires theoretically to include the unbound fraction of the drug in plasma (fu_p). Interestingly, several authors reported an underestimation of the predicted in vivo clearance by taking into account the fu_p alone.[29,38,39] A better correlation was observed between the actual and predicted clearance values when the free fraction in plasma was ignored in the equations describing hepatic extraction or when the nonspecific binding to microsomes (fu_{mic}) and plasma (fu_p) were both taken into account. Obach et al.[29,38] recommended to use both unbound fractions for the prediction of the in vivo hepatic clearance.

However, there are exceptions to this rule suggesting that other factors may influence the availability of the drug at the active site of the enzyme. This might be due to the use of isolated parameters (fu_p, fu_{mic}, CL_{int}), measured separately, without taking into account the various competitions occurring at the cellular level. In early metabolism programs, with the event of 96-well plate technologies and complete automation, all these individual parameters can be measured today on series of compounds.

In theory, a drug must first dissociate from the nonspecific binding sites before it can bind to, and be metabolized by enzymes.[29] The availability of a drug at the enzymatic site is therefore dependent on both the specific and nonspecific binding to proteins as well as on the affinity toward metabolizing enzymes.

Coincubation models, including the interaction between plasma, nonspecific proteins present in the classical in vitro metabolism models, and metabolizing enzymes have therefore been developed. Isolated hepatocytes suspended in serum have been found to give more accurate predictions of hepatic clearances.[40]

5.10.3.4   Organ and Species Differences

Early discovery programs have mainly been based on in vitro hepatic tools using microsomes as the simplest and fastest in vitro model to generate the first metabolism data. Now that metabolic stability screens have been performed successfully with microsomes for a number of years, we observe more and more non-CYP reactions, particularly soluble enzymes metabolizing the drugs. Microsomal information is therefore very often completed with hepatocyte experiments, but these enzymes are also present in other tissues such as the gut or the lung and can be of major importance in the bioavailability of the drug.

Even drugs stable on hepatic tools (CYP- and non-CYP-dependent reactions) can be oxidatively metabolized in other tissues such as the gut. This is probably not related to catalytic differences between hepatic and intestinal enzymes but is more related to the protein binding properties of the drug of interest. For a highly protein-bound drug, the free fraction available to the enzyme is low in the liver but much higher in the intestine, because the drug is assumed to be free in the intestine. Intestinal extraction can therefore be as important or even greater than the hepatic extraction. Midazolam, as a highly protein-bound and pure CYP3A4 substrate, is a good example of the importance of intestinal (0.4–0.6) versus liver extraction (0.3–0.5).[26]

With regard to animal species, the rat is probably the most frequently used comparative species in early programs, which are completed later with dog and monkey data. However, one very often sees, with hepatic tools, major differences in rate between rat (rapid) and human (slow). The validity of the metabolic bioavailability predicted in humans is then questioned. One has to remember that these results only suggest that the compound is stable in the liver. However, because in many cases the liver is not the only metabolic organ, these results will have to be completed with extrahepatic in vitro tools. A typical example is, with the advent of combinatorial chemistry and the frequent use of amide bonds, the differences in hydrolytic activities between rodent species and larger species including humans, as well as some tissue differences (plasma, liver, and gut).

An interesting rodent species to be added in early programs is the mouse. This is documented as regards to the enzymatic content of the intestine. The importance of CYP3A isoenzymes in the mouse intestine appears to mirror that in humans, whilst in rat this family of enzymes is negligible and the CYP2C family is of prime importance.[41] This correlation between human and mouse can also be observed for the liver with common pathways and similar rates on chemical series conversely to the rat.

5.10.4   Metabolic Profiling

Today drug research programs, including early screening studies on large series of compounds and more informative studies on selected candidates, allow a rapid understanding of the in vivo fate of a drug candidate. These studies are able to define systematically, at early stages, the metabolic stability, drug permeability, and drug solubility of the drugs tested, with the aim of selecting rapidly lead candidates.[42] These programs are then completed with in vivo studies in animals. But most of this early information is oriented toward predicting differences in rate of metabolism (clearances) with the ultimate goal of predicting the in vivo pharmacokinetic behavior of a drug. In comparing species only on the basis of their metabolic stability, one tends to forget sometimes the vast difference that can exist in the nature and amount of metabolite formed. Some phenomenon observed in animal species can be of little relevance to man as regards safety and efficacy profile of the NCE tested.

5.10.4.1   Biological Importance of Metabolites

In the actual research and development programs, knowledge on metabolites has been described as important in order to assist in the selection of the appropriate compound for development.[43] Moreover, the chemical nature of the metabolites is essential because they have to be evaluated for their biological activity. This includes the evaluation of their activity in pharmacology, safety in toxicology and both safety and efficacy aspects in clinical programs.

Before discussing how the in vitro tools can give, as early as possible, an insight into both the safety and efficacy aspects of an NCE, it is worth trying to understand what the future requirements will be at the registration stage.

For the pharmacological/clinical aspects, some authors have proposed that the metabolites participating to the overall pharmacological activity (at least 25% of the total activity) and which are major in human,[44] will have to be monitored in future animal and clinical studies.

Although it is difficult to give an absolute number as a definition for a major metabolite, because it can be very much drug-specific, a number of proposals has been made which can be grouped in two categories:

1. For drug interaction purposes, some authors have proposed to initiate CYP phenotyping only for major reactions namely if a sum total of 25–30% of compound clearance appears to be related to the formation of primary oxidative metabolites.[45,46]
2. For safety testing of metabolites, the recent draft guidelines[47] have focused on circulating metabolites representing more than 10% of the administered dose or circulating drug related material, whichever is the less. The guidance describes a unique metabolite as one produced only in humans or formed to a much greater extend in humans compared to animal species used in toxicology programs.

In any case, identification of major active metabolites is essential for the interpretation of the clinical response.[43] The same applies for the safety profile of an NCE in animals in support of human safety: one has to compare exposures to individual metabolites or global routes of metabolism obtained in humans to those obtained in mutagenicity tests and animal species of the toxicological safety program. Some authors have highlighted the potential safety concerns of metabolites specific to humans as well as the presence of structural alerts.[48]

The question remaining to be answered is what can be defined as an adequate coverage and safety ratio. An optimal '25 times rule' difference in exposure between animals and humans has been discussed for either the parent drug or metabolites.[44] In a number of cases the net exposure, defined as the sum of the exposure to the drug and its metabolites, will have to be used. In practice, for minor metabolites which can be of importance in the safety aspects, it is not always easy to compare chemical entity per chemical entity throughout the species because of possible major differences in rate (formation and elimination), e.g., between rodents and humans. The combination of both chemical entities and common metabolic routes, using urinary and fecal metabolites, is then more appropriate. Safety aspects mainly focus on phase I pathways even though conjugation reactions can be of importance when dealing with mutagenicity and carcinogenicity mechanisms.

When looking at the degree of precision required in the assessment of safety and efficacy profiles of the drug and its metabolites before the clinical programs, one can easily see the enormous expectations put on in vitro tools in early research programs. They are frequently seen as decision-making tools, and it tends to be forgotten, as for other fields such as pharmacology or in vitro toxicology, that they are only models able to bring new building blocks to our understanding of particular series or processes. A combination of in vitro approaches together with tools integrating this information into a more physiological context has in recent years be a major step to increase our knowledge in early screening phases and to better challenge and combine our in vitro results. In any case, the actual recommendations are that attempts should be made as early as possible in the drug development process to identify differences in drug metabolism between animals used in nonclinical safety assessments compared to humans.

5.10.4.2 Identification of Metabolites

For biotransformation investigations of in vitro and in vivo studies in development programs, radioactively labeled drugs provide one of the most important tools to completely track the metabolites in complex biological matrices, such as blood, urine, feces, and bile, and in in vitro samples. The most frequently used approach is to administer the drug containing a radioactive isotope such as ^{14}C or ^{3}H. In combination with separation methods, e.g., liquid chromatography (LC), radioactive labeling allows the highly selective, sensitive, and quantitative detection of unknown metabolites. Coupled with mass spectrometric and/or other spectroscopic data it allows to completely elucidate biotransformation pathways for an administered drug. Major progress in the on-line radio detection such as LC-accurate radioisotope counting (LC-ARC) and LC topcount have brought the detection level from 200–500 disintegration per minute (dpm) down to 5–20 dpm.[42] These techniques are nowadays essential for the quantification of circulating metabolites. The use of radioisotopes is however not a good alternative in early research programs.

Identification of the metabolites produced in in vitro experiments is generally not performed in a screening approach. At this stage it is a difficult and time-consuming exercise. It can be done and is very useful on a few selected compounds most of the time at a somewhat later stage. This identification is performed using mass spectrometry (MS), which is an extremely powerful technique in the detection and identification of unknown chemical entities. However one has to remember that MS remains (in the absence of synthesized reference metabolites), a qualitative technique. Non or poorly ionizable compounds will also be difficult to detect using LC/MS/MS. Overall, the metabolic pathways obtained from microsomal incubations give a good picture of the primary reactions occurring in the drug, or in other words major metabolic reactions but not necessarily the major metabolites. One of the reasons can be the absence of secondary and tertiary phase I metabolites that are simply not formed in sufficient amount in the in vitro conditions. In vitro profiles can also in many instances give a reasonable qualitative picture of the circulating metabolites of major importance in the biological profiles of the drug. With hepatocytes and the presence of phase I soluble enzymes and phase II conjugation enzymes, we can complete the picture (see 5.06 Principles of Drug Metabolism 2: Hydrolysis and Conjugation Reactions). Some in vitro/in vivo differences can also occur with regard to glucuronides predicted to be a major pathway in human hepatocytes but finally representing a minor metabolite in vivo. This can be the case for certain conjugation reactions of acids.

One of the pitfalls of early identification of metabolites is that most experiments are done on what we can call an a posteriori metabolite identification. The MS expert is searching for possible metabolites either on the basis of the chemical structure of the drug or making hypotheses on the sites of attack based on his knowledge of the metabolic reactions. This latter stage can be improved with the use of databases predicting potential pathway but the risk of missing an important metabolite or pathway exists (see 5.33 Comprehensive Expert Systems to Predict Drug Metabolism).

One interesting alternative could come from the treatment of data. One illustration can be derived from the metabolomic field (see 5.43 Metabonomics) where powerful statistical tools such as principal component analysis are used to follow the changes in urinary and plasma patterns of endogenous markers after treatment with a drug in comparison with a control. An automated treatment can enable to separate drug related from endogenous material in in vitro incubations as well as in vivo samples helping in the early identification of metabolites. This can be achieved with a reduced number of samples as compared with a classical metabolomic approach requiring statistical power. This approach has the advantage of being an a priori technique of metabolite identification. It gives information on the number of metabolites having an MS response (molecular ion) in the sample. Even if this technique is not quantitative in the absence of reference products, it allows one to monitor the change in concentration of these metabolites with the dose, comparing species, and can be applied to both in vitro and in vivo samples. Knowing the number of metabolites present in the sample allows a further identification to be performed to define the exact structure of these metabolite(s) using further LC/MS/MS techniques and LC-(NMR) nuclear magnetic resonance to position the chemical modification.[42]

5.10.4.3 Link between Metabolism and Safety

5.10.4.3.1 Activity of metabolites

Historically, the selection of the large toxicological species using the in vitro interspecies comparison study performed with liver microsomes or hepatocytes was the major link between metabolism studies and safety (see 5.39 Computational Models to Predict Toxicity). Today, with the early identification of human metabolites, the question of their potential positive (pharmacological) or negative (toxicological) activity can be asked (see 5.08 Mechanisms of Toxification and Detoxification which Challenge Drug Candidates and Drugs).

Even though toxicity of a metabolite is difficult to identify in research programs, it can be defined as either an extension of target pharmacology or mediated via other receptors. In this case the on-line combination of mass spectrometric characterization and biological screening based on ligand–receptor or antigen–antibody interactions can be attractive for some applications. An example of such an approach is on-line affinity capillary electrophoresis–MS. The receptor is present in the electrophoresis buffer and the metabolite mixture is injected as the sample. Metabolites showing strong binding to the receptor are retained and thus separated from compounds that do not interact. On-line MS detection allows direct characterization of the interesting ligands.[42]

5.10.4.3.2 Metabolites in genotoxicity tests

Genotoxicity is typically not an extension of the pharmacology and has to be studied separately. It represents today the earliest link with safety done during the early genotoxicity screens such as Ames II and in vitro micronucleus tests. When positive, many discussions on possible work around based on the nature of metabolites and/or degradation products as well as alternative compounds within a series are held within the research teams. With regard to genotoxicity screens, the presence of major human metabolites or a different metabolic pathway in comparison to those formed in the test is the prime information necessary.

The metabolic profile in the media used in these in vitro tests is essential for the understanding of the toxicological profile of the drug. This evaluation has therefore to be done in the strict conditions of the mutagenic test because of the hydrolytic enzymes present in certain media as well as in the light of the high concentrations used and the resulting frequent saturation of certain metabolic pathways. Because the activation media are most often from rat origin it would also be useful, when possible, to perform these tests in parallel with human activation in vitro media (S9 or microsomes).

5.10.4.3.3 Structural alerts and reactive metabolites

It is well known that reactive structures or substructures as well as reactive metabolites formed after activation of chemicals by metabolic enzymes can potentially covalently bind to cellular proteins. This binding leads to modified proteins and can trigger various organ toxicities. A comprehensive listing of compounds, the proposed metabolic activation process as well as the type of toxicity observed is given by Evans *et al.*[49] As a logical consequence of metabolic activation, the liver, as the main metabolizing organ, is the main target organ of these reactive metabolites.

Capturing reactive xenobiotic metabolites using glutathione (GSH) or potassium cyanide are well-described techniques. These tests are performed in the presence of liver microsomes or hepatocytes and in the case of GSH the loss of the gamma-glutamyl moiety in MS analysis[36] can point toward a potential GSH adduct which however needs to be identified. A highly efficient screening method has been developed mixing natural GSH and GSH labeled with two ^{13}C and one ^{15}N (+ 3 in mass) in equal molar ratios. With this technique a unique MS signature is obtained with the presence of an isotopic doublet in MS.[50] There is no consensus with regard to the minimum amount of adduct acceptable in terms of safety. The threshold amount proposed is $50 \, pmol^{-1} mg$ of proteins.[50]

Identifying potential structural alerts can be done with the appropriate databases containing more and more the potential structural alerts.

5.10.5   Drug Interactions

5.10.5.1   Importance of In Vitro Tools in Drug Interaction Studies

Drug interactions can be defined as the modification of the safety and efficacy profile of a medication following the coadministration of drugs, environmental pollutants, ingredients or additives present in the diet (*see* 5.35 Modeling and Simulation of Pharmacokinetic Aspects of Cytochrome P450-Based Metabolic Drug–Drug Interactions).

Drug interactions are usually defined as being either of pharmacodynamic or of pharmacokinetic origin. These two aspects are closely linked and need to be studied together to evaluate both safety and efficacy in the clinic. Only the latter category, defined also as metabolism-based interactions (i.e., the alteration of the metabolic clearance of a drug by another coadministered drug) will be further detailed in this chapter.

With the increasing prevalence of polypharmacy,[51,52] these drug interactions have become more frequent,[51,53,54] and can lead to serious side effects, e.g., Cerivastatin[55] and increased morbidity and mortality.[54]

It is well recognized that testing in the clinic all the possible combinations between an NCE and the potential interacting compounds on the market is impossible for obvious cost and time reasons, but also because it is nonrelevant. In vitro studies have therefore become essential to explore the molecular mechanisms involved in these drug interaction processes with the aim of first selecting a compound with a safe or a controlled drug interaction profile but also in order to help in the future design and optimization of drug interaction clinical studies.

In 2003, the Pharmaceutical Research and Manufactures of America (PhRMA) Drug Metabolism and Clinical Pharmacology Technical working Groups have defined a minimal in vitro and in vivo pharmacokinetic drug interaction package for registration purposes.[46] In October 2004, a preliminary concept paper of the US Food and Drug Administration (FDA) has focused more on study design, data analysis, as well as implications on dosing and labeling.[45]

CYP are involved in the metabolism of more than 90% of currently available drugs[56] and are recognized by regulatory authorities as an important cause of drug interactions.[57,58] Therefore, the guidelines focus essentially on those enzymes that can be inhibited, activated, or induced by concomitant drug treatments. These changes in enzyme activities may significantly increase or decrease the exposure of the body to the drug or metabolites at potentially toxic levels. The approaches described can however be adapted to other enzymes.

This chapter suggests the following general topics to be investigated during the drug development program:

- Drug metabolism enzyme identification
- Evaluation of CYP inhibition
- Evaluation of CYP induction.

A pragmatic approach is recommended to define the drug interaction strategy which will, at the registration level, be able to justify the selection of the interacting compounds used during the clinical trials, limit the number of clinical studies to be undertaken, and, finally, in the labeling, position the compound as regards future comedications as well as suggest dosage adjustments if required.[46]

5.10.5.2 Reaction Phenotyping

Reaction phenotyping consists in identifying the enzyme(s) involved in a metabolic reaction. Before initiating these phenotyping studies, one should ideally first determine the in vitro metabolic profile of a drug in humans (identity of the metabolites formed and their quantitative importance) and estimate the relative contribution of each enzyme to the total clearance.[45]

There is no rationale for performing CYP reaction phenotyping with an NCE if CYP-mediated reactions play only a minor role in the overall clearance, i.e., less than 25% as recommended recently.[45]

One has to keep in mind that in screening programs the use of in vitro tools most of the time only allows the assessment of the primary reactions of a drug but not necessarily the major in vivo pathways. The in vitro model selected at this stage will only be fully validated once the metabolic profile in humans is known. It is recommended that the metabolism of an NCE is examined in an, as complete as possible, in vitro system.[46] Freshly isolated hepatocytes are theoretically the best choice for conducting such experiments even though microsomes are preferred for technical reasons (characterization, availability, presence of membranes, etc.)

The enzymes involved in the drug clearance can be identified using four well-defined approaches:

1. Specific chemical inhibitors for each CYP (or other enzymes).
2. Specific antibodies for each CYP.
3. Individual human recombinant CYPs.
4. A bank of human liver microsomes characterized for CYP activities prepared from individual donor livers[45] (correlation studies).

At least two of these methods should be performed to identify the specific enzyme(s) responsible for drug metabolism.

The specific inhibition experiments should preferably be conducted in pools of human liver microsomes (except for correlation studies) to represent the average population. The in vitro incubations should be performed under linear rate conditions (linearity of metabolite production rates with respect to time and enzyme concentrations) and whenever possible, with pharmacologically relevant concentrations of drugs.[1,46] It is also recommended that each metabolite produced by individual CYP selected for identification be mentioned. By consensus, the pharmaceutical companies[46] have classified the CYPs as major (1A2, 2C9, 2C19, 2D6, and 3A4), of emerging importance (2C8, 2B6, and 3A5), and minor in drug metabolism (1A1, 1B1, 2A6, 2E1, 4A11, etc.).

No explicit recommendations are given for the number of CYPs to investigate so far in the current guidance[57] or the new draft paper.[45] However, the major CYPs can represent the minimum package to be investigated in screening programs completed, in later programs, and for registration purposes, with the other CYPs.

Table 3 Method for identification of pathways involved in the oxidative biotransformation of a drug

In vitro system	Condition	Tests[a]
Microsomes	+/− NADPH	CYP, flavine monooxygenases (FMO) versus other oxidases
Microsomes, hepatocytes	+/− 1-aminobenzotriazole + clotrimazole, 1-benzylimidazole, SKF-525A or anti-NADPH-CYP reductase antibodies (+ cofactor required)	Broad-specificity CYP-inactivator
Microsomes	45 °C pretreatment	Inactivates FMO
S9	+/− pargyline	Broad monoamine oxidase (MAO) inactivator
S9	+/− menadione, allopurinol	Mo-CO (oxidase) inhibitors

[a]See [45,46].

The FDA guidelines[45] encourage the identification of non-CYP enzymes such as glucuronyltransferases, sulfotransferases, and N-acetyltransferases, although there are only few documented cases of clinically significant drug interactions related to these classes of enzymes. These enzymes can be of importance in certain chemical series if they are involved in primary reactions and before initiating the phenotyping experiments, it is important to define the role of CYP and non-CYP enzymes in the metabolic clearance of the drug (**Table 3**). Taking a close look at the chemical structures will allow the selection of a more appropriate in vitro model and limit the number of systematic experiments at early stages.

Most of the early investigations are performed by monitoring the substrate only usually using a LC/MS/MS method; alternatively LC coupled with UV or fluorescence can be used but these methods give too low a sensitivity to monitor low drug concentration incubations. This is why the characterization of the enzyme kinetics, for each major pathway, is performed in later stages of the drug development plan in order to fine-tune the predictions of drug interactions generated in early stages.

5.10.5.2.1 Use of specific inhibitors

For phenotyping studies, the use of specific CYP inhibitors requires well-defined conditions to ensure a reliable result.[59] The substrate concentration should be lower than or equal to the K_m and the inhibitor concentration should ensure selectivity and adequate potency. A range of inhibitor concentrations may be used to aid the interpretation.[28] It is necessary that their conditions of use have been validated earlier in order to ensure a reliable answer. For example, some specific inhibitors have a very low metabolic stability (such as coumarin, a potential CYP2A6 inhibitor) and they do not show any inhibitory potential in typical incubation conditions because they are completely metabolized.

The contribution of a specific CYP can be given directly by the percentage of inhibition of the NCEs metabolism. The most common CYP specific inhibitors are given in **Table 4**.

In early screening, only the major CYPs (1A2, 2C9, 2C19, 2D6, and 3A4) are studied, with CYP3A4 and 2D6 frequently selected first, since they represent the CYP isozymes very often involved in the metabolism of the currently marketed drugs.[60,61] At a somewhat later stage, a larger panel of CYPs will be investigated and the effect of the CYP specific inhibitors can be studied on each metabolic reaction.

5.10.5.2.2 Use of specific antibodies

The problem of some of the specific inhibitors lies in their poor specificity and potency although they are preferred to antibodies in drug-interaction studies. However, the use of inhibiting antibodies has been recently promoted[62,63] in the quantitative assessment of a particular CYP in the metabolic pathway of a drug, in particular when several CYPs are involved.[64] Much improvement has been made in their selectivity, with reduced cross reactivity between homologous CYPs (e.g., 1A2 and 1A1).

It remains however difficult for some CYP antibodies to attain complete inhibition of the activity of the enzyme (at most 80%) even with large amounts of antibodies. Another drawback is their limited commercial availability, and antibodies raised against non-CYP enzymes are even less accessible. Finally, the use of inhibiting antibodies requires further optimization (titration curve) and they are not suitable for early screening programs.

Table 4   Most common specific CYP inhibitors used for phenotyping studies (see [45,66])

Enzyme	Specific CYP inhibitors	Range of K_i from the literature (μM)
CYP1A1	Alpha-naphthoflavone	0.01–10
CYP1A2	Furafylline[a]	0.6–4.7
CYP2A6	Tranylcypromine	0.02–0.2
CYP2A6	Methoxsalen	0.01–0.2
CYP2A6	Pilocarpine	1–4
CYP2B6	8-Methoxypsoralene (8-MOP)[a]	
CYP2B6	ThioTEPA[a] (triethylenethiophosphoramide)	4.8
CYP2C8	Quercetine[b]/Retinol	1.1/7
CYP2C9	Diclofenac/Sulfaphenazole	20/0.3–0.5
CYP2C19	N3-benzyl nirvanol	0.25–0.41
CYP2D6	Quinidine	<0.1–0.5
CYP2E1	Diethyldithiocarbamate (DDC)[a]	13–24
CYP2E1	Disulfiram	50
CYP3A4	Ketoconazole	0.015–0.15
CYP3A4	Itraconazole	0.27–2.3
All CYP	Clotrimazole	0.1–12
FMO	Methimazole	5700–8600

[a] Mechanism based inhibitor necessitating a preincubation period before adding substrate.
[b] CYP2C8 and 1A2 inhibitor.

5.10.5.2.3   Use of recombinant enzymes

The use of cDNA-expressed enzymes is now preferred to estimate the contribution of each CYP in the clearance of the NCE.[45] The great advantage of these recombinant human CYPs is that they contain one isolated enzyme ruling out the problem of selectivity of the chemical inhibitors or CYP inhibiting antibodies.

The experiments can be conducted in microplates (96 to 1536 wells)[65] and are suitable for a complete automation making them fully adapted to screening programs. However, because the turnover number of each recombinant is not representative of those found in human liver microsomes, their incubation with the NCE cannot alone directly give the relative contribution of the CYPs in the overall metabolism of the drug. Without scaling factors, it only gives the ability of a particular enzyme to metabolize an NCE.

The microsomal environment, i.e., lipids, apoproteins, level of both cytochrome b5 and NADPH cytochrome CYP reductase, is different in the cDNA-expressed CYP system from the one in human liver microsomes. These differences can affect the turnover number (V_{max}) for a given enzyme, although the affinity (K_m) of CYP is quite comparable between recombinant enzymes and human liver microsomes (**Table 5**). A thorough literature search performed with the aid of a drug interaction database (Aurquest database[66]) can illustrate this aspect (**Table 5**).

This link between the recombinant CYP and human liver microsomes can be obtained using a relative activity factor (RAF) for each CYP.[62] This RAF is a sort of extrapolation factor from a simple system expressing only one enzyme to the human microsomes containing a mixture of enzymes. These RAF can be calculated using the maximal velocity of the reaction (V_{max}) or better, by estimating of intrinsic clearance CL_{int}[67,68] in both systems using specific probe substrates of each CYP.

Table 5 Comparison of the metabolic affinities (K_m) for CYP-dependent reactions obtained in human liver microsomes and recombinant system (obtained from Aurquest database[66]). Results are expressed as median values (Min, Max)[a]

CYP	Substrate	Reaction	K_m HLM (μM)	K_m Recombinant (μM)
1A2	Phenacetin[b]	O-Deethylation	45.3 (2.7–7691)	29.9 (9.19–107)
1A2	Caffeine	N3-Demethylation	640 (190–3000)	310 (190–430)
1A2	Ethoxyresorufin	O-Deethylation	0.34 (0.061–2.9)	0.844 (0.05–3)
2A6	Coumarin[b]	7-Hydroxylation	1 (0.3–50)	2.45 (0.61–7.93)
2A6	Nicotine[b]	C-Oxidation	64.9 (39.6–95.3)	25.5 (11–47)
2B6	Efavirenz[b]	Hydroxylation	30.1 (20.2–39.9)	12.4 (6.4–20.2)
2C19	S-Mephenytoin[b]	Phenyl-4-hydroxylation	42 (29.1–903)	48.0 (21.7–72)
2C19	Omeprazole	5-Hydroxylation	6.4 (3.82–106)	12.2 (—)
2C8	Taxol[b]	6-Hydroxylation	15 (3.37–26)	6 (2.85–29.1)
2C9	S-Warfarin[b]	7-Hydroxylation	4.73 (3.79–27)	5.8 (0.78–30)
2C9	Diclofenac[b]	4'-Hydroxylation	12 (1.5–31)	4.8 (0.9–15.5)
2C9	Tolbutamide[b]	4-Methylhydroxylation	125 (18–2580)	132 (43–523)
2D6	(+/−)-Bufuralol[b]	1'-Hydroxylation	9.3 (3–250)	10 (1.5–54.5)
2D6	Dextromethorphan[b]	O-Demethylation	5.5 (1.5–282)	2.3 (0.3–8.5)
2E1	Chlorzoxazone[b]	6-Hydroxylation	70 (15.8–238)	132 (6–660)
3A4	Alprazolam	4-Hydroxylation	530 (234–764)	531 (238–743)
3A4	Alprazolam	α-Hydroxylation	409 (256–575)	178 (90–328)
3A4	Dextromethorphan[b]	N-Demethylation	286 (217–700)	232 (18–660)
3A4	Midazolam[b]	1'-Hydroxylation	5.6 (1.37–14)	2.35 (0.8–5)
3A4	Nifedipine	Oxidation	20 (13–30)	8.75 (2.35–20.9)
3A4	Quinidine	3-Hydroxylation	17.6 (4–85.5)	2.25 (1–88)
3A4	Testosterone[b]	6-β-Hydroxylation	63 (33.48–120)	60 (33–108)

[a] Biphasic kinetics explain the wide range of K_m obtained in HLM (high and low affinity enzymes). Recombinant CYP, containing only one enzyme, cannot display the lowest affinity enzymatic system (lower range of K_m).
[b] Preferred chemical substrates for in vitro experiments.[47]
HLM, human liver microsomes; Recomb., recombinant CYPs.

Individual CYP CL_{int} values for a new compound, obtained separately with the expressed systems, can then be multiplied by the corresponding RAF factors allowing to assess the CL_{int} of the corresponding CYP in human microsomes. This approach is integrating the relative abundance of each hepatic isoform[62] and allows the estimation of the relative contribution of an individual CYPs to the metabolism of an NCE.

This method is of course applicable for all types of enzymes once a recombinant form of the enzyme and a specific substrate are available, such as is the case for some UGTs.[69] However, this approach can be time consuming for screening programs on large series of compounds.

An approach derived from the model proposed by Crespi *et al.*[70] is in use in our laboratories. Using their relative activity factor, each major CYP (1A2, 2C9, 2C19, 2D6, and 3A4) is mixed with relative proportions allowing to mimic the metabolic capacity found in a pool of human liver microsomes. Another mix is prepared by omitting the CYP of interest. The velocity of the overall reaction or intrinsic clearance of the NCEs is then measured using a mix of all the major CYPs, the same mix missing one of the CYPs and the CYP alone. The comparison of the intrinsic clearances then allows the estimation of the relative contribution of the CYPs of interest. This experimental approach, limited in this example to the five major human CYPs, can be extended to other enzymes reconstituting and standardizing in a way human microsomes. This approach is particularly adapted when evaluating large series of compounds.

Although the RAF approach is now commonly used either indirectly in screening programs or directly in development programs, there are still uncertainties about this method since controversial results showed that this relative activity approach may or may not depend on the index reaction used to create the RAFs for each CYP isoform and the expression system employed.[62,68,71] A closely related method named relative substrate activity factor (RSF) is based on the ratio of activities of the recombinant CYP toward the NCE and the CYP specific substrates. The RSF and RAF approaches are similar and they suffer from the same limits.[72] An alternative approach (relative abundance) is based on the immunoquantification of CYPs, but is very dependent on the source of recombinant enzymes and very time-consuming.[62,72,73]

Despite these limits, regulatory agencies recommend the use of recombinant human CYPs as they consider that they can provide the relative contribution of each pathway if more than one CYP is involved.[45]

However, at early stages, the emphasis is more on identifying the drugs metabolized by a single enzyme. A contribution of more than 70% to the metabolic clearance can be given as a general rule at this stage. Other clearances (biliary and renal) can decrease the risk of drug interaction due to coadministered drugs.

In later programs, the impact of substrate concentrations toward the relative contribution of enzymes will have to be defined with proper enzyme kinetics (K_m, V_{max} of each enzyme). When a NCE is metabolized by multiple enzymes, the effect of multienzyme inhibitors should also be considered in later stages.

5.10.5.2.4 Correlation studies

The different rates for each primary pathway of an NCE are correlated with reference activities of each individual CYP in the same bank of liver microsomes. This bank of individual human liver microsomes should be constituted of at least 10 donors with a wide range of enzyme activities for each CYP and no cross-correlations between each individual enzyme.

This approach, more than the others, is not sufficient in itself due to its lack of accuracy. Only the major enzymes involved in a metabolic reaction can be characterized and no quantitative contribution can be deduced. The lack of power of this method means that it is definitively not suitable for early screening programs.

5.10.5.3 Inhibition

Assessment of the inhibition potential of a drug and the evaluation of its impact on plasma concentrations of coadministered drugs has been a major part of the early in vitro metabolism programs in the past decade. Fortunately, even though the potential consequences of inhibition on safety and efficacy are major, the basic understanding of the mechanisms involved as well as the in vitro tests available to address these issues are well known by metabolism scientists.

As a logical consequence regulatory agencies have required in vitro studies to be undertaken to determine the inhibitory potential as well as the mechanism of inhibition of an NCE toward CYPs.

These studies contain three major steps:

1. Kinetic experiments, determining IC_{50}s or K_is of NCE toward CYPs.
2. Type of inhibition (time-dependent experiments assessing a mechanism based inhibition).
3. Prediction of the risk of interaction (extrapolation to the in vivo situations).

The new FDA draft guidelines recommend to select specific substrates with preferentially a simple metabolic scheme (ideally no sequential metabolism), allowing to set up sensitive, rapid and simple assays with reasonable incubation times.[47] A list of CYP probe substrates proposed by this draft guidance is given in **Table 5**. The probe substrates listed covers the major CYPs as well as those of emerging importance. There is no specific recommendation on investigating non-CYPs.

Similarly to phenotyping studies, initial rate conditions need to be respected. No more than 10% substrate or inhibitor depletion should occur.[45] This is somewhat difficult to maintain for the substrate in order to obtain a reasonable detection of the CYP specific metabolite. Solvents should be used at low concentrations ($\leqslant 1\%$, v/v and preferably $<0.1\%$) because of their potential inhibiting or activating effects.[1,26] No particular recommendations are made for the human model to be used. Determinations of inhibition potentials, in development programs, are based on monitoring the changes in the metabolism of specific CYP substrates in human liver microsomes or recombinant CYPs in the presence of an adapted range of concentrations of the NCEs of interest. [52]

Enzyme inhibition can be divided into reversible and irreversible processes (discussed below).

5.10.5.3.1 Reversible inhibition

Reversible inhibition can be classified further as competitive, uncompetitive, mixed, or noncompetitive (**Table 2**).

For competitive inhibition, the binding of the inhibitor prevents binding of the substrate to the active site of the enzyme; for noncompetitive inhibition, the inhibitor binds to another site of the enzyme and the inhibitor has no effect on binding of substrate, but the enzyme-substrate-inhibitor complex is nonproductive. In the case of uncompetitive inhibition, the inhibitor does not bind the free enzyme but binds to the enzyme-substrate complex and again, the enzyme-substrate-inhibitor complex is nonproductive.[74] Finally in the case of mixed (competitive–noncompetitive) inhibition, the inhibitor binds to the active site as well as to another site on the enzyme, or the inhibitor binds to the active site but does not block the binding of the substrate.

The kinetics and the affinity with which an inhibitor binds to an enzyme are described by the inhibition constant K_i.

Transformations of the Michaelis–Menten equation (**Table 2**) are not only used for calculating K_i values but also for graphical depiction of the type of inhibition.[28] These equations allow to predict the risk of drug interaction as a change in substrate exposure (AUC) in presence of an inhibitor (AUC$_i$/AUC) (**Table 2**).

5.10.5.3.1.1 K_i Determination

An accurate appreciation of the inhibitory potential and the type of inhibition are given by the determination of the inhibition constant K_i. The standard approach consists in incubating varying concentrations of substrate (around the K_m) and inhibitor (around the K_i) with fixed amounts of enzyme for a constant period of time after having checked the linearity of the kinetics.[45]

An alternative technique, derived from the SDK method (*see* Section 5.10.3.1) requires only two-substrate and two-inhibitor concentrations.[33] However, even if this approach is more rapid than the regular Michaelis–Menten one, both experiments are too time-consuming to be included in early stages of drug development and will only be conducted if necessary and for a limited number of CYPs. Determination of K_i is too labor-intensive, so only the IC$_{50}$ is measured in early stages. This constant represents the concentration of inhibitor required to reach 50% inhibition under given experimental conditions.[28]

5.10.5.3.1.2 IC$_{50}$ determination

To determine IC$_{50}$, only one concentration of the probe substrate, equal to or below the K_m of the index reaction, is incubated with several inhibitor concentrations.

In screening programs, a simplified method using cDNA expressed systems and non-CYP selective fluorogenic probes (**Table 6**), able to be conducted in 96 and even 1536 well microplates, has been developed.[65] It offers the great advantage of being suitable for a full automation.[75,76] Because cDNA-expressed systems contain only one enzyme, there is no need to use a selective probe substrate for each CYP as would be the case with human liver microsomes.[18] The method is based on the use of low or nonfluorescent CYP substrates producing highly fluorescent metabolites able to be read using a fluorescence plate reader, which does away with the need laborious and time-consuming LC methods.[76]

At this early stage of development, the inhibitory potential of the NCE is studied toward the major CYPs only (i.e., 1A2, 2C9, 2C19, 2D6, and 3A4).[1] To avoid solubility problems, the NCEs are usually solubilized in DMSO (0.5% maximum final concentration) and incubated at relatively low concentrations (maximum 25 μM). A positive control (known CYP specific inhibitor) can be added to validate the assay. If several concentrations of NCE are incubated, an IC$_{50}$ can be calculated. Due to the low inhibitor concentrations incubated, only IC$_{50}$ below 25 μM can be assessed, enabling to screen potent (IC$_{50} < 1$ μM), marginal (1 μM $<$ IC$_{50} < 10$ μM), or weak (IC$_{50} > 10$ μM) inhibitors.[46]

However, this method is limited by the quality of expression of these isolated systems in the microsomal environment or the turnover number of these enzymes.

Table 6 List of the fluorimetric probe substrates (and their respective CYP-dependent index reaction) commonly used for screening the inhibitory potential of NCEs

Substrate	CYP	Metabolite
Resorufin benzyl ether (BzRes)	1A1/1B1/3A4	Resorufin
3-Cyano-7-ethoxycoumarin (CEC)	1A2/2C19	3-Cyano-7-hydroxycoumarin (CHC)
Coumarin	2A6	7-Hydroxycoumarin (7-HC)
7-Ethoxy-4-trifluoromethylcoumarin (EFC)	2B6	7-Hydroxy-4-trifluoromethylcoumarin (7-HFC)
Dibenzylfluorescein (DBF)	2C8/2C9/2C19	Fluorescein
7-Methoxy-4-trifluoromethylcoumarin (7-MFC)	2C9/2E1	7-Hydroxy-4-trifluoromethylcoumarin (7-HFC)
3-O-Methylfluorescein (OMF)	2C19	Fluorescein
7-Methoxy-4-(aminomethyl) coumarin (MAMC)	2D6	7-Hydroxy-4-(aminomethyl) coumarin (HAMC)
3-[2-(N,N-Diethyl-N-methyl-ammonium)ethyl]-7-methoxy-4-methylcoumarin (AMMC)	2D6	3-[2-(N,N-diethyl-amino)ethyl]-7-hydroxy-4-methylcoumarin (AHMC)
Benzyloxyquinoline (BQ)	3A4	Quinolinol
7-Benzyloxy-4-trifluoromethylcoumarin (BFC)	3A4/3A5	7-Hydroxy-4-trifluoromethylcoumarin (7-HFC)
Dibenzylfluorescein (DBF)	3A4	Fluorescein

When several CYPs are involved in the metabolism of a compound, the use of cDNA-expressed systems may not always properly estimate the inhibitory effects of a given drug. Contradictory results on the correlation level of inhibitory potential using different fluoroprobes have been shown in the literature.[65,76]

In any case it remains a very useful tool for preliminary screening stages, completed later on selected compounds with conventional probes in models showing a full metabolic competence.[99] When working on small series of compounds, human liver microsomes are the preferred test system to determine the inhibitory potential,[46,65] because CYP kinetic measurements are not confounded with other metabolic processes or cellular uptake, as can be the case in hepatocytes.

The major issue in using human liver microsomes is the selection of highly specific CYP probes or index reaction (**Table 5**). The kinetic parameters should be assessed and the conditions of incubation should be validated by testing and comparing the inhibitory potential of known inhibitors with literature values.

In the past, the index reactions were monitored using several analytical techniques namely, LC coupled with UV, fluorescence, radiochemical, or MS/MS detections. Today a single LC/MS/MS technique can follow most of the CYP substrates and their respective metabolites within the same run, supplanting the standard use of radiolabeled substrates.[42] A method is in use in our laboratory enabling us to follow 11 metabolic reactions, each specific for one human CYP[42] in a single incubate.

IC_{50} or K_i generated in vitro can be corrected for nonspecific microsomal binding (not including plasma binding) in order to avoid an underestimation of the drug interaction risk or even false negatives. Tran et al.[77] have shown that the inhibition of several CYP3A4 inhibitors on diazepam metabolism decreased when microsomal proteins increased. This highlights the interest of correcting inhibition constants for microsomal binding.

These IC_{50} determinations are sufficient in early stages to extrapolate a preliminary risk of drug interaction using a pragmatic safety approach. They do not, however, allow one to distinguish the type of inhibition that is normally required to select the right model for drug interaction predictions (**Table 2**).

5.10.5.3.1.3 In vivo risk prediction

It can be assumed, as a simplification, that the vast majority of inhibitors involved in drug interactions are competitive inhibitors. In this case when the substrate concentration is equal to the K_m, the K_i is theoretically equal to $IC_{50}/2$.

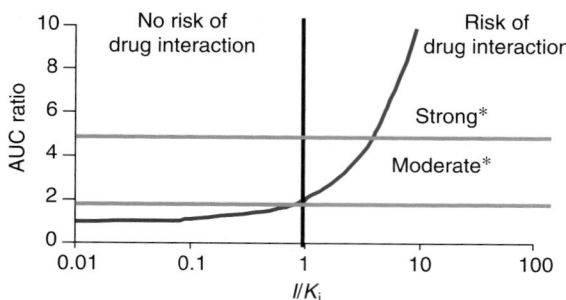

Figure 1 Influence of the inhibition index (I/K_i) of a drug on the drug interaction risk (based on CYP 3A4 substrates). AUC ratio: AUC_i/AUC represents the predicted ratio of the exposure to the drug undergoing the drug interaction in presence of the interacting compound (AUC_i) and the exposure to the drug without the interacting compound (AUC). I/K_i represents the inhibition index where I is the concentration of the inhibitor at the active site of the enzyme and K_i the constant of inhibition. *, with respect to the safety margins.

As shown in **Table 2**, the prediction of the risk of drug interaction is based on the inhibition index I/K_i (**Figure 1**) resulting from equation

$$AUC_i/AUC = 1 + I/K_i \qquad [9]$$

I is the concentration of inhibitor at the active site of the enzyme which is an unknown parameter somewhat difficult to figure out.

Mean steady-state C_{max} ($C_{max}(ss)$) values for total drug following administration of the highest proposed clinical dose have been recommended as surrogate estimates of the inhibitor concentration for in vitro–in vivo scaling.[45,46] As reviewed by Venkatakrishnan et al.[26] if some reasonably accurate projections of the magnitude of clinical interactions have been obtained using $C_{max}(ss)$ as an inhibitor concentration, underestimated predictions have been observed,[78] which implies the risk of developing a potential inhibitory drug.

In early screening programs the emphasis is on safety and the objective is to predict the highest possible risk as regards to drug interactions. This safety approach consists in using the maximum concentration reaching the liver (C_{in}) as the concentration at the site of the enzyme using the following equation:

$$C_{in} = \frac{k_a \times F_{abs} \times D_a}{Q} \qquad [10]$$

where C_{in} represents the concentration of the inhibitor entering the liver after its oral administration at the dose D_a; F_{abs} is the fraction absorbed, k_a the absorption rate constant, and Q the hepatic blood flow. This equation assumes that the NCE follows a first-order absorption kinetics.

Some authors recommend to use the C_{in} corrected by the unbound fraction in plasma (fu_p).[79] However, in respect to a safety approach, it is deemed better to use the total fraction to obtain the maximal inhibitor concentration at the active site of the enzyme (overestimation).

The therapeutic dose in humans can be estimated by allometry from the pharmacological dose in the rat making the assumption of linear kinetics.

The default value for the absorption rate constant[79] is 0.1 min^{-1}. In order to maximize the C_{in}, the NCE is assumed to be completely absorbed by the gastrointestinal tract ($F_{abs} = 1$).

At steady state, Ito et al.[79] considered the inhibitor concentration at the active site of the enzyme as the sum of C_{in} and C_{max} weighed by the plasma unbound fraction (fu_p). Successful predictions of clinical drug interactions have been obtained using this approach. It is however only applicable at clinical stages. A combined use of $C_{max}(ss)$ and $C_{in} + C_{max}$ can avoid a number of false negatives.[78] When using such a safety approach one has to accept the risk of false positives.

In addition, some compounds have been found to accumulate extensively in the hepatocytes (e.g., itraconazole, ketoconazole, diltiazem, verapamil[80]) so that even the C_{in} might underestimate the intrahepatic inhibitor concentrations. An alternative proposed by Yamano et al.[80] and Ito et al.[79] would be to measure this partition using hepatocytes in suspension. Such experiments are only adaptable to small series of compounds.

When using hepatocytes to measure apparent IC_{50}, this uptake is taken into account, even though it is only adaptable to small series of compounds.

When the intrahepatic enzyme-available inhibitor concentration has been estimated, the inhibition index I/K_i (with $K_i = IC_{50}/2$) and the ratio AUC_i/AUC can be calculated. The new FDA draft guidance[45] classifies a drug as a strong (AUC_i/AUC is ≥ 5) or moderate inhibitor $2 \leq AUC_i/AUC < 5$.

An AUC_i/AUC ratio of 2 or higher can be used as a trigger to plan further in vitro and in vivo drug interaction studies with different probe substrates and concomitant treatments.

5.10.5.3.2 Irreversible inhibition

In irreversible inactivation, the inhibitor first acts on the target enzyme before inhibition occurs. In many case, this involves conversion of the inhibitor into a chemically reactive intermediate that forms a covalent bond with the enzyme, inactivating it permanently. For CYP enzymes, irreversible inhibition can be noncovalent occurring through metabolite–intermediate complexes tightly bound to the iron of the heme prosthetic group (quasi-irreversible inhibition). Irreversible inactivation, whether it occurs through formation of covalent bonds or iron complexes, is the underlying mechanism of some high-magnitude drug interactions,[16] e.g., CYP2D6 inactivation by paroxetine,[81] CYP3A4 inactivation by diltiazem,[82] erythromycin, or troleandomycin,[26,74,83] CYP2C9 inactivation by tienilic acid,[84] etc.

It is crucial to distinguish time-dependent inhibition due to the reversible inhibitory effect of a metabolite formed during the incubation from the irreversible inactivation of the enzyme by a reactive intermediate formed from the drug itself (mechanism-based inhibition). The risk is an underestimation of drug interaction potential. This has been demonstrated by Mayhew et al.[85] with diltiazem, clarithromycine, and fluoxetine significantly increasing the exposure to CYP3A4 substrates, for which a simple competitive model failed to predict drug interactions.[85]

In vitro, a mechanism-based inhibition is characterized by: the irreversibility of the reaction, an NADPH-, time-, and concentration-dependent enzyme inactivation, substrate protection,[86] loss of the CYP content, and the incapability of superoxide dismutase, catalase or exogenous nucleophiles such as glutathione to reverse the inactivation.

The consequences of a mechanism-based inhibition are not only an alteration of the kinetic behavior of the drug involved by a decrease in its first pass clearance but also possible potent drug interactions. In this respect, this kind of inhibition has to be evaluated as early as possible, as recommended by the new FDA draft guidelines[45] even if the full set of experiments required to assess the mechanism involved can be quite time-consuming.

However, a simple approach can be used as a preliminary step. This consists in experiments in which the metabolism-dependent inhibitor is preincubated (usually for 15–60 min (30 min being recommended by FDA draft guidance[45])), allowing the metabolite(s) to be formed. The probe substrate is then added assessing the inhibition potential of both inhibitor and its metabolites in comparison to the substrate only in control samples. In order to minimize reversible inhibition due to metabolites, a dilution of the preincubate has been proposed (minimum $10 \times$ dilution) to discriminate between reversible and mechanism-based inhibition. This technique should, however, be handled with care for specific substrates with low turnovers.

To appreciate the potency of mechanism-based inhibition and predict the risk of drug interaction, the inactivation kinetic constant k_{inact} and inhibition constant K_i should be measured in vitro. The impact of a mechanism-based inhibition can now be better appreciated with the use of adapted models for the prediction of drug interaction.[85] In addition to K_i and k_{inact}, this model requires the rate of de novo biosynthesis of enzyme (k_{degrad}) estimated at $0.000825 \, min^{-1}$ by Mayhew et al.[85]; this probably constitutes the major limit of this model.

All this information is required to optimize the design of future clinical drug interaction studies. For example, in case of a mechanism-based inhibition, it may be important to administer the inhibitor prior to (e.g., 1 h) the administration of the substrate to maximize the effect.[45] It is interesting to notice that in case of mechanism-based inhibitions, a time-dependent deviation of the first order kinetics can lead to false predictions of intrinsic clearances using the in vitro half-life approach.[26]

In conclusion, even though in vitro based drug interaction predictions have been very successful in early metabolism programs, one has to keep in mind some of the pitfalls of this approach. It includes the percentage of solvent used, microsomal protein binding, substrate depletion, mechanism-based inhibition, probe selectivity, substrate susbstitution (particularly for CYP3A4), and uptake in the hepatocytes. The in vitro incubation conditions including buffers, ionic strength, pH; and addition of cytosol and/or albumin can also influence the metabolic properties of in vitro microsomal systems, thereby having implications on the predictions of the magnitude of drug interaction.[26]

5.10.5.4 Induction

Induction has been defined as an increased level of enzyme gene expression (see 5.08 Mechanisms of Toxification and Detoxification which Challenge Drug Candidates and Drugs), and hence an increase in enzyme activity, resulting from exposure to a xenobiotic or endogenous components.[87]

The consequences are modifications in exposure to the drug (decrease) and its metabolites (increase) leading to potential changes of the efficacy profile of the drug (duration and intensity of action) or to its safety profile (reduced therapeutic margin). This is the case of the well-known interaction between oral contraceptives and rifampicin where the bioavailability of the former decreased by 42% due to induction of CYP3A4 by rifampicin, resulting in the loss of contraceptive efficacy.[87] If both parent drug and major metabolites have the same pharmacological activity, then the pharmacological effect of the induction is reduced although the pharmacokinetics of both parent drug and metabolites have changed. This has been observed for the induction of alprenolol by pentobarbital.[88]

In addition, the induction may enhance the formation of reactive metabolites resulting in potential adverse reactions. The timescale of an induction and an inhibition is somewhat different. Induction processes, at the level of an organism, are not as immediate as inhibition mechanisms. They involve fewer isozymes, affecting only the inducible ones (principally CYP1A1–1A2, CYP2C9, CYP2E1, and CYP3A4).

The prime concern is the induction potential of an NCE able to affect the metabolism of coadministered compounds. On the other hand known inducers can also affect the metabolism of an NCE. For the latter case the information on the identity of the enzymes involved in the metabolism of an NCE is crucial to evaluate potential impacts.

Evaluation of the induction potential of a drug has been a weak point in early screening programs, carried out only for specific projects in which induction was known to represent a potential issue.

Strategies for evaluating the induction potential of drug candidates have often been, and still are, based on ex vivo animal liver analyses obtained from toxicological studies.[31] However, with regard to the number of animals, amount of compound, and time required, they have proven to be inadequate for early screening. In addition, scaling results from animals to humans is unreliable due to interspecies differences[89] in metabolizing enzymes and their regulation factors (such as nuclear receptor CAR, PXR, or AhR).[87]

5.10.5.4.1   Primary cultures of human hepatocytes

Primary cultures of human hepatocytes have been considered by most academic, industrial and regulatory scientists to be the current 'gold standard' for preclinical induction studies.[8,87]

Culture conditions have been optimized to increase the culture period duration, the parallel decrease in enzymatic activity over time being less an issue for induction than metabolic profiling purposes.[8–10] Hepatocytes can be cultured on collagen-coated 96-well plates, more adapted for screening purposes.

Nevertheless, no real standardized conditions exist between laboratories thus making some comparison quite difficult. A preliminary concept paper by the FDA is under discussion to propose some guidelines that could allow a homogenization of the techniques.[45]

Practically, the study design recommended consist in a 1 or 2 day culture period of fresh or cryopreserved hepatocytes (time of attachment) after which the cells are treated with the test compounds for an additional 2–4 day induction period. Finally, gene or protein expression or CYP specific activities are measured.[8,45,87] Usually, only CYP1A2 and 3A4, as the most frequently induced enzymes, are studied in screening programs. However, because CYP expression can be regulated by multiple pathways and that one activator of a nuclear receptor can cause the induction of several CYPs it is worth completing the panel of enzymes at latter stages of development.

Three concentrations of NCE should be tested and should span the expected human plasma drug concentrations at the therapeutic dose, with at least one concentration that is an order of magnitude greater than the average concentration.[45] It is clear that the highest concentration tested should be compatible with the solubility of the compound and should not cause cellular toxicity.

A known potent inducer (positive control causing more than twofold increase in enzyme activity) (**Table 7**), should be included in the test together with a negative control (solvent treated hepatocytes).[45] Once the CYP specific activities are measured, the final results may be expressed as induction potential (x-fold induction over control), EC_{50} (effective concentration for 50% maximal induction), or potency index (the ratio of induction response of the test compound compared to a reference inducer).[28,46] It is highly recommended to compare the results with a positive control given the high variability in induction response.[87] For example, in similar culture conditions, rifampicin can produce a 5- to 50-fold increase in the 6β-hydroxy-testosterone CYP3A4-dependent activity.[90,91]

A drug that produces a greater than twofold increase in activity or more than 40% of the positive control can be considered as an enzyme inducer and will have to be evaluated in vivo.[46]

Even if long-term cultures permit re-use of the same preparation for several additional induction studies following a washout period, the major issue of the fresh human hepatocytes still remains their availability. To avoid high variability in induction response it is recommended to use preparations from at least three individual donors[20,46] which further

Table 7 Typical chemical inducers for in vitro experiments[45]

CYP	Inducer	Inducer concentration (μM)	Induction
1A2	Omeprazole	25–100	14–24
	β-Naphthoflavone	33–50	4–23
	3-Methylcholanthrene	1.2	6–26
2A6	Dexamethasone	50	9.4
2B6	Phenobarbital	500–1000	5–10
2C8	Rifampicin	10	2–4
2C9	Rifampicin	10	3.7
2C19	Rifampicin	10	20
3A4	Rifampicin	10–50	4–31

complicates the issue of availability. This is why cryopreserved human hepatocytes have been proposed as an acceptable alternative for induction studies.[46] Some batches of these cryopreserved human hepatocytes commercially available are capable of attaching in culture and have shown to be as inducible as fresh hepatocytes in certain culture conditions.[15]

Other model such as the collagen sandwich model[92] has been considered by some authors as the most suitable for induction studies since it gets closer to reality in terms of cell polarity, structural and functional normal bile canalicular network, and cellular integrity. This model is however less suitable for early screening studies.

In addition to the more classical measurements of enzyme activity, other techniques allow the level of induction to be assessed. The increase in expression level of an enzyme can be evaluated by western blotting using enzyme selective antibodies, but this is not suitable for screening purposes.

Messenger RNA (mRNA), coding for metabolizing inducible enzymes, can be quantified by real-time techniques reverse-transcriptase polymerase chain reaction (RT-PCR) using either the Taq-Man or SIBERGREEN technologies.[11,93] Care must be taken in the interpretation of the results since an increase in the gene expression is not always mirrored by an increase in the protein activity (false positive); vice versa, a false negative can occur if the inducer affects a non-transcriptional mechanism such as activation of translation (e.g., isoniazid toward CYP2E1[94]) or protein stabilization (e.g., acetone toward rat hepatic N-nitrosodimethylamine demethylase[95]). However some rather good correlations have been obtained between real-time RT-PCR and enzyme activity results.[96]

DNA microarrays and more targeted low-density DNA microarrays dedicated to drug metabolizing enzymes and transporters are also useful tools for studying changes of CYP gene expression in response to drug treatment.[97] Such a low-density human hepatochip is in use in our laboratory allowing us to screen 151 genes (additionally including nuclear receptors, kinases, growth factors, etc.) after treatment of human fresh and cryopreserved hepatocytes (10^6 cells) with potential inducing drugs.[97] The possibility of analyzing the level of all of these genes in one shot makes these techniques nicely adapted to early research programs.

5.10.5.4.2 Genetically engineered tools

In order to overcome the problem of availability of hepatocytes and/or to make induction studies more adapted to screening programs, genetically engineered tools have become available. Even if they still necessitate further validation, they represent promising alternatives or preliminary techniques to evaluate the induction potential of an NCE in the earliest stage of drug development. These tools regroup new cell lines expressing a wider range of enzymes and transporters as well as reporter gene technologies and DNA constructs.

5.10.5.4.2.1 Hepatoma cell lines

Hepatoma cell lines have been proposed for studying CYP induction in vitro. Among the numerous cell lines developed, HepG2 is the most popular for use in induction studies. However, because of the discrepancy between this cell line and hepatocytes (*see* Section 5.10.2.2) care should be taken in the interpretation of induction results. The HepaRG line exhibits higher degrees of similarities with fresh hepatocytes and should represent a better alternative for in vitro induction studies.

In addition to HepaRG, new avenues of research have been opened, with in particular, human mesenchymal stem cells or peripheral blood monocytes that might be transformed into human hepatocytes in certain culture conditions (*see* Section 5.10.2.2).[24,25]

5.10.5.4.2.2   Reporter gene and other DNA constructs

Nuclear receptors constitute a large family of ligand-dependent DNA transcription factors. This family includes receptors such as AhR, PXR, CAR, and VDR which regulate xenobiotic metabolism through CYP.

Nuclear receptor activation assays, utilizing cell culture with transient transfection of reporter gene under the control of various nuclear receptors and subsequent assays of luciferase or µ-galactosidase activities are an alternative to assess the induction potential of a drug.[98]

These constructs have shown to be valuable tools for the assessment of CYP3A4 induction for example using PXR activation assays, with a good correlation with human hepatocytes in screening series (internal data). There are however false positives such as troleandomycin, tamoxifen, or ritonavir, which markedly activated human PXR in reporter gene assay whereas in primary cultures of human hepatocytes, pretreatment with these compounds did not significantly increase CYP3A4-dependent metabolic activity.[98]

For the other receptors, the correlation with in vivo CYP induction requires additional validation.[46] They can however be very useful for comparing activation profiles within series of compounds. The selected compounds can then be tested as regards to their induction profile with human hepatocytes. One also has to keep in mind that they do not address post-transcriptional mechanisms of induction.[87]

5.10.6   Discussion and Conclusion

Progress in optimizing the way to carry out in vitro metabolic studies in R&D has been constant over the last decade. In a way, xenobiotic metabolism represents a quite unique domain within R&D. It is effectively one of the sole disciplines present at every major step in R&D, and provides pivotal information to pharmacologists, chemists, toxicologists, and clinical scientists. It has therefore evolved toward the use of a large range of metabolic tools, allowing the rapid understanding of a drug's fate.

This nonexhaustive panel of in vitro tools has been very much used by pharmaceutical companies to better select drug candidates in high-throughput screening (HTS) programs. However, go/no-go decisions on simple parameters are somewhere an idealistic search for the (perfect) drug. The actual strategies have moved toward better evaluation, anticipation, and control of the potential safety and efficacy issues that one can, and probably will, encounter at a certain degree when administering a xenobiotic to a living organism.

These safety and efficacy aspects have been discussed in this chapter, trying to show how in vitro metabolism tools can better address, at very early stages, the nature and the importance of metabolites formed throughout species as well as the drug interactions risks.

These two issues have been recognized by regulatory authorities as major and the studies to be undertaken have been defined quite precisely for development programs and registration purposes in recent draft guidelines on safety testing of metabolites and predictions of drug interactions. These new requirements will have to be discussed and integrated by research teams. But one has to keep in mind that getting a similar level of understanding in research projects on series of drugs will only be possible if one can further adapt the tools using new technologies and apply different and more global interpretation strategies.

Screening programs have in a way broken down the important mechanisms involved in drug disposition (absorption, metabolism, etc.), trying to avoid their potential negative consequences in humans (variability, drug interactions, etc.). Today a large number of individual key parameters (solubility, permeability, rates of metabolism, inhibition and induction characteristics, etc.) or even surrogate markers (log P, parallel artificial membrane permeability assay (PAMPA), etc.) of these parameters exist to screen series of compounds.

Despite the large number of tests used routinely, a number of weak points remain to be addressed. Because these programs have been successful in selecting more stable compounds as regards CYP oxidative pathways, the role of non-CYPs has become more apparent and represents one important aspect to be developed by metabolism scientists.

Nonspecific binding to microsomes will have to be further integrated on a systematic basis in these early programs. The incubation of microsomes, in the presence of plasma or otherwise, should enable these binding issues to be simply addressed in routine tests.

The impact of gastrointestinal tract enzymes (endogenous enzymes from small intestine but also exogenous microflora enzymes from colon) on drug metabolism as well as on the risk of drug interaction is another aspect that has been neglected in early evaluations. In this respect, there are no real validated human tools as regards in vivo extrapolations. The CYP activity profile of human gut microsomes, in the presence of protease inhibitors, is variable with the time of storage with sometimes a dramatic loss in certain activities, hence a limited predictivity.

On the other hand the in vivo relevance of well-known human liver tools such as microsomes and hepatocytes is regularly questioned. The former, even though they are used as pools of several livers, are frequently not representative of the average population because these pools are difficult to constitute. An alternative, discussed in this chapter, would be to reconstruct human microsomes using expressed enzymes. This is a way to standardize microsomes, better representing the average population for all major and minor enzymes, or if needed any other type of subpopulation, i.e., integrating the polymorphic aspects of these enzymes. Hepatocytes, mainly because of their poor availability, are not really adapted to large screening programs. Cell lines as well as stem cells or medullar cells are probably going to resolve these aspects in the near future.

The domain having recently progressed the most in metabolism screening is probably the evaluation of the induction potential. Nuclear receptor activation assays, even though the exact correlation with the in vivo induction profile still has to be fully validated, are useful for evaluating the induction potential within series of compounds. Human hepatocytes, as a reference model for induction studies, can be used on smaller series of compounds at a somewhat later stage.

As regards the nature of metabolites, the identification of major metabolic pathways in drug discovery programs has always represented a significant analytical challenge. It seems that with the new technological advances, this information, very frequently asked for by chemists, pharmacologists, and toxicologists in research groups, will be available on small series of compounds. It will contribute to more precisely determine their positive (efficacy) and negative (safety) importance in nonclinical and clinical programs.

However, the future challenge is more in combining this early information, and thus reconstructing partially or totally these mechanisms at the level of the cell, the organ, or the organism.

At the cellular level coculture or coincubation systems (hepatocytes and plasma or microsomes and plasma) better integrate the different equilibrium occurring in and between the cells. Measuring individual parameters such as protein binding on one side and clearance of the drug on the other side and mathematically combining them to predict the bioavailability has proven to fail in a number of cases.

At the organ or the organism level, the use of physiologically-based pharmacokinetics (PBPK) models is a way to extrapolate in vitro data in order to get in vivo time concentration profiles. This approach allows one to challenge the multiple assumptions made when measuring in vitro parameters and extrapolating them to the entire organism (*see* 5.37 Physiologically-Based Models to Predict Human Pharmacokinetic Parameters).

On the other hand, the enormous amount of absorption, distribution, metabolism, and excretion (ADME) data generated over the last decade in HTS programs has been used to develop in silico prediction packages. Even though these in silico approaches need further refinements and validations, they will be of great value when linked to PBPK models and will probably in the future replace entire parts of HTS programs.

For drug interaction predictions, based on the inhibition and induction potential as well as knowledge of the enzyme(s) involved in the drug's metabolism, one can get a clear picture of the potential risks of coadministered drugs in the clinic, using drug interaction databases such as AurQuest,[66] DIDB,[99] drug interaction,[100] or Gentest websites.[101] Even though interlaboratory differences in kinetic parameters (K_m, V_{max}, CL_{int}, K_i, and IC_{50}) can be quite large in the literature, it is a useful source of in vitro data, especially with packages such as AurQuest integrating the calculation of drug interaction risks.

In conclusion, in vitro metabolism tools have been recognized by the recent guidelines on drug interactions as valid decision-making tools able to avoid unnecessary in vivo studies. In parallel, more information on the identity and the concentration of metabolites will potentially be available helping for a better interpretation of drug safety and efficacy programs. After almost a decade of selection of drug candidates on simple ADME parameters, the need for more global tools allowing to better combine this early information has emerged. New approaches, such as physiological models, integrating in vitro parameters, and emerging molecular modeling tools will, without doubt, be key elements in the future selection processes of drug candidates. They represent also valuable tools to optimize future clinical studies (*see* 5.34 Molecular Modeling and Quantitative Structure–Activity Relationship of Substrates and Inhibitors of Drug Metabolism Enzymes).

References

1. Riley, R. J.; Grime, K. *Drug Disc. Today* **2004**, *1*, 365–372.
2. Brandon, E. F. A.; Raap, E. F.; Meijerman, I.; Beijnen, H. H.; Schellens, J. H. *Toxicol. Appl. Pharmacol.* **2003**, *189*, 233–246.
3. Berry, M. N.; Barritt, G. J.; Edwards, A. M. High-Yield Preparation of Isolated Hepatocytes from Rat Liver. In *Laboratory Techniques in Biochemistry and Molecular Biology Isolated Hepatocytes Preparation Properties and Applications*; Burdon, R. H., van Knippenberg, P. H., Eds.; Elsevier: Amsterdam, 1991; Vol. 21, pp 24–32.

4. Puviani, A. C.; Ottolenghi, C.; Tassinari, B.; Pazzi, P.; Morsiani, E. *Comp. Biochem. Physiol. A. Mol. Integr. Physiol.* **1998**, *121*, 99–109.
5. Guyomard, C.; Chesne, C.; Meunier, B.; Fautrel, A.; Clerc, C.; Morel, F.; Rissel, M.; Campion, J. P.; Guillouzo, A. *Hepatology* **1990**, *12*, 1329–1336.
6. Gebhardt, R.; Hengstler, J. G.; Muller, D.; Glockner, R.; Buenning, P.; Laube, B.; Schmelzer, E.; Ullrich, M. et al. *Drug Metab. Rev.* **2003**, *35*, 145–213.
7. George, J.; Goodwin, B.; Liddle, C.; Tapner, M.; Farrell, G. C. *J. Lab. Clin. Med.* **1997**, *129*, 638–648.
8. LeCluyse, E. L. *Eur. J. Pharm. Sci.* **2001**, *13*, 343–368.
9. Meunier, V.; Bourrie, M.; Julian, B.; Marti, E.; Guillou, F.; Berger, Y.; Fabre, G. *Xenobiotica* **2000**, *30*, 589–607.
10. Ferrini, J. B.; Pichard, L.; Domergue, J.; Maurel, P. *Chem. Biol. Interact.* **1997**, *107*, 31–45.
11. LeCluyse, E. L.; Madan, A.; Hamilton, G.; Carroll, K.; DeHaan, R.; Parkinson, A. *J. Biochem. Mol. Toxicol.* **2000**, *14*, 177–188.
12. Treijtel, N.; Barendregt, A.; Freidig, A. P.; Blaauboer, B. J.; van Eijkeren, J. C. *Drug Metab. Dispos.* **2004**, *32*, 884–891.
13. McGinnity, D. F.; Soars, M. G.; Urbanowicz, R. A.; Riley, R. J. *Drug Metab. Dispos.* **2004**, *32*, 1247–1253.
14. Griffin, S. J.; Houston, J. B. *Drug Metab. Dispos.* **2004**, *32*, 552–558.
15. Hengstler, J. G.; Utesch, D.; Steinberg, P.; Platt, K. L.; Diener, B.; Ringel, M.; Swales, N.; Fischer, T.; Biefang, K.; Gerl, M. et al. *Drug Metab. Rev.* **2000**, *32*, 81–118.
16. Hewitt, N. J.; Hewitt, P. *Xenobiotica* **2004**, *34*, 243–256.
17. Takahashi, S.; Takahashi, T.; Mizobuchi, S.; Matsumi, M.; Morita, K.; Miyazaki, M.; Namba, M.; Akagi, R.; Hirakawa, M. *J. Int. Med. Res.* **2002**, *30*, 400–405.
18. Mace, K.; Aguilar, F.; Wang, J. S.; Vautravers, P.; Gomez-Lechon, M.; Gonzalez, F. J.; Groopman, J.; Harris, C. C.; Pfeifer, A. M. *Carcinogenesis* **1997**, *18*, 1291–1297.
19. Gomez-Lechon, M. J.; Donato, T.; Jover, R.; Rodriguez, C.; Ponsoda, X.; Glaise, D.; Castell, J. V.; Guguen-Guillouzo, C. *Eur. J. Biochem.* **2001**, *268*, 1448–1459.
20. Mills, J. B.; Rose, K. A.; Sadagopan, N.; Sahi, J.; de Morais, S. M. *J. Pharm. Exp. Ther.* **2003**, *309*, 303–309.
21. Gripon, P.; Rumin, S.; Urban, S.; Le Seyec, J.; Glaise, D.; Cannie, I.; Guyomard, C.; Lucas, J.; Trepo, C.; Guguen-Guillouzo, C. *Proc. Natl. Acad. Sci. USA* **2002**, *99*, 15655–15660.
22. Aninat, C. Expression and Regulation of Nuclear Factors, CYPs and Conjugating Enzymes in Human Hepatoma HepaRG cells. *European ISSX Meeting*, Nice, June 2005.
23. LeVee, M.; Jigorel, E.; Glaise, D.; Gripon, P.; Guguen-Guillouzo, C.; Fardel, O. Functional Expression of Drug Transporters in the Differentiated Human Hepatoma Cell Line HepaRG. *Pharmaceutical Sciences Fair and Exhibition*, Nice, June 2005.
24. Ruhnke, M.; Nussler, A. K.; Ungefroren, H.; Hengstler, J. G.; Kremer, B.; Hoeckh, W.; Gottwald, T.; Heeckt, P.; Fandrich, F. *Transplantation* **2005**, *79*, 1097–1103.
25. Lee, K. -D.; Kuo, T. K. -C.; Whang-Peng, J.; Chung, Y. -F.; Lin, C. -T.; Chou, S. -H.; Chen, J. -R.; Chen, Y. -P.; Lee, O. K. -S. *Hepatology* **2004**, *40*, 1275–1284.
26. Venkatakrishnan, K.; von Moltke, L. L.; Obach, R. S.; Greenblatt, D. J. *Curr. Drug Metab.* **2003**, *4*, 423–459.
27. Friedberg, T.; Pritchard, M. P.; Bandera, M.; Hanlon, S. P.; Yao, D.; McLaughlin, L. A.; Ding, S.; Burchell, B.; Wolf, C. R. *Drug Metab.* **1999**, *31*, 523–544.
28. Madan, A.; Usuki, E.; Burton, L. A.; Ogilvie, B. W.; Parkinson, A. Drugs and Pharmaceutical Sciences. In *Drug–Drug Interactions*; Rodriguez, A. D., Ed.; Marcel Dekker: New York, 2002; Vol. 116, pp 217–294.
29. Obach, R. S. *Drug Metab. Dispos.* **1999**, *27*, 1350–1359.
30. Obach, R. S.; Baxter, J. G.; Liston, T. E.; Silber, B. M.; Jones, B. C.; MacIntyre, F.; Rance, D. J.; Wastall, P. *J. Pharm. Exp. Ther.* **1997**, *283*, 46–58.
31. Bertrand, M.; Jackson, P.; Walther, B. *Eur. J. Pharm. Sci.* **2000**, *11* (Suppl 2.), S61–S72.
32. Obach, R. S.; Reed-Hagen, A. E. *Drug Metab. Dispos.* **2002**, *30*, 831–837.
33. Parmentier, Y.; Laisney, M.; Proust, L.; Cardona, H.; Bertrand, M.; Martinet, M.; Walther, B. The Substrate Disappearance Kinetic (SDK) Method in K_m, V_{max} and K_{ic} estimation: Limits and Benefits. European DMW Congress, Valencia: Spain, 2002.
34. Tracy, T. S. *Curr. Drug Metab.* **2003**, *4*, 341–346.
35. Hutzler, J. M.; Tracy, T. S. *Drug Metab. Dispos.* **2002**, *30*, 355–362.
36. Houston, J. B.; Kenworthy, K. E. *Drug Metab. Dispos.* **2000**, *28*, 246–254.
37. Galetin, A.; Clarke, S. E.; Houston, J. B. *Drug Metab. Dispos.* **2003**, *31*, 1108–1116.
38. Obach, R. S. *Drug Metab. Dispos.* **1997**, *25*, 1359–1369.
39. Mc Lure, J. A.; Miners, J. O.; Birkett, D. J. *Br. J. Clin. Pharmacol.* **2000**, *49*, 453–461.
40. Shibata, Y.; Takahashi, H.; Ishii, Y. *Drug Metab. Dispos.* **2000**, *28*, 1518–1523.
41. Perloff, M. D.; von Moltke, L. L.; Greenblatt, D. J. *Xenobiotica* **2003**, *33*, 365–377.
42. Neugnot, B.; Bossant, M. J.; Caradec, F.; Walther, B. Metabolic Studies in Research and Development. In *Pharmacokinetic Profiling in Drug Research: Biological, Physicochemical and Computational Strategies*; Testa, B., Kraemer, S., Wunderli-Allenspach, H., Folkers, G., Eds.; Verlag Helvetica Chimica Acta: Zürich, Switzerland, 2005, pp 79–92.
43. Pritchard, J. F.; Jurima-Romet, M.; Reimer, M. L.; Mortimer, E.; Rolfe, B.; Cayen, M. N. *Nat. Rev. Drug Disc.* **2003**, *2*, 542–553.
44. Baillie, T. A.; Cayen, M. N.; Fouda, H.; Gerson, R. J.; Green, J. D.; Grossman, S. J.; Klunk, L. J.; LeBlanc, B.; Perkins, D. G.; Shipley, L. A. *Toxicol. Appl. Pharm.* **2002**, *182*, 188–196.
45. FDA Preliminary concept paper: *Drug Interaction Studies:* Study Design, Data Analysis, and Implications for Dosing and Labeling (for discussion purposes only, 1 Oct, 2004). http://www.fda.gov/ (accessed May 2006).
46. Bjornsson, T. D.; Callaghan, J. T.; Einolf, H. J.; Fischer, V.; Gan, L.; Grimm, S.; Kao, J.; King, S. P.; Miwa, G.; Ni, L. et al. *Drug Metab. Dispos.* **2003**, *31*, 815–832.
47. FDA, *Safety Testing of Drug Metabolites*. Draft guidance for comment purposes only, June 2005, 1–11.
48. Hasting, K. L.; El-Hage, J.; Jacobs, A.; Leighton, J.; Morse, D.; Osterber, R. E. *Tox. Appl. Pharm.* **2003**, *190*, 91–92.
49. Evans, D. C.; Watt, A. P.; Nicoll-Griffith, D. A.; Baillie, T. A. *Chem. Res. Toxicol.* **2004**, *17*, 3–16.
50. Yan, Z.; Caldwell, G. W. *Anal. Chem.* **2004**, *76*, 6835–6847.
51. Fattinger, K.; Roos, M.; Vergeres, P.; Holenstein, C.; Kind, B.; Masche, U.; Stocker, D. N.; Braunschweig, S.; Kullak-Ublich, G. A.; Galeazzi, R. L. et al. *Br. J. Clin. Pharmacol.* **2000**, *49*, 158–167.
52. Routledge, P. A.; O'Mahony, M. S.; Woodhouse, K. W. *Br. J. Clin. Pharmacol.* **2004**, *57*, 121–126.

53. Lazarou, J.; Pomeranz, B. H.; Corey, P. N. *JAMA* **1998**, *279*, 1200–1205.
54. Guédon-Moreau, L.; Ducrocq, D.; Duc, M. F.; Quieureux, Y.; L'Hote, C.; Deligne, J.; Caron, J. *Eur. J. Clin. Pharmacol.* **2004**, *59*, 899–904.
55. FDA talk paper T01-34 Aug 8, **2001**, 1–2.
56. Bachmann, K. A.; Ghosh, R. *Curr. Drug Metab.* **2001**, *2*, 299–314.
57. FDA. Guidance for Industry: Drug Metabolism/Drug Interaction Studies in the Drug Development Process: Studies In Vitro. http://www.fda.gov/cder/guidance.htm (accessed May 2006).
58. Note for Guidance on the Investigation of Drug Interactions. Guidance CPMP/EWP/560/95, Dec. 1997. http://www.eudra.org/emea.html (accessed May 2006).
59. Pelkonen, O.; Maenpaa, J.; Taavitsainen, P.; Rautio, A.; Raunio, H. *Xenobiotica* **1998**, *28*, 1203–1253.
60. Wrighton, S. A.; Stevens, J. C. *Crit. Rev. Toxicol.* **1992**, *22*, 1–21.
61. Thummel, K. E.; Wilkinson, G. R *Annu. Rev. Pharmacol. Toxicol.* **1998**, *38*, 389–430.
62. Soars, M. G.; Gelboin, H. V.; Krausz, K. W.; Riley, R. J. *Br. J. Clin. Pharmacol.* **2003**, *55*, 75–181.
63. Shou, M.; Lu, T.; Krausz, K. W.; Sai, Y.; Yang, T.; Korzekwa, K. R.; Gonzalez, F. J.; Gelboin, H. V. *Eur. J. Pharmacol.* **2000**, *394*, 199–209.
64. Gelboin, H. V.; Krausz, K. W.; Gonzalez, F. J.; Yang, T. J. *Trends Phamacol. Sci.* **1999**, *20*, 432–438.
65. Trubetskoy, O. V.; Gibson, J. R.; Marks, B. D. *J. Biomol. Screen.* **2005**, *10*, 56–66.
66. Barberan, O.; Ijjaali, I.; Petitet, F.; Michel, A. Knowledge-Based Drug Profiling for ADME Properties: A Cytochrome 2D6 Application. *CHI Predictive ADME*, Jan 10–11, 2005, San Diego, CA.
67. Venkatakrishnan, K.; von Moltke, L. L.; Greenblatt, D. J. *J. Pharm. Sci.* **1998**, *87*, 845–853.
68. Nakajima, M.; Nakamura, S.; Tokudome, S.; Shimada, N.; Yamazaki, H.; Yokoi, T. *Drug Metab. Dispos.* **1999**, *27*, 1381–1391.
69. Toide, K.; Terauchi, Y.; Fujii, T.; Yamazaki, H.; Kamataki, T. *Biochem. Pharmacol.* **2004**, *67*, 1269–1278.
70. Crespi, C. L.; Miller, V. P. *Pharmacol. Ther.* **1999**, *84*, 121–131.
71. Venkatakrishnan, K.; von Moltke, L. L.; Court, M. H.; Harmatz, J. S.; Crespi, C. L.; Greenblatt, D. J. *Drug Metab. Dispos.* **2000**, *28*, 1493–1504.
72. Roy, P.; Yu, L. J.; Crespi, C. L.; Waxman, D. J. *Drug Metab. Dispos.* **1999**, *27*, 655–666.
73. Stormer, E.; von Moltke, L. L.; Greenblatt, D. J. *J. Pharm. Exp. Ther.* **2000**, *295*, 793–801.
74. Lin, J. H.; Lu, A. Y. *Clin. Pharmacokinet.* **1998**, *35*, 361–390.
75. Crespi, C. L.; Miller, V. P.; Stresser, D. M. *Methods Enzymol.* **2002**, *357*, 276–284.
76. Donato, M. T.; Jimenez, N.; Castell, J. V.; Gomez-Lechon, M. J. *Drug Metab. Dispos.* **2004**, *32*, 699–706.
77. Tran, T. H.; von Moltke, L. L.; Venkatakrishnan, K.; Granda, B. W.; Gibbs, M. A.; Obach, R. S.; Harmatz, J. S.; Greenblatt, D. J. *Drug Metab. Dispos.* **2002**, *30*, 1441–1445.
78. Ito, K.; Brown, H. S.; Houston, J. B. *Br. J. Clin. Pharmacol.* **2004**, *57*, 473–486.
79. Ito, K.; Iwatsubo, T.; Kanamitsu, S.; Ueda, K.; Suzuki, H.; Sugiyama, Y. *Pharmacol. Rev.* **1998**, *50*, 387–412.
80. Yamano, K.; Yamamoto, K.; Kotaki, H.; Takedomi, S.; Matsuo, H.; Sawada, Y.; Iga, T. *Drug Metab. Dispos.* **1999**, *27*, 1225–1231.
81. Bertelsen, K. M.; Venkatakrishnan, K.; von Moltke, L. L.; Obach, R. S.; Greenblatt, D. J. *Drug Metab Dispos.* **2003**, *31*, 289–293.
82. Jones, D. R.; Gorski, J. C.; Hamman, M. A.; Mayhew, B. S.; Rider, S.; Hall, S. D. *J. Pharmacol. Exp. Ther.* **1999**, *290*, 1116–1125.
83. Pessayre, D.; Descatoire, V.; Tinel, M.; Larrey, D. *J. Pharmacol. Exp. Ther.* **1982**, *221*, 215–221.
84. Lopez-Garcia, M. P.; Dansette, P. M.; Valadon, P.; Amar, C.; Beaune, P. H.; Guengerich, F. P.; Mansuy, D. *Eur. J. Biochem.* **1993**, *213*, 223–232.
85. Mayhew, B. S.; Jones, D. R.; Hall, S. D. *Drug Metab. Dispos.* **2000**, *28*, 1031–1037.
86. Zhou, S.; Chan, E.; Lim, L. Y.; Boelsterli, U. A.; Li, S. C.; Wang, J.; Zhang, Q.; Huang, M.; Xu, A. *Curr. Drug Metab.* **2004**, *5*, 415–442.
87. Luo, G.; Guenthner, T.; Gan, L. S.; Humphreys, W. G. *Curr. Drug Metab.* **2004**, *5*, 483–505.
88. Collste, P.; Seideman, P.; Borg, K. O.; Haglund, K.; von Bahr, C. *Clin. Pharmacol. Ther.* **1979**, *25*, 423–427.
89. Kocarek, T. A.; Schuetz, E. G.; Strom, S. C.; Fisher, R. A.; Guzelian, P. S. *Drug Metab. Dispos.* **1994**, *23*, 415–421.
90. Kostrubsky, V. E.; Ramachandran, V.; Venkataramanan, R.; Dorko, K.; Esplen, J. E.; Zhang, S.; Sinclair, J. F.; Wrighton, S. A.; Strom, S. C. *Drug Metab. Dispos.* **1999**, *27*, 887–894.
91. Wen, Y. H.; Sahi, J.; Urda, E.; Kulkarni, S.; Rose, K.; Zheng, X.; Sinclair, J. F.; Cai, H.; Strom, S. C.; Kostrubsky, V. E. *Drug Metab. Dispos.* **2002**, *30*, 977–984.
92. LeCluyse, E. L.; Madan, A.; Hamilton, G.; Caroll, K.; Dehaan, R.; Parkinson, A. *J. Biochem. Mol. Toxicol.* **2000**, *14*, 177–188.
93. Bowen, W. P.; Carey, J. E.; Miah, A.; McMurray, H. F.; Munday, P. W.; James, R. S.; Coleman, R. A.; Brown, A. M. *Drug Metab. Dispos.* **2000**, *28*, 781–788.
94. Park, K. S.; Sohn, D. H.; Veech, R. L.; Song, B. J. *Eur. J. Pharmacol.* **1993**, *248*, 7–14.
95. Song, B. J.; Veech, R. L.; Park, S. S.; Gelboin, H. V.; Gonzalez, F. J. *J. Biol. Chem.* **1989**, *264*, 3568–3572.
96. Burczynski, M. E.; McMillian, M.; Parker, J. B.; Bryant, S.; Leone, A.; Grant, E. R.; Thorne, J. M.; Zhong, Z.; Zivin, R. A.; Johnson, M. D. *Drug Metab Dispos.* **2001**, *29*, 1243–1250.
97. Meneses-Lorente, G.; de Longueville, F.; Dos Santos-Mendes, S.; Bonnert, T. P.; Jack, A.; Evrard, S.; Bertholet, V.; Pike, A.; Scott-Stevens, P.; Remacle, J.; Sohal, B. *Chem. Res. Toxicol.* **2003**, *16*, 1070–1077.
98. Luo, G.; Cunningham, M.; Kim, S.; Burn, T.; Lin, J.; Sinz, M.; Hamilton, G.; Rizzo, C.; Jolley, S.; Gilbert, D. et al. *Drug Metab. Dispos.* **2002**, *30*, 795–804.
99. Levy, R. H.; Hachad, H.; Yao, C.; Ragueneau-Majlessi, I. *Curr. Drug Metab.* **2003**, *4*, 371–380.
100. Drug interaction. http://www.drugs.com (accessed Sept 2006).
101. Gentest. http://bdbiosciences.com (accessed May 2006).

Biographies

Yannick Parmentier studied Pharmacy in Lille (Faculté des Sciences Pharmaceutiques et Biologiques – France) and molecular and cellular pharmacology in 'Ecole Normale Supérieure' of Paris VI.

He spent 1 year in the Institut de Recherches Servier at Croissy (France) to study the involvement of PPARγ nuclear receptors in the cellular differentiation of adipocytes.

In 1998, he worked 18 months in the Metabolism Department at Servier Research and Development (England) as research officer. He was in charge of the validation of the cytochrome P450 co-expressed functionally in *E. coli* with human NADPH-P450 reductase and engineered by Dundee University (Scotland).

He took on a position as Study Director in the Metabolism Department at Servier (France) in 1999 before being appointed as Section leader in the DMPK predevelopment Division in 2005 in charge of the supervision of all ADME predevelopment selection programs as well as the development of new in vitro models.

Marie-Jeanne Bossant studied Analytical Chemistry, Metabolism and Pharmacokinetics in Paris (Faculté des Sciences Pharmaceutiques Paris XI – France) and obtained his PhD on PAF-Acether: characterization and quantitation (collaboration between Paris XI and INSERM U200, Clamart – France – J Benveniste)

She spent 2 years as a postdoctoral fellow at the NIH (NIAAA Bethesda – USA – Norman Salem Jr – 1989–91) period during which, she was involved on metabolism of lipids in alcoholic patients, before taking up a position as Study Director in the Metabolism Department at Servier (France) in 1991.

She has since been involved in medical writing for registration purposes, design of metabolism databases and is currently following the predevelopment of New Chemical Entities for both their Pharmacokinetic and Metabolic properties.

Marc Bertrand studied Biology and Physiology in Clermont Ferrand (University of Clermont Ferrand – France) and then Biochemistry in Lyon (University Claude Bernard – Lyon – France) and obtained his PhD in 1987 at INSERM U171 (Lyon – Pr J-F Pujol) on central catecholaminergic systems developing an in vitro brain slice model.

He took up a position as Study Director in the Metabolism Department at Servier (Orléans – France) in June 1987, an important part of his role being to develop and implement in vitro models.

In 1991, he was appointed Head of the Metabolism Department (Orléans – France) managing both in vitro experiments such as interspecies comparisons, drug–drug interaction and in vitro based pharmacokinetic parameter predictions as well as in vivo animal and human metabolism studies for registration purposes.

From 1998 onward he took over the Drug Discovery Support Department, responsible for the generation and integration of biopharmaceutical parameters (physicochemical, metabolism, pharmacokinetics, and safety fields) within research programs as well as the development of new approaches in early discovery support (new in vitro based models, transcriptomics and in silico molecular modeling prediction tools).

Bernard Walther studied Pharmacology in Nancy (Faculté des Sciences Pharmaceutiques et Biologiques – France) and obtained his PhD at the Centre du Médicament (Nancy – Pr G Siest) on cytochrome P450: characterization and role in the brain.

He spent 3 years as a postdoctoral fellow at the School of Pharmacy, University of Lausanne (Switzerland – Pr B Testa – 1986–88) period during which, he was involved on QSAR and CNS uptake projects. He was also supervising a PhD on the hepatic and cerebral hydrolysis of nicotinic acid esters and the relation between structure and metabolism, before taking up a position as Study Director in the Metabolism Department at Servier (France) in 1988.

In 1991, he was appointed as General Manager of the Pharmacokinetic and Metabolism laboratory at Servier Research and Development (England).

From 1995 onward he took over the Pharmacokinetic and Metabolism Centre as Director of the Pharmacokinetics Centre at Technologie Servier in France (Orléans), involved in drug screening and development and more recently dedicated to early discovery and predevelopment programs as well as the implementation of new technologies.

Comprehensive Medicinal Chemistry II
ISBN (set): 0-08-044513-6

ISBN (Volume 5) 0-08-044518-7; pp. 231–257

5.11 Passive Permeability and Active Transport Models for the Prediction of Oral Absorption

P Artursson, S Neuhoff, and P Matsson, Uppsala University, Uppsala, Sweden
S Tavelin, Umeå University, Umeå, Sweden

© 2007 Elsevier Ltd. All Rights Reserved.

5.11.1 Introduction

A recent analysis of the success rates from first-in-human studies to registration during a 10 year period (1991–2000) for the 10 big pharma companies in Europe and the USA indicates an average success rate for all therapeutic classes in drug development of around 11%.[1] A closer examination identifies the major underlying causes of attrition as problems related to clinical safety, efficacy, formulation, pharmacokinetics/bioavailability, toxicology, cost of goods, and commercial reasons, and it also reveals how these underlying causes have developed from 1991 to 2000. Only in the case of pharmacokinetic/bioavailability issues can a positive development be observed, although these issues may also be incorporated within the categories of toxicology, clinical safety, and efficacy.[1] In 1991, pharmacokinetic/bioavailability issues were the most significant cause of attrition, accounting for about 40% of all attrition.[2] A decade later, the percentage of clinical drug failures that could be related to pharmacokinetic/bioavailability issues had been reduced to about 10%.[1] This development occurred in parallel with the development and implementation of cell culture-based in vitro routine tools for the prediction of ADMET properties in drug discovery, and, in particular, absorption and metabolism models. It is likely that these models have assisted in providing better grounds for making pharmacokinetics-related decisions for the selection of lead compounds in drug discovery.

What factors spurred the development of predictive cell culture models for the assessment of permeability and absorption in the ADMET area? Apart from the awareness that inferior pharmacokinetic properties, including inadequate drug absorption, were major reasons for the failure of candidate drugs in the clinical phase,[3] it was realized that drug absorption across biological barriers is a fairly complex process involving several pathways and that, because of this, it cannot easily be delineated in experimental animals.[4] Another, more concrete, reason was that the increasing output of compounds from combinatorial chemistry could not be accommodated by the conventional absorption assessment, which was based on low-throughput animal experiments. This was probably the most important reason for implementing cell culture-based ADMET models, including permeability models such as the Caco-2 model, in drug discovery settings.

About 15 years have passed since Caco-2 cells grown on permeable supports were introduced as an experimental tool for mechanistic studies of intestinal drug transport.[5–8] At that time it was also suggested that the Caco-2 model might be suitable for screening intestinal drug permeability and predicting the oral absorption potential of new drug compounds.[9] After some initial skepticism, Caco-2 cells were gradually accepted as a versatile in vitro tool for predictive and mechanistic studies of intestinal drug absorption. A period of critical evaluation followed, and nowadays the majority of researchers using Caco-2 cells and other similar models are aware not only of their advantages but also of their limitations.[4,10]

In this article, we review cell culture-based methods for the prediction of drug permeability, transport mechanisms, and oral absorption. First, we review the discrete barriers to drug absorption provided by the intestine, and discuss whether they need to be considered in in vitro permeability and transport models. Then we review the various drug transport routes across the intestinal epithelium, and discuss their relative importance. A number of cell culture models used for the in vitro prediction of intestinal permeability, transport, and absorption are discussed, as is the general methodology used in these studies and the limitations of this methodology. Many of the examples are taken from studies conducted in Caco-2 cells, simply because of the dominant position occupied by this cell line in drug permeability and transport studies for the prediction of oral absorption. However, the discussion is of a principal nature, and is, therefore, also applicable to emerging alternatives to Caco-2 cells.

5.11.2   The Role of Permeability, Transport, and Absorption Prediction in Drug Discovery

Ideally, assessment of drug permeability and transport should be used at all stages of drug discovery. In the early stages, when in silico screening of virtual or real combinatorial libraries is required, global rules or filters based on simple molecular properties, such as the 'rule of five'[11] and the polar surface area (PSA),[12–14] are the best choice of screening technique since they are cheap and rapid, and provide immediate feedback to the discovery scientists.[15,16] In contrast, during later stages of drug discovery when the number of compounds under investigation has been reduced, and the iterative process of lead optimization is taking place, the use of experimental permeability assessments should be increased in order to obtain more reliable predictions of the drug permeability after oral administration. Consequently, for maximum impact, drug permeability and transport assessment (and pharmaceutical profiling in general) should be performed simultaneously with the determination and optimization of pharmacological properties.[15–19]

5.11.3   Intestinal Drug Absorption – Solubility and Permeability Issues

In the intestine, a single layer of epithelial cells covers the interior of the intestinal wall and forms the rate-limiting barrier to the absorption of a dissolved drug. Although many other factors, and in particular the drug solubility in the intestinal fluids, may limit the bioavailability of a drug (**Figure 1**), the intestinal epithelial cell layer has proved to be the rate-limiting step to the absorption of many registered oral drug molecules (i.e., drugs for which solubility problems tend to have been sorted out during drug development). It is for this reason that it is possible to predict human intestinal absorption of druglike molecules from drug permeability studies in in vitro models consisting of a single layer of epithelial cells, such as Caco-2 cells. Whether drug permeation or drug solubility should be the assay of choice in pharmaceutical profiling should, ideally, be dependent on an educated guess of which of these two absorption barriers will be the rate-limiting step for the set of compounds of interest.[20] Simple rule-based systems based on molecular properties such as lipophilicity are generally used to this end. The physicochemical properties that influence the pharmacokinetic properties of drugs are discussed elsewhere in this volume (e.g., *see* 5.18 Lipophilicity, Polarity, and Hydrophobicity).

In many drug discovery programs, very lipophilic compounds are obtained in the effort to generate high-affinity ligands.[11] Such compounds often have severe solubility problems, and their absorption may therefore be solubility limited.[15,20,21] Because of this, in some drug discovery settings, such compounds are initially screened for solubility rather than for permeability. However, very lipophilic compounds may also distribute strongly to the lipophilic cell membranes, which may result in significantly reduced permeation.[22–24] Thus, for such drugs too, it will eventually be important to study drug permeability in in vitro models such as Caco-2. Methods for studies of drug solubility and dissolution are treated elsewhere in this volume (*see* 5.17 Dissolution and Solubility), and the specific problems associated with permeability studies of lipophilic compounds discussed elsewhere.[25]

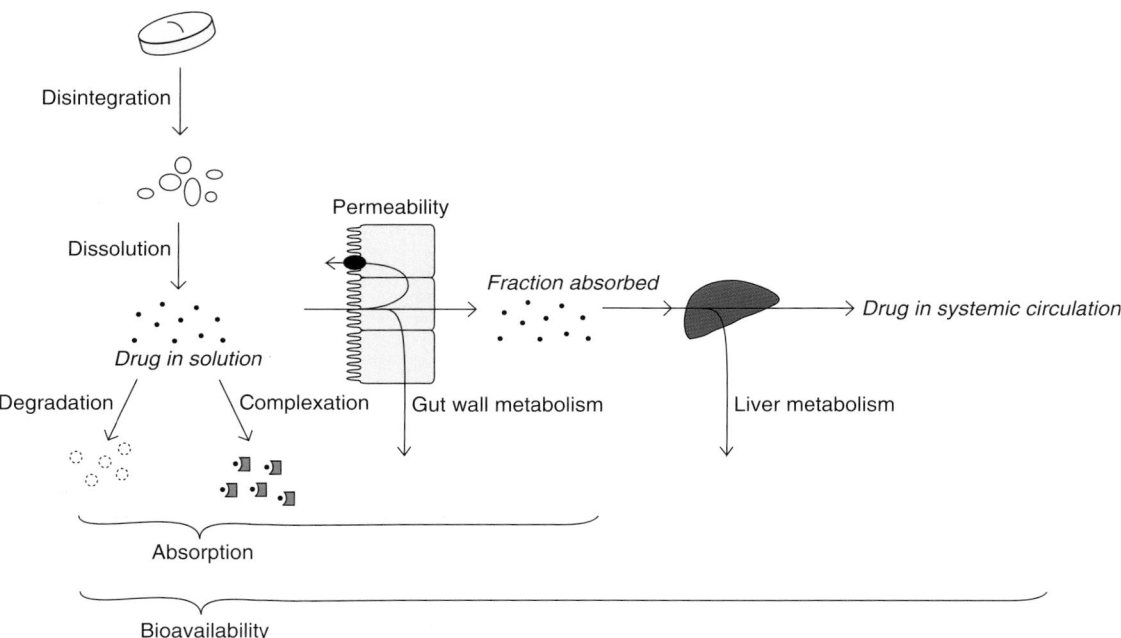

Figure 1 The sequential events during the transfer of a drug molecule from a solid dosage form in the gastrointestinal tract to the systemic circulation. The dissolution and permeability across the epithelium are the two most important determinants for the absorption process. Enzymatic and chemical degradation as well as complexation in the GI tract, active secretion, and enzymatic degradation in the intestinal epithelium, and enzymatic degradation in, and biliary excretion from, the liver may decrease the fraction of the drug entering the systemic circulation.

The relative importance of drug solubility and permeability for ensuring appropriate absorption will be dependent on the assumed dose of the druglike compound. For traditional drugs given in relatively high doses, the demands for good solubility and permeability are high. For high-affinity drugs, which can be administered in lower doses, the demands are less strict. Unfortunately, in early drug discovery the minimal dose that needs to be absorbed after oral administration in vivo is rarely known.

5.11.4 Physiological and In Vitro Barriers to Intestinal Drug Absorption

Knowledge of the physiological barriers to intestinal drug absorption is vital if one is to appreciate the opportunities and limitations that the cell culture models of the intestinal epithelium represent. The intestinal mucosa can be considered to act as a system of sequential barriers to drug absorption, the outermost barrier being the unstirred water layer (UWL) and the mucus layer.

In vivo, the luminal fluids in the intestine are considered to be well stirred.[26,27] As a result of efficient mixing, even rapidly absorbed drugs are available at the same concentration as in the bulk fluid of the intestinal lumen at the luminal surface of the mucus layer covering the intestinal epithelium. In in vitro screening models such as Caco-2, the situation is different. Without stirring, the UWL adjacent to the apical epithelial surface facing the luminal buffer compartment is rapidly depleted of high-permeability drugs, as they are quickly absorbed across the cell monolayer (**Figure 2**).[9,28,29] As a result, the drug permeability becomes limited by the (artificially thick) UWL (where the drug concentration is lower than in the bulk), rather than by the epithelial cell monolayer. Importantly, the drugs that are limited by the UWL are the ones with high-transcellular-permeability coefficients, and, therefore, in some screening settings, stirring is not considered to be necessary when the aim is binning into well and poorly absorbed compounds. This may be acceptable when it is sufficient to obtain qualitative rather than quantitative results.

However, when more accurate quantitative results are required, for example for ranking compounds, controlled stirring is mandatory.[30] When extrapolations of Caco-2 permeability data are desired to obtain a value for the permeation of, for example, the blood–brain barrier, a well-stirred system is necessary for the predictions to be of use.[18,31,32] This becomes clear if we consider that the UWL in the brain capillaries cannot, in the unstirred situation, be larger than the diameter of these capillaries (3 μm), and is probably much smaller because of the blood flow and cellular

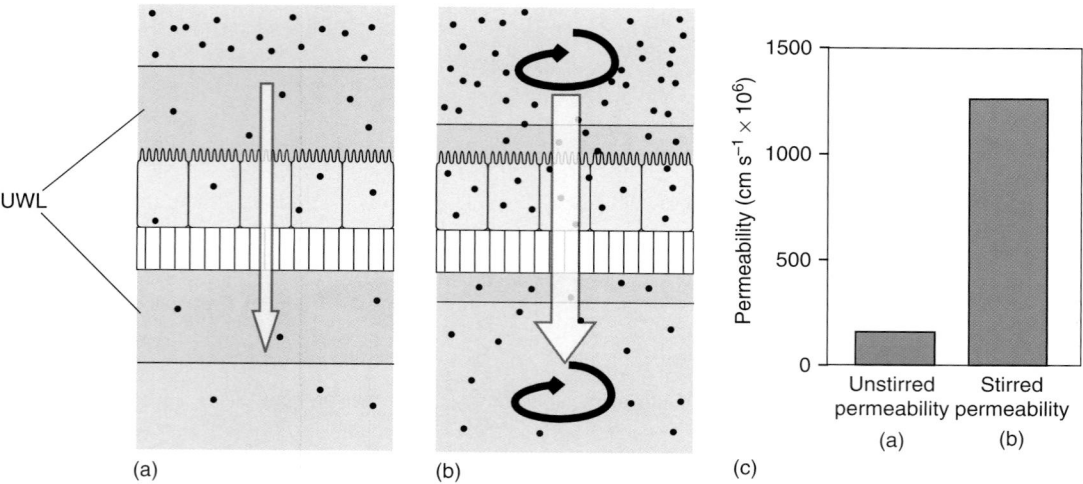

Figure 2 The effect of stirring in in vitro permeability measurements. In an unstirred system (a), drugs with high cellular permeability will be limited by their diffusion through the unstirred water layer (UWL). Under stirring conditions (b), the thickness of the UWL is reduced, and the bulk concentration will be available for diffusion through the cell monolayer. The permeability of verapamil, an example of a high-permeability drug, was measured at low and high stirring rates (c). Data redrawn from Avdeef, A.; Artursson, P.; Neuhoff, S.; Lazorova, L.; Gråsjö, J.; Tavelin, S. *Eur. J. Pharm. Sci.* **2005**, *24*, 333–349, with permission from Elsevier.

contents.[32] Another consequence of disregarding stirring is that computational models built from drug transport experiments performed under inadequate stirring conditions will be overly affected by the barrier of the UWL.

The next barrier that a permeating drug molecule encounters during its approach to the intestinal wall is the mucus layer. In general, the hydrogel-like structure of the mucus does not restrict the diffusion of normal-sized drugs.[33] However, the mucus is thought to present a barrier to the absorption of highly lipophilic drugs and some peptides because of the restrictions to diffusion in this matrix.[34] Probably, such interactions are saturable under in vivo conditions, and therefore of little significance. This conclusion is supported by the fact that the permeability of lipophilic drugs in general is high and the reduction in drug permeability caused by interactions with the mucus layer will in most cases not influence the absorbed fraction of the drug. Thus, in permeability screening of average-sized oral drugs,[35] the barrier of the mucus layer does not have to be considered. Cell culture models of the intestinal epithelial goblet cells that produce a mucus layer are available,[36,37] but their application is normally limited to specific investigations on the effects of molecular structures that are larger than those of normal drugs, such as macromolecules,[38–41] and nanoparticles.[42] Attempts to produce co-cultures of Caco-2 and HT29 goblet cells have been made, but no real advantages over standard Caco-2 cell cultures have been revealed.[43–45]

Independently of the bulk pH of the luminal fluids of the human intestine, which ranges between 5.5 and 7.5,[46] the mucus layer buffers an acid microclimate that keeps the pH adjacent to the apical surface of the enterocytes at the villus tips relatively constant. The acid microclimate is maintained by sodium–proton exchangers in combination with several HCO_3^--ion exchangers (e.g., the Cl^-/HCO_3^--exchanger) and other transporters.[47] The expression and activity of the transporters along the GI tract and along the crypt–villus axis will also influence the buffer capacity.[48] The pH adjacent to the human jejunal surface was determined to be 5.9,[49] and the surface pH of the normal human rectal mucosa was between 6.2 and 6.8.[50] The acid microclimate can easily be mimicked during permeability experiments across cell monolayers such as Caco-2, and will influence both the outcome and interpretation of such experiments.[25]

The absorptive epithelium lining the GI tract follows the folds and villi that increase the anatomical surface area of the mucosa several-fold in the small intestine.[52] The villi are interspaced with crypts in which the regeneration of intestinal cells occurs. In between the crypts and the tips of the villi are the more basal parts of the villi (**Figure 3a**). The properties that are relevant for drug absorption differ along the crypt–villus axis.

Owing to its enlarged surface area, the small intestine is considered to be the major site for passive transcellular drug absorption. The anatomical surface area of the intestinal epithelium is, however, not necessarily analogous to an effective absorptive surface area (**Figure 3b**).[53] For example, it has been calculated that the most rapidly absorbed compound known to date, glucose, has an absorptive surface area that covers only a small part of the total anatomical surface area of the small intestine (i.e., the villus tips).[27,54] By contrast, a low-permeability drug does not distribute rapidly into the cell membrane of the villus cells and therefore resides longer in the intestinal fluids.

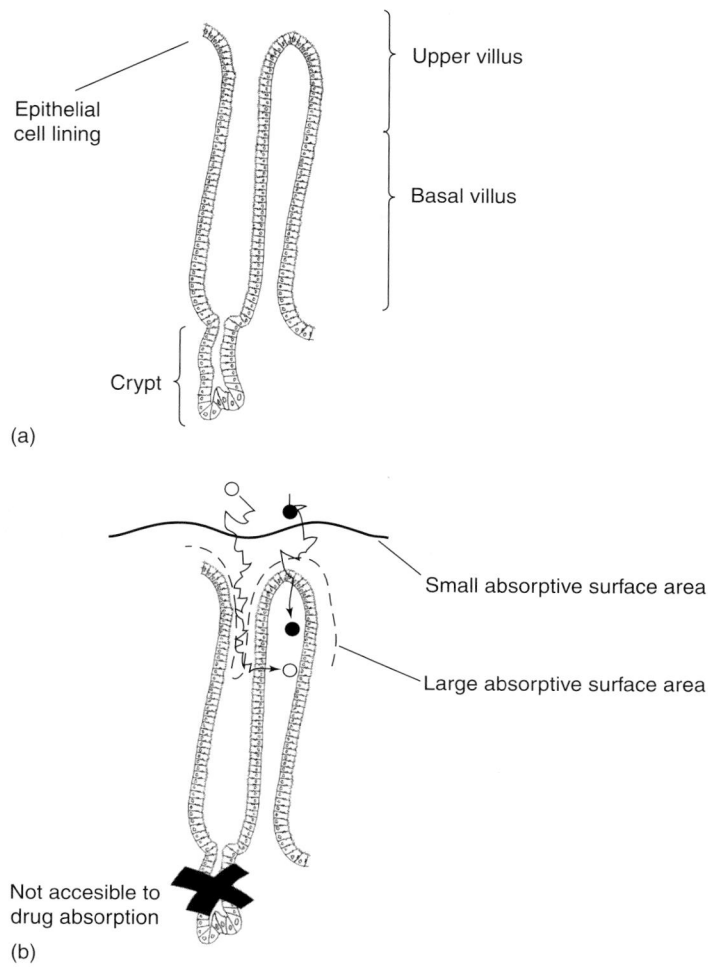

Figure 3 The crypt–villus axis and the properties of its various parts. (a) Schematic representation of the crypt–villus axis, divided into upper villus, basal villus, and crypt. (b) The effective intestinal surface area that the drug encounters will be determined by the epithelial permeability to the drug. Compounds with lower permeability coefficients (open circles) will have time to diffuse further down the crypt–villus axis, and can therefore be absorbed across a larger surface area (dashed line) than can the high-permeability compounds (solid circles) that are rapidly absorbed at the villus tips and therefore utilize a smaller absorptive surface area (solid line). The average size of the paracellular pores varies along the crypt–villus axis, with the pores becoming more narrow as the cells mature and move toward the villus tips.[51]

During this time the drug will diffuse down the length of the crypt–villus axis to be absorbed eventually over a larger absorptive surface area. Thus, the contribution of the absorptive surface area to the drug transport of the intestinal mucosa will be dependent on the permeability to the drug. The variability in absorptive surface area is an unknown parameter, and, thus, not considered in intestinal perfusion studies in animals or humans, where the absorptive surface area is treated as a constant and considered to be equal to a smooth tube of the intestinal inner diameter.[55,56]

The intestinal epithelium does not only restrict access, thereby protecting the body from harmful agents, but also allows selective absorption of nutrients and secretion of waste products and xenobiotics. The intestinal epithelium consists of several types of cells including the absorptive enterocytes and mucus-secreting goblet cells, of which the enterocytes are the most abundant.[52] At the villus tips, where the absorption of most orally administered drugs occurs, the epithelium is dominated by well-differentiated enterocytes. Caco-2 cells are unusually well differentiated, and therefore express many of the functions found in the mature small intestinal villus enterocytes, despite the fact that they originate from a human colorectal carcinoma.[57,58] In fact, so far all attempts to isolate and grow or immortalize normal intestinal epithelial cells have resulted in less differentiated phenotypes than that of Caco-2 cells.[59,60]

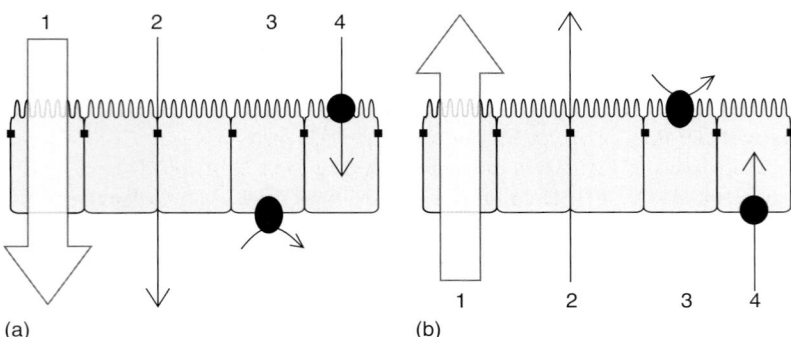

Figure 4 Transport routes across the intestinal epithelium that are relevant for the absorption of druglike compounds: 1, passive transcellular; 2, passive paracellular transport; 3, active efflux; 4, active uptake. Passive permeability can occur in both absorptive (a) and secretory (b) directions, depending on local drug concentrations. Likewise, because the membrane proteins responsible for efflux and uptake can be situated either in the apical or in the basolateral membrane, active transport can occur in both absorptive and secretory directions.

The barrier properties of the differentiated enterocyte at the villus tips is best described by an examination of the several parallel drug transport mechanisms across these cells (**Figure 4**). These mechanisms and their impact on drug absorption are discussed in the next section.

5.11.5   Transport Mechanisms Across the Intestinal Epithelial Barrier

There are two general pathways by which a drug molecule can cross the intestinal epithelium: the transcellular pathway, which requires that the drug can cross the lipophilic cell membranes, and the paracellular pathway, in which diffusion occurs through the water-filled pores of the tight junctions between the cells. Both passive and active transport may contribute to the permeability of drugs via the transcellular pathway (and also via the paracellular pathway). The passive and active transcellular pathways are inherently different, and therefore the molecular properties that influence drug transport by these routes are also different. The four principal drug transport pathways are presented in **Figure 4**. Note that all of the transport processes may occur in both directions, depending on local drug concentrations (in the case of passive transport) and the location and directionality of the involved transporter (in the case of active transport). The various transport mechanisms across the intestinal epithelium are discussed below.

5.11.5.1   Passive Transcellular Transport

The passive transcellular pathway is still by far the most important absorption pathway for drugs. Drug transport via the passive transcellular route requires that the solute permeates the apical cell membrane. The composition of the phospholipids and proteins of the cell membranes varies from cell type to cell type, and may, theoretically, give rise to different permeability properties depending on the cell type. In addition, monolayer-forming cells such as the intestinal enterocytes have a polarized cell membrane with distinct differences in membrane composition in the apical and the basolateral membrane.[61] It is generally believed that the apical membrane has a lower permeability than the basolateral membrane, and therefore the former is considered to be the rate-limiting barrier to passive transcellular drug transport.[62]

Early models of the transcellular pathway regarded the cell membrane as a homogeneous barrier, and thereby considered drug permeability to be a process of partitioning into the lipid bilayer followed by diffusion across the membrane.[63] This 'solubility–diffusion' model was later extended with the so-called pH-partitioning theory, which states that only the uncharged form of an electrolytic molecule will partition into the cell membrane.[64] One consequence of the assumption, that drug partitioning into the membrane is a one-step process, is that the cell membrane is considered to be an isotropic system. This assumption explains why drug partitioning in simple isotropic solvent systems, such as octanol–water, has been extensively used to predict passive membrane permeability (e.g., *see* 5.19 Artificial Membrane Technologies to Assess Transfer and Permeation of Drugs in Drug Discovery; 5.28 In Silico Models to Predict Oral Absorption).[65,66] However, the protein-containing lipid bilayer that constitutes the cell membrane is an anisotropic system.[67] Multiple-step permeability models have therefore been proposed to account for the anisotropic nature of the cell membrane.[68] Molecular dynamics simulations have shown that a drug molecule entering the cell membrane will experience different diffusion rates in different parts of the membrane,[69–71] and, perhaps more importantly, different inter-molecular forces will affect the permeating substance depending on the local environment in the lipid bilayer.[67,69–74]

Independently of whether the solubility–diffusion model or a more complex description of the cell membrane is used, passive transcellular drug permeability can usually be relatively well described using rather simple molecular properties such as size, charge, polarity, and hydrogen bonding (*see* 5.19 Artificial Membrane Technologies to Assess Transfer and Permeation of Drugs in Drug Discovery; 5.28 In Silico Models to Predict Oral Absorption).[15,17] Many experimental and theoretical predictive models of intestinal drug permeability have therefore been based on descriptors relating to these basic molecular properties (*see* 5.28 In Silico Models to Predict Oral Absorption).[15,17] Importantly, these models assume that the passive transcellular route dominates the drug permeation. However, when other transport routes, such as the paracellular or active transcellular routes, are involved to a significant extent, they will introduce a bias that reduces the predictivity of such models. Likewise, significant intestinal metabolism will be a complicating factor.

5.11.5.2   Paracellular Transport

Drugs of low to moderate molecular weight (MW) can permeate the intestinal epithelium through the water-filled pores between the cells. This process is known as paracellular transport, and is generally considered to be a passive process, even if this pathway appears to be selective for cationic rather than anionic and neutral drugs.[62,75,76] The paracellular pathway has also been shown to be saturable[77–79] by at least two independent mechanisms, one of which involves an intracellular process.[79] These examples illustrate the complexity and dynamics of the regulation of this pathway, which had previously been considered to be invariable.

The paracellular route is guarded by extracellular tight junction proteins that create narrow water-filled pores between the cells. The narrow pores of the tight junction complex, which form a seal at the most apical part of the intercellular space, constitute the rate-limiting step of the paracellular pathway. This route is, therefore, mainly accessible to smaller molecules, and is of significance only if these small drugs are too hydrophilic to distribute into the cell membranes at appreciable rates. If one considers the large difference in surface area between the membrane surface of intestinal epithelial cells and that of the intercellular spaces between the cells, it may be surprising to find that the paracellular route is of any importance at all. The issue remains controversial, and different opinions are held about its significance, ranging from it being a route that significantly contributes to drug absorption to it being of negligible importance.[75,80,81]

Drugs believed to be significantly absorbed by the paracellular route include the beta receptor antagonist atenolol (MW: 266),[82] the H_2 receptor antagonists cimetidine (MW: 252), ranitidine (MW: 314), and famotidine (MW: 337),[78,79] and the loop diuretic furosemide (MW: 331).[83] These drugs are all of moderate MW (around 250–340), and are relatively hydrophilic. Paracellularly absorbed drugs are usually incompletely absorbed since the absorption only occurs during the passage through the small intestine where the pores are sufficiently large to allow the passage of small solutes. In addition, the paracellular pores cover only 0.01–0.1% of the total surface area of the intestine,[84,85] and the size-restricting gate function of the paracellular pathway significantly limits the permeability.[86] The paracellular permeability is dynamically regulated, and varies both along the path of the intestine and along the crypt–villus axis.[51,87–89] Quite specific ways of bringing about pharmacological regulation of the paracellular pathway have been proposed, with the intention of enhancing the absorption of polar drugs, but fear of inducing nonspecific simultaneous absorption of toxic compounds from the gut has reduced interest in pursuing this approach.[77,90–94]

The paracellular pathway of the small intestine appears to be a dual-pore system, with larger pores further down the crypt–villus axis[51,95] Given this, it can be hypothesized that a hydrophilic drug that is partitioning slowly into the intestinal cell membrane will have time to diffuse further down the crypt–villus axis (**Figure 3b**) and eventually be absorbed via paracellular pores that are significantly larger than those available at the villus tips. Thus, such compounds may exhibit higher in vivo absorption than expected from their passive transcellular permeability.

5.11.5.3   Active Transport

Global gene expression studies on intestinal tissues show that more than 100 transport proteins and ion channels are expressed at the mRNA level in the gut.[96,97] While most of these transporters are highly specialized and mediate facilitated transport of essential nutrients, an increasing number have been found to have more general transport functions. These transporters, which have broader substrate specificity, are often referred to as polyspecific transport proteins, and have the capacity to transport drug molecules. Comparison with Caco-2 cells has revealed that, although

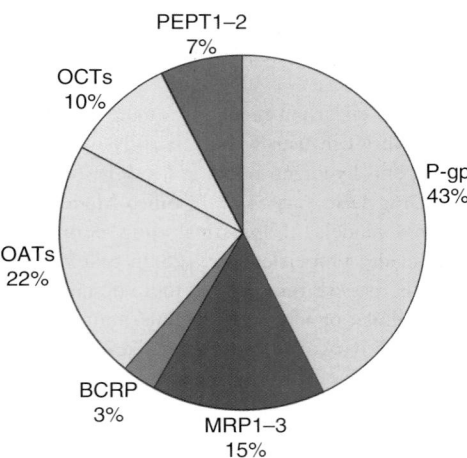

Figure 5 Comparison of the number of compounds reported to have affinity for the major drug transporters (OCT, organic cation transporter; OAT, organic anion transporter; PEPT, peptide transporter of the intestine; other transporters defined in the text). The data were extracted from the publicly available transporter database from Tokyo University.[99] Note that the numbers reflect the extent to which the transporters have been studied and not their relative importance as drug transporters. In total, 994 compounds were listed in the database in May 2005.

the majority of the transporter genes had detectable expression, the expression levels generally differed from those in human intestinal tissue. However, the relevance of the reported mRNA expression levels will remain uncertain until more data on protein expression and function has become available. In at least one case, a clear correlation between transporter mRNA expression, protein expression, and protein function was observed in Caco-2 cells.[98] Importantly, for the majority of drugs, it is still not known which transporters of all those possible are of significance for their intestinal absorption.

An examination of a database containing data assembled from the literature on substrates/inhibitors for various transporters indicates how the research community has scored their relevance for drug transport (**Figure 5**).[99] Clearly, the extensively studied efflux protein P-glycoprotein (P-gp) has the highest number of substrates/inhibitors, followed by other efflux transporters of the ATP-binding cassette (ABC) transporter family, various transporters for organic anions, cations, and the oligopeptide transporter PEPT1.[99] This rating can probably be explained by the role of ABC transporters in drug resistance,[100,101] brain uptake,[102,103] and drug–drug interactions,[104–106] and by the importance of ABC transporters and organic anion and cation transporters for drug transport and elimination in the liver, kidney, and, perhaps, the intestine,[107] and by the exploitation of PEPT1 as a (pro)drug target for enhanced absorption.[108]

Examples of successful exploitation of PEPT1, which is responsible for the uptake of dietary small peptides from the intestine, include the modification of the antiviral drug acyclovir to the PEPT1 substrate valacyclovir, which increased the bioavailability from 22% to 77%.[109] A potential drawback with the exploitation of drug transporters for increased drug absorption could be the occurrence of complex drug–drug interactions. For instance, the bioavailability of cefixime, which is another substrate for PEPT1, was increased by the calcium channel blocker nifedipine,[110] an effect attributed to calcium ion effects on proton-dependent transporters such as sodium–proton exchangers that maintain the apical acid microclimate necessary for PEPT1 function.[111,112] Further, successful improvement of drug uptake with the aid of a transporter such as PEPT1 will not necessarily result in increased absorption since the compound may be trapped and degraded inside the cell.[108,113]

In contrast to transport proteins that mediate uptake into the cells, other transport proteins expel their substrates out of the cells. These so-called efflux proteins belong to the ATP-binding cassette (ABC) super-family of membrane transporters. The ABC gene family has about 50 members, and includes the multidrug resistance 1 (MDR1, ABCB1) gene product P-gp and the multidrug resistance associated protein family (MRP, ABCC).[114,115] Fairly recently the homodimeric breast cancer resistance protein (BCRP, ABCG2) has been identified as a potential contributor to actively limiting the oral bioavailability of some drugs.[116,117] The function of the efflux proteins in the intestine may be to prevent the uptake of toxic substances, and also to eliminate such substances from the blood.[106,118,119]

While the significance of drug efflux proteins such as P-gp and BCRP are undisputed in barriers such as the endothelial blood–brain barrier,[120] which is exposed to comparably low and often nonsaturating concentrations of efflux substrates, their role in limiting drug absorption in the small intestine, which is exposed to much higher, often saturating drug concentrations, is disputed.[121–123] In one case, a polymorphism in P-gp was suggested to influence intestinal absorption of digoxin,[124] but the relevance of this finding has been questioned.[125] In particular, the effects of concomitantly administered drugs on the P-gp substrate digoxin has been studied.[105,106,126–128] Interestingly, the serum levels of orally administered digoxin have been shown to increase in a stepwise manner with the number of P-gp inhibitors co-administered to patients, suggesting that several weaker P-gp inhibitors may have additive effects in the clinical setting.[104] Thus, an emerging concern is that drug–drug interactions may be more pronounced in the clinical setting, where co-administration of several P-gp-interacting drugs is a reality.

5.11.5.4 Complexity of Drug Transport

Absorption processes involving transport proteins are generally saturable. Because of this, drugs that are substrates for an active transport protein can display a nonlinear dose–response relationship. In addition, many of these proteins transport nutrients, and therefore their capacity is likely to be influenced by food intake.[129] These factors may complicate the oral delivery of drugs that are absorbed by active mechanisms.[130]

A further complication arises because one drug may be a substrate for several transporters. For instance, the drug may be taken up into the cell by one transporter and effluxed by another.[131–133] Moreover, an interplay between drug transport and drug metabolism may occur.[134–138] Last, but not least, active transport components often occur in parallel with passive diffusion processes, the latter being the most significant process. For instance, the pharmacokinetics of the partly actively transported drug levodopa was dramatically improved by maximizing the passive diffusion of the drug across the intestinal wall.[139,140] Thus, in summary, an intricate picture has emerged where several transport processes may act in parallel or in series with each other in the transfer of a drug across the intestinal epithelium. It is very difficult to clarify such multifaceted processes in the complex intestinal tissue. They are more easily studied in the simple cell culture models of epithelial cells, where each transport process can be examined in isolation. Thus, it is possible to extract information about specific transport processes in Caco-2 cells and alternative epithelial cell cultures that would be difficult to obtain in more complex systems, such as models based on whole tissues from experimental animals.

5.11.6 Permeability Assay

Transport studies in Caco-2 cells and in other monolayer-forming cells are generally performed using an experimental setup in which the cells are grown on a porous filter insert, although some variants of the setup have been suggested. For instance, continuous-dissolution–Caco-2 permeability systems[141–143] and an automated continuous-perfusion system have been suggested.[144] In the standard setup, the porous support covers the bottom of a plastic cell culture insert, which forms a small compartment to which the cell culture medium is added. The insert is placed in a slightly larger cell culture well, to form a second compartment that is filled with cell culture medium. The medium is then changed to a buffer suitable for transport experiments, and the drug is added to one of the compartments, called the donor compartment. The amount of transported drug is determined from samples withdrawn from the other compartment, called the acceptor compartment. Detailed descriptions of the cell culture procedures for Caco-2 cells (and other cell lines) and more general descriptions of how to perform drug transport experiments in these cell cultures are available in handbooks on cell culture methodologies.[145–148]

This experimental setup has several unique features that are attractive for permeability assessment. If the filter is sufficiently porous, it will not form a rate-limiting barrier to drug permeation. This means that true intrinsic cellular permeability coefficients can be obtained.[149] If the system is well stirred, so that stagnant water layers are avoided,[150] the measured passive transcellular permeability coefficients are in quantitative agreement with those measured in the perfused human jejunum.[151] In addition, since the apical and basolateral surface areas are approximately equal, the passive transcellular permeability will be equal in the absorptive and secretory directions.[152] Similarly, the paracellular surface areas are comparable, and, hence, paracellular surface markers such as mannitol generally have the same permeability in both directions.

Thus, in principle, it can be revealed that an active transport process is involved if a clear difference is observed between the permeabilities in the two opposing directions. Therefore, a common choice in the screening of intestinal permeability is to measure the permeability in both directions across the cell monolayers, and determine the efflux or uptake ratio. The efflux ratio is calculated as the ratio of the secretory permeability to the absorptive permeability ($P_{app, B \to A}/P_{app, A \to B}$), and the uptake ratio as the inverse of the efflux ratio. An efflux or uptake ratio significantly larger or smaller than unity is generally taken to be an indication of an active transport process.[153,154]

A complicating issue in the assessment of permeability is that many investigators prefer to perform permeability studies using a slightly acid pH on the apical side of the monolayer. There are two explanations for this: first, this mimics the acid microclimate on the apical side of the enterocytes in the small intestine, and, second, independent studies have suggested that slightly better correlations were obtained with in vivo data under these conditions.[96,155,156] However, since most drugs have protonizable functionalities, the asymmetrical pH condition will result in a different concentration of the uncharged drug species on each side of the monolayer.[25] In turn, this will result in passive asymmetric transport, that is, the absorptive permeability will be different from the secretory permeability without any transporter being involved.[157] Another complication is that access to the transporter may differ between the apical and basolateral sides.[158,159]

The use of an effective inhibitor of the transporter may be an alternative to studies of asymmetrical transport. In this setup, the drug permeability in the absorptive direction is studied in the absence and presence of the inhibitor, and, if the resulting permeability ratio ($P_{app, -inhibitor}/P_{app, + inhibitor}$) is significantly different from 1, an active transport mechanism may be involved. Indeed, when such an inhibition ratio was used in vitro to predict the effect of P-gp on plasma drug levels in mice, a better in vivo prediction was obtained than for efflux ratios obtained from measurements of asymmetrical transport.[160]

5.11.6.1   Quality of Data and Data Sets

It is not possible to obtain good predictions of human intestinal permeability and absorption from cell culture models if the data generated are of a poor quality. Nevertheless, higher-throughput permeability data, often based on just one or a few samples per compound, are produced in many screening settings. If appropriately applied, though, such data are still useful for binary classification of groups of compounds or even whole chemical libraries in earlier stages of drug discovery. However, more recently, the value of such data sets has been questioned due to the improvements seen in in silico predictions of drug permeability.

Caco-2 data sets can also serve as reference data sets for the evaluation of computational permeability filters. Permeability data from Caco-2 cells have been successfully used to evaluate computational screening filters that extracted compound sets enriched in high-permeability compounds from chemical libraries.[161]

One result of the fact that the permeability data generated in drug discovery are often obtained from rapid, more or less binary screening assays is that the data are of insufficient quality to allow development of the good predictive relationships needed in later drug discovery phases. Unfortunately, the literature contains very few reliable data sets of sufficient size even though increased information sharing is predicted to be beneficial for academia, government, and industry alike.[162] This deficiency of high-quality published data sets has resulted in repetitive and sometimes uncritical use of poor data in predictions and computer modeling of drug absorption. Thus, in the absence of large data sets of high quality, relatively small data sets comprising reliable and reproducible experimental data have been used.[30,163]

One bias that is evident in in vitro models (and, indeed, in in silico models) intended to predict drug absorption is that the data sets are generally over-represented by completely absorbed compounds.[164] This is probably a consequence of the data sets being based on orally administered marketed drugs, most of which have favorable oral absorption properties. Despite their bias, models based on such skewed data sets can often distinguish between high- (>80% absorbed) and low-permeability compounds, and rapidly give useful information about the expected average permeability of compound libraries. However, compounds with moderate absorption properties may also be of interest in the early phases of drug discovery, and it is therefore advantageous to include such compounds in the evaluation of permeability models.[164] The need to include a significant number of incompletely absorbed compounds in the data set is emphasized during lead optimization, when ranking compounds according to their permeability or their predicted absorption properties is desirable. The fact that the data sets are generally based on passively transported compounds and that no consideration is given to the potential influence of active transport processes is indicative of another weakness. Although this approach gives better models of passive permeability, these models only generate selective information about the passive route, and their performance in the screening of data sets that include actively transported compounds should be evaluated.

5.11.6.2   Prediction of In Vivo Drug Permeability and Absorption Using Caco-2 Cells and Alternative Models

Initially, the good relationship between the passive drug transport across Caco-2 cells and the fraction absorbed after oral administration to humans[9] may be surprising, given that oral drug absorption is influenced by many factors besides drug permeability, such as drug solubility, dissolution, active transport, and, in some cases, presystemic metabolism.[165] The first study conducted with Caco-2 cells was performed under highly controlled conditions on registered drugs that were known to be free from solubility problems and which were mostly transported by passive transport.[9] Furthermore, their metabolism could be accounted for. Later investigations have revealed that results of a similar standard can be obtained when the same parameters are strictly controlled in expanded data sets (**Figure 6a**, solid symbols).[30]

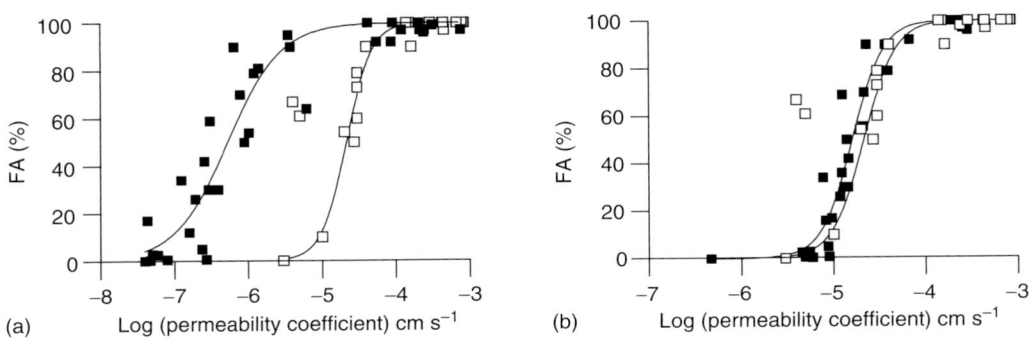

Figure 6 Comparison of permeability coefficients obtained from human jejunal perfusion experiments (open symbols)[166,167] and in Caco-2 cells (a, solid symbols)[30,168] and 2/4/A1 cells (b, solid symbols).[168] The 2/4/A1 cell line gives permeability coefficients that overlap the human jejunal permeability coefficients. FA is the fraction of a dose that is absorbed after oral administration. (Reproduced from Artursson, P.; Tavelin, S. Caco-2 and Emerging Alternatives for Prediction of Intestinal Drug Transport: A General Overview. In *Drug Bioavailability – Estimation of Solubility, Permeability, Absorption and Bioavailability*; Artursson, P., Ed.; Wiley-VCH: Weinheim, Germany, 2003, pp 72–89, with permission from Wiley-VCH.)

However, early on, many drug discovery scientists were disappointed when their experimental in-house compounds gave rise to relationships between FA and permeability with a much larger scatter than reported in the original publication. With hindsight, these results are not surprising considering the fact that the majority of the discovery compounds had neither been characterized nor optimized with regard to the chemical stability, metabolism, solubility, or (if available in crystalline form) dissolution rate. Another factor contributing to this scatter is related to the experimental design. For small data sets, Caco-2 permeability measurements are generally performed manually, with full attention being given to, for example, issues related to mass balance and the contribution from active transport. In contrast, permeability measurements for large data sets are normally established in automated systems, with little consideration being given to mass balance and active transport. Despite these simplifications, there are several reports that have indicated the usefulness of permeability data obtained in automated systems in predictions of oral absorption[161,169–172] and drug efflux.[173] The implementation of Caco-2 cells in the 96-well format was reported fairly recently.[174] Unfortunately, miniaturization to plate formats below 96-well plates has been technically difficult to achieve, because of the small surface area of the cell monolayers in such formats.[19] Also, discovery scientists have experienced larger problems with the mass balance in this format, in that larger fractions of lipophilic compounds are sometimes lost during transport experiments. In addition to automating the experimental runs, another way to increase the efficiency of permeability screening is to use cassette dosing (*n*-in-one). Independent investigations have shown that the dosing of several compounds to each cell monolayer gives fairly good permeability data, but concerns have been expressed about the impact of potential drug–drug interactions (e.g., on transporter inhibition), and, in one case, interactions between acidic compounds were observed.[175–177] Analytical methods for cassette dosing of sets of reference compounds sets have been developed.[178–180] An extreme but useful variant of cassette dosing is to study partly purified extracts from, for example, plants or food supplements in the cell monolayers.[181]

One example of the wide-spread acceptance of absorption prediction from in vitro permeability measurements in Caco-2 cells and similar cell lines is that the US Food and Drug Administration (FDA) has approved drug permeability measurements in these cell lines as substitutes for in vivo investigations in bioequivalence studies of well-absorbed drugs.[182,183] Thus, according to the FDA-implemented biopharmaceutics classification system (BCS), it is possible to waive FDA bioequivalence studies for high-solubility and high-permeability drugs (*see* 5.42 The Biopharmaceutics Classification System). Ways of expanding the BCS by allowing a broader definition of high-solubility/high-permeability drugs or of other classes of drugs have been discussed.[184–186] The need to consider the dose in the biopharmaceutical classification has also been investigated, and was found to be particularly important for solubility-limited drugs.[187] The classification of druglike compounds according to the BCS is increasingly applied in drug discovery, and modifications to the scheme have been suggested with this in mind.[163,188]

It is noteworthy that the Caco-2-based relationship in **Figure 6a** was established using a structurally diverse data set and that analogous series of compounds often give better relationships. Consequently, Caco-2 cells have often been used to rank series and libraries of analogous compounds. When such data sets are used, permeability measurements from Caco-2 cells and other cell monolayers provide the opportunity to establish structure–permeability relationships for quite different analogous series of drugs. Several examples of the latter have been published. For example, these include series of conventional drugs,[5,189,190] peptides, and peptide mimetics[74,191–193] as well as newer types of druglike compounds.[194–197] Although most of these structure–permeability relationships have been established for passive membrane permeability, there are also many examples of structure–transport relationships for series of drugs that are absorbed by means of an active transport mechanism.[198,199]

5.11.6.3   Alternatives to Caco-2 Cells

Over the years, several alternatives to Caco-2 cells have been suggested for the assessment of drug permeability. In fact, MDCK cells, which are epithelial cells derived from dog kidney, were the first cells grown as monolayers on a permeable support,[200] and were proposed as a cellular drug transport barrier as early as 1989.[201] However, while MDCK cells were exploited as host cells for transfected transport proteins,[202] it was not until later that their performance was compared with that of Caco-2 cells with regard to predictions of oral drug absorption in humans.[203] The results revealed that the two cell lines were comparable for the investigated data set, and the Papp values from the two models correlated relatively well ($r^2 = 0.79$).[204] Other alternative cell lines include the Caco-2 clone TC7, a cell line which, in contrast to parenteral Caco-2 cells, also expresses some drug-metabolizing enzymes.[205,206] Drug permeabilities determined in Caco-2TC7 cells result in fairly good correlations with the absorbed drug fraction after oral administration to humans, as do other monolayer-forming intestinal epithelial cell lines, such as the mucus-producing cell clone HT29-MTX.[207]

In contrast to high-permeability drugs, which partition into the cell membranes at comparable, rapid speeds in vitro and in vivo, compounds of intermediate or low permeability have a lower permeability in the Caco-2 model than in the human small intestine in vivo.[151,208,209] As can be seen in **Figure 6a**, this difference increases with a decrease in the compound permeability. One reason for this difference is that the paracellular route is tighter in Caco-2 cells than that in the small intestine in vivo. While the average pore radius of the tight junctions in the human small intestine is around 8–13 Å,[210] the corresponding radius in Caco-2 cells is about 4 Å.[168] When the paracellular pathway is narrower, the intrinsic permeability will be lower than in the in vivo situation. The contribution of the paracellular pathway to the overall transport of one lipophilic high-permeability drug (alfentanil) and one hydrophilic low-permeability drug (cimetidine) was investigated in Caco-2 cells.[81,211]

A more leaky cell culture model established from the rat fetal intestine, 2/4/A1, was suggested as a solution to this problem.[208,212] This cell line, which has a paracellular permeability comparable to that of the human small intestinal epithelium in vivo (pore radius 9 Å), gives a better quantitative relationship than Caco-2 cells with human permeability data generated in the Loc-I-Gut perfusion technique (**Figure 6b**).[164,168,208] Interestingly, the 2/4/A1 cell line does not seem to functionally express important drug transporters,[213] which makes it an interesting alternative in studies of passive permeability. Another advantage of the 2/4/A1 cell line is that relatively large amounts of low-permeability drugs are transported, thereby eliminating the need for expensive analytical equipment such as liquid chromatography with tandem mass spectrometry.

In summary, passive permeability measurements conducted not only in Caco-2 cells but also in other monolayer-forming epithelial cell lines seem to predict oral drug absorption of registered drugs. The question then arises if even simpler models, such as monolayers of the less differentiated and therefore less complex 2/4/A1 cells, or even artificial membranes consisting of phospholipids (parallel artificial membrane permeability assay, PAMPA[214]), or of organic solvents such as hexadecane membrane (HDM[215]), can rank compounds with regard to permeability or oral drug absorption equally well as the more complex cell culture models. A recent study investigated how the more leaky small intestinal-like 2/4/A1 cells compared with Caco-2 cells, which have tighter tight junctions and express many functional transport proteins, and with HDM, which lack tight junctions completely and have no transport proteins.[164] The completely absorbed compounds in this study (fraction absorbed >80%), were well predicted by all three models. This was most likely because all of these compounds utilized the transcellular route, which was available in all models. However, when incompletely absorbed compounds were also included in the data set (fraction absorbed = 30–80%), the two cell culture models, which in addition to the transcellular route also have a paracellular route, performed much better. For instance, the Spearman rank order correlation (r_s) with fraction absorbed was around 0.75 for the cell cultures and 0.47 for the artificial membranes when a data set of compounds evenly distributed with regard to intestinal absorption (FA = 0–100%) was used.[164] Thus, cellular permeability models that take the paracellular route into consideration seem to perform better than those that do not.

In the study comparing Caco-2, 2/4/A1, and HDM permeability data, the significance of active transport in permeability screening was challenged, since half of the compound data set of 30 compounds was reported to be at least partly actively transported.[164] Neither 2/4/A1 cells nor the artificial membranes exhibit significant transport activities, and, with the possible exception for some ABC transporters,[117] the transporter expression in Caco-2 cells does not correspond quantitatively to that found in the human intestine.[96] Thus, it was expected that compounds that were transported actively to a significant extent would be outliers. However, the influence of active transport was not clear-cut. For instance, only three actively transported compounds, digoxin, a substrate for MDR1, GlySar, a substrate for PEPT1, and methotrexate, a substrate for the reduced folate transporter, were clear outliers in the correlation based on 2/4/A1 cell permeabilities while all other actively transported compounds were predicted equally well as the passively transported compounds. When the three outliers were removed, the rank order correlation with the fraction absorbed increased to 0.95. While these results bring hope for the prediction of oral absorption from permeability data, they require an explanation. The most likely reason is that active transport is insignificant in comparison to the strongly dominant passive component and that passive permeability dominates even more at the early phase of absorption, when the oral dosage form has been dissolved and a very high concentration of the drug is presented to the intestinal wall.[139,140] Despite these results, which were obtained with registered oral drug products, it is clear that in many cases the impact of active drug transport is more pronounced for many current discovery compounds, which often have higher receptor affinities and, hence, are intended for administration at lower, non-saturating doses. In vitro tools for studies of active drug transport will therefore most likely become a more important tool in ADMET research.

5.11.6.4 Prediction of Active Drug Transport

The functional expression of many relevant drug transporters in Caco-2 cells may complicate the interpretation of the results. For instance, when druglike compounds are screened in Caco-2 cells, they are usually used at low, micromolar range, concentrations. At such low concentrations, many active drug transport processes might not be saturated. The results from Caco-2 cells may, for example, then suggest that, at micromolar range concentrations, efflux systems significantly limit the permeability of the drug at the same time that simple calculations may show that the drug when administered to humans will be presented to the epithelium of the upper small intestine at a much higher concentration (millimolar range) than that used in the screening.[15,216] Since the transport mechanism may be saturated at higher concentrations, the drug efflux may sometimes be considered as an in vitro artifact that will not be relevant in the in vivo situation, at least for compounds that are administered at higher doses or that have high protein binding.

Another complicating factor in performing in vitro prediction of the active transport in vivo is that there is no quantitative relationship between active drug transport in the cell culture models and in vivo.[151,217] The explanation for this may be that the level at which the transporter is expressed in Caco-2 cells is not comparable to that in vivo.[117,218] In addition, studies of active transport are further complicated by the absence of specific substrates. For example, most substrates for the drug efflux transporters can be transported by several transport proteins.[219] A look into the listed substrates/inhibitors in the transporter database provided by the University of Tokyo reveals that a large fraction of the listed compounds interact with several transporters; for example, taurocholic acid was reported to interact with 14 different transporters.[99] Whether the compounds listed as interacting with a single transporter are indeed specific for that particular transporter remains to be shown, since most compounds have only been studied with a single or a few transporters.

In the absence of specific substrates, epithelial cell culture models, such as LLC-PK1 (a monolayer-forming epithelial cell line originating from the pig kidney) and MDCK, which have a low background expression of endogenous transporters and which can sort many transfected transporters to the correct membrane, have been stably transfected to express high levels of a specific transporter, as discussed in numerous references.[220–224] These models have become increasingly popular for the characterization of active transport mechanisms, since, in theory, the requirement for specific substrates should be lower in these models. Unfortunately, epithelial cell lines such as MDCK and LLC-PK1 also in fact do express functional endogenous transporters at levels that can complicate the interpretation of the results.[223,225–227] A comparative study of the performance of Caco-2 cells and P-gp-transfected cell lines for evaluating the in vivo function of P-gp showed that all cell models performed well, although with slightly better results being achieved with the transfected cell lines.[228] Similarly as for passive permeability screening[174] the implementation of Caco-2 cells in the 96-well plate format has been reported for P-gp interaction studies.[19]

A further simplification in the prediction of active transport is to use adherent or suspension cell lines that are either transiently or stably transfected with the transporter of choice. A commonly used adherent cell line that is easily transfected with various transporter genes is HEK293. This cell line is derived from human embryonic kidney, and has been extensively used in the characterization of various uptake carriers.[229–231] One example of a commonly used suspension cell line is CHO (derived from Chinese hamster ovaries), which has been used to study the uptake characteristics for a variety of transporters.[232–234] Surprisingly, the endogenous 'background' expression of transporters in the various expression systems is generally unknown. An exception is the Saos-2 cell line (derived from a human osteosarcoma), which was shown to lack significant expression of at least a few important ABC drug transporters, and was therefore used as a relatively clean expression system for BCRP,[235] an ABC transporter of increasing significance.[119,236] Importantly, suspension cell lines, as well as adherent cell lines such as HEK293, cannot be used as models for drug transport across epithelial barriers, since they do not form monolayers. Rather, they have found their application in the characterization of active drug uptake into or efflux out of cells by means of the expressed transporter.

Other assays for transporter interaction, such as ATP consumption by transporters of the ABC transporter family, have been used to assess active drug transport.[237,238] In the case of P-gp, two studies compared the performance of assays based on active transport across cell monolayers, interaction studies using a fluorescent substrate (rhodamine 123 or calcein) and ATP consumption.[153,154] In the first study, about half of the investigated compounds exhibited concordance between the assays. The compounds that differed between the assays were divided into those that were and those that were not actively effluxed across MDCK cell monolayers expressing P-gp. In general, compounds that had lower permeabilities were actively effluxed, while those having higher permeabilities were not. In contrast, the calcein and ATPase assays revealed P-gp interactions for the more highly permeable compounds, while these assays were less responsive to compounds exhibiting lower permeabilities.[153] On the basis of these findings, the authors recommended that the efflux assay across MDCK-MDR1 cells should be used as the primary screen (despite its lower throughput), since P-gp-mediated efflux is more relevant for in vivo drug disposition when the passive permeability is low, and this assay was the only one responding properly to such compounds. It was also suggested that an efflux ratio as low as 1.5 should be taken as an indication of P-gp involvement. The second study focused on interaction assays using rhodamin 123 and calcein as fluorescent probes, and drug efflux studies across P-gp-expressing LLC-PK1 cells were only performed on a subset of the compounds.[154] The authors concluded that the more indirect fluorescence-based interaction assays should be used as the primary screen, and that these assays should be followed by an ATPase assay, but with the reservation that the results from the primary screens should be followed up by efflux studies across cell monolayers at later stages of the drug development. As in the first study, an efflux ratio of 1.5 was taken as an indication of a significant active efflux. The authors also used cell systems expressing P-gp homologs from different species, and observed clear differences in P-gp interactions between the species.

One problem that has emerged with the transgenic expression systems is that the active transport of some drugs can be mediated by two (or more) transporters simultaneously or in series.[239] This problem becomes particularly evident when polar compounds that are substrates for efflux transporters are to be studied, and may explain some of the unexpected findings in the comparative studies discussed above. Since the substrate-binding sites of important efflux proteins such as P-gp are suggested to be located in the inner leaflet of the plasma membrane or directly adjacent to the cytosolic compartment, more polar drugs may need an uptake transporter to gain access to the transporter. In this context, it was reported that the active transport of pravastatin, a substrate for both the basally located organic anion transporting polypeptide 2 (OATP2) and the apically located drug efflux transporter MRP2 in the liver, could only be studied in MDCK cells double transfected with these two transporters while no significant efflux of the drug was observed in MDCK cells expressing MRP2 only,[240] since the substrate was not able to reach the transporter-binding site. Another recent example is the compound CCK-8, which was found to be transported by both OATP8 and MRP2 across double-transfected MDCK cells.[241] Thus, transgenic cell lines overexpressing a single transporter may be inadequate when more complex drug transport mechanisms are to be clarified. Unfortunately, it is difficult to generate multiple-transfected cell lines that retain the required differentiated properties. It will also be very demanding to maintain a whole panel of double-transfected MDCK cells and the necessary control cell lines, including mock-transfected cells and single-transfected cells, although such efforts are underway in some laboratories. Caco-2 cells may, therefore, remain a viable alternative in the future in some of these situations, provided that sufficiently specific substrates or inhibitors can be identified.

5.11.7   Conclusions

We can conclude that technologies are now available for the study of drug permeability and transport in vitro with good precision. However, sound predictions of drug absorption rely on additional data, such as information on the drug solubility in the intestinal fluids and in the acid microclimate adjacent to the intestinal epithelium. Furthermore, better knowledge of the expression and function of the increasing number of drug-transporting proteins is required. Only through the combination of physicochemical and biological data will new and improved predictions of oral absorption, and even bioavailability eventually, be obtained.

Caco-2 cells remain a versatile and general model for the screening of drug permeability and for the prediction of oral absorption, provided that the absorption is not solubility limited. The comparable performance of similar cell culture models in the prediction of drug absorption suggests that fine tuning of permeability protocols may not be critical for all applications. Techniques have been developed to account for in vitro artifacts in the cell culture monolayers, as have ways to minimize and compensate for the variability in the permeability assays. Further optimization of the active transport studies is needed; for instance, with regard to the use of transport ratios or inhibitors in primary screening of active transport.

The rich selection of functional transporters in Caco-2 cells sometimes makes it difficult to study passive permeability or specific active transport processes in isolation. New, alternative cell models that exhibit fewer drug transport pathways may be preferable in situations where specific drug transport mechanisms are to be studied. Thus, one current trend is to develop alternative models to Caco-2 cells by adopting a reductionist approach, either by isolating the study of specific transporters to various transgenic cell lines or by isolating the study of passive transport using cell lines in which transporters are not functionally expressed.

Successful evaluation of cellular models for studies of passive permeability and active transport is dependent on data sets of high quality, obtained with compounds that represent the appropriate part of the chemical space. Unfortunately, such data sets are rare. It is the conviction of the authors that technology development would be driven to the benefit of all if such data sets were to be constructed through cooperation between academia, government, and the drug industry.

Acknowledgments

This work was supported by grant No. 9478 from the Swedish Research Council, the Knut and Alice Wallenberg Foundation, the Swedish Fund for Research without Animal Experiments, and the Swedish Animal Welfare Agency.

References

1. Kola, I.; Landis, J. *Nat. Rev. Drug Disc.* **2004**, *3*, 711–715.
2. Kennedy, T. *Drug Disc. Today* **1997**, *2*, 436–444.
3. Prentis, R. A.; Lis, Y.; Walker, S. R. *Br. J. Clin. Pharmacol.* **1988**, *25*, 387–396.
4. Artursson, P.; Borchardt, R. T. *Pharm. Res.* **1997**, *14*, 1655–1658.
5. Artursson, P. *J. Pharm. Sci.* **1990**, *79*, 476–482.
6. Hilgers, A. R.; Conradi, R. A.; Burton, P. S. *Pharm. Res.* **1990**, *7*, 902–910.
7. Hidalgo, I. J.; Raub, T. J.; Borchardt, R. T. *Gastroenterology* **1989**, *96*, 736–749.
8. Wilson, G.; Hassan, I. F.; Dix, C. J.; Williamson, I.; Shah, R.; Mackay, M.; Artursson, P. *J. Control. Release* **1990**, *11*, 25–40.
9. Artursson, P.; Karlsson, J. *Biochem. Biophys. Res. Commun.* **1991**, *175*, 880–885.
10. Gumbleton, M.; Audus, K. L. *J. Pharm. Sci.* **2001**, *90*, 1681–1698.
11. Lipinski, C. A.; Lombardo, F.; Dominy, B. W.; Feeney, P. J. *Adv. Drug Deliv. Rev.* **1997**, *23*, 3–25.
12. Palm, K.; Stenberg, P.; Luthman, K.; Artursson, P. *Pharm. Res.* **1997**, *14*, 568–571.
13. Kelder, J.; Grootenhuis, P. D.; Bayada, D. M.; Delbressine, L. P.; Ploemen, J. P. *Pharm. Res.* **1999**, *16*, 1514–1519.
14. Ertl, P.; Rohde, B.; Selzer, P. *J. Med. Chem.* **2000**, *43*, 3714–3717.
15. Stenberg, P.; Bergström, C. A.; Luthman, K.; Artursson, P. *Clin. Pharmacokinet.* **2002**, *41*, 877–899.
16. Kerns, E. H.; Di, L. *Drug Disc. Today* **2003**, *8*, 316–323.
17. Artursson, P.; Matsson, P. Absorption Prediction. In *Pharmaceutical Profiling in Drug Discovery for Lead Selection*; Wang, B., Ed.; AAPS Press: Arlington, VA, 2004, pp 3–26.
18. van de Waterbeemd, H. *Basic Clin. Pharmacol. Toxicol.* **2005**, *96*, 162–166.
19. Balimane, P. V.; Chong, S. *Drug Disc. Today* **2005**, *10*, 335–343.
20. Lipinski, C. A. *J. Pharmacol. Toxicol. Methods* **2000**, *44*, 235–249.
21. Bergström, C. A. *Basic Clin. Pharmacol. Toxicol.* **2005**, *96*, 156–161.
22. Aungst, B. J.; Nguyen, N. H.; Bulgarelli, J. P.; Oates-Lenz, K. *Pharm. Res.* **2000**, *17*, 1175–1180.
23. Sawada, G. A.; Barsuhn, C. L.; Lutzke, B. S.; Houghton, M. E.; Padbury, G. E.; Ho, N. F.; Raub, T. J. *J. Pharmacol. Exp. Ther.* **1999**, *288*, 1317–1326.
24. Wils, P.; Warnery, A.; Phung-Ba, V.; Legrain, S.; Scherman, D. *J. Pharmacol. Exp. Ther.* **1994**, *269*, 654–658.
25. Department of Pharmacy, Uppsala University, Uppsala, Sweden: Drug Delivery; http://www.farmfak.uu.se/farm/lmformul-web/ (accessed April 2006).
26. Lennernäs, H. *J. Pharm. Sci.* **1998**, *87*, 403–410.
27. Levitt, M. D.; Fine, C.; Furne, J. K.; Levitt, D. G. *J. Clin. Invest.* **1996**, *97*, 2308–2315.
28. Karlsson, J.; Artursson, P. *Biochim. Biophys. Acta* **1992**, *1111*, 204–210.
29. Naruhashi, K.; Tamai, I.; Li, Q.; Sai, Y.; Tsuji, A. *J. Pharm. Sci.* **2003**, *92*, 1502–1508.
30. Stenberg, P.; Norinder, U.; Luthman, K.; Artursson, P. *J. Med. Chem.* **2001**, *44*, 1927–1937.
31. Faassen, F.; Vogel, G.; Spanings, H.; Vromans, H. *Int. J. Pharm.* **2003**, *263*, 113–122.
32. Avdeef, A.; Nielsen, P. E.; Tsinman, O. *Eur. J. Pharm. Sci.* **2004**, *22*, 365–374.
33. A. Wikman Larhed, *Comprehensive Summaries of Uppsala Dissertations from the Faculty of Pharmacy*; Uppsala University: Uppsala, Sweden, 1997, No. 162.
34. Larhed, A. W.; Artursson, P.; Gråsjö, J.; Björk, E. *J. Pharm. Sci.* **1997**, *86*, 660–665.
35. Leeson, P. D.; Davis, A. M. *J. Med. Chem.* **2004**, *47*, 6338–6348.
36. Wikman, A.; Karlsson, J.; Carlstedt, I.; Artursson, P. *Pharm. Res.* **1993**, *10*, 843–852.
37. Behrens, I.; Stenberg, P.; Artursson, P.; Kissel, T. *Pharm. Res.* **2001**, *18*, 1138–1145.

38. Keely, S.; Rullay, A.; Wilson, C.; Carmichael, A.; Carrington, S.; Corfield, A.; Haddleton, D. M.; Brayden, D. J. *Pharm. Res.* **2005**, *22*, 38–49.
39. Schipper, N. G.; Varum, K. M.; Stenberg, P.; Ocklind, G.; Lennernäs, H.; Artursson, P. *Eur. J. Pharm. Sci.* **1999**, *8*, 335–343.
40. Lesuffleur, T.; Roche, F.; Hill, A. S.; Lacasa, M.; Fox, M.; Swallow, D. M.; Zweibaum, A.; Real, F. X. *J. Biol. Chem.* **1995**, *270*, 13665–13673.
41. Phillips, T. E.; Huet, C.; Bilbo, P. R.; Podolsky, D. K.; Louvard, D.; Neutra, M. R. *Gastroenterology* **1988**, *94*, 1390–1403.
42. Norris, D. A.; Puri, N.; Sinko, P. J. *Adv. Drug Deliv. Rev.* **1998**, *34*, 135–154.
43. Wikman Larhed, A.; Artursson, P. *Eur. J. Pharm. Sci.* **1995**, *3*, 171–183.
44. Walter, E.; Janich, S.; Roessler, B. J.; Hilfinger, J. M.; Amidon, G. L. *J. Pharm. Sci.* **1996**, *85*, 1070–1076.
45. Hilgendorf, C.; Spahn-Langguth, H.; Regårdh, C. G.; Lipka, E.; Amidon, G. L.; Langguth, P. *J. Pharm. Sci.* **2000**, *89*, 63–75.
46. Fallingborg, J.; Christensen, L. A.; Ingelman-Nielsen, M.; Jacobsen, B. A.; Abildgaard, K.; Rasmussen, H. H. *Aliment. Pharmacol. Ther.* **1989**, *3*, 605–613.
47. Chang, E. B.; Rao, M. C. Intestinal Water and Electrolyte Transport. Mechanism of Physiological and Adaptive Responses. In *Physiology of the Gastrointestinal Tract*; Johnson, R., Ed.; Raven Press: New York, 1994.
48. Daniel, H.; Neugebauer, B.; Kratz, A.; Rehner, G. *Am. J. Physiol.* **1985**, *248*, G293–G298.
49. Lucas, M. L.; Cooper, B. T.; Lei, F. H.; Johnson, I. T.; Holmes, G. K.; Blair, J. A.; Cooke, W. T. *Gut.* **1978**, *19*, 735–742.
50. McNeil, N. I.; Ling, K. L.; Wager, J. *Gut.* **1987**, *28*, 707–713.
51. Fihn, B. M.; Sjöqvist, A.; Jodal, M. *Gastroenterology* **2000**, *119*, 1029–1036.
52. Madara, J. L.; Trier, J. S. The Functional Morphology of the Mucosa of the Small Intestine. In *Physiology of the Gastrointestinal Tract*, 3rd ed.; Johnson, R., Ed.; Raven Press: New York, 1994, pp 1577–1622.
53. Strocchi, A.; Levitt, M. D. *Dig. Dis. Sci.* **1993**, *38*, 385–387.
54. Kawaguchi, A. L.; Dunn, J. C.; Lam, M.; O'Connor, T. P.; Diamond, J.; Fonkalsrud, E. W. *J. Pediatr. Surg.* **1998**, *33*, 1670–1673.
55. Komiya, I.; Park, J. K.; Kamani, A.; Ho, N. F. H.; Higuchi, W. I. *Int. J. Pharm.* **1980**, *4*, 249–262.
56. Lennernäs, H.; Ahrenstedt, O.; Hallgren, R.; Knutson, L.; Ryde, M.; Paalzow, L. K. *Pharm. Res.* **1992**, *9*, 1243–1251.
57. Zweibaum, A.; Laburthe, M.; Grasset, E.; Louvard, D. Use of Cultured Cell Lines in Studies of Intestinal Cell Differentiation and Function, In: *Handbook of Physiology, Section 6. The Gastrointestinal System*; Schultz, S., Ed.; American Physiological Society: Bethesda, MD, 1991, pp 223–255.
58. Fogh, J.; Fogh, J. M.; Orfeo, T. *J. Natl. Cancer Inst.* **1977**, *59*, 221–226.
59. Quaroni, A.; Beaulieu, J. F. *Gastroenterology* **1997**, *113*, 1198–1213.
60. Pageot, L. P.; Perreault, N.; Basora, N.; Francoeur, C.; Magny, P.; Beaulieu, J. F. *Microsc. Res. Tech.* **2000**, *49*, 394–406.
61. van Meer, G.; Simons, K. *EMBO J.* **1986**, *5*, 1455–1464.
62. Reuss, L. Tight Junction Permeability to Ions and Water. In *Tight Junctions*, 2nd ed.; Anderson, J., Ed.; CRC Press: Boca Raton, FL, 2001, pp 61–88.
63. Collander, R. *Physiol. Plant.* **1954**, *7*, 420–445.
64. Shore, P.; Brodie, B.; Hogben, C. *J. Pharmacol. Exp. Ther.* **1957**, *119*, 361–369.
65. Testa, B.; Carrupt, P. A.; Gaillard, P.; Billois, F.; Weber, P. *Pharm. Res.* **1996**, *13*, 335–343.
66. Avdeef, A.; Testa, B. *Cell. Mol. Life Sci.* **2002**, *59*, 1681–1689.
67. Marrink, S. J.; Berendsen, H. J. C. *J. Phys. Chem.* **1994**, *98*, 4155–4168.
68. Burton, P. S.; Conradi, R. A.; Hilgers, A. R.; Ho, N. F. H.; Maggiora, L. L. *J. Control. Release* **1992**, *19*, 87–97.
69. Stouch, T. R.; Alper, H. E.; Bassolino, D. *Comput. Aided Mol. Des.* **1995**, *589*, 127–138.
70. Bassolino, D.; Alper, H.; Stouch, T. R. *Drug Des. Disc.* **1996**, *13*, 135–141.
71. Bemporad, D.; Essex, J. W.; Luttmann, C. *J. Phys. Chem. B.* **2004**, *108*, 4875–4884.
72. Jacobs, R. E.; White, S. H. *Biochemistry* **1989**, *28*, 3421–3437.
73. Marrink, S. J.; Berendsen, H. J. C. *J. Phys. Chem.* **1996**, *100*, 16729–16738.
74. Goodwin, J. T.; Conradi, R. A.; Ho, N. F.; Burton, P. S. *J. Med. Chem.* **2001**, *44*, 3721–3729.
75. Adson, A.; Raub, T. J.; Burton, P. S.; Barsuhn, C. L.; Hilgers, A. R.; Audus, K. L.; Ho, N. F. H. *J. Pharm. Sci.* **1994**, *83*, 1529–1536.
76. Karlsson, J.; Ungell, A. L.; Gråsjö, J.; Artursson, P. *Eur. J. Pharm. Sci.* **1999**, *9*, 47–56.
77. Gan, L. -S. L.; Yanni, S.; Thakker, D. R. *Pharm. Res.* **1998**, *15*, 53–57.
78. Lee, K.; Thakker, D. R. *J. Pharm. Sci.* **1999**, *88*, 680–687.
79. Zhou, S. Y.; Piyapolrungroj, N.; Pao, L.; Li, C.; Liu, G.; Zimmermann, E.; Fleisher, D. *Pharm. Res.* **1999**, *16*, 1781–1785.
80. Lennernäs, H. *Pharm. Res.* **1995**, *12*, 1573–1582.
81. Nagahara, N.; Tavelin, S.; Artursson, P. *J. Pharm. Sci.* **2004**, *93*, 2972–2984.
82. Adson, A.; Burton, P. S.; Raub, T. J.; Barsuhn, C. L.; Audus, K. L.; Ho, N. F. *J. Pharm. Sci.* **1995**, *84*, 1197–1204.
83. Flanagan, S. D.; Takahashi, L. H.; Liu, X.; Benet, L. Z. *J. Pharm. Sci.* **2002**, *91*, 1169–1177.
84. Nellans, H. *Adv. Drug Deliv. Rev.* **1991**, *7*, 339–364.
85. Pappenheimer, J. R.; Reiss, K. Z. *J. Membr. Biol.* **1987**, *100*, 123–136.
86. Madara, J. L. *Ann. Rev. Physiol.* **1998**, *60*, 143–159.
87. Madara, J. L.; Marcial, M. A. *Kroc. Found. Ser.* **1984**, *17*, 77–100.
88. Artursson, P.; Ungell, A. L.; Löfroth, J. E. *Pharm. Res.* **1993**, *10*, 1123–1129.
89. Fihn, B. M.; Jodal, M. *Pflugers Arch.* **2001**, *441*, 656–662.
90. Wong, V.; Gumbiner, B. M. *J. Cell Biol.* **1997**, *136*, 399–409.
91. Ouyang, H.; Morris-Natschke, S. L.; Ishaq, K. S.; Ward, P.; Liu, D.; Leonard, S.; Thakker, D. R. *J. Med. Chem.* **2002**, *45*, 2857–2866.
92. Ward, P. D.; Klein, R. R.; Troutman, M. D.; Desai, S.; Thakker, D. R. *J. Biol. Chem.* **2002**, *277*, 35760–35765.
93. Tavelin, S.; Hashimoto, K.; Malkinson, J.; Lazorova, L.; Tóth, I.; Artursson, P. *Mol. Pharmacol.* **2003**, *64*, 1530–1540.
94. Kondoh, M.; Masuyama, A.; Takahashi, A.; Asano, N.; Mizuguchi, H.; Koizumi, N.; Fujii, M.; Hayakawa, T.; Horiguchi, Y.; Watanbe, Y. *Mol. Pharmacol.* **2005**, *67*, 749–756.
95. Söderholm, J. D.; Olaison, G.; Kald, A.; Tagesson, C.; Sjödahl, R. *Dig. Dis. Sci.* **1997**, *42*, 853–857.
96. Sun, D.; Lennernäs, H.; Welage, L. S.; Barnett, J. L.; Landowski, C. P.; Foster, D.; Fleisher, D.; Lee, K. D.; Amidon, G. L. *Pharm. Res.* **2002**, *19*, 1400–1416.
97. Bates, M. D.; Erwin, C. R.; Sanford, L. P.; Wiginton, D.; Bezerra, J. A.; Schatzman, L. C.; Jegga, A. G.; Ley-Ebert, C.; Williams, S. S.; Steinbrecher, K. A.; Warner, B. W.; Cohen, M. B.; Aronow, B. J. *Gastroenterology* **2002**, *122*, 1467–1482.
98. Taipalensuu, J.; Tavelin, S.; Lazorova, L.; Svensson, A. C.; Artursson, P. *Eur. J. Pharm. Sci.* **2004**, *21*, 69–75.

99. University of Tokyo, Tokyo, Japan: Transporter Database; http://www.tp-search.jp (accessed April 2006).

100. Huang, Y.; Anderle, P.; Bussey, K. J.; Barbacioru, C.; Shankavaram, U.; Dai, Z.; Reinhold, W. C.; Papp, A.; Weinstein, J. N.; Sadee, W. *Cancer Res.* **2004**, *64*, 4294–4301.

101. Annereau, J. P.; Szakacs, G.; Tucker, C. J.; Arciello, A.; Cardarelli, C.; Collins, J.; Grissom, S.; Zeeberg, B. R.; Reinhold, W.; Weinstein, J. N.; Pommier, Y.; Paules, R. S.; Gottesman, M. M. *Mol. Pharmacol.* **2004**, *66*, 1397–1405.

102. Schinkel, A. H.; Smit, J. J. M.; van Tellingen, O.; Beijnen, J. H.; Wagenaar, E.; van Deemter, L.; Mol, C. A. A. M.; van der Valk, M. A.; Robanus-Maandag, E. C.; te Riele, H. P. J.; Berns, A. J. M.; Borst, P. *Cell* **1994**, 77, 491–502.

103. Sadeque, A. J.; Wandel, C.; He, H.; Shah, S.; Wood, A. J. *Clin. Pharmacol. Ther.* **2000**, *68*, 231–237.

104. Englund, G.; Hallberg, P.; Artursson, P.; Michaelsson, K.; Melhus, H. *BMC Med.* **2004**, *2*, 8.

105. Fromm, M. F.; Kim, R. B.; Stein, C. M.; Wilkinson, G. R.; Roden, D. M. *Circulation* **1999**, *99*, 552–557.

106. Drescher, S.; Glaeser, H.; Murdter, T.; Hitzl, M.; Eichelbaum, M.; Fromm, M. F. *Clin. Pharmacol. Ther.* **2003**, *73*, 223–231.

107. Mizuno, N.; Niwa, T.; Yotsumoto, Y.; Sugiyama, Y. *Pharmacol. Rev.* **2003**, *55*, 425–461.

108. Våbenø, J.; Lejon, T.; Nielsen, C. U.; Steffansen, B.; Chen, W.; Ouyang, H.; Borchardt, R. T.; Luthman, K. *J. Med. Chem.* **2004**, *47*, 1060–1069.

109. Steingrimsdottir, H.; Gruber, A.; Palm, C.; Grimfors, G.; Kalin, M.; Eksborg, S. *Antimicrob. Agents Chemother.* **2000**, *44*, 207–209.

110. Duverne, C.; Bouten, A.; Deslandes, A.; Westphal, J. F.; Trouvin, J. H.; Farinotti, R.; Carbon, C. *Antimicrob. Agents Chemother.* **1992**, *36*, 2462–2467.

111. Wenzel, U.; Kuntz, S.; Diestel, S.; Daniel, H. *Antimicrob. Agents Chemother.* **2002**, *46*, 1375–1380.

112. Thwaites, D. T.; Kennedy, D. J.; Raldua, D.; Anderson, C. M.; Mendoza, M. E.; Bladen, C. L.; Simmons, N. L. *Gastroenterology* **2002**, *122*, 1322–1333.

113. Steffansen, B.; Nielsen, C. U.; Frokjær, S. *Eur. J. Pharm. Biopharm.* **2005**, *60*, 241–245.

114. Borst, P.; Evers, R.; Kool, M.; Wijnholds, J. *J. Natl. Cancer Inst.* **2000**, *92*, 1295–1302.

115. Tanigawara, Y. *Ther. Drug Monit.* **2000**, *22*, 137–140.

116. Jonker, J. W.; Smit, J. W.; Brinkhuis, R. F.; Maliepaard, M.; Beijnen, J. H.; Schellens, J. H.; Schinkel, A. H. *J. Natl. Cancer Inst.* **2000**, *92*, 1651–1656.

117. Taipalensuu, J.; Törnblom, H.; Lindberg, G.; Einarsson, C.; Sjöqvist, F.; Melhus, H.; Garberg, P.; Sjöström, B.; Lundgren, B.; Artursson, P. *J. Pharmacol. Exp. Ther.* **2001**, *299*, 164–170.

118. Hoffman, T. L.; Canziani, G.; Jia, L.; Rucker, J.; Doms, R. W. *Proc. Natl. Acad. Sci. USA* **2000**, *97*, 11215–11220.

119. Jonker, J. W.; Merino, G.; Musters, S.; van Herwaarden, A. E.; Bolscher, E.; Wagenaar, E.; Mesman, E.; Dale, T. C.; Schinkel, A. H. *Nat. Med.* **2005**, *11*, 127–129.

120. Eisenblatter, T.; Huwel, S.; Galla, H. J. *Brain Res.* **2003**, *971*, 221–231.

121. Chiou, W. L.; Chung, S. M.; Wu, T. C.; Ma, C. *Int. J. Clin. Pharmacol. Ther.* **2001**, *39*, 93–101.

122. Lin, J. H.; Yamazaki, M. *Clin. Pharmacokinet.* **2003**, *42*, 59–98.

123. Faassen, F.; Vromans, H. *Clin. Pharmacokinet.* **2004**, *43*, 1117–1126.

124. Hoffmeyer, S.; Burk, O.; von Richter, O.; Arnold, H. P.; Brockmoller, J.; Johne, A.; Cascorbi, I.; Gerloff, T.; Roots, I.; Eichelbaum, M.; Brinkmann, U. *Proc. Natl. Acad. Sci. USA* **2000**, *97*, 3473–3478.

125. Morita, Y.; Sakaeda, T.; Horinouchi, M.; Nakamura, T.; Kuroda, K.; Miki, I.; Yoshimura, K.; Sakai, T.; Shirasaka, D.; Tamura, T.; Aoyama, N.; Kasuga, M.; Okumura, K. *Pharm. Res.* **2003**, *20*, 552–556.

126. Hager, W. D.; Mayersohn, M.; Graves, P. E. *Clin. Pharmacol. Ther.* **1981**, *30*, 594–599.

127. Rodin, S. M.; Johnson, B. F.; Wilson, J.; Ritchie, P.; Johnson, J. *Clin. Pharmacol. Ther.* **1988**, *43*, 668–672.

128. Westphal, K.; Weinbrenner, A.; Giessmann, T.; Stuhr, M.; Franke, G.; Zschiesche, M.; Oertel, R.; Terhaag, B.; Kroemer, H. K.; Siegmund, W. *Clin. Pharmacol. Ther.* **2000**, *68*, 6–12.

129. Shiraga, T.; Miyamoto, K.; Tanaka, H.; Yamamoto, H.; Taketani, Y.; Morita, K.; Tamai, I.; Tsuji, A.; Takeda, E. *Gastroenterology* **1999**, *116*, 354–362.

130. Wu, C. Y.; Benet, L. Z. *Pharm. Res.* **2005**, *22*, 11–23.

131. Lalloo, A. K.; Luo, F. R.; Guo, A.; Paranjpe, P. V.; Lee, S. H.; Vyas, V.; Rubin, E.; Sinko, P. J. *BMC Med.* **2004**, *2*, 16.

132. Su, Y.; Zhang, X.; Sinko, P. J. *Mol. Pharm.* **2004**, *1*, 49–56.

133. Kamath, A. V.; Morrison, R. A.; Harper, T. W.; Lan, S. J.; Marino, A. M.; Chong, S. *J. Pharm. Sci.* **2005**, *94*, 1115–1123.

134. Gan, L. S.; Moseley, M. A.; Khosla, B.; Augustijns, P. F.; Bradshaw, T. P.; Hendren, R. W.; Thakker, D. R. *Drug Metab. Dispos.* **1996**, *24*, 344–349.

135. Fisher, J. M.; Wrighton, S. A.; Calamia, J. C.; Shen, D. D.; Kunze, K. L.; Thummel, K. E. *J. Pharmacol. Exp. Ther.* **1999**, *289*, 1143–1150.

136. Benet, L. Z.; Cummins, C. L.; Wu, C. Y. *Curr. Drug Metab.* **2003**, *4*, 393–398.

137. Hochman, J. H.; Chiba, M.; Nishime, J.; Yamazaki, M.; Lin, J. H. *J. Pharmacol. Exp. Ther.* **2000**, *292*, 310–318.

138. Wacher, V. J.; Wu, C. Y.; Benet, L. Z. *Mol. Carcinogen.* **1995**, *13*, 129–134.

139. Bredberg, E.; Nilsson, D.; Johansson, K.; Aquilonius, S. M.; Johnels, B.; Nyström, C.; Paalzow, L. *Eur. J. Clin. Pharmacol.* **1993**, *45*, 117–122.

140. Nyholm, D.; Askmark, H.; Gomes-Trolin, C.; Knutson, T.; Lennernäs, H.; Nyström, C.; Aquilonius, S. M. *Clin. Neuropharmacol.* **2003**, *26*, 156–163.

141. Kataoka, M.; Masaoka, Y.; Yamazaki, Y.; Sakane, T.; Sezaki, H.; Yamashita, S. *Pharm. Res.* **2003**, *20*, 1674–1680.

142. He, X.; Kadomura, S.; Takekuma, Y.; Sugawara, M.; Miyazaki, K. *J. Pharm. Sci.* **2004**, *93*, 71–77.

143. Ginski, M. J.; Taneja, R.; Polli, J. E. *AAPS Pharm. Sci.* **1999**, *1*, E3.

144. Masungi, C.; Borremans, C.; Willems, B.; Mensch, J.; Van Dijck, A.; Augustijns, P.; Brewster, M. E.; Noppe, M. *J. Pharm. Sci.* **2004**, *93*, 2507–2521.

145. Artursson, P.; Karlsson, J.; Ocklind, G.; Schipper, N. Studying Transport Processes in Absorptive Epithelia. In *Cell Culture Models of Epithelial Tissues – A Practical Approach*; Shaw, A., Ed.; Oxford University Press: New York, 1996, pp 111–133.

146. Ho, N. F. H.; Raub, T. J.; Burton, P. S.; Bausuhm, C. L.; Adson, A.; Audus, K. L.; Borchard, R. Quantitative Approaches to Delineate Passive Transport Mechanisms in Cell Culture Monolayers. In *Transport Processes in Pharmaceutical Systems*; Lee, P. I., Ed.; Marcel Dekker: New York, 2000, pp 219–316.

147. Tavelin, S.; Gråsjö, J.; Taipalensuu, J.; Ocklind, G.; Artursson, P. *Methods Mol. Biol.* **2002**, *188*, 233–272.

148. Lehr, C. M. *Cell Culture Models of Biological Barriers; In Vitro Test Systems for Drug Absorption and Delivery*; CRC Press: Boca Raton, FL, 2002.

149. Avdeef, A.; Artursson, P.; Neuhoff, S.; Lazorova, L.; Gråsjö, J.; Tavelin, S. *Eur. J. Pharm. Sci.* **2005**, *24*, 333–349.

150. Karlsson, J.; Artursson, P. *Int. J. Pharm.* **1991**, *71*, 55–64.

151. Lennernäs, H.; Palm, K.; Fagerholm, U.; Artursson, P. *Int. J. Pharm.* **1996**, *127*, 103–107.

152. van Meer, G. Polarity and polarized transport of membrane lipids in a cultured epithelium. In *Functional Epithelial Cells in Culture*; Valentich, J. D., Ed.; Alan R. Liss: New York, 1989, pp 43–69.
153. Polli, J. W.; Wring, S. A.; Humphreys, J. E.; Huang, L.; Morgan, J. B.; Webster, L. O.; Serabjit-Singh, C. S. *J. Pharmacol. Exp. Ther.* 2001, *299*, 620–628.
154. Schwab, D.; Fischer, H.; Tabatabaei, A.; Poli, S.; Huwyler, J. *J. Med. Chem.* 2003, *46*, 1716–1725.
155. Yee, S. *Pharm. Res.* 1997, *14*, 763–766.
156. Yamashita, S.; Furubayashi, T.; Kataoka, M.; Sakane, T.; Sezaki, H.; Tokuda, H. *Eur. J. Pharm. Sci.* 2000, *10*, 195–204.
157. Neuhoff, S.; Ungell, A. L.; Zamora, I.; Artursson, P. *Pharm. Res.* 2003, *20*, 1141–1148.
158. Troutman, M. D.; Thakker, D. R. *Pharm. Res.* 2003, *20*, 1200–1209.
159. Troutman, M. D.; Thakker, D. R. *Pharm. Res.* 2003, *20*, 1210–1224.
160. Collett, A.; Tanianis-Hughes, J.; Hallifax, D.; Warhurst, G. *Pharm. Res.* 2004, *21*, 819–826.
161. Pickett, S. D.; McLay, I. M.; Clark, D. E. *J. Chem. Inf. Comput. Sci.* 2000, *40*, 263–272.
162. Lipinski, C. A. *Drug Disc. Today Tech.* 2004, *1*, 337–341.
163. Bergström, C. A.; Strafford, M.; Lazorova, L.; Avdeef, A.; Luthman, K.; Artursson, P. *J. Med. Chem.* 2003, *46*, 558–570.
164. Matsson, P.; Bergström, C. A.; Nagahara, N.; Tavelin, S.; Norinder, U.; Artursson, P. *J. Med. Chem.* 2005, *48*, 604–613.
165. Rowland, M.; Tozer, T. N. *Clinical Pharmacokinetics. Concepts and Applications*, 3rd ed.; Lippincott, Williams and Wilkins: Philadelphia, PA, 1995.
166. Winiwarter, S.; Bonham, N. M.; Ax, F.; Hallberg, A.; Lennernäs, H.; Karlén, A. *J. Med. Chem.* 1998, *41*, 4939–4949.
167. Lennernäs, H.; Knutson, L.; Knutson, T.; Hussain, A.; Lesko, L.; Salmonson, T.; Amidon, G. L. *Eur. J. Pharm. Sci.* 2002, *15*, 271–277.
168. Tavelin, S.; Taipalensuu, J.; Söderberg, L.; Morrison, R.; Chong, S.; Artursson, P. *Pharm. Res.* 2003, *20*, 397–405.
169. Mandagere, A. K.; Thompson, T. N.; Hwang, K. K. *J. Med. Chem.* 2002, *45*, 304–311.
170. Garberg, P.; Eriksson, P.; Schipper, N.; Sjöström, B. *Pharm. Res.* 1999, *16*, 441–445.
171. Stevenson, C. L.; Augustijns, P. F.; Hendren, R. W. *Int. J. Pharm.* 1999, *177*, 103–115.
172. McKenna, J. M.; Halley, F.; Souness, J. E.; McLay, I. M.; Pickett, S. D.; Collis, A. J.; Page, K.; Ahmed, I. *J. Med. Chem.* 2002, *45*, 2173–2184.
173. Hugger, E. D.; Cole, C. J.; Raub, T. J.; Burton, P. S.; Borchardt, R. T. *J. Pharm. Sci.* 2003, *92*, 21–26.
174. Alsenz, J.; Haenel, E. *Pharm. Res.* 2003, *20*, 1961–1969.
175. Bu, H. Z.; Poglod, M.; Micetich, R. G.; Khan, J. K. *Rapid Commun. Mass Spectrom.* 2000, *14*, 523–528.
176. Tannergren, C.; Langguth, P.; Hoffmann, K. J. *Pharmazie.* 2001, *56*, 337–342.
177. Laitinen, L.; Kangas, H.; Kaukonen, A. M.; Hakala, K.; Kotiaho, T.; Kostiainen, R.; Hirvonen, J. *Pharm. Res.* 2003, *20*, 187–197.
178. Hakala, K. S.; Laitinen, L.; Kaukonen, A. M.; Hirvonen, J.; Kostiainen, R.; Kotiaho, T. *Anal. Chem.* 2003, *75*, 5969–5977.
179. Palmgren, J. J.; Mönkkönen, J.; Jukkola, E.; Niva, S.; Auriola, S. *Eur. J. Pharm. Biopharm.* 2004, *57*, 319–328.
180. Augustijns, P.; Mols, R. *J. Pharm. Biomed. Anal.* 2004, *34*, 971–978.
181. Laitinen, L. A.; Tammela, P. S.; Galkin, A.; Vuorela, H. J.; Marvola, M. L.; Vuorela, P. M. *Pharm. Res.* 2004, *21*, 1904–1916.
182. US Department of Health and Human Services, Center for Drug Evaluation Research, Food and Drug Administration, Washington, DC; http://www.fda.gov/cder/guidance/ (accessed April 2006).
183. Amidon, G. L.; Lennernäs, H.; Shah, V. P.; Crison, J. R. *Pharm. Res.* 1995, *12*, 413–420.
184. Polli, J. E.; Yu, L. X.; Cook, J. A.; Amidon, G. L.; Borchardt, R. T.; Burnside, B. A.; Burton, P. S.; Chen, M. L.; Conner, D. P.; Faustino, P. J. et al. *J. Pharm. Sci.* 2004, *93*, 1375–1381.
185. Rinaki, E.; Dokoumetzidis, A.; Valsami, G.; Macheras, P. *Pharm. Res.* 2004, *21*, 1567–1572.
186. Yu, L. X.; Amidon, G. L.; Polli, J. E.; Zhao, H.; Mehta, M. U.; Conner, D. P.; Shah, V. P.; Lesko, L. J.; Chen, M. L.; Lee, V. H. et al. *Pharm. Res.* 2002, *19*, 921–925.
187. Rinaki, E.; Valsami, G.; Macheras, P. *Pharm. Res.* 2003, *20*, 1917–1925.
188. Wu, C. -Y.; Benet, L. Z. *Pharm. Res.* 2005, *22*, 11–23.
189. Palm, K.; Luthman, K.; Ungell, A. L.; Strandlund, G.; Beigi, F.; Lundahl, P.; Artursson, P. *J. Med. Chem.* 1998, *41*, 5382–5392.
190. Liang, E.; Proudfoot, J.; Yazdanian, M. *Pharm. Res.* 2000, *17*, 1168–1174.
191. Burton, P. S.; Conradi, R. A.; Ho, N. F.; Hilgers, A. R.; Borchardt, R. T. *J. Pharm. Sci.* 1996, *85*, 1336–1340.
192. Stenberg, P.; Luthman, K.; Artursson, P. *Pharm. Res.* 1999, *16*, 205–212.
193. Werner, U.; Kissel, T.; Stuber, W. *Pharm. Res.* 1997, *14*, 246–250.
194. Stenberg, P.; Luthman, K.; Ellens, H.; Lee, C. P.; Smith, P. L.; Lago, A.; Elliott, J. D.; Artursson, P. *Pharm. Res.* 1999, *16*, 1520–1526.
195. Palanki, M. S.; Erdman, P. E.; Gayo-Fung, L. M.; Shevlin, G. I.; Sullivan, R. W.; Goldman, M. E.; Ransone, L. J.; Bennett, B. L.; Manning, A. M.; Suto, M. J. *J. Med. Chem.* 2000, *43*, 3995–4004.
196. Ekins, S.; Durst, G. L.; Stratford, R. E.; Thorner, D. A.; Lewis, R.; Loncharich, R. J.; Wikel, J. H. *J. Chem. Inf. Comput. Sci.* 2001, *41*, 1578–1586.
197. Schipper, N. G.; Österberg, T.; Wrange, U.; Westberg, C.; Sokolowski, A.; Rai, R.; Young, W.; Sjostrom, B. *Pharm. Res.* 2001, *18*, 1735–1741.
198. Cianchetta, G.; Singleton, R. W.; Zhang, M.; Wildgoose, M.; Giesing, D.; Fravolini, A.; Cruciani, G.; Vaz, R. J. *J. Med. Chem.* 2005, *48*, 2927–2935.
199. Våbenø, J.; Nielsen, C. U.; Ingebrigtsen, T.; Lejon, T.; Steffansen, B.; Luthman, K. *J. Med. Chem.* 2004, *47*, 4755–4765.
200. Cereijido, M.; Robbins, E. S.; Dolan, W. J.; Rotunno, C. A.; Sabatini, D. D. *J. Cell Biol.* 1978, *77*, 853–880.
201. Cho, M. J.; Thompson, D. P.; Cramer, C. T.; Vidmar, T. J.; Scieszka, J. F. *Pharm. Res.* 1989, *6*, 71–77.
202. Borst, P.; Evers, R.; Kool, M.; Wijnholds, J. *Biochim. Biophys. Acta.* 1999, *1461*, 347–357.
203. Braun, A.; Hammerle, S.; Suda, K.; Rothen-Rutishauser, B.; Gunthert, M.; Krämer, S. D.; Wunderli-Allenspach, H. *Eur. J. Pharm. Sci.* 2000, *11 Suppl 2*, S51–S60.
204. Irvine, J. D.; Takahashi, L.; Lockhart, K.; Cheong, J.; Tolan, J. W.; Selick, H. E.; Grove, J. R. *J. Pharm. Sci.* 1999, *88*, 28–33.
205. Carriere, V.; Lesuffleur, T.; Barbat, A.; Rousset, M.; Dussaulx, E.; Costet, P.; de Waziers, I.; Beaune, P.; Zweibaum, A. *FEBS Lett.* 1994, *355*, 247–250.
206. Raeissi, S. D.; Guo, Z.; Dobson, G. L.; Artursson, P.; Hidalgo, I. J. *Pharm. Res.* 1997, *14*, 1019–1025.
207. Pontier, C.; Pachot, J.; Botham, R.; Lenfant, B.; Arnaud, P. *J. Pharm. Sci.* 2001, *90*, 1608–1619.
208. Tavelin, S.; Milovic, V.; Ocklind, G.; Olsson, S.; Artursson, P. *J. Pharmacol. Exp. Ther.* 1999, *290*, 1212–1221.
209. Artursson, P.; Tavelin, S. Caco-2 and Emerging Alternatives for Prediction of Intestinal Drug Transport: A General Overview. In *Drug Bioavailability – Estimation of Solubility, Permeability, Absorption and Bioavailability*; Artursson, P., Ed.; Wiley-VCH: Weinheim, Germany, 2003, pp 72–89.
210. Fine, K. D.; Santa Ana, C. A.; Porter, J. L.; Fordtran, J. S. *Gastroenterology* 1995, *108*, 983–989.

211. Palm, K.; Luthman, K.; Ros, J.; Gråsjö, J.; Artursson, P. *J. Pharmacol. Exp. Ther.* **1999**, *291*, 435–443.
212. Paul, E. C.; Hochman, J.; Quaroni, A. *Am. J. Physiol.* **1993**, *265*, C266–C278.
213. Tavelin, S.; Taipalensuu, J.; Hallböök, F.; Vellonen, K. S.; Moore, V.; Artursson, P. *Pharm. Res.* **2003**, *20*, 373–381.
214. Kansy, M.; Senner, F.; Gubernator, K. *J. Med. Chem.* **1998**, *41*, 1007–1010.
215. Wohnsland, F.; Faller, B. *J. Med. Chem.* **2001**, *44*, 923.
216. Curatolo, W. *Pharm. Sci. Tech. Today* **1998**, *1*, 387–393.
217. Chong, S.; Dando, S. A.; Soucek, K. M.; Morrison, R. A. *Pharm. Res.* **1996**, *13*, 120–123.
218. Prime-Chapman, H. M.; Fearn, R. A.; Cooper, A. E.; Moore, V.; Hirst, B. H. *J. Pharmacol. Exp. Ther.* **2004**, *311*, 476–484.
219. Litman, T.; Druley, T. E.; Stein, W. D.; Bates, S. E. *Cell. Mol. Life Sci.* **2001**, *58*, 931–959.
220. Pastan, I.; Gottesman, M. M.; Ueda, K.; Lovelace, E.; Rutherford, A. V.; Willingham, M. C. *Proc. Natl. Acad. Sci. USA* **1988**, *85*, 4486–4490.
221. Horio, M.; Chin, K. V.; Currier, S. J.; Goldenberg, S.; Williams, C.; Pastan, I.; Gottesman, M. M.; Handler, J. *J. Biol. Chem.* **1989**, *264*, 14880–14884.
222. Evers, R.; Zaman, G. J.; van Deemter, L.; Jansen, H.; Calafat, J.; Oomen, L. C.; Oude Elferink, R. P.; Borst, P.; Schinkel, A. H. *J. Clin. Invest.* **1996**, *97*, 1211–1218.
223. Evers, R.; Kool, M.; van Deemter, L.; Janssen, H.; Calafat, J.; Oomen, L. C.; Paulusma, C. C.; Oude Elferink, R. P.; Baas, F.; Schinkel, A. H.; Borst, P. *J. Clin. Invest.* **1998**, *101*, 1310–1319.
224. Kinoshita, S.; Suzuki, H.; Ito, K.; Kume, K.; Shimizu, T.; Sugiyama, Y. *Pharm. Res.* **1998**, *15*, 1851–1856.
225. Tang, F.; Horie, K.; Borchardt, R. T. *Pharm. Res.* **2002**, *19*, 765–772.
226. van der Sandt, I. C.; Blom-Roosemalen, M. C.; de Boer, A. G.; Breimer, D. D. *Eur. J. Pharm. Sci.* **2000**, *11*, 207–214.
227. Tang, F.; Horie, K.; Borchardt, R. T. *Pharm. Res.* **2002**, *19*, 773–779.
228. Adachi, Y.; Suzuki, H.; Sugiyama, Y. *Pharm. Res.* **2001**, *18*, 1660–1668.
229. Nozawa, T.; Imai, K.; Nezu, J.; Tsuji, A.; Tamai, I. *J. Pharmacol. Exp. Ther.* **2004**, *308*, 438–445.
230. Morisaki, K.; Robey, R. W.; Özvegy-Laczka, C.; Honjo, Y.; Polgar, O.; Steadman, K.; Sarkadi, B.; Bates, S. E. *Cancer Chemother. Pharmacol.* **2005**, *56*, 161–172.
231. Gupta, A.; Zhang, Y.; Unadkat, J. D.; Mao, Q. *J. Pharmacol. Exp. Ther.* **2004**, *310*, 334–341.
232. Su, T. Z.; Feng, M. R.; Weber, M. L. *J. Pharmacol. Exp. Ther.* **2005**, *313*, 1406–1415.
233. Suhre, W. M.; Ekins, S.; Chang, C.; Swaan, P. W.; Wright, S. H. *Mol. Pharmacol.* **2005**, *67*, 1067–1077.
234. Covitz, K. M.; Amidon, G. L.; Sadee, W. *Pharm. Res.* **1996**, *13*, 1631–1634.
235. Wierdl, M.; Wall, A.; Morton, C. L.; Sampath, J.; Danks, M. K.; Schuetz, J. D.; Potter, P. M. *Mol. Pharmacol.* **2003**, *64*, 279–288.
236. van der Heijden, J.; de Jong, M. C.; Dijkmans, B. A.; Lems, W. F.; Oerlemans, R.; Kathmann, I.; Schalkwijk, C. G.; Scheffer, G. L.; Scheper, R. J.; Jansen, G. *Ann. Rheum. Dis.* **2004**, *63*, 138–143.
237. Litman, T.; Zeuthen, T.; Skovsgaard, T.; Stein, W. D. *Biochim. Biophys. Acta* **1997**, *1361*, 159–168.
238. Scarborough, G. A. *J. Bioenerg. Biomembr.* **1995**, *27*, 37–41.
239. Cui, Y.; Konig, J.; Keppler, D. *Mol. Pharmacol.* **2001**, *60*, 934–943.
240. Sasaki, M.; Suzuki, H.; Ito, K.; Abe, T.; Sugiyama, Y. *J. Biol. Chem.* **2002**, *277*, 6497–6503.
241. Letschert, K.; Komatsu, M.; Hummel-Eisenbeiss, J.; Keppler, D. *J. Pharmacol. Exp. Ther.* **2005**, *313*, 549–556.

Biographies

Per Artursson studied pharmacy at Uppsala University, Sweden, where he also obtained his thesis in 1985. He spent 1 year as a postdoctoral fellow at the Medical Products Agency, Uppsala (1986), and 1 year as a visiting scientist at Advanced Drug Delivery Research, Ciba-Geigy, UK (1987), before taking up a position as Assistant Professor in Pharmaceutics at Uppsala University. In 1992 he was appointed to his present post as Professor of Dosage Form Design at the Department of Pharmacy, Uppsala University. In 1996 he spent a sabbatical at GeneMedicine, the Woodlands, TX. Dr Artursson's research interests range from drug and gene delivery and targeting to drug absorption. He is on the editorial boards for the leading scientific journals in the pharmaceutical sciences, was Editor-in-Chief for the *European Journal of Pharmaceutical Sciences* (1998–2001), and has received several awards for his research. He is one of the worlds most cited scientists in *Pharmacology and Toxicology* (ISI, 2004). In 2004, he founded a new unit at his department, dedicated to pharmaceutical screening and informatics.

Sibylle Neuhoff graduated from the Johan Wolfgang Goethe University, Frankfurt am Main, Germany, with a BSc in food chemistry in 1992, and gained her MSc in chemistry in 1996. Between 2002 and 2005 she studied for her PhD at the Department of Pharmacy, Uppsala University, Sweden, with Professor Per Artursson as her supervisor, and partly at AstraZeneca R&D, Sweden, with Associate Professor Anna-Lena Ungell as her supervisor (2002–05). Her PhD thesis was entitled "Refined in vitro Models for Prediction of Intestinal Drug Transport".

Pär Matsson graduated from Uppsala University, Sweden, with an MSc in pharmacy. He started his PhD studies at the Department of Pharmacy, Uppsala University, with Professor Per Artursson as his supervisor in 2002.

Staffan Tavelin graduated from Uppsala University, Sweden, with an MSc in pharmacy in 1996. Between 1996 and 2003 he studied for his PhD at the Department of Pharmacy, Uppsala University, with Professor Per Artursson as his supervisor. His thesis was entitled "New Approaches to Studies of Paracellular Drug Transport in Intestinal Epithelial Cell Monolayers". He currently holds a position as a lecturer in pharmaceutics at Umeå University, Sweden.

Comprehensive Medicinal Chemistry II
ISBN (set): 0-08-044513-6

ISBN (Volume 5) 0-08-044518-7; pp. 259–278

5.12 Biological In Vitro Models for Absorption by Nonoral Routes

P Colombo, S Cagnani, F Sonvico, and P Santi, University of Parma, Parma, Italy
P Russo, University of Salerno, Fisciano, Italy
G Colombo, University of Ferrara, Ferrara, Italy

5.12.1 Introduction to Nonoral Routes for Drug Absorption

In recent years, pharmacotherapy has been used to search for novel routes of drug absorption as an alternative to the oral route or the invasive injection pathway. The main reason for this interest is the increasing number of therapeutic substances, in particular peptides and proteins, that can not be administered orally. Pharmacotherapeutic innovations in drug administration have been boosted by the development of controlled release preparations (CR).

The development of products for alternative administration routes continues to be based on strategic considerations that can be summarized as the three Cs of alternative drug delivery: clinic, compliance, and commercial. These highlight the need for assessing the bioavailability and safety of the product linked to a specific route, as well as the acceptability by the patient and the marketability of the product. Thus, these elements play a pivotal role in the introduction of products for alternative nonoral administration routes.

The evolution of drug release control has led to a great number of modified release preparations. Research in this area has resulted in new materials, structures, mechanisms, instruments, and techniques. Controlled release refers to the study of principles and devices for the control of drug availability in time and space. Initially, most CR products were given by mouth or by injection to obtain a systemic effect, whereas other routes were mainly explored for the treatment of local diseases. More recently, advances in the field of drug release science and technology have made it possible to attain a systemic effect by nonenteral and nonparenteral administration, such as buccal, nasal, pulmonary, vaginal, or transdermal. For example, the introduction and application of the bioadhesion concept developed in parallel with these alternative routes. The goal of controlled drug delivery based on bioadhesion is to localize the formulation in an accessible body area in order to enhance drug absorption in a site-specific way. The use of bioadhesion has been proposed for novel buccal, transmucosal, and vaginal drug delivery systems,[1] as localization of the drug formulation might enhance absorption of the released drug. Good mucoadhesion at the gel carrier/mucosa interface is obtained in the presence of molecular adhesion promoters, i.e., polymeric structures, which are free to diffuse across the interface in the mucus.[2]

The most extensively explored alternative absorption route for systemic effects has been the skin, which successfully fulfils the criteria of the three Cs of alternative drug delivery. In general, when a new route is proposed, the greatest concern is the safety of the application. Using the skin as a delivery site is relatively safe since the structure of skin makes it less vulnerable to applied foreign substances. In contrast, the mucosal epithelia typical of the other routes of administration exist for specific physiological functions and not as a transport pathway for substances of varying aggressiveness. Thus, safety in the case of mucosal involvement is very important and explains why one of the oldest alternative mucosal routes was the buccal route. In fact, as the first part of the GI tract, the mucosa of the mouth has a structure adapted for dealing with aggressive substances from food or the environment. Similarly, rectal administration has been used for a long time and was successful in some countries, although its use is limited by poor patient compliance and low therapeutic value. The safety aspects related to the use of the highly permeable nasal, pulmonary, and vaginal routes have only recently been tackled; the acquisition of a sufficient amount of knowledge as to the effect of the exposure to foreign substances at this level has opened up the possibility of using these routes for systemic effect. A few examples of marketed products for administration by nonoral routes are shown in **Table 1**.

An aspect that is common to all the alternative routes is the need to enhance the permeability of the tissue to compensate for the limited surface area involved. The use of penetration enhancers, i.e., substances that reorganize the epidermis or epithelial structures or open up the intercellular tight junctions, is very important in this area, but it adds further safety concerns and requires careful evaluation of both immediate and long-term effects on tissue integrity and functionality.

Finally, the studies of different routes of drug administration initiated the science of drug delivery, i.e., methods and techniques for efficiently administering a drug to a specific site. The strategies adopted for drug delivery to the nose, lungs, or mouth are the heritage of different technologies developed for CR preparations. At present, mucosal drug delivery only involves products employed for immediate release and prompt effect, whereas the conjugation of CR and drug delivery is classically obtained with the transdermal systems. Alternative drug delivery has evolved from the science of drug release control, since the products or systems developed are mainly designed for targeting a specific site of drug absorption. Alternative drug delivery routes that incorporate controlled or prolonged release are currently the subject of a great deal of research. Quality, efficacy, and safety issues must be considered very carefully when exploiting these ways of systemic application.

5.12.2   Mucosal Routes for Drug Absorption

5.12.2.1   Buccal Route

Buccal administration is intended for delivering drugs within/through the buccal mucosa in order to achieve a local or systemic effect. This route is particularly attractive since substances absorbed through the buccal mucosa bypass gastrointestinal enzymatic degradation and the hepatic first-pass effect. The mouth has a relatively large area for drug application and good accessibility compared to the nose, rectum, and vagina.[3] In addition, the mucosa is resistant to damage or irritation because of the rapid cell turnover[4] and frequent exposure to food. The buccal mucosa consists of a surface layer of stratified squamous epithelium linked to the underlying connective tissue (lamina propria and submucosa) by a basal lamina. In the connective tissue a network of blood capillaries is present where drugs that have permeated through the epithelium can enter the systemic circulation. The surface layers of the epithelium (approximately the uppermost 25–30%) are reported to be the main barrier to the penetration of substances.[5,6] Cell layers of the buccal mucosa contain membrane-coating granules (MCGs), which are spherical or oval organelles, 100–300 nm in diameter. MCGs appear to fuse with cell membranes and extrude their content, mainly lipids, into the intercellular space. By means of confocal laser scanning microscopy, the region corresponding to MCG in porcine buccal epithelium was visualized as the rate-limiting layer for diffusion of hydrophilic fluorescein isothiocyanate.[7] The intercellular lipids extruded by MCG (150–200 μm deep) represent the permeability barrier for hydrophilic compounds. A recent paper proposed a bilayer diffusion model to quantitatively describe the relative contributions of epithelium and connective tissue to the permeation barrier for 2′,3′-dideoxycytidine.[8] It was shown that the basal lamina layer within the buccal epithelium acted as an important barrier to drug permeation.

The drug transport pathways across buccal mucosa can be both transcellular and paracellular, but for many hydrophilic drugs the permeation is mainly due to passive diffusion via the paracellular route. Most transbuccal permeation studies have been conducted in vitro[9] using various diffusion cells (flow-through cells, vertical Franz cell, horizontal cells, Ussing chamber). A schematic representation of a common vertical Franz diffusion cell is reproduced in **Figure 1**.

Porcine buccal mucosa is the most common model barrier for in vitro experiments. The anatomy and metabolism of this nonkeratinized epithelium are similar to that of human buccal mucosa. In general, full-thickness buccal mucosa is used in permeation studies. Technically, pig buccal tissue (cheek) is obtained from a slaughterhouse and transported to

Table 1 Some examples of marketed products formulated for administration by nonoral routes

Proprietary name	Approved name	Uses	Route and dose
Buccal			
Nitrolingual Pumpspray (First Horizon Pharm., USA)	Nitroglycerin	Treatment of angina pectoris	Metered dose spray to apply onto or under the tongue; 400 µg of nitroglycerin per spray
Striant (Columbia Laboratories, USA)	Testosterone	Hormone replacement therapy in males	Buccal system; mucoadhesive tablet for the gum tissue; 30 mg of testosterone
Nasal			
Miacalcin (Novartis Pharm., USA)	Calcitonin-salmon	Treatment of postmenopausal osteoporosis	Spray pump; nasal solution; 200 IU of calcitonin per actuation
DDAVP (Aventis, USA)	Desmopressin acetate	Management of primary nocturnal enuresis and central cranial diabetes insipidus	Spray pump; nasal solution; 10 µg of desmopressin acetate per actuation
Synarel (Searle, USA)	Nafarenil acetate	Treatment of central precocious puberty	Spray pump; nasal solution; 200 µg of nafarenil per actuation
Pulmonary			
Intal Inhaler (Monarch, USA)	Cromolyn sodium	Prophylactic agent in the management of patients with bronchial asthma	Metered dose inhaler; delivered dose: 800 µg of cromolyn sodium
Atrovent (Boehringer Ingelheim, USA)	Ipratropium bromide	Bronchodilator for maintenance treatment of bronchospasm associated with COPD	Metered dose inhaler; delivered dose: 18 µg of ipratropium bromide
Serevent Discus (GlaxoSmithKline, USA)	Salmeterol xinafoate	Maintenance treatment of asthma and prevention of bronchospasm	Dry powder inhaler; delivered dose: 47 µg of salmeterol xinafoate
Clenil jet (Chiesi Farmaceutici, Italy)	Beclomethasone dipropionate	Control of asthmatic disease development and of bronchostenotic conditions	Metered dose inhaler; CFCs free; delivered dose: 50 µg of beclomethasone dipropionate
Exubera (Aventis/Pfizer, USA)	Insulin human Phase III completed	For the treatment of type 1 and type 2 diabetes	Dry powder inhaler; metered dose: 1–3 mg of insulin
Vaginal			
Nuvaring (Organon, USA)	Etonogestrel/ethinyl estradiol	Prevention of pregnancy	Nonbiodegradable, flexible vaginal ring releasing 0.120 mg day^{-1} of etonogestrel and 0.015 mg day^{-1} of ethinyl estradiol
Prochieve (Columbia Laboratories, USA)	Progesterone	Treatment of secondary amenorrhea and in assisted reproductive technology for infertile women	Bioadhesive vaginal gel; once daily application of 1.125 g of 4% w/w gel delivering 45 mg of progesterone
Cervidil (Forest, USA)	Dinoprostone	Cervical ripening in patients at or near term for the induction of labor	Thin polymeric slab for vaginal insertion, releasing 0.3 mg h^{-1} of dinoprostone

continued

Table 1 Continued

Proprietary name	Approved name	Uses	Route and dose
Transdermal			
Duragesic (Janssen, USA)	Fentanyl	Treatment of chronic pain that requires continuous opioid administration	Transdermal system releasing $25\,\mu g\,h^{-1}$ of fentanyl per $10\,cm^2$
Ortho evra (Ortho-Mcneil, USA)	Norelgestromin/ethinyl estradiol	Prevention of pregnancy	Transdermal system releasing $150\,\mu g\,day^{-1}$ of norelgestromin and $20\,\mu g$ of ethinylestradiol
Oxytrol (Watson, USA)	Oxybutynin	Treatment of overactive bladder with symptoms of urinary incontinence	Transdermal system ($39\,cm^2$) releasing $3.9\,mg\,day^{-1}$
Climara (Berlex, USA)	Estradiol	Relief of moderate to severe vasomotor symptoms associated with menopause	Trandermal system ($25\,cm^2$) releasing $0.1\,mg\,day^{-1}$ of estradiol

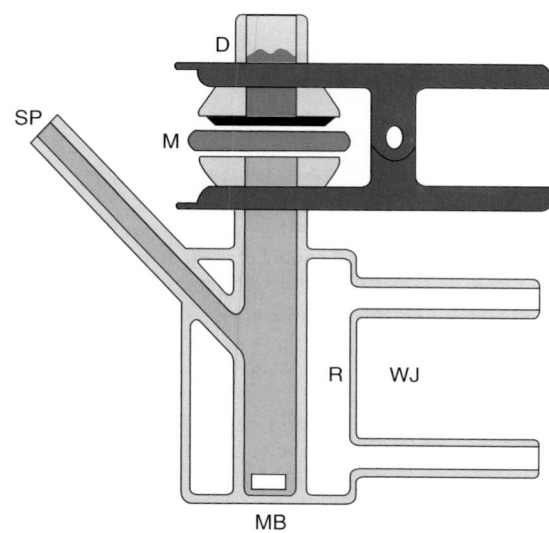

Figure 1 Schematic representation of an assembled vertical Franz diffusion cell: D, donor compartment; M, membrane; SP, sampling port; R, receptor compartment; MB, magnetic bar; WJ, water jacket.

the laboratory in cold Krebs buffer (pH 7.4). The buccal mucosa, along with a part of the submucosa, is carefully separated from fat and muscles using a scalpel. Then, by means of an electro-dermatome, the epithelium (including mucus layer and basal lamina) is isolated from the underlying tissue. The average thickness of samples is about $500\,\mu m$. Because of the time-dependent viability, the buccal epithelium must be used within 2 h of removal. The mucosa is mounted, for example, on an Ussing chamber with the mucosal side facing the donor compartment. Donor and acceptor compartments are filled with Krebs buffer (pH 7.4). Carbogen gas (95% O_2, 5% CO_2) is circulated through both compartments in order to maintain tissue viability and provide adequate mixing. Electrophysiological parameters are determined for assessing the integrity and viability of the biological sample. After a 1-h equilibration period at $34\pm0.5\,^\circ C$, the receptor is replaced with fresh Krebs buffer and the donor side is filled with the donor solution. A saturated solution of drug is used as donor to measure the permeability coefficient. Diffusion experiments are conducted for 3 h. In the case of determination of the drug concentration profile within the buccal epithelium, the thin-slicing technique has been applied after the transport experiments.[10] Buccal mucosa specimens were also snap-frozen in liquid nitrogen and stored at $-85\,^\circ C$ for periods of up to 6 months. Frozen specimens were then used for permeability studies: it was observed that the permeability characteristics of the mucosa were not adversely affected by freezing and storage.[11]

Data analysis for permeability coefficient determination through the mucosa is usually performed in the steady-state transport period of time.[8] For one-dimensional diffusion, the permeability of a diffusant through a solid membrane can be calculated as:

$$P_e = \frac{J_{SS}}{\Delta C} \qquad [1]$$

where P_e is the permeability coefficient of the diffusant (cm s^{-1}), ΔC is the concentration difference between the two surfaces of the membrane, and J_{SS} is the flux at steady state (mg s^{-1} cm^{-2}).

The steady-state flux is given by the following equation:

$$J_{SS} = \frac{\Delta M}{\Delta t \cdot A} \qquad [2]$$

where ΔM is the amount of diffusant transported through the membrane during the interval time Δt at steady state, and A is the diffusion area.

Buccal drug absorption studies were performed in vitro and in vivo with porcine buccal mucosa.[12] The distribution of fluorescein isothiocyanate (FITC)-labeled dextrans in the epithelium using confocal laser scanning microscopy allowed the permeation pathways to be visualized. At molecular weights lower than 20 kDa the passage of hydrophilic FITC-dextrans was impeded. The paracellular route was found to be the main pathway for these molecules. An in vivo delivery device, consisting of an application chamber containing a solution of FITC-labeled dextran 4400 (FD4) or buserelin, was attached to the buccal mucosa for 4 h with an adhesive. Steady-state plasma levels were rapidly achieved. Coadministration of 10 mM sodium glycodeoxycholate, an absorption enhancer, increased the absolute bioavailability of FD4.

In a more recent study,[13] the in vitro permeability of porcine buccal mucosa showed consistently lower values with different markers (arecoline, 17β-estradiol, water, and vasopressin) compared to porcine mouth floor mucosa. Porcine mouth floor mucosa was a good model of human buccal mucosa using a continuous flow-through perfusion system (20 °C, 24 h).

To enable in vitro studies of irritation, oral pathologies, and basic oral cavity phenomena, EpiOral and EpiGingival tissue models are available, consisting of normal, human-derived epithelial cells (MatTek corporation, Ashland, MA, USA). The cells have been cultured to form multilayered, highly differentiated models of the human buccal and gingival phenotypes.

The delivery of thiocolchicoside, a muscle-relaxant, through oral mucosa was studied by investigating its in vitro permeation through porcine oral mucosa and in vivo buccal transport in humans.[10] A bioadhesive disk and a fast dissolving disk for buccal and sublingual administration, respectively, were tested. The in vitro permeation of thiocolchicoside through porcine buccal mucosa from these dosage forms was compared with the in vivo drug absorption in humans. The buccal absorption test was performed on healthy volunteers in accordance with Rathbone.[14] Before each administration, the volunteers washed their mouths with 100 mL of distilled water. The dosage form (4 mg of thiocolchicoside) was then placed under the tongue (fast dissolving disk) or in contact with the gingival mucosa (bioadhesive disk) and kept in place avoiding swallowing during a fixed period of time. The residue of the dosage form was then expelled and the mouth was rinsed with water. The residue of dosage form and the washing solutions were combined and analyzed for the residual drug content. The fast dissolving form (sublingual) resulted in a quick uptake of 0.5 mg of thiocolchicoside within 15 min, whereas with the bioadhesive buccal form the same dose could be absorbed over an extended period of time. Despite the variability of the in vivo results, an interesting correlation between in vitro (porcine) and in vivo (human) data for both dosage forms was found.

Buccal delivery systems include mouthwashes, sprays, chewing gums, bioadhesive tablets, gels, and patches. Transbuccal delivery devices can be easily applied and removed. However, drug therapy within the oral cavity is subjected to a rapid elimination of drug due to the flushing action of saliva, and may require repeated and frequent doses. This aspect can affect the inter-patient variability and is likely strongly dependent on the system technology used for keeping the drug product in contact with the absorbing mucosa. In fact, significant absorption is obtained by a prolonged exposure of the drug to the mucosal surface. Lectins or bioadhesive substances have been proposed to prolong the system residence time and improve drug absorption through the oral mucosa.[15] A sublingual tablet for rapid drug absorption based on mixtures of carrier particles partially covered by fine drug particles was studied for fentanyl citrate buccal delivery.[16] Plasma concentrations of fentanyl were obtained within 10 min, with no second peak. The bioadhesive component prevented the fentanyl from being swallowed, without hindering its release and absorption. Chimera agglomerates, a formulation in powder form based on primary particles that are agglomerated in soft and porous clusters, represent a new dosage form for buccal insufflation.[17] These free-flowing powders can be used for generating a mouth aerosol or for direct introduction into the gingival space in order to achieve bioadhesion and prompt or delayed dissolution. The appropriate choice of drug product formulation and mode of application can improve the reproducibility of the administration and of the response. However, it must be remembered that the doses that can be applied remain in the order of a few tenths of milligrams and the lipid solubility of the drug improves the reliability of the administration. This last aspect was supported by the good absorption of fentanyl compared to the more hydrophilic morphine.[18]

Buccal immunization with films loaded with plasmid DNA (CMV-beta-gal) or beta-galactosidase protein has also been studied.[19] Bilayer films were developed using different polymers as the mucoadhesive layer and a pharmaceutical wax as the impermeable backing layer. These films were applied to the buccal pouch of rabbits and the immune response to beta-gal was determined. All rabbits were immunized with plasmid DNA administered via the buccal route, whereas none was by subcutaneous injection of the antigen protein. Different mucoadhesive films based on chitosan hydrochloride and polyacrylic acid sodium salt showed the possibility of achieving buccal absorption of problematic drugs such as acyclovir.[20]

Finally, another approach in buccal delivery was based on a bioadhesive device as a method for controlling delivery of cyanocobalamin to the gastrointestinal tract in male beagle dogs.[21] The novelty was in the use of bioadhesive controlled delivery in the mouth for improving gastrointestinal absorption of this actively transported drug. Significantly higher bioavailability was observed with the buccal bioadhesive device than with the oral immediate release capsule.

5.12.2.2   Nasal Route

Nasal administration is an alternative to injection for low- dose drugs when rapid onset of action is required.[22] This route is also useful for peptide substances, as demonstrated by the marketed products calcitonin, buserelin, desmopressin, and nafarelin. As for all inhalation products, nasal products are characterized by two features that are relevant to drug bioavailability: the formulation itself and the delivery device. In the pharmaceutical development, the nasal formulation has to be adapted to the delivery device, in order to provide complete deposition of the dose on the nasal mucosa. Nasal deposition is relevant for avoiding drug swallowing, which for certain drugs can be detrimental to the bioavailability due to instability in the GI tract. However, with nasal administration, due to the structure of the nasal cavity, swallowing of the drug product is minimal compared with the pulmonary delivery.

Nasal absorption can be investigated by in vitro studies on: (1) primary cell cultures of human nasal epithelium; (2) human nasal cell lines[23]; (3) excised nasal epithelium from different animals (rabbit, pig, cow, sheep); or (4) in vivo studies with animal models. The degree of epithelial perturbation in the in vitro models is higher than in vivo, where the tissue benefits from the protective mucus and mucociliary clearance. On the other hand, the mucus that normally lines the nasal mucosa and protects the epithelium can also affect the absorption of drugs. A comprehensive review of the cell models for nasal absorption, metabolism, and toxicity is available in the literature.[24]

Studies with primary cell cultures of human nasal epithelium allow for the exploration of drug permeation, metabolism, and toxicity while bypassing potential problems arising from interspecies diversity. Nasal cell specimens can be obtained from the vestibular area, atrium, or turbinates. The area of the nasal cavity should be selected depending on the zone where the formulation is expected to deposit. In many cases, the selection falls on the middle and inferior turbinate, which is covered by a pseudostratified columnar epithelium consisting of basal cells, goblet cells, and columnar cells. This respiratory area is lined with a moist mucous membrane that contains tall column-shaped cells. On the surface of these cells fine hair-like projections, known as cilia, collect and move debris toward the oropharynx. Using human nasal epithelial tissues obtained during elective surgery, the absorption enhancement of methionine enkephalin in the presence of protease inhibitors and absorption enhancers was investigated.[25] The nasal epithelium was dissociated with pronase, filtered, and preplated in a 95% O_2 and 5% CO_2 environment in order to reduce fibroblast contamination. Epithelial cells were then seeded at a density of 5×10^5 cells cm^{-2}, on a different support depending on the study. For transport studies, cells were cultured on collagen CD-24 inserts. A procedure for collecting cells with the brushing technique without using anesthesia has been described.[26] Cells were collected by means of plastic strips equipped with a sticky albumin surface. Independently of the method used, permeation studies are performed on viable cells grown to confluence. Cytotoxicity studies are therefore required, especially when the permeation pathway involves active transport or other viable cell functions.

Primary cell cultures of human nasal epithelium present several drawbacks, which could be overcome by culturing nasal cell lines that show extended life span, easier proliferation, and good homogeneity. Human nasal septum tumor cells (RPMI 2650) are an example of nasal cell lines having metabolic activity close to that of human nasal tissue. However, since they are poorly differentiated, not polarized, and unable to form a monolayer, RPMI cells are considered unsuitable for transport studies.[27] Alternatively, human lung adenocarcinoma cell lines (Calu-3) have been used to simulate nasal epithelium in drug permeation studies, owing to their similarity with the serous cells of the upper airway. Moreover, they are able to grow in a confluent polarized monolayer with tight junctions. Calu-3 cell lines were used to investigate the bioadhesive properties of formulations and the bovine serum albumin transport in vitro.[28]

Nasal mucosa excised from different animal species (rabbit, sheep, cow, and pig) is mounted as a barrier on diffusion cells in order to predict the intranasal transport. The method used to obtain the nasal mucosa specimens depends on the animal. In general, the nasal mucosa has to be dissected immediately after the animal's death. With rabbits, for example, the nasal bone is incised with a knife along the welding line from the frontal bone to the nose tip. After the opening of the nasal cavity, the nasal septum is separated by cutting it from its connections with the maxillary region, and extracted. The mucosa layers are then carefully detached from the septum cartilage by means of a pair of tweezers. The specimens removed (surface area 0.7–1 cm^2; thickness 100 μm) are rinsed with a saline solution and immediately inserted between the donor and the receptor compartments of a vertical or horizontal diffusion cell, with the mucosal surface facing the donor. In order to preserve the viability of the mucosa during the experiment, a gas mixture O_2–CO_2 (95:5) or oxygen alone can be bubbled through. Before starting the permeation experiment, the mucosa mounted on the cell is allowed to equilibrate at 37 °C and the integrity is assessed. The formulation to be tested is then introduced into the donor compartment and aliquots are sampled from the receiving medium. Usually, the diffusion cells have a receptor volume ranging between 1 and 12 mL; small cells are required when drug quantification is a critical issue. Compartments can be equipped with ports for the introduction of electrodes for continuous electrophysiological measurements. In the determination of permeability coefficients, good stirring of the solutions contained in both compartments is required to ensure sink conditions.

Permeation studies performed under the above conditions provide information on transport rates, quantity of permeated drug, metabolism, and effect of absorption enhancers. The cumulative amount of drug permeated per unit area (Q_n) is calculated by the following equation:

$$Q_n = \frac{V_R \times C_n + \sum_{i=0}^{n-1} V_P \times C_i}{A} \qquad [3]$$

where C_n is the drug concentration in the receiving compartment at time n, V_R is the volume of the receiving compartment, V_P is the sampling volume, C_i is the drug concentration at time i, and A is the permeation area.

Permeation data are generally plotted as quantity of drug permeated per unit of area versus time in order to verify the steady-state flux. **Figure 2** illustrates the results of a transport experiment through nasal mucosa analyzed in order to calculate the permeation parameters. The experiment compared the transport of desmopressin from powder and liquid formulations.

The permeability coefficient (P_e) across the mucosa has been calculated from the slope of the linear part of the curve relative to the desmopressin solution, according to eqn [1]; P_e is related to the diffusion coefficient D ($cm^2\,s^{-1}$) and mucosa-medium partition coefficient K by the following equation:

$$P_e = \frac{D \times K}{h} \qquad [4]$$

where h is the mucosa thickness (cm).

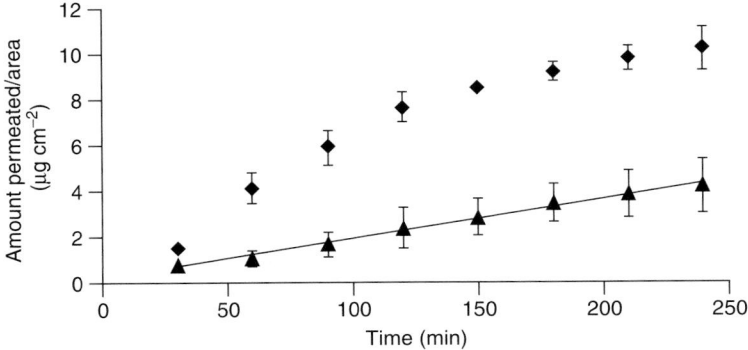

Figure 2 Desmopressin permeation through rabbit nasal mucosa from a solution (▲) and a nasal powder formulation (♦). Regression line of the solution steady-state permeation profile: $y = 0.0173x + 0.2559$ ($R^2 = 0.9919$). Permeation parameters according to eqn [1]: $J_{SS} = 0.0173\ \mu g\,min^{-1}\,cm^{-2}$; $\Delta = 100\ \mu g\,mL^{-1}$; $P_e = 2.88 \times 10^{-6}\,cm^2\,s^{-1}$ (s.d. 8.84×10^{-7}).

An in vitro model for permeation studies using porcine nasal mucosa has been described.[29] Three different nasal mucosal tissues were evaluated: full thickness septum mucosa, dermatomed septum mucosa, and nasal cavity mucosa. The dermatome was used for the removal of the underlying connective tissue, in order to achieve a more homogeneous thickness of septum mucosa. This study demonstrated that the drug permeability depended not only on the thickness of the mucosa, but also on the presence of connective tissue.

Permeation across porcine nasal mucosa of drugs with different physicochemical properties (insulin, lidocaine, nicotine, propranolol, sumatriptan) has also been measured.[30] Permeability coefficients determined with the Ussing chamber were correlated with the literature data on the fraction absorbed after nasal administration in humans. A weak correlation was found between excised porcine mucosa and nasal absorption in humans: among the different drugs, the closest correlation was found for drugs transported by passive diffusion.

An original ex vivo model not performed with classical diffusion cells has been reported in the literature.[31] Nasal turbinate mucosa of sheep was detached from the underlying bone and mounted at the bottom of a cylindrical support connected to the drive shaft of the basket dissolution apparatus (USP Apparatus 1). A metoclopramide formulation was introduced in the cylinder closed at the bottom by the mucosa. The system was rotated at 50 r.p.m. in the dissolution vessel keeping the mucosa just in contact with the surface of the receiving medium. The transport-enhancing properties of chitosan were evidenced. During the permeation test, transmission electron microscopy analysis was also performed in order to study the morphological changes of the epithelial cells. It was observed that chitosan/alginate microspheres were able to open the cell tight junctions.

Hussain[32] described three variants of a frequently used rat model for studying nasal drug absorption, from which he predicted the absorption profile after nasal administration in other species. They were an in situ method, an in vivo in situ method, and an in vivo method. In the in situ method, the rat is anesthetized and the trachea is cannulated. A tube inserted through the esophagus to the posterior part of the nasal cavity serves to circulate the perfusion solution into the nasal cavity. The absorption rate is determined by analyzing the remaining amount of drug in the perfusing solution. The in vivo in situ method is a variant of the previous in situ recirculation method to keep the solution in the nasal cavity. A small volume of drug is administered into the nasal cavity for absorption and hydrolysis studies. In the in vivo method, the drug is directly deposited on the nasal mucosa of the rat and blood samples are withdrawn at predetermined time points. Owing to the use of a close and confined system, the authors claimed to have reproducible and reliable data.

Nasal mucociliary clearance has implications in nasal drug absorption.[33] Drugs are rapidly cleared from the nose after intranasal administration, and several approaches have been proposed to prolong the residence time of formulations in the nasal cavity. It has already been emphasized that increasing the contact time between the formulation and the nasal mucosa can improve the absorption of drugs. Bioadhesive drug delivery systems can help to counteract clearance of the formulation. Intranasal in vivo administration of apomorphine powder in rabbits was performed in order to determine the effect of bioadhesion on drug absorption.[34] A sustained plasma level of apomorphine using a carboxymethylcellulose formulation was obtained, reaching a relative bioavailability equivalent to subcutaneous injection. The nasal formulations were insufflated with an air-filled syringe expelling the drug through a plastic tip inserted into the rabbit's nostril. Sometimes this procedure requires superficial local anesthesia of the external nostrils to prevent the animal sneezing.

Interspecies differences between rat, rabbit, and man have been described for the nasal administration of peptide and protein drugs with cyclodextrins as absorption enhancers.[35] In fact, the nasal absorption of drugs, in particular of polar compounds (buserelin, leuprolide, ACTH analog, LHRH analogs, calcitonin, glucagon, insulin) can be improved by combining drugs with the absorption enhancer to promote the permeation. The physical state of the formulation could also have an important effect on the absorption rate, as demonstrated by administration of a desmopressin solution and desmopressin nasal powder in a rabbit mucosa model (see **Figure 2**). The nasal absorption of insulin after its administration in the form of insulin/chitosan nanoparticles, insulin/chitosan solution, and insulin/chitosan powder formulations was evaluated in anesthetised rats and in conscious sheep.[36] In the sheep model, chitosan nasal powder was the most effective chitosan-based formulation for insulin absorption, with a bioavailability of 17%.

Finally, nasally administered drugs could reach the central nervous system directly from the nasal cavity, thus bypassing the blood–brain barrier. Studies in animal models and in man have shown this possibility with uptake of drugs into the cerebrospinal fluid and the brain tissue olfactory bulb depending on their molecular weight and lipophilicity.[37]

5.12.2.3   Pulmonary Route

The inhalation route has been developed primarily for the treatment of respiratory disorders that today affect two hundred million people worldwide. Pulmonary drug administration allows locally acting drugs to be administered directly to their site of action such as asthma and chronic obstructive pulmonary disease (COPD) treatments. More recently, pulmonary drug delivery has gained interest for the administration of systemic drugs, due to the large absorptive surface of the lung, the high permeability, and the fact that it bypasses the liver.

The respiratory tract is divided into the conducting and the respiratory zones.[38] An epithelial lining fluid, acting as a trap for inhaled particles, covers the conducting airways. The airways are lined by the secreting goblet cells (except in the bronchioles) and ciliated cells moving the mucus back to the pharynx. The respiratory zone comprising the alveolar region is lined with continuous sheets of epithelial cells, which differentiate in type and functions throughout the tracheobronchial tree. In the alveoli, type I cells cover almost 97% of the exposed surface area; the remaining surface area is made up of type II pneumocytes secreting surfactant. Type I pneumocytes are large (approx. 200 μm) and thin (less than 0.2 μm) cells, so their barrier to drug transport is at least one order of magnitude lower than typical mucosal or epithelial membranes. This allows the inhaled drug dose to be absorbed quickly into the blood circulation.

Passive transport and carrier-mediated active transport are the two major mechanisms of absorption. In the case of passive transport, drugs cross the cell membrane by travelling through small pores (bulk flow) or are absorbed by pinocytosis. Recently, it has been suggested that even caveolae (omega-shaped invaginations of the plasma membrane of type I cells) play a role in drug absorption.[39] The alveoli contain freely roaming macrophages that act as scavengers for inhaled particles, engulfing them and moving out of the respiratory tract with their payload.

Drugs can be easily absorbed from the alveolar site, provided that particles or droplets survive transit and clearance mechanisms. It is known that lung uptake of a formulated drug is affected in rate and extent by formulation pH, ionic strength, partition coefficient, and solubility. Macromolecules are absorbed in the lung to a degree that is inversely proportional to their molecular weight.[40]

All the phenomena affecting drug absorption in the GI tract (dissolution, permeation, stability) are also relevant for the pulmonary route, but an additional aspect has to be considered, i.e., the need for lung deposition. The pulmonary route is accessible for drugs only if they are inhaled as aerosols, i.e., airborne suspensions of fine particles, and deposited on the epithelium. It is essential to consider that the formulation and the device together constitute the inhalation product. The two components determine the performance of the administration, and a change in formulation or device will affect the bioavailability of the drug. Available devices for producing such fine aerosols are metered dose inhalers (MDIs), dry powder inhalers (DPIs), nebulizers and metered dose nebulizers (MDNs), or soft mist inhalers. The popular MDIs consist of a pressurized canister containing a mixture of micronized drug suspended or dissolved in a propellant and are devices that require coordination between dose delivery and patient inhalation during administration. DPIs were designed to deliver the drug with the minimum patient coordination. Nebulizers are predominantly employed for antibiotic therapy or in emergency situations. Metered dose nebulizers, the last generation devices, generate a slow-moving aerosol as a bolus dose to be subsequently inhaled.

Drug deposition and absorption from the lung can be assessed using in vitro and in vivo methods. Considering that a drug for pulmonary absorption requires entry into the lung by inhalation, the aerodynamic behavior of particles is decisive for the biopharmaceutical quality of the preparation. Inhaled asthma drugs act locally on receptors in the airway walls, while inhaled drugs intended for systemic delivery must be deposited in the alveolar regions to be absorbed optimally. Thus, the deposition in the lungs has a predictive role for the absorption of drugs and is the limiting step for the product's efficacy. Deposition is dependent on aerosol behavior. Particle size parameters of an inhaler, i.e., mass median aerodynamic diameter (MMAD), geometric standard deviation, and the fine particle dose (or fraction) have to be assessed. The fine particle dose has particular relevance, since it expresses the respirable mass of drug contained in an aerosol. Inertial separation methods are prescribed in order to measure the fine dose of a therapeutic aerosol. Aerosol particles are separated according to their aerodynamic diameter, which describes the particle tendency to fly along an air stream. European guidelines consider those particles with an aerodynamic diameter less than 5 μm to be respirable. **Figure 3** shows the Andersen cascade impactor (US and European Pharmacopoeia 5) for determination of aerodynamic particle size distribution and lung deposition prediction.

Methods used in the early stages of formulation involve the sizing of the aerosol particles using laser diffraction and particle time-of-flight methods. The first technique determines the volume diameter and the second the aerodynamic size directly. Some authors have compared results from impactors with laser scattering methods for simplifying the determination of the aerodynamic diameter.[41] Even though these tools are routine in the early development of a pulmonary formulation, further experiments must prove an effective correlation between in vivo and in vitro data.[42] Particle size data are essential in the quality control of the pharmaceutical product and for bioavailability expectations. The relationship between particle size data and lung deposition data is complex. The fine particle fraction or dose has been found to overestimate whole lung deposition.[43]

In vivo lung deposition is quantified by external imaging radiographic techniques employing a radiolabeled drug molecule (typically γ-emitting 6 h half-life nucleotide ^{99m}Tc) that is inhaled. Besides the commonly used gamma

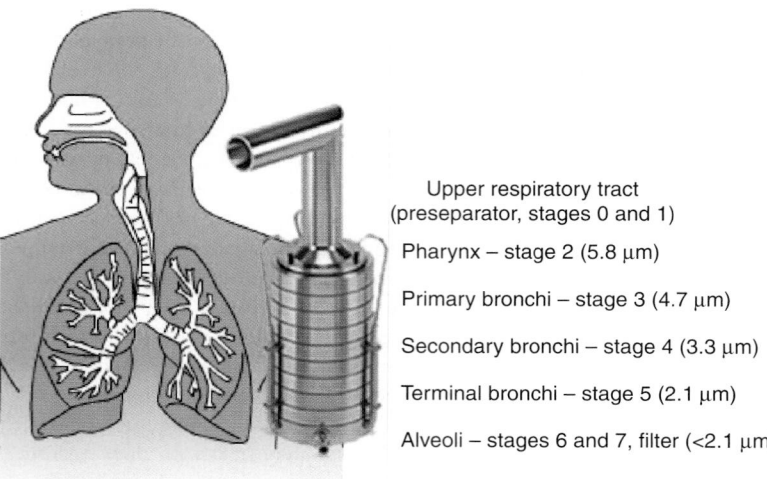

Upper respiratory tract
(preseparator, stages 0 and 1)

Pharynx – stage 2 (5.8 μm)

Primary bronchi – stage 3 (4.7 μm)

Secondary bronchi – stage 4 (3.3 μm)

Terminal bronchi – stage 5 (2.1 μm)

Alveoli – stages 6 and 7, filter (<2.1 μm)

Figure 3 Aerodynamic assessment of aerosol. Andersen Cascade Impactor: seven stages impactor able to separate the aerosol particles sizes in relation to the deposition at different levels of the human respiratory tract.

scintigraphy, other techniques include single photon emission computed tomography (SPECT), positron emission tomography (PET), and magnetic resonance imaging (MRI), the latter not employing radioactive substances.[44] Gamma scintigraphy has been used despite some concerns due to the fact that about 30% of photons detected are wrongly positioned.[45] In comparison with conventional γ-scintigraphy, SPECT has the advantage of constructing a three- dimensional image of the lung. PET does not need to chemically modify the structure of the drug to link the radioactive moiety. Results from imaging techniques are suitable for comparing efficiency of delivery and bioequivalence of pulmonary formulations. However, owing to the practical problems associated with the use of SPECT and PET, traditional planar gamma scintigraphy remains a standard.

For studying pulmonary drug absorption, in vitro models are used mainly as a formulation-screening tool, whereas in vivo analysis monitors the drug in plasma and tissues after inhalation. Cell lines and primary cultures of airways and alveolar epithelia from animals and humans have been developed and studied for pulmonary drug transport as a predictive tool to estimate the in vivo absorption.[46, 47] Examples of airway epithelial cell lines are the bronchial HBE4/E6/E7, 16HBE14o-, used for assessing β_2-adrenergic drug permeation, and the Calu-3 cells,[48] extensively investigated because of their ability to form tight junctions. Human cell line A549 is derived from type II cells that rapidly differentiate into monolayer morphologically and functionally similar to those of type I pneumocytes.[49] This cell line exhibits the in vivo enzyme activities, but it does not form tight junctions and lacks some clearance mechanisms typical of the alveolar epithelium. Monitoring the protein expression and permeability of Calu-3 cell lines grown in different conditions and placed in a liquid impinger has provided an estimation of the effect of particle impact. This interesting experimental set up combines an in vitro model with aerosol deposition.[50]

Isolated perfused organs from rats, guinea pigs, and rabbits have also been used to study deposition and transport phenomena within the lung.[51] The advantages range from the absence of interactions with other organs to the integrity of the anatomical structure and cellular organization. The role of respiration, i.e., the dynamic change in the lung tissue on drug transport across the air–blood barrier, has been investigated in a bullfrog model, in order to establish the quantitative relationship between the effect of respiration and the physicochemical properties of drugs.[52] It was suggested that respiration not only increases the surface area of lung membrane for permeation, but also dramatically affects the permeability of the lung membrane. Furthermore, the enhancement in permeation rate produced by respiration was observed to be in a linear correlation with the hydrophilicity of the substance.

Lung slices and homogenates have also been used for studying the drug transport. They are easier to prepare, but also undergo variations in enzyme activity and oxygen uptake.

Pharmacokinetic (PK) studies by body fluid analysis establish the amount absorbed of the inhaled drug, which depends on deposition and residence time.[53,54] This approach is direct for the systemic acting drugs, but is less stringent for locally acting substances. The different target, local or systemic, obliges to manipulate the drug particles formulation in order to modulate the pulmonary deposition. For example, drugs exhibiting a fast dissolution profile will be rapidly absorbed from the lung. For drugs acting locally, an increase in pulmonary selectivity has been reached by creating surface-modified microparticles. For high-molecular-weight drugs, such as peptides and proteins, an increase in the lung bioavailability could be obtained, since modifying the formulations can modulate the pulmonary targeting. For example, with insulin, large porous microparticles[55] and drug release control[56] led to an increased bioavailability in animals.

Pharmacokinetic methods for the clinical evaluation of pulmonary formulations must be considered as being highly drug specific, and provide no information about the regional deposition of the active ingredient. The assessment is obtained by comparing the plasma concentration-time profiles of the drug after inhalation (AUC_{inh}) and intravenous administration (AUC_{iv}). A problem relevant to the use of the PK approach in drug absorption studies is that a systemic drug level does not only result from lung exposure. In the case of corticosteroids delivered with MDI devices, drug passage and absorption in the GI tract is also possible, thus the AUC_{inh} will also account for this fraction of drug absorbed orally. In the case of terbutaline sulfate pulmonary bioavailability, in order to avoid concomitant oral absorption, the technique of charcoal blocking has been proposed[57]; this consists of mouth rinsing and swallowing a charcoal slurry immediately after inhalation to adsorb the drug that has entered the GI tract. The pulmonary bioavailability[54] can then be obtained by comparing AUCs:

$$F_{inh} = \frac{AUC_{inh,char}}{AUC_{iv}} \qquad [5]$$

Pulmonary availability of aerosol generated by a device (fine particle fraction) differs from one formulation to another, ranging from 10% for the older MDIs up to 40% for newer MDIs and DPIs. The pharmacokinetic approach has been used to highlight the differences in the bioavailability among different formulations, different devices, and even different patients. Then, not only the formulation differences contribute to the variability but also the respiratory effort linked to obstructive diseases, particularly where patient-driven devices are used, can add important variability to lung deposition.[58]

Pharmacodynamic (PD) data are considered to be the most reliable for proving the efficacy and safety of formulations for inhalation. Very recently, many PD studies have been carried out for bronchodilators and inhaled corticosteroids and several protocols have been developed for this use. For systemically acting drugs, PD analysis is conducted by monitoring changes in biochemical parameters, for example, the serum glucose and C-peptide in the case of inhaled insulin.

Correlations between clinical evidence and in vitro results have been reviewed by Newman[43] in a very comprehensive contribution for identifying the role of each method in inhalation drug absorption. No method for assessing inhaled drug delivery is perfect, and data from several sources are welcomed. Gamma scintigraphy is used to quantify lung deposition, but PK methods do not require the drug formulation to be manipulated during the radiolabeling process. Respirability experiments are valuable for quality assessment purposes, although their importance also extends to bioavailability.

5.12.2.4 Vaginal Route

The relatively large surface area of the vagina wall and the dense network of blood vessels make this organ suitable for systemic treatments. Vaginal administration of drugs circumvents hepatic first-pass metabolism but, in particular, opens up the possibility for prolonged drug delivery. In hormone therapies for contraception and postmenopause, the vaginal route offers greater efficacy with lower side effects than oral administration. From a histological point of view, the vaginal mucosa is formed by a lamina propria, consisting of collagen and elastin with an extensive vascular network, and a nonsecretory squamous epithelium (200–300 μm thick). In designing drug delivery systems for vaginal administration, the different physiological conditions occurring during the menstrual cycle or at different ages, along with the enzymatic activity of the vaginal microflora, should be taken into account. In particular, the pH and composition of vaginal fluid change from low pH values (3.5–4.5) during the ovulation phase to a higher pH during menstruation.[59] As for other mucosal routes, drug absorption may occur through the transcellular, paracellular, or receptor-mediated pathways.[60]

In vitro studies performed on human vaginal mucosa mounted on continuous flow-through diffusion cells allowed for the determination of the permeability of a series of molecules, in comparison with other tissues, such as buccal, small intestinal, and colonic mucosa. The vaginal mucosa specimens, obtained from postmenopausal patients after hysterectomies, had the connective and adipose tissue trimmed. They were either immediately used or snap-frozen in liquid nitrogen and stored at −85 °C for several months, without affecting their permeability.[11] Vaginal mucosa was shown to be more permeable than intestinal mucosa to various molecules and to a greater extent to lipophilic hormones such as 17β-estradiol.[61,62] In comparison with human buccal mucosa, a similar transport was shown in the case of vasopressin.[63] Recent studies on the dependence of the permeability of vaginal mucosa on temperature suggested that the main barrier to drug diffusion was lipoidal in nature and mainly linked to the presence of inter- and intracellular granules containing lipid lamellae.[64] Average thickness, surface keratinization, lipid composition, and the presence of intra- and intercellular MCGs make the vaginal mucosa similar to buccal mucosa.[65] From the histological and structural point of view, porcine vaginal mucosa is more representative of human mucosa than other animal models.[66] Sheep,[67] rabbit,[68] and cow vaginal mucosa[69] has also been employed as permeation barriers.

The use of cultures of vaginal epithelial cells has also been proposed for permeability studies. Human vaginal epithelial cells were obtained from healthy tissue specimens and grown in well plates for 2–3 weeks in an appropriate medium. At confluence, cells were trypsinized and seeded on collagen-coated membranes in a Transwell system. The flux of tritiated water through these vaginal epithelial cell layers was found to be three times higher than that measured for the intact mucosa.[70] Similar reconstructed tissues have been used for the in vitro study of mucosal irritation.[71] Highly differentiated multilayer vaginal tissue models are commercially available (EpiVaginal by MatTek, Ashland, MA, USA; Reconstituted Human Vaginal Epithelium by SkinEthic, Nice, France).

Traditional pharmaceutical forms for vaginal delivery include creams, ointments, gels, pessaries, tablets, and vaginal rings. In the case of semisolid formulations and conventional tablets, short residence time at the administration site and product leakage have been evidenced.[60] To circumvent these problems, attention has been devoted to the development of mucoadhesive formulations based on derivatives of cellulose and chitosan or on acrylic acid- based polymers. The goal was to prolong the residence time in order to enhance absorption.[72] In vitro tests for mucoadhesion properties of drug delivery systems have been proposed, determining the tensile strength of system/mucosa detachment.[69,73]

For the in vivo evaluation of vaginal formulations, several animal models have been exploited. Mice, rats, and rabbits are the most commonly used species Rhesus monkeys were employed for the development of a novel vaccine against HIV infection through vaginal administration.[74] In a recent study, the aminopeptidase activity of vaginal homogenates from rabbit, rat, guinea pig, and sheep have been compared with the activity of the same enzymes in human vaginal tissue. This could allow for the identification of the best animal model for vaginal administration of proteins and peptides. Rabbit and rat vaginal mucosa displayed aminopeptidase activity similar to human tissue.[75]

Nowadays, vaginal drug delivery is applied for local therapies with antibacterial, antifungal, antiviral, or labor-inducing drugs and for systemic treatments with hormones for contraception and postmenopausal therapies. However, the characteristics of the site and the improvements made in delivery systems indicate that vaginal administration shows promise in the delivery of peptides, proteins, and vaccines.[60,76]

5.12.2.5   Rectal Route

Rectal route for drug administration has been proposed to complement the classical GI way for systemic delivery of drugs either in cases of peculiar patient conditions, such as nausea, vomiting or convulsions, or, more importantly, for treating children.[77] From an anatomical point of view, the rectal mucosa differs slightly from other mucosal epithelia of the GI tract. However, the dense vascular network of the rectum potentially allows the systemic distribution of absorbed drugs while bypassing the first pass metabolism. In fact, even though the superior rectal vein drains into the portal vein, the inferior and middle rectal veins drain to inferior vena cava. However, the unpredictability of the exposition to the appropriate veins as well as others physiological parameters, such as pH, volume, and viscosity of the rectal fluid, are responsible for important interpatient variability in serum levels of rectally administered drugs.[78] Among the drugs administered via the rectal route anticonvulsants, analgesics, antiemetics, antipyretics, antibacterial agents, migraine medications, diuretics, and vaccines have been reported.[79–81]

The establishment of an in vitro system based on cell culture for the evaluation of rectal drug transport has been proposed.[82] However, no model system has been realized yet. In order to assess rectal formulations, in vivo experiments using rats[83] or rabbits[84] as animal models are usually performed.

In summary, despite the advantages of this administration route in pediatrics, until now the erratic drug absorption and low patient compliance have downgraded this delivery approach to line extension products.

5.12.3   Dermal and Transdermal Route for Drug Absorption

5.12.3.1   Biological Background

The skin has traditionally been used for the topical application of creams, pastes, or gels for local effect, although since the 1980s it has also been used for drug administration for systemic effect, using transdermal patches. The skin is the

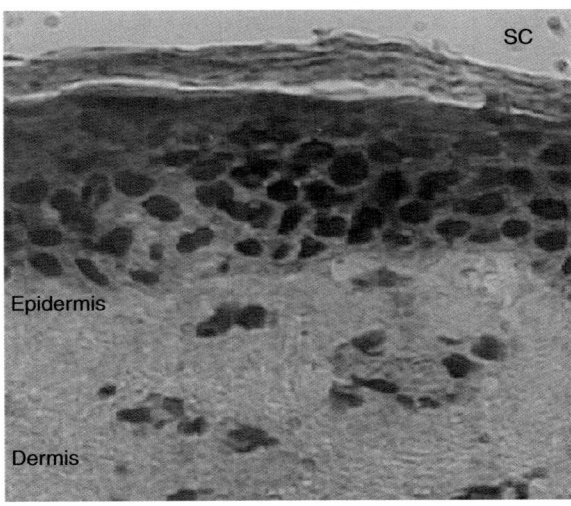

Figure 4 Microscopic image of a human skin section.

largest organ of the body, with an average weight of 7 kg in adults. With a surface area of approximately $2 m^2$, the skin offers a unique and easily accessible body surface across which drugs can be delivered. Transdermal administration is noninvasive, and allows for the attainment of constant plasma levels over extended periods of time. Additionally, the incidence of side effects after transdermal administration may be lower than after oral administration, such as in the case of estradiol.[85] Transdermal patches are discreet to wear and were preferred for nicotine replacement therapy over other administration devices (nasal sprays and pulmonary inhalers) that, although equivalent in efficacy, were considered embarrassing to use.[86]

The skin is a multilayered organ delimiting the body. **Figure 4** shows a section of human skin (haematoxylin-eosin stain) The main layers of the skin are the stratum corneum (10–15 μm thick), the viable epidermis (100–150 μm), and the dermis (1000 μm).

The structure of the stratum corneum can be described as a brick&mortar wall,[87] in which bricks represent the corneocytes and mortar the extracellular matrix. The particular composition and structure of the stratum corneum controls the diffusion and penetration of molecules in and across the skin. Cellular elements, the corneocytes, are embedded in an intercellular matrix organized as lamellar lipid layers.[88] The stratum corneum lamellar structure has been extensively investigated using different techniques, such as freeze- fracture electron microscopy followed by osmium or ruthenium tetroxide fixation[89,90] or wide and small angle x-ray diffraction.[91] Other techniques used to investigate stratum corneum structure and modifications are infrared spectroscopy, differential scanning calorimetry (DSC), and laser scanning confocal microscopy (LSCM).[92]

The stratum corneum lipids are mainly composed of ceramides, cholesterol, and free fatty acids. There are at least seven classes of ceramides, each containing various species, which have the same head group and different chain lengths. Ceramides play an important role in maintaining and structuring the water permeability barrier of the skin.

5.12.3.2   Structure–Permeation Relations

Drug absorption across the skin is considered as being composed of partition and diffusion processes, the limiting step being, in most cases, the stratum corneum. The active molecule included in the dosage form is released to the skin surface and then diffuses through the cells of the stratum corneum, eventually reaching the dermis, where blood capillaries are located. Two permeation routes across the skin are possible (**Figure 5**): annexial (i.e., through the annexes of the skin, namely pilosebaceous units and sweat glands) and transepidermal (i.e., across the stratum corneum). The latter involves diffusion through the intercellular lipid matrix and can be divided into polar and lipid routes. The annexial route is predominant for hydrophilic drugs, since skin pores have a hydrophilic character. However, since skin annexes constitute only 1% of the total surface area, the annexial route contributes to a limited extent to the total transport. The annexial route may be relevant in specific conditions, such as in the case of application of electric current (iontophoresis). Beside the use of confocal microscopy, an original approach for studying the relevance of the annexial route of skin penetration is the use of a skin sandwich.[93]

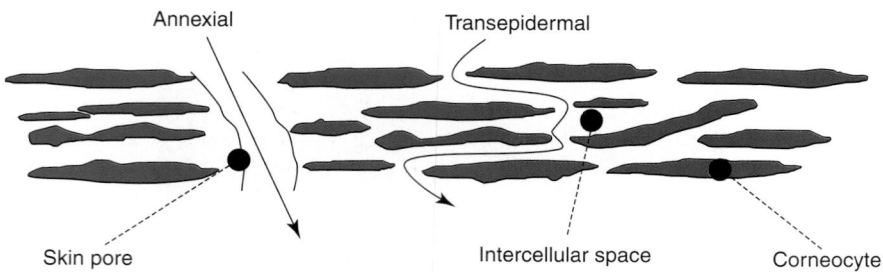

Figure 5 Permeation routes across the skin.

Owing to the excellent barrier properties of the skin, only molecules possessing the appropriate physicochemical properties can be used for this administration route.[94] In particular, water and lipid solubility should be such that the drug can cross the lipophilic barrier of the stratum corneum. On the other hand, the permeant should not be too lipophilic, otherwise the subsequent partitioning in the hydrophilic living epidermis would be hindered.

The molecular size and weight is another limiting parameter, the upper limit being around 500 Da in conditions of passive diffusion.[95] It is interesting to point out that the limit of molecular weight for penetration across the skin is identical to the limit set out by Lipinski for oral drug absorption.[96] It is well accepted that, without penetration enhancement, only relatively small and lipophilic molecules can cross the skin in therapeutically active amounts. Several attempts have been made to predict the permeability of a molecule across the skin from its physicochemical properties, the first being that of Potts and Guy,[97] who predicted the value of permeability coefficient (P_e) of a permeant on the basis of octanol/water partition coefficient (P) and molecular weight (MW):

$$\log P_e = -6.36(\pm 0.28) + 0.72(\pm 0.13)\log P - 0.0059(\pm 0.0014)\text{MW}$$
$$n = 93,\ r^2 = 0.67,\ q^2 = 0.65,\ s = 0.74,\ \text{and}\ F = 92 \tag{6}$$

More recently, solute hydrogen bond basicity and volume have been taken into consideration since they can influence permeability coefficient.[98] In a recent review, Geinoz and co-workers emphasized the limitations of such structure–permeation relationships, stressing that their applicability is restricted by the limited range of polarity and size of the permeants on which they are based.[99]

The possible approaches to increasing passive skin penetration, such as the use of penetration enhancers or supersaturated solutions, have recently been reviewed by Moser and colleagues.[100] Other alternatives are the use of physical techniques, such as iontophoresis, electroporation or sonophoresis,[94] or minimally invasive techniques such as microneedles.[101] In particular, iontophoresis consists of the application of a small electric current density ($<0.5\,\text{mA cm}^{-2}$) to the skin in order to facilitate the transdermal passage mainly of ionized molecules. The technique has been shown to promote the transdermal penetration of peptides and proteins, such as calcitonin in vivo in rabbits.[102]

5.12.3.3 In Vitro Models

In vitro permeation across the skin is usually studied using diffusion cells. Either horizontal or vertical diffusion cells can be used, the constituting material usually being glass.[103] The receptor compartment is maintained at approximately 37 °C, in such a way that the temperature of the skin surface is at its physiological value of 32 °C. Care should be taken to avoid the formation of air bubbles in the receptor compartment (which can reduce the area of contact) and to guarantee homogeneous mixing. The choice of the receptor medium, which should mimic body fluids, is conditioned by the solubility of the permeant. Isotonic pH 7.4-buffered saline can contain solubilizing additives, such as bovine serum albumin or cyclodextrins, which increase the solubility of the permeant without altering the permeability of the skin, and guarantee sink conditions to be respected. As an alternative, for low solubility permeants, flow-through diffusion cells can be used, although in this case the dilution of the receptor solution may limit the analytical sensitivity. Other special types of diffusion cells exist, most of them designed for use in specific cases, such as when iontophoresis is applied. One of them is the dual chamber diffusion cell, which allows the two electrodes to be placed on the same site of the skin – in exactly the same way as electrodes are placed in vivo.[104]

The analyte permeated across the skin is then quantified in the samples taken from the receptor compartment, using a validated analytical method. The use of radioactive tracers is in some respects quite convenient, for the rapidity of analysis, but does not allow for evaluating skin metabolism, as do chromatographic methods, since it is not specific.

High-performance liquid chromatography (HPLC) represents the method of choice in most cases. Microdialysis, a typical in vivo technique, has been proposed as an in vitro tool for assessing drug penetration through the skin, with the advantage of measuring local drug concentration at the skin level the probe was inserted.[105] Running a permeation experiment on static diffusion cells is time-consuming, and needs the presence of an operator to withdraw samples of receptor solution at each sampling interval. Automatic diffusion cell systems are available on the market containing up to 12 diffusion cells that can run simultaneously, while samples are collected automatically into HPLC or liquid scintillation vials. The Microette Plus Diffusion Cell Test System (Hanson Research, Chatsworth, CA, USA) has been validated for drug release from semisolid formulations across a synthetic membrane.[106] The use of a synthetic membrane as a barrier for separating the donor and receptor solutions has the advantage of easy availability and reproducibility. There are no membranes able to simulate the complex structure of the skin. Synthetic membranes can be useful for quality control purposes in topical formulations, but for skin permeation studies a biological barrier has to be used. The most obvious skin model used in in vitro penetration experiments is excised human skin. Cadaver skin or skin coming from surgery can be used as full thickness, i.e., with all layers present. However, drug absorption to the systemic circulation takes place at the level of the upper dermis, and so the use of full thickness skin can create an artificial barrier to drug penetration that is not present in vivo. For this reason, sections of the skin can be used to better mimic the in vivo situation. The use of a dermatome allows for the preparation of a sample with a predetermined and constant thickness. As an alternative, the full thickness skin can be heat-separated into epidermis and dermis. Although skin heating (at approximately 60 °C for 1 min) can produce a modification of viability,[107] this technique is widely used, and preferred over chemical separation.[108] Artusi and co-workers [109] used both full thickness and heat-separated epidermis for studying skin permeation of thiocolchicoside, and found that with epidermis the lag time was 4 h, compared to 12 h for full thickness skin. Another alternative is the use of isolated stratum corneum, which can be separated from epidermis through the enzymatic action of trypsin. The rationale behind the use of isolated stratum corneum is that, with polar drugs, it represents the rate limiting step to drug transport. Sekkat and co-workers[110] demonstrated that lidocaine permeation across pig ear skin increased when the stratum corneum was progressively removed.

Although the use of fresh skin has the basic advantage of preserving the metabolic activity of the skin, it is seldom available. Frozen skin (at − 20 °C) is more commonly used, although freezing can alter its metabolic activity.[107] There are several reports concerning the differences between fresh and frozen skin, most of them suggesting that freezing increases the permeability of the skin. However, for some molecules, such as 8-methoxypsoralen, fresh skin is more permeable than frozen skin, owing to alterations in the molecular arrangement of the skin components during freezing.[111]

Animal skin is often used as an alternative to human skin.[112] The use of hairless or bare species is preferred over hairy ones, because it has been shown that the presence of hair follicles produces shunts available for permeation that produce an increase in permeability. Additionally, shaving can damage the stratum corneum, leading to an increase in skin permeability. Mouse, rat, pig, and guinea pig are the most common species used, although their permeability is usually higher than that of human skin.[100] Pig skin has been shown to be most similar to human skin in terms of both biochemical composition of the stratum corneum lipids and of permeability.[113]

Perfused skin models use excised regions of the skin, complete with their associated microvasculature. They have the advantage of maintaining the viability of the skin, since the tissue is continuously perfused. Rabbit ear and pig skin flaps have been used for this purpose.[114,115]

Skin equivalents have recently been developed as alternatives to the use of excised skin. All of them are obtained from keratinocytes cultured in vitro on a substrate of collagen or fibroblast containing collagen gel. The use of air interface technique allows for epidermis differentiation, mimicking the natural differentiation of the skin in vivo. This procedure allows for the formation of a stratum corneum, if an appropriate medium is used, having a composition very close to human skin. Among commercially available products are EpiSkin developed by Imedex and Oreal and EpiDerm developed at MatTek. These skin equivalents can be successfully used for in vitro skin irritation and metabolism studies, since they are a living substrate. However, the permeability of skin equivalents is typically much higher than that of excised human skin. Wagner and coworkers examined the permeability to flufenamic acid of EpiDerm, isolated stratum corneum, epidermis, and full thickness human skin. The permeability obtained with the skin equivalent was 5 times higher than with isolated stratum corneum and epidermis and 30 times higher than with full thickness skin.[116]

The flux of a drug across a membrane (J) can be described in terms of Fick's law of diffusion[117]:

$$J = -D\frac{\mathrm{d}c}{\mathrm{d}x} \qquad [7]$$

where D is the diffusion coefficient and $\mathrm{d}c/\mathrm{d}x$ represents the concentration gradient across the barrier.

In the simplest case of an infinite donor solution at a constant concentration, the cumulative drug permeated is plotted versus time, and the steady-state permeation rate is calculated from the slope of the linear portion of the curve. However, in the case of skin penetration, the attainment of steady-state conditions may take too long to be reliably measured in a diffusion cell experiment. A better approach is to fit the entire curve to the appropriate solution of Fick's law of diffusion[118]:

$$Q = (K\,H)C_{veh}\left[\frac{D}{H^2}t - \frac{1}{6} - \frac{2}{p^2}\sum_{n=1}^{n}\frac{(-1)}{n^2}\exp\left(\frac{-Dn^2p^2t}{H^2}\right)\right] \qquad [8]$$

where Q is the cumulative amount of drug permeated per unit area at time t, C_{veh} is the concentration of the drug in the donor vehicle, K is the stratum corneum/vehicle partition coefficient, D is the diffusion coefficient, and H is the diffusion path length.

The permeability coefficient (P_e) can be calculated as:

$$P_e = K * H * \frac{D}{H^2} \qquad [9]$$

and steady-state flux (J_{SS}) as:

$$J_{SS} = P_e * C_{veh} \qquad [10]$$

This method can be used to gain indications as to the mechanism of action of different enhancers. Harrison and co-workers[119] used this method to compare the relative effect of Azone and Transcutol on 4-cyanophenol permeation and found that Azone reduces the diffusional resistance while Transcutol increases the solubility of the permeant in the stratum corneum.

Skin stripping is a technique that can be used either in vitro or in vivo. The rationale behind its use is the consideration that this layer is often the rate-limiting barrier for percutaneous penetration. It is reasonable then to expect that the concentration of a topically applied drug in the stratum corneum will be related to its concentration in deeper skin layers and in some cases also to the total amount absorbed.[120] Tape stripping consists of the progressive removal of stratum corneum by serial application of adhesive tape and subsequent determination of the amount of drug in each tape strip.[100] If the amount of stratum corneum removed by each tape strip is known, it is possible to calculate the stratum corneum thickness and to determine the permeation parameters of the permeants, using an appropriate solution of Fick's law of diffusion, which predicts the concentration of the drug as a function of position and time within the stratum corneum ($C(x)$)[121]:

$$C(x) = PC_{veh}\left[1 - \frac{x}{H} - \frac{2}{\pi}\sum_{n=1}^{n}\frac{1}{n}\sin\frac{n\pi x}{H}\exp\left(\frac{-Dn^2\pi^2t}{H^2}\right)\right] \qquad [11]$$

where C_{veh} is the concentration of the drug in the donor vehicle, D is the diffusion coefficient in the stratum corneum, P is the stratum corneum/vehicle partition coefficient, and H is the diffusional path length.

Attenuated total reflectance Fourier transform infrared spectrometry (ATR/FTIR) has been used as an alternative to the conventional diffusion cell to determine drug or solvent permeation across the stratum corneum.[119] The stratum corneum is placed on the ZnSe crystal of the ATR/FTIR spectrometer face-up and is covered with the drug donor solution. The diffusion of the permeant is monitored by the appearance and increase in the drug-specific infrared (IR) absorbance as a function of time. By appropriate calibration, the absorbance can be transformed into concentration. Pirot and coworkers[121] validated the ATR/FTIR method, using liquid scintillation counting as analytical reference method and 4-cyanophenol as a model permeant: they found a direct correlation between the amount of permeant determined by ATR/FTIR and liquid scintillation counting. Unfortunately, this technique is limited to IR-active permeants, showing absorption bands in the regions of the spectrum free of skin-related bands.

Finally, autoradiography and laser scanning confocal microscopy (LSCM) are used mostly to visualize and to quantify penetration through and distribution within the skin. Applications of LSCM in transdermal drug delivery, recently reviewed by Alvarez-Roman et al.,[122] include the determination of the distribution of fluoropores in hair follicles and the study of enzyme activity and localization in the skin.

5.12.4 **Conclusions**

Nonoral routes are considered novel routes for drug administration. The increased frequency in their application for systemic drug absorption is due to the increased understanding of the transport mechanisms across tissues as well as the development of new materials and devices. The introduction in therapy of new products for alternative routes of administration is mainly presented as line extension. Nevertheless, it is useful as a series of additional in vivo studies, particularly where the anatomic district involved in the administration has never been exposed to the drug formulation before. However, while showing the advantages of new ways of administering a drug, safe use of the drug product originally formulated for other routes requires not only in vitro demonstration, but also in particular additional clinical data. The potential benefits in terms of patient acceptance, therapy outcome, and pharmacoeconomic value ensure that the pharmaceutical development of drug products for alternative routes of administration continues to move forward.

References

1. Peppas, N. A.; Buri, P. A. *J. Control. Release* **1985**, *2*, 257–275.
2. Peppas, N. A. *Int. J. Pharm.* **2004**, *277*, 11–17.
3. Rathbone, M. J.; Drummond, B. K.; Tucker, I. G. *Adv. Drug Deliv. Rev.* **1994**, *13*, 1–22.
4. Squier, C. A.; Wertz, P. W. Structure and Function of the Oral Mucosa and Implications for Drug Delivery. In *Oral Mucosal Drug Delivery*; Rathbone, M. J., Ed.; Marcel Dekker: New York, 1996, pp 1–26.
5. Squier, C. A.; Hall, B. K. *Arch. Oral Biol.* **1985**, *30*, 485–491.
6. De Vries, M. E.; Bodde, H. E.; Verhoef, J. C.; Ponec, M.; Craane, W. I. H. M.; Junginger, H. E. *Int. J. Pharm.* **1991**, *76*, 25–35.
7. Hoogstraate, J. A.; Cullander, C.; Nagelkerke, J. F.; Spies, F.; Verhoef, J. C.; Schrijivers, A. H. G. J.; Junginger, H. E.; Bodde, H. E. *J. Control. Release* **1996**, *39*, 71–78.
8. Xiang, J.; Fang, X.; Li, X. *Int. J. Pharm.* **2002**, *231*, 57–66.
9. Zhang, H.; Robinson, J. R. In Vitro Methods for Measuring Permeability of the Oral Mucosa. In *Oral Mucosal Drug Delivery*; Rathbone, M. J., Ed.; Marcel Dekker: New York, 1996, pp 85–100.
10. Artusi, M.; Santi, P.; Colombo, P.; Junginger, H. E. *Int. J. Pharm.* **2003**, *250*, 203–213.
11. Van der Bijl, P.; Van Eyk, A. D.; Thompson, I. O *S. Afr. J. Sci.* **1998**, *94*, 499–502.
12. Junginger, H. E.; Hoogstraate, J. A.; Verhoef, J. C. *J. Control. Release* **1999**, *62*, 149–159.
13. Van Eyk, A. D.; Van der Bijl, P. *Arch. Oral Biol.* **2004**, *49*, 387–392.
14. Rathbone, M. J. *Int. J. Pharm.* **1991**, *69*, 103–108.
15. Smart, J. D.; Nantwi, P. K.; Rogers, D. J.; Green, K. L. *Eur. J. Pharm. Biopharm.* **2002**, *53*, 289–292.
16. Bredenberg, S.; Duberg, M.; Lennernas, B.; Lennernas, H.; Pettersson, A.; Westerberg, M.; Nystrom, C. *Eur. J. Pharm. Sci.* **2003**, *20*, 327–334.
17. Russo, P.; Buttini, F.; Sonvico, F.; Bettini, R.; Massimo, G.; Sacchetti, C.; Colombo, P.; Santi, P. *J. Drug Deliv. Sci. Tech.* **2004**, *14*, 449–454.
18. Weinberg, D. S.; Inturrisi, C. E.; Reidenberg, B.; Moulin, D. E.; Nip, T. J.; Wallenstein, S.; Houde, R. W.; Foley, K. M. *Clin. Pharmacol. Ther.* **1988**, *44*, 335–342.
19. Cui, Z.; Mumper, R. J. *Pharm. Res.* **2002**, *19*, 947–953.
20. Rossi, S.; Sandri, G.; Ferrari, F.; Bonferoni, M. C.; Caramella, C. *Pharm. Dev. Technol.* **2003**, *8*, 199–208.
21. Tiwari, D.; Goldman, D.; Town, C.; Sause, R.; Madan, P. L. *Pharm. Res.* **1999**, *16*, 1775–1780.
22. Illum, L. *Drug Disc. Today* **2002**, *7*, 1184–1189.
23. Dimova, S.; Brewster, M. E.; Noppe, M.; Jorissen, M.; Augustijns, P. *Toxicol. In Vitro* **2005**, *19*, 107–122.
24. Merkle, H. P.; Ditzinger, G.; Lang, S. R.; Peter, H.; Schmidt, M. C. *Adv. Drug Deliv. Rev.* **1998**, *29*, 51–79.
25. Agu, R. U.; Vu Dang, H.; Jorissen, M.; Kinget, R.; Verbeke, N. *Peptides* **2004**, *25*, 563–569.
26. Harris, C. M.; Mendes, F.; Dragomir, A.; Doull, I. J.; Carvalho-Oliveira, I.; Bebok, Z.; Clancy, J. P.; Eubanks, V.; Sorscher, E. J.; Roomans, G. M. et al. *J. Cyst. Fibros.* **2004**, *3*, 43–48.
27. Werner, U.; Kissel, T. *Pharm. Res.* **1996**, *13*, 978–988.
28. Witschi, C.; Mrsny, R. J. *Pharm. Res.* **1999**, *16*, 382–390.
29. Wadell, C.; Bjork, E.; Camber, O. *Eur. J. Pharm. Sci.* **1999**, *7*, 197–206.
30. Wadell, C.; Bjork, E.; Camber, O. *Eur. J. Pharm. Sci.* **2003**, *18*, 47–53.
31. Gavini, E.; Rassu, G.; Sanna, V.; Cossu, M.; Giunchedi, P. *J. Pharm. Pharmacol.* **2005**, *57*, 287–294.
32. Hussain, A. A. *Adv. Drug Deliv. Rev.* **1998**, *29*, 39–49.
33. Merkus, F. W.; Verhoef, J. C.; Schipper, N. G.; Marttin, E. *Adv. Drug Deliv. Rev.* **1998**, *29*, 13–38.
34. Ikechukwu Ugwoke, M.; Kaufmann, G.; Verbeke, N.; Kinget, R. *Int. J. Pharm.* **2000**, *202*, 125–131.
35. Merkus, F. W.; Verhoef, J. C.; Marttin, E.; Romeijn, S. G.; Van der Kuy, P. H.; Hermens, W. A.; Schipper, N. G. *Adv. Drug Deliv. Rev.* **1999**, *36*, 41–57.
36. Dyer, A. M.; Hinchcliffe, M.; Watts, P.; Castile, J.; Jabbal-Gill, I.; Nankervis, R.; Smith, A.; Illum, L. *Pharm. Res.* **2002**, *19*, 998–1008.
37. Illum, L. *J. Pharm. Pharmacol.* **2004**, *56*, 3–17.
38. Altiere, R. J.; Thompson, D. C. Physiology and Pharmacology of the Airways. In *Inhalation Aerosols*; Hickey, A. J., Ed.; Marcel Dekker: New York, 1996, pp 233–272.
39. Gumbleton, M.; Hollins, A. J.; Omidi, Y.; Campbell, L.; Taylor, G. *J. Control. Release* **2003**, *87*, 139–151.
40. Byron, P. R.; Patton, J. S. *J. Aerosol Med.* **1994**, *7*, 49–75.
41. Vecellio None, L.; Grimbert, D.; Becquemin, M. H.; Boissinot, E.; Le Pape, A.; Lemarie, E.; Diot, P. *J. Aerosol Med.* **2001**, *14*, 107–114.

42. Newman, S. P. *J Aerosol Med.* **1998**, *11*, S97–S104.
43. Newman, S. P.; Wilding, I. R.; Hirst, P. H. *Int. J Pharm* **2000**, *208*, 49–60.
44. Newman, S. P.; Pitcairn, G. R.; Hirst, P. H.; Rankin, L. *Adv. Drug Deliv. Rev.* **2003**, *55*, 851–867.
45. Fleming, J. S.; Conway, J. H.; Bolt, L.; Holgate, S. T. *J. Aerosol Med.* **2003**, *16*, 9–19.
46. Mathias, N. R.; Yamashita, F.; Lee, V. H. L. *Adv. Drug Deliv. Rev.* **1996**, *22*, 215–249.
47. Forbes, B. *Methods Mol. Biol.* **2002**, *188*, 65–75.
48. Florea, B. I.; Cassara, M. L.; Junginger, H. E.; Borchard, G. *J. Control. Release* **2003**, *87*, 131–138.
49. Yang, X.; Ma, J. K.; Malanga, C. J.; Rojanasakul, Y. *Int. J. Pharm.* **2000**, *195*, 93–101.
50. Fiegel, J.; Ehrhardt, C.; Schaefer, U. F.; Lehr, C. M.; Hanes, J. *Pharm. Res.* **2003**, *20*, 788–796.
51. Smith, B. R.; Bend, J. R. *Methods Enzymol.* **1981**, 77, 105–120.
52. Yu, J.; Chien, Y. W. *Pharm. Dev. Technol.* **2002**, 7, 215–225.
53. Gonda, I. *J. Pharm. Sci.* **1987**, 77, 340–346.
54. Mobley, C.; Hochhaus, G. *Drug Disc. Today* **2001**, *6*, 367–375.
55. Edwards, D. A.; Hanes, J.; Caponetti, G.; Hrkach, J.; Ben-Jebria, A.; Eskew, M. L.; Mintzes, J.; Deaver, D.; Lotan, N.; Langer, R. *Science* **1997**, *276*, 1868–1871.
56. Surendrakumar, K.; Martyn, G. P.; Hodgers, E. C.; Jansen, M.; Blair, J. A. *J. Control. Release* **2003**, *91*, 385–394.
57. Borgstrom, L.; Derom, E.; Stahl, E.; Wahlin-Boll, E.; Pauwels, R. *Am. J. Respir. Crit. Care Med.* **1996**, *153*, 1636–1640.
58. Borgstrom, L.; Bengtsson, T.; Derom, E.; Pauwels, R. *Int. J. Pharm.* **2000**, *193*, 227–230.
59. Alexander, N. J.; Baker, E.; Kaptein, M.; Karck, U.; Miller, L.; Zampaglione, E. *Fertil. Steril.* **2004**, *82*, 1–12.
60. Hussain, A.; Ahsan, F. *J. Control. Release* **2005**, *103*, 301–313.
61. Van der Bijl, P.; Thompson, I. O.; Squier, C. A. *Eur. J. Oral Sci.* **1997**, *105*, 571–575.
62. Van der Bijl, P.; Van Eyk, A. D.; Thompson, I. O. *Oral Surg. Oral Med. Oral Pathol. Oral Radiol. Endod.* **1998**, *85*, 393–398.
63. Van der Bijl, P.; Van Eyk, A. D.; Thompson, I. O.; Stander, I. A. *Eur. J. Oral Sci.* **1998**, *106*, 958–962.
64. Van der Bijl, P.; Van Eyk, A. D.; Kriel, J. *S. Afr. J. Sci.* **2003**, *58*, 95–101.
65. Thompson, I. O.; Van der Bijl, P.; Van Wyk, C. W.; Van Eyk, A. D. *Arch. Oral Biol.* **2001**, *46*, 1091–1098.
66. Davis, C. C.; Kremer, M. J.; Schlievert, P. M.; Squier, C. A. *Am. J. Obstet. Gynecol.* **2003**, *189*, 1785–1791.
67. Vermani, K.; Garg, S.; Zaneveld, L. J. *Drug Dev. Ind Pharm* **2002**, *28*, 1133–1146.
68. Acarturk, F.; Robinson, J. R. *Pharm. Res.* **1996**, *13*, 779–783.
69. Valenta, C.; Kast, C. E.; Harich, I.; Bernkop-Schnurch, A. *J. Control. Release* **2001**, 77, 323–332.
70. Van Eyk, A. D.; Van der Bijl, P. *S. Afr. J. Sci.* **1998**, *53*, 497–503.
71. D'Cruz, O. J.; Waurzyniak, B.; Uckun, F. M. *Toxicol. Pathol.* **2004**, *32*, 212–221.
72. Edsman, K.; Hagerstrom, H. *J. Pharm. Pharmacol.* **2005**, *57*, 3–22.
73. Sandri, G.; Rossi, S.; Ferrari, F.; Bonferoni, M. C.; Muzzarelli, C.; Caramella, C. *Eur. J. Pharm. Sci.* **2004**, *21*, 351–359.
74. Bogers, W. M.; Bergmeier, L. A.; Ma, J.; Oostermeijer, H.; Wang, Y.; Kelly, C. G.; Ten Haaft, P.; Singh, M.; Heeney, J. L.; Lehner, T. *Aids* **2004**, *18*, 25–36.
75. Acarturk, F.; Parlatan, Z. I.; Saracoglu, O. F. *J. Pharm. Pharmacol.* **2001**, *53*, 1499–1504.
76. Kwant, A.; Rosenthal, K. L. *Vaccine* **2004**, *22*, 3098–3104.
77. Van Hoogdalem, E.; de Boer, A. G.; Breimer, D. D. *Clin. Pharmacokinet.* **1991**, *21*, 11–26.
78. Song, Y.; Wang, Y.; Thakur, R.; Meidan, V. M.; Michniak, B. *Crit. Rev. Ther. Drug Carrier Syst.* **2004**, *21*, 195–256.
79. Worthington, I. *Can. Fam. Physician* **2001**, *47*, 322–329.
80. Regdon, G., Jr.; Deak, D.; Regdon, G., Sr.; Musko, Z.; Eros, I. *Pharmazie* **2001**, *56*, 70–73.
81. McCluskie, M. J.; Davis, H. L. *Vaccine* **2000**, *19*, 413–422.
82. Audus, K. L.; Bartel, R. L.; Hidalgo, I. J.; Borchardt, R. T. *Pharm. Res.* **1990**, 7, 435–451.
83. Sallai, J.; Vernyik, A.; Regdon, G.; Gombkoto, S.; Nemeth, J.; Regdon, G., Jr. *J. Pharm. Pharmacol.* **1997**, *49*, 496–499.
84. Yagi, N.; Taniuchi, Y.; Hamada, K.; Sudo, J.; Sekikawa, H. *Biol. Pharm. Bull.* **2002**, *25*, 1614–1618.
85. Ramachandran, C.; Fleisher, D. *Adv. Drug Deliv. Rev.* **2000**, *42*, 197–223.
86. Hajek, P.; West, R.; Foulds, J.; Nilsson, F.; Burrows, S.; Meadow, A. *Arch. Intern. Med.* **1999**, *159*, 2033–2038.
87. Harding, C. R. *Dermatol. Ther.* **2004**, *17*, 6–15.
88. Elias, P. M.; Goerke, J.; Friend, D. S. *J. Invest. Dermatol.* **1977**, *69*, 535–546.
89. Haftek, M.; Teillon, M. H.; Schmitt, D. *Microsc. Res. Tech.* **1998**, *43*, 242–249.
90. Van Hal, D. A.; Jeremiasse, E.; Junginger, H. E.; Spies, F.; Bouwstra, J. A. *J. Invest. Dermatol.* **1996**, *106*, 89–95.
91. Brinkmann, I.; Muller-Goymann, C. C. *Pharmazie* **2005**, *60*, 215–220.
92. Potts, R. O. Physical Characterization of the Stratum Corneum: The Relationship of Mechanical Barrier Properties to Lipid and Protein Structure. In *Transdermal Drug Delivery*; Hadgraft, J., Guy, R. H., Eds.; Marcel Dekker: New York, 1989, pp 23–57.
93. Essa, E. A.; Bonner, M. C.; Barry, B. W. *J. Pharm. Pharmacol.* **2002**, *54*, 1481–1490.
94. Naik, A.; Kalia, Y. N.; Guy, R. H. *Pharm. Sci. Technol. Today* **2000**, *3*, 318–326.
95. Bos, J. D.; Meinardi, M. M. *Exp. Dermatol.* **2000**, *9*, 165–169.
96. Lipinski, C. A.; Lombardo, F.; Dominy, B. W.; Feeney, P. J. *Adv. Drug Deliv. Rev.* **2001**, *46*, 3–26.
97. Potts, R. O.; Guy, R. H. *Pharm. Res.* **1992**, *9*, 663–669.
98. Abraham, M. H.; Martins, F. *J. Pharm. Sci.* **2004**, *93*, 1508–1523.
99. Geinoz, S.; Guy, R. H.; Testa, B.; Carrupt, P. A. *Pharm. Res.* **2004**, *21*, 83–92.
100. Moser, K.; Kriwet, K.; Naik, A.; Kalia, Y. N.; Guy, R. H. *Eur. J. Pharm. Biopharm.* **2001**, *52*, 103–112.
101. Guy, R. H. *B.T. Gattefossé* **2003**, *96*, 47–62.
102. Santi, P.; Colombo, P.; Bettini, R.; Catellani, P. L.; Minutello, A.; Volpato, N. M. *Pharm. Res.* **1997**, *14*, 63–66.
103. Brain, K. R.; Walters, K. A.; Watkinson, A. C. Investigation of Skin Permeation In Vitro. In *Dermal Absorption and Toxicity Assessment*; Roberts, M. S., Walters, K. A., Eds.; Marcel Dekker: New York, 1998, pp 161–188.
104. Glikfeld, P.; Cullander, C.; Hinz, R. S.; Guy, R. H. *Pharm. Res.* **1988**, *5*, 443–446.
105. Leveque, N.; Makki, S.; Hadgraft, J.; Humbert, P. *Int. J. Pharm.* **2004**, *269*, 323–328.
106. Rapedius, M.; Blanchard, J. *Pharm. Res.* **2001**, *18*, 1440–1447.
107. Wester, R. C.; Christoffel, J.; Hartway, T.; Poblete, N.; Maibach, H. I.; Forsell, J. *Pharm. Res.* **1998**, *15*, 82–84.

108. Diaz, L. A.; Heaphy, M. R.; Calvanico, N. J.; Tomasi, T. B.; Jordon, R. E. *J. Invest. Dermatol.* **1977**, *68*, 36–38.

109. Artusi, M.; Nicoli, S.; Colombo, P.; Bettini, R.; Sacchi, A.; Santi, P. *J. Pharm. Sci.* **2004**, *93*, 2431–2438.

110. Sekkat, N.; Kalia, Y. N.; Guy, R. H. *Pharm. Res.* **2004**, *21*, 1390–1397.

111. Shaikh, N. A.; Ademola, J. I.; Maibach, H. I. *Skin Pharmacol.* **1996**, *9*, 274–280.

112. Panchagnula, R.; Stemmer, K.; Ritschel, W. A. *Methods Find. Exp. Clin. Pharmacol.* **1997**, *19*, 335–341.

113. Meyer, W.; Schwarz, R.; Neurand, K. *Curr. Probl. Dermatol.* **1978**, 7, 39–52.

114. Williams, P. L.; Carver, M. P.; Riviere, J. E. *J. Pharm. Sci.* **1990**, *79*, 305–311.

115. Seki, T.; Hosoya, O.; Yamazaki, T.; Sato, T.; Saso, Y.; Juni, K.; Morimoto, K. *Int. J. Pharm.* **2004**, *276*, 29–40.

116. Wagner, H.; Kostka, K. H.; Lehr, C. M.; Schaefer, U. F. *J. Control. Release* **2001**, *75*, 283–295.

117. Cussler, E. L. Models for Diffusion. In *Diffusion – Mass Transfer in Fluid Systems*; Cussler, E. L., Ed.; Cambridge University Press: Cambridge, UK, 1984, pp 1–10.

118. Moser, K.; Kriwet, K.; Froehlich, C.; Kalia, Y. N.; Guy, R. H. *Pharm. Res.* **2001**, *18*, 1006–1011.

119. Harrison, J. E.; Watkinson, A. C.; Green, D. M.; Hadgraft, J.; Brain, K. *Pharm. Res.* **1996**, *13*, 542–546.

120. Shah, V. P.; Flynn, G. L.; Yacobi, A.; Maibach, H. I.; Bon, C.; Fleischer, N. M.; Franz, T. J.; Kaplan, S. A.; Kawamoto, J.; Lesko, L. J. et al. *Skin Pharmacol. Appl. Skin Physiol.* **1998**, *11*, 117–124.

121. Pirot, F.; Kalia, Y. N.; Stinchcomb, A. L.; Keating, G.; Bunge, A.; Guy, R. H. *Proc. Natl. Acad. Sci. USA* **1997**, *94*, 1562–1567.

122. Alvarez-Roman, R.; Naik, A.; Kalia, Y. N.; Fessi, H.; Guy, R. H. *Eur. J. Pharm. Biopharm.* **2004**, *58*, 301–316.

Biographies

Paolo Colombo, born in 1944 in Italy, was graduated in Pharmacy in 1968 at the University of Pavia, Italy. In 1986 he became Full Professor of Pharmaceutical Technology at the University of Parma, Italy, where he is now active. He is presently the Head of the Department of Pharmacy.

Paolo Colombo was acknowledged with the award for scientific production by University of Pavia in 1973, the Colorcon Award in 1991 and the AFI-Pharma Works 1992 Award. He was called as AAPS Fellow in 1996 and awarded with the 1999 Jorge Heller Journal of Controlled Release/CRS Outstanding Paper Award. From April 18 to July 30, 2001 he has been Associated Professor at the Faculty of Pharmacy of the University of Paris Sud. In 2003 the Academy of Pharmacy of France nominated Prof Colombo as foreign correspondent member. In 2004 he received the Maurice-Marie Janot Award as highly recognized drug research scientist. Professor Paolo Colombo has over 180 original papers and 30 patents on drugs and drug delivery systems.

His research interests are devoted to oral matrices delivery, inhalatory nasal and pulmonary powders and transdermal delivery. Many patents filed became products registered in various countries. One is the Geomatrix technology consisting in a hydrophilic matrix, partially coated with impermeable films that allows the modulation of drug delivery kinetics. Several products based on this technology are registered. The first one was Dilacor XR, containing diltiazem. Another marketed product is the new physical form of the antiulcer drug sucralfate. This new form named sucralfate gel, resulted much more active than the regular sucralfate powder, allowing the reduction of the therapy schedule of sucralfate from 4 to $2\,\mathrm{g\,day}^{-1}$.

Stefano Cagnani received his PharmD from University of Parma in 1998 and the PhD degree in Biopharmaceutics and Pharmacokinetics from the same University in 2004. His interests are focused on the pulmonary field with particular attention to the delivery of peptides and proteins. He is co-inventor of a formulation for the pulmonary administration of insulin.

Fabio Sonvico was born in Italy in 1974 and he graduated in Pharmaceutical Chemistry and Technology in 1999 at the University of Pavia (Italy) after spending a research stage at the laboratories of the Bayer Italia in Milan. He obtained a co-tutorial PhD at the Université de Paris Sud and the University of Parma. The subject of the research was 'Folate-targeted metallic nanoparticles fot the treatment of solid tumors.' Since 2004, he is funded from the National Institute for the Physics of the Matter (INFM) on a research project dealing with the development of biopolymeric nanosystems suitable for the controlled delivery of drugs.

Patrizia Santi received her Pharm D and subsequently PhD in Pharmaceutical Chemistry and Technology from the University of Parma (Italy), where, in 1991, she became Assistant Professor. From 1998 to 2002 she was Associate

Professor in the same University (Faculty of Pharmacy) where she is Full Professor of Pharmaceutical Technology since 2002. She has published 58 papers and she is co-inventor of nine patents. Her main research interests are transdermal drug delivery and iontophoresis.

Paola Russo was born in Italy in 1977. She graduated in Pharmaceutical Chemistry and Technology in 2001 at the University of Naples (Italy), where she also obtained her PhD. The subject of the research was 'Microparticulates for drug administration.' In 2004 she was granted of a fellowship from 'Consorzio SPINNER' working on a research project dealing with nasal drug delivery at the Pharmaceutical Department, University of Parma. Today, she has a research assistant position at the university of Salerno (Italy).

Gaia Colombo was born in Italy in 1976. In 2001 she graduated in Pharmaceutical Chemistry and Technology at the University of Parma (Italy), after an 8-month research stage at the faculty of Pharmacy, University of Paris XI (France). From 2002 to 2004 she was granted of a PhD scholarship from the University of Parma and spent an 18-month appointment at Massachusetts Institute of Technology (Cambridge, MA, USA) within Prof Robert Langer's research team, where she worked on new systems for prolonged local anesthesia. Since January 2004, she has a permanent research assistant position at the University of Ferrara (Italy), where she is involved in several projects about nasal, buccal, and oral drug delivery.

Comprehensive Medicinal Chemistry II
ISBN (set): 0-08-044513-6

ISBN (Volume 5) 0-08-044518-7; pp. 279–299

5.13 In Vitro Models for Examining and Predicting Brain Uptake of Drugs

N J Abbott, King's College London, London, UK

5.13.1   Introduction

5.13.1.1   Central Nervous System Barriers

The neurons of the brain communicate with each other using chemical and electrical signals that depend on fine control of the local ionic microenvironment; moreover such communication requires greater protection from circulating toxins than found in most other tissues of the body.[1–3] Barriers at three interfaces separate the blood from the brain interstitial fluid (ISF, also called extracellular fluid, ECF) (**Figure 1**): the blood–brain barrier (BBB), formed by the endothelial cells lining the microvessels (**Figure 2**); the blood–CSF barrier, formed by the choroid plexus epithelium, which also secretes cerebrospinal fluid (CSF); and the arachnoid epithelium, which forms part of the meningeal covering.[4] At each of these layers, tight junctions between cells form the 'physical' barrier, specific transport proteins/ transcytosis mechanisms mediate uptake and efflux ('transport' barrier), and enzymes add a 'metabolic' barrier. Together these mechanisms regulate molecular traffic between blood and brain, both inward and outward. The BBB, with the largest surface area and shortest diffusion distances to neurons, is the most important interface in regulating drug permeability to the brain and is the focus of this chapter; however, for some drugs and pathological conditions, the other two barriers may also be relevant.

5.13.1.2   Cellular Anatomy of the Blood–Brain Barrier In Vivo

A number of cell types are closely associated with the brain endothelium (**Figure 2**). These include pericytes, astrocytic glial cells, microglia, and several cells that may be present in the perivascular space, including derivatives from the meninges, and in certain pathological conditions such as inflammation, invading cells from the blood.[3] Neurons occasionally send processes toward the endothelium, but are generally separated from direct contact by the astrocytes. In arterioles, smooth muscle cells controlling microvessel diameter are present between the astrocytic sheath and the endothelium. There is growing evidence for a 'modular' organization by which clusters of neurons are surrounded by the processes of specific astrocytes, which in turn contact individual segments of the vascular bed; the resulting 'neurovascular unit' may be a mechanism by which groups of neurons regulate their own blood flow.[5] At the capillary level, a finer degree of modular organization may be present, with neurons signaling via individual astrocytes to short segments of the capillary endothelium, possibly regulating local metabolic interaction.[6,7] However, most studies of drug permeation in vivo using animal models examine the blood–brain interface without approaching the level of individual microvessels. Hence the BBB activity observed will reflect the average function over the region assayed, often confined to cortical regions of the cerebral hemispheres. Microdissection of brain regions and autoradiographic techniques can be used where regional distribution is likely to be important.

5.13.1.3   Routes Across the Blood–Brain Barrier

For drug discovery programs involving compounds designed to target or avoid the central nervous system (CNS), several potential routes for permeation across the BBB are relevant.[1,8,9] The brain endothelial tight junctions severely restrict the penetration of hydrophilic solutes via the intercellular cleft (paracellular pathway), forcing molecular traffic between blood and brain to take a largely transcellular route (**Figure 3**).[2,3] Gaseous molecules such as O_2 and CO_2, and small lipophilic agents, including many CNS drugs, can diffuse passively through the lumen-facing (apical) and brain-facing (basal, abluminal) endothelial cell membranes[10]; hence, compounds of greater lipophilicity generally show higher permeability (**Figure 4**). The endothelial membranes contain a number of specific transport (carrier) proteins that can shuttle necessary nutrients into the brain (uptake carriers, e.g., for glucose and amino acids)[2,3,11] or efflux waste products and exogenous compounds of potential toxicity (efflux carriers, e.g., P-glycoprotein, P-gp).[9,12–15] Molecules too large for carrier-mediated entry, such as peptides and proteins, may be able to cross the endothelium to a limited degree via a

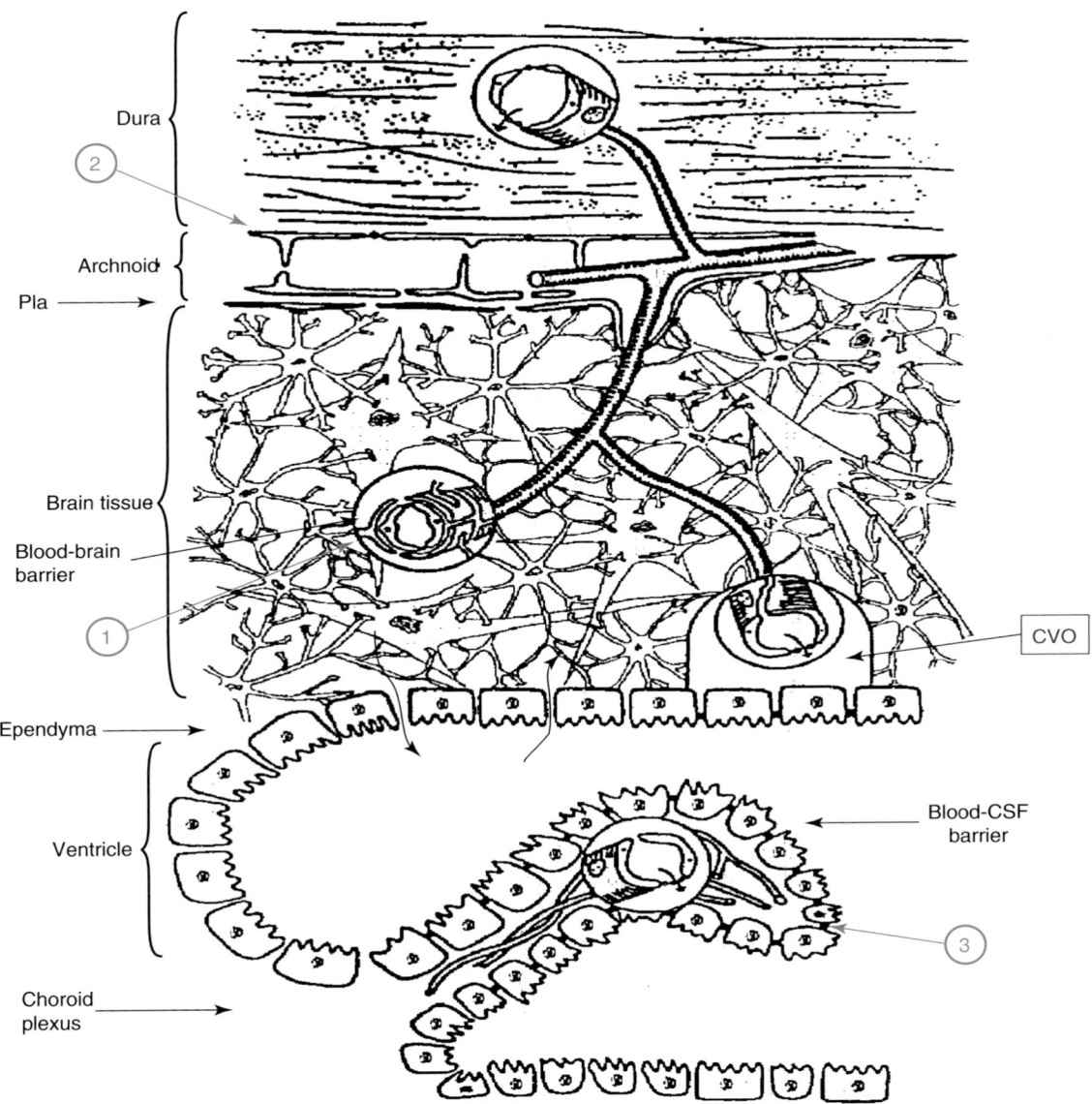

Figure 1 Location of barrier sites in the CNS. Barriers are present at three main sites: (1) the brain endothelium forming the blood–brain barrier (BBB), (2) the arachnoid epithelium forming the middle layer of the meninges, and (3) the choroid plexus epithelium which secretes cerebrospinal fluid (CSF). In each site, the physical barrier is caused by tight junctions that reduce the permeability of the paracellular (intercellular cleft) pathway. In circumventricular organs (CVO), containing neurons specialized for neurosecretion and/or chemosensitivity, the endothelium is leaky. This allows tissue–blood exchange, but as these sites are separated from the rest of the brain by an external glial barrier, and from CSF by a barrier at the ependyma, CVOs do not form a leak across the BBB. (Reproduced with permission from Abbott, N. J. *Drug Disc. Today: Technol.* **2004**, *1*, 407–416 (with permission from Elsevier), based on Figure 1.1 in Segal, M. B.; Zlokomic, B. V. *The Blood–Brain Barrier, Amino Acids and Peptides*; Kluwer: Dordrecht, 1990, with kind permission of Springer Science and Business Media.)

vesicular route, either by specific receptor-mediated transcytosis (RMT) or following nonspecific adsorption of cationic molecules to the membrane surface (adsorptive-mediated transcytosis or AMT).[9,16] The endothelial cells express a number of surface and intracellular enzymes (e.g., peptidases, monoamine oxidase, cytochrome P450 enzymes) that form a metabolic barrier.[3] Several aspects of the barrier phenotype have been shown to be induced by cell types associated with the endothelial cells in vivo (**Figure 2**) especially astrocytic glia, but including neurons, pericytes, and perivascular macrophages[17]; this complexity has implications for development of in vitro models of the BBB (see below).

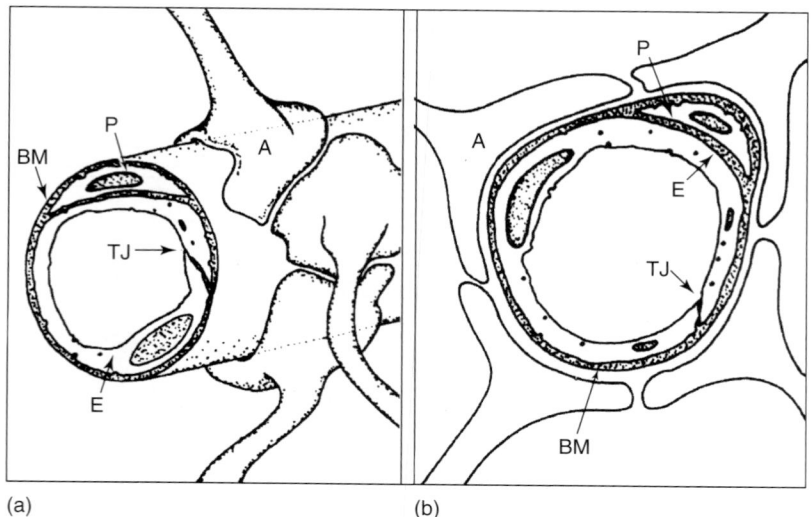

(a)   (b)

Figure 2 Cell types at the BBB, shown (a) in three dimensions, and (b) in transverse section. The endothelial layer coupled by tight junctions forms the barrier; perivascular astrocytes and pericytes have associated roles, including induction of BBB phenotype in the endothelium. Other associated cell types such as neurons and microglia are not shown. A, astrocyte; BM, basement membrane; E, endothelial cell; P, pericyte; TJ, tight junction. (Reproduced with permission from Abbott, N. J. *et al. Progr. Appl. Microcirc.* **1990**, *16*, 1–19.)

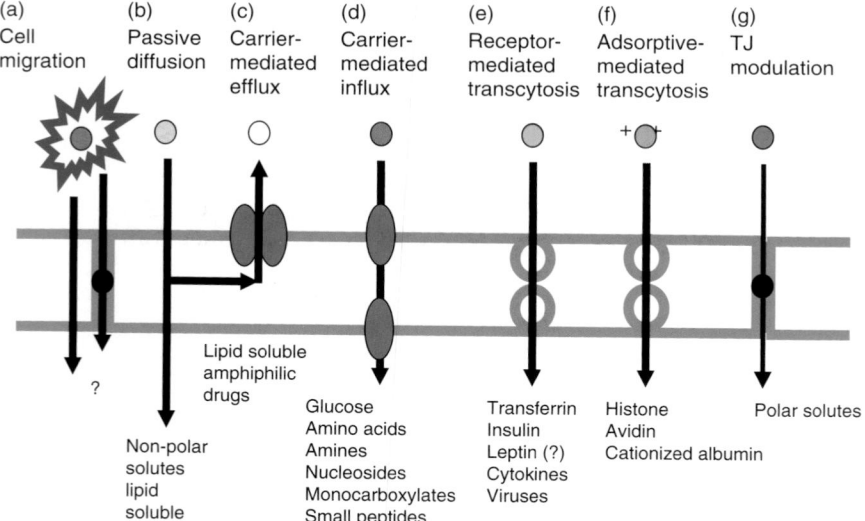

Figure 3 Potential routes for transport across the brain endothelium forming the BBB. (a) Leucocytes may cross the BBB adjacent to, or by modifying, the tight junctions. (b) Solutes may passively diffuse through the cell membrane and cross the endothelium. Greater lipid solubility favors this process. (c) Active efflux carriers may intercept some of these passively penetrating solutes and pump them out of the endothelial cell. (d) Carrier-mediated influx, which may be passive or secondarily active, can transport many essential polar molecules, such as glucose, amino acids and nucleosides, into the CNS. (e) Receptor-mediated transcytosis (RMT) can transport macromolecules such as peptides and proteins across the cerebral endothelium. (f) Adsorptive-mediated transcytosis (AMT) appears to be induced nonspecifically by negatively charged macromolecules and can also result in transport across the BBB. (g) Tight junction (TJ) modulation may occur, which 'relaxes' the junctions and wholly or partially opens the paracellular aqueous diffusional pathway. (Modified with permission from Begley, D. J.; Brightman, M. W. In *Progress in Drug Research*; Prokai, L., Prokai-Tatrai, K., Eds.; Birkhaüser: Basle, Switzerland, 2003; Vol. 61 © Birkhaüser Verlag, Switzerland.)

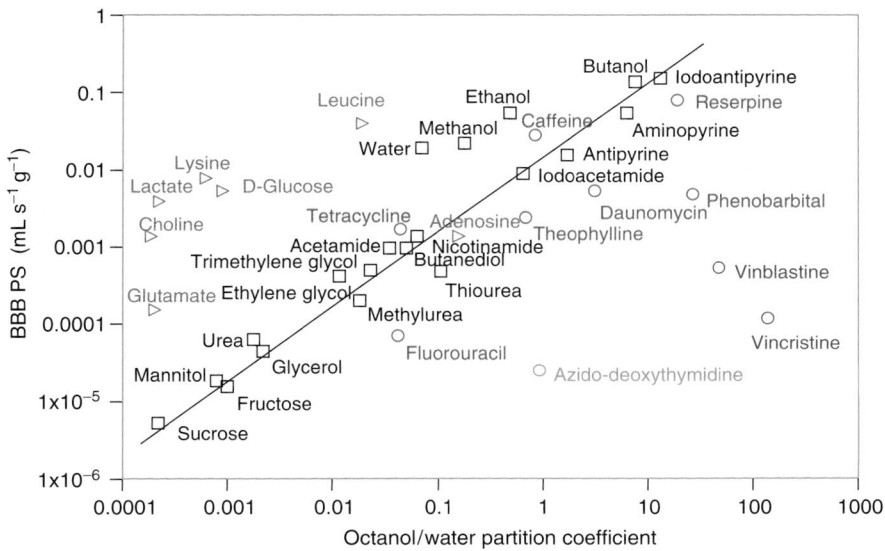

Figure 4 BBB permeability versus lipophilicity. Relation between measured BBB capillary permeability × surface area (BBB PS) and the octanol/water partition coefficient for reference tracers (squares), drugs (circles), and nutrients (triangles). BBB PS data are all from in situ brain perfusion in the rat the absence of plasma protein binding. Each point represents the mean of $n = 3$–8 perfusions. (Reproduced with permission from Smith, Q. R. A Review of Blood–Brain Barrier Transport Techniques. In *Methods in Molecular Medicine 89: The Blood–Brain Barrier: Biology and Research Protocols*; Nag, S., Ed.; Humana: Totowa, NJ, 2003, pp 193–208.)

5.13.1.4   Drug Concentration in Brain Interstitial Fluid

The concentration of compounds in the brain under steady-state conditions will depend on a number of factors[18]: the plasma concentration versus time curve (pharmacokinetics in blood), the degree of plasma protein binding (since the effective concentration determining BBB permeation is the free concentration in blood), the permeability across the BBB (whether by passive diffusion or some mediated mechanism), metabolic modification by barrier enzymes, and the 'sink effect' of the continual secretion and drainage of CSF and ISF, giving turnover of both fluids.[4] Methods are available for determining all of these, but for most compounds complete information is lacking; instead, a mixture of technologies can be used to estimate the likely behavior of compounds at the chief barrier layer, the BBB, on the understanding that this is generally the most critical step in determining CNS penetration and concentration (**Table 1**).[18]

5.13.1.5   Need for In Vitro Models

The earliest studies of the blood–brain interface were made more than 100 years ago, and involved in vivo preparations.[19] Injection of dyes and other chemical substances intravascularly in animal models, followed by assessment of their degree of penetration into the brain and CSF, gave evidence for the barrier layers indicated above (*see* Section 5.13.1.1), and demonstrated some general principles of BBB penetration. Thus BBB penetration was observed for certain lipophilic uncharged compounds, while larger, more hydrophilic and charged compounds showed lower permeability, and specific carrier-mediated transport mechanisms for nutrients and waste products were demonstrated. In vivo preparations preserve the normal relations of brain cellular and fluid compartments, as well as cerebral blood flow, and allow investigation of drug pharmacokinetics (*see* 5.02 Clinical Pharmacokinetic Criteria for Drug Research). However, the complexity of the cell types and processes occurring at the BBB mean that it is often difficult in vivo to separate the mechanisms that lead to the observed drug distribution. Cell-based in vitro methods were introduced in the 1970s, and have allowed rigorous investigation not only of the events occurring at the brain endothelial cell membranes, but also of the intracellular signal transduction pathways that regulate BBB function. Certain of the in vitro models introduced allow relatively rapid measurement of BBB permeability, and of drug–transporter interaction, hence have applications in drug screening. Information gained from both in vivo and in vitro studies is now being used to construct and validate in silico tools (*see* 5.31 In Silico Models to Predict Brain Uptake),

Table 1 Comparison of technologies for studying BBB permeability and transport

Technology	In vivo: log BB	In vivo: log PS	In vivo, knockout (k-o) animals	In vitro cell-based assays	Non-cell-based assays	In silico modeling
Summary, species	Brain:plasma ratio, ideally from AUC Rat, mouse	Permeability × surface area, from unidirectional uptake coefficient K_{in} Rat, mouse	Influence of P-gp – mdr1a/1b –/– or triple k-o together with MRP1 –/– (mouse) Effect of BCRP/ABCG2 and other efflux transporters using inhibitors	Choice of primary culture (1°), 1° + astrocyte influence, or cell line model Study of permeability, transport. Most useful: bovine, porcine, rat, mouse, human	Use of immobilized artificial membrane (IAM = HPLC technology) or membrane filter systems (PAMPA) mimicking aspects of BBB membranes	Based on correlations between compound permeation and physicochemical descriptors (see 5.31 In Silico Models to Predict Brain Uptake)
Pros	Reflects real behavior of compounds in blood: brain distribution	Accurate measure of BBB permeability step, better basis for predictive modeling of this interface Reliable 'gold standard' reference for other methods	Relatively clear-cut 'classification' system to distinguish efflux transporter substrates	Medium throughput; can select model according to information required (permeability, mechanism of transport)	PAMPA: Medium to high throughput (UV analysis) Moderately good predictor for passive permeation	Several models available, similar levels of predictive power for passively permeating compounds Some quite high throughput
Cons	Influenced by factors apart from BBB permeability, especially nonspecific binding Low throughput	Labor intensive Low throughput May need 'capillary depletion' technique to confirm compound reaches parenchyma	A 'classification' rather than ranking system – not well suited for deriving structure–activity relationships information or generating predictive models	No industry standard model. Primary cultures labour intensive Cell line models generally too leaky for permeability assays Very lipophilic compounds may stick to membranes Need to minimize unstirred water layers	Less suitable for compounds not detectable by UV – then need to use LC/MS/MS[a] Very lipophilic compounds may stick to membranes Need to correct for unstirred water layers	Few models can yet offer prediction for carrier-mediated transport. For predicting passive permeability, accuracy depends on training set used, and chemical space of test compound set
References	18, 107	24, 27	32, 34	37, 46, 58–71, 76, 96, 99, 108	38, 39	27, 107–116

[a] LC/MS/MS, liquid chromatography followed by tandem mass spectrometry and identification of product ions.

with the aim of conducting as much drug screening as possible without the need to use biological preparations. However, the complexities of the living system mean that prediction is likely to remain an inexact science for the foreseeable future, so that cell-based models will continue to play an important role.

5.13.1.6 Cell-Based In Vitro Models

A number of approaches have been tried to create in vitro models that can act as simpler systems for examining BBB function. Inevitably these will not reproduce all aspects of the in vivo BBB, but the aim is to develop models that will show features relevant to the specific application. Thus isolated brain capillaries may be useful for investigations of brain endothelial metabolism and transporters,[20–23] and brain slices make possible examination of the neurovascular unit.[6] However, by far the most useful models for understanding BBB mechanisms at the cellular and molecular level have been cell culture models, either using brain capillaries as a starting material to obtain endothelial cell cultures, or using cell lines that when cultured show some features of the in vivo barrier. Before these can be used as reliable models of the BBB, they need careful characterization, with comparison of their properties with those of the in vivo BBB. In the field of drug delivery to the brain, this characterization will include examination of the extent to which the in vitro model reproduces the tightness and transport properties of the in vivo BBB.

Section 5.13.2 below summarizes key technologies that help establish and predict BBB permeation using in vivo and in vitro models.

5.13.2 Key Technologies

5.13.2.1 Range of Technologies

This chapter focuses on in vitro techniques, while other technologies relevant to the BBB are covered in detail elsewhere in this volume (*see* 5.31 In Silico Models to Predict Brain Uptake). For completeness, methods that give valuable data for reference or comparison with the in vitro models[1,24,25] are summarized here. **Table 1** lists the six technologies that have contributed most to understanding compound behavior at the BBB over the last 5–10 years.

Designing drugs to treat brain disorders needs to take into account the properties of the BBB outlined above. Improved understanding has come from techniques for measuring BBB permeability in vivo, and for establishing the properties of both passive and mediated transport routes across the barrier. In vivo techniques still provide some of the most reliable reference information for testing and validating other models, but in vitro technologies are increasingly useful, especially to provide mechanistic information. For passively permeating compounds, non-cell-based assays may be valuable, and methods combining in vitro and physicochemical information may even be able to predict penetration for compounds using mediated transport systems. In silico (computer-based) modeling and BBB permeability prediction have been limited by the quality and quantity of the data (training set) used to construct the models, but more practical methods are becoming available. Although there is as yet no agreement on 'industry standard' models, the ability to choose from a range of technologies to generate a BBB 'permeability profile' for a drug provides valuable internal checks and prediction covering an expanding chemical space.

5.13.2.2 In Vivo and In Situ Methods

Of several methods for assessing BBB permeability in vivo,[24] two are widely used: determinations of brain:plasma ratio (log BB); and measurement of the permeability × surface area product (PS or log PS), from which permeability P_e can be derived provided vessel surface area (S) can be estimated.

5.13.2.2.1 Extent of brain penetration: brain:plasma ratio, log BB

Most pharmaceutical companies generate log BB data in animals (often rat) as part of standard pharmacokinetic profiling of compounds.[18,26] Single time-point log BB determinations are of limited value as they depend critically on the time chosen and the relation between the concentration in plasma and brain at this time point. The ratio of the areas under the curve (AUC) for brain and plasma concentrations is more useful. However, these measurements are generally made over several hours, with a number of animals required per data point, so are costly and labor intensive. Moreover, a number of factors, including metabolism and binding, affect the brain distribution, so that log BB is not a pure measure of BBB permeability.

5.13.2.2.2   Rate of brain penetration: unidirectional influx coefficient (K_{in}) and permeability–surface area product (PS)

Determining the unidirectional influx coefficient (K_{in}) using the in situ saline-based perfusion method (for rat, mouse, typically over 1–10 min, plotting the volume of distribution against time, from which slope = K_{in}) more accurately reflects the BBB permeation step, effectively isolating this 'kinetic' element of drug penetration.[24,26] The saline composition of the perfusate can be controlled, and plasma protein included if needed to test the effect of protein binding. If perfusate flow is also measured, K_{in} can be converted to PS (see above). These are absolute values that can be compared across preparations and tissues, and used to follow changes under altered conditions (e.g., in pathology). Because of the accurate quantitation, the K_{in} (or PS) measurement acts as a valuable 'gold standard' reference for other methods.[27]

5.13.2.2.3   Microdialysis: free drug concentration in brain

For drugs acting on the chief CNS target sites (membrane receptors, transporters), the critical concentration is the free concentration in brain ISF. Brain:plasma ratio measured especially at longer times and for more lipophilic agents will be affected by drug distribution into brain lipids and nonspecific binding. Measuring free concentration with a microdialysis probe is possible but technically difficult, and there is a particular problem of recovery of more lipophilic agents.[28,29] Measurement of CSF:plasma ratio has been used as a 'surrogate' measure of ISF concentration, especially in steady state.[30] However, depending on the balance between uptake and clearance mechanisms influencing drug concentration in ISF and CSF, including the 'sink effect' of CSF and ISF turnover and drainage, the CSF:ISF ratio may be greater or less than 1.0.[31]

5.13.2.2.4   Knockout and gene-deficient animals

Although there is a general relation between BBB permeability and lipophilicity (**Figure 4**), such plots show a number of outliers. Those considerably above the line generally prove to be substrates for uptake carriers at the BBB, while many below the line prove to be substrates for efflux transporters, such as P-gp. This was most strikingly demonstrated in P-gp knockout mice (initially mdr1a, later double-knockout mdr1a/1b –/–), when log BB increased up to 90-fold for certain compounds, identifying them as P-gp substrates.[32] A high proportion of established CNS-active drugs are proving to be substrates for P-gp[33] yet these are efficacious drugs; there are indications that if passive BBB permeability, and potency on the target site, are high enough, these factors can outweigh the effect of P-gp-mediated efflux at the BBB. Nevertheless, transport by P-gp needs to be taken into account in designing and optimizing new CNS drugs, since it will affect the free drug concentration in brain ISF concentration resulting from any particular plasma concentration versus time curve. Following the identification of further efflux transporters at the BBB (breast cancer resistance protein BCRP1/ABCG2; major vault protein MVP; the Ral-interacting protein RLIP-76),[34–36] knockout and gene-deficient animals will be valuable in assessing their role.

5.13.2.3   In Vitro Cell-Based Methods

All the animal-based studies discussed above are labor intensive and low throughput. Given the need for higher-throughput methods to deal with the increasing numbers of compounds generated by combinatorial chemistry, there has been great interest in in vitro methods.[37] Cell-based model systems and the techniques for their examination are the most widely applied and are discussed in more detail in Sections 5.13.3 and 5.13.4 below.

5.13.2.4   Non-Cell-Based Permeability Assays

In the 1980s, efforts were made to generate high-performance liquid chromatography (HPLC) columns, 'immobilized artificial membranes' (IAMs), that mimicked the properties of biological membranes to use as a permeability screen. Some were moderately successful, capable of ranking compounds according to BBB permeability,[38] but they have not proved suitable for medium- to high-throughput operation. A more promising technology is the parallel artificial membrane permeability assay (PAMPA) (*see* 5.19 Artificial Membrane Technologies to Assess Transfer and Permeation of Drugs in Drug Discovery), first developed as a surrogate for gastrointestinal absorption. PAMPA is suitable for passively permeating compounds, showing moderately good correlation with data from the human colonic epithelial cell line (Caco-2) and in vivo studies. By modifying the lipid composition of the artificial membranes, the system appears capable of predicting CNS permeability with reasonable accuracy.[39]

5.13.2.5   High-Throughput Assays

None of the in vitro tools for studying the BBB has so far led to high-throughput assays. At best, cell culture preparations have allowed medium-throughput studies of BBB features of interest, for example interaction of modulators and inhibitors with transporters expressed, such as the efflux transporter P-gp.[40,41]

5.13.3   Theoretical Basis: Permeability and Transport Measurements in Blood–Brain Barrier Models

The theory behind flux measurements in in vitro BBB models is the same as that for measurements of this kind in other cell and monolayer systems, such as Caco-2 and Madin–Darby canine kidney cell line (MDCK) epithelial cells (*see* 5.11 Passive Permeability and Active Transport Models for the Prediction of Oral Absorption), and the same basic equations apply.[42,43] The mathematical treatments frequently used for BBB models are described in this section.

5.13.3.1   Measurement of Cell Monolayer Permeability

Cell monolayers for BBB permeability measurement, as for epithelial models, are generated by growing the cells on permeable filters, mounted in a device that separates the apical (luminal) compartment and the basal (abluminal) compartment (e.g., Ussing chamber, Transwell system). Drug is added to one compartment (the donor) and the amount appearing in the other compartment (receiver), is determined over time. Permeability across the monolayers can be assessed using the equations governing chemical flux, based on Fick's first law:

$$J = P_e(C_1 - C_2) \qquad [1]$$

where J is flux ($mol\,s^{-1}cm^{-2}$ membrane), P_e is the permeability coefficient ($cm\,s^{-1}$), and $C_1 - C_2$ ($mol\,mL^{-1}$) is the concentration difference between compartments 1 (donor) and 2 (receiver). If the experimental conditions mean that the concentration in the receiver compartment is negligible compared with that in the donor compartment, then this can be simplified to:

$$J = P_e C_1 \qquad [2]$$

and hence $P_e = J/C_1$

In the pharmaceutical science field, it is often sufficient to gain a measure of permeability ranking within a compound series, hence the measure frequently used is apparent permeability (P_{app}) ($cm\,s^{-1}$):

$$P_{app} = (dQ/dt) * (1/c_0 S) \qquad [3]$$

where dQ/dt ($mol\,s^{-1}$) is the increase in the amount of drug in the receiver chamber per time interval, S (cm^2) is the growth area of the cell culture filter insert, and c_0 ($mol\,mL^{-1}$) is the initial drug concentration in the donor chamber. P_{app} is an empirical value, since there is generally no attempt to correct for the series barrier of the filter on which the cells are grown. However, if the filter is sufficiently porous, and the cell layer sufficiently tight (so that leakage through the tight junctions, the paracellular pathway, is small), and unstirred water layers are minimized by stirring or shaking, then useful ranking of drug permeabilities can be obtained. Refinement of the method to check for mass balance and allow for cell retention of the drug can be used where this is a problem, e.g., for very lipophilic drugs.[44]

An alternative approach is to make use of the clearance principle (widely used in physiology) to give a concentration-independent transport parameter. Here the 'cleared volume' is a calculated virtual volume equivalent to the smallest volume of donor solution necessary to deliver the amount of drug appearing in the receiver solution, i.e., the volume of donor solution from which that drug has been 'cleared.' It is calculated as described by Siflinger-Birboim *et al.*[45] by dividing the amount of compound in the receiver compartment by the drug concentration in the donor compartment. The normal direction for study is from luminal to abluminal compartment, but for studies of efflux transport, the opposite direction can be used.

For the most reliable analysis, filters with cells grown on, as well as blank filters, coated with any cell growth matrix but without cells, are studied. At time zero, an aliquot of the test compound in buffered saline is added to the donor compartment, and at set times (typically 10, 15, 20, 30, and 45 min), aliquots are removed from the receiver compartment, and the amount of compound determined. After removal of the aliquots, fresh buffer is added to the receiver solution to maintain equal volumes in both chambers, the dilution being taken into account in data analysis.

Alternatively, the filter insert is moved to a fresh receiver well at each time point, to minimize possible back-flux from the receiver compartment, and aliquots of each receiver well, and of the initial donor solution, are analyzed.[46]

The total volume cleared (V_{CL}) at each time point is calculated by summing the cleared volumes up to the given time point:

$$V_{CL} = A/C_d \qquad [4]$$

where A is the amount of test compound on the receiver side, and C_d is the concentration of test compound on the donor side at each time point. If permeability is constant during the experiment, then the total cleared volume will increase linearly with time, and the slope of the line (cleared volume divided by time) calculated by linear regression analysis will give the $P_e S_t$ (or total clearance), where $P_e S$ is the permeability $\times$ surface area product in $mL\,s^{-1}$. The slope of the clearance curve for control (blank) filters is denoted $P_e S_f$. Hence the $P_e S$ value for the endothelial or epithelial cell monolayer ($P_e S_e$) can be calculated from

$$1/P_e S_e = (1/P_e S_t) - (1/P_e S_f) \qquad [5]$$

$P_e S_e$ is then divided by the surface area of the filter S to give the monolayer permeability coefficient P_e ($cm\,s^{-1}$). The advantage of this parameter is that it represents the measured permeability of the test molecule through that cell system, and can be compared with measurements made across other biological interfaces (e.g., gastrointestinal tract, skin) and by in vivo or in situ techniques. Since the permeability coefficient is proportional to the mobility of the drug in the monolayer or membrane system, divided by the path length (often the cell or membrane thickness),[47] absolute values of P_e made in different tissues and membrane systems will generally differ. However, these differences can give valuable information about the nature of the barrier layers.

5.13.3.2   Measurement of Uptake Transport by Cells

Studies of brain endothelial cell uptake of a test substance typically involve incubation of the cells in a tracer molecule of interest, stopping the uptake process at set time points (e.g., by washing with cold buffer solution, or using a stopping solution containing specific inhibitors to block transport), lysing the cells and analyzing the contents for presence of the tracer (e.g., by radioactive counting, quantification of fluorescence, HPLC or liquid chromatography-mass spectrometry (LC/MS)), as well as total protein content. The volume of distribution of the test substance, V_d ($\mu L\,mg^{-1}$ protein), is determined from the ratio of counts (or amount) per milligram protein to counts (or amount) per microliter incubation medium. Where uptake is mediated by transporters, kinetic parameters of transport can be estimated using nonlinear regression analysis of the concentration (C) dependence of the influx, J_{in} ($nmol\,min^{-1}mg^{-1}$ protein) based on conventional Michaelis–Menten kinetics,[48] e.g.,

$$J_{in} = (V_{max}C)/(K_m + C) + K_D C \qquad [6]$$

where K_m is the half-saturation concentration, V_{max} is the maximum transport capacity ($nmol\,min^{-1}mg^{-1}$ protein), and K_D the constant of passive diffusion ($\mu L\,min^{-1}mg^{-1}$ protein).

5.13.4   Cell-Based In Vitro Tools for Studying Blood–Brain Barrier Function

5.13.4.1   Range of In Vitro Tools

A variety of in vitro tools have been developed to study the behavior of compounds at the BBB, and to predict their BBB permeability. The earliest studies compared in vivo permeation with compound behavior in brain slices, the latter designed to bypass the BBB. Later it became possible to study isolated brain microvessels, to examine brain capillary endothelium directly. Both these techniques have recently returned to the experimental repertoire, thanks to improvements in imaging technology. However, the introduction of brain endothelial cell culture was the most significant advance in the 1970s. This led to the isolation of cell membrane and organelle fractions, to examine function in more detail at the molecular level. Where drug distribution has proved to be largely dependent on physical chemistry, particularly the nature of the lipid bilayer cell membranes, attempts have been made to mimic these in non-cell-based in vitro systems. All these tools continue to be used, often in combination to obtain the most reliable information, and to make allowance for the artifacts associated with each method.

5.13.4.2   Brain Slices

In the early 1960s, the location of the BBB was still not established with certainty; one view was that the apparent restriction on compound entry to the brain was due to a limited extracellular space.[19] Brain slices were used to establish that the brain contained appreciable extracellular fluid, so that restriction on uptake in vivo must be due to a barrier at the capillary level.[49] In the 1970s brain slices were used to make significant advances in understanding of brain biochemistry and neurochemistry, for example for glucose transport.[50,51] However, they were not suitable for kinetic studies, since diffusion in the extracellular space rather than the transporter itself often proved the rate-limiting step, and cell swelling caused changes in the fluid compartments.[51]

More recently, improvements in techniques for isolation and maintenance of brain slices with well-preserved extracellular space,[52] together with new imaging techniques (such as 2-photon confocal microscopy) allowing examination of structures within the slice, have allowed studies of cell–cell interaction within the neurovascular unit, including influences of astrocytes on blood vessels.[6] However, as the vessels within the slice are unperfused, these preparations are not suitable for examining BBB permeability and transport. For studies of gene and protein expression, the powerful technique of laser capture microdissection (LCM) can be used not only to sample from individual blood vessels in defined regions of brain slices, but also to separate relatively cleanly the endothelial and perivascular cells.[53] This tool has significant applications in determining changes in phenotype of blood vessels in pathological human brain material (resected, biopsy, or post-mortem), to guide drug delivery strategies.

5.13.4.3   Isolated Cerebral Microvessels

In 1975 Goldstein *et al.*[21] reported isolation of metabolically active capillaries from rat brain, and this preparation was used for a number of studies on transport and metabolism, e.g., for hexoses[20] and small ions.[22] Since Na^+-dependent glutamate uptake by the BBB could not be detected in vivo, but was observed in isolated brain microvessels, where the bathing medium mainly has access to the outer vessel surface, an abluminal location for this transport was proposed.[54] The vessels were less suitable for studies of processes with high energy demand such as transcytosis, possibly because of the damage caused in vessel isolation and the resultant impairment in ATP production. Recent improvements in technique have led to more physiological preparations of isolated brain microvessels preserving normal polarized (apical-basal) function, and modern confocal imaging techniques using fluorescent substrates have allowed good resolution of real-time transport activity at the single vessel level.[55] The technique has been valuable in identifying location and specificity of transporters present at the BBB, and for examining drug interaction with these transporters,[56,57] but is not suitable for screening large numbers of compounds to establish whether they are substrates for these transporters.

5.13.4.4   Cell Culture Preparations

Brain microvessels can be readily isolated from brain; when seeded in culture medium, endothelial cells grow out to form monolayers suitable for experimental examination.[58] So far no single in vitro BBB model has been adopted as 'industry standard'; rather the choice depends on the application, convenience, and resource available.

5.13.4.4.1   Primary endothelial cultures
Primary endothelial cultures maintain a number of features of the in vivo BBB phenotype, though with some downregulation or altered expression: tight junctions, transporters, enzymes, and receptors are preserved to differing extents.[59,60] Coculture with astrocytic glial cells (or astrocytoma cells) generally leads to some upregulation of BBB phenotype, including tightness,[61,62] making monolayers suitable for permeability screening.[63]

5.13.4.4.2   Immortalized brain endothelial cell lines
These have proved useful for examining mechanisms of transport and cell–cell interaction, but generally make leakier monolayers than primary cells.[64,65] Some conditionally immortalized cells appear to express transporters at higher levels than in primary cultures.[66] Growing endothelial cells in porous tubes with luminal flow (dynamic in vitro BBB model) may help differentiation of a BBB phenotype.[67,68] Some of the models are suitable for examining permeability under conditions mimicking pathology, such as ischemia and hypoxia,[69,70] and infection.[71] Further detail is given in Section 5.13.5 below.

5.13.4.5   Membrane Vesicles

In vitro models based on whole living cells are often too complex to permit detailed examination of the interactions of compounds including drugs with individual receptors or transporters, so a number of techniques have been developed for isolating cell membranes, both the plasmalemma and membranes from organelles, and maintaining them for the few hours needed for experimental study.[72,73] By using marker enzymes or antibodies against specific membrane antigens it is possible to distinguish the membranes coming from apical and basolateral plasmalemma, and from different organelles. For study of ATP-dependent processes, 'inside-out' vesicles can be used, e.g., for examination of P-gp or breast cancer resistance protein (BCRP1 = ABCG2). These vesicles can be used to screen for compound interaction with transporters, but not in high-throughput mode for identifying transporter substrates.

5.13.5   Cell Culture Models of the Brain Endothelium

5.13.5.1   History of Brain Endothelial Cell Culture

The first successful isolation of viable brain microvessels[74] was soon followed by the observation that if the microvessel fragments were seeded in dishes in culture medium, cells would grow out, proliferate, and become confluent.[75] Several culture models have since been developed and characterized.[58,76]

5.13.5.2   Primary Cultures

All the methods for generating primary cultures of microvessel endothelial cells from brain tissue aim to maximize the yield and purity of the cells obtained, and to minimize or eliminate fast-growing contaminating cells, chiefly pericytes, fibroblasts, and smooth muscle cells. This can be achieved by various combinations of enzymatic digestion, filtration, centrifugation, differential adhesion on different matrices, and techniques to kill contaminants.[76] The last were initially based on use of complement with antibodies against antigens expressed on contaminating cells, such as the surface glycoprotein thymus protein-1 (Thy-1 or CD90) expressed on pericytes and fibrobasts. More recently, methods based on growth in medium containing a cytotoxic P-gp substrate such as puromycin have been introduced; the principle is that brain endothelial cells (expressing P-gp) will survive, while contaminating cells lacking this transporter will not.[77] It may be necessary to allow presence of some contaminants in the early growth phase of the culture, as they appear to produce important growth factors for the endothelium, but to remove contaminants later to encourage endothelial differentiation and formation of uniform monolayers. The nature of the extracellular matrix on which the cells are grown can influence differentiation.[78]

It is important to characterize the culture produced to demonstrate that important features of the in vivo BBB relevant to the study planned are expressed. A number of suitable biomarkers are available, to assess endothelial phenotype (e.g., von Willebrand factor, platelet endothelial adhesion molecule PECAM-1, alkaline phosphatase),[79–82] presence of tight junction proteins (e.g., occludin, claudins, zonula occludens protein-1 (ZO-1)) and apical–basal transport polarity (e.g., apical transferrin receptor, basolateral Na^+, K^+-ATPase).[83] The best established primary cultures of brain microvessel endothelial cells have been generated and characterized from bovine,[61] porcine,[59] rat,[76,77] mouse,[81] and human brain,[84] and methods have also been reported for ovine, goat, and monkey brain.

5.13.5.3   Immortalized Brain Endothelial Cell Lines

There are several drawbacks to primary cultures, the most important being their limited lifespan, and hence the need to obtain fresh brains as a source material on a regular basis. Other problems are lack of consistency resulting from variability in state and viability of the starting material, the need to cull large numbers of animals when working with small (e.g., rodent) brains, difficulty in removing contaminating cells, and batch-to-batch variation in cell growth. This last is often difficult to explain, but may relate to minor differences in cell isolation and growth conditions. As a result of these problems, a great deal of effort has been put into developing immortalized brain endothelial cell lines, by transfection of pure brain endothelial cells. The technology has been particularly applied to rat and mouse brain,[85] as these animals are widely used as in vivo models (for physiological, pharmacological, and pharmaceutical applications), so that a great deal of reference information is available for them, and useful probes (e.g., antibodies, oligonucleotides) have been developed. A drawback is that immortalized cell lines generally form leakier monolayers than primary cultures, although a recently reported mouse cell line (cEND) may be an encouraging exception.[82]

5.13.5.4   Methods to Upregulate Blood–Brain Barrier Features in Cell Culture Models

Many BBB features present in vivo, and still detectable in freshly isolated brain capillaries, are found to disappear or decline in expression over hours to days in culture.[86] These include presence and correct localization of tight junctional and associated proteins, certain enzymes such as alkaline phosphatase and gamma-glutamyl transpeptidase,[83] and presence and polarized membrane localization of efflux proteins such as P-gp. By contrast, certain proteins expressed in non-brain endothelium but apparently downregulated in BBB endothelium (MRP1, multidrug resistance associated protein-1,[87] aquaporin1 water channel[88]), begin to reappear in culture. Some of the phenotypic features of the BBB phenotype can be reinduced by coculture of the brain endothelium with astrocytes or astrocytic cell lines (e.g., C6 glioma),[89] suggesting that in vivo the perivascular astrocyte endfeet play an important role in inducing (sculpting) and maintaining the BBB phenotype.[79] Recent studies indicate that other cells associated with brain microvessels may also take part in induction, and that it is a two-way or even multiway process between the cell types involved.[90]

A number of candidate molecules for inducing agents have been identified, including basic fibroblast growth factor (bFGF), glial cell-derived neurotrophic factor (GDNF), and transforming growth factor beta (TGF-β),[17,79] but so far it has not been possible to mimic all aspects of the differentiated BBB phenotype by adding any one of them to the medium. It is useful to distinguish between 'permissive' influences that prime the endothelium (intraluminal flow,[67] corticosteroids,[59] elevation of intracellular cyclic AMP[86]), and release of specific 'inductive' influences from associated cell types that trigger the full differentiation.[90,91] Generally the combination of permissive and inductive influence leads to the highest expression of BBB features.[17,79]

This knowledge has led to development of a number of coculture models that generate a more differentiated BBB. The key features that are relevant to pharmaceutical studies are the tightness of the endothelial monolayer, and the presence and polarized location of the appropriate transporters.[83] For a number of years, the tightest in vitro models were obtained from primary bovine brain endothelial cells cocultured above primary rat astrocytes.[46] More recently it has been possible to generate tight porcine brain endothelial cell monolayers by exposure to hydrocortisone.[59] Both models have proved valuable in pharmaceutical applications. It has recently become possible to make tight rat[77] and mouse[81] primary coculture models, and although these are not so suitable for medium-throughput drug permeability screening, they will be valuable where correlation with animal studies is needed (e.g., drug pharmacokinetics in rat, studies in transgenic mice). Certain of the cell line models also retain responsiveness to glial factors and hydrocortisone,[82,85] suggesting that generating cell line models as tight as the best primary models may be possible in the future.

5.13.5.5   Models for Studying Specific Transport by Brain Endothelial Cells

Since production of the earliest primary cultures of brain endothelial cells, and then cell line models, these preparations have proved valuable for studying the mechanisms for specific transport, their kinetics, and mode of operation. Cell lines often express a reduced phenotype compared to primary culture (e.g., in range of transporters expressed), but this can be turned to advantage for examining the mechanisms that are expressed.[92] Where the main question concerns the mechanism and kinetics of transport, monolayer tightness is unimportant.[48] For transporters with polarized (apical–basal) expression, full polarization is likely to require the presence of effective tight junctions, which not only restrict paracellular permeability ('gate' function) but also limit lateral diffusion of membrane proteins such as transporters within the membrane and hence segregate apical and basolateral transporters ('fence' function).[92,93]

5.13.5.6   Models for Studying Permeability and Transport Across the Blood–Brain Barrier

Effective tight junctions are an important element of a fully differentiated BBB. This is largely because the physical barrier to hydrophilic molecules thus provided is one of the defining roles of the BBB. As mentioned above, polarization of transporters to apical and basolateral membrane domains also requires effective tight junctions. Hence the two features for pharmaceutical applications, tightness and polarized expression of transporters, depend on the tight junctions. This readily explains why the two models most commonly used in pharmaceutical assays are the bovine brain endothelial model (with or without astrocytes)[37,46,61] and the porcine model (with or without hydrocortisone).[59] Furthermore, scientific study of the mechanisms underlying polarized transendothelial transport, whether mediated by membrane carrier proteins or by vesicles (RMT or AMT), requires tight junctions restricting enough to allow resolution of the transport without excessive shunting (short-circuiting flux) via the junctions.[94]

The tightness of an endothelial or epithelial layer can be assessed by two measures of paracellular permeability, the transendothelial (or epithelial) electrical resistance (TEER, ohm cm^2) and the transmonolayer permeability to a hydrophilic tracer molecule with negligible cell uptake or binding. Suitable permeability markers of a size similar to small drug molecules (typically 100–400 Da mol. wt.) are sucrose and mannitol (used as radiolabeled probes) or fluorescent indicators such as fluorescein and Lucifer yellow.[76] As electrical resistance is the reciprocal of conductance, related to the permeability of small ions, TEER is inversely proportional to the permeability or conductance of Na^+ and Cl^- through the tight junctions, these being the major charge carriers in extracellular fluids.[92]

A plot of TEER versus small solute permeability through an epithelial or endothelial monolayer will give an asymptotic curve reflecting this relationship.[95] Net solute flux increases as the sum of fluxes across individual junctional clefts, while TEER is most affected by areas with the lowest resistance, which shunt current flow. Hence a small flaw in a monolayer can cause a large drop in TEER with little drop in sucrose or fluorescein permeability. Monolayers with TEER greater than $\sim$100–200 ohm cm^2 give the expected relation between TEER and solute permeability,[37,76,95] so TEER assessment is a useful way of checking the uniformity and tightness of a monolayer before starting a drug permeability assay. The ability of a monolayer to rank drug permeabilities (its dynamic range) will depend partly on its tightness, as this will determine the degree of paracellular leak of the drug, and hence the dynamic range of the model.

5.13.5.7   Models Incorporating Intraluminal Flow

As discussed above, when endothelial cells are exposed to fluid flow exerting a shear stress across their apical surface, and inducing factors from astrocytes on their abluminal surface, this should provide an optimal combination of conditions encouraging BBB differentiation. Indeed, in the three-dimensional 'dynamic' in vitro BBB model (DIV–BBB) in which endothelial cells are seeded inside porous tubes with astrocytes on the outside, and medium flow is maintained via the lumen, TEER of up to 2000 ohm cm^2 has been reported.[67,68] A number of transporters and receptors characteristic of the BBB are also upregulated in the cells. Although less easy to use for kinetic studies than flat monolayer cultures, as the control and sampling of apical and basolateral fluid is less direct, the model has been used to test drug permeability. Recent miniaturization of the system and connection to a control module will make it possible to run several sets in parallel for statistical treatments.

5.13.6   In Vitro Surrogate Blood–Brain Barrier Models

In spite of the documented value of cultured brain endothelial cells as BBB models, they have not been universally popular with pharmaceutical companies. This is largely due to the labor-intensive nature of their preparation, and the fact that the expertise needed to prepare reproducible cultures is greater than for commonly used epithelial models such as Caco-2 and MDCK. These are widely used as assay systems to predict gastrointestinal permeability of a compound or drug (*see* 5.11 Passive Permeability and Active Transport Models for the Prediction of Oral Absorption).

Since the earliest studies of BBB and blood–CSF barrier permeability in vivo, it has been clear that for many drug compounds, there is an approximate relation between permeability and lipid solubility, indicating a major role for lipid membranes of the barrier layers in determining drug delivery to the brain. As Caco-2 and MDCK cells are already used to give information about permeability across barrier layers, albeit for the gastrointestinal tract, many companies have applied these models as 'BBB surrogates' to estimate entry to brain. For compounds penetrating the brain by passive partition into brain endothelial membranes involving no interaction with transporters, this can give a reasonable prediction.[96,97] However, careful comparison using groups of druglike molecules shows that the correspondence is not exact,[63,98] indicating that for decisions affecting CNS-targeted drugs, a real BBB-based model gives better prediction (**Figure 5**).[63,99] Certainly the profile of transporter expression in the gastrointestinal tract and the BBB is different, and since a high percentage of current drugs appear to be substrates for efflux transporters, the expression profile and transporter density in the barrier layer (gastrointestinal tract or BBB) is likely to have major impact on drug penetration across the barrier.

It has recently become clear that the nature of the lipid membranes is different in the gastrointestinal tract and BBB.[39,100] This has implications for passive permeability and permeability ranking of drugs between the two systems. In addition, as more becomes known about the ways in which the membrane lipids influence the activity and transit time of substrates for uptake and efflux transporters,[101] it is likely that we will become aware of differences in drug delivery across the gut wall and across the BBB that can be attributed to the differences in lipid composition.

For studies where it is desirable to use human-derived cells, the ECV304/C6 coculture model is of interest.[58] The ECV304 cell line is apparently a clonal derivative of T24 human bladder carcinoma cells, and expresses a mixed endothelial–epithelial phenotype. It is widely used as a model of endothelial cell function, for example expressing a number of antigens recognized by anti-endothelial antibodies. When cocultured with C6 glioma cells

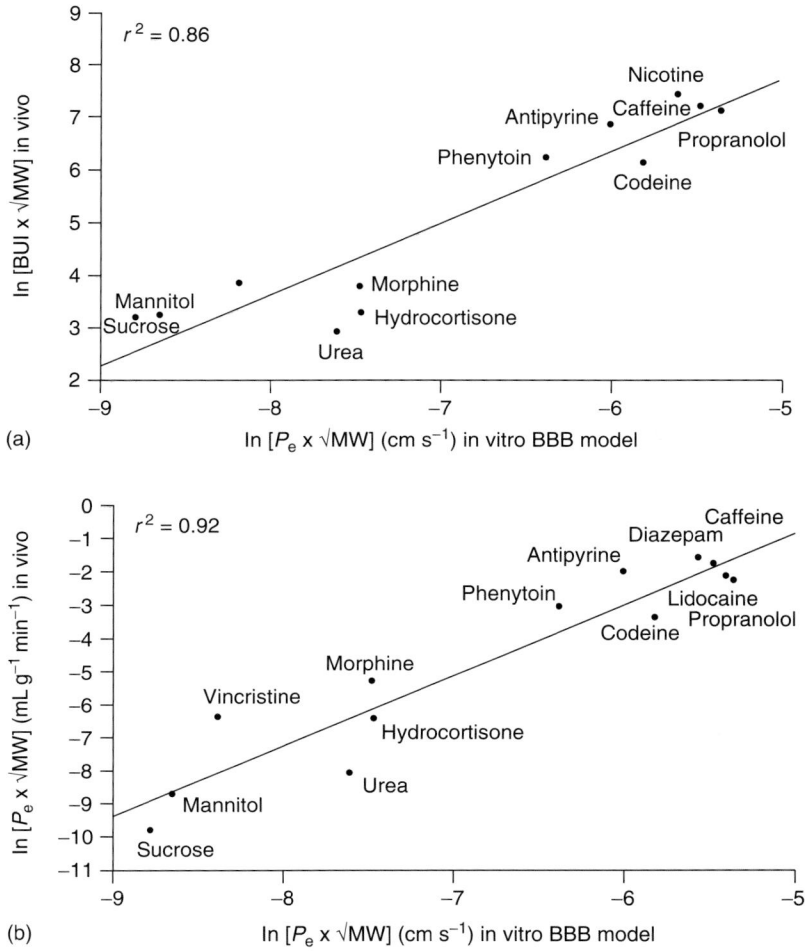

Figure 5 Ability of an in vitro brain endothelial model to predict BBB permeability in vivo. In vitro BBB permeability (bovine brain endothelium cocultured above astrocytes) as a function of (a) brain uptake index (BUI), $r^2 = 0.86$, and (b) in vivo BBB permeability from in situ brain perfusion ($r^2 = 0.92$). All parameters were normalized for molecular weight (MW). There was less good correlation for the plot of BUI versus permeability across Caco-2 monolayers, $r^2 = 0.46$ (not shown). (Reproduced with permission from Lundquist, S.; Renftel, M.; Brillault, J.; Fenart, L.; Cecchelli, R.; Denouck, M.-P. *Pharm. Res.* **2002**, *19*, 976–981.)

it shows upregulation of some BBB features, in particular tight junctional organization, generating TEER of typically 100–300 ohm cm^{-2}.[102] It has been used to screen BBB permeability of flavonoid compounds,[103] and performed well in the recent study sponsored by the European Centre for the Validation of Alternative Methods (ECVAM) comparing cell culture models as drug screening tools.[96] Although it appears to express P-gp, it does so at lower levels than the BBB in vivo, so should be used with caution.

5.13.7 Cell Lines Transfected to Express Transporters

Rather than try to create in vitro models that expresses all relevant BBB features, an alternative approach is to develop assays for particular features that can be used as parallel screening tools. If each has advantages of reproducibility and potential for automation, this may prove a better strategy overall. Thus cell lines expressing the major drug efflux transporters present at the BBB including P-gp, multidrug resistance associated proteins (MRPs), and organic anion and cation transporters are currently used.[104,105] There have so far been few studies assessing the reliability of this kind of approach; one problem is that it is difficult to allocate 'weighting factors' that reflect the relative importance of particular transporters in the in vivo BBB, since this information is itself lacking.

5.13.8   Comparison of In Vitro Tools: Cell-Based versus Non-Cell-Based

Non-cell-based tools for permeability measurement that have been used to mimic the BBB include IAM columns and PAMPA (*see* 5.19 Artificial Membrane Technologies to Assess Transfer and Permeation of Drugs in Drug Discovery). PAMPA using filters coated with brain lipids in solvent can give permeability results well correlated with those determined in in vivo models[39]; interestingly the correlation was less good using a filter coating appropriate for mimicking gastrointestinal membranes, confirming the differences in the cell membranes of these two barrier layers. So far no direct comparisons have been made between the performance of PAMPA and cell-based in vitro BBB models, although it would be predicted that PAMPA would do well with passively distributed compounds, less well with those subject to specific carrier-mediated transport. Interestingly, it has been possible using high-speed magnetic stirring to minimize unstirred water layers in the PAMPA system, something not so easy to do with the more vulnerable membranes of living cell cultures.

5.13.9   Relevance to Predictability in Humans

The best test of the validity of any in vitro or in silico model is to establish how well it can predict and rank order the permeability of drugs determined in an in vivo or in situ model (for methods, *see* Section 5.13.2.2 above; also 5.31 In Silico Models to Predict Brain Uptake). Surprisingly few studies have attempted such a comparison. One difficulty is that the bulk of the data from in vivo BBB models come from brain:blood ratio measurements, where other factors apart from BBB permeability influence the ratio. The P_{app} data from in vitro cell culture models are derived from a kinetic measure, the rate of solute flux. Some comparisons have used the kinetic brain uptake index (BUI) data,[46] but this has poor resolution for the more lipid soluble compounds. Small sets of compounds have been compared across the bovine brain endothelial/rat astrocyte coculture model, and in in situ brain perfusion (K_{in} or PS), with good correlations, indeed better than for Caco-2 cells.[63]

The recent ECVAM study compared the efficacy of a number of in vitro BBB models, and their ability to predict the rank order of permeabilities of a drug series determined in an in vivo preparation.[60,96] The best-performing models were Caco-2 cells and bovine brain endothelial cells cocultured with rat astrocytes; however, even for these models the correlation with the in vivo data was poor. MDCK cells expressing the human P-gp (MDCK-MDR1) gave the best separation of passively permeating and effluxed compounds. The conclusion is that further development and optimization of models was required, and testing with larger data sets, covering a greater 'drug-like' chemical space. The final test of predictability, the degree to which the model is able to predict drug permeability at the human BBB, must await improved noninvasive methods for determining human BBB permeability.

5.13.10   Conclusions and Strategy

The properties of the brain endothelium forming the BBB are important considerations when designing drugs to target or avoid the brain. Molecular traffic across the BBB is regulated by a combination of physical, transport, and metabolic barrier properties. Tight junctions restrict the paracellular pathway, and endothelial membrane lipids and other components determine the passive pathway across the cells. Specific uptake transport proteins and vesicular transcytotic mechanisms mediate the transcellular entry of necessary nutrients unable to use the lipid pathway, while efflux transporters help exclude potentially toxic (including lipophilic) agents, and contribute to clearance of waste products from the brain.

Given the complexity of BBB function, it is not surprising that in vivo models still provide some of the most reliable measurements of BBB drug permeation, as is necessary for the derivation and validation of predictive tools. Measurements of brain:plasma ratio, as routinely made by pharmaceutical companies, do not necessarily give a clear picture, either of the BBB entry step, or of the free drug concentration in brain ISF. Measurement of K_{in} or PS focuses on the kinetic (BBB entry) step and may be more useful in generating data for predictive modeling. Brain microdialysis is technically demanding and difficult to use with very lipophilic agents, but may be necessary for establishing the ISF concentration of 'anomalous' compounds that do not behave as predicted from simpler approaches. Knockout and gene-deficient mouse models allow testing of the role of specific transporters in regulating BBB permeation, under close to physiological conditions, but results may be affected by compensatory upregulation of other transporters.

Several practical in vitro models of the brain endothelium have been developed, which have helped to give valuable mechanistic insights (especially into transporter function). Unlike Caco-2 and MDCK cells, which are frequently used for predicting gastrointestinal absorption, no simple cell-line assay is available for BBB prediction, mainly because they are generally too leaky for permeability studies. However, the best primary cultures do give medium throughput and

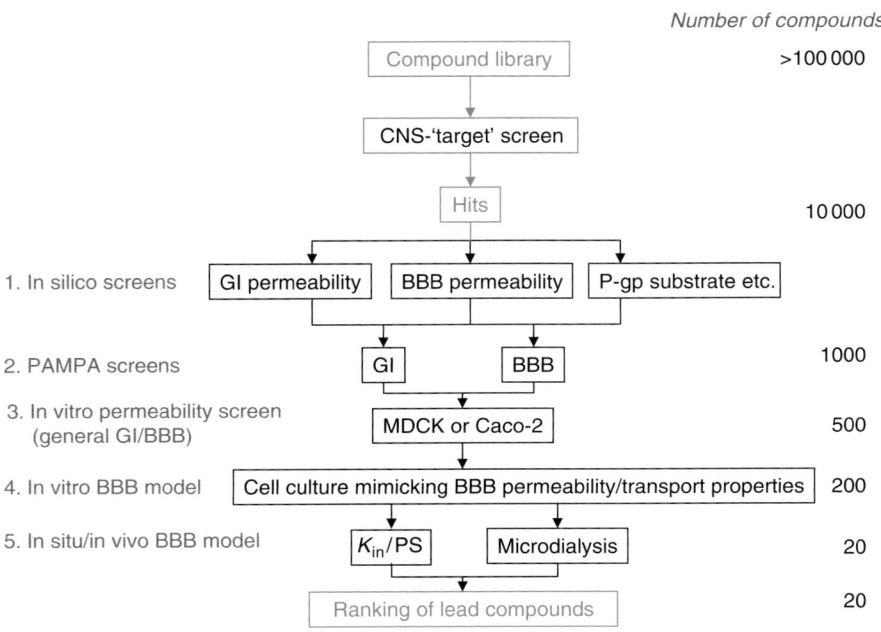

Figure 6 Suggested strategy for prediction of new drug permeability across the BBB. Flow chart showing suggested sequence of screens to handle 'hits' from high-throughout CNS-target screen, to facilitate ranking of lead compounds for optimal BBB penetration.

are used by companies with large CNS programmes. Higher-throughout methods ('surrogate BBB' cell-based; non-cell-based PAMPA; in silico modeling) can help in prediction, but a combination of techniques is still required to cover the range of drug entry and efflux mechanisms present at the BBB.

In summary, companies needing to predict BBB permeation data for novel compounds should consider a serial and parallel screening process that incorporates the best of current knowledge and understanding, selecting from 'hits' (observed activity in suitable target assay) to 'leads' (lead compounds with most potential for development as drugs) (**Figure 6**).

1. An initial high-throughput in silico screen of hits, using a number of computational tools, to predict passive gastrointestinal and BBB permeability, interaction with the main BBB efflux transporters, and chemical fit with known substrates for uptake transporters.
2. Selection of the most promising agents to test in a non-cell-based medium-throughput permeability screen (e.g., PAMPA), using artificial lipid membranes that mimic as closely as possible those of the gastrointestinal tract and BBB.
3. Use of data from MDCK (preferably expressing MDR1) or Caco-2 cells where available as check on PAMPA permeability data. Medium throughput.
4. Testing of permeability in validated in vitro BBB model (warranted for CNS drug program). Medium–low throughput.
5. In situ/in vivo screen (warranted for selected members of classes of new compounds, or for ranking lead compounds), preferably K_{in} (PS) from brain perfusion, or brain microdialysis; note problems in interpretation of log BB data (above). Low throughput.

Several pharmaceutical companies use in vitro tools together with information from in vivo and in situ studies to optimize the rate and extent of brain penetration of drugs. Increasing passive permeability, and reducing efflux transport and brain tissue binding are valuable strategies to increase drug concentration at CNS target sites.[26] Where microdialysis is judged too labor-intensive, equilibrium dialysis of drug concentration in brain homogenate can give an approximate measure of free drug concentration in brain,[106] usefully complementing the measures of BBB permeability.

There are still many challenges. These include the need for a simple, cost-effective industry-standard in vitro BBB model; difficulty in assessing the effects of all influx and efflux transporters and enzymes present at the BBB, since it is likely that more remain to be discovered; problems predicting drug concentration in brain ISF beyond the BBB; difficulty extrapolating from animal models to human, with growing evidence for species differences in BBB transport proteins, especially for efflux; and the need to mimic pathological effects on the BBB in model and predictive systems for greater relevance to clinical situations.

References

1. Abbott, N. J. *Drug Disc. Today: Technol.* **2004**, *1*, 407–416.
2. Abbott, N. J.; Romero, I. A. *Molec. Med. Today* **1996**, *2*, 106–113.
3. Begley, D. J.; Brightman, M. W. Structural and Functional Aspects of the Blood–Brain Barrier. In *Progress in Drug Research 61*; Prokai, L., Prokai-Tatrai, K., Eds.; Birkhäuser: Basle, 2003, pp 39–78.
4. Abbott, N. J. *Neurochem. Int.* **2004**, *45*, 545–552.
5. Iadecola, C. *Nat. Rev. Neurosci.* **2004**, *5*, 347–360.
6. Nedergaard, M.; Ransom, B.; Goldman, S. A. *Trends Neurosci.* **2003**, *26*, 523–530.
7. Leybaert, L. *Cerebr. Blood Flow Metab.* **2005**, *25*, 2–16.
8. Begley, D. J. *Pharmacol. Therapeutics* **2004**, *104*, 29–45.
9. Pardridge, W. M. *NeuroRx* **2005**, *2*, 3–14.
10. Bodor, N.; Buchwald, P. *Am. J. Drug Deliv.* **2003**, *1*, 13–26.
11. Tsuji, A. *NeuroRx* **2005**, *2*, 54–62.
12. Mahar Doan, K. M.; Humphreys, J. E.; Webster, L. O.; Wring, S. A.; Shampine, L. J.; Serabjit-Singh, C. J.; Adkison, K. K.; Polli, J. W. *J. Pharmacol. Exp. Ther.* **2002**, *303*, 1029–1037.
13. Begley, D. J. Efflux Mechanisms in the Central Nervous System: A Powerful Influence on Drug Distribution within the Brain. In *Blood–Spinal Cord and Brain Barriers in Health and Disease*; Sharma, H. S., Westman, J., Eds.; Elsevier/Academic Press: San Diego, CA, 2004, pp 83–97.
14. Löscher, W.; Potschka, H. *NeuroRx* **2005**, *2*, 86–98.
15. Terasaki, T.; Ohtsuki, S. *NeuroRx* **2005**, *2*, 63–72.
16. Pardridge, W. M. *Molec. Interventions* **2003**, *3*, 90–105.
17. Abbott, N. J. *J. Anat.* **2002**, *200*, 629–638.
18. Feng, M. R. *Curr. Drug Metab.* **2002**, *3*, 647–657.
19. Bradbury, M. *The Concept of a Blood–Brain Barrier*; Wiley: Chichester, UK, 1979.
20. Betz, A. L.; Csejtey, J.; Goldstein, G. W. *Am J. Physiol.* **1979**, *236*, C96–C102.
21. Goldstein, G. W.; Wolinsky, J. S.; Csejtey, J.; Diamond, I. *J. Neurochem.* **1975**, *25*, 715–717.
22. Goldstein, G. W. *J. Physiol.* **1979**, *286*, 185–195.
23. Goldstein, G. W.; Betz, A. L.; Bowman, P. D. *Fed. Proc.* **1984**, *43*, 191–195.
24. Smith, Q. R. A Review of Blood–Brain Barrier Transport Techniques. In *Methods in Molecular Medicine 89: The Blood–Brain Barrier: Biology and Research Protocols*; Nag, S., Ed.; Humana: Totowa, NJ, 2003, pp 193–208.
25. Bickel, U. *NeuroRx* **2005**, *2*, 15–26.
26. Liu, X.; Chen, C. *Curr. Opin. Drug Disc. Dev.* **2005**, *8*, 505–512.
27. Gratton, J. A.; Abraham, M. H.; Bradbury, M. W.; Chadha, H. S. *J. Pharm. Pharmacol.* **1997**, *49*, 1211–1216.
28. Hansen, D. K. *J. Pharm. Biomed. Anal.* **2002**, *27*, 945–958.
29. Dai, H.; Elmquist, W. F. Drug Transport Studies using Quantitative Microdialysis. In *Methods in Molecular Medicine 89: The Blood–Brain Barrier: Biology and Research Protocols*; Nag, S., Ed.; Humana: Totowa, NJ, 2003, pp 249–264.
30. Atkinson, F.; Cole, S.; Green, C.; van de Waterbeemd, H. *Curr. Med. Chem. – Central Nervous System Agents* **2002**, *2*, 229–240.
31. Shen, D. D.; Artru, A. A.; Adkison, K. K. *Adv. Drug Deliv. Rev.* **2004**, *56*, 1825–1857.
32. Schinkel, A. H. *Adv. Drug. Deliv. Rev.* **1999**, *36*, 179–194.
33. Doran, A.; Obach, R. S.; Smith, B. J.; Hosea, N. A.; Becker, S.; Callegari, E.; Chen, C.; Chen, X.; Choo, E.; Cianfrogna, J. et al. *Drug Metab. Dispos.* **2005**, *33*, 165–174.
34. Cisternino, S.; Mercier, C.; Bourasset, F.; Roux, F.; Scherrmann, J. M. *Cancer Res.* **2004**, *64*, 3296–3301.
35. Sisodiya, S. M.; Martinian, L.; Scheffer, G. L.; van der Valk, P.; Cross, J. H.; Scheper, R. J.; Harding, B. N.; Thom, M. *Epilepsia* **2003**, *44*, 1388–1396.
36. Awasthi, S.; Hallene, K.L.; Fazio, V.; Singhal, S.S.; Cucullo, L.; Awasthi, Y.C.; Janigro, D. *BMC Neurosci.* **2005**, Sep 27, 6:61.
37. Gumbleton, M.; Audus, K. L. *J. Pharm. Sci.* **2001**, *90*, 1681–1698.
38. Reichel, A.; Begley, D. J. *Pharm. Res.* **1998**, *15*, 1270–1274.
39. Di, L.; Kerns, E. H.; Fan, K.; McConnell, O. J.; Carter, G. T. *Eur. J. Med. Chem.* **2003**, *38*, 223–232.
40. Bauer, B.; Miller, D. S.; Fricker, G. *Pharm. Res.* **2003**, *20*, 1170–1176.
41. Bachmeier, C. J.; Miller, D. W. *Pharm. Res.* **2005**, *22*, 113–121.
42. Krämer, S. D.; Abbott, N. J.; Begley, D. J. In *Pharmacokinetic Optimization in Drug Research: Biological, Physicochemical and Computational Strategies*; Testa, B., Van der Waterbeemd, H., Folkers, G., Guy, R., Eds.; Wiley-VHCA: Zurich, Switzerland, 2000, pp 401–428.
43. Braun, A.; Hammerle, S.; Suda, K.; Rothen-Rutishauser, B.; Gunthert, M.; Kramer, S. D.; Wunderli-Allenspach, H. *Eur. J. Pharm. Sci.* **2000**, *11*, S51–S60.
44. Youdim, K. A.; Avdeef, A.; Abbott, N. J. *Drug Disc. Today* **2003**, *8*, 997–1003.
45. Siflinger-Birnboim, A.; Del Becchio, P. C.; Cooper, J. A.; Blumenstock, J. N.; Shepard, J. N.; Malik, A. B. *J. Cell. Physiol.* **1987**, *132*, 111–117.
46. Dehouck, M. P.; Dehouck, B.; Schluep, C.; Lemaire, M.; Cecchelli, R. *Eur. J. Pharm. Sci.* **1995**, *3*, 357–365.
47. Rapoport, S. I. *Blood–Brain Barrier in Physiology and Medicine*; Raven: New York, 1976; pp 17–42.
48. Chishty, M.; Begley, D. J.; Abbott, N. J.; Reichel, A. *NeuroReport* **2003**, *14*, 1087–1090.
49. Davson, H.; Spaziani, E. *J. Physiol.* **1959**, *149*, 135–143.
50. Bachelard, H. S. *J. Neurochem.* **1971**, *18*, 213–222.
51. Lund-Andersen, H. *Physiol. Rev.* **1979**, *59*, 305–352.
52. Nicholson, C. *J. Neu. Transmiss.* **2005**, *112*, 29–44.

53. Kinnecom, K.; Pachter, J. S. *Brain Res. Protocols* **2005**, *16*, 1–19.
54. Hutchison, H. T.; Eisenberg, H. M.; Haber, B. *Exp. Neurol.* **1985**, *87*, 260–269.
55. Miller, D. S.; Nobmann, S. N.; Gutmann, H.; Toeroek, M.; Drewe, J.; Fricker, G. *Mol. Pharmacol.* **2000**, *58*, 1357–1367.
56. Hartz, A. M.; Bauer, B.; Fricker, G.; Miller, D. S. *Mol. Pharmacol.* **2004**, *66*, 387–394.
57. Bauer, B.; Hartz, A. M.; Fricker, G.; Miller, D. S. *Mol. Pharmacol.* **2004**, *66*, 413–419.
58. Reichel, A.; Begley, D. J.; Abbott, N. J. An Overview of In Vitro Techniques for Blood–Brain Barrier Studies. In *Methods in Molecular Medicine 89: The Blood–Brain Barrier: Biology and Research Protocols*; Nag, S., Ed.; Humana: Totowa, NJ, 2003, pp 307–324.
59. Franke, H.; Galla, H.; Beuckmann, C. T. *Brain Res. Protocols* **2000**, *5*, 248–256.
60. Prieto, P.; Blaauboer, B. J.; de Boer, A. G.; Boveri, M.; Cecchelli, R.; Clemedson, C.; Coecke, S.; Forsby, A.; Galla, H. J.; Garberg, P. et al. *Altern. Lab. Anim.* **2004**, *32*, 37–50.
61. Dehouck, M. P.; Meresse, S.; Delorme, P.; Fruchart, J. C.; Cecchelli, R. *J. Neurochem.* **1990**, *54*, 1798–1801.
62. Török, M.; Huwyler, J.; Gutmann, H.; Fricker, G.; Drewe, J. *Exp. Brain Res.* **2003**, *153*, 356–365.
63. Lundquist, S.; Renftel, M.; Brillault, J.; Fenart, L.; Cecchelli, R.; Denouck, M.-P. *Pharm. Res.* **2002**, *19*, 976–981.
64. Toimela, T.; Maenpaa, H.; Mannerstrom, M.; Tähti, H. *Tox. Appl. Pharmacol.* **2004**, *195*, 73–82.
65. Lauer, R.; Bauer, R.; Linz, B.; Pittner, F.; Peaches, G. A.; Ecker, G.; Friedl, P.; Noe, C. R. *Farmaco* **2004**, *59*, 133–137.
66. Terasaki, T.; Ohtsuki, S.; Hori, S.; Takanaga, H.; Nakashima, E.; Hosoya, K. *Drug Disc. Today* **2003**, *8*, 944–954.
67. Stanness, K. A.; Westrum, L. E.; Fornaciari, E.; Mascagni, P.; Nelson, J. A.; Stenglein, S. G.; Myers, T.; Janigro, D. *Brain Res.* **1997**, *771*, 329–342.
68. Cucullo, L.; McAllister, M. S.; Kight, K.; Krizanac-Bengez, L.; Marroni, M.; Mayberg, M. R.; Stanness, K. A.; Janigro, D. *Brain Res.* **2002**, *951*, 243–254.
69. Dehouck, M. P.; Cecchelli, R.; Richard Green, A.; Renftel, M.; Lundquist, S. *Brain Res.* **2002**, *955*, 229–235.
70. Krizbai, I. A.; Bauer, H.; Bresgen, N. *Cell. Mol. Neurobiol.* **2005**, *25*, 129–139.
71. Toborek, M.; Lee, Y. W.; Flora, G.; Pu, H.; Andras, I. E.; Wylegala, E.; Hennig, B.; Nath, A. *Cell. Mol Neurobiol.* **2005**, *25*, 181–199.
72. Hawkins, R. A.; Peterson, D. R.; Vina, J. R. *IUBMB Life* **2002**, *54*, 101–107.
73. Peterson, D. R.; Hawkins, R. A. *Methods Mol. Med.* **2003**, *89*, 233–247.
74. Joo, F.; Karnushina, I. *Cytobios,* **1973**, *8*, 41–48.
75. Panula, P.; Joo, F.; Rechardt, L. *Experientia* **1978**, *34*, 95–97.
76. Deli, M. A.; Abraham, C. S.; Kataoka, Y.; Niwa, M. *Cell. Mol. Neurobiol.* **2005**, *25*, 59–127.
77. Perrière, N.; Demeuse, P.; Garcia, E.; Regina, A.; Debray, M.; Andreux, J. -P.; Couvreur, P.; Scherrmann, J. -M.; Temsamani, J.; Couraud, P. -O. et al. *J. Neurochem.* **2005**, *93*, 279–289.
78. Tilling, T.; Korte, D.; Hoheisel, D.; Galla, H. J. *J. Neurochem.* **1998**, *71*, 1151–1157.
79. Abbott, N. J. *Cell. Mol. Neurobiol.* **2005**, *25*, 5–23.
80. Weksler, B. B.; Subileau, E. A.; Perriere, N.; Charneau, P.; Holloway, K.; Leveque, M.; Tricoire-Leignel, H.; Nicotra, A.; Bourdoulous, S.; Turowski, P.; et al. *FASEB J.* **2005**, *19*, 1872–1874.
81. Coisne, C.; Dehouck, L.; Faveeuw, C.; Delplace, Y.; Miller, F.; Landry, C.; Morissette, C.; Fenart, L.; Cecchelli, R.; Tremblay, P.; Dehouck, B. *Lab. Invest.* **2005**, *85*, 734–746.
82. Förster, C.; Silwedel, C.; Golenhofen, N.; Burek, M.; Kietz, S.; Mankertz, J; Drenckhahn, D. *J. Physiol.* **2005**, *565*, 475–486.
83. Cornford, E. M.; Hyman, S. *NeuroRx* **2005**, *2*, 27–43.
84. Wong, D.; Dorovini-Zis, K.; Vincent, S. R. *Exp. Neurol.* **2004**, *190*, 446–455.
85. Roux, F.; Couraud, P. -O. *Cell. Mol. Neurobiol.* **2005**, *25*, 41–57.
86. Wolburg, H.; Neuhaus, J.; Kniesel, U.; Krauss, B.; Schmid, E. M.; Ocalan, M.; Farrell, C.; Risau, W. *J. Cell Sci.* **1994**, *107*, 1347–1357.
87. Regina, A.; Koman, A.; Piciotti, M.; El Hafny, B.; Center, M. S.; Bergmann, R.; Couraud, P. -O.; Roux, F. *J. Neurochem.* **1998**, *71*, 705–715.
88. Dolman, D. E. M.; Drndarski, S.; Abbott, N. J.; Rattray, M. *J. Neurochem.* **2005**, *93*, 825–833.
89. Boveri, M.; Berezowski, V.; Price, A.; Slupek, S.; Lenfant, A. M.; Benaud, C.; Hartung, T.; Cecchelli, R.; Prieto, P.; Dehouck, M. P. *Glia* **2005**, *51*, 187–198.
90. Abbott, N. J.; Rönnback, L.; Hansson, E. *Nat. Neurosci. Rev.* **2006**, *7*, 41–53.
91. Haseloff, R. F.; Blasig, I. E.; Bauer, H. -C.; Bauer, H. *Cell Mol. Neurobiol.* **2005**, *25*, 25–39.
92. Madara, J. L. *Annu. Rev. Physiol.* **1998**, *60*, 143–159.
93. Wolburg, H.; Lippoldt, A. *Vascul. Pharmacol.* **2002**, *38*, 323–337.
94. Dehouck, B.; Dehouck, M. P.; Fruchart, J. C.; Cecchelli, R. *J. Cell Biol.* **1994**, *126*, 465–473.
95. Gaillard, P.; de Boer, A. G. *Eur. J. Pharm. Sci.* **2000**, *12*, 95–102.
96. Garberg, P.; Ball, M.; Borg, N.; Cecchelli, R.; Fenart, L.; Hurst, R. D.; Lindmark, T.; Mabondzo, A.; Nilsson, J. E.; Raub, T. J. et al. *Toxicol. in vitro* **2005**, *19*, 299–334.
97. Polli, J. W.; Humphreys, J. E.; Wring, S. A.; Burnette, T. C.; Read, K. D.; Hersey, A.; Butina, D.; Bertolotti, L.; Pugtnaghi, F.; Serbjit-Singh, J. A Comparison of Madin–Darby Canine Kidney Cells and Bovine Brain Endothelial Cells as a Blood–Brain Barrier Screen in Early Drug Discovery. In *Progress in Reduction, Refinement and Replacement of Animal Experimentation*; Balls, M., van Zeller, A. -M., Halder, M. E., Eds.; Elsevier Sciences: Amsterdam, the Netherlands, 2000, pp 271–289.
98. Faassen, F.; Vogel, G.; Spanings, H.; Vromans, H. *Int. J. Pharmaceut.* **2003**, *263*, 113–122.
99. Lohmann, C.; Huwel, S.; Galla, H. J. *J. Drug Targ.* **2002**, *10*, 263–276.
100. Krämer, S. D.; Hurley, J.; Abbott, N. J.; Begley, D. J. *In vitro Cell. Dev. Biol.* **2002**, *38A*, 557–565.
101. Kamau, S. W.; Krämer, S. D.; Gunthert, M.; Wunderli-Allenspach, H. *In vitro Cell Dev. Biol. Anim.* **2005**, *41*, 207–216.
102. Hurst, R. D.; Fritz, I. B. *J. Cell Physiol.* **1996**, *167*, 81–88.
103. Youdim, K. A.; Dobbie, M. S.; Kuhnle, G.; Proteggente, A. R.; Abbott, N. J.; Rice-Evans, C. *J. Neurochem.* **2003**, *85*, 180–192.
104. Mertsch, K.; Maas, J. *Curr. Med. Chem. – Central Nervous System Agents* **2002**, *2*, 187–201.
105. Polli, J. W.; Wring, S. A.; Humphreys, J. E.; Huang, L.; Morgan, J. B.; Webster, I. O.; Serabjit-Singh, C. S. *J. Pharmacol. Exp. Ther.* **2001**, *299*, 620–628.
106. Maurer, T. S.; DeBartolo, D. B.; Tess, D. A.; Scott, D. O. *Drug Metab. Dispos.* **2005**, *33*, 175–181.
107. Platts, J. A.; Abraham, M. H.; Zhao, Y. H.; Hersey, A.; Ijaz, L.; Butina, D. *Eur. J. Med. Chem.* **2001**, *36*, 719–730.
108. Usansky, H. H.; Sinko, P. J. *Pharm. Res.* **2003**, *20*, 390–396.
109. van de Waterbeemd, H.; Camenisch, G.; Folkers, G.; Chretien, J. R.; Raevsky, O. A. *J. Drug Target.* **1998**, *6*, 151–165.

110. Kelder, J.; Grootenhuis, P. D.; Bayada, D. M.; Delbressine, L. P.; Ploemen, J. P. *Pharm. Res.* **1999**, *16*, 1514–1519.
111. Avdeef, A; Testa, B. *Cell. Mol. Life Sci.* **2002**, *59*, 1681–1689.
112. Rose, K.; Hall, L. H. *J. Chem. Inf. Comput. Sci.* **2002**, *42*, 651–666.
113. Sippl, W. *Curr. Med. Chem.* **2002**, *2*, 211–227.
114. Abraham, M. H.; Ibrahim, A.; Zissimos, A. M.; Zhao, Y. H.; Comer, J.; Reynolds, D. P. *Drug Disc. Today* **2002**, 7, 1056–1063.
115. Clark, D. E. *Drug Disc. Today* **2003**, *8*, 927–933.
116. Didziapetris, R.; Japertas, P.; Avdeef, A.; Petrauskas, A. *J. Drug Targ.* **2003**, *11*, 391–406.

Biography

N Joan Abbott, PhD, is professor of neuroscience at King's College London, and a member of the Wolfson Centre for age-related disease which specializes in neuroscience. With Dr David J Begley she co-directs the Blood–Brain Barrier (BBB) Research Group. She has worked on the BBB for more than three decades, recently focusing on in vitro models for study of the cell physiology, pharmacology, and toxicology of the brain endothelium. A particular interest is in the development, optimization, and characterization of in vitro BBB models suitable for studying drug delivery to the central nervous system, with emphasis on the nature of their lipid membranes, and specific transport functions.

5.14 In Vitro Models for Plasma Binding and Tissue Storage

P Barton, R P Austin, and R E Fessey, AstraZeneca R&D Charnwood, Loughborough, UK

5.14.1   Plasma Protein Binding

5.14.1.1   Overview

The role of plasma protein binding (PPB) in the discovery process and the impact this key parameter has on the discovery and clinical process is now becoming fully realized. The pharmacokinetic and pharmacodynamic properties of drugs are greatly influenced by the reversible binding to plasma proteins with the unbound fraction of the drug being responsible for the biological activity. Methods are now available for the rapid determination of free levels in plasma using a multitude of techniques from the traditional dialysis methods to the more recent surface plasmon resonance methodologies.

5.14.1.2   Biological Aspects

Human plasma is known to contain over 60 different proteins, the major component being albumin, which comprises approximately 60% of the total plasma protein.[1] The next most abundant and well-characterized protein is α_1-acid glycoprotein. Human serum albumin has a molecular weight of 66 kDa and contains 585 amino acid residues. At least 18 different mutations of human serum albumin have been identified and are primarily due to a single amino acid mutation, accounting for distinct protein–ligand binding.[2] The concentration of albumin in a normal healthy adult male is typically $43\,\mathrm{g\,L^{-1}}$, with a range of 35–$53\,\mathrm{g\,L^{-1}}$. Females have approximately a 9% lower concentration ($38\,\mathrm{g\,L^{-1}}$)[1,3] and this is argued to account for the gender difference in binding of chlorodiazepoxide and warfarin.

In diseased patients the albumin concentration can be significantly different. For patients with nephrotic syndrome, burns, or cirrhosis, the albumin concentration can be less than $10\,\mathrm{g\,L^{-1}}$, i.e., 20–30% of the normal concentration.[4] In contrast, the normal plasma concentration (human) of α_1-acid glycoprotein is 0.4–$1.0\,\mathrm{g\,L^{-1}}$ (10–$30\,\mu\mathrm{M}$) (MW = 44 kDa), and in patients with inflammatory diseases, it can be elevated by up to 4–5-fold.[4] Taking into account the significant variation in protein content and concentration in human plasma (and other species) it is useful to determine PPB using large plasma pools containing a statistically reasonable number of donors.[2]

Plasma contains many other globulins (the name of a family of proteins precipitated from plasma or serum by addition of ammonium sulfate). These can be separated into many subgroups, the main ones being alpha-, beta-, and gamma-globulins which differ with respect to the associated lipid or carbohydrate. Immunoglobulins (antibodies) are in the a and b fractions, lipoproteins are in the c and d fractions. Other substances in the globulin fractions include macroglobulin, plasminogen, prothrombin, euglobulin, antihemomorphic globulin, fibrinogen, and cryoglobulin. The lipoproteins in plasma, of which α_1-acid glycoprotein is one, can be further classified into very high-density (VHDL), high-density (HDL), low-density (LDL), and very low-density (VLDL). The higher the density the lower the lipid content. Lipoproteins are macromolecular complexes displaying characteristic sizes, densities, and compositions. All lipoproteins contain protein components, called apoproteins, and polar lipids (phospholipids) in a surface film surrounding a neutral core (free and esterified cholesterol, triglycerides). The plasma lipoproteins vary in composition with respect to the lipid component, because their principal physiological function is to transport lipids in a water-soluble form, but also vary with respect to the polypeptide chain composition. Lipoprotein plasma concentration may vary 5–10-fold. Gamma-globulins generally only marginally account for the plasma binding of drugs.[5] Often it is only when a drug is present at very high concentrations that binding to components other than albumin or α_1-acid glycoprotein occurs.[6] Considerable intersubject variability in the PPB of some compounds (4–5-fold variation in $f_{u,p}$ is not uncommon) and in the concentration of proteins exists even within healthy human volunteers.[3] Genetically determined variations in amino acid sequences of human serum albumin can also contribute to variability in binding, and cause higher variability in patients with highly bound drugs. The binding affinities of warfarin, salicylate, and diazepam to five known variants of human serum albumin have been studied.[7] The association constants for all three drugs to albumin decreased by a factor of 4–10-fold for the mutations relative to each other.

5.14.1.3   Effect of Plasma Protein Binding on Drug Disposition

The extent of plasma protein binding is extremely important in its influence on many pharmacokinetic parameters. The relationship between the in vitro intrinsic clearance, $\mathrm{CL_{int}}$ and the in vivo clearance, CL, may be understood by the use of the well-stirred model, which is given in its most simple form by eqn [1][8]:

$$\mathrm{CL} = \frac{(Q_\mathrm{H} \times f_{u,b} \times \mathrm{CL_{int}})}{(f_{u,b} \times \mathrm{CL_{int}} + Q_\mathrm{H})} \quad\quad [1]$$

where Q_H is the hepatic blood flow, with units of $\mathrm{mL\,min^{-1}\,kg^{-1}}$, and $f_{u,b}$ the fraction of compound unbound in blood as defined in Section 5.14.2.1.

For drugs with low hepatic clearance, compared to liver blood flow, the in vivo clearance can be approximated by[9]:

$$CL \cong f_{u,p} \times CL_{int} \qquad [2]$$

where CL_{int} is the intrinsic clearance of the drug and $f_{u,p}$ is the fraction of compound unbound in plasma. Under these conditions clearance is directly proportional to $f_{u,p}$. An example of this behavior is given by the clearance of warfarin in male Sprague–Dawley rats which is proportional to the free fraction of the drug within each individual rat,[10,11] as shown in **Figure 1a**. For drugs with high hepatic clearance, the clearance is largely controlled by the liver blood flow, Q_H, and is independent of the extent of plasma binding[9]:

$$CL \cong Q_H \qquad [3]$$

This type of behavior is illustrated by propranolol where the clearance in humans does not depend on the human free fraction of drug in plasma[12] (**Figure 1b**).

The degree of plasma protein binding cannot only influence metabolic clearance, but also renal clearance.[13] The renal clearance (CL_R) of a drug consists of four different processes; glomerular filtration (CL_f), active secretion (CL_{rs}),

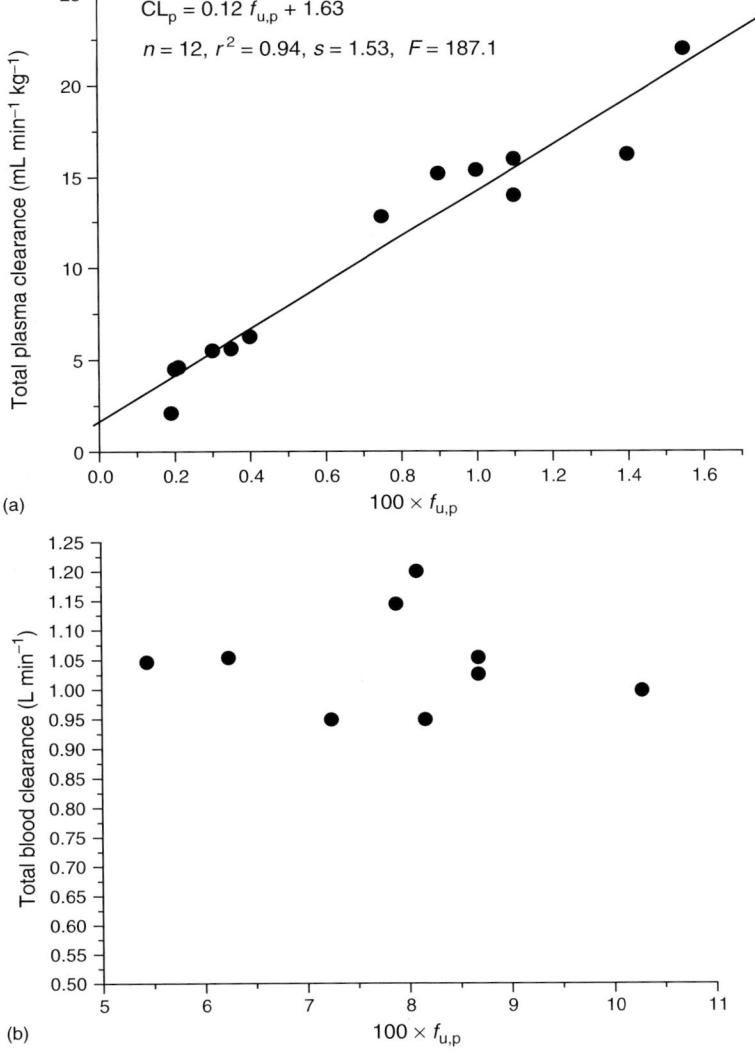

Figure 1 (a) Clearance of warfarin in male Sprague–Dawley rats showing dependence of the clearance on the plasma free fraction. (Reproduced with permission from Yacobi, A.; Levy, G. *J. Pharm. Sci.* **1974**, *63*, 805–806). (b) Clearance of propranolol in humans showing the lack of relationship between clearance and the plasma free fraction. (Reproduced with permission from Evans, G. H.; Nies, A. S.; Shand, D. G. *J. Pharmacol. Exp. Ther.* **1973**, *186*, 114–122.)

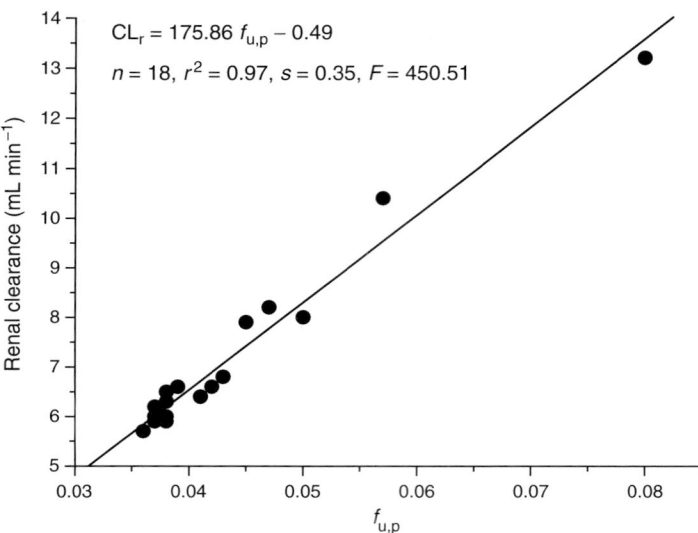

$$CL_r = 175.86 \, f_{u,p} - 0.49$$
$$n = 18, \, r^2 = 0.97, \, s = 0.35, \, F = 450.51$$

Figure 2 Human renal clearance data for ceftriaxone showing the dependence on free fraction.

passive reabsorption (F_r), and renal metabolism (CL_{rm}).[14] The glomerular clearance is related to $f_{u,p}$ by:

$$CL_f = f_{u,p} \times GFR \qquad [4]$$

where GFR is the rate at which plasma water is filtered through the glomerulus, which is approximately $125 \, mL \, min^{-1}$ in a 70-kg human.

Figure 2 demonstrates how the human renal clearance of ceftriaxone is proprtional to the plasma free fraction as shown by eqn [4], with higher plasma free fraction in individual patients leading to higher renal clearance of the drug.[13]

5.14.1.4   Influence of Blood:Plasma Ratio on Pharmacokinetic Parameters

Clearance and volume of distribution measurements depend on the reference body fluid, such as blood or plasma, in which drug concentrations are measured.[15,16] Blood clearance, CL_b, is the actual volume of blood cleared of a drug per unit time from the either the entire blood pool in the body (systemic blood clearance) or from the blood passing through the eliminating organs (organ blood clearance). The systemic blood clearance is the sum of all the organ blood clearances.

Plasma clearance, CL_p, does not represent the actual volume of plasma cleared of drug, rather it is the apparent volume of plasma cleared per unit time, reflecting the ratio between the rate of drug elimination from the entire body (or the organs) and the drug concentration in plasma. Plasma clearance and volume of distribution are more often measured, because sample preparation and analysis are easier in plasma than in blood.

The relationship between blood clearance and plasma clearance is given by eqn [5], and between blood volume of distribution ($V_{ss,b}$) and plasma volume of distribution ($V_{ss,p}$) by eqn [6]:

$$CL_p = CL_b \times \left(\frac{C_b}{C_p} \right) \qquad [5]$$

where C_b/C_p is the experimentally accessible blood:plasma ratio. Hence there can be substantial differences between blood and plasma clearances. If $C_b/C_p > 1$, plasma clearance is greater than blood clearance; if $C_b/C_p < 1$ then plasma clearance is lower than blood clearance. It is important therefore to ensure that it is known whether plasma or blood clearance is being considered. Likewise, differing blood:plasma ratios can lead to differences in the steady-state volume of distribution in blood and plasma.

$$V_{ss,p} = V_{ss,b} \times \left(\frac{C_b}{C_p} \right) \qquad [6]$$

Table 1 Drug to plasma ratios and pharmacokinetic parameters for marketed drugs in human patients

Drug	C_b/C_p	CL_p $(mL\,min^{-1}\,kg^{-1})$	$V_{ss,p}$ $(L\,kg^{-1})$	CL_b $(mL\,min^{-1}\,kg^{-1})$	$V_{ss,b}$ $(L\,kg^{-1})$
Pravastatin	0.55	13.5	0.46	24.5	0.84
Omeprazole	0.58	7.5	0.34	12.9	0.59
Valproic acid	0.64	0.11	0.22	0.17	0.34
Sufentanil	0.74	12.7	1.7	17.2	2.3
Trimethoprim	1.0	1.9	1.6	1.9	1.6
Metoprolol	1.0	15	4.2	15	4.2
Chloroquine	9	8.4	197	0.93	21.8
Tacrolimus	35	0.90	0.91	0.025	0.026

Data from Goodman, A. G.; Gilman, A. G. In *The Pharmacological Basis of Therapeutics*, 10th ed.; Harman, J. G., Limbird, L. E., Gilman, A. G., Eds.; McGraw-Hill: New York, 2001, pp 1924–2023.

If two drugs have equivalent properties except differences in C_b/C_p, clearance and volumes of distribution of the drugs will be different depending on the reference fluid (blood or plasma); however, there will be no differences between blood and plasma half-lives. The lower limit for a blood to plasma ratio is 0.5, which corresponds to a compound that has no affinity for the blood and is present totally in the plasma compartment. Blood:plasma ratios > 50 are possible. A list of drugs with a range of blood to plasma ratios and the corresponding blood and plasma pharmacokinetic parameters in human patients is given in **Table 1**.

5.14.1.5 Drug–Drug Interactions

The obvious question arises as to whether highly plasma bound drugs can displace each other from plasma proteins. The potential does exist, but as most drugs have therapeutic concentrations far below the plasma concentration of albumin and often below that of α_1-acid glycoprotein, proteins are rarely at the point of saturation and so competition for binding sites is much diminished.[9]

Although the literature is replete with examples of drug–drug interactions suggested to be due to PPB interactions, many of these have now been attributed to other mechanisms.[18] The drug–drug interactions are often due to cytochrome P450 inhibition, or to coadministered drugs acting as competitive substrates, although a few may still be attributable to pure displacement. **Table 2** lists some drugs for which other mechanisms are in fact responsible for drug–drug interactions, rather than the original suggestion of PPB displacement.

The fact that high drug concentrations are required in order to induce significant displacement interactions is illustrated by studies on the effect of salicylic acid on the PPB of ibuprofen[19,41] which show that very high concentrations of salicylic acid (1500 μM) have only a moderate effect on the free concentration of ibuprofen. The possibility of significant albumin displacement interaction in vivo is therefore only a reality when one of the interacting drugs has a very low potency (or short half-life) and consequently is dosed in such a way that very high plasma concentrations are achieved. Since plasma concentrations of α_1-acid glycoprotein are about 40-fold lower than that of albumin, displacement interactions at this protein could be possible at more typical therapeutic drug concentrations.

5.14.2 Experimental Aspects

5.14.2.1 Scales of Measurement of Plasma Protein Binding

The term protein binding normally refers to the reversible association of a drug with the proteins of the plasma compartment of blood, and this binding is due to electrostatic and hydrophobic forces between drug and protein.[3] Drug that is bound reversibly will be in equilibrium with the free (unbound) drug, with the amount bound being dependent on both the affinity of the drug for the various proteins and the binding capacity of each protein.

Upon entering plasma, most drugs bind rapidly to the plasma constituents, which principally include albumin and α_1-acid glycoprotein.[20,21] Measurement of the extent of this binding requires either a physical separation of free from

Table 2 Drug–drug interactions originally suggested being attributable to plasma protein displacement, and the actual mechanism responsible

Drug	Displacing drug	Mechanism responsible
Methotrexate	Salicylate	Inhibition of renal clearance
Phenytoin	Valproate	Inhibition of metabolism
Tolbutamide	Phenylbutazone	Inhibition of metabolism
	Salicylates	Pharmacodynamic
	Sulfonamides	Inhibition of metabolism
Warfarin	Phenylbutazone	Inhibition of metabolism
	Clofibrate	Pharmacodynamic
	Choral hydrate	Possible pure displacement
	Sulfinpyrazole	Inhibition of metabolism

Reproduced with permission from Rolan, P. E. *Br. J. Clin. Pharmacol.* **1994**, *37*, 125.

bound drug, or a technique that can distinguish some property of bound from unbound drug. Equation [7] defines the free fraction of a drug in plasma:

$$f_{u,p} = [\text{free drug in plasma}]/[\text{total drug in plasma}] \qquad [7]$$

The fraction of drug bound $f_{b,p}$ is given by:

$$f_{b,p} = 1 - f_{u,p} \qquad [8]$$

While plasma binding refers to the binding of a drug to the plasma compartment of blood, studies can also be carried out on binding to serum or to whole blood. The major difference between plasma and serum is the removal of fibrinogen from plasma, and since most drugs do not bind to fibrinogen, no differences in binding to plasma or serum are expected.[22] Plasma is the fluid that remains after blood cells have been removed by centrifugation, and is usually obtained using an anticoagulant so that all clotting factors (e.g., fibrinogen) are retained. Serum is the fluid that remains after clotting factors have been removed by first allowing clotting to take place. Proteins not involved in clotting, e.g., albumin, are still present in serum. In whole blood, drugs can bind to plasma proteins and to blood cells, and hence there can be significant differences between blood binding and plasma binding.[20] The whole blood free fraction of a compound, $f_{u,b}$, is related to the plasma free fraction by eqn [9]. Here C_b/C_p is the experimentally accessible blood to plasma ratio of the drug, i.e., the ratio of total concentration of drug in whole blood to the total concentration of drug present in the plasma compartment of the blood.

$$f_{u,b} = \frac{f_{u,p}}{C_b/C_p} \qquad [9]$$

Highly detailed studies of PPB allow derivation of several binding constants from the experimental data. Two different formulations may be used for this purpose, one being stoichiometric and the other being site-oriented.[23–25] However, most plasma binding data are simply quoted on a percentage scale as given by:

$$\%PPB = 100 f_{u,p} \qquad [10]$$

Because the percentage scale is bounded by 0 and 100, a pseudo-binding constant ($K^{B/F}$) for PPB is frequently used, and this is defined by[26,27]:

$$K^{B/F} = \frac{(1 - f_{u,p})}{f_{u,p}} = \frac{\%bound}{\%free} \qquad [11]$$

More detailed quantitative binding information can be obtained by carrying out a Scatchard analysis.[28,29] In this analysis, B/F is plotted against B for a compound, covering a range of compound concentrations at a chosen protein or plasma concentration. For binding of compound to a single site on the protein, this plot is linear; the slope is related to the equilibrium-binding constant of the compound and protein, K, and the intercept being equal to the concentration of the binding protein. For a situation where binding to a number of different sites on protein, or sites on a multitude of proteins, occurs over the compound concentration range studied, curved Scatchard plots are observed, and nonlinear regression curve fitting then allows the binding constants to each site to be determined.

5.14.2.2 Experimental Methods for Plasma Protein Binding Measurement

There are many experimental methods that have been used to determine the extent of plasma protein binding and these techniques are summarized in more detail in the following section.

5.14.2.2.1 Equilibrium dialysis

Equilibrium dialysis is still regarded as the 'gold standard' method by most researchers. In this technique two cells are separated by a semipermeable membrane which precludes high molecular weight compounds and plasma proteins from crossing from one cell to the other, but does allow transfer of low molecular weight compounds to occur.[30,31] In one cell, an appropriate volume of plasma is placed. In the other cell, an equivalent volume of an appropriate aqueous buffer solution is placed. The compound is spiked into either dialysis cell, more often the plasma containing cell.

The cells are then equilibrated at the desired temperature (often 37 °C) for the chosen dialysis time. The plasma dialysis cell contains compound which is bound to plasma proteins and compound which is unbound (free) the latter being able to cross the semipermeable membrane into the buffer side, and vice versa. If the dialysis time is sufficiently long then an equilibrium is reached, where the free concentration of compound is the same in both the plasma and buffer dialysis cells. An aliquot from the buffer cell is analyzed, which gives a measure of the free concentration of compound. An aliquot from the plasma cell is also analyzed and this gives a measure of the total concentration of compound (bound and free). The extent of plasma protein binding as measured by $f_{u,p}$ is then given by:

$$f_{u,p} = \frac{C_{buffer}}{C_{plasma}} \quad [12]$$

where C_{buffer} is the concentration of compound in the buffer dialysis cell and C_{plasma} the total concentration of compound in the plasma cell. A variant of equilibrium dialysis is dynamic dialysis.[32,33]

5.14.2.2.2 Fluid shifts during equilibrium dialysis

Osmotic volume shifts across the dialysis membrane can occur during equilibrium dialysis experiments.[34,35] These result in a dilution of the plasma with the buffer, leading to an overestimate of the free fraction, $f_{u,p}$. A correction can be made for the observed average volume shift for a given equilibrium dialysis apparatus and equilibration time as given by:

$$f_{u,p} = 1 - \frac{(C_{plasma} - C_{buffer})(V_{final}^{plasma}/V_{initial}^{plasma})}{(C_{plasma} - C_{buffer})(V_{final}^{plasma}/V_{initial}^{plasma}) + C_{buffer}} \quad [13]$$

where C_{plasma} and C_{buffer} refer to the observed concentrations in the plasma and buffer cells at equilibrium, V_{final}^{plasma} is the volume of the plasma at equilibrium (accounting for the fluid shift) and $V_{initial}^{plasma}$ the volume of the plasma at the start of the experiment.

5.14.2.2.3 The Donnan effect in equilibrium dialysis

One aspect of equilibrium dialysis that is sometimes raised as an issue is the so-called Donnan effect.[36–38] This results in an unequal distribution of ionic species in the dialysis cells during an equilibrium dialysis experiment, which causes a difference between the observed binding compared to the situation where no Donnan effect occurred. This arises because the plasma protein molecules themselves often possess a net charge, but they cannot diffuse between the two cells, owing to the presence of a semipermeable membrane. Small ionic species present in the buffer and plasma are free to diffuse across the membrane. Owing to the conditions of the balanced chemical potential of ionic species on both sides of the dialysis membrane and electroneutrality, the charge on the protein molecules on one side of the membrane is compensated for in part by a reduced concentration of small (diffusible) ions on that side of like charge, and in part by an increased concentration of small ions of the opposite charge. However, the Donnan effect can be almost completely suppressed if a sufficiently high concentration of electrolyte is used, i.e., a large buffer concentration is used (ionic strength >0.10 M), as in many equilibrium dialysis experiments.[36]

5.14.2.2.4   Ultrafiltration

Determination of the extent of plasma protein binding by ultrafiltration involves separation of the unbound drug present in an equilibrated solution of drug and plasma proteins by filtering it through a semipermeable membrane (no protein can pass through the membrane).[1,39] The separation is assisted either by application of a vacuum, positive pressure from compressed nitrogen, or by centrifugation. Ultrafiltration has the advantage of a fast equilibration time, typically 30 min at 37 °C. Equilibrium dialysis is rate limited by the time it takes the unbound compound to pass through the semipermeable membrane and reach equilibrium between the two dialysis cells; this often takes several hours. One issue with ultrafiltration is that significant nonspecific binding to the membrane can occur and this needs to be characterized. In addition, the plasma protein concentration increases during the filtration step, which can potentially cause a change in the binding of the drug and introduce a bias in the measured binding; however, only a small volume is filtered (typically 5–10% of the initial volume) so the disturbance of the equilibrium is minimal and the error introduced is quite small. It is also possible for the pore size of the membrane to be decreased by accumulation of protein at the membrane surface.

5.14.2.2.5   Ultracentrifugation

The ultracentrifugation method achieves separation of free and protein bound drugs by centrifugation ($10^5 g$ or higher) of a mixture of drug and plasma in a tube with no membrane.[1,40–42] Ultracentrifugation is an alternative to both equilibrium dialysis and ultrafiltration as it removes the problems associated with membrane effects (e.g., nonspecific binding) and osmotic volume shifts. However, ultracentrifugation takes 12–16 h and causes errors in the estimation of the free drug concentration by some physical phenomena such as sedimentation, back diffusion, and viscosity, and additionally requires expensive centrifugation equipment.[42]

5.14.2.2.6   Alternative methods for the determination of plasma protein binding

A number of alternative methods have been employed for the determination of plasma binding and to describe these all in any detail is beyond the scope of this chapter. However, the authors point the reader to key references for a number of other methods: chromatographic methods,[43–45,65] fluorescence spectroscopy,[46,47] ultraviolet spectroscopy,[46] derivative spectroscopy,[48–50] circular dichroism,[51] nuclear magnetic resonance spectroscopy,[52,53] surface plasmon resonance,[54–58] capillary electrophoresis,[59–63] microdialysis,[64,65] and high-throughput equilibrium dialysis methods.[66,67]

5.14.3   Tissue Binding

5.14.3.1   Overview

Although not studied as extensively as plasma binding, tissue binding can have as profound an effect upon the pharmacokinetics of a compound as can the plasma binding.[68–70] The binding of drugs to plasma and tissue constituents affects both the pharmacokinetics and pharmacodynamics of the compound. The pharmacokinetic parameters, steady-state volume of distribution, clearance, and half-life, all depend upon the free fraction in plasma that is in turn dependent upon the relative affinity between plasma and tissue constituents.

The methods for the determination of tissue binding can generally be divided into two rather distinct areas of in vivo methods, and in vitro methods each of which possess their own advantages and disadvantages based around the complexity of the experiment and the ability to monitor biologically relevant processes.

5.14.3.2   In Vitro Methods for the Determination of Fraction Unbound in Tissue

5.14.3.2.1   Binding to tissue homogenates

The binding to tissue homogenates can be determined by many of the methods described for protein binding. Generally, equilibrium dialysis is the preferred method but wide use has also been made of the ultrafiltration and ultracentrifugation methods. One clear disadvantage of the use of tissue homogenates is the complete disruption of the cellular integrity of the tissue homogenate. However, the tissue-to-plasma equilibrium constant, K_p, obtained from this method is not distorted by uptake into the tissue. Homogenization may release tissue constituents that the drug would otherwise not be exposed to and this may also distort the observed K_p value. Because of the fact that tissue homogenate is difficult to handle due to the formation of a paste like heterogeneous suspension, experiments are generally carried out in diluted homogenate. This then leads to the question as to whether the binding is linear with tissue concentration.[71] It was noted that for warfarin, phenytoin, and quinidine a linear relationship exists between B/F and the percent muscle in the homogenate, but for propranolol and imipramine, linear extrapolation from lower percent muscle in homogenate to higher levels would be incorrect and lead to an over estimate of the B/F term.

5.14.3.2.2 Binding to tissue slices

Determination of binding to tissue slices is a frequently used method and has the advantage of retaining the cellular integrity, a fact that is often a deciding factor for the use of tissue slices over tissue homogenates. Because of the intact nature of the tissue it is possible to observe accumulation of compound due to uptake processes although the accumulation rates may vary significantly from that observed in vivo. The use of tissue slices can also be extended to tissue pieces as it often difficult to generate tissue slices with certain tissue such as muscle and lung.

Tissue pieces have been used to predict total water and extracellular space using the marker compounds ^{14}C-urea and ^{3}H-inulin.[72] In this case the K_p values for ^{14}C-urea were greater than the total tissue water due to tissue swelling. The tissue swelling was driven by osmotic pressure to equilibrate the protein-containing tissue with the protein-free media. The apparent K_p value for ^{14}C-urea was corrected for the ^{14}C-urea in the absorbed media and the corrected K_p values were then considerable closer to the in vivo situation when the correction was made (**Table 3**).

The extent to which albumin diffusion from tissue into the surrounding medium during in vitro incubations has also been shown to have a significant affect on the K_p values obtained from tissue pieces.[73] In this study 12 different tissues were obtained from rats after an intravenous dose of ^{125}I-human albumin and tissue pieces incubated to determine the efflux of albumin over a 2 h and 4 h period. The authors derived a mathematical model eqn [14] and a correction factor eqn [15], with readily measurable or derived terms, to modify the apparent K_{pu} taking into account diffusion of albumin from the tissue:

$$\frac{K_{puapp}}{K_{pu}} = \frac{1 - (1 - f_r)f_p\left(\dfrac{1 - f_{u,p}}{f_{u,p}}\right)}{1 + f_p\left(\dfrac{1 - f_{u,p}}{f_{u,p}}\right) + \left(\dfrac{1}{\theta_E} - 1\right)/f_{ul}} \Bigg/ 1 + \theta_E\dfrac{V_T}{V_M}f_p\left(\dfrac{1 - f_{u,p}}{f_{u,p}}\right)(1 - f_r)$$

$$[14]$$

where $\theta_E = V_E/V_T$ (V_E and V_T are the extracellular and total tissue volumes respectively), V_M the volume of the media; f_p is the ratio of the plasma binding concentrations in the extracellular space to that in plasma[74]; $f_{u,p}$ the unbound fraction of compound in plasma; f_r, the fraction of extracellular binding protein remaining in tissue at the end of the incubation; f_{ul}, the unbound intracellular fraction of the compound.

Table 3 Tissue swelling, tissue water content, ^{14}C-urea partition coefficients with and without swelling

Tissue	% Swelling (sd)	Fraction water content (sd)	^{14}C-Urea (K_p)	^{14}C-Urea (K_{pcorr})	^{14}C-Urea ($K_{p\ in\ vivo}$)
Adipose	12 (1.0)	0.047 (0.002)	0.23 (0.008)	0.099 (0.011)	0.131
Bone	3.5 (1.7)	0.16 (0.0044)	0.37 (0.042)	0.33 (0.030)	0.559
Brain	34 (2.1)	0.62 (0.014)	1.2 (0.026)	0.61 (0.013)	0.137
Heart	8.5 (1.2)	0.79 (0.006)	0.92 (0.016)	0.77 (0.010)	0.837
Intestine	− 5.0 (2.8)	0.76 (0.002)	0.68 (0.026)	0.79 (0.064)	1.06
Kidney	39 (1.4)	0.73 (0.004)	1.0 (0.033)	0.46 (0.014)	2.27
Liver	13 (0.9)	0.65 (0.005)	0.80 (0.011)	0.59 (0.014)	0.69
Lung	19 (1.2)	0.70 (0.035)	1.0 (0.022)	0.70 (0.008)	0.625
Muscle, thigh	41 (3.8)	0.73 (0.005)	1.2 (0.043)	0.55 (0.011)	0.735
Skin	18 (1.1)	0.54 (0.008)	0.88 (0.011)	0.59 (0.011)	0.617
Spleen	6.7 (0.6)	0.74 (0.003)	0.75 (0.021)	0.64 (0.013)	0.735
Stomach	− 0.4 (2.4)	0.73 (0.023)	0.88 (0.022)	0.91 (0.037)	0.496
Testes	− 34 (2.8)	0.83 (0.024)	0.65 (0.023)	0.55 (0.028)	0.551
Thymus	3.9 (4.0)	0.74 (0.017)	0.98 (0.039)	0.93 (0.065)	ND

Reproduced from Ballard, P.; Leahy, D. E.; Rowland, M. *Pharm. Res.* **2000**, *17*, 660–663, with kind permission of Springer Science and Business Media.

A correction factor (CF) of the apparent tissue-to-medium partition coefficient to give the true K_{pu} when no albumin has diffused out of the tissue is given by:

$$\text{CF} = \frac{K_{pu}}{K_{puapp}} = \left(1 + \theta_E\left(\frac{V_t}{V_M}\right) \cdot \left(\frac{1 - f_{u,p}}{f_{u,p}}\right) f_p(1 - f_r)\right) + \frac{\theta_E(1 - f_r)(1 - f_{u,p}/f_{u,p})f_P}{K_{puapp}} \qquad [15]$$

From the models the authors concluded that albumin diffusion would have a minimal effect on the measured K_p unless the compound is highly plasma bound, in which case the measured K_p would be significantly underestimated.

The ability to determine in vivo tissue-to-unbound plasma distribution coefficients K_{pu} from in vitro data was examined for a series of 5-n-alkyl-5-ethyl barbituric acids.[75] Unbound tissue-to-unbound-plasma distribution coefficients were determined in 15 rat tissue/organs for a homologous series of nine barbiturates. The steady-state in vivo K_{pu} values were estimated from tissue and plasma concentrations following constant rate intravenous infusions for the homologous series. Correction of the in vitro tissue-to-unbound-plasma distribution coefficients for tissue swelling and barbiturate in the vasculature showed a good prediction of the in vivo values over a wide range of lipophilicity. However, it was noted that for the most lipophilic barbiturates the in vitro K_{pu} under predicted those that were observed in vivo. This under-prediction may have been associated with the combination of binding to albumin and partitioning into lipids that had diffused from the tissue into the medium or insufficient equilibration time.

5.14.3.3   In Vivo Methods for the Determination of Fraction Unbound in Tissue

In vivo methods generally involve the inference of tissue binding from the plasma or blood concentration time course or from the determination of tissue to plasma or blood concentration ratio. For example tissue to blood partition coefficients for lidocaine were calculated following constant rate infusion of the dug to marmoset monkeys.[76] Also tissue to plasma partition coefficients were obtained for digoxin in rats.[116] However, the tissue to plasma partition coefficients for adriamycin were calculated from the terminal elimination phase following an intravenous bolus dose of the drug. In general there is no consensus as to which method is the most appropriate for the determination of tissue to plasma partition coefficients. The volume of distribution can be expressed as:

$$V_{ss} = V_P + V_t \cdot \frac{f_{u,p}}{f_{u,t}} \qquad [16]$$

This can be rearranged to give:

$$f_{u,t} = \frac{V_t \cdot f_{u,p}}{V_{ss} - V_P} \qquad [17]$$

where $f_{u,t}$ is the fraction unbound in tissue, $f_{u,p}$ the fraction unbound in plasma, V_P the plasma volume, V_t the volume of other tissue in the body, and V_{ss} the apparent volume of distribution at steady state. Using this approach the steady-state volume of distribution is determined in the normal manner and plasma binding can be determined in vitro by any of the methods outlined earlier in the chapter. The $f_{u,t}$ determined in this way represents the average fraction unbound in all tissue. From these measurements it is not possible to draw any conclusions about the disposition of the drug in any particular tissue or organ. This methodology was used to estimate the $f_{u,t}$ values of phenytoin[77] in healthy volunteers. For compounds, which are highly bound to plasma proteins, it was pointed that the use of eqn [17] may be misleading[116] as this does not take into account the fact that plasma proteins are distributed throughout the extracellular fluids.

5.14.3.4   Tissue-to-Plasma Concentration Ratios In Vivo

The second approach to the determination the tissue binding is to determine tissue-to-plasma or tissue-to-blood partition ratio, which is defined below:

$$\frac{C_{TSS}}{C_{PSS}} = K_P = \frac{f_{u,p}}{f_{u,t}} \qquad [18]$$

where C_{TSS} is the concentration in the tissue of interest at steady state and C_{PSS} the steady-state plasma concentration. The ratio of the concentration in tissue at steady state to that in plasma at steady state is the tissue to plasma partition coefficient denoted by K_p. The assumption in the derivation of eqn [18] is that at steady state the free concentration between tissue and plasma are equal. In order to fully characterize K_p, knowledge of whether the compound is eliminated from the organ of interest is required.[76] For eliminating organs such as the liver and lung calculation of K_p is only possible if the extraction ratio for that organ is known and in this case the apparent $K_{p(app)}$ can be corrected using:

$$K_{p(app)} = K_p(1 - E)$$ [19]

It is worth noting that to unambiguously assign K_p to a given organ or tissue it should be nonelimination and steady state should be achieved by constant rate infusion. For eliminating organs and when using bolus dosing regimen K_p depends on the assumptions that the perfusion rate limits the distribution of the drug into the tissue and the tissue is viewed as well stirred. In this case the emergent blood is in equilibrium with the tissue.[76] If this is not the case then the relationship for K_p given in eqn [19] is not valid.

5.14.3.5   Binding to Adipose Tissue

Binding and disposition studies of drugs into adipose tissue have received considerably fewer investigations than into other tissue. However, adipose tissue must be considered an important storage tissue as it contains a high proportion of stored lipid in the form of triglycerides and the tissue mass of adipose in human is large.

Since many compounds that accumulate into adipose tissue have long half-lives, steady-state distribution measurements in adipose tissue are not practicable. This led to the introduction of the adipose storage index (ASI),[78–80] which is defined as:

$$ASI = \frac{C_{admax}}{D_{admax}}$$ [20]

where C_{admax} is the maximum concentration in adipose tissue after a single dose and D_{admax} the hypothetical (average) concentration of evenly distributed drug at $t = t_{max}$. This can be calculated from the mass balance of the kinetic distribution experiment, or if elimination is slow, the value can be approximated to $t = 0$. This has been suggested to be superior to the frequently used adipose/plasma concentration ratio.

A number of studies have been aimed at understanding the properties that control disposition into adipose tissue. It has been noted that there was a surprising lack of correlation between adipose tissue build up and lipophilicity in vivo. It was suggested the distribution is the binding competition between adipose tissue, lean tissue and plasma proteins. The binding to adipose tissue is driven mainly by log P (octanol/water partition coefficient) whereas blood and lean tissue binding is determined by log P and the drug's molecular structure.

5.14.4   Incubational Binding

5.14.4.1   Overview

The kinetics of in vitro metabolism are most frequently studied using liver microsomes or hepatocytes as the incubation medium, and from such kinetic experiments the in vitro intrinsic clearance of the compound under study may be derived. The intrinsic clearance data is then often used in the important field of in vitro–in vivo correlations for the purpose of predicting the in vivo clearance of the compound in animals or humans.[81–83] Nonspecific binding of compounds to the microsomes or hepatocytes in the incubation medium will cause a reduction of free drug available to interact with the drug metabolising enzymes, and hence a decrease in the observed intrinsic clearance. The scaling methods used in in vitro–in vivo correlation require the use of unbound intrinsic clearance ($CL_{int,u}$) rather than the observed intrinsic clearance (CL_{int}), and the two properties are related by[81]:

$$CL_{int,u} = \frac{V_{max}}{K_{M,U}} = \frac{V_{max}}{K_M f_{u,inc}} = \frac{CL_{int}}{f_{u,inc}}$$ [21]

where V_{max} is the maximum velocity, $K_{M,U}$ the Michaelis constant based on free substrate concentrations, K_M the observed Michaelis constant determined from plots of reaction rate against total substrate concentration, and $f_{u,inc}$ the free fraction of compound in the incubation medium.[84] The most widely used model for relating the observed CL_{int} to in vivo clearance is the well-stirred model[84,85]:

$$CL = \frac{(Q_H . A . B . f_{u,p} . CL_{int}/f_{u,inc})}{(Q_H + (A . B . f_{u,p} . CL_{int}/f_{u,inc}))}$$ [22]

where A is a constant representing the amount of microsomes or hepatocytes per gram of liver, B a constant describing the grams of liver per kilogram of body weight, Q_H represents liver blood flow, and $f_{u,p}$ the free fraction of the compound in plasma. Equation [22] indicates that knowledge of the extent of binding of drugs to in vitro incubations is required in order to make best use of in vitro–in vivo scaling methods, and this has been confirmed in studies of the prediction of human clearance of diverse sets of marketed drugs from in vitro data.[86,87]

5.14.4.2   Microsome Binding

5.14.4.2.1   Experimental methods
The binding of drugs to liver microsomes can in principle be determined using a variety of the techniques described for the measurement of plasma binding. However, the literature of microsome binding data is very much dominated by the equilibrium dialysis technique, with the larger compilations of data all using this method.[86,88–91]

5.14.4.2.2   Use of microsome binding to determine unbound intrinsic clearance
An important consequence of in vitro microsomal binding is that it causes the observed intrinsic clearance to be dependent on the concentration of microsomes, such that CL_{int} decreases with increasing microsomal concentration, whereas $CL_{int,u}$ should be independent of microsomal protein concentration. This has been clearly demonstrated in several studies.[90,92,93] In terms of the components of CL_{int} (eqn [21]), it is the observed K_M that is dependent on microsome concentration while V_{max} is not.[92,93] The concentration dependence can be strong, and some compounds can have very different values of CL_{int} at different microsomal protein concentrations. For example, the lipophilic base amiodarone has been shown to have $CL_{int} = 115$, 28, and $3 \, \mu L \, min^{-1} \, mg^{-1}$ protein at microsomal protein concentrations of 0.25, 1, and $4 \, mg \, mL^{-1}$, respectively.[90] The use of such data for making in vivo clearance predictions using eqn [22], but with ignorance of the $f_{u,inc}$ terms, would lead to three very different clearance predictions that are clearly erroneous. However, there are many examples of this mistreatment of data in the literature. Once data like that of amiodarone above have been corrected for the extent of microsomal binding, the unbound intrinsic clearance should be constant within experimental error,[90] yielding data that lead to better predictions of in vivo clearance.[86,87]

In addition to the important use of microsomal binding data for the correction of observed intrinsic clearance, the data also has a role in the correction of cytochrome P450 inhibition constants determined in microsomal systems. It has been shown that correction of the observed inhibition constants should lead to better predictions of in vivo drug interactions from in vitro data.[94,95]

5.14.4.3   Dependence of Binding on Drug and Microsome Concentration

The extent of microsomal binding will clearly increase with increasing microsomal protein concentration, and several studies have demonstrated this.[88,90–92] However, it is also possible that $f_{u,inc}$ of a compound will depend on compound concentration and that the binding is saturable. This area has not been widely studied experimentally, but it has been shown that the microsomal binding of amitriptyline, imipramine, propranolol, phenytoin, and tolbutamide is independent of compound concentration over the range of concentration studied (up to $100 \, \mu M$),[88,91,92] and weak concentration dependence has been demonstrated for nortriptyline and warfarin.[88,89] Microsomal binding is therefore not strongly dependent on compound concentration, and saturation does not occur at the low concentrations relevant to in vivo metabolic assays. It is therefore clear that microsomal binding, at concentrations relevant to in vitro metabolic assays, is not characterized by saturable binding to specific binding sites, but is instead controlled by rather nonspecific binding to various protein and lipid components of the microsomes. A model involving specific and saturable binding sites has been reported[89] for describing the concentration dependence of microsomal binding, but appears to be unnecessarily complex since a more simple phase equilibrium model has been successfully used to describe the binding of 12 diverse drugs as a function of microsomal concentration.[90] This binding model (eqn [23]) can generate an estimate of the extent of microsomal binding (f_{u_2}) at given microsomal concentration (C_2) from the observed extent of binding (f_{u_1}) at a different microsomal concentration (C_1):

$$f_{u_2} = \frac{1}{\frac{C_2}{C_1}\left(\frac{1-f_{u_1}}{f_{u_1}}\right) + 1}$$

[23]

Microsome binding could in principle show a dependence on the animal species from which the microsomes are derived. However, It has been shown that propranolol, imipramine, and warfarin have no species dependence in their binding to rat, dog, monkey, and human microsomes,[84] and another study on a set of eight drugs demonstrates no species dependence in binding to rat, dog, and human microsomes.[91] Species dependence in general is therefore likely to be weak or nonexistent, which is consistent with the finding that binding to other types of tissue such as muscle is independent of species for a wide variety of drugs.[93]

5.14.4.4 Relationship of Microsome Binding with Physicochemical Properties

The extent of nonspecific binding of drugs to liver microsomes is expected to be dominated by lipophilicity since microsomes contain about 60% phospholipid by mass.[97] A relationship of binding with lipophilicity was first shown in a semiquantitative way with a small set of acidic and basic drugs,[89] and this study also highlighted enhanced binding of the basic drugs compared with acidic drugs. This emerging trend was made clearer in later study which demonstrated a quantitative relationship of microsomal binding with lipophilicity using a diverse set of 37 acidic, basic, and neutral drugs.[90] It was found that when octanol/water $\log D_{7.4}$ was used as the index of lipophilicity, a broad trend of increasing binding with increasing lipophilicity was apparent, but with enhanced binding of basic compounds compared with neutral or acidic compounds of the same $\log D_{7.4}$. This was argued as being consistent with the enhanced affinity of basic compounds for phospholipid membranes compared with octanol, as previously demonstrated by liposome binding studies.[98,99] The phospholipid studies demonstrate that the affinity of compounds for model membranes at pH 7.4 is better modeled by $\log P$ than $\log D_{7.4}$ for bases, but by $\log D_{7.4}$ rather than $\log P$ for acids. The difference in behavior of the two classes is due to the ability of the ionized form of bases to bind to the membrane with similar affinity to the unionized form, whereas the ionised form of acids have much lower membrane affinity than the neutral form. To account for the different behavior of acids and bases in microsome binding, a combined lipophilicity descriptor, $\log(P/D)$ was used. This descriptor is equal to the $\log P$ of the molecule if it is a base (basic $pK_a > 7.4$) and equal to the $\log D_{7.4}$ of the molecule if it is an acid (acidic $pK_a < 7.4$). For neutral molecules $\log P$ and $\log D_{7.4}$ are equivalent, and that value is used. It was found that the extent of binding to rat liver microsomes with $1\,mg\,mL^{-1}$ microsomal protein concentration, expressed as $\log((1-f_u)/f_u)$, is highly correlated with the combined lipophilicity descriptor as shown in **Figure 3**.[90]

The least squares fit to the data in **Figure 3**, eqn [24], allows a prediction of $f_{u,inc}$ from readily available lipophilicity and ionization data, and was proposed as a useful method for estimating $CL_{int,u}$ from observed intrinsic clearance data,

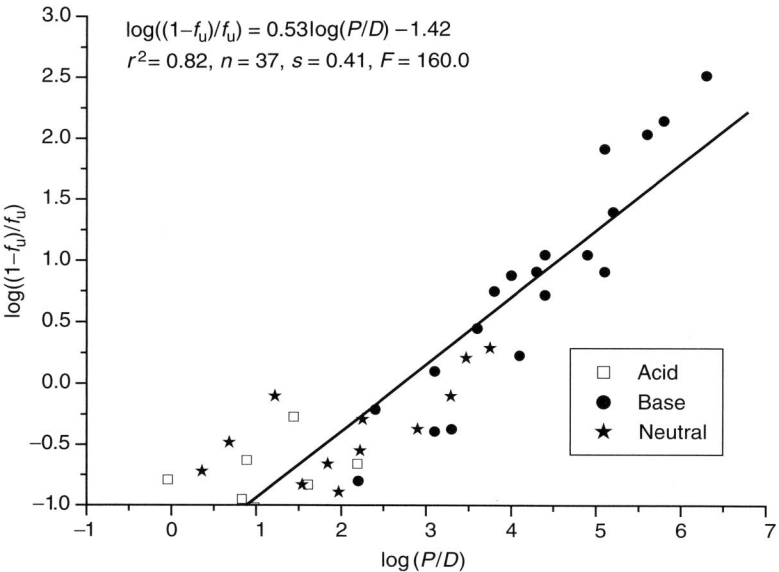

Figure 3 Plot of extent of microsomal binding against the combined lipophilicity descriptor log(P/D). (Reproduced with permission from Austin, R. P.; Barton, P.; Cockroft, S. L.; Wenlock, M. C.; Riley, R. J. *Drug Metab. Dispos.* **2002**, *30*, 1497–1503.)

which would be helpful in the prediction of in vivo clearance from in vitro data[90]:

$$\log((1 - f_u)/f_u) = 0.53 \log(P/D) - 1.42$$
$$n = 37 \quad r^2 = 0.82 \quad s = 0.41 \quad F = 160.0 \qquad [24]$$

5.14.5   Hepatocyte Binding

5.14.5.1   Experimental Methods

Most hepatocyte binding studies have used rapid centrifugation methods to separate drug in solution from that bound to or contained within the hepatocytes.[100–102] These methods generally involve centrifugation of a hepatocyte suspension through a layer of oil which allows a very rapid separation of the hepatocytes and bound compound from aqueous solution. These methods are rapid enough to allow measurement of the kinetics of uptake of compounds into hepatocytes, which has been the focus of many of the reports. A recent study has used an alternative method of using metabolic inhibitors along with the equilibrium dialysis technique.[103] The inhibitors effectively remove all metabolic turnover of the compounds, which then enables hepatocyte binding to be determined using the relatively slow equilibrium dialysis technique.

5.14.5.2   Mechanism of Drug Binding to Hepatocytes

The term binding to hepatocytes refers to the net effect of a number of different processes. There will be passive processes such as binding to the cell wall and other phospholipids. Another process is accumulation of basic compounds within regions of the hepatocyte where the pH is lower than the extracellular pH.[104,105] An active process for some compounds will be carrier mediated transport into hepatocytes.[101,102] All of these processes will affect the free concentration of compound that accesses the drug metabolizing enzymes, and hence will influence the observed intrinsic clearance in an in vitro assay.

Studies on some lipophilic pesticides have shown that the binding of these compounds to rat hepatocytes is a passive process with a well-defined equilibrium position that is reached rapidly.[106] The extent of hepatocyte binding was found to increase with increasing lipophilicity, although the extent of binding solely to hepatocytes was unclear due to the presence of bovine serum albumin in the buffer. In another study, the binding of parathion to rat hepatocytes has also been shown to be rapid and reversible, and to reach a defined equilibrium position.[107] Most other investigations relating to hepatocyte binding have focused on kinetic studies of uptake and metabolism of compounds by hepatocytes where active uptake processes were expected to be present, for example with drugs known to have extensive biliary clearance.[101,108–111] In these studies, the active and passive components of hepatocyte uptake were deconvoluted from the results of kinetic measurements as a function of compound concentration, and a variety of inhibitors of uptake transporters were shown to decrease the rate of uptake of those compounds with an active component.

5.14.5.3   Use of Hepatocyte Binding to Determine Unbound Intrinsic Clearance

Hepatocytes have historically been less widely utilized than liver microsomes for in vitro drug metabolism studies. However, the current easier availability of hepatocytes, for example with cryopreserved hepatocytes, is leading to their increased use. One study analyzed the influence of binding of a set of nine compounds to rat hepatocytes and found an improvement in the clearance prediction of all compounds when the $f_{u,inc}$ correction was applied.[100] The accuracy of the clearance prediction for some of the compounds was still quite poor, but the $f_{u,inc}$ correction did always move the clearance prediction closer to the correct value. Another study involved the prediction of human clearance of a large set of marketed drugs from human hepatocyte intrinsic clearance.[87] The extent of incubational binding was estimated from the lipophilicity of the compounds using a quantitative structure–activity relationship (QSAR) model, and incorporation of the predicted $f_{u,inc}$ values significantly improved the correlation of observed versus predicted human clearance.

5.14.5.4   Relationship of Hepatocyte Binding with Physicochemical Properties

Since an active transport component of hepatocyte binding exists for some compounds, and given that basic compounds can accumulate within certain regions of hepatocytes due to pH differences, it is likely that the dependence of hepatocyte binding on physicochemical properties will be more complex than for microsomal binding. One study reported hepatocyte binding data on a set of 17 drugs that had a diverse range of lipophilicity and ionization class.[103] It was found that the extent of binding of these compounds was related to lipophilicity. The correlation of binding with lipophilicity was best when log D was used for acidic and neutral compounds and log P was used for bases, in a similar way to the microsomal binding studies described previous. Despite the potential complexities of the hepatocyle binding process, this simple physicochemical model was able to predict $f_{u,inc}$ with significantly less than twofold error for 16 of the 17 drugs. The method therefore has promise as facile way of obtaining an initial estimate of $f_{u,inc}$ when performing in vitro–in vivo correlation using intrinsic clearance data from hepatocytes, as has been demonstrated with the human clearance prediction of a diverse set of marketed drugs.[87]

5.14.6 Influence of Plasma Binding and Tissue Binding on Volume of Distribution

5.14.6.1 Overview

Volume of distribution (V_{ss}) is a critical pharmacokinetic parameter that is an important component of drug in vivo half-life. The ability to generate reliable predictions of human V_{ss} from preclinical data will frequently be a key component of the success of a drug research project. Volume of distribution can be viewed as a property that arises from the relative affinity of a compound for plasma versus tissue. Consequently, most methods for understanding and predicting V_{ss} do so with considerations of absolute tissue affinity and/or plasma affinity, or interspecies differences in these affinities. Such methods include those based on the Oie–Tozer equation, and those that make use of tissue plasma partition coefficients. Another frequently used method for V_{ss} prediction is allometry, which does not directly consider relative affinities for tissue and plasma and instead, in its simplest form, assumes a relationship between V_{ss} in various species and the body weights of the animals.

5.14.6.2 Allometry

It is well known that there can be large interspecies variation in volume of distribution (V_{ss}), and a recent study on a set of 103 drugs showed that V_{ss} data from rat, dog, and monkey all show a similarly poor correlation with human V_{ss}.[112] Allometry has been a widely used method for the prediction of human V_{ss},[113,114] and the largest reported study describes its application to this 103 drug data set resulting in rather modest levels of predictivity.[112] Since it has been shown that interspecies differences in plasma affinity are much greater than for tissue affinity,[96] it should be expected that observed interspecies differences in V_{ss} will be strongly influenced by the interspecies differences in plasma affinity. It is significant therefore that for a data set of 14 drugs it has been shown that allometry using unbound volume of distribution ($V_{ss}/f_{u,p}$), which takes into account the interspecies differences in plasma binding, greatly outperforms simple allometry.[115] If, for certain compounds, interspecies differences in plasma affinity happen to be related to animal size, and hence are allometrically related, then simple allometry will as a consequence be able to model interspecies differences in V_{ss} in terms of interspecies differences in plasma affinity. In cases such as this, simple allometry would be expected give better results. However, a more reliable way of incorporating interspecies differences in plasma affinity into allometry is to work with unbound volume of distribution, and as more examples of this methodology are reported it expected that more accurate predictions of V_{ss} will generally be found than with the use of simple allometry.[112]

5.14.6.3 Methods Based on the Oie–Tozer Equation

Many nonallometric methods for predicting V_{ss} and for understanding the influence of plasma binding on V_{ss} have been based on the Oie–Tozer equation[116]:

$$V_{ss} = V_p(1 + R_{E/I}) + f_{u,p}V_p(V_E/V_p - R_{E/I}) + \frac{f_{u,p}}{f_{u,t}}V_R \qquad [25]$$

The parameters V_p, V_E, and $R_{E/I}$ are constants describing the plasma and extracellular fluid volumes ($L\,kg^{-1}$) and the ratio of extravascular to intravascular albumin respectively. V_R is the physical volume into which the drug distributes minus the extravascular space (often referred to as V_t), and $f_{u,t}$ the fraction unbound in tissues. This equation is often used in a simplified format eqn [16].[116] These equations clearly show how V_{ss} is largely controlled the relative affinity of a compound for plasma and tissue. The way these equations have been used involves rearranging them such that $V_t/f_{u,t}$ or $V_R/f_{u,t}$ can be evaluated in terms of measurements of V_{ss} and $f_{u,p}$ in a given animal species, and estimates of the other parameters from physiological data. It is then possible to obtain estimates of $V_t/f_{u,t}$ in several animal species. Several studies have shown that for a variety of drugs, there is often a significant interspecies variation in V_{ss}, while the variation in $V_t/f_{u,t}$ is much smaller.[117,118] It then follows that if $V_t/f_{u,t}$ has little dependence on species, the value of $V_t/f_{u,t}$ in humans can be assumed to be the same as that in an animal species, which can then be substituted into eqns [25] or [16] along with a measured value of human $f_{u,p}$ in order to obtain a prediction of human V_{ss}. This approach has been applied to a range of drugs resulting in predictions of human V_{ss} of very useful accuracy.[115,117-119] An advantage of this method of V_{ss} prediction compared with allometry is it is able to produce good predictions from in vivo data in a single animal species such as rat,[119] allowing generation of human V_{ss} predictions earlier in the drug discovery process and with less use of in vivo experiments.

A further simplification of the methods based on the Oie–Tozer equation is to assume in eqn [16] that V_p is small compared with $f_{u,p}V_t/f_{u,t}$, which is a reasonable approximation for drugs with $V_{ss} > 0.3 \, \text{L kg}^{-1}$. Equation [16] then simplifies to:

$$\frac{V_{ss}}{f_{u,p}} = \frac{V_t}{f_{u,t}} \qquad [26]$$

As described earlier, $V_t/f_{u,t}$ tends to be species independent, hence $V_{ss}/f_{u,p}$, known as the unbound volume of distribution, should also be approximately independent of species. This approach leads to the very simple equation below for prediction of human V_{ss}:

$$V_{ss,human} = V_{ss,animal} \frac{f_{u,p,human}}{f_{u,p,animal}} \qquad [27]$$

All that is needed for human V_{ss} prediction using eqn [28] is a measurement of V_{ss} in a single animal species and the extent of plasma binding in animal plasma and human plasma. This method has been shown to give very good predictivity for a set of 14 drugs using data from dog as the animal species.[115]

Equation [25] has been combined with a empirical QSAR approach in order to predict human V_{ss} without the use of any in vivo data.[120,121] In these studies, eqn [25] was rearranged to give $f_{u,t}$ in terms of a training set of human V_{ss} and $f_{u,p}$ data from firstly 64, and later 120 neutral and basic drugs. Regression analysis was then used to model $\log f_{u,t}$ in terms of $\log D_{7.4}$ and the fraction ionized at pH 7.4 (from pK_a) and fraction unbound at pH 7.4 (from PPB) of the drugs. It was found that $\log f_{u,t}$ has a good relationship with the physicochemical properties, and the equation derived from the later study of 120 drugs is[121]:

$$\log f_{u,t} = 0.23(\pm 0.04)\log D - 0.93(\pm 0.08) f_{i,7.4} + 0.89(\pm 0.10)\log f_u$$
$$n = 120; \quad r^2 = 0.87; \quad q^2 = 0.85; \quad \text{rmse} = 0.37; \quad F = 250.9 \qquad [28]$$

Equation [28] was then used to predict $f_{u,t}$ for a separate test set of 18 drugs. These predicted values of $f_{u,t}$ were then substituted into eqn [25] along with measured values of $f_{u,p}$ in order to generate predictions of human V_{ss}. This method produced fairly good predictions of human V_{ss} with 10 out 18 compounds in the separate test set having a predicted V_{ss} within a factor of 2 of the experimental value. Recently QSAR methods have been published to predict V_{ss} purely from molecular structure.[127,128]

5.14.6.4   Methods Based on Tissue Plasma Partition Coefficients

In the field of physiologically based pharmacokinetic modeling, V_{ss} is related to the volumes of the tissues into which a drug distributes multiplied by the corresponding tissue-to-plasma partition coefficients, in addition to the plasma volume.[122] This relationship is expressed quantitatively by[123]:

$$V_{ss} = \sum V_t P_{t:p} + V_e E{:}P + V_p \qquad [29]$$

where V_t and V_e are the volumes (L kg^{-1}) of tissue and erythrocytes respectively, $E{:}P$ is the erythrocyte to plasma concentration ratio, and $P_{t:p}$ the tissue-to-plasma partition coefficient. The sum is taken over all tissue types (muscle, liver, lung, etc.). The widespread use of equations such as eqn [29] has largely been prevented by the difficulties in the measurement of the tissue-to-plasma partition coefficients and the lack of availability of human tissue.[122] A useful simplification of eqn [29] arises from the fact that the tissue plasma partition coefficients from different tissues are highly correlated with one another. Linear regression has been used on a set of 65 drugs to derive empirical relationships between muscle-to-plasma partition coefficients and tissue-to-plasma coefficients of other tissues.[124] These relationships can be used to estimate the tissue-to-plasma partition coefficients of a compound from a measurement of only the muscle-to-plasma partition coefficient. Further developments that should facilitate more widespread use of eqn [29] involve prediction of tissue-to-plasma partition coefficients from readily available physicochemical properties. A QSAR approach has been applied to the prediction of tissue-to-plasma partition coefficients of a set of nine homologous barbituric acids[125] and to a set of ten lipophilic basic and neutral drugs.[126] Good relationships were found between the tissue-to-plasma partition coefficients and lipophilicity in both reports, but a limitation of this approach is that the derived equations are likely to only apply with useful accuracy to chemical structures that are similar to those used for building the QSAR models. A more recent tissue composition based mechanistic method[124] for predicting tissue-to-plasma partition coefficients from knowledge of solvent partition coefficients, pK_a, and $f_{u,p}$ is likely to be more widely applicable than the QSAR equations. The method was validated on a set of 65 drugs, where it was found that 70% of the predicted tissue-to-plasma partition coefficients were within a factor of 2 of the experimental value.[124] This method was later applied to the prediction of tissue-to-plasma partition coefficients of another set of 123 drugs followed by substitution of the predicted values in eqn [29] in order to predict human and rat V_{ss}.[123] It was found that 80% of the predicted V_{ss} values were within a factor of 2 of the experimental value, indicating good predictive power of a method that only uses in vitro data.

5.14.7 Outlook

Due to the high attrition rate in late stage development, the pharmacokinetic and pharmacodynamic profiles of new compounds are obtained at an earlier stage than ever before. In order to fully understand these data, plasma protein binding and tissue studies are in greater demand than ever before. Understanding of the properties controlling plasma and tissue binding is increasing dramatically. Studies are now being undertaken on the crystallization of albumin and α_1-acid glycoprotein with ligands of interest in order to further elucidate the mechanism of binding to these proteins and their ligands.

References

1. Lin, J. H.; Cocchetto, D. M.; Duggan, D. E. *Clin. Pharmacokinet.* **1987**, *12*, 402–432.
2. Kariv, I.; Cao, H.; Oldenburg, K. R. *J. Pharm. Sci.* **2001**, *200*, 580–587.
3. Wilkinson, G. R. *Drug Metab. Rev.* **1983**, *14*, 427–465.
4. Blaschke, T. F. *Clin. Pharmacokinet.* **1977**, *2*, 32–44.
5. Piafsky, K. M. *Clin. Pharmacokinet.* **1980**, *5*, 246–262.
6. Tillement, J. P.; Houin, G.; Zini, R.; Urien, S.; Albengres, E.; Barre, J.; Lecomte, M.; D'Athis, P.; Sebille, B. *Adv. Drug Res.* **1984**, *13*, 59–93.
7. Kragh-Hansen, U.; Brennan, S.; Galliano, M.; Sugita, O. *Mol. Pharmacol.* **1990**, *37*, 238–242.
8. Pang, K. S.; Rowland, M. *J. Pharmacokinet. Bio.* **1977**, *5*, 625–653.
9. Benet, L. Z.; Hoener, B. *Clin. Pharmacol. Ther.* **2002**, *71*, 115–121.
10. Levy, G.; Yacobi, A. *J. Pharm. Sci.* **1974**, *63*, 805–806.
11. Yacobi, A.; Levy, G. *J. Pharm. Sci.* **1975**, *64*, 1660–1664.
12. Evans, G. H.; Nies, A. S.; Shand, D. G. *J. Pharmacol. Exp. Ther.* **1973**, *186*, 114–122.
13. McNamara, P. J.; Stoeckel, K.; Ziegler, W. H. *Eur. J. Clin. Pharmacol.* **1982**, *22*, 71–75.
14. Kwon, Y. In *Handbook of Essential Pharmacokinetics, Pharmacodynamics, and Drug Metabolism for Industrial Scientists*; Kluwer Academic/Plenum Press: New York, 2001, pp 95–96.
15. Rowland, M.; Tozer, T. N. In *Clinical Pharmacokinetics: Concepts and Applications*, 3rd ed.; Balado, D., Klass, F., Stead, L., Forsyth, L., Magee, R. D., Eds.; Lippincott Williams & Wilkins: Baltimore, MD, 1995, pp 1924–2023.
16. Kwon, Y. In *Handbook of Essential Pharmacokinetics, Pharmacodynamics, and Drug Metabolism for Industrial Scientists*; Kluwer Academic/Plenum Publisher: New York, 2001, p 100.
17. Goodman, A. G.; Gilman, A. G. In *The Pharmacological Basis of Therapeutics*, 10th ed.; Harman, J. G., Limbird, L. E., Gilman, A. G., Eds.; McGraw-Hill: New York, 2001, pp 1924–2023.
18. Rolan, P. E. *Br. J. Clin. Pharmacol.* **1994**, *37*, 125.
19. Hartrick, C. T.; Dirkes, W. E.; Coyle, D. E.; Prithvi Raj, P.; Denson, D. D. *Clin. Pharmacol. Ther.* **1984**, *36*, 546.
20. Laznicek, M.; Laznickova, A. *J. Pharm. Biomed. Anal.* **1995**, *13*, 823–828.
21. Tillement, J. P.; Houin, G.; Zini, R.; Urien, S.; Albengres, E.; Barre, J.; Lecomte, M.; D'Athis, P.; Sebille, B. In *Advances in Drug Research*; Testa, B., Ed.; Academic Press: London, 1984; Vol. 13, pp 59–93.
22. Rowland, M.; Tozer, T. N. In *Clinical Pharmacokinetics – Concepts and Applications*, 3rd ed.; Balado, D., Klass, F., Stead, L., Forsyth, L., Magee, R. D., Eds.; Lippincott Williams & Wilkins: London, 1995.
23. Klotz, I. M. *Acc. Chem. Res.* **1974**, *7*, 162–168.
24. Klotz, I. M.; Hunston, D. L. *J. Biol. Chem.* **1975**, *250*, 3001–3009.
25. Klotz, I. M.; Hunston, D. L. *Proc. Natl. Acad. Sci. USA* **1977**, *74*, 4959–4963.
26. Valko, K.; Nunchuck, S.; Bevan, C.; Abraham, M. H.; Reynolds, D. *J. Pharm. Sci.* **2003**, *92*, 2236–2248.
27. Toon, S.; Rowland, M. *J. Pharmacol. Exp. Ther.* **1983**, *225*, 752–763.
28. Scatchard, G. *Ann. NY Acad. Sci.* **1949**, *51*, 660–672.
29. Scatchard, G.; Scheinberg, I. H.; Armstrong, S. H., Jr. *J. Am. Chem. Soc.* **1950**, *72*, 535–540.
30. Weder, H. G.; Schildknecnt, J; Kesselring, P. *Am. Lab.* **1971**, *3*, 15–21.
31. Klotz, I. M.; Walker, F. M.; Pivan, R. B. *J. Am. Chem. Soc.* **1946**, *68*, 1486–1490.
32. Colowick, S. P.; Womack, F. C. *J. Biol. Chem.* **1968**, *244*, 774–777.

33. Meyer, M. C.; Guttman, D. E. *J. Pharm. Sci.* **1968**, *57*, 895–918.
34. Lohman, J. J. H. M.; Hooymans, P. M.; Verhey, M. T. J. M.; Koten, M. L. P.; Meus, F. W. H. M. *Pharm. Res.* **1984**, *4*, 187–188.
35. Boudinot, F. G.; Jusko, W. J. *J. Pharm. Sci.* **1984**, *73*, 775–780.
36. Tanford, C. *Physical Chemistry of Macromolecules*; Wiley: New York, 1961, pp 225–227.
37. Suter, P.; Rosenbusch, J. P. *Anal. Biochem.* **1977**, *82*, 109–114.
38. Totsune, K. *Jinko Zoki.* **1988**, *17*, 69–72.
39. Rheberg, P. B. *Acta Phys. Scand.* **1943**, *5*, 305–310.
40. Hall, S.; Rowland, M. *J. Pharm. Exp. Ther.* **1983**, *227*, 174–179.
41. Aarons, L.; Grennan, D. M.; Siddiqui, M. *Eur. J. Clin. Pharmacol.* **1983**, *25*, 815–818.
42. Barre, J.; Chamourd, J. M.; Houin, G.; Tillement, J. P. *Clin. Chem.* **1985**, *31*, 60–64.
43. Aubry, A. F.; Markoglou, N.; Descorps, V.; Wainer, I. W.; Felix, G. *J. Chromatogr., A* **1994**, *685*, 1–6.
44. Hage, D. S.; Tweed, S. A. *J. Chromatogr., B* **1997**, *699*, 499–525.
45. Hage, D. S.; Austin, J. *J. Chromatogr., B* **2000**, *739*, 39–54.
46. Chignell, C. F. *Mol. Pharmacol.* **1969**, *5*, 244–252.
47. Rockey, J. H.; Li, W. *Ophthalm. Res.* **1982**, *14*, 416–427.
48. Rochas, M. A.; Tufenkji, A. E.; Levillain, P.; Houin, G. *Arzneim.-Forsch.* **1991**, *41*, 1286–1288.
49. Lachau, S.; Rochas, M. A.; Tufenkji, A. E.; Martin, N.; Levillain, P.; Houin, G. *J. Pharm. Sci.* **1992**, *81*, 287–289.
50. Maia, M. B. S.; Tufenkji, A. E.; Rochas, M. A.; Saivin, S.; Houin, G. *Fund. Clin. Pharmacol.* **1994**, *8*, 178–184.
51. Bertucci, C.; Salvadori, P.; Domenici, E. *The Impact of Stereochemistry on Drug Development and Use*; Wiley: New York, 1997, pp 521–543.
52. Jardetsky, O.; Wade-Jardetsky, N. *Mol. Pharmacol.* **1965**, *1*, 214–230.
53. Liu, M.; Nicholson, J. K.; Lindon, J. C. *Anal. Commun.* **1997**, *34*, 225–228.
54. Myszka, D. G.; Rich, R. L. *Pharm. Sci. Technol. Today* **2000**, *3*, 310–317.
55. Frostell-Karlsson, A.; Remaeus, A.; Roos, H.; Andersson, K.; Borg, P.; Hamalainen, M.; Karlsson, R. *J. Med. Chem.* **2000**, *43*, 1986–1992.
56. Rich, R. L.; Day, Y. S. N.; Morton, T. A.; Myszka, D. G. *Anal. Biochem.* **2001**, *296*, 197–207.
57. Dennis, M. S.; Zhang, M.; Meng, Y. G.; Kadkhodayan, M.; Kirchhofer, D.; Coms, D.; Samico, L. A. *J. Biol. Chem.* **2002**, *277*, 35035–35043.
58. Day, Y. S. N.; Myszka, D. G. *J. Pharm. Sci.* **2003**, *92*, 333–343.
59. Shibukawa, A.; Yoshimoto, Y.; Ohra, T.; Nakagawa, T. *J. Pharm. Sci.* **1994**, *83*, 616–619.
60. Quaglia, M. G.; Bossu, E.; Dell'Aquila, C.; Guidotti, M. *J. Pharm. Biomed. Anal.* **1997**, *15*, 1033–1039.
61. McDonnell, P. A.; Caldwell, G. W.; Masucci, J. A. *Electrophoresis* **1998**, *19*, 448–454.
62. Ishihama, Y.; Miwa, T.; Asakawa, N. *Electrophoresis* **2002**, *23*, 951–955.
63. Zhou, D.; Li, F. *J. Pharm. Biomed. Anal.* **2004**, *35*, 879–885.
64. Wang, H.; Zou, H.; Zhang, Y. *Biomed. Chromatogr.* **1998**, *12*, 4–7.
65. Liu, Z.; Li, F.; Huang, Y. *Biomed. Chromatogr.* **1999**, *13*, 262–266.
66. Kariv, I.; Cao, H.; Oldenburg, K. R. *J. Pharm. Sci.* **2001**, *200*, 580–587.
67. Banker, M. J.; Clark, T. H.; Williams, J. A. *J. Pharm. Sci.* **2002**, *92*, 967–974.
68. Curry, S. H. *J. Pharm. Pharmacol.* **1970**, *22*, 753–757.
69. Gillette, J. R. *J. Pharmcokin. Biopharm.* **1973**, *1*, 497–520.
70. Jusco, W. J.; Gretch, M. *Drug Metab. Rev.* **1976**, *5*, 43–140.
71. Kurz, H.; Fictl, B. *Drug Metab. Rev.* **1983**, *14*, 467–510.
72. Ballard, P.; Leahy, D. E.; Rowland, M. *Pharm. Res.* **2000**, *17*, 660–663.
73. Ballard, P.; Arundel, P. A.; Leahy, D. E.; Rowland, M. *Pharm. Res.* **2003**, *20*, 857–863.
74. Dewey, W. C. *Am. J. Physiol.* **1959**, *197*, 423–431.
75. Ballard, P.; Leahy, D. E.; Rowland, M. *Pharm. Res.* **2003**, *20*, 864–872.
76. Rowland, M. *Pharmacol. Ther.* **1985**, *29*, 49–68.
77. Gibaldi, M.; McNamara, P. J. T. N. *J. Pharm. Sci.* **1977**, *66*, 1211–1212.
78. Bickel, M. H. *Prog. Drug Res.* **1984**, *28*, 273.
79. Bickel, M. H. *Adv. Drug Res.* **1994**, *25*, 56.
80. Bickel, M. H.; Clausen, J. *J. Pharm. Sci.* **1993**, *82*, 345.
81. Watsubu, T.; Hirota, N.; Ooie, T.; Suzuki, H.; Shinimada, N.; Chiba, K.; Ishazaki, T.; Green, C. E.; Tyson, C. A.; Sugiyama, Y. *Pharmacol. Ther.* **1997**, *73*, 147–171.
82. Lave, T.; Coassolo, P.; Reigner, B. *Clin. Pharmacokinet.* **1999**, *36*, 211–231.
83. Houston, J. B. *Biochem. Pharmacol.* **1994**, *47*, 1469–1479.
84. Obach, R. S. *Drug Metab. Dispos.* **1996**, *24*, 1047–1049.
85. Pang, K. S.; Rowland, M. *J. Pharmacok. Biopharm.* **1997**, *5*, 625–653.
86. Obach, R. S. *Drug Metab. Dispos.* **1999**, *27*, 1350–1359.
87. Riley, R. J.; McGinnity, D.; Austin, R. P. *Drug Metab. Dispos.* **2005**, *33*, 1304–1311.
88. Obach, R. S. *Drug Metab. Dispos.* **1997**, *25*, 1359–1369.
89. Mclure, J. A.; Miners, J. O.; Birkett, D. J. *Br. J. Clin. Pharmacol.* **2000**, *49*, 453–461.
90. Austin, R. P.; Barton, P.; Cockroft, S. L.; Wenlock, M. C.; Riley, R. J. *Drug Metab. Dispos.* **2002**, *30*, 1497–1503.
91. Naritomi, Y.; Terashita, S.; Kimura, S.; Suzuki, A.; Kagayama, A.; Sugiyama, Y. *Drug Metab. Dispos.* **2001**, *29*, 1316–1324.
92. Venkatakrishnan, K.; Von Moltke, L. L.; Obach, R. S.; Greenblatt, D. J. *J. Pharmacol. Exp. Ther.* **2000**, *293*, 343–350.
93. Kalvass, J. C.; Tess, D. A.; Giragossian, C.; Linhares, M. C.; Maurer, T. S. *Drug Metab. Dispos.* **2001**, *29*, 1332–1336.
94. Ishigam, M.; Uchiyama, M.; Kondo, T.; Iwabuchi, I.; Inoue, S.; Takasaki, W.; Ikeda, T.; Komai, T.; Ito, K.; Sugiyama, Y. *Pharm. Res.* **2001**, *18*, 622–631.
95. Margolis, J. M.; Obach, R. S. *Drug Metab. Dispos.* **2003**, *31*, 606–611.
96. Fichtl, B.; Nieciecki, A. V.; Walter, K. *Adv. Drug Res.* **1991**, *20*, 115–166.
97. Ernster, L.; Siekevitz, P.; Palade, G. E. *J. Cell Biol.* **1962**, *15*, 541–562.
98. Austin, R. P.; Davis, A. M.; Manners, C. N. *J. Pharm. Sci.* **1995**, *84*, 1180–1183.
99. Krämer, S. D.; Braun, A.; Jakits-Deiser, C.; Wunderli-Allenspach, H. *Pharm. Res.* **1998**, *15*, 739–744.
100. Naritomi, Y.; Terashita, S.; Kagayama, A.; Sugiyama, Y. *Drug Metab. Dispos.* **2003**, *31*, 580–588.

101. Joppen, C.; Petzinger, E.; Frimmer, M. *Naunyn-Schmiedeberg's Arch. Pharmacol.* **1985**, *331*, 393–397.
102. Zhong, Z.-D.; Wattiaux-De Connick, S.; Wattiaux, R. *Biochim. Biophys. Acta* **1993**, *1176*, 77–82.
103. Austin, R. P.; Barton, P.; Mohmed, S.; Riley, R. J. *Drug Metab. Dispos.* **2005**, *33*, 419–425.
104. Siebert, G. A.; Hung, D. Y.; Chang, P.; Roberts, M. S. *J. Pharmacol. Exp. Ther.* **2004**, *308*, 228–235.
105. MacIntyre, A. C.; Cutler, D. J. *J. Pharm. Sci.* **1993**, *82*, 592–600.
106. Ichinose, I.; Kurihara, N. *Pestic. Biochem. Physiol.* **1985**, *14*, 116–122.
107. Nakatsugawa, T.; Bradford, W. L.; Usui, K. *Pestic. Biochem. Physiol.* **1980**, *14*, 13–25.
108. Tsuji, A.; Terasaki, T.; Takanosu, K.; Tamai, I.; Nakashima, E. *Biochem. Pharmacol.* **1986**, *35*, 151–158.
109. Ishida, S.; Sakiya, Y.; Ichikawa, T.; Taira, Z. *Biol. Pharm. Bull.* **1993**, *16*, 293–297.
110. Yamazaki, M.; Suzuki, H.; Hanano, M.; Tokui, T.; Komai, T.; Sugiyama, Y. *Am. J. Physiol.* **1993**, *264*, G36–G44.
111. Petzinger, E.; Müller, N.; Föllmann, W.; Deutscher, J.; Kinne, R. K. H. *Am. J. Physiol.* **1989**, *256*, G78–G86.
112. Ward, K. W.; Smith, B. R. *Drug Metab. Dispos.* **2004**, *32*, 612–619.
113. Mahmood, I. *J. Pharm. Pharmacol.* **1999**, *51*, 905–910.
114. Mahmood, I. *J. Pharm. Pharmacol.* **1998**, *50*, 493–499.
115. Obach, R. S.; Baxter, J. G.; Liston, T. E.; Silber, B. M.; Jones, B. C.; MacIntyre, F.; Rance, D. J.; Wastall, P. *J. Pharmacol. Exp. Ther.* **1997**, *283*, 46–58.
116. Oie, S.; Tozer, T. N. *J. Pharm. Sci.* **1979**, *68*, 1203–1205.
117. Sawada, Y.; Hanano, M.; Sugiyama, Y.; Harashima, H.; Iga, T. *J. Pharmacokinet. Biopharm.* **1984**, *12*, 587–596.
118. Sawada, Y.; Hanano, M.; Sugiyama, Y.; Iga, T. *J. Pharmacokinet. Biopharm.* **1984**, *12*, 241–261.
119. Sawada, Y.; Hanano, M.; Sugiyama, Y.; Iga, T. *J. Pharmacokinet. Biopharm.* **1985**, *13*, 477–492.
120. Lombardo, F.; Obach, R. S.; Shalaeva, M. Y.; Gao, F. *J. Med. Chem.* **2002**, *45*, 2867–2876.
121. Lombardo, F.; Obach, R. S.; Shalaeva, M. Y.; Gao, F. *J. Med. Chem.* **2004**, *47*, 1242–1250.
122. Theil, F.-P.; Guentert, T. W.; Haddad, S.; Poulin, P. *Toxicol. Lett.* **2003**, *138*, 29–49.
123. Poulin, P.; Theil, F.-P. *J. Pharm. Sci.* **2002**, *91*, 129–156.
124. Poulin, P.; Theil, F.-P. *J. Pharm. Sci.* **2000**, *89*, 16–35.
125. Nestorov, I.; Aarons, L.; Rowland, M. *J. Biopharm. Pharmacokinet.* **1998**, *26*, 521–545.
126. Yokogawa, K.; Nakashima, E.; Ishizaki, J.; Nagano, T.; Ichimura, F. *Pharm. Res.* **1990**, *7*, 691–696.
127. Gleeson, M. P.; Waters, N. J.; Paine, S. W.; Davis, A. M. *J. Med. Chem.* **2006**, *49*, 1953–1963.
128. Lombardo, F.; Obach, R. S.; DiCapua, F. M.; Bakken, G. A.; Lu, J.; Potter, D. M.; Gao, F.; Miller, M. D.; Zhang, Y. *J. Med. Chem.* **2006**, *49*, 2262–2267.

Biographies

Patrick Barton received his first degree in Applied Chemistry from the University of Huddersfield in 1989 followed by a PhD in the field of Physical Organic Chemistry from the same institution in 1993, under the supervision of Prof M I Page investigating the reaction catalyzed by Porcine Liver Esterase. He is now Team Leader of Physical Organic Chemistry at AstraZeneca R&D Charnwood. His interests are in physical property measurements, the influence of physical properties on drug metabolism and pharmacokinetics and QSAR methodologies.

Rupert P Austin received his first degree in Chemistry from University College London in 1990 followed by a PhD in the field of Physical Organic Chemistry from the same institution in 1993, studying under Prof J H Ridd. He joined what is now AstraZeneca in 1993 working in the Physical Organic Chemistry group. His interests include QSAR, the binding of molecules and to biological matrices such as plasma, phospholipids, and proteins and the structure activity relationships of these processes, and laboratory automation.

Roger E Fessey received his first degree in Chemistry from the University of Liverpool in 1993 followed by a PhD in the field of Physical Organic Chemistry from the same institution in 1997, under the supervision of Prof D Bethell. He joined what is now AstraZeneca in 1998 working in the Physical Organic Chemistry group. His interests include structure–activity relationships, QSAR, and binding of molecules to phospholipids and plasma proteins.

Comprehensive Medicinal Chemistry II
ISBN (set): 0-08-044513-6

ISBN (Volume 5) 0-08-044518-7; pp. 321–340

5.15 Progress in Bioanalytics and Automation Robotics for Absorption, Distribution, Metabolism, and Excretion Screening

J Wang, Novartis Institutes for Biomedical Research, Cambridge, MA, USA
B Faller, Novartis Institutes for Biomedical Research, Basel, Switzerland

5.15.1 The Automated Absorption, Distribution, Metabolism, and Excretion (ADME) Screening Era

The high costs involved in the discovery and development of a new drug and the high attrition rate of drug candidates in development have shifted the current drug discovery and development paradigm towards the parallel assessment of efficacy and comprehensive ADME (absorption, distribution, metabolism, and excretion) properties of drug candidates.[1–5]

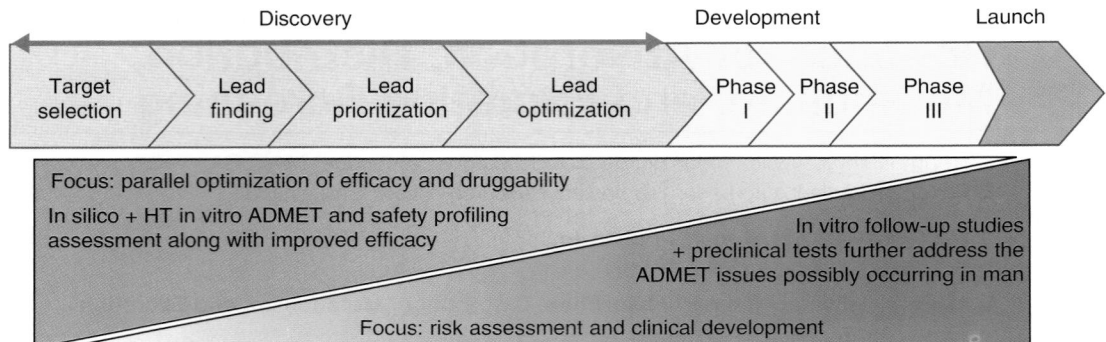

Figure 1 New strategy for drug discovery and development where direct interactions between discovery and development are greatly enhanced and the optimization of efficacy and druggability will be performed in parallel. The new strategy will require a comprehensive suite of predictive in silico and automated in vitro ADMET profiling tools in the early discovery phase.

Multiple predictive filters of ADME properties (such as in silico, high-throughput in vitro and in vivo pharmacology assays in animals, etc.) are established in the various stages of drug discovery and development process to flag the potential ADME issues of new chemical entities (NCEs) in man.[6–10] Of these, high-throughput (HT) in vitro ADME assays offer the first experimental tools to rank and prioritize NCEs in the early discovery phase in a faster and cost-effective fashion.

Such a new trend to evaluate therapeutic and drug-like features of NCEs together has led to a number of changes in the modern drug discovery and development process. First, a new mentality to break down the wall between drug discovery and development has appreciably encouraged the interdisciplinary and translational interactions to drive the projects effectively toward the same goal, the clinical launch of the medicine. As a result, industry tends to migrate from sequentially assessing efficacy and druggability to a parallel process in the new drug discovery and development strategy as shown in **Figure 1**. Hits, leads, and NCEs are being assessed at different milestones. The 'go' and 'no-go' decisions to advance drug candidates to the next stage should rely not only on a particular parameter (e.g., efficacy) but also on the drug's overall performance and their potential to become a commercial drug. In addition, industry has committed to investment in developing and implementing cascaded HT, miniaturized, cost-effective automated in vitro ADME profiling assays with good predictivity for in vivo druggability, for usage at various stages in the drug discovery and development process. This offers a solution in early discovery to modernize those in vitro ADME assays that were usually costly, labor-intensive, low throughput, and required a large amount of materials and were therefore available only in the late development stages. Finally, the follow-up 'pharmacoinformatic' studies are critical to establish reliable and predictive models to translate the in vitro ADME profiling data into useful druggability information for an NCE. It is essential to gradually build up the predictive in silico models for future ADME profiling. All of these can be implemented in the optimization cycle of chemical synthesis and thus can directly improve and accelerate the selection of early hits, leads, and NCEs in the early stages of discovery.

5.15.1.1 Is Automation Absolutely Necessary for High-Throughput In Vitro Absorption, Distribution, Metabolism, and Excretion Assays?

In the past decades, many novel in vitro ADME assays have been established to predict the potential pharmacokinetic (PK) and other toxicological or pharmacological behaviors of new drug candidates in animals and man. From proof of concept, in vitro assays were found to be quite useful to assess the respective in vivo activities. However, they used to be available only in the development phase due to limited capacity. In early ADME, those in vitro ADME assays are modernized in the early discovery phase by utilizing state-of-the-art automation and streamlined bioanalysis technologies that are widely applied in high-throughput screening (HTS) for a drug's efficacy. A number of review articles have summarized the latest developments in the automated ADME profiling field.[11–15] **Table 1** outlines the major differences between conventional in vitro ADME assays currently operating in development laboratories and those for HT ADME assays in the discovery phase.

Currently, HT ADME assays are an integral part of the early drug discovery process ranging from hit/lead nomination to lead optimization. Sometimes in vitro follow-up assays are required for the declaration of NCEs. Therefore, cascaded in vitro ADME assays are required with a greatly enhanced capacity (thousands to tens of thousands) to support such activities. In addition, the data must be available for periods lasting hours to days to cope with the designated decision tree used for each chemistry cycle in the lead optimization phase. Furthermore, a streamlined and efficient logistic

Table 1 Comparison of assay requirements for HT in vitro ADME assays with those for conventional in vitro ADME assays offered in development

	Conventional in vitro ADME assays	*HT in vitro ADME assays*
Capacity (yearly)	10s to 100s	1000s to 10 000s
Turn-around time	Days/weeks to months	Hours to days
Sample amounts	1 to 100s µmol	0.1 to 10s nmol
Cell culture/prep	Manual, individual labs	Automated (QC on wells)
Assay platform	Manual in vials/test tubes	Automated in 96/384-plate format
Bioanalytical format	Manual/semiautomated	HT program, automatic
Cost per sample	$100s to $1000s	$10s to $100s
Data tracking/mining	Not available mostly	Database/LIMS

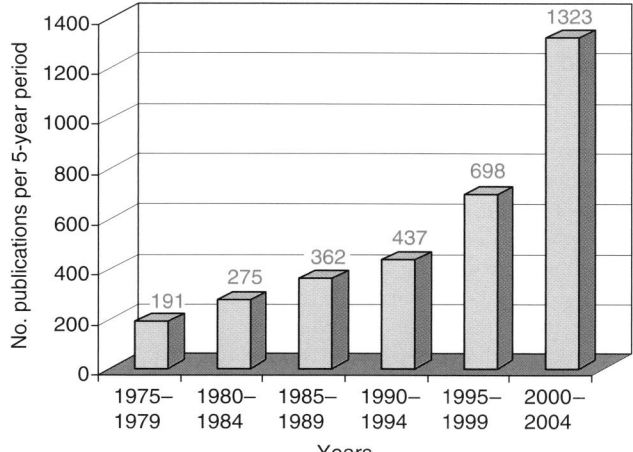

Figure 2 Number of publications on automated ADME assays (data source: SciFinder, using a combination of keywords 'automation' and 'absorption, distribution, metabolism, excretion').

process is a prerequisite for handling high demands with the limited sample amount available in the early phase. In practice, this cannot be achieved by simply multiplying the capacity of the conventional in vitro ADME assays by the additional number of full-time employees (FTEs) required. Most organizations consequently seek solutions via automation and streamlined bioanalysis. **Figure 2** demonstrates that the number of publications on ADME automation has risen steeply recently.

5.15.1.2   Automation Strategy

Implementation of efficient and user-friendly automated in vitro ADME assays is not a trivial matter, as the automation platforms required are more complex and thus far less standardized than those used in HTS systems. For instance, most of the in vitro permeability assays (using either an artificial membrane or a cell monolayer) utilize a sandwich-type of device comprising 96-well donor and receiver plates to mimic the in vivo permeation of test compounds across the gastrointestinal walls in animals or man. Another example is the automated HT-equilibrium solubility assay involving very complicated sample preparation steps that are difficult to automate. Complicating matters even further, the development of in vitro ADME assays has not reached a static stage and the constantly improved/revised assay protocols for the designated parameters (e.g., solubility, permeability, clearance, etc.) add additional challenges or variability to the design of the automated ADME assay platform.

5.15.1.2.1 Shared assay workstations

Sometimes, an automated assay workstation system can be integrated such that it can run a number of in vitro ADME assays sequentially. This is feasible and cost effective for small organizations or even business units (or satellite sites) of large organizations that have limited capacity and usage requirements for each of the in vitro ADME assays. This shared workstation solution is ideal to cope with a number of in vitro assays utilizing similar biological and analytical approaches by sharing the same labware, device(s), and readout equipment. For example, the determination of lipophilicity (log P)[35] via the microtiter-plate format[16] can be carried out on the same automated workstation that is used for the measurement of parallel artificial membrane permeability assay (PAMPA).[17,18]

5.15.1.2.2 Individual workstation dedicated to a specific assay

The latest trend in industry is to centralize the in vitro ADME screening assays in one or two hub laboratories within an organization. It offers a number of advantages such as: (1) reduced cost by consolidating the requests; (2) more easily managed internal logistics; and (3) efficiently maintained global assay standard and quality control (QC) within an organization. Therefore, the automation of a dedicated assay in a user-friendly, efficient, and reliable format will allow each hub laboratory to capture requests from the satellite sites without additional manpower. With the increased responsibility to support projects at multiple sites, the hub laboratories must acquire a favorable service contract with the major automation vendors to ensure the shortest downtime of the automated system (e.g., within 24–48 h). Contingency plans should always be available between the major hub laboratories to support each other under those conditions.

5.15.1.2.3 'Fully-automated' versus 'semi-automated' systems

The dedicated automated ADME systems can be integrated in distinct formats. Not every step needs to be automated and a balance between the complexity for automation/cost and the gain needs to be carefully evaluated. For example, it might not be necessary to automate loading of a high-performance liquid chromatography (HPLC) autosampler if it takes only a few minutes per day for an operator to do it manually. On the other hand, steps such as plate sealing might be worth automating if they allow for an unattended overnight run. In another example, the cell permeability assay can be assimilated either as 'pseudo-full automation' or as 'full automation.' Both approaches were attempted in our laboratories. Pseudo-full automation can separate the entire assay process into 3–4 segments such as: (1) sample preparation (from solid powders or DMSO stock solution); (2) cell culturing (of cell monolayers on 96-transwell plates in 21 days); (3) permeability assay (assemble donor/receiver and perform incubation); and (4) bioanalysis (quantification of samples in liquid chromatography/mass spectrometry (LC/MS). Individual workstations are integrated for each of the four dedicated tasks in an automated fashion with minimal personal involvement for sample/plate transfers between workstations (e.g., once a day). Alternatively, in the 'full automation' mode, those four self-automated workstations can be further integrated by a Thermo-CRS or Mitsubishi mobile robot so that no personal intervention is required.

Apparently both approaches have advantages and limitations. 'Full automation' has complex integration and interfacing among individual workstations, generally involving multiple vendors. In addition, the individual workstations may often be idle as required by the scheduling of shared components (e.g., CRS robot) and therefore the entire system is not effectively utilized. Despite there being no need for personnel to transfer plates between workstations, the system may still require qualified workers for maintenance and daily operation. In the worst-case scenario, the technical failure of one individual workstation may interrupt the operation of the entire automated system. The challenges for integration and maintenance have been significantly reduced in pseudo-full automation and all four workstations can be operational in parallel without interference. However, the limitation is a required level of assay monitoring between workstations. As most of the repetitive and tedious components of the assay are performed on the individual automated workstations and the manual interventions between workstations are limited, we found this to be the most efficient and effective solution for our laboratory.

5.15.1.2.4 Large-scale fully integrated automation systems versus automated workstations

The challenges in dealing with complicated interfacing among individual workstations are being tackled by automation- or profiling-specialized vendors such as Thermo Electron Corp. (Thermo), The Automation Partnership, and Cyprotex. This type of undertaking requires the corporations to have integration teams with a good understanding of the business needs and future trends for in vitro ADME screening. In addition, they often have strong business units in automation and bioanalytical equipment, as well as a software programing team. They intend to generalize the majority of user requirements to create a universal automation platform for multiple in vitro ADME screening assays in a streamlined format. Apparently, they will help users to address the interfacing between workstations, the compatibility between assays sharing the same workstations, and compound tracking and data processing/interpretation issues that are commonly encountered by scientists of in vitro ADME laboratories. This full integration is a new approach and its fate lies in the future successful resolution of the technical roadblocks such as: (1) the flexibility of a universal giant to address the specific requirements in assay protocols by an individual user; (2) the applicability or adaptability of the generic system to assays that are extremely challenging to automate (e.g., HT thermodynamic solubility); and (3) its financial advantage (capital investment, operating and personnel costs) over the individual automated workstations.

5.15.2   Automation of In Vitro Absorption, Distribution, Metabolism, and Excretion Assays in Users' Laboratories

At present, there appears to be a consensus in the early ADME field that automation is a must to enhance the throughput and productivity of in vitro ADME assays in early drug discovery. However, adoption of an appropriate implementation strategy is essential for a successful integration, timely launch, and smooth subsequent operation of the in vitro ADME screening assays.

5.15.2.1   Off-the-Shelf Workstation

Nowadays, fewer people utilize a sewing machine to make their own clothes. Construction of a customized house can be a similar luxury to meet specific cultural, architectural, or other personal preference, not necessarily for the economic benefit. Off-the-shelf products are generally welcome as long as they have the important features required by a majority of users. This is particularly crucial for HT ADME assays as they can alleviate scientists from automation, integration, interfacing, and programing problems. For a successful and well-received system, the cost of the investment by the vendors can be shared by users, which may lead to some financial benefits over an individual integration. Another advantage lies in the existence of tangible products with confirmed specifications, which translates to minimal risk in delayed implementation and in failure of deliverables. **Table 2** summarizes the off-shelf automated ADME products on the market.

Factors to be considered prior to purchase of an off-the-shelf automated system include:

- quality of deliverables (whether they meet the specific requirements of a user's organization);
- designed throughput (maximal and sustainable);
- user-friendliness of the system (automation, analysis, data processing, and quality of life for operators);
- reliability and technical support (in particular for the centralized laboratories);
- stability of the products and the vendors (to avoid discontinued products);
- cost effectiveness (both the upfront investment and the subsequent operational cost including personnel expenses);
- possibility to be upgraded in the future (to a higher or more advanced version); and
- compatibility to the infrastructure within a user's organization (e.g., information technology (IT), assay format, etc.).

Commonly, evaluations are carried out prior to purchase of an off-the-shelf product. Typically, the users have a set of reference compounds with known data serving as a 'gold standard' for the assessment. Sometimes inconsistent and unsatisfactory results are found after installation compared to those from evaluation, possibly because of overly optimistic data derived from evaluating well-behaved commercial drugs. In reality, the current drug discovery compounds are more challenging due to their properties being poorer than those of commercial drugs.[19] Therefore, it is essential to create the test set by combining commercial drugs and internal discovery compounds. Ideally, side-by-side comparisons should be made if there are multiple products available.

5.15.2.2   The Customized Workstation in Users' Laboratories

5.15.2.2.1   Automation planning and strategy

One of the largest advantages of integrating your own automation system is the high degree of flexibility, wherein the users can design a completely customized system based on specific requirements. It also allows users to combine favorite components and other special devices. Nonetheless, designing a customized system is a large project entailing significant commitments. Several key elements should be considered before an automation project is initiated: (1) necessity for customization (review existing off-shelf product availability); (2) urgency (generally 3–12 months are

Table 2   Summary of off-the-shelf automated ADME products

Assays	*Vendor/website*	*Main features*
Solubility	pION (www.pion-inc.com)	Start with DMSO solution on Beckman Tecan 8-probe liquid handlers, pH monitored, plate filtration Readout: UV (with HPLC & MS interface)
	Symyx (www.symyx.com)	Start with powders, single probe liquid-handler, temperature controlled, pH monitored, hot filtration Readout: UV or HPLC
	Analiza (www.analiza.com)	Start with powders/DMSO solution single probe liquid-handler, vial/plate filtration Readout: LC/CLND
Intrinsic solubility	pION (www.pion-inc.com)	Start with powders, intrinsic solubility and solubility as a function of pH (1–13) Readout: pH-metric titration
	Sirius (www.sirius-analytical.com)	Start with powders, intrinsic solubility Readout: pH-metric titration
Rate dissolution	Delphian Tech Inc. (www.delphian-tech.com)	On-line rate dissolution detection system using optic fibers Readout: UV (PDA)
PAMPA	pION (www.pion-inc.com)	Start with DMSO solution on Tecan/Beckman 8-probe liquid handlers Readout: UV (with LC/MS interface)
pK_a	Sirius (www.sirius-analytical.com)	Automated medium throughput pK_a determination Readout: pH-metric/on-line UV reader High-throughput pK_a analyzer (using fast mixing technology) Readout: on-line UV reader (multiwavelength)
	Combisep (www.combisep.com)	96-Capillary array pK_a analyzer (mobility versus pH) Readout: CE/UV
log P/D	Analiza (www.analiza.com)	Start with DMSO, on single-probe Hamilton liquid-handler, using 96-deep-well plate Readout: LC/CLND
	Sirius (www.sirius-analytical.com)	Automated medium throughput log P/D determination Readout: pH-metric/on-line UV reader Automated high-throughput log D determination Readout: HPLC
	Combisep (www.combisep.com)	96-Capillary array log P/D analyzer (mobility versus pH) Readout: CE/UV
BBB	Analiza (www.analiza.com)	Start with DMSO, on single-probe Hamilton liquid-handler, using 96-deep-well plate
Serum/plasma protein binding	Biacore (www.biacore.com)	Using Biacore's surface plasmon resonance technology

required to complete customized automation from scratch); (3) availability of a validated protocol for a manual assay that can feasibly be automated; and (4) availability of an internal team with all related knowledge and expertise needed (e.g., ADME assay, automation, bioanalysis, programing, etc.); and (5) the budget.

In summary, the users must have a clear idea of the scope of the project, with clearly defined but realistic time-lines and anticipated deliverables. In the cases where the assay to be automated has only partially been validated and the additional development is still underway, the proposed automation solution has to be flexible enough to cover most of the possible scopes of the project. Once the project is initiated, a project leader should be nominated to oversee it during its entire course and coordinate all activities. A detailed business plan and strategic vision should include an assessment of the anticipated demand for compound profiling, the allowable operating costs, the anticipated turnaround time, and the number of staff required.

5.15.2.2.2 Preparation of user requirement specifications (URS)

Typical issues involved in customized automation are: (1) missed completion time-lines; (2) substantially deteriorated assay quality in automated format; (3) overspending; (4) incompatibility or failed interfacing between components from different vendors; and (5) bespoke systems that are so inflexible that no new features can be accommodated. In these circumstances, most of the disputes between frustrated users and vendors result from the lack of a clearly defined user requirement specification (URS) document. Therefore, users are highly recommended to prepare a URS document to clearly describe their specifications (e.g., deliverables) and required time-line. This is an extremely useful exercise that allows the end users to compose different aspects of the assay and foresee the scope of the entire project. It is critical to include all possible changes of the protocol or future development of the assay so that contingency plans can be arranged upfront by the vendors. Normally, the selected vendors will convert this URS into a technical requirement. These documents serve as a vital guidance for the integration and implementation of the automated system. It is also important to establish guidelines for the project to accommodate changes in scope and modifications of requirements. The project team needs to balance delivering a quality system with schedule and budget.

5.15.2.2.3 Selection of proper components and vendors

Unlike the large-scale automated system or off-the-shelf integrated system described earlier (*see* Sections 5.15.1.2.3 or 5.15.2.1), most users also have the option to select individual components for the assembly. The challenge is that the users have to be very familiar with the latest technologies on the market for the critical components of the integration. A common and also efficient practice is to interview prospective vendors after they have looked over the URS. In general, users should break down the integrated system into several functional units requiring different types of vendors (e.g., special cell culturing facilities, liquid handlers, bioanalytical equipment, or mobile robots to incorporate all selected components together). Clearly, this is a time-consuming process, as, usually, there are a number of well-established suppliers in each of those fields. Here are some tips for the selection process:

1. Include someone in the team who is knowledgeable in those products and also familiar with the integration process of automated ADME systems. The involvement of an automation and/or IT specialist is very beneficial for any project involving intermediate to advanced complexity, as they can provide technical advice on instruments and can comment on design preferences thus aiding in selection.
2. Distribute the URS to all potential vendors with a given deadline (typically 2–4 weeks) for response, provided that a nondisclosure agreement is signed ahead of time by the vendors. For large projects a RFP (request for proposal) can take up to 8 weeks. Establishing clear rules during the RFP is important in these cases. Usually vendors are expected to give their presentation in person.
3. In addition to the features, specifications, and cost analysis provided by a vendor, take into consideration the instrument reliability, data quality, and customer service reputation as well as the stability of the vendor. For a large organization where the particular integration will be implemented in parallel in multiple sites, the above assessment has to be done on a global basis. Generally, there are minimal risks when choosing the most popular models from vendors with an excellent global sales and support network. Some users tend to retain no more than two brands of equipment for ease of training and maintenance within the laboratory.
4. Identify the crucial steps or specifications of your ADME assays for the evaluation. For instance, the expected percent coefficient of variation (CV%) for dilution curves (e.g., for IC_{50} determination of cytochrome P450 (CYP) inhibition and preparation of calibration curve for thermodynamic solubility, etc.) will guide the assessment for precision and accuracy of the respective liquid handling steps. Keep in mind that some of the specifications provided by vendors are reached under optimized conditions using ideal samples. It is essential to confirm the

performance of the products and the quality of the deliverables from daily samples (which normally are much more challenging than those used by vendors) under routine conditions. It is useful to prepare a set of validation samples, which can be tested in the demo experiments by each vendor. Sometimes vendors are willing to ship their demo equipment to users' laboratories to conduct such testing.

5. The evaluation criteria should be quite similar to those used for 'off-the-shelf' systems (*see* Section 5.15.2.1) except that interfacing and compatibility of each component of an integrated system become the user's responsibilities. In general, major vendors have experience of interfacing with peripheral equipment from divergent suppliers. It is wise and efficient to encourage direct communications between vendors of the different functions.

6. The users should be prepared to receive an alternative or revised proposal that may utilize a different technical approach than that originally described in the URS, normally due to the vendor's technical or budgetary limitations. Users should be flexible for noncritical items but should demand that key requirements are met to avoid sacrificing quality or capacity.

7. It is important to confirm if the proposed automated solution is feasible for the labware used in the assays. For example, we found the medium exchange protocol for 21-day cell culturing on the 96-well filter plates from some suppliers was perfectly fine in the manual operation. In the automated procedure, however, the robot failed to consistently remove the lid of the filter plates because of surface tension when taking the plate out of the cell-culturing incubator. In addition, the exchange of the fed media was incomplete because the robotic liquid handler could not remove all remaining fed media from the basolateral well of the plates. Although the suppliers for the filter plates eventually resolved these issues after 6–12 months, clearly this hurdle kept users from offering the designated capacity within the desired timeframe.

8. For a large and complex project, execution of the vendor selection process should be completed with a formal selection process where vendors are narrowed to 1–2 candidates. Once a vendor is selected, all vendors may be notified in writing of the outcome. The letters should be brief and courteous, to clearly communicate the outcome of the selection process and to strengthen the relationship between the vendor and the customer for future projects.

5.15.2.2.4 Keeping integration on track with system factory acceptance test and site acceptance test

The completion of integration projects can be delayed in many cases for various reasons. The project manager from the user side should closely monitor the project with predefined milestones. Typically, the time-line, any necessary URS revisions, including special requirements, and the criteria for site acceptance test (SAT) should be clearly defined in the quote and agreed among all parties. In addition, good communication is a must to avoid last minute changes or misunderstandings.

For a large automation project, it is important to consider a factory acceptance test (FAT) in addition to a SAT. The FAT requires the customer to be at the vendor's site while the test is run. A proper FAT will reduce the risk of the system arriving at the customer site unfit for its designated purpose. It is always cheaper, easier, and faster for a vendor to fix problems at the vendor site than at the customer site. A system shipped unfit for its defined purpose can lead to delays due to the 'out of sight, out of mind' mentality that ensues as the vendor moves on to the next project. The testing protocols used for the FAT and SAT should be outlined in detail and agreed upon by both parties. These documents should be signed prior to the start of testing and once the testing has been completed.

5.15.3 Early Absorption, Distribution, Metabolism, and Excretion Analytics

Bioanalytics is an important factor in early ADME screening as it is often the bottleneck in assay automation. In practice, almost all automated in vitro ADME assays were designed and integrated based on selection of the assay readout. In vitro assays used to assess ADME properties of drug candidates utilize a broad range of detection methods (**Table 3**). Direct optical detection without any chromatographic step has the advantage of speed and is relatively easy to incorporate in a robotic workstation. It also allows the user to follow reaction kinetics if valuable to an assay, for example, time-dependent CYP inhibition. One disadvantage of this approach is that the detection is not compound specific; therefore, one needs to have tight requirements for compound purity ($>95\%$).

Solubility and permeability measurements are the two prominent assays for which insufficient compound purity can cause the most severe problems in bioanalysis. In these cases, more compound-specific detection methods like HPLC/UV or LC/MS tolerate a lower compound purity but are more difficult to automate and generally more labor

Table 3 Assay technology and detection methods for in vitro ADME assays

Molecular property	Assay technology	Detection mode
Kinetic solubility	Nephelometric titration semi-shake flask	Light scattering, HPLC/UV/MS
Equilibrium solubility	Shake flask	UV, LC/UV, LC/MS, LC/CLND
	Potentiometric titration	pH electrode
Ionization	Rapid mixing + UV titration	UV/DAD
	Titration	pH electrode
	Ionization-related mobility	CE/HPLC
Permeability	PAMPA	UV plate reader, HPLC/UV, LC/MS (/MS)
	Caco-2 monolayers/P-gp-transferred MDCK cells	LC/MS/MS, LC/MS/MS
Protein binding	Dialysis/centrifugation	LC/MS/MS
Metabolic clearance	Liver microsomes/S9/hepatocytes	LC/MS/MS, LC/MS/MS
CYP inhibition	Recombinantly expressed CYP + fluorogenic substrates	Fluorescence
	Human liver microsomes + prototypical substrates	LC/MS/MS

intensive. While the data acquisition itself can be automated and multiparallel systems are used to increase throughput, data analysis remains labor intensive. Detection based on mass spectrometry has the advantage of a higher sensitivity, which can be particularly useful for the analysis of compounds with low solubility. However, it requires a higher capital investment.

5.15.3.1   Kinetic Solubility

Kinetic solubility is one of the approaches to estimating aqueous solubility of NCEs rapidly. It is mainly achieved by directly introducing a dimethyl sulfoxide (DMSO) solution of the test compound into the designated media, followed by an immediate readout using optical detection techniques such as turbidimetry,[20,21] nephelometry,[22,23] laser flow cytometry,[24] or direct-UV (using a UV plate reader).[25] In general, multiple samples are prepared with increasing concentrations of the test compound so that one can define the solubility by monitoring the concentration at which particles fall out of the solution. The advantage is that this is a fast and relatively universal detection mode. Sensitivity varies with particle size. Solubility limited by lipophilicity is usually more difficult to detect than solubility limited by other factors such as crystal packing, which usually produces a stronger signal.[26] In addition to the errors caused by the insufficient incubation time and coexistence of DMSO in kinetic solubility determinations, the readout is also interfered by impurities of low solubility, which can lead to a substantially underestimated solubility value of the parent compound.

The above kinetic solubility samples can also be quantified using the direct-UV tactic at a specific wavelength. It offers a fast analytical approach but requires high compound purity and a filtration procedure to remove precipitates from solutions. An alternative method based on full spectra analysis enhances sensitivity, while comparison of the shape of the reference with the analyte spectra can flag the presence of an impurity interfering with the measurement. However, this is successful only if the UV spectrum of the impurity is different from the parent compound, which may not be the case, especially if the impurity is a synthesis by-product that shares the same scaffold as the parent compound. From a process viewpoint, flagging mechanisms imply that the sample needs to be reprocessed using a different readout. The additional steps for this type of follow-up analysis, however, can add complexity to the automation process. It is also worth mentioning that the use of different detection systems for the same assay may affect assay robustness.

5.15.3.2   Equilibrium Solubility

There are essentially two techniques to measure equilibrium solubility. The 'classical' approach is based on the shake flask followed by separation of the solute from the precipitate and quantification of the amount of compound in filtrate.

Whereas various approaches are being used to quantify the compound concentration, the most common approach in the development laboratories is to use generic HPLC/UV. While the data capture itself is relatively easy, automated data analysis is challenging for different reasons: (1) automatic peak integration requires a good signal/noise ratio (S/N); (2) some compounds give more than one peak (diastereoisomers, some enantiomers); (3) retention time can shift with concentration; and (4) when impurities are more soluble than the parent compound, correct peak identification is not always straightforward (on-line qualitative MS can then help to identify which is the compound of interest). Currently, a number of vendors such as Agilent, Waters, and Nanostream have developed customized application software to automate the time-consuming data processing steps and to derive ADME data, such as equilibrium solubility, in a HT fashion. One of the key features of this type of software is its ability to review and process the HPLC data interactively and tackle only the problematic data that are flagged by the predefined criteria. It should be mentioned that some laboratories may utilize the direct-UV approach to quantify the equilibrium solubility samples to enhance the throughput but the assay using this approach will face the same analytical challenges as mentioned in Section 5.15.3.1.

The other accepted method to measure solubility is potentiometric titration, popularized by pION Inc.[27] In this case, there is no need to separate the filtrate from the precipitate as the pH electrode measures proton exchange of the fraction of compound that is in solution. However, this method is time consuming and not very amenable to HT as each sample requires a specific experimental design and data analysis requires a skilled operator and would be difficult to automate.

5.15.3.3   Ionization

Measurement of ionization constants via potentiometric titration is usually a lengthy process. In order to record the pH changes accurately (with a glass electrode) caused by introducing a known volume of titrants to the well-mixed solution of a drug candidate, one needs to wait until the equilibrium is reached after each titrant addition. This approach, albeit reliable, can only analyze 5–10 compounds daily and requires materials in the milligram scale; therefore, it is commonly utilized in the late discovery and early development stages. In addition, solubility is often an issue, as potentiometry requires concentrations in the 0.1–1 mM range in aqueous media. Cosolvents are usually used to circumvent solubility, but several titrations are required to back extrapolate to aqueous media. One additional problem with potentiometric titration is the increasing number of NCEs in early discovery that are delivered as salts with protogenic counter-ions, like acetate or fumarate.

Capillary electrophoresis (CE) can be utilized to monitor the migration or mobility of the compounds under electric potential. As mobility is highly dependent on the ionization process, pK_a data of NCEs can be assessed accordingly.[21,28,29] This tactic demonstrates the advantages of dealing with the common issues in early discovery such as interference of impurities and a limited amount of sample materials. However, the inability to handle poorly soluble compounds that are frequently analyzed in the early phase becomes the major hurdle for its applications in early discovery. A new approach utilizing an HPLC method may help accurately measure pK_a for sparingly soluble compounds.[30]

Another method based on a continuously flowing pH gradient and a UV-DAD, (diode array detector) was developed a few years ago and allows for a much higher throughput.[31] The method establishes a stable time-dependent pH gradient by rapidly mixing acidic and basic buffers, during which drug candidates predissolved in organic solvent are introduced at different pH conditions. The assay works effectively with poorly soluble NCEs by using cosolvent in the media. The pK_a data measured on this 'rapid-mixing' approach correlate very well with those from the potentiometric titration method. The method has shown a number of key advantages in measuring pK_a in early discovery as being high throughput, automatic, robust, and, most importantly, quite predictive to the reliable potentiometric titration method used in later stages. The limitation is that only ionizations that induce a change in the UV spectra can be detected. In other words, not only will a UV chromophore be required but also it may have to locate close enough to the ionization center within a NCE. According to our experience, this fast method allows full characterization of ionizable groups for ca. 70% of the possible pK_a data and usually gives high-quality data, that is, close to what is obtained by potentiometric titration. Nonetheless, this tactic can be nicely combined with the potentiometric approach. It should be mentioned that neither this 'rapid mixing' nor the potentiometric titration approach is compound specific, so compound purity >95% is required to get accurate data.

5.15.3.4   In Vitro Permeability Assays

A variety of approaches were developed to predict permeability of NCEs in early drug discovery.[5,32,33] The PAMPA[17,18,36–39] using a chemical membrane (e.g., dodecane or hexadecane) immobilized on a 96-well filter plate offers an avenue to quickly estimate permeability for NCEs with passive diffusion mechanisms and small molecular weights (e.g., <500). For the Madin–Darby canine kidney (MDCK) cell model that originates from dog kidney, the expression of transporters is quite different from that of human intestine.[32,40] As a result, the MDCK monolayer is commonly used for permeability evaluation of NCEs transported by passive diffusion mechanisms,[41] as does the PAMPA model. The latest development has extended the permeability studies using a P-glycoprotein (P-gp)-transferred MDCK model to estimate the contributions of efflux transporters.[42] The human colon adenocarcinoma (Caco-2) cell permeability model, exhibiting morphological (e.g., tight junction and brush-border) as well as functional (e.g., multiple transport mechanisms) similarities to human intestinal enterocytes, has been widely received in drug discovery and development.[41,43,44] Caco-2 cells extensively express a variety of transport systems beyond P-gp normally found in small intestinal enterocytes, which made it possible to investigate the interplay among different transport systems and differentiate the relative contributions from passive and active transport mechanisms to the overall permeability across the human gastrointestinal (GI) tract.[45] Despite the extensive insights offered by the Caco-2 model over other in vitro permeability models, its lengthy and complicated cell culturing has led to high cost for assay maintenance and daily operation. It is preferable to use PAMPA as a fast and HT prescreening tool for permeability ranking and apply Caco-2 assays only to the challenging compounds that failed in the PAMPA assay.

The permeability assays are carried out similarly for the above in vitro models. In Caco-2 or P-gp transferred MDCK models the bidirectional mode (apical to basolateral chambers and basolateral to apical compartments) is utilized to estimate the permeability in both directions and to differentiate the contribution of the active transporters in the permeation process of NCEs. For PAMPA, permeability can be assessed by either direction due to the lack of active transporters. Analytically, while the cellular permeability assays are commonly quantified by LC/MS/MS, PAMPA is analyzed by UV plate reader, HPLC, and/or LC/MS. Therefore, the same arguments used in Section 5.15.3.2 still apply here. For all permeability measurements, it is important to avoid precipitation in the donor compartment or it will lead to underestimated permeability values. In order to maximize the S/N ratio when UV is used as a detection method, one needs to optimize the loading concentration based on solubility of the compound. This extra step makes the compound management more complicated but increases the fraction of compounds that can be quantified using the UV method. A fraction of compounds cannot be measured in this way because they lack a UV chromophore or because their solubility prevents the use of a concentration suitable for UV detection. LC/MS can then be used as a complementary method. The use of formulation[46] or a cosolvent[47] can be used to prevent compound precipitation. The alternative is to work at a lower loading concentration also using LC/MS as a complementary method. The main issue with this strategy is that accurate UV measurements impose requirements on high compound purity (e.g., >95%) and a tight correlation between the two detection modes used. Data given in **Figure 3** show the assay reproducibility using UV or MS

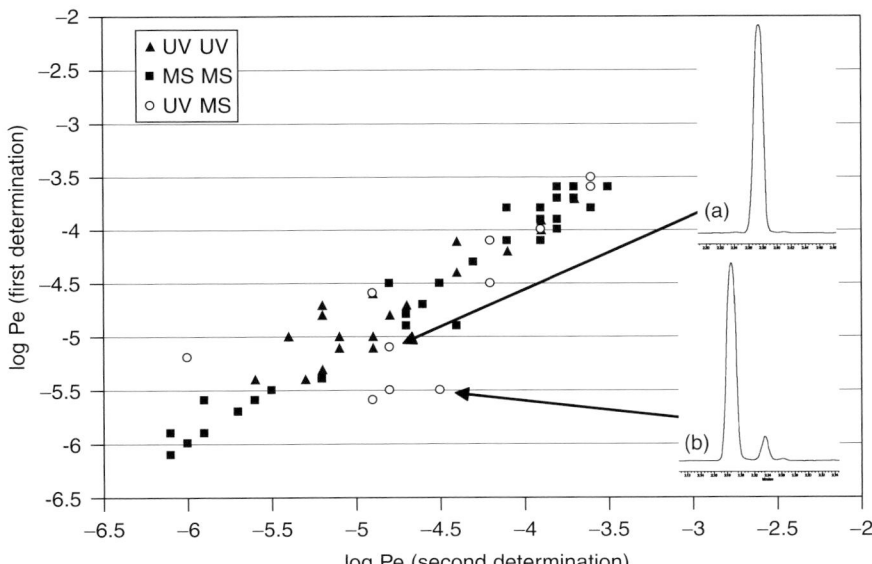

Figure 3 Measurements of effective permeability using UV and LC/MS detection. Filled triangles: both runs were analyzed by a UV plate reader; filled squares: both runs were measured by the LC/MS approach; open circles: one of the two determinations was quantified by LC/MS and the other by a UV plate reader.

detection. The two detection methods correlate fairly tightly for highly permeable compounds (-3.1 to -4.5 in **Figure 3**), but the situation becomes more complex when permeability of test compounds decreases where permeable impurities can interfere with the optical detection.

5.15.3.5   Metabolic Clearance

Metabolic clearance, particularly hepatic, is recognized as one of the main determinants of the concentration in blood as well as the volume of the distribution of NCEs, and therefore is used to help predict bioavailability and toxicokinetics.[48,49] A metabolically unstable NCE, albeit orally absorbed, might never reach the required therapeutic concentration. Alternatively, a certain degree of instability might be desirable in cases where a metabolite is more active than its parent (i.e., a prodrug).

Currently, in vitro approaches that are utilized to monitor the metabolic clearance as well as to predict the human clearance of drug candidates include the use of recombinant CYP enzymes, liver microsomes, S9 fraction (the 9000 g supernatant of a liver homogenate), isolated hepatocytes, and liver slices.[50–52] While liver microsomes and S9 remain most commonly used to measure clearance in early discovery, utilization of hepatocytes has been drastically increased due to the latest breakthrough in cryogenic technology in handling them.

Analytically, a generic LC/MS/MS approach is usually used to probe the disappearance of the test compounds and report the half-life and intrinsic clearance values. This is a very compound-specific detection method, which allows for a quick LC step. Small compound impurities are usually not an issue unless they inhibit CYP enzymes that are involved in the metabolism of the parent compound.

5.15.3.6   Cytochrome P450 Inhibition

Currently, two approaches are commonly utilized to monitor the metabolism-related drug–drug interactions through cytochrome CYP inhibition: LC/MS and fluorogenic methods.[53–55] The former uses clinically relevant substrates such as midazolam or testosterone (e.g., for CYP3A4) and human liver microsomes (HLM), allowing concurrent metabolic events as well as interplay among multiple CYP isozymes in HLM to be monitored. The latest study using the above substrate/enzyme/detection combination approach has demonstrated a good correlation to the in vivo drug–drug interaction issues.[56] In addition, this approach complies with the US Food and Drug Administration (FDA) requirement and is therefore considered a 'gold standard' in vitro assay. Despite the wide acceptance in the development phase, its application in early discovery, however, was somewhat hindered by the limited capacity of LC/MS quantification. In this case, the obtained metabolites are separated and analyzed by LC/MS (occasionally by LC/UV), which imposes major limitations to the throughput and cost effectiveness.

The fluorogenic approach, using a fluorescence plate reader, quantifies the fluorescent metabolites from a known substrate, generally used in combination with recombinant CYP enzymes, and monitors the fluorescent signal reduction as a result of the inhibition of CYP enzymes caused by the coexistence of drug candidate(s). The fluorogenic approach does not require metabolite separation by LC, substantially improving the throughput, turn-around time, and cost effectiveness. The downside of this approach, however, is its limited sensitivity and assay robustness for each enzyme/ substrate pair when the DMSO concentration reaches a specific level (e.g., $>0.1\%$).[57,58] It should be mentioned that although the DMSO-induced analytical issues also occur with the LC/MS/HLM approach, their impact on the fluorescent methodology is significantly greater due to the limited dynamic range of the detection in the fluorogenic method. It should be noted that, unlike the prototypical substrates that are commercial drugs used in the LC/MS/ HLM assay, the fluorescent substrates are not mimicking clinically relevant drugs, but rather are derived from known fluorescent compounds, such as coumarin or fluorescein. In general, those fluorogenic substrates are smaller molecules serving only as analytical markers. Therefore, their physiological relevance and predication to the in vivo drug – drug interactions are yet to be established. Colored, impure, unstable compounds and auto-fluorescent compounds (and/or metabolites) will interfere with the assay detection and need to be flagged. In some organizations, the fluorescent assay is used as a first filter followed by a confirmation using the LC/MS assay. The fluorescent assay is also practical to study time-dependent inhibition as kinetics are easily followed using the optical readout, provided that automation-compatible temperature control and/or reaction mixing can be properly addressed.

5.15.3.7   Impact of Dimethyl Sulfoxide on Early Absorption, Distribution, Metabolism, and Excretion Analytics

In order to make compound logistics efficient and keep compound consumption to a minimum, scientists usually prefer to handle solutions rather than dry powders. The most common compound format for in vitro ADME studies is to start

Table 4   Factors affecting ADME automation and analysis

Factors	Impacts	Potential assays/process	Suggestions
DMSO (other solvents)	Precision/accuracy	Liquid handling for dilution process, kinetic solubility, CYP inhibition	Evaluation of solvent class specification
DMSO	Fluorescence or LC/MS readout	CYP inhibition (inhibit enzyme & increase assay variability)	Minimize DMSO concentration if possible
			Try other solvents (e.g. acetonitrile (ACN), etc.)
Purity or chemical stability	Almost all readouts by plate readers	Solubility, PAMPA, CYP inhibition	Utilize LC or other online purification
			Verify purity prior to profiling assays
Solubility in DMSO	Accuracy for all assays	All	Ensure compounds fully dissolved in DMSO
			Include sonication for prep. if needed
			Monitor precipitation by light scattering
			Quantify the concentration of DMSO solution if needed
Solubility in assay media	Accuracy for all assays	All	Set up solubility surrogate in ADME assays
			Properly define assay dynamic range
			Perform assay in validated formulations
Colored samples	Accuracy for those using a laser or UV plate reader	Kinetic solubility or PAMPA	Utilize LC/MS for colored samples

with a 10 mM DMSO solution. Most compounds are fully dissolved at that concentration and this 10 mM concentration in the DMSO stock solution is high enough to get to a reasonably low amount of solvent at the final compound concentration in most assays. It is advisable to check whether the compound is truly in solution at the beginning of the process. Nephelometry easily detects the presence of microcrystals that are not visible to the eye. DMSO solutions are hygroscopic and as much as 10% of water can be taken up, thus diluting the sample and changing the physical properties of the solution. As many of the ADME assay endpoints (e.g., solubility and IC_{50} for CYP inhibition, etc.) are back-calculated from 10 mM DMSO stock, it is preferable to confirm the true concentration of 10 mM stock in DMSO. In addition, all DMSO stocks should be properly sealed to avoid adsorption of water moisture. It has also been shown that freeze/thaw cycling can cause significant issues in solubility and stability of the compounds in solution. To minimize the possible stability issues, the DMSO stocks should be used as quickly as possible subsequent to preparation and stored at room temperature only. **Table 4** summarizes all factors that affect ADME automation and analysis, as discussed throughout the chapter.

Keeping the solvent concentration low enough is critical for some assays like CYP inhibition or metabolic clearance because even 1% of DMSO can significantly inhibit the enzyme activities (unpublished data). In the kinetic solubility assay, solubility of test compounds was drastically enhanced by a few orders of magnitude when the DMSO concentration increased from 0.5 to 8% (unpublished data). DMSO can also directly interfere with detection methods where its signals overlap or suppress those from test compounds. For example, one tends to be careful to use UV detection at wavelengths above 260 nm to avoid high background contributions from DMSO. The presence of DMSO also prevents the use of direct flow injection in MS analyzers.

5.15.3.8   Laboratory Information Management System within Absorption, Distribution, Metabolism, and Excretion Laboratories

The most critical part of any automation system is the management of the sample and dataflow and the synchronization between the two. A proper laboratory information management system (LIMS) is designed to bring scientists the right information at the right time, keep track of samples and plates, monitor assay QC, and secure data storage. As all of these are used daily, the system stability is critical. LIMS can be fully developed in-house or partially with external vendors. In the latter case, the integration with the in-house IT infrastructure and the corporate databases needs to be carefully planned. LIMS is at the interface between laboratory-automated workstations and corporate databases where the final results are stored and made available to the scientific community. It is important to balance the system capabilities versus complexity so that the LIMS to be established will be relatively simple to use. Manual data input must be kept to a minimum in order to minimize risk of human errors as well as to improve the employee's quality of life. Typically, sample vials or plates are registered using bar code readers and the information flow between the different modules does not require manual editing. While LIMS has proven effective in assuring data integrity and increasing productivity in the cascaded automated ADME assays in particular for the hub laboratories within large organizations, their flexibility in comparison to the standalone instruments needs further improvement. In general, process changes and assay modifications require planning and need to be validated and integrated into the LIMS before productive implementation. A well-designed LIMS should include mechanisms allowing scientists to introduce small changes on their own at least in the pilot study, without relying on the input of IT engineers, who should be responsible for more profound changes and system maintenance.

5.15.4   Outlook for Absorption, Distribution, Metabolism, and Excretion Automation

In the next decade, the demands for automated in vitro ADME assays will continue to grow in major pharmaceutical and biotech companies. More ADME assessments will be shifted from development to the early discovery phase. Additional in vitro ADME profiling assays will be developed to predict the in vivo behaviors of NCEs. The high-quality in vivo and in vitro ADME profiling data derived from drug-like libraries[34] allow for the development of robust predictive in silico models.[59–61] These require industry to reshape and enhance the ADME profiling strategy by consolidating and streamlining the existing automated assays to one or two major sites within an organization to improve efficiency and return on investment (ROI) in both assay development and automation aspects. Increasing efforts are dedicated to the quality and robustness of the automated in vitro ADME assays to warrant the integrity and predictivity (to the next tier of ADME assays) of the data derived.

As automation technologies approach maturity, the challenges to integrate and operate automated in vitro ADME assays will shift away from the liquid handling aspect to the streamlined logistics, high-end quantification, management of the mass data derived and the interface among the above components. Despite the attractive concept and substantial improvements made over the past years, large-scale automated ADME systems will be accepted only after they can demonstrate the flexibility to address the individual assay requirements, the realistic operational scheme for multiple assays, the user-friendly interface between different functional modules, and the capability to accommodate the addition of new assays in the future. Nonetheless, the individually integrated automated in vitro ADME assays, either the off-the-shelf systems or custom-built units, will still play critical roles due to the divergent user groups and assay requirements.

References

1. Hodgson, J. *Nature Biotech.* **2001**, *19*, 722–726.
2. Li, A. P. *Drug Disc. Today* **2001**, *6*, 357–366.
3. Kerns, E. H.; Di, L. *Drug Disc. Today: Technol.* **2004**, *1*, 343–348.
4. Roberts, S. A. *Curr. Opin. Drug Disc. Dev.* **2003**, *6*, 66–80.
5. Wang, J.; Urban, L. *Drug Disc. World* **2004**, *5*, 73–86.
6. Biller, S. A.; Custer, L.; Dickinson, K. E.; Durham, S. K.; Gavai, A.; Hamann, L. G.; Josephs, J. L.; Moulin, F.; Pearl, G. M.; Flint, O. P. et al. In *Pharmaceutical Profiling in Drug Discovery for Lead Selection*; Borchardt, R. T., Kerns, E. H., Lipinski, C. A., Thakker, D. R., Wang, B., Eds.; AAPS Press: New York, 2004, pp 413–429.
7. Penzotti, J. E.; Landrum, G. A.; Putta, S. *Curr. Opin. Drug Disc. Dev.* **2004**, *7*, 49–60.
8. Kennedy, T. *Drug Disc. Tech.* **1997**, *2*, 436–441.
9. Yu, H.; Adedoyin, A. *Drug Disc. Today* **2003**, *8*, 852–861.
10. Kassel, D. B. *Curr. Opin. Chem. Biol.* **2004**, *8*, 339–345.
11. Rutherford, M. L.; Stinger, T. *Curr. Opin. Drug Disc. Dev.* **2001**, *4*, 343–346.
12. Van de Waterbeemd, H. *Curr. Opin. Drug Disc. Dev.* **2002**, *5*, 33–43.
13. O'Connor, D. *Curr. Opin. Drug Disc. Dev.* **2002**, *5*, 52–58.
14. Saunders, K. C. *Drug Disc. Today: Technol.* **2004**, *1*, 373–380.

15. Lesturgeon, C.; Wasselin, G.; Fossati, L.; Dechaume, R.; Chameroy, N.; Flageolet, A.; Hardillier, E.; Guillet, F.; Moissenet, S.; Chevillon, D.; Maubon, N. *BIOforum Europe* **2005**, *9*, 58–60.
16. Faller, B.; Grimm, H. P.; Loeuillet-Ritzler, F.; Arnold, S.; Briand, X. *J. Med. Chem.* **2005**, *48*, 2571–2576.
17. Kansy, M.; Senner, F.; Gubernator, K. *J. Med. Chem.* **1998**, *41*, 1007–1010.
18. Wohnsland, F.; Faller, B. *J. Med. Chem.* **2001**, *44*, 923–930.
19. Lipinski, C. A. *J. Pharmacol. Toxicol. Methods* **2000**, *44*, 235–249.
20. Lipinski, C. A.; Lombardo, L.; Dominy, B. W.; Feeney, P. J. *Adv. Drug Delivery Rev.* **1997**, *23*, 3–25.
21. Kibbey, C. E.; Poole, S. K.; Robinson, B.; Jackson, J. D.; Durham, D. *J. Pharm. Sci.* **2001**, *90*, 1164–1175.
22. Bevan, C. D.; Lloyd, R. S. *Anal. Chem.* **2000**, *72*, 1781–1787.
23. Kariv, I.; Rourick, R. A.; Kassel, D. B.; Chung, T. D. Y. *Combinat. Chem. High Throughput Screen.* **2002**, *5*, 459–472.
24. Stresser, D. M.; Broudy, M. I.; Ho, T.; Cargill, C. E.; Blanchard, A. P.; Sharma, R.; Dandeneau, A. A.; Goodwin, J. J.; Turner, S. D.; Erve, J. C. L. et al. *Drug Metab. Dispos.* **2004**, *32*, 105–112.
25. Pan, L.; Ho, Q.; Tsutsui, K.; Takahashi, L. *J. Pharm. Sci.* **2001**, *90*, 521–529.
26. Lipinski, C. L. Solubility in the Design of Combinatorial Libraries. In *Chemical Analysis 163* (*Analysis and Purification Methods in Combinatorial Chemistry*); John Wiley & Sons: New York, 2004, pp 407–434.
27. Avdeef, A. High-Throughput Measurements of Solubility Profiles. In *Pharmacokinetic Optimization in Drug Research*; Testa, B., van de Waterbeemd, H., Folkers, G., Guy, R., Eds.; Wiley-VCH: Weinheim, Germany, 2001, pp 305–325.
28. Ishihama, Y.; Nakamura, M.; Miwa, T.; Kajima, T.; Asakawa, N. *J. Pharm. Sci.* **2002**, *91*, 933–942.
29. Cleveland, J. A., Jr.; Benko, M. H.; Gluck, S. J.; Walbroehl, Y. M. *J. Chromatogr. A* **1993**, *652*, 301–308.
30. Oumada, F. Z.; Ràfols, C.; Rosés, M.; Bosch, E. *J. Pharm. Sci.* **2002**, *91*, 991–999.
31. Box, K.; Bevan, C.; Comer, J.; Hill, A.; Allen, R.; Reynolds, D. *Anal. Chem.* **2003**, *75*, 883–892.
32. Balimane, P. V.; Chong, S.; Morrison, R. A. *J. Pharmacol. Toxicol.* **2000**, *44*, 301–312.
33. Hämäläinen, M. D.; Frostell-Karlsson, A. *Drug Disc. Today: Technol.* **2004**, *1*, 397–406.
34. Lipinski, C. A. *Drug Disc. Today: Technol.* **2004**, *1*, 337–342.
35. Hartmann, T.; Schmitt, J. *Drug Disc. Today: Technol.* **2004**, *1*, 431–439.
36. Sugano, K.; Takata, N.; Machida, M.; Saitoh, K.; Terada, K. *Int. J. Pharm.* **2001**, *228*, 181–188.
37. Zhu, C.; Jiang, L.; Chen, T. M.; Hwang, K. K. *Eur. J. Med. Chem.* **2002**, *37*, 399–407.
38. Ruell, J. A.; Tsinman, K. L.; Avdeef, A. *Eur. J. Pharm. Sci.* **2003**, *20*, 393–402.
39. Kansy, M.; Avdeef, A.; Fischer, H. *Drug Disc. Today: Technol.* **2004**, *1*, 349–356.
40. Irvine, J. D.; Takahashi, L.; Lockhart, K.; Cheong, J.; Tolan, J. W.; Selick, H. E.; Grove, J. R. *J. Pharm. Sci.* **1999**, *88*, 28–33.
41. Ungell, A. L.; Karlsson, J. Cell Culture in Drug Discovery: An Industrial Perspective. In *Drug Bioavailability*; van de Waterbeemd, H., Lennernas, H., Artursson, P., Eds., Wiley-VCH: Weinheim, Germany, 2003, pp 90–131.
42. Bohets, H.; Annaert, P.; Mannens, G.; Van Beijsterveldt, L.; Anciaux, K.; Verboven, P.; Meuldermans, W.; Lavrijsen, K. *Curr. Top. Med. Chem.* **2001**, *1*, 367–383.
43. Artursson, P.; Tavelin, S. Caco-2 and Emerging Alternatives for Prediction of Intestinal Drug Transport: A General Overview. In *Drug Bioavailability*; Van de Waterbeemd, H., Lennernas, H., Artursson, P., Eds.; Wiley-VCH: Weinheim, Germany, 2003, pp 72–89.
44. Lennernäs, H.; Lundgren, E. *Drug Disc. Today: Technol.* **2004**, *1*, 417–422.
45. Ungell, A. -L. B. *Drug Disc. Today: Technol.* **2004**, *1*, 423–430.
46. Kansy, M.; Fisher, H.; Kratzat, K.; Senner, F.; Wagner, B.; Parilla, I. High Throughput Artificial Membrane Permeability Studies in Early Lead Discovery and Development. In *Pharmacokinetic Optimization in Drug Research*; Testa, B., Van de Waterbeemd, H., Folkers, G., Guy, R., Eds.; Wiley-VCH: Weinheim, Germany, 2001, pp 447–464.
47. Ruell, J. A.; Tsinman, O.; Avdeef, A. *Chem. Pharm. Bull.* **2004**, *52*, 561–565.
48. Obach, R. S. *Curr. Opin. Drug Disc. Dev.* **2001**, *4*, 36–44.
49. Keldenich, J. *Drug Disc. Today: Technol.* **2004**, *1*, 389–396.
50. Obach, R. S. *Drug Metab. Dispos.* **1999**, *27*, 1350–1359.
51. Obach, R. S.; Baxter, J. G.; Liston, T. E.; Silber, B. M.; Jones, B. C.; MacIntyre, F.; Rance, D. J.; Wastall, P. *J. Pharmacol. Exp. Ther.* **1997**, *283*, 46–58.
52. Lau, Y. Y.; Krishna, G.; Yumibe, N. P.; Grotz, D. E.; Sapidou, E.; Norton, L.; Chu, I.; Chen, C.; Soares, A. D.; Lin, C. C. *Pharm. Res.* **2002**, *19*, 1606–1610.
53. Crespi, C. L.; Miller, V. P.; Penman, B. W. *Anal. Biochem.* **1997**, *248*, 188–190.
54. Crespi, C. L.; Miller, V. P.; Stresser, D. M. *Methods Enzymol.* **2002**, *357*, 276–284.
55. Riley, R. J.; Grime, K. *Drug Disc. Today: Technol.* **2004**, *1*, 365–372.
56. Ito, K.; Hallifax, D. R.; Obach, S.; Houston, B. J. *Drug Metabol. Dispos.* **2005**, *33*, 837–844.
57. Cohen, L.; Remley, M.; Raunig, D.; Vaz, A. *Drug Metab. Dispos.* **2003**, *31*, 1005–1015.
58. Busby, Jr.; William, F.; Ackermann, J. M.; Crespi, C. L. *Drug Metabol. Dispos.* **1999**, *27*, 246–249.
59. Egan, W. J.; Zlokarnik, G.; Grootenhuis, P. D. J. *Drug Disc. Today: Technol.* **2004**, *1*, 381–388.
60. Rostami-Hodjegan, A.; Tucker, G. *Drug Disc. Today: Technol.* **2004**, *1*, 441–449.
61. Schmitt, W.; Willmann, S. *Drug Disc. Today: Technol.* **2004**, *1*, 449–456.

Biographies

Jianling Wang is a Sr Research Investigator in Novartis institutes for Biomedical Research at Cambridge, MA. He obtained his PhD in biophysical chemistry in US. He extended his career as a postdoctoral fellow at AT&T Bell laboratories and joined Novartis (formerly Ciba-Geigy) in 1995. Recently, he has devoted to the novel techniques in HT ADME profiling tools that can significantly improve the existing drug discovery and development process. Currently, he is heading the early ADME program in Novartis at Cambridge and responsible for the development and/or implementation of a number of automated ADME assays such as HT-equilibrium solubility, artificial and Caco-2 permeability, pK_a, log P/D, chemical and metabolic stability, CYP-450 inhibition, and rate of dissolution in Cambridge, in support of the drug discovery research throughout the company.

Bernard Faller graduated as a biochemist from the University of Strasbourg, France where he went on to obtain a PhD in 1991. His initial interest was in inhibitors of neutrophil elastase for the treatment of respiratory diseases. In 1991 he moved to Ciba-Geigy as a postdoctoral fellow and focused on enzyme kinetics of aromatase inhibitors. In 1995 he became head of laboratory at Ciba-Geigy (and then Novartis) responsible for physicochemical characterization of iron-chelating agents. He then focused on biochemical characterization of fatty-acid binding proteins and in 1999 moved to central technologies and established the foundations of the Novartis early ADME profiling platform, which investigates molecular properties governing drug disposition. In 2001, he became technology program head for physicochemical profiling in the Preclinical Compound Profiling Unit. Today, he leads the Profiling Group Basel within Discovery Technologies Department of Novartis Institutes for BioMedical Research.

Comprehensive Medicinal Chemistry II
ISBN (set): 0-08-044513-6

ISBN (Volume 5) 0-08-044518-7; pp. 341–356

5.16   Ionization Constants and Ionization Profiles

J E A Comer, Sirius Analytical Instruments Ltd, Forest Row, UK

© 2007 Elsevier Ltd. All Rights Reserved.

5.16.1   Terminology, pH Scale and pK_a Definitions

This section introduces the terminology of pH and ionization, as a well as equations for calculating the pH of solutions of weak acids and bases, and for plotting distribution of species profiles. It also introduces ampholytes and zwitterions, and the concept of protonation microconstants.

5.16.1.1   Solution Equilibria

This chapter is concerned only with acid–base ionizations of organic molecules involving the gain or loss of hydrogen ions; this is the main class of ionization that can occur in solution. The terminology for ionization processes developed over more than 100 years of solution chemistry, with the result that many different words and phrases have been used to describe the same things. Thus, hydrogen ions are written as H^+, but they are also referred to as protons. The acquisition of a hydrogen ion by a molecule is called protonation; the loss of a hydrogen ion is called deprotonation. Molecules that can gain or lose one proton may be called monobasic, though the word monoprotic is clearer; those that can gain or lose two are called diprotic, and so on. Molecules that can be protonated to create positively charged cations are called bases, and the protonated cations are called conjugate acids; molecules that can be deprotonated to create negatively charged anions are called acids, and the deprotonated anions are called conjugate bases. Some molecules can both gain and lose protons, existing as cations, anions, and unionized molecules, depending on pH. These are called either ampholytes or zwitterions; there is a tendency to use these terms interchangeably, but they represent distinctly different chemical forms, as explained later. The different forms of the molecule (unionized, ionized) are called species. Aside from mentioning these terms in this section, the remainder of the chapter will use a standardized set of terms that may be convenient for medicinal chemists, who are not on the whole concerned with the theories of solution chemistry but with the practical use of the numbers.

Ionizable drugs and drug-like molecules transform between their unionized and ionized states at pH values between 1 and 13. They are called weak acids and bases. The strength of the acid or base is denoted by its ionization constant, K_a. Besides ionization constant, this term has been called several other names in the literature, such as protonation constant, equilibrium constant, and dissociation constant. There is a related term, stability constant, which is equivalent to $1/K_a$. K_a is usually expressed logarithmically as a pK_a value, meaning $-\log_{10} K_a$; where possible, the term pK_a will be used throughout this chapter.

5.16.1.2   Definition of pH

Because pK_a values are numbers on the pH scale, it is important to understand the concept of pH. The term pH is used to express the acidity of aqueous solutions, and was first introduced by Sørensen in 1909. In two important papers published simultaneously in German and French,[1] he compared the usefulness of the degree of acidity with that of the total acidity, proposed the hydrogen ion exponent, set up standard methods for the determination of hydrogen ion concentrations by both electrometric and calorimetric means, with a description of suitable buffers and indicators, and discussed in detail the application of pH measurements to enzymatic studies.

Sørensen introduced a potential of hydrogen $P_H = -\log C_{H^+}$, which was subsequently modified to pH. Modern usage recognizes the following terms that are used in the so-called operational scale,

$$pH = pa_H = -\log a_H = -\log_{10} [H^+]\gamma_+ \qquad [1]$$

as well as a concentration pH, which is valid in solutions of constant ionic strength,

$$pC_{H^+} = -\log [H^+] = p[H^+] \qquad [2]$$

In the above expressions, a_H represents the activity of the hydrogen ion, γ_+ represents the activity coefficient, and C_{H^+} represents the concentration of hydrogen ions.

5.16.1.3   Concentration and Activity

Concentration expresses the amount of solute as a proportion of the amount of solvent or of solution, in units such as or moles per kilogram, moles per liter, grams per liter, etc. It is conventionally denoted by the use of square brackets in equilibrium expressions, so [S] means the concentration of S. The concept of activity was developed to relate concentration to thermodynamic activity as measured by solubility, vapor pressure, or electrochemical potential. For rough calculations, or in very dilute solutions that approach zero ionic strength, activity is more or less equal to concentration, but at higher concentrations activity (a) is shown as

$$a_S = [S]\gamma_S \tag{3}$$

where S denotes solute and γ_S is the activity coefficient of the solute. When considering molecules that ionize in solution, it would be easy to regard the ions as mere points of charge. However, this is not realistic, and the activity of ions is affected by a number of factors. One factor is temperature, and another is the dielectric of the solvent medium. Another significant factor is that hydrated ions occupy a volume of the solution in proportion to their size, and exclude other ions from this solution space. This competition for space inhibits the dissociation at high ionic strength. Ions self-associate to form ion pairs, so as the concentration of the solute increases, the activity of ions increases somewhat less. The equations to derive activity coefficients are complicated. However, two special cases are often used to avoid this complication. Results may be reported at zero ionic strength, where the activity coefficient is unity, though this introduces some practical difficulties. More usually, if the solute under study is dilute but the total ionic strength is high because an inert electrolyte has been added, then the activity coefficient of the solute ions will have a constant value at constant ionic strength.

Although the relationship between concentration and activity has been properly considered in fundamental work to establish the composition of primary buffer solutions, it is too complicated to apply in everyday measurement situations. For this reason, pH is measured and reported in units of activity. Values for p[H] can be obtained by measuring pH in solutions that have been adjusted to a constant ionic strength by adding a solution of an inert electrolyte; under these conditions, pH is proportional to p[H]. A constant ionic strength background of 0.15 M KCl is commonly used in the measurement of pK_a values. The ionic strength of this solution is close to that of human blood; KCl is used in preference to NaCl because high concentrations of sodium ions interfere with the response of the glass electrode used to measure pH.

5.16.1.4   pH of Water

Liquid water is composed of molecules of H_2O. Interactions between the molecules lead to complex metastable structures within liquid water, and a solvated ionized molecule ($H_9O_4^+$), as well as structures with additional solvation shells have been proposed.[2] It is too complicated to include these hydration structures in written acid–base reactions. The ionization of water can be expressed in terms of the hydrated hydrogen ion H_3O^+,

$$H_2O + H_2O \rightleftharpoons H_3O^+ + OH^- \tag{4}$$

or more simply using the proton H^+ itself (even though H^+ ions do not exist as free species in solution). Thus, the ionization of water can be written as

$$H_2O \rightleftharpoons H^+ + OH^- \tag{5}$$

In terms of ionic activities an equation can be written containing the equilibrium constant K^o:

$$a_H a_{OH} = K^o a_{H_2O} \tag{6}$$

The activities of H^+ and OH^- are proportional to their concentrations, but by convention the activity of H_2O is proportional to the mole fraction of water in the solution. Thus, a_{H_2O} is close to 1.000 in dilute solutions, and may be included in the equilibrium constant K_w^o:

$$a_H a_{OH} = [H^+]\gamma_+ [OH^-]\gamma_- = K_w^o \tag{7}$$

If the activity coefficients are also included in the constant, the ionic product of water K_W can be derived:

$$[H^+][OH^-] = K_w \tag{8}$$

Values for K_w have been calculated from conductance data measured in water at varying temperatures and pressures,[3] and a value of $10^{-13.997}$ (25 °C, 1 bar, zero ionic strength) is generally accepted.[4] The corresponding value of pK_w ($-\log_{10} K_w$) is 13.997 (i.e. close to 14.0 under standard conditions). The concentration of H^+ and OH^- must be equal when water dissociates. Therefore, if $[H^+] = [OH^-]$, and $[H^+][OH^-] = 10^{-14}$ then $[H^+]$ in pure water has a value of 10^{-7}, and the pH ($-\log_{10} 10^{-7}$) of pure water is 7.0 at 25 °C. In fact, the pH of water at 25 °C is usually about 5.5, because water in equilibrium with air becomes saturated with carbon dioxide, thus forming a dilute solution of carbonic acid.

Note that pH varies with temperature, especially for solutions at high pH. This is because temperature affects the activity coefficients involved in the calculation of K_w^o (eqn [7]). The values of pK_w vary from 14.96 at 0 °C to 13.26 at 50 °C.[5] Note also that the value of K_w varies as a function of ionic strength. The values of the pK_w of water at 25 °C vary between 13.997 at $0 \, mol \, L^{-1}$ to 14.18 at $3.0 \, mol \, L^{-1}$.[6,7]

5.16.1.5   pH of Strong Acids and Bases

Acids are molecules that are not ionized at low pH, but which become negatively charged ions at higher pH because they lose one or more hydrogen ions to the solution. Bases are not ionized at high pH, but become positively charged ions at lower pH because they gain one or more hydrogen ions from the solution. HCl and KOH are examples of a strong acid and base, respectively, meaning that they are fully ionized in aqueous solution; their unionized forms would only exist in solution outside the range of the aqueous pH scale. These definitions are in accordance with Brønsted's concept that an acid is a species having a tendency to lose a proton.

The pH values of solutions of strong acids and bases in water can be easily calculated. In a solution of HCl of concentration $C \, mol \, L^{-1}$, the following equations may be defined:

$$[Cl^-] = C \text{ (because dissociation is complete)} \tag{9}$$

$$[H^+][OH^-] = K_w = 10^{-14} \text{ (ionic product)} \tag{10}$$

$$[H^+] = [OH^-] + [Cl^-] \text{ (charge balance)} \tag{11}$$

Substituting for $[Cl^-]$ from eqn [9] and $[OH^-]$ from eqn [10] in eqn [11] gives the following equation, which has terms in units of $[H^+]$ alone:

$$[H^+] = \frac{K_w}{[H^+]} + C \tag{12}$$

which reduces to a quadratic equation of the form

$$[H^+]^2 - C[H^+] - K_w = 0 \tag{13}$$

which could be solved for an explicit value of $[H^+]$.

However, consider a solution of $10^{-1} M$ HCl in water. The concentration of hydrogen ions supplied by the dissociation of the HCl is 10^{-1}, and the term $K_w/[H^+]$ in eqn [12] is 10^{-13}, which is very small compared with 10^{-1} and can be neglected. Equation [12] therefore reduces to

$$[H^+] = C = 10^{-1} \tag{14}$$

which rearranges to

$$pH = -\log_{10}[H^+] = 1.00 \tag{15}$$

A similar series of equations can be defined for a strong base, such as a solution of KOH of $C \, mol \, L^{-1}$:

$$[K^+] = C \text{ (because dissociation is complete)} \tag{16}$$

$$[H^+][OH^-] = K_w = 10^{-14} \text{ (ionic product)} \tag{17}$$

$$[H^+] + [K^+] = [OH^-] \text{ (charge balance)} \tag{18}$$

Substituting for $[K^+]$ from eqn [16] and $[OH^-]$ from eqn [17] in eqn [18] gives the following equation that has terms in units of $[H^+]$ alone:

$$[H^+] + C = \frac{K_w}{[H^+]} \qquad [19]$$

which reduces to a quadratic equation of the form

$$[H^+]^2 + C[H^+] - K_w = 0 \qquad [20]$$

which can be solved for an explicit value of $[H^+]$.

However, in basic solutions the pH is higher than 7, and therefore $[H^+]$ is less than 10^{-7}. If C is much larger than 10^{-7} then $[H^+]^2$ is negligible compared with C, and eqn [19] could be simplified to

$$C[H^+] - K_w = 0 \qquad [21]$$

which rearranges to

$$p[H] = pK_w + \log C = 14.00 + \log C \qquad [22]$$

In a solution of 10^{-1} M KOH in water, $\log C = -1$, therefore pH = 13.

Solutions of HCl and KOH supply points at the extremes of the pH scale. On this logarithmic scale, a 0.1 M solution of HCl would have a pH of 1, and a solution of 0.1 M KOH would have a pH of 13. While solutions exist whose pH values are outside the range 1 to 13, they are not normally encountered in the human body, and these values represent the limits of the pH scale for most aspects of medicinal chemistry.

5.16.1.6   Weak Acids, Weak Bases, and Ionization Constants (pK_a)

Ionizable drugs are called weak acids and bases because their unionized forms exists in solution over at least a part of the pH range from about 1 to 13. A convenient nomenclature uses AH (and AH_2, AH_3, etc.) to represent weak acids, B to represent weak bases, and XH to represent ampholytes and zwitterions.

Consider a monoprotic (i.e., singly ionizing) acid AH that dissociates according to

$$AH \rightleftharpoons A^- + H^+ \qquad [23]$$

The term K_a is the ionization constant such that at equilibrium,

$$K_a = \frac{[A^-][H^+]}{[AH]} \qquad [24]$$

The ionization of a monoprotic base B may be represented in a similar way:

$$BH^+ \rightleftharpoons B + H^+ \qquad [25]$$

and

$$K_a = \frac{[B][H^+]}{[BH^+]} \qquad [26]$$

From the above equations, when $[A^-] = [AH]$ or $[B] = [BH^+]$, then $[H^+] = K_a$. Therefore, for monoprotic compounds the pK_a ($= -\log_{10} K_a$) is the pH at which equal concentrations of neutral and ionized forms of the sample are present in the solution.

Equations [24] and [26] can be written in the following form, which is frequently cited as the Henderson–Hasselbalch equation:

$$pH = pK_a - \log\frac{[AH]}{[A]} \qquad [27]$$

In a series of acids with increasing pK_a values, those with higher pK_a values are said to be weaker, while acids with lower pK_a values are stronger. In contrast, bases with higher pK_a values are stronger, while bases with lower pK_a values are weaker.

As with pH values, pK_a values vary with temperature and ionic strength, so it is helpful if these conditions are cited along with the pK_a value. Conditions of 25 °C and 0.15 M ionic strength are widely used for drugs. It is also important to know whether the pK_a refers to an acidic or basic group.

5.16.1.7   Distribution of Species

The pK_a and the type of compound (acid or base) provide a snapshot of the acid-base behavior of a molecule, and may be visualized graphically. For example, ibuprofen is a carboxylic acid with a pK_a of 4.35. The relative concentration of unionized and ionized species as a function of pH may be plotted as a distribution of species graph (**Figure 1**). The graph consists of two lines plotted on the same axes. While these graphs conventionally plot the relative concentrations as percentages, they are based on the following equations. At any pH, the fraction of compound in the form A⁻ is expressed as

$$[A^-] = \frac{K_a}{[H^+] + K_a} \qquad [28]$$

and the fraction in the form AH is expressed as

$$[AH] = \frac{[H^+]}{[H^+] + K_a} \qquad [29]$$

Figure 2 shows the distribution of species graph for carvedilol, a base with a pK_a of 7.97. The fraction in the form B is expressed as

$$[B] = \frac{K_a}{[H^+] + K_a} \qquad [30]$$

and the fraction in the form BH⁺ is expressed as

$$[BH^+] = \frac{[H^+]}{[H^+] + K_a} \qquad [31]$$

5.16.1.8   Ionization Equations for Samples with up to Six pK_a Values

Table 1 shows a general scheme for describing the ionization of compounds with up to six pK_a values, together with equations for plotting the distribution of species. The equations in the table, as well as those used to draw the graphs and profiles of lipophilicity and solubility appearing in this chapter, are derived from principles of mass balance and

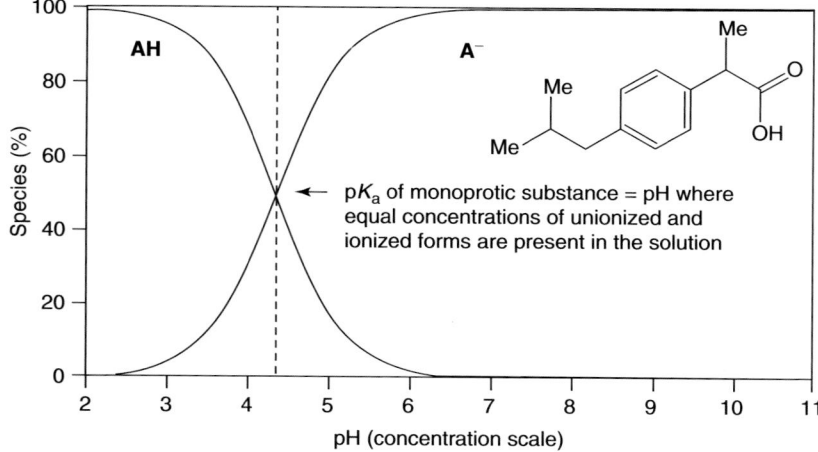

Figure 1   Distribution of species of ibuprofen, an acid with $pK_a = 4.35$. A total of 18% of ibuprofen is in unionized form at pH 5, which is the pH in the duodenum and jejunum in the fed state.

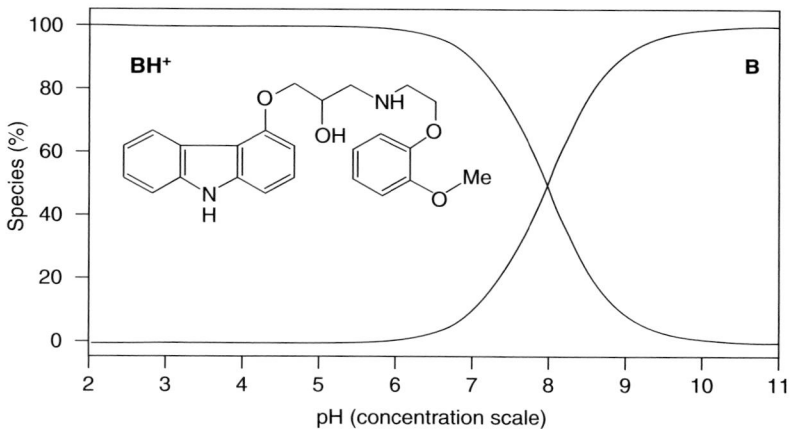

Figure 2   Distribution of species of carvedilol, a base with $pK_a = 7.97$. A total of 6.3% of carvedilol is in unionized form at pH 6.8, which is the pH in the duodenum and jejunum in the fasted state.

charge balance. Each class of molecule (monoprotic acid, diprotic acid, etc.) has its own set of equations for each type of graph. The general principles of mass balance and charge balance are explained in Butler's book on ionic equilibrium,[8] which guides readers in the derivation of new equations. Another source is Albert and Serjeant's book on the determination of ionization constants.[9] Unfortunately, the latter book is out of print, but as it provides an excellent introduction to the subject, it is worth trying to obtain a copy.

5.16.1.9   Ampholytes and Zwitterions

Ampholytes and zwitterions are molecules with at least two pK_a values, at least one of which is acidic and at least one is basic. **Figures 3** and **4** show the distribution of species graph for clioquinol (an ampholyte) and ampicillin (a zwitterion). Both molecules have two pK_a values. In clioquinol, the base pK_a is lower than the acid pK_a, so at pH values between the pK_a values, the molecule is neutral. However, the acid pK_a for ampicillin is lower than the base pK_a, and the molecule is charged across the entire pH scale. Ampholytes are most lipophilic and least soluble in their neutral form. However, zwitterions are always charged; they are rarely lipophilic, and they are often soluble in water at any pH. The following equations are used for the ionization and distribution of species of diprotic ampholytes and zwitterions. Because there are two ionizations, the ionization constants must be labeled K_{a1} and K_{a2}.

$$XH_2^+ \rightleftharpoons XH + H^+ \qquad [32]$$

$$K_{a1} = \frac{[XH][H^+]}{[XH_2^+]} \qquad [33]$$

and

$$XH \rightleftharpoons X^- + H^+ \qquad [34]$$

$$K_{a2} = \frac{[X^-][H^+]}{[XH]} \qquad [35]$$

The distribution of species equations for diprotic ampholytes and zwitterions are

$$[X^-] = \frac{K_{a1}K_{a2}}{[H^+]^2 + K_{a1}[H^+] + K_{a1}K_{a2}} \qquad [36]$$

Table 1 General ionization scheme

Equation	pK_a order	Ionization constant	Plot versus pH for distribution of species graph for species [X], [XH], etc.
Monoprotic (1 pK_a value)			
$X + H = XH$	–	$K_{a1} = \dfrac{[X][H]}{XH}$	$[X] = \dfrac{K_{a1}}{[H] + K_{a1}}$ Monoprotic bases: unionized species = X $[XH] = \dfrac{[H]}{[H] + K_{a1}}$ Monoprotic acids: unionized species = XH
Diprotic (2 pK_a values)			
$X + H = XH$	Highest	$K_{a2} = \dfrac{[X][H]}{[XH]}$	$[X] = \dfrac{K_{a1}K_{a2}}{[H]^2 + K_{a1}[H] + K_{a1}K_{a2}}$ Diprotic bases: unionized species = X
$XH + H = XH_2$	Lowest	$K_{a1} = \dfrac{[XH][H]}{[XH_2]}$	$[XH] = \dfrac{K_{a1}[H]}{[H]^2 + K_{a1}[H] + K_{a1}K_{a2}}$ Ampholytes, zwitterions: unionized species = XH $[XH_2] = \dfrac{[H]^2}{[H]^2 + K_{a1}[H] + K_{a1}K_{a2}}$ Diprotic acids: unionized species = XH_2
Triprotic (3 pK_a values)			
$X + H = XH$	Highest	$K_{a3} = \dfrac{[X][H]}{XH}$	$[X] = \dfrac{K_{a1}K_{a2}K_{a3}}{[H]^3 + K_{a1}[H]^2 + K_{a1}K_{a2}[H] + K_{a1}K_{a2}K_{a3}}$
$XH + H = XH_2$	→	$K_{a2} = \dfrac{[XH][H]}{[XH_2]}$	$[XH] = \dfrac{K_{a1}K_{a2}[H]}{[H]^3 + K_{a1}[H]^2 + K_{a1}K_{a2}[H] + K_{a1}K_{a2}K_{a3}}$
$XH_2 + H = XH_3$	Lowest	$K_{a1} = \dfrac{[XH_2][H]}{[XH_3]}$	$[XH_2] = \dfrac{K_{a1}[H]^2}{[H]^3 + K_{a1}[H]^2 + K_{a1}K_{a2}[H] + K_{a1}K_{a2}K_{a3}}$ $[XH_3] = \dfrac{[H]^3}{[H]^3 + K_{a1}[H]^2 + K_{a1}K_{a2}[H] + K_{a1}K_{a2}K_{a3}}$

4 pK_a values

Equation	pK_a order	Constant
$X + H = XH$	Highest	K_{a4}
$XH + H = XH_2$	→	K_{a3}
$XH_2 + H = XH_3$		K_{a2}
$XH_3 + H = XH_4$	Lowest	K_{a1}

5 pK_a values

Equation	pK_a order	Constant
$X + H = XH$	Highest	K_{a5}
$XH + H = XH_2$		K_{a4}
$XH_2 + H = XH_3$	→	K_{a3}
$XH_3 + H = XH_4$		K_{a2}
$XH_4 + H = XH_5$	Lowest	K_{a1}

6 pK_a values

Equation	pK_a order	Constant
$X + H = XH$	Highest	K_{a6}
$XH + H = XH_2$		K_{a5}
$XH_2 + H = XH_3$	→	K_{a4}
$XH_3 + H = XH_4$		K_{a3}
$XH_4 + H = XH_5$		K_{a2}
$XH_5 + H = XH_6$	Lowest	K_{a1}

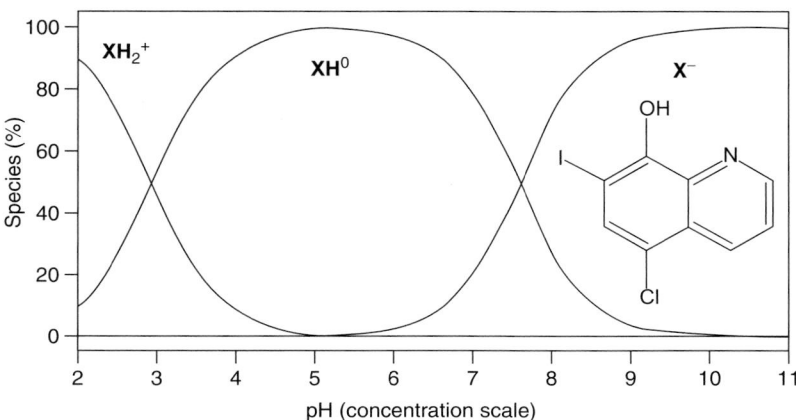

Figure 3 Distribution of species of clioquinol, an ampholyte with base $pK_a = 2.96$ and acid $pK_a = 7.60$. Clioquinol is unionized at intestinal pH.

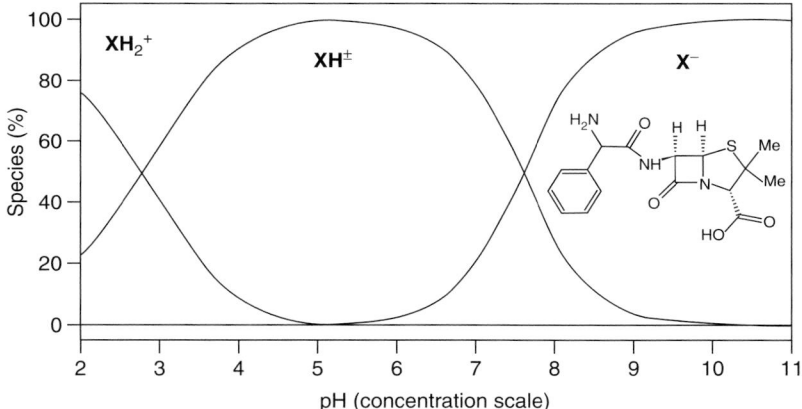

Figure 4 Distribution of species of ampicillin, a zwitterion with base $pK_a = 7.14$ and acid $pK_a = 2.55$. Ampicillin is ionized across the entire pH range.

$$[XH] = \frac{K_{a1}[H^+]}{[H^+]^2 + K_{a1}[H^+] + K_{a1}K_{a2}} \tag{37}$$

$$[XH_2^+] = \frac{[H^+]^2}{[H^+]^2 + K_{a1}[H^+] + K_{a1}K_{a2}} \tag{38}$$

Selected ampholytes and zwitterions are listed in **Table 2**. Diprotic molecules with one acidic and one basic pK_a that are far apart behave unequivocally as ampholytes or zwitterions. However, those whose pK_a values are separated by less than about three units may exhibit both zwitterionic and ampholytic properties. Note that the species XH in the equations above represents a composite of the neutral species XH^0 and the charged species $XH^\pm$, which can both exist at the same time. To properly describe the ionization of diprotic ampholytes and zwitterions with close pK_a values, a more detailed picture of the ionization equilibria is required. This must include microconstants, which are the pk_a values for the ionization of the microspecies (note the lower-case k in pk_a). This is illustrated in **Figure 5** for the ionization of a diprotic ampholyte or zwitterion. (Molecules with three or more pK_a values including at least one acidic pK_a that is lower than a basic pK_a may also be evaluated in terms of microconstants. Examples include tetracycline (discussed in Section 5.16.2.2.2) and cetirizine (discussed in Section 5.16.3.4).

Table 2 pK_a values of ampholytes and zwitterions. A or B written after each pK_a denotes whether the ionizable group is acidic or basic. The term A – B is the difference between the nearest acidic and basic pK_a values. Compounds for which A – B is less than 3 are shown in bold, as they are expected to exhibit both ampholytic or zwitterionic properties, depending on the solution conditions

Compound	Designation	pK_{a1}	pK_{a2}	pK_{a3}	A – B	Ref.
Sulfanilamide	Ampholyte	2.00 B	10.43 A		8.43	46
Nitrazepam	Ampholyte	3.02 B	10.37 A		7.35	46
Acyclovir	Ampholyte	2.27 B	9.25 A		6.98	46
Albendazole	Ampholyte	3.28 B	9.93 A		6.65	46
Carbendazim	Ampholyte	4.48 B	10.53 A		6.05	80
Luminol	Ampholyte	1.34 B	6.29 A		4.95	80
Clioquinol	Ampholyte	2.96 B	7.60 A		4.64	61
Famotidine	Ampholyte	6.74 B	11.19 A		4.45	46
Sulfacetamide	Ampholyte	1.76 B	5.52 A		3.76	46
5-Hydroxyquinoline	Ampholyte	5.22 B	8.54 A		3.32	46
Amodiaquine	Ampholyte	7.37 B	8.24 B	11.49 A	3.25	80
Piroxican	**Ampholyte**	**2.33. B**	**5.07 A**		**2.74**	46
Sildenafil	**Ampholyte**	**6.78 B**	**9.12 A**		**2.34**	32
Sotalol	**Ampholyte**	**8.28 B**	**9.72 A**		**1.44**	46
Sulpiride	**Ampholyte**	**9.00 B**	**10.19 A**		**1.19**	46
Morphine	**Ampholyte**	**8.18 B**	**9.26 A**		**1.08**	79
Serotonin	**Zwittterion**	**9.92 A**	**10.90 B**		**– 0.98**	46
Enrofloxacin	**Zwittterion**	**6.16 A**	**7.75 B**		**– 1.59**	46
Pefloxacin	**Zwittterion**	**6.03 A**	**7.80 B**		**– 1.77**	46
Labetalol	**Zwittterion**	**7.41 A**	**9.36 B**		**– 1.95**	46
Niflumic acid	**Zwittterion**	**2.24 A**	**4.44 B**		**– 2.2**	46
Norfloxacin	**Zwittterion**	**6.40 A**	**8.70 B**		**– 2.3**	46
Nicotinic acid	**Zwittterion**	**2.10 A**	**4.63 B**		**– 2.53**	46
Ciprofloxacin	**Zwittterion**	**6.16 A**	**8.62 B**		**– 2.46**	11
Ampicillin	Zwittterion	2.55 A	7.14 B		– 4.59	79
Cetirizine	Complex	2.12 B	2.90 A	7.98 B		38
Tetracycline	Complex	3.30 A	7.68 B	9.30 A		46

5.16.1.10 The Curious Behavior of Zwitterions

The ratio K_z between the concentrations of XH^0 and $XH^\pm$ changes with the solution conditions. **Figure 6** shows that the concentration of charged species of labetalol ($pK_a = 9.36$ (base), 7.41 (acid)) diminishes with increasing solvent content of aqueous solution.[10] This is because the base pK_a is lowered in the presence of solvent and the acid pK_a is raised. Together these changes make the molecule more ampholytic with increasing solvent. This observation could explain the relatively high bioavailability of some zwitterions, as the dielectric constant in water–solvent mixtures is much closer to the value at water–membrane interfaces (about 32) than to the value in water (about 78).

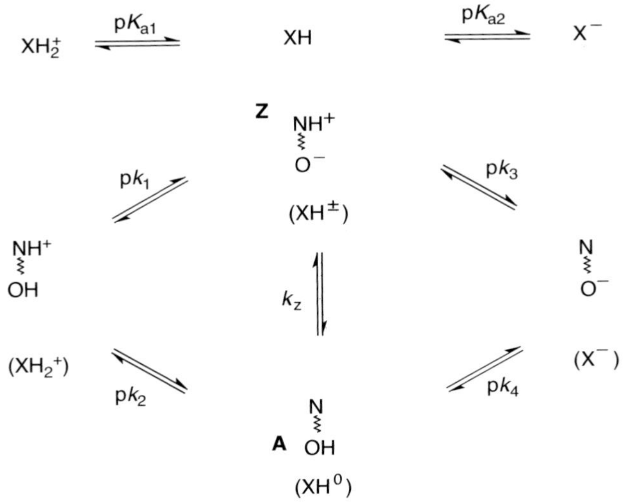

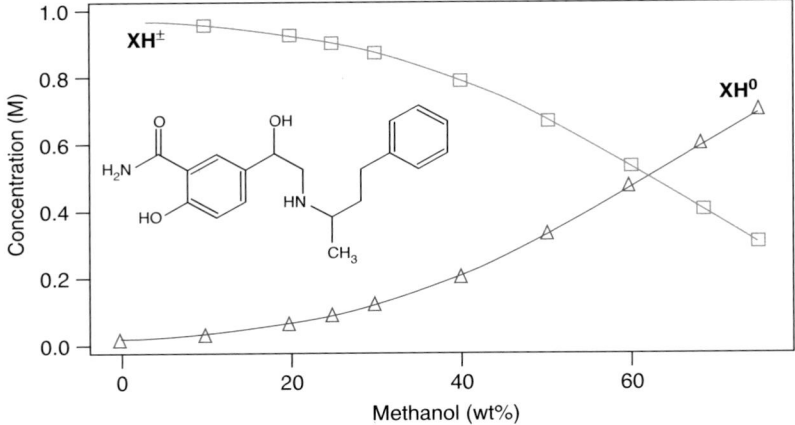

$$XH_2^+ \quad \xrightarrow{\text{p}K_{a1}} \quad XH \quad \xrightarrow{\text{p}K_{a2}} \quad X^-$$

Figure 5 Ionization of microspecies for ampholytes and zwitterions. If the acid pK_a is much lower than the base pK_a, the zwitterionic microspecies Z will predominate. If the base pK_a is much lower than the acid pK_a, the ampholytic microspecies A will predominate. If the pK_a values are close together, A and Z will coexist. The macrospecies XH is a mixture of the Z and A microspecies.

Figure 6 Zwitterion and neutral distribution of species for labetalol. The zwitterionic species predominates in water. In the presence of methanol, the acid pK_a rises and the base pK_a falls, causing labetalol to convert to the ampholytic species with increasing concentration of methanol.

Some properties of zwitterions are normally measured in terms of the composite species XH. However, the form XH^0 is much more lipophilic than $XH^{\pm}$. A value of log P(XH) for labetalol has been measured at 1.48, from which a log P(XH^0) of 2.70 has been calculated using equations derived from microconstants.[10]

Ciprofloxacin is a zwitterion with an acidic pK_a of 6.16 and a basic pK_a of 8.62.[11] Considering only its charge, the zwitterionic species $XH^{\pm}$ might be expected to be more soluble than XH_2^+ or X^-. However, ciprofloxacin can precipitate from solutions whose pH lies between its pK_a values. This clearly occurs in pH-metric titrations to measure solubility of ciprofloxacin, whose solubility–pH profile is shown in **Figure 15**. Presumably it is the ampholyte form XH^0 that precipitates, suggesting that once a precipitate has formed a few crystals, the precipitated state is energetically more favored than the dissolved zwitterionic state.

5.16.2 Ionizable Groups

This section reviews the most common ionizable groups, and the structural factors that influence pK_a values.

5.16.2.1   The Number of Ionizable Drugs

To evaluate their ionization properties, compounds can be divided into several classes. Some compounds are neutral (nonionizable); they will not be discussed in this chapter. Most ionizable compounds are acids or bases with one, two or more ionizable groups. Another class includes ampholytes and zwitterions, which are molecules with at least one acidic and one basic group. Finally, compounds such as quaternary ammonium salts contain species that are always charged; they rarely make successful drugs, so will not be discussed in this chapter. A study published in 1988 suggested that 75% of drugs at the time were bases, 20% were acids, and only 5% were nonionizable.[12] A search of the World Drug Index for 1999 suggested that 63% of the 51 600 listed drugs were ionizable, of which 67% were bases, 15% were acids, and 18% were ampholytes.[10] While it is difficult to ascertain the reliability of either set of figures, it is clear that most drug molecules are ionizable, and that most ionizable drugs are bases.

5.16.2.2   The Most Common Ionizable Groups

Ionization in drugs occurs at well-recognized ionizable groups. **Figures 7–13** give the structures of molecules containing the most common ionizable groups that are found in drugs. Some structures are simple organic molecules; others are drugs.

5.16.2.2.1   Carboxylic acids

The carboxylic acid group is the most common type of acidic group found in drugs. Their pK_a values typically occur between 3 and 5 for monobasic acids, unless other structural features have a significant effect on the ionization. Of the monobasic carboxylic acids in **Figure 7**, two have pK_a values outside this range. The pK_a of trifluoroacetic acid is very low because the CF_3 group is strongly electron-withdrawing; this property tends to lower the pK_a values of both acidic and basic groups. Other properties that modify pK_a values by electrostatic and inductive effects are listed in **Table 3**. The pK_a of flumequine is higher than typical carboxylic acids (6.27) because the proton is stabilized by hydrogen bonding to the adjacent $=O$ group, and is held on the molecule until relatively high pH. Carboxylic acids are larger than hydroxyl or primary amino groups, and are therefore more susceptible to steric interactions of this type.

The pK_a values of some dicarboxylic acids are shown in **Figure 7**. For diprotic acids, it is the first pK_a (that which describes the transition between the neutral molecule and the species with a charge of -1) that has the major influence on pK_a-dependent properties of pharmacokinetic interest such as lipophilicity and solubility.

Molecules with two identical ionizable groups tend to have two well-separated pK_a values. This can be illustrated by the ionization of fumaric acid, a dibasic carboxylic acid:

$$HOOCCH = CHCOOH \rightleftharpoons HOOCCH = CHCOO^- + H^+ \qquad [39]$$

$$HOOCCH = CHCOO^- \rightleftharpoons {}^-OOCCH = CHCOO^- + H^+ \qquad [40]$$

In the first stage of ionization, the equilibrium shown in eqn [39] can lose a proton from two positions, but it can add a proton from only one position. In the second stage of ionization shown in eqn [40], it can lose a proton from only one position but it can add to two positions. Thus equilibrium [39] is therefore statistically favored over equilibrium [40], with the result that the pK_a is lower for this ionization. The pK_a difference is further enhanced by electrostatic effects, because the negatively charged species in eqn [39] tends to repel the negatively charged hydroxide ions that are required to raise the pH for the second ionization.

The pK_a values of maleic acid (*cis*-isomer) are more widely spaced than those of fumaric acid (*trans*-isomer) because the proton involved in the higher pK_a is stabilized by hydrogen bonding; for steric reasons, this cannot occur in fumaric acid, whose pK_a values are therefore much closer together.

The carboxylate pK_a values of olsalazine are lower than expected, while the phenol pK_a values are higher. This may be because after the loss of the protons, stable hydrogen-bonded structures form between the phenol –OH groups and the carboxylate $=O$, which also serve to make it more difficult to remove the protons from the phenols.

Sometimes, large symmetrical molecules with identical ionizable groups that are well separated lose two protons simultaneously. The author has observed several instances of this behavior, but unfortunately not for any whose structures are in the public domain. However, this behavior has been reported for a dye, *meso*-tetra (3-hydroxyphenyl)porphine (see **Figure 13**), which appears to lose two protons at a common pK_a.[13]

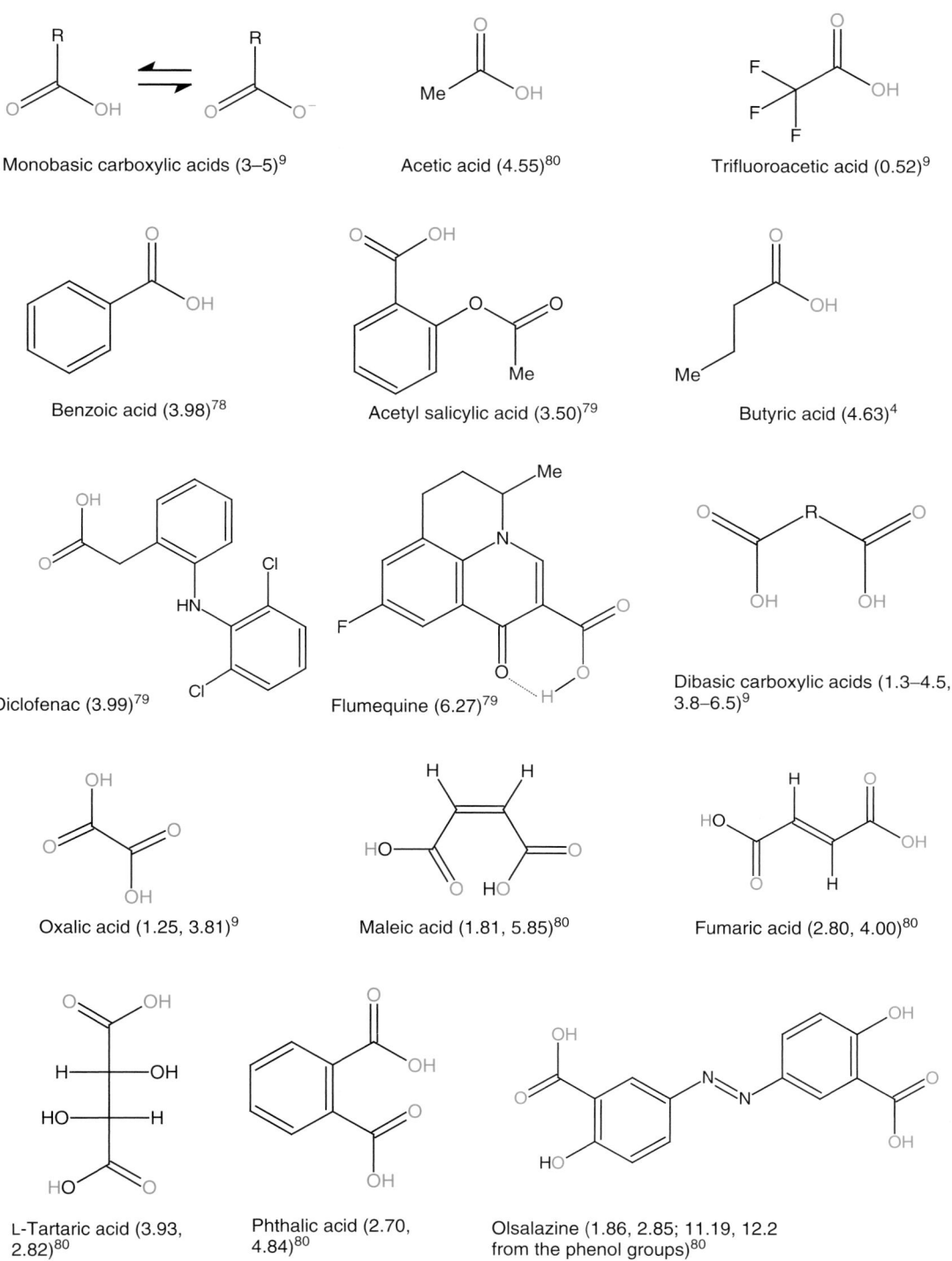

Monobasic carboxylic acids (3–5)[9]

Acetic acid (4.55)[80]

Trifluoroacetic acid (0.52)[9]

Benzoic acid (3.98)[78]

Acetyl salicylic acid (3.50)[79]

Butyric acid (4.63)[4]

Diclofenac (3.99)[79]

Flumequine (6.27)[79]

Dibasic carboxylic acids (1.3–4.5, 3.8–6.5)[9]

Oxalic acid (1.25, 3.81)[9]

Maleic acid (1.81, 5.85)[80]

Fumaric acid (2.80, 4.00)[80]

L-Tartaric acid (3.93, 2.82)[80]

Phthalic acid (2.70, 4.84)[80]

Olsalazine (1.86, 2.85; 11.19, 12.2 from the phenol groups)[80]

Figure 7 Structures of ionizable molecules: carboxylic acids. Here and in **Figures 8–13**, red atoms denote acidic groups, and blue atoms indicate basic groups (pK_a values are given in parentheses).

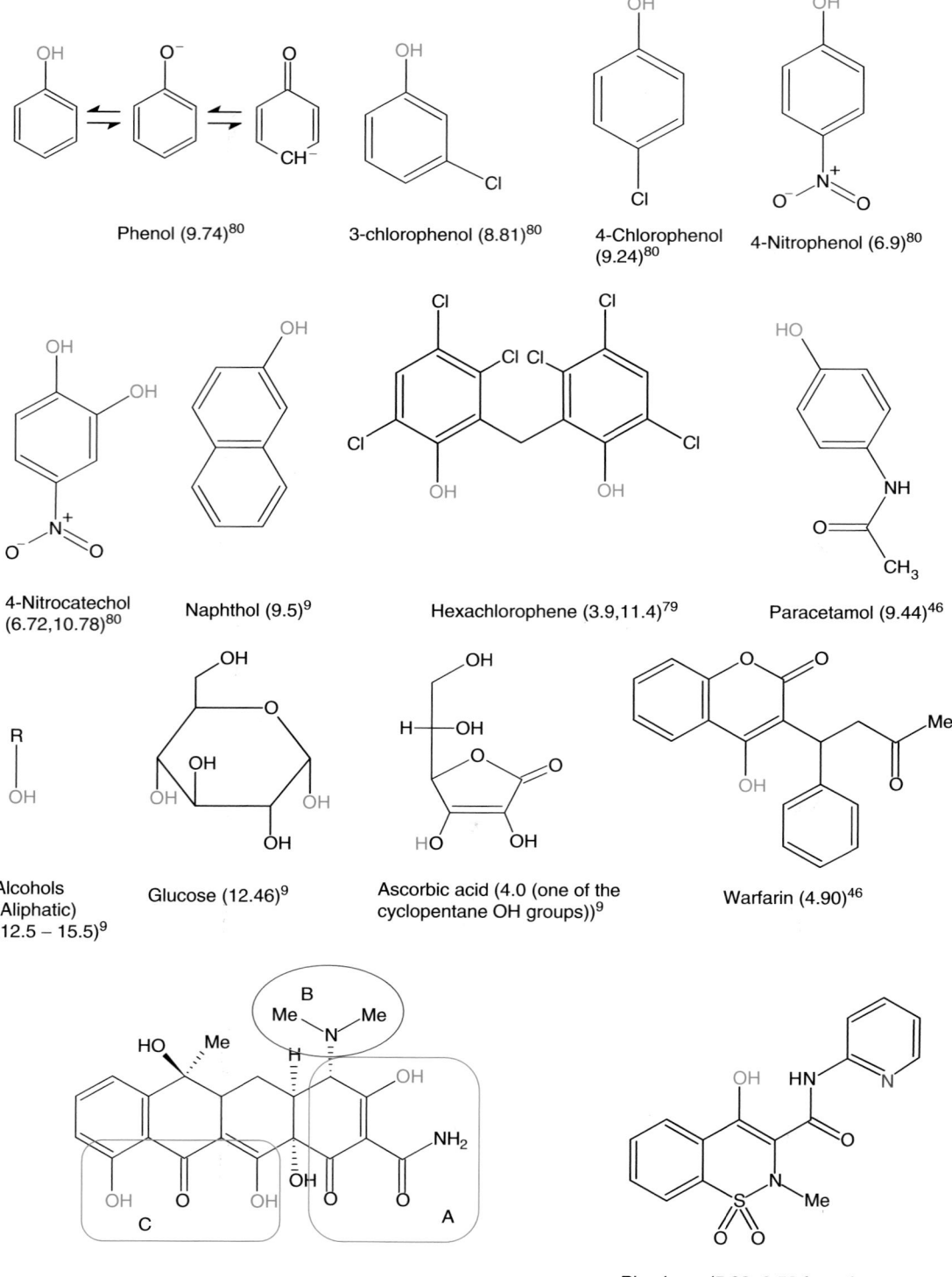

Figure 8 Structures of ionizable molecules: phenols, alcohols, and enols.

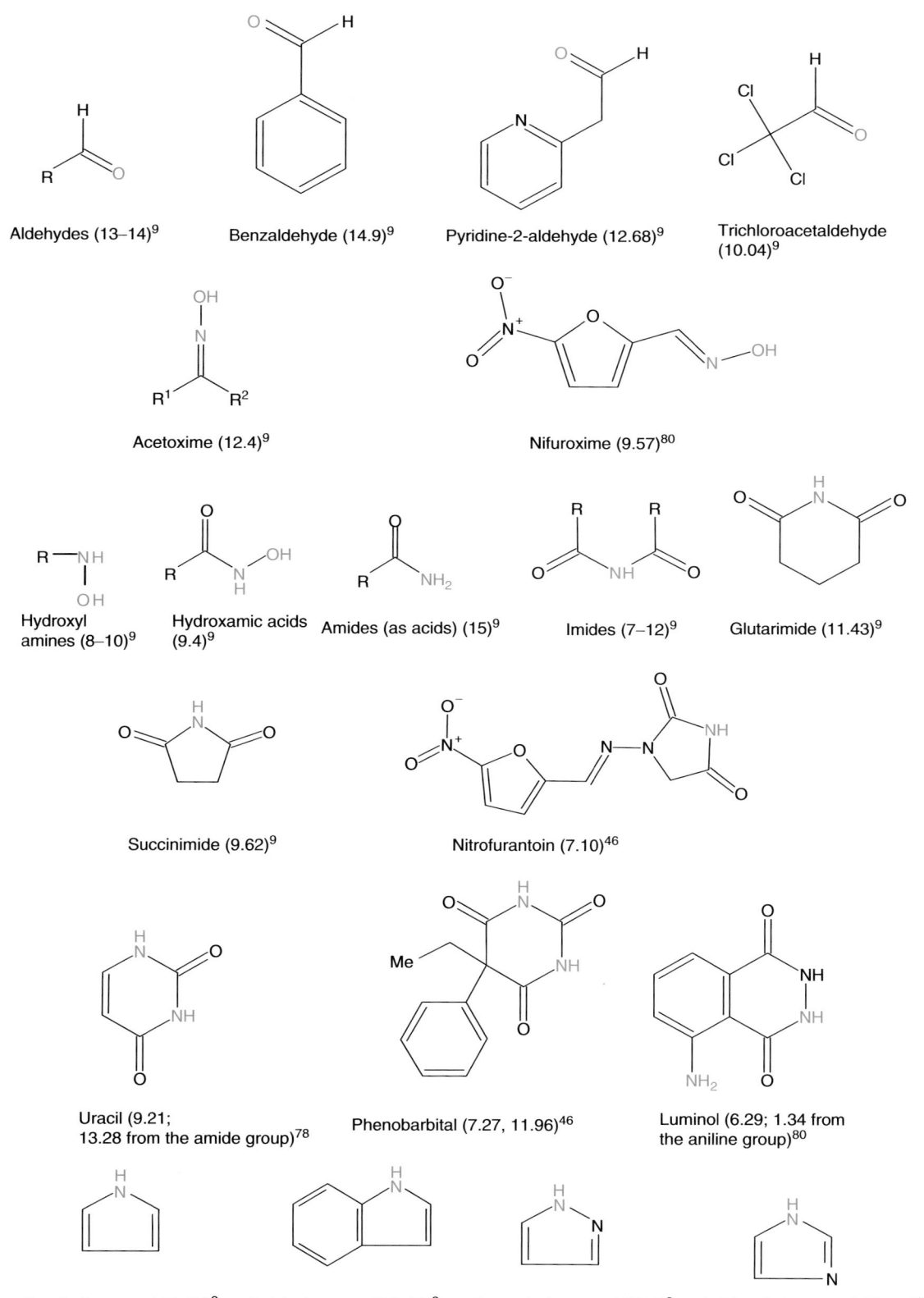

Aldehydes (13–14)[9]   Benzaldehyde (14.9)[9]   Pyridine-2-aldehyde (12.68)[9]   Trichloroacetaldehyde (10.04)[9]

Acetoxime (12.4)[9]   Nifuroxime (9.57)[80]

Hydroxyl amines (8–10)[9]   Hydroxamic acids (9.4)[9]   Amides (as acids) (15)[9]   Imides (7–12)[9]   Glutarimide (11.43)[9]

Succinimide (9.62)[9]   Nitrofurantoin (7.10)[46]

Uracil (9.21; 13.28 from the amide group)[78]   Phenobarbital (7.27, 11.96)[46]   Luminol (6.29; 1.34 from the aniline group)[80]

Pyrrole (as an acid)(>15)[9]   Indole (as an acid)(>12)[9]   Pyrazole (as an acid)(14)[9]   Imidazole (as an acid)(14.4)[9]

Figure 9 Structures of ionizable molecules: other acidic groups.

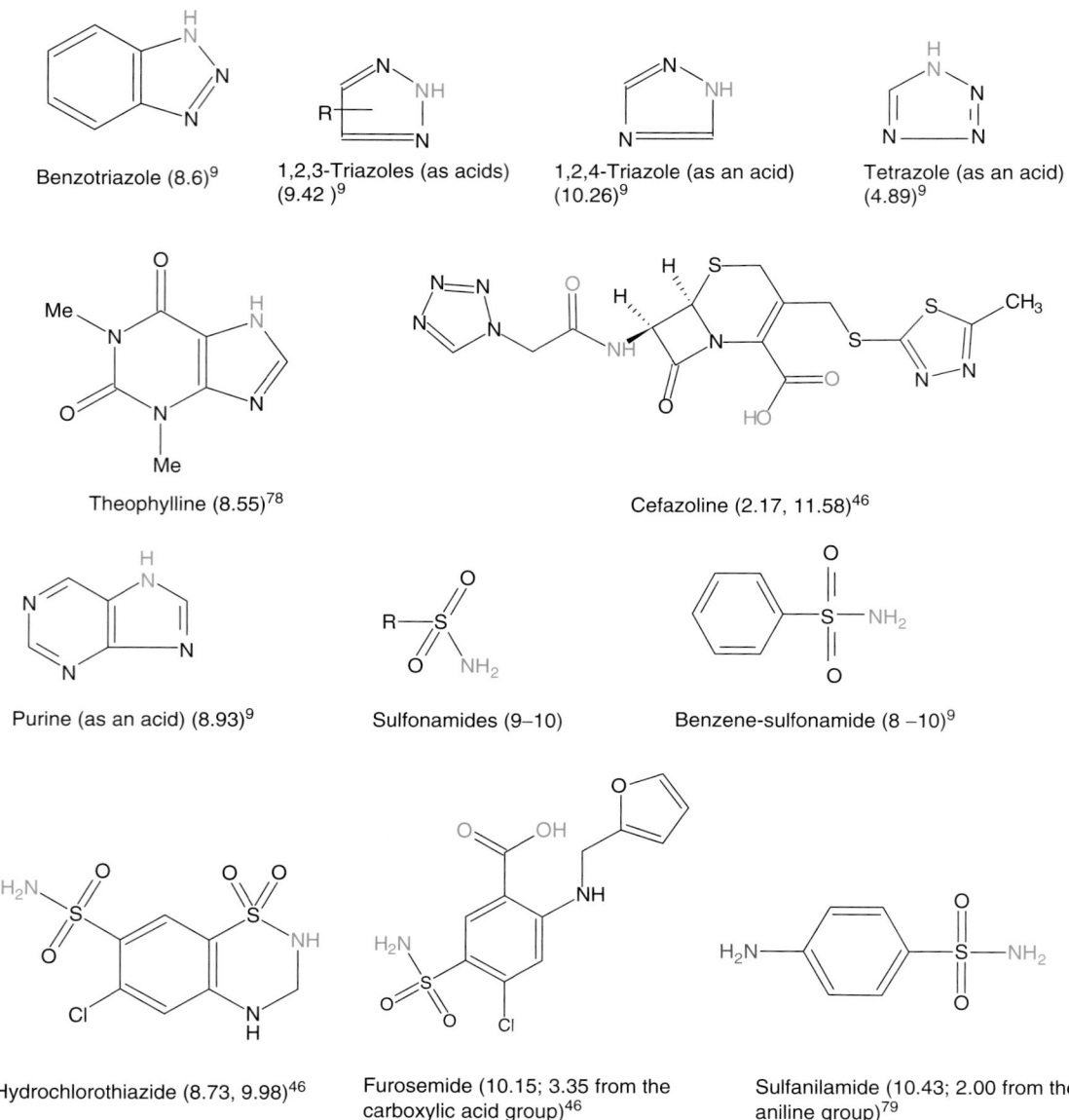

Benzotriazole (8.6)[9]

1,2,3-Triazoles (as acids) (9.42)[9]

1,2,4-Triazole (as an acid) (10.26)[9]

Tetrazole (as an acid) (4.89)[9]

Theophylline (8.55)[78]

Cefazoline (2.17, 11.58)[46]

Purine (as an acid) (8.93)[9]

Sulfonamides (9–10)

Benzene-sulfonamide (8 –10)[9]

Hydrochlorothiazide (8.73, 9.98)[46]

Furosemide (10.15; 3.35 from the carboxylic acid group)[46]

Sulfanilamide (10.43; 2.00 from the aniline group)[79]

Figure 9 Continued

5.16.2.2.2 Phenols, alcohols, and other molecules with –OH groups

The –OH group that occurs in phenols and alcohols is the other common acidic ionizable group found in drugs. While the pK_a of phenol itself is 9.74, the pK_a values of substituted and heterocyclic phenols are typically between 7 and 10.5 (see **Figure 8**). In contrast to phenols, the pK_a values of aliphatic alcohols are greater than 13 unless lowered by the effect of substituents. The pK_a of phenol is so much lower than that of alcohols because of the electron-withdrawing effect of the benzene ring coupled with the negative mesomeric effect in the anion. The phenol anion exists in equilibrium between two forms, as shown in **Figure 8**. The pK_a of 4-nitrophenol (6.9) is much lower than that of phenol because of the combined electron-withdrawing and mesomeric effects of the nitro group in the *para* position. A similar lowering of pK_a occurs in 4-nitrocatechol. The pK_a of 4-chlorophenol (9.24) is closer to that of phenol than the pK_a of 3-chlorophenol (8.81), because in 4-chlorophenol a positive mesomeric effect via the *para* substitution opposes the electron-withdrawing tendency of the Cl atom.

Hexachlorophene has two phenol groups, both with unusual pK_a values. With six Cl substituents, the molecule is strongly electron-withdrawing, leading to the very low pK_a (3.9) for the transition between neutral and ionized. Once the

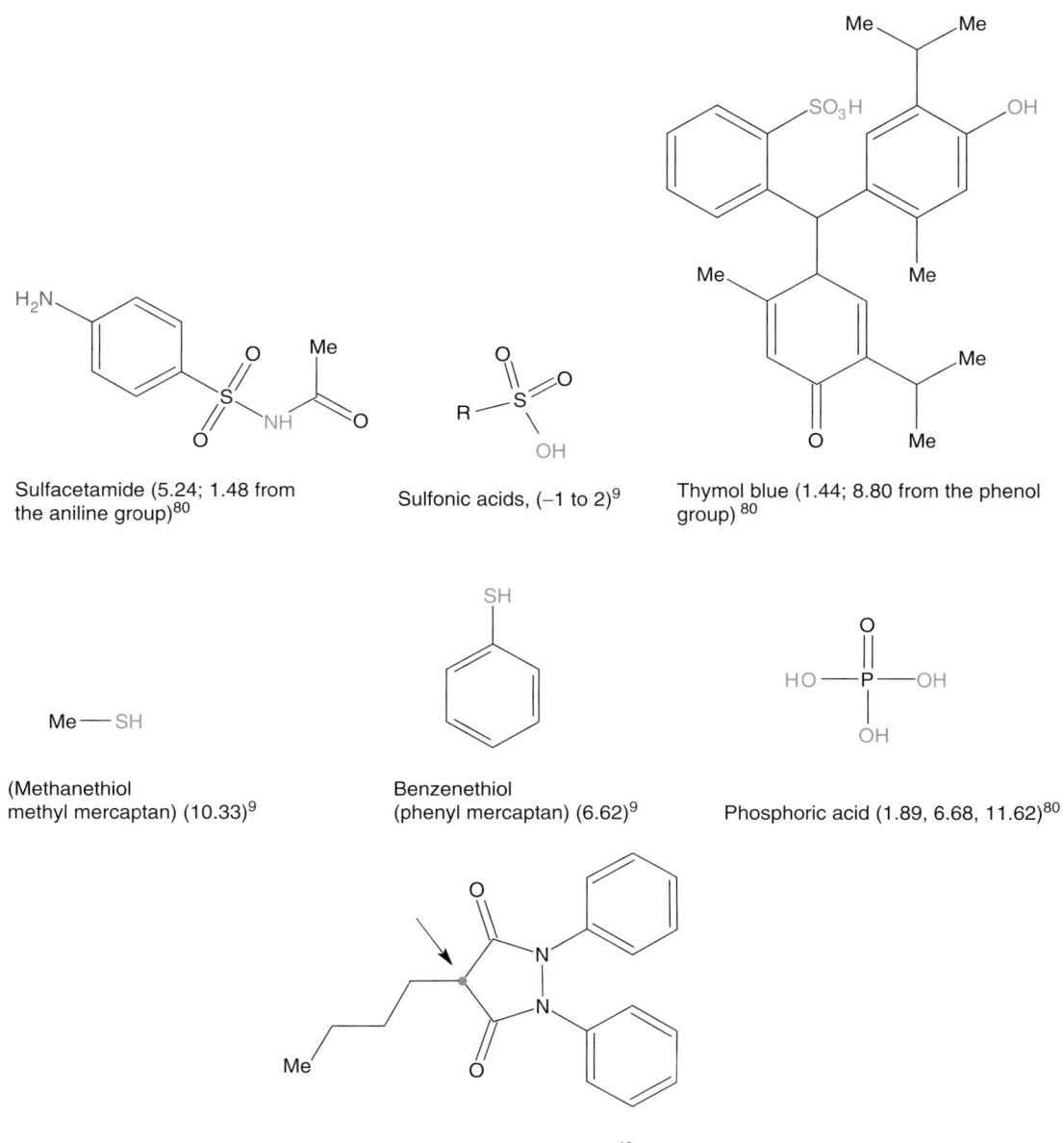

Sulfacetamide (5.24; 1.48 from the aniline group)[80]

Sulfonic acids, (−1 to 2)[9]

Thymol blue (1.44; 8.80 from the phenol group) [80]

Me—SH

(Methanethiol methyl mercaptan) (10.33)[9]

Benzenethiol (phenyl mercaptan) (6.62)[9]

Phosphoric acid (1.89, 6.68, 11.62)[80]

Phenylbutazone (4.23)[46]

Figure 9 Continued

first proton has been lost, however, it is likely that the second phenol forms a stable hydrogen bond with the first phenol in its ionized keto form, and the molecule loses the second proton only at very high pH, as denoted by the pK_a of 11.4.

While aliphatic alcohol pK_a values are very high, the pK_a values of −OH groups can be lower as a result of substitution. The −OH groups that result from keto–enol tautomerism usually have quite low pK_a values, as seen in warfarin (4.9), piroxicam (5.3), and ascorbic acid (4.0).

Tetracycline has three pK_a values, and it is known that two are acidic and one is basic. While the basic amine is easily identified, the acidic pK_a values are difficult to assign by casual inspection of the structure. Stevens *et al.* in 1956 measured pK_a values of tetracycline, substituted tetracyclines, and substructures of tetracycline, and deduced that the pK_a values could be assigned as shown in **Figure 8**, where A and C are acidic functions, and B is a base.[14] Nuclear magnetic resonance (NMR) spectroscopy has since been used to determine a protonation scheme for tetracycline, showing the possibility of eight microspecies, and determining microconstants for each equilibrium.[15]

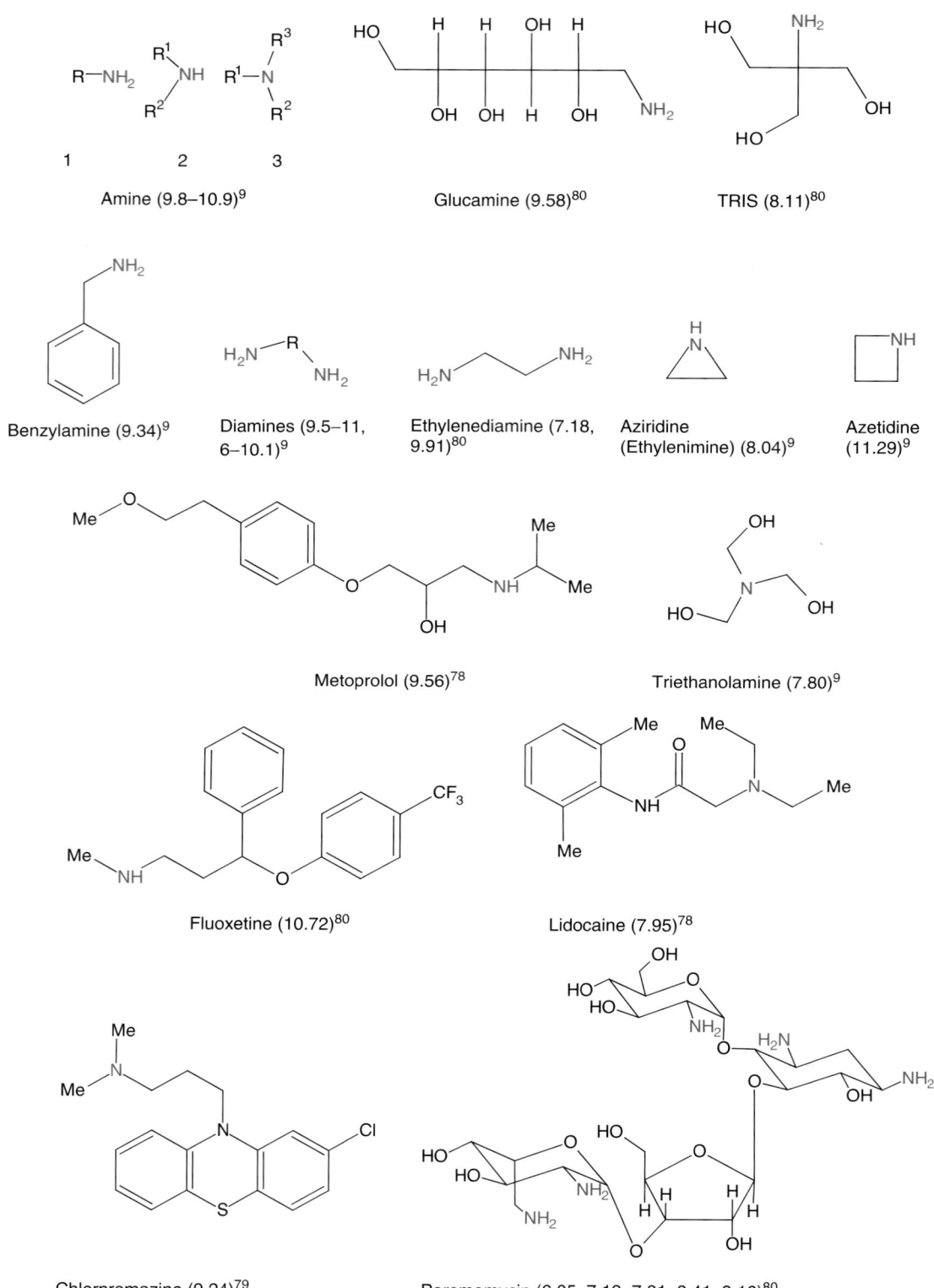

Figure 10 Structures of ionizable molecules: aliphatic and alicyclic amines.

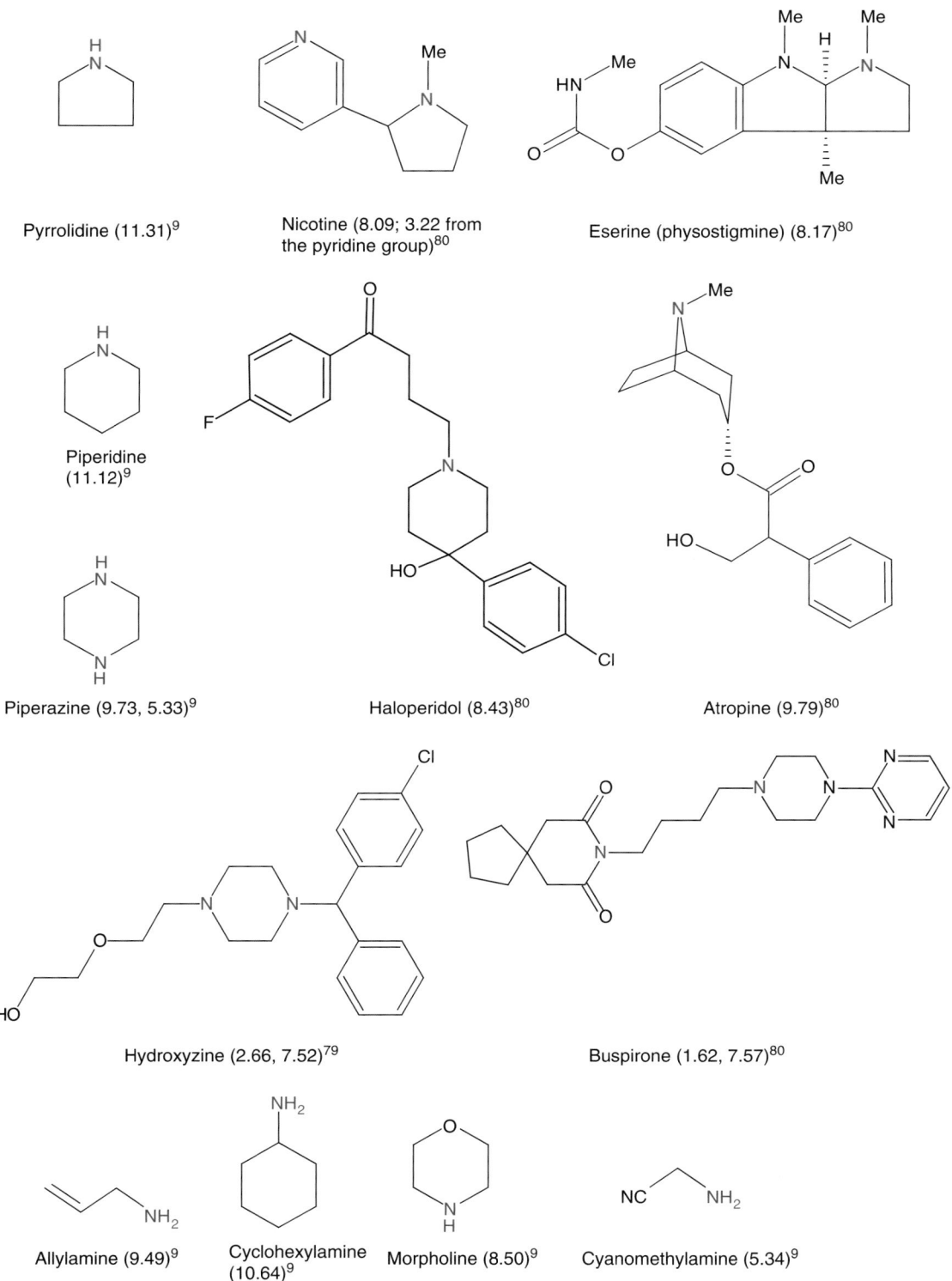

Pyrrolidine (11.31)[9]

Nicotine (8.09; 3.22 from the pyridine group)[80]

Eserine (physostigmine) (8.17)[80]

Piperidine (11.12)[9]

Piperazine (9.73, 5.33)[9]

Haloperidol (8.43)[80]

Atropine (9.79)[80]

Hydroxyzine (2.66, 7.52)[79]

Buspirone (1.62, 7.57)[80]

Allylamine (9.49)[9]

Cyclohexylamine (10.64)[9]

Morpholine (8.50)[9]

Cyanomethylamine (5.34)[9]

Figure 10   Continued

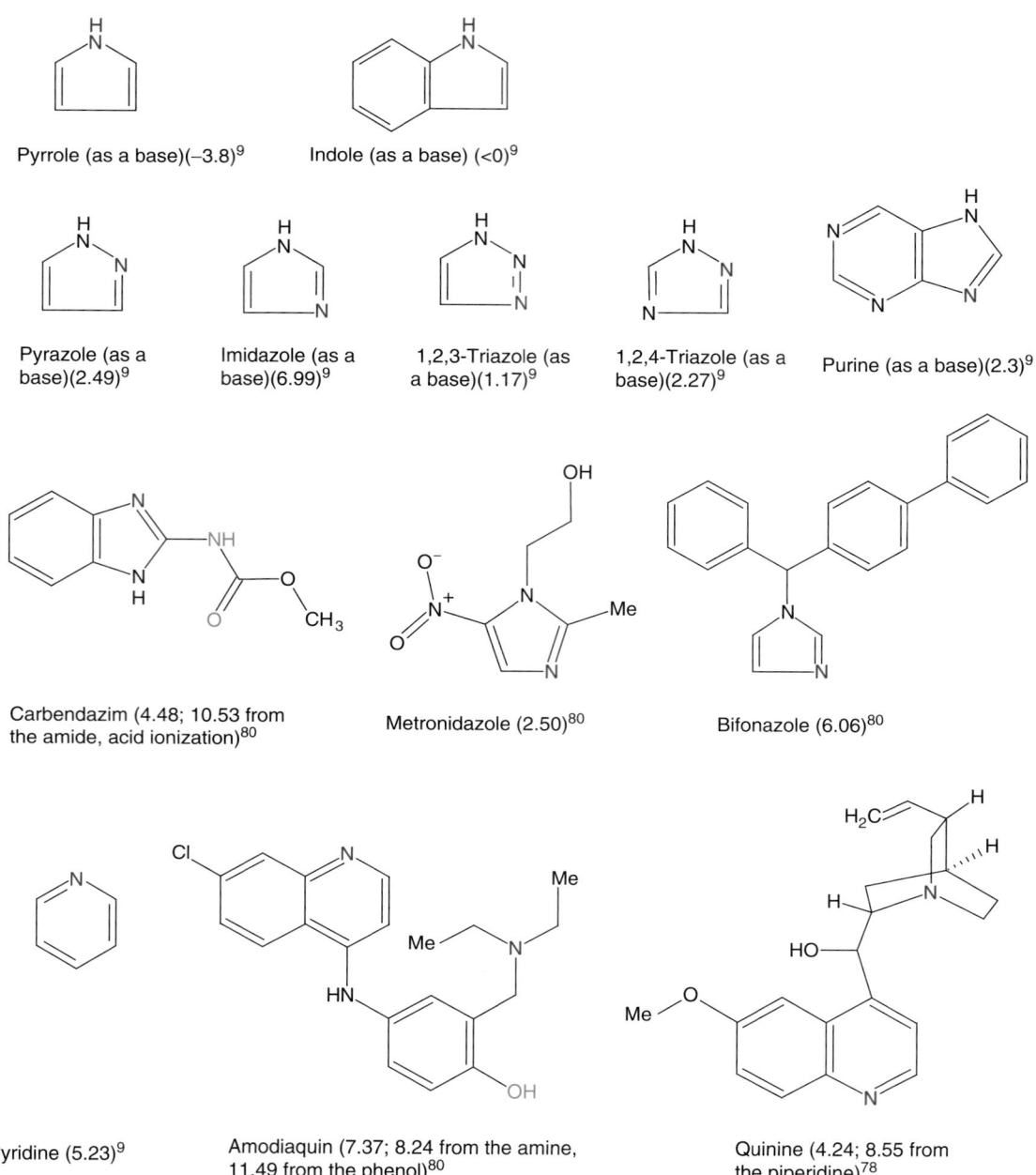

Figure 11 Structures of ionizable molecules: aromatic and heteroaromatic bases.

5.16.2.2.3 Other acidic ionizable groups

The structures shown in **Figure 9** illustrate aldehydes and other acidic ionizable groups, including molecules in which the dissociable proton is attached to nitrogen, sulfur, and phosphorus-containing groups. The pK_a values of aldehydes are generally above 13, but substitution of electron-withdrawing groups can lower them, as shown by the pK_a of trichloroacetaldehyde (10.1). There are many acidic drugs in which the dissociable proton is attached to a nitrogen atom, including oximes, hydroxylamines, sulfonamides, and rings containing N atoms. Some acidic nitrogen-containing groups can also be bases at different parts of the pH scale. For example, purine is an acid with a pK_a of 9 and a base with a pK_a of 2.3; and triazole is an acid with a pK_a of 10.3 and base with a pK_a of 2.3. Phenylbutazone is a carbon acid. Its pK_a of 4.2 corresponds with the loss of a proton from the carbon atom between the two $=O$ groups.

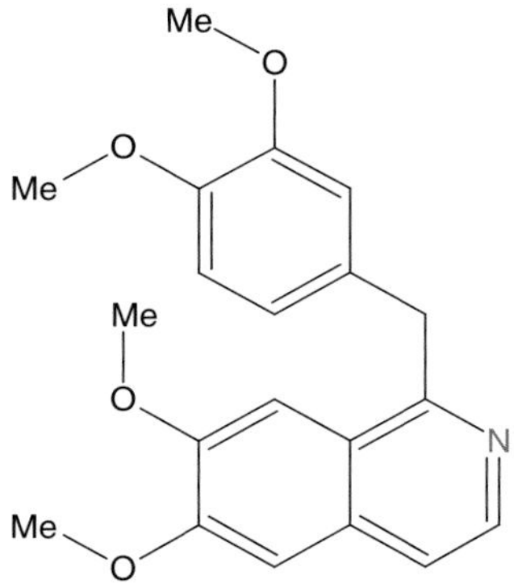

Papaverine (6.45)[46]

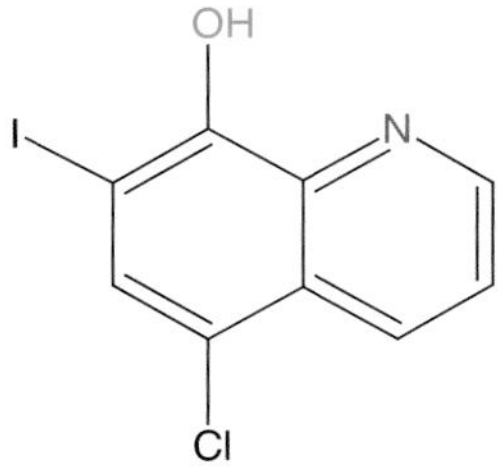

Clioquinol (2.81; 7.49 from the phenol)[80]

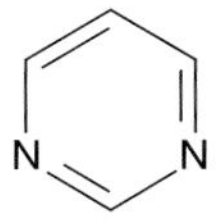

Pyrimidine (1.23)[9]

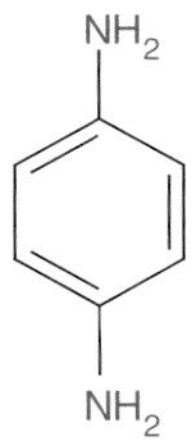

1,3-Diamino-benzene (5.11, 2.5)[9]

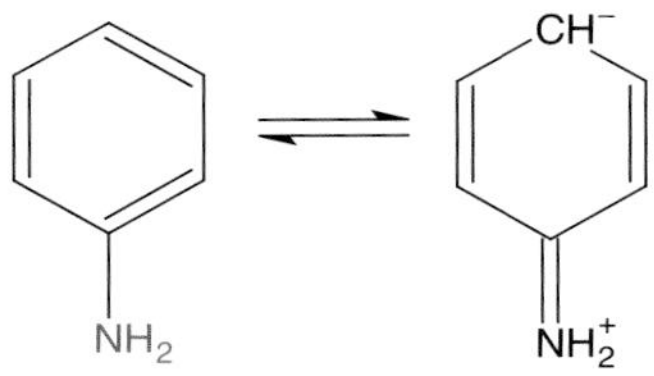

Aniline (4.87)[9]

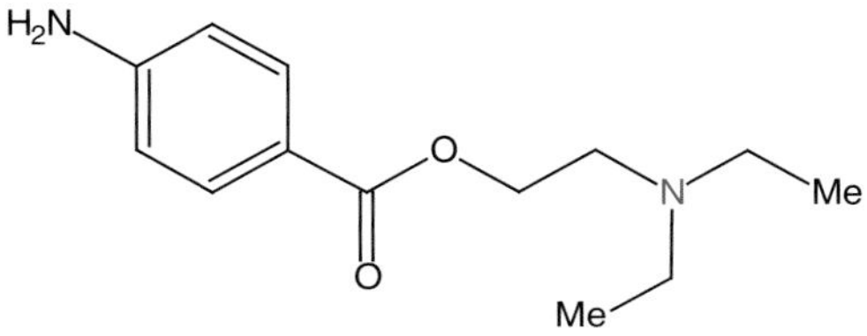

Procaine (2.29; 9.04 from the amine)[78]

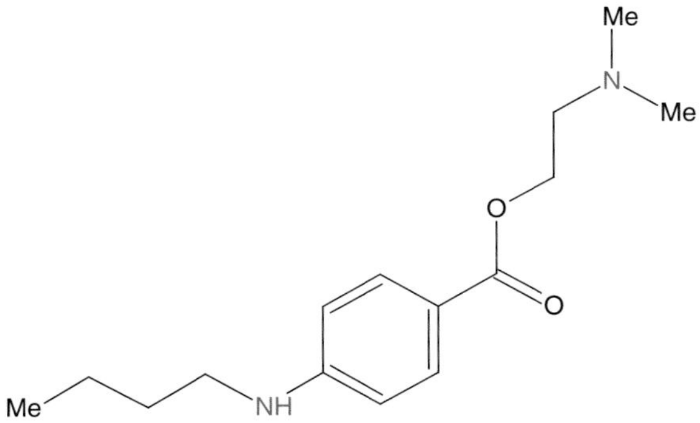

Tetracaine (2.39; 8.49 from the amine group)[78]

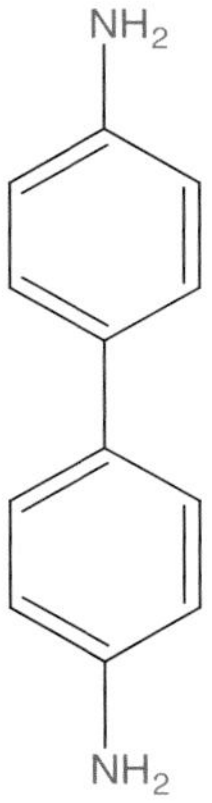

Benzidine (4.65, 3.43)[9]

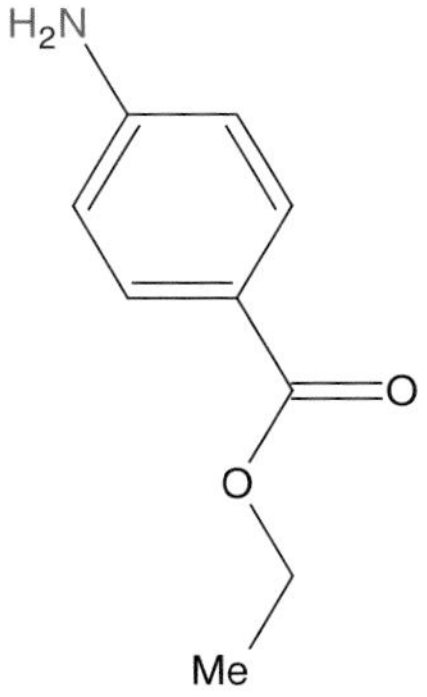

Benzocaine (2.39)[78]

Figure 11 Continued

5.16.2.2.4 Aliphatic bases

This class includes aliphatic amines or saturated nitrogen-containing rings such as piperazine, pyrollidine, or morpholine (see **Figure 10**). Most aliphatic bases are quite strong, typically with pK_a values of 9 and above. When incorporated into drugs, steps to reduce the pK_a values using electron-withdrawing substituents may be required, otherwise the molecules will be fully ionized at the pH of body tissues, and thus will be poorly permeable by passive diffusion.[16]

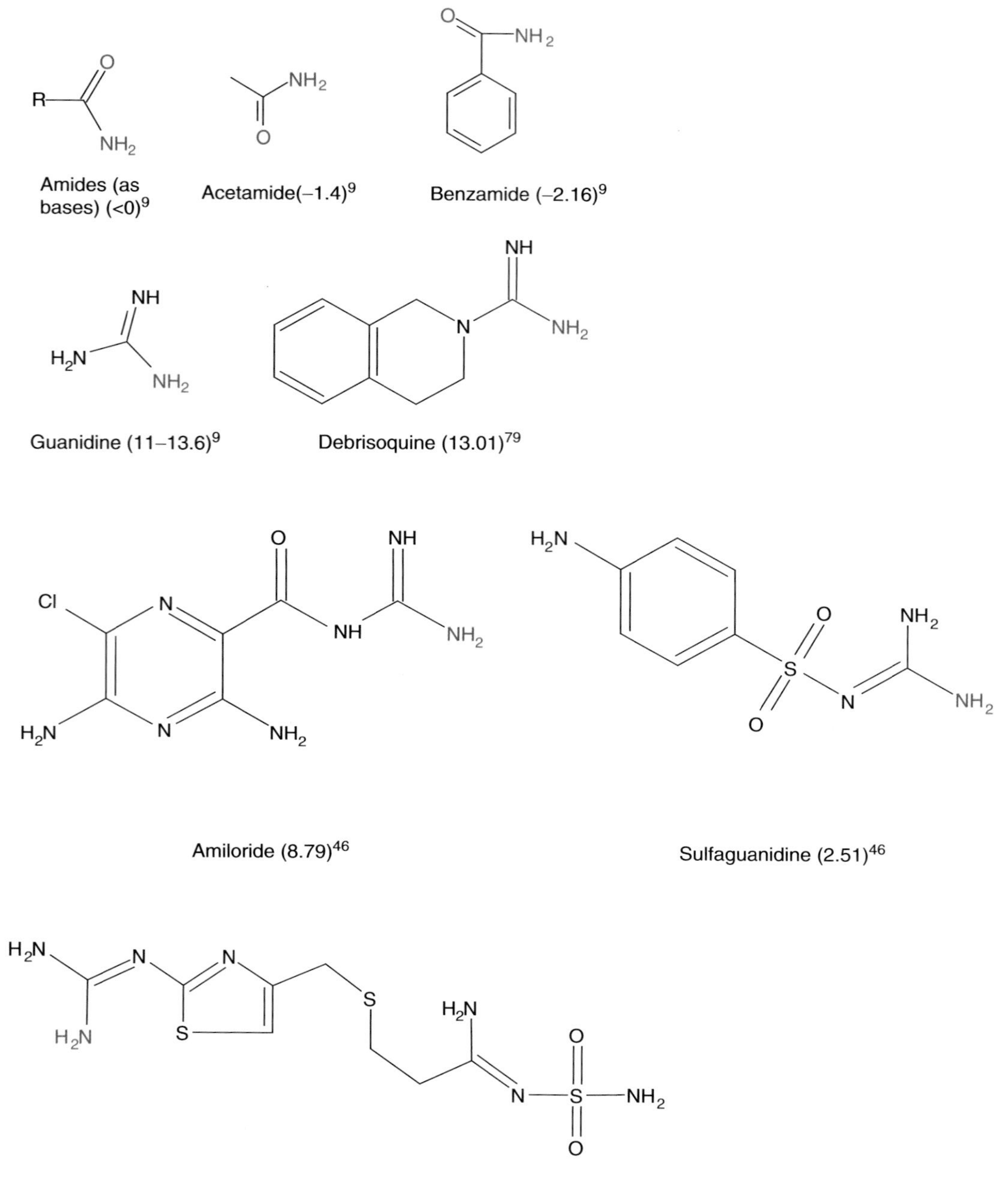

Amides (as bases) (<0)[9]   Acetamide(−1.4)[9]   Benzamide (−2.16)[9]

Guanidine (11–13.6)[9]   Debrisoquine (13.01)[79]

Amiloride (8.79)[46]   Sulfaguanidine (2.51)[46]

Famotidine (6.84; 11.32 from the sulfanilamide group)[46]

Figure 12 Structures of ionizable molecules: amides and guanidines.

5.16.2.2.5 Aromatic and heteroaromatic bases

The pK_a values of aromatic nitrogen bases are generally much lower than those of aliphatic bases (see **Figure 11**). When incorporated in drugs, they are often unionized at body pH, causing molecules to be poorly water-soluble. Aniline (4.87) is a much weaker base than its saturated analog cyclohexylamine (10.64) because of resonance in the neutral molecule that is not possible in the ion, as shown in **Figure 11**. Additional base weakening occurs in aniline because of the electron-withdrawing effect of the benzene ring, which adds to the effect of the resonance. The pK_a values of anilines that are substituted in the *ortho* position are generally higher than expected because their bulk twists the amino group out of the plane of the benzene ring, thus interfering with the resonance. *Meta*-substituted anilines show similar

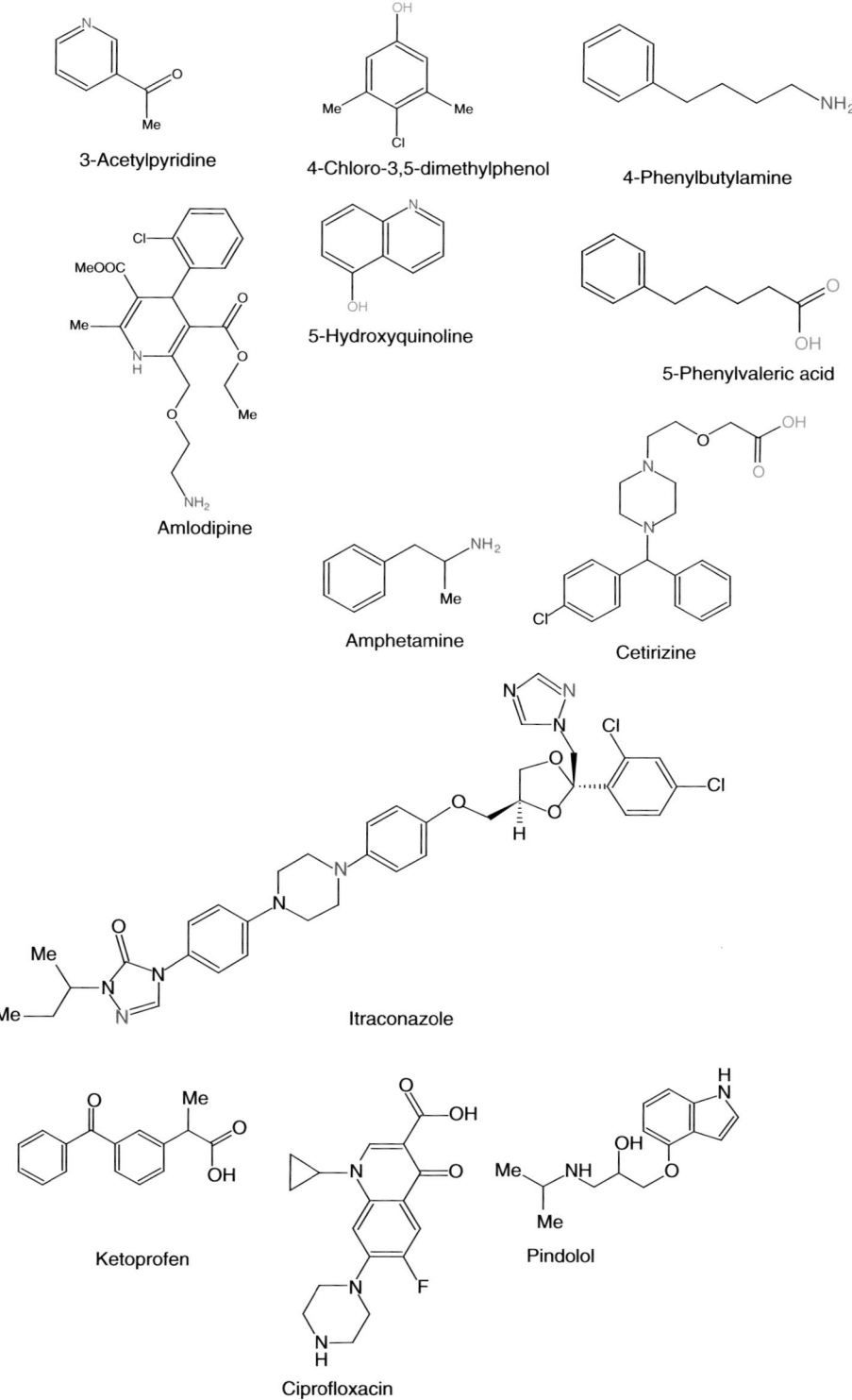

Figure 13 Structures of other ionizable molecules discussed in the text.

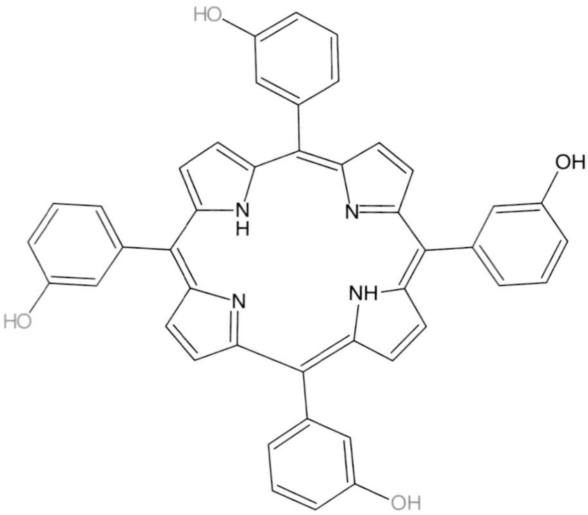

meso-Tetra(3-hydroxyphenyl)porphine

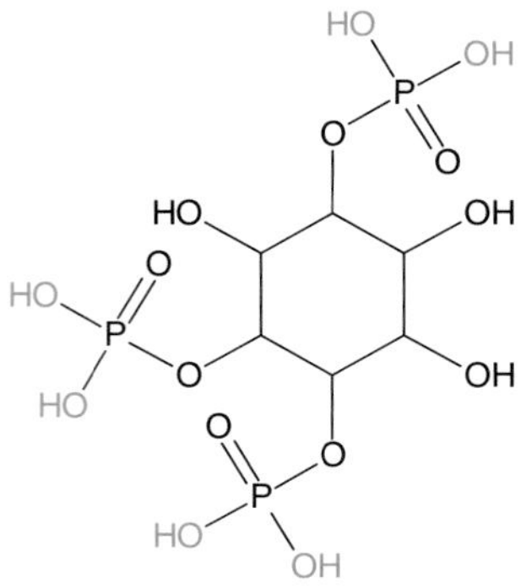

Myoinositol 1,4,5-triphosphate

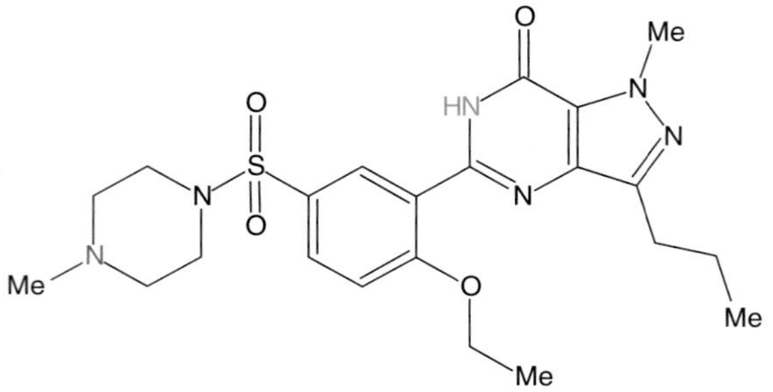

Sildenafil

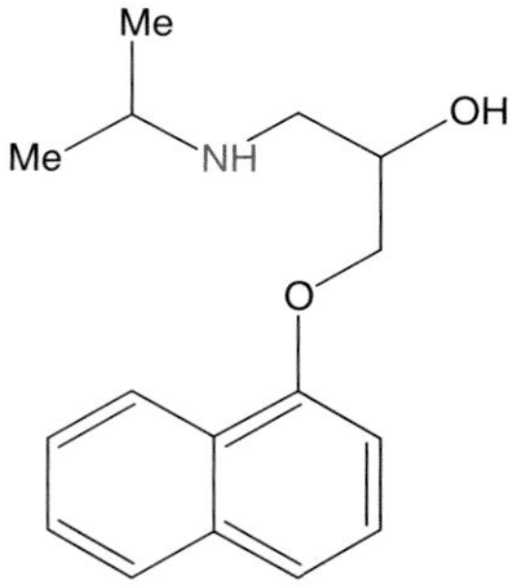

Propranolol

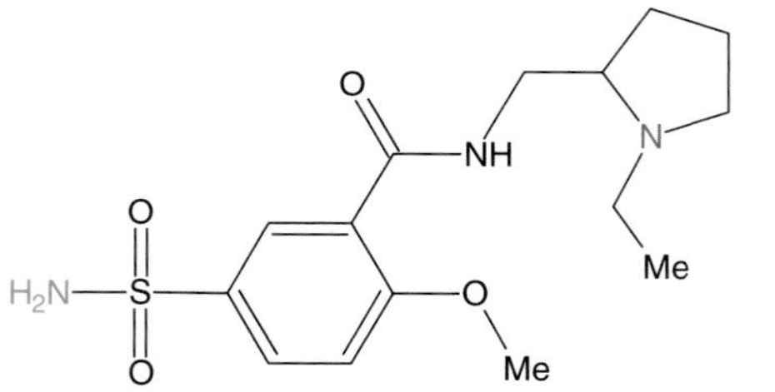

Sulpiride

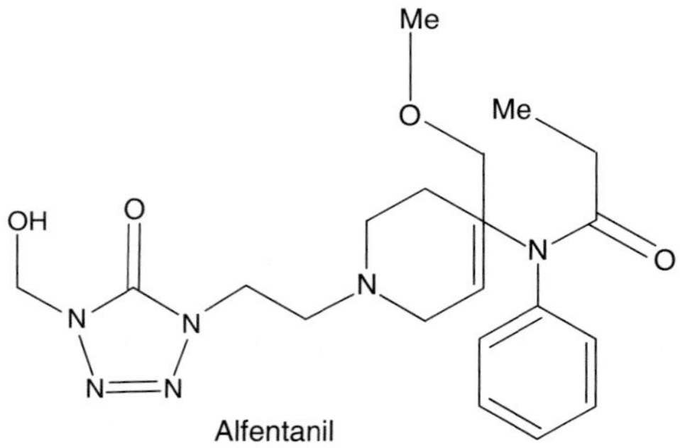

Alfentanil

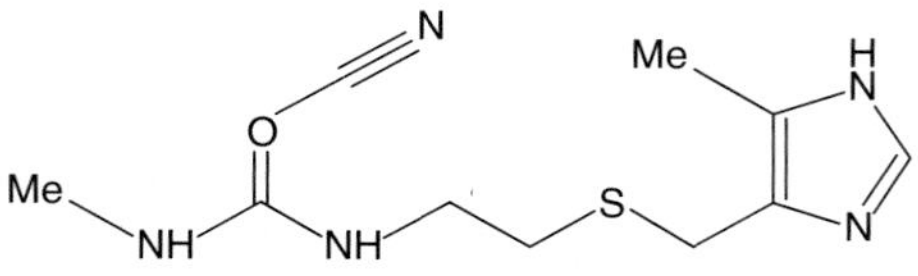

Cimetidine

Figure 13   Continued

Table 3 Structural factors that modify pK_a values

Name of effect	Symbol	What happens	Groups causing the effect
Negative inductive effect, caused by electron-withdrawing groups	− I	Lowers the pK_a values of both acids and bases	R_3N^+, NO_2, SO_2R, CN, F, Cl, Br, I, CF_3, COX (X = OH, NH_2, OR), COR, OR, SR, NH_2, C_6H_5
Positive inductive effect, caused by electron-donating groups	+ I	Raises the pK_a values of both acids and bases	CO_2^-, O^-, NH^-, alkyl
Negative mesomeric effect	− M	Lowers the pK_a values of both acids and bases	NO_2, CN, COR, SO_2R
Positive mesomeric effect	+ M	Raises the pK_a values of both acids and bases	F, Cl, Br, I, OCH_3 OH, NR_2, NH, COR, O^-, alkyl, aryl

R = alkyl.

inductive effects to those seen with carboxylic acids, which are reinforced by mesomeric effects in *para*-substituted anilines. The effects of substituents in pyridines is similar to the effect in anilines. In contrast to aniline, resonance in the cationic species occurs in the five-membered imidazole ring (6.95), which is a much stronger base than pyrrole (<0).

5.16.2.2.6 Amides and guanidines
Amides commonly occur in drugs, and have both basic and acidic possibilities (see **Figure 12**). However, their basic pK_a values are very low, and their acidic pK_a values are very high, and, unless affected by substitution, they are always unionized at physiological pH. Guanidine is a strong base with a pK_a > 13, and is therefore always ionized at physiological pH. Guanidine groups are sometimes attached to molecules to improve their solubility. As explained below, their pK_a values can change significantly in response to substitution.

5.16.2.3 Structural Factors that Modify pK_a Values

The pK_a values of ionizable groups are influenced by other structural factors in the molecule, as illustrated in **Table 3**. They comprise inductive, electrostatic, and electron delocalization (mesomeric) effects, together with contributions from hydrogen bonding, conformational differences, and steric factors.

Electrical work is required to remove a proton from an acidic center and transfer it to a solvent molecule, or to remove a proton from a solvent molecule and transfer it to a basic center. The amount of work is influenced by the locations and distributions of dipoles and electrical charges. The inductive effects of charge are transmitted through bonds in the molecule, while the electrostatic effects of charge operate across the low dielectric cavity provided by the solute or through the solvent. Inductive effects would be expected to remain more or less constant in *cis*- and *trans*-isomers, with the major difference in pK_a values values arising from differences in electrostatic field effects, but because inductive and electrostatic effects operate in similar directions, it is difficult to consider them separately.

If a substituent withdraws electrons, and therefore reduces the electron density in other parts of the molecule, it has a negative inductive effect. For example, the electrons forming the C–F bonds in trifluoroacetic acid are displaced toward the F atoms, making the C atom more electron-deficient than an ordinary methyl group. In turn, the carbonyl oxygen atoms are more electron-deficient, making it easier for a proton to be removed from the unionized acid. The consequence is that the pK_a of trifluoroacetic acid is significantly lower (0.32) than the pK_a of acetic acid (4.54). Electron-withdrawing groups make acids stronger and bases weaker.

If a substituent donates electrons, it increases the energy density at the ionizable group, making is harder for the group to become deprotonated. For example, the –$C(CH_3)_3$ group donates electrons, raising the pK_a of isobutyl acetic acid to 4.83. Electron-donating groups make acids weaker and bases stronger.

Mesomeric effects result from π-electron delocalization, and contribute significantly to changes in the strength of acids and bases caused by remote substituents, especially via double bonds in conjugation with the ionizable center, including *ortho* or *para* (but not *meta*) substituents in aromatic or heteroaromatic systems. Large free energy differences can result from charge delocalization, so that acids or bases may become much stronger or weaker than would otherwise be expected. Mesomeric effects may enhance or oppose inductive effects. Charge delocalization leads to the high basic pK_a of guanidine (>13), though pK_a values of substituted guanidines can be appreciably lower because of the electron-withdrawing effects of substituents (8.79 for amiloride, 2.51 for sulfaguanidine).

5.16.3   **Ionization and Drugs**

Over 700 publications since the first edition of *Comprehensive Medicinal Chemistry* in 1989 have mentioned pK_a, ionization constants, or related properties in connection with drugs. This section reviews a selection of these publications to cover the most popular topics, including drug–receptor binding, ionization and absorption, macromolecules and NMR. It also cites publications with compilations of pK_a values.

5.16.3.1   **Drug–Receptor Binding**

The comparison between the ionization state and the binding properties on brain membrane receptors of myo-inositol 1,4,5-trisphosphate (see **Figure 13**) has been studied by pH-metric titration, leading to the conclusion that the biologically active species may be either the monoprotonated or the fully deprotonated trisphosphate.[17]

It has been shown that the binding of sulpiride (see **Figure 13**) to D_2 dopamine receptors in the bovine brain is highly pH-dependent.[18] It is thought that the negatively charged sulpiride ion binds with an ionized residue at the active site. Similar behavior has been observed for other substituted benzamide drugs.[19]

In a study of the design, synthesis, and biological activity of a series of high-affinity basic ligands for the cholecystokinin-B receptor, the compounds (which incorporated a piperidin-2-yl or a homopiperidin-2-yl group attached to C-5 of a benzodiazepine core structure) had significantly higher pK_a values than previously reported antagonists based on 5-amino-1,4-benzodiazepines (e.g., 9.48 versus 7.1). In view of their high pK_a values, significant concentrations of ionized species would exist at lower pH, and it is thought that they bind to the CCK-B receptor in their protonated form.[20]

The binding of lidocaine (a base with pK_a = 7.95 – see **Figure 10**) and homologs to specific sites associated with the sodium channel were investigated. The lidocaine homologs differed in the length of the link between the arylamide and amine domains of the molecule and in the number of carbons attached to the terminal amine. Drug affinity was measured with a radio-ligand binding assay, and freshly isolated cardiac myocytes. Optimal binding was obtained when the link between the arylamide and amine domains was two carbons in length. The affinity of the drug for the receptor was optimal with four or more amino-terminal carbons; the precise arrangement of the carbons was not important. Each of the amino-terminal carbons independently contributed 1.26 kJ of free energy of binding, suggesting that the carbons dissolve in a hydrophobic pocket.[21]

To determine the binding site of quinine (a base with two pK_a values, 8.55 and 4.24, see **Figure 11**), the effect of quinine and a permanently charged quaternary derivative of quinine were investigated at various external and internal pH values of mammalian cells. Raising the external pH increases the concentration of the uncharged form and increases the block rate and potency, while increasing the internal pH reduced the cationic form in the cytoplasma. The uncharged drug crosses the membrane and binds to the receptor, probably in its ionized form. The results indicate that the binding site for quinine is intracellular, possibly within the pore.[22]

5.16.3.2   **Drug Binding with Membranes, Membrane Analogs, and Proteins**

Changes in the fluorescence properties of quinine after association with neutral and negatively charged small unilamellar lipid vesicles at pH 7 and 37 °C were used to obtain binding isotherms over a range of phospholipid compositions at different ionic strengths. All the findings suggested that the association of quinine to liposomes is controlled primarily through electrostatic attractions, and, to a lesser extent, by hydrophobic forces. Electrostatic and hydrophobic interactions play a crucial role in both the drug–membrane affinity and the location of the drug.[23]

The distribution of four ionizable molecules (amlodipine, 5-phenylvaleric acid, 4-phenylbutylamine, and 5-hydroxyquinoline – see **Figure 13**) between small unilamellar vesicles of dimyristoylphosphatidylcholine and aqueous buffers was studied as a function of pH using an ultrafiltration method. The results show that the charged forms of the molecules are able to interact with the phospholipid bilayer.[24]

The liposomal membrane/water partition coefficients of eight ionizable drugs (5-phenylvaleric acid and propranolol (see **Figure 13**), ibuprofen (see **Figure 1**), diclofenac (see **Figure 7**), warfarin (see **Figure 8**), lidocaine (see **Figure 10**), and tetracaine and procaine (see **Figure 11**)) were determined using a pH-metric technique, and it was shown that the liposomal membrane/water partition coefficients as derived from the pH-metric technique are consistent with those obtained from alternative methods such as ultrafiltration and dialysis. It was found that in the liposome system, partitioning of the ionized species is significant and is influenced by electrostatic interaction with the membranes.[25]

To gain insight into the molecular mechanisms governing drug binding to albumin, an ultracentrifugation method was used to measure the drug fraction bound to bovine serum albumin for a series of 14 structurally diverse drugs.

Values for log D at pH 7.4 were obtained from pK_a and log P values. The study shows good correlation between binding and log $D_{oct}(7.4)$ for neutral compounds and bases, but not for acids, for which bovine serum albumin binding is stronger, probably because of supplementary electrostatic interactions.[26]

5.16.3.3 Ionization and Absorption

According to the classic pH partition theory, only the uncharged species of ionizable drugs will partition into biological phases.[27] Thus molecules are more permeable by passive diffusion through lipid membranes when unionized, while ionized molecules remain in aqueous phases such as blood and are more readily excreted via the kidneys. The fraction of drug absorbed may be higher than expected from pH partition theory when biological phenomena relating to membranes are taken into account. Thus, the pH microclimate at the surface of the membrane may allow a molecule to become more (or less) ionized close to the membrane surface, with consequences for its lipophilicity. Drugs may bind to the membrane surface, or their transport may be retarded by an unstirred aqueous diffusion layer adjacent to the membrane surface.[28] After these phenomena are taken into account, pH partition theory adequately explains drug transport through membranes by passive diffusion, though the existence of other transport mechanisms (e.g., paracellular transport or active transport) must be noted.

The high- and low-permeability drugs alfentanil and cimetidine were investigated in three models expressing different drug permeability mechanisms: hexadecane membranes (HDMs), Caco-2, and 2/4/A1 cell monolayers. For both drugs, sigmoidal relationships between membrane permeability and pH were observed in all models. The ionized species of cimetidine was found to permeate through the cell membranes, but not through the HDMs. The results showed that the paracellular route has a significant role in the permeability of small basic hydrophilic drugs, such as cimetidine, in leaky small intestinal-like epithelia such as 2/4/A1. By contrast, in tighter epithelia such as Caco-2 and in artificial membranes such as HDM, the permeability of the ionized forms of the drugs and the paracellular permeability were lower or insignificant, respectively.[29]

Synthetic chemists often seek to modify the properties of molecules by modifying ionizable groups, for example to reduce their lipophilicity or to provide specific drug–receptor interactions. An example is a study to enhance the oral absorption of selective serotonin 5HT$_{1D}$ receptor agonists.[16] A lead compound with good efficacy and minimal side effects was identified; however, its bioavailability was low because of deficient oral absorption. The lead compound contained a basic piperidine group with a pK_a of 9.7, so it existed only in ionized form at the pH encountered in the small intestine, leading to poor absorption by passive diffusion. A series of fluorine-substituted analog structures was synthesized. The electron-withdrawing properties of fluorine lowered the pK_a values of the analogs to between 8.2 and 8.7, leading to increased oral absorption in rats. The authors do not comment on the mechanism of absorption. However, it can be deduced from the distribution of species that a base with a pK_a of 9.7 is only 0.2% unionized at pH 7, suggesting that passive transport of the unionized species would not be an effective mechanism for oral absorption. On the other hand, a base with a pK_a of 8.2 is 6% unionized at pH 7, and should be much better transported.

The gastrointestinal absorption of 3-amino-1-methyl-5H-pyrido[4,3-b]indole (Trp-P-2), a mutagen/carcinogen, was investigated in rats. Trp-P-2, a weak base with a pK_a of 8.2, was poorly lipophilic below pH 5, and therefore the gastric absorption was negligible at pH 1.1 and 3.0. On the other hand, this compound was rapidly absorbed from the small intestine even at pH 4.0, and the absorption was even faster at higher pH values, where the unionized fraction increases.[30]

Many basic drugs are absorbed in the intestine, where the pH may vary between 5 and 8 depending on which part of the intestine is considered, and whether the patient is in the fed or fasted state (food raises the pH of the stomach, but reduces the pH of the intestine). Itraconazole (marketed as Sporanox; see **Figure 13**) is an antifungal drug that may be taken orally. Patients are advised to take itraconazole with meals, when the pH of the intestine contents may go down to 5 or below. The wall of the intestine is shielded from the intestine contents by a layer of mucus. The pH at the wall itself is around 6, and not greatly affected by the pH of the intestine contents.[31] While pK_a and log P values have not been reported in the literature, itraconazole is a base, and at high pH it is not ionized and is almost insoluble. It ionizes at low pH, and would be rather more soluble at pH 5 than at pH 6. At pH 5 some compound can dissolve in the intestine contents. This enables it to approach the wall, where it will be transformed into the neutral form that is more easily absorbed.

Sildenafil is a vasodilator, and is marketed under the name Viagra (see **Figure 13**). This molecule is an ampholyte, with an acidic pK_a of 9.12 and a basic pK_a of 6.78.[32] While sildenafil is charged at low and high pH, it is neutral at mid-range pH, and therefore exists in its most lipophilic form (log P = 3.18) at intestinal pH; this enhances its absorbance in the gastrointestinal tract.

5.16.3.4   Ionization and Macromolecules; the Use of Nuclear Magnetic Resonance Spectroscopy

Macromolecules such as DNA, RNA, proteins, and receptors are so large and contain so many ionizable groups that it is not possible to measure the pK_a values of these groups by the methods generally used for small-molecule drugs. The technique of NMR spectroscopy, coupled with pH-metric titration[33] or buffers, is commonly used to study individual ionizable groups attached to macromolecules.

To understand why the RNA–RNA duplexes in general have a higher thermodynamic stability over the corresponding DNA–DNA duplexes, the pK_a values of nucleosides modeling as donors as well as acceptors of base pairs in duplexes have been measured by ^{1}H-NMR (at 500 MHz) with 20–33 different pH measurements for each compound. The study showed that monomeric DNA nucleobases are more basic than those of the corresponding RNA. The pK_a values of the monomeric nucleotide blocks, and differences between pK_a values of the donor as well as acceptor, can be used to understand the relative base-pairing strength in the oligomeric duplexes in the RNA and DNA series.[34]

To investigate the formation of aggregates in the peptide fragment of the non-amyloid-beta component, pK_a values were determined by pH titration experiments in NMR spectroscopy, indicating that the protonation of the carboxyl group of the N-terminal glutamic acid triggers the aggregation.[35]

The catalytic activity of serine proteases is due to a serine–histidine–aspartate triad of amino acids. The pK_a values for the histidine residue have been determined by NMR for several cases in which there is a negative charge installed at the serine residue to mimic the oxyanionic intermediate and the related transition state for the catalytic pathway. In every case studied, there was an elevation of the pK_a at the histidine residue when the negative charge is installed at the serine position.[36]

One benefit of pK_a measurement by NMR is the ability to assign pK_a values unambiguously to particular ionizable groups; this can be difficult for molecules with two or more pK_a values whose values are close together. An example is a study of the zwitterionic antihistamine cetirizine and its parent drug hydroxyzine (see **Figure 13**). The two basic pK_a values were attributed by NMR spectroscopy, and the most common conformations for each electrical species were determined by molecular dynamics simulations and confirmed by NMR measurements. For cetirizine, the results demonstrate that the zwitterion, which is the predominant species at physiological pH, exists as folded conformers able to partly mask polar groups. Extended and folded conformers of similar energy were also found for neutral hydroxyzine, whose monocationic species displayed folded conformers stabilized by intramolecular hydrogen bonds. These findings could explain the favorable pharmacokinetic properties of the drug.[37]

While NMR provides an unambiguous assignment of pK_a values, the technique is complicated and expensive. It is often possible to assign them by deduction from other experimental techniques such as pH-metric and pH/ultraviolet (UV) spectroscopy. These techniques were combined to elucidate the microconstants of cetirizine by a two step divide and conquer approach, in which values were determined for 12 pk microconstants and four tautomerization constants.[38] Other examples of deductive assignment are given in Section 5.16.5.1 (pH-metric assignment) and Section 5.16.5.3 (UV assignment). NMR, as well as Raman spectroscopy has been used to measure very low pK_a values (down to −7) for very weak bases and strong acids. Experiments are typically done in strong sulfuric acid solutions [9].

5.16.3.5   Other Connections between pK_a and Drugs

Ionization is important in the chemical analysis of drugs by high-performance liquid chromatography, because ionized and unionized forms of molecule behave quite differently in chromatographic systems.[39,40]

As well as its influence on gastrointestinal absorption, ionization is important in the transport and disposition of drugs in other organs and biological fluids. For example, it is noted that in the treatment of diseases of the prostate, the tissues are best penetrated by antimicrobial drugs with a high lipid solubility and high pK_a.[41] Irritation caused by 12 basic drugs with pK_a values between 1.4 and 11.2 applied to the skin increased significantly with increasing pK_a, the drugs which are ionized at skin pH causing the greatest irritation.[42] In forensic investigations of drugs of abuse, secretion into saliva is suggested to be by passive diffusion and to depend on lipid solubility, pK_a, plasma protein binding, and the pH of saliva.[43] Amphetamine (see **Figure 13**) and related compounds are bases, with pK_a values around 9.9, so they are ionized at physiological pH. However, their relatively low molecular weights allow amphetamine-type stimulants to diffuse easily across cell membranes and lipid layers and to those tissues or biological substrates with a more acidic pH than blood, facilitating their detection in alternative matrices at relatively high concentrations.[44]

5.16.3.6 Published Values for pK_a Values of Drugs

Wherever possible, original literature references have been provided for pK_a values cited in this chapter. While most publications report on just one or two measurements, lists and compilations have also been published. In 1978, Newton and Kluza published a compilation of pK_a values from the literature with references for about 400 drugs.[45] In his 2003 book, Avdeef[31] showed a compilation of about 250 measured pK_a values with references, including some values measured by the author. Box *et al.* measured pK_a values of 71 drugs in 2003.[46]

5.16.4 The Effect of Ionization on Other Physicochemical Properties

This section discusses the effect of ionization on the lipophilicity and solubility of drugs, and provides equations for calculating the shapes of lipophilicity and solubility profiles.

5.16.4.1 Ionization and Lipophilicity

Molecules are most lipophilic when they are unionized, and their lipophilicity decreases rapidly when they are ionized. This has an important implication for the absorption and transport of drugs as a result of their permeability by passive diffusion through membranes and lipid bilayers. **Figure 14** shows the lipophilicity (expressed as octanol/water partition) of three ionizable drugs as a function of pH. Significant variations in lipophilicity occur over the physiological pH range. The pH of blood and at the blood–brain barrier is 7.4. The pH in the stomach can vary between 1 and 7, but is usually between 1.4 and 2.1. In the gastrointestinal tract, where many drugs are absorbed, there is a pH gradient in the small intestine, with values gradually rising from about 5 to 7.4 between the duodenum and ileum. Also, the pH changes slightly according to the fed and fasted state, and the type of food ingested.[47] These factors can influence the conversion of drugs into their unionized form, and underline the importance of obtaining log D values at the appropriate pH.

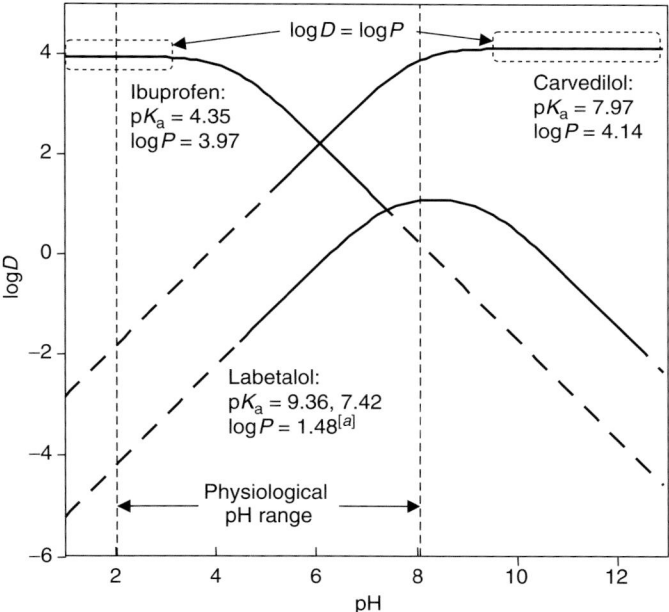

Figure 14 Lipophilicity profiles for ibuprofen (acid, see **Figure 1**), carvedilol (base, see **Figure 2**) and labetalol (ampholyte/zwitterion, see **Figure 6**). log D denotes the ratio between the concentration of sample in all forms dissolved in solvent and the concentration in water. The lines are drawn using eqns [41]–[43], which are based on the partition coefficient of the unionized species (P) and pK_a. Dashed lines indicate the pH range where the molecule may partition in its ionized form as an ion pair with a counter-ion (depends on conditions).

[a]The pK_a values of labetalol are close together, and it exhibits both ampholye and zwitterionic properties, depending on solution conditions. This graph represents the partition of the composite species XH. It has been shown that log P for the neutral species $XH^0 = 2.70$.[10]

Lipophilicity profiles complement the aqueous distribution of species graph, and are useful for assessing the lipophilicity at a given pH. Bases with low pK_a values and acids with high pK_a values are most lipophilic at gastrointestinal pH values. However, provided at least some of the unionized species is available at a given pH, the molecule will be able to partition from an aqueous solution into a lipid phase (e.g., from the intestine into the gut wall). As the unionized molecules partition, and thus have been removed from the aqueous phase, the aqueous acid–base equilibrium redistributes, and more unionized species will be produced in solution.

The following equations were used to obtain log D values in **Figure 14**:

- lipophilicity–pH profile of a monoprotic acid,

$$\log D_{AH} = \log_{10}\left(P\frac{[H^+]}{[H^+] + K_{a1}}\right) \tag{41}$$

- lipophilicity–pH profile of a monoprotic base,

$$\log D_{B} = \log_{10}\left(P\frac{K_{a1}}{[H^+] + K_{a1}}\right) \tag{42}$$

- lipophilicity–pH profile of an ampholyte,

$$\log D_{XH} = \log_{10}\left(P\frac{K_{a1}[H^+]}{[H^+]^2 + K_{a1}[H^+] + K_{a1}K_{a2}}\right) \tag{43}$$

In these equations, P represents the octanol/water partition coefficient of the unionized species, and D represents the ratio of the concentrations of all species (unionized plus ionized) partitioned between octanol and water at a given pH. K_{a1} and K_{a2} are taken from **Table 1**. To draw these graphs, it is assumed that only the unionized species partitions into the lipid phase. These equations represent the three most common classes of ionizable compound.

Table 1 illustrates a general ionization scheme for compounds with one to six pK_a values. As explained in the table, the charge on X can vary, depending on the class of compound. The form of the equations is quite general. Explicit equations are provided for for [X], [XH], [XH$_2$], and [XH$_3$]; equations for [XH$_4$] and higher can easily be deduced.

To calculate log D (the lipophilicity at a given pH) for a lipophilicity–pH profile for a compound with up to six pK_a values, identify the unionized species (X, XH, XH$_2$ etc.), obtain the log P value (partition coefficient of unionized species), and calculate P, then calculate $\log(P \times [X])$, etc., at each pH. For examples:

$$\log D = \log (P \times [XH]) \text{ (for a monoprotic acid)} \tag{44}$$

$$\log D = \log(P \times [X]) \text{ (for a monoprotic base)} \tag{45}$$

The equations in **Table 1** can only be used to plot the lipophilicity of the unionized species. Avdeef has published general equations for the lipophilicity–pH profiles of all classes of ionizable molecules, covering both the partition of unionized and ionized species.[48]

pK_a values are used in the pH-metric method for measuring log P of ionizable compounds. In this method, the shape of a titration curve of a sample titrated in a water–octanol mixture is compared with a curve based on calculated pH values; the pK_a is required to calculate the pH values used in the calculated titration curve.[49]

5.16.4.2 Ionization and Solubility

Ionizable molecules are least soluble when unionized, and their solubility increases rapidly when they are ionized. Drugs are permeable by passive diffusion when they are unionized, but they may be so poorly soluble at the pH where they are unionized that they cannot get into solution in order to be permeable. Poor solubility is a major problem in drug discovery. The solubility–pH profiles in **Figure 15** show how solubility increases when molecules are ionized. These profiles are calculated from the equations below. Superimposed on these profiles are lines suggesting the solubility of the drug in salt form. The positions of these lines vary according to the counter-ion. Some salts are so soluble that the entire substance remains in solution at all pH values where the substance is ionized. The solubility for different types of molecule is calculated as follows:

- solubility–pH profile of monoprotic acid,

$$\log S_{\text{pH}} = \log_{10}\left(S_0 \frac{[\text{H}^+] + K_{\text{a1}}}{[\text{H}^+]}\right) \qquad [46]$$

- solubility–pH profile of monoprotic base,

$$\log S_{\text{pH}} = \log_{10}\left(S_0 \frac{[\text{H}^+] + K_{\text{a1}}}{K_{\text{a1}}}\right) \qquad [47]$$

- solubility–pH profile of ampholyte,

$$\log S_{\text{pH}} = \log_{10}\left(S_0 \frac{[\text{H}^+]^2 + K_{\text{a1}}[\text{H}^+] + K_{\text{a1}}K_{\text{a2}}}{K_{\text{a1}}[\text{H}^+]}\right) \qquad [48]$$

In these equations, S_0 refers to the intrinsic solubility of the sample, which is the concentration of the free acid or base form of an ionizable compound at a pH where it is fully unionized in a saturated solution when excess solid is present, and the solution and the solid are at equilibrium; S_{pH} refers to the concentration of the substance in solution in

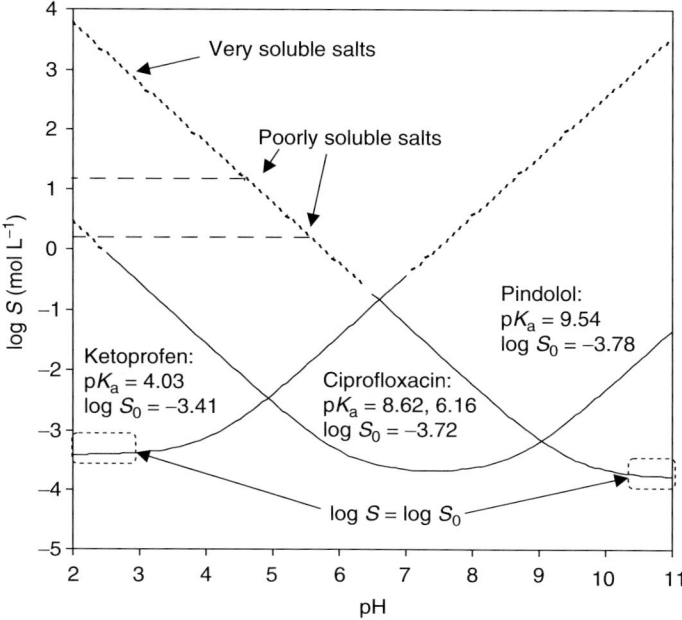

Figure 15 Solubility–pH profiles of ketoprofen (acid), pindolol (base), and ciprofloxacin (zwitterion). log S denotes the concentration of sample species that could dissolve at any pH. The lines are drawn using eqns [37]–[39], which are based on intrinsic solubility (S_0) and pK_a. The solubility would only follow the dashed lines if (1) a sufficient weight of the sample was available, and (2) if salts of the sample were soluble. See discussion of ciprofloxaxin in Section 5.16.1.10.

equilibrium with excess solid at a given pH; K_{a1} and K_{a2} are taken from **Table 1**. In the graphs, S_0 and S_{pH} are expressed in moles per liter.

The equations in **Table 1** can be used to calculate log S_{pH} (the solubility at a given pH) for compounds with up to six pK_a values. First, identify the unionized species (X, XH, XH$_2$, etc.), then obtain the log S_0 value (intrinsic solubility of unionized species), and calculate S_0, then follow the examples below:

$$\log S_{pH} = \log\frac{S_0}{[XH]} \text{ (for a monoprotic acid)} \qquad [49]$$

$$\log S_{pH} = \log\frac{S_0}{[X]} \text{ (for a monoprotic base)} \qquad [50]$$

Streng *et al.* have published equations for solubility–pH profiles of acids, bases, and ampholytes, including the effect of salts.[50,51] Ozturk *et al.* have published equations showing the dependence of the dissolution rate on pK_a and intrinsic solubility.[52]

pK_a values are used in pH-metric methods for measuring solubility of ionizable compounds. A method has been described in which the shape of a titration curve of a sample that precipitated during the titration is compared with a curve based on calculated pH values; the pK_a is required to calculate the pH values.[53] Another method calculates the solubility during a process of chasing equilibrium for compounds that form supersaturated solutions; this calculation also requires the pK_a value.[54]

5.16.5   Measuring pK_a Values

To measure pK_a values it is necessary to expose the compound to an environment of changing pH, and monitor a property that changes as a function of the ionization state of the molecule. Several considerations affect the choice of method. If the sample is poorly water-soluble in its unionized form, it must either be measured in very dilute solution, or in water–solvent mixtures. If only a very small quantity of sample is available (e.g. less than 1 mg), the only practical methods involve UV spectroscopy. If high throughput measurement is required, not all methods are suitable.

5.16.5.1   The Effect of Co-Solvents on pK_a

Methods of pK_a of measurement were developed using water-soluble samples. However, many drugs are poorly soluble in water alone, and require the presence of a water-miscible co-solvent to keep them in aqueous solution. The solvent affects the pK_a in two ways. It causes the pH scale to shift, and it causes the pK_a to shift. The consequence is that apparent pK_a values measured in the presence of solvent are different from aqueous values.

For the sake of improving throughput, it is tempting to ignore this effect, and to work always in water–solvent mixtures and report the apparent pK_a values, which may be considered 'good enough' for use during early stage discovery. As seen in **Figure 16**, however, the shifts can be quite high. An error in pK_a of 0.5 unit, when applied in a calculation of log D from a log P would cause an error in log D estimation of about 0.5 log D units. It causes a similar error in pH-metric measurement of solubility. When pK_a values are close to the physiological pH in the region where the drug is absorbed, errors pK_a in could have a significant effect on predictions of drug behavior.

The technique of extrapolating a series of p$_s$$K_a$ (apparent pK_a) values obtained in several ratios of water/solvent to obtain an aqueous value is well established, but three or more experiments at different water/solvent ratios are required. The Yasuda–Shedlovsky plot of (p$_s$$K_a$ + log[H$_2$O]) versus $1/\varepsilon$ (where ε is the dielectric constant of the water–solvent mixture) is widely used for methanol and 1,4-dioxane.[55,56] A linear plot of p$_s$$K_a$ versus the water/solvent ratio is preferred for dimethyl sulfoxide (DMSO), whose dielectric constant is similar to that of water. A method of calculating aqueous pK_a values for various classes of organic acids and bases from single apparent pK_a values obtained in water–solvent mixtures has been reported,[57] and looks promising as a way of speeding pK_a measurement.

The most widely used solvent for pK_a measurements is methanol. However, not all drugs dissolve in water–methanol mixtures. Other solvents have been used, notably 1,4-dioxane and DMSO, though the latter solvent has high UV absorbance at low wavelengths, and absorbs water at room temperature, causing pK_a measurement difficulties. Quite often, only small sample quantities are available and there is no time to do experiments to choose the best solvent. In this case it is desirable to use a water–solvent mixture in which the largest number of compounds will dissolve. Some mixtures of solvents with low UV absorbance have been described for pK_a measurement. A 1:1:1 mixture

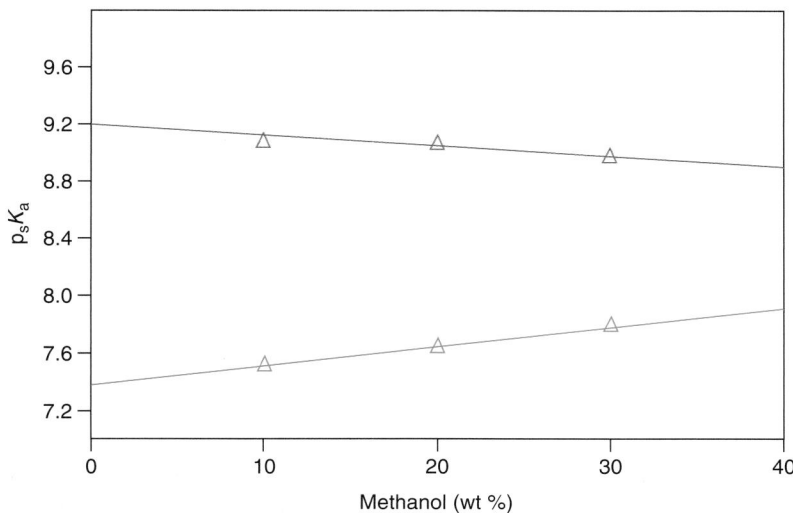

Figure 16 Apparent pK_a values of labetalol in increasing concentrations of methanol. The acid pK_a increases from 7.39 in water to 7.9 at 40% methanol.

of methanol, acetonitrile, and diethyleneglycol monoethyl ether, mixed 15% or 20% with an aqueous buffer has been used in the pH-gradient method [81]; to correct apparent pK_a values to their aqueous values, 0.3 was subtracted from acid values, and 0.2 was added to bases. A 1:1:1 mixture of methanol, 1,4-dioxane, and acetonitrile (called MDM) has been used in titrations at various water/solvent ratios, with pK_a values calculated by extrapolation. These mixtures contain both polar and nonpolar solvents, offering a good opportunity to solubilize the maximum number of drug molecules.

Figure 16 shows p$_s$$K_a$ values versus the water/methanol ratio extrapolated to the aqueous pK_a for labetalol (see Figure 6). Note how the three p$_s$$K_a$ values for each ionization extrapolate to the aqueous pK_a value. Note also the slopes of the two lines. The p$_s$$K_a$ values of acids increase with increasing solvent, and the p$_s$$K_a$ values of bases decrease, in both cases because the molecules form charged species with increasing difficulty under lower dielectric conditions. These slopes may be used to assign pK_a values to the corresponding acid and base ionizable groups.

5.16.5.2 pH-Metric Titration

The pH-metric method can be used to measure all pK_a values between 2 and 12, with or without a UV chromophore, provided the sample can be dissolved in water or water/co-solvent over the pH range of interest.

Albert and Serjeant[9] described a simple pH-metric method for measuring pK_a values, illustrated here for a sample of 30.9 mg of boric acid in 47.5 mL of water at 20 °C. Ten aliquots of NaOH titrant are added, and the pH measured after each addition. The results are calculated as shown in **Table 4**. They illustrated similar methods in which corrections were made for concentration of hydrogen ions and hydroxyl ions that are present in increasing quantity as the pH deviates from 7. These simple methods are suitable for high weights (> 30 mg) of water-soluble samples with a single pK_a in the mid-pH region (4–10), and they can be carried out with the simplest of apparatus and calculations. However, these calculations involve approximations, and for smaller weights, overlapping pK_a values (i.e., values separated by less than 3 units) and values outside the range 4–10, methods using charge and mass balance without approximations are required. These involve the use of automatic titration apparatus and computer programs, which must be able to deal with changes in ionic strength and their effect on ionic activity, the effect of carbon dioxide on the titration curve, and the volume changes in solutions as titrants are added. **Figure 17** shows some of the graphs that can be obtained using Sirius RefinementPro 2 software during the pH-metric measurement of pK_a of labetalol.

In an automated pH-metric titration for measuring pK_a, the pH of a solution of the drug is adjusted by adding acid or base until the sample is fully ionized, and then titrated with base or acid until the sample is fully unionized. The pH of the sample solution is monitored with a glass pH electrode, and the pK_a is calculated from the change in shape of the titration curve.[55] A sample concentration of at least about 5×10^{-4} M is required in order for any significant change in shape to be detectable between pH 3 and 11, and a higher concentration is required outside this range.[58] Although pH-metric pK_a titration in volumes as low as 100 μL has been reported using pH micro-electrodes,[59] the physical size of glass pH electrodes and stirrers generally dictates that the smallest practical volume of sample solution

Table 4 Method published by Albert and Serjeant[9] to measure the pK_a of boric acid. The sample is 30.9 mg of boric acid in 47.5 mL of water at 20 °C

1	2	3	4	5	6	7
Titrant 0.1 N KOH (mL)	Meter reading (pH)	C^0 (the original conc.) diminished by tenths	C^0 minus column 3	Column 3 divided by column 4	log of column 5	pK_a (columns 2 + 6)
0	6.16	0.01	0			
0.5	8.34	0.009	0.001	9/1	0.95	9.29
1	8.68	0.008	0.002	8/2	0.60	9.28
1.5	8.89	0.007	0.003	7/3	0.37	9.26
2	9.07	0.006	0.004	6/4	0.18	9.25
2.5	9.26	0.005	0.005	5/5	0.00	9.26
3	9.43	0.004	0.006	4/6	−0.18	9.25
3.5	9.62	0.003	0.007	3/7	−0.37	9.25
4	9.84	0.002	0.008	2/8	−0.60	9.24
4.5	10.14	0.001	0.009	1/9	−0.95	9.19
5	10.56	0	0.01		Mean	9.25

Reproduced from Albert, A.; Serjeant, E. P. *The Determination of Ionization Constants*, 3rd ed.; Chapman and Hall: London, UK, 1984, with kind permission of Springer Science and Business Media.

is about 5 mL. This requires at least 1 mg of sample for a compound with a molecular weight of 400 to achieve a concentration of 5×10^{-4} M. A typical titration between pH 3 and 11 requires about 45 measured pH values. The need to wait for equilibrium after each pH measurement means that automated titrations typically take 20–40 min to perform.

The pK_a result is calculated by comparing the experimental titration curve (consisting of measured pH versus the volume of titrant added) with a theoretical curve made up from calculated pH values. Each calculated pH requires a pK_a value for the sample. The pH values are calculated and recalculated iteratively using different pK_a values until the experimental and theoretical curves reach the closest agreement. The pK_a value giving best agreement is taken to be the true measured value.

The calculation of pH values of aqueous solutions of weak acids and bases is based on mass balance and charge balance. Consider a monoprotic acid AH of concentration C_{AH} moles per liter, which dissociates according to

$$AH \rightleftharpoons H^+ + A^- \tag{51}$$

Based on the pK_a of AH, we can write

$$[H^+][A^-] = K_a[AH] \tag{52}$$

The other equations required to calculate the pH of the solution are

$$[H^+][OH^-] = K_w = 10^{-14} \text{ (ionic product)} \tag{53}$$

$$C_{AH} = [A^-] + [AH] \text{ (mass balance)} \tag{54}$$

$$[H^+] = [A^-] + [OH^-] \text{ (charge balance)} \tag{55}$$

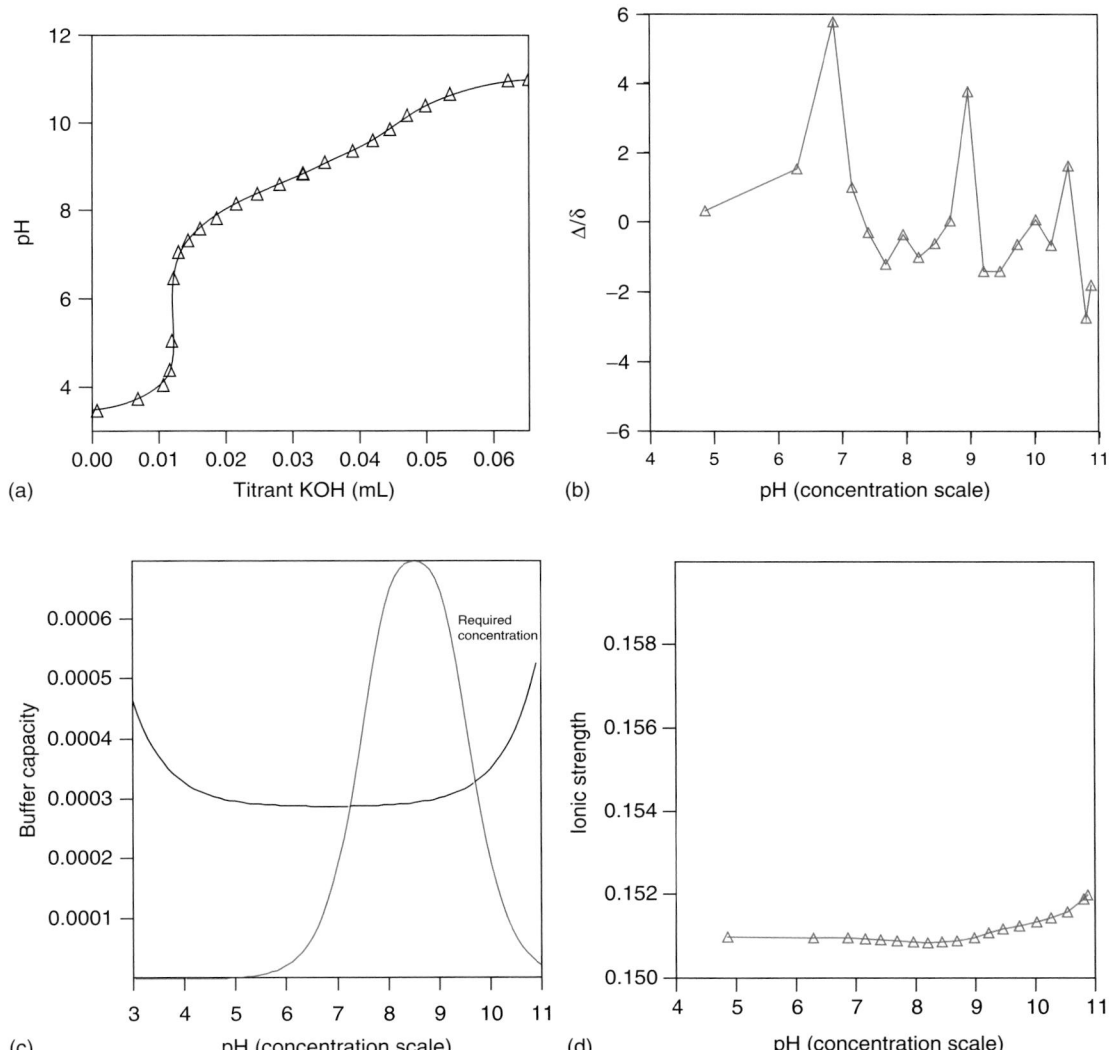

Figure 17 Selection of graphs obtained during the pH-metric measurement of p_sK_a of labetalol. The sample weight was 2.77 mg. The sample was dissolved in 10 mL of an acidified water–methanol mixture (49% methanol, 0.15 M KCl, 25 °C). The titrant was 0.5 M KOH. (a) The titration curve. (b) The residuals at each pH, representing the unexplained pH after comparing the measured and calculated pH for each point. (c) The buffer capacity of the water–solvent background, and the contribution from the buffer capacity of the sample. When the sample peak is higher than the background signal, there is sufficient signal for a good-quality pH-metric measurement. (d) The total ionic strength of the solution at each pH during the titration

Combining these equations, substituting C_A for $[A^-]$, and rearranging leads to a cubic polynomial equation:

$$[H^+]^3 + [H^+]^2(C_A + K_a) - [H^+](C_{AH}K_a + K_w) - K_wK_a = 0 \qquad [56]$$

Values for C_A and C_{AH} change with each point in the titration curve in response to the increase in solution volume caused by the addition of titrants. Note that the comparable equations will be different for di- or triprotic substances, and different again for bases and ampholytes. This class of equation can be solved explicitly for $[H^+]$ and thus for pH, but it is not easy to do so by simple algebraic techniques. A number of methods have been proposed in which results can be obtained graphically. The Newton–Raphson method is widely used. The pH-metric method may be used for samples with one, two, three, or more pK_a values, as well as for samples with titratable counter-ions. In these cases, the equations used to calculate the pH of each point must take account of each pK_a of the substance and the counter-ion.

5.16.5.3   Hybrid pH-Metric/Ultraviolet Method

Significantly smaller sample concentrations (10^{-5} M or below) are required for pK_a measurement by the hybrid pH-metric/UV method, in which multiwavelength UV absorbance of the sample solution is monitored throughout the titration. Samples must have a chromophore, and the absorbance must change as a function of ionization. Recent advances in this method have enabled pK_a measurement in volumes less than 8 mL by the use of a fiber-optic dip probe in a titration cell, and has been automated on the Sirius GLpKa with D-PAS attachment. During measurement, samples are acid–base titrated across a pH range that includes the pK_a values, and multiwavelength UV spectra are measured at each pH. The pK_a values are calculated using a technique based on target factor analysis (TFA).[60] The apparatus has been used for the determination of pK_a values of samples with multiple ionization centers,[61] and the determination of protonation microconstants of zwitterionic molecules.[38,62] Titrations in unbuffered solutions take up to 30 min because of the need to wait for electrode stability, but they can be done in less than 4 min in the presence of linear buffer solutions, taking advantage of the fast electrode equilibration time in buffered solutions. The ability to work at lower concentrations increases the scope for measuring pK_a values of poorly soluble samples by the Yasuda–Shedlovsky technique, but throughput will be lower if this technique is used.

   Figure 18 shows some of the graphs that can be obtained using Sirius RefinementPro 2 software during the pH/UV measurement of pK_a of labetalol. The structure of labetalol contains two ionizable groups: a phenol and an amine. The phenol is part of a strong UV-absorbing chromophore, while the base is distant from the chromophore, and will exert only a small influence on the UV absorbance. From the graphs, it is clear that the absorbance change associated with the equilibrium $X^- + H^+ \rightleftharpoons XH$ is much smaller than the change associated with $XH + H^+ \rightleftharpoons XH_2^+$. This provides another way to assign the pK_a values, as it can be deduced that $X^- + H^+ \rightleftharpoons XH$ refers to the ionization of the amine and $XH + H^+ \rightleftharpoons XH_2^+$ refers to the ionization of the phenol.

5.16.5.4   pH Gradient Titration

A technique called pH gradient titration can be used for the rapid measurement of pK_a. In this technique, samples are injected into a flowing pH gradient, which is created by mixing an acidified and a basified linear buffer together using syringe pumps running at inversely varying speeds. During the gradient the pH varies linearly, and the pH at any given time may be predicted from the time elapsed since the start of gradient generation. This eliminates the need for pH measurement, which is one of the rate-limiting steps in conventional titration. The gradient and sample pass through a diode array UV spectrophotometer, and pK_a values are determined from changes in absorption as a function of pH. The linear buffer is created by mixing several weak acids and bases in varying concentrations such that the titration curve is linear in terms of pH, and the ionic strength approximates to 0.15 M at any pH. The buffer components were chosen because of their low UV absorbance, and have a total of nine pK_a values spaced approximately equally over the 2–12 pH range. Experimental data consist of measured absorbance, wavelength, and pH (which is calculated from the time elapsed from the start of gradient generation). The pK_a values are obtained by differentiation or by TFA using methods described elsewhere.[63] A validation study has been published showing good correlation between pK_a values measured by this method and previously measured values for 71 pharmaceutical drugs with published structures.[46] The pH gradient technique is offered on the Sirius ProfilerSGA instrument, which accepts samples in 10 mM stock solutions in 96-well plates. Typical assay cycle is 4 min per sample, including a gradient time of 90 s, leading to a throughput of over 200 samples per day.

5.16.5.5   Capillary Electrophoresis

Measurement of the pK_a by UV absorption after separation by capillary electrophoresis has been described for acids[64] and bases.[65] The principle of pK_a determination by capillary electrophoresis is the measurement of the ionic effective mobility of the solute as a function of pH. Equations that relate effective electrophoretic mobility to pK_a for monoprotic and diprotic acids and bases have been published.[66] The measurement of pK_a by capillary electrophoresis requires small sample quantities, and the rate of throughput using single-channel instruments is currently about 20–35 samples per day.[67] A method using an array of 96 capillaries has been described for measuring pK_a values of samples supplied in solutions in 96-well plates.[82] A rate-limiting step in the capillary electrophoresis method is the need to make separate experimental runs at different buffered pH values to determine the relative concentration of ionized species at each pH; this limitation does not apply to titration methods, which are inherently faster than methods using buffers.

5.16.5.6   Other Methods

Albert and Serjeant[9] described a general method for measuring pK_a by conductivity. This was a widely used method up to about 1932, before pH-metric methods took over. The conductometric method is extremely temperature-sensitive,

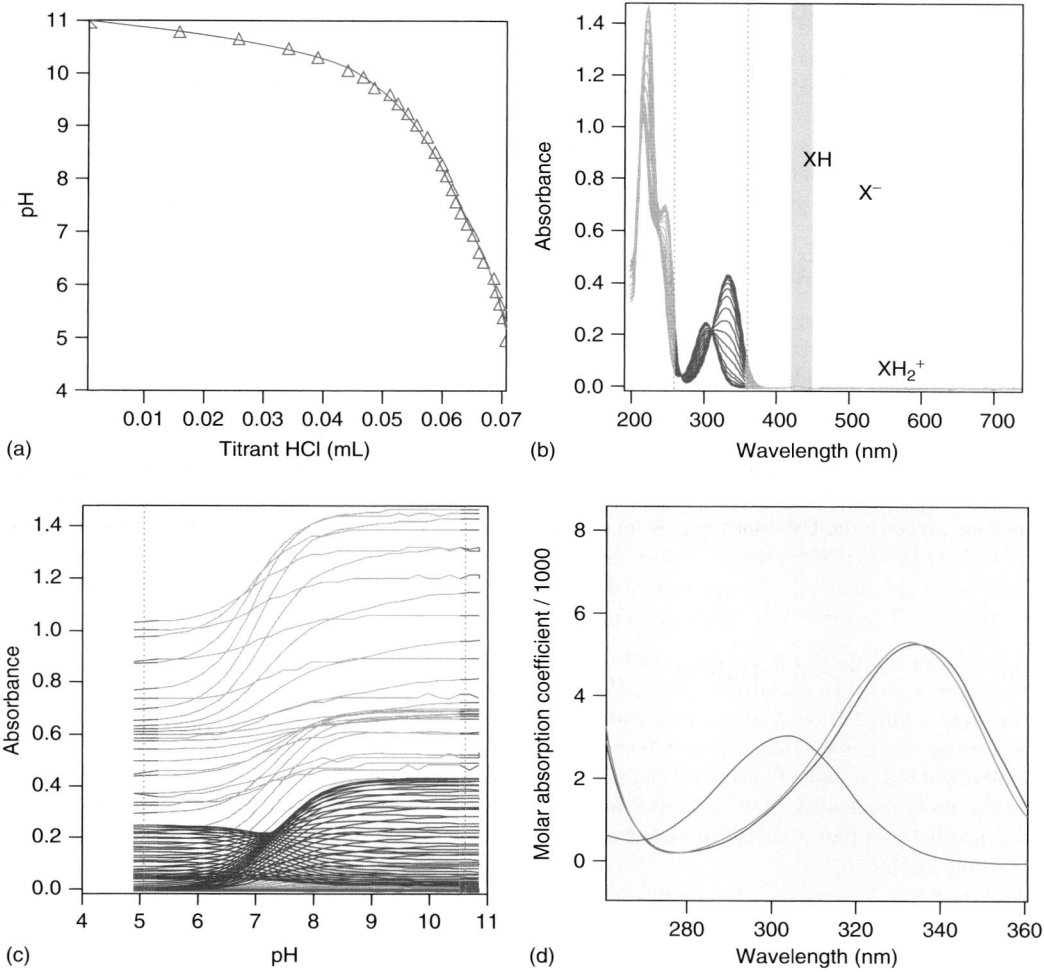

Figure 18 Selection of graphs obtained during the pH/UV measurement of pK_a of labetalol. The sample weight was 0.6 mg. The sample was dissolved in 20 mL of basified 0.15 M KCl solution in water containing 0.0001 M K_2HPO_4 buffer (25 °C). The titrant was 0.5 M HCl. (a) The titration curve. (b) the absorbance versus the wavelength. Each line represents a different pH. Only the absorbance between 260 nm and 360 nm are used in the calculation. (c) The absorbance versus the pH. The pK_a values of labetalol occur in those pH regions where the lines are crossing. TFA deconvolutes the data to find the exact pK_a values. (d) The absorption of each species of labetalol (XH_2^+, XH, X^-).

and cannot be used to measure overlapping pK_a values. A possible advantage for samples without a chromophore is that it can be used in very dilute solutions, down to 2.8×10^{-5} M. In recent years, conductometry has been used to measure the pK_a of lidocaine,[68] though the result obtained was 0.7 unit lower than the generally accepted pH-metric value.

Albert and Serjeant[9] also described a method for determining the pK_a from solubility measurements. As mentioned above, solubility is pK_a-dependent, so if the solubility is measured at several pH values on a curve, such as is shown in **Figure 15**, the pK_a could be deduced. Peck and Benet[69] described a general method for estimating pK_a values for monoprotic, diprotic, and amphoteric substances, given a set of solubility and pH measurements. Hansen and Hafliger[70] derived the pK_a of a sample that decomposed rapidly by hydrolysis from its initial dissolution rates as a function of pH in a rotating-disk apparatus. The result agreed well with a pH/UV result, but decomposition made the latter method difficult.

A hybrid pH/fluorometry method has been used to measure pK_a values of porphyrins[13] and ciprofloxacin.[71] The methodology and calculations required are similar to those used in pH/UV methods. Fluorescence methods have the advantage of using very dilute drug solutions, and may be applicable in cases where UV methods are restricted.

A flow injection method has been reported in which pK_a values are measured by UV absorbance after injection into a flowing phosphoric acid buffer.[72] No mention of throughput was made.

5.16.6   Calculating pK_a Values from Chemical Structures

The acidic and basic groups in organic molecules are easily recognized, and the structural factors that modify pK_a values are well known (see **Table 3**). Experienced chemists can estimate the pK_a values of molecules by inspecting their structures. Over the years, these rules have consolidated into a series of methods for predicting pK_a values, often with high accuracy, especially for molecules that are structurally similar to other molecules with measured pK_a values; they are summarized in Perrin's 1981 book.[73] This section cites examples from the book, and briefly reviews the currently available prediction software. More information is provided in 5.25 In Silico Prediction of Solubility, on pK_a prediction for.

5.16.6.1   Hammett and Taft Equations

Values can be predicted using linear free energy relationships, from the class of compound, and also by analogy and extrapolation. The change in free energy (ΔG) that is produced by adding a substituent leads to a corresponding change in pK_a, and the effects are additive, more or less. This observation forms the basis of the Hammett and Taft equations that are the most widely used methods of prediction. As seen in **Figures 7–13**, pK_a values change according to the class of compound. An example is the pK_a values of primary, secondary, and tertiary aliphatic amines, whose values fall into the ranges 10.6 ± 0.2, 11.1 ± 0.1, and 10.5 ± 0.2, respectively. Values like these can be used to provide the starting pK_a of the parent compound.

Hammett[74] demonstrated that the effects on of *meta*- and *para*-substituted benzoic acids were linear and additive. This led to the Hammett equation

$$pK_a = pK_a^0 - \rho(\Sigma\sigma) \qquad [57]$$

where ρ is a constant for the particular equilibrium (i.e., an acid or base functional group), σ is a constant that is characteristic of a particular substituent, and pK_a^0 is the pK_a of the unsubstituted acid or base.

In the original formula, two constants needed to be assigned to each substituent, σ_{meta} for *meta*-substitutions and σ_{para} for *para*-substitutions. This was extended to aliphatic systems by Taft,[75] who introduced the term σ^*.

Hammett and Taft σ and σ^* values for large numbers of substituents have been published, as have equations for different classes of compound. The best predictions of pK_a use the least number of substituent sigma constants and the least number of approximations. Ideally they should be based on a model compound as similar in structure as possible to the sample.

The method may be illustrated by the following examples:

- Calculation of the pK_a of 3-acetylpyridine (see **Figure 13**). First, find a Hammett or Taft equation for the pK_a of a substituted pyridine, such as the following:[76]

$$pK_a = 5.25 - 5.90\sum\sigma^* \qquad [58]$$

- Next, find a σ_{meta} value for the –COCH$_3$ substituent. According to Perrin, σ_{meta} for –COCH$_3$ is 0.36. Therefore,

$$pK_a = 5.25 - (5.90 \times 0.36) = 3.13 \qquad [59]$$

- Calculate the pK_a of 4-chloro-3,5-dimethylphenol (see **Figure 13**). The following Hammett equation has been published for phenol[77]:

$$pK_a^0 = 9.92 - 2.23\sum(\sigma^*) \qquad [60]$$

- σ_{meta} for –CH$_3$ is -0.06, and σ_{para} for –Cl is 0.24. Therefore,

$$pK_a = 9.92 - [2.23 \times (0.24 - 0.06 - 0.06)] = 9.65 \qquad [61]$$

A problem associated with pK_a prediction based on the use of Hammett and Taft equations is the difficulty of finding values for the substituent constants and for the acid/base functional group under consideration. While they may be obtained from books and publications, they are more easily handled by databases; and the calculation of pK_a values is more easily handled by software.

5.16.6.2 Software for Predicting pK_a Values

Several vendors, including Compudrug (Pallas), Simulations Plus, ACD, and Pharma Algorithms supply software for the calculation of pK_a values. Software predicts values very well for simple molecules, but for more complex or novel structures, the prediction may be suspect. Many software developers are keen to improve their predictions by taking advantage of published, measured pK_a values for new compounds, and it seems likely that predictions will improve considerably during the coming years (*see* 5.25 In Silico Prediction of Ionization).

References

1. Sørensen, S. P. L. *Biochem Z.* **1909**, 21–131.
2. Kusalik, P. G.; Svishchev, I. M. *Science* **1994**, *265*, 1219–1221.
3. Marshall, W. L.; Frank, E. U. *J. Phys. Chem. Ref. Data* **1981**, *10*, 295–304.
4. Smith, W.; Martell, A. E. *Critical Stability Constants*; Plenum Press: New York, 1974–1989; Vols 4–6.
5. Ackermann, T. Z. *Elektrochemie* **1958**, *62*, 411–419.
6. Sillén, L. G.; Martell, A. E. *Stability Constants. Special Publication No. 17*; The Chemical Society: London, UK, 1964.
7. Sillén, L. G.; Martell, A. E. *Stability Constants. Special Publication No. 25*; The Chemical Society: London, UK, 1971.
8. Butler, J. N. *Ionic Equilibrium, Solubility and pH Calculations*; John Wiley: New York, 1998.
9. Albert, A.; Serjeant, E. P. *The Determination of Ionization Constants*; 3rd ed.; Chapman and Hall: London, UK, 1984.
10. Comer, J.; Tam, K. Lipophilicity Profiles: Theory and Measurement. In *Pharmacokinetic Optimization in Drug Research: Biological, Physicochemical and Computational Strategies*; Testa, B., van de Waterbeemd, H., Folkers, G., Guy, R., Eds.; VHCA: Zurich, Switzerland, 2001, pp 275–304.
11. Avdeef, A. *Curr. Topics Med. Chem.* **2001**, *1*, 277–351.
12. Wells, J. I. *Pharmaceutical Preformulation*; Ellis Horwood: London, UK, 1988.
13. Cunderikov, B.; Kaalhus, O.; Cunderlik, R.; Mateasik, A.; Moan, J.; Kongshaug, M. *Photochem. Photobiol.* **2004**, *79*, 242–247.
14. Stephens, C. R.; Murai, K.; Brunings, K. J.; Woodward, R. B. *J. Am. Chem. Soc.* **1956**, *78*, 4155–4157.
15. Rigler, N. E.; Bag, S. P.; Leyden, D. E.; Sudmeier, J. L.; Reilley, C. N. *Anal. Chem.* **1965**, *37*, 872–875.
16. Castro, J. L.; Collins, I.; Russell, M. G. N.; Watt, A. P.; Sohal, B.; Rathbone, D.; Beer, M. S.; Stanton, J. A. *J. Med. Chem.* **1998**, *41*, 2667–2670.
17. Schmitt, L.; Schlewer, G.; Spiess, B. *Biochim. Biophys. Acta* **1991**, *1075*, 139–140.
18. Presland, J. P.; Strange, P. G. *Biochem. Pharmacol.* **1991**, *41*, R9–R12.
19. D'Souza, U. M.; Strange, P. G. *Biochem. Soc. Trans.* **1992**, *20*, 146S.
20. Castro, J. L.; Broughton, H. B.; Russell, M. G. N.; Rathbone, D.; Watt, A. P.; Ball, R. G.; Chapman, K. L.; Patel, S.; Smith, A. J.; Marshall, G. R.; Matassa, V. G. *J. Med. Chem.* **1997**, *40*, 2491–2501.
21. Sheldon, R. S.; Hill, R. J.; Taouis, M.; Wilson, L. M. *Mol. Pharmacol.* **1991**, *39*, 609–614.
22. Srinivas, M.; Hopperstad, M. G.; Spray, D. C. *Proc. Natl. Acad. Sci. USA* **2001**, *98*, 10942–10947.
23. Porcar, I.; Codoner, A.; Gomez, C. M.; Abad, C.; Campos, A. *J. Pharm. Sci.* **2003**, *92*, 45–57.
24. Austin, R. P.; Davis, A. M.; Manners, C. N. *J. Pharm. Sci.* **1995**, *84*, 1180–1183.
25. Avdeef, A.; Box, K. J.; Comer, J. E. A.; Hibbert, C.; Tam, K. Y. *Pharm. Res.* **1998**, *15*, 209–215.
26. Ermondi, G.; Lorenti, M.; Caron, G. *J. Med. Chem.* **2004**, *47*, 3949–3961.
27. Shore, P. A.; Brodie, B. B.; Hogben, C. A. *J. Pharmacol. Exp. Ther.* **1957**, *119*, 361–369.
28. Camenisch, G.; Folkers, G.; van de Waterbeemd, H. *Pharm. Acta Helv.* **1996**, *71*, 309–327.
29. Nagahara, N.; Tavelin, S.; Artursson, P. *J. Pharm. Sci.* **2004**, *93*, 2972–2984.
30. Kimura, T.; Nakayama, T.; Kurosaki, Y.; Suzuki, Y.; Arimoto, S.; Hayatsu, H. *Jpn. J. Cancer Res.* **1985**, *76*, 272–277.
31. Avdeef, A. *Absorption and Drug Development. Solubility, Permeability and Charge State*; Wiley-Interscience: Hoboken, NJ, 2003.
32. Gobry, V.; Bouchard, G.; Carrupt, P. A.; Testa, B.; Girault, H. H. *Helv. Chim. Acta* **2000**, *83*, 1465–1474.
33. Hägele, G.; Holzgrabe, U. pH-Dependent NMR Measurements. In *NMR Spectroscopy in Drug Development and Analysis*; Holzgrabe, U., Wawer, I., Diehl, B., Eds.; Wiley-VCH: Weinheim, Germany, 1999, pp 61–76.
34. Acharya, P.; Cheruku, P.; Chatterjee, S.; Acharya, S.; Chattopadhyaya, J. *J. Am. Chem. Soc.* **2004**, *126*, 2862–2869.
35. Abe, H.; Nakanishi, H. *J. Peptide Sci.* **2003**, *9*, 177–186.
36. Kahyaoglu, A.; Jordan, F. *Protein Sci.* **2002**, *11*, 965–973.
37. Ermondi, G.; Caron, G.; Bouchard, G.; van Balen, G. P.; Pagliara, A.; Grandi, T.; Carrupt, P. A.; Fruttero, R.; Testa, B. *Helv. Chim. Acta* **2001**, *84*, 360–374.
38. Tam, K. Y.; Quéré, L. *Anal. Sci.* **2001**, *17*, 1203–1208.
39. Espinosa, S.; Bosch, E.; Rosés, M. *J. Chromatogr. A.* **2002**, *947*, 47–58.
40. Espinosa, S.; Bosch, E.; Rosés, M. *J. Chromatogr. A.* **2002**, *945*, 83–96.
41. Shoskes, D. A. *Can. J. Urol.* **2001**, *8*, 24–28.
42. Nangia, A.; Andersen, P. H.; Berner, B.; Maibach, H. I. *Contact Dermatitis* **1996**, *34*, 237–242.
43. Skopp, G.; Potsch, L. *Int. J. Legal Med.* **1999**, *112*, 213–221.
44. de la Torre, R.; Farre, M.; Navarro, M.; Pacifici, R.; Zuccaro, P.; Pichini, S. *Clin. Pharmacokinet.* **2004**, *43*, 157–185.
45. Newton, D. W.; Kluza, R. B. *Drug Intell. Clin. Pharm.* **1978**, *12*, 546–554.
46. Box, K.; Bevan, C.; Comer, J.; Hill, A.; Allen, R.; Reynolds, D. *Anal. Chem.* **2003**, *75*, 883–892.
47. Dressman, J. B.; Amidon, G. L.; Reppas, C.; Shah, V. P. *Pharm. Res.* **1998**, *15*, 11–22.
48. Avdeef, A. Assessment of Distribution–pH Profiles. In *Lipophilicity in Drug Action*; Pliska, V., Testa, B., van de Waterbeemd, H., Eds.; VCH: Weinheim, Germany, 1995, pp 109–139.

49. Slater, B.; McCormack, A.; Avdeef, A.; Comer, J. E. *J. Pharm. Sci.* **1994**, *83*, 1280–1283.
50. Streng, W. H.; Hsi, S. K.; Helms, P. E.; Tan, H. G. *J. Pharm. Sci.* **1984**, *73*, 1679–1684.
51. Streng, W. H.; Tan, H. G. H. *Int. J. Pharm.* **1985**, *25*, 135–145.
52. Ozturk, S. S.; Palsson, B. O.; Dressman, J. B. *Pharm. Res.* **1988**, *5*, 272–282.
53. Avdeef, A. *Pharm. Pharmacol. Commun.* **1998**, *4*, 165–178.
54. Stuart, M.; Box, K. *Anal. Chem.* **2005**, 77, 983–990.
55. Avdeef, A.; Comer, J. E. A.; Thomson, S. J. *Anal. Chem.* **1993**, *65*, 42–49.
56. Takács-Novák, K.; Box, K. J.; Avdeef, A. *Int. J. Pharm.* **1997**, *151*, 235–248.
57. Rived, F.; Canals, I.; Bosch, E.; Rosés, M. *Anal. Chim. Acta* **2001**, *439*, 315–333.
58. Comer, J. E. A.; Avdeef, A.; Box, K. J. *Am. Lab.* **1995**, *4*, 36C–36I.
59. Morgan, M. E.; Liu, K.; Anderson, B. D. *J. Pharm. Sci.* **1998**, *87*, 238–245.
60. Allen, R. I.; Box, K. J.; Comer, J. E. A.; Peake, C.; Tam, K. Y. *J. Pharm. Biomed. Anal.* **1998**, *17*, 699–712.
61. Tam, K. Y.; Takács-Novák, K. *Anal. Chim. Acta* **2001**, *434*, 157–167.
62. Takács-Novák, K.; Tam, K. Y. *J. Pharm. Biomed. Anal.* **2000**, *21*, 1171–1182.
63. Comer, J. E. A. High Throughput Measurement of log *D* and p*K*ₐ. In *Drug Bioavailability. Estimation of Solubility, Permeability and Bioavailability*; Artursson, P., Lennernas, H., van de Waterbeemd, H., Eds.; Wiley-VCH: Weinheim, Germany, 2003, pp 21–45.
64. Gluck, S. J.; Steele, K. P.; Benko, M. H. *J. Chromatogr. A* **1996**, *745*, 117–125.
65. Caliaro, G. A.; Herbots, A. A. *J. Pharm. Biomed. Anal.* **2001**, *26*, 427–434.
66. Wan, H.; Holmen, A.; Nagard, M.; Lindberg, W. *J. Chromatogr. A* **2002**, *979*, 369–377.
67. Poole, S. K.; Patel, S.; Dehring, K.; Workman, H.; Poole, C. F. *J. Chromatogr. A* **2004**, *1037*, 445–454.
68. Sjöberg, H.; Karami, K.; Beronius, P.; Sundelöf, L. *Int. J. Pharm.* **1996**, *141*, 63–70.
69. Peck, C. C.; Benet, L. Z. *J. Pharm. Sci.* **1978**, *67*, 12–16.
70. Hansen, J. B.; Hafliger, O. *J. Pharm. Sci.* **1983**, *72*, 429–431.
71. Vazquez, J. L.; Berlanga, M.; Merino, S.; Domenech, O.; Vinas, M.; Montero, M. T.; Hernandez-Borrell, J. *Photochem. Photobiol.* **2001**, *73*, 14–19.
72. Saurina, J.; Hernandez-Cassou, S.; Tauler, R.; Izquierdo-Ridorsa, A. *Anal. Chim. Acta* **2000**, *408*, 135–143.
73. Perrin, D. D.; Dempsey, B.; Serjeant, E. P. *pK*ₐ *Prediction for Organic Acids and Bases*; Chapman and Hall: London, UK, 1981.
74. Hammett, L. P. *Chem. Rev.* **1935**, *17*, 125–136.
75. Taft, R. W., Jr. Separation of Polar, Steric and Resonance Effects in Reactivity. In *Steric Effects in Organic Chemistry*; Newman, M. S., Ed.; John Wiley: New York, 1956, pp 556–675.
76. Clark, J.; Perrin, D. D. *Q. Rev.* **1964**, *18*, 295–320.
77. Biggs, A. I.; Robinson, R. A.; *J. Chem. Soc. London* **1961**, 338–393.
78. Sirius Analytical Instruments *Sirius Technical Application Notes 1*; Sirius Analytical Instruments: Forest Row, UK, 1994.
79. Sirius Analytical Instruments *Sirius Technical Application Notes 2*; Sirius Analytical Instruments: Forest Row, UK, 1995.
80. Sirius Analytical Instruments. Unpublished data, 2005.
81. Loeuillet-Ritzler, F.; Faller, B. High-Throughput p*K*ₐ with the Sirius Profiler SGA Using a Co-Solvent Approach. In *Pharmacokinetic Profiling in Drug Research*; Testa, B., Krämer, S. D., Wunderli-Allenspach, H., Folkers, G., Eds.; VHCA: Zürich, Switzerland, 2006, Poster A19 (CD).
82. Kenseth, J.; Bastin, A.; Gossett, K.; Prog, H. High Throughput Determination of p*K*ₐ Values of Soluble and Insoluble Compounds Using Multiplexed Capillary Electrophoresis with Absorbance Detection. In *Pharmacokinetic Profiling in Drug Research*; Testa, B., Krämer, S. D., Wunderli-Allenspach, H., Folkers, G., Eds.; VHCA: Zürich, Switzerland, 2006, Poster Session A, 3 (on CD).

Biography

John E A Comer, after studying chemistry at the universities of Warwick and Newcastle upon Tyne, joined the analytical instrument industry, working for several years in technical and marketing roles with Orion Research Inc., before co-founding Sirius Analytical Instruments in 1990. Sirius Analytical Instruments develops and manufactures instruments for the measurement of drug pK_a, lipophilicity, and solubility, and provides analytical services. Currently the technical director, Comer leads a team of chemists and software engineers who are researching into novel aspects of the measurement of physicochemical properties of drugs and other small organic molecules. Achievements include the development and introduction of automated instrumentation for the measurement of pK_a by pH-metric and pH/ultraviolet techniques, together with several publications in the field of pK_a measurement. Current research is concerned with the behavior of ionizable drugs, looking at parallel artificial membrane permeability assay and its correlation with $\log P$ and $\log D$; solubility measurement by a novel, pH-metric approach; miniaturization of measurement apparatus; ultraviolet methods for pK_a; solubilization of drugs for analysis; and $\log P$ measurement using small sample quantities.

Comprehensive Medicinal Chemistry II
ISBN (set): 0-08-044513-6

ISBN (Volume 5) 0-08-044518-7; pp. 357–397

5.17 Dissolution and Solubility

A Avdeef, D Voloboy, and A Foreman, pION Inc., Woburn, MA, USA

5.17.1 Introduction

This chapter critically reviews the experimental strategies and theoretical basis of the pH-dependent measurement of solubility and dissolution of multiprotic ionizable drugs. Property assessment of ionizable molecules is more complicated than that of neutral molecules, and in this review, focus is primarily placed on the former class of molecules. Well-tested methods, suitable for application in discovery as well as preformulation, and appropriate for the analysis of solubility and dissolution of problematic molecules, are discussed. Medicinal chemists confronted by low-solubility compounds may find several of the discussed topics helpful. If the complexity of some parts of the presentation are of limited direct interest, the medicinal chemist will at least have some basis for asking critical questions of their physical chemistry counterparts.

Several carefully selected molecules that illustrate important concepts are considered in detail, with some literature data illustrated, using improved computational tools. Considered elsewhere are extensive compilations of experimental results.[1] Several excellent publications serve as important background material for the present review, stressing solubility,[2] salt formation,[3] and dissolution,[4] and covering the solubility analysis of complicated systems.[5]

The solubility treatment here should be of practical interest to discovery-support physical chemists intent on making fast, quality measurements of physical properties of test compounds in lead finding and optimization, and on effectively communicating these measured results to medicinal chemists, so that rational structural modifications leading to improved properties can be made.[6–10] Moreover, this treatment should be particularly useful to downstream preformulation scientists, sometimes charged with the daunting task of formulating poorly soluble molecules into drug products.[3,5]

The dissolution coverage is primarily focused on the disk intrinsic dissolution rate (DIDR) and powder sample dissolution rate measurement used in early preformulation, rather than late-stage highly regulated development and quality control applications, which have been covered very well in the recent literature.[11–17] The close relationship between solubility and dissolution is stressed throughout this review. The case study molecules have been selected in many instances to link applications in solubility to those in dissolution.

5.17.1.1 Solubility

The measurement of the solubility of multiprotic drugs can be very challenging.[18] Speciation characteristics (usually pH-dependent), such as multiple (often overlapping) ionization,[19] complexation,[20] aggregation,[21] micelle formation,[22,23] and common-ion effects (*see* Section 5.17.6.5.2)[24,25] can hamper the interpretation of the measurements. Potential experimental artifacts, such as adsorption to microporous filters, plastic (or glass) surfaces, 'promiscuous inhibitor' molecules forming very small (0.1–0.2 μm) particles that can pass through filters,[26] and a multitude of other confounding effects may escalate the interlaboratory variability of the measured values. Different polymorphic forms of a compound may produce subtle differences in solubility and dissolution behavior.[27,28] Unanticipated polymorphs isolated during the scale-up of synthesis can mislead interpretations of measurements.

Two types of solubility assays are considered in detail, based on using (1) just a pH electrode as a concentration sensor (the dissolution titration template (DTT) method),[29–31] and (2) ultraviolet (UV) microtitre plate technology.[8,9,19] In the potentiometric method, it is not necessary to separate the solid from the solution in order to deduce solubility. In the UV-metric methods, however, the solid usually needs to be separated from the solution, either by microfiltration or centrifugation, before concentrations are determined. The small-volume dissolution apparatus (*see* Sections 5.17.6.3 and 5.17.7.2), which is used also for solubility measurement, however, has a special algorithm to allow the direct concentration determination in a turbid solution. As special case studies, the apparent aggregation reactions of piroxicam, indomethacin, 2-naphthoic acid, and phenazopyridine are illustrated using the 'shift in the apparent pK_a' method.[1]

The early results of the DDT solubility–pH method have been successfully compared with those obtained by traditional shake-flask methods,[30,35,36] and an optimistic assessment has been suggested.[37] In the US Food and Drug Administration's biopharmaceutics classification system (BCS; *see* 5.42 The Biopharmaceutics Classification System), the new acid–base titration method is deemed suitable for the determination of solubility.[38]

5.17.1.2   Dissolution

The theory of dissolution kinetics of multiprotic ionizable molecules, based on the convective diffusion with simultaneous chemical reaction (CDR) model, explored extensively in the pioneering contributions of Higuchi *et al.*,[40] Mooney and co-workers,[41–44] McNamara and co-workers,[45–47] and Southard *et al.*,[48] and a number of others, is still evolving, and according to McNamara *et al.*[4]:

Today, the opportunity is still available, for those inclined to numerical analysis, to derive a robust numerical dissolution model capable of including contributions due to diffusion, convection, and reaction which can encompass the entire pH scale.

Our further analysis of the CDR model[39] has uncovered some limitations to the assumptions made in earlier treatments. With the new computational treatment, only briefly described in this article (*see* Section 5.17.5), we explore the insights gained from the improved model.

For example, in the new treatment, the electrostatic fields created in the aqueous boundary layer (ABL) (surrounding the dissolving solid) by the convection and diffusion of ions with different mobilities are explicitly considered, using the Vinograd–McBain[49] formalism. This is an important generalization of the earlier CDR treatments.

Concentration–time profiles as a function of pH, buffers, salt, and aggregation are considered. Concentration–position profiles of all species in the ABL, including solid species, are also considered under the above conditions. An intriguing 'fog of solid' in the ABL can be predicted under certain circumstances. The dissolution flux profiles of indomethacin, reported earlier,[41,42] are re-evaluated, taking into account suspected anionic aggregates formed during dissolution of this molecule.

The dissolution kinetics equations for two variants in the CDR model, (1) rotating disk[50] and (2) spherical particles,[51,52] are discussed. The new small-volume (2–20 mL) dip probe UV detection apparatus, developed for polymorph screening, stability assessment, and solubility determination, is applied to phenazopyridine hydrochloride dissolution as a case study.

5.17.1.3   The Relation between Solubility and Dissolution

Solubility and dissolution are closely linked. The thin ABL adjacent to the surface of the dissolving solid particles, separating the solid–liquid interface from the bulk medium, is the rate-limiting barrier in dissolution. When the solid is introduced to the dissolution medium, almost instantaneously (assuming 'wettability' is not an issue, a topic beyond the scope of the present coverage), the concentration of the drug at the solid–liquid interface becomes equal to its solubility under the conditions of the bulk medium. After a further short delay (typically <1 s), the steady state (when the dissolution rate is constant in time) is established. At that time, the sink condition is still in effect (since the amount of drug dissolved in the bulk medium is vanishingly low), and the linear concentration gradient which initially drives the dissolution process (Fick's law of diffusion) is equal to the solubility divided by the thickness of the ABL. As more compound dissolves (>10% of the solubility value), the process becomes more complicated to analyze (due to nonsink conditions), but the concentration of the drug at the solid–liquid interface still remains equal to the solubility. Dissolution will continue until the concentration of the drug in the bulk medium becomes equal to the solubility, or until all of the solid dissolves. In such a view, solubility drives dissolution.

5.17.1.4   Why Should Medicinal Chemists Care about Dissolution?

In discovery, compounds are usually stored in dimethyl sulfoxide (DMSO), and when compounds are initially isolated as a solid, their form is often amorphous. Early evaluation of the physical properties are made. But after a period of storage (e.g., 6 months or more), some amorphous compounds may undergo partial transformation into a crystalline form. Properties evaluated at that time can be very different from those observed earlier. To anticipate or check for such polymorphic changes, dissolution measurements of minute quantities of the powdered sample may be made, and compound-sparing small-volume dissolution screening for polymorphic changes can be practical.

As medicinal chemists scale up the synthesis of a particularly promising compound, it is not uncommon that the isolated solid is morphologically transformed in new batches from its earlier states, as a result of altered procedures. Dissolution measurements on small quantities of the compound can reveal such polymorphic changes. If the particle size distribution is not known, such measurements are only approximate, but still may be useful for comparative assessments.

In the earliest stages of development, where drug delivery issues are first explored and syntheses of promising compounds are scaled up further, there is often a shortage of the candidate material to use for some of the desirable testing. The traditional US Pharmacopeia (USP)-type dissolution methods[11-17] are based on 500 or 900 mL assay volumes. There simply is not enough of the active pharmaceutical ingredient (API) then, and API-sparing methods need to be considered. Excipient-sparing methods are also needed, especially when expensive media, such as simulated intestinal fluid (SIF) or human intestinal fluid, are used. However, at formulation-oriented conferences (e.g., the long-standing June Land-of-Lakes Conference), presentations on 'API-sparing' topics still describe the use of 900 mL vessels. The apparent reluctance among some of the practitioners to use smaller volumes stems from concerns that the familiar (USP) 'hydrodynamic conditions' may not be valid in small-volume settings, when employed volumes are as low as 2 mL. Were small-volume dissolution apparatus used, SIF media would be cost-effective to use in early development.

The powder sample dissolution rates measured in 2 mL volumes will very likely be different from those done in 900 mL apparatus. This may be a source of confusion in discovery projects. The discussion of scalability of measurement (e.g., 2–900 mL), done under non-USP conditions, both its limitations and its advantages, could be of use to medicinal chemists, before their compounds transfer into development. The present theoretical discussion of dissolution simulation can help to address such issues. The small-volume, small-sample consumption methods described here could allow critical tests to be performed at more upstream positions than conventional methods would allow.

Since the dip-probe UV method (see Section 5.17.6.2) does not depend on the separation of solid form the suspension, in situ measurement of concentration is possible. This enables the method to be applied to the assessment of the in situ stability of very small quantities of the API in various test media.

5.17.2   Theoretical Solubility

The idealized (Henderson–Hasselbalch type) relationships[1,2,19] between solubility and pH can be easily derived for any given equilibrium model, although this may not be a familiar process to many practicing pharmaceutical scientists. The model refers to a set of equilibrium equations and the associated equilibrium constants.

5.17.2.1   Monoprotic Weak Acid, HA

In a saturated solution, the two relevant equilibrium equations for the case of a monoprotic weak acid, HA, are the ionization reaction (defined by pK_a) and the solubility reaction (defined by S_0, the intrinsic solubility − the solubility of the uncharged molecular species, [HA]). The solubility, S, at a particular pH is defined as the mass balance sum of the concentrations of all of the species dissolved in the aqueous phase:

$$S = [A^-] + [HA] \qquad [1]$$

where the square brackets denote the concentration of species. It is useful to convert eqn [1] into an expression containing only constants and $[H^+]$ (as the only variable), by substituting the ionization and solubility equations into it[1,2,19]:

$$S = S_0(10^{-pK_a+pH} + 1) \qquad [2]$$

with the limiting linear solutions,

$$\log S = \log S_0 \qquad \text{for } pH \ll pK_a \qquad [3a]$$

$$\log S = \log S_0 - pK_a + pH \qquad \text{for } pH \gg pK_a \qquad [3b]$$

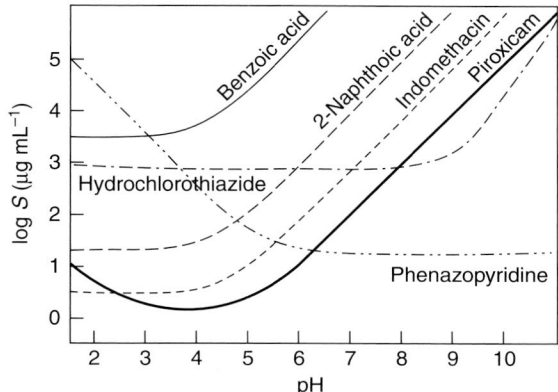

Figure 1 Log solubility–pH profiles for monoprotic acids (thin solid, long-dash, and short-dash curves), diprotic acid (dash-dot curve), a base (dash-dot-dot curve), and ampholyte (thick solid curve). For monoprotic acids, the tangent horizontal line segment denotes the intrinsic solubility, and the slope of +1 segment denotes the increase in the apparent solubility due to the formation of the anionic species. For diprotic acids, the ascending segment has a slope of +2. For monoprotic bases, the ascending portion of the curve has a slope of −1. An ampholyte has both acidic and basic behavior to the ascending portions of the curves. Salt precipitation is not indicated in these curves.

Figure 1 shows a plot of log S versus pH for several monoprotic acids (benzoic acid, 2-naphthoic acid, and indomethacin). Note that in eqn [3b], log S is defined by a straight line as a function of pH, exhibiting a slope of 1. At the bend in the curves in **Figure 1**, the pH equals the pK_a.

5.17.2.2   Diprotic Weak Acid, H_2A

The solubility for a diprotic weak acid is defined by (cf., eqn [1])

$$S = [A^{2-}] + [HA^-] + [H_2A] \qquad [4]$$

The conversions of all variables into expressions containing only constants and $[H^+]$ produces[1,19]

$$S = S_0(10^{-pK_{a1}-pK_{a2}+2pH} + 10^{-pK_{a1}+pH} + 1) \qquad [5]$$

where $S_0 = [H_2A]$, and $pK_{a1} < pK_{a2}$ are the two ionizations constants. The dash-dot curve in **Figure 1** shows the plot of log S versus pH for the diprotic molecule hydrochlorothiazide, which has overlapping pK_as (**Table 1**). For pH $\ll pK_{a1}$, the function reduces to a horizontal line, with log $S = \log S_0$. For pH values between pK_{a1} and pK_{a2}, log S is straight line as a function of pH, exhibiting a slope of 1 (provided the pK_a constants are not overlapping). For pH $\gg pK_{a2}$, log S is a straight line as a function of pH, exhibiting a slope of 2, as is evident above pH 10 in **Figure 1**.

5.17.2.3   Monoprotic Weak Base, B

For a monoprotic weak base,

$$S = [BH^+] + [B] \qquad [6a]$$

$$= S_0(10^{pH+pK_a} + 1) \qquad [6b]$$

where $S_0 = [B]$. The limiting linear solutions are

$$\log S = \log S_0 \qquad \text{for pH} \gg pK_a \qquad [7a]$$

$$\log S = \log S_0 + pK_a - pH \qquad \text{for pH} \ll pK_a \qquad [7b]$$

The dash-dot-dot curve in **Figure 1** shows the plot of log S versus pH for the weak base phenazopyridine. Note the sign reversal between eqns [7b] and [3b]. This shows up as the mirror relationship between this base curve and those of the acids.

Table 1 Equilibrium constants for selected compounds[a]

Compound	MW	pKa		−log S_0[b]	S_0 (μg mL^{-1})	−log K_{sp}[c]
Benzoic acid	122.1	3.98 4.03[d]		1.59 1.67[d]	3100 2600	
2-Naphthoic acid	172.2	4.16 4.31[i] 4.02[d] 4.30[e]		3.93 3.76[i] 3.89[d] 3.38[e]	20 30 22 72	
Indomethacin	357.8	4.42 4.17[d]		5.06 5.58[d]	3 1	
Hydrochlorothiazide	297.7	9.95	8.76	2.63	700	
Piroxicam	331.4	5.07 5.34[f]	2.3 2.3	5.48 4.18[g]	1 22	
Phenazopyridine hydrochloride	249.7	5.15 5.20[h]		4.24 3.83[h]	14 37	3.53[h]

[a] 25 °C, 0.15 M (KCl) ionic strength, solubility determined by the DTT method,[31] unless otherwise noted.
[b] Instrinsic solubility, based on molar units.
[c] $K_{sp} = [BH^+][Cl^-]$, the solubility product.
[d] 25 °C, 0.5 M (KCl).[42]
[e] 25 °C, 0.1 M (KCl).[46]
[f] 37 °C, 0.15 M (KCl) (pION, unpublished work).
[g] Re-analysis of 37 °C data of Jinno et al.[58]
[h] 37 °C, no ionic strength adjuster added.[53]
[i] 25 °C, 0.01 M ionic strength, determined by the UV method (pION, unpublished work).

Also shown in **Figure 1** is the case of the ampholyte piroxicam (thick line). An ampholyte is a molecule containing both acid and base functionality (*see* 5.16 Ionization Constants and Ionization Profiles). The curve shows features of both an acid (pH > 5) and a base (pH < 3).

5.17.2.4   Gibbs pK_a (Weak Base Example)

Although **Figure 1** properly conveys the shapes of solubility–pH curves in saturated solutions of uncharged ionizable species, the effect of salt precipitation is not shown. It has been observed, for example, that in 0.15 M NaCl solutions, when the solubility exceeds its lowest (intrinsic) value by about four orders of magnitude for a weak acid and about three orders of magnitude for a weak base, the sodium and chloride salts of the charged drugs, respectively, will likely precipitate, and under the right circumstances, co-precipitate with the free acid or free base.[30] This has been called the 'sdiff 3–4' effect.[1,19] Salt formation stunts further increases in equilibrium solubility, which is not represented in **Figure 1**.

Consider the case of a very concentrated solution of the weak base hypothetically titrated with HCl from high pH (well above its pK_a) down in pH to the point where the solubility product is exceeded. At the start, the saturated solution contains only the free base precipitate. As the pH is decreased below the pK_a, the solubility increases, as more of the free base ionizes, and a portion of the solid B dissolves, as indicated by the dash-dot-dot curve in **Figure 1**. When the concentration of BH$^+$ times the concentration of Cl$^-$ reaches the solubility product value, at the critical pH, solid BH$^+$Cl$^-$ starts to precipitate, but at the same time there may be remaining free base, B, precipitate. The simultaneous presence of the two solids invokes the Gibbs phase rule constraint, forcing the pH and the solubility to be precisely constant. (As shown by Gibbs, the thermodynamic phase rule describes the possible degrees of freedom in a closed system at equilibrium, which puts a limit on the number of independent reactants that describe the system uniquely, for a given number of separate phases in the system.) As long as B(s) still remains, the HCl titrant, in effect, is diverted entirely into the solid state, to transform the remaining B(s) into BH$^+$Cl$^-$(s). This process produces a 'perfect' pH buffer.[29] This special pH point has been called the Gibbs pK_a, indicated as pK_a^{Gibbs}.[1,29] The equilibrium equation associated with this phenomenon is

$$BH^+Cl^-(s) \rightleftharpoons B(s) + H^+ + Cl^- \qquad K_s = [H^+][Cl^-] \qquad [8]$$

where the constant, $K_a^{Gibbs} = [H^+] = K_s/[Cl^-]$. This is a conditional constant, inversely depending on the value of the background $[Cl^-]$, as is illustrated in **Figure 2**, for the profile of phenazopyridine hydrochloride at two different levels of added KCl, using the 37 °C data of Serajuddin and Jarowski[53] in the simulation (see **Table 1**). The decrease in the solubility at low pH is an example of the common-ion (chloride) effect.[24,25] The chloride contributed by the HCl, needed to lower the pH, drives down the solubility of the charged form of the drug.

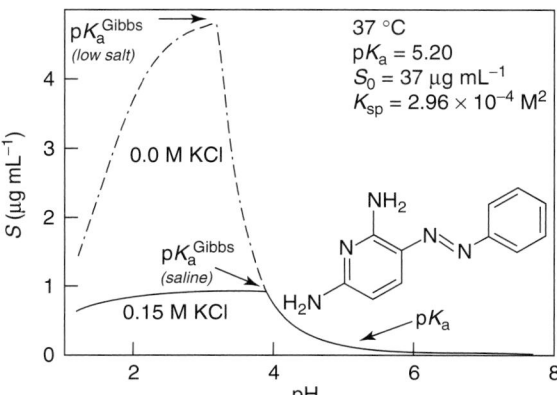

Figure 2   Log solubility–pH profiles of phenazopyridine,[53] indicating the formation of chloride salt precipitates. The solid curve is the predicted profile in the background of 0.15 M KCl, whereas the dash-dot curve is the predicted profile in the presence of no added KCl. In both cases, the simulation considered a titration of the drug suspension solution using concentrated HCl, from high pH to low pH. The 'common-ion' effect refers to the decrease in solubility of the salt below pH 2, due to the increased presence of chloride ions entering the solution from the HCl titrant used to lower the pH. The point where both the solid free base and the chloride salt of the conjugate acid coprecipitate is referred to as the Gibbs pK_a.

Since solubility is constant (the Gibbs phase rule constraint) during the two-solid interconversion process, one may set eqn [6b] equal to eqn [2], noting that for a weak base, $10^{-pK_a+pH} \ll 1$ and $10^{+pK_a-pH} \gg 1$, to get (in logarithmic form) the conditional expression[1]

$$\log S_i - \log S_0 = pK_a - pK_a^{Gibbs} \qquad [9]$$

where S_i is the solubility of the ionized form of the drug.

5.17.3   Solubility Measurement

5.17.3.1   Shake-Flask Method

The traditional solubility measurement[2] under equilibrium conditions is largely a manual, time-intensive process, taking many hours, sometimes as long as several days, at a given pH. It is often a simple procedure. The drug is added to a standard buffer solution until saturation occurs, indicated by undissolved excess drug. The saturated solution is shaken at constant temperature as equilibration between the two phases establishes. After centrifugation and/or microfiltration, the concentration of the substance in the supernatant solution is then determined. If a solubility–pH profile is required, then the measurement needs to be performed in parallel in several different pH buffers. The whole process can easily take a week to complete.

Many of the measurements by the shake-flask method use a pH 6.5 or 7.4 buffer solution (often 50 mM phosphate) as the medium. Alternatively, a biomimetic SIF medium can be used, which consists of phospholipids, bile salts, and other ingredients, in addition to buffer.[12,13] Many low-solubility drugs show elevated solubility in a SIF medium. Due to the complexity and relatively high cost of the SIF medium, it is used less often than simple buffers.

It is critical that the pH at the end of each assay be carefully measured, when working with ionizable drugs. Surprisingly, this is not always done.

5.17.3.2   Potentiometric Method

In acid–base dissolution titrations of ionizable compounds, the titrant perturbs the saturated system at equilibrium and causes a portion of the suspended solid to dissolve (taking the system out of equilibrium), a process which can be very slow, depending on a number of factors related to dissolution kinetics (see Section 5.17.5). In order to collect pH data suitable for the reliable determination of solubility constants, it is necessary to allow the titrated system to return to an equilibrium state following each titrant addition.

The DTT method consists of an unusual data collection regime, based on the use of precalculated titration curves serving as templates.[29–31] The instrument takes as input parameters the pK_a and the octanol/water partition coefficient (log P_{ow}) of the sample. The latter parameter is used to estimate the intrinsic solubility, S_0, according to the simple linear expression log $S_0 = 1.17 - 1.38 \log P_{ow}$.[54] Using the pK_a and the estimated S_0, the instrument calculates the entire titration curve before starting the assay. The curve serves as a template that the instrument uses as a guide on how to collect individual pH measurements in the course of the titration.

Enough sample is weighed to cause precipitation during some portion of the titration. The pH where precipitation takes place is estimated from the template simulation, and the data collection strategy is set accordingly.[31] Titrations of weak acids begin at low pH, and those of weak bases begin at high pH. KOH (or HCl) titrant is dispensed accurately and slowly into the slurry, to drive the pH of the solution in the direction of dissolution, eventually well past the point of complete dissolution.

As titrant is added, careful measurements of pH are made. The instrument dramatically slows down the rate of data taking as the point of complete dissolution approaches in the titration. It uses the Noyes–Whitney dissolution rate expression,[2] with the necessary reactant concentrations calculated on the basis of the template model, to estimate how long to wait for equilibrium to be reached following the addition of titrant. Consequently, some data are collected at moderate speeds and some at very slow speeds. Only after the precipitate completely dissolves (assessment based on the template prediction), does the instrument collect the remainder of the data rapidly, as would an ordinary titrator. Typically, 4–10 h is required for the entire equilibrium solubility data taking (20–50 pH points). The more insoluble the compound is anticipated to be (based on the template), the longer the recommended assay time.

5.17.3.3   High-Throughput Microtitre Plate Methods

5.17.3.3.1   Turbidimetric assays ('kinetic' solubility)
The turbidity method described by Lipinski et al.,[6] which assesses 'kinetic' solubility, although not thermodynamically rigorous, is an attempt to rank molecules according to expected solubilities. Versions of the method are used by several

pharmaceutical companies, with custom-built equipment. Nephelometers using 96-well microtitre plates have been adapted into these systems.[7] Often, single-pH (6.5 or 7.4) 50 mM buffer solutions are used. Note that phosphate at such high concentration can form complicated-stoichiometry salts with positively charged drugs.[35] Buffers such as MES and HEPES are less problematic.[1] Being a result of a 'kinetic' method, the measured solubility can depend dramatically on the timings used in the assay. In this regard, standard methodology is not well defined. For binning purposes, the turbidity analysis is fine. However, the turbidity-based measurements may not be of sufficient quality for medicinal chemistry structure–activity relation (SAR) applications.

5.17.3.3.2 High-performance liquid chromatography (HPLC) based assays

Several pharmaceutical companies have taken the classical saturation shake-flask method and transferred it to 96-well plate technology and a robotic liquid dispensing system. Analyses are performed with reverse-phase HPLC, often using fast generic gradient methods. In some laboratories, the DMSO is first eliminated by a freeze–drying procedure, before the aqueous buffers are added. Data handling is often the rate-limiting step in the customized operations.[7–9] Such methods can be very reliable,[35] but the lack of method standardization is potentially problematic.

5.17.3.3.3 Direct ultraviolet assay

A highly-automated direct-UV 96-well microtitre plate equilibrium solubility method has been described recently.[19] Samples are typically introduced as 10 mM DMSO solutions in a 96-well microtitre plate. The robotic liquid-handling system (e.g., the Beckman FX series) draws a 3–10 μL aliquot of the DMSO stock solution and mixes it into an aqueous universal buffer solution, so that the final (maximum) sample concentration is 50–150 μM and the DMSO concentration is 0.5–1.0% (v/v). The solutions are stirred magnetically, then filtered, and assayed by direct UV spectrophotometry. The direct UV measurement eliminates the need for method development, and easily lends itself to more complete automation and, hence, higher throughput. The buffers used in the assay are automatically prepared by the robotic system. The quality controls of the buffers and the pH electrode are performed by alkalimetric titration, incorporating the Avdeef and Bucher[55] procedure. This method has been applied to determine solubilities as low as 2 ng mL^{-1}.[73]

5.17.4 Common Complications in Solubility Measurement of Low-Solubility Compounds

5.17.4.1 Aggregation Reactions in Solubility Measurement

When a compound forms a dimer or a higher-order oligomer in aqueous solution, the characteristic log S–pH profile (e.g., see **Figure 1**) often indicates the observed pK_a (pH at the bend in the log S–pH profile in **Figure 1**), to be different from the true pK_a, as measured in dilute solutions. Besides aggregation, other anomalies can cause the pK_a to appear shifted, including compound adhesion to the filter material, DMSO binding, and excipient effects. This shift-in-the-pK_a effect has been discussed previously, although its use appears not to be widespread.[1]

If an uncharged molecule undergoes some speciation anomaly (aggregation, DMSO binding, filter retention, etc.), weak acids will indicate an apparent pK_a higher than the true pK_a, as indicated schematically in **Figure 3a**, and weak bases will indicate an apparent pK_a lower than the true pK_a, as indicated schematically in **Figure 3b**. If the observed shifts are opposite to those stated above, then the charged (rather than the neutral) species is involved in the anomaly (**Figures 3c** and **3d**). Anomalies involving charged species also reveal themselves by nonunit slopes in the diagonal portions of the log S–pH curves (see below). Although the precise mechanism of the anomaly may not be apparent, the shift combined with the apparent solubility will often reveal the intrinsic (unshifted) solubility, S_0. **Figure 3** illustrates these four cases schematically as 'stick' diagrams.

In this section we will present several examples of putative aggregation reactions, involving 2-naphthoic acid, indomethacin, phenazopyridine, and piroxicam. The medicinal chemist may wish to skip the mathematical treatment below, but an appreciation of the characteristic shapes of log S–pH profiles under conditions of 'anomaly' would be beneficial.

5.17.4.1.1 Indomethacin and 2-naphthoic acid anion aggregation

Roseman and Yalkowsky[22] described the solubility–pH behavior of prostaglandin $F_{2\alpha}$, and noted that anionic aggregates appeared to form, as in **Figure 3c**. No further characterization of the aggregation behavior was reported. There are numerous other literature examples of the phenomenon, which is often not fully assessed.

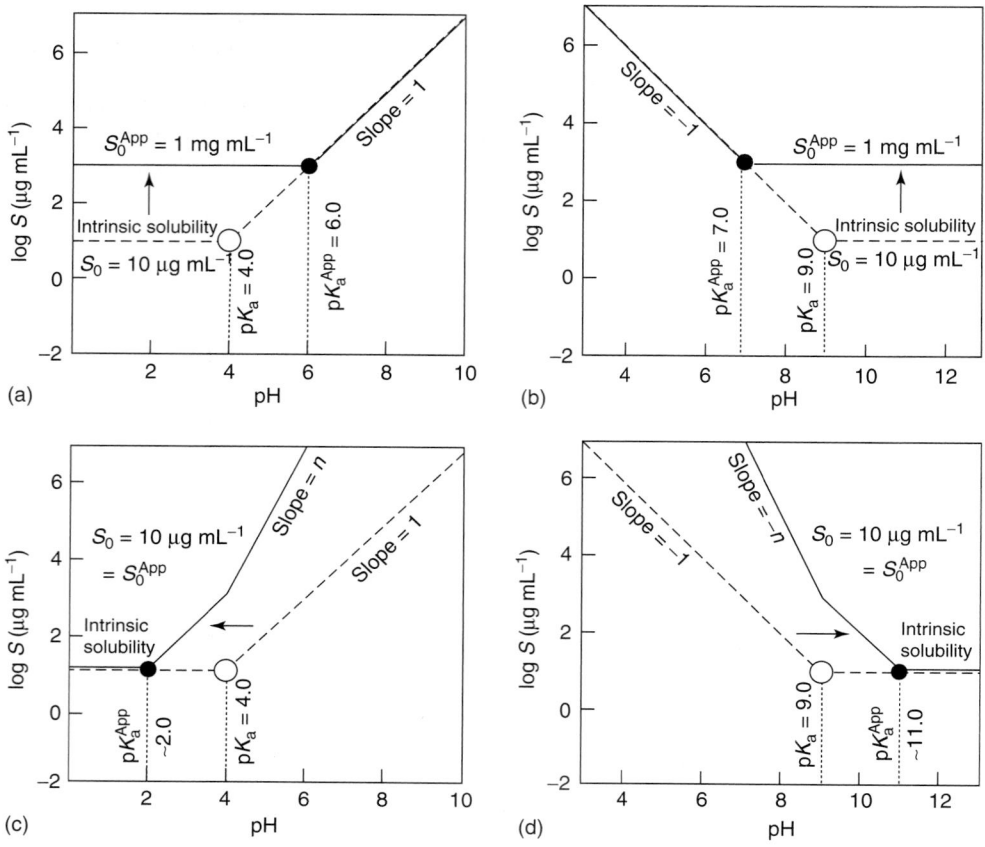

Figure 3 Stylized schematic log solubility–pH profiles, showing just the straight line segments, and not the curved regions linking them (as in **Figure 1**). (a) The solid segments are observed in the case of a weak acid, where neutral aggregates form. The dashed segments refer to the underlying curve that would have been observed had there been no aggregation. The difference between the apparent pK_a (solid circle) and the true pK_a (open circle), when subtracted from the apparent intrinsic log solubility (low pH horizontal region), reveals the hidden intrinsic log solubility. (c) The solid segments are observed in the case of a weak acid where anionic aggregates form. The expected slope in the right-most portion of the ascending solid plot is equal to the order of aggregation, n, if just a single aggregate stoichiometry exists. (b, d) The parallel cases to parts a and c, but concerning weak-base behavior (see the text).

Let us hypothesize that the oligomeric species A_n^{n-} forms. The required equilibrium equations and the associated concentration quotients to completely define the mass balance problem are

$$HA \rightleftharpoons H^+ + A^- \qquad K_a = [H^+][A^-]/[HA] \tag{10a}$$

$$HA(s) \rightleftharpoons HA \qquad S_0 = [HA] \tag{10b}$$

$$nA^- \rightleftharpoons A_n^{n-} \qquad K_n = [A_n^{n-}]/[A^-]^n \tag{11}$$

The expanded form of eqn [1] becomes

$$S = [A^-] + [HA] + n[A_n^{n-}] \tag{12}$$

Following the steps that led to eqn [2],[1,19] we now obtain[39]

$$\log S = \log S_0 + \log[1 + 10^{+(pH-pK_a)} + n10^{+\log K_n + n(pH-pK_a) + (n-1)\log S_0}] \tag{13}$$

Two limiting forms of eqn [13] may be posed as (see **Figure 3c**)

$$\log S = \log S_0 \quad \text{for pH} \ll pK_a \tag{14a}$$

$$\log S = \log n + \log K_n + n(\log S_0 - pK_a) + npH \quad \text{for pH} \gg pK_a \tag{14b}$$

Equation [14a] indicates that the formation of charged aggregates does not obscure the value of the intrinsic solubility in low-pH solutions. If high-quality data are available over an extensive range of pH, and aggregation takes place (as evidenced by an apparent pK_a significantly different from the true pK_a) then eqn [14b] may be useful in assessing the stoichiometry of aggregation, provided that aggregates can be described by a single value of n. A plot of log S versus pH (see eqn [14b] and **Figure 3c**) should indicate the value of n as the slope in the curve at high pH. The specialist reader is encouraged to explore other properties of eqn [13], using, for example, a spreadsheet.

Figure 4 contains log S versus pH plots of three 'anomaly' molecules, 2-naphthoic acid, indomethacin, and phenazopyridine, all measurements done by the UV microtitre plate method in 0.01 M ionic strength solutions. One

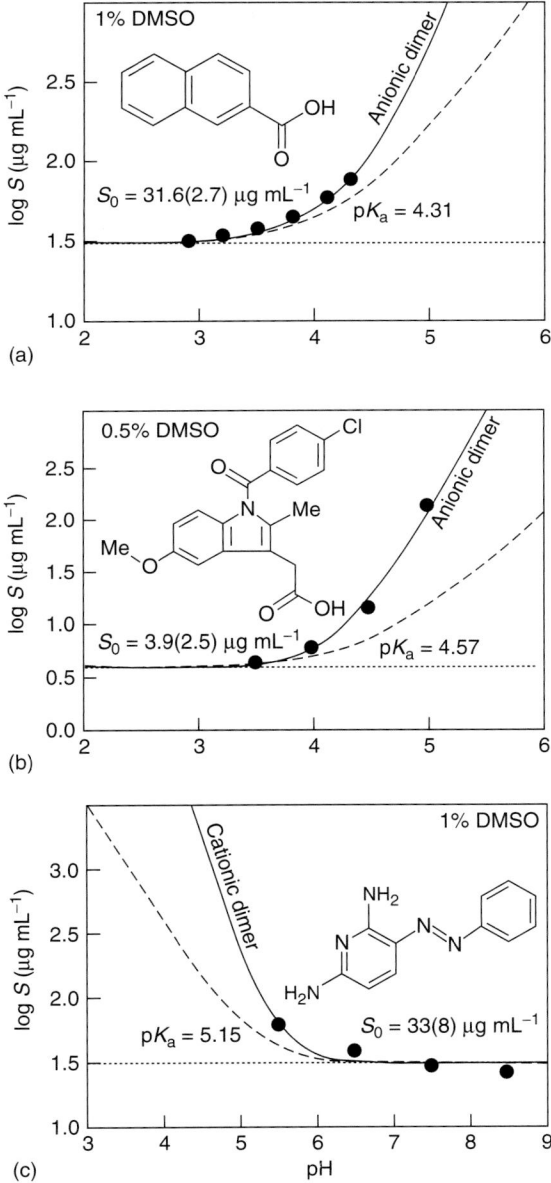

Figure 4 Specific examples of aggregation (cf. the schematic depictions in **Figure 3**). Solubility determined by the miniaturized 'self-calibrating' UV method.[19] (a) 2-Naphthoic acid, with an indication of anionic aggregation (0.2 μm filtration after 17 h, 1% DMSO). (b) Indomethacin, with an indication of extensive anionic aggregation (centrifugation after 23 h, 1.0% DMSO). (c) Phenazopyridine, with indications of cationic aggregation (0.2 μm filtration after 21 h, 1% DMSO). The dashed curves were calculated using the idealized Henderson–Hasselbalch equation, which neglects aggregation.

can see that the apparent pK_a appears to be different from the true pK_a. The direction of the shifts in the pK_a in the three cases suggest the presence of charged aggregates.

If one assumes that indomethacin forms a dimer ($n = 2$), then **Figure 4b** indicates that $K_2 = 1.8 \times 10^5\,M^{-1}$. For 2-naphthoic acid, a weaker dimerization constant is apparent ($K_2 = 1.5 \times 10^3\,M^{-1}$), consistent with the small shift in **Figure 4a**.

It is interesting to note that indomethacin solubility characterized in 0.3 M KCl solution indicates pK_a^{App} 4.81[56] (pK_a 4.17 in **Table 1**, 0.5 M KCl, from the same laboratory), and $-\log S_0^{App}$ 5.06 ($-\log S_0$ 5.58 in **Table 1**, from the same laboratory), indicating the predominant presence of neutral aggregates under the high-salt condition, an entirely different behavior from that observed in 0.01 M ionic strength solution, as shown in **Figure 4b**.

5.17.4.1.2 Phenazopyridine cation aggregation

Figure 4c show the log S–pH plot of phenazopyridine hydrochloride. As can be seen, the apparent pK_a is shifted to a higher value (cf. **Figure 3d**), consistent with the formation of cationic aggregates. An analysis along the lines of that described in Section 5.17.4.1.1 may be performed to obtain an equation similar to eqn [13].[39] The dimerization constant consistent with the data is $K_2 = 1.0 \times 10^4\,M^{-1}$. We have seen cationic aggregation for other weak bases as well: chlorprothixene, cyproheptadine, fendiline, imipramine, and mifepristone (unpublished data). With today's all too often low-solubility discovery compounds, such phenomena are probably more common than is actually realized by pharmaceutical scientists. If such effects are ignored, then later applications of the measured solubility to rationalizing dissolution curves may lead to erroneous conclusions (*see* Section 5.17.6.4).

5.17.4.1.3 Piroxicam neutral species aggregation effects

Shifts in the pK_a can also be expected if water-soluble aggregates form from the uncharged monomers (see **Figures 3a** and **3b**). This may be expected with stackable planar molecules, such as piroxicam.[57] Let us hypothesize that a specific neutral n-oligomer of piroxicam forms:

$$H_2X^+ \rightleftharpoons H^+ + HX \qquad K_{a1} = [H^+][HX]/[H_2X^+] \qquad [15a]$$

$$HX \rightleftharpoons H^+ + X^- \qquad K_{a2} = [H^+][X^-]/[HX] \qquad [15b]$$

$$HX(s) \rightleftharpoons HX \qquad S_0 = [HX] \qquad [15c]$$

$$nXH \rightleftharpoons H_nX_n \qquad K_n = [X_nH_n]/[XH]^n \qquad [16]$$

So,

$$S = [H_2X^+] + [XH] + [X^-] + n[X_nH_n] \qquad [17]$$

Let us for the moment consider only pK_{a2} of amphoteric piroxicam, so that the treatment simplifies to that of a monoprotic acid. Consequently,

$$\log S = \log S_0 + \log(1 + K_a/[H^+] + nK_nS_0^{n1}) \qquad [18]$$

At pH values well below pK_{a2} (but above pK_{a1}), eqn [18] reduces to (the horizontal solid line in **Figure 3a**)

$$\log S = \log S_0 + \log(1 + nK_nS_0^{n1}) \quad \text{for } pK_{a1} < pH < pK_{a2}$$
$$= \log S_0^{App} \qquad [19]$$

From the schematic representation in **Figure 3a**, eqn [19] suggests

$$\log S_0^{App} - \log S_0 = \log(1 + nK_nS_0^{n-1}) \qquad [20]$$

Aggregation of the neutral species elevates the apparent solubility of the neutral species (**Figure 3a**). However, if $pH \gg pK_{a2}$ (past the pK_{a2}^{App} value, thus $K_a/[H^+] \gg nK_nS_0^{n-1 \gg 1}$), eqn [18] reduces to eqn [3b].

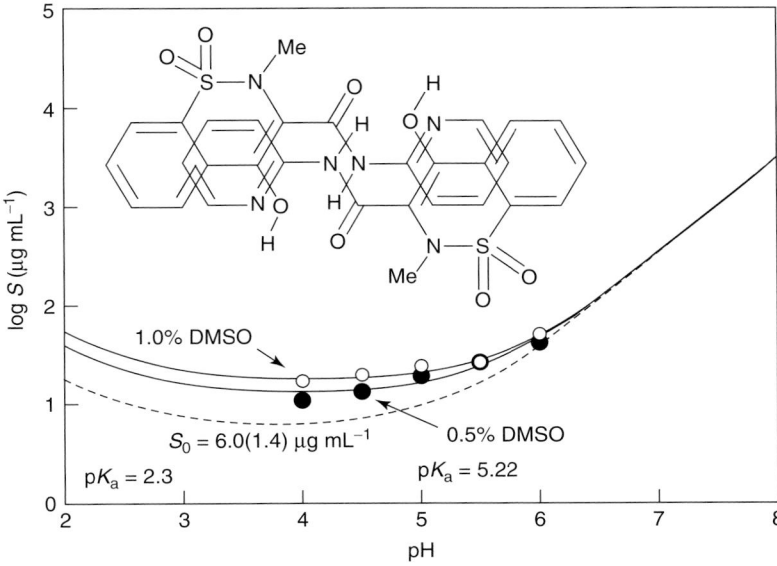

Figure 5 Log solubility–pH profiles for piroxicam, 25 °C, 0.01 M ionic strength, miniaturized 'self-calibrating' UV method (*see* Section 5.17.3.3.3),[19] centrifugation after 23 h, in the presence of 0.5% DMSO (solid circles) and 1% DMSO (open circles). The solid curves are fitted to the points with the refined dimerization constants listed in the figure. The nearly identical dashed curves were derived by the shift-in-the-pK_a procedure (*see* **Figure 3a**), with intrinsic solubility resulting in $6.0 \pm 1.4 \, \mu g \, mL^{-1}$. The structure shown is based on the known structure of the dimer present in the solid state.[57]

If the true pK_a is subtracted from pK_a^{App}, the difference then subtracted from the logarithm of the apparent solubility, S_0^{App}, yields the true aqueous solubility constant:

$$\log S_0 = \log S_0^{App} \pm (pK_a - pK_a^{App}) \tag{21}$$

with the ' + ' sign for ionizable acids (**Figure 3a**) and the ' − ' sign for ionizable bases (**Figure 3b**).

Figure 5 shows the log S–pH plots for piroxicam, at 25 °C and 0.01 M ionic strength, in the presence of 0.5% and 1.0% DMSO. The plots indicate neutral species anomaly. It was possible to fit the data with a dimerization ($n = 2$) model, with refined constants best fitting the elevated curves in **Figure 5** as $K_2 = 3.5 \times 10^4 \, M^{-1}$ (0.5% DMSO) and $6.4 \times 10^4 \, M^{-1}$ (1.0% DMSO). The underlying (hidden) intrinsic solubility, indicated by the dashed curve in **Figure 5**, is the same $6.0 \pm 1.1 \, \mu g \, mL^{-1}$, regardless of which DMSO level data are used in its evaluation. It is interesting to note that Rozou *et al.*,[57] using nuclear magnetic resonance, detected the presence of dimers in low-pH solution, although binding constants were not reported. The dimerization constant appears to be sensitive to the level of DMSO present, although the same intrinsic solubility is calculated from either of the two sets.

5.17.4.2   Effects of Excipients on Solubility

Jinno *et al.*[58] showed how excipient sodium lauryl sulfate (SLS) affected the solubility–pH profile of piroxicam, at 37 °C. For each level of SLS, the solubility–pH data were fitted by eqn [2], with pK_a refined as the apparent value. **Table 2** summarizes the shift-in-the-pK_a analysis of their data. Note that as the level of SLS increases from 0% to 2%, the pK_a^{App} value increases systematically. **Figure 6** shows the data and the best-fit results. The application of eqn [21] indicates that the underlying intrinsic solubility, S_0, remains remarkably invariant, averaging $21.3 \pm 1.6 \, \mu g \, mL^{-1}$, even though the apparent intrinsic solubility rises to about $500 \, \mu g \, mL^{-1}$ at 2% SLS.

5.17.4.3   What the Shift-in-the-pK_a Method Can and Cannot Do

The method based on shift in the pK_a, schematically illustrated in **Figure 3**, is best used as a 'quick-alert' tool. (As is implied, the molecule must have an ionization group within the accessible pH range, in order for the method to work.) When a log S versus pH plot is inspected, and the true pK_a is known, the investigator can quickly surmise whether aggregates (or 'anomalies') may be present, and whether these effects are due to the neutral or the charged

Table 2 Piroxicam intrinsic solubility[a]

SLS (%)	pK_a^{app}	S_0 ($\mu g\,mL^{-1}$)
0.0	5.44 ± 0.16	24.6 ± 8.9
0.5	6.32 ± 0.13	20.5 ± 6.1
1.0	6.48 ± 0.10	21.8 ± 5.0
2.0	6.77 ± 0.12	20.0 ± 5.6

[a] Data from Jinno, J.; Oh, D.-M.; Crison, J. R.; Amidon, G. L. *J. Pharm. Sci.* **2000**, *89*, 268–274 at 37 °C. The measured pK_{a2} of piroxicam is 5.34 at 37 °C (see **Table 1**). The calculated values are from the pDISS program.

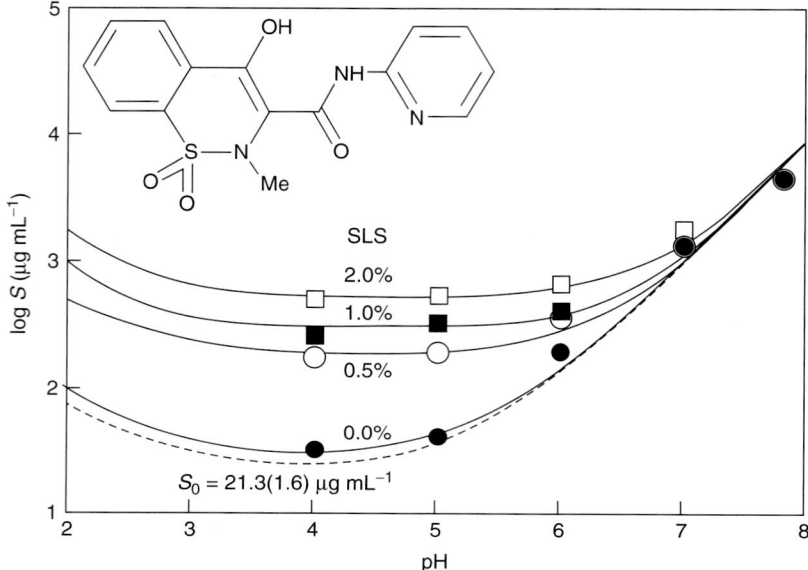

Figure 6 Log solubility–pH profiles for piroxicam as a function of excipient SLS, 37 °C, 0.1–0.2 M ionic strength, data (*see* Section 5.17.4.2) from Jinno, J.; Oh, D.-M.; Crison, J. R.; Amidon, G. L. *J. Pharm. Sci.* **2000**, *89*, 268–274. The solid curve for 0% SLS was fitted to the points with the refined dimerization constants listed in the figure. The other solid curves were fitted to the shift-in-the-pK_a analysis (see **Figure 3a**), with intrinsic solubility resulting in $21.3 \pm 1.6\,\mu g\,mL^{-1}$.

form of the species. Furthermore, the intrinsic solubility may be calculated from the magnitude or the direction of the pK_a shift. So, one has the opportunity to measure solubility in the presence of excipients and at the same time to assess the solubility that would have been evident in the absence of added excipients. This is the best use of the shift method. Caution is needed not to overinterpret the measurement data, however.

5.17.5 Convective Diffusion with Reaction: Rotating Disk Dissolution of Ionizable Drugs

The concentration gradients of all species present in the ABL determine how quickly the drug is transported from the solid state to the solution in the medium. When the solid drug is first placed in the dissolution medium, the concentration gradient that quickly establishes across the ABL is proportional to the solubility, according to the Noyes–Whitney equation.[59–62] The overall dissolution rates are dependent on how quickly the drug diffuses through the ABL, how strongly the drug interacts with the medium components in the ABL, such as excipients, and how pH gradients develop in the ABL, as a consequence of all ionizable components present. The pH gradient is a very important and subtle determinant of the rate of dissolution. For example, an often surprising observation (to a nonspecialist) is that the dissolution rate can double if the concentration of the medium buffer is doubled, even though the buffer pH stays the same (*see* Section 5.17.6.6). There are thus opportunities to improve drug dissolution properties by synthetic alterations of the molecules, to change solubility and excipient binding properties.

This section briefly summarizes the conceptual aspect of modern dissolution theory, applied to the rotating disk geometry (e.g., Wood's apparatus).[39] Dissolution of powders is covered in Section 5.17.7. The medicinal chemist more interested in dissolution applications than theory may safely skip to Section 5.17.6.

5.17.5.1   Equations in Convective Diffusion Mass Transport

In the rotating disk dissolution experiment, once an ionizable compound dissolves at the solid–liquid interface and begins to move across the ABL, which is in contact with the surface of the solid, the molecules may undergo chemical reactions in the ABL, such as ionization, complexation, micelle formation, and excipient interaction. (Also, the properties of the ABL are affected by non-API components, often critically affecting the pH gradient, which impacts on the overall dissolution process.) The dissolved molecules move through the ABL by the combined action of diffusion and the convective fluid flow generated by the action of the rotating disk. The convective flow equations due to the rotation in a viscous fluid were originally solved by von Kármán in 1921, and later substantially refined by Cochran in 1934. More recent discussions may be found in the classic books by Levich,[64] Cussler,[65] and Schlichting and Gersten,[66] and in research papers.[39,63,67,68]

The partial differential equation describing the convective diffusion process is

$$\partial C(x, t)/\partial t = \partial(D(x)\,\partial C(x, t)/\partial x)/\partial x + |V(x)|\,\partial C(x, t)/\partial x \qquad [22]$$

where $C(x, t)$ is the concentration of the reactant, at position x (distance from the surface of the rotating disk) and time t. $D(x)$ is the diffusivity at position x (which indirectly has a time dependence), and $|V(x)|$ is the absolute value of the axial velocity of the convective fluid flow, at the distance x from the rotating surface. Each different species in solution has its own form of eqn [22]. For example, for a diprotic weak acid, the different species are H_2A, HA^-, A^{2-}, H^+, and OH^-. Normally, eqn [22] needs to be solved for $C(x, t)$ of each of the species. (For the moment, the chemical reaction terms are left out in eqn [22]. We will come back to them later.)

In our treatment of the theoretical problem,[39] the diffusivity of species is not assumed to be constant throughout the ABL. We will show that in certain cases the diffusivity can vary by an order of magnitude (or even more) at various locations in the ABL. This can be explained by the Vinograd–McBain[49] electric field neutrality treatment, which addresses the spatial dependence of diffusivity, as a function of all ions present in solution.

5.17.5.2   Vinograd–McBain Treatment of the Diffusion of Electrolytes and of the Ions in Their Mixture

Electric fields, created in solutions containing ions migrating under concentration gradients, can act on each ion, to either retard or accelerate its diffusion velocity from the canonical value the ion would have (D^*) in the absence of the gradients. If a solution contains a highly mobile ion, such as H^+, and a sluggish ion, such as the drug molecule, A^-, and there is an imposed concentration gradient, the faster ion has the innate tendency to diffuse ahead of the slower ion, which creates a charge separation. This in turn lowers the diffusivity of the fast ion and raises that of the slow one, so that very quickly both ions travel at the same steady state velocity, in order to maintain spatial charge neutrality. For example, the diffusivity of H^+ in a dilute salt solution can be as low as $10 \times 10^{-6}\,cm^2\,s^{-1}$ (see below). This is a surprisingly low value, since, in a solution containing $0.1\,M\,KCl$ and no concentration gradients, the diffusivity of hydrogen ions is $93 \times 10^{-6}\,cm^2\,s^{-1}$. Such swings in diffusivity need to be properly accounted for in the analysis of dissolution experiments, when the ionic strength of the solution is low.

By imposing the electrical neutrality condition for the ion mixture, Vinograd–McBain[49] derived the expression for the resultant diffusivity at position x in the diffusion layer for a given ion, $D_j(x)$, as a fraction of the diffusivity that the ion would have in the absence of concentration gradients, D_j^* (position independent),

$$\frac{D_j(x)}{D_j^*} = 1 \mp \frac{|z_j|\,C_j(x, t)}{dC_j(x, t)/dx}\left(\frac{\sum_i(z_i/|z_i|)D_i^*(dC_i(x, t)/dx)}{\sum_i D_i^*|z_i|\,C_i(x, t)}\right) \qquad [23]$$

where the '$-$' sign in $\mp$ is used if the species j is a cation, and the '$+$' sign used if it is an anion. The summations are taken over all the ions present in the solution. The shifts in the diffusivity coefficients predicted by eqn [23] are less dramatic when an appreciable amount of background salt (e.g., KCl) is added. In the weak acid example below, the coupling between A^- and H^+ is relaxed by the K^+ and Cl^- ions from the 'swamping' level of background electrolyte. Such changes in diffusivities as a function of the ion concentrations and concentration gradients in solution are automatically compensated for in the computer program developed to solve eqn [22].[39]

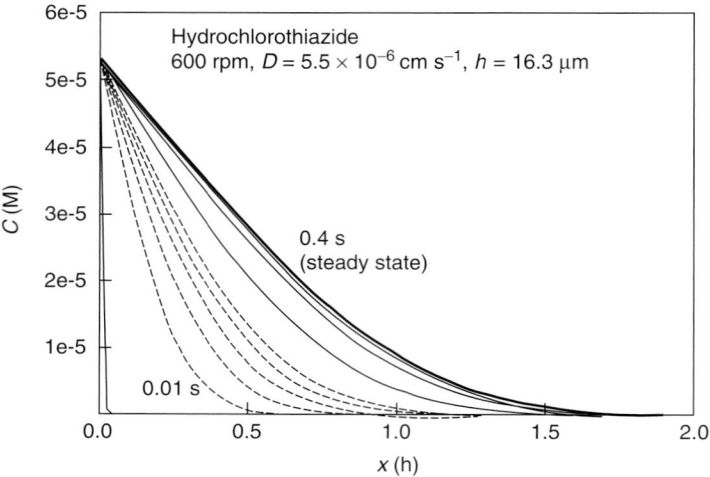

Figure 7 Concentration–time profile in the ABL until the time steady state is reached, calculated by the numerical integration technique (*see* Section 5.17.5.3) for diprotic hydrochlorothiazide, under the conditions identified in the figure. The dashed lines indicate $t = 0.01$, 0.02, ..., 0.05 s, while the solid curves above those lines indicate $t = 0.1$, ..., 0.4 s, the latter value being approximately the steady state time.

5.17.5.3   Chemical Reaction during Convective Diffusion

Once an ionizable substance dissolves and begins to diffuse across the ABL, it may undergo chemical reactions, as noted above. Yet, eqn [22] does not explicitly contain reaction terms. It was shown by Olander[69] that if the differential equations for each species, such as eqn [22] applied to H_2A, HA^-, A^{2-}, H^+, and OH^-, are combined to express total concentrations of weak acid drug, A, all reaction terms cancel out, provided the reactions are reversible and fast compared with diffusion.[69] It is simply not necessary to consider explicit chemical reaction rate constants, as Olander had demonstrated a long time ago. For example, the total partial differential expression for the diprotic weak acid becomes

$$\partial A/\partial t = \partial(D_{H2A}\,\partial C_{H2A}/\partial x)/\partial x + \partial(D_{HA}\,\partial C_{HA}/\partial x)/\partial x + \partial(D_A\,\partial C_A/\partial x)/\partial x + |V(x)|\,\partial A/\partial x \qquad [24]$$

Equation [24] can be solved for the total concentrations A. With A calculated, then the nonlinear computation of all reactant and associated species can proceed, provided all the relevant equilibrium constants are known.[1,32–34] A general and robust numerical method to solve eqn [24] has been described.[39] This was compared with an analytical solution in the same study.

Figure 7 shows the time progression of the numerical solutions[39] to eqn [24] for diprotic hydrochlorothiazide, as a function of the distance from the rotating disk surface, at 600 rpm. The (gradient-free) diffusivity of hydrochlorothiazide was taken to be 5.5×10^{-6} cm^2 s^{-1}. Under these circumstances, the thickness of the ABL, according to the Levich equation,[64] is $h = 0.00163$ cm. The dashed curves indicate $t = 0.01$, 0.02, ..., 0.05 s, while the solid curves above those lines indicate $t = 0.1$, ..., 0.4 s, the last value being approximately the steady state time. When the monoprotic indomethacin example[42] is considered, the resultant steady state curve is identical to that shown in **Figure 7**. The solution to the total concentration partial differential equations is entirely independent of the chemical reactions taking place, as originally recognized by Olander.

The numerical integration scheme is flexible and robust (albeit not fast). We demonstrated practical equivalence between the steady state concentration profile determined by the numerical procedure (which is very easy to manipulate for new boundary conditions and geometries) and that suggested by the much faster analytical equation procedure[65,70] (which is not easy to alter and manipulate). The program developed to do all these computations is called pDISS.[39]

5.17.6   Dissolution Apparatus and Measured Results

5.17.6.1   Wood's Rotating Disk Apparatus for Mechanistic Studies

Because of its well-characterized fluid hydrodynamics, the rotating disk method is a useful tool for mechanistic dissolution studies. The DIDR measurements have been used for many years to characterize solid drugs, including studies of dissolution–pH rate profiles in the presence of buffers, complexing agents, and various excipients.[2,41–47,53,58] It is currently debated whether the DIDR method can be used to determine solubility class membership in the Biopharmaceutics Classification System, with encouraging early indications.[71]

In the various modifications of the method, a pure drug is generally compressed in a die with a hole of known diameter to produce a pellet of known exposed surface area (e.g., see [50]). It is assumed that during the dissolution period, the exposed area of the pellet remains constant.

Often, traditional thermostated baths, such as those available from Sotax or a number of other manufacturers, are used, accommodating USP volumes of 500 and 900 mL. However, when the experiments are performed in early preformulation, typical volumes can be in the range of 175–250 mL, and single-sample jacketed beakers are used, thermostated with a circulating bath.[46] The temperature is usually maintained constant at 37 °C, but in research settings it is not uncommon to use 25 ± 1 °C.

About 100–700 mg of pure drug substance are compressed into disks with a punch and die, using a hydraulic pump (Carver, Inc.). The typical surface area ranges from 0.5 to 1.3 cm^2.[41,71] Typical pressures used are about 2000 lb in^{-2} (14 bar), applied for about 1 min.

Rotational speeds between 15 and 250 rpm are often used, with a preference for 100 rpm,[71] although values as high as 600–900 rpm have been reported.[41] Usually the speeds are elevated to increase the dissolution rates of very insoluble compounds.[46]

Buffered media are often made with 0.1 M HCl (pH 1.2), 0.2 M acetate buffer (pH 4.5), or 0.2 M phosphate buffer (pH 6.8). However, unbuffered solutions, with 0.1 M KCl,[45] 0.5 M KCl,[41] and without background salt[53] have been used, where the pH was controlled with a pH stat, by titrating the solution with standard 0.01–1 M NaOH or HCl to maintain constant pH. A combination pH electrode is used to measure the pH of the solution directly in the bath. One such apparatus is illustrated in **Figure 8a**.[41] Usually, optical density data are collected at the isosbestic wavelength, where the molar absorptivity is pH-independent. The turbid solutions are traditionally filtered in order to read the UV spectra reliably. (Newer in situ methods do not require filtration; *see* Section 5.17.6.2.) A plot of absorbance versus time is made. A series of standard solutions of the drug is prepared in advance, and the dissolution versus time profile is calibrated against the standards. Experiments can run from 5 min to several hours, depending on the solubility of the compound.

5.17.6.2　In Situ Fiber Optic Dip Probe Ultraviolet Spectrophotometry

As is seen from the above description, the dissolution testing method can be time-consuming and labor-intensive. The method can be made more efficient and quicker if the UV data-taking were performed directly in the assay vessel. To that end, Purdue Pharmaceuticals developed an in situ fiber optic dip probe UV apparatus, called the Rainbow Dynamic Dissolution Monitor (Delphian Technology), which eliminates external sampling of the solutions.[14] Since moving parts are eliminated, sampling errors due to sipper mechanism failure, sample contamination, filter clogging, adsorption to sampling tubing, etc., are not of concern. The UV system employs a photodiode array (Zeiss) for fast whole-spectrum

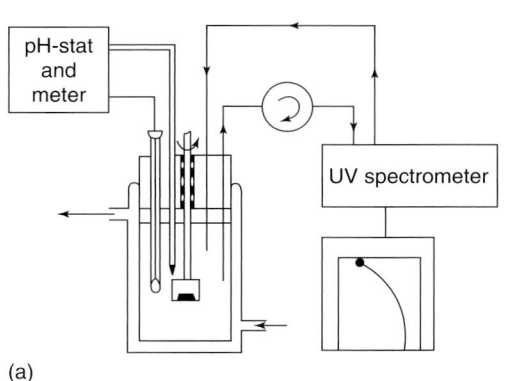

(a)　　　　　　　　　　　　　　　　　　(b)

Figure 8 (a) Apparatus for DIDR measurement used by Mooney and co-workers (*see* Section 5.17.6.1). (Reprinted from Mooney, K. G.; Mintun, M. A.; Himmelstein, K. J.; Stella, V. J. *J. Pharm. Sci.* **1981**, *70*, 13–22; Copyright © (1981) with permission of Wiley-Liss, Inc., a subsidiary of John Wiley & Sons, Inc.) (b) Small-volume dissolution apparatus (Delphian Technology) used for polymorph rank-order screening (*see* Section 5.17.6.3). Reproduced with permission from Delphian Technology.

UV data-taking (one complete-curve scan per second). Slow-release or rapidly dissolving products can be characterized. For example, vitamin B$_{12}$, whose tablet dissolution half-life is about 1 min, was successful characterized.[16] Interference due to excipient background turbidity is minimized by a proprietary spectral second-derivative method.

5.17.6.3   Small-Volume Dissolution Apparatus for Polymorph Rank-Order Screening

Because of the increasing interest in characterizing intrinsic dissolution in early preformulation, where the tested API is not available in large quantities, a small-volume dissolution instrument, called μDISS, has been developed, as shown in **Figure 8b**. The working volume can be as low as 2–3 mL. Pure powder suspensions of the API are tested, with sample weights as low as 50–100 μg (hence the name 'μDISS'), instead of the usual 100–700 mg. Currently, the apparatus is used for three purposes: (1) to rank-order compounds for dissolution and detect any polymorphic changes, a problem that can occur during scale-up in synthesis in early preformulation, (2) to measure API stability over time, and (3) to determine solubility very precisely. In the first instance, the working premise is that if a compound undergoes a polymorphic transformation during the various stages of synthesis scale-up, redetermined dissolution curves will indicate a departure from previous determinations. Also, in some instances (e.g., unstable organic solvates of the API), a polymorphic transformation may be observed during the actual dissolution interval. This could suggest to the medicinal chemist to modify the isolation and purification methods used. Solubility determination is possible when enough solid API is added so that at the end of sufficiently long time, the solution reaches equilibrium and a state of uniform saturation, where the indicated bulk concentration of the API is its solubility. Given today's low-solubility molecules, it is not really necessary to have very much API to produce a saturated solution.

Figure 9 shows an example of the dissolution profile of powdered phenazopyridine hydrochloride, with each of the six dissolution vessels containing 20 mL 0.05 M HEPES buffer with the pH adjusted to 7.4. The data are quantitatively analyzed for both dissolution and solubility in Section 5.17.7, using the spherical particle model of Wang and Flanagan.[51,52]

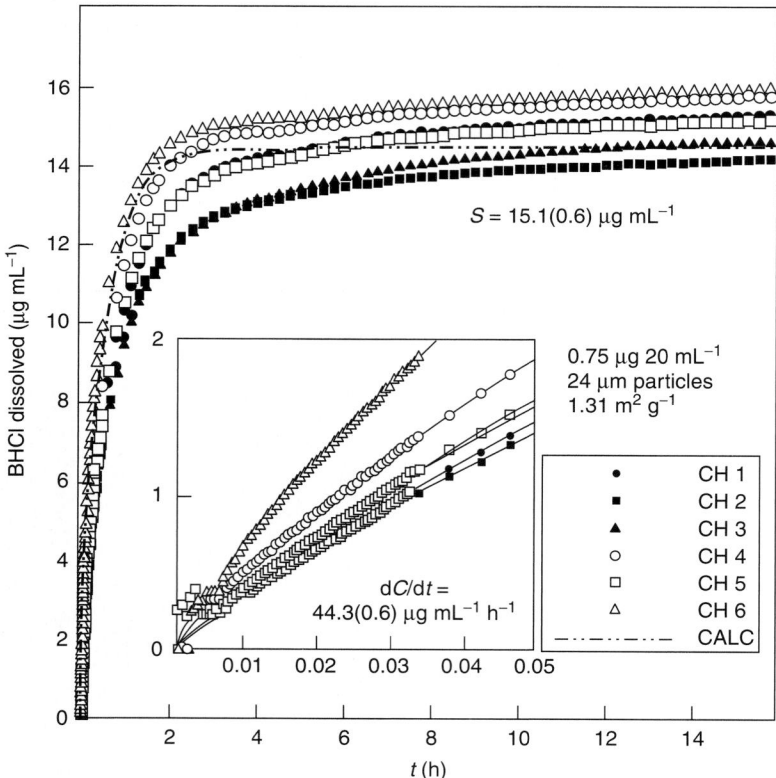

Figure 9   Powder dissolution–time profiles using the apparatus pictured in **Figure 8b**, in 0.05 M HEPES buffer at pH 7.4, stirred at 700 rpm, of phenazopyridine hydrochloride, 0.75 ± 0.04 mg in 20 mL of buffer. The dash-dot-dot line corresponds to the calculated dissolution profile using the Wang–Flanagan model, with an assumed 24 μm particle diameter, and the constants in **Table 1** corresponding to 25 °C, adjusted to 0.05 M ionic strength. The inset is the expanded-scale view. The average rate of dissolution, $dC/dt = 44.3 \pm 0.6 \, \mu g \, mL^{-1} \, h^{-1}$, corresponds to $9.9 \, (\pm 2.7) \times 10^{-10} \, mol \, s^{-1}$.

5.17.6.4   Example of a Weak Acid Dissolution Measurement by the Disk Intrinsic Dissolution Rate Method

In this section, the DIDR dissolution simulation computer program, pDISS, is applied to indomethacin.[42,46] **Table 3** shows the Mooney data for indomethacin, the least soluble of the weak acids considered in our examples. The calculated flux fits the observed data reasonably well, but only at low pH. It is suspected that the formation of anionic aggregates may be the reason for the poor prediction at high pH (see **Figure 4b**). The last column in **Table 3** lists three calculated values, which agree with the observed fluxes, where the calculation assumed higher solubility constants than the value reported by Mooney. To explain the data at pH 9, it was necessary to assume an intrinsic solubility about 10-fold higher than that of Mooney. **Figure 4b** shows that solubility measurements of indomethacin indicate an enhanced solubility due to anionic aggregate formation, of the order of magnitude needed to explain the observed DIDR data. These new predicted insights need to be verified with further experimentation.

5.17.6.5   Example of a Weak-Base Drug: Phenazopyridine

The well-designed DIDR studies of phenazopyridine by Serajuddin and Jarowski[53] show extraordinary complexity, which can most easily be grasped with the aid of simulations using the improved computational procedures[39] briefly described in this review. Serajuddin and Jarowski determined at 37 °C the physicochemical constants $pK_a = 5.20$, $-\log S_0 = 3.83$ (molarity scale), and $pK_{sp} = 3.53$ (chloride; see **Table 1**). A thorough dissolution–pH study was performed using Wood's rotating disk apparatus, operated at 200 rpm, with a pellet surface area of 0.95 cm². Particular emphasis was placed by the researchers on characterizing the common-ion and self-buffering effects. Pellets of the free base and those of the hydrochloride salt of phenazopyridine were studied independently, and the possible formation of dual-solid phases at certain pH conditions were explored by the investigators, with scrapings of the pellet surfaces analyzed by differential scanning calorimetry. Adjunct ionic strength adjusters (e.g., NaCl) were not used. (Hence, it was particularly important to apply the Vinograd–McBain formalism (*see* Section 5.17.5.2) in the simulations.) Low-pH solutions were prepared with 0.1, 0.01, and 0.001 M HCl. Other pH values of working solutions were prepared by adding small volumes of standardized 1 M NaOH or 1 M HCl. A pH stat was employed to adjust the pH of the bulk solution. **Table 4** summarizes some of the results of the dissolution study.

5.17.6.5.1   Diffusivity may reveal aggregation

The two right-most columns in **Table 4** are expected to be the experimental diffusivity. The value averages to 7.5×10^{-6} cm² s⁻¹. The theoretically expected diffusivity is 8.6×10^{-6} cm² s⁻¹ for a monomer and 6.3×10^{-6} cm² s⁻¹

Table 3 Indomethacin dissolution data

pH_{bulk}	pH_0	S ($\mu g\,mL^{-1}$)	$J_{max}^{Obs\,a}$ ($10^{-12}\,mol\,cm^{-2}\,s^{-1}$)	$J_{max}^{Calc\,a}$ ($10^{-12}\,mol\,cm^{-2}\,s^{-1}$)	J_{max}^{Calc} ($10^{-12}\,mol\,cm^{-2}\,s^{-1}$)
2.00	2.00	0.95	12	9	
3.00	3.00	1.00	27	10	
4.00	3.93	1.60	31	17	27[b]
5.00	4.50	3.50	81	37	
6.00	4.50	3.50	134	38	125[c]
7.00	4.50	3.50	151	38	
7.50	4.50	3.50	166	38	
8.00	4.50	3.50	181	38	
8.50	4.50	3.50	227	38	
8.70	4.75	5.50	253	59	
9.00	4.75	5.50	338	60	334[d]

[a] Mooney et al.,[42] 25 °C, 0.50 M KCl, pK_a 4.17, $-\log S_0 = 5.58$, 600 rpm, $D = 5.6 \times 10^{-6}$ cm² s⁻¹, pH_0 is the interfacial pH.
[b] Calculated with $-\log S_0 = 5.38$ (see **Figure 3**).
[c] Calculated with $-\log S_0 = 4.88$ (see **Figure 3**).
[d] Calculated with $-\log S_0 = 4.45$.

Table 4 Phenazopyridine dissolution case study[a]

pH of medium	J $(10^{-8} mol\, cm^{-2} s^{-1}\, units)^a$		$pH_{x=0}$		S $(mg\, mL^{-1})$		Jh/S $(10^{-6} cm^2 s^{-1}\, units)^a$	
	BHCl pellet	B pellet	BHCl pellet	B pellet	BHCl pellet	B pellet	BHCl pellet	B pellet
1.1	0.561	59.34[b]	1.2	3.3–3.6	0.68	63.5[b]	6.49	7.35[b]
2.05	–	6.274[b]	–	3.6	3.20	6.7[b]	–	7.36[b]
3.05	4.272	0.687	3.2	3.9	4.40	0.75	7.64	7.20
5.0	4.258	0.087	3.4	5.0	4.46	0.090	7.51	7.60
7.0	4.305	–	3.4	7.0	4.46	0.037	7.59	–

[a] Serajuddin and Jaworski.[53] B, phenazopyridine free base; $h = 0.00315\,cm$ (based on $D = 8.56 \times 10^{-6}\,cm^2 s^{-1}$ and 200 rpm), $A = 0.95\,cm^2$. Jh/S values are the experimentally determined diffusivity coefficients.
[b] Approximate (due to the presence of supersaturation).

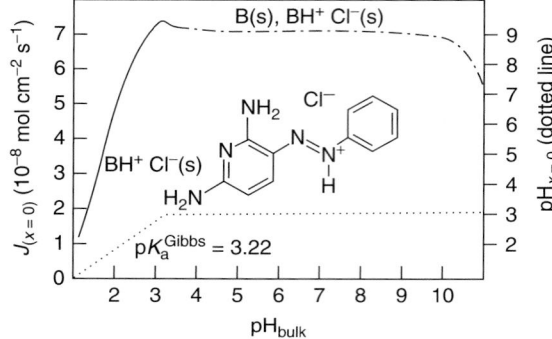

Figure 10 Simulated flux profiles of a 100 mg phenazopyridine hydrochloride pellet, as a function of bulk pH, at 37 °C, 200 rpm, no ionic strength adjuster added, ABL thickness $h = 0.00315\,cm$, $D_B = 8.56 \times 10^{-6}\,cm^2 s^{-1}$, $A = 0.95\,cm^2$, $V = 1\,cm^3$. The dotted curve represent the interfacial pH ($x = 0$). The region past the bulk pH 3.22 contains the coprecipitate of $BH^+ Cl^-$ (s) and B(s).

for a dimer, using a formula related to molecular weight.[1] **Figure 4c** indicates that phenazopyridine forms a cationic aggregate in low pH solution. In **Table 4**, at pH 1.1, the phenazopyridine hydrochloride pellet experiment shows an apparent diffusivity of $6.5 \times 10^{-6}\,cm^2 s^{-1}$. This suggests the possibility that the pH 1.1 dissolution data are indicative of some aggregation of the cationic species, perhaps a dimer – a new perspective on the interesting 1985 study.

5.17.6.5.2 Phenazopyridine hydrochloride dissolution as a function of pH: the self-buffering effect

Figure 10 shows plots of the calculated fluxes ($x = 0$) for a pellet made from the hydrochloride salt $BH^+ Cl^-$ (s), (solid and dash-dot curve) as a function of bulk pH. As the pH is lowered below 3.22, the flux decreases steadily, as more of the $BH^+ Cl^-$ (s) precipitates, and as solubility decreases. The HCl used to lower the pH below 3.22 introduces chloride ions, which interact with the positively charged form of the weak base to form the additional precipitate. This is known in solubility measurements as the common-ion effect. The interfacial pH ($x = 0$) also linearly decreases from pH 3.14 to 1.1 (dotted curve in **Figure 10**).

However, above pH 3.22, the conditions become just right for the free base, B(s), to co-precipitate on the surface of the $BH^+ Cl^-$ (s) pellet. The Gibbs phase rule then applies, and both the pH and the solubility on the surface of the pellet becomes absolutely constant. The dotted line indicates the interfacial pH as a function of the bulk pH values. As can be seen, the pH at $x = 0$ is fixed at 3.14, and remains so, as the bulk pH varies from 3.22 to 10.5 in that interval. This is the condition where the free base and the chloride salt of its conjugate acid co-precipitate, and the interfacial pH $= pK_a^{Gibbs}$. Furthermore, at the solid–liquid interface, the solubility becomes constant ($4.88\,mg\,mL^{-1}$). The flux also becomes nearly constant, as indicated in **Figure 10** by the dash-dot curve.

Above bulk pH 9.5, another very interesting phenomenon occurs: the free base not only precipitates on the surface of the $BH^+ Cl^-$ (s) pellet but is predicted also to precipitate throughout much of the ABL. **Figure 11** shows the

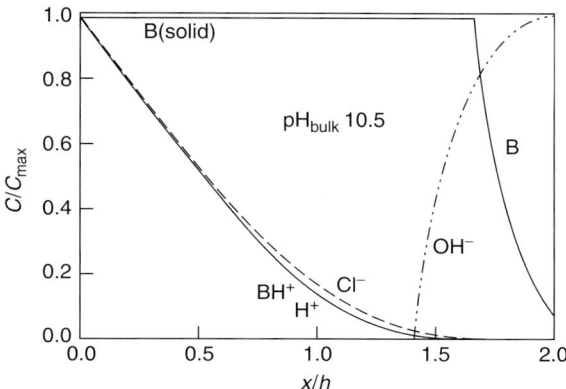

Figure 11 Calculated ABL cross-section for phenazopyridine hydrochloride, showing fractional concentration (C/C_{max}) profiles with the fractional distance across the ABL, up to $x/h = 2$, at bulk pH 10.5.

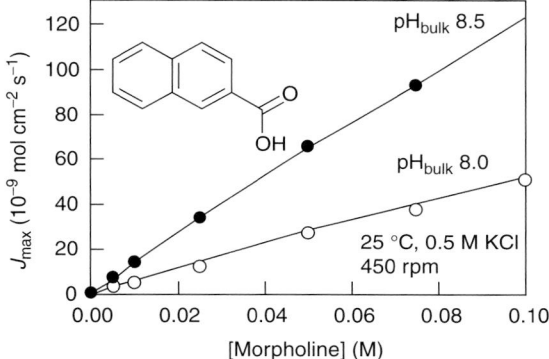

Figure 12 The predicted flux of 2-naphthoic acid dissolving at bulk pH 8.0 and 8.5, at various buffer concentrations.

relative concentrations of the species in the fractional coordinates of the ABL. Shown at bulk pH 10.5 is the region of the B(s) precipitate within the ABL, from $x/h = 0$ to $x/h = 1.7$. Since this solid is constant in 'concentration' throughout most of the ABL, and since the solid is uncharged, there is no driving force (other than convection) to move the solid out of the ABL. One can imagine a 'fog of solid' suspended in the ABL. Serajuddin and Jarowski did observe the surface of the pellet change from brick red to yellow in color as the free base loosely attached to the rotating pellet. DSC analysis of the shavings of the yellow solid confirmed it to be the free base.

Once the free base starts to precipitate in the ABL, the flux is predicted to decline, as is shown in **Figure 10** by the dash-dot curve for bulk pH > 9.5.

Even though the free base of phenazopyridine is quite insoluble at pH 7.4, the dissolution rate of a pellet of BH^+Cl^-(s) maintains a very high flux, due to the self-buffer effect mentioned above. This is a fascinating case of the Gibbs interfacial 'super buffer.'[29]

5.17.6.6   The Effect of Buffer on the Intrinsic Dissolution Rate

Figure 12 shows the measured fluxes for 2-naphthoic acid in pH_{bulk} 8.0 and 8.5 solutions, as a function of the concentration of the added buffer morpholine with measured data taken from Mooney *et al.*[43] The solid lines were calculated by pDISS.[39] The fit is excellent. What this illustrates, as has been well understood ever since the 1980s, with the studies of Mooney and co-workers, as well as those of McNamara and Amidon, is that the observed dissolution rates depend not only on the pH of the buffer but also on how much buffer is present in the dissolution medium. The amount of buffer affects the pH gradient in the ABL, and that impacts on the dissolution rate of the drug molecule, albeit indirectly.

5.17.7   Spherical Particle Dissolution (Powder Model)

5.17.7.1   Theory

So far, we have considered intrinsic dissolution of compounds from rotating planar surfaces, where, at steady state, the classical Nernst–Brünner thin-film model predicts a linear decrease in concentration of the dissolving species throughout the ABL. With a spherical particle, on the other hand, the concentration gradient is nonlinear (bowed) in the ABL, and decreases proportionately with the inverse distance from the surface of the particle. Dissolution of a spherical surface will not reach a conventional steady state, because the surface area continually changes with time, as the dissolving particles shrink in size. The most common expression used to describe the dissolution of a spherical particle follows the so-called cube root law.[72]

Wang and Flanagan[51,52] compared particle dissolution equations, and found that the classical equations in current use were only approximate, since assumptions made in their derivations were not valid under the broad range of practical conditions. These investigators proceeded to derive a general model for diffusion-controlled particle dissolution, the most comprehensive to date.

5.17.7.2   Application of the Wang–Flanagan Model to the Dissolution of Phenazopyridine Hydrochloride, as Powders of Known Particle Size Distribution

The small-volume dissolution apparatus pictured in **Figure 8b** (μDISS), was used to measure the dissolution rates of powders of phenazopyridine hydrochloride. By a different analysis, the specific surface area of phenazopyridine hydrochloride was determined to be $1.31 \, m^2 \, g^{-1}$, and the average size of particles was about $24 \, \mu m$.

About 0.75 mg of the hydrochloride salt of phenazopyridine were weighed into each of six test tubes of the μDISS apparatus. To each, 20 mL of 0.05 M HEPES buffer, pH 7.4, were added, and the stirring rate was set to 700 rpm. The thermodynamic calculation by pDISS[39] predicted that the solid, introduced as a powder in the form of the hydrochloride salt, should undergo a transformation into the free base solid at pH 7.4. This must be a very fast conversion, since the observed rate is relatively slow, and is comparable to that expected of the dissolution of free base solid.

In about 8 h, the dissolution reached a steady state. Simulations indicate that at pH 7.4, phenazopyridine is not completely dissolved. This was confirmed visually. **Figure 9** shows the six traces of the salt dissolution curves, as well as the calculated (Wang–Flanagan model) dissolution curve. The concentration–time profile was calculated by the μDISS processing software, based on the area under the second-derivative curves (optical density versus wavelength) in the wavelength interval 322–360 nm, a strategy minimizing the contributions due to the background scattering of the turbid solutions. The Wang–Flanagan model predicted the time course of the dissolution of the powder reasonably well, when the experimental surface area is calculated based on the average radius of assumed spherical particles.

After about 18 h, the concentrations in each of the six vessels leveled off to a constant value of $15.1 \pm 0.6 \, \mu g \, mL^{-1}$, very close to the $14.5 \, \mu g \, mL^{-1}$ solubility determined independently by the 'self-calibrating' microtitre plate UV method (see **Table 1**).[19]

It is interesting to note that if more sample had been used (e.g., 750 mg of phenazopyridine hydrochloride, rather than 0.75 mg), the chloride–phenazopyridine solubility product would have been exceeded. There would have been enough compound to lead to the formation of the co-precipitates of the free base and the hydrochloride salt. Under this condition of the Gibbs 'super buffer,' the interfacial pH is predicted to drop to 3.14 (from a value slightly below pH 7.4 in the 0.75 mg case), and the flux to increase by more than 100-fold, corresponding to the conditions associated with the dash-dot line in **Figure 10**. The equilibration time would drop from about 20 h to less than 1 h. However, this rapid rate of dissolution would only be sustained under an excessive amount of the compound. As the dissolution process decreases the amount of solid below the phenazopyridinium chloride solubility product limit, then the slower dissolution mechanism would take over (see **Figure 9**).

5.17.8   Outlook

The tools and methods described in this presentation are highly refined and are suitable for the measurement of solubility and dissolution of molecules which emerge out of today's lead-finding and optimization programs, molecules that sometimes are very low in solubility. These are pharmaceutically problematic molecules, requiring considerable care and analytical experience and insight to characterize reliably. If the earlier measurements of the physicochemical properties are done well, the structural fine-tuning by medicinal chemists can be made more efficient. The selection process can be expected to produce better behaving candidates, ones that formulation scientists can transform into drug products in a shorter time. Quality measurements can begin at the earlier stages, can be fast (made possible by thoughtful automation), and can benefit considerably from the state-of-the-art tools and strategies currently available.

References

1. Avdeef, A. *Absorption and Drug Development*; Wiley-Interscience: New York, 2003.
2. Grant, D. J. W.; Higuchi, T. *Solubility Behavior of Organic Compounds*; John Wiley: New York, 1990.
3. Anderson, B. D.; Flora, K. P. In *The Practice of Medicinal Chemistry*; Wermuth, C. G., Ed.; Academic Press: London, UK, 1996, pp 739–754.
4. McNamara, D. P.; Vieira, M. L.; Crison, J. R. In *Transport Processes in Pharmaceutical Systems*; Amidon, G. L., Lee, P. I., Topp, E. M., Eds.; Marcel Dekker: New York, 2000.
5. Streng, W. H. *Characterization o Compounds in Solution − Theory and Practice*; Kluwer Academic: New York, 2001.
6. Lipinski, C. A.; Lombardo, F.; Dominy, B. W.; Feeney, P. J. *Adv. Drug Deliv. Rev.* **1997**, *23*, 3–25.
7. Bevan, C. D.; Lloyd, R. S. *Anal. Chem.* **2000**, *72*, 1781–1787.
8. Pan, L.; Ho, Q.; Tsutsui, K.; Takahashi, L. *J. Pharm. Sci.* **2001**, *90*, 521–529.
9. Chen, T.-M.; Shen, H.; Zhu, C. *Combi. Chem. HTS* **2002**, *5*, 575–581.
10. Ruell, J.; Avdeef, A. *Modern Drug Disc.* **2003**, *June*, 47–49.
11. Amidon, G. L.; Lennernäs, H.; Shah, V. P.; Crison, J. R. *Pharm. Res.* **1995**, *12*, 413–420.
12. Dressman, J. B.; Amidon, G. L.; Reppas, C.; Shah, V. P. *Pharm. Res.* **1998**, *15*, 11–22.
13. Dressman, J. B. In *Oral Drug Absorption − Prediction and Assessment*; Dressman, J. B., Lennernäs, H., Eds.; Marcel Dekker: New York, 2000, pp 155–182.
14. Bynum, K.; Roinestad, K.; Kassis, A.; Pocreva, J.; Gehriein, L.; Cheng, F.; Palermo, P. *Dissol. Tech.* **2001**, *8*, 13–22.
15. Gray, V. A. *Am. Pharm. Rev.* **2003**, *6*, 26–30.
16. Toher, C. J.; Nielsen, P. E.; Foreman, A. S. *Dissolut. Tech.* **2003**, *Nov*, 20–25.
17. Kostewicz, E. S.; Wunderlich, M.; Brauns, U.; Becker, R.; Bock, T.; Dressman, J. B. *J. Pharm. Pharmacol.* **2004**, *56*, 43–51.
18. Avdeef, A.; Testa, B. *Cell. Mol. Life Sci.* **2003**, *59*, 1681–1689.
19. Avdeef, A. Pharmacokinetic Optimization in Drug Research. In *Verlag Helvatica Chimca Acta*; Testa, B., Van de Waterbeemd, H., Folkers, G., Guy, R., Eds.; Wiley-VCH: Zurich, Weinheim, Germany, 2001, pp 305–326.
20. Ritschel, W. A.; Alcorn, G. C.; Streng, W. H.; Zoglio, M. A. *Methods. Find. Exptl. Clin. Pharmacol.* **1983**, *5*, 55–58.
21. Streng, W. H.; Yu, D. H.-S.; Zhu, C. *Int. J. Pharm.* **1996**, *135*, 43–52.
22. Roseman, T. J.; Yalkowsky, S. H. *J. Pharm. Sci.* **1973**, *62*, 1680–1685.
23. Crison, J. R.; Shah, V. P.; Skelly, J. P.; Amidon, G. L. *J. Pharm. Sci.* **1996**, *85*, 1005–1011.
24. Lindenbaum, A. S.; Tiguchi, T. *J. Pharm. Sci.* **1976**, *65*, 747–749.
25. Miyazaki, S.; Oshiba, M.; Nadai, T. *J. Pharm. Sci.* **1981**, *70*, 594–596.
26. McGovern, S. L.; Caselli, E.; Grigorieff, N.; Shoichet, B. K. *J. Med. Chem.* **2002**, *45*, 1712–1722.
27. Brittain, H. G. *J. Pharm. Sci.* **1997**, *86*, 405–411.
28. Pudipeddi, M.; Serajuddin, A. T. M. *J. Pharm. Sci.* **2005**, *94*, 929–939.
29. Avdeef, A. *Pharm. Pharmacol. Commun.* **1998**, *4*, 165–178.
30. Avdeef, A.; Berger, C. M.; Brownell, C. *Pharm. Res.* **2000**, *17*, 85–89.
31. Avdeef, A.; Berger, C. M. *Eur. J. Pharm. Sci.* **2001**, *14*, 281–291.
32. Avdeef, A. *Anal. Chim. Acta* **1983**, *148*, 237–244.
33. Avdeef, A. In *Computational Methods for the Determination of Formation Constants*; Leggett, D. J., Ed.; Plenum Press: New York, 1985, pp 355–474.
34. Avdeef, A. *J. Pharm. Sci.* **1993**, *82*, 183–190.
35. Bergström, C.A.S. PhD dissertation, University of Uppsala, Uppsala, Sweden, 2003.
36. Bergström, C. A. S.; Strafford, M.; Lazarova, L.; Avdeef, A.; Luthman, K.; Artursson, P. *J. Med. Chem.* **2003**, *46*, 558–570.
37. Faller, B.; Wohnsland, F. In *Pharmacokinetic Optimization in Drug Research*; Testa, B., van de Waterbeemd, H., Folkers, G., Guy, R., Eds.; Verlag Helvetica Chimica Acta, Wiley-VCH: Zürich, Weinheim, 2001, pp 257–274.
38. US FDA. *Guidance for Industry, Waiver of In Vivo Bioavailability and Bioequivalence Studies for Immediate Release Solid Oral Dosage Forms Based on a Biopharmaceutics Classification System*; Food and Drugs Administration: Washington, DC, 2000.
39. Avdeef, A. pDISS: Simulation of Dissolution of Multiprotic Ionizable Molecules under the Influence of Chemical Reactions, Convection, and Diffusion, pION Inc., 2006.
40. Higuchi, W.; Parrot, E. L.; Wurster, D. E.; Higuchi, T. *J. Am. Pharm. Assoc.* **1958**, *47*, 376–383.
41. Mooney, K.G. PhD dissertation, University of Kansas, KS, 1980.
42. Mooney, K. G.; Mintun, M. A.; Himmelstein, K. J.; Stella, V. J. *J. Pharm. Sci.* **1981**, *70*, 13–22.
43. Mooney, K. G.; Mintun, M. A.; Himmelstein, K. J.; Stella, V. J. *J. Pharm. Sci.* **1981**, *70*, 22–32.
44. Mooney, K. G.; Rodriguez-Gaxiola, M.; Mintun, M.; Himmelstein, K. J.; Stella, V. J. *J. Pharm. Sci.* **1981**, *70*, 1358–1365.
45. McNamara, D.P. Ph.D. dissertation, University of Michigan, MI, 1986.
46. McNamara, D. P.; Amidon, G. L. *J. Pharm. Sci.* **1986**, *75*, 858–868.
47. McNamara, D. P.; Amidon, G. L. *J. Pharm. Sci.* **1988**, *77*, 511–517.
48. Southard, M. Z.; Green, D. W.; Stella, V. J.; Himmelstein, K. J. *Pharm. Res.* **1992**, *9*, 58–69.
49. Vinograd, J. R.; McBain, J. W. *J. Am. Chem. Soc.* **1941**, *63*, 2008–2015.
50. Wood, J. H.; Syarto, J. E.; Letterman, H. *J. Pharm. Sci.* **1965**, *54*, 1068.
51. Wang, J.; Flanagan, D. R. *J. Pharm. Sci.* **1999**, *88*, 731–738.
52. Wang, J.; Flanagan, D. R. *J. Pharm. Sci.* **2002**, *91*, 534–542.
53. Serajuddin, A. T. M.; Jarowski, C. I. *J. Pharm. Sci.* **1985**, *74*, 142–147.
54. Yalkowsky, S. H.; Banerjee, S. *Aqueous Solubility: Methods of Estimation for Organic Compounds*; Marcel Dekker: New York, 1992, pp 58–68.
55. Avdeef, A.; Bucher, J. J. *Anal. Chem.* **1978**, *50*, 2137–2142.
56. Okimoto, K.; Rajewski, R. A.; Uekama, K.; Jona, J. A.; Stella, V. J. *Pharm. Res.* **1996**, *13*, 256–264.

57. Rozou, S.; Voulgari, A.; Antoniadou-Vyza, E. *Eur. J. Pharm. Sci.* **2004**, *21*, 661–669.
58. Jinno, J.; Oh, D.-M.; Crison, J. R.; Amidon, G. L. *J. Pharm. Sci.* **2000**, *89*, 268–274.
59. Noyes, A. A.; Whitney, W. R. *J. Am. Chem. Soc.* **1897**, *19*, 930–934.
60. Brünner, E. *Z. Phys. Chem.* **1904**, *47*, 56–102.
61. Nernst, W. *Z. Phys. Chem.* **1904**, *47*, 52–55.
62. Weiss, T. F. *Cellular Biophysics*; MIT Press: Cambridge, MA, 1996; Vol. 1.
63. Litt, M.; Serad, G. *Chem. Eng. Sci.* **1964**, *19*, 867–884.
64. Levich, V. G. *Physiochemical Hydrodynamics*; Prentice-Hall: Englewood Cliffs, NJ, 1962.
65. Cussler, E. L. *Diffusion – Mass Transfer in Fluid Systems*, 2nd ed.; Cambridge University Press: Cambridge, UK, 1997.
66. Schlichting, H.; Gersten, K. *Boundary Layer Theory*, 8th ed.; Springer-Verlag: Berlin, Germany, 2000.
67. Kelson, N.; Desseaux, A. *ANZIAM J.* **2000**, *42*, C837–C855.
68. Miklavèiè, M.; Wang, C. Y. *Z. Angew. Math. Phys.* **2004**, *54*, 1–12.
69. Olander, D. R. *AIChE J.* **1960**, *6*, 233–239.
70. Pohl, P.; Saparov, S. M.; Antonenko, Y. N. *Biophys. J.* **1997**, *72*, 1711–1718.
71. Yu, L. X.; Carlin, A. S.; Amidon, G. L.; Hussain, A. S. *Int. J. Pharm.* **2004**, *270*, 221–227.
72. Hixson, A. W.; Crowell, J. H. *Ind. Eng. Chem.* **1931**, *23*, 923–931.
73. Bendels, S.; Tsinman, O.; Kansy, M.; Avdeef, A. *Pharm. Res.* Submitted.

Biographies

Alex Avdeef, PhD, is the cofounder of pION Inc. (USA) and Sirius Analytical Instruments Ltd (UK). Formerly, he was Assistant Professor of Chemistry at Syracuse University (USA). He is on the Advisiory Board of the *European Journal of Pharmaceutical Sciences*. He has authored nearly 100 technical papers in scientific journals and chapters in books, and has given nearly 300 invited talks. He recently authored *Absorption and Drug Development* (Wiley-Interscience, 2003). He has held various positions at pION inc. (USA), Sirius Analytical Instruments Ltd (UK), Orion Research (USA), Syracuse University (USA), the University of California (Berkeley, USA), and the California Institute of Technology (CALTECH) (Pasadena, USA).

Dmytro Voloboy, as a staff scientist, is primarily responsible for writing and maintaining scientific software code for high-throughput instrumentation at pION Inc. Prior to his employment at pION, Voloboy served as a developer for the SI Bridge of Dnipropetrovsk, developing scientific instruments for a broad range of applications. Dmytro Voloboy holds an MSc in physics from Dnipropetrovsk State University, Ukraine. Also, he finished 3 years of postgraduate study at Dnepropetrovsk State University in the Automation of Physical Experiments program.

Alexis Foreman is the product specialist for the fiber optic dissolution apparatus, and the designer of the μDISS small volume dissolution system. He holds a BSc in chemistry from Montana State University. He held positions at Montana State University and The Gillette Co. before joining pION Inc.

Comprehensive Medicinal Chemistry II
ISBN (set): 0-08-044513-6

ISBN (Volume 5) 0-08-044518-7; pp. 399–423

5.18 Lipophilicity, Polarity, and Hydrophobicity

G Caron and G Ermondi, University of Torino, Torino, Italy
R A Scherrer, BIOpK, White Bear Lake, MN, USA

5.18.1 General Introduction

In the first edition of *Comprehensive Medicinal Chemistry*, chapters on lipophilicity were included in the Quantitative Drug Design section. In this second edition, published about 15 years later, they appear in Physicochemical Tools in absorption, distribution, metabolism, excretion, and toxicity (ADMET). This change reflects the revolution in medicinal chemistry itself, undoubtedly due to the elucidation of the human genome, along with new automated experimental technologies and progress in computer sciences. These brought an enormous increase in the number of newly discovered active molecules, but not in the number of drug candidates surviving early clinical trials.

This limitation is also due to poor ADMET properties, in particular poor bioavailability. In fact it is now clearly recognized that a successful drug candidate requires not only potency and selectivity, but also a suitable pharmacokinetic profile.[1]

Lipophilicity is an important property of drugs with a bearing on absorption/permeability, volume of distribution, plasma protein binding, metabolism, and toxicity.[1-3] It is a cornerstone of the well-accepted rule-of-five derived by Lipinski and co-workers.[4] Based on an analysis of the key properties of compounds in the World Drug Index (WDI), this alert states that compounds are most likely to have poor absorption when the molecular weight (MW) is > 500, the calculated n-octanol/water partition coefficient (CLOGP) is > 5, the number of H-bond donors is > 5 and the number of H-bond acceptors is > 10 (substrates for active transporters and natural products or mimics of these may be exceptions to this alert). These simple rules and some succeeding studies clearly demonstrate the reasons drug candidates must have ad hoc lipophilic properties to meet the demands of clinical studies.[4-8]

The original guidelines focused on $\log P$ of the unionized species because of the ability to calculate that property for hundreds of marketed drugs. Today, attention is increasingly being given to $\log D$ at physiological pH ($\log D^{pH}$), including methods for its rapid measurement and its correlation with biological properties.

Some key areas for investigation today include finding solvents more membrane-like than n-octanol and the contribution of ionized forms to the overall molecular lipophilicity of drugs.

The lack of partitioning data for ion pairs (*see* Section 5.18.8 below for ion pair definition) has led to inaccurate calculation procedures, to assumptions that the ionized species is not important, and to compromises such as using $\log D$, a mixture of terms, as a single term. Certainly it is known that almost all drug candidates are either ionizable[9] or are converted in vivo to ionizable metabolites.[10-12] Ionized species are also involved throughout the ADMET pathways, so there is considerable incentive to understand their properties. Commercialization of equipment for automated measurements is making it easier to obtain specific data on the physicochemical properties of ionized species. Better $\log P^I$ data (*see* definition in Section 5.18.2 below) should allow more attention to be given to the ionized portion of $\log D$ (*see* Section 5.18.8.6).

Undoubtedly, an obvious consequence of the growing complexity of systems to measure lipophilicity is the increase in information content in related descriptors. This hidden information must be extracted and interpreted, but the debate is still open to understand which lipophilicity descriptors are really useful in ADMET profiling. Some determinants are largely superimposable and others are only relevant from a speculative point of view, but have poor application in drug discovery.

5.18.2   Lipophilicity Parameters and the Effect of Ionization

The most common lipophilicity descriptors are listed and defined in **Table 1**. Basically, we distinguish simple ($=$ one value) from combined ($=$ obtained from two $\log P/\log D/pK_a$ values) parameters.

Among simple descriptors, the partition coefficients, expressed as P, are valid for a single electrical species, to be specified (P^N for neutral forms and P^I for ionized species). Whereas the partitioning of neutral species has been widely studied and understood in the past, the same cannot be said for ions that have received attention only in recent years.

Table 1   Simple and combined lipophilicity descriptors

Category	Symbol	Definition
Simple	$\log P^X_{aaa}$	Logarithm of the partition coefficient P of the electrical species X in the aaa/water system
	$\log D^{pH}_{aaa}$	Logarithm of the distribution coefficient D at a given pH expressed as the concentration ratio of a solute present in more than one electrical state
	$\log P^N$	Logarithm of the partition coefficient of the neutral form (in n-octanol/water)
	$\log P^I$	Logarithm of the partition coefficient of an ionized form (in n-octanol/water); C and A may be used as a substitute for superscript I, denoting a cation or anion, respectively
Combined	$\Delta\log P_{aaa-bbb}$	Difference between the $\log P$ values of the same compound in the same electrical state, measured in the systems aaa/water and bbb/water
	$diff(\log P^{X-Y})$	Difference between the $\log P$ of X and Y obtained with the same organic solvent. X and Y can be two electrical forms of the same compound (in which case the aqueous phases were different), or two closely related compounds

The distribution coefficient, expressed as D^{pH}, is a pH-dependent simple descriptor for ionizable solutes and results from the weighted contributions of all electrical forms present at this pH, as illustrated by eqn [1]:

$$D^{pH} = f^N \times P^N + \sum (f^I \times P^I) \tag{1}$$

where f^N and f^I are the respective molar fractions of the neutral and ionized forms.

In practical terms, it is easily inferred from eqn [1] that $\log D^{pH}$ is: (1) either lower or equal to $\log P^N$; and (2) either higher or equal to $\log P^I$. Please note that the pH regions in which $\log P^N$ and $\log D^{pH}$ have the same value are opposite for acids and bases, since $\log D^{pH}$ is pK_a-dependent. In particular, $\log P^N = \log D^{pH}$ for acids when the pH is about 2 units below the pK_a and for bases when the pH is 2 units above the pK_a. A similar estimation for ampholites is more complex (*see* Section 5.18.8.5).

Equation [1] includes all species present at a given pH and is easily obtained from the definition of D as shown below for a generic acid HA:

$$D^{pH} = \frac{[HA]_{org} + [A^-]_{org}}{[HA] + [A^-]} \tag{2}$$

where [HA] is the weak acid concentration in the aqueous phase and $[HA]_{org}$ is the weak acid concentration in the nonaqueous phase, $[A^-]$ is the fully deprotonated acid concentration in the aqueous phase, and $[A^-]_{org}$ is the fully deprotonated acid concentration in the nonaqueous phase.

Dividing each term of eqn [2] by [HA], eqn [3] is obtained:

$$D^{pH} = \frac{\dfrac{[HA]_{org}}{[HA]} + \dfrac{[A^-]_{org}}{[HA]}}{1 + \dfrac{[A^-]}{[HA]}} \tag{3}$$

and substituting eqns [4], [5], and [6] in eqn [3], eqn [7] is found:

$$K_a = \frac{[A^-] \times [H^+]}{[HA]} \tag{4}$$

$$P^N = \frac{[HA]_{org}}{[HA]} \tag{5}$$

$$P^I = \frac{[A^-]_{org}}{[A^-]} \tag{6}$$

$$D^{pH} = \frac{P^N + P^I \times \dfrac{K_a}{[H^+]}}{1 + \dfrac{K_a}{[H^+]}} \tag{7}$$

which, rewritten by substituting the definition of pH and pK_a, becomes eqn [8], that corresponds to eqn [1] in the case of a generic acid HA:

$$D^{pH} = f^N \times P^N + f^I \times P^I \tag{8}$$

The fractionation of $\log D$ into terms representing the distribution of individual species is discussed in Section 5.18.8.6.

The relevance of ion partitioning, and thus of the P^I terms in eqn [1], has been emphasized in several papers[9,13–16] and will be discussed more fully in Section 5.18.8. Nevertheless, a method has been proposed in the literature whereby, in some cases, it is possible when using $\log D$ to dismiss the contribution of some ionized species. This can occur in conditions for which ion partitioning is not significant.[17]

Combined lipophilicity descriptors have been classified into two groups[15]: those derived from $\log P$ values obtained in the same solvent system (the *diff* descriptors), and those derived from $\log P$ values obtained in different solvent systems (the $\Delta \log P$ descriptors that have been used for a long time to investigate inter- and intramolecular hydrogen bonding properties that determine many pharmacokinetic processes[4,15,18–22]).

The *diff* parameters may refer either to two distinct states of a single solute (e.g., neutral and ionized), or to a pair of related chemicals in a given biphasic system (e.g., parent and metabolites). Another *diff* term refers to pK_a, *diff* $pK_{a\ (oct-W)}$ for acids and *diff* $pK_{a\ (W-oct)}$ for bases. The use of *diff* is more recent[15] and some examples will be discussed below.

The difference between the $\log P$ of a neutral compound and its ion pair ($diff(\log P^{N-I})$) can easily vary more than two log units, even within the same chemical class, depending on certain structural features. It is not correct to make broad generalizations that this difference for all carboxylic acids is the same, or for all phenols or all amines.

Anyone having trouble appreciating the concept of ionization in *n*-octanol can see the evidence of it in **Figure 1a** and **b**. These show the lipophilicity profile of the acid, ketoprofen, and the base propranolol as obtained by applying eqn [1].[9,14,23] The upper inflections in both cases occur at the aqueous pK_a. The lower inflections must occur at the corresponding octanol pK_a ($pK_{a\ oct}$).[14,24] The $pK_{a\ oct}$ is seen in **Figure 1a** at pH 7.78. The octanol pK_a is equal to the pH of the aqueous phase when the concentrations of neutral and ion pair species in the octanol phase are equal[24] (indicated in **Figure 1a**). The pH of the second phase is thermodynamically equivalent to the pH of the aqueous phase with which it is in equilibrium. Contrary to some thinking, when the ionization behavior of an acid or base changes as it enters *n*-octanol (or a biolipid phase), the pH has not changed; it is the pK_a of the solute that changes continuously as the compound progresses from water to octanol.

Figure 1c illustrates a valuable relationship between partitioning and ionization in water and *n*-octanol. These processes form a thermodynamic cycle. The Gibbs free energy of an equilibrium is given by the equation in **Figure 1c**. The net free energy around the cycle is zero. This provides the useful relationship that the difference between two of the equilibria, $\log P^N$ and $\log P^I$, is the same as the difference between the pK_a and $pK_{a\ oct}$ (eqn [9]).[24] An alternative expression that is gaining favor is to use *diff* terms (eqn [10]). These *diff* terms, when used to represent differences in

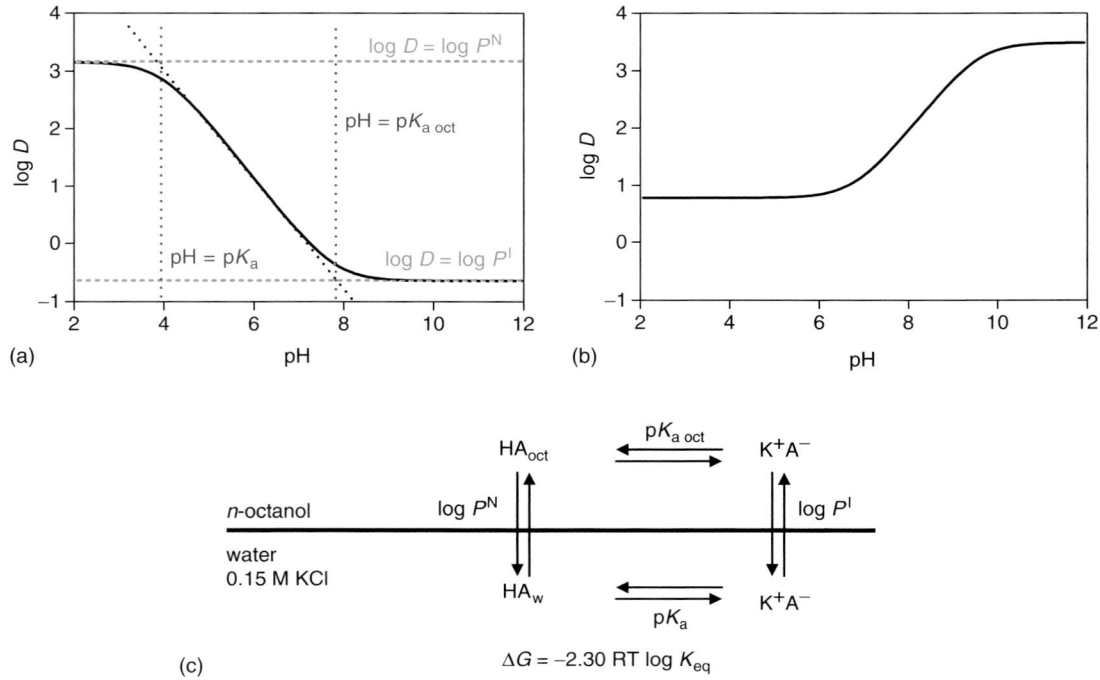

Figure 1 (a) Lipophilicity profile of an acid, ketoprofen, in the *n*-octanol/water system. The aqueous and octanol pK_a's can be read from this profile. The difference between these pK_a's is the same as the difference between $\log P^N$ and $\log P^I$ (pK_a 3.98; $\log P^N$ 3.16; $\log P^I$ –0.64; $pK_{a\ oct}$ 7.78). (b) Distribution profile of a base, propranolol, in the *n*-octanol/water system. Note that in these profiles, the aqueous pK_a always occurs at the upper inflection (pK_a 9.53; $\log P^N$ 3.48; $\log P^I$ 0.78; $pK_{a\ oct}$ 6.83). (c) The thermodynamic cycle of partitioning and ionization. The Gibbs free energy change associated with an equilibrium constant is given by the equation shown. The net free energy change around the cycle is zero.

equilibria, are free energy terms. One *diff* log *P* unit, or *diff*(pK_a), equals 1.36 kcal mol^{-1} at 25 °C.

$$\log P^N - \log P^I = \pm (pK_{a \; oct} - pK_a) \tag{9}$$

$$diff(\log P^{N-I}) = diff \; pK_a^{\pm (oct-W)} \tag{10}$$

The $\pm$ sign is + for acids and − for bases; oct and W stand for the *n*-octanol and aqueous phases. These equations apply to multifunctional molecules as well, in which case, each ionized species is incorporated into its own thermodynamic cycle(s).

In water an ion and its counterion are independently solvated and require little molecular interaction. A useful way to look at the structure of an ion pair is in a stepwise progression following ionization in *n*-octanol. First, mentally convert the ionizable group, in a fixed conformation of the molecule, to the corresponding ion pair. This high-energy form will seek all possible ways to lower the energy of this charge. Most importantly, it will attract water from the solvent to form hydrogen bonds around the ion pair, so the more access water has to the charge, the lower its energy (e.g., in primary versus tertiary amines). It is also important how closely the counterions can approach each other. The next most important change will be a reorientation of the molecule to place as many H-bond donating and accepting groups as possible in the vicinity of the ion pair. The task then becomes quantifying these factors. For example, what size ring is formed when a hydroxy group or ether oxygen coordinates with the ammonium or the carboxylate ion pair? Are there steric interactions that decrease the benefit? Not every interaction can occur simultaneously. In triethanolamine, for example, the first two hydroxy groups will contribute to a much greater extent than the third. In addition, each class of acid or base has its characteristic *diff*(log P^{N-I}) from which the above kinds of adjustments are made. An observation is that, within a given class, the more acidic (or more basic) the group, the lower the *diff*(log P^{N-I}). The positive message is that for a given ion pair and environment, a wide range of molecular structures are possible without affecting the difference between log P^N and log P^I.

Please note that there is another kind of pK_a, designated pK_a'' (double prime), obtained by direct titrations in KCl/water-saturated *n*-octanol. pK_a'' differs from p$K_{a \; oct}$ by a constant, (0.8 ± 0.1),[25] as discussed in Section 5.18.5, and thus can be used in eqns [9] and [10].[26]

Since ADMET prediction studies require the analysis of large series of compounds (a significant data set and the rules to obtain it are given in references [27–29]), it is often informative to visualize the data graphically by ad hoc plots. In particular, in a series of congeneric compounds it is useful to look at the plot log P^N versus log P^I. Please note that the value of the *Y*-intercept is the average *diff*(log P^{N-I}) of the series. Specific examples are discussed below and in Sections 5.18.8.1–5.18.8.1.4.

5.18.3 Chemical Forces Encoded in Lipophilicity

5.18.3.1 Intermolecular Forces

Two or more molecules interact when they affect each other's state or evolution. The 'non-covalent' forces, which encompass an enormous range of attractive and repulsive components, are responsible for the interactions between molecules and may be grouped into two main categories: (1) the electrostatic forces that arise from the forces between the charged entities that make up the molecules[30]; and (2) the hydrophobic forces that occur in a wide range of compounds.[31–33]

The intermolecular forces can be further classified into two main categories (**Table 2**): the fundamental and the combined ones. The fundamental electrostatic forces are usually considered pairwise interactions and can be represented by a potential function (that basically depends on the nature of the interacting charges, e.g., permanent, induced, etc.). The combined forces are mainly of electrostatic origin and can be explained as a mixture of the fundamental ones.

The precise mechanism of fundamental hydrophobic interactions has been under debate for many decades and all the details are still not well understood.[31,34] Hydrophobic interactions do not primarily depend on attractive intermolecular interactions between the involved species. Rather, they are driven by the tendency of water molecules to retain their own water–water hydrogen bond interactions as much as possible, leading to a tendency to arrange nonpolar entities such that the contact surface area between these and water is minimized.[34] Classically, the 'iceberg model'[32] assumes that during the hydration of a nonpolar compound a reduction in the number of hydrogen bonds between water molecules occurs, but that water molecules next to the interface form stronger hydrogen bonds than those in the bulk water phase. Whereas the enthalpic contribution almost cancels out for this process at room temperature (fewer but

Table 2 Main inter- and intramolecular forces in chemistry

A. Fundamental intermolecular forces

Category			Physical origin[a], pair-wise potential[b] and approximate energy of interaction	

A1. Electrostatic

Category			Potential	Energy
Ion	–	Ion	$\dfrac{1}{4\pi\varepsilon_0}\cdot\dfrac{q_1q_2}{r}$	$25\,\text{kJ mol}^{-1}$
Ion	–	Permanent dipole	$-\dfrac{1}{(4\pi\varepsilon_0)kT}\cdot\dfrac{q^2p^2}{r^4}$	$50\text{–}200\,\text{kJ mol}^{-1}$
Permanent dipole	–	Permanent dipole (Keesom Forces)	$-\dfrac{2}{3(4\pi\varepsilon_0)^2kT}\cdot\dfrac{p_1^2p_2^2}{r^6}$	$50\text{–}500\,\text{kJ mol}^{-1}$
Permanent dipole	–	Induced dipole	$-\dfrac{1}{(4\pi\varepsilon_0)^2}\cdot\dfrac{p^2\alpha_p}{r^6}$	$<5\,\text{kJ mol}^{-1}$
Induced dipole	–	Induced dipole (Dispersion forces)	$\sim a\dfrac{-1}{r^6}$	$<5\,\text{kJ mol}^{-1}$

van der Waals forces brackets the last four rows.

Always present
Quantum mechanical origin

Repulsive forces — Always present
Quantum mechanical origin
Repulsion derived from overlapping electron clouds

A2. Hydrophobic

B. Combined intermolecular forces

Category	Physical origin, forces involved and approximate energy of interaction	
Cation-π and π–π	Attractive interactions involving π-systems. The interaction energy depends on both the nature of the π-system and the nature of cation. When the ligand is a metal cation, electrostatic forces dominate the interaction. When the ligand is a non-polar molecule (hydrocarbons …) the dispersive interactions dominate. A combination of electrostatic and dispersive forces governs the interaction when the ligand is polar.	$5\text{–}80\,\text{kJ mol}^{-1}$ and $0\text{–}50\,\text{kJ mol}^{-1}$
Normal and reinforced hydrogen bond	The hydrogen bond is an intermediate range intermolecular interaction between an electron-deficient hydrogen and a region of high electron density. Hydrogen bonds result from an electrostatic attraction between a hydrogen atom bound to an electronegative atom X (usually N or O) and an additional electronegative atom Y or a π-electron system.	$4\text{–}120\,\text{kJ mol}^{-1}$

Table 2 Continued

<table><tr><td colspan="3" align="center">C. Intramolecular interactions</td></tr>
<tr><td>*Factor*</td><td>*Ionization influence*</td><td>*Comment*</td></tr>
<tr><td>Electronic conjugations</td><td>Indirect</td><td>They may occur in aromatic systems and across aliphatic segments. Indirect</td></tr>
<tr><td>Interactions involving polar groups</td><td>Direct, high. Depends on the pK_a and the chemical class of the ionizing group</td><td>Proximity effects between two neutral polar groups, internal H-bonds, hydrophilic collapse, proximity effects between polar and nonpolar groups.</td></tr>
<tr><td>Hydrophobic effects</td><td>Indirect</td><td>Mainly hydrophobic collapse</td></tr>
<tr><td>Steric factors around the charged center</td><td>Direct, medium.</td><td>Related to solvation with water and closeness of approach of the counter-ion</td></tr></table>

[a] One-color circles refer to ions; two-color ovals to permanent dipoles; two-color circles refer to induced dipoles; the arrows are used to indicate the random motion of the object responsible for the averaging of the interaction.

[b] q_i is the charge of the ion i; p_i is the dipole moment of the dipole i; α_p is the polarizability; r the distance between objects; k is the Boltzmann constant; ε_0 is the permittivity of the free space; T is the temperature; the subscript i is omitted where unnecessary; a and b constants are not represented in explicit form for brevity reasons.

stronger hydrogen bonds instead of many such bonds of medium strength), the entropy decreases because of a higher ordering of the water molecules. This step is entropically disfavored. This classical view, however, is not generally accepted. An alternative approach does not regard the structure of the water molecules as the main reason for hydrophobic interactions. Instead, it involves a positive enthalpy resulting from the rupture of several hydrogen bonds in order to create a cavity in the water structure that subsequently accommodates the nonpolar compounds.[31]

5.18.3.2   Intramolecular Interactions

Functional groups in solute molecules may interact with each other in a number of ways (*see* Section 5.18.3.1) depending on their own steric and electronic properties, on the number and nature of interconnecting bonds, and on intramolecular distances. It is thus evident that intramolecular interactions (summarized in **Table 2**) strongly affect the lipophilicity of compounds. In addition, since a polar group may exist in a neutral or charged state, the intramolecular interactions involving this group may dramatically vary according to its electrical state and the dielectric constant of the medium (*see* Section 5.18.8.1).

Testa *et al.* suggested that dichotomic classifications of intramolecular interactions are misleading since they tend to neglect overlaps and intermediate cases.[35,36]

Flexible compounds with suitable moieties may exhibit hydrophobic collapse in polar solvents, and hydrophilic folding in lipidic environments[35,36] (examples of hydrophobic collapse and hydrophilic folding are given in **Figure 2**). These conformational changes, which are postulated to allow flexible compounds to adapt and to mimic their environment, are difficult to predict but can strongly affect the fate of drugs in the body[17,18] and thus cannot be neglected as most commercial software does.

5.18.3.3   Factorization of Lipophilicity

Lipophilicity (expressed as $\log P$) is a molecular parameter encoding both electrostatic and hydrophobic intermolecular forces as well as intramolecular interactions (**Table 2**), as shown by eqn [11], discussed in its original form some years ago.[35,37]

$$\log P = vV - \Lambda + I + IE \qquad [11]$$

where v is a constant, V is the molar volume which assesses the solute's capacity to elicit nonpolar interactions (i.e., hydrophobic forces, and to some extent dispersive and repulsive forces), Λ accounts for the polarity of the molecule, I accounts for ionic interactions, and IE represents the intramolecular effects.

Equation [11] can be applied to all electrical species and all partitioning systems (a classification of systems is reported in Section 5.18.4).

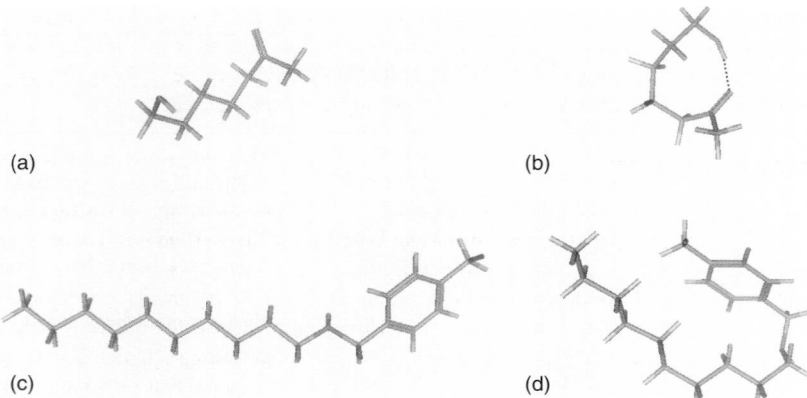

Figure 2 Examples of hydrophilic folding (upper plate) and hydrophobic collapse (lower plate): (a) and (b) represent two conformers for 7-hydroxy-heptan-2-one, of which (b) illustrates hydrophilic olding. Analogously, (c) and (d) show two conformers of *N*-decyl-*para*-methylbenzylamine, of which (d) illustrates the conformation in a hydrophobic collapse. (Reproduced with permission from Caron, G.; Ermondi, G. *Mini Rev. Medicinal Chem.* **2003**, *3*, 821–830.)

Please also note that eqn [11] clearly states that lipophilicity is not synonymous with hydrophobicity, as also indicated by their different International Union of Pure and Applied Chemistry (IUPAC) definitions.[38] Hydrophobicity is the association of nonpolar groups or molecules in an aqueous environment which arises from the tendency of water to exclude nonpolar molecules. Lipophilicity represents the affinity of a molecule or a moiety for a lipophilic environment.

To describe the intermolecular forces that govern the partitioning of simple, neutral solutes, the polar interactions encoded in lipophilicity are hydrogen bonds plus forces involving permanent dipoles, whereas the nonpolar interactions are the London (dispersion) forces and hydrophobic interactions (**Table 2**). This is expressed by the so-called solvatochromic approach,[39] conveyed by eqn [12] (a particular case of eqn [11])[40]:

$$\log P = v \times V/100 + p \times \pi^* + b \times \beta + a \times \alpha + c \qquad [12]$$

where $V/100$ is the calculated van der Waals volume; π^*, α, and β are the solvatochromic parameters (respectively, dipolarity/polarizability π^*, hydrogen bond donor acidity α and hydrogen bond acceptor basicity β) and v, p, a, and b are the regression coefficients reflecting the contribution of each parameter to $\log P$.[29,41]

More difficult is the definition of the I term of eqn [11] because little systematic work has been done on ionized species despite their particular importance. The same is true for the IE term. Work is in progress to address these concerns. In particular, two recent papers from Abraham and Zhao[42,43] report important advances in using the Abraham linear free-energy relationship approach to sort out the factors that correlate with the partitioning of ions into *n*-octanol. This work is in the very early stages and is focusing on the properties of the free ions by using tetra-phenyl-substituted counterions.

Finally, it is noteworthy that inter- and intramolecular interactions governing lipophilicity are of the same nature as those that govern drug recognition and binding to biological sites of action.[36,44]

5.18.4   Lipophilicity Systems

The number of types of lipophilicity systems under study has grown to enhance the possibility of finding good mimic biomembrane models. Equation [11] holds for all systems, but the balance of its terms varies with the system investigated (i.e., sometimes ionic interactions represent the predominant system/solute interaction force, whereas in other cases hydrophobicity and/or polar forces govern the interaction[45]). Great caution must therefore be used to assume that what we learn in a given system can be applied to a second system, particularly when ionized species are considered.

We call standard those biphasic systems in which the two phases are the only components of the system and are in direct contact with each other.

When the second phase is an organic solvent (e.g., *n*-octanol), we have an isotropic system; when the second phase is a suspension (e.g., liposomes), it is an anisotropic system.[45] Anisotropic systems resemble biological systems more than do isotropic ones. Conversely, isotropic systems are easier to use and interpret experimentally.

A search for systems that are markedly easier to obtain and use is always in progress. In particular, because of the ready availability of data, chromatographic systems are often used as an alternative to standard systems.[27,46–53] The presence of chromatographic media along with the one or two liquid phases causes this system to be governed by a balance of intermolecular forces different from the corresponding standard methods.

5.18.4.1   Isotropic Systems

Among biphasic systems, n-octanol/water remains the reference, and thus should always at least be used for comparison. Most of the lipophilicity profiles available to date were determined in that system, and even the shortened classification, $\log D_{\text{oct}}^{7.4}$ data, can give useful information on implications for drug discovery.[58]

Besides n-octanol/water, some alternatives have been proposed in the literature. In particular standard isotropic lipophilicity systems (the 'critical quartet') can be classified by the characteristics of the second phase[41,54,55]: (1) amphiprotic solvents such as n-octanol; (2) hydrogen bond acceptor solvents such as di-n-butylether (dbe); (3) hydrogen bond donor solvents such as chloroform; and (4) apolar aprotic inert solvents such as alkanes.

Solubility and toxicity are two key issues that must be kept in mind when using organic solvents in partitioning measurements. In fact, most solvents show lower solubilizing power than n-octanol and higher toxicity (see below).

In practice, $\log P/\log D^{\text{pH}}$ values will vary with the partitioning solvent. Collander[56,57] was the first to show that $\log P$ (read as $\log P^{\text{N}}$) values from different isotropic solvent systems are linearly correlated:

$$\log P_1 = a \times \log P_2 + b \qquad [13]$$

where subscripts 1 and 2 refer to solvent systems 1 and 2, respectively.

This equation is only valid when the organic solvents have similar physical properties, in particular, similar hydrogen bonding capacity. In fact, for the dbe/water system a satisfactory correlation exists between the $\log P_{\text{oct}}$ and $\log P_{\text{dbe}}$ (**Figure 3a**; $n = 70$; $r^2 = 0.92$).[29] Conversely, when 1,2-dichloroethane (dce) is used, a similar plot of $\log P_{\text{dce}}$ versus $\log P_{\text{oct}}$ splits the training set into two lines (**Figure 3b**)[28] due to the α properties of the solutes (α = hydrogen bond donor acidity). 1,2-dce is an organic solvent replacing alkanes, as explained below.

Alkane/water is a lipophilicity system largely employed in ADMET analyses. In fact, $\log P_{\text{alk}}^{\text{N}}$ is very important alone and in combination with $\log P_{\text{oct}}^{\text{N}}$ ($\Delta \log P_{\text{oct}-\text{alk}}^{\text{N}} = \log P_{\text{oct}}^{\text{N}} - \log P_{\text{alk}}^{\text{N}}$) to investigate those inter- and intramolecular hydrogen bonding properties that determine many pharmacokinetic processes.[4,15,18,22] In particular, in addition to the relevance of $\Delta \log P_{\text{oct}-\text{alk}}^{\text{N}}$ in determining brain penetration,[19–21] a close correlation has also been found between $\log P_{\text{alk}}^{\text{N}}$ and permeability, as measured through artificial membranes.[59,60] It should be noted that $\Delta \log D_{\text{oct}-\text{solv}}^{\text{pH}}$ is equivalent to $\Delta \log P_{\text{oct}-\text{solv}}$ over the pH range from where $\log D = \log P$, to within about 2 units of the pK_a in the more polar of the two solvents. This is evident by comparing the lipophilicity profiles for two solvents and water on the same plot. Beyond this point the ion pair begins to contribute and the relationship becomes much more complex. $\log D^{\text{N}}$ and $\log D^{\text{I}}$ terms are discussed in Section 5.18.8.6.

To overcome some experimental problems caused by the low alkane solubility of many compounds, dce[28] and, more recently, $ortho$-nitrophenyloctyl ether (o-NPOE)[61] have been proposed by a few academic labs as replacements. Unfortunately, dce is considered so toxic that its use in many industrial labs has been banned. On the other hand, o-NPOE shows a number of advantages over dce (higher viscosity, lower volatility, lower solubility in water, lower toxicity, and greater chemical stability[61]), but is too recent to have gained sufficient reliability among researchers. Alkanes (mainly n-dodecane and cyclohexane) remain the elective inert solvents applied in standard lipophilicity measurements.

From a practical standpoint it is convenient to use a graphical approach to extract information from $\log P_{\text{alk}}$ data. In the literature the plot of $\log P_{\text{oct}}^{\text{N}}$ versus $\log P_{\text{aaa}}^{\text{N}}$, where subscript aaa represents a generic biphasic system is often applied.[27–29] In the cases for which a linear trend is found, the Y-intercept represents the average $\Delta \log P_{\text{oct}-\text{aaa}}$ for the investigated series. In **Figure 3c**, aaa is dce, and the corresponding $\Delta \log P$ is about 1.

For studying the lipophilicity of ion pairs, n-octanol/water stands far above any other isotropic system. Since $\log P^{\text{I}}$ values are best determined by looking at the difference between the $\log P$ of the parent neutral compound and the $\log P$ of the ion pair, the library of known $\log P_{\text{oct}}^{\text{N}}$ values provides a valuable starting point for such calculations.

Unfortunately, to date no gold standard regarding which solvent systems should be routinely used in ADMET studies has been proposed. In Section 5.18.7 some ideas are given, but since a lot of experimental data (especially for ion pairs) is unavailable, a complete comparative study among all descriptors has never been done. We rather think that the choice of measuring one parameter or another is based on lab equipment and traditions rather than on other more realistic aspects. Finally, it should be pointed out that predictive tools are rather deficient in this field (see 5.27 Rule-Based Systems to Predict Lipophilicity).

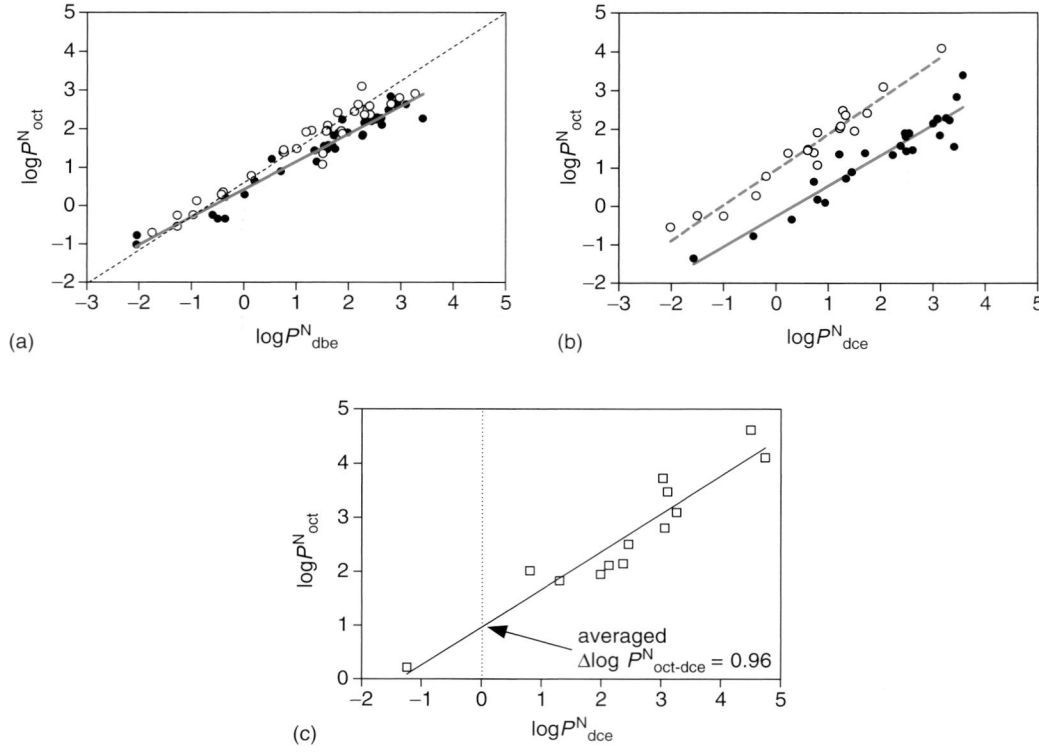

Figure 3 (a) Partition coefficients measured in *n*-octanol/water ($\log P_{oct}$) versus dbe/water ($\log P_{dbe}$). Open circles, solutes with $\alpha > 0$; filled circles, solutes with $\alpha = 0$.[29] The regression line is represented in red ($n = 70$; $r^2 = 0.92$); the ideal line of slope 1 and intercept 0 is in black. (b) Partition coefficients measured in *n*-octanol/water ($\log P_{oct}$) versus 1,2-dichloroethane/water ($\log P_{dce}$). Open circles, solutes with $\alpha > 0$; filled circles, solutes with $\alpha = 0$.[28] Regression lines for both data sets are represented in red. (c) Averaged $\Delta \log P^N_{oct-dce}$ (*n*-octanol and 1,2-dichloroethane) for a series of β-blockers[111]: a value of 0.96 is found (see text for details). It is the outliers from this fit that can provide clues to the role this difference might play in an ADMET process.

5.18.4.2 Anisotropic Systems

In contrast to the isotropic solvents traditionally used in lipophilicity studies, artificial (liposomes) and natural membranes are anisotropic media and their use in lipophilicity studies has therefore led to the concept of anisotropic lipophilicity that looks for good biomembrane models in the effort to reproduce drug–membrane interactions which are involved in both pharmacokinetic and pharmacodynamic processes.[45,62]

In the simplest pharmacokinetic case, membranes are a barrier between two aqueous compartments in which a drug is distributed. Isotropic lipophilicity parameters can model such a situation for either neutral compounds or chemicals that are ionizable but neutral at physiologically relevants pHs (*see* Section 5.18.6), but their combination with anisotropic lipophilicity data is required when the contribution of ionized species is important (*see* Section 5.18.8.3).

In an anisotropic system, if the component providing the anisotropy contains no ionized groups, one would expect little difference from isotropic systems. In other words, anisotropic systems provide the most important information when they contain (or provide) ionized groups and one is dealing with ionizable compounds since the location of charges (**Figure 4**) also plays a role in determining the presence of ionic interaction (eqn [11]) which is very strong in a low dielectric environment.[63]

The entropy of a solute (related to the number of rotatable bonds and their flexibility) may be more affected by binding to a surface or a large molecule than to a simple counterion. The closeness of approach (steric factors) will also be more important. These are attributes not of concern in isotropic systems. Another difference is illustrated below in which a 'negative' phospholipid component of membranes is proposed to form an HIP with a basic solute (*see* Section 5.18.8.3).

Positive and negatively charged substrates will be affected differently in anisotropic systems containing charges. Each anisotropic system is dependent on the nature of its charged components. For this reason, a model system may apply only narrowly to related biological systems.

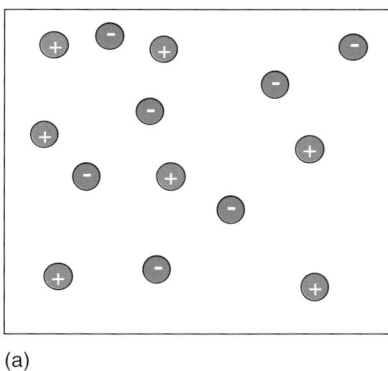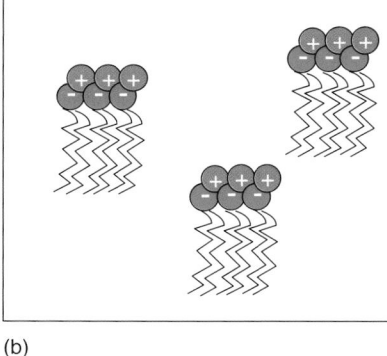

(a)   (b)

Figure 4 Charge location governs the absence/presence of ionic interactions in isotropic/anisotropic systems.[63] (a) In isotropic systems, charges are carried by salt buffers and solutes and thus have no defined location. As a result no 'specific' electrostatic interaction occurs. (b) In anisotropic systems charges are located on the lipidic phase (e.g., the polar head of phospholipids). As a result these systems provide the most important information when one is dealing with ionizable compounds. (Reproduced with permission from Caron, G.; Ermondi, G. New Insights into the Lipophilicity of Ionized Species. In *Pharmacokinetic Profiling in Drug Research: Biological, Physicochemical and Computational Strategies*; Testa, B., Kraemer, S., Wunderli-Allenspach, H., Folkers, G., Eds.; Verlag Helvetica Chimica Acta: Zürich, 2005, pp 165–185.)

An example of a pharmacodynamic process involves membranes as targets of drug action. Anisotropic systems may help to rationalize the behavior of drugs whose target is a membrane, especially when used with a pool of sophisticated experimental techniques (i.e., small-angle x-ray diffraction and small-angle neutron diffraction to support partition coefficients determined by centrifugation assays). These are discussed in detail by Herbette and co-workers.[62] In other cases, membranes provide the matrix in which enzymes or receptors are embedded. Binding assays, coupled with dose-response experiments, are needed to assess specific drug-receptor interactions.[45]

Anisotropic lipophilicity represents an intermediate case between partitioning and binding. As discussed elsewhere, energetic criteria allow one to distinguish partitioning from binding mechanisms,[63] but not the nature of the intermolecular interactions involved, since they are the same, as explained above.

Other anisotropic systems (mainly micelles/water) have also been presented in the literature,[64–70] but their use is still not common.

5.18.4.3   Chromatographic Systems

Chromatographic methods have recently been reviewed by Comer.[58] Most are HT methods and can be classified according to whether or not *n*-octanol is used. The presence of *n*-octanol tends to provide chromatographic systems that more closely reproduce the standard *n*-octanol lipophilicity values.

Although chromatographic retention parameters are good lipophilicity indices that can be used to generate predictive ADMET models, their transformation into $\log D_{\text{oct}}$ values should in principle be avoided. This is especially true for ionized species, since ionic interactions with the stationary phase could significantly affect the balance of intermolecular forces (eqn [11]).[27,53] Even for neutral species, subtle effects such as differences in hydrogen bonding capacity can perturb the correlation between $\log D_{\text{oct}}$ and chromatographic data.[71] In spite of these problems, some good methods are reported in the literature that, after a carefully analysis of the data, are able to convert chromatographic $\log k$ into $\log P/\log D$ parameters. Among them, $E\log D$ by Lombardo and co-workers[47,48] is one of the most appreciated. Please note that the authors clearly underline the limits of the method to achieve high accuracy in determining data for anions, probably because of specific interactions with the stationary phase.

An HT chromatographic system using an octadedecyl-poly(vinyl alcohol) column with a methanol/water gradient has been described by Donovan and Pescatore.[72] Their method has been applied to thousands of compounds and closely reproduces *n*-octanol $\log P$ values ($r^2 = 0.88$, $s = 0.43$ for 120 fungicides, herbicides, and drugs). Routine use of pH 2 and 10 eluents (up to pH 13 when necessary) insures that values are for unionized species.

The preparation and validation of liposomes make their use in partitioning studies time-consuming and tedious. The use of immobilized artificial membranes (IAMs), in contrast, combines the speed of high-performance liquid chromatography (HPLC) technology with a model of phospholipid partitioning.[73] In fact, capacity factors measured on IAMs[74–86] are expected to reflect the partitioning of compounds in the liposomes/water system. (IAM columns are prepared by covalent binding of a monolayer of phospholipids on silica particles.) Some examples of relationships between IAM data and other lipophilicity descriptors are discussed in Section 5.18.8.4.

In immobilized liposome chromatography[81,87] phospholipid-based stationary phases of different chemical composition can be easily and reversibly immobilized on suitable gel supports, and thus chromatographic retention on phospholipids is devoid of any effect caused by the presence of a silica matrix. Still, the possible influence of polymeric support on retention and reproducibility of column performance remains to be fully understood.

Solid-supported lipid membranes (SSLM), commercially available as Transil, are systems in which porous silica beads are covered by a unilamellar liposomal membrane which is noncovalently bound to the bead and which retains its fluid character.[88,89] Schematically, a bead suspension and a solution of the test compound are mixed and left to equilibrate. The beads are then separated by centrifugation, and finally the supernatant is analyzed for quantification of remaining solute. The SSLM permit an HT determination of membrane partitioning, but more published work is needed to assess their potential and limitations better.

5.18.5 Low-Throughput and High-Throughput Methods to Measure Lipophilicity in Non-Chromatographic Systems

Methods for obtaining physicochemical properties for profiling drug candidates have been widely reviewed.[15,58] Even if in principle an optimized experimental technique to measure lipophilicity were able to produce the whole lipophilicity profile (*see* Section 5.18.4.1) in any system, any method has its own experimental window. In practice, the HT revolution often demands a single lipophilicity value (often $\log D_{oct}^{7.4}$) and not the whole profile (*see* Section 5.18.6).

The strategy for measuring $\log P/\log D^{pH}$ is determined by the solubility of a compound. The compound must be soluble (and stable) during any procedure to ensure equilibrium is maintained. $\log P/\log D^{pH}$ determination of poorly soluble compounds is a problem. Provided the $\log P/\log D^{pH}$ is high enough, the value may be determined by titration, adding the sample to the octanol first. The compound will then back-partition into the aqueous layer.

Though accurate, the classic manual shake-flask method[90] of measuring lipophilicity can be very time-consuming and thus emphasis has turned to HT modifications involving multiple samples using a modified titer plate. Examples of this shake-plate method include reports by Kratochwil *et al.*,[91] Zhu *et al.*,[92] and Valko.[93]

The reference method for measuring the partition coefficients of the neutral and ionized forms of a compound (and thus $\log D^{pH}$ by eqn [1]) is the pH-metric method, namely a two-phase titration.[14] *n*-Octanol is the usual second phase,[45,94–97] but other organic solvents[61,96,98–100] and suspensions of liposomes[45,101–105] have been used. The aqueous phase is usually 0.15 M KCl or NaCl. The titration occurs in the aqueous phase. The method is based on the apparent shift in pK_a in the presence of the second phase due to partitioning of neutral and ionized species into that phase. The extent of the pK_a shift can be related to $\log P^N$ and $\log P^I$. At least two titrations with different *n*-octanol:water ratios are required to solve for the two partition coefficients. Advantages and drawbacks of this method, which more recently has been fully automated, are listed and commented upon in the review of Plemper Van Balen *et al.*[45]

Single-phase titrations in KCl/water-saturated *n*-octanol is an alternative procedure for obtaining lipophilicity data.[25,26] The principle is based on the thermodynamic cycle in **Figure 1c** and the relationships derived in eqns [9] and [10]. It appears that pK_a'' differs from $pK_{a\ oct}$ by a constant, the electrode correction factor discussed below, and this is the working hypothesis. Contrary to one's instinct, the curves are very smooth – almost like aqueous titrations. **Figure 5** shows the titration curve of pyridine-2-acetic acid. Calculation of pK_a'' is by curve fitting to a Henderson–Hasselbalch equation using the Solver program in Excel. A key to accurate calculations was determination of the pK_as of water in KCl/water-saturated 1-octanol: -1.31 ± 0.09 ($n = 7$) and 17.02 ± 0.05 ($n = 8$). The pK_as were determined by titrations of HCl and of triflic acid from pH 2 to 14, as well as by extending titrations of other acids to pH 14 and of bases to pH 2–3.

The titration procedure uses the same apparatus and general protocol as two-phase titrations. The few differences include preparing the titrants in ethanol, widening the criteria for equilibration at each point, and extending the stirring time after each pulse of titrant to compensate for a slower electrode response time. Typically, the titrant represents < 1% of the final volume. The procedure is automated for unattended multisample application. Details are reported on the CD accompanying the reference[26] and a full manuscript is in preparation.

The electrode correction factor (ecf), for the conversion of pK_a'' to $pK_{a\ oct}$ (eqn [14]) is 0.8 ± 0.1, based on comparison of two-phase and single-phase titrations in *n*-octanol.[25,26] A component of ecf undoubtedly accounts for the different background counterion concentration in the two methods.

$$pK_{a\ oct} = pK_a'' \pm 0.8 \quad (+\text{for acids}, -\text{for bases}) \qquad [14]$$

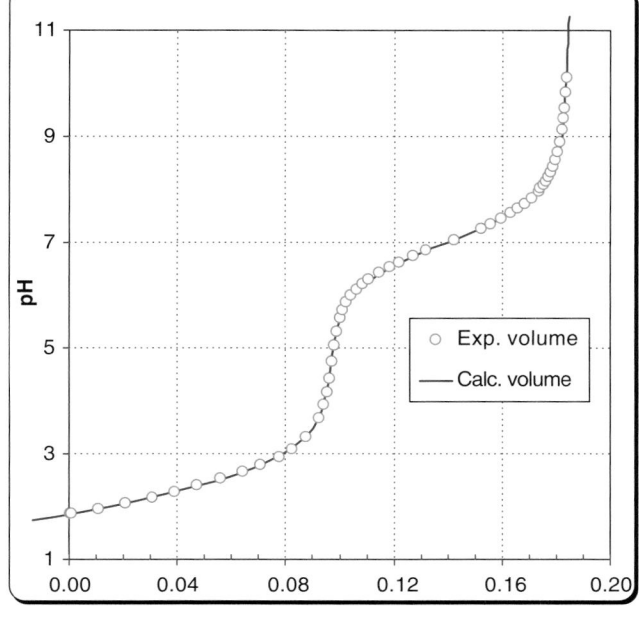

pK$_a$" (1), pK$_a$" (2)	2.309	6.994
SD	0.0038	0.0036
r^2, s	0.9999	0.0005
pK$_a$" water	−1.215	
SD	0.00363	
r^2, s	0.9999	0.0005

pH (y-axis), Volume (mL) (x-axis)

○ Exp. volume
— Calc. volume

Figure 5 The titration curve for pyridine-2-acetic acid hydrochloride in KCl/water-saturated octanol. The pK$_a$"'s are determined by curve-fitting to a Henderson–Hasselbalch equation using the Solver and Excel programs. There are enough data points to solve simultaneously for the pK$_a$" of water in n-octanol. The equivalent volume for pK$_{a1}$ was iteratively set equal to that for pK$_{a2}$. The appearance, statistics, and reproducibility of this method show a close approximation to titrations in water. See text for details.

Reymond *et al.*[16] indicate that cyclic voltammetry (CV) is the only method that can yield the intrinsic $\log P^{\mathrm{I}}$ (called $\log P^{0,\mathrm{i}}$) for the bare ion without a counterion. This contrasts with $\log P^{\mathrm{I}}$ values obtained by traditional methods (e.g., shake-flask and titration) which are strongly influenced by experimental conditions, and particularly by phase volumes and the nature of counterions. Even if the electrochemical approach is able to guarantee a very high number of reliable data,[99,106–112] it must be stressed that it cannot be routinely applied in medicinal chemistry because of its serious experimental limitations, particularly the very few usable organic solvents (dce and o-NPOE) and the resulting inaccessibility of most biphasic systems such as the standard n-octanol/water.

Faller and co-workers have adapted an HT permeation assay method (no stationary phase involved) to measure the partition coefficient of compounds in hexadecane[59] and now in n-octanol.[113] The method is based on the relation:

$$P_0 = P_{\mathrm{solvent}} \times \left(\frac{D}{h}\right) \qquad [15]$$

where P_0 is the intrinsic permeability, P_{solvent} is the partition coefficient in the solvent/water system, D the diffusion coefficient in the solvent, and h the thickness of the fluid membrane.

Finally, anisotropic lipophilicity can be measured by a number of direct methods, such as potentiometry (see above), dialysis, ultracentrifugation, ultrafiltration, calorimetry, and spectroscopic techniques.[45,114]

5.18.6   Strategies for Determining the Optimum Lipophilicity Parameters to be Used in Absorption, Distribution, Metabolism, Excretion, and Toxicity Studies

Most scientists accept $\log D_{oct}^{7.4}$ as one of the most relevant lipophilicity descriptors to be applied in ADMET studies. A proposed enhancement to $\log D$ is discussed in Section 5.18.8.6 which involves factoring $\log D$ into separate terms for the neutral and ionized species. Comer has given some general guidelines about $\log D_{oct}^{7.4}$ values and their implication for drug development.[58] Briefly, when $\log D_{oct}^{7.4}$ is below zero, the compounds have intestinal absorption and central nervous system (CNS) permeability problems and are susceptible to renal clearance; $0 < \log D_{oct}^{7.4} < 1$ generally indicates a good balance between solubility and permeability; $1 < \log D_{oct}^{7.4} < 3$ is an optimum range for CNS and non-CNS orally active drugs; $3 < \log D_{oct}^{7.4} < 5$ is a range in which metabolic liabilities tend to increase. Finally, $\log D_{oct}^{7.4} > 5$ is synonymous with poor oral bioavailability.

It is noteworthy that modern screening technology and the trend to combinatorial synthesis has resulted in hits that are generally more lipophilic than marketed drugs or compounds currently undergoing clinical evaluation. High lipophilicity is often accompanied by poor aqueous solubility. This can bring with it many challenges, often making development of a seemingly promising drug candidate very difficult.

The application of $\log D_{oct}^{7.4}$ in ADMET requires some alerts. Faced with the selected database(s) of drug candidates for which lipophilicity determinants are required, we can distinguish three classes of compounds: (1) chemicals that are neutral; (2) chemicals that are ionizable but neutral at physiologically relevants pHs; and (3) chemicals that are ionizable and partly or mostly charged at physiologically relevants pHs.

The first step of the study therefore consists in classifying each compound. To do that you have to know the pK_a(s) and its (their) acidic or basic nature. Then one has to determine the required $\log D_{oct}^{7.4}$.

Basically, dealing with compounds belonging to the first two classes is an easier task than working with compounds charged at physiologically relevant pH. In most cases, for (1) and (2) the estimation of $\log P_{oct}^{N}$ is sufficient since here $\log P_{oct}^{N}$ is equal to $\log D_{oct}^{7.4}$. Only when anomalies are evident, i.e., $\log P_{oct}^{N}$ is not in line with values found for congeneric chemicals or with the calculated value (*see* 5.27 Rule-Based Systems to Predict Lipophilicity; and Section 5.18.7), is it advisable to do something more. This means checking the experimental value by an additional experiment (better with a different method) and testing a second isotropic solvent system very different from *n*-octanol/water, i.e., alkane/water.

The presence of a charge (compounds belonging to class 3) complicates the situation considerably. One must note the dominant species at the considered pH before any determination. Then $\log D_{oct}^{7.4}$ can be measured by the selected method.

Once the $\log D_{oct}^{7.4}$ values are obtained, it is possible to check relationships with ADMET parameters (i.e., absorption/permeability, distribution volume, clearance) for all compounds belonging to the series of projects. If no model can be developed, or a significant correlation is found, but contains outliers, further studies should begin with members of class 3 and the application of molecular modeling tools is advisable.

A number of successful correlations between lipophilicity and ADMET descriptors have been reported in the literature. The correlation found between $\log D_{oct}^{7.4}$ and renal clearance ($mL\,min^{-1}\,kg^{-1}$) for a series of β-adrenergic antagonists in healthy human subjects[1] is an excellent example. It demonstrates the role of lipophilicity in renal excretion. Compounds with $\log D_{oct}^{7.4}$ above 0 will undergo near complete absorption in the kidney tube and then metabolism to render them more polar, while compounds with $\log D_{oct}^{7.4}$ below 0 will be poorly reabsorbed and thus undergo considerable renal clearance.

When ions are the dominant species (or the sole species) at the pHs of interest, we suggest obtaining additional $\log D^{pH}$ ($\log P^{I}$) in anisotropic systems and comparing these side by side with $\log D_{oct}^{pH}$. By such comparison, and a careful analysis of data, one can check the influence of the charge on ADMET phenomena.

This approach has been applied to investigate peculiar pharmacokinetic properties of amlodipine[62] ($pK_a = 9.07$ and $\log D^{7.4} = 1.55$; the chemical structure is reported in **Figure 7**, below) for which it has been demonstrated that ionic interactions are responsible for the long stay in membrane, which in turn is mandatory for the large volume of distribution, V_d ($25\,L\,kg^{-1}$) and the long plasma half-life (30 h). The extent of ionization and the balance between ionic and hydrophobic interactions are therefore the crucial features that differentiate the membrane-binding mode of amlodipine from other long-acting calcium antagonists (e.g., lacidipine and lercanidipine).[98]

When dealing with zwitterions it is advisable to estimate the whole lipophilicity profile (*see* Section 5.18.8).

Finally, we also recommend a combination of experimental data with ad hoc computational studies whose description is beyond the scope of this chapter.

5.18.7   Unraveling the Lipophilicity of Neutral Species

The combined descriptor $diff(\log P^{\exp-calc})$ represents the difference between an experimental $\log P$ and the corresponding calculated value (*see* 5.27 Rule-Based Systems to Predict Lipophilicity) by a favorite or best known method. As illustrated by Caron and Ermondi,[17] this *diff* descriptor (**Table 1**) has at least three applications of high utility in ADMET studies: (1) optimization of the adopted experimental procedure; (2) identification of the best predictive methods for a structurally related series of compounds; and (3) extraction of structural information from investigated drug candidates and from chemicals involved in hit characterization, lead profiling, and lead optimization.

A second approach to studying the lipophilicity of neutral species is based on graphs of $\log P^{N}_{oct}$ versus $\log P^{N}_{aaa}$[27–29] (*see* Section 5.18.4.1 and **Figure 3c**). Once the relationship for an optimized training set of chemicals is known, the position of new compounds on these graphs can give useful information on the presence or absence of intramolecular effects affecting their neutral forms.[98,111,115] This approach allowed the elucidation of the peculiar intramolecular interactions affecting amlodipine[98] and some β-blockers.[115]

The use of these plots in drug design is strongly recommended by academics, but unfortunately, to date, is poorly accepted by those in industry. The plots allow early identification of compounds with unusual intramolecular H-bonding, molecular tautomerism, or conformational flexibility. It is crucial to be aware when such features are present. Any drug candidate of this type will be endowed with a peculiar physicochemical profile, and thus is likely be an outlier in traditional classification schemes. These compounds could be excellent starting points for designing new chemical entities of pharmaceutical interest.

5.18.8   Unraveling the Lipophilicity of Ions and Zwitterions

It is widely assumed[25,116,117] that in partition phenomena the ionized species migrating into the nonaqueous phase is accompanied by a counterion, forming a charge-neutral ion pair.[118] According to IUPAC, an ion pair consists of a pair of oppositely charged ions held together by Coulomb attraction without formation of a covalent bond. Experimentally, an ion pair behaves as one unit in determining conductivity, kinetic behavior, and osmotic properties, etc.

The nature of the counterion and its concentration both influence $\log P^{I}$, so these have to be taken into account. Many feel that 0.15 M NaCl or KCl, corresponding to extra- and intracellular concentrations of these salts, respectively, is the logical choice for the aqueous phase in partitioning studies.[14,94]

5.18.8.1   The Use of $diff(\log P^{N-I})$ and $diff(pK_a'')$ to Calculate the Lipophilicity of Ion Pairs in *n*-Octanol/Water

Generally speaking, the change in lipophilicity of a molecule when it ionizes is determined by its structure in the near vicinity of the charge center. This simplifies the prediction process since, whatever the size or makeup of the remainder of the molecule, if important features around the charge center do not change, the difference between $\log P^{N}$ and $\log P^{I}$ will be about the same. Titrations in *n*-octanol have proven to be an ideal method for identifying these important features and quantifying their influence. Four primary factors affecting $\log P^{I}$[25] are illustrated in Sections 5.18.8.1.1–5.18.8.1.4 using data from **Table 3**. The titration method is based on eqns [9] and [14]. A particular advantage for current purposes is that, when looking at differences between $diff(pK_a'')$ values, the electrode correction factors cancel out, giving values sensitive to a few hundredths of a pK_a or $\log P^{I}$ unit. A distinctive feature of this method is that it is insensitive to the magnitude of $\log P$, so a wide range of simple compounds is available. A glance at the variation in the $diff(pK_a'')$ values in **Table 3** shows the limited value in the generalization that $diff(\log P^{N-I})$ is the same for all amines, and the same for all carboxylic acids.[119]

5.18.8.1.1   The influence of steric factors on $\log P^{I}$

The hindered König's base, *N, N*-diisopropylethylamine (**Table 3**), has a $diff(pK_a'')$ 0.62 units larger than for triethylamine. This means its $diff(\log P^{N-I})$ is also greater by the same amount. The more hindered amine has a lower $\log P^{I}$ with respect to its neutral form than does triethylamine. On the other hand, the much smaller $diff(pK_a'')$ for phenethylamine, compared to triethylamine, tells us that primary ammonium compounds are much more lipophilic compared to their neutral form than are simple tertiary amines, by about 1.4 log units in this case. This is reflected in their relative pK_as in water and *n*-octanol. In *n*-octanol, phenethylamine is a stronger base than triethylamine (pK_a'' column), just the opposite of their relative pK_as in water (pK_a column). These property differences can have a direct bearing on biological activity, or on ADMET properties, particularly if only one

Table 3 Titrations in KCl/water-saturated *n*-octanol illustrating structure–*diff* pK_a relationships. The lower the *diff*(pK_a''), the more lipophilic the ion pair in relation to its neutral form

		log P^{Na}	p$K_a{}^a$	p$K_a'' \pm$ SD $(n)^b$	*diff* (pK_a'') or (pK_a − pK_a'')^c
Bases					
1	(*N*, *N*-Diisopropyl)ethylamine	2.35^d	11.44^e	7.84 ± 0.00 (4)	**3.60**
2	Triethylamine	1.45	10.72	7.74 ± 0.03 (2)	**2.98**
3	Triethanolamine	−1.00	7.76	6.50 ± 0.05 (4)	**1.26**
4	Diethanolamine	−1.43	8.88	7.89 ± 0.03 (5)	**0.99**
5	Phenethylamine	1.41	9.83	8.24 ± 0.00 (2)	**1.59**
6	*t*-Butylamine	0.40	10.68	8.97 ± 0.00 (2)	**1.71**
7	Tris(hydroxymethyl)-aminomethane	−1.38^d	8.08	7.56 ± 0.01 (4)	**0.52**
8	*N*-Methyl-D-glucamine	−3.37^d	9.62^f	9.27 ± 0.03 (4)	**0.35**
Acids					*diff*(pK_a'') or (pK_a'' − pK_a)
9	Benzoic acid	1.87	4.20	7.60 ± 0.04 (9)	**3.40**
10	Hexanoic acid	1.92	4.88	8.10 ± 0.00 (2)	**3.22**
11	α-Hydroxyphenyl acetic acid	0.62	3.39	6.25 ± 0.01 (2)	**2.86**
12	3-Hydroxybutyric acid	−1.14^d	4.39	6.89 ± 0.04 (3)	**2.50**
13	Phenyltetrazole	1.65	4.38	6.43 ± 0.01 (2)	**2.05**
14	3,5-Dichlorophenol	3.52	8.22	11.16 ± 0.04 (4)	**2.94**
15	2,6-Dinitrophenol	1.37	3.56	5.29 ± 0.02 (5)	**1.73**
16	Diphenylphosphate	1.34^d	1.36^g	2.46 ± 0.01 (2)	**1.10**

^{a} From ACD database unless otherwise indicated.
^{b} Previously unpublished but implied by [26].
^{c} Taken from [26].
^{d} Calculated by ACD log P v8.02.
^{e} Taken from [152].
^{f} Taken from [153], p. 38.
^{g} Taken from [25].

of the species, the protonated or the neutral, is the active form. ('Active form' is the species involved in the rate-limiting step in the process being measured. It could be the species in equilibrium with the target biological receptor, the species that is the substrate of a metabolizing enzyme, or the form involved in active transport.) Whether one chooses to look at the lipophilicity of ionized species, or their *n*-octanol pK_a, these properties do not always parallel those of the neutral species. Such differences may provide an opportunity to identify the active species, but even if they do not, one can see the importance of giving specific consideration to the ionized form of compounds being modeled in ADMET studies.

5.18.8.1.2 The influence of hydrogen bond donors and acceptors on log P^I

The advantage of being able to use pK_a'' instead of log P values to study hydrogen bonding effects on log P^I is that one is able to use much simpler compounds to isolate the property to be studied from other potential factors and interactions. Polyhydroxy compounds would need to have quite high molecular weights for their log P^I values to be easily measured, for example. It would be difficult to isolate the influence of the hydroxy groups. Hydroxy groups can have a powerful influence on the lipophilicity of ionized acids and bases. The *diff*(log P^{N-1}) for triethanolamine is 1.7 units smaller than that for triethylamine (**Table 3**). Similar comparisons can be made for *tris*(hydroxymethyl)aminomethane (tromethamine), which has a *diff*(log P^{N-1}) 1.19 units lower than *t*-butylamine, and for the acids 3-hydroxybutyric acid and hexanoic acid, the two *diff*(log P^{N-1})s differ by 0.72 units. At the extreme is *N*-methyl-D-glucamine (meglumine), $CH_3NH-CH_2(CHOH)_4CH_2OH$, which has a $diff(pK_a'')$ of only 0.35, meaning the cationic form is almost as lipophilic as the free base. It is interesting that both tromethamine and meglumine have been used for the formulation of oral and injectable drugs and x-ray and magnetic resonance imaging contrast agents (e.g., carboprost tromethamine, flunixin megumine, tetrazolast meglumine, iodoxamate meglumine, diatrizoate meglumine, and gadoterate meglumine).

5.18.8.1.3 The influence of pK_a on $\log P^I$

It is difficult to find a uniform series of acids or bases in which the pK_a varies over a wide-enough range to make useful conclusions about the influence of pK_a on the lipophilicity of their ion pairs. There is one excellent series of 24 chloro- and nitrophenols reported by Escher and Schwarzenbach[120] in which the $\log P^I$ of each member was measured. The corresponding $pK_{a\ oct}$ values can be calculated using eqn [9]. The following regression equation obtained by Scherrer[25] showed the relation:

$$pK_{a\ oct} = 1.19(\pm 0.04)pK_a + 0.33(\pm 0.08)I_{Cl} + 1.92(\pm 0.27)$$
$$r^2 = 0.98;\ s = 0.32;\ n = 23;\ F = 401(4 - Cl\ omitted)$$ [16]

where I_{Cl} is an indicator term for the number of o-chloro groups. The coefficient of the pK_a term is > 1, which means that the more acidic the phenol, the less the difference between $pK_{a\ oct}$ and pK_a (and between $\log P^N$ and $\log P^I$; eqn [2]). The $diff(\log P^{N-1})$ decreases about 0.2 units for each unit decrease in pK_a. No special term is needed for nitrophenols. This equation validates the similar equation for 13 phenols based on pK_a'' values[25]:

$$pK_a'' = 1.16(\pm 0.02)pK_a + 0.28(\pm 0.07)I_{Cl} + 1.16(\pm 0.19)$$
$$r^2 = 0.99;\ s = 0.19;\ n = 13;\ F = 1135$$ [17]

The coefficient of the pK_a term was also > 1 for a series of 22 benzoic and salicylic acids,[25] (eqn [18]). It appears to be general that the more acidic and resonance-stabilized an acid, the lower the $diff(\log P^{N-1})$. It seems reasonable to believe that this will apply to bases as well.

For benzoic acids,

$$pK_a'' = 1.25(\pm 0.06)pK_a + 0.60(\pm 0.08)I_{Cl} - 1.18(\pm 0.09)I_{NO_2} + 2.47(\pm 0.17)$$
$$r^2 = 0.99;\ s = 0.17;\ n = 22;\ F = 467$$ [18]

for which I_{Cl} is an indicator variable for $ortho$-chloro groups and I_{NO2} an indicator for the presence of a nitro group $ortho$ or $para$ to the phenolic OH of salicylic acids.

In **Table 3**, the highly acidic diphenylphosphate, pK_a 1.36, is seen to form a relatively lipophilic ion pair. The $diff(pK_a'')$ is only 1.10. It would seem that many of the hydroxy phosphoric acid mono- and di-esters in nature and biochemistry, like phosphatidylinositol (PI), will form very stable low-energy ion pairs in low dielectric environments. These can be added to the list of reasons Westheimer gives for 'Why Nature chose phosphates.'[121]

5.18.8.1.4 The influence of the chemical class on $\log P^I$

Finally, various classes of acids and bases have characteristic $diff(\log P^{N-1})$ values. For example, compare benzoic acid and phenyltetrazole in **Table 3**. Even though they have about the same aqueous pK_a, their pK_a'''s differ much more, and therefore their $diff(\log P^{N-1})$ will differ as well. This category overlaps some of the preceding ones. For example, primary, secondary, and tertiary amines could be considered separate classes. There are many classes for which it would be difficult to predict $diff(\log P^{N-1})$ and for which specific data need to be obtained. Some examples are enolic acids, sulfonamides, acylsulfonamides, amidines, and nitroalkanes.

5.18.8.1.5 The influence of the immiscible solvent phase on $\log P^I$

Ionized compounds put a unique demand on a solvent, so their lipophilicity may vary more widely between solvents than neutral species. n-Octanol is able to hold water and counterions and has both hydrogen donor and acceptor properties.[122] dce and o-NPOE provide partitioning information different from n-octanol because they lack H-bond donor and acceptor activity.[13,61,107–109,111,123] In particular, Bouchard $et\ al.$ have proposed a $diff(\log P^{N-1})$ scale in dce/water that provides useful information about the behavior of ions in different systems.[13]

5.18.8.2   Hydrophobic Ion Pairs

The first use of the term 'hydrophobic ion pairing' appears in a review by Meyer and Manning.[124] The concept is further discussed in their subsequent patents and publications[125–128] on drug delivery. HIP is an ion-pairing phenomenon that refers to the replacement of simple inorganic counterions with ionic detergent-like compounds that dramatically alter their partition coefficient. Examples include α-chymotrypsin that can be solubilized in *iso*-octane with retention of activity, and insulin that can be transported through methylene chloride as an illustration of its unusual solubility properties.

The effect of HIPs on drugs can also be dramatic. A recent report[126] describes the preparation of a HIP of cisplatin with dioctyl sulfosuccinate (AOT), a US Food and Drug Administration-approved surfactant. The *bis* AOT ion pair had $\log P^{\mathrm{I}}_{\mathrm{oct}}$ 2.17, compared to $\log P^{\mathrm{I}}_{\mathrm{oct}}$ of −2.35 for cisplatin. When cells were exposed to equal concentrations of the HIP or the parent drug for 1 h, the HIP produced cellular platinum levels 4–5 times higher than the cisplatin. Studies on various tumor lines showed the AOT complex to be 1.7–5 times more potent.

HIPs are at one end of a continuum of salt forms of acids and bases with increasingly lipophilic counter ions. Their lipophilicity profile can be illustrated using the left or right half of **Figure 6a**.[151] Covering the left half of the figure leaves lipophilicity profiles that could represent a series of salts of a basic compound with pK_a 9.0. The lowest curve, **d**, could

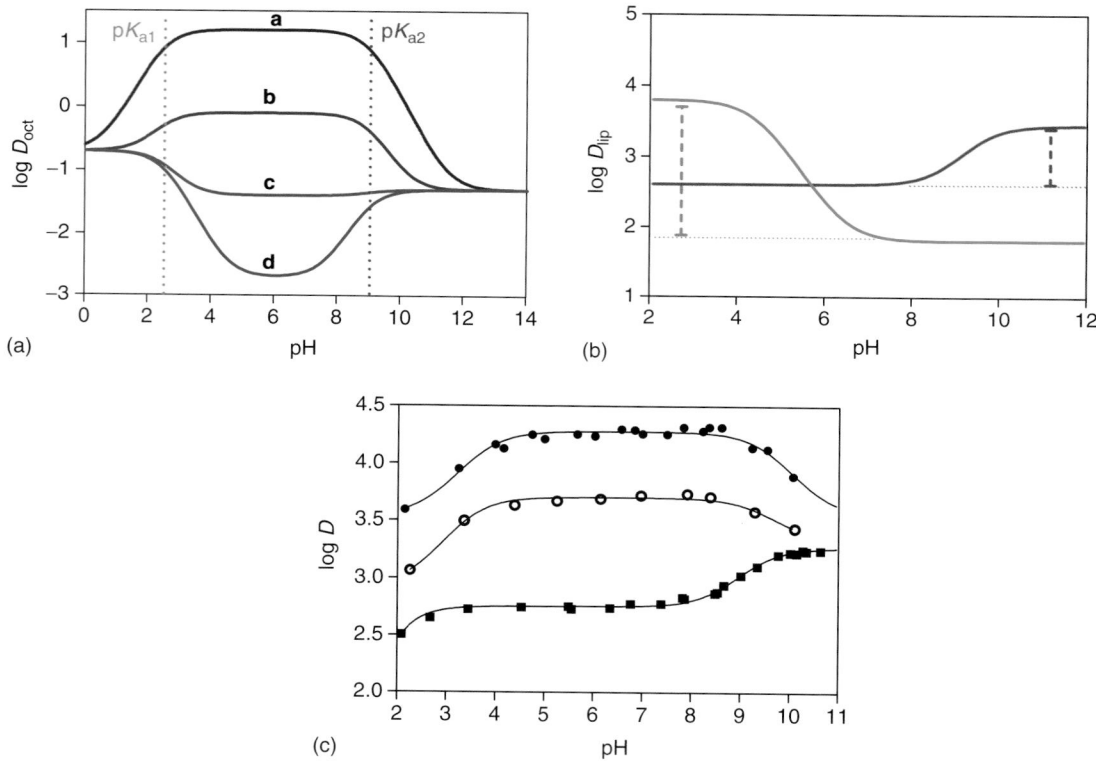

Figure 6 (a) The distribution profile of a series of ampholytes showing a progression from type 3 (**d**) to type 4 (**a**) ampholyte. For comparison purposes, all have the same partition coefficients for their cationic and anionic species at extreme pH values. Their acidic and basic groups also have the same respective pK_a values, 2.5 and 9.0. The only difference is that the $\log P^Z$ of their zwitterions varies. (Reproduced with permission from Berthod, A.; Carda-Broch, S.; Garcia-Alvarez-Cogue, M. C. *Anal. Chem.* **1999**, *71*, 879–888 © American Chemical Society.) Note how the p$K_{a \, \mathrm{oct}}$ at the lower inflection of each curve shifts through the series. (b) Partitioning in an anisotropic system. The distribution profile (data taken from reference[101]) of a monoacid drug (ibuprofen, in red) and of a monobasic drug (propranolol, in blue) in the PhC-liposome/water system. From the difference between $\log P^N$ and $\log P^I$ it is clear that charged forms of bases have larger affinity for liposomes than charged forms of acids. Comparison with **Figure 1b** shows that *diff*($\log P^{N - I}$) for propranolol is lower in liposomes than in octanol. (c) Partition profiles of (*RS*)-[³H]propranolol in: filled squares, P^C-; filled circles, P^I-, and open circles, P^C/P^I (7/3 mol/mol)-liposomes/SUBS systems. (Reproduced from Krämer, S. D.; Braun, A.; Jakits-Deiser, C.; Wunderli-Allenspach, H. *Pharm. Res.* **1998**, *15*, 739–744, with kind permission of Springer Science and Business Media.)

be that for the base with a chloride counter ion. That ion pair has a lower lipophilicity than the neutral amine, and the $pK_{a\,oct}$ 6.0 (lower inflection) is lower than the aqueous pK_a 9.0, which is also typical. If one were to determine the lipophilicity profiles of this amine with a series of alkanesulfonate counterions, they would look like the right half of **Figure 6a**. The $\log P$ of the ion pair will increase with chain length until it equals, and then exceeds, the $\log P$ of the neutral parent amine. This cross-over might be called the defining point for a hydrophobic ion pair. Ion pairs with $\log P$ values 2–4 units higher than the parent acid or base are readily obtainable.[124] Referring to **Figure 6a** again, for the upper curves one can see that the ion pair exists over an increased pH range and that the $pK_{a\,oct}$ is now higher than the aqueous pK_a. The shift reflects the lower free energy to form the ion pairs.

5.18.8.3 Lipophilicity of Ions in Anisotropic Systems

For neutral compounds, partition coefficients in isotropic and anisotropic media are known to be correlated.[74,101] In contrast, partition coefficients of cationic species are much enhanced in anisotropic media (liposomes and cell membranes) compared to n-octanol.[100,103,129] In **Figure 6b**, propranolol and ibuprofen are taken as examples to demonstrate this finding.

These data have been rationalized[62,98,129–132] by postulating that the cationic form of a drug interacts with the anionic oxygen of the phospholipid head group of the membrane bilayer, leading to an additional stabilization. In other words, when trying to model the partitioning of ionized compounds into membranes, the presence of phospholipids as potential counterions in the membrane is an important component missing from any isotropic medium.

A model for the interaction of basic compounds with the anion of phosphatidylcholine (PC) was described by Avdeef *et al.* as part of their 'pH-piston' hypothesis.[101] There is enough free energy gain from the phosphate/ammonium ion association that the choline of the zwitterion is displaced from the ion pair and extends outward to pair with chloride in the medium, resulting in the appearance of a positive charge on the surface. The field has progressed from proposals that basic drugs are present in membranes as unpaired cations,[39] to designing a series of long-acting dual D_2-receptor/β-adrenoceptor agonists based on their forming ion pairs with membrane phospholipids.[133] Austin *et al.*[39] have shown that there is little chloride ion pair in membranes, since, unlike the situation in octanol, increasing the aqueous chloride ion concentration does not appreciably increase the $\log P^l$ of basic compounds.

It is interesting to note that 'negative' phosphatidyl esters, such as PI, have perfect properties for forming hydrophobic ion pairs. They are very lipophilic, detergent-like strong acids. Unlike PC, they have no competing built-in counterion. The powerful solubilizing effects of hydrophobic ion pairs can be seen in membranes. Krämer *et al.*[134] compared the pH distribution profile of tritiated propranolol in 'negative' and 'neutral' liposomes prepared from PI, PC, and mixtures of the two. The profile in PI liposomes is markedly different from that in PC liposomes (**Figure 6c**). In PI liposomes the propranolol cation is more lipophilic than the free base ($\log P$ 4.24 versus 3.49), a clear sign of a HIP (*see* Section 5.18.8.2). PI has an additional beneficial property: the inositol hydroxy groups offer additional charge stabilization, just as in the examples in **Table 3**. HIP formation also appears to take place with desipramine in PI membranes.[135]

The difference between the pure liposomes in **Figure 6c** is dramatic. The $\log P^l$ of propranolol cation is 1.90 units higher in P^l than P^C liposomes. This is highly relevant to those building models of absorption and distribution, since Krämer showed that membrane phospholipids behave the same way in a mixed membrane as they do in pure liposomes.[134] Since membranes in nature contain on average about 10% negative phospholipids,[64] these could be major transporters of cations across cell membranes.

5.18.8.4 Chromatographic Systems

Many recent studies have used IAM columns to investigate the lipophilicity of ions, only some of which can be listed here.[78,81,85,136,137] The intense activity in this area in recent years comes from the advantage of this technique over traditional anisotropic systems in being suitable for HT screening. A perceptive review by Taillardat-Bertschinger and co-workers[137] indicates that the IAM capacity factors of neutral solutes often correlate with other lipophilicity parameters such as CLOGP, $\log P/\log D^{pH}$, capacity factors measured on reversed-phase HPLC columns, and liposomes/water partitioning. For ionized solutes, in contrast, IAM chromatography, liposomes/water, and n-octanol/water yield distinct lipophilicity scales, particularly for hydrophilic compounds. In fact, several studies have shown that the balance of recognition forces governing partitioning in liposomes and IAMs is not identical.[63,74,138] This may be explained by the marked structural differences existing between the two systems for which IAM cannot exactly emulate lipid dynamics and thus cannot replace liposomal systems.

444   **Lipophilicity, Polarity, and Hydrophobicity**

5.18.8.5   Ampholytes and Zwitterions

Compounds containing both acidic and basic groups are called ampholytes. An important subset is zwitterions in which, at some pH, the acidic and basic groups are simultaneously ionized (5.16 Ionization Constants and Ionization Profiles). The further classification of zwitterionic ampholytes is discussed below. Examples of a number of current and formerly marketed zwitterionic drugs are shown in **Figure 7**.

Many more examples of zwitterions occur in biology, including amino acids and peptides. Understanding the physicochemical properties of this class can sometimes be challenging. A useful review is provided by Pagliara *et al.*[139] A more recent analysis of ampholytes based on their pK_as in water and *n*-octanol simplifies classification and helps in understanding their physicochemical properties.[25] Often-asked questions about ampholytes include whether or not a zwitterion is present in the aqueous or *n*-octanol phases, over what pH range they exist, and to what extent a zwitterion exists as a neutral versus a doubly charged species. These questions can be answered quantitatively using p$K_{a\,oct}$ values.

The classification of ampholytes and zwitterions into four types is shown in **Table 4**. Type 1 is a simple ampholyte and the others are zwitterionic ampholytes. Types 3 and 4 cannot be distinguished by their aqueous pK_as. Both are strong acids and bases and capable of forming a zwitterion in both phases. What distinguishes them is whether or not the two charges can be internally compensated, either by two groups being physically able to come into contact, or through resonance forms. When they can be compensated, the compound has a dramatically different lipophilicity profile, as seen for type 4 ampholytes. Put another way, the type 4 profile is diagnostic that internal compensation is occurring in octanol.

Figure 6a shows lipophilicity profiles for a series of zwitterions providing a regular progression from type 3 to type 4 ampholytes. They have the same aqueous pK_a values, 2.5 and 9, the same cationic and anionic partition coefficients, 0.2 and 0.05, respectively, but with the P of the zwitterion (P^Z) varying from 0.002 to 16 in 20-fold intervals. The aqueous pK_a for each compound is found at the upper inflections of each curve (where the dashed lines intersect). The p$K_{a\,oct}$ values are always found at the lower inflections. One can see that the p$K_{a\,oct}$ values for the compound represented by curve **d** are 4.5 and 6.0. The compound represented by curve **a**, on the other hand, has p$K_{a\,oct}$ of 0.6 and 11.5. It is a much stronger acid and base in *n*-octanol than in water!

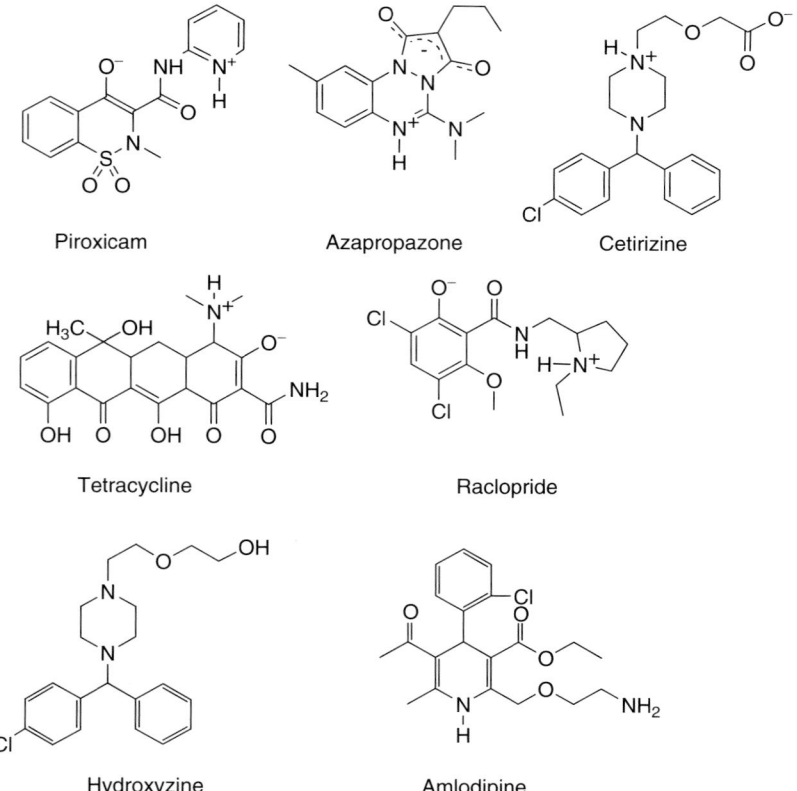

Figure 7 Chemical structures of investigated chemicals.

Table 4 Classification of simple and zwitterionic ampholytes

Type	Example	Lipophilicity profile	Water phase	n-Octanol phase	Charge interaction
1	m-Aminophenol OH pK_a 9.87 pK_a 4.17 NH_2		The base has a lower pK_a than the acid.	The acids and bases are even weaker in octanol.	No interaction in either phase. The compound is a simple ampholyte.
2	Labetalol NH_2 7.4 CH₃ R OH NH 9.4 O HO		The base has a higher pK_a than the acid.	In octanol, the base has a lower pK_a than the acid.	Zwitterion occurs in the aqueous phase only.
3	Acrivastine O O⁻ 2.2 N R 9.55 H—N⁺		A strong acid and base. Produces a zwitterion in water.	A weaker acid and weaker base in octanol, but the base still has a higher pK_a than the acid.	Zwitterion is formed in both phases. No internal compensation of charges.
4	Cetirizine (Figure 7). pK_a 2.9 (carboxyl) pK_a 8.0 (distal piperazine N)		A strong acid and base. Produces a zwitterion in water.	A stronger acid and stronger base in octanol than in water. A zwitterion over a greater pH range than in water.	Zwitterion in both phases. Internal compensation in octanol by contact.
	Azapropazone (Figure 7). pK_a 0.5 (endione) pK_a 6.9 (=N-C-NMe₂)				Interaction in octanol of adjacent tautomeric ions.

The curves in **Figure 6a** show a progression of profiles from type 3 zwitterions (**d**) to type 4 internally compensated zwitterions. The largest contribution to P^Z in this transition is the internal compensation of charges with the elimination of the counterions. Additional factors, such as the size of the ring, the nature and interaction of the charged groups, and steric interactions with axial substituents, will all contribute to the free energy of the ion pair, in a positive or negative way, and thus to $\log P^Z$. Examples of drugs of the type 4 zwitterion class include cetirizine, piroxicam, and azapropazone, and by their lipophilicity profiles (RA Scherrer, unpublished observations), should also include raclopride[140] and tetracycline[141] (chemical structures are shown in **Figure 7**).

The distribution profiles of hydroxyzine and cetirizine in **Figure 8**[142] provide an opportunity to illustrate the very large free energy change that takes place with the formation of internally compensated zwitterions. The pK_a change of the distal piperazine nitrogen of each tells the story. The pK_a of hydroxyzine (7.52) and cetirizine (8.00) are shifted in *n*-octanol by an amount equal to $diff(\log P^{N-1})$ (eqn [10]). The hydroxyzine shift is in the usual direction for amines, down 2.57 pK_a units. With cetirizine, the internal compensation of the cation by the carboxylate is so energy-favorable that the pK_a shift is up by 1.69 units (calculated again by $diff(\log P^{N-1})$). Comparing the actual $pK_{a\ oct}$ of cetirizine to what it would have been without internal compensation gives a net change of 4.26 pK_a units. This is equal to 5.8 kcal mol^{-1} at 25 °C (from the Gibbs free energy equation). The free energy difference is undoubtedly reflected in its physicochemical properties. The cetirizine ion pair has a seven-membered cyclic form, but is likely predominantly in the open form in water. This drug has unusual properties, one being a very small volume of distribution.[142] Cetirizine is known as a non-sedating antihistamine, but its low brain level is not due to an inability to penetrate the blood–brain barrier.[142] On the contrary, experiments with labeled drug show that it enters the brain extremely rapidly, but exits the brain even faster. Active transport presumably, but then, that is an interesting property, too.

The classification of zwitterions by their aqueous and *n*-octanol pK_a exposes a previously unnoticed type of stabilization.[25] It occurs in zwitterions that lack the obvious neutralization of charges through a cyclic interaction. The compounds may appear to be too weakly acidic and basic to produce zwitterions in *n*-octanol, yet they have the lipophilicity profile of a type 4 zwitterion. The interaction is by what one might call a localized interaction of resonance forms. Marketed drugs in this category include piroxicam and azapropazone. They partition without counterions. Tetracycline appears to fall into this category.[141] The energy gain from internal neutralization in *n*-octanol is high enough that one can find members of this class with ring sizes not usually seen in chemistry, such as ten-member ring interactions found with tripeptides.[25,143]

In addition to the work of Plemper van Balen *et al.* about cetirizine,[102] a few studies in the literature report the lipophilic behavior of zwitterions in systems other than *n*-octanol/water.[99,108,144,145]

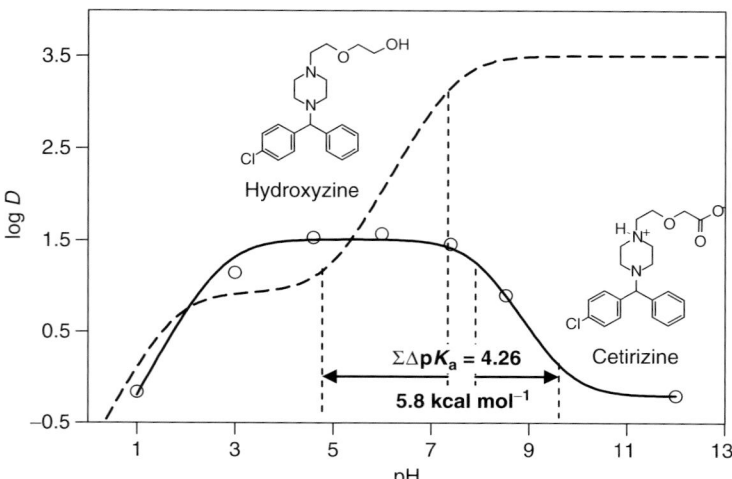

Figure 8 The lipophilicity profiles of cetirizine and hydroxyzine in *n*-octanol/water. Hydroxyzine: pK_a 7.52, $pK_{a\ oct}$, 4.95; cetirizine: pK_a 8.00, $pK_{a\ oct}$ 9.69. In this instance, the net difference between an internally compensated zwitterion and a similar non-compensated ion pair is 4.26 pK_a units, corresponding to a free-energy difference of about 5.8 kcal mol^{-1} at 25 °C. (Reproduced with permission from Pagliara, A.; Testa, B.; Carrupt, P. A.; Jolliet, P.; Morin, C.; Morin, D.; Urien, S.; Tillement, J. P.; Rihoux, J. P. *J. Med. Chem.* **1998**, *41*, 853–863 © American Chemical Society.)

5.18.8.6   Getting the Most out of logD^{pH}

Log$D$ has proven to be a useful physicochemical property for the analysis of ADMET-related activities. We see examples throughout this volume. When the use of log$D$ for QSAR was proposed in 1976,[146] it was to quantify the presence of the unionized species in the lipid phase at a particular pH. Partitioning of the ionized form was assumed to be insignificant. Today it is well accepted that the concentration of ionized species in the lipid phase can be significant and that log$D$ should account for all the forms of an agent in both phases. This is, of course, what is experimentally measured. Computations should only omit ionized species after deliberate consideration of their possible contribution and impact on the intended application (*see* Section 5.18.2).

This definition that the distribution coefficient, D, comprises neutral and ionic species presents a problem if the ionized and neutral species are not biologically equivalent for the activity or property being measured. One consequence is that regression analysis can find acids, bases, and neutral compounds falling into distinct groups, each correlating with log$D$, and these groups still may contain mixtures of species. Binding of drugs to serum albumin is an example.[147] This suggests that, if one actually fractionated log$D$ into its neutral and ionized species before analysis, better results would be obtained. Two approaches have been proposed to improve on using log$D$. One is to fractionate D into terms for its component species, D^{N} and D^{I},[25] illustrated in eqns [19]–[21].[147] D^{N} and D^{I} are called species-specific distribution coefficients. A disadvantage is that it requires knowing logP^{I}, p$K_{\mathrm{a\ oct}}$ or pK_{a}''.

The second approach, by Lombardo *et al.*,[3,148] uses the HT-determined Elog$D_{7.4}$ and an additional term, $f_{\mathrm{i\ (7.4)}}$, the fraction of ionized species in the aqueous phase. What is really needed is $f_{\mathrm{i\ oct}}$ instead. That way the exact representation of log$D$ could be obtained. The benefit of the practical approach of using the aqueous ionized fraction depends on the correlation between the fraction ionized in each phase. Regardless of theory, the Lombardo approach has led to the calculation of high-quality human volume of distribution values.[3,148]

As mentioned above, much more specific analysis of protein binding or biological activity is possible by incorporating physicochemical properties specific for cations, anions, and neutral species, namely, logD^{C}, logD^{A}, and logD^{N}, where C and A refer to cationic and anionic species. There are two requirements for this analysis: (1) access to good logP^{I} values; (2) a proper data set. To distinguish between the activity of an ion pair and the neutral form, *diff*(log$P^{\mathrm{N-I}}$) must vary within the series, otherwise, a correlation with logP^{N} is equivalent to a correlation with logP^{I}. In addition, the pK_{a} or p$K_{\mathrm{a\ oct}}$ of members of the series must cover a sufficient range, otherwise the percent ionized in equilibrium with the active site (the site which, when occupied, produces the benefit under study) will be a constant. A correlation with the neutral species will be equally valid for the ion pair. Small 'proper' data sets are difficult to find in the literature. Perhaps that is why log$D$ has been useful without (much) controversy, even as a mixture term. (One successful analysis is for uncoupling oxidative phosphorylation by phenols, which is highly correlated with logD^{I} and not log$D$ or logD^{N}.[25])

On the other hand, the large series of diverse compounds used in ADMET studies are perfect for sorting out the role of ionized and neutral species. All that is needed are good logP^{I} values from which to calculate logD^{I}. Unfortunately, that is where the bottleneck to rapid advances regarding ion pair lipophilicity lies – the scarcity of physicochemical data. Hopefully, titrations in KCl/water-saturated octanol will be adopted by others able to meet this need. Each of the areas represented in ADMET, and covered in detail in other chapters in this volume, brings questions about the significance of ionized species. The use of species-specific log$D$ terms, as calculated below, may help provide some answers.

The distribution coefficient of a compound is the sum of the distribution coefficients of each of various charge-related forms it is capable of assuming (eqn [19]). This is merely rewriting eqn [1]. The individual terms are pH-dependent distribution coefficients for each ionic species in which the D^{I} terms represent the n ionizable groups.

$$D = D^{\mathrm{N}} + D^{\mathrm{I(n)}} + D^{\mathrm{I(n-1)}} + \dots \qquad [19]$$

Formulas for calculating the individual terms can be derived from eqn [7] (by fractionating and converting to pH and pK_{a} terms) and are illustrated below for D^{N} and only one of the D^{I} terms with its individual pK_{a} in the case of acids.

$$D^{\mathrm{N}} = P^{\mathrm{N}}/(1 + 10^{(\mathrm{pH}-\mathrm{pK_a})}) \qquad [20]$$

and

$$D^{\mathrm{I}} = P^{\mathrm{I}} \times 10^{(\mathrm{pH}-\mathrm{pK_a})}/(1 + 10^{(\mathrm{pH}-\mathrm{pK_a})}) \qquad [21]$$

5.18.9   Conclusion: Lipophilicity in Absorption, Distribution, Metabolism, Excretion, and Toxicity Studies

The concepts and applications of lipophilicity have evolved enormously since the first edition of *Comprehensive Medicinal Chemistry*. Today there is a huge amount of information contained in lipophilicity data, but what part, applied to any given series of potential drug candidates, will give the most useful results: a single-column table? a multi-column table? a collection of plots and graphs? It is evident that the more data available, the more time will be required to evaluate it, but also the more information that can be obtained. Just what information is truly useful for ADMET studies? We think it is difficult to make generalizations, but throughout this chapter we demonstrate that in the presence of one or more ionizable centers, lipophilicity studies become more complicated. At the same time the significance of $\log P/\log D$ data goes beyond their numerical values.

Recently, Thomae and co-workers demonstrated that, in contrast to the expectations of the pH partition hypothesis, lipid bilayer permeation of an acidic compound can be completely controlled by the anion at physiological pH[149] (in the cited paper, liposomes have been used to describe not only membrane affinity but also membrane permeation kinetics). According to this and other papers we are led to believe that major progress in understanding and applying lipophilicity in ADMET areas will involve taking ionization into full account.

This area has long been held back by a lack of good physicochemical property values. To some extent, development has been slowed by the neglect of a fundamental physical property of all ionized compounds: their medium-associated pK_a. pK_a represents the quantifiable, analyzable free energy of ionization and allows a link to the free energy of partitioning. A number of advances in understanding and applying lipophilic properties have come from such a linking.

Up to now, in the absence of measured data on ionized species, one has been forced to use a crude estimate of properties, or to ignore the species. It should be a priority to accumulate more accurate and useable data. The shortcomings of commercial programs for calculating $\log D$ are discussed in a recent paper by Tetko and Poda[150] and in 5.27 Rule-Based Systems to Predict Lipophilicity.

It is easy to believe that advances will come from being able to fractionate $\log D$ into its component parts. Studies will be specifically designed to differentiate the biological activity of ionized and neutral species. In another area, it seems reasonable that the dramatic free-energy changes associated with the formation of internally compensated ampholytes will have an analogy in receptor pockets when a group on a drug finds an oppositely charged group on the receptor wall.

Given the theoretical complexity of these biological topics, a second big advance in understanding and applying lipophilicity in ADMET requires the combination of experimental data with ad hoc molecular modeling tools (not just software able to predict $\log P$ data!).

Ionized species play important roles in nature – in protein folding, in binding, in signal transduction, in metabolism to highly ionized species, in enzyme and receptor site specificity, and in the evolution of low-energy, lipid-soluble forms of ion pairs. Researchers in this area are in a position to advance science beyond the confines of ADMET.

References

1. Van de Waterbeemd, H.; Smith, D. A.; Beaumont, K.; Walker, D. K. *J. Med. Chem.* **2001**, *44*, 1313–1333.
2. Van de Waterbeemd, H.; Smith, D. A.; Jones, B. C. *J. Comput.-Aided Mol. Design* **2001**, *15*, 273–286.
3. Lombardo, F.; Obach, R.; Shalaeva, M. Y.; Gao, F. *J. Med. Chem.* **2002**, *45*, 2867–2876.
4. Lipinski, C. A.; Lombardo, F.; Dominy, B. W.; Feeney, P. J. *Adv. Drug Deliv. Rev.* **1997**, *23*, 3–25.
5. Wenlock, M. C.; Austin, R. P.; Barton, P.; Davis, A. M.; Leeson, P. D. *J. Med. Chem.* **2003**, *46*, 1250–1256.
6. Vieth, M.; Siegel, M. G.; Higgs, R. E.; Watson, I. A.; Robertson, D. H.; Savin, K. A.; Durst, G. L.; Hipskind, P. A. *J. Med. Chem.* **2004**, *47*, 224–232.
7. Veber, D. F.; Johnson, S. R.; Cheng, H. Y.; Smith, B. R.; Ward, K. W.; Kopple, K. D. *J. Med. Chem.* **2002**, *45*, 2615–2623.
8. Smith, P. A.; Sorich, M. J.; Low, L. S. C.; McKinnon, R. A.; Miners, J. O. *J. Mol. Graph. Model.* **2004**, *22*, 507–517.
9. Comer, J. E.; Tam, K. Y. Lipophilicity Profiles: Theory and Measurement. In *Pharmacokinetic Optimization in Drug Research*; Testa, B., Van de Waterbeemd, H., Folkers, G., Guy, R. H., Eds.; Wiley-VCH: Zürich, 2001, pp 275–304.
10. Walther, B.; Vis, P.; Taylor, A. Lipophilicity of Metabolites and its Role in Biotransformation. In *Lipophilicity in Drug Action and Toxicology*; Pliska, V., Testa, B., Van de Waterbeemd, H., Eds.; VCH: Weinheim, 1996, pp 253–261.
11. Testa, B.; Cruciani, G. Structure–Metabolism Relations, and the Challenge of Predicting Biotransformation. In *Pharmacokinetic Optimization in Drug Research: Biological, Physicochemical and Computational Chemistry*; Testa, B., Van de Waterbeemd, H., Folkers, G., Guy, R. H., Eds.; Wiley-Verlag Helvetica Chimica Acta: Zürich, 2001, pp 65–84.
12. Testa, B. Drug Metabolism. In *Burger's Medicinal Chemistry and Drug Discovery*, 5th ed.; Wolff, M. E., Ed.; Wiley: New York, 1995; Vol. 1, pp 129–180.
13. Bouchard, G.; Carrupt, P. A.; Testa, B.; Gobry, V.; Girault, H. H. *Chem. Eur. J.* **2002**, *8*, 3478–3484.
14. Avdeef, A. Assessment of Distribution–pH Profiles. In *Lipophilicity in Drug Action and Toxicology*; Pliska, V., Testa, B., Van de Waterbeemd, H., Eds.; VCH: Weinheim, 1996, pp 109–139.

15. Caron, G.; Reymond, F.; Carrupt, P. A.; Girault, H. H.; Testa, B. *PSTT* **1999**, *2*, 327–335.
16. Reymond, F.; Gobry, V.; Bouchard, G.; Girault, H. H. Electrochemical Aspects of Drug Partitioning. In *Pharmacokinetic Optimization in Drug Research*; Testa, B., Van de Waterbeemd, H., Folkers, G., Guy, R. H., Eds.; Verlag Helvetica Chimica Acta: Zürich, 2001, pp 327–349.
17. Caron, G.; Ermondi, G. *Mini Rev. Medicinal Chem.* **2003**, *3*, 821–830.
18. Pagliara, A.; Reist, M.; Geinoz, S.; Carrupt, P. A.; Testa, B. *J. Pharm. Pharmacol.* **1999**, *51*, 1339–1357.
19. Young, R. C.; Mitchell, R. C.; Brown, T. H.; Ganellin, C. R.; Griffiths, R.; Jones, M.; Rana, K. K.; Saunders, D.; Smith, I. R.; Sore, N. E. et al. *J. Med. Chem.* **1988**, *31*, 656–671.
20. Chadha, H. S.; Abraham, M. H.; Mitchell, R. C. *Bioorg. Med. Chem. Lett.* **1994**, *4*, 2511–2516.
21. Ashwood, V. A.; Field, M. J.; Horwell, D. C.; Julien-Larose, C.; Lewthwaite, R. A.; McCleary, S.; Pritchard, M. C.; Raphy, J.; Singh, L. *J. Med. Chem.* **2001**, *44*, 2276–2285.
22. Caron, G.; Ermondi, G. *J. Med. Chem.* **2005**, *48*, 3269–3279.
23. Avdeef, A.; Kearney, D. L.; Brown, J. A.; Chemotti, A. R., Jr. *Anal. Chem.* **1982**, *54*, 2322–2326.
24. Scherrer, R. A. The Treatment of Ionizable Compounds in QSAR Studies with Special Consideration to Ion [Pair] Partitioning. In *Pesticide Synthesis Through Rational Approaches*; Magee, P. S., Kohn, G. K., Menn, G. G., Eds.; American Chemical Society: Washington, DC, 1984, pp 225–246.
25. Scherrer, R. A. Biolipid pK_a Values and the Lipophilicity of Ampholites and Ion Pairs. In *Pharmacokinetic Optimization in Drug Research*; Testa, B., Van de Waterbeemd, H., Folkers, G., Guy, R. H., Eds.; Wiley-VCH: Zürich, 2001, pp 351–381.
26. Scherrer, R. A. Automated Titrations in KCl/Water-Saturated Octanol. In *Pharmacokinetic Profiling in Drug Research, Biological, Physicochemical and Computational Strategies*; Testa, B., Kraemer, S., Wunderli-Allenspach, H., Folkers, G., Eds.; Verlag Helvetica Chimica Acta: Zürich, 2004 (Presentation A35 included on CD).
27. Pagliara, A.; Khamis, E.; Trinh, A.; Carrupt, P. A.; Tsai, R. S.; Testa, B. *J. Liq. Chromatogr.* **1995**, *18*, 1721–1745.
28. Steyaert, G.; Lisa, G.; Gaillard, P.; Boss, G.; Reymond, F.; Girault, H. H.; Carrupt, P. A.; Testa, B. *J. Chem. Soc. Faraday Trans.* **1997**, *93*, 401–406.
29. Pagliara, A.; Caron, G.; Lisa, G.; Fan, W.; Gaillard, P.; Carrupt, P. A.; Testa, B.; Abraham, M. H. *J. Chem. Soc. Perkin Trans.* **1997**, *2*, 2639–2643.
30. Maitland, G. C.; Rigby, M.; Smith, E. B.; Wakeham, W. A. *Intermolecular Forces*, 2nd ed.; Breslow, R.; Goodenough, J. B.; Halpern, J.; Rowlinson, J. S., Eds.; International Series of Monographs on Chemistry; Oxford University Press: New York, 1987, pp 1–88.
31. Blokzijl, W.; Engberts, J. B. F. N. *Angew. Chem. Int. Edit.* **1993**, *32*, 1545–1579.
32. Gohlke, H.; Klebe, G. *Angew. Chem. Int. Edit.* **2002**, *41*, 2645–2676.
33. Chandler, D. *Nature* **2002**, *417*, 491.
34. Sijbren, O.; Engberts, J. B. F. N. *Org. Biomol. Chem.* **2003**, *1*, 2820.
35. Van de Waterbeemd, H.; Testa, B. The Parametrization of Lipophilicity and Other Structural Properties in Drug Design. In *Advances in Drug Research*; Testa, B., Ed.; Academic Press: London, 1987; Vol. 16, pp 87–227.
36. Testa, B.; Carrupt, P. A.; Gaillard, P.; Tsai, R. S. Intramolecular Interactions Encoded in Lipophilicity: Their Nature and Significance. In *Lipophilicity in Drug Action and Toxicology*; Pliska, V., Testa, B., Van de Waterbeemd, H., Eds.; VCH: Weinheim, 1996, Chapter 4, pp 49–71
37. Testa, B.; Seiler, P. *Arzneimittel-Forsch* **1981**, *31*, 1053–1058.
38. Van de Waterbeemd, H.; Carter, R. E.; Grassy, G.; Kubinyi, H.; Martin, Y. C.; Tute, M. S.; Willett, P. *Pure Appl. Chem.* **1997**, *69*, 1137–1152.
39. Austin, R. P.; Barton, P.; Davis, A. M.; Manners, C. N.; Stansfield, M. C. *J. Pharm. Sci.* **1998**, *87*, 599–607.
40. Abraham, M. H.; Chadha, H. S. Application of a Solvation Equation to Drug Transport Properties. In *Lipophilicity in Drug Action and Toxicology*; Pliska, V., Testa, B., Van de Waterbeemd, H., Eds.; VCH: Weinheim, 1996, pp 311–337.
41. El Tayar, N.; Tsai, R. S.; Testa, B.; Carrupt, P. A.; Leo, A. *J. Pharm. Sci.* **1991**, *80*, 590–598.
42. Abraham, M. H.; Zhao, Y. H. *J. Org. Chem.* **2004**, *69*, 4677–4685.
43. Abraham, M. H.; Zhao, Y. H. *J. Org. Chem.* **2005**, *70*, 2633–2640.
44. Testa, B. *Med. Chem. Res.* **1997**, *7*, 340–365.
45. Plemper van Balen, G.; a Marca Martinet, C.; Caron, G.; Bouchard, G.; Reist, M.; Carrupt, P. A.; Fruttero, R.; Gasco, A.; Testa, B. *Med. Res. Rev.* **2004**, *24*, 299–324.
46. Nasal, A.; Siluk, D.; Kaliszan, R. *Curr. Med. Chem.* **2003**, *10*, 381–426.
47. Lombardo, F.; Shalaeva, M. Y.; Tupper, K. A.; Gao, F.; Abraham, M. H. *J. Med. Chem.* **2000**, *43*, 2922–2928.
48. Lombardo, F.; Shalaeva, M. Y.; Tupper, K. A.; Gao, F. *J. Med. Chem.* **2001**, *44*, 2490–2497.
49. Yamagami, C. Recent Advances in Reversed-Phase HPLC Techniques to Determine Lipophilicity. In *Pharmacokinetic Optimization in Drug Research*; Testa, B., Van de Waterbeemd, H., Folkers, G., Guy, R. H., Eds.; Wiley-VCH: Zürich, 2001, pp 383–400.
50. Van de Waterbeemd, H.; Kansy, M.; Wagner, B.; Fischer, H. Lipophilicity Measurement by Reversed-Phase High Performance Liquid Chromatography. In *Lipophilicity in Drug Action and Toxicology*; Pliska, V., Testa, B., Van de Waterbeemd, H., Eds.; VCH: Weinheim, 1996, pp 73–87.
51. Kansy, M.; Fischer, H.; Van de Waterbeemd, H. *Correlations Between log P and Reverse Phase HPLC log k'_w-Values: Effect of Charge and Conformation*; Sanz, F.; Giraldo, J.; Manaut, F., Eds.; Prous Science: Barcelona, 1995, pp 45–48.
52. Valko, K.; Snyder, L. R.; Glachj, J. L. *J. Chromatogr. A* **1993**, *656*, 501–520.
53. Barbato, F.; Caliendo, G.; La Rotonda, M. I.; Morrica, P.; Silipo, C.; Vittoria, A. *Farmaco* **1990**, *45*, 647–663.
54. Maran, U.; Karelson, M.; Katritzky, A. R. *Quant. Struct.-Act. Rel.* **1999**, *18*, 3–10.
55. Schop, J.; Wiese, M.; Cordes, H. P.; Seydel, J. K. *Eur. J. Med. Chem.* **2000**, *35*, 619–634.
56. Collander, R. *Acta Chem. Scand.* **1950**, *4*, 1085.
57. Collander, R. *Acta Chem. Scand.* **1951**, *5*, 774.
58. Comer, J. E. High-Throughput Measurement of log D and pKa. In *Drug Bioavailability*; Van de Waterbeemd, H., Lennernaes, H., Artursson, P., Eds.; Wiley-VCH: Weinheim, 2003, pp 21–45.
59. Wohnsland, F.; Faller, B. *J. Med. Chem.* **2001**, *44*, 923–930.
60. Box, K. J.; Comer, J. E.; Huque, F. Correlations between PAMPA Permeability and log P. In *Pharmacokinetic Profiling in Drug Research: Biological, Physicochemical and Computational Strategies*; Testa, B., Kraemer, S., Wunderli-Allenspach, H., Folkers, G., Eds.; Verlag Helvetica Chimica Acta: Zürich, 2005, pp 243–257.
61. Liu, X.; Bouchard, G.; Müller, N.; Galland, A.; Girault, H. H.; Testa, B.; Carrupt, P. A. *Helv. Chim. Acta* **2003**, *86*, 3533–3547.
62. Herbette, L.; Rhodes, D. G.; Preston Mason, R. *Drug Design Deliv.* **1991**, *7*, 75–118.
63. Caron, G.; Ermondi, G. New Insights into the Lipophilicity of Ionized Species. In *Pharmacokinetic Profiling in Drug Research: Biological, Physicochemical and Computational Strategies*; Testa, B., Kraemer, S., Wunderli-Allenspach, H., Folkers, G., Eds.; Verlag Helvetica Chimica Acta: Zürich, 2005, pp 165–185.

64. Taillardat-Bertschinger, A.; Carrupt, P. A.; Testa, B. *Eur. J. Pharm. Sci.* **2002**, *15*, 225–234.
65. Abraham, M. H.; Chadha, H. S.; Dixon, J. P.; Rafols, C.; Treiner, C. *J. Chem. Soc., Perkin Trans.* **1995**, *2*, 887–894.
66. Abraham, M. H.; Chadha, H. S.; Dixon, J. P.; Rafols, C.; Treiner, C. *J. Chem. Soc., Perkin Trans.* **1997**, *2*, 19–24.
67. Escuder-Gilabert, L.; Sagrado, S.; Villaneuva-Camanas, R. M.; Medina-Hernandez, M. J. *Anal. Chem.* **1998**, *70*, 28–34.
68. Escuder-Gilabert, L.; Sanchis-Mallols, J. M.; Sagrado, S.; Medina-Hernandez, M. J.; Villaneuva-Camanas, R. M. *J. Chromatogr. A* **1998**, *823*, 549–559.
69. Khaledi, M. G.; Breyer, E. D. *Anal. Chem.* **1989**, *61*, 1040–1047.
70. Ishihama, Y.; Oda, Y.; Uchikawa, K.; Asakawa, N. *Chem. Pharm. Bull.* **1994**, *42*, 1525–1527.
71. Van de Waterbeemd, H.; Kansy, M.; Wagner, B.; Fischer, H. Lipophilicity Measurement by Reversed-Phase High Performance Liquid Chromatography. In *Lipophilicity in Drug Action and Toxicology*; Pliska, V., Testa, B., Van de Waterbeemd, H., Eds.; VCH: Weinheim, 1996, pp 73–87.
72. Donovan, S. F.; Pescatore, M. C. *J. Chromatogr. A* **2002**, *952*, 47–61.
73. Taillardat-Bertschinger, A.; Carrupt, P. A.; Barbato, F.; Testa, B. *J. Med. Chem.* **2003**, *46*, 655–665.
74. Taillardat-Bertschinger, A.; a Marca Martinet, C.; Carrupt, P. A.; Reist, M.; Caron, G.; Fruttero, R.; Testa, B. *Pharm. Res.* **2001**, *19*, 729–737.
75. Yang, C. Y.; Cai, S. J.; Liu, H.; Pidgeon, C. *Adv. Drug Deliv. Rev.* **1996**, *23*, 229–256.
76. Liu, H.; Ong, S.; Glunz, L.; Pidgeon, C. *Anal. Chem.* **1995**, *67*, 3550–3557.
77. Barbato, F.; di Martino, G.; Grumetto, L.; La Rotonda, M. I. *Eur. J. Pharm. Sci.* **2004**, *22*, 261–269.
78. Amato, M.; Barbato, F.; Morrica, P.; Quaglia, F.; La Rotonda, M. I. *Helv. Chim. Acta* **2000**, *83*, 2836–2847.
79. Taillardat-Bertschinger, A.; Barbato, F.; Quercia, M. T.; Carrupt, P. A.; Reist, M.; La Rotonda, M. I.; Testa, B. *Helv. Chim. Acta* **2002**, *85*, 519–532.
80. Valko, K.; My Du, C.; Bevan, C.; Reynolds, D.; Abraham, M. H. *J. Pharm. Sci.* **2000**, *89*, 1085–1095.
81. Alifrangis, L. H.; Christensen, I. T.; Berglund, A.; Sandberg, M.; Hovgaard, L.; Frokjaer, S. *J. Med. Chem.* **2000**, *43*, 103–113.
82. Ong, S.; Liu, H.; Qiu, X.; Bhat, G.; Pidgeon, C. *Anal. Chem.* **1995**, *67*, 755–762.
83. Stewart, B. H.; Chan, O. H. *J. Pharm. Sci.* **1998**, *87*, 1471–1478.
84. Ducarme, A.; Neuwels, M.; Goldstein, S.; Massingham, R. *Eur. J. Med. Chem.* **1998**, *33*, 215–223.
85. Barbato, F.; La Rotonda, M. I.; Quaglia, F. *Eur. J. Med. Chem.* **1996**, *31*, 311–318.
86. Barbato, F.; La Rotonda, M. I.; Quaglia, F. *J. Pharm. Sci.* **1997**, *86*, 225–229.
87. Österberg, T.; Svensson, M.; Lundhal, P. *Eur. J. Pharm. Sci.* **2001**, *12*, 427–439.
88. Loidl-Stahlhofen, A.; Eckert, A.; Hartmann, T.; Schöttner, M. *J. Pharm. Sci.* **2001**, *90*, 599–606.
89. Loidl-Stahlhofen, A.; Hartmann, T.; Schöttner, M.; Röhring, C.; Brodowsky, H.; Schmitt, J.; Keldenisch, J. *Pharm. Res.* **2001**, *18*, 1782–1788.
90. Dearden, J. C.; Bresnen, G. M. *Quant. Struct.-Act. Rel.* **1988**, 7, 133–144.
91. Kratochwil, N. A.; Huber, W.; Müller, F.; Kansy, M.; Gerber, P. R. *Biochem. Pharmacol.* **2002**, *64*, 1355–1374.
92. Zhu, C.; Jiang, L.; Chen, T.-M.; Hwang, K.-K. *Eur. J. Med. Chem.* **2002**, *37*, 399–407.
93. Valko, K. Separation Methods in Drug Synthesis and Purification. In *Handbook of Analytical Separations*; Smith, R. M., Ed.; Elsevier: Amsterdam, 2000, pp 539–542.
94. Clarke, F. H.; Cahoon, N. M. *J. Pharm. Sci* **1996**, *85*, 178–183.
95. Takacs-Novak, K.; Avdeef, A. *J. Pharm. Biomed. Anal.* **1996**, *14*, 1405–1413.
96. Avdeef, A.; Barrett, A.; Shaw, P. N.; Knaggs, R. D.; Davis, S. S. *J. Med. Chem.* **1996**, *39*, 4377–4381.
97. Slater, B.; McCormack, A.; Avdeef, A.; Comer, J. E. *J. Pharm. Sci.* **1994**, *83*, 1280–1283.
98. Caron, G.; Ermondi, G.; Damiano, A.; Novaroli, L.; Tsinman, O.; Ruell, J. A.; Avdeef, A. *Bioorg. Med. Chem.* **2004**, *12*, 6107–6118.
99. Bouchard, G.; Pagliara, A.; Plemper van Balen, G.; Carrupt, P. A.; Testa, B.; Gobry, V.; Girault, H. H.; Caron, G.; Ermondi, G.; Fruttero, R. *Helv. Chim. Acta* **2001**, *84*, 375–387.
100. Plemper van Balen, G.; Carrupt, P. A.; Morin, D.; Tillement, J. P.; Le Ridant, A.; Testa, B. *Biochem. Pharmacol.* **2002**, *63*, 1691–1697.
101. Avdeef, A.; Box, K. J.; Comer, J. E. A.; Hibbert, C.; Tam, K. Y. *Pharm. Res.* **1998**, *15*, 209–215.
102. Plemper van Balen, G.; Caron, G.; Ermondi, G.; Pagliara, A.; Grandi, T.; Bouchard, G.; Fruttero, R.; Carrupt, P. A.; Testa, B. *Pharm. Res.* **2001**, *18*, 694–701.
103. Fruttero, R.; Caron, G.; Fornatto, E.; Boschi, D.; Ermondi, G.; Gasco, A.; Carrupt, P. A.; Testa, B. *Pharm. Res.* **1998**, *15*, 1407–1413.
104. Caron, G.; Ermondi, G.; Boschi, D.; Carrupt, P. A.; Fruttero, R.; Testa, B.; Gasco, A. *Helv. Chim. Acta* **1999**, *82*, 1630–1639.
105. Caron, G.; Gaillard, P.; Carrupt, P. A.; Testa, B. *Helv. Chim. Acta* **1997**, *80*, 449–462.
106. Ulmeanu, S. M.; Jensen, H.; Bouchard, G.; Carrupt, P. A.; Girault, H. H. *Pharm. Res.* **2003**, *20*, 1317–1322.
107. Liu, X.; Bouchard, G.; Girault, H. H.; Testa, B.; Carrupt, P. A. *Anal. Chem.* **2003**, *75*, 7036–7039.
108. Bouchard, G.; Pagliara, A.; Carrupt, P. A.; Testa, B.; Gobry, V.; Girault, H. H. *Pharm. Res.* **2002**, *19*, 1150–1159.
109. Gobry, V.; Bouchard, G.; Carrupt, P. A.; Testa, B.; Girault, H. H. *Helv. Chim. Acta* **2000**, *83*, 1465–1474.
110. Chopineaux-Courtois, V.; Reymond, F.; Bouchard, G.; Carrupt, P. A.; Testa, B.; Girault, H. H. *JACS* **1999**, *121*, 1743–1747.
111. Caron, G.; Steyaert, G.; Pagliara, A.; Reymond, F.; Crivori, P.; Gaillard, P.; Carrupt, P. A.; Avdeef, A.; Comer, J. E.; Box, K. J. et al. *Helv. Chim. Acta* **1999**, *82*, 1211–1222.
112. Reymond, F.; Brevet, P. F.; Carrupt, P. A.; Girault, H. H. *J. Electroanal. Chem.* **1997**, *424*, 121–139.
113. Faller, B.; Grimm, H. P.; Loeuillet-Ritzler, F.; Arnold, S.; Briand, S. *J. Med. Chem.* **2005**, *48*, 2571–2576.
114. Krämer, S. D. Liposome/Water Partitioning: Theory, Techniques and Applications. In *Pharmacokinetic Optimization in Drug Research: Biological, Physicochemical and Computational Strategies*; Testa, B., Van de Waterbeemd, H., Folkers, G., Guy, R. H., Eds.; Wiley-Verlag Helvetica Chimica Acta: Zürich, 2001, pp 401–428.
115. Galland, A.; Bouchard, G.; Caron, G.; Plemper van Balen, G.; a Marca Martinet, C.; Geinoz, S.; Rey, S.; Ermondi, G.; Vacondio, F.; Mor, M. et al. *Chimia* **2002**, *56*, 373.
116. Avdeef, A. *Quant. Struct.-Act. Rel.* **1992**, *11*, 510–517.
117. Avdeef, A. *J. Pharm. Sci.* **1993**, *82*, 183–190.
118. Avdeef, A. *Absorption and Drug Development. Solubility, Permeability, and Charge State*; Wiley: New York, 2003.
119. Kubinyi, H. *QSAR: Hansch Analysis and Related Approaches*; VCH: Weinheim, 1993.
120. Escher, B. I.; Schwarzenbach, R. P. *Environ. Sci. Technol.* **1996**, *30*, 260–270.
121. Westheimer, F. W. *Science* **1987**, *235*, 1173–1178.
122. Abraham, M. H.; Takacs-Novak, K.; Mitchell, R. C. *J. Pharm. Sci.* **1997**, *86*, 310–315.
123. Gobry, V.; Ulmeanu, S. M.; Reymond, F.; Bouchard, G.; Carrupt, P. A.; Testa, B.; Girault, H. H. *JACS* **2001**, *123*, 10684–10690.

124. Meyer, J. D.; Manning, M. C. *Pharm. Res.* **1998**, *15*, 188–193.
125. Lengsfeld, C. S.; Pitera, D.; Manning, M. C.; Randolph, T. W. *Pharm. Res.* **2002**, *19*, 1572–1576.
126. Feng, L.; De Dille, A.; Jameson, V. J.; Smith, L.; Dernell, W. S.; Manning, M. C. *Cancer Chemother. Pharm.* **2004**, *54*, 441–448.
127. Falk, R.; Randolph, T. W.; Meyer, J. D.; Kelly, R. M.; Manning, M. C. *J. Control. Release* **1997**, *44*, 77–85.
128. Zhou, H.; Lengsfeld, G.; Claffey, D. J.; Ruth, J. A.; Hybertson, B.; Randolph, T. W.; Manning, M. C. *J. Pharm. Sci* **2002**, *91*, 1502–1511.
129. Austin, R. P.; Davis, A. M.; Manners, C. N. *J. Pharm. Sci.* **1995**, *84*, 1180–1183.
130. Bäuerle, H.-D.; Seelig, J. *Biochemistry* **1991**, *30*, 7203–7211.
131. Preston Mason, R.; Campbell, S. F.; Wang, S.-D.; Herbette, L. *Mol. Pharmacol.* **1989**, *36*, 634–640.
132. Preston Mason, R. *Biochem. Pharmacol.* **1993**, *45*, 2173–2183.
133. Austin, R. P.; Barton, P.; Bonnert, R. V.; Brown, R. C.; Cage, P. A.; Cheshire, D. R.; Davis, A. M.; Dougall, I. G.; Ince, F.; Pairaudeau, G. et al. *J. Med. Chem.* **2003**, *46*, 3210–3220.
134. Krämer, S. D.; Braun, A.; Jakits-Deiser, C.; Wunderli-Allenspach, H. *Pharm. Res.* **1998**, *15*, 739–744.
135. Marenchino, M.; Alpstäg-Wöhrle, A.; Christen, B.; Wunderli-Allenspach, H.; Krämer, S. D. *Eur. J. Pharm. Sci.* **2004**, *21*, 313–321.
136. Kaliszan, R.; Nasal, A.; Bucinski, A. *Eur. J. Med. Chem.* **1994**, *29*, 163–170.
137. Taillardat-Bertschinger, A.; Carrupt, P. A.; Barbato, F.; Testa, B. *J. Med. Chem.* **2003**, *46*, 655–665.
138. Ottiger, C.; Wunderli-Allenspach, H. *Pharm. Res.* **1999**, *16*, 643–650.
139. Pagliara, A.; Carrupt, P. A.; Caron, G.; Gaillard, P.; Testa, B. *Chem. Rev.* **1997**, *97*, 3385–3400.
140. Tsai, R. S.; Carrupt, P. A.; Testa, B.; Gaillard, P.; El Tayar, N.; Hoegberg, T. *J. Med. Chem.* **1993**, *36*, 196–204.
141. Colaizzi, J. L.; Klink, P. R. *J. Pharm. Sci.* **1969**, *58*, 1184–1189.
142. Pagliara, A.; Testa, B.; Carrupt, P. A.; Jolliet, P.; Morin, C.; Morin, D.; Urien, S.; Tillement, J. P.; Rihoux, J. P. *J. Med. Chem.* **1998**, *41*, 853–863.
143. Avdeef, A. *Sirius Technical Application Notes 2*; Sirius Analytical Instruments: East Sussex, 1995, pp 12–16a.
144. Tsai, R. S.; Carrupt, P. A.; El Tayar, N.; Testa, B.; Giroud, Y.; Andrade, P.; Brée, F.; Tillement, J. P. *Helv. Chim. Acta* **1993**, *76*, 842–854.
145. Caron, G.; Pagliara, A.; Carrupt, P. A.; Gaillard, P.; Testa, B. *Helv. Chim. Acta* **1996**, *79*, 1683–1685.
146. Scherrer, R. A.; Howard, S. M. *J. Med. Chem.* **1977**, *20*, 53–58.
147. Scherrer, R. A.; Donovan, S. F. *29th Species-Specific Log D in Drug Design and ADMET Analyses*; 29th National Medicinal Chemistry Symposium, Madison, WI, 27 June–1 July 2004, Abst. A70.
148. Lombardo, F.; Obach, R.; Shalaeva, M. Y.; Gao, F. *J. Med. Chem.* **2004**, *47*, 1242–1250.
149. Thomae, A. V.; Wunderli-Allenspach, H.; Krämer, S. D. *Biophys. J.* **2005**, *89*, 1802–1811.
150. Tetko, I. V.; Poda, G. I. *J. Med. Chem.* **2004**, *47*, 5601–5604.
151. Berthod, A.; Carda-Broch, S.; Garcia-Alvarez-Coque, M. C. *Anal. Chem.* **1999**, *71*, 879–888.
152. Fujii, T.; Nishida, H.; Abiru, Y.; Yamamoto, M.; Kise, M. *Chem. Pharm. Bull.* **1995**, *43*, 1872–1877.
153. Avdeef, A. *Sirius Technical Application Notes 1*; Sirius Analytical Instruments: East Sussex, 1994, pp 146–147.

Biographies

Giùlia Caron studied at the University of Torino (Italy) where she received a BSc in Pharmaceutical Chemistry and Technology in 1992, and a BSc in Pharmacy in 1994. She then moved to the University of Lausanne for doctoral studies, and was awarded a PhD in Pharmaceutical Sciences in 1997. Since 1999, she is assistant professor at the Faculty of Pharmacy of the University of Torino where she also teaches pharmaceutical analysis. Her primary interests were experimental and computational lipophilicity, recently moved to the in silico prediction of ADMET properties.

Giuseppe Ermondi graduated as a chemist and obtained a PhD from the University of Torino (Italy) under the supervision of Prof Silvio Aime. He was a postdoc at the University of Lausanne, then became assistant professor at the University of Piemonte Orientale and now he is associate professor at the University of Torino, where he teaches pharmaceutical analysis. Following early work in NMR of paramagnetic ions, he rapidly became interested in computational chemistry and physical chemistry with applications in molecular modeling techniques and physicochemical properties as applied to drug design and ADMET prediction.

Robert A Scherrer received a BS in chemistry at the University of California, Berkeley, in 1954, and a PhD in organic chemistry under E J Corey at the University of Illinois, Urbana, in 1958. He then began a 44-year laboratory career in medicinal chemistry; at Parke-Davis for 9 years, followed by 35 years with 3M Pharmaceuticals in St Paul. Over this time he was an inventor or coinventor on 78 patents, coedited a two-volume work on antiinflammatory agents, was a Senior Editor of the *J. Med. Chem.* for 2 years, an author of 9 book chapters, 9 journal articles, and 24 national or international presentations. Two marketed drugs, mefenamic and meclofenamic acids, have come from his patents. Other areas include antibacterials, antiasthmatics, antioxidants, viscoelastic collagen, transmucosal patches, and hydrophobic ion pairs. A keen interest in QSAR, and the lack of partition coefficients for ion pairs, led to the first single-phase titrations in octanol in 1977. Promising results, and later collaboration with Frank Clarke (ChemClarke) on two-phase titrations, spurred a 25-year '15%-time' project on the lipophilicity of ion pairs. Today, through BIOpK, he continues promoting the benefits of thinking about biolipid pK_a and its applications to QSAR and drug design.

Comprehensive Medicinal Chemistry II
ISBN (set): 0-08-044513-6

ISBN (Volume 5) 0-08-044518-7; pp. 425–452

5.19 Artificial Membrane Technologies to Assess Transfer and Permeation of Drugs in Drug Discovery

K Sugano, Pfizer Inc., Aichi, Japan

5.19.1 Permeability: An Intersection of Biology and Physicochemistry

Since the 1990s the paradigm of drug discovery has shifted from pharmacology-focused drug discovery to multidimensional drug discovery. In the multidimensional drug discovery paradigm, ADME, toxicology, and formulation suitability are simultaneously taken into account as well as pharmacological activity. Under this paradigm shift, the reductionism approach has been employed in ADME studies because it enables high-throughput screening and distinct structure–activity/property relationship studies.[1–4] In drug discovery today, many in vitro ADME assays are replacing in vivo assays. Cell-based assays (e.g., Caco-2 and MDCK cells) are widely used in drug discovery. However, cell-based assays still contain multiple mechanisms, so that they can be further reduced to individual enzyme level assays. Driven by advances in biotechnology, high-throughput enzyme level assays for each ADME process are becoming available for drug discovery.[5] Simultaneously, high-throughput assays for passive permeation across lipid bilayers have been required for drug discovery.

On the other hand, the relationship between the physicochemical properties of drugs and the ADME process has been extensively studied for more than 100 years.[6] The physicochemical properties of drugs affect the passive transfer and permeation processes in ADME. For example, intestinal brush border membrane (BBM) permeation, blood–brain barrier (BBB) permeation, the volume of distribution, uriniferous tubular reabsorption, and transdermal permeation are known to depend on the physicochemical properties of a chemical substance.[7] To evaluate the physicochemical properties required for passive ADME processes, isotropic solvent/water partition systems, such as octanol/water, have long been used in drug discovery. However, the in vivo predictability of isotropic solvent/water partition systems alone are insufficient for drug discovery, owing to the dissimilarity between the isotropic solvent and the biological membrane. Therefore, the construction of biomimetic physicochemical tools has been investigated.

The artificial membrane technologies can be mapped at the intersection of the biological reductionistic strategy and the physicochemical approach. In this article, artificial membrane technologies for transfer and permeation assessment in drug discovery are reviewed. The features of biological membranes will first be explained in detail, down to the molecular constituent level. Theories of passive permeation will then be briefly described. Finally, artificial membrane technologies for drug discovery will be reviewed in detail, mainly from the viewpoint of drug discovery utility and relevance to in vivo membrane permeation.

5.19.2   Biological Membranes: Constituents, Morphology, and Physicochemistry

There is marked variation between the biological membranes of the various organs. In this section, features of biological membranes will be reviewed from the chemical, morphological, and physical points of view. Active transporters and metabolisms are outside the of scope of this article, and are reviewed elsewhere (*see* 5.04 The Biology and Function of Transporters; 5.11 Passive Permeability and Active Transport Models for the Prediction of Oral Absorption).

5.19.2.1   Chemical Components

Typical lipid components of biological membranes are shown in **Figure 1**. Phosphatidylcholine (PC) and phosphatidylethanolamine (PE) are zwitter ionic phospholipids with zero net charge at neutral pH. The difference in chemical structure between these two phospholipids is only at the amine moiety (amine and trimethylammonium); however, they show different phase structures in water.[8] In addition, although PE is a zwitter ionic phospholipid, the

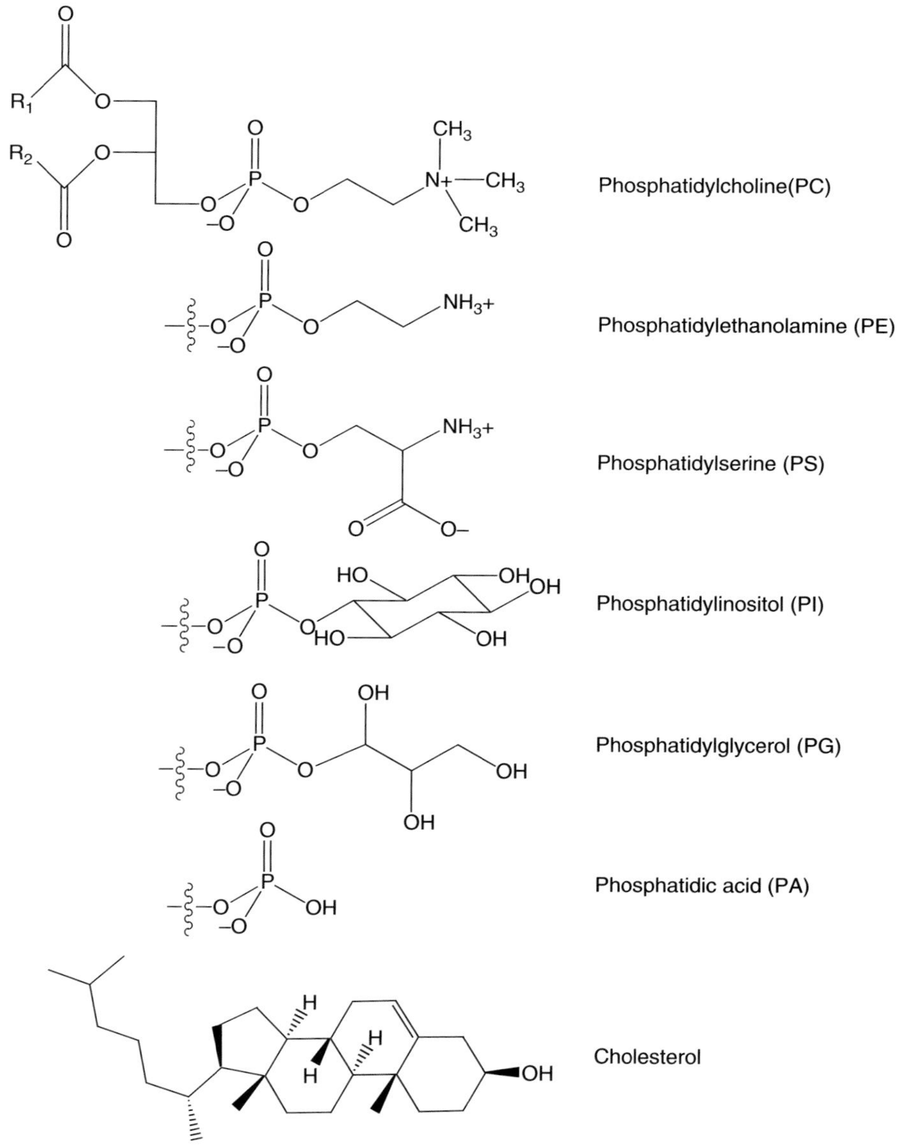

Figure 1   Chemical structures of phospholipids and cholesterol. R_1 and R_2 are fatty acid chains.

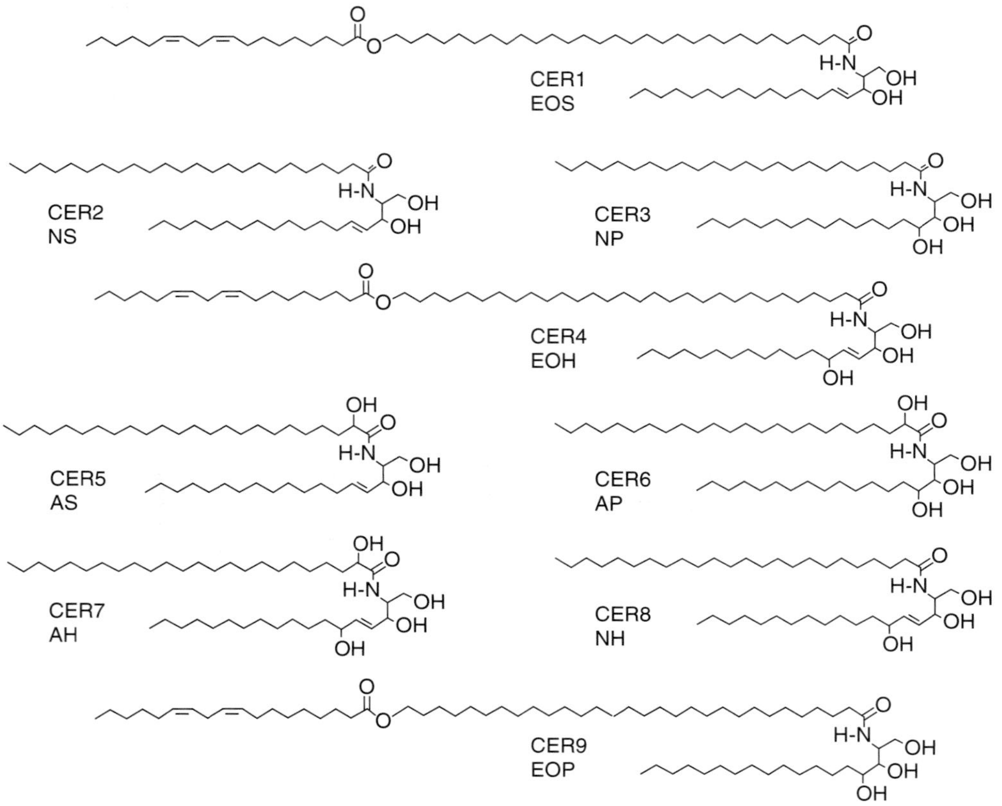

Figure 2 Chemical structures of human SC CERs. E, ester-linked fatty acids; O, –OH fatty acids; S, sphingosine; N, non–OH fatty acids; P, phytosphingosine; A, –OH fatty acids. (Reprinted from, De Jager, M. W.; Gooris, G. S.; Dolbnya, I. P.; Ponec, M.; Bouwstra, J. A. *Biochim. Biophys. Acta* **2004**, *1664*, 132–140, copyright (2004), with permission from Elsevier.)

incorporation of PE into PC vesicles has been found to slightly enhance the negative charge of the phosphate group.[9] Phosphatidylserine (PS) has two negatively charged moieties (pK_a on the membrane, carboxylate (pK_a = 5.5) and phosphate (pK_a < 1)) and one positively charged moiety (amine (pK_a = 11.3)), total − 1 net charge.[10] Phosphatidyl-linositol (PI) and phosphatidylglycerol (PG) have one negatively charged moiety (phosphate (pK_a = 2.7 (PI), 2.9 (PG))).

Ceramides (CERs) are important components of the skin permeation barrier, the stratum corneum (SC) (**Figure 2**).[11] The CER head group can form lateral hydrogen bonds with adjacent CER molecule. CERs are categorized into several subgroups; CER1, CER4, and CER9 have exceptionally long acyl chains that affect the interlamellar distance (*see* Section 5.19.2.5).

5.19.2.2   Physicochemical Properties of Lipid Bilayers

Lipid bilayers mainly composed of phospholipids are a heterogeneous system which can be roughly divided into four regions (see **Figure 3c**)[12,13]; the boundary of each region is not explicit:

- *Low-density head group region.* This region ranges from the point where the presence of the membrane begins to perturbate the bulk water structure to the point where the water density and head group density are comparable. This region can be large because the perturbation of water molecules can extend over a long range.
- *High-density head group region.* This region is ∼7.5 Å wide. In this region, bulk-like water no longer exists. It has a high dielectric constant ($\varepsilon = ∼30$)[14] and high viscosity, and is abundant in hydrogen bond acceptors.
- *High-density tail region.* This region is ∼7 Å wide. It has a low dielectric constant and high viscosity. The hydrocarbon tail has a high density and is highly ordered. The region is suggested to resemble a soft polymer.[15]
- *Low-density tail region.* This region is ∼11 Å wide (both halves of the bilayer). It has a low dielectric constant ($\varepsilon = ∼2$) and low viscosity. The region is suggested to resemble a low-density alkane fluid, such as dodecane or hexadecane.

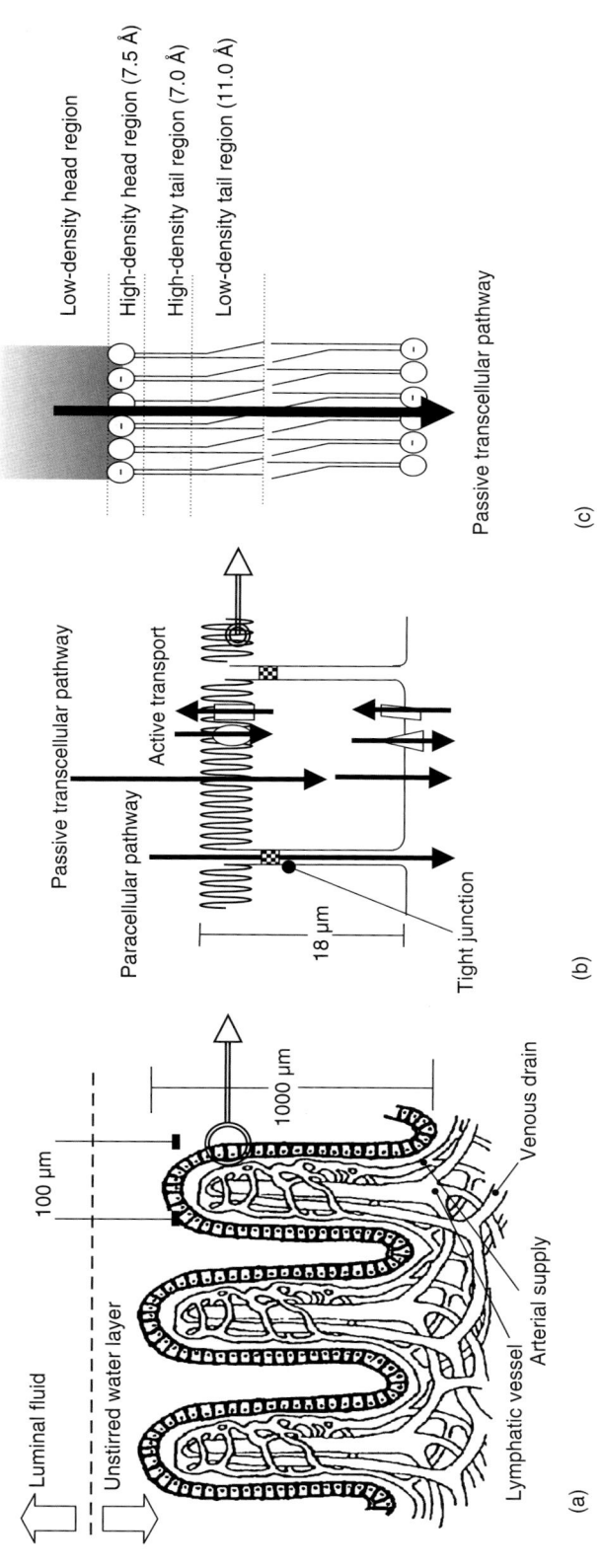

Figure 3 Intestinal membrane structure. The bold arrows in the figure illustrate the permeation pathways of compounds. (a) Villous structure of intestine. Unstirred water layer is on the top of villi. (Adapted from Avdeef, A. *Curr. Top. Med. Chem.* **2001**, *1*, 277–351, with permission from Bentham Science Publishers.) (b) Permeation pathways of compounds across the intestinal epithelial membrane. (c) Lipid bilayer. Four regions of the lipid bilayer are indicated: (1) low-density head group region (gray gradient); (2) high-density head group region; (3) high-density tail region; (4) low-density tail region.

5.19.2.3   Intestinal Membrane: Morphology, Physicochemistry, and Lipid Composition

5.19.2.3.1   Morphology

The human small intestine is about 6–7 m in length, and loosely divided into three sections, the duodenum, jejunum, and ileum.[16] The human small intestine membrane has a fractal-like structure, i.e., the ridges (oriented circumferentially around the lumen), villi, and microvilli (**Figures 3a** and **3b**).[14] The membrane surface is expanded approximately up to 600-fold by the villi and microvilli. Due to this large surface area, the small intestine is the main region of absorption for drugs.[16] The intestinal membrane has the mucus layer on the villi. This mucus layer is thought to maintain the unstirred water layer (UWL), about 30–100 μm in thickness, where agitation from intestinal movement does not reach.[17] The villous structure can cause a concentration gradient along with the trough, and affects the effective surface area and effective permeability (*see* Section 5.19.3.3).[18]

The permeation pathways of compounds across the intestinal epithelial membrane can be categorized into three groups: the passive transcellular pathway, the paracellular pathway, and active transport. The passive transcellular pathway is the permeation across the lipid bilayer of the intestinal epithelial cells. Intestinal epithelial cells tightly adhere to each other by intercellular adhesion molecules, and construct a tight junction.[19] However, small hydrophilic compounds can permeate through this junction. This permeation route is called the paracellular pathway. Permeability through the paracellular pathway differs in each part of the intestine. In the jejunum, the pore radius is about 7–9 Å, and in the ileum the pore radius is about 3–4 Å.[20] In addition, the paracellular pathway differs among animal species: dogs have a much larger pore size than humans and rats[21]; and Caco-2 cells have a tighter paracellular pathway than the in vivo human intestine.[22]

5.19.2.3.2   pH and transit time at each gastrointestinal site

The microclimate pH near the membrane surface is maintained at about pH 6.0–7.0, lower than that of the intestinal luminal fluid (**Table 1**).[23–25] The microclimate pH increases as the intestinal tract descends.[23] The transit time through each part of the intestine is an important factor that determines the fraction of a dose absorbed from each region. An administered drug passes through the stomach within about 0.5–1.5 h, through the small intestine in about 3–4 h, and through the large intestine in about 7–20 h.[26]

5.19.2.3.3   Lipid composition

The brush border (apical) membrane of the intestinal epithelial cell is mainly composed of PC, PE, PS, PI, cholesterol, and sphingomyelin (**Table 2**).[27] In contrast to red blood cells, negatively charged lipids distribute to both the inner and outer leaflets of the bilayer (**Table 3**).[28,29]

5.19.2.3.4   Other properties

Biles on the apical side of the membrane and plasma proteins on the basolateral side play an important role in the in vivo situation. Bile micelles can bind compounds and reduce the free concentration, resulting in a decreased flux.[34] On the other hand, biles increase the dissolution/solubility of a drug.[35] Plasma proteins may wash away the permeant from the basolateral side of the intestinal membrane, retaining a sink condition on the basolateral side.[34,36]

5.19.2.4   The Blood–Brain Barrier: Morphology

The structure of the BBB is shown in **Figure 4**.[37] In contrast to the intestinal membrane, the BBB has neither the mucus layer nor the UWL. The paracellular pathway is negligible for most compounds under physiological conditions.[32] The BBB has an estimated surface area of 12 m².[32]

Table 1 Variation of surface area, segment length, and pH in the gastrointestinal (GI) tract[16,22–26]

Position in the GI tract	Surface area (m²)[a]	Segment length (cm)	Luminal pH	Microclimate pH
Stomach	3.5	0.25	1.0–2.0	8
Duodenum	1.9	35	4.0–5.5	6.3–6.4
Jejunum	184	280	5.5–7.0	6.0–6.4
Ileum	276	420	7.0–7.5	6.6–6.9
Colon	1.3	150	7.0–7.5	6.9

[a]Expansion by the fold, villious, and microvillious structure was included in the calculation.

Table 2 Lipid composition of BBM,[27] porcine brain extract,[30] Caco-2 cells,[31] RBE4 rat endothelial immortalized cells (a BBB model),[32] egg lecithin,[33] and soybean lecitin[33]

Lipid[a]	BBM (rat)[b]	PBL (porcine)[c]	Caco-2[d]	RBE4[e]	Egg lecithin[f]	Soybean lecithin[g]
Percentage of total lipid						
PC	20	13	53	18	73	24
PE	18	33	19	23	11	18
PS	6	19	17	14	–	–
PI	7	4	8	6	1	12
Sph	7	–	3	8	–	–
Cholesterol + CE	37	–	–	26	–	–
TG	–	–	–	1	13	37
Percentage of phospholipid						
PC	39	18	54	30	87	44
PE	35	48	20	38	13	33
PS	12	27	17	23	–	–
PI	14	6	9	10	1	22

[a] CE, cholesterol esters; Sph, sphingomyelin; TG, triglycerides; see text for other definitions.
[b] Data from.[27] Reconstituted brush border membrane (BBM).
[c] PBL, porcine polar brain lipid (Avanti Polar Lipids, Alabaster, AL).
[d] Data from.[31] A model for oral absorption. Total lipid extract from Caco-2 cell homogenate.
[e] Data from.[32] A model for BBB. Total lipid extract from RBE4 rat endothelial immortalized cell line homogenate.
[f] Total extract from chicken egg (60% lecithin grade) (Avanti Polar Lipids, Alabaster, AL).
[g] Total extract from soy bean (20% lecithin grade) (Avanti Polar Lipids, Alabaster, AL).

Table 3 Lipid distribution on inner and outer leaflets (w/w%)[28,29]

Lipid	BBM (rabbit)[28]		Red blood cells (human)[29]	
	Outer	Inner	Outer	Inner
PC	32	68	76	24
PE	34	66	20	80
PS	44	56	0	100
PI	40	60	–	–
Sph	63	37	82	18

See **Table 2** for definitions of abbreviations.

5.19.2.5 The Skin Permeation Barrier (Stratum Corneum): Morphology and Lipid Composition

The transdermal permeation rate of most drugs is limited by the SC.[38,39] The thickness of the SC is different in each body part, about 15 μm in the abdominal skin and 10 μm in the dorsal skin. The pH near the surface of the skin is about 5. The structure of the SC is represented by the brick and mortar model, in which keratin-filled cells (corneocytes) (brick) are embedded within intercellular lipids (mortar) (**Figure 5**). The intercellular lipids form a lamellar structure, which orients parallel to the corneocyte surface. The primary transport pathway for most drugs traversing the SC is the intercellular lamellar lipid region. The path length relative to the thickness of the SC is about 13 due to the tortuous pathway.[40]

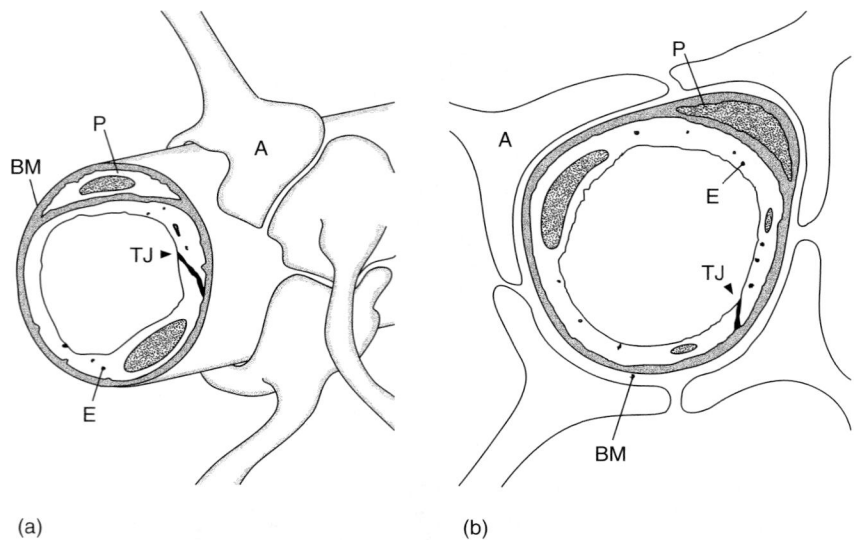

Figure 4 The brain capillary endothelium (i.e., the BBB), showing the cell types forming the blood–brain interface of a capillary in mammalian brain parenchyma: E, endothelium; BM, basement membrane; P, pericyte; A, astrocytic process. Arrowheads indicate the interendothelial clefts closed by tight junctions (TJs). (Reprinted from Abbott, N. J. *Drug Disc. Today* **2004**, *1*, 407–416, copyright (2004), with permission from Elsevier.)

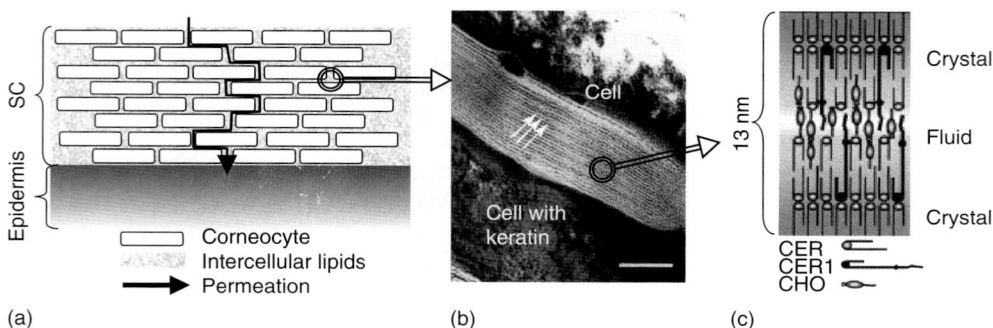

Figure 5 The SC. (a) The brick and mortar model. The arrow in the figure illustrate the primary permeation pathway for most drugs. (b, c) Lamellar structure of SC lipids. The interval of the lamellar structure is about 13 nm; CHO, cholesterol. A crystalline part is predominantly present, while most probably a subpopulation of lipids from a more fluidic part. (b, c: Reprinted from Bouwstra, J. A.; Honeywell-Nguyen, P. L.; Gooris, G. S.; Ponec, M. *Prog. Lipid Res.* **2003**, *42*, 1–36, copyright (2003), with permission from Elsevier.)

The intercellular lipids are mainly composed of free fatty acids (FFAs, 10–15%), cholesterol (25%), sterol esters (5%), and CERs (50%). The phase behavior of the lammellar lipid is different form that of the lipid bilayer mainly composed of phospholipids. In the SC, a crystalline part is predominantly present, while, most probably, a subpopulation of lipids form a more fluidic region (**Figure 5**).[38]

5.19.3   Permeation Theory and In Vivo Extrapolation

5.19.3.1   Transmembrane Permeation

5.19.3.1.1   The solubility–diffusion model for a homogeneous membrane

First approximation of a biological lipid membrane is a homogeneous organic solvent membrane (**Figure 6**).[41] Fick's first law can be applied to the homogeneous membrane permeation. The passive permeation across a membrane is a diffusion process, the driving force of which is the concentration gradient across the membrane (i.e., Fick's first law). If the interfacial resistance at the lipid–water interface is assumed to be negligible, flux (J, $mol\,cm^{-2}\,s^{-1}$) can be

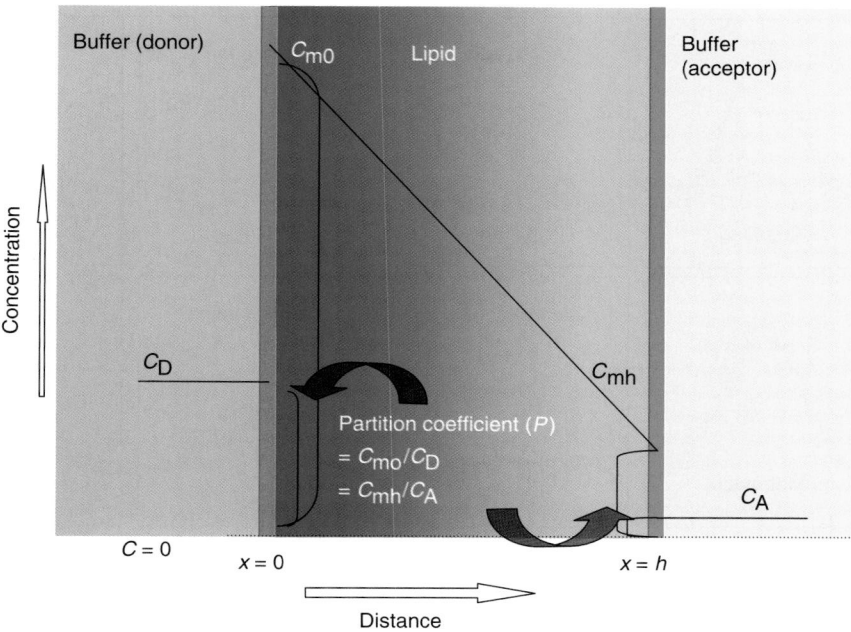

Figure 6 The solubility–diffusion model. The vertical axis and the blue gradient indicate the concentration. C_{m0}/C_D is the partition coefficient (P).

expressed as

$$J = \frac{D_m dC_m}{dx} = \frac{D_m(C_{m0} - C_{mh})}{h} \qquad [1]$$

where D_m is the diffusion coefficient in the membrane ($cm^2\,s^{-1}$), x is the position in the membrane, C_{m0} and C_{mh} are the concentrations at positions 0 and h in the membrane, respectively, and h is the thickness of the membrane (cm). C_{m0} and C_{mh} can be expressed by a partition coefficient of the organic solvent (P) and the concentration in the water phases of the donor and acceptor sides (C_D and C_A, respectively). Considering the sink condition, C_A is approximated by zero.

$$J = \frac{D_m P(C_D - C_A)}{h} \approx \frac{D_m P C_D}{h} = P_m C_D \qquad [2]$$

$$P_m = \frac{D_m P}{h} \qquad [3]$$

where P_m is membrane permeability ($cm\,s^{-1}$). Equation [3] indicates that the permeability is determined by the partition coefficient (a static parameter) and the diffusion coefficient (a kinetic parameter), and the thickness of the membrane.

5.19.3.1.2 The solubility–diffusion model for an inhomogeneous membrane

As shown in **Figure 3c**, the lipid bilayer is an inhomogeneous membrane. The solubility–diffusion model can be extrapolated to an inhomogeneous membrane. The permeability coefficient is the reciprocal of the permeation resistance, and the total permeation resistance connected in series is the sum of each resistance, similar to Ohm's law.

$$\frac{1}{P_m} = \int_0^h \frac{1}{P(x)} dx = \int_0^h \frac{1}{D(x)K(x)} dx \qquad [4]$$

where $P(x)$ is the local membrane permeability coefficient at position x, $D(x)$ is the local diffusion coefficient at position x, and $K(x)$ is the local partition coefficient between water and position x.

The barrier domain solubility–diffusion model was introduced by Anderson and co-workers.[42,43] According to eqn [4], the lowest-permeability region (barrier domain) limits the total permeability. Therefore, eqn [4] can be

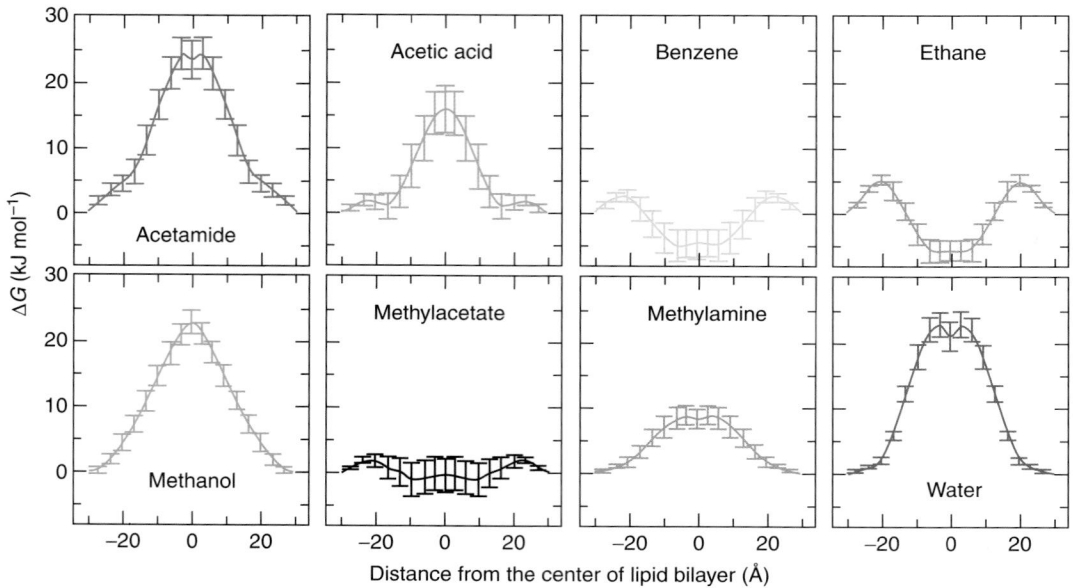

Figure 7 Free energy profiles of small molecules in a phospholipid bilayer estimated by a molecular dynamic simulation. Error bars are standard errors calculated from the difference in the force of the five individual simulations from their average. (Reprinted from Bemporad, D.; Essex, J. W.; Luttmann, C. *J. Phys. Chem.* **2004**, *108*, 4875–4884. Copyright (2004) American Chemical Society.)

simplified to

$$P_\mathrm{m} \approx \frac{D_\mathrm{barrier} K_\mathrm{barrier/water}}{h_\mathrm{barrier}} \qquad [5]$$

where h_barrier is the thickness of the barrier domain, D_barrier is the diffusion coefficient in the barrier domain, and $K_\mathrm{barrier/water}$ is the partition coefficient of a solute from water (not from the polar head group interface) to the barrier domain.

$K(x)$ can be expressed in terms of free energy as

$$K(x) = \exp\left(-\frac{\Delta G(x)}{RT}\right) \qquad [6]$$

where R is the gas constant and T is the absolute temperature. $\Delta G(x)$ is free energy difference between the water phase and position x in the membrane. Therefore, permeability can be expressed as

$$\frac{1}{P_\mathrm{m}} = \int_0^h \frac{1}{D(x)\exp[-\beta\,\Delta G(x)]}\mathrm{d}x \qquad [7]$$

where β is $1/RT$. The diffusion coefficient in the membrane is suggested to be lower than that in a nonpolar solvent such as hexadecane.[44] The ordered region of the hydrophobic core (high-density tail region in **Figure 3a**) is suggested to behave as a soft polymer, leading to a reduction of the diffusion coefficient in this region.

Owing to recent increases in computation speed, massive molecular dynamics simulations of lipid bilayers have become available.[45,46] In the case of hydrophilic compounds, the main permeation barrier was suggested to be the high tail density region of the membrane. In the case of hydrophobic compounds, the main barrier to permeation was suggested to be the head group region (*see* Section 5.19.2.2). **Figure 7** shows the change in free energy for eight neutral small molecules calculated by a molecular dynamic simulation.

5.19.3.1.3 The two-step model (flip-flop model)

The two-step model (or the flip-flop model) has been proposed to describe the transmembrane movement of a substrate, especially that of large amphiphilic molecules or peptide mimetic molecules (e.g., doxorubicin) (**Figure 8**).[47–50] The transmembrane movement can be described as (1) incorporation of a compound into one membrane leaflet, and (2) transfer (flip-flop) across the lipid core. In the case of fatty acids, the first step is much faster than the second flip-flop step, and the flip-flop rate decreases as the chain length increases.[50,51]

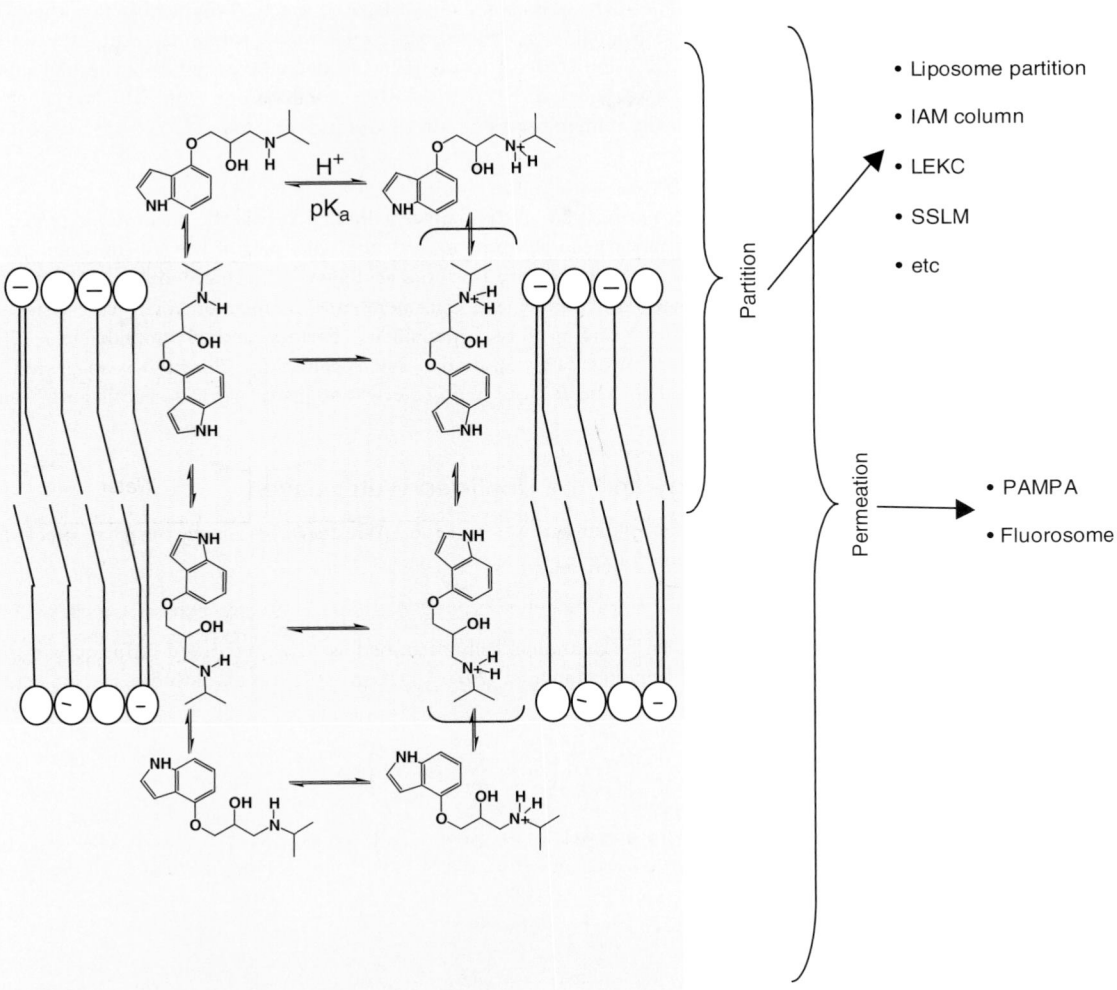

Figure 8 The two-step model, and categories of physicochemical tools. (Adapted from Sugano, K. *PharmStage* **2002**, *2*, 72–78.)

5.19.3.1.4 The relationship between permeation and physicochemical properties of a drug

5.19.3.1.4.1 The solubility–diffusion model

According to the solubility–diffusion model, the membrane permeability coefficient can be related to the partition coefficient between water and the highest ΔG region, which is the rate-limiting barrier. If a suitable organic solvent that resembles the rate-limiting barrier were chosen, the membrane permeability coefficient would be calculated from the partition coefficient between water and the organic solvent, the diffusion coefficient, and the thickness of the barrier (eqn [5]).

In the case of a lipid bilayer mainly composed of phospholipids, simple alkanes[44,52] or alkenes[53,54] were suggested to reflect the rate-limiting permeation barrier. Octanol was suggested to be less suitable.[53] The solubility–diffusion model suggests that the membrane partition measurement cannot directly represent the transmembrane permeation,[41,55] because the region of maximum partition in the membrane is the lowest ΔG region.

In the case of skin permeation, the permeability coefficient can be described by the solubility–diffusion model, so that by partition and diffusion coefficients of a drug in the permeation barrier.[56] Drugs permeate through the intercellular part of the SC mainly composed of CER, cholesterol, and FFAs, which organize a lamellar structure (*see* Section 5.19.2.5, the brick and mortar model). The octanol/water partition coefficient (K_{oct}) is most often used as a surrogate for the partition coefficient into the SC (K_{sc}).[56,57] K_{sc} was calculated as

$$\log K_{SC} = -0.024 + 0.59 \log K_{oct} \qquad [8]$$

($n = 45$, $r^2 = 0.84$, standard deviation $= 0.32$). The regression coefficient of $\log K_{oct}$ was 0.59, indicating that the skin membrane permeation barrier is more polar than octanol,[56] or a partial desolvation of the solute caused by the water associated with CER polar head group.[57] The diffusion coefficient was found to decrease as not only the molecular weight but also the hydrogen bond acidity/basicity increase.[57,58] This finding is consistent with diffusion along a nonpolar pathway hindered by interaction with the immobilized polar head group of SC lipids.[58]

5.19.3.1.4.2 The two-step model of lipid bilayer permeation

According to the two-step model of lipid bilayer permeation, in the first step, lipophilicity is the major driving force.[59] In the second step, interactions between the bilayer head group region and the polar part of the substrate are more important than lipophilicity.[60] In contradiction to the suggestion from the solubility–diffusion model, several reports show that the membrane partition measurement can predict the transmembrane permeation for a wide range of chemically diverse compounds (see Section 5.19.6 and see 5.18 Lipophilicity, Polarity, and Hydrophobicity). The membrane partition may represent the first step of the two-step model. The second step (flip-flop) is suggested to be more hydrogen bond-dependent, and related to $\Delta \log P$,[48] the heptane/ethylene glycol partition coefficient,[60] and the polar surface area.[61]

5.19.3.2 The Paracellular Pathway and the Unstirred Water Layer

In the case of the intestinal membrane, the paracellular pathway and the UWL are important factors that affect the total passive membrane permeability (see Section 5.19.2.3.1).

5.19.3.2.1 The paracellular pathway

The paracellular pathway permeability (P_{para}) can be mathematically described as a size-restricted diffusion within a negative electrostatic force field (eqn [9]).[62,63] Renkin's function ($F(r/r_{para})$, eqn [10]) was employed as the molecular size restrictor.

$$P_{para} = A \frac{1}{r} F\left(\frac{r}{r_{para}}\right)\left(f_0 + \sum^{z(z \neq 0)} f_z E(z)\right) \qquad [9]$$

$$F\left(\frac{r}{r_{para}}\right) = \left(1 - \frac{r}{r_{para}}\right)^2\left[1 - 2.104\frac{r}{r_{para}} + 2.09\left(\frac{r}{r_{para}}\right)^3 - 0.95\left(\frac{r}{r_{para}}\right)^5\right] \qquad [10]$$

$$E(z) = \frac{Cz}{1 - \exp(-Cz)} \qquad [11]$$

where f_z is the fraction of each charged species (z is the charge number) calculated from pK_a, r_{para} is the apparent pore radius of the paracellular pathway, r is the molecular radius of a permeant, and A is a fitting coefficient. C is in relation to the apparent electric potential of the paracellular pathway (for the intestine, 18–80 mV).[62,64] $F(r/r_{para})$ decreases as the molecular radius of a permeant increase. This paracellular pathway model equation is a first approximation. The paracellular pathway permeability was suggested to be affected by the lipophilicity of the substrate, which is not incorporated in this model.[65,66]

The contribution of the paracellular pathway to the total permeability depends on the physicochemical characteristics of a substrate. Small (molecular weight <200–300) hydrophilic cationic compounds can permeate via the paracellular pathway more readily than large, neutral, or anionic compounds. Cimetidine[67] and atenolol[63] are typical compounds for which the paracellular pathway permeation contributes significantly.

5.19.3.2.2 Unstirred water layer permeability

UWL permeability (P_{UWL}) can be mathematically modeled as a simple diffusion process in a water layer, according to the Einstein–Stokes equation,

$$P_{UWL} = A' \frac{1}{MW^{1/3}} \qquad [12]$$

where A' is a fitting coefficient and MW is the molecular weight. This UWL model equation is a first approximation. In the case of the intestinal membrane, the UWL is superimposed on the mucus layer, and the UWL permeability was suggested to be affected by the lipophilicity of a substrate, probably due to the interaction between the mucus and a substrate.[68]

5.19.3.3 The Effective Intestinal Membrane Permeability

The effective intestinal membrane permeability coefficient (P_{eff}) is an in vivo or in situ intestinal membrane permeability. P_{eff} in humans is measured using a technique based on the single-pass perfusion of the human jejunum segment between two inflated balloons.[17] The villous structure of the intestinal membrane is not considered in the P_{eff} calculation. As an approximation, P_{eff} can be expressed as

$$\frac{1}{P_{eff}} = \frac{1}{aP_m + bP_{para}} + \frac{1}{cP_{UWL}} \qquad [13]$$

The coefficients a, b, and c are fitting coefficients, which include the ratio of the available surface area. However, this equation assumes that the intestinal membrane is a flat. As noted in Section 5.19.2.3.1, the intestinal membrane has a fractal-like structure. This fractal structure should be taken into account for a quantitative in vitro/in vivo correlation (IVIVC) in a wide permeability range, since this fractal-like structure causes a concentration gradient down the villous structure. The magnitude of the concentration gradient depends on the membrane permeability. The concentration gradient changes the available surface area. It is suggested that a high-permeability compound is absorbed mainly from the top of the villi, while a low-permeability compound is absorbed from the whole villous surface.[18] A mathematical model to treat the villous structure was reported by Winne *et al.*[18]

5.19.3.4 The Link to the Fraction of a Dose Absorbed (A%)

The effective permeability and A% can be related by the parallel tube model, and the relationship is a hyperbolic curve.[69]

$$A\% = (1 - \exp(-GP_{eff})) \times 100 \qquad [14]$$

where G is the lump constant of the available intestinal surface area, the transit time through the absorption site of the gastrointestinal tract, and a fitting coefficient. The available intestinal surface area is estimated to be $800\,cm^2$ on P_{eff} bases.[70] To take into account the difference in permeability for each intestinal region, the compartment absorption transit (CAT) model has been proposed (*see* 5.38 Mechanism-Based Pharmacokinetic–Pharmacodynamic Modeling for the Prediction of In Vivo Drug Concentration–Effect Relationships – Application in Drug Candidate Selection and Lead Optimization).[71] The CAT model can be further connected to the physiologically based pharmacokinetic (PBPK) model.[72–74]

5.19.3.5 The pH–Partition Hypothesis and pH–Permeability Curves

A theoretical pH–permeability curve can be derived from the pH–partition hypothesis, allowing for the effect of the UWL in the calculation. According to the pH–partition hypothesis,[75] a molecule needs to be in its uncharged form to partition into (or flip-flop across) the lipophilic permeation barrier. The permeation of charged species is negligible. In addition, the UWL is adjacent to the membrane. Therefore, the apparent membrane permeability coefficient ($P_{m,\,app}$) of a monoprotic acid and base can be expressed as

$$\frac{1}{P_{m,\,app}} = \frac{1}{P_0 f_u} + \frac{1}{P_{UWL}} = \frac{1 + 10^{-pK_a + pH}}{P_0} + \frac{1}{P_{UWL}} \quad \text{(monoprotic acid)} \qquad [15]$$

$$\frac{1}{P_{m,\,app}} = \frac{1}{P_0 f_u} + \frac{1}{P_{UWL}} = \frac{1 + 10^{pK_a - pH}}{P_0} + \frac{1}{P_{UWL}} \quad \text{(monoprotic base)} \qquad [16]$$

where f_u is the fraction of uncharged species and P_0 is the intrinsic permeability coefficient of uncharged species. The fraction of uncharged species can be calculated from pK_a by the Henderson–Hasselbalch equation (*see* 5.16 Ionization Constants and Ionization Profiles) for a monoprotic acid and base.

A theoretical pH–permeability curve based on eqn [15] is show in **Figure 9**. The slope of the logarithmic plot is 1. In this slope region, the fraction of uncharged species changes as the pH of the buffer varies. One unit difference of pH or pK_a (a logarithmic scale) corresponds to a 10-fold change in permeability on a normal scale. Therefore, pH and pK_a are important determinants when in vitro and in silico systems are constructed. When the effect of the UWL is negligible, the horizontal line corresponds to P_0. The cross-over point of the slope line and the horizontal line is the pK_a. When the UWL limits the permeability, the horizontal line is lower than P_0, and the cross-over point (pK_a^{flux}) is not the same as the pK_a.

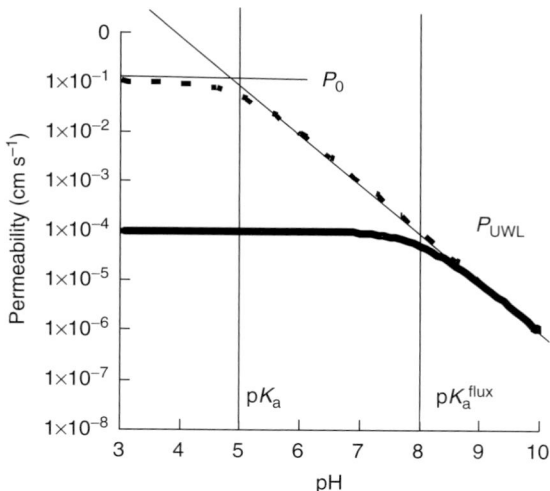

Figure 9   A theoretical pH–permeability curve. $P_0 = 0.1\,\mathrm{cm\,s^{-1}}$, $P_{UWL} = 1 \times 10^{-4}\,\mathrm{cm\,s^{-1}}$, $pK_a = 5$.

In the case of artificial membranes without surface charge, the pH–permeability profiles were found to follow the pH–partition hypothesis, and strictly adhere to the theoretical curve of eqns [15] and [16].[76,77] However, in the case of in vitro cells and the in vivo membranes, deviations from the theoretical curve are sometimes observed. These deviations can be due to (1) the effect of microclimate pH (see Section 5.19.2.3.2),[78] (2) active transport (see 5.04 The Biology and Function of Transporters; 5.11 Passive Permeability and Active Transport Models for the Prediction of Oral Absorption), (3) paracellular pathway permeation (see Section 5.19.3.2.1),[79] and (4) ion pair transport.[80,81] In the case of negatively charged artificial membranes, the deviations were observed in some basic compounds while not for acidic compounds (see Section 5.19.5.1.2.1).[10,82]

Equations [15] and [16] are often employed to reduce the apparent permeability into intrinsic parameters (P_0, pK_a, and P_{UWL}), similarly to how the octanol/buffer distribution coefficient (D_{oct}) is reduced to the intrinsic octanol/water partition coefficient of a neutral species (K_{oct}) and pK_a (see 5.18 Lipophilicity, Polarity, and Hydrophobicity). Intrinsic parameters can be more directly related to the chemical structure of the substrate than apparent parameters.

5.19.4   Comparing In Vivo Predictability Among Physicochemical Tools: Precautions

Before moving onto the introduction of each technology, precautions to compare in vivo predictability (IVIVC) are reviewed in this section. It is difficult to quantitatively compare the IVIVC of each method by literature surveys.

5.19.4.1   The Probe Compound Data Set

5.19.4.1.1   Inconsistency of the probe compound data set
To examine IVIVC, dozens to hundreds of compounds were employed as probe compounds with their in vivo data. However, a selection of probe compounds was not consistent in the literature for each IVIVC. In addition, the in vivo protocol was not necessarily consistent.

5.19.4.1.2   Structural diversity of the probe compound data set
In the literature, marketed drugs were most often employed as probes for the IVIVC study, because of (1) the availability of clinical ADME data, (2) the commercial availability of compounds, and (3) the confidentiality of in-house compounds. In addition, highly lipophilic compounds were not included in the probe compound set, probably because of experimental difficulties in obtaining in vivo data on intravenous administration, which is required to calculate the pharmacokinetic parameters. Therefore, the probe compound set was biased toward drug-like compounds, although structurally diverse compounds are often submitted for an assay in a drug discovery situation.[83,84] It is recommended that the IVIVC for structurally diverse compounds (or a chemotype series of each project) should be examined before implementation of an in vitro assay for practical use in drug discovery.

5.19.4.2 Complexity of the Biological Data for Oral Absorption and Blood–Brain Barrier Penetration

5.19.4.2.1 Oral absorption data

Regarding oral absorption prediction, the fraction of a dose absorbed in humans (A%),[85] P_{eff},[17] and the Caco-2 permeability coefficient (P_{app}) were often employed as biological data (*see* 5.02 Clinical Pharmacokinetic Criteria for Drug Research; 5.11 Passive Permeability and Active Transport Models for the Prediction of Oral Absorption, respectively). Typical A% values are summarized in **Table 4**. A% is determined by both solubility/dissolution and P_{eff}.[69] P_{eff} and P_{app} consist of passive transcellular permeation, paracellular permeation, and active transport (both influx and efflux). Artificial membrane technologies only reflect passive transcellular permeation. Therefore, those compounds which undergo solubility-limited absorption, paracellular pathway permeation, and active transport must be carefully excluded from the probe compound set. P-glycoprotein (P-gp) can reduce the oral absorption.[86] However, if the concentration of a drug in the intestine is high enough to saturate P-gp efflux, the effect of P-gp becomes less significant.[87] The K_m of P-gp substrates is relatively low, ranging from $4 \, \mu mol \, L^{-1}$ to $213 \, \mu mol \, L^{-1}$.[87]

5.19.4.2.2 Blood–brain barrier permeation data

The brain/plasma ratio in animal models has been often employed as probe data.[37] However, this parameter can be that of the steady state and unsuitable to relate directly to kinetic parameters. In addition, the effect of plasma protein binding was sometimes not corrected. Brain perfusion studies directly provide permeability data that can be a gold standard reference.[37,88] A binary expression of the central nerve system availability (CNS +/−) was often employed as an in vivo parameter.[89] This expression may include the in vivo results from (1) the perfusion method, (2) the brain/plasma ratio calculated from the area under the curve or at the steady state, and (3) the existence of pharmacological activity due to BBB penetration. In the case of BBB permeation, P-gp is well known to retard permeability.[86]

5.19.4.3 The Unstirred Water Layer of In Vitro Assays

The UWL thickness of in vitro planar membrane permeation assays (e.g., Caco-2[90]) and the parallel artificial membrane permeation assay (PAMPA),[91,92] can reach about $1500-4000 \, \mu m$ without stirring. This is much thicker than in vivo situations ($30-100 \, \mu m$ in the intestine, and much less in the BBB; *see* Sections 5.19.2.3 and 5.19.2.4). Nevertheless, without stirring, in vitro planar membrane permeation assays were found to be able to predict A%, suggesting that the UWL does not interfere with A% prediction. The UWL permeability of an in vitro planar membrane system was reported to be approximately $20-60 \times 10^{-6} \, cm \, s^{-1}$. A% < 90% corresponds to about $< 10 \times 10^{-6} \, cm \, s^{-1}$ in Caco-2 and PAMPA. Therefore, the contribution of the UWL is insignificant for A% prediction. The expansion of the effective surface area by the villous structure cancels out the reduction of the UWL in the in vivo situation (the UWL is on the top of villi).[17] However, the UWL of planar in vitro assays can interfere with the prediction of the high intestinal membrane permeability ($P_{eff} > \sim 1 \times 10^{-4} \, cm \, s^{-1}$), as well as BBB permeability. In addition, the effect of the UWL should be eliminated to obtain a correct structure membrane permeability relationship in a high membrane permeability range, since the apparent permeability may just reflect the diffusion in water when the UWL limits the permeation.

5.19.5 Permeation Assessment Tools for Drug Discovery

In the following sections, features and the in vivo relevance of artificial membrane tools for drug discovery will be discussed. These tools can be categorized into two types: (1) permeation across the membrane, and (2) partition into the membrane (interfacial region and/or interior hydrophobic region) (see **Figure 8**).[93] Tools to investigate permeation across the membrane (PAMPA and Fluorosome) are reviewed in this section. Tools to investigate partition into the membrane (chromatography-based artificial membrane-binding assays, the surface plasmon resonance (SPR)-based liposome-binding assay, solid-supported lipid membrane (SSLM) technologies, etc.) are reviewed in Section 5.19.6. In, addition, the surface activity assay is reviewed in Section 5.19.7.

5.19.5.1 Parallel Artificial Membrane Permeation Assay (PAMPA)

PAMPA has become widespread in drug discovery in recent years.[94–97] The first published study using PAMPA, by Kansy *et al.* in 1998,[94] attracted a lot of favorable attention. The first international symposium on PAMPA was held in 2002 in San Francisco.

PAMPA is an extension of the black lipid membrane (BLM), in which a planer membrane is constructed by brushing a lipid–organic solvent mixture over a small hole (0.5 mm diameter).[98] A serious drawback of BLM for use in drug

Table 4 Typical probe set compounds employed for IVIVC of A%

No.	Drug name	A%[a]	Reference	No.	Drug name	A%[a]	Reference
1	Acebutolol	80–90	64,152	37	Diclofenac	100	64
2	Acetaminophen	80	64	38	Dicloxacillin	35–76	64
3	Acetylsalicylic acid	84–100	94,152	39	Dilthiazem	80	64
4	Acyclovir	20–23	64,95,152	40	Diltiazem	92	94
5	Allopurinol	90	64	41	Dipyridamole	66	152
6	Alprenolol	93–96	64,94,95,128,152	42	Doxycycline	90–100	64
7	Amiloride	50	64,95,152	43	Enalapril	66	152
8	Ampicillin	62	152	44	Erythritol	90	64
9	Antipyrine	97	64	45	Ethambutol	80	64
10	Atenolol	50–54	64,94,95,128,152	46	Ethionamide	80	64
11	Aztreonam	1	64	47	Etoposide	50	152
12	Barbital	90	152	48	Famotidine	38–45	64,95
13	Bromocriptine	28	64	49	Fenoterol	60	64
14	Bupropion	87	64	50	Flecainide	81	64
15	Caffeine	100	152	51	Flucytosin	75–90	64
16	Carbamazepine	70–100	95,128,152	52	Foscarnet	17	128
17	Ceftriaxone	1	64,94,152	53	Furosemide	50–61	64,95,128,152
18	Cefuroxime	5	64	54	Ganciclovir	3	64
19	Cephalexin	0	94	55	Gentamycin	0	152
20	Chloramphenicol	90	64,94	56	Guanabenz	75–80	64,94,95,152
21	Chlorothiazide	13	64	57	HBED	5	64,95
22	Chlorpromazine	100	95	58	Hydrocortisone	55–91	64,94,95,128,152
23	Cidofovir	3	152	59	Imipramine	99–100	64,94,95,152
24	Cimetidine	64–95	64,95	60	Indomethacin	100	152
25	Ciprofloxacin	69	64,152	61	Isoniazid	80	64
26	Cloxacillin	37–60	64	62	Ketoprofen	100	64,128
27	Clozapine	100	95	63	Labetalol	90	95
28	Corticosterone	100	94	64	Lactulose	0.6	64,128
29	Coumarin	100	94	65	Lansoprazole	85	64
30	Creatinine	80	64	66	Lincomycin	28	64
31	Cymarin	47	64	67	Mannitol	16–26	64,128
32	Cytarabine	<20	64	68	Metaproterenol	44	64
33	Desferrioxamine	2	95	69	Metformin	86	64
34	Desipramine	100	95,128	70	Methotrexate	20	95
35	Dexamethasone	80–100	64,94,152	71	Methylprednisolone	82	64
36	Diazepam	100	152	72	Metolazone	64	95

Table 4 Continued

No.	Drug name	A%[a]	Reference	No.	Drug name	A%[a]	Reference
73	Metoprolol	95	64,94,95,128,152	97	Quinine	90	152
74	Nadolol	35–57	64,152	98	Raffinose	0.3	64,128
75	Naltrexone	96	64	99	Ranitidine	50–64	64,95,152
76	Naproxen	99–100	64,128	100	Ribavirin	33	152
77	Nicotinic acid	88	152	101	Salicylic acid	100	94,152
78	Norfloxacin	71	64	102	Sotalol	60	64
79	Olsalazine	2	64,94	103	Streptomycin	1	152
80	Oxacillin	30–35	64	104	Sulfasalazine	12–59	94,95,128,152
81	Oxprenolol	97	128	105	Sulindac	90	64,152
82	Oxybutynin	6	152	106	Sulpiride	35–44	64,94,95,128,152
83	Oxytetracycline	60	64	107	Sumatriptan	57	64
84	Phenobarbital	100	152	108	Terbutaline	62–73	64,94,95,128,152
85	Phenytoin	90	95	109	Testosterone	98–100	94,95
86	Pindolol	87–92	64,128,152	110	Tetracycline	75–80	64
87	Piroxicam	100	95	111	Theophylline	98–100	64,94,152
88	Practolol	100	64	112	Tiacrilast	99	94
89	Pravastatin	34	64	113	Timolol	72–95	64,95,152
90	Prazosin	77–95	64	114	Tolbutamide	85	64
91	Prednisolone	99	152	115	Tranexamic acid	55	64
92	Procainamide	75–95	64,152	116	Valsartan	55	64,95
93	Progesterone	91	95	117	Verapamil	95–100	64,94,128,152
94	Propranolol	90–100	64,94,95,128,152	118	Warfarin	93–98	64,94,95,152
95	Propylthiouracil	76	64	119	Zidovudine	100	64
96	Quinidine	81	152				

[a] The A% value employed for IVIVC in the literature.

discovery is that the membrane is extremely fragile. This drawback can be overcome by using a microporous filter as a scaffold for the lipid membrane. This filter-immobilized artificial membrane was introduced by Thompsom et al. in 1982.[99,100] In 1997, Camenisch et al. applied the filter-immobilized organic solvent membrane to compare the permeability of the organic solvent membrane and the Caco-2 membrane.[101] Kansy et al. applied the filter immobilized artificial membrane for 96-well plate formats to put it to practical use in drug discovery. The facileness and open accessibility of PAMPA are factors for its recent popularity. PAMPA does not require any special instruments, and is easily performed by standard laboratory instruments. Therefore, PAMPA is easily available regardless of the size of the budget of an organization.

5.19.5.1.1 Parallel artificial membrane permeation assay experimental operation

The PAMPA sandwich is constructed of a 96-well microplate and a filter plate (**Figure 10a**). The artificial membrane is formed by impregnating an organic solvent solution of lipids to the filter bottom of the filter plate. A dimethyl sulfoxide (DMSO) stock solution can be used as a sample resource. The incubation time depends on the experimental conditions (i.e., the lipid composition, pH, the stirring speed, additives to the donor and accepter solutions, etc.). After incubation, the concentrations of sample in the acceptor and/or donor plate ($C_{ac}(t)$ and $C_{do}(t)$, respectively) are

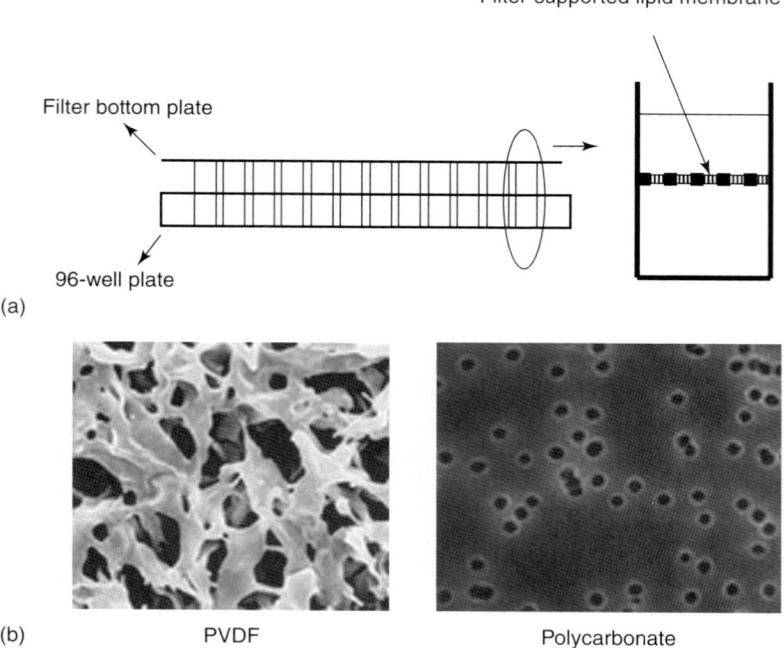

(a)

(b)   PVDF   Polycarbonate

Figure 10   (a) The PAMPA sandwich, constructed of a 96-well microplate (bottom) and a filter plate (top). The bottom of the well of the filter plate is a filter. (b) Structures of filter supports: polyvinylidene difluoride (PVDF) membrane and polycarbonate isopore membrane. (b: Adapted with permission from the Millipore Inc. website.)

measured, for example by ultraviolet spectroscopy (UV)[94] or liquid chromatography–mass spectrometry (LC/MS).[102] Usually, the concentration is measured at one time point (t, s). In the case of a no-gradient condition, the permeability (P_e, cm s^{-1}) and the membrane retention (R_e) are calculated as[14,95]

$$P_e = \frac{-0.203}{A_m t}\left(\frac{V_{ac} V_{do}}{V_{ac} + V_{do}}\right)\log\left[1 - \left(\frac{V_{ac} + V_{do}}{(1 - R_e)V_{do}}\right)\left(\frac{C_{do}(t)}{C_{do}(0)}\right)\right] \qquad [17]$$

$$R_e = \frac{C_{do}(t)}{C_{do}(0)} - \frac{V_{ac} C_{ac}(t)}{V_{do} C_{do}(0)} \qquad [18]$$

where A_m is the membrane area (cm^2), and V_{ac} and V_{do} are the volume of the donor and acceptor wells, respectively. Equations for gradient conditions have also been introduced.[10]

5.19.5.1.2   Effect of experimental conditions on parallel artificial membrane permeation assay permeability

The experimental conditions used for PAMPA vary; for example, lipid composition, the organic solvent used to dissolve the lipid, donor and acceptor buffer solutions, and stirring speed can all differ. These variations significantly affect permeability and in vivo predictability.

5.19.5.1.2.1   Lipid composition

Many PAMPA variations contain anionic (phospho-)lipids in the membrane to increase the in vivo predictability.[10,96] As described in Sections 5.19.2.3 and 5.19.2.4, the intestinal BBM and the BBB membrane contain anionic phospholipids. Anionic phospholipids significantly increased the PAMPA permeability of hydrophilic basic compounds, such as metoprolol, timolol, procainamide, and pindolol (**Figure 11**).[82,103,104] The PAMPA permeability of hydrophilic basic compounds is saturable and inhibited by the co-addition of cationic compounds. The pH–permeability profile deviates from the pH–partition hypothesis. Quaternary ammonium compounds permeate the membrane by the aid of anionic phospholipids.[82,104] It has been suggested that the PAMPA permeability of cationic species is mediated by anionic

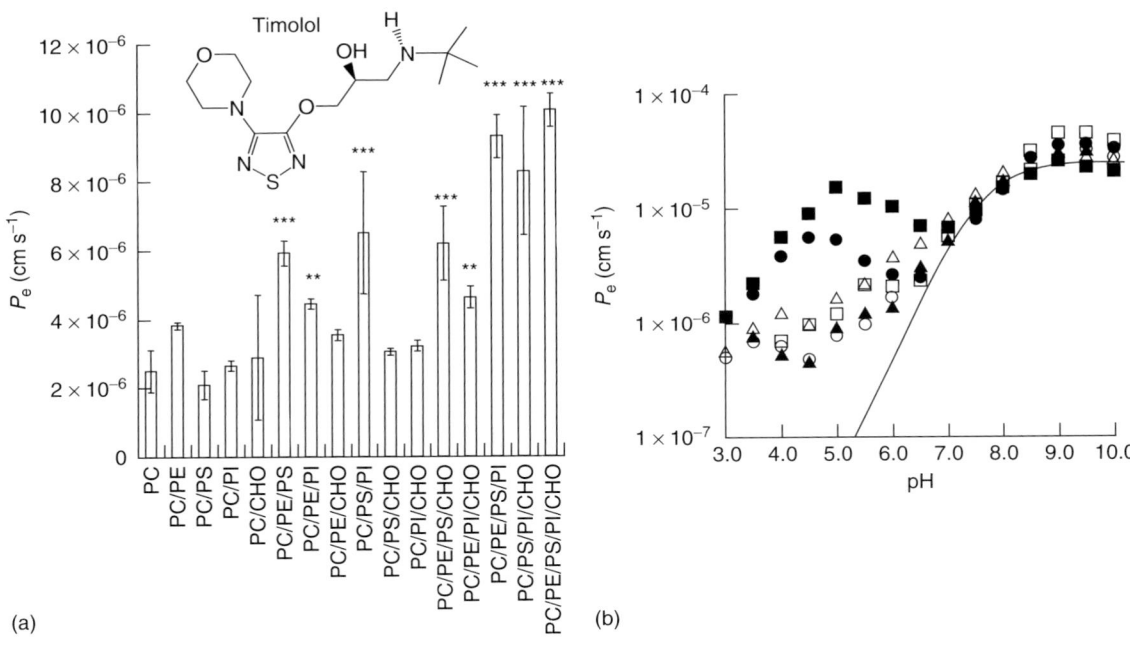

Figure 11 Effect of lipid composition on the PAMPA permeability (P_e) of timolol, a hydrophilic basic compound. Conditions: phospholipids (2%)/CHO (cholesterol) (1%)/1,7-octadiene (97%) (w/w) membrane, pH 6.0. PE, PS, and PI were added (0.8%, 0.2%, and 0.2%, respectively), and total phospholipids were adjusted to 2% by PC.[103] (a) Effect of various lipid compositions on the PAMPA permeation of timolol.[103] (b) pH–PAMPA permeability profile of timolol. Lipid composition: PC (○), PC/PE (△), PC/PS (□), PC/PI (●), PC/CHO (▲), and PC/PE/PS/PI/CHO (■). (Reprinted from Sugano, K.; Nabuchi, Y.; Machida, M.; Asoh, Y. *Int. J. Pharm.* **2004**, *275*, 271–278, copyright (2004) with permission.)

phospholipids. Some PAMPA variations employ egg or soybean extracted lecithins as the lipid source. These lecithins are not necessary 100% pure PC, and may contain anionic lipids to some extent, depending on the grade.

For neutral and acidic compounds, the reported results of the effects of phospholipids are varied. Avdeef reported that the phospholipid concentration in the membrane affected both the membrane permeability and the membrane retention of neutral and acidic compounds.[10] He varied the phospholipid concentration from 0% to 74%, and measured the permeability of structurally diverse compounds. On the other hand, Anderson *et al.* reported that, based on the barrier domain solubility diffusion theory (*see* Section 5.19.3.1.2), 1,9-decadiene can mimic the chemical selectivity of the phospholipid bilayer.[54] Wohnsland *et al.* reported that a PAMPA with the hexadecane membrane without phospholipids could adequately predict A% (*see* Section 5.19.5.1.3.2).[95]

5.19.5.1.2.2 Organic solvents in the parallel artificial membrane permeation assay membrane

In most cases, dodecane is employed as the organic solvent. It has been suggested that the organic solvent used to dissolve the lipids remains in the membrane, affecting membrane permeability. The permeability across an alkyldiene–phospholipid membrane was higher than that across an alkane–phospholipid membrane (**Figure 12a**).[105] The chain length of organic solvents also affects PAMPA permeability (**Figure 12b**).[96]

5.19.5.1.2.3 pH and solubilizers in the donor and acceptor buffers

The pH condition largely affects PAMPA permeability in accordance with the pH–partition hypothesis (*see* Section 5.19.3.5). To reflect the pH variations during the gastrointestinal transition (*see* Section 5.19.2.3.2), multiple pH conditions can be used for in vivo prediction.[94,95,106]

Co-solvents are often employed to dissolve poorly water-soluble compounds. DMSO, ethanol, PEG400,[107] 1-propanol,[10] and acetonitrile[108] were investigated for their effect on permeability. It was reported that the effect of acetonitrile can be corrected by hydrogen bond parameters.[109] The addition of co-solvents increases the affinity of hydrophobic compounds to the water phase, resulting in a decrease in the partitioning to the lipophilic part of the lipid membrane. Simultaneously, the co-solvent lowers the dielectric constant and increases the fraction of uncharged species. These phenomena affect the permeability oppositely, and may differ among each substrate.

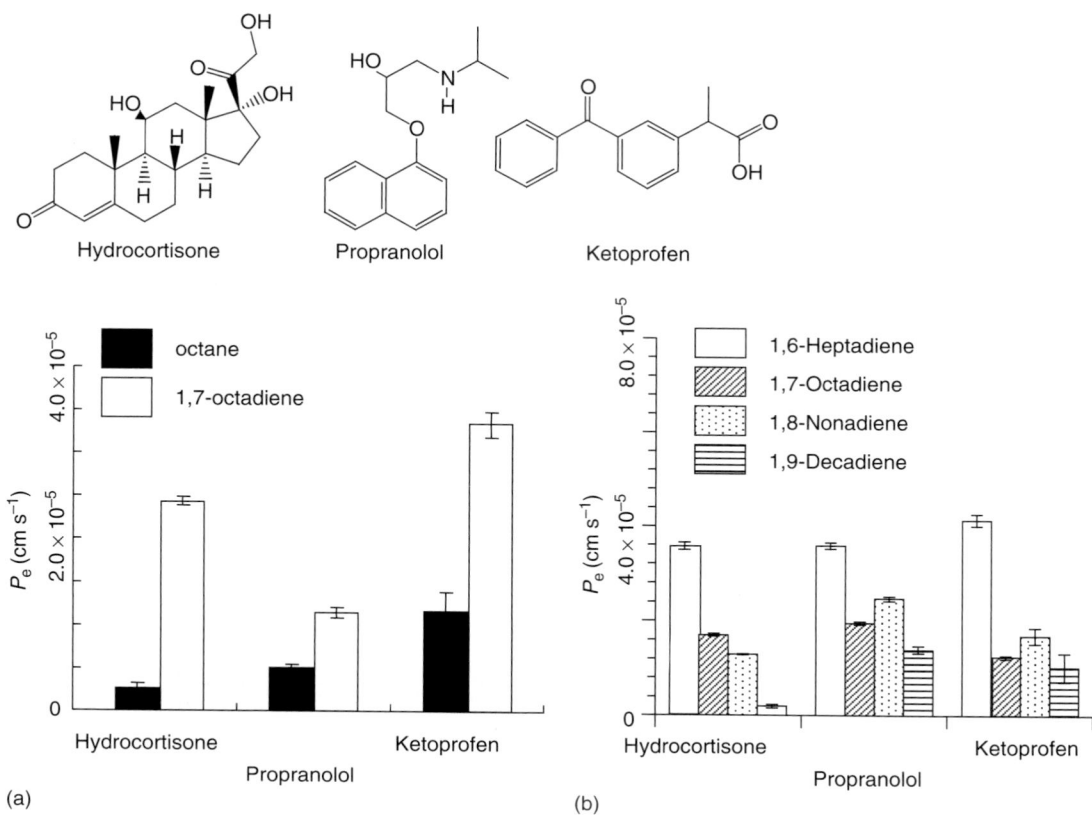

Figure 12 Effect of organic solvent remaining in the membrane on PAMPA permeability (P_e). (a) Octane versus octadiene (PC (0.8%)/PE (0.8%)/PS (0.2%)/PI (0.2%)/CHO (cholesterol) (1%)/organic solvent (97%) (w/w)).[105] (b) Chain length dependency (PC (2%)/organic solvent (98%) (w/w)). (Reprinted from Sugano, K.; Hamada, H.; Machida, M.; Ushio, H. *J. Biomol. Screen.* **2001**, *6*, 189–196, with permission from SAGE.)

Surfactants were also examined as the solubilizer. Glycocholic acid in the donor buffer was reported to have a minimal effect on permeability for most compounds.[109] On the other hand, taurocholic acid decreased the permeability of lipophilic compounds.[10] Tween 80 and Briji-35 at 0.5–5% (w/v) did not disrupt the PAMPA membrane, while Cremophor EL did at 0.5%.[110]

Surfactants in the acceptor buffer can increase the permeability of lipophilic compounds by facilitating the redistribution of lipophilic compounds from the PAMPA membrane to the acceptor buffer phase.[10]

5.19.5.1.2.4   Filters

Hydrophobic polyvinylidine difluoride (PVDF),[94] hydrophilic PVDF,[106] and polycarbonate filers[95] have been employed as a scaffold of the lipid membrane (see **Figure 10b**). The exact structure of the PAMPA membrane built on the PVDF filters is not known. PVDF filters have a sponge-like structure, and the thickness is about 125 μm. Polycarbonate filters have straightthrough pores, and the thickness is about 10 μm. Microscope observations suggest that the pore size of the polycarbonate filter affects the integrity of the lipid membrane.[67] In the case of a 0.45 μm pore size filter, the impregnated organic solvent (hexadecane) remains in the pore, and no defect (aqueous pores) was observed. In the case of a 3.0 μm pore size filter, defects were observed.

5.19.5.1.3   Variations in parallel artificial membrane permeation assay and their in vivo relevance

5.19.5.1.3.1   The original parallel artificial membrane permeation assay (egg lecithin parallel artificial membrane permeation assay)

Kansy *et al.* reported the original PAMPA in 1998.[94] The original PAMPA employed the hydrophobic PVDF filter as a membrane support. Detail of the lipid composition was not explained in this original article; however, it was later

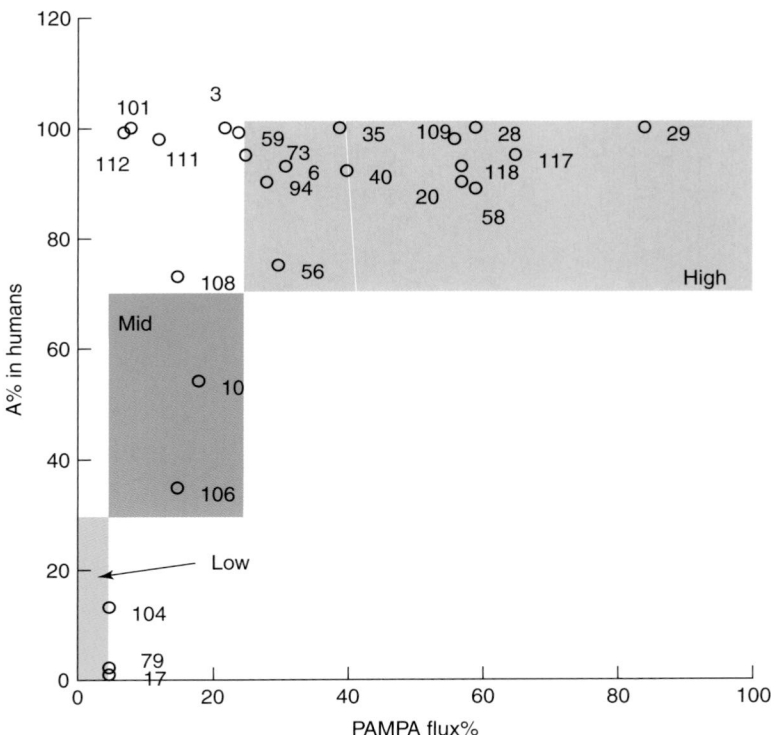

Figure 13 A% predictability of the original PAMPA. The PAMPA permeability with higher Flux% at pH 6.5 or 7.4 was employed. High: A% >70%, PAMPA flux >25%. Mid: A% = 30–70%, PAMPA flux 5–25%. Low: A% <30%, PAMPA flux <5%. See **Table 4** for assignment of compounds. (Reprinted from Kansy, M.; Senner, F.; Gubernator, K. *J. Med. Chem.* **1998**, *41*, 1007–1010 and modified to exclude the substrates of active transporters. Copyright (1998) American Chemical Society.)

published to be 10% (w/v) egg lecithin in dodecane.[104] The incubation time was 15 h. The sample buffer solution contained 5% DMSO. The permeability was reported as Flux%. Flux% was measured at pH 6.5 and 7.4, and the higher Flux% at pH 6.5 or 7.4 was used to predict A% in humans. The A% predictability was examined using 25 marketed drugs (**Figure 13**). The original PAMPA adequately categorized the A% classes to high (>70%), middle (30–70%), and low (30%). The number of probe compounds was relatively small, and A% data were biased to the high A% category.

Zhu *et al.* developed a similar PAMPA system, but employed a hydrophilic PVDF filter as a membrane support, resulting in a shorter incubation time (2 h).[106] The membrane lipid was 1% (w/v) egg lecithin in dodecane. The permeability was measured at pH 5.5 and 7.4. The higher permeability at pH 5.5 or 7.4 was used to correlate with oral absorption parameters. The A% predictability of PAMPA was examined using over 90 marketed drugs, and compared with that of Caco-2 permeability (pH 7.4), the octanol/buffer distribution coefficient at pH 7.4 ($D_{oct, 7.4}$, measured by a shake flask method), and the polar surface area (PSA). The A% prediction accuracy of PAMPA was comparable to Caco-2; however, it was significantly better than log $D_{oct, 7.4}$ and the PSA. The PAMPA permeability correlated well with the Caco-2 permeability at pH 7.4 ($r = 0.82$, $N = 49$).

5.19.5.1.3.2   Hexadecane membrane parallel artificial membrane permeation assay (HDM-PAMPA)

Faller *et al.* used a simplified membrane, the hexadecane membrane (HDM-PAMPA).[95] The rationale why HDM-PAMPA without phospholipids can predict the bio-membrane permeation is the assumption that a distinct permeation barrier for most drugs is the interior region of the lipid bilayer and that hexadecane can offer a similar environment (*see* Section 5.19.3.1.2 and 5.19.3.1.4.1, on the barrier domain solubility diffusion model). To construct a thin hexadecane membrane, a hexadecane–hexane mixture was added to the filter support, and the hexane was removed by evaporation. A polycarbonate filter (10 μm thickness and 3 μm pore size) was employed. Permeability was measured at pH 4–8, and the highest permeability was used to predict A% in humans. A% predictability was examined using 32 marketed drugs (**Figure 14**).

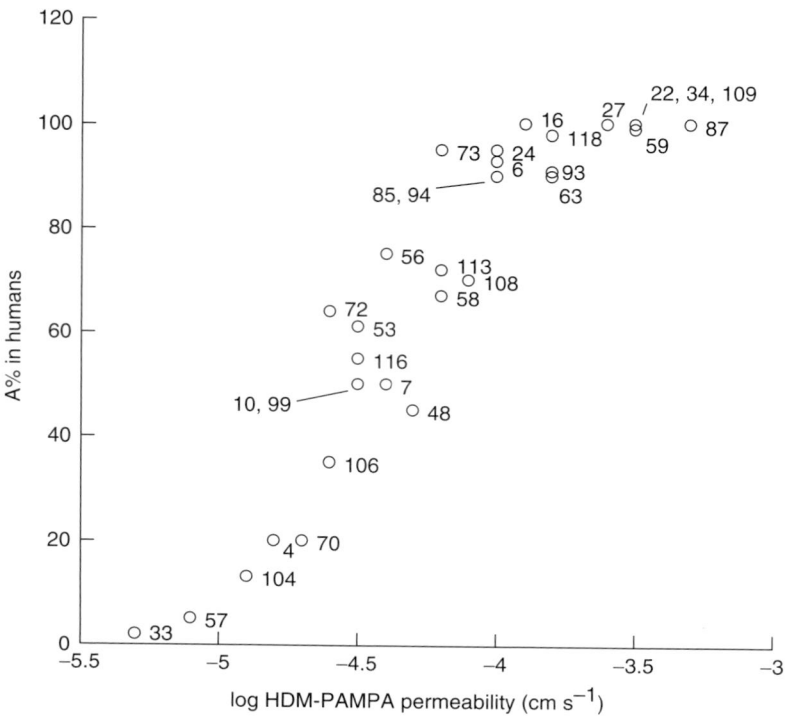

Figure 14   The A% predictability of HDM-PAMPA. PAMPA permeability with the highest permeability within the pH range 4–8 was employed. See **Table 4** for the assignment of compounds. (Reprinted from Wohnsland, and modified to exclude the substrates of active transporters and compound with MW <200. Copyright (2001) American Chemical Society.)

In the case of an isotropic organic solvent membrane, the membrane permeability coefficient and the partition coefficient correlate as eqn [3]. Therefore, PAMPA can also be used to measure the organic solvent/water partition coefficient.[95,111]

5.19.5.1.3.3   Biomimetic parallel artificial membrane permeation assay (BM-PAMPA)

Sugano *et al.* modified the original PAMPA to a more biomimetic assay (BM-PAMPA) (**Figure 15**).[64] The lipid composition was modified to PC (0.8%)/PE (0.8%)/PS (0.2%)/PI (0.2%)/cholesterol (1%) (w/w) to represent the lipid composition of the intestinal BBM. In addition, the organic solvent was changed from dodecane to 1,7-octadiene. The incubation time was 15 h. The sample solution contained 5% DMSO. In BM-PAMPA, permeability at pH 6.0–6.5 was found to be the best pH condition for A% prediction.[64,107] The A% and P_{eff} values of small and cationic drugs were underestimated by BM-PAMPA itself; however, they were adequately estimated when BM-PAMPA was combined with a paracellular pathway permeability model equation (A% prediction: root mean square error = 14%, N = 80) (*see* Section 5.19.3.2.1). [64,96] The Caco-2 permeability was also adequately predicted.[112] BM-PAMPA is not suitable for robotic automation in an open laboratory because 1,7-octadiene is an irritant.

5.19.5.1.3.4   Double-sink parallel artificial membrane permeation assay (DS-PAMPA)

After extensive examinations of assay conditions, Avdeef *et al.* developed the DS-PAMPA system. DS-PAMPA is commercially available from pION, Inc. As a source of phospholipids, Avdeef *et al.* employed a special lipid mixture similar to crude soybean lecithin, which is PI-rich (*see* **Table 2**).[33] The concentration of phospholipids was increased to 20%. To construct a double-sink condition, a pH gradient condition (pH 5.0–7.4 (donor) and pH 7.4 (acceptor)) and a lipophilicity gradient (surfactant in an acceptor compartment) were employed (sink conditions for acidic and lipophilic compounds, respectively). In addition, rapid stirring was introduced to adjust the UWL effect to that in the intestine.[92] These conditions reflect the in vivo situation. The in vivo predictability was examined by comparing with P_{eff} in humans (eight marketed drugs) and A% (21 marketed drugs).[10] In addition, the DS-PAMPA permeability was compared with the rat in situ intestinal membrane permeability using a series of fluoroquinolone derivatives (**Figure 16**).[113]

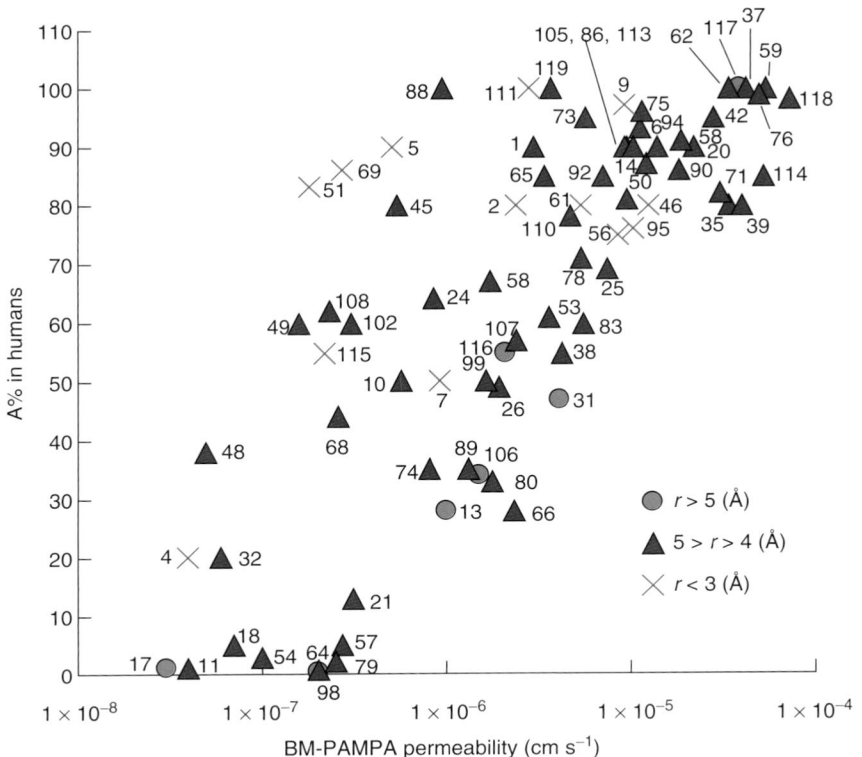

Figure 15 The A% predictability of BM-PAMPA. Compounds with MW 129–654 and −2 to +2 charge are plotted. BM-PAMPA permeability was measured at pH 6.0. See **Table 4** for the assignment of compounds. *r* is the molecular radius. (Reprinted from Sugano, K.; Takata, N.; Machida, M.; Saitoh, K.; Terada, K. *Int. J. Pharm.* **2002**, *241*, 241–251, copyright (2002), with permission from Elsevier.)

5.19.5.1.3.5   Blood–brain barrier parallel artificial membrane permeation assay (BBB-PAMPA)

Di *et al.* modified PAMPA for the prediction of the BBB penetration (BBB-PAMPA).[97] They employed a crude extract from the porcine brain as a lipid source (**Table 2**). BBB-PAMPA was found to predict CNS+/CNS− of 25 marketed drugs (100% correct), as well as their in-house compounds. The palmitoyloleilphosphatidylcholine (POPC)/dodecane membrane was found to have less predictability (88% correct, no false positives, false negatives: alprazolam, clonidine, oxazepam).

Recently, Avdeef *et al.* developed a new PAMPA system to predict the BBB permeability (**Figure 17**).[114] Brain perfusion data were used as probe data, and converted to the intrinsic permeability coefficient (P_0^{BBB}). The apparent PAMPA permeability of probe compounds was also converted to the intrinsic permeability coefficient (P_0^{PAMPA}) (*see* Section 5.19.3.5). A multiple linear regression model to calculate $\log P_0^{BBB}$ from $\log P_0^{PAMPA}$ in combination with other chemical descriptors (in combo PAMPA) was constructed with $\log P_0^{BBB}$ as the dependent variable, and $\log P_0^{PAMPA}$, along with Abraham solute property molecular descriptors: excess molar refraction, solute dipolarity/polarizability, overall hydrogen bond acidity, overall hydrogen bond basicity, and McGowan molecular volume as independent descriptors. P_0^{BBB} was adequately predicted by in combo PAMPA for a 10^7 order range.

5.19.5.1.4   Advantages, limitations, and utility of parallel artificial membrane permeation assay in drug discovery

PAMPA is especially advantageous in early drug discovery.[115] PAMPA is cost-effective, easy to automate, and compatible for a high percentage of solubilizers. In current drug discovery, we often encounter poorly water-soluble compounds.[84] PAMPA has good day-to-day reproducibility and small variability.[106] The cost of PAMPA is more than 10-fold lower than that of Caco-2 assay.[115] The PAMPA membrane can be instantly prepared at the time of use, while Caco-2 requires a 3–21 day culture before use. PAMPA enables studies of passive transcellular permeation without intervention by paracellular and active transports. This simplicity is advantageous for quantitative structure–permeability relationship (QSPR) studies. QSPR by Ano *et al.*[116] and Huque *et al.*[117] indicated that hydrogen bonds play a key role in PAMPA

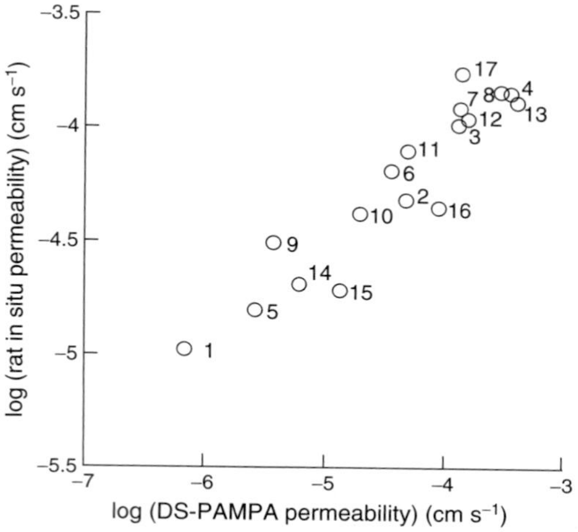

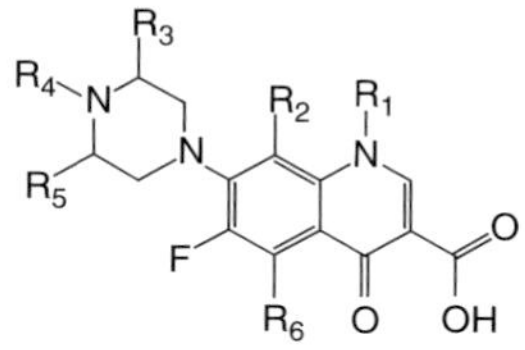

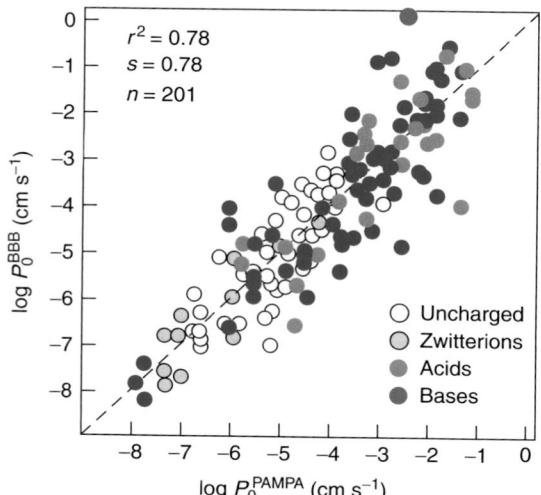

	R_1	R_2	R_3	R_4	R_5	R_6
1 (norfloxacin)	Et	H	H	H	H	H
2	Et	H	H	Me	H	H
3	Et	H	H	nPr	H	H
4	Et	H	H	nBu	H	H
5 (ciprofloxacin)	cPro	H	H	H	H	H
6	cPro	H	H	Me	H	H
7	cPro	H	H	nPr	H	H
8	cPro	H	H	nBu	H	H
9	cPro	H	H	H	Me	H
10	cPro	H	H	Me	Me	H
11	cPro	H	H	Et	Me	H
12	cPro	H	H	nPr	Me	H
13	cPro	H	H	nBu	Me	H
14 (ofloxacin)	$R_1, R_2 = -CH(CH_3)CH_2O-$		H	Me	H	H
15 (sarafloxacin)	4-F-phenyl	H	H	H	H	H
16 (sparfloxacin)	cPro	F	Me	H	Me	NH_2

17 (flumequine)

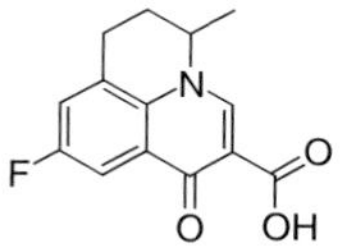

Figure 16 The logarithm of the apparent permeability coefficient for perfusion experiments in rats versus the logarithm of the intrinsic permeability coefficients based on DS-PAMPA. (Reprinted from Bermejo, M.; Avdeef, A.; Ruiz, A.; Nalda, R.; Ruell, J. A.; Tsinman, O.; Gonzalez, I.; Fernandez, C.; Sanchez, G.; Garrigues, T. M. *et al. Eur. J. Pharm. Sci.* **2004**, *21*, 429–441, with permission from Elsevier.)

Figure 17 BBB permeation predictability of PAMPA in combo approach. P_0^{BBB} is the intrinsic BBB permeability coefficient, and P_0^{PAMPA} is the intrinsic PAMPA permeability corrected by the Abraham solute property molecular descriptors. (Reproduced from Avdeef, A. EUFEPS/AAPS Meeting, Nice, France, June 2005, with permission from pION, Inc.)

permeation. Veber *et al.* measured PAMPA permeability of 3061 in-house compounds, and QSPR performed.[118] In addition to the hydrogen bond parameters, an increase of the number of rotatable bond was suggested to decrease permeability. Hwang *et al.* showed that PAMPA data can help permeation assessment of a metabolically unstable compound.[119]

PAMPA has a synergistic effect when combined with other tools. For example, combination with the in silico paracellular model was found to increase in vivo predictability.[64,120] Several reports suggested that the combination of PAMPA and Caco-2 could diagnose the participation of active transporters.[115,116,121] For this use, PAMPA permeability can be corrected by an in silico paracellular pathway model for the Caco-2 tight junction.[62,112] If Caco-2 permeability is different to that estimated by PAMPA permeability, the compound undergoes active transport. This diagnosis can trigger further studies to confirm the permeation mechanism. A combination of PAMPA and a high-throughput solubility assay enables biopharmaceutical classification in early drug discovery (*see* 5.42 The Biopharmaceutics Classification System).[122,123] This application is important for the fine tuning of a biopharmaceutical profile considering the target disease and clinical utility. To overcome the lack of active transporters and metabolisms, high-throughput molecular enzymological assays or in silico prediction for these processes can be used together. These kinds of in combo approaches are synergistically effective in various drug discovery situations.

One limitation of PAMPA is that the membrane is not exactly same as the biological membrane. The PAMPA membrane contains an organic solvent, possibly resulting in a nonbilayer membrane structure. In PC–dodecane membrane PAMPA and BM-PAMPA, parabolic relationships were reported between lipophilicity and PAMPA permeability,[109,124] similarly to the case of cell-based permeation assays.[125–127] In both PAMPA cases, maximum permeability was observed around the octanol/buffer distribution coefficient of 1.5. Since the probe data set for IVIVC is biased to low to moderate lipophilicity drugs, the in vivo relevance of PAMPA for high-lipophilicity compounds might not be assured. In addition, PAMPA has neither active transport systems nor a paracellular pathway. The in vivo prediction accuracy of PAMPA is comparable to or less than Caco-2[110,106,121]; however, it is good for compound ranking. Consequently, PAMPA is used as a complement, rather than an alternative, to Caco-2 assay in early drug discovery.[115,121]

5.19.5.2   Fluorosome

Fluorosome technologies are an application of liposomes.[128] In 1965, liposomes were introduced as a model for the cellular membrane, and have been used to study bilayer membrane permeation.[53,129] Liposomes almost exactly mimic the real biological lipid membrane. However, liposome permeability assays were difficult to implement in drug discovery because it was difficult to measure the concentration of a substrate in the aqueous phase inside the liposome cavity. Fluorosome technology overcomes this drawback by placing fluorescein-labeled proteins, which bind about 95% of all small molecules, in the liposome cavity. The advantage of Fluorosome technology is that it utilizes liposomes, an artificial membrane similar to the real biological membrane. In addition, in contrast to other liposome-based methods, membrane permeation is assessed separately from membrane binding. Fluorosome is commercially available from GLSynthesis, Inc.

5.19.5.2.1   Method

A test compound dissolved in DMSO is added to a Fluorosome suspension in a standard spectrofluorometer and measured to obtain a fluorescence–time curve. This time dependence is directly related to the rate of compound permeation across the bilayer. The data are curve fitted to give a first-order rate constant k, which is used in conjunction with the Fluorosome diameter to calculate permeability.

5.19.5.2.2   In vivo relevance

Fluorosome technology for assessing the membrane permeation has two variations, Fluorosome-trans-pc and Fluorosome-trans-ent. Fluorosome-trans-pc is composed of only phosphatidicholine, while Fluorosome-trans-ent is composed of a consensus mixture of the major lipids present in mammalian intestinal (enterocyte) membranes. The in vivo relevance was examined using 44–47 marketed drug A% values in humans (**Figure 18**). Similarly to PAMPA, a biomimetic lipid composition increased the in vivo predictability.

5.19.6   Partition Assessment Tools for Drug Discovery

In this section, artificial membrane partition assessment tools are reviewed. Traditionally, the artificial membrane partition was measured by an equilibrium dialysis method with liposomes. However, this method is not suitable for drug discovery. Several new technologies have been introduced to measure the artificial membrane partition in drug

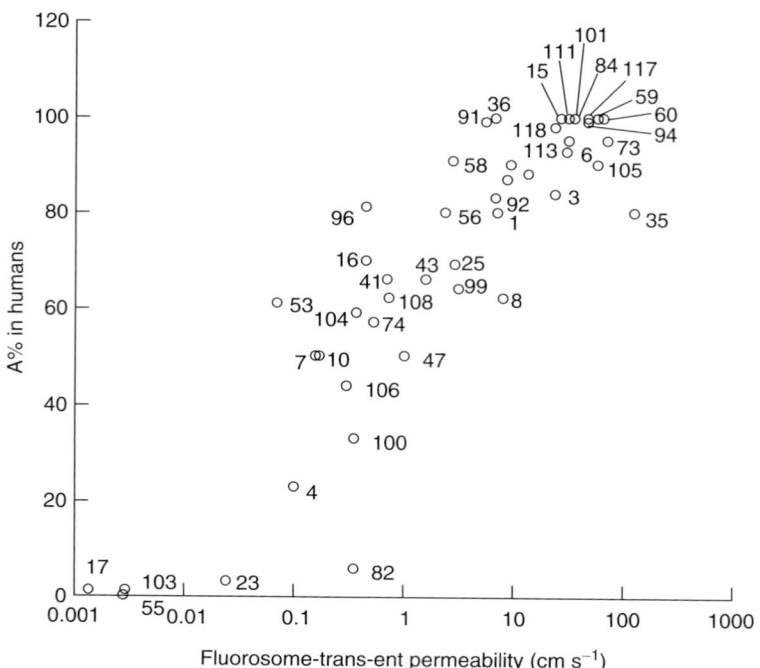

Figure 18   A% predictability of Fluorosome-trans-ent. Permeability was measured at pH 7.4. See **Table 4** for the assignment of compounds. (Reproduced from, GLSynthesis, Worcester, MA, with permission from GLSynthesis, Inc.)

discovery. In this section, immobilized artificial membrane (IAM) column chromatography, immobilized liposome chromatography (ILC), liposome electrokinetic chromatography (LEKC), pH titration, the surface plasmon resonance (SPR)-based method, and the solid-supported lipid membrane (SSLM) techniques are reviewed. Some of these methods are described from the viewpoint of physicochemical characteristics in another article (*see* 5.18 Lipophilicity, Polarity and Hydrophobicity). In this article, these methods are reviewed mainly from the viewpoint of drug discovery utility and relevance to in vivo membrane permeation.

5.19.6.1   Chromatography-Based Methods

When an artificial membrane is employed as the stationary phase of chromatography, retention indices correlate with the artificial membrane partition coefficient. An advantage of chromatography-based methods is that neither high sample purity nor establishment of quantitative analysis is required.

5.19.6.1.1   Immobilized artificial membrane column chromatography

IAM column chromatography was introduced by Pidgeon and co-workers,[130–132] and has been widely investigated for its physicochemical characteristics and its in vivo predictability.[133,134] The IAM column is commercially available from Regis technologies, Inc. Phospholipids are covalently bound to the silica particle at a monolayer level (**Figure 19a**). PC, PE, PG, PS, and phosphatidic acid (PA) versions have been reported.[135] The capacity factor of IAM column chromatography (k_{IAM}) correlated with liposome binding better than the octanol/water partition coefficient did.[132]

The retention time of a substrate on the IAM column can be measured utilizing conventional high-performance liquid chromatography (HPLC) instruments. To cover a wide range of compounds, organic solvents (e.g., methanol and acetonitril) can be added to accelerate elution.[136–138] The capacity factors obtained at a given percentage of organic solvents are then extrapolated to 100% aqueous mobile phase. k_{IAM} has been investigated for correlation with Caco-2 permeability,[131,133,139] and rat intestinal absorption (**Figure 19b**).[131,133,140] Correlation was improved when k_{IAM} was corrected by MW[131] and/or hydrogen bond factors.[139] The IAM column capacity factor has also investigated for BBB permeability prediction.[133,141–143]

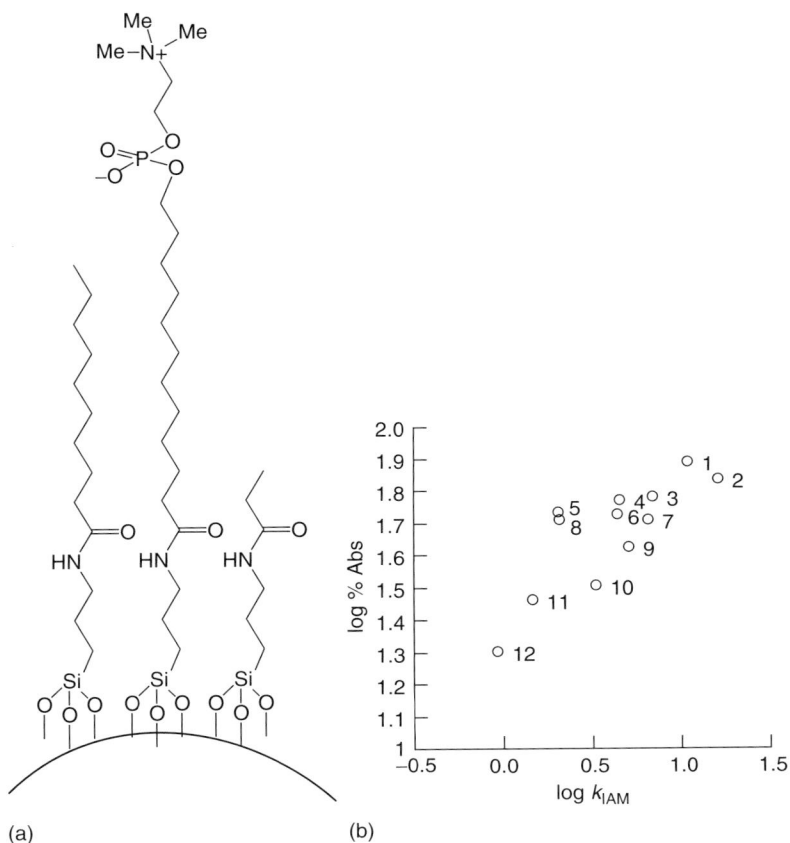

Figure 19 (a) Structure of IAM. PC C10/C3 column. (Reprinted from Ong, S.; Liu, H.; Qiu, X.; Bhat, G.; Pidgeon, C. *Anal. Chem.* **1995**, *67*, 755–762. Copyright (1995) American Chemical Society.) (b) Correlation of k_{IAM} with the rat intestinal drug absorption (Abs%). Retention time of substrate were measured at pH 5.4. 1, *m*-nitroaniline; 2, *p*-nitroaniline; 3, salicylic acid; 4, *p*-toluidine; 5, aniline; 6, *m*-nitrobenzoic acid; 7, phenol; 8, benzoic acid; 9, acetanilide; 10, antipyrine; 11, theophylline; 12, acetylsalicyclic acid. (Reprinted from Pidgeon, C.; Ong, S.; Liu, H.; Qiu, X.; Pidgeon, M.; Dantzig, A. H.; Munroe, J.; Hornback, W. J.; Kasher, J. S.; Glunz, L. *J. Med. Chem.* **1995**, *38*, 590–594. Copyright (1995) American Chemical Society.)

5.19.6.1.2 Immobilized liposome chromatography

ILC is a more biologically relevant in vitro model for the cell membrane than the IAM stationary phase, because liposomes are noncovalently immobilized to gel beads as a stationary phase of chromatography.[144–147] PC, a PC/PS mixture, an egg phospholipid mixture, human red blood cell extract (with or without proteins), and porcine intestinal BBM extract have been reported as lipids for ILC.

Liposomes are remixed with dry gel beads and immobilized by gel bead swelling followed by freezing–thawing to induce liposome fusion. During the freezing–thawing process, the liposomes grow in size and are entrapped in the gel beads pore. The immobilized liposome has a multilamellar structure.[148] The liposome immobilized gel is packed into a column. The retention time of the drugs is measured using an HPLC apparatus, and the capacity factor (K_s) calculated.

The correlation between A% in humans and K_s (PC and human red blood cell extract liposomes, pH 7.4) were examined using more than a dozen marketed drugs (**Figure 20**).[144,146] A bell-shaped correlation curve was observed between A% in humans and K_s. The correlation between P_{eff} in humans and K_s (egg yolk phospholipid and porcine intestinal brush border lipid) was also examined.[147,149]

5.19.6.1.3 Liposome electrokinetic chromatography

LEKC is a capillary electrophoresis method where liposomes are incorporated in the buffer as a pseudo-stationary phase for the separation of charged and uncharged molecules.[148,150] In order to determine retention indices for charged molecules, the migration time was measured with and without the presence of liposomes. Recently, liposomes that reflect the lipid composition of the Caco-2 membrane were examined for usage in LEKC.[151]

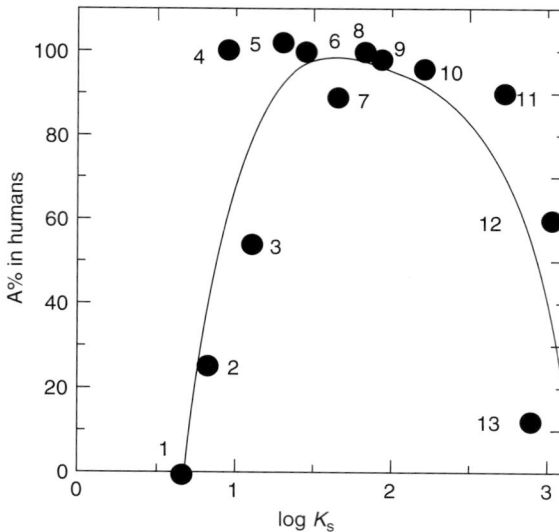

Figure 20 A% in humans versus $\log K_s$ values obtained by ILC on the PC vesicles. The drugs are: 1, polyethylene glycol (average MW 200); 2, mannitol; 3, atenolol; 4, acetylsalicylic acid; 5, metoprolol; 6, salicylic acid; 7, hydrocortisone; 8, corticosterone; 9, warfarin; 10, alprenolol; 11, propranolol; 12, terbutaline; 13, sulphasalazin. A single gel bed was used to determine the $\log K_s$ values. The equation for the curve is $y = -390.38 + 967.75x - 711.67x^2 + 232.93x^3 - 29.106x^4$. (Reprinted from Beigi, F.; Gottschalk, I.; Lagerquist Hagglund, C.; Haneskog, L.; Brekkan, E.; Zhang, Y.; Osterberg, T.; Lundahl, P. *Int. J. Pharm.* **1998**, *164*, 129–137, with permission from Elsevier.)

5.19.6.2   Other Partition Assessment Tools

5.19.6.2.1   Surface plasmon resonance

The phenomenon of SPR is sensitive to changes in refractive index at the sensor surface caused by change in mass. When liposomes are fixed on the sensor surface, the interaction between liposomes and drugs can be measured (**Figure 21a**).[152] Since SPR measures a change in mass on the sensor surface, concentration determination (i.e., UV, HPLC, LC/MS, etc.) is not required. In addition, the SPR instrument is medium-throughput adaptable. One 96-well plate can be analyzed in 24 h, with a total hands-on time of less than 3 h.[153] In addition, sample precipitation can be easily detected as an irregularly shaped sensorgram.[153] These features of this method are advantageous for drug discovery. Another distinctive advantage of the SPR method is that it can measure the kinetics of adsorption and desorption, which indicates differences in binding mode.[154] However, the SPR instrument is expensive.

Liposomes are attached to a sensor chip, the surface of which is coated with alkyl tails. During injection, the SPR signal shows the binding of a compound to the liposomes, and at the end of injection, it shows the release of a compound from the liposome surface. Steady state levels are rapidly achieved during injection.

The presence of 10% DMSO decreased the amount of liposome captured on the sensor surface by 30%, as well as the drug-binding affinity by twofold (desipramine, POPC liposome). The liposome diameter did not affect the binding affinity of each drug (0.05–2 µm, equivalent to a 64 000-fold difference in volume).[154]

In the first report of this method applied to oral absorption prediction, the correlation between A% in humans and the sensor response was examined using POPC and POPC/ganglioside GM1 liposomes. The POPC liposome surface has no net charge, while the POPC/GM1 liposomes have a net negative charge derived from the carboxylic group of GM1. Seventeen marketed drugs were employed as a probe set (**Figure 21b**).[152] The sensor response was measured at pH 7.4 with injection of a 500 µM sample solution containing 0.1% DMSO. POPC/GM1 liposomes were found to have slightly better A% predictability than POPC liposomes.

Frostell-Karlsson *et al.* reported that a two-dimensional plot of membrane binding (PC and PC/PE/PS (2:5:3) liposomes, pH 6.5) could classify A% into three ranges (high (>80%), medium (<80%, >20%), and low (>20%)).[153] The PC/PE/PS (2:5:3) system, which mimics the lipid composition of the intestinal membrane, enabled adequate A% prediction of basic compounds. In addition, the assay condition was adapted to a lower concentration (150 µM), a higher temperature (37 °C) and an increased DMSO concentration (3%). These conditions make this assay better suited for poorly water-soluble compounds.

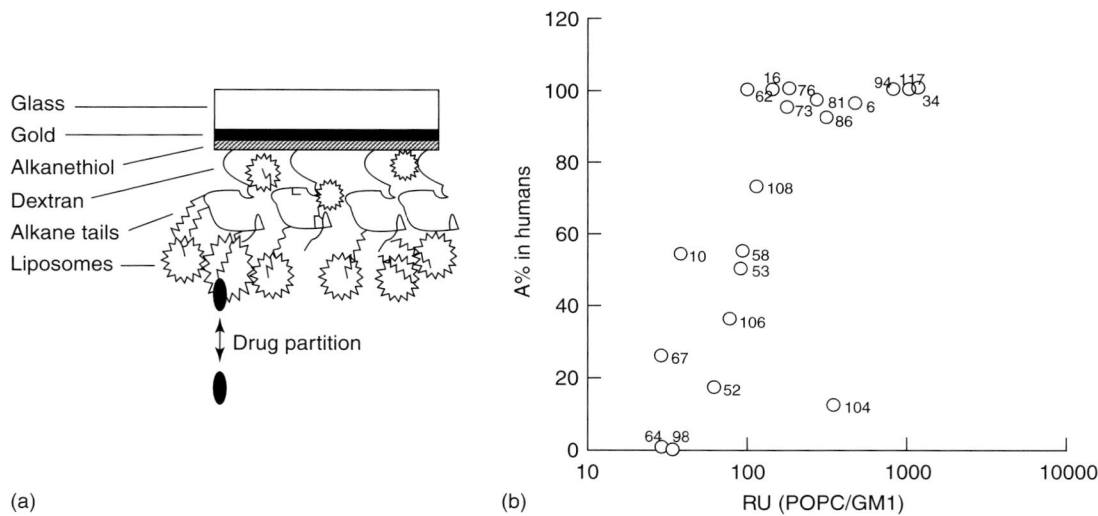

Figure 21 (a) SPR sensor chip and attachment of liposome. Liposomes are captured on alkane tails covalently attached to a dextran matrix. (b) Correlation between the signal from POPC/GM1 (in response units, RUs) and A% in humans. See **Table 4** for the assignment of compounds. (Reprinted from Danelian, E.; Karlen, A.; Karlsson, R.; Winiwarter, S.; Hansson, A.; Lofas, S.; Lennernas, H.; Hamalainen, M. D. *J. Med. Chem.* **2000**, *43*, 2083–2086. Copyright (2000) with permission from the American Chemical Society.)

5.19.6.2.2 pH titration

The liposome partition coefficient of dissociable compounds can be determined by the pH titration method.[155–157] A substrate solution is titrated with and without the coexistence of liposome, and the partition coefficient is calculated from the difference in the titration curves. This method has low to medium throughput, but can provide the pH–partition profile of both nonionized and ionized species.

The correlation between A% in humans and the liposome partition coefficient at pH 6.8 was examined using 21 marketed drugs. The liposomes employed were mainly composed of PC. After converting to the absorption potential in which solubility and the clinical dose were taken into account, a good correlation was observed.

5.19.6.2.3 Solid-supported lipid membrane

In the SSLM technique, a unilamellar liposomal membrane noncovalently surrounds porous silica beads.[158,159] The SSLM permits a high throughput determination of the membrane affinity. To measure SSLM binding, An SSLM is added to the sample solution ($\mu g^{-1} mL^{-1}$ range). A short mixing and incubation step of 2 min allows complete partitioning, and the beads are then separated by vacuum filtration or low-speed centrifugation. Quantitative analysis of the supernatant yields the membrane affinity. The SSLM membrane affinity has been related to A% in humans of nearly 30 compounds. The SSLM is commercially available from Nimbus Biotechnology as TRANSIL.

5.19.7 Surface Activity

Surface activity parameters may be classified as a family of purely physicochemical parameters (i.e., the bulk organic solvent water partition coefficient, hydrogen bond acidity/basicity, etc.) rather than as biomimetic physicochemical parameters. Nevertheless, surface activity parameters are closely related to membrane partition and permeation.

Surface activity parameters have been studied for their relevance to in vivo BBB penetration,[160–162] and K_m of the P-gp substrate.[163,164] Surface activity parameters are a quantitative measure of the tendency to move to the air–water interface. At the air–water interface, molecules are organized to expose the hydrophobic part of a molecule part to the air, while the hydrophilic part maximizes contact with the water. Surface activity parameters can be calculated form the Gibbs adsorption isotherm (**Figure 22**).

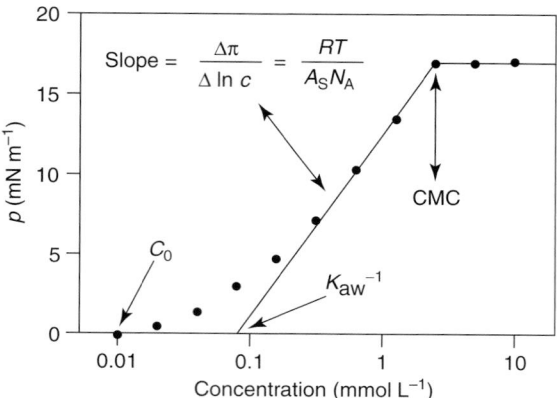

Figure 22 The Gibbs adsorption isotherm. Surface activity parameters are obtained from the adsorption isotherm: concentration (c), surface pressure (π), onset concentration (C_{on}), critical micelle concentration (CMC), air–water partitioning coefficient (K_{aw}), and molecular cross-sectional area (A_S). (Reprinted from Suomalainen, P.; Johans, C.; Soderlund, T.; Kinnunen, P. K. *J. Med. Chem.* **2004**, *47*, 1783–1788. Copyright (2004) American Chemical Society.)

5.19.7.1   Method

The surface tension is measured with a tensiometer (a multi-channel microtensiometer is commercially available). A probe hung down from the microbalance is soaked in the sample buffer solution. The maximum force exerted by the surface tension is recorded as the probe is lifted up from the solution.

The surface pressure (π) is related to the surface tension through

$$\pi = \gamma^0 - \gamma \tag{19}$$

where γ^0 is the surface tension of the bare interface and γ is that in the presence of surface-active substrates. A typical surface pressure–concentration plot (Gibbs adsorption isotherm) is shown in **Figure 22**.[161,162] From the pressure–concentration plot, the critical micelle concentration (CMC), the air/water partition coefficient (K_{aw}), the concentration of surface activity onset (C_{on}), and the molecular cross-sectional area (A_s) can be obtained.

The membrane partition coefficient (K_{memb}) is estimated as[161,162]

$$\log K_{memb} = \frac{RT \ln(K_{aw}) - \pi_{memb} A_s N_A}{2.302 RT} \tag{20}$$

where π_{memb} is the surface pressure of a lipid bilayer, and N_A is Avogadro's number.

5.19.7.2   In Vivo Relevance

5.19.7.2.1   Blood–brain barrier

Seelig *et al.* found that a combination of surface activity parameters could distinguish the CNS+/CNS− of model compounds (**Figure 23**).[160,161] Before the in vivo relevance study, possible P-gp substrates were excluded from the probe data set.[163,165] A three-dimensional plot using K_{aw}^{-1}, CMC and A_S was found to distinguish CNS+/CNS− (**Figure 23**). In addition, compounds with $pK_a < 4$ for acid and $pK_a > 10$ for base were found to be CNS−, even if they were predicted to be CNS+ from the surface activity parameters.

Suomalainen *et al.* performed a similar study, but using only K_{memb} as a parameter.[162] They employed a multi-channel microtensiometer, and demonstrated the adaptability of this method to drug discovery.

5.19.7.2.2   P-Glycoprotein

Seelig *et al.*[163] and Onishi *et al.*[164] suggested that surface activity parameters can predict P-gp activity. The rationale for this relationship is that a substrate accumulated in the membrane is transported by P-gp.[166] The $\log K_{aw}$–$\log 1/K_m$ plot yielded a straight line with the correlation coefficient of 0.95 (**Figure 24**).[163]

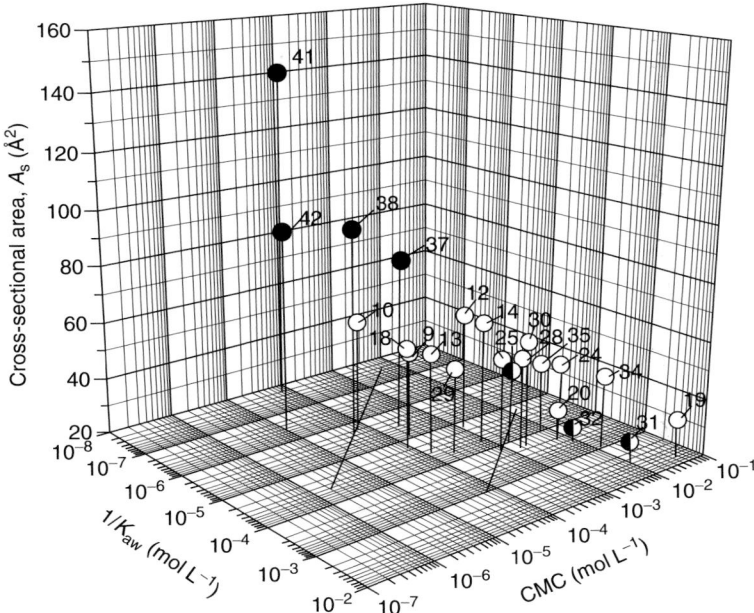

Figure 23 Three-dimensional plot using K_{aw}^{-1}, CMC, and A_s to classify CNS +/−: closed circles, CNS−; open circles, CNS +; half-filled circles, hydrophilic compounds which are pharmacologically applied at low concentration and therefore appear as CNS−. See [161] for the assignment of compounds. (Reproduced from Fischer, Fischer, H.; Gottschlich, R.; Seelig, A. *J. Membr. Biol.* **1998**, *165*, 201–211, with permission from SpringerLink.)

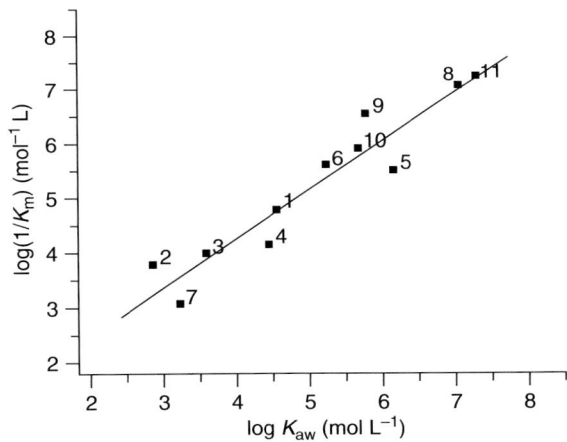

Figure 24 Correlation between the air–water partition coefficient (K_{aw}), determined from measurements of the surface pressure as a function of drug concentration (Gibbs adsorption isotherm) and the inverse of the Michaelis–Menten constant (K_m) of P-gp substrates in a buffer solution (50 mM Tris–HCl, containing 114 mM NaCl) at pH 7.4. Compounds measured: progesterone (1), propranolol (2), amitriptyline (3), diltiazem (4), amiodarone (5), racemic verapamil (6), colchicine (7), gramicidin S, (8), daunorubicin (9), vinblastine (10), and cyclosporin A (11). (Reprinted from Seelig, Seelig, A.; Landwojtowicz, E. *Eur. J. Pharm. Sci.* **2000**, *12*, 31–40, with permission from Elsevier.)

5.19.7.3   Advantages, Limitations, and Utility in Drug Discovery

Concentration determination (i.e., UV, HPLC, LC/MS, etc.) is not required for surface tension determination. A multichannnel microtensiometer compatible with the 96-well plate format is commercially available from Kibron, Inc. This system enables a theoretical throughput of over 1500 compounds per day.[162] These features of the method are advantageous for drug discovery. One limitation of this method might be the lack of surface activity for a subset of compounds.[162] Surface activity parameters have definitive physicochemical meaning related to membrane binding, therefore enabling straightforward interpretation.

5.19.8   Conclusion and Outlook

Since the mid-1990s, with the evolution of the drug discovery paradigm, several high-throughput technologies for passive membrane permeability assessment have been invented, accompanied by scientific progress in the field of the passive membrane permeation. Even though passive membrane permeation has been under investigation for over 300 years, since the time of Isaac Newton, the current state of the science and technology in this area indicates that there are still many questions to answer and technical issues to overcome. We have vast frontiers to explore in this area.

References

1. Eddershaw, P. J.; Beresford, A. P.; Bayliss, M. K. *Drug Disc. Today* **2000**, *5*, 409–414.
2. Kerns, E. H.; Di, L. *Drug Disc. Today* **2003**, *8*, 316–323.
3. Beresford, A. P.; Selick, H. E.; Tarbit, M. H. *Drug Disc. Today* **2002**, *7*, 109–116.
4. Van de Waterbeemd, H.; Gifford, E. *Nat. Rev. Drug Disc.* **2003**, *2*, 192–204.
5. Mizuno, N.; Niwa, T.; Yotsumoto, Y.; Sugiyama, Y. *Pharmacol. Rev.* **2003**, *55*, 425–461.
6. Tien, T. H.; Ottova, A. L. *J. Membr. Sci.* **2001**, *189*, 83–117.
7. Van de Waterbeemd, H.; Smith, D. A.; Beaumont, K.; Walker, D. K. *J. Med. Chem.* **2001**, *44*, 1313–1333.
8. Gennis R. B. *Biomembrane*; Springer-Verlag: Tokyo, Japan, 1990; Chapter 2, pp 32–78.
9. Roy, M. T.; Gallardo, M.; Estelrich, J. *J. Colloid Interface Sci.* **1998**, *206*, 512–517.
10. Avdeef, A. *Absorption and Drug Development: Solubility, Permeability and Charge State*; Wiley-Interscience: Englewood Cliffs, NJ, 2003; Chapter 7, pp 116–246.
11. De Jager, M. W.; Gooris, G. S.; Dolbnya, I. P.; Ponec, M.; Bouwstra, J. A. *Biochim. Biophys. Acta* **2004**, *1664*, 132–140.
12. Marrink, S. J.; Berendsen, H. J. C. *J. Phys. Chem.* **1994**, *98*, 4155–4168.
13. Marrink, S. J.; Berendsen, H. J. C. *J. Phys. Chem.* **1996**, *100*, 16729–16738.
14. Avdeef, A. *Curr. Top. Med. Chem.* **2001**, *1*, 277–351.
15. Leib, W. R.; Stein, W. D. *Nature* **1969**, *224*, 240–243.
16. Balimane, P. V.; Chong, S. *Drug Disc. Today* **2005**, *10*, 335–343.
17. Lennernas, H. *J. Pharm. Sci.* **1998**, *87*, 403–410.
18. Winne, D. *J. Math. Biol.* **1978**, *6*, 95–108.
19. Anderson, J. M. *News Physiol. Sci.* **2001**, *16*, 126–130.
20. Seorgel, K. H. *Gastroenterology* **1993**, *105*, 1247–1250.
21. He, Y.; Murby, S.; Warhurst, G.; Gifford, L.; Walker, D.; Ayrton, J.; Eastmond, R.; Rowland, M. *J. Pharm. Sci.* **1998**, *87*, 626–633.
22. Artursson, P.; Ungell, A. L.; Löfroth, J. E. *Pharm. Res.* **1993**, *10*, 1123–1129.
23. Said, H. M.; Blair, J. A.; Lucas, M. L.; Hilburn, M. E. *J. Lab. Clin. Med.* **1986**, *107*, 420–424.
24. Lucas, M. *Gut* **1983**, *24*, 734–739.
25. Maxwell, J. D.; Watson, W. C.; Watt, J. K.; Ferguson, A. *Gut* **1968**, *9*, 612–616.
26. Dressman, J. B.; Amidon, G. L.; Reppas, C.; Shah, V. P. *Pharm. Res.* **1998**, *15*, 11–22.
27. Proulx, P. *Biochim. Biophys. Acta* **1991**, *1071*, 255–271.
28. Lipka, G.; Op den Kamp, J. A.; Hauser, H. *Biochemistry* **1991**, *30*, 11828–11836.
29. Verkleij, A. J.; Zwaal, R. F.; Roelofsen, B.; Comfurius, P.; Kastelijn, D.; van Deenen, L. L. *Biochim. Biophys. Acta* **1973**, *323*, 178–193.
30. Kramer, S. D.; Hurley, J. A.; Abbott, N. J.; Begley, D. J. *In Vitro Cell Dev. Biol. Anim.* **2002**, *38*, 557–565.
31. Dias, V. C.; Wallace, J. L.; Parsons, H. G. *Gut* **1992**, *33*, 622–627.
32. Kramer, S. D.; Abbott, N. J.; Begley, D. J. A Biological Models to Study Blood Brain Barrier Permeation. In *Pharmacokinetic Optimization in Drug Research: Biological, Physicochemical and Computational Strategies*; Testa, B., van de Waterbeemd, H., Folkers, G., Guy, R., Eds.; Verlag Helvetica Chimica Acta: Zürich, Switzerland, 2001, pp 127–153.
33. Avanti Polar Lipid, Alabaster, AL http://www.avantilipids.com (accessed May 2006).
34. Yamashita, S.; Furubayashi, T.; Kataoka, M.; Sakane, T.; Sezaki, H.; Tokuda, H. *Eur. J. Pharm. Sci.* **2000**, *10*, 195–204.
35. Singh, B. N. *Clin. Pharmacokinet.* **1999**, *37*, 213–255.
36. Sawada, G. A.; Ho, N. F.; Williams, L. R.; Barsuhn, C. L.; Raub, T. J. *Pharm. Res.* **1994**, *11*, 665–673.
37. Abbott, N. J. *Drug Disc. Today* **2004**, *1*, 407–416.
38. Bouwstra, J. A.; Honeywell-Nguyen, P. L.; Gooris, G. S.; Ponec, M. *Prog. Lipid Res.* **2003**, *42*, 1–36.
39. Hadgraft, J. *Eur. J. Pharm. Biopharm.* **2004**, *58*, 291–299.
40. Talreja, P.; Kleene, N. K.; Pickens, W. L.; Wang, T. F.; Kasting, G. B. *AAPS PharmSci.* **2001**, *3*, E13.
41. Diamond, J. M.; Katz, Y. *J. Membr. Biol.* **1974**, *17*, 121–154.
42. Xiang, T. X.; Anderson, B. D. *Biophys. J.* **1997**, *72*, 223–237.
43. Xiang, T.; Xu, Y.; Anderson, B. D. *J. Membr. Biol.* **1998**, *165*, 77–90.
44. Walter, A.; Gutknecht, J. *J. Membr. Biol.* **1986**, *90*, 207–217.
45. Bemporad, D.; Essex, J. W.; Luttmann, C. *J. Phys. Chem.* **2004**, *108*, 4875–4884.
46. Bemporad, D.; Luttmann, C.; Essex, J. W. *Biophys. J.* **2004**, *87*, 1–13.
47. Burton, P. S.; Conradi, R. A.; Hilgers, A. R. *Adv. Drug Deliv. Rev.* **1991**, *7*, 365–386.

48. Parterson, D. A.; Conradi, R. A.; Hilgers, A. R.; Vidmar, T. J.; Burton, P. S. *Quant. Struct.–Act. Relat.* **1994**, *13*, 4–10.
49. Eylan, G. D.; Kuchel, P. W. *Int. Rev. Cyt.* **1999**, *190*, 175–250.
50. Kleinfeld, A. M.; Chu, P.; Storch, J. *Biochemistry* **1997**, *36*, 5702–5711.
51. Kamp, F.; Zakim, D.; Zhang, F.; Noy, N.; Hamilton, J. A. *Biochemistry* **1995**, *34*, 11928–11937.
52. Walter, A.; Gutknecht, J. *J. Membr. Biol.* **1984**, 77, 255–264.
53. Mayer, P. T.; Xiang, T. X.; Anderson, B. D. *AAPS PharmSci.* **2000**, *2*, E14.
54. Mayer, P. T.; Anderson, B. D. *J. Pharm. Sci.* **2002**, *91*, 640–646.
55. Malkia, A.; Murtomaki, L.; Urtti, A.; Kontturi, K. *Eur. J. Pharm. Sci.* **2004**, *23*, 13–47.
56. Potts, R. O.; Guy, R. H. *Pharm. Res.* **1992**, *9*, 663–669.
57. Pugh, W. J.; Roberts, M. S.; Hadgraft, J. *Int. J. Pharm.* **1996**, *138*, 149–165.
58. Pugh, W. J.; Degim, I. T.; Hadgraft, J. *Int. J. Pharm.* **2000**, *197*, 203–211.
59. Jacobs, R. E.; White, S. H. *Biochemistry* **1989**, *28*, 3421–3437.
60. Burton, P. S.; Conradi, R. A.; Hilgers, A. R.; Ho, N. F. H.; Maggiora, L. L. *J. Control. Release* **1992**, *19*, 87–97.
61. Stenberg, P.; Luthman, K.; Artursson, P. *Pharm. Res.* **1999**, *16*, 205–212.
62. Adson, A.; Ruab, T. J.; Burton, P. S.; Barsuhn, C. L.; Hilgers, A. R.; Audus, K. L.; Ho, N. F. H. *J. Pharm. Sci.* **1994**, *83*, 1529–1530.
63. Adson, A.; Burton, P. S.; Ruab, T. J.; Barsuhn, C. L.; Audus, L.; Ho, N. F. H. *Pharm. Res.* **1995**, *84*, 1197–1203.
64. Sugano, K.; Takata, N.; Machida, M.; Saitoh, K.; Terada, K. *Int. J. Pharm.* **2002**, *241*, 241–251.
65. Kristl, A.; Tukker, J. *J. Pharm. Res.* **1998**, *15*, 499–501.
66. Sugano, K.; Yoshida, S.; Takaku, M.; Haramura, M.; Saitoh, R.; Nabuchi, Y.; Ushio, H. *Bioorg. Med. Chem. Lett.* **2000**, *10*, 1939–1942.
67. Nagahara, N.; Tavelin, S.; Artursson, P. *J. Pharm. Sci.* **2004**, *93*, 2972–2984.
68. Larhed, A. W.; Artursson, P.; Grasjo, J.; Bjork, E. *J. Pharm. Sci.* **1997**, *86*, 660–665.
69. Amidon, G. L.; Sinko, P. J.; Fleisher, D. *Pharm. Res.* **1988**, *5*, 651–654.
70. Yu, L. X. *Pharm. Res.* **1999**, *16*, 1883–1887.
71. Yu, L. X.; Lipka, E.; Crison, J. R.; Amidon, G. L. *Adv. Drug Deliv. Rev.* **1996**, *19*, 359–376.
72. Parrott, N.; Lave, T. *Eur. J. Pharm. Sci.* **2002**, *17*, 51–61.
73. Poulin, P.; Theil, F.-P. *J. Pharm. Sci.* **2002**, *91*, 1358–1370.
74. Dickins, M.; van de Waterbeemd, H. *Drug Disc. Today* **2004**, *2*, 38–45.
75. Hogben, C. A. M.; Tacco, D. J.; Brodie, B. B.; Schanker, L. S. *J. Pharmacol. Exp. Ther.* **1959**, *269*, 275–282.
76. Walter, A.; Gutknecht, J. *J. Membr. Biol.* **1984**, 77, 255–264.
77. Ruell, J. A.; Tsinman, K. L.; Avdeef, A. *Eur. J. Pharm. Sci.* **2003**, *20*, 393–402.
78. Hogerle, M. L.; Winne, D. *Naunyn-Schmiedeberg Arch. Pharmacol.* **1983**, *322*, 249–255.
79. Avdeef, A.; Artursson, P.; Neuhoff, S.; Lazorova, L.; Grasjo, J.; Tavelin, S. *Eur. J. Pharm. Sci.* **2005**, *24*, 333–349.
80. Neubert, R. *Pharm. Res.* **1989**, *6*, 743–747.
81. Hadgraft, J.; Valenta, C. *Int. J. Pharm.* **2000**, *200*, 243–247.
82. Sugano, K.; Nabuchi, Y.; Machida, M.; Asoh, Y. *Int. J. Pharm.* **2004**, *275*, 271–278.
83. Lipinski, C. A.; Lombardo, F.; Dominy, B. W.; Feeney, P. J. *Adv. Drug Deliv. Rev.* **1997**, *23*, 3–25.
84. Lipinski, C. A. *J. Pharmacol. Toxicol. Methods* **2000**, *44*, 235–249.
85. Zhao, Y. H.; Le, J.; Abraham, M. H.; Hersey, A.; Eddershaw, P. J.; Luscombe, C. N.; Boutina, D.; Beck, G.; Sherborne, B.; Cooper, I. et al. *J. Pharm. Sci.* **2001**, *90*, 749–784.
86. Lin, J. H.; Yamazaki, M. *Clin. Pharmacokinet.* **2003**, *42*, 59–98.
87. Chiou, W. L.; Chung, S. M.; Wu, T. C.; Ma, C. *Int. J. Clin. Pharmacol. Ther.* **2001**, *39*, 93–101.
88. Abraham, M. H. *Eur. J. Med. Chem.* **2004**, *39*, 235–240.
89. Crivori, P.; Cruciani, G.; Carrupt, P. A.; Testa, B. *J. Med. Chem.* **2000**, *43*, 2204–2216.
90. Karlsson, J. P.; Artursson, P. *Int. J. Pharm.* **1991**, 7, 55–64.
91. Youdim, K. A.; Avdeef, A.; Abbott, N. J. *Drug Disc. Today* **2003**, *8*, 997–1003.
92. Avdeef, A.; Nielsen, P. E.; Tsinman, O. *Eur. J. Pharm. Sci.* **2004**, *22*, 365–374.
93. Sugano, K. *PharmStage* **2002**, *2*, 72–78.
94. Kansy, M.; Senner, F.; Gubernator, K. *J. Med. Chem.* **1998**, *41*, 1007–1010.
95. Wohnsland, F.; Faller, B. *J. Med. Chem.* **2001**, *44*, 923–930.
96. Sugano, K.; Hamada, H.; Machida, M.; Ushio, H. *J. Biomol. Screen.* **2001**, *6*, 189–196.
97. Di, L.; Kerns, E. H.; Fan, K.; McConnell, O. J.; Carter, G. T. *Eur. J. Med. Chem.* **2003**, *38*, 223–232.
98. Mueller, P.; Rudin, D. O.; Tien, H. T.; Wescott, W. C. *Nature* **1962**, *194*, 979–980.
99. Thompson, M.; Krull, U. J.; Worsfold, P. J. *Anal. Chim. Acta.* **1980**, *117*, 133–145.
100. Thompson, M.; Lennox, R. B.; McClelland, R. A. *Anal. Chem.* **1982**, *54*, 76–81.
101. Camenisch, G.; Folkers, G.; van de Waterbeemd, H. *Int. J. Pharm.* **1997**, *147*, 61–70.
102. Avdeef, A.; Strafford, M.; Block, E.; Balogh, M. P.; Chambliss, W.; Khan, I. *Eur. J. Pharm. Sci.* **2001**, *14*, 271–280.
103. Sugano, K. *J. Pharm. Sci. Tech. Jpn.* **2003**, *63*, 225.
104. Kansy, M.; Avdeef, A. *Drug Disc. Today* **2004**, *1*, 349–355.
105. Sugano, K. *Abstracts.* PAMPA 2002, San Francisco, 2002, pp 4–16.
106. Zhu, C.; Jiang, L.; Chen, T. M.; Hwang, K. K. *Eur. J. Med. Chem.* **2002**, *37*, 399–407.
107. Sugano, K.; Hamada, H.; Machida, M.; Ushio, H.; Saitoh, K.; Terada, K. *Int. J. Pharm.* **2001**, *228*, 181–188.
108. Ruell, J. A.; Tsinman, O.; Avdeef, A. *Chem. Pharm. Bull.* **2004**, *52*, 561–565.
109. Kansy, M.; Fischer, H.; Kratzat, K.; Senner, F.; Wanger, B.; Parrilla, I. High-Throughput Artificial Membrane Permeability Studies in Early Lead Discovery and Development. In *Pharmacokinetic Optimization in Drug Research: Biological, Physicochemical and Computational Strategies*; Testa, B., van de Waterbeemd, H., Folkers, G., Guy, R., Eds., Verlag Helvetica Chimica Acta: Zürich, Switzerland, 2001, pp 448–464.
110. Liu, H.; Sabus, C.; Carter, G. T.; Du, C.; Avdeef, A.; Tischler, M. *Pharm. Res.* **2003**, *20*, 1820–1826.
111. Faller, B.; Grimm, H. P.; Loeuillet-Ritzler, F.; Arnold, S.; Briand, X. *J. Med. Chem.* **2005**, *48*, 2571–2576.
112. Saitoh, R.; Sugano, K.; Takata, N.; Tachibana, T.; Higashida, A.; Nabuchi, Y.; Aso, Y. *Pharm. Res.* **2004**, *21*, 749–755.
113. Bermejo, M.; Avdeef, A.; Ruiz, A.; Nalda, R.; Ruell, J. A.; Tsinman, O.; Gonzalez, I.; Fernandez, C.; Sanchez, G.; Garrigues, T. M. et al. *Eur. J. Pharm. Sci.* **2004**, *21*, 429–441.

114. Avdeef, A. EUFEPS/AAPS Meeting, Nice, France, June 2005.
115. Kerns, E. H.; Di, L.; Petusky, S.; Farris, M.; Ley, R.; Jupp, P. *J. Pharm. Sci.* **2004**, *93*, 1440–1453.
116. Ano, R.; Kimura, Y.; Shima, M.; Matsuno, R.; Ueno, T.; Akamatsu, M. *Bioorg. Med. Chem.* **2004**, *12*, 257–264.
117. Huque, F. T.; Box, K.; Platts, J. A.; Comer, J. *Eur. J. Pharm. Sci.* **2004**, *23*, 223–232.
118. Veber, D. F.; Johnson, S. R.; Cheng, H.-Y.; Smith, B. R.; Ward, K. W.; Kopple, K. D. *J. Med. Chem.* **2002**, *45*, 2615–2623.
119. Hwang, K. K.; Martin, N. E.; Jiang, L.; Zhu, C. *J. Pharm. Pharm. Sci.* **2003**, *6*, 315–320.
120. Matsson, P.; Bergström, C. A. S.; Nagahara, N.; Tavelin, S.; Norinder, U.; Artursson, P. *J. Med. Chem.* **2005**, *48*, 604–613.
121. Miret, S.; Abrahamse, L.; de Groene, E. M. *J. Biomol. Screen.* **2004**, *9*, 598–606.
122. Obata, K.; Sugano, K.; Machida, M.; Aso, Y. *Drug Dev. Ind. Pharm.* **2004**, *30*, 181–185.
123. Amidon, G. L.; Lennernas, H.; Shah, V. P.; Crison, J. R. *Pharm. Res.* **1995**, *12*, 413–420.
124. Obata, K.; Sugano, K.; Saitoh, R.; Higashida, A.; Nabuchi, Y.; Machida, M.; Aso, Y. *Annual Meeting of the Pharmaceutical Society of Japan*; Poster: Osaka, 2005.
125. Wils, P.; Warnery, A.; Phung-Ba, V.; Legrain, S.; Scherman, D. *J. Pharmacol. Exp. Ther.* **1994**, *269*, 654–658.
126. Sawada, G. A.; Barsuhn, C. L.; Lutzke, B. S.; Houghton, M. E.; Padbury, G. E.; Ho, N. F.; Raub, T. J. *J. Pharmacol. Exp. Ther.* **1999**, *288*, 1317–1326.
127. Krishna, G.; Chen, K.; Lin, C.; Nomeir, A. A. *Int. J. Pharm.* **2001**, *222*, 77–89.
128. GLSynthesis, Worcester, MA. http://www.glsynthesis.com/fluoro1.html (accessed May 2006).
129. Bangham, A. D.; Horne, R. W. *J. Mol. Biol.* **1964**, *8*, 660–668.
130. Pidgeon, C.; Venkataram, U. V. *Anal. Biochem.* **1989**, *176*, 36–47.
131. Pidgeon, C.; Ong, S.; Liu, H.; Qiu, X.; Pidgeon, M.; Dantzig, A. H.; Munroe, J.; Hornback, W. J.; Kasher, J. S.; Glunz, L. *J. Med. Chem.* **1995**, *38*, 590–594.
132. Ong, S.; Liu, H.; Qiu, X.; Bhat, G.; Pidgeon, C. *Anal. Chem.* **1995**, *67*, 755–762.
133. Stewart, B. H.; Chan, O. H. *J. Pharm. Sci.* **1998**, *87*, 1471–1478.
134. Taillardat-Bertschinger, A.; Carrupt, P. A.; Barbato, F.; Testa, B. *J. Med. Chem.* **2003**, *46*, 655–665.
135. Yang, C. Y.; Cai, S. J.; Liu, H.; Pidgeon, C. *Adv. Drug Del. Rev.* **1997**, *23*, 229–256.
136. Lepont, C.; Poole, C. F. *J. Chromatogr. A* **2002**, *946*, 107–124.
137. Barbato, F.; Rotonda, M. L.; Quaglia, F. *Eur. J. Med. Chem.* **1996**, *21*, 311–318.
138. Taillardat-Ertschinger, A.; Galland, A.; Carrupt, P. A.; Testa, B. *J. Chromatogr. A* **2002**, *953*, 39–53.
139. Stewart, B. H.; Chung, F. Y.; Tait, B.; Blankley, C. J.; Chan, O. H. *Pharm. Res.* **1998**, *15*, 1401–1406.
140. Genty, M.; Gonzalez, G.; Clere, C.; Desangle-Gouty, V.; Legendre, J. Y. *Eur. J. Pharm. Sci.* **2001**, *12*, 223–229.
141. Salminen, T.; Pulli, A.; Taskinen, J. *J. Pharm. Biomed. Anal.* **1997**, *15*, 469–477.
142. Reichel, A.; Begley, D. J. *Pharm. Res.* **1998**, *15*, 1270–1274.
143. Pehourcq, F.; Matoga, M.; Bannwarth, B. *Fundam. Clin. Pharmacol.* **2004**, *18*, 65–70.
144. Beigi, F.; Yang, Q.; Lundahl, P. *J. Chromatogr. A* **1995**, *704*, 315–321.
145. Lundahl, P.; Beigi, F. *Adv. Drug Deliv. Rev.* **1997**, *23*, 221–227.
146. Beigi, F.; Gottschalk, I.; Lagerquist Hagglund, C.; Haneskog, L.; Brekkan, E.; Zhang, Y.; Osterberg, T.; Lundahl, P. *Int. J. Pharm.* **1998**, *164*, 129–137.
147. Engvall, C.; Lundahl, P. *J. Chromatogr. A* **2004**, *1031*, 107–112.
148. Wiedmer, S. K.; Riekkola, M.-L.; Jussila, M. S. *Trends Anal. Chem.* **2004**, *23*, 562–582.
149. Lagerquist, C.; Beigi, F.; Karlen, A.; Lennernas, H.; Lundahl, P. *J. Pharm. Pharmacol.* **2001**, *53*, 1477–1487.
150. Zhang, Y.; Zhang, R.; Hjerten, S.; Lundahl, P. *Electrophoresis* **1995**, *16*, 1519–1523.
151. Carrozzino, J. M.; Khaledi, M. G. *Pharm. Res.* **2004**, *21*, 2327–2335.
152. Danelian, E.; Karlen, A.; Karlsson, R.; Winiwarter, S.; Hansson, A.; Lofas, S.; Lennernas, H.; Hamalainen, M. D. *J. Med. Chem.* **2000**, *43*, 2083–2086.
153. Frostell-Karlsson, A.; Widegren, H.; Green, C. E.; Hamalainen, M. D.; Westerlund, L.; Karlsson, R.; Fenner, K.; van de Waterbeemd, H. *J. Pharm. Sci.* **2005**, *94*, 25–37.
154. Abdiche, Y. N.; Myszka, D. G. *Anal. Biochem.* **2004**, *328*, 233–243.
155. Balon, K.; Riebesehl, B. U.; Muller, B. W. *J. Pharm. Sci.* **1999**, *88*, 802–806.
156. Balon, K.; Riebesehl, B. U.; Muller, B. W. *Pharm. Res.* **1999**, *16*, 882–888.
157. Avdeef, A.; Box, K. J.; Comer, J. E.; Hibbert, C.; Tam, K. Y. *Pharm. Res.* **1998**, *15*, 209–215.
158. Loidl–Stahlhofen, A.; Eckert, A.; Hartmann, T.; Schottner, M. *J. Pharm. Sci.* **2001**, *90*, 599–606.
159. Loidl–Stahlhofen, A.; Hartmann, T.; Schottner, M.; Rohring, C.; Brodowsky, H.; Schmitt, J.; Keldenich, J. *Pharm. Res.* **2001**, *18*, 1782–1788.
160. Seelig, A.; Gottschlich, R.; Devant, R. M. *Proc. Natl. Acad. Sci. USA* **1994**, *91*, 68–72.
161. Fischer, H.; Gottschlich, R.; Seelig, A. *J. Membr. Biol.* **1998**, *165*, 201–211.
162. Suomalainen, P.; Johans, C.; Soderlund, T.; Kinnunen, P. K. *J. Med. Chem.* **2004**, *47*, 1783–1788.
163. Seelig, A.; Landwojtowicz, E. *Eur. J. Pharm. Sci.* **2000**, *12*, 31–40.
164. Onishi, Y.; Hirano, H.; Nakata, K.; Oosumi, K.; Nagakura, M.; Tarui, S.; Ishikawa, T. *Chem-Bio Inf. J.* **2003**, *4*, 175–193.
165. Seelig, A.; Blatter, X. L.; Wohnsland, F. *Int. J. Clin. Pharmacol. Ther.* **2000**, *38*, 111–121.
166. Sharom, F. J.; Liu, R.; Romsicki, Y.; Lu, P. *Biochim. Biophys. Acta* **1999**, *1461*, 327–345.

Biography

Kiyohiko Sugano graduated from Waseda University (Department of Chemistry) in 1995, studying under Prof Masaru Tada (Organic Chemistry), and joined Chugai Pharmaceutical Co., Ltd. In 2002, he received his PhD from Toho University (Department of Pharmaceutics) under the mentorship of Prof Katsuhide Terada. In 2004, he joined Pfizer Inc. His main interests as a pharmaceutical industrial researcher are physicochemical properties, especially in relation to pharmacokinetics, preformulation studies, and structure–pharmacokinetics relationships.

Comprehensive Medicinal Chemistry II
ISBN (set): 0-08-044513-6

ISBN (Volume 5) 0-08-044518-7; pp. 453–487

5.20 Chemical Stability

E H Kerns and L Di, Wyeth Research, Princeton, NJ, USA

5.20.1 Introduction: Needs and Scenarios

Chemical stability is not usually one of the highest priority issues on the minds of medicinal chemists during the effort to discover new drugs. Issues such as novelty, potency, and selectivity typically demand greater attention. Nevertheless, in recent years pharmaceutical research organizations have placed increasing emphasis on the study of compound properties. Physicochemical, metabolic, and pharmacokinetic properties have emerged as key determinants of success in drug discovery and development.[1–5] If the liabilities of a compound series can be identified and corrected during the lead optimization phase, then the development candidate is likely to be more successful. Chemical instability can result in significant liabilities for development candidates and discovery leads.

In development and clinical application, instability adds time and expense to the development of stable dosage forms. Decomposition causes problems with chemical processes and clinical batch release. It reduces the amount of drug delivered to the therapeutic target in living systems. Degradants may cause side effects and toxicity in patients.

In discovery, drugs are exposed to various conditions in the medicinal chemistry laboratory that can facilitate decomposition reactions. These include storage conditions in the solid or solution state, which expose the compound to oxygen and water from the air, light, elevated temperature, or various solution pHs. Basicity from glassware can catalyze reactions. Purification protocols may produce decomposition products, such as when high-performance liquid chromatography (HPLC) mobile phases that contain acid modifiers are evaporated. When compounds are placed in libraries for high-throughput screening (HTS), they are often plated in dimethyl sulfoxide (DMSO) solution and stored; exposure to oxygen or water from the air can induce decomposition and produce incorrect structure–activity relationship (SAR) during screening.

Instability can also occur in biological or property assay laboratories. This will misdirect the research efforts through inaccurate biological or property data, which leads to erroneous SAR for the chemical series. Compounds may be chemically unstable in the assay media by reacting with media components, or the pH of the media may catalyze hydrolysis. Reduction of the concentration of the putative compound will reduce the apparent activity. Decomposition products may have enhanced or reduced activity in the target activity assay or selectivity assays, compared to the putative compound. Series analogs can have different rates of decomposition due to their structural differences, which might be mistaken as real SAR. They can also produce false results from property assays. Compounds may not be active when dosed orally, owing to instability in the pHs or enzymes of the gastrointestinal system.

Several key chemical mechanisms lead to compound decomposition. These include hydrolysis, hydration, oxidation, and isomerization. As background, common destabilizing reactions are examined in this chapter. The potential reactions of a chemical series may be immediately obvious to the medicinal chemist, or they may be temporarily overlooked as higher priority issues take precedence in research.

To assist in screening for potential chemical instability mechanisms, several tools and methods have been developed and reported in the literature. These give the medicinal chemist tools to detect and study compound stability. At the simplest level, it is useful to be aware of common structural features that are susceptible to chemical instability. High to moderate throughput methods have been described for profiling a large number of compounds in a short period of time. These are consistent with the demands of modern drug discovery strategies. Customized methods may be designed to diagnose specific mechanisms for a compound series. Also, in-depth methods, which are adapted from development laboratories, are used to study the mechanisms and kinetics of reactions for a few selected late-stage compounds.

Having these tools in place allows medicinal chemists to efficiently and effectively deal with stability issues. A consistent pattern of modern drug research is to move the investigation of issues, which were traditionally studied during development (e.g., toxicity, metabolism, pharmacokinetics, stability), forward into the discovery period. This approach allows advancement of the highest quality candidates that have the lowest risk. Thus, modern discovery organizations have increasingly made the tools for these investigations available to medicinal chemists. With the many possibilities for chemical instability in drug discovery and development, vigilance and testing are reasonable considerations for medicinal chemists.

5.20.2   Chemical Aspects

5.20.2.1   Decreasing Chemical Instability

Care can be exercised in the chemical laboratory when handling compounds, in order to reduce chemical decomposition. Compounds can react with oxygen or water from the air. Solids are penetrated by gaseous molecules, especially at or near the surface of the particle and if the material is amorphous. Oxygen and water from the air dissolve in solutions and can react. To reduce these reactions, compounds can be stored under inert gas in sealed containers. Storage at reduced temperature decreases the rate of decomposition reactions. Reactions can also be induced by light in the laboratory. Therefore, they should be stored away from the light or in amber containers. Counter ions can react with the compounds, especially if they are strong acids or bases. Storage at reduced temperature may slow such reactions. Lipinski[6] has discussed the dissolution of water in DMSO solutions, which is also enhanced by freeze–thaw cycles that are common in biological testing laboratories. HTS scientists have, in recent years, taken steps to store DMSO testing plates for long periods of time at low temperature and under inert dry gas. However, when plates are transferred to the biological laboratory, they can be repeatedly taken in and out of the refrigerator, creating opportunities for dissolution of water from the air. Lipinski advises storing these plates at room temperature, if they are to be used within a few days. When compounds are purified by preparative HPLC, mobile phase components can react with compounds, especially if the solutions are concentrated on a rotary evaporator.

Compound testing for biological, physicochemical, metabolic, or pharmacokinetic properties always involves dissolution in aqueous solution. Care should be taken to insure that the compound is not unstable in the buffer. A prime consideration is the buffer pH. Decomposition can also occur from reaction with assay buffer components in assay media used for in vitro enzyme, receptor and cell-based assays or for nuclear magnetic resonance (NMR) binding studies.

Decomposition reactions of functional groups commonly found in drugs are treated in detail in several texts and reviews.[7–14]

5.20.2.2   Decomposition Reactions

Chemical decomposition reaction mechanisms, kinetics, and their minimization are a primary focus of development pharmaceutical scientists.[8–13] However, insights from these extensive studies are important for medicinal chemists to keep in mind during the discovery phase.

5.20.2.2.1 Hydrolysis

Hydrolysis has been observed for many drugs. This can occur for compounds that contain the functional groups: ester, amide, thioester, imide, imine, carbamate, acetal, alkyl chloride, nitrate, lactam, lactone, and sulfonamide.[7–9] Such reactions may be catalyzed by acidic solutions, basic solutions, or enzymes. General schemes for hydrolysis are shown in **Figure 1**. Two product fragments are formed unless the hydrolyzed group is part of a cyclic structure (e.g., lactam). Reaction rates are affected by the moieties attached to the hydrolyzable functional group[7] and they decrease with a decrease in dielectric constant.[8] Examples of hydrolysis include aspirin, which contains an ester and hydrolyzes to salicylic acid and acetic acid (**Figure 2**). Chloramphenicol, which contains an amide group, hydrolyzes to the corresponding amine and dichloroacetic acid[8] (**Figure 3**). Under basic conditions, penicillin G (I) hydrolyzes to benzylpenicilloic acid (II), which further decomposes to benzylpenilloic acid (III) and benzylpenicilloaldehyde (IV) plus carbon dioxide plus penicillamine (V)[8] (**Figure 4**).

Hydrolysis also occurs in vivo as compounds are exposed to endogenous hydrolase enzymes that occur in locations such as the intestine, blood, and liver. Hydrolytic metabolism is discussed elsewhere (*see* 5.06 Principles of Drug Metabolism 2: Hydrolysis and Conjugation Reactions). Hydrolases catalyze reactions of endogenous compounds and may have affinity for some drugs, thus increasing their hydrolysis rate. Testa and Mayer have extensively discussed hydrolytic drug metabolism in a volume that is very useful to medicinal chemists.[7]

The prodrug strategy makes use of in vivo hydrolases. Prodrugs (*see* 5.45 Drug–Polymer Conjugates) are derivatives of active drugs that are designed to improve a compound's liability, such as solubility or permeability, in order to improve oral absorption. After the functional group produces its desired effect (e.g., increasing intestinal solubility), in vivo enzymes cleave the functional group, typically by hydrolysis, to release the active drug.[7] In vitro assays are used to screen various prodrugs prior to in vivo study.

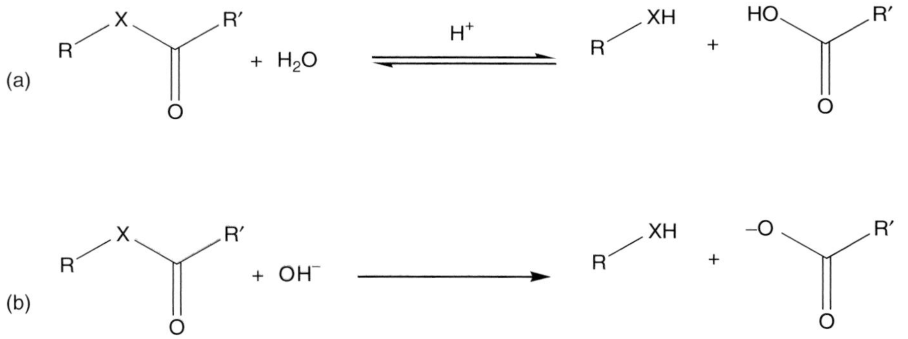

Figure 1   General schemes for (a) acid-catalyzed hydrolysis and (b) base hydrolysis, where X is O, NH, NR″, or S.

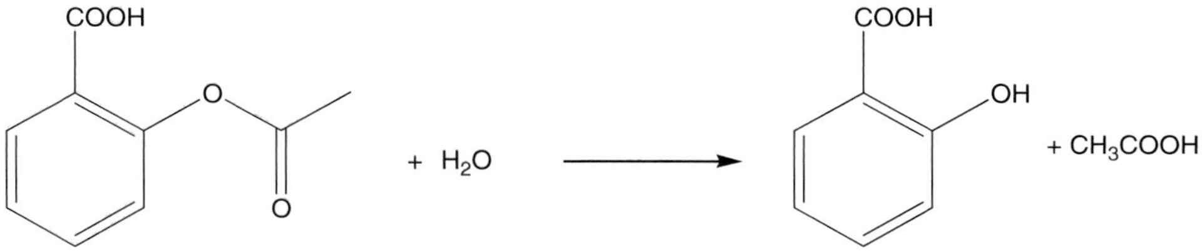

Figure 2   Hydrolysis example: acetylsalicyclic acid (Aspirin).[8]

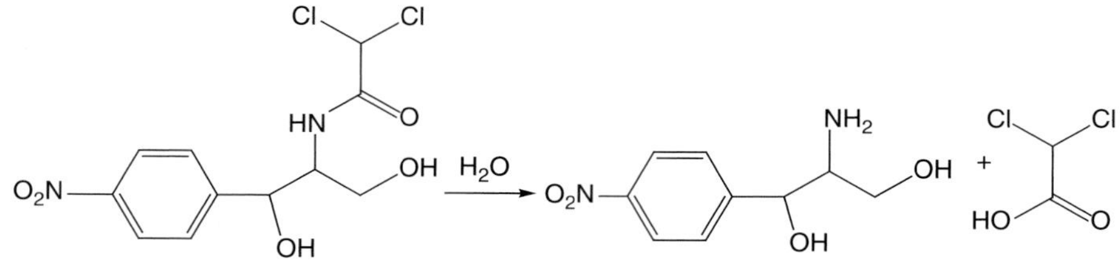

Figure 3 Hydrolysis example: choramphenicol.[8]

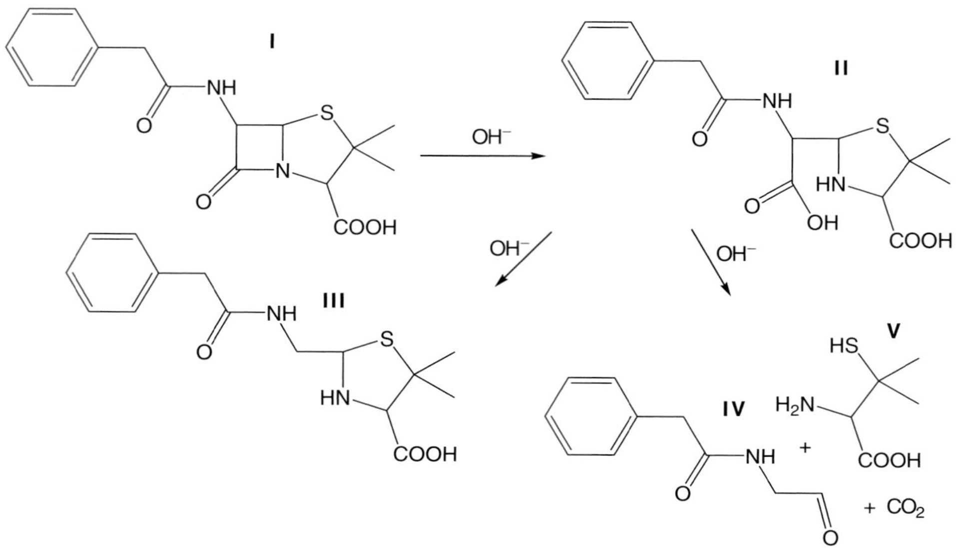

Figure 4 Hydrolysis example: penicillin G under basic conditions.[8]

5.20.2.2.2 Hydration

A molecule of water may also be added to a structure without cleavage of a fragment. This reaction is termed hydration. One example is the hydration of an epoxide to form a dihydrodiol. These reactions can occur in laboratory solutions and can be catalyzed by enzymes.

5.20.2.2.3 Oxidation

Oxidation occurs by way of free radical reactions.[8] Trace metal ions isolated with the compound material can catalyze oxidation. Light can initiate oxidation by free radical mechanism. In aqueous solution, pH and temperature affect the reaction rate. Many drugs are subject to autoxidation. Double bonds can be oxidized to hydroperoxides and then to aldehydes. Amines can form N-oxides and sulfides can form sulfoxides and sulfones. Benzylic sites, aldehydes, and ketones are also susceptible to oxidation. Dimerization can result from oxidation, such as for morphine at the phenolic oxygen, and captopril at the sulfide to form the disulfide (**Figure 5**). Such reactions primarily occur in aqueous solution. In the solid state, the presence of absorbed water at the solid surface can enhance oxidation. Oxidation mechanisms are complex. Furosemide, for example, is unstable in sunlight, forming several oxidized and reduced forms.[9] A major source of drug oxidation in vivo is metabolic enzymes (e.g., cytochrome P450). These greatly enhance the rate of oxidation (*see* 5.05 Principles of Drug Metabolism 1: Redox Reactions).

Photolysis involves the absorption of a photon of light to activate the molecule. This can lead to dissociation of the molecule or oxidation via a free radical mechanism. In solution, the photon may also be absorbed by a solution component, which in turn reacts with the drug compound. In the solid state, photolysis is confined to the surface of the particle. Photolytic reactions are possible for compounds having carbonyl, nitroaromatic, N-oxide, alkene, aryl chloride, sulfide, polyene, and weak C–H or O–H functional groups.[13] The photostability of drugs is treated in detail by Tonnesen *et al*.[14]

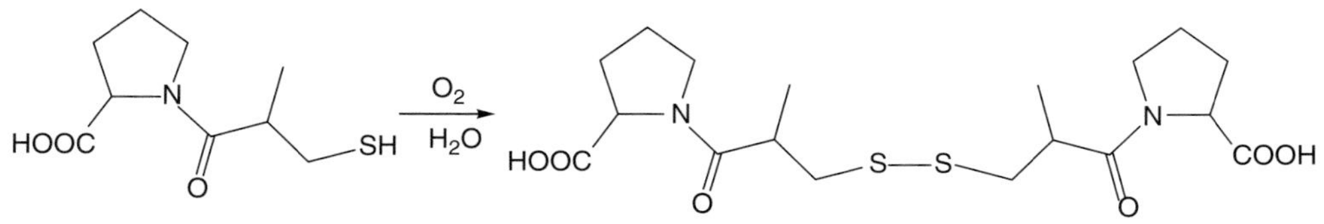

Figure 5 Oxidative dimerization of captopril.[8]

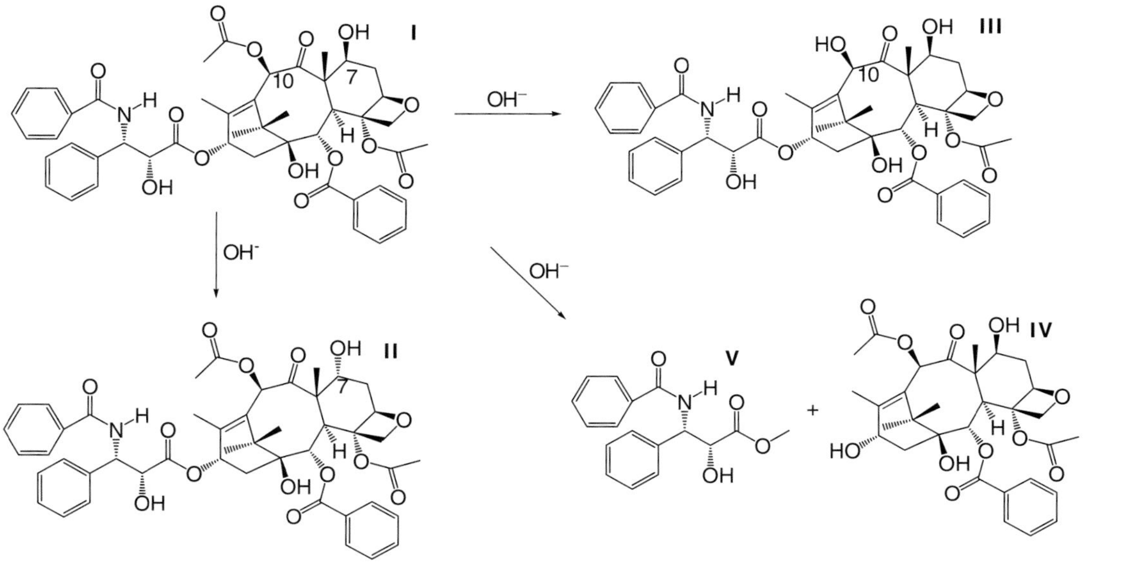

Figure 6 Base induced degradation of paclitaxel (I) in methanol to 7-epipaclitaxel (II), 10-deacetyl paclitaxel (III), baccatin III (IV), and side chain methyl ester (V).[19]

5.20.2.2.4 Stereochemical conversion

Stereoisomers having instability at a stereogenic center can react to produce stereochemically modified forms.[15–18] This can be triggered by an adjacent group that decreases stability at the chiral center.[15] An enantiomer can irreversibly 'racemize' and reach the racemic mixture at equilibrium. It can also 'enantiomerize,' or reversibly interconvert to the other enantiomer. Diastereomers can 'diastereomerize' or 'epimerize' by conversion of an unstable stereogenic center. For example, paclitaxel (I) epimerizes in organic solutions under basic conditions at the 7 position (II) in addition to other decomposition reactions (**Figure 6**).[19] Thalidomide enantiomerizes at basic pHs and in solution with human serum albumin (likely due to catalysis by Arg and Lys residues).[18]

5.20.3 Considerations on Analytical Methods for Chemical Stability

5.20.3.1 Choice of Stability Conditions to Monitor

The study of chemical stability requires analytical methodology. In considering what methods to implement or have access to, it is useful to consider the various stability challenges that compounds encounter in the laboratory, during in vitro testing and in vivo.[3] Even before testing, chemists can consider if these conditions could be anticipated to cause compound reaction.

In the laboratory, compounds are exposed to light, oxygen, and water. As compounds are stored in solid or solution form, they are exposed to fluctuations in temperature and relative humidity. These challenges are rigorously addressed when compounds reach development. Several references detail these stability conditions, including the US Food and Drug Administration (FDA) guidance documents to industry[20,21] and practical guides to their implementation.[22] FDA

Table 1 FDA guidance on stability testing[20,21]

Chemical stability question	*Stress test conditions*
Storage	12 months at 25°C/60% relative humidity
	6 months at 30°C/60% relative humidity
	6 months at 40°C/75% relative humidity
Oxidation	High oxygen atmosphere
Photolysis	D65/ID65 standard (or cool white fluorescent lamp), 1.2 million lux h and UV (320–400 nm); overall illumination $\geqslant 200$ watt hr m^{-2} (UV)
Hydrolysis	A wide pH range
Decomposition Structures	Identity and chemical structure for mechanism of formation

documents define regulatory tests, but they serve as useful guides to conditions that cause decomposition of drugs and also serve as guides for test methodology. A summary of FDA guidance on stability testing is listed in **Table 1**.[20,21] FDA tests usually focus on accelerated testing, which increases the rate of chemical decomposition using enhanced conditions (e.g., higher temperature) for the purpose of calculating kinetic parameters for the prediction of drug product expiration date. The terms accelerated testing and stress testing are used. FDA also emphasizes the importance of testing in open containers. This enhances air circulation. In discovery, the focus is on the drug substance in its various forms in short-term experiments. Development methods can be simplified for discovery stability studies.

During in vitro activity and selectivity testing, compounds encounter new challenges as they are exposed to aqueous solutions, mixed aqueous/organic solutions, and a range of pHs. Compounds can also assume various physical or equilibrium forms (e.g., aggregates,[23] micelles) that increase or decrease their apparent activity.

In vivo testing presents new challenges. Following oral dosing, compounds encounter the highly acidic environment of the stomach, which can be as low as pH 1–2. Compounds also encounter hydrolytic enzymes in the stomach, intestine, liver, and blood. These hydrolases have been used to advantage in the development of prodrugs.[24]

5.20.3.2   Quantitative and Qualitative Data Needs

The study of stability in support of medicinal chemistry utilizes two types of data, quantitative and qualitative. Quantitative data are derived from methods that measure the concentration of a compound at time points during incubation under a set of conditions that are designed to simulate a stability challenge. The data are used to compare analogs for their relative stability, resulting in structure–property relationships (SPRs), which is analogous to SAR. Medicinal chemists use this data to rank order compounds. It also assists with understanding whether stability is a major concern and with selecting compounds to go to the next stage, such as in vivo testing and advancement to development. Quantitative data are also used to determine the kinetics of the decomposition reaction. Kinetic rates are helpful in determining the extent of decomposition expected at key time points.

Additional insight is gained from qualitative data. This involves the identification of structures of the decomposition products of the reaction. Structures are very useful in determining the mechanism of the reaction. This in turn allows chemists to decide on strategies for structural modification to enhance chemical stability in subsequent analogs. In some cases, the decomposition products may be tested for activity and selectivity, in order to estimate potential efficacy complications. Toxicity testing may also be indicated in order to assess potential liabilities that may terminate the project if the decomposition cannot be eliminated. In these cases, material will need to be isolated for further testing. Structural information is obtained using HPLC, mass spectrometry (MS), and NMR techniques.

5.20.3.3   Appropriate Methods for Different Research Stages

While it is satisfying to have detailed data for every research question, modern drug discovery has continuously dealt with the need for cost efficiency. Therefore, a strategy has arisen: use an appropriate method for the question and stage of research. This requires tailoring the methodology to the situation. A three-tier strategy has proven be useful.

5.20.3.3.1 Tier one: high-throughput methods in early drug discovery

In early discovery, only milligram quantities of compounds are available for testing. Thus, methods must provide useful data from a small quantity of material. In addition, a typical research organization examines compounds from many research projects. These compounds have diverse structures, so the testing methods must be 'generic' and accommodate compounds of widely differing physicochemical properties. Discovery projects can produce thousands of compounds per year, so methods should have high capacity, often using robotics and well plate technology. Medicinal chemists also need data quickly. Current strategies call for property profiles to be produced at the same rate as biological data, which is typically a few days to 2 weeks. This allows the data to be included in decision-making along with activity data during the fast-paced stages of research. Finally, for efficiency, organizations need to select the highest priority properties to monitor at this stage. Thus, potential issues of stability must compete with other physicochemical and metabolic properties for inclusion in the property-profiling scheme.

5.20.3.3.2 Tier two: diagnostic methods to answer specific questions

When specific questions arise regarding a compound series or an unexpected observation, it is useful to design specific diagnostic methods for targeted information. These methods usually deal with a smaller set of compounds and can still utilize milligram amounts of compound. Assay conditions are more specific than the generic screening methods discussed above. Turnaround time for the data may be increased in order to accommodate the more detailed conditions or experiment time periods. More time-consuming procedures, such as incubation protocols, or advanced analytical instruments, may be used. In many cases, the questions addressed with diagnostic assays are not covered by high-throughput methods. They arise because of a specific concern about a structural moiety in the molecule(s) or a specific question. The questions are not usually sufficiently generic to warrant testing of all compounds with high-throughput methods. Another application of these assays is to diagnose more complicated processes, such as bioavailability in vivo or decomposition in solution, by breaking the complex phenomenon down into individual properties that can be measured to pinpoint the specific cause.

5.20.3.3.3 Tier three: in-depth methods to meet specific criteria

In late stages of discovery, organizations often impose criteria for advancement. For example, decision-makers may want to see a profile of the stability of a development candidate under a range of physical and physiological conditions. Specific methods have been devised in many organizations to obtain data under these specific conditions. Many of these methods are adapted from development methods, anticipating the testing and stability challenges that a drug will encounter in development. The methods often require a larger amount of material (tens to hundreds of milligrams for each test), longer times (weeks to months), and are performed in low throughput. Detailed quantitative, kinetic, structural, and mechanistic data and interpretations are provided.

Another useful application of this information, that is gathered during the drug discovery medicinal chemistry phase, is for drug development laboratories to anticipate issues they will face. For example, the chemical process research and development laboratory can anticipate conditions under which the compound will be unstable. They may need to avoid process steps that include exposure, of the compound to basic conditions, light exposure, or heating if the compound has been shown to be unstable under these respective conditions. The analytical research and development laboratory can anticipate the major degradants and conditions under which they will be formed, to assist their development of stability indicating HPLC methods. Pharmaceutical science laboratories can gain insights for initial studies on formulations for phase I clinical trials. Pharmacokinetics groups can learn whether compounds are unstable in plasma, which will affect the bioanalytical methods that are developed. Thus, stability data can be part of a comprehensive technology package that is transferred from the medicinal chemistry discovery department to development colleagues to accelerate the time-line leading to phase I studies.

5.20.4 High-Throughput and Diagnostic Chemical Stability Methods

In vitro assays are used to test compounds rapidly against key stability challenges. The following sections discuss methods to screen stability under physicochemical conditions in the chemical laboratory: solution pH, light, heat and humidity, and oxidation. Then methods for chemical stability in biological systems are discussed: biological assay matrices, and plasma and gastrointestinal.

5.20.4.1   pH Stability

Compounds are exposed to various pHs in the laboratory, in biological testing, and in vivo efficacy and pharmacokinetics studies. The most generally applicable pHs for stability in earlier stages of drug discovery are pH 1–2, indicating stability in the stomach, pH 6.6, indicating stability in the small intestine, and pH 7.4, indicating stability in neutral assay buffers and body fluids. At later stages, a fuller profile of pH stability assures that all cases are covered.

Several methods have been described for pH stability profiling. For low sample throughput, it is usually sufficient to conduct the studies in small vials, such as HPLC autosampler vials. For higher throughput, 96-well plates are quite useful for incubating larger numbers of compounds in small quantities. It is advisable to protect the samples from light and control the temperature, which are both factors that will affect the reaction rate.

pH stability can be screened as part of a high-throughput process for several physicochemical properties, in order to get an overview of each new compound. For example, Kibbey et al.[25] reported an efficient instrumental method for screening chemical stability at multiple pHs, along with log P and pK_a. Chemical stability was measured by overnight incubation at 100 µM in pH 2, 7, and 12 buffers, and in 3% hydrogen peroxide (see oxidation stability below). Samples were incubated in 96-well plates. In methods of this type, low solubility should always be monitored, because of the large number of low solubility compounds that occur in drug discovery. Low solubility can cause erroneous results if the compound precipitates from solution. In this case, solubility was enhanced by using a solution of 50% acetonitrile and 50% aqueous buffer. Because organic solvent is likely to have an effect on reaction rate, due to the reduced dielectric constant, an alternative would be to lower the test compound concentration so that less organic solvent content is needed to solubilize the low solubility compounds. The method is also an example of incorporating robotics (Gilson 215 robot) for unattended operation, including automated addition of reagents and sample injection into an HPLC running a short mobile phase gradient cycle of 5 min. As in many higher-throughput methods, a classification system was used for data reporting on a scale of 1 to 5. This is because high-throughput methods often have lower precision and accuracy than would be desired for in-depth methods, so it is prudent to use data in ranges (sometimes called 'bins'). A method of this type has a throughput of up to 350 compounds per week. This method approach allows a widely applicable screen for stability across a wide pH range.

pH stability studies may be targeted to specific research questions, such at the prediction of in vivo absorption. For example, Caldwell et al.[26] reported using a stability assay at pH 2 as part of a preclinical compound property evaluation scheme. An in vitro pH 2 stability assay was run for 75 min (the mean human stomach residence time) to model stomach conditions. This data was combined with data on solubility at pH 2 to produce a 'liberation ranking,' which combines the effects of dissolution and stability. When compounds had stability at pH 2 that was greater than 50% and solubility that was greater than 0.1 mg mL^{-1}, then a 'high' liberation ranking was predicted. This ranking was combined with other in vitro assays for absorption and metabolism to predict oral bioavailability classification prior to in vivo dosing.

pH stability studies can also be a part of the evaluation of prodrug designs, Nielsen et al.[27] performed pH stability studies over the range of pH 0.1–9.8 at 37 °C. The method description provides useful guidance on pH buffer preparation for pH stability studies. Buffers were prepared at 0.02 M concentration with: acetate (pH 5.0), phosphate (pH 3.0, 6.9, 7.4), borate (pH 8.5 and 9.75) and hydrochloric acid below pH 1.1. Constant ionic strength was maintained, because it affects reaction rate, by adding a calculated amount of potassium chloride. Several references are available for preparing buffers at different pH ranges, including The Chemist's Companion.[28] Test compound was added at 100 µM. The test was run in an HPLC autosampler vial and samples of the supernatant were withdrawn at selected time points and immediately injected into an HPLC.

Studies may be conducted in microtiter plates for higher throughput. Kerns et al.[29] incubated test compounds in 96-well plates with 0.1 mL of HCl (pH 1) for 2 h to simulate acidic conditions in the stomach. After 2 h, 0.2 mL of phosphate buffer at pH 7.4 was added to adjust the pH and samples were analyzed by LC/MS. One problem with this approach is that compounds continue to sit in aqueous buffer for an undetermined time prior to injection into the HPLC. Even at or near neutral pH, compounds may continue to decompose and there is a 2–4 h time gap between injection of the sample in the first well of the plate and injection of the 96th sample in the last well, thus exposing different samples to decomposition for different time periods.

The problems of inability to quench decomposition and the time lag between analysis of samples in the same set may be eliminated with the method of Di et al.[30] The programmable injection and temperature control features of modern HPLC autosamplers were used to perform the stability experiment completely within the instrument. Samples were placed in a well plate in solution and the plate was placed in the autosampler. The autosampler was programed to add samples to stability matrix (e.g., buffers at various pH) at a preprogramed time. The solution was

Table 2 pH Stability applications and methods

Tier	Conditions	Application	Method	Throughput per week	Reference
1	pH 2, 7, 12	New compound alerts	Robotics, well plate, HPLC	>300	25
1	pH 2	Stomach stability		>300	26, 29
2	pH 0.1–9.8	Prodrugs in vitro	Manual, 1.5 mL vial, HPLC	10	27
2	pH 1–10 37 °C	Diagnostic	Autosampler incubation, HPLC	50–300 (depending on time points)	30
3	pH 1–12	Predevelopment qualification	Manual, HPLC	10	
	37 °C				
	1–24 h				
	1N HCl 1N NaOH	Development	Heat and time sufficient for 10–20% degradation, up to 70 °C and 1 week		12

mixed using up- and downagitation of the injection needle and the sample was immediately injected for the time zero sample. The autosampler was maintained at a constant preset temperature between 0 and 40 °C and continued to inject aliquots of the sample at preprogramed time points. Multiple samples were analyzed simultaneously in the same manner, with the instrument tracking when to reinject each sample. In this manner, each sample was incubated for a specified time and did not sit for a variable time prior to injection. Furthermore, the experiment was conducted in an unattended mode. An overview of pH stability methods is found in **Table 2**.

5.20.4.2 Light Stability

Some compound series react when irradiated with light. This is a particular problem for topical drugs or drugs that accumulate in the skin. Outdoor light is the most intense, but indoor lighting in laboratories, pharmacies, clinics, and storage facilities can cause decomposition. Most organizations do not screen light stability for all compounds in high throughput. However, in selected cases there may be an interest in diagnosing light stability when it is suspected by an observation, or when the structure of the compound or historical information from analogous structures suggests the potential for instability in light. These tests are usually conducted under lights that simulate the wavelength spectrum and intensity of fluorescent lights. As indicated by the FDA guidance, the spectrum should cover the visible and near ultraviolet (UV) spectrum (**Table 1**). Care should be taken to control the temperature of the irradiated samples during the exposure, because they will tend to heat up under the lights and the rate of the photoinduced or other decomposition reactions may increase.

A traditional approach for studying photostability was described for RG 12915.[31] Compound in solution at $100 \, \mu g \, mL^{-1}$ in pH 6.8 phosphate buffer, with NaCl added for an ionic strength of 0.2, was placed in a 2 mL borosilicate glass vial. The vial was held at 22–25 °C in the open air and was exposed to 180-foot-candle light from a daylight fluorescent light source. Aliquots were withdrawn for HPLC analysis on a time course experiment.

For diagnostic or higher-throughput methods a light box with the brand name Rayonette, model RPR-200 from Southern New England Ultraviolet Co. has been used to generate high-intensity light for accelerated stability assays.[29] These instruments are often used as photochemical reactors in the synthetic chemistry laboratory. Light bulbs with various wavelength spectra can be selected according to need. A 96-well plate can be accommodated within the light chamber on a shelf that can be cut out to expose the under side of the plate in addition to illumination from above. This allows higher throughput or studies of different sets of solution conditions (e.g., pH, mix of aqueous and organic solvents, ionic strength). For studies in drug discovery, exposure for 24 h at high intensity is usually sufficient to determine if a major photostability liability exists or to diagnose if photostability is the cause of an unexplained observation.

Photostability chambers that meet the FDA guidelines can be obtained from several manufacturers (e.g., Harris Weathering Products, Caron Products, Lunaire). The units are typically larger than what is needed for discovery studies, but they are well designed and are reliable. Some groups have built their own smaller light chambers for late discovery applications. An engineering department can assist in measuring the intensity and spectrum of the light, because this is critical to the experiment and must be controlled. As light bulbs age, the intensity of their light tends to decrease and the spectrum of emission tends to shift; thus these should be monitored periodically. For these studies, the material is spread in a thin layer on an open dish to maximize surface exposure to light and air. International Conference on Harmonization (ICH) guidelines stipulate no more than a 3 mm depth. Alternatively, the samples can be tumbled in vials for uniform light exposure.[13]

Light-induced chemical reactions differ between the solid state and solution state and vary with solvent. Study conditions should best simulate the light exposure and solution state that the compound will receive in the discovery laboratory, process laboratory, or storage environment.

5.20.4.3   Heat and Humidity Stability

FDA guidelines endorse the use of elevated temperature to accelerate chemical decomposition of drugs in the solid phase in order to simulate long-term storage. This suggests that, in discovery, elevated temperatures may be used to accelerate decomposition reactions in air that would normally require much longer times to develop. Reactions such as unimolecular decomposition, hydrolysis or oxidation with water or oxygen from the air, respectively, may be observed. A rule of thumb for the effect of temperature on reaction rate comes from the Arrhenius equation. For an average compound, reaction rate doubles for each 10 °C of temperature increase.

FDA guidelines[20–22] point out the importance of maintaining humidity in studies of solid materials. Heating in an oven will lower the relative humidity unless additional water is introduced. A commercial chamber that is designed for heat and humidity testing can best maintain humidity at established levels with little variation during heating. Such studies should at least be conducted for compounds being considered for development, in order to reveal any liabilities. In the solid state, compounds should have full exposure to air in an open vessel, by having only a thin layer or continuous agitation.

During drug discovery, heat and humidity stability studies may be performed to diagnose the possible causes of observed compound degradation, but are uncommon in high throughput. In late discovery, they provide estimates of the stability of compounds prior to development.

For the in-depth study of temperature, pH, and dielectric strength on chemical stability in solution, Shah et al.[32] described an integrated system that controls all of these variables for kinetics studies. A thermostated/heated reaction vessel surrounds the sample solution. Samples of the solution are taken at regular time points by sampling with a microdialysis probe that is placed in the solution. The microdialysis samples are automatically introduced into an on-line HPLC with UV detection for quantitation of the compound. Decomposition products could also be readily measured. Temperature control allows good kinetic analysis.

5.20.4.4   Oxidation Stability

Oxidation experiments were performed by Won et al. by placing a 1 mg mL^{-1} aqueous solution in a 100 mL volumetric flask in a constant 60 °C water bath with protection from light.[31] Samples were withdrawn at time points and analyzed by HPLC. Subsequent experiments were conducted with the addition of cupric sulfate (0.1 to 1×10^{-5} M), to study the effect of metals, and at different pHs (2.1, 4.8, and 7.6), to study the effect of pH.

Oxidation has been studied by incubation with 3% hydrogen peroxide in pH 7.4 buffer for 10 min.[33] This method may be used in microtiter plates for high throughput. Hydrogen peroxide at 3% is a standard test that has been submitted to the FDA. It provides a convenient probe of labile substructures in the molecule. Some development laboratories prefer using an oxidation initiator and incubation under molecular oxygen at high pressure. This is because peroxide oxidation can follow a different pathway than oxygen oxidation, and there is no way to quench a peroxide oxidation, so it usually goes to completion.[13] Quantitation can be followed up with qualitative structural studies of the degradants, in order to better understand the oxidation site and mechanism.

Lombardo and Campos[34] recently discussed various oxidation-profiling methods and reported an instrumental method for rapidly assessing oxidative stability. The method was carefully studied and well developed and validated. Test compounds are diluted in aqueous buffer and injected into electrochemical cells (ESA, Inc.) of increasing voltage. The cells are designed to operate as HPLC detectors, so they are amenable to being on-line with HPLC separations. This allows individual sample components to be studied separately from impurities. The extent of oxidation at various voltages can be measured. Comparison of the oxidation potentials and extent of conversion permits rank ordering of compounds. Implementation of the method involves comparison to compounds of known oxidation liability. Oxidation stability methods are listed in **Table 3**.

Table 3 Oxidative stability applications and methods

Tier	Conditions	Application	Method	Throughput per week	Reference
1	3% H_2O_2, 10 min	Oxidation liability alert	Robotics, no preparation, LC/MS/MS	>300	33, 19
1,2	On-line redox cell and HPLC	Oxidation screen	Electrochemical cells, HPLC	>300	34
3	Solid, thin layer	Predevelopment	30–70 °C, 1–7 days	10	
3	100 mL flask, aqueous buffer	Development	50 h, 60 °C, pH 2.1–7.6, open to air	10	31

Other researchers have previously used electrochemical cells to qualitatively study oxidative reactions. Coupling to a tandem mass spectrometry allows on-line structure elucidation of the oxidative products. Volk et al.[35] studied endogenous oxidative biochemical pathways. Jurva et al.[36,37] used this methodology to study the structures of oxidized drugs in an attempt to produce phase I metabolites. One-electron oxidations were observed, but oxidations via direct hydrogen atom abstraction were not observed.

These studies indicate the potential for rapid structural identification of oxidative reaction products.

5.20.4.5 Buffer and Biological Assay Matrix Stability

Occasionally, unexpected or variable results are obtained from bioassays for enzyme/receptor binding or cell-based assays. Solubility may be a major cause of this, but, occasionally, stability in assay buffer may be the cause. Bioassay matrices may be at a pH at which some compounds are unstable, as discussed above, or they may contain components that react with some compounds. An assay using bioassay matrices is useful for diagnosing stability as a problem under these conditions.

The assay (above) of Di et al.[30] can also be used for this purpose. Instead of pH buffer, the sample can be added to biological assay matrix by the HPLC autosampler. This assay can be run with multiple time points to obtain kinetic data, or it can be operated with a single time point injection for high throughput.

It was previously shown[38] that a single time point method can be accurately run in drug discovery settings to assay large numbers of compounds when the reaction kinetics are first order. Three quantitative factors were examined: (a) half-life, (b) assay variability, and (c) the nonlinear relation of half-life to percent remaining for first-order kinetics. For short half-life (1–10 min) compounds, little compound remains unreacted when sampled at 30–60 min and half-life estimates would be of questionable value. In addition, resolution of significantly different quantitative values, which is needed for differentiating structure–stability relationships, is limited by the variability of the assay. There is insufficient resolution of unstable compounds when sampling is performed at late time points. Conversely, stable compounds with half-life greater than 30 min are not resolved using samples obtained at early time points (5–10 min). The method variability sets the resolution of calculated half-lives, based on statistical confidence. As a general guide, single-point sampling at 15 min permits resolution of compounds with low stability and allows statistically significant half-life calculation up to about 30 min. If the compounds are all more stable and the desired upper range of half-life resolution is higher than 30 min, then the single time point for sampling should be adjusted upward. This study provides a foundation for selecting a single time point for sampling in higher-throughput methods. At the higher-throughput level, single time point screening allows rapid throughput for a large number of compounds. When greater detail is needed, the method can be repeated for a smaller number of selected compounds and more time points can be obtained for more detail for making decisions. For high-throughput methods, instead of fitting a line by regression, the quantitative values from time zero and after incubation (e.g., time equals 3 h) can be used with the linear rate equation to calculate the half-life. More time points simply add more precision and accuracy to the measurement. The single time point approach is very effective for efficiently screening a large number of compounds in drug discovery.

5.20.4.6 Plasma and Gastrointestinal Stability

Hydrolytic enzymes in the blood can be responsible for compound degradation in vivo and result in higher clearance and reduced in vivo pharmacokinetic half-life. Incubation of test compounds with plasma is more convenient than incubation with blood and can reveal the potential for decomposition in the blood stream (*see* 5.10 In Vitro Studies of Drug Metabolism).

A moderate-throughput plasma stability assay was described by Wang *et al.*[39] It uses an HPLC autosampler to incubate the compounds in plasma. Plasma samples are much more complex than aqueous buffers and usually require some type of cleanup prior to MS analysis, because of interference by plasma to proper ionization of the test compound in the MS ion source. Wang demonstrated that this problem could be solved, by using a mixed function HPLC column (Capcell MF C8) to separate the major plasma components from the test compounds, prior to the mass spectrometer. The major plasma components, that would produce interference and rapidly build up in the ion source, causing reduction of signal generation, are diverted to waste as they elute early from the HPLC column. The MS/MS instrument can be operated in the multiple reaction monitoring (MRM) mode to selectively and sensitively detect the test compound.

Plasma is often used as a surrogate for blood in studying prodrugs in vitro prior to in vivo studies. Nielsen *et al.*[27] incubated prodrug compounds with human plasma and simulated intestinal fluid (SIF). Plasma was prepared at 4:1 (v/v) with 67 mM phosphate buffer at pH 7.4. Samples were incubated at 37 °C. At selected time points, sample was withdrawn and mixed with 2 volumes of acetonitrile to precipitate the protein, followed by centrifugation (14 500 rpm for 5 min), a common technique in the analysis of plasma samples. The centrifugate was diluted 1:1 with mobile phase and injected onto the HPLC.

An optimized high-throughput method was developed and reported by Di *et al.*[40] It was shown in earlier studies on liver microsome stability[41] that assay conditions can greatly affect the results. The present studies showed that plasma, instead, has a high catalytic capacity and can tolerate a range of incubation conditions. Approximately, the same rate of decomposition occurred at test compound concentrations from 0.5 μM to 20 μM, at dilutions of plasma in pH 7.4 buffer from undiluted to 1:4 (100% to 20% plasma), in the presence of DMSO from 0.5% to 5% and for incubation times up to 22 h. The final selected incubation conditions for plasma degradation studies were: 1 μM compound concentration, 50% plasma in pH 7.4 buffer, 2.5% DMSO and 3 h incubation. Dilution of plasma in buffer allowed pH stabilization of the plasma and reduced viscosity for improved liquid handling. Considerable plasma batch variation was observed, indicating the need for quality control (QC) samples and testing of each new batch before it is used in analysis. Conditions can be adjusted (e.g., time of incubation) to achieve the same QC results from batch to batch. Incubations were in 96-well plates and LC/MS/MS was used for quantitation, allowing a throughput of up to 10 samples h^{-1}. Incubates can also be analyzed qualitatively using LC/MS and LC/MS/MS to identify the reaction products. This aids in planning synthesis of more stable analogs. It was suggested that the method can be used to obtain an early alert to labile substructural motifs, prioritization of compounds for in vivo studies, screening prodrugs, and the structural identification of hydrolytic products. Plasma stability methods are listed in **Table 4**.

It is crucial to remember that plasma from different species can produce different rates of hydrolysis.[40] When performing in vitro experiments in order to predict in vivo hydrolysis, the same species and strain of plasma should be used as in the in vivo efficacy model.

Additional insights are gained from incubating compounds with simulated gastrointestinal fluids. These consist of simulated gastric fluid (SGF) and SIF. The components of these mixtures are specified in the United States Pharmacopoeia (USP). SGF and SIF contain hydrolytic enzymes from successive portions of the gastrointestinal tract to which drugs are exposed.

Table 4 Plasma stability applications and methods

Tier	Conditions	Application	Method	Throughput per week	Reference
1	Plasma	Blood stability	Autosampler incubation, mixed function HPLC, LC/MS	>300	39
1,2	1 μM compound, 1:1 plasma: buffer, 2.5% DMSO, 3 h	Blood stability	Well plates, acetonitrile precipitation, LC/UV/MS	>300	40
2	Plasma, simulated intestinal fluid	Prodrug study in vitro	1.5 mL vials, acetonitrile precipitation, HPLC	30	27
3	Human plasma, simulated intestinal fluids, simulated gastric fluid	Predevelopment	Vials, 1–24 h	10	

5.20.4.7 Stereochemical Methods

Many of the previous sets of conditions can produce stereochemical inversions. Owing to the different pharmacological and toxicological properties of stereoisomers, there will be cases when it is important to determine the stereochemical effects of conditions to which the compound is exposed. A number of chiral analytical methods are available for this purpose. Diastereomers can usually be separated using standard reversed phase HPLC methods. Enantiomers usually require special chiral HPLC phases. It is necessary to have the racemate material to be assured of resolution and it is necessary to have each purified enantiomer to be assured of the peak assignment. Chiral separation methods are also available that utilize chiral gas chromatography phase, supercritical fluid chromatography and capillary electrophoresis. Chiral shift reagents can be used with NMR. An 'indirect' NMR method uses deuterated water to substitute for the mechanism of the chiral interconversion,[18] which has been shown to follow the same mechanism. This has the advantage of not requiring purified enantiomers and can be used for reactions that are too rapid for other methods.

5.20.5 Low-Throughput In-Depth Chemical Stability Methods

5.20.5.1 Pharmaceutical Solution and Solid State Stability

Advancement of a compound to development requires an in-depth data package, which examines the major issues that indicate development success or potential liabilities. Chemical stability is a standard requirement of such reviews. The methods used need not be as rigorous as those used in development for regulatory filings, but they parallel the development protocols in order to indicate consistent trends. Examples of in-depth conditions for chemical stability examination may include the following:

- Aqueous buffers (37 °C, pH 1–12)
- Simulated intestinal fluid (USP, 37 °C, 1–24 h)
- Simulated gastric fluid (USP, 37 °C, 1–24 h)
- Simulated bile/lecithin mixture (37 °C, 1–24 h)
- Human plasma (37 °C, 1–24 h)
- High-intensity light (room temperature, 1–7 days)
- Heat (30–75 °C, 1–7 days).

Tests over a prolonged time reveal trends and kinetics of chemical decomposition. For solution-based studies, vials for each condition can be prepared and placed into an HPLC autosampler, which can be programed to inject samples a specific time points. This eliminates the need to quench the reaction, which can be difficult. Multiple time points are sampled in order to enhance the statistical accuracy, indicate the order of the reaction and calculate the kinetics. Cosolvent may be necessary to completely dissolve low-solubility compounds; however, it is important that these not be reactive, such as alcohols would be in hydrolysis of carboxylic acid derivatives. The information is also useful for providing development departments with initial data to start to their work.

5.20.5.2 Structure Elucidation of Chemical Decompostion Products

The identification of decomposition product structures can be considered a diagnostic or in-depth method. An example of degradants identification was described by Volk *et al.*,[19] in the study of degradants of paclitaxel when it was exposed to various conditions. Paclitaxel was treated with methanolic/aqueous Na_2CO_3 at pH 8 for 10 min to produce accelerated basic pH stress. It was also treated with HCl at 0.7 M for 4 h to simulate acidic stress. Hydrogen peroxide in buffer at 3% produced an oxidative stress. Finally, solid drug was exposed to 1000 foot-candle intensity light for 92 days to produce a light stress. These conditions were selected based on experience with paclitaxel and with common procedures in early development. LC/MS in the full scan mode was used to separate and determine the molecular weights of the decomposition products. Then, LC/MS/MS (tandem quadrupole instrument) in the product ion mode was used to obtain fragment ions from which the structures of the reaction products were identified. Nine major decomposition products were identified. Treatment of paclitaxel (I) with base in the presence of methanol (**Figure 6**) resulted in cleavage of the side chain to produce the sidechain methyl ester (V) and the baccatin III core (IV), as well as 7-epipaclitaxel (II) and 10-deacetyl paclitaxel (III). Acidic treatment produced 10-deacetyl paclitaxel. Under light (**Figure 7**), a C3–C11 bridged paclitaxel isomer (VI) and oxetane ring opened paclitaxel (VII) were formed. Isolation and NMR were necessary to determine the acyl migration. The structures of three minor degradants resulting from loss

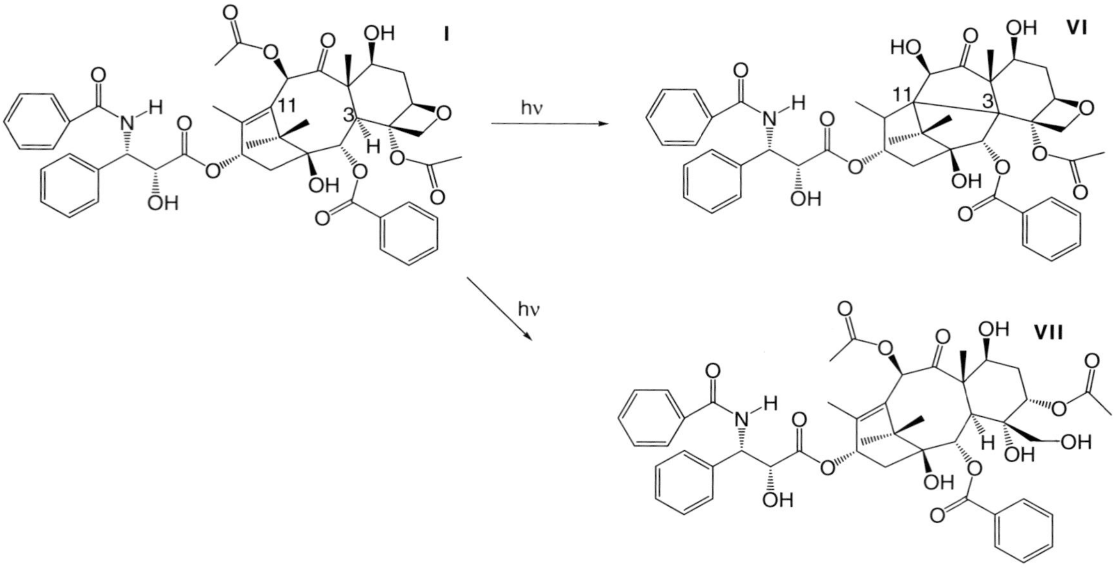

Figure 7   Light induced degradation of paclitaxel (I) to C3–C11 bridged paclitaxel (VI) and oxetane ring opened paclitaxel (VII).[19]

of CO from paclitaxel following light treatment could not be definitively determined using LC/MS/MS. The portion of the work involving LC/MS/MS analysis and structural interpretations was performed in less than one day. Other examples of drugs for which rapid identification of drug degradants by LC/MS/MS techniques were used include cefadroxil[42] and butorphanol.[43]

Following LC/MS studies, any unresolved structural details typically require NMR. Individual decomposition products can be readily isolated in sufficient quantity (20–50 µg) by fraction collection from HPLC. While LC/NMR can be used, isolation for NMR allows long-term NMR data acquisition and two-dimensional NMR spectroscopy for enhanced structural evidence.

In another example, two major degradants of risperidone were identified using LC/MS/MS (two-dimensional ion trap instrument), along with isolation and NMR.[44] Samples were prepared by incubating the compound in: acid (0.5 N HCl, reflux, 3 h), base (1:1 v/v of 0.1N NaOH:methanol, reflux, 3 h), oxidizer (3% H_2O_2 in methanol, 1 h). Two major degradants were identified: 9-hydroxy risperidone and resperidone N-oxide (piperidyl).

Identification of degradants provides several opportunities. Structural identities enable the modification of structures during lead optimization, in order to enhance chemical stability. The degradant compounds can be obtained through synthesis and used as quantitative standards in monitoring stability studies during development. Such material can also be used to check the toxicity or reactivity of the degradants. Knowing the identities of the decomposition products allows a kinetic study of the appearance and disappearance of individual products, so that the mechanisms of formation and further reaction may be better understood.

5.20.6   Applications of Stability Information in Medicinal Chemistry

Chemical stability data can assist medicinal chemistry in many ways. Overall, the data are part of the comprehensive data package that is used for decision-making that affects selection and redesign of structures.

The data can be effectively used in selecting compounds to move to the next level of investigation. For example, in one research program over 200 analogs were synthesized and screened for in vitro activity as well as physico-chemical and metabolic properties. The most significant differentiation among compounds was found in the in vitro activity and plasma stability. Compounds that were unstable contained peptide bonds. Thus, the plasma stability data were useful in selecting compounds that would be stable in vivo in the bloodstream during in vivo activity studies. In another program, compounds were screened for their stability in SIF and SGF, because of the presence of esters and amides. This provided predictions of compounds that were expected to be stable in vivo in the gastrointestinal tract.

Another application of stability data is in the optimization of structures. Structure–stability relationships can be developed that indicate how the structure may be modified to improve stability in plasma. **Figure 8** indicates the structure–stability relationships of a series from a human cytomegalovirus (HCMV) protease program.[45] It was observed that increasing the steric hindrance and increasing the electron withdrawal at the lactam result in improved stability. An analog was found that balanced activity and stability.

Figure 9 provides another structure–stability relationship example.[45] The stability against SN1 hydrolysis at pH 1 was improved by electron withdrawing R groups. Pharmacological activity was maintained, while the structural modification improved acid stability.

Stability data can also assist with understanding the kinetics of a reaction and the conditions to avoid when processing or storing the compound. **Figure 10** shows the pH–stability relationships of a beta lactam antibiotic using the HPLC autosampler method[30] described above. Such information can be used during lead optimization to diagnose the source of observed inconsistencies and during late discovery to anticipate storage stability of compounds.

Nielsen *et al.*[27] used pH, plasma, and SIF stabilities to screen *N*-acyloxymethyl prodrugs of bupivacaine and lidocaine. The prodrug moiety provided a quaternary amine to enhance solubility in the gastrointestinal tract. The prodrug was then cleaved at the brush border of the intestine, in order to produce a supersaturated solution at the surface of the gastrointestinal membrane to enhance permeability. Study of the rates of conversion in vitro provided a proof of the concept in advance of in vivo pharmacokinetic studies.

Compounds may also be screened for stability in biological assay buffer. An entire plate of compounds can be assayed in a matter of hours using a single time point experiment, in order to detect if there is a potential problem with stability

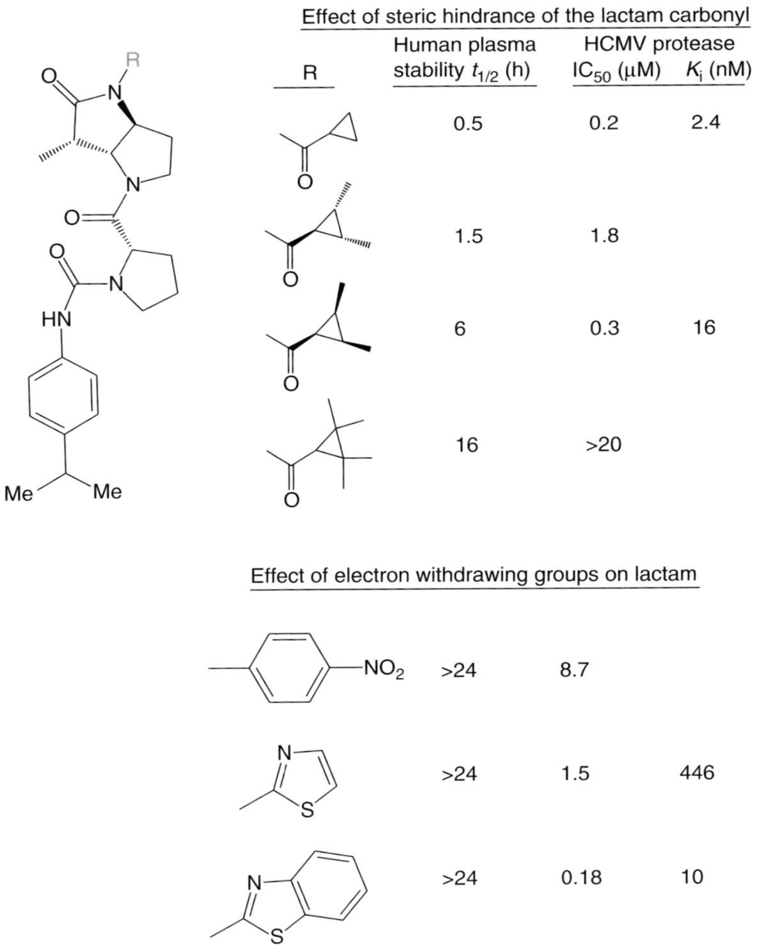

Figure 8 Structure–stability relationships in an HCMV protease program.[45]

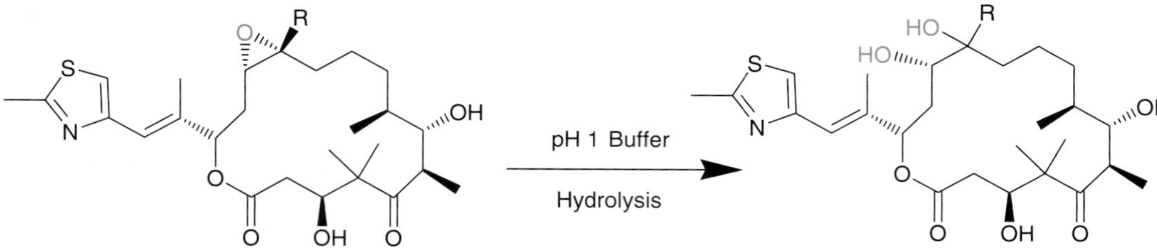

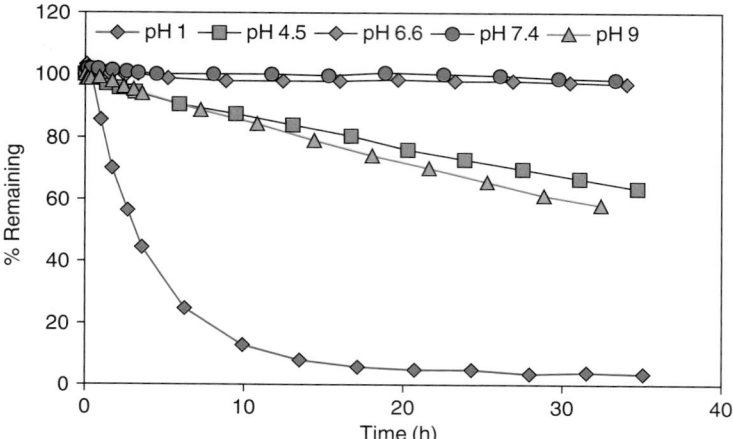

Figure 9 Electron–withdrawing R group slows down SN1 hydrolysis.[46] t_{95} is the time for the concentration of starting compound to decrease to 95% of its original level.

R	E	$C_{0.01}$ (tubuline)	IC_{50} (HCT-116)	t_{95}
Me	(Epothilone B)	2.2 μM	4.4 nM	< 0.2 h
CN		2.5 μM	4.1 nM	11 h

Figure 10 pH–stability profile for a beta lactam compound.[30]

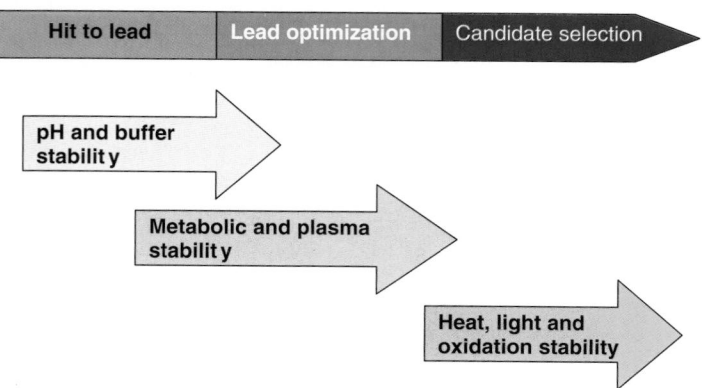

Figure 11 A possible strategy for stability profiling in drug discovery.

and to diagnose if poor activity was due to instability in the assay medium. For biological assays, inconsistencies can be diagnosed retrospectively using such an approach.

As projects approach development candidate selection, predevelopment pharmaceutics departments begin preparations by developing a 'stability indicating method.' This is an HPLC method that is likely to separate all of the chemical decomposition products that are likely to be observed in development stress testing. Rao *et al.*[46] described the development of a stability indicating method for rivastigmine. To provide test samples, they incubated the drug under the following conditions: light (254 nm, 10 days), heat (60 °C, 10 days), acid (0.5 N HCl, 48 h), base (0.5 N NaOH, 48 h), and oxidation (3% H_2O_2, 48 h). This insures that the method separates key potential impurities and degradants.

5.20.7 Conclusions

Chemical stability is important to consider in the comprehensive assessment of pharmaceutical properties, activity, and selectivity during drug discovery. There are many sources of guidance on what chemical stability issues are a concern: medicinal chemistry experience, FDA guidance, and issues that occur during the research project. Methods can be implemented or customized with high, moderate, or low throughput, to meet the needs for the number of samples, and depth of data needed. Both quantitative and qualitative data may be obtained for answering various questions.

For a drug discovery program, it may be useful to decide how and when to study different stability challenges. The scheme in **Figure 11** provides one approach. In early research, the focus is on finding a diverse set of structures that produce sufficient activity to indicate their future potential. It is important at this stage to insure that the compounds will be stable under laboratory handling and biological assays, thus, pH and bioassay buffer become important conditions to test for stability. After this stage, the series scaffold is locked in and these studies serve to insure that the scaffold is stable. During lead optimization, the focus is on in vivo stability as oral dosing and in vivo efficacy studies are planned. Thus, screening for metabolic and plasma stability are important during optimization. At this stage, stability may be modified by chemical modifications to the scaffold. During candidate selection for development, a series of stability challenges that are indicative of drug storage problems, can be employed. This insures that the candidate meets established criteria for advancement.

It is useful to establish the rate of crucial stability reactions. If the half-life is on the order of minutes or hours, then there will be an effect on the 'pharmacological timescale' and this affects activity in vivo. If the half-life is on the order of months or years, then there will be an effect on the 'pharmaceutical timescale' and this affects formulation and storage conditions.[15]

Chemical stability profiling as part of medicinal chemistry provides several opportunities:

- Select quality leads for optimization that have stable scaffolds and few liabilities
- Diagnose the cause(s) of unexpected observations or inconsistent results
- Modify structures to enhance stability of a lead series
- Better plan and interpret biological, metabolic, and physicochemical studies
- Select among compounds for crucial studies, such as in vivo efficacy and pharmacokinetics
- Provide data that accelerates early development activities
- Avoid candidates that shift a burden to development in time, expensive packaging and storage, or short expiration dates.

Integration of physicochemical properties, such as chemical stability, into the entire data package available to medicinal chemists, helps to make the drug discovery process more successful in producing high-quality clinical candidates.

References

1. Van de Waterbeemd, H.; Smith, D. A.; Beaumont, K. D.; Walker, K. *J. Med. Chem.* **2001**, *44*, 1313–1333.
2. Lipinski, C. A.; Lombardo, F.; Dominy, B. W.; Feeney, P. J. *Adv. Drug Deliv. Rev.* **1997**, *23*, 3–25.
3. Kerns, E. H.; Di, L. *Drug Disc. Today* **2003**, *8*, 316–323.
4. Di, L.; Kerns, E. H. *Curr. Opin. Chem. Biol.* **2003**, *7*, 402–408.
5. Venkatesh, S.; Lipper, R. A. *J. Pharm. Sci.* **2000**, *89*, 145–154.
6. Oldenburg, K.; Pooler, D.; Scudder, K.; Lipinski, C.; Kelly, M. *Comb. Chem. High Through. Screen.* **2005**, *8*, 499–512.
7. Testa, B.; Mayer, J. M. *Hydrolysis in Drug and Prodrug Metabolism: Chemistry, Biochemistry and Enzymology*; Verlag Helvetica Chimica Acta: Zurich, Switzerland, 2003.
8. Connors, K. A.; Amidon, G. L.; Stella, V. J. *Chemical Stability of Pharmaceuticals*; John Wiley: New York, 1986.
9. Martin, A. *Physical Pharmacy*; Lea & Febiger: Philadelphia, PA, **1993**; pp 305–316.
10. Carstensen, J. T. *Drug Stability, Principles and Practices*, 2nd ed.; Marcel Dekker: New York, 1995.

11. Waterman, K. C.; Adami, R. C.; Alsante, K. M.; Antipas, A. S.; Arenson, D. R.; Carrier, R.; Hong, J.; Landis, M. S.; Lombardo, F.; Shah, J. C. et al. *Pharm. Dev. Technol.* **2002**, 7, 113–146.
12. Waterman, K. C.; Adami, R. C.; Alsante, K. M.; Hong, J.; Landis, M. S.; Lombardo, F. *Pharm. Dev. Technol.* **2002**, 7, 1–32.
13. Alsante, K. M.; Friedmann, R. C.; Hatajik, T. D.; Lohr, L. L.; Sharp, T. R.; Snyder, K. D.; Szczesny, E. J. Degradation and Impurity Analysis for Pharmaceutical Drug Candidates. In *Handbook of Modern Pharmaceutical Analysis*; Ahuja, S., Scypinski, S., Eds.; Academic Press: San Diego, CA, 2001, pp 85–172.
14. Tonnesen, H. H., Ed. *Photostability of Drugs and Drug Formulations*, 2nd ed.; CRC Press: Boca Raton, FL, 2004.
15. Testa, B.; Carrupt, P. A.; Gal, J. *Chirality* **1993**, 5, 105–111.
16. Reist, M.; Testa, B.; Carrupt, P. A.; Jung, M.; Schurig, V. *Chirality* **1995**, 7, 396–400.
17. Reist, M.; Testa, B.; Carrupt, P. A. *Enantiomer* **1997**, 2, 147–155.
18. Reist, M.; Testa, B.; Carrupt, P. A. Drug Racemization and Its Significance in Pharmaceutical Research. In *Handbook of Experimental Pharmacology, Vol. 153, Stereochemical Aspects of Drug Action and Disposition*; Eichelbaum, M., Testa, B., Somogyi, A., Eds.; Springer Verlag: Berlin, Germany, 2003, pp 91–122.
19. Volk, K. J.; Klohr, S. E.; Rourick, R. A.; Kerns, E. H.; Lee, M. S. *J. Pharm. Biomed. Anal.* **1996**, 14, 1663–1671.
20. US Department of Health and Human Services, Food and Drug Adminstration, Center for Drug Evaluation and Research (CDER), Center for Biologics Evaluation and Research (CBER), ICH, August 2001, Revision 1, http:// www.fda.gov/cder/guidance/4282fnl.htm, Guidance for Industry, Q1A Stability Testing of New Drug Substances and Products.
21. US Department of Health and Human Services, Food and Drug Adminstration, Center for Drug Evaluation and Research (CDER), Center for Biologics Evaluation and Research (CBER), ICH, November 1996, http:// www.fda.gov/cder/guidance/1318.htm, Guidance for Industry, Q1B Photostability Testing of New Drug Substances and Products.
22. Cha, J.; Ranweiler, J. S.; Lane, P. A. Stability Studies. In *Handbook of Modern Pharmaceutical Analysis*; Ahuja, S., Scypinski, S., Eds.; Academic Press: San Diego, CA, 2001, pp 445–483.
23. McGovern, S. L.; Caselli, E.; Grigorieff, N.; Shoichet, B. K. *J. Med. Chem.* **2002**, 45, 1712–1722.
24. Stella, V. J. *Expert Opin. Ther. Patents* **2004**, 14, 277–280.
25. Kibbey, C. E.; Poole, S. K.; Robinson, B.; Jackson, D. L.; Durham, D. *J. Pharm. Sci.* **2001**, 90, 1164–1175.
26. Caldwell, G. W. *Curr. Opin. Drug Disc. Dev.* **2000**, 3, 30–41.
27. Nielsen, A. B.; Buur, A.; Larsen, C. *Eur. J. Pharm. Sci.* **2005**, 24, 433–440.
28. Gordon, A. J.; Ford, R. A. *The Chemists's Companion: A Handbook of Practical Data, Techniques, and References*; John Wiley: New York, 1973.
29. Kerns, E. H. Presented at 3rd Annual Symposium on Chemical and Pharmaceutical Structure Analysis, Princeton, NJ, Sept 26–28, **2000**.
30. Di, L.; Kerns, E. H.; Chen, H.; Petusky, S. L. *J. Biomol. Screen.* **2006**, 11, 40–47.
31. Won, C. M.; Tang, S. Y.; Strohbeck, C. L. *Int. J. Pharm.* **1995**, 121, 95–105.
32. Shah, K. P.; Zhou, J.; Lee, R.; Schowen, R. L.; Elsbernd, R.; Ault, J. M.; Stobaugh, J. F.; Slavik, M.; Riley, C. M. *J. Pharm. Biomed. Anal.* **1994**, 12, 993–1001.
33. Volk, K. J.; Hill, S. E.; Kerns, E. H.; Lee, M. S. *J. Chromatogr., B* **1997**, 696, 99–107.
34. Lombardo, F.; Campos, G. How Do We Study Oxidative Chemical Stability in Discovery? Some Ideas, Trials and Outcomes. In *Pharmaceutical Profiling in Drug Discovery for Lead Selection*; Borchardt, R., Kerns, E. H., Lipinski, C. A., Thakker, D. R., Wang, B., Eds.; AAPS Press: Arlington, VA, 2004, pp 183–194.
35. Volk, K. J.; Yost, R. A.; Bajter-Toth, A. *Anal. Chem.* **1992**, 64, 21A–30A.
36. Jurva, U.; Vikström, H. V.; Bruins, A. P. *Rapid Commun. Mass Spectrom.* **2000**, 14, 529–537.
37. Jurva, U.; Vikström, H. V.; Weidolf, L.; Bruins, A. P. *Rapid Commun. Mass Spectrom.* **2003**, 17, 800–808.
38. Di, L.; Kerns, E. H.; Gao, N.; Li, S. Q.; Huang, Y.; Bourassa, J. L.; Huryn, D. M.; Carter, G. T. *J. Pharm. Sci.* **2004**, 93, 1537–1545.
39. Wang, G.; Hsieh, Y.; Cheng, K. C.; Ng, K.; Korfmacher, W. A. *J. Chromatogr., B* **2002**, 780, 451–457.
40. Di, L.; Kerns, E. H.; Hong, Y.; Chen, H. *Int. J. Pharm.* **2005**, 297, 110–119.
41. Di, L.; Kerns, E. H.; Hong, Y.; Kleintop, T.; McConnell, O. J. *J. Biomol. Screen.* **2003**, 8, 453–463.
42. Rourick, R. A.; Volk, K. J.; Klohr, S. E.; Spears, T.; Kerns, E. H.; Lee, M. S. *J. Pharm. Biomed. Anal.* **1996**, 14, 1743–1752.
43. Volk, K. J.; Klohr, S. E.; Rourick, R. A.; Kerns, E. H.; Lee, M. S. *J. Pharm. Biomed. Anal.* **1996**, 14, 1663–1674.
44. Borthwick, A. D. *J. Med. Chem.* **2003**, 46, 4428–4449.
45. Regueiro-Ren, A. *J. Med. Chem.* **1999**, 42, 2162–2168.
46. Rao, B. M.; Srinivasu, M. K.; Kumar, K. P.; Bhradwaj, N.; Ravi, R.; Mohakhud, K.; Reddy, G. O.; Kuman, P. R. *J. Pharm. Biomed. Anal.* **2005**, 37, 57–63.

Biographies

Edward H Kerns is Associate Director, Chemical and Screening Sciences, Wyeth Research, where he leads the in vitro physicochemical and metabolic profiling group for medicinal chemistry in drug discovery. He was previously with the Bristol-Myers Squibb Pharmaceutical Research Institute in analytical research and development. His research interests include the development of methods for property assessment, the use of property data in medicinal chemistry and the application of advanced analytical technologies for research and development.

Li Di, PhD is Principal Research Scientist II, Chemical and Screening Sciences, Wyeth Research, where she leads the Princeton, NJ physicochemical and metabolic property profiling group. She was previously with Syntex Research in drug discovery and development. Her research interests include the development and application of methods for physicochemical and metabolic property measurement and the selection and optimization of discovery lead compounds, and solid state chemistry. Dr Di received her PhD, Chemistry from Brandies University in 1991 and completed a postdoctoral fellowship at Boston University, School of Medicine in Biophysics in 1993.

Comprehensive Medicinal Chemistry II
ISBN (set): 0-08-044513-6

ISBN (Volume 5) 0-08-044518-7; pp. 489–507

5.21 Solid-State Physicochemistry

D Giron, Novartis Pharma AG, Basel, Switzerland

5.21.1 Introduction

As a result of the advances in high-throughput technologies, it is now possible to evaluate a large number of drug candidates. Key data in the attrition of new molecules are ADME (absorption, distribution, metabolism, and excretion) data based on in silico, in vitro, and in vivo results. Physicochemical properties of a molecule such as its pK_a, lipophilicity, and chemical stability in solution are independent of the solid state. Most drugs are administered orally. Therefore, the absorption of the drug substance in the body can be limited by its solid-state properties such as dissolution rate and solubility.

It has been stated that most new molecules in late development and in newly marketed products are poorly soluble.[1] The interface between drug discovery and development is very critical to optimize the candidate selection in order to reduce considerably the number of projects rejected during development. The 'developability assessment' has become an increasingly important criteria in addition to traditional drug efficacy and toxicity evaluation.[2]

The physicochemical properties of acidic or basic compounds in the solid state can be modified by salt formation.[3] However, an organic compound can have several solid forms, crystalline or not, and its physical properties in the solid state will depend on these solid forms. Therefore, the choice of a salt form has to be done in correlation with its 'polymorphic' behavior. In addition, particle size reduction and pharmaceutical formulations can enhance dissolution and thereby absorption.

The pharmaceutical community has recognized the relevance of the polymorphic behavior of organic substances since the epoch-making papers of Haleblian and McCrone in the early 1970s.[4,5] According to McCrone,[4] for a compound "the number of forms known is proportional to the time and energy put in research on that compound." The market withdrawal of ritonarvir in 1998 1 year after its launch was caused by its far too slow dissolution rate due to an insoluble stable polymorph not found during development. The high number of efficacy and safety problems has led health authorities to set guidelines for the study of polymorphism and particle size for each new active ingredient, in connection with the properties of the drug product (ICH guideline Q6A[6]). The number of publications, review articles, and patents on the subject is constantly increasing since solid-state behavior is a major concern for drug safety and quality. There are a number of important reviews in the literature including strategic considerations at the interface of drug discovery and development.[7–39] For example, a recent review lists more than 500 drug substances exhibiting polymorphism.[34]

It is the objective of this chapter to summarize the thermodynamic rules that are the driving force of polymorphism, pseudo-polymorphism, and apomorphism in order to optimize the selection of salt and/or crystal form and to avoid the hurdles and problems involved in ADME and toxicology studies. This chapter also points to some strategies to overcome solubility problems by modification of salt form, crystal form, particle size, complexes, co-crystals, or formulations.

5.21.2   Basic Considerations

To introduce this section, **Figure 1** summarizes all possible organizations of organic substances in the solid state.

5.21.2.1   Definitions, Polymorphs, Solvates, Hydrates, Amorphous Forms

Polymorphism is the ability of a substance to crystallize into different crystalline forms. These crystalline forms are called polymorphs or crystalline modifications. Polymorphs have the same liquid or gaseous state but they behave differently in the solid state. There are two cases, enantiotropy, where at a given pressure and depending on temperature, each form has a domain of stability, and monotropy (only one form is stable whatever the temperature).

The expression pseudopolymorphism applies to hydrates and solvates. Solvent and also water may be part of the crystal lattice; a new compound is formed between a volatile phase and the drug substance. In addition solvates and hydrates of defined composition may have polymorphs.[23,34]

The amorphous state produced by precipitation, milling, drying, melting, lyophilization, and spray-drying is characterized by solidification in a disordered, random manner; it is structurally similar to the liquid state.

5.21.2.2   The Crystalline and Amorphous States: Basic Concepts

In crystalline substances atoms have defined positions and arrangements, and lattice energy plays a role in all solid-state reactions. This is not the case with amorphous substances, where it is difficult to know how and where counterions and excipients are bound. Methods used in polymer analysis may be used, e.g., thermal analysis, infrared (IR) and Raman spectroscopy, and nuclear magnetic resonance (NMR) spectroscopy.

On heating, crystalline substances undergo a first-order transition into the liquid or gaseous phase. If a physical property of a crystalline substance is plotted against temperature, a sharp discontinuity occurs at the melting point. For amorphous substances, there is no sharply defined melting point, and a change of slope occurs at the so-called glass transition temperature T_g. Below this temperature, the amorphous phase has certain properties of a crystalline solid (e.g., plastic deformation) and is termed 'glassy.' Above this temperature, the substance retains some of the properties of a liquid, e.g., molecular mobility, and is termed 'rubbery.'

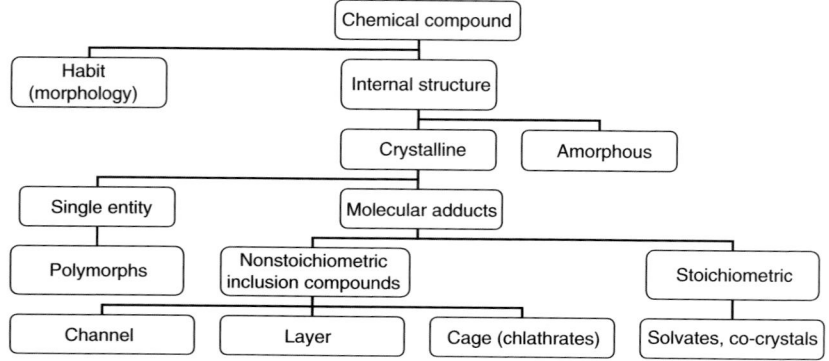

Figure 1   Possible solid phases of a chemical compound according to Haleblian. (Reprinted from Haleblian, J.; McCrone, W. *J. Pharm. Sci.* **1969**, *58*, 911–929; Copyright © (1969), with permission of Wiley-Liss, Inc., a subsidiary of John Wiley & Sons, Inc.)

5.21.2.3 Thermodynamic Background, Gibbs' Rules, and Phase Diagrams

The relationships between different phases are governed by the Gibbs' phase rule:

$$V = C + 2 - \Psi \tag{1}$$

where V is the variance, C the number of constituents, and Ψ the number of phases.

In the case of polymorphism, $C = 1$. If two solid phases are present and if both pressure and temperature vary, the variance is unity. If pressure is fixed, variance is zero. Phase diagrams of pressure versus temperature illustrate the different equilibrium curves for polymorphism (**Figure 2**).

For each solid form, there is a solid–liquid equilibrium curve and a solid–vapor equilibrium curve. In the case of enantiotropy, there is an equilibrium curve where both polymorphs are in equilibrium. In cases of 'monotropy' there is no thermodynamic transition between two phases since only one solid form is thermodynamically stable.

The ability of a system to perform work and to undergo a spontaneous change at constant pressure is measured by the Gibbs energy ΔG (formerly called free energy). ΔG has been defined in terms of enthalpy and entropy changes (ΔH and ΔS, respectively) at temperature T as:

$$\Delta G = \Delta H - T\Delta S \tag{2}$$

Gibbs energy (G), enthalpy (H), and entropy (S) are state functions and the term Δ can be omitted giving:

$$G = H - TS \tag{3}$$

At 0 K, $G = H$. Since the entropy S is always positive, G decreases with increasing temperature. The energy diagram G versus temperature at a given pressure reflects the transition observed between solid phases and between solid and liquid phases. If a transition between phases occurs, at this temperature both phases have the same Gibbs energy G.

In general, the thermodynamic relationship between two polymorphic phases is represented by plotting Gibbs energy as a function of temperature for each form. If the two curves intersect below the melting point of each polymorph, a reversible transition occurs at the temperature T_t of the intersection. At temperatures below T_t, polymorph A has the lower Gibbs energy and is therefore the thermodynamically stable form, while at temperatures above T_t polymorph B is stable. The transition point can be low − close to 40 °C in the case of tolbutamide or close to 100 °C in the case of propyphenazone or even higher than 200 °C.[34] In the case of monotropy, there is no intersection of the Gibbs energy of both forms and the higher melting form is always the thermodynamically stable form.

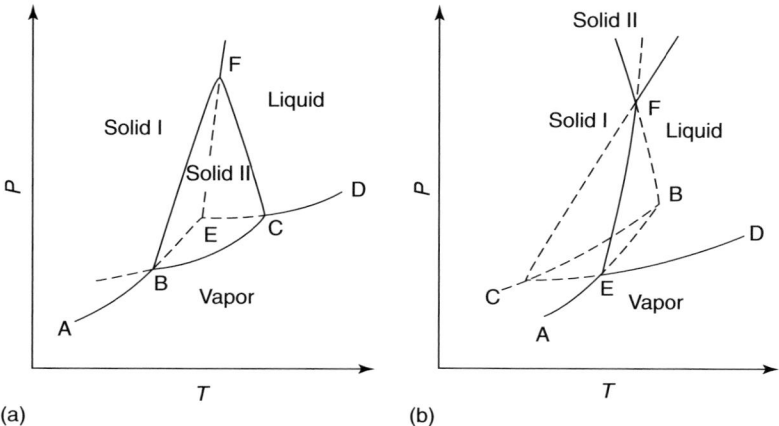

(a)   (b)

Figure 2 Phase diagrams of pressure versus temperature for a single crystalline compound with two crystalline forms showing: (a) enantiotropic behavior (AB, sublimation curve of solid I; BC, sublimation curve of solid II; BF, solid I ⇔ solid II equilibrium curve; CF, solid II melting curve); (b) monotropic behavior (AE, sublimation curve of solid II; EF, melting curve of solid II). Dashed lines represent metastable transformation curves.[7] (Reproduced with permission from Giron, D.; Grant, D. J. W. In *Handbook of Pharmaceutical Salts: Properties, Selection and Use*; Stahl, P. H., Wermuth, C. G., Eds.; IUPAC, Verlag Helvetica Chimica Acta: Zürich, Switzerland, 2002, pp 42–81.)

The relative position of the G-isobars ($\Delta G_{A \to B}$) of different modifications can be determined by solubility experiments in a given solvent (S_A is the saturation solubility of the modification A and S_B the saturation solubility of the modification B):

$$\Delta G_{A \to B} = RT \; \ln(S_A/S_B) \qquad [4]$$

Each crystal form has its own heat capacity, which is a function of the enthalpy H and the temperature. The heat capacities of solids at constant volume and constant pressure are about the same. The H isobars of two modifications are parallel. Their distance is the transition enthalpy ΔH_t.

Burger[35] proposed that energy/temperature diagrams of the Gibbs energy and the enthalpy should be plotted, as functions of temperature. This proposal is the fundamental tool for the solution of complex polymorphic systems.[36] As shown in **Figure 3a** and **b**, a notable difference between enantiotropy and monotropy is the melting enthalpy of the higher melting form. In the case of enantiotropy, the higher melting form has the lower melting enthalpy. In the case of monotropy, the higher melting form has the higher melting enthalpy.

The relationship between melting enthalpies of two solid phases A and B and the heat of transition is approximately:

$$\Delta H_t = \Delta H_A^f - \Delta H_B^f \qquad [5]$$

The phase diagrams of solvates and hydrates are more complex since binary mixtures are implied with different compositions. The new compound may have a congruent melting or a noncongruent melting (**Figure 4**). By heating, the melting of the solvate or hydrate may be observed followed by transformation to an anhydrous form, or the solvent is evolved with an endothermic transition into the anhydrous form. A series of such binary phase diagrams have to be considered if several compounds are formed. These diagrams are fundamental for the understanding of crystallization and drying steps.

The glass transition of amorphous solids is a second order transition and is characterized by a change of heat capacity. Above this temperature, the increase in molecular mobility facilitates spontaneous crystallization into the crystalline form with an exothermic enthalpy change after the glass transition. The use of amorphous forms is attractive, particularly for sparingly soluble compounds because of the enhanced solubility and dissolution rate over the crystalline state leading to increased bioavailability. Amorphous additives or excipients can give miscible phases with a new temperature of glass transition higher than the drug substance with the ability to stabilize it in the amorphous state. However, the amorphous state is thermodynamically unstable. The glass transition temperature, T_g, is lowered by water or other additives according to the Gordon–Taylor equation,[40] facilitating conversion to the rubbery state and hence facilitating crystallization. Moreover, uncontrolled crystallizations can occur at any time, and are accelerated by the presence of water. **Figure 5** shows the changes in Gibbs energy between crystalline state, amorphous state, and liquid state.[36]

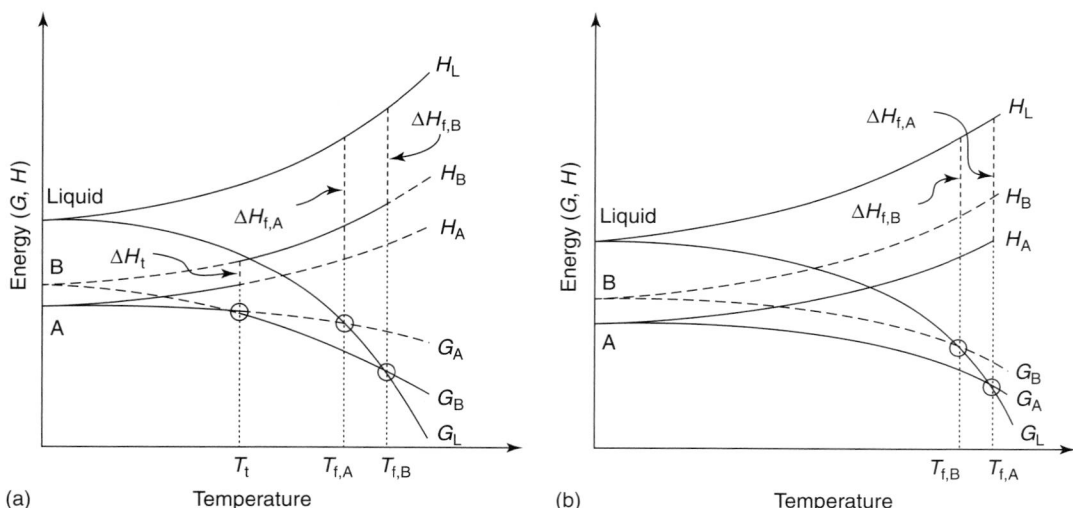

Figure 3 Energy diagrams according to Burger showing the plots of Gibbs energy G and enthalpy H versus temperature for the solid and liquid phases of a single compound with two crystalline forms showing (a) enantiotropy and (b) monotropy. (Reproduced with permission from Giron, D.; Grant, D. J. W. In *Handbook of Pharmaceutical Salts*: *Properties, Selection and Use*; Stahl, P. H., Wermuth, C. G., Eds.; IUPAC, Verlag Helvetica Chimica Acta: Zürich, Switzerland, 2002; pp 42–81.)

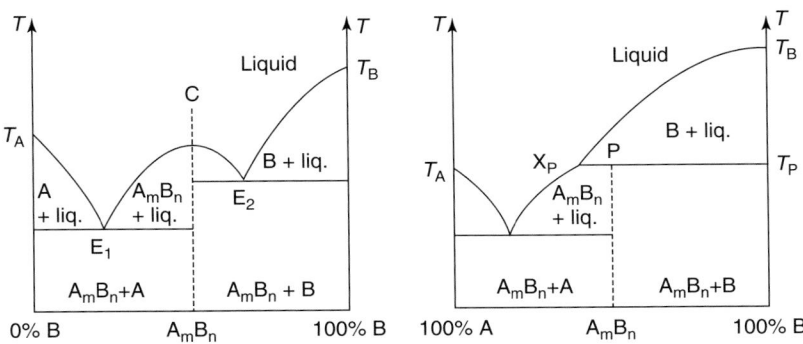

Figure 4 Phase diagrams of binary mixtures with formation of a new compound: (a) congruent melting point; and (b) incongruent melting point with dissociation of both components in liquid state (mostly occurring for solvates). (Reproduced with permission from Giron, D.; Grant, D. J. W. In *Handbook of Pharmaceutical Salts: Properties, Selection and Use*; Stahl, P. H., Wermuth, C. G., Eds.; IUPAC, Verlag Helvetica Chimica Acta: Zürich, Switzerland, 2002, pp 42–81.)

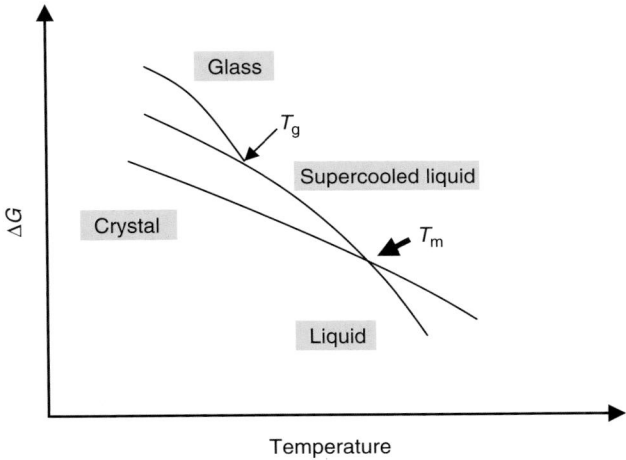

Figure 5 Graphic plot of the changes in Gibbs energy among crystalline state, amorphous state, and liquid state. (Reproduced with permission from Giron, D.; Grant, D. J. W. In *Handbook of Pharmaceutical Salts: Properties, Selection and Use*; Stahl, P. H., Wermuth, C. G., Eds.; IUPAC, Verlag Helvetica Chimica Acta: Zürich, Switzerland, 2002, pp 42–81.)

Liquid crystals are a state of order between crystals and liquids. They have imperfect long-range orders of orientation and position. Thus, they can be fluid like a liquid and they can have anisotropic properties like crystals (e.g., cholesterol ester, phenyl benzoates, paraffins, surfactants, glycol-lipids). Thermotropic phases such as fenoprofen Na, Ca, or K with water have been reviewed.[41] The solid properties of chiral substances depend on phase diagrams and have been discussed elsewhere.[42]

When the solid phases are in contact with a solvent, the Gibbs phase rule applies. Depending on the temperature, different domains of stability of each crystal form may exist. In the case of enantiotropic polymorphism and in the case of solvates there are points where the solubility of forms is the same. An example of solubility versus temperature in the case of pseudopolymorphism is given in **Figure 6** for an acidic compound N existing as anhydrous, monohydrate, and trihydrate forms.

5.21.2.4   Methods for Screening, Detection, Characterization, and Modeling

As demonstrated above, different solid phases can be stable in different environments. In the case of enantiotropy, for example, the low melting form is the thermodynamically stable form at temperatures below the transition point. A hydrate can crystallize from aqueous solution and an anhydrous phase from organic solutions, both crystalline forms

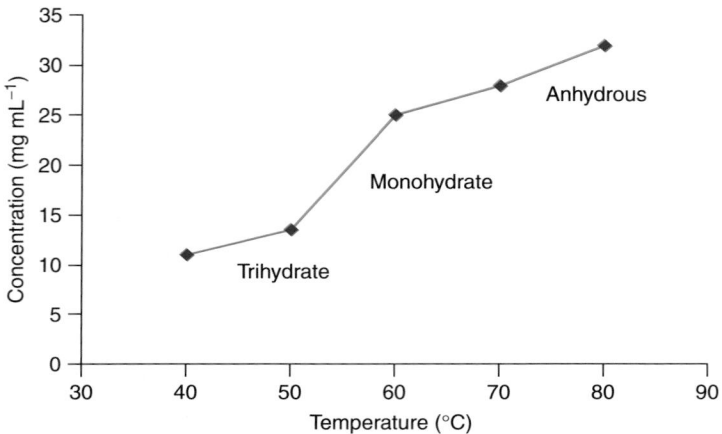

Figure 6 Solubility curves (expressed as measured concentration) versus temperature. Experimental curves for compound N exhibiting an anhydrous crystalline form, a monohydrate, and a trihydrate.

being stable. An anhydrous phase can be stable at low relative humidity (RH) and a hydrate at high relative humidity. But 'metastable' or 'unstable' forms may coexist resulting from supersaturation or from desolvation or dehydration. In early development, salt selection and polymorphism screening are strongly correlated.[16,17]

All investigations on polymorphism should consider crystallizations and parameters such as temperature, coprecipitation, and slurry in different solvents and under different conditions in order to explore a great number of parameters.[17,23,43–49] Relevant high-throughput systems are now used at the drug discovery stage and a number of companies offer specialist equipment or services. These include Tessella Inc., USA; Solvias, Basel, Switerland; RPD TOOL, Muttenz, Switerland; Zinsser Analytic GmbH, Eschborner, Germany; Advantium Technology BV, The Netherlands; SCCI Inc, West Lafayette, USA; Ricerca Bioscience LLC, USA; Pharmaterials Ltd, London, UK; Transform Pharmaceuticals, USA; and Symix, USA. A list of preferred solvents based on electronegativity has been proposed.[50]

Quenching from the melt, lyophilization, and precipitation by adding cosolvent can yield the amorphous state. Heating/cooling curves produced using differential scanning calorimetry (DSC) or the combined technique DSC–spectroscopy permit forms that are not obtained from solvents to be observed.[23,34] Screening methods may yield the drug candidate in crystalline form via salts if feasible, detect solvate formation, manufacture 'unstable' or 'metastable' forms, study hydrate formation by water sorption–desorption experiments (hygroscopicity), consider solvate formation by including solvent atmosphere studies to the crystallizations, determine the effect of pressure and milling, grinding, or granulating, and manufacture the amorphous phase.

In crystallization systems, the thermodynamic stable form(s) depends on temperature and pH, whereas in atmosphere systems they depend on temperature and relative humidity. The analytical method here is generally x-ray diffraction followed by thermal analysis or spectroscopy. Thereafter, pure forms have to be manufactured and characterized, and thermodynamic relationships understood. For this purpose, DSC and microcalorimetry are the best techniques.[23,34,51,52] The thermogravimetric analysis can also detect solvates and hydrates.

Very often the differences in melting energies of the crystalline forms are too small for fair characterization; therefore, it is convenient to mix the solid phases and to equilibrate saturated solutions as slurries in different solvents and to analyze the insoluble solid. The transformation of metastable phases in the thermodynamic stable phases takes place via solvent-mediated transition.

A knowledge of the temperature of transition in cases of enantiotropy or solvates (hydrates) is often crucial. It can be determined by equilibration at different temperatures, evaluation from DSC, or calculation from DSC data. However, using the relationship of solubility to temperature is not accurate enough according to Burger.[37] **Table 1** summarizes the data obtained for a neutral purine derivative MKS492[53] by using a combination of techniques.

X-ray diffraction is widely used for detection in high-throughput screening (HTS), calculation of the phase purity by calculating theoretical x-ray diffractions from the single crystal structure, and for quantification.[23] Modeling software allows crystal structures to be calculated from x-ray diffraction patterns.[54–56] The use of synchrotron is the most sophisticated technique. Polymorphism prediction software can contribute to such studies, but there are limitations.[57,58] The modeling of morphology can be successfully used to modify the crystal habit by changing the solvents of crystallization.

Table 1 Example of characterization of polymorphs of the neutral purine derivative MKS492

Characteristic	Form A	Form B	Form B'	Form C	Form D
DSC onset (°C)	111	–	128	118	109
Melting enthalpy (J g^{-1})	93	98^a	92	89	65
Transition heat (J g^{-1})	–	6		–	–
Temperature (°C)		108–112			
Density (g cm^{-3})	1.400	1.422		1.411	
Solubility water 20 °C in % (w/w)	0.27	0.17		0.2	0.18
Solubility ethanol 25 °C in % (w/w)	2.4	1.5		2.5	
Solubility isopropanol 25 °C in % (w/w)	1.3	0.9		1.5	
Solubility acetone 25 °C in % (w/w)	7.1	4.3		7.2	
Dissolution rate in water at 37 °C:					
Time in s for 50%	88	109		99	340
Time in s for 80%	219	315		330	972
Intrinsic dissolution rate, 37 °C in mg min^{-1} cm^{-2}:					
In water		0.12		0.13	
In HCl 0.1N		0.18		0.19	
X-ray angle 2θ in deg.	12.7	7.1		4.9	3.8

According to Giron, D.; Piechon, P.; Goldbronn, C.; Pfeffer, S. *J. Therm. Calorim.* 1999, *57*, 61–75.
^{a} Calculated.

5.21.3 Physicochemical Properties in the Solid State

5.21.3.1 Overview of Relevant Properties for Drug Formulation

All physicochemical characteristics of the solid state are involved in polymorphism and pseudopolymorphism. The main properties affected are color,[59–61] morphology (crystal shape), volume, density, melting and sublimation temperatures, heat capacity, conductivity, viscosity, crystal hardness, refractive index, solubility, dissolution rate, stability, hygroscopicity, processability, and solid-state reactions. The single crystal structure of crystalline substance delivers the density and the morphology.[23] The density can be measured by pycnometry. According to the polymorphism rule of Burger, the higher the density, the greater the stability. Some polymorphs may have differences of melting point smaller than 1°C or larger than 100 °C (e.g., riboflavin).[62] For ADME studies, solubility, dissolution in aqueous media as well as hygroscopicity and chemical stability are obviously of relevance.

5.21.3.2 Solubility

It is trivial knowledge that the pH-dependent solubility of an ionizable drug depends on its pK_a. The relation of solubility to temperature obeys the phase rule (eqn [1]). In cases of enantiotropic polymorphism and pseudopolymorphism, there is a thermodynamically fixed temperature at which the crystal forms in equilibrium have the same solubility (**Figure 6**). Measurements of solubility should take polymorphism into account. Immediate solubility can be different from equilibrium solubility and supersaturation can result from kinetic slow transformation into the stable form in the solvent at the temperature of measurement (*see* Section 5.21.4). The concentrations should be measured as a function of time when determining the solubility of the different forms because of possible transformation during the measurement. Furthermore, the temperature should be kept constant and should be accurately defined.

Solubility can be measured by phase solubility analysis,[63] where an excess of the solid remains in equilibrium with the saturated solution. The solution can be analyzed by gravimetry, high-performance liquid chromatography (HPLC), or spectroscopy. High-throughput methods of solubility determination have been developed by UV spectrometry or by nephelometry based on the appearance of the solid phase. In all solubility determinations, soluble impurities can significantly influence the measured solubility.

The 'equilibrium' solubility must be studied with good knowledge of the different phases and with an analysis of the residual solid. Since several thermodynamic stable species may exist depending on the solvent, pH, temperature, and relative humidity, it is more appropriate to define the 'equilibrium solubility' of the stable forms in the medium and at the temperature considered. In the case of salts, exchange of the counterion with the buffers has to be considered.[16,17] A recent review[67] compiled solubility ratios of polymorphs of 55 compounds. The study concluded that the typical ratio of solubility is less than 2. In addition to the experimental data, the calculations of solubility based on thermal data, melting point, and melting energy showed good agreement. The solubility of solvates in water is higher than the solubility of nonsolvated forms.[63] This is demonstrated by the example of a compound P in **Figure 7**. Hydrates are less soluble (compound Q in **Figure 8**). The solubility of amorphous solids is generally higher than that of crystalline solids. **Table 2** lists some published examples.

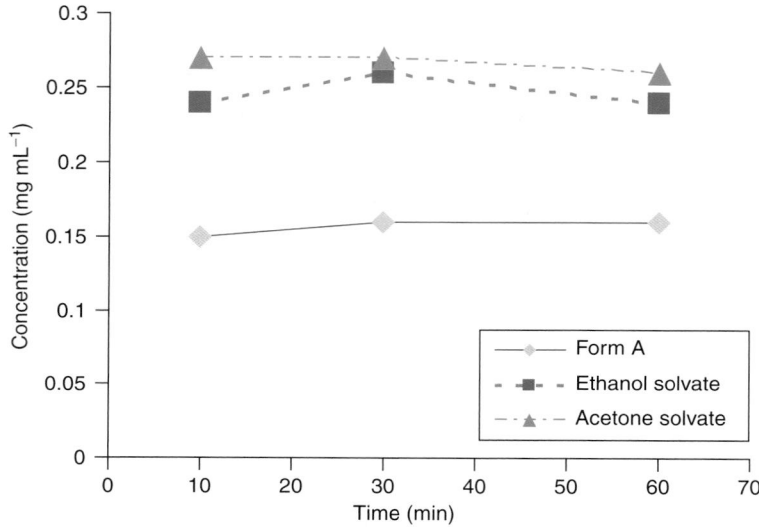

Figure 7 Water solubility of solvates of compound P compared to the crystalline form A.

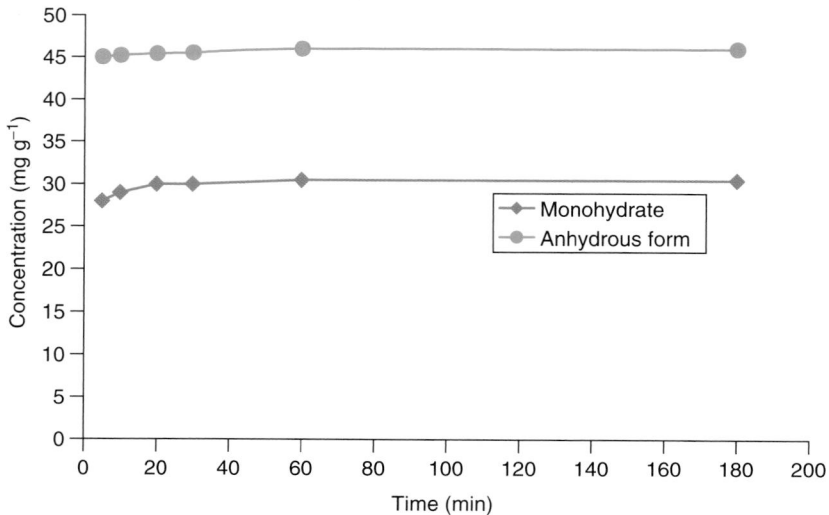

Figure 8 Water solubility of compound Q as a hydrate and an anhydrous form.

Table 2 Some examples of the impact of amorphous solid form on solubility

Substance	Forms	Solubility ratio	Reference
Glibenclamide	Amorphous/crystalline	14 (23 °C, buffer)	Mosharraf and Nystrom[68]
Griseofulvin	Amorphous/crystalline	1.4 (21 °C, water)	Flammin et al.[69]
Indometacin	Amorphous/gamma	4.5 in water (5 °C and 25 °C)	Hancock and Parks[70]
Iopanoic acid	Amorphous/crystalline I	3.7 (37 °C phosphate buffer)	Stagner and Guillory[71]
MK0591 sodium	Amorphous/crystalline	Approx. 1000	Clas et al.[72]

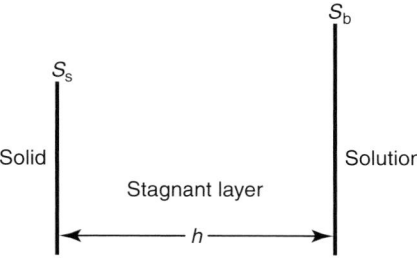

Figure 9 Scheme of interface during dissolution (S_s, solubility at interface; S_b, solubility in solution).

5.21.3.3   Dissolution Rate and Intrinsic Dissolution Rate

For poorly soluble compounds, the absorption rate may be limited by the dissolution rate.[64–66] Dissolution rate (DR) and intrinsic dissolution rate (IDR) are defined according to the Noyes–Whitney equation:

$$\text{DR} : \mathrm{d}M/\mathrm{d}t = (A \bullet D/h) \bullet (S_s - S_b(S_s - S_b)) = (AD/h) \bullet (S_s) \qquad [6]$$

$$\text{IDR} : (\mathrm{d}M/\mathrm{d}t) \bullet (1/A) = kS_s \qquad [7]$$

where A is the total surface area, D the diffusion coefficient, h the stagnant layer, S_s the solubility at the interface, and S_b the solubility in solution. **Figure 9** illustrates the dissolution interface.

The dissolution rates of powders are often measured by the flow-cell method. **Figure 10** illustrates the example of three polymorphic crystalline forms of the neutral purine derivative MKS492 (**Table 1**). The impact of polymorphs is not very high since the drug is dissolved within 20 min. The case of metolazone[7] shown in **Figure 11** is quite different. The stable form is so insoluble that even after 10 h 80% dissolution is not reached.[7]

The dissolution rate per unit surface area (the intrinsic dissolution rate, eqn [7]) is independent of particle size and is therefore very appropriate for polymorphic studies. In the 'disk' method, the powder is compressed to produce a compact disk or tablet. The method is described in the US Pharmacopoea.[73] Only one face of the disk is exposed to the dissolution medium and the cumulative amount dissolved per unit surface area is determined by ultraviolet spectrophotometry until 10% of the solid is dissolved. The slope of the plot of mass dissolved per unit surface area against time gives the intrinsic dissolution rate in appropriate units, e.g., $\text{mg min}^{-1}\text{cm}^{-2}$. If a change in the slope is observed during the course of the experiment, this implies a change in the solid phase exposed to the solvent occurs during the experiment. Dissociation of the unionized form as well as a polymorphic change can occur.[17] The ratio of intrinsic dissolution of two phases is the same ratio as their solubility. According to Kaplan,[74] there is no problem of dissolution for a compound if the IDR > 1. IDR is considered to have good predictability for new drugs.[75]

The pharmaceutical significance of polymorphism is exemplified in **Table 3** for poorly soluble substances. An amorphous form can improve behavior, but by a factor < 5. If different salt forms are possible with the same counterion, their IDR is also very relevant (**Table 3**). In the example given in the last two columns of **Table 3**, the monosodium salt is very soluble, the monohydrate somewhat less so, and the hemisalt is poorly soluble.

Drug solubility and dissolution rate are not constant throughout the gastrointestinal tract since S_s is pH dependent. **Figure 12** illustrates the pH-dependent IDR of a monobasic compound R as base with a pK_a of c. 7 in different hydrochloric acid (HCl) solutions. The effect of counterions on the IDR of this compound can be seen in **Figure 13**.[3,17,139]

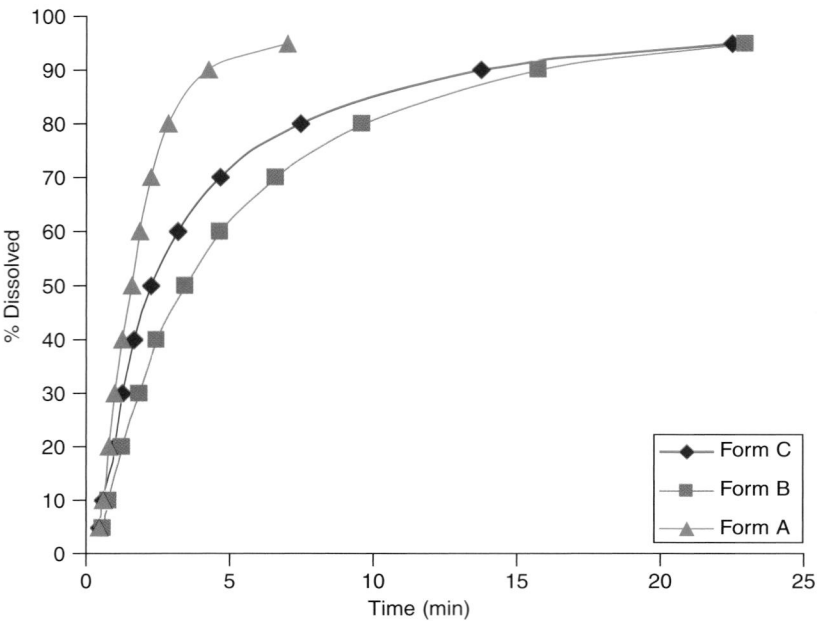

Figure 10 Dissolution rate of three crystalline forms of the neutral purine derivative MKS492.[53] All forms are quickly dissolved.

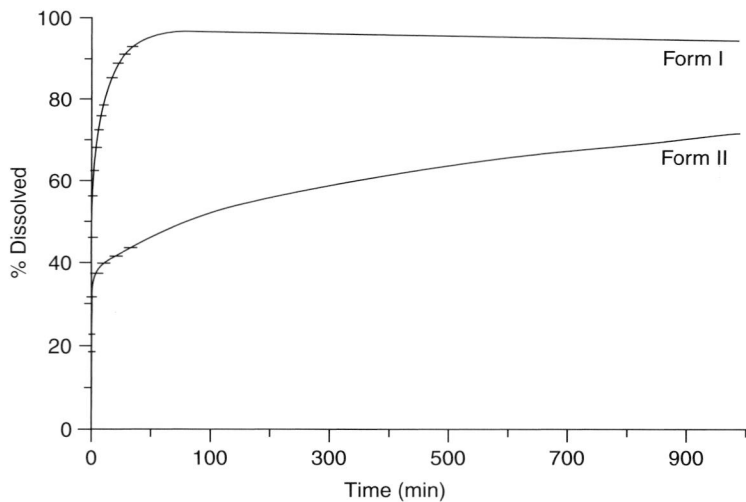

Figure 11 Dissolution rate of two forms of metolazone. Both samples were micronized and had the same particle size distribution. The difference in behavior is significant. After 900 min (15 h), form II had not reached 80% dissolution. (Modified from Giron, D. *Labo-Pharma-Probl. Techn.* **1981**, *307*, 151–160.)

5.21.3.4   Hygroscopicity and Interaction with Water Vapor Expressed by Sorption Isotherms

At a given temperature, the relative humidity (RH) is determined as the ratio of actual water vapor pressure over saturated water vapor pressure. The behavior of drugs at different temperatures and humidities for different climates is generally studied by gravimetry. The RH for gravimetric studies of water sorption and desorption can be controlled by saturated salt solutions[76] or by continuous humidification of a stream of air or nitrogen to which the solid sample is exposed (Dynamic Vapour Sorption (DVS), Surface Measurement Systems Ltd, UK). Solid-state hydration may occur leading to the formation of hydrates. X-ray diffractometers may be equipped with special sample cells for exposing the

Table 3 Examples of values of the intrinsic dissolution rate (IDR in mg min⁻¹ cm⁻²) for polymorphs

Drug as base[23]		Neutral drug[23]		Ipanoic acid[71]		Weak diacidic drug as sodium salt[23]	
Polymorph	IDR in water with 0.2% LDAO	Polymorph	IDR in water	Polymorph	IDR in water	Salt/form	IDR in water
Amorphous	0.048	Amorphous	0.269	Amorphous	0.0703	Monosalt Na	43.6
Form B	0.035	Form A	0.117	Form I	0.00739	Monosalt monohydrate	17.6
Form D	0.011	Form B	0.085	Form II	0.0117	Hemisalt	0.40

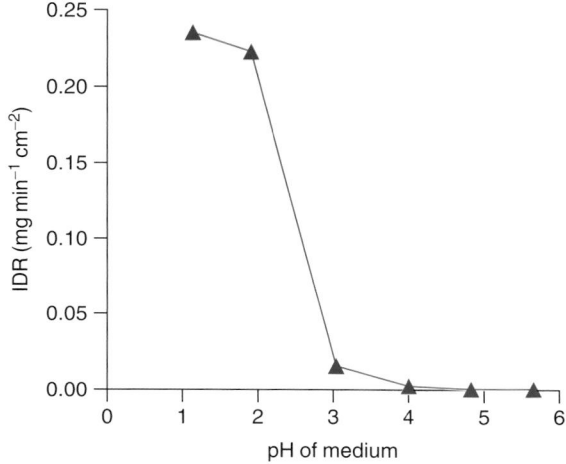

Figure 12 Intrinsic dissolution rate versus pH of the monobasic compound R (HCl solutions). (Reproduced with permission from Giron, D. Pharmacokinetic Profiling in Drug Research. In *Biological, Physicochemical and Computational Strategies*. Testa, B., Krämer, S., Wunderli-Alllenspach, H., Folkers, G., Eds.; Verlag Helvetica Chimica Acta: Zürich, Switzerland, 2006, 307–329.)

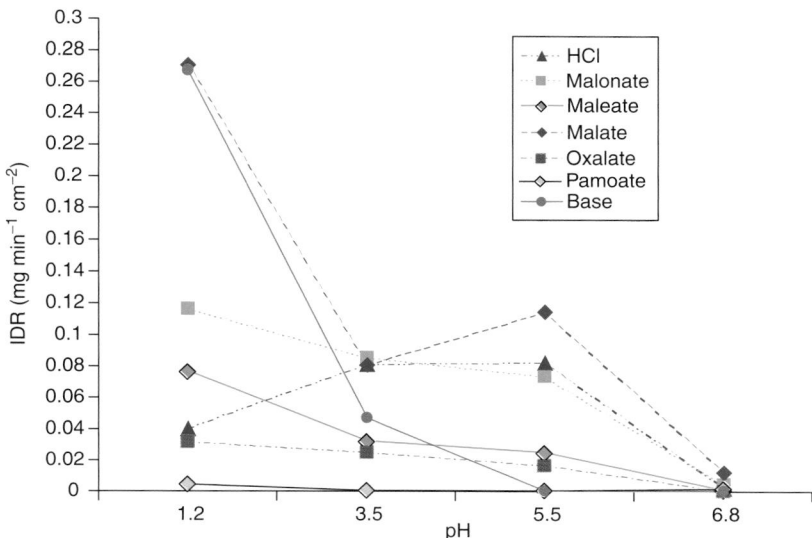

Figure 13 pH-dependent intrinsic dissolution rate of various salts of compound R in different buffer media. (Reproduced with permission from Giron, D. Pharmacokinetic Profiling in Drug Research. In *Biological, Physicochemical and Computational Strategies*. Testa, B., Krämer, S., Wunderli-Alllenspach, H., Folkers, G., Eds.; Verlag Helvetica Chimica Acta: Zürich, Switzerland, 2006, 307–329.)

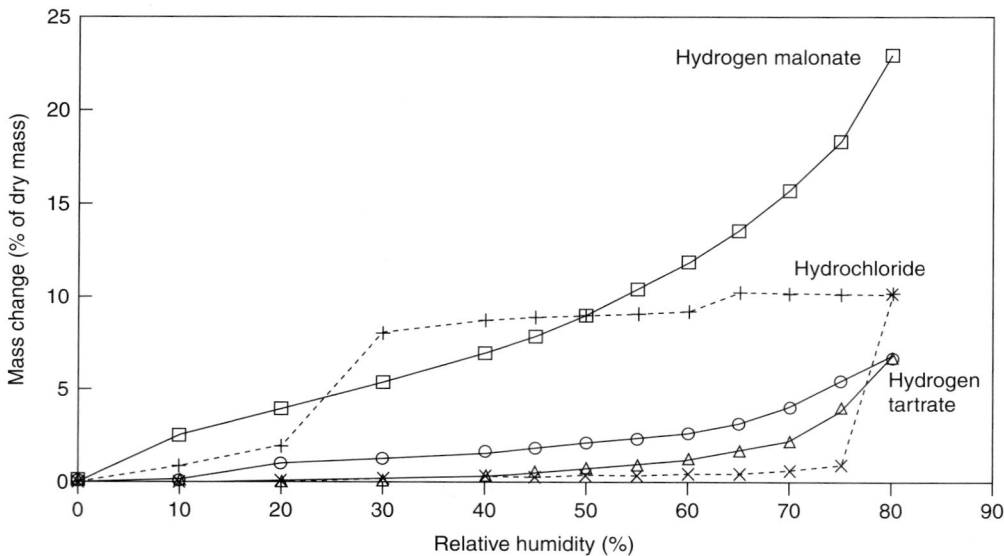

Figure 14 Hygroscopicity: water sorption–desorption isotherms of different salts of compound S. The hydrochloride transforms into a hydrate. The hydrogen malonate is hygroscopic until deliquescence. The hydrogen tartrate is reversibly hygroscopic. The base and the hydrogen maleate are not hygroscopic until 90% RH (not shown in the figure).

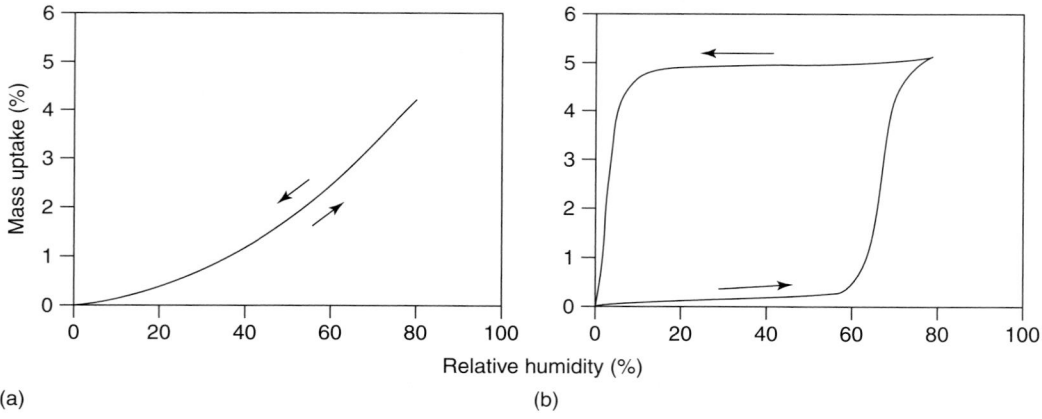

Figure 15 Examples of typical water sorption–desorption isotherms: (a) for amorphous state with reversibility and (b) for an anhydrous crystalline substance converted to a stable hydrate with hysteresis in the desorption. (Reproduced from Giron, D.; Grant, D. J. W. In *Handbook of Pharmaceutical Salts: Properties, Selection and Use*; Stahl, P. H., Wermuth, C. G., Eds.; IUPAC, Helvetica Chimica Acta: Zürich, Switzerland, 2002; pp 42–81.)

sample to controlled temperature and humidity. **Figure 14** shows the sorption–desorption isotherms of different salt forms of a compound S.[14] **Figure 15** shows typical curves: (a) with a complete reversibility, a phenomenon generally observed for amorphous solids; and (b) with a typical hysteresis for a defined hydrate formation.

5.21.3.5　Chemical Stability

Chemical reactivity in the solid state is correlated with the nature of the crystalline modifications. Walkling[77] found that the two crystalline modifications of fenretinide behave quite differently. After 4 weeks at 25 °C, the stable form showed no detectable degradation, whereas the unstable form showed 8% degradation.

The amorphous state is very reactive.[78] **Table 4** illustrates the difference in reactivity between the two crystalline forms and the amorphous form of a very insoluble compound T (example 1). The crystalline form A is more stable than the crystalline form B and the amorphous form. Also, for a small molecule such as compound U (example 2) that is very

Table 4 Influence of polymorphism on chemical stability in the solid state

	Degradation (HPLC)	
Example 1 (compound T, neutral, MW *c.* 1200, insoluble in water)	1 month at 80 °C (oxygen/water)	
Crystalline form A	No degradation	
Crystalline form B	0.5–1.5% degradation	
Amorphous form	2–3.5% degradation	
Example 2 (compound U, base pK_a 8, MW *c.* 300, very soluble in water)	1 month at 60 °C	
Crystalline form	No degradation	
Amorphous form	5% degradation	
Example 3 (compound V, neutral, MW *c.* 900)	2 weeks at 50 °C	Exposition 1200 klx h
Monohydrate A	No degradation	10%
Monohydrate B	12%	23%
Example 4 (peptide W, as base)	1 week at 70 °C	Exposition 300 klx h
Crystalline form	10%	2%
Amorphous form	80%	38%

soluble in water, the amorphous form is very unstable compared to the crystalline one. Large differences are observed for the two polymorphs, a dihydrate of compound V (example 3) with several chemical unstable functions. In the amorphous state of the peptide W (example 4), both the base and its hydrochloride were very unstable. However, the base could be obtained as a crystalline material with a substantial gain in stability.[23]

Polymorphs can behave differently under exposure to light as described for several drugs.[23,79–82] In the case of ethoxycinnamic acid, different photolytic degradation products were obtained for each form.[83] Generally, stability decreases as the particle size decreases.

5.21.4 Kinetic Aspects

The processing of drug substances and drug products involves solvent(s), temperature and pressure changes, as well as mechanical stress; different solid phases may coexist in the drug product. Organic substances show supersaturation behavior and unstable solid phases, which should not exist in defined temperature, pressure and humidity may behave like stable forms. These solid metastable phases obtained outside their domains of stability will convert to the thermodynamically stable forms at given temperatures, pressures, and relative humidities. These conversions driven by thermodynamics are governed by kinetics and are influenced by impurities, particle size, crystal defects, the presence of seeds, and temperature. An energy barrier has to be attained to start a transformation.[84] Many examples show that in early development metastable forms – often called 'disappearing polymorphs'[138] – or amorphous solids are first obtained and then later thermodynamically stable forms can appear unexpectedly. Low purity of research samples might hinder crystallization. In the case of ritonavir, this acute problem led to the withdrawal of a marketed formulation.[85] Basic thermodynamic relationships may be understood by considering the phase diagrams and the stability domains: melting enthalpy rule by DSC, solvent-mediated transformation studies, dissolution, and solubility rule. **Figure 16** illustrates the decrease of solubility due to transformation in the thermodynamically stable form in water for a compound X.

Figure 17 shows an old example of an abandoned indole derivative compound Y for which a slow transformation was initiated by the presence of seeds during storage.[83] Two batches of the base were manufactured as a metastable crystalline modification. Analysis of samples using formal stability studies showed transformation of the second batch while the first batch remained unchanged. The only difference between the two batches was the presence of a small amount of the insoluble stable form in the second batch.

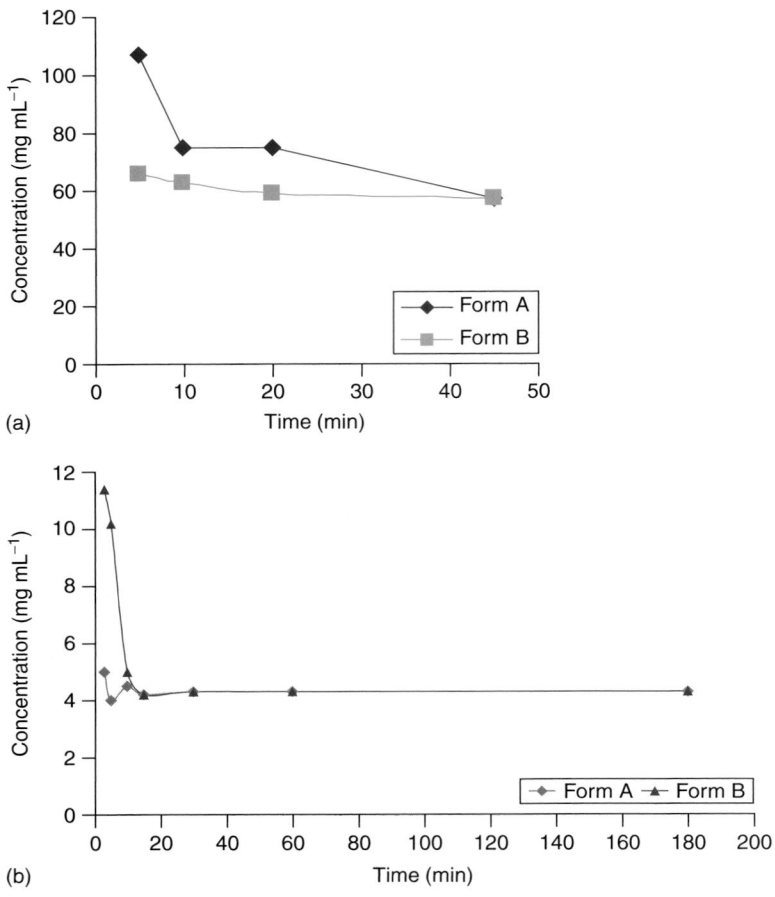

Figure 16   The case of compound X: (a) transformation of A into B; and (b) transformation of A and B into a monohydrate. The apparent solubility is the solubility of the metastable form. The thermodynamic solubility is the solubility of the stable form at saturation.

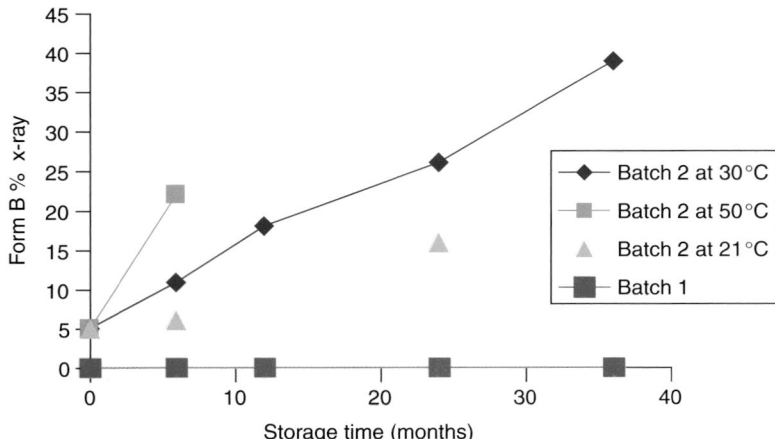

Figure 17   Role of seeds in transformation during storage (compound Y). Two batches put in a stability program showed different behavior. Batch 1 did not change; however, because of the presence of seeds of the stable form B, batch 2 was transformed into form B. The higher the temperature, the faster the transformation (comparison of data after 6 months at 21 °C, 30 °C, and 50 °C).[83]

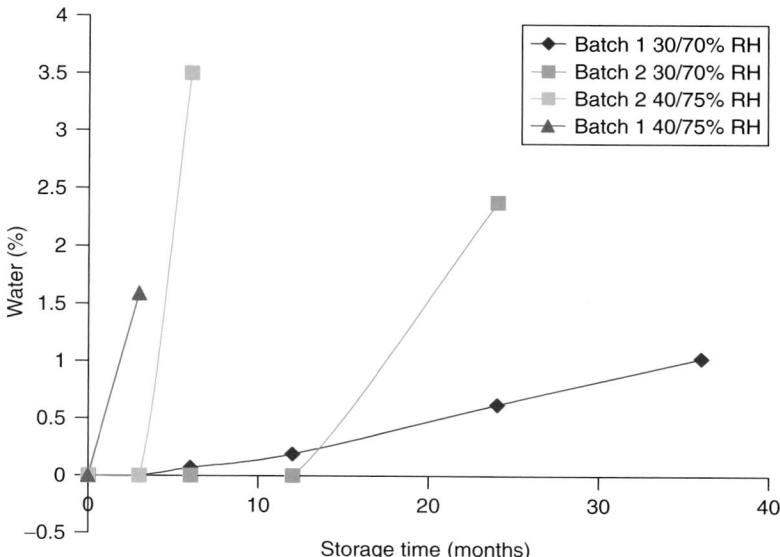

Figure 18 Kinetic effect, compound Z: slow transformation of a crystalline anhydrous form into a monohydrate when stored in a nonmoisture-protecting packaging. The transformation is followed by the measurement of water in %. The transformation rate is batch dependent as demonstrated for batch 1 and batch 2 after 3 months at 40 °C/75% RH and 24 months at 30 °C/70% RH. No transformation occurred in tight packaging.

Figure 18 shows the slow transformation of an anhydrous form of a compound Z into a monohydrate during storage in nonprotective packaging for several months at 40 °C/75% RH and 30 °C/70% RH. The two batches showed different kinetics of transformation.

Mechanical stress can induce transformation in metastable or amorphous forms.[86,87] Amorphous solids are very difficult to stabilize, give product variability and stability.[88,89] Prediction of the fragility of amorphous solids can be calculated from thermal data[90,91] and storage condition rules proposed.[91] However, water greatly reduces the temperature of the glass transition[92] and spontaneous crystallization may occur during storage. This recrystallization is the basis of the quantitative determination of amorphous content in drug substances.[93–95]

Kinetic effects have to be considered for ionizable drug substances since competition with the chloride counterion can play a role in the gastric fluid.[96,97] For acidic salts, dissociation into the base may slowly occur.[16,17]

5.21.5 Modulation of Physicochemical Properties

According to Johnson and Wolfgang,[98] assessment of the potential toxicity of new pharmaceuticals is also correlated to their bioavailability in toxicology studies. They recommend a scheme for optimizing oral bioavailability in toxicology studies, which is comparable to the recommendations for developability assessment[2] of poorly soluble substances. This section gives some suggestions for modifying the physicochemical properties of drug substances in order to enhance their bioavailability.

5.21.5.1 Effect of Salt Form

Low bioavailability is most common with oral dosage forms of poorly water-soluble, slowly absorbed drugs. More factors can affect bioavailability when absorption is slow or incomplete than when it is rapid and complete. The alteration of the pH of the formulation is the first way to solubilize acidic or basic compounds. Lin[99] described the example of L-735,524, a human immunodeficiency virus (HIV) protease inhibitor. When given orally as a suspension in 0.5% methylcellulose (MC, pH 6.5), its bioavailability was 18% in the rat and 16% in the dog. When the same dose was given in a citric acid solution (pH 2.5), bioavailability rose to 75% in the dog but remained at 20% in the rat.

For most substances limited absorption is due to solubility; therefore, for acidic and basic compounds the relationship between solubility and pH (or intrinsic dissolution rate and pH) has to be assessed (**Figure 12**). Salt formation can modify this behavior (**Figure 13**), but reprecipitation or dissociation may appear during absorption in living cells.

The fourth example given in **Table 3** (*see* Section 5.21.3.3) shows the impact of salt formation on the IDR of a weak diacidic drug. A factor of about 100 in IDRs was found between the sodium hemisalt and the monosodium salt.

For the basic protein kinase inhibitor LY333531, which was developed for the treatment of complications in diabetes, Engel and co-workers[100] selected the salt form based on a two-tier approach. The following salts were manufactured, hydrochloride, sulfate, mesylate, succinate, tartrate, acetate, and phosphate, and their physicochemical properties compared. In the first tier, selection was done based on crystallinity, solubility, and manufacturing purification. The remaining candidates were hydrochloride anhydrous and monohydrate and the mesylate monohydrate. The polymorphic behavior, stability, filterability, purification behavior, and bioavailability in dogs were compared. The mesylate monohydrate was selected due to a 2.6-fold higher bioavailability. This difference was presumed to be due to the fivefold higher solubility of mesylate in water; the common ion effect of hydrochloride in the GI tract was also discussed.

Piroxicam is a zwitterion with two pK_a values: 1.86 and 5.46. Its monoethanolamine salt was proven to have the highest bioavailability corresponding to the highest dissolution rate at pH 6.8.[101] Patent applications for the enhancement of bioavailability are increasing. For example, the arylsulfonic acid salts improve oral bioavailability of imidazoylpyrimidine antiviral agents as base by a factor greater than 2.[102] A factor of about 10 was also found between bendazac and its lysine salt.[103] A recent example demonstrating the importance of salt selection in obtaining in vivo exposure has been given for the weak acid PNU-103017 and its calcium, sodium, lysine, and arginine salts.[104] The solubility of codeine base at $8.3\,mg\,mL^{-1}$ in water was increased to $435\,mg\,mL^{-1}$.[105]

Enhanced solubility does not necessarily translate into better in vivo absorption. There are several reports of salts with differing solubilities behaving similarly in bioavailability studies. Better solubility may simply be a pH effect that is cancelled in the gastric or intestinal milieu, with solubility changing to reflect local pH. Conversely, it is also feasible that the pH engendered by a salt in its microenvironment facilitates dissolution, the salt acting as its own buffer. Once in the solvated state, the dynamics of transport or reprecipitation may be such that there is a net enhancement of amount dissolved and absorbed.

According to Davies, the motto "changing the salt, changing the drug"[106] should be taken into consideration. Once a drug has been marketed, there may be sound reasons for reformulating it in a different salt form to change its physicochemical properties. An example is provided by the analgesic propoxyphene, which was originally formulated as a hydrochloride salt. Propoxyphene was widely used in a fixed-dose combination with aspirin, but since aspirin proved to be unstable in close physical contact with propoxyphene hydrochloride, a reformulation was necessary. When propoxyphene was reformulated as a napsylate salt, there was no problem of aspirin instability. The relative insolubility of the napsylate salt form compared with the hydrochloride was also an advantage, as it reduced the potential for parenteral abuse of propoxyphene and its toxicity in oral form. The pamoate salt of pyrantel is reported to be three times as effective as the citrate against large bowel parasites, including resistant strains, because of its lower rate of absorption and consequently greater retention in the gastrointestinal tract. Some cations and anions are known to be associated with toxic effects and will contribute to the intrinsic toxicity of the salt form. For example, lithium cations have no toxic effect in small quantities but when ingested in large amounts can cause irreversible damage to the kidney. Similarly, tartrate anions, which are usually absorbed only minimally from the gastrointestinal tract, can cause renal damage if they reach the circulation in high concentrations. In addition, pravadoline maleate caused renal tubular lesions in the dog, as a result of maleic acid formed from the maleate anion.[106]

The relevance of salt selection and its impact on bioavailability were first emphasized by Berge and colleagues[107,108] who discussed its most important factors. Examples of this and discussions about the counterion have been reviewed recently,[3] with in-depth discussions on possible counterions and their physical and toxicological properties, the physicochemical properties of salts and polymorphs, the pH effect, as well as the toxicological impact. Examples of salt selection and manufacturing are presented and some patent issues discussed. **Table 5** offers a comparison of the bioavailability of the acid (3%), the monosodium salt (15%), and the disodium salt (50%) of PNU-140690.[109] **Table 5** also gives the results of the solubilities at saturation (maximal concentration in water) and the IDR (see [3]: p. 93). The ranking of bioavailability and of in vitro results are the same.

Counterions can affect all properties from melting point to solubility, hygroscopicity, stability, crystallinity, morphology, and feasibility. The increase in solubility can be obtained with strong acids for weak bases or with carboxylic or hydroxylic acids. Mono-, di-, or tri-functional groups may give competition between different salt forms. Kinetic factors are also important since reprecipitation into the base may occur for very weak acids or bases. Competition between salt-forming agents and buffers of measurement are often observed.[15]

Table 6 shows the results of a screening for salt selection of a new drug candidate. A compromise between solubility and stability was found and the hydrogen maleate salt (hml) chosen. **Figure 19** demonstrates the impact of salt forms on the morphology of crystals obtained for a drug candidate.

Table 5 Comparison of bioavailability in dogs and IDR in water for PNU-140690 (acid, pK_a 6.2 and 7.6, monosodium and disodium salts)

Form	Bioavailability (%)	IDR ($\mu g\,cm^{-2}\,s^{-1}$)	Max. concentration in water ($\mu g\,mL^{-1}$)
Acid	3	<0.2	60
Monosodium salt	15	115	530
Disodium salt	50	450	650

Table 6 Results of screening of salts of a drug

Property	Base	hfu	hml	ch	hta	hmo
Melting (°C)	98	196	161	251	122	72
DSC purity (%)	99	99	99.9	–	–	–
Hygroscopicity	No	No	No	Yes	Yes	No
X-ray	Cryst.	Cryst.	Cryst.	Cryst.	Cryst.	Cryst.
Polymorphic behavior	Mono	Mono	Mono	Poly	Poly	Poly
Feasibility	Good	Good	Good	Good	Good	Good
Solubility						
Water (%)	<0.01	0.5	0.8	0.8	>3	>3
HCl 0.1N (%)	0.3	0.3	0.3	0.3	0.3	0.3
Stability bulk						
96 h Xenon (%)	0	0	0	2–5	0	0
1 week 70 °C (%)	0	0	0	<2	0	10–20
1 week 70 °C/95% RH (%)	0	0	0	<2	>20	10–20
Methanol (%)	0	0	0	10–20	<2	0
Water (%)	>20	>90	>20	>20	>90	>20
Compatibility 1 week 70 °C						
Mixture 1 (%)	0	>20	0	<2	0	0
Mixture 1/95% RH (%)	>90	>20	>20	>20	>90	10–20
Mixture 2 (%)	2–5	<2	<2	0	10–20	2–5
Mixture 2/95% RH (%)	5–10	>20	5–10	10–20	>20	10–20
Particle size 99% (µm)	115	50	17	–	–	–

hfu, hydrogen fumarate; hml, hydrogen maleate; ch, hydrochloride; hta, hydrogen tartrate; hmo, hydrogen malonate.

The selection of the salt form is the first study to be done with new drug candidates. All high-throughput screening systems considered at drug discovery provide salt selection as well as polymorphism screening (*see* Section 5.21.2.4).

5.21.5.2 Effect of Polymorph, Solvate, and Amorphous Form

The absorption rates of many poorly water-soluble substances depend upon the rate of drug dissolution. Because of the effect of solid-state forms on solubility and dissolution rates the use of different amorphous polymorphs would be expected to affect the bioavailability. Some well-known examples are chloramphenicol palmitate[4] and carbamazepine.[110] Differences were also observed for chlortetracycline[111] and sulfamethoxazole.[112] Other examples are summarized in [113]. However, no impact in vivo was found for proglumetacin maleate[114] and phenobarbital.[115] The amorphous solid state is extremely attractive but its stabilization is a challenge.[116]

Figure 19 Surface electron microscopy of different salt forms of a drug candidate: (a) base; (b) hydrochoride; (c) hydrogen maleate; and (d) mesylate. (Reproduced with permission from Giron, D.; Grant, D. J. W. In *Handbook of Pharmaceutical Salts: Properties, Selection and Use*; Stahl, P. H., Wermuth, C. G., Eds.; IUPAC, Verlag Helvetica Chimica Acta: Zürich, Switzerland, 2002, pp 42–81.)

5.21.5.3   Effect of Particle Size

Figure 20 illustrates the impact of particle size on dissolution. As a consequence correlation with bioavailability has been described for a great number of poorly soluble drugs.[117–124] Bioavailability may be increased by micronization.

5.21.5.4   Complexation, Adducts, and Formulation

The excipients[125] used to solubilize drugs in oral and injectable dosage forms include pH modifiers, water-soluble and water-insoluble organic solvents, surfactants, medium-chain triglycerides, long-chain triglycerides, cyclodextrins, and phospholipids.

The group of solid dispersions comprises different physical systems such as simple eutectic mixtures (reduction of particle size), glass dispersions, amorphous precipitates, or complex formation. Solubility properties differ depending on the properties of the carrier, e.g., griseofulvin absorption increases by solid dispersion based on polyethyleneglycol formulation. Solid dispersions have been reviewed in the literature.[126–129] Certain solid dispersions are formed in order to maintain the drug substance in the amorphous state and are called 'solid-solutions.'

Polyvinylpyrrolidone polymers have been successfully used. For example, a fourfold increase in bioavailability was obtained for indometacin.[130]

Cyclodextrins are natural derivatives that are able to form complexes with drug substances with a considerable enhancement of bioavailability. Cyclodextrins do not permeate the gastrointestinal mucosa, so their action is to increase availability of dissolved drug substance at the aqueous mucosal surface. There are more than 600 publications on the use of cyclodextrins. Comprehensive reviews have been published by Thomson[131] and Loftsson.[132,133] **Table 7** gives the characteristics of the major cyclodextrins.

Co-crystals (multicomponent pharmaceutical phases in solid state) are able to overcome poor solubility by opening up a large space for nonionizable drug substances. By adjusting the nonactive component, enhancement in the solubility and bioavailability is expected.[134]

Micellar solubilization with surfactants giving colloidal clusters in solution known as micelles has the ability to increase the solubility of sparingly soluble drug substances.[135] Liposomes are self-forming lipid bilayers that are able to solubilize poorly soluble substances.[136] Microemulsions are clear isotropic mixtures of oil, water, and surfactants.[137]

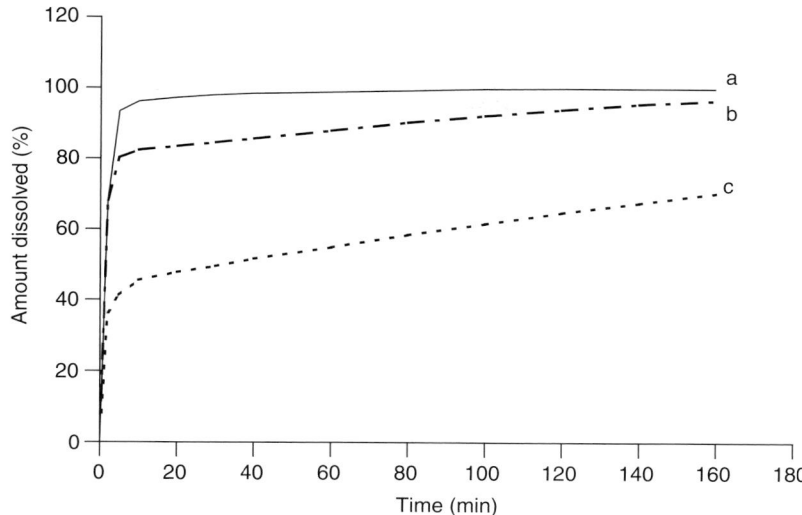

Figure 20 Influence of the particle size on the dissolution rate of reserpine hydrochloride in 0.1N HCl: (a) particle size 90% <40 µm; (b) particle size 90% <52 µm; and (c) particle size 90% <120 µm. (Reproduced with permission from Giron, D.; Grant, D. J. W. In *Handbook of Pharmaceutical Salts: Properties, Selection and Use*; Stahl, P. H., Wermuth, C. G.; Eds.; IUPAC, Verlag Helvetica Chimica Acta: Zürich, Switzerland, 2002; pp 42–81.)

Table 7 Cyclodextrin (CD) characteristics

	α-CD	β-CD	γ-CD	2-Hydroxy-propyl β-CD	2-Hydroxy-propyl γ-CD
Molecular masse	972	1135	1297	1400	1576
Glucose units	6	7	8	7	8
Water solubility (mg mL^{-1}, 20 °C)	145	18.5	232	>600	>500
Cavity diameter (Å)	4.7–5.3	6.0–6.5	7.5–8.3		

5.21.6 Conclusion

The solid-state properties of drug substances have a direct impact on the physicochemical properties relevant for ADME and early toxicity studies. An early selection of the salt and the crystalline form for the development of new active ingredients is recommended. The objective of this selection should be to achieve a soluble, less hygroscopic, reproducible, and stable crystalline form. This should result in a good bioavailability and a linear exposure in toxicity studies. In difficult cases, greater efforts are necessary to enhance solubility via special formulations within a 'developability assessment.' The thermodynamic rules show that depending on temperature, pressure, and concentration in solvents, different stable forms may exist. These forms may be incorrectly assessed if phase diagram rules are ignored. Special formulations produced in order to maintain drug substances in a metastable or unstable state should be used with caution in order to obtain reproducible results. Therefore, the thermodynamic principles described in this chapter are highly recommended to avoid misinterpretation of chemical and physical stability during ADME and early toxicity studies.

References

1. Leuner, C.; Dressmann, J. *Eur. J. Pharm. Biopharm.* **2000**, *50*, 47–60.
2. Lipper, R. A. *Mod. Drug Disc.* **1999**, *2*, 55–60.
3. Stahl, P. H.; Wermuth, C. G., Eds. *Handbook of Pharmaceutical Salts: Properties, Selection and Use*; IUPAC, Verlag Helvetica Chimica Acta: Zürich, Switzerland, 2002.
4. Haleblian, J.; McCrone, W. *J. Pharm. Sci.* **1969**, *58*, 911–929.
5. Haleblian, J. *J. Pharm. Sci.* **1975**, *64*, 1269–1288.

6. International Conference on Harmonization (ICH). Guideline Specification Q6A, October 1999.
7. Giron, D. *Labo-Pharma-Probl. Techn.* **1981**, *307*, 151–160.
8. Byrn, S. R.; Pfeiffer, R. R.; Ganey, M.; Hoiberg, C.; Poochikian, G. *Pharm. Res.* **1995**, *12*, 945–954.
9. Vitez, I. M.; Newman, A. W.; Davidovich, M.; Kiesnowski, C. *Thermochim. Acta* **1998**, *324*, 187–196.
10. Caira, M. R. *Top. Curr. Chem.* **1998**, *198*, 163–208.
11. Byrn, S. R.; Pfeiffer, R. R.; Stowell, J. G. *Solid-State Chemistry of Drugs*, 2nd ed.; SSCI Inc.: West Lafayette, IN, 1999.
12. Brittain, H. G., Ed. *Polymorphism in Pharmaceutical Solids*; Marcel Dekker: New York, 1999.
13. Bernstein, J. *Polymorphism in Molecular Crystals*; Oxford University Press: Oxford, UK, 2002.
14. Kibbey, C. E.; Poole, S. K.; Robinson, B.; Jackson, J. D.; Durham, D. *J. Pharm. Sci.* **2001**, *90*, 1164–1175.
15. Stephenson, G.; Forbes, R. A.; Reutzel-Edens, S. M. *Adv. Drug Deliv. Rev.* **2001**, *48*, 67–90.
16. Giron, D.; Grant, D. J. W. In *Handbook of Pharmaceutical Salts: Properties, Selection and Use*; Stahl, P. H., Wermuth, C. G., Eds.; IUPAC, Verlag Helvetica Chimica Acta: Zürich, Switzerland, 2002, pp 42–81.
17. Giron, D. *J. Therm. Anal. Calorim.* **2003**, *73*, 441–457.
18. Singhal, D.; Curatolo, W. *Adv. Drug. Deliv. Rev.* **2004**, *56*, 335–347.
19. Huang, L.; Tong, W. *Adv. Drug. Deliv. Rev.* **2004**, *56*, 321–334.
20. Grant, D. J. W.; Byrn, S. *Adv. Drug. Deliv. Rev.* **2004**, *56*, 237–239.
21. Yu, L. X; Furnes, M. S.; Raw, A.; Woodland Outlaw, K. P.; Nashed, N. E.; Ramos, E.; Miller, S. P. F.; Adams, R. C.; Fang, F.; Patel, R. M. et al. *Pharm. Res.* **2003**, *20*, 531–536.
22. Giron, D. *Eng. Life Sci.* **2003**, *3*, 103–112.
23. Giron, D.; Garnier, S.; Mutz, M. *J. Therm. Anal. Calorim.* **2004**, *77*, 709–747.
24. Gu, C.; Grant, D. J. W. *J. Pharm. Sci.* **2001**, *90*, 1277–1287.
25. Rasenack, N.; Müller, B. W. *Pharm. Ind.* **2005**, *67*, 323–326.
26. Chawla, G.; Bansal, A. K. *CRIPS* **2004**, *5*, 9–12.
27. Snider, D. A.; Addicks, W.; Owens, W. *Adv. Drug. Deliv. Rev.* **2004**, *56*, 391–395.
28. Storey, R.; Docherty, R.; Higginson, P.; Dallman, C.; Gilmore, C.; Barr, G.; Dong, W. *Crystallogr. Rev.* **2004**, *10*, 45–56.
29. Hursthouse, M. *Crystallogr. Rev.* **2004**, *10*, 85–96.
30. Blagden, N.; Davey, R. J. *Cryst. Growth Design* **2003**, *3*, 873–880.
31. Goho, A. *Sci. News* **2004**, *166*, 122–133.
32. Ware, E. C.; Lu, D. R. *Pharm. Res.* **2004**, *21*, 177–184.
33. Datta, S.; Grant, D. J. W. *Nat. Rev. Drug Disc.* **2004**, *3*, 42–57.
34. Giron, D. *Thermochim. Acta* **1995**, *248*, 1–59.
35. Burger, A.; Ramberger, R. *Mikrochim. Acta* **1989** *II*, 259–271, 273–316.
36. Grunenberg, A.; Henck, J.-O.; Siesler, H. W. *Int. J. Pharm.* **1996**, *129*, 147–158.
37. Burger, A. *Acta Pharm. Technol.* **1982**, *28*, 17–25.
38. Hancock, B. C.; Shamblin, S. L. *Thermochim. Acta* **2001**, *380*, 95–107.
39. Grant, D. J. W. In *Polymorphism in Pharmaceutical Solids*; Brittain, H. G., Ed.; Marcel Dekker: New York, 1999, pp 1–33.
40. Gordon, M.; Taylor, J. S. *J. Appl. Chem.* **1952**, *2*, 493–500.
41. Rades, T.; Müller-Goymann, C. *Eur. J. Biopharm.* **1994**, *40*, 277–282.
42. Li, Z. J.; Grant, D. J. W. *J. Pharm. Sci.* **1997**, *86*, 1073–1078.
43. Giron, D. *J. Therm. Anal. Calorim.* **2001**, *64*, 37–52.
44. Anderton, C. *Eur. Pharm. Rev.* **2004**, 68–74.
45. Morisette, S. L.; Almarsson, O.; Peterson, M. L.; Remenar, M. L.; Read, M. J.; Lemmo, A. V.; Ellis, S.; Cima, M. J.; Gardner, C. R. *Adv. Drug Deliv. Rev.* **2004**, *56*, 275–290.
46. Wu, R.; Papoutsakis, D.; Karpinski, P. Poster 437i, presented at the poster session of the Annual Meeting of Pharmaceutical Technology, Cincinatti, 2005.
47. Balbach, S.; Korn, C. *Int. J. Pharm.* **2004**, *275*, 1–12.
48. Hilfiker, R.; Berghausen, J.; Blatter, F.; Burkhard, A.; von Raumer, M. *J. Therm. Anal. Calorim.* **2003**, *73*, 429–440.
49. Barr, G.; Dong, W.; Gilmore, C. J. *J. App. Crystallogr.* **2004**, *37*, 658–664.
50. Mirmehrabi, M.; Rohani, S. *J. Pharm. Sci.* **2005**, *94*, 1560–1576.
51. Urakami, K. *Curr. Pharm. Biotechnol.* **2005**, *6*, 193–203.
52. David, W. I. F.; Shankland, K.; Shankland, N. *Chem. Commun.* **1998**, 931–932.
53. Giron, D.; Piechon, P.; Goldbronn, C.; Pfeffer, S. *J. Therm. Anal. Calorim.* **1999**, *57*, 61–75.
54. Stephenson, G. A. *Rigaku J.* **2005**, *22*, 2–15.
55. Harris, K. D. M.; Cheung, E. Y. *Chem. Soc. Rev.* **2004**, *33*, 426–538.
56. David, W. I. F.; Shankland, K.; Shankland, N. *Chem. Comm.* **1998**, 931–932.
57. Gavezotti, A.; Filippini, G. *J. Am. Chem. Soc.* **1996**, *118*, 7153.
58. Price, S. L. *Adv. Drug Deliv. Rev.* **2004**, *56*, 301–315.
59. Fukuda, H.; Amimoto, K.; Koyama, H.; Kawato, T. *Org. Biomol. Chem.* **2003**, *1*, 1578–1583.
60. Desiraju, G. R.; Paul, I. C.; Curtin, D. Y. *J. Am. Chem. Soc.* **1977**, *99*, 1594–1601.
61. Yatsenko, A. V.; Paseshnichenko, K. A. *Acta Crystallogr.* **2001**, *C57*, 961–964.
62. Wells, J. L., Ed. *Pharmaceutical Preformulation: The Physicochemical Properties of Drug Substances*; Ellis Horwood Ltd: Chichester, UK, 1988.
63. Giron, D.; Goldbronn, C. *Analusis* **1979**, *7*, 109–126.
64. Shefter, E.; Higuchi, T. *J. Pharm. Sci.* **1963**, *52*, 781–791.
65. Henwood, S. Q.; Liebenberg, W.; Tiedt, L. R.; Lötter, A. P.; de Villiers, M. M. *Drug. Dev. Ind. Pharm.* **2001**, *27*, 1017–1030.
66. Shah, J. C.; Chen, J. R.; Chow, D. *Drug Dev. Ind. Pharm.* **1999**, *25*, 63–67.
67. Pudipeddi, M.; Serajuddin, A. T. M. *J. Pharm. Sci.* **2005**, *94*, 929–939.
68. Mosharraf, M.; Nystrom, C. *Pharm. Sci.* **1998**, *1*, S268.
69. Flamin, A. A.; Ahlneck, C.; Alderbon, G.; Nystrom, C. *Int. J. Pharm.* **1994**, *111*, 159–164.
70. Hancock, B. C.; Parks, M. *Pharm. Res.* **2000**, *17*, 397–400.

71. Stagner, W. C.; Guillory, J. K. *J. Pharm. Sci.* **1979**, *68*, 1005–1009.
72. Clas, S. D.; Faizer, R.; O'Connor, R. E.; Vadas, E. B. *Int. J. Pharm.* **1995**, *121*, 73–80.
73. United States Pharmacopeia (USP) **2003**, *26*, 2333–2334.
74. Kaplan, S. A. *Drug. Metab. Rev.* **1972**, *1*, 15–25.
75. Amidon, G. *Int. J. Pharm.* **2004**, *270*, 221–227.
76. Nyqvist, H. *Int. J. Pharm. Tech. Prod. Mfr.* **1983**, *4*, 47–53.
77. Walkling, W. D.; Reynolds, B. E.; Fegely, B. Y.; Janicki, C. *Drug Dev. Ind. Pharm.* **1983**, *9*, 809–813.
78. Shalaev, E. Y.; Zografi, G. *J. Phys. Org. Chem.* **1998**, *9*, 729–738.
79. Teraoka, R.; Otsuka, M.; Matsuda, Y. *Int. J. Pharm.* **2004**, *286*, 1–8.
80. Nyqvist, H.; Wadsten, T. *Acta Pharm. Technol.* **1986**, *32*, 130–132.
81. Qin, X.; Frech, P. *J. Pharm. Sci.* **2001**, *90*, 833–844.
82. Cohen, M. D.; Green, B. S. *Chem. Br.* **1973**, *9*, 490–497.
83. Giron, D. *Mol. Cryst. Liq. Cryst.* **1988**, *161*, 77–100.
84. Masse, J.; Bauer, M.; Billot, P.; Broquaire, M.; Chauvet, A.; Doveze, J.; Garinot, O.; Giron, D.; Popoff, C. *STP Pharma Pratiques* **1997**, 7, 235–246.
85. Chemburkar, S. R.; Bauer, J.; Deming, K.; Spiwek, H.; Patel, K.; Morris, J.; Henry, R.; Spanton, S.; Dziki, W.; Porter, W. *Org. Process Res. Dev.* **2000**, *4*, 413.
86. Trask, A. V.; Shan, N.; Motherwell, W. D. S.; Jones, W.; Feng, S.; Tan, R. B. H.; Carpenter, K. J. *Chem. Commun.* **2005**, 880–882.
87. Chan, H. K.; Doelker, E. *Drug. Dev. Ind. Pharm.* **1985**, *11*, 315–332.
88. Wikllart, J. F.; De Gussene, A.; Hemon, S.; Odou, G.; Danede, F.; Descamps, M. *Solid State Commun.* **2001**, *119*, 501–505.
89. Roberts, C.; Debenedetti, P. G. *AIChE Journal* **2002**, *48*, 1140–1144.
90. Crowley, K. J.; Zografi, G. *Thermochim. Acta* **2001**, *380*, 79–93.
91. Hancock, B. C.; Shamblin, S. L. *Thermochim. Acta* **2001**, *380*, 95–107.
92. Hancock, B. C.; Zografi, G. *Pharm. Res.* **1994**, *11*, 471–477.
93. Anberg, M.; Nyström, C.; Castensson, S. *Int. J. Pharm.* **1992**, *81*, 153–162.
94. Giron, D.; Remy, P.; Thomas, S.; Vilette, E. *J. Therm. Anal.* **1997**, *48*, 465–472.
95. Buckton, G.; Darcy, P.; Greenleaf, D.; Holbrook, P. *Int. J. Pharm.* **1995**, *116*, 113–125.
96. Li, S.; Wong, S.; Sethia, S.; Almoazen, H.; Joshi, Y. M.; Serajuddin, A. T. M. *Pharm. Res.* **2005**, *22*, 628–635.
97. Yu, L.; Ng, K. *J. Pharm. Sci.* **2002**, *91*, 2367–2375.
98. Johnson, D. E.; Wolfgang, G. H. I. *Curr. Top. Med. Chem.* **2001**, *1*, 233–245.
99. Lin, J. H. *ISSX Proc.* **1994**, *6*, 31–33.
100. Engel, G. L.; Nagy, A. F.; Faul, M. M.; Richardson, L. A.; Winneroski, L. L. *Int. J. Pharm.* **2000**, *198*, 239–247.
101. Gwak, H.-S.; Choi, J.-S.; Choi, H.-K. *Int. J. Pharm.* **2005**, *297*, 156–161.
102. Powers, J. P. PCT Int. AppL. 2001 (2005 ACS on SciFinder).
103. Catenese, B.; Barillari, G.; Iorio, E.; Silvestrini, B.; Angeli, F. *Bolletino Chim. Farmac.* **1982**, *121*, 87–90.
104. Strong, L. G. Optimization of Drug-Like Properties During Lead Optimization. Presented at the AAPS Workshop, September 2004.
105. Higgins, J. D.; Rocco, W. L. *Today's Chemist at Work*, July 22–26, 2003.
106. Davies, G. *Pharm. J.* **2001**, *266*, 322–323.
107. Berge, S. M.; Bighley, L. D.; Monckhouse, D. C. *J. Pharm. Sci.* **1977**, *66*, 1–19.
108. Bighley, L. D.; Berge, S. M.; Monckhouse, D. C. Salt Forms of Drugs and Absorption. In *Encyclopedia of Pharmaceutical Technology*; Swarbrick, J., Boylan, J. C., Eds.; Marcel Dekker: New York, 1995; Vol. 13, pp 453–499.
109. Hawley, M.; Douglas, S.; Morozowich, W. Poster No. 2448, presented at the AAPS Annual Meeting, San Francisco, CA, USA, 1998 (Abstract published in *Pharm. Sci.* **1998**, *1*, S267).
110. Kabayashi, Y. *Int. J. Pharm.* **2000**, *193*, 137–146.
111. Miyazaki, S.; Arita, T.; Hori, R.; Ito, K. *Chem. Pharm. Bull.* **1974**, *22*, 638–642.
112. Nurasiyah, S.; Yuwono, T.; Farmasi, M. *Indonesia* **2001**, *12*, 205–211 AN: 2002: 222517.
113. Giron, D. *STP Pharma* **1988**, *4*, 330–340.
114. Yasutomo, T.; Shirai, S.; Sato, T.; Yada, I.; Matsuda, Y.; Otsuka, M. *Yakuzaigaku* **2001**, *61*, 97–108 CAN 137:83519.
115. Kato, Y.; Watanabe, F. *Yakugaku Zasshi* **1978**, *98*, 639–648.
116. Yu, L. *Adv. Drug Deliv. Rev.* **2001**, *48*, 27–42.
117. Aungst, B.; Nguyen, N. H.; Taylor, N. J.; Bindra, S. D. *J. Pharm. Sci.* **2002**, *91*, 1390–1395.
118. Jindal, K. C.; Chaudhary, R. S.; Singal, A. K.; Ganswal, S. S.; Khanna, S. *Indian Drugs* **1995**, *32*, 100–107.
119. Grau, M. J.; Kayser, O.; Müller, R. H. *Int. J. Pharm.* **1989**, *51*, 9–17.
120. Jounela, A. J.; Pentikainen, P. J.; Sothmann, A. *Eur. J. Clin. Pharmacol.* **1975**, *8*, 365–370.
121. McInnes, G. T.; Asbury, M. J.; Ramsay, L. E.; Shelton, J. R.; Harrison, I. R. *J. Clin. Pharmacol.* **1982**, *22*, 410–417.
122. Chakrabarti, S.; Moerman, E.; Belpaire, F. *Pharmazie* **1979**, *34*, 242–243.
123. Nimmerfall, F.; Rosenthaler, J. *J. Pharm. Sci.* **1980**, *69*, 605–610.
124. Rasenack, N.; Müller, B. W. *Pharm. Ind.* **2005**, *67*, 447–451.
125. Strickley, R. G. *Pharm. Res.* **2004**, *21*, 201–230.
126. Ford, J. L. *Pharm. Acta Helv.* **1986**, *61*, 69–88.
127. Leuner, C.; Dressman, J. *Eur. J. Pharm. Biopharm.* **2000**, *50*, 47–60.
128. Rasenack, N.; Müller, B. W. *Pharm. Ind.* **2005**, *67*, 583–591.
129. Serajuddin, A. T. M. *J. Pharm. Sci.* **1999**, *88*, 1058–1066.
130. Fujii, M.; Okada, H.; Shibata, Y.; Teramachi, H.; Kondoh, M.; Watanabe, Y. *Int. J. Pharm.* **2005**, *293*, 145–153.
131. Thompson, D. O. *Crit. Rev. Ther. Drug Carrier Syst.* **1997**, *14*, 1–104.
132. Loftsson, T.; Hreinsdottir, D.; Masson, M. *Int. J. Pharm.* **2005**, *302*, 18–28.
133. Loftsson, T.; Brewster, M. E.; Masson, M. *Am. J. Drug Deliv.* **2004**, *2*, 1–15.
134. Almarsson, O.; Zaworotko, M. *J. Chem. Commun.* **2004**, 1889–1896.
135. Rangel-Yagui, C. O.; Pessoa, A.; Tavares, L. C. *J. Pharm. Sci.* **2005**, *8*, 147–163.

136. Lawrence, M. J.; Rees, G. D. *Adv. Drug Deliv. Rev.* **2000**, *45*, 89–121.
137. Lian, T.; Ho, R. J. Y. *J. Pharm. Sci.* **2001**, *90*, 667–680.
138. Dunitz, J. D.; Bernstein, J. *Acc. Chem. Res.* **1995**, *28*, 193.
139. Giron, D. Pharmacokinetic Profiling in Drug Research. In *Biological, Physicochemical and Computational Strategies*; Testa, B., Krämer, S., Wunderli-Alllenspach, H., Folkers, G., Eds.; Verlag Helvetica Chimica Acta: Zürich, Switzerland, 2006, pp 307–329.

Biography

Danielle Giron studied engineering at the School of Industrial Chemistry and Physics in Lyon. She holds an MS and PhD in analytical chemistry from the University of Lyon. She joined Sandoz in 1971 to establish the laboratory of thermal analysis and polymorphism. With different positions in analytical and research development at Novartis, she gained a broad experience of the development process from research to production for drug substances and drug products. She has written more than 50 published journal articles and book chapters, and speaks regularly at conferences in Europe and the US. She is President of the Swiss Society for Thermal Analysis and Calorimetry (STK).

Dr Giron won the STK Award in 1992 and the Novartis Leading Scientist Award in 2000. She is presently Group Leader and Principal Fellow at Novartis and is also a private docent at the University of Basel.

Comprehensive Medicinal Chemistry II
ISBN (set): 0-08-044513-6

ISBN (Volume 5) 0-08-044518-7; pp. 509–530

5.22 Use of Molecular Descriptors for Absorption, Distribution, Metabolism, and Excretion Predictions

S Winiwarter, M Ridderström, and A-L Ungell, AstraZeneca R&D Mölndal, Mölndal, Sweden
T B Andersson, Institute of Environmental Medicine, Karolinska Institutet, Stockholm, Sweden
I Zamora, Universitat Pompeu Fabra, Barcelona, Spain

5.22.1 Introduction: Absorption, Distribution, Metabolism, and Excretion (ADME) in the Drug Discovery Process

The pharmacokinetic behavior of a drug has to be considered during the drug discovery process in order to find compounds that will reach their site of action and maintain their concentration long enough to be able to produce the desired effect. Specific properties are needed for an orally administered drug that is to be dosed once daily. Bioavailability of a compound depends on both absorption and first pass metabolism. To obtain the correct dose regimen, distribution and elimination (metabolism and excretion) properties are important. These ADME properties (absorption, distribution, metabolism, and excretion) are evaluated during the drug discovery process using in silico, in vitro, and/or in vivo systems.

In vitro and in vivo studies can only be performed on synthesized compounds and are capacity limited, even though high-throughput techniques have largely increased in vitro capacity during recent years. The complexity of the experimental data, usually higher for in vivo systems compared to in vitro, has to be considered as well. Great efforts have been made to develop in vitro models that may help to understand and interpret factors that influence the pharmacokinetic characteristics of a compound in vivo. However, even in vitro systems comprise various factors that influence the behavior of the compounds in different ways.

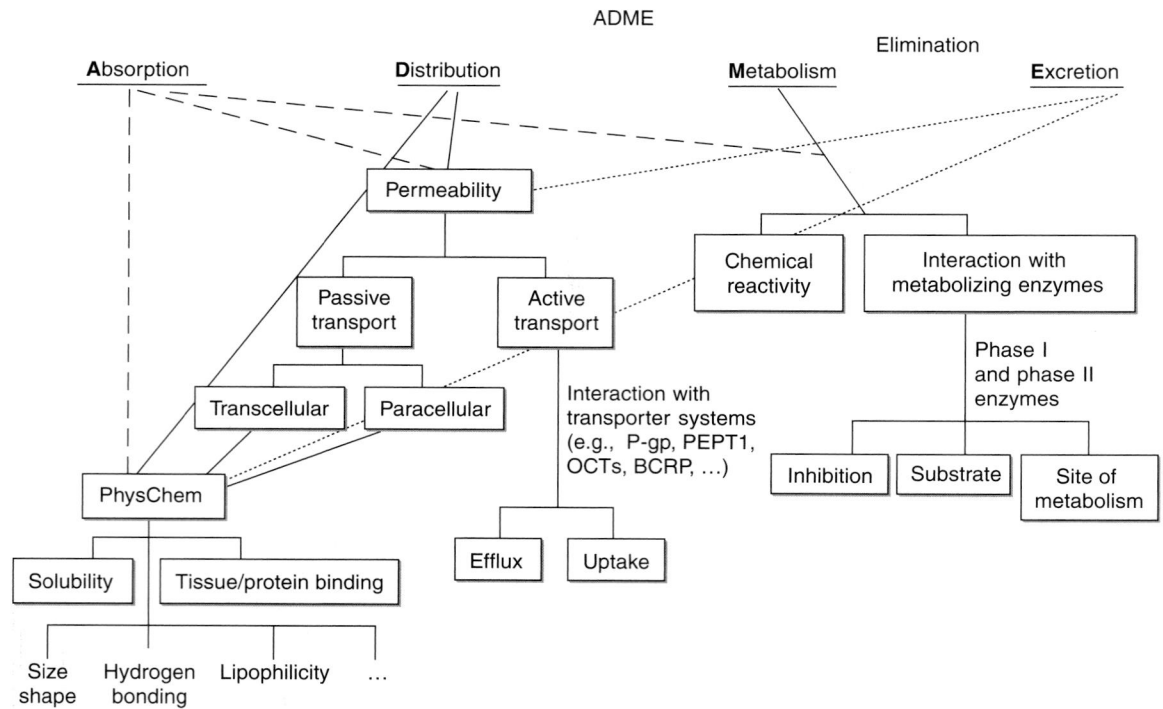

Figure 1 Deconvolution of ADME properties to partial effects.

In silico methods, in contrast, can also be used for virtual, i.e., not synthesized compounds and can, in general, be applied on very large compound libraries. Nevertheless, in silico methods are based on in vitro and/or in vivo measurements and their applicability is limited by the information that was extracted from these experiments. The quality of the models will depend on the quality, complexity, and understanding of the experimental data they are based on. In order to accelerate the discovery of new chemical entities (NCEs) with acceptable pharmacokinetic properties an increasing number of in silico ADME models has been developed. The number of reviews in this area published within the last few years shows the interest in these models.[1–8]

ADME properties are influenced by different factors (**Figure 1**). Absorption is determined by solubility, permeability (active and passive), and first pass metabolism across the gut wall. Distribution is controlled by permeability through various membranes (e.g., the blood–brain barrier) and by plasma protein or tissue binding. Metabolism is determined by the interaction with metabolizing enzymes and the reactivity of the compound. Excretion of a compound is governed by its physicochemical properties and the interaction with transporter systems. The more general term elimination comprises both metabolism and excretion of the compound.

Many of these factors are themselves affected by different molecular properties, like lipophilicity and hydrogen bonding, but also by structural features like protein binding site complementarity. The passive transcellular permeability of a compound, for instance, is often modeled on molecular size, lipophilicity, and/or hydrogen bonding. All three properties are correlated to each other and a simple, but sufficient model may be achieved by using only one or two of these properties.[9–11] However, if absorption is not primarily passive, i.e., if a carrier-mediated system is involved, the relationship between the molecular structure and the transport protein needs to be considered and therefore three-dimensional quantitative structure-activity relationship (3D-QSAR) techniques or pharmacophoric definitions have to be applied.

The scope of the analysis needs to be considered as well. Local models, on the one hand, are useful when a closely related family of compounds is studied. Global models, on the other hand, are needed if the entire chemical space is to be described in order to identify new interesting chemical series. In both scenarios the experimental data, the compounds, and the descriptors used are critical for building the model. In general, local models for congeneric series of compounds yield acceptable prediction capability within the series. Quantitative models can be obtained and are useful for closely related compounds. For global models, which by definition explore the chemical space, the experimental data most often involves several, possibly unknown factors. For such a diverse data set, a qualitative model that can discriminate between compounds that are active and nonactive (permeable and nonpermeable, inhibiting and noninhibiting of cytochrome P450 should be considered.

ADME predictions are often based on quantitative structure–activity (property) relationship (QSAR or QSPR) approaches. Several statistical techniques are available and have been used to correlate molecular descriptors to the ADME properties ranging from multiple or single regression analysis, multivariate data analysis like partial least squares analysis (PLS) to nonlinear methods like regression trees or neural networks.[9,12–16]

Molecular descriptors used for ADME modeling need to explain relevant properties such as size, lipophilic and hydrophilic profiles, hydrogen bonding patterns, and structural features specific for the property under analysis, e.g., pharmacophores for a metabolizing enzyme. Additionally, acidity and/or basicity (pK_a) of a compound and its (aqueous) solubility need to be considered for a proper description. Useful descriptors can range from simple atom counts, e.g., number of hydrogen bond donor and acceptor atoms, to more elaborate molecular structure description (like volume or polar surface area) or to quantum chemistry derived descriptors, e.g., atomic charges or HOMO and LUMO energies. One way to classify descriptors is to define them as (0D) 1D, 2D, and 3D (and even 4D) descriptors. However, the definition for each class is somewhat ambiguous.[17,18] 1D (0D) descriptors can be defined as those for which no structural information related to the connectivity of the atoms is necessary. Only the sum formula of the molecule is needed. Molecular weight and atom counts, number of atoms or number of a specific atom type, would fit into this definition. 'Bulk descriptors,' including log P, for which calculation of at least a 2D structure is necessary, have also been suggested as 1D descriptors.[18] 2D descriptors consider the 2D representation of the molecules. These descriptors are often referred to as topological descriptors. Typical 2D descriptors are connectivity indices and fragment counts, the latter are also suggested as 1D descriptors.[17] 3D descriptors are derived from the 3D structure and both molecular volume and polar surface area would fit into this category. However, polar surface area, which has become widely used in permeability models[11,19,20] can also be obtained from the 2D structure alone.[21] In this chapter we classify the descriptors according to the molecular property they are intended to describe, present descriptors derived from molecular interaction fields (MIFs) and add some other descriptor sets that have been used in ADME predictions.

5.22.2 Physicochemical Descriptors

Fairly simple descriptors have been shown to give good correlations to ADME properties. In 1997, Lipinski proposed the 'rule of 5,' an important, but not sufficient filter for oral drugs.[22] This 'rule of 5' comprises four rules (molecular weight MW < 500, Clog P < 5, number of hydrogen bond donor atoms HBD < 5, and number of hydrogen bond acceptor atoms HBA < 10) and states that good absorption is less likely for a compound if two or more of these rules are violated. However, this can only help for a first interpretation of a molecule – a compound that fulfills all criteria is more likely to be permeable and, thus, also more likely to be orally bioavailable. Nevertheless, not all compounds that do fulfill the criteria have good bioavailability or even good absorption (e.g., acyclovir, fluvastatin, and terbutaline all have bioavailability below 30%,[23] acyclovir also has a fraction absorbed below 30%,[24] and they do not violate any of Lipinski's rules). In addition, compounds that do not fulfill the criteria are successfully used as drugs (e.g., cyclosporine or digoxin have both bioavailability up to 60% and thus also reasonable high absorption[24]). Since the 'rule of 5' is based on the distribution of properties that were calculated for several thousand drugs, such outliers are to be expected. Surprisingly, most of these outliers belong to a few specific compound classes, some of which are known to be substrates to a transport system.[22]

The above study and other analyses show that descriptors for molecular size, lipophilicity, and hydrogen bonding are important for modeling ADME properties.[22,25,26] In the following the impact of these properties on ADME will be illustrated, showing how these properties can be described and explaining especially those descriptors that have been used in ADME predictions.

5.22.2.1 Molecular Size and Shape

Molecular size is generally important for the permeability of compounds. There are reports in the literature showing a correlation between transcellular and/or paracellular permeability and molecular size,[25,27] whereas others emphasize that molecular size determines which absorption route a drug can take: it is suggested that polar compounds with low molecular weight (below 200) may be able to use the paracellular route.[28,29] However, the size limit varies in different regions of the intestine and species differences in paracellular transport have also been shown.[30,31]

Molecular size can be assessed in different ways. The molecular weight is easily calculated from the molecular formula. Also a simple atom count can be seen as a crude measure of molecular size. Other descriptors often used are molecular volume and molecular surface. The molecular surface can be defined as the Van der Waals surface of the

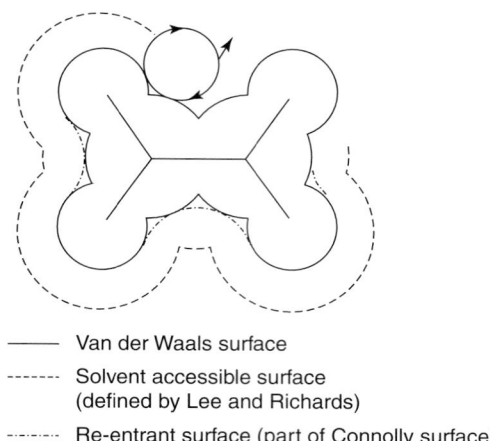

————— Van der Waals surface

- - - - - - Solvent accessible surface
(defined by Lee and Richards)

- - - - - - Re-entrant surface (part of Connolly surface)

Figure 2 Solvent-accessible surface area.

molecule, assuming all atoms to be spheres defined by their Van der Waals radii. Another possibility is to define a solvent-accessible surface area with the help of a spherical probe that is 'rolled' over the Van der Waals surface. This probe usually has a radius of about 1.4 or 1.5 Å, thus being the size of a water molecule. However, the solvent-accessible surface area can be defined in two ways, either by the center of the spherical probe[32] or by the contact and re-entrant surface of such a probe.[33,34] The latter type of solvent-accessible surface is often referred to as the Connolly surface, since Connolly *et al.* proposed a feasible and fast algorithm to obtain such a surface.[34] The computational method used will influence the results; in particular, the difference between the solvent-accessible surface area as defined by Lee and Richards and the Connolly surface should be noted (see **Figure 2**). Other molecular surfaces are contour surfaces that can be defined by molecular properties like electron isodensity, molecular electrostatic potential, or molecular orbitals.

The molecular volume can be defined according to the molecular surface, e.g., the Van der Waals volume. It can also be determined experimentally from the molecular weight, the liquid density, and the Avogadro number. McGowan's characteristic volume is defined by the sum of atomic volume parameters corrected by a factor for the number of atoms.[35] Another size descriptor is the molar refractivity, which is directly related to the molar volume (molecular weight divided by the liquid density) and a function of the refractive index of the liquid, thus containing information about polarizability. Refractivity is an additive property similar to the volume descriptors and can thus easily be calculated by a group contribution method. Molecular size descriptors are usually highly correlated,[10,17] which makes it unnecessary to use all available descriptors.

Shape descriptors are also often highly correlated to molecular size. Ovality is defined as the ratio between the surface of the molecule and the surface of a sphere with the same volume. Topological indices like connectivity[36,37] or shape indices[38] also give information about the molecular shape. Furthermore, flexibility measures, e.g., the number of rotatable bonds, can be interpreted as information about both size and variable shape of the molecule. It was suggested that less flexibility as measured by the number of rotatable bonds improves oral bioavailability.[39]

Molecular size and shape descriptors are usually part of every descriptor set used in ADME modeling, although they are not necessarily important in the final model. This does not imply that molecular size does not influence the ADME property in question, but rather that molecular size may be considered implicitly in other descriptors like lipophilicity or polar surface area.

5.22.2.2   Lipophilicity and Hydrophobicity

Lipophilicity, the 'love of fat,' and hydrophobicity, the 'fear of water,' are often taken as synonyms, but do not exactly describe the same property: hydrophobicity considers the interaction between the compound and water, whereas lipophilicity is a measure of the interaction with a lipid. It has been suggested that hydrophobicity may be a component of lipophilicity[40,41]:

$$\text{lipophilicity} = \text{hydrophobicity} - \text{polarity} \qquad [1]$$

and that lipophilicity may consist of a cavity (volume or size related) and a polarity term (the combination of hydrogen bonding and dipolarity/polarizability)[42]:

$$\log P = aV + \Lambda$$
$$(V = \text{molecular volume}, \ \Lambda = \text{polarity term}) \qquad [2]$$

It has long been assumed that molecular size in general and hydrogen bonding or polarity together are largely able to explain lipophilicity.[43,44] In summary, hydrophobicity can be seen as a (mostly) size- or cavity-related term, i.e., only dependent on the molecule itself, whereas lipophilicity includes a polarity factor and depends additionally on the lipid used. Lipids with different polar properties will give different lipophilicity values for the same compound, a phenomenon that is used for assessment of hydrogen bonding.[42,45,46] Lipophilicity, as the ability of a molecule to mix with an oily phase rather than with water, is usually measured as partition coefficient, P, between the two phases and is often expressed as log P.[47]

Lipophilicity was long ago found to correlate to drug potency – the first mention of such a correlation goes back to around 1900.[48] Lipophilicity has also been found to affect a number of pharmacokinetic parameters: higher lipophilicity gives, in general, lower solubility, higher permeability in the gastrointestinal tract, across the blood–brain barrier and other tissue membranes, higher affinity to metabolizing enzymes and efflux pumps, and higher protein binding.[49–52] Good correlations between lipophilicity and such ADME properties can often be obtained, at least within a congeneric series of compounds.[53] However, lipophilicity alone will not be able to predict any ADME property for structurally diverse compounds. For example, Yazdanian *et al.* state that there is no simple mathematical correlation between lipophilicity and Caco-2 permeability for a diverse set of more than 50 drugs, although a cut-off value for high permeability could be determined (log $D_{oct} > 0.5$).[54] Other workers have reported both linear, bilinear, sigmoidal, or parabolic relationships between permeability and lipophilicity.[12]

Due to the importance of lipophilicity, different ways to assess the value experimentally and to calculate log P have been developed: the shake flask method, chromatography,[55] or pH titration[56] are routinely used to obtain experimental lipophilicity values. Since it has been found that log P_{oct}, the logarithm of the partition coefficient between n-octanol and water, can be calculated from the sum of the contributions (π) of the molecular fragments,[57] various fragmental approaches for calculating log P have been developed. Either molecular fragments and correction factors,[58–60] or atomic contribution with rather complicated atomic types (Alog P)[61–64] or both approaches together are being used.[65] Other calculation methods include Abraham's linear free-energy relationship (LFER) approach[66] and quantum chemical methods.[67,68] However, quantum chemical methods are not as widely used since they are usually more computationally intensive and do not give better results than the empirical approaches.[58,69] In the mid-1990s, a 3D descriptor for lipophilicity became available, the molecular lipophilicity potential[70–72]: atomic lipophilicity values are projected onto the molecular surface to visualize the 3D property. Available programs and methods for log P calculations have been reviewed repeatedly (*see* 5.27 Rule-Based Systems to Predict Lipophilicity).[69,73–76]

The partition coefficient of a molecule in a system depends on the lipophilic phase used. n-Octanol has become the 'golden standard,' being easy to work with and having physicochemical properties similar to lipids. Furthermore, a lot of reference data exist by now.[49,76] Both experimentally obtained and calculated log P values usually refer to the partitioning between n-octanol and water, if nothing else is stated. This parameter is sometimes also termed log K_{OW}.[17] However, it is important to distinguish between the pH-independent partition coefficient, P, and the pH-dependent distribution coefficient, D. The former expresses the quotient of the concentrations of the neutral compound in both phases, whereas the latter expresses the quotient of the concentration of all ionized and unionized species of the compound in both phases. It is generally assumed that (almost) only the unionized species can partition into the lipid phase.[49] Although ionized species are probably able to partition into the lipid phase, at least together with a counter ion,[77] the partition coefficient of such an ionic species is usually about three orders of magnitude lower than the partition coefficient of the neutral species (i.e., if log $P_{neutral}$ is 4, log P_{ion} is about 1).[78] It is, thus, easily understandable that log D will become lower the more of the ionized species is present at the investigated pH. Experimental log D values are often transformed into the corresponding log P value by utilizing this theory, provided that the pK_a value of the investigated compound is known or can be obtained. However, possible ion pair partitioning is not always considered (*see* 5.18 Lipophilicity, Polarity, and Hydrophobicity). Another way to obtain log P is to ensure that (almost) only the unionized compound is present during the experiment by using a pH far from the pK_a. However, it is likely that log D at physiological pH is actually the more interesting parameter, since it gives the lipophilicity at a relevant pH. The pH to consider is 7.4 if blood/body tissue conditions are studied and around 6.0–8.0 in order to mimic the conditions along the gastrointestinal tract. pH varies from very acidic in the stomach, around pH 6.0–6.5 in the proximal small intestine to slightly basic in the colon, pH 8.0.[79] pH varies also with food intake.[80] Calculating log D

values has been found more troublesome than log P values since pK_a values are not as easy to predict (*see* 5.25 In Silico Prediction of Ionization).[81] A recent study showed that pK_a values predicted by ACDlabs[82] were found in some cases to be very different from those experimentally obtained, especially for new types of chemical structures.[83,84] This discrepancy will affect the quality of the log D prediction. Newer versions of this software allow training with in-house data sets to refine the predictions.

5.22.2.3   Hydrogen Bonding

Hydrogen bonding has been found to be an important part in structure permeation relationships.[9,19,85,86] Hydrogen bonds are also important for molecular recognition and thus will not only determine a compound's activity but also its metabolism or transport properties.

One way to assess hydrogen bonding strength experimentally is the measurement of hydrogen bonding equilibrium constants: hydrogen bond donors are measured against a common hydrogen bond acceptor and acceptors against a common donor. Abraham and co-workers found that data measured against different hydrogen bond acceptors/donors could be combined to give general hydrogen bonding scales for donors and acceptors, respectively, the α and β scales.[66] The applicability of these descriptors has been enhanced by an empirical calculation method based on a group contribution approach.[87,88] However, this approach can be erroneous for druglike molecules, which are usually larger compounds with various chemical functions.[10,89] Intramolecular hydrogen bonding may be difficult to account for by a group contribution approach. Hydrogen bond donor acidity and acceptor basicity scales can also be obtained by quantum chemistry methods.[90–92]

Another possible way to obtain hydrogen bond activities is to investigate the thermodynamics of hydrogen complex formation. Raevsky *et al.*[93] collected a thermodynamic, hydrogen bonding database of several thousand reactions and used these data to statistically determine C_a (proton acceptor free energy factor) and C_d (proton donor free energy factor) values that fitted the ΔH (change in enthalpy) and ΔG (change in free energy) values measured. Both C_a and C_d were found to correlate well with Abraham's hydrogen bonding scales and can be calculated for new compounds.[93]

Lipophilicity of a given compound is reported to be partly explained by its hydrogen bonding ability. This can be used to estimate hydrogen bonding ability by comparing the partition coefficients between water and different lipophilic phases, which need to be different in respect to their hydrogen bonding ability.[42,94,95] The lipid used as 'golden standard' for lipophilicity measurements, *n*-octanol, has some hydrogen bonding ability whereas a hydrocarbon solute, like cyclohexane or *n*-heptane does not. It has been suggested that Δlog P obtained from the difference between log P_{oct} and log P_{hept} will mainly capture the hydrogen bond donor capacity of a solute,[95] but also its polarity or polarizability may be encoded.[96,97] Δlog P (defined as difference between log P_{oct} and log P_{cyc}) was found to correlate to brain permeation for a set of H_2 receptor histamine antagonists,[45] showing the relevance of this descriptor. However, this approach is very difficult if not impossible for less lipophilic compounds, because of low partitioning into the heptane (or cyclohexane) phase.[10,19]

The data set mentioned above[45] was also studied by Van de Waterbeemd *et al.*,[19] who suggested that the calculated polar surface area (PSA) might be a more easily accessible descriptor of hydrogen bonding ability. Since then, various different definitions for polar surface area have been used.[19,20,98] The value for PSA will differ depending on what type of surface is calculated (e.g., Van der Waals surface, solvent-accessible surface,[32] or Connolly surface[33,34]) and which atoms are used to define the surface. Nitrogen, oxygen and attached hydrogen atoms usually define a polar surface area, although sulfur atoms have been suggested as well.[9] The type of the calculated surface is especially important if cut-offs from the literature are to be used in other settings: For example, a PSA less than 140 Å^2 for better chance of oral absorption is only appropriate if a Van der Waals or Connolly type solvent-accessible surface is used,[11] whereas an accessible surface as originally defined by Lee and Richards will give higher PSA values and needs a higher limit (see **Figure 3**). For flexible molecules it may also be of interest to obtain a dynamic polar surface area from all relevant, low-energy conformations weighted by the Boltzman distribution.[20,99,100] However, it has been shown that the static polar surface area derived from one reasonable conformation alone, in general, gives very similar results to the computationally much more costly dynamic PSA.[11,99–103] It has also been proposed that an indicative PSA can be derived from the 2D structure alone.[21] In some studies the relative PSA (PSA/total surface area × 100) is used as parameter.[104,105] However, the relative PSA is more related to lipophilicity than to hydrogen bonding,[10] whereas the absolute PSA represents the possible polar interactions in water and is, therefore, related to the energy necessary for desolvation.[106]

Another way to describe hydrogen-bonding properties theoretically is the combination of (calculated) atomic charges with other molecular properties. Dearden *et al.* suggested using the partial atomic charge of the most positive hydrogen atom within the molecule in combination with energy of the lowest unoccupied molecular orbital, obtained

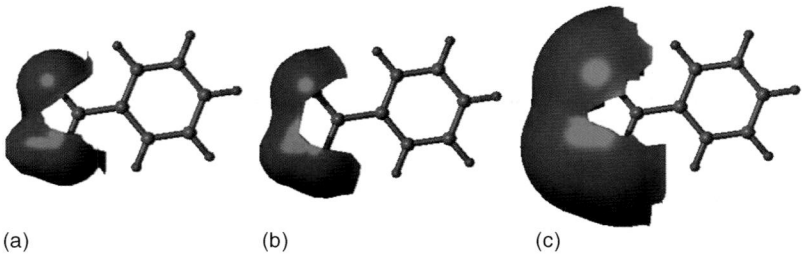

Figure 3   Polar surface area (PSA) for benzoic acid: (a) Van der Waals surface (PSA = 32 Å^2), (b) solvent accessible surface according to Connolly (PSA = 33 Å^2), (c) accessible surface according to Lee and Richards (PSA = 83 Å^2). Surfaces are drawn and calculated with Molcad within Sybyl[107] (the spherical probe used for surfaces (b) and (c) had a radius of 1.4 Å).

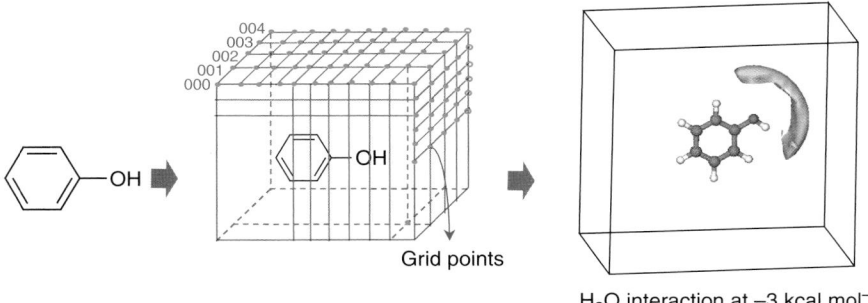

Figure 4   Molecular interaction field (MIF) computation.

from a quantum chemistry calculation as a measure for hydrogen bond donor activity.[96,108] A slightly changed descriptor, the sum of the partial charges of all hydrogen atoms connected to a noncarbon atom, was later found to correlate highly to other hydrogen bond descriptors for drug molecules.[10] Charge weighted molecular or partial surfaces, as introduced by Jurs *et al.*[109] can give similar information.

It has been found that simple counting descriptors may give a good enough correlation to interesting ADME properties, e.g., to permeability.[9] Such counting descriptors for hydrogen bonding are the number of hydrogen bond donors and acceptors, which can be defined either as donor and acceptor atoms[22] or as donor hydrogen atom and acceptor electron pair.[26]

5.22.3   Molecular Interaction Field Derived Descriptors

The calculation of the interaction energy between a small molecule or particle and a compound in a 3D regular grid yields MIFs (**Figure 4**) (*see* 5.25 In Silico Prediction of Ionization). At each grid point in the 3D field the potential energy of interaction between the probe and the compound is calculated. MIFs can represent one interaction type by describing the electrostatic, steric, hydrogen bonding, or hydrophobic interactions. For example, the electrostatic interaction field can be computed using a positive charge as particle. Other MIF types, like the GRID molecular interactions fields,[110] are calculated using probes with more than one interaction capability, such as a water molecule. Thereby, all interactions (electrostatic, steric, etc.) of this probe with the target molecule are summarized in one field. The interaction energies used in the MIF calculation can be derived both from quantum or molecular mechanics approaches. MIFs are sensitive to the orientation and the conformation of the compound investigated. Thus, a proper conformational analysis and alignment of the investigated compounds is necessary in advance when used directly as descriptor sets. Moreover, the use of MIFs in the analysis of structure activity relationships requires statistical projection tools that can cope with the immense number of highly correlated descriptors. Principal component analysis (PCA) and projection to latent structures (PLS) are the methods employed in various approaches.[111,112] The MIF approach has been extensively used within the ADME field in comparative molecular field analysis (CoMFA)[113–123] or selectivity analysis.[124,125] MIFs have further been used to compute molecular descriptors, that are less alignment and/or conformation dependent, such as VolSurf and grid-independent (GRIND) descriptors.[126–128]

5.22.3.1   VolSurf Descriptors

The program VolSurf derives molecular descriptors from the GRID molecular interaction fields, based on three different probes, the water (H_2O), the hydrophobic (DRY), and the hydrogen bond acceptor (O) probes (**Figure 5**). These probes were selected since properties such as shape, electrostatics, hydrogen bonding, and hydrophobicity are known to influence the interaction of molecules with biological membranes.[127] The descriptors from VolSurf are alignment independent and moderately dependent on the conformation ($\sim 10\%$ variation with the conformation).

5.22.3.1.1   A summary of VolSurf descriptors

1. Descriptors derived from the water interaction (H_2O probe):
   ○ Molecular surface (S) is computed as the isoenergy surface that encircles the compound at the interaction energy level of $0.2\,kcal\,mol^{-1}$. This descriptor is related to the size of the compound.
   ○ Molecular volume (V) is computed as the volume enclosed by the above defined surface, and is related to the size of the compound.
   ○ Rugosity (R) is the ratio between the molecular volume and the surface. This descriptor measures how wrinkled the compound is.
   ○ Molecular globularity (G) is the ratio of the molecular surface (S) and the surface of a sphere with the same molecular volume, i.e., the 'ovality' of the compound. This descriptor considers the similarity of the compound compared to a sphere.
   ○ Hydrophilic regions (W1–W8) are defined by the isoenergy surfaces of the interaction energy between molecule and water probe at eight different energy levels: -0.2, -0.5, -1.0, -2.0, -3.0, -4.0, -5.0, and $-6.0\,kcal\,mol^{-1}$. It is a measure of the compound's solvation properties and size, depending on the energy level analyzed. For example, a large hydrophilic region at $-6.0\,kcal\,mol^{-1}$ means that the compound has a strong interaction with water. The hydrophilic region parameters give information similar to the previously described PSA descriptor. However, the hydrophilic regions may be able to distinguish better between different polar groups. For example, the carboxylic acid and the nitro functional groups have similar PSA values ($40\,\text{Å}^2$ for nitrobenzene and $46\,\text{Å}^2$ for the benzoic acid anion, as calculated by the fast 2D method[21]), but they show different water interaction in the GRID field (**Figure 6**): At high energy levels (-0.2, -0.5, and $-1\,kcal\,mol^{-1}$) the volume of the interaction with the water is quite similar since it represents the steric interaction forces. However, at lower energy levels the interaction volume is bigger for benzoic acid than for nitrobenzene, emphasizing the difference between the two polar groups. The hydrophilic regions at lower energy levels depend on the hydrogen bond and electrostatic interactions.

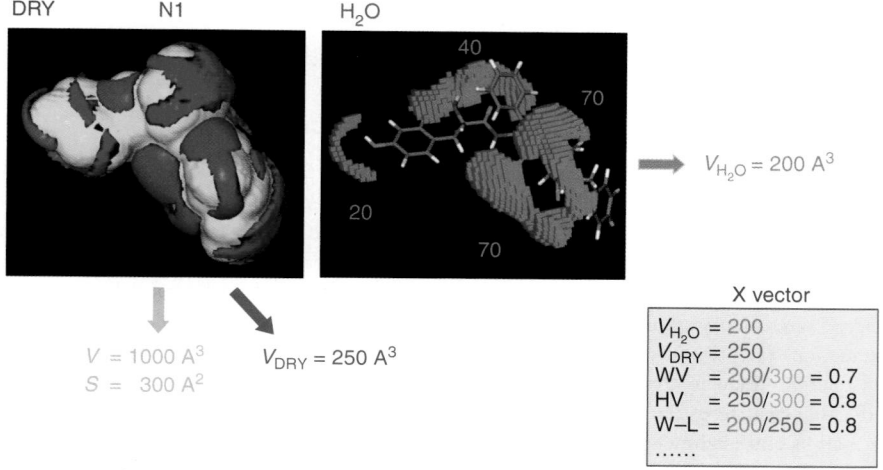

Figure 5   VolSurf descriptor visualization. Yellow: volume of the compound. Green: hydrophobic interaction (DRY probe) at $-1\,kcal\,mol^{-1}$. Blue: hydrophilic interaction (N1 and H_2O probe, as indicated) at $-4\,kcal\,mol^{-1}$; V, volume; S, Surface area; V_{DRY}, volume of the DRY probe at $-1\,kcal\,mol^{-1}$; V_{H_2O}, volume for the water probe at $-4\,kcal\,mol^{-1}$; WV, volume of the H_2O probe at $-1\,kcal\,mol^{-1}$ divided by the surface area of the compound (capacity factor); HV, volume of the DRY probe at $-1\,kcal\,mol^{-1}$ divided by the surface area; W–L, volume of the H_2O probe at $-4\,kcal\,mol^{-1}$ divided by the volume of the DRY probe at $-1\,kcal\,mol^{-1}$ (hydrophilic–lipophilic balance).

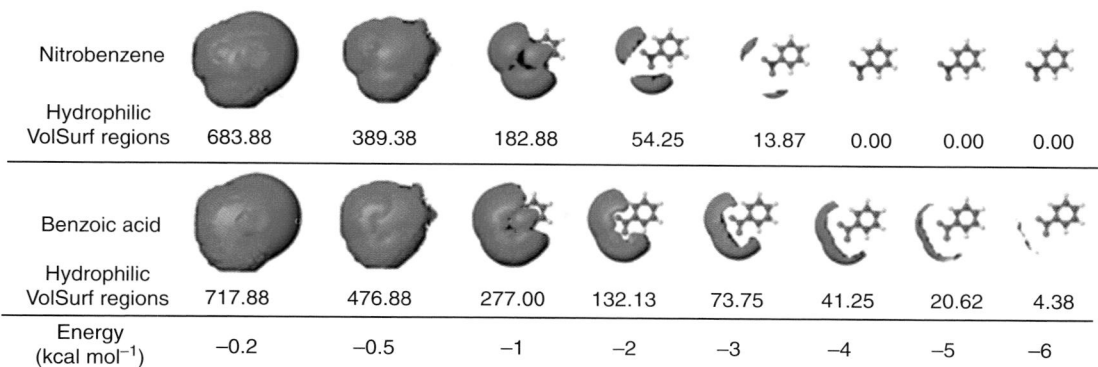

Nitrobenzene Hydrophilic VolSurf regions							
683.88	389.38	182.88	54.25	13.87	0.00	0.00	0.00

Benzoic acid Hydrophilic VolSurf regions							
717.88	476.88	277.00	132.13	73.75	41.25	20.62	4.38

Energy (kcal mol^{-1})							
−0.2	−0.5	−1	−2	−3	−4	−5	−6

Figure 6 Comparison of the water interaction and resulting VolSurf descriptors for nitrobenzene and benzoic acid as anion.

○ Best volumes are the largest three hydrophilic volumes generated by a water probe at the energy levels of − 1 (BV11, BV21, BV31) and − 3 kcal mol^{-1} (BV12, BV22, BV32). The descriptors represent the three largest isolated water interaction regions at a given energy level, indicating the distribution of the polar groups around the compound.

○ Integy moments (Iw1–Iw8) define the distances between the center of mass of a molecule and the centers of all hydrophilic regions around it at each of the eight energy levels. A higher integy moment value indicates a clearer separation between polar and hydrophobic interaction sites. Again the distribution of the polar groups in the molecule is described.

○ Capacity factors (Cw1–Cw8) are the ratios between the hydrophilic regions at each energy level and the molecular surface and show the concentration of polar interactions on the molecular surface.

2. VolSurf descriptors derived from the hydrophobic interaction (DRY probe):

○ Hydrophobic regions (D1–D8) are defined by the isoenergy surfaces of the interaction energy between the molecule and the DRY probe at energy levels of − 0.2, − 0.4, − 0.6, − 0.8, − 1.0, − 1.2, − 1.4, and − 1.6 kcal mol^{-1}. The definition is similar to the hydrophilic regions but the hydrophobic interactions are considered.

○ Hydrophobic integy moments (ID1–ID8) are the distances between the center of mass of a molecule and the centers of the hydrophobic regions around it at each of the eight energy levels. These descriptors show the distribution of the hydrophobic groups.

3. VolSurf descriptors from both the H$_2$O and the DRY probes:

○ Local interaction energy minima (Emin1–Emin3) are the interaction energy values for the lowest three local minima for both the H$_2$O and DRY probes (in kcal mol^{-1}).

○ Local interaction energy minima distances (d12, d13, d23) are the distances between the three lowest local energy minima.

○ Hydrophilic–lipophilic balances (HL1, HL2) are calculated as the ratios between the hydrophilic regions measured at − 3 and − 4 kcal mol^{-1} and the hydrophobic regions measured at − 0.6 and − 0.8 kcal mol^{-1}, respectively. These descriptors represent the balance between hydrophilic and lipophilic (hydrophobic) molecular parts.

○ Amphiphilic moment (A) is the distance between the centre of the hydrophobic domain and the center of the hydrophilic domain. This parameter measures the distribution of the polar and nonpolar groups in the molecule.

○ Critical packing parameter (CP) is the ratio between the maximum distance between the hydrophobic groups and the volume of the hydrophobic parts of a molecule, as defined by the DRY probe. In contrast to HL balance, CP refers just to the molecular shape.

4. VolSurf descriptors from the hydrogen bond acceptor interaction (O probe):

○ Polar regions (WO1–WO8) are defined by the isoenergy surfaces of the interaction energy between the molecule and the O probe at eight different energy levels: − 0.2, − 0.5, − 1.0, − 2.0, − 3.0, − 4.0, − 5.0, and − 6.0 kcal mol^{-1}. These descriptors consider the hydrogen bond acceptor capacity of the compounds.

○ Hydrogen bonding descriptors (HB1–HB8) give the differences between the hydrophilic volumes as defined by the water probe (W1–W8) and any other polar probe included, e.g., the O probe (WO1–WO8), for each level considered. They give additional information about the hydrogen bond capacity of the compounds.

5. Additional VolSurf descriptors:
   - ○ Polarizability (POL) is an estimate of the average molecular polarizability, calculated from atom contributions.[129]
   - ○ Elongation (Elong) is the ratio between the maximum extension of a molecule, considering the most extended conformation, and the extension of the rigid core of the molecule.
   - ○ Diffusivity (DIFF) is the estimation of diffusion of a compound in a water solution using the GRID-based VolSurf descriptors.[129]
   - ○ Lipophilicity (log P) is the estimation of the octanol–water partition coefficient using a model derived from the GRID-based VolSurf descriptors.[129]

VolSurf-derived descriptors can be used to characterize the interaction of a compound with different environments like water or a hydrophobic lipid membrane. Therefore, they are suitable for studying processes like passive permeability, solubility, unspecific binding to plasma proteins, and passive distribution of the compound to different tissues. Models based on VolSurf parameters show the differences in the lipophilic–hydrophilic balance important for the property in question. Properties that have been modeled successfully with VolSurf descriptors are, for example, solubility and passive permeability.

Solubility is a prerequisite for complete absorption of the compound from the gastrointestinal tract to the blood. Many different modeling efforts to predict solubility can be found in the literature (*see* 5.26 In Silico Predictions of Solubility).[130] The intrinsic solubility (thermodynamic solubility of the neutral species) for a set of 1028 compounds has been modeled using the VolSurf descriptors based on GRID and PLS multivariate analysis ($r^2 = 0.74$, $q^2 = 0.73$).[129] The model interpretation based on the PLS coefficients indicates that the ratio of the surface that has a positive interaction with the water probe and the total surface of the compound (the capacity factor) contribute positively to the solubility, while the hydrophobic interaction and log P have a negative contribution.

Passive transcellular transport is mainly determined by water desolvation of the compound before entering the membrane and solvation in a lipophilic environment. In order to account for these processes the interaction of the compound with water and a lipophilic environment has to be computed. This type of interaction is well described by VolSurf descriptors, which were used to model the passive absorption across the Caco-2 cell monolayer (*see* 5.28 In Silico Models to Predict Oral Absorption)[129,131] and the blood–brain barrier (*see* 5.31 In Silico Models to Predict Brain Uptake).[132] However, data from Caco-2 experiments do not only represent the passive pathway, but also a combination of the different absorption routes and mechanisms that affect each compound differently. Moreover, the combination of experimental data from different sources, possibly obtained under different experimental conditions, may fail to give a quantitatively reliable data set. A discriminative model may be more useful in such cases and was therefore used for both Caco-2 cell monolayer and blood–brain barrier permeability.[129,132] This approach was also used for modeling the permeability across an artificial membrane and produced similar results (*see* 5.28 In Silico Models to Predict Oral Absorption).[133]

5.22.3.1.2 Comparison between different experimental models for absorption

Different experimental techniques are used to predict a compound's permeability: transport experiments across a Caco-2 cell monolayer or rat intestinal tissue using the Ussing Chamber technique, physicochemical parameters like log D, or retention times in chromatographic systems. In order to find characteristic differences between such techniques permeability/lipophilicity values of a small set of adrenoreceptor antagonists obtained in six different systems were investigated: Caco-2 cell monolayers (Caco-2), rat ileum (Ileum), and rat colon (Colon) membrane in the Ussing chamber, distribution in octanol/water at pH 7.4 (log D), immobilized liposome chromatography (log k_{ILC}), and capillary electrophoresis (log k_{CE}).[20,100,134] A model for the experimental results for the small set of compounds in each of these systems was obtained using the VolSurf descriptors. The PLS coefficients for the models were combined in a single matrix and a PCA was performed. The resulting scores plot (**Figure 7**) shows the clustering of the different experimental models as they are seen by the descriptors.

The first principal component explains more than 42% of the variance and discriminates between the biological assays and the experimental physicochemical descriptors. This component is mainly dependent on the hydrophobic surface descriptors. The clear separation between the biological cell- and membrane-based methods on the one hand, and the physicochemical methods on the other hand, indicates that the molecular descriptors most important for distribution in octanol/water (log D), chromatography or capillary electrophoresis may not be as relevant for permeation through a biological cell membrane.

The second principal component (36% of the variance explained) divides the techniques further. The Ussing chamber experiments using rat colon segments is closer to the Caco-2 experiment than the experiment using rat ileum segments. In this principal component the hydrophilic volumes make the largest contribution. These descriptors

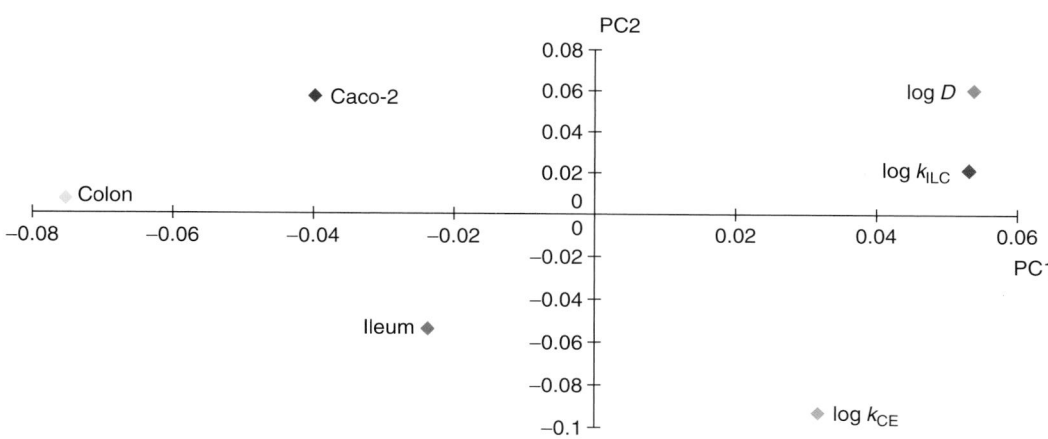

Figure 7 Score plot for the PCA of the PLS coefficients for a comparison between different experimental models for absorption/permeability.

correlate negatively to permeability in general. Their importance in distinguishing the Caco-2 cells and the colon membrane from the ileum membrane could be of physiological relevance with respect to lipid composition of the three biological membranes.

5.22.3.2   GRIND Descriptors

The GRIND descriptors were developed in order to keep the pharmacophoric information found in the MIF in an alignment-independent way. This pharmacophoric type of representation may be helpful when modeling active transport, metabolic stability or enzyme inhibition. The Almond software[126] transforms the MIF into distance-based descriptors of the molecule's interaction pattern (**Figure 8**). The resulting parameters describe the geometry of the interaction and QSAR models can be derived where the interaction with a protein is essential. Since the most relevant interactions are usually described as hydrogen bond acceptor, hydrogen bond donor and hydrophobic interactions, GRID probes that describe these types of interactions are used for the GRIND descriptor computation: N1 (amide nitrogen atom type with 1 hydrogen atom), O (oxygen atom from a carbonyl group), and DRY (hydrophobic group). In addition a description of shape is added using the curvature of the positive energy with the N1 or O probes.[135]

The GRIND analysis does not use all the points in the MIF, but extracts only the most relevant ones. Two criteria are used for the selection of the relevant grid points: the interaction energy, which should be high, and the distance between the selected points. For example, if there are two grid points with the same energy and one is 2 Å and the second one 10 Å away from a previously selected point, the methodology would select the second one. The user can define the balance between the two selection criteria and the number of selected points.

The relevant grid points are then used in the molecular descriptor calculation. Each selected grid point from any of the investigated fields is combined with another grid point from the same or another field. If the default conditions are applied, three fields plus the shape description are used and 10 pairs of interactions are obtained: DRY–DRY, N1–N1, O–O, Shape–Shape, DRY–N1, DRY–O, DRY–Shape, N1–O, N1–Shape, and O–Shape. The energy of the two points, scaled by the maximum energy of the respective probe, is multiplied and the distance between them determined. The distance is used to define the "bin" to which the energy value is to be associated. Only the highest energy value, i.e., the most important interaction, is kept for each distance bin. The use of the maximum energy of interaction in each bin allows saving the coordinates of each point that contributes to the distance. Thus, it is possible to decode the descriptors back into 3D space, which facilitates the model interpretation.

The transformation of the MIF into descriptors based on the distance between the selected points allows an alignment independent analysis of ligand–protein interactions. However, the descriptors still depend on the conformation used.

5.22.3.2.1   Active transport: P-glycoprotein

P-glycoprotein (P-gp) is an important efflux transporter, which was shown to influence both intestinal absorption and distribution into the brain for many drugs.[136] It is useful to apply a combination of descriptors based on the global properties of the compound and descriptors based on a pharmacophoric representation to describe the interaction with

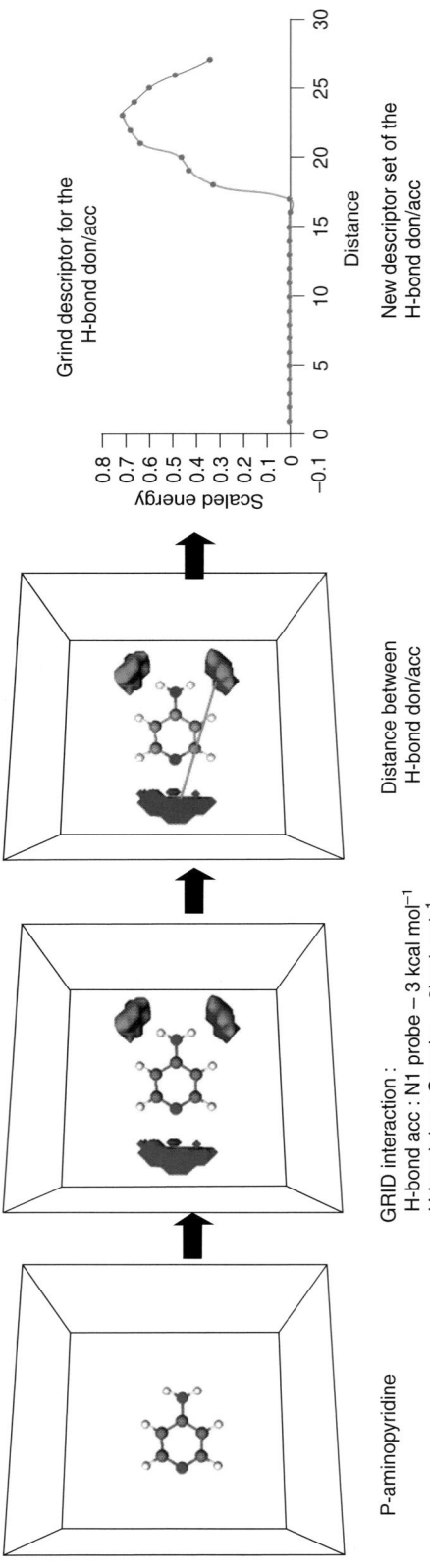

Figure 8 GRIND descriptors calculation. Step 1: GRID field computation using the DRY, N1, and O probes. Step 2: Selecting the most relevant grid points and distance computation. Step 3: Selecting the highest cross-correlation and autocorrelation values for each distance to obtain the descriptors.

this transporter correctly. Conceptually, the global properties would describe the initial passive membrane permeation required to reach the site of action. The specific ligand–protein interactions are explained by the pharmacophoric descriptors. This approach was successful in the P-gp case, where the two processes are important for transport; passive transport into the cell and active efflux transport from within the cell to the outside. Cianchetta *et al.*[137] obtained an initial model for the inhibition of the P-gp transport using the VolSurf descriptor set alone, with $r^2 = 0.71$ and $q^2 = 0.51$. This initial model was not, however, able to predict the most active compounds. When the authors expanded the model[137] with the GRIND descriptors they obtained better statistics ($r^2 = 0.81$ and $q^2 = 0.73$) and a similar prediction rate in the entire activity range.

5.22.3.2.2 Metabolic stability: the CYP3A4 case

The rate of metabolism is an important and well-studied factor that contributes to the elimination of the compound from the body. An easy, often used method to obtain metabolic stability data in screening is the determination of the percentage remaining after a certain period of time for a given compound at one concentration in an enzyme incubation. However, this metabolic data is not precise and can only be used to classify the compounds as metabolically stable or unstable. Such data has been used to model CYP3A4 metabolic stability,[138] an enzyme with broad substrate specificity and for which lipophilicity has been described as a major determining factor for affinity. Cruciani *et al.* showed that VolSurf descriptors could be used to obtain a model with reasonable statistics.[129] However, some compounds were not well classified. In order to improve the compound description and the model predictive power, GRIND descriptors were used additionally.[138] The resulting model showed better predictivity for the different compound families, although the overall prediction rates were similar.

5.22.3.3 Alternative Use of MIF-Derived Descriptors

MIFs can be used to investigate similarities and differences between various molecular structures using a new statistical technique, the consensus principal component analysis (cPCA).[139] This technique can analyze the GRID-MIFs of aligned protein structures and decomposes the typical PCA analysis to identify the contribution of each block of variables (i.e., each GRID probe). The data decomposition provides the opportunity to analyze the contribution made by different blocks of data. This statistical analysis technique is included in the software GOLPE.[112] It was used to study the different interaction patterns for CYP2C9, CYP2C19, CYP2D6, CYP3A4, and CYP1A2, a selectivity analysis. The obtained information is useful for predicting the selective sites of metabolism, i.e., those sites in a molecule that will be metabolized by one particular cytochrome. Moreover, the selectivity analysis has also been used to compare crystal structures to homology models for CYP2C5 and CYP2C9.[125]

5.22.3.3.1 Site of metabolism prediction

The elucidation of the site of metabolism in a molecule is one of the most time-consuming experimental tasks in the ADME field. Moreover, in some cases, experiments can only determine the region in the molecule that is metabolized, but not on which atom the metabolic reaction actually occurs. Nevertheless, such information is needed in order to explore the pharmacological activity and the toxic effects of the formed metabolites. Knowledge about the site of metabolism may also help in the chemical protection of the molecule, resulting in a compound that is less liable to metabolic reactions. A new technique called MetaSite[128] has been developed in order to consider the interaction with the CYP enzyme and the chemical reactivity of the compounds towards the oxidation. The protein recognition part compares the interaction profile of the enzyme based on GRID-MIFs and different conformations of the potential substrates. The prediction rates for the site of metabolism for five CYP enzymes (CYP1A2, CYP2C9, CYP2C19, CYP2D6, and CYP3A4) were validated using more than 100 metabolic reactions in each case (**Table 1**).

5.22.3.3.2 Selective site of metabolism

The aim of this study was to predict the site of metabolism that is due to one particular CYP enzyme. The investigated CYPs were aligned and the resulting GRID-MIFs used as input for a cPCA, in order to find the interactions specific for each enzyme. The loading values resulting from this analysis were used to scale the original GRID-MIFs. A protein recognition, like the one used in MetaSite, was based on these scaled MIFs in order to identify the selective sites of metabolism for the investigated compounds. In contrast, MetaSite suggests all possible sites, since it is based on the original, not scaled GRID-MIFs of the cytochromes.

The selective site of metabolism was studied for the CYP2C subfamily (CYP2C9, CYP2C8, CYP2C19, and CYP2C18). **Tables 2–5** summarize the experimental results. The selective site of metabolism investigation yielded in total 72% good predictions.

Table 1   Rate of prediction for the site of metabolism; 'Good prediction' means that the experimentally determined site of metabolism is among the first three sites predicted by MetaSite

Cytochrome	Number of reactions	Rate of good prediction
CYP1A2	120	73%
CYP2C9	125	83%
CYP2C19	109	75%
CYP2D6	130	85%
CYP3A4	220	77%

For all CYP2C9 cases analyzed the correct site of metabolism was predicted within the first three suggestions. Eight out of the 11 substrates tested for CYP2C19 were well predicted (the correct site of metabolism was found within the first three options presented). However, the following three compounds could not be predicted well by the selective site of metabolism analysis: (1) diazepam undergoes an N-demethylation catalyzed by CYP2C19, but the technique suggested a hydrogen atom abstraction from the phenyl ring; (2) R-mephobarbital is hydroxylated by CYP2C19 at the *para* position, while the method selected the *meta* position; and (3) ticlopedine is oxidized by CYP2C19 at two positions, but none of them was found within the three top solutions.

For three out of five CYP2C8 substrates the selective site of metabolism analysis was able to find the correct site of metabolism within the top three solutions. However, retinoic acid and R-troglitazone were not correctly predicted. R-Troglitazone undergoes a ring opening catalyzed by CYP2C8, but a hydrogen atom abstraction at the methyl groups in the phenyl ring was predicted. This may be due to the fact that the type of reaction, the ring opening, was not considered in the analysis.

There is only one reaction described in the literature as selective for the CYP2C18. The reported metabolic reaction requires the abstraction of the hydrogen at the carbon atom next to the sulphur in the thiophene ring (**Table 5**), which is close in space to the site selected by the selective site of metabolism analysis.

5.22.4   Miscellaneous

Molecular descriptors are available from various software – Sybyl, MOE, ACD, GRID, etc. – or can be calculated from in house software or simple scripts. Recently, web-based descriptor generation tools have been proposed as well.[169] A molecular descriptor generation software that calculates more than 1500 molecular descriptors is the Dragon software.[170] It includes many of those mentioned here, but also a lot of descriptors that have not yet been used too often in ADME modeling and therefore are beyond the scope of this chapter. The Dragon descriptors are explained in Todeschini's Handbook of Molecular Descriptors,[17] a general encyclopaedia on descriptors useful for QSAR and QSPR studies.

Below some examples of descriptor sets and other descriptors used in ADME modeling are presented. The descriptor sets usually include descriptors with information on molecular size, lipophilicity, and hydrogen bonding.

5.22.4.1   Abraham Descriptors

Abraham's LFER approach[66] uses five descriptors only: E, the excess molar refraction, S, the solute polarity/polarizability, A (α), the solute H-bond acidity, B (β), the solute H-bond basicity, and V, McGowans characteristic molecular volume (*see* 5.31 In Silico Models to Predict Brain Uptake).[35] It is easy to see the molecular size and hydrogen bonding descriptors in this set. Lipophilicity information is included as well, since it was shown that the descriptors could be used to estimate partition coefficients, like log P_{oct}.[66,171] The descriptors can be assessed experimentally, tabulated, and used for correlations to physicochemical properties. Often reasonable correlations can be found for which not all five descriptors need to be important. These descriptors can also be obtained theoretically, either based on fragments[87] or on quantum chemical calculations, e.g., the conductor-like screening model (COSMO) approach.[172] Such theoretically derived descriptors were successfully used in several QSPR studies, e.g., for the prediction of lipophilicity and cell permeation.[173,174]

Table 2 CYP2C9 substrates: the experimentally determined site of metabolism

Compound	Structure	Reaction	K_i (μM)	K_m (μM)	V_{max} (min^{-1})	Reference
S-Warfarin		7-Hydroxylation	20	4.1 ± 0.6/ 5.8 ± 0.8	0.421 ± 0.021/ 0.248 ± 0.018	140–142
Progesterone		21-Hydroxylation	5.5	17 ± 2	0.51 ± 0.02	140,143
Diclofenac		4′-Hydroxylation	2	3.6 ± 0.4/ 3.9 ± 0.3	$41/35.6 \pm 1.3$	140,142,144
S-Ibuprofen		3-Hydroxylation		35-82		145
R-Naproxen		O-Demethylation		126		146
Piroxicam		5′-Hydroxylation		40 ± 3	0.408 ± 0.026	142
Torasemide		Hydroxylation		40	5.2	147
Phenytoin		4′-Hydroxylation	6.0	15 ± 4	0.191 ± 0.018	140,142
Fluvastatin		6-Hydroxylation	2.2/3.3 0.3 ± 0.1/ 0.5 ± 0.1	0.9	0.078	140,148
Gemfibrozil		Hydroxylation	5.8			149
S-Miconazole		Hydroxylation	6.0			140
R-Fluoxetine		N-Demethylation and hydroxylation		13.6 ± 1.6	17.0 ± 0.7	150

Big dot (●) indicates site of metabolism by CYP2C9.

Table 3 CYP2C19 substrates: the experimentally determined site of metabolism

Compound	Structure	Reaction	K_i (μM)	K_m (μM)	V_{max} (min^{-1})	Reference
R-Lansoprazole		5-Hydroxylation				151
Moclobemide		Hydroxylation				152
Imipramine		N-Demethylation		24.7 ± 3.0	29.8 ± 4.7	153
R-Omeprazole		O-Demethylation		4.48 ± 2.1	0.60 ± 0.05	154,155
Proguanil		Hydroxylation				156
Bufuralol		1'-Hydroxylation		36	36.9	157
Ticlopedine		Hydroxylation	1.2 ± 0.5			158
S-Mephenytoin		N-Demethylation		54	2.1	159
R-Warfarin		7-Hydroxylation		55 ± 18	0.329 ± 0.062	160
Diazepam		N-Demethylation		21 ± 3	1.76 ± 0.07	161
R-Mephobarbital		4-Hydroxylation	34			162

5.22.4.2 Quantum Chemical Descriptors

The MolSurf program[68] uses quantum chemical calculations to derive descriptors that were found useful for ADME property predictions. The program calculates a constant electron density surface that represents the valence region of a given molecule. The electrostatic potential and the local ionization energy are calculated on evenly distributed points on this surface. This information is used to calculate chemically understandable descriptors for the molecular surface, Lewis acid and Lewis base properties, charge transfer characteristics, hydrogen bonding ability, polarizability, polarity, steric factors, and lipophilicity. Norinder et al. used these descriptors in a series of studies and found that ADME properties like Caco-2 cell permeability (17 drugs; $r^2 \sim 0.9$, $q^2 \sim 0.8$), human intestinal absorption (20 drugs; $r^2 \sim 0.9$, $q^2 \sim 0.7$), brain–blood partitioning (63 drugs; $r^2 \sim 0.8$, $q^2 \sim 0.78$), and P-gp interaction (22 drugs; $r^2 = 0.7$, $q^2 = 0.7$) could be modeled reasonably well.[175–178] However, in later studies they found that simpler descriptors could actually give reasonable correlations as well.[179]

Table 4 CYP2C8 substrates: the experimentally determined site of metabolism

Compound	Structure	Reaction	K_m (μM)	V_{max} (min^{-1})	Reference
Retinoic acid		4-Hydroxylation	100	0.38	163
R-Troglitazone		Quinone type	2.7 ± 2.5	4.2 ± 1.0	164
Rosiglitazone		Hydroxylation N-demethylation	44 ± 17 10 ± 1.3	174 ± 49 ($nmol^{-1}h^{-1}mg^{-1}$) 146 ± 8.4	165
Zopiclone		N-Demethylation(N-oxide)	71 ± 6	2.5 ± 0.1	166
Amiodarone		N-Deethylation	8.6 ± 2.5	2.3 ± 0.2	167

Table 5 CYP2C18 substrates: the experimentally determined site of metabolism

Compound	Structure	Reaction	K_m (μM)	K_{cat} (min^{-1})	K_{cat}/K_m ($min^{-1}\mu M^{-1}$)	Reference
TA derivative		5-Hydroxylation	9 ± 1	125 ± 25	13	168

Quantum chemically derived descriptors include also descriptors like highest occupied molecular orbital (HOMO), lowest unoccupied molecular orbital (LUMO), hardness (LUMO − HOMO)/2, dipole moment, atomic charges, polarizability, polarity, ionization potential, electrostatic potentials, molecular energy values, and others. Quantum chemical descriptors and their use in QSAR/QSPR studies were reviewed some time ago.[180] Descriptors based on quantum chemical calculations were successfully used in ADME studies by, e.g., Lombardo *et al.* who correlated free energies of the solvation in water as computed with the AMSOL program to log BB values.[181]

5.22.4.3 Surface-Related Descriptor Sets

Jurs *et al.* introduced the charged partial surface areas (CPSAs)[109] which primarily describe the polar interaction between molecules. The rationale behind these descriptors was the assumption that such interactions are mainly due to the charge distribution within the molecule and that the influence of one single atom depends on its exposure at the surface of the molecule (a solvent-accessible surface as defined by Lee and Richards[32]). Thus, a set of 25 descriptors was defined by summing the partial charges of all atoms with positive (or negative) partial charges and weighting them by multiplication with these partial charges in various combinations. They were found useful for correlation to physicochemical properties such as boiling point, and have also been used for ADME studies, together with other descriptors.[15] This set of descriptors has inspired other scientists to define charge weighted partial surface area descriptors, which have been shown to be hydrogen-bond related.[10]

Recently, an analogously defined set of 25 descriptors regarding the hydrophobic interactions, the hydrophobic surface area descriptors, was presented.[182] Instead of atomic charges, atomic hydrophobicity indices were used. These descriptors were found to be unique descriptors that can capture information regarding hydrophobic intermolecular interactions and should, therefore, be useful for ADME modeling. A first model on BBB data of a data set of 97 commercially available structurally diverse drugs was found to be slightly improved by addition of the hydrophobic surface area descriptors.

Simple partitioned total surface areas alone were shown to be able to classify oral drugs according to their absorption properties.[183] Bergström *et al.* conclude that surface areas describing the nonpolar part of the molecule yield good predictions of solubility, whereas those describing specific polar parts, like parts regarding only hydrogens or other polar atom types, can be used for predicting permeability.

Molecular hashkeys[184] describe a molecule in terms of surface similarity to a given basis set of common molecules, considering steric, polar positive, and polar negative features of the molecule. The hashkey for a given molecule comprises the surface similarities of each of the molecules in the basis set with the best fitting conformation of the molecule in question. It was shown, that such hashkeys were able to predict, the ranking order of fraction absorbed values of a diverse set of compounds.[184]

MS-WHIM descriptors[185] were developed by applying the weighted holistic invariant molecular (WHIM[186]) approach to molecular surface points. By centering the molecular surface coordinates with respect to their property specific weighted mean value and performing a weighted PLS, the resulting molecular descriptors become invariant to both translation and rotation. Thus, MS-WHIM descriptors are alignment independent, but depend on the conformation used for surface generation. Several weighting schemes can be used, e.g., the molecular electrostatic potential, hydrogen bonding capacity, and hydrophobicity. These descriptors were used successfully in a study on binding affinity of 16 coumarine-like molecules to CYP2A5,[185] and in a study on Caco-2 permeability of 17 structurally diverse compounds.[187]

5.22.4.4   Fingerprints

A molecular fingerprint[17,18] in the classical sense represents the molecule by a string of 0s and 1s, which correspond to the presence or absence of specific molecular features. According to the type of features used for the fingerprint, one can distinguish between 2D or 3D fingerprints. 2D fingerprints typically relate to the presence or absence of molecular fragments, functional groups, or atom types within the molecule in question, whereas 3D fingerprints collect three or four features including the corresponding interfeature distances. Fingerprints, where each bit is dedicated to a specific structural feature, are often called structural keys. Another way to define fingerprints is by a specific fingerprinting algorithm, like the one used by Daylight.[188] This algorithm generates several patterns for each molecule and combines them all together by a 'logical OR,' generating the fingerprint of the molecule. The patterns represent the atoms in the molecule, atoms and their closest neighbors, each group of atoms connected by two to four or more bonds. Each pattern is used as seed to a pseudorandom number generator, which results in a fingerprint string for this specific pattern, where typically four or five bits are set. The combination of all pattern fingerprints gives the molecular fingerprint, in which all single patterns are represented. Since there is no predefined set of patterns, it is not possible to assign a pattern to a specific bit. But all the bits of a specific subset's fingerprint are also set in the molecule's fingerprint (even though one specific bit may be contained in more than just one subset). This approach is more general, as the structural features investigated need not to be defined in advance; the resulting fingerprints can be smaller, since there need not to be a bit for each and every structural feature investigated.

Fingerprints are very fast and easy to calculate and compare with each other due to their data structure.[17] Thus, this type of descriptor is especially useful for quickly assessing the similarity of different molecules, for substructure searches, or for selecting the most diverse compounds from a given data set. Fingerprints can be used to check for a druglike score of investigated molecules,[189] which could be interesting in early drug design. The fingerprint approach can also be used in ADME studies by defining a specific metabolic fingerprint as was suggested by Keserü and Molnár.[190] In this case the fingerprints are not binary strings but bin arrays. Each bin is associated with a particular metabolic reaction (or a summary of the molecules and their metabolite properties) and holds a value that indicates how often a specific reaction occurs for the investigated structure. In this broader sense, the previously described molecular hashkeys or the GRID-based GRIND descriptors can also be seen as fingerprints.

5.22.5   Closing Remarks

We have here discussed a number of molecular descriptors that are currently used in the area of ADME modeling and predictions. Some of the descriptor sets, e.g., MIF-derived descriptors, can comprise a huge number of highly correlated descriptors and require statistical tools that can cope with such data, e.g., the PLS method.[12,191] Additionally, variable selection techniques can be used in order to find the most significant variables, which will help the model interpretation. These techniques are also useful, if several descriptor sets are used, since such combinations will often lead to highly redundant information. An alternative way to cope with this redundancy between different descriptor sets is to model the property using each data set on its own and combine the final models, for example, by hierarchical modeling as implemented in SIMCA.[192,193]

The main molecular properties found important for many ADME properties (lipophilicity, hydrogen bonding, and molecular size) are characterized by all descriptor sets, which is one reason for the data redundancy often found. Another reason is that they can be described in several and very different ways, e.g., by simple physicochemical descriptors or by descriptors based on quantum-chemical calculations or derived from molecular field interactions. Models for ADME properties such as permeability (passive), solubility, or unspecific protein binding can generally be described using only such general molecular descriptors. Other ADME properties, such as active transport or metabolism, which relate to interaction with a protein, need a pharmacophoric type of description, even though lipophilicity – at least for a congeneric series of compounds – may be very important. In this case descriptors based on molecular interaction fields are superior to the physicochemical descriptors. In addition, structural keys that recognize the absence or presence of specific essential molecular fragments can be useful. Furthermore, if the 3D structure of the protein in question, e.g., a CYP protein, is known or could be easily modeled by homology, a description of the ligand–protein interactions can be added.

Selection of the descriptor set to be used also depends on the scope of the model: if a set of closely related compounds is to be investigated, i.e., a local model to be developed, descriptors that have been tailored to this data set can be used. For example, descriptors for selected molecular fragments can be calculated. Local models should help in understanding the molecular properties required to produce certain specified ADME behavior. Therefore, molecular descriptors and statistical techniques that are easy to interpret should be chosen. These models can be used to suggest chemical changes in a molecule to obtain the desired ADME property, or even in the understanding of the complex ADME data.

For global models it is necessary to use descriptors that are available for as many molecules as possible – both for those that are used in the training set and possible molecules that should be predicted with the model. These models are useful for filtering out a great number of virtual (or synthesized) compounds and selecting those with the appropriate properties, but an exact interpretation of the influence of each variable may not be necessary.

Moreover, local and global models may differ in the impact of the complexity inherent in ADME data. Related compounds usually have similar factors that affect their pharmacokinetic behavior, while in a diverse set, which is required for a global model, the different factors could affect the property measured in various ways. For example, Caco-2 permeability data comprise both active and passive transport. In a set of similar compounds one can assume that all of them are either passively absorbed or have affinity to the same transporter(s) in varying degrees. In contrast, in a diverse data set one has to expect that some of the compounds are transported by different transporter proteins, whereas others are just passively absorbed. For example, if the passive route dominates in the data, it will be possible to obtain a reasonable model, and actively transported compounds will most probably be outliers.

Another aspect is the probable type of correlation: in many studies a linear correlation is assumed between the descriptors and the property in question, in others both linear and quadratic terms are used randomly. A good understanding of both the descriptors used and the property investigated will help in selecting the descriptors for which quadratic terms or other types of correlations may be useful, but is also a necessity for proper interpretation of the final model.

References

1. Butina, D.; Segall, M. D.; Frankcombe, K. *Drug Disc. Today* **2002**, *7*, S83–S88.
2. Ekins, S.; Rose, J. *J. Mol. Graph. Model.* **2002**, *20*, 305–309.
3. Lombardo, F.; Gifford, E.; Shalaeva, M. Y. *Mini Rev. Med. Chem.* **2003**, *3*, 861–875.
4. Stouch, T. R.; Kenyon, J. R.; Johnson, S. R.; Chen, X.-Q.; Doweyko, A.; Li, Y. *J. Comput.-Aided Mol. Des.* **2003**, *17*, 83–92.
5. Van de Waterbeemd, H.; Gifford, E. *Nat. Rev. Drug Disc.* **2003**, *2*, 192–204.
6. Beresford, A. P.; Segall, M.; Tarbit, M. H. *Curr. Opin. Drug Disc. Dev.* **2004**, *7*, 36–42.
7. Penzotti, J. E.; Landrum, G. A.; Putta, S. *Curr. Opin. Drug Disc. Dev.* **2004**, *7*, 49–61.
8. Votano, J. R. ; *Curr. Opin. Drug Disc. Dev.* **2005**, *8*, 32–37.
9. Winiwarter, S.; Bonham, N. M.; Ax, F.; Hallberg, A.; Lennernäs, H.; Karlén, A. *J. Med. Chem.* **1998**, *41*, 4939–4949.
10. Winiwarter, S.; Ax, F.; Lennernäs, H.; Hallberg, A.; Pettersson, C.; Karlén, A. *J. Mol. Graph. Model.* **2003**, *21*, 273–287.
11. Clark, D. E. *J. Pharm. Sci.* **1999**, *88*, 807–814.
12. Kubinyi, H. *QSAR: Hansch Analysis and Related Approaches*; Mannhold, R.; Krogsgaard-Larsen, P.; Timmerman, H.; Eds.; VCH Publishers: New York, 1993.
13. Devillers, J., Ed. *Neural Networks in QSAR and Drug Design*; Academic Press: London, 1996.

14. Bai, J. P. F.; Utis, A.; Crippen, G.; He, H.-D.; Fischer, V.; Tullman, R.; Yin, H.-Q.; Hsu, C.-P.; Jiang, L.; Hwang, K.-K. *J. Chem. Inf. Comput. Sci.* **2004**, *44*, 2061–2069.
15. Wessel, M. D.; Jurs, P. C.; Tolan, J. W.; Muskal, S. M. *J. Chem. Inf. Comput. Sci.* **1998**, *38*, 726–735.
16. Raevsky, O. A.; Fetisov, V. I.; Trepalina, E. P.; McFarland, J. W.; Schaper, K.-J. *Quant. Struct.-Act. Relat.* **2000**, *19*, 366–374.
17. Todeschini, R.; Consonni, V. *Handbook of Molecular Descriptors*; Mannhold, R., Kubinyi, H., Timmerman, H., Eds.; Wiley-VCH: New York, 2000.
18. Xue, L.; Bajorath, J. *Comb. Chem. High Throughput Screen.* **2000**, *3*, 363–372.
19. Van de Waterbeemd, H.; Kansy, M. *Chimia.* **1992**, *46*, 299–303.
20. Palm, K.; Luthman, K.; Ungell, A.-L.; Strandlund, G.; Artursson, P. *J. Pharm. Sci.* **1996**, *85*, 32–39.
21. Ertl, P.; Rohde, B.; Selzer, P. *J. Med. Chem.* **2000**, *43*, 3714–3717.
22. Lipinski, C. A.; Lombardo, F.; Dominy, B. W.; Feeney, P. *J. Adv. Drug Deliv. Rev.* **1997**, *23*, 3–25.
23. Turner, J. V.; Glass, B. D.; Agatonovic-Kustrin, S. *Analyt. Chim. Acta.* **2003**, *485*, 89–102.
24. Zhao, Y. H.; Le, J.; Abraham, M. H.; Hersey, A.; Eddershaw, P. J.; Luscombe, C. N.; Boutina, D.; Beck, G.; Sherborne, B.; Cooper, I.; Platts, J. A. *J. Pharm. Sci.* **2001**, *90*, 749–784.
25. Van de Waterbeemd, H.; Camenisch, G.; Folkers, G.; Raevsky, O. *Quant. Struct.-Act. Relat.* **1996**, *15*, 480–490.
26. Ren, S.; Das, A.; Lien, E. J. *J. Drug Target.* **1996**, *4*, 103–107.
27. Shah, M. V.; Audus, K. L.; Borchardt, R. T. *Pharm. Res.* **1989**, *6*, 624–627.
28. Artursson, P.; Ungell, A.-L.; Löfroth, J.-E. *Pharm. Res.* **1993**, *10*, 1123–1129.
29. Lennernäs, H. *J. Pharm. Sci.* **1998**, *87*, 403–410.
30. Ungell, A.-L.; Nylander, S.; Bergstrand, S.; Sjöberg, Å.; Lennernäs, H. *J. Pharm. Sci.* **1998**, *87*, 360–366.
31. He, Y.-L.; Murby, S.; Warhurst, G.; Gifford, L.; Walker, D.; Ayrton, J.; Eastmond, R.; Rowland, M. *J. Pharm. Sci.* **1998**, *87*, 626–633.
32. Lee, B.; Richards, F. M. *J. Mol. Biol.* **1971**, *55*, 379–400.
33. Richards, F. M. *Annu. Rev. Biophys. Bioeng.* **1977**, *6*, 151–176.
34. Connolly, M. L. *J. Appl. Crystallogr.* **1983**, *16*, 548–558.
35. Abraham, M. H.; McGowan, J. C. *Chromatographia* **1987**, *23*, 243–246.
36. Wiener, H. *J. Am. Chem. Soc.* **1947**, *69*, 17–20.
37. Randic, M. J. *J. Am. Chem. Soc.* **1975**, *97*, 6609–6615.
38. Kier, L. B. *Quant. Struct.-Act. Relat.* **1986**, *5*, 1–7.
39. Veber, D. F.; Johnson, S. R.; Cheng, H.-Y.; Smith, B. R.; Ward, K. W.; Kopple, K. D. *J. Med. Chem.* **2002**, *45*, 2615–2623.
40. Testa, B.; Carrupt, P.-A.; Gaillard, P.; Billois, F.; Weber, P. *Pharm. Res.* **1996**, *13*, 335–343.
41. Carrupt, P.-A.; Testa, B.; Gaillard, P. Computational Approaches to Lipophilicity Methods and Applications. In *Reviews in Computational Chemistry*; Lipkowitz, K. B., Boyd, D. B., Eds.; John Wiley: New York, 1997; Vol. 11, pp 241–315.
42. El Tayar, N.; Testa, B.; Carrupt, P. A. *J. Phys. Chem.* **1992**, *96*, 1455–1459.
43. Moriguchi, I.; Kanada, Y.; Komatsu, K. *Chem. Pharm. Bull.* **1976**, *24*, 1799–1806.
44. Testa, B.; Seiler, P. *Arzneim.-Forsch./Drug Res.* **1981**, *31*, 1053–1058.
45. Young, R. C.; Mitchell, R. C.; Brown, T. H.; Ganelling, C. R.; Griffiths, R.; Jones, M.; Rana, K. K.; Saunders, D.; Smith, I. R.; Sore, N. E.; Wilks, T. J. *J. Med. Chem.* **1988**, *31*, 656–671.
46. Burton, P. S.; Conradi, R. A.; Ho, N. F. H.; Hilgers, A. R.; Borchardt, R. T. *J. Pharm. Sci.* **1996**, *85*, 1336–1340.
47. Van de Waterbeemd, H.; Carter, R. E.; Grassy, G.; Kubinyi, H.; Martin, Y. C.; Tute, M.; Willett, P. *Pure Appl. Chem.* **1997**, *69*, 1137–1152.
48. Meyer, H. *Naunyn Schmiedebergs Arch. Exp. Path. Pharm.* **1899**, *42*, 109–118; and Overton, E. *Studien über die Narkose, zugleich ein Beitrag zur allgemeinen Pharmakologie*; G. Fischer: Jena, Germany 1901; cited in Kubinyi, H. *Quant. Struct.-Act. Relat.* **2002**, *21*, 348–356.
49. Leo, A.; Hansch, C.; Church, C. *J. Med. Chem.* **1969**, *12*, 766–771.
50. Scherrer, R. A.; Howard, S. *J. Med. Chem.* **1977**, *20*, 53–58.
51. Waterhouse, R. N. *Mol. Imag. Biol.* **2003**, *5*, 376–389.
52. Kratochwil, N. A.; Huber, W.; Müller, F.; Kansy, M.; Gerber, P. R. *Biochem. Pharmacol.* **2002**, *64*, 1355–1374.
53. Testa, B.; Crivori, P.; Reist, M.; Carrupt, P.-A. *Persp. Drug Disc. Des.* **2000**, *19*, 179–211.
54. Yazdanian, M.; Glynn, S. L. *Pharm. Res.* **1998**, *15*, 1490–1494.
55. Mirrlees, M. S.; Moulton, S. J.; Murphy, C. T.; Taylor, P. J. *J. Med. Chem.* **1976**, *19*, 615–619.
56. Avdeef, A. *Quant. Struct.-Act. Relat.* **1992**, *11*, 510–517.
57. Fujita, T.; Iwasa, J.; Hansch, C. *J. Am. Chem. Soc.* **1964**, *86*, 5175–5180.
58. Leo, A. J. *Chem. Rev.* **1993**, *93*, 1281–1306.
59. Mannhold, R.; Rekker, R. F. *Persp. Drug Disc. Des.* **2000**, *18*, 1–18.
60. Petrauskas, A. A.; Kolovanov, E. A. *Persp. Drug Disc. Des.* **2000**, *19*, 99–116.
61. Broto, P.; Moreau, G.; Vandycke, C. *Eur. J. Med. Chem. – Chim. Ther.* **1984**, *19*, 71–78.
62. Ghose, A. K.; Crippen, G. M. *J. Comput. Chem.* **1986**, *7*, 565–577.
63. Croizet, F.; Dubost, J. P.; Langlois, M. H.; Audry, E. *Quant. Struct.-Act. Relat.* **1991**, *10*, 211–215.
64. Ghose, A. K.; Viswanadhan, V. N.; Wendoloski, J. J. *J. Phys. Chem. A.* **1998**, *19*, 172–178.
65. Moriguchi, I.; Hirono, S.; Liu, Q.; Nakagome, I.; Matsushita, Y. *Chem. Pharm. Bull.* **1992**, *40*, 127–130.
66. Abraham, M. H. *Chem. Soc. Rev.* **1993**, *22*, 73–83.
67. Bodor, N.; Huang, M.-J. *J. Pharm. Sci.* **1992**, *81*, 272–281.
68. Sjöberg, P. MolSurf: A Generator of Chemical Descriptors for QSAR. In *Computer-Assisted Lead Finding and Optimization: Current Tools for Medicinal Chemistry*; Van de Waterbeemd, H., Testa, B., Folkers, G., Eds.; VHCA: Zurich, Switzerland, 1998, pp 83–92.
69. Mannhold, R.; Van de Waterbeemd, H. *J. Comp.-Aided Mol. Des.* **2001**, *15*, 337–354.
70. Audry, E.; Dubost, J.-P.; Colleter, J.-C.; Dallet, P. *Eur. J. Med. Chem. – Chim. Ther.* **1986**, *21*, 71–72.
71. Heiden, W.; Moeckel, G.; Brickmann, J. *J. Comput.-Aided Mol. Des.* **1993**, *7*, 504–514.
72. Gaillard, P.; Carrupt, P.-A.; Testa, B.; Boudon, A. *J. Comput.-Aided Mol. Des.* **1994**, *8*, 83–96.
73. Rekker, R. F.; ter Laak, A. M. *Quant. Struc.-Act. Relat.* **1993**, *12*, 152–157.
74. Mannhold, R.; Dross, K. *Quant. Struc.-Act. Relat.* **1996**, *15*, 403–409.
75. Van de Waterbeemd, H.; Mannhold, R. *Quant. Struc.-Act. Relat.* **1996**, *15*, 410–412.

76. Duban, M. E.; Bures, M. G.; DeLazzer, J.; Martin, Y. C. Virtual Screening of Molecular Properties: A Comparison of LogP Calculators. In *Pharmacokinetic Optimization in Drug Research*; Testa, B., Van de Waterbeemd, H., Folkers, G., Guy, R., Eds.; VHCA: Zurich, Switzerland, 2001, pp 485–497.
77. Clarke, F. H.; Cahoon, N. M. *J. Pharm. Sci.* **1987**, *76*, 611–620.
78. Avdeef, A. Assessment of Distribution-pH Profiles. In *Lipophilicity in Drug Action and Toxicology*; Pliska, V., Testa, B., Van de Waterbeemd, H., Eds.; VCH: Weinheim, Germany, 1996, pp 109–139.
79. Fallingborg, J.; Christensen, L. A.; Ingelman-Nielsen, M.; Jacobsen, B. A.; Abildgaard, K.; Rasmussen, H. H. *Aliment. Pharmacol. Ther.* **1989**, *3*, 605–613.
80. Dressman, J. B.; Berardi, R. R.; Dermentzoglou, L. C.; Russel, T. L.; Schmaltz, S. P.; Barnett, J. L.; Jarvenpaa, K. M. *Pharm. Res.* **1990**, *7*, 756–761.
81. Xing, L.; Glen, R. C.; Clark, R. D. *J. Chem. Inf. Comput. Sci.* **2003**, *43*, 870–879.
82. ACD/Labs 7.07, ACD pKa DB Advanced Chemistry Development Inc, 90 Adelaide Street, West Toronto, Ontario, M5H3V9, Canada.
83. Wan, H.; Holmén, A. G.; Wang, Y.; Lindberg, W.; Englund, M.; Någård, M.; Thompson, R. *Rapid Comm. Mass Spectr.* **2003**, *17*, 2639–2648.
84. Ulander, J.; Broo, A. *Int. J. Quant. Chem.* **2005**, *105*, 866–874.
85. Stein, W.D. The Molecular Basis of Diffusion Across Cell Membranes. In *The Movement of Molecules across Cell Membranes*; Academic Press: New York, 1967; pp 65–125.
86. Conradi, R. A.; Hilgers, A. R.; Ho, N. F. H.; Burton, P. S. *Pharm. Res.* **1991**, *8*, 1453–1460.
87. Platts, J. A.; Butina, D.; Abraham, M. H.; Hersey, A. *J. Chem. Inf. Comput. Sci.* **1999**, *39*, 835–845.
88. Abraham, M. H.; Ibrahim, A.; Zissimos, A. M.; Zhao, Y. H.; Comer, J.; Reynolds, D. R. *Drug Discov. Today* **2002**, *7*, 1056–1063.
89. Norinder, U.; Haeberlein, M. *Adv. Drug Deliv. Rev.* **2002**, *54*, 291–313.
90. Platts, J. A. *Phys. Chem. Chem. Phys.* **2000**, *2*, 973–980.
91. Platts, J. A. *Phys. Chem. Chem. Phys.* **2000**, *2*, 3115–3120.
92. Cacelli, I.; Campanile, S.; Giolitti, A.; Molin, D. *J. Chem. Inf. Model.* **2005**, *45*, 327–333.
93. Raevsky, O. A.; Grigorév, V. Y.; Kireev, D. B.; Zefirov, N. S. *Quant. Struct.–Act. Relat.* **1992**, *11*, 49–63.
94. Seiler, P. *Eur. J. Med. Chem.* **1974**, *9*, 473–479.
95. El Tayar, N.; Tsai, R.-S.; Testa, B.; Carrupt, P.-A.; Leo, A. *J. Pharm. Sci.* **1991**, *80*, 590–598.
96. Dearden, J. C.; Ghafourian, T. *J. Chem. Inf. Comput. Sci.* **1999**, *39*, 231–235.
97. Tsantili-Kakoulidou, A.; Varvaresou, A.; Siatra-Papastaikoudi, T.; Raevsky, O. A. *Quant. Struct.–Act. Relat.* **1999**, *18*, 482–489.
98. Feher, M.; Sourial, E.; Schmidt, J. M. *Int. J. Pharm.* **2000**, *201*, 239–247.
99. Palm, K.; Stenberg, P.; Luthman, K.; Artursson, P. *Pharm. Res.* **1997**, *14*, 568–571.
100. Palm, K.; Luthman, K.; Ungell, A.-L.; Strandlund, G.; Beigi, F.; Lundahl, P.; Artursson, P. *J. Med. Chem.* **1998**, *41*, 5382–5392.
101. Clark, D. E. *J. Pharm. Sci.* **1999**, *88*, 815–821.
102. Van de Waterbeemd, H.; Camenisch, G.; Folkers, G.; Chretien, J. R.; Raevsky, O. A. *J. Drug Target.;* **1998**, *6*, 151–165.
103. Stenberg, P.; Luthman, K.; Artursson, P. *J. Control. Rel.* **2000**, *65*, 231–243.
104. Stenberg, P.; Luthman, K.; Ellens, H.; Lee, C. P.; Smith, P. L.; Lago, A.; Elliott, J. D.; Artursson, P. *Pharm. Res.* **1999**, *16*, 1520–1526.
105. Stenberg, P.; Norinder, U.; Luthman, K.; Artursson, P. *J. Med. Chem.* **2001**, *44*, 1927–1937.
106. Van de Waterbeemd, H. Quantitative Structure–Absorption Relationships. In *Pharmacokinetic Optimization in Drug Research*; Testa, B., Van de Waterbeemd, H., Folkers, G., Guy, R., Eds.; VHCA: Zurich, Switzerland, 2001, pp 499–511.
107. SYBYL: Molecular Modeling Software, Tripos Inc., 1699 South Hanley Rd, St. Louis, MO 63144, USA.
108. Ghafourian, T.; Dearden, J. C. *J. Pharm. Pharmacol.* **2000**, *52*, 603–610.
109. Stanton, D. J.; Jurs, P. C. *Anal. Chem.* **1990**, *62*, 2323–2329.
110. Goodford, P. J. *J. Med. Chem.* **1985**, *28*, 849–857.
111. Cramer, R. D., III; Patterson, D. E.; Bunce, J. D. *J. Am. Chem. Soc.* **1988**, *110*, 5959–5967.
112. Baroni, M.; Clementi, S.; Cruciani, G.; Costantiono, G. *Quant. Struct.–Act. Relat.* **1993**, *12*, 9–20.
113. Gebauer, S.; Knutter, I.; Hartrodt, B.; Brandsch, M.; Neubert, K.; Thondorf, I. *J. Med. Chem.* **2003**, *46*, 5725–5734.
114. Swaan, P. W.; Koops, B. C.; Moret, E. E.; Tukker, J. J. *Receptors Channels* **1998**, *6*, 189–200.
115. Zamora, I.; Oprea, T. I.; Ungell, A.-L. Prediction of Oral Drug Permeability. In *Rational Approaches to Drug Design*; Höltje, H. D., Sippl, W., Eds.; Prous Science: Barcelona, Spain, 2001, pp 271–280.
116. Suhre, W. M.; Ekins, S.; Chang, C.; Swaan, P. W.; Wright, S. H. *Mol. Pharmacol.* **2005**, *67*, 1067–1077.
117. Locuson, C. W., II; Suzuki, H.; Rettie, A. E.; Jones, J. P. *J. Med. Chem.* **2004**, *47*, 6768–6776.
118. Suzuki, H.; Kneller, M. B.; Rock, D. A.; Jones, J. P.; Trager, W. F.; Rettie, A. E. *Arch. Biochem. Biophys.* **2004**, *429*, 1–15.
119. Haji-Momenian, S.; Rieger, J. M.; Macdonald, T. L.; Brown, M. L. *Bioorg. Med. Chem.* **2003**, *11*, 5545–5554.
120. Asikainen, A.; Tarhanen, J.; Poso, A.; Pasanen, M.; Alhava, E.; Juvonen, R. O. *Toxicol. In Vitro* **2003**, *17*, 449–455.
121. He, M.; Korzekwa, K. R.; Jones, J. P.; Rettie, A. E.; Trager, W. F. *Arch. Biochem. Biophys.* **1999**, *372*, 16–28.
122. Lozano, J. J.; Pastor, M.; Cruciani, G.; Gaedt, K.; Centeno, N. B.; Gago, F.; Sanz, F. *J. Comput.-Aided Mol. Des.* **2000**, *14*, 341–353.
123. Poso, A.; Gynther, J.; Juvonen, R. *J. Comput.-Aided Mol. Des.* **2001**, *15*, 195–202.
124. Ridderström, M.; Zamora, I.; Fjällström, O.; Andersson, T. B. *J. Med. Chem.* **2001**, *44*, 4072–4081.
125. Afzelius, L.; Raubacher, F.; Karlen, A.; Jorgensen, F. S.; Andersson, T. B.; Masimirembwa, C. M.; Zamora, I. *Drug Metab. Disp.* **2004**, *32*, 1218–1229.
126. Pastor, M.; Cruciani, G.; McLay, I.; Pickett, S.; Clementi, S. *J. Med. Chem.* **2000**, *43*, 3233–3243.
127. Cruciani, G.; Guba, W. Molecular Field-Derived Descriptors for the Multivariate Modeling of Pharmacokinetic Data. In *Molecular Modeling and Prediction of Bioactivity*; Gundertofte, K., Jørgensen, F. E., Eds.; Kluwer Academic/Plenum Press: New York, 2000, pp 89–94.
128. Zamora, I.; Afzelius, L.; Cruciani, G. *J. Med. Chem.* **2003**, *46*, 2313–2324.
129. Cruciani, G.; Meniconi, M.; Carosati, E.; Zamora, L.; Mannhold, R. VolSurf: A Tool for Drug ADME-Properties Prediction. In *Drug Bioavailability*; Van de Waterbeemd, H., Lennernäs, H., Artursson, P., Eds.; Wiley-VCH: New York, 2003; Vol. 18, pp 406–419.
130. Lobell, M.; Sivarajah, V. *Mol. Div.* **2003**, *7*, 69–87.
131. Cruciani, G.; Pastor, M.; Guba, W. *Eur. J. Pharm. Sci.* **2000**, *11*, S29–S39.
132. Crivori, P.; Cruciani, G.; Carrupt, P.-A.; Testa, B. *J. Med. Chem.,* **2000**, *43*, 2204–2216.

133. Ano, R.; Kimura, Y.; Shima, M.; Matsuno, R.; Tamio Ueno, T.; Akamatsu, M. *Bioorg. Med. Chem.* **2004**, *12*, 257–264.
134. Örnskov, E.; Gottfries, J.; Erickson, M.; Folestad, S. *J. Pharm. Pharmacol.* **2005**, 57, 435–442.
135. Fontaine, F.; Pastor, M.; Sanz, F. *J. Med. Chem.* **2004**, *47*, 2805–2815.
136. Sharom, F. Probing of Conformational Changes, Catalytic Cycle and ABC Transporter Function. In *ABC proteins: from bacteria to man*; Holland, I. B., Cole, S. P. C., Kuchler, K., Higgins, C. F., Eds.; Academic Press: San Diego, CA, 2003, pp 107–133.
137. Cianchetta, G.; Singleton, R. W.; Zhang, M.; Wildgoose, M.; Giesing, D.; Fravolini, A.; Cruciani, G.; Vaz, R. J. *J. Med. Chem.* **2005**, *48*, 2927–2935.
138. Crivori, P.; Zamora, I.; Speed, B.; Orrenius, C.; Poggesi, I. *J Comput. Aided Mol. Des.* **2004**, *18*, 155–166.
139. Kastenholz, M. A.; Pastor, M.; Clementi, M.; Cruciani, G. *J. Med. Chem.* **2000**, *43*, 3033–3044.
140. Afzelius, L.; Zamora, I.; Ridderström, M.; Andersson, T. B.; Karlén, A.; Masimirembwa, C. M. *Mol. Pharmacol.* **2001**, *59*, 909–919.
141. Rettie, A. E.; Korsekwa, K. R.; Kunze, K. L.; Lawrence, R. F.; Eddy, A. C.; Aoyama, T.; Gelboin, H. V.; Gonzalez, F. J.; Trager, W. F. *Chem. Res. Toxicol.* **1992**, *5*, 54–59.
142. Takanashi, K.; Tainaka, H.; Kobayashi, K.; Yasumori, T.; Hosakawa, M.; Chiba, K. *Pharmacogenetics* **2000**, *10*, 95–104.
143. Yamazaki, H.; Shimada, T. *Arch. Biochem. Biophys.* **1997**, *346*, 161–169.
144. Mancy, A.; Broto, P.; Dijols, S.; Dansette, P. M.; Mansuy, D. *Biochemistry* **1995**, *34*, 10365–10375.
145. Hamman, M. A.; Thompson, G. A.; Hall, S. D. *Biochem. Pharm.* **1997**, *54*, 33–41.
146. Miners, J. O.; Coulter, S.; Tukey, R. H.; Veronese, M. E.; Birkett, D. J. *Biochem. Pharm.* **1996**, *51*, 1003–1008.
147. Miners, J. O.; Coulter, S.; Birkett, D. J.; Goldstein, J. A. *Pharmacogenetics* **2000**, *10*, 267–270.
148. Fischer, V.; Johanson, L.; Heitz, F.; Tullman, R.; Graham, E.; Baldeck, J. P.; Robinson, W. T. *Drug Metab. Disp.* **1999**, *27*, 410–416.
149. Wen, X.; Wang, J. S.; Backman, J. T.; Kivistö, K. T.; Neuvonen, P. J. *Drug. Metab. Disp.* **2001**, *29*, 1359–1361.
150. Margolis, J. M.; O'Donnell, J. P.; Mankowski, D. C.; Ekins, S.; Obach, R. S. *Drug Metab. Disp.* **2000**, *28*, 1187–1191.
151. Pearce, R. E.; Rodrigues, A. D.; Goldstein, J. A.; Parkinson, A. *J. Pharmacol. Exp. Therap.* **1996**, *277*, 805–816.
152. Gram, L. F.; Guentert, T. W.; Grange, S.; Vistisen, K.; Brøsen, K. *Clin. Pharmacol. Ther.* **1995**, *57*, 670–677.
153. Koyama, E.; Chiba, K.; Tani, M.; Ishizaki, T. *J. Pharmacol. Exp. Ther.* **1997**, *281*, 1199–1210.
154. Äbelö, A.; Andersson, T. B.; Antonsson, M.; Knuts-Naudot, A.; Skånberg, I.; Weidolf, L. *Drug Metab. Disp.* **2000**, *28*, 966–972.
155. Li, X.-Q.; Andersson, T. B.; Ahlström, M.; Weidolf, L. *Drug Met. Disp.* **2004**, *32*, 821–827.
156. Setiabudy, R.; Kusaka, M.; Chiba, K.; Darmansjah, I.; Ishizaki, T. *Br. J. Clin. Pharmacol.* **1995**, *39*, 297–303.
157. Mankowski, D. C. *Drug Metab. Disp.* **1999**, *27*, 1024–1028.
158. Ko, J. W.; Desta, Z.; Soukhova, N. V.; Tracy, T.; Flockhart, D. A. *Br. J. Clin. Pharmacol.* **2000**, *49*, 343–351.
159. Wedlund, P. J.; Aslanian, W. S.; McAllister, C. B.; Wilkinson, G. R.; Branch, R. A. *Clin. Pharmacol. Ther.* **1984**, *36*, 773–780.
160. Jung, F.; Griffin, K. J.; Song, W.; Richardson, T. H.; Yang, M.; Johnson, E. F. *Biochemistry* **1998**, *37*, 16270–16279.
161. Jung, F.; Richardson, T. H.; Raucy, J. L.; Johnson, E. F. *Drug Metab. Disp.* **1997**, *25*, 133–139.
162. Kobayashi, K.; Kogo, M.; Tani, M.; Shimada, N.; Ishizaki, T.; Numazawa, S.; Yoshida, T.; Yamamoto, T.; Kuroiwa, Y.; Chiba, K. *Drug Metab. Disp.* **2001**, *29*, 36–40.
163. Nadin, L.; Murray, M. *Biochem. Pharmacol.* **1999**, *58*, 1201–1208.
164. Yamazaki, H.; Shibata, A.; Suzuki, M.; Nakajima, M.; Shimada, N.; Guengerich, F. P.; Yokoi, T. *Drug Metab. Disp.* **1999**, *27*, 1260–1266.
165. Baldwin, S. J.; Clarke, S. E.; Chenery, R. J. *Br. J. Clin. Pharmacol.* **1999**, *48*, 424–432.
166. Becquemont, L.; Mouajjah, S.; Escaffre, O.; Beaune, P.; Funck-Brentano, C.; Jaillon, P. *Drug Metab. Disp.* **1999**, *27*, 1068–1073.
167. Ohyama, K.; Nakajima, M.; Nakamura, S.; Shimada, N.; Yamazaki, H.; Yokoi, T. *Drug Metab. Disp.* **2000**, *28*, 1303–1310.
168. Minoletti, C.; Dijols, S.; Dansette, P. M.; Mansuy, D. *Biochemistry* **1999**, *38*, 7828–7836.
169. Tetko, I. V.; Tetko, I. V. *Drug Disc. Today* **2005**, *10*, 1497–1500.
170. Consonni, V.; Mauri, A.; Pavan, M. Dragon v. 5, 1997–2005 Talete srl – Milano, Italy, http://www.talete.mi.it/products/DRAGON_info.pdf (accessed May 2006).
171. Taft, R. W.; Abraham, M. H.; Famini, G. R.; Doherty, R. M.; Abboud, J.-L. M.; Kamlet, M. J. *J. Pharm. Sci.* **1985**, *74*, 807–814.
172. Zissimos, A. M.; Abraham, M. H.; Klamt, A.; Eckert, F.; Wood, J. *J. Chem. Inf. Comput. Sci.* **2002**, *42*, 1320–1331.
173. Platts, J. A.; Abraham, M. H.; Butina, D.; Hersey, A. *J. Chem. Inf. Comput. Sci.* **2000**, *40*, 71–80.
174. Platts, J. A.; Abraham, M. H.; Hersey, A.; Butina, D. *Pharm. Res.* **2000**, *17*, 1013–1018.
175. Norinder, U.; Österberg, T.; Artursson, P. *Pharm. Res.* **1997**, *14*, 1786–1791.
176. Norinder, U.; Österberg, T.; Artursson, P. *Eur. J. Pharm. Sci.* **1999**, *8*, 49–56.
177. Norinder, U.; Sjöberg, P.; Österberg, T. *J. Pharm. Sci.* **1998**, *87*, 952–959.
178. Österberg, T.; Norinder, U. *Eur. J. Pharm. Sci.* **2000**, *10*, 295–303.
179. Österberg, T.; Norinder, U. *Eur. J. Pharm. Sci.* **2001**, *12*, 327–337.
180. Karelson, M.; Lobanov, V. S. *Chem. Rev.* **1996**, *96*, 1027–1043.
181. Lombardo, F.; Blake, J. F.; Curatolo, W. J. *J. Med. Chem.* **1996**, *39*, 4750–4755.
182. Stanton, D. T.; Mattioni, B. E.; Knittel, J. J.; Jurs, P. C. *J. Chem. Inf. Comput. Sci.* **2004**, *44*, 1010–1023.
183. Bergström, C. A. S.; Strafford, M.; Lazorova, L.; Avdeef, A.; Luthman, K.; Artursson, P. *J. Med. Chem.* **2003**, *46*, 558–570.
184. Ghuloum, A. M.; Sage, C. R.; Jain, A. N. *J. Med. Chem.* **1999**, *42*, 1739–1748.
185. Bravi, G.; Wikel, J. H. *Quant. Struct.–Act. Relat.* **2000**, *19*, 29–38.
186. Todeschini, R.; Lasagni, M.; Marengo, E. *J. Chemometr.* **1994**, *8*, 263–272.
187. Bravi, G.; Wikel, J. H. *Quant. Struct.–Act. Relat.* **2000**, *19*, 39–49.
188. James, C.A.; Weininger, D.; Delany, J. Daylight Theory Manual Daylight Version 4.9, Release Date 04/08/05, http://www.daylight.com/dayhtml/doc/theory/theory.toc.html (accesses May 2006).
189. Oprea, T. I.; Davis, A. M.; Teague, S. J.; Leeson, P. D. *J. Chem. Inf. Comput. Sci.* **2001**, *41*, 1308–1315.
190. Keserü, G. M.; Molnár, L. *J. Chem. Inf. Comput. Sci.* **2002**, *42*, 437–444.
191. Wold, S.; Johansson, E.; Cocchi, M. PLS – Partial Least-Squares Projection to Latent Structures. In *3D-QSAR in Drug Design: Theory, Methods and Applications*; Kubinyi, H., Ed.; ESCOM Science Publishers: Leiden, the Netherlands, 1993, pp 523–550.
192. Wold, S.; Kettaneh, N.; Tjessem, K. *J. Chemometr.* **1996**, *10*, 463–482.
193. SIMCA-P, Umetri AB: SE-90719 Umeå, Sweden.

Biographies

Susanne Winiwarter studied pharmacy at the University of Vienna, Austria. She received her PhD from the same university in 1991 under the supervision of Dr P Wolschann and Prof G Buchbauer, using molecular modeling techniques to investigate structure–activity relationships. After one postdoctoral year with Prof H J Roth at the University of Tübingen, Germany, she spent several years at the University of Uppsala, Sweden, in the research group of Dr A Karlén, in the Department of Medicinal Chemistry. In 2003, she joined AstraZeneca in Mölndal, Sweden, as senior research scientist.

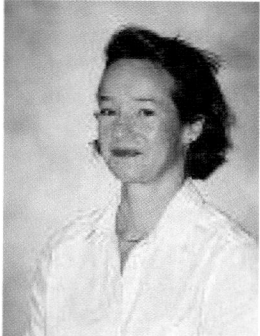

Marianne Ridderström received her PhD in biochemistry in Prof Bengt Mannervik's group at Uppsala University in 1997. Her thesis concerned structure–activity studies of the glutathione-linked enzymes glyoxalases I and II. She joined AstraZeneca R&D Mölndal on a postdoctorial position in 1999 and engaged in structure–activity studies and modeling of CYPs. Since 2001 she has been employed in the DMPK and Bioanalytical Chemistry department at AstraZeneca Mölndal, involved in the in vitro metabolism field and metabolism modeling in discovery projects.

Anna-Lena Ungell received her PhD degree in zoophysiology 1985 at the University of Gothenburg, Sweden, and after a postdoctorial fellowship in 1985–86 at the University of Würzburg in Germany she received a scholarship from the Swedish Royal Academy of Science for one year. She started 1987 at AstraZeneca R&D (former Astra Hässle AB), Mölndal, Sweden, and since 2004 has been an Associate Professor in Pharmaceutics at Uppsala University. She presently holds a position as a Team Leader and Principal Scientist at DMPK & Bioanalytical Chemistry, AstraZeneca R&D Mölndal, Sweden. Her group consists of 11 members working as support to preclinical, clinical, and pharmaceutical projects regarding drug absorption and transporter issues. She is responsible for developing models to study drug absorption, screening for structure–absorption relationship using automated screen facilities, and QSAR modeling, evaluating different absorption mechanisms and transporters, which link in vitro animal studies to in vivo pharmacokinetics and evaluate drug–drug interactions.

Tommy B Andersson is a Senior Principal Scientist at AstraZeneca R&D Mölndal Sweden. He received his PhD from the University of Gothenburg, Sweden, in 1986 and later became an assistant professor at the same university. Dr Andersson was appointed assistant professor in Drug Metabolism at the Karolinska Instutet, Stockholm in 1998 and was promoted to adjunct professor at the same institute in 2003. He joined AstraZeneca R&D Mölndal in1993 where he has been responsible for the in vitro metabolism work in drug discovery and development. He is the author and co-author of some 110 original research papers.

Ismael Zamora studied chemical engineering and organic chemistry at the Institut Químic de Sarriá in Barcelona (Spain). He received his PhD in organic chemistry in 1998 for studies on the synthesis of molecular modeling of natural products. After a 6-month postdoctoral period with Prof Gabriele Cruciani at the Univeristy of Perugia, Italy, he joined AstraZeneca R&D Mölndal, Sweden, in a postdoctoral position for 9 months, and stayed there as a research scientist for another 2 years. After that he founded the Lead Molecular Design Co., devoted to software development in collaboration with Molecular Discovery Ltd. Since 2002, he has been an associate professor at the Pompeu Fabra University, Barcelona, Spain.

Comprehensive Medicinal Chemistry II
ISBN (set): 0-08-044513-6

ISBN (Volume 5) 0-08-044518-7; pp. 531–554

5.23 Electrotopological State Indices to Assess Molecular and Absorption, Distribution, Metabolism, Excretion, and Toxicity Properties

L H Hall, Eastern Nazarene College, Quincy, MA, USA and Hall Associates Consulting, Quincy, MA, USA

L B Kier, Virginia Commonwealth University, Richmond, VA, USA

L M Hall, Hall Associates Consulting, Quincy, MA, USA

5.23.1 Introduction

In the study of drugs and the design of new drugs, the realization of the important role played by pharmacokinetic attributes of molecules has become of great importance. Of particular interest are the properties of absorption, distribution, metabolism, excretion, and toxicity, collectively called ADMET. Over recent years, in the drug design process, these properties of candidate drug molecules have become co-equal with biological activity in the goal of creating clinically useful drugs. Many investigators have been describing the serious problems that arise from poor pharmacokinetic and toxicity properties of drug candidates. Estimates have been made that indicate that the percentage of late stage drug candidates that fail to make it to market because of ADMET failures range from 20% to 40%.[1-3] Problems associated with ADMET performance must be identified experimentally early or predicted via testing, modeling, or other approaches, so that design changes can be made in the candidate molecules or decisions made to abandon the project before significant resources are squandered.

One approach to prediction is through procedures using in vitro experiments. These may take the form of simulations using tissue or actual measurements of an ADMET property in an animal. How successful are the predictions of ADMET properties with all of the current technologies? The general view is that they are not good enough to make predictions leading to a steady stream of safe, new drugs to come to market each year.[4] The alternative approach to in vitro testing is the process for the creation of models of various types, including structure-based approaches as well as physiologically based modeling and in silico models of ADMET properties using quantitative structure–activity relationship (QSAR) methods.[5,6]

These QSAR models are of two general types. One type uses physicochemical properties to describe molecules and then to develop a relationship with an ADMET property. Prominent descriptors in this type of approach are the partition coefficient (*see* 5.18 Lipophilicity, Polarity, and Hydrophobicity; 5.27 Rule-Based Systems to Predict Lipophilicity), pK_a (*see* 5.16 Ionization Constants and Ionization Profiles; 5.25 In Silico Prediction of Ionization), and the Hammett sigma (*see* 4.22 Topological Quantitative Structure–Activity Relationship Applications: Structure Information Representation in Drug Discovery; 5.22 Use of Molecular Descriptors for Absorption, Distribution, Metabolism, and Excretion Predictions; 5.25 In Silico Prediction of Ionization). The information derived from a successful property-based model is indirectly related to structure, and requires further processing to convert it into guidance that is meaningful in terms of molecular structure for chemical modification. The design of a new molecule requires additional development of molecular structure information.

The second type of QSAR in silico model is based on a numerical encoding of structure that yields information directly representing molecular structure. These models are derived from nonempirical indices such as the electrotopological state and molecular connectivity. This approach, structure information representation, is of direct value in analyzing the structure for design modification leading to the synthesis of a new molecule.[7,8]

To provide a basis for the presentation of several modeling studies on ADMET properties, we begin with a brief description of the methods for molecular structure description. We consider the structure of a molecule as a source of information. This approach proceeds not to the creation of a geometric distribution of points but to the systematic development of the useful information resident in the structure of a molecule and additional information that can be developed.

5.23.2   The Molecule as an Information Network

The initial encounter of two molecules has been conventionally dissected into two separate phenomena in a reductionist approach. Thus, the electronic structure and the topology of the molecule under study have usually been treated as distinct entities. The goal has been to quantify each attribute and enter these independently into the process aimed at acquiring a model of the structure engagement in the encounter. Success by this approach over the years is clear. But we may achieve greater insight if we accept the reality that molecular structure is a representation of the complex system that is made up of atoms and functional groups with internal patterns describable by information from both the electronic content and topological arrangement in an integrated manner.

One useful way to view a molecule is as an information network, with atoms comprising the nodes. Each node has three characteristics. It has a state; that is, it is an element with electrons and protons. Second, it has some relationship to other nodes (atoms) in the network through the bonds directly connecting neighboring atoms and beyond. The valence state is represented by electronic information associated with the set of directly bonded atoms and larger-scale patterns such as multiple bonding, aromaticity, and charge. Finally, each node (atom) has a specific influence on its (connected) bonded neighbors. Each node also manifests its effect on more distant neighbors in the network (molecule). This influence is summarized as the effect of electronegativity on the bonded neighbors through the bonds and on more remote network-connected neighbors as well. The electronegativity effect is a complex emergent property of the (valence) state of the node and its connectedness in the network (the electronic structure and the bonded relationships in the whole molecule). Viewed in this way, the development of structure information is a process of synthesis, leading to structure descriptors, rather than reduction, leading to atomic indices.

5.23.3   Structure Information Representation

Based on the concept described above, we can define two structure information systems that are a major part of what we refer to collectively as structure information representation.[7,8] A more complete description of this approach to structure representation is given elsewhere (*see* 4.22 Topological Quantitative Structure–Activity Relationship Applications: Structure Information Representation in Drug Discovery). We review these two approaches here.

5.23.3.1 The Electrotopological State (E-State) Index

Electronic structure, electronegativity, and topology cannot be meaningfully treated as separate aspects to be combined in some fashion, simple linear or otherwise, in order to create a predictive model in structure–activity analyses. Using the network analogy for a molecule, as outlined above, we can say that network structure and function are complex emergent properties of the node structure, including the influence of each node on bonded as well as remote neighbors. The recognition that structure is an emergent attribute casts the molecule into a class of complex systems. Accordingly, an alternative to the reductionist approach must be invoked to model such a system.

We begin by recognizing that the electronic structure of a molecule is a critical factor governing its interaction with another molecule. Based on the principle of logical depth, we accept the premise that it is not meaningful to say that a molecule engages any entity other than one at its own level, such as another molecule.[9] It is not very useful to speak of a molecule engaging a cell. What is happening, at a level that permits quantification, is that the molecule engages another molecule, or molecule-sized feature, that is part of the cell surface. Carrying this principle further, it is significant to consider for study only part of a molecule (an atom, functional group, or fragment) as it engages a comparable-level feature on another molecule. This principle of logical depth leads us to attempt to characterize these molecular features in order to model the possibility, extent, and specificities of a molecule engaging another molecule.

A model fulfilling these expectations must, therefore, address subfeatures of a molecule, encode electronic and topological attributes, and also encode the influence by these same features on neighboring features. This is the philosophy behind the creation of the E-state system over a decade ago.[10] The important features in the design of this system are described below.

In order to characterize the important subfeatures of a molecule that are implicated in the encounters of a molecule with other molecules, we begin by focusing attention on each atom (or hydride group) in the molecule as a distinct entity. Two characteristics of this atom are encoded into a uniform attribution value, both electronic and topological. This unification brings together the electron richness and the influence on electron richness (expressed conveniently through the electronegativity), together with local topology, expressed in a quantification of the relative adjacency of the atom in the molecular skeleton.[10]

Each of these attributes has been addressed and quantified by us in past work.[11,12] The sigma orbital electronegativity of an atom or hydride fragment has been shown by Kier and Hall to be described by an expression that includes counts of the number of valence electrons, the number of hydrogen atoms (in the case of a hydride group), and the number of sigma-bonded atoms other than hydrogen.[10,11] The expression contains the count of the extrajacent electrons on a sigma-bonded atom, which is the count of pi (π) and lone pair (n) electrons. The topology of an atom (or hydride group) in the molecular skeleton is characterized by the count (δ) of its sigma bonds (σ) to atoms other than hydrogen (h): $\delta = \sigma - h$. We have shown that the count of pi and lone pair electrons provides a strong relationship to valence state electronegativity.[11] The electron accessibility of the atom or hydride group is described as a ratio of the valence state electronegativity to the topological character, expressed as the δ value.

The provisional expression $(\pi + n)/\delta$ encodes the relative electron accessibility of the particular molecular fragment (atom, hydride group) in isolation from bonded atoms. This expression results in zero values for all $C(sp^3)$ atoms; so, the equation is modified with a constant value of one in the numerator, giving $(\pi + n + 1)/\delta$. Finally, to scale most values to be greater than 1, the expression is modified to give an intrinsic value of the noncovalent intermolecular interaction potential, I_i, for atom i[10]:

$$I_i = [(\pi_i + n_i + 1)/\delta_i] + 1 \qquad [1]$$

This expression holds for second-row atoms; an expanded form is used for higher rows of the periodic chart.[10]

The count of the pi and lone pair electrons on a bonded atom can be justified as a model of the electronegativity of the atom on both theoretical and experimental grounds. First, the work of Slater has shown that the pi and lone pair electrons have a high probability of being further from the atom core than the sigma electrons, resulting in less shielding. As a result of the presence of pi and lone pair electrons, the atomic core (nuclear protons and inner shell electrons) can exert a greater influence on the sigma-bonding electrons. This effect is the essence of electronegativity. Second, the Kier–Hall electronegativity is very successful in relating to the experimental Mulliken–Jaffe electronegativity derived from the ionization potential and electron affinity of atoms in their valence state.[11] In the correlation of pi and lone pair electron count with valence state electronegativity, the standard error is essentially equal to the known error in the valence state electronegativity.

The intrinsic state, I, of an atom or hydride group is an incomplete expression of the structure, because that feature is a part of the whole molecule and is affected by other atoms. Each atom in a molecule is in a field of the whole

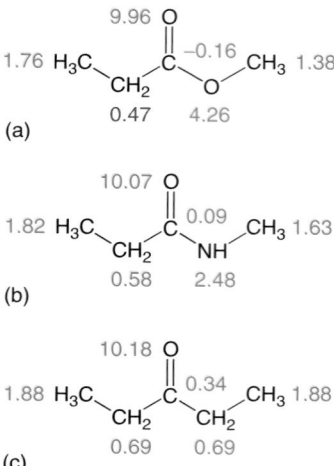

Figure 1   Computed E-state values for the atoms in a molecular series with iso-connective skeletons: (a) methylpropanoate, (b) *N*-methylpropanamide, and (c) 3-pentanone.

molecule, a field created by the combined effects of all atoms. The field is an information field, with the information derived from the influences as expressed through electronegativity differences. This influence is encoded into a series of perturbations in units of the intrinsic state of each atom. We have defined this perturbation on an atom as the difference between the intrinsic state of that atom and the intrinsic state of each other atom. The influence is mitigated by the distance through skeletal bonds, r_{ij}, between the atom and every atom in the molecule. These perturbations are summed and applied to the reference atom intrinsic state to yield the E-state value$_j$:

$$S_i = I_i + \sum_j (I_i - I_j)/r_{ij}^2 \qquad [2]$$

in which r_{ij} is the count of atoms in the shortest skeletal path of atoms connecting the reference atom to the perturbing atom. The sum of the intrinsic state and its perturbations is defined as the E-state, S_i, for atom i in the molecule under study.

Some aspects of the nature of E-state structure descriptors are represented in **Figure 1**. In 3-pentanone (**Figure 1c**), the carbonyl oxygen atom has the largest E-state value, representing both its electron-richness and topological accessibility. This E-state value encodes the electron accessibility of the carbonyl oxygen. On the other hand, the carbonyl carbon atom is both electron depleted and less topological accessible within the molecular structure. The two methyl groups have identical values, in keeping with molecular symmetry, as have the two methylene groups. Both methyl and methylene groups have E-state values less than their intrinsic state values (2.00 and 1.50, respectively) because of electron depletion arising from the high electronegativity of the carbonyl oxygen atom.

In *N*-methylpropanamide (**Figure 1b**), the E-state of the amide –NH– group is intermediate between the carbon and oxygen atoms because of its intermediate valence state electronegativity. The two methyl groups have different E-state values, being no longer in symmetric positions. Furthermore, the methyl group attached to the nitrogen atom has a smaller value because of the attraction of electrons to the nitrogen atom.

In the ester (**Figure 1c**), the carbonyl carbon atom has the smallest value among these three molecules. The carbonyl carbon is electron depleted because of the attraction of electrons from both of the oxygen atoms. The carbonyl oxygen atom also has the smallest E-state value (of the three molecules) because of the presence of the other oxygen atom. The methyl groups also have smaller values, as expected (*see* 4.22 Topological Quantitative Structure–Activity Relationship Applications: Structure Information Representation in Drug Discovery for more detailed information).

This simple illustration indicates the type of structure information resident in the E-state values and the trends that one can expect in a series of related structures. When E-state descriptors appear in a model, the chemist can draw upon this kind of information to interpret the model in terms of molecular structure for two purposes: (1) understanding the structure features that are important in the property and (2) guidance for structure modification of drug candidates.

The presentation above developed the formalism for the atom-level E-state value, the value computed for each atom in the structure. These values serve very well for investigations of biological activity in which the data are based on a common core skeleton structure. The atom level E-state values can be entered directly into the modeling as variables. Many such investigations have been reported.[10] For data sets most often encountered in ADMET studies, the data consist of very diverse structures with no common core skeleton. For these studies, the E-state formalism has been extended to the creation of the atom-type E-state. For the atom-type descriptors, all atoms in the molecule are classified into one of many atom types. These atom types are the classic types of organic chemistry, such as $-CH_3$, $-CH_2-$, $-NH-$, $=O$, $>C=$, $-Cl$, $-S-$, and aromatic C and CH. For the atom-type E-state value of atom type X, $S^T(X)$, the value is simply the sum of the E-state values for all atoms of that type, s_X, in the molecule:

$$S^T(X) = \sum_X s_X \qquad [3]$$

The atom-type E-state index encodes three different types of structure information:

(1) the electron accessibility of the atoms of that type;
(2) the presence/absence of the atom type; and
(3) the count of atoms of that type in the molecule.

This definition makes it clear that much more than feature count is included in the atom-type E-state. For that reason, the E-state is showing promising use in ADMET studies such as those described later in this article.

It is apparent that E-state descriptors encode information about both the valence electron structure of an atom and also its presence in the network making up the molecule. Interactions within the skeleton result in the perturbations to its intrinsic state. The E-state values are rich in information about the electron accessibilities of the atoms (hydride groups) in the molecule. These atom level descriptors of electron accessibility have been the basis for many published studies on biological activity and properties.[10] This formalism describes the E-state value for each atom (and hydride group) in a molecule. This concept has been extended to the description of other features that are significant for ADMET modeling, including a separate formalism for hydrogen atoms; development of E-state values for atom type; creation of the E-state for a bond; and development of E-state values for bond types.[10] Recently, this approach has been extended to groups or fragments in organic molecules (see 4.22 Topological Quantitative Structure–Activity Relationship Applications: Structure Information Representation in Drug Discovery for an overview of these developments).

5.23.3.2 Molecular Connectivity

The E-state formalism focuses on individual atoms, and provides descriptors of the atom within the molecular network. It is also useful and possible to encode structure information representing the whole molecule. In this system we create structure indices encoding information about the connectedness (connectivity) of whole-network depictions of a molecule. The objective is characterization of the whole molecule as a unified entity, that is, as a complex system. For simplicity we define the nodes in a molecular network as representations of chemical groups made up of atoms or of atom hydrides ($>N-$, $-NH-$, $-NH_2$, $-O-$, $-OH$, etc., as is done in the E-state formalism). The state of each node is defined by a number, the simple electron count, corresponding to the count of the neighbors for each node (the count of sigma bonds, not counting hydrogen, to each atom or atom hydride in the molecule). The network can be dissected into a set of fragments representing one or more nodes, that is, atoms. For a brief presentation of this approach, let us consider only the set of pairs of neighboring nodes. Each set is defined by the simple electron count for the two atoms, i and j, in the set; any property of the network is hypothesized to be a function of these sigma electron count values. The numerical value that characterizes the two-atom set, c_{ij}, is derived by taking the reciprocal square root of the product of the simple electron count values. This value, c_{ij}, for each fragment is combined by simple addition for all network sets to give a network index (molecular index) called the chi index, $^1\chi$.[13]

The nature of some of the structure information in the simple first-order molecular connectivity index $^1\chi$ is illustrated in **Figure 2**. The structures in **Figure 2a** show the increase in the index value as the number of atoms increases. In **Figure 2b** the heptane isomers (together with *n*-heptane in **Figure 2a**) clearly indicate that the $^1\chi$ index decreases in value as the degree of branching increases: *n*-heptane > 2-methylhexane > 2,3-dimethylpentane > 3,3-dimethylpentane > 2,2,3-trimethylbutane.[13] When this index is part of a model, the chemist can use this structure information in the molecular modification process for design of candidate drug molecules. If the $^1\chi$ index makes a positive contribution to predicted property values, structures with a decreased degree of branching will tend to have larger predicted property values (see 4.22 Topological Quantitative Structure–Activity Relationship Applications: Structure Information Representation in Drug Discovery for more detailed information).

This quantity is a first-order simple index for the network, encoding the sum of first-neighbor relationships throughout the skeleton of the molecule. This index represents the topology of the molecule as derived from the

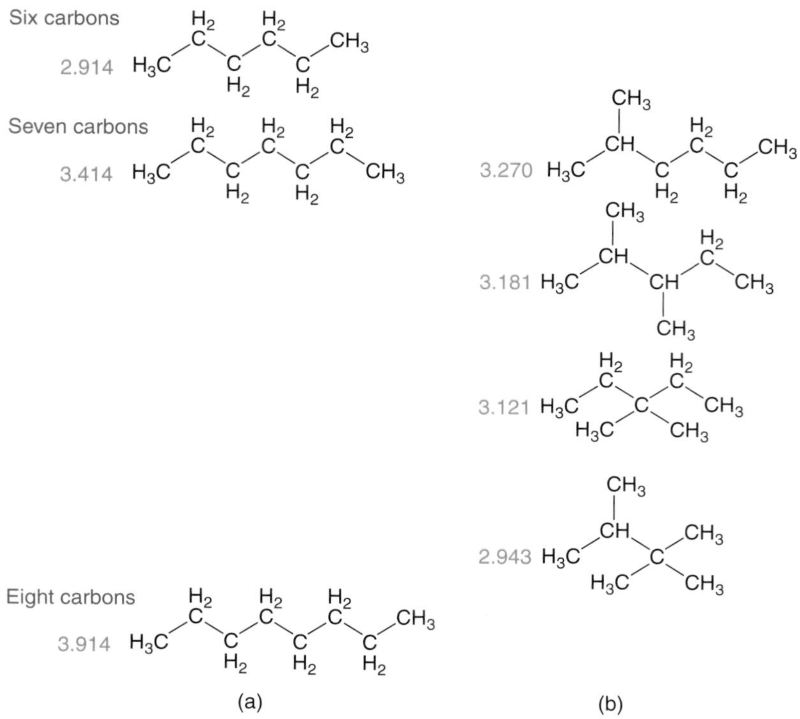

Figure 2 Structures of alkanes to illustrate variation of the simple molecular connectivity index 1χ (a) with molecular size and (b) with degree of skeletal branching.

undifferentiated connections (skeletal bonds) in that molecule, that is, the set of bonded relationships. This molecule index encodes the topological structure information of the pattern of bonded relationships within the network of the molecular skeleton.

This index is the prototype index in this structure information representation system that has been greatly expanded to describe multiple neighbors and atom states in a molecule.[7,8] Furthermore, it has been shown that this index encodes information about the topology of a molecule as it impacts on nearby molecules in intermolecular interactions.[14,15] Much more detail on the nature of molecular connectivity indices is available elsewhere.[13]

5.23.3.3   Nature of Topological Structure Descriptors

The description given above for the basis and formalism of both the E-state and molecular connectivity indices provides a basis for understanding their nature, which is the foundation for the structure information representation approach to molecular structure. This representation is derived from the atomic nature and skeletal connections within a molecule, as usually presented in the molecule connection table. In this sense, these indices represent relationships within the molecule.

It has become a common practice to include molecular structure descriptors in categories based on an assumed dimensionality. Using these rubrics, a three-dimensional descriptor is one that is derived from information built upon atom positions in a three-dimensional space. These positions are defined by their constant relationship to other atoms in a molecule. It is assumed that there is no variation in these dimensions for any specific value of the descriptor. This means that the conformation is constant for that particular descriptor value.

Structure descriptors derived from a molecular graph are not three-dimensional descriptors. They belong to a graph-based category that includes the E-state and molecular connectivity indices. They have been incorrectly described as two-dimensional descriptors. A two-dimensional descriptor would be based upon a statement of a pairwise separation value of two atoms in a molecule and knowledge of a specific angle between three adjacent atoms. Thus, the positions of each atom would be known precisely, in two-dimensional space. This information is not resident in the E-state or molecular connectivity descriptors.

These graph-based descriptors are derived from knowledge of the state of each atom in the molecule and the bonding relationship among them. It is a relational network that depicts the atoms and their neighbors in a model that describes some structure attributes of a molecule. It is an abstraction of the complete structure of a molecule. There is assumed to be sufficient information encoded in such a structure description that it has utility in correlating and predicting the properties of molecules. It is inappropriate to refer to these descriptors in a dimensional sense. It is appropriate to refer this category of molecular structure descriptors as relational network descriptors or molecular network descriptors.

This article focuses on models of ADMET properties, as discussed in the sections that follow. Each model captures the parallel between the structure of the molecules in the data set and the corresponding property. The molecules are three-dimensional entities engaging with other three-dimensional molecules; the properties arise from those engagements. Furthermore, the models are capable of prediction of the property values. Therefore, the descriptors in the model must contain implicit information about the structure of the molecules, including something of their three-dimensional nature. More detail is given on this important issue is given elsewhere (*see* 4.22 Topological Quantitative Structure–Activity Relationship Applications: Structure Information Representation in Drug Discovery).

5.23.4   Summary of Recent Studies on Absorption, Distribution, Metabolism, Excretion, and Toxicity Properties Using the Electrotopological State with Other Topological Indices

The combination of the E-state and the molecular connectivity chi indices together with other topological indices as structure representation has proven a useful approach to modeling ADMET properties. The combination of the electrotopological state, molecular connectivity, and related representations of molecular structure has been called the structure information representation.[7,8] The foundation of this approach rests on the extraction of basic salient structure information from the molecular connection table with the resultant development of expressions for valence state electronegativity and volume. As described elsewhere (*see* 4.22 Topological Quantitative Structure–Activity Relationship Applications: Structure Information Representation in Drug Discovery), this information is the basis for the development of several classes of structure descriptors and their applications.[10,13,16–31,35–39] These studies are aimed at the development of structure information that can lead to improved pharmacokinetic properties and greater safety of drug design candidates. Their focus has been the development of in silico methods for validated prediction that can be employed early in the drug design process as well as in the drug maturation stage. Furthermore, high priority is placed on the need for models whose interpretation can be directly implemented in the overall drug design process.

Two threads of thought, validated prediction and direct structure interpretation, converge on the idea that a broad formalism can be built on the structure information representation approach, together with appropriate statistical techniques for model development. We have found the formalism for structure information representation to be a solid foundation for the construction of approaches to ADMET property modeling and structure interpretation in a broad sense, including the development of new structure descriptors and new approaches to descriptor creation.

Several studies illustrate aspects of the modeling approach, and will be described briefly. ADMET investigations are reported here in order of perceived increasing complexity:

- aqueous solubility;
- human intestinal absorption;
- protein binding;
- blood–brain barrier partitioning;
- metabolic stability;
- genotoxicity (Ames mutagenicity);
- risk of human carcinogenicity; and
- hepatotoxicity

5.23.4.1   Procedures and Statistical Methods

In Chapter 4.22 on modeling ADMET properties, several terms, symbols, and quantities of statistical nature are regularly used. These terms are summarized, and a very brief description of their nature and use is given. For simplicity, the observed values (dependent statistical variables) will be referred to as the property values (ADMET).

5.23.4.1.1 Training and test sets

To provide for both the separate development of a model and its testing or validation, the original data set is usually partitioned between a training set and a test set, often called the external validation test set. The test set is put aside and not used in any way for model development. The test set is predicted by the model after the model is developed, providing evaluation statistics. Depending on the size of the data set (number of observations or compounds), 10–20% of the whole set may be set aside for testing. Some form of random selection is used to create the test set. In our approach, the whole data set is clustered in a structure space composed of atom-type E-state descriptors and simple molecular connectivity chi indices, providing clusters of structurally similar compounds. Standard clustering techniques, such as Ward's hierarchical method, are used. Compounds for the test set are then selected from each cluster so as to cover the property range, without selecting compounds with the extreme property values in the overall data set. In this manner, the external validation test set is a representative of the whole data set, suitable for testing the model. The training set is used exclusively for the development of the model, including descriptor selection and parameter estimation.

5.23.4.1.2 Modeling methods

Most of the terms used for indicating statistical significance originated in regression analysis with multiple linear regression (MLR) models. These terms are now widely used in other forms of modeling such as artificial neural networks (ANNs), k-nearest neighbor analysis (kNN), and partial least squares (PLS). Most recently, support vector machines (SVMs) have been adapted from their early use with binary data to the analysis of continuous data.

5.23.4.1.3 Statistical terms and symbols

To express statistical results the commonly used terms and symbols are employed. A summary of the symbols is given in **Table 1**.

Some investigators create many subsets with a model for each, and then use the average of the predictions from all the subset models as a form of consensus or ensemble model prediction. Statistics on the training subsets from such a process are also called cross-validation statistics.

5.23.4.1.4 Randomization tests

Another approach to assessing the quality of the model is to compare the modeling statistics to those obtained from the use of random numbers in place of variables in the data set. The objective of this approach is to determine how much better the model is than one based on random numbers. Some authors have replaced the whole set of independent variables with random numbers and repeated the modeling process, a very time-consuming process. A typically used alternative technique is to randomize only the observed property values many times and repeat the computation of the correlation coefficient. The most commonly reported statistics include the average r^2 for 100 random trials $\left(r^2_{\text{ave}}\right)$ and the maximum r^2 $\left(r^2_{\text{max}}\right)$ obtained in the process. The average value provides some indication of the statistical significance; however, the r^2_{max} value gives a better estimate of the probability that the model is better than a random one, that is, the chance (1 in 100) that random numbers will provide a correlation with $r^2 = r^2_{\text{max}}$.

Table 1 Terms and symbols used to express statistical information on the models discussed in this article

Term	Symbol/abbreviation	Use
Correlation coefficient	r^2	Statistic calculated on the whole training set
Correlation coefficient	q^2, r^2_{press}	Statistic calculated for cross-validation or the external validation set
Standard error	s	Statistic calculated on the whole or the standard deviation training set
Standard error or standard deviation	s	Statistic calculated on the whole training set
Press		Predicted residual sum squared error
Mean absolute error	MAE	Statistic for the external test set

5.23.4.1.5 Acceptable measure of criteria

The measure of statistical quality that indicates an acceptable level for predictive use depends heavily upon the experimental quality of the data set. For excellent physical and physicochemical data, very high r^2 (>0.90) and low standard error s values can be expected. However, for in vitro biological data, such as IC_{50} for binding, lower values of r^2 (>0.80) are usually obtained. For in vivo data on a population of subjects, the range of experiment values for a single compound is much larger; hence, much lower r^2 values are usually obtained. At present, no specified level of correlation has been universally accepted for these studies. In this article, different levels of r^2 are reported as acceptable depending upon the quality of the experimental data. Data taken from public sources, based on several information sources, often reveal a significant spread in values from different experimental protocols.

5.23.4.2 Aqueous Solubility

Early studies by several investigators on small data sets successfully used both E-state and molecular connectivity structure descriptors.[29,30] Recently, larger data sets were investigated with more complex methods to create sound models based on the structure information representation approach.[31] Cross-validation techniques were applied before testing the model on external validation test sets, data that were not used in the development of the models. In this case, the training set was divided into 10 randomly selected sets, and an ANN method was employed to create 10 models as the basis for a consensus or ensemble model. Prediction of the external test set is obtained as the average of the predictions from the 10 separate ANN models.

External validation is considered a necessary part of the determination of the predictive capabilities of the model. In addition, we and our collaborators have sought what we term independent validation. In this process, a group independent from the modelers is sought that can test the model predictions on independently measured data. For example, Lobell has investigated nine available commercial models for aqueous solubility, including one based on the structure information representation approach.[31] Lobell found that the model based on the structure information representation approach gave the best results (among commercially available models) on an independent set of data obtained in his laboratory.[32] **Table 2** contains data extracted from Lobell's study, and presents the comparative results from Lobell's work on solubility of 442 predominantly uncharged organic compounds. The model designated as CSLogWS is based on the structure information representation approach.[33]

The results in **Table 2** are rank ordered on model quality. At the head of the table is the CSLogWS model with the highest r^2 value and the smallest MAE value. This independent work indicates that the model based on the structure information representation approach is superior to the results of the other commercial models reported. Of particular interest is the much smaller percentage of predictions with large errors, that is, errors greater than 1.0 log unit (last column). These results tend to indicate the quality of the modeling approach, the combination of the structure information representation and nonlinear modeling. Software development continues, and it is expected that this progress will create software with greater accuracy in prediction than reported here.

In a second test of the solubility model, predictions were made on the solubility at pH 7.4 of a set of 80 charged aromatic compounds, including 31 cations, 24 anions, and 25 zwitterions.[31,32] The pK_a values were estimated by CSpKa,[31] so that solubility could be computed from the intrinsic solubility (CSLogWS) and pK_a using a relation between log S and pK_a based on standard mass action relations. For this external test set, $r^2 = 0.77$ and MAE $= 0.57$.[31] This result, currently considered as preliminary, is significantly better than those obtained from the only other commercial software for estimating aqueous solubility of charged compounds.

5.23.4.3 Human Intestinal Absorption

The effectiveness of an oral drug is dependent upon sufficient absorption through the human intestines (HIA). Estimates of HIA are very valuable in the early stages of drug design in order to prevent development of candidates with inadequate HIA values.

In this preliminary study, the collected data set consists of 417 therapeutic drugs in which HIA is expressed as the percentage absorbed (%OA).[34] We used the structure information representation modeling approach, implemented with ANN modeling techniques , to develop the final model. The predictive quality of the model was tested on an external validation test set of 195 drugs; 92% of the compounds were predicted within 25% (%OA) of their reported experimental values. **Figure 3** provides a plot of predicted %OA values. The ANN model was based on two data sets, one for low molecular weight (MW < 251) compounds thought to be involved in passive paracellular transport, and a second set for higher molecular weight compounds ($252 < $ MW < 1450) thought to use transcellular transport. The MAE (11% (%OA)) for the test set indicates useful results for this model. However, several compounds appear as

Table 2 Summary for comparison of models for aqueous solubility prediction

Method to predict log(WS_o)[a]	r[2]	MAE	AE > 1 (%)[b]
CSLogWS (ChemSilico)[c]	0.58	0.70	21
ws2 (Novartis)[d]	0.46	0.91	36
ABB[a]	0.50	1.04	45
ACD 6.0 (ACD Labs)[e]	0.65	1.06	43
Tetko LogS[f]	0.46	1.17	50
QikProp 2.0 (Schrödinger)[g]	0.33	1.36	52
C2-ADME (Accelrys)[h]	0.19	1.35	54
PreADME[i]	0.68	1.49	58
Syracuse Research Corporation[j]	0.58	1.85	62

[a] For prediction of intrinsic aqueous solubility, WS_0; statistical results supplied by Mario Lobell, OSI Pharmaceuticals, Oxford, UK.[32]
[b] Percentage of compounds with predicted absolute error (AE) greater than 1 log unit.
[c] Model from ChemSilico, 48 Baldwin Street, Tewksbury, MA 01876, http://www.chemsilico.com.
[d] Model from Novartis, Novartis International AG, CH-4002 Basel, Switzerland, http://www.novartis.com.
[e] Model from ACD Labs, 90 Adelaide Street West, Suite 600, Toronto, ON M5H 3V9, Canada, http://www.acdlabs.com.
[f] Tetko model.[29]
[g] Model from Schrödinger, 120 West Forty-Fifth Street, 32nd Floor, Tower 45, New York, NY 10036-4041, http://www.schrodinger.com.
[h] Model from Accelrys Inc., 9685 Scranton Road, San Diego, CA 92121-3752, http://www.accelrys.com.
[i] Model from PreADME, http://preadme.bmdrc.org.
[j] Physical Chemical Property Database (PHYSPROP), Syracuse Research Corporation, SRC Environmental Research Center, Syracuse, NY, 1999.

outliers based on residuals being greater than 25 (%OA). This study further underscores the need for a larger data set in which the data are more uniform from the experimental point of view, based on the same experimental protocol.

To indicate further the nature of the results, we divided the validation test set predictions into three categories: low, 0–19%OA ($n = 16$); moderate, 20–59%OA ($n = 21$); and high, 60–100%OA ($n = 158$). The percentages predicted in the correct category are as follows: 0–19, 88%; 20–59, 67%; and 60–100, 94%.[33] **Figure 4** displays the predictions by bin. Overall, 91% of the compounds were predicted in the correct category. No compound in the low category was predicted to be in the high category, and vice versa, further underscoring the usefulness of the predictions.

These HIA results are usable for both the prediction of categorical results (low, medium, high) or for a continuous value of the percentage absorbed. Important structure descriptors used in this model include atom-type E-state indices for secondary aniline amines and for ether oxygen. In addition, the model includes the E-state descriptor for the $>C=O$ bond type, the E-state for the topological polar surface area (TPSA), and a measure of internal hydrogen bonding, the E-state for a hydrogen bond donor and a hydrogen bond acceptor separated by four skeletal bonds (SHBint4). The model indicates that large TPSA values make a negative contribution to the calculated HIA whereas SHBint4 makes a positive contribution. Both of these indications agree with the understanding of the factors that affect HIA. Use of the model structure descriptors has the combined advantages of being exactly calculated by the software, and do not require any specific mechanism for their interpretation. Therefore, these structure interpretations, drawn from the model, are useful for design to improve absorption. When HIA predictions from the model indicate poor absorption, structure interpretations from the model can guide the chemist in making synthetic modifications to improve absorption.

5.23.4.4   Protein Binding

The binding of a drug to serum proteins in human plasma has a powerful influence on its pharmacodynamic behavior, and can affect the systemic distribution of the drug in several ways. Estimation of this significant ADME property is one of the important activities carried out in the lead optimization stage of drug design.

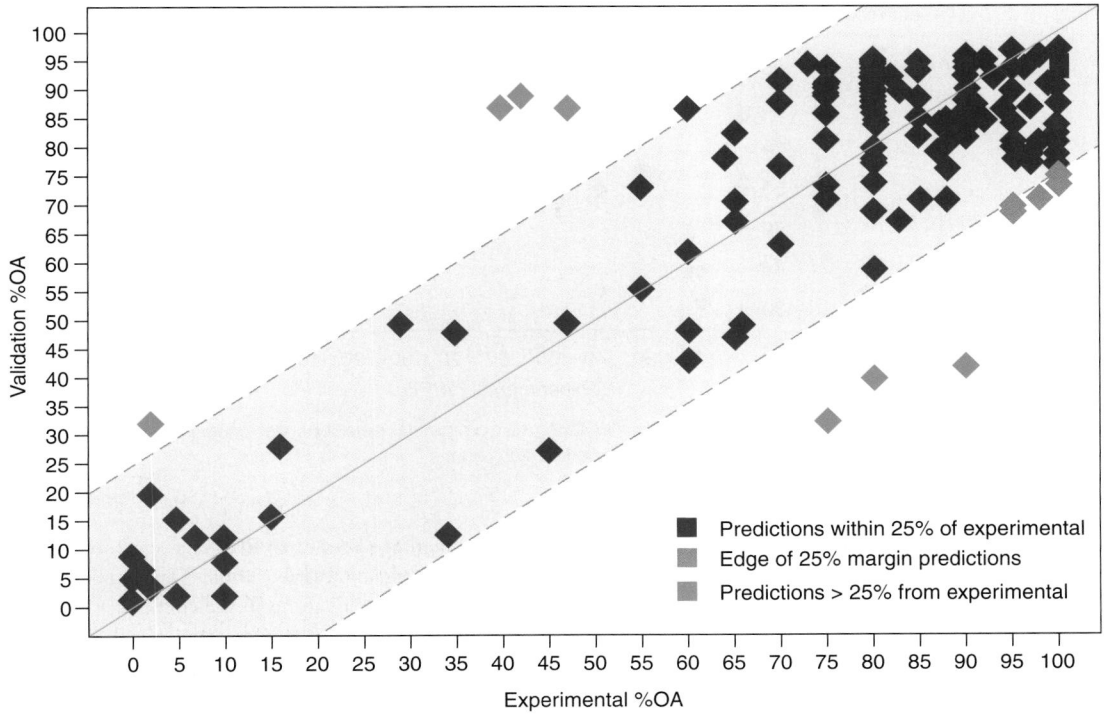

Figure 3 External validation test set predictions of HIA on 195 drugs, showing the %OA predicted by the model versus the experimental value. Ninety-two percent of the predictions are within 25% of the experimental values.

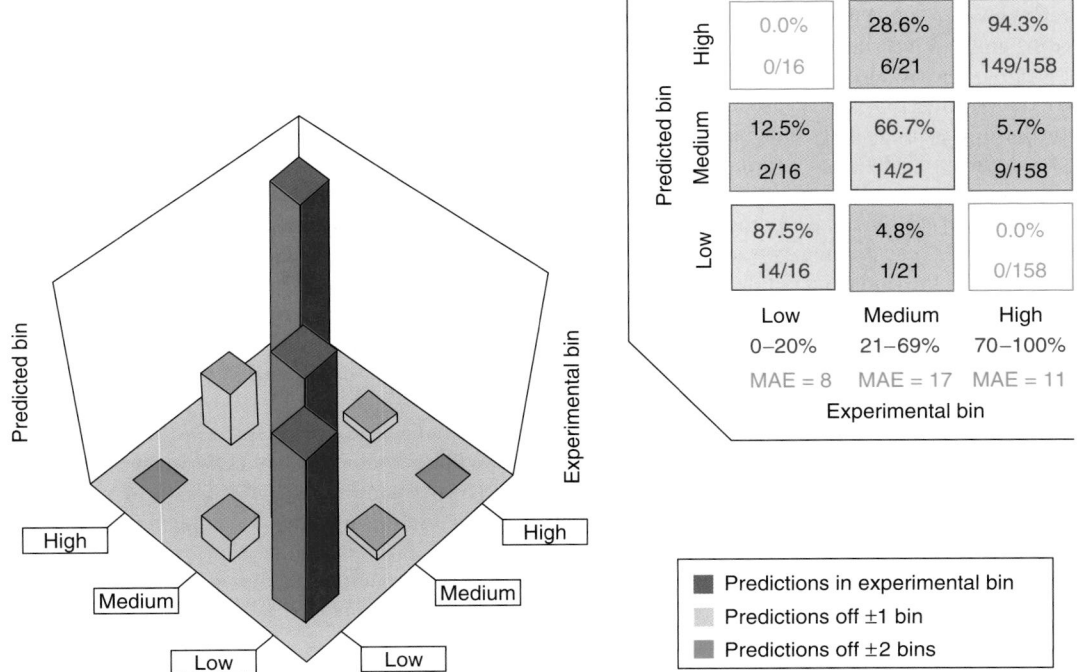

Figure 4 External validation test set predictions on 195 drugs, given in three bins, indicating that the predictions of %HIA by the model yield 91% correct placement in the high, medium, and low categories.

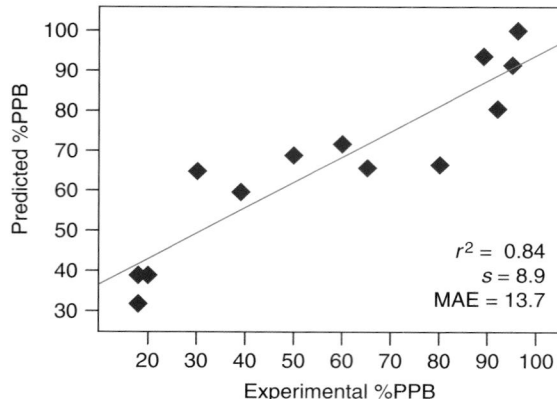

Figure 5 External validation test set predictions for %PPB of 13 commercial penicillins not used in model development, showing the %PPB predicted by the model versus the experimental value.

In a first study, protein-binding data were obtained (given as the percentage bound, %PPB) for a set of 74 penicillins from a published drug design series.[25] The structures among the substituents are highly varied. The number of atoms in the substituents on the penicillin skeleton range from 1 to 22, compared with 17 atoms in the common scaffold. The data set included penicillins with acyclic, monocyclic, and bicyclic substituents; heteroatoms included nitrogen, oxygen, sulfur, fluorine, chlorine, and bromine. Protein-binding values range from 7.2%PPB to 96.0%PPB. The modeling method, using MLR, employed the structure information representation, and led to a successful (cross-validated) model, with sound direct statistics: $r^2 = 0.80$ and $s = 12.1$ (i.e., 12.1% bound, in the units of the experimental data, not necessarily 12.1% of the range or average value). The model is based on five molecular structure descriptors that include E-state indices for aromatic carbons, methylene groups, amines, and halogens; also included is a molecular connectivity descriptor for the degree of skeletal branching and the size of the substituents, designated as R_1.

In a second study, the predictive capability of the model was successfully tested by the prediction of protein binding for 13 commercial penicillins, an external validation test set not used in model development. The protein-binding values for these 13 compounds effectively covered the training set range from 18%PPB to 92%PPB. The statistics for prediction on this external validation test set are quite acceptable, and compare favorably to the training set statistics: $r^2_{press}(= q^2) = 0.84$ and MAE = 12.7. **Figure 5** shows a plot of the predictions on this external test set. This test of the model strongly indicates that results from a model based on a drug design series can be used to predict a set of structures (for protein binding) under consideration for further synthesis and testing. This significant result indicates the quality that is needed in any stage of drug design.

Furthermore, in the third part of this investigation, when 28 cephalosporins were included in the training set, a new satisfactory model was developed for the full data set of 115 β-lactams.[25] The substituents contain from six to 21 atoms in the R_1 position of the β-lactam core skeleton. In the R_2 position, substituents contain from one to 12 atoms. Protein-binding values for the cephalosporins range from 2.0% to 98.0%. The full model for 115 β-lactams produced satisfactory statistics for prediction, based on cross-validation. For this purpose, 10% of the data was left out and predicted from the remaining 90%. This process was repeated until each compound was left out once. Then, that process was repeated 10 times (10 × 10 cross-validation), yielding 1150 residual values and the following statistics: $r^2_{press} = 0.79$, $s_{press} = 14.0$, and MAE = 10.9. In the y randomization test, $r^2_{max} = 0.16$.

This β-lactam model is based on eight structure descriptors, including the five obtained in the penicillin model. The three additional descriptors represent features present in the cephalosporins, including E-state descriptors for the carbonyl oxygen and $-N=$ nitrogen groups. Also included is a molecular connectivity index for the degree of skeletal branching and size of the substituents located in the second position of substitution, R_2. Structure interpretation of the model was also described in detail in the study.[25] For each descriptor in the model, detailed statements were made about the variation in the predicted protein binding as a result of the variation in the descriptor value. Based on this information, a chemist can design modifications of molecules to increase (or decrease) protein binding, a feature necessary for effective drug design.

In subsequent work, the modeling approach (based on the structure information representation descriptors) was used to develop an ANN model on a larger and significantly more diverse data set of 1000 drugs, including the prediction of an external validation test set.[33] The MAE for the external validation set of 208 drugs is 12.7 (percentage

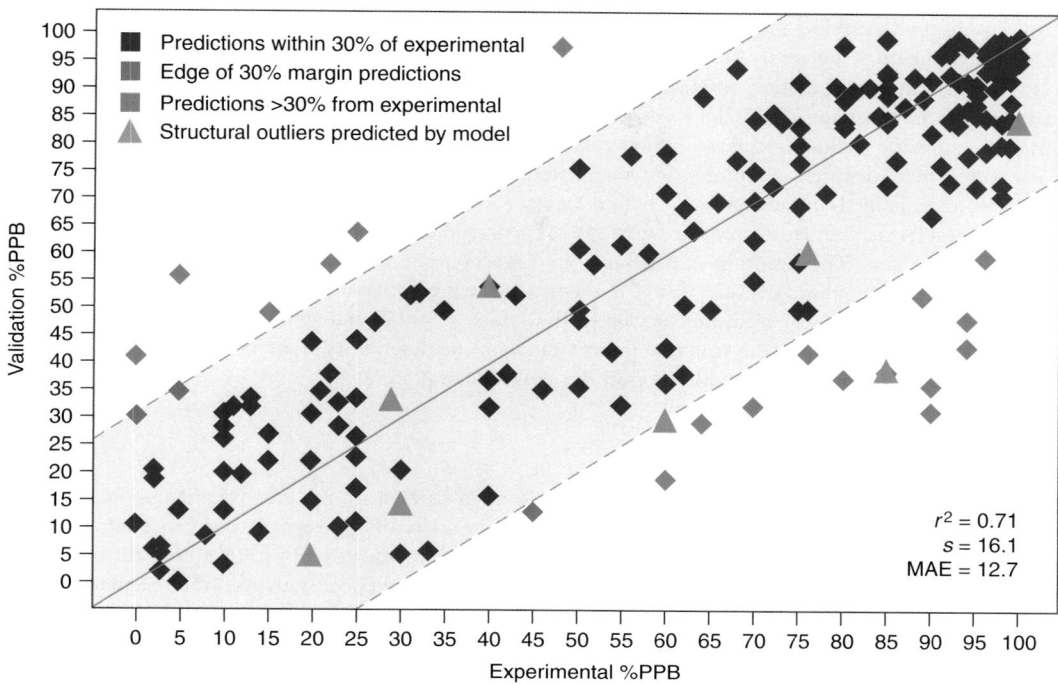

Figure 6 External validation test set predictions for %PPB of 208 compounds, showing the %PPB predicted by the model versus the experimental value. Included are eight compounds (green triangles) determined to be extremely nonsimilar to the training set (linkage distance > 1500 in hierarchical clustering). Ninety percent of the predictions are within 30% of the experimental values.

bound). Furthermore, 92% of these predicted values lie within 30% (percentage bound) of the experimental value. The external validation r^2 is 0.71. **Figure 6** provides a plot of the predicted values for this external test set. During this analysis, several errors were discovered in the original input data: for several compounds the value (percentage bound) in the compendium was actually found in the original literature to be reported as the percentage unbound. For a few other compounds, the experimental value is actually for a metabolite. Data were also found where the endpoint had been measured on nonhuman plasma, such as bovine serum. A clustering technique based on the atom-type E-state descriptors was used to facilitate this error analysis by identifying compounds with an experimental value that significantly differed from structurally similar compounds, suggesting further examination to confirm whether a data error was present.

In an extension of the method, eight compounds had been found to be significantly nonsimilar to the structures of the compounds in the training set. In the hierarchical clustering method (Ward's) used to create the external test set, these eight compounds were found at a large linkage distance from the rest of the data set. These eight compounds were removed from the training set. Subsequent to the testing of the model on the external validation test set of 208 compounds, predictions were made on these eight compounds. Their predictions are shown as green triangles in **Figure 6**. Seven of the eight predictions fall within the same range of predictive quality as for the test set. This finding may indicate that use of topological descriptors may provide the basis for the prediction of compounds that are not highly similar to those of the training set. Further investigations of this issue are underway.

The investigation on serum protein binding was extended to consider the use of data collected as the high-performance liquid chromatography retention time on immobilized albumin.[26] Retention time has been interpreted as the binding affinity of a compound to albumin. Some investigators have suggested that this highly automated experimental technique might be a suitable substitute for other methods of measuring protein binding. In order to investigate whether the modeling approach used for serum protein binding might also be useful for this alternative type of affinity data, the method was applied to a data set of 94 drugs. The number of (nonhydrogen) atoms in the molecules in the data set ranged from 10 (salicylic acid) to 57 (digitoxin). The molecules contain from one to six rings. The experimental data were modeled as the logarithm of the retention time, log k(HSA). For the training set of 84 drugs, direct statistics are quite acceptable: $r^2 = 0.77$ and $s = 0.29$. These modeling results are better than those reported using other published modeling methods.[35] In fact, Colmenarejo (*see* 5.36 In Silico Prediction of Plasma and Tissue Protein Binding) found it necessary to remove five compounds deemed to be outliers. The model, based on the structure information representation described above, gave quite acceptable results for 84 compounds, including all five excluded by Colmenarejo. The model is based on six molecular structure descriptors, including atom type E-state descriptors for aromatic carbons, saturated carbon atoms, fluorine and chlorine, and hydroxyl groups. In addition, two molecular connectivity descriptors represent six-membered rings.

Models were examined in several ways before being accepted as predictive tools. In the first test, the y randomization results are very satisfactory. In an extensive cross-validation process, each compound was left out once in 10 unique leave-out sets. The resulting statistics for this process are reported in the study, and further support the validity of the model. The most stringent test of the model is the prediction of 10 drugs in an external validation test set. For those 10 compounds, which covered the log k(HSA) range, $r^2 = 0.74$ and MAE = 0.31. Finally, our investigation presented detailed information on the structure interpretation of the descriptors found in the model. This information can serve as the basis for molecular modification in the drug design process.[26]

5.23.4.5 Blood–Brain Partitioning

Whether drug design criteria require significant brain penetration or demand minimal partitioning into brain tissue, estimation of the brain–blood ratio, BB = [brain]/[blood], is essential in drug design. In our first study, we organized experimental data, given as the logarithm, log BB, from 11 literature sources to create a very diverse set of 106 compounds.[28] MLR methods led to models, but the linear technique yielded poor statistics. When nonlinear terms were developed for specific structure descriptors, an adequate model was obtained: $r^2 = 0.66$ and $s = 0.45$. Two issues arose that appear typical of ADMET modeling. First, it was necessary to introduce nonlinear terms into the model in order to achieve useable results. Second, four compounds appeared to be outliers and were removed. Although the exact reasons for outlier status are not clear, two explanations were considered: (1) the eclectic nature of the experimental data, which include several types of experimental protocols, and (2) the possibility that salient structure features are not adequately encoded in the descriptor set. Because this data set is eclectic from the experimental protocol point of view, no reliable estimate of the actual experimental error for the data set is available. However, statistics from cross-validation indicate the model to be useful for drug design purposes.

This model for blood–brain barrier partitioning was also directly interpretable in terms of molecular structure, leading to statements about the manner in which calculated blood–brain barrier partitioning varies with molecular structure as a guide to molecular modification. For example, one of the descriptors in the model is the hydrogen E-state value for hydrogen bond donors (*see* 4.22 Topological Quantitative Structure–Activity Relationship Applications: Structure Information Representation in Drug Discovery). The model indicates that as the descriptor for hydrogen bond donor strength increases, log BB decreases, in agreement with the assumed nature of membrane passage across the blood–brain barrier. Furthermore, the model indicates that the contribution of an hydrogen bond donor group covers a range of values, depending on the bonding environment of the donor group. The range in contribution is traced directly to the fact the E-state value for an atom or group encodes its molecular bonding environment. This bonding environment aspect of E-state descriptors highlights their significant advantage over a simple fragment count, which is insensitive to molecular environment.

In one additional study with the model for blood–brain barrier partitioning, log BB values were predicted for a set of 20 039 organic molecules from the Pomona MedChem database.[28] The experimental log BB values were not available for these compounds, but the range of predicted values could indicate the potential for realistic predictions. In this test, 99.3% of the predicted values fell within the same log BB range as the training set. No unreasonable values were computed by the model.

Subsequent unpublished work has extended the range of applicability of the BB model. To deal with the nonlinear nature of the property–structure relation, an ANN model was developed for 103 compounds, and tested by the prediction of an external validation test set.[33] **Figure 7** shows the predictions for this external test set. One significant aspect of the new validation test set is that it consists of data on human brain and human cerebrospinal fluid (CSF) as well as data on rat and monkey. Data in the training set are entirely based on rat brain, but the external validation test set includes human brain data as well as CSF data in addition to monkey brain. The validation test results are statistically satisfactory, $r^2_{press} = 0.62$ and MAE = 0.39, even though the experimental nature of the test data arises from different sources than the training set. To provide a different view of the predictions, the data were divided into three bins, corresponding to low, medium, and high partitioning. These results are presented in **Figure 8**. Note that 83% of the predictions fall into the correct bin. Furthermore, no prediction on a compound in the high bin falls into the low bin, and vice versa.

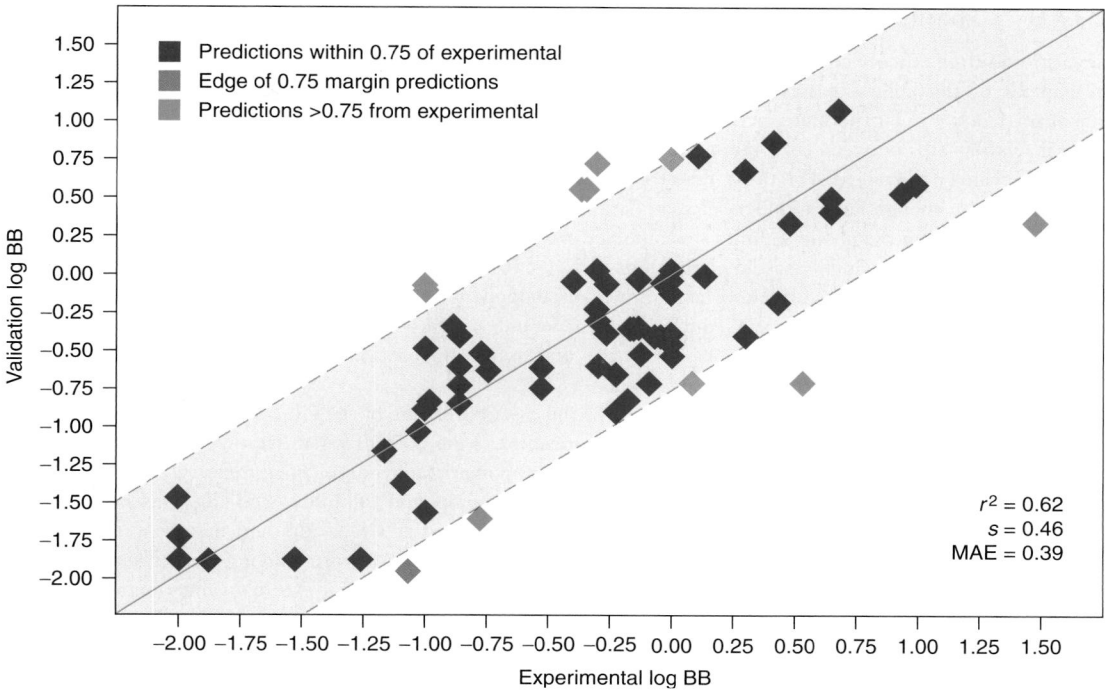

Figure 7 External validation test set predictions on 74 commercial drugs for blood–brain barrier partitioning, showing log BB predicted by the model versus the experimental value. The model is based entirely on data from rat brain. This validation set consists of data from several types of measurements: 18, rat brain; 44, human CSF; 5, human brain; 2, monkey brain; 3, other animals. Eighty-eight percent of the predictions are within 0.75 log units of the experimental values.

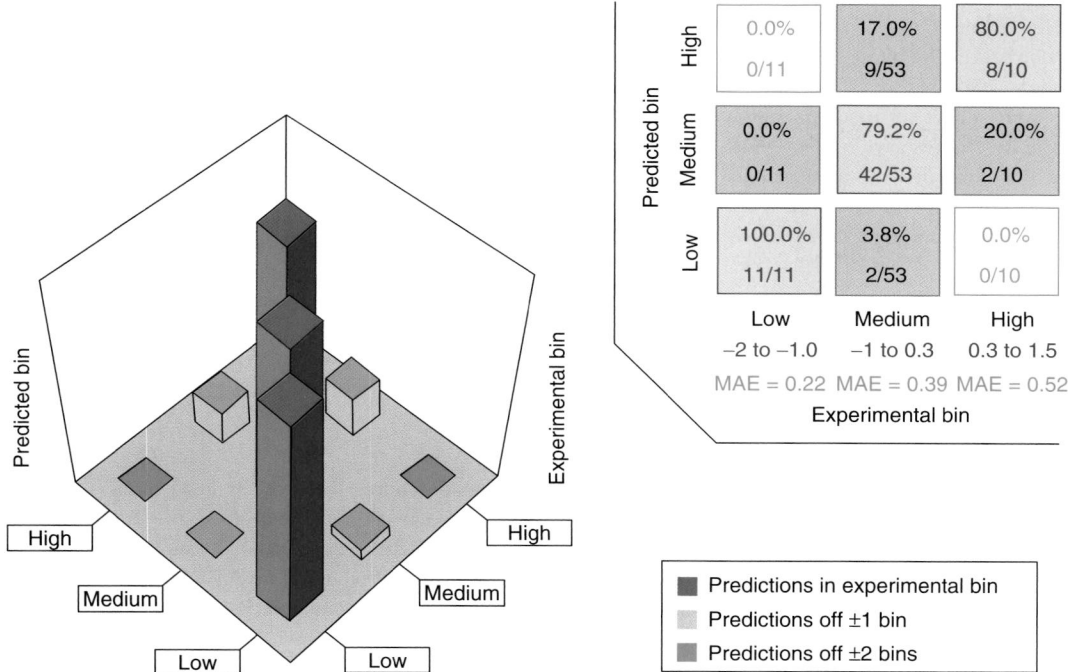

Figure 8 External validation test set predictions on 74 commercial drugs for blood–brain barrier partitioning as log BB given in three bins, indicating that the predictions of the model yield 83% correct placement in the high, medium, and low categories. The model is based entirely on data from rat brain. This validation set consists of data from several types of measurements: 18, human brain; 44, human CSF; 5, human brain; 2, monkey brain; 3, other animals.

5.23.4.6   Human Metabolic Stability

A key factor in drug efficacy and toxicity is the metabolic stability of the drug. Part of the approach to this issue is the development of the ability to recognize atom types and/or fragments involved in metabolism in phase I and II metabolism reactions. Furthermore, it is important to establish the ability of QSAR methods to account for the nonlinear relationships between compound structure and the toxic property, effects that arise from the multiplicity of possible mechanisms involved. For these reasons, we engaged in a preliminary investigation of modeling metabolic stability data even though large numbers of such data measured in a consistent protocol are not readily available.

The data for human metabolic stability was obtained from Cyprotex, based on in-house data as clearance, CL_{int}.[49] These values ranged from highly stable compounds ($CL_{int} < 10$) to unstable ($CL_{int} > 100$). The chemical structure diversity of the data set was investigated using principal component analysis (PCA) using 115 topological structure descriptors. The first two components, T_1 and T_2, account for only 17.2% of the total structure space variance, and only 43.3% of structure space variance was explained by the first nine components. These results indicate high chemical structure diversity among the 769 data set compounds.

A brief description of the experimental data indicates the essential nature of the CL_{int} values. The solution of a test compound was incubated with human liver microsomes. Reactions were terminated at five time points to provide time development data; analysis was carried out by liquid chromatography–mass spectrometry. Kinetic solubility measurements were made on all compounds to ensure adequate compound solubility. The CL_{int} ranged from ~ 0 to 500 µL min per milligram of protein. Metabolic stability was indicated as $CL_{int} < 25$; and metabolic instability was indicated as $CL_{int} > 100$. From the data set of 769 compounds, 693 compounds were randomly selected as the training set for ANN model development; the 76 remaining compounds were set aside for external validation of the model.

An ANN model was indicated in previous publications.[35,41] The developed ANN model is based on 19 topological descriptors, mainly atom-type, bond-type, and group-type E-state indices. The actual ANN model is called a consensus or ensemble model. In this technique, the final prediction is obtained as the average of the predictions of 10 models. The consensus model yielded correlation statistics, as follows: $r^2 = 0.77$ and $MAE = 19$ for a training set of 693 compounds. The validation test set ($n = 76$ compounds) yielded $r^2 = 0.56$ and $MAE = 22$; 11 compounds have absolute error > 35. The y randomization test also yielded satisfactory results; all the randomized sets yielded $r^2 < 0.05$ for the train and validation sets for this ANN model, indicating that the model is significantly different from a random selection of input vectors.

These preliminary results on 769 compounds indicate that the structure information representation approach to structure coupled to the consensus/ensemble ANN modeling technique is able to provide an approximate model for metabolic centers for phase I reactions as represented by this data set. The quality of the test set r^2 value is in keeping with the eclectic nature of both the structures and the methods of data measurement. More satisfactory statistical results await a data set based on consistent measurement. Another study using the kNN technique with topological structure descriptors gave similar QSAR findings on metabolic stability of 639 drug-like compounds.[50] Based on these results, it appears that this approach, using the ensemble ANN modeling technique or other nonlinear methods based on topological structure descriptors, is capable of yielding useful results. However, we continue to gather information with the hope of crating a database with more accurate experimental data as the basis for improved models.

5.23.4.7   Toxicity

This section consists of a brief presentation of toxicity studies as follows: fish toxicity, Ames test genotoxicity, risk of human carcinogenicity, and hepatotoxicity.

5.23.4.7.1   Fish toxicity

In earlier studies, models were developed for the toxicity of organic chemicals to the fathead minnow (*Pomephales promelas*), based entirely on molecular connectivity.[36–38] Later, Gough and Hall examined the toxicity of amide herbicides[21] using a combination of E-state and molecular connectivity. Recently, using the E-state formalism, Rose and Hall re-examined fish toxicity data (*Poecilia reticulata*), to create an externally validated model.[39] The data set consisted of 92 compounds characterized as phenols, anilines, aromatic hydrocarbons, nitro-aromatics, and halogenated aromatic hydrocarbons. Toxicity values (LC_{50}) ranged over four orders of magnitude.

Three E-state atom type descriptors (for chlorine, ether oxygen, and the hydrogen atom with maximum hydrogen E-state value) and one molecular connectivity index ($^1\chi^v$) provided the basis for an excellent model: $r^2 = 0.87$ and $s = 0.25$. An external validation test set consisted of data on three species other than that in the training set, as described in detail by Rose and Hall.[39]

In this study, the E-state model was compared to one based on sophisticated quantum chemical indices[40]; the E-state model was found to be significantly better in statistical quality.[39] Furthermore, direct structure interpretation was given for the model based on structure information representation. These descriptions of structure interpretation can be used directly to determine how toxicity is influenced by changes in the molecular structure of candidate molecules. No structure interpretation was given for the model based on quantum mechanical computations because of the complex nature of the principal components analysis that was used to construct the model.

An important feature of these studies is their applicability to an allied area of molecule design, environmental toxicity, and pollution. As part of the manufacturing process, chemists must consider the potential for toxicity and pollution. The structure information representation approach has shown its ability to assist this process through the development of models that can be predictive of toxicities to fish and other animals. What is learned in the arena of drug design can be useful in human health as well as in the environment.

5.23.4.7.2 Ames genotoxicity

This toxicity study illustrates the application of our method to a standard test required in drug submission to regulatory bodies, the Ames test for mutagenicity, based on mutations developed by chemicals in the bacterial system *Salmonella typhimurium*. The use of the Ames test is based on the assumption that any substance that is mutagenic (for the bacteria used in the test) may also turn out to cause cancer. Although, in fact, some substances that cause cancer in laboratory animals do not give a positive Ames test (and vice versa), the ease and low cost of the test make it invaluable for screening substances for possible carcinogenicity.

Recently, we applied the structure information representation approach to an Ames test data set that included 290 drugs.[41] The data set consisted of 3363 very diverse compounds with ratio of mutagens to nonmutagens as 60/40. For modeling, compounds were designated as mutagen or nonmutagen. An ANN model, developed to deal with binary data, yielded 89% correct classification in the training set and 83% correct in the external validation test set with a low rate of false positives. For 30 drugs in the external test set, false positives were 7%. **Figure 9** provides the model statistics along with a comparison to predictions produced by other methods.

In this investigation, using the structure information representation approach, three types of statistical modeling were examined: a nonlinear learning method, an ANN; a nonparametric method, the kNN technique, and a classification method (the decision forest). The results of the three approaches were carefully compared with an emphasis on the use of external validation test sets that included drugs. For this data set, the ANN method provided the best statistical results on the external test set. All of the other methods available for comparison also yielded higher false-positive percentages.

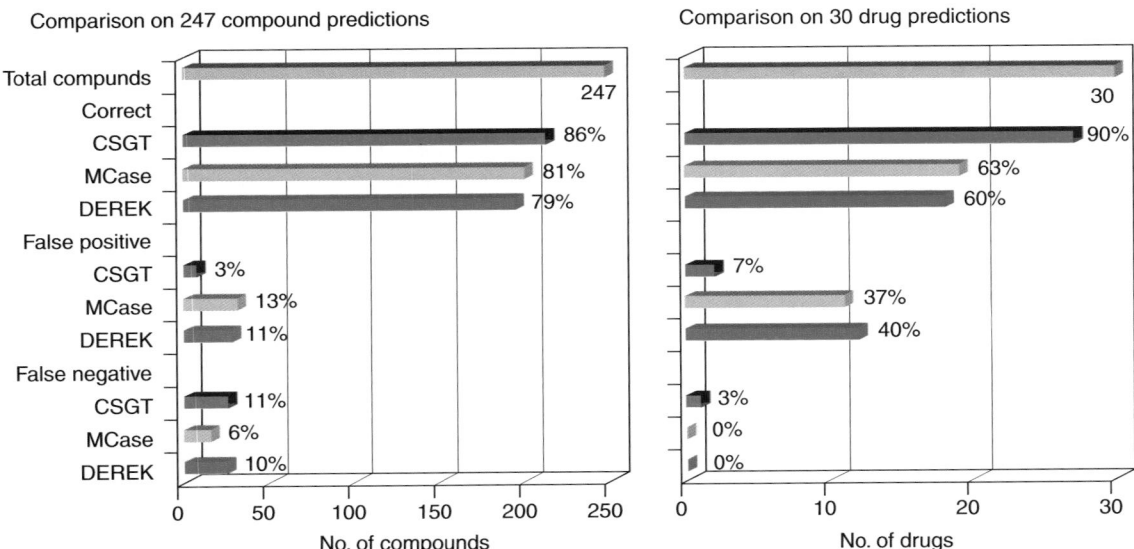

Figure 9 Comparison of overall predictive accuracy, false positives, and false negatives for the validation of Ames mutagenicity models from three sources; ChemSilico Predict 2.0.3.1 (CSGT), MULTICASE (MCase PCA2i), and DEREK (version 3.4). Validation predictions were made on 217 organic compounds, including 30 drugs. The complete validation set for CSGT consisted of 400 compounds, but comparison was available for only 247 of the 400 because 183 compounds were also in the MULTICASE training set.

Of significant importance for the results of this investigation is the combination of structure descriptors found to be important in the model. First, the whole molecule descriptors, such as molecular connectivity indices, encode significant information on the general structure features such as size, shape, and skeletal variation and complexity. These descriptors appear to provide useful discrimination among various structure classes present in the data set. Second, the atom-type, bond-type, and group-type E-state descriptors encode important structure features such as atoms and bonds associated with important functional groups. Of particular significance is the occurrence of E-state descriptors that correspond to known structure alerts. Several examples are described in the study.[41]

5.23.4.7.3   Risk of human carcinogenicity

A significantly expensive part of the drug approval protocol is the test for risk of human carcinogenicity based on a US Food and Drug Administration (FDA) protocol, performed on rats and mice for 2 years or more.[42] In recent years, otherwise successful drug candidates have failed this test. This kind of late failure is very costly and causes delays in the drug development process. An in silico predictive method that would give reasonably valid predictions of human risk would be invaluable in the early stages of drug design. Such a method would give early warnings so that alternative candidates could be developed to have estimated low risk.

The earlier studies on toxicity suggested that the structure information representation approach might be usefully applied to the prediction of the potential for human carcinogenicity. In conjunction with the FDA (Center for Drug Evaluation and Research), MDL Information Systems used a combination of E-state and molecular connectivity indices to create a model for the human risk of carcinogenicity.[43] Nonparametric discriminant analysis was used on a training set of 1022 drugs. The carcinogenicity risk was indicated as 'high' or 'low,' based on FDA criteria.[42] A form of nonparametric discriminant analysis was used to establish the model, which was validated on an external test set of 50 drugs. For the external test set, this model yielded 76% correct prediction for a 'high' risk of carcinogenicity and 84% correct for a 'low' risk. Of particular importance is the low percentage of false positives, 16%.

5.23.4.7.4   Hepatotoxicity dose-dependent modeling of therapeutic drugs

Drug-induced toxicity and adverse drug reactions are among the primary causes of postmarket withdrawal of drugs. These effects also account for over 50% of the acute liver cases in the USA.[1,3,46,47] Approximately 50% of new pharmaceuticals that show hepatotoxicity effects in clinical drug development also have corresponding effects in animal toxicity studies.[2,45]

A preliminary study was undertaken to assess the feasibility for modeling the hepatotoxicity of drugs. Although data measured by a consistent set of protocols are not available in large numbers, it is believed useful to examine the possibility for modeling this important type of toxicity, even on a small data set. For this reason we consider this a preliminary study. The sources for the data were *Mosby's GenRx* and *Therapeutic Drugs*.[44a,b]

Initially, 366 compounds were obtained from the data source and compiled in the database. Compounds were eliminated on the following basis: duplications (same drug with different associated salts), intravenously or topically administered drugs, and drugs containing coordinated Ca^{2+} or Mg^{2+} ions. The final number of orally administered drugs used in this study was 291. The dosage was given as milligrams, and assumed to be administered daily. With an assumed average body weight of 60 kg, the assumed daily dosage was converted to millimoles per kilogram. The dosage range covered 41 000 in $mmol\,kg^{-1}$ units for 291 compounds. For modeling purposes, the data were converted to the logarithm scale as $log(mmol\,kg^{-1})$, to be used as the dependent variable in modeling. No information was provided on age, gender, or the duration of oral consumption.

The final set of 291 drugs was rechecked for structure duplications.[48] This final set contained 63 of the most prescribed drugs from the 1995 to 2003 Rx list.[44b] An assessment of the chemical diversity of 291 drugs was conducted to establish the diversity of these 291 drugs. First, principal component analysis was performed, using 115 topological descriptors. A second similarity analysis was performed by computing the Tanimoto coefficient between every pair of molecules in the data set.[17] Both methods indicated the significant diversity of the structures in the data set.

As a basis for modeling, the total available set of structure descriptors was trimmed before modeling to an initial set of 210 structure information representation descriptors. Trimming was accomplished by removal of descriptors that only count features and those with a nonzero population less than 3%. An ANN ensemble model was built with a training set of 259 compounds. The average predicted value $(log(mmol\,kg^{-1}))$ from the 10 nets produced by the ensemble model was used as the predictive value for the external validation test set of 33 compounds. The dose-dependent range for the training set was 4.6 log units, and 3.4 log units for the 33 randomly selected drugs used for the external validation. The r^2 value for the training set was 0.64, and 0.42 for the external test set. The MAEs for the training and test sets were 0.39 and 0.44, respectively, amounting to about 10% of the range of data in the two sets.

To test whether the 10 descriptors obtained in the ANN model were accidental, two randomizations of the experimental data were made. The correlation coefficients obtained were given as $r^2 = 0.03$ and 0.07, respectively, for the training set. Similar values were found for the validation set. MLR analysis was also performed, yielding poor results: $r^2 = 0.42$ for the training set and 0.03 for the validation set, using a 10-fold ensemble model with 16 topological descriptors.

Given the lack of any specific information on age, gender, population size, or total dose, these results are very encouraging for this eclectic data set. The MAE for the training set was 0.39. Furthermore, 76% of the calculated values lie within 0.6 log units of the experimental values. Likewise, the 33 drug new chemical entities in the validation test set gave an MAE of 0.44; and 67% of the predicted values were within 0.6 log units of the experimental values. These results indicate that ensemble ANN modeling techniques using only 10 structure descriptors did an adequate job in predicting dose-dependent hepatotoxicity effects for this rather small and eclectic data set.

5.23.5 Software

Computation of the structure information representation descriptors is readily available in software from EduSoft, LC.[52] Furthermore, models for ADMET properties based on these structure descriptors are commercially available from ChemSilico LLC.[33]

5.23.6 Conclusions

This article summarizes a small number of QSPR investigations on ADMET properties that indicate the wide applicability of the structure information representation approach that includes the E-state. Each model is validated by prediction on an external validation test set, not used in model development. These test sets are diverse and generally among the largest yet published. The focus of this article is on modeling ADMET properties, but many more investigations have been reported elsewhere.[10,12,14–31,35–39,51]

The E-state description of molecular structure has been presented and shown to be of significant value in encoding information relevant to pharmacodynamic and biological properties. As derived, the E-state indices are based on electron counts and orbital distributions. Their relationship to chemical structure is verified by the ability to relate these indices to electronegativity, molecular volume, molecular orbital free valence, and nuclear magnetic resonance chemical shifts, among other properties.

These atom-based indices, along with whole-molecule topological descriptors, the molecular connectivity indices, have been used to model the pharmacodynamic and biological behavior of molecules of interest as drugs. In particular, we have shown that a class of properties known as ADMET can be successfully modeled using these descriptors.

The models can be used to predict activity within these categories. Since the E-state and molecular connectivity indices are derived from structure, the models created with them can be interpreted directly as molecular structure. Property prediction and drug design are thus derivatives of the models we have discussed here.

References

1. Dearden, J. C. *J. Comput.-Aided Mol. Des.* **2003**, *17*, 1–9.
2. Kola, I.; Landis, J. *Nat. Rev. Drug Disc.* **2004**, 7, 11–725.
3. Van de Waterbeemd, H.; Gifford, E. *Nat. Rev. Drug Disc.* **2003**, *2*, 192–204.
4. Beresford, A. P.; Segall, M.; Tarbit, M. H. *Curr. Opin. Drug Disc. Dev.* **2004**, 7, 36–42.
5. Banik, G. *Curr. Drug Disc.* **2004**, May, 31–34.
6. Ekins, S.; Boulanger, B.; Swaan, P. W.; Hupcey, M. A. *J. Comput.-Aided Mol. Des.* **2002**, *16*, 381–401.
7. Hall, L. H.; Hall, L. M. *SAR QSAR Environ. Res.* **2005**, *16*, 13–41.
8. Hall, L. H. *Chem. Biodivers.* **2004**, *1*, 183–201.
9. Kier, L. B.; Testa, B. *Adv. Drug Res.* **1995**, *26*, 1–35.
10. Kier, L. B.; Hall, L. H. *Molecular Structure Description: The Electrotopological State*; Academic Press: San Diego, CA, 1999.
11. Kier, L. B.; Hall, L. H. *J. Pharm. Sci.* **1981**, *70*, 583–589.
12. Hall, L. H.; Kier, L. B. *J. Chem. Inf. Comput. Sci.* **1995**, *35*, 1039–1045.
13. Kier, L. B.; Hall, L. H. *Molecular Connectivity in Chemistry and Drug Research*; Academic Press: New York, 1976; Kier, L. B.; Hall, L. H. *Molecular Connectivity in Structure–Activity Analysis*; John Wiley: New York, 1986.
14. Kier, L. B.; Hall, L. H. *J. Chem. Inf. Comput. Sci.* **2000**, *40*, 792–795.
15. Hall, L. H.; Kier, L. B. *J. Mol. Graph. Model.* **2001**, *20*, 4–18.
16. Kier, L. B.; Hall, L. H. *SAR QSAR Environ. Sci.* **2001**, *12*, 55–74.
17. Kier, L. B.; Hall, L. H. *J. Chem. Inf. Comput. Sci.* **2000**, *40*, 784–791.

18. Kier, L. B.; Hall, L. H. *J. Chem. Inf. Comput. Sci.* **1997**, *37*, 548–552.
19. Kier, L. B.; Hall, L. H. *Med. Chem. Res.* **1992**, *2*, 497–502.
20. Hall, L. H.; Mohney, B. M.; Kier, L. B. *J. Chem. Inf. Comput. Sci.* **1991**, *31*, 76–82.
21. Gough, J. D.; Hall, L. H. *Environ. Toxicol. Chem.* **1999**, *18*, 1069–1075.
22. Maw, H. H.; Hall, L. H. *J. Chem. Inf. Comput. Sci.* **2000**, *40*, 1270–1275.
23. Maw, H. H.; Hall, L. H. *J. Chem. Inf. Comput. Sci.* **2001**, *41*, 1248–1254.
24. Hall, L. H.; Kier, L. B. *Med. Chem. Res.* **1992**, *2*, 497–502.
25. Hall, L. M.; Hall, L. H.; Kier, L. B. *J. Comput.-Aided Mol. Des.* **2003**, *17*, 103–118.
26. Hall, L. M.; Hall, L. H.; Kier, L. B. *J. Chem. Inf. Comput. Sci.* **2003**, *43*, 2120–2128.
27. Hall, L. H.; Kier, L. B. *J. Chem. Inf. Comput. Sci.* **1995**, *35*, 1039–1045.
28. Rose, K.; Hall, L. H. *J. Chem. Inf. Comput. Sci.,* **2002**, *42*, 651–666.
29. Huuskonen, J.; Livingstone, D. J.; Tetko, I. V. *J. Comp. Inf. Comput. Sci.* **2000**, *40*, 773–777.
30. Livingstone, D. J.; Ford, M. G.; Huuskonen, J; Salt, D. *J. Comput.-Aided Mol. Des.* **2001**, *15*, 741–752.
31. Votano, J. R.; Parham, M. E.; Hall, L. H.; Kier, L. B.; Hall, L. M. *Chem. Biodivers.* **2004**, *1*, 1829–1841.
32. Lobell, M.; Sivarajah, V. *Mol. Divers.* **2003**, *7*, 69–87.
33. See the web site for model information at ChemSilico. www.chemsilico.com (accessed May 2006).
34. Votano, J. R.; Parham, M. E.; Hall, L. H.; Kier, L. B. *Mol. Divers.* **2004**, *8*, 379–391.
35. Colmenarejo, G.; Alvarez-Pedraglio, A.; Lavandera, J.-L. *J. Med. Chem.* **2001**, *44*, 4370–4378.
36. Hall, L. H.; Maynard, E. L.; Kier, L. B. *Environ. Toxicol. Chem.* **1989**, *8*, 431–436.
37. Hall, L. H.; Maynard, E. L.; Kier, L. B. *Environ. Toxicol. Chem.* **1989**, *8*, 783–788.
38. Hall, L. H.; Kier, L. B. *Environ. Toxicol. Chem.* **1986**, *5*, 333–337.
39. Rose, K.; Hall, L. H. *SAR QSAR Environ. Res.* **2003**, *14*, 113–129.
40. Robert, D.; Carbo-Dorca, R. *SAR QSAR Environ. Sci.* **1999**, *10*, 401–422.
41. Votano, J. R.; Parham, M. E.; Hall, L. H.; Kier, L. B.; Oloff, S.; Tropsha, A.; Xie, Q.; Tong, W. *Mutagenesis* **2004**, *19*, 365–378.
42. Tennant, R. W. *Mutat. Res.* **1993**, *286*, 111–118.
43. Contrera, J. F.; Hall, L. H.; Kier, L. B.; MacLaughlin, P. *Curr. Drug Disc. Technol.* **2005**, *2*, 55–67.
44a. Schrefer, J., Ed. *Mosby's GenRx*; Harcourt Health Sciences Company: Elsevier, 2000.
44b. Dollery, C., Ed. *Therapeutic Drugs*, 2nd ed.; Churchill Livingstone, Harcourt Brace and Company Ltd, Elsevier, 1999.
45. www.rxlist.com (accessed May 2006).
46. Xu, J. J.; Diaz, D.; O'Brien, P. J. *Chemico-Biol. Interact.* **2004**, *150*, 115–128.
47. Aldrige, J. E.; Gibbons, J. A.; Flaherty, M. M.; Kreider, M. L.; Romano, J. A.; Levin, E. D. *Toxicol. Sci.* **2003**, *76*, 3–20.
48. Kier, L. B.; Hall, L. H. *Chem. Biodivers.* **2005**, *2*, 1428–1437.
49. Private communication: Intrinsic Clearance (CL$_{int}$) data and compound structures were supplied by Cyprotex Ltd, Manchester, UK.
50. Shen, M.; Xiao, Y.; Golbraikh, A.; Gombar, V. J.; Tropsha, A. *J. Med. Chem.* **2003**, *46*, 3013–3020.
51. www.eslc.vabiotech.com/malconn/mconpubs.html (accessed May 2006).
52. EduSoft LC, PO Box 1811, Ashland, VA 23005 and www.eslc.vabiotech.com (accessed May 2006).

Biographies

Lowell H Hall is a physical chemist who has pioneered methods of molecular structure representation for over 30 years. He received the PhD in physical chemistry from The Johns Hopkins University and pursued postdoctoral work at The National Bureau of Standards and Oak Ridge National Laboratory in single-crystal x-ray crystallography. He has been Professor of Chemistry at Eastern Nazarene College, a liberal arts college in Quincy, MA, since 1967. He began his research into quantitative structure–activity relationship (QSAR) models during a sabbatical leave with Lemont B Kier at Massachusetts College of Pharmacy in 1974. Together, Hall and Kier have written four books, eight book chapters, and more than 100 papers. Their most recent book, '*Molecular Structure Description: The Electrotopological State*' (Academic Press, 1999) presents their E-state approach to molecular structure representation. Hall and Kier are the co-developers of the structure information representation method as well as topological QSAR modeling. Professor Hall has been a consultant to pharmaceutical companies, a regular participant in the Gordon Conference on QSAR/CADD, and is currently a consultant with MDL Information Systems. He is the creator of the Molconn software, and also runs Hall Associates Consulting. He is a partner in ChemSilico LLC, a producer of model predictors for properties of interest in the pharmaceutical area.

Lemont B Kier received a BS in pharmacy from Ohio State University, and a PhD in medicinal chemistry from The University of Minnesota. He has been a pioneer in the development of several theoretical methods used for rational drug design, based on molecular structure. He introduced the use of molecular orbital theory to compute the preferred conformation of molecules. New insight into the pharmacophores of several neurotransmitters and drug molecules were predicted by Kier using these calculations. In another application, Kier pioneered the use of molecular orbital calculated energies to predict intermolecular interactions, now called molecular docking. In the mid-1970s, Kier and Hall developed nonempirical molecular descriptors named molecular connectivity indices. Information derived from these indices is a nonempirical estimate of the valence state electronegativity, called the Kier–Hall electronegativity, closely correlated with the Mulliken–Jaffe electronegativity values. In the mid-1980s, Kier introduced the kappa indices, encoding information about several aspects of molecular shape and flexibility. Another approach to structure description was developed by Kier and Hall, in the early 1990s. This set of descriptors is called the electrotopological state (E-state), and is based on the Kier–Hall electronegativity and the topological characteristics of an atom or group in a molecule.

L Mark Hall received a BS in biology with minor degrees in chemistry and writing from Eastern Nazarene College, where he did undergraduate research developing artificial neural network (ANN) quantitative structure–activity relationship (QSAR) models on pesticide toxicity and human plasma protein binding. He has worked as a consultant to vendors that develop computational chemistry software tools. This consulting has included the development of product user guides, software specifications, and quality assurance reports. He has also worked in product development for ChemSilico, helping to develop commercial ANN predictors for human intestinal absorption, human plasma protein binding, and blood–brain barrier partition. He has published QSAR studies for human intestinal absorption, human plasma protein binding, albumin binding affinity, aqueous solubility, and Ames genotoxicity. He has worked for Hall Associates Consulting on the continuing development of the Molconn molecular descriptor software. These projects include the development of a new class of descriptors designed for use in modeling the acid ionization constant of drugs, the development of a system for recognizing and correcting aromaticity-encoding errors in computer-based molecular structure files, and the development of new approaches to error analysis of QSAR models using novel applications of hierarchical clustering. In 2005, Hall was elected to serve as the Communications Director for the Boston area Group for Informatics and Modeling.

Comprehensive Medicinal Chemistry II
ISBN (set): 0-08-044513-6

ISBN (Volume 5) 0-08-044518-7; pp. 555–576

5.24 Molecular Fields to Assess Recognition Forces and Property Spaces

G Vistoli and A Pedretti, Institute of Medicinal Chemistry, University of Milan, Milan, Italy

5.24.1   Setting the Scene

5.24.1.1   The Dimensionality of Molecular Information

It is a general assumption that molecular properties/attributes can be classified according to their dimensionality corresponding to that of the molecular representation from which the property was derived. As seen in **Figure 1a**, molecular representations can be divided into elementary (1D), symbolic (2D), spatial (3D), dynamic (4D), and social (5D), and multidimensional, nD).[1]

For the elementary 1D depiction, molecules are represented by their chemical formulae. This is the constitution of the molecule in terms of atom content and count. The 1D representation is so poorly informative that it does not unambiguously define a molecule, since all positional isomers have the same molecular formula.

At the 2D symbolic level, molecules are represented by illustrations depicting the presence and the adjacency of atoms of various types. The code letters and connecting lines in these representations are the familiar structures that we normally draw on paper. The main information encoded is the connectivity (i.e., the bonding relationship). The connectivity allows us to dissect a molecule according to its functional groups, facilitating a qualitative interpretation of most chemical phenomena (e.g., reactivity or acid/base equilibria).

The 3D spatial representation includes all the information encoded at the 2D level plus the rich information deriving from the fact that a molecule is an object in 3D space. This information includes the spatial arrangement of atoms or groups (i.e., the conformation), the mantle or buried status of an atom, and the approximate spatial occupancy of parts or the whole of a molecule. Geometric 3D information can be graphically represented using molecular surfaces that give a good approximation of steric requirements (*see* Section 5.24.2.1). Physicochemical properties of great pharmacological and biological relevance, including hydrophobicity, hydrogen bond donor and acceptor capacity, and lipophilicity depend strongly on the 3D geometry of a molecule. A powerful method to study these recognition forces is the computation of molecular fields, for example molecular electrostatic potentials (MEPs, *see* Section 5.24.3.1), which encode electrostatic forces, molecular lipophilic potentials (MLPs, *see* Section 5.24.3.2), which encode hydrophobicity, hydrogen-bonding capacity and polarizability, the recent molecular hydrogen-bonding potentials (MHBPs, *see* Section 5.24.3.3), and, more generally, molecular interaction fields (MIFs, *see* Section 5.24.3.1), which can encode a wide range of recognition forces depending on the probes used (e.g., the fields computed with the GRID program, *see* Section 5.24.3.1).

When time is added to the 3D description, the dynamic 4D profile of a molecule is revealed. This permits consideration of conformational flexibility and prototropic behavior. The dynamic behavior of molecules can be expressed in conformational hypersurfaces, and the ensemble of all conformers of a given compound is often taken as

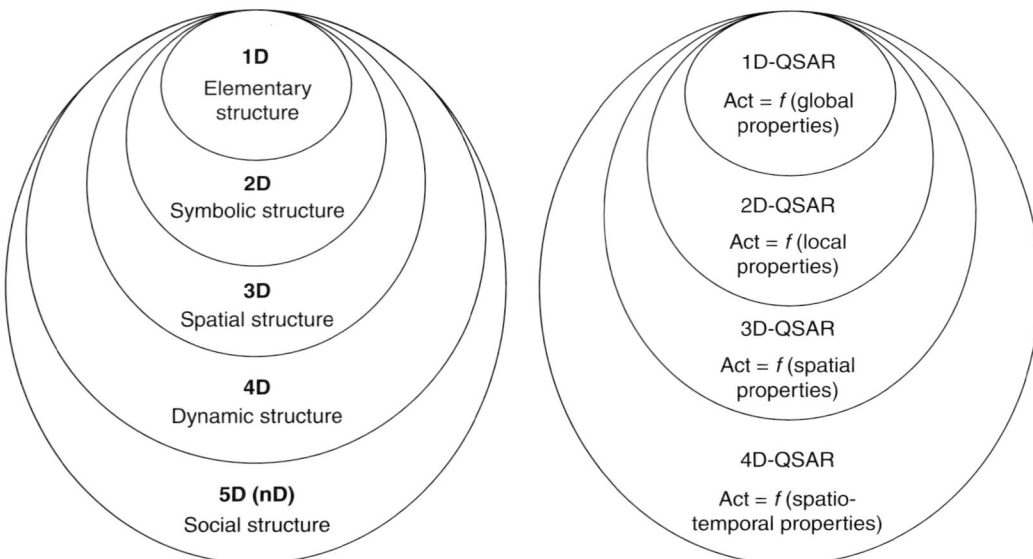

Figure 1   (a) A multilevel description of molecular structures and properties. (b) A multilevel classification of QSAR approaches. 1D, one-dimensional; 2D, two-dimensional; 3D, three-dimensional; 4D, four-dimensional; nD, multidimensional; QSAR, quantitative structure–activity relationships. (Reproduced from Testa, B.; Kier, L.; Carrupt, P. A. *Med. Res. Rev.* **1997**, *17*, 303–326, with permission from John Wiley & Sons, Inc.)

defining a conformational space. Also, the conformer-dependent properties will exhibit variability over time, since there is a one-to-one correspondence between the conformational space and the property values.

Thus far, molecules have been considered to be in isolation. This, however, is only partly correct, since, as Helmholtz repeatedly stated,[2] there exists no intrinsic property, since a property is always the ability of a subject to exert some effects on an object even if the latter is not defined or remains implicit. Such an implicit object is well illustrated by the environment or medium surrounding the subject. This model underlines the interactive nature of a property and implies a dynamic profile, since a property requires a process by which it can emerge and a discrete duration for its manifestation.[3]

Attributes and properties corresponding to the elementary and symbolic levels of description (mainly 1D and 2D properties) are environment-invariant, whereas stereo-electronic properties (mainly 3D and 4D properties) are influenced by the molecular environment. At this level of interaction, the description of molecular structure and properties becomes highly complex, and the molecular environment (e.g., solvent, bulk liquid, or crystal) must be explicitly considered.

A molecule may be viewed as a social entity, exhibiting emergent multidimensional properties from nonlinear interactions with other molecules (which may belong to the same compound as in a crystal, to other chemicals, or to a solvent). These properties are novel with respect to those which emerge at lower levels of description, and in a rigorous sense they can be exhibited not by individual molecules, but by a chemical compound.[4] This concept is also reflected in the social nature of living organisms, which are currently seen as dynamic complex systems that develop and evolve in relation to other organisms and complex environments.[5]

It is interesting to observe that the quantitative structure–activity relationship (QSAR) techniques can be similarly classified according to the dimensionality of molecular properties/descriptors used as independent variables (**Figure 1b**). Thus, the 1D-QSAR approaches correlate global properties (computable or measurable) with biological data, while the 2D-QSAR approaches use, as independent variables, fragmental increments conceptually derived from the dissection of a molecule according to its functional groups (as seen in multivariate Hansch analysis). 2D-QSARs rely on the concept that functional groups influence the biological activity to differing extents, and the independent variables must be focused on most determinant groups, neglecting the unimportant or invariant ones that represent useless 'noise.'

The 3D-QSAR techniques exploit conformer-dependent variables obtained considering the variation of molecular properties in the space around molecules. These variables are often computed using the molecular field formalism (as in the comparative molecular field analysis (CoMFA) method, *see* Section 5.24.4.1) or by projecting the molecular properties onto maps (as in the VolSurf approach, *see* Section 5.24.3.1).

The recent 4D-QSAR approaches attempt to include the property variability in the independent variables used to predict the biological data. These methods are still being developed, and can be divided into those based on flexible fields and dynamic occupancy (*see* Section 5.24.5.1) and those based on the property space concept (*see* Section 5.24.5.2).

This chapter is mainly devoted to molecular fields as prototypes of conformer-dependent 3D properties classified according to the form–function–fluctuation triad (*see* Section 5.24.1.2). Attention is also focused on the use of molecular fields in 3D- and 4D-QSAR approaches and, in particular, on their applications in absorption, distribution, metabolism, and excretion (ADME) predictions. More extended treatments of these computational methods can be found in Volumes 3 and 4.

5.24.1.2 The Form–Function–Fluctuation Triad

The growing computational power available to researchers is proving to be an invaluable tool with which to investigate the dynamic behavior of molecular systems, showing that a molecule cannot be considered simply as a static object but must instead be seen as an object whose conformational changes may significantly affect the profile of any of its properties.

The dynamic behavior of molecular systems depends mainly on molecular fluctuations that influence both form (what a molecule is) and function (what a molecule does), delineating the ensemble of all probabilistic transitions a molecule can undergo in form and function. This generates a very large number of molecular states, which are snapshots of the molecule at a given moment in time. Each state is characterized by a unique combination of form (geometric descriptors) and function (associated properties). Reciprocally, any function (property) exhibited by a compound will have a distinct value for each molecular state occupied by that compound.

Fluctuation, form, and function represent the interrelated elements of a triad that schematizes the behavior of any molecular entity. These components cannot be ordered causally or hierarchically. Rather, they are viewed as being of equal importance and feeding on each other. Fluctuation, form, and function can be quantitatively analyzed using appropriate molecular descriptors. Molecular form and its attributes (*see* Section 5.24.2) can be defined using geometrical (distance, angle, torsion, etc.) or structural descriptors (Van der Waals surface (vdWS), solvent accessible surface, volume, radius of gyration, etc.), while function can be Section 5.24.3). Molecular fluctuations can be monitored by analyzing the variability of properties as defined, for example, by their corresponding property spaces (*see* Section 5.24.5). A property space can also be conceived as the basin of attraction of the property states of a compound.

A physical representation of a basin of attraction of all molecular states is afforded by an energy landscape, namely a hypersurface whose dimensions are the energy of the system, plus all its other variables. Usually, the more probable states of a molecule (i.e., its states of lowest energy) are represented as valleys in the energy landscape, whereas the states of highest energy are represented by peaks, and the transition states as mountain passes.

5.24.1.3   Theoretical Bases of Molecular Fields

A molecular field is a classical example of a scalar field, namely a 3D region where a scalar (1D) value is assigned to each point in 3D space.[6] The concept of molecular fields is strictly related to that of molecular grids, a computational approach that divides the space around a molecule in cubic cells (also called voxels), whose edge length defines the grid resolution and, thus, the precision of results. Using the 3D grid approach, molecular field analysis is reduced to the calculation of a field value for each unoccupied grid voxel. The 3D grid approach finds various applications in computational chemistry; for example, the molecular volume can easily be calculated by multiplying the voxel volume by the number of occupied voxels (a voxel is considered occupied if there is at least an atom whose distance to the voxel center is less than the corresponding atomic radius).[7]

Molecular fields fall into two main classes. The first type is represented by MIFs, namely fields whose values are obtained by calculating the interaction energy between a target molecule and a probe located in each unoccupied grid cell. The second type involves fields obtained by projecting on unoccupied cells molecular properties for which atomic contributions have been parameterized.

MIFs are probably the most intuitive molecular fields. The probe can be monoatomic or polyatomic, being a real molecule (e.g., water), a fragment or an atom encoding specific interactions (thus, the neutral oxygen atom probe encodes the ability of a molecule to donate hydrogen bonds). The energy calculation[8] can describe total noncovalent energy or specific contributions – namely electrostatic (E_c), Van der Waals (E_{vdW}), and hydrogen bonding (E_{HB}):

$$E_{tot} = \sum_p \sum_m (E_c + E_{vdW} + E_{HB}) \qquad [1]$$

The first sum (p) concerns the probe, and the second (m) the molecule atoms.

Molecular fields derived from atomic contributions can be computed for all molecular properties for which it is possible to determine atomic increments (namely lipophilicity, polarizability, hydrogen bonding, and so on). The general formula to compute this type of molecular field is represented by

$$P_{xyz} = \sum_{i=1}^{i=n} a_i f(d_{xyz,i}) \qquad [2]$$

where P_{xyz} is the molecular field value for the point xyz, a_i is the local atomic contribution for atom i, and $f(d_{xyz,i})$ is a distance function between atom i and the field point xyz. The sum runs over the n atoms (or fragments) of the considered molecule.

Field visualization is useful to gain insight into the data and to highlight relations between values that may not be apparent from the raw data. The most obvious visualization is color: a set of colors is used to represent the range of values the data can assume. The colors should be chosen to make the values intuitive. A common technique is to use 'cold' colors such as blue for low values and 'hot' colors such as red for high values; green tones are used for mid-range values. Furthermore, the colored points cannot be represented as such because most would be masked, and relationships in the field values lost. A possible solution is to render the points transparent, increasing their strength from front to back, to render the obscured points more visible (the so-called alpha-blending technique).[9]

In fact, the commonly adopted solution to visualize molecular fields is the projection of field values on surfaces. These visualizations fall in two main classes. The first class involves the projection of field data onto a molecular surface, which yields a colored visualization of the local molecular properties (as seen later in **Figures 4** and **5**). But one must note that the surface itself can be considered a particular field where all grid points not on the surface are nil. A second method involves isosurfaces,[10] where grid cells with similar values are connected together by a tessellated surface. Each isosurface can have a different color to make it easier to distinguish, and the isosurfaces should be drawn partially transparent to render them clearer.

Existing MIF-based tools are introduced in the following sections in a progressive and logical manner. A selection of currently available MIF-based tools is presented in **Table 1**.

Table 1 A selection of currently available QSAR tools based on MIFs

Program/approach	Type	Internet resource	Ref.
CoMFA	3D-QSAR	www.tripos.com	30
CoMSIA	3D-QSAR	www.tripos.com	31
GRID	MIF calculation	www.moldiscovery.com	58
GOLPE	3D-QSAR	www.miasrl.com	60
VolSurf	3D-QSAR	www.moldiscovery.com	62
HINT	Lipophilic field	www.eslc.vabiotech.com/hint/	66
MLP	Lipophilic field	129.194.54.222/site_lct_gp/body.php	67
HYBOT	Hydrogen bond calculation	software.timtec.net/hybot-plus.htm	86
WHIM	3D-QSAR	www.talete.mi.it	98
FieldPrint and FieldScreen	Virtual screening	www.cresset-bmd.com	100
EVA	3D-QSAR	www.tripos.com	101
CoMMA	3D-QSAR	www.twcbiosearch.com/DOCS/1/QSARIS.htm	102
Catalyst	Pharmacophore	www.accelrys.com/products/catalyst/	104

5.24.2   Description of Molecular Form

5.24.2.1   Molecular Surfaces: Van der Waals, Solvent-Accessible, and Richards Surface

Although molecules do not have defined surfaces in a strict quantum mechanical sense, formal molecular surfaces have become important tools when analyzing both molecular shapes and local physicochemical properties, allowing fertile applications in rational drug design. Molecular surfaces are based on the modeling of atoms using hard spheres of defined radius, as suggested by the sharp rise of potential energy during the collision of two atoms. The molecular surface can be defined by the union of atomic surfaces for all atoms constituting a given molecule. Moreover, since the bond length is always less than the sum of atomic radii, the surface is modeled using fused spheres by allowing them to interpenetrate.[11,12] The molecular surface depends on the choice of atomic radii: for example, using the Van der Waals radii, one obtains a vdWS of fused spheres, which often represents a good approximation of steric requirements (**Figure 2a**).

Since in solution only accessible functional groups can interact with the solvent, it becomes important to assess which parts of a solute are indeed accessible. Despite its complexity, this problem can be approximated by rolling a sphere representing a molecule of solvent onto vdWS of the solute, thus obtaining a new smoothed surface defined by the centroid of the sphere, called the solvent accessible surface (SAS).[13] This is equivalent to increasing the atomic radii of vdWS by the radius of the solvent probe. A sphere of 1.4–1.5 Å radius is often taken to model a water molecule (**Figure 2b**).

A third type of surface representation tries to minimize the excessive extension of SAS by collecting points traced out by the lower face of the probe sphere. This surface is often referred to as a Richards surface or as a solvent excluded surface (SES).[14] It is the union of the 're-entrant' surface, generated by the underside of the probe when it is in contact with two or more spheres and the 'contact' surface that is the part of probe touching the vdWS (**Figure 2c**). While SAS and vdWS calculations are not very complicated, the SES calculation is a difficult and demanding problem.

A surface can be represented using either a collection of points (as seen later in **Figure 5**, a dot surface) or a solid rendering (as seen later in **Figure 4**, surface tessellation). In the second case, the surface is modeled into spherical polygons, permitting a more precise calculation of both the molecular surface and the enclosed volume.[15] Furthermore, the surface tessellation allows the calculation of surface derivatives with respect to the atomic coordinates.[16] The analytical derivatives can be used to evaluate better molecular shape and electrostatic interactions and find successful applications in the analysis of the solvent effects (implicit solvent continuum models), protein structure, enzymatic catalysis, and biomolecular recognition.[17]

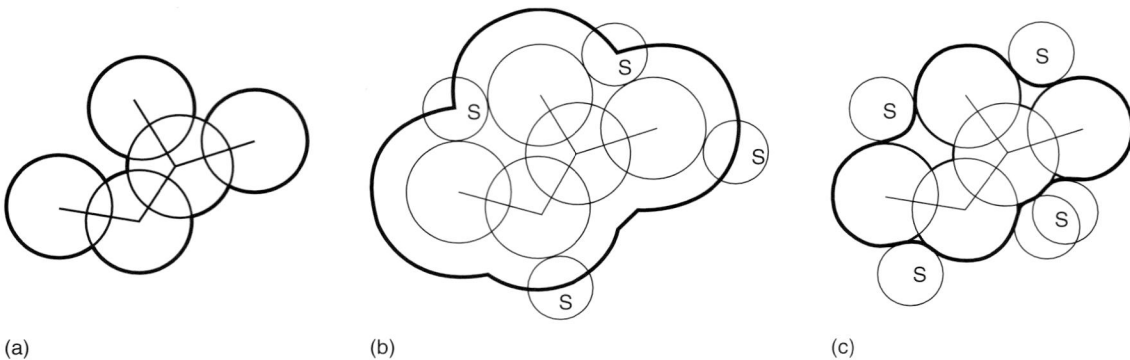

(a)   (b)   (c)

Figure 2 Examples of (a) a vdWS (Van der Waals surface), (b) an SAS (Connolly surface), and (c) an SES (Richards surface). (Reproduced from Pascual-Ahuir, J. L.; Silla, E.; Tunon, I. *J. Comput. Chem.* **1994**, *15*, 1127–1138, with permission from John Wiley & Sons, Inc.)

A precise alignment of molecules is necessary to compare molecular surfaces for similarity or complementarity.[18] This can be done manually by selecting the matching atoms, or automatically by calculating the rototranslational matrix (including three translation and three rotation values) to be used to align two or more rigid molecules. Several approaches to align molecules have been developed. Such methods include the matching of knobs and holes (which is especially useful in complementarity analysis through the computation of solid angles[19]), least squares methods (where a pattern of points is taken on molecular surfaces[20]), methods using surfaces with thickness,[21] and surface comparison by (1) projection onto a plane, (2) gnomonic projection (i.e., onto the surface of a sphere[22]), (3) projection onto a cylinder, and (4) Fourier analysis.[23]

5.24.2.2   The Polar Surface Area (PSA) and Derived Surfaces

The PSA is obtained by subtracting from the molecular surface the area of carbon atoms, halogens, and hydrogen atoms bonded to carbon atoms (i.e., nonpolar hydrogen atoms). In other words, the PSA is the surface associated with heteroatoms (namely oxygen, nitrogen, and phosphorous atoms) and polar hydrogen atoms.[24,25] Palm and co-workers[26] underlined that the polar area is sensitive to 3D conformation for a given molecule, and is thus better described using a weighted dynamic average, which considers all significant conformers, rather than by a single static PSA value. Ertl and co-workers[27] proposed an approach for the fast calculation of the PSA as the sum of fragment-based increments, which allows the rapid analysis of large data sets. An extensive study of PSA values highlights that the polar area correlates better with hydrogen bonding (both donor and acceptor groups) than with lipophilicity ($r^2 = 0.76$ versus $r^2 = 0.30$).[28]

Conversely, the apolar surface area (ASA, also called the nonpolar surface area) can be computed by considering only carbons, halogens, and nonpolar hydrogen atoms. ASA values are well related to experimentally observed thermodynamic changes in heat capacity and enthalpy for protein-folding processes, protein–solvent interactions, and ligand–receptor binding.[29]

5.24.2.3   Steric Fields

In the first 3D-QSAR approach (the CoMFA method[30]), the steric field was computed by means of the Lennard-Jones (LJ) 6–12 potential energy between the ligand and a Csp^3 probe. This field is characterized by a very sharp increase at short distances around the vdWS, thus requiring a cut-off value (usually $+125 \, kJ \, mol^{-1}$), above which it must be truncated. The hardness of the LJ potential makes CoMFA results very sensitive to the relative orientation and precise alignment of molecules. Moreover, this steric field also makes the CoMFA method very dependent on selected conformers, which are treated as rigid bodies despite the fact that both the ligand and receptor can co-adapt with low energy expenditure.

Different approaches have been developed to overcome these pitfalls. Some alternative steric fields, as used in comparative molecular similarity indices analysis (CoMSIA),[31] replace the 6–12 distance function of LJ energy with Gaussian-type functions, such as $\exp(-\alpha r^2)$ where the α term provides a smoothing effect. The steric fields so obtained do not require cut-off thresholds, and are less sensitive to conformation and superposition of ligands. Different steric fields are based on the analysis of similarity indices (e.g., the minimum topological difference method[32]) or Van der Waals intersection volumes.[33]

equations to evaluate the hydrogen bond interaction energy and accurate atomic parameterization from experimental hydrogen-bonding data. The conformer-dependent description of hydrogen bonding can be obtained using three main approaches: (1) quantum mechanical calculations, which are more rigorous but very time-consuming; (2) MIFs obtained by empirical force field calculations, which are quite useful when using precise equations for the hydrogen-bonding energy and suitable probes (e.g., the hydroxy group or a neutral oxygen atom); and (3) projection of atomic increments, mainly taken from solvatochromic analyses. All three methods can give a conformer-specific hydrogen-bonding description, and so are suitable in 3D-QSAR analyses. Furthermore, such methods allow discrimination between donor and acceptor ability, while other hydrogen bond descriptors such as the PSA or $\Delta \log P$ (namely the difference between $\log P$ values in octanol and cyclohexane) cannot distinguish between these two components.

Focusing on quantum mechanical approaches,[72] the donor ability can be modeled by using either the energy of the LUMO or the polarizability (or the superdelocalizability) of the heavy atom attached to the donor hydrogen. In contrast, the acceptor ability can be described by using either the energy of the HOMO or the MEP (see Section 5.24.3.1).

MIF calculations require an accurate equation to evaluate hydrogen-bonding contributions. Indeed, several force fields do not include specific terms for hydrogen bonds, but consider it to be implicit in the electrostatic and Van der Waals terms. Conversely, the master energy equation of GRID includes specific terms for hydrogen bonding, making this method useful to evaluate hydrogen bond interaction fields, by selecting suitable probes able to evaluate the hydrogen-bonding capacity (as seen in Section 5.24.3.1).

The third type of hydrogen-bonding field is based on an atomic parameterization that discriminates between donor and acceptor ability. Various methods exist to parameterize the hydrogen-bonding ability of molecules. The simpler descriptors (such as the number of potential hydrogen bond acceptor (HBA) and donor (HBD) groups or the sum of both[28]) are inadequate for atomic parameterization since they account neither for the strength of hydrogen bonds nor for the possibility of intramolecular hydrogen bonds. More sophisticated parameters based on solvatochromic analysis were developed by Taft and co-workers[73,74] in the 1970s, and can be considered as an application of the extrathermodynamic correlations (linear free energy relationships) elaborated by Hammett for chemical equilibria.[75] Abraham[76] and Raevsky[77] have developed quantitative scales of hydrogen-bonding acidity (α) and basicity (β) based on lipophilicity and thermodynamic data (see 5.31 In Silico Models to Predict Brain Uptake). The difference between two different $\log P$ scales is another experimental hydrogen-bonding descriptor first proposed by Seiler, who successfully correlated hydrogen bonding capacity with $\Delta \log P$ for 195 heterogeneous molecules.[78] El Tayar and co-workers[79] have developed the Λ descriptor obtained from experimental lipophilicity data and from calculated molecular volumes, to yield an indirect estimate of hydrogen-bonding properties.

On these grounds, hydrogen-bonding fields called MHBPs have recently been developed.[80] They consist of an HBD potential and an HBA potential. MHBPs use a fragmental system based on literature donor (α) and acceptor (β) values, and yield field values with an algorithm similar to that of MLPs. The value for the xyz point in a field (MHBP_{xyz}) is calculated by

$$\text{MHBP}_{xyz} = \sum_i \sum_j F_{ij}\, f(d_{j,xyz})\, g(U_{j,xyz}) \qquad [4]$$

where the first sum defines the number of molecular fragments in the compound, and the second sum the number of polar atoms in the molecular fragment; F_{ij} is the α and/or β value of atom j in fragment i, $f(d_{j,xyz})$ is the distance function between the polar atom j and the point xyz, and $g(U_{j,xyz})$ is the angular function of the angle defined by the point xyz, the polar atom j, and the polar hydrogen or the lone pair belonging to the polar atom j. It is worth noting that eqn [4] includes two geometric functions because hydrogen bonds are directional and the interaction energy decreases when moving away from the hydrogen bond axis.

5.24.3.4 Overview of Absorption, Distribution, Metabolism, and Excretion Correlations with Descriptors of Function

Although the relationships between pharmacokinetic parameters and physicochemical properties go beyond the objectives of this chapter, it appears useful here to describe some significant trends in these relations while avoiding a systematic analysis of such studies.[81]

The noteworthy role of electrostatic factors in molecular recognition finds some significant evidence also in ADME correlations, even if other factors such hydrophobicity also play a major role in pharmacokinetic processes. For example, recent studies have correlated the blood–brain permeation (assessed by log BB, see 5.31 In Silico Models to Predict Brain Uptake) with quantum chemical descriptors obtained from MEP surfaces.[46] Perhaps a main drawback of MEP analyses is that they require time-consuming quantum mechanical calculations and are therefore rarely applicable to large chemical libraries.

In contrast, it is well known that lipophilicity plays a key role in determining pharmacokinetic behavior[82]; the opportunity to compute virtual log P values from MLPs for flexible molecules has therefore found numerous applications in ADME studies. A classic example concerns the differing abilities of the two morphine glucuronides to cross the blood–brain barrier.[83] The MLP approach has demonstrated that these two metabolites have markedly different virtual log P values, existing in extended, more hydrophilic and folded, more lipophilic conformers. This indicates that the peculiar pharmacokinetic profiles can be attributed to the so-called chameleon effect (**Figure 6**).

Despite the significant role of log P in ADME processes, it is worth noting that predictive equations that successfully correlate lipophilicity with ADME parameters are not as numerous as one would expect.[84] This inherent difficulty can be explained considering that pharmacokinetic parameters result from a multiplicity of interactions with a variety of biological constituents that can be seen as fuzzy targets due to their fluctuating diversity. Moreover, a second difficulty is due to the heterogeneity of pharmacokinetic parameters, which, being obtained from complex biological systems, shows a decreased relatedness to molecular properties. These problems become apparent when examining correlations between lipophilicity and skin permeability; for example, it was possible to derive parabolic equations with a significant predictive power when considering separately a homogeneous set of phenols and of nonsteroidal anti-inflammatory agents, but no correlation was obtained for the set of all permeants.[79,85] Similar examples exist for membrane permeation, gastrointestinal absorption and blood–brain permeation. Overall, one can conclude that most correlations have a limited predictive capacity, yet can suggest mechanistic trends, and, as such, are of interest.

The relationships between hydrogen bonding and permeability processes have been validated on numerous occasions. First, as already outlined in Section 5.24.2.4, the PSA, a simple, but effective descriptor of hydrogen bonding, finds various successful correlations with pharmacokinetic parameters. Raevsky and co-workers used hydrogen bond thermodynamics descriptors to predict human absorption data through a nonlinear equation based on HBA and HBD

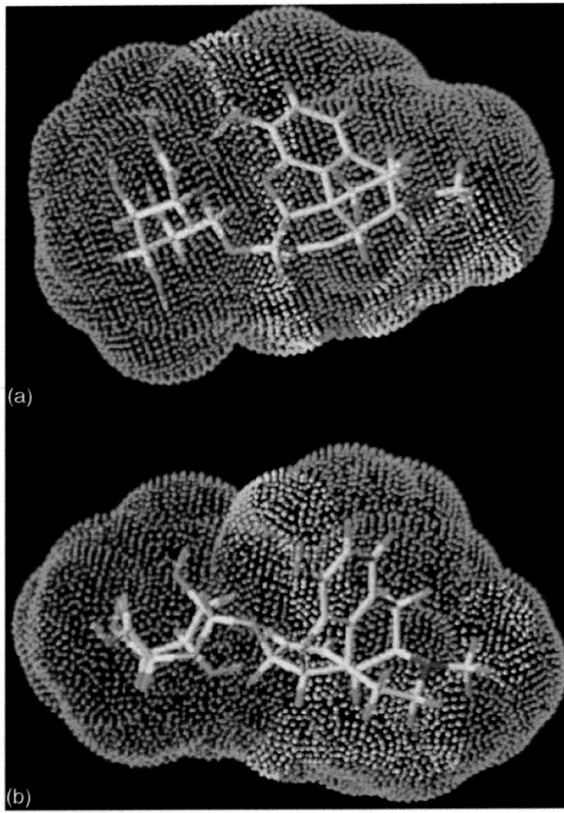

Figure 6 MLP of morphine 6-O-glucuronide (M6G) as a (a) folded neutral conformer and (b) as an extended neutral conformer. The MLP is displayed on the water-accessible surface area of the molecule. (Reproduced from Testa, B.; Kier, L.; Carrupt, P. A. *Med. Res. Rev.* **1997**, *17*, 303–326, with permission from John Wiley & Sons, Inc.)

thermodynamic potentials.[86] Moreover, when analyzing the correlations between ADME and experimental hydrogen-bonding descriptors, it is worth remembering the pioneering study of Young and co-workers, who[87] in 1988 showed that the difference between log P in octanol and in cyclohexane is inversely related to brain penetration for a set of centrally acting histamine H_2 antagonists.

5.24.4   Molecular Fields and Three-Dimensional Quantitative Structure–Activity Relationship

5.24.4.1   Methodological Bases: CoMFA and Derived Methods

In essence, 3D-QSAR methods correlate differences in conformer-dependent properties with biological data. This section will be focused on 3D-QSAR approaches involving molecular field calculations. To compute 3D fields, these approaches require the manual or automatic selection of biologically relevant conformations for all modeled compounds, followed by the manual or automatic superimposition of all selected conformers. These first steps represent a problematic issue because the bioactive conformation is seldom known experimentally. In the absence of such information, it is very useful to assume that, when ligands bind to a target, their steric and physicochemical patterns will be such as to maximize the fit. This is the pharmacophore idea, which defines the basic 3D arrangement of functional groups responsible for biological activity. Nevertheless, one must remember that this is a static concept since it does not consider ligand flexibility nor multiple binding modes to the target.

All 3D-QSAR approaches are based on assumptions such as (1) a unique bioactive conformation is responsible for the biological effect; (2) the binding site is exactly the same for all ligands considered; (3) the affinity is mainly due to enthalpic factors, while the entropic ones are constant and negligible; and (4) other aspects such as flexibility, solvent effects, and kinetic factors can be ignored.

From a historical point of view, 3D-QSAR approaches were born in 1986, when Wold and co-workers proposed use of the partial least squares (PLS) analyses to correlate field values with biological properties.[88,89] They found their precise definition in 1988 with CoMFA,[31] whose diffusion in the scientific community was accompanied by the availability of patented commercial software (**Figure 7**).

Clearly, the improvement of statistical techniques has played a relevant role in 3D-QSAR developments.[90] In particular, PLS generalizes and combines features from PCA and multiple regression, reducing the large amount of field data to a few significant orthogonal descriptors (also called latent variables or principal components).[91] The correct number of descriptors is chosen for the highest statistical significance. It is worth underlining that r^2 values are not very indicative of significance in 3D-QSARs due to the large number of independent variables. It is therefore necessary to compute cross-validated r^2 values (also written as q^2 or r^2_{CV}, an estimate of model predictivity) by iteratively eliminating one or more compounds from the data set, deriving a model equation with the remaining set, and predicting the biological value of the omitted compounds. For models with similar cross-validated r^2 values, the model with the smallest standard error of prediction (PRESS) is preferred.

In the CoMFA method, the biological responses are predicted using steric (Lennard-Jones, as seen in Section 5.24.2.3) and electrostatic field values computed for each molecule in the set. Attempts have been made to include other properties in CoMFA; for example, lipophilicity fields (HINT and MLP approaches, see Section 5.24.3.2), MIFs (e.g., as computed with the GRID program, see Section 5.24.3.1), hydrogen-bonding fields (see Section 5.24.3.3), or reactivity fields obtained by quantum mechanical calculations (e.g., HOMO/LUMO fields, see Section 5.24.3.1).

The results of CoMFA correspond to a regression analysis with thousands of coefficients. The common way to present the results is by means of colored contour maps, showing favorable or unfavorable steric and electrostatic regions around the molecules. In detail, green contours indicate regions where an increase in steric bulk will enhance activity, the opposite being the case for yellow contours. For electrostatic fields, blue and red contours usually correspond to regions where an increase of positive or negative charge, respectively, will enhance affinity (as seen later in **Figure 9**). The results can be used in lead optimization by designing new derivatives based on contour maps or calculating the corresponding fields and incorporating them in predictive models.

Despite the considerable success of CoMFA, some serious problems can arise during analysis and when interpreting results. The most obvious problem concerns the choice of bioactive conformations and the rules for mutual alignment.[92] The comparison can be based on either atom matching or property similarity, but the analysis of highly flexible compounds is always quite difficult and time-consuming. When at least one of the considered molecules is conformationally rigid, the choice of bioactive conformers and the alignment rules are easier, as other compounds have to mimic the geometry of the rigid molecule. The first step can be facilitated if at least one experimentally resolved ligand–target complex can be used as a template, and any source of error can be eliminated when such a complex is

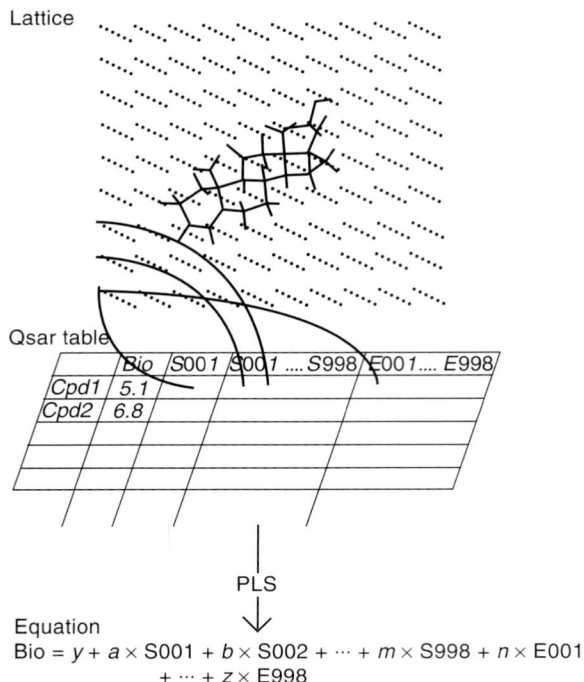

Figure 7 Historical diagram from the pioneering work of Cramer and co-workers, depicting the main logical steps of the CoMFA approach. The field values obtained inserting molecules in 3D grids (lattice) are used as independent variables to derive predictive models through a PLS technique. (Reprinted from Cramer, R. D.; Patterson, D. E.; Bunce, J. D. *J. Am. Chem. Soc.* **1988**, *110*, 5959–5967 © American Chemical Society.)

known for all molecules; but the question arises of whether CoMFA is the best choice in such cases. A simple iterative approach is to refine CoMFA models, modifying the conformation and/or alignment of molecules that do not fit the model, using preliminary results as a guide to improve the model.

Another source of errors is the hardness of field values, especially the steric ones, which are truncated when exceeding cut-off thresholds. Consequently, contour maps become fragmented and often difficult to interpret. As seen in Section 5.24.2.3, some attempts have been made to resolve this problem by using Gaussian-type distance functions that smooth the field values and do not require cut-off.

Finally, CoMFA results show a significant dependence on the overall orientation of ligands in the grid box.[93] Several studies on the stability of CoMFA have shown that the quality of models can be increased by displacing the aligned molecules in the grid box. For example, the systematic rotation of ligands in the box changed CoMFA q^2 values by as much as 0.5.[94] This can happen when the fields are computed on such a coarse grid that the field values are inadequately distributed. As a rule, a decrease in grid spacing increased the probability of placing the probe atom in a region where the steric and electrostatic fields best correlated with biological activity. Finally, these studies suggested that users have to pay attention to box properties when generating the field values, even if the choice of CoMFA parameters based on the highest q^2 is questionable, as it can bias the results.

Considering these pitfalls, it comes as no surprise that several alternative 3D-QSAR approaches have been proposed in recent years.[95,96] One of the most important is CoMSIA, proposed by Klebe and co-workers.[31] It uses a PLS technique to analyze fields of different physicochemical properties based on similarity indices of aligned molecules. The fields use a Gaussian-type distance function not requiring a cut-off. The results are statistically significant, and the maps obtained are not fragmented, highlighting regions within the area occupied by ligands that require a particular physicochemical property for activity (as illustrated later in **Figure 10**). Alternative 3D-QSAR approaches give up the use of grid calculation and project the computed property fields onto molecular surfaces. The use of molecular surfaces should simplify calculations, making the results more realistic, as the interactions between the ligand and the receptor occur near the molecular surface. These nongrid methods (e.g., comparative molecular surface analysis, CoMSA),[97] exploit other statistical techniques such as neural networks or genetic algorithms to improve the statistical goodness of results.

A third type of 3D-QSAR approach tries to resolve the problem of alignment by calculating coordinate-invariant 3D descriptors. For example, the weighted holistic invariant molecular descriptor (WHIM) approach[98] calculates property values using PCA. This procedure assures invariance to translation and to rotation, and molecules do not need to be aligned before computing WHIM parameters. A second method to compute 3D descriptors that does not require alignment of molecules is the VolSurf approach, which has found several applications in ADME predictions (*see* Section 5.24.3.1).[62] Stiefl and Baumann[99] recently reported a new approach to map property distribution on molecular surfaces based on radial distribution functions (distance-dependent count statistics). The calculated descriptors are invariant to roto-translation, and are well correlated with biological data through chemometric regression techniques in combination with a variable selection to identify highly relevant variables. Vinter and co-workers[100] proposed the FieldScreen approach, which reduces molecular fields to just a few key field points, obtaining a field pattern that represents the ligand from the viewpoint of the protein. This method, being alignment-independent, markedly simplifies field treatment, and is well suited for database screening. Other alignment-independent methods have been proposed recently (e.g., eigenvalues analysis (EVA) descriptors derived from the vibrational frequencies, which are calculated or extracted from experimental infrared spectra,[101] and comparative molecular moment analysis (CoMMA), based on the moments of the mass and charge distribution.[102] However, it is worth remembering that these methods are conformation-dependent. Since the best way to find similar bioactive conformations in a set of molecules is to superimpose them, these methods do not really escape the alignment problem.

Finally, methods based on pharmacophore mapping, even if they do not involve field analysis, can be considered as 3D-QSAR approaches.[103] For example, the program Catalyst analyzes the position in space of chemical functions (namely hydrogen-bonding groups, polar functions, and hydrophobic moieties) of molecules, and generates pharmacophore hypotheses in terms of the 3D arrangement of chemical functions to explain the differences in activity. The activity of a new compound can be quantitatively predicted by means of its ability to fit the pharmacophore model.[104] Clearly, pharmacophore analysis and CoMFA-like approaches are strictly related; indeed, pharmacophore hypotheses can be used to begin 3D-QSAR analyses, by suggesting alignment rules, whereas 3D-QSAR results can be used to validate and explain pharmacophore hypotheses.

5.24.4.2 Applications of Three-Dimensional Quantitative Structure–Activity Relationship Approaches to Absorption, Distribution, Metabolism, and Excretion Predictions

Since 3D-QSAR approaches can account for specific interactions between ligands and a target, and suggest pharmacophore hypotheses, they have found several applications in metabolism prediction. Analogously, 3D-QSAR has been applied extensively to predict the interactions between ligands and transporters, since such targets, like enzymes, are biomacromolecules. Conversely, the prediction of permeation processes finds rarer applications in 3D-QSAR, as the behavior of fuzzy targets is less suitable for these computational methods. This section will consider, in turn, drug-metabolizing enzymes (Section 5.24.4.2.1), transporter proteins (Section 5.24.4.2.2), and permeation processes (Section 5.24.4.2.3).[105]

5.24.4.2.1 Applications of three-dimensional quantitative structure–activity relationship approaches to drug-metabolizing enzymes

5.24.4.2.1.1 Oxidoreductase

The cytochrome P450 enzymes (CYPs, EC 1.14.14.1) are membrane-bound proteins that catalyze oxidative reactions of xenobiotics and endobiotics with broad substrate specificity, playing a key role in drug metabolism. Briefly, CYP catalysis involves reduction of the heme iron, with resulting cleavage of the oxygen–oxygen bond. The electrons are donated by reductases, creating an 'electron transfer chain.'[106,107]

The CYP proteins are classified into families and subfamilies according to their sequence homology. More than 3000 CYP gene sequences are known. The first three families (CYP1, CYP2, and CYP3) are enzymes involved in drug metabolism, being responsible for about 90% of drug oxidations in humans, while other families play specific roles (e.g., CYP11, CYP17, CYP19, and CYP21 are implicated in steroid biosynthesis).

The 3D-QSAR techniques are suitable for studying CYP-catalyzed oxidations as they can analyze how ligands may bind to these enzymes as substrates and/or inhibitors. It is worth noting that in these studies, 3D-QSAR approaches are mainly used to build pharmacophore models rather than to yield predictive equations of limited interest, due to the complexity of these enzymes and the heterogeneity of their substrates. The soundness of the results was often confirmed by other computational approaches (e.g., homology modeling and molecular docking) and by experimental studies (x-ray crystallography and/or site-directed mutagenesis). This section will consider some relevant examples of 3D-QSAR

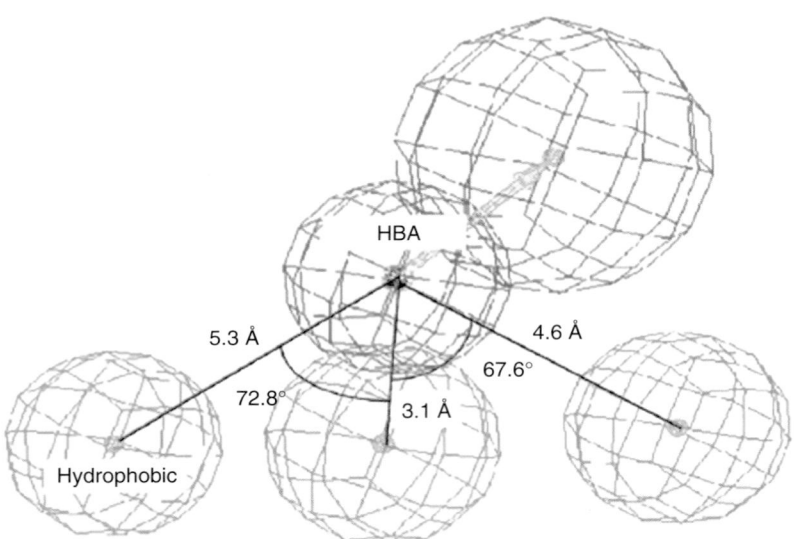

Figure 8 The CYP2B6 substrate pharmacophore illustrating three hydrophobic areas (cyan) and a HBA feature (green), with a vector in the direction of the putative hydrogen bond. (Reproduced from Ekins, S.; Bravi, G.; Ring, B. J.; Gillespie, T. A.; Gillespie, J. S.; VandenBranden, M.; Wrighton, S. A.; Wikel, J. H. *J. Pharmacol. Exp. Ther.* **1999**, *288*, 21–29, with permission from the American Society for Pharmacology and Experimental Therapeutics.)

studies on CYP ligands. The relevance of these studies lies in the major role of CYP enzymes in human metabolism; understanding the basis of their activity is important to determine their role and predict effects on new substrates.[108,109]

Extensive analyses have been performed on CYP2B6 substrates using two 3D-QSAR approaches, namely WHIM and pharmacophore mapping.[110] Both methods suggest the crucial relevance of three hydrophobic regions and one HBA, located at defined distances (**Figure 8**). These results are in agreement with classic QSAR studies that correlated binding affinities to CYP2B6 with log *P* and hydrogen-bonding descriptors.[111] It is interesting to observe that both 3D-QSAR methods, even if conceptually different, yielded very similar results, suggesting some degree of mutual validation, although both methods failed to predict molecules not included in the training set. Docking analyses on homology-modeled CYP2B6 revealed that its binding site consists of three well-defined hydrophobic binding pockets adjacent to the catalytic heme. The size, shape, and position of these hydrophobic pockets are in agreement with pharmacophore models.[112,113]

The inability to predict CYP activity for substrates not included in training sets is explained by considering the broad specificity of these enzymes, which renders difficult or impossible the creation of a unique pharmacophore hypothesis. For example, the CYP2C family can bind both neutral and anionic ligands, and different models were developed to accounts for these two classes of molecules. All models show a hydrophobic/aromatic region between the hydroxylation site and an electron-rich moiety, which interacts with the heme iron. Extensive analyses of CYP2C9 substrates show that the highest affinity is shown by molecules having two negative regions equally separated from a central aromatic group.[114] Homology models of CYP2C9 have indicated that Phe110 and Phe114 may be involved in aromatic interactions,[115] while mutagenesis experiments have suggested that Arg97 and Arg108 may constitute the cationic binding sites.[116] The key roles of aromatic and electrostatic interactions are also confirmed by classic QSAR analyses, which correlate binding affinities to CYP2C9 with pK_a and log *P* values.[111] A CoMFA analysis of a set of hydantoin and barbiturate substrates revealed that electrostatic features seem unimportant, while steric features and lipophilic properties are well correlated with binding data for these enzymes.[117] This does not imply that electrostatic features of the hydantoin and barbiturate rings are not important in binding, simply that these features are not significant in describing the differences in binding of these compounds.

CYP2D6 is a polymorphic CYP member well known for its absence in 5–9% of the Caucasian population. 3D-QSAR studies on CYP2D6 substrates showed that these molecules are characterized by a basic group at about 5–7 Å from the hydroxylation site, and coplanar aromatic rings.[118] These observations are strongly confirmed by homology modeling and mutagenesis experiments, which showed that two acidic residues (Glu216 and Asp301) interact with the basic group in the ligand.[119] The key role of a basic group was also confirmed by classic QSAR analyses, which correlated binding affinities to CYP2D6 with pK_a values.[111] Comparable analyses with CYP2D6 inhibitors underlined the major role of hydrogen bonding (as well as features already described for substrates) in accounting for differences in binding affinity.

The complexity and the very broad specificity of CYP enzymes find their greatest expression in the CYP3A subfamily, especially in CYP3A4.[120] Indeed, many data suggest two or more binding modes due to the presence of at least two binding subpockets and one 'effector-binding' region. A more recent hypothesis proposes that the CYP3A4 can exist in multiple conformations regulated by allosteric effects.[121] The CYP3A4 promiscuity has produced a number of theoretical models to predict its enzymatic activity, even if the predictions are often of modest statistical quality. To date, all QSAR analyses indicate the relevance of hydrophobic interactions and at least one hydrogen bond, and the crystal structure of CYP3A4 revealed the main residues involved in these interactions.[122] Docking analyses confirmed the key role of multiple hydrophobic domains in the active site of CYP3A4.[123]

These few examples indicate the increasing interest in modeling human CYPs using 3D-QSAR analyses. Despite their modest ability to predict the enzymatic parameters for heterogeneous molecules, such models can predict qualitatively the ability of molecules to bind to CYPs. As such, they can be useful tools even if they fail to distinguish substrates from inhibitors. Furthermore, the increasing availability of crystal structures or homology models of relevant CYPs should facilitate docking analysis on these enzymatic targets.

To date, 3D-QSAR analyses have accounted for inhibitors and substrates, while the ability of molecules to induce CYPs are just beginning to be evaluated by such methods, due to the lack of sufficiently broad experimental data sets.[124]

Among the other oxidative enzymes involved in metabolism, monoamine oxidases (MAOs, EC 1.4.3.4), particularly MAO-B, have attracted great interest since their inhibitors play diverse roles in the pharmacological management of some neurological disorders. Recent studies applied CoMFA and GOLPE procedures to a large data set of MAO inhibitors, revealing that hydrophobicity and steric hindrance are the main factors involved in binding. Interestingly, docking studies on the crystal structure of human MAO-B have confirmed the CoMFA results and highlighted the role, essential for MAO-B activity and selectivity, of a hydrophobic cavity (the so-called entrance cavity) connecting the surface of the protein to the catalytic cavity.[125] Other studies devoted to specific series of compounds (namely coumarins[126] and indolylmethylamine[127] derivatives) have led to similar results, and underlined the role of lipophilicity in both the affinity and selectivity of such inhibitors.

5.24.4.2.1.2 Transferases

UDP-glucoronosyltransferases (UGTs) are among the most important conjugating enzymes involved in drug-metabolism. Like CYPs, UGTs exist in multiple enzyme isoforms, differing in substrate specificity.[128]

The first studies applied 3D-QSAR techniques to inhibitor and substrates of rat and human UDPs.[129] These studies highlighted the key role of electrostatic factors in inhibition and metabolism. The results are in agreement with those obtained with classic QSAR analyses, which correlated the rate of glucuronidation with the log P and pK_a values of ligands.[130] The log P term may reflect diffusion to the active site, and pK_a the apparent lipophilicity.

More recent studies have applied the pharmacophore analysis to specific UGT isoforms. In particular, the enzymes in the UGT1A subfamily share common pharmacophore features, with the site of glucuronidation invariably adjacent to a hydrophobic region, and another hydrophobic domain located 6–8 Å from the site of conjugation.[131] Unfortunately, these pharmacophore models are useless to predict rates of glucuronidation, perhaps due to multiple binding modes. Indeed, a more comprehensive study, including substrates of 12 human UGTs and multiple pharmacophore models, has generated good predictability for all considered isoforms.[132]

A second class of conjugating enzymes that attracted great interest are the catechol O-methyltransferases (COMTs). These enzymes catalyze the methylation of various catechol derivatives such as catecholamines, catechol estrogens, and their metabolites, and several drug metabolites. There are two forms of COMT coded by a single gene: the soluble, cytosolic form (s-COMT) and the membrane-bound form (mb-COMT), found in the rough endoplasmic reticulum. COMT inhibitors are used as drugs in the treatment of Parkinson's disease. CoMFA analyses[133,134] have revealed the key role played by the electrostatic field in predicting the enzyme kinetic parameters of s-COMT. These studies underline how CoMFA results are sensitive to the method used to calculate atomic charge calculation. Semi-empirical charge calculations performed clearly better than fully empirical ones. The CoMFA results are in agreement with docking analyses of the s-COMT crystal structure, which have revealed the interaction between the catechol oxygen atoms and a Mg^{2+} ion involved in catalysis (**Figure 9**).

Catecholamine sulfotransferase (SULT1A3) is also involved in catechol metabolism. An extensive CoMFA study with 95 substrates has revealed the remarkable role that the electronic and steric fields play in enzyme interaction, as also confirmed by docking analysis.[135] In particular, CoMFA results revealed the importance of an electrostatic interaction near Glu146 in SULT1A3, plus the fact that bulky substituents *para* to the reacting hydroxy group are unfavorable.

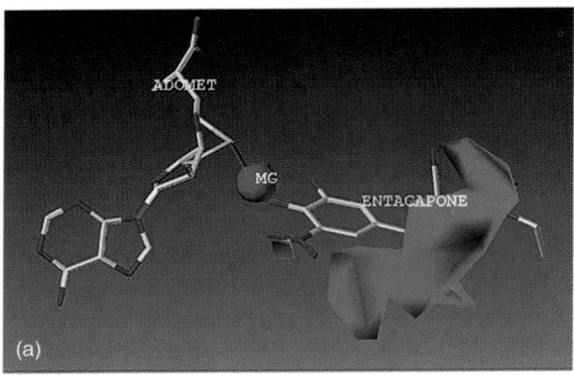

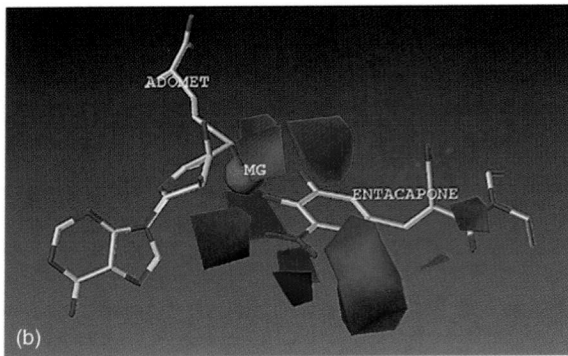

Figure 9 Contour maps of steric (a) and electronic (b) interactions in a CoMFA model obtained to predict the V_{max} values of s-COMT ligands. The maps are displayed with the inhibitor entacapone, Mg^{2+}, and the cosubstrate AdoMet. Green and yellow denote favorable and unfavorable regions for bulky substituents, respectively. The positive charge is favored near the blue regions, and the negative charge near the red regions. (Reprinted from Sipila, J.; Taskinen, J. *J. Chem. Inf. Comput. Sci.* **2004**, *44*, 97–104 © American Chemical Society.)

5.24.4.2.2   Applications of three-dimensional quantitative structure–activity relationship to transporters

Among known transporter proteins, monoamine transporters such as the dopamine (DAT) and the serotonin transporter (SERT) have been extensively studied. It has been shown that cocaine and other drugs of abuse can bind to monoamine transporters. A number of 3D-QSAR studies were performed to uncover the details of their binding site. CoMFA studies based on phenyltropane derivatives[136] have suggested that an increased negative electrostatic potential near the 3β substituent of the tropane ring and the *para* position of the phenyl ring favored inhibition of the monoamine transporters. Recent studies on aryltropanes and piperidinols suggested that DAT and SERT have a large cavity that can accommodate bulky C-2 substituents of tropane, and that the size of substituents in the *para* position in both phenyl rings of piperidinols is important for the inhibition of dopamine reuptake.[137] The CoMFA model of Wright and co-workers[138] indicates that hydrophobic interactions make a dominant contribution to binding. Newman and co-workers[139] also studied *N*-substituted tropanes, concluding that the steric interaction of the *N*-substituent with the DAT is a pivotal factor for the binding affinity. An extensive analysis of piperidine-based analogs of cocaine suggested two possible binding modes at the DAT, in both of which steric and electrostatic fields play important roles.[140]

P-glycoprotein (P-gp), which is encoded by highly conserved MDR (multidrug resistance) genes, is a member of the ATP-binding cassette superfamily of membrane transporters. P-gp transports a wide variety of xenobiotic and cytotoxic endogenous chemical agents out of the cell at the expense of ATP hydrolysis. The effect of P-gp-mediated drug efflux on limiting intestinal absorption, oral bioavailability, and tissue distribution has important implications for the efficacy of pharmacological therapies. 3D-QSAR studies have highlighted the crucial role of hydrophobic interactions and hydrogen bonds to promote the P-gp ATPase activity. Pharmacophore analyses have identified two electron donor groups, 2.5 Å or 4.6 Å apart.[141] Multiple pharmacophore models showed a promising capacity to predict

P-gp substrates.[142] Several 3D-QSAR models have been developed for propaphenone-type molecules[143] and phenothiazines.[144] Extensive analyses of steroid derivatives that act as substrates and inhibitors on P-gp allowed to recognize the basic differences between them. Electrostatic factors are more important for substrates, while strong hydrophobicity is more essential for inhibitors. The steric field appears crucial in both classes, but with different requirements; bulky substituents surrounding C-6 are not tolerated for substrates, while bulky groups around C-3 decrease inhibitory activity.[145] A recent 3D-QSAR study developed a pharmacophore hypothesis for P-gp substrates. The model includes two hydrophobic areas and two HBA groups, and the molecular size plays a major role in the interaction.[146] A more extensive treatment of transporters and their computational approaches can be found elsewhere (see 5.04 The Biology and Function of Transporters; 5.32 In Silico Models for Interactions with Transporters, respectively).

Proton-coupled peptide transporters, which are localized at brush border membranes of intestinal and renal epithelial cells, play important roles among the intestinal carriers in peptide absorption. These transporters also have significant pharmacological and pharmacokinetic relevance to the transport of various peptide-like drugs such as beta-lactam antibiotics.[147] CoMFA/CoMSIA analyses of small intestinal peptide carrier (oligo)peptide transporter 1 (PEPT1) substrates indicated that six structural elements favor the binding to PEPT1, namely (1) the presence of bulky side chains (**Figure 10**), (2) a positively charged N-terminus and a region of high-electron density at the C-terminus, (3) two hydrophilic regions, (4) a preferred hydrophobic region at the C-terminus, (5) an HBD region at the N-terminus, and (6) an HBA region crucial for differentiation between stereoisomers.[148]

A second relevant intestinal carrier is the apical sodium-dependent bile acid transporter (ASBT), which plays a pivotal role in cholesterol transport and represents an interesting target for drug delivery, enhancing the intestinal absorption of poorly absorbable drugs.[149] Pharmacophore mapping revealed the features essential for ASBT affinity, namely one HBD, one HBA, and three hydrophobic moieties suitably arranged. In natural bile acids, ring D, methyl-18, and methyl-21 represent the hydrophobic moieties, the hydroxy group in position 7 or 12 represents the HBD, and the HBA is afforded by the negatively charged side chain. The ASBT pharmacophore hypothesis is in good agreement with the 3D-QSAR models developed using a series of inhibitors and substrates. These models should facilitate the rational design of molecules targeted to the ASBT.[150]

5.24.4.2.3 Applications of three-dimensional quantitative structure–activity relationship approaches to permeation processes

The MOLSURF approach has found several applications in permeation modeling.[151] It is based on surface projection of electronic properties derived from semi-empirical or ab initio methods. Concerning the prediction of blood–brain distribution, Norinder and co-workers[152] used a strategy combining MOLSURF and PLS, and obtained a good predictive model ($q^2 = 0.78$). The most important factors influencing the models were associated with polarity and Lewis base strength. Properties associated with hydrogen bonding were also found to be important, and should be kept to a minimum to facilitate high brain–blood partitioning. High lipophilicity was also identified as favorable for high partitioning. Furthermore, polarizable electrons are also beneficial for the compounds to penetrate the brain, suggesting the positive role of charge–transfer interactions. Similar results were obtained using MOLSURF for intestinal absorption[153] and Caco-2 cell permeability.[154]

To account for the high heterogeneity of their compounds, Wolohan and Clark[155] recently combined field calculations with soft independent modeling of class analogy to predict pharmacokinetic properties. This approach involves molecular alignment using idiotropic field orientation (IFO), a generalization of inertial field orientation. The method incorporates electrostatic and the other molecular fields directly into alignment. Models were presented for human intestinal absorption, blood–brain permeation, and oral bioavailability, showing that this tool can be useful to predict pharmacokinetic behavior and explore the influence of orientation and molecular fields.

Several models have been built using the VolSurf approach. Crivori and co-workers[156] reported a qualitative analysis of blood–brain permeation considering 229 compounds and 72 VolSurf descriptors, analyzed by PLS method. The obtained model predicts the correct blood–brain permeation for over 90% of the compounds (**Figure 11**). The most important descriptors were the polar ones, which were inversely correlated with permeation. Overall, blood–brain permeation seems governed by a balance of all descriptors rather than a single descriptor type.

Ekins and co-workers[157] compared different 3D-QSAR approaches (namely CoMFA, VolSurf, molecular surface (MS)-WHIM, and Catalyst) to predict Caco-2 permeability using a set of 2-aminobenzimidazole derivatives. The study showed that VolSurf and Catalyst yielded better models, while MS-WHIM and CoMFA were less predictive. This contrasts with previous studies where MS-WHIM was used to predict Caco-2 permeability of more structurally diverse compounds. These observations suggest the value of assessing multiple approaches for permeability prediction, as it

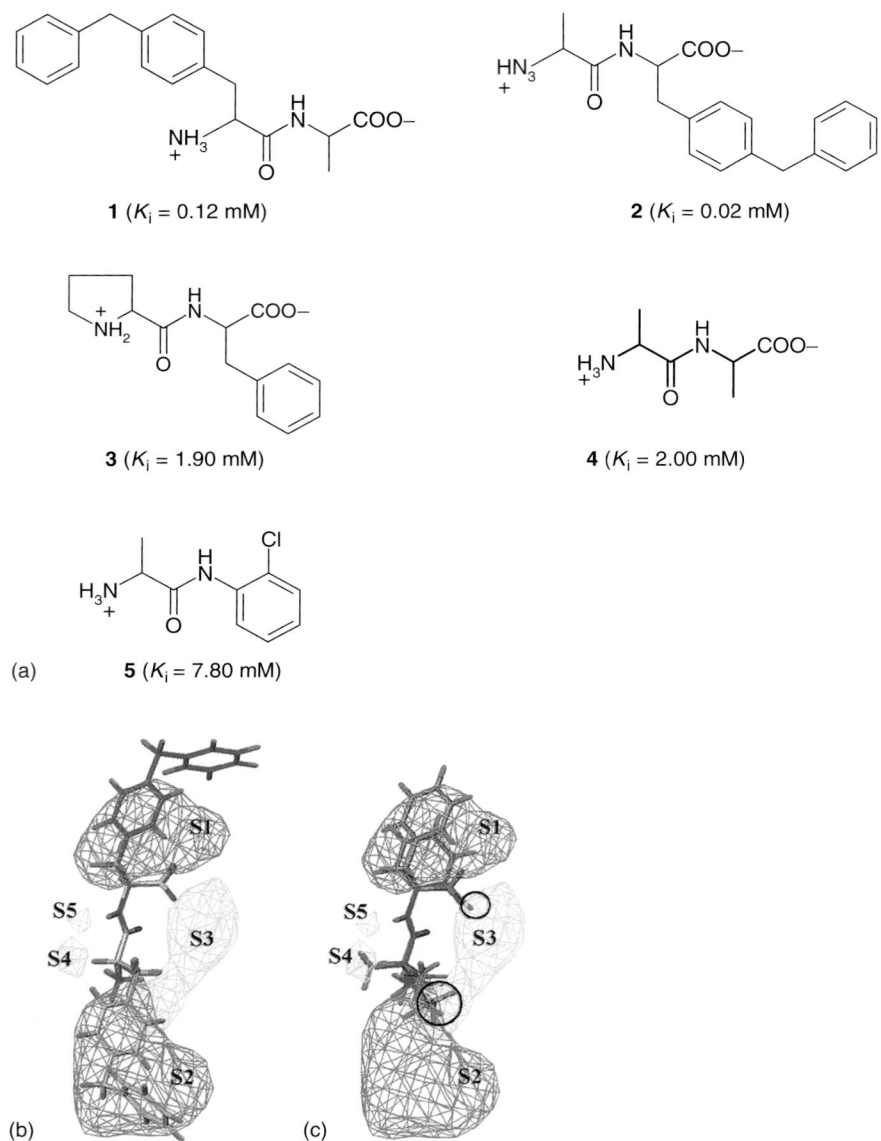

Figure 10 Contour plots for steric properties in a CoMSIA model obtained from an analysis of peptide substrates of the mammalian H^+/peptide co-transporter PEPT1. Green regions indicate regions where bulky groups enhance affinity. Yellow regions should be kept unoccupied. (a) Molecular structure of ligands displayed in parts (b) (**1** and **2**) and (c) (**3**, **4**, and **5**). (b) High-affinity substrates **1** (gray) and **2** (purple). (c) Low-affinity substrates **3** (orange), **4** (green), and **5** (purple). Circles in S3 denote substituents in disfavored regions. (Reprinted from Gebauer, S.; Knutter, I.; Hartrodt, B.; Brandsch, M.; Neubert, K.; Thondorf, I. *J. Med. Chem.* **2003**, *46*, 5725–5734 © American Chemical Society.)

appears that some methods do better than others for particular data sets. Moreover, the CoMFA model was not improved using the automatic conformer alignment in Catalyst, indicating that the modest CoMFA permeability prediction was due to lack of efficacy of molecular fields, rather than to an uncorrected alignment. Finally, the study demonstrated that 3D-QSAR techniques have considerable value in predicting passive permeability, at least for congeneric series of derivatives, and represent a valuable asset in drug discovery.

The recently proposed MHBPs (as outlined in Section 5.24.3.3) have found significant applications in ADME predictions.[80] One study showed that the hydrogen-bonding donor capacity ($MHBP_{do}$), but not the acceptor capacity ($MHBP_{acc}$), appears able to predict oral absorption. This is in agreement with the ratio of HBD and HBA groups in the 'rule-of-five.'

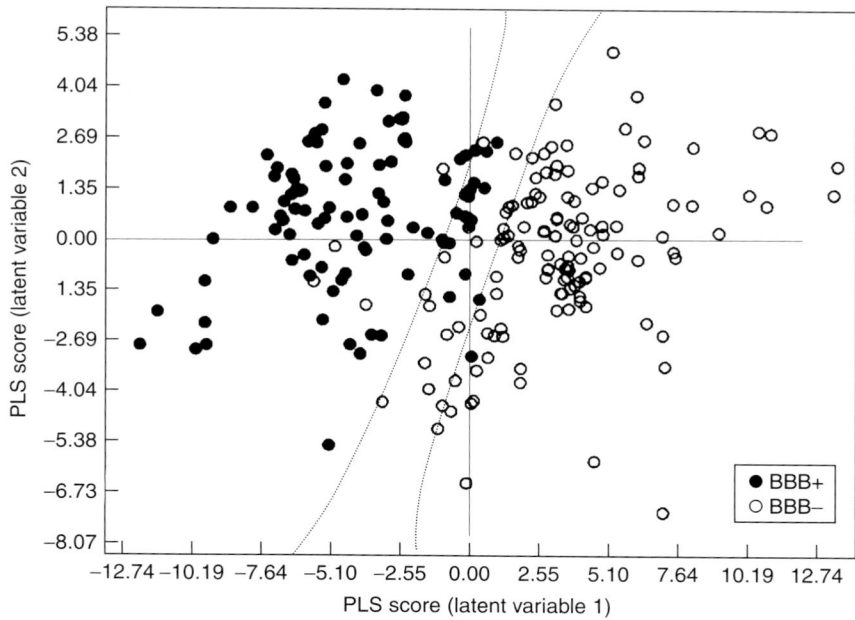

Figure 11 A discriminant PLS score plot for a model predicting blood–brain barrier penetration (BBB). The model offers good discrimination between the permeants (BBB+) and nonpermeants (BBB–), since it assigned a correct BBB profile to >90% of the compounds in the data set. (Reprinted from Crivori, P.; Cruciani, G.; Carrupt, P. A.; Testa, B. *J. Med. Chem.* **2000**, *43*, 2204–2216 © American Chemical Society.)

5.24.5 Description of Molecular Fluctuation (Four-Dimensional Quantitative Structure–Activity Relationship)

5.24.5.1 Approaches Based on Multidimensional Quantitative Structure–Activity Relationship

One of the main drawbacks of 3D-QSAR approaches is their inability to account for molecular flexibility, restricting the range of compounds that can be tested and introducing assumptions concerning the choice of alignment rules and hypothetical bioactive conformations. Hence, Hopfinger and co-workers[158] proposed a 4D-QSAR analysis formalism based on a Boltzmann sampling of the conformations available to each molecule. This approach aligns the molecules by combining the conformers obtained, and uses PLS techniques as in COMFA to analyze for each alignment the grid cell occupancy values taken as molecular descriptors. A general finding of this method is that a relatively small number of grid cell descriptors is significant in the relationships, providing compact models that are easy to explore and very informative. Finally, the evaluation of all models permits the analysis of the results without respect to alignment and conformational sampling, suggesting a hypothetical bioactive conformation for each ligand. This formalism was also exploited to calculate alignment-independent molecular similarity indices.[159] A recent comparison between such a 4D-QSAR method and a classic CoMFA analysis on a set of ecdysteroids showed that the two best 4D models have the smallest average residuals of prediction and the best cross-validated correlation coefficients.[160]

A second type of multidimensional QSAR approach considers the receptor-to-ligand adaptation, which can influence biological response, especially when this effect is a significant consequence of the ligand-induced triggering mechanism. Vedani and co-workers[161] proposed the Quasar approach, which accounts for induced fit by mapping a 'mean envelope' (which comprises the surfaces of all ligands) onto an 'inner envelope' (which accommodates the individual ligand). The energetic expenditure associated with this adaptation can be estimated from the mean envelope to individual envelope root mean square shift. Furthermore, such an approach accounts for unknown receptor flexibility by considering several protocols, to adapt the mean envelope to the individual envelope. This variability is measured using an induced-fit varying from 0.0 (no fit) to 1.0 (perfect fit). This variable represents the fifth dimension, leading to a so-called 5D-QSAR. Recently, Martinek and co-workers[162] accounted for ligand flexibility by calculating molecular descriptors based on the free energy encountered during the shift of the ligand conformational ensemble to the active conformation. This descriptor can be combined with traditional 3D-QSAR approaches giving a '3 + 3D-QSAR' methodology. Furthermore, Lukacova and Balaz[163] described an approach to incorporate multimode ligand binding in CoMFA analyses. Since experimental binding constants can be considered as the weighted average of specific constants for different binding modes, they are predictable as weighted combinations of the field values used to predict specific constants.

5.24.5.2   Approaches Based on the Concept of Property Space

The concept of property space has been applied to large series of compounds to give a quantitative description of chemical space, by characterizing each considered property of each compound with a single value. Applications of this approach include the evaluation of chemical libraries obtained by combinatorial chemistry, or the estimation of the optimal value a given molecular property should exhibit for biological relevance (e.g., indices of drug-likeness).

Whereas this approach has proven very successful in comparing chemical libraries and designing 'combichem' series, it is nevertheless based on the assumption that the molecular properties are discrete, invariant ones. This assumption derives from the restrictions imposed by the handling of huge databases, but like many assumptions it tends to fade in the background and be taken as fact. Yet as chemistry progresses, so does the understanding of molecular structure taken in its broadest sense, namely the mutual interdependence between geometric features and physicochemical properties.

The concept of property space is beginning to be used to gain a comprehensive understanding of the dynamic behavior of a single compound. In this dynamic vision, a molecular property can be described either by an average value or by descriptors defining its property space.[70]

The average value of a property, and especially a weighted average, contains more information than a conformer-specific value (even if of the lowest-energy conformer or of the hypothetical bioactive conformer). However, this average value does not yield information on the property space itself. To this end, one should use descriptors specifying the property range and distribution in relation to conformational changes and other property profiles.

A property space can be defined using two classes of descriptors. The first class includes descriptors quantifying the variability (spread) of values, and the range is probably the most intuitive one. The second class of descriptors relates the dynamic behavior of a given property with other geometric or physicochemical properties. Such correlations can reveal if and how two molecular properties change in a coherent manner.

The relationships between physicochemical properties and geometric descriptors describe the ability of a physicochemical property to fluctuate when the 3D geometry fluctuates. These relationships also lead to the concept of molecular sensitivity, since there will be sensitive molecules whose property values are markedly influenced by small geometric changes, and insensitive molecules whose properties change little even during major geometric fluctuations. One can assume that molecular sensitivity can affect biological properties, as the latter are dynamic properties in themselves whose emergence will depend on the ability of a molecule to fit into and interact with an active site. Work in progress indicates that explorations of the property spaces of bioactive molecules may lead to new and promising tools in dynamic QSARs.[164]

5.24.5.3   Absorption, Distribution, Metabolism, and Excretion Applications of Four-Dimensional Descriptors

The key role of molecular fluctuations in ADME predictions finds compelling evidence in the significant relation between molecular flexibility and oral bioavailability. For instances, Veber and co-workers[165] analyzed 1100 drug candidates, and found that there is an inverse relation between the number of rotatable bonds and oral bioavailability. Furthermore, they showed that a molecule must have less than seven rotors to possess a significant bioavailability (irrespective of the molecular weight). Similar results were obtained by Lu and co-workers,[166] who showed that a compound will have a good bioavailability if it possesses less than 10 rotors and a PSA value smaller than 140Å^2.

Therefore, it comes as no surprise that 4D-QSAR approaches have also found successful applications in ADME predictions. Among them, we single out the remarkable and recent study by Hopfinger and co-workers,[167] who predicted the blood–brain permeation of 150 heterogeneous compounds using a combination of 4D-QSAR approaches and cluster analysis. This study divided the data set into clusters based on 4D-QSAR results, and determined predictive QSAR models for each cluster. The results indicate that the properties governing blood–brain permeation may vary across diverse classes of compounds, and that it is essential to cluster compounds into chemically similar classes in order to derive predictive models. 4D-based molecular similarity indices seem more efficient for this clustering than traditional similarity descriptors.

4D-QSAR analysis was also used to study a series of azole CYP51 inhibitors.[168] Ten different alignments were used to build models, and one alignment yielded a significantly better model with higher predictivity than the others. The active site mapped out in this 4D-QSAR study suggests that hydrogen bonding is not prevalent in the interaction, as confirmed by a crystal structure that does not indicate any possible hydrogen bond.

Even without considering the property space concept in its full potentiality, Ekins and co-workers tentatively used the range and standard deviation of MS-WHIM descriptors (obtained considering a set of significant conformers for each compound) as independent variables to predict the biological data of CYP2C9, CYP2D6, and CYP3A4 inhibitors.[169–171] These studies highlight that 4D-QSAR approaches can increase model predictivity compared with 3D-QSAR models, confirming that molecular fluctuations do play a role in biological profiling. The limited role of 4D descriptors in these models may be explained if it is considered that activity and affinity depend mainly on the ability of the ligand to assume well-defined property values, a type of information not encoded in property ranges or standard deviations.

5.24.6   Conclusion

MIFs are mathematical models depicting the physical reality of recognition forces surrounding molecules, or, stated in more vivid language, 'projected' by molecules around themselves. Such recognition forces account for the long-distance (ion–ion, ion–dipole) and short-distance (dipole–dipole, hydrophobic) interactions between chemical compounds, and they are the physicochemical substratum of all biochemical processes. Their essential significance in the interaction of drugs with small and large biomolecules does not need to be stressed.

It thus follows that understanding, describing and parameterizing MIFs may well prove to be the most fruitful approach to obtaining broad and solid QSAR and quantitative structure–disposition relationship models. As this overview has tried to show, this approach is finding ever increasing applications in relating molecular structures to biological properties.[172,173] But current successes should not hide our conviction that the MIF approach is still in its infancy despite impressive advances in recent years due to theoretical progress and the continuous increase in computer power. It is our belief, which we hope to share with readers, that a huge potential for progress lies in the development of MIF-based instruments and their application to drug research.

References

1. Testa, B.; Kier, L.; Carrupt, P. A. *Med. Res. Rev.* **1997**, *17*, 303–326.
2. Helmholtz, H. *Gesammelte Abhandlungen* **1887**, *3*, 356–391.
3. Kim, J. *Events as Property Exemplification in Action Theory*; Reidel: Dordrecht: Germany, 1976, pp 196–215.
4. Testa, B.; Kier, L. B. *Entropy* **2000**, *2*, 1–25.
5. Van de Vijver, G.; Van Speybroeck, L.; Vandevyvere, W. *Acta Biotheor.* **2003**, *51*, 101–140.
6. Morse, P. M.; Feshbach, H. *Methods of Theoretical Physics*, Part I ; McGraw-Hill: New York, NY, 1953, pp 4–8.
7. Stouch, T. R.; Jurs, P. C. *J. Chem. Inf. Comput. Sci.* **1986**, *26*, 4–12.
8. Wang, W.; Donini, O.; Reyes, C. M.; Kollman, P. A. *Annu. Rev. Biophys. Biomol. Struct.* **2001**, *30*, 211–243.
9. Foley, J. D.; Van Dam, A.; Feiner, S. K.; Hughes, J. F. *Computer Graphics: Principles and Practice*; Addison-Wesley: Boston, MA, 1990.
10. O'Rourke, J. *Computational Geometry in C*; Cambridge University Press: Oxford, UK, 1994.
11. Mezej, P. G. *Molecular Surfaces*. In *Reviews in Computational Chemistry*; Lipkowitz, K. B., Boyd, D. B., Eds.; John Wiley: Chichester, UK, 1990, Vol. 1, pp 263–294.
12. Lewis, R. A. *Methods Enzymol.* **1991**, *202*, 126–156.
13. Lee, B.; Richards, F. M. *J. Mol. Biol.* **1971**, *55*, 379–400.
14. Richards, F. M. *Annu. Rev. Biophys. Bioeng.* **1977**, *6*, 151–176.
15. Pascual-Ahuir, J. L.; Silla, E.; Tunon, I. *J. Comput. Chem.* **1994**, *15*, 1127–1138.
16. Cossi, M.; Mennucci, B.; Cammi, R. *J. Comput. Chem.* **1996**, *17*, 57–73.
17. Fogolari, F.; Brigo, A.; Molinari, H. *J. Mol. Recog.* **2002**, *15*, 377–392.
18. Masek, B. B. Molecular surface comparison. In *Molecular Similarity in Drug Design*; Dean, P. M., Ed.; Blackie: London, UK, 1995, pp 163–186.
19. Connolly, M. L. *J. Mol. Graph.* **1986**, *4*, 3–6.
20. Bacon, D. J.; Moult, J. *J. Mol. Biol.* **1992**, *255*, 849–858.
21. Wang, H. *J. Comput. Chem.* **1991**, *12*, 746–750.
22. Chau, P. L.; Dean, P. M. *J. Mol. Graph.* **1987**, *5*, 97–100.
23. Leicester, S. E.; Finney, J. L.; Bywater, R. P. *J. Mol. Graph.* **1988**, *6*, 104–108.
24. McCracken, . R. O.; Lipkowitz, K. B. *J. Parasitol.* **1990**, *76*, 180–185.
25. Van de Waterbeemd, H.; Kansy, M. *Chimia* **1992**, *46*, 299–303.
26. Palm, K.; Luthman, K.; Ungell, A.; Strandlund, G.; Artursson, P. *J. Pharm. Sci.* **1996**, *85*, 32–39.
27. Ertl, P.; Rohde, B.; Selzer, P. *J. Med. Chem.* **2000**, *43*, 3714–3717.
28. Winiwarter, S.; Ax, F.; Lennernäs, H.; Hallberg, A.; Pettersson, C.; Karlén, A. *J. Mol. Graph. Model.* **2003**, *21*, 273–287.
29. Baker, B. M.; Murphy, K. P. *Methods Enzymol.* **1998**, *295*, 294–314.
30. Cramer, R. D.; Patterson, D. E.; Bunce, J. D. *J. Am. Chem. Soc.* **1988**, *110*, 5959–5967.

31. Klebe, G.; Abraham, U.; Mietzner, T. *J. Med. Chem.* **1994**, *37*, 4130–4146.
32. Oprea, T. I.; Kurunczi, L.; Olah, M.; Simon, Z. *SAR QSAR Environ. Res.* **2001**, *12*, 75–92.
33. Sulea, T.; Oprea, T. I.; Muresan, S.; Chan, S. L. *J. Chem. Inf. Comput. Sci.* **1997**, *37*, 1162–1170.
34. Lipinski, C. A.; Lombardo, F.; Dominy, B. W.; Feeney, P. J. *Adv. Drug Deliv. Rev.* **2001**, *46*, 3–26.
35. Norinder, U.; Haeberlein, M. *Adv. Drug Deliv. Rev.* **2002**, *54*, 291–313.
36. Ecker, G. F.; Noe, C. R. *Curr. Med. Chem.* **2004**, *11*, 1617–1628.
37. Stenberg, P.; Norinder, U.; Luthman, K.; Artursson, P. *J. Med. Chem.* **2001**, *44*, 1927–1937.
38. Egan, W. J.; Lauri, G. *Adv. Drug Deliv. Rev.* **2002**, *54*, 273–289.
39. Hou, T. J.; Zhang, W.; Xia, K.; Qiao, X. B.; Xu, X. J. *J. Chem. Inf. Comput. Sci.* **2004**, *44*, 1585–1600.
40. Mälkiä, A.; Murtomäki, L.; Urtti, A.; Kontturi, K. *Eur. J. Pharm. Sci.* **2004**, *23*, 13–47.
41. Calder, J. A. D.; Ganellin, C. R. *Drug Design Disc.* **1994**, *11*, 259–268.
42. Bergström, C. A. S.; Strafford, M.; Lazorova, L.; Avdeef, A.; Luthman, K.; Artursson, P. *J. Med. Chem.* **2003**, *46*, 558–570.
43. Naray-Szabo, G.; Ferenczy, G. G. *Chem. Rev.* **1995**, *95*, 829–847.
44. Scrocco, E.; Tomasi, J. *Top. Curr. Chem.* **1973**, *42*, 95–170.
45. Pagliara, A.; Testa, B.; Carrupt, P. A.; Jolliet, P.; Morin, C.; Morin, D.; Saïk, U.; Tillement, J. P.; Rihoux, J. P. *J. Med. Chem.* **1997**, *41*, 853–863.
46. Hutter, M. C. *J. Comput.-Aided Mol. Des.* **2003**, *17*, 415–433.
47. Carbo, R.; Leyda, L.; Arnau, M. *Int. J. Quantum Chem.* **1980**, *17*, 1185–1189.
48. Hodgkin, E. E.; Richards, W. G. *Chem. Br.* **1988** 1141–1144.
49. Besalu, E.; Girones, X.; Amat, L.; Carbo-Dorca, R. *Acc. Chem. Res.* **2002**, *35*, 289–295.
50. Gasteiger, J.; Li, X.; Uschold, A. *J. Mol. Graph.* **1994**, *12*, 90–97.
51. Chessari, G.; Hunter, C. A.; Low, C. M.; Packer, M. J.; Vinter, J. G.; Zonta, C. *Chemistry* **2002**, *8*, 2860–2867.
52. Kellog, G. E.; Phatak, S.; Nicholls, A.; Grant, J. A. *QSAR Comb. Sci.* **2003**, *22*, 959–964.
53. Rahnasto, M.; Raunio, H.; Poso, A.; Wittekindt, C.; Juvonen, R. O. *J. Med. Chem.* **2005**, *48*, 440–449.
54. Kellogg, G. E.; Kier, L. B.; Gaillard, P.; Hall, L. H. *J. Comput.-Aided Mol. Des.* **1996**, *10*, 513–520.
55. Bradley, M.; Waller, C. L. *J. Chem. Inf. Comput. Sci.* **2001**, *41*, 1301–1307.
56. Stone, A. J. *The Theory of Intermolecular Forces*; Clarendon Press: Oxford, UK, 1996.
57. Kantola, A.; Villar, H. O.; Loew, G. H. *J. Comput. Chem.* **1991**, *12*, 681–689.
58. Goodford, P. J. *J. Med. Chem.* **1985**, *28*, 849–857.
59. Brooks, B. R.; Bruccoleri, R. E.; Olafson, B. D.; States, D. J.; Swaminathan, S.; Karplus, M. *J. Comput. Chem.* **1983**, *4*, 187–217.
60. Baroni, M.; Costantino, G.; Cruciani, G.; Riganelli, D.; Valigi, R.; Clementi, S. *Quant. Struct.-Act. Relat.* **1993**, *12*, 9–20.
61. Cruciani, G.; Watson, K. A. *J. Med. Chem.* **1994**, *37*, 2589–2601.
62. Cruciani, G.; Crivori, P.; Carrupt, P. A.; Testa, B. *J. Mol. Struct. (Theochem)* **2000**, *503*, 17–30.
63. Carrupt, P. A.; Testa, B.; Gaillard, P. *Rev. Comp. Chem.* **1997**, *11*, 241–315.
64. Audry, E.; Dubost, J. P.; Colleter, J. C.; Dallet, P. *Eur. J. Med. Chem.* **1986**, *21*, 71–72.
65. Rekker, R. F. *The Hydrophobic Fragmental Constant*, Nauta, W. T.; Rekker, R. F., Eds.; Elsevier: Amsterdam, UK, 1977.
66. Kellogg, G. E.; Semus, S. F.; Abraham, D. J. *J. Comput.-Aided Mol. Des.* **1991**, *5*, 545–552.
67. Gaillard, P.; Carrupt, P. A.; Testa, B.; Boudon, A. *J. Comput.-Aided Mol. Des.* **1994**, *8*, 83–96.
68. Heiden, W.; Schlenkrich, M.; Brickmann, J. *J. Comput.-Aided Mol. Des.* **1990**, *4*, 255–269.
69. Kraszni, M.; Banyai, I.; Noszal, B. *J. Med. Chem.* **2003**, *46*, 2241–2245.
70. Vistoli, G.; Pedretti, A.; Villa, L.; Testa, B. *J. Med. Chem.* **2005**, *48*, 1759–1767.
71. Vistoli, G.; Pedretti, A.; Villa, L.; Testa, B. *J. Med. Chem.* **2005**, *48*, 6926–6935.
72. Gancia, E.; Montana, J. G.; Manallack, D. T. *J. Mol. Graph. Model.* **2001**, *19*, 349–362.
73. Kamlet, M. J.; Taft, R. W. *J. Am. Chem. Soc.* **1976**, *98*, 377–383.
74. Kamlet, M. J.; Taft, R. W. *J. Am. Chem. Soc.* **1976**, *98*, 2886–2894.
75. Hammet, L. P. *Physical Organic Chemistry: Reaction Rates, Equilibria and Mechanism*; McGraw-Hill: New York, NY, 1970.
76. Abraham, M. H. *Chem. Soc. Rev.* **1993**, *22*, 73–83.
77. Raevsky, O. A.; Grigorev, V. Y.; Kireev, D. B.; Zefirov, N. S. *QSAR* **1992**, *11*, 49–63.
78. Seiler, P. *Eur. J. Med. Chem.* **1974**, *9*, 473–479.
79. El Tayar, N.; Tsai, R. S.; Testa, B.; Carrupt, P. A.; Leo, A. *J. Pharm. Sci.* **1991**, *80*, 590–598.
80. Rey, S.; Caron, G.; Ermondi, G.; Gaillard, P.; Pagliara, A.; Carrupt, P. A.; Testa, B. *J. Mol. Graph. Model.* **2001**, *19*, 521–535.
81. Lombardo, F.; Gifford, E.; Shalaeva, M. Y. *Mini Rev. Med. Chem.* **2003**, *3*, 861–875.
82. Testa, B.; Crivori, P.; Reist, M.; Carrupt, P. A. *Perspect. Drug. Disc. Des.* **2000**, *19*, 179–211.
83. Carrupt, P. A.; Testa, B.; Bechalany, A.; El Tayar, N.; Descas, P.; Perrissoud, D. *J. Med. Chem.* **1991**, *34*, 1271–1275.
84. Van de Waterbeemd, H.; Smith, D. A.; Jones, B. C. *J. Comput.-Aided Mol. Des.* **2001**, *15*, 273–286.
85. Potts, R. O.; Guy, R. H. *Pharm. Res.* **1993**, *10*, 635–637.
86. Raevsky, O. A.; Skvortsov, V. S. *J. Comput.-Aided Mol. Des.* **2002**, *16*, 1–10.
87. Young, R. C.; Mitchell, R. C.; Brown, T. H.; Ganellin, C. R.; Griffiths, R.; Jones, M.; Rana, K. K.; Saunders, D.; Smith, I. R.; Sore, N. E.; Wilks, T. K. *J. Med. Chem.* **1988**, *31*, 656–671.
88. Wold, H. Partial least squares. In *Encyclopedia of Statistical Sciences*; Kotz, S., Johnson, N. L., Eds.; Wiley: New York, NY, 1985; Vol. 6, pp 581–591.
89. Stahle, L.; Wold, S. *J. Pharmacol. Methods* **1986**, *16*, 91–110.
90. Lindgren, F.; Rännar, S. *Perspect. Drug. Disc. Des.* **1998**, *12–14*, 105–113.
91. Geladi, P.; Kowlaski, B. *Anal. Chim. Acta* **1986**, *35*, 1–17.
92. Lemmen, C.; Lengauer, T. *J. Comput.-Aided Mol. Des.* **2000**, *14*, 215–232.
93. Melville, J. L.; Hirst, J. D. *J. Chem. Inf. Comput. Sci.* **2004**, *44*, 1294–1300.
94. Cho, S. J.; Tropsha, A. *J. Med. Chem.* **1995**, *38*, 656–671, 1060–1066.
95. Norinder, U. *Perspect. Drug Disc. Des.* **1998**, *12–14*, 25–39.
96. Akamatsu, M. *Curr. Top. Med. Chem.* **2002**, *2*, 1381–1394.
97. Polanski, J.; Walczak, B. *Comp. Chem.* **2000**, *24*, 615–625.
98. Todeschini, R.; Gramatica, P. *SAR QSAR Environ. Res.* **1997**, *7*, 89–115.

99. Stiefl, N.; Baumann, K. *J. Med. Chem.* **2003**, *46*, 1390–1407.

100. Cheeseright, T.; Mackey, M.; Vinter, A. *Drug Disc. Today: Biosilico* **2004**, *2*, 57–60.

101. Turner, D. B.; Willett, P. *Eur. J. Med. Chem.* **2000**, *35*, 367–375.

102. Silvermann, B. D.; Platt, D. E. *J. Med. Chem.* **1996**, *39*, 2129–2140.

103. Ghose, A. K.; Wendoloski, J. J. *Perspect. Drug. Disc. Des.* **1998**, *9–11*, 253–271.

104. Kurogi, Y.; Guner, O. F. *Curr. Med. Chem.* **2001**, *8*, 1035–1055.

105. Ekins, S.; Waller, C. L.; Swaan, P. W.; Cruciani, G.; Wrighton, S. A.; Wikel, J. H. *J. Pharmacol. Toxicol. Methods* **2000**, *44*, 251–272.

106. Denisov, I. G.; Makris, T. M.; Sligar, S. G.; Schlichting, I. *Chem. Rev.* **2005**, *105*, 2253–2278.

107. Shaik, S.; Kumar, D.; de Visser, S. P.; Altun, A.; Thiel, W. *Chem. Rev.* **2005**, *105*, 2279–2328.

108. Ekins, S.; De Groot, M. J.; Jones, J. P. *Drug Metab. Dispos.* **2001**, *29*, 936–944.

109. De Groot, M. J.; Ekins, S. *Adv. Drug Deliv. Rev.* **2002**, *54*, 367–383.

110. Ekins, S.; Bravi, G.; Ring, B. J.; Gillespie, T. A.; Gillespie, J. S.; VandenBranden, M.; Wrighton, S. A.; Wikel, J. H. *J. Pharmacol. Exp. Ther.* **1999**, *288*, 21–29.

111. Lewis, D. F. V. *Toxicol. In Vitro* **2004**, *18*, 89–97.

112. Bathelt, C.; Schmid, R. D.; Pleiss, J. *J. Mol. Model.* **2002**, *8*, 327–335.

113. Wang, Q.; Halpert, J. R. *Drug Metab. Dispos.* **2002**, *30*, 86–95.

114. Jones, J. P.; He, M.; Trager, W. F.; Rettie, A. *Drug Metab. Dispos.* **1996**, *24*, 1–6.

115. Haining, R. L.; Jones, J. P.; Henne, K. R.; Fisher, M. B.; Koop, D. R.; Trager, W. F.; Rettie, A. E. *Biochemistry* **1999**, *38*, 3285–3292.

116. Ridderstrom, M.; Masimirembwa, C.; Trump-Kallmeyer, S.; Ahlefelt, M.; Otter, C.; Andersson, T. B. *Biochem. Biophys. Res. Commun.* **2000**, *270*, 983–987.

117. Suzuki, H.; Kneller, M. B.; Rock, D. A.; Jones, J. P.; Trager, W. F.; Rettie, A. E. *Arch. Biochem. Biophys.* **2004**, *429*, 1–15.

118. De Groot, M. J.; Ackland, M. J.; Horne, V. A.; Alex, A. A.; Jones, B. C. *J. Med. Chem.* **1999**, *42*, 1515–1524.

119. De Groot, M. J.; Bijloo, G. J.; Martens, B. J.; Van Acker, F. A. A.; Vermeulen, N. P. E. *Chem. Res. Toxicol.* **1997**, *10*, 41–48.

120. Ekins, S.; Stresser, D. M.; Williams, J. A. *Trends Pharm. Sci.* **2003**, *24*, 161–165.

121. Scott, E. E.; Halpert, J. R. *Trends Biochem. Sci.* **2005**, *30*, 5–7.

122. Yano, J. K.; Wester, M. R.; Schoch, G. A.; Griffin, K. J.; Stout, C. D.; Johnson, E. F. *J. Biol. Chem.* **2004**, *279*, 38091–38094.

123. Tanaka, T.; Okuda, T.; Yamamoto, Y. *Chem. Pharm. Bull.* **2004**, *52*, 830–835.

124. Ekins, S.; Wrighton, S. A. *J. Pharmacol. Toxicol. Methods* **2001**, *45*, 65–69.

125. Carrieri, A.; Carotti, A.; Barreca, M. L.; Altomare, C. *J. Comput.-Aided Mol. Des.* **2002**, *16*, 769–778.

126. Guerre, C.; Catto, M.; Leonetti, F.; Weber, P.; Carrupt, P. A.; Altomare, C.; Carotti, A.; Testa, B. *J. Med. Chem.* **2000**, *43*, 4747–4758.

127. Moron, J. A.; Campillo, M.; Perez, V.; Unzeta, M.; Pardo, L. *J. Med. Chem.* **2000**, *43*, 1684–1691.

128. Smith, P. A.; Sorich, M. J.; Low, L. S. C.; McKinnon, R. A.; Miners, J. O. *J. Mol. Graph. Model.* **2004**, *22*, 507–517.

129. Said, M.; Ziegler, J. C.; Magdalou, J.; Elass, A.; Vergoten, G. *Quant. Struct. Act. Relat.* **1996**, *15*, 382–388.

130. Cupid, B. C.; Holmes, E.; Wilson, I. D.; Lindon, J. C.; Nicholson, J. K. *Xenobiotica* **1999**, *29*, 27–42.

131. Smith, P. A.; Sorich, M. J.; McKinnon, R. A.; Miners, J. O. *Clin. Exp. Pharmacol. Physiol.* **2003**, *30*, 836–840.

132. Sorich, M. J.; Miners, J. O.; McKinnon, R. A.; Smith, P. A. *Mol. Pharm.* **2004**, *65*, 301–308.

133. Tervo, A. J.; Nyronen, T. H.; Ronkko, T.; Poso, A. *J. Comput.-Aided Mol. Des.* **2003**, *17*, 797–810.

134. Sipila, J.; Taskinen, J. *J. Chem. Inf. Comput. Sci.* **2004**, *44*, 97–104.

135. Sipila, J.; Hood, A. M.; Coughtrie, M. W.; Taskinen, J. *J. Chem. Inf. Comput. Sci.* **2003**, *43*, 1563–1569.

136. Muszynski, I. C.; Scapozza, L.; Kovar, K. A.; Folkers, G. *Quant. Struct.-Act. Relat.* **1999**, *18*, 342–353.

137. Appell, M.; Dunn, W. J., 3rd; Reith, M. E.; Miller, L.; Flippen-Anderson, J. L. *Bioorg. Med. Chem.* **2002**, *10*, 1197–1206.

138. Lieske, S. F.; Yang, B.; Eldefrawi, M. E.; MacKerell, A. D., Jr.; Wright, J. *J. Med. Chem.* **1998**, *41*, 864–876.

139. Robarge, M. J.; Agoston, G. E.; Izenwasser, S.; Kopajtic, T.; George, C.; Katz, J. L.; Newman, A. H. *J. Med. Chem.* **2000**, *43*, 1085–1093.

140. Yuan, H.; Kozikowski, A. P.; Petukhov, P. A. *J. Med. Chem.* **2004**, *47*, 6137–6143.

141. Seelig, A. *Eur. J. Biochem.* **1998**, *251*, 252–261.

142. Penzotti, J. E.; Lamb, M. L.; Evensen, E.; Grootenhuis, P. D. J. *J. Med. Chem.* **2002**, *45*, 1737–1740.

143. Fleischer, R.; Wiese, M. *J. Med. Chem.* **2003**, *46*, 4988–5004.

144. Pajeva, I. K.; Wiese, M. *J. Med. Chem.* **1998**, *41*, 1815–1826.

145. Yates, C. R.; Chang, C.; Kearbey, J. D.; Yasuda, K.; Schuetz, E. G.; Miller, D. D.; Dalton, J. T.; Swaan, P. W. *Pharm. Res.* **2003**, *20*, 1794–1803.

146. Cianchetta, G.; Singleton, R. W.; Zhang, M.; Wildgoose, M.; Giesing, D.; Fravolini, A.; Cruciani, G.; Vaz, R. J. *J. Med. Chem.* **2005**, *48*, 2927–2935.

147. Terada, T.; Inui, K. *Curr. Drug Metab.* **2004**, *5*, 85–94.

148. Gebauer, S.; Knutter, I.; Hartrodt, B.; Brandsch, M.; Neubert, K.; Thondorf, I. *J. Med. Chem.* **2003**, *46*, 5725–5734.

149. Meier, P. J.; Stieger, B. *Annu. Rev. Physiol.* **2002**, *64*, 635–661.

150. Baringhaus, K. H.; Matter, H.; Stengelin, S.; Kramer, W. *J. Lipid Res.* **1999**, *40*, 2158–2168.

151. Sjoberg, P. MolSurf – a generator of chemical descriptors for QSAR. In *Computer-Assisted Lead Finding and Optimization*; Van de Waterbeemd, H., Testa, B., Folkers, G., Eds.; Verlag Helvetica Chimi Acta: Basel, Switzerland, 1997, pp 81–92.

152. Norinder, U.; Sjoberg, P.; Osterberg, T. *J. Pharm. Sci.* **1998**, *87*, 952–959.

153. Norinder, U.; Osterberg, T.; Artursson, P. *Eur. J. Pharm. Sci.* **1999**, *8*, 49–56.

154. Norinder, U.; Osterberg, T.; Artursson, P. *Pharm. Res.* **1997**, *14*, 1786–1791.

155. Wolohan, P. R. N.; Clark, R. D. *J. Comput.-Aided Mol. Des.* **2003**, *17*, 65–76.

156. Crivori, P.; Cruciani, G.; Carrupt, P. A.; Testa, B. *J. Med. Chem.* **2000**, *43*, 2204–2216.

157. Ekins, S.; Durst, G. L.; Stratford, R. E.; Thorner, D. A.; Lewis, R.; Loncharich, R. J.; Wikel, J. H. *J. Chem. Inf. Comput. Sci.* **2001**, *41*, 1578–1586.

158. Hopfinger, A. J.; Wang, S.; Tokarski, J. S.; Jin, B.; Albuquerque, M.; Madhav, P. J.; Duraiswami, C. *J. Am. Chem. Soc.* **1997**, *119*, 10509–10524.

159. Duca, J. S.; Hopfinger, A. J. *J. Chem. Inf. Comput. Sci.* **2001**, *41*, 1367–1387.

160. Ravi, M.; Hopfinger, A. J.; Hormann, R. E.; Dinan, L. *J. Chem. Inf. Comput. Sci.* **2001**, *41*, 1587–1604.

161. Vedani, A.; Dobler, M. *J. Med. Chem.* **2002**, *45*, 2139–2149.

162. Martinek, T. A.; Ötvös, F.; Dervarics, M.; Tóth, G.; Fülöp, F. *J. Med. Chem.* **2005**, *48*, 3239–3250.

163. Lukacova, V.; Balaz, S. *J. Chem. Inf. Comput. Sci.* **2003**, *43*, 2093–2105.

164. Vistoli, G.; Pedretti, A.; Villa, L.; Testa, B. *J. Med. Chem.* **2005**, *48*, 4947–4952.

165. Veber, D. F.; Johnson, S. R.; Cheng, H.-Y.; Smith, B. R.; Ward, K. W.; Kopple, K. D. *J. Med. Chem.* **2002**, *45*, 2615–2623.

166. Lu, J. J.; Crimin, K.; Goodwin, J. T.; Crivori, P.; Orrenius, C.; Xing, L.; Tandler, P. J.; Vidmar, T. J.; Amore, B. M.; Wilson, A. G. E.; Stouten, P. F. W.; Burton, P. S. *J. Med. Chem.* **2004**, *47*, 6104–6107.
167. Pan, D.; Iyer, M.; Liu, J.; Li, Y.; Hopfinger, A. J. *J. Chem. Inf. Comput. Sci.* **2004**, *44*, 2083–2098.
168. Liu, J; Pan, D.; Tseng, Y.; Hopfinger, A. J. *J. Chem. Inf. Comput. Sci.* **2003**, *43*, 2170–2179.
169. Ekins, S.; Bravi, G.; Binkley, S.; Gillespie, J. S.; Ring, B. J.; Wikel, J. H.; Wrighton, S. A. *J. Pharmacol. Exp. Ther.* **1999**, *290*, 429–438.
170. Ekins, S.; Bravi, G.; Binkley, S.; Gillespie, J. S.; Ring, B. J.; Wikel, J. H.; Wrighton, S. A. *Pharmacogenetics* **1999**, *9*, 477–489.
171. Ekins, S.; Bravi, G.; Binkley, S.; Gillespie, J. S.; Ring, B. J.; Wikel, J. H.; Wrighton, S. A. *Drug Metab. Dispos.* **2000**, *28*, 994–1002.
172. Moro, S.; Bacilieri, M.; Cacciari, B.; Bolcato, C.; Cusan, C.; Pastorin, G.; Klotz, K.-N.; Spalluto, G. *Bioorg. Med. Chem.* **2006**, *14*, 4923–4932.
173. Cheeseright, T.; Mackey, M.; Rose, S.; Vinter, A. *J. Chem. Inform. Mod.* **2006**, *46*, 665–676.

Biographies

Giulio Vistoli was born in 1968. He received his Laurea degree in medicinal chemistry at University of Milan in 1994. During his PhD studies with Prof L Villa, he spent a period in Lausanne under the supervision of Prof B Testa with whom he has fruitfully collaborated since 1996. In 1999, he became assistant professor in medicinal chemistry at University of Milan. His recent research focuses on developing the property space concept to explore the dynamic profile of molecular fields, deriving fertile descriptors for dynamic 4D-QSAR analyses.

Alessandro Pedretti was born in 1970. He received his Laurea degree in medicinal chemistry at University of Milan, in 1995. After PhD studies with Prof L Villa, he became assistant professor in medicinal chemistry at University of Milan, in 2001. His interests deal mainly with computer programming applied to computational chemistry, realizing novel software tools for molecular modeling and docking analysis. In particular, he developed VEGA (available at www.ddl.unimi.it), a program able to compute several molecular fields analyzing their dynamic profiles during the simulation time.

Comprehensive Medicinal Chemistry II
ISBN (set): 0-08-044513-6

ISBN (Volume 5) 0-08-044518-7; pp. 577–602

5.25 In Silico Prediction of Ionization

R Fraczkiewicz, Simulations Plus, Inc., Lancaster, CA, USA

5.25.1 Introduction

The major component of the human body is water – over 70% by weight.[1] The liver, our main 'biochemical factory,' is an amazing 96% water! Many vital biochemical processes either directly involve, or depend on, acid/base ionization of chemical compounds in water. Generally, ionizable molecules exist in aqueous environment as a population of species: one electrically neutral (on each ionization center) and multiple ionized forms; their exact number depends on the ionization center count. Ionized species may have very different properties from the neutral form. From a physiologist's point of view, this may mean preserving a relatively constant pH of the organism's fluids via biological buffers, pH-dependent reactivity of biomolecules that have charged functional groups, pH-dependent solubility, and the ability of compounds to distribute through an organism. To a pharmaceutical scientist, the importance of ionization cannot be overstated. It influences almost all the subjects of pharmaceutical science: drug absorption, transport, distribution, and elimination, solubility, formulation, chemical stability, pharmacodynamics, drug analysis, etc. Crossing cellular membranes, for example, is strongly related to molecular lipophilicity understood in terms of the n-octanol/water partition coefficient (log P). It has been observed that the lipophilicity of ionized forms of a chemical compound is significantly lower than that of the neutral form. This phenomenon is of course related to the energetic cost of removing the molecule from an aqueous environment and placing it in an organic solvent – removing the hydration layer is in general more difficult in the case of charged forms. Therefore, the neutral form of the same compound will permeate more easily than any of its charged forms. In principle, this trend should extend to relating permeabilities of predominantly neutral versus predominantly ionized drugs at physiological pH and holds true in many cases, indeed. However, one must remember that the observed lipophilicities of ionizable compounds are population averages over all species, i.e., the pH-dependent distribution coefficients (log D). Ionization and pH determine the relative contents of all the mentioned forms in water. It is possible for one ionizable drug to have log D in certain pH range equal to or even higher than log $D = \log P$ of another drug possessing no ionizable groups. Hence, one must be cautious in permeability comparisons of different compounds. An opposite general trend is observed with regards to the drug's water solubility – a very important factor for solid dosage forms. Ionized (salt) forms will dissolve better than the neutral form due to enhanced hydration. Indeed, a plot of solubility versus log P/log D shows strong, negative correlation between the two. A pharmaceutical formulation scientist must always carefully consider the interplay between these opposite effects and strive for their balance. These issues are specifically addressed elsewhere (*see* 5.11 Passive Permeability and Active Transport Models for the Prediction of Oral Absorption; 5.16 Ionization Constants and Ionization Profiles; 5.17 Dissolution and Solubility; 5.18 Lipophilicity, Polarity, and Hydrophobicity; 5.19 Artificial Membrane Technologies to Assess Transfer and Permeation of Drugs in Drug Discovery; 5.28 In Silico Models to Predict Oral Absorption) the reader is encouraged to consult these chapters for more details.

The acid/base properties of a chemical compound are quantified by its ionization constants (pK_a) that can be either measured or calculated. Experimental aspects of ionization, as well as elementary definitions, are described elsewhere (*see* 5.16 Ionization Constants and Ionization Profiles) in this book and will not be repeated here. Measurement of ionization constants has become more easy and convenient over recent years. However, it requires either purchase, or synthesis and purification, of chemical compounds. This may not always be cost-effective, particularly for early drug discovery scientists who wish to screen up to millions of compounds in virtual libraries. In situations like this, rapid in silico estimation of pK_a is the most practical way to obtain results.

This chapter will be focused on pK_a prediction by empirically based computational methods, although first-principles methods of prediction, as well as methods derived from statistical thermodynamics, will also be mentioned. Sections 5.25.2 and 5.25.3 will discuss the theory and algorithmic approaches to the prediction. Currently available software packages will be listed in Section 5.25.4. Available sources of data for pK_a modeling will be the subject of Section 5.25.5. Finally, Section 5.25.6 will discuss the future of ionization prediction.

5.25.2   Theoretical Basis of Ionization Prediction

Before discussing individual predictive algorithms in detail, it is imperative to understand the phenomenon of protolytic dissociation, as well as the factors influencing it. Simple at the first sight, the process of proton detachment from an organic molecule in an aqueous environment is actually quite complex, as it depends on a delicate balance of sizable opposing forces. A discussion of these forces is the subject of the first two subsections. The mutual interactions of individual ionization sites in multiprotic molecules introduce another complication leading to microequilibria theory – the subject of the third subsection.

5.25.2.1   Thermodynamic Forces Driving Ionization

Complex phenomena may become tractable after dissecting them into simpler components. Dissociation of a monoprotic organic acid at constant temperature in water is thus considered in terms of a thermodynamic cycle used by quantum chemists to calculate ionization constants[2-4] and depicted in **Figure 1**.

It is immediately apparent that the process of aqueous dissociation, shown by the bottom reaction, should be understood as proton exchange between an acid, HA, and a base, H_2O, rather than the simple process of protolysis. The

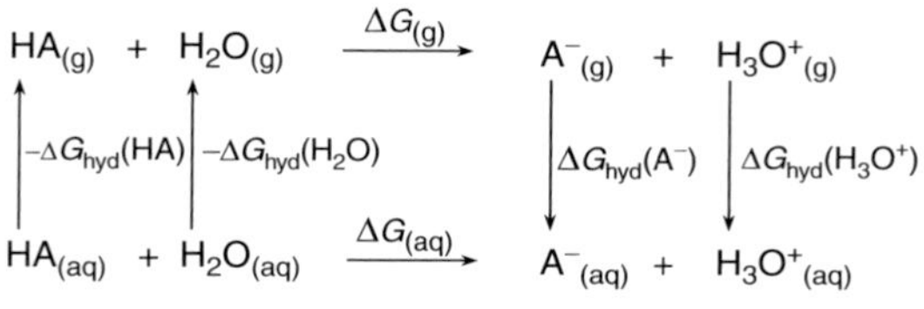

Figure 1 Thermodynamic cycle used to calculate the Gibbs free energy change upon acid dissociation in water. Gas phase is symbolized by the '(g)' subscript, aqueous phase by the '(aq)' subscript, while 'hyd' denotes solvation.

negative logarithm of the equilibrium constant for this exchange, the pK_a, is directly related to the free energy change upon ionization[3]:

$$pK_a = \frac{\Delta G_{aq}}{2.303RT} + \text{const} \qquad [1]$$

where R is the universal gas constant and T stands for temperature. The top reaction in **Figure 1**, conducted in the gas phase, is free of solvation components and its ΔG_g can be further expressed in terms of free energies of bond breaking:

$$\Delta G_g = \Delta G_{diss}(H-A) - \Delta G_{diss}(H-OH_2^+) \qquad [2]$$

Since the net free energy change of a closed cycle is zero, the aqueous free energy of dissociation can be decomposed as follows:

$$\begin{aligned} \Delta G_{aq} = &\Delta G_{diss}(H-A) + [\Delta G_{hyd}(A^-) - \Delta G_{hyd}(HA)] \\ &+ [\Delta G_{hyd}(H_3O^+) - \Delta G_{hyd}(H_2O) - \Delta G_{diss}(H-OH_2^+)] \end{aligned} \qquad [3]$$

The last three terms, pertaining to water (labeled further as $\Delta\Delta G(\text{water})$), are constant for all acids and their net effect strongly favors ionization. The $\Delta\Delta G(\text{water})$ can be defined as the free energy difference for between relative hydration and protonation of the hydronium ion. For the purpose of comparing relative acid strength, it is then sufficient to consider the remaining components: (1) the free energy of breaking the H–A bond in the gas phase, and (2) the difference in solvation between protonated and deprotonated acid. The magnitude of solvation is much greater for ions than for neutral molecules.[5,6] For example, the free energy of solvation for the acetate ion is about 10 times greater than the free energy of solvation of the acetic acid.[5] In the case of weak organic acids, solvation forces cooperate with $\Delta\Delta G(\text{water})$ driving ionization forward, while the process of breaking the H–A bond opposes it. The case of weak monoprotic bases is quite different in the solvation aspect. Replacing protonated acid HA by protonated base BH^+, and anion A^- by free base B in **Figure 1** results in a thermodynamic cycle for base ionization. The aqueous free energy of base dissociation is equal to:

$$\Delta G_{aq} = \Delta G_{diss}(B-H^+) + [\Delta G_{hyd}(B) - \Delta G_{hyd}(BH^+)] + \Delta\Delta G(\text{water}) \qquad [4]$$

Unlike acid ionization, which produces two ions out of neutral substrates, the deprotonation of one positively charged base on the left-hand side results in one positive hydronium ion on the right-hand side. For this reason, the sensitivity of pK_a to solvation is much lower for bases than that for acids. For example, acid strength drops rapidly with the solvent's dielectric constant, while base pK_a is affected only weakly.[7]

5.25.2.2   Molecular Factors Affecting Ionization

Being thermodynamic variables, ionization constants are obviously influenced by 'environmental' factors such as temperature, solvent, ionic strength, etc. For the purpose of predictive modeling these variables are assumed fixed, e.g., this chapter considers pK_a measured in water only. The main reasons for intermolecular pK_a variability are thus molecular factors that, in the light of aforementioned thermodynamic considerations, influence the stability of the H–A (or B–H$^+$) bond and the solvation differential between a product and a substrate of the ionization reaction. For a proton dissociating from the same functional group, the question is how much the rest of the molecule favors (stabilizes) the ionic form over the neutral one. These effects can be either electronic/electrostatic, or steric in nature. The first group can be further subdivided into electrostatic field (through-space), inductive (through-bond), and mesomeric (arising from π-electron delocalization) effects. Since Perrin *et al.* provide an excellent, detailed explanation of these molecular factors,[7] this chapter will only summarize the most important ideas.

Electrostatic interactions between a charged ionization center and electric charges in other parts of the molecule, permanent or induced, may stabilize or destabilize the center depending on whether the prevailing interactions are attractive or repulsive. Of course, polar solvent acting as dielectric usually dampens the magnitude of these field effects. In addition, field interactions determine the amount of electrical work required to remove a proton from the ionization center.

The electronegativity difference between hydrogen and chlorine atoms in, e.g., an HCl molecule causes a net shift of electron density toward the chlorine atom, resulting in charge polarization. This inductive effect acting through chemical bonds is intuitively understood in terms of 'electron-withdrawing' and 'electron-donating' atoms and groups.

This effect has short range in terms of the number of bonds between the acting group and the ionization center. Naturally, the electron density present in the A–H bond determines its strength; hence density withdrawal will promote ionization and vice versa. Of course, inductive influence is not only limited to bond breaking. It is also strongly related to the relative stabilization of the charged reaction center.

Unlike inductive effects acting through localized bonds, mesomeric influences are long-ranged and depend on existing π-conjugation between remote substituents and the reaction center. If the conjugation exists, then the electric charge resulting from ionization may delocalize, resulting in lower free energy than that expected from electrostatic/inductive effects alone. For example, the pK_a of simple aliphatic alcohols ranges above 14, whereas that of phenols is around 10. More information on electronic effects can be found in representative publications.[8–46]

Steric factors are much more specific and diverse. Stereochemical constraints may impose isomer-specific electronic/electrostatic interactions. For example, *cis/trans* isomerism is responsible for different electrostatic repulsion in fumaric and maleic acids leading to markedly different pK_a values.

Conformational effects influence the ionization constants of axial versus equatorial cyclohexane-carboxylic acids. Similar geometrical effects may also induce different solvation patterns, internal hydrogen bonding, and changes in π-electron delocalization. Internal hydrogen bonding provides an extra stabilization of the carboxylate anion in salicylic acid, resulting in lower primary and higher secondary pK_a in comparison to *p*-hydroxy benzoic acid. Local crowding hinders solvation and thus ion stabilization, leading to raised acid pK_a and lowered base pK_a.

This effect is particularly important for larger reacting groups. Steric hindrance to resonance effects is responsible for as much as a 10-fold difference in ionization of N,N-dimethylaniline and N,N-diethylaniline, as well as blocking the nitro group from achieving anion-stabilizing resonance with the ring in the 3,5-dimethyl-4-nitrophenol.

Some organic molecules isomerize by migrating a proton within the same molecule from one group to another. A well-known example of this tautomerism is keton–enol equilibrium:

$$-CH_2-C{=}O \Leftrightarrow -CH{=}C-OH$$

In most cases the equilibrium is strongly shifted to the left, but in some systems the enol form may become significant. This has profound consequences for pK_a prediction as the migration of one proton from carbon to oxygen increases the number of potentially ionizable groups by one. Since pK_a values of tautomers may be markedly different, they should be treated as separate molecules for predictive purposes.

Effects of the solvent and solvation have been discussed in previous subsection from thermodynamic point of view. However, their importance warrants more detailed consideration. For example, the expected and observed in gas phase order of basicity of methylamines, $Me_3N > Me_2NH > MeNH_2 > NH_3$ is changed by water to $Me_2NH > MeNH_2 > Me_3N > NH_3$.[47] Early treatment of solvent as a dielectric continuum influencing only through-space field interactions[8] turned out to be a crude approximation. The effects of solvent (especially water) are much more complex and subtle. Specific interactions mediated through hydrogen bonding affect not only a reaction center, but also electronic properties of other molecular groups that interact with it. In turn, neutral and ionized forms of an organic molecule have different influence on the structure of liquid water. Krishnan and Friedman saw four distinct components of solvation: cavity formation, hydrogen bonding, van der Waals dispersion forces, and structure modifying term almost entirely associated with the effect of nonpolar groups on the self-organization of water molecules.[48]

Topsom performed a detailed theoretical study of explicit water molecules interacting with carboxylic acids and amines in gas phase.[6] In the case of substituted acetic acids, XCH_2COOH, it takes just one water molecule hydrogen-bonded to the –OH group in substrate, but as many as three molecules directly bonded to the $–COO^-$ anion in the ionization product to explain all experimental substituent effects observed in aqueous solution. Similarly, one water molecule attached to $–NH_2$ and three water molecules attached to $–NH_3^+$ explain aqueous basicities of simple methylamines. In both cases it is the charged species that differentiates pK_a. However, explicit hydration with 1–3 waters of just the reaction site in larger molecules like substituted pyridines, quinuclidines, and phenols does not explain their aqueous basicity and acidity adequately. According to Topsom, at larger distances the bulk water influence on the 'lines of force' between reaction center and a substituent becomes more significant. Detailed studies by Arnett on literature data of hydration enthalpies of pyridines and pyridinium ions, however, do suggest dominance of the hydrogen bonding term related to the bulk water interacting with these molecules.[49] Bulk water was found to be twice as good hydrogen bond acceptor and hydrogen bond donor as monomeric water toward the pyridinium ions and pyridines, respectively.

Hydration of a substituent may affect its electronic influence on the reaction site: field, resonance, electronegativity, and polarizability. Based on hydration energy alone, the field effect of substituents like –H, alkyl, $–CF_3$, $–NO_2$, –F, and –CN is not influenced significantly by the specific hydration.[6] At the other end of hydration spectrum are substituents $–NH_2$, $–NMe_2$, –OH, –OMe, –CHO, –COMe, and $–CO_2Me$ whose effective dipole moment, thus field effect, is significantly changed by the first solvation shell. Calculations also show a significant decrease of π-electron donation by $–NH_2$, $–NMe_2$, and –OMe upon hydration as the lone pair becomes less involved in conjugation. By contrast, π-electron withdrawal by –COH and –COMe (but not –COOR) is enhanced by hydration.

The decades-old quest toward understanding the aforementioned 'anomalous' ordering of aqueous basicities of methylamines was comprehensively summarized by Caskey et al.[47] The authors contributed to this knowledge pool by performing effective fragment potential (EFP) calculations of methylamine– and methylammonium–water complexes with the number of water molecules increasing from 0 to 14. The structural, as well as energetic, differences between amines and corresponding ammonium ions are stark. Methylamine molecules are attached to hydrogen-bonded water clusters almost as an afterthought, making one acceptor N–HOH hydrogen bond and only occasionally one more donor $NH–OH_2$ hydrogen bond when applicable. On the other hand, ammonium ions are the 'center of attention' of surrounding water clusters, always utilizing all the available protons for donor $NH–OH_2$ bonding. These observations are commensurate with stabilization energetics due to hydration.[47] Methylamines are stabilized to about the same extent as the number of methyl groups increases. In contrast, relative methylammonium ion stabilization decreases with the number of methyl groups. (In concordance with the work of Topsom, the first few water molecules contribute the bulk of hydration enthalpy to ammonium ions.) This is consistent with the results of previous research that the electrostatic term dominates hydration enthalpy of methylammonium ions and decreases with the ion size, thus explaining the observed ordering of methylamine basicities in water.

The above examples are only a tip of an iceberg of research work done on organic ion solvation, but well illustrate the great complexity of this problem.

Finally, statistical factors and microscopic constants mentioned by Perrin and coauthors are manifestations of the same set of important phenomena to which the entire next subsection is dedicated.

5.25.2.3 Microequilibria of Acid/Base Reactions

Microequilibria, pK_a microconstants, microdissociation, microspecies, and microspeciation – these terms appearing in the chemical literature may mislead the reader into believing that there exists some mysterious microscopic-level theory of proton dissociation, or associate these with quantum chemical description of the said reactions. Nothing can be further from the truth. The terminology simply applies to an accurate description of acid/base equilibria of multiprotic molecules. A brief survey of randomly chosen college chemistry textbooks and scientists in the pharmaceutical industry shows that the microequilibria theory is rarely taught, if at all, even at the graduate level. Instead, an approximate picture of sequential dissociation is presented, where each pK_a determined by titration can be 'assigned' to specific ionizable functional group in a multiprotic molecule. The accurate description, already noticed by Bjerrum in 1923,[50] is that all protonation sites participate at every stage of multiprotic dissociation.[51–53]

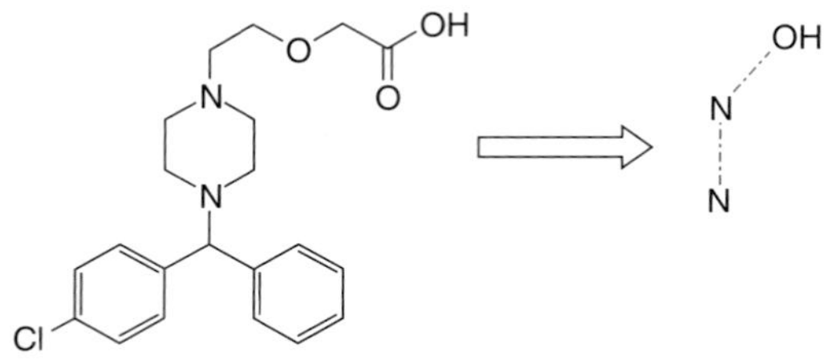

Figure 2 Molecule of cetirizine and schematic representation of its protonation sites.

The experimentally observed pK_a constants, called macroconstants from now on, are thus an effect of statistical averaging of individual microdissociation events.

The concepts presented above are best illustrated by an example. The drug cetirizine (ZYRTEC manufactured by Pfizer) (**Figure 2**) seems to be an ideal candidate. All 12 microconstants and 8 microspecies (protonation states) of this triprotic molecule have been determined experimentally,[54] along with three pK_a macroconstants: 7.98, 2.90, and 2.12. To interpret the latter, an average organic chemist would immediately say: 'in a zwitterionic compound like this the 7.98 constant reflects deprotonation of one of the tertiary amine groups, while 2.90 can be assigned to dissociation of the carboxylic acid.' That leaves the 2.12 constant which must be 'assigned' to the remaining amine group. Which one? Our chemist hesitates: "Well, maybe it is 2.12 that represents carboxylic acid." Of course, that makes 2.90 still "unassigned"...

The difficulties associated with the sequential ionization point of view are apparent and only multiply with the number of protonation sites in the molecule. Let us examine titration experiments closer. Placing cetirizine in a water solution at very high pH eliminates all dissociable protons leaving a negative ion. This fully deprotonated state (FDS) will be our reference state. Addition of a strong acid lowers pH and makes more protons available for binding by cetirizine. Adding positively charged proton to the negative FDS is easier than adding it to the electrically neutral state, while adding a proton to a positive ion is most difficult of all. These simple arguments allow us for a tentative ordering of macroconstants corresponding to the increasing protonation count on cetirizine:

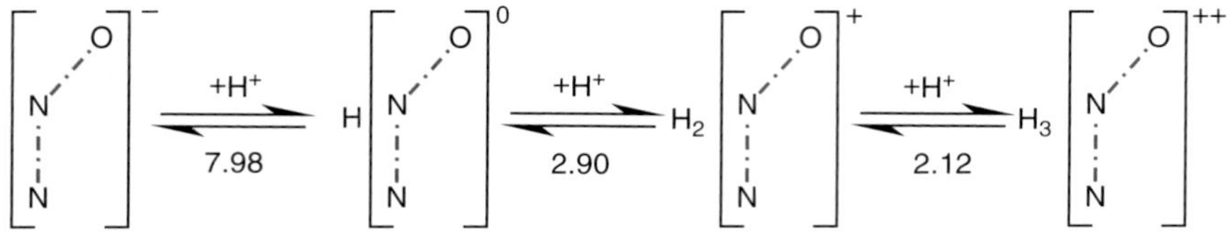

In principle, the overall proton count in each of the macrostates shown above is all that can be deduced from titration experiments. We know that, e.g., in neutral species there is one proton bound somewhere. There is no physicochemical reason to strictly determine which site the proton binds to. In fact, the proton may bind to any available site with nonzero probability. In the neutral macrostate there are three available sites leading to three neutral microstates. There are three different ways of distributing two protons among three sites providing another three microstates in the $+1$ macrostate. Together with FDS and the fully protonated microstate, there are 8 possible protonation microstates of cetirizine bound together by a network of 12 microequilibria as shown in **Figure 3**.

The existence of all the depicted microstates has been confirmed experimentally.[54] The neutral macrostate is strongly dominated by a zwitterionic microstate with proton bound to the middle nitrogen. Therefore, the idea of the 7.98 macroconstant being 'assigned' to this amine group is a good approximation, justifying the aforesaid sequential protonation point of view. However, 'assigning' the remaining two constants, 2.90 and 2.12, to any functional group cannot be done, since the components of $+1$ macrostate contribute with probabilities of the same order of magnitude. In general, the closer the microconstants are to each other, the less accurate the sequential ionization picture is. It is also noteworthy that so-called 'statistical factors,' artificial constructs mentioned in the Perrin's book[7] as significant correction parameters to pK_a prediction for symmetric molecules, are quite a natural consequence of protonation microequilibria if identical ionization centers are treated as discernible.

The analysis presented above has profound consequences for in silico prediction of ionization constants. Since macroscopic pK_a are predicted from molecular structure, having uniquely defined structure with unambiguously positioned protons is a necessary condition of a successful prediction. Unfortunately, unique structures are characteristic of microstates, not macrostates. In conclusion, macroconstants cannot be derived directly from the

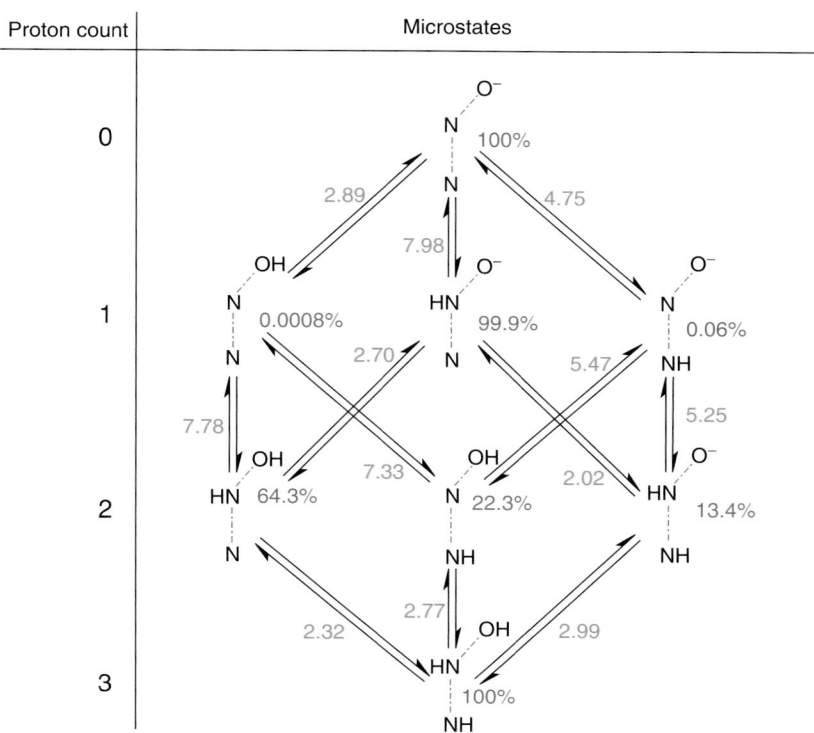

Proton count	Microstates
0	
1	
2	
3	

Figure 3 Protonation microequilibria of cetirizine. Arrows are labeled by microconstants measured experimentally.[54] Protonation microstates are labeled by their relative contributions (probabilities) to the respective parent macrostate.

structure. One way of solving this dilemma would be building predictive models from measured microconstants, then combining the results to form macroconstants. Although microconstants have been measured in a few impressive experiments,[54–84] they are available only for a handful of compounds and the applied methodologies are difficult and limited up to triprotic molecules when no symmetry constraints are present.[53] In an iterative approach predictive models may be gradually extended from monoprotic to multiprotic molecules with increasing degree of complexity. Finally, predictive models may be indirectly optimized against observed macroconstants, which were measured for tens of thousands of chemicals, via rigorous application of the microequilibria mathematics. Another difficulty is exponentially growing number of microconstants and microstates as a function of the number of ionizable groups per molecule. Generally, for n protonation sites there are as many as 2^n microstates and $n \cdot 2^{n-1}$ microconstants. In the case of a polypeptide containing 20 ionizable groups, the respective numbers are 1 048 576 microstates and 10 485 760 microconstants! Clearly, this presents significant challenge to programmers of predictive software. Many prediction methods avoid this challenge by neglecting microspeciation altogether or by using various approximation schemes. One possible way of overcoming the performance barrier would be guessing the dominant microstates and presenting them as the only representatives of the respective macrostates, thus restoring sequential ionization. Hence, macroscopic pK_a would be approximated by suitable dominant microconstants at a cost of neglecting lesser microstates. For example, microconstants collected along the dominant protonation path in **Figure 3** (7.98, 2.70, and 2.32) approximate the observed macroconstants with the maximum absolute error of only 0.2. Choice of other protonation paths, however, would result in a catastrophic loss of accuracy. Furthermore, the error of this approach becomes larger with the number of ionization centers, as the chance of equivalent and nearly equivalent microstates increases. The quality of prediction in this case strongly depends on the quality of guessing algorithm, which faces an arduous task of a priori determination of prevailing microstates from microequilibria diagrams, such as the one in **Figure 3**, but with all the numerical labels removed.

Fortunately, in drug discovery and development applications the investigated molecules rarely possess high numbers of ionizable groups. Therefore, exact resolution of complete sets of ionization microequilibria is computationally feasible and there is no need to introduce additional sources of error by using approximate algorithms. In Section 5.25.4, software packages for pK_a prediction will be labeled by their approach to the microequilibria problem.

5.25.3   Algorithms to Predict Ionization Constants

Computational methods of predicting pK_a can be divided into three major groups. Small molecules are amenable to approaches based on the power of modern quantum chemistry and starting from first principles; this first group will be labeled as quantum chemical methods. On the other end of the molecular size spectrum are proteins and nucleic acids – molecules with large numbers, but with limited variety, of ionization sites. These biopolymers are treatable with the methods derived from statistical thermodynamics, the second group. Methods that use empirical predictive models to predict pK_a belong to the third group.

5.25.3.1   Empirical Methods for Small Molecules

Empirical methods of small molecule pK_a estimation can be tentatively divided into three groups based on the approach used:

1. Linear free-energy relationships (LFER) – methods utilizing empirical relations of Hammett and Taft.
2. Quantitative structure–property relationships (QSPR) – methods correlating calculated structural descriptors with pK_a.
3. Database lookup – methods searching of similar structures in a predetermined database of molecules with known experimental pK_a.

In this section, we describe scientific principles and review relevant research pertaining to these approaches. Description of individual software packages will be postponed to Section 5.25.4.

5.25.3.1.1   Methods based on linear free-energy relationships
5.25.3.1.1.1   Hammett–Taft constants

Historically, LFER-based methods are the oldest. In as early as 1935, Hammett first discovered a simple linear relationship between the ionization constants, K_a's, of benzoic acid (0) and its *para*-substituted derivative (S)[85]:

$$\log\left(\frac{K_a^S}{K_a^0}\right) = \sigma_S \qquad [5]$$

where σ_S is a Hammett constant characteristic to a given substituent S. Taft observed similar relations in aliphatic systems.[86] The logarithm of the ionization constant, as shown in Section 5.25.2.1, is proportional to the Gibbs free energy of ionization, hence the LFER acronym. The above relation was later extended to multiple substituents and other classes of compounds:

$$pK_a^S = pK_a^0 - \rho \sum \sigma_S \qquad [6]$$

In the above, pK_a^0 symbolizes ionization constant for the parent molecule, while ρ is a characteristic constant for the entire reaction class considered (and equal to 1 for *para*-substituted benzoic acids) multiplied by the sum of substituent contributions. In principle, the reaction constant is considered independent of individual molecules, whereas substituent constants are assumed to be additive and independent of individual reactions. With the exception of experimentally determined pK_a^0, values of all other parameters are found by regression.

It should be duly noted that all the LFER methods belong to a general class of perturbational approaches. In short, these methods are valid as long as substituents are small chemical perturbations of the parent molecule. This concept is well illustrated by the following trivial example: when *p*-nitrobenzoic acid is considered as a benzoic acid perturbed by a nitro group, the calculated acidic $pK_a = 3.42$ agrees very well with experimentally determined[87] value of 3.44 (see **Figure 4**). However, when the molecule is split into a formic acid 'parent' and *p*-nitrophenyl 'substituent,' the agreement becomes much worse with calculated $pK_a = 2.62$.

Obviously, the parent molecule must carry the bulk of the chemical information (aromatic carboxyl as opposed to aliphatic carboxyl in this case) pertaining to a given reaction series. In this case, the application of substituent corrections to the parent's experimental pK_a^0 may result in highly accurate estimation. Indeed, in many simple systems, the prediction results turned out to be quite precise. For more complex molecules the parent/substituents may not be so clear, leading to ambiguity in selecting an appropriate equation and set of constants.

By the early 1980s, LFER methods had been studied in depth and developed into an elaborate prediction system, now firmly established in both academia and industry. Thousands of Hammett and Taft constants were determined

$$pK_a = 4.66 - 1.62\Sigma\sigma^*$$
$$\sigma^* (p\text{-}C_6H_4NO_2) = 1.26$$
$$pK_a = 2.62$$

$$pK_a = 4.20 - 1.0\Sigma\sigma$$
$$\sigma(p\text{-}NO_2) = 0.78$$
$$pK_a = 3.42$$

Figure 4 Two different ways of splitting p-nitrobenzoic acid into a 'parent molecule' and a 'substituent.' Values of pK_a are calculated for each split using the LFER approach. Constants and equations were taken from Perrin, D. D.; Dempsey, B.; Serjeant, E. P. pK_a Prediction for Organic Acids and Bases; Chapman and Hall: London and New York, 1981.

experimentally for many systems. Cases where LFER breaks down have been identified.[88–90] Once again, the author is vastly grateful to D D Perrin, B Dempsey, and E P Serjeant for writing such an excellent reference book[7] on the subject, where principles and variations of LFER-based pK_a prediction have been explained and richly illustrated with a multitude of examples. This enables us to concentrate this chapter on significant methodological developments that occurred after the book was published.

5.25.3.1.1.2 SPARC method: a step beyond Hammett–Taft approach

In the early 1990s Karickhoff, Carreira, and co-workers announced SPARC – an expert system for predicting chemical reactivity from structures.[91–95] The method was a blend of the LFER approach used to compute thermodynamic properties, and Dewar's perturbed molecular orbital (PMO) theory[96] to describe quantum effects such as electron delocalization energies or π-polarizabilities. The basic philosophy of SPARC is fairly in tune with that described above. A molecule is divided into a reaction center C (equivalent to parent molecule) and a set of one or more perturbers P (substituents). With '0' and 'f' describing the initial and final state in a chemical reaction, the energy difference, $\Delta E \equiv E_f - E_0$, between products and reactants is expressed as:

$$\Delta E = \Delta E_C + \delta_P(\Delta E_C) \tag{7}$$

where the first term is an intrinsic energy change of the reaction center inferred from experiment and the second describes perturbational corrections facilitated by perturbers. At first glance, eqns [6] and [7] are essentially the same. However, the perturbation term in eqn [7] is not a simple linear function of substituent constants, but a sum of terms corresponding to a number of potential mechanisms for interactions of P and C. For ionization reactions in water, the expansion is:

$$\delta_P(pK_a^0) = \delta_{ele} + \delta_{res} + \delta_{sol} + \delta_{H\text{-}bond} + \delta_{stat} \tag{8}$$

The individual terms in this expansion describe electrostatic, resonance, solvation, hydrogen bonding, and statistical effects of the perturber(s) (*see* Section 5.25.2.2). The electrostatic term is further split into three terms:

$$\delta_{ele} = \delta_{fixed} + \delta_{mesomeric} + \delta_{induction} \tag{9}$$

reflecting interactions of permanent charges and dipoles, and induced charges through mesomeric and inductive effects, respectively. An effort is made to express the δ-terms for each perturber P as a product of three portable structural components:

$$\delta_x(C, P) = \rho_C B_{P-C} F_P \tag{10}$$

where F_P describes the perturber's 'strength' to create the effect x, ρ_C is the susceptibility of the reaction center to effect x, and B_{P-C} is the 'conduction descriptor' of the intervening bond network between P and C. This expression is obviously reminiscent of the 'classic LFER' reaction constants and Hammett–Taft constants. The reader should keep in mind that the σ_S term in eqn [6] is a simple constant wholly describing substituent S, while terms in eqn [10] may be complex functions of the molecular structure, themselves depending on many parameters.

5.25.3.1.1.3   MCASE methodology

At about the same time Klopman and Fercu applied the MCASE expert system to study the primary acidity of thousands of organic acids.[97] Although the title of their paper contains the 'QSAR' phrase which seemingly qualifies it for the next section, most of Klopman's approach is essentially based on the perturbational LFER principles. The MCASE program has been specifically designed for studies of biological or physicochemical activity of chemical compounds.[98] All investigated molecular structures are fragmented into all possible 2–10 heavy atom connected substructures. These fragments are then labeled as 'active' or 'inactive' depending on the parent molecule activity and some arbitrary threshold. Fragments of highest activity can be used as descriptors (binary indicator variables) in a multivariate regression analysis. Since activity of compounds may be due to different mechanisms, MCASE searches the whole database for a fragment (called 'biophore') accounting for the activity of the largest portion of the database. All the compounds associated with this biophore are assumed to possess the same mode of action and are subsequently removed. The process of searching the next biophore is then restarted on the remainder of the data until all the biophores are found. The 2464-acid database used to build a model of primary pK_a provided 22 distinct biophores, with a carboxylic group being the most import one. All the remaining molecular fragments are called "modulators." For each database subset characterized by a common biophore a local multivariate linear regression model is built using modulators as descriptors. The equation used by Klopman for pK_a was:

$$pK_a = \text{const} + \sum r_i\, n_i\, M_i \qquad [11]$$

where M_i is a binary {0,1} indicator for the presence of modulator i, n_i is the number of modulators i in a compound, and r_i stands for the regression coefficient. Because of the locality of the regression models built for distinct classes of compounds, there are formal similarities between the Klopman method and the classical Hammett–Taft approach. It can be seen by the association of concepts: a biophore can be equivalent to a parent molecule, modulators can be perceived as perturbing substituents, the constant parameter in eqn [11] would resemble the biophore's intrinsic susceptibility to ionization, like pK_a^0, and regression coefficients could encode products of reaction and substituent constants. The difference is the scope of the models involved. The equations of classical LFER apply to very narrow classes of compounds; e.g., substituted acetic acids need different equations from those for substituted formic acids. On the other hand, Klopman's equation for the –COOH biophore applies to all carboxylic acids; it is much less 'localized.' Nevertheless, the basic principles of the perturbational approach, including fragmentation of molecules, are still preserved.

The real novelty in Klopman's work is inclusion of additional molecular descriptors in the linear regression equations as 'modulators,' but only for selected biophores. The descriptors were physicochemical (log P, water solubility, molecular weight) and quantum chemical (Hückel MO charge densities, HOMO and LUMO coefficients, absolute electronegativity, and hardness). Whole-molecule descriptors are clearly beyond the concepts of perturbational LFER. Regression coefficients of these descriptors, however, were significantly overshadowed by contributions from fragmental modulators.

5.25.3.1.1.4   Concluding remarks

Since 1981 many researchers have proposed numerous schemes for improving and refining the existing LFER methodologies to predict chemical reactivity. Regretfully, space limitations preclude this author from providing an exhaustive review of these techniques. The interested reader is referred to the original literature, taking [23,25,26,28–33,46,99–101] as a good starting point.

The prediction of all macroscopic pK_a for molecules with multiple ionizable groups using microspeciation is possible within the LFER framework, but at the expense of adding more substituent parameters. For example, to calculate all microconstants proper to an amine group in cetirizine (Section 5.25.2.3), one would need to know Taft σ^* parameters not only for –COOH and >N– groups, but also for their ionized forms, –COO$^-$ and >NH$^+$–. Thus, the number of required parameters for ionizable substituents doubles. Other challenging issues mentioned in Section 5.25.2.3 still remain. To the author's knowledge, only the SPARC method has been demonstrated as fully capable of computing multiprotic pK_a through a complete microspeciation.[93,95]

As a prediction system, LFER provides good estimations of pK_a, but even in its simplest form it requires vast numbers of parameters (many of which must be measured) and highly diverse, huge databases of experimental results. It is therefore very expensive to develop. Of course, it is the reflection of a very large number of possible parent molecules multiplied by various substituent combinations. Mathematically, a complete LFER system consists of thousands highly localized linear regression models plus complex rules of relating model parameters to ever-changing molecular topologies. Again, if a molecule in question falls within the scope of one of these models, the estimate of its ionization may be of high accuracy. From the point of view of an innovative pharmaceutical chemist, the perturbational nature of the LFER approach may be the source of its weakness when investigated drug candidates are so novel that suitable parent molecules cannot be found in existing databases. It is then necessary to conduct an appropriate series of experiments to arrive at new parameterization of, e.g., Hammett–Taft equations applicable to the studied compounds.

5.25.3.1.2 Methods based on quantitative structure–property relationships

Others chapters in this book (see all the chapters of 'In Silico Tools in ADMET' in this volume) well illustrate the success of predicting whole molecule properties from calculated structural descriptors using a QSAR/QSPR approach. In its most general form, the quantity of interest, y, is a function, sometimes very complex, of molecular descriptors, $\{x_1 \ldots x_N\}$:

$$y = \sum_j a_j f_j (x_1 \cdots x_N; b_1 \cdots b_M) + \text{const} \qquad [12]$$

where a_j and $\{b_1 \ldots b_M\}$ denote linear and nonlinear parameters of the model. If all the functions f_j are simple identities, i.e., $f(x) = x$, then the model is a simple multiple linear regression, most common in the literature. But the above equation may also denote one of the machine learning methods, such as artificial neural networks.[102-120] The main problem in applying eqn [12] to pK_a prediction is the choice of proper descriptors. Clearly, unlike lipophilicity or solubility, which can be modeled by molecular descriptors, ionization constants localized on specific atoms require a different kind of descriptors.

Similarly to LFER approaches, QSPR models may or may not take microspeciation into account. Microspecies are distinguished naturally if model descriptors take into account the protonation state of other groups. Although quite challenging to model developers, exact and complete inclusion of microspeciation is rewarded with much better pK_a prediction for multiprotic systems.

5.25.3.1.2.1 Atomic descriptors

Already in 1938 Kirkwood and Westheimer used classical electrostatics of spherical and ellipsoidal molecules immersed in continuous dielectric medium to predict pK_a shifts of organic diacids.[8-10] Extension of this theory to the electrostatic influence of any monopolar or dipolar substituent at a distance r from the reaction center results in the following expressions:

$$\Delta pK_a = \frac{q_1 q_2}{2.303 kT \varepsilon r}$$

$$\Delta pK_a = \frac{q_1 \mu_2 \cos\theta_{12}}{2.303 kT \varepsilon r^2} \qquad [13]$$

where q_1 and q_2 are electric charges on the reaction center and substituent, respectively, μ_2 is the dipole moment of the substituent, and θ_{12} is the angle between dipole moment vector and the axis joining the reaction center and substituent. The symbol ε denotes an effective dielectric constant, whereas kT is the absolute temperature scaled by the Boltzmann constant. Please note that values of these expressions depend both on a specific reaction center atom and its molecular environment. Hence, eqn [13] can be viewed as the first literature example of an atomic descriptor defined as a function describing an atom in a molecule. It seems natural that atomic descriptors can serve as inputs to QSPR models describing localized properties, like ionization constants, similarly to molecular descriptors used in modeling whole molecule properties.

Examples of atomic descriptors are scarce in the scientific literature. To author's best knowledge, the following paragraphs will be the first review of this subject. Catalan *et al.* expressed gas-phase proton affinities of oxygen, nitrogen, and carbon bases as multivariate linear functions of 1s orbital binding energies and first ionization potentials.[121] Many other types of quantum-mechanical atomic descriptors were also used by other research groups in developing QSPR correlations for either gas-phase proton affinities, or aqueous pK_a of narrow classes of compounds.[34,35,39,40,122-128] Unfortunately, quantum descriptors are (still) computationally too expensive to be used in high-throughput in silico predictions. Their review is postponed until the next section. An excellent example of empirical atomic descriptors is described in the 1984 paper of Gasteiger and Hutchings.[129] These authors used residual electronegativities and effective polarizabilities (serving as charge stabilizers) of a protonated group to predict gas-phase proton affinities of alcohols, ethers, their thio analogs, and carbonyl compounds.[130] Noteworthy, these descriptors do include the influence on the ionization center exerted by type and distance of the remaining atoms present in concentric topological spheres. This approach was later extended to ionization in aqueous phase.[131] Recently, Gasteiger reported on an extended set of atomic descriptors, both empirical and quantum, fed into artificial neural networks that successfully predicted pK_a of monoprotic –C–OH and –C–SH acids and monoprotic >N– bases.[132] Empirically calculated inductive partial atomic charges by an electronegativity-based algorithm were also utilized as atomic descriptors in QSPR models of pK_a of organic –OH acids.[133] Gross *et al.* critically reviewed different partial charge algorithms, most of them quantum-mechanical, with respect to the quality of correlation with pK_a of benzoic acids[34] as well as phenols and anilines.[134] Partial charges alone, however, may not be sufficient to build a universal pK_a model applicable to a wide class of compounds. For example, in a series of alkylamines NR_xH_{3-x} (R = Me,Et) the partial charge on the N atom increases with x implying smaller electron density and diminishing proton affinity.[27] The observed proton affinity trend, however, is quite the opposite. Mercier and co-workers list both empirical and quantum atomic descriptors used along with molecular descriptors in the predictive modeling system OASIS to model pK_a of purines: atomic topological and steric indices.[135] Rajasekaran revisited the Kirkwood–Westheimer descriptor of electrostatic repulsion between charged groups in dicarboxylic acids,[20] while Charton has analyzed its distance dependence in detail.[25]

All of the above examples are quite limited in scope, although they comprise important stepping stones toward a general predictive model. It is the universality of pK_a prediction that is the most alluring promise of QSPR models using calculated descriptors. If properly chosen and parameterized, atomic descriptors, calculable for any input molecule, may account for all of the factors influencing protonic ionization and may lead to predictive models that are strong, where LFER approaches are weak in their heavy reliance on experimental constants and their limited scope. Hence, global predictive models will be the focus of the following discussion.

5.25.3.1.2.2 The method of Xing

A predictive model with wide scope was the leading ambition of Xing and coworkers when they applied molecular tree structured fingerprints (MTSF), a novel class of universal atomic descriptors based on SYBYL atom types (Tripos, Inc.[169]), to build QSPR models of pK_a for 645 acids and 384 bases.[136] The idea of MTSF is very simple: progressive breadth-first traversal of a molecular graph rooted in an ionizable atom and counting specific atom types and functional groups at each search depth level.

This is best explained by an example, shown in **Figure 5**. Potentially, the ionizable atom type can be any one of the 24 defined in SYBYL (search level 0). There are 33 atomic descriptors (24 atom types and 9 types of functional groups) defined at each subsequent search level. The breadth-first search is truncated on depth level 5. Putting the numbers together, each ionizable atom can be associated with a $24 + 5*33 = 189$-dimensional vector of atomic descriptors. The exact number is smaller since 'impossible' atom types were eliminated at certain levels. Still, there are many descriptors, which is a concern given the relatively small size of training data sets. Xing *et al.* reduced the independent variable dimensionality by training their model using a partial least squares (PLS) algorithm.[137] The resulting model was able to predict only 'single' (i.e., monoprotic, while assuming all other ionizable groups in their neutral state) pK_a values, but for a wide array of diverse organic acids and bases. The model was tested on an external set of 25 molecules taken from Perrin's book[7] that were absent from the training sets. It 'correctly predicted the order of the ionizing strength of individual groups in multiprotic molecules' from the predicted single pK_a although in quantitative terms its performance was worse compared to the results of the Hammett–Taft approach used by Perrin. These results well illustrate the serious difficulties in building a universal model for all types of acids or bases, especially when the model uses a single class of descriptors. Overall, the intriguing approach of Xing was probably the first published QSPR model of pK_a with such a wide scope of applicability and great potential for future development. In their next work Xing and

	N.3	C.3	C.ar	Cl	O.3	H
Level 0	1	0	0	0	0	0
Level 1	0	1	0	0	0	2
Level 2	0	1	1	0	0	1
Level 3	0	0	2	0	0	3
Level 4	0	0	2	0	1	1
Level 5	0	1	1	1	0	1

| 1 | 0 | 0 | 0 | 0 | 0 | 0 | 1 | 0 | 0 | 0 | 2 | 0 | 1 | 1 | 0 | 0 | 1 | 0 | 0 | 2 | 0 | 0 | 3 | 0 | 0 | 2 | 0 | 1 | 1 | 0 | 1 | 1 | 1 | 0 | 1 |

Figure 5 An example of MTSF descriptor vector created for the amine atom in the molecule shown. The sp³ nitrogen atom (type = N.3) constitutes the entire 0th search level by itself. Subsequent levels 1–5 contain counts of any of the shown six atom types leading to a 36-dimensional vector of descriptors.

coworkers managed to significantly improve predictive performance by incorporating the idea of building separate QSPR models for narrower classes of compounds[138]:

1. Aromatic carboxylic, sulfonic and sulfinic acids.
2. Phenols and thiophenols.
3. Aliphatic and alicyclic carboxylic, sulfonic, sulfinic acids plus alcohols and thiols.
4. Acidic nitrogens and carbons.
5. Pyridines (defined here as six-member aromatic rings with at least one nitrogen).
6. Anilines.
7. Imidazoles.
8. Alkylamines.

At the same time, however, the researchers drifted off the modeling universality afforded by their previous approach. Namely, the eight classes were further divided into narrower subclasses depending on chemical characteristics of member molecules. Ionization constants were then modeled as follows:

$$\mathrm{p}K_a = \mathrm{p}K_a^0 + \sum_i a_i x_i + \sum_j g_j y_j + \sum_k q_k z_k \qquad [14]$$

where $\mathrm{p}K_a^0$ is a constant characteristic for each subclass, x_i and y_j are atom type and group type MTSF descriptors after PLS analysis, z_k are class-specific indicator variables, while a, g, and q symbolize regression parameters. For example, in the first class there were three distinct $\mathrm{p}K_a^0$ constants, one per each ionizing group type. The indicator variables, z_k, for the same class consisted of MTSF descriptors calculated for specific atoms at *ortho*, *meta*, and *para* positions on the six-member ring with respect to the acidic group. This definition by itself greatly limits the scope of the model since it excludes, e.g., a carboxyl group attached to a five-membered heteroaromatic ring, fused rings, etc. Characteristic constants, $\mathrm{p}K_a^0$, and indicator variables, z_k, used here are quite reminiscent of the 'parent molecule' $\mathrm{p}K_a$ and Hammett substituent constants shown in eqn [6]. Furthermore, multiprotic molecules were assumed to (de)protonate sequentially ignoring microspeciation. Individual protonation states were guessed from the predicted single $\mathrm{p}K_a$. The overall performance of this model was greatly improved in comparison to the preliminary model of Xing *et al.* but at a serious cost of scope limitation. The latter is proven by the fact that due to 'missing parameters' the new model could not process as many as 4 out of the 25 test molecules.[138] The same test set reported in the first Xing's paper was processed in its entirety.

5.25.3.1.2.3 Commercial models

Commercial researchers working for various chemistry software companies developed other global QSPR models for $\mathrm{p}K_a$ prediction. Unfortunately, due to competitive concerns, details of these models were not published. Nevertheless, some information is available from the respective companies' websites (URL addresses are summarized in **Table 1**) and posters presented at conferences. CSpKa from ChemSilico is a composite of 12 artificial neural network models for 12 ionization center types using topological and electrotopological descriptors (*see* 5.23 Electrotopological State Indices to Assess Molecular and Absorption, Distribution, Metabolism, Excretion, and Toxicity Properties). It is not clear whether these descriptors are calculated at the atomic level. On the other end of the spectrum exists the ChemAxon's 'pK_a Plugin' to the program Marvin with a complete treatment of microspeciation. Hence, multiprotic ionization is treated accurately, at least in theory. However, the model uses just three simple atomic descriptors (partial charge, atomic polarizability, and occasionally intramolecular hydrogen bond indicator) in simple nonlinear relations for microconstants.[139]

One of the recently developed commercial models can be described in much greater detail, though. While working at Simulations Plus, Inc., the author of this chapter (with a substantial help from coworkers) developed a comprehensive QSPR model for the prediction of multiprotic $\mathrm{p}K_a$, now a part of the ADMET Predictor software. For this purpose, we have created a unique and, most importantly, diverse set of 44 empirical atomic descriptors encompassing all the factors that influence ionization (*see* Section 5.25.2.2). A raw data set of 8590 diverse organic compounds with measured 10 674 multiprotic ionization constants was purchased from BioByte, Inc. (*see* Section 5.25.5). Each of these $\mathrm{p}K_a$ values was assigned to the macroscopic protonation states of the respective molecules, mostly through referral to the original literature. Of these, 2143 $\mathrm{p}K_a$ values (complete compounds) were sequestered as an external test set, while 8531 of the remaining constants were used to train the model. We have chosen to construct artificial neural network ensembles (ANNEs) optimized with an early stopping criterion[140] to prevent overtraining. Neural networks are well known for their excellent ability to encode nonlinear relationships between descriptors and a predicted property. The author has developed a special training algorithm that optimizes ANNE weights against measured $\mathrm{p}K_a$ macroconstants.

Table 1 Software for empirical pK_a prediction

Software product	Vendor	Method	Treatment of microstates	URL
ACD/pK_a DB	Advanced Chemistry Development	LFER	Partial[a]	www.acdlabs.com
Pallas	CompuDrug	LFER	None	www.compudrug.com
SPARC	University of Georgia	LFER + PMO[b]	Complete	ibmlc2.chem.uga.edu/sparc/index.cfm
Marvin	ChemAxon	QSPR	Complete	www.chemaxon.com
ADMET Predictor	Simulations Plus	QSPR	Complete	www.simulations-plus.com
CSpKa	ChemSilico	QSPR[c]	None	www.chemsilico.com
ADME Boxes	Pharma Algorithms	Database[d]	None	www.ap-algorithms.com
DISCON	Russian Academy of Medical Sciences	Database	None	www.ibmh.msk.su/molpro/discon.html

[a] Limited to diprotic and perfectly symmetric systems only ('apparent exact' constants in ACD terminology).
[b] Blend of LFER and perturbed molecular orbital (PMO) methods.
[c] Limited to five ionization centers per molecule only.
[d] Estimates only principal acid and base pK_a.

The artificial neural network outputs in an ensemble correspond to the ionization of individual atoms, while protonation states of all other ionizable atoms are strictly determined, i.e., to microconstants. This, plus the fact that this pK_a routine can process about 537 complete microstates per second on average (version 1.1.4, Pentium IV/512 MB RAM), allows the model to completely calculate all the 2^n microspecies for an n-protic molecule, and hence to treat microspeciation exactly. Moreover, we assumed that the main factor determining an atom's ionization is its type, followed by its molecular environment. Consequently, we have built separate ANNEs for the following eight atom types:

1. O in all hydroxyacids (–OH group; alcohols, phenols, carboxyls, sulfo acids, phosphoric acids, etc.).
2. O in basic N-oxides (O=N– group; both aromatic and tertiary aliphatic).
3. S in thioacids (–SH group; thiols, thiocarboxyls, etc.).
4. C in carboacids (–CH2– and >CH– groups; barbiturates, acetylacetone, nitroalkanes, etc.).
5. N in acidic aliphatic amides (–NH2 and –NH– groups; carboxyamides, sulfonamides, etc.)
6. N in acidic aromatic NH (–NH– incorporated into an aromatic ring; uracils, imidazoles, etc.).
7. N in basic amines (–NH2, –NH–, >N– groups; aliphatic amines, aromatic anilines, etc.).
8. N in basic aromatic N (–NH– and –N– incorporated into an aromatic ring; pyridines, pyrroles, etc.).

The above types encompass a vast majority of organic acids and bases and reflect the diversity of BioByte database. The corresponding ANNEs are combined into one predictive model within the microequilibria framework. The actual numbers of selected descriptors, as well as network complexity, are commensurate with the representation in the above eight subsets. The fact that it was at all possible to build one ANNE covering, e.g., such a multitude of markedly different aromatic and aliphatic hydroxyacids spelled out in type 1 accounts for the diversity of the employed atomic descriptors. There are no structural limitations and the model is global in the sense of covered atom types. However, the model would not predict ionization of some 'exotic' groups like, e.g., hydroselenides, –SeH. To the author's knowledge, no other LFER or QSPR method would process such acids, either. Of course, if such experimental data becomes available, there is no obstacle to building additional ANNEs.

5.25.3.1.3 Methods rooted in database lookups

Roger Sayle has noticed that direct assignment of 'typical' pK_a values according to detected ionizable atom types predicts the pK_a of simple molecules with a standard deviation of 0.95 log units.[141] The observation that similar molecules, or similar molecular fragments, have similar pK_a is the basis of the last class of methods. A database containing either complete structures or structural fragments and experimental pK_a is searched for the nearest neighbors of the compound in question. Correction factors may be eventually applied to found pK_a values. Of course, treatment of microspeciation is not possible in this approach.

5.25.3.2   Quantum Chemical Methods

This section starts with the continuation of the QSPR approach, but this time with quantum mechanical descriptors. A brief description of the first principles method is presented thereafter.

5.25.3.2.1   Predictive quantitative structure–property relationship correlations using quantum mechanical descriptors

The diversity, universality, and level of detail represented by quantum mechanical atomic descriptors vastly surpass that offered by empirical atomic descriptors. Therefore, high-accuracy QSPR predictions of ionization are possible in this case. A global model of this kind including multiprotic molecules has not yet been built. The main drawback is the relatively high computational cost of solving the Schrödinger's equation, mandating long processing times per molecule. Numerous authors have studied a multitude of local correlative models. Longuet-Higgins found out that pK_a differences in the class of heteroaromatic amines are explained by a product of the atomic π-density and local perturbation of the Coulomb integral.[15] Study of the basicity of heteroaromatic nitrogens was continued by Spanget-Larsen who correlated pK_a and proton affinities with shifts of valence atomic orbital energy occurring upon placing a neutral nitrogen atom in a molecule.[126] The energy shifts were in turn expressed as simple functions of atomic charges and electronic repulsion terms. Semiempirical partial charges, highest occupied molecular orbital (HOMO) energies of the anion, as well as the energy difference between anionic and neutral forms were correlated with pK_a of phenols and carboxylic acids.[122] The same group of acids plus aliphatic alcohols were revisited again by Citra, who added O–H bond orders to the semiempirical partial charge descriptors.[123] Atomic charges were a frequently recurring type of descriptors. Gross et al. evaluated different means of computing partial atomic charges for the purpose of predicting ionization variations in anilines and phenols.[134] The same group later extended the list of studied descriptors to geometrical properties (bond lengths and angles), rotation barriers around C–O bonds, orbital energies, relative proton transfer enthalpies,[39] as well as minimum local properties on molecular surface: ionization energy and electrostatic potential.[40] The quality of these correlations equaled and is some cases surpassed that afforded by empirical Hammett constants. Molecular electrostatic potential was also applied to study the basicity of amines,[44] later extended by computation of hydration effects.[142] Tehan et al. embarked on using parameters obtained from frontier orbital theory: electrophilic and nucleophilic frontier electron densities, electrophilic, nucleophilic, and radical superdelocalizabilities, and atom self-polarizabilities to predict pK_a of, once again, phenols and carboxylic acids,[143] but also amines, anilines, and heterocyclics.[144] Impressive accuracy has been achieved by Adam who correlated ionization of carboxylic acids, phenols and pyridines with the Atoms in Molecules energy of the dissociating proton.[145] Based on the same theory and the same classes of molecules, quantum topological molecular similarity descriptors were tested for pK_a estimation.[128] Multiple or single linear regression was the modeling method chosen in all of the above treatments, except the last one that invoked partial least squares.

An interesting group of approaches applied comparative molecular field analysis (CoMFA) to pK_a prediction of benzoic acids and imidazoles,[124,125] and nucleic acid components.[127] CoMFA is usually used to correlate biological activity with molecular structures of very narrow classes of compounds having a common scaffold. In brief, three-dimensional structures of these molecules are superimposed using scaffold atoms, then immersed in a discrete cubic lattice. Next, electrostatic potential energy is calculated at each lattice point with the aid of specific probe (e.g., H^+ ion). The electrostatic field is generated by partial atomic charges calculated by quantum methods. Energy values are then placed in a matrix whose columns are labeled by individual lattice points and rows by participating molecules. Partial least squares transformation reduces usually large row vector dimensionality and leads to much smaller number of latent components – the CoMFA descriptors. A related technique, comparative molecular surface analysis (CoMSA) was also applied to benzoic and alkanoic acids.[146] Finally, many authors pursued an indirect pK_a estimation by correlation of Hammett and Taft constants with quantum mechanical descriptors.[11,13,14,17,34,35,147–151]

The above are mere examples of the amount of important research work done in this field. In spite of the large number of papers a general QSPR model utilizing quantum mechanical descriptors has not emerged so far. Up to date research efforts in this arena should therefore be classified as exploratory.

5.25.3.2.2   Ab initio calculation of ionization constants

The most universal computational methods for pK_a prediction based on first principles of quantum mechanics are also most time-consuming of all. The underlying paradigm here is thermodynamic description of ionization already explained in Section 5.25.2.1. Quantum ab initio methods allow for calculation of the free energy terms present in eqns [3] and [4]. The inherent difficulty is that the road to the final, relatively small free energy of ionization leads through subtraction of sizable terms. In one anecdote it is said that the problem is equivalent to determining the weight of a transoceanic ship's captain by weighing the entire ship with the captain aboard, then weighing the ship without him and subtracting the two numbers. Consequently, 'ship weights' in pK_a prediction must be determined with excruciating accuracy. An error of just $5.7 \, kJ \, mol^{-1}$ in calculated free energy corresponds to the pK_a shift of 1 log unit.[145] Although high-precision calculations of a single molecule energy in gas phase have recently become possible, hydration terms remain a bottleneck hindering progress in this area. Classical molecular mechanics models hydration by surrounding the molecule in question with many water molecules and solving Newton's equations in time domain. Calculations of this kind in quantum realm are still beyond the reach of existing computers. Therefore, many different approximation schemes have been proposed for aqueous hydration. The common method for calculating hydration terms is the polarized continuum method (PCM). In there, solvent is approximated by a homogenous continuum polarized by the solute placed in a cavity. Solvent interactions are then included as one-electron interaction potential terms in the system Hamiltonian. This and other methods have been reviewed by Tomasi[152] and Cramer.[153]

Quite often pK_a values are calculated indirectly, i.e., instead of aqueous dissociation shown in **Figure 1** one calculates the free energy of proton exchange between the molecule in question, HAR_1, and its close analog, HAR_2:

$$HAR_1 + [AR_2]^- \Leftrightarrow [AR_1]^- + HAR_2$$

Calculations for this isodesmic (i.e., where the number of bonds remains invariant) process are more accurate due to cancelation of errors in the entropy and hydration terms. Still, use of finite basis sets introduces systematic bias in all the energy terms that can result in calculated pK_a values twice as large as observed.[154]

In principle, any ab initio software package can be used to calculate ionization constants. Program Jaguar has been tailored specifically for this purpose and includes linear empirical correction factors applied to the 'raw' ab initio results (Jaguar 4.2: Schrödinger, Inc.[170]). Each chemical class has its own equation derived from a direct fit to experimental data. It seems then that the Jaguar's 'raw pK_a' play a factual role of rather costly descriptors fed into linear QSPR models.

5.25.3.3 pK_a Estimation for Biopolymers

Biopolymers (polymeric molecules of biological origin: proteins and nucleic acids) pose a special challenge for pK_a predictors. Large number of ionization centers, varying solvation environments, and conformational flexibility greatly increase complexity of the problem in comparison to small molecules. In spite of this difficulty, solving this puzzle received significant attention by scores of research groups around the world, since key properties of biopolymers, such as structural folding, stability, ligand binding, enzyme activity, etc., are strongly dependent on pH. Current mainstream methodologies for pK_a prediction in proteins and nucleic acids are based on classical molecular mechanics and statistical thermodynamics. Since these methods have been thoroughly described in a recent review,[155] we will only summarize major points to help in their understanding.

Unlike small molecules whose topological diversity is almost unlimited, naturally occurring biopolymers are built of monomeric units coming from a limited set of molecules. Ionization properties of, e.g., amino acids, the building blocks of proteins, are well known, and so are the nearest local environments of all possible ionization centers. In tune with LFER principles, the pK_a of, for example, the $-NH_2$ group in lysine's side chain can be decomposed into two terms:

$$pK_a^P = pK_a^S + \Delta pK_a^{S \to P} \tag{15}$$

where the superscript P indicates protein environment, S stands for the bulk solvent, and the ΔpK_a term indicates ionization constant shift on going from solvent to protein environment. The above expression is understood in terms of a model compound undergoing the thermodynamic cycle similar to the one depicted in **Figure 1**. Since protein researchers think in terms of amino acid residues, most often the model compound is a free amino acid whose alpha amino and the carboxyl ends have been substituted by blocking groups, but it can also be reduced to a rigid part of the molecule around the given ionization center.[156] The pK_a^S term is then the ionization constant of the free model compound in bulk water, in analogy to the gas phase ionization shown in **Figure 1**. The bottom part of **Figure 1** corresponds to the dissociation of the same group while attached to the protein molecule. Ideally, the pK_a shift would cover an entire set of complex interactions turned on by incorporating the model compound onto the protein backbone. The following equation illustrates key simplifying assumptions made at this point:

$$\Delta pK_a^{S \to P} = \Delta pK_a^{(pc)} + \Delta pK_a^{(ic)} + \Delta pK_a^{(desolv)} \tag{16}$$

In other words, only the following effects are taken into account: an electrostatic interaction with protein's permanent charges (pc), an electrostatic interaction with other ionization centers (ic), and a desolvation penalty (desolv). It must be stressed out that all of the above terms strongly depend on the protein's conformational flexibility. Other ionization centers are distinguished from permanent charges because their charge state depends on pH. Desolvation energy is a cost of losing a part of the hydration shell when the model compound is placed in the protein environment (compare to vertical arrows in **Figure 1**). Equations [15] and [16] are then combined and rearranged into pH-independent and pH-dependent terms:

$$pK_a^p = pK_a^{(int)} + \Delta pK_a^{(ic)}(pH) \qquad [17]$$

where superscript (int) stands for the intrinsic pK_a in protein attained when all other ionization centers are neutral. Both terms are calculated with the aid of molecular mechanics force fields and the solution of the Poisson–Boltzmann equation for electrostatic potential, plus the assumed solvation model (continuum dielectric, explicit water molecules, etc.).[155] While $pK_a^{(int)}$ needs to be calculated only once per given conformation, the mutually dependent interactions of the multiple ionization centers cannot be calculated directly. Instead, Bashford and Karplus proposed microequilibria approach covering all n ionizable centers in the protein molecule,[157] where the probability of a given ionization site i to be protonated is given by the Boltzmann weighted sum:

$$p_i = \frac{\sum_{\{\vec{s}\}} s_i \exp(-\Delta G(\vec{s})/RT)}{\sum_{\{\vec{s}\}} \exp(-\Delta G(\vec{s})/RT)} \qquad [18]$$

where both summations are performed over the set of all 2^n microstates, $s_i = 1$ when site i is protonated and 0 when deprotonated, and $\vec{s}$ represent an n-dimensional vector of $\{0,1\}$ protonation states for n sites (microstate vector). The free energy of the entire system in a given ionization microstate, $\Delta G(\vec{s})$, is given by:

$$\Delta G(\vec{s}) = \sum_{i=1}^{n} 2.303 RT (pK_{a,i}^{(int)} - pH) + 1/2 \sum_{i,j \neq i}^{n} W_{s_i s_j} \qquad [19]$$

where $W_{s_i s_j}$ represents electrostatic interaction energy between sites i and j in their respective protonation states. Equation [18] allows for calculating the expected protonation state in the protein molecule at a given pH, which in turn can be used to calculate the pK_a shift present in eqn [17]. Because of the exponential growth of the number of summation terms in eqn [18], it becomes unrealistically time consuming for $n > 30$ sites. In one approximation scheme, distant parts of the protein molecule are summed independently at much lower cost, then their interactions are calculated using averaged electrostatic potentials (mean field approximation). Another is based on Monte Carlo samplings of the microstates. The conformational dependence of the above equations should not be forgotten, either. One conformation is not good enough to obtain realistic results. Calculations are performed over an ensemble of conformations collected experimentally, or via molecular dynamics simulations.

Overall, pK_a calculations of the ionizable sites in biopolymers are a time consuming process. Protein and nucleic acid researchers, however, are rarely interested in the individual pK_a values of specific sites (which may not be very accurate), but rather in the expected protonation microstate distribution as a function of pH, or simply in the total number of ionizable protons attached in function of pH (the titration curve).

5.25.3.4   A Word about Accuracy

In most cases, modern in silico prediction methods allow for a quick and convenient generation of pK_a estimates. Although individual features of different prediction methods are directly comparable, a universal global method of assessing the accuracy of these methods does not exist. Test results regarding new compounds have been published, but the numerical values of prediction errors always depend on a particular data set chosen for the test. This observation is valid for any predictive model. Moreover, the reader should keep in mind that all empirical predictive models are trained on an available set of chemicals with measured data (*see* 5.22 Use of Molecular Descriptors for Absorption, Distribution, Metabolism, and Excretion Predictions). Hence, the amount of interpolated chemical space covered by a model is limited to this training set. Beyond that corner of chemical space, the model extrapolates with diminishing accuracy. Errors ranging from 0.1 up to 2.0 log units are possible. For the above reasons, test results using different databases, or databases that are too small, should not be compared directly. Instead, a user is advised to perform a local validation, i.e., to test each method on his/her particular set of compounds with known pK_a. It is quite possible that different users will find different methods to be superior. Additionally, different users will be satisfied with different levels of the prediction error, depending on a particular application. For example, estimated pK_a is often a stepping stone toward in silico prediction of the complete pH-dependent profiles of drug solubility (log S) and lipophilicity (log D) (*see* 5.16 Ionization Constants and Ionization Profiles; 5.17 Dissolution and Solubility, for detailed explanations). Taking a monoprotic compound as an example, its log S and log D profiles are relatively insensitive to pH except a narrow region in the vicinity of pH $= pK_a$ where the profile curve is very steep. Errors in pK_a estimation may shift the steep part horizontally. Therefore, for users studying drug behavior in blood at pH $= 7.4$ the accuracy of a pK_a predicted at, e.g., 7.1 is much more critical than the accuracy of a $pK_a = 2.3$ estimate. But the latter is much more important for users studying, e.g., drug solubility in stomach. Of course, the same picture for polyprotic compounds is more complicated. It is critical to understand these dependencies, since errors in pK_a prediction may compound with the errors committed during the modeling of intrinsic solubility and log P, respectively.

5.25.4   Software Packages for pK_a Prediction

Currently available software packages for empirical prediction of ionization constants ordered by prediction method are summarized in **Table 1**. Three computer programs have been around for many years, all are based on the LFER approach: ACD/pK_a DB from Advanced Chemistry Development, Pallas from CompuDrug, and SPARC from the University of Georgia. Newer software uses either QSPR methodology (Marvin from ChemAxon, ADMET Predictor from Simulations Plus, and CSpKa from ChemSilico), or database lookups (ADME Boxes from Pharma Algorithms and DISCON from the Russian Academy of Medical Sciences). Information presented below has been taken from the respective company websites and publications, where applicable, with the exception of ADMET Predictor personally developed by author. Interested readers are encouraged to contact companies directly.

5.25.4.1   ACD/pK_a DB and ACD/pK_a Batch

The pK_a prediction software from Advanced Chemistry Development comes in two flavors: DB integrated with ChemSketch graphical interface allows for pK_a predictions one molecule at a time and contains an internal database of nearly 16 000 structures with measured pK_a, while nongraphical Batch form is designed for large sets of compounds at once (version 8.0). The program uses a large number of Hammett-type equations for the popular substituents and ionization centers. When the required substituent constants are not available from the experimental database for a given substitution position they are estimated from related constants via the calculation of transmission effects. The model is extendable if a user has own database of measured pK_a values. The DB version displays method of calculations and literature references, if available. Although the software allows for pK_a microconstants prediction for specific microstates (microconstants of the current form), complete microspeciation is limited to diprotic systems, most likely due to computational performance concerns. Otherwise, 'apparent approximate macroconstants' are calculated from an assumed dominant protonation form. Other limitations involve molecular size (no more than 255 atoms) and the number of ionizable groups (no more than 20). Accepted input formats include internal ACD database, SMILES strings, MDL SDfile, and MDL database file (MDL, Inc.[171]).

5.25.4.2   pKalc Module of Pallas

Not much information about pKalc is available from the CompuDrug website. The program's algorithm uses a set of almost 1000 Hammett- and Taft-like equations. Only the lowest acidic and highest basic pK_a are considered. Ionized substituents are taken into account, but there is no mention about microspeciation. Development of new Hammett–Taft equations is enabled from user's data. A database of experimental ionization constants is included in the program suite.

5.25.4.3   SPARC

Methods used by academic software SPARC[91–95], partially funded by the US Environmental Protection Agency (EPA), have been described in detail in Section 5.25.3.1.1. The program runs on a University of Georgia web server and is accessible through a web browser interface. It is probably the only program that, along with water pK_a, also estimates gas-phase and nonaqueous pK_a in a variety of solvents. Molecular structure must be submitted as a SMILES string,[158] one molecule at a time. A searchable database of measured pK_a is also available. Full microspeciation is supported, but quite limited by the code performance. Submission of viomycin, a cyclic peptide with 16 ionizable groups, in full speciation mode results in the following message: "This will result in ~ 96468992 calculations and may take as long as 964690.59 minutes" (about 1.8 years). Output includes list of dominant species, microscopic and macroscopic pK_a, and pH profiles. For some reason, macroscopic pK_a are not calculated accurately by microequilibria theory equations, but estimated from cross points on the drawn pH profile graphs.

5.25.4.4   Marvin

Similarly to SPARC, ChemAxon's Marvin is a Java-based software application accessible through a web browser. It includes a nice on-line molecular drawing tool. Prediction of pK_a is performed by one of the 'Calculator Plugins.' Molecular structures can be imported in a multitude of formats, but calculations are performed one molecule at a time. Program's method is strong on microspeciation providing a complete microstate distribution of molecules containing no more than 14 ionizable groups making it suitable for multiprotic dissociation. According to an animated demo example, processing time for a 12-group cyclic peptide was 1020 s (17 min, Pentium III/256 MB RAM). On output dominant microspecies are displayed graphically together with their pH-dependent concentration profiles. However, the underlying algorithm for microconstant calculation is extremely simplified: only three atomic descriptors (empirical partial charge, polarizability, and internal hydrogen bonding correction) take into account only selected factors influencing ionization.[139]

5.25.4.5   ADMET Predictor

Methods and models used by ADMET Predictor to predict pK_a have been described in Section 5.25.3.1.2. Since its inception, the program has been coded with high-throughput batch prediction of a variety of molecular properties in mind. Even though full microspeciation puts exponential demands on the computer's CPU, it is calculated completely for each compound. In a recent test, all 65 536 microstates of the 16-group viomycin were processed in 2 min 2 s (version 1.1.4, Pentium IV/512 MB RAM). In the same test, the 12-group amphotericin B was processed in 8 s. Of course, these are extreme examples as drug molecules have smaller number of ionizable groups on average, resulting in a rate of ~ 69 compounds per second including the prediction of the full pK_a spectrum, 244 descriptors, and 50 ADMET properties per molecule. The program accepts inputs in SMILES, SYBYL MAC, MDL MOL, SDF, and RDF formats and can export predicted data in the same form. The atom count is limited to 256 and the ionizable group count is limited to 20 per molecule. Calculation of ionization of isolated aliphatic –OH and aliphatic –C($=$O)NH– groups are ignored by default, but can be turned back on when needed. The output includes predicted macroscopic pK_a and graphical display of contributing microstates with relative percentages and individual proton dissociation probabilities (see **Figure 6**) as well as micro- and macrostate distributions as a function of pH.

5.25.4.6   CSpKa

CSpKa from ChemSilico recognizes the following 12 ionization centers:

Aromatic N	COOH	NH_2	OH
Unsaturated ring N	$O=P(OH)_2$	RNR	$O=POH$
Sulfonamides	Amides	NHR	SH

The corresponding 12 artificial neural network models were built upon 16 000 experimental pK_a for 11 000 compounds. A set of proprietary topological descriptors plus 350 well-known topological and E-state indices are used as input variables.[159] The program accepts structure input in MDL MOL and SDF formats. It is the software's good point that only prediction is returned on output, not a substituted experimental value if the submitted compound was used in the method construction. Prediction is limited up to only five apparent/macroscopic pK_a per molecule assigned to corresponding groups, thus ignoring microspeciation. According to ChemSilico's web page, "[CSpKa] does not treat microscopic constants due to the inability to calculate such constants accurately except for simple organic compounds."

5.25.4.7   ADME Boxes

ADME Boxes from Pharma Algorithms includes the Ionization Filter: a model that counts the total number of ionizable groups and estimates one or two primary pK_a values for the strongest acid and the strongest base, where applicable. The program uses database lookup approach recognizing over 100 electrolyte groups and over 1300 generic skeleton

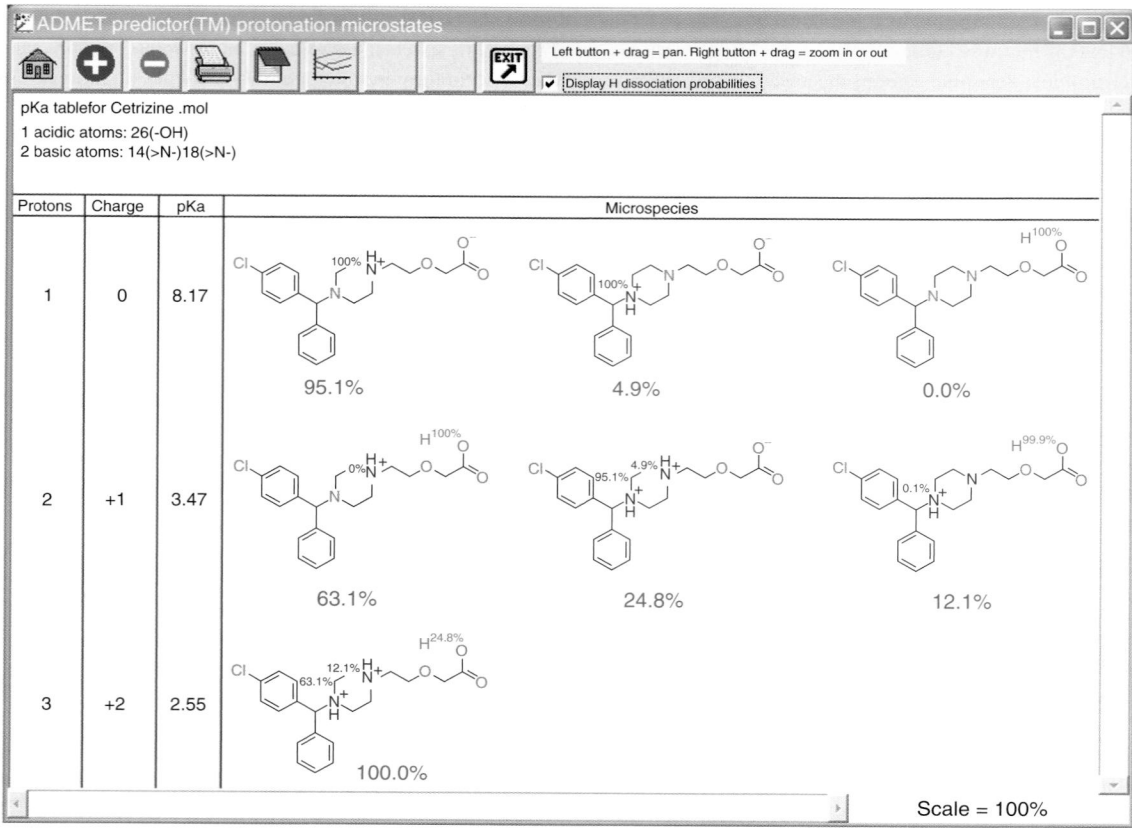

Figure 6   Microspecies distribution for cetirizine calculated by ADMET Predictor.

(fragment) classes. On average a skeleton has been derived from five compounds with known pK_a values. Nearest neighbors are found based on fragment similarity. The novelty is an application of correction factors that 'account for new interactions within generic skeletons.' Over 1000 structures can be processed per minute ($\sim 16\,s^{-1}$). Along with estimated pK_a values fractions of different ionic forms are displayed graphically as a function of pH.

5.25.4.8   DISCON

Little information can be discerned from the DISCON web page at the Russian Academy of Medical Sciences. Estimation of pK_a values is based on the search of the nearest neighbors in the database containing 2471 'chemicals' and 296 drugs. Calculation results include minimal and maximal pK_a values of neighbors and their dispersion.

5.25.5   Sources of pK_a Data

A lot of experimental data on aqueous ionization of organic chemicals has been collected over the last 150 years. Many dissociation constants have been included in literature compilations[7,160–166] and chemical indexes.[167,168] Examples of electronic resources, free and commercial, are summarized in **Table 2**.

5.25.6   Outlook

Undoubtedly, as long as inquisitive curiosity remains a part of human nature, in silico methods of ionization prediction will continue to be developed. Since chemical science became an integral part of many commercial enterprises, strong marketing forces will only add an extra motivation toward creation of more and more sophisticated computer models of not only ionization, but also chemical reactivity in general. Definitely, these trends are observed today. LFER methodology has quickly risen from the vast amount of experimental data collected in the twentieth century to the

Table 2 Electronic sources of pK_a data

Database	Vendor	URL
ACD/pK_a DB	Advanced Chemistry Development	www.acdlabs.com
Pallas	CompuDrug	www.compudrug.com
LOGKOW	Sangster Research Laboratories	logkow.cisti.nrc.ca/logkow/index.jsp
PHYSPROP	Syracuse Research, Inc.	www.syrres.com
BioLoom Database	BioByte, Inc.	www.biobyte.com
pK Database	University of Tartu, Estonia	mega.chem.ut.ee/tktool/teadus/pkdb/
HSDB	National Institutes of Health	toxnet.nlm.nih.gov
SPARC	University of Georgia	ibmlc2.chem.uga.edu/sparc/index.cfm
ADME INDEX	Lighthouse Data Solutions	www.lighthousedatasolutions.com
MolSuite DB	ChemSW	www.chemsw.com

point of becoming a mature methodology. The amount of available new data continues to be its bottleneck. Unless the large vaults of corporate data (especially in the pharmaceutical industry) are open, no significant new developments are expected in this area. On the other hand, empirical QSPR methods for pK_a prediction face wide-open possibilities for further development, especially considering the creation of new atomic descriptors. Quantum QSPR approaches still are in their early development stage, but with suitable progress in computational technology they are destined to catch up with empirical methods. It is author's belief that in the future a distinction between 'empirical' and 'quantum' QSPR will blur out. The net result will be one class of methods generating highly effective atomic descriptors fast enough to be usable for high-throughput prediction of high accuracy. First principles ab initio calculations of ionization will unquestionably benefit from advances in computer technology, albeit at a slower pace. Difficulties in proper treatment of solvation effects will hinder further progress, until computational molecular quantum dynamics becomes a reality. One point remains certain regardless of approach: to be effective no method should discard the underlying physical chemistry of ionization, particularly all the factors influencing pK_a including microequilibria theory.

References

1. Orten, J. M.; Neuhaus, O. W. *Human Biochemistry*, 10th ed.; C. V. Mosby Co: St. Louis, MO, 1982.
2. da Silva, C. O.; da Silva, E. C.; Nascimento, M. A. C. *Chem. Phys. Lett.* **2003**, *381*, 244–245.
3. Pliego, J. R., Jr. *Chem. Phys. Lett.* **2003**, *367*, 145–149.
4. Pliego, J. R., Jr. *Chem. Phys. Lett.* **2003**, *381*, 246–247.
5. Liptak, M. D.; Shields, G. C. *Int. J. Quantum Chem.* **2001**, *85*, 727–741.
6. Topsom, R. D. *Prog. Phys. Org. Chem.* **1990**, *17*, 107–120.
7. Perrin, D. D.; Dempsey, B.; Serjeant, E. P. *pK_a Prediction for Organic Acids and Bases*; Chapman and Hall: London and New York, 1981.
8. Kirkwood, J. G.; Westheimer, F. H. *J. Chem. Phys.* **1938**, *6*, 506.
9. Westheimer, F. H.; Kirkwood, J. G. *J. Chem. Phys.* **1938**, *6*, 513.
10. Westheimer, F. H.; Shookhoff, M. W. *J. Am. Chem. Soc.* **1939**, *61*, 555.
11. Ponec, R.; Girones, X.; Carbo-Dorca, R. *J. Chem. Inf. Comp. Sci.* **2002**, *42*, 564–570.
12. Jaffe, H. H. *Chem. Rev.* **1953**, *53*, 191.
13. Dewar, M. J. S.; Grisdale, P. J. *J. Am. Chem. Soc.* **1962**, *84*, 3539.
14. Dewar, M. J. S.; Grisdale, P. J. *J. Am. Chem. Soc.* **1962**, *84*, 3548.
15. Longuet-Higgins, H. C. *J. Chem. Phys.* **1950**, *18*, 265, 275 and 283.
16. Thomas, T. D. *J. Chem. Soc. Perkin Trans.* **1994**, *2*, 1945–1948.
17. Kuthan, J.; Danihel, I.; Skela, V. *Collect. Czech. Chem. Commun.* **1978**, *43*, 447.
18. Reynolds, W. F.; Mezey, P. G.; Hehre, W. J.; Topsom, R. D.; Taft, R. W. *J. Am. Chem. Soc.* **1977**, *99*, 5821–5822.
19. Houk, K. N. Theory of Acids and Bases. I. Relationship between Structure and Acidity and Basicity. II. Theory of Medium Effects on Acidity and Basicity. In *Acidity and Basicity of Solids: Theory, Assessments and Utility*; Fraissard, J., Petrakis, L., Eds.; NATO ASI Ser., Ser. C; Kluwer Academic Publishers: Dordrecht, 1994, pp 33–51.
20. Rajasekaran, E.; Jayaram, B.; Honig, B. *J. Am. Chem. Soc.* **1994**, *116*, 8238.
21. Yanez, O. M. M.; Esseffar, M. *J. Phys. Org. Chem.* **1994**, *7*, 685–695.
22. Exner, O.; Böhm, S. *J. Org. Chem.* **2002**, *67*, 6320–6327.
23. Exner, O.; Böhm, S. *Phys. Chem. Chem. Phys.* **2004**, *6*, 3864–3871.
24. Exner, O.; Fiedler, P. *Coll. Czech. Chem. Commun.* **1980**, *45*, 1251–1268.

25. Charton, M. J. *Chem. Soc. Perkin Trans.* **1999**, *2*, 2203–2211.
26. Charton, M. J. *Am. Chem. Soc.* **1969**, *91*, 615–618.
27. Maksic, Z. B.; Vianello, R. *J. Phys. Chem. A* **2002**, *106*, 419–430.
28. Pytela, O. *Collect. Czech. Chem. Commun.* **1994**, *59*, 381–390.
29. Pytela, O. *Collect. Czech. Chem. Commun.* **1994**, *59*, 159.
30. Pytela, O. *Collect. Czech. Chem. Commun.* **2002**, *67*, 596–608.
31. Pytela, O.; Otyepka, M.; Kulhanek, J.; Otyepkova, E.; Nevecna, T. *J. Phys. Chem. A* **2003**, *107*, 11489–11496.
32. Brändström, A. *J. Chem. Soc. Perkin Trans.* **1999**, *2*, 1855–1857.
33. Brändström, A. *J. Chem. Soc. Perkin Trans.* **1999**, *2*, 1847–1853.
34. Hollingsworth, B. R.; Seybold, P. G.; Hadad, C. M. *Int. J. Quantum Chem.* **2002**, *90*, 1396–1403.
35. Karaman, R.; Huang, J.-T. L.; Fry, J. L. *J. Comp. Chem.* **1990**, *11*, 1009–1016.
36. Lee, I.; Rhee, S. K.; Kim, C. K.; Chung, D. S.; Kim, C.-K. *Bull. Korean Chem. Soc.* **2000**, *21*, 882–890.
37. Sohn, C. K.; Chun, Y. I.; Rhee, S. K.; Kim, C. K.; Kim, C. K.; Lee, I. *Bull. Korean Chem. Soc.* **2000**, *21*, 1202–1206.
38. Sohn, C. K.; Lim, S. H.; Rhee, S. K.; Kim, C. K.; Kim, C. K.; Lee, I. *Bull. Korean Chem. Soc.* **2000**, *21*, 891–895.
39. Gross, K. C.; Seybold, P. G. *Int. J. Quantum Chem.* **2001**, *85*, 569–579.
40. Gross, K. C.; Seybold, P. G.; Peralta-Inga, Z.; Murray, J. S.; Politzer, P. *J. Org. Chem.* **2001**, *66*, 6919–6925.
41. Babij, C.; Poë, A. J. *J. Phys. Org. Chem.* **2004**, *17*, 162–167.
42. Edward, J. T.; Farrell, P. G.; Halle, J.-C.; Kirchnerova, J.; Schaal, R.; Terrier, F. *J. Org. Chem.* **1979**, *44*, 615–619.
43. Howard, S. T.; Platts, J. A.; Coogan, M. P. *J. Chem. Soc. Perkin Trans.* **2002**, *2*, 899–905.
44. Nagy, P.; Novak, K.; Szasz, G. *J. Mol. Struct. (THEOCHEM)* **1989**, *201*, 257–270.
45. Ohwada, T.; Hirao, H.; Ogawa, A. *J. Org. Chem.* **2004**, *69*, 7486–7494.
46. Swain, C. G.; Unger, S. H.; Rosenquist, N. R.; Swain, M. S. *J. Am. Chem. Soc.* **1983**, *105*, 492.
47. Caskey, D. C.; Damrauer, R.; McGoff, D. *J. Org. Chem.* **2002**, *67*, 5098–5105.
48. Krishnan, C. V.; Friedman, H. L. *J. Phys. Chem.* **1969**, *73*, 1572.
49. Arnett, E. M.; Chawla, B.; Bell, L.; Taagepera, M.; Hehre, W. J.; Taft, R. W. *J. Am. Chem. Soc.* **1977**, *99*, 5729–5738.
50. Bjerrum, N. *Z. Phys. Chem., Stoechiom. Verwandtschaftsl.* **1923**, *106*, 209.
51. Borkovec, M.; Brynda, M.; Koper, G. J. M.; Spiess, B. *Chimia* **2002**, *56*, 695–701.
52. Noszal, B. *J. Phys. Chem.* **1986**, *90*, 4104–4110.
53. Szakacs, Z.; Noszal, B. *J. Math. Chem.* **1999**, *26*, 139–155.
54. Tam, K. Y.; Quere, L. *Anal. Sci.* **2001**, *17*, 1203–1208.
55. Allen, R. I.; Box, K. J.; Comer, J. E. A.; Peake, C.; Tam, K. Y. *J. Pharm. Biomed. Anal.* **1998**, *17*, 699–712.
56. Tam, K. Y. *Anal. Lett.* **2000**, *33*, 145–161.
57. Tam, K. Y. *Mikrochim. Acta* **2001**, *136*, 91–97.
58. Tam, K. Y.; Takacs-Novak, K. *Pharm. Res.* **1999**, *16*, 374–381.
59. Tam, K. Y.; Takacs-Novak, K. *Anal. Chim. Acta* **2001**, *434*, 157–167.
60. Takacs-Novak, K.; Tam, K. Y. *J. Pharm. Biomed. Anal.* **2000**, *21*, 1171–1182.
61. Takacs-Novak, K.; Noszal, B.; Tokes-Kovesdi, M.; Szasz, G. *J. Pharm. Pharmacol.* **1995**, *47*, 431–435.
62. Takacs-Novak, K.; Noszal, B.; Hermecz, I.; Kereszturi, G.; Podanyi, B.; Szasz, G. *J. Pharm. Sci.* **1990**, *79*, 1023–1028.
63. Takacs-Novak, K.; Kökösi, J.; Podanyi, B.; Noszal, B.; Tsai, R. S.; Lisa, G.; Carrupt, P.-A.; Testa, B. *Helvet. Chim. Acta* **1995**, *78*, 553.
64. Mitchell, R. C.; Salter, C. J.; Tam, K. Y. *J. Pharm. Biomed. Anal.* **1999**, *20*, 289–295.
65. Sturgeon, R. J.; Schulman, S. G. *J. Pharm. Sci.* **1977**, *66*, 958–961.
66. Streng, W. H. *J. Pharm. Sci.* **1978**, *67*, 666–669.
67. Streng, W. H.; Huber, H. E.; DeYoung, J. L.; Zoglio, M. A. *J. Pharm. Sci.* **1976**, *65*, 1034–1038.
68. Noszal, B. *J. Phys. Chem.* **1986**, *90*, 6345–6349.
69. Noszal, B. *J. Phys. Chem. B* **2003**, *107*, 5074–5080.
70. Noszal, B.; Guo, W.; Rabenstein, D. L. *J. Phys. Chem.* **1991**, *95*, 9609–9614.
71. Noszal, B.; Kassai-Tanczos, R. *Talanta* **1991**, *38*, 1439–1444.
72. Noszal, B.; Kassai-Tanczos, R.; Nyiri, J.; Nyeki, O.; Schon, I. *Int. J. Pept. Protein Res.* **1991**, *38*, 139–145.
73. Noszal, B.; Osztas, E. *Int. J. Pept. Protein Res.* **1989**, *33*, 162–166.
74. Noszal, B.; Rabenstein, D. L. *J. Phys. Chem.* **1991**, *95*, 4761–4765.
75. Noszal, B.; Sandor, P. *Anal. Chem.* **1989**, *61*, 2631–2637.
76. Noszal, B.; Visky, D.; Kraszni, M. *J. Med. Chem.* **2000**, *43*, 2176–2182.
77. Nyeki, O.; Osztas, E.; Noszal, B.; Burger, K. *Int. J. Pept. Protein Res.* **1990**, *35*, 424–427.
78. Nagy, P. I.; Takacs-Novak, K. *J. Am. Chem. Soc.* **1997**, *119*, 4999.
79. Almasi, J.; Takacs-Novak, K.; Kokosi, J.; Noszal, B. *Int. J. Pharm.* **1999**, *180*, 1–11.
80. Szakacs, Z.; Kraszni, M.; Noszal, B. *Anal. Bioanal. Chem.* **2004**, *378*, 1428–1448.
81. Niebergall, P. J.; Schnaare, R. L.; Sugita, E. T. *J. Pharm. Sci.* **1972**, *61*, 232–234.
82. Edsall, J. T.; Martin, R. B.; Hollingworth, B. R. *Proc. Natl. Acad. Sci. USA* **1958**, *44*, 505–518.
83. Peinhardt, G.; Wiese, M. *Int. J. Pharm.* **2001**, *215*, 83–89.
84. Zekri, O.; Boudeville, P.; Genay, P.; Perly, B.; Braquet, P.; Jouenne, P.; Burgot, J.-L. *Anal. Chem.* **1996**, *68*, 2598–2604.
85. Hammett, L. P. *Physical Organic Chemistry*; McGraw-Hill: New York, 1940.
86. Taft, R. W.; Lewis, I. C. *J. Am. Chem. Soc.* **1959**, *81*, 5343.
87. Willi, A. V.; Meier, W. *Helvet. Chim. Acta* **1956**, *39*, 318.
88. Sjöström, M.; Wold, S. *Chem. Scr.* **1976**, *9*, 200.
89. Exner, O. In *Advances in Linear Free Energy Relationships*; Shorter, J., Ed.; Plenum Press: New York, 1972, pp 1–69.
90. Exner, O. Critical Compilation of Substituent Constants. In *Correlation Analysis in Chemistry*; Chapman, N. B., Shorter, J., Eds.; Plenum Press: New York, 1978, pp 439–540.
91. Hilal, S. H.; Carreira, L. A.; Karickhoff, S. W. Estimation of Chemical Reactivity Parameters and Physical Properties of Organic Molecuels Using SPARC. In *Theoretical and Computational Chemistry*; Politzer, P., Murray, J. S., Eds.; Elsevier Science: Amsterdam, 1994, pp 291–353.
92. Karickhoff, S. W.; McDaniel, D. H.; Melton, C.; Vellino, A. N.; Nute, D. E.; Carreira, L. *Envir. Toxicol. Chem.* **1991**, *10*, 1405.

93. Hilal, S. H.; Carreira, L. A.; Karickhoff, S. W. *Talanta* **1999**, *50*, 827.
94. Hilal, S. H.; Karickhoff, S. W.; Carreira, L. A. *Quant. Struct.-Act. Relat.* **1995**, *14*, 348.
95. Hilal, S. H.; El-Shabrawy, Y.; Carreira, L. A.; Karickhoff, S. W.; Toubar, S. S.; Rizk, M. *Talanta* **1996**, *43*, 607–619.
96. Dewar, M. J. S.; Dougherty, R. C. *The PMO Theory of Organic Chemistry*; Plenum Press: New York, 1975.
97. Klopman, G.; Fercu, D. *J. Comp. Chem.* **1994**, *15*, 1041–1050.
98. Klopman, G. *Quant. Struct.-Act. Relat.* **1992**, *11*, 176.
99. Spanjer, M. C.; Ligny, C. L. D.; Houwelingen, H. C. V.; Weesie, J. M. *J. Chem. Soc. Perkin Trans.* **1985**, *2*, 1401–1412.
100. Pytela, O. *Collect. Czech. Chem. Commun.* **1997**, *62*, 645–655.
101. Bosch, E.; Rived, F.; Roses, M.; Sales, J. *J. Chem. Soc. Perkin Trans.* **1999**, *2*, 1953–1958.
102. Agatonovic-Kustrin, S.; Beresford, A. P. *J. Pharm. Biomed. Anal.* **2000**, *22*, 717–727.
103. Anzali, S.; Gasteiger, J.; Holzgrabe, U.; Polanski, J.; Sadowski, J.; Teckentrup, A.; Wagener, M. *Perspect. Drug Disc. Design* **1998**, *9/10/11*, 273–299.
104. Bodor, N.; Huang, M. J.; Harget, A. *J. Mol. Struct.* **1994**, *309*, 259–266.
105. Burden, F. R.; Winkler, D. A. *J. Med. Chem.* **1999**, *42*, 3183–3187.
106. Fujiwara, S.; Yamashita, F.; Hashida, M. *Int. J. Pharm.* **2002**, *237*, 95–105.
107. Gasteiger, J.; Teckentrup, A.; Terfloth, L.; Spycher, S. *J. Phys. Org. Chem.* **2003**, *16*, 232–245.
108. Huuskonen, J.; Salo, M.; Taskinen, J. *J. Chem. Inf. Comput. Sci.* **1998**, *38*, 450–456.
109. Gobburu, J. V.; Shelver, W. H. *J. Pharm. Sci.* **1995**, *84*, 862–865.
110. Huuskonen, J. J.; Livingstone, D. J.; Tetko, I. V. *J. Chem. Inf. Comp. Sci.* **2000**, *40*, 947–955.
111. Kvasnicka, V.; Sklenak, S.; Pospichal, J. *J. Am. Chem. Soc.* **1993**, *115*, 1495–1500.
112. Kvasnicka, V.; Sklenak, S.; Pospichal, J. *Theor. Chim. Acta* **1993**, *86*, 257.
113. Livingstone, D. J.; Manallack, D. T.; Tetko, I. V. *J. Comput.-Aided. Mol. Design* **1997**, *11*, 135–142.
114. Manallack, D. T.; Livingstone, D. J. Neural Networks: A Tool for Drug Design. In *Advanced Computer-Assisted Techniques in Drug Discovery. Methods and Principles in Medicinal Chemistry*; van de Waterbeemd, H., Ed.; VCH: Weinheim, Germany, 1995, pp 293–318.
115. Peterson, K. L. Artificial Neural Networks and Their Use in Chemistry. In *Reviews in Computational Chemistry*; Lipkowitz, K. B., Boyd, D. B., Eds.; Wiley-VCH: New York, 2000, pp 53–140.
116. Tetko, I. V.; Luik, A. I.; Poda, G. I. *J. Med. Chem.* **1993**, *36*, 811–814.
117. Tetko, I. V.; Villa, A. E. P. *Neural Networks* **1997**, *10*, 1361–1374.
118. Tetko, I. V.; Tanchuk, V. Y.; Villa, A. E. *J. Chem. Inf. Comp. Sci.* **2001**, *41*, 1407–1421.
119. Sadowski, J.; Wagener, M.; Gasteiger, J. *Angew. Chem. Int. Ed. Engl.* **1995**, *34*, 2674–2677.
120. Wan, C.; Harrington, P. D. B. *J. Chem. Inf. Comp. Sci.* **1999**, *39*, 1049–1056.
121. Catalan, J.; Mo, O.; Perez, P.; Yanez, M. *J. Chem. Soc. Perkin Trans.* **1982**, *2*, 1409–1418.
122. Grüber, C.; Buss, V. *Chemosphere* **1989**, *19*, 1595–1609.
123. Citra, M. J. *Chemosphere* **1999**, *38*, 191–206.
124. Kim, K. H.; Martin, Y. C. *J. Med. Chem.* **1991**, *34*, 2056–2060.
125. Kim, K. H.; Martin, Y. C. *J. Org. Chem.* **1991**, *56*, 2723–2729.
126. Spanget-Larsen, J. *J. Phys. Org. Chem.* **1995**, *8*, 496–505.
127. Gargallo, R.; Sotriffer, C. A.; Liedl, K. R.; Rode, B. M. *J. Comput.-Aided Mol. Design* **1999**, *13*, 611–623.
128. Chaudry, U. A.; Popelier, P. L. A. *J. Org. Chem.* **2004**, *69*, 233–241.
129. Gasteiger, J.; Hutchings, M. G. *J. Am. Chem. Soc.* **1984**, *106*, 6489–6495.
130. Hutchings, M. G.; Gasteiger, J. *J. Chem. Soc. Perkin Trans.* **1986**, *2*, 447–454.
131. Hutchings, M. G.; Gasteiger, J. *J. Chem. Soc. Perkin Trans.* **1986**, *2*, 455–462.
132. Gasteiger, J.; Yan, A.-X.; Kleinöder, T. The Prediction of Water Solubility and of pK_a values by Physicochemical Descriptors. In *Chemical Information and Computation 2002*, 223rd American Chemical Society National Meeting and Exposition, Orlando, FL, Apr 7–11, **2002**, pp 98-102.
133. Dixon, S. L.; Jurs, P. C. *J. Comp. Chem.* **1992**, *13*, 492.
134. Gross, K. C.; Seybold, P. G.; Hadad, C. M. *Int. J. Quantum Chem.* **2002**, *90*, 445–458.
135. Mercier, C.; Mekenyan, O.; Dubois, J. E.; Bonchev, D. *Eur. J. Med. Chem. Chim. Ther.* **1991**, *26*, 575–592.
136. Xing, L.; Glen, R. C. *J. Chem. Inf. Comp. Sci.* **2002**, *42*, 796–805.
137. Glen, W. G.; Dunn, W. J.; Scott, D. R. *Tetrahedron Comput. Methdol.* **1989**, *2*, 349.
138. Xing, L.; Glen, R. C.; Clark, R. D. *J. Chem. Inf. Comp. Sci.* **2003**, *43*, 870–879.
139. Szegezdi, J.; Csizmadia, F. Prediction of Dissociation Constants using Microconstants. In Conference Abstracts, 227th National Meeting of the American Chemical Society, Anaheim, CA, Mar 28–Apr 1, **2004**.
140. *Cheminformatics: A Textbook*; Gasteiger, J., Engel, T., Eds.; Wiley-VCH: Weinheim, Germany, **2003**.
141. Sayle, R. Physiological Ionization and pK_a Prediction. Metaphorics LLC, 2000, http://www.daylight.com/meetings/emug00/Sayle/pkapredict.html (accessed May 2006).
142. Nagy, P. *J. Mol. Struct. (THEOCHEM)* **1989**, *201*, 271–286.
143. Tehan, B. G.; Lloyd, E. J.; Wong, M. G.; Pitt, W. R.; Montana, J. G.; Manallack, D. T.; Gancia, E. *Quant. Struct. –Act. Relat.* **2002**, *21*, 457–472.
144. Tehan, B. G.; Lloyd, E. J.; Wong, M. G.; Pitt, W. R.; Gancia, E.; Manallack, D. T. *Quant. Struct. –Act. Relat.* **2002**, *21*, 473–485.
145. Adam, K. R. *J. Phys. Chem. A* **2002**, *106*, 11963–11972.
146. Polanski, J.; Gieleciak, R.; Bak, A. *J. Chem. Inf. Comp. Sci.* **2002**, *42*, 184–191.
147. Dewar, M. J. S. *J. Am. Chem. Soc.* **1962**, *84*, 3541.
148. Dewar, M. J. S. *J. Am. Chem. Soc.* **1962**, *84*, 3546.
149. Ertl, P. *Quant. Struct. –Act. Relat.* **1997**, *16*, 377–382.
150. Girones, X.; Carbo-Dorca, R.; Ponec, R. *J. Chem. Inf. Comput. Sci.* **2003**, *43*, 2033–2038.
151. Cherkasov, A.; Sprous, D. G.; Chen, R. *J. Phys. Chem. A* **2003**, *107*, 9695–9704.
152. Tomasi, J.; Persico, M. *Chem. Rev.* **1994**, *94*, 2027.
153. Cramer, C. J.; Truhlar, D. G. *Chem. Rev.* **1999**, *99*, 2161.
154. Tran, N. L.; Colvin, M. E. *J. Mol. Struct. (THEOCHEM)* **2000**, *532*, 127–137.

155. Karshikoff, A. Calculations of Ionization Equilibria in Proteins. In *Enzyme Functionality*; Svendsen, A., Ed.; Marcel Dekker: New York, 2004, pp 149–183.
156. Grycuk, T. *J. Phys. Chem. B* **2002**, *106*, 1434–1445.
157. Bashford, D.; Karplus, M. *Biochemistry* **1990**, *29*, 10219–10225.
158. Weininger, D. *J. Chem. Inf. Comp. Sci.* **1988**, *28*, 31–36.
159. Todeschini, R.; Consonni, V. *Handbook of Molecular Descriptors*; Wiley-VCH: Weinheim, Germany, 2000.
160. Albert, A. Ionization Constants of Heterocyclic Substances. In *Physical Methods in Heterocyclic Chemistry*; Katritzky, A. R., Ed.; Academic Press: New York, 1963.
161. Kortüm, G.; Vogel, W.; Andrussow, K. *Dissociation Constants of Organic Acids and Bases*; Butterworth: London, UK, 1961.
162. Martell, A. E.; Motekaitis, R. J. *The Determination and Use of Stability Constants*; VCH Publishers: New York, 1988.
163. Martell, A. E.; Smith, R. M. *Critical Stability Constants*; Plenum Press: New York, 1982.
164. Perrin, D. D. *Dissociation Constants of Organic Bases in Aqueous Solution*; Butterworth: London, UK, 1965.
165. Perrin, D. D. *Dissociation Constants of Organic Bases in Aqueous Solution*; Butterworth: London, UK, **1972**; Supplement.
166. Serjeant, E. P.; Boyd, D. *Ionization Constants of Organic Acids in Aqueous Solution*; Pergamon Press: Oxford, UK, 1979.
167. *The Merck Index: An Encyclopedia of Chemicals, Drugs and Biologicals*, 13th ed.; CRC Press: Boca Raton, FL, **2002**.
168. *Handbook of Chemistry and Physics*, 84th ed.; CRC Press: Boca Raton, FL, 2003.
169. Tripos, Inc. http://www.tripos.com/ (accessed May 2006).
170. Schrödinger, Inc. http://www.schrodinger.com/ (accessed May 2006).
171. MDL, Inc. http://www.mdli.com/ (accessed May 2006).

Biography

Robert Fraczkiewicz has been Product Manager at Simulations Plus, Inc., since August 1998. His primary role at Simulations Plus is the development of a commercial computer program ADMET Predictor for prediction of crucial properties from molecular structure. Dr Fraczkiewicz was also instrumental in developing computational routines present in other software products: GastroPlus, ADMET Modeler, and DDDPlus. Prior to joining Simulations Plus, he completed a postdoctoral fellowship at the University of Texas Medical Branch (1998, Prof Werner Braun), where he developed new algorithms for molecular modeling of proteins in dihedral space including solvent effects (program FANTOM). Dr Fraczkiewicz received his PhD in computational chemistry from the University of Houston in 1996. His thesis included building a computer program for calculating vibrational spectra of molecules (RAMVIB) and its application to study of active centers in metalloproteins. The University of Wrocław, Poland, had awarded the MSc degree in theoretical chemistry and spectroscopy to Robert Fraczkiewicz in 1988. For the next three years, prior to receiving an invitation from the University of Houston, he worked as a staff scientist in the Radioisotope Laboratory, Technical University of Wrocław, Poland, where he also completed a graduate course in computer applications in chemistry.

5.26 In Silico Predictions of Solubility

J Taskinen, University of Helsinki, Helsinki, Finland
U Norinder, AstraZeneca R&D, Södertälje, Sweden

5.26.1 Background

5.26.1.1 Relevance of Solubility for Drug Applications

The importance of addressing physicochemical properties early in the drug discovery process is generally recognized. Inappropriate physicochemical properties are likely to result in poor pharmacokinetics or other problems in drug development.[1] Properties affecting oral absorption are especially important. The flux by passive diffusion through intestinal membranes depends on the effective permeability of the solute and its concentration at the site of absorption, which is limited by solubility.[2] Very high solubility, on the other hand, may be accompanied by poor permeability. Solubility required for good absorption depends on the balance between the solubility, permeability, and potency of the drug molecule. Adequate solubility may vary from $1\,\mu g\,mL^{-1}$ for a high-permeability, high-potency compound (clinical dose of $0.1\,mg\,kg^{-1}$) to $2\,mg\,mL^{-1}$ for a low-permeability, low-potency compound (clinical dose of $10\,mg\,kg^{-1}$).[3,4] The molar solubility (log S) of typical orally available drugs spans the range from -6 to 2 log units. For comparison, the lipophilicity of typical drug compounds varies in the log P range of -2 to $+6$ (where P is the octanol/water partition coefficient). Aqueous solubility can vary considerably depending on the environment (pure water, buffer, gastrointestinal fluid, blood). Solubility in organic solvents may be relevant in different stages of the drug development process. Dimethyl sulfoxide (DMSO) has become especially important for bioactivity screening because of its excellent solvent properties, for instance combinatorial libraries are usually stored in DMSO stock solutions.

5.26.1.2   What is Solubility?

The solubility of a solid nondissociating organic compound B is proportional to the equilibrium constant K_s in the partitioning of the compound between a solid phase and a dissolved phase:

$$B_{(solid)} \overset{K_s}{\rightleftharpoons} B_{(dissolved)} \qquad [1]$$

Solubility depends on temperature, pressure, and the polymorphic form of the solid. Thermodynamic solubility is the concentration of the solute in saturated solution in equilibrium with the most stable crystal form of the solid compound.[2,5] Less stable crystal forms and amorphous forms with noncrystalline structure have higher solubility. Solubility reflects the equilibrium competition of solute molecules between themselves (crystal energy) and the solvent (solvation energy).

In aqueous solution, ionizable compounds may be present as various charged species depending on the pK_a values of the respective ionizable groups and the pH of the solution. Solubility of the uncharged species is defined as the intrinsic solubility S_0. The effective solubility (S_{tot}), at a particular pH, is the sum of the concentrations of all the compound species dissolved in the aqueous medium.[2] The pH-dependent solubility is usually calculated using Henderson–Hasselbalch-type equations, such as the following that gives the solubility–pH profile for monoprotic acids in a saturated solution of the uncharged species:

$$pK_a = pH - \log[(S_{tot} - S_0)/S_0] \qquad [2]$$

The pH change is expected to raise the solubility according to eqn [2] with increasing concentration of the ionized species, until at some pH value the solubility product of the salt will be reached and the salt will precipitate. As a rule of thumb, the solubility plateau is assumed to be reached at a log S level 3–4 log units above log S_0.[2] The specific response of a compound to counter-ion effects may cause large deviations from the expected pH–solubility profile. Bergström et al.[6] found that both the solubility range and the slope for the change of solubility with pH varied by several log units for basic drugs in divalent buffer systems.

5.26.1.3   Kinetic and Thermodynamic Aspects of Solubility

Solubility is an equilibrium property. In practical situations, the equilibrium may not be attained, and the kinetically controlled apparent solubility may differ considerably from the equilibrium thermodynamic solubility. A slow dissolution rate may result in low apparent solubility, especially for sparingly soluble solids, because the dissolution rate depends on the equilibrium solubility.[5] The apparent solubility may exceed the thermodynamic solubility, if it is kinetically controlled by slow formation of the most stable crystal form. This may be the case if a solute is added to water in an organic solvent. High apparent solubility may be useful in the early discovery phase, allowing very insoluble compounds to be tested for efficient characterization of structure–activity relationships.[4] However, the thermodynamic aqueous solubility is the most important solubility measure for evaluation of a molecule as a drug candidate.

Thermodynamically, the transfer of solid solute to the solvent can be thought of as occurring by fluidization of the solid and mixing of the supercooled liquid solute with the solvent. The free energy changes of the processes govern solubility. The mole fraction solubility of solute B in solvent S (x_S^B) is given by

$$\ln x_S^B = (1/RT)\Delta G_{fus}^B - \ln \gamma_S^B \qquad [3]$$

where ΔG_{fus}^B is the free energy of fusion and γ_S^B is the activity coefficient accounting for liquid state interactions.

Approximate expressions based on thermodynamic considerations have been derived to allow the estimation of solubility from accessible experimental quantities. The ΔG_{fus} term can be approximated from the entropy of fusion (ΔS_{fus}) and the absolute melting point (T_m) as

$$\ln x_S^B = -(\Delta S_{fus}/RT)(T_m - T) - \ln \gamma^B \qquad [4]$$

where R is the gas constant and T is the temperature of the solution.

Assuming a constant value for the entropy of fusion, the effect of fluidization can be estimated from the melting point alone. This approximation was first used by Irmann[7] in his fragmental model for aqueous solubility of hydrocarbons and

Table 1 Expression for the terms of the MOD solubility model[11,12]: $\ln \Phi_B = A + B + D + F + O + OH$

Term	Expression
A	$-0.02278(T_m - 298.15)$ or $-\Delta H_{fus}/R(1/T - 1/T_m)$
B	$0.5\Phi_S(V_B/V_S - 1) + 0.5\ln(\Phi_B + \Phi_S^2 V_B/V_S)$
D	$-[1.0 + \max(K_{Oi}, K_{OHi})\Phi_S/V_S]^{-1}\Phi_S^2 V_B/RT(\delta_B' - \delta_S')$
F	$-r_s\Phi_S V_B/V_S + \sum \upsilon_{OHi}\Phi_S(r_s + b_i)$
O	$\sum \upsilon_{Oi} \ln[1 + K_{Oi}(\Phi_S/V_S - \upsilon_{Oi}\,\Phi_B/V_B)]$
OH	$\sum \upsilon_{OHi}\,[\ln(1 + K_{OHi}\Phi_S/V_S + K_{BBi}\Phi_B/V_B) - \ln(1 + K_{BBi}/V_B)]$

Φ, volume fraction solubility; T_m, melting point; V, molar volume, δ', modified nonspecific cohesion parameter; K_{Oi}, K_{OHi}, association constants for hydrogen bonding between the solvent and proton acceptor and donor sites of the solute, respectively; r_S, structure factor of the solvent; υ_{Oi}, υ_{OHi}, number of active proton acceptor or donor groups of type i on the solute; b_i, constant accounting for the type of hydroxyl group; K_{BBi}, stability constant for solute self-association.

halogenated hydrocarbons. Hansch *et al.*[8] found that the aqueous solubility of liquids (and γ) was correlated with the octanol/water partition coefficient log P. Yalkowsky and co-workers[9,10] then derived a model for the prediction of the aqueous solubility from the melting point and log P:

$$\log S_w = 0.5 - 0.01(mp - 25) - \log P \qquad [5]$$

where S_w is the molar solubility in water and mp is the Celsius melting point. The term $(mp - 25)$ is equal to zero for solutes that are liquid below 25 °C. The following approximations were used in the derivation of the model: the entropy of fusion was approximated by $\Delta S_{fus} = 56.5\,J\,K^{-1}\,mol^{-1}$ according to Walden's rule for rigid aromatic compounds; approximation of the partition coefficient by the solubility ratio $P = S_{oct}/S_w$ gave the relationship $\log S_w = \log S_{oct} - \log P$ for the aqueous solubility of a liquid solute; and assuming complete miscibility of organic solutes in octanol gave a constant value of 0.5 for log S_{oct}.

Ruelle and co-workers[11,12] developed a model that presents a more involved treatment of the solution phase effects. The general form of the model for the volume fraction solubility of solute B in solvent S is given by

$$\ln \Phi_S^B = A + B + D + F + O + OH \qquad [6]$$

where A represents the effects of crystallinity, and the other terms account for the various solute–solvent interactions (entropy of mixing, nonspecific cohesion forces, hydrophobic effect, and interactions of hydrogen bond acceptor and donor sites, respectively) that contribute to the activity coefficient. The A term can be estimated by Yalkowsky's approximation. For evaluation of the other terms, Ruelle and co-workers derived approximate expressions based on the mobile order and disorder (MOD) thermodynamics (**Table 1**).

The Yalkowsky and Ruelle models were derived by thermodynamics-based approximations of the terms of eqn [3], and contain no coefficients obtained by fitting experimental solubilities. Both models have been shown to give reasonable predictions, but since experimental parameters are required, they cannot be considered in silico tools. However, the models contribute to the theoretical background useful for hypothesis design in quantitative structure–property relationship (QSPR) model development.

5.26.1.4   Causes of Errors in the Literature

A limit to the accuracy of in silico methods is set by the accuracy of the experimental solubility data. Several types of error may contribute to the uncertainty of solubility values. Pure experimental error is caused by, essentially, random variation in experimental variables (purity of compounds, experimental protocols, analytical methods, etc.). A study of the AstraZeneca in-house database showed an average standard deviation 0.49 log units for repeated measurements on different batches of the same compounds.[13] Katrizky *et al.* calculated an average standard deviation of 0.58 for solubility data from different references.[14] Occasionally, the solubility values reported for one compound may differ by 2–3 log units; this large difference may originate from different experimental protocols, and may reflect the effects of kinetic control or ionization. In the case of sparingly soluble compounds, slow kinetics may result in values that do not represent thermodynamic solubility, because the system studied was far from equilibrium. The aqueous solubility of ionizable compounds is often measured at pH 7 or in unbuffered water. Inconsistent data sets may be compiled, if the effect of ionization is not taken into account. If solubility is determined in unbuffered water, the pH of the aqueous solution depends both on the pK_a of the solute and on the total concentration of the solute, and hence its solubility. The error caused by ionization then depends on solubility, and may be negligible for highly soluble acids or bases, but 1 or 2 log unit for sparingly soluble compounds.[15]

5.26.2   Relevance and Applicability of In Silico Tools

5.26.2.1   When Do We Use an In Silico Model?

There are situations when it is desirable to use an in silico model. One such instance is when there is a need to understand the underlying properties involved in governing the level of solubility. From the derived model it is possible to gain insight into which properties are particularly important and how these properties influences solubility (e.g., whether an increase in a molecular property will promote or impair solubility). Another situation when it is desirable to use an in silico model is for predictive purposes. In this case, the solubilities of a (possibly large) number of (virtual) compounds are to be estimated. Two scenarios are possible: the first is relevant for a small number of compounds for which the solubilities are to be predicted, while the other concerns the opposite situation where a large number of predictions are to be performed. In the former case, perhaps as part of a research project in the lead optimization phase, a more local single model based on physicochemical parameters that are easy to interpret (e.g., log P and hydrogen-bonding properties) is probably the preferred choice. Such a model more directly answers questions such as "Which is the next molecule to make?" in order to improve or maintain the solubility for a particular series of compounds. For the latter scenario (i.e., predictions for large sets of virtual compounds), robustness and speed of prediction are more important than interpretability of the model. Robustness may be increased by consensus or ensemble modeling, where several models, possibly using different statistical techniques and/or descriptors, are used. The reason for focusing on robustness is that subsequent to the predictions of a large set of compounds there is usually a selection procedure where only a small portion of the compounds will be identified for further work (e.g., synthesis or testing). The project team does not want to find out after further work has been performed on this selected subset that too many of these compounds are poorly soluble; solubility prediction will help prevent this scenario.

In silico models can also be utilized in intellectual property situations to either avoid the intrusion of existing patents and still develop compounds with desirable properties or make more educated decisions on how to meet the claims of a particular patent.

As with all other modeling activities, the scope and limits (i.e., the applicability domain) of the derived model in question should be identified to the users of the model (*see* Section 5.26.2.3).

5.26.2.2   Statistical Methods and Their Meaning

The practical methods for in silico prediction of solubility are not thermodynamic or other theoretical models; rather, they are empirical models obtained by statistical fitting of a mathematical model to experimental solubility data. This type of approach is termed QSPR modeling. A QSPR model for solubility typically has the form

$$\log S = f(x_i) \tag{7}$$

where the logarithm of the molar solubility S is a function of structural variables x_i calculated for each compound from the molecular structure. The structural variables may be whole-molecule properties (e.g., molecular weight), or topological, geometric, or electronic descriptors calculated from the two- or three-dimensional structure. The function may be a simple polynomial or a complex model such as a neural network (NN). The optimal model is found by fitting candidate models using a regression method. Multiple linear regression (MLR), partial least-squares regression (PLS), or NN regression are the methods used for fitting in most solubility modeling studies.

Development of a QSPR model involves four key issues: (1) selection of a set of compounds with known chemical structure and measured solubility (the training set); (2) selection of the models to be tested and the calculation descriptors; (3) fitting the model to the data with the regression method; and (4) validation of the prediction ability of the model with test sets.

There are a number of statistical terms with which to judge the quality of a derived in silico model:

- The squared correlation coefficient of the training set (r^2) is a measure of the variability explained by, or due to regression, and assumes a value between 0 and 1, where values closer to 0 indicate a random model.
- The squared correlation coefficient of the test set (q^2) is an estimate of the predictive capability of the derived model. It is either calculated by cross-validation within the training set or on the external test set. q^2 is similar to r^2 but can for poor models assume negative values.
- The average error (AE) and the average absolute error (AAE) are two additional ways to measure the quality of a model. While AE can assume both positive and negative values, AAE can only assume positive numbers, and a value close to 0 for the test set indicates a good predictive capability for the derived model. Also, since positive and negative errors may cancel out in AE and result in an overestimation of the quality of the derived model, a large discrepancy between AE and AAE indicates some systematic error in the model.
- The root mean squared error (RMSE) is a measure of the total error defined as the square root of the mean of the squared sum of errors,

$$\sqrt{\sum (y_{i,\text{measured}} - y_{\text{predicted}})^2 / n} \qquad [8]$$

where n is the number of observations.

- The standard deviation (s) is a measure of the degree of dispersion of the data from the mean value and the term is defined as

$$\sqrt{\sum (y_i - y_{\text{mean}})^2 / (n-1)} \qquad [9]$$

Furthermore, validation of the derived model is essential in order to determine its predictive capability on an external test set.

Thorough model validation is an instrumental part of the successful development of any statistical model. Without proper validation the predictive ability of the derived model cannot be estimated. Likewise, the derived model may equally well be nothing more than a random model. Techniques that should be employed to ensure proper validation will now be discussed.

Cross-validation is one method for the internal validation of a proposed model. In cross-validation the training set is divided into groups, usually 4–7, and one group is removed from the set. The model is then derived using the rest of the training set. The developed model then predicts the solubilities of the compounds of the left-out group. Each group is successively left out and predicted in the same manner as just described. The predicted residual error sum of squares (PRESS) is computed from all the predictions. The PRESS value is then compared with the sums of squares for the solubility dependent variable y (SSY):

$$\sum (y_{i,\text{measured}} - y_{\text{mean}})^2 \qquad [10]$$

The squared correlation coefficient q^2 is then defined as

$$q^2 = 1 - \text{PRESS/SSY} \qquad [11]$$

q^2 should be >0.5 for the model to be considered to have reasonable practical predictive performance.

An external validation set should be used as an independent test of the predictive ability of a derived model.

Randomization of solubility values is another technique used to validate a model. Here, the values are randomly redistributed among the compounds. A model is then derived based on the redistributed values, and checked for its predictive performance using cross-validation as well as external validation. This procedure is repeated a number of times, typically between 50 and 100. There should exist a clear separation in predictive ability between the model based on the 'true' dependent values compared with models based on redistributed values.

The problem of overfitting may not be revealed by the standard statistical tests. Overfitting means using a model that is more flexible than needed, or includes irrelevant components.[16] Decisions about overfitting need comparison. A model is overfitting if its predictions are not better than those of another, simpler, model.

5.26.2.3   **Applicability Domain**

Another important issue, far too little discussed, is the applicability domain for a particular in silico model. All models in use should have some way of describing the descriptor space in which they operate. This is especially important for models that are mounted and accessed through inter- and intranet web services. It is very important that the users of such models receive feedback in terms of a clear indication if the prediction made by the model is to be regarded as an interpolation or extrapolation to the model. If a derived statistical model is to be regarded as poor from a predictive point of view, this should be done based on valid reasons, namely that the model truly has poor predictive ability and not because the model cannot estimate outliers to the model with acceptable accuracy. In many cases it is difficult, if not impossible, to find out about the compounds used as training set and/or the chemical description used in the model. Thus, many compounds outside the applicability domain of the model will be submitted. The outlier information, and possibly also how far from the model the compound in question is, may in many cases be utilized in a more proactive way than just realizing that a number of compounds submitted to the model for prediction are, in fact, outliers to the present model. Thus, by analyzing the outliers, perhaps virtual compounds, from various points of views (e.g., structural or synthetic), some of these compounds may later be synthesized and tested experimentally. The same compounds may then be incorporated into a revised model that will have a broader applicability domain.

There are different methods for determining whether a particular compound is to be labeled as an outlier or not. Two such measures are the Mahalanobis distance[17] and the residual standard deviation (RSD),[18] that is, the remaining amount of information present in the variables used to describe the compound that has not been utilized by the model. Recently, Sheridan *et al.* investigated different measures that could serve as good discriminators for prediction accuracy in statistical models.[19] They suggested, based on retrospective analysis of QSAR investigations, that the similarity of a molecule to be predicted to the nearest molecule in the training set and/or the number of neighbors in the training set are two good measures for this purpose. Another useful approach using a combination of the convex hull method and uncertainty estimation has been proposed by Fernandez Pierna *et al.*[20] as a practical way for detecting outliers in prediction.

5.26.3   **In Silico Methods for the Prediction of Aqueous Solubility**

Since the turn of the millenium, about half a dozen new QSPR methods have been published yearly for the in silico prediction of aqueous solubility. The published methods represent greatly varying approaches as to the selection of the experimental database, description of chemical structures, model selection and fitting, and validation of the prediction power. In the following discussion the methods are grouped together based on the type of structure representation, and representative examples provided. Most of the methods published since 2000, and selected earlier methods, are summarized in **Table 2**.

5.26.3.1   **Data Sets**

The original results of experimental solubility measurements are dispersed in the scientific literature or stored in the confidential files of private companies. The data set for solubility modeling may be collected from the original sources, but more often it is obtained from secondary sources, such as reference books, commercial databases, or published compilations of earlier modelers. Models from private companies are usually based on public data combined with proprietary data from one company. Only in exceptional cases has the research group developing a solubility model any control over solubility measurements.

At present, two commercially available databases are the prime sources of published solubilities. The AQUASOL database of the University of Arizona[71] contains almost 20 000 solubility records for almost 6000 compounds extracted from over 1800 references, and the database is updated continuously. The PHYSPROP database of the Syracuse Research Corporation[72] contains more than 6300 experimental solubility records.

Certain smaller published compilations extracted from the large databases or from original literature have been repeatedly used in solubility modeling or used as data sources for new compilations. Huuskonen[21] extracted from the AQUASOL and PHYSPROP databases a set of 1297 compounds representing diverse structures. The data set of Mitchell and Jurs[22] ($n = 332$) was extracted from AQUASOL, and that of Abraham and Le[15] ($n = 664$) was collected from various literature sources. All these compilations are of a general type, containing many small and liquid compounds, several homolog or analog series of simple compounds, and some drugs.

Several small data sets containing a major proportion of druglike compounds have also been compiled and published. Huuskonen *et al.*[23] collected from literature data 211 druglike compounds. Jorgensen and Duffy[24] used this data and

Table 2 In silico models for the prediction of aqueous solubility

Proponent	Solubility data	Validation	Type of model and descriptors
Fragmental models			
Clark (2005)[31]	2657 compounds from PHYSPROP	230 internal test set RMSE = 1.1 1297 Huuskonen set RMSE = 0.82	PLS group contribution 257 fragments
Hou et al. (2004)[39]	1290 from the Huuskonen set (revised by Tetko)	120 Klopman test set RMSE = 0.79	MLR atom contribution 76 atom types, 2 correction factors
Sun (2004)[30]	1297 from the Huuskonen set 211 from Huuskonen et al. Drug-like (25–35%)	Cross-validation $q^2 = 0.81$	PLS atom contribution 218 atom types, 26 correction factors
Klopman and Zhu (2001)[29]	1168 from the literature Drug-like (<10%)	120 from the literature Drug-like (<10%) RMSE = 0.79	MLR group contribution 118 fragments
Models based on atom type E-state indices			
Tetko et al. (2001)[35]	1291 from the Huuskonen set (revised)	412 internal test set RMSE = 0.62	NN 33-4-1 32 atom type E-state indices and MW
Livingstone et al. (2001)[37]	900 from AQUASOL	68 random internal test set RMSE = 0.63	NN 30-10-2 30 atom type E-state indices
Huuskonen (2000)[21]	1297 from AQUASOL and PHYSPROP	413 internal test set RMSE = 0.60	NN 30-12-1 24 atom type E-state, 6 other topological indices
Huuskonen et al. (1998)[23]	211 from the literature Drug-like	51 random internal test set RMSE = 0.53	NN 23-5-1 14 atom type E-state indices 9 other topological indices
Wanchana et al. (2002)[44]	211 from Huuskonen et al.	Cross-validation $q^2 = 0.79$ RMSE = 0.68	PLS with 19 principal components 29 topological indices
Models based on solvation properties			
Abraham and Le (1999)[15]	664 from various literature sources	65 random internal test set $s = 0.50$	MLR 5 solvation descriptors
Jorgensen and Duffy (2000)[24]	150 from the literature Drug-like (40%)	Cross-validation $q^2 = 0.87$ RMSE = 0.72	MLR 4 descriptors from Monte Carlo simulations, 2 functional group indicators
Models based on log P			
Lobell and Sivarajah (2003)[50]	202 from OSI Pharmaceuticals in-house Drug-like Nonionized at pH 7	442 from the *Journal of Medicinal Chemistry* nonionized at pH 7 AAE = 0.66	Linear regression on calculated log P
Delaney (2004)[51]	1144 from the literature 485 pesticides 1245 from Syngenta in-house	528 from Syngenta in-house RMSE = 0.96 150 Jorgensen and Duffy set AAE = 0.71	MLR log P, MW, rotatable bonds, aromatic proportion

continued

Table 2   Continued

Proponent	Solubility data	Validation	Type of model and descriptors
Liu and So (2001)[42]	1312 from the Huuskonen set	258 internal test set RMSE = 0.71	NN 7-2-1 log P, polar surface area (PSA), MW, rotatable bonds, 3 topological indices
Yan et al. (2004)[56]	2084 from the Merck KGaA compiled from Beilstein, the Merck catalog, and the literature	936 internal test set RMSE = 0.62 799 Huuskonen set RMSE = 0.72	NN 18-11-1 log P, 17 topological descriptors
McFarland et al. (2001)[26]	22 in-house measurements Drug-like Intrinsic thermodynamic	Cross-validation $q^2 = 0.64$	MLR log P, partial charge, and H bonding descriptors
Meylan and co-workers (1996, 2000)[54,55]	1450 from AQUASOL, PHYSPROP, and other sources	3000 compounds RMSE = 0.90	MLR log P, MW, 15 group indicators
Cheng and Merz (2003)[53]	809 from AQUASOL and PHYSPROP Unbuffered, thermodynamic Drug-like (7%)	34 random internal test set, RMSE = 0.62 61 drugs from the PDR, RMSE = 0.95 161 drug-like from the CMC, RMSE = 1.15 1404 from PHYSPROP, RMSE = 1.10	MLR log P, HBD*HBA, HBD, rotatable bonds, 4 topological descriptors
Bergström et al. (2004)[27]	85 compounds from AstraZeneca and the pharmaceutical industry Drug-like Intrinsic thermodynamic	29 random internal test set RMSE = 0.86 207 from Huuskonen et al., and Jorgensen and Duffy RMSE = 0.80	PLS, 3 components log P and six 2D descriptors
Butina and Gola (2003)[58]	3328 from PHYSPROP	640 random internal test set AAE = 0.68	MLR, rule-based four equations log P, 51 counts of atom type fragments and functional groups
Engvist and Wrede (2002)[41]	1318 from the Huuskonen set	2767 from PHYSPROP RMSE = 1.18	NN 63-5-1 log P and 62 2D descriptors properties, topological, atom and group counts
Engkvist and Wrede (2002)[41]	3351 from PHYSPROP	307 from the Huuskonen set RMSE = 0.80	NN 63-5-1, 384 weights log P and 62 2D descriptors
Catana and Gao (2005)[57]	473 from AQUASOL 307 from Pfizer in-house 130 from the literature Drug-like (40–50%)	177 internal test set RMSE = 0.48	PLS, 40 components log P, 22 MOE descriptors, 65 ISIS keys
Stahura et al. (2002)[59]	650 from the Jurs, Huuskonen et al., and Huuskonen sets	100 internal test set	Binary QSAR log P 33 1D and 2D descriptors
Lind and Maltseva (2003)[38]	1295 from the Huuskonen set	412 random internal test set RMSE = 0.68	Support vector regression Molecular fingerprints

Table 2 Continued

Proponent	Solubility data	Validation	Type of model and descriptors
Wegner and Zell (2003)[40]	1269 from the Huuskonen set	21 from the Yalkowsky test set RMSE = 0.79	NN 9-15-1 log P, XlogP, 7 other descriptors
Bruneau (2001)[13]	522 from Astra-Zeneca Nonionizable at pH 7, equilibrium 1038 from the literature Drug-like (40–50%)	261 from AstraZeneca in-house 673 from the Huuskonen set RMSE = 0.81	Bayesian NN log P and 15 2D and 3D descriptors
Other property-based models			
Yan and Gasteiger (2003)[43]	1293 from the Huuskonen set	496 internal test set RMSE = 0.59 1587 from the Merck KGaA RMSE = 0.93	NN 40-8-1 32 3D and 8 other descriptors
Chen *et al.* (2002)[25]	321 from literature sources Solid drug-like Intrinsic thermodynamic	54 random internal test set RMSE = 0.86	MLR model of 3 equations 8 physicochemical descriptors
Yaffe *et al.* (2001)[61]	515 from the literature Simple compounds	78 random internal test set AAE = 0.14 (Fuzzy ARTMAP) AAE = 0.28 (BPNN)	Fuzzy ARTMAP and NN 11-13-1 Quantum chemical and topological descriptors
Mitchell and Jurs (1998)[22]	332 from AQUASOL	32 random internal test set RMSE = 0.34	NN 9-6-1 9 2D and 3D descriptors
Mosier and Jurs (2002)[62]	399 from Michel and Jurs and other sources Diverse	51 internal test set RMSE = 0.83	Generalized regression NN 5 2D and 3D descriptors
McElroy and Jurs (2001)[60]	399 from Michel and Jurs and other sources Diverse	50 internal test set RMSE = 0.69	NN 11-5-1 11 2D and 3D descriptors
Katrizky *et al.* (1998)[14]	411 from PHYSPROP	Cross-validation $q^2 = 0.87$	MLR 6 3D descriptors

1D, 2D, 3D, one-, two-, and three-dimensional; HBA, hydrogen bond acceptor; HBD, hydrogen bond donor; MW, molecular weight.

the compilations of Jurs and Abraham to construct a data set of 150 compounds where one-half of the structures could be considered druglike. Venkatesh and co-workers[25] compiled data for 321 druglike compounds from the Analytical Profile of Drug Substances, the Merck Index, and original literature. Only intrinsic thermodynamic solubilities for solid druglike compounds at or around 25 °C were accepted. McFarland *et al.*[26] used intrinsic thermodynamic solubilities measured in-house for 22 drugs. Bergström *et al.*[27] compiled in-house solubility data with other data of confirmed quality. Intrinsic thermodynamic solubilities with known accuracy were obtained for 85 druglike compounds. A majority of the compounds were bases and a minority was nonproteolytes, a distribution in accordance to that of registered drugs. Principal component analysis (PCA) showed that the compounds evenly covered a large volume of the druglike space (**Figure 1a**). The authors compiled an external test set of 207 druglike compounds from the sets of Huuskonen *et al.* and the set of Jorgensen and Duffy. PCA showed that these compounds were clustered in one quarter of the PCA plot defined by their set of 85 compounds (**Figure 1b**).

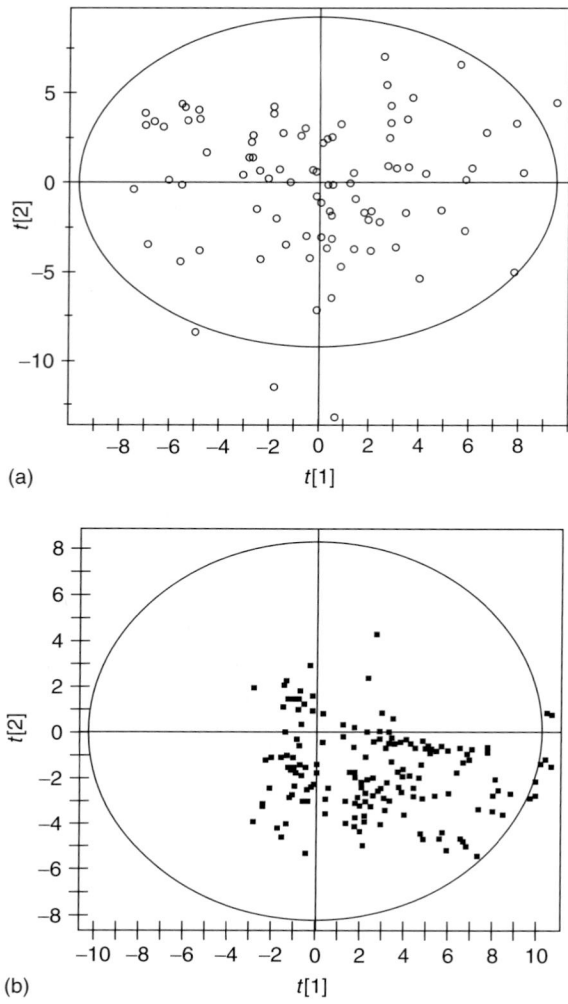

Figure 1 Score plots of the PCA (first two principal components) performed on all calculated, nonskewed descriptors for two druglike compound sets: (a) the set of Bergström et al.[27]; (b) the set compiled from Huuskonen et al.[23] and Jorgensen and Duffy.[24] (Reprinted with permission from Bergström, C. A. S.; Wassvik, C. M.; Norinder, U.; Luthman, K.; Artursson, P. *J. Chem. Inf. Comput. Sci.* **2004**, *44*, 1477–1488 © American Chemical Society.)

The origins of the data sets used for model building are given in **Table 2**. In most cases, inadequate documentation makes it difficult to assess whether a data set represents consistent values regarding thermodynamic equilibrium or ionization. This problem is common to data sets compiled from public sources and from proprietary sources. Most models are probably intended to predict the intrinsic solubility, although this may not be explicitly stated. Whenever the authors specify ionization (pH of measurement or intrinsic solubility), the information is given in **Table 2**.

Another concerns regarding a data set compiled for model building is the distribution of the compounds in the chemical structure space. Although skewness of the data set affects the applicability domain and accuracy of a model, systematic analysis of the distribution of the compounds in the structure space was reported only for one of the data sets listed in **Table 2**.

5.26.3.2 Fragmental Methods

Fragmental methods, also called atom or group contribution methods, are based on the idea that structural fragments have a constant contribution to solubility. A large number of fragments and correction factors usually have to be defined for modeling a diverse set of complex structures. A typical equation for calculating the aqueous solubility in a fragmental model is

$$\log S_w = a_0 + \sum a_i n_i + \sum b_j m_j \qquad [12]$$

where a_i and b_j are the contribution coefficients of the ith group or atom type, and the jth correction factor, respectively, and n_i and m_i are the frequencies of occurrence of the respective fragments or corrections in a molecule. The contribution coefficients are determined by MLR or another regression technique.

The first pure group contribution model for solubility prediction, without additional experimental parameters, was published by Klopman et al.[28] The model was based on 52 basic organic atom and functional groups. The model has been refined by defining 118 group parameters.[29] Even more extensive fragmentation schemes have been presented recently. For instance, the atom type classification scheme of Sun[30] defines 234 atom types and correction factors, including 88 types of carbon atom and 55 types of nitrogen atom.

The fragmental method of Clark[31] represents a new approach for decomposing molecules into a series of overlapping fragments, instead of conventionally used discrete fragments. Two classes of fragments were used, one representing basic chemical functionalities, the other representing larger biorelevant fragments. The overlapping substructures were planned to provide more variation than counting discrete fragments, and to allow distinction, for instance, between positional isomers. A large data set of 2427 compounds was extracted from the PHYSPROP database, excluding compounds with extreme log S or log P values. The data set represents the log S range of -8.8 to 1.2, and solubility values measured between 20 and 30 °C. The author does not discuss the basis for selection of the solubility values for ionizable compounds.

A predictive PLS model was built using 257 fragments, giving $r^2 = 0.73$ and an RMSE of 1.01 for the training set. The fragments with highest positive impact on solubility were found to be pyridine, $S=O(=O)$, acetone, amine, and ROH, while highest negative effects were due to naphthalene, bromine, iodine, and sulfur. Correlation of the experimental and calculated log S values is shown in **Figure 2**. The data set represents high structural diversity, and there are a number of compounds with large fitting errors (>3 log units). In fact, the r^2 value for the log S range of typical drugs (-2 to -6) is not higher than 0.42. The predictive power was validated with an internal test set of 230 compounds and with the set of Huuskonen ($n = 1297$), giving rms errors of 1.1 and 0.82 log units, respectively. Interestingly, the larger set of Huuskonen was predicted with better accuracy.

The modeling approach using the atom type E-state indices of Hall and Kier[32] can be considered as a special case of the atom contribution approach. The first solubility model of this type was developed by Huuskonen et al.,[23] using a small set of druglike compounds to train a predictive NN. Subsequently, Huuskonen used the same approach, with a larger data set of a general type.[21] Aqueous solubilities at 20–25 °C were extracted from the AQUASOL and PHYSPROP databases for 1297 compounds, representing diverse structures including some drugs (15–20%), but also a large number of simple compounds, such as unsubstituted hydrocarbons or halogenated hydrocarbons. The criteria for selecting solubility values for ionizable compounds were not stated. The log S values ranged from -11.62 to $+1.58$. The whole data set was used to build NN models, one part as the training set (884 compounds), the other part (413 compounds) as the test set for controlling the training endpoint. The final network, with a 30-12-1 architecture,

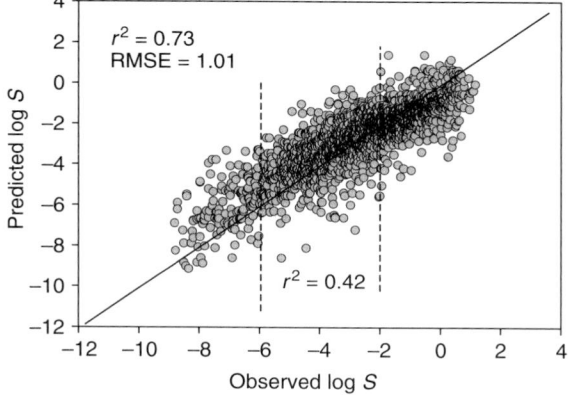

Figure 2 Correlation between the observed log S and the log S predicted by the fragmental model of Clark[31] for the training set.

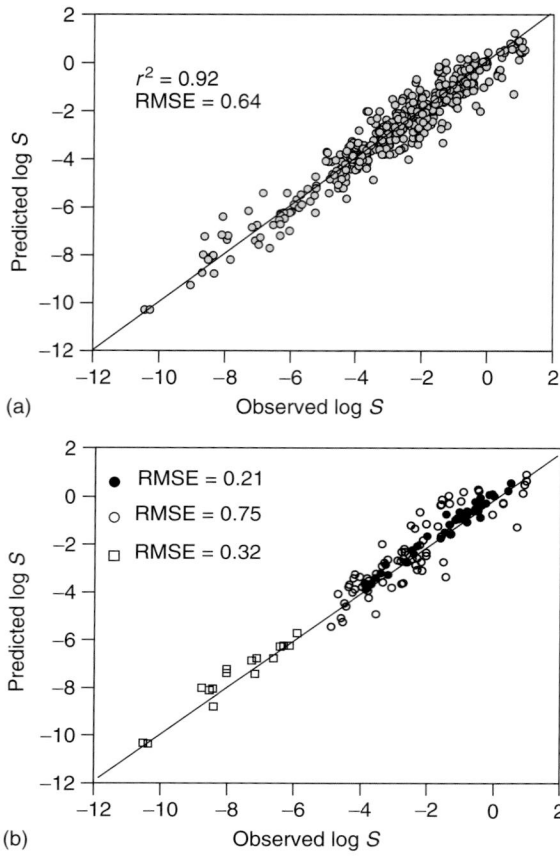

Figure 3 (a) Correlation between the observed log S and log S predicted by Huuskonen's model[21] for the test set. (b) Correlation plot for three subgroups of the test set. ●, aliphatic hydrocarbons and monofunctional oxygen compounds; □, polychlorinated biphenyls; ○, nitrogen heterocycles.

contained 24 atom type E-state indices and six other topological descriptors as input variables. The model was trained to best prediction accuracy for the internal test set corresponding to rms errors of 0.47 and 0.60 for the training and test sets, respectively. The low error values are probably related to the flexibility of the nonlinear model with 385 adjustable weights, and may give an overoptimistic view about the generalization ability of the model regarding druglike compounds. Correlation of the predicted and observed solubilities for the test set are shown in **Figure 3**. **Figure 3b** shows that the prediction accuracy varies depending on compound type. Very small errors were found for simple compounds, such as the aliphatic cluster (RMSE of 0.21), whereas a more diverse group of nitrogen heterocycles gave an RMSE of 0.75.

The Huuskonen model contains a considerably smaller number of atom types than conventional atom/group contribution models, such as the Sun model, which was based on the same data set. It is usually assumed that the atom type E-state indices contain more relevant information than mere counting of atom types. However, it has been reported that comparable prediction ability can be obtained using counts of the corresponding atom types instead of E-state indices.[33]

The modeling approach of Huuskonen or his data set has been used to develop several other solubility models, including commercial programs and free Internet methods.[30,34–44] It has been pointed out that the data set is limited in structural diversity, resulting in models that may give disappointing predictions, when challenged with an external test set.[41,43]

5.26.3.3 Property-Based Methods

In property-based models, the molecular structures of the solutes are represented by physicochemical parameters or molecular descriptors calculated from two- or three-dimensional molecular structures. Selection of the model variables may be hypothesis based or computer assisted, or may combine these two approaches. Hypothesis-based modeling typically means that a small number of presumably relevant descriptors are selected for statistical testing. The resulting models are typically straightforward to interpret in terms of medicinal chemistry. The computer-assisted model building typically means calculation of hundreds of descriptors, and the selection of the best descriptor combination and model form from a large number of possibilities. Various computational methods, such as genetic algorithms or entropic based descriptor selection, are used to make the process more efficient. The approach may lead to models that are rather obscure. It is common that a property-based solubility model includes a few terms representing indicators or counts of atom types or functional groups. Some recent models based on large and diverse data sets can be considered to be hybrids of property-based and fragmental models.

5.26.3.3.1 Models based on solvation properties

A group of hypothesis-based solubility models is related to Ruelle's MOD model in that interest is focused on the solvation process. The linear solvation energy relationship model, originally presented by Taft and Kamlet and their co-workers,[45,46] gives an empirical relationship between a solvation-dependent property and five experimental parameters related to solute–solvent interactions. The equation derived by Abraham and Le for aqueous solubility includes an additional hydrogen-bonding cross-term:

$$\log S = 0.52 - 1.00R_2 + 0.77\pi_2^H + 2.17\sum \alpha_2^H + 4.24\sum \beta_2^H - 3.3.6\sum \alpha_2^H \sum \beta_2^H - 3.99V_x \qquad [13]$$

$$n = 659, \quad r^2 = 0.92, \quad s = 0.56, \quad F = 1256$$

where R_2, π_2^H, α_2^H, $\sum \beta_2^H$, and V_x are the excess molar refraction, the dipolarity/polarizability, the hydrogen bond acidity, the hydrogen bond basicity, and McGowan's characteristic molecular volume, respectively.[15] The equation is based on solubility data for 664 compounds compiled from various literature sources, representing the log S range from -9 to $+2$. The compound set contained a large proportion of liquids, several homolog series of simple organic compounds, and some druglike compounds. Five compounds (cyclopropyl-5-spirobarbituric acid, uracil, chlorphenioramine, fentanyl, and adenine) were found to be outliers. The correlation between observed and predicted solubilities is shown in **Figure 4**.

Abraham and Le suggest that the cross-term accounts for solid state effects. However, the model still predicts systematically too high solubility for high-melting-point compounds, resulting in an average absolute error of -0.85 log units for the subset of 39 compounds with mp $>200\,^\circ$C.

The model aims to predict intrinsic solubility, although the experimental solubilities for ionizable compounds correspond to values observed at the pH of the saturated solution. The effect of ionization is discussed thoroughly by the authors, and is concluded to have a minor contribution to the prediction error.

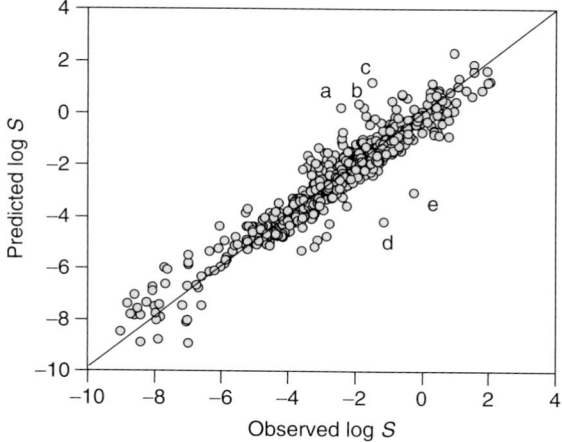

Figure 4 Correlation between the observed log S and the log S predicted by the solvation model of Abraham and Le[15] for the combined training and test sets. Outliers to the model: a, adenine; b, cyclopropyl-5-spirobarbituric acid; c, uracil; d, fentanyl; e, chlorpheniramine.

The descriptor values used to obtain eqn [13] stem from experimental parameters. However, a method for calculation of the solvation descriptors from the molecular structure has been published,[47] and a commercial in silico model, Absolv (see **Table 4**), has been developed based on Abraham's model.

Jorgensen and Duffy[48] tackled the solvation problem by running Monte Carlo simulation for the solute in water using a small data set of 150 compounds appropriate for the computational approach. Eleven descriptors related to solute–water interaction energies, surface areas, and hydrogen bonding were calculated for the solutes. A five-term predictive MLR model was derived, including four Monte Carlo descriptors and two functional group counts. The calculations, however, take too much time to be practical for predicting larger sets of compounds. The authors subsequently developed algorithms for the rapid estimation of the descriptors from a three-dimensional molecular structure, leading to a more practical method, which has been incorporated in the commercial QikProp program.[49]

5.26.3.3.2 Models based on log *P*

Since the early work of Hansch *et al.*[8] it has been known that the aqueous solubility of liquids is strongly correlated with the octanol/water partition coefficient log *P*, which is also a message of Yalkowsky's model. The solubility of the compounds compiled by Abraham and Le, or Jorgensen and Duffy, is also remarkably well explained by ClogP alone ($r^2 = 0.78$ and 0.70, respectively). This is true for several other data sets containing a large proportion of liquids and relatively small and simple compounds. Lobell and Sivarajah[50] even found high correlation ($r^2 = 0.71$) between aqueous solubility and calculated log *P* values for 442 druglike compounds that are predominantly uncharged at pH 7.

A group of solubility models is based on the hypothesis that log *P* is the key descriptor, and a few other physically meaningful properties can be found to account for the solid phase effects and inadequacies of log *P*. Typically, a small number of other preselected two-dimensional descriptors has been tested along with log *P*. A representative example, explicitly using Yalkowsky's model as the starting point, is the ESOL model of Delaney.[51]

The initial parameter set of the ESOL model included nine descriptors: calculated log *P* (ClogP), molecular weight (MW), the number of rotatable bonds (RB), the proportion of heavy atoms in a molecule that are located in aromatic rings (aromatic proportion), the noncarbon proportion, the hydrogen bond donor count, the hydrogen bond acceptor count, and the polar surface area. In addition to ClogP, only three parameters were found significant in the MLR analysis of a large training set, leading to the ESOL model:

$$\log S_w = 0.16 - 0.63\text{ClogP} - 0.0062\text{MW} + 0.066\text{RB} - 0.74\text{AP} \qquad [14]$$

$$n = 2874, \quad r^2 = 0.72, \quad s = 0.97, \quad F = 1865$$

The training data originated from three sources: one subset consisted of 1144 general organic compounds from the literature with a log *S* range of -11.6 to $+1.6$ and an average MW of 205, the second subset consisted of 485 pesticide products (average MW = 294), and the third subset consisted of 1245 compounds from the Syngenta proprietary database (average MW = 341). The criterion for selecting values for ionizable compounds was not given. Correlation of the calculated and observed solubilities for the first subset is shown in **Figure 5**. The model was validated using an independent test set of 528 proprietary compounds, a small set ($n = 21$) designed by Yalkowsky and the set of Jorgensen and Duffy ($n = 150$).[24,52] The average absolute errors for the three sets were 0.96, 0.69, and 0.71. The RMSE for prediction of the proprietary test set was 0.96.

The correlation plot in **Figure 5** shows some points with large deviations. None of these represent druglike compounds. Besides a few individual outliers, there are also classes of compounds with large systematic errors. For instance, small pyridine derivatives are predicted too insoluble (AE = +2.3) and polyaromatic hydrocarbons too soluble (AE = −1.7), while the error for 1-alkanols changes with the chain length from +1 to −2.8. Obviously, this type of error could be minimized by adding correction terms to the model.

The ESOL model contains only two more parameters than Yalkowky's model. Delaney compared the performance of the two models for 1305 training set compounds with measured melting points, and found AAE = 0.75 and 0.81 for the ESOL and Yalkowsky models, respectively. It was concluded that the non-log *P* terms of the ESOL model probably provide an enhanced estimate of ΔS_{fus} and T_{m}.

Other researchers have combined heuristic and computer-assisted approaches to find a few physically meaningful descriptors in addition to calculated log *P*. The MLR model of Cheng and Merz involves six additional descriptors and a hydrogen-bonding cross-term.[53] Liu and Sho developed a 7-2-1 NN with log *P* and six other descriptors.[42] The log *P*-based models seem to have a consensus about the molecular size and flexibility/rigidity being important complementing

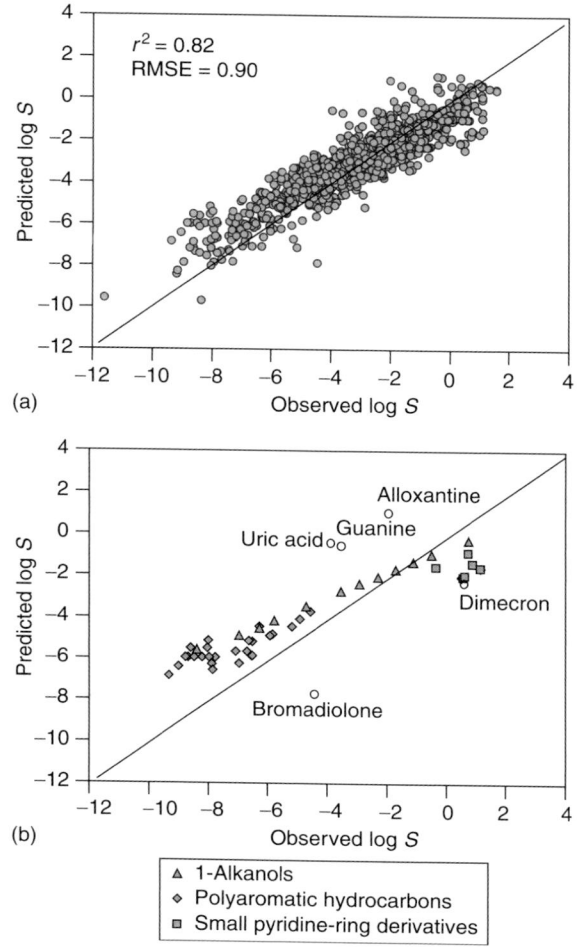

Figure 5 (a) Correlation between the observed log S and log S predicted by Delaney's ESOL model[51] for a training subset. (b) Predicted versus observed log S for five compounds showing the largest fitting errors and three compound groups with large systematic errors.

properties, but the role of hydrogen-bonding descriptors is less clear. Three hydrogen-bonding related terms are presented in the model of Cheng and Merz, while no such term was found significant in the work of Delaney, based on a much larger training set.

The example of Jorgensen's model and the analysis of Delaney's data show that the fit of simple property-based models can be easily improved by adding functional group-specific correction terms to the model. Meylan and co-workers used 15 group indicators besides the molecular weight as additional descriptors in their log P-based solubility model.[54,55]

Another approach is to include log P in a large descriptors set, which may comprise two- and three-dimensional structure-dependent properties, topological indices, and various counts typical for fragmental methods. Although no hypothesis about the model is made, the computer-assisted model building typically finds log P as the most important descriptor. Depending on the data set and the procedure used, the resulting model may resemble a hypothesis-based model with a rather small number of parameters,[13,27,40,56] or resemble a hybrid of property-based and fragmental models.[41,57–59] Examples of the last type are the models of Butina and Gola[58] and of Engkvist and Wrede.[41] Both used a large data set extracted from the PHYSPROP database for more than 3300 structurally diverse compounds. The strategy of Butina and Gola was to combine the calculated log P with a two-dimensional descriptor set comprising various counts including atom-based fragments, functional groups, and hydrogen bond donors and acceptors. The best results were obtained with a Cubist model that consisted of separate MLR equations for four rule-based subsets. The equations included ClogP and a maximum of 35 other descriptors – 52 descriptors in total. Engkvist and Wrede developed an NN model with 63-5-1 architecture involving 320 adjusted weights. The 63 descriptors included atom, bond, and ring type counts, and various topological indices, in addition to log P.

A few studies compared the power of two- and three-dimensional dependent descriptors in combination with log P. It was found that a combination log P with two-dimensional descriptors gave a better model than a combination with three-dimensional descriptors,[27,56] or including of three-dimensional descriptors in a model did not lead to significant improvement.[53]

5.26.3.3.3 Other property-based models

One group of solubility models are built without any preselected descriptors or hypotheses about the relationship between structure and solubility, by using computer-assisted selection of an optimal combination of variables from a large pool of calculated descriptors. Hundreds of topological, geometric, electronic, and combination descriptors may be calculated from two- and three-dimensional structures. Semiempirical quantum chemical methods are frequently used for these kinds of computation. A key problem is the effective selection of the best descriptor combination. The work of McElroy and Jurs[60] is a representative example of this approach. They compiled, from the literature, a solubility data set of 399 compounds containing at least one oxygen or nitrogen atom with a molecular weight range of 53–959. Solubility values ranged from -7.41 to $+0.96$ log units. The three-dimensional structures were optimized, and charge distributions calculated using semiempirical quantum chemical PM3 and AM1 methods. For each compound, 229 descriptors were calculated, including topological, geometric, and electronic descriptors, and combinations of these descriptors. Descriptors with little or redundant information were removed, leaving a reduced pool of 98 descriptors. Two NN models were developed using different procedures to select an optimal subset of descriptors. A feature selection routine based on generalized simulated annealing and MLR resulted in a set of 11 descriptors. The other procedure was fully nonlinear, using genetic algorithm and NN fitness evaluation to select the optimal set of 11 descriptors. The two procedures resulted in remarkably different selections of topological, geometric, and combination descriptors, with only two descriptors in common. Still, the ability of the two models to predict solubility for a random internal test set ($n = 51$) was similar, with an RMSE of about 0.7 log units after removing one outlier. Correlation between the observed solubilities and those predicted by the fully nonlinear model with 11-5-1 architecture is shown in **Figure 6**. Several other solubility models based on a similar modeling strategy have been published.[14,25,43,60–62] The applicability of this type of models is limited to rather small compound sets due to high computational cost.

5.26.3.4 Consensus or Ensemble Modeling

Consensus modeling, or ensemble modeling means that more than one model is used for prediction and the results are averaged. By using this approach, the weakness of one particular model is compensated by another model, thus obtaining a more robust behavior for the ensemble of models.

Support for the general usefulness of consensus modeling has been provided by the computer simulation of Wang and Wang on consensus scoring for virtual library screening.[63] The simulation suggested that three or four methods are

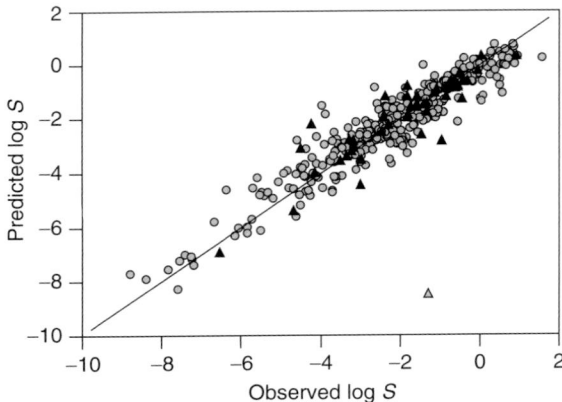

Figure 6 Correlation between the observed log S and the log S predicted by the neural network model of McElroy and Jurs.[60] ○, training set; ▲, test set; △, octacholorodibenzo-p-dioxin.

Table 3 Prediction of aqueous solubility using rule-based systems (Rule Discovery System, RDS, www.compumine.com) and ensemble modeling (U. Norinder, P. Lidén, and H. Boström, unpublished results)

Model type	Training set			Test set 1			Test set 2		
	n	r^2	s	n	r^2	s	n	r^2	s
PLS	800	0.87	0.69	497	0.93	0.58	21	0.80	0.82
RDS/ensemble	800	0.97	0.35	497	0.95	0.51	21	0.87	0.67

sufficient to improve the results significantly. Examples of fortuitous consensus modeling appear in the solubility-modeling literature. Yalkowsky and co-workers[64] found in a study comparing the performance of their model with the Abraham model that the average of predicted log S_w values using the two independent methods gives better prediction than either method used alone. Bergström et al.[27] developed three different PLS models based on either two- or three-dimensional descriptors, or a combination of both. They found that best results were obtained by averaging the predictions of the three models.

There are commercial software packages such as KnowItAll[73] and the Rule Discovery System[74] that employ consensus modeling. An example of systematic ensemble modeling to obtain more robust predictions instead of a single model is exemplified by a rule-based ensemble model using two-dimensional parameters on the Huuskonen data set (U. Norinder, P. Lidén, and H. Boström, unpublished results). The model consists of hierarchically organized rules by means of recursive partitioning with sets of if–then rules, where the condition part of each rule puts restrictions on some of the variables. In the case of a regression, a numeric value is assigned to the dependent variable (e.g., log S) for examples covered by a particular rule. The results of the ensemble modeling (50 models) are compared in **Table 3** with a single model using traditional PLS methodology.

5.26.3.5 Commercial Software

A number of commercial programs for predicting aqueous solubility have been recently introduced (**Table 4**). The models used in these programs represent many of the modeling approaches discussed in the previous sections, and some of them are described in the literature[34,49,65] CSlogWS and COSMO-RSol are representative examples of the diversity of the methods. CSlogWS is an NN model predicting intrinsic thermodynamic aqueous solubility using E-state indices and other topological descriptors.[34] The model was developed using solubility data for almost 6000 compounds. COSMO-RSol combines a theoretical model to calculate the liquid phase contribution and a QSPR model for the solid phase contribution.[65] The method can be used to predict solubility in any solvent, but due to the density of the functional quantum chemical calculations required, it is not practical for prediction of large compound sets. In addition to commercial programs, there are free programs available on the Internet for prediction of aqueous solubility (**Table 4**).

5.26.3.6 Accuracy and Applicability of In Silico Tools for Druglike Compounds

The prediction accuracy of a solubility model for druglike compounds may not be straightforward to evaluate from the published validation data. The majority of the data for training and test sets may represent compounds that are very far from drugs regarding structure and physicochemical properties, while regions of druglike structure space are poorly represented. The quality of a model is usually demonstrated by a plot showing the correlation between experimental and predicted log S values and by giving the corresponding r^2 value. High correlation may be shown for the whole solubility range, while only modest correlation is true for the most relevant range. The effect is demonstrated in **Figure 2**.

Using the overall error (rms, AAE, etc.) obtained for a test set as the measure of prediction accuracy may be deceptive as well. As demonstrated in **Figures 3** and **5**, the magnitude and sign of error may vary with structural type. Tetko et al.[36] found that the prediction error of their method increased approximately linearly with molecular size. It has also been found that the error is typically smaller for liquids than for solids, and may increase with melting point. Druglike compounds tend to be complex, large, solid compounds prone to larger prediction errors.

Root mean square errors in the range of 0.5–0.8 log units have been observed for most of the methods shown in **Table 2**, when tested with a set selected randomly from the data compiled for model development. Some models are

Table 4 Commercial software and free programs

Method	Supplier	Solubility data	Type of model and descriptors
Prediction of aqueous solubility			
QikProp	Schödinger Inc. www.schrodinger.com	281 from the Jorgensen and Duffy set and others	MLR 8 descriptors, SASA, HBD, HBA, etc.
Rule Discovery System	Compumine AB www.compumine.com	1297 from the Huuskonen set	
CSlogWS	ChemSilico www.chemsilico.com	5964 diverse compounds from AQUASOL, PHYPROP, and other sources	Two NNs with atom type E-state other topological indices
Absolv/ADME Boxes	Pharma Algorithms www.ap-algorithms.com		Based on the Abraham solvation model
ACD/Solubility DB	Advanced Chemistry Development www.acdlabs.com		
WSKOWWIN	Syracuse Research Corporation www.syrres.com/epi.htm		MLR, Meylan *et al.* log P, MW, and fragments
COSMO-RS	Cosmologic GmbH & Co. www.cosmologic.de	127 solid compounds from Jorgensen and Duffy for ΔG_{fus} modeling	Combination of a theoretical model for solvation and a QSPR model for ΔG_{fus}
ALOGPS	Free program: Virtual Computational Chemistry Laboratory (VCC-LAB) (http://www.vcclab.org)	1291 from the Huuskonen set (revised by Tetko)	NN 33-4-1 32 atom type E-state indices and MW
IAlogS	Free program: Interactive Analysis (http://www.logP.com)		NN Atom type E-state indices
Prediction of DMSO solubility			
AB/DMSO	Pharma Algorithms www.ap-algorithms.com	> 2000 DMSO solubilities	

See **Table 2** for definitions of abbreviations.

based on large and diverse sets containing probably most of the applicable data in PHYSPROP.[31,41,58] In these cases, RMSEs larger than 1 log unit have been observed for random test sets.

Lobell and Sivarajah[50] collected solubilities for 442 druglike compounds cited in the *Journal of Medicinal Chemistry* between 1982 and 2000. The set represented compounds that were classified as predominantly uncharged at the pH of measurement (MW = 129–903, with an average of 523). This set was used to compare the prediction accuracy of nine commercial or Internet methods with the log *P*-based equation of the authors. CSlogWS along with the log *P* equation showed an AAE of ~0.7. The other eight methods had AAE values between 0.9 and 1.9.

Cheng and Merz[53] tested their model with more than 220 drugs and druglike compounds from the Physician's Desk References and the Comprehensive Medicinal Chemistry database, observing RMSEs of 1–1.2 log units, almost twice the error found for the test set selected randomly from their working data set.

The data set of 85 compounds compiled by Bergström *et al.*[27] was proposed by the authors to represent a balanced coverage of the largest volume of druglike space published so far, and contain only high-quality intrinsic solubility values. We used part of this set (79 compounds with publicly available structures) to evaluate the prediction power of seven methods. All methods predicted an absolute error larger than − 3 log units for structure SKF105657. The model of Bergström *et al.* also predicted this compound poorly. The solubility of SKF105657 (log $S = -8.76$) is outside of the typical range for drugs. The RMSE for the other 78 compounds varied from 1.06 log units for ESOL to 1.54 log units for IALOGS (**Table 5**). It is noteworthy that the complex models did not work better than the ClogP model of Lobell and Sivarajah. Consensus modeling improved the accuracy to an RMSE of 1.00 log units. The clearly lowest RMSE obtained for the model for Bergström *et al.* is not comparable, because most of the compounds were present in the training set. The RMSE they report for their external test set is 1.01 log units, and several compounds showed large errors. It can be seen in **Figure 1** that part of the structure space occupied by the external test set is poorly covered by the set of Bergström *et al.*

Table 5 Prediction accuracy for a set of drug-like compounds compiled by Bergström et al.[a]

Method	r^2	RMSE	Compounds with AAE >2.5 log units
ESOL[51]	0.63 (0.61)	1.06 (1.12)	1
Absolv[b]	0.54 (0.54)	1.12 (1.17)	3
Lobell and Sivarajah[50] (ClogP model)	0.51 (0.52)	1.15 (1.19)	5
ALOGPS[c]	0.54 (0.51)	1.16 (1.21)	3
Compumine[d]	0.46 (0.45)	1.21 (1.28)	3
QikProp[e]	0.53 (0.51)	1.29 (1.35)	4
IALOGS[f]	0.25 (0.23)	1.54 (1.62)	8
Consensus prediction[g]	0.63 (0.62)	1.00 (1.07)	1
Bergström et al.[27] (two-dimensional model)	0.72 (0.72)	0.89 (0.94)	0

[a] Compounds with public structures from the set of Bergström et al.[27] excluded SKF105657 ($n = 78$). SKF105657 is included for values in parentheses ($n = 79$).
[b] Pharma Algorithms, www.ap-algorithms.com.
[c] VCC-LAB, www.vcclab.org.
[d] Compumine AB, www.compumine.com.
[e] QikProp 1.6, Schrödinger, Inc.
[f] Interactive Analysis, www.logP.com.
[g] Average prediction of seven models.

Considering the results discussed above, an RMSE in the range of 1–1.5 log units seems likely for a test set representing broadly druglike compounds. There are probably several reasons for the modest performance: inconsistencies in experimental solubility data, skewed or nonrelevant distribution of compounds in the training set, and inadequate structure representation. The domain of applicability of most solubility models covers only part of the druglike structure domain. Accuracy also varies within the descriptor range of a model. Some parts of the response surface are better covered by training set, or described by more appropriate structural parameters. A description of crystallinity effects is a special challenge for solubility modeling.

5.26.4   Methods for the Prediction of Solubility in Organic Solvents

The thermodynamics based MOD model of Ruelle is, in principle, applicable to the calculation of solubility in any solvent, providing that the necessary physicochemical parameters are available. The model has been tested for the prediction of solubility for hydroxysteroids and related compounds in 24 organic solvents.[12] The COSMO-RS method of Klamt[65,66] is also a general model, and, in principle, allows the prediction of solubility in any solvent without any experimental data. The predictive ability has been studied by Ikeda et al.[67] for 15 drugs and druglike compounds in three organic solvents. The rms errors for ethanol, acetone, and chloroform were 0.61, 0.84, and 0.56, respectively. A major limitation of the applicability of the method is the long time required for quantum chemical computations using the density functional theory.

A few QSPR studies have been carried out to model solubility in organic solvents. Acree and Abraham[68] investigated the applicability of Abraham's solvation model using a large number of solvents, and anthracene, phenanthrene, and hexachlorobenzene as model solutes. The solvents included aliphatic and aromatic hydrocarbons, alkanols, acetonitrile, and ethyl acetate. The model predicted solubility within an average absolute deviation of 35%. For comparison, the solubilities were calculated also using the MOD model, which gave an average absolute deviation of 110%. The most extensive work on solubility models for organic solvents has been carried out by Katritzky et al., who have modeled around 500 compounds in various series in different solvents, such as various aliphatic and aromatic hydrocarbons, alcohols, and ethers, as well as some dipolar aprotic solvents.[69,70] They used the CODESSA PRO software both for generating the chemical descriptors for the investigated compounds as well as for the subsequent statistical QSPR analysis. They developed a large number of equations for various solvents with high statistical quality. Some general trends were observed with respect to descriptor occurrence in the various models. A frequency analysis showed that electrostatic, topological, and hydrogen-bonding parameters were the most important, while descriptors related to geometry or derived from quantum chemical calculations were less important, and thermodynamic descriptors were of low importance. The solubility in aliphatic and aromatic hydrocarbon solvents is well described by electrostatic and topological parameters, while solubility in inert solvents (e.g., chlorine-containing solvents such as chloroalkanes or chloroaromatics) are governed, to a large extent, by cavity formation and other nonspecific terms. Thus, variables related to Randic topological indices occur frequently in these models. For protic solvents, parameters related to hydrogen bonding are of major importance. Variables related both to counts of donor and acceptor sites but also to parameters associated with charge-weighted molecular surfaces have a significant impact in these models. Polarizability and polarity as well as Lewis acid basicity are important parameters for describing solubility in polar aprotic solvents.

Despite the importance of DMSO for bioactivity screening, published QSPR modeling studies in are sparse, probably due to the lack of large publicly available experimental databases. The extensive QSPR work of Katritzky *et al.*[70] includes a model for DMSO solubility. A commercial program for predicting DMSO solubility has been recently introduced (**Table 4**).

5.26.5   Conclusion

There are presently a multitude of in silico methods available for the prediction of solubility. Methods representing very different modeling approaches apparently perform well within their domain of applicability, which, however, in most cases covers only a limited volume of the druglike structure space. Improving accuracy and applicability seems to require more consideration of the consistency of the experimental solubility data and the training set composition, although advances in structure representation to account for solid state effects may be the most critical aspect in improving prediction accuracy.

References

1. Van de Waterbeemd, H.; Smith, D. A.; Beaumont, K.; Walker, D. K. *J. Med. Chem.* **2001**, *44*, 1313–1333.
2. Avdeef, A. *Curr. Top. Med. Chem.* **2001**, *1*, 277–351.
3. Lipinski, C. A. *J. Pharmacol. Toxicol. Methods* **2000**, *44*, 235–249.
4. Lipinski, C. A.; Lombardo, F.; Dominy, B. W.; Feeney, P. J. *Adv. Drug Deliv. Rev.* **1997**, *23*, 3–25.
5. Grant, D. J. W.; Higuchi, T. *Solubility Behavior of Organic Compounds*; John Wiley: New York, NY, 1990.
6. Bergström, C. A. S.; Luthman, K.; Artursson, P. *Eur. J. Pharm. Sci.* **2004**, *22*, 387–398.
7. Irmann, F. *Chem. Ing. Technol.* **1965**, *37*, 789–798.
8. Hansch, C.; Quinlan, J. E.; Lawrence, G. L. *J. Org. Chem.* **1968**, *33*, 347–350.
9. Yalkowsky, S. H.; Valvani, S. C. *J. Pharm. Sci.* **1980**, *69*, 912–922.
10. Jain, N.; Yalkowsky, S. H. *J. Pharm. Sci.* **2001**, *90*, 234–252.
11. Ruelle, P.; Rey-Mermet, C.; Buchmann, M.; Nam-Tran, H.; Kesselring, U. W.; Huyskens, P. L. *Pharm. Res.* **1991**, *8*, 840–850.
12. Ruelle, P.; Farina-Cuendet, A.; Kesselring, U. W. *Perspect. Drug Disc. Des.* **2000**, *18*, 61–112.
13. Bruneau, P. *J. Chem. Inf. Comput. Sci.* **2001**, *41*, 1605–1616.
14. Katritzky, A. R.; Wang, Y.; Sild, S.; Tamm, T.; Karelson, M. *J. Chem. Inf. Comput. Sci.* **1998**, *38*, 720–725.
15. Abraham, M. H.; Le, J. *J. Pharm. Sci.* **1999**, *88*, 868–880.
16. Hawkins, D. M. *J. Chem. Inf. Comput. Sci.* **2004**, *44*, 1–12.
17. De Maesschalck, R.; Jouan-Rimbaud, D.; Massart, D. L. *Chemom. Intell. Lab. Syst.* **2000**, *50*, 1–18.
18. Eriksson, L.; Johansson, E.; Kettaneh-Wold, N.; Wold, S. *Multi- and Megavariate Data Analysis: Principles and Applications*; Umetrics: Umeå, Sweden, 2001.
19. Sheridan, R. P.; Feuston, B. P.; Maiorov, V. N.; Kearsley, S. K. *J. Chem. Inf. Comput. Sci.* **2004**, *44*, 1912–1928.
20. Fernandez Pierna, J. A.; Wahl, F.; de Noord, O. E.; Massart, D. L. *Chemom. Intell. Lab. Syst.* **2002**, *63*, 27–39.
21. Huuskonen, J. *J. Chem. Inf. Comput. Sci.* **2000**, *40*, 773–777.
22. Mitchell, B. E.; Jurs, P. C. *J. Chem. Inf. Comput. Sci.* **1998**, *38*, 489–496.
23. Huuskonen, J.; Salo, M.; Taskinen, J. *J. Chem. Inf. Comput. Sci.* **1998**, *38*, 450–456.
24. Jorgensen, W. L.; Duffy, E. M. *Bioorg. Med. Chem. Lett.* **2000**, *10*, 1155–1158.
25. Chen, X.-Q.; Cho, S. J.; Li, Y.; Venkatesh, S. *J. Pharm. Sci.* **2002**, *91*, 1838–1852.
26. McFarland, J. W.; Avdeef, A.; Berger, C. M.; Raevsky, O. A. *J. Chem. Inf. Comput. Sci.* **2001**, *41*, 1355–1359.
27. Bergström, C. A. S.; Wassvik, C. M.; Norinder, U.; Luthman, K.; Artursson, P. *J. Chem. Inf. Comput. Sci.* **2004**, *44*, 1477–1488.
28. Klopman, G.; Wang, S.; Balthasar, D. M. *J. Chem. Inf. Comput. Sci.* **1992**, *32*, 474–482.
29. Klopman, G.; Zhu, H. *J. Chem. Inf. Comput. Sci.* **2001**, *41*, 439–445.
30. Sun, H. *J. Chem. Inf. Comput. Sci.* **2004**, *44*, 748–757.
31. Clark, M. *J. Chem. Inf. Model.* **2005**, *45*, 30–38.
32. Hall, L. H.; Kier, L. B. *J. Chem. Inf. Comput. Sci.* **1995**, *35*, 1039–1045.
33. Butina, D. *Molecules* **2004**, *9*, 1004–1009.
34. Votano, J. R.; Parham, M.; Hall, L. H.; Kier, L. B.; Hall, L. M. *Chem. Biodivers.* **2004**, *1*, 1829–1841.

35. Tetko, I. V.; Tanchuk, V. Y.; Kasheva, T. N.; Villa, A. E. *J. Chem. Inf. Comput. Sci.* **2001**, *41*, 246–252.
36. Tetko, I. V.; Tanchuk, V. Y.; Kasheva, T. N.; Villa, A. E. *J. Chem. Inf. Comput. Sci.* **2001**, *41*, 1488–1493.
37. Livingstone, D. J.; Ford, M. G.; Huuskonen, J. J.; Salt, D. W. *J. Comput. Aided Mol. Des.* **2001**, *15*, 741–752.
38. Lind, P.; Maltseva, T. *J. Chem. Inf. Comput. Sci.* **2003**, *43*, 1855–1859.
39. Hou, T. J.; Xia, K.; Zhang, W.; Xu, X. J. *J. Chem. Inf. Comput. Sci.* **2004**, *44*, 266–275.
40. Wegner, J. K.; Zell, A. *J. Chem. Inf. Comput. Sci.* **2003**, *43*, 1077–1084.
41. Engkvist, O.; Wrede, P. *J. Chem. Inf. Comput. Sci.* **2002**, *42*, 1247–1249.
42. Liu, R.; So, S. S. *J. Chem. Inf. Comput. Sci.* **2001**, *41*, 1633–1639.
43. Yan, A.; Gasteiger, J. *J. Chem. Inf. Comput. Sci.* **2003**, *43*, 429–434.
44. Wanchana, S.; Yamashita, F.; Hashida, M. *Pharmazie* **2002**, *57*, 127–129.
45. Taft, R. W.; Abraham, M. H.; Doherty, R. M.; Kamlet, M. J. *Nature* **1985**, *313*, 384–386.
46. Kamlet, M. J.; Doherty, R. M.; Abboud, J.-L. M.; Abraham, M. H.; Taft, R. W. *Chemtech* **1986**, *16*, 566–576.
47. Platts, J. A.; Butina, D.; Abraham, M. H.; Hersey, A. *J. Chem. Inf. Comput. Sci.* **1999**, *39*, 835–845.
48. Duffy, E. M.; Jorgensen, W. L. *J. Am. Chem. Soc.* **2000**, *122*, 2878–2888.
49. Jorgensen, W. L.; Duffy, E. M. *Adv. Drug Deliv. Rev.* **2002**, *54*, 355–366.
50. Lobell, M.; Sivarajah, V. *Mol. Divers.* **2003**, *7*, 69–87.
51. Delaney, J. S. *J. Chem. Inf. Comput. Sci.* **2004**, *44*, 1000–1005.
52. Yalkowsky, S. H.; Bajernee, S. *Aqueous Solubility. Methods of Estimation for Organic Compounds*; Marcel Dekker: New York, NY, 1992.
53. Cheng, A.; Merz, K. M., Jr. *J. Med. Chem.* **2003**, *46*, 3572–3580.
54. Meylan, W. M.; Howard, P. H. *Perspect. Drug Disc. Des.* **2000**, *19*, 67–84.
55. Meylan, W. M.; Howard, P. H.; Boethling, R. S. *Environ. Tox. Chem.* **1996**, *15*, 100–106.
56. Yan, A.; Gasteiger, J.; Krug, M.; Anzali, S. *J. Comput. Aided Mol. Des.* **2004**, *18*, 75–87.
57. Catana, C.; Gao, H. *J. Chem. Inf. Model.* **2005**, *45*, 170–176.
58. Butina, D.; Gola, J. M. R. *J. Chem. Inf. Comput. Sci.* **2003**, *43*, 837–841.
59. Stahura, F.; Godden, J. W.; Bajorath, J. *J. Chem. Inf. Comput. Sci.* **2002**, *42*, 550–558.
60. McElroy, N. R.; Jurs, P. C. *J. Chem. Inf. Comput. Sci.* **2001**, *41*, 1237–1247.
61. Yaffe, D.; Cohen, Y.; Espinosa, G.; Arenas, A.; Giralt, F. *J. Chem. Inf. Comput. Sci.* **2001**, *41*, 1177–1207.
62. Mosier, P. D.; Jurs, P. C. *J. Chem. Inf. Comput. Sci.* **2002**, *42*, 1460–1470.
63. Wang, R.; Wang, S. *J. Chem. Inf. Comput. Sci.* **2001**, *41*, 1422–1426.
64. Yang, G.; Ran, Y.; Yalkowsky, S. H. *J. Pharm. Sci.* **2002**, *91*, 517–533.
65. Klamt, A.; Eckert, F.; Hornig, M.; Beck, M. E.; Burger, T. *J. Comput. Chem.* **2002**, *23*, 275–281.
66. Klamt, A.; Eckert, F.; Hornig, M. *J. Comput. Aided Mol. Des.* **2001**, *15*, 355–365.
67. Ikeda, H.; Chiba, K.; Kanou, A.; Hirayama, N. *Chem. Pharm. Bull.* **2005**, *53*, 253–255.
68. Acree, W. E.; Abraham, M. H. *Can. J. Chem.* **2001**, *79*, 1466–1476.
69. Katritzky, A. R.; Oliferenko, A. A.; Oliferenko, P. V.; Petrukhin, R.; Tatham, D. B. *J. Chem. Inf. Comput. Sci.* **2003**, *43*, 1806–1814.
70. Katritzky, A. R.; Oliferenko, A. A.; Oliferenko, P. V.; Petrukhin, R.; Tatham, D. B. *J. Chem. Inf. Comput. Sci.* **2003**, *43*, 1794–1805.
71. College of Pharmacy. www.pharmacy.arizona.edu (accessed June 2006).
72. Syracuse Research Corporation. www.syrres.com (accessed June 2006).
73. Bio-Rad Laboratories. www.bio-rad.com/B2B/Bio-Rad/ (accessed June 2006).
74. CM Compumine. www.compumine.com (accessed June 2006).

Biographies

Jyrki Taskinen (born 1942) graduated and received his PhD in organic chemistry from Helsinki University of Technology, Department of Chemistry, Finland. He was a professor of pharmaceutical chemistry at the Faculty of Pharmacy, University of Helsinki in 1994–2005, and retired in 2005. Previously, he has worked at the Research Laboratories of the State Alcohol Monopoly and in the R&D Department of Orion Pharma. His research interests include computer-assisted drug design and drug metabolism.

Ulf Norinder was born in Sweden in 1956. He received his MS in chemical engineering and PhD in organic chemistry from Chalmers University of Technology, Gothenburg, Sweden, in 1981 and 1984, respectively. He is currently a principal scientist at AstraZeneca R&D Södertälje, and is also an associate professor at the Department of Organic Chemistry, Chalmers University of Technology, as well as an adjunct professor in computational pharmaceutics at the Department of Pharmacy, Uppsala University, Sweden. His research interests include computer-assisted drug design and pattern recognition, with special emphasis on multivariate data analysis.

Comprehensive Medicinal Chemistry II
ISBN (set): 0-08-044513-6

ISBN (Volume 5) 0-08-044518-7; pp. 627–648

5.27 Rule-Based Systems to Predict Lipophilicity

I V Tetko, Institute of Bioorganic and Petrochemistry, Kiev, Ukraine
D J Livingstone, ChemQuest, Sandown, Isle of Wight, UK

5.27.1 Introduction

Physicochemical properties were recognized as important determinants of biological activity more than 130 years ago. Richardson showed that the toxicities of ethers and alcohols were inversely related to their water solubility[1] and Richet demonstrated a relationship between the narcotic effect of alcohols and their molecular weight.[2] The activity of local anesthetics was independently described by Overton[3] and Meyer[4] in terms of oil/water partition coefficients and the distribution ratio itself was first defined by Berthelot and Jungfleisch in 1872.[5] It was not until the 1960s, however, that the octanol/water partition coefficient system (log P) became recognized as a standard system for the description of lipophilicity. Corwin Hansch first proposed the use of octanol/water[6] for a number of theoretical and practical reasons:

- Octanol should be a good mimic for the long hydrocarbon chains with a polar headgroup found in membranes.
- Octanol dissolves water thus emulating the aqueous component of biological hydrophobic regions such as membranes.
- Octanol is cheap, easy to purify and lacks a chromophore, which would interfere with the spectroscopic determination of compound concentrations.

Hansch defined a substituent constant, π, as the difference between the partition coefficients for a parent compound and substituted derivatives:

$$\pi_X = \log P(R_X) - \log P(R_H) \qquad [1]$$

In eqn [1] the parent is indicated by the subscript H and the substituent by X. The first series for which this parameter was derived was a set of monosubstituted phenoxyacetic acids and measurements on polysubstituted compounds showed that π values were additive. As more measurements were made, however, it soon became clear that π values were not strictly additive across different parent series, due principally to electronic interactions, and it became necessary to measure π values in other series such as substituted phenols, benzoic acids, anilines, and so on.[7] This proliferation of different π series actually became a major problem in the use of hydrophobic descriptors since it required the selection of the 'correct' scale for some quite complex molecules. A further complication arises in the choice of substitution position or even parent structure. Most real drug molecules are much more complex than the simple compounds used in the chemical model systems to characterize particular effects. These compounds are selected so that the assignment of effects to particular chemical fragments is unambiguous but it is rarely obvious how these values should be assigned to bioactive structures.

The problem of the selection of an appropriate π scale has been resolved by the increasing use of either whole molecule $\log P$ values or of 'fragmental' values to describe portions of molecules. The fragmental values have often been produced as by-products of the process of creating expert systems that could be used to predict $\log P$ values or in attempts to account for hydrophobic effects in molecular modeling systems. In fact, as the 'traditional' QSAR approach to drug design became increasingly combined with molecular modeling techniques in the 1980s, more and more diverse sets of compounds were the subject of study and the 'standard' substituent constants used to describe hydrophobic, steric, and electronic effects were found unsuitable.

In order to measure a partition coefficient using a method such as the traditional 'shake flask' technique[8–10] it is necessary to have a reasonable idea of the expected $\log P$ value so as to choose the appropriate volumes of octanol and water phases for the partitioning experiment. If a model of biological activity has been constructed using $\log P$ values then, in order to use the model predictively, it is necessary to be able to calculate $\log P$ values for compounds that have not yet been synthesized. Finally, as mentioned above, when a set of compounds contains reasonably complex or diverse molecules it is often not possible to use substituent constants such as π and values of $\log P$ for the whole molecule or molecular fragments are needed. For all these reasons it became obvious that some form of calculation procedure was needed for partition coefficients as discussed in the next section.

5.27.2 Early Systems for log P Calculation

The fragmental approach of Nys and Rekker was one of the earliest systems to be developed for $\log P$ calculation and was based on a statistical analysis of a large number of measured partition coefficient values to give the 'best' values for particular molecular fragments.[11–13] In actual fact these are not 'best' values but average values and so although they work well in most situations there are certain molecules that require correction factors for some of the fragments. This approach was termed 'reductionist' since it involved the breakdown of measured values into contributions from their component fragments and resulted in a large number of fragments with a small number of correction factors. Many of the correction factors involved multiples of a single term, which in the original reports was called a 'magic' factor. $\log P$ is calculated in this approach as shown in eqn [2]:

$$\log P = \sum_{i=1}^{n} a_i f_i + \sum_{1}^{m} k_n \text{CM} \qquad [2]$$

In this equation a_i is the number of occurrences of fragment f_i and the second term is a summation of the number of times (k_n) that the correction factor (CM), which has a value of 0.289, has to be applied to each of the m fragments that need correction. In practice the second term in eqn [2] turns into a simple multiple of CM.

The Rekker system was quickly followed by a 'constructionist' approach due to Hansch and Leo.[14,15] This method was based on the use of a small number of fundamental fragments, derived from very precise $\log P$ measurements, with a correspondingly larger set of correction factors. The equation for the calculation of $\log P$ values using this system is very similar to eqn [2] for the Rekker method with the major difference being that correction factors can take on many values and are thus not simple multiples of a constant correction term. The fragments used in the Rekker method are mainly recognizable 'chunks' of a molecule such as functional groups and rings, although they also include some heteroatoms. The fragments used in the Hansch and Leo system, on the other hand, are single atoms apart from polar fragments, which are considered as multiatom groups[16] Both of these approaches have their advantages and disadvantages. The advantage of Rekker's fragments is that it is easy to make chemical 'sense' of the effect of any change in chemical structure on $\log P$; a disadvantage is that different results may be obtained if the structure is fragmented in different ways. The advantage of the Hansch and Leo method is that it has a set of rules, which will uniquely fragment any structure; a disadvantage is that this fragmentation for some molecules may result in missing fragments. A recent modification to the CLOGP program claims to have reduced the incidence of missing fragments by creating a set of rules to estimate them.[17] Comparisons of the two techniques have concluded, perhaps not surprisingly, that each method gives better results for some sets of compounds than others.[18–21]

At first, the two methods were used to calculate $\log P$ values manually and this had major drawbacks since not only was it labor intensive but it was also difficult to achieve consistency since even a relatively simple molecule may be broken into fragments in a number of different ways. The Hansch and Leo technique was the first to be made available as an automated system in the program CLOGP, which used the elegantly simple SMILES (*see* 3.13 Chemical Information Systems and Databases) notation for chemical structure input.[22,23] Since then the SMILES system has been widely adopted as the basis of database systems and as a chemical structure entry system for a variety of chemical modeling programs. An example of a breakdown of the calculation of log P for three different molecules by the two techniques is shown in **Figure 1**. These three compounds were chosen because they show that both methods work well for hydroxyacetic acid, the Hansch and Leo system works better than the Rekker method for 2-phenoxyethanol, and the Rekker system is better for 1,2-methylenedioxybenzene. log P calculation systems are now available on-line (see later), and these three molecules were submitted to the Virtual Computational Chemistry Laboratory[24,25] website[117] for calculation as shown in **Table 1**. This website not only calculates log P using its own method (ALOGPS) but also links to several other websites to call for calculations. As **Table 1** shows, the calculation by the Hansch and Leo system (ClogP) for 1,2-methylenedioxybenzene has been considerably improved to 2.11 compared with 1.34 given by the original system.

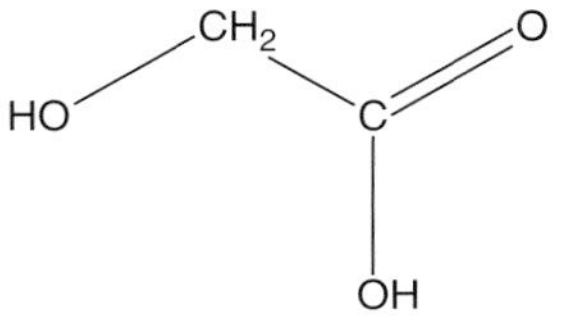

Measured log $P = -1.11$

Rekker:

$$\log P = f_{\text{OH,al}} + f_{\text{CH}_2} + f_{\text{COOH,al}} + 3\text{CM}$$
$$= -1.022 \qquad \Delta = 0.09$$

Hansch & Leo:

$$\log P = f_{\text{OH}} + f_{\text{CH}_2} + f_{\text{COOH}} + (2-1)F_b + F_{P1}$$
$$= -1.06 \qquad \Delta = 0.05$$

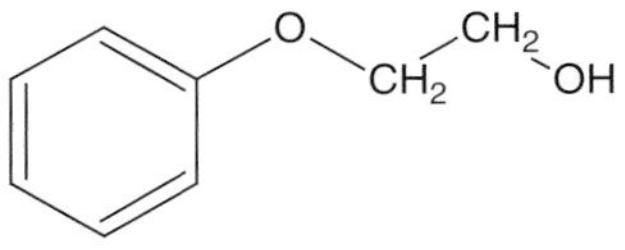

Measured log $P = 1.16$

Rekker:

$$\log P = f_{\text{C}_6\text{H}_5} + f_{\text{O,ar}} + 2f_{\text{CH}_2} + F_{\text{OH,al}} + 2\text{CM}$$
$$= 1.55 \qquad \Delta = 0.39$$

Hansch & Leo:

$$\log P = f_{\text{C}_6\text{H}_5} + f_{\text{O}}^{\Phi} + 2\,f_{\text{CH}_2} + F_{\text{OH}} + (4-1)F_b + F_{P2}$$
$$= 1.20 \qquad \Delta = 0.04$$

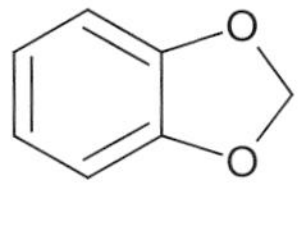

Measured log $P = 2.08$

Rekker:

$$\log P = f_{\text{C}_6\text{H}_4} + 2\,f_{\text{O,ar}} + f_{\text{CH}_2} + 3\text{CM}$$
$$= 2.17 \qquad \Delta = 0.09$$

Hansch & Leo:

$$\log P = 4f_{\text{CH}} + 2f_{\text{C}}^{\Phi} + 2\,f_{\text{O}}^{\Phi} + F_{\text{CH}_2} + 3F_b + F_{P1} + F_{P2}^{\Phi}$$
$$= 1.34 \qquad \Delta = -0.74$$

Figure 1 Examples of the calculation of log P by the Rekker and Hansch and Leo systems. (Reprinted with permission from Mayer, J. M.; van de Waterbeemd, H.; Testa, B. *Eur. J. Med. Chem.* **1982**, *17*, 17–25, with permission from Elsevier.)

Table 1 Calculated log P values from the VCCLAB website for the three compounds shown in **Figure 1**

Compound	$logP^a$	$ALOGPS^b$	IA_logP	$ClogP$	$MlogP$	$KOWWIN$	$XlogP$
I	−1.11	−1.01	−1.63	−1.04	−1.35	−1.07	−1.12
II	1.16	1.22	1.32	1.19	1.45	1.10	1.23
III	2.08	1.71	1.71	2.11	1.82	2.05	1.78

See **Table 3** for program details.
[a] Measured value.
[b] Visit the website (www.vcclab.org) for details of calculations.

5.27.3   Partition in Other Solvent–Water Systems

Partition coefficients and π values have been shown to correlate with measures of biological activity in a very wide variety of experimental systems, ranging from simple protein binding to animal and human effects in vivo. This is presumably because hydrophobic effects are important not only in the intermolecular interactions that occur between a drug and its target site but also in the distribution of a compound within a biosystem, its interaction with competing binding sites, passage across and into membranes, and its interaction with metabolizing enzymes. It may be questioned, however, whether octanol/water is the 'right' model system for hydrophobic effects. That it has been successful is without question but might not another model system be more successful? Part of the answer to this is the fact that partition coefficients from many different solvent systems may be modeled by the use of Collander[26] equations:

$$\log P_2 = a \log P_1 + c \qquad [3]$$

Here $\log P_2$ and $\log P_1$ represent partition coefficients measured in two different organic solvent–water systems with the coefficients a and c estimated by least squares fit. This might explain how a single partition coefficient could be applicable to so many different mechanisms since the weighting coefficient implicit in the Collander relationship would be estimated as part of the mathematical modeling process. It has been shown, however, that partition coefficients from other systems do contain extra information, which is useful in the description of biological properties. Young and co-workers demonstrated[27] that the difference between octanol/water and cyclohexane/water log P values ($\Delta \log P$) could be used to explain brain penetration as shown in eqn [4] and **Figure 2**. It was suggested that this parameter, first introduced by Seiler,[28] might be a useful general descriptor for brain penetration:

$$\log(C_{brain}/C_{blood}) = -0.49(\pm 0.16)\Delta \log P + 0.89(\pm 0.5)$$
$$n = 20 \ \ r^2 = 0.83 \ \ s = 0.44 \ \ F = 40.2 \qquad [4]$$

Extending this concept of the utility of other partitioning systems, Leahy and co-workers suggested that four model systems might be required in order to describe the properties of real membranes.[29] These models consist of water combined with:

- an amphiprotic solvent (e.g., octanol);
- an inert solvent (e.g., any alkane);
- a pure proton donor (e.g., chloroform); and
- a pure proton acceptor.

Propyleneglycol dipelargonate (PGPD) was proposed as a suitable compound for the pure proton acceptor and 216 partition coefficient values were reported along with a calculation scheme for other values. The lack of a wider range of measured values or of any automated calculation procedures is presumably the reason why this approach has not received wider attention.

Partition coefficients in other solvent systems have been reported in the literature but, in general, there have been few reports of attempts to create calculation schemes for them. One notable exception to this is a method known as linear solvation energy relationships (LSER) developed by Kamlet, Taft, Abraham, and co-workers.[30,31] Equations have been developed[32] for a number of solvent–solvent and solvent–gas systems and calculations can be made using the program Absolv.[118] A recent publication describes the calculation of 'virtual' log P values for alkanes and octanol using molecular modeling techniques.[33] The term virtual is used here because the calculations are carried out for individual

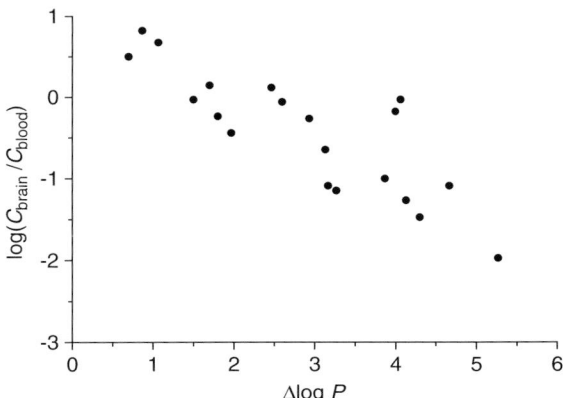

Figure 2 Relationship between blood – brain barrier uptake and log P for 20 structurally diverse compounds. (Reprinted with permission from Young, R. C.; Mitchell, R. C.; Brown, T. H.; Ganellin, C. R.; Griffiths, R.; Jones, M.; Rana, K. K.; Saunders, D.; Smith, I. R.; Sore, N. E. *et al. J. Med. Chem.* **1988**, *31*, 656–671. Copyright (1988) American Chemical Society.)

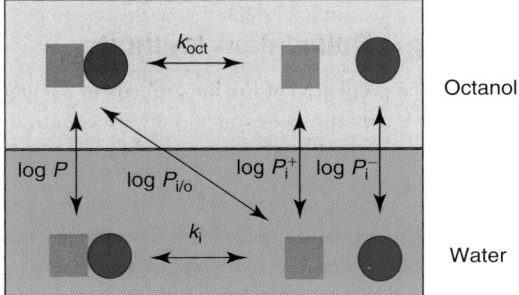

Figure 3 Octanol–water partition of a partially ionized compound. The partition of ions, log $P_i^{+/-} = \log P_i^+ + \log P_i^-$, can significantly contribute to lipophilicity of charged compounds.[116]

conformers and the authors explain how to calculate $\Delta\log P$ values (octanol/alkane) and even $\log D$ (see below) values for the water/alkane system. Thus, this approach deals with two separate issues, which affect partition in any solvent system, flexibility, and ionization.

5.27.4   Distribution Coefficient

The partition coefficient is defined as the ratio of the concentration of a solute in the organic phase to its concentration in the water phase. This definition applies to a neutral species and, for that matter, the same species. Ionization will clearly affect the distribution equilibrium between the two phases, as will other phenomena such as self-association for whatever reason. The distribution coefficient, $\log D$, applies to the measured value of partition for an ionized compound at a particular pH. Assuming that only the neutral form of a molecule will partition into the organic phase then the observed $\log D$ may be related to the $\log P$ and pK_a of the compound, at the pH of the measurement, using an equation such as that shown for monoprotic basic compounds below:

$$\log D = \log P - \log\left(1 + 10^{pK_a - pH}\right) \qquad [5]$$

The assumption that only the neutral form will partition into octanol is, at first sight, reasonable, but unfortunately octanol dissolves quite a large amount of water (water saturated octanol contains around 5 M water) and thus charged compounds will partition into it as ion pairs (*see* 5.18 Lipophilicity, Polarity, and Hydrophobicity). A more complete system of $\log D$ calculations should consider all possible combinations of partition coefficients as shown in **Figure 3**. The $\log D$ of a compound depends not only on the pH but also the concentration and nature of the counter-ions.[34] Having said this, the distribution coefficient is an important quantity in many applications and thus procedures for its calculation are useful, despite flaws in their theoretical basis. The calculation of $\log D$ is discussed in a later section.

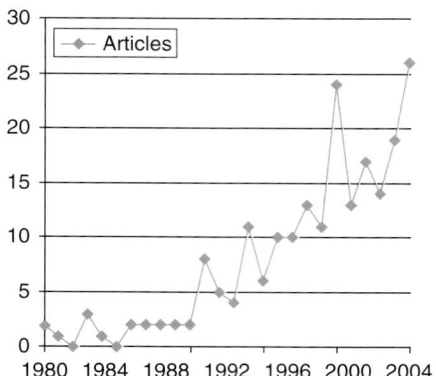

Figure 4   The number of methodological publications for computational $\log P/D$ prediction as a function of time (for 2005 it is a projected number). The publications were collected searching PubMed, ISI, and Google scholar servers. The 2000-year peak is partially explained by publication of the dedicated issue of *Perspectives in Drug Design and Discovery* that covered main $\log P$ prediction methods. (Hydrophobicity and Solvation in Drug Design, Parts I and II, *Perspect. Drug Disc. Des.* **2000**, *17* & *18*.)

5.27.5   The Explosion of $\log P$ Calculation Methods

The interest of pharmaceutical companies in prediction of the lipophilicity of compounds and multiple success stories about the correlation of this property with important biological and physicochemical properties of molecules caused a growth in the number of methods developed to predict this property. **Figure 4** indicates the number of publications with new $\log P$ calculation methods as a function of time.

The developed methods can be grouped according to two main criteria: according to the type of descriptor sets used, i.e., substructure and whole molecule approaches,[35] and according to the machine learning methods, e.g., linear or nonlinear approaches, used to establish structure–$\log P$ relationships. A simple analysis of the major methods indicates that linear methods tend to use fragment-based descriptors, while the nonlinear methods can exploit more efficiently whole molecular approaches.

Several reviews have examined the developed methods from a point of view of the representation of indices.[35–37] A different strategy is followed here and an analysis is provided according to the type of models used. Since the performance of the methods depends on the number of compounds in the training set, number of indices, machine learning method used as well as composition and diversity of the training and test sets (if the latter was employed), a comparison of the approaches according to their published results, in our opinion, does not really make sense. Thus, instead, a description of the philosophy of the developed approaches is focused on here, including their unique features that make them interesting for the scientific community. A comparison of several methods will be performed using some test sets in Section 5.27.9.

5.27.6   Linear Models

5.27.6.1 $\log P$ Prediction Using Fragmental Descriptors

As was mentioned above, the substructure approaches, such as fragmental and atom contribution methods, cut the analyzed molecule into fragments or atoms (as degenerate case of fragments), attribute a particular value for each group, and calculate the $\log P$ value as:

$$\log P = a + \sum_{i=1}^{N} b_i G_i \qquad [6]$$

where G_i is the number of occurrences of the group i and a and b_i are regression coefficients. This formula, however, may require additional coefficients, F_j, known as correction factors:

$$\log P = a + \sum_{i=1}^{N} b_i G_i + \sum_{i=1}^{K} c_i F_i \qquad [7]$$

The Rekker method, as shown in eqn [2], can be seen to conform to this general equation. A large number of group contribution methods have been developed during the last few years. The most popular methods include those that are fragment based such as ClogP,[17,38] ACD/logP,[39] KOWWIN,[40,41] Σf-SYBYL,[42,43] KlogP,[44] HlogP,[45] AB/logP,[46] and techniques based on atom-contribution approaches such as AlogP,[47–49] XlogP,[50,51] and SMILOGP.[52] The fragments used in these approaches range from 68 in AlogP[48] to about 2000 in HlogP[45] while the number of correction factors varies from 0 in Alog P to several thousand in the ACD/log P method.

Many of the fragment-based methods have been reviewed previously.[35,37,53] Only the main features and conceptual differences among the methods are pointed out below.

The poor performance of group contribution methods for the prediction of new compounds can be attributed to the presence of some new groups/correction factors that are not covered in the training set. As mentioned above, a new version of the CLOGP program (v. 4.0 and higher) includes an algorithm for the 'ab initio' calculation of the contribution of fragments if they are not found for the training set. Thus, new fragments are easily calculated and included to estimate $\log P$ for unusual molecules.[17]

ACD/$\log P$ also has to deal with the missing fragment problem. This program estimates them using a fragmental increment equation similar to the Ghose/Crippen approach[54] and a multilinear equation similar to the Hammett–Palm equation.[39] The result is calculated as a sum of all the fragments. If new experimental data are available, the contribution of some of the fragments, particularly missed ones, can be recalculated from the new data. This option is claimed to significantly improve the accuracy of the program and constitutes the so-called 'user-training' feature of ACD/$\log P$.[55] The practical application, such as the analysis of a few thousand compounds, however, indicated severe limitations of the method in speed, memory, and disk usage while the resulting improvement of accuracy was only marginal and increased r^2 between predicted and calculated values from 0.3 to 0.5.[56]

The KOWWIN program includes an interesting methodology, experimental value adjusted approach,[40] which evaluates the contribution of fragments by comparing closely related analogs. Thus, if a new compound has to be predicted, experimental value adjusted approach identifies the closest analog from the training set and calculates the $\log P$ of the new molecule by adding or subtracting the contribution of the groups required to transform the query structure into its analog:

$$\log P = \log P_{\text{exp,analog}} + \Delta \log P_{\text{target} \rightarrow \text{analog}}$$

$$\Delta \log P_{\text{target} \rightarrow \text{analog}} = \left(\sum b_{i,\text{target}}\, G_{i,\text{target}} + \sum c_{i,\text{target}}\, F_{i,\text{target}} \right) - \left(\sum b_{i,\text{analog}}\, G_{i,\text{analog}} + \sum c_{i,\text{analog}}\, F_{i,\text{analog}} \right) \quad [8]$$

where $\Delta \log P_{\text{target} \rightarrow \text{analog}}$ corresponds to the sum of contributions of fragments and correction factors required to change the target molecule into its analog. **Table 2** demonstrates fragment and correction factors for atropine and cocaine. The sum of contributions to map cocaine to atropine is $\Delta P_{\text{cocaine} \rightarrow \text{atropine}} = 0.26$. Thus, considering that the experimental atropine value is 1.83, the cocaine value estimated by the experimental value adjusted method is $\log P_{\text{cocaine}} = 1.83 + 0.26 = 2.09$; this is quite a good estimation of the experimental value of this compound, which is 2.30 log units. In a similar way the $\log P$ value of atropine can be estimated from the experimental value of cocaine as $\log P_{\text{atropine}} = 2.30 - 0.26 = 2.04$. Note that the above equation can also be rewritten (it just requires that several terms in the equation are swapped and that the summation fragments and correction factors that are common for target and analog compounds are also included) as:

$$\log P_{\text{target}} = \log P_{\text{predicted, target}} + \left(\log P_{\text{exp, analog}} - \log P_{\text{predicted, analog}} \right)$$

$$= \log P_{\text{predicted, target}} - \Delta \log P_{\text{error, analog}} \quad [9]$$

where $\log P_{\text{predicted}}$ refers to the value predicted using solely the fragment-based approach. This simple mathematical trick gives another treatment of the experimental value adjusted approach – the predicted value of the target compound is corrected using the error of its nearest neighbor. This simple but fundamental equation represents the basis of two other recently and independently proposed methods, SLIPPER[57] and ALOGPS.[58,59]

AlogP[47–49] is a pure atom-based contribution method based on a parameter set of only 68 atom type descriptors. The simplicity and as a result robustness of the method is probably the most remarkable feature of this approach. Because of simple implementation, the AlogP method can be easily reproduced and is implemented in several software packages, e.g., in DRAGON.[60]

Table 2 Fragment contributions calculated for atropine and cocaine by KOWWIN

Cocaine   Atropine

No.	Type	Fragment description	Coefficient	Number of groups	
				Atropine	Cocaine[a]
1	Fragment	–CH$_3$ (aliphatic carbon)	0.5473	1	2 (+1)
2	Fragment	–CH$_2$- (aliphatic carbon)	0.4911	5	3 (−2)
3	Fragment	–CH (aliphatic carbon)	0.3614	4	4
4	Fragment	–OH (hydroxy, aliphatic attach)	−1.4086	1	0 (−1)
5	Fragment	–N< (aliphatic attach)	−1.8323	1	1
6	Fragment	Aromatic carbon	0.2940	6	6
7	Fragment	–C(=O)O (ester, aliphatic attach)	−0.9505	1	1
8	Fragment	C(=O)O (ester, aromatic attach)	−0.7121	0	1 (+1)
9	Correction factor	Fused aliphatic ring unit correction	−0.3421	1	1
10	Constant	Equation constant	0.2290	1	1
Experimental values of compounds, $\log P_{exp}$				1.83	2.30
Cocaine $\log P$ value estimated from the $\log P_{exp}$ value of atropine					2.09

[a] The number of fragments/correction factors required to change atropine to cocaine is indicated in parentheses.

HlogP[45] uses an extended representation of 2D fragments based on structural keys (i.e., presence or absence of certain groups). The representation of a molecule using these fragments is called a 'molecular hologram.' It is possible to control the length of fragments and thus the 'fragment collision,' when different fragment types occupy the same bin. This happens when the hologram length is smaller than the number of distinct fragments. The data analysis method used in HlogP is partial least squares regression.

A new version of the KlogP program[61] is also based on fragmental indices but it profits from knowledge of the 3D structure of molecules by means of steric hindrance indices, H, proposed by Cherkasov.[62] The modified eqn [6], i.e.:

$$\log P = a + \sum_{i=1}^{N} b_i (1 - H_i) G_i \tag{10}$$

takes into account the hindrance H_i of the atoms in the fragment G_i as $H_i = \Sigma_{j \neq i}^{n} R_j^2 / A \cdot r_{ij}^2$, where A is a constant, R_j the atomic radius of the jth atom, and r_{ij} the distance between the ith and jth atoms.[62] The H_i index weights the contribution of different fragments according to their availability to solvent. The use of eqn [10] instead of [7] remarkably increased the performance of the method for drugs and decreased the standard deviation from 1.08 to 0.78 log units for a test set of 137 drugs.

The substructural molecular fragment (SFM) method[63] splits a molecule into fragments of two different types, 'sequences' (I) and 'augmented atoms' (II). For each type of fragment one can define indices of three subtypes AB, A, and B. For example I(B, 2–6) considers only types of bonds from 2 to 6 atoms. In a similar way to HlogP the resolution of the data can be easily controlled by including different numbers of atoms for the generation of fragments. The data analysis in this method is performed using the singular value decomposition.

5.27.6.2 log P Models Based on Nonfragmental Descriptors

5.27.6.2.1 Methods based on a few theoretically justified descriptors

While fragmental descriptors are one of the most frequently used descriptor systems for the prediction of lipophilicity of compounds, there are theoretical justifications for the use of property-based descriptors.

For two relatively immiscible solvents $\log P$ can be considered[64] proportional to the molar Gibbs free energy ($\Delta G^0_{o \to w}$) of transfer between octanol and water:

$$\log P = \frac{1}{RT} \Delta G^0_{o \to w} \qquad [11]$$

The solvation theory[65] indicates that the major contribution to this energy is the cavity term, usually considered proportional to the volume (or surface) of the solute. Thus, the partition coefficient depends on some molecular properties, which contribute to this energy term.

Mobile order and disorder (MOD) theory provides another nice framework for theoretical analysis of the lipophilicity of compounds.[66] The theory was challenged by Einstein who first proposed to express the equilibrium as time fractions rather than classical ensemble fractions in the Boltzmann model and thus eqn [11].[67] The MOD also uses mobile molecular domains for the calculation of the entropy of mixing compared to quasi-lattice models used in the classical Boltzmann model. Thermodynamically, the partition coefficient can be regarded not only as the ratio of concentrations at the equilibrium but also as the ratio of saturation concentrations or solubilities. Thus, it can be obtained from the differences of its volume fraction solubilities between less polar (octanol) and more polar (water) phases:

$$\log P = (\ln 10)^{-1}[\ln \Phi^o_B - \ln \Phi^w_B] \qquad [12]$$

The universal predictive equation for $\log P$ is given by the MOD theory[68,69] as the sum of Gibbs free energy contributions:

$$\log P = \Delta B_{o/w} + \Delta F_{o/w} + \Delta (O + OH)_{o/w} + \Delta D_{o/w} \qquad [13]$$

Here, $\Delta B_{o/w}$ corresponds to the differences between the two phases in the entropy of the solute–solvent exchange, $\Delta F_{o/w}$ is the hydrophobic effect-related term accounting for the differences in the propensity to squeeze the solute out of the solution, $\Delta(O + OH)_{o/w}$ expresses the differences in the strength of the H-bonds that bind the solute and the solvent in both phases, and the last term $\Delta D_{o/w}$ is similar to the previous one but accounts for the nonspecific forces only.

The general form of eqn [13] reduces to a very simple linear equation relating lipophilicity of molecules and their molar volume (V_B):

$$\log P = \Delta B_{o/w} + \Delta F_{o/w} = -0.48 + 0.03328\, V_B \qquad [14]$$

when the differences in the changes of nonspecific forces ($\Delta D_{o/w} = 0$) are negligible and when no solute–solvent specific interaction takes place in either phase ($\Delta O + OH)_{o/w} = 0$). Thus, despite different basic assumptions, the main conclusion of both classical solvation and MOD theory are basically the same and provide a direct basis for the empirical correlation of $\log P$ values to their molar volume or any related property (surface area, parachor, molar refraction, etc.). Indeed, there were multiple $\log P$ calculation models in-line with the conclusions of these studies For example, the QlogP model:

$$\log P = 0.032(\pm 0.0002)\, V - 0.723(\pm 0.007)\, N + 0.01(\pm 0.0007)\, I$$
$$n = 320 \text{ alkanes},\ r^2 = 0.98,\ s = 0.21 \qquad [15]$$

utilizes the molecular volume (V) as its central descriptor as well as a correction parameter to account for the hydrogen bonding effect between the solvent and oxygen–nitrogen-containing functional groups (N) of the solute molecules and an alkane indicator variable (I).[70] The solute size (favors octanol) together with solute hydrogen-bond basicity (favors water) were named as the main parameters of the Abraham's general linear solvation energy equation.[71] Xing and Glen[72] also calculated a significant model for $\log P$ prediction using just three parameters, polarizability and partial atomic charges on nitrogen and oxygen:

$$\log P = 0.29(\pm 0.07) + 0.199(\pm 0.004)\alpha - 14.9(\pm 0.5)q_N^2 - 8.4(\pm 0.2)q_O^2$$
$$n = 592, \; r^2 = 0.89, \; s = 0.65 \quad\quad [16]$$

where α is the molecular polarizability and q_N^2 and q_O^2 are the total squared partial charges on nitrogen and oxygen atoms, respectively. The model predicts higher lipophilicity values for molecules with bigger polarizability, i.e., molecules that require a bigger cavity are predisposed to move into the 1-octanol.

The use of eqn [13] requires knowledge of solubility data of monofunctional systems that are, to some extent, similar to fragment contributions. A nice feature of the MOD approach is that essentially the same equation can also be applied to derive partition coefficients in other two-phase systems made of two largely immiscible solvents. An application of eqn [13] to a wide set of chemicals in 16 different two-phase systems calculated a standard deviation error of 0.48 log units for 2263 predicted values.[69] However, in its current form the model can only be applied to essentially nonfunctional or monofunctional compounds but not to complex polyfunctional molecules and, in particular, to those with conjugated or internally H-bonded structures.[69]

There were several studies attempting to empirically correlate the $\log P$ values of chemicals from molecular properties contributing to Gibbs free energy pioneered as early as 1969 by Rogers.[73] The quantum-chemical calculations (MINDO/3 and Hückel-type calculation based primarily on topology) were challenged to predict the partition coefficient by Klopman.[74] The BlogP method, involving 18 parameters, was developed using AM1 methodology.[75]

Despite a great educational influence and explanatory power, the above articles could not be considered as important practical methods. Indeed, the number of compounds used in those studies was usually in the order of several hundreds and the molecules were from structurally simple classes (i.e., alkanes). However, even for such molecular series with limited diversity there was sometimes a need for correction factors and indicator variables (QlogP[70]) to account for nonadditive effects. The descriptors used in the aforementioned articles could be very useful for correlations in homogeneous series of compounds to build local lipophilicity models.

Nevertheless, in recent years new techniques based on quantum-chemical calculations have appeared. A well-known method, QikProp, is based on a study[76] that used statistical Monte Carlo simulations to calculate 11 parameters, including solvent accessible solvent area (SASA), solute–solvent energies, solute dipole moment, number of solute–solvent interactions $< -2.75\,\text{kcal mol}^{-1}$ (INME), number of solute as donor/hydrogen bonds (HBDN/HBAC), and some of their variations. These parameters made it possible to estimate a number of free energies of solvation of chemicals in hexadecane, octanol, water, as well as octanol/water distribution coefficients. The equation calculated for octanol/water coefficient is:

$$\log P = 0.015\,(\text{SASA}) - 0.58\,(\text{HBAC}) - 1.09\,(\text{no. of amines})$$
$$+ 1.10\,(\text{no. of nitro acid groups}) - 0.102\,(\text{INME}) - 1.81$$
$$n = 200, \; r^2 = 0.91, \; s = 0.55 \quad\quad [17]$$

The dominating term in the equation is the total solvent area of the molecule (SASA). Compounds with larger SASA favor solvation in octanol, which is in accordance with the importance of the size of molecules for lipophilicity of chemicals predicted by both solvation and MOD theory.

Another interesting approach to calculate partition coefficients using quantum-chemical calculations was developed by the Klamt group.[119] The COSMOtherm program describes the interactions in a fluid as local contact interactions of molecular surfaces.[77,78] This makes it possible to derive models for different solvent–water partition systems, including octanol, benzene, hexane, etc.

5.27.6.2.2 Large-scale property-based models

A popular method for calculation of $\log P$, MlogP, was developed by Moriguchi and co-workers.[79] MlogP uses the sum of lipophilic (carbons and halogens) atoms and the sum of hydrophilic (nitrogen and oxygen) atoms as two basic descriptors. These two descriptors were able to explain 73% of the variance in the experimental $\log P$ values for a database of 1230 compounds. The use of 11 correction factors covered 91% of the variance. Because of its simplicity of implementation, the MlogP method was widely used as a calculation and reference approach for many years.

SLIPPER[57,80] calculates lipophilicity as:

$$\log P = 0.267\alpha - \sum C_a \qquad [18]$$

which includes only two terms, polarizability (α) and the hydrogen bond acceptor strength ($\sum C_a$) of the molecule. Using only these two descriptors the authors calculated good results for a database of 2850 simple compounds ($n = 2850$, $r^2 = 0.94$, $s = 0.23$). However, they pointed out that the problem of predicting lipophilicity for compounds with several functional groups is much more difficult. Therefore, they proposed to correct the $\log P$ prediction of the target compound according to the lipophilicity values of the nearest neighbors as:

$$\log P_{\text{target}} = \left(0.267a - \sum C_{a,\text{target}}\right) + \frac{1}{K}\sum_{j=1}^{K}\left(\log P_{\text{exp},j} - \frac{1}{K}\sum_{j=1}^{K} 0.267\alpha_j - \sum C_{a,j}\right) \equiv$$

$$\equiv \log P_{\text{predicted,target}} + \frac{1}{K}\sum_{j=1}^{K}\left(\log P_{\text{exp},j} - \log P_{\text{predicted},j}\right), \text{ where}$$

$$\log P_{\text{predicted}} = 0.267a - \sum C_a \qquad [19]$$

where the K-nearest neighbors are determined using cosine similarity measure for representation of molecules by molecular fragments. Thus, the actual predicted value is corrected by the average error of its nearest neighbors. It is easy to notice that experimental value adjusted approach of KOWWIN uses exactly the same equation with the exception that correction in the former method is done using only one nearest neighbor and, of course, the predicted values are calculated using a different method. Using this approach the authors significantly improved the statistical results for a large database of 10937 chemicals compared to the use of the original eqn [18].

5.27.7 Nonlinear Methods

The dependency of $\log P$ on chemical descriptors clearly has a nonlinear character. This fact is taken into account in linear methods by introducing correction factors, F_j, in fragment-based approaches (eqn [7]), use of indicator variables, or development of methods to predict lipophilicity departing from the nearest neighbor analog (KOWWIN and SLIPPER). The use of nonlinear approaches, such as neural networks, makes it possible to more easily incorporate the nonlinear effects in the model and can, generally, result in models with higher prediction ability. This explains the appearance of a large number of studies performed to provide nonlinear modeling of $\log P$ as a function of molecular parameters.

Some of the first attempts to predict $\log P$ values using artificial neural networks (ANNs) were published in 1994 and 1995.[81–83] These works analyzed rather small sets of compounds and mainly introduced a new methodology and demonstrated its performance compared to traditional multiple linear regression (MLR) analysis. It is interesting that all of these studies used parameters derived from quantum-chemical calculations and thus were targeted to develop very general models that could predict compounds in the whole chemical universe. The use of advanced methodology for descriptor selection[84] made it possible to significantly decrease the number of required descriptors for the data set of Bodor[75,81] and to calculate improved statistical results for the independent test set used in the original study. Thus, this study demonstrated that the ANN design has a significant impact on the quality of calculated results. Similar conclusions and the importance of variable selection to construct reliable neural network models were also reported elsewhere.[85,86]

The development of $\log P$ prediction methods based on quantum-chemical parameters was continued by the group of Clark.[87,88] Both these reports were based on 1085 molecules and 36 descriptors were calculated following structure optimization and electron density calculation with the AM1 method. The descriptors selected with an MLR model were used as an initial set of descriptors, which was further optimized by trial-and-error variation. The new analysis also proposed to estimate the reliability of neural network prediction by analysis of the standard deviation error of an ensemble of 11 networks trained on different randomly selected subsets of the initial training set.[88]

The AUTOLOGP program[89,90] was developed using 2D autocorrelation descriptors.[91] The autocorrelation descriptors consider a molecule as a graph with the distance between nodes determined as the smallest number of edges between them. Any atomic contributions AC_i can be used to calculate products, $AC_i \times AC_j$, $i,j = 1,\ldots,n$, where n is the maximal distance in the molecule. The sum of these products for the same distance in the graph gives a component of the autocorrelation vector for the selected property. The authors used autocorrelation vectors encoding lipophilicity, molar refractivity, and hydrogen bond acceptors/donors. Only 35 autocorrelation indices were required to describe the molecules correctly and model log P.

The E-state indices[92,93] were developed to cover both topological and valence states of atoms. These indices were successfully used in many studies[92] and new applications of this methodology are extensively reviewed in this book (see 5.23 Electotopological State Indices to Assess Molecular and Absorption, Distribution, Metabolism, Excretion, and Toxicity Properties). Several articles by different authors demonstrated the applicability of the indices for lipophilicity predictions.[94–99] The ALOGPS[100] program was developed using 12 908 compounds from the PHYSPROP database,[101] which is one of the largest data sets used to predict lipophilicity of chemicals. The neural networks were trained using 73 E-state descriptors and number of hydrogen and nonhydrogen atoms and produced a significant improvement compared to MLR and several other methods, e.g., ClogP,[38] XLOGP,[51] etc., compared in the study. The authors warned that if molecules in training and test sets have different chemical diversity, the prediction ability of programs developed using different methodology but using the same training set is very similar and unacceptably low.[100] Following this observation, a feature to improve the prediction ability of the ALOGPS program was developed using the Associative Neural Networks approach.[58,59,102] This technique made it possible to incorporate new data into the memory of neural networks without their retraining. The basic equation is the same used in experimental value adjusted approach of KOWWIN[40] and SLIPPER-2001,[80] i.e., the prediction of a new compound is corrected by the average error of its nearest neighbors. The principal difference is in the definition of similarity between the molecules, which is defined as the rank correlation in the space of trained models.[59,102] This definition is claimed to be more accurate since it includes the features normalized by the neural network according to the target property, i.e., lipophilicity. The applicability of the method was successfully demonstrated in a number of studies for 'in-house' data of pharmaceutical firms.[103,104]

The question of whether one should use diverse or large libraries of compounds was challenged in another study.[105] It was argued that the use of huge libraries containing nondrug compounds may overfit the programs to such series of compounds and thus the developed programs could not predict the drug-like compounds. The authors selected a set of 78 compounds that were outliers for several studies, e.g., KOWWIN,[40] ClogP,[17,38] as well as additional drug-like compounds to give a database of 625 molecules. After the development of their approach, AutoQSAR models, using this set of compounds the authors were able to predict another set of compounds missed from the public databases. It is interesting that contrary to previous studies they calculated the best results for the test set with MLRA but not with ANN and PLS methods. However, since the final test set included just 18 compounds, their conclusions should not be over-generalized.

There have also been studies attempting to correlate log P with connection matrices[106] or fragmental indices.[107] The authors calculated improved models and argued that the advantage of the fragment-based descriptors is that they are more easily interpreted. It is interesting that the same group of authors mathematically proved that any topological index can be replaced with a set of fragmental descriptors provided that the structure-property data set is sufficiently large to build statistically significant models.[108] This result is not surprising, considering the fact that as low as 1 bits/atom are required to encode a molecule. Thus, theoretically as low as one to two topological indices calculated with float precision can be sufficient to encode/decode a molecule.[109] Any representation that utilizes a larger number of bits, i.e., fragmental indices, can be used for the same purposes and thus provide equivalent mapping between different representations. Of course, the differences may arise in the number of molecules required to make each model (e.g., based on topological or fragmental indices) statistically significant.

5.27.8   1-Octanol/Water Partition Coefficient Calculation Programs

5.27.8.1   log P

Since log P is such an important property for so many aspects of drug design it is not surprising that there has been a rapid increase in the availability of calculation programs to accompany the 'explosion of calculation techniques' Individual programs have been mentioned repeatedly in this chapter so a list of the free, commercial, and web accessible calculation routines are gathered together here. Table 3 is not an exhaustive listing of the available programs and web addresses were correct at the time of writing, of course.

5.27.8.2   log D

Calculation of the distribution coefficient, log D, as discussed in Section 5.27.4 is complicated by the fact that this requires knowledge of both log P and pK_a(s). In fact, log P values are only true log P values if they are measured at a pH at which the ionization of any ionizable groups is suppressed. It is unusual to find measured log P values quoted with a

Table 3 A list of log P calculation programs

Program[a]	Calculation method[b]	Method	Supplier
Commercial			
AB/logP	Fragmental-A/f	Linear	www.ap-algorithms.com
ACD/LogP	Fragmental-A/F	Linear	www.acdlabs.com
AutologP	Properties	Neural networks	j.devillers@ctis.fr
CERIUS[2*]	Atomic values		www.accelrys.com
CLOGP[#]	Fragmental-HL	Linear	www.biobyte.com
CslogP	Topological descriptors	Neural networks	www.chemsilico.com
HINT	Properties	Linear	www.eslc.vabiotech.com/hint
K-Pro	Fragmental-C	Linear	www.multicase.com
MlogP	Number of lipo- and hydrophilic groups	Linear	www.tripos.com
PCMODELS	Fragmental-HL	Linear	www.daylight.com
PrologP	Fragmental-R	Linear	www.compudrug.com
S + logP	Topological	Neural networks	www.simulations-plus.com
SLIPPER	Properties	Linear	www.timtec.net
SYBYL*	Fragmental-R	Linear	www.tripos.com
TerraQSAR-logP	?	Neural networks	www.terrabase-inc.com
TlogP	Topological and substructure coding	?	www.upstream.ch
TSAR*	Atomic values	?	www.accelrys.com
VLOGP*	Topological descriptors	Linear	www.accelrys.com
Free			
CHEMICALC-2	Atomic values	Linear	www.osc.edu/ccl/qcpe
KOWWIN[#]	Fragmental-A/F	Linear	www.epa.gov/opptintr/exposure/docs/episuite.htm
XLOGP[#]	Atomic values	Linear	mdl.ipc.pku.edu.cn/drug_design/work/xlogp.html
Via the Web			
Osiris	Atomic values	Linear	www.organic-chemistry.org/prog/peo
ALOGPS[#]	Topological descriptors	Neural networks	www.vcclab.org
IA_logP[#]	Topological descriptors	Neural networks	www.logp.com
MiLogP[#]	Group contributions	Linear	www.molinspiration.com
SklogP	Topological descriptors	Neural networks	preadme.bmdrc.org

[a] These are stand-alone programs except those marked with *. Results of programs marked with # are accessible from the Java applet at www.vcclab.org site.

[b] The fragmental methods refer to the system of Hansch and Leo (HL), Rekker (R), computer identified (C), and atom/fragment contributions (A/F). 'Properties' means that various molecular properties are used in the calculations. 'Atomic values' means that tables of atom-based values are used. 'Topological descriptors' means (usually) electrotopological descriptors.

measurement pH and the assumption, perhaps often wrong, is that the experimentalist chose an appropriate pH. Of course the casual user of log P may find it confusing to see measurements reported at pH 2 or 11 when the expected important pH of biological systems is 7.4.

When pK_a values are known then the calculation of log D from either measured or calculated log P values is trivial. When the pK_a values are unknown, though, the process of log D calculation becomes problematic since calculation of the ionization constant of an acid or base is arguably more difficult than the calculation of log P. The first step in the calculation of pK_a values for any molecule is the recognition of the ionizable group or groups and this needs an algorithm, which has some chemical 'sense.' Having recognized the groups there are several ways in which pK_a values

Table 4 Programs for the calculation of log D

Program	Calculation method	Supplier
ACDlogD	Fragmental-A/F	www.acdlabs.com
ALOGPS[a]	Topological descriptors	www.vcclab.org
CSlogP	Topological descriptors	www.chemsilico.com
PrologD	Hammett/Rekker	www.compudrug.com
SLIPPER	Properties	www.timtec.net
Plug-in		www.chemaxon.com
ADME/Tox Web, Tox Boxes		www.ap-algorithms.com

[a] log D calculations are available as the user-training feature.

may be estimated including methods based on: the Hammett equation, atomic charges, fragments, semiempirical and ab initio molecular orbital calculations, 3D QSAR, and hybrid (combined methods) systems (*see* 5.25 In Silico Prediction of Ionization). There is no space to discuss them here but the difficulty of pK_a prediction means that there are far fewer programs available for log D calculation as shown in **Table 4**.

5.27.9   Assessment of Performance

An assessment of the performance of the models is crucial for their application and development of new approaches. There were numerous studies published with a detailed analysis of different log P calculation methods attempting to derive an objective opinion about the relative performance of different strategies (i.e., fragmental and whole-molecule approaches) and machine learning methods (i.e., linear versus nonlinear).[37,53,110,111] The model performances in such studies are usually compared on published data or some relatively small test set of publicly available compounds that are normally selected to be diverse and 'drug-like' (i.e., drugs or drug-like molecules). The main assumption of such studies is that the analyzed approaches tested on the 'drug-like' set will have a similar performance for new unseen 'drug-like' molecules. Unfortunately, such assessment of performance of the methods cannot be truly objective. In general, the quality of the models to be compared critically depends on three main parts: (1) molecular descriptors; (2) machine learning approach; and (3) the diversity and the size of the training sets. While the first two items correspond to an intuitive understanding of the 'quality of the model' and 'predictive performance,' the third item can actually dominate in the method performance. Indeed, numerous studies suggest that the use of only a single compound per series or scaffold of compounds in the training set may improve the prediction ability of the method for the whole series by several times.[55,58,100] Thus, in the absence of knowledge whether the test molecules and/or their analogs were used or not to develop the tested model an objective comparison of the programs is difficult.

We compared several programs, ALOGPS, KOWWIN, IA_logP, CLOGP, XLOGP, miLogP available at the VCCLAB website[24,117] using 20 series of compounds that were published in leading chemical journals during 2003–2004. Six of the series were contributed by pharmaceutical companies or were the result of collaboration between industry and academia. The molecules were downloaded from the LOGKOW database[120] supported by Dr J. Sangster and were checked to eliminate possible typing errors.

The first analysis of the series indicates that most reported 1-octanol/water partition values are log D rather than log P values. This can be explained logically for at least two reasons, experimental and biological. First, it is considerably easier and cheaper to make measurements of lipophilicity using, for example, phosphate buffer, under fixed pH. The fixed pH can also be efficiently used with cheap experimental methods, such as high-performance liquid chromatography (HPLC), and is most suitable for high-throughput measurement systems. In contrast to that, the identification of log P is more complex and requires several steps of titration toward the direction of its neutral form. This may dramatically decrease its solubility and makes measurements of log P inaccurate. Second, an argument goes that the compounds will perform their action on the biological target in the organism under physiological pH (7.4) or a range of pH values in the gut when it should be absorbed by the organism. Although this has appeal, it should be remembered that this 'physiological pH' is a bulk pH, not the pH of the microenvironment where the molecules interact with their receptor.

We used predicted pK_a values of the ACD and Pallas programs to decide if the analyzed compounds are predicted to be ionized or neutral under the pH of measurements. Eight series were predicted to be neutral at the measurement pH and, interestingly, these series had considerably lower prediction errors with all methods (**Figure 5**). The mean absolute error (MAE) for these series averaged over all methods was about 0.6 units as shown in **Figure 6a**. The lowest errors for these sets were calculated using XLOGP (MAE = 0.53).

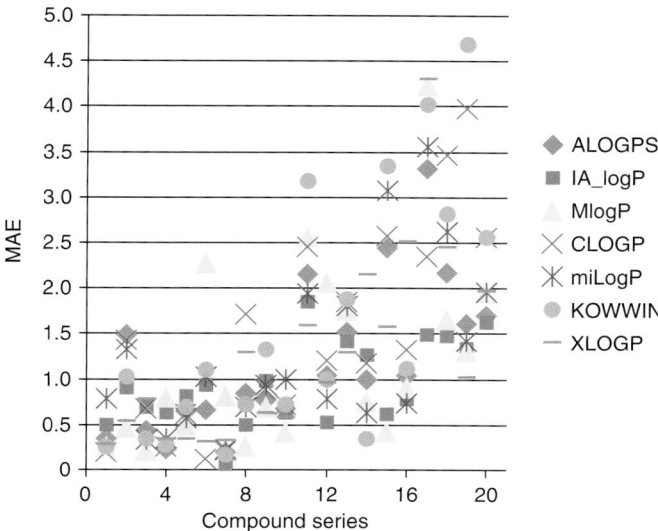

Figure 5 The mean absolute error (MAE) of the methods for prediction of different series of compounds. The first eight series are compounds predicted to be neutral at pH of measurements according to the ACD Laboratories and Pallas programs.

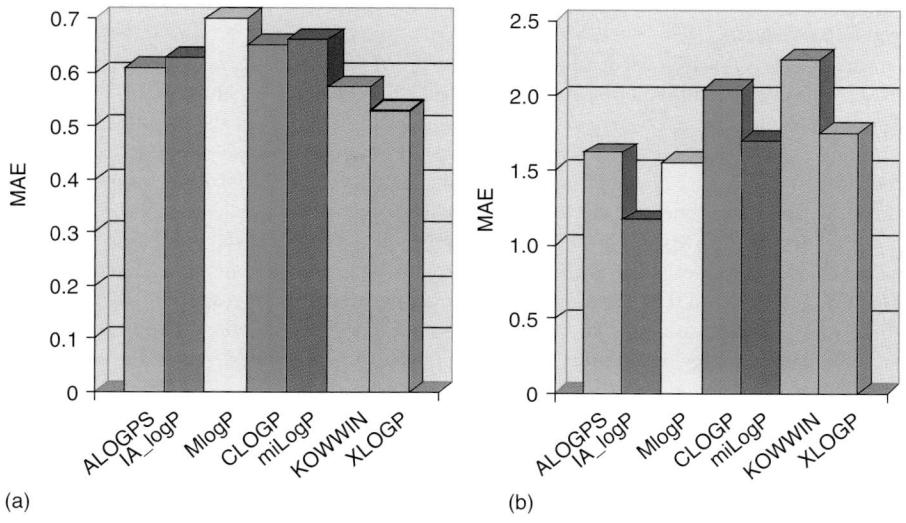

(a) (b)

Figure 6 Average mean absolute error (MAE) of different programs for 8 neutral (a) and 12 charged (b) data sets.

Table 5 ACD laboratories and Pallas log$P$ and log$D$ values calculated for pyrazolopyridine derivatives[112] using experimental pK_a values

Compound number	X	log D	pK_a	log P		log D 7.4 values estimated using eqn [5], calculated log P, and experimental pK_a	
				ACD labs	Pallas Prolog	ACD labs	Pallas Prolog
4	H	0.96	8.23	4.53	3.83	3.64	2.94
6		1.14	9.98	4.39	3.3	1.80	0.72
8	4'-F	0.70	8.14	4.9	4.04	4.08	3.22
11	3'-Cl	1.67	9.05	3.9	3.25	2.24	1.59
13	4'-Cl	1.32	7.98	4.70	3.57	4.01	2.88
16	3'-CH$_3$	2.27	7.06	5.43	4.57	5.27	4.41
			MAE	3.30	2.42	2.17	1.45

On the contrary, the accuracy of all methods dropped dramatically to MAE = 1.84 if predictions were made for compounds predicted to be charged at the pH of measurements (**Figure 6b**). This result may suggest that the accuracy of the pK_a prediction is very important for the total accuracy of the programs prediction. However, this is not always the case. Let us consider the series of pyrazolopyridine[112] derivatives that have experimental pK_a and log$D$ values for some compounds investigated in the article (**Table 5**). The ACD labs and Pallas predicted these compounds as uncharged at the pH (7.4) used in the experiment, while, in fact, the compounds were partially ionized at this pH according to the measurements.[112]

Using experimental pK_a values to calculate log$D$ using eqn [5] increased the accuracy of both these programs as shown in the table. However, their prediction accuracy remains unacceptably low Thus, in this example the accuracy of prediction of lipophilicity for uncharged compounds, log$P$, but not the accuracy of the pK_a prediction dominates in the total accuracy of log$D$ prediction. Of course, it is also possible that in this case eqn [5] does not apply since the solvation of ionized compounds in 1-octanol should not be neglected. A poor performance of programs to predict log$D$ measured by Pfizer[103] and AstraZeneca[104] could be of the same origin.

The self-learning user-training feature of the ALOGPS program can increase the prediction ability of this program. The MAE error of the ALOGPS method after the user-training with molecules from the analyzed series decreased from 0.58 and 1.50 log units to 0.22 and 0.51 log units for neutral and charged series, respectively. For series of compounds from references[113] and[114] there was a 10-fold increase in the accuracy of prediction. The overall performance of the self-training is very similar to previous results.[58,59,102–104] Thus, a few measured values can be used to create very precise models to predict compounds from a similar series. As was mentioned above, KOWWIN, SLIPPER, and ACD Labs provide conceptually similar methods for the user-training feature but they were not tested in this study. On the other hand, the calculated results do not allow a particular statement to be made in favor of any analyzed method.

5.27.10 Conclusions and Guidelines

That log$P$ is an important property for 'drug design' is without doubt and that it can be calculated with reasonable accuracy for many different chemical structures is also without doubt. The importance of lipophilicity prediction remains the focus of interest of many researchers who are concerned with its measurement and calculation and this can be seen by the series of conferences on $\log P$ held in Lausanne and Zurich in 1995, 2000, and 2004. Similarly, the steadily rising number of publications on $\log P$ calculation as shown in **Figure 3** testifies to its importance.

In our comparison of neutral series of compounds all investigated methods performed quite well. However, such a result should not always be expected, particularly if prediction of in-house pharmaceutical data is involved.[59,104] Another interesting question is 'how well should the prediction work?' A review by Morris and Bruneau[115] compares the performance of several $\log P$ prediction routines on some 1300 proprietary AstraZeneca compounds that have in-house measured values. The correlation between predicted and measured values ranged from as low as 0.26 to the best fit of 0.75, which is still not a very good result. The overall errors were large compared with the results reported for the sets used to develop the methods but this was a 'real' test set and thus the toughest test that any prediction technique could face. The conclusion from this comparison was that the errors seen were quite typical for all the methods when applied to novel, diverse structures and that there was no 'best' method. An important recommendation was to consider what any predicted $\log P$ value was needed for. If all that is required is to know whether $\log P$ is 3 or 4 then these calculation techniques may be adequate.

One of the problems with the prediction of almost any property for a new chemical structure lies in the answer to the question "how similar is this structure to the compounds used to develop the property prediction model" and "how reliable is my prediction"? There are ways to answer this question,[46,98,99,109] but the quality of such estimations need yet to be verified in large-scale experiments. Given the fast increase in the number of new methods to predict $\log P$ and physicochemical and biological properties, the approaches that can cover the largest number of compounds with reasonable error and provide a reliable estimation of their applicability domain will have the most practical value in the pharmaceutical industry To this extent sharing of data measured in major pharmaceutical firms can have a great impact on the development of the field.[109]

The prediction of lipophilicity of charged compounds is more problematic as has been demonstrated here and elsewhere.[103,104] This may be due to problems in the prediction of pK_a or $\log P$ or partition of ions (**Figure 3**) that is often neglected in modeling. The recent studies to investigate the octanol/water partition coefficients' ionic species $(\log P_i^+, \log P_i^-)$ were challenged by Zhao and Abraham.[116] This may provide a starting point for development of new approaches for $\log D$ prediction.

A piece of simple and obvious advice in the use of any predicted value is to compare it with a measured value for a similar structure. With the development of high-throughput methods for physicochemical property measurement it should not be too much of a problem to obtain measured values for new series of interest. These new values may then also be used to improve the predictions using models that have already been built. This can be done using the user-training option provided by several programs, e.g., ACD labs, ALOGPS, SLIPPER, KOWWIN, etc., and fairly good local models for new accurate predictions within a similar series of compounds can be derived. However, a global method to provide reliable global $\log P$ and $\log D$ models has yet to be designed.

References

1. Richardson, B. J. *Medical Times and Gazette* **1868**, *2*, 703.
2. Richet, C. R. *Seances Soc. Biol.* **1893**, *9*, 775.
3. Overton, E. *Phys. Chem.* **1897**, *22*, 189–209.
4. Meyer, H. *Arch. Exp. Path. Pharm.* **1899**, *42*, 109–118.
5. Berthelot, M.; Jungfleisch, E. *Ann. Chim. Phys.* **1872**, *4*, 396–407.
6. Hansch, C.; Maloney, P. P.; Fujita, T.; Muir, R. M. *Nature* **1962**, *194*, 178–180.
7. Fujita, T.; Iwasa, J.; Hansch, C. *J. Am. Chem. Soc.* **1964**, *86*, 5175–5180.
8. Hersey, A.; Hill, A. P.; Hyde, R. M.; Livingstone, D. J. *Quant. Struct.-Act. Relat.* **1989**, *8*, 288–296.
9. Leo, A. J. *Methods Enzymol.* **1991**, *202*, 544–591.
10. Dearden, J. C.; Bresnen, G. M. *Quant. Struct.-Act. Relat.* **1998**, 7, 133–144.
11. Rekker, R. F. *The Hydrophobic Fragmental Constant*; Elsevier: Amsterdam, Netherlands, 1977.
12. Nys, G. G.; Rekker, R. F. *Chim. Ther.* **1973**, *8*, 521–535.
13. Rekker, R. F.; Mannhold, R. *Calculation of Drug Lipophilicity. The Hydrophobic Fragmental Constant Approach*; VCH: Weinheim, Germany, 1992.
14. Leo, A.; Jow, P. Y.; Silipo, C.; Hansch, C. *J. Med. Chem.* **1975**, *18*, 865–868.
15. Hansch, C.; Leo, A. *Substituent Constants for Correlation Analysis in Chemistry and Biology*; Wiley: New York, 1979.
16. Hansch, C.; Leo, A. In *Exploring QSAR Fundamentals and Applications in Chemistry and Biology*; American Chemical Society: Washington, DC, 1995, pp 125–168.
17. Leo, A. J.; Hoekman, D. *Perspect. Drug Disc. Des.* **2000**, *18*, 19–38.
18. Mayer, J. M.; van de Waterbeemd, H.; Testa, B. *Eur. J. Med. Chem.* **1982**, *17*, 17–25.
19. Van de Waterbeemd, H.; Testa, B. *Adv. Drug Res.* **1987**, *16*, 85–225.
20. Mannhold, R.; Dross, K. P.; Rekker, R. F. *Quant. Struct. Act.-Relat.* **1990**, *9*, 21–28.
21. Rekker, R. F.; ter Laak, A. M.; Mannhold, M. *Quant. Struct.-Act. Relat.* **1993**, *12*, 152–157.

22. Weininger, D. *J. Chem. Inf. Comput. Sci.* **1988**, *28*, 31–36.
23. Weininger, D.; Weininger, A.; Weininger, J. L. *J. Chem. Inf. Comput. Sci.* **1989**, *29*, 97–101.
24. Tetko, I. V.; Gasteiger, J.; Todeschini, R.; Mauri, A.; Livingstone, D.; Ertl, P.; Palyulin, V. A.; Radchenko, E. V.; Zefirov, N. S.; Makarenko, A. S. et. al. *J. Comput.-Aided. Mol. Des.* **2005**, *19*, 453–463.
25. Tetko, I. V. *Drug Disc. Today* **2005**, *10*, 1497–1500.
26. Collander, R. *Acta Chem. Scand.* **1951**, *5*, 774–780.
27. Young, R. C.; Mitchell, R. C.; Brown, T. H.; Ganellin, C. R.; Griffiths, R.; Jones, M.; Rana, K. K.; Saunders, D.; Smith, I. R.; Sore, N. E. et al. *J. Med. Chem.* **1988**, *31*, 656–671.
28. Seiler, P. *Eur. J. Med. Chem.* **1974**, *9*, 473–479.
29. Leahy, D. E.; Taylor, P. J.; Wait, A. R. *Quant. Struct.-Act. Relat.* **1989**, *8*, 17–31.
30. Abraham, M. H.; Doherty, R. M.; Kamlet, M. J.; Taft, R. W. *Chem. Br.* **1986**, *22*, 551–554.
31. Kamlet, M. J.; Abboud, J.-L. M.; Abraham, M. H.; Taft, R. W. *J. Org. Chem.* **1983**, *48*, 2877–2887.
32. Platts, J. A.; Abraham, M. H.; Butina, D.; Hersey, A. *J. Chem. Inf. Comput. Sci.* **2000**, *40*, 71–80.
33. Caron, G.; Ermondi, G. *J. Med. Chem.* **2005**, *48*, 3269–3279.
34. Wang, P. H.; Lien, E. J. *J. Pharm. Sci.* **1980**, *69*, 662–668.
35. Mannhold, R.; van de Waterbeemd, H. *J. Comput.-Aided Mol. Des.* **2001**, *15*, 337–354.
36. Japertas, P.; Didziapetris, R.; Petrauskas, A. *Mini Rev. Med. Chem.* **2003**, *3*, 797–808.
37. Klopman, G.; Zhu, H. *Mini Rev. Med. Chem.* **2005**, *5*, 127–133.
38. Leo, A. J. *Chem. Rev.* **1993**, *93*, 1281–1306.
39. Petrauskas, A. A.; Kolovanov, E. A. *Perspect. Drug Disc. Des.* **2000**, *19*, 99–116.
40. Meylan, W. M.; Howard, P. H. *Perspect. Drug Disc. Des.* **2000**, *19*, 67–84.
41. Meylan, W. M.; Howard, P. H. *J. Pharm. Sci.* **1995**, *84*, 83–92.
42. Mannhold, R.; Rekker, R. F. *Perspect. Drug Disc. Des.* **2000**, *18*, 1–18.
43. Mannhold, R.; Rekker, R. F.; Dross, K.; Bijloo, G.; de Vries, G. *Quant. Struct.-Act. Relat.* **1998**, *17*, 517–536.
44. Klopman, G.; Li, J.-Y.; Wang, S.; Dimayuga, M. *J. Chem. Inf. Comput. Sci.* **1994**, *34*, 752–781.
45. Viswanadhan, V. N.; Ghose, A. K.; Wendoloski, J. J. *Perspect. Drug Disc. Des.* **2000**, *19*, 85–98.
46. Japertas, P.; Didziapetris, R.; Petrauskas, A. *Quant. Struct.-Act. Relat.* **2002**, *21*, 23–37.
47. Viswanadhan, V. N.; Ghose, A. K.; Revankar, G. R.; Robins, R. K. *J. Chem. Inf. Comput. Sci.* **1989**, *9*, 163–172.
48. Ghose, A. K.; Viswanadhan, V. N.; Wendoloski, J. J. *J. Phys. Chem. A* **1998**, *102*, 3762–3772.
49. Wildman, S. A.; Crippen, G. M. *J. Chem. Inf. Comput. Sci.* **1999**, *39*, 868–873.
50. Wang, R. X.; Fu, Y.; Lai, L. H. *J. Chem. Inf. Comput. Sci.* **1997**, *37*, 615–621.
51. Wang, R. X.; Gao, Y.; Lai, L. H. *Perspect. Drug Disc. Des.* **2000**, *19*, 47–66.
52. Convard, T.; Dubost, J. P.; Le Solleu, H.; Kummer, E. *Quant. Struct.-Act. Relat.* **1994**, *13*, 34–37.
53. Mannhold, R.; Petrauskas, A. *QSAR Comb. Sci.* **2003**, *22*, 466–475.
54. Ghose, A. K.; Crippen, G. M. *J. Chem. Inf. Comput. Sci.* **1987**, *27*, 21–35.
55. Petrauskas, A.; Kolovanov, E. A. In *LogP 2000 – The Second LogP Symposium*; Testa, B., Ed.; University of Lausanne Press: Lausanne, Switzerland, 2000.
56. Walker, M. J. *QSAR Comb. Sci.* **2004**, *23*, 515–520.
57. Raevsky, O. A.; Trepalin, S. V.; Trepalina, H. P.; Gerasimenko, V. A.; Raevskaja, O. E. *J. Chem. Inf. Comput. Sci.* **2002**, *42*, 540–549.
58. Tetko, I. V.; Tanchuk, V. Y. *J. Chem. Inf. Comput. Sci.* **2002**, *42*, 1136–1145.
59. Tetko, I. V. *J. Chem. Inf. Comput. Sci.* **2002**, *42*, 717–728.
60. Todeschini, R.; Consonni, V. *Handbook of Molecular Descriptors*; Wiley-VCH: Weinheim, Germany, 2000.
61. Zhu, H.; Sedykh, A.; Chakravarti, S. K.; Klopman, G. *Curr. Comp.-Aid. Drug Des.* **2005**, *1*, 3–9.
62. Cherkasov, A.; Jonsson, M. *J. Chem. Inf. Comput. Sci.* **1998**, *38*, 1151–1156.
63. Solov'ev, V. P.; Varnek, A.; Wipff, G. *J. Chem. Inf. Comput. Sci.* **2000**, *40*, 847–858.
64. Sangster, J. *J. Phys. Chem. Ref. Data* **1989**, *18*, 1111–1229.
65. Grant, D. J. W.; Higuchi, T. *Solubility Behavior of Organic Compounds (Techniques of Chemistry)*; Wiley: New York, USA, 1990.
66. Huyskens, P. L. *J. Mol. Struct.* **1992**, *274*, 223–246.
67. Pais, A. *Subtle is the Lord, The Science and the Life of Albert Einstein*; Oxford University Press: Oxford, UK, 1982.
68. Ruelle, P. *Chemosphere* **2000**, *40*, 457–512.
69. Ruelle, P. *J. Chem. Inf. Comput. Sci.* **2000**, *40*, 681–700.
70. Bodor, N.; Buchwald, P. *J. Phys. Chem. B* **1997**, *101*, 3404–3412.
71. Abraham, M. H.; Chadha, H. S.; Whiting, G. S.; Mitchell, R. C. *J. Pharm. Sci.* **1994**, *83*, 1085–1100.
72. Xing, L.; Glen, R. C. *J. Chem. Inf. Comput. Sci.* **2002**, *42*, 796–805.
73. Rogers, K. S.; Cammarata, A. *Biochim. Biophys. Acta* **1969**, *193*, 22–29.
74. Klopman, G.; Iroff, L. D. *J. Comput. Chem.* **1981**, *2*, 157–160.
75. Bodor, N.; Huang, M. J. *J. Pharm. Sci.* **1992**, *81*, 272–281.
76. Duffy, E. M.; Jorgensen, W. L. *J. Am. Chem. Soc.* **2000**, *122*, 2878–2888.
77. Klamt, A.; Eckert, F. In *Rational Approaches to Drug Design*; Höltje, H.-D., Sippl, W., Eds.; Prous Science S.A.: Barcelona, Spain, 2001, pp 195–205.
78. Eckert, F.; Klamt, A. *Aiche J.* **2002**, *48*, 369–385.
79. Moriguchi, I.; Hirono, S.; Liu, Q.; Nakagome, I.; Matsushita, Y. *Chem. Pharm. Bull.* **1992**, *40*, 127–130.
80. Raevsky, O. A. *SAR QSAR Environ. Res.* **2001**, *12*, 367–381.
81. Bodor, N.; Huang, M. J.; Harget, A. *J. Mol. Struct. (THEOCHEM)* **1994**, *309*, 259–266.
82. Cense, J. M.; Diawara, B.; Legendre, J. J.; Roullet, G. *Chem. Intell. Lab. System.* **1994**, *23*, 301–308.
83. Grunenberg, J.; Herges, R. *J. Chem. Inf. Comput. Sci.* **1995**, *35*, 905–911.
84. Duprat, A. F.; Huynh, T.; Dreyfus, G. *J. Chem. Inf. Comput. Sci.* **1998**, *38*, 586–594.
85. Tetko, I. V.; Villa, A. E.; Livingstone, D. J. *J. Chem. Inf. Comput. Sci.* **1996**, *36*, 794–803.
86. Tetko, I. V.; Livingstone, D. J.; Luik, A. I. *J. Chem. Inf. Comput. Sci.* **1995**, *35*, 826–833.
87. Breindl, A.; Beck, B.; Clark, T.; Glen, R. C. *J. Mol. Model.* **1997**, *3*, 142–155.

88. Beck, B.; Breindl, A.; Clark, T. *J. Chem. Inf. Comput. Sci.* **2000**, *40*, 1046–1051.
89. Devillers, J.; Domine, D.; Karcher, W. *Polycyclic Aromat. Compd.* **1996**, *11*, 211–217.
90. Devillers, J.; Domine, D.; Guillon, C.; Bintein, S.; Karcher, W. *SAR QSAR Environ. Res.* **1997**, 7, 151–172.
91. Moreau, G.; Broto, P. *Nouv. J. Chim.* **1980**, *4*, 359–360.
92. Kier, L. B.; Hall, L. H. *Molecular Structure Description: The Electrotopological State*; Academic Press: London, UK, 1999.
93. Hall, L. H.; Kier, L. B. *J. Chem. Inf. Comput. Sci.* **1995**, *35*, 1039–1045.
94. Parham, M. E.; Hall, L. H.; Kier, L. B. *Abstracts of Papers*, 220th National Meeting of the American Chemical Society, August 20–24, 2000; American Chemical Society: Washington, DC, 2000; U288.
95. Huuskonen, J. J.; Livingstone, D. J.; Tetko, I. V. *J. Chem. Inf. Comput. Sci.* **2000**, *40*, 947–955.
96. Huuskonen, J. J.; Villa, A. E. P.; Tetko, I. V. *J. Pharm. Sci.* **1999**, *88*, 229–233.
97. Livingstone, D. J.; Ford, M. G.; Huuskonen, J. J.; Salt, D. W. *J. Comput.-Aided Mol. Des.* **2001**, *15*, 741–752.
98. Gombar, V. K.; Enslein, K. *J. Chem. Inf. Comput. Sci.* **1996**, *36*, 1127–1134.
99. Gombar, V. K. *SAR QSAR Environ. Res.* **1999**, *10*, 371–380.
100. Tetko, I. V.; Tanchuk, V. Y.; Villa, A. E. *J. Chem. Inf. Comput. Sci.* **2001**, *41*, 1407–1421.
101. PHYSPROP. The Physical Properties Database (PHYSPROP) is a trademark of Syracuse Research Corporation. www.syrres.com (accessed June 2006).
102. Tetko, I. V. *Neur. Proc. Lett.* **2002**, *16*, 187–199.
103. Tetko, I. V.; Poda, G. I. *J. Med. Chem.* **2004**, *47*, 5601–5604.
104. Tetko, I. V.; Bruneau, P. *J. Pharm. Sci.* **2004**, *93*, 3103–3110.
105. Eros, D.; Kovesdi, I.; Orfi, L.; Takacs-Novak, K.; Acsady, G.; Keri, G. *Curr. Med. Chem.* **2002**, *9*, 1819–1829.
106. Schaper, K. J.; Samitier, M. L. R. *Quant. Struct.-Act. Relat.* **1997**, *16*, 224–230.
107. Artemenko, N. V.; Palyulin, V. A.; Zefirov, N. S. *Dokl. Chem.* **2002**, *383*, 114–116.
108. Zefirov, N. S.; Palyulin, V. A. *J. Chem. Inf. Comput. Sci.* **2002**, *42*, 1112–1122.
109. Tetko, I. V.; Abagyan, R.; Oprea, T. I. *J. Comput.-Aided. Mol. Des.* **2005**, *19*, 749–764.
110. Taskinen, J.; Yliruusi, J. *Adv. Drug. Deliv. Rev.* **2003**, *55*, 1163–1183.
111. Winkler, D. A. *Drugs Future* **2004**, *29*, 1043–1057.
112. de Mello, H.; Echevarria, A.; Bernardino, A. M.; Canto-Cavalheiro, M.; Leon, L. L. *J. Med. Chem.* **2004**, *47*, 5427–5432.
113. Hutchinson, J. H.; Halczenko, W.; Brashear, K. M.; Breslin, M. J.; Coleman, P. J.; Duong le, T.; Fernandez-Metzler, C.; Gentile, M. A.; Fisher, J. E.; Hartman, G. D. *J. Med. Chem.* **2003**, *46*, 4790–4798.
114. Leisen, C.; Langguth, P.; Herbert, B.; Dressler, C.; Koggel, A.; Spahn-Langguth, H. *Pharm. Res.* **2003**, *20*, 772–778.
115. Morris, J. J.; Bruneau, P. P. In *Virtual Screening for Bioactive Molecules*; Bohm, H. J., Schneider, G., Eds.; Wiley-VCH: Chichester, UK, 2000, pp 33–58.
116. Zhao, Y. H.; Abraham, M. H. *J. Org. Chem.* **2005**, *70*, 2633–2640.
117. Virtual Computational Chemistry Laboratory. www.vcclab.org (accessed April 2006).
118. Absolv Computer Program. www.ap-algorithms.com/absolv.htm (accessed April 2006).
119. Klamt Group. www.cosmologic.de (accessed April 2006).
120. LOGKOW Database. logkow.cisti.nrc.ca (accessed April 2006).

Biographies

Igor V Tetko graduated (cum laude) from the Faculty of Physical & Chemical Biology, Moscow Institute of Physics and Technology in 1989. He received a PhD in Chemistry (application of artificial neural networks in structure–activity relationship studies) from the Biomedical Department, Institute of Bioorganic & Petroleum Chemistry, Kiev, Ukraine under the supervision of Prof A I Luik in 1994. Currently, he is a senior research scientist at this institution. He was a recipient of the Human Frontier Program Organization (HFSPO) fellowship in 1996 at the Institute of Physiology, University of Lausanne, Switzerland where he continued his postdoctoral work, as assistant diplômé and premier assistant, between 1996 and 2001. In 2001 he became a senior research scientist at the Institute for Bioinformatics, GSF-National Research Centre for Environment and Health, Neuherberg, Germany. Igor is coauthor of more than 80 publications in peer-reviewed scientific journals. His main interests include development and application of artificial neural networks and nonlinear methods of data analysis in chemistry, development of new methods for robust prediction of physicochemical properties of compounds, functional annotation of proteins, and data mining in bioinformatics. He maintains the Virtual Computational Chemistry Laboratory site (www.vcclab.org).

David J Livingstone After working for almost twenty years in industrial pharmaceutical research at Wellcome and SmithKline Beecham, David Livingstone set up the ChemQuest consultancy business in 1995 offering training, advice and contract research to the chemical industry.

Dr Livingstone obtained his PhD in physical organic chemistry from the University of Dundee. He joined the Department of Biophysics and Biochemistry at Wellcome research where he was involved in the design and testing of compounds which improved the ability of hemoglobin to deliver oxygen. By working closely with the computational chemists at Wellcome he was responsible for some of the earliest routine applications of computationally derived properties in QSAR and was a pioneer in the use of pattern recognition methods in drug design.

He is now a visiting professor at the Centre for Molecular Design at the University of Portsmouth. He is a member of the editorial board of three journals and has published more than 80 papers, 12 book chapters, a textbook, and has coedited three books. His recent research interests include: the use of neural networks and other AI methods in QSAR; the prediction of toxic effects by QSAR; the characterization of intermolecular interactions; the development and application of multivariate techniques in QSAR; and the analysis of multiple response data.

5.28 In Silico Models to Predict Oral Absorption

H Van de Waterbeemd, AstraZeneca, Macclesfield, UK

5.28.1   Introduction

Most drugs are preferably given by the oral route, since this is the most convenient one for the patient, and contributes to high compliance. Good oral drugs therefore have to achieve high bioavailability ($F\%$), defined as the percentage available in the systemic circulation (**Figure 1**). Bioavailability is a very complex absorption, distribution, metabolism, and excretion (ADME) property, and often incorrectly confused with absorption ($A\%$).[1,2] Many of the contributing properties to bioavailability are now reasonably well characterized (**Figure 2**), although their interplay is less well understood. The key contributors to bioavailability include dissolution/solubility, permeability/absorption, interactions with transporters and gut wall, and liver metabolism.[3]

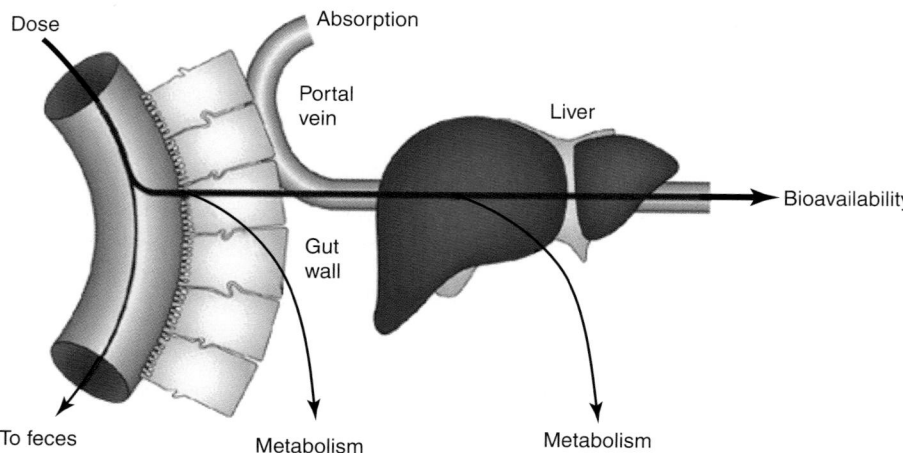

Figure 1   Schematic definition of oral GI absorption (percentage of dose reaching portal vein) and bioavailability (percentage of dose reaching system circulation). (Reproduced from Van de Waterbeemd, H.; Gifford, E. *Nature Rev. Drug Disc.* **2003**, *2*, 192–204, by permission from Nature Reviews Drug Discovery, copyright (2003) Macmillan Magazines Ltd.)

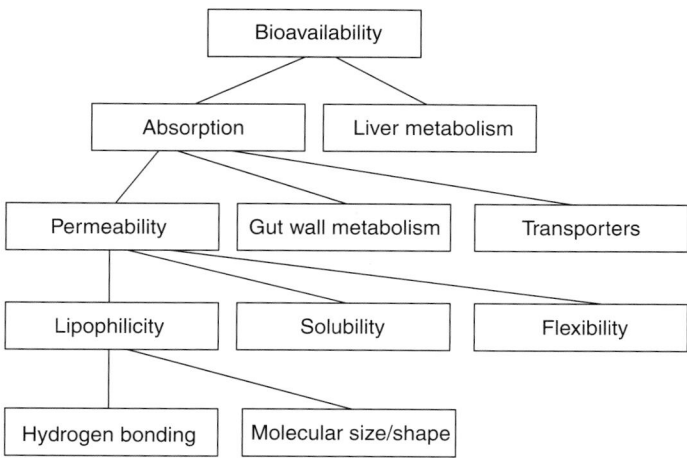

Figure 2 Bioavailability and absorption are complex properties that can be the unraveled into more basic contributions that offer chemists guidance in the optimization of their series and compounds.

Absorption can be defined and measured in different ways. A direct way is to measure the percentage of dose reaching the portal vein before entering the liver. Since such studies are not often performed in humans, more indirect methods are generally applied[4]:

- bioavailability measurement;
- excretion in urine and feces following oral administration; and
- the ratio of cumulative urinary excretion of drug-related material following oral and intravenous administration.

A number of high-throughput techniques have been developed in recent years to identify compounds with good oral absorption; as discussed elsewhere in detail (*see* 5.11 Passive Permeability and Active Transport Models for the Prediction of Oral Absorption; 5.19 Artificial Membrane Technologies to Assess Transfer and Permeation of Drugs in Drug Discovery). These include cell-based methods such as Caco-2, and artificial membranes, such as parallel artificial membrane permeation assay (PAMPA).[5] These methods are resource-intensive in terms of personnel and cost, and the in vivo predictions not always reliable. For both reasons, there is increasing interest in developing the best possible in silico models for human absorption. However, it is extremely important to understand the strengths and weaknesses of such computational models, and apply these with scientific judgment in the discovery process. Also, in silico models are not perfect due to the limitations and accuracy in measured data to build the model, coverage of sufficient chemical space, suitability of the descriptors, and choice of the mathematical tool.

Two types of predictive modeling of human oral absorption can be distinguished. One is based on physiological principles, and attempts to model details of the dissolution/solubility, including dose and formulation variations, and permeability processes, as well as taking into account the effects of transporters and metabolizing enzymes. Others are simpler but not necessarily less successful approaches that ignore effects of solubility, dose, formulation, active transport, and metabolism. This chapter will discuss both of these modeling efforts. An excellent historical account on theoretical transport and absorption models can be found elsewhere.[6] Older models see a membrane as an organic compartment with aqueous compartment on either side. Based on a critical analysis of the processes involved in passive membrane crossing,[6,7] in newer models an aqueous pore pathway has been added to better reflect current views of a membrane.[7,8]

A good understanding of the key properties contributing to oral absorption provides the medicinal chemist with a powerful tool to design new molecules.

Lipinski[9] and others[10–12] were among the pioneers who pointed out that a limited set of descriptors can be defined to distinguish poor and good oral absorption. As early as 1971, in a study of red cell membranes, it was observed that "within the membrane, lipophilic permeability is modified both by steric factors and by the formation of hydrogen bonds with membrane components."[13]

A quantitative rule which can be used for a qualitative prediction of oral absorption, is known as the rule-of-five,[9] and states that good oral absorption can be expected for compounds with a calculated log $P < 5$, number of hydrogen bond donors (defined as the sum of –OH and –NH groups) < 5, the number of hydrogen bond acceptors (expressed as the sum of N and O atoms) < 10 and a molecular weight below 500. This simple concept and its modified variants are now widely used in library and singleton design. The general approach to optimized ADME properties has been called property-based design,[14,15] as a complement to structure-based design. These simple rules are also often cited in the context of defining drug- or lead-likeness of compounds.[16]

5.28.2   In Vivo Permeability, Absorption, and Bioavailability

5.28.2.1   Biological Permeation

5.28.2.1.1   Permeability principles

Molecules can cross the intestinal epithelium via two important routes for permeation, namely passive diffusion and carrier-mediated influx via transporter proteins (*see* 5.04 The Biology and Function of Transporters). For passive diffusion we can distinguish transport via the paracellular and the transcellular routes. For most drugs the passive transcellular route will be the only or major pathway from the gastrointestinal (GI) lumen to the portal vein blood. Transporter proteins may promote (influx) or hinder (efflux) the uptake of drugs.[17] Drugs can also be metabolized in the gut wall, in particular by cytochrome P450 3A4 (CYP3A4),[18,19] which obviously affects their absorption.

Some simple rules are that compounds with a molecular weight of < 200 may use the paracellular pathway to cross a membrane, and are potential candidates for active transport mechanisms, while compounds with a molecular weight above 500 have poorer membrane diffusion characteristics, and may be more susceptible to interactions with P-glycoprotein, both limiting membrane permeation and absorption. For most drugs with a molecular weight in the range of 200–500, the permeation will be driven by passive diffusion.

It has been argued in the literature that most of the absorption takes place in the small intestine (the absorption window concept).[20] The influence of pH, solubility, paracellular and/or transcellular membrane transfer, transporters, and metabolizing enzymes may vary along the GI tract depending on the character (neutral, acid, base, zwitterions, lipophilic, hydrophilic) of the drug.

Neutral compounds tend to cross a membrane more easily than charged species following the pH–partition hypothesis at various pH values. However, deviations have been found. These can be explained by the presence of an unstirred water layer (UWL), also called the aqueous boundary layer. A microclimate pH will prevail in this layer, changing the ionization state of the permeating solute. Ionization is in some cases less susceptible, since, although the molecule may be formally charged, this charge may be considerably delocalized, and therefore less 'hard.'

5.28.2.1.2   Human perfusion studies

Intestinal effective permeability (P_{eff}) across the intestinal membrane can be measured by regional perfusion, as performed by Lennernäs's group at the University of Uppsala.[21] Typical P_{eff} values are reported in **Table 1**.[22] A sigmoidal relationship is observed between the percentage absorbed in human (*A%*) and these P_{eff} values, which can be described by

$$A\% = 100 \times [1 - \exp(-P_{eff})] \qquad [1]$$

5.28.2.2   Barrier Differences: Comparison of the Gastrointestinal with the Blood–Brain Barrier, Skin, and Other Membranes

Membranes differ from organ to organ in relation to the function they are fulfilling.[23] The basic constitution is of course similar, and consists of phospholipids. However, the concentration of, for example, cholesterol varies between membrane types. Also, the number and distribution of various influx and efflux transporters varies. In addition, cell membranes are equipped with a wide range of transporters and ion channels. In summary, cell membranes are, in principle, similar but not identical. The key molecular descriptors to describe permeability through any type of membrane (e.g., gut wall, skin, blood–brain barrier, spinal meninges),[24] or hepatocytes, are therefore also quite similar. This can be seen by comparing the properties used in the predictive models described in other chapters (*see* 5.28 In Silico Models to Predict Oral Absorption; 5.30 In Silico Models to Predict Passage through the Skin and Other Barriers; 5.31 In Silico Models to Predict Brain Uptake).

5.28.2.3   Effect of Dose, Particle Size, Metabolism, Transporter Proteins, Absorption Enhancers, and Food

The role of transporters and metabolizing enzymes in oral absorption is now much better understood, and these can have an important effect on the percentage absorbed.[17–19,25] Dose also plays a critical role, since at higher dosage,

Table 1 Human effective permeability $(n = 28)$[22]

Compound	P_{eff} ($\times 10^{-4}$ cm s^{-1})
Amiloride	1.6
Amocillin	0.3
Antipyrine	4.5
Atenolol	0.2
Benserazide	3.8
Carbamazepine	4.3
Cimetidine	0.3
Creatinine	0.3
Desipramine	4.4
Enalaprilat	0.2
Furosemide	0.05
Hydrochlorothiazide	0.04
Ketoprofen	8.4
Levodopa	3.4
L-Leucine	6.2
Lisinopril	0.33
α-Methyldopa	0.2
Metoprolol	1.3
Naproxen	8.3
PEG400	0.56
Phenoxymethylpenicillin	0.26
Phenylalanine	3.4
Piroxicam	7.8
Propranolol	2.9
Ranitidine	0.27
Terbutaline	0.3
Urea	1.4
Verapamil	6.7

Reproduced from Alsenz, J.; Haenel, E. *Pharm. Res.* **2003**, *20*, 1961–1969, with kind permission of Springer Science and Business Media.

transporters and metabolizing enzymes can be saturated, and their effect is less or not visible compared with (much) lower doses. In addition, pharmaceutical factors such as particle size and the addition of absorption enhancers can modify absorption levels, as well as intake with or without food and disease state, but this is beyond the scope of this chapter and more relevant to late-stage clinical development.

5.28.2.4   Species Differences

The rat is a widely used preclinical animal species for pharmacokinetics studies. Despite differences in clearance and metabolism, and often bioavailability, the volume of distribution and oral absorption appear to correlate well between human and rat. For a set of 64 compounds the following relationship was published[26]:

$$A\%(\text{human}) = 1.01 \ A\%(\text{rat}) - 0.2$$
$$n = 64, \ r^2 = 0.97, \ s = 5.6; \ F = 2176 \qquad [2]$$

(r, s, and F are defined in the Nomenclature). Equation [2] is particularly valid in the dose-independent absorption range. In a later study comprising 98 compounds, a nearly identical equation was derived[27]:

$$A\%(\text{human}) = 0.91 \ A\%(\text{rat}) + 6.9$$
$$n = 98, \ r^2 = 0.88, \ s = 11, \ F = 675 \qquad [3]$$

A similar good relationship was found for monkey oral absorption data[28] for a set of 43 compounds:

$$A\%(\text{human}) = 1.04 \ A\%(\text{monkey}) - 2.9$$
$$n = 43, \ r^2 = 0.97 \qquad [4]$$

Marked differences between human and dog in the nonlinear absorption profiles were found for some drugs. Caution is thus needed in the interpretation of dog absorption data. Therefore, a further study comparing dog absorption with that of human was less successful[29]:

$$A\%(\text{human}) = 1.58 \ A\%(\text{dog}) - 55.7$$
$$n = 43, \ r^2 = 0.51 \qquad [5]$$

Another way of studying species differences is by use of the Ussing chamber, in which drug transport across small sections of the GI tract can be evaluated.[30] This has been done for rat, dog, monkey, and human tissue. This technique appears to be useful for poorly water-soluble drugs,[31] but is of course low throughput, and therefore mainly suited for a development setting.

5.28.3   In Vitro Permeability/Absorption Approaches

5.28.3.1   Physicochemical Properties and Models

Models that address the pH-dependent solubility, permeability, and absorption behavior have been developed, and will be briefly presented in this section. More can be found elsewhere (see 5.37 Physiologically-Based Models to Predict Human Pharmacokinetic Parameters).

5.28.3.1.1   Physicochemical profiling

Physicochemical profiling has a long tradition, but only recently became more automated and widely used in the pharmaceutical industry. Excellent reviews on current techniques and technologies have been published.[3,32–35] An overview of typical in vitro screens is given in **Table 2**. Among the physicochemical properties, solubility, lipophilicity, and ionization (pK_a) are considered the most important.

5.28.3.1.2   The biopharmaceutics classification system (BCS)

The BCS is a concept combining permeability and solubility, and is discussed in full detail elsewhere (see 5.42 The Biopharmaceutics Classification System).[36] The BCS classifies compounds into four classes according to their permeability and solubility characteristics: class I, high solubility and high permeability; class II, low solubility and high permeability; class III, high solubility and low permeability; and class IV, low solubility and low permeability. Therefore, measurement and/or prediction of both solubility and permeability give a good estimate of developability of a new compound, class I being the ideal profile. This has been called rational drug delivery.[37] A proposal has been made to extend the BCS to six classes.[38]

It has also been argued that, in fact, molecular size and hydrogen bonding should be the basic properties of the BCS,[39] since these are both strongly determinant for solubility and permeability.

Table 2   In vitro screening for oral absorption/permeability[173]

Physicochemical profiling
Solubility
Ionization (pK_a)
Lipophilicity
Octanol/water distribution – partitioning
Cyclohexane/water distribution – partitioning
Heptane/ethylene glycol distribution – partitioning
Immobilized artificial membranes
Immobilized liposome chromatography
Micellar electrokinetic chromatography
Biopartitioning micellar chromatography
Hydrogen bonding
$\Delta \log P = \log P_{oct} - \log P_{alkane}$

Noncell-based in vitro models
Phospholipid vesicles
Liposome partitioning
Impregnated artificial membranes
Parallel artificial membrane permeability assay (PAMPA)
Hexadecane-coated polycarbonate filters/membrane (HDM)
Transil particles
Surface plasmon resonance biosensor (SPR)
Surface activity

Cell-based in vitro models
Caco-2
MDCK
TC7
HT29
2/4/A1

Transporters and transfected cell lines
MDR1 (P-glycoprotein)
Calcein AM
Cycling
ATPase
MDCK-MDR1
MDCK-MRP2-OATP2
LLC-PK1-CYP3A

Ex vivo
Ussing chamber
Everted rings

In situ
Intestinal perfusion

Reproduced from Van de Waterbeemd, H. Property-Based Lead Optimization. In *Biological and Physicochemical Profiling in Drug Research*; Testa, B., Krämer, S. D., Wunderli-Allenspach, H., Folkers, G., Eds.; VHCA: Zürich, Switzerland; Wiley-VCH: Weinheim, Germany, 2006, pp 25–45, with permission from Verlag Helvetica Chimica Acta AG.

5.28.3.1.3 Absorption potential

Absorption is not dissolution limited for most drugs, and the main permeability mechanism is by passive diffusion. However, dissolution and solubility can be an issue for some classes of drugs (*see* 5.17 Dissolution and Solubility). The absorption potential (AP) and maximum absorbable dose (MAD) approach discussed below, have been developed to deal with such compounds and get early estimates of potential limited oral absorption.

The AP of a compound is defined as[40]

$$AP = \log[Pf_{non}(S_w V_L)/X_0] \qquad [6]$$

where P is the octanol/water partition coefficient, f_{non} the fraction of un-ionized species at pH 6.5, S_w the solubility of the un-ionized species in water, V_L the volume of the lumen contents, and X_0 the dose administered. The authors found a sigmoidal relationship between AP and the fraction absorbed in humans for a small set of seven chemically diverse drugs. The general applicability has not been proven.

5.28.3.1.4 The maximum absorbable dose concept

The MAD concept of has been introduced,[41,42] and relates drug absorption to solubility via

$$MAD = S \times k_a \times SIWV \times SITT \qquad [7]$$

where S is the solubility ($mg\,mL^{-1}$) at pH 6.5, k_a is the transintestinal absorption rate constant (min^{-1}), SIWV is the small intestinal water volume (mL), assumed to be ca 250 mL, and SITT is the small intestinal transit time (min), assumed to be 4.5 h (270 min).

5.28.3.1.5 Three-dimensional (3D) solubility parameters

Hansen's 3D solubility parameters related to contributions of dispersion forces, polar interactions, and hydrogen bond formation have been used to predict permeation rates for human skin and absorption rates of drugs from the rectum, and to predict GI absorption sites and absorption duration.[43] Such studies might help in the rational choice of dosage forms.

5.28.3.1.6 The absorption parameter

A model has been proposed to predict either high or low fraction absorbed for an orally administered, passively absorbed drug on the basis of a new absorption parameter π.[44] The model requires two inputs: the octanol/water partition coefficient P and the dimensionless oversaturation number O_{lumen}, where

$$\pi = P/O_{lumen} \qquad [8]$$

O_{lumen} is the ratio of the concentration of drug delivered to the GI fluid to the solubility of the drug in that environment, and is equal to the dose-normalized solubility for suspensions, and is unity for solutions. For a list of 98 drugs, 88% were correctly predicted solely based upon whether their π value is greater than or less than unity.[44] Since this approach relies on dose estimates, it is a late-stage candidate selection tool only.

5.28.3.1.7 How much solubility and permeability are required?

The MAD analysis also provides an indication to the question of how soluble a drug candidate should be, by rewriting eqn [7] as[41]

$$minimum\ acceptable\ solubility = MAD/(k_a \times SIWV \times SITT) \qquad [9]$$

The minimum required solubility ($S_{w(min)}$ in $mg\,L^{-1}$) of a drug has also been proposed[45] to satisfy the equation

$$dose \times (1 - FA) - 0.25 S_{w(min)} = 0 \qquad [10]$$

where 0.25 is the GI volume in liters and FA is the fraction absorbed. Thus, for a 100 mg dose the solubility should be between $400\,mg\,L^{-1}$ (FA = 0) and $4\,mg\,L^{-1}$ (FA = 1). In reality, these numbers need to be higher, since dissolution is a kinetic process.[45]

The required level of permeability and solubility needed for oral absorption is related to potency.[46] **Table 3** gives an indication of the levels of the required solubility, in relation to the projected dose.

Table 3 Minimum acceptable solubility in $\mu g\,mL^{-1}$ for low-, medium-, and high-permeability compounds at clinical dose[46]

Projected Dose (mg kg^{-1})	Permeability		
	Low	Medium	High
0.1	21	5	1
1.0	207	52	10
10	2100	520	100

Reproduced from Lipinski, C. *J. Pharmacol. Toxicol. Methods* **2000**, *44*, 235–249, with permission from Elsevier.

5.28.3.2   Artificial Membranes

Table 2 lists a number of techniques in use to estimate the in vivo permeability/AP of compounds. Artificial membranes, such as immobilized artificial membranes and the PAMPA approach, have been studied in various forms (*see* 5.19 Artificial Membrane Technologies to Assess Transfer and Permeation of Drugs in Drug Discovery). Other systems[47] include surface plasma resonance[48,49] and surface activity.[50] Membrane affinity can be measured in high throughput using solid-supported lipid membranes.[51]

Based on observations that transport through impregnated artificial membranes is comparable, but not identical, to Caco-2 flux,[52] the PAMPA method was developed.[53–55] Its simplicity, low cost, and high throughput makes it potentially an attractive permeability screen for drug discovery. It is not clear, however, what is the in vivo equivalent of this 'permeability.'

In one study it was demonstrated that the percentage transport (%T) across the PAMPA lipid bilayer correlates to the percentage absorbed in humans ($A\%$) via the following equation, derived from a training set of 50 compounds:

$$A\% = 100 \times [1 - \exp(-0.85 \times \%T)] \qquad [11]$$

A test set of 42 compounds was poorly predicted with $r^2 = 0.46$, or $r^2 = 0.66$ after removing two outliers. These outliers are actively transported or have a paracellular component. However, in a screening setting, this information may in general not be available. It was concluded that PAMPA might not be sensitive enough to predict human absorption, but that it might rank compound permeability.[56] In another study it was found that intrinsic PAMPA values correlate very well with log D measured in a dodecane/water solvent system.[57] This causes doubt about the role of the phospholipids addition in PAMPA. It also shows that PAMPA is in fact simply another type of lipophilicity scale, and its value in predicting human absorption remains unclear.

5.28.3.3   Cell-Based Models

Intestinal epithelial cell lines, such as Caco-2, MDCK, and HT29, are widely used to assess the AP of new compounds (**Table 2**).[58,59] These assays can now be run in a 96-well format. (For a detailed discussion, *see* 5.10 In Vitro Studies of Drug Metabolism.)

5.28.3.4   In Situ and Ex Vivo Models

Biological techniques involving isolated tissues (ex vivo) and perfused organs (in situ), including the GI tract, have been used to discriminate between drug candidates as models for intestinal absorption.[60] Since these techniques are very resource-intensive and low throughput, they are nowadays only found in a development departments, and have been replaced in discovery by higher throughput techniques discussed briefly in Sections 5.28.3.1–5.28.3.3.

5.28.4   Physiologically-Based Models for the Fraction Absorbed

5.28.4.1   Rate Constants of Absorption

Drug distribution is a dynamic process, with only pseudo-equilibrium states in certain organs and membranes in the body. Hence, rates of transport might be more important to consider than equilibrium constants such as permeability constants and partition/distribution coefficients.[61] Membrane permeability in centimeters per second (e.g., in Caco-2 cells) would be the typical measure for in vitro permeability, and the percentage or fraction absorbed is the measure for human absorption. Although it is widely agreed that most drugs cross a membrane by passive diffusion as the major driving force, it is less trivial which parameter describes this process best. Membrane permeability, absorption rate constant, or absorption clearance can a priori all be used.[62] First-order permeation rate constants (k_p) are related to the permeability coefficient (P_e) via

$$k_p = (A/V)P_e \qquad [12]$$

where A and V are the surface and volume of the human GI tract, respectively.[63] Drug absorption kinetics have also been studied in Caco-2 cells,[63] and it was demonstrated that rate constants for cellular uptake into cerebral capillary endothelial cell monolayers might be a better predictor for brain uptake.[64] Absorption rates are often the preferred parameter from in situ single-pass techniques using rat.[65–67]

5.28.4.2   Biophysical Compartment Models

The role of lipophilicity and molecular weight in absorption mechanisms has been studied by, for example, the Plá-Delfina's group, and mathematically described by biophysical compartment models. For a review, see Camenisch et al.[6] A biophysical drug absorption model termed PATQSAR (population analysis by topology-based quantitative structure–activity relationship) has been proposed, based upon the topological DARC/PELCO methodology.[66] The model considers the absorption process from the intestinal lumen as the sum of two resistances in series, namely an aqueous diffusional barrier and a lipoidal membrane. Following this approach, it was shown that lipophilicity plays a major role in a sigmoidal relationship with absorption rate constants (k_a) obtained from the in situ rat gut technique[67]:

$$k_a = cD^d/(1 + b \times \sqrt{MW \times D^d}) \qquad [13]$$

where D is the distribution coefficient (measured at pH 7), MW is the molecular weight, and b, c, and d are regression coefficients. These k_a values can then be related to the percentage absorbed ($A\%$) as follows:

$$A\% = 100(1 - e^{-\alpha k_a}) \qquad [14]$$

5.28.4.3   Advanced Compartmental Absorption and Transit (ACAT) Modeling

Much effort went into the detailed description of the absorption processes in so-called ACAT models.[25,68–70] Such oral absorption simulators are based on physiological models of the GI tract, divided into several sections. The effect of compound properties such as log P, solubility, pK_a, and formulation properties, such as particle size, particle density, and diffusion, can be simulated to make estimates on the fraction absorbed and its effect on plasma concentration–time curves.[71] Attempts have also been made to include first-pass metabolism and the effect of transporters in these models.[25,70,72] Eventually, this may well lead to modeling of bioavailability.

Software has become available to perform simulations of the absorption process, for example GastroPlus (Simulations Plus),[70] or the iDEA approach (Lion Bioscience, formerly from Navicyte or Trega),[73] as well as newer programs including Simcyp, Cloe PK,[74] and PK-Sim[75,76] (see **Table 4**).

5.28.4.4   Physiologically-Based Pharmacokinetic (PBPK) Modeling

PBPK simulation modeling (see 5.37 Physiologically-Based Models to Predict Human Pharmacokinetic Parameters; 5.35 Modeling and Simulation of Pharmacokinetic Aspects of Cytochrome P450-Based Metabolic Drug–Drug Interactions) has developed considerably in recent years.[73,74,77,78] In combination with ACAT models it is therefore now possible to simulate the whole body pharmacokinetic events and to predict the human or other species most important pharmacokinetic properties, including estimates of oral absorption. The model parameters are separated into physiological parameters for the species studied and compound-specific parameters, which can be measured or estimated computationally.

Commercial programs (**Table 4**) include Simcyp, PK-Sim, and Cloe PK. In the future, simulation of pharmacokinetics in virtual humans will no doubt play a greater role in compound design and evaluation.

Table 4 Commercial software and database used to predict human oral absorption

Software	Vendor	Contact
AbSolv	Pharma Algorithms	www.ap-algorithms.com
	Sirius Analytical Instruments	www.sirius-analytical.com
ADME Boxes	Pharma Algorithms	www.ap-algorithms.com
	Sirius Analytical Instruments	www.sirius-analytical.com
ADME Index	Lighthouse Data Solutions	www.lighthousedatasolutions.com
Admensa	Inpharmatica	www.inpharmatica.com
ADME Works	Fujitsu/FQSPoland	www.fqspl.com.pl
BioPrint	CEREP	www.cerep.fr
Cerius2	Accelrys	www.accelrys.com
Cloe PK	Cyprotex	www.cyprotex.com
GastroPlus	SimulationsPlus	www.simultations-plus.com
KnowItAll	BioRad	www.biorad.com
		www.knowitall.com
OraSpotter	ZyxBio	www.zyxbio.com
Pipeline Pilot	SciTegic	www.scitegic.com
PK-Sim	Bayer Technology Services	www.pksim.com
Simcyp	Simcyp Ltd.	www.simcyp.com
SLIPPER 2001	Raevsky	raevsky@ipac.ac.ru
	MolPro	www.ibmh.msk.su/molpro/
TruPK	Strand Genomics	www.strandgenomics.com
VolSurf	Cruciani	gabri@chemiome.chm.unipg.it
	Molecular Discovery	www.moldiscovery.com
	Tripos	www.tripos.com
WB-PK	Sunset Molecular Discovery	www.sunsetmolecular.com

5.28.5 Data Sets and Molecular Descriptors for In Silico Prediction of Human Oral Absorption

5.28.5.1 Data Sets

The maximum number of compounds in a data set used thus far in modeling human oral absorption is 1260.[79] Earlier studies were done on much smaller data sets (**Table 5**).[4,56] Care needs to be taken not to mix data obtained via different methods of experimental estimation of oral absorption, as this dilutes the quality of the data. Another problem is that most data sets are skewed toward high absorption. Therefore, unlike $\log P$ or solubility data, there is only a relatively small data set available for predictive modeling of oral absorption. Fortunately, drugs from various chemical classes are included that cover a reasonable chemical space and diversity. Another set of data consists of human effective permeability data, which were measured for a relatively small set ($n = 23$) of selected compounds using the Gut-O-Loc balloon technique[80] (*see* Section 5.28.2.1.2). A compilation of this data has been made[22] (see **Table 1**).

Table 5   Human oral absorption $(n = 169)$[4]

Compound	Absorption (%)	Compound	Absorption (%)
Acarbose	2	Famotidine	38
Acebutolol	80	Felbamate	90
Acetaminophen	80	Felodipine	88
Acetylsalicylic acid	84	Fenclofenac	100
Adefovir	16	Fenoterol	60
Alprazolam	90	Flecainide	81
Alprenolol	93	Fluconazole	95
Amiloride	50	Flumazenil	95
Aminopyrine	100	Fluvastatin	100
Amphetamine	90	Foscarnet	17
Amrinone	93	Fosfomycin	3
Antipyrine	97	Fosmidomycin	30
Ascorbic acid	35	Furosemide	61
Atenolol	50	Gallopamil	100
Atropine	98	Ganciclovir	3
Betaxolol	90	Glyburide	100
Bromazepam	84	Granisetron	100
Bumetanide	96	Guanabenz	80
Bupropion	87	Guanoxon	50
Caffeine	100	Hydrochlorothiazide	65
Camazepam	100	Hydrocortisone	91
Captopril	84	Ibuprofen	95
Carfecillin	99	Imipramine	100
Chloramphenicol	90	Indomethacin	100
Cicaprost	100	Isoniazid	80
Cidofovir	3	Isoxicam	100
Cimetidine	64	Isradepine	92
Cisapride	100	Kanamycin	1
Clofibrate	97	Ketoprofen	92
Codeine	95	Ketorolac	90
Corticosterone	100	Labetalol	95
Cycloserine	73	Lactulose	0.6
Cymarin	47	Lamivudine	87
Cyproterone acetate	100	Lamotrigine	98
Desipramine	100	Lansoprazole	8
Dexamethasone	80	Levonorgestrel	100
Diazepam	100	Lincomycin	28
Diclofenac	100	Lormetazepam	100
Digoxin	81	Lornoxicam	100
Dihydrocodeine	89	Mannitol	16
Disulfiram	97	Meloxicam	90
Ethambutol	80	Mercaptoethane	77
Ethinylestradiol	100	Metaproterenol	44
Famciclovir	77	Metformin	53

Table 5 Continued

Compound	Absorption (%)	Compound	Absorption (%)
Methadone	80	Raffinose	0.3
Methylprednisolone	82	Recainam	71
Metolazone	64	Reproterol	60
Metoprolol	95	Rimiterol	48
Mexiletine	100	Saccharin	88
Mifobate	82	Salicylic acid	100
Minoxidil	98	Scopolamine	95
Morphine	85	Sorivudine	82
Moxonidine	88	Sormodren	100
Nadolol	57	Sotalol	95
Naloxone	91	Stavudine	100
Naproxen	99	Streptomycin	1
Nefazodone	100	K-Strophanthoside	16
Neomycin	1	Sudoxicam	100
Netivudine	28	Sulindac	90
Nicotine	100	Sulpiride	44
Nisoldipine	90	Sultopride	89
Nitrendipine	88	Sumatriptan	57
Nizatidine	90	Tenidap	89
Nordiazepam	99	Tenoxicam	100
Omeprazole	80	Terazosin	90
Ondansetron	100	Terbutaline	62
Ouabain	1.4	Testosterone	100
Oxatomide	100	Theophylline	100
Oxazepam	89	Timolol	95
Oxprenolol	95	Tolbutamide	85
Oxyfedrine	85	Tolmesoxide	98
Phenglutamide	100	Topiramate	8
Phenytoin	90	Torasemide	96
Pindolol	87	Toremifene	100
Pirbuterol	60	Tramadol	90
Piroxicam	100	Trapidil	96
Piroximone	81	Trimethoprim	97
Practolol	95	Urapidil	78
Praziquantel	100	Valproic acid	100
Prednisolone	99	Venlafaxine	97
Progesterone	100	Verapamil	100
Propiverine	84	Viloxazine	98
Propranolol	99	Warfarin	98
Propylthiouracil	76	Ziprasidone	60
Quinidine	81		

Reprinted from Zhao, Y. H.; Le, J.; Abraham, M. H.; Hersey, A.; Eddershaw, P. J.; Luscombe, C. N.; Boutina, D.; Beck, G.; Sherborne, B.; Cooper, I.; Platts, J. A. *J. Pharm. Sci.* **2001**, *90*, 749–784, with permission of Wiley-Liss, Inc., a subsidiary of John Wiley & Sons, Inc.

5.28.5.2   Molecular Descriptors for Oral Absorption Prediction

The calculation of many different 1D, 2D, and 3D descriptors is possible using a range of commercially available software packages, such as Sybyl, Cerius2, MOE, Dragon, Molconn-Z, Hybot, BCI, and BCUT. A more extensive discussion on descriptors used in ADME prediction can be found elsewhere (*see* 5.22 Use of Molecular Descriptors for Absorption, Distribution, Metabolism, and Excretion Predictions). Several new descriptor sets are based on quantification of 3D molecular surface properties and these have been explored for the prediction of, for example, Caco-2 permeability and oral absorption. The problem with some of these approaches is that non-intuitive descriptors are used, such as UNI and PMEP in MS-WHIM,[81] or the integy moment in VolSurf.[82] Modelers have therefore the choice of using complex molecular descriptors and/or more intuitive, simple and obvious molecular descriptors.

The most important properties for absorption and permeability appear to be solubility, and those related to hydrogen-bonding capacity and molecular size of the drug, rather than lipophilicity alone. However, solubility is only used in approaches such as the BCS (*see* 5.42 The Biopharmaceutics Classification System) and physiologically-based models, and is described elsewhere (*see* 5.17 Dissolution and Solubility; 5.26 In Silico Prediction of Solubility). In structure-based QSAR approaches solubility is not explicitly taken into account, but many of the descriptors used in fact do encode for solubility as well (*see* Section 5.28.9 and **Figure 6**). A short summary of the key descriptors is given below.

5.28.5.2.1   Lipophilicity
Lipophilicity is a key property in transport processes, including intestinal absorption, membrane permeability, protein binding, and distribution to different tissues and organs, including the brain.[83] Octanol/water partitioning ($\log P$) (and distribution ($\log D$)) has established itself as the reference solvent system, although other systems have been studied as potential alternatives. Alkane/water partitioning is being used as a model for the blood–brain barrier. Of particular interest is also the nonaqueous heptane/glycol partitioning,[84] although this system is practically quite demanding. A number of comprehensive reviews on the estimation of lipophilicity have been published and are recommended for further reading[85,86] (*see* 5.18 Lipophilicity, Polarity, and Hydrophobicity). Due to its key importance, there is continued interest in developing good $\log P$ calculation programs. Some computational $\log P$ approaches are limited by a lack of parameterization of certain fragments. For the widely used CLOGP program (*see* 5.27 Rule-Based Systems to Predict Lipophilicity), missing fragments are estimated, so that a result is always produced, albeit of varying quality.[87] Most predictive $\log P$ programs refer to the octanol/water partition system. Others, such as $\log P$ for alkane/water have been discussed, or can be found in the Absolv (ADME Boxes) software (see **Table 4**).

5.28.5.2.2   Hydrogen bonding
A simple measure of the hydrogen-bonding capacity of a molecule is to calculate its polar surface area (PSA), which is the sum of the fractional contributions to the surface area of all nitrogen and oxygen atoms.[88] The physical explanation is that the polar groups are involved in desolvation when they move from an aqueous extracellular environment to the more lipophilic interior of membranes. PSA thus represents, at least in part, the energy involved in the membrane transport of a compound. PSA is dependent upon the conformation of the molecule, but the original method was based on a single minimum energy conformation.[88] Later work has taken into account conformational flexibility, and coined the term 'dynamic PSA,' in which a Boltzmann-weighted average PSA is computed.[89] However, in most cases it was demonstrated that PSA calculated for a single minimum energy conformation is sufficient to produce an approximate sigmoidal relationship to intestinal absorption, and differed very little from the dynamic PSA described above.[90,91] A fast calculation of the PSA as the sum of fragment-based contributions has been published,[92] called the topological PSA (TPSA) (**Figure 3**), allowing the application of these calculations to large compound sets, such as combinatorial or virtual libraries. Other descriptors based on the PSA have been suggested. One example is the partitioned total surface area (PTSA).[93] More recently, the scaled PSA[94] was presented, which assigns scaling factors according to their known hydrogen-bonding characteristics. Another extension is called the high-charged PSA (HCPSA), which only counts contributions from polar atoms with high charge densities.[95] Alternative descriptors for hydrogen bonding include the HYBOT descriptors,[96] or partial atomic charges. Also, the non-polar surface area (NPSA) has been evaluated.[97] A further descriptor is the molecular hydrogen-bonding potential (MHBP), based on the 3D shape of a molecule.[98] It appears that, in particular, the hydrogen-bonding donor capacity ($MHBP_{do}$) correlates better with oral absorption than the hydrogen-bonding acceptor capacity ($MHBP_{ac}$).

5.28.5.2.3   Size descriptors
Molecular weight is often taken as the size descriptor of choice, as it is easy to calculate and is in the chemist's mind. However, other size and shape properties are equally simple to calculate, and may offer a better estimate of the potential for permeability. These include molecular volume, total surface area, and gyration.[95] Thus far, no detailed,

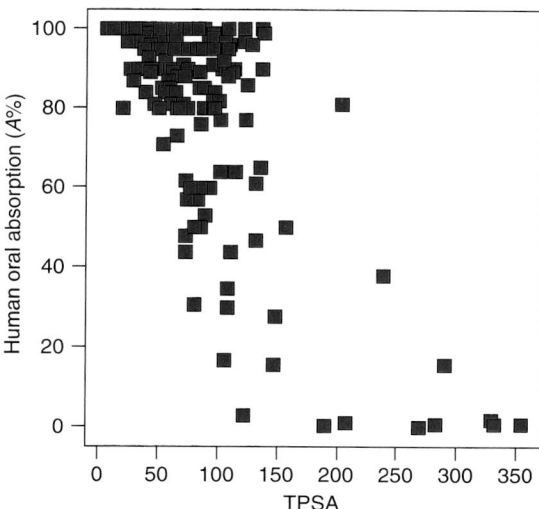

Figure 3 Approximate sigmoidal relationship between human oral absorption (data from Zhao *et al.*[4]) and topological polar surface area (TPSA).[92]

systematic investigation has been reported which size descriptor is most relevant. The cross-sectional area A_D, obtained from surface activity measurements, has been reported as a useful size descriptor to discriminate compounds that can access the brain $(A_D < 80 \text{ Å}^2)$ from those that are too large to cross the blood–brain barrier.[99] Similar studies have been performed to define a cut-off for oral absorption.[100]

5.28.5.2.4 Topological descriptors

There are many ways to use molecular topology as a basis to define molecular topological descriptors (*see* 5.22 Use of Molecular Descriptors for Absorption, Distribution, Metabolism, and Excretion Predictions). The advantage is that these are easy to calculate from the 2D structure. The disadvantage is that the direct relationship to structure is not intuitive, and their use in QSAR approaches is therefore often debated. It has been demonstrated that so-called electrotopological state indices (*see* 5.23 Electrotopological State Indices to Assess Molecular and Absorption, Distribution, Metabolism, Excretion and Toxicity Properties) can be useful to derive predictive models for oral absorption.[101–103] Simple counts of atom types are another easy way to derive useful topology descriptors.[104]

5.28.5.2.5 Other descriptors

Many other molecular descriptors have been tried out in quantitative structure–absorption relationships, as will be discussed in the sections below.

Often, AP can be rapidly estimated using a simple set of descriptors. An evaluation of the use of properties generated with the well-known ACD package (Advanced Chemistry Development, Toronto, Canada) was reported.[105] These ACD properties are derived from 2D graphs, and need no 3D optimization. These simple methods appear to be predictive, with correlation coefficients in the range of 0.73–0.95, depending on the predicted property. This can be considered as sufficient to screen virtual libraries, but insufficient for lead optimization.

Flexibility of a molecule, expressed as the number of free rotatable bonds, has been proposed as an important factor for bioavailability/absorption,[106] but should be used with caution.[107]

Molecular hashkeys have been proposed as a descriptor encoding the surface properties of a molecule, which can be used to predict, for example, properties such as $\log P$ and oral absorption, applying neural networks or machine learning.[108]

5.28.6 Computational Prediction of Human Oral Absorption

5.28.6.1 Types of Models

ADME modeling and, in particular, the prediction of intestinal absorption has been discussed extensively in the literature.[5,16,23,109–118]

For the medicinal chemist the most elegant approach to optimizing compounds for high oral absorption is the modification of a single property that easily relates to molecular structure. We therefore will first discuss the relationship between oral absorption and some of the most popular descriptors. However, complex ADME properties such as oral absorption are more likely to be dependent on a subset of fundamental properties. In Abraham's linear solvation energy relation method, a set of five descriptors has been used to predict a range of solvation-related properties such as partition coefficients in various solvents, blood–brain barrier penetration, and oral absorption. The serious limitation of this approach is that the five descriptors might be intercorrelated, and therefore the equation is statistically not robust.

The most general approach to derive a quantitative structure–property relationship is to select the appropriate subset of descriptors to build a predictive model from a large number (hundreds) of possible descriptors. This might indeed lead to statistically sound models, but suffer from interpretability ('black box' models). In quantitative structure–activity relationship (QSAR) studies as presented here to predict human oral absorption, a key question is: what are good and appropriate descriptors to build the model? The simplest approach is to look at correlations with individual descriptors. Another approach is to make an educated guess, as is done in the Abraham equation. The most common approach is to use a multivariate technique, and let the best descriptors emerge from a larger pool. There are basically two main approaches to deal with the descriptor selection problem, the filter and the wrapper approaches,[119,120] but further discussion is beyond the scope of this chapter.

Good introductory texts to QSAR modeling can be found in the literature.[121,122] The key elements of a QSAR analysis are descriptors and a model. In the remainder of the chapter we have chosen a mixed approach in presenting the various attempts published in the literature. Some headings refer to a particular set of descriptors, while other headings point to a mathematical technique used to derive the model.

The statistics for the equations given below have the following meaning: n is the number of compounds, r the correlation coefficient, q the cross-validated correlation coefficient using the leave-one or leave group out method, s the standard error of the regression, rms(e) the root mean square (of the standard error), and F is the Fisher statistic.[122] If reported in the original source, the regression coefficients include either their standard deviation or 95% confidence interval.

5.28.6.2   Correlations with Single Descriptors

5.28.6.2.1   log P/log D and Δlog P

Most of the membrane models developed in the 1970s used partition coefficients (log P) as the key descriptor, and it was later demonstrated that molecular size should be considered as an additional factor. It was suggested that the relationship between lipophilicity and Caco-2 flux could best be described by a set of sigmoidal curves.[123] This study underlined how complex these relationships are, and why one rarely finds good (linear) correlations using single properties alone.

We mention here also the Δlog P approach,[124,125] although this descriptor is in fact a measure for hydrogen bonding and thus similar to the PSA. For a series of azole endothelin antagonists, Δlog P was used to improve their oral absorption.[126]

5.28.6.2.2   Molecular weight

It has frequently been observed that permeability, and therefore absorption and bioavailability, decreases with increasing molecular weight. Furthermore, biliary excretion tends to increase with increasing molecular size. Both principles have been nicely illustrated in studies on the peptide-like renin inhibitors.[11] In the analysis leading to the 'rule-of-five,' it was found that 90% of all drugs in the World Drug Index have a molecular weight below 500 da.[9] This observation can therefore be taken as a guide in the design of orally active compounds.

5.28.6.2.3   Polar surface area

Following the first report on the use of the PSA to predict central nervous system (CNS) uptake,[88] this also became a popular descriptor for membrane permeability, for example in the prediction of the oral absorption of beta-blockers[89] and the design of peptide analogs.[127] There is a clear sigmoidal trend when plotting human oral absorption data against PSA (see **Figure 3**). The sigmoidal relationship can be described by

$$A\% = 100/[1 + (PSA/PSA_{50})^\gamma] \tag{15}$$

where $A\%$ is percentage of orally absorbed drug, PSA_{50} is the PSA at the 50% absorption level, and γ is a regression coefficient.[93] Also, a Boltzmann sigmoidal (general form given in eqn [16]) curve has been used[90]:

$$y = bottom + (top - bottom)/[1 + \exp(x_{50} - x)/slope)]$$ [16]

$$A\% = 100/[1 + \exp(PSA_{50} - PSA)/slope]$$ [17]

As a guide, orally active drugs that are transported by the transcellular route should not exceed a PSA of about $120\,\text{Å}^2$.[128,129] Similarly, for good brain penetration of CNS drugs, this number should even be tailored to $PSA < 100\,\text{Å}^2$ [129] or even smaller, $<60–70\,\text{Å}^2$.[128]

Poorly absorbed compounds have been identified as those with a $PSA > 140\,\text{Å}^2$. Many compounds do not show only simple passive diffusion but are affected by active carriers, efflux mechanisms involving P-glycoprotein and other transporter proteins, and gut wall metabolism. A further refinement in the PSA approach is expected to come from taking into account the strength of the hydrogen bonds, which in principle is the basis of the HYBOT approach described later.

An absorption model based on the combination of the PSA and lipophilicity (AlogP98) has been suggested,[112,130] distinguishing between compounds that are well absorbed ($A\% > 90\%$) or poorly absorbed ($A\% < 30\%$).

5.28.6.2.4 Nonpolar surface area

The Caco-2 cell membrane permeability of three series of peptides and endothelin antagonists could be predicted by a theoretical model that considered both the polar (PSA_d) and nonpolar ($NPSA_d$) parts of the dynamic molecular surface area of the molecules.[97,131] The three peptide series were AcHN-X-phenethylamides, AcHN-XD-Phe-NHMe derivatives and D-Phe-oligomers. Experimental log D (octanol/water) values gave a rank order of permeability within each series, but failed across the three series. Possibly, some of the compounds are substrates for one or more transporters present in Caco-2 cells, but this needs further investigation. A strong correlation was found between log D and $NPSA_d$ ($r^2 = 0.96$). A good sigmoidal correlation was obtained when P_{app} (Caco-2 permeability) was plotted against a linear combination of PSA_d and $NPSA_d$. Thus, this model predicts permeability based on a combination of hydrogen-bonding capacity and hydrophobicity. The latter is thought to be related to the transport of a compound from the aqueous environment into the polar head group region of the membrane, while hydrogen bonding is detrimental to transport into the nonpolar interior of the membrane.[131]

5.28.6.2.5 Percentage polar surface area

The percentage polar surface area (%PSA) has been investigated as another surface property,[127] but without success.[97,131] %PSA is a number between 0 and 100, and is not related to more fundamental phenomena such as solvation energy, as is the case with the PSA.

5.28.6.2.6 Sigmoidal surface fit

Single-descriptor approaches often assume a sigmoidal relationship between a property and oral absorption. These relationships have been studied in more detail[132] and extended to a surface fit taking into account two properties simultaneously. Both symmetric and asymmetric sigmoidal relationships were compared, and often found to perform equally. Models were built for 93 compounds and tested for 31 compounds using descriptors calculated with the MOE software. It was found that the total hydrogen bond descriptor (a_T) and the calculated log P (SlogP, see 5.27 Rule-Based Systems to Predict Lipophilicity) were most important in the models. The best symmetric surface was given by

$$A\% = 33.1(\pm 0.1)\{\pi/2 + \arctan[-0.42(\pm 0.09)a_T + 0.39(\pm 0.09)SlogP + 3.9(\pm 0.9)]\}$$
$$n = 124, \quad r^2 = 0.80, \quad q^2 = 0.79, \quad s = 13.4$$ [18]

where $A\%$ is the percentage human intestinal absorption (**Figure 4**).

5.28.6.3 Solvation Equation

Instead of looking at only one or two descriptors, a linear free-energy relationship approach based on the 'solvation equation' has been used to predict solution related properties such as solubility, partitioning, blood–brain transport, and GI absorption. These properties are correlated with a set of five molecular descriptors, called Abraham descriptors,

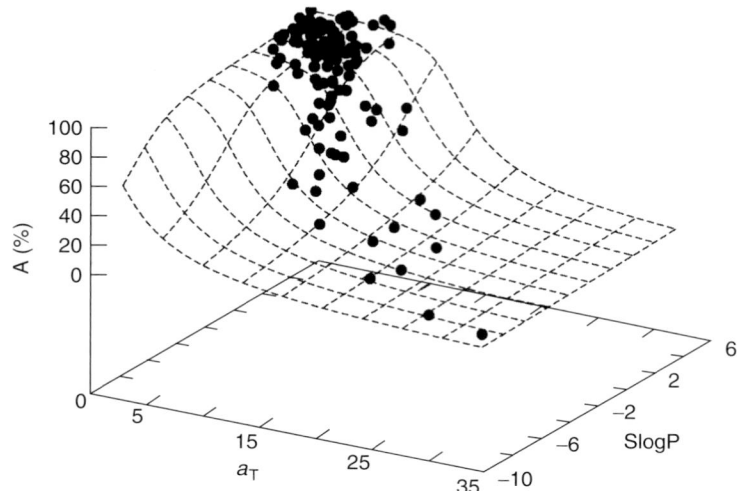

Figure 4 Surface fit of experimental human intestinal absorption data ($n = 93$) using the descriptors a_T and SlogP.[132] (Reproduced from Deretey, E.; Feher, M.; Schmidt, J. M. *Quant. Struct.-Act. Relat.* **2002**, *21*, 493–506, with permission of Wiley-VCH Verlag GmbH.)

which are excess molar refraction (E), solute polarity/polarizability (S), solute overall acidity (A) and basicity (B), and the McGowan characteristic volume (V).

In a detailed study, the human intestinal absorption of 241 drugs was evaluated.[4] After careful consideration of the data, only 169 compounds (**Table 5**) could be used to develop the absorption model.

$$A\% = 92 + 2.94E + 4.10S + 10.6V - 21.7A - 21.1B$$
$$n = 169, \quad r^2 = 0.74, \quad s = 14, \quad F = 93 \qquad [19]$$

This equation shows that size is beneficial and hydrogen bonding is detrimental to oral absorption. The quality and error of the original data are such that the final model accounts for 74% of the variance and that the predictions have a 14% standard error.

By transforming the data to the effective rate of absorption (GI k_{eff}), the following equation was obtained[133]:

$$\log \text{GI } k_{\text{eff}} = 0.54 - 0.03E + 0.01E + 0.14S - 0.41A - 0.51B + 0.20V$$
$$n = 127, \quad r^2 = 0.79, \quad s = 0.29, \quad F = 0.84 \qquad [20]$$

The Abraham descriptors can be calculated by the AbSolv program (available via ADME Boxes; see **Table 4**). It is important to note that in the Abraham approach, as exemplified by eqns [19] and [20], the selected five descriptors are always incorporated into the equation, despite the fact that some may be intercorrelated.[134] This is not good statistical practice, and may lead to an overestimation of the relevance of the equation, and therefore its general applicability.

5.28.6.4   Multivariate Models

5.28.6.4.1   ADAPT

Over the years, Jurs's group has developed a range of molecular descriptors, which have been implemented in the ADAPT software and used to predict properties such as water solubility and human intestinal absorption.[135] From a larger pool of descriptors, a neural network model selected six key descriptors. Of these six descriptors, three encode for size (cube root of gravitational index), shape (SHDW-6, the normalized 2D projection of the molecule on the YZ plane), and flexibility (NSB, the number of single bonds), while the three others are related to hydrogen-bonding properties (CHDH-1, the charge on donatable hydrogen atoms; SCAA-2, the surface area multiplied by the charge of hydrogen bond acceptor atoms; and SAAA-2, the surface area of hydrogen bond acceptor atoms). A 16% rms error was observed in an external test set.

In Silico Models to Predict Oral Absorption 687

5.28.6.4.2 MolSurf

The program MolSurf offers a number of descriptors related to physicochemical properties, such as lipophilicity, polarity, polarizability, and hydrogen-bonding ability. Their relevance for predicting oral absorption was investigated using a partial least squares (PLS) analysis.[136] Good statistical models were obtained, revealing that properties associated with hydrogen bonding had the largest impact on absorption and should be kept to a minimum. However, MolSurf analysis is detailed, computationally expensive, and not automated.

5.28.6.4.3 VolSurf

The VolSurf descriptors are a set of descriptors related to the surface properties of a molecule, and are calculated using an H2O and a DRY probe in the program GRID. These descriptors have been evaluated in correlations with human absorption.[82,137] A descriptor called 'integy moment' was defined, analogous to the dipole moment, and describes the distance from the center of mass to the barycenter of polar interaction sites at a given energy level. A high integy moment reflects a clear separation between polar and nonpolar parts of a molecule. Hydrophobicity and high integy moments are positively correlated with human intestinal absorption, whereas polarity and a high concentration of polar interaction sites on the molecular surface are detrimental to absorption.

VolSurf has also been used in a successful in silico prediction of intestinal lymphatic transfer of a series of compounds co-administered with a long-chain triglyceride.[138]

5.28.6.4.4 HYBOT

Based on experimental thermodynamic data, hydrogen bond donor and acceptor descriptors have been developed, which have been correlated to permeability and absorption data.[96] It was concluded that both hydrogen bond donors (ΣCd) and acceptors (ΣCa), often in combination with a steric descriptor, are important physicochemical properties for permeation processes. However, due to the frequently observed intercorrelation between donors and acceptors, only the more significant one can be used in multiple linear regression equations, though this problem can be avoided using other statistical tools such as the PLS or neural networks. It may also be more sensible to use the combined acceptor plus donor term.[139,140]

Raevsky[139,141] and others[45,142] pointed out that data presented as the percentage effect or fractional effect such as oral absorption are problematic in linear regression techniques, because predictions can be negative or $>100\%$. This difficulty may be overcome by transforming the values into logit values (e.g., for the fraction absorbed, FA):

$$\text{logit } FA = \log[FA/(1 - FA)] \qquad [21]$$

These logit FA values can be correlated to a set of independent variables or descriptors. Thus a nonlinear equation can be derived as follows:

$$Z = \text{logit } FA = \log[FA/(1 - FA)] \qquad [22]$$

$$FA/(1 - FA) = 10^Z \qquad [23]$$

$$FA = 1/(1 + 10^{-Z}) \qquad [24]$$

where

$$Z = f(Xi) \qquad [25]$$

where Xi is a range of relevant physicochemical descriptors.

For a set of 31 compounds, the following nonlinear sigmoidal equation was found[131]:

$$A\%/100 = FA = 1/(1 + 10^{-[5.05 - 0.36\Sigma Ca. + 0.26\Sigma Cd]})$$
$$n = 31, \quad r^2 = 0.95, \quad q^2 = 0.92, \quad s = 0.09 \qquad [26]$$

Addition of other descriptors had only little influence. This equation demonstrates the key importance of hydrogen bonding for absorption processes. Such a relationship is very similar to the one obtained using the PSA as another descriptor for hydrogen bonding. However, it appears that these equations do not perform satisfactorily with a test set.[143] In more recent work, QSARs based on the HYBOT descriptors have been combined with a similarity approach,[141,142,144] leading to eqn [27].[141] The prediction is obtained using a model derived from a training set and the difference of properties between the molecule of interest, virtual or real, and a similar drug with known oral absorption.

$$A\%/100 = FA = 1/(1 + 10^{\{(-\log[FA_{nrs}/(1-FA_{nrs})]-0.36\Delta\Sigma Ca(\text{drug-nrs})+0.26\Delta\Sigma Cd(\text{drug-nrs})\}})$$
$$n = 100, \quad r^2 = 0.89, \quad q^2 = 0.89, \quad s = 0.11$$

[27]

where FA_{nrs} is the fraction absorbed for a nearest related structure, $\Sigma Ca(\text{drug-nrs})$ is the difference in calculated Ca (HYBOT hydrogen bond acceptor factors), and $\Sigma Cd(\text{drug-nrs})$ the difference in calculated Cd (HYBOT hydrogen bond donor factors).

5.28.6.4.5 Other multivariate studies

In a study of several data sets,[130] the most appropriate factors to consider predicting passive absorption from structure were lipophilicity, hydrophilicity, and size. The authors chose AlogP98 (*see* 5.27 Rule-Based Systems to Predict Lipophilicity), PSA, and molecular weight as variables to measure these factors. It was concluded that molecular weight was a redundant descriptor since it was seen as a component of both PSA and AlogP98. Sharp, nonlinear drops in permeability were observed below certain values for both descriptors, in line with the sigmoidal relationships observed by many others. The authors stressed the need for large experimental data sets to advance ADME models in general and, in particular, prediction of oral absorption and bioavailability.

Topological descriptors have been used to develop intestinal absorption models. Examples include electrotopological state indices.[101,102] Also, structural descriptors using the CASE program led to successful models.[145] An external test set of 50 compounds had a standard deviation of 12.3% using the model based on 417 drugs.[145]

Using a set of 219 atom-type descriptors and PLS discriminant analysis (PLS-DA) a human intestinal absorption predictive model was derived.[146] The data set[4] of 169 drugs was divided into three classes: >80% absorption, 20–80% absorption, and <20% absorption. A five-component PLS-DA model separated the compounds into these classes with an $r^2 = 0.92$ and $q^2 = 0.79$.

5.28.6.5   Other Approaches

5.28.6.5.1 GRID

Using the program GRID (*see* 4.11 Characterization of Protein-Binding Sites and Ligands Using Molecular Interaction Fields), hydrogen-bonding capacity could be quantified using an NH amide probe to explore the hydrogen bond acceptor regions, a carbonyl probe to detect hydrogen bond donor areas, and a water probe to characterize both.[147] The water surface interaction map generated by the water probe appears to be a good descriptor in the prediction of drug permeability, although no improvement over previously reported methods was obtained.

5.28.6.5.2 Neural networks

Neural networks have the attractive property that they model nonlinear relationships. However, they have the disadvantage that they look more like a 'black box' approach. Various approaches and different architectures of neural nets have been used in the prediction of biological properties, the most useful of which appear to be the back-propagation and Kohonen nets.

Through the combination of simulated annealing and a neural network approach, a seven-descriptor model for a set of 120 compounds was derived to estimate human intestinal absorption.[148] The chosen descriptors are easy and fast to compute, and therefore would allow screening of larger libraries. However, the overall error of prediction is about 25%, which probably reflects the error on the input data. Clearly, an extension of the database, particularly if populated over the whole 0–100% absorption range, would help to improve the model.

In another study, a set of 86 drug compounds[135] was investigated.[149] Although 57 global molecular descriptors were initially used, in the final model, 15 input descriptors were chosen with a genetic algorithm. The most significant descriptors relate to lipophilicity, conformational stability and inter-molecular interactions reflected by polarity and hydrogen bonding. The rms error for an external prediction set was 16.0%, which is comparable to experimental error in the human absorption data.

The same 86 compounds[135] were studied using only 2D topological descriptors.[150] The data were modeled with a general regression neural network (GRNN) and a probabilistic neural network (PNN), which are variants of normalized radial basis function networks. Both performed well, with an rmse of 22.8% for the external prediction set for a GRNN model, and 80% of the external prediction set were correctly classified for the PNN model.

A set of 141 topological descriptors were used to build an artificial neural network (ANN) model for a set of 417 therapeutic drugs.[102] This was validated by testing 195 new drugs not used in model building; 92% of the compounds were predicted within 25% of their reported value. The PSA and log P appear to be important additional properties in the ANN model.

A small set of 28 compounds was studied using a neural network and so-called CODES topological descriptors.[151] This set is too small to give meaningful predictions.

5.28.6.5.3 Decision trees and other rule-based approaches

Recursive classification regression trees (CARTs) are a data-mining tool suitable for exploration of relatively small data sets, as is the case with human oral absorption data. The original CART method is used to bin compounds in a number of predefined classes. A modified version is capable of nonlinear regression, and can produce numerical outputs.[79] A set of 1260 structures was used to build recursive partitioning models for an 899-compound training set and a 362-compound validation set. The fraction of the dose absorbed was divided into six classes of equal size. A set of 28 descriptors was used for the modeling of oral absorption. Two test sets of unpublished proprietary compounds showed 79–86% prediction plus/minus one class. This error level is comparable to current in vitro/in vivo correlations. This in silico approach also appears to be superior too predictions based on data obtained with artificial membranes.[79]

A sequence of recursive partitioning analyses based on multiple physicochemical and structural descriptors for a set of over 1000 druglike compounds using the Algorithm Builder software led to a rule-based algorithm for intestinal permeability.[152] The fragment builder in this software has some unique properties different from other approaches. Interestingly, the model can easily be retrained as compounds with new fragments are added to the training set. The model is currently a two-class model, where poor absorption is <10% and good absorption is >15%. The general applicability of this approach using, for example, different cut-offs for poor, medium, and good absorption needs to be demonstrated.

5.28.6.5.4 Support vector machines (SVMs)

The relatively new QSAR method of SVMs, a machine learning technique, has been used on human intestinal absorption data for 164 compounds.[119] A new descriptor selection algorithm picked out the 50 most relevant descriptors from a pool of in total 2929 descriptors. The high relevance of the TPSA was confirmed in this work. A 12.7% error rate in the predictions was found.

5.28.6.6 Hybrid Models Using In Silico and In Vitro Data (In Combo Approach)

QSAR descriptors can be purely computational or experimental. If a mixture of both is required, this can be called an in combo approach.[78] An example is given below.

Relationships with the PSA and several other descriptors using the MOLCAD module within SYBYL have been studied for effective permeability data in humans, based on a single minimum energy conformation.[153,154] For the 13 passively transported compounds in the data set, a linear correlation with the PSA was obtained with $r^2 = 0.76$. A plot of these data shows that the trend is possibly sigmoidal, but with some scatter. Even more scatter is observed when all the compounds in the study ($n = 22$) are plotted against the PSA. Clearly, the PSA alone is insufficient to account for effective permeability or absorption. In this case, the cut-off for poor absorption seems to be at the lower PSA values of around 100Å^2. This may be due to a scaling difference between methods used here and by others. The outliers, identified as glucose, levodopa, and amoxicillin, are believed to have active uptake mechanisms, and therefore are better absorbed than predicted by the PSA. The following equations have been obtained[153]:

$$\log P_{\text{eff}} = -0.01\text{PSA} + 0.19 \ \log D_{5.5} - 0.24\text{HBD} - 2.88$$
$$n = 13, \ r^2 = 0.93, \ q^2 = 0.90 \tag{28}$$

$$\log P_{\text{eff}} = -0.01\text{PSA} + 0.16\text{CLOGP} - 0.24\text{HBD} - 3.07$$
$$n = 13, \ r^2 = 0.88, \ q^2 = 0.85 \tag{29}$$

where P_{eff} is the in vivo permeability measured with a single-pass perfusion technique.[21,80] $\log D_{5.5}$ is the octanol/water distribution coefficient measured at pH 5.5, believed by the authors to be the most relevant value for absorption as it reflects the pH in the unstirred mucus layer adjacent to the intestinal wall. HBD is the number of hydrogen atoms connected to nitrogen and oxygen atoms (i.e., the total potential hydrogen-donating capacity). Since these models are based on only 13 compounds, the three-parameter eqns [28] and [29] have limited statistical significance. No definitive conclusions can be drawn on the role of a lipophilicity descriptor. The best result was obtained by combining two hydrogen bond descriptors (PSA and HBD) and the experimental log D at pH 5.5. However, the partial correlation between the PSA and HBD is 0.82. This sheds serious doubt on eqns [28] and [29], despite the fact that it was derived using PLS. A larger data set is required to more fully explore this approach. The continuous variable PSA is strongly correlated with a simple count of hydrogen bonds. However, the PSA is probably a better reflection of hydrogen-bonding capacity, since it takes conformational behavior into account. Various alternatives for hydrogen-bonding descriptors were explored.[154] Combinations of a general hydrogen-bonding descriptor and a lipophilicity descriptor give the best models.[154]

5.28.6.7   Commercial Softwares

In the discussion presented above, a number of commercial packages have been mentioned, and which are listed in **Table 4**. Some of these can be used to generate descriptors and/or to generate the models. A number of these products have a preprogrammed human oral absorption model 'off the shelf.' Also included in this overview are sources for oral absorption data.

5.28.7   Prediction of Caco-2 Flux

One in vitro approach to predict human intestinal absorption is to use Caco-2 cell permeability data. A more indirect method consists of predicting Caco-2 flux from physicochemical and/or structural properties. These are models of a model for oral absorption, and one might question the predictive power of such an approach. Among others,[155] a number of predictive Caco-2 models are discussed here.

Initially, it was thought that only poor linear correlation between Caco-2 flux and log D values is observed.[156] However, the relationship between permeability and lipophilicity is in fact thought to be sigmoidal,[6,7,123,140] and can be described as

$$A\% = (0 - 100)/[1 + (\log D/\log D_{50\%})^k] + 100 \qquad [30]$$

where $A\%$ is the percentage of orally absorbed drug, log $D_{50\%}$ is the log D at the 50% absorption level, and k is a slope factor[93] (compare also to eqns [14]–[17]). Using human absorption data,[157] a sigmoidal correlation between Caco-2 data and octanol/water log D values was observed for compounds within the molecular weight range of 200–500.[12] It was also suggested that, in fact, such a sigmoidal relationship exists for each group of compounds with identical molecular weight.[123] Thus, a plot of log P_e (permeability) against log D (lipophilicity) becomes meaningful when adding the molecular weight (size) contour plots.

However, such relationships to lipophilicity are not always clear-cut. For the solvents octanol, hexadecane, and propyleneglycol dipelargonate (PGDP) as lipophilicity scales, no simple relationships were found for a set of 51 structurally diverse low molecular weight compounds and their Caco-2 permeability coefficients.[158] Furthermore, in the same study no significant correlation was found using $\Delta\log D$ values, derived from the difference between log D_{hex} and either log D_{oct} or log D_{PGDP} as a measure for hydrogen bonding.

Prediction of Caco-2 permeability using computed molecular properties has been studied by several groups using multiple linear regression (MLR),[95,140] principal component analysis, cluster analysis, linear discriminant analysis (LDA),[159] PLS,[160–163] 3D-QSAR,[164] and membrane-interaction QSAR.[165]

Simple MLR equations can be derived combining a size and hydrogen bond descriptor, such as in

$$\log P_{app} = 0.008(\pm 0.002)\text{MW} - 0.043(\pm 0.008)\text{PSA} - 5.17(\pm 0.61)$$
$$n = 17, \quad r^2 = 0.69 \qquad [31]$$

where P_{app} is the permeability constant across Caco-2 cells and MW is the molecular weight of the compound.[132] Interestingly, this equation contains no lipophilicity term such as log P or log D.

In a study of 100 compounds for which Caco-2 data were collected from the literature, cellular permeability was primarily dependent upon the experimental distribution coefficient (log D), the high-charged PSA (HCPSA), and the radius of gyration (r_{gyr}).[95] The fraction of rotatable bonds (f_{rotb}) had only a minor contribution.

$$\log P_{app} = 0.25[-1.8 < \log D < 2.0] - 0.0048\text{HCPSA} - 0.19r_{gyr} + 1.06f_{rotb} - 4.39 \qquad [32]$$

The authors used a spline function for the log D term. This term equals the experimental log D if within the range -1.8 to 2.0, otherwise it will be either -1.8 for compounds with a log D below that value or 2.0 for compounds with a higher log D. A regression with splines does allow incorporating features that do not have a linear effect over the full property range. The HCPSA takes into account that only highly charged polar atoms will be involved in hydrogen bonding. The radius of gyration is a molecular shape descriptor believed to be more relevant for membrane permeability than molecular weight. Using a test set of 23 compounds, $r^2 = 0.78$ was found, which is not impressive.

In a PLS study, descriptors including polarization, solvent accessible surface area, hydration energy, heat of formation, and dipole moment have been explored.[161] The 51 compounds in this study were from Yazdanian et al.[158] Increased log D and hydration energy facilitate permeability, and an increased dipole moment of the molecules has a negative effect on the Caco-2 flux. Hydrogen bonding is one of the important factors associated with permeability.

Using 70 descriptors, including lipophilicity, hydrogen bonding, polar surface area, size and charge descriptors, and some nonlinear terms, a PLS model for Caco-2 permeability was derived ($n = 46$, $r^2 = 0.79$, $q^2 = 0.65$) and tested on two external test sets.[162] For a small test set ($n = 5$) the rmse was 0.45 log P_{app} units. Overall, 82% of the compounds were correctly classified as low, medium, or high permeability.

A model predicting Caco-2 cell permeability from 2D topological descriptors was developed using a combined genetic algorithm and a PLS method for 73 structurally diverse compounds.[163] The final PLS model consisted of 12 principal components of 24 descriptors, and gave a correlation coefficient of 0.89. However, this must be considered as a very high number of principal components and therefore unreliable for prediction.

A novel set of topological descriptors obtained from graph theoretical methods and LDA as well as MLR was used for 33 structurally diverse drugs and another 18 drugs to test the models.[159] The LDA model discriminates between good and poor–moderate permeability with a global classification of 87.9%.

3D-QSAR approaches have been used to predict permeability through Caco-2 cells for a series of close structural analogs.[166] A comparison was made between Catalyst, CoMFA, VolSurf, and genetic function approximation (GFA) with MS-WHIM descriptors.[81] The Catalyst pharmacophore model appears to give the best predictions (test set versus observed permeability $r^2 = 0.94$). The GFA/WHIM approach gave the poorest results.

Using a data set of 712 compounds for which permeability data were measured using Caco-2 and MDCK cells, a two-class model (above and below 4×10^{-6} cm s^{-1}) was developed using nearest-neighbor classification.[167] The key descriptors were the molecular surface, the number of flexible bonds, the number of hydrogen bond acceptors, the number of hydrogen bond donors, and the PSA. In an external test set of 112 compounds, 93% of the compounds could be classified in one of two permeability classes. The misclassification rate was 15%, and no compounds were misclassified in the nonpermeable group.

One of the problems of developing a general Caco-2 model is the interlaboratory differences in the data. To overcome this problem, a 'latent membrane permeability' concept was developed.[168] In this approach, Caco-2 data from different sources are treated by a computational algorithm to extract the common relationship between the membrane permeability and properties of the compounds.

5.28.8   Prediction of Parallel Artificial Membrane Permeation Assay (PAMPA) Data

PAMPA[53] is discussed in more detail elsewhere (see 5.19 Artificial Membrane Technologies to Assess Transfer and Permeation of Drugs in Drug Discovery). For a set of 43 ionizable compounds and using the Abraham linear free-energy relation method and using calculated Abraham descriptors, a predictive PLS model has been derived for the intrinsic PAMPA permeability ($P_{0\ PAMPA}$)[169]:

$$\log P_{0PAMPA} = -4.03 + 1.06E - 0.95S - 2.26A - 2.84B + 2.58V$$
$$n = 43, \quad r^2 = 0.78, \quad q^2 = 0.69, \quad \text{rmse} = 0.92 \tag{33}$$

In this study, membranes were made from a 2% solution of dioleylphosphatidylcholine in dodecane. Rigorous stirring compensated for the UWL near the membrane, and to derive the intrinsic permeability for the neutral species. The model clearly demonstrates the relative contributions of various factors to PAMPA permeability. It is unclear how this relates to in vivo permeability. It should also be noted, however, that about 20% remains unexplained, which is considerable for a relatively simple physicochemical model. This number is also of the same order as the coefficient of variance of the assay.

In a subsequent study, the Abraham descriptors were measured, leading to a statistically improved equation[170]:

$$\log P_{0PAMPA} = -3.45 + 0.43E - 1.42S - 3.37A - 4.79B + 4.30V$$
$$n = 41, \ r^2 = 0.94, \ \text{rmse} = 0.49$$

[34]

This equation is very similar to the expressions obtained for alkane/water partitioning. This therefore demonstrates that $\log P_{\text{alkane/water}}$ has the same information content as a PAMPA P_e value.

5.28.9   Outlook

A growing consensus is that in silico absorption predictions are as predictable as data from in vitro tests, with the advantage that much less investment in resources, technology, and time is needed. Additionally, it is possible to screen before synthesis.[167] The in silico approach also appears to be superior over predictions based on data obtained with artificial membranes.[79]

In comparing various predictive approaches for human intestinal absorption (**Table 6**), one should consider the drug discovery stage and the required precision. In early stages, predictions serve to decide on the value of making a library, or at the high-throughput screening stage on making choices between promising series. In later stages, effects of formulation and even patient populations may be of greater interest.

Current purely structure-based in silico predictions of absorption are most likely to be reliable only for compounds that are transported mainly by passive diffusion. The most dominant descriptors for oral absorption are hydrogen bonding, molecular size, and lipophilicity. A minimal set of descriptors can frequently give reasonable models.[142] Probably the best estimations of oral absorption are obtained using an in combo approach (i.e., a hybrid in vitro/in silico approach (see 5.41 The Adapative In Combo Strategy and **Figure 5**). This could in a simple form (e.g., mean to take experimental solubility into consideration), as in the BCS approach.

Oral absorption is a complex property (see **Figure 2**), with contributions from solubility, lipophilicity, permeability, transporters, and metabolism (**Figure 6**). Therefore, a more realistic predictive absorption model should account for

Table 6 Approaches to absorption prediction

Compounds	In vivo	In vitro	In silico
Virtual			Human oral absorption models (**Table 4**)
Synthesized		Solubility	MAD
		Caco-2/MDCK	BCS
		$\log D$/$\log P$	PBPK
		PAMPA	
		(see **Table 2**)	
Synthesized	Rat, dog, and monkey oral absorption		Allometric scaling

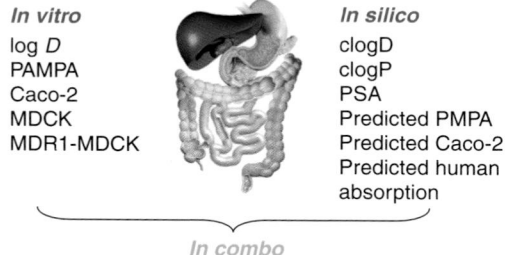

In vitro
log D
PAMPA
Caco-2
MDCK
MDR1-MDCK

In silico
clogD
clogP
PSA
Predicted PMPA
Predicted Caco-2
Predicted human
absorption

In combo

Figure 5 The in combo approach to the prediction of oral absorption: combination of the in vitro and in silico methods to obtain the best possible estimate of human intestinal absorption.

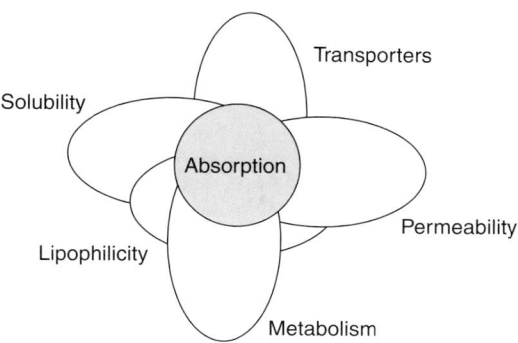

Solubility = A1 + A3 + A6
Lipophilicity = A2 + A3 + A5
Permeability = A1 + A4 + A5 + A6
Transporters = A2 + A3 + A5
Metabolism = A1 + A4 + A7
Absorption = A1 + A3 + A4 + A5 + A6

Figure 6 Opening the 'black box' of oral absorption prediction. Mechanistic insight is obtained by studying the properties/descriptors (schematically represented here as A1 to A7) influencing the individual contributions of solubility, lipophilicity, permeability, transporters, and metabolism to the overall property absorption (see also **Figure 2**). (Reproduced from Van de Waterbeemd, H. Property-Based Lead Optimization. In *Biological and Physicochemical Profiling in Drug Research*; Testa, B., Krämer, S. D., Wunderli-Allenspach, H., Folkers, G., Eds.; VHCA: Zürich, Switzerland; Wiley-VCH: Weinheim, Germany, 2006, pp 25–45, with the permission of Verlag Helvetica Chimica Acta AG.)

these basic contributions. More needs to be learned about how to account for transporters involved in active uptake and secretive efflux, and the role of gut wall metabolism[18,19] in such predictions. Some of this learning is already being used in physiologically-based models (*see* Section 5.28.4.3). The physiological role of transporters in oral absorption is discussed in Chapter 5.04, while predictive modeling of transporters is reviewed in Chapter 5.33.

For any ADME endpoint, more data of the highest possible quality covering much chemical space will further improve the predictive power of in silico ADME models. As discussed earlier, the data spread over the entire range (0–100% in the case of the amount of oral absorption) within the data set is important.

Combining several approaches gives a consensus prediction that often performs better than the individual models. This has been shown in, for example, log *P* predictions (see the KnowItAll approach). The CEREP/BioPrint models, including their human absorption model, are also typically based on the consensus of two approaches, namely a neural network and a predictive neighborhood behavior (nearest neighbor) model. It is expected that many in silico ADME models, including oral absorption, will in the near future be based on some form of consensus modeling,[171] such as the decision forest.[172]

References

1. Sietsema, W. K. *Int. J. Clin. Pharmacol. Ther. Toxicol.* **1989**, *27*, 179–211.
2. Burton, P. S.; Goodwin, J. T.; Vidmar, T. J.; Amore, B. M. *J. Pharmacol. Exp. Ther.* **2002**, *303*, 889–895.
3. Van de Waterbeemd, H. In *Physicochemical Concepts*. Van de Waterbeemd, H., Lennernäs, H., Artursson, P., Eds.; Wiley-VCH: Weinheim, Germany, 2003; Vol.18, Chapter 1, pp 3–20
4. Zhao, Y. H.; Le, J.; Abraham, M. H.; Hersey, A.; Eddershaw, P. J.; Luscombe, C. N.; Boutina, D.; Beck, G.; Sherborne, B.; Cooper, I.; Platts, J. A. *J. Pharm. Sci.* **2001**, *90*, 749–784.
5. Mälkiä, A.; Murtomäki, L.; Urtii, A.; Kontturi, K. *Eur. J. Pharm. Sci.* **2004**, *23*, 13–47.
6. Camenisch, G.; Van de Waterbeemd, H.; Folkers, G. *Pharm. Acta Helv.* **1996**, *71*, 309–327.
7. Camenisch, G.; Folkers, G.; Van de Waterbeemd, H. *Eur. J. Pharm. Sci.* **1998**, *6*, 321–329.
8. Sugano, K.; Takata, N.; Machida, M.; Saitoh, K.; Terada, K. *Int. J. Pharm.* **2002**, *241*, 241–251.
9. Lipinski, C. A.; Lombardo, F.; Dominy, B. W.; Feeney, P. J. *Adv. Drug Deliv. Rev.* **1997**, *23*, 3–25.
10. Chan, O. H.; Stewart, B. H. *Drug Disc. Today* **1996**, *1*, 461–473.
11. Stewart, B. H.; Chan, O. H.; Lu, R. H.; Reyner, E. L.; Schmid, H. L.; Hamilton, H. W.; Steinbaugh, B. A.; Taylor, M. D. *Pharm. Res.* **1995**, *12*, 693–699.
12. Van de Waterbeemd, H. Intestinal Permeability: Prediction from Theory. In *Oral Drug Absorption*; Dressman, J. B., Lennernäs, H., Eds.; Marcel Dekker: New York, 2000, pp 31–49.

13. Sha'afi, R. I.; Gary-Bobo, C. M.; Solomon, A. K. *J. Gen. Physiol.* **1971**, *58*, 238–258.
14. Van de Waterbeemd, H.; Smith, D. A.; Beaumont, K.; Walker, D. K. *J. Med. Chem.* **2001**, *44*, 1313–1333.
15. Abraham, M. H.; Ibrahim, A.; Zissimos, A. M.; Zhao, Y. H.; Comer, J.; Reynolds, D. P. *Drug Disc. Today* **2002**, *7*, 1056–1063.
16. Matter, H.; Baringhaus, K. H.; Naumann, T.; Klabunde, T.; Pirard, B. *Comb. Chem. High Throughput Screen* **2001**, *4*, 453–475.
17. Dantzig, A. H.; Hillgren, K. M.; De Alwis, D. P. *Annu. Rep. Med. Chem.* **2004**, *39*, 279–291.
18. Benet, L. Z.; Wu, C.-Y.; Hebert, M. F.; Wacher, V. J. *J. Control. Release* **1996**, *39*, 139–143.
19. Wacher, V. J.; Salphati, L.; Benet, L. Z. *Adv. Drug Deliv. Rev.* **1996**, *20*, 99–112.
20. Lacombe, O.; Woodley, J.; Solleux, C.; Delbos, J.-M.; Boursier-Neyret, C.; Houin, G. *Eur. J. Pharm. Sci.* **2004**, *23*, 385–391.
21. Petri, N.; Lennernäs, H. In Vivo Permeability Studies in the Gastrointestinal Tract of Sumans. In *Methods and Principles in Medicinal Chemistry*; Van de Waterbeemd, H., Lennernäs, H., Artursson, P., Eds.; Wiley-VCH: Weinheim, Germany, 2003; Vol. 18, Chapter 7, pp 155–188.
22. Alsenz, J.; Haenel, E. *Pharm. Res.* **2003**, *20*, 1961–1969.
23. Pagliara, A.; Reist, M.; Geinoz, S.; Carrupt, P.-A.; Testa, B. *J. Pharm. Pharmacol.* **1999**, *51*, 1339–1357.
24. Bernards, C. M.; Hill, H. F. *Anesthesiology* **1992**, *77*, 750–756.
25. Pang, K. S. *Drug Metab. Dispos.* **2003**, *31*, 1507–1519.
26. Chiou, W. L.; Barve, A. *Pharm. Res.* **1998**, *15*, 1792–1795.
27. Zhao, Y. H.; Abraham, M. H.; Le, J.; Hershey, A.; Luscombe, C. N.; Beck, G.; Sherborne, B.; Cooper, I. *Eur. J. Med. Chem.* **2003**, *38*, 233–243.
28. Chiou, W. L.; Buehler, P. W. *Pharm. Res.* **2002**, *19*, 868–874.
29. Chiou, W. L.; Jeong, H. Y.; Chung, S. M.; Wu, T. C. *Pharm. Res.* **2000**, *17*, 135–139.
30. Lee, C.-P.; De Vrueh, R. L. A.; Smith, P. L. *Adv. Drug Deliv. Rev.* **1997**, *23*, 47–62.
31. Watanabe, E.; Takahashi, M.; Hayashi, M. *Eur. J. Pharm. Biopharm.* **2004**, *58*, 659–665.
32. Avdeef, A. *Curr. Top. Med. Chem.* **2001**, *1*, 277–351.
33. Avdeef, A.; Testa, B. *Cell. Mol. Life Sci.* **2002**, *59*, 1681–1689.
34. Krämer, S. D. *Pharm. Sci. Technol. Today* **1999**, *2*, 373–380.
35. Van de Waterbeemd, H. *Curr. Opin. Drug Disc. Devel.* **2002**, *5*, 33–43.
36. Amidon, G. L.; Lennernäs, H.; Shah, V. P.; Crison, J. R. *Pharm. Res.* **1995**, *12*, 413–420.
37. Varma, M. V. S.; Khandavilli, S.; Askokraj, Y.; Jain, A.; Dhanikula, A.; Sood, A.; Thomas, N. S.; Pillai, O.; Sharma, P.; Gandhi, R. et al. *Curr. Drug Metab.* **2004**, *5*, 375–388.
38. Bergström, C. A. S.; Strafford, M.; Lazorova, L.; Avdeef, A.; Luthman, K.; Artursson, P. *J. Med. Chem.* **2003**, *46*, 558–570.
39. Van de Waterbeemd, H. *Eur. J. Pharm. Sci.* **1998**, *6*, 1–3.
40. Dressman, J. B.; Amidon, G. L.; Fleisher, D. *J. Pharm. Sci.* **1985**, *74*, 588–589.
41. Curatolo, W. *Pharm. Sci. Technol. Today* **1998**, *1*, 387–393.
42. Johnson, K.; Swindell, A. *Pharm. Res.* **1996**, *13*, 1795–1798.
43. Breitkreutz, J. *Pharm. Res.* **1998**, *15*, 1370–1375.
44. Sanghvi, T.; Ni, N.; Mayersohn, M.; Yalkowsky, S. H. *QSAR Comb. Sci.* **2003**, *22*, 247–257.
45. Zhao, Y. H.; Abraham, M. H.; Le, J.; Hersey, A.; Luscombe, C. N.; Beck, G.; Sherborne, B.; Cooper, I. *Pharm. Res.* **2002**, *19*, 1446–1457.
46. Lipinski, C. *J. Pharmacol. Toxicol. Methods* **2000**, *44*, 235–249.
47. Balimane, P. V.; Chong, S.; Morrison, R. A. *J. Pharmacol. Toxicol. Methods* **2000**, *44*, 301–312.
48. Kim, K.; Cho, S.; Park, J. H.; Byon, Y.; Chung, H.; Kwon, I. C.; Jeong, S. Y. *Pharm. Res.* **2004**, *21*, 1233–1238.
49. Frostell-Karlsson, A.; Widegren, H.; Green, C. E.; Hämäläinen, M. D.; Westerlund, L.; Karlsson, R.; Fenner, K.; Van de Waterbeemd, H. *J. Pharm. Sci.* **2005**, *94*, 25–37.
50. Suomalainen, P.; Johans, C.; Soderlund, T.; Kinnunen, P. K. *J. Med. Chem.* **2004**, *47*, 1783–1788.
51. Loidl-Stadelhofen, A.; Eckert, A.; Hartmann, T.; Schöttner, M. *J. Pharm. Sci.* **2001**, *90*, 597–604.
52. Camenisch, G.; Folkers, G.; Van de Waterbeemd, H. *Int. J. Pharma.* **1997**, *147*, 61–70.
53. Kansy, M.; Senner, F.; Gubernator, K. *J. Med. Chem.* **1998**, *41*, 1007–1010.
54. Avdeef, A. *Absorption and Drug Development*; John Wiley: Hoboken, Germany, 2003.
55. Kansy, M.; Avdeef, A.; Fischer, H. *Drug Disc. Today* **2004**, *1*, 349–355.
56. Hwang, K.-K.; Martin, N. E.; Jiang, L.; Zhu, C. *J. Pharm. Pharma. Sci.* **2003**, *6*, 315–320.
57. Box, K.; Comer, J.; Huque, F. Correlations between PAMPA Permeability and log P. In *Biological and Physicochemical Profiling in Drug Research*; Testa, B., Krämer, S. D., Wunderli-Allensbach, H., Folkers, G., Eds.; VHCA: Zurich, Germany; Wiley-VCH: Weinheim, Germany, 2006, pp 243–257.
58. Artursson, P.; Palm, K.; Luthman, K. *Adv. Drug Deliv. Rev.* **2001**, *46*, 27–43.
59. Ungell, A.-L.; Karlsson, J. Cell Cultures in Drug Discovery: An Industrial Perspective. In *Drug Bioavailability*; Van de Waterbeemd, H., Lennernäs, H., Artursson, P., Eds.; Wiley-VCH: Weinheim, Germany, 2003; Vol. 18, Chapter 5, pp 90–131
60. Stewart, B. H.; Chan, O. H.; Jezyk, N.; Fleisher, D. *Adv. Drug Deliv. Rev.* **1997**, *23*, 27–45.
61. Van de Waterbeemd, H.; Smith, D. A. Relations of Molecular Properties with Drug Disposition: The Cases of Gastrointestinal Absorption and Brain Penetration. In *Pharmacokinetic Optimization in Drug Research; Biological, Physicochemical and Computational Strategies*; Testa, B., Van de Waterbeemd, H., Folkers, G., Guy, R., Eds.; HCA: Basel, Switzerland, 2001, p 5164.
62. Yamashita, S.; Yoshida, M.; Taki, Y.; Sakane, Y.; Nadai, T. *Pharm. Res.* **1994**, *11*, 1646–1651.
63. Polli, J. E.; Ginski, M. J. *Pharm. Res.* **1998**, *15*, 47–52.
64. Johnson, M. D.; Anderson, B. D. *J. Pharm. Sci.* **1999**, *88*, 620–625.
65. Sugawara, M.; Takekuma, Y.; Yamada, H.; Kobayashi, M.; Iseki, K.; Myazaki, K. *J. Pharm. Sci.* **1998**, *87*, 960–966.
66. Bermejo, M.; Merino, V.; Garrigues, T. M.; Pla Delfina, J. M.; Mulet, A.; Vizet, P.; Trouiller, G.; Mercier, C. *J. Pharm. Sci.* **1999**, *88*, 398–405.
67. Merino, V.; Freixas, J.; Bermejo, M.; Garrigues, T. M.; Moreno, J.; Plá-Delfina, J. M. *J. Pharm. Sci.* **1996**, *84*, 777–782.
68. Yu, L. X.; Amidon, G. L. *Int. J. Pharm.* **1999**, *186*, 119–125.
69. Yu, L. X.; Gatlin, L.; Amidon, G. L. *Drugs Pharm. Sci.* **2000**, *102*, 377–409.
70. Agoram, B.; Woltosz, W. S.; Bolger, M. B. *Adv. Drug Deliv. Rev.* **2001**, *50*, S41–S67.
71. Parrott, N.; Lavé, T. *Eur. J. Pharm. Sci.* **2002**, *17*, 51–61.
72. Tam, D.; Tirona, R. G.; Pang, K. S. *Drug Metab. Dispos.* **2003**, *31*, 373–383.
73. Grass, G. M.; Sinko, P. J. *Adv. Drug Deliv. Rev.* **2002**, *54*, 433–451.

74. Leahy, D. E. *Drug Disc. Today* **2004**, *2*, 78–84.
75. Willmann, S.; Lippert, J.; Severstre, M.; Solodenk, J.; Schmitt, W. *BioSilico* **2003**, *1*, 121–124.
76. Willmann, S.; Schmitt, W.; Keldenich, J.; Lippert, J.; Dressmann, J. B. *J. Med. Chem.* **2004**, *47*, 4022–4031.
77. Schmitt, W.; Willmann, S. *Drug Disc. Today* **2004**, *1*, 449–456.
78. Dickins, M.; Van de Waterbeemd, H. *BioSilico* **2004**, *2*, 38–45.
79. Bai, J. P. F.; Utis, A.; Crippen, G.; He, H.-D.; Fischer, V.; Tullman, R.; Yin, H.-Q.; Hsu, C.-P.; Jiang, L.; Hwang, K.-K. J. *Chem. Inf. Comput. Sci.* **2004**, *44*, 2061–2069.
80. Lennernäs, H.; Knutson, L.; Hussain, A.; Tesko, L; Salmonson, T.; Amidon, G. *Eur. J. Pharm. Sci.* **2002**, *15*, 271–277.
81. Bravi, G.; Wikel, J. H. *Quant. Struct.-Act. Relat.* **2000**, *19*, 39–49.
82. Cruciani, G.; Crivori, P.; Carrupt, P.-A.; Testa, B. *Theochem* **2000**, *503*, 17–30.
83. Testa, B.; Crivori, P.; Reist, M.; Carrupt, P.-A. *Perspect. Drug Disc. Des.* **2000**, *19*, 179–211.
84. Goodwin, J. T.; Conradi, R. A.; Ho, N. F. H.; Burton, P. S. *J. Med. Chem.* **2001**, *44*, 3721–3729.
85. Buchwald, P.; Bodor, N. *Curr. Med. Chem.* **1998**, *5*, 353–380.
86. Carrupt, P. A.; Testa, B.; Gaillard, P. *Rev. Comput. Chem.* **1997**, *11*, 241–315.
87. Leo, A. J.; Hoekman, D. *Perspect. Drug Disc. Des.* **2000**, *18*, 19–38.
88. Van de Waterbeemd, H.; Kansy, M. *Chimia* **1992**, *46*, 299–303.
89. Palm, K.; Luthman, K.; Ungell, A. L.; Strandlund, G.; Beigi, F.; Lundahl, P.; Artursson, P. *J. Med. Chem.* **1998**, *41*, 5382–5392.
90. Clark, D. E. *J. Pharm. Sci.* **1999**, *88*, 807–814.
91. Österberg, Th.; Norinder, U. *J. Chem. Inf. Comput. Sci.* **2000**, *40*, 1408–1411.
92. Ertl, P.; Rohde, B.; Selzer, P. *J. Med. Chem.* **2000**, *43*, 3714–3717.
93. Stenberg, P.; Norinder, U.; Luthman, K.; Artursson, P. *J. Med. Chem.* **2001**, *44*, 1927–1937.
94. Saunders, R. A.; Platts, J. A. *New J. Chem.* **2004**, *28*, 166–172.
95. Hou, T. J.; Zhang, W.; Xia, K.; Qiao, X. B.; Xu, X. J. *J. Chem. Inf. Comput. Sci.* **2004**, *44*, 1585–1600.
96. Raevsky, O. A.; Schaper, K.-J. *Eur. J. Med. Chem.* **1998**, *33*, 799–807.
97. Stenberg, P.; Luthman, K.; Ellens, H.; Lee, C. P.; Smith, P. L.; Lago, A.; Elliott, J. D.; Artursson, P. *Pharm. Res.* **1999**, *16*, 1520–1526.
98. Rey, S.; Caron, G.; Ermondi, G.; Gaillard, P.; Pagliara, A.; Carrupt, P.-A.; Testa, B. *J. Mol. Graph. Mol.* **2001**, *19*, 521–535.
99. Fischer, H.; Gottschlich, R.; Seelig, A. *J. Membr. Biol.* **1998**, *165*, 201–211.
100. Fischer, H. PhD Thesis, University of Basel, Switzerland, 1998.
101. Norinder, U.; Österberg, Th. *J. Pharm. Sci.* **2001**, *90*, 1076–1985.
102. Votano, J. R.; Parham, M.; Hall, L. H.; Kier, L. B. *Mol. Divers.* **2004**, *8*, 379–391.
103. Hall, L. H.; Hall, L. M. *SAR QSAR Environ. Res.* **2005**, *16*, 13–41.
104. Butina, D. *Molecules* **2004**, *9*, 1004–1009.
105. Österberg, Th.; Norinder, U. *Eur. J. Pharm. Sci.* **2001**, *12*, 327–337.
106. Veber, D. F.; Johnson, S. R.; Cheng, H.-Y.; Smith, B. R.; Ward, K. W.; Kopple, K. D. *J. Med. Chem.* **2002**, *45*, 2615.
107. Lu, J. J.; Crimin, K.; Goodwin, J. t.; Crivori, P.; Orrenius, C.; Xing, L.; Tandler, P. J.; Vidmar, T. J.; Amore, B. M.; Wilson, A. G. E. et al. *J. Med. Chem.* **2004**, *47*, 6104–6107.
108. Ghuloum, A. M.; Sage, C. R.; Jain, A. N. *J. Med. Chem.* **1999**, *42*, 1739–1748.
109. Van de Waterbeemd, H.; Gifford, E. *Nature Rev. Drug Disc.* **2003**, *2*, 192–204.
110. Yamashita, F.; Hashida, M. *Drug Metab. Pharmacokin.* **2004**, *19*, 327–338.
111. Clark, D. E.; Grootenhuis, P. D. J. *Curr. Top. Med. Chem.* **2003**, *3*, 1193–1203.
112. Egan, W. J.; Lauri, G. *Adv. Drug Deliv. Rev.* **2002**, *54*, 273–289.
113. Van de Waterbeemd, H. Quantitative Structure–Absorption Relationships. In *Pharmacokinetic Optimization in Drug Research; Biological, Physicochemical and Computational Strategies*; Testa, B., Van de Waterbeemd, H., Folkers, G., Guy, R., Eds.; HCA: Basel, Switzerland, 2001, pp 499–511.
114. Clark, D. E.; Grootenhuis, P. D. J. *Curr. Opin. Drug Disc. Devel.* **2002**, *5*, 382–390.
115. Stenberg, P.; Luthman, K.; Artursson, P. *J. Control. Release* **2000**, *65*, 231–243.
116. Stenberg, P.; Bergström, C. A. S.; Luthman, K.; Artursson, P. *Clin. Pharmacokinet.* **2002**, *41*, 877–899.
117. Van de Waterbeemd, H.; Jones, B. C. *Progr. Med. Chem.* **2003**, *41*, 1–59.
118. Boobis, A. R.; Gundert-Remy, U.; Kremers, P.; Machers, P.; Pelkonen, O. *Eur. J. Pharm. Sci.* **2002**, *17*, 183–193.
119. Fröhlich, H.; Wegner, J. K.; Zell, A. *QSAR Comb. Sci.* **2004**, *23*, 311–318.
120. Wegner, J. K.; Fröhlich, H.; Zell, A. *J. Chem. Inf. Comput. Sci.* **2004**, *44*, 931–939.
121. Van de Waterbeemd, H.; Rose, S. Quantitative Approaches to Structure–Activity Relationships. In *The Practice of Medicinal Chemistry*, 2nd ed.; Wermuth, C. G., Ed.; Academic Press: London, UK, 2003, pp 351–369.
122. Livingstone, D. *Data Analysis for Chemists.*; Oxford Science Publications: Oxford, UK, 1995.
123. Camenisch, G.; Alsenz, J.; Van de Waterbeemd, H.; Folkers, G. *Eur. J. Pharm. Sci.* **1998**, *6*, 313–319.
124. Seiler, P. *Eur. J. Med. Chem.* **1974**, *9*, 473–479.
125. Young, R. C.; Mitchell, R. C.; Brown, Th. H.; Ganellin, C. R.; Griffiths, R.; Jones, M.; Rana, K. K.; Saunders, D.; Smith, I. R.; Sore, N. E. et al. *J. Med. Chem.* **1988**, *31*, 656–671.
126. Von Geldern, T. W.; Hoffmann, D. J.; Kester, J. A.; Nellans, H. N.; Dayton, B. D.; Calzadilla, S. V.; Marsch, K. C.; Hernandez, L.; Chiou, W.; Dixon, D. B. et al. *J. Med. Chem.* **1996**, *39*, 982–991.
127. Barlow, D.; Satoh, T. *J. Control. Release* **1994**, *29*, 283–291.
128. Clark, D. E. *J. Pharm. Sci.* **1999**, *88*, 815–821.
129. Kelder, J.; Grootenhuis, P. D. J.; Bayada, D. M.; Delbressine, L. P. C.; Ploemen, J.-P. *Pharm. Res.* **1999**, *16*, 1514–1519.
130. Egan, W. J.; Merz, K. M.; Baldwin, J. J. *J. Med. Chem.* **2000**, *43*, 3867–3877.
131. Stenberg, P.; Luthman, K.; Artursson, P. *Pharm. Res.* **1999**, *16*, 205–212.
132. Deretey, E.; Feher, M.; Schmidt, J. M. *Quant. Struct.-Act. Relat.* **2002**, *21*, 493–506.
133. Abraham, M.; Zhao, Y. H.; Hersey, A.; Luscombe, C. N.; Reynolds, D. P.; Beck, G.; Sherborne, B.; Cooper, I. *Eur. J. Med. Chem.* **2002**, *37*, 595–605.
134. Abraham, M. H.; Chadha, H. S. Applications of a Solvation Equation to Drug Transport Properties. In *Lipophilicity in Drug Action and Toxicology*; Pliska, V., Testa, B., Van de Waterbeemd, H., Eds.; VCH: Weinheim, Germany, 1996, pp 311–337.

135. Wessel, M. D.; Jurs, P. C.; Tolan, J. W.; Muskal, S. M. *J. Chem. Inf. Comput. Sci* **1998**, *38*, 726–735.
136. Norinder, U.; Österberg, Th.; Artursson, P. *Eur. J. Pharm. Sci.* **1999**, *8*, 49–56.
137. Guba, W.; Cruciani, G. Molecular Field-Derived Descriptors for the Multivariate Modeling of Pharmacokinetic Data. In *Molecular Modelling and Prediction of Bioreactivity*; Gundertofte, K., Jorgensen, F. S., Eds.; Plenum Press: New York, NY, 2000, pp 89–94.
138. Holm, R.; Hoest, J. *Int. J. Pharm.* **2004**, *272*, 189–193.
139. Raevsky, O. A.; Fetisov, V. I.; Trepalina, E. P.; McFarland, J. W.; Schaper, K. J. *Quant. Struct.-Act. Relat.* **2000**, *19*, 366–374.
140. Van de Waterbeemd, H.; Camenisch, G.; Folkers, G.; Raevsky, O. A. *Quant. Struct.-Act. Relat.* **1996**, *15*, 480–490.
141. Raevsky, O. A. *Mini-Rev. Med. Chem.* **2004**, *4*, 1041–1052.
142. Oprea, T. J.; Gottfries, J. *J. Mol. Graphics. Mod.* **2000**, *17*, 261–274.
143. Raevsky, O. A.; Schaper, K.-J.; Artursson, P.; McFarland, J. W. *Quant. Struct.-Act. Relat.* **2002**, *20*, 402–413.
144. Raevsky, O. A.; Raevsky, O. A. *SAR QSAR Environ. Res.* **2001**, *12*, 367–381.
145. Klopman, G.; Stefan, L. R.; Saiakhov, R. D. *Eur. J. Pharm. Sci.* **2002**, *17*, 253–263.
146. Sun, H. *J. Chem. Inf. Comput. Sci.* **2004**, *44*, 748–757.
147. Segarra, V.; López, M.; Ryder, H.; Placios, J. M. *Quant. Struct.-Act. Relat.* **1999**, *18*, 474–481.
148. Gohlke, H.; Dullweber, F.; Kamm, W.; März, J.; Kissel, Th.; Klebe, G. Prediction of Human Intestinal Absorption Using a Combined Simulated Annealing/Backpropagation Neural Network Approach. In *Rational Approaches to Drug Design*; Höltje, H. D., Sippl, W., Eds.; Prous Science: Barcelona, Spain, 2001, pp 261–270.
149. Agatonovic-Kustrin, S.; Beresford, R.; Yusof, A. P. M. *J. Pharm. Biomed. Anal.* **2001**, *25*, 227–237.
150. Niwa, T. *J. Chem. Inf. Comput. Sci.* **2003**, *43*, 113–119.
151. Dorronsoro, I.; Chana, A.; Abasolo, M. I.; Castro, A.; Gil, C.; Stud, M.; Martinez, A. *QSAR Comb. Sci.* **2004**, *23*, 89–98.
152. Zmuidinavicius, D.; Didziapetris, R.; Japertas, P.; Avdeef, A.; Petrauskas, A. *J. Pharm. Sci.* **2003**, *92*, 621–633.
153. Winiwarter, S.; Bonham, N. M; Ax, F.; Hallberg, A.; Lennernäs, H.; Karlén, A. *J. Med. Chem.* **1998**, *41*, 4939–4949.
154. Winiwarter, S.; Ax, F.; Lennernäs, H.; Hallberg, A.; Pettersson, C.; Karlén, A. *J. Mol. Graph. Mod.* **2003**, *21*, 273–287.
155. Ren, S.; Lien, E. J. *Prog. Drug Res.* **2000**, *54*, 1–23.
156. Artursson, P.; Karlsson, J. *Biochem. Biophys. Res. Commun.* **1991**, *175*, 880–885.
157. Yee, S. *Pharm. Res.* **1997**, *14*, 763–766.
158. Yazdanian, M.; Glynn, S. L.; Wright, J. L.; Hawi, A. *Pharm. Res.* **1998**, *15*, 1490–1494.
159. Ponce, Y. M.; Cabrera Perez, M. A.; Zaldivar, V. R.; Diaz, H. G.; Torrens, F. J. *Pharm. Pharm. Sci.* **2004**, *7*, 186–199.
160. Norinder, U.; Österberg, Th.; Artursson, P. *Pharm. Res.* **1997**, *14*, 1785–1791.
161. Tantishaiyakul, V. *Pharmazie* **2001**, *56*, 407–411.
162. Nordqvist, A.; Nilsson, J.; Lindmark, T.; Eriksson, A.; Garberg, P.; Kihlén, M. *QSAR Comb. Sci.* **2004**, *23*, 303–310.
163. Yamashita, F.; Wanchana, S.; Hashida, M. *J. Pharm. Sci.* **2002**, *91*, 2230–2239.
164. Ekins, S.; Waller, C. L.; Swaan, P. W.; Cruciani, G.; Wrighton, S. A.; Wikel, J. H. *J. Pharmacol. Toxicol. Methods* **2000**, *44*, 251–272.
165. Kulkarni, A.; Han, Y.; Hopfinger, J. *J. Chem. Inf. Comput. Sci.* **2002**, *42*, 331–342.
166. Ekins, S.; Durst, G. L.; Stratford, R. E.; Thorner, D. A.; Lewis, R.; Loncharich, R. J.; Wikel, J. *J. Chem. Inf. Comput. Sci.* **2001**, *41*, 1578–1586.
167. Refsgaard, H. H. F.; Jensen, B. F.; Brockhoff, P. B.; Padkjaer, S. B.; Guldbrandt, M.; Christensen, M. S. *J. Med. Chem.* **2005**, *48*, 805–811.
168. Yamashita, F.; Fujiwara, S.; Hashida, M. *J. Chem. Inf. Comput. Sci.* **2002**, *42*, 408–413.
169. Huque, F. T. T.; Box, K.; Platts, J. A.; Comer, J. *Eur. J. Pharm. Sci.* **2004**, *23*, 223–232.
170. Box, K.; Comer, J. Personal Communication.
171. O'Brien, S. E.; De Groot, M. J. *J. Med. Chem.* **2005**, *48*, 1287–1291.
172. Tong, W.; Hong, H.; Fang, H.; Xie, Q; Perkins, R. *J. Chem. Inf. Comp. Sci.* **2003**, *43*, 525–531.
173. Van de Waterbeemd, H. Property-Based Lead Optimization. In *Biological and Physicochemical Profiling in Drug Research*; Testa, B., Krämer, S. D., Wunderli-Allenspach, H., Folkers, G., Eds.; VHCA: Zurich, Switzerland; Wiley-VCH: Weinheim, Germany, 2006, pp 25–45.

Biography

Han Van de Waterbeemd studied physical organic chemistry at the Technical University of Eindhoven, The Netherlands, and did a PhD in medicinal chemistry at the University of Leiden, The Netherlands. After post-doctoral research with Bernard Testa at the School of Pharmacy of the University of Lausanne, Switzerland, he held a 5 year faculty position at the same institution. He also taught medicinal chemistry to pharmacy students at the universities of Berne and Basel in Switzerland from 1987 to 1997. In 1988 he joined F Hoffmann-La Roche Ltd in Basel as head of the Molecular Properties Group. He then moved in 1997 to Pfizer Central Research UK, later Pfizer Global Research and Development, and held various positions in the Department of Drug Metabolism, later called PDM (pharmacokinetics, dynamics and metabolism), including Head of Discovery, and Head of Automation and In Silico ADME Technologies. Van de Waterbeemd has published more than 135 peer-reviewed papers and book chapters, and co-edited 11 books. His research interests include physicochemical and structural molecular properties and their role in drug disposition, as well as the in silico modeling of ADMET properties. He was secretary of The QSAR and Modelling Society from 1995 to 2005.

5.29 In Silico Prediction of Oral Bioavailability

J V Turner, The University of Queensland, Brisbane, Qld, Australia
S Agatonovic-Kustrin, The University of Western Australia, Perth, WA, Australia

5.29.1 Introduction

Oral dosing is the preferred route of administration for most drugs due to ease of administration and patient compliance. Orally delivered pharmacologically active compounds must have favorable absorption and clearance properties and satisfactory metabolic stability to provide adequate systemic exposure to elicit a pharmacodynamic response. Once ingested, the drug must undergo absorption, distribution, and hepatic metabolism prior to reaching the systemic circulation and becoming bioavailable (*see* 5.02 Clinical Pharmacokinetic Criteria for Drug Research). A bioavailable compound is then able to bind to target receptors and exert a pharmacological effect. Poor bioavailability limits oral administration of drugs, while high clearance limits desirable dosing regimens.

In order to assess oral pharmacokinetic parameters such as bioavailability, it is necessary to administer the drug both intravenously and by the oral route. Bioavailability is then extracted from the combined intravenous and oral concentration–time data. Due to practical, ethical, legal, and safety constraints, pharmacokinetic characterization is often attended to late in the developmental process. Clearly, if pharmacokinetic input is to be available early in drug discovery and development, more viable predictive methodologies are necessary.

The quantitative structure–activity relationship (QSAR) approach developed last century[1] has recently been adapted to construct quantitative structure–pharmacokinetic relationships (QSPkRs).[2] In addition to exploring physicochemical parameters including partitioning, solubility, and charge characteristics, increasing attention is being paid to the extraction of useful information from the molecular structure of a compound for QSPkRs.[3]

QSPkR modeling of bioavailability must take into consideration structural influences on absorption and metabolism of a compound, as well as the variability inherent in human/biological systems. Various computational techniques have been developed, some of which utilize experimental data for model generation while others rely on theoretically derived parameters including those that relate to molecular physicochemical properties. In silico methods depend heavily on the nature and quality of data used to construct models. Model validity, therefore, is influenced by issues of data variability, diversity, and experimental quality. These concepts and the theory behind them will now be discussed in relation to the current status of in silico bioavailability modeling.

5.29.2   Theoretical and Biological Basis of Bioavailability Prediction

5.29.2.1   Drug-Likeness

Druglikeness may be defined as a complex balance of various molecular properties and structural features which determine whether a particular molecule is drug or nondrug. These properties, mainly lipophilicity, electronic distribution, hydrogen bonding characteristics, molecule size and flexibility, and presence of various pharmacophoric features influence the behavior of a molecule in a living organism, including characteristics such as transport, affinity to proteins, reactivity, toxicity, metabolic stability, and many others. Extensive work has been performed to examine known druglike and nondruglike molecules and to assess what structural motifs are associated with druglike molecules (these concepts are explored in more detail in Volume 4 of this series).

Although absorption and bioavailability are separate parameters, information about absorption can be of benefit for bioavailability modeling. One way to screen out compounds with probable absorption problems is known as Lipinski's 'rule of five.'[4] The rule states that poor absorption or permeation of a drug is more probable when the chemical structure fulfils two or more of the following criteria:

1. Molecular weight (MW) is greater than 500.
2. The calculated log P value is above five.
3. There are more than five hydrogen bond donors (–NH–, –OH).
4. The number of hydrogen bond acceptors (–N $=$, –O–) is greater than ten.

It is important to note that the rule of five does not estimate oral absorption in a quantitative manner, nor are compounds that do not break any of the rule of five criteria necessarily orally bioavailable. The rule of five does not definitively categorize all well and poorly absorbed compounds, although it is simple, fast, and provides a reasonable degree of classification (see 5.28 In Silico Models to Predict Oral Absorption). It follows then, that there is a clear need for quantitative models able to screen large chemical databases to identify new chemical entities (NCEs) with good solubility and oral absorption. Although well-absorbed compounds can have low bioavailability due to high presystemic clearance, low bioavailability for marketed drugs is more common with oral dosage forms of poorly water-soluble, slowly absorbed compounds.

5.29.2.2   Pharmacokinetic Properties

5.29.2.2.1   Bioavailability versus absorption

Oral bioavailability is particularly important for lead optimization considering the importance of developing orally administered commercial products. The absolute oral bioavailability (F) of a particular compound is defined as the fraction of that compound reaching the bloodstream after oral administration. The terms absorption and bioavailability are often incorrectly and interchangeably used. The confusion occurs when attempts are made to correlate oral bioavailability with intestinal permeation. Oral absorption can be defined as the amount of drug that passes through the intestinal tissue and enters into the portal vein. In order for a drug to be orally bioavailable, it must survive:

1. Gastrointestinal absorption and metabolism.
2. Blood metabolism (mainly hydrolytic).
3. Hepatic clearance (metabolism and biliary excretion) before reaching the general circulation.[5]

Oral bioavailability can be regarded as a superposition of the major processes of absorption in the gastrointestinal tract and first-pass metabolism in the liver. The percentage bioavailability should be determined primarily by both absorption and metabolism, and any attempt to model bioavailability would require that both properties be taken into account (*see* 5.03 In Vivo Absorption, Distribution, Metabolism, and Excretion Studies in Discovery and Development; 5.05 Principles of Drug Metabolism 1: Redox Reactions; 5.06 Principles of Drug Metabolism 2: Hydrolysis and Conjugation Reactions; 5.07 Principles of Drug Metabolism 3: Enzymes and Tissues).

A common method of determining bioavailability is by using area under the curve (AUC) calculations for oral and intravenous doses:

$$F = \frac{\text{AUC}_{\text{oral}} \times \text{dose}_{\text{IV}}}{\text{AUC}_{\text{IV}} \times \text{dose}_{\text{oral}}} \qquad [1]$$

Other methods include amount of a drug absorbed into the bloodstream and urinary excretion studies or cumulative excretion of drug following oral and intravenous administration.[6]

5.29.2.2.2 Factors implicated in bioavailability

There are multiple factors affecting oral bioavailability, including disintegration of the dosage form, dissolution of the drug molecule, metabolism in the intestinal tract or mucosa, permeation across the intestinal mucosa, and hepatic first-pass metabolism and/or drug elimination (**Figure 1**). Most of these effects are class-specific and cannot be described collectively by known physicochemical parameters, adding to the difficulty of bioavailability modeling. Physicochemical properties currently utilized in drug discovery programs include solubility, lipophilicity, molecular size, hydrogen

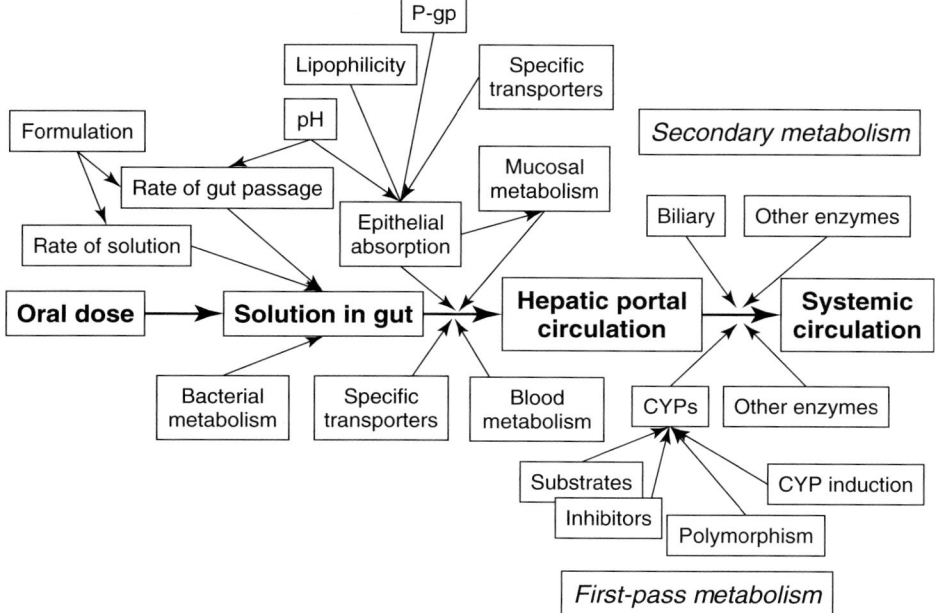

Figure 1 Schematic of some of the logical relationships between components of an oral bioavailability calculation. The model can be divided into pharmaceutics (formulation), gut physiology and hepatic metabolism (first-pass clearance). An arrow indicates a logical or algorithmic dependence: the model is not complete. CYP, cytochrome P450; P-gp, P-glycoprotein. (Reproduced with permission from Bains, W.; Gilbert, R.; Sviridenko, L.; Gascon, J. M.; Scoffin, R.; Birchall, K.; Hetherington, L.; Harvey, I.; Caldwell, J. *Curr. Opin. Drug Disc. Dev.* **2005**, 5, 44–51 © The Thomson Corporation and Current Opinion in Drug Discovery & Development.)

bonding capacity, and charge, all of which are considered important molecular features relating to various aspects of drug disposition.[7,8] However, there are no universal rules for their application nor means to foretell which the dominant property in any given situation will be. The search for better molecular descriptors has not decreased the use of established descriptors, nor have recently developed theoretical descriptors been found applicable to all areas of computational pharmacology and pharmaceutics.[9] The current state of molecular descriptor use is reviewed elsewhere (*see* 5.22 Use of Molecular Descriptors for Absorption, Distribution, Metabolism, and Excretion Predictions).

In order to be bioavailable, oral drugs must comply with the following requirements:

1. Solubility (dissolve in the stomach or intestine under variable pH: $2 < pH < 8$).
2. Chemical stability (withstand acid hydrolysis in stomach at $pH < 2$).
3. Permeability (permeate through intestinal membrane by paracellular, transcellular, or active transport).
4. Withstand P-glycoprotein (P-gp) efflux (backward transport through intestinal membrane).
5. Withstand first-pass metabolism in liver and gut wall (susceptibility to metabolic transformations catalyzed by enzymes in liver and intestine).

The gastrointestinal tract can be seen as both a physical and biochemical barrier to oral drug absorption. Absorption across the intestinal epithelium may be passive or active. Passive transport is governed by physicochemical properties whereas active transport involves specific binding of a molecule to a binding site on a transport protein. The former can be modeled with physicochemical and structural properties, whereas modeling the properties of biological membranes still remains a major challenge. This is due to secretory intestinal processes, such as P-gp transport, and gut wall metabolism including by the cytochrome P450 (CYP) isoform 3A4 (CYP3A4).[10] Further information on these systems can be found in elsewhere (*see* 5.04 The Biology and Function of Transporters; 5.07 Principles of Drug Metabolism 3: Enzymes and Tissues).

5.29.2.3   Absorption

Intestinal drug absorption, controlled by a number of factors, can be represented by various molecular descriptors (**Table 1**). These factors include: (1) dissolution rate and solubility, determining how fast a drug reaches a maximum concentration in the luminal intestinal fluid; and (2) permeability, which relates to the rate at which a dissolved drug will cross the intestinal wall to reach the portal blood circulation. Following oral administration, the drug must be capable of moving across cell membranes (transcellular absorption) or between tight gaps in between cells of the gastrointestinal mucosa (paracellular absorption). Both transcellular and paracellular diffusion depend upon the properties of drug molecules, such as molecular size, polarity, and lipophilicity.[11] For most drugs, the predominant mechanism of absorption is via the transcellular route (*see* 5.28 In Silico Models to Predict Oral Absorption).

Determination of the dissolution, solubility at different pH values, and permeability properties of drug candidates can thus provide information about their absorption potential. This also allows evaluation of the compounds according to the biopharmaceutics classification system (BCS) (*see* 5.42 The Biopharmaceutics Classification System) which distinguishes four classes relating to various degrees of absorption potential (**Table 2**).[12]

Since drug absorption potential has become a more important criterion for decisions about candidate progression early in the drug discovery process, screening methods are needed to assess drug dissolution and permeability. In silico modeling of ionization, solubility, lipophilicity, and absorption are covered elsewhere (*see* 5.25 In Silico Prediction of Ionization; 5.26 In Silico Predictions of Solubility; 5.27 Rule-Based Systems to Predict Lipophilicity; 5.28 In Silico Models to Predict Oral Absorption, respectively).

Table 1   Factors influencing absorption

Characteristic	*Sample descriptors*
Solubility	Solubility parameters, hydrophilic–lipophilic balance, hydrophilic surface area
Polarity	Dipole moment, polar surface area
Lipophilicity	log P, log D
Molecular size	Molecular mass, molar volume
Molecular shape	Molecular width, length, depth, connectivity indices

Table 2 Biopharmaceutics classification system (BCS) classification matrix

	High solubility	*Low solubility*
High permeability	Class I	Class II
Low permeability	Class III	Class IV

Different carriers and transporters have also been implicated in drug passage across the intestinal mucosa. P-gp is a membrane transporter which functions as a drug efflux pump with a broad spectrum of substrates. P-gp is located on the apical side of the intestinal membrane and is involved in the secretory transport of a wide range of drugs. As well as P-gp itself affecting drug permeability, hydrogen bonding capacity has been implicated in drug efflux since it slows membrane crossing and may therefore increase the affinity of a compound for P-gp.[13]

Other absorption/permeability parameters of interest include membrane permeation measured by hydrocarbon films[14] and or parallel artificial membrane permeability assay (PAMPA)[15] (*see* 5.19 Artificial Membrane Technologies to Assess Transfer and Permeation of Drugs in Drug Discovery), P-gp efflux (*see* 5.32 In Silico Models for Interactions with Transporters) and hepatic cell metabolism,[16] and transport mechanisms.[17] These experimentally derived parameters may provide useful data for in silico predictive models in the future. Further discussion may also be found elsewhere (*see* 5.10 In Vitro Studies of Drug Metabolism; 5.11 Passive Permeability and Active Transport Models for the Prediction of Oral Absorption).

5.29.2.3.1 Effects of formulation and dose

A number of drugs are not administered orally as the free base or acid due to their poor water solubility, stability, hygroscopicity, crystallinity, or purity. These drugs are usually combined with acids or bases to form a salt or formulated with a lipophilic solvent, a hydrophilic solvent, and a surfactant that interact to assist in dispersion and emulsification. The bioavailability of these drugs depends on their formulation.[18]

Another consideration is the dose-dependent absorption of certain drugs: the percentage of the oral dose absorbed and its resultant bioavailability and excretion may decline with increasing dose.[19] For example, low oral bioavailability of propranolol is due in part to first-pass hepatic extraction. Bioavailability of propranolol is dose dependent, most likely due to saturation of the CYP enzymes. On the other hand atenolol is incompletely bioavailable due to incomplete intestinal absorption with no contribution of hepatic first-pass metabolism.[20] It is difficult to give a definition of such dose dependency since many factors can affect solubility in intestinal fluid and absorption in humans and, therefore, bioavailability. These include absorption mechanisms,[21] drug formulation, food composition, chemical composition, pH of the intestinal secretions, gastric emptying time, intestinal motility, and blood flow.[22]

5.29.2.3.2 Solubility

Solubility of the drug molecule in the specific biological fluid at the site of absorption is an important prerequisite for passive membrane permeation. Solubility depends on the physicochemical properties of the fluid and the lipophilicity and ionization status of the drug. Knowledge and understanding of the dissociation constants (pK_a values) of a drug may also be helpful for the prediction of its behavior in vivo. The ionization state of a drug compound in a specific body compartment can be calculated by considering its dissociation constants relative to the pH in the body compartment of interest, for example, acidic for stomach, basic for intestine, pH 7.4 for plasma, pH 4.5–8.0 for urine. The ionization state of a drug may affect its solubility, lipophilicity, and suitability for passive transcellular diffusion and/or suitability as a substrate for active transport mechanisms (*see* 5.16 Ionization Constants and Ionization Profiles; 5.17 Dissolution and Solubility); predictive models are discussed elsewhere (*see* 5.25 In Silico Prediction of Ionization; 5.26 In Silico Predictions of Solubility). The pH partition model has long been used to correlate pK_a and partition coefficients with absorption through membranes by passive diffusion.[23]

For compounds that are poorly soluble in the gastrointestinal tract, appropriate pharmaceutical formulations may provide a solution.[24] Formulation design including particle size, co-solvents, and immediate and controlled release formulations, can allow control of release rate into solution[25] and, therefore, rate of transfer into the systemic circulation and bioavailability.

5.29.2.3.3 Lipophilicity

The best-known physicochemical approach to predict membrane permeation is the determination of lipophilicity using the logarithm of the partition coefficient between an octanol and a water phase (log *P*).[26] The membrane/water partition coefficient is, therefore, often assumed to resemble the oil/water partition coefficient. Although relationships between log P values and intestinal drug absorption have been established, log P does not necessarily quantitatively reflect the membrane/water partitioning behavior of a solute.[15] For example, membrane/water partition coefficients for some substituted benzenes and phenols were found to be up to six times higher than octane/water partitioning coefficients,[27] indicating that QSARs based on log P can contain a degree of uncertainty in their predictions. Also, polar compounds with hydrogen donor or acceptor groups partition significantly more readily into membranes than predicted by log P.[28] Relationships between log P and absorption can be drawn, however, by correlation algorithms such as the Collander equation.[29] Lipophilicity is a critical variable implicated in intestinal absorption and membrane permeability in general and has proven useful in predictive bioavailability models.[30] In silico approaches for log P calculation include both atomic/substructure contribution methods[31] and whole model approaches which make use of molecular lipophilic potentials (MLP),[32] topological indices, and other molecular properties such as charge density and electrostatic potentials for log P calculation.[33] Detailed discussion of the theoretical and practical aspects of lipophilicity and log P calculation can be else where found elsewhere (*see* 5.18 Lipophilicity, Polarity, and Hydrophobicity; 5.27 Rule-Based Systems to Predict Lipophilicity).

The log D (octanol/water distribution coefficient) value takes into account the extent of ionization as well as the partitioning of various microspecies for each compound in both aqueous and organic phases. Many metabolic and other pharmacokinetic processes can be related to distribution and partition coefficients.[34] For example, the human intestine at pH 6.5 is where oral absorption predominantly occurs, while distribution can be related to the physiological pH value of 7.4. Log D is thus a powerful descriptor since it contains a correction of the degree of ionization as well as combining aspects of size and ability to form hydrogen bonds.[7]

5.29.2.3.4   Polarity and hydrogen bonding

A molecule that crosses cell membranes via the transcellular pathway needs to penetrate both hydrophilic and hydrophobic regions since cell membranes are composed of a lipid bilayer. As a result, both hydrophilic and lipophilic properties of a molecule should be taken into account when predicting drug permeability. This is encompassed in the hydrophilic-lipophilic balance (HLB) value which is a numerical measure of the overall proportion of mass of a molecule that is hydrophilic. HLB has recently been found to be important in the prediction of oral bioavailability.[35] It is difficult for a drug molecule with a mainly hydrophilic structure to penetrate the outer phospholipid layer of the cell membrane by transcellular diffusion. Hence, the intestinal absorption decreases as the HLB value increases. Amphiphilicity is a more specific parameter than HLB and is determined by the distinction between a polar (hydrophilic) head region and nonpolar (hydrophobic) tail region of a molecule. Amphiphilicity can affect processes such as dissolution and partitioning as well as permeability.[36] It is worthwhile to note that HLB, amphiphilicity, and log P are all ratios and indicate relative rather than absolute values of hydrophilicity and hydrophobicity of a molecule. Such ratios can be further utilized, for example in the calculation of MLP which is a whole-molecule value determined by summation of fragmental values of log P.[32] The balance of molecular surface properties such as polar surface area (PSA) and nonpolar surface area have also been implicated in human intestinal absorption of drug molecules.[37]

Dipole moment and PSA[38] are both measures of the polarity of a molecule. Dipole moment describes the intramolecular electronic effect which may be related to molecular reactivity.[39] PSA has been found to be related to drug permeability such that drugs with high polarity are less likely to be absorbed by the small intestine and have correspondingly low bioavailability.[40] PSA is a surface descriptor, defined as a part of the surface area contributed by nitrogen, oxygen, and connected hydrogen atoms.[41] As such, it is clearly related to the capacity of a drug to form hydrogen bonds and is, therefore, a hydrogen-bonding descriptor. Molecules with many hydrogen bond donors and a large PSA tend to have low permeability values.[42]

5.29.2.3.5   Flexibility

An analysis of the measured oral availability in rats has revealed the unexpected positive influence of increasing molecular rigidity, measured by the rotatable bond count, and the negative impact of increasing PSA.[43] The commonly held association of increased absorption for compounds with molecular weight less than 500 was proposed to result from greater molecular rigidity rather than lower molecular weight per se. It was found that molecules possessing fewer than 10 rotatable bonds and having a PSA less than 140 Å^2 after oral administration generally showed bioavailability in the rat exceeding 20%. These results suggested that such criteria could be used as filters in early discovery to identify drug candidates with minimally acceptable oral bioavailability. These findings, however, were contested in a more recent publication which cautioned the used of such generalizations.[44] It was described that molecular rigidity is a much more complex issue than the simple counting of rotatable bonds, while rotatable bond count has a much less sensitive influence on oral bioavailability than PSA.[44]

5.29.2.3.6 Size and shape

It is well known that the size and shape of a drug can determine its orientation and the rate of movement across biological membranes.[45] Molecular dimensions and shape can be described using molecular width, length, and depth, and using connectivity indices.[46] Compounds with larger molecular width, length, and depth are not well absorbed from the intestinal epithelium due to the large molecular size limiting passage through the paracellular pores. The paracellular route, therefore, restricts passage to mainly small molecules[7] which corresponds to the rule of five indicating good absorption for compounds of MW less than 500.[4] With increasing lipophilicity, the transcellular pathway becomes important but also involves an efflux component which itself can be affected by the rate of membrane transfer.

5.29.2.4 Metabolism

In addition to possible intestinal metabolism by the gut wall enzymes, metabolic reactions in the gastrointestinal tract occur predominantly via CYPs, the major enzyme being CYP3A. More than 50% of drugs may be substrates for CYP3A, potentially resulting in low oral bioavailability due to extensive metabolism in the gastrointestinal tract.[47,48]

CYPs may be induced by a structurally diverse range of xenobiotics including rifampicin, omeprazole, and phenobarbital, as well as cigarette smoking and alcohol.[49] The induction of CYPs leads to two different problems: effects on self, and coadministered drug–drug interactions. On chronic administration ritonavir induces CYP3A leading to a decrease of its own AUC.[50] Regarding the latter effect, an inducible drug can influence the pharmacokinetics of a coadministered drug which shares the same metabolic pathway. For example, rifampicin induces metabolism of ethynylestradiol which can lead to the failure of contraceptive therapy.[49] More sophisticated approaches accounting for drug interactions with transporter proteins[51,52] and metabolizing enzymes are required to enable accurate predictive models to be developed.

It is important to determine not only the stability of a potential drug but also the route of its metabolism. Structural determinants of metabolism, including calculated molecular descriptors, can then be elucidated and incorporated into rational drug design to improve stability and consequent bioavailability and efficacy. Principles of drug metabolism are further discussed elsewhere (see 5.05 Principles of Drug Metabolism 1: Redox Reactions; 5.06 Principles of Drug Metabolism 2: Hydrolysis and Conjugation Reactions; 5.07 Principles of Drug Metabolism 3: Enzymes and Tissues).

5.29.2.5 Impact of Interindividual Variability

Drug concentration is amongst the most important determinants of clinical response to a drug. Variability in pharmacokinetic profiles makes drug concentrations unpredictable: the greater the variability, the greater the magnitude of this problem.

Bioavailability of a drug may vary within the same person over time as well as between different people in a population. The cause of large interpatient pharmacokinetic variability is multifactorial and includes differences in drug absorption, metabolism, or distribution, and complex drug–drug or drug–food interactions.[53] Population subgroups such as infants, pregnant women, and other groups with underlying traits or disease states are also likely to exhibit variable bioavailability and absorption, distribution, metabolism, and excretion (ADME) characteristics.[54] Population variability in clinical data is a reality and should be incorporated into the modeling process.

Although predictive models may be constructed with good accuracy based on a given data set, the quality of such models ultimately reflects the quality of the data: poor-quality bioavailability data with wide uncertainty will result in poor models regardless of apparent accuracy. Care must also be taken when extrapolating outside the limits of model training data.

5.29.2.6 Summary of Determinants of Oral Bioavailability

Theoretical determination of bioavailability relies on derivation of information relating to solubility and permeability which drive passive absorption rates, and metabolic stability including rates and routes of metabolism. The complexity of metabolic routes may necessitate knowledge of information relating to different metabolic pathways for accurate bioavailability prediction. Existing ADME models are structure-based, physicochemical property-based, or incorporate a mixture of both structural and physicochemical descriptors. Numerous descriptors implicated in bioavailability have been described but no universally or common descriptor/s have yet been defined. This is not surprising considering the multiplicity of factors that contribute to oral bioavailability.

5.29.3   Data Usage for Bioavailability Prediction

During drug discovery compounds are first screened to find hits, which are then turned into leads from which a clinical candidate hopefully emerges. The challenge is the prediction of human pharmacokinetic characteristics of drugs from preclinical data: accurate prediction would avoid unnecessary clinical evaluation of compounds with poor pharmacokinetic performance in humans.

A number of difficulties should not be overlooked when using computational and physicochemical properties to assess bioavailability. Some relevant properties such as $\log D$ and hydrogen bonding are in fact conformation-dependent.[55] Further progress in the analysis of molecular states including conformation and ionization, and the development of more appropriate descriptors may lead to improved models in the future.

Numerous methods have been used to predict human pharmacokinetics from in vivo animal preclinical pharmacokinetic data[56] including physiologically-based pharmacokinetic (PBPK) (see 5.37 Physiologically-Based Models to Predict Human Pharmacokinetic Parameters). Another well-described technique is allometric scaling.[57]

5.29.3.1   Allometric Scaling for Bioavailability

The word allometry refers to bodies that are not isometric but which change with size according to a particular rule. Allometric scaling is an empirical technique and was developed to explain the observed relationships between organ size and body mass of mammals.[58] Relationships revealed between body size and physiological parameters consequently led to the application of allometric scaling in correlating human pharmacokinetics with pharmacokinetic parameters obtained in animal studies.[59,60] According to allometry a number of body processes can be described in a very simple way:

$$Y = A \times X^k \qquad\qquad [2]$$

where Y is a body process such as metabolic rate, X a measure of the size of the organism, for example body surface area or mass, and A and k are constants. The constant A scales across species and is expected to be different for each species.

Numerous studies have attempted to correlate preclinical pharmacokinetics in animal species to pharmacokinetics in humans.[61] Both absolute bioavailability prediction in humans from animal data using interspecies scaling as well as indirect approaches have been described.[62] However, a comparison of absolute bioavailability prediction in humans using allometric scaling (absolute bioavailability versus body mass) with four indirect interspecies scaling approaches suggested different degrees of accuracy of these methods. It was concluded that accurate prediction of absolute bioavailability in humans from animal data was unreliable. Interspecies scaling for bioavailability prediction, therefore, has not become a popular approach.

The major drawback in allometric scaling is its empirical nature. It is very sensitive to both the species in which testing is performed and how/if extrapolations are made. Traditional allometric scaling introduced a large degree of uncertainty into data extrapolations and did not take into account species differences in physiology that could modify drug uptake, metabolism, or physiological responsiveness. Recent methods have combined allometric scaling with in vitro human metabolism data in order to overcome species differences in metabolism.[63,64] A rough estimate of categorical bioavailability was successful for a small number of compounds for which in vivo data were available[65] despite the fact that only hepatic microsomal metabolism was considered a limitation.

5.29.3.2   In Vivo Bioavailability from In Vitro Data

The methodologies and mathematics behind approaches to predict in vivo clearance from intrinsic clearance data have been summarized in a recent review.[66] In vitro/in vivo correlations that have included data from both human and preclinical animal species have been achieved in practice.[67] A number of in vitro systems can be used to obtain hepatic intrinsic clearance data. The most commonly used systems are liver microsomes, hepatocytes, and cut liver slices. Each system possesses unique advantages and disadvantages. In general, for kinetic experiments, the body of data available suggests that hepatocytes are a superior method with regard to accurate predictions of in vivo data, with microsomes also providing good results.[68–71]

As examples, one graphic model utilized permeability in Caco-2 cells and liver microsomal clearance to estimate oral bioavailability[72] while a different mathematical model integrated Caco-2 permeability and hepatic extraction based on microsomal clearance to project oral bioavailability.[73] It was also demonstrated in a retrospective study that oral bioavailability could be predicted by human liver microsomal clearance only when the hepatic first pass was considered to be the limiting factor.[65]

5.29.3.3   Input Data Quality and Type

Pharmacokinetics presents a number of unique problems for statistical modeling (**Table 3**). Data derived from literature sources, particularly in vitro data, are usually quite reliable from a qualitative point of view but much less reliable from a quantitative perspective. The quality of such experimental bioavailability data varies enormously and is often restricted in quantity and chemical diversity[74] due to economic and ethical considerations. Another limiting factor is the variability of pharmacokinetic results. Traditional pharmacokinetic studies have placed a priority on minimizing variability through careful control over study participants and other aspects of the experiment, and thus obtain an understanding of bioavailability in only a narrowly defined group.

It is also important to distinguish between uncertainty and inherent biological variability. In practice, the distinction is usually less clear, as illustrated by the often large degrees of error present in available bioavailability data. One study pooled literature values of 49 compounds for which a reliable, numerical range of human oral bioavailability was available (**Figure 2**)[75] and found that a large proportion of ranges were wide in both absolute and relative terms. Perhaps the most pressing need at this time is for larger, high-quality experimental data sets to provide a sound basis for model building.[76]

The QSAR approach was initially developed from and applied to sets of congeneric chemicals. Development of successful QSARs require the chemical space for training and testing sets to be analyzed followed by identification and removal of structural outliers. QSAR models will then have an applicability domain centered within the range of parameter values used in modeling. Beyond this range, any relationship will fade in an unpredictable manner.

In contrast, models based on large heterologous chemical data sets are less dependent on structural similarity and therefore place less emphasis on eliminating outliers. Excluding outliers subsequent to model construction may render such models statistically invalid if excluded compounds were utilized during model construction or training.[75] Removal of outliers then retraining of predictive models may avoid this problem but introduces another problem of further reducing the size of the data set and, therefore, the chemical space and extrapolative capability of the final model. Small data sets, especially those composed of congeneric compounds, often require the use of leave-*n*-out

Table 3   Problems with human bioavailability data

Within studies	*Between studies*
Limited number of compounds	Quality of data/study: methodology and analysis
Limited diversity of compounds	Difference in experimental conditions
Narrow range of human subjects	Quantitative versus qualitative data
	Variability between studies

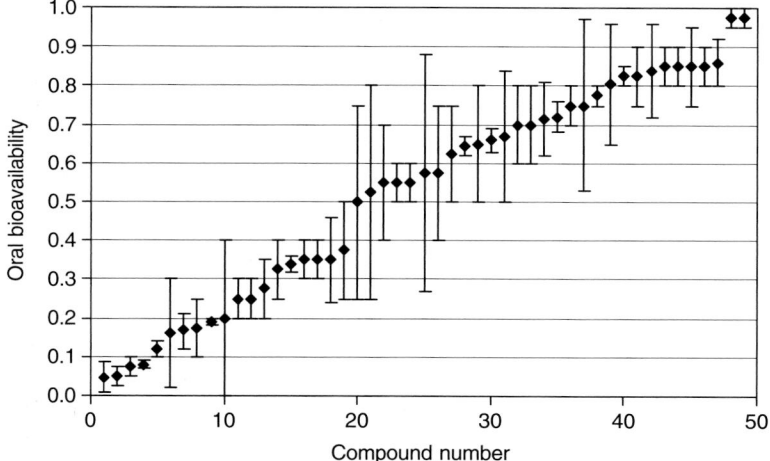

Figure 2   Variability of human oral bioavailability data. (Reproduced with permission from Bains, W.; Gilbert, R.; Sviridenko, L.; Gascon, J. M.; Scoffin, R.; Birchall, K.; Hetherington, L.; Harvey, I.; Caldwell, J. *Curr. Opin. Drug Disc. Dev.* **2005**, *5*, 44–51 © The Thomson Corporation and Current Opinion in Drug Discovery & Development.)

cross-validation or a separate testing set to gauge model performance. These testing subsets are, in effect, considered statistically part of the training set of compounds.

For generalizability, large structurally dissimilar data sets cover a broad chemical space and may better reflect the complex aspects involved in predicting bioavailability. Although they may enable the use of a statistically independent validation set of compounds to test true predictive ability[77] such heterologous data sets also face greater data quality assurance problems.

One relevant consideration often overlooked in data sets is the inclusion of a range of compounds including both developmental failures as well as successes. Data sets with only successful compounds will be biased and likely unable to detect potential failures.[78]

5.29.4 Commercial Tools and Software

A number of commercial entities produce proprietary software capable of predicting a range of ADMET properties. Molecular structures can be input via SMILES (Simplified Molecular Input Line Entry System), MOL files, or other industry standard notation. For the most part, these describe two-dimensional (2D) structure wherein spatial positioning of atoms is of little importance.

Attention must be paid to the three-dimensional (3D) structure of molecules when accounting for energy interactions across space. Energy minimization routines vary depending on the application used while conformation endpoints can depend on the selected solvent such as aqua or octanol, or whether minimization is done in vacuo. Appropriate 3D conformation and energy minimization parameters are essential to extract meaningful information from structural descriptors.

5.29.4.1 Available Data Sources

The scope of computational programs and models continues to increase as more compounds are added to bioavailability databases. Several companies specialize in providing experimentally derived libraries of information for bioavailability prediction in addition to supplying data based on high-throughput screening (HTS) (**Table 4**).

For example, the AurSCOPE database contains drug information for over 7000 compounds collated from the literature. Bioavailability data is available for over 10% of compounds with human bioavailability data currently available for 140 compounds. AurSCOPE is updated regularly and is subject to rigorous internal quality control to ensure data reliability. Another example is the BioPrint database which contains substantially fewer compounds but which contains pharmacologic, ADME screen, and literature data. HTS resources are also available such as the ECLiPS library of over 7.5 million small molecules which is claimed to contain a majority of compounds expected to have good oral absorption due to property calculation and selection criteria.

Table 4 Commercial computational bioavailability prediction tools

Product/module	Supplier/developer	Capability
Predictor		
ADME Boxes	Pharma Algorithms www.ap-algorithms.com	Prediction of bioavailability, physicochemical parameters, aqueous solubility, solvation, pK_a, P-gp substrate specificity
truPK Bioavailability predictor	Strand Genomics trupk.strandgenomics.com	Bioavailability, plasma protein binding, volume of distribution, elimination half-life and rate of absorption prediction models; based on 2D structure and small validation set
KnowItAll ADME-Tox	Bio-Rad Laboratories www.knowitall.com www.bio-rad.com	ADME prediction including bioavailability, analytical informatics, database building and mining, structure drawing, reporting
Database/library		
AurSCOPE	Aureus Pharma www.aureus-pharma.com	ADME/drug–drug interaction, GPCR, ion channel, hERG channel and kinase components
BioPrint	Cerep www.cerep.fr	Pharmaco-informatics platform containing ADME database
ECLiPS libraries	Pharmacopeia www.pharmacopeia.com	High-throughput screening/combinatorial libraries available for bioavailability prediction

A significant problem is the degree of human bias in selecting compounds to be included particularly in smaller libraries. This may be due to availability and costs of reagents, pharmacokinetic considerations, and preferences of medicinal chemists themselves in producing samples of such compounds.[79] Compound databases should ideally contain experimentally derived data from marketed drugs, failed drug candidates, bioactive natural substances, and withdrawn drugs in order to avoid bias towards established drug like molecules.

5.29.4.2 Bioavailability Prediction Software

There are few commercial programs that specifically predict bioavailability of compounds (**Table 4**). The majority come with bundled modules that allow bioavailability estimation in addition to deriving other parameters.

ADME Boxes[80] predicts bioavailability using a combination of probabilistic and mechanistic methods. Experimental data and proprietary algorithms take into consideration water solubility, acid stability, passive and active intestinal transport characteristics, P-gp efflux susceptibility, and first-pass metabolism to classify compounds into low ($<30\%$), moderate (30–70%) and good ($>70\%$) bioavailability classes. Although classification provides less information than quantitative prediction, this may be sufficient simply for screening purposes. The probability or certainty of classification is also provided in ADME Boxes to give a measure of confidence. The database contains 723 compounds for which bioavailability data is provided as averages of experimental values and ranges found in clinical trial and review publications. As discussed previously, drug information extracted from the literature may not comprise data generated under the same experimental conditions. The relatively small number of compounds in the database also limits the chemical space represented.

Strand Genomics utilized an even smaller dataset of 227 compounds from mostly commercially available preparations to develop a specific prediction model for bioavailability. Their truPK Bioavailability predictor[81] incorporated the use of artificial neural networks, decision trees, and support vector machine technology to identify relevant descriptors from which a predictive model was developed. Strand Genomics offers other customizable models using focused training and validation data sets which may counter some of the potential bias in their data. Clinical data supplied by manufacturers need also be treated with caution since these may potentially be skewed towards favorable results, in a similar manner to publication bias in the literature.[82]

Validation of models in KnowItAll is achieved with consensus modeling for both continuous and classification models. In consensus modeling, a number of individual models are generated from which a weighted average is calculated to best match experimental data. The compound database taken from the literature[30] contains 232 compounds and their bioavailability classification into one of five categories: $\leqslant 20\%$, 20–49%, 50–79%, and $\geqslant 80\%$. One advantage of the database is its widespread use which enables comparisons with other studies to be performed. Database size is again, however, a limiting factor. The ADME-Tox module[83] for bioavailability prediction provides a distinct advantage over static models since it enables data to be added or removed when needed, thus allowing models to be updated and expanded upon without new models required to be generated. The suitability of commercial applications for bioavailability predictions needs to weigh the relative importance of chemical space representation, postmodel construction flexibility, purpose (quantitative or qualitative predictions), and cost.

Generic software and modeling approaches enables a great deal of flexibility to be adopted for in silico bioavailability prediction. Programs such as artificial neural networks (ANN), genetic algorithms (GA), and expert systems are readily available both commercially and as open-access code, and can be customized for modeling tasks and to the data available. Several recent applications of these programs for bioavailability prediction have been published[84,85] with reasonable success for the small data sets used. Custom operations using other proprietary software such as Equbits Foresight[86] in KnowItAll and the QSAR module of SYBYL[87] can also be performed for construction of predictive models. Comparative molecular field analysis (CoMFA) and soft-independent modeling of class analogy (SIMCA) methodologies have been utilized for bioavailability prediction.[88] These both require that appropriate 3D structure of molecules be obtained using appropriate energy-minimization routines to extract relevant information from molecular fields. It should be noted though, that conformation-independent techniques can deliver better correlative models than by using CoMFA.[89]

The various commercial computational resources available for in silico bioavailability prediction allow a flexible approach to be taken. The continued increase in availability of quality data will see commercial databases advancing in scope and size. In addition to ever-improving algorithms and computing capabilities, this will doubtless enable both proprietary and generic models to become better validated and more accurate in the future.

5.29.5   Current Models

5.29.5.1   Absorption and Bioavailability Prediction

Several approaches have focused on predicting the required physicochemical properties for drug-like pharmacokinetics for NCEs.[90,91] Studies combining MW and PSA have revealed that recently marketed orally bioavailable drugs mostly have MW < 500 and PSA < 120 Å^2.[92] From data on 20 selected compounds, it has been shown that PSA > 140 Å^2 results in incomplete (<10%) oral absorption whereas PSA < 60 Å^2 results in excellent (>90%) absorption.[93] Other computational models including dynamic PSA have been reported[41,94] although these have no advantage over PSA calculated from the minimum energy conformation.

Assuming good metabolic stability, the two key parameters controlling drug absorption are membrane permeability and dose/solubility ratio.[95] It has recently been shown that when the aqueous solubility of drugs is less than 100 µg mL^{-1} dissolution is the limiting factor for oral absorption.[95] Oral bioavailability of drugs that are sparingly soluble and do not have significant first-pass metabolism are, therefore, dissolution-limited and may have dose-dependent bioavailability. Modeling dose-dependency of ganciclovir demonstrated the utility of experimentally determined physicochemical parameters in one study.[96] Permeability, surface area, transit time, solubility, and dissolution rate data obtained from a consortium of pharmaceutical companies and the literature and were used to construct a physiological simulation model to predict the fraction of drug absorbed into the portal vein (FD$_p$%). These models, for both human and dog, were tested with 12 compounds exhibiting a wide range amongst the physicochemical parameters included and also for a further three commercial drugs. The model qualitatively delineated compounds with low and high absorption values as evidenced by good FD$_p$% prediction. Although limited by the small number of compounds and FD$_p$% not being a true measure of bioavailability as claimed, the methodology and results obtained were encouraging. Another study utilized calculated FD$_p$% and other experimentally derived parameters to classify human and rat bioavailability of 140 compounds.[73] Results suggested that low in vitro bioavailability was reasonably well correlated with low in vivo bioavailability. Misclassification rates corresponding to large numbers of false positives and false negative were minimized around in vitro bioavailability of 15% suggesting that the utility of the model was limited to identifying compounds with low in vivo bioavailability only.

5.29.5.2   Classification Models

Qualitative prediction or classification of oral bioavailability may be a less demanding goal than quantitative prediction. In one early study, fuzzy adaptive least-squares (FALS) analysis was used to develop a structure–bioavailability relationship for 188 known drugs.[97] Three separate models based on this expert system approach were developed for the three classes of compounds: aromatics, nonaromatics, and heteroaromatics. A number of physicochemical and structural descriptors were generated including log P, MW, and functional group counts. Interestingly, squared transformations of both log P and MW were performed to account for nonlinearity amongst the data but only the MW^2 variable was included in the final aromatic model. The negative influence of MW on bioavailability found in this study has been corroborated in more recent, extensive work.[43] Compounds were classified into one of three classes, for bioavailability values <50%, between 50% and 89%, and >89%. Although representing a wide variety of structures, there was no discussion of the range of actual bioavailability values used and, since this data set consisted of marketed drugs, there may have been some bias towards compounds with high clinical bioavailability. Model performance was gauged by leave-one-out cross-validation which is not a true measure of predictive ability unlike predictions for an independent validation set of compounds.[77] Nevertheless, the reasonably accurate classifications of the wide range of chemical structures enabled some discussion of the influence of various descriptors on bioavailability. The complexity inherent in bioavailability prediction[85] was demonstrated by the opposite effects of –OH groups on bioavailability for the different classes of compounds.

Structural and physicochemical descriptors similar to those in the study described above were used in combination with the ordered multicategorical classification method using the simplex technique (ORMUCS), an adaptive least squares (ALS) approach, to categorize bioavailability of a total of 272 drugs.[30,98] In addition to being the most extensive study yet conducted in this field, the authors introduced a new parameter, Δlog D (**Figure 3**). This was defined as the difference between the logarithm of the distribution coefficient at pH 6.5 (intestine) versus pH 7.4 (blood) for an ionizable species. The purpose of this descriptor was to account for the apparent higher bioavailability observed for many acidic compounds.

Bioavailability data for these commercial compounds were extracted from the literature and placed into one of four categories based on their oral bioavailability: ≤20%, 20–49%, 50–79%, and ≥80% (**Table 5**). Classification of both training and testing data was found to be more accurate for compounds with higher bioavailability for which there were

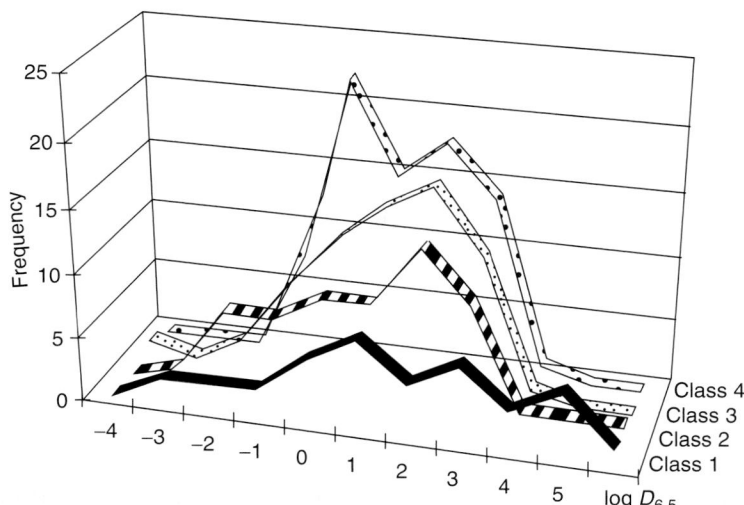

Figure 3 Distribution of $\log D_{6.5}$ within each bioavailability class showing broader distribution as bioavailability decreases. (Reproduced with permission from Yoshida, F.; Topliss, J. G. *J. Med. Chem.* **2000**, *43*, 2575–2585. Copyright year (2000) American Chemical Society.)

greater numbers of available compounds. Poor predictions for compounds with low bioavailability may be due to the complexity introduced by extensive enzymatic metabolism of these compounds.[85] An overall classification rate of 71% was achieved, with 97% correct being obtained within one class. Furthermore, 60% (95% within one class) of the 40 independent testing compounds were also correctly classified using this linear QSAR equation.

Of the various descriptors in the final models, $\log D_{6.5}$ was shown to exert a significant negative influence on bioavailability while similar effects were seen for phenolic –OH and $-SO_2NH_2$ groups. The results indicated the feasibility of obtaining reasonable estimates of oral bioavailability from molecular structures when physically and biologically meaningful descriptors are employed. These conclusions must be taken in light of the accuracy of the model which has since been improved upon.[84]

A recent adaptive fuzzy partitioning (AFP) study[84] employed the same data and classification system as described above[30] as well as including a further set of 235 compounds extracted from elsewhere in the literature. This study exploited a number of innovative approaches including fuzzy logic (FL), GA, and self-organizing map (SOM) techniques to select descriptors and construct models (the success of GAs has been reported elsewhere[75] although results have not been subject to review). Classification results for independent testing compounds (true predictive ability) were improved over the previous study[30] with similarly better classification achieved for compounds with higher recorded bioavailability. The descriptors used were 2D topological and electrotopological descriptors as well as $\log D$ and $\log P$ variables, all generated in silico from molecular structure. After selection of an optimal descriptor subset, it was found that compounds in the independent testing set were not fully representative of the training set. The authors therefore reselected testing compounds that better represented the training set in the descriptor hyperspace. While adequate representation is strongly desirable to validate models, bias introduced by nonrandomly selecting testing compounds from a relatively small data set can generate potentially artificial results.

A higher level of descriptor complexity, involving derivation of information from 3D drug structure using CoMFA, has been described.[88] Following descriptor calculation from drug structure and conformation, principal component analysis (PCA) was used to determine independent latent variables that best characterized descriptor class and combinations. Human bioavailability data taken from two literature sources for 606 compounds[5,30] involved nominally continuous values but, due to the large errors in determination, values were considered of a qualitative nature. Categories were conveniently divided into distinct classes (0–15%, 25–36%, 45–76%, and 80–100%) which reduced the effects of poor data quality and borderline predictions. Separate and combined models for the two data sets were constructed and it was found that the large, combined data set did not necessarily achieve superior predictions. This emphasized the effects of data variability since original values were collated from a variety of sources initially.[5,30] The two separate models were validated with an independent set of compounds with only 8 from 20 and 9 from 41 compounds correctly classified (**Table 6**). Better results (16 from 18 and 31 from 41, respectively) were claimed when correct classification was extended to plus or minus one category level. The literature data used was biased towards

Table 5 Bioavailability class and log D values

Number	Compound	Type[a]	$\log D_{6.5}$	$\Delta \log D$[b]	pK_a	Class[f]
1	Adrenaline	B	− 3.49	− 0.90	9.9	1
2	Alprenolol	B	0.40	− 0.89	9.2	1
3	Clomethiazole	N	2.12	0.00		1
4	Coumarin	N	1.39	0.00		1
5	Dobutamine	B	− 0.48[d]	− 0.90	9.5	1
6	Domperidone	B	2.83	− 0.75	7.7	1
7	Dopamine	B	− 3.37	− 0.90	8.9	1
8	Epanolol	B	− 0.40	− 0.77	7.8[e]	1
9	Estradiol	N	4.01	0.00		1
10	Felodipine	N	3.22[d]	0.00		1
11	Hydralazine	B	0.14	− 0.61	7.3	1
12	Isradipine	N	1.72[d]	0.00		1
13	Ketamine	B	1.14	− 0.69	7.5	1
14	Lofepramine	B	4.58[d]	− 0.32	6.7[e]	1
15	Lovastatin	N	4.26	0.00		1
16	Mebendazole	N	2.83	0.00		1
17	Meptazinol	B	0.71[d]	− 0.88	8.7	1
18	Mercaptopurine	B	− 1.12	− 0.72	7.6	1
19	Nabumetone	N	2.83[c]	0.00		1
20	Nalbuphine	B	− 0.63	− 0.88	8.7	1
21	Naloxone	B	0.64	− 0.81	7.9	1
22	Nimodipine	N	0.73	0.00		1
23	Nisoldipine	N	1.58	0.00		1
24	Nitrendipine	N	0.97	0.00		1
25	Phenolphthalein	N	2.41	0.00		1
26	Probucol	N	7.29[d]	0.00		1
27	Prochlorperazine	B	2.51	− 0.69	7.5	1
28	Progesterone	N	3.87	0.00		1
29	Selegiline	B	0.62[d]	− 0.89	9.2[e]	1
30	Simvastatin	N	4.68	0.00		1
31	Sumatriptan	B	− 2.07	− 0.90	9.4[e]	1
32	Tacrine	B	− 0.59	− 0.90	9.8	1
33	Terbutaline	B	− 1.82[e]	− 0.89	8.8	1
34	Testosterone	N	3.32	0.00		1
35	Tetrahydrocannabinol	N	4.49[d]	0.00		1
36	Venlafaxine	B	− 2.29	− 0.89	9.2	1

Table 5 Continued

Number	Compound	Type^a	log $D_{6.5}$	$\Delta log D^b$	pK_a	Class^f
37	Xamoterol	B	-0.80^c	-0.78	7.8^e	1
38	Acebutolol	B	-0.99	-0.89	9.2	2
39	Alprazolam	B	1.10	-0.16	6.2^e	2
40	Amitriptyline	B	2.14	-0.90	9.4	2
41	Chlorpromazine	B	2.49	-0.89	9.2	2
42	Cisapride	B	1.79^d	-0.75	7.7^e	2
43	Clemastin	B	1.65	-0.89	9.2^e	2
44	Chlorothiazide	A	-0.45	0.57	6.7	2
45	Desipramine	B	1.20	-0.90	10.2	2
46	Dextropropoxyphene	B	2.20	-0.89	9.2^e	2
47	Diltiazem	B	1.47	-0.75	7.7	2
48	Diprafenone	B	0.59	-0.89	9.2^e	2
49	Doxepin	B	1.55^d	-0.82	8.0	2
50	Encainide	B	-0.34^d	-0.90	10.2	2
51	Ethinyl estradiol	N	3.67	0.00		2
52	Etilefrine	B	-2.86^c	-0.90	9.8^e	2
53	Famotidine	B	-1.27	-0.52	7.1	2
54	Fluorouracil	A	-0.89	0.10	8.0	2
55	Imipramine	B	1.80	-0.90	9.5	2
56	Indoramine	B	1.56	-0.75	7.7	2
57	Isoprenaline	B	-2.76	-0.88	8.7	2
58	Isotretinoin	A	4.30	0.90	4.5^e	2
59	Labetalol	B	0.35	-0.90	9.3	2
60	Lidocaine	B	0.88	-0.79	7.9	2
61	Lorcainide	B	1.44	-0.88	8.7^e	2
62	Medifoxamine	B	-0.43^d	-0.89	9.2^e	2
63	Metoprolol	B	-1.32	-0.90	9.7	2
64	Mianserin	B	2.78^d	-0.52	7.1	2
65	Midazolam	B	1.37	-0.14	6.1	2
66	Moricizine	B	2.73	-0.21	6.4	2
67	Morphine	B	-2.34	-0.90	9.6	2
68	Nadolol	B	-2.46	-0.90	9.7	2
69	Nafcillin	A	-0.84^d	0.90	2.7	2
70	Naltrexone	B	0.84	-0.84	8.1	2
71	Nicardipine	B	1.18	-0.72	7.6^e	2

continued

Table 5 Continued

Number	Compound	Type[a]	$\log D_{6.5}$	$\Delta \log D^{b}$	pK_a	Class[f]
72	Nicotine	B	− 0.34	− 0.82	8.0	2
73	Oxacillin	A	− 1.32	0.90	2.8	2
74	Oxprenolol	B	− 0.72	− 0.90	9.3	2
75	Pentazocine	B	1.05	− 0.88	8.8	2
76	Pentoxifylline	N	0.29	0.00		2
77	Phenylephrine	B	− 3.02	− 0.90	9.8	2
78	Pimozide	B	5.42	− 0.62	7.3	2
79	Pirenzepine	B	− 1.51	− 0.83	8.1	2
80	Prenalterol	B	− 1.61^c	− 0.89	9.2^e	2
81	Promethazine	B	2.21	− 0.89	9.1	2
82	Propafenone	B	0.15^d	− 0.89	9.2^e	2
83	Propranolol	B	− 0.02	− 0.90	9.5	2
84	Ritodrine	B	− 1.00^c	− 0.89	9.0	2
85	Salbutamol	B	− 2.69^c	− 0.90	9.3	2
86	Scopolamine	B	0.11	− 0.72	7.6	2
87	Spironolactone	N	2.78	0.00		2
88	Sulpiride	B	− 2.04	− 0.89	9.0	2
89	Thioridazine	B	2.90	− 0.90	9.5	2
90	Triazolam	B	2.26	− 0.14	6.2^e	2
91	Verapamil	B	1.39	− 0.89	8.9	2
92	Acetaminophen	N	0.51	0.00		3
93	Acetylsalicylic acid	A	− 1.81	0.90	3.5	3
94	Amantadine	B	− 1.86	− 0.90	10.8	3
95	Amiodarone	B	2.29	− 0.29	6.6	3
96	Amlodipine	B	0.78	− 0.90	9.5^e	3
97	Atenolol	B	− 2.66	− 0.90	9.3	3
98	Atropine	B	− 1.37	− 0.90	9.7	3
99	Bepridil	B	1.10	− 0.90	9.3^e	3
100	Betamethasone	N	1.94	0.00		3
101	Bevantolol	B	1.11	− 0.87	8.4	3
102	Brotizolam	B	2.63	− 0.14	6.2^e	3
103	Bufuralol	B	1.00	− 0.89	9.0	3
104	Captopril	A	− 1.78^c	0.90	3.7	3
105	Chloramphenicol	N	1.14	0.00		3
106	Chlorpheniramine	B	0.69	− 0.89	9.2	3
107	Chlorthalidone	N	0.22^c	0.00		3

Table 5 Continued

Number	Compound	Type[a]	$\log D_{6.5}$	$\Delta \log D$[b]	pK_a	Class[f]
108	Cimetidine	B	-0.08	-0.38	6.8	3
109	Clomipramine	B	2.49	-0.89	9.2[e]	3
110	Clopenthixol	B	2.71	-0.83	8.1[e]	3
111	Clozapine	B	1.81	-0.81	8.0	3
112	Codeine	B	-0.28	-0.80	7.9	3
113	Dexamethasone	N	1.83	0.00		3
114	Diclofenac	A	2.69	0.89	4.8	3
115	Dicloxacillin	A	-0.79	0.90	2.8	3
116	Diphenhydramine	B	0.77	-0.89	9.0	3
117	Doxazosin	B	1.14[d]	-0.25	6.5	3
118	Enoximone	N	2.32[d]	0.00		3
119	Ethambutol	B	-4.02	-0.90	9.5	3
120	Finasteride	N	3.03	0.00		3
121	Fluoxetine	B	0.92	-0.90	10.2[e]	3
122	Flupenthixol	B	2.88	-0.83	8.1	3
123	Fluvoxamine	B	-0.13[d]	-0.90	9.5[e]	3
124	Furosemide	A	-0.02	0.90	4.7	3
125	Haloperidol	B	1.42	-0.86	8.3	3
126	Hydrochlorothiazide	A	-0.19	0.42	7.0	3
127	Levobunolol	B	-0.40	-0.90	9.3	3
128	Levomepromazine	B	1.98	-0.89	9.2[e]	3
129	Maprotiline	B	0.52	-0.90	10.5	3
130	Meperidine	B	0.25	-0.88	8.7	3
131	Metoclopramide	B	-0.18	-0.90	9.3	3
132	Moclobemide	B	1.26[d]	-0.15	6.2	3
133	Nifedipine	N	0.42	0.00		3
134	Nifurtimox	N	0.08	0.00		3
135	Nitrazepam	N	2.25	0.00		3
136	Norethisterone	N	2.97	0.00		3
137	Nortriptyline	B	1.08	-0.90	9.7	3
138	Omeprazole	B	2.21	-0.02	5.2[e]	3
139	Ondansetron	B	1.13[d]	-0.61	7.3[e]	3
140	Paroxetine	B	-1.96[d]	-0.90	11.2[e]	3
141	Penbutolol	B	1.35	-0.90	9.3	3
142	Perphenazine	B	2.88	-0.77	7.8	3

continued

Table 5 Continued

Number	Compound	Type^a	$\log D_{6.5}$	$\Delta \log D^b$	pK_a	Class^f
143	Pindolol	B	−0.55	−0.88	8.8	3
144	Prazosin	B	1.04^d	−0.25	6.5	3
145	Primaquine	B	−2.19^d	−0.90	10.3^e	3
146	Procainamide	B	−1.82	−0.89	9.2	3
147	Procyclidine	B	1.30^d	−0.88	8.8^e	3
148	Quinidine	B	1.08	−0.86	8.3	3
149	Raclopride	B	0.43	−0.89	9.2	3
150	Ranitidine	B	−1.44	−0.84	8.2	3
151	Timolol	B	−0.99	−0.90	9.3	3
152	Triamterene	B	0.80	−0.15	6.2	3
153	Urapidil	B	0.90	−0.52	7.1	3
154	Zofenoprilat	A	−0.64^d	0.90	3.7^e	3
155	Allopurinol	N	−0.55	0.00		4
156	Amobarbital	A	2.05	0.13	7.8	4
157	Amrinone	B	−0.72	−0.02	5.2^e	4
158	Betaxolol	B	−0.07	−0.90	9.4	4
159	Bisoprolol	B	−0.83	−0.89	9.2^e	4
160	Bumetanide	A	−1.84^d	0.90	4.0	4
161	Caffeine	N	−0.07	0.00		4
162	Carbamazepine	N	2.45	0.00		4
163	Carteolol	B	−1.35	−0.89	9.2^e	4
164	Chlorambucil	A	2.29	0.82	5.8	4
165	Chlordiazepoxide	B	2.43	−0.01	4.8	4
166	Chloroquine	B	0.33	−0.90	9.9	4
167	Chlorpropamide	A	0.76	0.89	5.0	4
168	Cibenzoline	B	−0.37^d	−0.90	10.3	4
169	Clobazam	N	0.95	0.00		4
170	Clonazepam	N	2.41	0.00		4
171	Clonidine	B	0.06	−0.82	8.0	4
172	Cyclophosphamide	N	0.63	0.00		4
173	Desmethyldiazepam	N	2.93	0.00		4
174	Diazepam	N	2.99	0.00		4
175	Diazoxide	A	1.20	0.03	8.5	4
176	Diflunisal	A	0.94	0.90	3.0	4
177	Disopyramide	B	−1.08	−0.90	10.4	4
178	Ethanol	N	−0.31	0.00		4

Table 5 Continued

Number	Compound	Type[a]	$\log D_{6.5}$	$\Delta \log D$[b]	pK_a	Class[f]
179	Ethosuximide	N	-0.33[c]	0.00		4
180	Flecaninide	B	0.24	-0.90	9.3	4
181	Flurbiprofen	A	1.81	0.90	4.2	4
182	Fluconazole	N	-0.11[c]	0.00		4
183	Flucytosine	N	-1.65[c]	0.00		4
184	Flunitrazepam	N	2.06	0.00		4
185	Gemfibrozil	A	1.47[d]	0.89	4.8[e]	4
186	Glipizide	A	1.31	0.81	5.9	4
187	Glyburide	A	1.85	0.87	5.3	4
188	Hexobarbital	A	1.48	0.04	8.3	4
189	Ibuprofen	A	2.18	0.88	5.2	4
190	Indomethacin	A	2.27	0.90	4.5	4
191	Isoniazide	N	-0.70	0.00		4
192	Isosorbide 2-nitrate	N	-0.40	0.00		4
193	Isosorbide 5-nitrate	N	-0.15	0.00		4
194	Ketoprofen	A	1.21	0.89	4.6	4
195	Ketolorac	A	-1.34[c]	0.90	3.5	4
196	Lorazepam	N	2.51	0.00		4
197	Mabuterol	B	-0.03[d]	-0.90	9.7[e]	4
198	Methadone	B	1.16	-0.90	9.3	4
199	Methylprednisolone	N	1.66[c]	0.00		4
200	Metronidazole	N	-0.02	0.00		4
201	Mexiletine	B	-0.55	-0.89	9.2	4
202	Naproxen	A	1.04	0.90	4.2	4
203	Nitrofurantoin	N	-0.47	0.00		4
204	Nizatidine	B	-0.08	-0.38	6.8	4
205	Oxaprozin	A	3.41	0.84	5.8[e]	4
206	Oxazepam	N	2.24	0.00		4
207	Phenobarbital	A	1.43	0.21	7.5	4
208	Phenylbutazone	A	1.06	0.90	4.4	4
209	Phenylethylmalonamide	N	0.13	0.00		4
210	Phenytoin	A	2.47	0.10	8.0	4
211	Prednisolone	N	1.62	0.00		4
212	Prednisone	N	1.46[d]	0.00		4
213	Primidone	N	0.91	0.00		4

continued

Table 5 Continued

Number	Compound	Type[a]	$\log D_{6.5}$	$\Delta \log D$[b]	pK_a	Class[f]
214	Probenecid	A	0.11	0.90	3.4	4
215	Protriptyline	B	0.34	−0.84	8.2	4
216	Quinine	B	0.34	−0.88	8.8	4
217	Salicylic acid	A	−1.24	0.90	3.0	4
218	Sulfadiazine	A	−0.39	0.65	6.5	4
219	Sulfamethoxazole	A	0.11	0.83	5.8	4
220	Sulfinpyrazone	A	−1.40	0.90	2.8	4
221	Sulfisoxazole	A	−0.50	0.89	5.0	4
222	Temazepam	N	2.19	0.00		4
223	Tenoxicam	A	0.56	0.88	5.1[c]	4
224	Theophylline	A	−0.02	0.02	8.8	4
225	Tocainide	B	−0.56	−0.78	7.8	4
226	Tolbutamide	A	1.11	0.87	5.3	4
227	Tolmetin	A	−0.21	0.90	3.5	4
228	Trazodone	B	1.47	−0.33	6.7	4
229	Trimethoprim	B	0.13	−0.57	7.2	4
230	Valproic acid	A	1.04	0.89	4.8	4
231	Warfarin	A	1.28	0.88	5.1	4
232	Zalcitabine	N	−1.33	0.00		4

Reproduced with permission from Yoshida, F.; Topliss, J. G. *J. Med. Chem.* **2000**, *43*, 2575–2585. Copyright year (2000) American Chemical Society.
[a] A, acid; B, base; N, neutral. Compounds not significantly ionized at pH 6.5 and 7.4 ($\Delta \log D = 0.00$) are designated as neutral.
[b] $\log D_{6.5}$–$\log D_{7.4}$.
[c] Derived from the CLOGP value.
[d] Derived from MLOGP value.
[e] Estimated.
[f] Bioavailability $\leqslant 20\%$ (class I), 20–49% (class II), 50–79% (class III), $\geqslant 80\%$ (class IV).

compounds with high bioavailability which may also have skewed predictions to the higher categories. Furthermore, although a novel approach was described, given the absolute margins between categories these models may not be sufficiently accurate in today's highly demanding drug discovery arena.

5.29.5.3 Quantitative Models

5.29.5.3.1 Regression models

The ability to accurately predict bioavailability in a quantitative manner for virtual compounds would be of immense value in drug discovery and development. The first published study to develop a quantitative structure–bioavailability relationship[99] made use of data for 591 compounds obtained from the literature. The multiple linear regression (MLR) model was optimized using recursive partitioning to analyze descriptor interactions. Only simple one-dimensional (1D) descriptors were required and consisted of 608 substructure counts for each molecule. From the final regression equation the relative influence of individual descriptors was able to be characterized. This information may be applied in library screening although should be considered along with the effects of all significant descriptors included in the final model. The 85 descriptors in the final model corresponded to approximately seven observations per descriptor, a

Table 6 Bioavailability class predictions using idiotropic field orientation (IFO)[88]

	Yoshida and Topliss data set[30]				Sietsema data set[5]			
	Literature class	n	Correct	Correct ±1	Literature class	n	Correct	Correct ±1
Steric IFO	1	38	36	36	1	64	19	34
	2	43	6	38	2	67	43	63
	3	52	46	48	3	131	84	128
	4	65	35	47	4	146	92	128
	Total	198	123	169	Total	408	238	353
	Percent		62%	85%	Percent		58%	87%
Electrostatic IFO	1	38	34	35	1	64	35	40
	2	43	37	41	2	67	51	64
	3	52	44	50	3	131	80	124
	4	65	46	56	4	146	104	125
	Total	198	161	182	Total	408	270	353
	Percent		81%	92%	Percent		66%	87%

Adapted from Wolohan, P. R.; Clark, R. D. *J. Comput.-Aided Mol. Des.* **2003**, *17*, 65–76.

fact which the authors claimed negated the chance of the model being overfitted. Nevertheless, the descriptor hyperspace was not analyzed for correlations or overlap and, based on the large number of descriptors in the final model and the minor structural variations between these descriptors, it is likely that some redundancy existed in the model. The authors also offered a comparison with benchmark rule of five[4] classification. This was performed by applying a good/poor bioavailability cutoff value of 20% to the leave-one-out predictions of the quantitative model. Comparisons were made between absorption and bioavailability which are separate pharmacokinetic parameters. Since oral bioavailability cannot exceed human intestinal absorption, compounds with low absorption may be assumed to have low bioavailability. The same is not necessarily true in reverse, however, since bioavailability may be limited by presystemic metabolism rather than being absorption-dependent.

Model performance was consistent with r^2 values of 0.71, 0.63, and 0.58 reported for the regression equation, leave-one-out cross-validation, and leave-n-out cross validation respectively. Even though both leave-one-out and leave-n-out validations were carried out, more rigorous model testing was not performed, for example, validation of the final model by an independent set of compounds.

A smaller MLR study of 169 compounds used leave-10-out testing to validate a quantitative model containing eight theoretical descriptors.[35] Regression results for this model were poorer than those in the previous study.[99] True predictive ability, however, was encouraging as evidenced by absolute predictions and differentiation of compounds with low and high observed bioavailability. Complexity of descriptors was increased to include numerous topological 2D variables as well as quantum chemical descriptors. The questionable benefits gained from using the more complex descriptors may not necessarily justify their use in this case. Nevertheless, useful information concerning the influence of calculated parameters was described. This included the positive influences of electron affinity, aromatic ring count, and hydrogen bonding on bioavailability as well as the negative influences of highest occupied molecular orbital (HOMO), log P, molar volume, HLB, and water solubility. Such regression models imply a linear relationship between descriptors and bioavailability but this may not always be the case.

5.29.5.3.2 Artificial neural network prediction

Since oral bioavailability is an inherently nonlinear process, the most appropriate modeling techniques may be those which are nonlinear themselves. ANNs are one such soft-computing paradigm of which there are several varieties. The utility of a radial basis function (RBF) ANN was demonstrated in a recent a quantitative structure–bioavailability relationship study.[85] The data set of 167 compounds obtained from the literature was divided into separate training, testing, and independent validation subsets. Even though the validated model made acceptable predictions, it was better able to qualitatively differentiate between poorly and highly bioavailable compounds. Naloxone, naltrexone, idarubicin, and losartan with low bioavailabilities were predicted as such but the model incorrectly assigned labetalol a high value. Predictions for metoclopramide, ranitidine, paracetamol, methyldopa, and nifedipine with somewhat higher bioavailabilities were well predicted but hydrochlorothiazide was given a low value. Propylthiouracil, cephalexin, and lamotrigine with high absolute bioavailabilities all had acceptable predicted values.

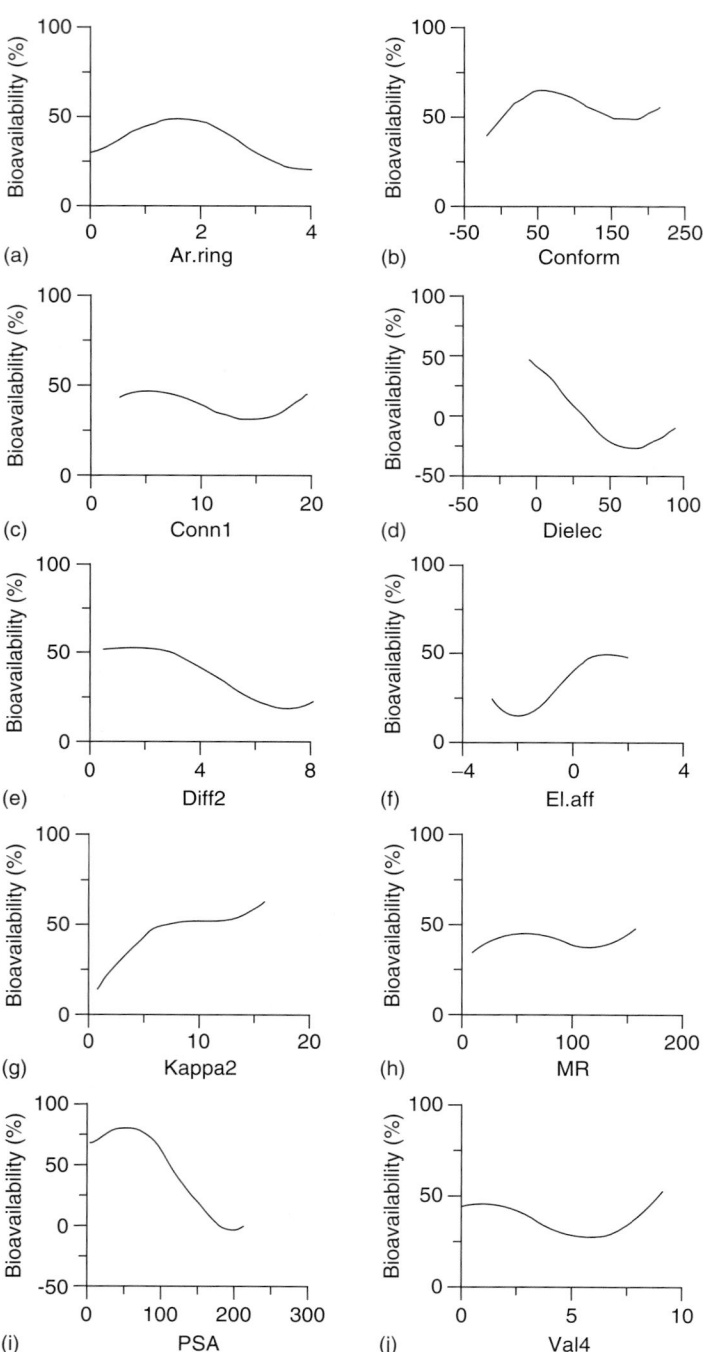

Figure 4 Descriptor response graphs for optimum ANN bioavailability model for structurally diverse dataset. (Reproduced from Turner, J. V.; Maddalena, D. J.; Agatonovic-Kustrin, S. *Pharm. Res.* **2004**, *21*, 68–82, with kind permission of Springer Science and Business Media.)

The descriptor hyperspace consisted of 1D constitutional, 2D topological, and 3D quantum chemical variables all of which were represented in the final model. Of particular note was the finding that individual descriptors influenced compound bioavailability in a nonlinear fashion (**Figure 4**). This was more consistent with the complex nature of bioavailability being determined by numerous biophysical and physicochemical processes rather than the simple relationships reported elsewhere.[30,97] Important descriptors found were aromatic rings (Ar.ring), conformational minimum energy (Conform), connectivity index order 1 (Conn1), dielectric energy (Dielec), connectivity difference index 2 (Diff2), electron affinity (El.aff), Kappa shape index 2 (Kappa2), molar refractivity (MR), polar surface area (PSA), and valence connectivity index order 4 (Val4). RBF ANNs are better able to correlate model inputs with the output parameter than conventional multilayer perceptron ANNs,[100] hence this model provides good confidence in the influence these descriptors have on bioavailability.

5.29.5.4 A Bioavailability Score

One recent study described a bioavailability scoring system based on analysis of 553 compounds containing 99 different rings and 180 side chains with experimentally determined rat bioavailability data.[101] Validation of the results was provided with human bioavailability values from the literature. It was shown that none of the rule of five, log P, log D, or the combination of the number of rotatable bonds and PSA could be used to successfully predict a compound having bioavailability greater than 10% in the rat. Instead, different properties of compounds depending on their predominant charge at biological pH were shown to govern bioavailability. It was concluded, therefore, that different physical properties are required for negatively charged compounds at pH 6–7 than for uncharged or positively charged compounds. The scoring system described was useful for classifying bioavailability in rat, Caco-2 permeability, and intestinal absorption in humans. Further work in this field may see development of a similar scoring system with greater success in human bioavailability characterization.

5.29.6 General Discussion and Conclusion

Data-driven modeling has been actively applied within drug discovery for several decades and now has the potential to impact all stages of the drug discovery process. While early efforts were focused on target affinity optimization, applications to lead finding and ADMET property prediction have more recently emerged. The dramatic increase in size of compound libraries and ultra-HTS has resulted in greatly increased rates at which biological activity data can be obtained. Nevertheless, research and development time-lines have remained virtually unchanged.[102] Data from HTS and other in vitro experiments have now been used to develop a first generation of ADMET predictive in silico tools which can be used in the early stages of the drug discovery process including library design and compound screening. The progress of qualitative and quantitative models using literature data and purely in silico methods is also encouraging.

Computational models, by their very nature, are dependent on the data used to generate them and on the understanding and insight of the scientists who created those models. This is especially true for bioavailability prediction considering the multiplicity of factors influencing drug translocation from the gastrointestinal tract into the systemic circulation. It follows that the availability of sufficiently good quality databases is another limiting factor for in silico bioavailability prediction.

Indication of potential bioavailability should ideally be obtained during drug discovery prior to clinical trials. Currently, early screening for absorption may be performed using simple methods such as the rule of five which rely on structural considerations only. This principle should be applied to bioavailability since this pharmacokinetic parameter has additional clinical and developmental implications including pharmacodynamic effects and dosing regimens. The complex nature of bioavailability renders this a considerable challenge. Similarly, in vitro/in vivo correlations using experimentally derived data have established utility in absorption prediction but are of limited use for bioavailability. Recent progress in the field is encouraging as predictive models become more refined and broader in scope. Purely in silico tools for bioavailability prediction have the potential to contribute significantly to drug discovery and development and, as they improve, will become a key component in the early stages.

References

1. Hansch, C.; Lien, E. J. *J. Pharm. Sci.* **1968**, *57*, 1027–1028.
2. Gobburu, J. V. S.; Shelver, W. H. *J. Pharm. Sci.* **1995**, *84*, 862–865.
3. Turner, J. V.; Maddalena, D. J.; Cutler, D. J. *Int. J. Pharm.* **2004**, *270*, 209–219.
4. Lipinski, C. A.; Lombardo, F.; Dominy, B. W.; Feeney, P. J. *Adv. Drug Deliv. Rev.* **2001**, *46*, 3–26.

5. Sietsema, W. K. *Int. J. Clin. Pharmacol. Ther. Toxicol.* **1989**, *27*, 179–211.
6. Schoenwald, R. D. Basic Principles. In *Pharmacokinetics in Drug Discovery and Development*; Schoenwald, R. D., Ed.; CRC Press: Boca Raton, FL, 2002, pp 3–30.
7. Van de Waterbeemd, H. Intestinal Permeability: Prediction from Theory. In: *Oral Drug Absorption: Prediction and Assessment*; Dressman, J.B., Lennerhäs, H. Eds.; Drugs and the Pharmaceutical Sciences 106; Marcel Dekker: New York, 2000, pp 31–49.
8. Van de Waterbeemd, H.; Smith, D. A.; Beaumont, K.; Walker, D. K. *J. Med. Chem.* **2001**, *44*, 1313–1333.
9. Devillers, J. No-Free-Lunch Molecular Descriptors in QSAR and QSPR. In *Topological Indices and Related Descriptors in QSAR and QSPR*; Balaban, A. T., Devillers, J., Eds.; Gordon and Breach: Amsterdam, The Netherlands, 1999, pp 1–20.
10. Lin, J. H.; Yamazaki, M. *Clin. Pharmacokinet.* **2003**, *42*, 59–98.
11. Balimane, P. V.; Chong, S.; Morrison, R. A. *J. Pharmacol. Toxicol. Methods* **2000**, *44*, 301–312.
12. Amidon, G. L.; Lennernäs, H.; Shah, V. P.; Crison, J. R. *Pharm. Res.* **1995**, *12*, 413–420.
13. Schmid, D.; Ecker, G.; Kopp, S.; Hitzler, M.; Chiba, P. *Biochem. Pharmacol.* **1999**, *58*, 1447–1456.
14. Wohnsland, F.; Faller, B. *J. Med. Chem.* **2001**, *44*, 923–930.
15. Kansy, M.; Senner, F.; Gubernator, K. *J. Med. Chem.* **1998**, *41*, 1007–1010.
16. Matheny, C. J.; Lamb, M. W.; Brouwer, K. R.; Pollack, G. M. *Pharmacotherapy* **2001**, *21*, 778–796.
17. Gao, J.; Sudoh, M.; Aube, J.; Borchardt, R. T. *J. Peptide. Res.* **2001**, *57*, 316–329.
18. Vincent, J.; Meredith, P. A.; Reid, J. L.; Elliott, H. L.; Rubin, P. C. *Clin. Pharmacokinet.* **1985**, *10*, 144–154.
19. Stein, G. E. *Am. J. Med.* **1987**, *82*, 18–21.
20. Letendre, L.; Scott, M.; Dobson, G.; Hidalgo, I.; Aungst, B. *Pharm. Res.* **2004**, *21*, 1457–1462.
21. Kramer, S. D. *Pharm. Sci. Technol. Today* **1999**, *2*, 373–380.
22. Loebstein, R.; Lalkin, A.; Koren, G. *Clin. Pharmacokinet.* **1997**, *33*, 328–343.
23. Yu, L. X.; Lipka, E.; Crison, J. R.; Amidon, G. L. *Adv. Drug Deliv. Rev.* **1996**, *19*, 359–376.
24. Aungst, B. J. *J. Pharm. Sci.* **1993**, *82*, 979–987.
25. Takayama, K.; Morva, A.; Fujikawa, M.; Hattori, Y.; Obata, Y.; Nagai, T. *J. Control. Release* **2000**, *68*, 175–186.
26. Van de Waterbeemd, H.; Smith, D. A.; Jones, B. C. *J. Comput.-Aided Mol. Des.* **2001**, *15*, 273–286.
27. Vaes, W. H.; Ramos, E. U.; Hamwijk, C.; Van Holsteijn, I.; Blaauboer, B. J.; Seinen, W.; Verhaar, H. J.; Hermens, J. L. *Chem. Res. Toxicol.* **1997**, *10*, 1067–1072.
28. Vaes, W. H.; Ramos, E. U.; Verhaar, H. J.; Cramer, C. J.; Hermens, J. L. *Chem. Res. Toxicol.* **1998**, *11*, 847–854.
29. Larsen, D. B.; Parshad, H.; Fredholt, K.; Larsen, C. *Int. J. Pharm.* **2002**, *232*, 107–117.
30. Yoshida, F.; Topliss, J. G. *J. Med. Chem.* **2000**, *43*, 2575–2585.
31. Wildman, S. A.; Crippen, G. M. *J. Chem. Inf. Comput. Sci.* **1999**, *39*, 868–873.
32. Testa, B.; Carrupt, P. A.; Gaillard, P.; Billois, F.; Weber, P. *Pharm. Res.* **1996**, *13*, 335–343.
33. Mannhold, R.; Van de Waterbeemd, H. *J. Comput.-Aided Mol. Des.* **2001**, *15*, 337–354.
34. Caron, G.; Ermondi, G. *J. Med. Chem.* **2005**, *48*, 3269–3279.
35. Turner, J. V.; Glass, B. D.; Agatonovic-Kustrin, S. *Anal. Chim. Acta* **2003**, *485*, 89–102.
36. Balaz, S.; Lukacova, V. *J. Mol. Graph. Model.* **2002**, *20*, 479–490.
37. Fu, X. C.; Chen, C. X.; Wang, G. P.; Liang, W. Q.; Yu, Q. S. *Pharmazie* **2005**, *60*, 674–676.
38. Van de Waterbeemd, H.; Kansy, M. *Chimia* **1992**, *46*, 299–303.
39. Agatonovic-Kustrin, S.; Beresford, R.; Yusof, A. P. M. *J. Pharm. Biomed. Anal.* **2001**, *25*, 227–237.
40. Refsgaard, H. H.; Jensen, B. F.; Brockhoff, P. B.; Padkjaer, S. B.; Guldbrandt, M.; Christensen, M. S. *J. Med. Chem.* **2005**, *48*, 805–811.
41. Ertl, P.; Rohde, B.; Selzer, P. *J. Med. Chem.* **2000**, *43*, 3714–3717.
42. Clark, D. E. *J. Pharm. Sci.* **1999**, *88*, 807–814.
43. Veber, D. F.; Johnson, S. R.; Cheng, H. Y.; Smith, B. R.; Ward, K. W.; Kopple, K. D. *J. Med. Chem.* **2002**, *45*, 2615–2623.
44. Lu, J. J.; Crimin, K.; Goodwin, J. T.; Crivori, P.; Orrenius, C.; Xing, L.; Tandler, P. J.; Vidmar, T. J.; Amore, B. M.; Wilson, A. G. et al. *J. Med. Chem.* **2004**, *47*, 6104–6107.
45. Agatonovic-Kustrin, S.; Ling, L. H.; Tham, S. Y.; Alany, R. G. *J. Pharm. Biomed. Anal.* **2002**, *29*, 103–119.
46. Estrada, E. *J Phys. Chem. A* **2002**, *106*, 9085–9091.
47. Benet, L. Z.; Wu, C. Y.; Hebert, M. F.; Wacher, V. J. *J. Control. Release* **1996**, *39*, 139–143.
48. Wacher, V. J.; Silverman, J. A.; Zhang, Y.; Benet, L. Z. *J. Pharm. Sci.* **1998**, *87*, 1322–1330.
49. Park, B. K.; Kitteringham, N. R.; Pirmohamed, M.; Tucker, G. T. *Br. J. Clin. Pharmacol.* **1996**, *41*, 477–491.
50. Hsu, A.; Granneman, G. R.; Bertz, R. J. *Clin. Pharmacokinet.* **1998**, *35*, 275–291.
51. Tsuji, A.; Tamai, I. *Pharm. Res.* **1996**, *13*, 963–977.
52. Yang, C. Y.; Dantzig, A. H.; Pidgeon, C. *Pharm. Res.* **1999**, *16*, 1331–1343.
53. Andrade, A.; Flexner, C. *Hopkins HIV Rep.* **2001**, *13*, 12–13.
54. Tandon, V. Bioavailability and Bioequivalence. In *Pharmaceutics in Drug Discovery and Development*; Schoenwald, R. D., Ed.; CRC Press: Boca Raton, FL, 2002, pp 97–112.
55. Testa, B. *Med. Chem. Res.* **1997**, *7*, 340–365.
56. Ings, R. M. *Xenobiotica* **1990**, *20*, 1201–1231.
57. Ritschel, W. A.; Vachharajani, N. N.; Johnson, R. D.; Hussain, A. S. *Comp. Biochem. Physiol. C* **1992**, *103*, 249–253.
58. Mordenti, J. *J. Pharm. Sci.* **1986**, *75*, 1028–1040.
59. Boxenbaum, H. *J. Pharmacokinet. Biopharm.* **1982**, *10*, 201–227.
60. Boxenbaum, H. *Drug Metab. Rev.* **1984**, *15*, 1071–1121.
61. Van de Waterbeemd, H.; Jones, B. C. *Prog. Med. Chem.* **2003**, *41*, 1–59.
62. Mahmood, I. *Drug Metab. Drug Interact.* **2000**, *16*, 143–155.
63. Lave, T.; Coassolo, P.; Ubeaud, G.; Brandt, R.; Schmitt, C.; Dupin, S.; Jaeck, D.; Chou, R. C. *Pharm. Res.* **1996**, *13*, 97–101.
64. Lave, T.; Dupin, S.; Schmitt, M.; Kapps, M.; Meyer, J.; Morgenroth, B.; Chou, R. C.; Jaeck, D.; Coassolo, P. *Xenobiotica* **1996**, *26*, 839–851.
65. Obach, R. S.; Baxter, J. G.; Liston, T. E.; Silber, B. M.; Jones, B. C.; MacIntyre, F.; Rance, D. J.; Wastall, P. *J. Pharmacol. Exp. Ther.* **1997**, *283*, 46–58.
66. Shou, M. *Curr. Opin. Drug Disc. Dev.* **2005**, *8*, 66–77.
67. Baarnhielm, C.; Dahlback, H.; Skanberg, I. *Acta Pharmacol. Toxicol.* **1986**, *59*, 113–122.

68. Ashforth, E. I.; Carlile, D. J.; Chenery, R.; Houston, J. B. *J. Pharmacol. Exp. Ther.* **1995**, *274*, 761–766.
69. Hayes, K. A.; Brennan, B.; Chenery, R.; Houston, J. B. *Drug Metab. Dispos.* **1995**, *23*, 349–353.
70. Vickers, A. E.; Connors, S.; Zollinger, M.; Biggi, W. A.; Larrauri, A.; Vogelaar, J. P.; Brendel, K. *Drug Metab. Dispos.* **1993**, *21*, 454–459.
71. Zomorodi, K.; Carlile, D. J.; Houston, J. B. *Xenobiotica* **1995**, *25*, 907–916.
72. Mandagere, A. K.; Thompson, T. N.; Hwang, K. K. *J. Med. Chem.* **2002**, *45*, 304–311.
73. Stoner, C. L.; Cleton, A.; Johnson, K.; Oh, D. M.; Hallak, H.; Brodfuehrer, J.; Surendran, N.; Han, H. K. *Int. J. Pharm.* **2004**, *269*, 241–249.
74. Poggesi, I. *Curr. Opin. Drug Disc. Dev.* **2004**, *7*, 100–111.
75. Bains, W.; Gilbert, R.; Sviridenko, L.; Gascon, J. M.; Scoffin, R.; Birchall, K.; Hetherington, L.; Harvey, I.; Caldwell, J. *Curr. Opin. Drug Disc. Dev.* **2002**, *5*, 44–51.
76. Clark, D. E.; Grootenhuis, P. D. *Curr. Opin. Drug Disc. Dev.* **2002**, *5*, 382–390.
77. Butina, D.; Segall, M. D.; Frankcombe, K. *Drug Disc. Today* **2002**, *7*, S83–S88.
78. Janssen, D. *Drug Disc. Dev.* **2002**, *5*, 38–40.
79. Diller, D. J.; Hobbs, D. W. *J. Med. Chem.* **2004**, *47*, 6373–6383.
80. *ADME Boxes*, v2.2; Pharma Algorithms: Toronto, Canada, 2004.
81. *truPK Bioavailability Predictor*; Strand Genomics: Redwood City, CA, 2004.
82. Sutton, A. J.; Duval, S. J.; Tweedie, R. L.; Abrams, K. R.; Jones, D. R. *Br. Med. J.* **2000**, *320*, 1574–1577.
83. *KnowItAll ADME/Tox*; Bio-Rad Laboratories: Philadelphia, PA, 2003.
84. Pintore, M.; Van de Waterbeemd, H.; Piclin, N.; Chretien, J. R. *Eur. J. Med. Chem.* **2003**, *38*, 427–431.
85. Turner, J. V.; Maddalena, D. J.; Agatonovic-Kustrin, S. *Pharm. Res.* **2004**, *21*, 68–82.
86. *KnowItAll Eqbits Foresight*; Bio-Rad Laboratories: Philadelphia, PA, 2003.
87. *SYBYL*, v7.1; Tripos: St Louis, MO, 2005.
88. Wolohan, P. R.; Clark, R. D. *J. Comput.-Aided Mol. Des.* **2003**, *17*, 65–76.
89. Wanchana, S.; Yamashita, F.; Hara, H.; Fujiwara, S.; Akamatsu, M.; Hashida, M. *J. Pharm. Sci.* **2004**, *93*, 3057–3065.
90. Lipinski, C. A. *J. Pharmacol. Toxicol. Methods* **2000**, *44*, 235–249.
91. Ajay, A.; Walters, W. P.; Murcko, M. A. *J. Med. Chem.* **1998**, *41*, 3314–3324.
92. Van de Waterbeemd, H.; Camenisch, G.; Folkers, G.; Raevsky, O. A. *Quant. Struct.-Act. Relat.* **1996**, *15*, 480–490.
93. Palm, K.; Stenberg, P.; Luthman, K.; Artursson, P. *Pharm. Res.* **1997**, *14*, 568–571.
94. Stenberg, P.; Luthman, K.; Artursson, P. *Pharm. Res.* **1999**, *16*, 205–212.
95. Amidon, G. L.; Sinko, P. J.; Fleisher, D. *Pharm. Res.* **1988**, *5*, 651–654.
96. Norris, D. A.; Leesman, G. D.; Sinko, P. J.; Grass, G. M. *J. Control. Release* **2000**, *65*, 55–62.
97. Hirono, S.; Nakagome, I.; Hirano, H.; Yoshii, F.; Moriguchi, I. *Biol. Pharm. Bull.* **1994**, *17*, 686–690.
98. Yoshida, F.; Topliss, J. G. *J. Med. Chem.* **2000**, *43*, 4723.
99. Andrews, C. W.; Bennett, L.; Yu, L. X. *Pharm. Res.* **2000**, *17*, 639–644.
100. Derks, E.; Pastor, M. S. S.; Buydens, L. M. C. *Chemometr. Intell. Lab. Syst.* **1995**, *28*, 49–60.
101. Martin, Y. C. *J. Med. Chem.* **2005**, *48*, 3164–3170.
102. Hecht, P. *Curr. Drug Disc.* **2002**, Jan, 21–14.

Biographies

Joseph V Turner, BMedSc (Hons), MBBS, PhD, is currently affiliated with the Rural Clinical Division of the School of Medicine, The University of Queensland, Australia. He hails from Armidale in country New South Wales and gained his Bachelor of Medical Science (Honours) from The University of Sydney in 1999. Joseph subsequently undertook postgraduate research and teaching in pharmaceutics at the Faculty of Pharmacy, The University of Sydney, and was awarded his PhD in 2004. His research focused on the use of artificial intelligence systems for the prediction of human pharmacokinetic parameters, a field in which he has a number of refereed publications and maintains an active interest.

Joseph is an altruist and has a passion for rural and remote healthcare provision, evident throughout and beyond his medical training at The University of Queensland.

Snezana Agatonovic-Kustrin, after being awarded a Bachelor of Pharmacy, accepted an assistant lectureship position at the Faculty of Pharmacy in Belgrade, Yugoslavia, where she was also involved in drug quality control with the Department of Drug Analysis. During this time she completed her Master of Pharmaceutical Science and PhD in Pharmaceutical Chemistry. From 1995 she held a number of academic positions in different pharmacy disciplines internationally, teaching in pharmaceutical chemistry, pharmaceutics, and medicinal chemistry. She has also been involved in development of major pharmacy curriculum components of the Bachelor of Pharmacy and related postgraduate courses.

Snezana recently joined the School of Biomedical, Biomolecular and Chemical Sciences at The University of Western Australia as a Senior Lecturer in Pharmaceutical Technology and Medicinal Chemistry, to assist in curriculum development and establishment of the Master of Pharmacy programme.

She has a diverse range of research interests including quantitative structure–activity relationship analysis, dosage form design, development and optimization of novel drug delivery systems especially of liquid dosage forms for pediatric patients, and quality evaluation of generic drugs including identification and quantitative control of their crystal or enantiomeric purity. Other research has included the application of statistical design, computer optimization techniques, and artificial neural network modeling.

Comprehensive Medicinal Chemistry II
ISBN (set): 0-08-044513-6

ISBN (Volume 5) 0-08-044518-7; pp. 699–724

5.30 In Silico Models to Predict Passage through the Skin and Other Barriers

M T D Cronin and M Hewitt, Liverpool John Moores University, Liverpool, UK

© 2007 Elsevier Ltd. All Rights Reserved.

5.30.1 Introduction

Membranes form biological barriers controlling and restricting the movement of molecules in vivo. **Table 1** summarizes the main routes by which chemicals may be taken up or excreted by an organism. Different molecules will cross membranes at different rates, depending on their physicochemical properties and attributes. Understanding, quantifying, and formalizing the relationship between physicochemical properties and membrane permeability will allow for the development of predictive models. Such models are often termed quantitative structure–activity relationships (QSARs) or, more specifically, quantitative structure-property relationships (QSPRs). They are encompassed in what is popularly termed 'in silico predictive science.'

The modeling of the transfer of molecules across biological membranes in this chapter will assume that passive diffusion. Such a process is thermodynamically driven, whereby a concentration gradient is required to achieve flow across a membrane. Thermodynamics drives the process such that the concentration gradient is diminished. Such transfer can be described in terms of the amount M of a chemical flowing through unit area (cross-section) S of a

Table 1 A summary of the main barriers to the uptake and excretion of compounds in vertebrate organisms, with an emphasis on the delivery of drugs

Uptake	*Excretion*
Skin	Renal
Eye	Breast milk
Placental blood	Semen
Fish gills	Bile
Lungs	Feces
Gastrointestinal tract[a]	
Blood–brain barrier[b]	
Buccal	
Nasal	
Rectal	

[a] *See* 5.28 In Silico Models to Predict Oral Absorption.
[b] *See* 5.31 In Silico Models to Predict Brain Uptake.

membrane in a particular time t. The transfer across the membrane, in these terms, is known as the flux J. These parameters are related by Fick's first law[1]:

$$J = \frac{\mathrm{d}M}{S \cdot \mathrm{d}t} \quad [1]$$

The flux of a compound through a membrane is also related to its intrinsic ability to diffuse through a membrane, i.e., its diffusivity. The diffusivity can be quantified in terms of a diffusion coefficient D. Other essential properties are the concentration C, and the thickness of the membrane (or the distance of movement perpendicular to the membrane) x. Thus, these parameters are related by:

$$J = -D \frac{\mathrm{d}C}{\mathrm{d}x} \quad [2]$$

With regard to biological membranes, there are many factors that can affect D, notably the concentration at the membrane, ionization, temperature, and effect of co-solvents (e.g., in formulation). Other physical properties can also affect D, though most, e.g., pressure, are less relevant for the modeling of flux across biological membranes. Modeling the membrane transfer of molecules assumes these properties to be constant, and an experimental test protocol is designed to achieve this. Thus, differences in flux relates between molecules should be related directly to their properties in a well-designed and executed test procedure.

Therefore, even a basic thermodynamic model of membrane passive diffusion provides a number of parameters that may be amenable to modeling. Key amongst these are the flux and diffusion coefficient. However, in practical terms a number of other quantities are also measured and have been considered for modeling. These are often dependent on the experimental protocol and requirements for the measurement.

There is great interest in measuring the ability of compounds to cross a large number of membranes. Readers may well wish to concentrate on optimizing drug activity, but they should also be aware of the efforts made in risk assessment. For instance, the experimental measurement, or prediction, of skin permeability can be useful in assessing the possible topical delivery of drugs. In addition, QSPRs for skin permeability are also very useful in risk assessment in a variety of areas, e.g., occupational or occasional exposure. Within these areas it is important to be able to quantify uptake for different scenarios, for instance, for a worker continually exposed to a pesticide spray, or an accidental splash. For other membranes, e.g., the blood–placenta barrier, most uses will be for risk assessment purposes. In terms of the evaluation of risk, there is a clear need for this information, as a potentially hazardous compound will have a reduced risk if it is not able to be absorbed.

In terms of predicting membrane permeability, QSARs, or other models, should ideally be mechanistic in nature.[2] What this means in practical terms is that the descriptors used in the model should be capable of some form of interpretation. Thus, it is expected that molecular features such as lipophilicity will generally control the passage across a lipid bilayer. Hydrogen bonding with phospholipids may retard transfer. In addition, molecular size and ionization may influence the ability to permeate through a membrane. Therefore, descriptors for these features can be used with confidence when modeling permeability. Inevitably, however, this is not an exhaustive list of physicochemical effects and other descriptors may be appropriate in certain QSPRs. In terms of an interpretable QSAR, preference is usually given to a transparent model with a 'clearly defined algorithm.' Regression analysis, in its various forms, meets these criteria. At the other end of the spectrum more multivariate and nonlinear techniques such as neural networks are perceived to be less interpretable. This does not exclude the use of such techniques, but it should be noted that transparency, which is most cherished, can be more difficult to ascertain from such models!

The aim of this chapter is to describe some of the approaches to predict, using a variety of in silico approaches, membrane permeability. The membranes considered in this chapter include the skin, the cornea, the placenta, the testis, as well as blood–(breast) milk transfer and renal excretion. Uptake across the gastrointestinal tract and blood–brain barrier (BBB) is not dealt with in this chapter as it is considered extensively elsewhere in this volume (*see* 5.28 In Silico Models to Predict Oral Absorption; 5.31 In Silico Models to Predict Brain Uptake). There is an emphasis on practical models for these membranes and an illustration of application, applicability, and future research requirements.

5.30.2 Dermal Penetration

The skin is the largest organ in the human body and provides an excellent biological protection against most chemicals. While the skin does have barrier properties, it is entirely possible for chemicals to permeate the skin. An appreciation of this process is beneficial for the increasingly popular use of transdermal drug delivery as well as for risk assessment following topical exposure to xenobiotics.

There are many methods to determine how much of a chemical will be transported across the skin, and many measurements that can be made and parameters that can be determined. These may be generalized into in vivo and in vitro measurements. From a practical point of view, in vitro measurements are easier to obtain and more widely available than in vivo measurements. They are also more likely to present the modeler with a greater selection of parameters (e.g., maximal flux, permeability coefficient etc.) to model. It should, however, be remembered that many experimental difficulties remain with measuring skin permeability either in vivo or in vitro. It is beyond the remit of this chapter to discuss these further[3–5] but these will impact on the precision of experimental data and thus the resultant accuracy of the models. Two guidelines for skin absorption have recently been adopted by the Organization of Economic Co-operation and Development (OECD), both on April 13, 2004, for in vivo measurement (OECD Guideline 427[78]) and for in vitro measurement (OECD Guideline 428[77]). It is hoped that use of these guidelines will provide reliable data for QSAR modeling in the future.

5.30.2.1 Quantitative Structure–Activity Relationships for Skin Permeability

5.30.2.1.1 Flynn database of skin permeability coefficients

There are a number of excellent reviews of QSARs that may be used to predict the ability of chemicals to cross the skin.[3–10] As such, this chapter will be restricted to the more useable and general models, or those with a particular interest or novelty. As with all QSAR modeling, predictivity is dictated by the data available to model. Key amongst the databases (see **Table 2** for a full listing) has been the Flynn[11] data set. This is a collation of historical skin permeability coefficients (K_p) for 94 compounds taken from literature sources. Although not originally intended for modeling, there have been numerous attempts to develop models from this compilation.

5.30.2.1.2 Modeling skin permeability coefficients with lipophilicity and molecular size

First amongst the modelers of the Flynn database were Potts and Guy,[12] who postulated that partition into the skin (K_p) is strongly related to lipophilicity, which may be quantified by the logarithm of the octanol/water partition coefficient (log P) and to diffusion through the skin, which is related to the size of the molecule, which in turn is easily quantified by molecular weight (MW). Thus, a general equation can be formulated to relate K_p to physicochemical properties:

$$\log K_p = \text{lipophilicity} - \text{molecular size} + \text{constant} \qquad [3]$$

or, putting the terms log P and MW into eqn [3]:

$$\log K_p = a \log P - b\, \text{MW} + c \qquad [4]$$

where a and b are the coefficients on the variables; and c is the constant.

Table 2   A nonexhaustive list of permeability databases suitable for modeling

Measurement and membrane	Number of compounds	Type of compounds	Comments	Reference
Permeability coefficients through human skin in vitro	94	Miscellaneous organic compounds, drugs, steroids	Historical database now updated with more reliable values (see Cronin[6])	Flynn[11]
Permeability coefficients through human skin in vitro	94	Miscellaneous organic compounds, drugs, steroids	Updated Flynn[11] database with new measurements	Cronin[6]
Permeability coefficients through human skin in vitro	99	Miscellaneous organic compounds, drugs, steroids	A reliable database of K_p values	Wilschut et al.[33]
Permeability coefficients through human skin in vitro	98	Miscellaneous organic compounds, drugs, steroids	Modified and updated Flynn[11] database	Buchwald and Bodor[20]
Miscellaneous in vitro and in vivo permeability endpoints for human skin	Over 300	Miscellaneous organic compounds, drugs, steroids	A web-based and fully searchable database with over 1600 in vitro and 800 in vivo results	Fitzpatrick et al.[22]
Permeability coefficients through human skin in vitro	119	Miscellaneous organic compounds and drugs	Reliable database corrected for temperature and ionization	Abrahams and Martin[24]
Human skin/water partition coefficients	45	Miscellaneous organic compounds and steroids	Reliable database	Abrahams and Martin[24]
Permeability coefficients through human skin in vitro	112	Miscellaneous organic compounds, drugs, steroids	Modified and updated (cleaned) database from Wilschut et al.[33] Available electronically	ten Berge database[32]
Permeability coefficients through human skin in vitro	127	Miscellaneous organic compounds, drugs, steroids	A fully validated database in terms of quality assurance, in addition to much additional information	Vecchia and Bunge[43]
Stratum corneum/water partition coefficient	76	Miscellaneous organic compounds, drugs, steroids	A fully validated database in terms of quality assurance, in addition to much additional information	Vecchia and Bunge[44]
Various databases of permeability through the cornea, sclera, conjunctiva, corneal epithelium, stroma, and endothelium	Over 150 compounds; the corneal permeability database is the largest (over 110 compounds)	Miscellaneous, mainly drugs	A number of different databases of varying quality	Prausnitz and Noonan[45]
Blood to breast milk (milk-to-plasma (M/P) concentration ratio)	60	Drugs	A compilation of historical data of varying quality	Agatonovic-Kustrin et al.[58]
Blood to breast milk (M/P concentration ratio)	115	Drugs	A compilation of historical data of varying quality	Katritzky et al.[61]

Table 2 Continued

Measurement and membrane	Number of compounds	Type of compounds	Comments	Reference
Various measures of blood to placenta transfer	Over 100	Miscellaneous, mainly drugs	A number of different databases of varying quality and size	Pacifici and Nottoli[51]
The unchanged drug excreted in the urine, expressed as a percentage of the intravenous dose	160	Drugs	Historical data compiled from Brunton et al.[54]	Manga et al.[55]
General pharmacokinetic properties of drugs	Hundreds	Drugs	Many and varied data	Brunton et al.[54]
Polydimethylsiloxane (PDMS)	256	Aromatic organic compounds	A consistent and high-quality database	Chen et al.,[69,70] as compiled by Cronin et al.[72]

Fitting the 94 K_p values to eqn [4] reveals the following, well-established equation:

$$\log K_p = 0.71(\pm 0.06)\log P - 0.0061(\pm 0.006)\,\mathrm{MW} - 6.3(\pm 0.8)$$
$$n = 93,\ r^2 = 0.67 \tag{5}$$

where the numbers in parentheses are the standard deviation on the regression coefficients; n is the number of observations; and r^2 is the square of the correlation coefficient.

Equation [5] is seminal in the development of QSPRs for skin permeability and is referred to frequently in this chapter, and by other authors, as the 'Potts and Guy equation.' Since its publication, more accurate $\log P$ data have become available and a more thorough QSAR analysis in statistical terms is warranted. To this end, Cronin[6] performed a new regression analysis of the complete Flynn[11] data set, resulting in the following:

$$\log K_p = 0.60\log P - 0.0058\,\mathrm{MW} - 2.64$$
$$n = 94,\ r^2 = 0.64,\ s = 0.79,\ F = 84 \tag{6}$$

where s is the standard error on the estimate; and F is the Fisher statistic.

Numerous workers were enthused by the Potts and Guy model to attempt to develop models for skin permeability coefficient further, in particular by improving statistical fit through the use of different physicochemical and structural properties. These mainly accounted for a subgroup of permeability coefficients for the steroids, which some workers also removed from the data set and modeled separately. Later studies e.g.,[1–4,13] showed that the original steroid permeability coefficients, which dated back to the 1960s, were not concordant with more recent measurements, indicating possible experimental error. When the original permeability coefficient data were substituted for the modern data, a better fit was observed. This would appear to be a salutary lesson for all QSAR modelers. The model, redeveloped by Cronin[6] with the new data, was:

$$\log K_p = 0.60\log P - 0.0053\,\mathrm{MW} - 2.64$$
$$n = 94,\ r^2 = 0.70,\ s = 0.68,\ F = 112 \tag{7}$$

Further analysis of the standard residuals from eqn [7] shows that some compounds, typically the steroids, are significant outliers. A number of approaches can be taken to this problem: (1) a 'let-it-be' attitude of doing nothing; (2) adding further descriptors to account for the steroids; or (3) removing the possibly erroneous data. As there is significant evidence that the permeability coefficients for the steroids were poor quality, Cronin[6] elected to remove 19 compounds, with steroid-type structures and for which there are no new data. Their removal results in the following QSAR, which has similar regression coefficients to eqn [7] and a better statistical fit:

$$\log K_p = 0.61\log P - 0.0058\,\mathrm{MW} - 2.54$$
$$n = 75,\ r^2 = 0.75,\ s = 0.63,\ F = 112 \tag{8}$$

Some compounds remain outliers, e.g., sucrose, although the model is not developed further. It is up to model users how they wish to use any of eqns [5–8] (or the QSARs) described below. Hopefully, future validation exercises for regulatory purposes will help resolve these issues.

The value of Potts and Guy approach was confirmed by Fujiwara et al.,[15] who extracted an 'essential' QSAR from those developed on a number of data sets. Ten skin permeability data sets were collected from the literature, including a total of 111 permeability coefficients in human, hairless mouse, or hairless rat skin for 94 structurally diverse compounds (the data are not reported in the manuscript but are almost certainly analogous to the Flynn[11] data set). QSARs based on log P and MW were developed for all 10 data sets. These QSARs were analyzed simultaneously, assuming that all sets shared a latent, common factor as far as the structure/permeability relationship is concerned. Using a novel statistical technique, the so-called latent score variable or latent membrane permeability, the analysis confirms the role of log P and MW, with approximately equal contributions to the prediction of skin permeability.

A database of over 100 physicochemical properties and skin permeation was compiled by Health Canada.[16] Analyses were performed using linear regression and QSARs for K_p based on log P were improved by grouping the compounds according to their respective molar volumes. This data set was reanalyzed by Cronin et al.,[17] who confirmed the role of the Potts and Guy approach, as well as the influence of a number of outliers (as noted above) on the data set. While the data set presented was larger than the original Flynn database, the work on these data has been criticized, correctly, as some of the data appear to have been predicted by the Potts and Guy equation (or something similar to this). This, of course, is not good practice in QSAR development. The problems of the Kirchner et al.[16] data set and the models developed by Cronin et al.[17] were exemplified by Poda et al.[18] and later by Frasch and Landsittel.[19]

5.30.2.1.3 Other approaches to modeling skin permeability coefficients

Buchwald and Bodor[20] collated data for the permeability coefficients of 98 compounds. This data set is, in effect, based on the Flynn database but, similar to the analysis of Cronin[6] (eqn [7]), updated to include more recently published data and also some of the corrected historical values (e.g., from Johnson et al.[21]). For the 98 compounds a relationship similar to the Potts and Guy[12] QSPR (eqn [8]) was obtained:

$$\log K_p = 0.60(\pm 0.04) \log P - 0.0052(\pm 0.0042) \text{ MW} - 6.02(\pm 0.12)$$
$$n = 98, \ r^2 = 0.75, \ s = 0.59, \ F = 144 \tag{9}$$

It is interesting to note that, while the data set has been 'cleaned' in terms of the steroid data, some of the outliers found by Cronin et al.[17] (e.g., digitoxin, ouabain, sucrose, etc.), which have legitimate physicochemical reasons for being removed, remained in the data set. This may explain the slightly lower r^2 than might reasonably be hoped for.

The analysis of the enlarged data set was extended by Buchwald and Bodor,[20] who correctly noted that log P and MW are relatively collinear for this data set. Collinearity of variables in a linear regression analysis is associated with instability, which may lead to false relationships being formed. To counteract this, log P and MW were replaced by terms relating hydrogen bond acceptor ability (N) and the effective van der Waals volume (V_e):

$$\log K_p = 0.013(\pm 0.0014) V_e - 0.49(\pm 0.036) N - 5.94(\pm 0.12)$$
$$n = 98, \ r^2 = 0.72, \ s = 0.60, \ F = 124 \tag{10}$$

In order to account for changes in hydrogen-bonding between the octanol and water phases, eqn [10] was rearranged to give:

$$\log K_p + 0.72 N = 0.021(\pm 0.0007) V_e - 6.25(\pm 0.14)$$
$$n = 98, \ r^2 = 0.91, \ s = 0.74, \ F = 1010 \tag{11}$$

This latter model (eqn [11]) requires further investigation to clarify its meaning and application. While the statistical fit of eqn [10] was slightly lower than that for eqn [9], the authors claim that these parameters negate the reliance on log P and V_e is a better estimate of molecular size than MW. Despite the lack of physicochemical similarity, Geinoz et al.[8] note that the two parameters N and V_e are highly correlated statistically, which many consider to be a drawback in regression-based QSARs.

A European Union Fifth Framework project, entitled Evaluation and Predictions of Dermal Absorption of Toxic Chemicals, more commonly known as the EDETOX project (project number QLKA-2000-00196), has confirmed the significance of the descriptors and mechanistic interpretation of the Potts and Guy equation by the measurement of further data. At the present time the best place to obtain information on the EDETOX project is from its website. The website also contains an online database that comprises data produced by in vitro and in vivo percutaneous penetration studies. These have been compiled from the published literature, and data generated during the EDETOX project have also been entered. The database has allowed for the Potts and Guy data set to be extended from 94 chemicals published by Flynn to over 140 (S. C. Wilkinson, 2004, personal communication). Further information on this database and modeling is available from Fitzpatrick *et al.*[22]

The effect of the experimental design of the measurement of skin permeability has, until recently, been insufficiently studied. For instance, Martinez-Pla *et al.*[23] investigated the effect of pH of formulations on the dermal absorption of chemicals. These authors confirmed using biopartitioning micellar chromatography (BMC) that the pH at which the measurement is taken will influence the modeling, and especially the parameterization of lipophilicity. The hypothesis that ionization, and also temperature of the test system, are important for modeling skin permeability was confirmed by Abraham and Martins.[24] These authors adjusted literature values of the permeability coefficients through human skin from water for ionization in water and for the temperature of the test system. The $\log K_p$ values obtained for 119 solutes, which were adjusted for temperature (to 37 °C) and for ionization, were correlated with Abraham descriptors to yield the following equation:

$$\log K_p = -0.11(\pm 0.12)\, E - 0.47(\pm 0.09)\, S - 0.47(\pm 0.09) A$$
$$- 3.00(\pm 0.15)\, B + 2.30(\pm 0.14)\, V - 5.43(\pm 0.10)$$
$$n = 119,\ r^2 = 0.83,\ s = 0.46,\ F = 112 \qquad [12]$$

where E is the solute excess molar refractivity (in units of $(cm^3\,mol^{-1})/10$); S is the solute dipolarity/ polarizability; A and B are the overall or summation hydrogen-bonding acidity and basicity respectively; and V is the McGowan characteristic volume (in units of $(cm^3\,mol^{-1})/10$).

Evaluating eqn [12] further, three separate test sets of 60 compounds had $\log K_p$ predicted with an s of 0.48 log units. The main factors shown to influence $\log K_p$ are solute hydrogen bond basicity, which lowers the permeability coefficient, and solute volume, which increases it.

Human skin/water partition coefficients ($\log K_{sc}$) were collected for 45 compounds and yielded the following equation [24]:

$$\log K^{SC} = 0.34(\pm 0.13)\, E - 0.21(\pm 0.10)\, S - 0.024(\pm 0.14)\, A$$
$$- 2.18(\pm 0.16)\, B + 1.85\, V(\pm 0.11) - 0.34(\pm 0.09)$$
$$n = 45,\ r^2 = 0.93,\ s = 0.22,\ F = 97 \qquad [13]$$

The TOPological Sub-structural MOlecular DEsign, topological substructural approach (the so-called TOPS-MODE methodology) was used by Estrada *et al.*[25] to develop a model for the permeability of a series of 12 commercial solvents through living human skin (for more details on the TOPS-MODE methodology, *see* Estrada *et al.*[25]). This model accounted for more than 95% of the variance in the experimental permeability of these solvents. Using the derived model, the structural factors responsible for the permeability of this series of solvents through living human skin were identified. Methyl groups bonded to heteroatoms or to CH_2 groups resulted in the greatest contributions to skin permeability. In contrast, groups of the type $X=O$ ($X = S, C$) were found to inhibit penetration. While there are only a small number of compounds utilized, this is an interesting paper as it makes a good attempt to relate structure back to activity (in this case, permeability) using a topological approach, something that is sadly missing from most topological QSAR studies. The applicability of the TOPS-MODE methodology should be demonstrated to a larger data set.

There may also be some value in the use of nonlinear models to predict skin permeability coefficients. For instance, Degim *et al.*[26] used artificial neural network (ANN) analysis to predict the skin permeability of selected xenobiotics. Log K_p, for 40 compounds, were obtained from various literature sources. A regression equation, based on $\log P$ and a cross-multiplied term for MW and partial charges of the penetrants (MW × charge) for the skin permeability for the set of 40 compounds had reasonable statistical fit (and similar to that for the Potts and Guy equation):

$$\log K_p = 0.59 \log P - 0.0028\, MW \times charge - 2.64$$
$$n = 40,\ r^2 = 0.67 \qquad [14]$$

An ANN was developed for the log K_p values for the same 40 compounds. It had $r^2 = 0.997$. While this seems to be a good model, further investigation is required to ascertain what a realistic level of predictivity is from such a model. In addition, these techniques are considered to lack transparency (i.e., no formal model is presented), are complex, and do not provide a easily portable algorithm as compared to, for instance, regression analysis.[2]

Nonlinear methods were also used in a similar fashion by Pannier *et al.*[27] These authors applied three fuzzy inference models using subtractive clustering to define natural structures within the data and assign subsequent rules. Each model was evaluated using the Flynn[11] data set. Fuzzy inference models successfully predicted skin permeability coefficients, with r^2 ranging from 0.83 to 0.97. The lowest correlation coefficient resulted from a model using log P and MW as inputs with two input membership functions evaluated by two fuzzy rules. The correlation coefficient of 0.97 occurred when log P and hydrogen bond donor activity were used as inputs with three input membership functions evaluated by three fuzzy rules. Again, there is a suspicion of overfitting in these models, and a realistic evaluation of predictivity is required.

A completely novel approach to modeling skin penetration has been applied by Santos-Filho *et al.*[28] A set of 152 compounds was collated from the literature; of these, 40 were selected to be representative and were those on which modeling was performed. This study combined so-called membrane-interaction (MI) descriptors with more traditional QSAR parameters. Molecular dynamics simulations were employed to determine how each test compound (solute) interacts with a model dimyristoylphosphatidylcholine (DMPC) monolayer membrane model. The authors claim that MI-QSAR models may capture features of cellular membrane lateral transverse transport involved in the overall skin penetration process by organic compounds. The MI methodology has previously been applied to model eye irritation and Caco-2 cell permeation coefficients, and is described in detail elsewhere.[29] A large number of regression-based models are reported; the best have been found to be those non-MI-QSAR and MI-QSAR descriptors. For instance:

$$\log K_p = 0.22 \log P - 0.14\, E_{ss}(\text{tor}) - 0.05\, E_{\text{inter}}(\text{vdW}) - 2.97$$
$$n = 39,\ r^2 = 0.80,\ q^2 = 0.77 \qquad [15]$$

where $E_{SS}(\text{tor})$ is the torsional energy at the total intermolecular system minimum energy for the solute; $E_{\text{inter}}(\text{vdW})$ is the energy of the van der Waals penetrant–membrane interaction; and q^2 is the leave-one-out cross-validated coefficient of determination.

Further improvements to these models are reported by increasing the number of descriptors up to six, and also by including quadratic terms. The use of these MI parameters is a novel and mechanistically appealing approach. It is hoped that the technique can be applied to all 152 compounds in the data set, and also that recommendations could be made for which is best of the 16 QSPRs published by Santos-Filho *et al.*[28] in terms of simplicity, robustness, and ease of application.

5.30.2.1.4 Modeling of maximal flux

As noted in the introduction, a number of other measurements are available from the experimental determination of skin absorption, some of which may be amenable for modeling. For instance, Magnusson *et al.*[30] have investigated the modeling of maximal flux ($J_{\max}$). These authors used values from an aqueous solution across human skin; these values were acquired or estimated from experimental data and correlated with solute physicochemical properties. Whereas epidermal permeability coefficients K_p are optimally correlated to log P, MW was found to be the dominant determinant of $J_{\max}$ for these literature data:

$$\log J_{\max} = -0.019\, \text{MW} - 3.90$$
$$n = 87,\ r^2 = 0.85 \qquad [16]$$

Addition of other physicochemical parameters to MW by forward stepwise regression only marginally improved the regression with a melting point (mp) term ($r^2 = 0.88$) and then hydrogen-bonding acceptor capability (H_a; $r^2 = 0.92$) being significant. Evaluation of eqn [16] was performed with a number of other data sets, including an aqueous vehicle with full- and split-thickness skin ($r^2 = 0.78$, $n = 56$). An analysis of the entire database gave the equation:

$$\log J_{\max} = -0.014\, \text{MW} - 4.52$$
$$n = 278,\ r^2 = 0.69 \qquad [17]$$

with inclusion of mp and H_a increasing r^2 to 0.76 ($n = 269$).

There is considerable motivation to develop simple qualitative rules for predicting skin penetration. Such a scheme was proposed by Flynn[11] for skin permeability coefficients. More recently Magnusson *et al.*[31] have proposed simple rules to define the potential of compounds to cross the skin. In vitro maximal flux values ($J_{\max}$) across human skin were collected for 87 compounds. Penetrants were assigned as being 'good' ($J_{\max} > 10^{-5.52}\,\text{mol cm}^{-2}\,\text{h}^{-1}$), 'bad' ($J_{\max} < 10^{-84}\,\text{mol cm}^{-2}\,\text{h}^{-1}$) or 'intermediate' based on mean ± 1 standard deviation. The study examined the

Table 3 Simple structural rules for the potential of compounds to cross the skin based on in vitro maximal flux values (J_{max}), as defined by Magnusson et al.[31]

Property	Good penetrants	Bad penetrants
MW	$\leqslant 152$	> 213
log S	> -2.3	< -1.6
HB	$\leqslant 5$	$\geqslant 4$
log P	< 2.6	> 1.2
mp	$\leqslant 432$	$\geqslant 223$

possibility of using fundamental and easily obtainable physicochemical properties, such as MW, mp (K), log P, water solubility (S, molarity), and the number of atoms available for hydrogen-bonding (HB). Structural rules were developed for good penetrants and are summarized in **Table 3**. Discriminant analysis using MW, HB and log P correctly assigned 70% of compounds, with no good penetrants being misclassified as bad or vice versa. This approach confirms the role of these fundamental physicochemical properties and provides very simple rules for their application.

5.30.2.2 Freely Available Expert Systems and other Formalized Models for the Prediction of Skin Permeability

Expert systems are useful as they have the capability of formalizing QSARs, QSPRs, and structure–activity relationships into user-friendly tools for the prediction of an endpoint directly from structural input. Several expert systems for the prediction of skin permeability are available. The best presented is the dermal permeability coefficient program (DERMWIN). This program is freely available from the US Environmental Protection Agency (EPA) through the EPISuite software and can be downloaded from its website. DERMWIN estimates the dermal permeability coefficient (K_p) and the dermally absorbed dose per event (DAevent) of organic compounds. Structures are entered into DERMWIN through the Simplified Molecular Input Line Entry System (SMILES) notation or can be retrieved for over 103 000 compounds from their Chemical Abstract Service (CAS) registry numbers. DERMWIN estimates a log P value for every SMILES string using the KOWWIN program and will automatically revert to an experimental log P value should one be available. DERMWIN uses a version of the Potts and Guy[12] general equation to predict K_p for all structures:

$$\log K_p = 0.71 \log P - 0.0061 \, MW - 2.72 \quad [18]$$

DERMWIN also makes K_p estimates for the alcohol, phenol, and steroid chemical classes using the following class-specific equations:

(Aliphatic) alcohols

$$\log K_p = 0.544 \log P - 2.88 \quad [19]$$

The aliphatic alcohol equation applies to any structure (except steroid-types) containing the –OH functional group connected to an aliphatic carbon.

Phenols

$$\log K_p = 2.39 \log P - 0.39 \, (\log P)^2 - 5.2 \quad [20]$$

The phenol equation applies to any structure containing the –OH functional group connected to an aromatic carbon.

Steroids

$$\log K_p = 1.01 \log P - 5.33 \quad [21]$$

DERMWIN also uses two methods to estimate the dermally absorbed dose per unit area per event (DAevent).

A calculation method is also provided by Ten Berge,[32] and is well described on his website. The website gives a good indication of the science behind the prediction. Skin permeation coefficients have been expressed mathematically as a complex of permeation coefficients, presenting the permeation through the subparts of the skin:

$$K_{p(\text{skin-water})} = \frac{1}{(1/(K_{\text{lip}} + K_{\text{pol}})) + (1/K_{\text{aq}})} \text{ cm h}^{-1} \tag{22}$$

$$\log K_{\text{lip}} = b1 + (b2 \times \log P) + b3 \times \text{MW}^{0.7} \tag{23}$$

$$K_{\text{pol}} = \frac{b4}{\text{MW}^{0.7}} \tag{24}$$

$$K_{\text{aq}} = \frac{b5}{\text{MW}^{0.7}} \tag{25}$$

where K_{lip} is the partition coefficient into the lipid layer; K_{pol} is the partition coefficient into the protein layer; K_{aq} is the partition coefficient into the aqueous layer; and $b1$, $b2$, $b3$, $b4$, $b5$ are regression coefficients.

Ten Berge[32] postulated that if the mathematical model is really a good description of the permeation process, it should be possible to estimate the regression coefficients from experimental results. To illustrate this, the paper of Wilschut et al.[33] provides an extensive database of measured dermal permeation coefficients for human skin. A slightly modified (cleaned) database was used to estimate the regression coefficients of eqns [22–25]. The results for the regression coefficients are: $b1 = -1.74$; $b2 = 0.72$; $b3 = -0.060$; $b4 = 0.00030$; and $b5 = 4.21$.

Regression coefficients $b1$, $b2$, and $b3$, accounting for the lipid permeation, are all statistically significant. However, $b4$ and $b5$ were not significant, which ten Berge[32] explained with pragmatic and empirically based considerations. If a compound has a $\log P$ less than 0, K_p is increasingly controlled by the permeation coefficient through the protein layer (regression coefficient $b4$). Conversely, if a compound has a $\log P$ greater than 4, the K_p coefficient is controlled by the aqueous permeation coefficient through the water layer below the stratum corneum (regression coefficient $b5$). The database of Wilschut et al.[33] is underpopulated with compounds with $\log P$ less than 0 and greater than 4, thus $b4$ and $b5$ were less accurately estimated. Moreover, $b4$ and $b5$ are only controlled by MW. These, and other conclusions regarding the regression coefficients $b1$–$b5$, illustrate the beauty of breaking the permeability process down into its component parts and also, importantly for modeling, that passive diffusion through the skin is not a simple single-event phenomenon. It can also be observed that water solubility is important, as an increasing skin permeation coefficient does not always result in a higher skin permeation rate of the pure substance in contact with the skin. This is because the substance has to permeate the postulated water layer below the stratum corneum.

Water solubility may also be modeled (by an inverse relationship) with $\log P$, and may explain why some models contain a quadratic function with $\log P$ and the relative success of nonlinear approaches (see 5.26 In Silico Predictions of Solubility).

Ten Berge[32] has also prepared some online software for use. A Windows version of the SKINPERM program is available from his website and formalizes eqns [22–25] for ease of use. Also available from this website is a Microsoft Excel file containing the database taken from Wilschut et al.[33] with slight modifications.

5.30.2.3   Other Issues in Modeling Skin Permeability

Current opinion is that percutaneous absorption can only be predicted for pure substances in an aqueous mixture, at an infinite dose.[3] While this is acceptable for modeling, it is simplistic and lacks an appreciation of the effects of formulations and mixtures, and the assumption of infinite dose is not realistic of most accidental and occupational exposure scenarios (although this may be what is strived for in dermal delivery of drugs).

Of these areas, there is certainly interest in predicting the relative enhancing effect of a solvent in a solution. There is much background knowledge regarding skin penetration enhancement[34,35]; however, currently no methods to 'predict' conclusively the effect of vehicles and solvents exist. Such an ability would be of particular benefit, for instance, in the assessment of toxicological endpoints such as skin sensitization.[36] An excellent review of structure–activity relationships for skin penetration enhancement, and compilation of data for this endpoint, is available from Kanikkannan et al.[37]

Some recent advances in the prediction of the effect of skin penetration enhancers are worthy of further consideration. Warner et al.[38] reported the findings of studies on the influence of n-alkanols, 1-alkyl-2-pyrrolidones, N,N-dimethlyalkanamides, and 1,2-alkanediols as skin permeation enhancers on the transport of a model permeant, corticosterone. The effects of sodium lauryl sulfate (SLS) on the stratum corneum partitioning and permeability in porcine skin of 10 agricultural and industrial chemicals in water, ethanol, and propylene glycol was investigated by van der Merwe and Riviere.[39] Some quantitative studies have also been performed to relate the degree of enhancement to structure. Borras-Blasco *et al.*[40] studied several concentrations of SLS on skin permeability. These effects were investigated using seven drugs, selected to cover a broad range in log P (from -0.95 to 4.2). Skin pretreatment with aqueous solutions of SLS does not increase K_p for the lipophilic compounds (log $P \geqslant 3$). For the other drugs, the increase in K_p depended on the concentration of SLS used in the skin pretreatment, and on the lipophilicity of the compounds tested. A QSAR was developed between SLS enhancer efficacy (ER) and log P.

QSAR techniques were also used by Ghafourian *et al.*[41] to investigate the activities of naturally occurring terpenes, pyrrolidinone, and *N*-acetylprolinate derivatives on the skin penetration of 5-fluorouracil, diclofenac sodium, hydrocortisone, estradiol, and benazepril. The resulting QSARs indicated that, for 5-fluorouracil and diclofenac sodium, less hydrophobic enhancers were the most active. For instance, for 26 terpene enhancers:

$$\log \mathrm{ER} = -5.79(\pm 0.95)\, q^- - 0.46(\pm 0.13)\, \mathrm{E_v}/10^4 + 0.14(\pm 0.26)$$
$$n = 26,\ r^2 = 0.63,\ s = 0.33,\ F = 19 \tag{26}$$

where q^- is the lowest atomic charge in the molecule; and E_v is the free energy of vaporization.

Although q^- is an electrostatic parameter explaining electrostatic interactions, it has been shown that it can also model hydrogen bonding in QSAR equations, with 'low' q^- values (high negative charges) leading to 'high' ability to accept hydrogens in hydrogen-bonding interactions. Thus it may be concluded that hydrogen-bonding has some controlling influence in skin permeability enhancement.

In contrast to eqn [26], skin penetration enhancement of hydrocortisone, estradiol, and benazepril showed a linear relationship with log P. For instance, the enhancement ratio of terpenes towards hydrocortisone was:

$$\log \mathrm{ER} = 0.15(\pm 0.03)\log P + 0.72(\pm 0.09)$$
$$n = 12,\ r^2 = 0.76,\ s = 0.089,\ F = 32 \tag{27}$$

There is clearly a lack of consistent information for the prediction of the skin penetration enhancement of chemicals. Some of the studies indicate that lipophilicity, as well as hydrogen-bonding, of enhancers is, at least partially, a driving force. There is a great requirement for a coherent set of systematic studies of varying enhancers on a suitable selection of permeants.

5.30.2.4   Data Sources for Skin Permeability Modeling

Reading this chapter, recent reviews, and reference to the recent EDETOX project should confirm that there are a reasonable number of data for skin absorption openly available (some of the major data sources are listed in **Table 2**). Data sourcing and quality assessment are important concerns, and these items are described in detail with regard to predictive modeling by Cronin.[42] There are a number of documented quality issues with the skin permeability data.[13,14] Despite that, some of the databases listed in **Table 2** are of particular use for modeling. When studying the data sources, the attention of the reader should be drawn to a number of issues. Firstly, the classic Flynn database[11] was never intended for the level of scrutiny, manipulation, and interpretation that it has received. Secondly, later databases, e.g., those presented by Wilschut *et al.*[33]; Vecchia and Bunge,[43,44] and the EDETOX project,[22] are of higher quality and have been better scrutinized in terms of suitability for modeling.

5.30.3   Corneal Permeability

An appreciation of the penetration of a compound through the corneal membrane has a number of potential applications. Most research is focused towards the ocular delivery of drugs. However, there could also be important implications for risk assessment, i.e., if a compound is known to be able to cross the cornea then there may be possible harmful effects to the eye. Conversely, if it is not able to cross the membrane then toxicity assessment may not be required. As described below, data for transcorneal permeability are available, and some modeling of the data has been performed, but generally there is much less interest than for other permeability endpoints. The lack of work in this area as compared to, for instance, dermal penetration may reflect the smaller market for the ocular delivery of drugs and that a full suite of validated in vitro assays is available for endpoints such as eye irritation and corrosion.

Measured corneal permeability coefficients are available from the open literature. For instance, Prausnitz and Noonan[45] have listed over 300 permeability measurements for the cornea, sclera, and conjunctiva, as well as from the corneal epithelium, stroma, and epithelium for nearly 150 compounds taken from more than 40 different studies. Inevitably this number should have increased in the best part of a decade since publication. For the reasons noted above, the database is made up mainly from druglike molecules.

The structural basis of the permeability of the membranes in the eye to chemicals was considered by Prausnitz and Noonan,[45] and latterly their data were formalized by Worth and Cronin[46] to develop some general models. Using corneal permeability coefficients ($K_{\text{p-corneal}}$) for 112 compounds taken from Prausnitz and Noonan,[45] Worth and Cronin[46] illustrated that, analogous to dermal penetration, a simple and mechanistic model could be developed on descriptors for lipophilicity (log P) and molecular size (MW). To model the data set, seven compounds were removed as being outside the acceptable range of the properties or as being outliers:

$$\log K_{\text{p-corneal}} = 0.39 \log P - 0.0033 \, \text{MW} - 4.77$$
$$n = 105, \; r^2 = 0.51, \; q^2 = 0.47, \; s = 0.46, \; F = 53 \qquad [28]$$

A more thorough QSAR analysis was performed using other established QSAR descriptors. A significant equation was derived based on descriptors for hydrogen-bonding (number of potential hydrogen-bonding a molecule can form; n_{H}), lipophilicity (log P) and molecular shape (the topological third-order kappa index: $\kappa3$). In order to develop models, a total of 17 compounds were removed from the data as being outliers in one or more of the three variables:

$$\log K_{\text{p-corneal}} = 0.23 \log P - 0.068 \, \eta_{\text{H}} - 0.17 \, \kappa3 - 4.35$$
$$n = 95, \; r^2 = 0.52, \; q^2 = 0.47, \; s = 0.46, \; F = 33 \qquad [29]$$

Under no circumstances could eqns [28] and [29] be considered as 'good' models in terms of statistical fit. However, it must be remembered that their fit probably reflects, and is realistic of, the measurement of corneal permeability, i.e., $s = 0.46$ in eqn [29], which is approximately equivalent to expected experimental error. Further evidence for the relevance of these equations is provided by Yoshida and Topliss,[47] who, analyzing smaller data sets (of approximately 20 compounds) confirmed the role of lipophilicity, hydrogen-bonding, and ionization. It should be noted that the effects of ionization were not considered by Worth and Cronin[46] and may provide some improvement in the models. The importance of lipophilicity in the prediction of transcorneal permeability was also noted by Kishida and Otori.[48] Edwards and Prausnitz[49] also modeled the passive, steady-state permeability of cornea and its component layers (epithelium, stroma, and endothelium) as a function of drug size and membrane-to-water distribution coefficient (ϕ). The approach provided strategies to enhance corneal permeability by targeting epithelial paracellular pathways for hydrophilic compounds ($\phi < 0.1 - 1$), epithelial transcellular pathways for intermediate compounds, and stromal pathways for hydrophobic compounds ($\phi > 10 - 100$). The effects of changing corneal physical properties (e.g., to mimic disease states or animals models) were also examined. This last point is of considerable interest, for instance, in understanding permeability for patients when a drug is not being applied to a healthy membrane (as would be expected in a disease state).

Agatonovic-Kustrin et al.[50] used a neural network approach, based on experimentally derived values of corneal permeability (log C), for 45 compounds. Using a very large descriptor pool (1194 calculated molecular structure descriptors) a genetic algorithm was used to select a suitable subset of parameters to model log C. A supervised network with a radial basis transfer function was applied, resulting in the best model having an architecture with four input descriptors and 12 hidden neurons. Inevitably, good correlations were obtained with these networks, with the correlation coefficient greater than 0.87 and 0.83 for the training and testing data sets respectively. Even with this good statistical fit, the small number of compounds in the model and complex architecture mean that the model must be treated with caution. The authors are pragmatic and recommend that expectations must be reasonable and large error limits accepted for predicted values from the models.

5.30.4 Transfer from Blood to Placenta

There is great and obvious concern about the effects of drugs and other xenobiotics on the developing fetus. This is exacerbated by the complexity and resource-hungry (in terms of time, finances, and animals) nature of reproductive toxicology. There are considerable efforts to find alternatives to the traditional reproductive toxicology tests, although there is a growing appreciation that a (possibly large) number of effects will need to be modeled. Part of this will obviously be the pharmacokinetic effects, and uptake is a simple place to start. It would be of great benefit to know whether a compound will be taken up across the placenta or into the testis (see Section 5.30.7). With regard to the development of new pharmaceuticals, benefits include reduction in the cost of development, the ability to identify toxic compounds early in the development pipeline, and a reduction in animal usage. Non governmental organizations are in favor of techniques such as the isolated dually perfused human placenta assay (see below), as it does not require animals and fresh placentas are constantly available. Regulatory agencies desire improved techniques to assess the safety of medicines more rapidly. An additional bonus could be that the understanding of the placental barrier may allow drugs to be developed to treat the unborn fetus.

Predictions of placental transfer could have a specific application in the risk assessment of pharmaceuticals, as well as other chemicals. If it may be predicted with some certainty that a compound will not be absorbed, then there should be no risk associated with it, despite any hazards that may be identified. Thus, the accurate modeling of the transfer of molecules across the placenta is desirable in terms of risk assessment at the very least.

It should, however, be remembered that, of mammalian membranes, the placenta is unique in that it separates the blood of two individuals and allows for the delivery of oxygen and nutrients to the fetus. As a starting place to learn about the assessment of placental transfer of drugs, the reader should consider the review from Pacifici and Nottoli.[51] This review describes not only the experimental measurement, but also the role of gross physicochemical properties important in placental transfer. There are a variety of methods to measure absorption across the placental membrane. A search of the available literature indicates that there is a reasonable amount of data available, some of which may be suitable for modeling. These data are given in many units (from a variety of assays) including: level in the placenta ($pg\,g^{-1}$); percentage transfer across the placenta; plasma enrichment (%); transplacental pulse fluxes ($\mu mol\,min^{-1}\,kg^{-1}$); permeability ($cm^3\,min^{-1}\,g^{-1}$); and the clearance index standardized to antipyrine:

$$\text{Clearance index (Cl)} = \frac{\text{Test compound clearance}}{\text{Antipyrine clearance}} \qquad [30]$$

Clearance towards the fetus is calculated as follows:

$$\text{Clearance index (Cl)} = \frac{(F_v - F_a) \times Q_f}{M_a} \qquad [31]$$

where F_v, fetal vein concentration; F_a, fetal artery concentration; Q_f, fetal flow rate; and M_a, maternal artery concentration.

Antipyrine was chosen for comparison as it is a small lipophilic molecule that is known to be transported across the placenta via passive diffusion. Its use as an internal standard is beneficial as it corrects for experimental variabilities such as placenta weight, blood flow, and placental exchange surface area. This then allows clearance index data from different sources to be compared without having compatibility issues. The use of the clearance index leads to a more consistent measure of transfer rate. A database has been created by the authors of this chapter (as part of the European Union 6th Framework ReProTect Integrated Project) containing clearance index data for 109 structurally diverse compounds and is due for publication in 2006 (please contact the authors for further details).

Despite the data being openly available, due to the structural differences in placenta from different species, extrapolation from other species to humans is not possible except for primates. Therefore an in vitro technique is now employed that uses fresh human placental tissue in order to determine experimentally placental transfer characteristics. This is known as the isolated dually in vitro perfused placenta assay.

There have been relatively few attempts to develop models for the transfer of molecules across the human placental barrier. As with the modeling of other membranes, these assume that a passive diffusion process is to be quantified. Thus, the capabilities to influence transfer of the active transporters known to be present in the placenta, such as P-glycoprotein (P-gp), multidrug resistance protein-1 and -2 (MRP1, MRP2), and breast cancer resistance protein (BCRP),[52] are ignored. In addition, it is appreciated that the placenta has considerable metabolism capacity, although this is not taken account of in the modeling.

The models that are available indicate the role of physicochemical properties such as lipophilicity. For instance, Akbaraly et al.[53] derived a QSAR using the placental transfer ratio (TR) with respect to antipyrine, the so-called clearance index described above. Akbaraly et al.[53] discovered a significant parabolic (as well as biphasic) relationship with $\log P$, following the removal of two outliers:

$$\log \text{TR} = 0.21(\pm 0.016) \log P - 0.074(\pm 0.0089)\,(\log P)^2 - 0.133(\pm 0.029)$$
$$n = 19,\ r^2 = 0.92,\ s = 0.95,\ F = 93 \qquad [32]$$

where the numbers in parentheses are 95% confidence intervals on the regression coefficients.

It should be noted that the $\log P$ values used were measured at $37\,^\circ C$ using a buffer at pH 7.4. Work in the laboratory of the authors of this chapter has not been able to repeat the success of the above equation due to differences in the reported measured $\log P$ values and those calculated using more up-to-date calculation methods.

Since the publication by Akbaraly et al.[53] in the mid-1980s there has been little subsequent development of QSARs for placental transfer. There have been a number of papers exploring the placental barrier and further data have been generated, but these have not been developed into a new QSAR. Although no quantitative studies have been published, Pacifini and Nottoli[51] summarized a number of possible factors determining the extent of placental transfer. For instance, drugs with MWs greater than 500 Da have an incomplete transfer across the human placenta. This is in line with similar findings relating to molecular size and passage across many, if not all, human membranes. Most (but not all) strongly dissociated acidic drug molecules have incomplete transfer. However, the extent of drug-binding to plasma protein does not appear to influence transfer of drugs across the human placenta. While these simple physicochemical rules may give some clues for risk assessment, they are not wholly suitable for discounting molecules as not being able to cross the placenta.

5.30.5   Renal Excretion

Excretion is an essential property to model if drug bioavailability is to be predicted successfully. Renal excretion of drugs and their metabolites is one of the key routes of drug clearance; the role of renal excretion in the prediction of pharmacokinetic parameters is well reviewed by Duffy.[7] It is also one of the more difficult properties to assess, as there is no direct measure of it, and those data that are available are not easy (i.e., often they are in the form of a percentage) to model. Typically data are obtained from literature sources, such as the ever-popular compilation of drug absorption, distribution, metabolism, and excretion data published by Brunton et al.,[54] commonly known as *Goodman and Gilman* (at the time of preparation of the manuscript, up to its 11th edition). Despite renal clearance being a useful parameter, there appear to be few models for it in the literature. An available model is described below.

Manga et al.[55] attempted to model the unchanged drug excreted in the urine, expressed as a percentage of the intravenous dose, administered for 160 drugs. This was an indirect attempt to model drug biotransformation. The data were categorized into classes according to excretion ranges. The cut-off values between those ranges were defined so as to enable optimal modeling. Modeling of the drug metabolism data was attempted utilizing a hierarchical approach comprising a set of rules combining both linear discriminant analysis and recursive partitioning. The model developed into a decision tree involving the following descriptors: the logarithm of the distribution coefficient at pH 6.5 ($\log D_{6.5}$), counts of H-bond donors, ionization potential, total energy, electronic energy, counts of OH groups and COOH groups and the sum of the total net charges. Overall, this model assigned 90% of the compounds correctly. The model was successfully validated using an external test set of 40 compounds. This approach indicates a number of issues in the modeling of these endpoints, namely that nonlinear techniques are required, and also the probable role of ionization (i.e., the use of $\log D$ as opposed to $\log P$) and hydrogen-bonding.

5.30.6   Transfer from Blood into Breast Milk

Xenobiotics, drugs, and other chemicals have the capability to pass from a mother's blood into breast milk. This could clearly be of great concern for a breast-feeding mother to prevent (accidental) exposure to a baby. There is no formal guideline to assess breast milk transfer, although it is accepted that the most straightforward method is to measure directly concentrations in the milk and blood. These may be extrapolated to others through the use of the milk-to-plasma (M/P) ratio. There are clear difficulties in obtaining these values, not least the practical and ethical problems associated with sourcing nursing women to take part in trials. Thus there are few M/P data, and it does present a greater emphasis on the modeling of this endpoint.

In terms of modeling this endpoint, again relatively little has been performed in comparison to, for instance, dermal penetration. We are fortunate to have an excellent review into models and methods for predicting drug transfer into human milk by Fleishaker[56] and the reader is referred to this paper as a starting point for information. Meskin and Lien[57] provided the most fundamental models of M/P based on regression analysis, and $\log P$ and MW for acidic drugs, and $\log P$ and a parameter based on ionization for basic drugs. The fit for these models was only moderate, but probably realistic of the data being modeled.

Agatonovic-Kustrin et al.[58] extended a previous study[59] using a neural network. The latter study expanded the previous data set of M/P values for 60 drugs to that for experimentally derived M/P values, taken from the literature, for 123 drug compounds. Descriptors were selected with a genetic algorithm and modeled using an ANN. A nine-descriptor neural network model was developed. The descriptors applied included the percentage of oxygen, parachor, density, highest occupied molecular orbital energy, and a number of topological indices. These were considered to indicate that molecular size, shape, and electronic properties are important to predict drug transfer into breast milk. The data collected by Agatonovic-Kustrin et al.[58] were reanalyzed by Yap and Chen.[60] These authors removed one erroneous value, norfluxexetine, and split the data set into a training set (102 compounds) and test set (20 compounds). Yap and Chen[60] also used a genetic algorithm to select seven descriptors possibly representing polarizability, electronegativity, and steric properties. A neural network approach was found to model the data much better than regression analysis on the basis of the prediction of values for the test set.

Katritzky et al.[61] have also searched the literature to obtain experimentally derived M/P ratio values for 115 widely used pharmaceuticals. Based on the data set, for 100 commonly used drugs, a seven-parameter regression-based QSAR model was derived showing satisfactory fit. While definitive mechanistic interpretation is not provided, the descriptors applied in the model encode possible information on hydrogen-bonding and molecular shape and symmetry. The commonality between the Agatonovic-Kustrin et al.[58] and Katritzky et al.[61] data sets has not been assessed at this time.

5.30.7 Blood–Testis

Related to the interest in predicting transfer across the placenta is interest in the modeling of the blood–testis barrier. If a chemical is able to cross this barrier from the blood, then it may damage the testes themselves, or the spermatosa contained therein. Ideally these effects will be found by reproductive toxicity testing, but it is likely that many are still unquantified, and may only be obtained by large-scale epidemiological studies (for a full description of the function of the blood–testis barrier, see Bart et al.[62]). The ongoing concerns about endocrine disruption, and its possible relationship to falling sperm counts in men, are a good example. The possible place of predictions of blood–testis transfer in integrated testing strategies can be considered to be similar to those for placental transfer (see above). Namely, if a compound is observed or predicted not to be able to be taken up into the testes, then it will be of lower concern for risk assessment. This has all the possibilities for reductions in costs and animal usage as noted previously. In addition, an understanding of the role of blood–testis transfer may assist in the optimization of drug delivery for the treatment of testicular tumors and other disease states.

Of all the membranes considered in this chapter, there are probably fewer openly available experimental data for blood–testis transfer than any other membranes. This probably reflects the subtlety, as well as the experimental complexity, of this endpoint. Inevitably, in terms of modeling this means that very limited (if any) models are available, and only a limited comprehension of the structural basis of transfer. A number of parameters from experimentation may be useful for modeling purposes, including: transfer rates ($\log k$ (min^{-1})); absolute concentrations ($ng\,g^{-1}$ (ml)); concentration in the testis ($\mu g\,g^{-1}$); and the testis uptake index. The latter index is calculated according to Sakiyama et al.[63]:

$$\text{Testis uptake index (TUI)} = \frac{\text{Extraction of the test compound}}{\text{Extraction of the reference compound}} \qquad [33]$$

To obtain these values, drug compounds are administered via injection along with radiolabeled butanol (which acts as a highly diffusible reference compound).

The literature contains some basic information about the role of structural properties to determine transfer from the blood to the testis. Johnson and Setchell[64] determined that the blood–testis barrier excluded high-molecular-weight immunoglobulins from seminiferous tubules. Okumura et al.[65] demonstrated that the permeability of the blood–testis barrier to nonelectrolytes was dependent on their molecular size. These studies are in further agreement with the known role of molecular size to influence uptake and suggest bulk flow through water-filled pores.

Due to the paucity of data, and the fact that there are few consistent data sets with more than five compounds, there are few QSARs to review for blood–testis transfer. Of these, Lien[66] modeled the transfer rate constant (k, min^{-1}) of some acidic drugs through the blood–testis barrier and determined a strong relationship with the chloroform/buffer distribution coefficient ($D_{CHCl_3/B}$):

$$\log k = \log D_{CHCl_3/B} - 1.59$$
$$n = 7, \ r^2 = 0.86, \ s = 0.19 \qquad [34]$$

The mechanistic interpretation of $\log D_{CHCl_3/B}$ is not clear, although it may emphasize the role of partitioning. Clearly, however, such a QSAR, based on only seven compounds and a relatively unfamiliar descriptor and with no proper documentation, is of limited practical use.

5.30.8 Other Membranes

5.30.8.1 Fish Gills

For environmental risk assessment it is useful to be able to estimate uptake of compounds, although these experiments are seldom performed, presumably due to their complexity and ease of measurement of fish lethality. Despite these practical difficulties some values for uptake across the fish gill are available and a limited amount of modeling is possible.

McKim et al.[67] published the percentage uptake efficiency of the trout (%uptake) for 14 chemicals. A limited QSPR analysis (unpublished results) reveals that one compound (mirex) is a significant outlier in the relationship with lipophilicity. The reason for this is unclear, though it is noted that the published (measured) log P of this compound was 5.28, whereas the value reported by McKim et al.[67] was 7.50. The latter value would appear to be a more reliable estimate of the hydrophobicity of mirex. Removal of this compound reveals a statistically significant quadratic relationship with log P (see 5.11 Passive Permeability and Active Transport Models for the Prediction of Oral Absorption):

$$\text{\%uptake} = 25 \log P - 2.5(\log P)^2 - 7.67$$
$$n = 13, \ r^2 = 0.93, \ s = 6.4, \ F = 62, \ \log P(\text{optimal}) = 5.14 \quad [35]$$

A quadratic relationship with log P was also seen for rate of absorption (primarily across the gills) in the guppy (k_{gill}) for the 16 compounds reported by Saarikoski et al.[68]:

$$\log k_{\text{gill}} = 0.80 \log P - 0.073(\log P)^2 + 0.017$$
$$n = 16, \ r^2 = 0.90, \ s = 0.17, \ F = 56, \ \log P(\text{optimal}) = 5.51 \quad [36]$$

5.30.8.2   Quantitative Structure–Permeability Relationships for other Routes of Uptake and Excretion

Table 1 clearly identifies a number of routes of exposure and clearance. This chapter has emphasized those where there is significant QSAR and QSPR analysis. It is with some surprise that few, if any, models exist for important routes such as uptake across the lung and excretion into bile.

5.30.9   Artificial Membranes

One of the major bottlenecks in developing QSPRs further is the paucity of available data. This has certainly held back the development of models for all the biological barriers discussed in this chapter. In addition, much of the variability in data quality and reliability can be accounted for by variations in the membranes themselves. It should not be ignored by modelers that most excised biological membranes are very delicate, and obtaining a source of a suitable membrane is difficult for a number of practical and ethical reasons. With regard to humans, as an example, it can only be obtained from cadavers, or following cosmetic (voluntary) surgery or other procedures such as amputations. It is well recognized that skin sourced from different parts of the body has different permeabilities, therefore obtaining consistent measured values may be difficult. Obtaining animal skin may also be fraught with ethical and legal reasons (e.g., in the UK, legislation requires pig skin to be boiled before distribution, thus destroying its integrity). Because of these and other issues, there has been considerable interest in the use of 'artificial' membranes. The area where this has been most commonly applied is in the assessment of skin penetration (particularly for product development of topical formulations). Most artificial membranes are some type of porous plastic, such as silastic and dialysis bags. This brief section will concentrate on the modeling of membrane transfer and the information it should give us. Consideration of other types of in vitro barrier systems, e.g., Caco-2, is provided elsewhere in this volume.

5.30.9.1   Polydimethylsiloxane

An excellent database of flux values through a PDMS membrane has been developed by Chen et al.[69,70] and Matheson et al.[71] The authors report flux through PDMS for over 250 compounds: the one drawback is that all compounds are aromatic or cyclic. There are no aliphatic or noncyclic compounds in the database. Despite this, the database has good structural heterogeneity. Flux values for 103 compounds from 15 ring classes, including benzene, quinoline, naphthalene, pyridine, naphthyridine, furan, benzofuran, imidazole, benzimidazole, indole, thiophene, pyrrole, pyrazole, pyridazine, and pyrazine were reported by Chen et al.[69] Maximum steady-state flux was measured using isopropyl alcohol as solvent. In a simple QSAR study, partial charge calculations combined with solubility and MW provided a good estimation of flux for all 15 classes of compounds. The flux of the imidazoles was found to be systematically slower than expected, while that of aliphatic amines was faster. Indicator variables for these types of compound were included in order to improve the model. Chen et al.[70] subsequently reported the maximum steady-state flux of 171 compounds through PDMS membranes, using isopropyl alcohol as solvent, as a demonstration of the former QSAR approach. The new data included a variety of substituted benzenes, naphthalenes, thiophenes, benzimidazoles, pyridines, quinolines, isoquinolines, pyrimidines, triazoles, and other heterocyclic classes. Chen et al.[70]

concluded that permeability decreases significantly as the atomic charge of diffusant is increased. It was considered that atomic charges can be used to represent polarity and is related to this property for an electrically neutral compound. In similar studies, Matheson and Chen[71] reported the release of 52 compounds (half benzene and half pyridine derivatives) into water, loaded at their solubility limits in a filter-supported PDMS matrix. Solubility of the solid compounds in the matrix was related to the mp, MW, partial atomic charge of the solutes, and an indicator variable for pyridines. Release of the compounds from the matrix into water was controlled by the matrix itself, with the initial release following the square root of time relationship. The release coefficient was defined as the slope of the linear portion of the Q-versus-t plot and was well predicted by mp, hydrogen-bonding energy group contribution, partial atomic charge in combination with an indicator variable as predictors (which could suggest it is solubility-dependent).

The data sets reported by Chen et al.[69,70] do provide a surprisingly good and consistent data set for modeling purposes. Cronin et al.[72] combined these two data sets and, after removing some duplicate compounds, developed QSPRs for maximum steady-state flux values for 256 compounds through a PDMS membrane (J_{PDMS}). Forty-three physicochemical parameters were calculated for each compound and their significance to flux determined. Removal of 14 outliers enabled derivation of a significant three-parameter QSPR based on the number of hydrogen bond acceptor and donor groups (HA and HD, respectively) and sixth-order path molecular connectivity ($^6\chi$):

$$\log J_{PDMS} = -0.56\,HA - 0.67\,HD - 0.801\,^6\chi - 0.38$$
$$n = 242,\ r^2 = 0.81,\ s = 0.46,\ F = 338 \tag{37}$$

In contrast to the findings of Chen et al.,[70] Cronin et al.[72] concluded that the mechanism of flux across a PDMS membrane is based mainly on hydrogen-bonding effects. In addition, Cronin et al.[72] investigated whether PDMS could act as a surrogate for skin flux measurements. There were no comparative flux measurements for both PDMS and skin, so an exercise was performed to compare the QSARs. Log P and MW (the parameters from the Potts and Guy equation (eqn [5])) were used to model PDMS flux. The coefficients of the QSPR based on these two parameters were not the same as those for skin, and a much reduced correlation coefficient was observed ($r^2 = 0.55$) as compared to eqn [37]. The authors concluded that permeation through PDMS occurs via a mechanism of action different from that of penetration of the skin in humans and that the artificial membrane may not be a good surrogate for skin. It must, however, be remembered that the data for J_{PDMS} were measured using isopropyl alcohol as solvent, and skin flux measurements use aqueous solutions. The effect of the solvent is not known at this time.

The data set measured by Chen et al.[69,70] and organized and modeled by Cronin et al.[72] was also modeled by Agatonovic-Kustrin et al.[73] A QSPR for the maximum steady-state flux values through PDMS was developed using an ANN. A total of 42 molecular descriptors were calculated for each compound and a genetic algorithm was used to select the most relevant. Again, in contrast to the findings of the previous studies, the model developed indicated that molecular shape and size, intermolecular interactions, hydrogen-bonding capacity of drugs, and conformational stability were important. A 12-descriptor neural network model has been developed for the estimation of log J values. The root mean square error was found to be 0.36 for the training set and slightly higher, at 0.59, for the test set.

5.30.9.2   Quantitative Structure–Property Relationships for other Artificial Membranes

There have been a small number of other QSPR studies on artificial membranes in the past few years. Cross et al.[74] examined the relationship between solvent uptake into a model membrane (silicone) with the physical properties of the solvents (e.g., solubility parameter, mp, MW). The topical penetration and retention kinetics of hydrocortisone from various solvents were also assessed. The sorption of solvents into the membrane (V_F) was related to differences in solubility parameters (the difference between the solubility parameter for the vehicle (δ_v) and the membrane (δ_m)), MW and hydrogen-bonding acceptor (α) and donor (β) ability:

$$V_F = 0.735 - 0.0015MW - 0.10|\delta_v - \delta_m| - 0.512\alpha + 0.087\beta$$
$$n = 28,\ r^2 = 0.76 \tag{38}$$

The authors concluded that a simple QSPR can predict the sorption of solvents into silicone membranes. Changes in solute diffusivity and partitioning appeared to contribute to the increased hydrocortisone flux observed with the various solvent vehicles.

The molecular properties that influence solute permeation across silicone membranes were assessed by Geinoz et al.[75] Permeability coefficients ($K_{p(sil)}$) of a series of model solutes across silicone membranes were determined from simple transport experiments using a pseudo-steady-state mathematical model of the diffusion process. QSPRs were developed utilizing the difference between octanol/water and 1,2-dichloroethane/water partition coefficients (Δlog $P_{oct/dce}$). This parameter is associated with H-bond donor activity and was supplemented by the computationally derived molecular hydrogen-bonding potential. Hydrogen bond donor acidity and lipophilicity were shown to influence permeation across silicone membranes greatly:

$$\log K_{p(sil)} = 0.56(\pm 0.37) \log P - 0.011(\pm 0.0067) \sum MHBP_{do} - 1.16(\pm 0.72)$$
$$n = 16, \ r^2 = 0.77, \ q^2 = 0.61, \ s = 0.35, \ F = 21 \qquad [39]$$

where $\sum MHBP_{do}$ is the sum of the calculated molecular hydrogen bond donor potential and the numbers in parentheses are 95% confidence intervals on the regression coefficients.

Furthermore, for a limited data set, a significant relationship was determined for solute permeation across the silicone membranes and through human epidermis (log $K_{p(epi)}$).

$$\log K_{p(sil)} = 1.15(\pm 0.36) \log K_{p(epi)} + 1.29(\pm 0.58)$$
$$n = 7, \ r^2 = 0.90, \ q^2 = 0.83, \ s = 0.19, \ F = 46 \qquad [40]$$

where the numbers in parentheses are 95% confidence intervals on the regression coefficients.

Geinoz *et al.*[75] concluded that they had determined the key molecular properties controlling solute permeation across silicone membranes. In contrast to the findings of Cronin *et al.*[72] who used a lateral approach to determine there was not a significant relationship between PDMS flux and that across skin, Geinoz *et al.*[75] showed that there is a good relationship between silicone and human skin K_p by the construction of a carefully chosen data set.

The modeling of permeability coefficients of various compounds through low-density polyethylene at 0 °C was investigated by Gonzalez and Helguera[76] using the TOPS-MODE approach. A model ($r^2 > 0.92$) was developed for the experimental permeability of 38 organic compounds using the TOPS-MODE parameters. The TOPS-MODE methodology outperformed other QSAR techniques and gives some interpretation as to the contribution of different fragments to the permeability coefficients.

5.30.10   Recommendations

There is a great deal of work in the development of QSAR and QSPRs for membrane permeability. This has varied greatly in terms of the approach and philosophy taken, and the quality of the resultant model. Some recommendations for use and the further development of the models are given below.

- QSAR modeling of membrane transfer of any sort works best (or, in most cases, requires) steady-state, infinite-dose data such as a permeability coefficient.
- QSAR modeling of membrane transfer works best for passive diffusion. Much more effort is required to understand, quantify, and model the effects of active transport and metabolism for many of the membranes described in this chapter.
- More effort is required (especially for risk assessment scenarios) to resolve issues for nonsteady-state application such as finite-dose exposures.
- More work is required on the effects of formulation, solvents, and mixtures through all membranes.
- There must be realistic expectations for predictions from models of membrane transfer taking into account experimental limitations.
- Evaluation of models, with the possible long-term aim of validation for regulatory purposes, is recommended.
- The development of simple and, preferably, freely available software (such as DermWin and SKINPERM) for the prediction of membrane transfer would be useful, but must be used with caution due to the limitations of the models.
- With the possible exception of skin, more high-quality, consistent data are required.
- A greater assessment of the role of artificial membranes to act as surrogates for biological membranes is required.

References

1. Martin, A.; Bustamonte, P.; *Physical Pharmacy*; Lippincott, Williams and Wilkins: Philadelphia, PA, USA, 1993.
2. Cronin, M. T. D.; Schultz, T. W. *J. Mol. Struct. (Theochem.)* **2003**, *622*, 39–51.

3. Jones, A. D.; Dick, I. P.; Cherrie, J. W.; Cronin, M. T. D.; van de Sandt, J. J. M.; Esdaile, D. J.; Iyengar, S.; ten Berge, W.; Wilkinson, S. C.; Roper, C. S. et al. *CEFIC Workshop on Methods to Determine Dermal Permeation for Human Risk Assessment*, Research report TM/04/0. Institute of Occupational Medicine: Edinburgh, Scotland, 2004.

4. Moss, G. P.; Dearden, J. C.; Patel, H.; Cronin, M. T. D. *Toxicol. In Vitro* **2002**, *16*, 299–317.

5. Patel, H.; Cronin, M. T. D. *Determination of the Optimal Physicochemical Parameters to Use in a QSAR-Approach to Predict Skin Permeation Rate. Final report. CEFIC-LRI project no. NMALRI-A2.2UNJM-0007*; Liverpool John Moores University: Liverpool, England, 2001.

6. Cronin, M. T. D. The Prediction of Skin Permeability using Quantitative Structure–Activity Relationships (QSARs). In *Dermal Absorption Models in Toxicology and Pharmacology*; Riviere, J. E., Ed.; CRC Press: Boca Raton, FL, USA, 2005, pp 113–134.

7. Duffy, J. C. Prediction of Pharmacokinetic Parameters in Drug Design and Toxicology. In *Predicting Chemical Toxicity and Fate*; Cronin, M. T. D., Livingstone, D. J., Eds.; CRC Press: Boca Raton, FL, 2004, pp 229–261.

8. Geinoz, S.; Guy, R. H.; Testa, B.; Carrupt, P.-A. *Pharm. Res.* **2004**, *21*, 83–92.

9. Walker, J. D.; Rodford, R.; Patlewicz, G. *Environ. Toxicol. Chem.* **2003**, *22*, 1870–1884.

10. Yamashita, F.; Hashida, M. *Adv. Drug Dev. Rev.* **2003**, *55*, 1185–1199.

11. Flynn, G. L. Physicochemical Determinants of Skin Absorption. In *Principles of Route-to-Route Extrapolation for Risk Assessment*; Gerrity, T. R., Henry, C. J., Eds.; Elsevier: New York, 1990, pp 93–127.

12. Potts, R. O.; Guy, R. H. *Pharm. Res.* **1992**, *9*, 663–669.

13. Degim, T.; Pugh, W. J.; Hadgraft, J. *Int. J. Pharm.* **1998**, *170*, 129–133.

14. Johnson, M. E.; Blankschtein, D.; Langer, R. *J. Pharm. Sci.* **1995**, *84*, 1144–1146.

15. Fujiwara, S. I.; Yamashita, F.; Hashida, M. *J. Pharm. Sci.* **2003**, *92*, 1939–1946.

16. Kirchner, L. A.; Moody, R. P.; Doyle, E.; Bose, R.; Jeffery, J.; Chu, I. *ATLA* **1997**, *25*, 359–370.

17. Cronin, M. T. D.; Dearden, J. C.; Moss, G. P.; Murray-Dickson, G. *Eur. J. Pharm. Sci.* **1999**, *7*, 325–330.

18. Poda, G. I.; Landsittel, D. P.; Brumbaugh, K.; Sharp, D. S.; Frasch, H. F.; Demchuk, E. *Eur. J. Pharm. Sci.* **2001**, *14*, 197–200.

19. Frasch, H. F.; Landsittel, D. P. *Eur. J. Pharm. Sci.* **2002**, *15*, 399–402.

20. Buchwald, P.; Bodor, N. *J. Pharm. Pharmacol.* **2001**, *53*, 1087–1098.

21. Johnson, M. E.; Blankschtein, D.; Langer, R. *J. Pharm. Sci.* **1997**, *86*, 1162–1172.

22. Fitzpatrick, D.; Corish, J.; Hayes, B. *Chemosphere* **2004**, *55*, 1309–1314.

23. Martinez-Pla, J.; Martin-Biosca, Y.; Sagrado, S.; Villanueva-Camanas, R. M.; Medina-Hernandez, M. J. *J. Chromatogr. A* **2004**, *1047*, 255–262.

24. Abraham, M. H.; Martins, F. *J. Pharm. Sci.* **2004**, *93*, 1508–1523.

25. Estrada, E.; Uriarte, E.; Gutierrez, Y.; Gonzalez, H. *SAR QSAR Environ. Res.* **2003**, *14*, 145–163.

26. Degim, T.; Hadgraft, J.; Ilbasmis, S.; Ozkan, Y. *J. Pharm. Sci.* **2003**, *92*, 656–664.

27. Pannier, A. K.; Brand, R. M.; Jones, D. D. *Pharm. Res.* **2003**, *20*, 143–148.

28. Santos-Filho, O. A.; Hopfinger, A. J.; Zheng, T. *Mol. Pharm.* **2004**, *1*, 466–476.

29. Kulkarni, A. S.; Han, Y.; Hopfinger, A. J. *J. Chem. Inf. Comput. Sci.* **2002**, *42*, 331–342.

30. Magnusson, B. M.; Anissimov, Y. G.; Cross, S. E.; Roberts, M. S. *J. Invest. Dermatol.* **2004**, *122*, 993–999.

31. Magnusson, B. M.; Pugh, W. J.; Roberts, M. S. *Pharm. Res.* **2004**, *21*, 1047–1054.

32. ten Berge, W. F. *Modeling Dermal Exposure and Absorption Through the Skin*. Available online at: http://home.planet.nl/~wtberge/skinperm.html (accessed May 2006).

33. Wilschut, A.; ten Berge, W. F.; Robinson, P. J.; McKone, T. E. *Chemosphere* **1995**, *30*, 1275–1296.

34. Walters, K. A.; Hadgraft, J., Eds. *Pharmaceutical Skin Penetration Enhancement*; New York: Marcel Dekker, p 448.

35. Williams, A. C.; Barry, B. W. *Adv. Drug Dev. Rev.* **2004**, *56*, 603–618.

36. Felter, S. P.; Robinson, M. K.; Basketter, D. A.; Gerberick, G. F. *Contact Dermatitis* **2002**, *47*, 257–266.

37. Kanikkannan, N.; Kandimalla, K.; Lamba, S. S.; Singh, M. *Curr. Med. Chem.* **2000**, *7*, 593–608.

38. Warner, K. S.; Li, S. K.; He, N.; Suhonen, T. M.; Chantasart, D.; Bolikal, D.; Higuchi, W. I. *J. Pharm. Sci.* **2003**, *92*, 1305–1322.

39. van der Merwe, D.; Riviere, J. E. *Toxicol.* **2005**, *206*, 325–335.

40. Borras-Blasco, J.; Diez-Sales, O.; Lopez, A.; Herraez-Dominguez, M. *Int. J. Pharm.* **2004**, *269*, 121–129.

41. Ghafourian, T.; Zandasrar, P.; Hamishekar, H.; Nokhodchi, A. *J. Cont. Rel.* **2004**, *99*, 113–125.

42. Cronin, M. T. D. Toxicological Information for Use in Predictive Modeling: Quality, Sources, and Databases. In *Predictive Toxicology*; Helma, C., Ed.; Taylor and Francis: Boca Raton, FL, USA, 2005, pp 93–133.

43. Vecchia, B. E.; Bunge, A. L. Skin Absorption Databases and Predictive Equations. In *Transdermal Drug Delivery*, 2nd ed.; Guy, R. H., Hadgraft, J., Eds.; Marcel Dekker: New York, 2003, pp 57–141.

44. Vecchia, B. E.; Bunge, A. L. Partitioning of Chemicals into Skin: Results and Predictions. In *Transdermal Drug Delivery*, 2nd ed.; Guy, R. H., Hadgraft, J., Eds.; Marcel Dekker: New York, 2003, pp 143–198.

45. Prausnitz, M. R.; Noonan, J. S. *J. Pharm. Sci.* **1998**, *87*, 1479–1488.

46. Worth, A. P.; Cronin, M. T. D. *ATLA* **2000**, *28*, 403–413.

47. Yoshida, F.; Topliss, J. G. *J. Pharm. Sci.* **1996**, *85*, 819–823.

48. Kishida, K.; Otori, T. *Jpn J. Ophthalmol.* **1980**, *24*, 251–259.

49. Edwards, A.; Prausnitz, M. R. *Pharm. Res.* **2001**, *18*, 1497–1508.

50. Agatonovic-Kustrin, S.; Evans, A.; Alany, R. G. *Pharmazie* **2003**, *58*, 725–729.

51. Pacifici, G. M.; Nottoli, R. *Clin. Pharmacokin.* **1995**, *28*, 235–269.

52. Leslie, E. M.; Deeley, R. G.; Cole, S. P. C. *Toxicol. Appl. Pharmacol.* **2005**, *204*, 216–237.

53. Akbaraly, J. P.; Leng, J. J.; Bozler, G.; Seydel, J. K. Quantitative Relationship Between Trans-Placental Transfer and Physicochemical Properties of a Series of Heterogeneous Drugs. In *QSAR and Strategies in the Design of Bioactive Compounds*; Seydel, J. K., Ed.; VCH: Weinheim, 1985, pp 313–317.

54. Brunton, L. L.; Lazo, J. S.; Parker, K. L.; *Goodman and Gilman's The Pharmacological Basis of Therapeutics* 11th ed.; MaGraw-Hill: New York, 2005.

55. Manga, N.; Duffy, J. C.; Rowe, P. H.; Cronin, M. T. D. *QSAR Comb. Sci.* **2003**, *22*, 263–273.

56. Fleishaker, J. C. *Adv. Drug Deliv. Rev.* **2003**, *55*, 643–652.

57. Meskin, M. S.; Lien, E. J. *J. Clin. Hosp. Pharm.* **1985**, *10*, 269–278.

58. Agatonovic-Kustrin, S.; Ling, L. H.; Tham, S. Y.; Alany, R. G. *J. Pharm. Biomed. Anal.* **2002**, *29*, 103–119.

59. Agatonovic-Kustrin, S.; Tucker, I. G.; Zecevic, M.; Zivanovic, L. J. *Anal. Chim. Acta* **2000**, *418*, 181–195.

60. Yap, C. W.; Chen, Y. Z. *J. Pharm. Sci.* **2005**, *94*, 153–168.

61. Katritzky, A. R.; Dobchev, D. A.; Hur, E.; Fara, D. C.; Karelson, M. *Bioorg. Med. Chem.* **2005**, *13*, 1623–1632.
62. Bart, J.; Groen, H. J. M.; van der Graaf, W. T. A.; Hollema, H.; Hendrikse, N. H.; Vaalburg W.; Sleijfer, D. T.; de Vries, E. G. E. *Lancet Oncol.* **2002**, *3*, 357–363.
63. Sakiyama, R.; Pardridge, W. M.; Musto, N. A. *J. Clin. Endocrinol. Metab.* **1988**, *67*, 98–103.
64. Johnson, M. H.; Setchell, B. P. *J. Reprod. Fertil.* **1968**, *17*, 403–408.
65. Okumura, K.; Lee, I. P.; Dixon, R. L. *J. Pharmacol. Exp. Ther.* **1975**, *194*, 89–94.
66. Lien, E. J. *Prog. Drug Res.* **1985**, *29*, 67–95.
67. McKim, J.; Schmieder, P.; Veith, G. *Toxicol. Appl. Pharmacol.* **1985**, *77*, 1–10.
68. Saarikoski, J.; Lindström, R.; Tyynelä, M.; Viluksela, M. *Ecotox. Environ. Safe.* **1986**, *11*, 158–173.
69. Chen, Y. S.; Yang, W. L.; Matheson, L. E. *Int. J. Pharmacol.* **1993**, *94*, 81–88.
70. Chen, Y. S.; Vayumhasuwan, P.; Matheson, L. E. *Int. J. Pharmacol.* **1996**, *137*, 149–158.
71. Matheson, L. E.; Chen, Y. S. *Int. J. Pharmacol.* **1995**, *125*, 297–307.
72. Cronin, M. T. D.; Dearden, J. C.; Gupta, R.; Moss, G. P. *J. Pharm. Pharmacol.* **1998**, *50*, 143–152.
73. Agatonovic-Kustrin, S.; Beresford, R.; Pauzi, A.; Yusof, M. *J. Pharmacol. Biomed. Anal.* **2001**, *26*, 241–254.
74. Cross, S. E.; Pugh, W. J.; Hadgraft, J.; Roberts, M. S. *Pharm. Res.* **2001**, *18*, 999–1005.
75. Geinoz, S.; Rey, S.; Boss, G.; Bunge, A. L.; Guy, R. H.; Carrupt, P. A.; Reist, M.; Testa, B. *Pharm. Res.* **2002**, *19*, 1622–1629.
76. Gonzalez, M. P.; Helguera, A. M. *J. Comp. Aid. Mol. Des.* **2003**, *17*, 665–672.
77. OECD. *Guideline for the Testing of Chemicals. Skin Absorption: In Vitro Method*, TG 428; 2004.
78. OECD. *Guideline for the Testing of Chemicals. Skin Absorption: In Vitro Method*, TG 427; 2004.

Biographies

Mark T D Cronin is Professor of Predictive Toxicology in the School of Pharmacy and Chemistry at Liverpool John Moores University, UK. He has over 15 years' expertise in the use of quantitative structure–activity relationships (QSARs) to predict toxicity and fate.

Mark Hewitt is a Research Assistant in the School of Pharmacy and Chemistry at Liverpool John Moores University, UK. He is developing alternative approaches to predict the mammalian toxicity of chemicals.

Comprehensive Medicinal Chemistry II
ISBN (set): 0-08-044513-6

ISBN (Volume 5) 0-08-044518-7; pp. 725–744

5.31 In Silico Models to Predict Brain Uptake

M H Abraham, University College London, London, UK
A Hersey, GSK Medicines Research Centre, Stevenage, UK

5.31.1 Introduction

One of the key pieces of information needed in designing novel therapeutic agents is the ability of a compound to cross the blood–brain barrier. This information is needed for different reasons depending on the therapeutic endpoint of the drug. For neurological and psychiatric disorders the molecular targets reside in the brain and it is necessary for compounds to achieve efficacious concentrations in the vicinity of the target. However, equally important today is the desire to prevent undesirable side effects in peripherally acting drugs and in many cases this includes the necessity to limit the brain penetration of compounds. An example of this would be the antihistamine compounds where the first-generation compounds penetrated the brain and caused sedation. In more recent years it has become desirable for new therapeutic agents in this class to have minimal brain penetration and consequently no sedative effect.[1] The nature of the blood–brain barrier[2] is such that many exogenous substances are unable to penetrate it. The junctions between the endothelial cells of the barrier are tight (transendothelial electrical resistance, TEER, value $\sim 8000\,\Omega$ cm),[2,3] eliminating the ability of compounds to enter the brain by a paracellular route, a route available for some compounds to be absorbed from the intestinal tract.[4] In addition, many potential therapeutic agents are substrates for the efflux protein P-glycoprotein (P-gp).[5,6] P-gp is present on the luminal side of endothelial cells of brain capillaries and has the effect of binding to compounds and pumping them out of the brain (*see* 5.32 In Silico Models for Interactions with Transporters). In consequence, the brain penetration for these compounds is severely limited. The effect of P-gp can be confirmed by studies that compare the uptake into the brain for compounds in wild-type (wt) mice and knockout (ko) mice (genetically modified mice where P-gp is absent). These studies show that the brain penetration is greatly enhanced in ko mice if a compound is a P-gp substrate.[5,6] In vitro models of the blood–brain barrier, in which the transport of compounds through cell monolayers is measured, have been set up and can be used as permeation screens. They 'mimic' the blood–brain barrier and hence provide the means for scientists to 'screen' compounds for their potential to penetrate the blood–brain barrier. The difficulty of maintaining the tight junctions between cells and in developing robust cell-based systems that express a full complement of transport proteins has meant that, to date, these in vitro methods have had limited utility in screening pharmaceutical agents although they have their place in specific mechanistic studies (*see* 5.12 Biological In Vitro Models for Absorption by Nonoral Routes). This means that in vivo experiments are still the main method of assessing the brain uptake of compounds. These experiments are costly both in terms of quantity of compound needed and number of animals required. Hence, there is a drive within the pharmaceutical industry to limit these experiments for both ethical and economic reasons. In silico models of brain penetration are one technique that can help in this respect.

Early in the drug discovery process large numbers of compounds are proposed for synthesis using combinatorial design methods. Here the use of in silico methods is particularly applicable if the calculations are sufficiently fast that they can profile thousands of compounds in a few minutes. In the later stages of lead optimization the project team is likely to have smaller numbers of compounds they wish to profile and will accept a slower calculation time per compound, particularly if it results in a more reliable prediction.

There is a wealth of information published in the literature and in the databases of pharmaceutical companies on compounds for which brain penetration has been measured[7–13] or on compounds that are known to exhibit CNS (central nervous system) mediated pharmacological effects.[14–17] By combining this information with knowledge about the molecular properties of compounds, it is possible to develop in silico models of brain penetration. Scientists are now able to use these models to make reasonably accurate predictions about which compounds are likely to have good brain penetration and which are not. This results in an ability to bias both the synthetic effort and prioritization for in vivo testing toward the compounds most likely to have the desired brain penetration profile.

Details of many in silico methods that attempt to predict a variety of measures of brain uptake have been published and reviewed.[7,18–21] Here, an attempt has been made to bring all this work together and analyze the common features responsible for brain uptake (or lack of it) and to consider their application in the delivery of therapeutic agents to the brain.

5.31.2   Measures of Brain Uptake

The terms 'brain uptake' or 'brain penetration' are frequently used to describe entry of compounds into the brain, but without any precise definition. They are in fact 'catch-all' phrases that encompass a variety of processes in which chemicals are transferred to the brain. The term 'brain uptake' is used here in the sense of a general term that includes various ways in which chemicals are transferred.

One of the first measures of brain uptake was that of biological activity. Hansch and co-workers[22] studied the hypnotic activity of a series of depressants and noted that the activity reached a maximum when the water/octanol partition coefficient, as $\log P_{oct}$, was near two. This seems to be the origin of the 'rule-of-two.' Timmermans and colleagues[23] appeared to confirm a parabolic relationship between brain uptake and $\log P_{oct}$, but on inspection it is very difficult to distinguish a parabola at all. Waterhouse[24] claimed that Dishino and co-workers[25] had shown a parabolic relationship between brain uptake and lipophilicity as measured by $\log P_{oct}$. The measure of brain uptake was the per cent extraction, as determined by a single pass method involving sequential injection of ^{15}O-labeled water and ^{11}C-labeled compound in an ethanol–saline bolus. This measure of per cent extraction would be expected to lead to a sigmoid curve. Dishino and co-workers[25] specifically explained the downturn for the more lipophilic compounds as being due to mixing of the ethanol–saline bolus with blood – the more lipophilic compounds are retained by blood and hence their per cent extraction is less than expected.

Oldendorf[26,27] described a much more rigorous measure of brain uptake, the brain uptake index (BUI). A mixture of a ^{14}C-labeled compound and ^3H-labeled water in saline is injected into a rat, and the radioactivity in the brain recorded 15 s after administration. The BUI is then defined as follows, where BUI = 100 for water:

$$\text{BUI} = 100(^{14}C/^3H)_{brain}/(^{14}C/^3H)_{saline} \qquad [1]$$

The BUI is a useful rank order of brain uptake,[26–28] but is difficult to analyze by physicochemical methods, since it is neither a rate nor an equilibrium quantity. Other measures of brain uptake are now much more commonly used.

The first such measure was introduced by Rappoport and colleagues[29,30] and involved intravenous injection. This was later developed[31,32] into the in situ perfusion technique, in which an aliquot portion of whole blood or plasma or saline containing a radiolabeled compound is injected into an artery and the amount that perfuses through the blood–brain barrier is determined at a given time, typically 1–2 min after perfusion. Many workers have developed this technique, including Deane and Bradbury,[33] and Pardridge and colleagues.[34] Results are obtained as a 'one point' kinetic experiment, from which a perfusion coefficient (PC; in units of $cm\,s^{-1}$) may be obtained. More usually, permeation is expressed as a product of permeability and surface area (PS; in $cm^3\,s^{-1}\,g^{-1}$). As mentioned, the perfusate can be whole blood, plasma, or various saline solutions, since there is no one generally accepted perfusate. There is no reason why the factors that influence perfusion should be quantitatively the same for the various perfusates, and so the different perfusates must of necessity be considered to be a different system, as regards physicochemical analysis. In all cases, however, the time scale of the experiment is very short, i.e., usually of the order of 1–2 min.

The second measure was described by Young and colleagues,[8,9] in two key papers that have formed the foundation of in vivo data on blood–brain distribution. A compound is administered to a rat and after a given time, the concentration of the compound in blood and in brain is determined. Usually, several rats are sacrificed in this way, in order to ascertain if the blood to brain concentration ratio reaches a steady state. If so, then the blood–brain distribution coefficient (BB) is defined as:

$$BB = [\text{conc. of compound in brain}]/[\text{conc. of compound in blood}] \qquad [2]$$

Typically, the distribution coefficient is obtained at intervals of time up to several hours, in contrast to the short time scale of the perfusion technique.

In addition to the in vivo technique developed by Young and colleagues,[8,9] environmental scientists have obtained distribution coefficients for volatile environmental pollutants by an in vitro method. The partition of a volatile compound between the gas phase and blood (K_{blood}) and between the gas phase and homogenized brain (K_{brain}) is determined as:[35,36]

$$K_{\text{blood}} = [\text{conc. of compound in blood}]/[\text{conc. of compound in air}] \qquad [3]$$

$$K_{\text{brain}} = [\text{conc. of compound in brain}]/[\text{conc. of compound in air}] \qquad [4]$$

Then,

$$BB = K_{\text{brain}}/K_{\text{blood}} \qquad [5]$$

Concentrations are usually taken as mol dm^{-3} in the gas phase and blood, and mol kg^{-1} in brain.

Abraham and co-workers[10] combined the in vivo BB values of Young and colleagues[8,9] for drug compounds with a compilation of in vitro values for environmental volatile organic compounds given by Abraham and Weathersby[37] to yield a set of BB values that has provided the basis of subsequent data sets on blood–brain distribution. It should be noted that the in vivo BB values of Young and colleagues[8,9] refer to rat blood and rat brain, but the in vitro values for environmental volatile organic compounds refer to both human and rat blood and brain.

Pardridge[38] has recently criticized the use of blood–brain distribution by the pharmaceutical industry, and has strongly argued that only the perfusion method is meaningful. However, this argument overlooks the very different time scales of the two methods. In some cases, it might be appropriate to consider perfusion results that indicate how quickly a drug will diffuse through the blood–brain barrier. In other cases, it might be more useful to consider the blood–brain ratio after several hours. In addition, there are complications regarding perfusion from saline of ionizable species, which are discussed later. Most recent work on physicochemical analyses has focused on blood–brain distribution, but both the perfusion and the distribution methods will be considered here.

A third measure of brain uptake is referred to as CNS penetration or CNS activity. Although this seems to be more specific than 'brain uptake' it is not a very well defined parameter, and is usually recorded just as CNS + or CNS −. In the review by Seelig and colleagues[14] CNS +/− are assigned using various processes. Later workers have expanded this compilation.[15,16] Ajay and colleagues[17] assigned CNS + on the basis of therapeutic activity of various classes of drugs, but pointed out that the assignment of CNS − is difficult on this basis because absence of therapeutic activity cannot be taken as synonymous with inability to cross the blood–brain barrier. The assignment of CNS + and CNS − will be considered later.

5.31.3   Blood–Brain Distribution

5.31.3.1   Calculations

By far the majority of papers on in silico calculations of 'brain uptake' have dealt with blood–brain distribution,[20,39–66] defined by eqn [2] for the direct determination of drugs, and through eqns [3] to [5] for the indirect determination of volatile pollutants. It should be pointed out that all these in silico calculations attempt to explain passive transport from blood to brain, and do not include processes such as efflux mechanisms or active transport.

If in silico calculations are to be used for screening of large numbers of candidate drugs, it is important that the data set covers as wide a chemical space as possible, especially as the chemical space covered by experimental methods is small. Furthermore, it is essential to provide some predictive assessment of any calculation. Very often, validation methods such as 'leave-one-out' are used, but these provide only an assessment of internal consistency. The only way to assess the predictive power of a calculation is to separate the initial data set into a training set and a test set.

The former is used to construct an equation and to predict values of the test set. Since the test set has not been used to obtain the equation, a comparison of predicted and experimental values for the test set yields an assessment of predictive capability. It is common to compare the sets of predictive and experimental values by plotting one set against the other. This is not correct, and it is better simply to obtain statistics on the two sets of data, such as the average error (AE), which is a measure of bias, and the absolute average error (AAE) and standard deviation (s), which are measures of goodness-of-fit.

The predictive assessment for a test set refers only to the chemical space of the compounds that make up the test set, and not to the chemical space of the training set. This is why it is important for the test set to cover as wide a range of chemical space as possible, and why test sets of only a handful of compounds yield no real general assessment of predictability.

In **Table 1**, a summary of the various in silico calculations to date is given. The number of compounds in the training set and the test set (n), and the standard deviation found for the training set and the test set are given. In some cases, the descriptor used is listed. ΔG_h^o is the standard free energy of hydration of the gaseous compound and PSA is the polar surface area; ANN means that an artificial neural network was used. As can be seen, many investigators give no results for test sets at all, and so no predictive assessment can be made in these cases.

In **Table 2**, the 'best' predictive methods are listed, in terms of the s value for the test set. There are several calculations that yield predictions with an s-value of around 0.4 log units for a reasonable number of compounds. We first concentrate on these particular procedures, and then draw some general conclusions from all the calculations listed in **Table 1**.

The simplest calculation is that of Keserü and Molnár[47] who used only one descriptor – the standard free energy of hydration of the gaseous compound (ΔG_h^o) in kJ mol^{-1}. This was related to values of log BB through the equation:

$$\log BB = 0.26 + 0.035 \, \Delta G_h^o \, (\text{kJ mol}^{-1})$$
$$n = 55, \; r^2 = 0.72, \; s = 0.37 \tag{6}$$

Here and elsewhere, n is the number of data points (usually the number of compounds), r is the correlation coefficient, and s is the standard deviation. Some years before this work, Lombardo and colleagues[40] also used ΔG_h^o as a predictor of log BB, although their calculation by AMSOL 5.0 with AM1-SM2 hydration model is considerably more time consuming than that of Keserü and Molnár. Even the latter calculation, at 6 s per molecule, is not very fast in the context of high-throughput screening. Lombardo and colleagues obtained eqn [7], but with ΔG_h^o in kcal mol^{-1}. The corresponding equation in the same units as eqn [6] is given as eqn [8], and it can be seen that the slopes of eqns [6] and [8] are considerably different.

$$\log BB = 0.43 + 0.054 \, \Delta G_h^o \, (\text{kcal mol}^{-1})$$
$$n = 55, \; r^2 = 0.67, \; s = 0.41, \; F = 108.3 \tag{7}$$

$$\log BB = 0.43 + 0.013 \, \Delta G_h^o \, (\text{kJ mol}^{-1}) \tag{8}$$

Not only do the slopes of the equations differ by a factor of 2.7, but also the actual numerical values of ΔG_h^o (in the same units) differ greatly between the two methods of calculation. Some comparisons are given in **Table 3**, mostly for simple compounds for which experimental values are available.[67] Differences between the results of Keserü and Molnár and experimental values are so large that clearly great care must be taken over predictions for compounds that are even slightly outside the chemical space used. One reason why the method of Keserü and Molnár[47] works as well as it does is that the slope in eqn [6] is so small that even large errors in ΔG_h^o are transformed into small errors in the calculated log BB values.

The method of Luco[44] also yields statistics for a test set of 37 compounds ($s = 0.37$) that show the predictive capability of the method. Luco uses 18 descriptors from an original set of 25 that include connectivity indices, a shape factor, molecular surface, and counts of hydrogen bond acid and hydrogen bond base groups. A partial least squares (PLS) analysis reduces the variables to a three-component model that correlates log BB for 55 compounds with $n = 55$, $r^2 = 0.85$, $s = 0.32$, and $F = 102$; however, note that the F-statistic is calculated on the basis of three variables and not 18. For an independent test set of 37 compounds, the s-value was only 0.37 log units, so that the method of Luco is one of the best, as regards predictability of log BB. However, it not clear how the method can be implemented; Luco[44] calculates the connectivity-topological descriptors with a computer software program obtained privately, and similarly for the calculation of charge and geometrical indexes.

Norinder and Österberg[50] also used a PLS method but this was based on only six descriptors. Four electrotopological state indices were calculated using Molconn-Z, a commercial software package, and the water/octanol partition coefficient ClogP and molar refraction were calculated through the software package Sybyl. Hence, all six descriptors

Table 1 Summary of recent in silico models of log BB

Equation	Desc.[a]	Training set		Test set		Reference
		n^b	s^c	n	s	
Brewster et al.	3	60	0.38			39
Lombardo et al.	$1(\Delta G_h^o)^d$	55	0.41	6	0.62	40
Kelder et al.	$1(\text{PSA})^e$	45	0.36			41
Clark	2	55	0.35	10^f	0.25	42
Norinder et al.	14	56	0.31	6	0.55	43
Luco	18	58	0.32	37	0.37	44
Feher et al.	3	61	0.42	14, 25	0.76, 0.80	45
Ertl et al.	1(PSA)	45	$r^2 = 0.84$			46
Keserü and Molnár	$1(\text{Gw})^d$	55	0.37	5, 25	0.14, 0.37	47
Liu et al.	2	55	0.35	11	0.43	48
Liu et al.	ANN	55	0.30	11	0.39	48
Kaznessis et al.	5	76	0.17	4	0.48	49
Norinder and Osterberg	6	28	0.37	30	0.35, 0.38	50
Norinder and Osterberg	6	58	0.34			50
Platts et al.	6	148	0.34			51
Platts et al.	6	74	0.34	74	0.38	51
Rose et al.	3	102	0.45	21^g	0.47	52
Subramanian and Kitchen	8	58	0.31	39	0.52	53
Ooms et al.	$72(4)^h$	79	0.78			54
Lobell et al.	5	48	(0.26)	17	(0.41)	55
Klamt et al.	5	65?	0.43			56
Norinder et al.	2	55	0.34			20
Salminen et al.	3	23	0.32			57
Kaliszan et al.	2	20	0.27			58
Iyer et al.	5	56		7	0.48^i	59
Sun	Fragment	57	0.26	13^j	0.67	60
Hou and Xu	3	72	0.36	35	0.41^k	61
Hou and Xu	3	78	0.37	35	0.44^k	61
Cabrera et al.	3	114	0.42			62
Cabrera et al.	3	81		33	0.43	62
Pan et al.	2	150				63
Winkler and Burden	7	85	0.37	21	0.54	64
Yap and Chen	9, MLR	125		30	0.40	65
Yap and Chen	9, MLFN	125		30	0.39	65

continued

Table 1 Continued

Equation	Desc.[a]	Training set		Test set		Reference
		n^b	s^f	n	s	
Yap and Chen	7, GRRN	125		30	0.37	65
Narayanan and Gunturi	3	88	0.42			66
Narayanan and Gunturi	4	88	0.39			66

[a] Desc. is the number of descriptors used.
[b] n is the number of data points (compounds).
[c] s is the standard deviation.
[d] Standard free energy of hydration.
[e] Polar surface area.
[f] Combined test set, calculated in this work.
[g] An external validation set $(+/-)$ was used; 27/28 compounds were correctly predicted.
[h] Thirty-one descriptors, in a four-component PLS analysis; $r^2 = 0.78$.
[i] Calculated from data given by Iyer et al.
[j] A second 'test' set was constructed by swapping compounds from the original training and test sets, and yielded $s = 0.33$; however, this is no longer an independent test set.
[k] From eqn [21] in the text. This is the corrected equation in the errata.[61] For the combined test sets, AE = 0.04, AAE = 0.29, and $s = 0.41$ calculated in this work.

Table 2 Best predictions of log BB (test sets)

Equation	Desc.	n	s
Luco	18	37	0.37
Keserü and Molnár	$1(\Delta G_h^o)$	25	0.37
Norinder and Osterberg	6	30	0.35
Platts et al.	6	74	0.38
Subramanian and Kitchen	8	39	0.52
Feher et al.	3	14, 25	0.76, 0.80
Hou and Xu	3	35	0.44
Cabrera et al.	3	33	0.43
Winkler and Burden	7	21	0.54
Yap and Chen	9	30	0.40
Yap and Chen	9	30	0.39
Yap and Chen	7	30	0.37

Desc. is the number of descriptors used.

can be calculated with available software. Two training sets of 28 compounds yielded r^2-values of 0.75 and 0.78 with corresponding s-values of 0.39 and 0.37 log units, respectively. The test sets of 30 compounds had s-values of 0.35 and 0.38 log units. The method of Norinder and Osterberg[50] thus seems to be one of the easiest to implement, and provides a predictive capability unlikely to be greatly exceeded. The only caveat is that the data set is restricted to a total of 58 compounds.

The method of Platts and co-workers[51] is quite different. Five descriptors plus an indicator variable are used in a multiple linear regression analysis. The descriptors are the well-known solvation parameters used in the general equation[68,68a]:

$$SP = c + e \cdot E + s \cdot S + a \cdot A + b \cdot B + v \cdot V \qquad [9]$$

Table 3 Calculated and experimental values of standard free energies of hydration, ΔG_h^0, in kJ mol^{-1}

Compound	Calc.[47]	Calc.[40]	Exp.[66]
Clonidine	−15.78	−41.92	N/A
Benzene	6.91	−2.89	−3.60
Butanone	−15.19	−13.22	−15.52
Ethanol	−1.71	−20.92	−20.95
1,1,1-Trichloroethane	2.28	−3.97	−0.80
Isoflurane	−6.08	2.51	0.40
Pentane	9.37	7.66	9.70

where SP is the dependent variable such as values of log BB for a series of compounds and the independent variables are solute properties or descriptors as follows. E is the solute excess molar refractivity in units of (cm^3 mol^{-1})/10, S is the solute dipolarity/polarizability, A and B are the overall or summation hydrogen bond acidity and basicity, respectively, and V is the McGowan characteristic volume in units of (cm^3 mol^{-1})/100. Platts and co-workers[51] constructed a large data base of 148 compounds, and used a training set of 74 compounds to predict the remaining log BB values in the test set with an s-value of 0.38 log units. They used an indicator variable (I) for carboxylic acids such that $I = 1$ for carboxylic acids and $I = 0$ for all other compounds. For the total set of 148 compounds Platts and co-workers found:

$$\log BB = 0.06 + 0.47\,E - 0.86\,S - 0.59\,A - 0.71\,B + 0.90\,V - 0.56\,I$$
$$n = 148,\ r^2 = 0.74,\ s = 0.34,\ F = 83 \qquad [10]$$

The indicator variable shows that carboxylic acids are less likely to distribute into the brain than expected from the usual descriptors. Equation [10] is about as good, statistically, as other equations for log BB, but has a decided advantage in pointing out the chemical factors that influence log BB. This will be returned to later. An initial disadvantage of eqn [10] was that descriptors had to be obtained from various experimental values. However, Platts and colleagues[69] showed that a fragmentation scheme could be constructed for the very rapid calculation of descriptors, and hence of log BB. More recently, a fragmentation scheme for the calculation of the descriptors has become available commercially,[70] and so the initial disadvantage no longer applies.

The largest data set used to date is that compiled by Yap and Chen,[65] who set out log BB values for 175 compounds. Of these, 12 were considered as outliers from previous work, and descriptors could not be calculated for the rare gases, leaving 159 compounds for the analysis; 125 were used to construct models, and 30 compounds were left as an independent test set. Yap and Chen set up three methods, all based on descriptors calculated by the DRAGON software. A simple multiple linear regression (MLR) used nine descriptors. A neural network, MLFN, also used nine descriptors and a Baysian-type neural network, GRNN, employed seven descriptors. Results are summarized in **Tables 1** and **2** and show that, for the test set, s-values of 0.37 to 0.40 (calculated in this work) could be achieved.

The above methods can all predict test sets of log BB to about 0.37 log units, so there is not much to choose between them in terms of predictability. The method of Luco seems difficult to implement, but the methods of Keserü and Molnár,[47] Norinder and Osterberg,[50] Platts and co-workers,[51] and Yap and Chen[65] can all be carried out using commercial software.

In addition to these methods, a number of others can be used to obtain log BB values, though to less accuracy as regards predictions. Of these, the calculations of Feher and colleagues[45] are very simple indeed. Only three descriptors are used: a calculated log P_{oct} value, the polar surface area, and the number of hydrogen bond acceptors in an aqueous medium, n_a. For a training set of 61 compounds, Feher and colleagues[45] obtained the equation:

$$\log BB = 0.43 - 0.39\,n_a + 0.11 \log P_{oct}(\text{calc}) - 0.0027\,\text{PSA}$$
$$n = 61,\ r^2 = 0.73,\ \text{RMSE} = 0.42,\ F = 51 \qquad [11]$$

The root mean square error (RMSE) is almost the same as the standard deviation (s) except that RMSE has one more degree of freedom (hence, RMSE is trivially less than s). The results of predictions through test sets are a little disappointing, with s-values of 0.76 and 0.81 for test sets with 14 and 25 compounds. These s-values have been calculated in this work. Feher and colleagues give RMSE values of 0.63 and 0.79 but these seem to be calculated from plots of predicted versus experimental values, which is not quite correct.

A number of other methods lead to reasonable results, but have no advantage over those mentioned above. Pan and co-workers[63] employed a very simple method that uses only two descriptors, the calculated log P_{oct} value and the polar surface area. Although they used a large data set of 150 compounds, they give few statistics for the full data set, only $r^2 = 0.69$ and $q^2 = 0.59$, and restrict further details to rather small subsets. One of the more recent models is that of Winkler and Burden[64] who used a Bayesian neural network to examine a number of sets of descriptors. They found that a set of seven quite simple molecular descriptors yielded the best results. However, the obtained statistics (see **Tables 1** and **2**) are not as good as those of several other models. Narayanan and Guntari[66] examined a data set of 88 log BB values, and calculated 324 descriptors by in-house software. The best three-descriptor model had $s = 0.42$ and the best four-descriptor model had $s = 0.39$, which are no better than many other models. Since the descriptors are very difficult to interpret in terms of chemical structure, the method offers no advantages.

Hou and Xu[61] use only three descriptors, a calculated log P_{oct} value, SlogP, based on their own methodology,[71] the high-charged polar surface area (HCPSA), and a molecular weight (MW) term that takes the value MW − 360 for compounds with MW > 360 and zero for other compounds A linear combination yields the equation:

$$\log BB = 0.008 + 0.20 \text{ SlogP} - 0.014 \text{ HCPSA} - 0.014 \text{ (MW} - 360)$$
$$n = 72, \ r^2 = 0.79, \ s = 0.36, \ F = 82.6 \tag{12}$$

Two test sets were predicted through eqn [12]. For the combined sets of 35 compounds we calculate that AE = 0.04, AAE = 0.29, and $s = 0.41$, using the equation and results listed in the errata of Hou and Xu.[61] Following the success of eqn [12], Hou and Xu[61] re-parametrized the calculation of SlogP, and replaced HCPSA with the more easily calculatable high-charged topological polar surface area (HCTPSA) and derived the equation:

$$\log BB = 0.13 + 0.16 \text{ SlogP} - 0.013 \text{ HCTPSA} - 0.015 \text{ (MW} - 360)$$
$$n = 78, \ r^2 = 0.74, \ s = 0.38, \ F = 66.9 \tag{13}$$

Note that in their eqn [24] given as our eqn [13], Hou and Xu[61] use HCPSA, but this must be an error. Also, it seems as though eqn [13] should have $n = 72$ but this was not mentioned in the errata.[61] When used to predict a combined test set of 35 compounds, an absolute mean error of 0.40 was found, and we calculate an s-value of 0.44 log units. The advantage of eqn [13] over eqn [12] is that the former can be calculated with greater speed, and is more suitable for high-throughput screening.

Cabrera and co-workers[62] use only three descriptors in their calculation:

$$\log BB = -0.03 - 0.05 \times 10^{-3} \text{ PS} \times \text{AM} + 0.23 \text{ H}$$
$$n = 114, \ r^2 = 0.70, \ s = 0.42, \ F = 127.8 \tag{14}$$

We use a simple notation for the descriptors: PS is related to the polar surface, AM is the atomic mass (which appears to be the mass of atoms in a molecule excluding hydrogen atoms), and H is related to hydrophobicity. The descriptors are calculated by an in-house program, TOPS-MODE. Thus, although eqn [14] is very good, and so is the predictive power, with $s = 0.43$ for a test set, the method will not be very easy to implement.

5.31.3.2   Factors that Influence Blood–Brain Distribution

Although a wide variety of descriptors have been used in calculations of log BB values, they share some common factors. In general, they fall into two classes: (a) descriptors of size of the compound; and (b) descriptors of the polar nature of the compound. In class (a) are molar refraction, connectivity and topological indices, molecular mass, and surface area. In class (b) are descriptors such as polar surface area, partial charges, and functions of hydrogen bond acid or hydrogen bond base groups. In general, class (a) descriptors drive compounds into brain and class (b) descriptors drive compounds into blood. This is, indeed, what would be expected from known properties of blood and brain. According to Abraham and Weathersby[37] and from the tissue compositions given by Poulin and Theil,[72] brain is considerably more lipophilic than blood, and nonpolar compounds will tend to prefer brain and polar compounds prefer blood.

This can be put on a more quantitative basis using eqn [10] of Platts and co-workers. The descriptors E and L are nonpolar or size-related, and an increase of either of these leads to an increase of log BB. The descriptors S, A, and B are all polar quantities, and an increase in any of these leads to a decrease in log BB. Many descriptors, including PSA, and counts of hydrogen bond acid groups and hydrogen bond basic groups, are approximations to A or B (or to A plus B).

We can also investigate the use of ΔG_h^o the standard free energy of hydration, as a single descriptor in the calculations of Lombardo and co-workers[40] and of Keserü and Molnár[47] (see eqns [6] and [8]). Equation [9] has been applied[66] to the partition of compounds between air and water, as log K_w:

$$\log K_w = -0.99 + 0.58\ E + 2.55\ S + 3.81\ A + 4.84\ B - 0.87\ V \qquad [15]$$

The descriptor ΔG_h^o can be converted into log K_w to obtain another version of eqn [7]:

$$\log BB = 0.43 - 0.07\ \log K_w \qquad [16]$$

Then insertion of log K_w from eqn [15] into eqn [16] yields eqn [17]:

$$\log BB = 0.50 - 0.04\ E - 0.19\ S - 0.28\ A - 0.36\ B + 0.06\ V \qquad [17]$$

Thus, indirectly, the use of ΔG_h^o as a descriptor implies that dipolarity/polarizability, hydrogen bond acidity, and hydrogen bond basicity all favor blood, and volume favors brain. The only anomaly is that eqn [17] has a small negative e-coefficient instead of a positive e-coefficient.

A similar analysis can be carried out for log P_{oct} as a predictor of log BB. Abraham and co-workers have shown that[73]:

$$\log P_{oct} = 0.09 + 0.56\ E - 1.05\ S + 0.034\ A - 3.46\ B + 3.81\ V \qquad [18]$$

So if log BB $= c + m$ log P_{oct}, then the best that can be done is to divide eqn [18] by 4.5 to make the b-coefficient and the v-coefficient agree with eqn [10]:

$$\log BB = 0.02 + 0.13\ E - 0.23\ S + 0.01\ A - 0.77\ B + 0.85\ V \qquad [19]$$

There is now one glaring difference between eqn [19] derived from log P_{oct} and eqn [10], namely that the a-coefficient in eqn [19] is almost zero. Thus, because log P_{oct} does not involve hydrogen bond acidity, and log BB depends quite markedly on hydrogen bond acidity, log P_{oct} cannot be used as a sole descriptor for log BB. Of course, there is nothing to prevent log P_{oct} from being one out of a number of descriptors.

Abraham and co-workers[74] took advantage of this in their log Pplus method, which simply uses log P_{oct} as a descriptor and adds the required parameters, to yield eqn [20]. It would be interesting to apply the method to a much larger data set of compounds:

$$\log BB = 0.06 + 0.20\ \log P_{oct} - 0.51\ A - 0.50\ B$$
$$n = 57,\ r^2 = 0.90,\ s = 0.20,\ F = 136.1 \qquad [20]$$

The Δlog P parameter of Seiler,[75] is defined through eqn [21], where P_{cyc} is the water/cyclohexane partition coefficient, and has been suggested[9] as a possible predictor of log BB:

$$\Delta\log P = \log P_{oct} - \log P_{cyc} \qquad [21]$$

The factors that influence Δlog P have been established as shown in eqn [22], where the sign has been changed to show more clearly the connection between Δlog P and log P_{oct}:

$$-\Delta\log P = 0.03 + 0.25\ E - 0.68\ S - 3.82\ A - 1.45\ B + 0.83\ V \qquad [22]$$

Although the signs of the coefficients in the equation for $-\Delta$log P match those in eqn [10] for log BB exactly, the relative magnitude of the coefficients demonstrates that Δlog P cannot be used to predict log BB to any useful degree of accuracy.

In conclusion, there are a number of in silico calculations of blood–brain distribution by passive transport that yield predictions of log BB to around 0.35 to 0.45 log units that could be used for screening purposes. It should be pointed out that the chemical space covered by the test sets in this work is limited by the small number of compounds studied, that is, between 25 and 74. Indeed, only one set of predictions[51] uses a test set with more than 40 compounds.

One point of interest that seems to have been overlooked is that the various calculations of log BB all use descriptors for the neutral species of proton acids and proton bases. Blood has a pH of 7.4 and so on the usual pH/pK_a calculation, a base with a pK_a of 9.4, for example propranolol, will be 99.9% ionized at this pH. So either the ionized form of strong bases can partition across the blood–brain barrier almost exactly as the neutral form, or the usual pH/pK_a calculation does not apply. The latter might well be the case. The pK_a we quote for propranolol is that for aqueous solution, and there is no reason why it should be the same in blood. In the latter, a compound will be partly in an aqueous environment and partly bound to plasma proteins, etc. The free fraction in the aqueous environment could ionize according to the aqueous pK_a, but the bound fraction will ionize according to the appropriate pK_a for the microenvironment.

Quite recently, Rodgers and colleagues[76,77] measured and set up calculations for plasma water/brain partition coefficients. Their results cannot be compared to those surveyed above, but it is of interest that in their calculations they use pK_a values of drugs that refer to aqueous solution. Although, following Rodgers and colleagues[76,77] the fraction of a drug in the aqueous environment ionizes according to the aqueous pK_a, this may not apply to the bound fraction.

5.31.4   Brain Perfusion

5.31.4.1   Calculations

The perfusion technique has been used on numerous occasions to obtain PCs or PSs. It has proved very difficult to reach any general conclusions in terms of correlations or predictions of either PC or PS values. One major problem is that different workers have used different perfusate solutions, usually made up in whole blood or in saline. Clearly, diffusion from whole blood will not be the same as diffusion from saline, and so data sets from whole blood or saline experiments cannot be combined. The second problem is that, for the most part, PC or PS values have been obtained only for small data sets.

Much of the early work correlated log PC with various combinations of $\log P_{oct}$ and molecular weight. Levin[78] showed a linear connection between log PC for perfusion from blood and the function $\log [P_{oct} (MW)^{-1.5}]$ for 22 compounds but Pardridge and Triguero[34] obtained only a poor plot of $\log [PC (MW)^{-1.5}]$ against $\log P_{oct}$. Chikhale and co-workers[79] obtained a reasonable correlation of log PS for seven peptides against $\Delta \log P$ but a poor correlation against $\log P_{oct}$. Begley[80] collected a number of log PC values and showed that there was a rather scattered plot against $\log P_{oct}$ for 24 compounds. Several other drugs were considerable outliers and included substrates for the P-glycoprotein efflux pump as well as those transported by carrier-mediated mechanisms. Bodor and Buchwald[81] carried out a similar analysis and 'omitting the obvious outliers' obtained a reasonable correlation of log PC. However, the PC values used appear to include permeation from both saline and plasma (corrected for protein binding), and for some compounds log D_{oct} at pH 7.4 was used instead of $\log P_{oct}$. Unfortunately, no numerical details of the PC values are given. In addition to this use of $\log P_{oct}$ as a descriptor, Liu and Hider[82] showed that for a series of hydroxypyridinones, a plot of log PS against the % polar surface area (%PSA) was an excellent straight line.

The first logical attempt to correlate log PC or log PS values was that of Gratton and co-workers.[11] These workers measured log PS values from saline at pH 7.4 and correlated them through the Abraham equation. They used the pH/pK_a hypothesis to correct for ionization of compounds at pH 7.4; unlike the case for transfer from blood, this is straightforward for perfusion from saline. Abraham[12] later questioned whether a full correction for ionization was necessary. He collected data on perfusion from saline at pH 7.4, and for 30 neutral compounds found the correlation in eqn [23] where PS is in units of $cm^3 s^{-1} g^{-1}$

$$\log PS = -0.64 + 0.31\,E - 1.01\,S - 1.90\,A - 1.64\,B + 1.71\,V$$
$$n = 30,\ r^2 = 0.87,\ s = 0.52,\ F = 32.2 \tag{23}$$

There were two proton acids and five proton bases that were left out of the correlation, as shown in **Table 4**. Also given in the table are observed values corrected for ionization in saline at pH 7.4, using the given pK_a values, on the assumption that only the neutral species permeates. Both the observed and the ion-corrected log PS values can be compared to the calculated values on eqn [23] for the neutral species. In some cases (1A, 4B, and 6B) the calculated values are quite close to the observed (uncorrected) values, implying that no correction need be carried out. In other cases (2A, 3B, and 5B) it seems as though only a 'partial' correction is needed, and for 7B a full correction is needed. Correlations of the 30 neutral compounds against $\log P_{oct}$ and against %PSA showed little more than random effects, with $r^2 = 0.15$ and $r^2 = 0.17$, respectively.

The only other recent analysis of log PS values is that by Liu and co-workers.[13] They measured PS values in units of $cm^3 s^{-1} g^{-1}$ for permeation from saline for 28 compounds. Of these, 20 compounds underwent passive perfusion.

Table 4 log PS values for perfusion from saline at pH 7.4 for proton acids and proton bases

Name[a]	pK_a	log PS		
		Obs.	Full correction	eqn [23]
22001 (1A)	5.30	−2.92	−0.82	−2.87
12002 (2A)	3.20	−3.29	0.91	−1.64
Propranolol (3B)	9.40	−1.02	0.98	−0.47
11003 (4B)	8.10	−1.53	−0.75	−1.71
13007 (5B)	10.10	−1.57	1.13	−0.74
95005 (6B)	7.95	−2.61	−1.95	−2.28
26006 (7B)	8.91	−3.31	−1.79	−1.48

[a] Numbers as in [12]. A denotes a proton acid; B denotes a proton base.

Liu and co-workers[13] started by calculating 50 descriptors for each compound, but reduced these to three descriptors in their preferred equation:

$$\log PS = -2.09 + 0.26 \log D_{\text{oct}} + 0.058 \text{ vas\_bas} - 0.009 \text{ TPSA}$$
$$n = 23, \ r^2 = 0.74, \ s = 0.50, \ F = 18.2 \qquad [24]$$

Here, D_{oct} is the water/octanol partition coefficient at pH 7.4, vas_bas is the van der Waals surface area of the basic atoms, and TPSA is the topological van der Waals polar surface area. Equation [24] includes five proton acids and three proton bases. For all 23 compounds, a very poor correlation of log PS was found against log P_{oct}.

The results of Liu and co-workers[13] are not incompatible with the findings of Abraham.[12] In eqn [24] the small coefficient in the log D_{oct} term means, in effect, that only a partial correction for ionization is made. The results in **Table 4** suggest that for six out of the seven compounds a partial correction is necessary in order to comply with eqn [23].

Clearly, as regards any sort of medium- to high-throughput screening, considerably more work will have to be carried out on the perfusion technique in order to ascertain the role of strong proton acids and proton bases. This is by no means straightforward. Wohnsland and Faller[83] have already shown that permeation through a hexadecane membrane is complicated by the presence of an unstirred water layer (UWL) adjacent to the membrane. This leads to permeation–pH profiles for strong acids and strong bases that are not those calculated on the usual pH–pK_a hypothesis. The effect of strong acids and strong bases on perfusion from saline suggests that an UWL is present in this system. Unfortunately, perfusion–pH profiles cannot be obtained in the brain perfusion system because the pH must be maintained at 7.4 in order not to compromise the integrity of the blood–brain barrier. In view of the difficulty of understanding the perfusion mechanism and of predicting perfusion of strong acids and strong bases, the suggestion of Pardridge[38] that blood–brain distribution be abandoned in favor of the perfusion method seems premature, to say the least. In addition, it should be noted that the most recent work on perfusion has used saline as the perfusate, rather than plasma or whole blood. It is not clear what the connection is between perfusion from saline and perfusion from plasma or whole blood, particularly with respect to perfusion of proton acids and proton bases.

5.31.4.2 Comparison of Distribution and Perfusion

Just as we can use eqn [9] to compare blood–brain distribution with, for example, water to octanol partition, so we can compare brain perfusion with various other processes. We list the coefficients in eqn [9] in **Table 5**. In order to compare coefficients, the form of the equations must be exactly the same, and so we use the equation of Platts and colleagues[51] without the extra indicator variable:

$$\log BB = 0.04 + 0.51 \ E - 0.89 \ S - 0.72 \ A - 0.67 \ B + 0.86 \ V \qquad [25]$$

Included also are coefficients for partition in the ethylene glycol to heptane system[84] suggested to be a model system for permeation of peptides in a number of systems.[85,86]

Table 5 Coefficients in eqn [9] for various processes

Process	e	s	a	b	v
$\log P_{oct}$	0.56	− 1.05	0.03	− 3.46	3.81
− $\Delta \log P_{cyc}$	0.25	− 0.68	− 3.82	− 1.45	0.83
$\log P_{eg/hept}$	− 0.08	− 1.20	− 3.79	− 2.20	2.09
$\log$ BB eqn [25]	0.51	− 0.89	− 0.72	− 0.67	0.86
$\log$ PS	0.31	− 1.10	− 1.90	− 1.64	1.71

Table 6 Comparisons between processes

Process	cos θ		Distance between the points	
$\log P_{oct}$	0.77	0.79	4.14	3.40
− $\Delta \log P$	0.74	0.89	3.21	2.15
$\log P_{eg/hept}$	0.85	0.97	3.70	2.06
$\log$ BB	1.00	0.94	0.00	1.76
$\log$ PS	0.94	1.00	1.76	0.00

By inspection of **Table 5**, it can be seen that $\log P_{oct}$ cannot be a good model for either $\log$ BB or $\log$ PS, because it has no real term in solute hydrogen bond acidity. The relationships between some of the other sets of coefficients are not so clear, but can be recognized by two particular methods. In the first method,[87,88] the five coefficients are regarded as points in five-dimensional space. It is simple to calculate distance between the points, and the smaller the distance the closer are the corresponding processes in chemical terms. In the second method, due to Ishihama and Asakawa,[89] the five coefficients define a line passing through the origin. Then the smaller the angle θ between any two lines, the nearer are the processes in mathematical or correlational terms.

Values of the distance between the points and cos θ are given in **Table 6**. Note that cos θ is not directly connected to the correlation coefficient, r, and that for there to be a reasonable correlation between two processes, cos θ must be very close to unity. As for the distance between the points, for there to be a strong chemical similarity between two systems, then this distance must be less than about 0.5 (on the arbitrary scale in **Table 6**).

On both measures of similarity, the processes that refer to $\log$ BB and $\log$ PS (the latter for permeation from saline) are not closely related. Log BB and log PS are therefore two different measures of 'brain uptake.' It can be seen also from **Table 6** that various parameters such as $\log P_{oct}$, $\Delta \log P$, and $\log P_{eg/hept}$ that have been put forward as models for either $\log$ BB or $\log$ PS are not at all good models. This is one of the reasons why so much effort has been put into devising in silico models for $\log$ BB, and why further efforts to obtain more data on perfusion should be expected.

5.31.5   Central Nervous System Uptake

5.31.5.1   Assessment of Central Nervous System Uptake

The lack of data on measured compound concentration in the brain or on the rate of uptake into the brain has limited the generalizability of in silico models built on these endpoints. This is because the 'chemical space' for brain uptake is not necessarily well represented by so few data points. There are only several hundred available in the literature for example. This has resulted in many workers attempting to build in silico models of brain penetration using data for which it is just known whether a compound exhibits a CNS effect (either a therapeutically beneficial effect or a side effect) or that it does not exhibit such an effect. These categories are generally referred to as CNS + and CNS–, respectively. This approach has the advantage that much more data is available (thousands of data points); however, while there is a good degree of certainty that compounds classified as CNS + will penetrate the blood–brain barrier, there is considerable uncertainty about whether a CNS– compound really has negligible brain uptake or just that it is pharmacologically inactive at CNS targets. The use of this type of data can be compared to the now well accepted 'Lipinski rule of five.'[90] Here, the analysis was performed on a compound set assumed to be orally bioavailable because the compounds in the set had United States Adopted Names (USANs). Hence, they were thought to have been progressed to the later stages of compound development, as these would be the only compounds to have had such names assigned. It was also assumed that compounds that had reached this phase of development would have had good bioavailability. The success of the 'Lipinski rule of 5' at removing undesirable compounds from large virtual libraries or screening sets perhaps suggests that despite the uncertainty in the assignments, the approach can be applied equally well to filtering CNS+ and CNS− compounds.

5.31.5.2 Factors Affecting Central Nervous System Uptake

As discussed in Section 5.31.2, many papers were published in the second half of the twentieth century that reported quantitative structure–activity relationships (QSARs) between centrally mediated pharmacological activity and lipophilicity.[91] The data were primarily gathered only on small data sets, some as small as five compounds, and are not particularly useful in furthering our general understanding of the relationship between chemical structure and brain uptake. It was not until the 1990s that analysis of larger data sets was undertaken.

Seelig and colleagues[14,92] identified a set of 28 (later expanded to 53) CNS+ and CNS− compounds and used measurements of surface-active properties (such as critical micelle concentration, air/water partition coefficient, and surface area of compound at the air–water interface) to classify the compounds. Interestingly, they used an in silico model[93] that identified which compounds were likely to be substrates for the efflux protein P-gp and eliminated them from their analysis set. This allowed them to have more confidence that they were just measuring properties for compounds for which brain penetration was limited by passive diffusion. These experiments resulted in some interesting observations about the optimal properties of molecules for brain penetration, e.g., the molecular cross-sectional area should be $<80\,\text{Å}^2$, and acceptable ranges for air/water partition coefficient, pK_a values and amphiphilicity. Whilst these relationships perhaps provide some insight into the properties needed for uptake into the brain this approach does not constitute a high-throughput screen that would enable scientists to assess the likely brain uptake of large sets of compounds.

Van de Waterbeemd and co-workers[16] expanded the original Seelig set to a total of 125 compounds and investigated the relationship between calculated physicochemical descriptors and categorical CNS activity. In particular, they looked at ways of calculating properties related to the measured cross-sectional area of Seelig. For the Seelig set they found that the use of molecular weight and a count of the total hydrogen bonds (acceptors and donors) gave as good a separation of the CNS and non-CNS drugs as when the cross-sectional area was used. On the bigger data set, they were able to identify that for compounds to penetrate the CNS they should have MW <450 and polar surface area $<90\,\text{Å}^2$. However, this does not mean that all compounds fulfilling these criteria will have good brain penetration. They described the shape of molecules in terms of the principal axes ratio (length/width) and found with only three exceptions CNS+ compounds had this ratio <5, i.e., CNS+ compounds are not long and thin but are more globular in shape.

In 1999 Ajay and co-workers[17] took the approach of trying to separate CNS+ and CNS− compounds a stage further by categorizing ∼65 K compounds as CNS active or inactive from the CMC[94] and MDDR[95] databases. This resulted in a set of 15 K CNS+ and 50 K CNS− compounds. They used seven simple descriptors such as MW, log P_{oct}, hydrogen bond acceptors and donors, and 166 functional group based descriptors to build a neural network that was capable of correctly predicting 92% of the actives and 71% of the inactives in a test set of 275 compounds. In addition, they used a further validation set (80 compounds) for which log BB was known and correctly predicted 97% of the actives and 71% of the inactives. Given the difficulties of correctly assigning the CNS− compounds this methodology gives a remarkably good method for selecting the compounds most likely to penetrate the blood–brain barrier when starting from large virtual libraries of compounds. In their more detailed analysis of the models built using the seven simple descriptors (correctly predicting 70% of the actives and 60% of the inactives in their test set) they concluded that increasing the number of hydrogen bond acceptors, MW, branching in a molecule, or the number of rotatable bonds will result in a compound being less likely to be CNS+. Increasing the aromatic density, log P, or the number of hydrogen bond donors results in a compound being more likely to be CNS+. It is worth noting that the observed effect of hydrogen bond donors is in direct contrast with that observed in other studies.[10] The usual observation is that the presence of hydrogen bonding groups (both acceptors and donors) results in a decrease in brain penetration. The latter seems the more easily understandable conclusion as passage through the blood–brain barrier requires a compound to cross a membrane, which is likely to be more difficult for compounds with polar or hydrogen-bonding groups.

Although a much smaller data set was used in the Crivori analysis,[15] the use of the Volsurf descriptors[96] has suggested further insights into the more detailed nature of polarity of molecules that affect their ability to cross the blood–brain barrier. The data set was comprised of 44 compounds (110 when stereoisomers were considered) for which antinociceptive potency was available from an in vivo test. Compounds that had no effect were assumed to have no brain penetration. They validated their model on a set of 120 compounds for which experimental brain penetration data was available. Their partial least squares discriminant analysis (PLS-DA) gave a correct classification of 90% on the CNS+ compounds and 65% on the CNS− compounds. This again highlights the difficulty in predicting CNS− compounds, which could in large part be due to an incorrect assignment in the training data set. Crivori and colleagues observed the usual trends of hydrophobicity favoring brain penetration whereas polarity and hydrogen bonding ability result in less favorable brain penetration. However, in addition they noted that the nature of the polar regions could affect the brain penetration in that diffuse polar regions in molecules allowed them to be CNS penetrant but localized polar regions did not. They further concluded that the distribution of hydrogen bonds and the hydrophilic–lipophilic balance in a molecule could also affect its brain penetration. In contrast to the Seelig and Van de Waterbeemd analysis they observe no correlation between size or shape and brain penetration.

Large data sets (>3 K compounds in each of the CNS+ and CNS− categories) have also been subjected to analysis by other workers.[97–99] The sets and their classifications have also been abstracted from a variety of databases although the problem always remains as to the correct assignment of CNS−. All these workers used fingerprint or fragment-based descriptors and a neural network to separate the classes. Although the classifications are reasonable the models can only be used as black boxes as these types of descriptors do not lend themselves to easy interpretation.

Engkvist and co-workers[99] additionally used a tool (SUBSTRUCT) that fragments molecules and then calculates the molecular sum over all fragments (MSOF). This is calculated as:

$$\sum F_{CNS+}/(F_{CNS+} + F_{CNS-}) \qquad [26]$$

where F_{CNS+} and F_{CNS-} are the fragment frequencies for each fragment in the molecule for each set. If MSOF >0.5 a molecule is classified as CNS+ and if MSOF <0.5 it is classified as CNS−. Engkvist suggests that certain fragments are more prevalent in CNS+ compounds than CNS− compounds. For example, carboxyl, hydroxyl, and alkoxy groups occur more frequently in CNS− compounds and halo groups and amines in CNS+ compounds. An alternative explanation could be (as reported previously by many workers) that hydrogen bonding tends to keep compounds out of the brain (carboxyl, hydroxyl, and alkoxy groups) and lipophilicity favors brain penetration (halo groups). The observation that amines occur more in CNS+ compounds could also just be a result of the functionality needed for activity at many CNS targets (lipophilic and basic). For example, the tricyclic antidepressants and opiate agonists and antagonists would be compounds that fall into this category.

Doniger and colleagues[100] achieved comparable predictive power on their set of only 324 compounds and 9 physicochemical descriptors using neural network and support vector machines, the latter giving the better predictability. Adenot and Lahana[101] started with a set of 62 K compounds from the World Drug Index (WDI) and used the anatomical therapeutic chemical classification (ATC)[102] to identify ~1.3 K CNS+ compounds and 360 CNS− compounds. This latter set was further divided into a PGP+ set (91 compounds), i.e, compounds known to be P-gp substrates. They also propose the remaining compounds to be assigned the class PGP−, although there is considerable uncertainty as to the validity of this assignment. They built two PLS-DA models to distinguish CNS+ from CNS− and PGP+ from PGP−, respectively. The PGP_pred and BBB_pred values are then plotted against each other to create a 'CNS drug map.' Visually at least, this gives distinct regions in which the CNS+, CNS−, and PGP+ compounds lie and could provide a useful way of classifying compounds according to their interaction at the blood–brain barrier. Adenot and Lahana[101] investigated a variety of descriptor types including topological, physicochemical, and 3D geometric and electronic descriptors. They conclude, however, that something as simple as the number of heteroatoms is sufficient to distinguish CNS+ and CNS− compounds. CNS− compounds have more heteroatoms but as with the Doninger analysis[100] this is only likely to be reflecting the overall hydrogen bonding capacity of the CNS− relative to the CNS+ set.

A recent paper by Li and colleagues[103] compared various statistical methods in terms of their ability to classify CNS+ and CNS− compounds. Their data set was made up of 415 compounds and the methods compared were linear regression, linear discriminant analysis, decision tree, probabilistic neural network, k-nearest neighbors, and support vector machines. As with the work of Doninger and co-workers[100] the support vector machine approach yielded the best discrimination. They also found that the use of recursive feature elimination (RFE)[104] showed improved predictability particularly in predicting the CNS− compounds. They started with a large set of descriptors ranging in complexity from simple counts of hydrogen bonds to quantum mechanical and geometric properties and used RFE to reduce the descriptors. As with other workers they highlight the importance of hydrophobicity, hydrogen bond donors and acceptors, and electrostatic descriptors in describing CNS penetration.

5.31.6   General Rules for Brain Uptake

Several groups[2,16,20,41,105] have identified what are now generally referred to as 'rules of thumb' for brain penetration. These are not models in that a statistically significant relationship has been obtained between brain penetration and a molecular property. They are better described as observations that the CNS + compounds fall into different descriptor ranges to CNS − compounds. These 'rules' generally follow the same descriptor trends described above for the statistical models of CNS activity.

The most frequently used of these 'rules' are PSA 'cut-offs' for brain penetration. Kelder and co-workers[41] analyzed the frequency of CNS + and CNS − compounds for different PSA ranges and concluded that CNS + compounds were most likely to have PSA < 60–$70 \text{\AA}^2$. Van de Waterbeemd and colleagues[16] put the limit at $90 \text{\AA}^2$. There are several explanations as to why different workers have chosen different cut-offs. First, the method of calculation is not identical and it has been shown that different calculation methods give different values for PSA.[46] Also, the data sets of Van de Waterbeemd and Kelder are different and in the latter case not particularly large (125 compounds), which could skew the results. Additionally, it is probably the case that both are right and the cut-off chosen depends on whether one wants to classify all the CNS + compounds correctly and accept some misclassifications of CNS − as CNS + (most likely for the $90 \text{\AA}^2$ cut-off) or one wants to classify CNS + compounds correctly but does not want any misclassifications of low CNS as high (most likely for the lower 60–$70 \text{\AA}^2$ cut-off). The effect of using different PSA cut-offs to predict CNS + and CNS − compounds can be seen from **Figure 1**. This shows the percentage of compounds with measured low, medium, and high CNS penetration that fall into various PSA classes for a set of compounds measured at GlaxoSmithKline.[106]

Bodor and Buchwald's analysis[2] of CNS + , but not CNS − compounds, concluded that CNS + compounds had log P_{oct} in the range 0.5 to 5.5, MW 150–450, 2–4 hydrogen bond acceptors, and 0–2 hydrogen bond donors. This is consistent with the observations of other studies. However, by not using a CNS − set for comparison it could be that Bodor's 'rules' are just reflecting the properties needed in drug molecules in general regardless of their target.

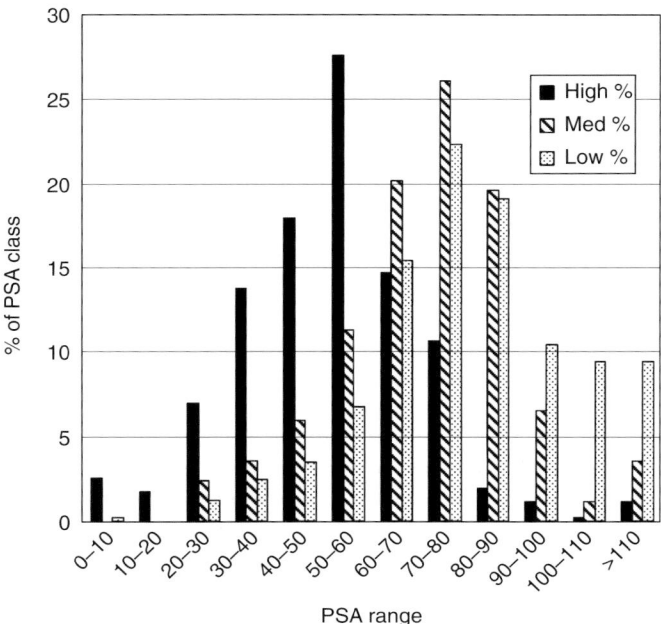

Figure 1   The distribution of ∼1000 compounds with measured high, medium, and low CNS penetration within specified PSA categories. Definitions of CNS penetration classes are: low log brain/blood $\leqslant -0.7$ [brain/blood $\leqslant 0.2$]; medium log brain/blood > -0.7 to < -0.4 [brain/blood > 0.2 to < 0.4]; and high log brain/blood $\geqslant -0.4$ [brain/blood $\geqslant 0.4$]. PSA is calculated as in [46] but phosphorus and sulfur are not included as polar atoms.

The more detailed study of Mahar Doan and co-workers,[105] albeit on a small data set (93 compounds), attempted to separate CNS + and CNS − compounds on the basis of both calculated properties and experimental measures of membrane permeability and P-gp profile. In addition to the observations that CNS + drugs had fewer hydrogen bond donors, greater lipophilicity, lower PSA and reduced flexibility, they also concluded that CNS + compounds should have an in vitro passive permeability $> 150 \, nm \, s^{-1}$ and should not be a good P-gp substrate.

A review of the methods for predicting brain penetration by Norinder[20] showed that a very simple rule of thumb for brain penetration was that brain penetration will be high if:

$$N + O \leqslant 5 \quad (\text{Rule 1}) \quad or \quad \log P - (N + O) > 0 \ (\text{Rule 2})$$

where $N + O$ is the number of nitrogen + oxygen atoms in a molecule. This is consistent with the many other observations of the effect of hydrogen bonding on brain penetration namely that brain penetration is favored by few hydrogen-bonding groups and high lipophilicity. This 'rule' has the attraction that it is extremely simple and quick to calculate yet it is reported to give good predictability on test compound sets (85% for rule 1 and 92% for rule 2). The effect of CNS activity on the descriptors used to determine relationships in these studies are summarized in **Table 7**.[14–17,92–101,108–114] The findings are no different from those observed on the quantitative [brain]/[blood] data. Most simply, the extent of the hydrogen bonding seems to be the most frequently reported determinant of the likelihood of CNS activity. Low hydrogen bonding favors brain penetration. Hydrogen bonding can be expressed as many descriptor types such as the counts of hydrogen bonds, the Abraham A and B descriptors, and the more complex descriptors of Crivori that describe the nature of the polar regions in a molecule. It can also be argued that high lipophilicity, which is also observed to be a factor favoring CNS + compounds, is just reflecting the lack of hydrogen-bonding groups in highly lipophilic molecules.

It has been regarded as a puzzle as to why models of brain penetration seem to work so well when very few workers have made any effort to remove P-gp substrates from their training and test sets and yet P-gp is well known to be an important factor in limiting brain penetration[5,6] and many known drug molecules are substrates for P-gp. It seems likely that the reason that this does not appear to matter lies in the observation that compounds with large numbers of hydrogen bonding groups are more likely to be P-gp substrates.[93,115] Didziapetris and co-workers[115] suggest the 'rule of fours' to separate P-gp substrates and nonsubstrates. Compounds with $N + O \geqslant 8$, $MW > 400$, and acid $pK_a > 4$ are likely to be P-gp substrates and compounds with $N + O \leqslant 4$, $MW < 400$, and basic $pK_a < 8$ are likely to be nonsubstrates. Hence, a compound with many hydrogen-bonding groups is likely to be excluded from the brain, not only because the presence of hydrogen bonds limits brain penetration via passive diffusion but also it makes the compound more likely to be a P-gp substrate and so the compound is effluxed out of the brain by that mechanism.

As regards quantitative in silico calculations, the position with regard to blood–brain distribution, as log BB, seems relatively straightforward. The calculations are for passive transport only, and effects such as efflux mechanisms and, indeed, any active transport are not included. Within this set of conditions, there are now several in silico calculations that yield fits of data with errors close to, or even less than, experimental error. Of course, these are only fits, and more weight should be given to calculations that include an assessment of predictive ability through the use of training sets and independent test sets. As shown in **Table 2**, several in silico calculations can predict log BB to a reasonable degree of accuracy (0.35 to 0.45 log units). All these calculations use descriptors for the neutral molecules and, apparently, no correction is needed for strong proton acids or strong proton bases. Interestingly, goodness of fits and goodness of predictions do not seem to depend either on the descriptors used or on the methods used. Simple descriptors such as polar surface area, molecular weight, volume, hydrogen bond acidity, and hydrogen bond basicity lead to similar fits to those achieved through the use of more complicated descriptors such as electrotopological state indices. Methods used range from multiple linear regression to partial least squares to Baysian neural networks, but results from the complicated methods appear to be no better than those from the simpler methods. It seems possible that the predictive capability of any calculation depends more on the experimental error in log BB values than on the descriptors or calculational methods used. Therefore, it is likely that future in silico calculations based on the same data sets that have already been used will not lead to predictive capability better than the 0.35–0.45 log units already achieved.

The situation with passive brain perfusion is quite different. There are very few in silico calculations of permeability coefficients with a reasonable number of compounds in the data set, and none with sufficient data to be able to assess predictive capability. Furthermore, the role of strong proton acids and strong proton bases is not well known, and a great deal more data on these types of compound is needed before any reliable in silico calculation of brain permeability can be constructed.

The descriptors identified for separating CNS + and CNS − compounds are largely similar to those that are able to rank the quantitative measure of brain to blood concentration. It is likely that workers will continue to develop models

Table 7 Summary of data sets, validation, and descriptor trends for models built using CNS uptake data

Reference	Data source	Data set size	Descriptors	Modeling technique	Test set prediction		Descriptor trends (effect of increasing descriptor value on CNS+)
					CNS+	CNS−	
14, 92	Not defined	53 (train)	Measured surface activity	Visual discrimination	na	na	Cross-sectional area $<80\,\text{Å}^2$
							K_{aw} $105–103\,\text{M}^{-1}$
							A(acid) $pK_a > 4$
							B(base) $pK_a < 10$
							Amphiphilicity $(\Delta\Delta G_{am} > -3\,\text{kJ mol}^{-1})$
16	(train)[14]	37 (train)	Calculated physchem	Visual discrimination	na	na	MW <450
	Not defined (test)	88 (test)					PSA $<90\,\text{Å}^2$
							Principal axes ratio <5
17	CMC/MDDR (train)	65 000 (train)	Physchem	NN	92%	71%	Low HB acceptors
	Not defined (test)	275 (test)	ISIS fingerprints				Low MW
							Low branching
							High aromatic density
							High log P
							High HB donors
15	(train)[114]	110 (train)	Volsurf descriptors	PLS-DA	90%	65%	High hydrophobicity
	Measured log BB (test)	120 (test)					Low polarity
							Low hydrogen bonding
							Diffuse polar regions
							Hydrogen bond distribution
							Hydrophilic/lipophilic balance
							(size and shape – no effect)

continued

Table 7 Continued

Reference	Data source	Data set size	Descriptors	Modeling technique	Test set prediction		Descriptor trends (effect of increasing descriptor value on CNS+)
					CNS+	CNS−	
97	CDSA and CPS (train/test)	7000 (train) 7000 (test)	2D unity fingerprints	NN	84%	87%	No analysis
98	MDDR (train and test)	10000 (train)	Ghose and Crippen atom types	NN	72%	76%	No analysis
	CMC (test)	CMC (test)			68%	35%	
100	PDR,[16,17,92] psychotropic DB (train/test)	324 (train)	9 physchem descriptors	NN	81.5%	69.9%	log P and hydrogen bond donors and acceptors.
		50 (test) but subsets of above		SVM	82.7%	80.2%	Note: no directional effect is assigned to these descriptors in respect of CNS penetration
99	WDI (train)	9000 (train)	SUBSTRUCT (substructural analysis)	MSOF (comparison of fragment frequency in CNS+ and CNS− set)	100%	67%	Protonated N, Aromatic rings, Cl, F more common in CNS+
	(test)[14,110,111]	20 (test)					O more common in CNS−
101	WDI (train)	1700 (train)	Ghose and Crippen atom types	NN	91%	78%	NN difficult to interpret
	(test)[15]	82 (test)	Topological, physchem and 3D descriptors	PLS-DA	90%	92%	Heteroatoms <9

train, training set; test, test set. Test set prediction is the percentage of CNS+ and CNS− compounds correctly predicted by model on an independent test set of compounds. NN, neural network; SVM, support vector machine; PLS-DA, partial least squares discriminant analysis. Databases: CMC,[94] MDDR,[95] PDR,[107] Psychotropic DB,[108] WDI,[109] CPS,[112] and CDSA.[113]

for CNS penetration based on CNS+ and CNS− classification in the attempt to better describe the chemical space in which CNS penetrant compounds lie. The challenges in this area will be in obtaining better assessments of CNS − and in improved molecular descriptors.

5.31.7   The Future of Brain Uptake Models

Over the last few years several groups of workers have started to question whether the use of the log BB (total compound concentration in brain/total compound concentration in blood) is the best measure of brain penetration to assess the likelihood of a compound having efficacy at a CNS receptor.[38,116,117] It is now thought that the compound concentration in brain extracellular fluid (ECF) is likely to be a better predictor of efficacy. This is sometimes referred to as free brain concentration and a surrogate for its measurement is the cerebrospinal fluid (CSF) compound concentration. Maurer and colleagues[118] recently undertook a study in mice on 33 compounds known to have CNS activity and investigated the relationship between the ratio of the unbound fractions in brain and plasma measured in vitro, the total brain/plasma ratio, and the CSF/brain ratio. In many cases, the fraction unbound in brain was predictive of the CSF/brain ratio. For this compound set there was a large variation in the total brain/plasma ratios but the free compound concentrations in brain and plasma were very similar. Maurer and co-workers[118] suggest these differences in brain/plasma ratios are therefore due to differences in nonspecific binding to plasma and brain tissue. The implication for in silico models is that designing compounds with properties to maximize the brain/blood ratio only increases the nonspecific or 'not useful' binding in the brain. It is anticipated that over the next few years these data will be subjected to in silico analyses and in silico models developed for fraction unbound in brain, for example, which is likely to show very different trends with lipophilicity, for example, than total brain/blood ratio. This does not mean that brain/blood ratio will become a redundant parameter for modeling. It is likely that it will be the balance of brain penetration and fraction unbound that will be the goal for therapeutic agents targeting receptors in the central nervous system.

References

1. Royal Pharmaceutical Society. *Martindale: The Complete Drug Reference*; Pharmaceutical Press: London, UK, 2005.
2. Bodor, N.; Buchwald, P. *Am. J. Drug Deliv.* **2003**, *1*, 13–26.
3. Feng, M. R. *Curr. Drug Metab.* **2002**, *3*, 647–657.
4. Yan-Ling, H.; Murby, S.; Warhurst, G.; Gifford, L.; Walker, D.; Ayrton, J.; Eastmond, R.; Rowland, M. *J. Pharm. Sci.* **1998**, *87*, 626–633.
5. Ayrton, A.; Morgan, P. *Xenobiotica* **2001**, *31*, 469–4976.
6. Wiese, M.; Pajeva, I. K. *Curr. Med. Chem.* **2001**, *8*, 685–713.
7. Clark, D. E. *Comb. Chem. High Throughput Screen.* **2001**, *4*, 467–496.
8. Young, R. C.; Mitchell, R. C.; Brown, T. H.; Ganellin, C. R.; Griffiths, R.; Jones, M.; Rana, K. K.; Saunders, D.; Smith, I. R.; Sore, N. E. et al. *J. Med. Chem.* **1988**, *31*, 656–671.
9. Young, R. C.; Ganellin, C. R.; Griffiths, R.; Mitchell, R. C.; Parsons, M. E.; Saunders, D.; Sore, N. E. *Eur. J. Med. Chem.* **1993**, *28*, 201–211.
10. Abraham, M. H.; Chadha, H. S.; Mitchell, R. C. *J. Pharm. Sci.* **1994**, *83*, 1257–1268.
11. Gratton, J. A.; Abraham, M. H.; Bradbury, M. W.; Chadha, H. S. *J. Pharm. Pharmacol.* **1997**, *49*, 1211–1216.
12. Abraham, M. H. *Eur. J. Med. Chem.* **2004**, *39*, 235–240.
13. Liu, X.; Tu, M.; Kelly, R. S.; Chen, C.; Smith, B. J. *Drug Metab. Dispos.* **2004**, *32*, 132–139.
14. Seelig, A.; Gottschlich, R.; Dervent, R. M. *Proc. Natl. Acad. Sci. USA* **1994**, *91*, 68–72.
15. Crivori, P.; Cruciani, G.; Carrupt, P.-A.; Testa, B. *J. Med. Chem.* **2000**, *43*, 2204–2216.
16. Van de Waterbeemd, H.; Caminisch, G.; Folkers, G.; Chretien, J.; Raevsky, O *J. Drug. Target.* **1998**, *6*, 151–165.
17. Ajay; Bemis, G. W.; Murcko, M. A. *J. Med. Chem.* **1999**, *42*, 4942–4951.
18. Clark, D. E. *Drug Disc. Today* **2003**, *8*, 927–933.
19. Kaznessis, Y. N. *Curr. Med. Chem. – Central Nervous System Agents* **2005**, *5*, 185–191.
20. Norinder, U.; Haeberlein, M. *Adv. Drug Deliv. Rev.* **2002**, *54*, 291–313.
21. Atkinson, F.; Cole, S.; Green, C.; Van de Waterbeemd, H. *Curr. Med. Chem. – Central Nervous System Agents* **2002**, *2*, 229–240.
22. Hansch, C.; Steward, A. R.; Anderson, S. M.; Bentley, D. L. *J. Med. Chem.* **1968**, *11*, 1–11.
23. Timmermans, P. B. M. W. M.; Brands, A.; van Zwieten, P. A. *Arch. Pharmacol.* **1977**, *300*, 217–226.
24. Waterhouse, R. N. *Mol. Imag. Biol.* **2003**, *5*, 376–389.
25. Dishino, D. D.; Welch, M. J.; Kilbourn, M. R.; Raichle, M. E. *J. Nucl. Med.* **1983**, *24*, 1030–1038.
26. Oldendorf, W. H. *Brain Res.* **1970**, *24*, 372–376.
27. Oldendorf, W. H. *Am. J. Physiol.* **1971**, *221*, 1629–1639.
28. Pardridge, W. M.; Mietus, L. J. *J. Clin. Invest.* **1979**, *54*, 145–154.
29. Ohno, K.; Pettigrew, K. D.; Rapoport, S. I. *Am. J. Physiol.* **1978**, *235*, H299–H307.
30. Rapoport, S. I.; Ohno, K.; Pettigrew, K. D. *Brain Res.* **1979**, *172*, 354–359.
31. Takasato, Y.; Rapoport, S. I.; Smith, Q. R. *Am. J. Physiol.* **1984**, *247*, H484–H493.
32. Smith, Q. R.; Yakasato, Y. *Ann. NY Acad. Sci.* **1986**, *481*, 186–201.
33. Deane, R.; Bradbury, M. W. B. *J. Neurochem.* **1990**, *54*, 905–914.

34. Pardridge, W. M.; Triguero, D. *J. Pharmacol. Exp. Ther.* **1990**, *253*, 884–891.
35. Fiserova-Bergerova, V.; Diaz, M. L. *Arch. Occup. Environ. Health* **1986**, *58*, 75–87.
36. Gargas, M. L.; Burgess, R. J.; Voisard, D. E.; Cason, G. H.; Anderson, M. E. *Toxicol. Appl. Pharmacol.* **1989**, *98*, 87–99.
37. Abraham, M. H.; Weathersby, P. K. *J. Pharm. Sci.* **1994**, *83*, 1450–1456.
38. Pardridge, W. M. *Drug. Disc. Today* **2004**, *9*, 392–393.
39. Brewster, M. E.; Pop, E.; Huang, M. J.; Bodor, N. *Int. J. Quantum Chem.* **1996**, *23*, 1775–1787.
40. Lombardo, F.; Blake, J. F.; Curatolo, W. J. *J. Med. Chem.* **1996**, *39*, 4750–4755.
41. Kelder, J.; Grootenhuis, P. D. J.; Bayada, D. M.; Delbressine, L. P. C.; Ploemen, J.-P. *Pharm. Res.* **1999**, *16*, 1514–1519.
42. Clark, D. C. *J. Pharm. Sci.* **1999**, *88*, 815–821.
43. Norinder, U.; Sjöberg, P.; Österberg, T. *J. Pharm. Sci* **1998**, *87*, 952–959.
44. Luco, J. M. *J. Chem. Inf. Comput. Sci.* **1999**, *39*, 396–404.
45. Feher, M.; Sourial, E.; Schmidt, J. M. *Int. J. Pharm.* **2000**, *201*, 239–247.
46. Ertl, P.; Rohde, B.; Seizer, P. *J. Med. Chem* **2000**, *43*, 3714–3717.
47. Keserü, G. M.; Molnar, L. *J. Chem. Inf. Comput. Sci.* **2001**, *41*, 120–128.
48. Liu, R.; Sun, H.; So, S.-S. *J. Chem. Inf. Comput. Sci.* **2001**, *41*, 1623–1632.
49. Kaznessis, Y. N.; Snow, M. E.; Blankley, C. J. *J. Comput.-Aided Mol. Des.* **2001**, *15*, 697–708.
50. Norinder, U.; Österberg, T. *J. Pharm. Sci.* **2001**, *90*, 1076–1085.
51. Platts, J. A.; Abraham, M. H.; Zhao, Y. H.; Hersey, A.; Ijaz, L.; Butina, D. *Eur. J. Med. Chem.* **2001**, *36*, 719–730.
52. Rose, K.; Hall, L. H.; Kier, L. B. *J. Chem. Inf. Comput. Sci.* **2002**, *42*, 651–666.
53. Subramanian, G.; Kitchen, D. B. *J. Comput.-Aided Mol. Des.* **2003**, *17*, 643–664.
54. Ooms, F.; Weber, P.; Carupt, P.-A.; Testa, B. *Biochim. Biophys. Acta* **2002**, *1587*, 118–125.
55. Lobell, M.; Molnar, L.; Keserü, M. *J. Pharm. Sci.* **2003**, *92*, 360–370, errors are given as 'mean absolute'.
56. Klamt, A.; Eckert, F.; Hornig, M. *J Comput.-Aided Mol. Des.* **2001**, *15*, 355–365.
57. Salminen, T.; Pulli, A.; Taskinen, J. *J. Pharm. Biomed. Anal.* **1997**, *15*, 469–477.
58. Kaliszan, R.; Markuszewski, M. *Int. J. Pharm.* **1996**, *145*, 9–16.
59. Iyer, M.; Mishra, R.; Hopfinger, A. *J. Pharm. Res.* **2002**, *19*, 1611–1621.
60. Sun, H. *J. Chem. Inf. Comput. Sci.* **2004**, *44*, 748–757.
61. Hou, T. J.; Xu, X. *J. Chem. Inf. Comput. Sci.* **2003**, *43*, 2137–2152; errata in **2004**, *44*, 766–770.
62. Cabrea, M. A.; Bermejo, M.; Perez, M.; Ramos, R. *J. Pharm. Sci.* **2004**, *93*, 1701–1717.
63. Pan, D.; Liu, J.; Li, Y.; Hopfinger, A. *J. Chem. Inf. Comput. Sci.* **2004**, *44*, 2083–2098.
64. Winkler, D. A.; Burden, F. R. *J. Mol. Graph. Mod.* **2004**, *22*, 499–505.
65. Yap, C. W.; Chen, Y. Z. *J. Pharm. Sci.* **2005**, *94*, 153–168.
66. Narayanan, R.; Gunturi, S. B. *Bioorg. Med. Chem.* **2005**, *13*, 3017–3028.
67. Abraham, M. H.; Andonian-Haftvan, J.; Whiting, G. S.; Leo, A.; Taft, R. W. *J. Chem. Soc. Perkin Trans.* **1994**, *2*, 1777–1791.
68. Abraham, M. H. *Chem. Soc. Rev.* **1993**, *22*, 73–83.
68a. Abraham, M. H.; Ibrahim, A.; Zissimos, A. M. *J. Chromatogr. A* **2004**, *1037*, 29–47.
69. Platts, J. A.; Butina, D.; Abraham, M. H.; Hersey, A. *J. Chem. Inf. Comput. Sci.* **1999**, *39*, 835–845.
70. PharmaAlgorithms ADME Boxes, Version 2.2, PharmaAlgorithms Inc., 591 Indian Road, Toronto, ON M6P 2C4, Canada.
71. Hou, T. J.; Xu, X. *J. Chem. Inf. Comput. Sci.* **2003**, *43*, 1058–1067.
72. Poulin, P.; Theil, F. P. *J. Pharm. Sci.* **2000**, *89*, 16–35.
73. Abraham, M. H.; Chadha, H. S.; Whiting, G. S.; Mitchell, R. C. *J. Pharm. Sci.* **1994**, *83*, 1085–1100.
74. Abraham, M. H.; Chadha, H. S.; Mitchell, R. C. *Drug Des. Disc.* **1995**, *13*, 123–131.
75. Seiler, P. *Eur. J. Med. Chem.* **1974**, *9*, 473–479.
76. Rodgers, T.; Leahy, D.; Rowland, M. *J. Pharm. Sci.* **2005**, *94*, 1237–1248.
77. Rodgers, T.; Leahy, D.; Rowland, M. *J. Pharm. Sci.* **2005**, *94*, 1259–1276.
78. Levin, V. A. *J. Med. Chem.* **1980**, *23*, 682–684.
79. Chickhale, E. G.; Burton, P. S.; Borchardt, R. A. *J. Pharmacol. Exp. Ther.* **1995**, *273*, 298–303.
80. Begley, D. J. *J. Pharm. Pharmacol.* **1996**, *48*, 136–146.
81. Bodor, N.; Buchwald, P. *Adv. Drug Deliv. Rev.* **1999**, *36*, 229–254.
82. Liu, Z. D.; Hider, R. C. *Med. Res. Rev.* **2002**, *22*, 26–64.
83. Wohnsland, F.; Faller, B. *J. Med. Chem.* **2001**, *44*, 923–930.
84. Abraham, M. H.; Martins, F.; Mitchell, R. C.; Salter, C. J. *J. Pharm. Sci.* **1999**, *88*, 241–247.
85. Burton, P. S.; Conradi, R. A.; Hilgers, A. R.; Ho, N. F. H.; Maggiora, L. L. *J. Control. Release* **1992**, *19*, 87–92.
86. Chikale, E. G.; Ng, K.-Y.; Burton, P. S.; Borchardt, R. T. *Pharm. Res.* **1994**, *11*, 412–419.
87. Abraham, M. H. Can We Identify Models for Intestinal Absorption, Blood–Brain Barrier Distribution and Skin Permeation? In *EuroQSAR 2002. Designing Drugs and Crop Protectants: Processes, Problems and Solutions*; Ford, M., Livingstone, D., Dearden, J., Van de Waterbeemd, H., Eds.; Blackwell: Oxford, UK, 2003, pp 5–7.
88. Abraham, M. H.; Martins, F. *J. Pharm. Sci.* **2004**, *93*, 1508–1523.
89. Ishihama, Y.; Asakawa, N. *J. Pharm. Sci.* **1999**, *88*, 1305–1312.
90. Lipinski, C.; Lombardo, F.; Dominy, B. W.; Feeney, P. J. *Adv. Drug Deliv. Rev.* **1997**, *23*, 3–25.
91. Hansch, C.; Bjorkroth, J. P.; Leo, A. *J. Pharm. Sci.* **1987**, *76*, 663–687, and references quoted therein.
92. Fischer, H.; Gottschlich, R.; Seelig, A. *J. Membr. Biol.* **1998**, *165*, 201–211.
93. Seelig, A. *Eur. J. Biochem.* **1998**, *251*, 252–261.
94. CMC (Comprehensive Medicinal Chemistry), Release 94, MDL Information Systems Inc., San Leandro, CA 94577, USA.
95. MDDR (MACCS Drug Data Report), MDL Information Systems Inc., San Leandro, CA 94577, USA.
96. Cruciani, C.; Crivori, P.; Carupt, P. A.; Testa, B. *J. Mol. Struct. (Theochem)* **2000**, *503*, 17–30.
97. Keserü, G. M.; Molnar, L.; Greiner, I. *Comb. Chem. High Throughput Screen.* **2000**, *3*, 535–540.
98. Podlogar, B. I.; Muegge, I. *Curr. Top. Med. Chem.* **2001**, *1*, 257–275.
99. Engkvist, O.; Wrede, P.; Rester, U. *J. Chem. Inf. Comput. Sci.* **2003**, *43*, 155–160.
100. Doniger, S.; Hofmann, T.; Yeh, J. *J. Comp. Biol.* **2002**, *9*, 849–864.

101. Adenot, M.; Lahana, R. *J. Chem. Inf. Comput. Sci.* **2004**, *44*, 239–248.
102. Guidelines for ATC Classification and DDD Assignment WHO Collaborating Centre for Drug Statistics Methodology, Oslo, Norway, 2001. http://www.whocc.no/atcdd (accessed April 2006).
103. Li, H.; Yap, C. W.; Ung, C. Y.; Xue, Y.; Cao, Z. W.; Chen, Y. Z. *J. Chem. Inf. Model.* **2005**, *45*, 1376–1384.
104. Xue, Y.; Yap, C. W.; Sun, L. Z.; Cao, Z. W.; Wang, J. F.; Chen, Y. Z. *J. Chem. Inf. Comput. Sci.* **2004**, *44*, 1497–1505.
105. Mahar Doan, K. M.; Humphreys, J.; Webster, L. O.; Wring, S. A.; Shampine, L. J.; Serabjit-Singh, C. J.; K Adkinson, K.; Polli, J. W. *J. Pharmacol. Exp. Ther.* **2002**, *303*, 1029–1037.
106. Hersey, A. Unpublished work; Computational Chemistry Dept, GlaxoSmithKline, 2006.
107. Lundbeck, H. Psychotropics Database 2000–2001.
108. *PDR (Physicians' Desk Reference)*; Medical Economics Company, 2005. http://www.pdr.com (accessed June 2006).
109. WDI (Derwent World Drug Index), Version 2001/01 Derwent Information Ltd., 2001. http://www.derwent.com (accessed June 2006).
110. Chang, M.; Sood, V. K.; Wilson, G. J.; Kloosterman, D. A.; Sanders, P. E.; Hauer, M. J.; Zhang, W.; Branstetter, D. G. *Drug Metab. Dispos.* **1997**, *24*, 828–839.
111. Crooks, P. A.; Li, M.; Dwoskin, L. P. *Drug Metab. Dispos.* **1997**, *25*, 47–54.
112. CPS (Cipsline Database) Prous Science, Barcelona, Spain.
113. CDSA (Chemical Directory – Sigma-Aldrich) Sigma-Aldrich Co., Milwaukee, WI 53223, USA.
114. Gardina, G.; Clarke, G. D.; Grugni, M.; Sbacchi, M.; Vecchietti, V. *Farmaco* **1995**, *50*, 405–418.
115. Didziapetris, R.; Japertas, P.; Avdeef, A.; Petrauskas, A. *J. Drug Target* **2003**, *11*, 391–406.
116. Martin, I. *Drug Disc. Today* **2004**, *9*, 161–162.
117. Liu, X.; Chen, C. *Curr. Opin. Drug Disc. Dev.* **2005**, *8*, 505–512.
118. Maurer, T. S.; DeBartolo, D. B.; Tess, D. A.; Scott, D. O. *Drug Metab. Dispos.* **2005**, *33*, 175–181.

Biographies

Michael H Abraham is Professor of Chemistry at University College London. His scientific interests include molecular properties that influence solvation and the transport of compounds from one phase to another. He has investigated various processes including gas to solvent and water to solvent partitions, skin permeation, intestinal absorption, brain perfusion and blood–brain distribution, as well as thresholds for odor, eye irritation, and nasal irritation of volatile organic compounds. Professor Abraham was awarded the American Pharmaceutical Association 1992 Ebert Prize for work on general anesthesia and the 2002 Ebert Prize for work on human intestinal absorption. He has published a book, numerous reviews, over 400 research papers, and is rated in the top 100 of the world's most cited chemists.

Anne Hersey is Section Head of the ADMET Modelling Group within the Department of Computational Chemistry at GlaxoSmithKline in Stevenage. After obtaining her PhD from the University of Kent at Canterbury, she worked at the Wellcome Research Laboratories in Beckenham, primarily on physicochemical property measurements and their applications. In 1996, she moved to GlaxoWellcome, Ware, to work on the in silico modeling of ADME properties. Her interest is in developing and using in silico ADMET models to aid the design of molecules with good developability profiles.

5.32 In Silico Models for Interactions with Transporters

M Wiese, University of Bonn, Bonn, Germany
I K Pajeva, Bulgarian Academy of Sciences, Sofia, Bulgaria

5.32.1 Introduction

During the past decade the extensive involvement of ATP-binding cassette (ABC) transporters in the pharmacokinetics and pharmacodynamics of drugs has been revealed. These findings established transport proteins as key determinants for drug transport and disposition. Their ubiquitous expression in various organs, particularly those involved in drug excretion (e.g., the liver and kidney) and tissue protection (e.g., the blood–brain barrier, BBB) suggest important physiological implications. Moreover, common pathways involving the same transporters have been identified for many drugs and essential nutrients. These findings point to the involvement of transport proteins in fundamental cellular processes, thus making them extremely significant subjects of investigation. ABC transporters are, furthermore, implicated in the development of de novo or acquired multidrug resistance (MDR) in many human malignancies. Through high overexpression in tumor cells, ABC transporters actively extrude chemotherapeutic drug substrates and reduce their concentration to subeffective intracellular levels, which in many cases results in failure of the cancer therapy. Investigational efforts have two significant aims: (1) the better understanding of the basic biological mechanisms involved and (2) the promotion of therapeutic interventions in medical treatments through identification of novel targets.

Independently of the progress achieved in recent years, the structure–function relationships of most transport proteins are unsolved, and their interactions with various ligands are poorly characterized at the molecular level. This makes it necessary to apply a more operative strategy of investigation that effectively combines experimental and in silico approaches.

This chapter presents an overview of the current approaches for molecular modeling of different transport proteins. As P-glycoprotein (P-gp) is the best characterized and the most significant MDR transport system, it will be in the main focus of discussion.

5.32.2   ATP-Binding Cassette Transporter Family

5.32.2.1   Physiological Significance and Role in Drug Distribution

The importance of drug transporters for intracellular drug concentration was discovered in tumor cells showing an MDR phenotype.[1] The first MDR transporter, P-gp, initially discovered in resistant CHO cells in 1976,[2] was found to be expressed in various resistant tumor cell lines of animal and human origin.[3] To date, P-gp overexpression is thought to be mainly responsible for MDR in cancer drug therapy. The protein is, furthermore, naturally expressed in many tissues, especially those with barrier functions such as liver, BBB, kidney, intestine, and placenta (*see* 5.04 The Biology and Function of Transporters).

P-gp (the product of the human gene *ABCB1*) belongs to the highly conserved superfamily of ATP-binding cassette (ABC) transporters. Several hundred compounds have been reported to be its substrates and/or inhibitors. They belong to various chemical classes, and generally do not share structural homology. The characteristic features that relate a P-gp compound to the group of the substrates or inhibitors are currently not well defined.

Table 1 lists well-known P-gp substrates and inhibitors. Among the compounds listed, verapamil, trifluoperazine, and propafenone represent ones of the best studied first-generation MDR modulators. Verapamil is used as a reference compound in most MDR studies, a fact that can be explained by historical rather than scientific reasons. The intensive search for more specific and less toxic MDR modulators led to the development of second- and third-generation MDR modulators, some of which are shown in **Figure 1**.

Similarly to P-gp, several proteins encoded by genes of the subfamily C of the ABC transporters also transport a large variety of substrates. These transporters are named MDR-associated proteins (MRPs). The first subfamily member MRP1 (ABCC1) was discovered in 1987.[4] In small quantities it is present in almost all mammalian cells, and it is expressed in the sinusoidal membrane of liver hepatocytes. MRP1 functions as a multispecific organic anion transporter. Both P-gp and MRP1 confer resistance to a similar but not identical spectrum of cytotoxic drugs.[5,6] MRP1 transports cytotoxic drugs and other hydrophobic compounds only in the presence of glutathione. MRP2 (ABCC2) is expressed on the canalicular (apical) side of the hepatocyte, and promotes biliary transport of anionic conjugates with glutathione, sulfate, or glucuronic acid on one hand, and anticancer drugs such as vinblastine on the other hand. Overexpression of MRP2 is associated with cisplatin resistance in tumor cells.[7] MRP3 is expressed in the intestine and liver, but little is known about its importance for drug distribution. MRP3 transports glucuronated metabolites of drugs.[8] Its substrate spectrum appears to be much narrower than those of MRP1 and MRP2. The human MRP4 and MRP5 proteins are organic anion transporters that transport cyclic nucleotides and some nucleoside monophosphate analogs, including nucleoside-based antiviral drugs. MRP4 also transports prostaglandins.[9]

Table 2 summarizes compounds shown to interact with MRP family members that are expressed at the BBB and therefore could have implications for brain penetration of central nervous system (CNS)-active drugs.

A novel member of the ABC transporter superfamily is breast cancer resistance protein (BCRP) (ABCG2).[54] It is expressed in the intestine, the bile canicular membrane, and the placenta.[55] BCRP has been shown to play a role in the absorption and secretion of clinically important drugs. **Table 3** lists substrates and inhibitors of BCRP. This transporter shows wide substrate recognition properties, including neutral, positively, and negatively charged compounds.

In summary, several ABC transporters, alone or in combination, are responsible for MDR and for the absorption, distribution, and elimination of drugs in humans. The next section presents a summary of experimental results on the structure and functioning of different transporters. The results are discussed in relation to their usefulness for in silico studies of the transport proteins, particularly P-gp.

5.32.2.2   Experimental Studies of P-Glycoprotein and Related Transporters

5.32.2.2.1   Structural data

High-resolution structures of human MDR proteins from the ABC family have not yet been resolved. However, x-ray structures of several other transporters have recently become available. These structures include MsbA from *Escherichia coli*[90] and *Vibrio cholerae*,[91] obtained in the absence of nucleotide at 4.5 Å and 3.8 Å resolution, respectively, and *E. coli* BtuCD resolved at 3.2 Å.[92]

The lipid transporter MsbA is phylogenetically close to P-gp and the bacterial multidrug transporter LmrA from *Lactococcus lactis*; all believed to act as flipases. MsbA is formed of two monomers, each containing six α helical transmembrane domains (TMDs) and a nucleotide-binding domain (NBD) linked by an intracellular domain. The homodimer has 12 transmembrane α helices (six per monomer). MsbA closely resembles the topological organization of P-gp and the degree of sequence homology between both proteins approaches 30%.

Table 1 Main groups of P-gp substrates and inhibitors

Class	P-gp substrate/inhibitor
Anticancer drugs	Anthracyclines (doxorubicin, daunorubicine, epirubicin)
	Epipodophyllotoxins (etoposide, teniposide)
	Taxanes (paclitaxel, docetaxel)
	Vinca alkaloids (vincristine, vinblastine)
	Actinomycin D
	Topotecan
Cytotoxic agents	Colchicine
	Puromycin
Calcium channel blockers	Dilthiazem
	Nicardipin
	Verapamil
HIV protease inhibitors	Indinavir
	Ritonavir
	Saquinavir
Steroids	Dexamethasone
	Methylprednisolon
	Progesterone
	Testosterone
Hydrophobic peptides	Cyclosporin A
	FK506
	Gramicidin D
	Rapamycin
Other compounds	Calcein AM
	Digoxin
	Dipyridamole
	Hoechst 33342
	Loperamide
	Morphin
	Propafenone
	Quetiapine
	Rhodamine 123
	Trifluoperazine

The vitamin B_{12} transporter BtuCD differs structurally from P-gp: it contains 20 TMDs (versus 12 TMDs), a low degree of homology to the P-gp primary sequence and does not have a domain between the TMDs and NBDs. Additionally to these structural differences BtuCD acts as an influx pump.

The first experimental insight into the three-dimensional (3D) structure of P-gp was obtained by electron microscopy of two-dimensional protein crystals and image analysis at 2.5 nm resolution.[93] The image revealed a large

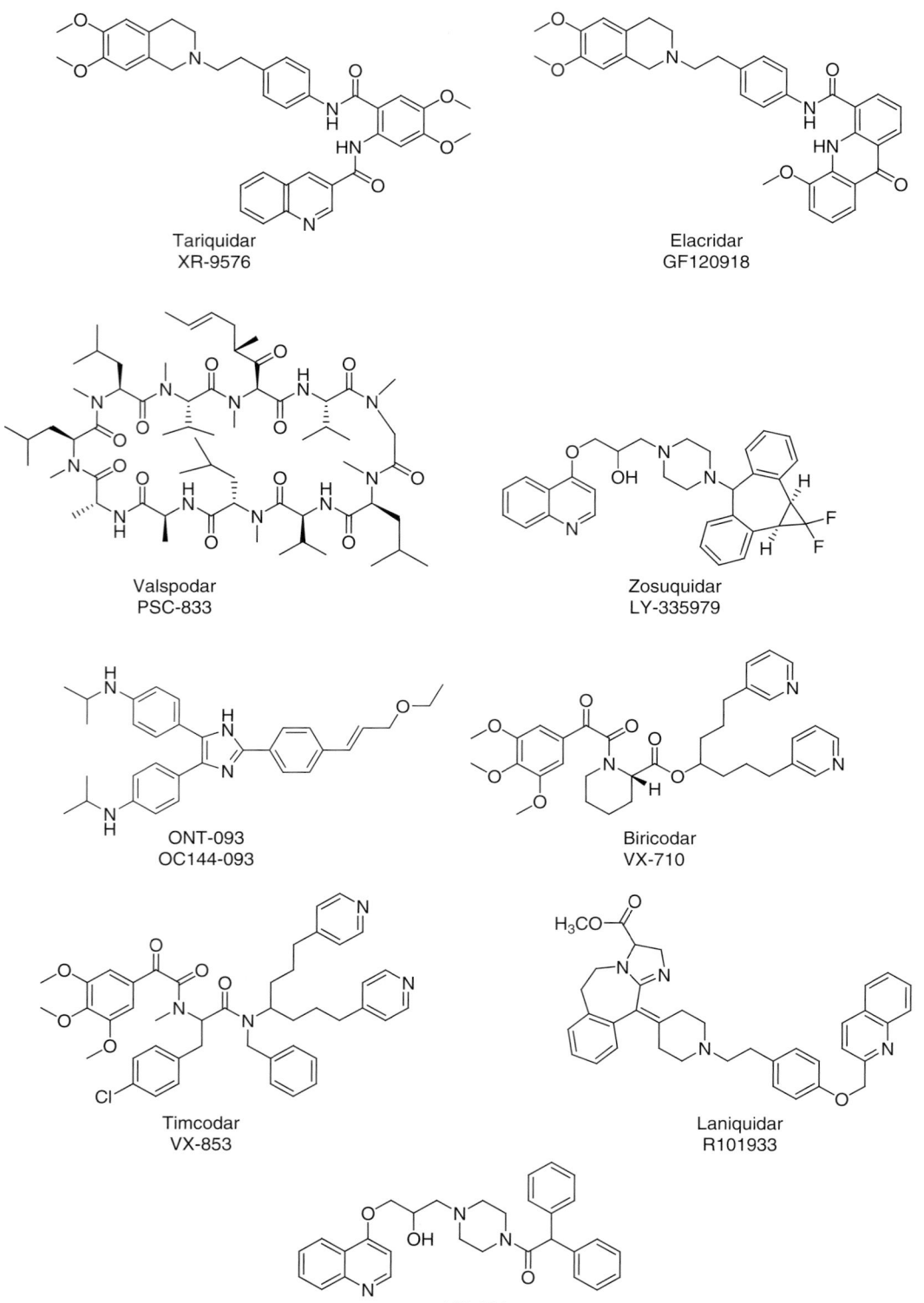

Figure 1 Structures of third-generation inhibitors of P-gp that are currently undergoing or have been investigated in clinical trials.

Table 2 Substrates and inhibitors interacting with transporters of the MRP family

Compound	Ref.
MRP1	
Bilirubin	10
Bilirubin glucoronides	11
Leukotriene C4, D4, E4	12–14
Glutathione disulfide	15
17β-Estradiol glucuronide	13, 16
S-Glutathionyl aflatoxin B1	17
S-(2,4-dinitrophenyl)glutathione	13
S-Glutathionyl ethacrynic acid	18
S-Glutathionyl prostaglandin A2	19
Vincristine plus reduced glutathione[a]	20–22
Daunorubicin plus reduced glutathione[a]	22
Etoposide plus reduced glutathione[a]	23
Etoposide glucuronide	13
3a-Sulfatolithocholyltaurine	13
Fluo-3	24
Fluorescein	25
Folic acid	26
Methotrexate	27
Saquinavir	28, 29
Leukotriene receptor antagonist MK 571	30
Curcumin	31
Phenethyl isothiocyanate	32
Indomethacin and derivatives	33, 34
Pyrrolopyrimidine-type inhibitors	35
Glibenclamide	36
Lonafarnib (SCH66336)	37
Lipophilic flavonoids	38
Glutathione-conjugate analogs	39
MRP2	
Etoposide	40
Bilirubin glucuronides	11
Estradiol 3-glucuronide	41
Ethinylestradiol-3-glucuronide	42
Indinavir[b]	43
Saquinavir[b]	28, 43

continued

Table 2 Continued

Compound	Ref.
Ritonavir[b]	43
Azithromycin	44
Curcumin	31
Lonafarnib (SCH66336)	37
MRP3	
9-(2-Phosphonylmethoxyethyl) adenine	45
Azidothymidine monophosphate	45
p-Aminohippurate	46
Prostaglandines	9
Urate	47
MRP4	
5-Chloromethylfluorescein diacetate	48
Fluorescein diacetate	48
cAMP	49
cGMP	49
6-Mercaptopurine	50
Thioguanine	50
9-(2-Phosphonylmethoxyethyl) adenine	50
S-(2,4-Dinitrophenyl)-glutathione	50
Sildenafil	49
Trequinsin	49
MRP5	
Leukotriene C4	51, 52
S-(2,4-Dinitrophenyl)glutathione	52
Endothelin receptor antagonist BQ-123	52, 53

[a] Requires the presence of glutathione; co-transport assumed.
[b] Transport increased by the nonspecific MRP inhibitors probenecid and sulfinpyrazon.

central pore, which was closed at the inner cytoplasmic face of the membrane. A ~10 Å resolution structure of P-gp in the absence of bound nucleotide was also reported, and significant conformational changes were observed upon interaction of nucleotides with the NBDs.[94] Comparing the 3D structure of P-gp in the absence of nucleotide and in the presence of the nonhydrolyzable ATP analog adenosine 5'- (B, γ-imino)triphosphate (AMP-PNP), Rosenberg *et al.*[95] demonstrated that in the ATP-free state the two transmembrane halves form a single barrel with a central pore being open to the extracellular side; in the bound state the TMDs reorganize in a manner that could allow entry of hydrophobic drugs from the lipid bilayer to the central pore of P-gp. Recently, the same group reported an asymmetric configuration of TMDs in the nucleotide-bound state of P-gp at ~8 Å resolution.[96]

Comparison of this first 3D structure for an intact eukaryotic ABC transporter with the structures of the prokaryotic MsbA and BtuCD showed significant differences in the packing of the TMD α helices within this protein family in the absence of bound nucleotide. The structures of the nucleotide-free MsbA from *E. coli*[90] and *Vibrio cholerae*[91] differed in the relative orientations of their NBDs, but both showed a chamber closed at the extracellular side, in contrast to the

Table 3 Substrates and inhibitors of BCRP

Compound	Ref.
Substrates	
Mitoxantrone	56, 57
Daunorubicin	56
Doxorubicin	56
Etoposide	58
Topotecan	59
Flavopiridol	60
Methotrexate polyglutamate	61
7-Ethyl-10-hydroxycamptothecin (SN-38)	62, 63
2-Amino-1-methyl-6-phenylimidazo[4,5-*b*]pyridine (PhIP)	64
Rhodamin 123	56
Hoechst 33342	65, 66
Cimetidine	67
Oxfendazole	68
Nitrofurantoin	69–71
Estrogen sulfate	72
Inhibitors	
Corticosterone	67
Beclomethasone	67
Digoxin	67
Methylprednisolone	67
Dexamethasone	67
Triamcinolone	67
GF120918	73, 74
Fumitremorgin C	75
Chrysin	76, 77
Benzoflavone	76
Biochanin A	77
Genistein	78
Acacetin	78
Kaempferol	78
Naringenin	78
Naringenin-7-glucoside	78
Silymarin	79
Hesperetin	79

continued

Table 3 Continued

Compound	Ref.
Quercetin	79
Daidzein	79
Resveratrol	79
Novobiocin	80
Nelfinavir	81
Ritonavir	81
Saquinavir	81
Biricodar (VX-710)	82
Taxane derivatives	83
Imatinib	84, 85
Gefitinib	85, 86
Tyrosine kinase inhibitor EKI-785	85
Diethylstilbestrol	87
Tamoxifen	87, 87
Toremifene	87, 87
Antiestrogen derivatives	87
Estrone	87, 88
17β-Estradiol	88
Reserpine	89

open chamber of the unbound P-gp.[95] This basic difference in the structural organization of P-gp and MsbA remains unclear; moreover, the P-gp functional homolog LmrA displays a chamber open to the intracellular site,[97] as observed in the original MsbA structure. The domain organization of the *E. coli* BtuCD resembles an inverted portal opened to the extracellular solution.[92] Although more similar to the unbound P-gp configuration, BtuCD could also not be readily modeled to the P-gp structure. The different number of TMDs (20 versus 12) and the lack of sequence similarity do not allow the reliable determination of which α helices, if any, correspond to the P-gp TMDs.[95]

Thus, the high-resolution 3D structures of transporters related to P-gp should be used with caution, considering the low degree of homology within the primary sequences of the TMDs, the differences in the helix packing, and protein functions.

5.32.2.2.2 P-glycoprotein functioning

Despite the significant progress achieved in recent years, the interaction of P-gp with its substrates and inhibitors remains a poorly understood process, and the details of the particular mechanisms of action are still subjects of controversy.

Biochemical and structural data agree that the protein undergoes large conformational changes during the transport cycle. P-gp shows a basal ATPase activity in the absence of any added transport substrate. Al-Shawi *et al.* suggested that this activity is an intrinsic mechanistic property of the protein rather than a consequence of transport of an unknown endogenous substrate.[98] According to the ATP switch model for the transport cycle of ABC transporters[99] the transport substrate binds to its high-affinity binding site on the TMDs from the inner leaflet of the membrane. This binding transmits a conformational change to the NBDs, thus enhancing the binding of ATP. The bound ATP molecules generate a closed NBD dimer that itself induces conformational changes in TMDs in such a way that the drug-binding site is exposed extracellularly. The affinity decreases, allowing drug release. During the subsequent steps of the transport cycle the transporter restores its basal configuration.

While the high-affinity binding site is mostly associated with the unbound state of the protein, there is no united understanding of the functional state of the protein in which the binding site adopts its low-affinity conformational state.[100,101] In a study reporting the 3D structure of P-gp[95] it was directly demonstrated that major conformational changes occurred upon nucleotide binding and did not require ATP hydrolysis. Similar results could furthermore be shown for MRP1. A more limited conformational change also occurs upon ATP hydrolysis.[94] This is thought to facilitate the release of the bound drug to the extracellular environment and/or to represent an intermediate step toward the resetting of the transporter to the conformation that corresponds to the high-affinity binding site.

The available structural and biochemical data on P-gp lead to two important conclusions for the modeling of drug–protein interactions: (1) the 3D structural data on MDR transporters obtained to date do not permit a reliable definition of the 3D arrangement of P-gp TMDs and (2) it is very likely that the high-affinity binding site of transported substrates is located on TMDs in conformations that correspond to the ATP-free functional state of the protein.

5.32.2.2.3 Drug-binding sites on P-glycoprotein

Initially, a large common binding site on P-gp was assumed, confirmed by results from ATPase activity assays.[102] Later, a more complex behavior of P-gp became apparent, when cooperative, competitive, and noncompetitive interactions between modulators were observed, suggesting a minimum of two binding sites.[103] These findings were confirmed by different groups.[104,105]

From experiments with proteoliposomes containing P-gp, Shapiro and Ling[106] postulated two distinct binding sites, the rhodamine and Hoechst 33342 sites, which interacted in a positively cooperative manner. Later a third, regulatory, binding site was reported for progesterone and prazosin.[107]

From radioligand-binding experiments, at least four different binding sites were inferred that were able to allosterically communicate.[100]

In a recent review, Safa summarized data on binding sites obtained by competition with photolabeling drugs.[108] Seven binding sites were proposed that partly interacted with each other.

Novel results on the interaction and transport characterization of P-gp have been reported by Loo and Clarke from cross-linking experiments.[109–113] Differences in the cross-linking patterns observed upon the binding of different drugs support the suggestion that substrate binding produces rotation and/or lateral movement of transmembrane (TM) helices. Thus, P-gp could accommodate varying substrates through an induced-fit mechanism. Recently, the same authors reported that TMD5 and TMD8, on one hand, and TMD2 and TMD11, on the other hand, are located close together on the cytoplasmic side of the membrane.[114,115] Their results indicate that TMD2/TMD11 and TMD5/TMD8 between the two halves enclose the drug-binding pocket on the cytoplasmic side of P-gp.

Ecker and co-workers also studied ligand-binding sites, using photoaffinity labeling and mass spectrometry.[116] They reported that TMD3, TMD5, TMD8, and TMD11 were mostly labeled by benzophenone-type photoaffinity ligands, suggesting the involvement of the interfaces between the two halves of the protein.[117]

In summary, P-gp is likely to possess multiple (at least two) distinct drug-binding sites. The binding sites of substrates and inhibitors can be identical, can overlap, or be distinct, and certainly are affected by induced-fit mechanisms. Unfortunately, no 3D structural data on protein–substrate or protein–inhibitor complexes are available to date. Thus, the exact localization and structure of the drug-binding sites remain to be specified. Therefore, in silico approaches have a significant role in better understanding the protein structure–function relationships and rational design of P-gp inhibitors.

5.32.3   In Silico Approaches

5.32.3.1   In Silico Tools for Modeling Interactions with the Transporters

The application of in silico approaches for the modeling of drug–protein interactions has several stages, depending on the level of knowledge available on the structure and function of the modeled protein. Lack of detailed information about the 3D structure of the multidrug transporters and the binding sites involved in the drug interactions precludes the use of in silico tools that are based on the so-called structure-based drug design approach (e.g., docking and, related to it, virtual screening). Instead, they are commonly restricted to ligand-based drug design methods. Among the mostly used are different versions of the classical quantitative structure–activity relationship (QSAR) analyses by Hansch[118] and Free–Wilson[119]; three-dimensional quantitative structure–activity relationship (3D-QSAR), comparative molecular field analysis (CoMFA),[120] and comparative molecular similarity indices analysis (CoMSIA).[121] Furthermore, correlations with molecular descriptors can be generated from 3D calculated GRID interaction energies,[122,123] or converted to a two-dimensional form, such as the VolSurf[124] descriptors. Depending on the number and type of the structural descriptors, different multivariate data techniques are utilized: multiple linear regression (MLR), principle component analysis, the partial least squares (PLS) method,[125] genetic algorithms (GAs),[126,127] and artificial neural networks.[128,129] Pattern recognition methods based on different classification rules are also used.

Earlier in silico applications mostly used QSARs and 3D-QSARs. In recent years, pharmacophore and homology models have rapidly developed in parallel with increasingly detailed information about drug-binding sites and additional 3D structural data on different transport proteins.

5.32.3.2   Case Studies of P-Glycoprotein Interactions

5.32.3.2.1   P-glycoprotein assays and data sets for in silico studies

Most experimental data available in the literature relate to drug–protein interactions in cancer MDR, particularly P-gp-associated MDR. In silico methods, when applied to these data, have some limitations. The problems arising from a variety of screening assays used for activity estimation have already been discussed in detail.[130] The main assays for the determination of P-gp interactions are discussed briefly below.

Among the most widely used indirect measurements is the so-called MDR ratio or fold reversal. It is defined as the ratio between the IC_{50} value of the cytotoxic drug in the absence and presence of a relatively nontoxic concentration of the modulator[131]:

$$MDR \ \ ratio = IC_{50 \ cytotoxic \ drug}/IC_{50 \ cytotoxic \ drug+modulator} \quad [1]$$

The reverse relationship, the increase in cytotoxicity of a subinhibitory concentration of cytotoxic drug in the presence of increasing modulator concentrations is also applied[132]:

$$MDR \ \ ratio = IC_{50 \ modulator}/IC_{50 \ modulator+cytotoxic \ drug} \quad [2]$$

Both equations reflect a general MDR reversal effect; however, eqn [1] measures the potentiation of drug cytotoxicity in the presence of the modulator and eqn [2] estimates the potentiation of modulator cytotoxicity in the presence of the cytotoxic drug.

MDR ratios can vary substantially for the same modulator, depending on the tumor cell line, the antitumor drug used for preselection, the antitumor drug whose cytotoxic effect is modulated, and, finally, the concentrations of the cytotoxic drug and modulator.[130] Therefore, the comparison of MDR ratios obtained from different cell lines is not straightforward. The ratios register the entire MDR reversing effect of the drug without accounting for a particular modulation mechanism.

Inhibition of the transport activity of the protein is a widely used assay for direct measurement of drug–protein interactions. The decreased intracellular accumulation of a drug or the increased drug efflux can be used to characterize the transport activity. For these experiments, either radiolabeled antitumor drugs (e.g., [3H]vinblastine) or fluorescent substrates are used.

The calcein AM influx assay is another assay for measuring P-gp activity. The dye calcein AM is a nonfluorescent P-gp substrate that is intracellularly hydrolyzed to the fluorescent calcein. As calcein is not a P-gp substrate and its permeability through the cell membrane is very low, the increase in intracellular fluorescence over time is an indirect measure of P-gp activity.[133]

The effectiveness of P-gp to expel a substrate depends on its activity and on the uptake rate by passive diffusion. This was shown by kinetic studies using different anthracycline antitumor drugs.[134,135] The balance of drug influx through passive diffusion and its active efflux by the transporter ultimately determines the fraction of the drug entering a particular compartment in the presence of the transporter. It was shown that drug affinity toward P-gp does not affect the distribution of a drug into tissue compartments in cases of high passive permeability.[136] This observation confirms the importance of true affinity measurements for modeling drug–transporter interactions.

The efflux inhibition expressed as IC_{50} is considered to be a measure of actual drug-binding affinity, and thus is preferable when modeling drug–protein interactions. IC_{50} values can be used to estimate the relative contribution of the protein-mediated transport to the whole MDR modulating effect.[130] In some cases, a high degree of correlation can be observed between the efflux inhibition and the MDR ratio.[137] But efflux experiments can also have drawbacks. Assays using fluorescent anthracyclines could not reveal the true intracellular drug concentration due to drug accumulation in the nuclei accompanied by quenching of drug fluorescence.[138] Additionally, for anthracyclines, dissociation from DNA can be rate limiting, and thus obscure the true efflux capacity of P-gp.[139]

Another problem occurring in both accumulation and efflux assays is associated with the measurement of fluorescence. As an example, the fluorescence of the P-gp substrate Hoechst 33342 is greatly enhanced when bound to DNA or incorporated in a lipophilic environment such as the cell membrane. A recent study demonstrated that daunorubicin and rhodamine 123 decreased the fluorescence of membrane-bound Hoechst 33342 considerably in liposomes containing no P-gp. Thus, either fluorescence quenching or drug–membrane interactions leading to a displacement of Hoechst 33342 from the membrane can interfere with functional assays using Hoechst 33342 as the substrate.[140]

Another assay measures ATPase activity modulation caused by P-gp drugs, and makes use of EC_{50} values for stimulation/inhibition of ATPase activity.[141] Although initially suggested to be an appropriate test for P-gp substrates, both substrates and inhibitors were shown to induce ATPase activity.[105] Furthermore, some substrates such as etoposide,[142] daunorubicin, or vincristine inhibit ATPase activity instead of stimulating it,[143] while other substrates and inhibitors have negligible effects.[102,105] Controversial results have been published for some drugs, for example colchicine. Differences in ATPase stimulation by the same drug (e.g., verapamil and vinblastine) have been observed,[141,144,145] suggesting that the ATPase activity assay is sensitive to experimental conditions and may not correlate well with drug affinity and transport activity of P-gp.

Transport assays using a cell line expressing P-gp in a polarized way have become popular when attempting to distinguish between substrates and nonsubstrates. Either the transport of the investigated compound itself or its influence on the transport of a fluorescent marker such as rhodamine 123 is determined. Some studies indicate that the use of rhodamine 123 as a fluorescent marker in monolayer transport assays is rather problematic.[146,147]

A comparison of ATPase activity assays using rhodamine 123 uptake or calcein AM influx led to the suggestion that indirect fluorescence indicator assays (rhodamine 123 or calcein AM) should be used for a primary screening, followed by an ATPase or transcellular transport assay to distinguish between substrates and inhibitors.[142] Polli *et al.* compared 65 different compounds classified either as substrates or nonsubstrates in the literature (**Table 5**). They used three functional assays: ATPase stimulation, transport across adenocarcinoma cell line derived from human colon and used for estimation of human absorption (Caco-2) monolayers, and increase in calcein fluorescence.[148] Besides unequivocal substrates and nonsubstrates, which showed activity in all or in none of these assays, respectively, a number of compounds were not active in the transport assay, while others were inactive in the ATPase assay. A good example is doxorubicin, a well-known substrate of P-gp, which showed no activity in all three assays. For some compounds such as progesterone and PSC 833, interaction with P-gp is well documented, but no transport by the protein has been registered. Are these compounds substrates of P-gp or not? The answer can be different, depending on the specific property the scientist is looking for. If a CNS-active drug is being developed, transport is of crucial importance. In the case of an inhibitor, tight interactions are desired, and transport abilities are of minor importance.

In summary, a variety of assays can be used to determine drug–transporter interactions, and none of these assays is faultless. The problems mentioned above suggest that a preliminary critical evaluation is absolutely necessary when deciding on the right data to be used for in silico studies.

5.32.3.2.2 Classical quantitative structure–activity relationships: Hansch and Free–Wilson approaches

Different versions of the Hansch analysis, utilizing physicochemical parameters, and of the Free–Wilson approach, making use of indicator variables for the presence/absence of particular structural features, have been used to relate the changes in the chemical structures of various classes of MDR modulators to changes in activity.

The class of propafenone-type MDR modulators has been intensively investigated by Chiba, Ecker, and co-workers. These authors have synthesized a large number of derivatives, and derived several QSARs for this type of MDR modulator.[149–157] In the earlier investigations, fold reversal was determined to quantify anti-MDR activity using a resistant T lymphoblast cell line.[149–151] Later, rhodamine 123[150,151] or daunorubicin[152–159] functional efflux inhibition assays were used.

The results obtained for derivatives with different substituents and modifications in the parental propafenone structure (**Figure 2**) demonstrated that the structural changes correlate with activity through changes in lipophilicity (log *P*) if only strongly homologous series are considered.[150–153] In a QSAR analysis of a larger data set of 48 propafenone derivatives, molar refractivity (MR) was found to be slightly superior to log *P*.[154] Both parameters, however, could only explain 56% and 55%, respectively, of the variance in anti-MDR activity. To improve the correlation, several structural features were included as indicator variables, yielding a mixed Hansch-type Free–Wilson equation.

The length of the spacer between the central aromatic ring and the nitrogen atom was also varied, and a bilinear dependency of anti-MDR activity on the chain length was observed.[155] Up to a length of five carbon atoms, the activity increased with increasing length of the spacer, and then remained constant up to a chain length of eight carbons.

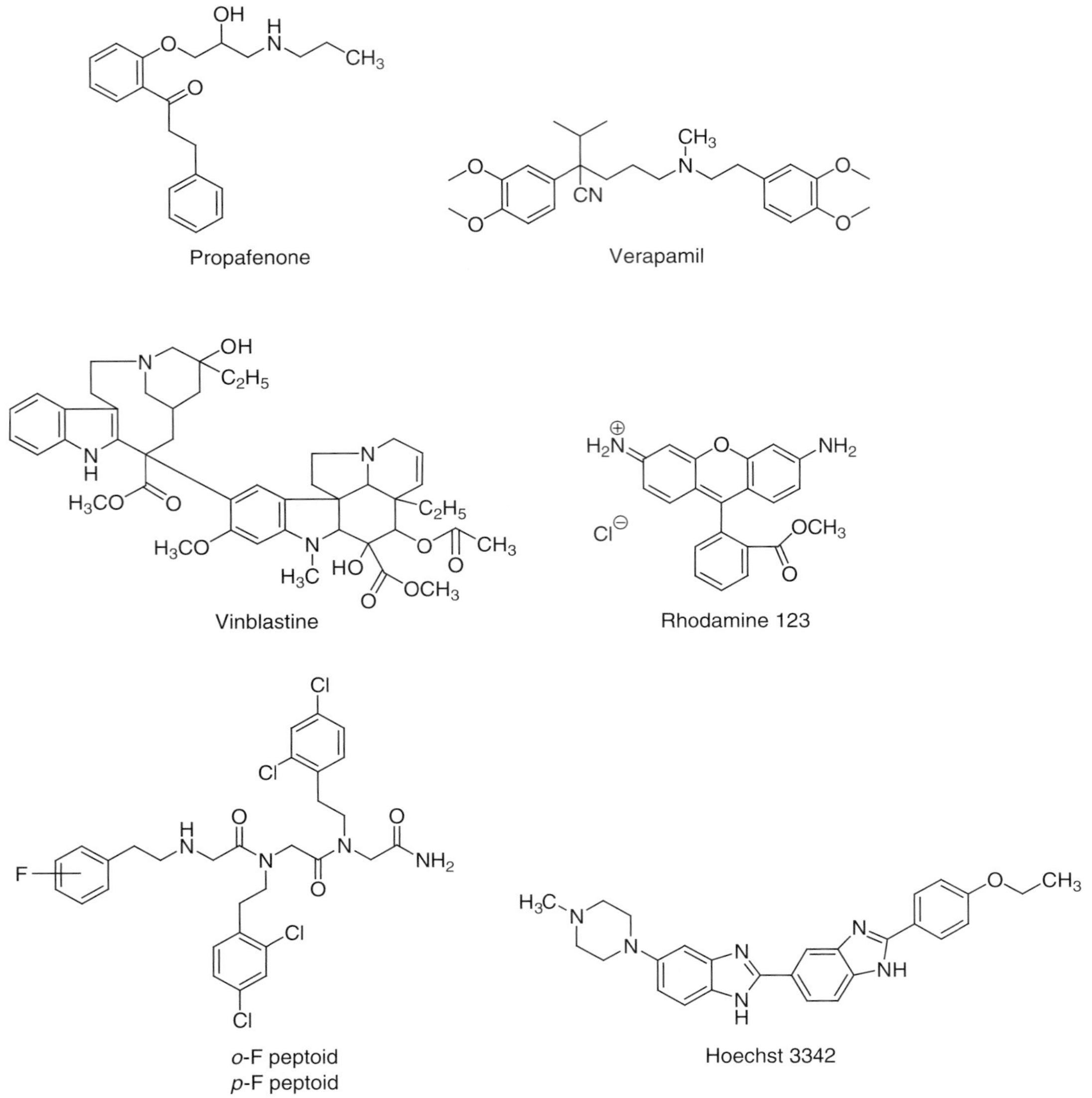

Figure 2 Structures of representative compounds used in the development of pharmacophore models for recognition/inhibition of P-gp.

The role of the carbonyl oxygen atom, which can act as a hydrogen bond (HB) acceptor, was further investigated using indanon analogs.[156] No statistically significant difference in anti-MDR activity was found for indanones differing in the position of the carbonyl group. However, inclusion of the charge of the acyl carbonyl oxygen atom in the regression improved the QSAR model of 15 structurally related 4-acylpyrazole derivatives, pointing to the HB acceptor ability of the oxygen atom as a feature contributing to MDR reversal activity.[158]

Further, the HB acceptor strength of 12 amines, amides, and anilines and one ester of the propafenone type was correlated with their P-gp efflux inhibition effect.[157] The HB acceptor strength yielded a model with high predictivity, in contrast to that using log P as the descriptor. The authors concluded that the nitrogen atom did not interact with P-gp in a charged form but instead functioned as an electron donor group.

In a recent study, Ecker and co-workers attempted to check the hypothesis about the role of the space-directed hydrophobicity suggested previously in a 3D-QSAR study of propafenone-type modulators[137] (discussed in more detail in Section 5.32.3.2.4). The authors systematically varied hydrophobicity distribution within a series of 32 propafenone derivatives, using different substituents on the nitrogen atom and varying the position of the side chain with the carbonyl group.[160] The results showed that with increasing lipophilicity of the substituent on the nitrogen atom, the importance of the position of the side chain increased, and the dependency on total lipophilicity decreased. This indicates that with increasing lipophilicity on the amino terminus the influence of the position of the substituent on the central aromatic ring gains increasing relevance, and that the hydrophobicity distribution within the molecule is important for P-gp inhibitory potency, confirming the results from the 3D-QSAR study.[137]

The same group investigated also series of tetrahydroquinoline-, dihydrobenzopyran-, and thienothiazine-type MDR modulators. Restrictions arising form the smaller data sets studied did not allow such a detailed analysis as for the propafenone-type modulators.[159,161] In all series, the best correlations were found with molecular refractivity.

Derivatives based on the verapamil (**Figure 2**) and related skeletons were investigated as MDR modulators by Teodori *et al*.[162] The compounds were designed by combining the skeletons with potent substituents that had been identified in other studies of different groups of MDR modulators. Attempts to analyze the structure–activity relationships in both a qualitative and quantitative manner led to no general pattern, with the exception that the core moiety of verapamil appeared to be the most useful.

For thioxanthenes and phenothiazines tested as MDR modulators by Ford *et al*.,[131,163] QSAR models were developed employing the Free–Wilson-type analysis.[164] Besides classical MLR, a GA-based MLR was applied, to test the predictivity of the models derived. The length of the chain, the type of the substituent on the basic nitrogen atom, and stereoisomery were found to be significant for the anti-MDR activity of thioxanthenes. For both data sets the activity was additionally dependent on the type of ring system and the substituent in position 2 of the tricyclic ring system. These results pointed to the importance of the relative disposition of the aromatic ring system and the basic nitrogen atom, in agreement with the QSAR studies of Pearce *et al*.[165] and, later, Suzuki *et al*.[166]

The data of Litman *et al*. on the stimulation of ATPase activity[143] was explored to model P-gp–drug interactions using MolSurf parameters and PLS.[167] Twenty-two compounds were selected, omitting large structures such as vinblastine or cyclosporine A. MolSurf parameters included 15 global descriptors, characterizing mostly HB properties. In the final model, the molecular surface, polarizability, Lewis acid property, the number of HB acceptor oxygen atoms, and the acceptor strength of the HB oxygen atoms were selected as parameters. The contribution of the five parameters, accounted for by one significant PLS variable, were almost equal, with the exception of the Lewis acid descriptor, which had a lower regression coefficient in the PLS models (**Table 4**). Once again, the HB acceptor properties were found to contribute predominantly to P-gp affinity.

A series of taxane-based MDR modulators was investigated by Brooks *et al*.,[168] using a Free–Wilson-type approach. In this study, the influence of the modulators on P-gp was investigated, and, additionally, on MRP1 (ABCC1) and BCRP (ABCG2) using the substrates daunorubicin and mitoxantrone. The observed patterns of the substituent contributions varied between the two substrate structures and the three investigated transport proteins.

A series of 25 xanthine derivatives, analogs of pentoxifylline, were studied as MDR modulators.[169] The best correlation between IC_{50} and the descriptors was obtained for a combination of molecular refractivity, crystal density, and calculated log *P*.

Generally, in the QSAR models derived, the descriptors varied between different chemical classes, and most of them held only for homologous series of compounds. In the Hansch analyses, the partition coefficient log *P*, molecular refractivity, and HB acceptor descriptors were the most important characteristics for the MDR modulating activity of the studied drugs. In the Free–Wilson and subsequent 3D-QSAR analyses, a role for the relative disposition of the hydrophilic and the hydrophobic parts was suggested, and confirmed for the propafenone-type MDR modulators.

5.32.3.2.3 Classification studies

In classification studies, structurally different compounds are used as training sets to identify structural features as 'decision rules' for compounds belonging to a particular group. Large data sets evaluated by multiple assays and widely

Table 4 A PLS model generated by MolSurf parameters[167]: $n = 21$, $r^2 = 0.72$, $q^2 = 0.70$, $s = 0.48$, $F = 48.4$

Parameter	Scaled PLS regression coefficient
Surface	0.260
Polarizability	0.247
No. of HB acceptors (O atoms)	0.232
HB acceptor strength (O atoms)	0.221
Lewis acid	0.138

varying in their activity values are commonly used. Therefore classifications are mostly performed on either functional estimates (e.g., substrates/nonsubstrates), or activity interval estimates (e.g., high/medium/low activity).

The correctness of the classification rules, however, is highly dependent on the correct division of the compounds into different groups. As discussed in Section 5.32.3.2.1, depending on the assay and the experimental conditions, the same compound can be assigned to different functional or activity groups. **Table 5** shows some compounds used in classification studies, for which different assignments have been reported.

Klopman and co-workers performed one of the most extensive classification studies. First, they applied a substructural analysis approach to 137 diverse MDR modifiers, and identified a number of relatively small structural features (biophores) that contributed in a statistically significant way to anti-MDR activity.[170] Data from different tumor cell lines and different measures of the MDR reversal effect were merged in this analysis. Later, the same approach employing the MULTICASE software was extended to 609 structurally and functionally diverse MDR reversal agents whose activities had been determined under the same experimental conditions.[171] The program was able to identify 35 biophores, among them 25 relevant to activity. The biophore contributing the most was C–C–X–C–C, with X = N, NH, or O, where a tertiary nitrogen atom linked to two unsubstituted alkyl groups was preferred. A partially positively charged nitrogen atom was found to be superior, and a quaternary nitrogen atom detrimental, for activity. In **Table 5**, some of the compounds classified according to the study of Klopman *et al.* are shown. As seen from the table, there is a partial correspondence between the experimental and classification results for many of the drugs shown; for some of them, such as prazosine and yohimbine, totally opposite assignments in the experiment and the classification are recorded.

The extended MULTICASE approach, called MCASE, was also applied to 130 propafenone-type MDR modulators.[172] In total, 10 biophores were identified, from which six gave a statistically significant contribution. The largest and top-scoring biophore contained a large part of the propafenone backbone, whereas the lowest-scoring biophore contained a *meta*-substituted aromatic ring. The predictive capability of this model was tested on structurally related compounds that were not propafenones. A correlation coefficient of $r^2 = 0.75$ between calculated and observed MDR reversal activity was obtained, suggesting more general applicability of the model.

Bakken and Jurs applied several classification methods to the same data set as Klopman *et al.*, but characterized molecular properties of the compounds by structural descriptors.[173] The potential variables were selected by a GA from a pool of 100 geometric, electronic, polar surface, and topological descriptors. When only topological descriptors were considered, the best models were obtained with nine or six variables, and achieved the highest classification success in the range 83.1–91.1%. These results suggest the potential of topological descriptors for high-throughput screening, as they can be easily generated from the two-dimensional structural formula of the compounds. As in other investigations of structurally dissimilar compounds, inclusion of log *P* did not improve the results.

A set of 44 compounds containing several pesticides was investigated by Bain *et al.*,[174] using a rule-based system to differentiate between 11 transported ligands, 18 inhibitors, and 15 noninteracting compounds. First, the transported substrates were selected if they fulfilled the following four criteria: (1) at least one ≥six-membered ring system; (2) molecular weight >399; (3) lipophilicity expressed as log K_{ow}<2; and (4) HB potential >8. Next, from the remaining compounds, the inhibitors were classified if they fulfilled the following three criteria: (1) at least one ≥six-membered ring system; (2) molecular weight >247; and (3) dipole moment ≥3.3. Compounds that fulfilled neither of the two rules were categorized as noninteracting. Using these rules, transported substrates, inhibitors, and noninteracting compounds were classified with 82%, 72%, and 89% accuracy, respectively. To test their rules, the authors selected six pesticides; five were correctly predicted, leading to an accuracy of 83%. In contrast to the results of other studies, the HB donor, rather than HB acceptor, potential was found to be important.

Gombar *et al.* used linear discriminant analysis to predict whether a given compound was a P-gp substrate or not.[175] The data set consisted of 95 compounds, classified either as substrates or nonsubstrates, based on the results from cell monolayer transport assays. The descriptors included log *P*, molecular refractivity, the electrotopological values of 13 substructures, and counts of 12 substructures. With 'leave-one-out' cross-validation, all substrates were correctly predicted, and only three out of 32 nonsubstrates were incorrectly classified as substrates. For a test set of 58 compounds, 50 (86.2%) were correctly predicted as either substrates or nonsubstrates. In **Table 5**, the correspondence between the assigned classification and the transport assay through Caco-2 cells can be seen; however, the classification does not completely match the other experimental assays.

A simple rule, termed the 'Gombar–Polli' rule, was developed, which was based on the molecular electrotopological state. According to this rule, compounds with a molecular electrotopological state value >110 are likely to be substrates, while compounds with a value <49 are nonsubstrates. When this rule was applied to a test set, 14 compounds were predicted to be substrates, and, indeed, all of them were confirmed to be substrates. Five compounds with a molecular electrotopological state value <49 were correctly predicted to be nonsubstrates, while the remaining 39 compounds fell into the undefined interval between 49 and 110.

Table 5 Classification of some P-gp-related compounds according to the results from experimental assays of P-gp activity and classification and modeling studies

Compound	P-gp activity assays				Classification and pharmacophore studies			
	Transport through Caco-2 cells[148]	*Increase of Calcein fluorescence[148]*	*Stimulation of ATPase activity[148]*	*Stimulation of ATPase activity[102]*	*Klopman et al.[171]*	*Gombar et al.[175]*	*Xue et al.[177]*	*Seelig[196]*
BIBW22	N							S
Chlorpromazine	N	S	S		S	N	N	B
Colchicine				N	N	S	S	I
Corticosterone					N		S	I
Daunorubicin						S	S	
Diltiazem	S	S	S		S	N	S	S
Doxorubicin	N	N	N	N		N	S	I
Estriol	N				N		S	
GF120918	N	S	N				S	
Hydrocortisone	N				N		S	S
Ketoconazole	N	S	S			N		
Mebendazole	N	S	S			N		
Methotrexate	N	N	N	N		N	S	B
Midazolam	N	S	S			N	N	I
Nicardipine	N	S	S		S	N	S	S
Nifedipine	N	S	S		N	N	S	I
Nitrendipine	N	S	S			N		
Prazosine	S	S	S		N	S	S	S
Progesterone				S	S		N	S
Propranolol	N	N	N		S	N		
Tamoxifen					N		N	S
Testosterone	N	S	N		N	N		
Trimethoprim					N	S		
Verapamil	N	S	S	S	S	N	S	S
Yohimbine	N	N	N		S	N	S	S

B, borderline substrate; I, inducer and substrate; N, nonsubstrate; S, substrate/modulator.

Didziapetris *et al.* developed a set of simple rules to discriminate P-gp substrates from nonsubstrates using a data set of 1000 compounds and a stepwise classification method.[176] From the initially selected molecular descriptors, Abraham's H-accepting β, followed by the molar weight or volume, had the highest discriminating power, and log *P* had the lowest. A compound is predicted to be a substrate if: (1) molecular weight > 400 and (2) $β > 1.7$. As the sum of oxygen and nitrogen atoms is a crude estimate for β, the substrate properties can be roughly estimated by the 'rule of fours': compounds with $(N + O) \geqslant 8$, molecular weight > 400, and acidic $pK_a > 4$ are likely to be P-gp substrates; whereas compounds with $(N + O) \leqslant 4$, molecular weight < 400, and base $pK_a < 8$ are likely to be nonsubstrates. Class-specific rules of high discrimination power were developed for structural subclasses with slightly varying cut-off values. The obtained results led to the view that P-gp possesses a large binding site with fuzzy specificity, rather than distinct binding sites. Recently, Xue *et al.* applied the support vector machine learning technique, to derive a model capable of discriminating substrates from nonsubstrates among 201 compounds.[177] From an initial pool of 59 descriptors, 22 remained significant after applying feature selection. Doxorubicin, one of the training set compounds, was erroneously included in the validation set under the name adriamycin, another name used to denote the same compound. In **Table 5**, the assigned classifications for some compounds used in this study are given for comparison with the other in silico classifications and experimental assays.

In summary, classification studies are suitable if the best candidates among a large number of potential ones are to be selected for further experimental screening. They are a worthy alternative for the in silico screening of large numbers of substrates and inhibitors of transport proteins, at least until more sophisticated methods such as docking and virtual screening become feasible. In addition, the classification rules may help identification of important structural discriminants for functionally different drugs, thus providing more in-depth knowledge on drug–protein interactions.

5.32.3.2.4   Three-dimensional quantitative structure–activity relationship studies

The phenothiazine-type MDR modulators were the first to be investigated using a 3D-QSAR approach.[178] The data set was collected from the works of Ford *et al.*,[131,163] and included both stereoisomers of thioxanthenes. To better decide on a possible alignment of the stereoisomers, a detailed molecular modeling study of *trans*- and *cis*-flupentixol was performed, and the 3D properties of the drugs were compared.[179] When both stereoisomers were aligned to be similar in shape, a good agreement of the hydrophobic HINT[180] fields was found. Therefore, it was suggested that *trans*- and *cis*-flupentixol could have a mirror-like orientation when interacting with P-gp. Interestingly, an opposite effect of *cis*- and *trans*-flupentixol on ATP hydrolysis and substrate recognition by P-gp has been reported,[181,182] suggesting a steroespecific interaction of the isomers with P-gp, in agreement with the modeling results. CoMFA models of 40 phenothiazines, thioxanthenes, and related drugs were further derived, involving steric, electrostatic, and hydrophobic fields alone and in combination. The best models reached high internal and external predictivity, and combined steric and hydrophobic fields.

Phenothiazine type MDR modulators, taken from Ramu and Ramu,[132] were also studied using CoMFA and CoMSIA.[183,184] Once more, the role of hydrophobicity expressed as CoMSIA similarity indices was confirmed. Additionally, the role of the HB acceptor indices for MDR reversal was demonstrated, confirming the contribution of HB acceptor descriptors identified in classical QSAR and classification studies.

Propafenone-type MDR modulators were investigated in several studies. The first 3D-QSAR using CoMFA[137] involved 28 compounds for which P-gp inhibition had already been determined.[150,151] The best models were obtained either with the HINT hydrophobic field alone or in combination with steric and electrostatic fields. Results of this study led to the conclusion that the standard steric and electrostatic fields could not fully represent the main forces driving MDR reversal and directed to the role of the hydrophobicity distribution as an important structural determinant for inhibition of the P-gp dependent efflux. This was further confirmed by comparing activity predictions employing log *P* and hydrophobic fields.[185] Later, a larger set of 70 propafenone-type compounds was also investigated using both CoMFA and CoMSIA.[186] The hydrophobic field alone or in combination with the steric field resulted in the best models again. As no HB donor and acceptor fields were included, the importance of HB acceptor properties was not investigated.

A series of imidazole derivatives that had led to the clinical candidate ONT-093 was investigated using both CoMFA and CoMSIA.[187] The best CoMFA model for 46 compounds was a two-component model in which the electrostatic and steric fields contributed nearly equally. Applying CoMSIA, a two-component model was also obtained with similar, but slightly worse, statistics. Analyzing the contribution of the five CoMSIA fields (steric, electrostatic, H acceptor, H donor, and hydrophobic), only the first four where found to contribute significantly, while the hydrophobic field did not, and was excluded.

The hydrophobic field models, calculated by HINT and CoMSIA hydrophobic indices, were compared in a series of triazine-type MDR modulators.[188] The data set contained 30 derivatives with MDR reversing activity measured by

Dhainaut *et al.*[189] Both CoMFA and CoMSIA models showed comparable results. The role of the hydrophobic fields was confirmed, and the electrostatic field was additionally found to contribute to the MDR reversal effect of the studied compounds.

Thirty sesquiterpenes of the dihydro-β-agarofuran type have been investigated with CoMSIA using wild-type and MDR *Leishmania tropica* expressing a P-gp-like transporter.[190] Activity was measured by percentage growth inhibition, and a fixed modulator concentration was used. All five CoMSIA fields were included as descriptors, and additionally the solvation energy of the compounds was considered. Separate models were generated, both for wild-type and MDR *Leishmania tropica*. The best models were obtained with an unusually high number of latent variables – 19 for MDR and 18 for the wild type. This result could be related to the use of the percentage growth inhibition as the dependent variable, as this introduces nonlinearity into the biological response. The electrostatic field had the highest contribution by far, of more than 50%, followed by the HB donor and hydrophobic fields. The contributions of the steric field and solvation energy were marginal.

Although different molecular fields and similarity indices were found to contribute to MDR reversal by different classes of modulators, in general, most of the 3D-QSAR models agreed on the role of the hydrophobic fields for their interactions with P-gp. The steric and HB acceptor fields also were found to contribute significantly. The role of the particular field seems, however, to vary, depending on the particular class studied.

5.32.3.2.5 Pharmacophore modeling

The term 'MDR pharmacophore' was first introduced by Pearce *et al.* in pioneering studies on structurally different MDR modulators, including verapamil, vinblastine, vindoline, chlorpromazine, chloroquine, and several reserpine analogs.[191,192] The existence of 'a conserved element of molecular recognition in modulator binding by P-gp,' was postulated, which was composed of two aromatic domains and a basic nitrogen atom. Although far from any particular protein-binding site associated with the drugs studied, this first pharmacophore model already presumed some common pharmacophore features of substrates and inhibitors when interacting with the transporter.

Based on analogies to binding of antipsychotic drugs to calmoduline, Hait and Aftab[193] proposed a hypothetical drug-binding site on P-gp for phenothiazines and related drugs. This model involved the same structural domains as proposed by Pearce *et al.*, assuming overlapping of the π orbitals of the aromatic rings of phenylalanine residues in the protein TMDs and of the phenothiazine and thioxanthene ring systems.

Another type of drug–protein interaction was suggested by Suzuki *et al.*[166] In a study of 24 quinoline-type MDR-reversing compounds, including MS-209 (**Figure 1**), the authors postulated that a nonplanar arrangement of the pharmacophoric aromatic rings was essential for high activity. Additionally, a distance of at least 5 Å between the basic nitrogen atom and the center of the hydrophobic part was set as a prerequisite for high activity within the series.

Later, Etievant *et al.* formulated a distance requirement for P-gp recognition based on structure–activity relationships of a podophyllotoxin series.[194] The authors demonstrated that the presence of an amido group in the fourth position of podophyllotoxin promotes P-gp recognition, and that a distance of 5 Å to the centroid of the trimethoxy aromatic ring appeared to be a predictive criterion for P-gp recognition.

While the interactions between P-gp and the aromatic rings of the drugs, regardless of ambiguity in the type of interactions, were well recognized from the very beginning, the role of the tertiary nitrogen atom remained unclear. Investigating MDR reversal activity of 232 phenothiazines and related drugs[132] and 311 structurally diverse compounds with two or more phenyl rings,[195] Ramu and Ramu came to the conclusion that a permanent charge, as in quaternary amines, abolished activity, and that a carbonyl group could participate in the formation of intra- or intermolecular HBs.

Several studies were undertaken to further elucidate the role of the nitrogen atom and the carbonyl group in the interaction of compounds with P-gp. Chiba *et al.*[150] reported that reduction of the carbonyl group as well as its conversion to a methylether moiety in the propafenon nucleus (**Figure 2**) led to a remarkable decrease in activity, whereby lipophilicity lost its predictive character as the main determinant for modulator potency. Similarly, the relative positioning of the acyl- and propanolamine side chains also influenced activity, so that the distance between the carbonyl group and the nitrogen atom seemed important. The role of the carbonyl group as an HB acceptor was demonstrated more clearly in another QSAR study of 12 propafenone analogs.[157]

In parallel with the elucidation of the nature of the drug–protein interactions, studies designed to differentiate between the pharmacophore patterns of P-gp substrates and inhibitors have been initiated. Seelig proposed specific recognition patterns for P-gp substrates and inhibitors based on the proximity of HB acceptors in space.[196,197] A general pattern for P-gp substrate recognition was proposed, and considered two types of patterns. Pattern I was formed by two electron donor groups with a spatial separation of 2.5 ± 0.3 Å. Pattern II was formed either by two electron donor groups with a spatial separation of 4.6 ± 0.6 Å or by three electron donor groups with a spatial separation of the two outer groups of 4.6 ± 0.6 Å. Substrates with a high affinity for P-gp should have at least one type I or one type II pattern. Binding would increase with the strength and with the number of electron-donor groups involved in type I and type II patterns. For transport, at least two type I patterns or one type I and one type II patterns were necessary. If two substrates are applied simultaneously to P-gp, the compound with the higher potential to form HBs generally acts as an inhibitor. Partitioning into the lipid membrane was proposed to be the rate-limiting step for the protein–substrate interaction, and dissociation of the substrate–protein complex was proposed to depend on the number and strength of the HBs formed.[198] The same substrate recognition principles were, furthermore, proposed for the transport protein MRP1.[199]

The above patterns were derived employing data from different experimental sources. As shown in **Table 5**, the substrates classified may possess different functionalities in the experiments, thus confirming the importance of the source data used to generate the pharmacophore model.

Using Seelig's pattern approach, Penzotti *et al.* proposed a computational ensemble pharmacophore model to identify substrates of P-gp.[200] A binary classification scheme was applied to a set of 195 compounds (144 for training and 51 for test): compounds were classified as substrates if transported by P-gp, and as nonsubstrates if not transported. This ranking scheme was thought to reduce the error associated with the collection of data from different laboratories, assays, cell lines, and experimental conditions. The top-scoring pharmacopore consisted of one HB acceptor, one HB donor, and two hydrophobic features. Depending on the conformation, different functional groups of the same compound could form the pharmacophore. Although the model of Penzotti *et al.* overcomes some of the drawbacks of Seelig's patterns (data amalgamation and conformational dependence), it only achieves a relatively low correct classification for the substrates: 64% for the training set and 53% for the test set. Interestingly, half of the pharmacophores derived contained a combination of HB acceptor, HB donor, and one or two hydrophobic groups. The presence of an HB donor was highlighted in particular by the authors, in contrast to the patterns found by Seelig.

In a series of publications, Doeppenschmitt *et al.*[201–203] and Neuhoff *et al.*[204] reported the affinity of structurally different compounds for the P-gp verapamil-binding site. These and other data on the interactions of structurally divergent drugs with P-gp[205] initiated a number of more sophisticated 3D pharmacophore modeling studies applying programs developed for the automatic identification of pharmacophore models, such as DISCO,[206] Catalyst,[207] and GASP.[208]

Ekins *et al.*[209] used Catalyst to generate 3D pharmacophore models of diverse series of P-gp inhibitors. Several training sets of compounds taken from the literature[204,205] and tested in vitro in different assays were studied. Four pharmacophore models were generated, and used to predict the inhibition of P-gp-mediated transport of digoxin, calcein accumulation, vinblastine accumulation, and vinblastine binding. All models contained at least one hydrophobic feature corresponding to either an aromatic ring or an aliphatic hydrophobe. Based on the good correlation between the models for digoxin transport and vinblastine binding, the authors hypothesized that vinblastine and digoxin share the same binding site. As the pharmacophores differed slightly in feature content, angles, and distances, it was suggested that the pharmacophores may correspond to multiple regions within the same binding site. In a further study,[210] a novel P-gp inhibition model was constructed using 16 compounds from the data set for the verapamil-binding site of P-gp by Neuhoff *et al.*[204] The model consisted of one HB-acceptor, one aromatic feature, and two hydrophobes. Comparing the degree of similarity in rank order prediction of previously generated data by this model and other models mentioned above, the authors suggested that verapamil was likely to bind to the same site, or an overlapping site, as digoxin and vinblastine.

Pajeva and Wiese[211] also used the data of Neuhoff *et al.*[204] to develop a pharmacophore model. Representatives of all reported structural classes were included, and the GASP program was applied. A general pharmacophore pattern was proposed for the verapamil-binding site of P-gp that involved two hydrophobic aromatic centers, three HB acceptors, and one HB donor. The distances were calculated as the averages of those found in the most active compounds. The same pharmacophore patterns were identified for enantiomers, in agreement with their binding affinity. It could be demonstrated that the binding affinity of the drugs depended on the number of pharmacophore points simultaneously involved in the interaction with the protein. The results suggest the verapamil-binding site of P-gp to have several points that can participate in hydrophobic and HB interactions, and that different drugs can interact with different receptor points in different binding modes.

Using (*R*)-verapamil, rhodamine 123, vinblastine, colchicine, and calcein AM as a training set, and Catalyst software, Masip *et al.*[212] also generated a pharmacophore model that contained two hydrophobic groups (aromatic or aliphatic chains), one aromatic ring, and two HB acceptors. The model was used to test two peptoids (**Figure 2**), identified through screening of a library of *N*-alkylglycine trimers that had been shown to cause a higher intracellular accumulation of daunorubicin than verapamil. Some of the conformations of both peptoids with energies close to the minimum conformers were consistent with the pharmacophore model.

A different 3D pharmacophore modeling approach was used by Garrigues et al.,[213] who studied a series of cyclic peptides, peptide-like compounds, verapamil, progesterone, and vinblastine for their effects, alone and in combination, on P-gp ATPase activity. The results of manual superimpositions on the MOLCAD surfaces were expressed as distances between selected consensus groups of hydrophobic and polar elements. Two pharmacophores were outlined. 'Pharmacophore 1' was produced from overlays of compounds such as verapamil, cyclosporine A, and actinomycin D, and consisted of one aromatic and two alkyl areas and one HB acceptor group. 'Pharmacophore 2' consisted of one aromatic and three alkyl areas and one HB-acceptor, and involved vinblastine and tentoxin. In order to explain why vinblastine displayed a competitive inhibitory effect with the ligands of the first pharmacophore, the authors suggested that the receptor counterparts of the pharmacophores are closely localized in the same multispecific binding pocket of P-gp. These results were considered to agree with the results by Sharom et al.[214] for different binding sites of verapamil, cyclosporine A, and vinblastine.

The group of Ecker et al. also used the Catalyst program for the identification of pharmacophore patterns of structurally similar series of benzofuran- and benzopyran-type MDR modulators[215] and propafenones.[216] In their first study, a pharmacophore model consisting of one HB acceptor, one hydrophobe, one aromatic hydrophobic, and one positive ionizable feature was obtained. The pharmacophore model of propafenones additionally involved one more aromatic hydrophobic feature. The positive ionizable group corresponded to the basic nitrogen atom, which could not be defined as an HB acceptor due to limitations of the software. When the authors excluded the ionization function, an HB acceptor was obtained in close proximity to the positively charged center. Although limited to a congeneric series of P-gp inhibitors, these pharmacophore models agree with models derived from structurally divergent P-gp drugs on features related to HB acceptor and hydrophobic interactions.

A pharmacophore model for the Hoechst 33342-binding site (**Figure 2**) on P-gp was also proposed.[217] The authors used experimental data on small molecules presumed to interact with the H-site of P-gp.[218] GASP produced a pharmacophore model consisting of a minimum of three hydrophobic features and one HB acceptor with the potential involvement of one additional HB donor.

For other MDR-related proteins, far fewer pharmacophore models have been developed. Recent studies demonstrate that the orphan nuclear receptor PXR can regulate the expression of CYP3A4 and ABCB1 simultaneously.[219,220] In human hepatocytes the expression of MRP2 can be upregulated by PXR ligands such as rifampicin.[221,222] Based on EC_{50} values for PXR activation, 12 PXR ligands were examined with Catalyst, and a pharmacophore could be generated that was able to distinguish between poor and the most potent activators.[223,224]

In summary, depending on the sets and software used, different combinations of pharmacophore elements were generated. Independent of the variety within the models reported, most of them suggest that a pharmacophore of a P-gp drug should combine at least one HB acceptor and two hydrophobic elements in a particular space arrangement. This combination can be termed 'a conserved pharmacophore pattern,' similarly to the term used by Pearce et al.[165] The hydrophobic elements can be either aromatic or aliphatic, or both. Additional HB acceptor and HB donor interactions can also be involved, which is entirely reasonable considering that proteins possess a large number of functional groups that allow both acceptor and donor interactions. Derivation of more sophisticated pharmacophore models related to particular binding sites of the protein will depend on both the use of more relevant experimental data and the application of more sophisticated program tools.

5.32.3.2.6 Homology modeling

Several studies deal with the homology modeling of transport proteins. Among them, P-gp is the most intensively modeled transporter, but others have also been studied (see Section 5.32.3.3).

Soon after the x-ray structure of E. coli MsbA in the absence of nucleotide was reported,[90] Seigneuret and Garnier-Suillerot published a structural model for the 'open' conformation of P-gp.[225] Comparative sequence analysis, motif search, and secondary structure prediction indicated a high structural similarity between each of the two P-gp halves and the MsbA monomer. The domain organization in the modeled P-gp structure, however, yielded a chamber that was closed on the extracellular side of the TMDs, in disagreement with the low-resolution 3D structure of the unbound P-gp.[93] As most of the residues that were known to be associated with drug binding were not found inside the chamber of the modeled P-gp, the authors suggested that P-gp could exist as open and closed structures.

Stenham et al.[226] generated two models of P-gp, refraining from modeling the P-gp domain organization reported by Seigneuret and Garnier-Suillerot.[225] They believed that the inverted V-shaped MsbA dimer, as observed crystallographically,[90] did not represent the physiologically relevant state of the transporter. The first model maintained the NBD:TMD interface suggested by the MsbA template, and the interface between the two halves of P-gp relied on the consensus interaction between the two NBDs reported by Smith et al.[227] Although V shaped, the resulting model had only a limited concordance with the published cross-linking data concerning the intracellular ends of TMDs, and was inconsistent with cross-linking data describing the TMD:TMD interface on the extracellular face of the membrane.[228–230] In the second model, each half of P-gp was modeled on the MsbA structure, and oriented as in the MsbA crystal. The two halves were then placed closer to each other by rotation of TMDs with respect to the NBDs so that the angle between the planes of the TMD:TMD and the NBD:NBD interfaces resembled the equivalent angle in the crystal structure of BtuCD.[92] This model was expected to embody the electron microscopy structure and to correspond to experimental cross-linking data. The correspondence was estimated based on cross-linkers; however, different lengths of cross-linkers were considered. Furthermore, broad intervals instead of particular distances were mostly used to validate the model, and distances measured in different functional states of P-gp (e.g., ATP-bound and vanadate trapped) were simultaneously considered, thus hampering the correct estimation of the reliability of the model.

Pajeva et al.[217] also proposed a homology model of P-gp reconstructed from the same E. coli MsbA x-ray structure.[90] The amino acids were aligned as in the model of Seigneuret and Garnier-Suillerot.[225] A rearrangement of both halves was performed to achieve an orientation of the NBDs that corresponded to the cross-linking data reported by Lee et al.[231] A model of P-gp, called the cross-linking model, was also generated, based on the cross-linking constrains defined by Loo and Clarke.[111,230] TMDs were built as α helices, the cross-linked amino acids were mutated to cysteine residues, and the helices rotated in such a way that the cross-linked residues faced each other. The distance constrains between the cross-linked cysteine residues were then incorporated into the model. In the following, the orientations of two sets of amino acid side chains were compared in the two models. These included those suggested to be involved in the binding of methanethiosulfonate–rhodamine, as identified in the Loo and Clarke cross-linking experiments, and also those assumed to be the binding site of Hoechst 33342, as defined from pharmacophore modeling (see Section 5.32.3.2.5). The authors found that the amino acid residues had an opposite orientation: in the homology model they faced the membrane, whereas in the cross-linking model the residues were directed toward the pore. They suggested that both models might describe two different functional states of P-gp. Generally, the comparable distances in the cross-linking model by Pajeva et al. were shorter than the ranges reported for the second P-gp model by Stenham et al.[226] The authors attributed this to the fact that, in contrast to the model of Stenham et al., where cross-linkers of different length were applied, the shortest possible cross-linkers were used as distance constrains.

Pleban et al.[117] generated a homology model of P-gp based on the nucleotide-free Vibrio cholerae MsbA structure.[91] Although the dimer position in the Vibrio cholerae MsbA was closer to the low-resolution electron microscopy data of P-gp, the homology model of Pleban et al. had a central cavity that was narrower at the extracytoplasmic face, in disagreement with the funnel-shaped form of P-gp with TMDs closer at their cytoplasmic ends. To fit the experimental P-gp structure, the authors repositioned the halves, moving the extracellular ends apart from each other and narrowing them at the cytoplasmic face. The rearranged model was reported to satisfy the majority of distance constrains that were similar to those used by Stenham et al.[226]

Very recently the first report on molecular dynamics simulation of an atomic resolution structural model of P-gp together with the lipid bilayer in the presence of transport substrates was published.[239] Similarly to the previously reported P-gp models 3D structural data from different transporters were combined to obtain a model corresponding to the physiologically relevant form of the protein. The so-called 'solvation exchange mechanism' was further proposed to explain the broad transport-substrate specificity that based on the P-gp selectivity to positively charged compounds.

The homology models of P-gp generated to date have a common drawback: neither of them copes with the problem of correct modeling of the TMD/NBD:TMD/NBD interface. The repositioning of the P-gp halves in order to achieve a funnel-like form, as in the low-resolution 3D structure of P-gp,[95] is performed manually. These manipulations aim to bring the orientation of the halves in accordance to the cross-linking constrains between amino acid residues reported in experimental studies. To date, there is no clear presentation about the functional state of the protein in which cross-linking distances have been defined. This results in the application of constrains that may hold true only for specific stages of the protein transport cycle. This could explain the fact that the correspondence between the distances achieved in the proposed homology models and the experimental distances reported is not met for most of the cross-linked domains.[117,226]

Recent results by Loo and Clarke on oxidative cross-linking with copper phenanthroline of mutants of human P-gp showed that TMD2–TMD11 and TMD5–TMD8 could be directly cross-linked, and that this linking was inhibited by vanadate trapping or the presence of certain drug substrates.[114,115] **Figure 3** shows the location of the mutated residues in the homology model obtained by alignment onto the x-ray structure of Vibrio cholerae MsbA[91] using an approach similar to that previously published.[217] As can be observed from the figure, the mutated residues are

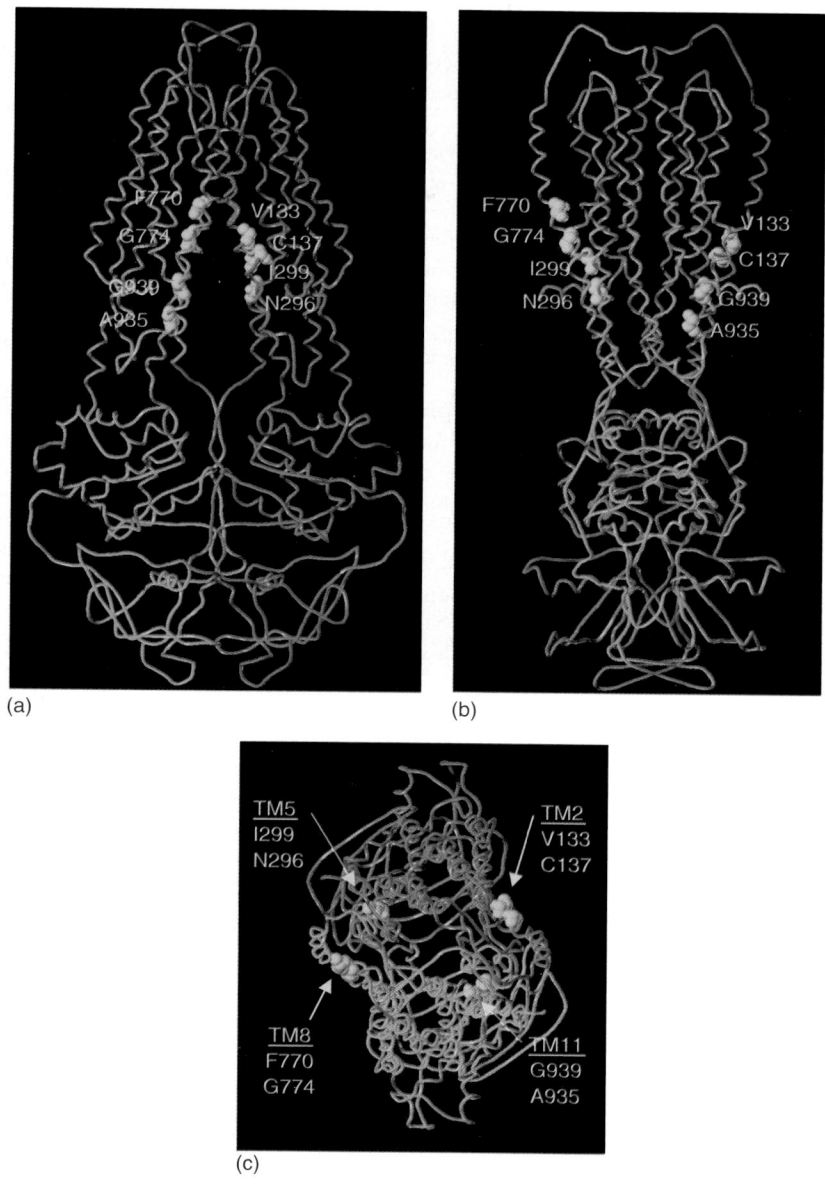

Figure 3 Homology model of P-gp generated from the x-ray structure of *Vibrio cholerae* MsbA[91] with the cross-linked residues (in yellow) in the N-terminal (in purple) and the C-terminal (in magenta): (a) front view, (b) side view, (c) top view.

arranged apart from each other (**Figure 3a** and **3b**), and, additionally, they are oriented oppositely, so that they cannot be directly cross-linked (**Figure 3c**). The distances between the mutated residues in the homology model are between 15 Å and 25 Å, which is far larger than the required distance of 5–7 Å. This implies that both halves of the protein should be further tilted toward each other, to allow the mutated residues to come closer. Additionally, to be cross-linked, the mutated residues should face each other. This, however, cannot be achieved by tilting only. Obviously, challenges such as finding the correct positioning of the protein domains and the correct correspondence of the P-gp structural models to the functional states remain to be solved. More precise constraining distances, higher-resolution protein and protein–ligand complex structural data, and extensive modeling studies involving molecular dynamics simulations could further throw light on the structure–function relationships of P-gp and its interaction with drugs.

5.32.3.3   Modeling of Other Multidrug Transporters

Compared with P-gp, other ABC transporters have not been investigated to the same extent. For several of them, only a few inhibitors with low potency are known. Examples include the MRP1 and MRP2 transporters, for which only very recently have more potent inhibitors been described.

Van Zanden *et al.* investigated a large series of flavonoids as inhibitors of MRP1 and MRP2 in transfected MDCKII cells, and performed the first MRP1-related structure–activity relationship study.[232] Three structural characteristics were found to be important for inhibition of MRP1: the number of methoxylated moieties, the number of hydroxyl groups, and the dihedral angle between the B and C rings. For inhibition of MRP2, only the presence of a flavonol B ring pyrogallol group was identified as important for inhibitory activity.

Hirono and co-workers also tried to identify structural features related to MRP2 recognition using 16 structurally diverse compounds.[233] Two kinds of binding conformations were extracted by a flexible alignment of ligands. For one of them, CoMFA yielded a statistically significant model in which the steric contribution was highest and log *P* contributed only marginally.

Similarly to P-gp, no high-resolution structures of MRPs are available to date. Rosenberg *et al.* were able to obtain structural information for MRP1 to approximately 22 Å resolution.[234] Their data provide evidence for a core structure consisting of two membrane-spanning domains, each followed by an NBD. Unlike P-gp and most other ABC superfamily members, MRP1 contains a third membrane-spanning domain with five predicted TMDs and an extracytosolic amino terminus.

The higher resolution x-ray structure of the *Vibrio cholerae* MsbA[91] initiated homology modeling studies of the bacterial MDR protein LmrA. This protein is a half transporter, and functions as a homodimer. Ecker *et al.* developed a 3D model of the LmrA substrate-binding domain, based on homology modeling and experimental data.[235,236] Photoaffinity labeling with propafenone-type compounds showed that mostly TMD3, TMD5, and TMD6, but also the NBD, were labeled. Linking the experimental data with the homology model, it was suggested that two symmetry-related ligand-binding domains are formed by TMD3 of one monomer and TMD5 and TMD6 of the other monomer. It was concluded that these domains remain close to each other during the entire transport cycle.

Recently, research interest has also focused on organic anion transporting polypeptides (rodent Oatps and human OATPs) that, similarly to P-gp and MRPs, constitute part of the organism detoxification system, and show broad substrate specificity.[237] Yarim *et al.*[238] modeled Oatp1a5–substrate interactions combining a Hansch-type analysis, pharmacophore (using GASP), and CoMFA approaches. The pharmacophore model outlined three important regions: HB donor, hydrophobic, and negatively charged areas; the QSAR and 3D-QSAR models yielded similar predictivity ($q^2 = 0.71$). Using the models, the authors identified potentially new substrate structures that could serve as templates for designing new Oatp1a5–substrates.

5.32.4   Conclusions

As outlined by many examples, remarkable progress has been made in recent years in the accurate prediction of interactions with transport proteins, and especially with P-gp. Both experimental and in silico studies have contributed to this progress. QSAR models and classification systems, sufficient to confidently predict drug activity, have been elaborated that allow high-throughput screening of drug candidates. Although a variety of significant descriptors have been identified, most of these studies define size and hydrophobicity as important properties for the interaction with P-gp either as a transported substrate or as an inhibitor. It is worth noting that hydrophobic fields rather than log *P* produce better models, thus pointing to the importance of the space location of the hydrophobic and hydrophilic parts in the drug structure. Additionally, the drug capacity for hydrogen bonding is shown to be an essential contribution to interactions with transporters. Even though the quality of some of the classification models does not permit wholly correct identification of compounds (non)interacting with P-gp, they nevertheless offer a preselection tool for experimental testing.

Several pharmacophore models have been developed that are capable of predicting the recognition of compounds by P-gp at the molecular level. In general, hydrophobic and HB acceptor interactions are the two molecular properties which appear in these models. It seems that at least one HB acceptor and two hydrophobic (either aromatic or aliphatic or both) pharmacophore points in a particular space arrangement are required for recognition by P-gp.

The first homology models contributed to a better understanding of the structure–function relationships of transport proteins; their reliability, however, was not sufficient. With the generation and collection of more precise structural data, further efforts can be made to generate more reliable 3D models.

The in silico models are valuable tools for elucidating the basis of drug–transporter interactions. Their improvement will allow the development of potent and highly specific inhibitors of ABC transporters, potentially giving rise to a new class of effective therapeutic agents in clinical applications such as the treatment of MDR tumors or pharmacokinetic optimizations.

Acknowledgments

We are grateful to the Alexander von Humboldt Foundation, DFG, DAAD, and the National Science Fund of Bulgaria for the financial support of our studies.

References

1. Dano, K. *Biochim. Biophys. Acta.* **1973**, *323*, 466–483.
2. Juliano, R. L.; Ling, V. *Biochim. Biophys. Acta.* **1976**, *455*, 152–162.
3. Gottesmann, M. M.; Pastan, I. *Annu. Rev. Biochem.* **1993**, *62*, 385–427.
4. Mirski, S. E. L.; Gerlach, J. H.; Cole, S. P. C. *Cancer Res.* **1987**, *47*, 2594–2598.
5. Hipfner, D. R.; Deeley, R. G.; Cole, S. P. C. *Biochim. Biophys. Acta.* **1999**, *1461*, 359–376.
6. Leslie, E. M.; Deeley, R. G.; Cole, S. P. C. *Toxicology* **2001**, *167*, 3–23.
7. Liedert, B.; Materna, V.; Schadendorf, D.; Thomale, J.; Lage, H. *J. Invest. Dermatol.* **2003**, *121*, 172–176.
8. Xiong, H.; Turner, K. C.; Ward, E. S.; Jansen, P. L. M.; Brouwer, K. L. *J. Pharmacol. Exp. Ther.* **2000**, *295*, 512–518.
9. Reid, G.; Wielinga, P.; Zelcer, N.; van der Heijden, I.; Kuil, A.; de Haas, M.; Wijnholds, J.; Borst, P. *Proc. Natl. Acad. Sci. USA* **2003**, *100*, 9244–9249.
10. Rigato, I.; Pascolo, L.; Fernetti, C.; Ostrow, J. D.; Tiribelli, C. *Biochem. J.* **2004**, *383*, 335–341.
11. Jedlitschky, G.; Leier, I.; Buchholz, U.; Hummel-Eisenbeiss, J.; Burchell, B.; Keppler, D. *Biochem. J.* **1997**, *327*, 305–310.
12. Leier, I.; Jedlitschky, G.; Buchholz, U.; Cole, S. P.; Deeley, R. G.; Keppler, D. *J. Biol. Chem.* **1994**, *269*, 27807–27810.
13. Jedlitschky, G.; Leier, I.; Buchholz, U.; Barnouin, K.; Kurz, G.; Keppler, D. *Cancer Res.* **1996**, *56*, 988–994.
14. Loe, D. W.; Almquist, K. C.; Deeley, R. G.; Cole, S. P. *J. Biol. Chem.* **1996**, *271*, 9675–9682.
15. Leier, I.; Jedlitschky, G.; Buchholz, U.; Center, M.; Cole, S. P.; Deeley, R. G.; Keppler, D. *Biochem. J.* **1996**, *314*, 433–437.
16. Loe, D. W.; Almquist, K. C.; Cole, S. P.; Deeley, R. G. *J. Biol. Chem.* **1996**, *271*, 9683–9689.
17. Loe, D. W.; Stewart, R. K.; Massey, T. E.; Deeley, R. G.; Cole, S. P. *Mol. Pharmacol.* **1997**, *51*, 1034–1041.
18. Zaman, G. J.; Cnubben, N. H.; van Bladeren, P. J.; Evers, R.; Borst, P. *FEBS Lett.* **1996**, *391*, 126–130.
19. Evers, R.; Cnubben, N. H.; Wijnholds, J.; van Deemter, L.; van Bladeren, P. J.; Borst, P. *FEBS Lett.* **1997**, *419*, 112–116.
20. Loe, D. W.; Almquist, K. C.; Deeley, R. G.; Cole, S. P. *J. Biol. Chem.* **1996**, *271*, 9675–9682.
21. Loe, D. W.; Deeley, R. G.; Cole, S. P. *Cancer Res.* **1998**, *58*, 5130–5136.
22. Renes, J.; de Vries, E. G.; Nienhuis, E. F.; Jansen, P. L.; Muller, M. *Br. J. Pharmacol.* **1999**, *126*, 681–688.
23. Leslie, E. M.; Deeley, R. G.; Cole, S. P. *Toxicology* **2001**, *167*, 3–23.
24. Keppler, D.; Cui, Y.; Konig, J.; Leier, I.; Nies, A. *Adv. Enzyme Regul.* **1999**, *39*, 237–246.
25. Sun, H.; Johnson, D. R.; Finch, R. A.; Sartorelli, A. C.; Miller, D. W.; Elmquist, W. F. *Biochem. Biophys. Res. Commun.* **2001**, *284*, 863–869.
26. Hooijberg, J. H.; Jansen, G.; Assaraf, Y. G.; Kathmann, I.; Pieters, R.; Laan, A. C.; Veerman, A. J.; Kaspers, G. J.; Peters, G. J. *Biochem. Pharmacol.* **2004**, *67*, 1541–1548.
27. Hooijberg, J. H.; Broxterman, H. J.; Kool, M.; Assaraf, Y. G.; Peters, G. J.; Noordhuis, P.; Scheper, R. J.; Borst, P.; Pinedo, H. M.; Jansen, G. *Cancer Res.* **1999**, *59*, 2532–2535.
28. Williams, G. C.; Liu, A.; Knipp, G.; Sinko, P. J. *Antimicrob. Agents Chemother.* **2002**, *46*, 3456–3462.
29. Dallas, S.; Ronaldson, P. T.; Bendayan, M.; Bendayan, R. *Neuroreport* **2004**, *15*, 1183–1186.
30. Gekeler, V.; Ise, W.; Sanders, K. H.; Ulrich, W. R.; Beck, J. *Biochem. Biophys. Res. Commun.* **1995**, *208*, 345–352.
31. Wortelboer, H. M.; Usta, M.; van der Velde, A. E.; Boersma, M. G.; Spenkelink, B.; van Zanden, J. J.; Rietjens, I. M.; van Bladeren, P. J.; Cnubben, N. H. *Chem. Res. Toxicol.* **2003**, *16*, 1642–1651.
32. Hu, K.; Morris, M. E. *J. Pharm. Sci.* **2004**, *93*, 1901–1911.
33. Benyahia, B.; Huguet, S.; Decleves, X.; Mokhtari, K.; Criniere, E.; Bernaudin, J. F.; Scherrmann, J. M.; Delattre, J. Y. *J. Neurooncol.* **2004**, *66*, 65–70.
34. Rosenbaum, C.; Rohrs, S.; Muller, O.; Waldmann, H. *J. Med. Chem.* **2005**, *48*, 1179–1187.
35. Wang, S.; Folkes, A.; Chuckowree, I.; Cockcroft, X.; Sohal, S.; Miller, W.; Milton, J.; Wren, S. P.; Vicker, N.; Depledge, P. et al. *J. Med. Chem.* **2004**, *47*, 1329–1338.
36. Payen, L.; Delugin, L.; Courtois, A.; Trinquart, Y.; Guillouzo, A.; Fardel, O. *Br. J. Pharmacol.* **2001**, *132*, 778–784.
37. Wang, E. J.; Johnson, W. W. *Chemotherapy* **2003**, *49*, 303–308.
38. Bobrowska-Hagerstrand, M.; Wrobel, A.; Mrowczynska, L.; Soderstrom, T.; Shirataki, Y.; Motohashi, N.; Molnar, J.; Michalak, K.; Hagerstrand, H. *Oncol. Res.* **2003**, *13*, 463–469.
39. Burg, D.; Wielinga, P.; Zelcer, N.; Saeki, T.; Mulder, G. J.; Borst, P. *Mol. Pharmacol.* **2002**, *62*, 1160–1166.
40. Matsumoto, Y.; Tamiya, T.; Nagao, S. *J. Med. Invest.* **2005**, *52*, 41–48.
41. Gerk, P. M.; Li, W.; Vore, M. *Drug Metab. Dispos.* **2004**, *32*, 1139–1145.
42. Chu, X. Y.; Huskey, S. E.; Braun, M. P.; Sarkadi, B.; Evans, D. C.; Evers, R. *J. Pharmacol. Exp. Ther.* **2004**, *309*, 156–164.
43. Huisman, M. T.; Smit, J. W.; Crommentuyn, M. L.; Zelcer, N.; Wiltshire, H. R.; Beijnen, J. H.; Schinkel, A. H. *AIDS* **2002**, *16*, 2295–2301.
44. Sugie, M.; Asakura, E.; Zhao, Y. L.; Torita, S.; Nadai, M.; Baba, K.; Kitaichi, K.; Takagi, K.; Takagi, K.; Hasegawa, T. *Antimicrob. Agents Chemother.* **2004**, *48*, 809–814.
45. Schuetz, J. D.; Connelly, M. C.; Sun, D.; Paibir, S. G.; Flynn, P. M.; Srinivas, R. V.; Kumar, A.; Fridland, A. *Nature Med.* **1999**, *5*, 1048–1051.

46. Smeets, P. H.; van Aubel, R. A.; Wouterse, A. C.; van den Heuvel, J. J.; Russel, F. G. *J. Am. Soc. Nephrol.* **2004**, *15*, 2828–2835.
47. Van Aubel, R. A.; Smeets, P. H.; van den Heuvel, J. J.; Russel, F. G. *Am. J. Physiol. Renal Physiol.* **2005**, *288*, F327–F333.
48. McAleer, M. A.; Breen, M. A.; White, N. L.; Mattews, N. *J. Biol. Chem.* **1999**, *274*, 23541–23548.
49. Jedlitschky, G.; Burchell, B.; Keppler, D. *J. Biol. Chem.* **2000**, *275*, 30069–30074.
50. Wijnholds, J.; Mol, C. A. A. M.; van Deemter, L.; de Haas, M.; Scheffer, G. L.; Baas, F.; Beijnen, J. H.; Scheper, R. J.; Hatse, S.; Clercq, E. D. et al. *Proc. Natl. Acad. Sci. USA* **2000**, *97*, 7476–7481.
51. Ilias, A.; Urban, Z.; Seidl, T. L.; Le Saux, O.; Sinko, E.; Boyd, C. D.; Sarkadi, B.; Varadi, A. *J. Biol. Chem.* **2002**, *277*, 16860–16867.
52. Belinsky, M. G.; Chen, Z.; Shchaveleva, I.; Zeng, H.; Kruh, G. D. *Cancer Res.* **2002**, *62*, 6172–6177.
53. Madon, J.; Hagenbuch, B.; Landmann, L.; Meier, P. J.; Stieger, B. *Mol. Pharmacol.* **2000**, *57*, 634–641.
54. Doyle, L. A.; Yang, W.; Abruzzo, L. V.; Krogmann, T.; Gao, Y.; Rishi, A. K.; Ross, D. D. *Proc. Natl. Acad. Sci. USA* **1998**, *95*, 15665–15670.
55. Maliepaard, M.; Scheffer, G. L.; Faneyte, I. F.; van Gastelen, M. A.; Pijnenborg, A. C. L. M.; Schinkel, A. H.; van de Vijver, M. J.; Scheper, R. J.; Schellens, J. H. M. *Cancer Res.* **2001**, *61*, 3458–3464.
56. Doyle, L. A.; Yang, W.; Abruzzo, L. V.; Krogmann, T.; Gao, Y.; Rishi, A. K.; Ross, D. D. *Proc. Natl. Acad. Sci. USA* **1998**, *95*, 15665–15670.
57. Miyake, K.; Mickley, L.; Litman, T.; Zhan, Z.; Robey, R.; Cristensen, B.; Brangi, M.; Greenberger, L.; Dean, M.; Fojo, T.; Bates, S. E. *Cancer Res.* **1999**, *59*, 8–13.
58. Allen, J. D.; Van Dort, S. C.; Buitelaar, M.; van Tellingen, O.; Schinkel, A. H. *Cancer Res.* **2003**, *63*, 1339–1344.
59. Maliepaard, M.; van Gastelen, M. A.; de Jong, L. A.; Pluim, D.; van Waardenburg, R. C.; Ruevekamp-Helmers, M. C.; Floot, B. G.; Schellens, J. H. *Cancer Res.* **1999**, *59*, 4559–4563.
60. Robey, R. W.; Medina-Perez, W. Y.; Nishiyama, K.; Lahusen, T.; Miyake, K.; Litman, T.; Senderowicz, A. M.; Ross, D. D.; Bates, S. E. *Clin. Cancer Res.* **2001**, *7*, 145–152.
61. Volk, E. L.; Schneider, E. *Cancer Res.* **2003**, *63*, 5538–5543.
62. Nakatomi, K.; Yoshikawa, M.; Oka, M.; Ikegami, Y.; Hayasaka, S.; Sano, K.; Shiozawa, K.; Kawabata, S.; Soda, H.; Ishikawa, T. et al. *Biochem. Biophys. Res. Commun.* **2001**, *288*, 827–832.
63. Kawabata, S.; Oka, M.; Shiozawa, K.; Tsukamoto, K.; Nakatomi, K.; Soda, H.; Fukuda, M.; Ikegami, Y.; Sugahara, K.; Yamada, Y. et al. *Biochem. Biophys. Res. Commun.* **2001**, *280*, 1216–1223.
64. Jonker, J. W.; Merino, G.; Musters, S.; van Herwaarden, A. E.; Bolscher, E.; Wagenaar, E.; Mesman, E.; Dale, T. C.; Schinkel, A. H. *Nat. Med.* **2005**, *11*, 127–129.
65. Kim, M.; Turnquist, H.; Jackson, J.; Sgagias, M.; Yan, Y.; Gong, M.; Dean, M.; Sharp, J. G.; Cowan, K. *Clin. Cancer Res.* **2002**, *8*, 22–28.
66. Ozvegy, C.; Varadi, A.; Sarkadi, B. *J. Biol. Chem.* **2002**, *277*, 47980–47990.
67. Pavek, P.; Merino, G.; Wagenaar, E.; Bolscher, E.; Novotna, M.; Jonker, J. W.; Schinkel, A. H. *J. Pharmacol. Exp. Ther.* **2005**, *312*, 144–152.
68. Merino, G.; Jonker, J. W.; Wagenaar, E.; Pulido, M. M.; Molina, A. J.; Alvarez, A. I.; Schinkel, A. H. *Drug Metab. Dispos.* **2005**, *33*, 614–618.
69. Kari, F. W.; Weaver, R.; Neville, M. C. *J. Pharmacol. Exp. Ther.* **1997**, *280*, 664–668.
70. Oo, C. Y.; Paxton, E. W.; McNamara, P. J. *Adv. Exp. Med. Biol.* **2001**, *501*, 547–552.
71. Gerk, P. M.; Kuhn, R. J.; Desai, N. S.; McNamara, P. J. *Pharmacotherapy* **2001**, *21*, 669–675.
72. Imai, Y.; Asada, S.; Tsukahara, S.; Ishikawa, E.; Tsuruo, T.; Sugimoto, Y. *Mol. Pharmacol.* **2003**, *64*, 610–618.
73. de Bruin, M.; Miyake, K.; Litman, T.; Robey, R.; Bates, S. E. *Cancer Lett.* **1999**, *146*, 117–126.
74. Maliepaard, M.; van Gastelen, M. A.; Tohgo, A.; Hausheer, F. H.; van Waardenburg, R. C.; de Jong, L. A.; Pluim, D.; Beijnen, J. H.; Schellens, J. H. *Clin. Cancer Res.* **2001**, *7*, 935–941.
75. Rabindran, S. K.; He, H.; Singh, M.; Brown, E.; Collins, K. I.; Annable, T.; Greenberger, L. M. *Cancer Res.* **1998**, *58*, 5850–5858.
76. Zhang, S.; Wang, X.; Sagawa, K.; Morris, M. E. *Drug Metab. Dispos.* **2005**, *33*, 341–348.
77. Zhang, S.; Yang, X.; Morris, M. E. *Mol. Pharmacol.* **2004**, *65*, 1208–1216.
78. Imai, Y.; Tsukahara, S.; Asada, S.; Sugimoto, Y. *Cancer Res.* **2004**, *64*, 4346–4352.
79. Cooray, H. C.; Janvilisri, T.; van Veen, H. W.; Hladky, S. B.; Barrand, M. A. *Biochem. Biophys. Res. Commun.* **2004**, *317*, 269–275.
80. Shiozawa, K.; Oka, M.; Soda, H.; Yoshikawa, M.; Ikegami, Y.; Tsurutani, J.; Nakatomi, K.; Nakamura, Y.; Doi, S.; Kitazaki, T. et al. *Int. J. Cancer* **2004**, *108*, 146–151.
81. Gupta, A.; Zhang, Y.; Unadkat, J. D.; Mao, Q. *J. Pharmacol. Exp. Ther.* **2004**, *310*, 334–341.
82. Minderman, H.; O'Loughlin, K. L.; Pendyala, L.; Baer, M. R. *Clin. Cancer Res.* **2004**, *10*, 1826–1834.
83. Minderman, H.; Brooks, T. A.; O'Loughlin, K. L.; Ojima, I.; Bernacki, R. J.; Baer, M. R. *Cancer Chemother. Pharmacol.* **2004**, *53*, 363–369.
84. Burger, H.; van Tol, H.; Boersma, A. W.; Brok, M.; Wiemer, E. A.; Stoter, G.; Nooter, K. *Blood* **2004**, *104*, 2940–2942.
85. Özvegy-Laczka, C.; Hegedüs, T.; Varady, G.; Ujhelly, O.; Schuetz, J. D.; Varadi, A.; Keri, G.; Örfi, L.; Nemet, K.; Sarkadi, B. *Mol. Pharmacol.* **2004**, *65*, 1485–1495.
86. Yanase, K.; Tsukahara, S.; Asada, S.; Ishikawa, E.; Imai, Y.; Sugimoto, Y. *Mol. Cancer Ther.* **2004**, *3*, 1119–1125.
87. Sugimoto, Y.; Tsukahara, S.; Imai, Y.; Sugimoto, Y.; Ueda, K.; Tsuruo, T. *Mol. Cancer Ther.* **2003**, *2*, 105–112.
88. Imai, Y.; Tsukahara, S.; Ishikawa, E.; Tsuruo, T.; Sugimoto, Y. *Jpn. J. Cancer Res.* **2002**, *93*, 231–235.
89. Zhou, S.; Schuetz, J. D.; Bunting, K. D.; Colapietro, A. M.; Sampath, J.; Morris, J. J.; Lagutina, I.; Grosveld, G. C.; Osawa, M.; Nakauchi, H. et al. *Nat. Med.* **2001**, *7*, 1028–1034.
90. Chang, G.; Roth, C. B. *Science* **2001**, *293*, 1793–1800.
91. Chang, G. *J. Mol. Biol.* **2003**, *330*, 419–430.
92. Locher, K. P.; Lee, A. T.; Rees, D. C. *Science* **2002**, *296*, 1091–1098.
93. Rosenberg, M. F.; Callaghan, R.; Ford, R. C.; Higgins, C. F. *J. Biol. Chem.* **1997**, *272*, 10685–10694.
94. Rosenberg, M. F.; Velarde, G.; Ford, R. C.; Martin, C.; Berridge, G.; Kerr, I. D.; Callaghan, R.; Schmidlin, A.; Wooding, C.; Linton, K. J. et al. *EMBO J.* **2001**, *20*, 5615–5625.
95. Rosenberg, M. F.; Kamis, A. B.; Callaghan, R.; Higgins, C. F.; Ford, R. C. *J. Biol. Chem.* **2003**, *278*, 8294–8299.
96. Rosenberg, M. F.; Callaghan, R.; Modok, S.; Higgins, C. F.; Ford, R. C. *J. Biol. Chem.* **2005**, *280*, 2857–2862.
97. Poelarends, G. J.; Konings, W. N. *J. Biol. Chem.* **2002**, *277*, 42891–42898.
98. Al-Shawi, M.; Polar, M. K.; Omote, H.; Figler, R. A. *J. Biol. Chem.* **2003**, *278*, 52629–52640.
99. Higgins, C. F.; Linton, K. J. *Nat. Struct. Mol. Biol.* **2004**, *11*, 918–926.
100. Martin, C.; Berridge, G.; Higgins, C. F.; Mistry, P.; Charlton, P.; Callaghan, R. *Mol. Pharmacol.* **2000**, *58*, 624–632.
101. Martin, C.; Higgins, C. F.; Callaghan, R. *Biochemistry* **2001**, *40*, 15733–15742.
102. Borgnia, M. J.; Eytan, G. D.; Assaraf, Y. G. *J. Biol. Chem.* **1996**, *271*, 3163–3171.

103. Ayesh, S.; Shao, Y. M.; Stein, W. D. *Biochim. Biophys. Acta.* **1996**, *1316*, 8–18.
104. Dey, S.; Ramachandra, M.; Pastan, I.; Gottesman, M. M.; Ambudkar, S. V. *Proc. Natl. Acad. Sci. USA* **1997**, *94*, 10594–10599.
105. Scala, S.; Akhmed, N.; Rao, U. S.; Paull, K.; Lan, L.-B.; Dickstein, B.; Lee, J.-S.; Elgemeie, G. H.; Stein, W. D.; Bates, S. E. *Mol. Pharmacol.* **1997**, *51*, 1024–1033.
106. Shapiro, A. B.; Ling, V. *Eur. J. Biochem.* **1997**, *250*, 130–137.
107. Shapiro, A. B.; Fox, K.; Lam, P.; Ling, V. *Eur. J. Biochem.* **1999**, *259*, 841–850.
108. Safa, A. R. *Curr. Med. Chem. Anti-Cancer Agents* **2004**, *4*, 1–17.
109. Loo, T. W.; Clarke, D. M. *J. Biol. Chem.* **2001**, *276*, 14972–14979.
110. Loo, T. W.; Clarke, D. M. *J. Biol. Chem.* **2001**, *276*, 31800–31805.
111. Loo, T. W.; Clarke, D. M. *J. Biol. Chem.* **2001**, *276*, 36877–36880.
112. Loo, T. W.; Clarke, D. M. *J. Biol. Chem.* **2002**, *277*, 44332–44338.
113. Loo, T. W.; Bartlett, M. C.; Clarke, D. M. *J. Biol. Chem.* **2003**, *278*, 50136–50141.
114. Loo, T. W.; Bartlett, M. C.; Clarke, D. M. *J. Biol. Chem.* **2004**, *279*, 7692–7697.
115. Loo, T. W.; Bartlett, M. C.; Clarke, D. M. *J. Biol. Chem.* **2004**, *279*, 18232–18238.
116. Ecker, G. F.; Csaszar, E.; Kopp, S.; Plagens, B.; Holzer, W.; Ernst, W.; Chiba, P. *Mol. Pharmacol.* **2002**, *61*, 637–648.
117. Pleban, K.; Kopp, S.; Csaszar, E.; Peer, M.; Hrebicek, T.; Rizzi, A.; Ecker, G. F.; Chiba, P. *Mol. Pharmacol.* **2005**, *67*, 365–374.
118. Kubinyi, H. The Extrathermodynamic Approach (Hansch Analysis). In *QSAR Hansch Analysis and Related Approaches*; Mannhold, R., Kroogsgard-Larsen, P., Timmerman, H., Eds.; VCH: Weinheim, Germany, 1993; Vol. 1, pp 57–85.
119. Kubinyi, H. The Free–Wilson Method and Its Relationship to the Extrathermodynamic Approach. In *Quantitative Drug design*; Ramsden, C. A., Ed.; Pergamon Press: Oxford, UK, 1990; Vol. 4, pp 589–643.
120. Cramer, R. D., III; DePriest, S. A.; Patterson, D. E.; Hecht, P. The Developing Practice of Comparative Molecular Field Analysis. In *3D QSAR in Drug Design. Theory Methods and Applications*; Kubinyi, H., Ed.; Escom: Leiden, Germany, 1993, pp 443–486.
121. Klebe, G. Comparative Molecular Similarity Indices Analysis: CoMSIA. In *3D-QSAR in Drug Design. Recent Advances*; Kubinyi, H., Klebe, G., Martin, Y., Eds.; Kluwer/Escom: Dordrecht, Germany, 1998; Vol. 3, pp 87–104.
122. Wade, R. C.; Goodford, P. J. *Prog. Clin. Biol. Res.* **1989**, *289*, 433–444.
123. Cruciani, G.; Pastor, M.; Clementi, S.; Clementi, S. GRIND Grid Independent Descriptors in 3D Structure–Metabolism Relationships. In *Rational Approaches to Drug Design, Proceedings of the 13th European Symposium on Quantitative Structure–Activity Relationships*; Hoeltje, H.-D., Sippl, W., Eds.; Prous Science: Barcelona, Spain, 2001, pp 251–260.
124. Cruciania, G.; Crivorib, P.; Carruptb, P.-A.; Testa, B. *J. Mol. Struct. Theochem.* **2000**, *503*, 17–30.
125. Leach, A. R.; Gillet, V. J. *An Introduction to Chemoinformatics*; Kluwer Academic Publishers: Dordrecht, Germany, 2003.
126. Leardi, R. *J. Chemom.* **2001**, *15*, 559–569.
127. Clark, D. E. Evolutionary Algorithms in Rational Drug Design: A Review of Current Applications and a Look to the Future. In *Rational Drug Design: Novel Methodology and Practical Applications. ACS Symposium Series.*; Parrill, A. L., Reddy, M. R., Eds.; American Chemical Society: Washington, DC, USA, 1999; Vol. 719, pp 255–270.
128. Zupan, J.; Gasteiger, J. *Neural Networks for Chemists: An Introduction*, 2nd ed; Wiley-VCH: Weinheim, Germany, 1999.
129. Anzali, S.; Gasteiger, J.; Holzgrabe, U.; Polanski, J.; Sadowski, J.; Teckentrup, A.; Wagener, M. The Use of Self-Organizing Neural Networks in Drug Design. In *3D QSAR in Drug Design*; Kubinyi, H., Folkers, G., Martin, Y. C., Eds.; Kluwer/ESCOM: Dordrecht, Germany, 1998; Vol. 2, pp 273–299.
130. Wiese, M.; Pajeva, I. K. *Curr. Med. Chem.* **2001**, *8*, 685–713.
131. Ford, J. M.; Prozialeck, W. C.; Hait, W. N. *Mol. Pharmacol.* **1989**, *35*, 105–115.
132. Ramu, A.; Ramu, N. *Cancer Chemother. Pharmacol.* **1992**, *30*, 165–173.
133. Köhler, S.; Stein, W. D. *Biotechnol. Bioeng.* **2003**, *81*, 507–517.
134. Mankhetkorn, S.; Dubru, F.; Hesschenbrouck, J.; Fiallo, M.; Garnier-Suillerot, A. *Mol. Pharmacol.* **1996**, *49*, 532–539.
135. Mankhetkorn, S.; Garnier-Suillerot, A. *Eur. J. Pharmacol.* **1998**, *343*, 313–321.
136. Leisen, C.; Langguth, P.; Herbert, B.; Dressler, C.; Koggel, A.; Spahn-Langguth, H. *Pharm. Res.* **2003**, *20*, 772–778.
137. Pajeva, I. K.; Wiese, M. *Quant. Struct.-Act. Relat.* **1998**, *17*, 301–312.
138. Schaich, M.; Neu, S.; Beck, J.; Gekeler, V.; Schuler, U.; Ehninger, G. *Leuk. Res.* **1997**, *21*, 933–940.
139. Stein, W. D. *Phys. Rev.* **1997**, *77*, 545–590.
140. Tang, F.; Ouyang, H.; Yang, J. Z.; Borchardt, R. T. *J. Pharm. Sci.* **2004**, *93*, 1185–1194.
141. Hamada, H.; Tsuruo, T. *Cancer Res.* **1988**, *48*, 4926–4932.
142. Schwab, D.; Fischer, H.; Tabatabaei, A.; Ploi, S.; Huwyler, J. *J. Med. Chem.* **2003**, *36*, 1716–1725.
143. Litman, T.; Zeuthen, T.; Skovsgaard, T.; Stein, W. D. *Biochim. Biophys. Acta.* **1997**, *1361*, 159–168.
144. Ambudkar, S. V.; Lelong, I. H.; Zhang, J.; Cardarelli, C. O.; Gottesman, M. M.; Pastan, I. *Proc. Natl. Acad. Sci. USA* **1992**, *89*, 8472–8476.
145. Shapiro, A. B.; Ling, V. *J. Biol. Chem.* **1994**, *269*, 3745–3754.
146. Troutman, M. D.; Thakker, D. R. *Pharm. Res.* **2003**, *20*, 1192–1199.
147. Troutman, M. D.; Thakker, D. R. *Pharm. Res.* **2003**, *20*, 1200–1209.
148. Polli, J. W.; Wring, S. A.; Humphreys, J. E.; Huang, L.; Morgan, J. B.; Webszerr, L. O.; Serabjit-Singh, C. S. *J. Pharmacol. Exp. Ther.* **2001**, *299*, 620–628.
149. Chiba, P.; Burghofer, S.; Richter, E.; Tell, B.; Moser, A.; Ecker, G. *J. Med. Chem.* **1995**, *38*, 2789–2793.
150. Chiba, P.; Ecker, G.; Schmid, D.; Drach, J.; Tell, B.; Goldenberg, S.; Gekeler, V. *Mol. Pharmacol.* **1996**, *49*, 1122–1130.
151. Ecker, G.; Chiba, P.; Hitzler, M.; Schmid, D.; Visser, K.; Cordes, H. P.; Csöllei, J.; Seydel, J. K.; Schaper, K.-J. *J. Med. Chem.* **1996**, *39*, 4767–4774.
152. Chiba, P.; Hitzler, M.; Richter, E.; Huber, M.; Tmej, C.; Giovagnoni, E.; Ecker, G. *Quant. Struct.-Act. Relat.* **1997**, *16*, 361–366.
153. Chiba, P.; Tell, B.; Jäger, W.; Richter, E.; Hitzler, M.; Ecker, G. *Arch. Pharm. Pharm. Med. Chem.* **1997**, *330*, 343–347.
154. Tmej, C.; Chiba, P.; Huber, M.; Richter, E.; Hitzler, M.; Schaper, K.-J.; Ecker, G. *Arch. Pharm. Pharm. Med. Chem.* **1998**, *331*, 233–240.
155. Chiba, P.; Annibali, D.; Hitzler, M.; Richter, E.; Ecker, G. *Farmaco* **1998**, *53*, 357–364.
156. Salem, M.; Richter, E.; Hitzler, M.; Chiba, P.; Ecker, G. *Sci. Pharm.* **1998**, *66*, 147–158.
157. Ecker, G.; Huber, M.; Schmid, D.; Chiba, P. *Mol. Pharmacol.* **1999**, *56*, 791–796.
158. Chiba, P.; Holzer, W.; Landau, M.; Bechmann, G.; Lorenz, K.; Plagens, B.; Hitzler, M.; Richter, E.; Ecker, G. *J. Med. Chem.* **1998**, *41*, 4001–4011.

159. Hiessböck, R.; Wolf, C.; Richter, E.; Hitzler, M.; Chiba, P.; Kratzel, M.; Ecker, G. *J. Med. Chem.* **1999**, *42*, 1921–1926.
160. Pleban, K.; Hoffer, C.; Kopp, S.; Peer, M.; Chiba, P.; Ecker, G. F. *Arch. Pharm. Pharm. Med. Chem.* **2004**, *337*, 328–334.
161. Chiba, P.; Erker, T.; Galanski, M.; Hitzler, M.; Ecker, G. F. *Arch. Pharm. Pharm. Med. Chem.* **2002**, *335*, 223–228.
162. Teodori, E.; Dei, S.; Quidu, P.; Budriesi, R.; Chiarini, A.; Garnier-Suillerot, A.; Gualtieri, F.; Manetti, D.; Romanelli, M. N.; Scapecchi, S. *J. Med. Chem.* **1999**, *42*, 1687–1697.
163. Ford, J. M.; Bruggeman, E. P.; Pastan, I.; Gottesmann, M.; Hait, W. T. *Cancer Res.* **1990**, *50*, 1748–1756.
164. Pajeva, I. K.; Wiese, M. *Quant. Struct.-Act. Relat.* **1997**, *16*, 1–10.
165. Pearce, H. L.; Safa, A. R.; Bach, N. J.; Winter, M. A.; Cirtain, M. C.; Beck, W. T. *Proc. Natl. Acad. Sci. USA* **1989**, *86*, 5128–5132.
166. Suzuki, T.; Fukazawa, N.; San-nohe, K.; Sato, W.; Yano, O.; Tsuruo, T. *J. Med. Chem.* **1997**, *40*, 2047–2052.
167. Österberg, T.; Norinder, U. *Eur. J. Pharm. Sci.* **2000**, *10*, 295–303.
168. Brooks, T. A.; Kennedy, D. R.; Gruol, D. J.; Ojima, I.; Baer, M. B.; Bernacki, R. J. *Anticancer Res.* **2004**, *24*, 409–416.
169. Kupsakova, I.; Rybar, A.; Docolomansky, P.; Drobna, Z.; Stein, U.; Walther, W.; Barancik, M.; Breier, A. *Eur. J. Pharm. Sci.* **2004**, *21*, 283–293.
170. Klopman, G.; Srivastava, S.; Kolossvary, I.; Epand, R. F.; Ahmed, N.; Epand, R. M. *Cancer Res.* **1992**, *52*, 4121–4129.
171. Klopman, G.; Shi, L. M.; Ramu, A. *Mol. Pharmacol.* **1997**, *52*, 323–334.
172. Klopman, G.; Zhu, H.; Ecker, G.; Chiba, P. *J. Comput.-Aided Mol. Des.* **2003**, *17*, 291–297.
173. Bakken, G. A.; Jurs, P. C. *J. Med. Chem.* **2000**, *43*, 4534–4541.
174. Bain, L.; McLachlan, J.; LeBlanc, G. *Environ. Health Perspect.* **1997**, *105*, 812–818.
175. Gombar, V. K.; Polli, J. W.; Humphreys, J. E.; Wring, S. A.; Serabjit-Singh, C. S. *J. Pharm. Sci.* **2004**, *93*, 957–968.
176. Didziapetris, R.; Japertas, P.; Avdeef, A.; Petrauskas, A. *J. Drug Target.* **2003**, *11*, 391–406.
177. Xue, Y.; Yap, C. W.; Sun, L. Z.; Cao, Z. W.; Wang, J. F.; Chen, Y. Z. *J. Chem. Inf. Comput. Sci.* **2004**, *44*, 1497–1505.
178. Pajeva, I. K.; Wiese, M. *J. Med. Chem.* **1998**, *41*, 1815–1826.
179. Wiese, M.; Pajeva, I. K. *Pharmazie* **1997**, *52*, 679–685.
180. Kellogg, G. E.; Semus, S. F.; Abraham, D. J. *J. Comput.-Aided Mol. Des.* **1991**, *5*, 545–552.
181. Safa, A. R.; Agresti, M.; Bryk, D.; Tamai, I. *Biochemistry* **1994**, *33*, 256–265.
182. Dey, S.; Hafkemeyer, P.; Pastan, I.; Gottesman, M. M. *Biochemistry* **1999**, *38*, 6630–6639.
183. Tsakovska, I. M. *Bioorg. Med. Chem.* **2003**, *11*, 2889–2899.
184. Tsakovska, I. M.; Wiese, M.; Pajeva, I. *Biotechnol. Biotechn. Eq.* **2003**, *17*, 163–169.
185. Pajeva, I. K.; Wiese, M. Comparative Molecular Field Analysis of Multidrug Resistance Modifiers. In *Molecular Modeling and Prediction of Bioactivity*; Gundertofte, K., Jorgensen, F. S., Eds.; Kluwer Academic/Plenum Publishers: New York, NY, 2000, pp 414–416.
186. Fleischer, R.; Wiese, M. *J. Med. Chem.* **2003**, *46*, 4988–5004.
187. Kim, K. H. *Bioorg. Med. Chem.* **2001**, *9*, 1517–1523.
188. Tsakovska, I. M.; Pajeva, I. K. *SAR QSAR Environ. Res.* **2002**, *13*, 473–484.
189. Dhainaut, A.; Régnier, G.; Atassi, Gh.; Pierré, A.; Léonce, St.; Kraus-Berthier, L.; Prost, J.-F. *J. Med. Chem.* **1992**, *35*, 2481–2496.
190. Cortes-Selva, F.; Campillo, M.; Reyes, C. P.; Jimenez, I. A.; Castanys, S.; Bazzocchi, I. L.; Pardo, L.; Gamarro, F.; Ravelo, A. G. *J. Med. Chem.* **2004**, *47*, 576–587.
191. Pearce, H. L.; Safa, A. R.; Bach, N. J.; Winter, M. A.; Cirtain, M. C.; Beck, W. T. *Proc. Natl. Acad. Sci. USA* **1989**, *86*, 5128–5132.
192. Zamora, J. M.; Pearce, H. L.; Beck, W. T. *Biochem. Pharmacol.* **1988**, *33*, 454–462.
193. Hait, N. W.; Aftab, D. T. *Biochem. Pharmacol.* **1992**, *43*, 103–107.
194. Etievant, C.; Schambel, P.; Guminski, Y.; Barret, J.-M.; Imbert, T.; Hill, B. T. *Anti-Cancer Drug Des.* **1998**, *13*, 317–336.
195. Ramu, A.; Ramu, N. *Cancer Chemother. Pharmacol.* **1994**, *34*, 423–430.
196. Seelig, A. *Eur. J. Biochem.* **1998**, *251*, 252–261.
197. Seelig, A. *Int. J. Clin. Pharmacol. Ther.* **1998**, *36*, 50–54.
198. Seelig, A.; Landwojtowicz, E. *Eur. J. Pharmaceut. Sci.* **2000**, *12*, 31–40.
199. Seelig, A.; Li Blatter, X.; Wohnsland, F. *Int. J. Clin. Pharmacol. Ther.* **2000**, *38*, 111–121.
200. Penzotti, J. E.; Lamb, M. L.; Evensen, E.; Grootenhuis, P. D. *J. Med. Chem.* **2002**, *45*, 1737–1740.
201. Doeppenschmitt, S.; Spahn-Langguth, H.; Regard, C. G.; Langguth, P. *Pharm. Res.* **1998**, *15*, 1001–1006.
202. Doeppenschmitt, S.; Langguth, P.; Regardh, C. G.; Andersson, T. B.; Hilgendorf, C.; Spahn-Langguth, H. *J. Pharmacol. Exp. Ther.* **1999**, *288*, 348–357.
203. Doeppenschmitt, S.; Spahn-Langguth, H.; Regardh, C. G.; Langguth, P. *J. Pharm. Sci.* **1999**, *88*, 1067–1072.
204. Neuhoff, S.; Langguth, P.; Dressler, C.; Andersson, T. B.; Regardh, C. G.; Spahn-Langguth, H. *J. Clin. Pharm. Ther.* **2000**, *38*, 168–179.
205. Wandel, C.; Kim, R. B.; Kajiji, S.; Guengerich, P.; Wilkinson, G. R.; Wood, A. J. *Cancer Res.* **1999**, *59*, 3944–3948.
206. Martin, Y. C.; Bures, M. G.; Danaher, E. A.; De Lazzer, J.; Lico, I.; Pavlik, P. A. *J. Comput.-Aided Mol. Des.* **1993**, *7*, 83–102.
207. Barnum, D.; Greene, J.; Smellie, A.; Sprague, P. *J. Chem. Inf. Comput. Sci.* **1996**, *36*, 563–571.
208. Jones, G.; Willett, P.; Glen, R. C. *J. Comput.-Aided Mol. Des.* **1995**, *9*, 532–549.
209. Ekins, S.; Kim, R. B.; Leake, B. F.; Dantzig, A. H.; Schuetz, E. G.; Lan, L. B.; Yasuda, K.; Shepard, R. L.; Winter, M. A.; Schuetz, J. D. et al. *Mol. Pharmacol.* **2002**, *61*, 964–973.
210. Ekins, S.; Kim, R. B.; Leake, B. F.; Dantzig, A. H.; Schuetz, E. G.; Lan, L. B.; Yasuda, K.; Shepard, R. L.; Winter, M. A.; Schuetz, J. D. et al. *Mol. Pharmacol.* **2002**, *61*, 974–981.
211. Pajeva, I.; Wiese, M. *J. Med. Chem.* **2002**, *45*, 5671–5686.
212. Masip, I.; Cortes, N.; Abad, M. J.; Guardiola, M.; Perez-Paya, E.; Ferragut, J.; Ferrer-Montiel, A.; Messeguer, A. *Bioorg. Med. Chem.* **2005**, *13*, 1923–1929.
213. Garrigues, A.; Loiseau, N.; Delaforge, M.; Ferte, J.; Garrigos, M.; Andre, F.; Orlowski, S. *Mol. Pharmacol.* **2002**, *62*, 1288–1298.
214. Sharom, F. J.; Yu, X.; Di Diodato, G.; Chu, J. W. *Biochem. J.* **1996**, *320*, 421–428.
215. Rebitzer, S.; Annibali, D.; Kopp, S.; Eder, M.; Langer, T.; Chiba, P.; Ecker, G. F.; Noe, C. R. *Farmaco* **2003**, *58*, 185–191.
216. Langer, T.; Eder, M.; Hoffmann, R. D.; Chiba, P.; Ecker, G. F. *Arch. Pharm. Pharm. Med. Chem.* **2004**, *337*, 317–327.
217. Pajeva, I. K.; Globisch, C.; Wiese, M. *J. Med. Chem.* **2004**, *47*, 2523–2533.
218. Kondratov, R. V.; Komarov, P. G.; Becker, Y.; Ewenson, A.; Gudkov, A. V. *Proc. Natl. Acad. Sci. USA* **2001**, *98*, 14078–14083.
219. Geick, A.; Eichelbaum, M.; Burk, O. *J. Biol. Chem.* **2001**, *276*, 14581–14587.
220. Synold, T. W.; Dussault, I.; Forman, B. M. *Nat. Med.* **2001**, *7*, 584–590.

221. Kast, H. R.; Goodwin, B.; Tarr, P. T.; Jones, S. A.; Anisfeld, A. M.; Stoltz, C. M.; Tontonoz, P.; Kliewer, S.; Willson, T. M.; Edwards, P. A. *J. Biol. Chem.* **2002**, *277*, 2908–2915.
222. Kauffmann, H. M.; Pfannschmidt, S.; Zoller, H.; Benz, A.; Vorderstemann, B.; Webster, J. I.; Schrenk, D. *Toxicology* **2002**, *171*, 137–146.
223. Ekins, S.; Erickson, J. A. *Drug Metab. Dispos.* **2002**, *30*, 96–99.
224. Ekins, S.; Mirny, L.; Schuetz, E. G. *Pharm. Res.* **2002**, *19*, 1788–1800.
225. Seigneuret, M.; Garnier-Suillerot, A. *J. Biol. Chem.* **2003**, *278*, 30115–30124.
226. Stenham, D. R.; Campbell, J. D.; Sansom, M. S.; Higgins, C. F.; Kerr, I. D.; Linton, K. J. *FASEB J.* **2003**, *17*, 2287–2289.
227. Smith, P. C.; Karpowich, N.; Millen, L.; Moody, J. E.; Rosen, J.; Thomas, P. J.; Hunt, J. F. *Mol. Cell* **2002**, *10*, 139–149.
228. Loo, T. W.; Clarke, D. M. *J. Biol. Chem.* **1997**, *272*, 20986–20989.
229. Loo, T. W.; Clarke, D. M. *J. Biol. Chem.* **2000**, *275*, 5253–5256.
230. Loo, T. W.; Clarke, D. M. *Proc. Natl. Acad. Sci. USA* **2002**, *99*, 3511–3516.
231. Lee, J. Y.; Urbatsch, I. L.; Senior, A. E.; Wilkens, S. *J. Biol. Chem.* **2002**, *277*, 40125–40131.
232. van Zanden, J. J.; Wortelboer, H. M.; Bijlsma, S.; Punt, A.; Usta, M.; Bladeren, P. J.; Rietjens, I. M.; Cnubben, N. H. *Biochem. Pharmacol.* **2005**, *69*, 699–708.
233. Hirono, S.; Nakagome, I.; Imai, R.; Maeda, K.; Kusuhara, H.; Sugiyama, Y. *Pharm. Res.* **2005**, *22*, 260–269.
234. Rosenberg, M. F.; Mao, Q.; Holzenburg, A.; Ford, R. C.; Deeley, R. G.; Cole, S. P. *J. Biol. Chem.* **2001**, *276*, 16076–16082.
235. Pleban, K.; Macchiarulo, A.; Costantino, G.; Pellicciari, R.; Chiba, P.; Ecker, G. F. *Bioorg. Med. Chem. Lett.* **2004**, *14*, 5823–5826.
236. Ecker, G. F.; Pleban, K.; Kopp, S.; Csaszar, E.; Poelarends, G. J.; Putman, D.; Kaiser, D.; Konings, W. N.; Chiba, P. *Mol. Pharmacol.* **2004**, *66*, 1169–1179.
237. Mikkaichi, T.; Suzuki, T.; Tanemoto, M.; Ito, S.; Abe, T. *Drug Metab. Pharmacokinet.* **2004**, *19*, 171–179.
238. Yarim, M.; Moro, S.; Huber, R.; Meier, P. J.; Kaseda, C.; Kashima, T.; Hagenbuch, B.; Folkers, G. *Bioorg. Med. Chem.* **2005**, *13*, 463–471.
239. Omote, H.; Al-Shawi, M. K. *Biophys. J.* **2006**, *90*, 4046–4059.

Biographies

Michael Wiese born in Neumünster, Germany, studied chemistry and pharmacy at Kiel University, where he obtained his PhD in pharmaceutical chemistry in 1985 under the direction of Prof J K Seydel. Afterward he worked at the research center Borstel as a Postdoctoral Research Fellow. In 1993, he finished his Habilitation and subsequently, he was awarded a Senior Professor at University of Halle in 1994 and took up his present position as Senior Professor and Head of Department at Bonn University in September 2000. His scientific interests include all aspects of molecular modeling and quantitative structure–activity relationships, in particular, transport proteins and G protein-coupled receptors and their ligands.

Ilza K Pajeva born in Bulgaria, studied at Mendeleev Institute of Chemical Technology in Moscow, where she obtained a MSc in chemical cybernetics; in 1989 obtained a PhD in biological science from Bulgarian Academy of Sciences in Sofia; in 1992 awarded a fellowship of the Alexander von Humboldt Foundation and spent 19 months in Borstel Research Center (Germany) under the direction of Prof J K Seydel. Returning to Sofia, she took her present position as an Associate Professor in Pharmacology at Centre of Biomedical Engineering, Bulgarian Academy of Sciences. The main scientific interests include QSAR and molecular modeling, multidrug resistance, and transport proteins.

Comprehensive Medicinal Chemistry II
ISBN (set): 0-08-044513-6

ISBN (Volume 5) 0-08-044518-7; pp. 767–794

5.33 Comprehensive Expert Systems to Predict Drug Metabolism

D R Hawkins, Huntingdon Life Sciences, Alconbury, UK

5.33.1 Introduction

Metabolism initially became an established part of drug development from a recognition of its importance to pharmacological and toxicological profiles. Thus, it was incorporated as part of the safety evaluation process. A widely expressed concept was that metabolism could be used to select the most appropriate species for toxicological evaluation of a drug candidate. This concept is rather simplistic, and clear opportunities to make decisions based on this comparison are seldom encountered, since the choice of species is limited, and species comparisons almost always consists of elaborating degrees of difference. Besides, a small difference in one aspect may prove to have greater significance than larger ones, and other variables such as pharmacokinetics may also be of some importance.

However, as the scientific knowledge of metabolism has developed, there has been an increasingly greater appreciation of the importance of considering metabolism predevelopment during drug discovery and the selection of development candidates, often called lead optimization. This provides the opportunity to incorporate structural features into a compound that can contribute to creating a drug that possesses the desirable pharmacokinetic properties, and avoids possible adverse properties such as toxicity or idiosyncratic drug reactions. Since these objectives can potentially reduce the number of drugs that fail during development and clinical evaluation, it is a compelling concept that has been readily adopted.

While relatively rapid generation of compound-specific metabolic information is possible using in vitro systems and modern analytical techniques, there are resource implications and some limitations to the extent of information that can be obtained. An alternative or complementary approach is to use in silico techniques, which consists of the application of existing knowledge to make predictions based on a consideration of the substrate structure alone. Since this is a theoretical exercise, it is possible to apply this approach as part of a process for designing new structures for modeling, synthesis, and biological screening. Thus, at the earliest stage it is possible to extend the chemical space of a particular compound series by devising modifications that may enhance metabolic stability, provide soft targets for metabolism, and also divert from the formation of potentially toxic metabolites or reactive intermediates.

In summary, there are several areas where the prediction of metabolites can be valuable:

- identifying sites of metabolism, and hence providing input into defining where structural modifications could be made to modulate metabolic stability;
- providing alerts to metabolites that could have potential pharmacological activity;

- providing alerts to the formation of reactive intermediates;
- providing alerts to species differences in metabolism that may be significant during the drug safety evaluation process;
- acting as a tool to aid in structural assignment during the identification of experimentally produced metabolites (e.g., by mass spectrometry and nuclear magnetic resonance); and
- highlighting specific structures to look for as metabolites in experimental investigations and aid identification of unknown metabolites.

Hence, the ability to make any prediction of metabolism can add value to drug discovery and development processes. From their very nature, predictions cannot be expected to always be correct, but performance will improve as knowledge increases. However, the applications of prediction mean that incorrect predictions will not have critical effects, and with intelligent use there is potentially a lot to be gained. The objectives of a prediction system should encompass the following:

- aim to predict the metabolites most likely to be formed;
- generally omit the prediction of minor metabolites (e.g., <5% dose or total metabolites) unless of special importance, such as where reactive intermediates are involved;
- alert to species differences;
- add value and be fit for the purpose; and
- capable of improvement by taking account of advances in knowledge.

5.33.2 The Challenge

With broad up-to-date knowledge of metabolic chemistry it is relatively easy to construct the pathways for major possible routes of metabolism for any novel structure. Some examples of metabolites involving more complex intramolecular rearrangements are less easy, but often these represent minor metabolites. However, the real challenge is to define which pathways will actually occur. There are many factors that will contribute to making some substructures and functional groups more or less susceptible, which include steric and electronic effects, the spatial interaction with enzyme active sites, and the physicochemical properties of the whole molecule. In the first instance, prediction requires the ability to decide which of the various options for metabolism will occur. The first option is that no metabolism will be necessary, and the drug is excreted unchanged. There are few examples of drugs that behave in this way, but the prediction of this outcome would be important, and one example is an analog of glutamic acid (1).[1] Clearly, compounds like this, which have low lipophilicity, have the requisite properties for efficient excretion, and metabolism is not necessary. A simple prediction rule based on log P could be applied, and the influence of lipophilicity on the pharmacokinetic behavior of drugs, which is a consequence of metabolism, has been described.[2]

The second type could be that no phase I metabolism is predicted (see 5.05 Principles of Drug Metabolism 1: Redox Reactions), only a direct phase II reaction with an existing functional group (see 5.06 Principles of Drug Metabolism 2: Hydrolysis and Conjugation Reactions). This process produces a readily excreted conjugate that is not uncommon, an example being ezetimibe (2) where the only major metabolite in plasma and urine was a glucuronide of the phenolic function. Oxidation of the benzylic alcohol to a ketone was the only minor phase I metabolite.[3]

When phase I reactions are the only choice, there will invariably be several options for the first-stage reactions, which in different combinations can be part of a sequence. In most cases, not all reactions will occur, at least to give significant amounts of metabolites. There are therefore two challenges, first to be able to assess probabilities of alternatives, and, secondly, to have criteria to terminate reaction sequences on the basis that a metabolite ceases to be a viable substrate for further metabolism and possesses the prerequisite properties to be excreted. Beyond this, there is also the challenge to predict notable species differences and to achieve the sophistication of being able to consider three-dimensional structures.

Species differences are primarily due to variation in the nature, number, and activities of enzymes that catalyze the individual reactions. These differences can be very large, differing by several orders of magnitude when comparing the rate of a reaction in the same organ. By far the greatest amount of knowledge exists for transformations involving the cytochrome P450 isozymes, but our understanding of differences is not generally well understood across the commonly encountered species (*see* 5.07 Principles of Drug Metabolism 3: Enzymes and Tissues; 5.10 In Vitro Studies of Drug Metabolism).

Consideration of the three-dimensional structure of compounds will take account of stereoselectivity in substrate metabolism and stereospecificity of product formation. The former would allow differences in metabolism of enantiomers and diastereoisomers (i.e., substrate specificity) to be predicted, and the latter would define the stereochemistry of reactions where a chiral center is generated (i.e., product stereoselectivity). These are long-term objectives.

5.33.3 Use of Knowledge Databases

A fundamental starting point for prediction is the ability to utilize existing knowledge, therefore collating metabolism data into a searchable format is strategically important. There are two such active databases containing information on drug metabolism, namely Metabolite, produced by MDL Information Systems, Inc. (San Leandro, CA, USA), and Metabolism, produced by Accelrys Ltd (Cambridge, UK) (**Table 1**). The latter was originally established using data compiled in the book series *Biotransformations: A Survey of the Biotransformation of Drugs and Chemicals in Animals*,[4] but it has continued to be developed by the addition of both historical and more recent literature data. Products of this type contain information that can be used to assist prediction in various ways. A database of reactions ensures that as a minimum there can be an awareness of all the types of reaction that are known to occur for functional groups and substructures. While most of the reactions are well established, new reactions continue to be found. These include oxidation of an oxime to a nitro group,[5] *S*-glucuronidation of a thiourea-type function,[6,7] and rearrangement of a substituted piperazine to an imidazoline.[8] The importance of a database is that it contains information that is searchable using the chemical structure as the key parameter. Hence, it should be possible to find all information on the metabolites of compounds containing defined substructures, and, if necessary, this could be selected further using in vivo, in vitro, and/or species criteria. Other selection filters could include parameters such as molecular weight and physicochemical properties such as log *P*. Searches can therefore be customized according to the compound of interest and purpose.

Information extractable from the databases can be used by an individual expert to formulate a prediction or as a basis for developing rules that can be incorporated into a software prediction program. How an individual might use a database can be illustrated by investigating what is known about the metabolic vulnerability and extent of degradation of two related heterocyclic ring systems, piperazine and morpholine (**Table 2**). A search of the Metabolism database showed that cleavage of the ring occurred for 7% of the total compounds (1043) that contained a piperazine group and for 30% of those (349) containing a morpholine group.

Table 1 Structure-searchable databases for metabolic pathways

Database	Content	Producer and website
Metabolite	Metabolites of primarily drugs	MDL Information Systems Inc. www.mdli.com
Metabolism	Metabolites of drugs and chemicals	Accelrys Ltd www.accelrys.com
BioPath	Biochemical pathways of endogenous compounds	University of Erlangen/Molecular Networks GmbH www2.uni-erlangen.de/services/biopath www.mol-net.de/databases/biopath

Table 2 Occurrence of analogous types of metabolites from piperazines and morpholines

Metabolite structure		Number of examples	
Piperazines	Morpholines	Piperazines	Morpholines
		11	16
		4	4
		0	13

Evaluation of the examples indicates that morpholines are more susceptible to extensive metabolism than piperazines. It is always important to ensure that it is valid to use a given example to make conclusions. For instance, the following need consideration:

- confidence in the structural assignments (speculative evidence should be downgraded);
- quantitative importance (minor metabolites should be excluded or given less importance); and
- completeness of the metabolic pathway (the importance of a particular pathway may not be fully estimated if a large proportion of the metabolites are unknown).

The input of human expertise is necessary at this stage to formulate rules that could be used for an expert prediction system.

Databases can also be used in very simple ways to give some basis for making predictions simply by finding out what is known about compounds with a high degree of structural similarity. For instance, there are a number of oral antidiabetic drugs that all contain a benzylthiazolidindione substructure (**Figure 1**). Clearly this could be a potential site of metabolism, but in many cases it appears that the main metabolites occur by oxidation and conjugation at the other end of the molecule when there are sites available.[9–12] Some of these metabolites also appear to have pharmacological activity. It is also of interest that for the compound MK-0767, where there is a blocking substituent on the terminal aromatic ring, the major metabolites are formed by oxidation in the thiazolindione ring.[13] On this basis it would therefore be reasonable to make some prediction for other compounds with similar structural features.

Analysis of a database can give some idea about the probability of a reaction type, and, importantly, highlight factors that may prevent its occurrence. Oxidation of arylmethyl groups was shown to occur for 89 of 106 compounds containing this group.[14] Where it did not occur, this could be ascribed to two reasons, one being that metabolism occurred at other alternative soft sites such as ester and dimethylamino functions, and the other due to steric hindrance from bulky *ortho* substituents. This type of analysis indicates how rules for basic reactions can be refined to expand the ability to make correct predictions. The use of these metabolism databases has been further reviewed in other publications.[15,16] An automated fragment-based searching and scoring method has been described that can be used in conjunction with any structure-searchable database.[17] The procedure consists of the analysis of the frequency of reactions for a defined structural fragment by mining all examples of specified reactions. The different options for metabolism are compared by calculating the frequency rates of the reactions, and normalizing to the most frequently occurring reaction. This process could be applied to all the key substructures in a drug, but the challenge is then to provide a composite view for the whole molecule.

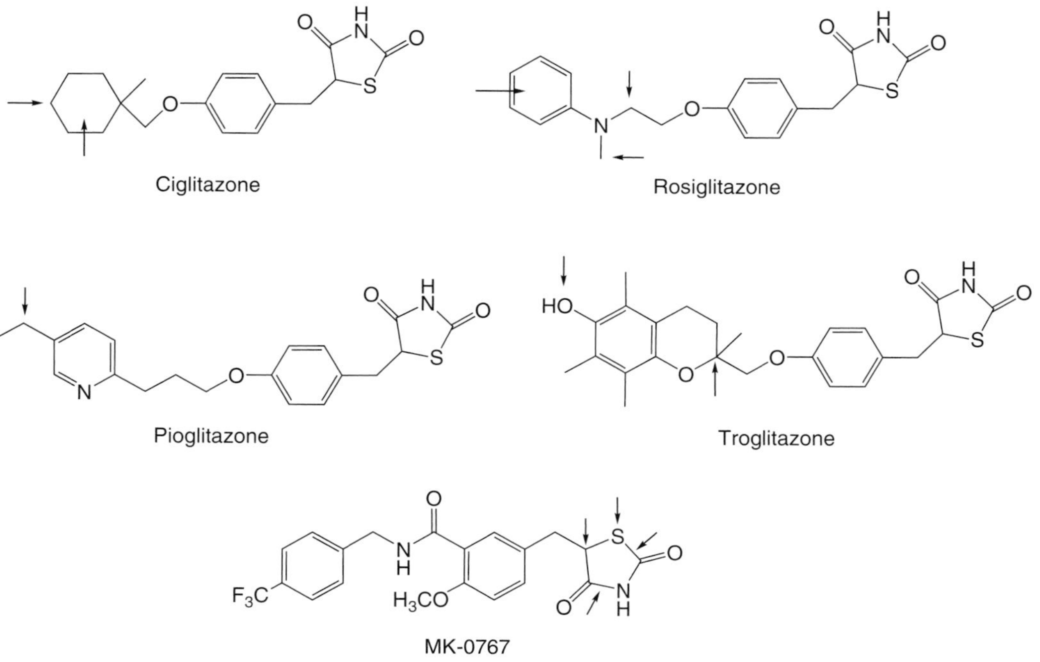

Figure 1   Structures of related thiazolidinedione antidiabetic agents, indicating sites of metabolism.

Another somewhat different database is BioPath, which contains the information on the well-known Biochemical Pathways Wall Chart, originally distributed by Boehringer Mannheim in a structure searchable format (see **Table 1**). It is possible to search for all the reactions in which a given biochemical, or those containing a certain substructure, occurs either as a substrate or a product.[18] Since many of the enzymes that participate in these processes may also accept drugs as substrates, the ability to search for this information could be a valuable tool in metabolite prediction.

5.33.4   Prediction Based on Reaction Mechanisms

If the mechanism of a particular biotransformation is understood, then it becomes easier to define a rule to predict the influence of changes in substituents on a functional group or substructure, on this reaction. The cytochrome P450-mediated cleavage of isoxazole rings involving breaking the N–O bond is a well-established reaction with many examples.[10] Whether this process will form a dominant pathway when there are other options will depend upon the ease of the reaction. In a series of substituted benzoxazoles of the general formula **3**, ring cleavage was enhanced when R^3 was electron-withdrawing, while an alternative reaction involving cleavage of the isoxazole indole bond (C–N bond) was enhanced when R^1 and R^2 were electron-withdrawing.[19]

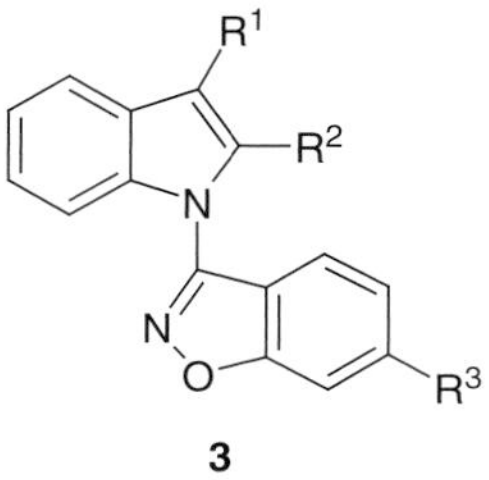

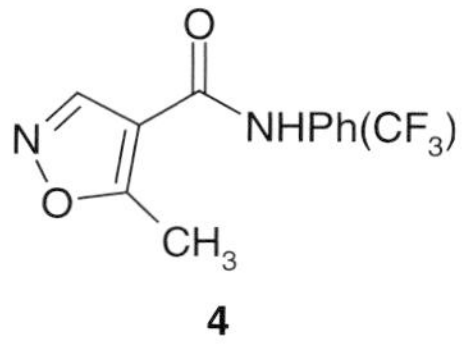

Similarly, for an isoxazole in a different structural environment such as **4**, other factors are important. Thus, when R is H, ring cleavage was very facile in rat/human plasma and liver microsomes; however, when R is CH_3, this pathway appeared to be blocked, and the ring methyl group was hydroxylated.[20] A mechanism of ring opening was postulated that was dependent upon the presence of a proton on the carbon atom adjacent to the nitrogen atom. These examples illustrate the huge challenge in attempting to predict metabolism and the level of sophistication that may be necessary. Thus, it is easy to define a basic rule for cleavage of isoxazoles and describe the structures of products, but isoxazoles will inevitably occur in complex structural environments with a wide diversity in ring substituents; the challenge is to predict the situations when ring cleavage will not occur. As an example, a report on the metabolism of valdecoxib (**5**), which has a fully substituted ring, showed that the main metabolite was formed by hydroxylation and conjugation of the ring methyl group, and that there was only one very minor metabolite resulting from ring cleavage.[21] This outcome could have been predicted based on the above knowledge of the likely mechanism.

5

5.33.5   Reactive Metabolites and Intermediates

The importance of trying to identify and avoid structures that may subsequently lead to unacceptable toxicity during preclinical development or idiosyncratic drug reactions during clinical trials or clinical use has become paramount to prevent late-stage failure of development candidates or even clinical withdrawal of marketed drugs. The impact of the above on the discovery and development of new drugs is critical and of key concern to the industry and human health. Many companies are now evolving strategies to deal with these issues, as identified in recent publications.[22,23] These considerations need to be incorporated at early stages of drug discovery rather than only at the lead optimization and selection stage, since it will be more efficient to include candidates in the potency optimization phase that have been designed on the basis of metabolism. One strategy to adopt is to have a mechanism for identifying substructures in substrates or their metabolites that have the potential to cause toxicity (*see* 5.39 Computational Models to Predict Toxicity). The first stage is to capture existing relevant knowledge, and to make sure this is used intelligently. There are some well-known structural groups associated with potential toxicity, such as quinones, which may be formed as part of an oxidative pathway, but there are certainly other less well-known indicators. A recent review has attempted to catalog all the structural groups where bioactivation to reactive intermediates can occur.[24]

Two examples of structures that could be highlighted as having toxicity potential are thiophenes and 3-methyleneindoles. The four thiophene compounds in **Figure 2** have all been identified as causing some toxicity.[25] There is evidence that oxidative metabolism is associated with the toxicity, and there are various proposals[26] for the nature of the toxic intermediate(s), including:

- epoxides analogous to those proposed for furans such as ipomeanol, a 3-alkylketofuran[27];
- *S*-oxidation, since thiophene sulfoxides are known to be very reactive; and
- unsaturated ring-cleaved products, such as unsaturated dialdehydes.

Hence, while it is known that some oxidative metabolism to a reactive intermediate is involved, the exact nature and structure of the intermediate is not known. It is noteworthy that these compounds have keto groups in the 2- or 3-position, so this could form the basis for a structural alert. This makes the alert more selective and therefore more valuable by not raising too many possibly false positives. In proposing a structural alert, it may appropriate to expand the horizon by including structures that are related by simple metabolic reactions. These would include the alcohols and their simple derivatives such as esters and ethers.

3-Methylindole (**6**) is a well-known pneumotoxin, and the mechanism is postulated to involve methyl oxidation, generating a reactive intermediate.[28] Evidence for this intermediate is provided by the identification of a 3-methylene-mercapturic acid metabolite. Pyrrolizidine alkaloids with the general structure **7** are known hepato- and pneumotoxins, and also form glutathione conjugates at the exocyclic methylene group, following formation of a 3-hydroxypyrrole substructure.[29]

There are also reports on the metabolism of two compounds (**8** and **9**) containing methyleneindole-related groups, both of which form reactive intermediates that are indicated by subsequent formation of glutathione conjugates.[30,31] For the former, an oxidative *N*-dealkylation reaction results in cleavage of the molecule, formation of a pyridine analog of hydroxymethylindole, and a mercapturic acid of this intermediate as a terminal metabolite. The last compound, zafirlukast, is a CYP3A4 inhibitor that may be mechanism based since that is the isozyme that catalyzes formation of the intermediate. The above evidence could be used for defining a structural alert that could be based on 3-methylenepyrrole and precursors (**10**).

The application of in silico techniques to reactive intermediates has a somewhat different objective than prediction of the structures of metabolites, and the performance criteria are not the same. Thus, it is not the objective to accurately predict the structures of metabolites but to raise an awareness of metabolism that could occur to produce reactive metabolites. In some cases the exact nature of the reactive intermediate will not be known. The objective

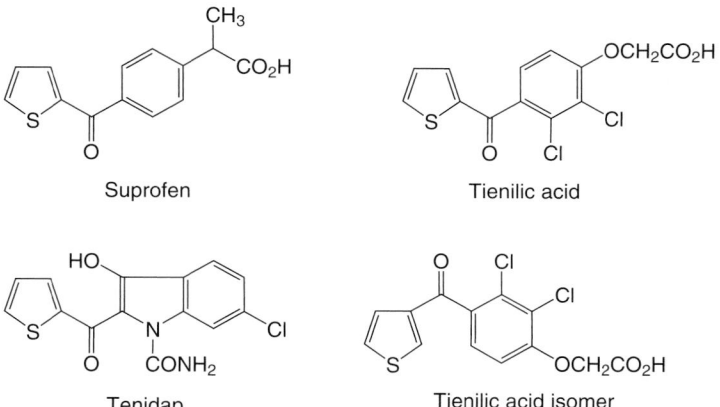

Figure 2 Structures of thiophene-containing compounds associated with toxicity.

should be to raise an alert for certain substructures that may have the potential to form the above, but not to predict that they will be formed or even if they are that they will lead to adverse effects in vivo. These substructures may be present in the parent molecule or formed in a metabolite. With this knowledge, the options are to consider including new compounds by making structural modifications that could preclude either the presence or the formation of the substructure alert, delete the inclusion of the alerted compounds as lead candidates, or begin to make some experimental investigations if they are strong candidates for other reasons. At the very least it provides a warning that can be kept in mind if any preclinical safety evaluation studies are conducted.

5.33.6 Development of Software to Predict Metabolism

5.33.6.1 Enzyme Active Site/Substrate Models

Any systems that can be used to predict specific aspects of metabolism could be used to contribute to a global prediction program. The most obvious is a system to predict the affinity of substrates for cytochrome P450 isozymes and the distribution of products as a consequence of alignment at the active site and steric/electronic components. An approach to predicting aliphatic hydroxylation reactions has been reported by Alhambra and Boyer[32] and Korzekwa et al.[33] These have been further extended to derive a general electronic model for cytochrome P450-mediated aromatic and aliphatic hydroxylation.[34] The model was constructed for small molecules where steric orienting effects at the enzyme active site are not of major importance. A more versatile model needs to combine electronic and steric effects. The importance of steric effects on the regioselectivity of quinoline and isoquinoline oxidation has been demonstrated.[35] Quinoline was hydroxylated at the 3-position, the position with the highest activation energy, whereas position 8 with the lowest activation energy was a relatively minor site for hydroxylation. The N-oxide was a major product for the isozymes CYP3A4 and CYP2A6, but it was not formed by CYP2E1 or CYP1A2. For isoquinoline the N-oxide was formed more predominantly than for quinoline. The conclusion from these investigations was that CYP3A4 and CYP2A6 are less sterically demanding than CYP2E1, CYP2B4, and CYP1A2.

Clearly, if models of the spatial interaction of substrates with enzyme active sites can be constructed and account taken of electronic effects when this is pertinent, they can lead to a prediction of the structures of metabolites. One such program has been developed called MetaSite (Molecular Discovery Ltd),[50] to predict the site of metabolism for substrates of CYP2C9, 2D6, 3A4, 1A2, and 2C9. It uses three-dimensional homology models for the enzymes as well as the three-dimensional structure of the substrates. The method uses GRID flexible molecular interaction field representations of the isozymes combined with different conformational structures of substrate molecules.[36] The predicted site of metabolism is reported as a ranking list of all the hydrogen atoms of each substrate. For CYP2C9, 87 oxidation reactions with 43 substrates only reported to be catalyzed by this enzyme were used to validate the methodology. These substrates displayed reactions consisting of metabolism at one to four different sites. It was concluded that in more than 90% of cases the method correctly predicted the three hydrogen atoms at sites where metabolism would occur. The conformation of the molecules had the biggest impact in the case of flexible molecules with multiple sites of oxidation.[36] A modeling approach for the other important isozymes listed above has also been reported,[37] stating that the first and second predicted options for sites of metabolism were correct to the following extents CYP2C9, 86%; CYP2C19, 81%; CYP2D6, 86%; CYP3A4, 78%; and CYP1A2, 75%. This program does not necessarily predict which isozymes will actually metabolize the substrates. A homology model for CYP2C9 has also been used to predict the sites of reaction for substrates ranked according to likelihood. Of 87 reactions reported to be catalyzed exclusively by CYP2C9, the model predicted the site of metabolism for more than 90% of the substrates within the first three ranked sites. The real test for this approach is whether for compounds with the complexity of structures now associated with drug research, there is sufficient understanding of the nature and behavior of active sites to predict the substrate interaction and when electronic and steric factors play a major role in the process.

Homology models for various cytochrome P450 isozymes have been developed by Lewis, initially based on bacterial CYP102,[38] and more recently using the crystal structure of the first mammalian cytochrome P450, namely CYP2C5.[39] While these models have not been developed into programs specifically to predict metabolism, part of their validation has involved showing that selective substrates fit within the active sites in such a manner that is consistent with the reported experimental data for known pathways of metabolism. Structure–activity relationships have been developed for estimations of reaction rates and clearances of substrates.[40,41] Lipophilicity has been shown to be important for substrate specificity since it combines hydrophobicity associated with molecular size, volume, or surface area with polarity due to the presence of hydrogen bond donor/acceptors in the structure. Log $D_{7.4}$ becomes important when the substrate contains ionizable groups. Typical log P ranges have been collated for various isozymes, although not surprisingly there is considerable overlap. Besides the above, the relative levels of the isozymes are also important in determining the extent and rate of metabolism and the degree of involvement of a particular isozyme. As the crystal structures of other mammalian isozymes are obtained, the models should improve. Incorporating these structural relationships into a composite metabolite prediction model could allow significant advances in performance.

5.33.6.2 Metabolite Prediction Programs

Metabolism is essentially a series of chemical reactions, each with a substrate and products. A computer system capable of recognizing chemical structures and portraying the products of metabolism is an ideal tool to support drug research and development. The requirements of such a system can be defined as follows:

- the ability to enter novel chemical structures;
- contains the available knowledge of all possible metabolic reactions supported by examples;
- contains rules or supporting intelligence that enables distinction between probable, possible, and unlikely metabolites;
- able to alert to reactions where reactive intermediates that have toxicity potential may be formed; and
- alert to possible species differences in probable reactions or species-specific reactions.

Various programs have been developed and made available, but all are still at a relatively low stage of sophistication. These include METEOR,[42] MetabolExpert,[43] META,[44] and the newly introduced MetaDrug[45] (**Table 3**). All nominally operate in a similar way mechanistically, although there are differences in the knowledge and rules used to arrive at the prediction.

METEOR is a program that was designed to predict the structures of likely metabolites from an evaluation of inbuilt rules on structure–metabolism relationships.[46] During its development it was recognized that it was necessary to have some basis to predict the most likely metabolites rather than all those theoretically possible. This has been approached by using an integrated reasoning engine based on a system of non-numerical argumentation.[47] According to preassigned levels of absolute and relative categorization of the likelihood of reactions and a link to log P calculation, metabolites can be predicted for five different levels, namely 'probable,' 'plausible,' 'equivocal,' 'doubted,' and 'improbable,' which can be preselected. In addition, the number of sequential phase I reaction stages can be selected, which reduces the combinatorial explosion of metabolites that can ensue when there are many different potential sites of metabolism. Query compounds can be constructed using a linked structure-drawing program or an imported molfile. Results can be reviewed that show the structure of the metabolite, highlighting the reaction site. Where cleavage of a molecule occurs, both products are shown. A recent evaluation of its performance has been reported for 10 substrates where the major metabolites were known.[48] For these substrates, 130 first-generation metabolites were predicted and/or identified experimentally. Of these, 30% were correct predictions, 62% apparently false positives, and 8% false negatives. Hence, 70% could be considered incorrect, consisting of 20% false positives due to molecular constraints not being taken into account, 3% false negatives due to missing reactions in the database, and the remainder ascribed to the complexity of biological factors. Utilization of this type of evaluation can help to make improvements to the system.

MetabolExpert uses a knowledge base that, for each transformation, contains a substructure, the product structure, a list of substructures that must be present for the reaction to occur, and a list of substructures whose presence

Table 3 Currently available metabolite prediction programs

Software program	Producer and website
META	MultiCASE Inc. www.multicase.com
MetabolExpert	CompuDrug Inc. www.compudrug.com
METEOR	Lhasa Ltd www.lhasalimited.org
MetaSite	Molecular Discovery Ltd www.moldiscovery.com
MetaDrug	GeneCo Inc. www.geneco.com

prevents the transformation from occurring. The number of levels of phase I transformations can be selected, and when a phase II metabolite is formed, no further transformations are performed. It operates in two ways, namely compiling a metabolic tree by matching known biotransformations for the respective substructures and functional groups in the molecule, or by making an analogy with information in the database on structurally similar compounds. The analogs available for comparison are listed, and the one used can be selected. This way of working is very similar to using one of the knowledge databases, although these will inevitably contain more extensive data. The basic biotransformation database of version 10.0 is reported to contain 179 transformations. There is also a version called RetroMex, which aims to predict substrates that would generate a defined compound by metabolism.

META is reported to have been assembled by starting with a compilation of the major enzymes involved in metabolism, and assessing the structural sites and their activities for each of these. Rules guiding their application to specific structures is stored in a series of logically linked dictionaries. The program operates on the basis of recognizing targets and subsequent transformations from the input structure, which may be a SMILES code or, more likely, a molfile. Each target/transformation receives a priority value. A stability module is incorporated, which checks if the primary molecular entity is stable or whether it will spontaneously degrade to a stable terminal product. The program can be used to proceed to sequential levels of metabolism. The dictionary of reactions is reported to contain more than 1000 transformations, covering the activity of 30 different enzyme systems, and more than 300 transformations in the stability module.

MetaDrug combines ruled-based metabolite prediction, quantitative structure–metabolism models for major drug-metabolizing enzymes, and gene network tools. Given its novelty, independent validations are required before its potential can be appreciated.

It is difficult to judge the merits of these existing systems, and neither is it appropriate to compare their performance in this chapter. It is inappropriate to compare the number of transformations each contains, since the substructure/functional groups pertaining to a reaction may be defined very specifically or more generally. A fair evaluation of their relative performance would be a rigorous exercise, comparing prediction against experimental data for a wide range of compounds. There is no doubt that it would be relatively easy to show that each performed well for one compound and equally badly for another. But no system should be condemned on the basis of one poor prediction. The overall performance is dependent on the quality and depth of knowledge each contains, which can only be assessed by an extensive evaluation. The question arises of how to quantify performance of a metabolite prediction system. The goal is to predict only those metabolites that are formed. To propose the structures of 20 metabolites in order to predict four that are actually produced is in most cases not helpful, and conversely to propose four metabolites and only predict one of four that are produced is also not a high success rate. Thus, there are two criteria that need to be considered and possible definitions are as follows:

$$\text{prediction coefficient} = C/A$$

$$\text{efficiency index} = C/B$$

where A is the total number of metabolites experimentally formed, B is the total number of metabolites predicted, and C is the total number of metabolites correctly predicted. The target is for both of these parameters is unity.

It could be argued that the prediction coefficient is the most important criterion. Namely, that false positives are less important than the omission of true positives. Clearly, we cannot expect this to be achieved at this stage, but any program can be continuously improved if up-to-date knowledge is incorporated and new rules developed. However, even the above analysis is not straightforward when metabolites are being predicted at different levels of probability, as with METEOR. The value a prediction system adds will depend on the knowledge of the user, and it is likely at this stage that the best human experts will be able to exceed the performance of any software program. There is extensive scope for improvement of the existing systems, since it is fair to comment that none of them yet achieve all the requirements and objectives defined above.

5.33.7   Future Development

5.33.7.1   Identifying Key Parameters

There is no doubt that our ability to predict metabolism and the processes to use are at an early stage of development. It is undeniably a huge challenge, and one easily dismissed as unachievable. This would be a mistake as long as some positive contribution can be made to drug research, and there is no doubt, as described in this chapter, that this is the case. As long as critical decisions are not made based solely on these predictions, it is a worthwhile exercise. As with any predictive exercise, it will never be entirely accurate, but we need to function with instructive minds that allow us to be comfortable with the degree of precision that the nature of the subject permits. The software programs that have been developed provide a good basis for evolving to a greater level of sophistication in performance. However, it is likely that this can only be achieved by combining some different approaches to prediction that complement the existing rule-based systems. The factors that contribute to structure–metabolism relationships have been reviewed by Testa and Cruciani,[49] with suggestions about how to take account of these and combine them into a prediction system based on rules and probabilities.

Besides compilation of a dictionary of all possible reactions, the key sequential stages that could be identified in prediction are as follows:

- ascertain that phase I metabolism is likely, based on physicochemical properties (log P and log D);
- ascertain whether metabolism could be restricted to phase II reactions;
- identify soft sites for metabolism according to hierarchically defined reactions;
- consider the influence of steric/electronic factors on the above, which may downgrade their probability; and
- link to models of cytochrome P450 isozymes active site/substrate interactions to assist in assigning relative probabilities of reactions.

These aspects need to be developed and supported by various initiatives in order to add value to the evolving products.

5.33.7.2   Capture and Collation of Knowledge

One key to improving and keeping systems up to date depends upon using as much information as possible on the metabolism of new compounds. The traditional sources available are the published scientific journals and presentations/posters at scientific meetings. But there is without doubt considerably more information not in the public domain but which could add immensely to the knowledge base. This includes data residing with regulatory authorities and also company archives. There have been some opportunities to utilize data from the authorities, more particularly in the USA, but there must be further possibilities to exploit this process more widely. But the greatest single source of existing data must be in the archives of research-based companies. While a certain amount of this may be considered confidential, there must also be a significant amount that could readily be released for utilization in the interests of scientific progress. The scientific community should support these initiatives, and the challenge is to find the best forum for catalyzing and implementing action.

5.33.7.3   Development of Structure–Metabolism Rules

Some of the aspects of metabolism that could be incorporated to enhance prediction have been referred to in this chapter. These include devising rules based on a detailed understanding of reaction mechanisms, which enables the structural requirements for reactions to be more precisely defined. Knowledge on species differences can be included to add value, and more detailed analysis of existing data might allow the development of more rules for this area. As knowledge of the interaction of substrates with enzyme active sites develops, this provides another evaluation tool to incorporate into a prediction process that can complement other rules. Some additional use of physicochemical properties could be appropriate to develop rules for the termination of phase I reaction sequences.

5.33.7.4   Incorporation of Knowledge into Prediction Programs

The final challenge is to incorporate all of the above into a seamless logical system that can operate reasoning according to rules and output the predictions clearly and concisely. When this can be achieved there will be no doubt that it will outperform the most knowledgeable scientist. The skill at this stage is to use our existing knowledge for predicting metabolism, and to use it in a way that means we gain from any positive contribution that can be taken, while avoiding any negative impact of incorrect predictions.

References

1. Johnson, J. T.; Mattiuz, E. L.; Chay, S. H.; Herman, J. L.; Wheeler, W. J.; Kassahun, K.; Swanson, S. P.; Phillips, D. L. *Drug Metab. Dispos.* **2002**, *30*, 27–33.
2. Testa, B.; Crivori, P.; Reist, M.; Carrupt, P.-E. *Perspect. Drug Disc. Des.* **2000**, *19*, 179–211.

3. Patrick, J. E.; Kosoglou, T.; Stauber, K. L.; Alton, K. B.; Maxwell, S. E.; Zhu, Y.; Statkevich, P.; Iannucci, R.; Chowdhury, S.; Affrime, M. et al. *Drug Metab. Dispos.* **2002**, *30*, 430–437.
4. Hawkins, D. R., Ed. *Biotransformations: A Survey of the Biotransformations of Drugs and Chemicals in Animals*; Royal Society of Chemistry: Cambridge, UK, 1988–1996, Vols 1–7.
5. Hainzl, D.; Loureiro, A. I.; Parada, A.; Soares-Da-Silva, P. *Xenobiotica* **2002**, *32*, 131–140.
6. Martin, I. J.; Lewis, R. J.; Bonnert, R. V.; Cage, P.; Moody, G. C. *Drug Metab. Dispos.* **2003**, *31*, 694–696.
7. Ethell, B. T.; Riedel, J.; Englert, H.; Jantz, H.; Oekonomopulos, R.; Burchell, B. *Drug Metab. Dispos.* **2003**, *31*, 1027–1043.
8. Doss, G. A.; Miller, R. R.; Zhang, Z.; Teffera, Y.; Nargund, R. P.; Palucki, B.; Park, M. K.; Tang, Y. S.; Evans, D. C.; Baillie, T. A. et al. *Chem. Res. Toxicol.* **2005**, *18*, 271–276.
9. Kawai, K.; Kawasaki-Tokui, Y.; Odaka, F.; Tsuruta, F.; Kazui, M.; Iwabuchi, H.; Nakamura, T.; Kinoshita, T.; Ikeda, T.; Yoshioka, T. et al. *Arzneim-Fosch* **1997**, *47*, 356–368.
10. Bolton, G. C.; Keogh, J. P.; East, P. D.; Hollis, F. J.; Shore, A. D. *Xenobiotica* **1996**, *26*, 627–636.
11. Krieter, P. A.; Colletti, A. E.; Doss, G. A.; Miller, R. R. *Drug Metab. Dispos.* **1994**, *22*, 625–630.
12. Barman Balfour, J. A.; Plosker, G. L. *Drugs* **1999**, *57*, 921–930.
13. Liu, D. Q.; Karanam, B. V.; Doss, G. A.; Sidler, R. R.; Vincent, S. H.; Hop, C. E. C. A. *Drug Metab. Dispos.* **2004**, *32*, 1023–1031.
14. Hawkins, D. R. *Drug Disc. Today* **1999**, *4*, 469–471.
15. Hawkins, D. R. *Med. Chem. Res.* **1998**, *8*, 434–443.
16. Erhardt, P. W., Ed. *Metabolism Databases and High-Throughput Testing During Drug Design and Development*; International Union of Pure and Applied Chemistry and Blackwell Science: Oxford, UK, 1999.
17. Boyer, S.; Zamora, I. J. *Comput.-Aided Mol. Des.* **2002**, *16*, 403–413.
18. Reitz, M.; Sacher, O.; Tarkhov, A.; Trumbach, D.; Gasteiger, J. *Org. Biomol. Chem.* **2004**, *2*, 3226–3237.
19. Tschirret-Guth, R. A.; Wood, H. B. *Drug Metab. Dispos.* **2003**, *31*, 999–1004.
20. Kalgutkar, A. M.; Hang, T. N.; Vaz, A. D. N.; Doan, A.; Dalvie, D. K.; Mcleod, D. G.; Murray, J. C. *Drug Metab. Dispos.* **2003**, *31*, 1240–1250.
21. Yuan, J. J.; Yang, D.-C.; Zhang, J. Y.; Bible, R.; Karim, A.; Findlay, J. W. A. *Drug Metab. Dispos.* **2002**, *30*, 1013–1021.
22. Nassar, A.-E. F.; Kamel, A. M.; Clarimont, C. *Drug Disc. Today* **2004**, *9*, 1055–1064.
23. Evans, D. C.; Watt, A. P.; Nicol-Griffith, D. A.; Baillie, T. A. *Chem. Res. Toxicol.* **2004**, *17*, 3–16.
24. Kalgutkar, A. S.; Gardner, I.; Obach, R. S.; Shaffer, C. L.; Callegari, E.; Henne, K. R.; Mutlib, A. E.; Dalvie, D. K.; Lee, J. S.; Nakai, Y. et al. *Curr. Drug Metab.* **2005**, *6*, 161–225.
25. Nelson, S. D. Structure Toxicity Relationships – How Useful Are They in Predicting Toxicities of New Drugs. In *Biological Reactive Intermediates VI*; Dansette, P. M., Ed.; Kluwer Academic/Plenum Press: New York, 2001; pp 33–43.
26. Dalvie, D. K.; Kalgutkar, A. S.; Khojasteh-Bakht, S. C.; Obach, R. S.; O'Donnell, J. P. *Chem. Res. Toxicol.* **2002**, *15*, 269–299.
27. Alvarez-Diez, T. M.; Zheng, J. *Chem. Res. Toxicol.* **2004**, *17*, 150–157.
28. Skiles, G. L.; Smith, D. J.; Appleton, M. L.; Carlson, J. R.; Yost, G. S. *Toxicol. Appl. Pharmacol.* **1991**, *108*, 531–537.
29. Mattocks, A. R. *Chemistry and Toxicology of Pyrrolizidine Alkaloids*; Academic Press: New York, NY, 1986.
30. Zhang, K. E.; Kari, P. H.; Davis, M. R.; Doss, G.; Baillie, T. A.; Vyas, K. P. *Drug Metab. Dispos.* **2000**, *28*, 633–642.
31. Skordos, K. W.; Yost, G. S. *Drug Metab. Rev.* **2003**, *35*, 49.
32. Alhambra, C.; Boyer, S. *Drug Metab. Rev.* **2002**, *34*, 109.
33. Korzekwa, K. R.; Jones, J. P.; Gillette, J. R. *J. Am. Chem. Soc.* **1990**, *112*, 7042–7046.
34. Jones, J. P.; Mysinger, M.; Korzekwa, K. R. *Drug Metab. Dispos.* **2002**, *30*, 7–12.
35. Dowers, T. S.; Rock, D. A.; Rock, D. A.; Perkins, B. N. S.; Jones, J. P. *Drug Metab. Dispos.* **2004**, *32*, 328–332.
36. Zamora, I.; Afzelius, L.; Cruciani, G. *J. Med. Chem.* **2003**, *46*, 2313–2324.
37. Cruciani, C.; Aristei, Y.; Vianello, R.; Baroni, M. In *Grid-Based Molecular Interaction Field in Cheminformatics and Drug Design*; Cruciani, G., Ed.; Wiley-VCH: Chichester, UK, 2005.
38. Lewis, D. F. V. *Drug Metab. Rev.* **2002**, *34*, 55–67.
39. Lewis, D. F. V.; Lake, B. G.; Dickens, M.; Goldfarb, P. S. *Xenobiotica* **2004**, *34*, 549–569.
40. Lewis, D. F. V. *Toxicology* **2000**, *144*, 197–203.
41. Lewis, D. F. V.; Dickens, M. *Toxicology* **2002**, *170*, 45–53.
42. Greene, N. Knowledge Based Expert Systems for Toxicity and Metabolism Prediction. In *Metabolism Databases and High-Throughput Testing During Drug Design and Development*; Erhardt, P. W., Ed.; International Union of Pure and Applied Chemistry and Blackwell Science: Oxford, UK, 1999; pp 289–296.
43. Darvas, P. MetabolExpert: Its Use in Metabolism Research and in Combinatorial Chemistry. In *Metabolism Databases and High-Throughput Testing During Drug Design and Development*; Erhardt, P. W., Ed.; International Union of Pure and Applied Chemistry and Blackwell Science: Oxford, UK, 1999; pp 237–270.
44. Klopman, G.; Tu, M. A Program for the Prediction of the Products of Mammal Metabolism of Xenobiotics. In *Metabolism Databases and High-Throughput Testing During Drug Design and Development*; Erhardt, P. W., Ed.; International Union of Pure and Applied Chemistry and Blackwell Science: Oxford, UK, 1999; pp 271–276.
45. Ekins, S.; Andreyev, S.; Ryabov, A.; Kirillov, E.; Rakhmatulin, E. A.; Bugrim, A.; Nikolskaya, T. *Exp. Opin. Drug Metab. Toxicol.* **2005**, *1*, 304–324.
46. Langowski, J. J.; Long, A. *Adv. Drug Deliv. Rev.* **2002**, *54*, 407–415.
47. Button, W. G.; Judson, P. N.; Long, A.; Vessey, J. D. *J. Chem. Inform. Comput. Sci.* **2003**, *43*, 1371–1377.
48. Testa, B.; Balmat, A.-L.; Long, A.; Judson, P. *Chem. Biodiversity* **2005**, *2*, 955–973.
49. Testa, B.; Cruciani, G. In *Pharmacokinetic Optimisation in Drug Research: Biological Physicochemical and Computational Strategies*; Testa, B., van de Waterbeemd, H., Folkers, G., Guy, R., Eds.; Wiley-Verlag Helvetica Chimica Acta: Zurich, Switzerland, 2001; pp 65–84.
50. www.moldiscovery.com (accessed July 2006).

Biography

David R Hawkins is a graduate in chemistry, and obtained a PhD in organic chemistry from the University of Birmingham in the UK. He began his career as a synthetic organic chemist in drug research for a pharmaceutical company. Subsequently, he has spent more than 30 years in the area of metabolism, and been responsible for directing the work of scientists as a head of department in a contract research organization. He has published more than 100 scientific papers and reviews, covering a range of different aspects of xenobiotic metabolism, and was the founding editor of the book series *Biotransformations*, published by the Royal Society of Chemistry. More recently, he has become involved in the development of methodologies for the in silico prediction of toxicity and metabolism based on structure–activity relationships, and serves as a board director and consultant to a software company in this area.

Comprehensive Medicinal Chemistry II
ISBN (set): 0-08-044513-6

ISBN (Volume 5) 0-08-044518-7; pp. 795–807

5.34 Molecular Modeling and Quantitative Structure–Activity Relationship of Substrates and Inhibitors of Drug Metabolism Enzymes

M J De Groot, Pfizer Global Research and Development, Sandwich, UK
D F V Lewis, University of Surrey, Guildford, UK
S Modi, GlaxoSmithKline Research and Development, Stevenage, UK

5.34.1 Introduction

Drug metabolism is an immense area of study where drugs undergo a range of enzyme-mediated chemical reactions, such as oxidation, reduction, hydrolysis, hydration, conjugation, and migration. Drug metabolism is normally divided into two phases depending on the reaction, phase 1 (functionalization reactions, e.g., oxidation, reduction, hydration, hydrolysis, or isomerization) and phase 2 (conjugation reactions). In general, the products of phase 1 reactions are more chemically reactive than the parent compound, and phase 1 is seen as the creation of a reactive species prior to true detoxification by phase 2 metabolism. In phase 2 metabolism the substrate is conjugated to a polar hydrophobic group making it more water soluble and suitable for excretion.[1,2]

Cytochrome P450 (also termed heme-thiolate protein P450) is the most important phase 1 enzyme. The cytochromes P450 constitute a superfamily containing over 70 families of membrane-bound heme-proteins that catalyse the oxidative (and sometimes reductive) metabolism of structurally diverse compounds.[3,4]

Phase 2 metabolism is carried out by a diverse group of enzymes acting on diverse types of compounds, generally leading to a water-soluble product that can be excreted in bile or urine. Major enzymes involved in this are glutathione *S*-transferase (GST), methyltransferase (MT), *N*-acetyltransferase (NAT), sulfotransferase (SULT), and UDP-glucuronosyltransferase (UGT).

Due to the extensive impact of drug-metabolizing enzymes on the availability of administrated drugs, it would be of considerable value, both in the design of new drugs and also in the understanding of environmental toxins, to be able to predict the metabolic fate of these compounds. A major advance in drug development will be the availability of a tool that can predict whether a given compound will interact with these drug-metabolizing enzymes, and, if so, which isoform of these it will preferentially interact with. Prediction of $IC_{50}/K_m/V_{max}$ values, as well as prediction of the site of metabolism of any given compound, will be valuable information to take into consideration during the drug development process. These predictions could significantly reduce the failure rate in clinical trials, by identifying potential problems at the early stages of development. A combination of these in silico models/tools can offer great advantages in improving the odds of success in a discovery programme. In the early stages, this can be done by eliminating the guesswork and decreasing the experimental load, while, in later stages, more precise information for a limited number of compounds can be predicted.

Drug metabolism by an enzyme can be thought of as three phases: recognition, accessibility, and reactivity. Early recognition takes place at long distance (i.e., at the protein surface), and is mainly an electrostatic interaction between the substrate and the enzyme. Accessibility is determined by the ease by which the substrates can reach the active site of the drug-metabolizing enzyme; for example, the overall dimensions of the channel leading to the active site (which extends from the surface of the protein to the active site) is in the case of a buried active site very important in substrate discrimination. The topography and size of the active site plays an important role in metabolism of a compound. Reactivity (oxidation, conjugation, etc.) occurs when the substrate has reached the active site.[5] Accessibility, conformation, and orientation of the substrates (leading to multiple sites of metabolism) and thermodynamics are the key determinants in this phase.

The three-dimensional (3D) structure of enzymes can provide a valuable insight into the function, and help derive structure–activity relationships. X-ray crystal structures are desirable, but due to solubility issues and membrane association this has not always been possible for all drug-metabolizing enzymes (e.g., cytochromes P450[6]). In the absence of crystal structures, alternative methods (such as homology models) have been applied. Homology models are based on the theory that proteins with similar amino acid sequences have a tendency to adopt similar 3D structures. This enables prediction of the 3D structures of these enzymes based solely on their amino acid sequences and the 3D structures of proteins with similar amino acid sequences.

Another methodology used in the absence of crystal structures has been the pharmacophore approach. Pharmacophore models utilize the common structural features associated with biological activity in a variety of substrates for a specific enzyme (in this case, binding and metabolism by a particular isoform of these enzymes). Ideally, a pharmacophore should be unique for a particular enzyme, and knowledge of the common 'core' features associated with each isoform of these enzymes can be very helpful. Due to lack of similarity and overlap of the common features between different isoforms (e.g., cytochromes P450), the utility of these pharmacophores can, however, be limited.

5.34.2   Modeling of Phase 1 Metabolism – Cytochromes P450

Cytochrome P450 is the most important phase 1 enzyme, and belongs to a superfamily containing over 70 families of membrane-bound heme proteins that catalyze the oxidative metabolism of structurally diverse chemicals.[3,4] All members of a particular cytochrome P450 subfamily have a 55% amino acid sequence homology with one another, while members of the same family show an identity of more than 40%.[7] The isoenzymes differ from each other in amino acid sequence, in control by inhibitors and inducing agents, in the substrates they act upon, and the reactions they catalyze. Some isoenzymes have overlapping substrate specificities, acting on the same substrates as each other but at differing rates. Each cytochrome P450 has a unique system of substrate recognition, leading to the observed substrate specificities; for example, CYP2C9 prefers acidic and neutral but not basic substrates, CYP2D6 prefers basic substrates, CYP3A4 metabolizes both acidic and basic substrates, but log P plays a role, and CYP1A2 metabolism is governed by size.

Seven of the 57 known human isoforms of cytochromes P450 (CYP1A2, CYP2C9, CYP2C18, CYP2C19, CYP2D6, CYP2E1, and CYP3A4) are responsible for more than 90% of the metabolism of all drugs in current clinical use.[8] Some of these enzymes (especially CYP2D6 and CYP2C9) show polymorphisms, which can either result in rapid or very poor metabolism.[9–11] For example, debrisoquine is extensively metabolized in normal individuals to 4-hydroxydebrisoquine, but poor metabolizers can develop high levels of the parent compound, which may be toxic.[12] The impaired metabolic oxidation of at least 30 drugs with diverse structures and pharmacological actions has been associated with the phenotype of poor debrisoquine metabolism. In a few cases, poor metabolizers may be unable to bioactivate a parent drug, such as encainide, to its therapeutically active metabolites.[13,14] Binding of any drug to these cytochromes P450 can cause drug–drug interactions, which can lead to severe side effects. This has in the last few years resulted in early termination of development, refusal of approval, severe prescribing restrictions, and withdrawal of drugs from the market. Regulators, including the US Food and Drug Administration, have therefore issued guidance for in vitro and in vivo drug interaction studies to be conducted during development.[15]

5.34.2.1   Structural Modeling of Cytochromes P450

The major human cytochromes P450 involved in xenobiotic transformations are from families 1, 2, and 3, with CYP3A4 representing the most important enzyme for phase 1 drug metabolism, which is broadly commensurate with its average percentage of the human hepatic microsomal cytochrome P450 complement.[16] **Table 1** summarizes the more important human cytochromes P450 involved in exogenous metabolism, and indicates typical substrates, inducers, and inhibitors of these enzymes, showing the structural diversity of such compounds. Apart from the determination of structural characteristics of these cytochromes P450 via x-ray crystallography,[17] where six mammalian structures (rabbit 2B4,[18] rabbit 2C5,[19–21] human 2C8,[22] human 2C9,[23,24] human CYP2D6,[25] and human 3A4[26,27]) have been resolved, the challenges for computational chemistry in this area have focused on homology modeling, requirements for substrate selectivity, estimation of binding affinity, and the search for descriptors that may correlate with kinetic data (rates or clearance) for cytochrome P450-mediated metabolism and other absorption, distribution, metabolism, excretion, and toxicology properties.[28,29]

Recently published x-ray crystal structures[30] of human cytochromes P450 (**Table 2**) provide an opportunity for comparing previously reported models of these enzymes, and to establish a degree of validity for homology modeling. For CYP2C8 and CYP2C9, the models have been shown to exhibit α-carbon root mean square distances of 1.2 Å and 1.5 Å, respectively, thus demonstrating the utility of, in this case, the CYP2C5 crystal structure as a template for generating three-dimensional models for other CYP2C subfamily enzymes.[31–33] An example of a CYP2C9 homology model is shown in **Figure 1**. Furthermore, there may be wider applicability of such crystallographic templates toward the CYP2 family in general, although the recent inhibitor-bound CYP2B4 crystal structure can be expected to represent an ideal template for the homology modeling of CYP2B6. Information from the human cytochrome P450 crystal structures is proving useful for the rationalization of substrate selectivity, which can lead to predictive tools for the high-throughput screening of new chemical entities (NCEs). For example, the active-site volumes of CYP2C8, CYP2C9, and CYP3A4 indicate the likely maximum sizes of substrates and their superimposed templates; also, the topography of these active sites helps to explain, in part, substrate structural preferences (summarized in **Table 3**). For example, there are clear topographical differences between CYP2C8 and CYP3A4 active sites that are in agreement with the known substrate selectivity between these two enzymes, despite evidence for some degree of overlap in this respect. More details with regard to the field of cytochrome P450 modeling can be found in a recent review. [34]

Table 1   Human cytochromes P450 from families CYP1, CYP2, CYP3, and CYP4[16,123–125]

CYP	Hepatic CYP (%)	Drug oxidations (%)	Typical substrates	Typical inhibitors	Typical inducers	Receptor/regulatory factor
1A1	1	3	Benzo[a]pyrene	9-Hydroxyellipticine	PAHs	AHR/ARNT
1A2	13	10	Caffeine	Furafylline	PAHs	AHR/ARNT
1B1	1	1	Estradiol	Resveratrol	PAHs	AHR/ARNT
2A6	4	3	Coumarin	Pilocarpine	Androgens	HNF4
2B6	1	4	Bupropion	Orphenadrine	Phenobarbital	CAR
2C8			Amodiaquine	Sulfinpyrazone		
2C9	19	25	(S)-Warfarin	Sulfaphenazole	Barbiturates	HNF4, RAR/RXR
2C19			(S)-Mephenytoin	Fluconazole		
2D6	3	15	Propranolol	Quinidine	None	Constitutively regulated
2E1	7	3	Chlorzoxazone	Pyridine	Ethanol	Post-translational modification
3A4/5	28	36	Nifedipine	Ketoconazole	Rifampicin	PXR, GR
4A11	NA	~0	Lauric acid	11-Imidazolyl decanoic acid	Clofibrate	PPAR

AHR, arylhydrocarbon receptor; ARNT, AHR nuclear transporter; CAR, constitutive androstane receptor; GR, glucocorticoid receptor; HNF4, heptocyte nuclear factor 4; NA, data not available; PAHs, polycyclic aromatic hydrocarbons; PPAR, peroxisome proliferator-activated receptor; PXR, pregnane X-receptor; RAR, retonic acid receptor.

Table 2 Mammalian cytochrome P450 crystal structures[30]

CYP	Resolution (Å)	PDB	Substrate/inhibitor present
2B4 (rabbit)	1.60	1ps5[18]	None
	1.90	1su0[18]	4-(4-Chlorophenyl)imidazole
2C5 (rabbit)	3.00	1dt6[19]	None
	2.30	1n66[20]	Dimethyl-2-phenyl-2*H*-pyrazol-3-ylbenzenesulfonamide (DMZ)
	2.10	1nr6[21]	Diclofenac
2C8 (human)	2.70	1pq2[22]	None
2C9 (human)	2.60	1og2[23]	None
	2.55	1og5[23]	(*S*)-Warfarin
	2.00	1r90[24]	Flurbiprofen
2D6 (human)	3.00	2f9q[25]	None
3A4 (human)	2.05	1tqn[26]	None
	2.74	1w0 g[27]	Metyrapone

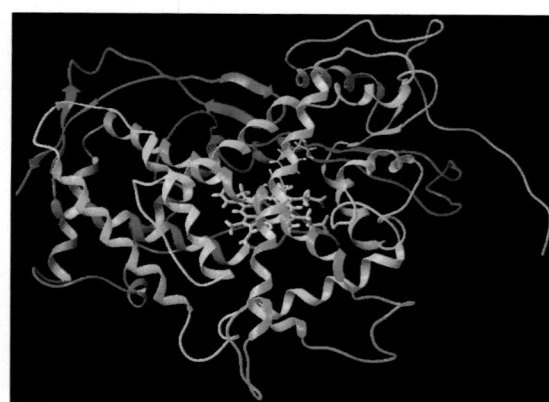

Figure 1 Homology model for human CYP2C9[31] based on the crystal structure of rabbit CYP2C5.[19] α-Helices and β-sheets are shown.

Modeling of the 3D structure and docking of substrates within these models can be very useful, but is complicated by another factor. Several years ago, evidence started to emerge that CYP3A4 could bind multiple substrates simultaneously.[35] This was later extended to various other cytochromes P450.[36] Further work suggested at least three different binding sites for CYP3A4, with testosterone, α-naphthoflavone, and midazolam occupying these respective sites,[37] which eventually led to all three of these substrates to be used in CYP3A4 inhibition experiments. For CYP3A4, the existence of multiple binding sites was further confirmed with spectroscopic[38] and, ultimately, crystallographic[26] evidence.

5.34.2.2   Molecular Dynamics (MD) Applied to Cytochromes P450

Although there is likely to be conformational flexibility in both enzymes and their substrates, the overall conformational space can be explored using MD simulations of the enzyme–substrate complex. In fact, it would appear that MD may be able to derive physiologically realistic structures for these complexes irrespective of whether the original 3D coordinates stem from a homology model or a crystal structure, as these probably represent different conformations of the same system that could converge energetically following a dynamics simulation.

Table 3 Information on CYP substrate characteristics and active site volumes from crystal structures[18,20–22,24,26,126,127]

CYP	Volume of active site ($\mathring{A}^3$)	Typical substrates/inhibitors	General substrate characteristics
2A6	250	Coumarin, nicotine, losigamone	Relatively small, compact molecules with hydrogen bond donor/acceptor atoms
2B4	260	Benzphetamine, testosterone, 4-chlorophenylimidazole	Relatively small/medium-sized molecules
2C5	645	Progesterone, DMZ, diclofenac	Small/medium-sized molecules with hydrogen bond donor/acceptor atoms
2C8	1386	Cerivastatin, taxol, rosiglitazone	Relatively large, S-shaped molecules
2C9	1137	Flurbiprofen, (S)-warfarin, diclofenac	Medium-sized molecules possessing an acidic group
3A4	1438	Erythromycin, cyclosporin, nifedipine	Relatively large, globular molecules

In order to prepare an enzyme–substrate complex for an MD run, a likely initial geometry needs to be established. This naturally requires some form of docking facility to derive an appropriate starting point prior to energy minimization and geometry optimization via a combination of molecular mechanics and dynamics procedures. A suggested sequence of stages in the process of generating a homology model from a suitable crystal structure can be outlined as follows:

(1) molecular mechanics energy minimization calculations on the initial crystal structure;
(2) changing the appropriate amino acid residues in accordance with the alignment;
(3) searching the protein databank (PDB) for suitable protein loops;
(4) checking the side chain torsional angles and modification as required;
(5) molecular mechanics energy calculations on the raw model;
(6) docking of the desired substrate or inhibitor; and
(7) MD simulations on the enzyme–substrate complex.

Of course, various groups of researchers will have different alternatives to the above multistage scheme for producing cytochrome P450 models by homology, and it is recognized that within each step there are several possibilities and a variety of starting conditions.[33,39–41] Nevertheless, the preferred methodology to employ is likely to arise from an experience of comparison between techniques, and of validation against experimental observations.

The choice of crystal structure template for homology model building will, in general, be dictated by primary sequence identity and similarity, following a multiple sequence alignment that may involve several cytochrome P450 from within the same family and/or subfamily. For the human CYP2 family there is a relatively high homology with the rabbit CYP2C5 crystal structure,[19–21] which displays sequence identities of over 40%, and greater than 70% for CYP2C subfamily enzymes (although the human CYP2C9 crystal structures[23,24] are better templates in this case). Once a satisfactory alignment has been produced, it should be an essentially straightforward procedure to generate a raw model by replacement of the necessary amino acid residues, although some residue deletions and insertions may also be required. The latter can be readily achieved via loop searching of the PDB and selection of the most appropriate section of peptide. It is advisable to check the torsional angles of the amino acid side chains in the resultant raw model, and unfavorable steric interactions caused by the above procedures can then be corrected prior to relaxation of the entire structure via molecular mechanics energy minimization.

MD calculations can improve these homology models further, and, indeed, enable a means of full convergence with x-ray geometries, since both are likely to represent alternative points in the conformational space of the overall enzyme. Using the example of CYP2C9 where the original crystal structure shows the substrate (S)-warfarin bound preferentially about 10 Å from the heme, it is found that 1–2 ns of MD will bring the substrate approximately 5 Å closer to the heme iron, and at an orientation which is consistent with 7-hydroxylation, the experimentally observed result. Such substrate movements have been reported for bacterial CYP102, where the bound substrate, lauric acid, moves about 6 Å nearer to the heme, following reduction of the enzyme.[42] Moreover, it is noted that, under dynamics simulation, the overall enzyme conformation changes significantly during the 2 ns time frame with specific motions of the structure, especially helical trajectories, cooperatively assisting the translation of (S)-warfarin towards the heme moiety. Consequently, it should be appreciated that crystal structures may not necessarily represent the paradigm for investigating cytochrome P450–substrate interactions and for estimating energies or for deriving other structural information from them. However, a combination of x-ray crystallography, molecular modeling, and dynamics simulation would appear to constitute an appropriate way forward in this, at times, inexact science. The important roles played by the energy calculations, both molecular mechanics and dynamics, in this respect are to relax the entire enzyme structure from its initially constrained solid state form, which is unlikely to be physiologically realistic.

It should also be recognized that some cytochrome P450 crystal structures do not conform to that of the wild-type enzyme, as certain residues have been deliberately modified to assist the crystallization process. For example, there are two particular mutations, S220P and P221A, in the original CYP2C9 crystal structures (PDB codes: 1og2 and 1og5[23]), that have resulted in a conformational feature of the tertiary fold that causes the (S)-warfarin substrate to bind on the periphery of the active site rather than at the heme locus. However, the recently published flurbiprofen-bound CYP2C9 crystal structure (PDB code: 1r90[24]) shows the substrate present within the active site at a position and orientation relative to the heme iron, which is consistent with its known route of metabolism. In this case, the enzyme is closer to the wild type with no mutations at the 220 and 221 locations, thus resulting in a more satisfactory conformation. As CYP2C19 has a primary sequence containing a proline residue at position 220, and an otherwise overall high homology (91%) with CYP2C9, then it is reasonable to expect that the 1og2 or 1og5 crystal structures will represent preferred templates for producing CYP2C19 models. Following homology modeling and minimization, however, it is recommended that the structure is validated (e.g., using ProCheck[43,44] or WhatIf?[45]), to ensure that there are no unacceptable protein geometries and that the whole 3D structure conforms to the now well-characterized features of protein structures in terms of backbone and side chain angles.

5.34.2.3   Substrate Docking and Binding Interactions of Cytochromes P450

Once a satisfactory geometry has been obtained following energy minimization using molecular mechanics, it is feasible to investigate the docking of substrates. There are several programs available that may be used for this process. One of the most accurate appears to be AutoDock[46,47] as, for cytochrome P450–substrate binding, the energies calculated via this program can be shown to agree well with experimental data for the change in binding free energy (ΔG_{bind}) obtained from the Michaelis rate constants (K_m).[48] This is, however, not the only available method. An extensive overview of models generated for cytochromes P450 and the software used has been published recently.[49] **Table 4** shows that, for a set of human cytochromes P450 and their selective substrates, there is a very good concordance ($r^2 = 0.95$) between AutoDock-calculated binding energies and those derived from the apparent K_m data published in the literature (despite the uncertainties in both the experimental and computational methodologies). Plotting these data shows that the calculated enzyme–substrate binding energies lie close to the 1:1 correspondence line with respect to ΔG_{bind}, derived from

$$\Delta G_{bind} = RT \ln K_m \qquad [1]$$

where R is the gas constant and T is the absolute temperature, taken as 310 K for data measured at physiological temperature.

Table 4   Comparison between calculated and experimental binding energies

CYP	Substrate	K_m (μM)	$\Delta G_{bind}^{expt\ a}$	$\Delta G_{bind}^{calc\ a,b}$
1A2	Caffeine	180	−5.31	−5.88
2C8	Amodiaquine	1.2	−8.40	−8.16
2C9	(S)-Warfarin	4	−7.66	−7.81
2C19	(S)-Mephenytoin	54	−6.05	−5.71
2D6	Propranolol	1.02	−8.50	−8.56
2E1	Chlorzoxazone	65.1	−5.94	−5.87
3A4	Nifedipine	15	−6.84	−6.76

[a] The correlation between the experimental and calculated ΔG_{bind} shows $r^2 = 0.95$ and a slope of 0.999, indicating a very close correspondence between the data.
[b] ΔG_{bind}^{calc} is the enzyme–substrate binding energy calculated using AutoDock 3.05.[47]

Interestingly, these AutoDock binding energy values were calculated prior to MD runs after the raw structures had been minimized using molecular mechanics. It is possible that the success of AutoDock in this respect lies in its particular scoring function, which takes hydrophobic interactions into account as well as the usual linear combination of hydrogen-bonded, ionic, and π–π stacking energies. Moreover, the local dielectric constant can be applied as a correction to the various terms. Other scoring functions are able to account for this factor as well, however, and some may even allow for user-defined scoring functions to be applied. It is possible to estimate binding energies quite satisfactorily using average values for the typical interactions and include a log P-derived partitioning energy as an approximation for desolvation effects. In such descriptions, the total binding energy can be represented as a linear combination of individual components,

$$\Delta G_{\text{bind}} = \Delta G_{\text{hb}} + \Delta G_{\text{ionic}} + \Delta G_{\pi-\pi} + \Delta G_{\text{desol}} + \Delta G_{\text{rotors}} \qquad [2]$$

where the subscripts refer to hydrogen bond, ionic, π–π stacking, desolvation, and loss of bond rotation upon binding.

5.34.2.4 Quantitative Structure–Activity Relationship Approaches and Cytochromes P450: Lipophilicity and Substrate Selectivity

The desolvation term in the above linear combination of contributions to ΔG_{bind} (eqn [2]) is equated with the octanol/water partitioning energy, ΔG_{part}, in accordance with

$$\Delta G_{\text{part}} = -RT \ln P \qquad [3]$$

where P is the octanol/water partition coefficient, R and T are as defined for eqn [1].

In fact, it is possible to demonstrate that there are so-called lipophilicity relationships for selective substrates of several major human drug-metabolizing cytochromes P450. These can be expressed as $-\log K_m$ versus $\log P$ or in terms of the corresponding free energy changes, ΔG_{bind} and ΔG_{part}, respectively. The use of the latter enables one to estimate the total intermolecular interaction energies common to a set of substrates from the value of the intercept on the y axis for ΔG_{bind} itself. For CYP2B6 substrates, the lipophilicity relationship is one of the most clear,[50] with an intercept corresponding to the average value of a hydrogen bond and π–π stack interaction, which is, in fact, found in practice when the substrate interactions are investigated using modeling.

It is also possible to derive a quantitative structure–activity relationship (QSAR) for human cytochrome P450–substrate binding based on the number of active site interactions (i.e., hydrogen bonds, ion pairs, and π–π stacks) and log P values of 90 compounds that are fairly selective substrates of the major human enzymes.[51] The various coefficients for each contribution agree closely with the average values reported in the literature, and the log P coefficient is in harmony with the factor employed to convert this quantity into the corresponding desolvation free energy. Neural network analysis[52] of a similar set of selective cytochrome P450 substrates reveals the major descriptors likely to be involved in enzyme selectivity for the individual cytochromes P450 themselves.

A combined loading of only three parameters: pK_a, number of hydrogen bond donors/acceptors and molecular mass, produced a 94% concordance with the known selectivity of 60 compounds, and the small number of outliers identified were those substrates capable of metabolism by more than one cytochrome P450.[51] The three descriptors selected by the neural network can all be rationalized in terms of typical characteristics exhibited by cytochrome P450 substrates because, for example, it would be expected that basic compounds are CYP2D6-selective whereas acidic groups represent a common feature in substrates of CYP2C9.[52]

5.34.2.5 Cytochromes P450: Molecular Size and Shape

The size and shape of molecules, together with the number and disposition of hydrogen bond donor/acceptor atoms and aromatic rings are important determinants of P450 selectivity.[51] To an extent, these factors are included in the neural net analysis,[52] although modeling the substrates within the active site region or, possibly, building up a template of superimposed substrates can reveal more information on the structural requirements for occupancy and binding in the heme pocket of the cytochrome P450. This is mirrored by the topography of the active site itself, and the substrate recognition site (SRS) regions,[53] including residues identified by site-directed mutagenesis or known to contact substrates in crystal structures, are presented in **Table 5**. The SRS regions were identified in the early 1990s as those regions responsible for substrate interactions in the 2-family of cytochromes P450, and later expanded to cover all cytochromes P450.[53]

Table 5 SRS regions[53] of human cytochrome P450s showing likely contact residues with substrates for various cytochrome P450s[a,b]

CYP SRS1

CYP																		
1A2	R_{108}	P	D	L	Y	T	S	T	L	I	T	D	G	Q	S	L	T	F_{125}
1B1	R_{117}	P	A_{119}	F	A	S	F	R	V	V	S	G	G	R	S	M	A	F_{134}
2A6	R_{101}	G	E	Q_{104}	A	T	F	D	W	L	F	K	G	Y	A	V	V	F_{118}
2B6	R_{98}	G	K	I_{101}	A	M	V	D	P	F_{107}	F	R	G	Y	G	V	I_{114}	F_{115}
2C8	R_{97}	G	N_{99}	S_{100}	P	I	S	Q	R	I	T	K	G	L	G	I	I	S_{114}
2C9	R_{97}	G	I_{99}	F	P	L	A	E	R	A	N	R_{108}	G	F	G	I	V_{113}	F_{114}
2C19	R_{97}	G	H_{99}	F	P	L	A	E	R	A	N	R	G	F	G	I	V	F_{114}
2D6	R_{101}	P	P	V	P	I	T_{107}	Q	I	L	G	F	G^c	Q	G	V	F_{120}	L_{121}
2E1	R_{100}	G	E	I	P	A	F	–	R	E	F	K	D	K	G	I	I	F_{116}
3A4	T_{103}	N	R_{105}	R	P	F_{108}	G	P	V	G	F	M	K	S	A	I	S_{119}	I_{120}

CYP SRS2

CYP													
1A2	T_{223}	H	E_{225}	F_{226}	V	E	T	A	S	S	G	N	P_{235}
1B1	N_{228}	E	E	F	G	R	T	V	G	A	G	S	L_{240}
2A6	L_{206}	G	I	F_{209}	Q	F	T	S	T	S	T	G	Q_{218}
2B6	T_{203}	Q	T	F_{206}	S	L	I	S	S	V	F	G	Q_{215}
2C8	N_{202}	E	N	F	R	I	L	N	S	P	W_{212}	I_{213}	Q_{214}
2C9	N_{202}	E	N_{204}	I_{205}	K	I	L_{208}	S	S	P	W_{212}	I	Q_{214}
2C19	N_{202}	E	N	I	R	I	V	S	T	P	W	I	Q_{214}
2D6	Q_{210}	E	G	L	K	E	E_{216}	S	G	F	L	R	E_{222}
2E1	N_{204}	E	N	F	Y	L	L	S	T	P	W	L	Q_{216}
3A4	K_{208}	K	L_{210}	L	R_{212}	F_{213}	D	F_{215}	L	D	P	F	F_{220}

CYP SRS4

CYP										
1A2	N_{312}	D	I	F_{315}	G	A	G	F	D_{320}	T_{321}
1B1	T_{325}	D	I	F	G	A	S	Q	D	T_{334}
2A6	L_{296}	N_{297}	L	F	I	G	G	T	E	T_{305}
2B6	L_{293}	S_{294}	L	F	F_{297}	A	G	T	E_{301}	T_{302}
2C8	A_{292}	D	L	F	V	A	G	T	E	T_{301}
2C9	V_{292}	D	L_{294}	F	G_{296}	A_{297}	G	T	E	T_{301}
2C19	A_{292}	D	L	L	G	A	G	T	E	T_{301}
2D6	A_{300}	D_{301}	L	F	S	A	G	M	V	T_{309}
2E1	A_{294}	D	M	F	F	A	G	T	E	T_{303}
3A4	I_{300}	I_{301}	F	I	F_{304}	A_{305}	G	Y	E	T_{309}

CYP SRS5

CYP							
1A2	F_{381}	L	P	F	T_{385}	I_{386}	P_{387}
1B1	F_{394}	V	P	V	T	I	P_{400}
2A6	V_{365}	I	P	M_{368}	S	L	A_{371}
2B6	L_{362}	L_{363}	P	M	G	V_{367}	P_{368}
2C8	L_{361}	V	P	T_{364}	G	V	P_{367}
2C9	L_{361}	L	P_{363}	T	S_{365}	L_{366}	P_{367}
2C19	L_{361}	I	P	T	S	L	P_{367}
2D6	I_{369}	V	P	L	G	V_{374}	T_{375}
2E1	L_{363}	V	P	S	N	L	P_{369}
3A4	I_{369}	A_{370}	–	M_{371}	R_{372}	L	E_{374}

CYP SRS6

CYP				
1A2	G_{496}	L	T_{498}	M_{499}
1B1	G_{508}	L	T	I_{511}
2A6	V_{478}	G	F	A_{481}
2B6	C_{475}	G	V_{477}	G_{478}
2C8	K_{474}	G	I_{476}	V_{477}
2C9	N_{474}	G	F_{476}	A_{477}

Table 5 Continued

2C19	N	G	F	A_{477}
2D6	$\mathbf{F_{481}}$	A	F_{483}	L_{484}
2E1	V_{476}	G	F	G_{479}
3A4	$\mathbf{L_{480}}$	G	$\mathbf{L_{482}}$	L_{483}

[a] Residues shown in **bold** have been reported by x-ray crystallographic and/or site-directed mutagenesis studies to have a relevance for substrate binding and/or selectivity.
[b] SRS3 is not thought to play a major role in substrate binding and orientation for metabolism.
[c] An additional tripeptide (PRS) is present at this point.

Table 6 Statistical learning techniques used to explore cytochromes P450[a]

Cytochrome P450	Support vector machine	Recursive partitioning (decision tree)	Bayesian	Logistic Regression	k-nearest neighbours	Partial least squares	Multiple linear regression	Combination
CYP1A2 inhibition[64]	–	X	X	–	–	X	X	√ (consensus)
CYP2D6 inhibition[65]	–	X	X	–	–	–	–	√ (combination, non-consensus)
CYP3A4 inhibition[128]	√	√	X	X	X	–	–	–
CYP3A4 inhibition[66]	√	–	–	–	–	–	–	–

[a] –, method not used; X, method used; √, method giving best results for this study.

Electronic structure calculations using molecular orbital (MO) methods may often reveal useful information relating to preferred sites of metabolism and the rate or clearance data for cytochrome P450 mediated oxidations frequently correlate with ionization energy. However, other approaches can also yield important information about the energetics of the metabolism process[54] and lead to a prediction of the metabolic route involved.[55] Neural network analyses are proving useful in this respect, as has been shown for CYP3A4 substrates and N-dealkylation reactions.[56,57] Furthermore, the regioselectivity of metabolism by cytochrome P450 enzymes can be explored using MO methods[58] or via a combination of molecular modeling techniques, as has been reported recently for CYP2B6.[39] In fact, the whole area of cytochrome P450 enzyme–substrate selectivity may be explored via specific QSAR approaches,[59] such as that afforded by pharmacophore modeling using such systems as Catalyst,[49,60,61] where even the overlapping specificities (due to high similarities in the substrate-binding site) of CYP3A4, CYP3A5, and CYP3A7 can be examined.[62] The field of cytochrome P450 modeling has been reviewed recently,[34] and the interested reader is referred to this review for further information.

5.34.2.6   Cytochromes P450: Statistical Learning Methods

Recent advances in computer sciences, combined with larger data sets for cytochromes P450 due to high-throughput screening techniques, have led recently to a large number of statistical approaches to predict the involvement of specific cytochromes P450 in the metabolism of compounds and inhibition of cytochromes P450 by specific inhibitors. Compared with QSAR techniques, where a descriptive relationship is sought to describe the inhibition or metabolism or specific energy terms describing these processes,[63] statistical learning methods (or machine-learning algorithms) take a large number of descriptors related to structure, etc., and derive models that can be used to predict new compounds but are not necessarily 'human readable.' It is not the intention of this chapter to describe these computational techniques in detail, as this is covered in Volume 4. **Table 6** shows that the approaches have been applied successfully to the inhibition of specific cytochromes P450 such as CYP1A2,[64] CYP2D6,[65] and CYP3A4.[32,66]

5.34.3   Modeling of Phase 2 Metabolism Enzymes

Whereas Phase 1 metabolism involves hydroxylation, oxidation, and reduction pathways, phase 2 metabolism involves primarily conjugation (of the phase 1 metabolites) with a variety of groups (e.g., sulfation, glucuronidation, glutathione conjugation, acetylation, amino acid conjugation, and methylation).[67] With the exceptions of methylation and acetylation, phase 2 biotransformation reactions result in a large increase in hydrophilicity.[68] Enzymes involved in phase 2 metabolism include, GST, MT, NAT, SULT, and UGT.[67,68] Modeling of phase 2 metabolism has not received as much attention as the modeling of phase 1 metabolism for various reasons, as described below. A short overview of 3D modeling (homology modeling, pharmacophore modeling, and 3D-QSAR) will be presented in the following sections for the main enzymes involved in phase 2 metabolism.

5.34.3.1   Glutathione *S*-Transferases

GSTs catalyze the nucleophilic attack of the thiol of the tripeptide glutathione (GSH) on electrophilic substrates, a reaction usually resulting in addition or substitution, depending on the nature of the substrate.[68] Many diverse compounds, including toxic xenobiotics and reactive products of intracellular processes such as lipid oxidation, act as GST substrates.[69] The primary function of GST isoenzymes is generally considered to be detoxification of both endogenous and xenobiotic compounds.[70,71] However, GSTs can also lead to the formation of more reactive intermediates, either directly or following conversion of glutathione conjugates by other biotransformation enzymes. In some cases the glutathione conjugates formed are labile, and dissociate to the parent electrophile in other tissues.[71] Substrates for GSTs share three common features: they are hydrophobic, contain an electrophilic atom, and also react nonenzymatically with GSH.[68]

Mammalian cytosolic GST isoenzymes have been grouped in eight families, alpha, mu, pi, theta, kappa, sigma, zeta, and omega,[68,72–75] which are distinguished by molecular masses, isoelectric points, and other properties. The primary sequence homology between subunits of the same family is >70%, while interfamily subunit homologies are approximately 30–40%.[68,70,76,77] In humans, multiple members of each subfamily class are now known.[68] Cytosolic GSTs are found as dimeric proteins comprised of two subunits.[78] Each isoenzyme subunit contains an active site consisting of two binding sites, one for the cofactor GSH (G-site) and one for the electrophilic substrate (H-site).[78,79] Furthermore, one nonsubstrate-binding site (binding compounds that are not conjugated but transported by GSTs) is present in the dimer.

Comparison of cytosolic GST crystal structures reveals that all cytosolic GST isoenzyme subunits share a number of general structural features, though they differ considerably in detail. The tertiary structure and a large part of the secondary structure are generally maintained between cytosolic GSTs from different families and different species.

Microsomal GSTs are membrane bound, and, in spite of the fact that their catalytic function is similar to soluble cytosolic GSTs, they share no sequence identity with the latter GSTs, as was shown by a two-dimensional projection structure of a microsomal membrane-bound GST determined using electron crystallography.[80] Microsomal GSTs occur as a trimer, which forms a pore-like structure with six α-helices delineating a region with the local symmetry axis in the center, suggesting a membrane transport function for these GSTs.[80] This indicates that for comparable catalytic functions of microsomal and cytosolic GSTs, structurally different enzymes exist.[80]

The first models for a small-molecule bound to a GST isoenzyme suggested both hydrophobic interactions and hypothetical hydrogen bonds to polar groups in the active site to be important for binding and orientation of (α-bromoisovaleryl)urea for conjugation.[81]

Another small-molecule model was derived for rat GST 4-4 (rGST M2-2; nomenclature adapted from Mannervik *et al.*[82] – the first letter indicates the species (h, human; m, mouse; p, pig; r, rat), the capital following 'GST' indicates the class, and the numbers indicate the specific GST), as a surrogate for the polymorphic hGST M1-1.[83] The derived substrate model incorporated information on regio- and stereoselective product formation of 20 substrates covering three chemically and structurally different classes (aromatic diol epoxides, aromatic chlorides, pyrene oxides, and (aza)phenanthrene oxides). The model highlighted three interaction sites responsible for Lewis acid–Lewis base interactions as well as a region responsible for aromatic interactions.[83] This substrate molecule model successfully predicted the conjugation to GSH of 11 substrates of GST 4-4 (representing three classes of compounds) that were not used to construct the model.[83] This small-molecule model was subsequently improved with another series of substrates, which extend beyond the original model in one specific area.[84] Based on these data, additional steric restrictions imposed by the protein in this region have been incorporated into the original small-molecule model.[83]

A small-molecule model based upon GSH analogs was constructed for hGST P1-1, which is expressed at elevated levels in tumor cells. A large number of GSH analogs were used to construct a pharmacophore for the G-site of GST P1-1, overlapping charged and hydrophobic centers.[85] The pharmacophore was consequently docked into the crystal structure of GST P1-1,[86] which indicated complementarity of all the major motifs.[85]

In the 1990s, when crystal structures were not available for all GSTs, several homology models were constructed. As there was generally at least one crystal structure available for each GST family, alignment between amino acid

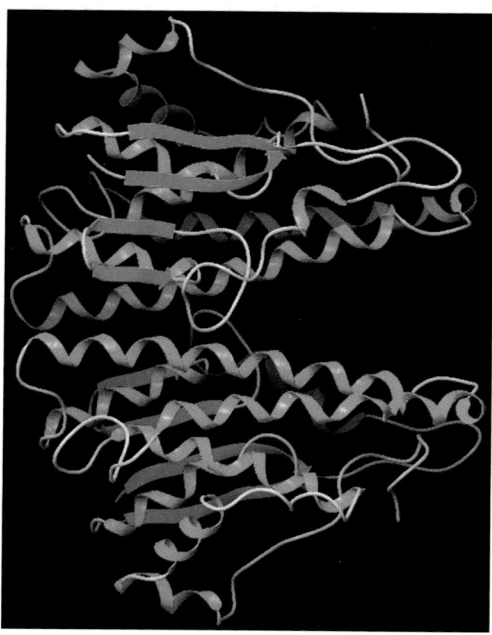

Figure 2 Homology model of cytosolic rGST M2-2 showing the dimeric (top/bottom) character of the enzyme.[89]

sequences of the crystal structure and a GST to be modeled was relatively easy when using a GST from the same family as a template for the model, due to the high homology. Furthermore, as the overall tertiary and secondary structures are largely maintained in all GSTs, the quality of the resulting alignments was generally excellent.

Cachau et al.,[87] Hsiao et al.,[88] and De Groot et al.,[89] used the crystal structure of rat mu class GST 3-3 (rGST M1-1)[90] as the template for the construction of homology models for, respectively, human mu class GSTs M1b-1b, M2-2, and M3-3,[87] chicken theta class GST,[88] and rat mu class GST 4-4 (rGST M2-2, shown in **Figure 2**).[89] All models closely resembled the crystal structures upon which the models were based. The modeled GST M2-2 structure[87] agreed remarkably well with the crystallographic structure of human muscle GST M2-2.[91] The orientation of four GSH conjugates docked into the GST 4-4 model appeared to be similar to the orientation of these substrates in a small-molecule model for GST 4-4.[83] The homology model of GST 4-4 (combined with the small-molecule model[83]) was used successfully to identify amino acids that could be involved in binding or activation of substrates in the active site.[89]

Several homology models have been generated for the GST theta family, for example a model for hGST T1,[92] hGST T2,[93] and for the mouse isoforms mGST T1, mGST T2, and mGST T3.[94] The model for hGST T2[93] was used to consolidate the role of specific amino acids in the active site. The mouse models used the crystal structures of hGST T2 as a template, and were used to compare the active sites and explain the inability of mGST T1 and mGST T3 to metabolize 1-menaphthyl sulfate.[94]

The crystal structure of hGST A1-1 has been used to generate homology models for rat GSTs of the alpha family, rGST A3 and rGST A5.[95] The resulting models were again used to investigate the importance of specific amino acids in the active site for activity toward conjugation of Aflatoxin B_1, supported by site-directed mutagenesis.[95]

GSTs from the pi family have so far only been used for docking studies (hGST P1a, hGST P1b, and hGST P1c)[96] and docking algorithm evaluation.[97]

Due to the number of crystal structures currently available for GSTs from all families, pharmacophore and homology modeling for these enzymes is not receiving as much attention as that for the cytochromes P450, compared to the efforts expended in the 1990s.

5.34.3.2 Methyltransferases

Methylation is a common but minor pathway for phase 2 metabolism of xenobiotics, and is catalyzed by various enzymes, such as thioether methyltransferase, phenyl O-methyltransferase, catechol O-methyltransferase (COMT), phenylethanolamine N-methyltransferase (PNMT), and histamine N-methyltransferase (HNMT).[68] Several modeling studies have been performed on MTs involved in endogenous pathways (e.g., DNA and RNA methylation), as well as studies in various plants. However, modeling of the human counterparts involved in drug metabolism has received relatively little attention.

A homology model of human HNMT was developed, and used to determine the location of the amino acid linked to polymorphism for this enzyme. This model supports experimental observations with regards to protein stability.[98] A set of homology models for rabbit and human indolethylamine N-methyltransferases was obtained, and used successfully to explain the observed species differences and polymorphisms.[99] These homology models were based on the rat crystal structure of COMT.[100] This crystal structure was also used for a docking study that, in combination with experiments using the human enzyme, resulted in various suggestions for mechanisms controlling the methylation by human COMT.[101]

A series of comparative molecular field analysis (CoMFA) studies on PNMT and the α_2-adrenoceptor was performed in order to investigate the factors that determine potency and selectivity for a series of PNMT inhibitors.[102,103] The results of the CoMFA studies were similar to those of previous QSAR analysis.[102] COMT inhibitors are used in the treatment of Parkinson's disease, but their application is limited due to adverse side effects. Docking in the rat crystal structure, CoMFA analysis, and GRID/GOLPE models were used to examine a series of COMT inhibitors.[104] The three models were in agreement with each other, and suggested that an interaction between the catechol oxygen atoms and the magnesium ion in COMT is important, while several hydrogen-bonding and hydrophobic contacts influence inhibitor binding.[104]

5.34.3.3   *N*-Acetyltransferases

N-Acetylation is a major route of metabolism for compounds containing an aromatic amine or hydrazine group, which are converted to aromatic amides and hydrazides, respectively.[68] The reaction is catalyzed by NATs, which are found in most mammals, and require the cofactor acetyl-CoA.[68] Humans, rabbits, and hamsters express only two NATs (NAT1 and NAT2), whereas mice express a third (NAT3).[68] NAT1 and NAT2 have different but overlapping substrate specificities, and both show genetic polymorphisms.[68] Homology models have been constructed for both human NAT1[105] and NAT2,[106] based on the crystal structure of NAT from *Salmonella typhimurium* (*St*NAT). The model for NAT1 was used to predict that human NATs have adopted a common catalytic mechanism from cysteine proteases to accommodate acetyl transfer reactions,[105] while the NAT2 model was used to explain the polymorphism of this enzyme.[106] The crystal structures of *S. typhimurium* and *Mycobacterium smegmatis* NATs were used to generate a homology model of *Mycobacterium tuberculosis* NAT (*Mt*NAT), and were used for specific inhibitor design against *Mt*NAT (which is responsible for resistance against the antitubercular drug isoniazid).[107]

5.34.3.4   Sulfotransferases

Sulfate conjugation generally produces a highly water-soluble sulfuric acid ester. The reaction is catalyzed by SULTs, a large gene family of soluble (cytosolic) enzymes, and involves transfer of a sulfonate (SO_3^-). The required cofactor for the reaction is 3'-phosphoadenosine 5'-phosphosulfate (PAPS).[68,108] Although the most common substrates for this enzyme are phenols and aliphatic alcohols, metabolism is not limited to those classes.[68] In mammals, sulfation is involved in the detoxification of therapeutic, dietary, and environmental xenobiotics, and contributes to the homeostasis and regulation of numerous biologically active endogenous chemicals such as steroids, iodothyronines, bile acids, and neurotransmitters.[109,110] In addition, for a large number of procarcinogens sulfation is the terminal step in the bioactivation pathway, and is necessary to reveal their mutagenic/carcinogenic activity. Multiple SULTs have been identified in all mammalian species, and are members of five gene families (SULT1 to SULT5).[68] Crystal structures are now available for seven human SULTs: the cytosolic SULT1A1, SULT1A3, SULT1E1, SULT2A1, SULT2B1A, and SULT2B1B and the Golgi-resident NDST-1.[108] Docking can be used to predict binding modes within these crystal structures, as was done for dopamine in SULT1A3.[111] The PAPS binding pocket is highly conserved, while the substrate-binding site of each SULT reflects the differences in substrate specificity in these crystal structures: the cytosolic SULTs typically recognize small hydrophobic molecules, whereas the Golgi-resident SULTs recognize hydrophilic carbohydrates and tyrosine residues in peptides.[108] The design of SULT inhibitors has focused on exploring environmental toxins and dietary agents, rather than drug development.[108] The crystal structure of mouse Sult1e1 has been used to build a homology model of human SULT1E1 before the human crystal structure was solved, to identify the potential (allosteric) binding pocket using a series of hydroxylated polychlorinated biphenyls.[112]

Similarly, a model of rat aryl sulfotransferase IV (AST IV) was constructed, based on mouse Sult1e1.[113] Docking of a series of N-hydroxyarylamines in the homology model of AST IV highlighted specific steric constraints within the active site, which could explain the change from substrate to competitive inhibitor within this series of compounds.[113] CoMFA has been performed on K_m values of a series of 35 substrates of rat AST IV, and could explain the activities of six more substrates.[114] The CoMFA results[114] showed a good fit with the predicted active site in the homology model of AST IV.[113]

Four CoMFA models were derived from the K_m values for 95 substrates of SULT1A3. The different models used subsets of the data set or the entire data set and different molecular alignment rules.[115] All four models were statistically significant, and highlighted factors affecting binding in the SULT1A3 binding site.

Recently, the crystal structures of numerous human SULTs have been obtained by the Structural Genomics Consortium.[116]

5.34.3.5   Uridine Diphosphate (UDP)-Glucuronosyltransferases

Glucuronidation is a major pathway of xenobiotic biotransformation in most mammalian species, and requires the cofactor uridine diphosphate-glucuronic acid.[68,117] The reaction is metabolized by UGTs (also called glucuronyl-transferases), which are present in many tissues.[68,117] The site of glucuronidation is generally an electron-rich nucleophilic heteroatom (oxygen, nitrogen, or sulfur).[68] Human UGTs are a family of enzymes that detoxify many hundreds of compounds by their conjugation to glucuronic acid, rendering them harmless, more water-soluble, and, hence, excretable. Genetic inheritance, age, and environmental factors largely determine the different profiles of the inducible hepatic UGTs. Variation in the complement of these UGTs may result in dramatic differences in the safe elimination of toxic metabolites.

UGTs are classified into two distinct gene families: UGT1 and UGT2, the latter showing genetic polymorphisms.[68] In humans, up to 16 different functional UGT isoforms belonging to subfamilies 1A and 2B have been characterized, showing that these isoforms possess distinct but frequently overlapping substrate specificities.[117] Certain unexpected adverse reactions and decreased efficiency of drug therapy may be due to interindividual variations in the expression of UGTs.

Human UGTs have been studied extensively using pharmacophore and 3D-QSMR approaches.[69,118] Pharmacophores for UGT1A1,[69,119–121] UGT1A3,[121] UGT1A4,[118,119,121,122] UGT1A6,[121] UGT1A7,[121] UGT1A8,[121] UGT1A9,[69,121] UGT1A10,[121] UGT2B4,[121] UGT2B7,[121] UGT2B15,[121] and UGT2B17[121] have been reported. Pharmacophore modeling of UGTs is hampered by the ability of UGT isoforms to accommodate multiple substrate-binding modes, and requires the construction of a 'glucuronidation feature' for use in the computational algorithm.[69,122]

Substrates of UGT1A1[120] and UGT1A4[122] share two key hydrophobic regions 3 Å and 6–7 Å from the site of glucuronidation.[119,120,122] These results were confirmed by pharmacophore-based 3D-QSAR and molecular field-based 3D-QSAR (self-organizing molecular field analysis) methods.[118,120,122] An aromatic ring attached to the nucleophilic group was found to increase the likelihood of glucuronidation by UGT1A6, UGT1A7, and UGT1A9.[121] The pharmacophore of UGT1A9 consists of two hydrophobic regions at similar distances to the site of glucuronidation, as for UGT1A1 and UGT1A4,[119,120,122] but also contains a hydrogen bond acceptor feature close to the most distant hydrophobic feature.[118] A large hydrophobic region close to the site of glucuronidation and a hydrogen bond acceptor 10 Å from the site of glucuronidation were common in most UGT2B4 substrates.

5.34.4   Conclusions

It is possible to make use of a considerable amount of accumulated data from experimental determinations and theoretical calculations to understand more about the way in which the selectivity of drug substrates for drug-metabolizing enzymes is encoded in their molecular structures. These features are likely to be complementary to the binding sites themselves, and, therefore, one can make predictions about the likely metabolic fate of new compounds (NCEs) from the currently available knowledge in this area (either using the 3D structures of the enzymes themselves or the accumulated features of known substrates for model building), thus leading to improved lead times for NCE development via high-throughput screening procedures.

So far, most attention has been paid to the prediction of cytochrome P450- and GST-catalyzed metabolism. As we start to understand how to design out metabolism by, for example, cytochromes P450, other metabolic pathways will take over, which will drive the need to understand these other pathways more thoroughly. With the drive to design out drug–drug interactions in an earlier phase of drug development, modeling of other enzymes involved in metabolism will therefore become more and more important. This is, of course, limited by the availability of experimental data to base these models on, like the in vivo or in vitro data described elsewhere in this volume. Combining in silico methods with in vitro screening (as described in 5.38 Mechanism-Based Pharmacokinetic–Pharmacodynamic Modeling for the Prediction of In Vivo Drug Concentration–Effect Relationships – Application in Drug Candidate Selection and Lead Optimization) will be a very powerful tool in the near future.

Acknowledgments

The financial support of GlaxoSmithKline Research and Development Limited, Merck, Sharp & Dohme Limited, British Technology Group Limited, the Kyushu Institute of Technology, and the University of Surrey Foundation Fund is gratefully acknowledged by DFVL.

References

1. Woolf, T. F. *Handbook of Drug Metabolism*; Marcel Dekker: New York, NY, 1999.
2. Coleman, M. *Human Drug Metabolism: An Introduction*; Wiley: Chichester, UK, 2005.
3. Ortiz de Montellano, P. R. *Cytochrome P-450: Structure, Mechanism, and Biochemistry*; Plenum Press: New York, NY, 1995.
4. Lu, A. Y. H.; Lee, J. S.; Obach, R. S.; Fisher, M. B. *Drug Metabolizing Enzymes: Cytochrome P450 and Other Enzymes in Drug Discovery and Development*; Marcel Dekker: New York, 1999.
5. Kemp, C. A.; Maréchal, J.-D.; Sutcliffe, M. J. *Arch. Biochem. Biophys.* **2005**, *433*, 361–368.
6. Cosme, J.; Johnson, E. F. *J. Biol. Chem.* **2000**, *275*, 2545–2553.
7. Nelson, D. R.; Koymans, L.; Kamataki, T.; Stegeman, J. J.; Feyereisen, R.; Waxman, D. J.; Waterman, M. R.; Gotoh, O.; Coon, M. J.; Estabrook, R. W. et al. *Pharmacogenetics* **1996**, *6*, 1–42.
8. Smith, D. A.; Jones, B. C. *Biochem. Pharmacol.* **1992**, *44*, 2089–2098.
9. Sullivan-Klose, T. H.; Ghanayem, B. I.; Bell, D. A.; Zhang, Z. Y.; Kaminsky, L. S.; Shenfield, G. M.; Miners, J. O.; Birkett, D. J.; Goldstein, J. A. *Pharmacogenetics* **1996**, *6*, 341–349.
10. Aithal, G. P.; Day, C. P.; Kesteven, P. J.; Daly, A. K. *Lancet* **1999**, *353*, 717–719.
11. Kidd, R. S.; Curry, T. B.; Gallagher, S.; Edeki, T.; Blaisdell, J.; Goldstein, J. A. *Pharmacogenetics* **2001**, *11*, 803–808.
12. Dayer, P.; Gasser, R.; Gut, J.; Kronbach, T.; Robertz, G. M.; Eichelbaum, M.; Meyer, U. A. *Biochem. Biophys. Res. Commun.* **1984**, *125*, 374–380.
13. Meyer, U. A.; Skoda, R. C.; Zanger, U. M. *Pharmacol. Ther.* **1990**, *46*, 297–308.
14. Daly, A. K.; Leahurt, J. B.; London, S. J.; Idle, J. R. *Human Genetics* **1995**, *95*, 337–341.
15. Bjornsson, T. D.; Callaghan, J. T.; Einolf, H. J.; Fischer, V.; Gan, L.; Grimm, S.; Kao, J.; King, S. P.; Miwa, G.; Ni, L. et al. *Drug Metab. Dispos.* **2003**, *31*, 815–832.
16. Rendic, S. *Drug Metab. Rev.* **2002**, *34*, 83–448.
17. Johnson, E. F. *Drug Metab. Disp.* **2003**, *31*, 1532–1540.
18. Scott, E. E.; He, Y. A.; Wester, M. R.; White, M. A.; Chin, C. C.; Halpert, J. R.; Johnson, E. F.; Stout, C. D. *Proc. Natl. Acad. Sci. USA* **2003**, *100*, 13196–13201.
19. Williams, P. A.; Cosme, J.; Sridhar, V.; Johnson, E. F.; McRee, D. E. *Mol. Cell* **2000**, *5*, 121–131.
20. Wester, M. R.; Johnson, E. F.; Marques-Soares, C.; Dansette, P. M.; Mansuy, D.; Stout, C. D. *Biochemistry* **2003**, *42*, 6370–6379.
21. Wester, M. R.; Johnson, E. F.; Marques-Soares, C.; Dijols, S.; Dansette, P. M.; Mansuy, D.; Stout, C. D. *Biochemistry* **2003**, *42*, 9335–9345.
22. Schoch, G. A.; Yano, J. K.; Wester, M. R.; Griffin, K. J.; Stout, C. D.; Johnson, E. F. *J. Biol. Chem.* **2004**, *279*, 9497–9503.
23. Williams, P. A.; Cosme, J.; Ward, A.; Angrove, H. C.; Vinkovic, D. M.; Jhoti, H. *Nature* **2003**, *424*, 464–468.
24. Wester, M. R.; Yano, J. K.; Schoch, G. A.; Yang, C.; Griffin, K. J.; Stout, C. D.; Johnson, E. F. *J. Biol. Chem.* **2004**, *279*, 35630–35637.
25. Rowland, P.; Blaney, F. E.; Smyth, M. G.; Jones, J. J.; Leydon, V. R.; Oxbrow, A. K.; Lewis, C. J.; Tennant, M. G.; Modi, S.; Eggleston, D. S. et al. *J. Biol. Chem.* **2006**, *281*, 7614–7622.
26. Yano, J. K.; Wester, M. R.; Schoch, G. A.; Griffin, K. J.; Stout, C. D.; Johnson, E. F. *J. Biol. Chem.* **2004**, *279*, 38091–38104.
27. Williams, P. A.; Cosme, J.; Vinkovic, D. M.; Ward, A.; Angove, H. C.; Day, P. J.; Vonrhein, C.; Tickle, I. J.; Jhoti, H. *Science* **2004**, *305*, 683–686.
28. Modi, S. *Drug Disc. Today* **2003**, *8*, 621–623.
29. Modi, S. *Drug Disc. Today* **2004**, *9*, 14–15.
30. Li, H.; Poulos, T. L. *Curr. Topics Med. Chem.* **2004**, *4*, 1789–1802.
31. De Groot, M. J.; Alex, A. A.; Jones, B. C. J. *Med. Chem.* **2002**, *45*, 1983–1993.
32. Afzelius, L.; Zamora, I.; Masimirembwa, C. M.; Karlén, A.; Andersson, T. B.; Mecucci, S.; Baroni, M.; Cruciani, G. *J. Med. Chem.* **2004**, *47*, 907–914.
33. Afzelius, L.; Raubacher, F.; Karlén, A.; Steen, F.; Jorgensen, S.; Andersson, T. B.; Masimirembwa, C. M.; Zamora, I. *Drug Met. Disp.* **2004**, *32*, 1218–1229.
34. De Groot, M. J.; Kirton, S. B.; Sutcliffe, M. J. *Curr. Topics Med. Chem.* **2004**, *4*, 1803–1824.
35. Shou, M.; Grogan, J.; Mancewicz, J. A.; Krausz, K. W.; Gonzalez, F. J.; Gelboin, H. V.; Korzekwa, K. R. *Biochemistry* **1994**, *33*, 6450–6455.
36. Korzekwa, K. R.; Krishnamachary, N.; Shou, M.; Ogai, A.; Parise, R. A.; Rettie, A. E.; Gonzalez, F. J.; Tracy, T. S. *Biochemistry* **1998**, *37*, 4137–4147.
37. Hosea, N. A.; Miller, G. P.; Guengerich, F. P. *Biochemistry* **2000**, *39*, 5929–5939.
38. Dabrowski, M. J.; Schrag, M. L.; Wienkers, L. C.; Atkins, W. M. *J. Am. Chem. Soc.* **2002**, *124*, 11866–11867.
39. Bathelt, C.; Schmid, R. D.; Pleiss, J. *J. Mol. Model.* **2002**, *8*, 327–335.
40. Kirton, S. B.; Baxter, C. A.; Sutcliffe, M. J. *Adv. Drug Devel. Rev.* **2002**, *54*, 385–406.
41. Modi, S.; Paine, M. J.; Sutcliffe, M. J.; Lian, L. Y.; Primrose, W. U.; Wolf, C. R.; Roberts, G. C. K. *Biochemistry* **1996**, *35*, 4540–4550.
42. Modi, S.; Sutcliffe, M. J.; Primrose, W. U.; Lian, L.-Y.; Roberts, G. C. K. *Nature Struct. Biol.* **1996**, *3*, 414–417.
43. Laskowski, R. A.; McArthur, M. W.; Moss, D. S.; Thornton, J. M. *J. Appl. Crystallogr.* **1993**, *26*, 283–291.
44. Laskowski, R. A.; Moss, D. S.; Thornton, J. M. *J. Mol. Biol.* **1993**, *231*, 1049–1067.
45. Rodriguez, R.; Chinea, G.; Lopez, N.; Pons, T.; Vriend, G. *Bioinformatics* **1998**, *14*, 523–528.

46. Morris, G. M.; Goodsell, D. S.; Huey, R.; Olson, A. J. *AutoDock. Automated Docking of Flexible Ligands to Receptors. Version 2.4.*; Scripps Research Institute: La Jolla, CA, 1996.

47. Morris, G. M.; Goodsell, D. S.; Halliday, R. S.; Huey, R.; Hart, W. E.; Belew, R. K.; Olson, A. J. *J. Comp. Chem.* **1998**, *19*, 1639–1662.

48. Bauer, C.; Osman, A. M.; Cercignani, G.; Gialluca, N.; Paolini, M. *Biochem. Pharmacol.* **2001**, *61*, 1049–1055.

49. De Groot, M. J.; Ekins, S. *Adv. Drug Devel. Rev.* **2002**, *54*, 367–383.

50. Lewis, D. F. V.; Jacobs, M. N.; Dickins, M. *Drug Disc. Today* **2004**, *9*, 530–537.

51. Lewis, D. F. V. *Curr. Drug Metab.* **2003**, *4*, 331–340.

52. Lewis, D. F. V. Unpublished results.

53. Gotoh, O. *J. Biol. Chem.* **1992**, *267*, 83–90.

54. Park, J.-Y.; Harris, D. *J. Med. Chem.* **2003**, *46*, 1645–1660.

55. Singh, S. B.; Shen, L. Q.; Walker, M. J.; Sheridan, R. P. *J. Med. Chem.* **2003**, *46*, 1330–1336.

56. Balakin, K. V.; Ekins, S.; Bugrim, A.; Ivanenkov, Y. A.; Korolev, D.; Nikolsky, Y. V.; Skorenko, A. V.; Ivashchenko, A. A.; Savchuk, N. P.; Nikoskaya, T. *Drug Metab. Disp.* **2004**, *32*, 1183–1189.

57. Balakin, K. V.; Ekins, S.; Bugrim, A.; Ivanenkov, Y. A.; Korolev, D.; Nikolsky, Y. V.; Ivashchenko, A. A.; Savchuk, N. P.; Nikoskaya, T. *Drug Metab. Disp.* **2004**, *32*, 1111–1120.

58. Lightfoot, T.; Ellis, S. W.; Mahling, J.; Ackland, M. J.; Blaney, F. E.; Bijloo, G. J.; De Groot, M. J.; Vermeulen, N. P. E.; Blackburn, G. M.; Lennard, M. S. et al. *Xenobiotica* **2000**, *30*, 219–233.

59. Ekins, S.; Wrighton, S. A. *J. Pharmacol. Toxicol. Methods* **2001**, *45*, 65–69.

60. Ekins, S.; Waller, C. L.; Swaan, P. W.; Cruciani, G.; Wrighton, S. A.; Wikel, J. H. *J. Pharmacol. Toxicol. Methods* **2001**, *44*, 251–272.

61. Ekins, S.; De Groot, M. J.; Jones, J. P. *Drug Metab. Dispos.* **2001**, *29*, 936–944.

62. Ekins, S.; Stresser, D. M.; Williams, J. A. *Trends Pharm. Sci.* **2003**, *24*, 161–166.

63. Ekins, S.; Andreyev, S.; Ryabov, A.; Kirillov, E.; Rakhmatulin, E. A.; Bugrim, A.; Nikolskaya, T. *Exp. Opin. Drug Metabol. Toxicol.* **2005**, *1*, 303–324.

64. Chohan, K. K.; Paine, S. W.; Mistry, J.; Barton, P.; Davis, A. M. *J. Med. Chem.* **2005**, *48*, 5154–5161.

65. O'Brien, S. E.; De Groot, M. J. *J. Med. Chem.* **2005**, *48*, 1287–1291.

66. Kriegl, J. M.; Arnhold, T.; Beck, B.; Fox, T. *J. Comput.-Aided Mol. Design* **2005**, *19*, 189–201.

67. Timbrell, J. A. *Factors Affecting Toxic Responses: Metabolism*; Taylor & Francis: London, UK, 1991.

68. Parkinson, A. *Biotransformation of Xenobiotics*; McGraw-Hill: New York, NY, 2001.

69. Miners, J. O.; Smith, P. A.; Sorich, M. J.; McKinnon, R. A.; Mackenzie, P. *J. Annu. Rev. Pharmacol. Toxicol.* **2004**, *44*, 1–25.

70. Armstrong, R. N. *Chem. Res. Toxicol.* **1991**, *4*, 131–140.

71. Commandeur, J. N. M.; Stijntjes, G. J.; Vermeulen, N. P. E. *Pharmacol. Rev.* **1995**, *47*, 271–330.

72. Board, P. G.; Coggan, M.; Chelvanayagam, G.; Easteal, S.; Jermiin, L. S.; Schulte, G. K.; Danley, D. E.; Hoth, L. R.; Griffor, M. C.; Kamath, A. V. *J. Biol. Chem.* **2000**, *275*, 24798–24806.

73. Board, P. G.; Coggan, M.; Chelvanayagam, G.; Jermiin, L. S. *Biochem. J.* **1997**, *328*, 929–935.

74. Mannervik, B.; Widerstern, M. *Human Glutathione Transferases: Classification, Tissue Distribution, Structure and Functional Properties*; European Commission: Brussels, Belgium, 1995.

75. Hayes, J. D.; Pulford, D. *J. Crit. Rev. Biochem. Mol. Biol.* **1995**, *30*, 445–600.

76. Sinning, I.; Kleynwegt, G. J.; Cowan, S. W.; Reinemer, P.; Dirr, H. W.; Huber, R.; Gilliland, G. L.; Armstrong, R. N.; Ji, X.; Board, P. G. et al. *J. Mol. Biol.* **1993**, *232*, 192–212.

77. Wilce, M. C. J.; Board, P. G.; Feil, S. C.; Parker, M. W. *EMBO J.* **1995**, *14*, 2133–2143.

78. Danielson, U. H.; Mannervik, B. *Biochem. J.* **1985**, *231*, 263–267.

79. Mannervik, B.; Guthenberg, C.; Jakobson, I.; Warholm, M. *Glutathione Conjugation: Reaction Mechanism of Glutathione S-Transferase A*; Elsevier/North-Holland Biomedical Press: Amsterdam, 1978.

80. Hebert, H.; Schmidt-Krey, I.; Morgenstern, R. *EMBO J.* **1995**, *14*, 3864–3869.

81. te Koppele, J. M.; Esajas, S. W.; Brussee, J.; Van der Gen, A.; Mulder, G. *J. Biochem. Pharmacol.* **1988**, *37*, 29–35.

82. Mannervik, M.; Awasthi, Y. C.; Board, P. G.; Hayes, J. D.; De Ilio, C.; Ketterer, B.; Listowsky, I.; Morgenstern, R.; Muramatsu, M.; Pearson, W. R. et al. *Biochem. Lett.* **1992**, *282*, 305–306.

83. De Groot, M. J.; Van der Aar, E. M.; Nieuwenhuizen, P. J.; Van der Plas, R. M.; Donné-Op den Kelder, G. M.; Commandeur, J. N. M.; Vermeulen, N. P. E. *Chem. Res. Toxicol.* **1995**, *8*, 649–658.

84. van der Aar, E. M.; De Groot, M. J.; Bouwman, T.; Bijloo, G. J.; Commandeur, J. N. M.; Vermeulen, N. P. E. *Chem. Res. Toxicol.* **1997**, *10*, 439–449.

85. Kauvar, L. M. *GST-Targeted Drug Candidates*; Taylor & Francis: London, UK, 1996.

86. Reinemer, P.; Dirr, H. W.; Ladenstein, R.; Huber, R.; Lo Bello, M.; Federici, G.; Parker, M. W. *J. Mol. Biol.* **1992**, *227*, 214–226.

87. Cachau, R. E.; Erickson, J. W.; Villar, H. O. *Protein Eng.* **1994**, *7*, 831–839.

88. Hsiao, C. D.; Martsen, E. O.; Lee, J. Y.; Tam, M. F. *Biochem. J.* **1995**, *312*, 91–98.

89. De Groot, M. J.; Vermeulen, N. P. E.; Mullenders, D. L. J.; Donné-Op den Kelder, G. M. *Chem. Res. Toxicol.* **1996**, *9*, 28–40.

90. Ji, X.; Zhang, P.; Armstrong, R. N.; Gilliland, G. L. *Biochemistry* **1992**, *31*, 10169–10184.

91. Raghunathan, S.; Chandross, R. J.; Kretsinger, R. H.; Allison, T. J.; Penington, C. J.; Rule, G. S. *J. Mol. Biol.* **1994**, *238*, 815–832.

92. Flanagan, J. U.; Rossjohn, J.; Parker, M. W.; Board, P. G.; Chelvanayagam, G. *Proteins Struct. Funct. Genet.* **1998**, *33*, 444–454.

93. Chelvanayagam, G.; Wilce, M. C. J.; Parker, M. W.; Tan, K. L.; Board, P. G. *Proteins Struct. Funct. Genet.* **1997**, *27*, 118–130.

94. Coggan, M.; Flanagan, J. U.; Parker, M. W.; Vichai, V.; Pearson, W. R.; Board, P. G. *Biochem. J.* **2002**, *366*, 323–332.

95. McDonagh, P. D.; Judah, D. J.; Hayes, J. D.; Lian, L.-Y.; Neal, G. E.; Wolf, C. R.; Roberts, G. C. K. *Biochem. J.* **1999**, *339*, 95–101.

96. Buolamwini, J. K.; Akande, O.; Antoun, G.; Ali-Osman, F. Docking study of 1-chloro-2,4-dinitrobenzene (CDNB) binding at the putative H-site of human glutathione-S-transferase pi (GST-π) polymorphic proteins. Proceedings of the 214th ACS National Meeting, 1997. American Chemical Society: Las Vegas, NV.

97. Zavodsky, M. I.; Sanschagrin, P. C.; Korde, R. S.; Kuhn, L. A. *J. Comput.-Aided Mol. Design* **2002**, *6*, 883–902.

98. Pang, Y.-P.; Zheng, X.-E.; Weinshilboum, R. M. *Biochem. Biophys. Res. Commun.* **2001**, *287*, 204–208.

99. Thompson, M. A.; Weinshilboum, R. M.; Yazal, J. E.; Wood, T. C.; Pang, Y.-P. *J. Mol. Model.* **2001**, *7*, 324–333.

100. Vidgen, J.; Svensson, L. A.; Liljas, A. *Nature* **1994**, *368*, 354–358.

101. Lautala, P.; Ulmanen, I.; Taskinen, J. *Mol. Pharmacol.* **2001**, *59*, 393–402.

102. Grunewald, G. L.; Dahanukar, V. H.; Jalluri, R. K.; Criscione, K. R. *J. Med. Chem.* **1999**, *42*, 118–134.
103. Grunewald, G. L.; Caldwell, T. M.; Dahanukar, V. H.; Jalluri, R. K.; Criscione, K. R. *Bioorg. Med. Chem. Lett.* **1999**, *9*, 481–486.
104. Tervo, A. J.; Nyroenen, T. H.; Roenkkoe, T.; Poso, A. *J. Comput.-Aided Mol. Design* **2004**, *17*, 797–810.
105. Rodriguez-Lima, F.; Deloménie, C.; Goodfellow, G. H.; Grant, D. M.; Dupret, J.-M. *Biochem. J.* **2001**, *356*, 327–334.
106. Rodriguez-Lima, F.; Dupret, J.-M. *Biochem. Biophys. Res. Commun.* **2002**, *291*, 116–123.
107. Sandy, J.; Mushtaq, A.; Kawamura, A.; Sinclair, J.; Sim, E.; Noble, M. *J. Mol. Biol.* **2002**, *318*, 1071–1083.
108. Rath, V. L.; Verdugo, D.; Hemmerich, S. *Drug Disc. Today* **2004**, *9*, 1003–1011.
109. Coughtrie, M. W. H. *Hum. Exp. Toxicol.* **1996**, *15*, 547–555.
110. Coughtrie, M. W. H.; Sharp, S.; Maxwell, K.; Innes, N. P. *Chem.-Biol. Interact.* **1998**, *109*, 3–27.
111. Dajani, R.; Cleasby, A.; Neu, M.; Wonacott, A. J.; Jhoti, H.; Hood, A. M.; Modi, S.; Hersey, A.; Taskinen, J.; Cooke, R. M. et al. *J. Biol. Chem.* **1999**, *274*, 37862–37868.
112. Heimstad, E. S.; Andersson, P. L. *Quant. Struct.-Act. Relat.* **2002**, *21*, 257–266.
113. King, R. S.; Sharma, V.; Pedersen, L. C.; Kakuta, Y.; Negishi, M.; Duffel, M. W. *Chem. Res. Toxicol.* **2000**, *13*, 1251–1258.
114. Sharma, V.; Duffel, M. W. *J. Med. Chem.* **2002**, *45*, 5514–5522.
115. Sipilä, J.; Hood, A. M.; Coughtrie, M. W. H.; Taskinen, J. *J. Chem. Inf. Comput. Sci.* **2003**, *43*, 1563–1569.
116. Structural Genomics Consortium: http://www.sgc.utoronto.ca.
117. Ouzzine, M.; Barré, L.; Netter, P.; Magdalou, J.; Fournel-Gigleux, S. *Drug Metab. Rev.* **2003**, *35*, 287–303.
118. Smith, P. A.; Sorich, M. J.; Low, L. S. C.; McKinnon, R. A.; Miners, J. O. *J. Mol. Graph. Model.* **2004**, *22*, 507–517.
119. Smith, P. A.; Sorich, M. J.; McKinnon, R. A.; Miners, J. O. *Clin. Exp. Pharmacol. Physiol.* **2003**, *30*, 836–840.
120. Sorich, M. J.; Smith, P. A.; McKinnon, R. A.; Miners, J. O. *Pharmacogenetics* **2002**, *12*, 635–645.
121. Sorich, M. J.; Miners, J. O.; McKinnon, R. A.; Smith, P. A. *Mol. Pharmacol.* **2004**, *65*, 301–308.
122. Smith, P. A.; Sorich, M. J.; McKinnon, R. A.; Miners, J. O. *J. Med. Chem.* **2003**, *46*, 1617–1626.
123. Rendic, S.; Di Carlo, F. *J. Drug Metab. Rev.* **1997**, *19*, 413–580.
124. Honkakoski, P.; Negishi, M. *Biochem. J.* **2000**, *347*, 321–337.
125. Dickins, M. *Curr. Topics Med. Chem.* **2004**, *4*, 1745–1766.
126. Scott, E. E.; White, M. A.; He, Y. A.; Johnson, E. F.; Stout, C. D.; Halpert, J. R. *J. Biol. Chem.* **2004**, *279*, 27294–27301.
127. Johnson, E. F.; Yano, J. Personal communication.
128. Arimoto, R.; Prasad, M.-A.; Gifford, E. M. *J. Biomol. Screen.* **2005**, *10*, 197–205.

Biographies

Marcel J De Groot, after graduating from Utrecht University (the Netherlands, MSc in Chemistry) finished a PhD in Computational Toxicology at the Vrije Universiteit Amsterdam, developing computational models for biotransformation enzymes. In 1997, he worked for half a year as Assistant professor in Computational Chemistry at the Vrije Universiteit Amsterdam, after which he joined the Pfizer Global Research and Development in Sandwich (UK) as a computational chemist. Within Pfizer, Marcel De Groot continues to focus on computational models for biotransformation enzymes and the prediction of ADME and toxicological related issues and is an author of over 40 publications, mainly in the area of computational ADMET.

David F V Lewis, after graduating in Chemistry at the University of Bath, gained an MSc in Spectroscopy and, subsequently, a PhD in Theoretical Chemistry from the University of Surrey. He remained at Surrey as a Research Fellow, working with Professor Dennis Parke on the cytochromes P450 and, in 2003, was made Professor of Structural Biology, also receiving a DSc from the University of Bath the same year. David Lewis is the author of two books on P450, and has over 200 publications in peer-reviewed journals. His main interest lies in the molecular modeling of human P450-substrate interactions.

Sandeep Modi, after doing, MSc in Chemistry from Delhi University, India, did PhD in Chemical Physics from Tata Institute of Fundamental Research, Bombay, India. He also did postdoc in Cambridge University (Biochemistry Dept) and also in Leicester University (Biochemistry Dept). He joined Glaxo Wellcome in 1997 as computational chemist. Currently, he is working on the development of various ADMET models and their applications in several lead optimization programs.

Comprehensive Medicinal Chemistry II
ISBN (set): 0-08-044513-6

ISBN (Volume 5) 0-08-044518-7; pp. 809–825

5.35 Modeling and Simulation of Pharmacokinetic Aspects of Cytochrome P450-Based Metabolic Drug–Drug Interactions

M Dickins, Pfizer Global Research and Development, Sandwich, UK

A Galetin, University of Manchester, Manchester, UK

N Proctor, Simcyp, Sheffield, UK

5.35.1 Introduction

Knowledge of drug–drug interactions (DDIs) has become a part of the process of enabling new drugs to be introduced to the market, and there is increasing need for the prediction of such interactions.[1] DDIs not only give potential new drugs a competitive disadvantage, but could lead to labeling restrictions and in extreme cases can lead to the regulatory authorities refusing drug approval or in market withdrawal.[2]

It is known that there are a number of mechanisms for DDIs. The majority of these interactions involve effects at the level of drug-metabolizing enzymes, particularly cytochrome P450 (CYP), and will be the focus of this chapter. However, not every DDI is metabolism-based but may arise from changes in pharmacokinetics caused by absorption, tissue, and/or plasma binding, distribution, and excretion interactions. Drug interactions involving transporter proteins are now being recognized as important factors, with a number of drugs being substrates and/or inhibitors/inducers of both drug-metabolizing enzymes and transporter proteins. In addition, DDIs may alter pharmacokinetic/pharmacodynamic relationships.

In recent years, a number of drugs have been withdrawn from the market because of problems relating to metabolic DDIs (mDDI). These compounds include drugs such as terfenadine, astemizole, cisapride, mibefradil, and cerivastatin which are metabolized by CYP enzymes, expressed primarily in the liver. The majority of mDDIs result from perturbation of the CYP enzyme system, with inhibition of the respective enzymes being the major reason for this type of interaction. The major CYPs affected by mDDIs involving inhibition are CYPs 1A2, 2C9, 2C19, 2D6, and 3A4,[3] although appreciation of the involvement of CYPs 2B6, 2C8, and 3A5 is increasing.[2]

Drug interactions resulting from CYP enzyme induction, resulting in an increase of enzyme content, occur far less frequently than inhibition.[2,4] CYP3A4 is the major enzyme affected by enzyme induction in humans and drugs such as rifampicin produce marked effects on CYP3A4 substrates[5] such as midazolam.[6] In these cases, the mDDI results in increased clearance of the affected drug, giving a reduced therapeutic effect or even a loss of efficacy. However, inducing agents such as some antiepileptic drugs (phenobarbital and carbamazepine) also upregulate other drug-metabolizing enzymes in addition to CYPs (e.g., conjugation enzymes such as uridine diphosphate-glucuronyl-transferases, UGTs).[7,8]

Although there are examples of clinically relevant DDIs being mediated at the level of transporters,[9,10] these are relatively few in number, and it is often difficult to dissociate CYP inhibition and inhibition of transport processes.[2] Nevertheless, despite the fact that some drug interactions may be due to effects at the level of CYPs, other drug-metabolizing enzymes, and also transporters, CYP inhibition remains the predominant mechanism both in terms of frequency and of the magnitude of the resultant pharmacokinetic effect.

The availability of in vitro metabolism data in early drug discovery has created the opportunity to evaluate better the magnitude of mDDIs. In fact, the potential for mDDIs is continuously measured using in vitro technology throughout the development of a new drug, with increasingly quantitative measurement and the use of clinically relevant drugs. At this stage, the magnitude of a drug interaction can be estimated by incorporating the I_u/K_i ratio, that is, the ratio of unbound inhibitor concentration to the inhibition constant for a particular CYP enzyme.[11,12]

However, improved prediction of mDDI requires an understanding of the physiological variability and variation in the drug-metabolizing enzyme content encountered in the human population.[1] By incorporating demographic, physiological, pathological, and genetic variation together with the range of CYP contents, it is possible to generate virtual populations that can be used in the estimation of mDDIs. It is this integration of population parameters with drug-specific parameters such as intrinsic clearance (CL_{int}) or K_i which makes this approach a powerful tool in mDDI prediction. In this way, simulation of mDDIs should be able to generate the range of values seen in the clinic and thus be more representative of the true situation.[13] The typical mDDI study carried out in the pharmaceutical industry uses only a small number of healthy volunteers (typically up to 20), whereas a simulation of this type of study can generate a large population, encountered when a drug has been marketed and dosed to a patient population. The influence of the disease status such as hepatic/renal impairment or populations which are predominantly male (heart attack) or female (rheumatoid arthritis) can also be simulated. Thus, simulation techniques can be extremely useful in assessing the impact of drug interaction potential in such populations. This chapter will discuss the main principles for clearance prediction for both classical Michaelis–Menten (K_m) and atypical in vitro kinetic data. Additionally, a number of substrate and inhibitor-related characteristics associated with the prediction of both reversible and irreversible DDIs will be reviewed. A number of drugs such as diltiazem and erythromycin are very weak reversible inhibitors of CYPs and would not act as inhibitors of drug-metabolizing enzyme activity in vivo. However, a number of compounds including these drugs act as time-dependent (irreversible) inhibitors of CYPs by inactivating CYPs and thus depleting the pool of active enzyme. This effect is more profound than reversible inhibition because new protein synthesis of the affected enzyme is necessary to remove the inhibitory effect.

Regulatory authorities such as the US Food and Drug Administration (FDA) and the European Agency for the Evaluation of Medicinal Products (EMEA) require investigation of the potential for mDDIs. In vitro studies can often provide important indications of the likelihood of a mDDI occurring in vivo and guidances have been issued for these studies to be conducted during drug development. These include the establishment of a set of model CYP substrates and inhibitors which are recommended for use in these investigations. However, in vivo DDI studies remain the definitive method for determining the magnitude of an mDDI. The studies address a number of issues, such as the likely effects on drugs in the marketplace and whether dosage adjustments are necessary for coadministered drugs. Simulations of mDDI using in vitro and in vivo inputs can provide valuable projections of the likelihood and magnitude of an mDDI and the use of such methods are set out in subsequent sections.

5.35.2   In Vitro Models for Studying Metabolic Drug–Drug Interaction Potential

5.35.2.1   In Vitro Metabolizing Systems

In order to determine the DDI potential accurately the metabolic clearance of the 'victim' compound must be fully characterized.[1,14] Once this is performed it should then be a relatively simple matter either experimentally or theoretically to block some of these metabolic pathways, thereby determining the predicted degree of interaction.

Traditionally, human liver microsomes (HLM) have been used for these purposes. HLMs represent a solution of endoplasmic reticular membranes commonly isolated from liver homogenate by differential centrifugation up to 100 000 g. Since these membranes are the major intracellular location of the CYPs, the resultant solution is a highly concentrated reaction environment for studying phase I metabolic reactions, particularly those involving CYPs. The major advantages of using hepatic microsomes include the relative ease of preparation and the ability to store the solutions for long periods of time with little or no reduction in activity.[15]

Typically, an arbitrary concentration (around 1 µM) of the new chemical entity (NCE) is first incubated with an arbitrary amount of HLM, and then a series of selective CYP inhibitors[16,17] can be used to examine the reduction in measured CL_{int} to derive an impression as to which CYPs are most important in the clearance of this compound. There are several fundamental assumptions to this approach (such as the substrate concentration being significantly below the lowest metabolite K_m and that of negligible nonspecific binding); however, advances in analytical sensitivity have allowed sufficient limits of quantification so that these can often be accounted for in the experimental design. It has been recognized that an NCE must be assessed as both a potential 'victim' and 'perpetrator' of DDIs. In parallel to the above experimental approach, where the 'victim' potential is explored, another approach is often used, whereby fluorescent CYP-selective probes (resorufin, coumarin, and fluorescein derivatives) are used in the presence of varying NCE concentrations to examine the 'perpetrator' potential. Later in the drug development process, as major metabolic pathways are identified and characterized, each particular metabolic route can be explored for liability to DDI.

The advent of recombinant human CYPs (rhCYPs)[18] has provided the opportunity to combine several of these steps and obviate some of the reliance upon the 'selective' chemical CYP inhibitors. The enzyme preparations are available from commercial suppliers such as BD BioSciences, Invitrogen, and Cypex. Since these rhCYP systems express only one CYP it is theoretically possible merely to determine the CL_{int} in these and scale them to in vivo both pre- and postmetabolite identification stages (eqn [1]).

$$CL_{int,met} = \sum_j CL_{int,j} \times A_j \qquad [1]$$

where $CL_{int,j}$ represents the intrinsic clearance (CL_{int}) from a single CYP enzyme and A_j the abundance of that enzyme in the human liver. The rhCYP systems in common use include those expressed in mammalian (lymphoblast), insect (baculovirus transfected *Trichoplusia ni*), bacterial (*Escherichia coli*), and yeast (*Saccharomyces cerevisiae*) systems.

Discrepancies between CL_{int} derived from HLM and rhCYP have been observed and may be due to several factors, including the extent of nonspecific microsomal binding.[19,20] Although differences between the systems in such binding per amount of microsomal protein are minimal,[21] experiments with rhCYP systems commonly employ lower protein concentrations than those using liver microsomes, which may cause disparity between the extent of nonspecific binding in the two systems, especially for lipophilic basic drugs. The CL_{int} per unit amount of CYP (intrinsic activity or turnover number) may also vary between HLM and rhCYP enzymes: this has been attributed to differences in the concentrations of accessory proteins (cytochrome b_5, NADPH:cytochrome P450 oxidoreductase) and the lipid microenvironment of the enzyme.[22–24] Several attempts have been made to correct for this discrepancy.[25–27] The initial approach was to use a relative activity factor (RAF) which would account for both this discrepancy and also the relative abundances of CYPs in HLM and rhCYP solutions. Although this approach has been used successfully on several occasions, it fails to address adequately the issues of interindividual variability in CYP expression and the apparent substrate specificity of RAFs. Indeed, the RAF merely demonstrates the amount of rhCYP required to give an equivalent reaction velocity to that of the particular HLM sample used (eqn [2]).

$$RAF_{h,j} \text{ (pmol rhCYP mg}^{-1} \text{ HLM)} = \frac{V_{max}HLM_{h,j} \text{ (nmol min}^{-1}\text{mg}^{-1} \text{ HLM)}}{V_{max}rhCYP_j \text{ (nmol min}^{-1}\text{pmol}^{-1} \text{ rhCYP)}} \qquad [2]$$

An alternative approach would be merely to compare the intrinsic activities of rhCYP versus HLM and provide CYP abundance scaling by simply mathematical means. This approach is described in detail elsewhere,[10] but briefly it is the above RAF approach adjusted for the actual amount of HLM CYP present rather than a theoretical amount (eqn [3]).

$$\text{VISEF}_j = \frac{V_{\max}\text{HLM}_j \ (\text{nmol min}^{-1}\text{mg}^{-1} \ \text{HLM})}{V_{\max}\text{rhCYP}_j \ (\text{nmol min}^{-1}\text{pmol}^{-1} \ \text{rhCYP}) \times \text{HLM CYP}_j \ \text{abundance} \ (\text{pmol mg}^{-1} \ \text{HLM})} \quad [3]$$

Since the ISEF uses the actual CYP abundance of that HLM sample, it is readily scalable to in vivo without the added complications of intersample uncertainty. Although maximum rate of metabolism ($V_{\max}$) values have been used in eqns [2] and [3], CL_{int} values for the relevant activities can also be used.[26,27]

In order to provide accurate and reproducible estimates of DDI potential at an early stage in drug development, the above issues must be taken into account when using both HLM and rhCYP. Human hepatocytes may also be used for the determination of DDI potential, but their practical application is hampered by the plethora of biochemical systems in place. HLM remain the experimental system of choice for providing quantitative predictions of DDI, although this can only be representative if the HLM batch in question has been thoroughly characterized with respect to CYP abundance.

5.35.2.2   In Vitro Systems for Assessment of Cytochrome P450 Induction and Transporter Activity

Human hepatocyte cultures are frequently used as an in vitro system to detect potential CYP-inducing agents in humans.[28–32] A number of clinically relevant inducing agents have been shown to induce CYPs known to be affected in vivo at both the mRNA and the catalytically active protein level.[33] More recently, immortalized human hepatocytes, which offer the potential of a more useful screening procedure for CYP inducers, have also been used.[34] Estimation of hepatic enzyme induction involves incubating the cultured cells for a 48–72-h period with a range of concentrations of the test drug (based on anticipated human plasma drug concentrations). A positive control compound, e.g., a known human clinical enzyme inducer such as rifampicin, is also incubated under the same conditions. A drug that produces a greater than twofold increase in enzyme activity (measured with a probe substrate such as midazolam for CYP3A4 activity) or a fold change in enzyme activity that is 40% of the positive control data is considered to be an enzyme inducer. However, although there are data relating the extrapolation of the in vitro measured effects to the in vivo situation, knowledge of the extent of increased enzyme content in vivo due to induction is scarce and therefore difficult to incorporate in simulations.

In addition, the role of efflux and uptake transporters cannot be ignored, although the level of our understanding of these proteins in terms of their quantitation and interplay lags behind that of the CYPs.[10,35,36] Exposure of a drug to enzymes (particularly CYP3A4) during its transit from the gut lumen through the enterocyte and into the hepatic portal system depends on the interplay of transporters, passive membrane permeability, and enterocytic blood flow. A quantitative approach to incorporate these processes in the gut has been described for midazolam, a highly permeable drug.[37] It is possible to estimate the effect of efflux transporters for other compounds by comparing the flux of midazolam in an in vitro system (such as Caco-2 cells) with that of other compounds where these data are reported.[38] For a drug with a high permeability such as midazolam, the flux value approximates enterocytic blood flow but for substrates of efflux transporters flux from the gut into the hepatic portal vein is reduced. For active uptake transporters, studies using drug in the presence of cells incubated at 4 and 37 °C or with selective uptake inhibitors can be used to assess the degree of uptake into liver cells.[39]

5.35.2.3   In Vitro Parameters to Measure

A physiological approach to the simulation of mDDIs requires a range of data on tissue blood flow rates and tissue uptake, plasma protein binding, and nonspecific binding of compounds to the in vitro metabolizing system (hepatocytes, liver microsomes, or rhCYPs, enzyme kinetic data (in the form of CL_{int} for substrates and K_i values for inhibitors). CL_{int} is the enzyme-catalyzed metabolic clearance of a drug which is not influenced by other physiological parameters such as hepatic blood flow. The CL_{int} value is a fundamental link between enzyme kinetics and pharmacokinetics.

Many of the approaches used to calculate the degree of an mDDI rely on estimation of the change of CL_{int} resulting from the interaction of an inhibitor at a particular enzyme (CYP). The determination of this value for a CYP substrate is frequently determined using HLMs by measuring the half-life (in vitro $t_{1/2}$) of disappearance of the compound at low ($\leqslant 1\,\mu M$) substrate concentrations,[40] as shown in eqn [4].

$$CL_{int} = \frac{0.693}{\text{in vitro } t_{1/2} \times \text{microsomal protein concentration}} \qquad [4]$$

The assumption is made that the total CL_{int} is mediated by CYP enzymes. In order to establish which CYPs are responsible for the metabolic process, selective chemical inhibitors are available which can inhibit individual CYP isoforms. Alternatively, rhCYPs can be incubated with the substrate to identify the CYP isoforms involved in metabolism, although adjustment of the rates with recombinant enzymes is necessary, as previously described (*see* Section 5.35.2.1).

For compounds where metabolites have been identified, the K_m and V_{max} parameters for the pathways can be generated. In these cases, the value of CL_{int} for compounds which obey Michaelis–Menten kinetics is obtained as the V_{max}/K_m ratio. The incorporation of CL_{int} for compounds which show atypical (non-Michaelis–Menten) kinetic features are described in Section 5.35.3.

It is important to recognize that the value of CL_{int} generated in an in vitro system may be affected by nonspecific binding of the test compound to the in vitro metabolizing system. This binding is referred to in the literature as $f_{u,mic}$[41] or $f_{u,inc}$[42] and is applied as a correction factor to CL_{int} or K_m values obtained in in vitro incubations (eqn [5]).

$$CL_{int,u} = \frac{CL_{int}}{f_{u,mic}} = \frac{V_{max}}{K_m \times f_{u,mic}} \qquad [5]$$

As shown in eqn [5], estimates of K_m in vitro are apparent values without appropriate correction for the fraction unbound in the assay system.[21,41–44] Knowledge of $f_{u,mic}$ has also been shown to be important for correction of the inhibition constant, K_i, used for the prediction of mDDI.[45–48] Substrates and inhibitors which are lipophilic bases are particularly affected by nonspecific binding, whereas acidic and neutral compounds are typically bound to a significant extent only if they are highly lipophilic ($\log D_{7.4} > 3$).[49]

Thus, it is only by incorporating such variability in simulations of CL prediction from CL_{int} (eqn [6]), and of mDDI prediction from reduction in CL_{int} due to inhibition, that the variability of the magnitude of the interaction seen in the clinic can be captured.

5.35.2.4 In Vitro to In Vivo Scaling Factors

It has previously been mentioned that the predictive accuracy of in vitro to in vivo extrapolation (IVIVE) approaches can depend greatly upon the in vitro systems of choice. It is also important to examine the extrapolative scaling factors used for sources of uncertainty and variability. The classical scaling approach (as described by Houston,[50] eqn [6]) has been modified to reflect the challenges in scaling from both HLM (eqn [7]) and rhCYPs (eqn [8]), whilst taking into account established measures of interindividual variability. It is important to note that the units in both latter strategies account for the extreme variability of CYP expression in tissues by expressing the rates per unit CYP isoform. The strategy used for scaling total hepatic intrinsic clearance ($CL_{H,int}$) from in vitro data is outlined below. Central to these strategies are the amount of microsomal protein per gram of human liver (MPPGL) and the weight of the liver to be projected.

$$CL_{H,int} = HLM\ CL_{int} \times MPPGL \times \text{liver weight} \qquad [6]$$

$$CL_{H,int} = \left[\sum_j HLM\ CL_{int,j}\right] \times CYP_j\ \text{abundance} \times MPPGL \times \text{liver weight} \qquad [7]$$

where hepatic intrinsic clearances are summed for each contributing CYP (j).

$$CL_{H,int} = \left[\sum_j rhCYP\ CL_{int,j}\right] \times ISEF_j \times CYP_j\ \text{abundance} \times MPPGL \times \text{liver weight} \qquad [8]$$

It has been demonstrated that human MPPGL has considerable interindividual variability and indeed can vary dramatically between investigating laboratories.[51–58] The experimental technique of differential centrifugation used for preparing HLM can be highly variable and it is therefore essential to determine a recovery factor for each preparation in order to quantify reliably the true MPPGL rather than merely the experimental yield.[58] The most common values in use of 52.5^{54} and 45^{59} mg g^{-1} would appear to owe more to their assistance in overcoming underpredictions than any valid human physiological basis.[60] A recent publication has attempted to define a value of MPPGL and also assess the true interindividual variability present.[58] The authors have incorporated adjustments of both yield and assay variability by the means of recovery factor and cross-validated measurements. A further update[61] to this work has established MPPGL in a total of 53 human liver samples and work is ongoing to examine relationships between MPPGL and age, disease state, or medication status of the donors. An overall geometric mean value of 28 mg g^{-1} (range 13–54 mg g^{-1}) has been obtained which appears to encompass other, smaller, data sets. Such a large degree of interindividual variation can obviously be significant when extrapolating to a human population.

The abundance of CYP isozymes in the liver and other tissues is also an issue of great importance. In addition to helping contextualize the results of in vitro experiments from a particular liver (or pool of livers), these values are essential in the successful scaling of rhCYP data. A recent exhaustive metaanalysis of the literature,[62] including the most commonly cited source of abundance data,[63] has shown a large degree of interindividual variability for many of the CYPs most commonly involved in drug metabolism. For example, reported CYP3A (3A4) content varies from 37^{64} to 248 pmol mg^{-1}[65] CYP3A4 is the major isoform of the CYP3A subfamily, both in terms of absolute amount and metabolic capacity.

A similar strategy has been pursued for the scaling of hepatocyte CL$_{int}$ (eqn [9]), with slightly greater success, although this system has particular disadvantages for predicting DDI liability, as previously discussed.

$$CL_{H,int} = \left[\sum_j Hepatocyte\ CL_{int,j} \right] \times HPGL \times liver\ weight \qquad [9]$$

The approach to defining hepatocellularity (or HPGL, quantity of hepatocytes per gram of liver) is slightly more challenging due in part to the difficulties involved in sourcing, isolating, and handling primary human hepatocytes. The approach has been well described,[7] although published literature has been unclear on the source of the most commonly accepted value of 120×10^6 cells g^{-1}. Although widely used,[54,59,66–69] the original methodologies used for determining this number have not been verified, although the original reference value[70] is identical to that verified several years earlier in the rat.[71] Initial estimates[58] ($n = 7$) have given a geometric mean value of 107×10^6 cells g^{-1} (range 65–185×10^6 cells g^{-1}), although more recent data[61] ($n = 24$) suggest that the true population will be somewhat lower than this.

5.35.2.5   Liver Models

Well-stirred, parallel-tube, and dispersion models are liver models used for the prediction of clearance. All three models assume homogeneous distribution of enzymes, no diffusion barriers in the liver, and that only unbound drug is able to cross the membrane and be metabolized. The well-stirred model is a simple and the most commonly used liver model, based on the assumption that the liver is a well-stirred compartment with the drug concentration in the liver being in equilibrium with the blood.[54] From eqn [6] the value for the total CL$_{int}$ of the liver, CL$_{H,int}$ (calculated from the in vitro HLM CL$_{int}$) can be entered into the well-stirred model to estimate total hepatic blood CL in vivo, CL$_H$.

$$CL_H = \cfrac{\dfrac{Q_H \times f_{u,b} \times CL_{H,int}}{f_{u,mic}}}{Q_H + \dfrac{f_{u,b} \times CL_{H,int}}{f_{u,mic}}} \qquad [10]$$

where Q_H is hepatic blood flow, $f_{u,b}$ is free fraction of the compound in the blood, CL$_{H,int}$ is the measured intrinsic clearance, and $f_{u,mic}$ is the free fraction of the compound in the in vitro incubation system (e.g., human hepatic microsomes or rhCYP).

5.35.3   Prediction of Clearance by CYP3A Enzymes: Incorporation of CL$_{max}$ and Allosteric Kinetics to Measurement of CL$_{int}$

5.35.3.1   Theoretical Considerations of Nonlinear Enzyme Kinetics

For many years the Michaelis–Menten model, and the existence of a single active site for the interaction of substrate with drug-metabolizing enzyme, has been used to describe in vitro kinetic data for both clearance and inhibition

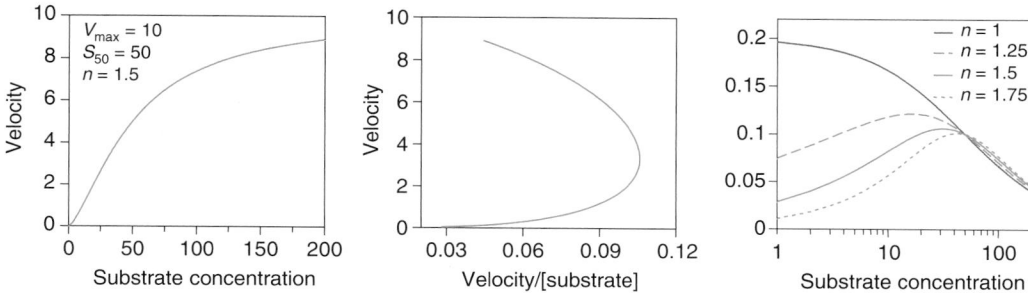

Figure 1 Kinetic profile simulations with the resulting Eadie–Hofstee and CL$_{int}$ plots for autoactivation.

determination. However, it has become increasingly common that atypical (non-Michaelis–Menten) kinetic features are observed.[22,72,73] The phenomena of auto- and heteroactivation; partial, cooperative and substrate inhibition; concentration-dependent effector responses (activation/inhibition); limited substrate substitution and inhibitory reciprocity cannot be readily interpreted using single active-site models. While most of these atypical properties are associated with CYP3A4, recent studies indicate similar kinetic behavior for some other human enzymes, particularly CYP2C9[74] and certain UGTs.[75,76] Atypical Michaelis–Menten kinetics are characterized by two particular types of rate-substrate concentration curves: (1) sigmoidal, believed to result from autoactivation; and (2) convex, resulting from substrate inhibition. Both give characteristic curved Eadie–Hofstee plots that deviate from the linear relationship expected from the Michaelis–Menten model and are useful diagnostic plots for identifying such behavior. These two types of homotropic effects (i.e., when two molecules of the same substrate bind to the active site) are of importance as neither allows the estimation of CL$_{int}$ in vitro by the standard method previously described.[77]

Autoactivation (positive cooperativity) has been observed for various CYP3A4 substrates, and in various in vitro systems. In order to explain and obtain the appropriate fit for the sigmoidal data, several models have been proposed, from the empirical Hill equation to more complex mechanistic ones.[22,78,79] Autoactivation is characterized by the dependence of clearance on the substrate concentration (**Figure 1**). As no linear region equivalent to Michaelis–Menten exists, an alternative to CL$_{int}$ for scaling of in vitro data is required. Maximum clearance, CL$_{max}$ (eqn [11]) has been suggested as an alternative parameter and provides an estimate for the maximum clearance when the enzyme is fully activated before the saturation occurs.[80]

$$CL_{max} = \frac{V_{max}}{S_{50}} \times \frac{(n-1)}{n(n-1)^{1/n}}$$ [11]

5.35.3.2 Practical Implications

The significance of nonstandard Michaelis–Menten data in vitro, and their correlation with the in vivo situation remains ambiguous,[73] since to date there are few confirmatory studies in vivo[81,82] and in certain cases their relevance could be questioned.[83,84] The introduction of CL$_{max}$ as an alternative for CL$_{int}$[80] represents one attempt to introduce autoactivation into the in vitro–in vivo scaling strategy. The dependence of clearance on substrate concentrations below the K_s is associated with positive cooperativity and indicates the possibility of clearance underestimation in rapid in vitro screening procedures based on only one substrate concentration.[77] How precisely the clearance estimate will be altered by the model misspecification will vary from case to case and will be dependent on the number and quality of the data points.[80] In terms of predicting DDIs in vivo, the use of multiple substrates in vitro at various substrate concentrations is recommended to explore the range of possible consequences of a heterotropic interaction.[2,16]

5.35.4 Use of Various Inhibitor Concentrations for the Prediction of Drug–Drug Interactions

5.35.4.1 Metabolic Drug–Drug Interaction Risk Evaluation

The use of in vitro data to identify the inhibition of CYP-mediated metabolic reaction and hence potential DDIs is attractive. The rapid and simple experimental procedures developed have allowed substantial technological advances in the conduct of in vitro studies. The studies are designed based on the steady-state kinetic approach, i.e., that the inhibitor concentration should be much higher than the enzyme concentration, such that inhibitor binding to the enzyme does not alter the free concentration of the inhibitor. In an ideal in vitro experimental design the inhibitor concentrations should span from 0 K_i (control), 1/3 K_i, K_i, 3 K_i, to up to 10 K_i values. In case of IC_{50} determinations the range of inhibition should range from virtually no inhibition to virtually complete inhibition in concentrations evenly spread on a log scale (e.g., 0, 0.1, 0.3, 1, 3, 10, 30, and 100 mM). The substrate concentration range should be in a similar range as that for determination of V_{max} and K_m. However, the interpretation of the parameters generated remains problematic due to lack of a quantitative framework for the in vitro–in vivo relationship for DDIs.[1]

The most promising approach to quantitative prediction of DDIs from in vitro data is based on the ratio between the concentration of the inhibitor in vivo at the enzyme active site $[I]$ and inhibition constant (K_i), assuming reversible single-site inhibition. The CL_{int} of substrate in the presence of inhibitor is reduced by a factor related to the inhibitor concentration available to the enzyme $[I]$ and the inhibition constant, K_i. The distinction between competitive and noncompetitive inhibition mechanisms is not relevant when the substrate concentration is much lower than the K_m value, the commonly encountered in vivo situation that results in linear kinetics.

$$CL_{int,inhibited} = \frac{CL_{int}}{1 + [I]/K_i} \qquad [12]$$

The major assumptions for this in vitro–in vivo extrapolation are reversible Michaelis–Menten type of inhibition, applicability of the 'well-stirred' liver model, and linear pharmacokinetics for the drug. The metric for the degree of DDI in vivo is the area under the plasma concentration–time curve (AUC) ratio for the plasma concentration in the presence and absence of the inhibitor after multiple oral dosing.[2,16] If the substrate is exclusively eliminated by liver by a single metabolic pathway which is subject to inhibition, the AUC ratio of orally administered substrate in the presence and absence of inhibitor reflects the ratio of clearances, as shown by eqn [13], where f_a is the fraction absorbed from gut into the portal vein, D is the dose, and $f_{u,b}$ is the unbound fraction in blood.

$$\frac{AUC_{inhibited}}{AUC_{control}} = \frac{f_{a,I} \times D_I}{f_a \times D} \times \frac{f_{u,b} \times CL_{int}}{f_{u,b,I} \times CL_{int,I}} = \frac{CL_{int}}{CL_{int,I}} \qquad [13]$$

This relationship is based on the assumption that there is no impact of inhibitor on either the intestinal absorption or plasma protein binding of the substrate. Therefore, the AUC ratio is dependent solely on the $[I]/K_i$ as shown in eqn [14]:

$$\frac{AUC_{inhibited}}{AUC_{control}} = 1 + \frac{[I]}{K_i} \qquad [14]$$

According to eqn [14], interactions can be regarded to be low-risk if the estimated $[I]/K_i$ ratio is < 0.1 and high-risk with a ratio of > 1. **Figure 2** shows the relationship of the AUC ratio and $[I]/K_i$ and classifications of the predictions into four zones: (1) true positives (AUC ratio > 2, $[I]/K_i > 1$); (2) true negatives (AUC ratio < 2, $[I]/K_i < 1$); (3) false positives (AUC ratio < 2, $[I]/K_i > 1$); or (4) false negatives (AUC ratio > 2, $[I]/K_i < 1$).[16,85]

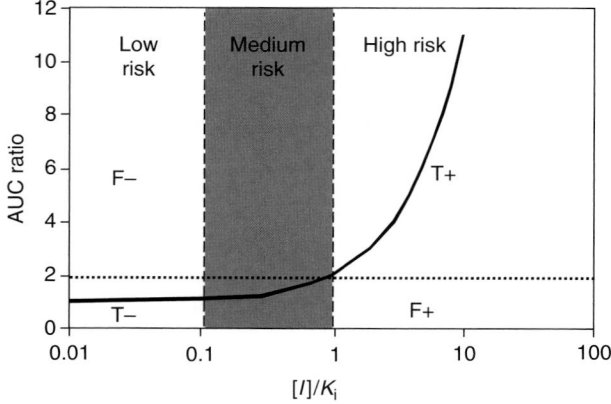

Figure 2   Qualitative zoning for the prediction of drug–drug interactions (AUC ratio) involving CYP inhibition using the $[I]/K_i$ ratio. The theoretical curve is based on eqn [14] where F − represents false-negative, T − true-negative, F + false-positive, and T + true-positive predictions. (Reproduced with permission from Houston, J. B.; Galetin, A. *Drug Metab. Rev.* **2003**, *35*, 393–415.)

In recent years there has been much interest in the use of $[I]/K_i$ to describe the degree of in vivo interaction between two drugs.[12,85–87] The accurate determination of $[I]$ is problematic as direct measurement is not possible and there is no generally accepted approach for the extrapolation of inhibitor concentration in the plasma to that at the enzyme site. A number of predictions of DDIs have been attempted, with varying degrees of success, using a range of $[I]$ values in eqn [14] including the average plasma total or unbound concentration or hepatic input concentration of the inhibitor.[3,85,87–89]

The average systemic plasma concentration after repeated oral administration ($[I]_{av}$), and the maximum hepatic input concentration ($[I]_{in}$) can be calculated as illustrated in eqns [15] and [16], respectively.[88] The $[I]_{in}$ value represents the combination of circulating systemic plasma concentration and the additional concentration occurring during the absorption phase.

$$[I]_{av} = \frac{D/\tau}{CL/F} \qquad [15]$$

$$[I]_{in} = [I]_{av} + \frac{k_a \times f_a \times D}{Q_H} \qquad [16]$$

where F and τ represent the fraction of dose systematically available and dosing interval, respectively, of the inhibitor used in the in vivo interaction study, Q_H is the hepatic blood flow ($1610\,mL\,min^{-1}$), D is the dose of an inhibitor, f_a is the fraction absorbed from the gastrointestinal tract, and k_a is the absorption rate constant.

Comprehensive analysis of 193 DDIs involving inhibition of CYP2C9, CYP2D6, and CYP3A4[85] has shown that hepatic input concentration as a surrogate for $[I]$ was the most successful for categorizing CYP inhibitors and for identifying true-negative DDIs. Although false-negative predictions were eliminated in this approach, a significant number of false positives was evident and most true positives were markedly overpredicted (**Figure 3**). The $[I]_{in}$ should be considered as an initial discriminator and is recommended, as true negatives can be identified. In addition, this approach would be particularly valuable in drug-screening processes, where false-negative predictions are to be avoided. However, this simple generic approach ignores specific substrate- or inhibitor-related properties that contribute to a number of overpredictions of true-positive interactions. The importance of these properties is evident by the number of true positives that are quantitatively overpredicted. These factors (e.g., the role of hepatic uptake transporters, plasma protein binding, appropriate absorption rate constant, existence of more than one metabolic/elimination pathway, impact of multisite kinetics for CYP3A4/5) require consideration in addition to the $[I]_{in}/K_i$ ratio in order to progress the prediction of CYP inhibition interactions toward a quantitative basis.

The absorption rate constant (k_a) values are frequently not reported in clinical studies; in the absence of this information and in order to avoid false-negative prediction and obtain the largest $[I]_{in}$, maximum value of $0.1\,min^{-1}$ was suggested as useful, assuming the gastric emptying is the rate-limiting step for absorption.[85,88] Recent studies by Brown et al.[87] have shown the impact of appropriate k_a values for 10 inhibitors for CYP2C9, CYP2D6, and CYP3A4, including azoles, quinidine, fluoxetine, fluvoxamine, etc. Refinement of this parameter resulted in the k_a values

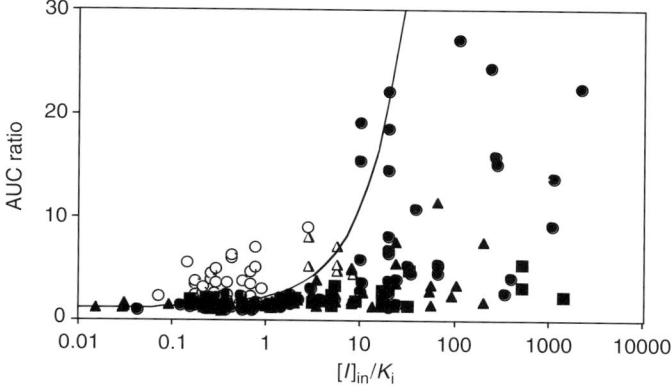

Figure 3   The graph shows 193 studies involving drug–drug interactions for CYP3A4 (filled circle), CYP2D6 (filled triangle), and CYP2C9 (filled square) substrates. Open symbols represent mechanism-based inhibitions. (Reproduced with permission from Houston, J. B.; Galetin, A. *Drug Metab. Rev.* **2003**, *35*, 393–415.)

2–14-fold lower than the initial estimates and reduction of the relative contribution of the absorption term in comparison with the systemic term to the $[I]_{in}$ value up to 13-fold in case of itraconazole. In addition, the k_a value may vary with the dose of inhibitor and the food intake (e.g., ketoconazole[90]), affecting the $[I]_{in}$ estimate and consequently the predicted AUC ratio.

5.35.4.2   Impact of Parallel Elimination Pathways on the Drug–Drug Interaction Assessment

Predictions of DDIs based on eqn [14] assume that the fraction of victim drug metabolized by the inhibited CYP pathway (fm_{CYP}) equals 1. However, parallel pathways of metabolism and renal clearance of unchanged drug will affect the fm_{CYP} estimate and consequently the predicted degree of interaction and therefore need to be incorporated into the prediction,[91] as shown in eqn [17]. Even minor changes in the fm_{CYP} value (e.g., from 1 to 0.98) may alter prediction accuracy significantly, as was illustrated in recent studies.[14,87,92]

$$\frac{CL_{control}}{CL_{inhibited}} = \frac{AUC_{inhibited}}{AUC_{control}} = \frac{1}{\left(fm_{CYP} \middle/ \left(1 + \left(\frac{[I]_{in\ vivo}}{K_i} \right) \right) \right) + (1 - fm_{CYP})} \tag{17}$$

A number of approaches can be employed to obtain fm_{CYP} value. The most explicit method is based on the comparison of phenotyping data in extensive and poor metabolizers, as shown for CYP2D6.[14] A good alternative to this approach is 'phenocopying,' based on the difference between the urinary recovery of metabolites in the presence and absence of a selective inhibitor (e.g., tolbutamide in the presence of sulfaphenazole as an inhibitor). CYP3A4 is the most problematic enzyme and the combined information on the urinary recovery of metabolites, biliary excretion, and the recovery of unchanged drug has recently been used.[87,92] As shown in the same systematic analysis, the use of fm_{CYP} data in the assessment of AUC ratio corrected several false-positive predictions, as well as significantly reduced the extent of overpredictions of true positives, with a minimal bias and high precision observed.

5.35.4.3   Complex Behavior of CYP3A4

Prediction of a potential DDI with CYP3A4 is particularly challenging due to a number of in vitro and in vivo factors, namely complexity of the in vitro kinetics observed for some of the CYP3A4 probes,[22,77] CYP3A interindividual variability in the abundance and activity in both liver and small intestine (with the variable contribution of polymorphic CYP3A5),[93,94] and overlapping substrate specificity with P-glycoprotein.[95]

In order to assess the importance of cooperativity and predict changes in the in vivo plasma concentration–time profile from CYP3A4 in vitro data, an equation (eqn [18]) was derived based on the same rapid equilibrium and steady-state assumptions as the single-site model.[92] In addition to simple $[I]/K_i$ ratio, this two-site model equation also incorporates changes in the catalytic efficacy (γ) and binding affinity (δ) in the presence of the inhibitor. In cases when $\gamma/\delta = 1$, the two-site prediction equation is reduced to the simple $1 + [I]/K_i$ relationship.

$$AUC\ ratio = \frac{\left(1 + \frac{[I]}{K_i} \right)^2}{1 + \frac{\gamma[I]}{\delta K_i}} \tag{18}$$

where I represents the in vivo inhibitor concentration (either $[I]_{in}$ – input plasma concentration – or $[I]_{av}$ – average plasma concentration during the dosing interval, whereas K_i estimates are obtained applying the generic two-site model). When the γ/δ ratio is either > or < 1 there is a change in both the position and the shape of the prediction curve, as shown in **Figure 4**. These simulations were performed for the case of inhibition ($\gamma \leqslant 0.5$) with a range of different binding affinities for the inhibitor in the presence of substrate. The analysis shows that the inclusion of CYP3A4 kinetic complexities (in the form of the γ/δ ratio) into the prediction strategy provides an explanation for the occurrence of some of the false negatives and reduces the overestimation of true-positive predictions (e.g., 28-fold for ketoconazole–triazolam interaction).[77]

However, a limitation of the multisite kinetic model approach to the analysis of DDIs lies in its complexity and the number of data points required for full characterization of the in vitro phenomenon observed. A recent study by Galetin et al.[92] explored the possibility of alternative and/or pragmatic approaches and substrates and their validity for reliable in vitro–in vivo prediction. Four different CYP3A4 probes, midazolam, testosterone, quinidine, and nifedipine,

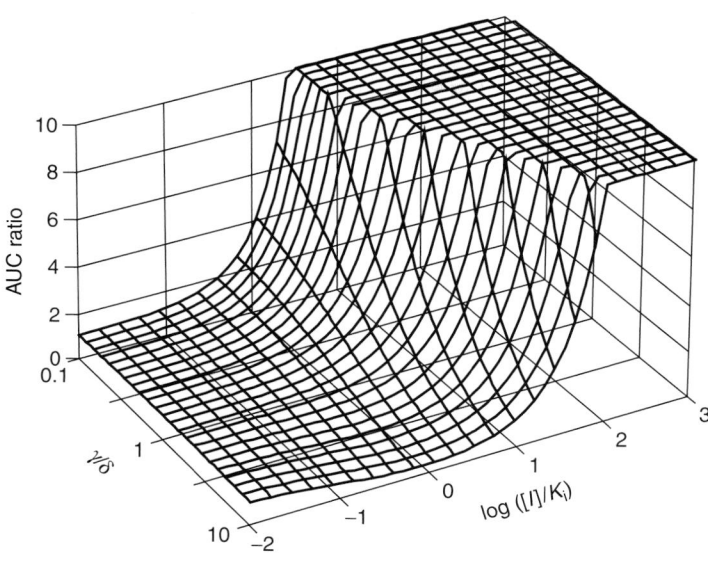

Figure 4 Practical implications of CYP3A4 multisite kinetics in predicting drug–drug interactions from in vitro parameters. The effect of interaction factors (shown as the γ/δ ratio) on the relationship between AUC ratio and $[I]/K_i$ ratio where $\gamma =$ catalytic efficacy and $\delta =$ binding affinity. (Reproduced with permission from Houston, J. B.; Galetin, A. *Drug Metab. Rev.* **2003**, *35*, 393–415.)

were assessed from the prospect of their in vitro complexities and possibility of substrate substitution in prediction of DDI potential. Midazolam and quinidine provided the best assessment of a range of selected 26 CYP3A4 drug interactions with azole inhibitors. Hyperbolic kinetics displayed in vitro for both of the substrates adds to their 'pragmatic appeal' in the evaluation of potential DDI compared to nifedipine and testosterone; however, the kinetic complexities of nifedipine and testosterone can be overcome by the appropriate mechanistic analysis. In addition, the study has shown that the extrapolation from one substrate to another (generally problematic in the case of CYP3A4) was successful, yet highly dependent on the incorporation of the detailed inhibitor- and substrate-related information in the prediction.

5.35.5   Prediction of Time-Dependent Drug–Drug Interactions

As the most abundant human P450 enzyme in both liver and intestine, CYP3A4 is susceptible to a number of reversible and irreversible metabolic DDIs. Irreversible, often referred to as mechanism-based, inhibition interactions, involve the metabolism of an inhibitor by CYP3A4 (i.e., require NADPH) to a reactive metabolite form which irreversibly inactivates the catalyzing enzyme (therefore removing it from the pool of active enzyme) in a concentration- and time-dependent manner.[96] A key characteristic of this type of inhibition is that inactivation occurs without the release of the reaction product from the catalytic site.[97] The interaction between the inactivating species and enzyme can be either covalent or noncovalent, involving binding to protein or heme moiety, respectively. Compounds such as bergamottin, ethinyl-estradiol, and mifepristone are indicated as covalent binders, whereas nitrosoalkanes formed by N-demethylation of tertiary amine group of macrolides (erythromycin) bind to the heme and inactivate the enzyme.[98]

The two major kinetic parameters that characterize time-dependent inhibition (TDI) interactions are k_{inact} and K_i, the maximal inactivation rate constant and the inhibitor concentration leading to 50% of k_{inact}, respectively.[96] The k_{inact}/K_I ratio is commonly taken as an indicator of the in vitro potency of a mechanism-based inhibitor. The established method of analysis of mechanism-based inhibitors[2,96] is based on the measurement of the residual enzyme activity using a probe substrate for the enzyme being investigated. This is achieved by preincubation of the range of inhibitor concentrations with the enzyme and cofactors. At set time points an aliquot of this mixture is taken and diluted into a secondary incubation containing the probe substrate and further cofactors; the dilution factor is necessary to reduce the occurrence of competitive inhibition between the inhibitor and marker substrate in the second incubation.

The methods used to obtain k_{inact} and K_I estimates generally vary across studies[99,100] from the CYP3A4 probes used (midazolam, testosterone, or triazolam), their concentration (from below the K_m to the concentrations equivalent to V_{max}), preincubation-to-incubation time ratio and data analysis method employed, all of which may affect the estimates of inhibitor potency. Ideally, preincubation-to-incubation time ratio should be at least >1, and a dilution factor from the preincubation to the incubation stage 1:10 should be applied to reduce the occurrence of competitive inhibition; in addition, the use of high substrate concentrations (equivalent to V_{max}) and nonlinear regression analysis are recommended for obtaining the k_{inact} and K_I parameters. A recent correlation analysis of k_{inact}/K_I estimates for five human immunodeficiency virus (HIV)-protease inhibitors by Ernest et al.[101] indicated higher potency estimates in recombinant enzymes in comparison to the liver microsomes. However, standardization of in vitro methodology and the use of the CYP3A4 probe at high substrate concentration (equivalent to the V_{max}) in the study by Ernest et al.[101] reduced the difference between the in vitro systems (with the exception of saquinavir) to only twofold.

Quantitative in vitro–in vivo drug interaction prediction work has mainly focused on reversible interactions, whereas TDIs were assessed on an individual case-by-case basis.[102,103] In contrast to reversible inhibition, enzyme activity can only be restored by synthesis of a new enzyme, often resulting in significant clinical interactions and a potential for idiosyncratic toxicities. Prediction of TDI using the $[I]/K_i$ approach with the assumption of reversible inhibition generally results in underprediction of the interaction observed, with a significant number of predictions being classified as false negatives.[1,85] The extent of TDI interaction, incorporating the contribution of parallel elimination pathways (other P450 enzymes or renal clearance, defined by $1 - fm_{CYP3A4}$), is defined by the eqn [19][103]:

$$\frac{AUC_{inhibited}}{AUC_{control}} = \frac{F'_G}{F_G} \times \frac{1}{\dfrac{fm_{CYP3A4}}{1 + \sum_{i=1}^{n} \dfrac{k_{inact,i} \times I_{u,i}}{k_{deg} \times (K_{I,u} + I_u)_i}} + (1 - fm_{CYP3A4})} \qquad [19]$$

The magnitude of interaction is assessed by the increase in the AUCi/AUC ratio for the plasma concentration–time profiles in the presence and absence of the inhibitor, respectively, as used previously.[16] In eqn [19], k_{inact} represents the maximal inactivation rate constant, K_I is the inhibitor concentration at 50% of k_{inact}, I_u is the unbound inhibitor concentration (either the average systemic plasma concentration after repeated oral administration ($[I]_{av}$), or the maximum hepatic input concentration ($[I]_{in}$), fm_{CYP3A4} is the fraction of victim drug metabolized by CYP3A4, k_{deg} is the endogenous degradation rate constant of the enzyme, and $F_{G'}$ and F_G are the intestinal wall availability in the presence and absence of inhibitor, respectively.

The rate of change of active enzyme concentration is determined by the equilibrium between the rates of de novo synthesis and degradation of the enzyme. Previous predictions of TDI[102,103] used CYP3A4 degradation half-life estimates ($t_{1/2deg}$) obtained in either rat or Caco-2 cells,[104,105] resulting in $t_{1/2deg}$ of 14–35 h. Recently, the estimates of human CYP3A4 k_{deg} from both induction studies and in vitro investigations in liver slices have been collated,[99,100] as shown in Table 1. The estimated decay of CYP3A4 activity is twofold longer in comparison with most other cytochrome P450 enzymes,[100] but comparable to CYP2D6.[24,106] Differential degradation half-lives reported for CYP3A4 and CYP3A5 in vitro (79 versus 35 h, respectively[106]) and less susceptibility to inhibition observed for CYP3A5[107] may contribute to the extent of interindividual variability observed in the magnitude of interactions.

Venkatakrishnan and Obach[24] have indicated the importance of accurate enzyme degradation estimates for the prediction of CYP2D6 inactivation by paroxetine; a recent comparable study was reported for CYP3A4.[99] The impact of interindividual variability of human CYP3A4 $t_{1/2deg}$ (20–146 h) on the assessment of TDI potential was assessed in addition to the fraction of the victim drug metabolized by CYP34 (defined by fm_{CYP3A4}), as this parameter has been shown to have a significant impact on the DDI prediction accuracy.[14,87,92] The sensitivity of the predicted extent of interaction to the differential CYP3A4 degradation rate was dependent on fm_{CYP3A4} (as illustrated in Figure 5). The prediction accuracy was very sensitive to the CYP3A4 degradation rate for substrates mainly eliminated by this enzyme

Table 1   CYP degradation rate constants and the corresponding half-lives in humans

CYP	k_{deg} (min^{-1})	$t_{1/2deg}$ range (h)	Method	Reference
CYP3A4	0.00016	20–146	Carbamazepine induction	108
	0.00015	17–no decay	In vitro – liver slices	106
CYP3A5	0.00032	15–70	In vitro – liver slices	
CYP2D6	0.00022	51	Paroxetine inhibition	24
CYP1A2	0.0003	27–54	Smoking induction	109

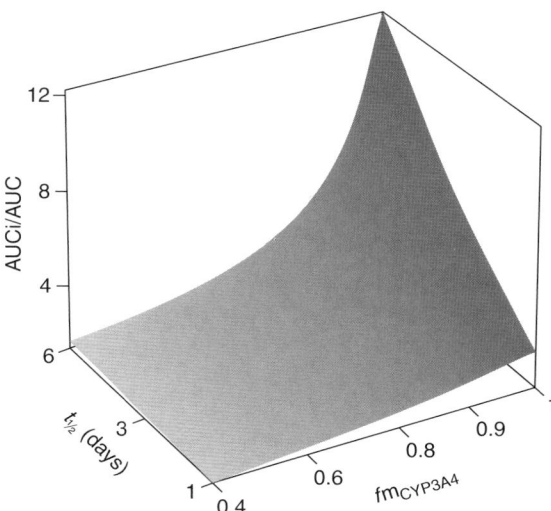

Figure 5 Effect of fm_{CYP3A4} and CYP3A4 $t_{1/2deg}$ on the predicted AUC ratio. Surface simulated for erythromycin k_{inact} and K_I values (0.025 min^{-1} and 13 µM, respectively) for [I] of 0.5 µM over the range of $fm_{CYP3A4} = 0.4–1$ and CYP3A4 $t_{1/2deg} = 1–6$ days; eqn [19] was used with $F_{G'}/F_G = 1$. (Reproduced with permission from Galetin, A.; Burt, H.; Gibbons, L.; Houston, J. B. *Drug Metab. Dispos.* **2006**, *34*, 166–175.)

($fm_{CYP3A4} \geqslant 0.9$); minimal effects are observed when CYP3A4 contributes less than 50% to the overall elimination in cases when the parallel elimination pathway is not subject to inhibition. The study also indicated the suitability of the mean CYP3A4 $t_{1/2deg}$ of 3 days in the assessment of time-dependent interaction potential.

5.35.6   Impact of the Intestinal Metabolic Interactions

Another confounding factor in the prediction of DDI is the possibility of the interaction at the level of the gut wall. Significant intestinal first-pass metabolism may contribute to the interindividual variability observed in bioavailability and extent of DDI for some CYP3A4 substrates. The impact of metabolic intestinal interaction has only recently been considered as part of the prediction strategy.[1,3,99,103]

A certain degree of interaction at the level of the gut wall is to be expected for some CYP3A4 substrates as the small intestine represents the first in the series of the first-pass eliminating organs and is exposed to the orally administered inhibitors or inducers. A number of studies indicated a more pronounced inhibition and induction of intestinal CYP3A4 in comparison to the liver.[110–112] The interpretation of such events must be cautious for compounds that are mutual substrates for P-glycoprotein and CYP3A4 (e.g., ciclosporin, tacrolimus) due to their interactive nature and inverse relationship in the intestine and liver.[113] Intestinal interactions are rarely incorporated in the assessment of interaction potential, whether for reversible or irreversible types,[1] which may represent a limitation for some CYP3A4 compounds. For example, a 2.6-fold greater increase in the AUC ratio after oral dose of midazolam in comparison to the intravenous (i.v.) administration was observed in the presence of clarithromycin, indicating the contribution of intestinal interaction.[114]

The contribution of the enzyme interaction in the intestinal wall can be evaluated by incorporating the ratio of intestinal wall availability in the presence and absence of inhibitor, $F_{G'}$ and F_G, respectively, into the hepatic prediction,[99,103] as shown in eqn [19]. The F_G control values for the corresponding substrates are obtained indirectly from oral and i.v. administration data assuming complete absorption, negligible intestinal metabolism after i.v. administration, and that the systemic clearance of a drug after i.v. dose reflects only hepatic elimination (eqns [20] and [21])[115]:

$$F = f_a(1 - E_G)(1 - E_H) \tag{20}$$

$$E_H = 1 - CL_H/Q_H \tag{21}$$

where F represents the overall oral bioavailability, f_a is the fraction of the oral dose absorbed intact across the apical membrane of the epithelial layer, E_G and E_H are the intestinal and hepatic extraction ratios, respectively, CL_H is the hepatic blood clearance and Q_H is the hepatic blood flow. **Figure 6** shows the general relationship between $F_{G'}/F_G$ ratio

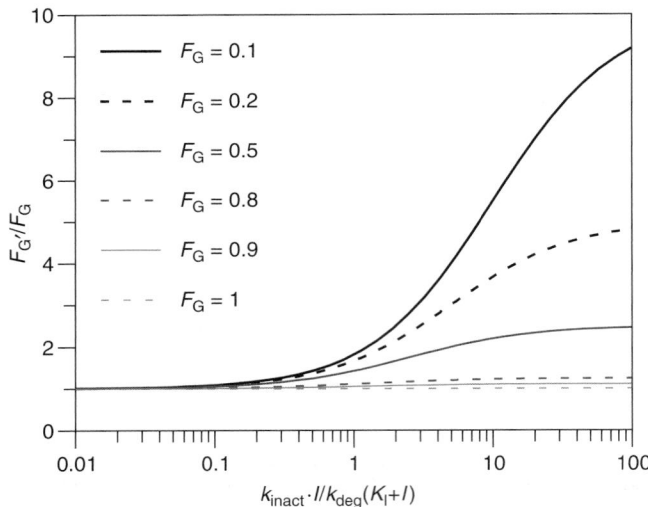

Figure 6   Relationship between $F_{G'}/F_G$ and inactivation index $k_{inact} \cdot I/k_{deg}(K_I + I)$ for varying degrees of intestinal extraction (0 – 1). $F_{G'}/F_G$ was simulated from the relative change in the intestinal intrinsic clearance caused by the inhibitor; this is estimated from the inactivation index parameters. (Reproduced with permission from Galetin, A.; Burt, H.; Gibbons, L.; Houston, J. B. *Drug Metab. Dispos.* **2006**, *34*, 166–175.)

and the relative change in the intestinal clearance caused by an inhibitor. The effect of F_G ratio is most pronounced for compounds with high intestinal extraction (e.g., buspirone, tacrolimus), whereas it is minimal for the drugs with intestinal extraction of around and below 50% (e.g., indinavir, alprazolam).

The $F_{G'}/F_G$ ratio can be determined experimentally from the relative change in the intestinal clearance of a compound in the presence of an inhibitor. However, when this is not available, an initial estimate of the extent of intestinal inhibition can be obtained by assuming the 'worst-case scenario,' i.e., the maximum inhibition of intestinal CYP3A4 ($F_{G'} = 1$). **Table 2** indicates a fivefold range of maximum $F_{G'}/F_G$ ratios for the nine CYP3A4 substrates.[99] A limitation of this approach is that the predicted values of the maximal $F_{G'}/F_G$ ratio are directly related to the accuracy of the initial estimates of the intestinal extraction and are based on the assumption that the extent of intestinal metabolism after i.v. administration is negligible and that absorption is complete, which may not be correct. For certain high-permeability/low-solubility compounds (tacrolimus, ciclosporin, and lovastatin),[36] possible contribution of the efflux transporters and their interplay with CYP3A4 will affect the initial significance of the intestinal metabolism. For these compounds the relationship between the F_G ratio and the inhibitor potency will be more complex in comparison to the substrates mainly cleared by metabolism (e.g., midazolam, nifedipine).

5.35.7   Concomitant Inhibition and Induction of Cytochrome P450

In some cases, a drug may be both an inhibitor and an inducer of drug-metabolizing enzymes and consequently the effects of such a drug will vary with single and multiple administration. Acute dosing will produce an inhibitory effect, whereas long-term administration will lead to induction of metabolism, usually by increasing the rate of protein synthesis. Thus the net effect of such a compound at any given time point during the course of administration will be a result of both mechanisms. Inhibition and induction of CYP3A4 can occur not only at the liver but also at the gut and the magnitude of the mDDI will depend on the relative concentrations of the interacting compounds in the hepatocytes and the enterocytes.

Ritonavir, the HIV-protease inhibitor, is an example of a compound which is both a potent inhibitor of CYP3A4[123,124] and also an inducer of CYP3A4-mediated metabolism.[125] Furthermore, ritonavir acts not only as a competitive inhibitor of CYP3A4 but also as a mechanism-based inhibitor of this enzyme,[101,126] resulting in the generation of catalytically inactive enzyme in the system.

Human hepatocytes treated with ritonavir showed a marked inhibition of the CYP3A4 marker activity testosterone 6β-hydroxylase but induction of CYP3A4 protein was also detected by immunoblotting techniques.[33] These results can be explained by the potent inhibitory effects of ritonavir on CYP3A4 in the cultures in combination with an inductive effect.

Table 2 List of F_G values and the maximum F_G ratio values for nine CYP3A4 substrates. Complete intestinal inhibition is assumed ($F_{G'} = 1$)

CYP3A4 substrate	F_G	Maximum $F_{G'}/F_G$	Reference
Buspirone	0.21	4.76	3
Ciclosporin	0.39	2.6	116
Felodipine	0.45	2.2	117
Midazolam	0.57	1.75	118
Triazolam	0.6	1.67	119
Simvastatin	0.66	1.51	3
Nifedipine	0.74	1.35	120
Quinidine	0.9	1.11	121
Alprazolam	0.98	1.02	122

Reproduced with permission from Galetin, A.; Burt, H.; Gibbons, L.; Houston, J. B. *Drug Metab. Dispos.* **2006**, *34*, 166–175.

Simulation methodology can be applied to take into account the time course of these effects and the differing mechanisms of inhibition. Thus values for the competitive inhibition constant (K_i) and the constants for mechanism-based inactivation (K_I, k_{inact}) can be used together with the value for k_{deg}, the endogenous degradation rate constant of the enzyme, and incorporated together with the increased synthetic rate of the enzyme (induction). A simulation of the interaction between ritonavir and methadone (a CYP3A4 substrate) has been generated which shows the shift in methadone kinetics from an acute inhibitory effect to an offset of this effect as induction due to ritonavir is manifested.[1] In the absence of clinical data, such simulations may be the only tool that can assess the impact of such a combination of effects over a time course of drug administration.

5.35.8 Simulation Software Available for Drug–Drug Interactions

A wide variety of computer software environments exists for the examination of potential DDI. These broadly fall into two categories: high-level computing software adapted for physiologically based simulation and specific biomathematical tools which have been developed for particular uses. Some examples of the former include programs based on the programming languages Fortran and C++, which, although very powerful and flexible, do require some advanced knowledge. Other more general tools include Matlab/Simulink,[138] MathCad,[139] Berkeley Madonna,[140] STELLA,[141] MLAB,[142] ModelMaker,[143] and GNU Octave.[144] All of these packages have been used for pharmacokinetic and/or pharmacodynamic model building and simulation; however, few examples can be found of their application in the predictive determination of DDIs. ACSLXtreme[145] has developed several modules for both pharmaco- and toxicokinetic applications: each of these could in theory be used for the prediction of DDIs. Precompiled modules for some of these applications are available for physiologically-based pharmacokinetic (PBPK) modeling; however, it is unclear whether they have been applied to the task of either individual or population-based DDI prediction. Although undoubtedly powerful tools, they do require a certain degree of specialist knowledge in model construction and mathematical simulation techniques.

Several packages have been developed specifically with pharmacokinetic simulation in mind and come in varying degrees of applicability to the drug metabolism scientist. These include NONMEM[146] and Pharsight's Trial Simulator and WinNonlin.[147] All of these products could be adjusted to perform predictions of DDIs in individuals, although this would require the construction and validation of novel models, again necessitating some degree of mathematical knowledge. Gastroplus[148] is primarily used for the simulation of gastrointestinal absorption but can also be used to investigate DDIs at the level of intestinal CYPs with some minor modifications. Another PBPK software, PK-Sim,[149] may also be used in this manner, although intimate knowledge of the existing exposure model would be necessary. Simcyp Clearance and Interaction Trial Simulator[150] has been developed in association with a consortium of pharmaceutical companies and regulatory authorities to simulate and predict pharmacokinetics and DDIs in populations of individuals. It combines common in vitro and preclinical data with information on human demography, anatomy, physiology, and genetics, resulting in simulations of population pharmacokinetics and DDIs involving both gastrointestinal and liver CYPs. In addition to simple, competitive interactions, enzyme induction and mechanism-based inhibition can also be simulated either alone or in combination with other effects.

Although many software tools are available for the physiologically-based prediction of DDIs, few offer ease of use to the nonstatistician. The flexibility and computational efficiency offered by programming languages such as C++ are largely offset by their relative difficulty of use. However, specialist PBPK modeling programs are often not designed for DDI simulation and must therefore be adapted for this purpose.

5.35.9   Success of Predictions

Some of the various difficulties in assessing the in vitro parameters required for the extrapolation of in vivo human DDIs have already been addressed (see below). The importance of determining the accurate concentrations of both 'victim' and 'perpetrator' drugs at the active site of the in vitro enzyme has been well described.[12,89] The potentially confounding extrapolation factors such as the free fraction of drug in blood, plasma, and the in vitro incubation of choice combine to demonstrate the robustness of current extrapolative models and their ability to provide reasonable predictions at all. Since all of the models presented for the extrapolation of DDIs require fixed in vitro inputs, the issue of uncertainty in these predictions must be addressed. In order to appreciate fully the impact of uncertainty on predictions of in vivo DDIs, this must be separated from biological variability, which is inherent in many of the in vitro systems of choice.

Standard experimental approaches of repeating both incubations and analyses and assigning confidence limits to nonlinear fits can be used to dissect out the impact of uncertainty. In a drug discovery setting these are of dubious value as the predictions are largely of a qualitative or categorical nature. It is relatively simple to substitute a range of values for each parameter and determine the sensitivity of the model to this variance. A more thorough approach would involve the use of fuzzy simulations in a similar manner to that already performed for certain PBPK models.[127] Perhaps the most confounding factor in any such sensitivity analyses would be the fact that the sensitivity to each in vitro parameter would be entirely drug-specific and therefore at best a set of general guidelines could be derived. A thorough examination of the experimental biases involved in standard experimental designs[128] has revealed that extreme care must be taken in the intelligent design of in vitro experiments in order to minimize the error passed into any extrapolative model.

Both uncharacterized HLM and human hepatocytes possess an inherent biological variability which has contributed to the lack of efficacy of some simple IVIVE approaches. When combined with experimental uncertainty, the biological variability of these in vitro systems increases the attractiveness of homogeneous rhCYPs.

The uncertainty and variabilities present in in vitro experimental systems can also be complemented by a potentially large degree of this observed in vivo. Thus it is essential a priori to define the target of any extrapolative attempt. Various differences have been observed between the typical healthy volunteer of early clinical studies and the patient for whom the drug has been designed. Differences have been observed in both CYPs[129,130] in addition to the disease-specific differences in demography, and other factors such as plasma protein concentrations.[91] Using the example of ritonavir and sildenafil,[131] it has been demonstrated that the individuals most likely to suffer extreme DDIs are those who would not usually be captured by early clinical investigation.

To a great extent the issue of CYP genetic polymorphisms can be counted as combinations of interindividual variability and DDIs. Once fm_{CYP} has been defined for a particular drug, the implications for the abolition of a polymorphic CYP has much the same impact as a potent inhibitor of that pathway. Indeed, the phenomenon of 'phenocopying'[132] has been observed in individuals taking ecstasy (methylenedioxymethamphetamine, MDMA) where they exhibit the metabolic characteristics of CYP2D6 poor metabolizers due to the potent mechanism-based inhibition of the drug. CYP2D6 has been particularly characterized in this respect as 5–10% of European populations are poor metabolizers.[133]

To carry out successful extrapolations, the a priori knowledge of the clinical situation should be as complete as possible. Knowledge of the clinical demographics, disease, and genetic status are important, as demonstrated in the example above. Equally important are the issues of formulation and proposed dosage regimen: this has been investigated both clinically[134,135] and by simulation,[136,137] using various pairs of drugs which have been dosed at various intervals relative to each other. The results demonstrated that even altering the relative dosage intervals by as little as 2 h could change the observed DDI (as measured by AUC ratio) by up to sixfold.[137]

To summarize, there are many potentially confounding factors to the successful extrapolation of DDIs from in vitro to in vivo, those related to the experimental system of choice, the assumptions inherent in the model, and the in vivo scenario to be predicted. In order to predict DDI potential accurately in humans, all of these factors must be considered individually and systematically before such strategies can be considered reliable. The prediction of DDIs involves the assessment of multiple factors, including clearance mechanisms for both substrates and inhibitors and the fraction metabolized by specific enzymes. Valid mechanistic approaches described here, based on incorporating population variability (both physiological and genetic), will allow the maximum information on the interindividual variation in the magnitude of a DDI to be obtained.

References

1. Rostami-Hodjegan, A.; Tucker, G. T. *Drug Disc. Today: Technologies* **2004**, *1*, 441–448.
2. Bjornsson, T. D.; Callaghan, J. T.; Einolf, H. J.; Fischer, V.; Gan, L.; Grimm, S.; Kao, J.; King, S. P.; Miwa, G.; Ni, L. et al. *Drug Metab. Dispos.* **2003**, *31*, 815–832.
3. Obach, R. S.; Walsky, R. L.; Venkatakrishnan, K.; Gaman, E. A.; Houston, J. B.; Tremaine, L. M. *J. Pharmacol. Exp. Ther.* **2006**, *316*, 336–348.
4. Dickins, M. *Curr. Top. Med. Chem.* **2004**, *4*, 1745–1766.
5. Niemi, M.; Backman, J. T.; Fromm, M. F.; Neuvonen, P. J.; Kivisto, K. T. *Clin. Pharmacokinet.* **2003**, *42*, 819–850.
6. Backman, J. T.; Kivisto, K. T.; Olkkola, K. T.; Neuvonen, P. J. *Eur. J. Clin. Pharmacol.* **1998**, *54*, 53–58.
7. Ebert, U.; Thong, N. Q.; Oertel, R.; Kirch, W. *Eur. J. Clin. Pharmacol.* **2000**, *56*, 299–304.
8. Hachad, H.; Ragueneau-Majlessi, I.; Levy, R. H. *Ther. Drug Monit.* **2002**, *24*, 91–103.
9. Ayrton, A.; Morgan, P. *Xenobiotica* **2001**, *31*, 469–497.
10. Shitara, Y.; Sato, H.; Sugiyama, Y. *Annu. Rev. Pharmacol. Toxicol.* **2005**, *45*, 689–723.
11. Bertz, R. J.; Granneman, G. R. *Clin. Pharmacokinet.* **1997**, *32*, 210–258.
12. Lin, J. H. *Curr. Drug Metab.* **2000**, *1*, 305–331.
13. Chien, J. Y.; Mohutsky, M. A.; Wrighton, S. A. *Curr. Drug Metab.* **2003**, *4*, 347–356.
14. Ito, K.; Hallifax, D.; Obach, R. S.; Houston, J. B. *Drug Metab. Dispos.* **2005**, *33*, 837–844.
15. Pearce, R. E.; McIntyre, C. J.; Madan, A.; Sanzgiri, U.; Draper, A. J.; Bullock, P. L.; Cook, D. C.; Burton, L. A.; Latham, J.; Nevins, C. et al. *Arch. Biochem. Biophys.* **1996**, *331*, 145–169.
16. Tucker, G. T.; Houston, J. B.; Huang, S. M. *Clin. Pharmacol. Ther.* **2001**, *70*, 103–114.
17. Williams, J. A.; Hurst, S. I.; Bauman, J.; Jones, B. C.; Hyland, R.; Gibbs, J. P.; Obach, R. S.; Ball, S. E. *Curr. Drug Metab.* **2003**, *4*, 527–534.
18. Tang, W.; Wang, R. W.; Lu, A. Y. *Curr. Drug Metab.* **2005**, *6*, 503–517.
19. Tucker, G. T. *Int. J. Clin. Pharmacol. Ther. Toxicol.* **1992**, *30*, 550–553.
20. Venkatakrishnan, K.; Von Moltke, L. L.; Greenblatt, D. J. *Biopharm. Drug Dispos.* **2002**, *23*, 183–190.
21. Venkatakrishnan, K.; von Moltke, L. L.; Obach, R. S.; Greenblatt, D. J. *J. Pharmacol. Exp. Ther.* **2000**, *293*, 343–350.
22. Tang, W.; Stearns, R. A. *Curr. Drug Metab.* **2001**, *2*, 185–198.
23. Nakajima, M.; Tane, K.; Nakamura, S.; Shimada, N.; Yamazaki, H.; Yokoi, T. *J. Pharm. Sci.* **2002**, *91*, 952–963.
24. Venkatakrishnan, K.; Obach, R. S. *Drug Metab. Dispos.* **2005**, *33*, 845–852.
25. Crespi, C. L. *Adv. Drug Res.* **1995**, *26*, 179–235.
26. Nakajima, M.; Nakamura, S.; Tokudome, S.; Shimada, N.; Yamazaki, H.; Yokoi, T. *Drug Metab. Dispos.* **1999**, *27*, 1381–1391.
27. Proctor, N. J.; Tucker, G. T.; Rostami-Hodjegan, A. *Xenobiotica* **2004**, *34*, 151–178.
28. LeCluyse, E.; Madan, A.; Hamilton, G.; Carroll, K.; DeHaan, R.; Parkinson, A. *J. Biochem. Mol. Toxicol.* **2000**, *14*, 177–188.
29. Drocourt, L.; Pascussi, J. M.; Assenat, E.; Fabre, J. M.; Maurel, P.; Vilarem, M. J. *Drug Metab. Dispos.* **2001**, *29*, 1325–1331.
30. Runge, D.; Kohler, C.; Kostrubsky, V. E.; Jager, D.; Lehmann, T.; Runge, D. M.; May, U.; Stolz, D. B.; Strom, S. C.; Fleig, W. E. et al. *Biochem. Biophys. Res. Commun.* **2000**, *273*, 333–341.
31. Kocarek, T. A.; Dahn, M. S.; Cai, H.; Strom, S. C.; Mercer-Haines, N. A. *Drug Metab. Dispos.* **2002**, *30*, 1400–1405.
32. Madan, A.; Graham, R. A.; Carroll, K. M.; Mudra, D. R.; Burton, L. A.; Krueger, L. A.; Downey, A. D.; Czerwinski, M.; Forster, J.; Ribadeneira, M. D. et al. *Drug Metab. Dispos.* **2003**, *31*, 421–431.
33. Luo, G.; Cunningham, M.; Kim, S.; Burn, T.; Lin, J.; Sinz, M.; Hamilton, G.; Rizzo, C.; Jolley, S.; Gilbert, D. et al. *Drug Metab. Dispos.* **2002**, *30*, 795–804.
34. Mills, J. B.; Rose, K. A.; Sadagopan, N.; Sahi, J.; de Morais, S. M. *J. Pharmacol. Exp. Ther.* **2004**, *309*, 303–309.
35. Williams, J. A.; Bauman, J.; Cai, H.; Conlon, K.; Hansel, S.; Hurst, S.; Sadagopan, N.; Tugnait, M.; Zhang, L.; Sahi, J. *Curr. Opin. Drug Disc. Dev.* **2005**, *8*, 78–88.
36. Wu, C. Y.; Benet, L. Z. *Pharm. Res.* **2005**, *22*, 11–23.
37. Rostami-Hodjegan, A.; Tucker, G. T. *Hepatology* **2002**, *35*, 1549–1550, author reply 1550–1551.
38. Polli, J. W.; Wring, S. A.; Humphreys, J. E.; Huang, L.; Morgan, J. B.; Webster, L. O.; Serabjit-Singh, C. S. *J. Pharmacol. Exp. Ther.* **2001**, *299*, 620–628.
39. Su, Y.; Zhang, X.; Sinko, P. J. *Mol. Pharm.* **2004**, *1*, 49–56.
40. Obach, R. S. *Curr. Opin. Drug Disc. Dev.* **2001**, *4*, 36–44.
41. McLure, J. A.; Miners, J. O.; Birkett, D. J. *J. Clin. Pharmacol. Br. J. Clin. Pharmacol.* **2000**, *49*, 453–461.
42. Austin, R. P.; Barton, P.; Cockroft, S. L.; Wenlock, M. C.; Riley, R. J. *Drug Metab. Dispos.* **2002**, *30*, 1497–1503.
43. Obach, R. S. *Drug Metab. Dispos.* **1999**, *27*, 1350–1359.
44. Kalvass, J. C.; Tess, D. A.; Giragossian, C.; Linhares, M. C.; Maurer, T. S. *Drug Metab. Dispos.* **2001**, *29*, 1332–1336.
45. Yao, C.; Kunze, K. L.; Kharasch, E. D.; Wang, Y.; Trager, W. F.; Ragueneau, I.; Levy, R. H. *Clin. Pharmacol. Ther.* **2001**, *70*, 415–424.
46. Ishigam, M.; Uchiyama, M.; Kondo, T.; Iwabuchi, H.; Inoue, S.; Takasaki, W.; Ikeda, T.; Komai, T.; Ito, K.; Sugiyama, Y. *Pharm. Res.* **2001**, *18*, 622–631.
47. Tran, T. H.; Von Moltke, L. L.; Venkatakrishnan, K.; Granda, B. W.; Gibbs, M. A.; Obach, R. S.; Harmatz, J. S.; Greenblatt, D. J. *Drug Metab. Dispos.* **2002**, *30*, 1441–1445.
48. Margolis, J. M.; Obach, R. S. *Drug Metab. Dispos.* **2003**, *31*, 606–611.
49. Venkatakrishnan, K.; von Moltke, L. L.; Obach, R. S.; Greenblatt, D. J. *Curr. Drug Metab.* **2003**, *4*, 423–459.
50. Houston, J. B. *Biochem. Pharmacol.* **1994**, *47*, 1469–1479.
51. Pelkonen, O.; Jouppila, P.; Karki, N. T. *Toxicol. Appl. Pharmacol.* **1972**, *23*, 399–407.
52. Schoene, B.; Fleischmann, R. A.; Remmer, H.; von Oldershausen, H. F. *Eur. J. Clin. Pharmacol.* **1972**, *4*, 65–73.
53. Baarnhielm, C.; Dahlback, H.; Skanberg, I. *Acta Pharmacol. Toxicol. (Cophenh.)* **1986**, *59*, 113–122.
54. Iwatsubo, T.; Hirota, N.; Ooie, T.; Suzuki, H.; Shimada, N.; Chiba, K.; Ishizaki, T.; Green, C. E.; Tyson, C. A.; Sugiyama, Y. *Pharmacol. Ther.* **1997**, *73*, 147–171.
55. Lipscomb, J. C.; Fisher, J. W.; Confer, P. D.; Byczkowski, J. Z. *Toxicol. Appl. Pharmacol.* **1998**, *152*, 376–387.
56. Snawder, J. E.; Lipscomb, J. C. *Regul. Toxicol. Pharmacol.* **2000**, *32*, 200–209.

57. Lipscomb, J. C.; Kedderis, G. L. *Sci. Total Environ.* **2002**, *288*, 13–21.
58. Wilson, Z. E.; Rostami-Hodjegan, A.; Burn, J. L.; Tooley, A.; Boyle, J.; Ellis, S. W.; Tucker, G. T. *J. Clin. Pharmacol. Br. J. Clin. Pharmacol.* **2003**, *56*, 433–440.
59. Soars, M. G.; Burchell, B.; Riley, R. J. *J. Pharmacol. Exp. Ther.* **2002**, *301*, 382–390.
60. Ito, K.; Houston, J. B. *Pharm. Res.* **2005**, *22*, 103–112.
61. Barter, Z. E.; PhD thesis. Sheffield, 2005.
62. Yeo, K. R.; Rostami-Hodjegan, A.; Tucker, G. T. *Br. J. Clin. Pharmacol.* **2004**, 57, 687–688.
63. Shimada, T.; Yamazaki, H.; Mimura, M.; Inui, Y.; Guengerich, F. P. *J. Pharmacol. Exp. Ther.* **1994**, *270*, 414–423.
64. Wang, Y. H.; Jones, D. R.; Hall, S. D. *Drug Metab. Dispos.* **2005**, *33*, 664–671.
65. Guengerich, F. P.; Turvy, C. G. *J. Pharmacol. Exp. Ther.* **1991**, *256*, 1189–1194.
66. Bayliss, M. K.; Bell, J. A.; Jenner, W. N.; Wilson, K. *Biochem. Soc. Trans.* **1990**, *18*, 1198–1199.
67. Bayliss, M. K.; Bell, J. A.; Jenner, W. N.; Park, G. R.; Wilson, K. *Xenobiotica* **1999**, *29*, 253–268.
68. Zuegge, J.; Schneider, G.; Coassolo, P.; Lave, T. *Clin. Pharmacokinet.* **2001**, *40*, 553–563.
69. McGinnity, D. F.; Soars, M. G.; Urbanowicz, R. A.; Riley, R. J. *Drug Metab. Dispos.* **2004**, *32*, 1247–1253.
70. Arias, I. M.; Popper, H.; Jakoby, W. B.; Schachter, D.; Shafritz, D. A. *The Liver Biology and Pathology*, 2nd ed.; Raven Press: New York, 1988.
71. Zahlten, R. N.; Stratman, F. W. *Arch. Biochem. Biophys.* **1974**, *163*, 600–608.
72. Galetin, A.; Clarke, S. E.; Houston, J. B. *Drug Metab. Dispos.* **2003**, *31*, 1108–1116.
73. Atkins, W. M. *Annu. Rev. Pharmacol. Toxicol.* **2005**, *45*, 291–310.
74. Hutzler, J. M.; Hauer, M. J.; Tracy, T. S. *Drug Metab. Dispos.* **2001**, *29*, 1029–1034.
75. Williams, J. A.; Ring, B. J.; Cantrell, V. E.; Campanale, K.; Jones, D. R.; Hall, S. D.; Wrighton, S. A. *Drug Metab. Dispos.* **2002**, *30*, 1266–1273.
76. Uchaipichat, V.; Mackenzie, P. I.; Guo, X. H.; Gardner-Stephen, D.; Galetin, A.; Houston, J. B.; Miners, J. O. *Drug Metab. Dispos.* **2004**, *32*, 413–423.
77. Houston, J. B.; Galetin, A. *1Drug Metab. Rev.* **2003**, *35*, 393–415.
78. Shou, M.; Lin, Y.; Lu, P.; Tang, C.; Mei, Q.; Cui, D.; Tang, W.; Ngui, J. S.; Lin, C. C.; Singh, R. et al. *Curr. Drug Metab.* **2001**, *2*, 17–36.
79. Galetin, A.; Clarke, S. E.; Houston, J. B. *Drug Metab. Dispos.* **2002**, *30*, 1512–1522.
80. Houston, J. B.; Kenworthy, K. E. *Drug Metab. Dispos.* **2000**, *28*, 246–254.
81. Tang, W.; Stearns, R. A.; Kwei, G. Y.; Iliff, S. A.; Miller, R. R.; Egan, M. A.; Yu, N. X.; Dean, D. C.; Kumar, S.; Shou, M. et al. *J. Pharmacol. Exp. Ther.* **1999**, *291*, 1068–1074.
82. Egnell, A. C.; Houston, B.; Boyer, S. *J. Pharmacol. Exp. Ther.* **2003**, *305*, 1251–1262.
83. Ngui, J. S.; Chen, Q.; Shou, M.; Wang, R. W.; Stearns, R. A.; Baillie, T. A.; Tang, W. *Drug Metab. Dispos.* **2001**, *29*, 877–886.
84. Hutzler, J. M.; Frye, R. F.; Korzekwa, K. R.; Branch, R. A.; Huang, S. M.; Tracy, T. S. *Eur J. Pharm. Sci.* **2001**, *14*, 47–52.
85. Ito, K.; Brown, H. S.; Houston, J. B. *J. Clin. Pharmacol. Br. J. Clin. Pharmacol.* **2004**, *57*, 473–486.
86. Rodrigues, A. D.; Winchell, G. A.; Dobrinska, M. R. *J. Clin. Pharmacol.* **2001**, *41*, 368–373.
87. Brown, H. S.; Ito, K.; Galetin, A.; Houston, J. B. *Br. J. Clin. Pharmacol.* **2005**, *60*, 508–518.
88. Kanamitsu, S.; Ito, K.; Sugiyama, Y. *Pharm. Res.* **2000**, *17*, 336–343.
89. Blanchard, N.; Richert, L.; Coassolo, P.; Lave, T. *Curr. Drug Metab.* **2004**, *5*, 147–156.
90. Daneshmend, T. K.; Warnock, D. W.; Turner, A.; Roberts, C. J. *J. Antimicrob. Chemother.* **1981**, *8*, 299–304.
91. Rowland, M.; Tozer, T. N. *Clinical Pharmacokinetics: Concepts and Applications*, 3rd ed; Williams and Wilkins: Baltimore, 1995.
92. Galetin, A.; Ito, K.; Hallifax, D.; Houston, J. B. *J. Pharmacol. Exp. Ther.* **2005**, *314*, 180–190.
93. Lin, Y. S.; Dowling, A. L.; Quigley, S. D.; Farin, F. M.; Zhang, J.; Lamba, J.; Schuetz, E. G.; Thummel, K. E. *Mol. Pharmacol.* **2002**, *62*, 162–172.
94. Xie, H. G.; Wood, A. J.; Kim, R. B.; Stein, C. M.; Wilkinson, G. R. *Pharmacogenomics* **2004**, *5*, 243–272.
95. Zhang, Y.; Benet, L. Z. *Clin. Pharmacokinet.* **2001**, *40*, 159–168.
96. Silverman, R. B. *Methods Enzymol* **1995**, *249*, 240–283.
97. Kent, U. M.; Juschyshyn, M. I.; Hollenberg, P. F. *Curr. Drug Metab.* **2001**, *2*, 215–243.
98. Zhou, S.; Yung Chan, S.; Cher Goh, B.; Chan, E.; Duan, W.; Huang, M.; McLeod, H. L. *Clin. Pharmacokinet.* **2005**, *44*, 279–304.
99. Galetin, A.; Burt, H.; Gibbons, L.; Houston, J. B. *Drug Metab. Dispos.* **2006**, *34*, 166–175.
100. Ghanbari, F.; Rowland-Yeo, K.; Bloomer, J.; Clarke, S. E.; Lennard, M. S.; Tucker, G. T.; Rostami-Hodjegan, A. *Curr. Drug Metab.* **2006**, 7, 315–334.
101. Ernest, C. S., 2nd; Hall, S. D.; Jones, D. R. *J. Pharmacol. Exp. Ther.* **2005**, *312*, 583–591.
102. Ito, K.; Ogihara, K.; Kanamitsu, S.; Itoh, T. *Drug Metab. Dispos.* **2003**, *31*, 945–954.
103. Wang, Y. H.; Jones, D. R.; Hall, S. D. *Drug Metab. Dispos.* **2004**, *32*, 259–266.
104. Correia, M. A. *Methods Enzymol* **1991**, *206*, 315–325.
105. Malhotra, S.; Schmiedlin-Ren, P.; Paine, M. F.; Criss, A. B.; Watkins, P. B. *Drug Metab. Rev.* **2001**, *33*, 97.
106. Renwick, A. B.; Watts, P. S.; Edwards, R. J.; Barton, P. T.; Guyonnet, I.; Price, R. J.; Tredger, J. M.; Pelkonen, O.; Boobis, A. R.; Lake, B. G. *Drug Metab. Dispos.* **2000**, *28*, 1202–1209.
107. McConn, D. J., 2nd; Lin, Y. S.; Allen, K.; Kunze, K. L.; Thummel, K. E. *Drug Metab. Dispos.* **2004**, *32*, 1083–1091.
108. Lai, A. A.; Levy, R. H.; Cutler, R. E. *Clin. Pharmacol. Ther.* **1978**, *24*, 316–323.
109. Faber, M. S.; Fuhr, U. *Clin. Pharmacol. Ther.* **2004**, *76*, 178–184.
110. Gomez, D. Y.; Wacher, V. J.; Tomlanovich, S. J.; Hebert, M. F.; Benet, L. Z. *Clin. Pharmacol. Ther.* **1995**, *58*, 15–19.
111. Floren, L. C.; Bekersky, I.; Benet, L. Z.; Mekki, Q.; Dressler, D.; Lee, J. W.; Roberts, J. P.; Hebert, M. F. *Clin. Pharmacol. Ther.* **1997**, *62*, 41–49.
112. Tsunoda, S. M.; Velez, R. L.; von Moltke, L. L.; Greenblatt, D. J. *Clin. Pharmacol. Ther.* **1999**, *66*, 461–4671.
113. Benet, L. Z.; Cummins, C. L.; Wu, C. Y. *Int. J. Pharm.* **2004**, *277*, 3–9.
114. Gorski, J. C.; Jones, D. R.; Haehner-Daniels, B. D.; Hamman, M. A.; O'Mara, E. M.; Hall, S. D. *Clin. Pharmacol. Ther.* **1998**, *64*, 133–143.
115. Hall, S. D.; Thummel, K. E.; Watkins, P. B.; Lown, K. S.; Benet, L. Z.; Paine, M. F.; Mayo, R. R.; Turgeon, D. K.; Bailey, D. G.; Fontana, R. J. et al. *Drug Metab. Dispos.* **1999**, *27*, 161–166.
116. Hebert, M. F.; Roberts, J. P.; Prueksaritanont, T.; Benet, L. Z. *Clin. Pharmacol. Ther.* **1992**, *52*, 453–457.
117. Lundahl, J.; Regardh, C. G.; Edgar, B.; Johnsson, G. *Eur. J. Clin. Pharmacol.* **1997**, *52*, 139–145.

118. Thummel, K. E.; O'Shea, D.; Paine, M. F.; Shen, D. D.; Kunze, K. L.; Perkins, J. D.; Wilkinson, G. R. *Clin. Pharmacol. Ther.* **1996**, *59*, 491–502.
119. Masica, A. L.; Mayo, G.; Wilkinson, G. R. *Clin. Pharmacol. Ther.* **2004**, *76*, 341–349.
120. Holtbecker, N.; Fromm, M. F.; Kroemer, H. K.; Ohnhaus, E. E.; Heidemann, H. *Drug Metab. Dispos.* **1996**, *24*, 1121–1123.
121. Damkier, P.; Hansen, L. L.; Brosen, K. *Eur. J. Clin. Pharmacol.* **1999**, *55*, 451–456.
122. Hirota, N.; Ito, K.; Iwatsubo, T.; Green, C. E.; Tyson, C. A.; Shimada, N.; Suzuki, H.; Sugiyama, Y. *Biopharm. Drug Dispos.* **2001**, *22*, 53–71.
123. Iribarne, C.; Berthou, F.; Carlhant, D.; Dreano, Y.; Picart, D.; Lohezic, F.; Riche, C. *Drug Metab. Dispos.* **1998**, *26*, 257–260.
124. Greenblatt, D. J.; von Moltke, L. L.; Harmatz, J. S.; Durol, A. L.; Daily, J. P.; Graf, J. A.; Mertzanis, P.; Hoffman, J. L.; Shader, R. I. *Clin. Pharmacol. Ther.* **2000**, *67*, 335–341.
125. Geletko, S. M.; Erickson, A. D. *Pharmacotherapy* **2000**, *20*, 93–94.
126. von Moltke, L. L.; Durol, A. L.; Duan, S. X.; Greenblatt, D. J. *Eur. J. Clin. Pharmacol.* **2000**, *56*, 259–261.
127. Gueorguieva, I. I.; Nestorov, I. A.; Rowland, M. *J. Pharmacokinet. Pharmacodyn.* **2004**, *31*, 185–213.
128. Yang, J.; Jamei, M.; Yeo, K. R.; Tucker, G. T.; Rostami-Hodjegan, A. *Eur. J. Pharm. Sci.* **2005**, *26*, 334–340.
129. Forsyth, J. T.; Grunewald, R. A.; Rostami-Hodjegan, A.; Lennard, M. S.; Sagar, H. J.; Tucker, G. T. *Br. J. Clin. Pharmacol.* **2000**, *50*, 303–309.
130. Rostami-Hodjegan, A.; Amin, A. M.; Spencer, E. P.; Lennard, M. S.; Tucker, G. T.; Flanagan, R. J. *J. Clin. Psychopharmacol.* **2004**, *24*, 70–78.
131. Yang, J.; Rostami-Hodjegan, A.; Tucker, G. T. *Br. J. Clin. Pharmacol.* **2002**, *53*, 438–439.
132. Heydari, A. *Br. J. Clin. Pharmacol.* **2003**, *55*, 430.
133. Alvan, G.; Bechtel, P.; Iselius, L.; Gundert-Remy, U. *Eur. J. Clin. Pharmacol.* **1990**, *39*, 533–537.
134. Neuvonen, P. J.; Varhe, A.; Olkkola, K. T. *Clin. Pharmacol. Ther.* **1996**, *60*, 326–331.
135. Seidegard, J. *Clin. Pharmacol. Ther.* **2000**, *68*, 13–17.
136. Fang, J.; McKay, G.; Hubbard, J. W.; Hawes, E. M.; Midha, K. K. *Biopharm. Drug Dispos.* **2000**, *21*, 249–259.
137. Yang, J.; Kjellsson, M.; Rostami-Hodjegan, A.; Tucker, G. T. *Eur J. Pharm. Sci.* **2003**, *20*, 223–232.
138. Matlab/Simulink. http://www.mathworks.com (accessed Aug 2006).
139. MathCad. http://www.mathsoft.com (accessed Aug 2006).
140. Berkeley Madonna. http://www.berkeleymadonna.com (accessed Aug 2006).
141. STELLA. http://www.hps-inc.com/ (accessed Aug 2006).
142. MLAB. http://www.civilized.com/ (accessed Aug 2006).
143. ModelMaker. http://www.modelkinetix.com (accessed Aug 2006).
144. GNU Octave. http://www.octave.org/ (accessed Aug 2006).
145. ACSLXtreme. http://www.aegisxcellon.com (accessed Aug 2006).
146. NONMEM. http://www.globomax.com/ (accessed Aug 2006).
147. Pharsight's Trial Simulator and WinNonlin. http://www.pharsight.com (accessed Aug 2006).
148. Gastroplus. http://www.simulations-plus.com (accessed Aug 2006).
149. PK-Sim. http://www.bayertechnology.com (accessed Aug 2006).
150. Simcyp Clearance and Interaction Trial Simulator. http://www.simcyp.com (accessed Aug 2006).

Biographies

Maurice Dickins has worked in drug metabolism in the pharmaceutical industry since 1984, following the award of his PhD from the University of Surrey, UK. He has held successive positions at Wellcome Research, Glaxo Wellcome, and GlaxoSmithKline and joined Pfizer in 2002.

His current role is to understand the enzymology of biotransformations in human in vitro systems and to assess their relevance in the clinic. He is particularly interested in the prediction of metabolism and potential drug interactions using in vitro and in silico approaches.

Aleksandra Galetin is the Pfizer Lecturer in Drug Metabolism and Pharmacokinetics in the School of Pharmacy and Pharmaceutical Sciences, University of Manchester, UK. Prior to her current position she was a postdoctoral research associate in the Centre for Applied Pharmacokinetic Research at the University of Manchester.

Her recent publications centred on the incorporation of homo/heterotropic cooperativity phenomena associated with CYP3A4 (and other enzymes) into the in vitro–in vivo prediction of either clearance or drug–drug interactions. Her current research activities focus on the mechanism-based prediction of human pharmacokinetics, in particular the contribution of the intestinal CYP and UGT enzymes to the clearance and drug interactions and investigations of time-dependent inhibition interactions.

Nick Proctor received his PhD from the University of Manchester, UK, under the supervision of Prof Brian Houston, investigating the impact of drug binding in the predictions of human pharmacokinetics and metabolic drug interactions. Nick was the Dr Hadwen Trust Postdoctoral Fellow at the University of Sheffield, UK, and worked on the refinement of automated in vitro–in vivo extrapolation techniques before becoming Business Development Manager for Simcyp in 2004.

Comprehensive Medicinal Chemistry II
ISBN (set): 0-08-044513-6

ISBN (Volume 5) 0-08-044518-7; pp. 827–846

5.36 In Silico Prediction of Plasma and Tissue Protein Binding

G Colmenarejo, GlaxoSmithKline, Madrid, Spain

5.36.1 Introduction

The reversible strong binding of a drug to its target (whether in the cell membrane, and therefore accessible to the extracellular fluid, or within the cell) is in competition with the reversible binding to multitude of low-affinity sites in the blood and the tissues: plasma proteins, membranes, intracellular proteins and nucleic acids, fat, cellular organelles like microsomes and lysosomes.[1–3] This distribution of the drug through the body conditions its pharmacokinetics. For instance, consider an orally administered drug. After being (partially) absorbed in the intestine, it travels in the circulation through the portal vein to the liver, and there it suffers some initial elimination by hepatic metabolism (the so-called first-pass metabolism). The fraction remaining (bioavailable) is distributed throughout the body, first in the circulation partially bound to plasma proteins, and later to the different tissues, until an equilibrium tissue/plasma distribution is established. Some drugs then remain in the extracellular fluid and do not enter the cells, while some others are able to cross the cell membranes and get inside the cell. With time, the fraction of the drug remaining in blood plasma is eliminated as it passes through the eliminating organs, mainly kidney (renal excretion) and liver (hepatic metabolism), while the tissue/plasma distribution equilibrium is reestablished at the expense of drug molecules leaving the tissues. At any moment, it is only the free drug at the extracellular fluid that is able to bind the target: directly in the case of targets in the cell membrane, or after (partially) crossing the cell membrane in the case of intracellular targets. Although of lower affinity, given the overwhelmingly larger amount of binding sites in plasma proteins and tissues compared to those of the target, the concentration of free drug in the extracellular fluid is determined by plasma protein and tissue binding.

There is an important pharmacokinetic concept that models the distribution of drugs within the body, the so-called volume of distribution, V_D (also $V_{D,ss}$), which is defined as[1]:

$$V_D = \frac{Dose}{C_o} \qquad [1]$$

where *Dose* is the administered dose, and C_o is the initial concentration of the drug in plasma, before elimination starts; here complete bioavailability is assumed. The volume of distribution corresponds to the theoretical volume where the dose should be distributed, if it had the same concentration as in plasma. It is easy to see that:

$$V_D = V_P + V_T \frac{f_{u,P}}{f_{u,T}} \qquad [2]$$

where V_P is the plasma volume, V_T is the tissue volume outside plasma where the drug distributes, and $f_{u,P}$ and $f_{u,T}$ are the fractions of unbound drug in plasma and tissue, respectively. Thus, the volume of distribution increases when the tissue binding is increased ($f_{u,T}$ decreases) while it decreases when the plasma binding is increased ($f_{u,P}$ decreases). The importance of the volume of distribution stems from the fact that, in combination with the clearance of the drug (CL, the volume of plasma with the amount of drug eliminated by unit time), it determines the half-life of the drug (the time required to metabolize and/or eliminate half of the administered drug):

$$t_{1/2} = 0.693 V_D / CL = 0.693\left(V_P + V_T \frac{f_{u,P}}{f_{u,T}}\right) / CL \qquad [3]$$

Thus, the half-life of the drug is linear with the volume of distribution.

All these considerations stress the fundamental role that the plasma and tissue protein binding of a drug have in determining both its free concentration in the extracellular fluid and its half-life. They must be considered in order to develop drugs with appropriate pharmacokinetic behavior. Modifications of molecular properties to fix absorption and clearance issues must in this way consider their impact in drug distribution by altering the plasma and/or tissue-binding (and therefore the volume of distribution) properties. In silico models to predict these properties are thus expected to be of much help in drug design. They are probably most useful at intermediate stages of drug discovery, e.g., in the lead optimization and preclinical phases, in order to guide the fine-tuning of the structure of the lead or candidate molecule to improve its pharmacokinetics and therapeutic windows in combination with clearance and toxicity data, as well as to provide first estimates for doses for in vivo assays in animals and humans (in phase I clinical assays). In very early stages like hit selection and the hit-to-lead phase they can be useful to detect extreme values for these parameters: for instance, Kratochwil et al.[4,5] found in a set of 408 therapeutic drugs a wide range of plasma protein-binding affinities, although no drugs were found with submicromolar binding. These extreme binding affinities (also applicable to tissue binding) should force the use of very large doses of compound in order to achieve an acceptable saturation of the target, with very negative consequences, such as toxicity, drug–drug interactions, impossibility of oral administration of such a dose given the physicochemistry of the compound, etc.

The present chapter reviews the current models for plasma and tissue protein-binding prediction. The models to predict the volume of distribution are briefly reviewed as well.

5.36.2   Biological Background

5.36.2.1   Blood, Plasma, and Serum

Human blood is composed of 45% cells, plus 55% plasma. In turn, plasma is an aqueous solution composed of 92% water, 7% proteins, and 1% other solutes such as inorganic ions. Serum is the aqueous solution that remains after precipitating and removing the clotting factors in plasma. Most of the plasma proteins are synthesized in the liver. They are grouped as follows: 60% albumin, 35% globulins, 4% clotting proteins, and 1% regulatory proteins. From the point of view of drug distribution, the most important proteins are human serum albumin (HSA) and α_1-acid glycoprotein (AGP)[6,7]; there are other less relevant proteins, like lipoproteins. Drug binding to blood cells is also possible, for instance in the case of drugs lipophilic enough to cross the membrane, or positively charged drugs that can bind the anionic phospholipids in the membrane; binding to blood cells is frequently weak and less relevant, though. Binding of drugs to plasma proteins influences the concentration of free drug in the extracellular space, and is especially important as the fraction of drug remaining in plasma is the one that is mainly subject to metabolism and excretion by the eliminating organs. In the next section, the structure and function of the plasma proteins relevant to drug distribution are described briefly.

5.36.2.1.1 Human serum albumin

HSA is by far the most abundant protein in serum, with a concentration of about $40\,mg\,mL^{-1}$ ($\sim 0.6\,mmol\,L^{-1}$).[8] Its main function is the transport of nonsterified fatty acids in the extracellular compartment, as well as of multiple additional endogenous ligands like bilirubin, bile salts, steroids, hemin, thyroxine, some vitamins, and metal ions.[8,9] In addition, this protein has the striking ability to bind a wide variety of structurally dissimilar drugs, and this is the reason for its importance in drug pharmacokinetics. Two main binding sites for drugs were proposed after biochemical studies: site I, or warfarin site, and site II, or indole-benzodiazepine site.[10,11]

HSA is a member of a multigene family of proteins, including vitamin D-binding protein and α-fetoprotein.[6,8] It is synthesized by hepatocytes in the liver as preproalbumin, which is further processed intracellularly to remove the corresponding pre- and propeptide sequences, and finally secreted into the blood.[12] It is composed of a single polypeptide chain 585 amino acids long, with 17 disulfide bridges and 66 400 molecular weight.[8] Its primary structure is organized in three consecutive homologous domains (I, II, and III) that evolved from the repetition of an ancestor gene.[13] Each domain is, in turn, composed of two partially homologous subdomains, named A and B. **Figure 1** displays a secondary structure scheme of HSA.

Several crystal structures of HSA, alone or complexed with fatty acids and/or other ligands, have been solved.[14–24] A query in the Protein Data Bank for 'human serum albumin' results in 26 entries as of May 2005. These are shown in **Table 1**, together with the ligands (if applicable), resolution, original reference and year of deposition.

All of them display a heart-shaped structure of the protein, with each of the three lobes of the heart corresponding to one of the homologous domains in the primary structure (**Figure 2**). Some 67% of the protein is α-helical, and each domain comprises 10 α-helices (h1–h10) separated by disordered loops, with six helices in the A subdomains (h1–h6), and four in the B subdomains (h7–h10). Consecutive helices in contiguous domains (h10(I)–h1(II); h10(II)–h1(III)) are aligned, forming long helices.

The crystal structures with fatty acids display multiple sites for the binding of these molecules, some of them overlapping with the drug-binding sites. Fatty acid binding drives a large conformational change in the protein in which domains I and III are rotated around domain II, and the long h10(I)–h1(II) helix is bent (**Figure 2b**).[15,17,18] The crystal structures with drugs have allowed to localize in the three-dimensional structure of the protein the two drug-binding sites described by Sudlow et al.,[10,11] at subdomains IIA and IIIA, respectively. **Table 2** lists the binding sites of multiple drugs as observed from crystal structures of HSA–drug complexes. In the cases of warfarin[19] and the two general anesthetics halothane and propofol,[16] the crystals have enough resolution to permit the fit of the drug structures in atomic detail.

All these structures highlight the promiscuity of this protein, that binds multiple and structurally different ligands at different binding sites.

5.36.2.1.2 α₁-Acid glycoprotein

Another abundant protein in serum is AGP, with a concentration of about $1\,mg\,mL^{-1}$ ($\sim 0.02\,mmol\,L^{-1}$), that can increase several times in acute-phase inflammation reactions.[25] AGP has a role as a natural antiinflammatory and immunomodulatory agent. It is synthesized and secreted to the bloodstream mainly in the liver by hepatocytes, although some extrahepatic synthesis has also been detected.

This acidic glycoprotein is composed of a single polypeptide chain of 183 amino acids and two disulfide bridges. It displays five N-linked glycans of high complexity and variable composition (12–20 different glycoforms are observed in normal human serum). The molecular weight thus varies between 41 and 43 kDa, with about 45% of the molecular weight corresponding to the carbohydrate. The polypeptide appears as two or three main forms, namely the A form and the F1 and/or S forms, encoded by three clustered genes: AGP-A, AGP-B, and AGP-B'.[25,26] The A form is encoded by the AGP-B and AGP-B' duplicated genes, while the F1 and S forms are encoded by two different alleles of the gene AGP-A. There is a difference in 22 bases between A and F1 encoding genes, and of a few bases between F1 and S genes.

AGP is another important nonspecific drug binder in plasma. It binds different drugs such as tamoxifen, propranolol, vanilloids, and phenobarbital, and also endogenous compounds, such as retinoic acid, steroids, platelet-activating factor, and heparin.[27] It is normally stated that HSA binds acidic drugs, while AGP binds basic or neutral drugs; this seems an oversimplification, as many basic drugs bind strongly to HSA,[8] and AGP binds acidic drugs like phenobarbital.[28] Given that its molar concentration is about 30 times lower than HSA, in principle its influence in whole plasma protein binding should be negligible compared to HSA.

5.36.2.1.3 Lipoproteins

Lipoproteins have been reported to contribute to the plasma binding of very lipophilic drugs.[28] Lipoproteins consist of a central core of lipids (triglycerides and cholesteryl esters) surrounded by a shell of apoproteins and polar lipids

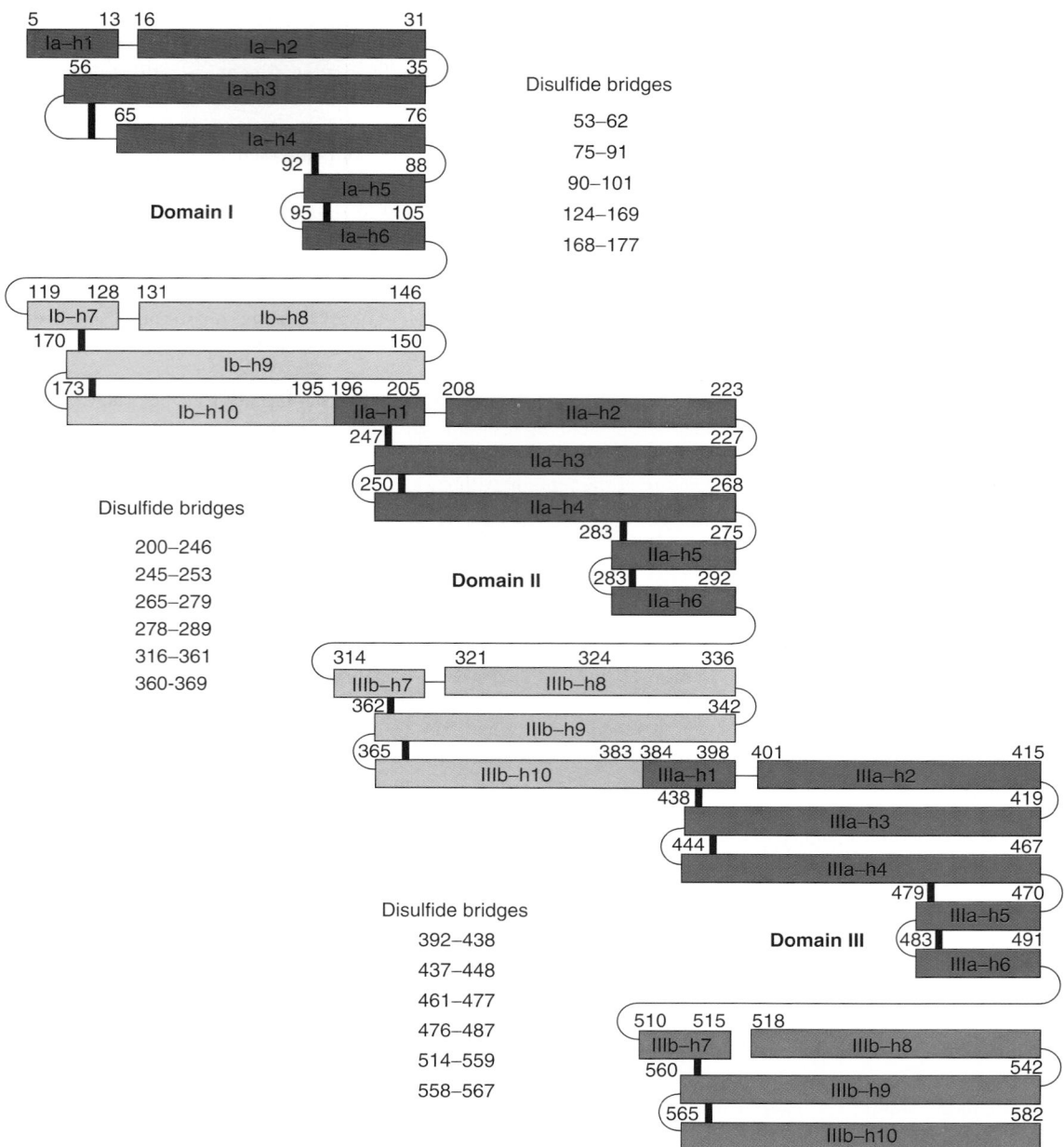

Figure 1 Secondary structure scheme of HSA, after Sugio *et al.*[14] The domains are color-coded as follows: I, red; II, green; III, blue. Within each domain, A and B subdomains are displayed in dark and light shades, respectively.

(phospholipids and cholesterol).[29] Their physiological role is the transport of cholesterol and triglycerides. Four classes of lipoproteins have been described, differing in size and density: high-density lipoproteins (HDL), low-density lipoproteins (LDL), very-low-density lipoproteins (VLDL), and chylomicrons; the density increases and the size decreases as the proportion of triglycerides of the particles decreases.

5.36.2.2 Tissues and Drug Distribution

As plasma represents about 4.5% of total body weight, the 'tissue' compartment should be in general more influential in determining the drug concentration in the extracellular fluid, as well as the volume of distribution (eqn [2]) and

Table 1 Structures of HSA currently present in the protein data bank (PDB) (as of May 2005)

PDB entry	Ligands	Resolution/Å	Reference	Year
1ao6	–	2.5	14	1997
1bj5	Myristic acid	2.5	15	1998
1bke	Myristic acid	3.15	15	1998
	Tri-iodobenzoic acid			
1bm0	–	2.5	14	1998
1e78	–	2.6	16	2000
1e7a	Propofol	2.2	16	2000
1e7b	Halothane	2.38	16	2000
1e7c	Myristic acid	2.4	16	2000
	Halothane			
1e7e	Decanoic acid	2.5	17	2000
1e7f	Dodecanoic acid	2.43	17	2000
1e7g	Tetradecanoic acid	2.5	17	2000
1e7h	Hexadecanoic acid	2.43	17	2000
1e7i	Octadecanoic acid	2.7	17	2000
1gni	cis-9-octadecenoic acid	2.4	18	2001
1gnj	cis-5,8,11,14-Eicosatetraenoic acid	2.6	18	2001
1h9z	Myristic acid	2.5	19	2001
	R-(+)-warfarin			
1ha2	Myristic acid	2.5	19	2001
	S-(−)-warfarin			
1hk1	Thyroxine	2.65	20	2003
1hk2	Thyroxine	2.8	20	2003
1hk3	Thyroxine	2.8	20	2003
1hk4	Myristic acid	2.4	20	2003
	Thyroxine			
1hk5	Myristic acid	2.7	20	2003
	Thyroxine			
1n5u	Myristic acid	1.9	21	2002
	Hemin			
1o9x	Myristic acid	3.2	22	2002
	Hemin			
1tf0	GA module	2.7	23	2004
	Decanoic acid			
	Citric acid			
1uor	–	2.8	24	1998

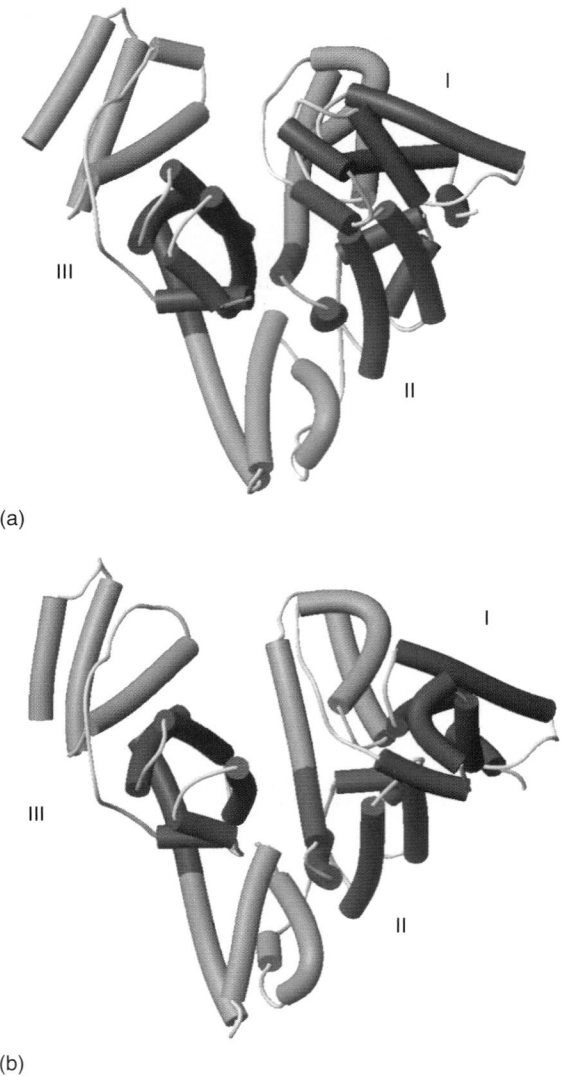

(a)

(b)

Figure 2 Three-dimensional structure of HSA. α-Helical segments are represented by rods, connected by disordered loops. The domains and subdomains are colored following the secondary structure schema in **Figure 1**. (a) HSA structure without fatty acids (PDB entry 1AO6). (b) HSA structure with (not displayed) fatty acids (PDB entry 1BJ5). Both structures were superimposed through their α-carbons at domain II. Notice the large conformational rearrangement and rotation of domains I and III upon binding of myristic acid.

therefore the half-life (eqn [3]).[1] Tissues are composed of cells (the water content, the intracellular fluid, is on average about 35% of body weight) and interstitial fluid (about 16%).[30] The interstitial fluid composition resembles that of plasma, including the presence of HSA and AGP,[31] and there is free exchange of small molecules between the two through pores in the capillary walls (an exception to this is the blood–brain barrier, where the capillary endothelial cells are joined by tight junctions and there are no pores). Drugs can bind to tissues by binding to proteins in the interstitial fluids, and/or by binding to cell components, such as membranes, proteins, and nucleic acids; there are also cellular organelles that have a remarkable capacity to store drugs (e.g., microsomes in hepatocytes and lysosomes of low internal pH in several cellular types that trap basic drugs).

The human body has different types of tissues, and **Table 3** lists the fractions of body weights for the most important ones.[32] Muscle is by far the most abundant tissue (about 40% of body weight), followed by adipose tissue (about 12%). Some experimental data suggest a strong correlation between the binding ratios in different tissues, indicating a similar mechanism of binding, but not between tissue and plasma.[1]

Table 2 Binding sites of different drugs to HSA observed at crystal structures of HSA–drug complexes

Ligand	Binding site(s)	Reference
Aspirin	IIA, IIIA	24
Warfarin	IIA	24
Diazepam	IIIA	24
Digitoxin	IIIA	24
Clofibrate	IIIA	24
Ibuprofen	IIIA	24
AZT	IIA, IIIA	24
5-Iodosalicilic acid	IIA, IIIA	24
3,5-Diiodosalicilic acid	IIA, IIIA	24
Propofol	IIIA, IIIB	16
Halothane (no myristic acid)	2 × IIA/IIB interface, IIIA	16
Halothane (with myristic acid)	3 × IIA/IIB interface, IIIA, 2 × IIA, IA/IIA interface	16
R-(+)-Warfarin (with myristic acid)	IIA	19
S-(−)-Warfarin (with myristic acid)	IIA	19

Table 3 Fractions of body weights for different tissues in the human body

Tissue	Fraction of body weight per $L\,kg^{-1}$
Adipose	0.1196
Bone	0.0856
Brain	0.02
Gut	0.0171
Heart	0.0047
Kidney	0.0044
Liver	0.026
Lung	0.0076
Muscle	0.40
Skin	0.0371
Spleen	0.0026
Plasma	0.0424
Erythrocytes	0.0347

Data taken from Poulin, P.; Theil, F.-P. *J. Pharm. Sci.* **2002**, *91*, 129–156.

Binding to adipose tissue has received much attention as fat can dissolve organic solvents and pollutants displaying negligible protein binding, that should result in very high adipose tissue/plasma distribution coefficients[33]; however, this seems not to be the case with lipophilic drugs, because these molecules also display a larger plasma protein binding that cancels out the stronger binding to fat and results in moderate distribution coefficients.

5.36.3   Prediction of Plasma Protein Binding of Drugs

In our description of in silico models for plasma protein binding of drugs, we will focus on global models developed using large and diverse data sets of drugs, and obtained with human proteins. On the one hand, there are many local models derived from small data sets of structurally related molecules. However, these models do not allow generalizations to the whole medicinal chemical space, as is the case with global models, and describing them here is not possible given the restricted space; they have been reviewed recently elsewhere.[34] On the other hand, large differences between binding data obtained from plasma proteins of different species[5] cast doubt on the validity of models obtained with data sets of nonhuman plasma proteins to predict human plasma protein binding, which is what we are really interested in.

Three main types of global models with human proteins have been derived: (1) models for whole plasma protein binding; (2) models for HSA binding; (3) and one model for AGP binding (although the importance of lipoproteins in the distribution of some lipophilic drugs has been described,[28] no global model has been derived for this type of plasma proteins). The source of data sets in the first case is equilibrium dialysis experiments, ultrafiltration, or ultracentrifugation experiments. HSA data sets are mainly derived from chromatographic techniques, although there is also one work that uses NMR titrations, and another that uses a compilation of HSA association constants gathered from the published works. Finally, the model for AGP uses a data set obtained by means of competition experiments using equilibrium dialysis. All these models are described below. Previous reviews of this area can be found in [5,34].

5.36.3.1   Literature Models to Predict Whole Plasma Protein Binding

Probably the first model of this type derived using a large and diverse data set was that of Saiakhov *et al.*[35] These authors used a set of 154 drugs with known protein affinities in the form of percentage of drug bound in plasma (%ppb). The source of the data set was mainly Goodman and Gilman's textbook.[36] For the analysis, the 154 %ppb values were categorized into three classes: (1) bound (%ppb > 32%); (2) marginally bound ($32\% \geqslant \%\text{ppb} \geqslant 19\%$); and (3) unbound (%ppb < 19%) drugs. The MCASE program of Klopman[37] was used to build the model. The program did not find a good correlation of the whole data set with log P, the logarithm of the octanol/water partition coefficient. Instead, the program identified eight main biophores, that is, substructures appearing with higher frequencies in the bound and marginally bound categories. For seven of these biophores it was possible to derive local quantitative structure–activity relationships (QSARs) where other substructures and/or physicochemical properties worked as positive or negative modulators. It is interesting to note that, although log P did not correlate well with %ppb for the whole data set, in almost all of the local QSARs it appeared as a positive modulator. This probably means that, in general, higher lipophilicity enhances plasma protein binding, although the way and extent of the enhancement depend on the particular structural class of molecules for which the local QSAR is developed. The resulting set of local QSARs displayed good statistics (concordance of 96% for a coverage of 92%). It was validated internally by three runs of cross-validation, and the resulting statistics indicated a good predictive power (average concordance of 85% for a coverage of 81%). Besides, the model was used to predict the %ppb of two penicillins outside the data set, giving predictions very close to the experimental value.

Lobell and Sivarajah[38] built models for the prediction of the following binding parameter:

$$\log\left(\frac{1 - f_{u,P}}{f_{u,P}}\right) = \log\left(\frac{\%\text{ppb}}{100 - \%\text{ppb}}\right) \qquad [4]$$

The authors collected from the literature a training set of 226 compounds, and a test set of 94 molecules. The sets were classified in subsets according to their predominant charge state: negatively charged, zwitterionic, uncharged, positively charged, and permanently positively charged. Equations were derived by fitting for each training subset a linear univariate model having as descriptor $A \log P$ (a calculated log P obtained with $A \log P$); for some cases, however, a correlation was not found and a constant binding value was postulated instead. **Table 4** collects all the resulting models and their statistics with both the training and test sets.

The statistics show reasonable, but not very good, predictions. All these models displayed similar slopes, although very different intercepts. For the same lipophilicity ($A \log P$), the charge state determines the binding affinity in this order: negatively charged compounds > uncharged compounds > positively charged compounds > permanently positively charged compounds. Zwitterionic compounds seem always to have low plasma protein binding.

Table 4 Models by Lobell and Sivarajah for the prediction of plasma protein binding

Charge state pH 7.4	Predicted $\log((1 - f_{u,P})/f_{u,P})$	Training set			Test set		
		n	r^2	MAE	n	r^2_{ext}	MAE
Negatively charged	$0.36 A\log P + 0.42$	52	0.50	0.56	25	0.46	0.51
Zwitterionic	$-0.48\ (f_{u,P} = 0.75)$	21		0.26			
Uncharged	$0.45 A\log P - 0.48$	78	0.63	0.51	38	0.56	0.54
Positively charged and $A\log P \leqslant 0.2$	$-0.95\ (f_{u,P} = 0.90)$	7		0.30			
Positively charged and $A\log P > 0.2$	$0.46 A\log P - 1.10$	63	0.67	0.38	33	0.51	0.51
Permanently positively charged	$0.38 A\log P - 2.10$	5	0.98	0.07			
All		226	0.68	0.45	94	0.51	0.53

Data taken from Lobell, M.; Sivarajah, V. *Mol. Divers.* **2003**, *7*, 69–87.

Yamakazi and Kanaoka[39] derived models for the prediction of %ppb. The authors used a data set obtained from the literature of 302 %ppb values for a wide variety of drugs. The following expression was used to model the data:

$$\%\text{ppb} = \frac{k_1 \exp(\log X) + k_2}{k_1 \exp(\log X) + k_2 + 1} \quad [5]$$

where, for $\log X$, $\log P$, or its version that takes into account the ionization state at different pHs, $\log D$, was tested. When fitting the model, neutral and basic drugs (143 drugs, including 53 zwitterions) were analyzed separately from acidic drugs (159 drugs, including 71 zwitterions). The best model for neutral and basic drugs used $\log D_{7.4}$ as an independent variable, and was obtained without zwitterions in the regression ($n = 84$, $r^2 = 0.80$, mean absolute error (MAE) $= 0.10$, six outliers). The prediction of zwitterions with this model was not good. Similarly, the regression for acidic drugs was performed without zwitterions, and only with a reduced subset of 44 compounds displaying a particular two-center pharmacophore. The best model was obtained with $\log P$ as descriptor ($r^2 = 0.79$, MAE $= 0.05$). The prediction with this model of the zwitterions or the nonzwitterions without two-center pharmacophore was bad. The model was validated with an external set of 20 drugs, none of which was a zwitterion, and with all acidic drugs displaying two-center pharmacophore, giving in this case good predictions ($r^2_{ext} = 0.83$).

5.36.3.2   Commercial Models to Predict Whole Plasma Protein Binding

Besides the literature models described in the previous section, there are several commercial models to predict whole plasma protein binding. **Table 5** gives details of the most relevant ones and gives some information about them.

The MCASE/ADME software implements the model by Saiakhov *et al.*[35] described in the previous section. For the other models the information available is limited; for those displaying more information (ADMET Predictor, truPK, and CSPB) the size of the data set and the quality of the statistics are similar to the literature models.

5.36.3.3   Models to Predict Human Serum Albumin Binding

Colmenarejo *et al.*[40] used a data set of binding constants to HSA for 94 diverse drugs obtained through high-performance affinity chromatography (HPAC) in a column with immobilized HSA (*see* 5.14 In Vitro Models for Plasma Binding and Tissue Storage). The binding parameter modeled was therefore $\log K'_{HSA} = \log((t - t_0)/t_0)$, where t and t_0 are the retention time of the drug and the dead time of the column, respectively. A genetic algorithm was used to search for model equations with good statistics, by trying 53 different molecular descriptors and equation terms of linear, quadratic, spline, offset-quadratic, and quadratic-spline type. Eighty-four compounds were

Table 5 Commercial in silico models to predict plasma protein binding

Software name	Company	URL	n^a	Comments	Reference
Cerius2.ADME/Tox	Accelrys	http://www.accelrys.com		Based on 2D descriptors	
MCASE/ADME	Multicase	http://www.multicase.com	402	Based on MCASE algorithm	35
Admensa	Inpharmatica	http://www.inpharmatica.co.uk			
ADMET Predictor	Simulations Plus	http://www.simulations-plus.com	298	Based on neural networks	
VolSurf	Tripos	http://www.tripos.com		Uses VolSurf approach	
TruPK	Strand Genomics	http://trupk.strandgenomics.com	306	Based on machine learning algorithms	
CSPB	ChemSilico	http://chemsilico.com	345	Combination of neural networks and multiple linear regression	

^{a} Size of data set.

selected to train the model, and 10 were used as an external prediction set. Initially, the following nonlinear global model was derived:

$$\log K'_{HSA} = 2.01 \times 10^{-2} + 5.54 \times 10^{-2} AM1\text{dip} - 1.22 JursRPSA$$
$$- 2.83 \times 10^{-2} (E_{HOMO} + 7.41)^2 + 0.15 C \log P - 3.48 \left[0.18 - {}^6\chi_{ring} \right]$$
$$n = 84; \ r^2 = 0.78; \ q^2 = 0.73 \tag{6}$$

where $AM1$dip is a calculated dipole moment, $JursRPSA$ is a Jurs polar surface descriptor, E_{HOMO} is the HOMO energy, $C\log P$ is a calculated $\log P$, and ${}^6\chi_{ring}$ is the sixth-order, ring-type Kier and Hall topological index (no confidence intervals were provided for the correlation coefficients). The statistics of both the regression ($r^2 = 0.78$, lack-of-fit statistic (LOF) = 0.12, five outliers) and the internal validations by leave-out-one cross-validation runs and randomization tests were good ($q^2 = 0.73$), indicating a good predictive power. The external test set was well predicted too ($r^2_{ext} = 0.88$). A series of 84 additional runs of the genetic algorithm was used to determine the descriptors most relevant to this system, by counting the frequencies of descriptors in the 84 winning equations. It turned out that just 12 descriptors appeared in these equations, with $C\log P$ being the most important (it was present in all the equations), followed by ${}^6\chi_{ring}$, and descriptors of other types (quantum mechanical, structural, Jurs-type). By starting with only the 12 relevant descriptors, the genetic algorithm selected this second model:

$$\log K'_{HSA} = -0.61 + 6.78 \times 10^{-2} [HBDon - 3]^2 - 9 \times 10^{-6} JursTPSA^2$$
$$- 2.83 \times 10^{-2} (E_{HOMO} + 7.41)^2 + 5.70 \times 10^{-3} AM1\text{dip}^2$$
$$+ 0.18 C \log P + 2.33 \ {}^6\chi_{ring}$$
$$n = 84; \ r^2 = 0.83; \ q^2 = 0.79 \tag{7}$$

where $HBDon$ is the number of hydrogen bond donors, and $JursTPSA$ is another Jurs polar surface descriptor. The statistics for the regression ($r^2 = 0.83$, LOF = 0.10) and internal validation ($q^2 = 0.79$) were slightly better than those of eqn [6], although the external set prediction was slightly worse ($r^2_{ext} = 0.82$). The interpretation given to the model was that HSA binding is driven to a large extent by lipophilicity, which is modulated by other additional factors of different nature (such as structural, quantum mechanical). As we will see later, the data set by Colmenarejo *et al.* has been used in several other studies.

Kratochwil *et al.*[4] used a data set comprising the logarithm of the association constant to HSA ($\log K_{HSA}$) in the literature for 138 diverse drugs, plus 25 in-house determined binding constants for Roche compounds. Obviously, $\log K_{HSA}$ is related to %ppb by:

$$\log K_{HSA} = \log \left(\frac{\% ppb}{100 - \% ppb} \right) - \log [HSA] \tag{8}$$

where [HSA] is approximated by the concentration of albumin in plasma (0.6 mmol L^{-1}). These association constants correlated well with the ones calculated from %ppb values available in the literature in the low-to-medium range, but for high association constants (corresponding to submicromolar binding or %ppb above 99%) there was poor correlation, probably due to the difficulty in measuring accurately %ppb values above 99%. An initial attempt to correlate log K_{HSA} with log $D_{7.4}$ for a subset of 76 compounds failed to yield a statistically sound model, indicating that lipophilicity alone cannot be used to predict HSA binding. Thus, in order to derive a global model, the authors used two sets of molecular descriptors, vectors in bases obtained from the diagonalization of the similarity matrices of the whole data set. In turn, the similarities are referred to a topological pharmacophore representation (TPR) of the molecules, or to Daylight fingerprints (DFP), respectively. These descriptors were fitted with a partial least squares (PLS) algorithm in order to derive two models. The models predicted well the training set ($r^2 = 0.72$ and 0.76 for the TRP and DFP models, respectively), and randomization tests indicated that they were not the result of chance correlations. However, cross-validation tests did not show good predictive powers ($q^2 = 0.48$ and 0.37 for the TPR and DFP models, respectively). In addition, when the TPR model was used to predict an external data set of 76 HPAC-obtained binding constants for glycine/N-methyl-D-aspartate (NMDA) receptor antagonists, the results were very bad ($r^2_{\text{ext}} = -0.70$). The authors attributed this to several factors: poor diversity of the external set, low similarity between the test set and the training set, and the possible absence of linear correlation of HPAC constants with association constants in the whole range of binding constants of the external set.

Hajduk et al.[41] developed a model to predict the logarithm of dissociation constants for binding to domain IIIA of HSA (log $K^D_{HSA-IIIA}$). The authors used a data set of dissociation constants of 889 diverse molecules to the isolated domain IIIA of HSA obtained through nuclear magnetic resonance and fluorescence titrations (where 657 molecules displayed negligible binding and were assigned the same dissociation constant of 2 mmol L^{-1}). log $K^D_{HSA-IIIA}$ displayed some correlation with Clog P, but in order to derive a statistically sound model the authors adopted a group contribution approach using the following equation:

$$\log K^D_{HSA-IIIA} = \sum_i w_i x_i$$
$$n = 889; \quad r^2 = 0.94; \quad q^2 = 0.90 \tag{9}$$

where w_i is the weighting coefficient of structural descriptor i, representing a particular chemical group, and x_i is the number of times i appears in the molecule. A total of 74 chemical groups were used, obtained by fragmenting the molecules in the training data set. The equation was fitted to the data set, and the resulting model displayed good statistics ($r^2 = 0.94$, average error = 0.11) and good predictive power in leave-group-out cross-validation tests and randomization tests ($q^2 = 0.90$). No external validation of the model was performed. However, it was seen that, for a series of antibacterials, a large proportion of those showing a higher than eightfold loss of activity in the presence of serum were predicted to have large affinity for the domain IIIA of HSA. Analysis of the weighting coefficients revealed some trends. For instance, positively charged groups like cyclic or acyclic amines had positive weighting coefficients; that is, they are detrimental for binding; the opposite is seen for negatively charged groups like carboxylic and sulfonic acids. Five- and six-membered rings and chlorine and fluorine also improve binding. More interestingly, there is a general trend for more hydrophobic chemical fragments to enhance binding.

Valko et al.[42] modeled the logarithm of an apparent association constant to HSA (log K^{app}_{HSA}), obtained from a fast-gradient HPAC technique developed by the authors. The log K^{app}_{HSA} values were determined after calibrating a column parameter (log t) against the log(%ppb/(101 − %ppb)) values for nine drugs, where the %ppb values were taken from the literature. In this way, the %ppb values for an independent set of 68 drugs, obtained by back-calculation from the log K^{app}_{HSA} values estimated from the column retention times, displayed an acceptable correlation with those from the literature. A combination of the previous 68 drugs with 52 compounds comprising the so-called Abraham set (containing some simple nondrug molecules) showed some correlation between log K^{app}_{HSA} and Clog P, and worse correlation between log K^{app}_{HSA} and ACDlog D at pH 7.4 (a calculated log D following the model of Advanced Chemistry Development/Labs, usually considered a not very reliable model). This indicates a major role of lipophilicity in binding to HSA, but also the need for additional factors in order to obtain a good model. The Abraham equation was thus fitted to the Abraham set, resulting in this best model:

$$\log K = -1.28 + 0.82(\pm 0.15)R_2 - 0.36(\pm 0.15)\pi^H_2 + 0.18(\pm 0.14)\alpha^H_2$$
$$- 1.97(\pm 0.15)\beta^H_2 + 1.62(\pm 0.21)\log V_x$$
$$n = 52; \quad r^2 = 0.83; \quad s = 0.33; \quad F = 44 \tag{10}$$

where R_2 is the drug excess molar refraction; π_2^H is the dipolarity/polarizability; α_2^H and β_2^H are the hydrogen bond acidity and basicity, respectively; and V_x is the McGowan volume. The model gave good statistics with the training set ($n = 52$, $r^2 = 0.83$, $s = 0.33$, $F = 44$), but when used to predict the set of 68 drugs, many drugs were observed to display much higher binding than predicted (no statistics were given). The authors argued that eqn [10] predicts a 'baseline' affinity based mainly on lipophilicity, and from there one compound could have increased affinity if it has good complementarity with the binding site.

Hall et al.[43] used topological and electrotopological descriptors to build a model using the data set of Colmenarejo et al. Multiple linear regression (MLR) was used, testing all the possible multivariate linear models up to a maximum number of terms. The best model is this:

$$
\begin{aligned}
\log K'_{HSA} = {}& 5.03 \times 10^{-2} (\pm 5.0 \times 10^{-3}) S_{arom}^T + 7.87 \times 10^{-2} (\pm 8.3 \times 10^{-3}) S_{CHsat}^T \\
& + 2.91 \times 10^{-2} (\pm 5.9 \times 10^{-3}) S_{-F,Cl}^T + 8.71 \times 10^{-3} (\pm 3.2 \times 10^{-3}) S_{-OH}^T \\
& + 2.38 (\pm 0.95)\, {}^6\chi_{CH}^V + 2.96 (\pm 0.99)\, {}^5\chi_{CH}^V - 1.18 \\
& n = 84;\ r^2 = 0.77;\ s = 0.29;\ q^2 = 0.70;\ F = 43
\end{aligned}
\qquad [11]
$$

where S_{arom}^T, S_{CHsat}^T, $S_{-F,Cl}^T$, and S_{-OH}^T are the sums of atom type E-state descriptors for aromatic carbon atoms, saturated carbon atoms (with hydrogens), fluorine and chlorine, and hydroxyl groups, respectively; and ${}^6\chi_{CH}^V$ and ${}^5\chi_{CH}^V$ are the six- and fifth-order valence molecular connectivity chi-chain indexes, respectively. The model displayed good statistics and predictive power ($r^2 = 0.77$, $s = 0.29$, $F = 43$, $q^2 = 0.70$, no outliers). The model was validated with randomization and leave-group-out cross-validation tests ($q^2 = 0.68$). It also predicted well the external set, although slightly worse than the model of Colmenarejo et al. ($r_{ext}^2 = 0.74$, MAE = 0.31). The authors interpreted the model as follows: electron accessibility and/or count of aromatic rings, chlorine and fluorine atoms, nonsubstituted aliphatic chains, electron accessibility and/or number of –OH groups, and six-membered heteroaromatic rings are all features that enhance HSA binding; in contrast, aliphatic chains with electronegative substituents and five-membered heteroaromatic rings all weaken the binding to HSA.

Xue et al.[44] derived two models with the data set of Colmenarejo et al. The first one was developed using MLR with a heuristic method for descriptor selection. An initial pool of 243 descriptors was tested to search for models while optimizing several statistics. The resulting best model was:

$$
\begin{aligned}
\log K'_{HSA} = {}& -2.51 (\pm 0.39) - 0.40 (\pm 0.08) HDCA2 \\
& + 7 \times 10^{-3} (\pm 1 \times 10^{-3}) MSA - 0.15 (\pm 0.02) NO \\
& + 9.21 (\pm 1.39) RNR - 3.94 (\pm 0.66) RNN \\
& + 0.40 (\pm 0.1) BI - 4.5 \times 10^{-2} (\pm 0.01) RNCS \\
& n = 84;\ r^2 = 0.86;\ s = 0.22;\ \text{RMSE} = 0.21;\ q^2 = 0.63;\ F = 63.9
\end{aligned}
\qquad [12]
$$

where HDCA2 is the area-weighted surface charge of hydrogen-bonding donor atoms, MSA is the molecular surface area, NO is the number of oxygens, RNR is the relative number of rings, RNN is the relative number of nitrogens, BI is the Balaban index, and RNCS is the relative negative-charged surface area. The statistics of this model were slightly better than those of Colmenarejo et al. models. However, the leave-out-one cross-validation test ($q^2 = 0.63$) and the external set prediction ($r_{ext}^2 = 0.71$, $F = 19.3$, $s = 0.47$) were worse. The second model was derived by training a support vector machine equipped with Gaussian kernels, and using the seven descriptors selected in eqn [12]. Very good statistics were returned from the regression ($r^2 = 0.94$, RMSE = 0.13), and the external set prediction was very good as well ($r_{ext}^2 = 0.89$, RMSE = 0.22), better than with the models of Colmenarejo et al. The authors interpreted the models as follows: both hydrogen bonding and electrostatic interactions should decrease HSA binding, while hydrophobic interactions, five- and six-membered rings, and molecular branching should enhance it.

Yap and Chen[45] modeled by means of general regression neural network (GRNN) the data set of Colmenarejo et al. For comparison purposes, additional models were derived by using MLR and multilayer feedforward neural network (MFNN). A very large pool of 1497 descriptors of different nature was used in the derivation, and they were selected by means of a genetic algorithm. The absence of chance correlations was estimated using randomization tests. The analyzed data set was split into a training set of 75 compounds and an external prediction set of 18 compounds. The best GRNN model obtained had six descriptors: SRW07 (self-returning walk count of order 07), GATS8e (Geary autocorrelation-lag 8/weighted by atomic Sanderson electronegativities), RDF040 m (radial distribution function-4.0/weighted by atomic masses), Mor20p (3D-MoRSE-signal 20/weighted by atomic polarizabilities), C-040 (an atom-centered fragment corresponding to R-C(=X)-X/R-C#X/X-=C=X), and H-050 (an atom-centered fragment corresponding to hydrogen attached to a heteroatom). The GRNN model predicted well the external set ($r^2 = 0.85$, Spearman rho coefficient (R_s) $= 0.82$, and RMSE $= 0.2$) and was seen not to appear as chance correlation; worse models were obtained with MLR and MFNN. The interpretation given to the model is that hydrogen-bonding ability decreases HSA binding, while polarizability, shape of the molecule, and electronegativity of molecular substituents all influence the binding to HSA, although in a complex way.

5.36.3.4   Models to Predict α₁-Acid Glycoprotein Binding

The only global model found for the prediction of AGP binding is that of Hervé et al.[26] The authors used a data set of 35 molecules with pIC₅₀ values obtained by means of equilibrium dialysis experiments, with both the A and the F1/S variants of AGP. Comparative molecular field analysis (CoMFA) was carried out with each variant, selecting the models displaying the largest q^2. For the A variant the best model ($q^2 = 0.57$, $r^2 = 0.95$, $F = 263$) had three PLS components and was derived from a CoMFA model using only the steric field. This model field could be approximated by a simple haptophoric model (**Figure 3**) where the strongly bound ligands should have a nitrogen atom, plus a ring placed in a hydrophobic pocket, and another ring interacting with a hydrophobic area, all separated by well-defined distances. On the other hand, for the F1/S variants no statistically sound model could be derived.

5.36.3.5   Conclusions

The models described in the previous section present some regularities worth mentioning. On the one hand, it seems that in general the binding to whole plasma proteins is well approximated by the binding to HSA alone. For instance, in the work of Kratochwil et al.[4] a good correlation was obtained between the association constants for HSA and those calculated from %ppb values for whole plasma proteins, for all the ranges of binding strengths except for very strong binding; the bad correlation in this range seems to be due to experimental unreliability of the %ppb values. Similarly, Valko et al.[42] could calibrate a column with immobilized HSA in order to obtain reasonable predictions for %ppb. That plasma protein binding can be approximated by HSA binding is not surprising, as HSA is a promiscuous protein, being by far the most abundant of all the plasma proteins (about 60%). This implies that in general, AGP, with a molar concentration approximately 30-fold lower than HSA, should have a secondary role in determining %ppb. It must be taken into account, however, that in acute-phase inflammation reactions its concentration can increase several times, and this can result in underpredictions for %ppb for drugs with strong binding to AGP if they are administered under these conditions. It is therefore necessary to do more work on the modeling of drug binding to AGP, with larger data sets than that of Hervé et al.,[26] and in comparison with whole plasma protein-binding data, in order to clarify the relevance of AGP binding in this area.

On the other hand, the works and models described seem to converge on a similar picture: although statistically sound models cannot be derived where plasma protein binding depends only on lipophilicity (e.g., log P, log D, or their calculated versions), this property can account for a large percentage of the variance. Saiakhov et al.,[35] Kratochwil et al.,[4] Hajduk et al.,[41] and Valko et al.[42] calculated such simple correlations between the binding parameter and a lipophilicity descriptor and, although in some cases some correlation was observed, in no case was it enough to derive an acceptable model. However, whether explicitly or implicitly, in almost all of the models lipophilicity appears as a dependent variable where its increase always results in an increased binding. Statistically reasonable models are then achieved in two different ways: a global model for the whole data set is derived by adding more descriptors, or several 'local' simpler

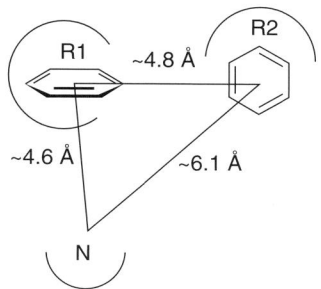

Figure 3 Haptophoric model proposed by Hervé et al.[26] for the binding to the A variant of AGP.

models are derived for subsets of similar molecules, where lipophilicity is the only one or one of the descriptors. Examples of the first case are the models by Colmenarejo et al.,[40] and Valko et al.,[42] while the models by Saiakhov et al.,[35] Lobell and Sivarajah,[38] and Yamakazi and Kanaoka[39] are examples of the second case. The models by Hajduk et al.,[41] Hall et al.,[43] Xue et al.,[44] and Yap and Chen[45] probably belong to the first case, but in them lipophilicity is implicitly included by means of other types of descriptor: structural or topological descriptors for lipophilic groups, and molecular suface area. In the model by Kratochwil et al.[4] the assignment of HSA binding constants is based on the similarity of the molecule to a set of topological pharmacophores; as these are not made explicit in this paper, it is difficult to judge the role of lipophilicity in their model.

Besides lipophilicity, there seem to be other consistent features in the models: negatively charged groups enhance binding, while positively charged groups tend to decrease it[38,39,41]; this is in agreement with the traditional (but oversimplified) idea that HSA binds predominantly acidic drugs.[46] Also the presence of six-membered rings[40,41,43,44] as well as chlorine and fluorine atoms[41,43] should enhance binding. No complete agreement is observed for five-membered rings: in two models[41,44] they appear to enhance binding, but in one model[43] they decrease it. In addition there is a variety of molecular properties with different modulating effects that are specific to each model.

In summary, prediction of plasma protein binding requires relatively complex models where lipophilicity repeatedly displays a prevailing role, but where other additional descriptors of different type are necessary to achieve statistical significance, such as molecular shape, charge, and electrostatic potential.

5.36.4 Prediction of Drug Tissue Binding

Tissue binding used to be described by means of tissue/blood or tissue/plasma distribution coefficients. In the literature these coefficients are frequently called tissue/blood or tissue/plasma partition coefficients, but following a suggestion from one editor the term distribution coefficients will be used instead, as they are actually ratios of the total concentrations of the drug (including all the ionization subspecies) in two phases.

The determination of tissue/plasma or tissue/blood distribution coefficients is much more difficult than plasma protein-binding data, and this has largely precluded the derivation of models to predict these important parameters. Initial models were mainly based on data sets of blood–air and tissue–air partition coefficients of low-molecular-weight, nonionizable chemicals obtained from tissue homogenates or blood.[47–50] Obviously these chemicals are devoid of drug-like properties, but at least the corresponding models provided good starting points for more complex models. In time different models were developed for drugs, using data determined both in vivo (measurements of concentrations in tissue and blood/plasma after administration to animals) and in vitro (equilibrium dialysis experiments with tissue homogenates).

In what follows, a review of the outstanding models to predict tissue/plasma or tissue/blood distribution coefficients of drugs is presented. The specific models for brain/blood distribution, related to the problem of blood–brain barrier permeation of drugs, are not discussed here; they are described in Chapter 5.31 (In Silico Models to Predict Brain Uptake). We focus instead on models developed to predict simultaneously the distribution of drugs through multiple tissues, and again consider only global models derived from large data sets of diverse drugs. The data sets used were obtained from different mammal species besides human (rat, rabbit, mouse), but, as we will see, the models can be adapted to any species and tissues by providing appropriate tissue-specific parameters to the equations.

5.36.4.1 Models to Predict Tissue/Plasma Distribution Coefficients

Poulin and Theil[51] developed mechanistic models to predict tissue/plasma distribution coefficients ($D_{T/P}$) that had as independent variables $P_{vo/w}$ (the vegetable oil/water partition coefficient of the molecule) and/or $f_{u,P}$. In the case of drugs displaying unrestricted homogeneous distribution by passive diffusion, the following equation was proposed:

$$D_{T/P} = \frac{\left[P_{vo/w}(v_{nl}^T + 0.3v_{pl}^T) + v_w^T + 0.7v_{pl}^T\right]f_{u,P}}{\left[P_{vo/w}(v_{nl}^P + 0.3v_{pl}^P) + v_w^P + 0.7v_{pl}^P\right]f_{u,T}}$$ [13]

where the v-type coefficients are the volume fractions of neutral lipids, phospholipids, or water (subscripts nl, pl, and w, respectively) in the tissue or plasma (superscripts T and P, respectively). For drugs with distribution restricted to the extracellular fluid, the corresponding model becomes:

$$D_{T/P} = \frac{F_T f_{u,T}}{F_P f_{u,P}}$$ [14]

where F_T and F_P are the fractional content of extracellular fluid in tissue and plasma, respectively. The model was used to predict the $D_{T/P}$ values for 65 basic drugs in eight rabbit, rat, and mouse nonadipose and nonexcretory tissues (269 data points). The experimental volume fractions and $f_{u,P}$ values were obtained from the literature. The $f_{u,T}$ values were approximated from the $f_{u,P}$ values by assuming a concentration of protein in plasma that was double that seen in tissues. The predictions were fairly good ($r^2 = 0.81$, $s = 1.40$), and 85% of the predicted values were within a factor of three of the experimental values. Larger discrepancies were ascribed to additional mechanisms not included in the model (e.g., lysosomal ionic trapping) or errors in the estimated fractions of unbound drug. It is worth noting that these equations have no adjustable parameters, but instead they attempt the prediction of $D_{T/P}$ from purely mechanistic reasoning.

In a later paper, Poulin et al.[52] developed a variation of eqn [13] to predict the $D_{T/P}$ for adipose tissue. The rationale used was that, on the one hand, protein binding should be negligible in adipose tissue given its low percentage; thus $f_{u,T} \approx 1$. On the other hand, the ionization state of ionizable compounds should also strongly affect the partition coefficient in this tissue of low water content, because it is expected that the nonionized species should partition in both the lipid and water fractions, while the ionized species should only be present in the water fraction. Therefore, $P_{vo/w}$ should be substituted by $D_{vo/w}$, the vegetable oil/water distribution coefficient, taking into account the ionization state of the drug. The resulting equation is thus:

$$D_{T/P} = \frac{\left[D_{vo/w}(v_{nl}^T + 0.3v_{pl}^T) + v_w^T + 0.7v_{pl}^T\right]f_{u,P}}{\left[D_{vo/w}(v_{nl}^P + 0.3v_{pl}^P) + v_w^P + 0.7v_{pl}^P\right]}$$
$$n = 16;\ r^2 = 0.90;\ s = 0.44;\ \text{RMSE} = 1.08 \quad [15]$$

The model (again with no adjustable parameters) was validated with a data set of 14 drugs with known adipose/plasma distribution coefficients for rats, rabbits, and humans (16 data points). The predictions with eqn [15] were good ($r^2 = 0.90$, $s = 0.44$, $\text{RMSE} = 1.08$). When octanol/water partition coefficients were used instead of the vegetable oil/water distribution coefficients, the prediction error increased significantly, indicating that vegetable oil is a much better surrogate of the lipid fraction in tissues than octanol.

A third mechanistic model is that of Rodgers et al.,[53] in this case attempting to predict the tissue/plasma water distribution coefficient ($D_{T/u,P}$): this is the ratio of the concentration of total drug in the tissue to the concentration of unbound drug in plasma. The model is specialized in the prediction of moderate-to-strong bases, and assumes that electrostatic and hydrophobic interactions dominate and the drugs distribute by passive diffusion. The tissues are modeled as composed of extracellular and intracellular water, plus neutral lipids and both neutral and acid phospholipids. The model equation (with no adjustable parameters once again) is as follows:

$$D_{T/u,P} = \frac{D_{T/P}}{f_{u,P}} = v_{ew}^T + \frac{1 + 10^{pK_a - pH_{iw}}}{1 + 10^{pK_a - pH_p}}v_{iw}^T + \frac{K_a[AP^-]_T 10^{pK_a - pH_{iw}}}{1 + 10^{pK_a - pH_p}}$$
$$+ \frac{Pv_{nl}^T + (0.3P + 0.7)v_{np}^T}{1 + 10^{pK_a - pH_p}} \quad [16]$$

where the v-type coefficients are the volume fractions of extracellular water, intracellular water, neutral lipids, and neutral phospholipids (subscripts ew, iw, nl, and np, respectively) in the tissue (superscript T); pH_{iw} and pH_p are the pHs of intracellular water and plasma, respectively; $[AP^-]$ is the concentration of acid phospholipids in tissues; and K_a is the affinity constant for acid phospholipids of the drug. The model was used to predict the $D_{T/u,P}$ values for 28 basic drugs in 13 rat tissues, including adipose and excretory ones (261 data points). The prediction was good (85.1% of the values within a factor of three to the experimental ones), and better than the predictions obtained using the models by Poulin and Theil[51] and Poulin et al.[52] (eqns [13]–[15]). The use of calculated pK_a's and $\log P$'s instead of experimental ones did not decrease the statistical quality of the model.

5.36.4.2 Models to Predict Tissue/Blood Distribution Coefficients

Zhang[54] developed a predictive model for tissue/blood distribution coefficients. Starting with a mechanistic reasoning, where tissues (but not blood) are described as composed of lipids, proteins, and water, and after assuming that the concentration of compound in the tissue water fraction equals that of blood, the author came up with this equation:

$$\log D_{T/B} = \log\left(\sum_{i,j} f_j 10^{\log P^j_{i/B} + \log v_i}\right)$$

$$n = 201; \quad r^2 = 0.82; \quad s = 0.29; \quad q^2 = 0.79 \qquad [17]$$

where f_j is the fraction of the compound at ionization state j (nonionized, positively ionized, negatively ionized), $P^j_{i/B}$ is some sort of tissue composition/blood partition coefficient for tissue composition i (lipids, proteins, water) at ionization state j, and v_i is the volume fraction of tissue composition i (approximated by the corresponding weight fraction). The exponents in eqn [17] are approximated as linear combinations of molecular descriptors for polarizability, charge, and hydrogen bonding. The data set analyzed included 36 volatile neutral organic compounds plus 10 basic drugs with known $P_{T/B}$ for rabbit fat, brain, muscle, lung, and heart (46 compounds and 265 data points). The data set was split into a training set of 35 compounds and 201 data points, and an external validation set of 11 compounds and 64 data points. eqn [17] was fitted to the training set, using experimental weight fractions from the literature and trying different linear combinations of molecular descriptors, resulting in three final models. The results indicate a good fit to the training set ($r^2 = 0.82$, $s = 0.29$, $q^2 = 0.79$ for the best model), and a good prediction of the external set ($r^2_{\mathrm{ext}} = 0.92$, $s = 0.25$). As regards the molecular descriptors in the best-fitting model, it is observed that the partition of neutral compounds into lipids depends on polarizability, hydrogen bonding, and charge, while the partition into proteins depends on polarizability; the partition of ionizable compounds into both lipids and proteins depends on charge. An interesting feature of this model is that it can deal with all types of tissue, including adipose and excretory ones.

5.36.4.3 Conclusions

All the models[51–54] described in the previous section adopt a mechanistic or semimechanistic approach where each tissue is described as a mixture of different extents of components (lipids of different types, water, and proteins) and the binding of a drug to a tissue is determined by its affinity for these components and their volume fraction in the tissue. The volume fractions of the components are taken directly from the literature instead of being estimated from a fit of the model to the data. The drug is described physicochemically by its octanol/water or olive oil/water partition coefficients, and (explicitly or implicitly) by its pK_a; in addition, its unbound fraction in plasma (and tissue) is also used as input in the models and taken from experiments. An additional parameter in the model by Rodgers et al.[53] is the binding constant to acid phospholipids. The only exception to this trend is the model by Zhang,[54] where, although it is based on mechanistic reasoning and uses the tissue component weight fractions from the literature, the drugs are described by linear combinations of molecular descriptors and the final models are obtained by fits to the data. Thus, currently no purely in silico model (not requiring any experimental input, and based only on molecular descriptors) has yet been developed to predict tissue/plasma or tissue/blood partition coefficients for diverse data sets of drugs.

5.36.5 Prediction of the Volume of Distribution of Drugs

As was seen in the introduction, the volume of distribution is a parameter that describes the balance between plasma and tissue binding (eqn [2]). In particular, the volume of distribution of total drug increases when tissue binding increases, and decreases when plasma protein binding increases.[1] The volume of distribution is linear with the half-life of the drug (eqn [3]), an extremely important pharmacokinetic parameter: as (free) drug is less available in plasma, it is less amenable to elimination by hepatic metabolism and/or renal excretion, and the drug lasts longer. This has raised interest in the development of models to predict the volume of distribution, because, in combination with experimental or predicted clearances, this should allow the prediction of drug half-lives. In this section, recent predictive models for the volume of distribution are reviewed.

5.36.5.1 Models to Predict Volumes of Distribution

Poulin and Theil[55] developed a mechanism-based model for $V_{D,ss}$, the volume of distribution at steady state:

$$V_{D,ss} = \sum_i V_{T_i} D_{T_i/P} + V_E R_{E/P} + V_P$$

$$n = 118; \quad r^2 = 0.86; \quad s = 039 \qquad [18]$$

where the V coefficients are the body volumes of tissue T_i, erythrocyte, and plasma (subscripts T_i, E, and P, respectively), and $R_{E/P}$ is the erythrocyte/plasma ratio. All these parameters are obtained or estimated from the literature, except the $D_{T_i/P}$ parameters, that are approximated by eqns [13]–[15]. Again, the model has no adjustable parameters and is valid for different species by just providing the appropriate species-specific parameters. A data set of 123 drugs with known rat and human $V_{D,ss}$ was reasonably predicted by the model (i.e., 80% of the predicted values were within a factor of two of the experimental values; $n = 147$, $r^2 = 0.61$, $s = 0.82$). However, a subset of 98 drugs within this data set displayed especially good agreement with the experimental data ($n = 118$, $r^2 = 0.86$, $s = 0.39$). For these drugs an agreement can be deduced with the assumptions of the model: tissues are composed of different proportions of cellular water and lipids, plus extracellular proteins; drugs bind to the tissue to an extent depending on their lipophilicity and their protein binding. The remaining 25 drugs showed a poorer agreement with the experimental values ($n = 29$, $r^2 = 0.18$, $s = 1.74$), which indicates additional factors not taken into account in the model: binding to cellular macromolecules, ionic drug–membrane interactions, lysosomal ion trapping, and active uptake processes. In general, this model predicts a larger $V_{D,ss}$ for more lipophilic drugs, and a reduced $V_{D,ss}$ for drugs that bind strongly to plasma proteins.

A slightly different mechanistic approach is that used by Lombardo et al. in two works.[56,57] These authors derived a model to predict $\log f_{u,T}$ for neutral and basic drugs. The $f_{u,T}$ so estimated could then be used, together with the experimental $f_{u,P}$, to predict the volume of distribution using the Oie–Tozer equation[58]:

$$V_{D,ss} = V_P(1 + R_{E/I}) + f_{u,P}V_P(V_{EF}/V_P - R_{E/I}) + \frac{V_R f_{u,P}}{f_{u,T}} \qquad [19]$$

where V_{EF} is the volume of the extracellular fluid (not including plasma), V_R is the volume of the remainder of the body, and $R_{E/I}$ is the ratio of extravascular to intravascular proteins. The authors used a data set of 64 basic and neutral structurally different drugs, extended to 120 drugs in the second work. The best model for the extended set was:

$$\log f_{u,T} = 8 \times 10^{-3}(\pm 0.07) - 0.23(\pm 0.04)E\log D$$
$$- 0.93(\pm 0.08)f_{i(7.4)} + 0.89(\pm 0.09)\log f_{u,P}$$
$$n = 120; \quad r^2 = 0.87; \quad \text{RMSE} = 0.37; \quad q^2 = 0.85; \quad F_{3,116} = 250.9 \qquad [20]$$

where $E\log D$ is a $\log D_{7.4}$ determined through a high-performance liquid chromatography method developed by the authors, and $f_{i(7.4)}$ is the fraction of drug ionized at pH 7.4. The fit was good ($r^2 = 0.87$, RMSE $= 0.37$, $q^2 = 0.85$, $F_{3,116} = 250.9$, $P < 0.0001$, mean-fold error $= 2.08$). The model was internally validated with randomization tests and leave-class-validation tests, resulting in a good predictive power (mean-fold errors between 1.26 and 2.51). It was externally validated with a set of 18 proprietary compounds, giving similar statistics as the previous model (mean-fold error of 2.26). The coefficients indicate that lipophilicity favors tissue binding, as well as electrostatic interactions with membrane phospholipids ($f_{i(7.4)}$ term); however, the strongest positive correlation is observed with the fraction of unbound drug in plasma, and this is rationalized by the authors from the large amount of proteins in the extravascular compartment.

Lobell and Sivirajah[38] created a fully in silico model. In a similar way to the model that the authors derived for plasma protein binding already described above, the training set was divided into subsets based on their main charge states at pH 7.4, and for each subset the correlation between $\log V_{D,ss}$ and $A \log P$ was investigated. In this case, the training set comprised 204 molecules, while the test set had 124 drugs. The resulting models and the corresponding statistics are collected in **Table 6**.

In general, it is observed that the $V_{D,ss}$ of compounds with positive $A\log P$ decreases in the order: positively charged > uncharged > permanently positively charged > negatively charged; while for compounds with negative $A\log P$, the trend is: uncharged > positively charged > negatively charged = zwitterionic. Negatively charged compounds always have a very low $V_{D,ss}$, while the largest $V_{D,ss}$ is observed for lipophilic positively charged and uncharged compounds.

5.36.5.2 Conclusions

The models by Poulin and Theil[55] and Lombardo et al.[56,57] show a strong dependence of the volume of distribution on the unbound fraction in plasma $f_{u,P}$. As the experimental unbound fraction in tissues $f_{u,T}$ is not used explicitly in their models, although it is required to determine the volume of distribution (eqn [2]), this probably indicates that it can be easily modeled from $f_{u,P}$. As a matter of fact, Poulin and Theil[55] approximate the former from the $f_{u,P}$ values by assuming an average concentration of protein in plasma that is double that seen in tissues, while Lombardo et al.[56,57]

Table 6   Models for the prediction of volumes of distribution

Charge state at pH 7.4	A log P range	Predicted $V_{D,ss}$	Training sets			Test sets			Average fold error
			n	r^2	MAE	n	r^2_{ext}	MAE	
Negatively charged		0.2	42		0.17	29		0.31	2.5
Zwitterionic	$\leqslant 0.1$	0.2	2						1.2
Zwitterionic	> 0.1	$0.44 A \log P - 0.75$	10	0.68	0.13	11	0.35	0.30	1.8
Uncharged	< 5	1	70		0.29	43		0.34	2.8
Uncharged	$\geqslant 5$	10	3		0.30	2			4.3
Positively charged	$\leqslant -2$	0.3	4		0.06				1.1
Positively charged	> -2 and < 5	$0.23 A \log P - 0.05$	59	0.53	0.26	36	0.02	0.51	3.5
Positively charged	$\geqslant 5$	20	4		0.84	3		0.47	6.1
Permanently positively charged		0.5	10		0.35				6.6
All			204			190		124	2.96

Data taken from Lobell, M.; Sivarajah, V. *Mol. Divers.* **2003**, *7*, 69–87.

approximate it as a linear function of $f_{u,P}$, $E \log D$ and $f_{i(7.4)}$. This is not surprising, taking into account the fact that the composition of the interstitial fluid resembles that of plasma, including the presence of HSA and AGP.[31] This opens the interesting possibility of developing a fully in silico model for the volume of distribution as a 'hypermodel' that includes a model for the prediction of plasma protein binding, like those described above, as one independent variable. Another possibility is a development purely from scratch; this is the approach taken by Lobell and Sivirajah.[38] Their model is indeed a set of local models, and the particular model to apply depends on the main charge state of the molecule to predict.

5.36.6   Future Directions

The distribution of drugs throughout the body is a complex problem, but of extreme importance in the design of drugs for therapeutic usage. It is no use having a very potent and selective drug if it has too short or long a half-life, if the dosage and concentrations in the extracellular fluid needed to obtain a therapeutic effect result in toxic concentrations at the tissues, or if most of it remains bound to plasma proteins, being unable to reach its target. So, in the same way that appropriate bioavailability and metabolism features are required for a drug to be successful, appropriate plasma and tissue protein-binding properties are also essential for the safe and effective usage of the drug.

Pharmacokinetic considerations like these are being taken into account at earlier stages in the drug discovery process in order to reduce the attrition rate as far as possible. The sooner an issue in this area is discovered, the better the researchers can react to fix it, and if no solution is possible, move the resources to an alternative candidate. In this way, in silico tools like the ones reviewed in this chapter are of great use. On the one hand, they provide predictions of the modeled parameters without the need to perform an experiment; this is very useful in areas like virtual screening, design of combinatorial libraries, and prioritization of compounds to assay, synthesize, or purchase. On the other hand, they rationalize large amounts of previous experimental data, and therefore can be used to suggest sound modifications of problematic molecules in order to resolve distribution-related issues.

The set of works reviewed in this chapter indicates that there are good models to predict plasma and tissue protein binding of drugs. Most of the models display acceptable predictive power and seem to agree in similar basic conclusions. There are, however, several areas of improvement. For instance, there is the need for more improved models for binding to AGP, with larger data sets and for the different variants of the protein, in order to ascertain the role of this protein in whole plasma protein binding, especially under acute-phase inflammation reactions. The models for HSA binding can be improved by taking into account structural information from x-ray structures of complexes with drugs, probably from high-throughput x-ray crystallography; currently, no model for HSA has used the known structure of HSA. Another area of improvement is the development of fully in silico models (not requiring experimental input and based only on molecular descriptors) for tissue binding and volume of distribution using large data sets of drugs; right now, the models derived with data sets of drugs use experimental plasma protein-binding data as input; the only exception should be the set of models for volume of distribution by Lobell and Sivirajah.[38] The development of models for tissue binding and volume of distribution is largely prevented by the difficulty in the generation of reliable data for tissue binding. These models – and in general the whole area of modeling of distribution-related properties – will benefit from the development of in vitro screens, as well as new noninvasive techniques for in vivo measurements and molecular imaging.

References

1. Fichtl, B.; Von Nieciecki, A.; Walter, K. *Adv. Drug Res.* **1991**, *20*, 117–166.
2. Smith, D. A.; Van de Waterbeemd, H.; Walker, D. K. Pharmacokinetics. In *Pharmacokinetics and Metabolism in Drug Design. Methods and Principles in Medicinal Chemistry*; Smith, D. A., Van de Waterbeemd, H., Walker, D. K., Mannhold, R., Kubinyi, H., Timmerman, H., Eds.; Wiley-VCH: Weinheim, 2001, pp 15–34.
3. Smith, D. A.; Van de Waterbeemd, H.; Walker, D. K. Distribution. In *Pharmacokinetics and Metabolism in Drug Design. Methods and Principles in Medicinal Chemistry*; Smith, D. A., Van de Waterbeemd, H., Walker, D. K., Mannhold, R., Kubinyi, H., Timmerman, H., Eds.; Wiley-VCH: Weinheim, 2001, pp 47–57.
4. Kratochwil, N. A.; Huber, W.; Mueller, F.; Kansy, M.; Gerber, P. R. *Biochem. Pharmacol.* **2002**, *64*, 1355–1374.
5. Kratochwil, N. A.; Huber, W.; Müller, F.; Kansy, M.; Gerber, P. R. *Curr. Opin. Drug Disc. Dev.* **2004**, 7, 507–512.
6. Hervé, F.; Urien, S.; Albengres, E.; Duché, J. C.; Tillement, J. *Clin. Pharmacokinet.* **1994**, *26*, 44–58.
7. Olson, R. E.; Christ, D. D. *Ann. Rep. Med. Chem.* **1996**, *31*, 327–336.
8. Carter, D. C.; Ho, J. X. *Adv. Protein Chem.* **1994**, *45*, 152–203.
9. Cistola, D. *Nat. Struct. Biol.* **1998**, *5*, 751–753.
10. Sudlow, G.; Birkett, D. J.; Wade, D. N. *Mol. Pharmacol.* **1975**, *11*, 824–832.
11. Sudlow, G; Birkett, D. J.; Wade, D. N. *Mol. Pharmacol.* **1976**, *12*, 1052–1061.
12. Galliano, M.; Kragh-Hansen, U.; Tarnoky, A. L.; Chapman, J. C.; Campagnoli, M.; Minchiotti, L. *Clin. Chim. Acta* **1999**, *289*, 45–55.
13. McLachlan, A. D.; Walker, J. E. *J. Mol. Biol.* **1977**, *112*, 543–558.
14. Sugio, S.; Kashima, A.; Mochizuki, S.; Noda, M.; Kobayashi, K. *Protein Eng.* **1999**, *12*, 439–446.
15. Curry, S.; Mankelkow, H.; Brick, P.; Franks, N. *Nat. Struct. Biol.* **1998**, *5*, 827–835.
16. Bhattacharya, A. A.; Curry, S.; Franks, N. P. *J. Biol. Chem.* **2000**, *275*, 38731–38738.
17. Bhattacharya, A. A.; Gruene, T.; Curry, S. *J. Mol. Biol.* **2000**, *303*, 721–732.
18. Petitpas, I.; Grüne, T.; Bhattacharya, A. A.; Curry, S. *J. Mol. Biol.* **2001**, *314*, 955–960.
19. Petitpas, I.; Bhattacharya, A. A.; Twine, S.; East, M.; Curry, S. *J. Biol. Chem.* **2001**, *276*, 22804–22809.
20. Petitpas, I.; Petersen, C. E.; Ha, C.-E.; Bhattacharya, A. A.; Zunszain, P. A.; Ghuman, J.; Bhagavan, N. V.; Curry, S. *Proc. Natl. Acad. Sci.* **2003**, *100*, 6440–6445.
21. Wardell, M.; Wang, Z.; Ho, J. X.; Robert, J.; Ruker, F.; Ruble, J.; Carter, D. C. *Biochem. Biophys. Res. Commun.* **2002**, *291*, 813–819.
22. Zunszain, P. A.; Ghuman, J.; Komatsu, T.; Tsuchida, E.; Curry, S. *BMC Struct. Biol.* **2003**, *3*, 6.
23. Lejon, S.; Frick, I.-M.; Björck, L.; Wikström, M.; Svensson, S. *J. Biol. Chem.* **2004**, *279*, 42924–42928.
24. He, X. M.; Carter, D. C. *Nature* **1992**, *358*, 209–215.
25. Fournier, T.; Medjoubi, N. N.; Porquet, D. *Biochim. Biophys. Acta* **2000**, *1482*, 157–171.
26. Hervé, F.; Caron, G.; Duché, J.-C.; Gaillard, P.; Rahman, N. A.; Tsantili-Kakoulidou, A.; Carrupt, P.-A.; D'Athis, P.; Tillement, J.-P.; Testa, B. *Mol. Pharmacol.* **1998**, *54*, 129–138.
27. Kremer, J. M. H.; Wilting, J.; Janssen, L. M. H. *Pharmacol. Rev.* **1988**, *40*, 1–47.
28. Urien, S.; Nguyen, P.; Bastian, G.; Lucas, C.; Tillement, J.-P. *Invest. New Drugs* **1995**, *13*, 37–41.
29. Stryer, L. *Biochemistry*, 4th ed.; W. H. Freeman: New York, 1995, pp 695–697.
30. Rang, H. P.; Dale, M. M.; Ritter, J. M. *Pharmacology*; Churchill Livingstone: Edinburgh, 1999.
31. Brée, F.; Houin, G.; Barré, J.; Moretti, J.-L.; Wirquin, V.; Tillement, J.-P. *Clin. Pharmacokinet.* **1986**, *11*, 336–342.
32. Poulin, P.; Theil, F.-P. *J. Pharm. Sci.* **2002**, *91*, 129–156.
33. Bickel, M. H. *Adv. Drug Res.* **1994**, *25*, 55–86.
34. Colmenarejo, G. *Med. Res. Rev.* **2003**, *23*, 275–301.
35. Saiakhov, R. D.; Stefan, L. R.; Klopman, G. *Perspect. Drug Disc. Des.* **2000**, *19*, 133–155.
36. Hardman, J. G.; Limbird, L. E.; Gilman, A. G. *Goodman & Gilman's The Pharmacological Basis of Therapeutics*; McGraw-Hill: New York, 1996.
37. Klopman, G. *Quant. Struct.–Act. Relat.* **1992**, *11*, 176–184.
38. Lobell, M.; Sivarajah, V. *Mol. Divers.* **2003**, *7*, 69–87.
39. Yamazaki, K.; Kanaoka, M. *J. Pharm. Sci.* **2004**, *93*, 1480–1494.
40. Colmenarejo, G.; Alvarez-Pedraglio, A.; Lavandera, J. L. *J. Med. Chem.* **2001**, *44*, 4370–4378.
41. Hajduk, P. J.; Mendoza, R.; Petros, A. M.; Huth, J. R.; Bures, M.; Fesik, S. W.; Martin, Y. C. *J. Comput.-Aided Mol. Des.* **2003**, *17*, 93–102.
42. Valko, K.; Nunhuck, S.; Bevan, C.; Abraham, M. H.; Reynolds, D. P. *J. Pharm. Sci.* **2003**, *92*, 2236–2248.
43. Hall, L. M.; Hall, L. H.; Kier, L. B. *J. Chem. Inf. Comput. Sci.* **2003**, *43*, 2120–2128.
44. Xue, C. X.; Zhang, R. S.; Liu, H. X.; Yao, X. J.; Liu, M. C.; Hu, Z. D.; Fan, B. T. *J. Chem. Inf. Comput. Sci.* **2004**, *44*, 1693–1700.
45. Yap, C. W.; Chen, Y. Z. *J. Pharm. Sci.* **2005**, *94*, 153–168.
46. Van de Waterbeemd, H.; Smith, D. A.; Jones, B. C. *J. Comput.-Aided Mol. Des.* **2001**, *15*, 273–286.
47. Abraham, M. H.; Kamlet, M. J.; Taft, R. W.; Doherty, R. M.; Weathersby, P. K. *J. Med. Chem.* **1985**, *28*, 865–870.
48. Abraham, M. H.; Weathersby, P. K. *J. Pharm. Sci.* **1994**, *83*, 1450–1456.
49. DeJongh, J.; Verhaar, H. J. M.; Hermens, J. L. M. *Arch. Toxicol.* **1997**, *72*, 17–25.
50. Baláž, Š.; Viera Lukáčová, V. *Quant. Struct.–Act. Relat.* **1999**, *18*, 361–368.
51. Poulin, P.; Theil, F.-P. *J. Pharm. Sci.* **2000**, *89*, 16–35.
52. Poulin, P.; Schoenlein, K.; Theil, F.-P. *J. Pharm. Sci.* **2001**, *90*, 436–447.

53. Rodgers, T.; Leahy, D.; Rowland, M. *J. Pharm. Sci.* **2005**, *94*, 1259–1276.
54. Zhang, H. *J. Chem. Inf. Model.* **2005**, *45*, 121–127.
55. Poulin, P.; Theil, F.-P. *J. Pharm. Sci.* **2002**, *91*, 129–156.
56. Lombardo, F.; Obach, R. S.; Shalaeva, M. Y.; Gao, F. *J. Med. Chem.* **2002**, *45*, 2867–2876.
57. Lombardo, F.; Obach, R. S.; Shalaeva, M. Y.; Gao, F. *J. Med. Chem.* **2004**, *47*, 1242–1250.
58. Oie, S.; Tozer, T. N. *J. Pharm. Sci.* **1979**, *68*, 1203–1205.

Biography

Gonzalo Colmenarejo received a degree in biology from the Universidad Complutense de Madrid (UCM). After that, he received an MSc in biophysics (UCM) by working in the statistical mechanics of nonspecific ligand binding to DNA. He then received his PhD in biophysics (UCM), with a thesis dealing with the interaction, molecular modeling, and photochemistry of photoactive drugs with nucleic acids. He was a postdoctoral researcher for 2 years in the Department of Chemistry of the University of California at Berkeley, and in the Structural Biology Department of the Lawrence Berkeley Laboratory, in the laboratory of Ignacio Tinoco, Jr, where he solved 3D ribozime structures by nuclear magnetic resonance and molecular dynamics. Then in 1999 he joined the GlaxoWellcome R&D center in Tres Cantos, Spain, for a position in Cheminformatics. He is currently an Investigator in the Cheminformatics Department of GlaxoSmithKline at Tres Cantos.

5.37 Physiologically-Based Models to Predict Human Pharmacokinetic Parameters

T Lavé, N Parrott, and H Jones, F. Hoffmann-La Roche Ltd, Basel, Switzerland

5.37.1 Introduction

In order to reduce failures in clinic related to poor absorption, distribution, metabolism, and excretion (ADME) properties, it is important to predict human pharmacokinetics (PK) as early as possible to help select the best candidates for development and reject those with a low probability of success. Predicting human PK can also assist in choosing the first dose for a clinical trial; in testing the suitability of the compound for the intended dosing regimen; in predicting any interactions and in predicting the expected variability in man. Furthermore, predicting human PK prior to phase I studies has been shown to result in significant time savings ranging from 1 to 6 months.[1] Previously the use of physiologically-based pharmacokinetic (PBPK) models was limited in drug development mainly due to the mathematical complexity of the models and the labor-intensive input data required. However advances in the prediction of hepatic metabolism and tissue distribution from in vitro and in silico data have made these models more attractive, providing the opportunity to integrate key input parameters from different sources to not only estimate PK parameters and predict plasma and tissue concentration time profiles, but also to gain mechanistic insight into compounds properties.

There is recent evidence of growing interest in applying PBPK for the discovery and development of compounds.[2–4] Thus, strategies for PBPK application in compound research are being published[5] and several commercial PBPK simulation tools have recently become available.

This chapter describes the usefulness of PBPK in predicting human pharmacokinetics during drug discovery and nonclinical development within pharmaceutical research. The potential and limitations as well as the predictive power of the PBPK approaches are also discussed. The specific tools and techniques employed are briefly described and illustrated with a number of examples. The examples chosen illustrate the utility and the potential of PBPK from early discovery to early development.

5.37.2   Methods

5.37.2.1   Construction of Physiologically-Based Pharmacokinetic Models

Physiologically-based pharmacokinetic models divide the body into compartments, including the eliminating organs (e.g., kidney and liver) and noneliminating tissue compartments (such as fat, muscle, brain, etc.) which are connected by the circulatory system (**Figure 1**).

The models use physiological and species specific parameters (such as blood flow rates and tissue volumes) to describe the pharmacokinetic processes. A list of pharmacokinetic parameters with their corresponding definition is provided in **Table 1**. These physiological parameters are coupled with compound specific parameters such as physicochemical and biochemical parameters (e.g., tissue/blood distribution coefficients and metabolic clearance) to predict the plasma and/or tissue concentration versus time profiles of a compound in an in vivo (animal and/or human) system.

Once a model has been developed, the concentrations in the various tissues can be determined by using the mass balance equation below:

$$(\text{Drug concentration in tissue}) = (\text{Rate of drug distribution into tissue}) - (\text{Rate of drug distribution out}) - (\text{Rate of drug elimination within the tissue}) \qquad [1]$$

Depending on the drug and tissue, the distribution is perfusion-rate or diffusion-rate limited. Perfusion-rate limited kinetics tend to occur with relatively low-molecular-weight, hydrophobic drugs which have no problem crossing the lipid barrier of the cell wall. In this case the process limiting the penetration of the drug into the cells is the rate at which it is delivered to the tissue, so that blood flow is the limiting process. By contrast diffusion-rate limited kinetics occurs with more polar drugs that do not freely dissolve in the lipid of the cell membrane and, therefore, have difficulty

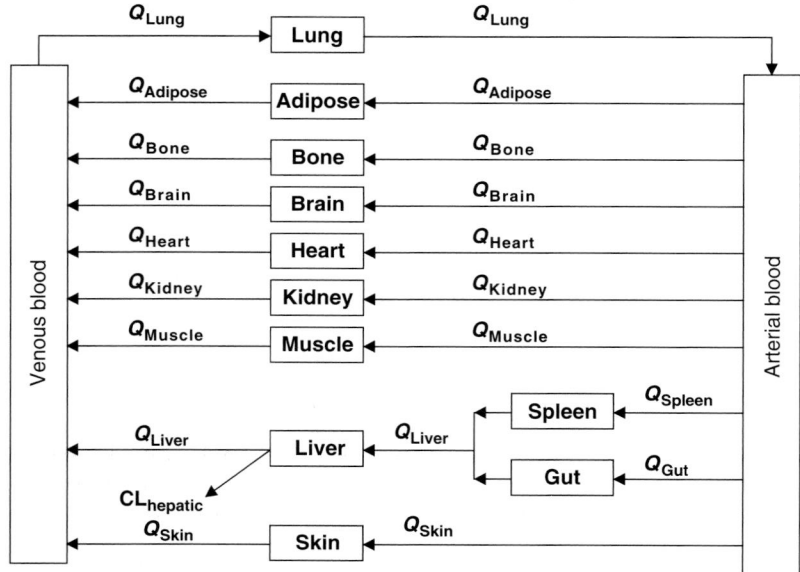

Figure 1   A PBPK model. Q, blood flow; CL, clearance.

Table 1 Definition of main pharmacokinetic parameters

PK parameter	Definition	Range of values (human)
AUC	Area under the plasma concentration-time curve following intra or extravascular administration of the compound	
CL	Total clearance This parameter represents the elimination (metabolism and excretion) of the compound from the body	
CL_H	Hepatic clearance This parameter represents the elimination of the compound through metabolism in the liver and/or biliary excretion	$0–20\,\mathrm{mL\,min^{-1}\,kg^{-1}}$ Low: $<6\,\mathrm{mL\,min^{-1}\,kg^{-1}}$ Intermediate: between 6 and $14\,\mathrm{mL\,min^{-1}\,kg^{-1}}$ High: $>14\,\mathrm{mL\,min^{-1}\,kg^{-1}}$
V_D and $V_{D,ss}$	Volume of distribution The volume of distribution (V_D) corresponds to the theoretical volume in which the drug would need to distribute to be at plasma concentration $V_D = \mathrm{dose}/C_0$ (iv) $V_{D,ss}$ corresponds to volume of distribution at steady state	From $0.1\,\mathrm{L\,kg^{-1}}$ (distribution restricted to vascular space) to $>50\,\mathrm{L\,kg^{-1}}$ (high volume due to extensive binding)
$t_{1/2}$	Half-life This parameter represents the time taken for the drug concentration in the plasma to decrease by 50% $t_{1/2}$ is a function of the clearance and volume of distribution, and reflects how often a drug needs to be administered $t_{1/2} = 0.693\,V_D/CL$	Few minutes to several months, depending on the drug
$A\%$	Fraction absorbed It is the fraction or percentage of the dose administered which becomes available in the portal vein after intestinal absorption $A\% = F\%(1 - CL_H/Q_H)$ where Q_H represents hepatic blood flow	0–100%
$F\%$	Bioavailability It is the fraction or percentage of the dose administered which becomes available in the systemic circulation Oral bioavailability is obtained by comparison of dose normalized exposure obtained after oral versus intravenous administration: $F\% = (AUC_{po}/AUC_{iv}) \times (Dose_{iv}/Dose_{po})$	0–100%

in penetrating into the cell. In this case the diffusion of drug across the membrane, which is independent of blood flow, becomes the limiting process.

For perfusion limitation, the rate of change of drug concentration in a tissue where no elimination occurs can be described as:

$$V\frac{dC}{dt} = Q\left(C_{in} - \frac{C_{in}}{P}\right) \qquad [2]$$

where V is the physical volume of the tissue, C is the drug concentration in tissue, Q is the blood flow to the tissue, C_{in} is the drug concentration entering the tissue, and P_{bt} is the partition coefficient of the drug between blood and tissue.

When diffusion rate limitation occurs, diffusion to and from the extra cellular space must be taken into account:

$$V_e\frac{dC_e}{dt} = Q(C_{in} - C_e) - P_m(C_e - C_i) \qquad [3]$$

where V_e is the anatomical extracellular volume, P_m is the membrane permeability coefficient, C_e is the free extracellular drug concentration, and C_i is the free intracellular drug concentration.

With organs such as liver, where elimination can occur, the rate of elimination must be included in the mass-balance equation. The rate of elimination is described by:

$$\text{Rate of elimination} = \text{CL}_H C_t \qquad [4]$$

where CL_H is the hepatic clearance of the drug and C_t is the concentration in tissue (in this case the liver) at time t. This term can then be inserted into the mass balance equation, such as that for perfusion limitation:

$$V\frac{dC}{dt} = Q\left(C_{in} - \frac{C_{in}}{P}\right) - \text{CL}_H C_t \qquad [5]$$

A recent excellent review of whole body physiological models is provided in.[6] The whole-body model applied in a drug discovery context is described more fully in.[7]

Physiologically-based components for prediction of distribution and metabolism may be combined in a whole-body model and linked to an oral absorption model. Thus, plasma and tissue concentration versus time profiles after intravenous and oral administration may be simulated based upon limited in data and/or in silico estimates.[7] In addition, the whole-body model provides a framework for incorporation of additional processes when these can be estimated.

5.37.2.2   Models to Predict Distribution

For each organ included in the PBPK model, estimates of tissue/plasma partitioning are required to quantify the distribution for the compound under study. The tissue plasma partitioning can be estimated in silico using mechanistic tissue composition models[8,9] or can be determined experimentally. Semiempirical methods to estimate tissue plasma partitioning based on in vivo $V_{D,ss}$ have also been reported.[10] These methods are reviewed below.

Models based on tissue composition have been developed to estimate tissue distribution based on in vitro and in silico data.[8,9] These models have greatly extended the applicability of the PBPK approach to early compound research and development. Briefly, these equations assume that the compound distributes homogenously into the tissue and plasma by passive diffusion where two processes are accounted for:

1. Nonspecific binding to lipids estimated from compound lipophilicity data (*see* 5.18 Lipophilicity, Polarity, and Hydrophobicity), namely $\log P$ (i.e., octanol/water partition coefficient of the unionized drug) and $\log D$ (octanol/water distribution coefficient of the ionized + unionized drug measured at a given pH).
2. Specific reversible binding to common proteins present in plasma and tissue estimated from the plasma f_u (f_{up}, fraction unbound in plasma). These models have been reviewed in detail.[7,11,12]

An alternative to the tissue composition model is provided with a semiempirical method originally proposed by Arundel.[10] In contrast to the tissue composition model, this approach requires in vivo input data for predicting tissue distribution. In brief, this approach allows the partitioning of a compound into different tissues to be estimated from the observed in vivo $V_{D,ss}$, by assuming an empirical relationship between this parameter and the disappearance rate constant (k_t) for a particular tissue (except adipose). This approach has been reviewed in detail elsewhere and was applied with some success to predicting tissue distribution in both rats and humans.[10,13]

A number of other approaches have been proposed to predict tissue distribution. These methods include in vitro–in vivo correlations of tissue/plasma partitioning data.[14–20] Other types of prediction methods are of computational nature and consist of mathematical models built in an empirical, semiempirical, or mechanistic framework. The two types of empirical models are (1) regression analyses made between data on in vivo $V_{D,ss}$ (and tissue/plasma partitioning) and common compound properties (e.g., lipophilicity, ionization, and binding to plasma proteins or lipids),[21–31] and (2) regression analyses made between muscle/plasma partitioning determined in vivo and tissue/plasma partitioning of other nonadipose tissues.[11,32] Another type of semiempirical model is based on the Oie–Tozer equation, which uses physiological and physicochemical input data to estimate $V_{D,ss}$ in vivo.[22,33]

5.37.2.3   Models to Predict Clearance

5.37.2.3.1   Hepatic metabolic clearance

Both empirical and physiologically-based approaches have been developed to predict the in vivo hepatic metabolic clearance in animals and humans.[13,34–39] Due to the improved availability of animal and human liver samples, the number of predictions that include in vitro data and are based on in vitro to in vivo physiologically-based direct scaling approaches have markedly increased during the past few years. By applying pharmacokinetic principles, it has recently been shown that human hepatic metabolic clearance can be predicted reasonably well from in vitro data.[37,38,40–43] Excellent reviews that cover all aspects of the in vitro to in vivo scaling have recently been provided.[37,38,41,44]

The strategy allowing this extrapolation of in vitro clearance to the in vivo situation is depicted in **Figure 2**.

In summary, the first step is to obtain CL_{int} (hepatic intrinsic clearance) from in vitro data. Various in vitro systems (e.g., hepatocytes, liver microsomes) are available for these metabolism studies. Thus, in vitro CL_{int} values determined from hepatocyte or microsomal substrate depletion or kinetic assays are normalized for cell or microsomal protein concentration to obtain CL_{int} in units of $\mu L\,min^{-1}\,10^{-6}$ cells or $\mu L\,min^{-1}\,mg^{-1}$ and corrected for any nonspecific binding. The second step consists of scaling the activity measured in vitro to the whole liver. Depending on the test system, this is achieved by using microsomal recovery factors and hepatocellularity values. Thus, in vitro CL_{int} is scaled to in vivo CL_{int} using a formal scaling procedure:

$$CL_{int}\ in\ vivo = CL_{int}\ in\ vitro \times SF \qquad [6]$$

where SF represents the milligrams of microsomal protein or million cells per gram of liver multiplied by the grams of liver weight as described by Houston.[41] Microsomal recovery and hepatocellularity scaling factors have been reported in the literature for the rat[44] and human.[45] The third stage involves the use of a liver model which incorporates the effects of liver blood flow and blood binding, to convert the estimated CL_{int} into a CL_H.

The main liver models used are: the venous equilibrium (well-stirred) model, the undistributed sinusoidal (parallel-tube) model, and the dispersion model. These models share three assumptions: Firstly, the distribution of the drug into the liver is perfusion rate rather than diffusion rate limited; secondly, that only the unbound drug in the blood crosses the cell membrane and occupies the enzyme site; and thirdly that there is a homogeneous distribution of drug metabolizing enzymes within the liver. The latter is not true, as it has been shown that cytochrome P450 (CYP) isoforms are distributed heterogeneously across the liver acinus. Therefore, these liver models do not truly represent the complex physiological nature of the in vivo liver situation but represent an approximation. The main difference between these three models is their description of the drug concentration profile within the liver. However, overall there is usually no significant difference between the models and as the well-stirred model is the most mathematically simple it is usually recommended for the prediction of CL_H from CL_{int}. In the well-stirred model the liver is conceived to be a single well-stirred compartment with the distribution of the unbound drug being instantaneous and the concentration of unbound drug in the hepatic compartment is assumed to be uniform and equivalent to the hepatic outflow concentration. This model can be successfully used for low clearance compounds where only a relatively small amount of the drug is eliminated in one pass through the liver. However, for high-clearance compounds there will be a higher concentration of drug entering the liver than leaving it, so the predictions are less accurate.[37,38,46,47] With the well-stirred model, CL_H is calculated as follows:

$$CL_H = Q_H f_u\ CL_{int}/(Q_H + f_u\ CL_{int}) \qquad [7]$$

where Q_H is the hepatic blood flow and f_u is the unbound fraction of drug in the blood.

5.37.2.3.2   Renal clearance

The prediction of renal clearance is based mainly on species scaling as no in vitro model is available to scale renal clearance from in vitro to in vivo. A number of examples support the conclusion that the allometric approach can be

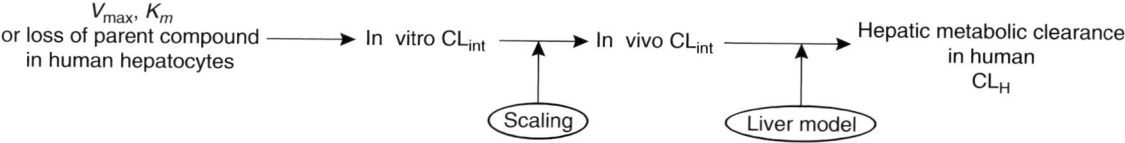

Figure 2   In vitro–in vivo scaling strategy for hepatic metabolic clearance. The arrows link the different steps of the scaling procedure.

used to predict the renal clearance of a compound in humans.[48–50] In these studies, the best predictions of human renal clearance were achieved when the intrinsic parameters were scaled across species. For example, a successful prediction was achieved for a lactamylvinylcephalosporin, when unbound renal clearance was used for interspecies scaling.[49]

Correlation analysis has also been used successfully to predict the human renal clearance from animal data. Sawada obtained good correlation coefficients for six beta-lactam antibiotics, with good predictions of clearance in humans based on monkey data.[50] As with allometric scaling, the use of an intrinsic parameter (e.g., unbound renal clearance) gave a better prediction and correlation than the corresponding hybrid parameter (bound + unbound renal clearance). Such results were confirmed by a study with 11 cephalosporins,[49] in which correcting for protein binding again improved the correlations between monkey and human.

The glomerular filtration rate (GFR) ratio approach was proposed by Lin.[51] With this approach, renal clearance is estimated using the ratio of the GFR between rats and humans. This approach can be used when minimal preclinical data is available (i.e., when only rat data are available). This method is based on the observation that the ratio of unbound renal clearance for a range of compounds is approximately equal to the ratio of GFR between rat and human (4.8); therefore, the human renal clearance can be estimated from information on the renal clearance in the rat as well as protein binding data in rats and humans.

5.37.2.3.3 Biliary clearance
A battery of in vitro models is available for predicting biliary clearance in vivo. They include: membrane vesicles, cultured hepatocytes, freshly isolated hepatocyte suspensions, hepatocyte couplets, overexpressed cell lines, and transfected cell lines. Overall, the ability of such models to predict in vivo data is qualitative rather than quantitative. Recently, however, the biliary secretion kinetics of selected compounds have been successfully predicted in some initial studies with sandwich hepatocyte cultures,[52] a model which conserves the integrity of the hepatocytes and seems to retain the capacity for biliary secretion.[53]

Attempts to apply allometric scaling techniques to biliary excretion have resulted in poor predictions of human clearance. Thus, the biliary clearances of susalimod[54] and napsagatran[55] were overestimated by 20- and 7-fold, respectively, and other similar failures to predict the human pharmacokinetics from preclinical data have also been reported.[56] These failures are probably due to species differences in the rate of biliary secretion. For example, the MRP2-mediated hepatobiliary transport of temocaprilat displays large differences between human, mouse, rat, guinea pig, rabbit, and dog.[57]

5.37.2.4 Models to Predict Oral Absorption
Oral absorption is a complex process determined by the interplay of physiological and biochemical processes, physicochemical properties of the compound, and formulation factors. Physiologically-based models to predict oral absorption in animals and humans have recently been reviewed[58–60] and several models are now developed to a degree that they are commercially available. The commercial models have not been published in detail for proprietary reasons, but in essence they are transit models segmenting the gastrointestinal tract into different compartments with the kinetics of transit, dissolution, and uptake described by differential equations. For example, the model underlying GastroPlus is known as the advanced compartmental absorption and transit (ACAT) model[58,59] (**Figure 3**) and is a semiphysiologically-based transit model, consisting of nine compartments corresponding to different segments of the digestive tract. Oral absorption simulations take in vitro and in silico input data such as solubility, permeability, particle size, log P, pK_a (ionization constant), and dose.

5.37.2.5 Physiologically-Based Pharmacokinetic Modeling Strategy for Prediction to Humans
A PBPK modeling strategy for drug discovery and early development has been proposed recently by Theil et al.[5] and is illustrated in **Figure 4**.

The proposed modeling strategy attempts to evaluate the data generated in drug discovery and early development in a more integrated way, by combining in silico and in vitro prediction methods for absorption, distribution and hepatic clearance to estimate the oral PK profile in humans. The strategy consists essentially of two major steps. The first step is to predict specific absorption, distribution, metabolism, and excretion (ADME) parameters, and then combining their input in a PBPK model to estimate, in a second step, the overall PK after oral and intravenous

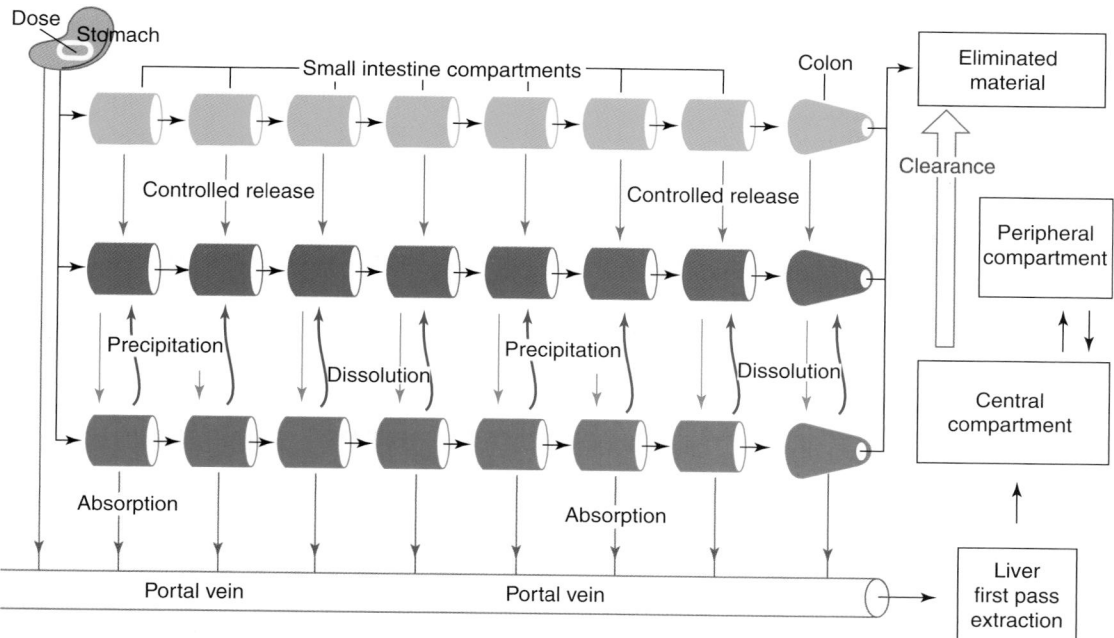

Figure 3 Advanced compartmental absorption transit (ACAT) model. The arrows represent the movement of the compound through the various compartments of the model. (Reprinted from Agoram, B. W. W.; Bolger, M. B. *Adv. Drug Deliv. Rev.* **2001**, *50*, S41–S67, with permission from Elsevier.)

administration. Before a prediction to humans is attempted, a comparison of the predicted PK parameters and the concentration/time profiles with in vivo data from animals (rat, dog, or monkey) is performed as a model validation. The rat is usually used, because it is by far the most commonly used animal species for the ADME characterization of compound candidates and because tissue composition models are available in rat and human. A satisfactory prediction for the rat suggests that the prediction for humans might be successful, although this cannot be considered as a full validation.

Thus, the PK in the rat is used to confirm that the PBPK model selected can be used for prediction of human PK. In a first step, disposition PK in the rat after intravenous administration is predicted with a PBPK model by using estimates of tissue: plasma partition coefficients, and hence the derived $V_{D,ss}$ as well as hepatic clearance as the assumed major component of clearance. If the predicted plasma concentration/time profile in rat describes the experimental data adequately, the intravenous PK in humans is simulated in a second step. In case of major deviations between predicted and experimental concentration/time profile in the rat, model refinements are necessary prior to prediction of human PK.

During the drug discovery and development process, more and more data become available and the data obtained are characterized by higher quality. This constantly increasing data quantity and quality potentially improves the model predictions as shown in **Figure 5**.

The value added by PBPK at the different stages of the discovery and development process, namely for PK simulations prior to in vivo studies, clinical candidate selection extrapolation to humans during the entry into human enabling phase is illustrated in **Figure 6** and discussed more in detail in the following section.

5.37.2.6   Commercial Tools Available

Several commercial software tools for physiology-based modeling of pharmacokinetic processes exist on the market. Generally these include (1) general simulation software packages, (2) software that simulates separate PK processes such as absorption, metabolism, and interaction, and (3) PBPK software with generic whole-body models.

A program developed to predict intestinal absorption is GastroPlus (SimulationsPlus[72]). In the most recent version, GastroPlus[59] includes a physiological intestinal model for humans in fed and fasted states as well as models for rat, dog, cynomolgus monkey, and mouse. Several additional modules are available for parameter optimization and for

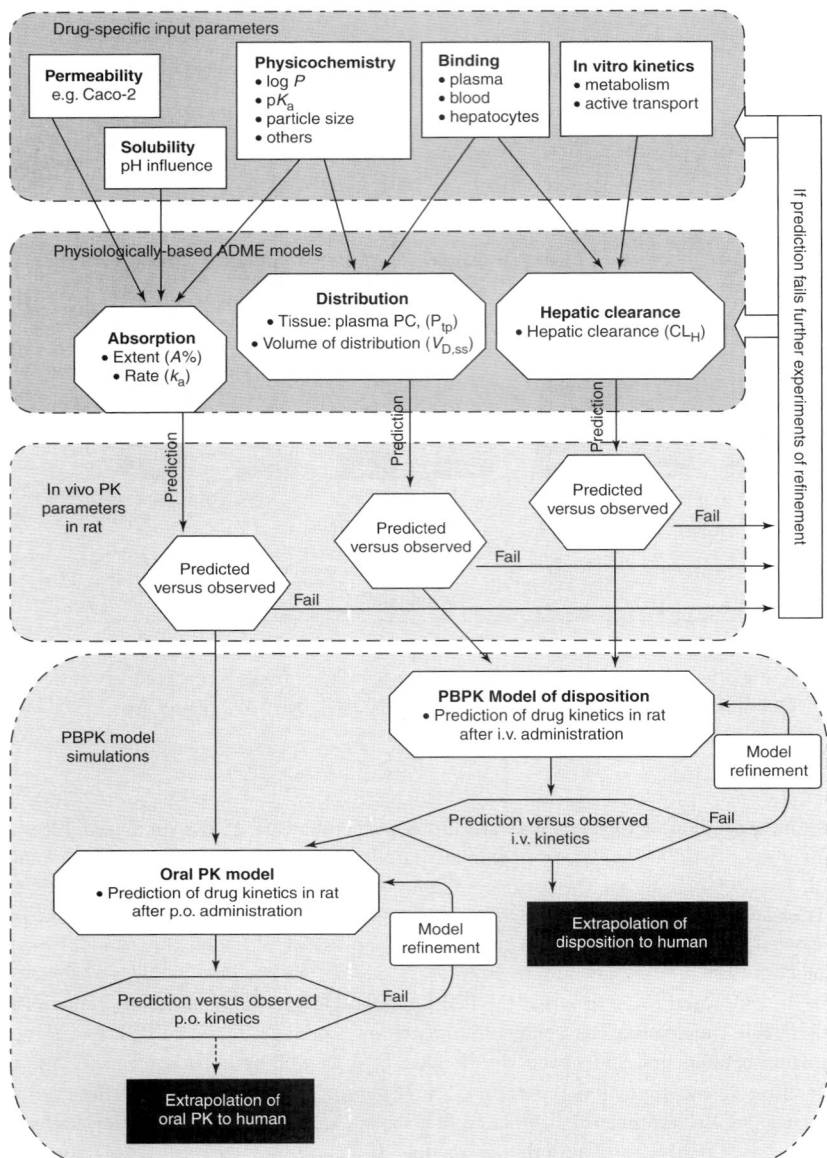

Figure 4   Strategy for PBPK predictions. (Reprinted from Theil, F. P.; Guentert, T. W.; Haddad, S.; Poulin, P. *Toxicol. Lett.* **2003**, *138*, 29–49, with permission from Elsevier.)

combining the absorption model with physiological and compartmental PK models as well as with pharmacodynamic models. The performance of GastroPlus in predicting absorption was evaluated by Parrott and Lavé.[61] In this study, GastroPlus was compared to iDEA another software to predict absorption which is no longer commercially available.

SIMCYP is a software suite specialized in the description of metabolic processes (Simcyp Ltd).[62] SIMCYP utilizes fundamental scaling procedures, described by Houston[41] for the prediction of in vivo hepatic clearance (CL_H) from in vitro metabolism data. These in vitro metabolism data are obtained from individual CYP in human-expressed recombinant systems. These are used for predicting not only the mean behavior in individuals, but also in whole populations or in special subpopulations. In order to predict drug–drug interactions involving CYP, SIMCYP utilizes the relationship between the inhibitor concentration at the active site in vivo and the K_i determined in vitro. Competitive inhibition, induction, and/or mechanism-based inactivation mechanisms can be investigated within this software, according to the principles described by Ito *et al.*[63] and Tucker *et al.*[64] SIMCYP not only predicts the mean value, but also simulates the extremes in the population by applying a Monte Carlo approach.

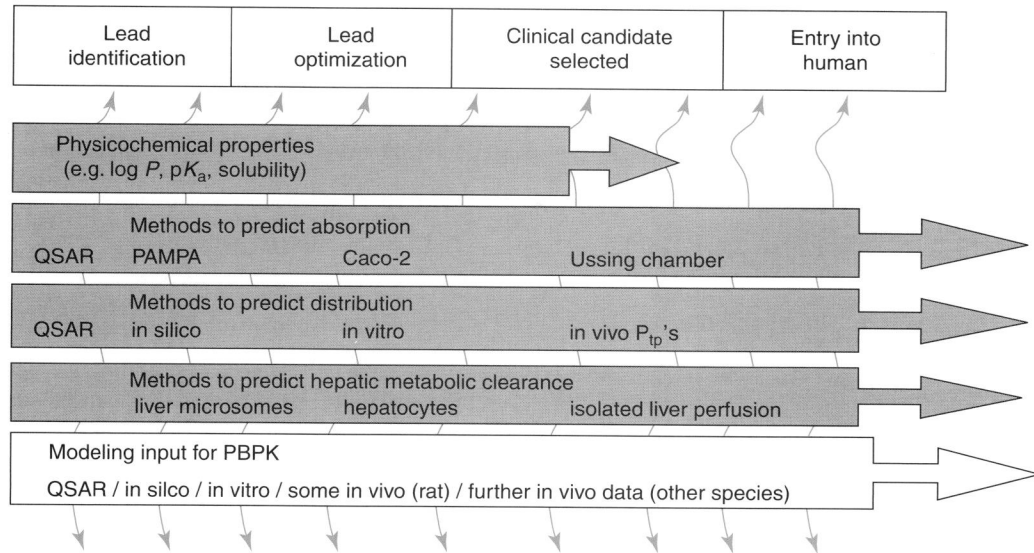

Figure 5 Modeling approach within drug discovery and early nonclinical development. (Reprinted from Theil, F. P.; Guentert, T. W.; Haddad, S.; Poulin, P. *Toxicol. Lett.* **2003**, *138*, 29–49, with permission from Elsevier.)

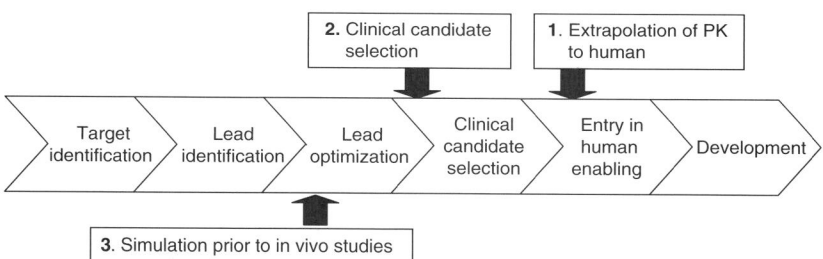

Figure 6 Application of PBPK during the drug discovery and development processes. (Reproduced with permission from Parrott, N.; Jones, H.; Paquereau, N.; Lavé, T. *Basic Clin. Pharmacol. Toxicol.* **2005**, *96*, 193–199.)

Commercial products proposing physiology-based whole-body models allowing the simulation of the whole pharmacokinetic behavior including absorption, distribution and elimination include PK-Sim[4] (Bayer Technology Services)[73] Cloe PK (Cyprotex),[74] and pkEXPRESSTM (Lion Bioscience).[75]

The different technologies for PBPK modeling discussed have all different strengths making them useful for different purposes within the research and development process in the pharmaceutical industry (see **Table 2**).

5.37.3   Applications and Limitations of the Current Methodology

5.37.3.1   Physiologically-Based Pharmacokinetic for Extrapolation of Pharmacokinetics to Humans

A number of publications illustrate the potential of PBPK, both for predicting human pharmacokinetics and for examining species differences in pharmacokinetics. For example, Sawada predicted the human plasma concentration versus time profiles of seven compounds (phenobarbital, phenytoin, hexobarbital, quinidine, tollbutamide, valproate, and diazepam).[65] Using a physiologically-based approach he incorporated the intrinsic (unbound compound) clearance and unbound tissue-to-plasma concentration ratios that were extrapolated from rat data, plus the human plasma protein binding and blood-to-plasma concentration ratios. The concentration–time curves that were predicted for phenytoin, hexobarbital, quinidine, and phenobarbital showed comparatively good agreement with the values observed in human plasma. By contrast, for tolbutamide, valproate, and diazepam, because of poor clearance predictions the areas under the concentration–time curve were poorly estimated.[65]

Table 2　Commercially available softwares

Process simulated	Absorption	Metabolism and drug–drug interaction	Whole-body simulations
Software/company website	GastroPlus (SimulationsPlus, www.simulationsplus.com/) IDEA[a] (Lion Bioscience, www.lionbioscience.com/)	SIMCYP (SimCyp, www.simcyp.com/)	Cloe PK (Cyprotex, www.cyprotex.com/) pkEXPRESS:[a] (Lion Bioscience, www.lionbioscience.com/) PK-Sim (Bayer Technology Services, www.pk-sim.com) SIMCYP (SimCyp, www.simcyp.com/)
Description	Semiphysiological absorption models Input data: structural descriptors, permeability, solubility, in vitro ADME data	Predictions in whole populations or in special subpopulations Input data: in vitro metabolism data on recombinant CYP	Physiological models including permeability limitations for PKSim Input data: physicochemical and in vitro ADME properties The coming versions of Gastroplus and SIMCYP will include PBPK into their new release

[a]No longer commercially available.

Successful predictions were also reported in the case of epiroprim.[66] This was achieved by combining PBPK and allometric scaling. Epiroprim, an antimicrobial agent from the diaminopyrimidine family, was considered a challenging compound for the current extrapolation methods because of its large interspecies differences in its pharmacokinetic properties mainly related to mixed elimination pathways, namely hepatic metabolism and biliary excretion of unchanged compound. For the prediction of distribution in humans, the tissue/plasma partition coefficients were predicted using the human tissue composition model based on in silico input parameters. The human clearance was predicted by using allometric scaling combining animal intrinsic in vivo blood clearance and in vitro hepatocyte data in animals and humans. Under these circumstances, PBPK led to reasonable predictions of the in vivo disposition profile of epiroprim in rat and human.

Lumped physiologically-based models (in which the tissues are subdivided into groups, according to their kinetic properties) have been used to predict human plasma concentration–time curves for the melatonin agonist S20098.[67] By combining in vitro metabolism data obtained from human CYP enzymes with partition coefficients in fat and liver, the range of plasma concentrations observed in human could be successfully predicted. In another study[13] a lumped physiologically-based pharmacokinetic model originally developed by Arundel for the rat[10] was extended and applied to human for ten extensively metabolized compounds. Overall, the results showed that this model might be used to predict concentration versus time profiles in human. Irrespective of the compounds' pharmacokinetic and physicochemical characteristics, the average error for the predicted pharmacokinetic parameters (CL (total clearance), $V_{D,ss}$ (volume of distribution at steady state), $t_{1/2}$ (half-life)) of the 10 compounds was less than twofold.[13]

Recently, empirical and PBPK approaches were compared for the prediction of human pharmacokinetics using 19 diverse compounds taken from recent clinical development projects at Roche.[68] The original set of compounds included in this analysis represented all the compounds developed at Roche that went into clinical development between 1998 and 2002. From this set biological compounds, prodrugs, and compounds that were not absorbed were excluded. The remaining compounds ($n = 19$) were used in the analysis. The compounds selected covered a wide range of physicochemical and PK properties. All compounds were lipophilic with log P values ranging between 1.2 and 6.6, with a mean of 4.0. Five compounds were acids with a $pK_a < 8$, seven were bases with a $pK_a > 6$ and seven were neutral or weakly ionized at physiological pH. Half-life in human ranged from short (0.50 h) to long (200 h). Elimination pathways included hepatic metabolism, renal excretion, biliary excretion, or a combination of these. Fraction unbound in plasma (f_u) values ranged from extensive (0.00060) to low (0.89). Distribution consisted of both passive and active processes and varied from limited to widespread. In the rat, absorption ranged from 15% to 100%. Predicted values (pharmacokinetic parameters and plasma concentrations) were compared to observed values in order to assess the accuracy of the prediction methods.

Based on the strategy proposed (*see* Section 5.37.2.5), a prediction would have been made prospectively for approximately 70% of the compounds. The accuracy for these compounds in terms of the percentage of compounds with an average-fold error of less than twofold was 83%, 50%, 75%, 67%, 92%, and 100% for clearance, volume, half-life, C_{max} (maximal (peak) plasma concentration), AUC (area under the plasma concentration–time curve) and t_{max}, respectively. For the other 30% unacceptable prediction accuracy was obtained in animals; therefore a prospective prediction would not have been made using PBPK. The accuracy for these compounds in terms of the percentage of compounds with an average-fold error of less than twofold was 40%, 40%, 60%, 0%, 40%, and 80% for clearance, volume, half-life, C_{max}, AUC, and t_{max}, respectively. In general, for compounds that were cleared by hepatic metabolism or renal excretion, and whose absorption and distribution were governed by passive processes, the prediction accuracy was very good (average-fold error less than twofold) using PBPK. Significant mispredictions were achieved when other elimination processes (e.g., biliary elimination) or active processes were involved. PBPK prediction accuracy was significantly improved when any nonlinearities in clearance were incorporated; when both microsomal and plasma binding were included in the scaling of microsomal data; when intestinal metabolism was incorporated for CYP3A4 substrates and when solubility was measured in physiological media for low-solubility compounds. The poorer predictions or the compounds for which a prediction was not attempted with PBPK were often as a result of processes that were not incorporated into the model due to the lack of in vitro input parameters to describe processes such as biliary excretion or enterohepatic recirculation.

The prospective use of PBPK approaches was described recently for one compound.[69] This basic (pK_a 10.3) and lipophilic (log P 3) compound showed good aqueous solubility ($9\,mg\,mL^{-1}$ at pH 5) and permeability measured in Caco-2 cells was high. Animal pharmacokinetics showed high blood clearance in rats ($50\,mL\,min^{-1}\,kg^{-1}$) but low clearance in dogs and monkeys (8 and $13\,mL\,min^{-1}\,kg^{-1}$, respectively) and a high volume of distribution (between 4 and $9\,L\,kg^{-1}$) and moderate binding to plasma proteins (approx. 25%) in all three species.

The volume of distribution predicted using tissue composition based equations was first compared to the observed volume in rat and was found to under predict by threefold. Therefore the predicted partition coefficients were replaced by experimentally determined values from a quantitative whole-body autoradiography study, although caution is needed in using such estimates of tissue partitioning based on total radioactivity since they measure the distribution of both parent compound and metabolites. However, in this case the volume in rat was more accurately predicted using such estimates and so the distribution in human was simulated using these values in a human PBPK model.

Direct physiological scaling of intrinsic clearance in rat and dog hepatocytes gave underpredictions of the observed clearance in both species and this underprediction could not be fully explained by renal or biliary clearance since these were known to be relatively minor (each <10% of total clearance). As this compound is mainly metabolized by CYP3A4, intestinal metabolism may also be important. However in vitro clearance obtained with gut microsomes was not available. Therefore the estimated human clearance was based on direct scaling of intrinsic clearance from human hepatocytes with inclusion of a predicted renal clearance calculated from the unbound renal clearance in rat corrected for the ratio of the GFR in rat to that in man. Finally the absorption in rat and dog had been shown to be well predicted with GastroPlus based on the experimental permeability and solubility data and so the same inputs were used to simulate the human oral profile which is shown in **Figure 7** together with the observed data.

All in all, there is growing evidence that physiologically-based models show improved prediction accuracy for extrapolation of human pharmacokinetics over more empirical methods. The PBPK approach is based upon solid physiological principles and so can often be extended in a rational way to include additional relevant information and data. PBPK models offer also a mechanistic framework which contributes to an improved understanding of the compound's properties which ultimately should result in better prediction accuracy. The variability and uncertainty associated with the predictions requires also some special attention. As is apparent from the literature and available clinical data for the compounds studied, there is a large degree of variability in the population both in terms of physiology (tissue volumes, blood flows, transit times, etc.) and biochemistry (plasma binding, CYP expression and activity, etc.). In order to produce predictions that are more realistic of the target population, this variability together with any uncertainty (variations due to assumptions, hypotheses, handling of system, etc.) ideally must be accounted for. The incorporation of these factors can be translated into a measure of confidence in the prediction.

5.37.3.2 Early Pharmacokinetic–Pharmacodynamic (PK/PD) Predictions in Human Based on Physiologically-Based Pharmacokinetic and Preclinical Data

In a recent study a mechanistic PBPK and pharmacokinetic/pharmacodynamic approach was reported for the prediction of the human response using the example of a immunomodulator[70] (*see* 5.38 Mechanism-Based Pharmacokinetic–Pharmacodynamic Modeling for the Prediction of In Vivo Drug Concentration–Effect Relationships – Application in Drug Candidate Selection and Lead Optimization). The example was based on the immunomodulator FTY720 (the first

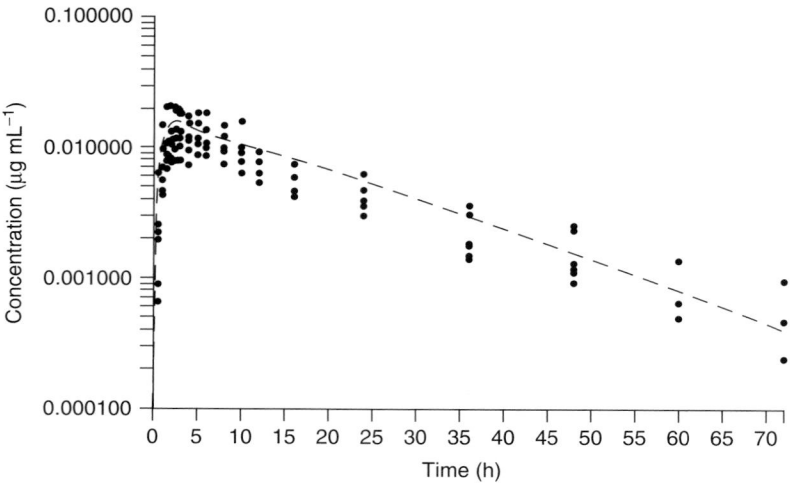

Figure 7 Observed data and physiologically-based predicted (oral administration) plasma concentration–time profiles in human for a 10 mg dose. (Reproduced with permission from Parrott, N.; Jones, H.; Paquereau, N.; Lavé, T. *Basic Clin. Pharmacol. Toxicol.* **2005**, *96*, 193–199.)

sphingosine-1-phosphate receptor agonist) acting by stimulating the sequestration of lymphocytes into lymph nodes in order to remove cells from blood circulation. A PBPK model incorporating permeability limitations in thymus, brain, and lymph nodes was connected to an indirect response model of the lymphocyte system to characterize the cell trafficking effects. Permeability was obtained after fitting of in vivo tissue data in rat. The IC_{50} of FTY720 was estimated from monkey; human was assumed to be similar to the monkey. To make predictions of the pharmacodynamic behavior for humans, systemic exposure was obtained from a physiology based model, and an estimate of lymphocyte turnover in human was obtained from preclinical species using allometric scaling. Predictions compared well with clinical results.

5.37.3.3   Physiologically-Based Pharmacokinetics Applied to Clinical Candidate Selection

In contrast to empirical techniques, an approach based upon a mechanistic framework can lead to greater insights into the behavior of compounds and offers more potential in the early stages of drug development since it does not require animal in vivo data in several species as is needed for allometric scaling. Therefore, PBPK has a great potential to assist clinical candidate selection where numerous factors need to be considered and data related to the pharmacokinetics and pharmacodynamics of the compounds needs to be combined and compared in a rational way.

An example showing the utility of such an approach was provided recently.[69] In this example in vitro and in vivo data for five potential clinical candidates were combined in PK/PD models to make estimate of the effective human doses and associated exposures and aid in the selection of the most promising compound. To ensure that the decision was based upon significant differences between the compounds, estimates of variability and/or uncertainty were carried through in the modeling of the PK and PD. The five compounds were from the same structural class and showed similar physicochemical properties. Molecular weights were in the range 406–472 g and all compounds were largely unionized at physiological pH. Values of log P ranged from 2.1 to 2.9 and permeability was good for all compounds while solubility ranged from 2.2 to 0.009 mg mL^{-1}. In vitro receptor binding and plasma protein binding in rat and human and in vivo pharmacokinetic and pharmacodynamic data in rat were also available for all compounds.

In a first step the scaling of intrinsic clearances determined in rat hepatocytes was compared to in vivo clearance. When nonlinearity was accounted for the estimated hepatic metabolic clearance values were in reasonable agreement with observed total clearances which ranged from 7 to 35 mL min^{-1} kg^{-1} and it was considered appropriate to estimate the expected clearances in human by similar scaling of human hepatocyte data. The error around the mean predicted human clearance was based on the variability seen in different batches of human hepatocytes.

Tissue composition based prediction of volume based on lipophilicity and plasma protein binding showed an average 2.2-fold error and the correlation of predicted versus observed volume for the five compounds was poor. For the prediction of volume of distribution it was assumed that the volume in L kg^{-1} in human was the same as the observed

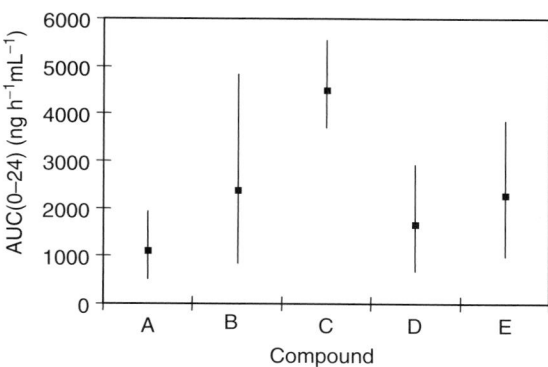

Figure 8 Simulated AUC at steady state for five potential clinical candidate compounds. Symbols give the mean prediction while lines indicate a 95% confidence range. (Reproduced with permission from Parrott, N.; Jones, H.; Paquereau, N.; Lavé, T. *Basic Clin. Pharmacol. Toxicol.* **2005**, *96*, 193–199.)

volume in the rat (ranging from 0.9 to 2.8 L kg^{-1} for the five compounds). Due to the uncertainty in the prediction of volume the error range associated with this parameter was set as a uniform distribution over a twofold range.

The effect versus plasma concentration relationship in the rat pharmacodynamic data was shown to be a direct one and was well described by a simple Emax model. Based on preclinical models for efficacy, a 90% effect was considered as the target for therapeutic effect. The effect versus plasma concentration relationship in the rat pharmacodynamic data was shown to be a direct one and was well described by a simple Emax model. A 90% effect was considered as the target for therapeutic effect. Finally the human C_{90} (human concentration corresponding to 90% effect) was estimated by accounting for the different affinities of each compound for the rat and human receptors and also for the different free fraction unbound in rat and human. Thus:

$$C_{90} \text{ human} = C_{90} \text{ rat} \cdot (K_i \text{ human}/K_i \text{ rat}) \cdot (f_u \text{ rat}/f_u \text{ human}) \qquad [8]$$

Finally, absorption in the rat was shown to be well predicted by GastroPlus since good agreement was seen between simulated and observed oral profiles and so the predicted human pharmacokinetics and pharmacodynamics were combined in a GastroPlus human model to allow for estimation of the effective steady state doses and exposures after repeated oral dosing. Incorporation of variability and uncertainty was achieved using some stochastic simulations. The predicted effective doses, and associated exposures including error ranges on the predictions were then provided for the various potential clinical candidates (see **Figure 8**).

This example showed that the PBPK approach assists a sound decision on the selection of the optimal molecule to be progressed by integrating the available information and focusing attention onto the expected properties in human. Importantly, the method is also able to include estimates of variability and uncertainty in the predictions.

5.37.3.4 Physiologically-Based Pharmacokinetics to Prioritize Compounds Prior to In Vivo Experiment

Drug discovery is increasingly 'data rich' with high-throughput chemistry generating numerous compounds that are rapidly screened for pharmacological and pharmacokinetic properties. Determination of in vivo PK is considerably more costly than in vitro screening and so there is interest in optimizing resources by using simulation to prioritize compounds. However, acceptance of this approach requires extensive validation. The practical value of generic physiologically-based models of pharmacokinetics at the early stage of drug discovery was evaluated.[69,71] In the study reported below, the validation was carried out with rat data but the ultimate aim is to run these predictions in humans.

In the study by Parrott *et al.*, a generic PBPK model was applied to predict plasma profiles after intravenous and oral dosing to the rat for a set of 68 compounds from six different chemical classes. The compounds were selected without particular bias and so are considered representative of current discovery compounds. The physicochemical properties of the compounds are rather different from those of marketed compounds; in particular they have high lipophilicity (mean log *P* of 4) and low aqueous solubility as well as a tendency to neutrality at physiological pH. The more extreme property values can present measurement difficulties and so for consistency all predictions were made on the basis of calculated lipophilicity and protein binding while in vitro measurements of intrinsic clearance in hepatocytes, ionization, solubility, and permeability were used for all compounds.

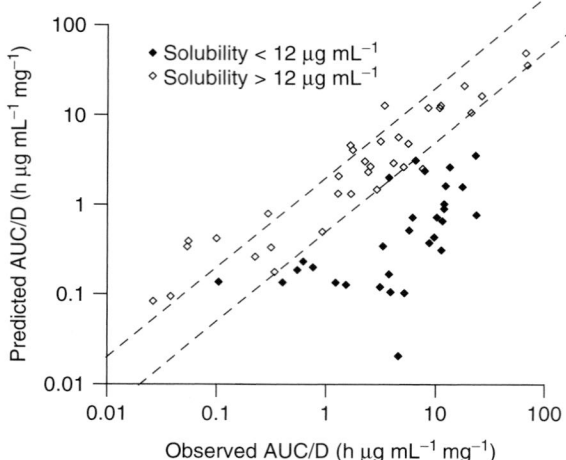

Figure 9 Predicted versus observed AUC/dose for 68 Roche compounds. (Reproduced with permission from Parrott, N.; Jones, H.; Paquereau, N.; Lavé, T. *Basic Clin. Pharmacol. Toxicol.* **2005**, *96*, 193–199.)

In a first stage distribution was predicted with tissue composition based equations and the estimated tissue partition coefficients were combined with clearance estimated by direct scaling of hepatocyte intrinsic clearance in a PBPK model as described earlier.

In a second stage GastroPlus was used to simulate oral absorption and oral profiles were produced by feeding this predicted input into a compartmental disposition model fitted to the mean observed (intravenous) iv data. For intravenous dosing simulations 60% of the clearance and volume predictions were within twofold of the observed values and the ranking of compounds by predicted versus observed parameters showed correlation coefficients of 0.8 and 0.6 for clearance and volume, respectively. For oral dosing simulations, 40% of the predicted AUC values were within twofold of observed values and the mean fold-error was 4.1. For compounds with measured solubility of less than $0.012\,\mathrm{mg\,mL^{-1}}$ only 10% of predictions were within twofold while for compounds with higher solubility 70% were within twofold[69] (**Figure 9**).

Overall, this study indicated that generic simulation of pharmacokinetics at the lead optimization stage could be useful to predict differences in pharmacokinetic parameters of threefold or more based upon minimal measured input data. Fine discrimination of pharmacokinetics (less than twofold) should not be expected due to the uncertainty in the input data at the early stages. It is also apparent that verification of simulations with in vivo data for a few compounds of each new compound class was required to allow an assessment of the error in prediction and to identify invalid model assumptions.

The value of PBPK for simulating the first-time-in-animal study was illustrated also by Germani *et al.*[71] In this study, the model was used to predict the plasma pharmacokinetics obtained in 24 rat and two mouse pharmacokinetic studies. This study was restricted to the prediction of disposition profiles (i.e., profiles observed after intravenous administration of the compounds). The administered compounds were synthesized as part of a number of different discovery programs. The simulated profiles were generally in agreement with the observed ones. Plasma clearance and mean residence times were well predicted: average fold errors were 2.7 and 2.5, respectively, with approximately one half of the cases predicted within a twofold of observed. Some over prediction of plasma clearance was observed in the range of low clearance values, in case of compounds with $\log P = 5$ and protein binding of 99%. On the other hand, slight underestimates were observed for molecules with $\log P$ near zero and protein binding of 99.9%. The error in the volume of distribution at steady state ($V_{\mathrm{D,ss}}$) was higher, but still acceptable (average fold error was 3.7).

5.37.3.5   Inclusion of Transporters and Gut Wall Metabolism into Physiologically-Based Pharmacokinetics

Although remarkable progress has been made in the identification and functional characterization of drug transporters, quantitative predictions of pharmacokinetic processes involving membrane transport is still challenging because of the lack of quantitative in vitro models and the sheer number of issues that need to be considered for the modeling approaches to allow an accurate description of the in vivo situation.

Mathematical models incorporating all nonlinear processes in a systematic manner must allow for discrimination of several distinct kinetic processes, such as uptake by passive and active transport, distribution, metabolism (e.g., oxidations and conjugations), and excretion (e.g., biliary excretion). Furthermore, each of the above processes must be linked to the appropriate drug concentration within the relevant medium (i.e., free concentration in plasma for the uptake processes, and the intracellular free drug concentration for the efflux, metabolism, and excretion processes). PBPK models offer the framework to potentially address and link all the foregoing aspects in a mechanistic fashion. Thus, these models should, theoretically, enable the prediction of the in vivo kinetics even for compounds for which such complex processes are involved.

Attempts were made recently to model drug absorption of compounds subject to gut efflux and/or metabolism.[59] Thus, using the ACAT model available in Gastroplus based on the assumed P-glycoprotein (P-gp) distribution in the gatrointestinal tract, and optimized V_{max} and K_m for substrate P-gp interactions, reasonable estimates of the in vivo plasma concentration profiles and bioavailabilities were obtained for oral digoxin administered alone and after rifampin treatment.[59]

The absorption of saquinavir, a drug metabolized and effluxed in the gut, was also successfully modeled in this study. Saquinavir was the first clinically available human immunodeficiency virus (HIV) protease inhibitor in the USA. It has a very low bioavailability (0.5–2%) due to extensive first pass metabolism by CYP3A4 in the gut and in the liver. It is also a gut P-gp substrate. The bioavailability of saquinavir was found to double when administered with food or with grapefruit juice. The interaction has been attributed to increased hepatic blood flow due to food intake and suppression of intestinal CYP3A4 and P-gp by grapefruit juice respectively. In order to model the absorption of saquinavir, in vitro and/or optimized kinetic constants (V_{max} and K_m) were used for both gut and liver metabolism and the amounts of CYP3A4 in the small intestine and liver were taken into consideration as well as the regional distribution of P-gp in the gut. Based on this model, the relative impact of solubility, passive permeability, efflux, and metabolism in gut and liver on the overall absorption of saquinavir could be estimated.

5.37.4   Discussion

For extrapolation of human pharmacokinetics, physiologically-based models show improved prediction accuracy over more empirical methods. The PBPK approach is based upon solid physiological principles and can be extended to include additional relevant processes. Also, besides satisfactory prediction capabilities, an approach based upon a mechanistic framework can lead to greater insights into the behavior of compounds and offers more potential in the early stages of drug development since it does not require animal in vivo data in several species as is needed for allometric scaling.

When applied at the clinical candidate selection phase the PBPK approach assists a sound decision on the optimal molecule to be progressed by integrating the available information and focusing the attention onto the expected properties in human. Importantly, the method is also able to include estimates of variability and uncertainty in the predictions, which should be considered as an important piece of information for decision-making.

The results on the use of generic simulation prior to in vivo studies indicate that some caution is required since certain chemical classes were poorly predicted and it is recommended that generic PBPK models should only be applied for prioritization after verification of the simulations with in vivo pharmacokinetics for a few compounds of a given chemical class. Such verification will help to identify invalid model assumptions or missing processes where additional data is needed. In addition, for poorly soluble compounds, the use of aqueous solubility is shown to be inadequate for reliable prediction of oral absorption in physiologically-based models.

Important limitations of the generic PBPK approach are realized for compounds that have significant active distribution/absorption processes, where biliary elimination is a major component of the elimination process or where the assumptions of flow limited distribution and well-mixed compartments are not valid and permeability limited distribution is apparent. These drawbacks could be addressed by the addition of permeability barriers for some tissues and by the incorporation of a more complex liver model which addresses active uptake into the liver, active efflux into the bile, biliary elimination, and enterohepatic recirculation. However, this will require the availability of the appropriate input data for quantification of the various processes involved as well as validation of the corresponding in vitro to in vivo scaling approaches.

Overall, these three examples illustrate how PBPK models can already add value at various stages of the preclinical compound research and development process although one has to keep in mind a number of limitations. Their use is growing and the potential will be fully exploited as powerful and userfriendly software continues to make PBPK models accessible to nonspecialists.

References

1. Reigner, B. G.; Williams, P. E. O.; Patel, J. H.; Steimer, J. L.; Peck, C.; van Brummelen, P. *Clin. Pharmacokinet.* **1997**, *33*, 142–152.
2. Dickins, M.; Van de Waterbeemd, H. *Drug Disc. Today: Biosilico* **2004**, *2*, 38–45.
3. Leahy, D. E. *Drug Disc. Today: Biosilico* **2004**, *2*, 78–84.
4. Willmann, S.; Lippert, J.; Sevestre, M.; Solodenko, J.; Fois, F.; Schmitt, W. *Drug Disc. Today: Biosilico* **2003**, *1*, 121–124.
5. Theil, F. P.; Guentert, T. W.; Haddad, S.; Poulin, P. *Toxicol. Lett.* **2003**, *138*, 29–49.
6. Nestorov, I. *Clin. Pharmacokinet.* **2003**, *42*, 883–908.
7. Poulin, P.; Theil, F. P. *J. Pharm. Sci.* **2002**, *91*, 1358–1370.
8. Poulin, P.; Theil, F. P. *J. Pharm. Sci.* **2000**, *89*, 16–35.
9. Poulin, P.; Schoenlein, K.; Theil, F. *J. Pharm. Sci.* **2001**, *90*, 436–447.
10. Arundel, P. A. The third IFAC symposium, Modeling and control in biomedical systems, University of Warwick, March 23–26, 1997.
11. Poulin, P.; Theil, F. P. *J. Pharm. Sci.* **2000**, *89*, 16–35.
12. Poulin, P; Theil, F. P. *J. Pharm. Sci.* **2002**, *91*, 129–156.
13. Lavé, T.; Luttringer, O.; Zuegge, J.; Schneider, G.; Coassolo, P.; Theil, F. P. *Ernst Schering Res. Found. Workshop* **2002**, *37*, 81–104.
14. Lin, J. H.; Sugiyama, Y.; Awazu, S.; Hanano, M. *J. Pharmacokinet. Biopharm.* **1982**, *10*, 637–647.
15. Schuhmann, G.; Fichtl, B. *Biopharm. Drug Dispos.* **1987**, *8*, 73–86.
16. Clausen, J.; Bickel, M. H. *J. Pharm. Sci.* **1993**, *82*, 345–349.
17. Bickel, M. H.; Raaflaub, R. M.; Hellmuller, M.; Stauffer, E. J. *J. Pharm. Sci.* **1987**, *76*, 68–74.
18. Ballard, P.; Arundel, P. A.; Leahy, D. E.; Rowland, M. *Pharm. Res.* **2003**, *20*, 857–863.
19. Ballard, P.; Leahy, D. E.; Rowland, M. *Pharm. Res.* **2000**, *17*, 660–663.
20. Reinoso, R. F.; Telfer, B. A.; Rowland, M. *Eur. J. Pharm. Sci.* **1998**, *6*, 145–152.
21. Ritschel, W. A.; Hammer, H. G. *Int. J. Clin. Pharmacol. Toxicol.* **1980**, *18*, 298–316.
22. Lombardo, F.; Obach, R. S.; Shalaeva, M. Y.; Gao, F. *J. Med. Chem.* **2002**, *45*, 2867–2876.
23. Wajima, T.; Fukumura, K.; Yano, Y.; Oguma, T. *J. Pharm. Pharmacol.* **2003**, *55*, 939–949.
24. Yokogawa, K.; Nakashima, E.; Ishizaki, J.; Maeda, H.; Nagano, T.; Ichimura, F. *Pharm. Res.* **1990**, 7, 691–696.
25. Yata, N.; Toyoda, T.; Murakami, T.; Nishiura, A.; Higashi, Y. *Pharm. Res.* **1990**, 7, 1019–1025.
26. Kaliszan, R. M. M. *Int. J. Pharm.* **1996**, *145*, 9–16.
27. Abraham, M. H.; Chadha, H. S.; Mitchell, R. C. *J. Pharm. Sci.* **1994**, *83*, 1257–1268.
28. Abraham, M. H.; Chadha, H. S.; Whiting, G. S.; Mitchell, R. C. *J. Pharm. Sci.* **1994**, *83*, 1085–1100.
29. Lombardo, F.; Blake, J. F.; Curatolo, W. J. *J. Med. Chem.* **1996**, *39*, 4750–4755.
30. Luco, J. M. *J. Chem. Inf. Comput. Sci.* **1999**, *39*, 396–404.
31. Nestorov, I.; Aarons, L.; Rowland, M. *J. Pharmacokinet. Biopharm.* **1998**, *26*, 521–545.
32. Bjorkman, S. *J. Pharm. Pharmacol.* **2002**, *54*, 1237–1245.
33. Sawada, Y.; Hanano, M.; Sugiyama, Y.; Harashima, H.; Iga, T. *J. Pharm. Sci.* **1984**, *12*, 5875–5896.
34. Zuegge, J.; Schneider, G.; Coassolo, P.; Lavé, T. *Clin. Pharmacokinet.* **2001**, *40*, 553–563.
35. Lavé, T.; Coassolo, P.; Reigner, B. *Clin. Pharmacokinet.* **1999**, *36*, 211–231.
36. Lavé, T.; Dupin, S.; Schmitt, C.; Chou, R. C.; Jaeck, D.; Coassolo, P. *J. Pharm. Sci.* **1997**, *86*, 584–590.
37. Houston, J. B. *Toxicol. In Vitro* **1994**, *8*, 507–512.
38. Houston, J. B.; Carlile, D. J. *Drug Metab. Rev.* **1997**, *29*, 891–922.
39. Mahmood, I.; Balian, J. D. *Xenobiotica* **1996**, *26*, 887–895.
40. Hoener, B. A. *Biopharm. Drug Dispos.* **1994**, *15*, 295–304.
41. Houston, J. B. *Biochem. Pharmacol.* **1994**, *47*, 1469–1479.
42. Houston, J. B.; Kenworthy, K. E. *Drug Metab. Dispos.* **2000**, *28*, 246–254.
43. Lavé, T.; Dupin, S.; Schmitt, C.; Valles, B.; Ubeaud, G.; Chou, R. C.; Jaeck, D.; Coassolo, P. *Pharm. Res.* **1997**, *14*, 152–155.
44. Carlile, D. J.; Zomorodi, K.; Houston, J. B. *Drug Metab. Dispos.* **1997**, *25*, 903–911.
45. Naritomi, Y.; Terashita, S.; Kimura, S.; Suzuki, A.; Kagayama, A.; Sugiyama, Y. *Drug Metab. Dispos.* **2001**, *29*, 1316–1324.
46. Wilkinson, G. R. *Pharmacol. Rev.* **1987**, *39*, 1–47.
47. Iwatsubo, T.; Hirota, N.; Ooie, T.; Suzuki, H.; Shimada, N.; Chiba, K.; Ishizaki, T.; Green, C. E.; Tyson, C. A.; Sugiyama, Y. *Pharmacol. Ther.* **1997**, *73*, 147–171.
48. Efthymiopoulos, C.; Battaglia, R.; Strolin-Benedetti, M. *J. Antimicrob. Chemother.* **1991**, *27*, 517–526.
49. Richter, W. F.; Heizmann, P.; Meyer, J.; Starke, V.; Lavé, T. *J. Pharm. Sci.* **1998**, *87*, 496–500.
50. Sawada, Y.; Hanano, M.; Sugiyama, Y.; Iga, T. *J. Pharmacokinet. Biopharm.* **1984**, *12*, 241–261.
51. Lin, J. H. *Drug Metab. Dispos.* **1998**, *26*, 1202–1212.
52. Liu, X.; Chism, J. P.; LeCluyse, E. L.; Brouwer, K. R.; Brouwer, K. L. *Drug Metab. Dispos.* **1999**, *27*, 637–644.
53. LeCluyse, E. L.; Audus, K. L.; Hochman, J. H. *Am. J. Physiol.* **1994**, *266*, C1764–C1774.
54. Pahlman, I.; Andersson, S.; Gunnarsson, K.; Odell, M. L.; Wilen, M. *Pharmacol. Toxicol.* **1999**, *85*, 123–129.
55. Lavé, T.; Portmann, R.; Schenker, G.; Gianni, A.; Guenzi, A.; Girometta, M. A.; Schmitt, M. *J. Pharm. Pharmacol.* **1999**, *51*, 85–91.
56. Lin, J. H. *Drug Metab. Dispos.* **1995**, *23*, 1008–1021.
57. Ishizuka, H.; Konno, K.; Shiina, T.; Naganuma, H.; Nishimura, K.; Ito, K.; Suzuki, H.; Sugiyama, Y. *J. Pharmacol. Exp. Ther.* **1999**, *290*, 1324–1330.
58. Grass, G. M.; Sinko, P. J. *Adv. Drug Deliv. Rev.* **2002**, *54*, 433–451.
59. Agoram, B. W. W.; Bolger, M. B. *Adv. Drug Deliv. Rev.* **2001**, *50*, S41–S67.
60. Norris, D. A.; Leesman, G. D.; Sinko, P. J.; Grass, G. M. *J. Control. Release* **2000**, *65*, 55–62.
61. Parrott, N. J.; Lavé, T. *Eur. J. Pharm. Sci.* **2002**, *17*, 51–61.
62. Rostami-Hodjegan, A.; Tucker, G. *Drug Disc. Today: Biosilico* **2004**, *1*, 441–448.
63. Ito, K.; Iwatsubo, T.; Kanamitsu, S.; Nakajima, Y.; Sugiyama, Y. *Annu. Rev. Pharmacol. Toxicol.* **1998**, *38*, 461–499.
64. Tucker, G. T.; Houston, J. B.; Huang, S. M. *Pharm. Res.* **2001**, *18*, 1071–1080.
65. Sawada, Y.; Harashima, H.; Hanano, M.; Sugiyama, Y.; Iga, T. *J. Pharm. Sci.* **1985**, *8*, 757–766.
66. Luttringer, O.; Theil, F. P.; Poulin, P.; Schmitt-Hoffmann, A. H.; Guentert, T. W.; Lavé, T. *J. Pharm. Sci.* **2003**, *92*, 1990–2007.

67. Bogaards, J. J.; Hissink, E. M.; Briggs, M.; Weaver, R.; Jochemsen, R.; Jackson, P.; Bertrand, M.; van Bladeren, P. J. *Eur. J. Pharm. Sci.* **2000**, *12*, 117–124.

68. Jones, H. R.; Parrott, N.; Jorga, K.; Lavé, T. *Clin. Pharmacokinet.* **2006**, in press.

69. Parrott, N.; Jones, H.; Paquereau, N.; Lavé, T. *Basic Clin. Pharmacol. Toxicol.* **2005**, *96*, 193–199.

70. Meno-Tetang, G. M. L.; Lowe, P. J. *Basic Clin. Pharmacol. Toxicol.* **2005**, *96*, 182–192.

71. Germani, M.; Crivori, P.; Rocchetti, M.; Burton, P. S.; Wilson, A. G. E.; Smith, M. E.; Poggesi, I. *Basic Clin. Pharmacol. Toxicol.* **2005**, *96*, 254–256.

72. Simulations Plus. www.simulationsplus.com/ (accessed May 2006).

73. PK-Sim. Bayer Technology Services. www.pksim.com/ (accessed May 2006).

74. Cloe PK. Cyprotex. www.cyprotex.com/ (accessed May 2006).

75. pkEXPRESSTM Lion Bioscience. www.lionbioscience.com/ (accessed May 2006).

Biographies

Thierry Lavé received his PharmD and PhD in pharmaceutical sciences from the University of Strasbourg in 1992 and 1993. He also received a statistical degree from the University of Paris in 1992 and did an internship in hospital pharmacy from 1988 to 1992 in Strasbourg. Thierry Lavé joined Roche's Department of Preclinical Pharmacokinetics and Drug Metabolism in 1992 and worked in this department as a postdoctoral fellow for 2 years. Thierry Lavé is currently a Scientific Expert at Hoffmann-La Roche and took responsbility for the global modeling and simulation group in preclinical research in 2002. His main area of research is in preclinical modeling of safety, pharmacokinetics, and formulation, physiologically-based pharmacokinetics, interspecies scaling, drug–drug interaction, and all aspects related to early predictions of pharmacokinetics/pharmacodynamics in humans during drug discovery and early drug development. He has published a number of papers in these areas. Thierry Lavé also teaches pharmacokinetics at the Universities of Pharmacy in Basel, Strasbourg, Helsinki, and Besançon.

Neil Parrott graduated in physics from the University of Bristol, UK, then specialized in information technology and medical imaging at the University of Aberdeen, UK. In 1988, he started work with Sandoz Pharma as a developer of chemistry and biology databases and search algorithms. Later with Sandoz he joined the combinatorial chemistry group developing software and automation for high-throughput solid-phase synthesis. In 1998, he joined the Informatics group in Roche Pharma Research where he developed data analysis tools for pharmacokinetic data and led a project to integrate DMPK data into the Roche Research global data warehouse. While working in the DMPK area he became interested in physiologically-based pharmacokinetics and transferred to the Research Modeling and Simulation group where he is now involved in preclinical project support with pharmacokinetic and absorption modeling.

Hannah Jones is a Research Scientist within the DMPK department at F Hoffmann-La Roche, where she is responsible for the simulation of human pharmacokinetics for discovery and early development compounds before entry into human. She received a BSc (Hons) in pharmacology and a PhD in drug metabolism from the School of Biological Sciences and the School of Pharmacy at the University of Manchester, UK, respectively. Hannah is interested in the prediction of human pharmacokinetics, including metabolism, distribution, absorption, drug–drug interactions, and the incorporation of variability and uncertainty. She has several research publications in this area.

5.38 Mechanism-Based Pharmacokinetic–Pharmacodynamic Modeling for the Prediction of In Vivo Drug Concentration–Effect Relationships – Application in Drug Candidate Selection and Lead Optimization

M Danhof, Leiden University, Leiden, The Netherlands

P H Van der Graaf, Pfizer Global Research and Development, Sandwich, UK

D M Jonker, Novo Nordisk A/S, Bagsværd, Denmark

S A G Visser, AstraZeneca R&D, Södertälje, Sweden

K P Zuideveld, F. Hoffmann-La Roche Ltd, Basel, Switzerland

5.38.1 Introduction

The application of pharmacokinetic–pharmacodynamic (PK/PD) modeling in drug development is well established. The primary objective of PK/PD modeling is prediction of the time course of the drug effect in vivo (intensity and duration) in health and disease.[1] As such PK/PD constitutes the scientific basis for optimization of the dosing and delivery profile of drugs in phase II clinical trials. Furthermore PK/PD modeling is also widely applied as the basis for the design (by clinical trial simulation) and the evaluation of phase III clinical trials.[1,2]

In the meantime PK/PD modeling is increasingly applied in drug discovery and preclinical development. Here the objective is the prediction of the PK/PD properties of novel drugs in humans on the basis of information from (high-throughput) in vitro bioassays and in vivo animal studies. Specific applications in this context include (1) the selection of drug candidates with a predicted optimal PK/PD profile in humans, (2) acquisition of pertinent information for lead optimization, and (3) the use of preclinical PK/PD information in optimizing early clinical trials.

The use of PK/PD modeling in drug discovery and early development relies on the prediction, in a strictly quantitative manner, of PK/PD properties in humans. Not surprisingly there is a clear trend toward mechanism-based PK/PD models, which have much improved properties for extrapolation and prediction. Mechanism-based PK/PD models differ from empirical, descriptive models in that they contain specific expressions for processes on the causal path between drug administration and effect. These include (1) target site distribution, (2) target binding and activation, (3) transduction, and (4) effect of homeostatic feedback mechanisms. Ultimately effects on (5) disease processes are also considered. Mechanism-based PK/PD modeling relies on the use of biomarkers to characterize quantitatively processes on the causal path.[3]

In this chapter a general overview of the basic principles of mechanism-based PK/PD modeling is presented. Next specific models and pertinent issues with regard to the prediction of PK/PD properties in humans are discussed on the basis of examples, illustrating the utility of preclinical PK/PD modeling in drug candidate selection and lead optimization.

5.38.2   Mechanism-Based Pharmacokinetic–Pharmacodynamic Modeling: Concepts

5.38.2.1   Pharmacokinetic Considerations: The Rationale of Obtaining Drug Concentrations in In Vivo Pharmacodynamic Investigations

In PK/PD modeling the time course of the drug effect in vivo is typically considered in conjunction with the time course of the drug concentration in plasma (i.e., the pharmacokinetics of the drug). In a compartmental model, drug disposition is described as the transfer of drug between interconnected compartments, serving to mimic drug absorption and elimination processes. This type of analysis of pharmacokinetics is often considered routine and straightforward.[4]

An important factor, however, is that typically wide differences between drugs in the time course of the concentration are observed. Strikingly, this is also the case for structurally similar compounds with nearly identical receptor binding properties. As result the pharmacokinetic properties of new chemical entities constitute an important selection criterion in drug development.[5] In order to predict the pharmacokinetics properties of new chemical entities in humans, the discipline of physiological pharmacokinetic modeling has been developed.[6] A discussion of this type of modeling is beyond the scope of this chapter, which focuses on the prediction of drug effects rather than concentrations.

The observation of wide differences in pharmacokinetics between structurally similar compounds has important implication for the design of preclinical in vivo pharmacodynamic investigations. As was outlined above, due to the wide differences in pharmacokinetics, identical doses of different drugs may result in widely different in vivo drug concentration profiles. This wide difference in pharmacokinetics makes the comparison of in vivo potency estimates on the basis of dose a meaningless exercise. Furthermore, pharmacokinetic properties typically also differ widely between species, which complicates the interspecies extrapolation of drug potency on the basis of doses. For the above-mentioned reasons is imperative that quantitative in vivo pharmacodynamic investigations be based on drug concentration as a measure of internal exposure, rather than dose. To this end in recent years several chronically instrumented rat models have been developed in which, following the (intravenous) administration of a single dose, the time course of the drug effect can be determined in conjunction with drug concentrations.[7–10] It has been amply demonstrated that by PK/PD modeling of these high-density data unique in vivo concentration–effect relationships can be obtained, which may be readily extrapolated from laboratory animals to humans.[11,12]

Recently the concept of 'population' pharmacokinetic analysis has been successfully introduced in preclinical PK/PD modeling. In population analysis, data from several experiments are simultaneously analyzed allowing estimation of the typical pharmacokinetic parameters in conjunction with their intra- and interindividual variability. Population pharmacokinetic analysis offers the advantage that it enables detailed description of the time course of the drug concentration in individual rats on the basis of sparse data or information obtained in separate experiments. This is particularly important in, e.g., behavioral pharmacology studies where blood sampling may interfere with the behavioral readout.[13]

5.38.2.2   Target Site Distribution: The Estimation of Drug Concentrations in the Biophase In Vivo

PK/PD modeling studies typically utilize drug concentrations in plasma as a measure of internal exposure to the drug. Most drugs however have a target site in peripheral tissues. For drugs producing their biological effect in a peripheral tissue, distribution to the site of action may represent a rate-limiting step for producing the biological effect. Typically this is reflected in a delay in the time course of the pharmacological effect relative to the plasma concentration, which is commonly referred to as 'hysteresis' (**Figure 1**).

In order to account for hysteresis in PK/PD investigations the so-called effect compartment model has been introduced.[14] In this model the distribution of drug to a hypothetical effect site is described by the following differential equation:

$$\frac{\mathrm{d}C_e}{\mathrm{d}t} = k_{1e} \cdot C_p - k_{e0} \cdot C_e \qquad [1]$$

where k_{1e} and k_{e0} are the first-order rate constants distribution into and out of the hypothetical effect compartment and C_p and C_e are the drug concentrations in plasma and the hypothetical effect compartment, respectively. In this model the amount of drug entering the effect compartment is considered to be negligible and therefore not reflected in the pharmacokinetics of the drug. Furthermore, for reasons of identifiability, the values of k_{1e} and k_{e0} are usually set equal to each other. In this manner it is inherently assumed that in steady state the drug concentration in the biophase is identical to the (free) plasma concentration.

The effect compartment model has been shown to be highly useful in describing delayed drug effects owing to distributional processes. The model has also been successfully applied in preclinical studies to derive meaningful in vivo drug concentration–effect relationships.[15] However, although for many drugs the assumption that in steady state the drug concentration in the effect compartment is identical to the (free) plasma concentration is plausible, this may not always be the case. Particularly for relatively large hydrophilic molecules and for compounds that are substrates for specific transporters distribution to the target site be restricted. This especially concerns drugs with an intracellular

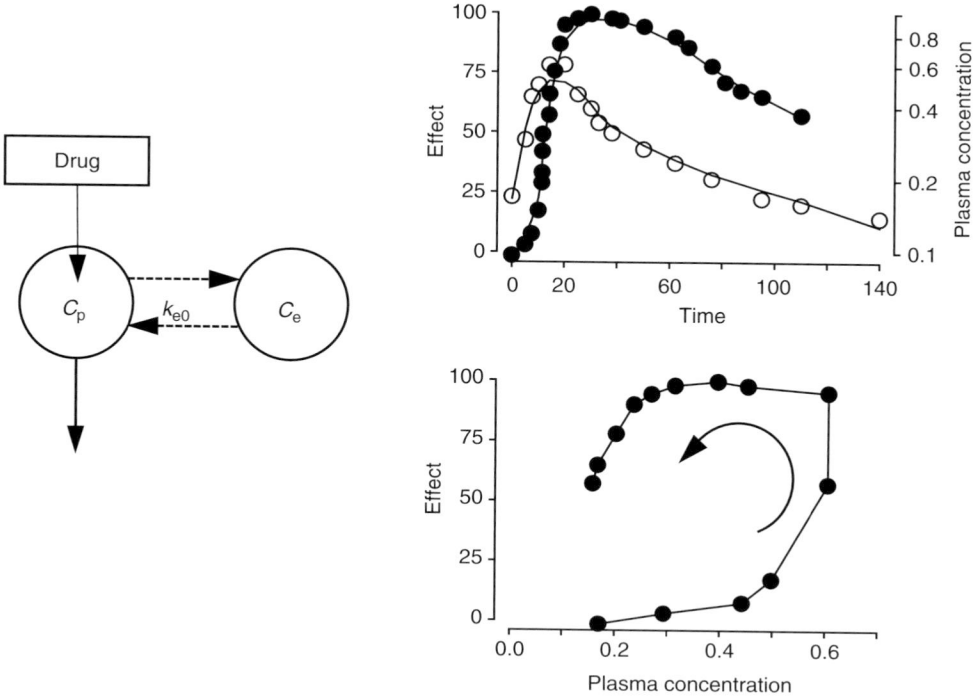

Figure 1   Schematic representation of the effect compartment model to account for delays (hysteresis) in the time course of the drug effect relative to the plasma concentration. By estimation of the rate constant k_{e0} the relationship between the drug concentration in plasma (C_p) and the hypothetical effect site (C_e) is characterized. In this manner target site drug concentration–effect relationships can be derived from plasma concentrations.

target (e.g., cytostatic drugs) and drugs that act in tissues which are protected by specific barriers (e.g., the central nervous system).[3] Recently several specific transporters have been discovered which may restrict the access of a drug to the site of action.[16] For drugs acting in the central nervous system measures of target exposure can be indispensable in PK/PD modeling. A novel technique for obtaining information on target site exposure in the central nervous system is intracerebral microdialysis. An example of the application of intracerebral microdialysis is the investigations on the PK/PD correlation of morphine, where P-glycoprotein and possibly other transporters restrict distribution into the central nervous system.[17]

5.38.2.3   Target Site Interaction and Activation: The Pharmacodynamic Basis of Predicting In Vivo Drug Concentration–Effect Relationships

In PK/PD modeling nonlinear drug concentration–effect relationships are commonly described on the basis of a hyperbolic function (the Hill equation or sigmoid E_{max} model):

$$E = \frac{\alpha \cdot C^{n_{\text{H}}}}{\text{EC}_{50}^{n_{\text{H}}} + C^{n_{\text{H}}}}$$

[2]

where E is the drug effect intensity, α is the maximum drug effect, C is the drug concentration, EC_{50} is the drug concentration at half-maximal effect, and n_{H} is the Hill factor, a parameter which reflects the steepness of the concentration–effect curve. Although very useful for descriptive purposes, a limitation is that the Hill equation provides no insight in the underlying factors that determine the shape and the location of the concentration–effect relationship. As such it does not constitute a scientific basis for the prediction of in vivo concentration–effect relationships on the basis of information from in vitro receptor binding assays.

According to receptor theory the in vivo potency (i.e., the EC_{50}) and intrinsic activity (i.e., α) are dependent on multiple factors related to the drug (receptor affinity, intrinsic efficacy) and the biological system (receptor density, transducer function relating receptor occupancy to the pharmacological effect) (**Figure 2**). The prediction of in vivo drug concentration–effect relationships therefore requires a strict separation between drug-specific properties (which can be estimated in in vitro receptor assays) and biological system-specific parameters (which can only be estimated on the basis of in vivo biological systems analysis).[18] Therefore we have developed a mechanism-based PK/PD strategy, which is based on concepts from receptor theory and which constitutes a scientific basis for the prediction of in vivo concentration–effect relationships.[19]

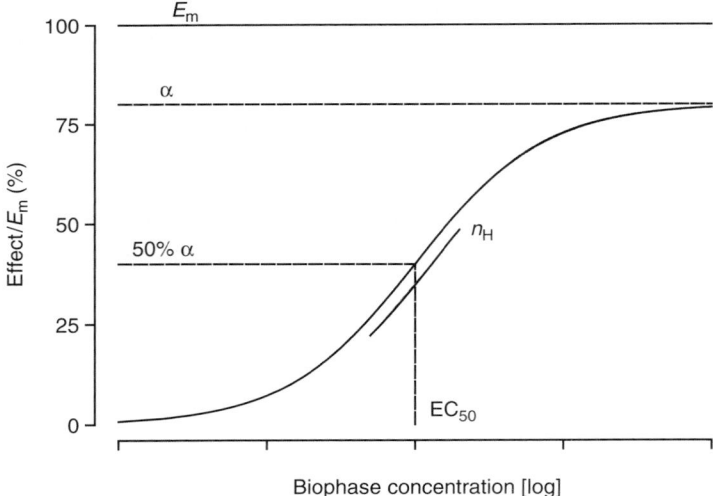

Biophase concentration [log]

Figure 2   The shape and location of an in vivo drug concentration–effect relationship depends on the properties of the drug and the properties of the biological system. Mechanism-based PK/PD models describe drug effects in 'drug-specific' and 'system-specific' parameters. This enables the prediction of in vivo drug concentration–effect relationships in humans on basis of information from in vitro receptor assays and in vivo animal studies.

5.38.2.3.1 Full parametric approach to the incorporation of receptor theory: the operational model of agonism

Classical receptor theory combines two independent parts to describe drug action: an agonist-dependent part, incorporating agonist affinity and intrinsic efficacy, and a tissue-dependent part, determined by the receptor concentration and the nature of the stimulus–response relation. The basic form of Black and Leff's[18] operational model of agonism is based on the observation that most agonist concentration–effect curves are hyperbolic in nature. The authors proved algebraically that if the law of mass action governs the binding of an agonist to the receptor:

$$[AR] = \frac{[R_0] \cdot [A]}{K_A + [A]} \qquad [3]$$

where $[AR]$ is the concentration of agonist–receptor complex, R_0 is the total receptor concentration, $[A]$ the concentration of the agonist, and K_A is the dissociation equilibrium constant of AR, then the function relating the concentration of agonist–receptor complex and response can only be a straight line or another hyperbolic. In order to allow for receptor reserve, the linear alternative needs to be rejected in favor of the hyperbolic transducer function, which can be represented as follows:

$$\frac{E}{E_m} = \frac{[AR]^n}{K_E^n + [AR]^n} \qquad [4]$$

where E is the response, E_m is the maximum system response, K_E is the value midpoint location of the transducer function, that is $[AR]$, that elicits half-maximal effect, and n is the slope of the transducer function. Combining eqns [3] and [4] yields an explicit model of agonism describing the pharmacological effect in terms of agonist concentration:

$$E = \frac{E_m \cdot \tau^n \cdot [A]^n}{(K_A + [A])^n + \tau^n \cdot [A]^n} \qquad [5]$$

where the transducer ratio τ is defined as R_0/K_E as a measure of the efficiency of the occupied receptors into pharmacological effect.

Leff et al.[20] have shown that the operational model of agonism can be employed to obtain estimates of affinity and intrinsic efficacy of a partial agonist by comparison with a full agonist. This so-called 'comparative method'[21] is based on the idea that for a full agonist the Hill equation parameters, intrinsic activity (α) and Hill slope (n_H), are identical to the operational model parameters E_m and n respectively. Therefore, it has been shown previously that when E_m and n are constrained to the estimates of α and n_H for the full agonist, respectively, K_A and τ for a partial agonist can be estimated by directly fitting the concentration–effect data to eqn [5]. To date the application of the comparative method has been confined mainly to in vitro studies. Recently, however, the operational model of agonism has been successfully applied in a series of investigations on the PK/PD correlations of A_1 adenosine receptor agonists,[22,23] μ opioid receptor (MOP) agonists,[15,24] and 5HT$_{1A}$ agonists.[25] Simultaneous analysis of in vivo concentration–effect relationships of different compounds within the nonlinear mixed effects modeling software NONMEM has been a key feature in these applications.

5.38.2.3.2 Semiparametric approach to the incorporation of receptor theory

A limitation of the operational model of agonism is that it contains a strict assumption on the shape of the transducer function. In theory a transducer function can take any shape. For this reason a semiparametric approach to the incorporation of receptor theory in PK/PD modeling has been proposed. In this approach the interaction between the drug and its receptor is still described by a hyperbolic function (eqn [1]), but no specific assumptions are made on the shape of the transducer function. The approach is based on Stephenson's concept of a stimulus to the biological system as result of the drug–receptor interaction.[26] This stimulus is defined as:

$$S = \frac{E \cdot [AR]}{[R_0]} = \frac{e \cdot [A]}{[A] + K_A} \qquad [6]$$

where S is the stimulus to the biological system and e is a dimensionless proportionality factor denoting the power of a drug to produce a response in a tissue. An important feature is that the intensity of the drug response is not directly proportional to the stimulus. Instead the response is assumed to be a function f of the stimulus that is monotonically increasing and continuous:

$$\frac{E_A}{E_{\max}} = f(S) = f\left[\frac{e \cdot [A]}{[A] + K_A}\right] \qquad [7]$$

in which $E_{\max}$ is the maximum effect of the hyperbolic concentration–effect relationship. Thus the introduction of the function f dissociates receptor stimulus and tissue response as directly proportional quantities. This also eliminates the necessity for the assumption that a maximal tissue response requires maximal receptor occupancy. This model was further refined by Furchgott,[27] who defined the concept of intrinsic efficacy ε as

$$\varepsilon = \frac{e}{[R_0]} \qquad [8]$$

thereby directly incorporating the receptor density R_0 into the model as major determinant of the effect. The factor ε characterizes the capacity of a drug to initiate a stimulus per receptor. Thus ε is a strictly drug-specific parameter. The incorporation of the term ε into eqn [5] results in

$$\frac{E_A}{E_{\max}} = f(S) = f\left[\frac{\varepsilon \cdot [R_0] \cdot [A]}{[A] + K_A}\right] \qquad [9]$$

Recently, a semiparametric approach to the incorporation of receptor theory in PK/PD modeling has been successfully applied in investigations on γ-aminobutyric acid A (GABA$_A$) receptor agonists.[28] A parameterized transduction function to describe GABA$_A$ receptor activation is presented below (*see* Section 5.38.3.3).

5.38.2.4   Transduction: The Modeling of Time-Dependent Pharmacodynamics to Derive In Vivo Drug Concentration–Effect Relationships

Transduction refers to the process of receptor activation in the ultimate pharmacological response. Specifically, binding of a drug molecule to a biological target initiates a cascade of biochemical and/or electrophysiological events resulting in the observable biological response. For most receptors (e.g., G protein-coupled receptors) second messengers such as phospholipases (e.g., 1,4,5-inositol triphosphate, diacylglycerol) and nucleotide cyclases (e.g., cAMP) serve as second messengers. For other receptors (e.g., glucocorticoid receptors) transduction is mediated through interaction with DNA thus regulating the expression of second messengers, proteins, or enzymes.

Large differences exist in the rates at which the various transduction processes occur in vivo. In many instances transduction is fast (i.e., operating with rate constants in the range of milliseconds to seconds) relative to the rate constants governing the disposition processes (typically minutes to hours). In that situation the transduction process determines the shape and the location of the in vivo concentration–effect relationship,[18,19] but it does not influence the time course of the drug effect relative to the plasma concentration. In contrast, transduction in vivo can also be slow, operating with rate constants in the order of hours to days, in which case transduction can become an important determinant of the time course of drug action.

As an approach to account for delays in between the drug concentration and effect, Dayneka *et al.*[29] have proposed a family of four in direct response models on the basis of the following equation

$$\frac{dR}{dt} = k_{in} - k_{out} \cdot R \qquad [10]$$

where R is a physiological entity, which is constantly being produced and eliminated in time, k_{in} is the zero-order rate constant for production of the physiological entity and k_{out} is the first-order rate constant for its loss. In this model the drug effect, $f(C)$, is described as stimulation ($[1 + f(C)]$) or inhibition ($[1 - f(C)]$) of the factors controlling either the input or the dissipation of drug response (k_{in} or k_{out}) in a direct concentration-dependent manner. The indirect response model has been successfully applied in preclinical investigations to derive in concentration–effect relationships of drugs with an indirect mechanism of action.[30]

For the modeling of complex transduction mechanisms, transduction can be modeled mechanistically on basis of intermediary processes between pharmacokinetics and response. In mathematical terms, the so-called transit compartment model has been proposed. This model relies on a series of differential equations to describe the cascade of events between receptor activation and final response.[31] Well-known examples of applications of this type of modeling are the modeling of the genomic effects of corticosteroids[32] and the modeling of hematologic toxicity in cancer.[33] The transit compartment model is attractive because of its flexibility, but for it to become fully mechanistic, pertinent information on the processes on the causal path is required. This underscores the need for biomarkers to characterize transduction mechanisms.[3]

5.38.2.5   Functional Adaptation: Modeling of Homeostatic Feedback to Derive In Vivo Concentration–Effect Relationships

Apart from a delay in the pharmacological response relative to the drug concentration in plasma, complex pharmacological effects versus time profiles may be observed when drug exposure leads to tolerance/sensitization or when homeostatic feedback mechanisms are operative. In those situations modeling of the homeostatic feedback is imperative for deriving in vivo drug concentration–effect relationships. Over the years several mathematical models for the characterization of homeostatic feedback have been proposed.

A useful model to describe complex effect versus time profiles is the so-called 'push-and-pull' model.

$$\begin{cases} \dfrac{dR}{dt} = k_{in} \cdot [1 - f(C)] - k_{out} \cdot R \cdot M \\ \dfrac{dM}{dt} = k_m \cdot R - k_m \cdot M \end{cases} \qquad [11]$$

where C the drug concentration, M is a response modifier value, and k_m is the rate constant for formation and dissipation of the response modifier. Due to its plasticity the push-and-pull model could be successfully applied to describe tolerance to the diuretic response upon repeated administration of furosemide.[34]

Another example of a PK/PD model describing tolerance is the 'precursor pool' model.

$$\begin{cases} \dfrac{dPool}{dt} = k_{in} - k_{loss} \cdot Pool \cdot [1 - f(C)] \\ \dfrac{dR}{dt} = k_{loss} \cdot Pool \cdot [1 - f(C)] - k_{out} \cdot R \end{cases} \qquad [12]$$

where $Pool$ is the precursor pool value and k_{loss} is the first-order rate constant for release of precursor into the central compartment. The precursor pool model can conceptually be considered a description of a tachyphylactic system and has been successfully applied to describe the effects of neuroleptic drugs on prolactin balance.[35]

Attempts to model physiological counterregulatory mechanisms have resulted in a series of advanced models describing complex behavior. These models are in part based on the work by Ekblad and Licko.[36] An example is the model proposed by Bauer *et al.* to characterize tolerance to the hemodynamic effects of nitroglycerin in experimental heart failure.[37] In the meantime this type of model of physiological counterregulatory effect has been successfully applied to describe tolerance and rebound to the effects of drugs such as alfentanil and omeprazole.[38–40] Moreover, a dynamic systems model has been proposed, which can account for the complex hemodynamic effects of arterial vasodilators (e.g., nifedipine) for which rate of administration is a major determinant of the effects.[41,42] The most recent development in the incorporation of dynamic systems analysis in PK/PD modeling has been the conceptualization of the so-called 'set-point' model.[43] This model was designed to describe complex effects versus time profiles of the hypothermic response following the administration of $5HT_{1A}$ receptor agonists to rats. In this model the indirect physiological response model is combined with a thermostat-like regulation of body temperature. Specifically in the model, body temperature and set-point temperature are interdependent through a feedback loop. This model has been successfully applied in preclinical investigations aiming at the characterization of the in vivo concentration–effect relationships of $5HT_{1A}$ receptor partial agonists.[25]

5.38.2.6   Disease: Modeling of Disease Processes to Derive In Vivo Drug Concentration–Effect Relationships

The latest development in PK/PD concerns the modeling of disease processes. Disease models are particularly important for drugs that interact in a highly specific manner with the disease process and that may have no direct observable effects in healthy subjects. Modeling of disease processes is imperative for drugs that are specifically designed to modify disease progression.

Chan and Holford[44] and Holford and Peace[45] were among the first to propose disease progression models for clinical rating scales. In these models the signs and/or symptoms of disease and their response to treatment are modeled directly, without consideration of the underlying biological system. Recently a theoretical framework for mechanism-based disease progression modeling has been proposed.[46] In the meantime steps have been taken toward the application of such mechanistic models for the effects of thiazolidinedione insulin-sensitizing agents (e.g., pioglitazone) in type 2 diabetes mellitus, using biochemical indices such fasting plasma glucose concentration (FPG), plasma insulin, and split proinsulin concentrations and % glycosylated hemoglobin (HbA$_{1C}$) as biomarkers.[47,48]

To date there are no reports of preclinical investigations with the aim to derive in vivo drug concentration effect relationships by modeling disease processes.

5.38.3   Application of Pharmacokinetic–Pharmacodynamic Modeling in Drug Design and Lead Optimization

5.38.3.1   Adenosine A$_1$ Receptor Agonists

The endogenous nucleoside adenosine exerts a variety of physiological effects via interactions with a family of G protein-coupled receptors consisting of four subtypes: the adenosine A$_1$, A$_{2A}$, A$_{2B}$, and A$_3$ receptors.[49] The A$_1$ receptor is the most comprehensively studied adenosine receptor subtype and has been cloned from a number of species, including humans. A$_1$ receptors are believed to act mainly via the G$_{i/o}$ family of G proteins and are widely expressed in the brain, spinal cord and a variety of peripheral tissues such as adipose tissue, heart, eye, adrenal gland, liver, kidney, salivary glands, and GI tract.[49] Despite this widespread distribution, A$_1$ knockout mice develop normally and display relatively subtle differences compared to wild-type animals, suggesting that adenosine A$_1$ receptors are particularly important under pathophysiological conditions.[50]

There has been considerable interest in the adenosine A$_1$ receptor as a potential target for therapeutic intervention for a variety of indications, such as cardiac arrhythmias, metabolic diseases, neurodegenerative diseases, neuropathic pain, and renal disorders.[51,52] Stimulation of the A$_1$ receptors in the heart produces negative dromo-, chrono-, and inotropic effects and adenosine itself is used clinically to terminate paroxysmal supraventricular arrhythmias and for myocardial perfusion imaging.[51,52] In addition, an N^6-substituted derivative of adenosine, tecadenoson (CVT-510) is a selective A$_1$ receptor agonist in phase III clinical trials for the treatment of supraventricular tachycardia.[138] On the other hand, the pronounced cardiodepressant effects induced by stimulating the A$_1$ receptor have been a major impediment to the research into the development of selective A$_1$ receptor ligands for other potential indications. One potential strategy to overcome this issue is based on the concept that low-efficacy (partial) agonists may display greater tissue selectivity compared to high-efficacy (full) agonists,[53] and during the last decade significant efforts have been put into medicinal chemistry programs targeting the development of selective, partial agonists for the adenosine A$_1$ receptor.[52,54,55]

An example of how such tools have been used to develop novel, mechanism-based PK/PD models is a project exploring the potential of a series of deoxribose and 8-alkylamino analogs of N^6-cyclopentyladenosine (CPA), which were identified as adenosine A$_1$ receptor agonists with reduced intrinsic efficacy[54,55] as antilipolytic agents for the treatment of non insulin-dependent diabetes mellitus. Increased levels of non-esterified fatty acids (NEFAs) are a characteristic of non insulin-dependent diabetes and are believed to exacerbate insulin resistance and hyperglycemia. Therefore, selective A$_1$ receptor agonists may provide a novel therapeutic strategy for the treatment of diabetes, since stimulation of adenosine A$_1$ receptors in adipose tissue has been shown to decrease NEFA levels in rat and human.[51]

A key first step in this project was the development of an efficient and quality animal model and sensitive analytical assays for simultaneous and detailed characterization of the pharmacokinetics and time course of effect of efficacy and safety biomarkers. For this purpose, conscious, freely moving rats were used with several arterial and venous cannulae which were implanted under anesthesia several days prior to the experiment, which in a single animal allowed for continuous hemodynamic recordings and repeated arterial blood sampling for measurement of drug and NEFA levels following controlled intravenous drug infusion.[56,57]

The CPA analogs that were studied in this in vivo model had previously been demonstrated to display lower 'GTP shifts' in an in vitro binding assay of the adenosine A$_1$ receptor than CPA itself. The guanosine triphosphate (GTP) shift is defined as the ratio between the apparent affinity in the presence and absence of GTP and is considered to be an in vitro measure of efficacy of A$_1$ receptor ligands.[54] By applying an integrated PK/PD modeling approach, it was found that the CPA analogs with a lower in vitro GTP shift displayed a reduced cardiovascular response and did indeed behave as partial agonists in vivo as evidenced by a significantly lower estimate of intrinsic activity (α) for the effect on heart rate.[56,57] Subsequently, the cardiovascular data for 10 analogs were analyzed simultaneously using the operational model of agonism (eqn [5]) and the 'comparative method' (see Section 5.38.2.3.1), assuming a direct link between drug exposure and pharmacodynamic effect. The key finding of this analysis, which provided the first 'proof of principle' for this mechanism-based PK/PD approach, was that the estimates of in vivo affinity (K_A) and efficacy (τ) were highly

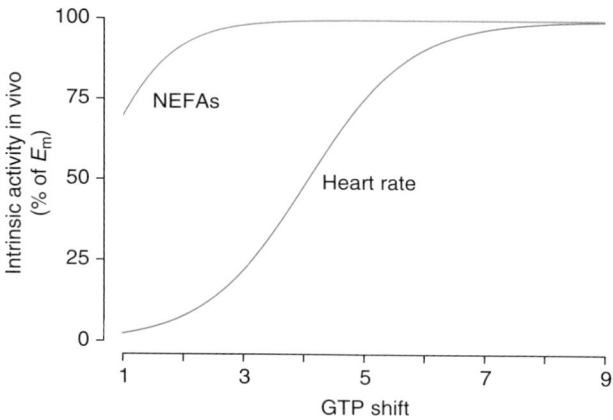

Figure 3 Relationship between in vitro GTP shift at the adenosine A_1 receptor and intrinsic activity (α) for the in vivo effect on NEFAs and heart rate in rats.

consistent with the results obtained from in vitro studies, i.e., in vivo estimates of K_A correlated very well with the in vitro estimates of K_i and the in vivo estimates of τ with the in vitro estimates of GTP shift.[22] A main utility of this approach compared to more empirical modeling strategies is that it provides a more robust basis for prediction of drug effects between different systems, since it explicitly separates drug and system properties in the model. This is exemplified in **Figure 3**, which shows the nonlinear relationship between the in vitro GTP shift and maximum effect on heart rate in vivo predicted by the mechanistic model.[22] Information like this not only provides novel mechanistic insights but can also be used to design effective and predictive in vitro and in silico screening strategies in drug discovery programs.

In order to validate the hypothesis that reducing intrinsic efficacy within the CPA series could enhance the cardiovascular safety window, the next step in this program was to compare the effects on heart rate with those on NEFA levels. It was indeed found that several analogs displayed profoundly reduced cardiovascular effects but still produced near-maximal inhibition of NEFAs in the rat model.[58] Once again, the operational model of agonism (eqn [5]) was employed to simultaneously analyze the NEFA effects of all ligands, this time incorporated into an indirect effect model (eqn [10]) to account for hysteresis between drug concentration and pharmacodynamic response. This approach yielded estimates of in vivo efficacy and affinity for the adenosine A_1 receptor-mediated effect on NEFAs that could be compared directly with those obtained for the effect on heart rate. The main outcome of this analysis was that on average all compounds displayed an $\sim$38-fold higher efficacy (τ in eqn [5]) for the NEFA effect compared to the cardiovascular effect, consistent with findings that the density of adenosine A_1 receptors is $\sim$25-fold higher in adipocytes compared to cardiac tissue.[23] Finally, this integrated PK/PD model for efficacy and safety was linked back to the in vitro data as described above and **Figure 3** shows the final model which can predict the therapeutic window in vivo for any compound from in vitro GTP shift data. An important conclusion from the model in **Figure 3** is that it indicates that even compounds with GTP shift values close to unity (i.e., ligands that appear to behave as antagonists in vitro) may still produce significant inhibition of lipolysis in vivo, whereas they are expected to be without cardiodepressant side effects. This illustrates our view that mechanism-based PK/PD modeling can yield insights that can provide new directions for drug discovery program in the search for compounds with improved efficacy and/or safety profiles.

5.38.3.2   μ Opioid Receptor Agonists

Opioid receptors are activated by both endogenously produced peptides (known as endorphins or enkephalins)[59,60] and exogenously administered opioid drugs which are among the most effective analgesics known. The opium alkaloid morphine and synthetic analogs are widely used in clinical practice for the treatment of severe pain. The opioid receptors belong to the family of G protein-coupled receptors (GPCRs), which are predominantly signaling through the $G_{i/o}$ transduction pathway. Four subtypes have been identified: μ, δ, κ, and N/OFQ opioid receptors (MOP, DOP, KOP, and NOP receptors, respectively).[61] An alternative nomenclature proposed by the International Union of Pharmacology to replace the Greek symbols μ, δ, and κ by OP3, OP1, and OP2, respectively has not gained wide acceptance in the literature and has now been withdrawn.[62]

The MOP receptor has been the most widely studies member of the opioid receptor family and it is generally believed that the main clinical effects of currently used opioid drugs are mediated predominantly via agonism at this subtype.[63] Despite the fact that morphine and synthetic analogs such as fentanyl are well established in clinical analgesia and anesthesia, there is still considerable interest in the development of novel agents that may address some of the key limitations of exiting drugs, in particular the occurrence of prolonged respiratory depression following anesthesia.[64] Another limitation to the long-term use of current opioid drugs as analgesics is the development of tolerance, although the significance of the clinical manifestation of this in some patient populations remains unclear.[61,65]

During the last decade, exciting new insights have been obtained in the molecular and pharmacological properties of the MOP receptor that may provide the basis for the development of the next generation of opioid drugs. For example, although only a single gene for the MOP receptor has been identified, it has been suggested that pharmacologically it may manifest itself as a heterogeneous phenotype, due to for example alternative splicing and heterodimerization.[66–71] In addition, considerable progress has been made in elucidating the molecular mechanisms that may underlie tolerance development to opioid drugs and one of the main hypotheses has been that this phenomenon is due to MOP receptor downregulation and/or desensitization in response to agonism.[61] However, although at the cellular level MOP receptor downregulation, uncoupling, and desensitization can be readily demonstrated in in vitro systems, the relevance of these findings in the context of clinical tolerance development to opioid analgesics remains controversial.[61] A number of integrated PK/PD models have been proposed to characterize functional adaptation to the effect of opioids in vivo. For example, some models describe tolerance development on the basis of distribution to a hypothetical tolerance compartment[72] whereas others postulate a physiological counter-regulatory mechanism to characterize functional adaptation.[39] A limitation to these models is that they are largely empirical in nature and that they do not incorporate MOP receptor regulation as a main mechanism of functional adaptation. Therefore, more recently focus has shifted toward the development of mechanism-based model to describe, explain, and predict opioid tolerance development.

An example of such an approach is based on the use of quantitative electroencephalogram (EEG) analysis, which has been used successfully as a biomarker for MOP receptor activity in preclinical models and humans.[73] Using the synthetic opioid alfentanil as a reference compound, a rat model was developed in which the amplitudes in the 0.5–4.5 Hz frequency band of the EEG spectrum were used to describe the time course of effect and derive PK/PD relationships for MOP receptor agonists.[73] In order to develop an initial mechanism-based PK/PD model, alfentanil and two, related 4-anilidopiperidine analogs, fentanyl and sufentanil, were studied. In vitro studies had demonstrated that these three ligands display very different levels of intrinsic efficacy at the MOP receptor, as judged by the 'sodium shift' which is the equivalent of the GTP shift described in Section 5.38.3.1 and is defined as the ratio between a ligand's MOP receptor affinity in the presence and absence of high concentrations of Na^+.[9] Despite the fact that in vitro sufentanil behaved as a low efficacy agonist (sodium shift = 2.8) compared to fentanyl and alfentanil (sodium shift = 13.3 and 19.1, respectively), a comparison of the concentration–effect relationships for the EEG effect revealed that in vivo the three compounds displayed the same maximal response (intrinsic activity). In order to explain this finding, the operational model of agonism (eqn [5]) was employed to integrate the in vitro and in vivo data.[9] This analysis showed that the sodium shift in vitro provided an accurate prediction of the expression of agonism in vivo, that is the in vivo efficacy parameter τ could be expressed as the product of sodium shift and an agonist-independent constant (4.3). In the operational model of agonism, τ is defined as the ratio of total receptor concentration ($[R_0]$) and the midpoint location of the transducer function (K_E) which relates agonist-occupied receptors to pharmacological effect (see Section 5.38.2.3.1). Because $[R_0]$ is specific for a particular pharmacological system and therefore ligand-independent, differences in τ are likely to reflect differences in K_E values between opioid ligands. Importantly, the model predicted that on the basis of the in vitro sodium shift it will not be possible to detect ligands that will behave as partial agonists in the EEG model, since a value of the sodium shift close to unity would still be expected to result in the expression of maximal possible intrinsic activity (**Figure 4**). This is consistent with findings described for adenosine A_1 agonists in Section 5.38.3.1, which suggests that amplification of agonist properties may often result in the expression of higher levels of efficacy in vivo compared to in vitro, which calls into question the reliability of some (high-throughput) in vitro screens and underscores the important role mechanism-based PK/PD modeling can play in drug discovery programs, in particular ones targeting receptor agonists.

This mechanism-based PK/PD model was successfully applied to predict the in vivo potency and intrinsic activity of the novel synthetic opioid remifentanil and its active metabolite GR90291 in humans on the basis of the results of preclinical investigations in rats.[11] The mechanism-based PK/PD model was subsequently used to describe the rapid functional adaptation observed with alfentanil following repeated administration in the rat EEG model. It was shown that a second administration of alfentanil resulted in a parallel shift of the in vivo concentration–effect relationship with an almost twofold decrease in potency (EC_{50}) without affecting the maximal EEG response. In terms of the

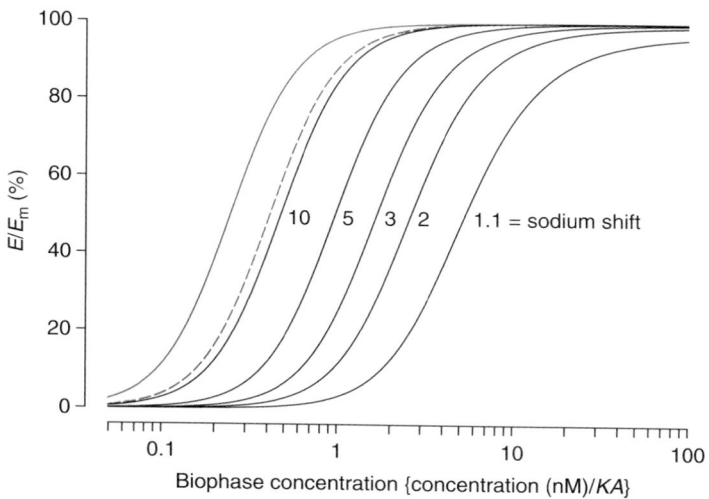

Figure 4 Operational model of agonism simulations showing the predicted in vivo concentration–EEG effect relationships on the basis of in vitro radioligand binding studies for MOP receptor agonists with constant affinity and different sodium shifts. The model predictions were obtained using eqn [5] with the following parameter values: $E_m = 100\%$, $n = 2.3$, $K_A = 1$, $\tau = 4.3 \times$ sodium shift. The solid and dashed green lines show the model predictions for alfentanil (sodium shift $= 19.1$) in an intact system and one with a 40% reduction in τ due to receptor desensitization or loss.

operational model, this could be explained by the presence of a high 'receptor reserve' and a decrease in τ reflecting a ~40% loss of functional MOP receptors without changes in receptor affinity.[74] Interestingly, this model-derived estimate of fractional receptor loss is very similar to ones obtained using molecular techniques following chronic opioid exposure.[61] Further validation of the mechanism-based PK/PD model was obtained using a novel experimental approach of MOP receptor 'knockdown' with the irreversible MOP antagonist β-funaltrexamine (β-FNA). It was shown that pretreatment of rats with β-FNA prior to the PK/PD experiment could produce a gradual reduction in the number of MOP receptors in rat brain of 40–60% without affecting the affinity of alfentanil, as judged by radioligand binding studies. Consistent with previous findings which had suggested the presence of a MOP receptor reserve for the EEG effect of opioids in this model, this loss of receptors did not result in a reduced intrinsic activity of alfentanil but in a two to threefold loss of potency in β-FNA-treated animals, again in line with expectations of a reduction in τ in the mechanism-based model.[24] An interesting feature of the pharmacological receptor knockdown approach, which differentiates it from more conventional molecular biology 'knockout' techniques, is that it provides a dynamic in vivo model system for the investigation of receptor turnover. Thus, it was observed that the effect on the concentration–EEG effect relationship for alfentanil was the same when β-FNA pretreatment was given 35 min or 24 h prior to the PK/PD investigation. In contrast, the marked effects of alfentanil on respiration depression and muscle rigidity appeared to be unchanged following 24 h washout of β-FNA compared to vehicle treatment but were not observed at all following 35 min washout, which could suggest that turnover of receptors mediating respiratory depression and muscle rigidity is faster than that of receptors mediating EEG effects. Whether or not this observation is related to for example differential expression and turnover of MOP splice variants in different tissues remains to be investigated further.[64]

Recently, it has been demonstrated that the in vivo knockdown approach can also be applied to other GPCR systems. For example, it was shown that the irreversible adenosine A_1 receptor antagonist, 8-cyclopentyl-3-N-[3-(4-(fluorosulphonyl)(benzoyl)-oxy)-propyl]-1-N-propyl-xanthine (FSCPX)[75] produces dose-dependent, irreversible antagonism of the cardiovascular effects of CPA in the conscious rat,[76] an approach which could be used to further validate the mechanistic PK/PD model developed for adenosine A_1 receptor agonists described in Section 5.38.3.1.

5.38.3.3   GABA$_A$ Receptor Agonists

The GABA$_A$ receptor is a ligand-gated ion channel, with five homologous membrane-spanning subunits that form an integral Cl^- channel.[77–80] Upon binding to the receptor, the neurotransmitter GABA facilitates the opening of the chloride channel, resulting in hyperpolarization of the neuron.[81,82] GABA$_A$ receptors can be formed from a pool of different subunits. Depending on their subunit composition, receptors exhibit distinct pharmacological and electrophysiological properties. There is a wide variation in the expression of different subtypes in various neuronal populations and in different regions of the CNS. To date, at least 18 human $GABA_A$ receptor proteins have been described. Furthermore at least eight $GABA_A$-receptor subunit types can be distinguished: α, β, γ, δ, θ, ε, π, and ρ.[79,83,84]

The $GABA_A$ receptor is an important target for the development of new drugs for a wide array of CNS disorders such as epilepsy, sleep disorders, anxiety, stress, depression, and cognitive failure.[85–87] The development of novel $GABA_A$ receptor ligands as innovative drugs is specifically directed toward the design of partial agonists with an improved selectivity of action, inverse agonists for the treatment of cognitive failure, and receptor subtype selective ligands.

Quantitative EEG parameters, specifically power or amplitudes in the β-frequency range, have proven valuable measures of $GABA_A$ receptor-mediated drug effects in vivo.[88,89] The EEG parameters appear to fulfill many of the characteristics of ideal pharmacodynamic measures, being continuous, sensitive, and reproducible.[90] An important feature of EEG effect measurements is that they can be obtained in both laboratory animals and humans. This enables the investigation of interspecies correlations of the pharmacodynamics.

Increase in β-activity has been used by several investigators to characterize the concentration–effect relationships of a variety of benzodiazepines in both experimental animals[7,12,91,92,101] and in humans.[93–95] In addition, for barbiturates, propofol, and neuroactive steroids, the changes in the β-frequency range has been successfully used as a biomarker in PK/PD investigations.[92,95–97] Interestingly, the GABA reuptake inhibitor tiagabine gives a characteristic change in the EEG comparable to benzodiazepines.[98] Using the β-frequency in the EEG as biomarker, benzodiazepines exhibit nonlinear concentration–effect relationships, which are readily described by the sigmoid E_{max} pharmacodynamic model (eqn [2]). Between benzodiazepines wide differences in potency (EC_{50}) and intrinsic activity (E_{max}) have been observed. Furthermore, inverse agonists display a negative EEG effect.[91,99,101] Interestingly, neuroactive steroids exhibit biphasic concentration–effect relationships in vivo. At low concentrations the EEG effect increases from baseline to a maximum value, which is approximately two to three times higher than the maximum observed for the benzodiazepine displaying the highest intrinsic activity (diazepam). At higher concentrations the effect decreases below the baseline toward isoelectric EEG. Based on these observations, a novel mechanism-based PK/PD modeling approach for $GABA_A$ receptor agonists has been proposed, which features a parameterized biphasic stimulus–response relationship.[96] Specifically, this model contains separate expressions to characterize the drug–receptor interaction and the transduction processes. The drug receptor interaction is characterized in terms of in vivo affinity and intrinsic efficacy using a hyperbolic function (eqn [6]). The transduction f (in eqn [7]) is described by a parabolic function according to:

$$E = E_{\text{top}} - a \cdot \left(S^d - b\right)^2 \qquad [13]$$

where S represents the stimulus to the biological system resulting from $GABA_A$ receptor activation, E_{top} represents the top of the parabola, a is a constant reflecting the slope of the parabola; $b^{1/d}$ is the stimulus for which the top of the parabola (i.e., the maximal effect, E_{top}) is reached and the exponent d determines the asymmetry of the parabola. When no drug is present the EEG is equal to its baseline value (E_0). Equation [13] then reduces to:

$$E_0 = E_{\text{top}} - a \cdot b^2 \qquad [14]$$

Substituting eqn [13] in eqn [14], and rearranging yields:

$$E = E_0 - a \cdot \left[(S^d)^2 - 2 \cdot b \cdot S^d\right] \qquad [15]$$

This parabolic function is determined by three parameters: a, b, and d. Parameter a determines the height of the maximal achievable response (E_{top}) and the steepness of the increasing and decreasing wing. Parameter b determines the location of E_{top} and d determines the asymmetry of the parabola.

The mechanism-based PK/PD model contains two drug-related and three system-related parameters that cannot be estimated independently from a single biphasic drug concentration–effect relationship. Similar to the analysis with the operational model of agonism, the efficacy for drugs can only be estimated compared to the drug with an efficacy that produces the maximal effects possible in the system. This comparative method, originally proposed by Barlow,[21] is based on the idea that the intrinsic activity of a full agonist is identical to the maximum system response. In the present analysis the neuroactive steroid alphaxalone was assumed to be a full agonist and its in vivo efficacy was fixed to the value of 1, resulting in a possible stimulus range between 0 and 1 for positive modulators. The in vivo efficacy for other drugs was estimated relative to the maximal efficacy of alphaxalone.

The model was applied to the concentration–effect relationships of neuroactive steroids.[97] Simultaneous analysis of the concentration–effect relationships of alphaxalone, pregnanolone, ORG 20599, and ORG 21465 on the basis of the mechanism-based model showed that all observations can be described with a single unique transducer function with parameter $d = 3.4$. This indicates that the obtained parameter is indeed system-specific and independent of the drug that was used. With respect to the receptor activation model, it was observed that all neuroactive steroids acted as high-efficacy modulators at the $GABA_A$ receptor with an intrinsic efficacy equal to that of alphaxalone. Wide differences in in vivo potencies were observed which were consistent with the differences in affinity in an in vitro bioassay.[100] Next the model was applied to the concentration–effect relationships of benzodiazepines, imidazopyridines, cyclopyrrolones, and a β-carboline.[101] These compounds differ in intrinsic efficacy at the $GABA_A$ receptor displaying the entire spectrum from full benzodiazepines receptor agonists, through partial and silent to inverse agonists.[102–104] Simultaneous analysis of the data on the basis of the mechanism-based model allowed estimation of the in vivo efficacy (e_{PD}) and affinity (K_{PD}) for each of the compounds utilizing the single unique transducer function that had previously been identified for neuroactive steroids. This analysis revealed that the $GABA_A$ receptor modulators differ from the neuroactive steroids solely in their in vivo efficacy (**Figure 5**). A highly significant correlation between the in vitro estimates for efficacy (GABA shift) and affinity (K_i) and the corresponding in vivo estimates was observed, confirming the validity of the proposed mechanism-based PK/PD model.

An important feature of mechanism-based PK/PD models is that they have much improved properties for extrapolation and prediction. This concerns especially the prediction of in vivo concentration–effect relationships from in vitro receptor assays, and the interspecies extrapolation from laboratory animals to humans of these concentration–effect relationships. The high correlation between the in vitro estimates for efficacy (GABA shift) and affinity (K_i) and corresponding in vivo estimates shows that in vivo concentration–effect relationships can be predicted on the basis of in vitro bioassays. **Figure 6** shows the complex relationship between the observed maximum EEG effect of $GABA_A$ receptor ligands and the GABA shift as a measure of intrinsic efficacy at the $GABA_A$ receptor in vitro. This relationship is rather shallow at low GABA shifts. This shows that the EEG effect is not a very sensitive pharmacodynamic endpoint for the identification of low-efficacy and near-silent agonists. This is important, since partial and near-silent agonists are of considerable interest therapeutically as they may display a much-improved selectivity of action. In contrast, large differences in EEG effects are observed for compounds with intermediate efficacy (benzodiazepines). Finally, the steepness of the stimulus response relationship at high stimulus intensities makes it difficult to identify differences in efficacy for neuroactive steroids.

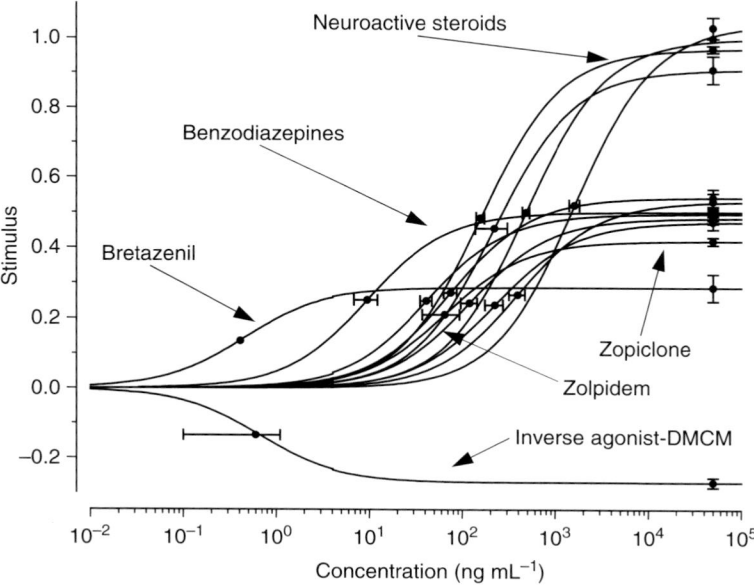

Figure 5 The relationship between drug concentration and stimulus at the $GABA_A$ receptor for the $GABA_A$ receptor modulators. Concentration (ng mL^{-1}) is depicted on the x-axis in logarithmic scale and the stimulus is depicted on the y-axis. The results of the mechanism-based PK/PD analysis show that functionally benzodiazepines, imidazopyridines, cyclopyrrolones and β-carbolines behave as partial agonists relative to neurosteroids.

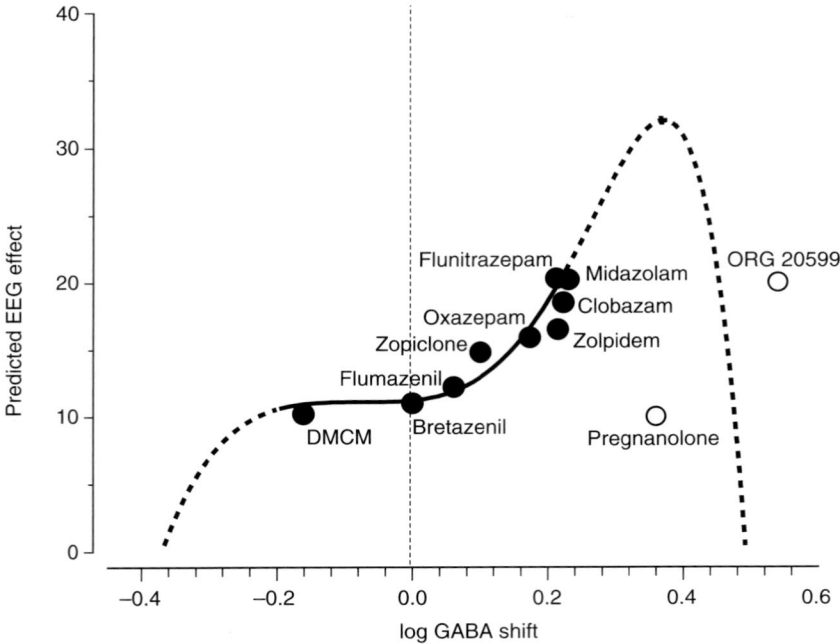

Figure 6 Prediction of in vivo concentration–EEG effect relationships of GABA$_A$ receptor modulators from in vitro GABA shift. On the y-axis the observed maximal EEG effect is plotted. Compounds with GABA shifts lower than 1 are negative modulators and give negative EEG response compared to baseline. Compounds with GABA shifts higher than 1 are positive modulators and give positive EEG response compared to baseline. It is predicted that compounds with GABA shifts higher than 2 will give biphasic concentration–effect relationships. Note that the GABA shift is depicted in a logarithmic scale.

For the extrapolation from animal to humans, it is important to obtain information about the shape of the stimulus–response relationship in humans. The exact shape of the transducer function in humans remains to be determined. There are indications however that the transducer function is quite similar, since nearly identical concentration–EEG effect relationships in rats and humans have been observed for a number of benzodiazepines.[12]

5.38.3.4   Serotonin (5HT$_{1A}$) Receptor Agonists

A wide array of selective and less selective 5HT$_{1A}$ receptor ligands has been developed. It has been shown that these compounds are of value in the treatment of anxiety disorders and depression. In addition they may be of value in therapeutic indications such as aggression, addiction, seizures, nausea, and neuroprotection. The mechanism of action of 5HT$_{1A}$ receptor ligands is not entirely resolved. It is believed however that differences in intrinsic efficacy at 5HT$_{1A}$ receptors in different areas in the brain are an important determinant of the therapeutic effect. Specifically so-called near-silent agonists are considered of great interest for a number of indications.

The regional distribution of 5HT$_{1A}$ receptors within the brain has been studied in many animal species including the rat, mouse, guinea pig, calf, cat, pig, monkey, and human. It has been shown to be similar in these species.[105] In short, 5HT$_{1A}$ receptors are distributed throughout the CNS, but high concentrations are found in the hippocampus, septum, and amygdala, areas that are typically thought to be associated with the control of mood.[106] Interestingly, differences in receptor reserve in various parts of the brain have been demonstrated. For example it has been shown that higher levels of receptor reserve are present in the raphe nuclei (mainly, presynaptic somatodendritic autoreceptors), compared to the hippocampus and hypothalamus (mainly postsynaptic receptors).[107] Hence a ligand may act as a full agonist for a presynaptically mediated response, whereas at the same time it may behave as a partial agonist for a postsynaptically mediated response.

Upon administration of 5HT$_{1A}$ receptor agonists a series of responses is observed which are all potentially suitable for PK/PD characterization. These in vivo effects can be classified into four groups: physiological (e.g., induction of hypothermia), endocrinological (e.g., stimulation of corticosterone release), behavioral (e.g., induction of flat body posture or forepaw treading), and therapeutic-like effects (e.g., reduction of fear-induced ultrasonic vocalizations or immobility in the forced-swimming test).[108] The hypothermic response is a continuous, reproducible, objective, sensitive, and selective $5HT_{1A}$ receptor mediated response, making it an attractive pharmacodynamic endpoint in PK/PD investigations. It is well established that the hypothermic effect of $5HT_{1A}$ receptor agonists is caused by a direct effect on the body's set-point temperature.[109] The hypothermic effect of $5HT_{1A}$ receptor agonists has been studied extensively and is considered one of the most robust $5HT_{1A}$ receptor-mediated responses.[110] It is considered the response of choice for the differentiation between full, partial, and silent agonists at the $5HT_{1A}$ receptor.[111] Furthermore, this effect can be observed both in rodents and in humans[111] thus enabling the investigation of animal to human extrapolation.

In PK/PD experiments in chronically instrumented rats the time course of the hypothermic response was determined in conjunction with plasma concentrations for R-8-OH-DPAT, S-8-OH-DPAT, flesinoxan, buspirone, buspirone's active (and but nonselective) metabolite 1-PP, WAY-100,135, and WAY-100,635.[43,112–114] Briefly, 8 days prior to the experiment, the rats were operated upon. Indwelling cannulae for drug administration and blood sampling as well as a telemetric transmitter in the abdominal cavity for the measurement of core body temperature were implanted. In the PK/PD experiments, conscious freely moving rats received an intravenous infusion of vehicle (saline) or active drug. The $5HT_{1A}$ receptor ligands were administered in a wide range of doses (high, middle, and low), infusion rates (short, intermediate, and computer controlled) and combinations (R-8-OH-DPAT with WAY-100,635). In each experiment from each individual rat approximately serial blood samples were taken and subsequently analyzed to determine the time course of the drug concentration.[43,112–114] Body temperature was measured continuously throughout the experiment using the telemetric system. The affinity and the in vitro efficacy (agonist ratio) of the various $5HT_{1A}$ receptor agonists have been determined in a series of receptor binding assays.[25]

In investigations where the hypothermic response is studied in detail, complex effect versus time patterns have been observed suggesting the involvement of homeostatic control mechanisms.[115,116] In order to characterize $5HT_{1A}$-agonist induced hypothermia in a mechanistic manner a mathematical model which describes the hypothermic effect based on the concept of a set-point[116–118] and a general physiological response model[29] has been proposed.[43] Briefly, as an agonist binds to the $5HT_{1A}$ receptor a stimulus, S is generated, which in turn drives physiological processes that lower the temperature. This stimulus, which is determined by the drug receptor interaction and hence the drug's affinity and efficacy, can be described by a sigmoid function where $f(C)$ for example equals eqn [2] or [5]. As S is assumed to be inhibitory, it is defined as $S = 1 - f(C)$. As the drug concentration C changes with time, S changes as well, governing the first timescale of the model. The second timescale on which the model operates is governed by physiological principles. The model that describes the hypothermic response utilizes the concepts of the indirect physiological response model as described by Dayneka et al. (eqn [10]).[29] In this model the change in temperature (T) is described as an indirect response to either the inhibition of the production of body heat or the stimulation of its loss, where k_{in} represents the zero-order fractional turnover rate constant associated with the warming of the body and k_{out} a first-order rate constant associated with the cooling of the body. The indirect physiological response model is combined with the thermostat-like regulation of body temperature. This regulation is implemented as a continuous process in which the body temperature is compared with a reference or set-point temperature (T_{SP}). It is accepted that $5HT_{1A}$ agonists elicit hypothermia by decreasing the value of the set-point temperature T_{SP}, and hence T_{SP} depends on the drug concentration C: $T_{SP} = T_{SP}(C)$. It is assumed that T_{SP} is controlled by the drug concentration C through eqn [16]:

$$T_{SP} = T_0[1 - f(C)] \tag{16}$$

where T_0 is the set-point value in the absence of any drug: $T_0 = T_{SP}(0)$. Combining the indirect physiological response model with the thermostat-like regulation then yields:

$$\begin{cases} \dfrac{\mathrm{d}T}{\mathrm{d}t} = k_{in} - k_{out} \cdot T \cdot X^{-\gamma} \\[2mm] \dfrac{\mathrm{d}X}{\mathrm{d}t} = a(T_0 \cdot [1 - f(C)] - T) \end{cases} \tag{17}$$

in which X denotes the thermostat signal. The change in X is driven by the difference between the body temperature T and the set-point temperature T_{SP} on a timescale that is governed by a. Hence when the set-point value is lowered, the body temperature is perceived as too high and X is lowered. To relate this decreasing signal to the drop in body temperature, an effector function $X^{-\gamma}$ was designed, in which γ determines the amplification. Raising this function to the loss term $k_{out} \cdot T$ therefore facilitates the loss of heat. In eqn [17], body temperature and set-point temperature are interdependent and a feedback loop is created that can give rise to oscillatory behavior, as has been shown.[43] The model is able to reproduce the observed complex effect versus time profiles, which are typically observed upon the administration of $5HT_{1A}$ receptor partial agonists.

A population approach[119] was utilized to quantify both the pharmacokinetics and pharmacodynamics of the $5HT_{1A}$ receptor agonists. This enabled the successful characterization of the time course of the hypothermic effect in terms of physiological parameters and drug specific parameters such as potency and intrinsic efficacy.[43,112–114] In the initial version of the model, the sigmoid–E_{max} model (eqn [2]) was used to describe the direct concentration effect relationships of the $5HT_{1A}$ receptor agonists at the receptor in terms of potency and intrinsic activity. In a subsequent step, which aimed at the development of a fully mechanistic model, the sigmoid–E_{max} model was replaced by the operational model of agonism[18] (eqn [5]). In this analysis the value of the system maximum E_{max} was constrained to the observed maximum effect for a full agonist R-8-OH-DPAT. The values of K_A and τ for the various partial agonists were estimated by directly fitting the operational model of agonism to the combined concentration–effect data. The values of K_A and τ are shown in **Table 1** together with the corresponding estimates of affinity (K_i) and intrinsic efficacy (log[Agonist ratio]) in a receptor binding assay. The correlation between pK_A and pK_i based on [^{3}H]-WAY-100,635 was rather poor ($P > 0.05$) compared to similar in vivo/in vitro correlations observed for adenosine A_1 agonists, synthetic opiates, and $GABA_A$ receptor agonists (see previous sections). Close inspection of this correlation showed that flesinoxan deviated from the line of identity. In fact, the correlation between the pK_A and pK_i became statistically significant when flesinoxan was excluded from the analysis ($P < 0.05$). Recently Van der Sandt et al.[120] have shown that active transport mechanisms (i.e., P-glycoprotein) at the blood–brain barrier are an important determinant for the brain distribution for flesinoxan. Thus it appears that complexities at the level of blood–brain distribution of flesinoxan explain the observed lack of correlation between in vitro and in vivo receptor affinity estimates.[25] Between the in vivo and in vitro efficacy (log τ and log[agonist ratio]) a significant correlation was found ($P < 0.05$). The correlation between log τ and log[agonist ratio] showed further that the in vivo 'test assay' was more sensitive for detecting $5HT_{1A}$ activity then the agonist ratio. For example, the significant in vivo agonist activity demonstrated for WAY-100,135 was not detected in vitro.

Thus by combining the semimechanistic PK/PD model for the hypothermic effect of $5HT_{1A}$ agonists with the operational model of agonism, a full mechanistic PK/PD model was obtained, which proved to be highly predictive of the in vivo intrinsic activity of ligands at this receptor. This is important, since the pharmacological and therapeutic properties of $5HT_{1A}$ agonists are closely related to the degree of intrinsic activity at $5HT_{1A}$ receptors. The ability of the in vivo assay to detect weak partial agonism underscores the importance of the use of in vivo models in the development of $5HT_{1A}$ agonists as clinical agents.

An important advantage of mechanism-based PK/PD models is their optimal properties for quantitative extrapolation of pharmacological responses across species. To this end allometric scaling of the rat PD parameters of the set-point model to predict the response in humans for buspirone and flesinoxan has been applied. Parameters were extrapolated based on body weight (M) following simple power laws, according to

$$Y = a \cdot M^b \qquad [18]$$

Table 1 Estimates of in vitro and in vivo affinity and efficacy for the $5HT_{1A}$ receptor ligands

Drug	In vitro[a]		Agonist ratio	In vivo[b]	
	[^{3}H]-R-OH-DPAT pK_i	[^{3}H]-WAY-100,635 pK_i		pK_A	log τ
R-8-OH-DPAT	8.36 (0.11)	7.35 (0.10)	10.13	7.35[c] (32%)	0.62 (53%)
S-8-OH-DPAT	7.95 (0.04)	7.22 (0.11)	5.34	6.68 (5%)	0.0523 (123%)
Flesinoxan	7.91 (0.04)	7.15 (0.09)	4.14	5.67 (15%)	0.206 (46%)
Buspirone	7.42 (0.16)	6.40 (0.09)	10.53	7.03 (32%)	−0.0684 (32%)
1-PP	5.38 (0.03)	4.76 (0.02)	5.33	5.68 (32%)	−0.291 (70%)
WAY-100,135	7.30 (0.04)	7.03 (0.07)	1.88	7.74 (130%)	−1.25 (88%)
WAY-100,635	8.73 (0.05)	8.94 (0.08)	0.61	8.63[d] (3%)	n.a.

[a] In vitro: the estimates of affinity (pK_i) are determined in the presence of [^{3}H]-R-OH-DPAT and [^{3}H]-WAY-100,635. The ratio between the two K_is serves as a measure for efficacy (agonist ratio). The parameters are expressed as mean (SD).

[b] In vivo: estimates of in vivo affinity and efficacy for the $5HT_{1A}$ receptor ligands on body temperature in the rat. In vivo estimates (population mean + coefficient of variation (%)) of affinity (pK_A) and efficacy (log τ) were obtained by fitting the data to the operational model of agonism.

[c] R-8-OH-DPAT's value has been fixed to the one obtained in the in vitro binding assay.[27]

[d] WAY-100,635 has been determined in a previous analysis.[24]

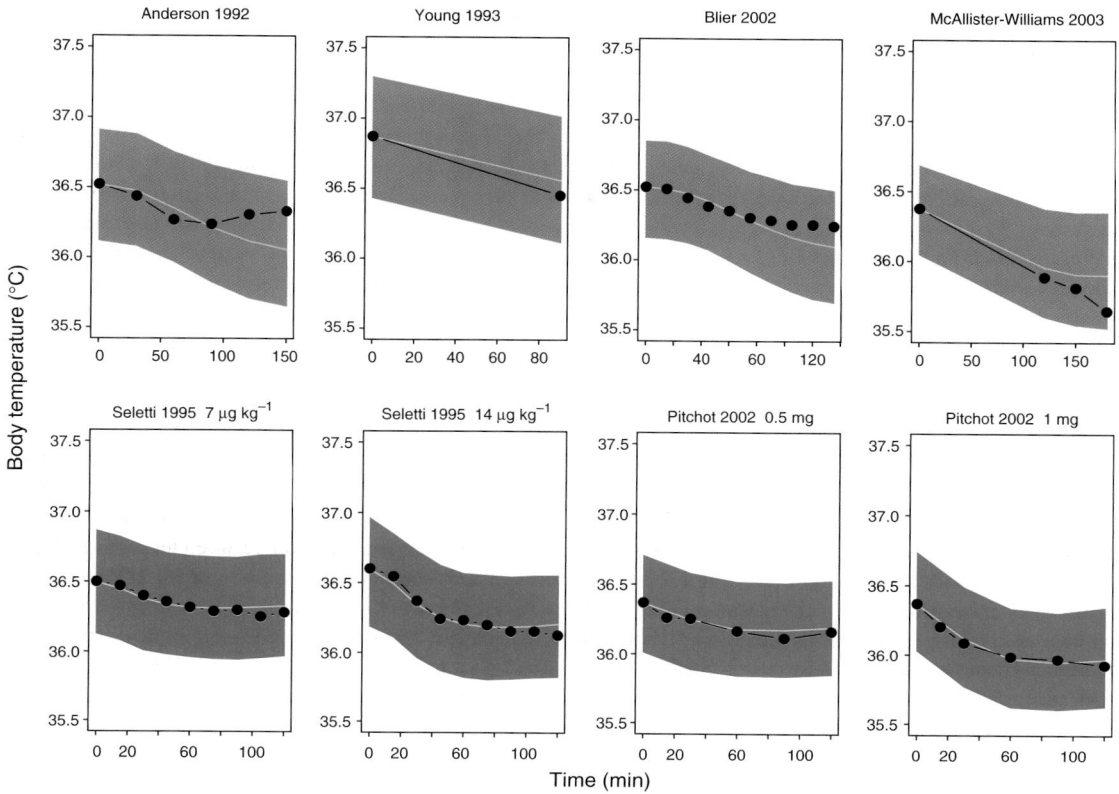

Figure 7 Simulated 90th percentile prediction intervals for the hypothermic responses in humans for buspirone (blue)[121–124] and flesinoxan (green) by study[125,126] and dose (only the active doses are depicted and the title of each graph denotes refers to the applied study design). The observed hypothermic response is denoted by the filled circles and connecting line.

where the exponent b assumes multiples of 1/4, Y is the parameter of interest, and a the weight-independent parameter. Typically b assumes values of $-1/4$ for growth rates and values of 3/4 for rate constants. Considering the fact that that the receptor systems are similar across species drug-specific parameters were not scaled. Furthermore, the interindividual variability in response was assumed to be identical between species. **Figure 7** depicts the simulated 90th percentile prediction intervals for the hypothermic responses in humans for buspirone[121–124] and flesinoxan.[125,126] Overall the time course of both responses is captured well for both drugs. The results of these preliminary analyses indicate that allometric scaling of mechanistic PK/PD models can be used to predict the pharmacodynamic response in humans. This approach combined with the operational model of agonism provides for a novel way of interpreting in vitro and in vivo pharmacological responses and will ultimately facilitate candidate selections and lead optimization.

5.38.3.5   Drug-Induced QT Interval Prolongation

The possibility of torsade de pointes (TdP) is recognized as a potential risk during treatment with over 100 antiarrhythmic and nonantiarrhythmic drugs.[127] TdP is a life-threatening form of ventricular tachycardia and is a documented side effect of over 40 nonantiarrhythmic drugs across all therapeutic drug classes. An unacceptably high incidence of TdP has led to withdrawal from the market of several drugs, including amongst others cisapride, astemizole, and sertindole. For a comprehensive overview of the accumulated evidence for an association between individual drugs and TdP, the reader is referred to [127,128].

The majority of drugs that promote TdP arrhythmias inhibit the rapidly activating, delayed rectifier potassium current (I_{Kr}) channels in cardiac cells. Through that mechanism, they effectively slow down cardiac repolarization, resulting in a measurable prolongation of the QT interval in the surface electrocardiogram (ECG). In clinical safety studies, the QT interval in the ECG is principally used to monitor cardiac repolarization. In the in vitro setting, inhibition of I_{Kr} conduction in potassium channels encoded by the human ether-a-go-go related gene (hERG) is the most extensively studied parameter for assessing the torsadogenic potential of drugs. Drug-induced TdP arrhythmias are believed to be triggered by ventricular extra beats, which can be induced by early depolarizations in cells with relatively long repolarization phases.

Although I_{Kr} inhibition has been identified as a major explanatory factor for drug-induced TdP arrhythmias, the torsadogenic properties of a drug may also depend on drug action on other cardiac ion channels that are involved in repolarization, such as the slowly activating, delayed rectifier potassium channels and L-type calcium channels. Drugs which block the hERG K^+ channel with a relatively high selectivity, are likely to do so with a potency that is similar to the potency for prolonging the QT interval in dog and human.[127] Conversely, a discrepancy is found between in vitro and in vivo conditions for many drugs with mixed ion channel properties. Further data suggest that, irrespective of prolongation of the action potential duration, drug-induced changes in the shape and stability of the action potential and increased dispersion of repolarization across the heart may be critical for arrhythmogenesis, and presumably for the risk of TdP.[129] The conclusion from this must be that the I_{Kr} current and QT interval prolongation are, in isolation or together, not unambiguously indicative of the risk of TdP arrhythmias.[128] These surrogate outcomes give only probabilistic answers, with the possibility of false positive and false negative findings.

The clinical frequency of TdP requiring a detailed risk–benefit analysis is very low (above 1 in 10^5 to 1 in 10^6 exposures)[130] and for that reason, TdP is not likely to be observed during clinical development programs. The torsadogenic risk for a candidate drug must therefore be assessed on the basis of surrogate endpoints, which poses an enormous challenge in drug development. In the absence of alternative outcomes that have been studied sufficiently well to replace the QT interval as a measure for torsadogenic toxicity, regulatory agencies require clinical safety studies to be performed with highly precise QT interval measurements. The finding of a prolongation of the QT interval with <10 ms may currently be sufficient reason for denial or withdrawal of marketing approval.[128] It is believed that a drug that does not cause even a small prolongation of the QT interval will probably not be torsadogenic at all, or so rarely that the perceived benefits of the drug outweigh the risk associated with its use.

In the preclinical setting, measurement of QT interval prolongation in the dog is considered to be the most predictive for cardiotoxicity in humans.[131,132] The canine hERG channel has been shown to be biophysically and pharmacologically similar to the human channel[133] and also the other ionic channels that are involved in cardiac repolarization have a similar distribution and activity between human and dog.[131] With regard to the correction of QT intervals for heart rate, the same considerations apply as described above for humans and deserve special attention since the morphology of ECGs is highly variable in dogs.[128]

Prediction of QT interval prolongation by candidate drugs remains a major challenge in drug development. In vitro drug properties and responses measured in animal models are believed to be predictive for the risk of TdP. In this context quantitative PK/PD modeling techniques can be used to establish the quality and limitations of such predictions to the clinic. At present a provisional 30-fold safety margin between the maximally reached free plasma concentration and the in vitro IC_{50} for I_{Kr} inhibition is proposed as a general safety guideline for the interpretation of clinical QT studies.[127] Although useful as a screening method, this safety margin does not allow to quantitatively predict the actual level of QT interval prolongation occurring in vivo. Drug-induced QT interval prolongation depends on several other factors, for example a difference between the observed free plasma concentration of a drug and the drug concentration at the effect site. By using an effect compartment model to characterize hysteresis between plasma concentration and QT prolongation, an estimate of the effect site concentration can be obtained even in small studies.[132,134,137]

However, the effect site concentration is not the only factor that needs to be taken into account for predicting QT interval prolongation from in vitro drug properties. The formation of metabolites affecting cardiac repolarization can bias the predictions from in vitro studies[128] and such effects have been observed with active metabolites of terfenadine, astemizole, and cisapride.[135] Despite efforts to perform in vitro studies under physiological conditions, for a range of compounds a systematic 10-fold difference was observed between the in vitro potency and unbound concentrations associated with QT interval prolongation or TdP in clinical studies.[132] The operational model of pharmacological agonism offers the opportunity to summarize such differences into the parameter τ defined in eqn [5] while simultaneously including in the model factors such as an effect compartment or the response to active metabolites.

The operational model was recently used to estimate τ on the basis of five clinical studies with the class III antiarrhythmic drug dofetilide and estimates of dofetilide potency in the in vitro hERG assay.[136] Dofetilide is a highly selective I_{Kr} blocker without known active metabolites that distinctly prolongs the QT interval and constitutes a suitable benchmark compound. A population PK/PD model including an effect compartment was developed by combining data from these five clinical studies. Introduction of the in vitro estimate of dofetilide potency for hERG current inhibition resulted in an estimate of τ at a value of 6.2. This estimate was used to relate the in vitro current inhibition in a strictly quantitative manner to clinical QT interval prolongation. For example, 10% inhibition of hERG

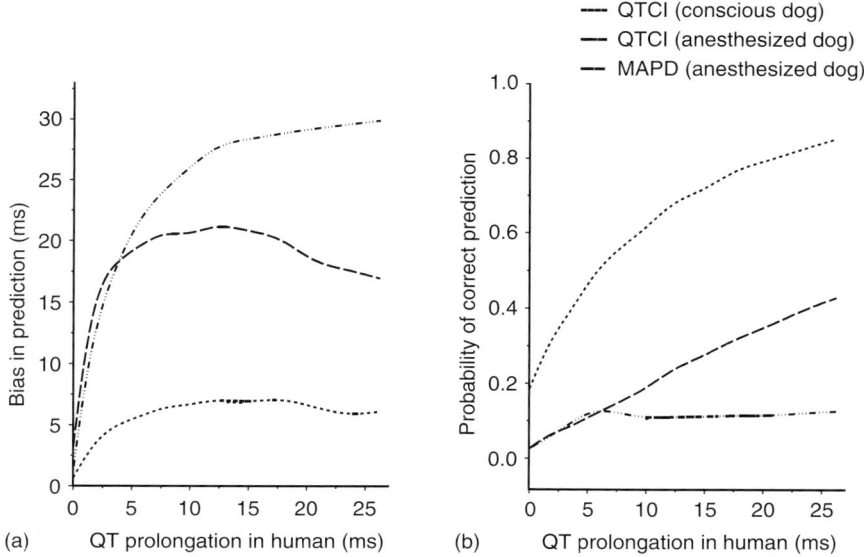

Figure 8 (a) The bias in the degree of QT interval prolongation predicted for humans from three measures of cardiac repolarization in dogs as a function of the predicted QT interval prolongation in human; (b) the probability that the prediction from the dog is not different from the true QT response in human.

currents by dofetilide corresponded to 20 ms of QT interval prolongation (95% confidence interval: 12–31 ms). Conversely, 20 ms QT interval prolongation was exceeded with statistical significance at 13% blockade of hERG currents.

In a second study, the operational model was used once more to determine the value of τ for dofetilide in the beagle dog. The aim of this study was to determine the predictiveness of experiments in the beagle dog under different conditions. QT interval prolongation was measured in conscious and isoflurane-anesthetized dogs; in the anesthetized dogs also the effects of dofetilide on monophasic action potenial duration (MAPD) was determined. A population PK/PD model was developed yielding estimates of τ ranging from 5.7 to 10. The left-hand panel in **Figure 8** shows the bias in predictions that was observed by directly superimposing the concentration–response relationships of dofetilide in human on top of the curves obtained in the dog models. This figure shows that in the clinically relevant range of 0–20 ms QT interval prolongation, a relative large bias was found in the data on anesthetized dogs compared to conscious dogs. The population model allowed considering in addition to the bias (based on the population mean prediction), also uncertainty and variability in the predictions. The precision of the predictions can be derived from these magnitudes and gives insight into the probability that the responses in the respective models are equal to each other. The right-hand panel in **Figure 8** shows that the highest precision was found for prediction from the conscious dog, and that this precision increases with the QT response. This result underscores that relatively large QT intervals can be predicted with reasonable precision from the conscious dog, but that this precision decreases rapidly when attempting to predict <10 ms QT interval prolongation.

5.38.4   **Conclusion**

This chapter presents an overview of the current status of PK/PD modeling concepts. It is shown that PK/PD modeling has progressed from an empirical descriptive discipline into a mechanistic science. Specific highlights in this development have been the recent incorporation of the principles of receptor theory and dynamic systems analysis in mechanism-based PK/PD modeling. This has resulted in a novel class of mechanism-based PK/PD models with much improved properties for extrapolation and prediction.

In the meantime PK/PD modeling is being widely applied in drug development. This concerns specifically its application in the design and analysis of phase I/II proof-of-concept and dose-finding clinical trials. In such trials it is of importance to maximize the information gathered as to better understand the in vivo safety and efficacy profile of novel drugs. Such analyses in turn facilitate the dose selection, design, and evaluation of confirmatory phase III clinical trials. In this chapter the application of PK/PD in preclinical investigations is illustrated for different drugs and an array of pharmacodynamic endpoints. It is shown that with mechanism-based PK/PD models in vivo drug concentration–effect relationships can be predicted on the basis of in vitro bioassays as was illustrated for adenosine A_1 receptor agonists (Section 5.38.3.1) and $GABA_A$ receptor agonists (Section 5.38.3.3). Furthermore, mechanism-based PK/PD modeling enables the prediction, in a strictly quantitative manner, of drugs effects in humans as was shown for MOP receptor agonists (Section 5.38.3.2), $GABA_A$ receptor agonists (Section 5.38.3.3), $5HT_{1A}$ receptor agonists (Section 5.38.3.4), and drug-induced QT interval prolongation (Section 5.38.3.5). This opens new opportunities in drug development, specifically in drug candidate selection and lead optimization. Thus, mechanism-based PK/PD modeling constitutes a novel scientific basis for the prediction of drug effects in humans, with regard to both efficacy and safety, enabling the selection of drug candidates with an optimal efficacy–safety balance.

To date mechanism-based PK/PD modeling has been mainly applied to ad hoc pharmacological effects of drugs (i.e., effects that can be observed in animals, healthy subjects, and patients at a particular point in time). A new challenge will be the modeling of drug effects on disease processes and disease progression in time. The development of novel animal models of disease and the identification of novel biomarkers, characterizing in a strictly quantitative manner disease processes in vivo, are indispensable in this regard.

References

1. Peck, C. C.; Barr, W. H.; Benet, L. Z.; Collins, J.; Desjardins, R. E.; Furst, D. E.; Levy, G.; Ludden, T.; Rodman, J. H. *J. Clin. Pharmacol.* **1994**, *34*, 111–119.
2. Breimer, D. D.; Danhof, M. *Clin. Pharmacokinet.* **1997**, *32*, 259–267.
3. Danhof, M.; Alvan, G.; Dahl, S. G.; Kuhlmann, J.; Paintaud, G. *Pharm. Res.* **2005**, *22*, 1432–1437.
4. Mager, D. E.; Wyska, E.; Jusko, W. J. *Drug. Metab. Disp.* **2003**, *31*, 510–519.
5. Jochemsen, R.; Van Boxtel, C. J.; Hermans, J.; Breimer, D. D. *Clin. Pharmacol. Ther.* **1983**, *34*, 42–47.
6. Rowland, M.; Balant, L.; Peck, C. C. *AAPS Pharm. Sci.* **2004**, *6*, E6.
7. Mandema, J. W.; Tukker, E.; Danhof, M. *Br. J. Pharmacol.* **1991**, *102*, 663–668.
8. Mathôt, R. A. A.; Van Schaick, E. A.; Langemeijer, M. W. E.; Soudijn, W.; Breimer, D. D.; Ijzerman, A. P.; Danhof, M. *J. Pharmacol. Exp. Ther.* **1994**, *268*, 616–624.
9. Cox, E. H.; Kerbusch, T.; Van der Graaf, P. H.; Danhof, M. *J. Pharmacol. Exp. Ther.* **1998**, *284*, 1095–1103.
10. Zuideveld, K. P.; Maas, H. J.; Treijtel, N.; Van der Graaf, P. H.; Peletier, L. A.; Danhof, M. *Am. J. Physiol.* **2001**, *281*, R2059–R2071.
11. Cox, E. H.; Langemeijer, M. W. E.; Gubbens-Stibbe, J. M.; Muir, K. T.; Danhof, M. *Anesthesiology* **1999**, *90*, 535–544.
12. Tuk, B.; Van Oostenbruggen, M. F.; Herben, V. M. M.; Mandema, J. W.; Danhof, M. *J. Pharmacol. Exp. Ther.* **1999**, *189*, 1067–1074.
13. Geldof, M.; Van Beijsterveldt, L., Danhof, M. *Psychopharmacology* **2005**, unpublished observations.
14. Sheiner, L. B.; Stanski, D. R.; Voseh, S.; Miller, R. D.; Ham, J. *Clin. Pharmacol. Ther.* **1979**, *25*, 358–371.
15. Cox, E. H.; Kerbusch, T.; Van der Graaf, P. H.; Danhof, M. *J. Pharmacol. Exp. Ther.* **1998**, *284*, 1095–1103.
16. Jonker, J. W.; Schinkel, A. H. *J. Pharmacol. Exp. Ther.* **2004**, *308*, 2–9.
17. Bouw, M. R.; Xie, R.; Tunblad, K.; Hammarlund-Udenaes, M. *Br. J. Pharamcol.* **2001**, *134*, 1796–1804.
18. Black, J. W.; Leff, P. *Proc. R. Soc. London B* **1983**, *220*, 141–162.
19. Van der Graaf, P. H.; Danhof, M. *Int. J. Clin. Pharmacol. Ther.* **1997**, *35*, 442–446.
20. Leff, P.; Prentice, D. J.; Giles, H.; Martin, G. R.; Wood, J. *J. Pharmacol. Toxicol. Methods* **1990**, *23*, 225–237.
21. Barlow, R. B.; Scott, N. C.; Stephenson, R. P. *Br. J. Pharmacol. Chemother.* **1967**, *31*, 188–196.
22. Van der Graaf, P. H.; Van Schaick, E. A.; Mathôt, R. A. A.; Ijzerman, A. P.; Danhof, M. *J. Pharmacol. Exp. Ther.* **1997**, *283*, 809–816.
23. Van der Graaf, P. H.; Van Schaick, E. A.; Visser, S. A. G.; De Greef, H. J. M. M.; Ijzerman, A. P.; Danhof, M. *J. Pharmacol. Exp. Ther.* **1999**, *29*, 702–709.
24. Garrido, M.; Gubbens-Stibbe, J. M.; Tukker, H. J.; Cox, E. H.; Drabbe von Freitag Drabbe Künzel, J.; Ijzerman, A. P.; Danhof, M.; Van der Graaf, P. H. *Pharm. Res.* **2000**, *17*, 653–659.
25. Zuideveld, K. P.; Van der Graaf, P. H.; Newgreen, D.; Thurlow, R.; Petty, N.; Jordan, P.; Peletier, L. A.; Danhof, M. *J. Pharmacol. Exp. Ther.* **2004**, 308, 1012–1020.
26. Stephenson, R. P. *Br. J. Pharmacol.* **1956**, *11*, 379–393.
27. Furchgott, R. F. In *Advances in Drug Research*; Harper, N. J., Simmonds, A. B., Eds.; Academic Press: New York, 1966, pp 21–55.
28. Tuk, B.; Van Oostenbruggen, M. F.; Herben, V. M. M.; Mandema, J. W.; Danhof, M. *J. Pharmacol. Exp. Ther.* **1999**, *289*, 1067–1074.
29. Dayneka, N. L.; Garg, V.; Jusko, W. J. *J. Pharmacokinet. Biopharm.* **1993**, *21*, 457–478.
30. Van Schaick, E. A.; De Greef, H. J. M. M.; Langemeijer, M. W. E.; Sheehan, M. J.; Ijzerman, A. P.; Danhof, M. *Br. J. Pharmacol.* **1997**, *122*, 525–533.
31. Sun, Y. N.; Jusko, W. J. *J. Pharm. Sci.* **1998**, *87*, 732–737.
32. Ramakrishnan, R.; DuBois, D. C.; Almon, R. R.; Pyszczynski, N. A.; Jusko, W. J. *J. Pharmacokin. Pharmacodyn.* **2002**, *29*, 1–24.
33. Sandström, M.; Lindman, H.; Nygren, P.; Lidbrink, E.; Bergh, J.; Karlsson, M. O. *J. Clin. Oncol.* **2005**, *23*, 413–421.
34. Wakelkamp, M.; Alvan, G.; Gabrielsson, J.; Paintaud, G. *Clin. Pharmacol. Ther.* **1996**, *60*, 75–88.
35. Movin-Osswald, M.; Hammarlund-Udenaes, M. *J. Pharmacol. Exp. Ther.* **1995**, *274*, 921–927.
36. Eckblad, E. B.; Licko, V. *Am. J. Physiol.* **1984**, *264*, R114–R121.
37. Bauer, J. A.; Balthasar, J. P.; Fung, H. L. *Pharm. Res.* **1997**, *14*, 114–145.
38. Åbelö, A.; Eriksson, U. G.; Karlsson, M. O.; Larsson, H.; Gabrielsson, J. *J. Pharmacol. Exp. Ther.* **2000**, *295*, 662–669.
39. Mandema, J. W.; Wada, D. R. *J. Pharmacol. Exp. Ther.* **1995**, *279*, 1035–1042.
40. Veng Pedersen, P.; Modi, N. B. *J. Pharm. Sci.* **1993**, *82*, 266–272.
41. Kleinbloesem, C. H.; Van Brummelen, P.; Van Harten, J.; Danhof, M.; Faber, H.; Urquhart, J.; Breimer, D. D. *Clin. Pharmacol. Ther.* **1987**, *41*, 26–30.

42. Francheteau, P.; Steimer, J. L.; Merdjan, H.; Guerret, M.; Dubray, C. *J. Pharmacokin. Biopharm.* **1993**, *21*, 489–514.
43. Zuideveld, K. P.; Maas, H. J.; Treijtel, N.; Van der Graaf, P. H.; Peletier, L. A.; Danhof, M. *Am. J. Physiol.* **2001**, *281*, R2059–R2071.
44. Chan, P. L.; Holford, N. H. G. *Ann. Rev. Pharmacol. Toxicol.* **2001**, *41*, 625–659.
45. Holford, N. H. G.; Peace, K. E. *Proc. Natl. Acad. Sci. USA* **1992**, *89*, 11466–11470.
46. Post, T. M.; Freijer, J. I.; De Jongh, J.; Danhof, M. *Pharm. Res.* **2005**, *22*, 1035–1049.
47. De Winter, W.; DeJong, J.; Ploeger, B. A.; Danhof, M. *J. Pharmacokinet. Pharmacodyn.* **2005**, PMID 16682968 [Epub ahead of print].
48. De Winter, W.; DeJong, J.; Ploeger, B. A.; Danhof, M. *J. Pharmacokinet. Pharmacodyn.* **2005**, in review.
49. Fredholm, B. B.; Ijzerman, A. P.; Jacobson, K. A.; Klotz, K.-N.; Linden, J. *Pharmacol. Rev.* **2001**, *53*, 527–552.
50. Fredholm, B. B.; Chen, J.-F.; Masino, S. A.; Vaugeois, J.-M. *Annu. Rev. Pharmacol. Toxicol.* **2005**, *45*, 385–412.
51. Dhalla, A. K.; Shryock, J. C. *Curr. Topics Med. Chem.* **2003**, *3*, 369–385.
52. Hutchinson, S. A.; Scammells, P. J. *Curr. Pharm. Des.* **2004**, *10*, 2021–2039.
53. Kenakin, T. *Pharmacologic Analysis of Drug–Receptor Interactions*, 2nd ed.; Raven Press: New York, 1993, pp 441–468.
54. De Ligt, R. A. F.; Ijzerman, A. P. *Curr. Pharm. Des.* **2002**, *8*, 2333–2344.
55. Soudijn, W.; Van Wijngaarden, I.; Ijzerman, A. P. *Curr. Topics Med. Chem.* **2003**, *3*, 355–367.
56. Mathôt, R. A. A.; Van Schaick, E. A.; Langemeijer, M. W.; Soudijn, W.; Breimer, D. D.; Ijzerman, A. P.; Danhof, M. *J. Pharmacol. Exp. Ther.* **1994**, *268*, 616–624.
57. Van Schaick, E. A.; De Greef, H. J. M. M.; Langemeijer, M. W.; Sheenan, M. J.; Ijzerman, A. P.; Danhof, M. *Br. J. Pharmacol.* **1997**, *122*, 525–533.
58. Van Schaick, E. A.; Tukker, E. E.; Roelen, H. C. P. F.; Ijzerman, A. P.; Danhof, M. *Br. J. Pharmacol.* **1998**, *124*, 607–618.
59. Snyder, S. H.; Pasternak, G. W. *Trends Pharmacol. Sci.* **2003**, *24*, 198–205.
60. Snyder, S. H. *Neuropharmacology* **2004**, *47*, 274–285.
61. Walhoer, M.; Bartlett, S.; Whistler, J. L. *Annu. Rev. Biochem.* **2004**, *73*, 953–990.
62. Foord, S. M.; Bonner, T. I.; Neubig, R. R.; Rosser, E. M.; Pin, J.-P.; Davenport, A. P.; Spedding, M.; Harmar, A. J. *Pharmacol. Rev.* **2005**, *57*, 279–288.
63. Gutstein, H. B.; Akil, H. In *Goodman & Gilman's The Pharmacological Basis for Therapeutics*, 10th ed.; Hardman, J. G., Limbird, L. E., Eds.; McGraw-Hill: New York, 2001, pp 569–619.
64. Dahan, A.; Yassen, A.; Bijl, H.; Romberg, R.; Sarton, E.; Teppema, L.; Olofson, E.; Danhof, M. *Br. J. Anaesth.* **2005**, *94*, 825–834.
65. Ossipov, M. H.; Lai, J.; King, T.; Vanderah, T. W.; Porreca, F. *Biopolymers (Pept. Sci.)* **2005**, *80*, 319–324.
66. Cadet, P. *Med. Sci. Monit.* **2004**, *10*, MS28–MS32.
67. Pasternak, G. W. *Neuropharmacology* **2004**, *47*, 312–323.
68. Pasternak, D. A.; Pan, L.; Xu, J.; Yu, R.; Xu, M.-M.; Pasternak, G. W.; Pan, Y.-X. *J. Neurochem.* **2004**, *91*, 881–890.
69. Pan, L.; Xu, J.; Yu, R.; Xu, M.-M.; Pan, Y.-X.; Pasternak, G. W. *Neuroscience* **2005**, *133*, 209–220.
70. Pan, Y.-X.; Xu, J.; Bolan, E.; Moskowitz, H. S.; Xu, M.; Pasternak, G. W. *Mol. Pharmacol.* **2005**, *68*, 866–875.
71. Walhoer, M.; Fong, J.; Jones, R. M.; Lunzer, M. M.; Sharma, S. K.; Kostenis, E.; Portoghese, P. S.; Whistler, J. L. *Proc. Natl. Acad. Sci. USA* **2005**, *102*, 9050–9055.
72. Ekblom, M.; Hammerlund-Udenaes, M. *J. Pharmacol. Exp. Ther.* **1993**, *266*, 244–252.
73. Cox, E. H.; Van Hemert, J. G. N.; Tukker, H. E.; Danof, M. *J. Pharmacol. Toxicol. Methods* **1997**, *38*, 99–108.
74. Cox, E. H.; Kuipers, J. A.; Danhof, M. *Br. J. Pharmacol.* **1998**, *124*, 1534–1540.
75. Lorenzen, A.; Beukers, M. W.; Van der Graaf, P. H.; Lang, H.; Van Muijlwijk-Koezen, J.; De Groote, M.; Menge, W.; Schwabe, U.; Ijzerman, A. P. *Biochem. Pharmacol.* **2002**, *64*, 1251–1265.
76. Van Muijlwijk-Koezen, J. E.; Timmerman, H.; Van der Sluis, R. P.; Van de Stolpe, A. C.; Menge, W. M. P. B.; Beukers, M.; Van der Graaf, P. H.; De Groote, M.; Ijzerman, A. P. *Bioorg. Med. Chem. Lett.* **2001**, *11*, 815–818.
77. Olsen, R. W.; Tobin, A. J. *FASEB J.* **1990**, *4*, 1469–1480.
78. Johnston, G. A. *Pharmacol. Ther.* **1996**, *69*, 173–198.
79. Barnard, E. A.; Skolnick, P.; Olsen, R. W.; Mohler, H.; Sieghart, W.; Biggio, G.; Braestrup, C.; Bateson, A. N.; Langer, S. Z. *Pharmacol. Rev.* **1998**, *50*, 291–313.
80. Costa, E. *Annu. Rev. Pharmacol. Toxicol.* **1998**, *38*, 321–350.
81. Lan, N. C.; Gee, K. W. *Horm. Behav.* **1994**, *28*, 537–544.
82. Olsen, R. W.; Sapp, D. W. *Adv. Biochem. Psychopharmacol.* **1995**, *48*, 57–74.
83. Rudolph, U.; Crestani, F.; Mohler, H. *Trends Pharmacol. Sci.* **2001**, *22*, 188–194.
84. Whiting, P. J.; Bonnert, T. P.; McKernan, R. M.; Farrar, S.; le Bourdelles, B.; Heavens, R. P.; Smith, D. W.; Hewson, L.; Rigby, M. R.; Sirinathsinghji, D. J. et al. *Ann. NY Acad. Sci.* **1999**, *868*, 645–653.
85. Costa, E.; Guidotti, A. *Trends. Pharmacol. Sci.* **1996**, *17*, 192–200.
86. Gasior, M.; Carter, R. B.; Witkin, J. M. *Trends Pharmacol. Sci.* **1999**, *20*, 107–112.
87. Rupprecht, R.; Holsboer, F. *Steroids* **1999**, *64*, 83–91.
88. Fink, M. *Annu. Rev. Pharmacol.* **1969**, *9*, 241–258.
89. Saletu, B.; Anderer, P.; Kinsperger, K.; Grunberger, J.; Sieghart, W. *Int. Clin. Psychopharmacol.* **1988**, *3*, 287–323.
90. Dingemanse, J.; Danhof, M.; Breimer, D. D. *Pharmacol. Ther.* **1988**, *38*, 1–52.
91. Mandema, J. W.; Sansom, L. N.; Dios-Vieitez, M. C.; Hollander-Jansen, M.; Danhof, M. *J. Pharmacol. Exp. Ther.* **1991**, *257*, 472–478.
92. Mandema, J. W.; Danhof, M. *Clin. Pharmacokinet.* **1992**, *23*, 191–215.
93. Mandema, J. W.; Tuk, B.; van Steveninck, A. L.; Breimer, D. D.; Cohen, A. F.; Danhof, M. *Clin. Pharmacol. Ther.* **1992**, *51*, 715–728.
94. Laurijssens, B. E.; Greenblatt, D. J. *Clin. Pharmacokinet.* **1996**, *30*, 52–76.
95. Cox, E. H.; Knibbe, C. A.; Koster, V. S.; Langemeijer, M. W.; Tukker, E. E.; Lange, R.; Kuks, P. F.; Langemeijer, H. E.; Lie, A.-H.-L.; Danhof, M. *Pharm. Res.* **1998**, *15*, 442–448.
96. Visser, S. A. G.; Smulders, C. J. G. M.; Reijers, B. P. R.; Van der Graaf, P. H.; Peletier, L. A.; Danhof, M. *J. Pharmacol. Exp. Ther.* **2002**, *302*, 1158–1167.
97. Visser, S. A. G.; Gladdines, W. W. F. T.; Van der Graaf, P. H.; Peletier, L. A.; Danhof, M. *J. Pharmacol. Exp. Ther.* **2002**, *303*, 616–626.
98. Cleton, A.; De Greef, H. J.; Edelbroek, P. M.; Voskuyl, R. A.; Danhof, M. *J. Pharmacokinet. Biopharm.* **1999**, *27*, 301–323.
99. Mandema, J. W.; Kuck, M. T.; Danhof, M. *J. Pharmacol. Exp. Ther.* **1992**, *261*, 56–61.
100. Anderson, A.; Boyd, A. C.; Byford, A.; Campbell, A. C.; Gemmell, D. K.; Hamilton, N. M.; Hill, D. R.; Hill-Venning, C.; Lambert, J. J.; Maidment, M. S. et al. *J. Med. Chem.* **1997**, *40*, 1668–1681.

101. Visser, S. A. G.; Wolters, F. L. C.; Gubbens-Stibbe, J.; Tukker, E.; Van der Graaf, P. H.; Peletier, L. A.; Danhof, M. *J. Pharmacol. Exp. Ther.* **2003**, *304*, 88–101.
102. Haefely, W. E. *Adv. Biochem. Psychopharmacol.* **1988**, *45*, 275–292.
103. Haefely, W. E.; Martin, J. R.; Richards, J. G.; Schoch, P. *Can. J. Psychiat.* **1993**, *38* (Suppl. 4), S102–S108.
104. Sieghart, W. *Trends Pharmacol. Sci.* **1992**, *13*, 446–450.
105. Hillegaart, V. *Acta Physiol. Scand Suppl.* **1991**, *598*, 1–54.
106. Albert, P. R.; Zhou, Q. Y.; Van Tol, H. H.; Bunzow, J. R.; Civelli, O. *J. Biol. Chem.* **1990**, *265*, 5825–5832.
107. Meller, E.; Bohmaker, K. *J. Pharmacol. Exp. Ther.* **1994**, *271*, 1246–1252.
108. Olivier, B.; Mos, J.; Van der Heyden, J. A. M.; Molewijk, H. E.; Van Dijken, H. H.; Zethof, T.; Van Vest, A.; Tulp, M. T. M.; Slangen, J. L. Functional Correlates of 5-HT Receptors Clinical Applications Possibilities of Serotonergic Drugs. In *Trends in Receptor Research*; Claassen, V., Ed.; Elsevier Science Publishers: Amsterdam, The Netherlands, 1993, pp 97–122.
109. Gudelsky, G. A.; Koenig, J. I.; Meltzer, H. Y. *Neuropharmacology* **1986**, *25*, 1307–1313.
110. Cryan, J. F.; Kelliher, P.; Kelly, J. P.; Leonard, B. E. *J. Psychopharmacol.* **1999**, *13*, 278–283.
111. Millan, M. J.; Rivet, J. M.; Canton, H.; Le Marouille Girardon, S.; Gobert, A. *J. Pharmacol. Exp. Ther.* **1993**, *264*, 1364–1376.
112. Zuideveld, K. P.; Treijtel, N.; Maas, H. J.; Gubbens-Stibbe, J. M.; Peletier, L. A.; Van der Graaf, P. H.; Danhof, M. *J. Pharmacol. Exp. Ther.* **2002**, *300*, 330–338.
113. Zuideveld, K. P.; Van Gestel, A.; Peletier, L. A.; Van der Graaf, P. H.; Danhof, M. *Eur. J. Pharmacol.* **2002**, *445*, 43–54.
114. Zuideveld, K. P.; Rusiç-Pavletiç, J.; Maas, H. J.; Peletier, L. A.; Van der Graaf, P. H.; Danhof, M. *J. Pharmacol. Exp. Ther.* **2002**, *303*, 1130–1137.
115. Yu, H.; Liu, Y.; Malmberg, A.; Mohell, N.; Hacksell, U.; Lewer, T. *Eur. J. Pharmacol.* **1996**, *303*, 151–162.
116. Zeisberger, E. *Prog. Brain Res.* **1998**, *115*, 159–176.
117. Bligh, J. In *Recent Studies of Hypothalamic Function*; Lederis, K., Cooper, K. E., Eds.; S. Karger: Calgary, Canada, 1974, pp 315–327.
118. Cabanac, M. *Annu. Rev. Physiol.* **1975**, *37*, 415–439.
119. Boeckman, A.; Sheiner, L. B.; Beal, S. L. *NONMEM User's Guide*; NONMEM Project Group, University of California: San Franciso, CA, 1992.
120. Van der Sandt, I. C. J.; Smolders, R.; Nabulsi, L.; Zuideveld, K. P.; De Boer, A. G.; Breimer, D. D. *Eur. J. Pharm. Sci.* **2001**, *14*, 81–86.
121. Anderson, I. M.; Cowen, P. J. *Psychopharmacology* **1992**, *106*, 428–432.
122. Young, A. H.; McShane, R.; Park, S. B.; Cowen, P. J. *Biol. Psychiat.* **1993**, *34*, 665–666.
123. Blier, P.; Seletti, B.; Gilbert, F.; Young, S. N.; Benkelfat, C. *Neuropsychopharmacology* **2002**, *27*, 301–308.
124. McAllister-Williams, R. H.; Massey, A. E. *Psychopharmacology* **2003**, *166*, 284–293.
125. Seletti, B.; Benkelfat, C.; Blier, P.; Annable, L.; Gilbert, F.; de Montigny, C. *Neuropsychopharmacology* **1995**, *13*, 93–104.
126. Pitchot, W.; Wauthy, J.; Hansenne, M.; Pinto, E.; Fuchs, S.; Reggers, J.; Legros, J. J.; Ansseau, M. *Psychopharmacology* **2002**, *164*, 27–32.
127. Redfern, W. S.; Carlsson, L.; Davis, A. S.; Lynch, W. G.; MacKenzie, I.; Palethorpe, S.; Siegl, P. K.; Strang, I.; Sullivan, A. T.; Wallis, R. et al. *Cardiovasc. Res.* **2003**, *58*, 32–45.
128. De Ponti, F.; Poluzzi, E.; Cavalli, A.; Recanatini, M.; Montanaro, N. *Drug Saf.* **2002**, *25*, 263–286.
129. Belardinelli, L.; Antzelevitch, C.; Vos, M. A. *Trends Pharmacol. Sci.* **2003**, *24*, 619–625.
130. Malik, M. *Clin. Pharmacol. Ther.* **2005**, 77, 241–246.
131. Gralinski, M. R. *Toxicol. Pathol.* **2003**, *31*, S11–S16.
132. Webster, R.; Leishman, D.; Walker, D. *Curr. Opin. Drug Disc. Devel.* **2002**, *5*, 116–126.
133. Wang, J.; Della Penna, K.; Wang, H.; Karczewski, J.; Connolly, T. M.; Koblan, K. S.; Bennett, P. B.; Salata, J. J. *Am. J. Physiol. Heart Circ. Physiol.* **2003**, *284*, H256–H267.
134. Le Coz, F.; Funck-Brentano, C.; Morell, T.; Ghadanfar, M. M.; Jaillon, P. *Clin. Pharmacol. Ther.* **1995**, *57*, 533–542.
135. Shah, R. R. *Br. J. Clin. Pharmacol.* **2002**, *54*, 188–202.
136. Jonker, D. M.; Kenna, L.; Leishman, D.; Wallis, R.; Milligan, P. A.; Jonsson, E. *Clin. Pharmacol. Ther.* **2005**, 77, 572–582.
137. Ollerstam, A.; Visser, S. A.; Persson, A. H.; Eklund, G.; Nilsson, L. B.; Forsberg, T.; Wiklund, S. J.; Gabrielsson, J.; Duker, G.; Al-Saffar, A. *J. Pharmacol. Toxicol. Methods* **2006**, *53*, 174–183.
138. Peterman, C.; Sanoski, C. A. *Cardiol. Rev.* **2005**, *13*, 315–321.

Biographies

Meindert Danhof is Professor of Pharmacology and Director of Research of the Leiden/Amsterdam Center for Drug Research at Leiden University in the Netherlands. His research interest is in the development of new theoretical concepts in pharmacokinetic/pharmacodynamic (PK/PD) modeling. Important contributions in his research have been on the incorporation of (i) receptor theory and (ii) dynamical systems analysis. This has resulted in a new class of 'mechanism-based' PK/PD models with considerably improved properties for extrapolation and prediction. His most recent contribution is the development of the concept of 'disease system analysis' as the basis for the prediction of drug effects on disease progression.

Meindert Danhof is also Senior Scientific Advisor of LAP&P Consultants BV, which provides a professional infrastructure for consulting on the application of advanced PK/PD modeling concepts to the international pharmaceutical industry.

Piet Hein van der Graaf studied Pharmacy in The Netherlands and obtained a PhD in Clinical Medicine at King's College London (United Kingdom) under supervision of Sir James Black for his work on quantitative pharmacological characterisation of α_1-adrenoceptors. He subsequently worked as a Post-Doc at the Division of Pharmacology at the Erasmus University in Rotterdam (The Netherlands) with Professor Pramod Saxena and as Team Leader Pharmacology in the Urology Group of Synthelabo Recherche in Paris (France). After this, he moved back to The Netherlands and worked as a Fellow of the Royal Netherlands Academy of Sciences at the Division of Pharmacology of the Leiden/ Amsterdam Center for Drug Research with Prof Meindert Danhof on the development of mechanism-based PK/PD models. Since 1999, he has been working at Pfizer Global Research & Development in Sandwich (UK) as Senior Director in Discovery Biology and Head of Sexual Health and currently as Head of Pharmacokinetics & PK/PD in the Department of Pharmacokinetics, Dynamics and Metabolism.

Daniël M Jonker obtained his PhD in Mathematics and Natural Sciences at Leiden University (The Netherlands) under supervision of Prof Meindert Danhof and Dr Rob Voskuyl for his work on the mechanism-based analysis of pharmacodynamic drug–drug interactions between antiepileptic drugs. He worked between 2003 and 2005 as a Postdoctoral Fellow at Uppsala University in Sweden with Dr Niclas Jonsson on PK/PD models for predicting QT interval prolongation. Since then he has been working as a Senior Scientist PK/PD at Grünenthal GmbH Aachen in Germany and as a Biomodeling Scientist at Novo Nordisk A/S Bagsværd in Denmark.

Sandra A G Visser obtained her PhD in Mathematics and Natural Sciences cum laude at Leiden University (The Netherlands) with her thesis on the mechanism-based pharmacokinetic–pharmacodynamic modeling of the $GABA_A$ receptor response in vivo. This work was done under supervision of Prof Meindert Danhof, Prof Bert Peletier and Dr Piet Hein van der Graaf. Between 2002 and 2004, she worked as a Postdoctoral Fellow at AstraZeneca R&D Södertälje in Sweden on modeling of tolerance development in temperature regulation and QT prolongation with Dr Johan Gabrielsson. From 2004 onwards she has been working as a Senior Research Scientist specialized in PK/PD analysis within both discovery and development DMPK at AstraZeneca R&D Södertalje.

Klaas P Zuideveid obtained his PhD in Mathematics and Natural Sciences cum laude at Leiden University (The Netherlands) under supervision of Prof Meindert Danhof, Prof Bert Peletier, and Dr Piet Hein van der Graaf for his work on mechanism-based pharmacokinetics–pharmacodynamic modeling: application to $5HT_{1A}$ receptor mediated responses. He then worked as a scientific consultant for Pharsight Corporation in the European strategic services group, working with numerous pharmaceutical companies on computer assisted trial design. Since 2002, he has been working at F Hoffman-La Roche Ltd. (Basel, Switzerland) in the clinical modeling & simulation group as a Senior Pharmacometrician. In this position he is responsible for the oncology disease area.

5.39 Computational Models to Predict Toxicity

N Greene, Pfizer Global Research and Development, Groton, CT, USA

5.39.1   Introduction

5.39.1.1   Definitions

In order to avoid any potential confusion it is important to first define the terms most often used in describing toxicity and the factors that influence decisions:

- Toxicity is a measure of the degree to which something is toxic or poisonous. The study of poisons is known as toxicology. Toxicity can refer to the effect on a whole organism, such as a human or a bacterium or a plant, or to a substructure, such as the liver.
- Hazard is the general term for anything that has the ability to cause injury or the potential to cause injury. The hazard associated with a potentially toxic substance is a function of its toxicity and the potential for exposure to the substance.
- Risk is defined as the prediction of the likelihood and extent of harm (negative consequences) resulting from a given action: risk = hazard × exposure.
- Risk assessment is the identification and quantification of the risk resulting from a specific use or occurrence of a chemical, taking into account the possible harmful effects on individual people or on society of using the chemical in the amount and manner proposed and all the possible routes of exposure. Quantification ideally requires the establishment of dose–effect and dose–response relationships in likely target individuals and populations.
- Safety is the practical certainty that injury will not result from exposure to a hazard under defined conditions: in other words, the high probability that injury will not result.

5.39.1.2   Why is Toxicity an Important Consideration in Medicinal Chemistry?

Recent trend analysis in the pharmaceutical industry has shown that the average cost of producing a new drug has risen to beyond US$800 million, and takes between 12 and 15 years to reach the marketplace. Moreover, fewer than 1 in 10 new drug candidates will reach the market once they enter clinical trials. Toxicity is one of the major causes of attrition, accounting for approximately 30% of the withdrawal of drugs from the market or from a company's development pipeline.[1] Late-stage development failures or withdrawals from the market are very costly, and create an unsustainable business model for the pharmaceutical industry.

Efforts to address this attrition problem require the early consideration of toxicity in the design and synthesis of new drug molecules. Once a chemical has been synthesized, its efficacy and safety profiles are, for the most part, fixed, although different formulations can have an impact on the exposure and toxicity profile of a compound. All that remains is to identify what those biological properties are and to manage the risks presented by the molecule. The ability to accurately predict the potential adverse effects of a novel compound in advance of entering in vivo testing would provide a distinct advantage in the fight to reduce attrition.

The high-profile market withdrawals of compounds for adverse safety profiles (e.g., Vioxx (rofecoxib), a drug used for the treatment of arthritis and acute pain, withdrawn in September 2004 due to an increased risk of adverse cardiovascular events after long-term dosing[2]) has created a social climate where there is increased scrutiny and pressure on the pharmaceutical industry to make 'safer' medicines and to publicly disclose the risks of taking the medicines that they sell. The safety of new medicines is now of paramount importance to the survival of the pharmaceutical industry.

5.39.1.3   The Challenges of Predicting Toxicity

There are many challenges that have to be overcome before one can accurately predict the toxicity of a novel chemical structure. Fundamental issues, such as the complex nature of biological systems and the ability of chemicals to interact with a biological system in multiple ways that could all potentially manifest themselves as similar toxic events, inevitably mean that the problem of predicting toxicity is not straightforward.

The diverse types of information that relate to the characterization of toxicological risk of chemicals[3] can be represented schematically as layers of an onion,[4] with the chemical of concern at the center, and the human health risk assessment in the outermost layer (**Figure 1**). The intervening layers range from calculated or measured physical properties, to mechanistically defined biochemical interactions, to in vitro cell culture bioassays, to in vivo responses in whole animals and populations. The radius of each layer increases according to the level of biological complexity and proximity to human health risk assessment, whereas the boundaries between the layers separate physically and conceptually distinct types of information. Structure–activity relationships (SARs) are models that attempt to extrapolate from the center of the onion, i.e., the chemical structure, to different types of biological endpoints.

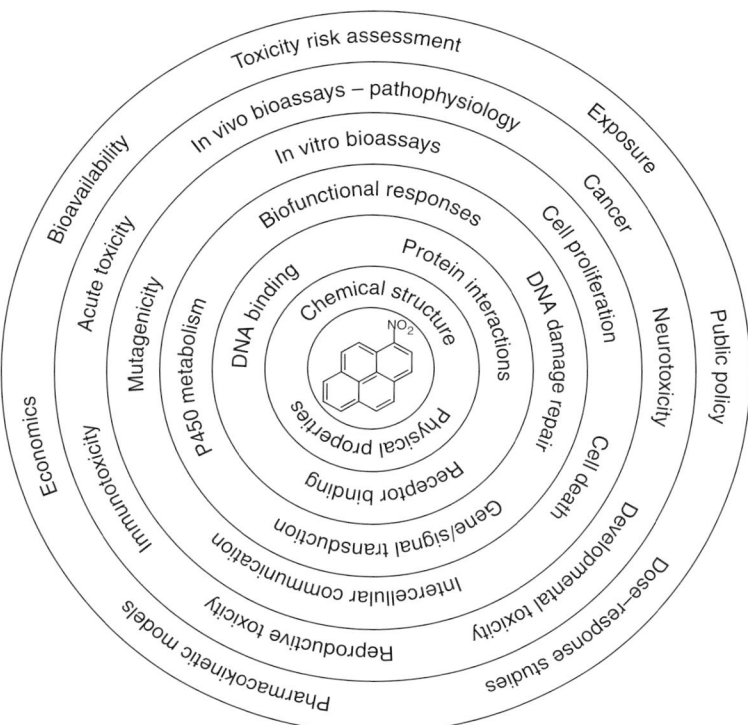

Figure 1 Layers of chemical and biological organization relevant to the toxicity prediction problem. (Adapted from Richard, A. M. *Knowledge Eng. Rev.* **1999**, *14*, 1–12.)

5.39.1.3.1 The nature of toxicity data

Most toxicity data consist of observations of the biological system under treatment in relation to a control group. Rarely are these observations done at a molecular level, making relationships to chemical structure difficult to identify. Add to this, species differences and genetic variations within a single population and between populations themselves, making the issues even more complex.

5.39.1.3.2 Factors affecting the determination of mechanisms of toxicity

The dose, metabolic fate, and exposure of a compound are also factors in determining the potential for toxicity that need to be considered in any mechanistic evaluation. The relationship between dose and response is one of the most fundamental concepts of toxicology, which is often forgotten in the public debate about safety of new medicines.

"What is there that is not a poison? All things are poison and nothing is without poison. Solely the dose determines that a thing is not a poison." (Paracelsus, 1493–1541.)

The metabolic fate of a chemical, or the susceptibility of a compound to undergo biological transformations, can have a profound effect on its ability to cause toxicity. Often a drug compound itself can be benign, but can be metabolized to a reactive intermediate that can elicit a toxic response following exposure to the drug. In other instances, metabolism can lead to detoxification of a drug molecule. Kalgutkar *et al.*[5] provide an extensive review of typical substructures that have been shown to be metabolized to reactive intermediates and have been associated with the expression of adverse drug reactions (ADRs).

One classic example, where excessive dose and metabolic activation can combine to cause toxicity, is paracetamol. When a person takes paracetamol, metabolism in the liver produces small amounts of *N*-acetyl-*p*-benzoquinone imine (NAPQI) (**Figure 2**). Under normal dose levels, this potentially toxic compound can quickly form conjugates with glutathione, thus neutralizing the toxic effects of NAPQI.

The problem occurs when the body runs out of glutathione. As glutathione stores are diminished, NAPQI is not detoxified, and covalently binds to the lipid bilayer of hepatocytes, causing centrilobular necrosis, resulting in hepatotoxicity. The maximum daily dose of paracetamol is 4 g in adults and 90 mg kg^{-1} in children. A single ingestion

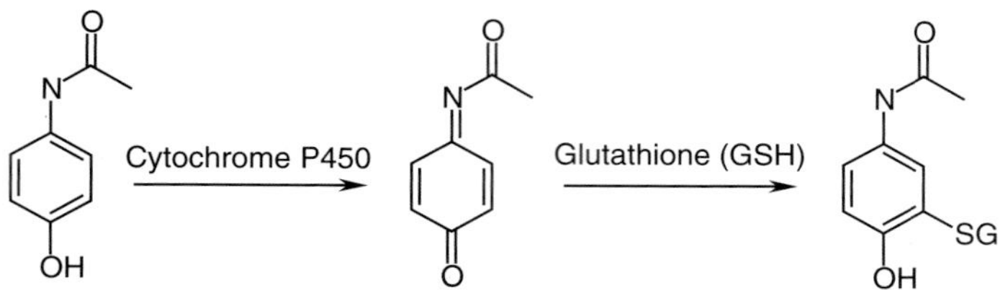

Figure 2 Metabolic formation of NAPQI and subsequent reaction with glutathione.

of 7.5 g in an adult or more than 150 mg kg^{-1} in a child is a potentially toxic dose of paracetamol. Paracetamol toxicity is the number one cause of hospital admission for acute liver failure in the USA, accounting for approximately 40% of all liver transplant cases.[6]

While dose is an important consideration in determining the exposure of a subject to a chemical, it does not always have a direct correlation to the quantity of drug circulating in the bloodstream. The absorption, distribution and excretion properties of a compound will also have a bearing on the observed toxicity resulting from the drug.

5.39.1.3.3 The available data pool for toxicity modeling

Consideration for the toxic effects of a compound is primarily driven by either social concerns for human health and safety via intended or accidental exposure to the chemical, or, in more recent years, it is the potential impact of a chemical on the environment if released in large quantities into the ecosystem. Therefore, the available data pool for modeling chemical toxicity tends to be both structurally and mechanistically diverse.[7] The primary goal of chemical toxicity prediction is to distinguish between toxicologically active and inactive compounds. Typically, multiple mechanisms can lead to the same toxicological endpoint, and therefore predictive models need to be able to distinguish multiple regions of activity amidst a mass of inactive chemical structures.

Toxicity tests, unlike those designed to study the pharmacological interactions of a compound, are often low-throughput and expensive with few in vitro models that have a high correlation with in vivo outcomes. These limitations result in practical and economic barriers to the generation of focused data sets that could be used for toxicity modeling applications.

5.39.1.4   Other Factors Influencing the Need to be Able to Predict Toxicity

In recent years, legislation in both the USA and EU stipulated that companies producing chemicals in quantities greater than 1000 tonnes per annum, often referred to as high-production volume (HPV) chemicals, should provide data on the human and environmental safety of these substances. Current estimates are that less than 10% of the 2500–3000 chemicals that fall under this legislation have sufficient data available for them.[8] In addition to this, other legislation, such as the Safe Drinking Water Act Amendment of 1996 require the US Environmental Protection Agency to assess thousands of contaminants believed to be present in drinking water and to develop and maintain a Contaminant Candidate List for further investigation.[9]

Social pressures to reduce the number of animals used in laboratory testing, particularly in the cosmetic and personal products industries, have resulted in greater interest and need for the ability to predict the human and environmental hazards posed by chemicals. Computer-based or in silico prediction of toxicity is seen as an ideal solution to the problem, as it is fast, cheap, and requires no additional resources. While the overall reliability of any single method for prediction of every possible toxic event is not currently adequate to replace laboratory experiments, at the very least it is hoped that in silico tools could be used to help to identify and prioritize those potential 'bad actors' in need of further exploration from those of lesser concern.

While environmental impact is an important consideration for any chemical produced in large quantities, the focus of this chapter will be the use of in silico methods to predict the potential adverse effects of a compound in humans. There are many reports in the literature concerning the prediction of the environmental fate of chemicals.[10,11]

5.39.2   Current Methods for In Silico Prediction of Toxicity

Early approaches developed by Hansch & Fujita and Free & Wilson using linear free energy have provided the basic scientific framework for the quantitative correlation of chemical structure with biological activity, and have spurred many developments in the field of quantitative SARs (QSAR).[12,13] These techniques have been extensively applied to the modeling of medicinal properties, where they are generally used to optimize the pharmacological activity within a series of chemically similar compounds.

However, for many years SARs and QSAR techniques have also been applied to the prediction and characterization of chemical toxicity.[7] These computational techniques have been applied to a wide variety of toxicological endpoints, from the prediction of LD_{50} (the dose at which 50% of a population exposed to the chemical will die) and maximum tolerated dose values to *Salmonella typhimurium* (Ames) assay results, carcinogenic potential and developmental toxicity effects. Generally, however, toxicological endpoints such as carcinogenicity, reproductive effects, and hepatotoxicity are mechanistically complex when compared with the prediction of pharmacological activity, and hence this leads to added difficulty when trying to predict these endpoints.[14]

In general terms, in silico methods can be categorized into three main classes:

- statistical methods;
- rule-based methods; and
- three-dimensional (3D) molecular modeling techniques.

Each of these approaches has advantages and disadvantages. As with the application of all models, the limitations and trends of the model should be recognized and considered by the user when using them to guide decisions.

5.39.2.1 Statistical Approaches to Predicting Toxicity

Statistical approaches use a variety of mathematical algorithms to explore and define the relationships between properties of the compounds and their respective activities. This requires a training set of chemicals from which a multitude of descriptors can be generated, usually calculated from the chemical structure of each compound. These can range from simple structural fragments of two or more atoms and bonds to more complex quantum mechanics calculations of molecular orbital (HOMO/LUMO) ratios.

These descriptors, along with a measure of the activity of each compound are used as substrate for mathematical learning algorithms that derive a statistical correlation between the descriptors and the observed activity. Many different algorithms have been developed in the scientific community to address problems of this type, ranging from the calculation of simplistic statistical frequency (e.g., descriptor X appears more frequently in active compounds than inactive ones) to sophisticated neural networks[15–17] or genetic algorithms.[18,19]

All statistical methods used to derive these relationships can suffer from over-fitting, which ultimately means that the resulting model becomes too specific to the training set of compounds, and reduces the predictive power of the model for compounds not included in the original set. To avoid over-training of the algorithm, the developer will often hold back a representative test set of structures from the training set that can be used to measure the predictive power of the resulting model. Other statistical techniques such as leave-one-out cross-validation are also used to ensure that the model is suitable for use in prediction. If the model generation is considered successful, it can then be used to predict the activity of a novel compound based on the respective values of its descriptors.

5.39.2.1.1 Advantages of statistical approaches

- The algorithms do not need to know the underlying biological mechanisms in order to resolve relationships between the descriptors and the observed activity; the relationships are based solely on mathematical correlations.
- The descriptors identified as being highly correlated to biological activity can potentially lead to a better understanding of the biological mechanism.
- The descriptors used in the model can be a combination of structural, physicochemical, and quantum mechanics-derived values that can lead to a more comprehensive assessment of the complex factors involved in the expression of toxicity.
- Where toxicity is measured quantitatively, these methods can be used to rank order compounds on their relative potency.

5.39.2.1.2 Disadvantages of statistical approaches

- Statistical methods work best if all molecules covered within a given class all act by the same underlying biological mechanism. Variation in mechanisms of action will cause deviations in the relative importance of descriptors within the algorithm, which ultimately lowers the ability of the model to predict accurately. This can be partially overcome by using more sophisticated mathematics.

- The quality of data used to develop the model is critical to the success or failure of the system to predict toxicity.
- The choice of descriptors is fundamental to the success of the model to predict the biological effect. However, the more descriptors used, the higher the likelihood that correlations of chance are discovered that have no real predictive power.
- Descriptors can often be based on theoretical concepts and calculations that are difficult to interpret in terms of their relationships to the chemical structure. This can often leave medicinal chemists at a loss when seeking structures that either maximize or minimize the value of a particular descriptor.
- Confusion can arise where models provide a probabilistic measure of effect. This is a measure of confidence in the classification by the model of toxic versus nontoxic, and should not be mistaken for a measure of relative potency in causing the toxicological effect for the compound.
- While statistical methods require the conversion of the biological effects of the compound into a numerical representation, toxicology is predominantly a descriptive science that relies on observation rather than quantifiable measurements. Even where numerical measurements are made, for example in liver function tests and other clinical chemistry parameters, there exists a natural variation between individual subjects that can result in an imprecise classification in terms of the biological effect.

Given the aforementioned issues, it is not surprising that statistical methods have had limited success in predicting the toxicological properties of novel chemical structures that are not representative of the training set used in developing the model. This can be attributed in part to the quality and quantity of the available toxicology data, but also to the complexity of the toxicological endpoint and breadth of mechanisms that they are attempting to predict.

5.39.2.2 Rule-Based Approaches to Predicting Toxicity

Rule-based approaches, sometimes referred to as knowledge-based approaches, rely on the human interpretation of data to derive relationships between chemical structure and/or physicochemical properties and the biological effects of a compound. While not always necessary, it is useful to have some basic idea of the mechanism by which toxicity is manifested, as this will help establish the chemical properties that could potentially play a role in the expression of toxicity. The development of these relationships or correlations requires a reasonably sized set of chemicals that contains both active and inactive examples, although the size of this data set will often be dependent on the suspected mechanism of action. Where mechanisms are well understood, the number of compounds required to derive a relationship are often lower than where the mechanisms are complex or ill defined.

For example, one mechanism of toxicity is through direct reaction with endogenous biomolecules (e.g., DNA or skin proteins). This mechanism of direct interaction with DNA or proteins has been implicated as one of the major mechanisms involved in the expression of mutagenicity and skin sensitization, respectively. Other types of toxicity can occur through the inhibition of enzyme or cellular activity, such as blockage of the cardiac potassium channel. This channel is sometimes referred to as the human ether-a-go-go-related gene (hERG) channel. hERG inhibition has been implicated in causing prolonged cardiac action potential duration, which is a potential risk factor in torsade de pointes, a cardiac arrhythmia.

The rule development process involves the formation of associations between the properties of a chemical and its observed effects by human experts. The resulting relationships derived from the training set are often expressed as a series of rules that describe the necessary components for the biological effects to manifest themselves on exposure of a biological system to the compound. These rules can be encoded electronically as structural alerts or decision trees, or combinations of the two. The resulting model can then be used to predict the biological properties of novel chemicals.

During the development of a rule-based model, it is unnecessary and often unwise to withhold data from evaluation, as this can lead to gaps in the generation of hypotheses and result in shortcomings in the predictive model. The downside to this, of course, is that the only way to test the performance of the resulting model is to conduct prospective testing of novel chemicals.

The predictions from rule-based approaches are often qualitative or semi-quantitative, using a classification system to indicate a level of concern. However, more often than not, rule-based systems will provide a textual rationale to the end user to explain the basis of its prediction, and aid the user in interpretation and the subsequent direction to take.

5.39.2.2.1 Advantages of rule-based approaches

- Predictions are often based on mechanistic understanding, and usually provide a rationale to the user to help in interpretation.
- Experts can assess data quality during the development of the model, and therefore the system is able to cope with and use imperfect data points.
- The properties used to build the model are usually readily identifiable with the structure, and give the user a clear indication on the component of the structure that is suspected of causing the toxicity.
- The use of textual phrases to classify concern levels allows the user to group chemicals into those of highest and lowest priority.

5.39.2.2.2 Disadvantages of rule-based approaches

- These systems are often labor-intensive and require individuals to possess1 a working knowledge of both toxicology and chemistry.
- While the developers can uncover or propose novel mechanisms during the analysis, the system itself is incapable of identifying potentially novel relationships. This often results in rule-based systems being biased toward or limited by current knowledge.
- Rule-based systems generally use descriptors that are easily identifiable with properties of the chemical structure, for example, the lipophilicity of a compound as described by the octanol/water partition coefficient often referred to as log P. These structure-based descriptors might not adequately capture or describe the toxicological activity of compounds, leading to false-positive or -negative predictions.
- By their very nature, rule-based systems often do not attempt to predict a continuous value of toxicity. Where toxicology endpoints have a continuous measure of activity, as with the inhibition of the hERG potassium channel, they are usually reduced to a simple classification system, for example, low, medium, or high; or positive or negative.

With all predictive models there is often a need to balance false-positive predictions (i.e., predicting a compound to be active when in fact it is not) and false negatives where the system predicts active compounds to be inactive. Rule-based systems often struggle with this balance, as developer have to rationalize their confidence in the mechanism of action with the available data. Inconsistencies among rules have to be guarded against when they are implemented in these systems. When using any predictive model it is important that the user understands the limitations of the system and its relative bias toward false negatives or false positives.

5.39.2.3 Three-Dimensional Approaches to Predicting Toxicity

3D approaches use knowledge about the shape and volume of either the submitted compound or the binding site thought to be involved in the mechanism of toxicity, or both. They often require knowledge that the chemicals act via a common mechanism or biological event. One example where 3D methods have been applied to predict toxicity is the binding or blockage of the potassium ion channel in the prediction of prolonged QT syndrome, another is the identification of substrates for one or more of the P450 cytochromes and their ability to cause induction or peroxisome proliferation. P450 induction where a patient is taking multiple therapeutic medicines has been linked to causing adverse events through prolonged or increased exposure to one of the drugs or decreased efficacy by increased metabolic clearance. This type of effect is often referred to as a drug–drug interaction (DDI).

As with the prediction of efficacy using 3D models, development of a model for toxicity requires, at a minimum, a set of chemicals with a range of known activities for a particular endpoint. If one assumes that these compounds all act via common interactions with proteins it is possible to develop a pharmacophore that describes the essential arrangement of structural components for a compound to express similar biological activity. Other techniques such as volume or molecular planarity estimates can also be used where there is little knowledge of the protein or its binding site.

Similarly, techniques such as comparative molecular field analysis (CoMFA),[20,21] developed by Tripos Inc., can also be applied in these types of situations. CoMFA models use estimates of steric and electrostatic interactions to describe the activities of molecules. The primary limitation in the development of CoMFA models is the need for good structural alignment between not only the molecules used to build the model but also those molecules for which predictions are made.

It is often preferable, although not necessary, to know which protein is involved in the expression of toxicity and its active binding site. This additional knowledge enables the developer to include more accurate estimates of volume restrictions and alternative conformations that may further enhance the prediction of protein–ligand interactions. Techniques such as ligand docking and scoring can be used to rank order chemicals according to their ability to bind to the receptor site. These techniques often have to compensate for flexibility not only in the ligand being docked but also the protein-binding site itself. This can be accomplished by either using multiple conformations of the compound or using energy minimization algorithms after the docking has been performed.

As with statistical methods, 3D models can suffer from over-fitting, which results in the model being too specific to the training set of compounds, reducing its predictive power. It is therefore common for developers to use separate training and test sets to ensure that over-fitting does not occur.

5.39.2.3.1 Advantages of three-dimensional approaches

- Predictions from these 3D approaches are usually based on mechanistic hypotheses of toxicity, and therefore usually provide reliable predictions where the molecule in question acts via this mechanism and is adequately represented in the model training set.
- These approaches take a whole-molecule view to predicting their activity, and can compensate for subtle variations in electronic or steric factors within the structure.
- The predictions made by 3D methods are often numerical estimates of the activity of a molecule, and therefore these approaches are capable of providing a rank ordering of novel chemical structures.

5.39.2.3.2 Disadvantages of three-dimensional approaches

- They rely on the assumption that the compounds in question cause toxicity through their interaction with specific proteins or receptors. Toxicity can also occur as a result of direct covalent binding to proteins or as a result of altering the physical state of the biological system for example by increasing or decreasing intracellular pH levels.
- 3D methods are often computationally intensive and require extensive calculation times to generate predictions. While this can be compensated for by parallelization, it ultimately leads to some level of approximation in the calculation, which can lead to errors in the predictions from the models.
- Often it is difficult to reliably confirm that all the molecules in the training set are acting through a common mechanism or even through interactions at the same binding site within the protein complex. These assumptions can lead to errors in the predictions from these models.
- Most types of toxicity studies do not yield data that are amenable to this type of model development, as often experiments are conducted with a view to observing changes at an organism level and not at the molecular level.

Due to the nature of toxicology studies there are few endpoints where specific proteins have been identified as the primary cause of the adverse effect or where a molecular shape has demonstrated good correlations with the data generated. This has naturally limited the application of 3D approaches to yield truly predictive models for toxicity. As knowledge around the mechanisms of toxicity expand and new technology facilitates more precise measurements of the interactions between molecules and a biological system, these types of approaches will most likely yield the best predictive models. However, the scope of each model needs to be adequately assessed and applied appropriately.

5.39.3   In Silico Models for Specific Toxicological Endpoints
5.39.3.1   In Silico Prediction of Genotoxicity

Genotoxicity results from the adverse effects of chemicals on DNA and the genetic processes of living cells.[22] These effects can be in the form of point mutations in the DNA (mutagenicity), DNA strand breaks, or chromosome alterations. For a long time now there has been a strong association between genetic damage and carcinogenicity, especially in biological systems that have the requisite metabolic activation.

In vitro assays for the detection of gene mutations have been in use for over two decades, with the Ames assay[23] being adopted as the de facto gold standard. The standard Ames assay consists of testing a compound against five different strains of *S. typhimurium* in both the presence and absence of S9 metabolic activation in concentrations up to

5000 µg per plate. An increase of greater than twofold in the number of revertant colonies per plate when compared with the control is considered a positive response for that compound. A single positive response for any of the strain/S9 conditions results in an overall positive mutagenicity call for the compound being tested.

While not considered 'high-throughput' by today's standards, the Ames assay is still regarded as one of the most rapid and reproducible assays available in toxicology. An analysis of the inter- and intralaboratory reproducibility of *Salmonella* test results yielded a strict positive-versus-negative concordance of 85%.[24] The speed, cost, and ease of use of this assay has resulted in a large number of unique chemicals being tested and the subsequent publication of their results in the public domain.

This rich source of data combined with the regulatory implications of having an Ames-positive result make mutagenicity an obvious target for the development of in silico models to predict the outcome of this assay. Because of the relatively simple mechanisms by which point mutations can occur, this toxicological endpoint renders itself amenable to all types of model development. Indeed, there are many commercial systems that claim to be able to predict mutagenicity (*see* Section 5.39.4 on commercial systems for more details).

In the late 1980s and early 1990s, Ashby and Tennant and their co-workers published reviews of compounds tested in genotoxicity assays as part of the US National Toxicology Program (NTP).[25–27] This analysis included an assessment of the relationship between chemical structure and its ability to cause genotoxicity. They identified that there were many chemical functional groups (**Figure 3**) that had a strong association with causing genetic toxicity.

These structural alerts have formed the basis of many predictive models for this endpoint, and been used extensively within the pharmaceutical and agrochemical industry to screen out and avoid potentially genotoxic compounds. The unfortunate consequence of this is that any training set based solely on pharmaceuticals or agrochemicals is going to be heavily biased toward Ames negatives, making the development of robust models difficult for these types of compounds. In a review of some 394 pharmaceuticals currently on the market, only 7.2% of compounds were reported to give a positive response in a bacterial mutation assay.[28] Even taking into account that most Ames-positive compounds are not pursued as viable drug candidates, reports suggest that the percentage of compounds synthesized and tested by the industry that are in fact Ames-positive is only around 10%.[14,29] In contrast, the data set assessed by Ashby and Tennant and their co-workers contained approximately 37% Ames positive compounds.

Given that there are substantial differences between the types of chemicals tested by the NTP and those synthesized by the pharmaceutical industry, models based solely on the NTP data set are unlikely to perform well against pharmaceuticals. This observation has been reported on numerous occasions in reference to the commercially available systems for toxicity prediction.[14,28–30]

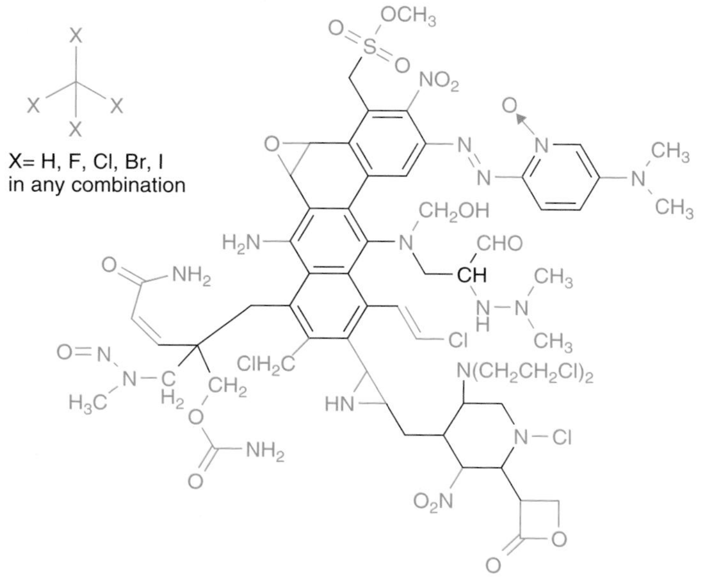

Figure 3 Composite model structure for DNA reactivity.[26]

Some in silico models for mutagenicity prediction take the overall Ames result as a simple binary classification in the development of the model (i.e., either positive or negative). However, other methods have sought to make the results of this assay a more continuous variable and hence amenable to statistical analysis. These methods have introduced the concept of mutagenic potency, and several methods for calculating mutagenic potency have been proposed, each with advantages and disadvantages:

- Method 1 uses the maximum increase in revertants as a measure of mutagenic potency, that is, the potency of a compound increases in relation to the maximum increase in revertants observed in the assay. This method provides an absolute measure of the potency, but does not take into account the potential for cytotoxicity that would limit the number of revertants at higher concentrations.
- Method 2 assumes that compounds that elicit positive responses in multiple-strain/S9 combinations are more potent mutagens than those that only show positive responses under a single condition. This method provides a more objective view of potency, but can become a subjective measure where compounds illicit a reproducible and robust high increase in the number of revertants but only under one or two specific conditions.
- Method 3 uses the drug concentration (e.g., µg per plate) that induces a 2.0-fold increase in the number of revertants, and therefore the more potent the mutagen the lower the concentration required to yield a positive response. Although this method offers the advantage that the concentration of compound is used, consideration of molar concentrations would further improve this approach. However, this method does not take into account the potential for cytotoxicity to impact on the number of revertants.

Often these methods will focus on particular classes of chemicals, developing mathematical relationships between the mutagenic potency and the chemical structures. One important class of potential mutagens – aromatic amines – has been the focus of attention for many researchers working in this area. This is primarily as a result of their suspected link with carcinogenicity through their presence in cooked foods, diesel fumes, and common environmental contaminants.

5.39.3.1.1   Mutagenicity of aromatic amines

The metabolic activation of aromatic amines is complex[31] (**Figure 4**). They can be converted to aromatic amides in a reaction catalyzed by an acetyl coenzyme A (CoA)-dependent acetylase. The acetylation phenotype varies among the population: persons with the rapid acetylator phenotype are at higher risk of colon cancer,[32,33] whereas those who are slow acetylators are at increased risk of bladder cancer.[34] This latter association may result from the fact that activation of aromatic amines by N-oxidation is a competing pathway for aromatic amine metabolism. Also, the N-hydroxylation products when protonated under the acid conditions in the urinary bladder form reactive electrophiles that bind covalently with DNA or proteins to produce macromolecular damage.

An initial activation step for both aromatic amines and amides is N-oxidation by CYP1A2. This cytochrome P450 is also responsible for the 3-demethylation of 1,3,7-trimethylxanthine (i.e., caffeine), the distribution of metabolic phenotypes in the population, as well as the disposition of an individual with respect to CYP1A2 metabolism, is relatively easy to determine.[35] The reaction of N-hydroxy-arylamines with DNA appears to be acid catalyzed, but they can be further activated by either an acetyl CoA-dependent O-acetylase or a 3'-phosphoadenosine-5'phosphosulfate-dependent O-sulfotransferase. The N-arylhydroxamic acids, which arise from the acetylation of N-hydroxy-arylamines or N-hydroxylation of aromatic amides, are not electrophilic; therefore, they require further activation. The predominant pathway for this occurs through acetyltransferase-catalyzed rearrangement to a reactive N-acetoxy-arylamine. Sulfotransferase catalysis results in the formation of N-sulfonyloxy-arylamides. This complex pathway results in two major adduct types: amides (i.e., acetylated) and amines (i.e., nonacetylated).

The heterocyclic amines are formed during the preparation of cooked food, primarily from the pyrolysis (> 150 °C) of amino acids, creatinine, and glucose. Heterocyclic amines have been recognized as food mutagens,[36] and they have been shown to form DNA adducts and cause liver tumors in primates.[37] Compared with other carcinogens, their metabolism is less well understood, but N-hydroxylation is considered to be a necessary step. Because they are similar in structure to the aromatic amines, it is not surprising that they too can be activated by CYP1A2. The N-hydroxy metabolites of 3-amino-1-methyl-5H-pyrido[4,3-b]indole (Trp-P-1), 2-amino-6-methyldipyrido[1,2-a:39,29-d]imidazole (Glu-P-1), and 2-amino-3-methyl-imidazo-[4,5-f]quinoline (IQ) can react directly with DNA. Unlike aromatic amines, however, this reaction is not facilitated by acid pH. Enzymic O-esterification of the N-hydroxy metabolite is important in the activation of these food mutagens, and the N-hydroxy metabolite is also a good substrate for transacetylases. This suggests a possible role for these chemicals in the etiology of colorectal cancer in combination with the rapid acetylator phenotype.

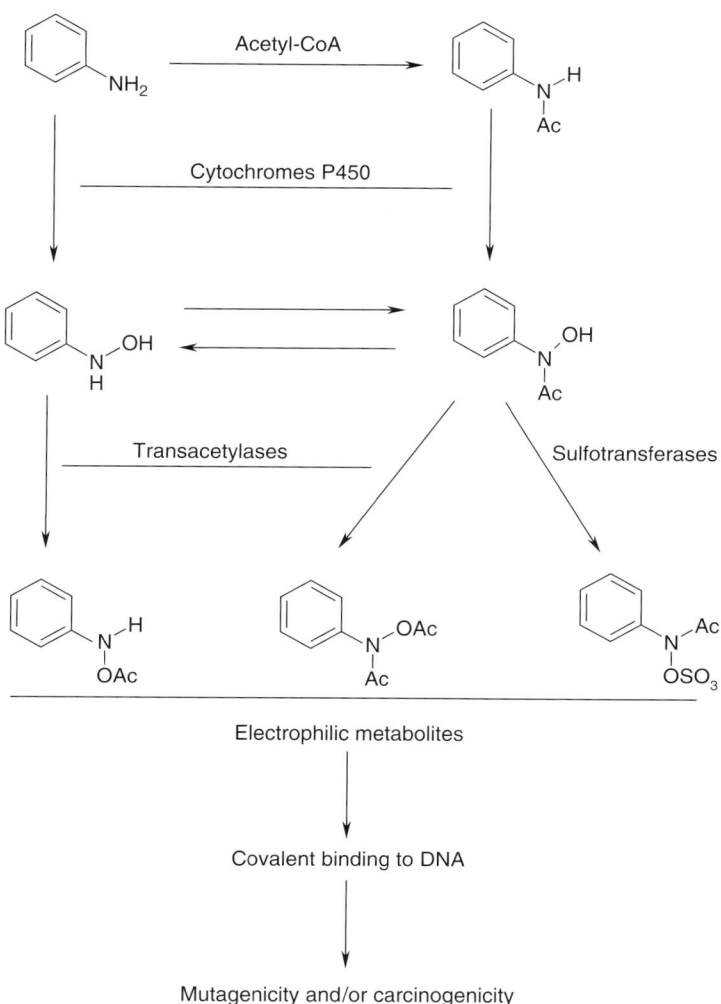

Figure 4 Metabolic activation pathways of aromatic amines.[38]

Through the use of various QSAR techniques, the mutagenic potency of aromatic amines has been correlated to a number of structural descriptors including lipophilicity as represented by the log P value of a chemical,[38] the relative stability of the nitrenium ion,[39] and the HOMO/LUMO calculations for class. However, upon further investigation, these methods, while providing an acceptable mechanism for predicting mutagenic potency, were poor at predicting the activity of nonmutagenic aromatic amines. Work conducted by Benigni et al.[40] showed that both electronic and steric factors at the amine site also play a role in the crucial metabolic activation of this class of compounds. For a comprehensive review of QSARs for aromatic amines, refer to Benigni et al.[41]

5.39.3.1.2 Predictive models for chromosomal damage

In contrast to mutagenicity, very few in silico models have been developed to predict the potential of chemicals to cause chromosomal damage. Chromosomal damage can take various forms, including structural aberrations such as clastogenicity, and/or numerical aberrations, including aneuploidy.[42] There are many assays developed for the detection of these endpoints, including the mouse lymphoma assay, in vitro cytogenetics assays using human lymphocyte cell lines, and in vivo micronucleus assays. Due to their low throughput and higher cost, these assays are often run only when compounds have already been shown to be negative in the Ames assay. Some companies also use higher-throughput screening assays, such as the in vitro micronucleus assay, to eliminate potential bad actors earlier in the drug discovery process. Often these company-generated data remain within corporate databases, and as such have not been subjected to the same level of scrutiny as that for Ames mutagens.

Some data on chromosome damaging agents have been published by Ishidate *et al.*[43] and, more recently, by Ojiama *et al.*[44] Despite this, only a few attempts have been made to develop predictive models for this endpoint. One reason for this could be the diversity and complexity of mechanisms that are involved in chromosomal damage. Another possibility is that comparing results from the different assays makes combining data from these difficult and can cause problems when applying statistical algorithms to develop QSAR models.

Serra *et al.*[45] report some success using topological descriptors and six-descriptor *k*-nearest-neighbor model to predict structural chromosomal aberrations. This model was built using a training set of some 245 chemicals, and was capable of predicting a test set comprised of 37 chemicals – 26 negatives and 11 positives – with an overall accuracy of 86.5%. Use of support vector machine algorithms[46] on the same training and test sets yielded models with an overall accuracy of 83.8%.

Other QSAR models for chromosomal aberrations have been developed by Tafazoli *et al.*,[47] but these were focused on a specific chemical class of compounds, namely short-chain chlorinated hydrocarbons. As such, the application of this model to noncongeneric data sets is inappropriate.

5.39.3.2   In Silico Prediction of Carcinogenicity

Cancer is a disorder in which the mechanisms that control proliferation of cells no longer function adequately. Cancer is considered to be an all-or-none effect (i.e., a tumor is or is not present). Because most cancers are considered irreversible life-threatening changes, chemical carcinogenesis (chemically induced cancer development) has hitherto been perceived by many to be different from other forms of chemical toxicity.

Currently there are no good in vitro assays that can be used to identify carcinogens. Most testing needs to be carried out in vivo, with bioassays for carcinogenicity consisting of exposing adequate numbers of animals to the test substance for a significant proportion of their normal life span. Typically, this is performed using both male and female rats and male and female mice over a 2 year test period. This combination helps to distinguish species and/or gender specificity for the chemical being tested. This test is expensive and time-consuming to conduct, and therefore alternatives or surrogate tests are often employed in advance of carcinogenicity testing, which thereby reduces the number of chemicals for which carcinogenicity data are generated.

One of these surrogate paradigms is genetic toxicity testing, as discussed in the previous section. However, some chemicals can cause cancer by mechanisms other than through direct interaction with DNA,[48,49] and are therefore not detected through the use of genotoxicity tests. These classes of compounds are commonly referred to as nongenotoxic or epigenetic carcinogens. This diversity of mechanisms and high cost of experimental testing has been a significant driving force for the development of in silico models for carcinogenicity.

Despite the high cost to test chemicals, there are in fact many sources of data for carcinogenicity. For example, the International Agency for Cancer Research (IARC) has been compiling and reviewing data on chemicals suspected of causing carcinogenicity in humans for over three decades.[50] Surprisingly there are only 68 unique substances that have been classified as IARC Group 1 Carcinogens (known human carcinogens) and only a further 56 substances categorized as IARC Group 2A Carcinogens (probable human carcinogens). The US NTP has been conducting and reviewing animal toxicity data for chemicals, and to date has compiled carcinogenicity data on over 530 chemicals and mixtures. Other sources of data include the Carcinogenic Potency Database (CPDB), compiled by Gold *et al.*,[51,52] which contains over 1300 chemicals and mixtures, both naturally occurring and synthetic, compiled from public literature sources. It is worth noting that there is significant overlap between the NTP and Gold databases.

A review of the CPDB shows that around 50% of the compounds analyzed were in fact considered to be carcinogenic in rodents.[53] The explanation for this probably lies in the fact that the rodents are given the chemical the maximum tolerated dose (i.e., at chronic, near-toxic doses). Evidence is accumulating to suggest that the high dose itself, rather than the chemical per se, can contribute toward the expression of cancer.[53] The relevance therefore of rodent carcinogenicity testing to the situation where humans are exposed to relatively low levels of these chemicals has to be questioned. The high proportion of positive chemicals has also been attributed to the selective testing of chemicals suspected to cause cancer. One advantage that this bias has yielded is that it makes the data set much more amenable to the development of in silico models.

Many methods have been applied to create predictive models of carcinogenicity with varying degrees of success. While the carcinogenicity endpoint is usually expressed as a binary classification (i.e., positive or negative), some methods have used estimates of carcinogenic potency to differentiate between those that cause cancer at relatively low exposures versus those that require high doses and prolonged exposures. One common method used for determining potency is the use of the TD_{50} approach. The TD_{50} approach was described by Peto *et al.* in 1984.[54] Put simply, the TD_{50} is the calculated daily dose per unit body weight causing a 50% incidence above control of tumors at the most sensitive site.

As with the prediction of genotoxicity, some developers of in silico models have focused their attention on specific classes of chemicals such as aromatic amines, nitroarenes, or nitrosamines. For a comprehensive review of structure-based approaches for predicting mutagenicity and carcinogenicity, refer to Benigni.[55]

5.39.3.2.1 Prospective challenges for carcinogenicity prediction

While most developers use a test set to validate the performance of a model, the true test of its predictivity comes from a prospective analysis, that is, where predictions are made in advance of the experimental tests being conducted. The NTP has issued two such prospective challenges to date for chemicals it planned on testing for rodent carcinogenicity. The first of these exercises took place in 1990, when Tennant et al.[56] published predictions for 44 chemicals that were due to be tested by the NTP, and invited others to do the same.[57] Seven developers in addition to Tenant et al. published their respective predictions prior to completion to testing.[58–64]

In 1993, once testing had been completed, the US National Institute of Environmental Health Sciences conducted an international workshop to evaluate what had been learned from the first predictive exercise. The workshop reached two main conclusions.[65] First, SAR-based models do not perform as accurately as models that utilize biological attributes and, second, models that used multiple attributes to represent the chemical carcinogenicity endpoint performed better than models that were based on one or two attributes.

The comparison between predicted and actual results has been the subject for many discussions and publications.[66,67] Benigni concludes that all systems currently suffer from an inability to distinguish those chemicals that contain a structural alert and are carcinogenic from those that also contain a structural alert but are inactive with the predictions made by the human experts, Ashby and Tennant being the most accurate, getting 75% of the predictions correct (**Table 1**).

In 1996, the NTP issued a second challenge to system developers to predict the outcome for a further 30 chemicals due to be tested for rodent carcinogenicity.[68] As before, most predictions were published prior to the completion of testing.[69–82] In this exercise there was a greater variety in the systems used to predict the outcome of the rodent carcinogenicity studies. They included several human expert opinions, the use of existing or new in vitro assays, as well as the use of computational approaches. This second set of chemicals presented developers with a greater challenge by avoiding chemicals with obvious modes of action or structural alerts. Many chemicals in this second exercise, in contrast to the first, were Ames-negative, implying that carcinogenicity would not be induced through direct interactions with DNA. To compensate for this additional challenge, developers were allowed access to the in vitro and short-term in vivo test results to aid them in making their predictions.

Benigni in his analysis of the results from this second prospective exercise concluded that at present the best systems for prediction still rely to some extent on human judgment and that the upper limit for accuracy from computational approaches is 65%.[55] Overall accuracy results for each system used for prediction are summarized in **Table 2**.

In general, the models for noncongeneric chemicals are inferior in their predictive power when compared with the classical QSARs for congeneric series. This is not entirely surprising, since a QSAR for congeners is aimed at modeling only one mechanistic phenomenon, whereas the general models for the noncongeneric sets try to model several mechanisms of action at the same time, each relative to a whole class of carcinogens. Obviously, this broad-sweeping

Table 1 Concordance between predictions and rodent results

System	Accuracy
Tennant et al.[56]	0.75
Jones and Easterly; RASH[62]	0.68
Bakale and McCreary[64]	0.65
Sanderson and Earnshaw; DEREK[63]	0.59
Benigni[61]	0.58
Sanderson and Earnshaw; DEREK hybrid[63]	0.57
Enslein et al.; TOPKAT[58]	0.57
Lewis et al.; COMPACT[60]	0.54
Rosenkranz and Klopman; MCASE[59]	0.49

Table 2 Accuracy of the QSAR predictions compared with the rodent bioassay results

System	Accuracy
OncoLogic[82]	0.65
SHE[77]	0.65
R1[75]	0.64
Huff et al.[81]	0.62
R2[75]	0.61
Benigni et al.[72]	0.61
Tennant et al.[79]	0.60
Ashby[80]	0.57
Bootman[83]	0.53
FALS[74]	0.50
RASH[70]	0.45
COMPACT[69]	0.43
DEREK[76]	0.43
S. typhimurium[68]	0.33
Purdy[78]	0.32
Progol[71]	0.29
MCASE[73]	0.25

approach is a much more difficult task. It should be noted that the successful modeling of noncongeners requires the availability of an adequate number of representatives for each chemical class or mechanism of action.

When assessing the potential for toxicity, human experts are more likely to be able to make use of a variety of different sources of information, including the general principles of chemistry and biochemistry, and adapt better to issues that require the simultaneous use of knowledge at different hierarchical levels. Obviously, the opposite is true in the case of the individual chemical classes, where the potency of the QSAR methods can model subtle gradations in a much more efficient way than can the human expert.

A further confirmation of the difficulty in crystallizing the SARs into automatic prediction tools comes from the experience of the Predictive Toxicology Challenge 2000–2001. In this exercise, developers of artificial intelligence algorithms were invited to make predictions on the 30 chemicals proposed by the NTP after training their algorithms on standardized training set. Models were developed, and the results submitted to the Predictive Toxicology Challenge website. Out of the 111 original attempts made by these sophisticated artificial intelligence methods to grasp the information contained implicitly in the carcinogenicity databases that were proposed by the artificial intelligence community, only five sets of predictions were generated that were considered to be better than random.[55]

5.39.3.3   In Silico Prediction of Skin Sensitization

Skin sensitization, sometimes referred to as allergic contact dermatitis (ACD), is considered to be one of the largest causes of occupational injury in the USA.[84] It often develops as a result of repeated exposure to a sensitizing chemical agent, and can cause severe and sometimes fatal reactions on subsequent exposure to the chemical. It is thought that skin sensitization occurs when a chemical first reacts with skin proteins either directly or after metabolic activation. In order for this reaction to occur, the chemical must first penetrate the skin at least as far as the viable epidermis. The resulting conjugate is then recognized as a foreign body, and elicits an immune reaction, producing an inflammatory response in the skin.[85] Due to the long-lasting memory of the immune system, once an individual is sensitized, he or she can remain sensitive to further exposure of a chemical for a long period of time.

Experimental techniques for the detection of skin sensitization include the standard guinea pig maximization test, the Buehler test (also performed using guinea pigs), and the more recently developed murine local lymph node assay (LLNA). With the exception of the LLNA, these assays only provide a dichotomous conclusion, that is, the presence or absence of an effect with no measure of the dose–response relationship. The LLNA, however, provides an experimental measure of stimulation at each particular dose that when compared with control animals can be expressed as a stimulation index (SI). For example, SI = 3 is the dose required to give a threefold increase in stimulation between treated and control groups in the LLNA. Indeed, use of the calculated or estimated concentration at which SI = 3 (more often referred to as EC3) has become a common method to report LLNA data in the literature rather than reporting the experimental concentrations and isotope readings. The EC3 can be taken as a quantifier of the sensitization potential that makes this assay more amenable to the application of QSAR techniques.[86]

Despite the fact that none of these assays is considered high throughput, regulatory requirements for occupational health and safety, as well as the need to demonstrate consumer safety for personal products and cosmetics, has meant that many chemicals have been tested in one or more of these experimental assays. In addition, strong public and political opinion on the use of laboratory animals for testing of personal products and cosmetics has put immense pressure on these industries to develop and adopt alternative methods for identifying skin sensitizers. This pressure has resulted in the release into the public domain of large quantities of data that were traditionally held in the internal archives of private corporations. This release of data along with public pressure was the primary driving force for the development of in silico models for the prediction of ACD.

Recent reviews[85,87] of the QSAR models for skin sensitization discuss the many approaches that have been applied in this area. These models range from the application of structural alerts as in the case of the DEREK for Windows (DfW) system[87–90] to more complex structural, topological and E-state descriptors.[86,91,92] Rodford et al.[85] conclude that while a variety of methods have been employed, each model is constrained by a lack of high-quality data that are sufficient to provide a truly global model for the prediction of skin sensitization. However the adoption of the LLNA will go some way to addressing this need, as this assay is less costly and produces quantifiable results compared to those methods utilizing guinea pigs.

In more recently publications, Li et al. have used methods such as random forests, and applied these to the prediction of skin sensitization.[93] The random forest algorithm, originally developed by Breiman,[94] grows a collection, called a forest, of classification trees, and uses these for classifying a data point into one of the classes (see Volume 4). Growing a forest of trees and using randomness in building each classification tree in the forest leads to better predictions compared with a single classification tree, and helps to make the algorithm robust to noise in the data set. Using this approach, Li et al. claim to get an overall accuracy of 83.5% with a sensitivity and specificity of 84% and 83%, respectively.

Other recent developments by Miller et al. have used quantum mechanical calculations to develop SARs for predicting the outcome from the LLNA.[96] This system classifies chemicals as strong, moderate, or weak sensitizers, and the authors claim to be able to correctly classify over 80% of the test set of chemicals. However, it is noted by the authors that this prediction method is not broadly applicable to chemicals that fall outside of its training set. In addition, literature reports on chemicals tested in the LLNA for sensitization have shown that there is often variation in their reported EC3 values, suggesting that a classification scheme may improve the accuracy of in silico approaches.

In developing predictive models for skin sensitization it is also important that the ability of a compound to penetrate the skin be taken into account in the model. Permeability coefficients are often expressed in the form log K_p, where the units are in centimeters per hour. In vitro permeability coefficients for human skin are calculated by measuring the flux across a known area of isolated human skin under a known concentration. Subsequently, many QSAR models have been developed for the calculation of this parameter. For an excellent review of these models refer to Walker et al.[97]

In general, skin absorption seems to be primarily dictated by the lipophilicity, molecular size, and solubility of the compound. Often, the log P value is used to estimate the lipophilicity, and molecular weight or volume can be used as a measure of size. Some researchers have used the melting point as an estimate of the solubility; however, to date there are no good methods for the structure-based calculation of melting point. Some commercial rule-based systems, such as DfW (see Section 5.39.4), have incorporated skin permeability estimates to help improve the performance of the models for skin sensitization, but to date there have been no reports on the success or failure of this technique.[14]

5.39.3.4   In Silico Prediction of Other Toxicological Endpoints

5.39.3.4.1   QT prolongation

QT prolongation, in of itself, is not considered to be a toxicity; however, this phenomenon has been observed in human subjects just prior to the onset of cardiac arrhythmias. This co-occurrence has led to a postulated link between QT and these often-fatal cardiac events, and, for this reason, drug-induced QT prolongation or long QT syndrome has recently been a major cause of drug withdrawals.

One of the primary causes of QT prolongation is thought to be blockage of the hERG potassium channel in cardiac myocytes. While the x-ray crystal structure of the hERG channel has not yet been elucidated, there are examples of bacterial potassium channel structures in the public domain. These structures have been used as the basis for some 3D predictive models in an attempt to identify chemicals capable of causing QT prolongation.[98–101] For a more detailed explanation of models for QT prolongation and its impact on human health (*see* 5.40 In Silico Models to Predict QT Prolongation).

While hERG plays a significant role in the expression of QT prolongation, there are some 14 other ion channels that can also impact on the adverse event in vivo. The complexity has driven some developers to apply a more holistic systems biology approach to the prediction of QT prolongation, creating complex mathematical models of the processes at a cellular level.

5.39.3.4.2 Respiratory sensitization

Respiratory sensitization is often characterized by episodes of wheezing, coughing, and chest tightness.[102] The condition can be severe and sometimes fatal.[103] The mechanisms by which chemicals can induce this adverse response are uncertain, but are assumed to be immune mediated with some pharmacological and neurological involvement. Currently there are no accepted, well-validated in vitro or in vivo models for the detection of respiratory sensitizers.[85] However, preliminary work by Dearman and co-workers has shown, for a limited number of respiratory sensitizers at least, that these chemicals all elicit a response in the LLNA used for the detection of allergic contact dermatitis.[104,105] Therefore, the suggestion can be made that chemicals that do not cause ACD will be unlikely to cause respiratory sensitization. However, there are significant differences between the mechanisms of these adverse effects that mean that not all contact allergens will necessarily be respiratory allergens.[85]

Despite a lack of accepted models, a review of the available data by Payne *et al.* did yield some structural features (e.g., isocyanates, halogenated diazines and triazines, β-lactams, and diazonium salts) that are thought to be important in the expression of respiratory sensitization.[106]

5.39.3.4.3 Phospholipidosis

Phospholipidosis, a phospholipid storage disorder, is defined by an excessive accumulation of intracellular phospholipids and the concurrent development of concentric lamellar bodies. Phospholipids are structural components of mammalian cytoskeleton and cell membranes. The metabolism of this essential cell component is regulated by the individual cell, and may be altered by drugs that interact with phospholipids or the enzymes that affect their metabolism.[107] Phospholipidosis can occur in multiple target organs, including the liver, kidney, brain, and lungs. While it is considered a disruption of normal cellular function, the long-term implications for human health remain an ongoing debate in the scientific community.[108] Drugs known to induce phospholipidosis all share several common physiochemical properties: a hydrophobic ring structure on the molecule and a hydrophilic side chain with a charged cationic amine group, hence the class term 'cationic amphiphilic drugs' (CADs).[107] Amiodarone (**Figure 5**) is a classic example of a CAD that has been well studied in the public literature and is known to cause phospholipidosis. To date, in silico models for the prediction of this toxicological endpoint have focused primarily on the use of physicochemical descriptors such as pK_a and log P to distinguish between positive and negative structures. Ploemen *et al.* have demonstrated that these two simple parameters are capable of distinguishing between inducers and noninducers.[109] The heuristics used in this exercise were that for a compound to cause phospholipidosis the calculated log P (ClogP)

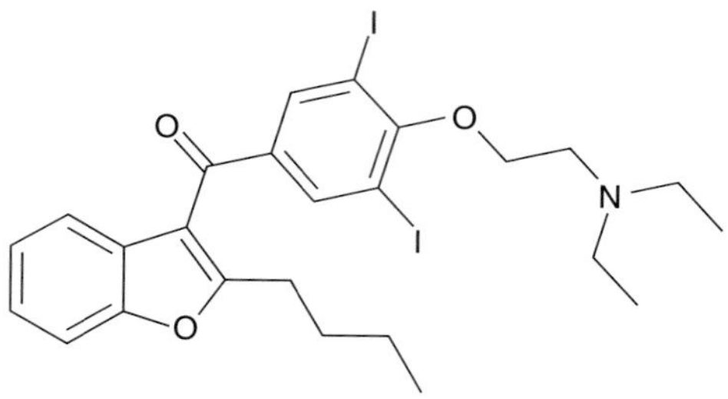

Figure 5 Structure of amiodarone.

must be greater than 1, and the first basic pK_a must be greater than 8. In addition, the combined sum of the squares of each parameter must be greater than 90 for the compound to be classed as positive (see eqn [1]):

$$\text{If} \ (\text{ClogP} > 1) \ \text{AND} \ (\text{p}K_a > 8) \ \text{AND} \ (((\text{ClogP})^2 + (\text{p}K_a)^2) > 90)$$
$$\text{then} \ \text{phospholipidosis} = \text{TRUE} \qquad\qquad [1]$$

An independent validation of this proposed model has not been published, and so independent verification is not available. As with all models, it is advisable that validation studies by third parties independent of the original developers be conducted prior to their broad implementation in a research environment.

5.39.3.4.4 Hepatotoxicity

Despite hepatotoxicity being a primary cause of market withdrawals, few developers have turned their attention to predicting this endpoint. Reasons for this could be because of the insufficient volume of high-quality data required to address the problem or because hepatotoxicity covers a multitude of complex endpoints that make relationships to chemical properties difficult to tease out from the available data. To date, only two developers, Lhasa Limited (the developers of DfW) and the Informatics and Computational Safety Analysis Staff (ICSAS) at the US Food and Drug Administration (FDA) have publicly acknowledged that they are working on this toxicological endpoint.[110,111]

Note et al. acknowledge that no systematic attempt has been made to provide comprehensive coverage of hepatotoxicity within the DfW knowledge base. However, efforts are underway to improve the predictive performance of DfW for this endpoint through the compilation and analysis of relevant toxicity data from a range of sources. To date, data have been collected for 1139 compounds, for which at least some evidence of hepatotoxicity has been reported either in humans and/or experimental animals. Efforts to derive new structural alerts or make modifications to existing alerts for hepatotoxicity based on the compound classes in the data set are currently ongoing.[110]

The ICSAS group has compiled a database on human liver adverse effects based on reports of ADRs reported to the regulatory authority. Of the 631 compounds contained in this data set, 490 have ADR data available for one or more of the 47 liver effects COSTAR term endpoints. The liver adverse effect data from the five liver enzyme increase endpoints were used to optimize the database for identification of a significant adverse effect. Matthews et al. have used the MultiCASE (MCASE) algorithms to develop predictive models based on these data.[111]

5.39.3.4.5 Reproductive and developmental toxicity

Reproductive toxicology is where exposure to a toxin causes adverse effects on either the male or female reproductive system. Developmental toxicity is where exposure to a chemical before conception (by either parent), during prenatal development, or postnatally until the time of puberty, causes the impairment of the normal growth and development of the offspring. In addition, another consideration is teratogenicity, where exposure to a chemical between conception and birth causes malformations in the offspring.

While this type of toxicity is not a common cause for withdrawal of drugs from the market, it can severely limit the population to which drugs can be administered, and hence impacts on their profitability. This fact, along with increasing concern over environmental exposure to chemicals, is driving the industry and regulators to seek ways to rapidly identify compounds that demonstrate this liability.

To date, only a few attempts have been made to develop in silico models for these toxicological endpoints. Gombar et al. compiled a database of 273 compounds that had been tested in rodents and reported to cause maternal toxicity or fetal anomalies that passed their criteria for data quality and structural inclusion.[112] These compounds were further subdivided into three chemical classes, aliphatic, carboaromatic, and heteroaromatic compounds, with a balanced number of positive and negatives totaling around 90 chemicals in each. The TOPKAT models use electrotopological descriptors as defined by Hall and Kier[113] to relate the adverse effects to the chemical structure. Using this approach, the developers claim, using a leave-one-out jackknife test, to get models with specificities ranging from 86% to 97%, and sensitivities between 86% and 89% for these chemical classes. The authors go on to say that in an evaluation of 18 pharmaceuticals only five had developmental toxicity data available, of which only three were considered to be inside the predictive space of the models although all three were correctly identified. It is worth noting that the training sets for these models are relatively small and that the independent test set is insufficient to accurately gauge their broad applicability or the reliability of the predictions from these models.

Ghanooni et al. developed a model using the MCASE technology[114] (see Section 5.39.4 on commercial systems for more details). They used a data set of some 323 compounds retrieved from the Teratogen Information System and a database used by the FDA.[115,116] Using an n-fold cross-validation, by withholding 10% of the data from the training set, yielded models with a pooled concordance of 72.6%. Gómez *et al.* also used MCASE to look at compounds shown to cause developmental toxicity in hamsters.[117] They assembled a data set of 192 chemicals taken from Schardein[118] and an additional 42 physiological chemicals that were assumed to not cause toxicity at physiological concentrations. The authors claim that the overall predictability of the model is 74%, although there is an insufficient number of compounds available for a truly external validation of the model.

In a report from an International Life Science Institute working party, Julien *et al.*[9] conclude that after considering two SAR systems (TOPKAT and MCASE) that have been applied to developmental toxicity illustrates the difficulties in predictive modeling of this type of toxicity. The difficulties stem from the fact that developmental toxicity encompasses a wide range of endpoints, yet there is little knowledge of the mechanisms underlying these diverse endpoints. Thus, in silico models for developmental toxicity must rely primarily on using a statistical approach. With this type of methodology, the activity (or inactivity) of each compound in the training set must be captured as a score that can be correlated with the presence or absence of chemical structural features. This requirement poses at least two major challenges: (1) defining the effect on which compounds are scored and (2) having an objective, rational, reproducible and transparent process for scoring compounds on that activity. Given the wide range of 'effects' encompassed by developmental toxicity and the complexity in evaluating the data, it is very difficult to score activities in a way that makes biological sense and also allows statistical analysis.

5.39.4　Commercially Available Systems for Toxicity Prediction

5.39.4.1　Major Systems and Suppliers

There are many commercial systems for the prediction of toxicity available on the market. To date, these systems, in general, fall into one or other of the three categories of models discussed earlier (statistics based, rule based, and 3D). In all cases, each system has advantages and disadvantages, and it is important that the user understands the limitations of each system when using them to help guide decision-making. **Table 3** lists the main systems available with their respective suppliers, system category, and key references for additional information.

For a more comprehensive review of DfW, MCASE, TOPKAT, and HazardExpert, refer to the review by Greene.[14]

5.39.4.2　Comparisons of System Performances

Most evaluations published on commercial systems to date have focused primarily on one or more of the four market-leading systems, TOPKAT, MCASE, DfW, and Hazard Expert. At the Environmental Mutagen Society meeting in 2003, Greene presented a comparison between all four of these systems, based on a proprietary data set of pharmaceutical-like compounds that had been tested in the Ames assay.[128] The versions of each system used and the criteria used to interpret the results are listed in **Table 4**.

The compounds used in the evaluation consisted of some 974 legacy Pfizer compounds, of which approximately 10% gave a positive response in the Ames assay, and 268 legacy Warner Lambert compounds, of which approximately 15% were Ames-positive. It should be noted that with an evaluation set of compounds that have a high bias toward

Table 3 Major commercial systems for toxicity prediction

System	Developer	Category	Reference
DEREK for Windows (DfW)	Lhasa Limited (www.lhasalimited.org)	Rule based	14,63,88,119
MCASE	MultiCase Inc. (www.multicase.com/)	Statistics based	59,73,120–122
TOPKAT	Accelrys Inc. (www.accelrys.com)	Statistics based	58,91,123
ToxBox	Pharma Algorithms (ap-algorithms.com/)	Hybrid[a]	124
CSGenoTox	ChemSilico (www.chemsilico.com/)	Statistics based	125
HazardExpert	CompuDrug (www.compudrug.com/)	Rule based	126
Know-It-All[b]	BioRad (www.bio-rad.com/)	Hybrid consensus	127

[a] Hybrid systems combine statistical and rule-based approaches.
[b] Know-It-All combines CSGenTox and HazardExpert, to give a consensus model.

Table 4 Commercial systems, versions, and interpretation criteria used in a cross-platform comparison

System	Version	Criteria
TOPKAT	Main software version 6.1 Ames mutagenicity module version 3.1	$p > 0.7$ = positive, $p < 0.3$ = negative, $0.3 < p < 0.7$ = indeterminate
DfW	Version 6.0	*Salmonella* species selected: plausible = positive, equivocal = open, no alert = negative
HazardExpert	Version 5.0	Used default probabilities to determine call
MCASE[a]	Version 3.45 module A2I	CASE units > 19 = positive, CASE units < 19 = negative

[a] No compounds were excluded for unknown fragments or other reasons.

Table 5 Results (as percentages) for DfW, TOPKAT, HazardExpert, and MCASE for the combined set of 1242 compounds

System	Sensitivity	Specificity	Concordance	Predictive value (postive)	Predictive value (negative)	Coverage
DfW	48.8	81.9	78.7	22.5	93.7	100.0
TOPKAT	27.4	83.4	77.6	16.1	90.8	87.7
HazardExpert	75.2	65.1	66.1	18.9	96.1	100.0
MCASE	33.1	84.0	79.0	18.3	92.1	99.9

negatives, models that have a tendency toward false negatives (i.e., predict positive compounds as being negative) will naturally have a high concordance measure. Indeed, in this particular case, predicting all compounds to be negative will give an overall concordance of nearly 90%. However, this type of model provides no value in terms of enabling decisions or prioritizing compounds for development. The combined results from both sets of compounds for each system are listed in **Table 5**.

In **Table 5**, sensitivity is the percentage of Ames-positive compounds correctly predicted; specificity is the percentage of Ames-negative compounds correctly predicted; concordance is the total percentage of compounds correctly predicted; predicted value of a positive is the percentage of compounds predicted positive that were positive; predicted value of a negative is the percentage of compounds predicted negative that were negative; and coverage is the percentage of compounds for which predictions could be made by the system.

While HazardExpert predicted the greatest number of positive compounds correctly, it also gave the highest number of false positives, predicting almost 35% (over 400 compounds) of the Ames-negative compounds to be positive. There was little difference between TOPKAT and MCASE, the two statistical models, in terms of their predictive power, and DfW was not significantly better, although its ability to predict Ames-positive compounds was better than the statistics-based methods.

Comparisons by other researchers have found similar results when comparing the predictions for mutagenicity against the actual Ames assay results for compounds. Cariello *et al.* used a set of some 414 compounds to compare the capabilities of the DfW and TOPKAT systems.[129] Of these 414 compounds, approximately 20% gave a positive response in the Ames assay. Their results indicated that DfW achieved overall concordance of 64.5% whereas TOPKAT achieved a concordance of 73.3%. The authors noted, however, that 26% of the 414 compounds were considered to be outside the prediction space for the TOPKAT model, and therefore the predictions for these compounds were considered unreliable and removed from the analysis.

Similarly, Pearl *et al.* looked at a set of 123 pharmaceutical compounds, of which 49 were Ames-positive, to compare the abilities of DfW, TOPKAT, and MCASE at predicting the results for compounds in the Ames assay.[130] The authors found the overall concordance for each system was 61%, 67%, and 72%, respectively, with DfW having the lowest false-negative rate at 8%. The authors suggest that by combining one or more of these predictive models it is possible to improve upon the concordance of the individual systems. The authors comment that using MCASE to confirm a DfW-positive call produced a combined system with a high overall concordance at 75% while maintaining coverage at 97%. While using all three systems did result in the highest overall concordance at 86%, the coverage of this approached dropped dramatically to 46%.

Table 6 Summary of the performance characteristics of MCASE, DfW, and TOPKAT

	Sensitivity	*Specificity*	*Concordance*
MCASE			
Ames	13/27 (48.1%)	307/330 (93.0%)	320/357 (89.6%)
In vitro cytogenetics	10/47 (21.3%)	174/184 (95%)	184/231 (79.7%)
MLA	7/23 (30.4%)	78/82 (95.1%)	85/105 (80.9%)
In vivo cytogenetics	6/25 (24.0%)	245/267 (91.7%)	251/292 (85.9%)
DfW			
Ames	14/27 (51.9%)	260/346 (75.1%)	274/372 (73.6%)
In vitro cytogenetics	15/47 (31.9%)	139/206 (67.5%)	154/253 (60.8%)
MLA	8/24 (33.3%)	50/85 (58.8%)	58/109 (53.2%)
In vivo cytogenetics	10/30 (33.3%)	194/276 (70.2%)	204/306 (66.7%)
TOPKAT			
Ames	10/23 (43.4%)	267/316 (84.5%)	277/339 (81.7%)
In vitro cytogenetics	11/48 (22.9%)	150/172 (87.2%)	161/220 (73.1%)
MLA	3/24 (12.5%)	67/76 (88.1%)	70/100 (70%)
In vivo cytogenetics	8/28 (28.5%)	219/260 (84.2%)	228/288 (79.2%)

White *et al.* in their comparison of TOPKAT, DfW, and MCASE used a set of some 520 proprietary pharmaceutical compounds that had been tested in the Ames assay.[29] This test set contained approximately 15% Ames-positive compounds, and therefore would favor systems that had a high false-negative rate. In their analysis they found that TOPKAT had an overall concordance of 74%, MCASE, 81%, and DfW, 72%. Interestingly, though, MCASE could only make predictions on 213 compounds (41%). Similarly, TOPKAT could only make predictions for 319 (61%) of the compounds, whereas DfW made predictions for all compounds in the set. The authors note, however, that when the predictions of all three systems are combined and all three agree, the concordance with the Ames assay rises to 95% but the number of compounds for which predictions are made drops to only 15% of the data set.

In their analysis, Snyder *et al.* looked at the genotoxicity of a set of 394 pharmaceutical compounds that are currently on the market.[28] The data for these compounds were collected from the Physicians Desktop Reference (PDR). The final set comprised 7.2% bacterial reverse mutagenesis (Ames) positives (27/375), 18.7% in vitro cytogenetics (primarily chromosome aberration) positives (47/251), 22% MLA positives (24/109), and 10.1% in vivo cytogenetics (primarily rodent micronucleus) positives (31/305). DfW, MCASE, and TOPKAT were used to make predictions, and their respective results are listed in **Table 6**.

Snyder concludes that the relative insensitivity of these computational models to predict Ames and, for that matter, genotoxicity in general may be due to both our incomplete understanding of genotoxic mechanisms and an insufficiency of structural coverage in these test systems. It is important, therefore, to consider the limitations of each system and make appropriate decisions when using the commercial in silico models.

5.39.5   Summary

The use of in silico models to predict toxicity offers some significant advantages over traditional in vitro and in vivo approaches. The models are generally much cheaper and faster on a per compound basis, and capable of providing an assessment of toxicity within a matter of seconds. They also do not require any physical chemical matter, and can therefore be employed even before the compounds have been synthesized. These advantages make the use of computational approaches very attractive to the chemical industry and regulators alike.

However, the identification of toxicity is a complex task, often not being revealed until humans are exposed to a compound. Nevertheless computational methods are able to predict individual toxic endpoints, which, when combined with information on likely exposure (e.g., duration of treatment, dose, and route), can provide useful information to assist in risk assessment.[131]

There are a number of areas in the drug discovery process where predictive techniques may be applied, but the nature of the method will normally define its appropriate use. For instance, methods for genotoxicity prediction may be applicable to screen out compounds from a virtual library with a high probability of toxicity. However, prediction of other toxicological endpoints, such as hepatotoxicity, may be more appropriately placed in the drug optimization stages of the discovery process. Needless to say, methods that are available for large-scale screening must be rapid, computer-based, and able to screen large numbers of compounds in a single run.

It is, however, crucial to know the limitations of in silico techniques and avoid applying them blindly to every instance. In this respect, it is advantageous to use techniques that present the rationales for their predictions in a traceable manner, and that indicate the limitation of their predictions clearly (e.g., compounds that fall beyond the applicability domain of the model or inconclusive evidence for predictions). In the case of unreliable predictions, it is better to perform additional biological experiments to clarify these issues than to trust in silico predictions implicitly.[132]

Significant improvements in the predictive ability of computational models for toxicological endpoints are unlikely to come from improvements in methodology alone, such as the use of better descriptors or new algorithms. The reality is that there is not a simple set of 'Lipinski-like' rules that will answer all questions concerning toxicity. An adverse event can be the culmination of an incredibly complex series of physical, biochemical, and physiological processes. This mechanistic complexity of toxicity coupled with incomplete knowledge is too fundamentally limiting to allow accurate predictions across all chemical space through the use of a single model. Simplistic modeling of such phenomena will only result in trivial models with limited utility.

All computational systems for toxicity prediction, whether rule or QSAR based, are limited by the availability of toxicity data on which to develop, evaluate, and validate models. Indeed, the whole area of predictive toxicology is blighted by the paucity of publicly available, high-quality data for modeling. Much toxicity data are, of course, held within industry archives, and while it is always easy to call for the release of corporate data, unless forced by legislation, large-scale public data release is highly unlikely. However, it should be remembered that commercially sensitive data might be used internally within individual companies to assist in the evaluation, validation, and further development of models.

Improvements in the prediction of toxicity will ultimately come from the more challenging and difficult task of improving our understanding of the mechanisms, and better integration of quantitative modeling QSAR, empirical association, biological mechanisms, and chemical reactivity considerations toward the goal of building mechanistically relevant and valid models across a wide range of mechanisms and chemical classes.[133]

In the future, the use of computer-aided toxicity prediction looks set to expand in the chemical industries and their regulation by government agencies. However, the use of predictive techniques must be tempered by the realization and appreciation of their limitations, and these points require transparency of models and expertise in their application. A more holistic approach to toxicity prediction is required that may include an integrated assessment of absorption, distribution, metabolism, and excretion (ADME), route of exposure, dose, etc. Data used to make an informed assessment may come from a variety of sources, including QSARs, expert systems, and knowledge of the effects of 'similar' chemicals (however similarity may be defined), as well as genomic level in vitro and in vivo experiments. To best address the challenge of structure-based toxicity prediction, we should seek a convergence of modeling approaches toward a more unified approach for the future, one that includes elements of hypothesis generation, data base exploration, statistical association, and existing knowledge. The ideal toxicity prediction method will ultimately be one that builds upon existing data and knowledge using the full range of human and machine capabilities for processing relevant information.[134]

References

1. Kola, I.; Landis, J. *Nature Rev. Drug Disc.* **2004**, *3*, 711–716.
2. Merck. Merck Announces Voluntary Worldwide Withdrawal of VIOXX®. Press release, Sept. 30, **2004**. Merck: Whitehouse Station, NJ.
3. Johannsen, F. R. *Crit. Rev. Toxicol.* **1990**, *20*, 341–367.
4. Richard, A. M. *Knowledge Eng. Rev.* **1999**, *14*, 1–12.
5. Kalgutkar, A. S.; Gardner, I.; Obach, R. S.; Shaffer, C. L.; Callegari, E.; Henne, K. R.; Mutlib, A. E.; Dalvie, D. K.; Lee, J. S.; Nakai, Y. et al. *Curr. Drug Metab.* **2005**, *6*, 161–225.
6. Wallace, J. L. *Br. J. Pharmacol.* **2004**, *143*, 1–2.
7. Benigni, R.; Richard, A. *Methods* **1998**, *14*, 264–276.

8. Brown, V. J. *Environ. Health Perspect.* **2003**, *111*, A766–A769.
9. Julien, E.; White, C. C.; Richard, A. M.; DeSesso, J. M. *Birth Defects Res.* **2004**, *70*, 902–911.
10. Schultz, T. W.; Cronin, M. T. D. *Environ. Toxicol. Chem.* **2003**, *22*, 599–607.
11. Cronin, M. T. D.; Dearden, J. C. *Quant. Struct.–Act. Relat.* **1995**, *14*, 1–7.
12. Fujita, T. In *Comprehensive Medicinal Chemistry*; Hansch, C. Ed.; Pergamon: Oxford, 1990, pp 497–560.
13. Franke, R. *Theoretical Drug Design Methods*; Elsevier: Amsterdam, The Netherlands, 1984.
14. Greene, N. *Adv. Drug. Deliv. Rev.* **2002**, *54*, 417–431.
15. Fausett, L. *Fundamentals of Neural Networks*; Prentice-Hall: Englewood Cliffs, NJ, 1994.
16. Gurney, K. *An Introduction to Neural Networks*; UCL Press: London, 1997.
17. Haykin, S. *Neural Networks,* 2nd edn.; Prentice Hall: Englewood Cliffs, NJ, 1999.
18. Holland, J. H. *Adaptation in Natural and Artificial System*; University of Michigan Press: Ann Arbor, 1975.
19. Goldberg, D. *Genetic Algorithms*; Addison Wesley: Boston, MA, 1988.
20. Cramer, R. D., III.; Patterson, D. E.; Bunce, J. D. *J. Am. Chem. Soc.* **1988**, *110*, 5959.
21. Kim, K. H.; Greco, G.; Novellino, E. In *3D QSAR in Drug Design*; Kubinyi, H., Folkers, G., Martin, Y. C., Eds.; Kluwer: Dordrecht, The Netherlands, 1998; Vol. 3, p 257.
22. Preston, R. J.; Hoffman, G. R. Genetic Toxicology. In *Casarett & Doull's Toxicology: The Basic Science of Poisons*, 6th ed.; Klassen, C. D. Ed.; McGraw-Hill: New York, 2001, Chapter 9, 321 pp.
23. Maron, D. M.; Ames, B. N. *Mutat. Res.* **1983**, *113*, 173–215.
24. Zeiger, E.; Ashby, J.; Bakale, G.; Enslein, K.; Klopman, G.; Rosenkranz, H. S. *Mutagenesis* **1996**, *11*, 471–484.
25. Ashby, J.; Tennant, R. W. *Mutat. Res.* **1988**, *204*, 17–115.
26. Ashby, J.; Tennant, R. W.; Zeiger, E.; Stasiewicz, S. *Mutat. Res.* **1989**, *223*, 73–103.
27. Ashby, J.; Tennant, R. W. *Mutat. Res.* **1991**, *257*, 229–306.
28. Snyder, R. D.; Pearl, G. S.; Mandakas, G.; Choy, W. N.; Goodsaid, F.; Rosenblum, I. Y. *Environ. Mol. Mutag.* **2004**, *43*, 143–158.
29. White, A. C.; Mueller, R. A.; Gallavan, R.; Aaron, H. S.; Wilson, A. G. E. *Mutation Res.* **2003**, *539*, 77–89.
30. Pearl, G. M.; Livingston-Carr, S.; Durham, S. K. *Curr. Topics Med. Chem.* **2001**, *1*, 247–255.
31. Beland, F. A.; Poirier, M. C. DNA Adducts and Carcinogenesis. In *The Pathobiology of Neoplasia*; Sirica, A. E., Ed.; Plenum Press: New York, 1989, pp 57–80.
32. Ilett, K. F.; David, B. M.; Detchon, P.; Castleden, W. M.; Kwa, R. *Cancer Res.* **1987**, *47*, 1466–1469.
33. Lang, N. P.; Chu, D. Z.; Hunter, C. F.; Kendall, D. C.; Flammang, T. J.; Kadlubar, F. F. *Arch. Surg.* **1986**, *121*, 1259–1261.
34. Cartwright, R. A.; Rogers, H. J.; Barham-Hall, D.; Glashan, R. W.; Ahmad, R. A.; Higgins, E.; Kahn, M. A. *Lancet* **1982**, *ii*, 842–845.
35. Butler, M. A.; Iwasaki, M.; Guengerich, F. P.; Kadlubar, F. F. *Proc. Natl. Acad. Sci. USA* **1989**, *86*, 7696–7700.
36. Felton, J. S.; Malfatti, M. A.; Knize, M. G.; Salmon, C. P.; Hopmans, E. C.; Wu, R. W. *Mutat. Res.* **1997**, *376*, 37–41.
37. Adamson, R. H. *Cancer Detect Prev.* **1989**, *14*, 215–219.
38. Trieff, N. M.; Biagi, G. L.; Sadagopa Ramanujam, V. M.; Connor, T. H.; Cantelli-Forti, G.; Guerra, M. C.; Bunce, H., III.; Legator, M. S. *Mol. Toxicol.* **1989**, *2*, 53–65.
39. Ford, G. P.; Griffin, G. R. *Chem. Biol. Interact.* **1992**, *81*, 19–33.
40. Benigni, R.; Andreoli, C.; Giuliani, A. *Environ. Mol. Mutag.* **1994**, *24*, 208–219.
41. Benigni, R.; Giuliani, A.; Franke, R.; Gruska, A. *Chem. Rev.* **2000**, *100*, 3697–3714.
42. Committee on Mutagenicity of Chemicals in Food, Consumer Products and the Environment. *Guidance on a Strategy for Testing of Chemicals for Mutagenicity*; Department of Health: London, UK, 2000.
43. Ishidate, M., Jr.; Harnois, M. C.; Sofuni, T. *Mutat. Res.* **1988**, *195*, 151–213.
44. Ojiama, T.; Hayashi, S.; Matsuoka, A.; *Compilation of Chromosomal Mutation Test Data*; Life Science Information Center: Tokyo, Japan, 1998.
45. Serra, J. R.; Thompson, E. D.; Jurs, P. C. *Chem. Res. Toxicol.* **2003**, *16*, 153–163.
46. Cristianini, N.; Shawe-Taylor, J. *Support Vector Machines and Other Kernel-Based Learning Methods*; Cambridge University Press: Cambridge, UK, 2000.
47. Tafazoli, M.; Baeten, A.; Geerlings, P.; Kirsch-Volders, M. *Mutagenesis* **1998**, *13*, 115–126.
48. Butterworth, B. E. *Mutat. Res.* **1990**, *239*, 117–132.
49. Berry, C. L. Epigenetic Carcinogenesis. In *General and Applied Toxicology*; Ballantyne, B., Marrs, T. C., Turner, P., Eds.; Macmillan: Basingstoke, UK, 1993, pp 979–987.
50. IARC. *IARC Monographs of the Evaluation of Carcinogenic Risks to Humans*; International Agency for Research on Cancer, Lyons, France, **1972–2005**; Vols. 1–91.
51. Gold, L. S.; Manley, N. B.; Slone, T. H.; Rohrbach, L. *Environ. Health Perspect.* **1999**, *107*, 527–600.
52. Gold, L. S.; Manley, N. B.; Slone, T. H.; Rohrbach, L.; Garfinkel, G. B. *Toxicol. Sci.* **2005**, *85*, 747–800.
53. Ames, B. N.; Swirsky Gold, L. *Mutat. Res.* **2000**, *447*, 3–13.
54. Peto, R.; Pike, M. C.; Bernstein, L.; Gold, L. S.; Ames, B. N. *Environ. Health Perspect.* **1984**, *581*, 1–8.
55. Benigni, R. *Chem. Rev.* **2005**, *105*, 1767–1800.
56. Tennant, R. W.; Spalding, J.; Stasiewicz, S.; Ashby, J. *Mutagenesis* **1990**, *5*, 3–14.
57. Parry, J. M. *Mutagenesis* **1990**, *5*, 89.
58. Enslein, K.; Blake, B. W.; Borgstedt, H. H. *Mutagenesis* **1990**, *5*, 305–306.
59. Rosenkranz, H. S.; Klopman, G. *Mutagenesis* **1990**, *5*, 425–432.
60. Lewis, D. F. V.; Ioannides, C.; Parke, D. V. *Mutagenesis* **1990**, *5*, 433–435.
61. Benigni, R. *Mutagenesis* **1991**, *6*, 423–425.
62. Jones, T. D.; Easterly, C. E. *Mutagenesis* **1991**, *6*, 507–514.
63. Sanderson, D. M.; Earnshaw, C. G. *Hum. Exp. Toxicol.* **1991**, *10*, 261–273.
64. Bakale, G.; McCreary, R. D. *Mutagenesis* **1992**, 7, 91–94.
65. Wachsman, J. T.; Bristol, D. W.; Spalding, J.; Shelby, M.; Tennant, R. W. *Environ. Health Perspect.* **1993**, *101*, 444–445.
66. Ashby, J.; Tennant, R. W. *Mutagenesis* **1994**, *9*, 7–15.
67. Benigni, R. *Mutat. Res.* **1997**, *387*, 35–45.
68. Bristol, D. W.; Wachsman, J. T.; Greenwell, A. *Environ. Health Perspect.* **1996**, *104*, 1001–1010.
69. David, F.; Lewis, V.; Ioannides, C.; Parke, D. V. *Environ. Health Perspect.* **1996**, *104*, 1011–1016.

70. Jones, T. D.; Easterly, C. E. *Environ. Health Perspect.* **1996**, *104*, 1017–1030.
71. King, R. D.; Srinivasan, A. *Environ. Health Perspect.* **1996**, *104*, 1031–1040.
72. Benigni, R.; Andreoli, C.; Zito, R. *Environ. Health Perspect.* **1996**, *104*, 1041–1044.
73. Zhang, Y. P.; Sussman, N.; Macina, O. T.; Rosenkranz, H. S.; Klopman, G. *Environ. Health Perspect.* **1996**, *104*, 1045–1050.
74. Moriguchi, I.; Hirano, H.; Hirono, S. *Environ. Health Perspect.* **1996**, *104*, 1051–1058.
75. Lee, Y.; Buchanan, B. G.; Rosenkranz, H. S. *Environ. Health Perspect.* **1996**, *104*, 1059–1064.
76. Marchant, C. A. *DEREK Collaborative Group. Environ. Health Perspect.* **1996**, *104*, 1065–1074.
77. Kerckaert, G. A.; Brauninger, R.; LeBoeuf, R. A.; Isfort, R. J. *Environ. Health Perspect.* **1996**, *104*, 1075–1084.
78. Purdy, R. *Environ. Health Perspect.* **1996**, *104*, 1085–1094.
79. Tennant, R. W.; Spalding, J. *Environ. Health Perspect.* **1996**, *104*, 1095–1100.
80. Ashby, J. *Environ. Health Perspect.* **1996**, *104*, 1101–1105.
81. Huff, J.; Weisburger, E.; Fung, V. A. *Environ. Health Perspect.* **1996**, *104*, 1105–1112.
82. Woo, Y. T.; Lai, D. Y.; Arcos, J. C.; Argus, M. F.; Cimino, M. C.; DeVito, S.; Keifer, L. *J. Environ. Sci. Health* **1997**, *C15*, 139.
83. Bootman, J. *Environ. Mol. Mutag.* **1996**, *27*, 237.
84. NIOSH. Worker Health Chartbook, 2000. Nonfatal Illness DHHS (NIOSH) Publication No 2002–120.; National Institute for Occupational Safety and Health: Atlanta GA, 2002.
85. Rodford, R.; Patlewicz, G.; Walker, J. D.; Payne, M. P. *Environ. Toxicol. Chem.* **2003**, *22*, 1855–1861.
86. Estrada, E.; Patlewicz, G.; Chamberlain, M.; Basketter, D.; Larbey, S. *Chem. Res. Toxicol.* **2003**, *16*, 1226–1235.
87. Hostynek, J. J.; Magee, P. S.; Maibach, H. I. *Curr. Probl. Dermatol.* **1996**, *25*, 18–27.
88. Ridings, J. E.; Barratt, M. D.; Cary, R.; Earnshaw, C. G.; Eggington, C. E.; Ellis, M. K.; Judson, P. N.; Langowski, J. J.; Marchant, C. A.; Payne, M. P. et al. *Toxicology* **1996**, *106*, 267–279.
89. Barratt, M. D.; Langowski, J. J. *J. Chem. Inf. Comput. Sci.* **1999**, *39*, 294–298.
90. Gerner, I.; Barratt, M. D.; Zinke, S.; Schlegel, K.; Schlede, E. *ATLA* **2004**, *32*, 487–509.
91. Enslein, K.; Gombar, V. K.; Blake, B. W.; Maibach, H. I.; Hostynek, J. J.; Sigman, C. C.; Bagheri, D. *Food Chem. Toxicol.* **1997**, *35*, 1091–1098.
92. Fedorowicz, A.; Zheng, L.; Singh, H.; Demchuk, E. *Int. J. Mol. Sci.* **2004**, *5*, 56–66.
93. Li, S.; Fedorowicz, A.; Singh, H.; Soderholm, S. C. *J. Chem. Inf. Model.* **2005**, *45*, 952–964.
94. Breiman, L. *Machine Learn.* **2001**, *45*, 5–32.
96. Miller, M. D.; Yourtee, D. M.; Glaros, A. G.; Chappelow, C. C.; Eick, J. D.; Holder, A. J. *J. Chem. Inf. Model.* **2005**, *45*, 924–929.
97. Walker, J. D.; Rodford, R.; Patlewicz, G. *Environ. Toxicol. Chem.* **2003**, *22*, 1870–1884.
98. Pearlstein, R. A.; Vaz, R. J.; Kang, J.; Chen, X.-L.; Preobrazhenskaya, M.; Shchekotikhin, A. E.; Korolev, A. M.; Lysenkova, L. N.; Miroshnikova, O. V.; Hendrixa, J. et al. *Bioorg. Med. Chem. Lett.* **2003**, *13*, 1829–1835.
99. Aronov, A. M.; Goldman, B. B. *Bioorg. Med. Chem.* **2004**, *12*, 2307–2315.
100. Aptula, A. O.; Cronin, M. T. *Sar Qsar Environ. Res.* **2004**, *15*, 399–411.
101. Ekins, S. *Biochem. Soc. Trans.* **2003**, *31* (Pt 3), 611–614.
102. Dearman, R. J.; Basketter, D. A.; Kimber, I. *J. Appl. Toxicol.* **1992**, *12*, 317–322.
103. Fabbri, L. M.; Danieli, D.; Crescioli, S.; Bevilacqua, P.; Meli, S.; Saetta, M.; Mapp, C. E. *Am. Rev. Respir. Dis.* **1988**, *137*, 1494–1498.
104. Dearman, R. J.; Kimber, I.; Crevel, R. W. R.; Basketter, D. A. *Contact Dermatitis* **2002**, *46*, 42.
105. Dearman, R. J.; Kimber, I. *J. Appl. Toxicol.* **2001**, *21*, 153–163.
106. Payne, M. P.; Walsh, P. T. Proceedings of the British Toxicological Society Meeting, York, UK, April **1996**.
107. Halliwell, W. H. *Toxicol. Pathol.* **1997**, *25*, 53–60.
108. Reasor, M. J.; Kacew, S. *Exp. Biol. Med.* **2001**, *226*, 825–830.
109. Ploemen, J. P.; Kelder, J.; Hafmans, T.; van de Sandt, H.; van Burgsteden, J. A.; Saleminki, P. J.; van Esch, E. *Exp. Toxicol. Pathol.* **2004**, *55*, 347–355.
110. Note, R.; Patel, M.; Marchant, C.; Greene, N. *Toxicol. Pathol.* **2005**, *33*, 185–203.
111. Matthews, E. J.; Kruhlak, N. L.; Weaver, J. L.; Benz, R. D.; Contrera, J. F. *Curr. Drug Disc. Technol.* **2004**, *1*, 243–254.
112. Gombar, V. K.; Enslein, K.; Blake, B. W. *Chemosphere* **1995**, *31*, 2499–2510.
113. Hall, H. L.; Mohney, B.; Kier, L. B. *J. Chem. Inf. Comput. Sci.* **1991**, *31*, 76–82.
114. Ghanooni, M.; Mattison, D. R.; Zhang, Y. P.; Macina, O. T.; Rosenkranz, H. S.; Klopman, G. *Am. J. Obstet. Gynecol.* **1997**, *176*, 799–805.
115. Friedman, J. M.; Little, B. B.; Brent, R. I.; Cordero, J. F.; Hanson, J. W.; Shepard, T. H. *Obstet. Gynecol.* **1990**, *75*, 594–599.
116. Briggs, G. G.; Freeman, R. K.; Yaffe, S. J. *Drugs in Pregnancy and Lactation*, 3rd ed.; Williams & Wilkins: Baltimore, 1990.
117. Gomez, J.; Macina, O. T.; Mattison, D. R.; Zhang, Y. P.; Klopman, G.; Rosenkranz, H. S. *Teratology* **1999**, *60*, 190–205.
118. Schardein, J. L. *Chemically Induced Birth Defects*; Marcel Dekker: New York, 1993.
119. Greene, N.; Judson, P. N.; Langowski, J. J.; Marchant, C. A. *SAR QSAR Environ. Res.* **1998**, *10*, 299–314.
120. Klopman, G. *J. Am. Chem. Soc.* **1984**, *106*, 7315–7324.
121. Klopman, G. *Quant. Struct. –Act. Relat.* **1992**, *11*, 176–184.
122. Rosenkranz, H. S.; Klopman, G. *Environ. Mol. Mutag.* **1993**, *21*, 193–206.
123. Enslein, K. *In vitro Toxicol.* **1993**, *6*, 163–169.
124. Zmuidinavicius, D.; Japertas, P.; Didziapetris, R.; Petrauskas, A. *Curr. Topics Med. Chem.* **2003**, *3*, 1301–1314.
125. Votano, J. R.; Parham, M.; Hall, L. H.; Kier, L. B.; Oloff, S.; Tropsha, A.; Xie, Q.; Tong, W. *Mutagenesis* **2004**, *19*, 365–377.
126. Smithing, M. P.; Darvas, F. HazardExpert: An Expert System For Predicting Chemical Toxicity. In *Food Safety Assessment*; Finlay, J. W., Robinson, S. F., Armstrong, D. J., Eds.; American Chemical Society: Washington, DC, 1992, pp 191–200.
127. Banik, G. *Curr. Drug Disc.* **2004**, May, 31–34.
128. Greene, N. *Environ. Mol. Mutagen.* **2003**, *41*, 218.
129. Cariello, N. F.; Wilson, J. D.; Britt, B. H.; Wedd, D. J.; Burlinson, B.; Gombar, V. *Mutagenesis* **2002**, *17*, 321–329.
130. Pearl, G. M.; Livingston-Carr, S.; Durham, S. K. *Curr. Topics Med. Chem.* **2001**, *1*, 247–255.
131. Cronin, M. *Business Brief. Pharmagenerics* **2003**, 50–52.
132. Helma, C. *Curr. Opin. Drug Disc. Devel.* **2005**, *8*, 27–31.
133. Richard, A. M. *Toxicol. Lett.* **1998**, *102–103*, 611–616.
134. Richard, A. M. *Mutat. Res.* **1998**, *400*, 493–507.

Biography

Nigel Greene received his BSc in chemistry and computational science from the University of Leeds in the UK in 1991. He was awarded his PhD also from the University of Leeds in 1994, where his work focused on the synthesis of organophosphorus ligands and transition metal complexes for the catalysis of acetic acid production. Following on from his PhD, Greene worked at Lhasa Ltd in the UK, promoting the use and development of the DEREK for Windows computer program for toxicity prediction and METEOR for metabolite prediction. In 1999, after 5 years with Lhasa Ltd, he then moved to Tripos Inc., where he promoted the use of the SYBYL molecular modeling software suite within the pharmaceutical industry. Since February 2001, Greene has worked for Pfizer Inc. in their World Wide Safety Sciences Department, heading a computational toxicology group whose aim is to use in silico methods for the early identification and screening of new molecular entities in drug discovery and development.

Comprehensive Medicinal Chemistry II
ISBN (set): 0-08-044513-6

ISBN (Volume 5) 0-08-044518-7; pp. 909–932

5.40 In Silico Models to Predict QT Prolongation

A M Aronov, Vertex Pharmaceuticals Inc., Cambridge, MA, USA

5.40.1 Introduction

Recent years have witnessed a revolution in philosophical approaches to drug discovery. Spiraling R&D costs along with higher failure rates of clinical candidates have contributed to the growing recognition by the drug discovery community that undesirable drug properties and toxicity are two of the major contributors to pipeline attrition. Owing to this awareness, many organizations have moved to incorporate predictive absorption, distribution, metabolism, excretion, and toxicity (ADMET) into early discovery programs under the slogan of 'fail fast, fail cheap.' By identifying and removing undesirable leads early, the research organizations should be able to prevent more costly failures downstream. Wider availability of commercial libraries, internal combinatorial chemistry capabilities, and ever-increasing throughput of high-throughput screening (HTS) campaigns have been providing medicinal chemists with multiple potential starting points, and the application of predictive ADMET is seen as a way to narrow the scope of lead exploration.

Originally the domain of experimentalists, ADMET began to change in the late 1990s, embracing the use of computational tools for predicting ADMET properties of compounds. As broader and higher quality proprietary data sets became available, computational approaches have advanced from 'first-generation' models, such as Lipinski's Rule of Five,[1] and in a broader sense drug-likeness[2,3] and lead-likeness,[4,5] to 'second-generation' models that focus primarily on predicting pharmacokinetic parameters, such as solubility, intestinal permeability, oral bioavailability, and blood–brain barrier penetration.[6,7] The 'third-generation' models[6] have taken aim at major determinants of metabolic fate of xenobiotics (cytochrome P450s)[8] and factors that induce them (pregnane X-receptor),[9] drug transport proteins affecting compound PK profile (P-glycoprotein),[10] and, more recently, ion channels implicated in QT interval prolongation (hERG).[11] It is this predictive cardiotoxicity screening that is the focus of this chapter.

It would be impossible to incorporate all of the information regarding QT prolongation in a single review chapter, and this was not the intended goal. The primary focus in this review is on in silico tools for predicting drug-induced QT prolongation, and structural lessons that have been derived in an effort to bias newly synthesized ligands away from the undesirable side effect of cardiotoxicity. The field of hERG modeling is very young and active, with most articles on the topic published since 2002. The most recent reviews published on the general topic of QT prolongation are by De Ponti et al.,[12] and Abriel et al.,[13] while Redfern et al.[14] is required reading for anyone interested in the acceptable historical safety windows for QT prolongation. Reviews by Mitcheson and Perry[15] and Sanguinetti and Mitcheson[16] cover the biomolecular underpinnings of hERG blockade. Finally, the chemical nature of hERG blockers is discussed by Pearlstein et al.,[17] Aronov,[18] and Recanatini et al.[19]

5.40.2 Human Ether-a-go-go Related Gene (hERG)-Related Cardiotoxicity

Sudden death as a side effect of the action of non-antiarrhythmic drugs is a major pharmacological safety concern facing the pharmaceutical industry and the health regulatory authorities.[20] In recent years, at least five blockbuster drugs (**Figure 1**) have been withdrawn from the market due to reports of sudden cardiac death, and a number of others were forced to carry strong 'black box' warning labels.[11,17,20] In all cases, long QT syndrome (LQTS), an abnormality of cardiac muscle repolarization that is characterized by the prolongation of the QT interval in the electrocardiogram, was implicated as a predisposing factor for torsades de pointes, a polymorphic ventricular tachycardia that can spontaneously degenerate to ventricular fibrillation and cause sudden death.

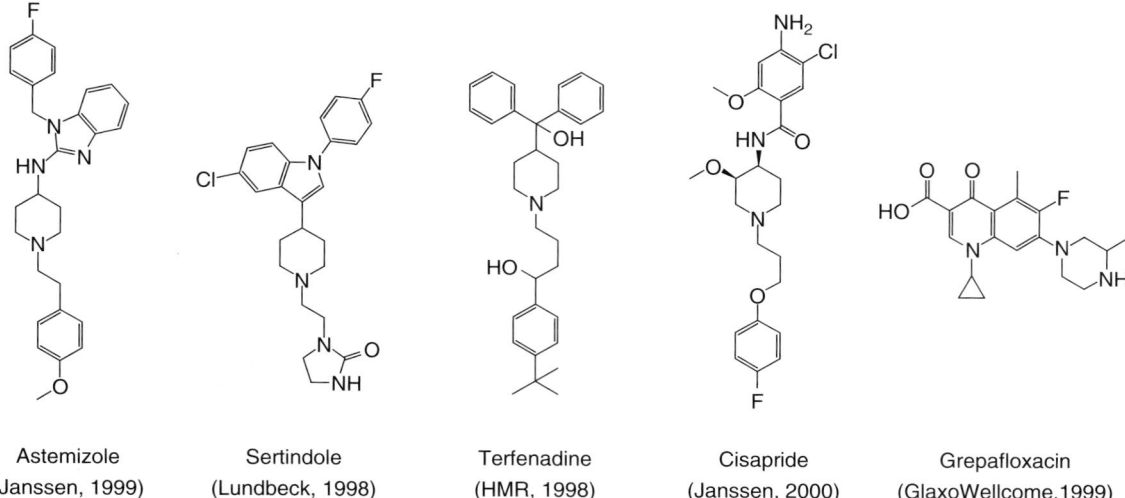

| Astemizole | Sertindole | Terfenadine | Cisapride | Grepafloxacin |
| (Janssen, 1999) | (Lundbeck, 1998) | (HMR, 1998) | (Janssen, 2000) | (GlaxoWellcome,1999) |

Figure 1 Drugs withdrawn due to QT prolongation concerns. (Reproduced with permission from Aronov, A. M. *Drug Disc. Today* **2005**, *10*, 149–155 © Elsevier.)

5.40.2.1   QT Interval and QT Prolongation

Cardiac action potential consists of four distinct phases (**Figure 2a**). In phase 0, upstroke occurs due to rapid transient influx of Na^+. Later, Na^+ channels are inactivated, combined with a transient efflux of K^+. In phase 2, also known as the plateau phase, the efflux of K^+ and the influx of Ca^{2+} are counterbalanced. At the end of the plateau, sustained repolarization occurs due to K^+ efflux via the delayed rectifier K^+ channels exceeding Ca^{2+} influx; this constitutes phase 3 of the action potential. Finally, as part of phase 4, resting potential in myocytes is maintained.

In the clinical setting, the QT interval is measured from the beginning of the QRS complex to the end of the T wave (**Figure 2b**). The QRS complex of the electrocardiogram corresponds to the action potential depolarization, while the T wave is associated with ventricular repolarization. Torsades de pointes is associated with the twisting of the QRS complex around the isoelectric line on the electrocardiogram. Since QT interval depends to a large degree on the heart rate, it is typically reported as QTc, a value normalized for a heart rate of 60 bpm. Bazzett's formula[21] is frequently used to introduce the heart rate correction. For reasons that are as yet unknown, QTc in adult females is about 20 ms longer than that in males. The normal limits of QTc in adults are 430 ms in males and 450 ms in females.[13] An increase of up to 20 ms is considered borderline, while longer QTc values correspond to prolonged QT interval.

5.40.2.2   Molecular Basis of QT Prolongation

Congenital LQTS can be traced back to a number of possible mutations resulting in defects in sodium channels, and two different potassium channels – the rapidly activating delayed rectifier I_{Kr} and the slowly activating delayed rectifier I_{Ks}.[22,23] The primary effect the mutations have on the length of the QT interval is due to the role many K^+ channels play in terminating the plateau phase of the action potential. Any delay in phase 3 repolarization manifests itself in QT interval prolongation. Mutations occur both in genes encoding α-subunits (KvLQT1, LQT1; hERG, LQT2; SCN5A, LQT3, etc.) and β-subunits (KCNE1, LQT5; KCNE2, LQT6) of ion channels.[13,16,23,24] Mutations in these genes are estimated to occur in ~ 1 in 5000 individuals worldwide and cause an estimated 3000 deaths annually in the US.[16]

5.40.2.3   Role of Human Ether-a-go-go Related Gene in QT Prolongation

Virtually every case of a prolonged duration of cardiac action potential related to drug exposure (acquired LQTS) can be traced to one specific mechanism – blockade of I_{Kr} current in the heart.[17] This current, a major contributor to phase 3 repolarization at the end of the QT interval, is conducted by tetrameric pores with the individual subunits encoded by the human ether-a-go-go related gene (hERG).[15] Not all hERG channel blockers induce torsades de pointes as they can also modulate other channels that counteract the effect of hERG blockade. However, the blockade of hERG K^+ channels is widely regarded as the predominant cause of drug-induced QT prolongation and an important indicator of potential pro-arrhythmic liability, making early detection of compounds with this undesirable side effect an important objective in the pharmaceutical industry.

Perhaps owing to the unique shape of the ligand-binding site and its hydrophobic character, the hERG channel has been shown to interact with pharmaceuticals of widely varying structure, often at concentrations similar to the levels of on-target activity of the respective compounds. While risk tolerance for QT prolongation may vary significantly depending on indication, development stage, etc., the documented hERG-blocking activity reduces the intrinsic value of the molecule, as it increases risk of clinical failure. Among the indications least tolerant of hERG blockade by the candidate compound are antivirals and antibacterials where high plasma concentrations of the drug are necessary to suppress resistance, and pain management drugs and antipsychotics where overdosing is likely. In addition, drug–drug interactions (e.g., via inhibition of P450 metabolism) may lead to unexpectedly high plasma levels of a drug, which are sufficient to prolong the QT interval. As a general rule, a safety window of greater than 30-fold has been recommended for the ratio of hERG IC_{50} to the expected compound C_{max} adjusted for its unbound fraction.[14] This ratio could be used as a 'go-no go' decision point in early discovery.

In recent years, a number of in silico approaches have attempted to mitigate the safety concern represented by hERG channel blockade. Some of these approaches have been aimed primarily at filtering out potential hERG blockers in the context of virtual libraries, while others have involved understanding structure–activity relationships governing hERG–drug interactions. This chapter summarizes the most recent efforts in this emerging field.

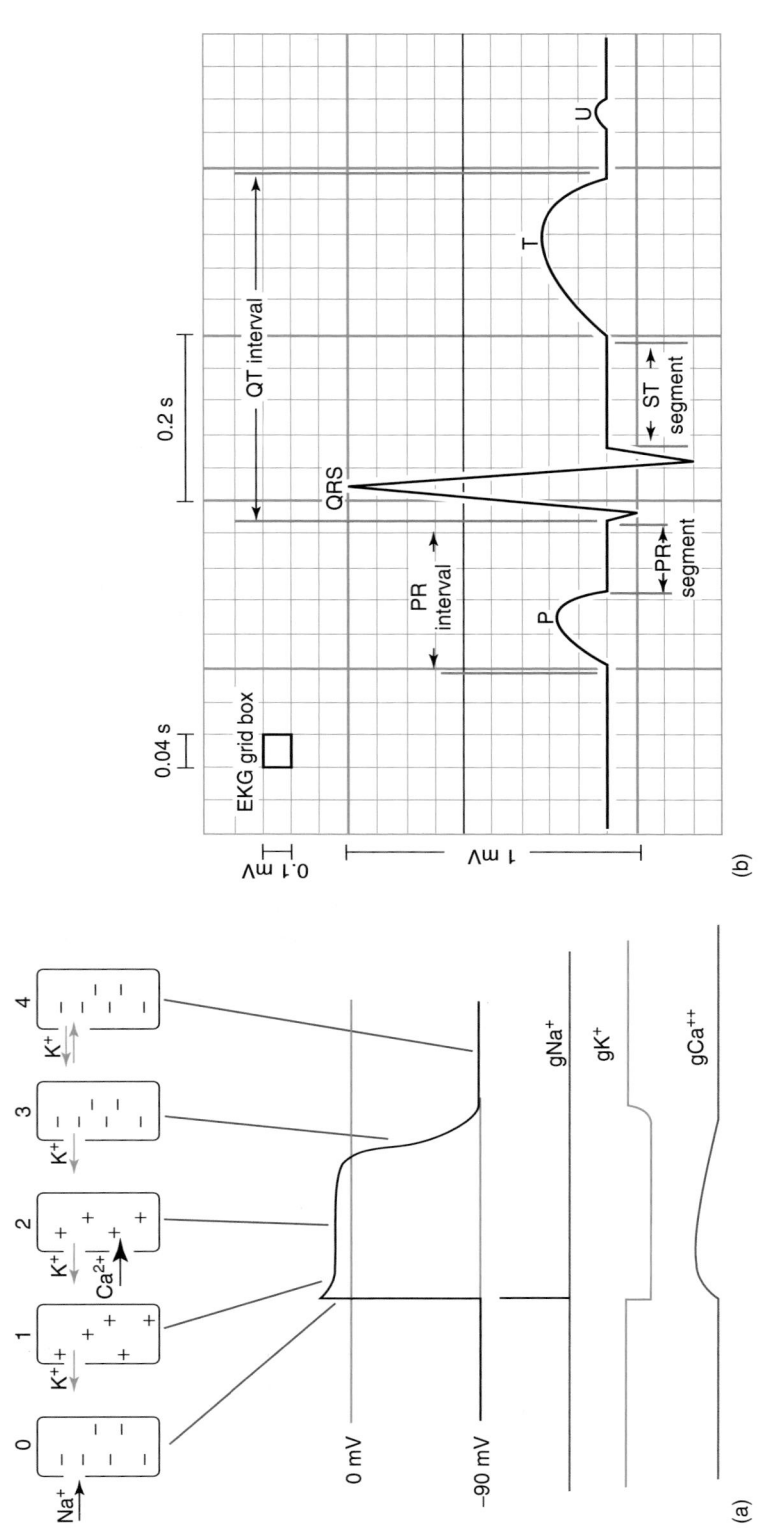

Figure 2 Cardiac action potential. (a) Voltage changes in the heart as a function of changes in ion currents into and out of the cell. (b) Schematic of an ECG tracing. Parameters derived from the ECG are depicted. QT interval is determined as the length of time separating the beginning of the QRS complex and the end of the T wave.

5.40.3   Screening for Human Ether-a-go-go Related Gene Blockers

Any discussion of predictive modeling approaches would be incomplete without mention of the type of data used to build predictive in silico models. Two excellent reviews in *Drug Discovery Today* have summarized the types of test systems and assays currently available for cardiotoxicity screening.[25,26]

5.40.3.1   In Vitro Telemetry

In vivo telemetry experiments in nonrodents (typically in conscious or anesthetized dogs) are the ultimate preclinical test for cardiotoxicity, with data generated under physiological conditions and related to the pharmacokinetic profile of the drug. However, its high cost severely limits its use at the earlier discovery stage.[25] In vitro electrophysiology in primary cardiac tissue, such as Purkinje fibers, is sometimes used, but is relatively low throughput and variability precludes its widespread application.

5.40.3.2   Patch Clamp Techniques

Voltage clamp techniques represent the 'gold standard' in the field and provide real time mechanistic information on ion channels.[26] The experiments are performed in mammalian cells (e.g., Chinese hamster ovary (CHO) cell line or human HEK293) transfected with the gene for hERG. The overwhelming majority of predictive hERG models have been built using mammalian patch clamp data. Electrophysiology data from experiments performed in *Xenopus* oocytes can be found in the literature. Unfortunately, the highly lipophilic environment in the oocytes appears to limit access of the drug to its site of action, leading to IC_{50} values being increased by as much as 100-fold.[26,27] While this could be tolerated in a classification model, quantitative structure–activity relationship (QSAR) approaches have tended to avoid using data collected in the oocytes.

5.40.3.3   High-Throughput Assays

Owing to the moderate throughput of patch clamp experiments, many companies have been willing to compromise on data quality to increase assay throughput. Techniques such as fluorescence-based assays with cells stably transfected with hERG and radioligand binding assays[25] have been successfully used to tune out hERG channel activity in lead series.[28,29] Radioligand binding assays typically utilize known hERG blockers, such as [^{3}H]dofetilide, [^{35}S]MK-499, or [^{3}H]astemizole to measure the extent of radiolabel displacement by compounds in question.[30] While it is thought that most hERG blockers bind at the intracellular region of the pore,[15,17] radioligand binding assays have a potential to produce false negatives in case a ligand binds to hERG elsewhere.[30] Fluorescence-based assays utilize voltage-sensitive dyes to monitor the membrane potential of the cell in response to a changing electrochemical gradient across the membrane. Another option is a rubidium flux assay, in which the rubidium gradient is characterized with atomic absorbance spectrometry. Unfortunately, the variation in potency obtained in these higher throughput experiments from patch clamp data and the frequent occurrence of false negatives due to a lower sensitivity of the assays[26,30] makes the data sets derived by these techniques less amenable to consistent model building.

5.40.3.4   Planar Patch Clamp

A new and exciting development in ion channel screening has been the recent introduction of automated high-throughput patch clamp machines (e.g., planar patch technology).[26] Most planar patch instruments utilize silicon-based plates with holes etched into them. Suction brings cells close to the hole, thus removing the need for a manually operated patch pipette. Several correlations to manual patch clamp data have been reported, but further validation is necessary. The approach suffers from problems typically associated with automated screening procedures, such as compound insolubility in buffer leading to false-negative results. If successful, this approach could revolutionize cardiac safety testing by increasing the rate of data acquisition by approximately 100-fold, ultimately providing larger and more diverse data sets for in silico modeling.

5.40.3.5   Whole Organism Approaches

Another recent innovation in cardiotoxicity screening is the use of zebrafish for whole organism compound testing.[31] Zebrafish provide scalability due to their rapid growth, making assays in multiwell plate format possible. In addition, they are tolerant to dimethyl sulfoxide (DMSO), typically used to store liquid compound samples in screening collections, and readily absorb compounds from the water. In the case of hERG gene, 99% amino acid identity between zebrafish and human sequences has been shown for the key S6 and pore domains.[32] Several known QT-prolonging drugs were shown to elicit arrhythmia in zebrafish embryos.[32] More validation of the zebrafish system is ongoing, but preliminary results appear to indicate its potential use in early toxicity screening. Zebrafish can be viewed as a powerful sieve, and while further compound attrition upon going from zebrafish to humans is expected, the attrition rate should be significantly lower.[31]

5.40.4   Predictive Modeling of Human Ether-a-go-go Related Gene Blockers

Of the variety of hERG modeling approaches that have appeared in the literature, most can be broadly divided into three categories: homology modeling, QSAR models, and classification methods.

5.40.4.1   Homology Modeling of Human Ether-a-go-go Related Gene

At least two groups[33,34] have reported hERG homology models using available atomic resolution structures of bacterial K^+ channels KcsA[35] (closed) and MthK[36] (open). These channels contain only two transmembrane domains (equivalent to helices S5/S6 in **Figure 3**), and the models therefore only cover the predicted structure of the hERG pore. The basic architecture of hERG channel is expected to be similar to that of other voltage-gated K^+ channels, such as KvAP (**Figure 3**).[37] The channel pore domain is formed by tetramerization from helices S5 and S6, as well as the pore helix P and the selectivity filter loop. The selectivity filter lies on the extracellular side of the membrane. The movement of S6 helices with respect to each other in a crossover fashion renders the channel closed, with the water-filled cavity isolated from cytosol. The voltage-sensing paddles formed by helices S3b and S4 are responsible for the voltage dependence exhibited by KvAP. The debate about the exact position of the S4 segment and the details of voltage sensing at the molecular level is ongoing.[38,39] While the structure and location of the paddles with respect to the membrane may be different in hERG, the basic structure of the pore is likely to be reasonably conserved.

Two bands of aromatic residues are predicted to line the cavity, with each monomer contributing Phe656 and Tyr652 (**Figure 4**).[15,33,34] These residues are both located on the S6 helix, with the tetrad of Phe656 situated closer to the mouth of the channel, and the four Tyr652 residues further toward the pore helix. The homology model is corroborated by earlier mutagenesis data. Sanguinetti and co-workers used alanine scanning to identify key residues responsible for hERG blockade by potent inhibitors terfenadine, cisapride, and MK-499.[33] Phe656 and Tyr652 appeared to be the primary interaction points. The current consensus implicates Phe656 in π-stacking interactions with the ligands, while Tyr652 is thought to participate in a cation–π interaction with the protonated basic nitrogen present in most of the reported hERG blockers. Recently, the potency for hERG blockade by these three drugs was shown by systematic mutagenesis to correlate well with measures of hydrophobicity of residue 656, such as its side chain van der Waals

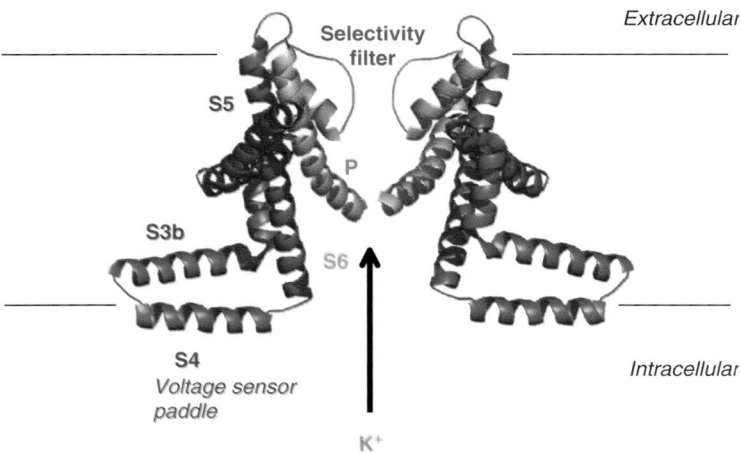

Figure 3 Structural model of hERG channels. Key elements of hERG channel topology illustrated using the x-ray structure of KvAP.[37] Two of the four subunits comprising the tetrameric channel are shown. (Reproduced with permission from Aronov, A. M. *Drug Disc. Today* **2005**, *10*, 149–155 © Elsevier.)

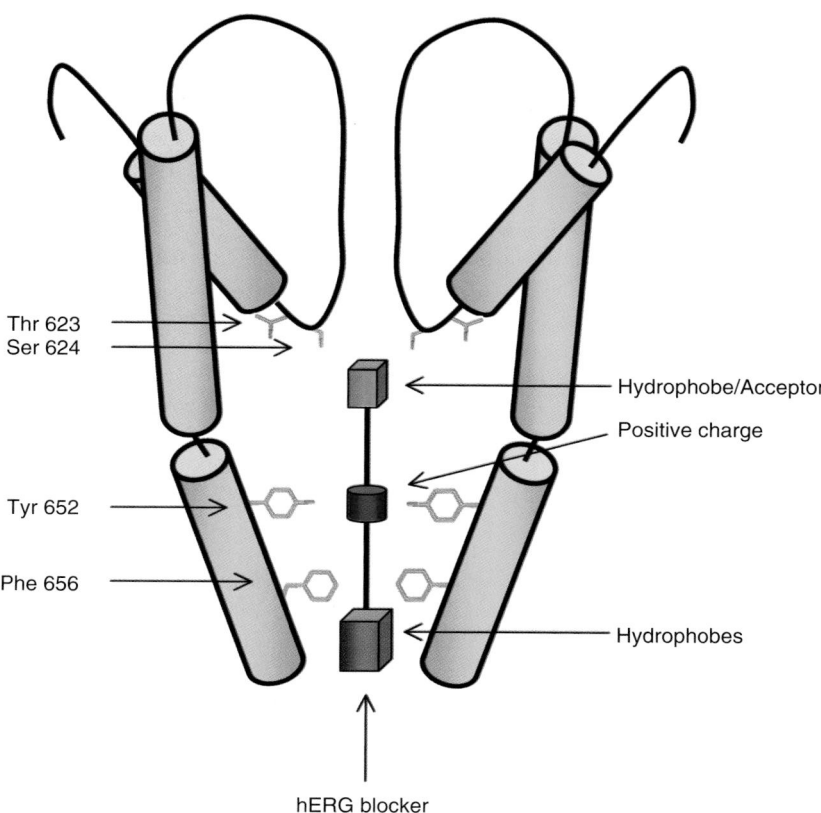

Thr 623
Ser 624

Hydrophobe/Acceptor

Positive charge

Tyr 652

Phe 656

Hydrophobes

hERG blocker

Figure 4 Model of the pore portion of hERG channel. The P-S6 fragment is shown for a dimer. Aromatic residues Phe656 and Tyr652 are critical for hERG block by most known small molecule ligands. Polar residues Thr623 and Ser624 modulate the binding potency for a number of reported hERG blockers. An hERG blocker is represented schematically based on published evidence. One or two hydrophobes interact with Phe656, a positive charge is stabilized by cation–π interaction with Tyr652, and the generally hydrophobic tail contains an acceptor able to interact with the polar residues on the loop that connects the pore helix to the selectivity filter. (Reproduced with permission from Aronov, A. M. *Drug Disc. Today* **2005**, *10*, 149–155 © Elsevier.)

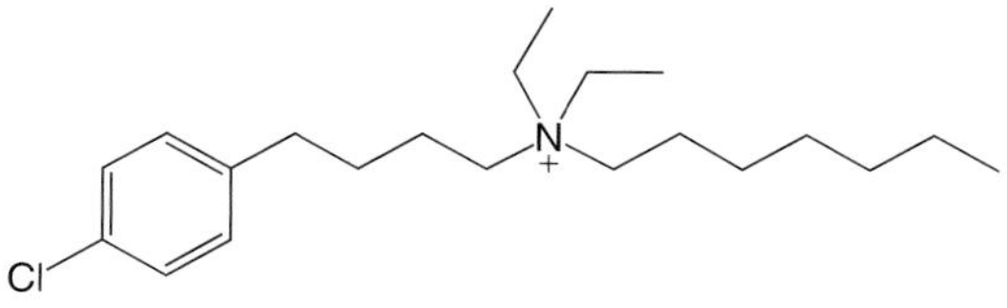

Figure 5 Structure of ibutilide.

hydrophobic surface area.[24] In the case of residue 652, the presence of an aromatic residue in this position is required for high-affinity hERG blockade, consistent with the importance of the cation–π interaction predicted by ligand-based models.

Additionally, residues Thr623 and Val625, located near the pore helix, were implicated in hERG binding to MK-499, but the effect was moderate in the cases of terfenadine and cisapride.[15] Both Thr623 and neighboring Ser624 (**Figure 4**) have been shown to have pronounced effects on hERG block by vesnarinone, clofilium, and ibutilide (**Figure 5**). These polar residues may be able to interact with the polar tails present in many of the potent hERG blockers.[15,17,34] A recent study of the structural determinants of hERG channel block by clofilium and ibutilide provided further evidence that residues at the base of the pore helices, which also face into the central cavity, can form interactions with drugs that may facilitate high-affinity binding.[40] Mutation of T623A and S624A resulted in approximately 90- and 60-fold increases in ibutilide IC$_{50}$ values, respectively. It is the specific interaction with Ser624 that appears to be responsible for the slow-off rates that characterize clofilium block of both wild-type and mutant D540K hERG channels. Based on the docking model of clofilium within the inner cavity of the hERG homology model, Perry and co-workers invoke polar interactions between the side chain hydroxyl of Ser624 and the chlorine atom in clofilium as the interaction that contributes to stabilization of clofilium binding.[40] The proposed orientation of both clofilium and ibutilide within the hERG channel includes aryl groups pointing toward the selectivity filter, a cation–π interaction of Tyr652 with the positively charged amine, and a hydrophobic contact of the fatty side chains of the drugs with Phe656. The role of Val625 in interactions with hERG blockers is less well understood, primarily because according to K$^+$ channel crystal structures,[35,36] the corresponding valine side chain is buried in the hydrophobic core surrounding the selectivity filter, and does not point into the inner cavity of the channel. Thus, the reason for the sensitivity of hERG block to the V625A mutation could lie in the allosteric effect of this residue on the conformation of neighboring side chains of Thr623 and Ser624.

A combination of structural features is thought to be responsible for the 'binding promiscuity' of hERG relative to other K$^+$ channels. The aromatic residues critical for binding to structurally diverse drugs are missing in channels of the Kv1–4 families, replaced by Ile or Val in positions equivalent to Tyr652 and Phe656.[15,17,24] The exact positioning of the aromatic residues is also of critical importance for binding. Structurally related EAG channels were made sensitive to the hERG blocker cisapride by moving the Tyr residue by one position along the S6 helix.[41] Finally, the hERG channel lacks the Pro-Val(Ile)-Pro motif on the S6 helix, which is present in Kv1–4 channels and is thought to decrease the size of the inner cavity by inserting a kink in the inner helices.[15,24]

The only published example of the use of homology models for quantitative prediction of hERG blockade is the study by Rajamani and colleagues.[42] The method described uses a two-state homology model to represent the flexibility of the channel. Two hERG homology models were built based on the crystal structure of the closed state of KcsA.[43] The lower resolution structure of MthK[36,37] in the open state was used as a guide providing insight into the movement of the S6 helix in going from the open to the closed states of the channel. As a result, two models of hERG channel were built – the partially open state (10° translation away from the KscA reference state) and the fully open state (19° translation, closely corresponding to MthK). IC$_{50}$ values for 32 ligands assembled by Cavalli and co-workers[44] (hERG pIC$_{50}$ ranging between 4 and 9) were used to derive the linear interaction energy (LIE) correlation. The ligand set was docked to the homology models, the best poses (one for each ligand) were energy minimized, and used to derive the van der Waals and electrostatic contributions to the binding energy. Models derived for the two homology models separately resulted in poor fit (e.g., $r^2 = 0.24$ for the partially open state). However, when a preference of each compound for a particular state was established by comparing the estimated interaction energy for each ligand in both states, separate better fits were obtained, with 21 ligands preferring the open state and 11 ligands the partially closed state of the channel. The final combined model:

$$pIC_{50} = -0.163(\Delta vdw) + 0.0009(\Delta ele) \quad [1]$$

produced $r^2 = 0.82$ (rmse = 0.56) for 27 ligands, with five compounds identified as potential outliers.[42] The study highlighted the difficulty of using hERG homology models for quantitative predictions of hERG blockade and the need for modeling techniques that can capture the flexibility of the K$^+$ channel.

5.40.4.2   Quantitative Structure–Activity Relationships

Ligand-based approaches have been extensively applied to understanding SARs of hERG channel blockers. One of the earliest SAR studies performed on a series of compounds that cause QT prolongation is by Morgan and Sullivan.[45] In an overview of Class III antiarrhythmic agents, the authors proposed a general structure of Q-Phenyl-A-NR$_1$R$_2$. R$_1$ and R$_2$ are preferentially hydrophobes, R$_1$ as alkyl or phenylalkyl (R$_1$ = H is also acceptable for inclusion in the class), and R$_2$ as alkyl, arylalkyl, or heteroalkyl. Linker A is a 1–4 atom chain that may contain heteroatoms or be part of a ring. This linker chain was noted as one of the more variable regions in the molecule. Finally, para-substituent Q is an electron-withdrawing group, such as nitro, cyano, N-imidazole, or N-methylsulfonamide.

The first example of hERG pharmacophore modeling appears to be the work by Matyus and co-workers[46] on elucidation of the pharmacophore for blockers of I_{Kr} current as Class III antiarrhythmic agents. Eleven ligands were divided into two sets. The first set included the six most active agents, spanning a 1.5 log activity range, and the second set contained the less active compounds. Starting structures were obtained by energy minimization in the Tripos force field.[47] Sets of conformations were generated by the Multisearch option in DISCO interface (see 4.06 Pharmacophore Modeling: 1 - Methods; 4.19 Virtual Screening),[47] up to a maximum of 100 conformations per ligand. As part of the conformation generation process, various energy minima were sampled by random perturbation of torsions, followed by

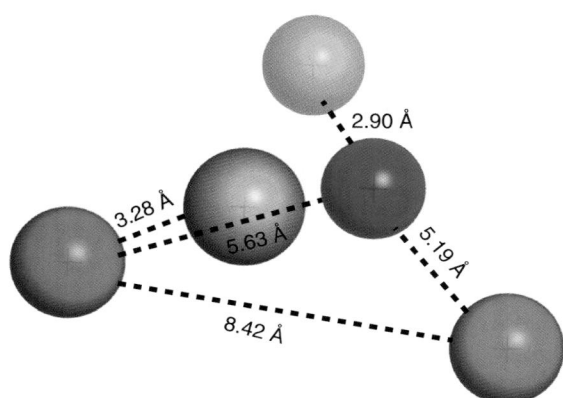

Figure 6 Pharmacophore for hERG blockers. (Reprinted from Matyus, P.; Borosy, A. P.; Varro, A.; Papp, J. G.; Barlocco, D.; Cignarella, G. *Int. J. Quant. Chem.* **1998**, *69*, 21–30, wtih permission of John Wiley & Sons, Inc.) Colored features correspond to donor atom (blue), receptor-associated acceptor site (red), aromatic moieties (green), and aliphatic chain (orange).

minimization and elimination of duplicates. Perturbed torsions were limited to those that have the potential to change the internal geometry of atoms that determine the DISCO features. A new feature definition was added to the hydrophobic class in DISCO to recognize hydrophobic aliphatic chains. DISCO was forced to consider five-point pharmacophores containing one donor atom, one receptor-associated acceptor site, and three hydrophobes, with dofetilide chosen as the reference. Fifty-three pharmacophores from DISCO were analyzed, and the best model was chosen (**Figure 6**). The basic nitrogen donor was within 5.19, 5.63, and 2.37 Å from the two aromatic features and the aliphatic feature, respectively. The less potent five-compound set was shown to satisfy only four points in the five-point pharmacophore, with one of the three hydrophobes missing.

The first hERG pharmacophore published since the renewed interest in predicting I_{Kr} blockade was described by Ekins and co-workers using 15 molecules from the literature.[48] The conformers were generated using BEST mode in Catalyst, up to a maximum of 255 conformations per ligand. Hydrophobic, ring aromatic, donor, acceptor, and positive ionizable features were selected for possible inclusion. Ten Catalyst hypotheses were assessed, and the lowest energy cost hypothesis was deemed the best. It contains four hydrophobes, one positive ionizable feature, and produced an r^2 value of 0.90. Proposed distances between the positive center and the hydrophobes are 5.2, 6.2, 6.8, and 7.5 Å. The model was further applied to predict IC_{50} values for a test set of 22 mostly antipsychotic compounds known to inhibit hERG ($r^2 = 0.83$). More recently, the initial training set was expanded to include 66 molecules, resulting in an observed-versus-predicted correlation of $r^2 = 0.86$.[49] This model produced a correlation of $r^2 = 0.67$ on an additional 25 molecules from the literature.

Cavalli and colleagues[44] constructed a hERG pharmacophore based on a training set of 31 carefully selected QT-prolonging drugs from the literature. The conformational space was sampled by means of Monte Carlo analysis as implemented in the MacroModel software.[50,51] All of the dihedral angles of single linear bonds were allowed to move freely, and an unusually large $100\,kJ\,mol^{-1}$ energy window was used to filter conformers. The procedure generated hundreds of conformers per ligand, which were then clustered using 1 Å root mean square displacement cutoff. A 'constructionist' approach to pharmacophore generation was employed, whereby astemizole was used as the template onto which other ligands were sequentially superimposed by means of the commonality of geometric and spacial characteristics. The proposed pharmacophore contains three aromatic moieties connected through a nitrogen function that is a tertiary amine throughout the whole set of molecules. The nitrogen and the aromatic moieties are separated by distances of 5.2–9.1, 5.7–7.3, and 4.6–7.6 Å (**Figure 7**). Comparative molecular field analysis (CoMFA) performed on the training set produced a correlation with $r^2 = 0.95$ ($q^2 = 0.77$), and its predictive ability was tested on a set of six additional compounds ($r^2_{pred} = 0.74$).

Pearstein and co-workers[34] reported a comparative molecular similarity indices analysis (CoMSIA) (*see* 4.23 Three-Dimensional Quantitative Structure–Activity Relationship: The State of the Art) model built using in-house patch clamp data for 28 compounds, 18 of them sertindole analogs ($q^2 = 0.57$). Conformational searching was performed using the MMFF94 force field. Ring centroids and the basic nitrogen were used as landmarks for superimposing the compounds onto sertindole using least-squares fitting (*see* 4.05 Ligand-Based Approaches: Core Molecular Modeling). According to the model, decreasing the positive charge on the central nitrogen and increasing the steric bulk on the hydrophobic end of the molecule are two potential ways to reduce hERG blocking activity. The model was tested on a holdout set

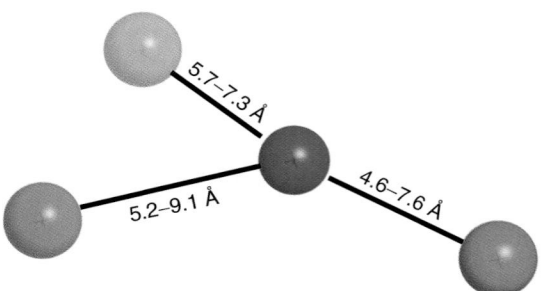

Figure 7 Four-point pharmacophore for hERG blockers. (Reprinted with permission from Cavalli, A.; Poluzzi, E.; De Ponti, F.; Recanatini, M. *J. Med. Chem.* **2002**, *45*, 3844–3853. Copyright (2002) American Chemical Society.) Colored features correspond to positively charged nitrogen (blue) and aromatic rings (green).

containing four sertindole analogs with widely varying potency for hERG. In addition, the authors constructed a homology model of the tetrameric pore region of hERG from the MthK template[36] and qualitatively docked the set of aligned inhibitor structures into the inner cavity of the channel (*see* Section 5.40.5.1 for the ligand-binding mode).

Keseru[52] used literature data on 55 compounds to train a QSAR model based on calculated descriptors. Five descriptors were used: ClogP, calculated molar refractivity (CMR), partial negative surface area, and the Volsurf W2 (polarizability) and D3 (hydrophobicity) descriptors. A model of acceptable quality was obtained ($r^2 = 0.94$, SSE = 0.82) and tested on a 13 compound holdout set ($r^2 = 0.56$, SSE = 0.98). A hologram QSAR (HQSAR; *see* 4.23 Three-Dimensional Quantitative Structure–Activity Relationship: The State of the Art)[53] model was then created that made use of 2D fragment fingerprints (threshold hERG $IC_{50} = 1 \mu M$).[52] In HQSAR, each molecule is divided into structural fragments that are counted in the bins of a fixed-length array to form a molecular hologram. These structural descriptors encoding compositional and topological molecular information are then used to derive a partial least squares regression model that correlates variation in structural composition with variation in experimental data. The best HQSAR model was validated on a holdout set of 13 compounds ($r^2 = 0.81$, SSE = 0.67).

In a study of Class III antiarrhythmics, Du and co-workers[54] selected a set of 34 compounds from the literature spanning a broad range of IC_{50} values. Using the features previously proposed to be important in characterizing hERG blockers,[46] the authors turned to HypoGen module in Catalyst. Models were allowed to contain at least one and at most two instances of every feature. The best pharmacophore hypothesis contained a positive ionizable feature, two aromatic rings, and a hydrophobic group. It was then applied to a test set of 21 compounds, which was split into three groups based on the hERG activity level (<1, 1–100, and $>100 \mu M$). The model predicted the activity of test ligands with $r^2 = 0.713$, with all highly active ($<1 \mu M$) compounds predicted correctly.

A simple two-component QSAR model for hERG blocking potency was proposed by Aptula and Cronin.[55] A set of 150 descriptors were calculated for 19 structurally diverse hERG blockers from the literature. The calculated variables included physicochemical parameters, topological indexes, and quantum chemical descriptors. Multiple linear regression produced a relationship between hERG blocking potency and two descriptors – $\log D$ (at pH = 7.4) and D_{max} (the maximum diameter of molecules):

$$pIC_{50}(hERG) = 0.58 \log D + 0.30 D_{max} - 0.36 \qquad [2]$$

with a reasonable correlation ($r^2 = 0.87$, $q^2 = 0.81$). A further analysis of 81 chemicals from Redfern and colleagues[14] was described, but no results were shown. The authors suggest the maximum diameter cutoff of 18 Å for cases where hERG blocking activity is undesirable, and $D_{max} > 18$ Å for antiarrhythmics. The relationship described[55] is rather intuitive, with hERG activity correlating with both hydrophobicity and ligand size. More lipophilic ligands that have a large diameter are capable of binding in the hERG channel and engaging most of the residues implicated in hERG blockade.[15,18] While the overall findings with regard to lipophilicity are generally in agreement with earlier studies,[56] the QSAR equation based on a study of only 19 ligands that contains no additional size constraints (e.g., a reasonable D_{max} limit beyond which the compound would be unable to enter the channel pore) appears a gross generalization.

One of the larger hurdles for building QSAR models using literature data has been the large discrepancy observed for hERG IC_{50} values determined in different laboratories. Interlaboratory variability of greater than 10-fold is not uncommon, even in cases where inhibition was measured using the same cell line. Additional efforts in generating internally consistent hERG data sets that could be made available to the broad scientific community are sorely needed to propel the field forward.

5.40.4.3 Classification Methods

While QSAR methods aim to predict absolute compound activity, classification methods attempt to bin compounds by their potential hERG inhibition. The earliest example of a hERG-based classification was reported by Roche and co-workers.[57] A total of 244 compounds representing the extremes of the data set (<1 and $>10\,\mu M$ for actives and inactives, respectively) were modeled with a variety of techniques such as substructure analysis, self-organizing maps, partial least squares, and supervised neural networks. The descriptors chosen included pK_a, Ghose-Crippen,[58] TSAR,[59] CATS,[60] Volsurf,[61] and Dragon[62] descriptors. The most accurate classification was based on an artificial neural network. In the validation set containing 95 compounds (57 in-house and 38 literature IC_{50} values) 93% of inactives and 71% of actives were predicted correctly.

In a decision tree-based approach to constructing a hERG model using calculated physicochemical descriptors, Buyck and co-workers[56] used three descriptors – ClogP, calculated molar refractivity (CMR), and the pK_a of the most basic nitrogen – to identify hERG blockers within an in-house data set. With $IC_{50} = 130\,nM$ as a cutoff, factors suggestive of hERG activity were determined to be $ClogP \geqslant 3.7$, $110 \leqslant CMR < 176$, and pK_a max $\geqslant 7.3$.

A combined 2D/3D procedure for identification of hERG blockers was proposed by Aronov and Goldman.[63] A 2D topological similarity screen utilizing atom pair[64] descriptors and an amalgamated similarity metric termed TOPO was combined with a 3D pharmacophore ensemble procedure in a 'veto' format to provide a single binary hERG classification model. A molecule flagged by either component of the method was considered a hERG active. In the course of 50-fold cross-validation of the model on a literature data set containing 85 actives (threshold HERG $IC_{50} = 40\,\mu M$) and 329 inactives, 71% of hERG actives and 85% of hERG inactives were correctly identified. The model utilizing the TOPO metric was shown to be superior to a number of other 2D models using the receiver operating characteristic metric. Additionally, five of eight (62.5%) hERG blockers were identified correctly in a 15 compound in-house validation set. Most of the statistically significant pharmacophores from the ensemble procedure were three-feature [aromatic]–[positive charge]–[hydrophobe] combinations (**Figure 8a–b**) similar to those reported by Cavalli and colleagues[44]; however, a novel three-point pharmacophore containing a hydrogen bond acceptor was also proposed (**Figure 8c**). The presence in hERG blockers of the acceptor functionality pointing toward the selectivity pore agrees with the previous observations[15,40,44] of a potential for polar interactions with the side chains of Thr623 and Ser624 to stabilize the hERG–ligand complex.

Testai and co-workers[65] evaluated a set of 17 antipsychotic drugs, all of them associated with reports of torsadogenic cardiotoxicity. The search for a common molecular feature that may be a requirement for hERG blockers focused on measuring several different distances between atoms that could constitute a hERG-active template. The authors hypothesized that such a template for hERG-active ligands consists of a hydrocarbon chain, three or four atoms long, serving as a spacer between a basic sterically hindered nitrogen atom, and a second, more variable, moiety. Focusing on the distance between the basic nitrogen and this second moiety, Testai and co-workers observed that for all of the compounds in the data set the distance converged in the range between 4.32 and 5.50 Å (average $= 4.87$ Å).

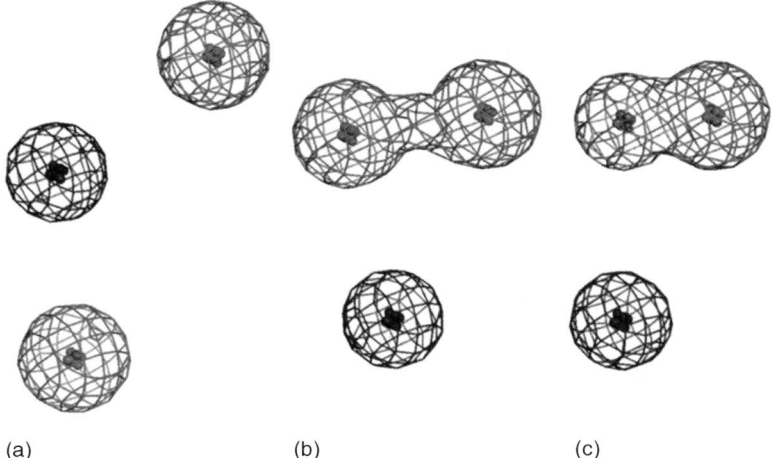

(a)   (b)   (c)

Figure 8 Three three-point pharmacophores (a–c) for in silico hERG block prediction (from Aronov, A. M.; Goldman, B. B. *Bioorg. Med. Chem.* **2004**, *12*, 2307–2315). Colored features correspond to positive charge (blue), hydrogen bond acceptor (red), and hydrophobic (green). (Reproduced with permission from Aronov, A. M. *Drug Disc. Today* **2005**, *10*, 149–155 © Elsevier.)

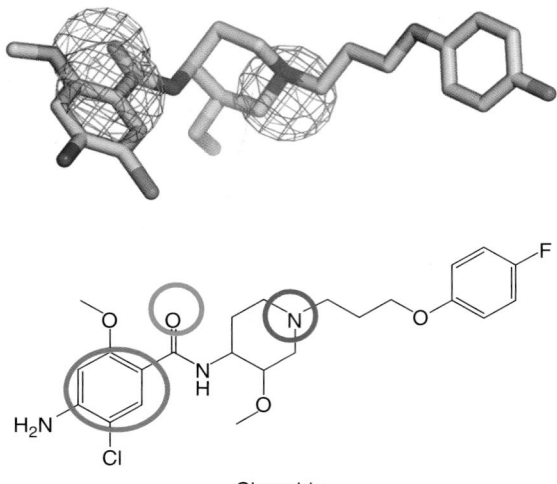

Cisapride

Figure 9 Acceptor-containing three-point pharmacophore mapped onto the structure of cisapride, a known potent hERG blocker. (Reproduced with permission from Aronov, A. M.; Goldman, B. B. *Bioorg. Med. Chem.* **2004**, *12*, 2307–2315 © Elsevier.) Feature colors are same as in **Figure 8**. Matching features are also shown mapped onto the 2D structure of cisapride.

Three-dimensional structures of the antipsychotics were generated starting from known small molecule x-ray structures followed by solvation and minimization. Interestingly, the variable moiety present in 14 of the 17 ligands is a hydrogen bond acceptor, either a carbonyl oxygen or a heteroaromatic nitrogen.[65] This is in agreement with the high information content seen for the acceptor-containing pharmacophore by Aronov and Goldman.[63] Indeed, the acceptor functionality can be seen not only in antipsychotic agents, such as risperidone and droperidol,[63] but also for other known hERG blockers, such as prokinetic agent cisapride (**Figure 9**). In the case of risperidone, the two groups pointed to different hydrogen bond acceptors that, incidentally, are located approximately the same distance from the basic nitrogen – the oxygen of the benzisoxazole[63] and the carbonyl of the pyrimidone.[65]

Bains and colleagues[66] described the application of genetic programming (GP) to building a predictive hERG model. GP, an evolutionary computing technique, 'evolves' an algorithm in silico that matches input variables with desired output, and performs both selection of relevant descriptors and algorithm building without human intervention. The data set comprised endpoints from the public domain with proprietary data, totaling 124 compounds. Three types of descriptors were utilized: general molecular descriptors (such as molecular weight), topological fragments, and experimental parameter descriptors. The largest set, topological fragments, was generated automatically and exhaustively from the data set by application of a maximum common subgraph-based algorithm. Only fragments containing at least four heavy atoms and present in two or more ligands were kept. Experimental parameters, a rather unusual addition, were abstracted from experimental sections in the literature. A total of 618 descriptors were used in the GP runs. GP algorithms were trained on a training set, selected for their ability to make predictions on a generalization set, and finally were tested on an independent validation set, their performance judged using Akaike fitness criterion and receiver operating characteristic. The best models achieved 85–90% accuracy in predicting on the validation set when $IC_{50} < 1 \mu M$ was used as a threshold for hERG-active ligands, which is consistent with the results of previously published classification studies.[57,63] Analysis of correlations between topological descriptors and hERG IC_{50} revealed that IC_{50} is:

- positively correlated with the presence of secondary or tertiary basic amines;
- positively correlated with the presence of at least one aromatic ring system;
- positively correlated with the presence of a five-membered nitrogen-containing heterocycle (including fused heterocycles, e.g., benzimidazole);
- negatively correlated with the presence of carboxylates; and
- negatively correlated with the presence of hydrogen bond acceptors.

While it is hardly possible to organize the topological fragment observations noted above into a stand-alone pharmacophore model, the authors pointed out that their results are broadly consistent with earlier publications.[15,17,33,34]

5.40.4.4 Incorporating Human Ether-a-go-go Related Gene Models into the Drug Discovery Process

With a variety of modeling approaches to the hERG problem that are aimed at various endpoints, the issue of incorporating the pertinent model(s) into the drug discovery process becomes essential. It is readily apparent that model utility is as much a function of how a model is being used as of how accurate the predictions are. For example, a model that is good at predicting hERG nonblockers (high ratio of true negatives to false negatives) may be useful in pointing out compounds that almost certainly do not block hERG. Low recovery of true hERG blockers would mean that it should be applied before the compounds are selected for in vitro screening, while a high recovery rate would make the model potentially competitive with the in vitro methods.

Another aspect critical to model deployment is its incorporation into the corporate database. Ideally, a link exists that allows for the model to make predictions on both the existing corporate collection and prophetic (virtual) compounds. The new hERG screening data made available in the course of medicinal chemistry programs can then be fed into the cycle, enabling constant expansion of the original training sets, as well as continuous model validation on an ongoing basis, which is made accessible to the medicinal chemists on the project. The data feedback loop helps address the malaise that most global models tend to suffer from – inadequate performance in the context of local predictions. Constant replenishment of the training set contributes to continuous model improvement in a local, as well as in a global sense, which can be monitored as new data arrives.

An example of one such approach is shown in **Figure 10**. The hERG model based on the publication by Aronov and Goldman[63] was incorporated into the VERDI chemoinformatics suite at Vertex Pharmaceuticals, Inc. A compound loaded as part of a search (**Figure 10a**) can be submitted for hERG prediction, which produces a set of most similar compounds based on which such a prediction is made (**Figure 10b**).

5.40.5 Understanding Structure–Activity Relationship of Human Ether-a-go-go Related Gene Blockers

As various in silico methods are being brought to bear on the problem of hERG, some understanding of the SAR relevant to hERG block is starting to emerge.

5.40.5.1 Ligand-Binding Mode

The current consensus[34,36,66] view of the proposed binding mode of hERG blockers within the channel pore is shown in **Figure 4**. The inhibitor orients itself along the pore axis, with the lipophilic end facing the opening to the cytosol and the polar tail facing the selectivity filter. One or two hydrophobic moieties interact with Phe656 side chains, likely via π-stacking. The basic nitrogen is involved in cation–π interaction with Tyr652 residues. While most known hERG blockers contain a basic nitrogen center that is expected to be protonated under biological conditions, some possess an aromatic ring in its place. The aromatic linker may be able to participate in favorable π-stacking with Tyr652 in lieu of a cation–π interaction. The tail end of the molecule appears in many of the potent hERG blockers. It is thought to extend deep into the pore and form a hydrogen bond between the acceptor feature[63] (**Figure 8c**) and the side chain hydroxyls of hERG, which may be able to explain ligand-dependent attenuation of hERG binding affinity by T623A and S624A mutations.

An important caveat to this discussion is a clear possibility that multiple ligand binding sites exist on hERG that are capable of modulating channel activity. The ability to locate and describe these alternative sites is likely to come from a concerted effort in radioligand competition assays, mutagenesis, homology modeling, and ultimately x-ray crystallography.

5.40.5.2 Structural Lessons Learned

Mapping the chemical space of known hERG blockers has shed some light on the structural aspects of ligands consistent with an hERG blocker profile:

1. Physicochemical property profiling of hERG actives and inactives clearly points to the fact that the likelihood of hERG block decreases significantly for polar low-molecular-weight ligands. As seen in **Figure 11**, few of the known hERG blockers have ClogP <1 or MW<250. This is in agreement with the ClogP$\geqslant 3.7$ limit for potent hERG blockers from Buyck and colleagues,[56] and may be due to the large pore size in hERG as well as the lipophilic character of the pore lining. Unfortunately, for a variety of reasons, which include target-binding

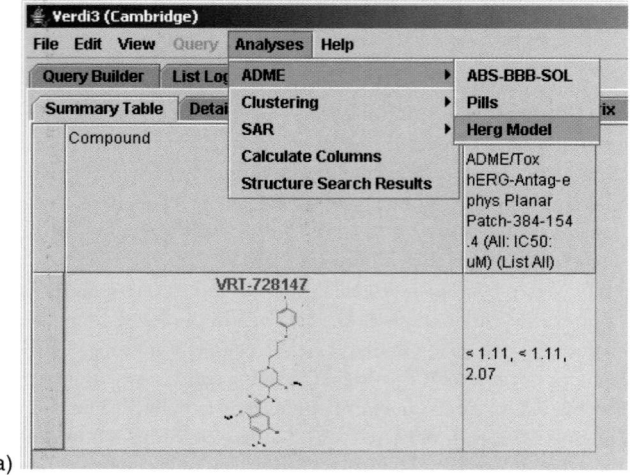

(a)

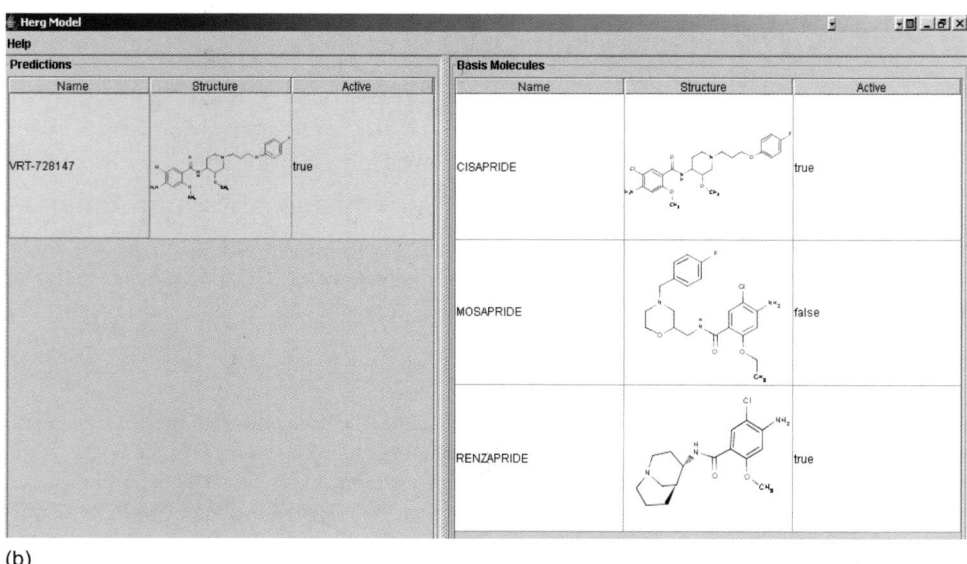

(b)

Figure 10 hERG model as incorporated into the VERDI chemoinformatics suite at Vertex Pharmaceuticals, Inc. (a) A compound (cisapride) loaded as part of a search in the process of being submitted for hERG prediction. (b) A set of most similar compounds produced based on which the prediction is made are displayed. Predictions on both real compounds from the corporate database and virtual compounds are enabled.

affinity, absorption, and clearance considerations, few of the discovery compounds that move into the clinic fall into this category.

2. The presence of a basic solubilizing group or a positively charged nitrogen in general increases the likelihood of hERG block. Conformational analysis of a number of hERG blockers from the literature led to the observation that shielding of the protonated form of the ligands decreases the amount of deprotonation, thus contributing to the increased potency for blocking hERG.[67] Tertiary amines, which form ammonium ions shielded by two structural fragments, block hERG more potently than compounds containing amines at the molecular periphery. Decreasing the pK_a of the basic amine may lower hERG activity by destabilizing the protonated species. Another useful strategy for attaching solubilizing groups to a molecule of interest may involve identification of multiple suitable substitution sites, thus increasing the odds of finding ligands devoid of hERG activity.

3. A large number of hERG blockers, both basic and neutral, contain flexible linkers that connect various molecular fragments. Flexibility appears to help the ligands find conformations compatible with the large binding site on hERG. Ligand rigidification may lead to removal of undesired hERG activity.

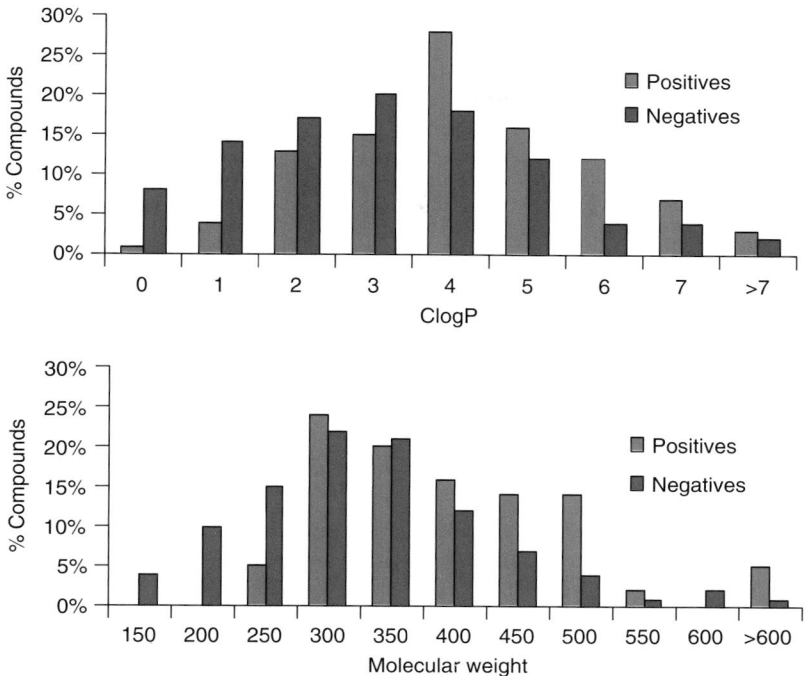

Figure 11 Reducing the risk of hERG interaction. ClogP and molecular weight distributions for hERG actives and inactives (data set from Aronov, A. M.; Goldman, B. B. *Bioorg. Med. Chem.* **2004**, *12*, 2307–2315 and references therein). Compounds with MW < 250 and ClogP < 1 are significantly less likely to block hERG. (Reproduced with permission from Aronov, A. M. *Drug Disc. Today* **2005**, *10*, 149–155 © Elsevier.)

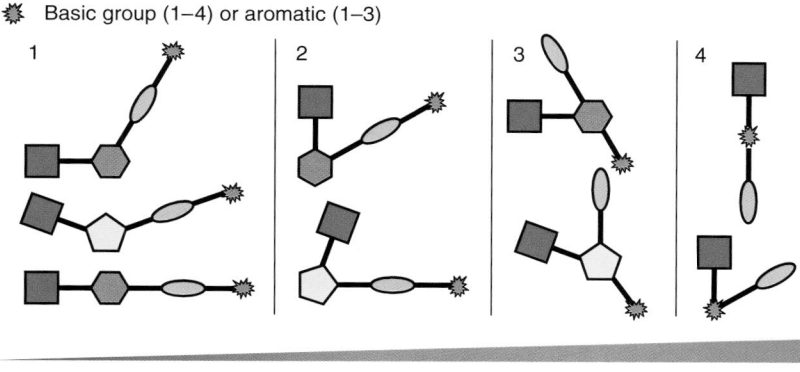

Figure 12 Molecular shapes more likely to result in a potent hERG interaction. Ellipses and squares symbolize rings of any size. Pentagons and hexagons symbolize five- and six-membered ring systems, respectively. Linear topology consistent with *meta-/para-* attachments is seen less frequently in hERG active sets than V-shaped geometry stemming from *ortho*-substitution patterns. The role of the basic group in the ligand–hERG interaction can be played by an aromatic ring. The placement of the solubilizing group, if present, may have a strong impact on the hERG profile of the molecule. (Reproduced with permission from Aronov, A. M. *Drug Disc. Today* **2005**, *10*, 149–155 © Elsevier.)

4. Some general observations of the topology of ring-linker arrangements versus the potential for hERG blockade are shown in **Figure 12**. Linear topology consistent with *meta-/para-* attachments is seen less frequently in hERG active sets than V-shaped geometry stemming from *ortho*-substitution patterns. The role of the basic group in the ligand–hERG interaction can be played by an aromatic ring. The placement of the solubilizing group, if present, may have a strong impact on the hERG profile of the molecule.

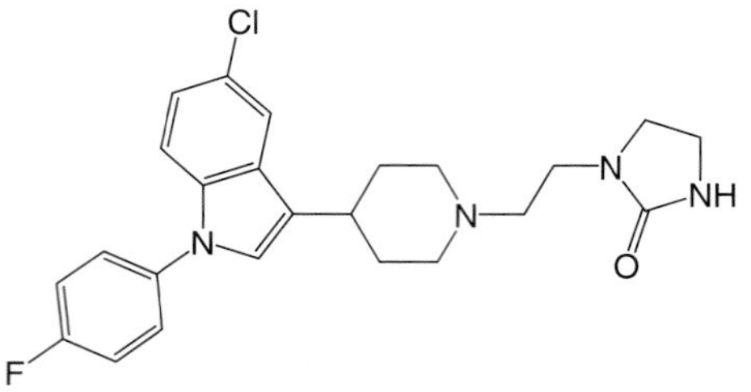

Figure 13 Structure of sertindole. See Pearlstein *et al.*[34] for modeling of sertindole analogs.

5.40.5.3   Successful Approaches to Reducing Human Ether-a-go-go Related Gene Activity within a Chemical Series

5.40.5.3.1   Reducing lipophilicity

One of the earliest observations regarding ligand propensity to inhibit I_{Kr} was that the likelihood of hERG block decreases significantly for polar low-molecular-weight ligands. This effect can be achieved through either introduction of polar functional groups or elimination of hydrophobic functionality. The challenge in log$P$ management is due to the fact that for most receptor binding sites compound potency often correlates with ligand lipophilicity, and the search for an optimal balance of on-target activity versus hERG potency is typically not trivial.

Pearlstein *et al.*[34] published a rather comprehensive hERG SAR for a series of sertindole analogs that provides a glimpse into the arsenal of modifications that can lead to a drop in hERG activity of a lead compound. For sertindole (**Figure 13**), the following modifications were successful:

- removal of either of the two halogen atoms;
- introduction of a polar substituent in the phenyl ring;
- saturating the phenyl moiety to cyclohexyl;
- saturating the indole to indoline;
- removal of the alkyl urea substituent in the piperidine ring; and
- removal of the alkylpiperidine.

A series of selective 5HT$_{2A}$ receptor antagonists possessing hERG activity was described by Rowley and colleagues.[68] The original 2-phenyl-3-piperidylindoles were potent hERG blockers as measured by the radioactive dofetilide displacement assay (**Figure 14a**). Removal of the *N*-phenethyl moiety led to a ~70-fold decrease in hERG binding, and the subsequent modification of 4-piperidine to 3-piperidine (not shown in **Figure 14a**) restored target activity.

Replacement of a phenyl ring with a methyl group, coupled with a switch from a difluoromethyl substituent to a cyclobutyl, and from pyridine to thiazole (**Figure 14b**) decreased the lipophilicity of the molecule by nearly a log unit, resulting in a greater than 50-fold drop in hERG blocking activity in a tertiary alcohol series of phosphodiesterase-4 inhibitors.[28]

A drop of nearly 10-fold in hERG blocking activity as determined by an in vitro electrophysiological assay using Cos7 cells was achieved in a series of 2-cyclohexyl-4-phenyl-1*H*-imidazoles designed as potent and selective neuropeptide Y Y5-receptor antagonists.[69] This result came from introducing a *gem*-dimethylglycinol amide functionality into the lead molecule (**Figure 14c**), which led to a large drop in lipophilicity of the compound with a concurrent nearly threefold increase in its polar surface area.

A phenolic hydroxyl was incorporated into the benzylpiperidine benzimidazoles (**Figure 14d**) active as antagonists of the NR2B subtype of the *N*-methyl-D-aspartate (NMDA) receptor.[29] A concomitant decrease in hERG activity of the ligands as measured in the MK-499 displacement assay was a modest twofold.

5.40.5.3.2   Reducing lipophilicity and basicity

In cases when lowering lipophilicity alone may not result in a sufficient drop in hERG blocking activity, an accompanying reduction in basicity (typically lowering the pK_a of the basic nitrogen) may prove successful. Bilodeau *et al.*[70] have recently disclosed a series of aminothiazole KDR inhibitors (**Figure 15a**), which contain a basic amine solubilizing group likely responsible for observed hERG blockade. The pK_a of the basic nitrogen was reduced by approximately three log units by converting pyrrolidine to *N*-acetylpiperazine, providing a 10-fold drop in hERG

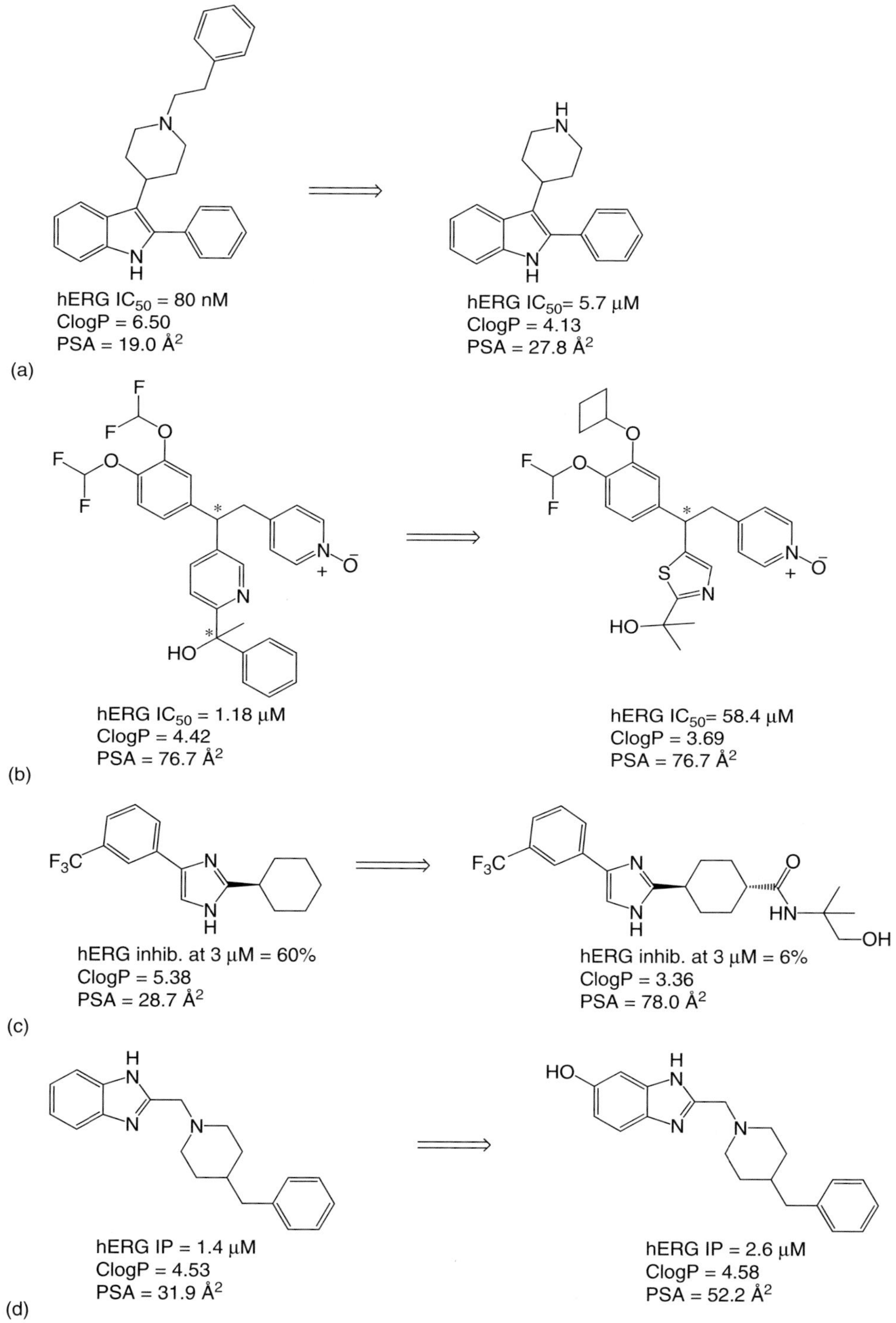

Figure 14 Lowering hERG liability : reduction in lipophilicity. (a) From Rowley, M.; Hallett, D. J.; Goodacre, S.; Moyes, C.; Crawforth, J.; Sparey, T. J.; Patel, S.; Marwood, R.; Thomas, S.; Hitzel, L. *et al. J. Med. Chem.* **2001**, *44*, 1603–1614 © American Chemical Society; (b) from Friesen, R. W.; Ducharme, Y.; Ball, R. G.; Blouin, M.; Boulet, L.; Cote, B.; Frenette, R.; Girard, M.; Guay, D.; Huang, Z. *et al. J. Med. Chem.* **2003**, *46*, 2413–2426 © American Chemical Society; (c) from Blum, C. A.; Zheng, X.; De Lombaert, S. *J. Med. Chem.* **2004**, *47*, 2318–2325 © American Chemical Society; (d) from McCauley, J. A.; Theberge, C. R.; Romano, J. J.; Billings, S. B.; Anderson, K. D.; Claremon, D. A.; Freidinger, R. M.; Bednar, R. A.; Mosser, S. D.; Gaul, S. L. *et al. J. Med. Chem.* **2004**, *47*, 2089–2096 © American Chemical Society.

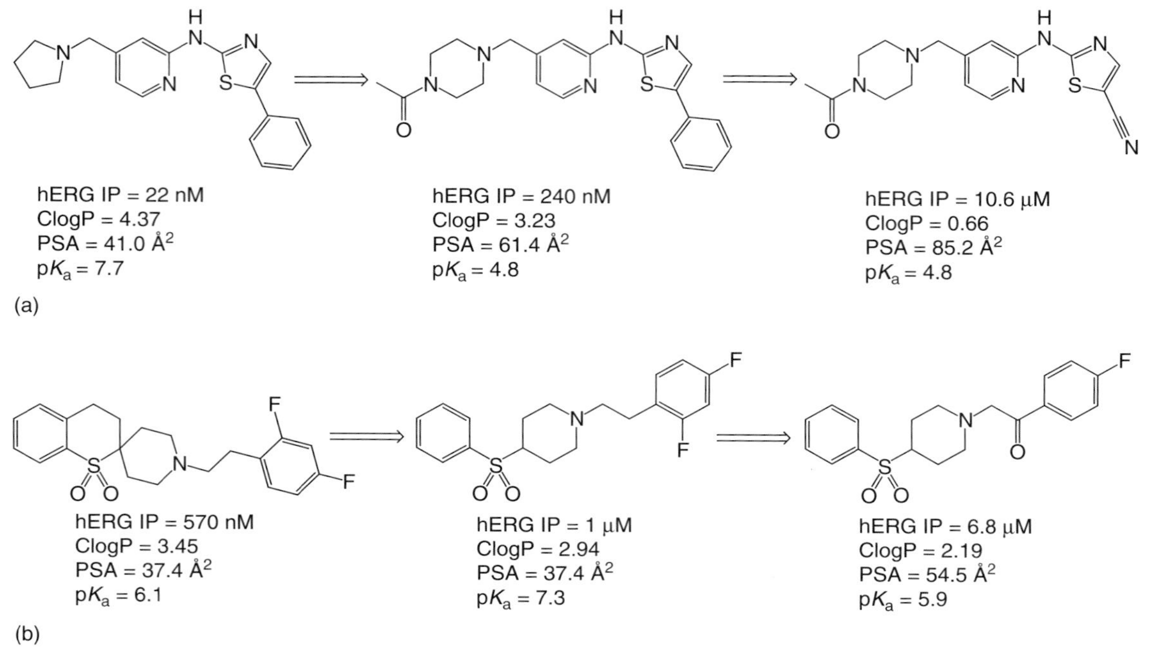

(a)

(b)

Figure 15 Lowering hERG liability: reduction in lipophilicity and basicity. (a) From Bilodeau, M. T.; Balitza, A. E.; Koester, T. J.; Manley, P. J.; Rodman, L. D.; Buser-Doepner, C.; Coll, K. E.; Fernandes, C.; Gibbs, J. B.; Heimbrook, D. C. *et al. J. Med. Chem.* **2004**, *47*, 6363–6372 © American Chemical Society; (b) from Fletcher, S. R.; Burkamp, F.; Blurton, P.; Cheng, S. K.; Clarkson, R.; O'Connor, D.; Spinks, D.; Tudge, M.; van Niel, M. B.; Patel, S. *et al. J. Med. Chem.* **2002**, *45*, 492–503 © American Chemical Society.

activity, measured as radioactive MK-499 displacement. A further reduction in lipophilicity of the molecule by substituting a cyano group for the phenyl ring resulted in an additional nearly 40-fold drop in hERG binding.

In order to lower hERG activity in a series of 4-(phenylsulfonyl)piperidines[71] as selective 5HT$_{2A}$ receptor antagonists, the spirocyclic sulfone was initially replaced with an acyclic sulfone moiety (**Figure 15b**). While lowering the logP, this structural change increased the pK_a of the basic amine, leading to a modest twofold drop in hERG binding, as measured in the dofetilide displacement assay in HEK293 cells, and later confirmed in vivo in the anesthetized ferret model. However, when the phenethyl fragment was converted to the acetophenone, the increase in polar surface area resulted in a lower logP, and the piperidine pK_a fell 1.4 log units due to the electron-withdrawing effect from the carbonyl. The result was a further sevenfold change in hERG binding.

5.40.5.3.3 Reducing flexibility

In addition to managing lipophilicity and basicity of the ligand, another potential avenue to modulating hERG activity within a series is by reducing the flexibility of a molecule. It appears that more flexible ligands as defined by the higher rotatable bond count tend to show higher hERG blocking ability as compared to their more rigid counterparts. The reason for this could lie in the greater ability of flexible ligands to adapt to the spacial constraints of the otherwise promiscuous binding site of hERG.

Drugs that activate the serotonin receptor subtype 5HT$_4$ are widely used to facilitate or restore motility in the gastrointestinal tract.[72] Cisapride, one such 5HT$_4$ agonist, was marketed in the US from 1993 until its withdrawal in 2000 following multiple reports of heart rhythm disturbances including QT prolongation, syncope, torsades de pointes, and 80 reports of death. Potet and co-workers[72] have reported comparative studies of K$^+$ currents in hERG-transfected COS-7 cells as affected by structurally similar 5HT$_4$ agonists cisapride, renzapride, prucalopride, and mosapride. While cisapride was confirmed as a submicromolar hERG blocker, mosapride was shown to be devoid of the hERG side effect. Currently marketed for gastritis in the Far East, mosapride has the potential to be widely utilized as a safe alternative to cisapride. The lower hERG activity in mosapride was partially achieved by decreasing its lipophilicity and basicity in

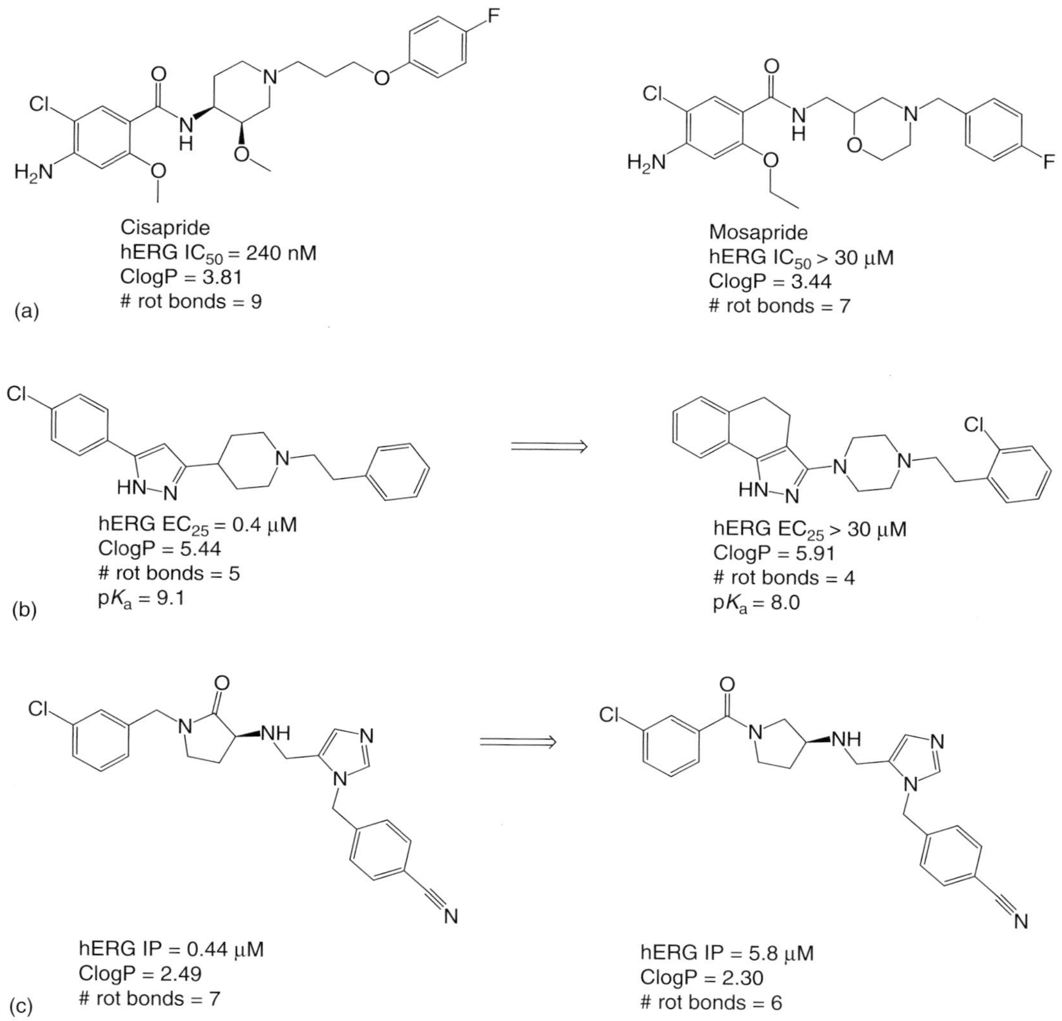

Figure 16 Lowering hERG liability: reduction in flexibility. (a) From Potet, F.; Bouyssou, T.; Escande, D.; Baro, I. *J. Pharmacol. Exp. Ther.* **2001**, *299*, 1007–1012; (b) from Collins, I. *et al. J. Med. Chem.* **2001**, *44*, 2933–2949 © Elsevier; (c) from Bell, I. M.; Gallicchio, S. N.; Abrams, M.; Beshore, D. C.; Buser, C. A.; Culberson, J. C.; Davide, J.; Ellis-Hutchings, M.; Fernandes, C.; Gibbs, J. B. *J. Med. Chem.* **2001**, *44*, 2933–2949 © American Chemical Society.

utilizing a morpholine core in place of piperidine (**Figure 16a**). ClogP decreased by 0.37 log units, and the pK_a of the basic nitrogen dropped by 0.5 log units. Another pronounced difference is that the flexibility of cisapride is partially addressed in mosapride, which has two fewer rotatable bonds. Following the replacement of the highly flexible propoxyphenyl moiety, mosapride affinity for hERG relative to cisapride decreased by more than 100-fold.

A similar approach to engineering out hERG activity was taken by Collins and co-workers[73] in a series of human dopamine D_4 receptor antagonists (**Figure 16b**). The original 3-(4-piperidinyl)-5-arylpyrazole scaffold was rigidified to produce fused partially saturated indazole analogs, and the piperidine moiety was replaced with a piperazine to lower the pK_a of the molecule. Indeed, the EC_{25}, estimated in vitro by the measurements made in ferret papillary muscle, increased by greater than 75-fold.

In the 3-aminopyrrolidinone series of farnesyltransferase inhibitors, the original *N*-benzyl-pyrrolidinones were shown to have I_{Kr} blocking activity in the submicromolar range.[74] By moving the carbonyl out of the ring, the team was able to reduce hERG activity by 13-fold (**Figure 16c**). The resulting *N*-benzamidopyrrolidines were more rigid than the starting series.

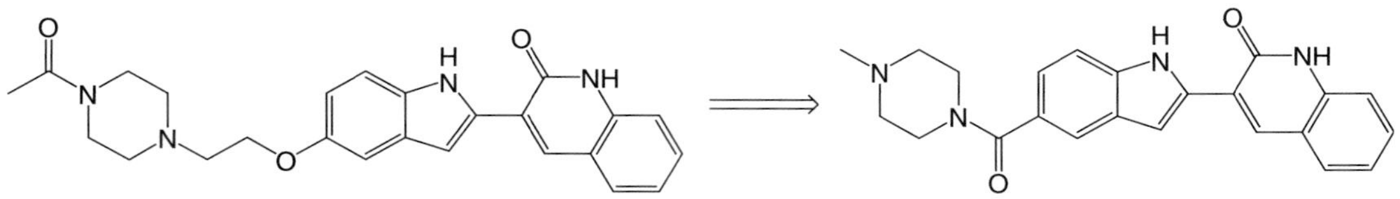

Figure 17 Lowering hERG liability: moving the basic center. (Reproduced with permission from Fraley, M. E.; Arrington, K. L.; Buser, C. A.; Ciecko, P. A.; Coll, K. E.; Fernandes, C.; Hartman, G. D.; Hoffman, W. F.; Lynch, J. J.; McFall, R. C. *et al. Bioorg. Med. Chem. Lett.* **2004**, *14*, 351–355 © Elsevier.)

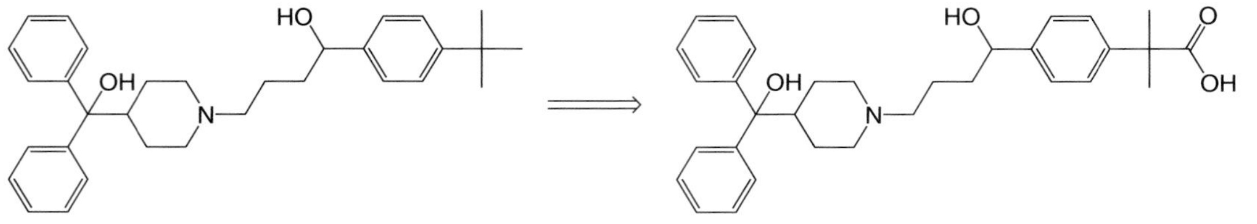

Figure 18 Lowering hERG liability: introducing a zwitterion. (Reproduced with permission from Paakkari, I. *Toxicol. Lett.* **2002**, *127*, 279–284 © Elsevier.)

5.40.5.3.4 Moving the basic center

One of the approaches to tuning out hERG activity is to evaluate a number of different attachment sites for a basic solubilizing group (*see* Section 5.40.5.2). The same result can often be achieved through variations in linker length, which effectively moves the charged center with respect to the rest of the ligand. A recent example of the success of this strategy is the optimization of solubilized indolyl quinolinones as KDR kinase inhibitors.[75] Owing to the presence of an internal hydrogen bond between the quinolinone carbonyl and the NH of the indole moiety, the scaffold is essentially flat, and requires a pendant-solubilizing group to improve its properties (**Figure 17**). One of the early compounds in the series, a linked acetylpiperazine, showed an inflection point of 520 nM in the MK-499/hERG displacement assay. Shortening of the linker and the transfer of the basicity to the terminal nitrogen of the piperazine produced the desired effect, reducing hERG inflection point to 10.8 µM, or greater than 20-fold.

5.40.5.3.5 Zwitterions

The best known example of the use of zwitterions to reduce hERG activity is the case of terfenadine.[76] A potent antihistamine, it was first reported to be cardiotoxic in 1989, and was later shown to block the hERG channel K^+ current. The QT prolongation effect was most pronounced when terfenadine was co-administered with CYP3A4 blockers such as ketoconazole. In contrast, fexofenadine, terfenadine's main metabolite, is devoid of hERG-blocking activity (**Figure 18**). The carboxylate in fexofenadine renders the molecule uncharged in solution, decreasing the potency of hERG blockade by over three orders of magnitude, and making fexofenadine (Allegra) a best-selling antihistamine.

5.40.6   Conclusions

How useful have most of the in silico methods for predicting hERG blockade been so far? Indeed, the publications described in this review have spawned a variety of commercially available models aimed at predicting which ligands may interact with hERG channel. Unfortunately, the impact of these predictive methods on drug discovery to date has been rather limited. Part of the challenge lies in the fact that none of the approaches tried so far can be considered universal. Homology modeling has been able to narrow the scope of mutational data required to characterize hERG, and has been rather successful in providing the rational for the effect these mutations have on channel function. Despite that, docking of prealigned hERG blockers into the homology models has so far only led to qualitative predictions of the importance of certain structural features present in potent hERG blockers. While useful to overall understanding of the problem, the utility of the 3D receptor-based approaches for rapid prediction of hERG liability in a set of ligands is currently rather low. The emergence of QSAR approaches as applied to hERG liability has produced models better suited to make ligand predictions. The drawback has been that, as is rather typical for QSAR, these models have been trained on rather small sets of known hERG blockers, and while highly predictive in local models, they tend to fail when confronted with broad SAR that is more global in nature. Yet another trend, ligand-based machine learning methods, has shown promise in the ability to discriminate between known hERG blockers and compounds devoid of hERG liability. However, this approach typically yields the prediction as a 'black box' calculation, making efforts to interpret the prediction difficult. Ultimately a widely adopted predictive tool for hERG should be able to make reliable and rationalizable predictions linked to interpretable features, such as presence or absence of certain molecular fragments.

At this point it appears that various hERG predictive methods have contributed to better understanding of the general SAR features implicated in hERG interactions. Many can be used as tools for the prioritization of compounds to be tested for hERG block due to their better performance in identifying hERG inactives. The main reason for that may lie in the data sets available for modeling. Naturally, they have tended to be global snapshots of structures associated with hERG activity, covering "broad but shallow regions of chemical space."[17] In other words, there is more known about nonblockers than blockers. This has rendered current in silico approaches less useful in making accurate predictions for each molecule, such as in the case of an ongoing medicinal chemistry optimization program. The drive toward better computational models may get a boost in the near future from increasing availability of high-throughput hERG channel assays, such as planar patch. While more work is needed to validate the results from these assays versus the 'gold standard' of the patch clamp, these techniques are providing a window into local structure-activity data sets not available previously. Current computational methods do, however, identify trends, which may be sufficient to permit in silico identification of potential scaffold-based QT liabilities in the course of whole library screens, internal as well as commercial, and real as well as virtual. It has been hypothesized that anywhere up to 25–30% of all compounds proceeding from screening to medicinal chemistry optimization have hERG blocking activity.[66] This observation may provide a further impetus to in silico approaches to identifying hERG blockers. These approaches have the potential to provide cost-saving tools that will be applied in combination with in vitro assays to facilitate the discovery of QT prolongation potential early in the drug discovery process, thus improving downstream attrition rates and ultimately the cardiac safety of medications.

References

1. Lipinski, C. A.; Lombardo, F.; Dominy, B. W.; Feeney, P. J. *Adv. Drug Deliv. Rev.* **1997**, *23*, 3–25.
2. Ajay; Walters, W. P.; Murcko, M. A. *J. Med. Chem.* **1998**, *41*, 3314–3324.
3. Egan, W. J.; Walters, W. P.; Murcko, M. A. *Curr. Opin. Drug Disc. Dev.* **2002**, *5*, 540–549.
4. Teague, S. J.; Davis, A. M.; Leeson, P. D.; Oprea, T. *Angew. Chem. Int. Ed. Engl.* **1999**, *38*, 3743–3748.
5. Oprea, T. I.; Davis, A. M.; Teague, S. J.; Leeson, P. D. *J. Chem. Inf. Comput. Sci.* **2001**, *41*, 1308–1315.
6. Beresford, A. P.; Segall, M.; Tarbit, M. H. *Curr. Opin. Drug Disc. Dev.* **2004**, 7, 36–42.
7. Clark, D. E. *Drug Disc. Today* **2003**, *8*, 927–933.
8. Jalaie, M.; Arimoto, R.; Gifford, E.; Schefzick, S.; Waller, C. L. *Methods Mol. Biol.* **2004**, *275*, 449–520.
9. Ekins, S.; Erickson, J. A. *Drug Metab. Dispos.* **2002**, *30*, 96–99.
10. Penzotti, J. E.; Lamb, M. L.; Evensen, E.; Grootenhuis, P. D. *J. Med. Chem.* **2002**, *45*, 1737–1740.
11. Ekins, S. *Drug Disc. Today* **2004**, *9*, 276–285.
12. De Ponti, F.; Poluzzi, E.; Cavalli, A.; Recanatini, M.; Montanaro, N. *Drug Safety* **2002**, *25*, 263–286.
13. Abriel, H.; Schlapfer, J.; Keller, D. I.; Gavillet, B.; Buclin, T.; Biollaz, J.; Stoller, R.; Kappenberger, L. *Swiss Med. Wkly* **2004**, *134*, 685–694.
14. Redfern, W. S.; Carlsson, L.; Davis, A. S.; Lynch, W. G.; MacKenzie, I.; Palethorpe, S.; Siegl, P. K.; Strang, I.; Sullivan, A. T.; Wallis, R. et al. *Cardiovasc. Res.* **2003**, *58*, 32–45.
15. Mitcheson, J. S.; Perry, M. D. *Curr. Opin. Drug Disc. Dev.* **2003**, *6*, 667–674.
16. Sanguinetti, M. C.; Mitcheson, J. S. *Trends Pharmacol. Sci.* **2005**, *26*, 119–124.
17. Pearlstein, R.; Vaz, R.; Rampe, D. *J. Med. Chem.* **2003**, *46*, 2017–2022.
18. Aronov, A. M. *Drug Disc. Today* **2005**, *10*, 149–155.
19. Recanatini, M.; Poluzzi, E.; Masetti, M.; Cavalli, A.; De Ponti, F. *Med. Res. Rev.* **2005**, *25*, 133–166.
20. Brown, A. M. *Cell Calcium* **2004**, *35*, 543–547.
21. Bazett, H. C. *Heart* **1920**, 7, 353–370.
22. Towbin, J. A.; Wang, Z.; Li, H. *Drug Metab. Dispos.* **2001**, *29*, 574–579.
23. Antzelevitch, C.; Shimizu, W. *Curr. Opin. Cardiol.* **2002**, *17*, 43–51.
24. Fernandez, D.; Ghanta, A.; Kauffman, G. W.; Sanguinetti, M. C. *J. Biol. Chem.* **2004**, *279*, 10120–10127.
25. Netzer, R.; Ebneth, A.; Bischoff, U.; Pongs, O. *Drug Disc. Today* **2001**, *6*, 78–84.
26. Wood, C.; Williams, C.; Waldron, G. J. *Drug Disc. Today* **2004**, *9*, 434–441.
27. Cavero, I.; Mestre, M.; Guillon, J. M.; Crumb, W. *Expert Opin. Pharmacother.* **2000**, *1*, 947–973.
28. Friesen, R. W.; Ducharme, Y.; Ball, R. G.; Blouin, M.; Boulet, L.; Cote, B.; Frenette, R.; Girard, M.; Guay, D.; Huang, Z. et al. *J. Med. Chem.* **2003**, *46*, 2413–2426.

29. McCauley, J. A.; Theberge, C. R.; Romano, J. J.; Billings, S. B.; Anderson, K. D.; Claremon, D. A.; Freidinger, R. M.; Bednar, R. A.; Mosser, S. D.; Gaul, S. L. et al. *J. Med. Chem.* **2004**, *47*, 2089–2096.
30. Finlayson, K.; Witchel, H. J.; McCulloch, J.; Sharkey, J. *Eur. J. Pharmacol.* **2004**, *500*, 129–142.
31. Goldsmith, P. *Curr. Opin. Pharmacol.* **2004**, *4*, 504–512.
32. Langheinrich, U.; Vacun, G.; Wagner, T. *Toxicol. Appl. Pharmacol.* **2003**, *193*, 370–382.
33. Mitcheson, J. S.; Chen, J.; Lin, M.; Culberson, C.; Sanguinetti, M. C. *Proc. Natl. Acad. Sci. USA* **2000**, *97*, 12329–12333.
34. Pearlstein, R. A.; Vaz, R. J.; Kang, J.; Chen, X. L.; Preobrazhenskaya, M.; Shchekotikhin, A. E.; Korolev, A. M.; Lysenkova, L. N.; Miroshnikova, O. V.; Hendrix, J. et al. *Bioorg. Med. Chem. Lett.* **2003**, *13*, 1829–1835.
35. Doyle, D. A.; Morais Cabral, J.; Pfuetzner, R. A.; Kuo, A.; Gulbis, J. M.; Cohen, S. L.; Chait, B. T.; MacKinnon, R. *Science* **1998**, *280*, 69–77.
36. Jiang, Y.; Lee, A.; Chen, J.; Cadene, M.; Chait, B. T.; MacKinnon, R. *Nature* **2002**, *417*, 515–522.
37. Jiang, Y.; Lee, A.; Chen, J.; Ruta, V.; Cadene, M.; Chait, B. T.; MacKinnon, R. *Nature* **2003**, *423*, 33–41.
38. Blaustein, R. O.; Miller, C. *Nature* **2004**, *427*, 499–500.
39. Cohen, B. E.; Grabe, M.; Jan, L. Y. *Neuron* **2003**, *39*, 395–400.
40. Perry, M.; de Groot, M. J.; Helliwell, R.; Leishman, D.; Tristani-Firouzi, M.; Sanguinetti, M. C.; Mitcheson, J. *Mol. Pharmacol.* **2004**, *66*, 240–249.
41. Chen, J.; Seebohm, G.; Sanguinetti, M. C. *Proc. Natl. Acad. Sci. USA* **2002**, *99*, 12461–12466.
42. Rajamani, R.; Tounge, B. A.; Li, J.; Reynolds, C. H. *Bioorg. Med. Chem. Lett.* **2005**, *15*, 1737–1741.
43. Zhou, Y.; Morais-Cabral, J. H.; Kaufman, A.; MacKinnon, R. *Nature* **2001**, *414*, 43–48.
44. Cavalli, A.; Poluzzi, E.; De Ponti, F.; Recanatini, M. *J. Med. Chem.* **2002**, *45*, 3844–3853.
45. Morgan, T. K., Jr.; Sullivan, M. E. *Prog. Med. Chem.* **1992**, *29*, 65–108.
46. Matyus, P.; Borosy, A. P.; Varro, A.; Papp, J. G.; Barlocco, D.; Cignarella, G. *Int. J. Quant. Chem.* **1998**, *69*, 21–30.
47. SYBYL 6.22 ed.; Tripos Inc.: St. Louis, MO, 1997.
48. Ekins, S.; Crumb, W. J.; Sarazan, R. D.; Wikel, J. H.; Wrighton, S. A. *J. Pharmacol. Exp. Ther.* **2002**, *301*, 427–434.
49. Ekins, S. *Biochem. Soc. Trans.* **2003**, *31*, 611–614.
50. Chang, G.; Guida, W. C.; Still, W. C. *J. Am. Chem. Soc.* **1989**, *111*, 4379–4386.
51. Mohamadi, F.; Richards, N. G. J.; Guida, W. C.; Liskamp, R.; Lipton, M.; Caufield, C.; Chang, G.; Hendrickson, T.; Still, W. C. *J. Comput. Chem.* **1990**, *1*, 440–467.
52. Keseru, G. M. *Bioorg. Med. Chem. Lett.* **2003**, *13*, 2773–2775.
53. Tong, W.; Lowis, D. R.; Perkins, R.; Chen, Y.; Welsh, W. J.; Goddette, D. W.; Heritage, T. W.; Sheehan, D. M. *J. Chem. Inf. Comput. Sci.* **1998**, *38*, 669–677.
54. Du, L. P.; Tsai, K. C.; Li, M. Y.; You, Q. D.; Xia, L. *Bioorg. Med. Chem. Lett.* **2004**, *14*, 4771–4777.
55. Aptula, A. O.; Cronin, M. T. *SAR QSAR Environ. Res.* **2004**, *15*, 399–411.
56. Buyck, C. In *EuroQSAR 2002. Designing Drugs and Crop Protectants: Processes, Problems, and Solutions*; Ford, M., Livingstone, D., Dearden, J., Van de Waterbeemd, H., Eds.; Blackwell Publishing: Oxford, UK, 2003, pp 86–89.
57. Roche, O.; Trube, G.; Zuegge, J.; Pflimlin, P.; Alanine, A.; Schneider, G. *Chembiochem* **2002**, *3*, 455–459.
58. Ghose, A. K.; Crippen, G. M. *J. Comput. Chem.* **1986**, *7*, 565–577.
59. TSAR 3.21 ed.; Oxford Molecular Ltd.: Oxford, UK.
60. Schneider, G.; Neidhart, W.; Giller, T.; Schmid, G. *Angew. Chem. Int. Ed. Engl.* **1999**, *38*, 2894–2896.
61. Volsurf 2.0.6 ed.; Multivariate Infometric Analysis S.r.l.: Perugia, Italy.
62. Dragon 1.11 ed.; Milano Chemometrics and QSAR group: Milano, Italy.
63. Aronov, A. M.; Goldman, B. B. *Bioorg. Med. Chem.* **2004**, *12*, 2307–2315.
64. Carhart, R. E.; Smith, D. H.; Venkataraghavan, R. *J. Chem. Inf. Comput. Sci.* **1985**, *25*, 64–73.
65. Testai, L.; Bianucci, A. M.; Massarelli, I.; Breschi, M. C.; Martinotti, E.; Calderone, V. *Curr. Med. Chem.* **2004**, *11*, 2691–2706.
66. Bains, W.; Basman, A.; White, C. *Prog. Biophys. Mol. Biol.* **2004**, *86*, 205–233.
67. Zolotoy, A. B.; Plouvier, B. P.; Beatch, G. B.; Hayes, E. S.; Wall, R. A.; Walker, M. J. A. *Curr. Med. Chem. – Cardiovasc. Hematolog. Agents* **2003**, *1*, 225–241.
68. Rowley, M.; Hallett, D. J.; Goodacre, S.; Moyes, C.; Crawforth, J.; Sparey, T. J.; Patel, S.; Marwood, R.; Thomas, S.; Hitzel, L. et al. *J. Med. Chem.* **2001**, *44*, 1603–1614.
69. Blum, C. A.; Zheng, X.; De Lombaert, S. *J. Med. Chem.* **2004**, *47*, 2318–2325.
70. Bilodeau, M. T.; Balitza, A. E.; Koester, T. J.; Manley, P. J.; Rodman, L. D.; Buser-Doepner, C.; Coll, K. E.; Fernandes, C.; Gibbs, J. B.; Heimbrook, D. C. et al. *J. Med. Chem.* **2004**, *47*, 6363–6372.
71. Fletcher, S. R.; Burkamp, F.; Blurton, P.; Cheng, S. K.; Clarkson, R.; O'Connor, D.; Spinks, D.; Tudge, M.; van Niel, M. B.; Patel, S. et al. *J. Med. Chem.* **2002**, *45*, 492–503.
72. Potet, F.; Bouyssou, T.; Escande, D.; Baro, I. *J. Pharmacol. Exp. Ther.* **2001**, *299*, 1007–1012.
73. Collins, I.; Rowley, M.; Davey, W. B.; Emms, F.; Marwood, R.; Patel, S.; Fletcher, A.; Ragan, I. C.; Leeson, P. D.; Scott, A. L.; Broten, T. *Bioorg. Med. Chem.* **1998**, *6*, 743–753.
74. Bell, I. M.; Gallicchio, S. N.; Abrams, M.; Beshore, D. C.; Buser, C. A.; Culberson, J. C.; Davide, J.; Ellis-Hutchings, M.; Fernandes, C.; Gibbs, J. B. *J. Med. Chem.* **2001**, *44*, 2933–2949.
75. Fraley, M. E.; Arrington, K. L.; Buser, C. A.; Ciecko, P. A.; Coll, K. E.; Fernandes, C.; Hartman, G. D.; Hoffman, W. F.; Lynch, J. J.; McFall, R. C. et al. *Bioorg. Med. Chem. Lett.* **2004**, *14*, 351–355.
76. Paakkari, I. *Toxicol. Lett.* **2002**, *127*, 279–284.

Biography

Alex M Aronov is Senior Research Scientist in the Applications Modeling Group at Vertex Pharmaceuticals, Cambridge, MA. His research interests include ways to capture chemical intuition as it relates to drug discovery, structure-based design, and ADMET modeling. Prior to coming to Vertex in early 2001, he was Postdoctoral Fellow in the laboratories of Irwin 'Tack' Kuntz, Paul Ortiz de Montellano, and Ching Ching Wang at the University of California–San Francisco, working on chemogenomic target validation of parasitic targets through structure-based design of novel antiparasitic agents. Originally trained in bioorganic and medicinal chemistry, he received his Doctor of Philosophy degree in 1998 from the University of Washington under the direction of Mike Gelb and Wim Hol, where he was involved in the design and synthesis of novel antitrypanosomal agents.

5.41 The Adaptive In Combo Strategy

D A Smith and L Cucurull-Sanchez, Pfizer Global Research and Development, Sandwich, UK

5.41.1 Introduction

In a recent article, Lombardino and Lowe[1] described changes in the way medicinal chemistry and indeed drug discovery are performed today compared to a decade or more earlier. They describe the much wider range of tools to help overcome the numerous hurdles in the drug discovery process (advances in synthetic, analytical, and purification technology, such as transition metal-catalyzed carbon–carbon bond-forming reactions, high-field nuclear magnetic resonance) as well as computer-assisted literature and data retrieval and analysis. The radical changes mentioned are combinatorial chemistry and high-throughput screening (HTS). Combinatorial chemistry allows chemists to generate rational, focused libraries of compounds that allow access to lead material and then potentially define structure–activity relationships (SAR) rapidly. They also identify an Achilles heel with the technology with the statement: "many companies are struggling to triage the large number of screening hits to viable lead compounds that can support a successful drug discovery project. In this struggle, costs can escalate significantly as the generation of large amounts of data is not the same as generating viable, quality leads." Taking this further, when combinatorial chemistry is used to generate SAR and aid lead development, the screening costs are multiplied even further. The logic of human genome-derived targets and subsequent in vitro screening of compounds against these specific molecularly defined targets is sound to derive specific agonists or antagonists. This is particularly true now as compound files have become more drug-like and project targets are defined by knowledge about the potential binding site as druggable. Selectivity is also readily and rapidly screened for against close neighbours or proteins associated with adverse events (e.g., the rapidly activating delayed rectifier potassium current channel (I_{Kr}) in cardiac tissues (hERGK$^+$, K$_v$11.1)). In vivo the slowing of repolarization, by compounds inhibiting this channel, leads to QT interval prolongation and the appearance of potentially fatal arrhythmias (torsade de pointes). Again, reflecting on a problem often encountered in the past, where discovery programmes concentrated on target potency, with the exclusion of other properties, humanized in vitro screens have been developed that relate to human pharmacokinetic performance. Such screens are termed generically ADME screens and generally address the key pillars of a compound's in vivo performance (drug-like). The pillars are illustrated in **Figure 1**.

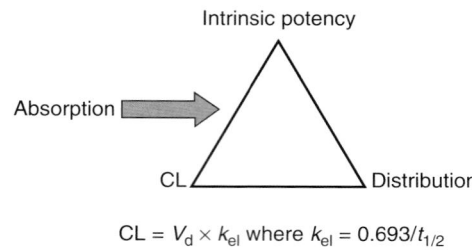

$$CL = V_d \times k_{el} \text{ where } k_{el} = 0.693/t_{1/2}$$

Figure 1   Schematic illustrating the key pillars that influence the dynamics of drugs.

5.41.1.1   In Vitro Screens for Oral Absorption

Absorption can be divided into dissolution and permeability phases. Solubility is a key parameter for dissolution of compounds following oral administration. The process depends on the surface area of the dissolving solid and the solubility of the drug at the surface of the dissolving solid. Solubility is inversely proportional to the number and type of lipophilic functions within the molecule and the tightness of the crystal packing of the molecule. Rapid, robust methods reliant on turbidimetry to measure solubility have been developed.[2,3]

Currently, in vitro methods such as Caco-2 or Madin–Darby canine kidney (MDCK) cell monolayers are widely used to make permeability estimates.[4-6] Both these cell lines form monolayers of polarized cells. These cell lines express drug transporters like P-glycoprotein (P-gp). They can be used in 24- and 96-well plates, but this is possibly the limit for these cell-based screens in their current form. The main advantage of MDCK cells is their shorter culture time. Artificial membranes such as parallel artificial membrane permeability assay (PAMPA) have been advanced for high-throughput permeability assessment. The original PAMPA method[7] involved creating a filter-immobilized artificial membrane by infusing a lipophilic microfilter with 10% (weight/volume) egg lecithin dissolved in n-dodecane. The filter membrane separated an aqueous solution containing the compound of interest from aqueous buffer. Microtiter plate technology allows rapid screening and the capacity can exceed the 96-well limit of cell-based assays. The use of nonbiological material and microtiter plates greatly increases speed and lowers cost.

The artificial membranes obviously lack transporters and other proteins, and the results generated are analogous to intrinsic membrane permeability. This produces a dilemma since ideally both pure membrane permeability and the influence of drug transporters would be desired to provide data that could be deconvoluted. Reliance on the cell-based systems generates complex SAR output that can be partially addressed by measuring flux in both directions or over a range of concentrations. Ultimately, these solutions lead to even more pressure on the in vitro screening systems.

5.41.1.2   In Vitro Screens for Clearance

The metabolic stability of compounds can be assayed in a high-throughput or semi-high-throughput screening system using recombinant enzymes, human liver microsomes, or human hepatocytes.[8] The use of mass spectrometry provides near-universal detectors for many of these in vitro systems, and separation systems are continually evolving to allow more and more direct introduction of sample. The choice of reagent governs the breadth of metabolic processes examined. The recombinant enzyme, normally CYP3A4, obviously only studies reactions performed by that enzyme. However some 60–70% of all drugs are cleared predominantly by CYP3A4, so screening against this isoenzyme will provide useful SAR. Microsomal systems with the appropriate cofactors provide a comprehensive screening reagent for oxidative metabolism by the liver. The cytosolic oxidation systems are absent such as aldehyde oxidase but these generally play a minor role. In most cases, screening is run using the microsomes fortified with NADPH (via a regenerating system) to study P450 metabolism (and flavin monooxygenases). Broad screening for metabolic stability is best accomplished with hepatocytes, which provide a system containing all the enzyme systems: oxidative, conjugative, and hydrolytic. In terms of ease of use the hierarchy is reversed as hepatocyte systems are more difficult to obtain, difficult to cryopreserve, and generally show limited linearity against cell concentration. This means, in terms of measuring stability, hepatocytes have a lower dynamic range. Human systems also suffer from the inter-subject variability of the donors and even when pooled into fairly large batches show differences in metabolic rate across batches. Screening results are normally percentage remaining or described by disappearance half-life. This is convertible into intrinsic clearance using appropriate scaling, a parameter that has direct pharmacokinetic significance in terms of in vivo pharmacokinetics. Intrinsic clearance (CL_i) relates to hepatic extraction (E) and hence systemic clearance CL_s and hepatic first-pass effect (F):

$$E = f_u \cdot CL_i / (f_u \cdot CL_i + Q_H) \qquad [1]$$

where f_u is fraction unbound in blood and Q_H is liver blood flow.

$$CL_s = Q_H \cdot E \qquad [2]$$

Bioavailability is also dependent on the fraction absorbed through the gut (permeability and gut metabolism, partially estimated by screens in Section 5.41.1.1 above).

$$F = 1 - E \qquad [3]$$

5.41.1.3   In Vitro Screens for Volume of Distribution

Various methods of prediction of the volume of distribution of a compound have been advanced. Very few are utilized in a high-throughput mode. These rely on looking at partitioning into tissue sections or pieces and then extrapolating that data using known values such as tissue volume and weight. The number of tissue types examined may be broad, like that described by Ballard *et al.*[9] (adipose, brain, heart, intestine, kidney, liver, lungs, muscle, pancreas, skin, spleen, stomach, testes, and thymus). Other prediction methods[10] have simplified the number of tissue-partitioning experiments, which need to be conducted and suggest that only two tissue types (adipose and muscle) are required.

5.41.2   The Obvious Solution

All these screens and others are high-value, but, as stated above, are very costly to maintain and run. If a number of discovery programmes are advanced using combinatorial chemistry, and there are compelling reasons to do this, the burden on compound control, formatting multiwell plates, robotic screening equipment, and reagent provision becomes unsupportable.

The immediate answer to the above is the use of in silico techniques. This approach offers none of the restrictions referred to once a model or system is developed and validated. These models could be used for prioritization: selecting which synthesized compounds to put into the expensive screening cascade referred to above or, more attractively, actually to guide synthesis. The latter could lower the number of compounds needed to develop the SAR nescessary to achieve a drug candidate and save both chemistry and screening resources.

There are numerous in silico approaches to attempt to model both the output of screens and actual human pharmacokinetic performance. The number of publications for the last decade describing aspects of ADME screening are shown graphically in **Figure 2**. There is an obvious upward trend for publications around in silico approaches and many reviews appear on this topic, summarizing the various attempts and approaches.[11,12]

The vast majority of these papers conclude with apparently successful methods that should be directly applicable to solving the pillars shown in **Figure 1** to an acceptable level of accuracy. Despite this huge and clear strategic drive to achieve the obvious solution, screening is still the method of choice and in silico methods have done little to change this.

5.41.3   The Solution Is Murky

Stouch and co-workers[13] produced a somewhat unusually titled paper that begins to unravel why the in silico revolution has not happened despite the promise of the literature. Their paper, 'In silico ADME/Tox: why models fail' describes several case examples which actually apply very often generically.

5.41.3.1   Data Quality

To gain a large enough data set literature data are employed. Such data will span many laboratories and will have been collected under varying experimental protocols and conditions. The data set therefore is worthless. Even in a single laboratory, data collected over time will often show variation. An example would be human microsomal or hepatocyte clearance. It is common practice to pool these reagents to try to lower variations; individual human samples show huge variation in certain enzymes.[14] For example, the major human P450 (CYP3A4) shows up to 80-fold variation across a

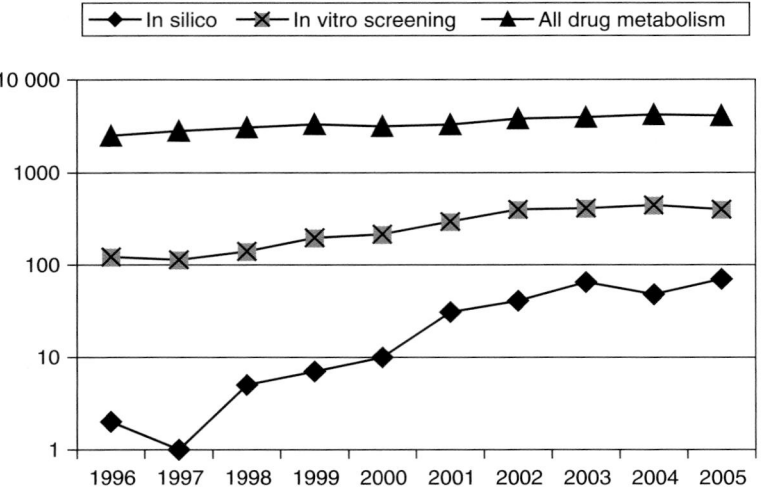

Figure 2 Graphical representation of the increasing publication of in silico approaches compared to in vitro screening and 'drug metabolism' in general.

series of human livers ($n = 45$).[15] Despite pooling, substantial variation can be seen between pooled batches. Even with the appropriate positive controls it is difficult to build a large data set from such screens. The commonly used Caco-2 cell line system shows variation in result due to cell passage numbers, culture time, type of support, and medium.[4,6] Hou *et al.*[5] in describing the correlation of Caco-2 permeation with simple molecular properties, highlight the variation in laboratory data. In their training data set the P_{eff} for mannitol, often used as a quality standard, showed 10-fold variation across referenced studies (-6.75 versus -5.49). Their approach in 'curating' the published data indicates the difficulty in literature-derived training sets. The authors also point out the somewhat more surprising ranges encountered in measured lipophilicity descriptors such as $\log D_{7.4}$. They note again a near-10-fold variation in values for compounds, as exemplified by atenolol (-1.29 versus -2.14).

Other aspects can also be added in that certain data are very hard to obtain. Whilst human bioavailability data are readily available, these data are a composite of fraction of the drug absorbed, gut first-pass metabolism, and liver first-pass metabolism. To obtain human absorption per se as a true permeability value and relate that to Caco-2 or replace such screens is thus highly problematic. Likewise, most of the data on blood–brain penetration is derived from animal experiments comparing whole blood or plasma with total brain. Such data are only of limited use compared to cerebrospinal fluid or extracellular fluid concentration data. Whole-brain partitioning actually represents partitioning into the lipid of the brain, and not actually access to drug receptors. The apparent dramatic differences in brain distribution described for total brain, as shown above (3–4 orders of magnitude), collapse to a small ratio when free (unbound) concentration of drug in plasma is compared to cerebrospinal fluid or extracellular fluid concentration.[16]

5.41.3.2   Compound Space

Almost invariably, any model will fail. No matter how many compounds are employed to build a data set, chemists will innovate and make structures dissimilar to those used to build the model. In the example quoted by Stouch *et al.*,[13] the model was based on 800 marketed drugs. These compounds showed only a very low similarity (less than 0.3 on a scale of 0–1) to those being investigated as putative drugs.

5.41.3.3   Accuracy around Critical Decision Points

Many screening cascades have a cut-off point to either test the compound further or to seek an alternative compound. These cut-off points are based on project wisdom and sometimes pragmatism. Many models can show a general trend, but fail to have any predictive value around a go/no-go value. Stouch *et al.*'s example[13] is a decision of CYP2D6 inhibition. Although the model was 90% predictive, the go/no-go point was 10 μM. Here the model predicted correctly 60% of the time and falsely 40% of the time.

In a similar vein, the paper includes a description of a solubility model where, even if accuracy had been achieved, the dynamic range did not address the key areas of interest. In this model the lowest limit was 1 μm for poorly soluble compounds. With potent molecules solubility below this value may be adequate. Moreover, progress in improving solubility from very low levels could not be predicted.

Lombardo and colleagues[17] also question the progress made with in silico modeling. They concluded from a critical review of work in the field of in silico ADME published during the period 2000–2002 that not much progress had been made in developing robust and predictive models. In agreement with the statements above, they also concluded that the lack of accurate data, together with the use of questionable modeling endpoints, had greatly hindered real progress in defining generally applicable models. They also refer to the complex nature of ADME phenomena, in which a single desired property such as absorption is actually a complex outcome of several biological processes. Local models refer to the concept of small more focused data sets that could be a single project, therapeutic area, or even a research site (as discussed later, in adaptive in combo models, in practice other definitions of 'local' can be applied). 'Local' will probably mean that the training set is more likely to reflect future compounds for which predictions are needed and also that the data used for training are comparable and consistent for each compound. They also state that the empirical nature of quantitative structure–activity relationship (QSAR)/quantitative structure–permeability relationship (QSPeR) approaches may ultimately dictate the use of 'local' models, since it would be hard to extrapolate predictions outside the parameters dictated by the training set.

5.41.4 Physicochemistry Provides a Model and a Link

Despite the apparent lack of progress, intrinsically ADME data have a strong relationship with the physicochemistry of the compound. Physicochemical properties should be predictable from consideration of the chemical structure. Many of the screens provide data that can only be rationalized by an understanding of the physicochemical properties. This relationship is illustrated in **Table 1**.

Lipophilicity is the key physicochemical parameter linking solubility, membrane permeability, and hence drug absorption and distribution with route and rate of clearance (metabolic or renal). Measured or calculated lipophilicity of a compound is readily amenable to automation at low cost and high throughput. Many of these calculation approaches rely on fragment values, where the model is broken down into fragments. The existing measured values of compounds form the basis for the values assigned to each fragment as well as various correction factors that correct for the proximity of certain groupings. These measured values determine to a large degree the accuracy of the method. De novo calculations often have inaccuracies due to missing fragments or that the core template expresses a combination of properties that the fragmental approach misses (e.g., internal or external hydrogen bonding). The accuracy of the calculation can be significantly improved by actual measurement of some representatives of the core template and/or compound containing or missing the fragment not in the original measured values. This can be likened to the exact approach we will discuss in more depth in Sections 5.41.5–5.41.7: that is, predictions on compounds that have properties (sufficiently) different to those studied previously not to allow the investigator to predict their properties from previous values (can be likened to a training set).

To understand when properties are sufficiently different to change the behavior of a molecule in a biological system, the concept of chemical space has been advanced and developed.[18] The huge number of compounds that can be synthesized represents the universe. Within that compounds can be mapped (chemography) against chemical descriptors of physicochemical and topological properties creating two-dimensional (2D), three-dimensional (3D), or

Table 1 Physicochemical properties and the relationship to key disposition (pillars) processes. Number of ticks indicates relative importance and the arrow indicates how an increase in the physicochemical property affects the ADME property, e.g., dissolution is decreased by increasing lipophilicity

	Lipophilicity	Molecular size	Hydrogen bonding	Ionization
Dissolution	√√√ ↓			√√
Absorption gastrointestinal tract permation, lipoidal	√√√ ↑	√√	√√	√
Absorption gastrointestinal tract permation, aqueous		√√√ ↓		
Metabolism	√√√ ↑			√
Renal clearance	√√√ ↓			
Volume of distribution	√√√ ↑			√√

multidimensional space. The hydrogen-bonding capacity of a drug solute is now recognized as an important constituent of the concept of lipophilicity. Initially $\Delta \log P$, the difference between octanol/water and alkane/water partitioning was used as a measure for solute hydrogen bonding, but this technique is limited by the poor solubility of many compounds in the alkane phase. Computational approaches to this range from simple heteroatom counts (O and N), division into acceptors and donors, and more sophisticated measures such as free energy factors[19] (used in programme Hybot), to polar surface area.[20]

5.41.5   Fixed or Adaptive in Combo Models

In Section 5.41.3, the statement was made that invariably any model will fail. The often suggested answer in computational terms is, to overcome failure, a bigger and better data set is required or that more or better descriptors are employed.

Despite the much larger data set used for training, new compounds explore chemistry space that has not been explored in the original training set. The chemical space is almost infinite. It has been calculated that 10^{60} different molecules could be made from a combination of just 30 C, N, O, and S atoms. It seems highly unlikely that a training set will ever exist that covers the 'space' of this dimension.[21] Whilst this may sound dramatic in terms of structural differences, creating an internal hydrogen bond in a molecule that masks a previously 'exposed' hydrogen bond can have dramatic effects on the biological behavior of a molecule (e.g., gut or cell penetration). The much larger data set allows a model to be built but the lack of precision in estimating parameters for a new molecule makes the use of the model of little value in filtering compounds for screening or guiding synthesis.

An alternative is to turn the cause of failure into an alternative strategy. The simple concept described in Section 5.41.7 – that of making the model adaptive, that is, it grows with new chemical space – should address a major cause of failure. Once a predictive model is constructed it is, by definition, restricted to the chemical space which built the model. To include new chemical space, new data are needed. This is provided by further data from in vitro screens. This interactive process is illustrated as **Figure 3**. Such an approach requires that all new compounds are examined for similarity to the original model training set. Those that fall outside chemical space need to be screened in vitro (e.g., singleton compounds). Those compounds that do fall within the chemical space of the model can be screened in silico, and then those that are predicted to be highly positive or negative of the screening scale results (e.g., metabolically

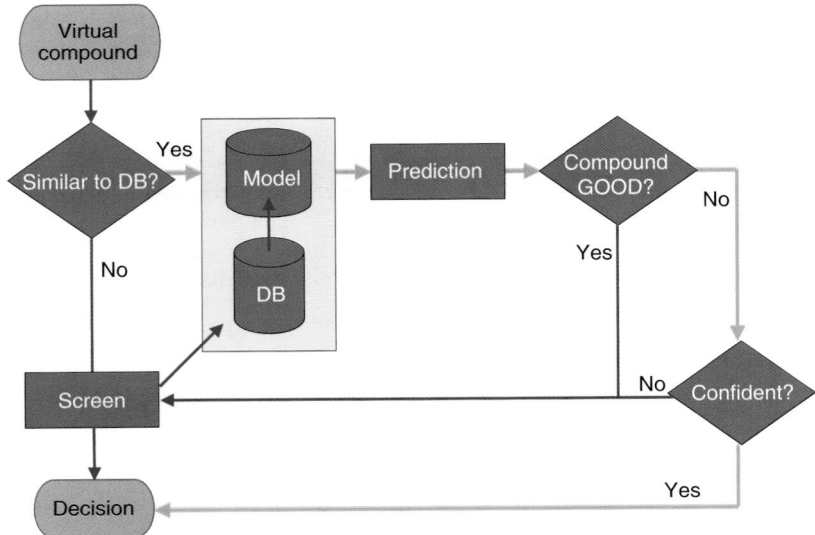

Figure 3 Scheme of an ideal adaptive in combo approach. The process starts with the design of a virtual compound. 'DB' represents the database used to train the in silico model and, if it exists within the compound space of the data base, a prediction on its ADME properties is made. Synthesis of the compound is triggered by a prediction of good ADME properties or it being outside the compound space of the training set, resulting in a low confidence in prediction. The compounds with low confidence in prediction are screened in the appropriate ADME screens and the data then add to the database. The combination of in silico and in vitro techniques provides the necessary information to make the decision faster and at a lower cost than a purely in vitro approach, and with more reliability than a purely in silico derivation of information. Since 'DB' is constantly growing in this combined approach, the model is constantly being updated to cover the evolving chemical space.

stable or labile, *see* Section 5.41.6) can be withdrawn from the in vitro screening routine with a high degree of confidence. If a more aggressive strategy needs to be addressed in order to reduce costs, then the number of screened compounds could be limited to those that are not within the chemical space of the model. The earlier this method is applied in the drug discovery and development pipeline, the further the costs are reduced – if a compound is identified by the in silico model as a failure and discarded when it is still virtual, then the savings derived from an unnecessary synthesis can be added on top of the screening savings. However, in order to adopt this strategy, the error associated with the in silico predictions should compare the error of the in vitro experiments, and a very robust method to measure the similarity of the compounds to the chemical space of the model training set must be put in place. The screened data are obviously available to medicinal chemists to guide synthesis. More importantly, perhaps the new data allow the model to be retrained with a data set that now covers the new chemical space (represented by 'DB' in **Figure 3**).

It is appropriate at this point to define further the concept of a compound's similarity or its position in chemical space, since it is key to the success of the in combo approach. Several recent examples in the literature use this concept to evaluate the quality of in silico QSAR models, either to study systematically the best approach for different kinds of training sets,[22] or to introduce new concepts like the dimension-related distance (DRD),[23] or to explore alternative approaches to linking similarity measures and model quality based on a residuals classification.[24] The bottom line is that the measure of similarity or diversity is context-dependent. Despite the intuitively simple idea of placing a compound in a chemical space and measuring its proximity to its nearest neighbors, there are several aspects of this process that need to be considered:

1. The kind of molecular descriptors that will be used to describe the chemical space and how many should there be: should they be the same descriptors used to develop the in silico model? Or should they invariably be structure-related (e.g., fingerprints)? Can we reduce the number of descriptors to latent variables (e.g., via principal component analysis, PCA) without losing significance in the measure of similarity?
2. How do we measure the distances in the chemical space? The most widely used metric for fingerprint-based chemical spaces is the Tanimoto coefficient, but when other types of descriptors are involved, then the choice of distance metrics expands: Euclidian, Manhattan, Mahalanobis, Pearson, etc.[25]
3. How many neighbors should our molecule have in the training set chemical space so that we can apply the model with confidence? It is clear in the case of a singleton that the number of neighbors will be zero or one at the most, but to choose a criterion for compounds embedded in the whole training set region is not trivial.

Eventually, any chosen strategy to evaluate a compound's similarity to the modeling chemical space will require optimization, so that the best descriptor space, distance metrics, and number of nearest neighbors can provide the best measure of the in silico model reliability. Recent examples that relate specifically to ADME are discussed in Section 5.41.8.

The adaptive in combo concept becomes even more attractive in today's environment. Combinational chemistry, applied these days to medicinal chemistry, relies on a common scaffold (or linking group). This scaffold is amenable to rapid parallel reactions at a number of sites to allow the rapid creation of new molecules. These scaffolds tend to be designed around core-privileged structures that begin to match the knowledge surrounding the various drug targets (e.g., aminergic and nonaminergic G protein-coupled receptors). Active compounds from a high-throughput screen can be rapidly exploited, as described above. Such programmes by definition occupy similar chemical space. Indeed, even in the absence of combinational techniques, the final compound selected in development or successful drug programmes shows high similarity to the original lead matter. An adaptive model, therefore, built on a smaller data set from a common scaffold will not fail due to being dissimilar in 2D or 3D chemical space, when new compounds from the same series are tested. The difference with combinational chemistry is that enough molecules are rapidly synthesized early in the programme to form a training set. As new scaffolds are adopted, so the model can be retrained, growing with the increasing chemical space.

Adaptive in combo models should have the maximum chance of fulfilling the three basic requirements of the training data set: (1) accuracy of the data; (2) diversity in terms of occupying the chemical space of the compounds for which predictions are made; and (3) quantity, in that there is sufficient density around the chemical space of the compounds for which predictions are made.

An adaptive in combo model, thus created, should have advantages beyond covering the expected chemical space. It is again conceivable that such models can be based on simplified descriptors, chosen not simply to 'fit' the data, but ones that can be directly applied to target new chemical synthesis, that is, to guide drug design. The section on physicochemistry links the pillars of drug behavior to properties that are directly in control of the medicinal chemist. They are easy to understand and to visualize. For instance, albeit with literature-derived training and test sets, Hou *et al.*[5] describe a model based on simple molecular properties correlating well ($r^2 = 0.67$) with Caco-2 permeation. Four molecular descriptors were found crucial to Caco-2 permeation: $\log D_{7.4}$, high-charged polar surface areas based on Gasteiger partial changes, the radius of gyration, and the fraction of rotatable bonds. All of these easily relate to medicinal chemistry and modifications that can be performed on subsequent molecules. Moreover, as the authors describe, these four descriptors give a meaningful physical picture of the mechanism of Caco-2 permeation. A lipophilic molecule penetrates the Caco-2 monolayer easier, but above a $\log D_{7.4}$ value of 2.0, the influence of lipophilicity reaches a maximum. Larger polar surface areas make a negative contribution to permeability, but this is mainly limited to those atoms with high-charge densities. In addition, larger and more rigid molecules have lower permeability than smaller or flexible ones. Other descriptors of lesser value include number of all atoms, number of aromatic atoms, number of hydrophobic atoms, number of all bonds, fraction of rotable bonds, mass density, dipole moment, heat of formation, Wiener index, Zagreb index, Hosoya index, and solvation free energy.

Definitely the data that are used to generate the model will not suffer from data quality, although by definition the data will be limited to the in vitro absorption, distribution, metabolism, excretion, and toxicity (ADMET) screens (e.g., **Figure 1**). Data quality will be ensured as the assays will be from a single source in a single laboratory with standardized reagents and conditions. The limitations are that these are in vitro results which then have to be extrapolated to the clinical situation, rather than a model based on human data. However, as outlined above, such global models invariably fail with new compound series that are removed in chemical space.

5.41.6   Limitations on What Can Be Achieved

An additional problem with many ADME parameters is the degree of precision attached to them. For instance, if in silico approaches allowed the in vitro pharmacological potency of a molecule to be determined from the structure of the ligand and some information on the target, then one or two orders of magnitude error is permissible. Such predictions would be of use in areas such as lead-finding or even compound optimization. Such error is not useful in ADME prediction, since it would often span the entire range of properties a molecule can possess.

A key difficulty with adaptive models is that the prediction is limited to in vitro properties: few compounds will progress to in vivo testing, and even fewer will be advanced to the clinic. In vitro tests are very good at predicting the extremes of the ranges of compound properties, particularly the positive end (the term 'positive' refers to the test and not the potential of the compound). Thus, high Caco-2 flux will invariably mean good absorption. Likewise, very fast metabolism by human liver microsomes or hepatocytes will lead to low bioavailability and high clearance. When the results move from the extremes, the systems are less predictive about eventual in vivo performance. This uncertainty has led to the 'binning approach,' where the ranges of the assay are divided into three or four segments. This limitation of the in vitro tests invariably means a compounded error in predicting in vivo performance by in silico methods. An example of this is provided by Shen et al.[26] The authors used a k-nearest neighbor approach, which relies on the assumption that chemically similar compounds possess similar physical and biological properties; a hypothesis strongly supported by **Table 1**. A four-'bin' model was developed to predict metabolic stability in human liver homogenate. The model was reasonably predictive of stable compounds; however, it was much less predictive of compounds in the intermediate stability 'bins'. Considering that the prediction of in vivo performance is most problematic from in vitro data in the intermediate range, the in silico output is likely to be even more uncertain.

Another variation of some of the problems on data quality and drift in assays is exemplified by solubility. Chemists expect a single right answer rather than a range. Even something as apparently simple as solubility does not have a single answer. A particular problem in drug discovery is that solubility is not constant. When a newly synthesized compound is first isolated, it is often as the amorphous form. In this form solubility is invariably at its highest. This is usually the stage at which solubility assays are run to guide new compound synthesis. Subsequent isolation will be in a crystalline form and a decrease in solubility is seen. As purification (crystallization) conditions are further improved, polymorphs are isolated, each representing the thermodynamically most stable form. Polymorphs are different crystalline forms which, although chemically identical, result from a different ordered arrangement of molecules within the crystalline lattice.

5.41.7   Adaptive in Combo Models in Practice

There should be no particular limit to the amount of chemical space covered in any particular model provided it has adequate distribution of structural types and data on the training set. Whilst we have stressed that combinational chemistry scaffolds around privileged structures may be a powerful tool in confirming the need for wide chemical space, we recognize the narrowness of this approach. This is particularly true in the early period of a project, when insufficient data will be available to train a model and early impact on the formative period, including behaviors (in silico–in project), will be lost. To begin to examine when the SAR of a few projects would be diluted by the diversity of many, we have examined the performance of a metabolic stability model constructed using data derived from projects on two sites. All the projects had diverse targets but some overlap in therapeutic area. One of the sites has excelled in traditional medicinal chemistry whilst the other is renowned for structure-based drug design. Both sites synthesize compounds using singleton and library approaches.

The results of this analysis are summarized in **Table 2**.

The models used experimental data derived from human microsomal systems optimized for oxidative metabolism to train a naive Bayesian classifier for each site (La Jolla (LJ) and Sandwich (SND) models) and for both sites (Global model). Several descriptor sets were used to build several models, and the best performers were kept as final models for each data set. The procedure to derive and evaluate the models was identical for both sites: 80% of the data available was used to train the model, whilst a test set was derived with the remaining 20% and used to evaluate the performance of the model (1 and 4 in **Table 2**). The chemical space of the test set is most likely covered by the training set. Each of these test sets was then run with the model developed for the other site to investigate the probable impact of compounds being outside the chemical space of the model (2 and 5 in **Table 2**). As expected, the performance of these models fell, although they still carried some predictive value. A further dimension of deviation from chemical space was examined by the 'external set' (3 and 6 in **Table 2**). These comprised the data points from newly synthesized compounds at each site. The global model was built with a merge of both training sets, and also evaluated with a merged test set (7 in **Table 2**). This represents an adaptive model since it has been trained to cover the compound space with the new data points from both sites.

The concordance values shown in **Table 2** represent the percentage of compounds correctly predicted. Sensitivity and specificity are the ratios of correctly predicted 'stable' and 'unstable' compounds, respectively. The positive and negative predictive values indicate respectively which proportions of 'stable' and 'unstable' predictions are consistent with the experimental data. It can be seen that the combination of both data sets yields a good model, since all the parameters relative to the global test set are ⩾80%. This can be contrasted with the performance of the site model's performance on compounds from the other sites. However, the most remarkable result is the higher prediction power of the global model with respect to the external validation sets (8 and 9 in **Table 2**): the percentage of captured positives amongst the La Jolla data set increases from 5% to 23%, and both the positive and the negative predictive values also increase for this site. At the same time, the predictions for the Sandwich compounds do not seem to improve or worsen with the inclusion of the La Jolla compounds into the training set, since the sensitivity increases from 16% to 55% at the expense of the negative predictive value, which decreases from 57% to 29%.

Table 2 Comparative performances, in the prediction of metabolic stability, of two local models (La Jolla (LJ) and Sandwich (SND)) versus a global model that combines La Jolla (LJ) and Sandwich (SND) data. The models are built from training sets exclusive to these research sites; these training sets are then combined to create the global model. The concordance values represent the percentage of compounds correctly predicted. Sensitivity and specificity are the ratios of correctly predicted 'stable' and 'unstable' compounds, respectively

	Model	Validation set (number of molecules)	Concordance (%)	Sensitivity (%)	Specificity (%)	Positive predictive value (%)	Negative predictive value (%)
1	La Jolla	LJ test set (408)	86	89	85	49	98
2	La Jolla	SND test set (446)	50	32	81	74	41
3	La Jolla	LJ external set (117)	81	5	99	50	82
4	Sandwich	SND test set (446)	86	87	85	91	80
5	Sandwich	LJ test set (408)	46	42	72	90	17
6	Sandwich	SND external set (80)	60	16	100	100	57
7	Global	GLB test set (854)	85	83	86	80	89
8	Global	LJ external set (117)	83	23	97	63	84
9	Global	SND external set (80)	61	55	92	97	29
10	Sandwich with 80% SND external set	20% SND external set (13)	77	40	100	100	73

These results lead to the conclusion that, as more chemical space is covered, the larger model performs better than the smaller dedicated models in dealing with new compounds, even from a local environment. The probable explanation is that, as soon as the two local models are combined, the chances of a new data set being predicted by interpolation rather than extrapolation are higher, hence the higher chances of obtaining correct predictions.

Another way of achieving interpolating rather than extrapolating predictions is to enrich the models with data from a chemical space adjacent in time, rather than in space, to the original training set of the model. That would be the case for the adaptive model represented in **Figure 3**. To illustrate this, 80% of the SND external set was randomly selected to generate an enriched Sandwich model, and the remaining 20% was used to test its performance. The results, included in row 10 of **Table 2**, show how the predictions for the SND external set have significantly improved when a representative fraction of its chemical space was incorporated in the Sandwich model: all the statistical indicators have either improved or remained unchanged relative to the results shown in row 6.

Of course, this result might apply to the particular endpoint modeled, or to the particular external tests used for the evaluation, but at least it provides evidence that the adaptive model is a sound strategy and can incorporate considerable chemical space and still be predictive.

5.41.8 Mapping Chemical Space and Absorption, Distribution, Metabolism, and Excretion

Sun[27] has focused attention on the data sets used in the models rather than the models themselves, in a manner analogous to our analysis of the problem, paying particular attention to their size and diversity and to the intrinsic quality. His belief is that data-driven methods, such as quantitative structure–activity/property relationships (QSA/PR), project complicated properties like ADME prediction on to the chemical space defined by compounds in the training set and that this projection simplifies the problem by avoiding the need to define and apply specific mechanisms. Thus, absorption or metabolism processes are complex but can be simplified in these models to a single output. This output is only useful if the training set is relevant in terms of chemical space to the chemical space occupied by the compounds for which predictions are desired.

How to define the need to retrain or adapt the model requires a knowledge of the chemical space occupied by the compounds used to train it. This map of chemical space needs to be specific enough to identify when new compounds, for which predictions are required, fall outside that chemical space.

There are a number of methods for estimating drug space and mapping it. One method (ChemGPS),[28] which has already been applied to ADME,[29] is to borrow from geography and utilize a system like the Mercator convention. Rules are established that are equivalent to dimensions such as longitude and latitude, while structures are equivalent to objects (e.g., cities and countries). Selected rules include size, lipophilicity, polarizability, charge, flexibility, rigidity, and hydrogen bond capacity. Selected objects include a set of 'satellite' structures and a set of representative drugs ('core' structures). Satellites, intentionally placed outside drug space, have extreme values in one or several of the desired properties, while containing drug-like chemical fragments found in drugs. ChemGPS (chemical global positioning system) is a tool that combines these predefined rules and objects to provide a global drug space map. Novel structures are placed in drug space via principal components analysis score prediction, providing a mapping device for the chemical space. ChemGPS scores are comparable across a large number of chemicals and do not change as new structures are predicted, allowing new compounds to be compared with existing compounds used in the model. Moreover, a record is provided for all previously explored regions of the chemical space, which may allow existing or new models to be assembled with excellent chemical space coverage and thus high predictivity for novel compounds. The principal properties can be translated into interpretable chemical descriptors, as shown in **Figure 4**.

5.41.9 In Silico–In Project

The examples provided above by Stouch *et al.*[13] highlighted critical decision points of projects. If in silico approaches are to succeed, they have to be seen as an integral part of the project decision process. This is a chicken-and-egg situation, since the in silico approach will not be trusted unless highly validated and trusted – to date something it has failed to do. Screens are often constructed to give decision point readouts such as the cut-off outlined for CYP2D6 inhibition (10 μm). Similar cut-off values are often used for metabolic liability (percentage remaining in a microsome or hepatocyte system after a given time) and gut permeability (flux value above a certain value, efflux ratio below a certain value). Being on the correct side of such values allows the compound to progress. This is an easy way of evaluating compounds as the screens can have a single assay point and the data lead to clear decisions. Such data may not be the best way either to construct or to use in silico models.

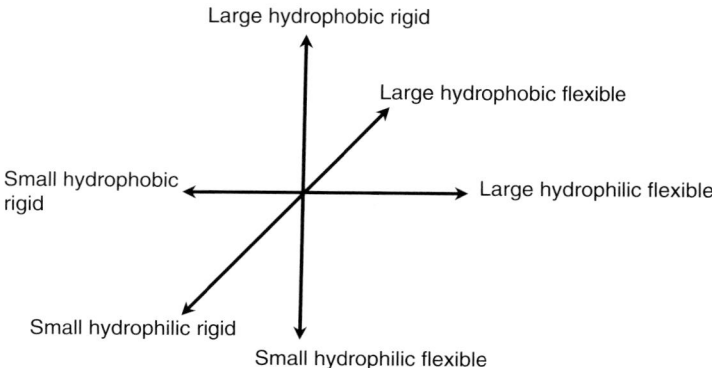

Figure 4 Principal properties translated into chemical descriptors, illustrating the three most significant dimensions of chemical space.

In fact the application of in silico techniques to drug discovery projects[30] has suffered from a disconnect between those doing the in silico modeling and those synthesizing the compounds. The disconnect can be summarized by the three lacks: (1) lack of alignment; (2) lack of inclusion; and (3) lack of enthusiasm to make change happen. What is needed is:

1. the project at the outset to adopt an in combo philosophy
2. to ensure the data from screens is in an appropriate form for in silico model development (e.g., dynamic range addresses all or the vast majority of compounds with real values and not greater or less than)
3. to ensure in synthesis plans that new chemical space is adequately addressed early to provide modeling substrate
4. the output of the model is in a form that the project team can use and will guide synthesis
5. that the in silico approach of the in combo paradigm is used fully to guide synthesis and that the project does not gradually move back to full reliance on in vitro data.

5.41.10 Conclusion

The adaptive model represents a future refinement of in silico techniques. What is needed is not better modeling techniques of the ADME data, but better ways of defining chemical space. Robust definition of chemical space will allow retraining of the model in a timely manner, triggered by the design of new compounds outside the chemical space and their subsequent screening. Pragmatic methods, such as limiting each model to a lead series, an individual project, or a cluster of projects around similar enzymes or receptors may offer a simple 'entry' solution, but, unless the models are adaptive, will begin to fail at some stage as new chemical templates are explored. Ultimately, with better measurement of ADME space, models that encompass many projects of diverse nature may be achievable.

References

1. Lombardino, J. G.; Lowe, J. A. *Nat. Rev. Drug Disc.* **2004**, *3*, 853–862.
2. Avdeef, A. *Pharm. Pharmacol. Commun.* **1998**, *4*, 165–178.
3. Lipinski, C. A.; Lombardo, F.; Dominy, B. W.; Feeney, P. J. *Adv. Drug Del. Rev.* **1997**, *23*, 3–25.
4. Artusson, P.; Palm, K.; Luthman, K. *Adv. Drug Del. Rev.* **1996**, *22*, 67–84.
5. Hou, T. J.; Zhang, W.; Xia, K.; Quiao, X. B.; Xu, X. J. *J. Chem. Info. Sci.* **2004**, *44*, 1585–1600.
6. Hidalgo, I. J. *Curr. Topics Med. Chem.* **2001**, *1*, 385–401.
7. Kansy, M.; Avdeef, A.; Fischer, H. *Drug Disc. Today: Technol.* **2004**, *1*, 349–355.
8. Ansede, J. H.; Thakker, D. R. *J. Pharm. Sci.* **2004**, *93*, 239–255.
9. Ballard, P.; Leahy, D. E.; Rowland, M. *Pharm. Res.* **2003**, *20*, 864–872.
10. Bjorkman, S. *J. Pharm. Pharmacol.* **2002**, *54*, 1237–1245.
11. Ekins, S.; Nikolsky, Y.; Nikolskaya, T. *T. I. P. S.* **2005**, *26*, 202–209.
12. Norinder, U. *SAR QSAR Environ. Res.* **2005**, *16*, 1–11.
13. Stouch, T. R.; Kenyon, J. R.; Johnson, S. R.; Chen, X. Q.; Doweyko, A.; Li, Y. *J. Comput.-Aided Mol. Des.* **2003**, *17*, 83–92.
14. Smith, D. A.; Jones, B. *Curr. Opin. Drug Disc. Dev.* **1999**, *2*, 33–41.
15. Sy, S. K. B.; Ciaccia, A.; Li, W.; Roberts, E. A.; Okey, A.; Kalow, W. B-K. Tang,. *Eur. J. Clin. Pharmacol.* **2002**, *58*, 357–365.
16. Van de Waterbeemd, H.; Smith, D. A.; Jones, B. C. *J. Comput.-Aided Mol. Des.* **2001**, *15*, 273–286.
17. Lombardo, F.; Gifford, E.; Shalaeva, M. Y. *Rev. Med. Chem.* **2003**, *3*, 861–875.

18. Lipinski, C.; Hopkins, A. *Nature* **2004**, *432*, 855–861.
19. Raevsky, O. A.; Skvortsov, V. S. *J. Comput.-Aided Mol. Des.* **2002**, *16*, 1–10.
20. Atkinson, F.; Cole, S.; Green, C.; Van de Waterbeemd, H. *Curr. Med. Chem.: C.N.S. Agents* **2002**, *2*, 229–240.
21. Bohacek, R. S. M.; Guida, W. C. C. *Med. Res. Rev.* **1996**, *16*, 3–50.
22. Sheridan, R. P.; Feuston, B. P.; Maiorov, V. N.; Kearsley, S. K. *J. Chem. Inf. Comput. Sci.* **2004**, *44*, 1912–1928.
23. Xu, Y.-J.; Gao, H. *QSAR Comb. Sci.* **2003**, *22*, 422–429.
24. Guha, R.; Jurs, P. C. *J. Chem. Inf. Model.* **2005**, *45*, 65–73.
25. Migliavacca, E. *M. R. M. C.* **2003**, *3*, 831–843.
26. Shen, M.; Xiao, Y.; Golbraikh, A.; Gombar, V. K.; Tropsha, A. *J. Med. Chem.* **2003**, *46*, 3013–3020.
27. Sun, H. *Curr. Comput.-Aided Drug Des.* **2005**, *1*, 179–193.
28. Oprea, T. I.; Gottfries, J. *J. Comb. Chem.* **2001**, *3*, 157–166.
29. Oprea, T. I.; Zamora, I.; Ungel, A.-L. *J. Comb. Chem.* **2002**, *4*, 258–266.
30. Smith, D. A. *Drug Disc. Today* **2002**, *7*, 1080–1081.

Biographies

Dennis A Smith has worked in the pharmaceutical industry for the past 30 years since gaining his PhD from the University of Manchester, UK, following research on the metabolism of diacetylmorphine. For the last 18 years he has been at Pfizer Global Research and Development, Sandwich, where he is Vice President – Pharmacokinetics, Dynamics and Metabolism. Prior to Pfizer he was head of the Metabolism and Pharmacokinetics Unit at Fisons Pharmaceuticals, Loughborough. During this 30-year span he has helped in the discovery and development of seven marketed NCEs, with hopefully several more to come. His 90 research publications include a number of seminal review papers on the impact of pharmacokinetics and drug metabolism on drug discovery, the role of physicochemical measurement in drug metabolism, species differences in metabolism, the substrate structure requirements of cytochrome P450 enzymes, the interpretation of animal pharmacokinetic and pharmacodynamic data in drug safety assessment, the role of high-throughput screening, and metabolites in safety and efficacy testing. He is active in a teaching role, holding appointments as Visiting Professor at the University of Liverpool and Honorary Senior Lecturer at the University of Aberdeen.

Lourdes Cucurull-Sanchez joined the in silico ADME modeling group in Pfizer in 2004, where she currently develops several in silico ADME models for end-points relevant to pharmacokinetics, dynamics, and metabolism. She was previously appointed as a postdoctoral research associate at the University of Cambridge, UK. Sponsored by

Unilever, she applied in silico modeling techniques to several ADME problems, including oral bioavailability and absorption mechanisms. She obtained her PhD in Chemistry in 2000 at the Universitat Autònoma de Barcelona, Spain, where she carried out several ab initio and molecular calculations to assist the bioinorganic chemical modeling of copper proteins. In 1995 she obtained an MPhil in Chemistry at the University of Newcastle upon Tyne, UK, after 1 year of research on x-ray structures of metal complexes with thiolates and related ligands.

Comprehensive Medicinal Chemistry II
ISBN (set): 0-08-044513-6

ISBN (Volume 5) 0-08-044518-7; pp. 957–969

5.42 The Biopharmaceutics Classification System

H Lennernäs, Uppsala University, Uppsala, Sweden
B Abrahamsson, AstraZeneca, Mölndal, Sweden

© 2007 Elsevier Ltd. All Rights Reserved.

5.42.1 Introduction

In modern pharmaceutical product development bioavailability (BA) and bioequivalence (BE) have been the subject of repeated studies. Presently, BE studies are being conducted for New Drug Applications (NDAs) of new chemical entity, in supplementary NDAs for new medical indications and product line extensions, in Abbreviated New Drug Applications (ANDAs) of generic products, and in applications for scale-up and post-approval changes. For example, NDA BE studies may be required comparing different formulations in pivotal clinical trials and products intended for market. The complexity and number of studies required are often driven by the fact that several dose strengths may be included in the development process. In addition, BE documentation may also be needed comparing blind and original comparator products in clinical trials. Thus, an NDA typically contains a multitude of in vivo BE studies and the biopharmaceutics classification system (BCS) may be a scientific and a regulatory tool to decrease the number of in vivo BE studies and replace them with BCS based biowaiver (as schematically described in **Figure 1**) which also includes nonregulatory applications.

The BCS has been developed to provide a scientific approach for classifying drug compounds based on solubility as related to dose and intestinal permeability in combination with the dissolution properties of the oral immediate release dosage form.[1–3] Additional considerations are pH, chemical and enzymatic stability in the gastrointestinal lumen, membrane transport mechanism, and therapeutic index. The aim of BCS is to provide a regulatory tool for replacing certain BE studies by accurate in vitro dissolution tests. It should be noted that permeability and solubility are substance-specific properties whereas dissolution is product-specific. Thus, it is only dissolution testing that is relevant as surrogate for in vivo BE tests while the BCS classification is a tool to select the compounds suitable for this approach. Presently, it is only high-solubility/high-permeability drugs (class I) for which this approach is acceptable. This is a minimal risk approach because BA of immediate release formulations of class I drugs is insensitive to product dissolution whereas for drugs with low permeability/solubility (classes II–IV) subtle differences in in vitro dissolution could translate into significant differences in BA. Thus, in the regulatory context the BCS classification is used as a risk management tool.

The BCS will certainly reduce costs and time in the drug development process, both directly and indirectly, and reduce unnecessary drug exposure in healthy volunteers, which is normally the study population in BE studies. For example, it has been reported that an application of a BCS strategy in drug development will lead to significant direct and indirect savings for pharmaceutical companies. The BCS is today only intended for oral immediate release products

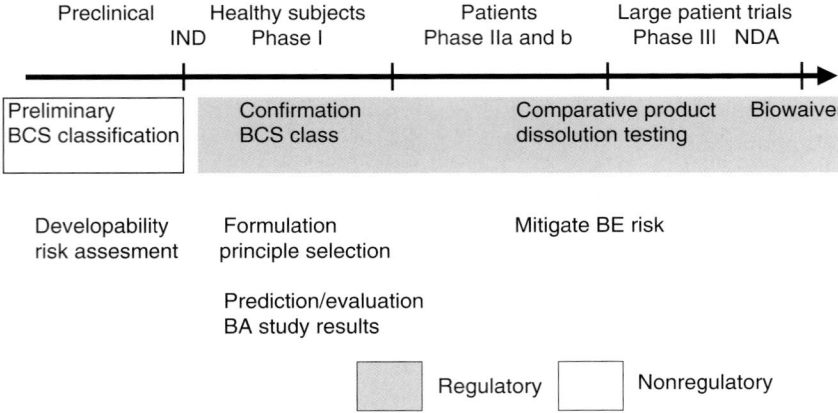

Figure 1 Where in the drug development process is the biopharmaceutics classification system (BCS) applied? BCS characterizations are *primarily determined* during preclinical phase and confirmed in phase I. Comparative dissolution tests of drug products as surrogate for in vivo BE studies are mainly performed during phase III and a biowaiver can be obtained. Several nonregulatory applications of BCS in drug development are also shown.

Biopharmaceutics Classification System
(BCS)

Class I High solubility High permeability	**Class II** Low solubility High permeability
Class III High solubility Low permeability	**Class IV** Low solubility Low permeability

Permeability (human intestinal absorption) — vertical axis

Solubility: Volume (mL) of water required to dissolve the highest dose strength over pH 1–7.5 range

Figure 2 The BCS provides a scientific basis for predicting intestinal drug absorption and for identifying the rate-limiting step based on primary biopharmaceutical properties such as solubility and effective intestinal permeability (P_{eff}). The BCS divides drugs into four different classes based on their solubility and intestinal permeability (defined as more than 90% of the dose absorbed). Drug regulation aspects related to in vivo performance of pharmaceutical dosage forms have been the driving force in the development of BCS. Guidance for industry based on BCS mainly clarifies when bioavailability/bioequivalence (BA/BE) studies can be replaced by in vitro bioequivalence testing.[127]

that are completely absorbed throughout the intestinal tract (>90% of the given dose), and accordingly have an absorption kinetics similar to an orally given solution with the same compound. Furthermore, the BCS has been shown to be very useful for identifying the rate-limiting step and predicting intestinal drug absorption based on primary biopharmaceutical properties such as solubility and effective intestinal permeability (P_{eff}).[1–3] A schematic drawing of the biopharmaceutics classification system (BCS) is shown in **Figure 2**.

BCS has primarily been developed for regulatory applications of oral immediate release products. However, it has also several other implications in the drug discovery, preclinical, and clinical drug development processes and has gained wide recognition within the research-based industry.[3–5] The importance of drug dissolution in the gastrointestinal (GI) tract and intestinal permeability in the oral absorption process has been well known since the 1960s, but the research carried out to constitute the BCS has provided new quantitative data of importance for modern drug development, especially within the area of clinical investigations and determinations of drug permeability with a single-pass perfusion of the human jejunum.[6–8] Another merit of the BCS is that it provides very clear and easily applied rules for determining the rate-limiting factor in the sometimes very complex GI drug absorption process. In this way the BCS framework can provide guidance in many situations during drug development, for example, to aid in lead optimization and developability risk assessment at clinical candidate drug (CD) selection, to assess feasibility of animal models for in vivo absorption studies, to predict and elucidate food interactions, to support determination of transport mechanism, to aid choice of formulation principle, including suitability for oral extended release (ER) administrations, and to open the possibility of in vitro/in vivo correlations in dissolution testing of solid formulations and risk assessment prior to in vivo BE studies.[5,9] BCS could thereby contribute to more rational decision-making both in screening for candidate drugs and in full development.

The aim of the present chapter is to discuss the past, the present, and the future application of BCS in pharmaceutical drug discovery, preclinical, and clinical product development for oral pharmaceutical products. It is important to realize that oral dosing is the most preferred drug administration route in the clinical usage of drugs: the majority (84%) of the 50 most-sold pharmaceutical products in the USA and European markets are given orally.[4] An extension of the BCS regulatory guideline is considered to be essential for widespread use in all development phases as the current version limits the broad use of BCS since class I drug substances today belong to a minority in pharmaceutical pipelines. For instance, the proportion of class I compounds in the development phase for oral immediate release formulations at AstraZeneca was less than 10% in 2001.[4] By contrast, among well-established drugs on the market, as represented by the 123 WHO oral drugs in immediate release dosage forms, about 25–30% were class I drugs.[10] An example list of drug BCS classification based on human permeability and solubility are given in **Table 1**.

Table 1 BCS classification based on human permeability[6] and dose number[10]

Drug	Human permeability ($\times 10^4$ cm s^{-1})	Dose number[a]	BCS class
α-Methyldopa	0.10	0.1	III
Amoxicillin	0.30	0.9	III
Antipyrine	5.60	0.20	I
Atenolol	0.20	0.02	III
Carbamazepine	4.30	80	II
Cephalexin	1.56	2	II
Cimetidine	0.26	3	III
Enalapril maleate	1.57	0.003	(I)[b]
Furosemide	0.05	30	IV
Hydrochlorothiazide	0.04	0.2	III
L-Dopa	3.40	1.0	(I)[b]
Lisinopril	0.33	0.002	III
Losartan potassium	1.15	0.004	III
Metoprolol tartate	1.34	0.0004	I
Naproxen sodium	8.50	0.06	I
Propanolol hydrochloride	2.91	0.01	I
Ranitidine hydrochloride	0.27	0.01	III
Terbutaline	0.30	0.01	III
Valacyclovir hydrochloride	1.66	0.02	I
Verapamil	6.80	0.004	I

[a] Dose number = (highest dose strength/gastric volume (250 mL))/solubility.
[b] High permeability due to carrier-mediated absorption; currently not included in BCS class I.

5.42.2   Definition of Relevant Pharmacokinetic Variables

Bioavailability (F) is the most useful pharmacokinetic variable for describing the overall quantitative role of all processes influencing the absorption (f_a) and first-pass metabolism and excretion (E_G and E_H) in the gut and liver.[11] This pharmacokinetic variable is used to illustrate the fraction of the dose that reaches the systemic circulation, and relate it to pharmacological and safety for oral pharmaceutical products from several different pharmacological classes. To summarize, bioavailability is mainly dependent on three general processes, the fraction dose absorbed (f_a), and the first-pass extraction of the drug in the gut wall (E_G) and by the liver (E_H)[1,6,12,13]:

$$F = f_a \cdot (1 - E_G) \cdot (1 - E_H) \tag{1}$$

There are different factors that might affect the fraction dose absorbed (i.e., f_a) and gut wall metabolism (i.e., E_G) of drugs. In general, these can be divided into three general categories: (1) pharmaceutical factors such as choice of excipients and method of manufacturing, (2) physiochemical factors of the drug molecule itself such as solubility and lipophilicity; and (3) physiological, genetic, biochemical, and pathophysiological factors in the intestine.[1,6,12,13] The f_a is the fraction of the dose transported (absorbed) across the apical cell membrane into the cellular space of the enterocyte according to scientific and regulatory definitions.[1,2,6,12–14] Accordingly, the rate (mass/time) and extent of drug absorption (f_a = mass/dose) from the intestinal lumen in vivo are influenced by: dose/dissolution ratio, chemical degradation or metabolism in the lumen, luminal complex binding, intestinal transit, and effective permeability (P_{eff}) across the intestinal mucosa. The fraction dose absorbed ($M(t)$/dose), i.e., the fraction of drug disappeared from the intestinal lumen during a certain residence time, assuming no luminal reactions, at any time t is[1]

$$\frac{M(t)}{dose} = \int_0^t \int \int A \cdot P_{eff} \cdot C_{lumen} \cdot dA \, dt \tag{2}$$

where A is the available intestinal surface area, P_{eff} is the average value of the effective intestinal permeability along the intestinal region where absorption occurs, and C_{lumen} is the free reference concentration of the drug in the intestinal lumen.[1,4,6] From eqn [2] it is obvious that the P_{eff} and the dissolved and free drug concentration are the key variables controlling the overall absorption rate and extent, and it is possible to use this regardless of the transport mechanism of the drug.[1,6] Once the drug has reached the intracellular site in the enterocyte, it may be subjected to cytochrome P450 metabolism, predominantly CYP3A4, as well as other enzymatic step such as phase II conjugation by glucuronides, sulfates, and/or glutathione. The enzymatic capacity of the small intestine to metabolize drugs can in pharmacokinetic terms be expressed as the extraction ratio of the intestine (E_G).[1,6,12,13] It is important to realize that CYP3A4 is not expressed in the colon.[15–17] The heterogeneity of CYP3A4-mediated gut wall metabolism is crucial to consider, especially for the assessment of interactions with other drugs and/or diet as well as development of ER dosage forms. Instead, drug metabolism by colonic microflora may play a crucial role in colonic drug absorption, especially with regard to drugs given in ER dosage forms, which may be subjected to predominantly hydrolytic and other reductive reactions.[18] The fraction that escapes metabolism in the small intestine ($1 - E_G$) may undergo additional metabolism and/or biliary secretion in the liver (E_H) before reaching the systemic circulation. The E_H is dependent on the blood flow (QH), the protein binding (f_u), and the intrinsic clearance of the enzymes and/or transporters (CL_{int}).[11] Recently, it has also been recognized that membrane transport into as well out of the hepatocyte has to be included in the models for predicting and explaining liver extraction.[19,20] These membrane transporters might influence uptake into the hepatocyte and canalicular transport into the bile. These processes are expected to play a crucial role for the local concentration–time profile in the vicinity of the intracellularly located enzyme in the liver.

5.42.3   The Biopharmaceutics Classification System

5.42.3.1   Present Regulatory Applications

Presently, the BCS is primarily used as a regulatory tool for identifying which drug substances in oral immediate release formulations are suitable for in vitro BE testing and a possible biowaiver. Additional criteria that must be verified for a biowaiver besides solubility, product dissolution, and permeability include drug stability in the GI fluids, therapeutic index classification (narrow or non-narrow), absence of effect of pharmaceutical excipients on the rate and extent of intestinal permeability, and lack of absorption of the parent drug from the oral cavity. If these criteria are met, test and reference products can be compared by in vitro dissolution testing and deemed bioequivalent if sufficiently similar results are obtained. In vitro dissolution testing should be done at three different pH values within the physiological range, typically pH 1.2, 4.5, and 6.8. Product dissolution must be complete ($>85\%$) within 30 min in order to utilize the in vitro BE route. The underlying rationale for this requirement for product performance is to ascertain that drug dissolution is fast enough so as not to become the rate-limiting step in the overall intestinal absorption. It is assumed that gastric emptying will control the onset and rate of intestinal absorption for class I drug substances in products with such a rapid dissolution and no effect on BA will be obtained for different in vitro dissolution profiles within acceptance limits. This has also been verified in vivo by studying metoprolol tablets (a class I drug) with different in vitro dissolution release profiles.[21]

The difference in dissolution rate between a test (T) and a reference (R) product should be investigated by use of the f_2-test (see eqn [3] below) where $f_2 > 50$ is the required limit for equivalence. This limit corresponds to an average difference in amount dissolved at different times (t) of less than 10%. If the dissolution is very rapid, i.e., complete dissolution within 15 min, the f_2 testing is not necessary.

$$f_2 = 50 \log \left[\left(1 + \frac{1}{n} \sum_{t=1}^{n} (R_t - T_t)^2 \right)^{-0.5} \cdot 100 \right] \qquad [3]$$

In addition to in vitro testing, the test and reference products must not contain excipients that could modify drug absorption in any way except for dissolution effects. For example, the potential for permeability-enhancing effects by surface-active agents, sometimes included in solid formulations, has been identified as one potential concern. However, the reported effects of formulation excipients on efflux inhibitors are not clearly defined, especially not in vivo. [5] Just recently it was reported by Bogman *et al.* that an in vivo intraduodenal perfusion study showed for the first time that low luminal concentrations of the dissolution enhancing vehicle tocopheryl polyethylene glycol succinate (TPGS) slightly enhanced (39%) the intestinal absorption of the P-glycoprotein (P-gp) substrate (ABCB1) talinolol. The concentration of TPGS used was close to the in vitro IC_{50} value.[22] We have also in an in vivo perfusion study in humans investigated the influence of a penetration enhancer, sodium caprate (C10), on the rectal absorption of phenoxymethylpenicillin (PcV) and antipyrine (**Figure 3**). Based on the plasma pharmacokinetics of PcV we showed a slightly increased rectal absorption at pH 7.4 in subjects where C10 was transported into the rectal tissue (absorbed) (**Figure 4**). However, the increased rectal P_{eff} for PcV was too small to detect from the outlet perfusate, which suggests that C10 alone has a limited effect on the in vivo permeability across the rectal epithelium when it is exposed in a solution.[23] These two studies provide an example of the maximum possibility for an interaction since they are perfused through a specific segment. Finally, the effect on GI transit by large amounts of sugars has been highlighted as another issue, especially for liquid formulations where large amounts of such excipients may be included.[3,24,25]

The present regulatory implications of a class I drug are not limited only to BE bridging for generic versus originator products or to different clinical trials versus commercial formulations but are also applied in postapproval manufacturing changes and in the justification of product quality control dissolution methods.[26,27]

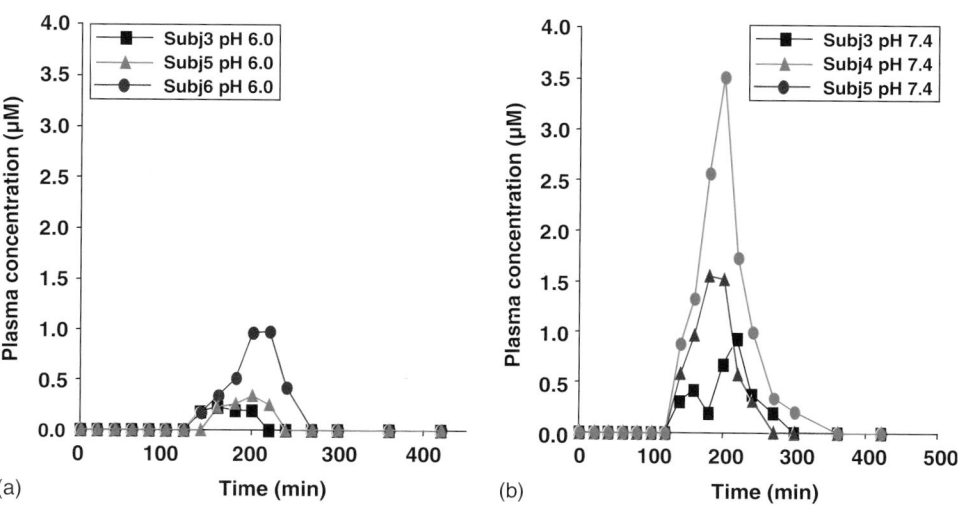

Figure 3 The plasma concentration of phenoxymethylpenicillin during an in vivo rectal perfusion in humans at (a) pH 6.0 and (b) pH 7.4. During 0–100 min and 100–200 min the rectum was perfused without and with caprate respectively at both treatment 1 (a) and treatment 2 (b).[23] Caprate was added to the perfusion solution at 100 min in both treatments 1 and 2.

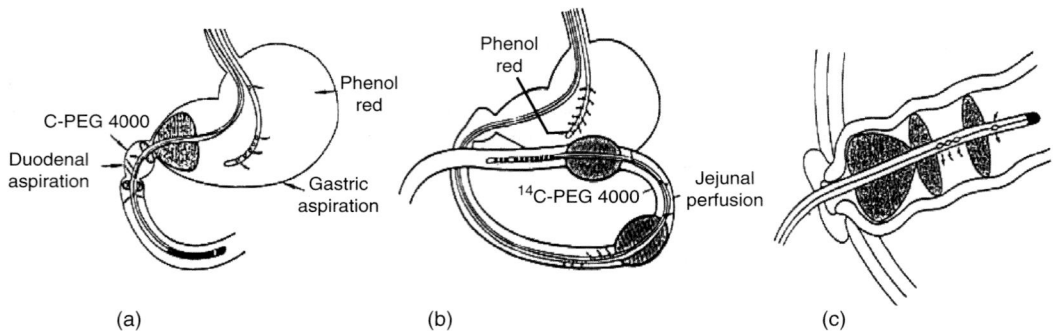

Figure 4 Single-pass intestinal perfusion of human duodenum, jejunum, and rectum in vivo. (a) The total Loc-I-Gut concept. To the left is shown a perfusion system of the duodenal segment. The tube system with double balloons (in the middle) allows a segmental single-pass perfusion of jejunum. To the right a perfusion system of the small intestinal stomi is presented. (b) Loc-I-Gut is a perfusion technique for the proximal region of the human jejunum.[7,8] The multichannel tube is 175 cm long and is made of polyvinyl chloride with an external diameter of 5.3 mm. It contains six channels and is provided distally with two 40 mm long, elongated latex balloons, placed 10 cm apart each separately connected to one of the smaller channels. The two wider channels in the center of the tube are for infusion and aspiration of perfusate. The two remaining peripheral smaller channels are used for the administration of marker substances and/or for drainage. At the distal end of the tube is a tungsten weight attached in order to facilitate passage of the tube into the jejunum. The balloons are filled with air when the proximal balloon has passed the ligament of Treitz. Gastric suction is obtained by a separate tube. ^{14}C-PEG 4000 is used as a volume marker to detect water flux across the intestinal barrier. (c) A schematic presentation of the perfusion tube with three balloons delimiting the segment located in the mid–upper part of the rectum or in a stomi patient.[23] Two balloons located on either side of the anal canal control the position of the tube. The central channel is used for the introduction of the tube by the endoscope, and decompression of gas and fluids during perfusion. The smaller channels are used to deliver air into the balloons and perfusion fluid to and from the rectal segment. The length of the perfused rectal segment is 8 cm.

5.42.3.2   Biopharmaceutics Classification System in Drug Discovery and Preclinical Development

Pharmacokinetic (absorption, distribution, metabolism, and excretion – ADME) parameters are today considered to have a crucial role in the selection process of oral candidate drugs for product development.[28–30] This understanding has also most probably contributed to significantly reduce the number of developmental failure due to pharmacokinetic factors during recent years.[31] Fundamental BCS parameters, such as permeability, solubility, and fraction dose absorbed, are among those ADME parameters and therefore it has been suggested that these fundamental BCS parameters should be useful in both the discovery and early development process.[1,32] For instance, it is clear that new compounds with a very low permeability and/or solubility/dissolution will certainly result in low and highly variable bioavailability, which may limit the possibilities that a clinically useful product can be developed. It is obvious that a selection of candidates that fulfil the BCS requirement of high permeability/high solubility (class I) almost guarantees the absence of failures due to incomplete and highly variable GI absorption. However, these BCS limits are generally too conservative to use as acceptance criteria in drug screening since many useful drugs can be found in classes II–III and even class IV.[10]

First of all, a class I drug is expected to provide complete absorption whereas a certain reduction in bioavailability due to permeability or solubility, as well as due to other reasons (e.g., first-pass metabolism), is generally acceptable. A summary of the different factors that have to be taken into account when defining more relevant acceptance criteria follows:

1. Acceptability of a low and highly variable bioavailability depending on
   ○ medical need
   ○ width of therapeutic window
   ○ potency
   ○ substance manufacturing costs.
2. Potential for poor in vivo predictability of early permeability and solubility characterizations, due to, e.g.,
   ○ active transport across the gut wall
   ○ high paracellular transport through gut wall
   ○ in vivo solubilization by bile salt micelles.
3. Opportunity to use formulation approaches that improve bioavailability, e.g.,
   ○ dissolution and solubility enhancement
   ○ permeation enhancers.

Thus, BCS points out some important variables in the screening of drug candidates whereas the proposed limits are less useful as acceptance criteria in a drug discovery context.

Experimental methods and relevant acceptance criteria regarding permeability and solubility are needed in the early drug discovery process.[33–36] Such procedures have also been introduced in the industry, including solubility screens using turbidimetric measurements and automatic permeability screens in cell-based systems (such as Caco-2 cell model). There appears to be a good correlation between in vitro and in vivo permeability for drugs with passive diffusion as the main transport mechanism, but there is a significant deviation for drugs absorbed through transporters.[36–39] It is also considered that the interpretation of the importance of efflux carrier on intestinal absorption process is overrated based on results obtained in tissue cell cultures (e.g., Caco-2 cells).[3,4,6,20,40,41] For instance, based on only cell culture data (Caco-2 cells), it was shown that the in vitro permeability of fexofenadine in the absorptive direction increased by approximately 200–300% in the presence of various P-gp inhibitors (such as verapamil, ketoconazole, and GF 120918) and that low passive diffusion was the main reason for the incomplete and variable intestinal absorption.[41] Interestingly, an in vivo perfusion of the proximal part of jejunum with our Loc-I-Gut technique showed that the in vivo permeability was affected by neither ketoconazole nor verapamil at clinical doses (**Figure 3**).[20,42] It clearly shows that the cell monolayer is too poorly defined to quantitatively predict drug–drug interactions at the transporter level. These cell models can only provide quantitative in vivo predictions of drug transport if an extensive mapping of the expression of functional activity of these membrane transporters is done. Therefore, there is a need to develop in vitro techniques with functional expressed transporter activities that are better correlated to the various regions of the human intestinal tract.[36,39,41,43,44] These in vitro permeability data need to be accurately compared with the corresponding in vivo data. We have over the past 15 years determined the human in vivo jejunal permeability with a single-pass perfusion technique (**Figure 3**) for a number of structurally different drugs, as illustrated in **Figure 5**. In addition, a critical interpretation and extrapolation of the in vivo relevance data of intestinal transport are crucial for a meaningful use of these improved cell monolayers.

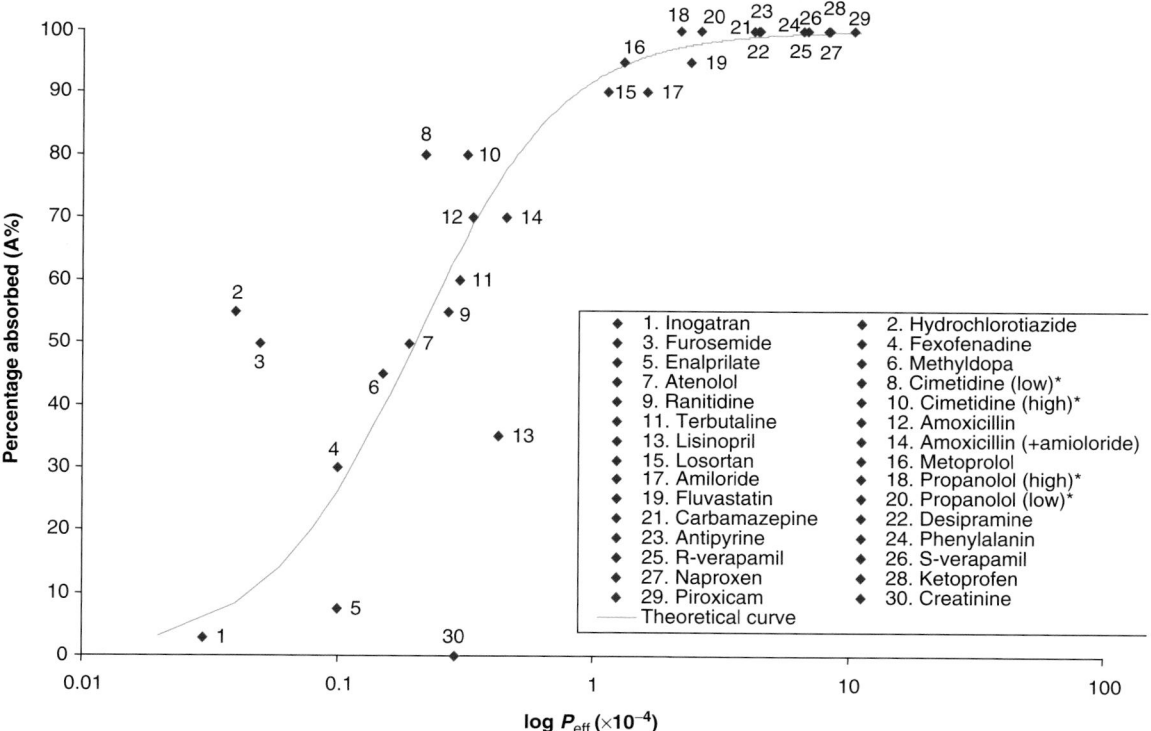

Figure 5 Human in vivo permeabilitity values (P_{eff}) can be determined by the use of a single-pass perfusion technique (Loc-I-Gut) in the human jejunum.[7,8] These human jejunal P_{eff} values have an excellent correlation to the percentage absorbed (A%). The human P_{eff}-values were originated from.[6–7,37,42,56,58,89,111,123–125]

It is well known that the intestinal absorption potential of drugs that are mainly transported by passive diffusion maybe be predicted from molecular properties, such as polar surface area, hydrogen bonding, $\log P$, $\log D$, and molecular weight.[33] Accordingly, the fundamental BCS parameters are suggested to be based on theoretical descriptors in the future and computational approaches have also been developed for permeability and solubility determinations.[1,10,32,45–49] If further refinement can be achieved for such methods, such as quantitative structure–activity relationship (QSAR) models to optimize ADME properties, it may be possible in the future to displace cell-based permeability screens and solubility estimates in screening for new drug molecules.[33,36] However, simple physicochemical descriptors of solubility and permeability must probably be assembled into and validated in more physiologically based models to be able to accurately predict human intestinal absorption.[50–52] Recently Lindahl *et al.* demonstrated that in vitro permeability values could be used to simulate the absorption from various intestinal segments in humans.[53] An intermediate approach for permeability determinations, between physicochemical and physiologically based methods, is the use of artificial membranes.[54,55] Such techniques provide a pure estimate of transcellular permeability without influence from transporters or the unstirred water layer present in cell-based models, which could provide useful feedback to chemists in the screening phase. Finally, it is important to understand that even if absorption properties are acceptable the drug may have a limited clinical use due to an extensive first-pass extraction in the gut and/or liver that is outside the topic of the present chapter.

In drug discovery there is a strong demand for early information on drug properties such as ADME variables. Several in vitro ADME screens have significantly increased the amount of experimental data generated in the early drug discovery process. In addition to these experimental techniques, there is a need for different in silico methods that can predict ADME variables with fairly high accuracy. For instance, ADME filters that can sort out drugs with undesirable ADME properties can be applied in virtual screening or drug design so as to reduce compound attrition rates. The basis for constructing these predictive models is the existence of high-quality experimental data. Several data sets exist in the literature and have been used for building models that predict ADME-related variables such as P_{eff}, f_a, and solubility.[29,33,45–50] Unfortunately, many of these data sets have been compiled from different sources and not with the same experimental conditions, making a direct validation of the different models difficult. Another shortcoming with the existing data sets is that different sets of drugs have been used for modeling different properties. This makes it difficult to construct robust and more complex quantitative models, of for example, drug absorption that would be based on $\log P/\log D$, solubility, intestinal permeability, metabolic stability, and carrier-mediated membrane transport data. In this case a relevant and common data set is essential that represents the available ADME space. New experimental tools are also constantly being developed that aim to complement in vitro and/or in vivo experiments. Many data sets in the literature contain high-quality experimental ADME-related data. Unfortunately, many of these include compounds that are not commercially available. Others are not structurally diverse, which is an important property when attempting to build predictive models of ADME properties.

In the later phases of preclinical development, various animal models are often used to evaluate the absorption potential of new drugs. The predictive value for humans of various models can be guided by BCS. For example, the rat is considered an appropriate model that predicts permeability-limited human intestinal absorption well.[56–63] In contrast, another very common model, the dog, has been shown to provide faster intestinal permeability and increased absorption compared to humans of low-permeability drugs whereas high-permeability drugs seem to be rapidly and completely absorbed in dog similarly to human.[36,64] In case of solubility limitations of drug absorption, the similarity in basic GI physiological conditions will be the primary determinant for the usefulness of an animal model. The key aspects are GI volumes, pH, and presence of solubilizing agents like bile acids. Several comparisons of these properties have been published for dog, rat, and human as well as other animal models.[65–67] Generally, the dog has been proposed to a reasonable model for BCS class II drugs due to the similarities in relevant aspects of GI physiology. Fairly similar drug solubility has also been shown (**Figure 6**) for a number of drugs in human and dog intestinal fluid obtained both under fasting and fed conditions in a recent study.[68] Thus, usefulness of animal models in drug absorption studies is dependent on the BCS characteristics of the drug compound.

5.42.3.3   Biopharmaceutics Classification System in the Product Development Phase

5.42.3.3.1   Approaches to overcome absorption limitations

BCS may be used to make the normally time-constrained pharmaceutical development process more efficient. If a drug is classified as having low solubility, it is obvious that the rate and extent of absorption may be improved by the use of formulation principles that increase the dissolution rate and/or drug solubility. There are several different formulation principles of varying complexity, ranging from selecting a suitable solid-state form or salt to the use of technologically more advanced formulation principles. Although their application could be limited by several practical factors, such as

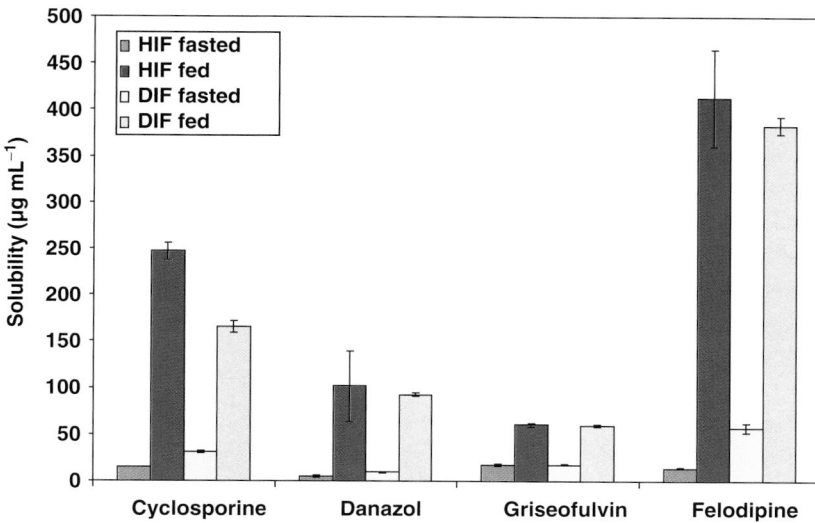

Figure 6 Solubility of four poorly soluble drugs in small intestinal fluid from human (HIF) and dog (DIF) obtained under fasting and fed conditions. Similar drug solubility was observed for human and dog intestinal fluid obtained both under fasting and fed conditions.[68]

poor drug stability, excessive size due to the need for large amounts of excipients in relation to the dose, technical manufacturing problems, and the high cost of goods, it is believed that many poorly soluble compounds with good pharmacological properties could be 'saved' by such approaches. One of the most recent advances in this area is the utilization of nanoparticles that can be obtained both by milling or precipitation methods.[69,70] One favorable feature of this approach is the relatively small amount of additional excipients needed, and very significant improvements in oral BA has been reported.[71] Microemulsions and cyclodextrin complexes are examples of other formulation approaches that have proven to increase BA several times for low-solubility drugs.[72]

Low-permeability compounds are less suitable for formulation improvements of BA. Despite extensive research in the area of permeation enhancers, very few products using such principles have entered the market. Significant increases in drug permeability found in in vitro models have often not provided the corresponding in vivo responses.[23,73–77] Another limitation of the enhancers used has been their unspecific nature, leading to a risk of increased permeability of toxins and other undesirable molecules.[78]

The use of prodrugs has proven to be a more successful approach to increase BA of low-permeability compounds.[79] Melagatran, a direct thrombin inhibitor, is a recent and illustrative example of how intestinal permeability has been increased and with it the developability of a drug.[80,81] It is produced by adding protecting chemical groups to its prodrug xi-melagatran (H376/95), which results in greater BA and less interindividual variability in the pharmacokinetics of melagatran. The two protecting groups change the pK_a values, producing a prodrug which is uncharged when its pH is above 6.2. Accordingly, the partition coefficient between octanol and water and in vitro permeability increased by 170 and 80 times, respectively.[80,81] The BA of melagatran increased from 5% to 20% with significant less interindividual variability when xi-melagatran was given orally instead of melagatran itself.

Targeting the carrier-mediated transport process is another successful prodrug strategy that has been shown to enhance the intestinal absorption.[79,82,83] For example, oligopeptide transporters are responsible for the active absorption of β-lactam antibiotics, the angiotensin-converting enzyme inhibitor enalapril, valaciclovir, and valganciclovir. The BA of valaciclovir was five times higher than that of aciclovir itself due to transport by the intestinal oligopeptide transporter PEPT1.[84–86]

5.42.3.3.2 Feasibility of extended release drug delivery

A successful development of oral ER formulations requires a strategy where BCS today could provide guidance. It is well known that not all drugs would clinically benefit from being given in an ER formulation due to unfavorable absorption properties along the human intestine. For clinical advantages to be obtained, the ER formulation most often requires drug release between 12 and 24 h. Thus, since the small intestinal transit of formulations is only 3–4 h, a significant part of the dose in an ER product will be delivered to the colon, which means that the absorption has to be

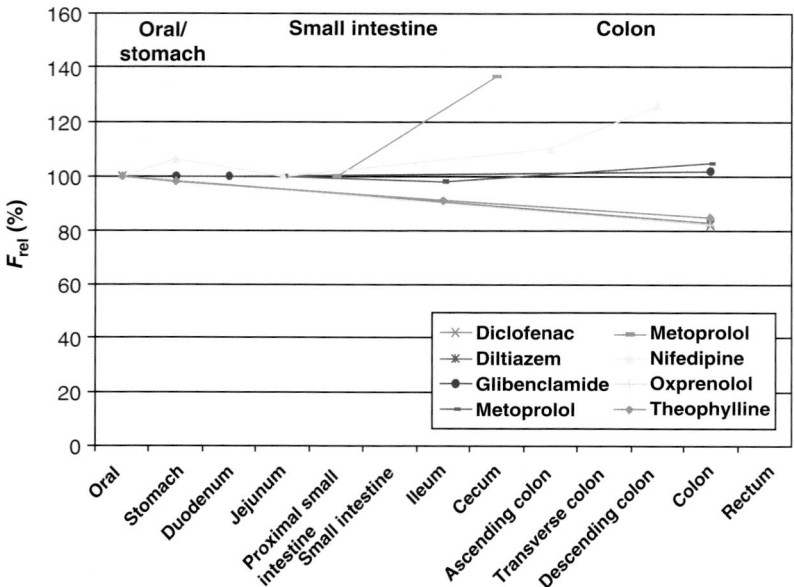

Figure 7 Mean relative bioavailability (F_{rel}) after administration to different sites in the GI tract of different high-permeability drugs to humans by use of intubations or remote control capsules. The amount of drug absorbed was similar over the entire GI tract also including the colon.

classified as high in the entire intestine. Class I drugs should therefore be the best candidates for ER development, which also is confirmed as several ER formulations are based on class I compounds.[62] The regional drug absorption in humans has been studied for several drugs by use of specialized techniques such as intubations or remote control capsules; results from such studies for highly permeable drugs are shown in **Figure 7**. The amount of drug absorbed was similar over the entire GI tract including the colon. Consequently, complete fraction absorbed seems to be maintained also after administration to the more distal parts for highly permeable drugs. It is our experience that if passive diffusion is the dominating membrane transport mechanism, an intestinal permeability equal to or higher than for metoprolol is an indication of a useful ER candidate. On the contrary, a low-permeability drug will not be completely absorbed in the small intestine given as a solution or an immediate release tablet, and accordingly the amount absorbed in the colon region will be even lower. This statement is based on experimental data demonstrating that the colonic membrane has an even lower permeability for drugs already classified as low-permeability drugs in the small intestine. For instance, human and rat intestinal specimens mounted in the Ussing chamber model (in vitro) have shown that the permeability of class III–IV drugs is even slower in the colon than the small intestine, whereas class I–II drugs show a slightly higher permeability in the colon when passive diffusion is the dominating membrane transport mechanism.[16,59,62,87] Consequently, it will not be possible to control the rate of absorption by an ER formulation for low-permeability drugs. In addition, a large part of the dose will not be absorbed, leading to a low and uneven variability, and no increased effect duration will be achieved. This is exemplified in ER tablets of amoxicillin. This high-solubility drug ($4.0 \, mg \, mL^{-1}$) is classified as a low P_{eff}, even if it is transported across the intestinal barrier via the oligopeptide carrier (PEPT1).[88,89] Essentially, no absorption occurred for an ER tablet when it entered the colon as determined by intubation technique and gamma scintigraphy, which is also in accordance with its slow passive permeability due to hydrophilic properties (polar surface area (PSA) $= 154.4 \, \mathring{A}^2$, $\log D_{6.5} = -1.7$).[45,90,91] However, recent advances in the area of gastric retention formulations hold some promise that ER formulations of low-permeability compounds may be possible in the future, i.e., the need for colonic drug absorption is no longer critical since the product stays in the upper GI tract during the entire period of drug release.[92]

Class II drugs are also frequently used in ER formulations. Felodipine ER tablets provide an example where a low-solublility compound is included in a useful ER product. It has been possible to administer felodipine, which has water solubility in the physiological pH range of about $1 \, \mu g \, mL^{-1}$, as a once-daily product in doses up to at least 20 mg without any reduced BA compared to an oral solution.[93] This successful absorption is indeed dependent on the use of a dissolution-enhancing formulation principle where the drug is given in a solubilized form. Other class II compounds, such as nifedipine ($10 \, \mu g \, mL^{-1}$/doses up to about 100 mg), is absorbed from the colon when released as a micronized drug.[94]

A special consideration may be made regarding classification of low-solublility compounds in ER forms. The standard classification is based on the idea that the drug should be completely dissolved in the gastric fluids, which has been estimated to be 250 mL[1]; this way of classifying drugs may be less relevant for ER formulations since only a very small part of the dose is made available for dissolution in the stomach. The dose is generally spread over the entire GI tract, making the effective water volume available as a dissolution medium for the drug probably larger than the 250 mL used in the original BCS. Furthermore, the drug permeability of these compounds is often much faster than the drug release, further preventing solubility limitations for class II drugs in ER formulations.[95] Thus, increased understanding of total available intestinal fluid volumes and influence of factors other than pH on colonic drug solubility/dissolution are needed before scientifically based solubility criteria can be proposed in the future. The dynamic processing of lipids and bile acids along the small intestine is also a challenging area, which needs to be further considered for understanding and prediction of drug solubility limitation of absorption along the intestine.[96,97]

5.42.3.3.3 In vitro/in vivo correlation (IVIVC)

In vitro dissolution tests are important for the successful development of solid pharmaceutical products and in batch quality controls. BCS is useful for predictions when in vitro/in vivo correction (IVIVC) could be expected for solid immediate release products. These in vitro dissolution tests can only model the release and dissolution rates of the drug and it is only when these processes are rate-limiting in the overall absorption that IVIVC can be established. In the case of class I drugs, the complete dose will already be dissolved in the stomach and, provided that the absorption in the stomach is negligible, the gastric emptying will be rate-limiting and therefore IVIVC is not expected. Thus, in vitro dissolution testing can be expected to be 'overdiscriminating' for those drugs, i.e., tablets showing different in vitro dissolution profiles will provide the same rate and extent of BA.[21]

Class II drugs are expected to have a dissolution-limited absorption and an IVIVC should be possible to establish using a well-designed in vitro dissolution test. One way to investigate and establish such a correlation is to study formulations containing drug particles with different surface areas. For example, the different peak plasma levels in a human BA study obtained between candersartan cilexitil (water solubility $< 1 \mu g \, mL^{-1}$) tablets (8 mg), containing drug particles with an average diameter of 4 and 9 μm, respectively, was predicted by a difference in in vitro dissolution in the standard US Pharmacopeia II (USP II) paddle method containing a buffer medium with addition of a surfactant. Examples of other composition/process factors of relevance for establishing IVIVC for immediate release formulations of class II drugs could be variations in solid-state form, effects of granulation conditions leading to different disintegration times of primary particles or tablets, variations in mixing leading to differences in wetting properties, and variations in pH of excipients.

Two cases could be identified for class II drugs when the establishment of simple IVIVCs is not feasible. First, there are a number of formulation principles that could enhance the dissolution rate and solubility of low-solublility compounds. It may be possible to achieve such a rapid and complete dissolution of a class II drug that the gastric emptying becomes rate-limiting, i.e., the BA of the solid dosage forms equals that of an oral solution. Thus, in such a case, the prerequisites for IVIVC will be identical to the situation for class I drugs, i.e., no correlation will be obtained as long as the dissolution rate is significantly faster than the gastric emptying. It has also been discussed that these drug and drug products should fall into a special intermediate class II, where a biowaiver should be applicable. This issue has also recently been discussed.[3,5] The second case in which IVIVC is not likely for class II drugs is the situation where the absorption is limited by the saturation solubility in the GI tract rather than the dissolution rate. In this situation, the drug concentration in the GI tract will be close to the saturation solubility, and changes of the dissolution rate will not affect the plasma concentration profile and the in vivo BA. Standard in vitro dissolution tests are carried out under 'sink conditions,' i.e., at concentrations well below the saturation solubility. Thus, only effects related to the dissolution rate can be predicted in vitro. If more physiologically relevant dissolution media are used, which do not necessarily provide sink conditions, the possibility for IVIVC could be improved as indicated by recent work using a simulated intestinal medium.[98] It has also been reported that, provided an appropriate composition is chosen for the dissolution test, the USP paddle apparatus can be used to reflect variations in hydrodynamic conditions in the upper GI tract.[99]

The absorption of class III drugs is limited by their intestinal permeability and no IVIVC should be expected. However, when the drug dissolution becomes slower than the gastric emptying, a reduction of the extent of BA will be found at slower dissolution rates since the time when the drug is available for transport across the small intestinal barrier will then be reduced.

5.42.3.3.4 Food–drug interactions

Food intake may affect the plasma exposure of a drug, which might influence the efficacy of a drug, and it is therefore crucial from a safety aspect to predict such effects at an early stage. However, an accurate prediction model is a challenge since several factors are involved in food–drug interactions including physicochemical effects such as increased solubility, binding to secretory or food components; physiological effects in the GI tract such as altered flow rates and gastric emptying; mechanical effects on formulations due to different motility patterns; permeability effects due to interactions with active transporters or effects on the membrane and altered first-pass metabolism. An extensive review of different mechanisms for food–drug interactions can be found elsewhere.[9]

The BCS can be used to predict the mechanisms and magnitude for food effects related to solubility, dissolution, gastric emptying, and permeability. The clinical important interaction with grapefruit juices and other food components that affect the first-pass metabolism is not possible to predict and describe with the use of BCS.[100] The most severe cases of food–drug interactions due to factors considered in BCS are generally found in the group of poorly soluble compounds (class II and IV drugs) given in high doses, as they approach the saturation solubility in the GI lumen. The BA of griseofulvin (class II or class IV) has been reported to be increased up to five times by food, and dosing recommendations requiring concomitant intake of the drug with a meal is often used in these cases.[101] The saturation solubility will be significantly improved by food due to solubilization in mixed micelles, including bile acids, lecithin and monoglycerides obtained from the dietary fat intake, and dissolution into emulsified nutritional lipids, and accordingly the amount of drug available for absorption will significantly increase.[97] An additional effect to the increased plasma exposure might be the relatively large fluid volume available in the stomach after a meal. The effect of just increasing the volume of water administered with a tablet from 200 to 800 mL increased the extent of BA by 50% in a human study of danazol (water solubility $0.001 \, \mathrm{mg \, L^{-1}}$) tablets 250 mg.[102]

Protolytic drugs with a pK_a within the physiological pH range will also be affected by the food-mediated pH changes in the stomach. A protein-rich meal could increase the pH from the fasting level of pH about 1.5–3 to a close to neutral pH.[103,104] Basic drugs, which are often freely dissolved in the acidic stomach, could thereby experience a reduced BA with food whereas it may be the other way around for low-solubility acids.

For class II drugs given in lower doses, the dissolution rate rather than the saturation solubility is the limiting factor. An increase in dissolution rate due to in vivo solubilization mediated by food intake could theoretically be obtained but this is not always found in vivo. For example, food does not affect the rate and extent of BA for candersartan cilexitil, a very poorly soluble compound given in a low dose.[105] An in vitro dissolution and solubility study of this compound in simulated intestinal media provided a potential explanation. It was revealed that the solubility was increased as a function of bile concentration as expected whereas the dissolution rate was not increased by the higher bile concentrations being representative for the fed state. The slower gastric emptying of dissolved drug might also counterbalance increases in dissolution rate in the fed state. Thus, although intestinal solubility most often will be increased in the fed state for class II drugs, this will not always lead to a more rapid dissolution.

For class I drugs, a slower rate of absorption could be expected after concomitant food intake (reduced C_{max} and prolonged t_{max}) due to the decreased gastric emptying rate induced by a meal. Gastric emptying is totally controlled by the two patterns of upper GI motility, i.e., the interdigestive and the digestive motility pattern. The interdigestive pattern, termed the migrating motor complex (MMC), dominates in the fasted state and is organized into alternating phases of activity and quiescence.[106] Gastric emptying in the fed state varies significantly depending on the motility pattern and meal composition, including factors such as energy content, osmolality, and pH. A gastric emptying half-life of about 10 min is obtained in the fasting state whereas a half-life of approximately 45 min has been reported for fluids when measured under nonfasting conditions.[107,108] It was also clearly shown that a smaller volume (50 mL) was more sensitive to the motility phase than large volumes (200 mL). Even during phase I of the MMC, the emptying rate was faster for the 200 mL volume.[107] In addition, it is important to understand that drugs with a shorter half-life will be more sensitive to changes in gastric emptying. Recently, it was reported that in vivo results appear to support the hypothesis that rapidly dissolving immediate release solid oral products containing a BCS class I drug are likely to be bioequivalent under fed conditions.[109]

Class III drugs are perhaps the least sensitive to food intake. None of the effects induced by food such as slower gastric emptying and increased solubilization capacity will be of any relevance. It is also well established that the transit from the main absorption region for a class III drug, i.e., the small intestine, would not be affected by food and, generally, no effect on drug permeability is expected by food.[16,59,110] For example, no interaction with food was obtained for the low permeability compound ximelagatran.[80] In addition, Yu et al. have confirmed this by suggesting that BCS class III drugs may have the potential to be bioequivalent under fed conditions.[109]

A plausible mechanism for interactions between food and drugs could occur for drugs that have carrier-mediated transport in the intestine, especially if nutritional carriers are involved. The two most important nutrient absorption carriers for drugs are the oligo-peptide carrier (hPEPT1) and the amino acid transport family. These carrier proteins have a high transport capacity in the human small intestine, and they seem less likely to be involved in direct food–drug interactions, unless high doses are given together with a protein-rich meal. For instance, it has been reported that a protein-rich meal does not affect the BA of levodopa, a drug that is transported by the several amino acid carriers in the human intestine, and this has been interpreted that the amino acids are absorbed rapidly and the risk for interaction is reduced and that the nutritional transporter has a high capacity.[111] Intestinal function depends on the presence of luminal nutrients and is altered during starvation and refeeding. The nutritional status could also cause transcriptional activation of the PEPT1 gene by selective amino acids and dipeptides in the diet.[112] It has also been reported that the integrated response to a certain stimulus may increase PEPT1 activity by translocation from a preformed cytoplasm pool.[113] This short-term regulation of drug transport requires further research. Amino acids are essential for enterocytes, but the luminal supply is compromised with changes in dietary intake. Howard *et al.* found that the removal of luminal nutrition increases the expression of amino acid and peptide transporter messenger RNAs (mRNAs) in the small intestine.[114]

5.42.4   Potential Future Extensions

Class I drug substances today are quite rare in pharmaceutical development, which limits the broad use of the BCS in the drug development. It has also been recognized that the present application represents a deliberately conservative approach, and proposals for extensions have been discussed since original publication of BCS.[1,3] For example, in a recent paper it was suggested that the requirement of highest pH for the solubility measurements could be changed from 7.5 to 6.8 since the latter is more relevant for the pH in the upper GI tract.[3] This revision would thus somewhat relax the requirements for basic drugs. It has also recently been suggested that for acidic drugs the boundaries for solubility are too restrictive (pH 1.0–7.5) and might be narrowed down to pH 5.0–7.4.[115] Yazdanian *et al.* reported that, based on the current definition, 15 of 18 acidic nonsteroidal anti-inflammatory drugs (NSAIDs) were classified as class II compounds. If only a neutral pH was used, 15 of the NSAIDs were classified as class I drugs, which may suggest the boundaries are too restrictive for solubility. Based on these findings, the authors suggested that there might be a need for an intermediate solubility classification for highly permeable ionizable compounds, especially when given in low doses.[115] Another proposal in the paper by Yu *et al.* was to reduce the high-permeability definition from 90% to 85% f_a based on observations that many drugs that are considered completely absorbed provide experimental values below 90%, i.e., 90% seems to be a too rigid criterion considering the precision of the experimental methods.[3] Other more radical relaxations of criteria, such as including class III drugs or class II drugs using simulated intestinal media with bile acids or increase volumes and further reducing the pH interval in solubility measurements, will most probably require further research and analysis of past experiences.[95,97,116,117] For instance, to better predict food effects on the BA/BE of drugs and drug products from in vitro data, a dissolution medium that simulates the initial composition of the postprandial stomach has been developed.[118] Furthermore, it has been suggested by Blume and Schug that the waiver of in vivo BE studies be extended to class III drugs.[119] For immediate release formulations of class III drugs (high solubility–low permeability), it is considered that the handling of the dosage form in vivo be similar to an oral solution. Accordingly, intestinal membrane permeability is expected to be rate-limiting in the overall drug absorption process. Therefore, the rate and extent of intestinal absorption are expected to be controlled by drug molecule properties and physiologic factors, rather than by properties related to the pharmaceutical formulation, given that none of the pharmaceutical excipients used in the formulation has a relevant influence on gastrointestinal transit and permeability of drug.[3] However, what kind of in vitro dissolution requirements should be set to ensure that the drug release has no significant impact on in vivo BA is still unknown, and data available for supporting biowaivers of class III drugs are still limited.[120] In a recent study by Cheng *et al.*, a biowaiver was suggested by the in vitro dissolution profiles and was justified by the in vivo BE data. The demonstration of BE between two immediate release tablets of class III drug metformin in healthy Chinese subjects serves as an example for supporting biowaivers for such cases.[121] Yu *et al.* also showed that a BCS class III drug, hydrochlortiazide, had a much narrower 90% confidence interval than BCS class II drugs and therefore may be a better candidate for biowaiver. However, important concerns for BCS class III drugs are the effect of excipients on GI transit and permeability (both passive and carrier-mediated uptake and/or efflux transport).[109] Further support for biowaiver of BCS class III drugs is available from Vogelpoel *et al.*, who showed recently that atenolol is a candidate for biowaiver, provided that the formulation contains well-known excipients.[122]

It has also been proposed that dissolution rate should be a more relevant variable than saturation solubility for BCS since the in vivo absorption process is rather a dynamic, rate-controlled process than a process at equilibrium conditions.[109] The rotating disk method, where the pure drug is compressed into a rotating disk holder to provide a constant area exposed to the dissolution medium, is the method of choice for such investigations (USP). It was shown in a limited set of drug substances that at a rotation speed of 100 rpm class I and III drugs had a dissolution rate faster than $0.1\,\text{mg}\,\text{min}^{-1}\text{cm}^{-2}$.[120]

Extensions of BCS beyond the oral immediate release area have also been suggested, e.g., to apply BCS in the ER area. However, this will provide a major challenge since the release from different formulations will interact in different ways with in vitro test conditions and the physiological milieu in the GI tract. For example, the plasma concentration–time profile differed for two felodipine ER tablets for which very similar in vitro profiles had been obtained, despite the fact that both tablets were of the hydrophilic matrix type based on cellulose derivates.[93] This misleading result in vitro was due to interactions between the test medium and the matrix forming polymers of no in vivo relevance. The situation for ER formulations would be further complicated by the need to predict potential food effects on the drug release in vivo. Although the present use of BCS is limited, extensions of applications should clearly be made without jeopardizing the quality of products on the market. This would more likely be achieved within the area of oral immediate release than for ER formulations.

The crucial role of membrane transporters in the ADME of drugs has only been fully recognized in the past decade. A new classification system having its origin in BCS has also recently been proposed that should be useful not only in predicting effects of dissolution and permeability on drug absorption but also allow for predictions of overall drug disposition, including routes of drug elimination.[126] It was noted that high-permeability compounds, i.e., BCS class I and II, are highly metabolized whereas low-permeability drugs are preferentially excreted unchanged through the biliary and/or renal route. It was also proposed that a relationship between dispostion and permeability determination would be possible to establish, and would form the basis for a Biopharmaceutic Drug Disposition Classification System (BDDCS). However, the proposed application of BDDCS as a tool to predict the effect of carrier-mediated membrane transport on absorption, first-pass liver extraction, distribution, and elimination is not established, since the exact in vivo transport mechanisms and their relevance for many drugs that are transport substrates are far from completely understood. In order to establish this, further validation of their in vivo relevance is needed as well as a significantly improved knowledge of the correlation between the in vitro and in vivo charecteristics. In the present genomics and proteomics era, a wealth of detailed information has become available on their structure, function, and topological expression. At the same time, a wide variety of advanced cell biological methodologies have been introduced to study their substrate specificity, the driving forces for membrane translocation, and potential interactions with other drugs and/or endogenous agents during cellular transport. Yet, although our knowledge in this field has been much deepened and our instrumentation amazingly sharpened, somehow an integral view on the practical implications of the related findings for ADME research seems to have been blurred by the multitude of molecular data, the uncritical use of, especially, in vitro methods, and the relative lack of human in vivo data. In other words, the 'black box' was filled up so rapidly that we did not even begin to perceive its functional organization. In particular, clinical data in various diseases are scarce and one seems to be at a loss as to which transporters to address first. Despite these difficulties lying ahead of this research, the proposed BDDCS may be developed into a thinking tool and classification system that should be very useful in the multidisciplinary field that is characteristic of drug discovery and development.

5.42.5 Conclusion

In the present chapter the importance of the fundamental variables in BCS, solubility and intestinal permeability for oral drug absorption has been scrutinized. The main regulatory impact today is the use of BCS as a framework for identifying drugs for which in vitro dissolution testing could replace in vivo studies to determine BE. Extensions of this approach to cases other than immediate release formulations of the relative rare class I drugs would significantly enhance the impact of BCS. Some drugs that fall into class II and III could be regulated by BCS rather than traditional BE testing.

The future application of BCS is most likely increasingly important when the present framework gains increase recognition, which will probably be the case if the BCS borders for certain class II and III drugs are extended. The future revision of the BCS guidelines by the regulatory agencies in communication with academic and industrial scientists is exciting and will hopefully result in an increased applicability in drug development.

Finally, we emphazise the great use of BCS as a simple tool in early drug development to determine the rate-limiting step in the oral absorption process, which has facilitated communication between different experts involved in the overall drug development process. This increased awareness of a proper biopharmaceutical characterization of new drugs may in the future result in drug molecules with sufficiently high permeability, solubility, and dissolution rates, and that will automatically increase the importance of BCS as a regulatory tool over time.

References

1. Amidon, G. L.; Lennernas, H.; Shah, V. P.; Crison, J. R. *Pharm. Res.* **1995**, *12*, 413–420.
2. CDER *Waiver of In Vivo Bioavailability and Bioequivalence Studies for Immediate Release Solid Oral Dosage Forms Based on a Biopharmaceutics Classification System*; Food and Drug Administration: Rockville, MD, USA, 2000.
3. Yu, L. X.; Amidon, G. L.; Polli, J. E.; Zhao, H.; Mehta, M. U.; Conner, D. P.; Shah, V. P.; Lesko, L. J.; Chen, M. L.; Lee, V. H.; Hussain, A. S. *Pharm. Res.* **2002**, *19*, 921–925.
4. Abrahamsson, B.; Lennernäs, H. Application of the Biopharmaceutic Classification System Now and in the Future. In *Drug Bioavailability, Estimation of Solubility, Permeability, Absorption and Bioavailability*; van de Waterbeemd, H., Lennernäs, H., Artursson, P., Eds.; Wiley-VCH: Berlin, 2003, pp 493–531.
5. Polli, J. E.; Yu, L. X.; Cook, J. A.; Amidon, G. L.; Borchardt, R. T.; Burnside, B. A.; Burton, P. S.; Chen, M. L.; Conner, D. P.; Faustino, P. J. et al. *J. Pharm. Sci.* **2004**, *93*, 1375–1381.
6. Lennernäs, H. *J. Pharm. Sci.* **1998**, *87*, 403–410.
7. Lennernäs, H.; Ahrenstedt, Ö.; Hällgren, R.; Knutson, L.; Ryde, M.; Paalzow, L. K. *Pharm. Res.* **1992**, *9*, 1243–1251.
8. Knutson, L.; Odlind, B.; Hallgren, R. *Am. J. Gastroenterol.* **1989**, *84*, 1278–1284.
9. Fleisher, D.; Li, C.; Zhou, Y.; Pao, L. H.; Karim, A. *Clin. Pharmacokinet.* **1999**, *36*, 233–254.
10. Kasim, N. A.; Whitehouse, M.; Ramachandran, C.; Bermejo, M.; Lennernäs, H.; Hussain, A. S.; Amidon, G. L. *Mol. Pharmaceut.* **2004**, *1*, 85–96.
11. Rowland, M.; Tozer, T. N. *Clinical Pharmacokinetics: Concepts and Applications*, 3rd ed.; Williams & Wilkins: Philadelphia, PA, 1995, pp 119–134.
12. von Richter, O.; Greiner, B.; Fromm, M. F.; Fraser, R.; Omari, T.; Barclay, M. L.; Dent, J.; Somogyi, A. A.; Eichelbaum, M. *Clin. Pharmacol. Ther.* **2001**, *70*, 217–227.
13. Wu, C. Y.; Benet, L. Z.; Hebert, M. F.; Gupta, S. K.; Rowland, M.; Gomez, D. Y.; Wacher, V. J. *Clin. Pharmacol. Ther.* **1995**, *58*, 492–497.
14. CPMP: *Note for Guidance on the Investigation of Bioavailability and Bioequivalence* (CPMP/EWP/QWP/1401/98); The European Agency for the Evaluation of Medicinal Products: London, UK, 2001.
15. Nakamura, T.; Sakaeda, T.; Ohmoto, N.; Tamura, T.; Aoyama, N.; Shirakawa, T.; Kamigaki, T.; Nakamura, T.; Kim, K. I.; Kim, S. R. et al. *Drug. Metab. Dispos.* **2002**, *30*, 4–6.
16. Berggren, S.; Lennernäs, P.; Ekelund, M.; Weström, B.; Hoogstraate, J.; Lennernäs, H. *J. Pharm. Pharmacol.* **2003**, *55*, 963–972.
17. de Waziers, I.; Cugnenc, P. H.; Yang, C. S.; Leroux, J. P.; Beaune, P. H. *J. Pharmacol. Exp. Ther.* **1990**, *253*, 387–394.
18. Goldin, B. R. *Ann. Med.* **1990**, *22*, 43–48.
19. Chandra, P.; Brouwer, K. L. *Pharm. Res.* **2004**, *21*, 719–735.
20. Tannergren, C.; Petri, N.; Knutson, L.; Hedeland, M.; Bondesson, U.; Lennernäs, H. *Clin. Pharmacol. Ther.* **2003**, *74*, 423–436.
21. Rekhi, G. S.; Eddington, N. D.; Fossler, M. J.; Schwartz, P.; Lesko, L. J.; Augsburger, L. L. *Pharm. Devel. Technol.* **1997**, *2*, 11–24.
22. Bogman, K.; Zysset, Y.; Degen, L.; Hopfgartner, G.; Gutmann, H.; Alsenz, J.; Drewe, J. *Clin. Pharmacol. Ther.* **2005**, *77*, 24–32.
23. Lennernäs, H.; Gjellan, K.; Hällgren, R.; Graffner, C. *J. Pharm. Pharmacol.* **2002**, *54*, 499–508.
24. Adkin, D. A.; Davis, S. S.; Sparrow, R. A.; Huckle, P. D.; Phillips, A. J.; Wilding, I. R. *Br. J. Clin. Pharmacol.* **1995**, *39*, 381–387.
25. Basit, A. W.; Podczeck, F.; Newton, J. M.; Waddington, W. A.; Ell, P. J.; Lacey, L. F. *Eur. J. Pharm. Sci.* **2004**, *21*, 179–189.
26. CDER: *Immediate-Release Solid Oral Dosage Forms: Scale-Up and Post-Approval Changes: Chemistry, Manufacturing and Controls, In vitro Dissolution Testing, and In vivo Bioequivalence Documentation*; Food and Drug Administration: Rockville, MD, USA, 1995.
27. Q6A International Conference on Harmonisation; *Guidance on Q6A Specifications: Test Procedures and Acceptance Criteria for New Drug Substances and New Drug Products: Chemical Substances.* Rockville, MD, USA, 2000.
28. Van de Waterbeemd, H.; Smith, D. A.; Beaumont, K.; Walker, D. K. *J. Med. Chem.* **2001**, *44*, 1313–1333.
29. Van de Waterbeemd, H.; Gifford, E. *Nat. Rev. Drug. Disc.* **2003**, *2*, 192–204.
30. Lajiness, M. S.; Vieth, M.; Erickson, J. *Curr. Opin. Drug. Disc. Devel.* **2004**, *7*, 457–470.
31. Kola, I.; Landis, J. *Nat. Rev. Drug Disc.* **2004**, *3*, 711–716.
32. van de Waterbeemd, H. *Eur. J. Pharm. Sci.* **1998**, *7*, 1–3.
33. Lipinski, C. A.; Lombardo, F.; Dominy, B. W.; Feeney, P. J. *Adv. Drug Deliv. Rev.* **2001**, *46*, 3–26.
34. Egan, W. J.; Lauri, G. *Adv. Drug. Deliv. Rev.* **2002**, *31*, 273–289.
35. Volpe, D. A. *Pharm. Sci.* **2004**, *6*, 1–6.
36. Sun, D.; Lennernäs, H.; Welage, L. S.; Barnett, J. L.; Landowski, C. P.; Foster, D.; Fleisher, D.; Lee, K.-D.; Amidon, G. L. *Pharm. Res.* **2002**, *19*, 1400–1416.
37. Lennernäs, H.; Palm, K.; Fagerholm, U.; Artursson, P. *Int. J. Pharm.* **1996**, *127*, 103–107.
38. Lennernäs, H. *J. Pharm. Pharmacol.* **1997**, *49*, 627–638.
39. Lennernäs, H. *J. Pharm. Pharmacol.* **2003**, *55*, 429–433.
40. Chiou, W. L.; Chung, S. M.; Wu, T. C.; Ma, C. *Int. J. Clin. Pharmacol. Ther.* **2001**, *39*, 93–101.
41. Petri, N.; Tannergren, C.; Rungstad, D.; Lennernäs, H. *Pharm. Res.* **2004**, *21*, 1398–1404.
42. Tannergren, C.; Knutson, T.; Knutson, L.; Lennernäs, H. *Br. J. Clin. Pharmacol.* **2003**, *55*, 182–190.
43. van Montfoort, J. E.; Hagenbuch, B.; Groothuis, G. M.; Koepsell, H.; Meier, P. J.; Meijer, D. F. *Curr. Drug Metab.* **2003**, *4*, 185–211.
44. Mizuno, N.; Niwa, T.; Yotsumoto, Y.; Sugiyama, Y. *Pharmacol. Rev.* **2003**, *55*, 425–461.
45. Winiwarter, S.; Bonham, N.; Hallberg, A.; Lennernäs, H.; Karlén, A. *J. Med. Chem.* **1999**, *41*, 4939–4949.
46. Winiwarter, S.; Ax, F.; Lennernäs, H.; Hallberg, A.; Pettersson, J.; Karlén, A. *J. Mol. Graph. Model.* **2003**, *21*, 283–287.
47. Bergstrom, C. A.; Norinder, U.; Luthman, K.; Artursson, P. *J. Chem. Inf. Comput. Sci.* **2003**, *43*, 1177–1185.
48. Fichert, T.; Yazdanian, M.; Proudfoot, J. R. *Bioorg. Med. Chem. Lett.* **2003**, *13*, 719–722.
49. Sanghvi, T.; Jain, N.; Yang, G.; Yalkowsky, S. H. *QSAR Combin. Sci.* **2003**, *22*, 258–262.
50. Agoram, B.; Woltosz, W. S.; Bolger, M. B. *Adv. Drug Deliv. Rev.* **2001**, *50*, S41–S67.
51. Grass, G. M.; Sinko, P. J. *Adv. Drug Deliv. Rev.* **2002**, *54*, 433–451.
52. Rowland, M.; Balant, L.; Peck, C., *PharmSci.* **2004**, *09*; *6*, E6.
53. Lindahl, A.; Sjöberg, Å.; Bredberg, U.; Toreson, H.; Ungell, A. L.; Lennernäs, H. *Mol. Pharmaceut.* **2004**, *1*, 347–356.
54. Bermejo, M.; Avdeef, A.; Ruiz, A.; Nalda, R.; Ruell, J. A.; Tsinman, O.; Gonzalez, I.; Fernandez, C.; Sanchez, G.; Garrigues, T. M.; Merino, V. *Eur. J. Pharm. Sci.* **2004**, *21*, 429–441.
55. Obata, K. M.; Sugano, K.; Machida, M.; Aso, Y. *Drug Dev. Ind. Pharm.* **2004**, *30*, 181–185.
56. Fagerholm, U.; Johansson, M.; Lennernäs, H. *Pharm. Res.* **1996**, *13*, 1335–1341.

57. Fagerholm, U.; Lindahl, A.; Lennernäs, H. *J. Pharm. Pharmacol.* **1997**, *49*, 687–690.
58. Fagerholm, U.; Nilsson, D.; Knutson, L.; Lennernäs, H. *Acta Physiol. Scand.* **1999**, *165*, 315–324.
59. Ungell, A. L.; Nylander, S.; Bergstrand, S.; Sjöberg, Å.; Lennernäs, H. *J. Pharm. Sci.* **1998**, *87*, 360–366.
60. Chiou, W. L.; Barve, A. *Pharm. Res.* **1998**, *11*, 1792–1795.
61. Berggren, S.; Hoogstraate, J.; Fagerholm, U.; Lennernäs, H. *Eur. J. Pharm. Sci.* **2004**, *21*, 553–560.
62. Corrigan, O. I. *Adv. Exp. Med. Biol.* **1997**, *423*, 111–128.
63. Anderson, C. M.; Grenade, D. S.; Boll, M.; Foltz, M.; Wake, K. A.; Kennedy, D. J.; Munck, L. K.; Miyauchi, S.; Taylor, P. M.; Campbell, F. C. et al. *Gastroenterol.* **2004**, *127*, 1410–1422.
64. Chiou, W. L.; Jeong, H. Y.; Chung, S. M.; Wu, T. C. *Pharm. Res.* **2000**, *17*, 135–140.
65. Dressman, J. B.; Yamada, K. Animal Models for Oral Drug Absorption. In *Pharmaceutical bioequivalence*; Welling, P. G., Tse, F. L. S., Dighe, S. V., Eds.; Marcel Dekker: New York, 1991, pp 235–266.
66. Kararli, T. T. *Biopharm. Drug Disp.* **1995**, *16*, 351–380.
67. Ritschel, W. A. *Sci. Tech. Prat. Pharm.* **1987**, *3*, 125–141.
68. Persson, E.; Carlsson, A.; Gustafsson, A. S.; Knutsson, L.; Lennernäs, H.; Abrahamsson B. *Pharm. Res.* **2005**, *22*, 2141–2151.
69. Hu, J.; Johnston, K. P.; Williams, R. O., III. *Drug Dev. Indust. Pharm.* **2004**, *30*, 233–245.
70. Patravale, V. B.; Abhijit, A; Kulkarni, R. M. *J. Pharm. Pharmacol.* **2004**, *56*, 827–840.
71. Wu, Y.; Loper, A.; Landis, E.; Hettrick, L.; Novak, L.; Lynn, K.; Chen, C.; Thompson, K.; Higgins, R.; Batra, U. et al. *Int. J. Pharm.* **2004**, *285*, 135–146.
72. Kim, C.-K.; Park, J.-S. *Am. J. Drug. Deliv.* **2004**, *2*, 113–130.
73. Fagerholm, U.; Sjöström, B.; Sroka-Markovic, J.; Wijk, A.; Svensson, M.; Lennernäs, H. *J. Pharm. Pharmacol.* **1998**, *50*, 467–473.
74. Schipper, N. G. M.; Vårum, K. M.; Stenberg, P.; Ocklind, G.; Lennernäs, H.; Artursson, P. *Eur. J. Pharm. Sci.* **1999**, *8*, 335–343.
75. Kato, Y.; Onishi, H.; Machida, Y. *Curr. Pharm. Biotechnol.* **2003**, *4*, 303–309.
76. Mahato, R. I.; Narang, A. S.; Thoma, L.; Miller, D. D. *Crit. Rev. Ther. Drug Carrier. Syst.* **2003**, *20*, 153–214.
77. Mrestani, Y.; Bretschneider, B.; Hartl, A.; Neubert, R. H. *J. Pharm. Pharmacol.* **2003**, *55*, 1601–1606.
78. Uchiyama, T.; Sugiyama, T.; Quan, Y. S.; Kotani, A.; Okada, N.; Fujita, T.; Muranishi, S.; Yamamoto, A. *J. Pharm. Pharmacol.* **1999**, *51*, 1241–1250.
79. Ettmayer, P.; Amidon, G. L.; Clement, B.; Testa, B. *J. Med. Chem.* **2004**, *47*, 2393–2404.
80. Eriksson, B. I.; Arfwidsson, A. C.; Frison, L.; Eriksson, U. G.; Bylock, A.; Kälebo, P.; Fager, G.; Gustafsson, D. *Thromb. Haemost.* **2002**, *87*, 231–237.
81. Eriksson, U. G.; Bredberg, U.; Hoffmann, K.-J.; Thuresson, A.; Gabrielsson, M.; Ericsson, H.; Ahnoff, M.; Gislen, K.; Fager, G.; Gustafsson, D. *Drug Metab. Disp.* **2003**, *31*, 294–305.
82. Steffansen, B.; Nielsen, C. U.; Brodin, B.; Eriksson, A. H.; Andersen, R.; Frokjaer, S. *Eur. J. Pharm. Sci.* **2004**, *21*, 3–16.
83. Thomsen, A. E.; Friedrichsen, G. M.; Sorensen, A. M.; Nielsen, C. U.; Brodin, B.; Begtrup, M.; Frokjaer, S.; Steffansen, B. *J. Contr. Release* **2003**, *86*, 279–292. Erratum in: *J. Contr. Release* **2003**, *88*, 343.
84. Beauchamp, L. M.; Orr, G. F.; De Miranda, P.; Burnette, T.; Krenitsky, T. A. *Antiviral Chem. Chemother.* **1992**, *3*, 157–164.
85. Han, H. K.; De Vrueh, R. L. A.; Rhie, J. K.; Covitz, K.-M. Y.; Smith, P. L.; Lee, C.-P.; Oh, D.-M.; Sadee, W.; Amidon, G. L. *Pharm. Res.* **1998**, *15*, 1154–1159.
86. Swaan, P. W.; Tukker, J. J. *J. Pharm. Sci.* **1997**, *86*, 596–602.
87. Connor, A. L.; Wray, H.; Cottrell, J.; Wilding, I. R. *Eur. J. Pharm. Sci.* **2001**, *13*, 369–374.
88. Oh, D. M.; Sinko, P. J.; Amidon, G. L. *J. Pharm. Sci.* **1993**, *82*, 897–900.
89. Lennernäs, H.; Knutson, L.; Knutson, T.; Hussain, A.; Lesko, L.; Salomonsson, T.; Amidon, G. L. *Eur. J. Pharm. Sci.* **2002**, *15*, 271–277.
90. Barr, W. H.; Zola, E. M.; Candler, E. L.; Hwang, S. M.; Tendolkar, A. V.; Shamburek, R.; Parker, B.; Hilty, M. D. *Clin. Pharmacol. Ther.* **1994**, *56*, 279–285.
91. Gottfries, J.; Svenheden, A.; Alpsten, M.; Bake, B.; Larsson, A.; Idström, J.-P. *Scand. J. Gastroenterol.* **1996**, *31*, 49–53.
92. Klausner, E. A.; Lavy, E.; Friedman, M.; Hoffman, A. *J. Contr. Release* **2003**, *90*, 143–162.
93. Abrahamsson, B.; Johansson, D.; Torstensson, A.; Wingstrand, K. *Pharm. Res.* **1994**, *11*, 1093–1097.
94. Grunda, J. S.; Foster, R. T. *Clin. Pharmacokinet.* **1996**, *30*, 28–51.
95. Bonlokke, L.; Hovgaard, L.; Kristensen, H. G.; Knutson, L.; Lennernäs, H. *Eur. J. Pharm. Sci.* **2001**, *12*, 239–250.
96. Kossena, G. A.; Charman, W. N.; Boyd, B. J.; Dunstan, D. E.; Porter, C. J. *J. Pharm. Sci.* **2004**, *93*, 332–348.
97. Porter, C. J.; Kaukonen, A. M.; Taillardat-Bertschinger, A.; Boyd, B. J.; O'Connor, J. M.; Edwards, G. A.; Charman, W. N. *J. Pharm. Sci.* **2004**, *93*, 1110–1121.
98. Kostewicz, E. S.; Brauns, U.; Becker, R.; Dressman, J. B. *Pharm. Res.* **2002**, *19*, 345–349.
99. Scholz, A.; Kostewicz, E.; Abrahamsson, B.; Dressman, J. B. *J. Pharm. Pharmacol.* **2003**, *55*, 443–451.
100. Edgar, B.; Bailey, D.; Bergstrand, R.; Johnsson, G.; Regardh, C. G. *Eur. J. Clin. Pharmacol.* **1992**, *42*, 313–317.
101. Charman, W. N.; Porter, C. J.; Mithani, S.; Dressman, J. B. *J. Pharm. Sci.* **1997**, *86*, 269–282.
102. Sunesen, V. H.; Vedelsdal, R.; Kristensen, H. G.; Christrup, L.; Mullertz, A. *Eur. J. Pharm. Sci.* **2005**, *24*, 297–303.
103. Dressman, J. B.; Berardi, R. R.; Dermentzoglou, L. C.; Russell, T. L.; Schmaltz, S. P.; Barnett, J. L.; Jarvenpaa, K. M. *Pharm. Res.* **1990**, 7, 756–761.
104. Lindahl, A.; Ungell, A.-L.; Lennernäs, H. *Pharm. Res.* **1997**, *14*, 497–502.
105. Gleiter, C. H.; Morike, K. E. *Clin. Pharmacokinet.* **2002**, *41*, 7–17.
106. Sarna, S. K. *Gastroenterology* **1985**, *89*, 894–913.
107. Oberle, R. L.; Chen, T. S.; Lloyd, C.; Barnett, J. L.; Owyang, C.; Meyerand, J.; Amidon, G. L. *Gastroenterology* **1990**, *99*, 1275–1282.
108. Ziessman, H. A.; Fahey, F. H.; Collen, M. J. *Dig. Dis. Sci.* **1992**, *37*, 744–750.
109. Yu, L. X.; Straughn, A. B.; Faustino, P. J.; Yang, Y.; Parekh, A.; Ciavarella, A. B.; Asafu-Adjaye, E.; Mehta, M. U.; Conner, D. P.; Lesko, L. J.; Hussain, A. S. *Mol. Pharmaceut.* **2004**, *5*, 357–362.
110. Davis, S. S.; Hardy, J. G.; Fara, J. W. *Gut* **1986**, *27*, 886–892.
111. Lennernäs, H.; Nilsson, D.; Aquilonius, S.-M.; Ahrenstedt, Ö.; Knutson, L.; Paalzow, L. K. *Br. J. Clin. Pharmacol.* **1993**, *35*, 243–250.

112. Shiraga, T.; Miyamoto, K.; Tanaka, H.; Yamamoto, H.; Taketani, Y.; Morita, K.; Tamai, I.; Tsuji, A.; Takeda, E. *Gastroenterology* **1999**, *116*, 354–362.
113. Thamotharan, M.; Bawani, S. Z.; Zhou, X.; Adibi, S. A. *Am. J. Physiol.* **1999**, *276*, C821–C826.
114. Howard, A.; Goodlad, R. A.; Walters, J. R.; Ford, D.; Hirst, B. H *J. Nutr.* **2004**, *134*, 2957–2964.
115. Yazdanian, M.; Briggs, K.; Jankovsky, C.; Hawi, A. *Pharm. Res.* **2004**, *21*, 293–299.
116. Kostewicz, E. S.; Wunderlich, M.; Brauns, U.; Becker, R.; Bock, T.; Dressman, J. B. *J. Pharm. Pharmacol.* **2004**, *56*, 43–51.
117. Rinaki, E.; Dokoumetzidis, A.; Valsami, G.; Macheras, P. *Pharm. Res.* **2004**, *21*, 1567–1572.
118. Klein, S.; Butler, J.; Hempenstall, J. M.; Reppas, C.; Dressman, J. B. *J. Pharm. Pharmacol.* **2004**, *56*, 605–610.
119. Blume, H. H.; Schug, B. S. *Eur. J. Pharm. Sci.* **1999**, *9*, 117–121.
120. Yu, L. X.; Ellison, C. D.; Conner, D. P.; Lesko, L. J.; Hussain, A. S. *PharmSci.* **2001**, *3*, E24.
121. Cheng, C. L.; Yu, L. X.; Lee, H. L.; Yang, C. Y.; Lue, C. S.; Chou, C. H. *Eur. J. Pharm. Sci.* **2004**, *22*, 297–304.
122. Vogelpoel, H.; Welink, J.; Amidon, G. L.; Junginger, H. E.; Midha, K. K.; Moller, H.; Olling, M.; Shah, V. P.; Barends, D. M. *J. Pharm. Sci.* **2004**, *93*, 1945–1956.
123. Lindahl, A.; Sandström, R.; Ungell, A.-L.; Abrahamsson, B.; Knutson, L.; Knutson, T.; Lennernäs, H. *Clin. Pharm. Ther.* **1996**, *60*, 493–503.
124. Petri, N.; Lennernäs, H. In Vivo Permeability Studies in the GI Tract. In *Drug Bioavailability: Estimation of Solubility, Permeability, Absorption and Bioavailability*; van de Waterbeemd, H., Lennernäs, H., Artursson, P., Eds.; Wiley-VCH: Weinheim, Germany, 2004, pp 345–386.
125. Sandström, R.; Karlsson, A.; Knutson, L.; Lennernäs, H. *Pharm. Res.* **1998**, *15*, 856–862.
126. Wu, C.-Y.; Benet, L. Z. *Pharm. Res.* **2005**, *22*, 11–23.
127. Food and Drug Administration. http://www.fda.gov/cder/guidance/3618fnl.htm (accessed July 2006).

Biographies

Hans Lennernäs has been a full professor of biopharmaceutics at Uppsala University, Sweden, since, 1 July 2000, and been adjunct professor of Biopharmaceutics at the Royal Danish School of Pharmacy in Copenhagen, Denmark, since 2000. His research interest is the clinical significance of mechanisms and regulation of membrane transport and metabolism of drugs/metabolites in the gastrointestinal tract, hepatobiliary system, and cancer tissues. Based on these findings, he is trying to develop novel strategies of tissue drug targeting and delivery in disease states, such as metabolic and cancer diseases. He has been the Principal Investigator in an extensive collaboration between the US Food and Drug Administration, the University of Michigan, USA, and the Medical Product Agency, Sweden, over several years to develop new guidelines for the Biopharmaceutics Classification System. He has established an extensive human permeability database (45 compounds) that is today widely used in academia and the pharmaceutical industry. He serves as reviewer for several scientific journals. His work has led to more 125 publications, 145 invited lectures, and more than 250 submitted presentations at scientific meetings. He has supervised 12 doctoral theses and acted as cosupervisor for two neurologists. He has obtained several national and international research grants. Dr Lennernäs has received the Glaxo Wellcome Achievement Award 1997, the Annual Award 1998 from the Industrial Pharmacy Section, Fédération Internationale Pharmaceutique (FIP), an Honourable Mentions at Eurand Award 2000. He was elected Fellow of the AAPS in 2004 and received the AAPS Meritorious Manuscript Award 2004. He is on the board of the nonprofit Drug Delivery Foundation, which promotes research and education in this area all around the world. He is also one of the innovators of a novel sublingual drug delivery system for the treatment of various acute pain conditions. He has invented five patent/patent applications, which have resulted in drug products in preclinical phase and in phase I and III clinical trials. His research team is currently composed of seven PhD students, and he has an extensive national and international interdisciplinary collaboration.

Bertil Abrahamsson has a broad experience of industrial drug development for more than 20 years within Astra and AstraZeneca. During this period he has held various project management positions, including overall responsibility for all aspects of a project during launch and life cycle management phases. Presently he is managing the biopharm group at AstraZeneca in Mölndal, Sweden, and is, as a Senior Principal Scientist, the most senior biopharmacist globally within AstraZeneca. His responsibilities include biopharmaceutical assessment prior to candidate drug selection as basis for the decision to progress compounds into development. In addition, his group provides biopharmaceutical input and testing, in vitro and in vivo, during product development as well as for biopharmaceutical NDA documentation. His expertise includes absorption pharmacokinetics in vivo predictive/in vitro dissolution and other gastrointestinal processes and functions, oral controlled-release drug delivery, in vivo imaging, and regulatory biopharmaceutics.

Bertil Abrahamsson has published 22 original papers, 6 reviews or book chapters, and more than 50 abstracts in the biopharmaceutical area. He has been invited as a speaker 15 times to international pharmaceutical and biopharmaceutical conferences and workshops. He is presently an adjunct Associate Professor in Biopharmaceutics at Uppsala University, Sweden. He has a large network globally within the scientific community and has managed research collaborations with several key universities in Europe and the USA. He also teaches at the university and supervises PhD students. Finally, he is frequently a referee for the most established journals in pharmaceutical science.

Bertil Abrahamsson has been active for many years in the board of Pharmaceutics and Biopharmaceutics department within Swedish Academy of Pharmaceutical Science, and he has organized and chaired several national and international scientific meetings.

Comprehensive Medicinal Chemistry II
ISBN (set): 0-08-044513-6

ISBN (Volume 5) 0-08-044518-7; pp. 971–988

5.43 Metabonomics

I D Wilson, AstraZeneca, Macclesfield, UK
J K Nicholson, Imperial College, London, UK

5.43.1 Introduction

Genomics, proteomics, and metabonomics represent a triumvirate of approaches that may aid our understanding of biological processes in cells, organs, and the whole organism at the biomolecular level. Recent advances in analytical technologies, coupled with the increasing availability of genome sequences, have led to intense interest in the application of these 'omics' technologies to obtain a greater understanding of the biological events leading to the onset and development of disease. Equally important to those engaged in pharmacological research is the response, both pharmacological and toxicological, of animals or human subjects to the therapeutic agents designed to combat or prevent these diseases. Outside this area there is also a need to assess the effects of the so-called functional foods as well as pesticides and other chemicals where there is likely to be environmental or industrial exposure to human populations.

A better understanding of the effects of xenobiotics, including drugs, on any biological system requires information on events and processes occurring at all levels of biomolecular organization within the organism, including genes, proteins, and metabolism. So while knowledge at the genome level will be of great value in studying disease or toxicity, it must be recognized that there are many environmental factors that will modulate the potential outcomes encoded in the genome. These factors include things such as diet, age, gender, strain, lifestyle, and even the gut microflora that live symbiotically within higher organisms but whose presence and contribution cannot be deduced from the genome sequence of the host. All of these, in combination, will affect the phenotype, and can have a very significant influence on outcomes. As result of this complexity the complementary 'omics' platforms of proteomics and metabonomics are also becoming widely employed (either singly or in combination) in so-called systems biology approaches to the study of disease and toxicity. Metabonomics, defined as "the quantitative measurement of the dynamic multiparametric response of a living system to pathophysiological stimuli or genetic modification,"[1,2] thus determines changes in the organisms complement of low-molecular-weight organic metabolites in biofluids and organs (in the related field of metabolomics, the aim is to determine the total small-molecule complement of the cell[1]). As such, metabonomics provides information on the endpoints of the various processes going on within the organism in response to change. Like all 'omics' approaches, metabonomics is a 'hypothesis-free' means of investigating metabolic systems and discovering biomarkers (insofar as there is a hypothesis, it is that "something metabolic may have changed; let's see if we can find out what!"). These biomarkers may then be used to monitor the responses of the system under examination and, perhaps more importantly, for the development of hypotheses that can advance mechanistic understanding.

5.43.2   Analytical Platforms for Metabonomics

The ideal characteristics of analytical techniques for metabonomic studies are that they should provide as comprehensive a metabolic fingerprint as possible in a reasonably short analysis time (so as to enable moderate to high throughput). Such an ideal technique would be unbiased toward particular classes of metabolites and be equally sensitive to all of the components in mixture. It should, in addition, possess a wide dynamic range, in order to cover the whole range of concentrations of endogenous molecules that might be present, and be able to provide quantitative data on all of the components present. Equally important, the technique should provide sufficient structural data to enable the investigator rapidly and unambiguously to identify the marker, or markers, detected. Not surprisingly, no such technique is currently available that can fulfill all of these criteria, and it is rather difficult to envisage any single methodology that could match up to this analytical challenge. Currently, the two major analytical methods used for obtaining metabolic profiles data are based on either high-resolution proton nuclear magnetic resonance spectroscopy (^{1}H NMR)[3–5] or, more recently, high-performance liquid chromatography coupled with mass spectrometry (HPLC-MS),[6] with a few examples of the use of gas chromatography (GC-MS)[7] and capillary electrophoresis-MS (CE-MS).[8,9] As well as these methodologies, others have advocated the use of less discriminatory techniques, such as HPLC-ultraviolet (UV),[10] HPLC-electrochemical detection (EC),[11] or infrared-spectroscopy,[12] but it is difficult to support the application of these approaches as they provide only a limited coverage of the molecules likely to be present in any biological sample, with minimal chances of full characterization and identification. So, while not entirely fulfilling the characteristics of the ideal detection and identification system described above, the NMR and separation MS-based methods provide the best compromise currently available and are described in more detail below.

5.43.2.1   Sample Types and Sampling

No matter how sophisticated the analytical technique used in the study, the key to any metabonomics investigation is the sample, and careful sample collection and treatment are essential if good results are to be obtained. The sample types that can be analyzed by the current techniques encompass all of those that might be required for biomedical or toxicological analysis, including urine, bile, blood plasma, intact tissues, and tissue extracts. When combined with chemometric techniques such as principal component analysis (PCA), particular metabolites, or groups of metabolites that provide specific markers for a particular condition (e.g., toxicity, disease, physiological variation) can be identified. In many ways urine provides an ideal method for the noninvasive study of the effects of such conditions on endogenous metabolic pathways. Samples can be taken over the duration of the study and provide a time course of effects that can be used to pinpoint onset and severity of toxicity, and determine the best times for other more invasive investigations. In addition, unless small rodents such as mice are involved, there are usually few restrictions on the size of the sample obtained. Blood plasma provides a more direct 'window' on the organism under study, but clearly requires more invasive procedures. There are also well-defined limits on the amounts of sample (and the number of sampling times) that can be taken in any given study. Tissue samples obtained from target organs clearly require surgical intervention, which in animal studies are usually only obtained on autopsy. In the case of humans the removal of, e.g., tumors or diseased organs as part of therapy affords the possibility of the direct study of these tissues.

However, when considering sampling, great care must be taken in all metabonomic studies to ensure both integrity and validity of the study design as there are numerous pitfalls for the unwary. Many factors can result in changes to sample composition and, for good results to be obtained, these must be controlled. Perhaps the most obvious is that biofluid samples provide ideal growth media for bacteria and, unless steps are taken, for example, to preserve urine being collected from animals in metabolism cages, the metabolic profile observed may be more indicative of fermentation than a response to an experimental treatment. More subtle factors such as the time of day of collection, and the gender, age, strain, and diet of the animals (or humans), as well as stress, exercise, and physical activity, can have very significant effects on global metabolite profiles.[13–17] If not carefully controlled there will be apparent study-related changes in metabolite profiles that are simply artefacts due to poor experimental design and no amount of advanced analytical technology can compensate for a badly designed study.

5.43.2.2 Nuclear Magnetic Resonance (NMR) Spectroscopy

5.43.2.2.1 Liquid samples

NMR spectroscopy has many of the ideal characteristics required for the nontargeted analysis of liquid samples for endogenous metabolites. There is thus no need for the preselection of the analytical conditions in the case of biofluid samples such as plasma, urine, and bile, as these can be analyzed without the need for any form of sample pretreatment (other than adding *c.* 10% by volume of D_2O to act as a field frequency lock for the spectrometer and buffering to minimize chemical shift variation). Currently, analysis at 600 MHz (for ^{1}H NMR) is the most popular method, requiring about 600 μL of sample in total, and taking *c.* 5–10 min per sample depending upon the sample and technique used. Higher-field-strength spectrometers are available, and probes with smaller sample requirements can be used. More recently the so-called cryoprobes have been introduced which combine high sensitivity with even more modest sample requirements. The resulting NMR spectra have very high information content, enabling the rapid detection and identification of analytes present in the sample. Another favorable feature of NMR spectroscopy is that it is nondestructive, permitting the subsequent reanalysis of the sample by other methods (e.g., HPLC-MS, GC-MS). A unique feature of NMR is that it readily allows equilibria between molecules in the sample to be observed. Against NMR spectroscopy as a bioanalytical tool for metabonomics is the criticism that it is relatively insensitive compared to mass spectrometry, for example. In answer to this criticism there is the argument that this disadvantage is more than compensated for by the fact that, unlike techniques such as MS and UV spectroscopy, NMR is equally sensitive for all proton-containing analytes. Also, as indicated above, the sensitivity of NMR spectrometers is constantly improving as a result of advances in field strength and probe design and currently lies in the ng range. The subject of the practice of NMR spectroscopy for metabonomics has been reviewed.[3–5] Typical spectra for control human plasma, bile, and urine samples are shown in **Figure 1**.

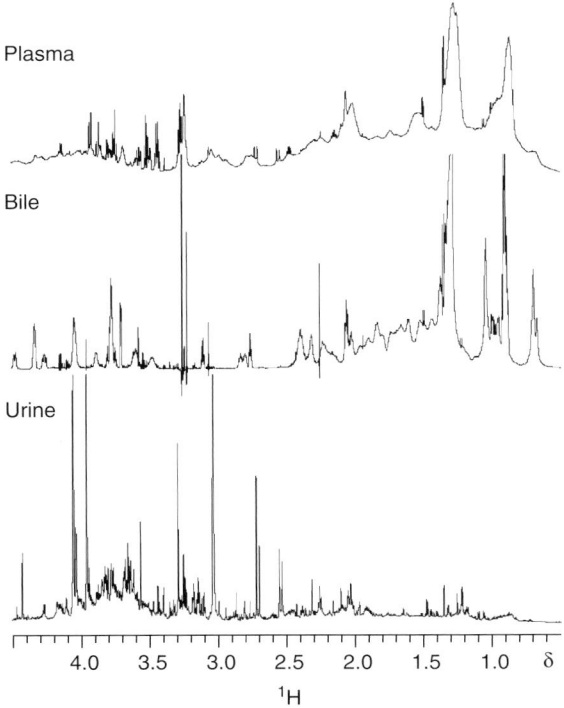

Figure 1 Partial ^{1}H NMR spectra (600 MHz) of control human plasma, bile, and urine.

5.43.2.2.2 Solid and semisolid samples

NMR spectroscopy can, in addition to biofluid samples, also be used to obtain metabolic profiles of solid and semisolid tissue samples by employing the technique known as magic angle spinning (MAS). High-resolution (HR) MAS has been used for the investigation of a range of tissue types, including kidney,[18] lymph node,[19] and prostate[20] tumors, as well as liver tissue,[21] and has recently been reviewed.[22] HRMAS provides a complementary method for the analysis of tissues compared to making extracts, with the advantage that intracellular compartmentation, which is lost during tissue homogenization and solvent extraction, is maintained. In a study of the toxicity of acetaminophen (paracetamol), liver samples from mice were analyzed both as extracts using conventional solution ^{1}H NMR spectroscopy and intact via HR ^{1}H MAS spectroscopy.[22,34] The MAS results clearly demonstrated significant perturbations in the profile of the liver tissue, with a rapid loss of glucose and glycogen combined with increased lipid content. These studies were complemented by transcriptomic[24] and proteomic[25] studies and the combination revealed a picture showing that the drug caused a global energy failure in the livers of mice receiving a toxic dose. More recent studies[26] on the hepatotoxin methapyrilene, administered to the rat at 0, 50, and $150\,mg\,kg^{-1}\,day^{-1}$ show a similar pattern of changes in the liver, as illustrated in the HRMAS NMR spectra shown in **Figure 2**. HRMAS NMR spectroscopy is not limited to tissues but can also be used to study the effects of various treatments on isolated organelles such as mitochondria.[27]

5.43.2.3   Mass Spectrometry

As discussed above, NMR spectroscopy can be performed directly on biofluid samples, but this approach is much less practicable for MS-based systems. This is because of problems due to ion suppression, which result in irreproducible and highly variable responses for analytes, while often ignored, currently present insurmountable difficulties for complex

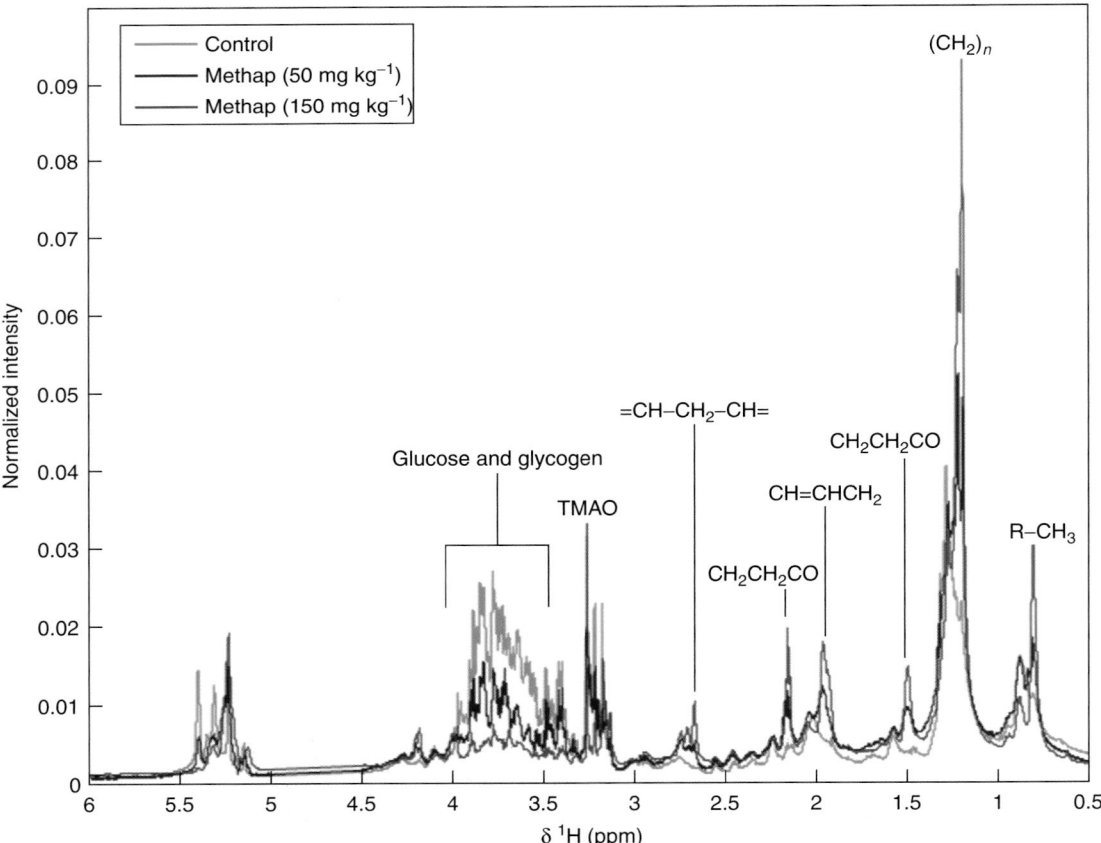

Figure 2   Average standard ^{1}H HRMAS NMR spectra of liver samples obtained from male rats dosed with 0, 50, and $150\,mg\,kg^{-1}\,day^{-1}$ of methapyrilene, showing dose-related changes in fatty acids, trimethylamine N-oxide (TMAO), glucose, and glycogen levels as a result of hepatotoxicity.[26]

samples such as urine and bile. This view is based on our unpublished studies comparing data obtained by the direct infusion of urine into the ion source of the MS versus HPLC-MS, which showed advantages for the latter. In our opinion, therefore, approaches employing MS for metabonomics that use a combination of MS with a separation technique, such as HPLC, GC, or CE, will give the best results by spreading out the components of a biological sample mixture, thus reducing the potential for ion suppression. These 'hyphenated techniques' are described in more detail below.

5.43.2.3.1 Gas chromatography-mass spectrometry

Although GC-MS, with HR capillary columns, has been widely applied to the metabolomic analysis of microorganisms and plants,[7] there are fewer published applications in mammalian systems. There is, however, considerable potential for GC-MS in this area, and it would be very surprising if many more examples did not appear in the future. An illustration of the potential of GC-MS for metabonomic analysis is shown by the example of the analysis of plasma from Zucker fa/fa and normal Wistar-derived animals given in **Figure 3**. The total ion current (TIC) traces shown are for both GC-MS with electron impact ionization (EI) (**Figure 3**) and chemical ionization (CI) (**Figure 4**). EI techniques provide mass spectra that contain diagnostic fragments that can enable much structural identification to be carried out (especially when combined with the extensive searchable databases). CI, on the other hand, gives mostly molecular ion information, enabling the confirmation of the molecular mass of the unknown. GC provides a highly developed, stable, selective, sensitive, and HR separation system. This capability is continually being enhanced and, with the introduction of GC-GC separations, combined with ever more powerful MS detectors, including time-of-flight (ToF) and ToF-ToF instruments, the comprehensive analysis of very complex samples should be possible. The use of ToF enables accurate masses to be obtained, with the benefit that atomic compositions can be deduced, providing further useful information for structure determination. The major disadvantage of

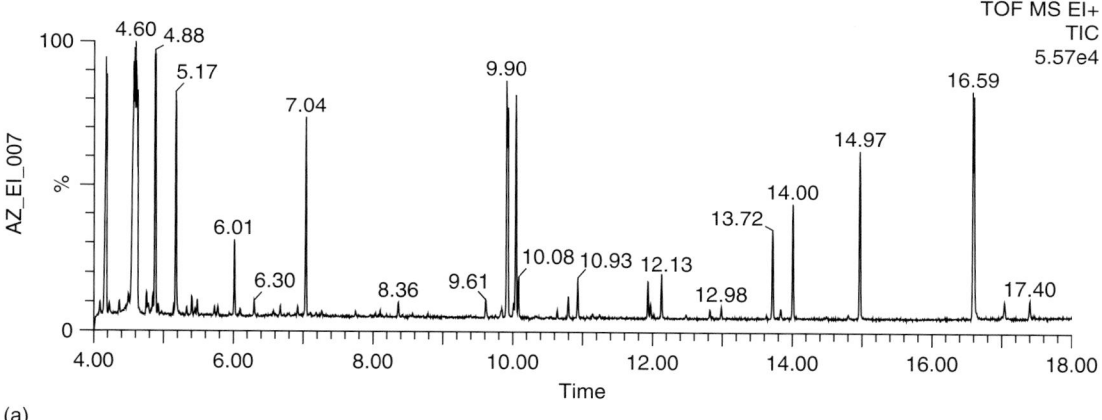

(a)

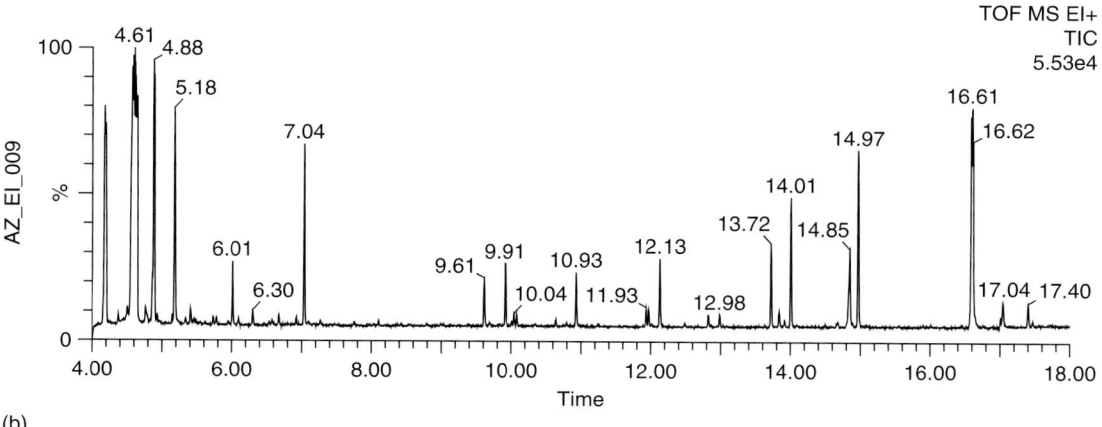

(b)

Figure 3 Typical TICs obtained from GC-EIMS analysis of plasma obtained from 20-week-old male animals corresponding to (a) Wistar-derived Alderley Park and (b) Zucker fa/fa rats.[106] (Reproduced by permission of the Royal Society of Chemistry.)

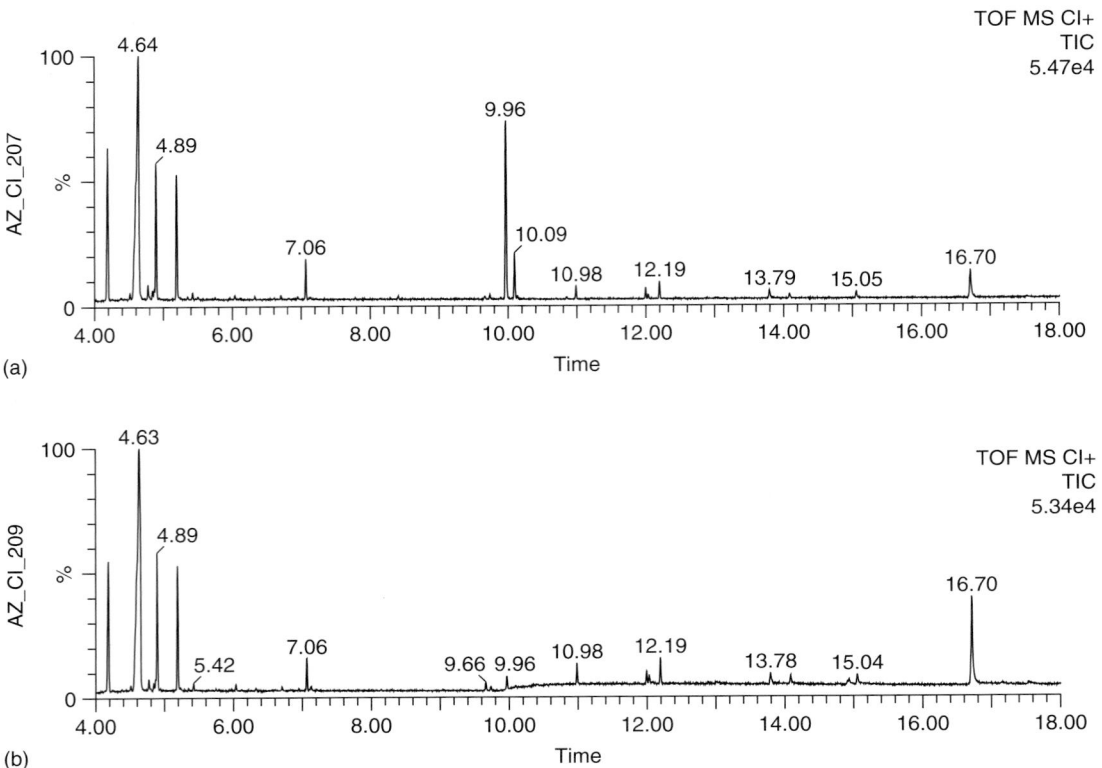

Figure 4 Typical TICs obtained from GC-CIMS analysis of plasma obtained from 20-week-old male animals corresponding to (a) Wistar-derived (AP) and (b) Zucker ($+/+$) rats.[106] (Reproduced by permission of the Royal Society of Chemistry.)

GC-based techniques is the need for a fair amount of sample processing prior to analysis. Thus, the direct injection of biofluids samples on to capillary GC columns is not technically viable and, even if it were, the bulk of the components in most biological fluids is relatively involatile. It is normal to analyze extracts of the sample after derivatization to provide volatile analytes.[7,28] The degree of sample preparation required, together with the relatively long run times currently associated with GC, means that the technique is relatively low-throughput compared to some other technologies.

5.43.2.3.2 High-performance liquid chromatography-mass spectrometry

HPLC-MS has begun to be employed in metabonomic studies only relatively recently.[6] However, given the very widespread availability of such systems in laboratories, and the compatibility of reversed-phase separations with many of the components in biofluids, a rapid increase in applications can be anticipated. Studies published to date cover areas such as rodent toxicology,[29–35] metabotyping (metabolic phenotyping) such as the study of strain, gender, and diurnal variation in mice,[14] and disease models.[36] A particular advantage of HPLC-MS over GC-based methods is that, for samples such as urine, it requires relatively little sample preparation, other than the removal of particulates, prior to analysis. For samples such as plasma or serum, proteins must be removed if the integrity of the HPLC column is to be protected, but this is easily done by precipitation with 2 or 3 volumes of acetonitrile. In general, gradient reversed-phase HPLC separations have been performed in preference to isocratic ones as these allow the separation of the maximum number of components in a given analysis time. The samples can be analyzed using a variety of ionization methods (e.g., electrospray ionization (ESI) or CI), with the current practice being to perform HPLC with ESI using both positive and negative ionization modes (usually in separate analytical runs).[7] In **Figure 5**, typical TICs are shown for the gradient HPLC of mouse urine in both positive and negative ionization mode. As alluded to above for GC-MS, using HPLC in combination with a ToF instrument to obtain accurate mass data from which atomic compositions can be determined greatly enhances the utility of the technique for the identification of unknowns. As well as HPLC-MS on conventional systems using 4.6 or 3.0 mm internal diameter (i.d.) columns packed with 3–5 μm packing materials, alternatives such as narrow-bore c. 2 mm i.d.), microbore (0.5, 1.0 mm i.d.) and capillary HPLC column formats can be employed.[7] More recently chromatography on 1.7 μm columns has been introduced in the form of ultra-performance

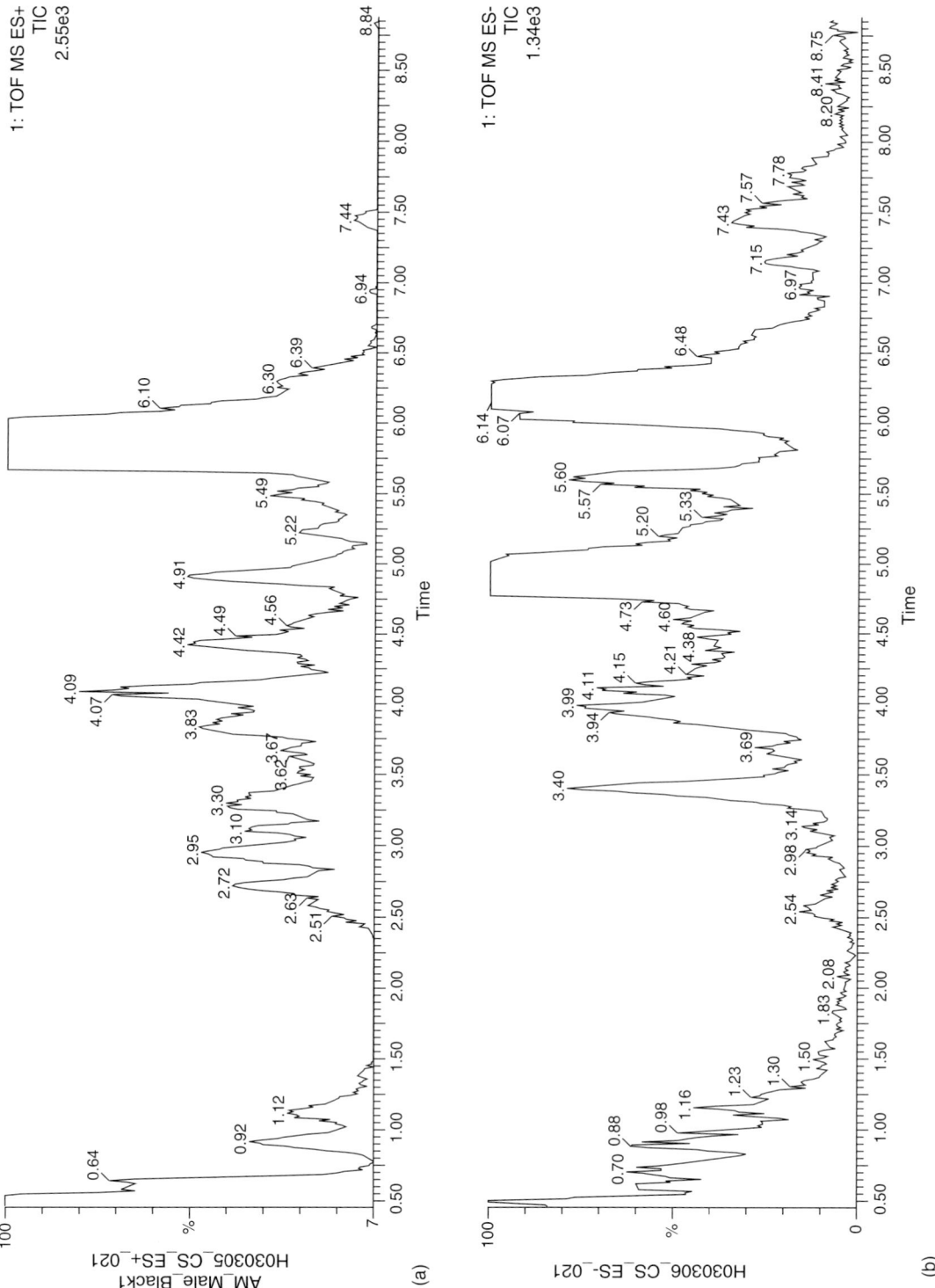

Figure 5 Reversed-phase gradient HPLC-MS analysis of a sample urine obtained from a male black mouse using (a) positive and (b) negative ESI modes.

LC (UPLC)-MS,[33,78] offering a substantial improvement in separation performance over conventional HPLC for complex mixture analysis. A typical example of the UPLC-MS of mouse urine is shown in the TIC in **Figure 6**. The improved resolution and increased number of peaks detected in UPLC-MS compared to conventional HPLC-MS are clear when these results are compared to those shown in **Figure 5**.

5.43.2.3.3 Capillary electrophoresis-mass spectrometry

The other major separation technique currently used to obtain metabolite profiles is capillary zone electrophoresis (CZE or, more commonly, CE), usually coupled with MS. To date the bulk of the applications in this area have been to bacterial samples using 'targeted' analyses against a panel of up to c. 1700 metabolite standards.[8,9] Like HPLC, CE methods have the advantage of requiring little or no sample preparation for samples such as urine. As an electrophoretic method it provides a different separation mechanism to HPLC or GC, based on charge, and CE-MS enables the HR separation and detection of metabolites. As a microbore separation technique CE requires only very small amounts of sample.

5.43.2.4   Data Processing and Model Building

The various analytical techniques described above all have one thing in common. They generate vast amounts of data, with typical metabolite profiles containing hundreds to thousands of components. While manual examination of such data sets is possible in a few cases, it is extremely time-consuming and cannot be advocated as a strategy. Instead, various multiparametic statistical approaches have been developed, of which the most widely applied is PCA. This is a well-known 'unsupervised' approach whereby the data from the various groups in the study are examined for differences without the introduction of bias by the investigator.[39] A typical example of a PCA scores and loadings plot for the ^{1}H NMR spectroscopic data obtained from normal and Zucker rats is given in **Figure 7**. The scores plot reveals the differences between the two experimental groups while the signals responsible for these differences can be found in the loadings plot. Any initial examination of the data generated by a metabolic profiling investigation should be based on such an unsupervised approach in the first instance. Supervised methods, where the investigator uses prior knowledge, such as which animals or subjects were controls and which dosed/diseased, can be used to construct models for the prediction of the class to which a sample belongs. One such supervised approach is 'projection to latent structures' by means of partial least squares (PLS), which can be described as the regression extension of PCA.[40,41] With PLS, unlike PCA, instead of describing the maximum variation in the data (X), what is attempted is to derive latent variables, which are analogous to principal components, that maximize the co-variation between the measured data (X) and the response variable (Y) regressed against. PLS discriminant analysis (PLS-DA) applies the PLS algorithm to classification of the data using a 'dummy' Y matrix that comprises an orthogonal unit vector for each class.[42] Other methods of supervised data analysis, such as soft independent modeling of class analogies (SIMCA),[43] have also been widely used for determining class in metabonomics studies. The various chemometric methods used for metabonomics have been reviewed.[44,45]

5.43.2.5   Metabolite Identification

Once detected, the biomarker, or biomarkers, have to be identified and strategies for this type of investigation have been described.[46] In the case of NMR-detected metabolites the simplest, and least time-consuming, method for accomplishing this is to use the structural information contained in the spectrum of the biofluid sample itself. Most of the common metabolites found in urine, for example, are now known, and many have characteristic ^{1}H NMR spectra.[47] Confirmation of identity can then be performed by overspiking with an authentic standard (this is particularly important where resonances can be affected by small variations in sample pH).

However, where it is not possible to assign unknowns directly from the standard one-dimensional (1D) ^{1}H NMR spectrum, then two-dimensional (2D) spectra can be obtained to aid the identification. Total correlation spectroscopy (TOCSY) is widely used for this, as protons within a spin system, especially when there are overlapping multiplets or there is extensive second-order coupling, are readily observed as off-diagonal peaks in the TOCSY spectrum. TOCSY gives both long-range and short-range correlations and is especially useful when coupling constants are small. Clearly all of the other standard NMR-based structure determination techniques can also be used.

The use of HPLC-MS for metabolite profiling is still in a state of rapid development and exhaustive databases of metabolites and retention time data are still being produced: similar comments can be made about CE-MS. In the case of GC-MS, extensive databases are currently available, although the coverage of analytes likely to be detected in metabonomics experiments is by no means complete. Typically the retention time and mass spectral data obtained for the unknown marker can be compared with those of known metabolites and, as with NMR spectroscopy, overspiking can be used to confirm identity. For complete unknowns a detailed examination of the mass spectral data and, where accurate

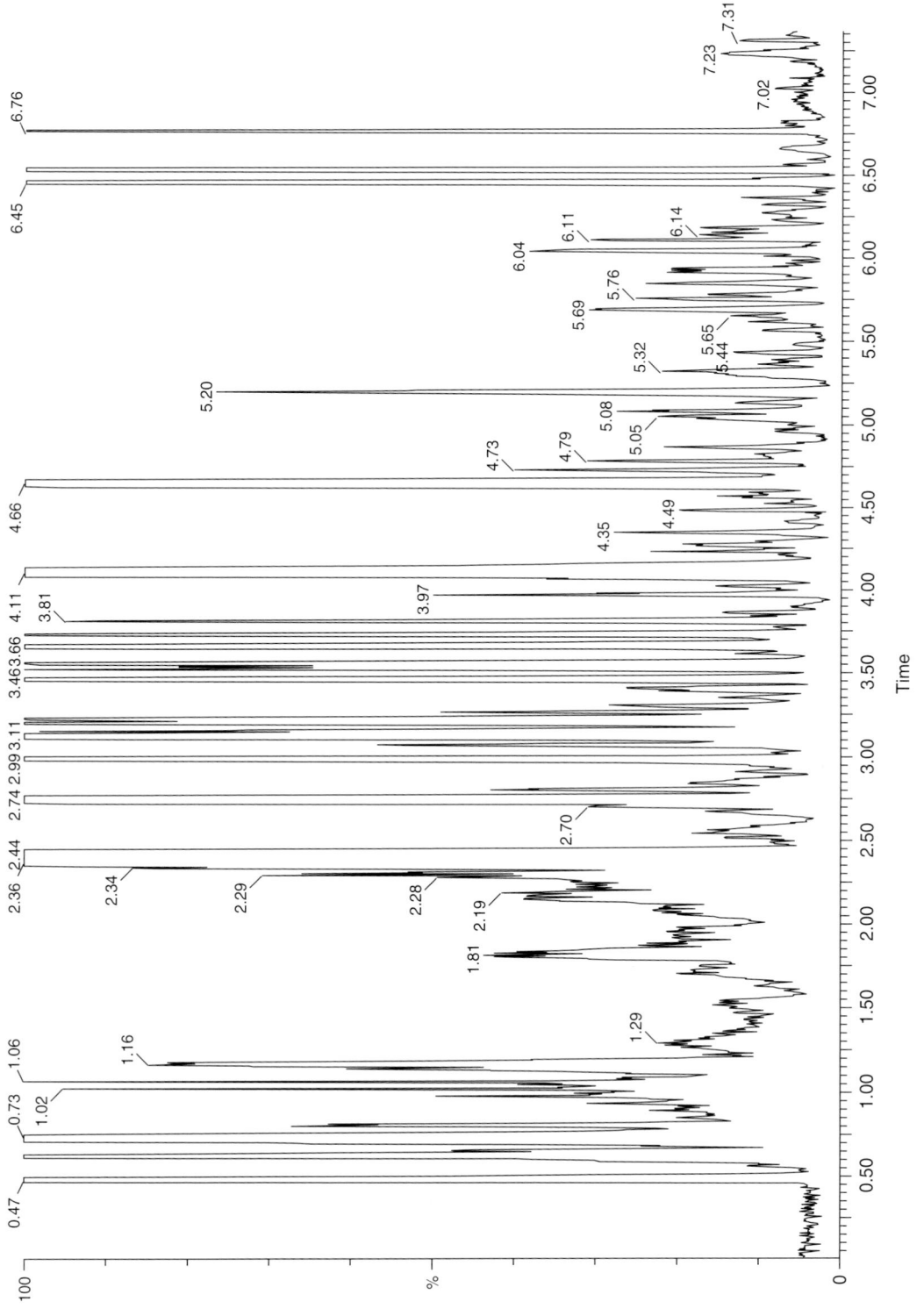

Figure 6 Reversed-phase UPLC-MS (positive ESI) of urine obtained from a sample of urine obtained from a male black mouse.

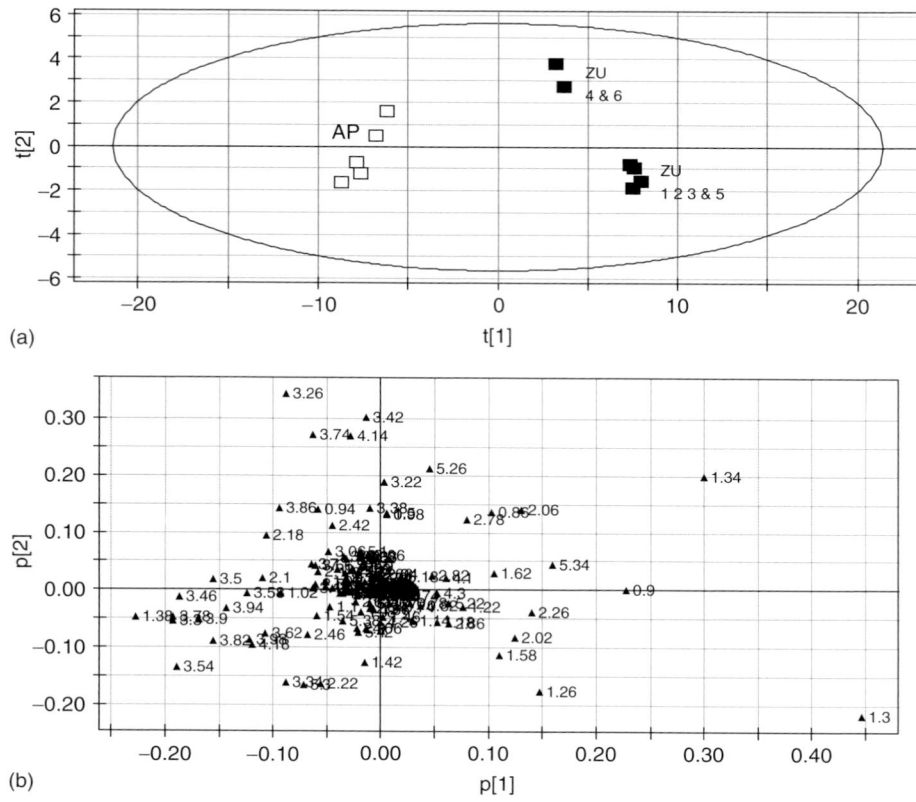

Figure 7 (a) Scores and (b) loading plots obtained following PCA of the NMR spectra obtained for the plasma of 20-week-old male Wistar-derived (AP, open squares) and Zucker fa/fa (ZU, solid squares) rats.[106] (Reproduced by permission of the Royal Society of Chemistry.)

mass data are acquired, the probable atomic composition, may provide clues to the identity of the compound that can be confirmed if an authentic standard can be procured. If derivatization was required in order to obtain the GC-MS data in the first place (i.e., the unknown was not volatile), then the chemical modifications introduced during derivatization can provide valuable clues as to the functional groups present on the unknown, which can further aid identification. If direct identification of the metabolite(s) of interest is not possible, then its isolation in a purified form will probably be required to enable detailed spectroscopic information to be obtained. A variety of methods are available to this.

5.43.2.5.1 Solid phase extraction chromatography (SPEC) nuclear magnetic resonance spectroscopy

Probably the least demanding method for the extraction and concentration of analytes from biofluids is solid phase extraction (SPE), where samples are passed through a chromatographic stationary phase contained within a syringe barrel (an SPE 'cartridge'). This results in the extraction of the analyte (together with many other components) which can then be recovered by elution with an organic solvent such as methanol. Using SPE several milliliters of sample (depending upon the amount of sorbent used) may be rapidly extracted, desalted, and eluted in a few hundred microliters of a volatile solvent. Such a procedure is sufficient for obtaining a concentrate but, if the extraction is followed by stepwise gradient elution with water/solvent mixtures ranging in proportion from 0 through 20, 40, 60, 80, and 100% organic solvent, a low-resolution chromatographic separation is obtained from which a fraction enriched in the target analyte can be garnered. Indeed, it has proved possible to isolate spectroscopically pure metabolites in this way.[48–51] Even if SPEC does not provide sufficiently pure analytes by itself, a feel is obtained for the chromatographic properties of the compounds under study that can help to guide further workup.

Once the fractions have been obtained, analysis can be performed on them by NMR and MS (if appropriate) or by HPLC-MS in order to locate the fraction(s) containing the metabolite of interest. When present in sufficient purity and quantity the identification is usually straightforward. An illustration of the use of this simple SPEC-based approach to isolation and identification is provided by the case of 5-oxoproline present in rats in urine following the administration of acetaminophen at 1% by weight of the diet.[49] Based on the chemical shifts and relative intensities of the signals detected in the urine it was possible to deduce the presence of a methylene group, adjacent to a carbonyl function, coupled to two other, strongly coupled highly nonequivalent methylene protons. These protons were also coupled to a single methine proton that had a chemical shift similar to that of an α-CH proton of an amino acid that formed the X of an ABX spin system. Despite these data the compound remained unidentified. However, SPEC,[49] followed by ^{1}H NMR to monitor the fractions, and then fast atom bombardment (FAB)-MS obtained the essential information that the unknown had a molecular mass of 129 Da and therefore contained, in addition to two methylene and a methine, at least 1 nitrogen. Based on these findings the unknown was tentatively identified as 5-oxoproline, which was then confirmed by comparison with an authentic standard. Similarly, SPEC was used on samples of urine from rats for the identification of 3-hydroxyphenylpropionic acid (3-HPPA),[50] derived from dietary chlorogenic acid via the gut microflora. Once again the 3-HPPA was partially characterized from the urinary ^{1}H NMR spectrum which showed four aromatic multiplets between 7.3 and 6.7 ppm and two triplets at 2.84 and 2.48 ppm integrating to two methylene and four aromatic protons respectively. This was consistent with a 1,3-disubstitution pattern on the aromatic ring, while the chemical shifts suggested the presence of a phenolic OH and the chemical shifts for the methylene groups suggested the presence of a carboxylic acid. SPEC was used to obtain a purified concentrate for ^{13}C NMR, which indicated that the unknown contained nine carbon atoms. These data led to a putative structure which was then confirmed as 3-HPPA by comparison to a standard.

5.43.2.5.2 High-performance liquid chromatography-nuclear magnetic resonance (HPLC-NMR) spectroscopy and high-performance liquid chromatography-mass spectrometry-based methods of identification

When low-resolution techniques prove to be unsuited to providing sufficiently pure material, then either preparative chromatographic isolation followed by spectroscopy or online methodologies such as HPLC-NMR and HPLC-MS must be performed.[46]

Although an efficient use of resources, HPLC-NMR is relatively insensitive and, where low-concentration analytes are encountered, the identification strategy must be formulated accordingly. Currently, a number of different modes of HPLC-NMR can be employed depending upon the problem. Where the sample (or extract) is concentrated, analysis can be performed on-flow, with the required spectra obtained as the sample elutes through the probe. If the analytes are present at lower concentrations, then stopped-flow techniques can be used where the peak corresponding to the analyte (observed using UV, MS, or some other conventional HPLC detector) is held stationary in an NMR-flow probe until a suitable spectrum is obtained. We have found this to be quite efficient in practice and shown that stopped-flow HPLC-NMR can be performed on several peaks in a single run without degrading the separation.

As the analyte is stationary in the flow probe quite complex (and time-consuming) NMR experiments can be performed in addition to simple 1D NMR spectroscopy, including 2D experiments such as TOCSY and correlation spectroscopy (COSY). Even if the peak contains several incompletely separated metabolites, techniques such as 'time slicing' can be used to obtain good-quality spectra. Here flow is restored for a few seconds to edge the peak a little further through the flow probe and this is then followed by the acquisition of a further spectrum. By taking a number of such spectra across the peak it is often possible to obtain spectra of the individual components. An extremely powerful approach, though somewhat time-consuming, is when the whole separation is subject to examination using the time-slicing method (alternatively, very low flow rates can be used).

As well as stopped-flow HPLC-NMR, the peaks of interest can be collected as they elute from the column in sample loops ('peak parking' or 'peak picking'), after which they can then be transferred from the sample collection loops into the NMR flow probe for spectroscopy. It is also possible to collect peaks via an online SPE where, if required, several runs can be performed, collecting and combining several samples of the unknown to provide a concentrate. The analyte may then be recovered using a fully deuterated solvent and directed into the NMR flow probe for spectroscopy.

An example where HPLC-NMR has been used in metabonomics studies concerns the identification of certain aromatic metabolites present in rat urine that separated two separate populations of the same strain.[51] Reversed-phase gradient HPLC combined with ^{1}H NMR spectroscopy allowed the identification of three of these peaks as hippuric acid, 3-HPPA, and 3-hydroxycinnamic acid. This enabled the conclusion to be reached that the observed differences between the two groups of rats resulted from differences in the respective populations of gut flora and this affected the metabolism of dietary aromatics such as chlorogenic acid.

Similarly, HPLC-MS-based approaches can be very useful in the identification of markers and, for example, xanthurenic and kynurenic acid (both metabolites of tryptophan) were detected as being correlated with nephrotoxicity in the rat.[32] Following detection by HPLC-Orthogonal acceleration (oa)ToF-MS a tentative identification was made based on the use of accurate mass and atomic composition data which was confirmed by the use of authentic standards.

5.43.3   Applications of Metabonomics

The applications of metabonomics have covered a very wide range of investigations, from basic studies in biology through more specific applications such as the investigation of disease models, toxicology, clinical investigations, and epidemiology. Typical examples of these studies are described below.

5.43.3.1   Metabotyping

The 'metabotype'[15] is a term coined to describe the metabolic phenotype presented by samples obtained from a particular population. An example of this might be the differences observed between the metabonomic profiles of, e.g., urine samples of different stains of rodents, or resulting from diurnal variation, diet, age, and gut microflora. Such differences are often by no means trivial, and if not properly understood can represent a confounding factor in metabonomic (or indeed other 'omic') investigations as changes may be seen in the profile that have little to do with the experimental treatment and everything to do with normal biology. Thus, using either ^{1}H NMR spectroscopy or HPLC-MS, it is relatively easy to distinguish between male and female rats or indeed mice of the same strain,[13–15,52] based simply on urine. Similarly, in the case of rodents, the age of an animal significantly modifies the urinary profile,[53,54] and this can be true over even relatively short periods. No doubt similar effects will be seen with other species. In **Figure 8** the effect of age on the profile of male Wistar-derived rats for a 16-week period, from 4 to 20 weeks of age, is shown.[54] The time of day when samples are collected is also important and if 24-h collections of urine are not obtained then the importance of carefully controlling sampling time to ensure that the same period is used for collection can not be overemphasized, as sample composition can change dramatically.[13–15] Similarly, diet can affect the urinary metabolite profile, as can the composition of the gut microflora, which can change during the study.[50,55–57] Thus every effort should be made to use a single batch of diet for the duration of a study, something which may not be easy if dosing over several years is contemplated. Clearly, before undertaking any series of metabonomic experiments it is important to try to establish the baseline conditions for the system under study and obtain an idea of the inherent biological variability in the system. However, as these studies show, metabonomics can provide useful and informative means of investigating the influence of normal physiological processes on the composition of biofluids.[58]

5.43.3.2   Metabonomics and the Study of Toxicity

A major area for the exploitation of metabonomics has been the study of toxicity and this area has been the subject of many publications and reviews (see [59–62] for some recent examples). An example of this type of application is provided by the work of the consortium for metabonomic toxicology (COMET), where a number of pharmaceutical companies and Imperial College have combined to look at a wide range of toxins in rodents.[63] This work was performed with the

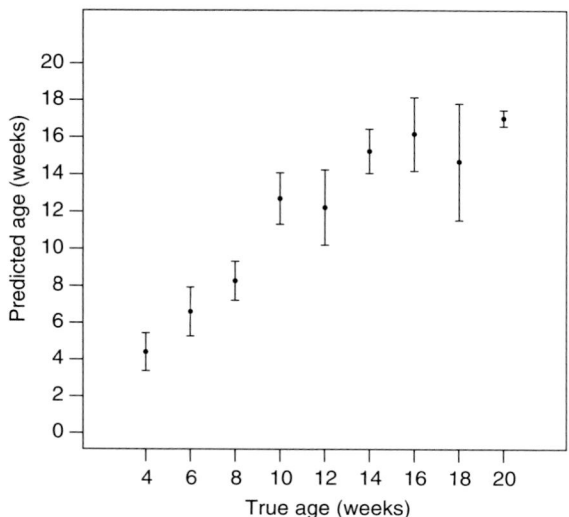

Figure 8   Age profile obtained from the urine of male Wistar-derived rats using ^{1}H NMR spectroscopy and PLS regression: cross-validated age predictions were obtained by a PLS model with three components. Data expressed as mean ± standard deviation.[54] (Reproduced by permission of the Royal Society of Chemistry.)

aim of building metabolic databases of responses to well-characterized 'typical' toxins to which the effects of new chemical entities could be compared to better predict their organ-specific toxicity and mechanism of action. In rodent studies, largely using ^{1}H-NMR-based sample analysis of urine or plasma, metabonomics has been demonstrated to be capable of classifying organ-specific toxicity in animals in tissues such as kidney,[64–80] liver[23,24,81–92] (including peroxisome proliferation[93,94]), testes,[95–97] phospholipidosis,[30,98,99] and vasculitis.[100,101] Of course, while it is possible to detect toxicity that is directed specifically to a particular organ, or indeed to a specific region within it, it is often the case that more than one organ can be affected,[92] and that the focus of the toxicity can change with time, so results from such studies need careful interpretation. In addition, as toxicity progresses, feeding may be reduced and this will also affect the metabolite profile.[102]

While the bulk of the metabonomics studies in toxicology have been conducted using NMR spectroscopic methods, more recently HPLC-MS-based analysis, either alone or together with NMR spectroscopy, has begun to be performed.[29–38] The combination of NMR and HPLC-MS techniques is particularly powerful as the different sensitivities and specificities of the two spectrometers enable a more complete metabolic profile to be generated. Indeed, a series of studies on a number of model nephrotoxins has shown the complementary nature of NMR and HPLC-MS.[32–34] In the first of these studies the effect of the administration of a single dose of mercuric chloride to male Wistar-derived rats on the urinary metabolite profiles of a range of endogenous metabolites was studied.[32] Urine was collected for the 9 days of the study and both HPLC-oa-TOF/MS and ^{1}H NMR spectroscopy revealed marked changes in the pattern of endogenous metabolites as a result of the $HgCl_2$-induced nephrotoxicity. The greatest disturbance in the urinary metabolite profiles was detected at 3 days postdose, after which the metabolite profile gradually returned to a more normal composition. The urinary markers of toxicity detected using ^{1}H NMR spectroscopy included increases in lactate, alanine, acetate, succinate, trimethylamine, and glucose, together with reductions in the amounts of citrate and α-ketoglutarate. In contrast, the HPLC-MS-detected markers (in positive ESI) included decreased kynurenic acid, xanthurenic acid, pantothenic acid, and 7-methylguanine concentrations, while an ion at *m/z* 188, possibly 3-amino-2-naphthoic acid, was observed to increase. In addition, unidentified ions at *m/z* 297 and 267 also decreased after dosing. Negative ESI revealed a number of sulfated compounds such as phenol sulfate and benzene diol sulfate, both of which appeared to decrease in concentration in response to dosing, together with an unidentified glucuronide (*m/z* 326). One conclusion from this study was that both NMR and HPLC-MS (positive and negative ESI) gave similar time courses for the onset of toxicity and recovery. However, the markers seen were quite different for each technique, clearly suggesting a role for both types of analysis. Similar conclusions about the complementary nature of NMR and HPLC-MS were confirmed in an investigation of the nephrotoxicity of the immunosuppressant cyclosporin A.[33] In this instance HPLC-MS analysis was complicated by the presence of ions derived from cyclosporin, its metabolites, and the dosing vehicle which had to be eliminated from the data prior to analysis by PCA. There was, however, once again, excellent concordance between the observed time course of toxicity whichever technique was used. However, as with the mercuric chloride example given above, the markers were different depending upon whether NMR or HPLC-MS was examined. A similar conclusion was reached when the effects of gentamicin on urinary metabolite profiles were examined by HPLC-MS and NMR spectroscopy.[35]

The complementary data provided by the combination of ^{1}H NMR and HPLC-MS suggest that, wherever possible, both techniques should be used to analyze samples. Similarly, a future role for GC-MS metabonomic studies of toxicity seems highly likely.

5.43.3.3 The Investigation of Disease Models

As well as metabotyping and providing organ-specific biomarkers of toxicity, metabonomics has great potential to aid in the characterization of animal disease models. This characterization is important as the extent to which the model actually mirrors human disease is, self-evidently, critical to the success of the drug discovery process. In addition, if the animal model does have some concordance with human disease, a detailed metabolic analysis may result in the detection of novel biomarkers that can be extended to the clinic to monitor efficacy.

An example of such studies examining disease models is provided by investigations of the urinary and plasma[103–106] metabolite profiles of Zucker (fa/fa) obese rats, which are used as a model of type 2 diabetes. Some of this work involved the collection of samples of urine from male rats at fortnightly intervals from weaning (4 weeks of age) until the animals were 20 weeks of age. These samples were compared with those of a control group of Wistar-derived animals. The urinary NMR and HPLC-MS profiles showed obvious age-related differences between conventional 20-week-old Wistar-derived and Zucker rats. In the ^{1}H NMR spectra these differences were evident as, e.g., decreased taurine and increased urinary glucose in the urine of Zucker compared to Wistar-derived rats. These differences are illustrated in **Figure 9** for 20-week-old animls. While the appearance of glucose in the urine of diabetic rats might easily have been

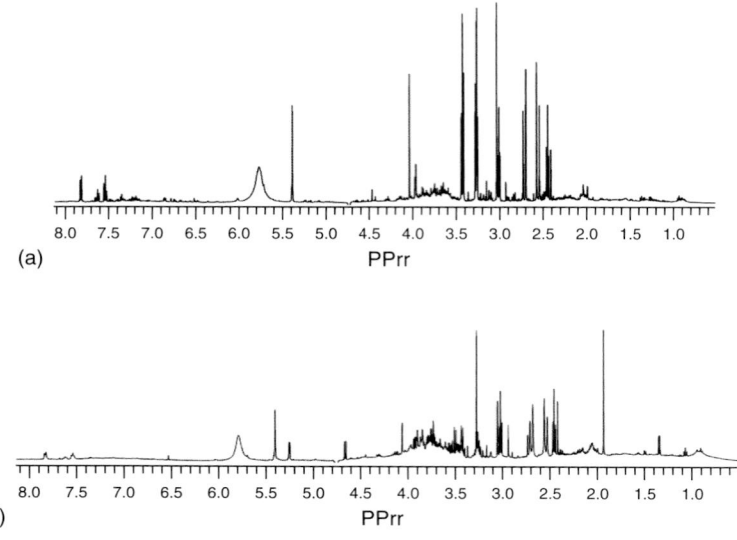

Figure 9 ¹H NMR spectra for urine samples obtained from (a) 20-week-old male Wistar-derived and (b) 20-week-old Zucker fa/fa rats.[106] (Reproduced by permission of the Royal Society of Chemistry.)

predicted, the meaning of the reduction in taurine excretion is not at all clear. The HPLC-MS data also revealed a similar progression with time. A mean trajectory for these animals, comparing the Zucker (fa/fa) obese and the control group, is shown in **Figure 10**. The two groups begin to diverge at *c.* 10 weeks of age, and by 20 weeks they occupy quite different areas of metabolic space.

5.43.3.4   Metabonomic Studies in Humans

The use of ¹H NMR spectroscopy for obtaining metabolic profiles for the diagnosis of metabolic diseases and inborn errors of metabolism is a long-standing application of the technique, with examples including such diseases as 'maple syrup urine' and 5-oxoprolinuria.[107–111] ¹H NMR spectroscopy has also been used to monitor toxicity in the case of drugs such as the anticancer agent ifosphamide. In this study the maximum kidney toxicity effects of ifosphamide were observed by the fourth treatment cycle.[112] ¹H NMR has also been used to examine the effects of phenol poisoning.[113]

However, studies undertaken to aid clinical development in human subjects, either to monitor treatment or discover new biomarkers, pose much greater difficulties compared to similar investigations in inbred strains of rodents. In particular, humans are much more diverse, and live in much less well-controlled environments than the laboratory animals used for toxicological studies. Thus the animals in metabonomic studies are housed in uniform, and carefully controlled, environmental conditions, and are similar genetically, the same age, usually the same gender, in the same weight range, and eat a controlled diet. In comparison humans, especially patients, in addition to not being housed in uniform environmental conditions are, by their very nature, often ill and subject to considerable variability in virtually all of the things that are carefully controlled in animal studies, and may also be on a range of medications. Even when such factors are controlled to some extent by performing the study on a small number of volunteers of the same gender in a clinical trials unit, there is large interindividual variability.[114] The effects of diet can be very significant and we have found unexpected dietary components (ethanol and ethyl glucoside) associated with the probable use of rice wine in cooking (or saki consumption) appearing in the urine.[115] Similarly, we have detected differences in betaine concentrations in urine between Swedish and British subjects that were sufficient to separate the two groups. This difference was most likely due to a higher consumption of fish by the Swedish population rather than any underlying genetic difference.[116] In the same study we noted dramatic effects on the urinary metabolite profile as a result of dietary changes by a female subject. Two samples were supplied by this individual, some months apart, the first of which was obtained when the subject was following a high-meat Atkins-style diet. This diet resulted in the excretion of large quantities of the amino acid taurine (often seen as a result of hepatotoxicity in rodents). However, when the second sample was analyzed the subject had reverted to a 'normal' pattern of food consumption, with the consequent absence of large amounts of taurine (and instead evidence of ethanol consumption).

where he received both BSc (1974) and MSc (1975) degrees. Subsequently, he obtained a PhD (1978) and more recently a DSc (1998) from Keele University. He is the author or co-author of some 300 papers or reviews, and has received a number of awards in separation and analytical science from the Royal Society of Chemistry, including the SAC Silver Medal for Analytical Chemistry (1990), the Analytical Separations Medal (1996), the Analysis and Instrumentation medal (2002) and the SAC Gold Medal for Analytical Chemistry (2006). He has also received the Jubilee Medal of the Chromatographic Society (1994) and gave the inaugural Desty memorial lecture for innovation in separation science (1996). As well as being on the editorial boards of a number of journals, he is Editor of the *Journal of Pharmaceutical and Biomedical Analysis* and Editor in Chief of the *Encyclopaedia of Separation Science*. He is currently visiting professor at the Universities of York, Keele, Sheffield Hallam, Manchester, and Imperial College. His research is presently directed toward the further development of hyphenated techniques in chromatography and their application to problems in drug metabolism and metabonomics.

Jeremy K Nicholson, BSc, PhD, CBiol, FIBiol, FRSA, FRC Path CChem, FRSC, obtained his BSc from Liverpool University (1977) and his PhD from London University (1980) in biochemistry working on the application of analytical electron microscopy and the applications of energy dispersive x-ray microanalysis in molecular toxicology and inorganic biochemistry. He was appointed Lecturer in Chemistry (Birkbeck College, London University, 1981–83) and Lecturer in Experimental Pathology at the London School of Pharmacy (1983–85) returning to Birkbeck as a Senior Lecturer in Chemistry, then Reader (1989) and Professor of Biological Chemistry (1992). Since 1998 he has been Professor and Head of Biological Chemistry at Imperial College London. Professor Nicholson is the author of over 500 scientific papers, patents and articles on the development and application of novel spectroscopic and chemometric approaches to the investigation of disturbed metabolic and physicochemical processes in cells and biofluids and their relationship to disease processes. His work has been recognized by the award of several scientific prizes, including: the 1992 Royal Society of Chemistry SAC Medal for Analytical Science; the 1997 Royal Society of Chemistry Gold Medal for Analytical Chemistry; the 1994 Chromatographic Society Jubilee Silver Medal; the 2002 Pfizer Prize for Chemical and Medicinal Technologies and the 2003 Royal Society of Chemistry Medal for Chemical Biology. These awards cover various aspects of the development of NMR, LC-NMR, and LC-NMR-MS approaches for biofluid analysis, drug metabolism studies, and for the development and application of metabonomic technologies for toxicological and clinical diagnostics. Research interests include: biological NMR spectroscopy, novel LC-MS and chemometric approaches to bioanalysis, metabolic modeling and studies leading to understanding the molecular basis of disease and toxic processes. Professor Nicholson holds visiting and honorary professorships in several countries and is on the editorial board of 10 international science journals, including Consulting Editor of the *Journal of Proteome Research*. He is also an honorary professor at six universities and a consultant to many pharmaceutical companies in the UK, continental Europe, and the US and is a founder director of Metabometrix, an Imperial College spin-off company specializing in molecular phenotyping, clinical diagnostics, and toxicological screening via metabonomics.

Comprehensive Medicinal Chemistry II
ISBN (set): 0-08-044513-6

ISBN (Volume 5) 0-08-044518-7; pp. 989–1007

5.44 Prodrug Objectives and Design

B Testa, University Hospital Centre, Lausanne, Switzerland

5.44.1 Introduction

This chapter aims at illustrating the various objectives of a prodrug strategy, as well as the biochemical pathways involved in prodrug activation (i.e., hydrolysis, oxidation or reduction, or even conjugation). More detailed information can be found in a number of reviews.[1–15] What makes prodrugs different from other drugs is the fact that they are devoid of intrinsic pharmacological activity. Thus, the simplest and clearest definition, in this writer's view, is that given by Adrien Albert in 1958,[16] who coined the term. In modified form, the definition reads:

Prodrugs are chemicals with little or no pharmacological activity, undergoing biotransformation to a therapeutically active metabolite.

Another reasonable but abandoned proposal is that of 'latent drugs' by Harper.[15]

The complete opposites of prodrugs are thus drugs that have no active metabolite (e.g., paracetamol). However, prodrugs should not be confused (as is too often the case), with drugs that are intrinsically active, yet are transformed into one or more active metabolites. In this case, two or more active agents will contribute to the observed clinical response in proportions that depend on differences in pharmacological activities, in compartmentalization, and in time profiles. Examples include cisplatin (which is chemically transformed to the monoaqua and diaqua species), morphine and its 6-O-glucuronide, diazepam (which is N-demethylated to nordiazepam), and codeine (which is O-demethylated to morphine). The search for pharmacologically active metabolites is thus an active topic in drug discovery.[17]

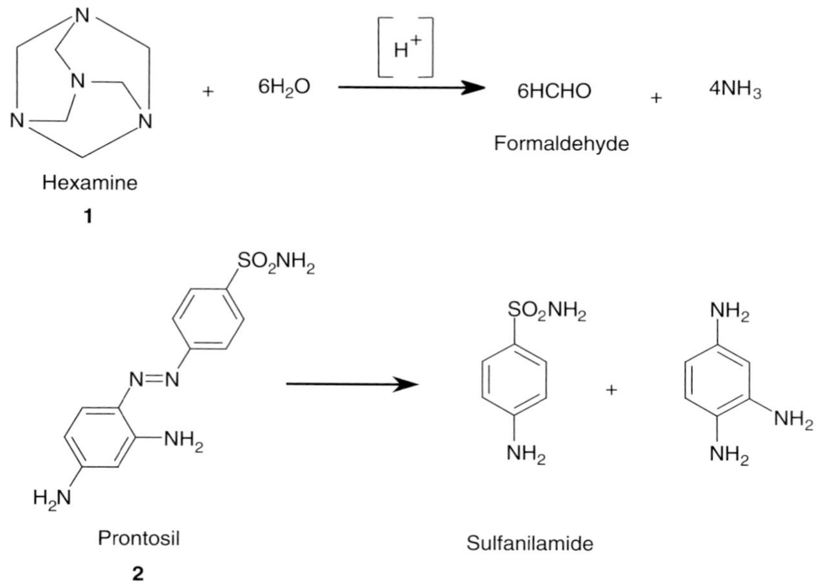

Figure 1 Two historical examples of prodrugs once in use, namely hexamine (**1**), an intentional prodrug of formaldehyde, and Prontosil (**2**), a fortuitous prodrug of sulfanilamide. (Reprinted from Testa, B. *Biochem. Pharmacol.* **2004**, *68*, 2097–2106, copyright (2004), with the permission from Elsevier.)

A number of criteria are available to classify the various types and subtypes of prodrugs. Chemical criteria (e.g., carrier-linked prodrugs versus bioprecursors) are central in our presentation, and will be covered extensively in Section 5.44.4. A historical discrimination can be made between intentional and fortuitous prodrugs. As the name indicates, the former are designed prodrugs obtained by chemical derivatization or modification of a known active agent. Many prodrugs in clinical use are of this type, and they were developed to improve the pharmaceutical and/or pharmacokinetic properties of an active agent, be it a lead candidate, a lead compound, a clinical candidate, a drug candidate, or a marketed drug. A historically interesting example is that of hexamine (**1, Figure 1**), an intentional prodrug of formaldehyde introduced over a century ago and used for decades as a throat desinfectant.

Whereas fortuitous prodrugs are few compared to intentional prodrugs, the discovery of the active agent they generate will contribute significantly to the understanding of the mechanism of action of the (pro)drug, and may even lead to the discovery of a new therapeutic class. The antibacterial medicine Prontosil (sulfamidochrysoidine) (**2, Figure 1**) offers a dramatic illustration of this phenomenon, since the discovery in 1935 of sulfanilamide as its active metabolite[18] was the milestone that led to the creation of all antibacterial sulfonamides.

5.44.2   A Look at Soft Drugs

It is essential to distinguish prodrugs from soft drugs, which are defined as "biologically active compounds (drugs) characterized by a predictable and fast in vivo metabolism to inactive and nontoxic moieties, after they have achieved their therapeutic role."[19,20] On the one hand, soft drug design resembles prodrug design since they both involve the introduction of a functional group of predictable metabolic reactivity, for example an ester linkage, to achieve metabolic promotion. On the other hand, prodrugs must be contrasted with soft drugs, which are active per se and yield inactive metabolites. And in a more global perspective, prodrugs and soft drugs appear as the two extremes of a continuum of possibilities where both the parent compound and the metabolite(s) contribute in a large or small proportion to the observed therapeutic response.

In practical terms, soft drugs (1) are bioisosteres of known drugs, (2) contain a labile bridge (often an ester group), and (3) are cleaved to metabolites known to lack activity and toxicity. In other words, soft drugs bear a close stereoelectronic analogy to the target drugs, but a rapid breakdown to inactive metabolites is programmed into their chemical structure.[21]

A typical example is succinylcholine (**3, Figure 2**), although the discovery of this agent predates by decades the concept of soft drugs. In most individuals, this curarimimetic agent is very rapidly hydrolyzed to choline and succinic acid by plasma cholinesterase, with a half-life ($t_{1/2}$) of about 4 min.[22]

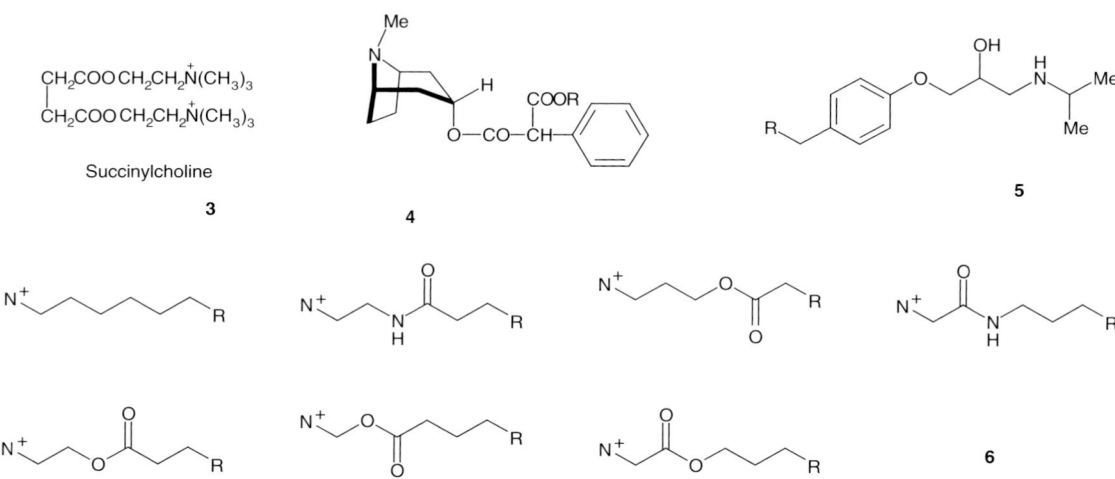

Figure 2 A few illustrative soft drugs, namely succinylcholine (**3**), whose hydrolysis yields choline and succinic acid; bioisosteres of atropine (**4**), whose hydrolysis of the –COOR moiety yields inactive acids; soft β-blockers (**5**), having an ester bridge in the R substituent, as exemplified by esmolol (R = CH₂COOCH₃), whose hydrolysis renders the side-chain too polar for receptor affinity; and quaternary ammonium antimicrobial agents (**6**), shown in order of decreasing stability under neutral conditions, with N⁺ being the quaternary ammonium group, and R being an alkyl chain (for references see the text).

Ad hoc designed soft drugs may be found in a variety of therapeutic classes, e.g., soft anticholinergics, soft β-blockers and soft antimicrobials. Thus, a number of potential soft anticholinergic agents (**4**, **Figure 2**) were prepared as bioisosteres of atropine, the –CH₂OH group being replaced by a labile –COOR group.[23] In vitro, the most active analog in the series was the isopropyl ester. The rates of chemical and enzymatic hydrolysis were found to decrease with the size of the R substituent. For example, the $t_{1/2}$ values in rat liver homogenates at 37 °C were approximately 3.5, 7, and 19 min for R = ethyl, isopropyl, and cyclohexyl, respectively, while the corresponding $t_{1/2}$ values in human plasma were much longer (about 80, 190, and 270 h, respectively), suggesting preferential and rapid hydrolysis by carboxylesterases.

A valuable class of soft β-blockers are aryloxypropanolamines, having an ester group in the *para* position (**5**, **Figure 2**).[24,25] In this series, the *para* substituent has an intermediate polarity compatible with good receptor affinity, but hydrolysis unmasks a carboxylic group whose polarity is too high for receptor binding. A promising compound in this series appears to be esmolol (**Figure 2**), whose ester bond is hydrolyzed by an esterase in the cytosol of red blood cells. The $t_{1/2}$ of esmolol in whole blood is about 1, 2, 13, and 23–27 min in guinea pigs, rats, dogs, and humans, respectively. In humans in vivo, the plasma $t_{1/2}$ of the drug is approximately 9 min, the recovery from beta blockade beginning 2 min after discontinuation of infusion and being complete at 18 min.[26–28]

Recent results suggest promising applications of soft drugs in the field of topical corticoids, calcium antagonists, and antimicrobial agents.[4,10] Antimicrobial agents deserve a special mention in the context of environmental safety ('green' chemistry). Thus, an extensive quantitative structure–activity relationship study was undertaken to assess the compared activity and stability of various quaternary ammonium antibacterials of interest as disinfectants and in environmental sanitation.[29] Whereas antimicrobial activities were comparable, large differences were seen in stability at neutral pH, allowing the authors to derive a safety index (hydrolytic rate constant divided by minimum inhibitory concentration) of clear environmental interest. As illustrated in **Figure 2** (**6**), stability in the four ester series decreased with increasing proximity between the carbonyl group and the quaternary nitrogen; the same was true in the two amide series. In fact, the effect appears so marked that the ester series with the carbonyl at the ε position had a higher stability than the more labile amide series. Such simple rules should be of clear interest to soft drugs and prodrugs designers.

For the sake of fairness, it must also be mentioned that the design of soft drugs is not without limitations. Indeed, caution is needed with the definition given at the beginning of this section, specifically with such terms as 'predictable manner' and 'controllable rate' used in the definition. What can indeed be predicted and controlled is, in qualitative terms, the site of cleavage, and, in semi-quantitative terms, the approximate lability of this site. What cannot be predicted and controlled in a quantitative manner is the actual in vivo rate of breakdown. This is due to the many biological factors that influence their biotransformation, in particular the nature, distribution, and levels of the inactivating enzymes, as well as the disposition of the soft drugs themselves, for example the extent of their binding to plasma proteins. A similar limitation also applies to many prodrugs, as discussed in the following sections.

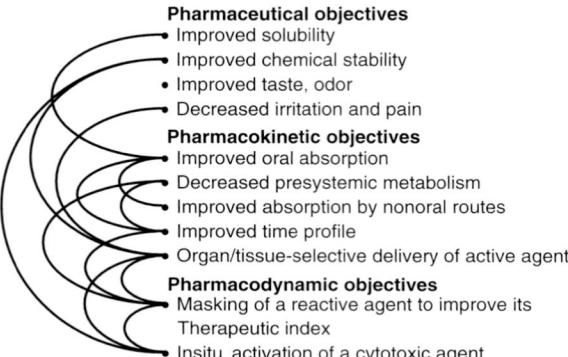

Figure 3 A schematic list of objectives in prodrug research, together with their overlaps. (Reprinted from Testa, B. *Biochem. Pharmacol.* **2004**, *68*, 2097–2106, copyright (2004), with the permission from Elsevier.)

5.44.3   Objectives of Prodrug Design

The prodrug concept has found a number of useful applications in drug research and development, as summarized in **Figure 3** and discussed in this section. However, it should be clear from the onset that such a view is too schematic, prodrug objectives being intertwined. Thus, an improved solubility can greatly facilitate oral absorption, while improving the chemical stability of an active agent can allow tissue-selective delivery and even lead to its in situ activation.

5.44.3.1   Pharmaceutical Objectives

Pharmaceutical scientists are often faced with serious formulation problems resulting from poor solubility, insufficient chemical stability, or poor organoleptic properties. While pharmaceutical technology can solve such problems in favorable cases (e.g., by improving the solubility of cyclosporine), success is not guaranteed and may be time-consuming to achieve. Rather than to wait for an uncertain and delayed pharmaceutical solution to a problem of solubility or stability, project leaders may prefer to take advantage of a prodrug strategy, and hope on an early solution.

A representative example of a prodrug solution to a solubility problem is afforded by dapsone, a drug still used in combination to fight lepra. In a project aimed at improving the water solubility of this drug, a number of prodrugs were prepared by creating an amido bridge between dapsone and an amino acid (**Table 1**).[30] Good water solubilities were indeed contributed by the more polar pro-moieties (i.e., glycyl, alanyl, and lysyl). The half-lives of hydrolysis in human blood and plasma were very promising.

Two recent examples illustrate different chemical strategies to achieve better water solubility. Thus, buparvaquone (**7**, **Figure 4**) is an effective antileishmania drug belonging to the class of hydroxynaphthoquinones, but its oral and topical availability suffers from poor water solubility. Two prodrugs were therefore prepared and examined for their solubility, rates of hydrolysis, and permeation properties.[31] As shown in **Table 2**, water solubility was markedly increased in the neutral and acidic pH range. The rates of chemical hydrolysis were rather slow in the acidic pH range, whereas enzymatic hydrolysis was fast except for compound **8** in human skin homogenates ($t_{1/2}$ about 9 h). Of great pharmacokinetic relevance are also the fluxes through human skin measured under steady state conditions, which suggest an excellent potential for topical (and oral) bioavailability. As a note of caution, however, it must be stated that the hydrolysis of prodrug **9** will liberate formaldehyde, a potentially toxic breakdown product (*see* Sections 5.44.4 and 5.44.5).

A second example of chemical interest is that of isotaxel (**10**, **Figure 4**), a prodrug of the potent antitumor agent paclitaxel (**11**).[32] Paclitaxel has a very poor water solubility (about 0.25 mg L^{-1}), which necessitates co-injection of a detergent. Taking advantage of the well-known mechanism of intramolecular nucleophilic transacylation, Hayashi *et al.*[32] prepared an analog of paclitaxel with the benzoyl group located on the α-hydroxy group rather than on the amino group (estimated pK_a about 8). As a result, what is an amido function in paclitaxel becomes a basic amino group in isotaxel, and solubility is greatly improved by protonation (water solubility of **10** · HCl is 0.45 g L^{-1}). *O,N*-Migration of the benzoyl moiety yields the active paclitaxel; as expected, the rate of *O,N-trans*-benzoylation increases with increasing pH (i.e., decreasing proportion of protonation). Thus, transbenzoylation occurred at pH 7.4 and 37 °C with a $t_{1/2}$ of 15.1 ± 1.3 min, but with a $t_{1/2}$ of 250 min at pH 4.9. In other words, the nonenzymatic activation of isotaxel to paclitaxel was fast under physiological conditions but slower under pharmaceutically compatible conditions. This study has recently been extended to other *O,N*-migrating acyl groups, leading to valuable structure–reactivity relations.[33]

Table 1 Solubility and enzymatic lability of amino acid derivatives of dapsone as water-soluble prodrugs[30]

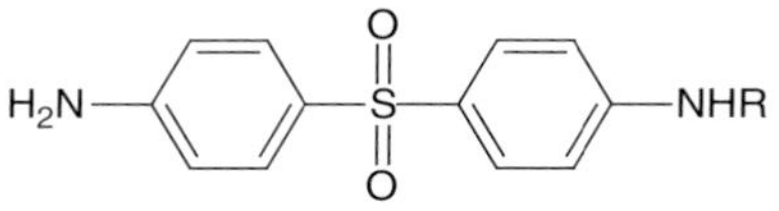	Solubility at 25 °C (mg dapsone equivalent per mL)		Stability at 37 °C in human blood, $t_{1/2} \pm SD$ (min)
	In water as HCl salt	In pH 7.4 buffer	
R = H (dapsone)	0.16	0.14	
R = –COCH$_2$NH$_2$ (glycyl)	> 15	0.87	14.6 ± 0.9
R = –COCH(CH$_3$)NH$_2$ (alanyl)	> 30	6.6	20.5 ± 0.7
R = –COCH(NH$_2$)CH$_2$CH(CH$_3$)$_2$ (leucyl)	> 25	0.31	1.7 ± 0.2
R = –COCH(NH$_2$)(CH$_2$)$_4$NH$_2$ (lysyl)	> 65	> 65	10.9 ± 0.7
R = –COCH(NH$_2$)CH$_2$C$_6$H$_5$ (phenylalanyl)	1.3	0.002	8.8 ± 0.7

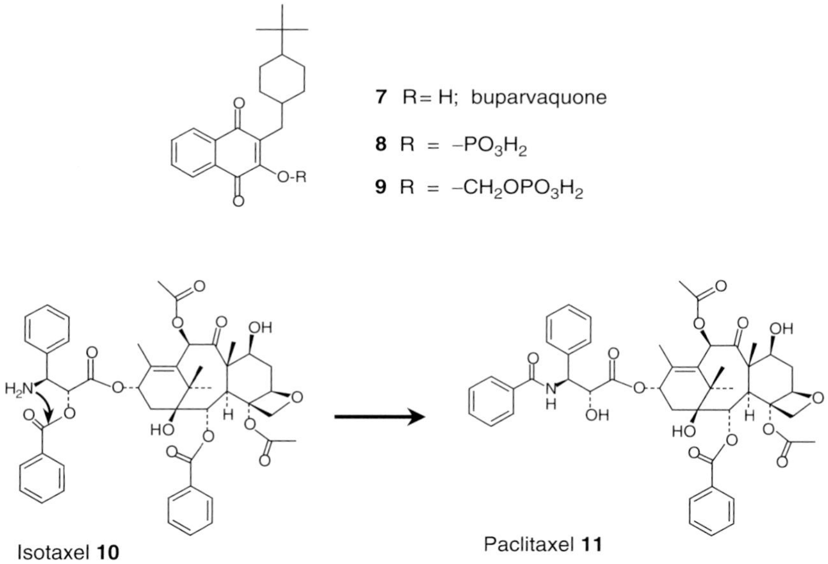

7 R = H; buparvaquone

8 R = –PO$_3$H$_2$

9 R = –CH$_2$OPO$_3$H$_2$

Isotaxel **10**   Paclitaxel **11**

Figure 4 Illustrative prodrugs developed for improved water solubility, namely two phosphate prodrugs (**8** and **9**) of buparvaquone (**7**), and isotaxel (**10**), a successful prodrug of paclitaxel (**11**).

As will be discussed later, increasing solubility is not only a pharmaceutical but also a pharmacokinetic objective. Indeed, and as made explicit in the Biopharmaceutical Classification Scheme,[34] solubility is one of the main factors influencing oral absorption (*see* 5.42 The Biopharmaceutics Classification System).

Improving bad taste is well exemplified by erythromycin A, an important antibiotic administered as enteric coated tablets or capsules to adults, but as a suspension to children. To prevent its bad taste causing rejection in children, erythromycin A is normally administered to children as its 2′-ethyl succinate, a taste-free prodrug, which, however, undergoes slow but non-negligible hydrolysis in the medicine bottle. A far more stable pro-prodrug has recently been developed, consisting in 9-enol ether of erythromycin B 2′-ethyl succinate.[35] This double prodrug is very poorly soluble in water, and thus remains stable in solution. In the acidic stomach, however, it converts rapidly to erythromycin B 2′-ethyl succinate, to be further hydrolyzed to erythromycin B (which is equiactive with erythromycin A) in the neutral condition of the small intestine.

Table 2 Properties of two phosphate prodrugs of buparvaquone[31]

		Buparvaquone (7)	*Prodrug (8)*	*Prodrug (9)*
Aqueous solubility at pH 7.4		$0.03 \pm 0.01 \, mg \, L^{-1}$	$> 3.5 \, g \, L^{-1}$	$> 3.5 \, g \, L^{-1}$
$t_{1/2}$ of hydrolysis at 37 °C	Borate buffer, pH 7.4	NA[a]	17 h	0.5 h
	Acetate buffer, pH 5.0	NA	50 h	8 h
	Alkaline phosphatase, pH 7.4	NA	1.2 min	3.8 min
	Human skin homogenate	NA	522 min	10.5 min
Steady state fluxes through human skin at 37 °C	Acetate buffer, pH 5.0	No permeation observed at 72 h	$0.13 \pm 0.04 \, nmol \, cm^{-2} \, h^{-1}$	$1.50 \pm 0.12 \, nmol \, cm^{-2} \, h^{-1}$
	Acetate buffer, pH 5.0/ethanol (50/50 v/v)	No permeation observed at 72 h	$0.11 \pm 0.08 \, nmol \, cm^{-2} \, h^{-1}$	$0.13 \pm 0.05 \, nmol \, cm^{-2} \, h^{-1}$

[a] NA = not applicable.

5.44.3.2   Pharmacokinetic Objectives

Pharmacokinetic objectives are currently the most important ones in prodrug research. Foremost among these is (1) a need to improve oral bioavailability, be it by improving the oral absorption of the drug, and/or by decreasing its presystemic metabolism (see **Figure 3**). Other objectives are (2) to improve absorption by parenteral (nonenteral, e.g., dermal, ocular) routes, (3) to lengthen the duration of action of the drug by slow metabolic release, and, finally, (4) to achieve the organ/tissue-selective delivery of an active agent. Some of these objectives are exemplified below with clinically successful prodrugs.

Achieving improved oral absorption by a prodrug strategy is a frequent rationale in marketed prodrugs.[5] An apt illustration is found within the neuraminidase inhibitors of therapeutic value against type A and B influenza in humans.[36–38] Here, target-oriented rational design has led to highly hydrophilic agents that are not absorbed orally. One of the two drugs in current clinical use is zanamivir (**12, Figure 5**), a highly hydrophilic drug administered in aerosol form. The other active agent is Ro-64-0802 (**14**), which also shows very high in vitro inhibitory efficacy toward the enzyme but low oral bioavailability due to its high polarity. To circumvent this problem, the active agent was derivatized to its ethyl ester prodrug, known as oseltamivir (**13**). Following intestinal absorption, oseltamivir undergoes rapid enzymatic hydrolysis, and produces high and sustained plasma levels of the active agent. As demonstrated by this example, the prodrug concept may thus be a valuable alternative to disentangle pharmacokinetic and pharmacodynamic optimization (see the conclusion, Section 5.44.6).

Absorption by nonoral routes includes topical, pulmonary, nasal, rectal, and vaginal delivery, but these routes are of rather limited impact compared with oral administration. This may explain why so few prodrugs have been marketed for nonoral routes. Thus, the many examples listed in an extensive review on prodrugs for topical delivery[39] are rich in scientific information but of modest clinical interest.

Pharmaceutical formulation is the most frequent method used to achieve slow release and prolong the duration of action of a given drug. However, a prodrug strategy can also be useful, as exemplified by depot formulation of esters of steroid hormones. A conceptually different and particularly elegant approach to slow metabolic release has been achieved with bambuterol (**15, Figure 6**), a prodrug of the β_2-adrenoreceptor agonist terbutaline (**16**).[40,41] Compared with terbutaline 5 mg taken three times daily, bambuterol 20 mg taken once daily provides smooth and sustained plasma levels of terbutaline, and a greater symptomatic relief of asthma with a lower incidence of side effects.[42] Similar results were obtained when comparing a slow-release formulation of terbutaline 10 mg with bambuterol $0.085 \, mg \, kg^{-1}$.[43] The two preparations elicited a comparable bronchodilating effect, but that caused by bambuterol appeared to be more sustained. The plasma levels of terbutaline were markedly more constant after bambuterol than terbutaline, but they were also significantly lower. This higher effect-to-plasma concentration ratio is another valuable feature of this prodrug, and it results from a marked uptake of bambuterol by the lung, where activation occurs in part.[41]

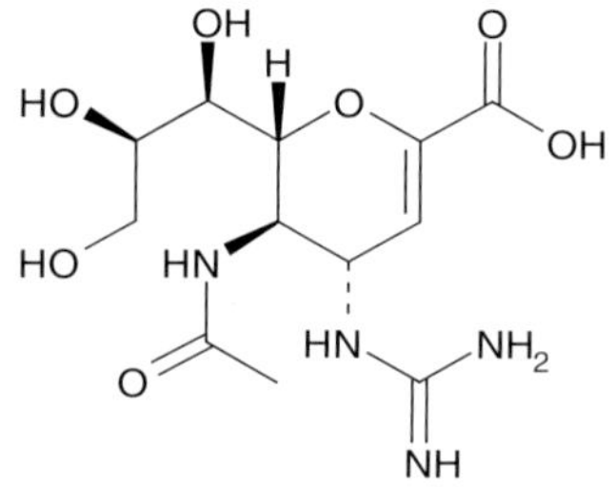

Zanamivir   **12**

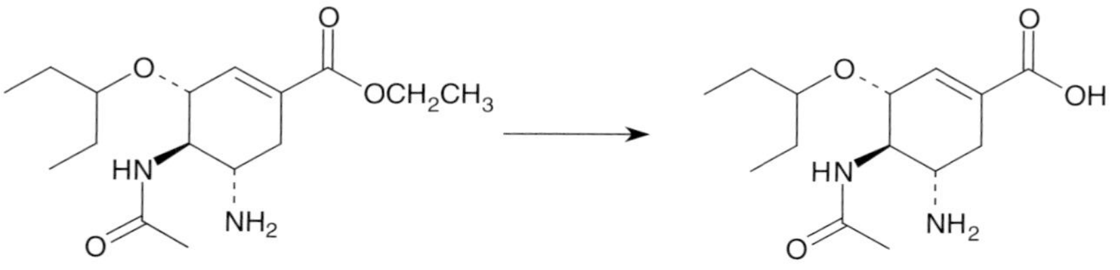

Oseltamivir   **13**

Ro-64-0802   **14**

Figure 5 Two neuraminidase inhibitors in current use against type A and B influenza in humans, namely the drug zanamivir (**12**) and the prodrug oseltamivir (**13**), whose active agent is Ro-64-0802 (**14**).

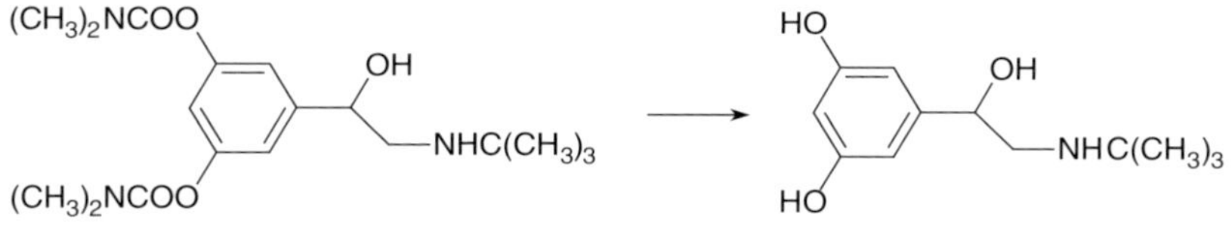

Bambuterol   **15**

Terbutaline   **16**

Figure 6 Structure of the useful prodrug bambuterol (**15**) and its active metabolite terbutaline (**16**). (Reprinted from Testa, B. *Biochem. Pharmacol.* **2004**, *68*, 2097–2106, copyright (2004), with the permission from Elsevier.)

Indeed, bambuterol is activated to terbutaline by cholinesterase (butyrylcholinesterase, EC 3.1.1.8) in blood serum, and by monooxygenase-catalyzed oxidation in the liver, lung, and other tissues. Following a first burst of terbutaline release, cholinesterase is inhibited by covalent attachment of the dimethylcarbamate moiety (Me_2NCO-), resulting in a high inhibitory capacity ($IC_{50} = 17\,nM$) with very slow reactivation of the enzyme. As a result, the inhibition of cholinesterase by bambuterol is of long duration but fully reversible, 2 days or more being necessary for complete recovery.

The last pharmacokinetic objective discussed here is the organ- or tissue-selective delivery of a given drug, also known as the search for the 'magic bullet.' This strategy is receiving ever-increasing interest in the tumor-selective delivery of anticancer agents.[44,45] A recent and clinically very significant example is that of capecitabine (**17**, **Figure 7**), a multistep, orally active prodrug of 5-fluorouracil (**20**).[46–48] Capecitabine is well absorbed orally, and undergoes three activation steps, resulting in high tumor concentrations of the active drug. It is first hydrolyzed by liver carboxylesterase, the resulting metabolite being a carbamic acid that spontaneously decarboxylates to 5'-deoxy-5-fluorocytidine (**18**). The enzyme cytidine deaminase, which is present in the liver and tumors, then transforms 5'-deoxy-5-fluorocytidine into 5'-deoxy-5-fluorouridine (**19**). Transformation into 5-fluorouracil is catalyzed by thymidine phosphorylase, and occurs selectively in tumor cells.

Capecitabine is of great interest in the context of this chapter. Clinically, it was first approved for the co-treatment of refractory metastatic breast cancer. Its therapeutic spectrum now includes metastatic colorectal cancer, and there are hopes that it might broaden further as positive results of new clinical trials become available. Capecitabine thus affords an impressive gain in therapeutic benefit compared with 5-fluorouracil, due to its oral bioavailability and a relatively

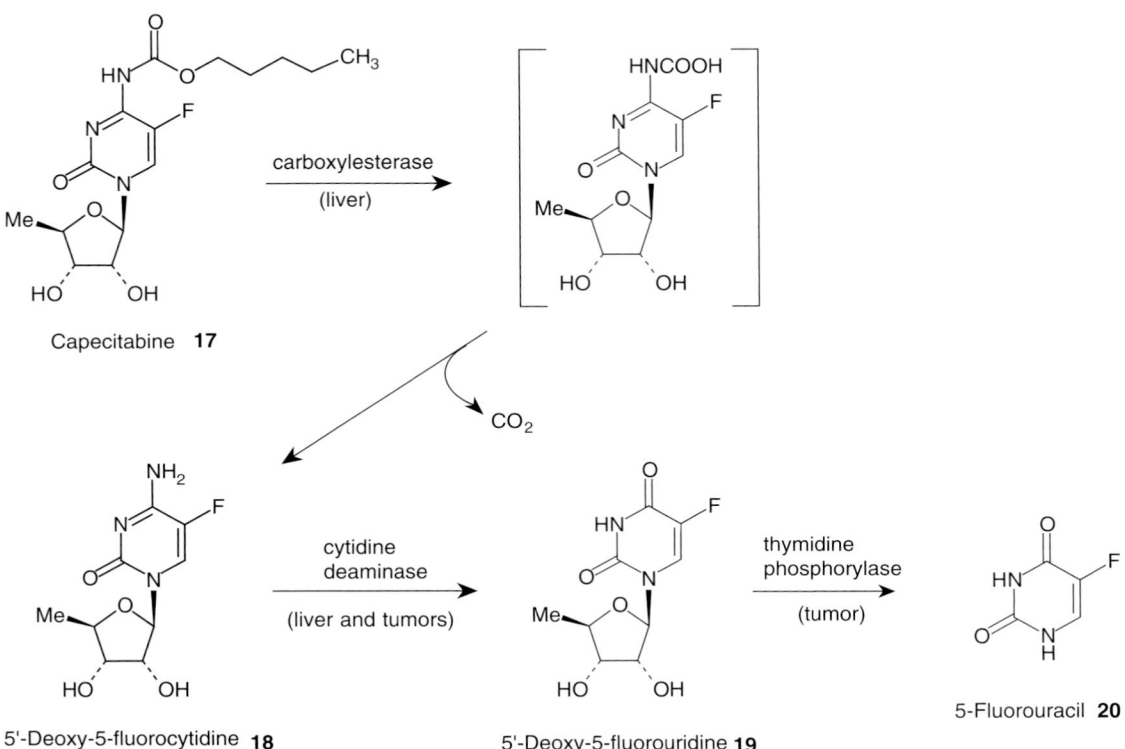

Figure 7 The stepwise activation of the successful prodrug capecitabine (**17**) to the antitumor drug 5-fluorouracil (**20**), intermediate metabolites being 5′-deoxy-5-fluorocytidine (**18**) and 5′-deoxy-5-fluorouridine (**19**). Note that the last activation step is tumor-selective. (Reprinted from Testa, B. *Biochem. Pharmacol.* **2004**, *68*, 2097–2106, copyright (2004), with the permission from Elsevier.)

selective activation in and delivery to tumors. This has been demonstrated in a number of studies, which have shown that the levels of 5-fluorouracil in tumors are clearly much higher after oral or intravenous capecitabine administration than after 5-fluorouracil intravenous administration. But even more impressive are the tumor-to-blood and tumor-to-gastrointestinal tract ratios achieved after capecitabine administration.[48]

5.44.3.3   Pharmacodynamic Objectives

Pharmacodynamic objectives can be understood as being synonymous with decreasing systemic toxicity. Two major cases are illustrated here, namely the masking of a reactive agent to improve its therapeutic index, and the in situ activation of a cytotoxic agent.

The masking of a reactive agent to improve its therapeutic index is aptly exemplified by the successful anti-aggregating agent clopidogrel (**21, Figure 8**). This compound, whose molecular mechanism of action was poorly understood for years, is now known to be a prodrug. However, it is of interest among prodrugs in that its major metabolic route in humans (about 85% of a dose) is indeed one of hydrolysis, but this reaction leads to the inactive acid. In contrast, clopidogrel is activated by cytochrome P450 3A in a two-step sequence. First, the cytochrome P450-catalyzed reaction oxidizes clopidogrel to 2-oxoclopidogrel (**22**). This is followed by a spontaneous and rapid hydrolysis of the cyclic thioester to a highly reactive thiol metabolite (**23**), which irreversibly antagonizes platelet ADP receptors via a covalent S–S bridge.[49,50] Interestingly, the same activation mechanism appears to account for the potent and irreversible inhibition of human cytochrome P450 2B6 by clopidogrel,[51] again demonstrating the high reactivity of the thiol metabolite.

In situ activation to a cytotoxic agent is part of the well-known mechanism of action of the antibacterial and antiparasitic nitroarenes such as metronidazol. Here, we examine this concept as currently intensively applied in the search for more selective antitumor agents.[44,52] Given that tumor cells have a greater reductive capacity that normal cells, various chemical strategies are being explored to design bioprecursors activated by reductive enzymes to cytotoxic agents, such as platinum(IV) complexes or *N*-oxides.

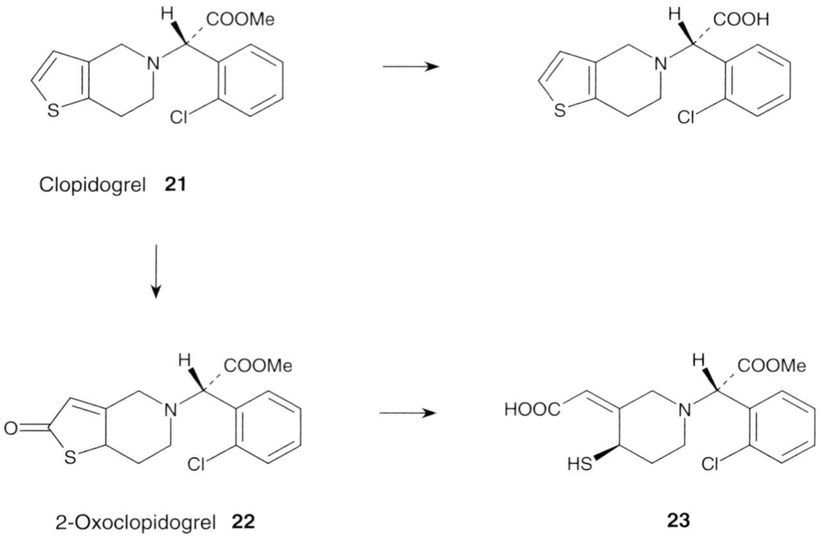

Figure 8 Major metabolic reactions of the successful prodrug clopidogrel (**21**) in humans. Most of a dose is inactivated by hydrolysis to the free carboxylic acid, whereas a smaller part is activated by cytochrome P450 3A to 2-oxoclopidogrel (**22**), followed by spontaneous hydrolytic ring opening to the active agent, a highly reactive thiol metabolite (**23**). (Reprinted from Testa, B. *Biochem. Pharmacol.* **2004**, *68*, 2097–2106, copyright (2004), with the permission from Elsevier.)

The latter case is aptly illustrated by the hypoxia-activated, bioreductive antitumor agent tirapazamine (**24**, **Figure 9**),[53] seemingly the best-studied drug candidate in this class.[52] Early evidence has shown that tirapazamine is activated to a cytotoxic nitroxide (**25**) by a one-electron reduction catalyzed by NADPH-cytochrome P450 reductase. More recent evidence, however, has implicated this activation to multiple reductases in the nucleus under hypoxia conditions, resulting in DNA single-strand and double-strand breaks.[54] The inactivation of tirapazamine is by two-electron reduction steps catalyzed by quinone reductase, and perhaps mainly by back-oxidation of the nitroxide radical by molecular oxygen in aerobic cells.[55] The metabolism of this bioprecursor is further complicated by spontaneous dismutation. In other words, cellular toxicity will depend first and foremost on oxygen levels in the cells. Of interest is also the fact that simple monosubstitution of tirapazamine can alter its lipophilicity enough to markedly improve extravascular transport and activity against target cells.

5.44.4 Prodrug Classes and their Metabolic Reactions of Activation

5.44.4.1 Chemical Criteria

Besides the criteria of classification discussed above, it is also useful from a pharmacochemistry perspective to examine prodrugs using chemical arguments. Thus, medicinal chemists find it useful to distinguish between four major classes of prodrugs, namely:

- carrier-linked prodrugs (*see* Section 5.44.4.2), where the active agent (the drug) is linked to a carrier (also known as a promoiety), and whose activation occurs by hydrolysis (esters, amides, imines, etc.), oxidation, or reduction (e.g., prontosil (**2**) see **Figure 1**);
- bioprecursors (*see* Section 5.44.4.3), which do not contain a promoiety yet are activated by hydration, oxidation, or reduction[7];
- drug–polymer conjugates (*see* 5.45 Drug–Polymer Conjugates), where the carrier is a macromolecule such as a polyethyleneglycol[56,57]; and
- drug–antibody conjugates and enzyme-directed prodrug therapies (*see* Section 5.44.4.4), where the carrier is an antibody raised against tumor cells.[4,58,59]

Here, we begin with carrier-linked prodrugs and bioprecursors, which remain by far the largest groups of prodrugs in use. Indeed, of 1562 different active substances marketed in Germany in 2002, 6.9% are prodrugs, with one-half of

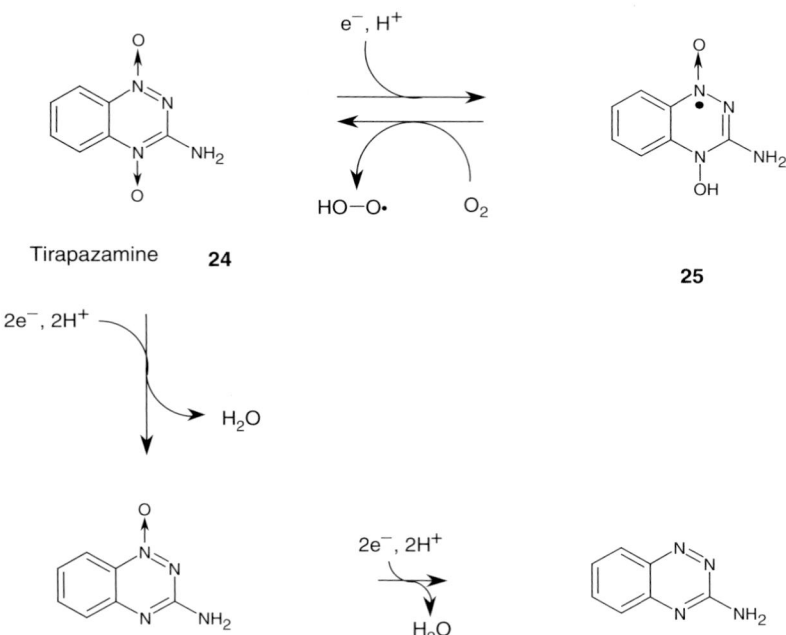

Figure 9 Pathways of activation and inactivation of the hypoxia-selective antitumor agent tirapazamine (**24**). Activation is by a one-electron reduction step catalyzed by cytochrome P450 reductase, occurs selectively in tumor cells, and produces a cytotoxic nitroxide (**25**). Inactivation is by two-electron reduction steps catalyzed by quinone reductase.

these being activated by hydrolytic cleavage of a promoiety, and one-quarter being bioprecursors.[2] Readers interested in a far more comprehensive treatment of carrier-linked prodrugs and bioprecursors are invited to consult the several relevant chapters in a recent book.[4]

5.44.4.2   Carrier-Linked Prodrugs

The carrier strategy remains the most common in prodrug design, although reductive bioprecursors are attracting an ever increasing interest, as Section 5.44.4.3 will show. Here, we illustrate and discuss a number of common and innovative promoieties classified according to the functional group in the active drug to which they are linked.

5.44.4.2.1   Prodrugs of acids

A carboxylic acid group, being ionized in the physiological pH range, will contribute significantly to reducing the lipophilicity of compounds containing this moiety. As a result, a large number of pharmacologically active carboxylic acids may display unfavorable pharmacokinetic properties such as low bioavailability, a problem of particular concern for compounds containing other moieties or a skeleton of high polarity. Thus, a major aim of prodrug design is to improve the pharmacokinetic behavior of active carboxylic acids, explaining why the majority of marketed or experimental prodrugs are derived from such acids. **Table 3** lists a number of common and less common carrier groups linked to a carboxylic group in drugs of the type R–COOH. In some cases, prodrugs of this type were developed to correct some pharmacokinetic defect(s) of established carboxylic drugs (post hoc prodrugs). In other cases, the problem was identified during development and ad hoc prodrugs were designed.

5.44.4.2.1.1   Ester prodrugs of carboxylic acids

The simplest carriers in this group are obviously alkyl promoieties, as exemplified by oseltamivir (**13, Figure 5**) discussed above. Another important therapeutic group illustrating ad hoc prodrug design is that of angiotensin-converting enzyme (ACE) inhibitors.[60] Captopril, the first marketed drug in this class, contains one carboxylate group, while more recent agents contain two carboxylate groups as important features of the pharmacophore. However, the polarity of these groups prevents good gastrointestinal absorption following oral administration. As a result, most dicarboxylate ACE inhibitors in the literature have been developed as prodrugs, being in fact ethyl monoesters, for example benazepril (**26, Figure 10**) and enalapril (**27**).

Table 3 Examples of common and less common carrier groups linked to a carboxylic group in drugs of the type R-COOH

Prodrug types	Generic structure(s)
Esters of simple alcohols or phenols	R^1COOR^2
Esters of alcohols containing an amino or amido group, or another functionality	e.g., $R^1COO(CH_2)_nNR^2R^3$ or $R^1COO(CH_2)_nCONR^2R^3$ or $R^1COO(CH_2)_nNHCOR^2$
(Acyloxy)methyl or (acyloxy)ethyl esters	$R^1COOCH_2OCOR^2$ or $R^1COOCH(CH_3)OCOR^2$
Hybrid glycerides formed from diacylglycerols	e.g., $R^1COOCH(CH_2OCOR^2)_2$
Esters of diacylaminopropan-2-ols	e.g., $R^1COOCH(CH_2NHCOR^2)_2$
N,N-Dialkyl hydroxylamine derivatives	$R^1COONR^2R^3$
Amides of amino acids	$R^1CONHCH(R^2)COOH$

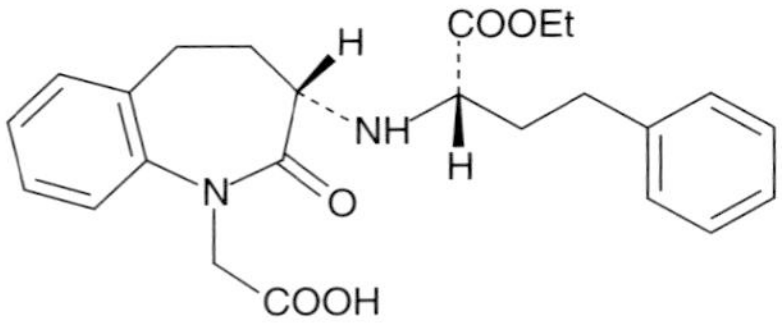

Benazepril **26**

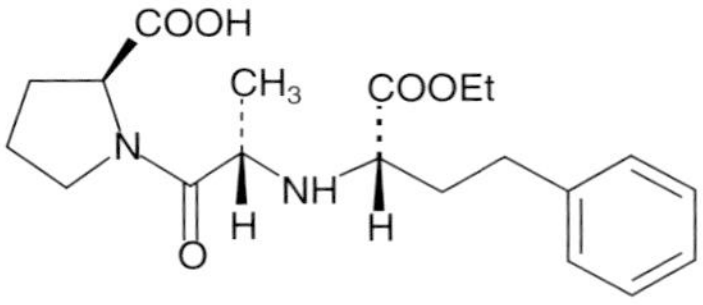

Enalapril **27**

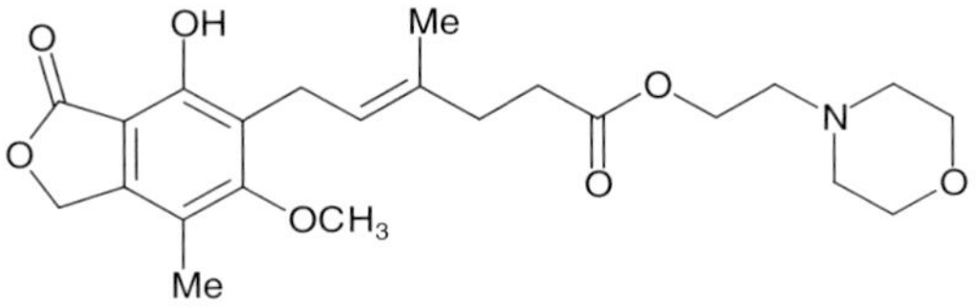

Mycophenolate mofetil **28**

Figure 10 The structure of ester prodrugs in use, namely benazepril (**26**), enalapril (**27**), and mycophenolate mofetil (**28**).

Benazepril is illustrative of the metabolism and disposition of ACE inhibitor prodrugs.[61] The compound is rapidly and well absorbed perorally in rats, dogs, baboons, and humans, in contrast to its active metabolite known as benazeprilate. The hydrolysis of benazepril to benazeprilate occurs very rapidly, and is catalyzed by carboxylesterases (EC 3.1.1.1). The latter enzymes are found in mammalian liver and in rat plasma, but are almost absent from the plasma of humans and other mammals. This difference might explain some interspecies variations in the rate of activation of ACE inhibitors.

Needless to say, many different n-alkyl, branched alkyl, cycloalkyl, alkylaryl, and aryl promoieties have been used in prodrug design. Here, we present a few functionalized alkyl groups whose properties are of interest. A comparison between the rates of activation of two types of prodrugs of nonsteroidal anti-inflammatory drugs (NSAIDs) will illustrate the influence of both the drug and carrier moieties.[62] As shown in **Table 4**, the (N,N-diethylcarbamoyl)-methyl esters were hydrolyzed in human plasma about two to four orders of magnitude faster than the methyl esters. When considering the influence of the acyl (i.e., active) moiety, differences of between 30- and 500-fold are apparent. In other words, both the drug and (mainly) the promoiety influence the rate of activation by human plasma hydrolases, but no structure–metabolism relation is apparent.

Basic promoieties are of interest since they maintain a sufficient water solubility and often afford improved oral absorption. This is nicely illustrated with mycophenolate mofetil (**28**, **Figure 10**), an orally effective prodrug of the immunosuppressant mycophenolic acid.[63] The later is of value in the suppression of acute allograft rejection following cadaver renal transplantation, but it is not absorbed orally. Esterification of the acid with the moderately basic (pK_a estimated around 8) 2-morpholinoethyl carrier resulted in a rapid and full oral absorption, followed by extensive presystemic de-esterification, presumably by serum cholinesterase. Whereas numerous basic promoieties have been

Table 4 Activation of prodrug esters of NSAIDs in human plasma under physiological conditions[62]

Drug (RCO–)	$t_{1/2}$ (80% human plasma at pH 7.4 and 37 °C) (min)	
	(N,N-Diethylcarbamoyl)methyl ester ($-OCH_2CON(Et)_2$)	Methyl ester ($-OCH_3$)
Fenbufen	3.8	280
Flurbiprofen	4.7	ND
Ibuprofen	4.0	ND
Indomethacin	25	9000
Ketoprofen	0.5	>1200
Naproxen	0.6	1200
Salicylic acid	0.05	1050
Sulindac	26	ND
Tolfenamic acid	5.0	6000

ND, not determined.

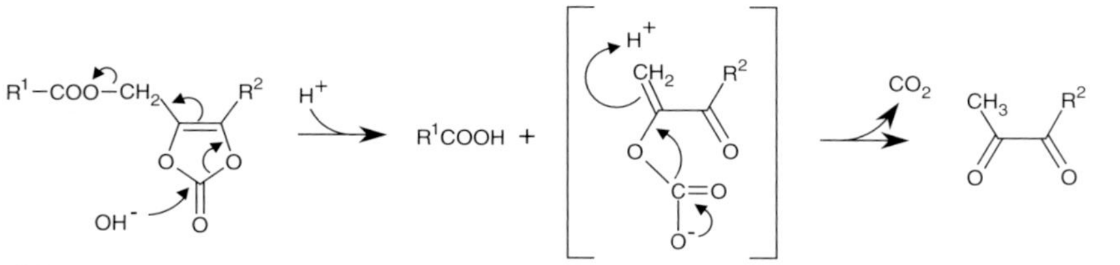

Figure 11 The general structure of (oxodioxolyl)methyl esters (daloxates) (**29**) and the mechanism of their base-catalyzed hydrolysis.

used in prodrug design, it is worth remembering that a too close proximity between the ester and amino groups may results in very fast chemical hydrolysis due to intramolecular catalysis.[4,64]

Fragmentable promoieties are another type of carriers, as exemplified by acyloxymethyl ($-CH_2OCOR$) and alkoxycarbonyloxyalkyl ($-CH_2OCOO-$alkyl) groups. The hydrolytic pathways by which these prodrugs are activated cleave their promoiety in two or three fragments, respectively.[4] The release of formaldehyde may account for some side effects of such prodrugs. This is certainly the reason why a 1-ethyl bridge should be preferred by medicinal chemists over a $-CH_2-$ bridge, despite somewhat greater synthetic difficulties

A fragmentable promoiety of value in prodrug design is the (2-oxo-1,3-dioxol-4-yl)methyl group, yielding prodrugs known as daloxates (**29**, **Figure 11**). The interest of this moiety was first documented with prodrugs of antibiotics. Available evidence supports the clean reaction of the base-catalyzed hydrolysis shown in **Figure 11**, although the actual mechanism may be more complex and/or condition-dependent.[4] Briefly, hydrolysis liberates the pro-moiety as an unstable carbonate monoester, which decarboxylates and generates a diketone. Enzymatic hydrolysis may also be involved, at least in some cases.

5.44.4.2.1.2 Amide prodrugs of carboxylic acids

Compared with esters, few amide prodrugs of carboxylic acids have been investigated. A well-known prodrug of this type is the tuberculostatic agent pyrazinamide (**30**, **Figure 12**). In vivo hydrolysis results in its inactivation, with the rate of the reaction determining its apparent half-life. Yet the tuberculostatic activity of pyrazinamide also implies its hydrolysis, since only the prodrug can penetrate into *Mycobacterium tuberculosis*, where it is hydrolyzed by pyrazinamide amidohydrolase to ammonia and the active pyrazinoic acid. The resistance to pyrazinamide can in most cases be explained by mutation(s) in pyrazinamide hydrolase and/or a pyrazinamide-selective transporter.[65]

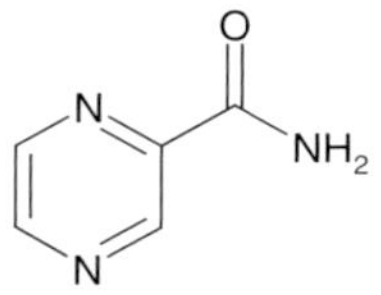

Pyrazinamide **30** 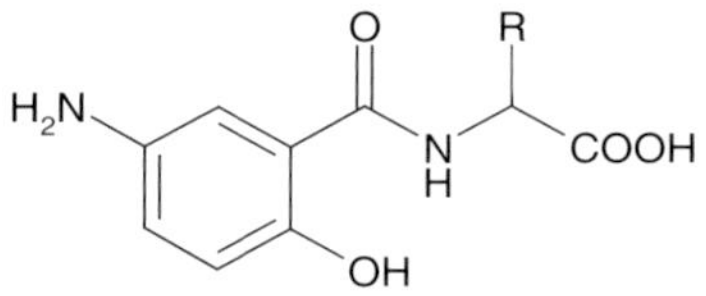   Amino acid amides of 5-aminosalicylic acid **31**

Figure 12 The structure of pyrazinamide (**30**) and amino acid amides of 5-aminosalicyclic acid (**31**).

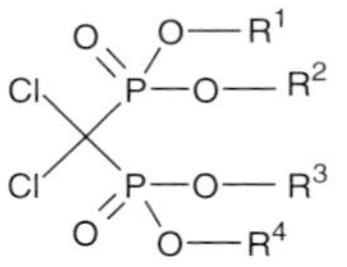

Prodrugs of clodronic acid   Prodrugs of etidronic acid   Adefovir (R = H)   **35**
                                                             Tenofovir (R = Me)
        32   **33**   **34**

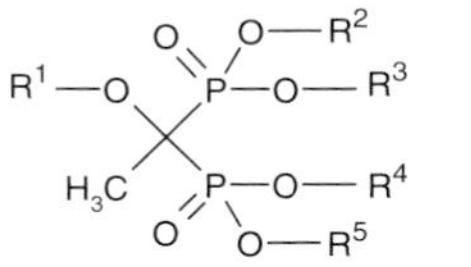

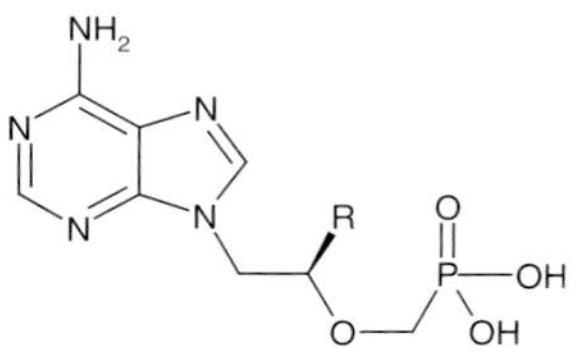

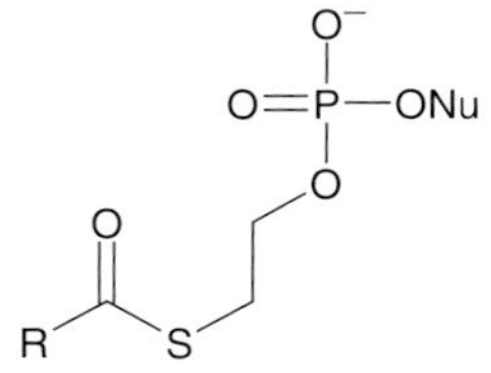

Figure 13 The structure of various phosphonate drugs (and prodrugs) discussed in the text: clodronic acid (**32**, $R^1 = R^2 = R^3 = R^4 = H$), etidronic acid (**33**, $R^1 = R^2 = R^3 = R^4 = R^5 = H$), adefovir (**34**, R = H), tenofovir (**34**, R = Me), and S-acyl-2-thioethyl phosphate esters of mononucleotides (**35**).

The drug 5-aminosalicylic acid (5-ASA) is of value in long-term maintenance therapy to prevent relapses of Crohn's disease and ulcerative colitis, yet it is absorbed rapidly in the upper intestinal tract and hardly reaches the colon. Prodrugs have therefore been designed with a view to achieve colon-specific delivery (i.e., activation in the colon), for example amino acid amides (**31**, **Figure 12**). When incubated with rat cecal content, the glycine amide released 80% of 5-ASA, whereas the aspartic and glutamic amides released 37% and 8%, respectively.[66,67] Oral administration to rats showed that most of 5-ASA-Asp was delivered to the small intestine, and about half of the administered dose was activated to liberate 5-ASA. In contrast, more than 80% of a dose of 5-ASA was recovered from urine, indicating extensive gastrointestinal absorption and negligible colon delivery.

5.44.4.2.1.3   Prodrugs of phosphoric acids

A number of phosphates and phosphonates are of value as drugs, for example bisphophonates used in the treatment of various bone diseases, and antiviral mononucleotides.[3] Because of the very high polarity of their ionized phosphate moiety, such drugs tend to suffer from limited oral bioavailability, and efforts are being directed toward prodrugs with improved pharmacokinetic properties.

Two typical bisphosphonates are clodronic acid (**32**, $R^1 = R^2 = R^3 = R^4 = H$, **Figure 13**) and etidronic acid (**33**, $R^1 = R^2 = R^3 = R^4 = R^5 = H$), with oral bioavailabilities in the range of 1–2% and 3–7%, respectively. Two types of acyloxyalkyl esters of clodronic acid were prepared and evaluated for their rate of release of the drug, namely pivaloyloxymethyl esters (**32**, $R^1 = R^2 = R^3 = R^4 = -CH_2OCOC(CH_3)_3$) and benzoyloxypropyl esters (**32**, $R^1 = R^2 = R^3 = R^4 = -CH_2CH_2CH_2OCOPhe$).[68] In vitro, the di-, tri-, and tetrabenzoyloxypropyl esters of clodronic acid did undergo some enzymatic hydrolysis at the carboxylate ester, but the resulting 3-hydroxypropyl phosphate esters were stable, and did not release clodronic acid. In contrast, the di-, tri-, and tetrapivaloyloxymethyl esters readily released clodronic acid by enzymatic hydrolysis of the carboxylate ester groups and spontaneous hydrolysis of the hydroxymethyl phosphate groups. It should be noted that the latter reaction liberated formaldehyde, an unwanted feature.

Etidronic acid (**33**, $R^1 = R^2 = R^3 = R^4 = R^5 = H$, **Figure 13**) represents a more complex case due to the presence of a hydroxy group on the bridging carbon atom. A variety of potential prodrugs were prepared and examined, with the most promising results being shown by the two derivatives having $R^1 = -COCH_3$, plus three or four pivaloyloxymethyl promoieties on the phosphonate groups.[69] The postulated mechanism involved enzymatic hydrolysis of the pivaloyloxymethyl promoities, and chemical hydrolysis of the acetyl ester group. But here again, liberation of formaldehyde occurred during hydrolytic cleavage of the pivaloyloxymethyl promoieties.

Analogous approaches also appear of interest in the search for prodrugs of antiviral agents such as the pseudonucleotides adefovir (**34**, R = H, **Figure 13**) and tenofovir (**34** , R = Me).[3,70] A different strategy is based on the use of SATE (*S*-acyl-2-thioethyl) promoieties (**35**), as applied in the preparation of prodrugs of medicinal phosphates and phosphonates. Mono- and bis-SATE derivatives have shown promising intracellular delivery, rates of activation, and antiviral activities in cell cultures.[71,72]

5.44.4.2.2 Prodrugs of alcohols and phenols

Prodrugs of active alcohols and phenols have received practically as much attention in the literature as prodrugs of active carboxylic acids. This symmetry may not extend to prodrugs in clinical use, since the improvement in pharmacokinetic properties is generally greater when masking a highly polar carboxylate group than a hydroxy group. However, it is clear that a number of active alcohols and phenols are gainfully used in therapy as ester prodrugs. This section considers a selection of the common and less common promoieties that are being used or have been examined in the design of prodrugs of active alcohols or phenols (**Table 5**).

5.44.4.2.2.1   Simple and functionalized acyl esters

The simplest and most common strategy to reversibly derivatize hydroxy groups is obviously with acyl groups, be they simple or functionalized ones. Thus, a number of acetate ester prodrugs are characterized by favorable pharmacokinetic properties and rates of activation. A large number of linear, branched, and cyclic alkanoic acids, as well as aromatic acids, can be found in the literature.[4]

The results with esters of di-acids appear difficult to rationalize. This can be seen with metronidazole, a drug which has attracted marked attention for prodrug design (**36**, R = H, **Figure 14**). At 37 °C and pH 7.4, the hemisuccinate ester (R = –COCH$_2$CH$_2$COOH) hydrolyzed with $t_{1/2}$ of 600–700 h in phosphate buffer and in human plasma, indicating the absence of enzymatic catalysis in this biological medium. Under the same conditions, similar results were obtained with the hemimaleinate (R = –COCH – CHCOOH), with $t_{1/2}$ of approximately 250–350 h. In contrast, the hemiglutarate (R = –COCH$_2$CH$_2$CH$_2$COOH) was hydrolyzed much faster in human plasma ($t_{1/2}$ about 16 h) than in phosphate buffer ($t_{1/2}$ about 800 h). This provides valuable information on the substrate selectivity of human plasma hydrolases, but further studies on a variety of other hemiglutarates as compared to hemisuccinates are required.[73]

5.44.4.2.2.2   Esters of amino acids

Most published examples of esters of amino acids contain an α-amino acyl moiety. A number of reasons may explain this fact, such as the lack of toxicity of these natural compounds, the large differences in lipophilicity and other properties between amino acids, and the variability afforded by N-substituents. Interesting examples are provided by salicylic acid and metronidazole. Thus, a number of α-amino acyl prodrugs of metronidazole (**36**, R = –COCH(R′)NH$_2$, **Figure 14**) were compared for their chemical and serum-catalyzed hydrolysis.[74] The amino acids used for esterification included

Table 5   Examples of common and less common carrier groups linked to a hydroxy group in drugs of the type R–OH

Prodrug types	*Generic structure(s)*
Esters of simple or functionalized aliphatic or alicyclic carboxylic acids	R^{1}OCOR^2
Esters of ring-substituted aromatic acids	ROCO–aryl
Esters of diacids	e.g., ROCO(CH$_2$)$_n$COOH
Esters of α-amino acids or other amino acids	e.g., R^{1}OCOCH(NH$_2$)R^2 or R^{1}OCO(CH$_2$)$_n$NR^{2}R^3
Carbonates or carbamates	R^{1}OCOOR^2 or R^{1}OCONR^{2}R^3
(Acyloxy)methyl or (acyloxy)ethyl ethers	e.g., R^{1}OCH$_2$OCOR^2 or R^{1}OCH(CH$_3$)OCOR^2
(Alkoxycarbonyloxy)methyl or (alkoxycarbonyloxy)ethyl ethers	e.g., R^{1}OCH$_2$OCOOR^2 or R^{1}OCH(CH$_3$)OCOOR^2
O-Glycosides	RO–sugar
Phosphates or sulfates	R^{1}OPO(OR^2)(OR^3) or ROSO$_3$H
Phosphoramidates	R^{1}OPO(OR^2)NHR^3

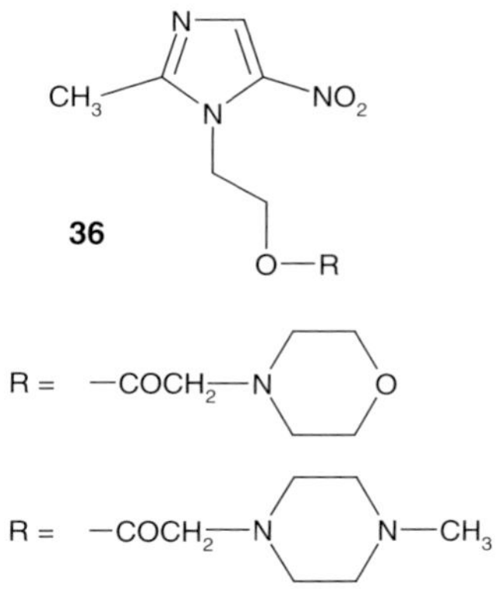

36

37

38

Prodrugs of metronidazole

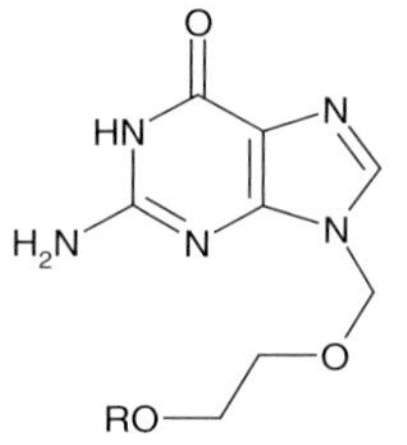

39

Prodrugs of acyclovir

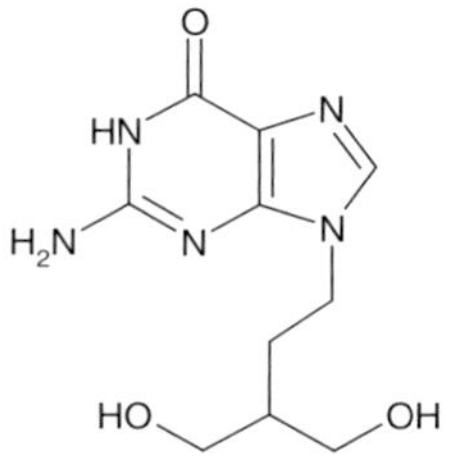

Penciclovir 40

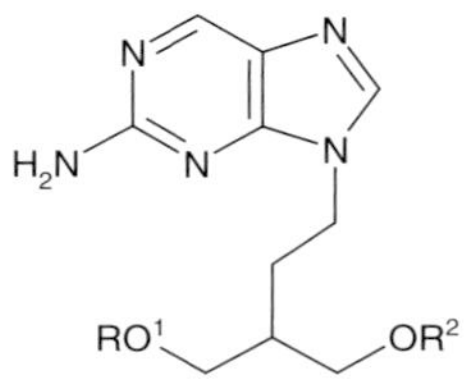

41

Pro-prodrugs of penciclovir

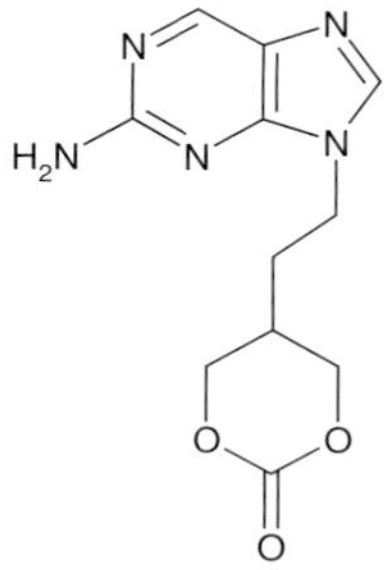

42

2-Amino-9-(2-(2-oxo-1,3-dioxan -5-yl)-ethyl)purine, a pro-prodrug of penciclovir

Figure 14 The structure of anti-infective agents whose hydroxy group has been derivatized to yield acyl ester or carbonate ester prodrugs.

alanine, glycine, isoleucine, leucine, lysine, phenylalanine, and valine. Under physiological conditions of pH and temperature, $t_{1/2}$ of hydrolysis in human serum ranged from 4.5 min for the Phe ester to 96 h for the Ile ester. A good linear relationship was established between the log of the rate constant of enzymatic hydrolysis and the log of the rate constant of hydroxyl-catalyzed hydrolysis of the protonated prodrugs. In other words, the rate of chemical hydrolysis was a good predictor of the rate of enzymatic hydrolysis. This renders the involvement of a plasmatic enzyme such as cholinesterase unlikely, but could suggest an albumin-mediated reaction.

Other types of amino acids have also been reported, as again exemplified for metronidazole by N-substituted glycine derivatives, 3-aminopropanoic acid (β-alanine), γ-amino-butyric acid (GABA), and the two promoieties **37** and **38** shown in **Figure 14**. $t_{1/2}$ for hydrolysis in buffer under physiological conditions ranged from 1 to 31 h, and between 8 min and 9 h in 80% human plasma.[75]

The use of amino acyl promoieties is of particular interest in view of recent results demonstrating that prodrugs of this type are substrates of intestinal peptide transporters such as PEPT1, resulting in enhanced intestinal absorption. Such results are of particular value for HIV protease inhibitors and antiviral nucleosides.[76,77] Thus, valaciclovir, the L-valyl ester of acyclovir (**39**, R = H, **Figure 14**), showed encouraging pharmacokinetic results in the rat and monkey. Indeed, the oral bioavailability of acyclovir following valaciclovir administration was 67 ± 13% in the monkey, a significant improvement over the limited availability of the drug administered as such.

5.44.4.2.2.3 Carbonate and carbamate esters

Carbonate esters (alkoxycarbonyl derivatives) are diesters of general formula R^1OCOOR^2. A single mechanism operates in the hydroxyl anion-catalyzed (and presumably also in the enzyme-catalyzed) hydrolysis of carbonate esters, namely a rate-determining addition of the base to the carbonyl carbon, forming an intermediate whose breakdown yields the drug (R^1OH), CO_2, and an alcohol (R^2OH). A number of carbonate prodrugs have been reported for the opioid antagonist naltrexone, whose phenolic group was esterified to yield a small series of alkyl *O*-carbonates (alkyl = methyl,

ethyl, *n*-propyl, isopropyl, *n*-butyl, and *n*-pentyl).[78] All prodrugs hydrolyzed to the parent drug when passing through sheets of human stratum corneum, with the methyl carbonate providing the highest drug flux, apparent permeability coefficient, and stratum corneum vehicle partition coefficient.

To return to antivirals, promising applications of the carbonate prodrug strategy have been reported for zidovudine and penciclovir, and their analogs. Encouraging results have thus emerged for prodrugs of penciclovir (**40, Figure 14**). This antiherpes agent has a poor oral bioavailability, which has led to the development and US Food and Drugs Administration approval of famciclovir (**41**, $R^1 = R^2 = -COCH_3$), a pro-prodrug of penciclovir activated by hydrolysis of its acetate groups and xanthine oxidase-mediated oxidation at the 6 position. Attempts to further improve the bioavailability of penciclovir have led to a variety of alkyl monocarbonate esters of 2-amino-9-(3-hydroxymethyl-but-1-yl)purine (**41**, $R^1 = H$, $R^2 = -COO-$alkyl) together with the cyclic carbonate **42**.[79] The rates of chemical hydrolysis in a pH 7.4 phosphate buffer at 37 °C ranged from about 3 h to 60 days, and they decreased in the order **42** > methyl > ethyl = *n*-pentyl > *n*-propyl > *n*-butyl = isopentyl > isobutyl > isopropyl. Interestingly, urinary recovery of penciclovir in mice and rats ranged from about one-third to one-half after oral pro-prodrug administration, whereas urinary recovery was 3–9% after penciclovir administration.

A variety of carbamate promoieties have also been attached to a hydroxy group and, particularly, a phenolic group, a good example being bambuterol, the prodrug of terbutaline (**15 and 16**, see **Figure 6**). The potential of carbamate ester prodrugs ($R^1OCONR^2R^3$) is well established by a number of studies on their stability and metabolism.[4] Upon hydrolysis, they liberate the active agent ROH and the carbamic acid R^2R^3NCOOH which, being unstable, breaks down to the amine R^2R^3NH and CO_2. The mechanism of hydroxyl-catalyzed hydrolysis of carbamate esters is more complex than that of carbonate esters. In the case of *N*,*N*-disubstituted carbamates ($R^1OCONR^2R^3$), the only possible mechanism is as for carbonates (nucleophilic attack at carbonyl carbon by hydroxyl anion to for a tetrahedral intermediate), which is presumably also the mechanism of enzyme-catalyzed hydrolysis of all carbamates.

In the case of *N*-monosubstituted carbamates ($R^1OCONHR^2$), where RO− is a good leaving group (i.e., having a $pK_a < c.12$, in practical terms a phenol), N-deprotonation does occur. The conjugate base then splits spontaneously in a rate-determining step to give the phenolate RO− and an isocyanate $R^2N = C = O$; the latter hydrates rapidly to the carbamic acid, which in turn breaks down to an amine and CO_2. This mechanism is thus characteristic for the nonenzymatic hydrolysis of aromatic *N*-monosubstituted carbamate, and it is usually several orders of magnitude faster than the mechanism of hydroxyl addition discussed above.

Thus, there will be a major difference in the design of carbamate prodrugs of phenols and alcohols. In the case of alcohols, both *N*-monosubstituted and *N*,*N*-disubstituted carbamates will a priori be chemically stable. Why so few carbamates of alcohols were reported as potential prodrugs might be due to enzymatic stability, which can be found to be uselessly stable.

For phenols, their *N*,*N*-disubstituted carbamate esters will also be high toward chemical hydrolysis, whereas their *N*-monosubstituted carbamates will as a rule be more labile. In addition, many other factors beside chemical stability can strongly influence the bioavailability and pharmacokinetic behavior of a prodrug, mainly enzymatic hydrolysis and lipophilicity. In concrete terms, this implies that careful design (and some luck) may lead to useful carbamates. That bambuterol is not an isolated example of success is demonstrated by the successful antitumor prodrug irinotecan (**43, Figure 15**). The story begins with the isolation of the natural compound camptothecin. Its antitumor activity proved highly promising, but its use was hampered by poor water solubility. Systematic studies led to the discovery of irinotecan, which combined high activity and good water solubility, and which was later shown to be a prodrug whose activity depends on the release of its active metabolite 7-ethyl-10-hydroxycaptothecin (**44**). Several human and rodent carboxylesterases were shown to hydrolyze irinotecan, but the overall activity varies from enzyme to enzyme. A highly active carboxylesterase was isolated from mouse liver and kidney.[80]

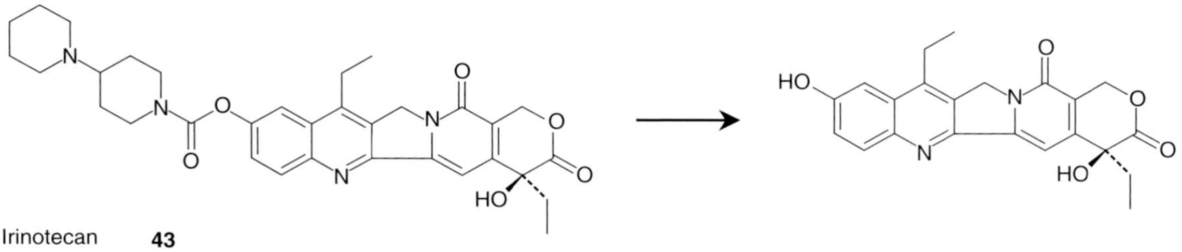

Irinotecan   **43**

7-Ethyl-10-hydroxycamptothecin   **44**

Figure 15 The structure of the successful antitumor prodrug irinotecan **43** and its active metabolite **44**.

5.44.4.2.2.4   Glycosides as prodrugs

Most glycosides reported as potential prodrugs are glucuronides, as illustrated below. Yet other sugars have also been used as promoieties, and it is worthwhile noting here that some natural glycosides are in fact prodrugs. This is the case for arbutin, the O-glucoside of hydroquinone. This conjugate is contained in some herbal medicines used as urinary antiseptics; its hydrolysis under the relatively acidic conditions prevailing in urine liberates the active hydroquinone. In other words, arbutin is a natural prodrug undergoing site-selective activation. Another example is afforded by the sennosides, which have been used since ages as laxatives; intact sennosides taken orally are hydrolyzed by colonic microflora, to liberate the active rhein anthrones.

Given the high activity of the gut microflora in various glycosidases, a number of synthetic glycosides were examined as potential colonic delivery systems, in particular corticoids.[66] The same is true for synthetic β-D-glucuronides. Note that glucuronides produced as conjugates ('phase II' metabolites) of drugs and excreted in the bile may also undergo intestinal hydrolysis, resulting in enterohepatic cycling. Hence, it is not surprising that the most promising application of glucuronide prodrugs appears to be in intestinal targeting, for example to treat ulcerative colitis and the irritable bowel syndrome. The same strategy was applied to the colon-specific delivery of corticosteroids used to treat inflammatory bowel disease.[81] Indeed, budesonide-β-D-glucuronide and dexamethasone-β-D-glucuronide underwent ready hydrolysis in the luminal contents of rat colon and cecum. Rat mucosal homogenates were less active, and hydrolysis in human fecal samples was quite low. Based on these and other studies, the prodrugs were found to be suitable candidates to deliver corticosteroids to the large intestine.

5.44.4.2.2.5   Phosphates and sulfates

A number of phosphate monoesters have been examined as potential prodrugs of active alcohols or phenols, the added phosphate promoiety contributing water solubility and being cleaved with relative ease by enzymatic reactions. A recent example of phosphorylation in prodrug design involves the endocannabinoid noladin ether (**45**, $R^1 = R^2 = H$, **Figure 16**). This cannabinoid CB1 receptor agonist reduces intraocular pressure, but its pharmacological profiling and pharmaceutical development are hindered by a poor aqueous solubility. Its mono- and diphosphate esters increased the water solubility of noladin ether more than 40 000-fold. They showed high stability against chemical hydrolysis, yet were rapidly hydrolyzed by alkaline phosphatase and liver homogenates ($t_{1/2}$ of a few minutes).[82] Hydrolysis in 4% cornea homogenates at 37 °C were also fast ($t_{1/2}$ of 17–18 min). When tested in vivo in rabbits, the monophosphate ester was very effective in reducing intraocular pressure.

Aryl monophosphates can also be found in the literature, as the phosphorylation of a phenolic group to produce potential prodrugs appears as a strategy of continued interest in medicinal chemistry. For example, the phenolic group of the chemically complex antitumor agent etoposide was monophosphorylated to increase water solubility and chemical stability.[83] After parenteral administration to patients, rapid and quantitative conversion to the parent drug was observed. Besides its pharmaceutical advantages, the prodrug proved to be bioequivalent to etoposide, and its administration resulted in a comparable efficacy.

A note of warning appears appropriate here, since Heimbach and colleagues have shown that a number of phosphate prodrugs may fail to be well absorbed orally due to the liberation (followed by precipitation) of the parent drug by intestinal alkaline phosphatase. Enzyme-mediated precipitation was found to depend on apparent supersaturation ratios, parent drug dose, solubility, and solubilization by the prodrug.[84]

Sulfate esters have but a modest importance as prodrugs, yet a few examples can be found in the literature. Such examples pertain to sulfate esters of phenols and secondary alcohols. A medicinal example is found with 5-ASA O-sulfate (**46**, **Figure 16**). As discussed earlier, 5-ASA is an agent for the treatment of ulcerative colitis and Crohn's disease of the large intestine, but it is unstable in the gastric juice. 5-ASA O-sulfate was therefore developed as a prodrug able to reach its site of action (the colon) following oral application.[85] In healthy human subjects, the prodrug

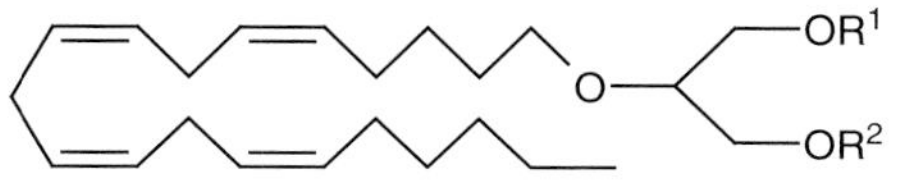

Prodrugs of noladin ether   **45** 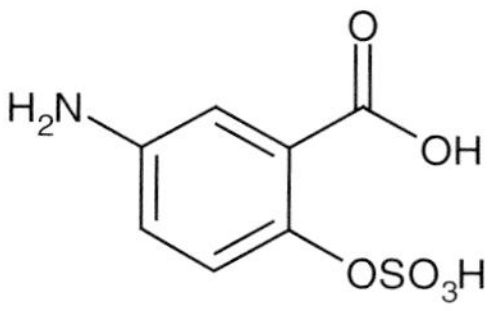   5-Aminosalicylic acid O-sulfate   **46**

Figure 16   The structure of phosphate (**45**; R^1 = phosphate, R^2 = H; or $R^1 = R^2$ = phosphate) and sulfate ester prodrugs (**46**).

was almost completely metabolized in the colon to the active agent. A high fecal and a low urinary excretion of the active metabolite 5-ASA was observed. Taken together, literature results suggest that sulfate esters can serve as valuable prodrugs for colon-specific delivery, since they avoid absorption in the small intestine and release the active compound by the action of microbial sulfatase in the colon.

5.44.4.2.3 Prodrugs of amines and amides

A number of active amines and amides have shown defects as such poor solubility or poor absorption. As a result, various prodrug strategies have been investigated, several of which are summarized in **Table 6**.

5.44.4.2.3.1 Amides of active amines

Derivatizing active amines to amides results in loss of basicity and increased lipophilicity. This is may be desirable in some cases, yet it is not a common strategy in the literature, perhaps because amides are inherently more stable than esters toward hydrolases. Yet valuable or promising prodrug amides do exist, particularly having an amino acid or short peptide as promoiety.

A fit example of the potential interest of prodrugs of this type is provided by dapsone derivatives presented in Section 5.44.3.1 and **Table 1**, which as explained were aimed at improving water solubility. An example of targeting a peptide transport system is found in recent carbapenem antibiotics.[86] A series of aminomethyl tetrahydrofuranyl-1β-methylcarbapenems (**47**, R = H, aminomethyl side-chain in position 2 or 5 (**Figure 17**)) having excellent in vitro broad-spectrum antibacterial activity exhibited only modest efficacy against acute lethal infections in mice following oral administration. In an effort to improve oral efficacy, a peptide-mediated transport strategy was investigated by substituting the amino group with a variety of amino acids. The peptidic prodrugs containing an L-amino acid were well absorbed and rapidly hydrolyzed, producing high plasmatic levels of the parent drug. Simultaneously, a 3–10-fold increased efficacy was shown by the Ala-, Val-, Ile-, and Phe-substituted prodrugs against acute lethal infections in mice. Median effective doses of $<1\,\mathrm{mg\,kg^{-1}}$ against infections caused by *Staphylococcus aureus*, *Escherichia coli*, *Enterobacter cloacae*, or penicillin-susceptible *Streptococcus pneumoniae* were obtained after the administration of single oral doses. In sharp contrast to these results, the D-forms were consistently less active than the parent drug. These results suggested an active absorption process with marked stereoselectivity for the prodrugs prepared from an L-amino acid. This hypothesis is further supported by the fact that the parent drugs demonstrated greater efficacy than the prodrugs following subcutaneous administration.

The oral absorption of an entirely different class of drugs, namely bisphosphonates, has also been improved by a peptide prodrug strategy.[87] The drugs pamidronate (**48**, $n = 2$, **Figure 17**) and alendronate (**48**, $n = 3$) were derivatized with the Pro-Phe dipeptidyl unit to yield the prodrugs Pro-Phe-pamidronate and Pro-Phe-alendronate (**49**, $n = 2$ and 3, respectively). The objective of this investigation was to target carrier systems in the intestine, particularly the intestinal peptide carrier system human oligo-peptide carrier for di- and tripeptide (hPEPT1), as well as cytosolic peptidases. In situ single-pass perfusion studies revealed competitive inhibition of transport by Pro-Phe, suggesting

Table 6 Examples of common and less common carrier groups linked to an amino or amido group in drugs of the type RNH_2 or R^1R^2-NH or $R^1R^2R^3N$

Prodrug types	Generic structure(s)
Amides formed from simple or functionalized acyl groups	$R^1R^2NCOR^3$
Amides of amino acids or peptides	$R^1R^2NCOCHR^3NHR^4$
Simple or structurally complex carbamates	R^1R^2NCOOR'' or, e.g., $R^1R^2NCOOCH(R^3)OCOR^4$ or $R^1R^2NCOOCH_2OPO_3H_2$
N-Mannich bases	$R^1R^2NCH_2NR^3R^4$
O-Mannich bases	e.g., $R^1R^2NCH_2OCOR^3$ or $R^1R^2NCH(CH_3)OCOR^3$
Cyclic Mannich bases	e.g., oxazolidines or imidazolidin-4-ones
Imines (Schiff bases)	$R^1N=CHNR^2R^3$
Phosphoramidates	$R^1R^2NPO_3H_2$
Azo conjugates	$R^1-N=NR^2$

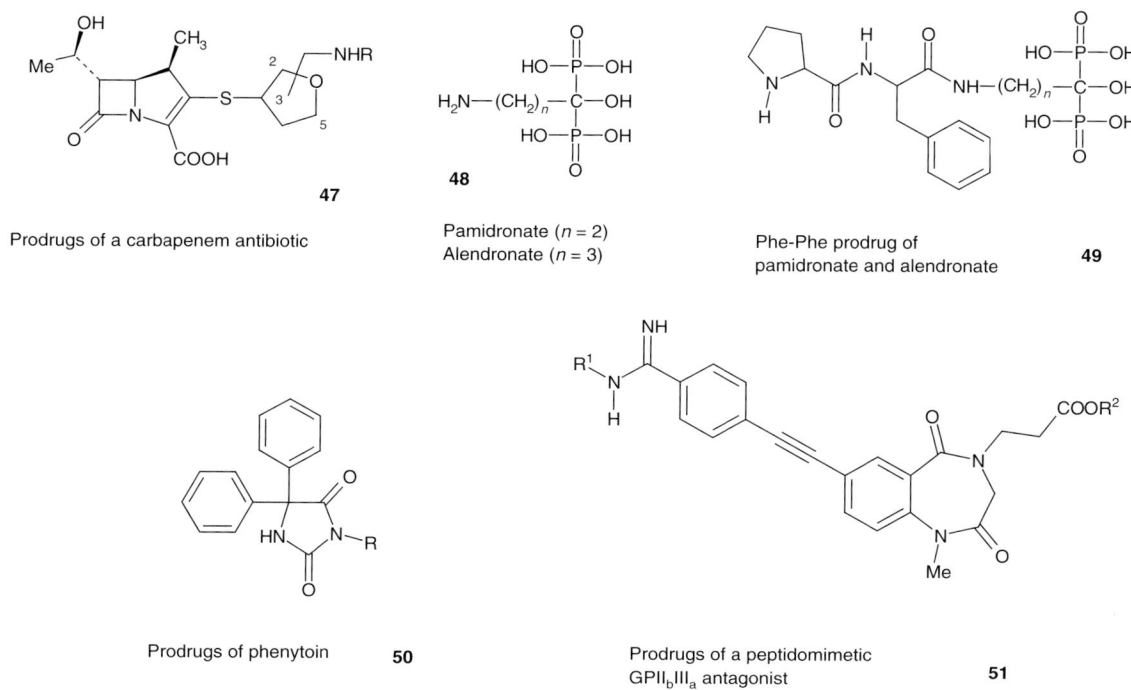

Figure 17 The structure of various amines, amides and prodrugs thereof.

peptide carrier-mediated transport. Prodrug transport in the Caco-2 cell line was significantly better than that of the parent drugs, and the prodrugs exhibited high affinity to the intestinal tissue. Oral administration of the dipeptidyl prodrugs in rats produced a threefold increase in drug absorption, and the bioavailability of Pro-Phe-alendronate was 2–3 times higher than that of the parent drug. The results indicate that the oral absorption of bisphosphonates can indeed be improved by a peptidyl prodrug strategy targeting hPEPT1 and other transporters.

5.44.4.2.3.2 Carbamates of active amines and amides

Carbamate prodrugs of active phenols and alcohols have been discussed in Section 5.44.4.2.2.3. The same principles operate when an active amine or amide is derivatized to a carbamate prodrug of the general formula $R^1R^2NCOOR^3$ (*N*-alkyloxycarbonyl derivatives), where R^3 can be an alkyl, arylalkyl, aryl, or a more complex promoiety. Interestingly, this strategy can operate for cyclic amides as well as for amines of strong to weak basicity. A successful example has already been presented earlier in the text, namely capecitabine (**17**, see **Figure 7**).

Phenytoin (**50**, R = H, **Figure 17**) is an antiepileptic drug whose poor water solubility and irregular bioavailability justify the development of prodrug forms. The simplest among these are alkylcarbamates. Thus, the ethyl- and isopropyl-*N*3-carbamates (**50**, R = –COOEt and –COOPr^i) were extensively investigated for their pharmacokinetic behavior in rats.[88] The two compounds appeared stable in buffers of pH 2 and 7.4; in rat plasma at 37 °C, they yielded phenytoin with a $t_{1/2}$ of 2.5 and 4.3 min, respectively. The oral bioavailability of phenytoin increased 8.5- and 6-fold, respectively, after administration of the two prodrugs. Such results, together with the lack of toxicity of the fragments, render these carbamates quite promising.

The derivatization of a basic function is exemplified by the amidine peptidomimetic **51** ($R^1 = R^2 = H$, **Figure 17**), an antagonist of the platelet membrane glycoprotein GPII$_b$III$_a$ aimed at decreasing thrombotic vascular stenosis and myocardial infarction.[89] The compound is a zwitterion at physiological pH, and showed limited oral bioavailability. A double prodrug approach was examined, with R^1 = alkoxycarbonyl and R^2 = Et. The rates of hydrolysis of the carbamate function in buffers increased only slightly from pH 2 to 7, and then sharply in the alkaline range. In the small series investigated, the rate of hydrolysis increased sharply from R^1 = –COOEt to R^1 = –COSEt to R^1 = –COO–phenyl, that is, with increasing acidity of the leaving group (ethyl –OH < ethyl –SH < phenyl –OH). These results are of interest since they suggest that the nonenzymatic activation of carbamate prodrugs of basic compounds may be tailored to desired values.

5.44.4.2.3.3   Mannich bases

Mannich bases are compounds containing the NCH_2X moiety, where the central carbon atom is rendered electrophilic by two flanking heteroatoms, one of which is nitrogen. Mannich bases examined as potential prodrugs are generally *O*-Mannich bases ($>NCH_2OR$) or *N*-Mannich bases ($>NCH_2NR^1R^2$). Such moieties usually undergo rapid nonenzymatic hydrolysis, for example

$$R^1R^2NCH_2NR^3R^4 + H_2O \rightarrow R^1R^2NH + HCHO + R^3R^4NH$$

As shown, the reaction releases formaldehyde. This is an unfavorable feature, given that formaldehyde may be a health hazard, depending on the amounts produced, and on the location and kinetics of its release.

A useful *N*-Mannich base is rolitetracycline, i.e., *N*-pyrrolidinemethyltetracycline.[90] This highly water-soluble prodrug of tetracycline is of value for parenteral administration following ex tempore dissolution. The rate of breakdown of the prodrug was found to vary only slightly in the pH range of 1–10. At 21 °C, the limit of stability (10% breakdown) was 1.6 h, whereas the $t_{1/2}$ at 37 °C and pH 7.4 was 43 min.

O-Mannich bases are exemplified by a number of prodrugs of phenytoin.[4] Thus, the 3-(*N*,*N*-dimethylglycyloxymethyl) and 3-(diethyl-β-alanyloxymethyl) derivatives of phenytoin (**50**, R = $-CH_2OCOCH_2N(CH_3)_2$ and $-CH-OCOCH_2CH_2N(CH_2CH_3)_2$, respectively (**Figure 17**)) appear of interest. At neutral pH, their solubilities were about 4000 and 40 times higher, respectively, than that of phenytoin. The $t_{1/2}$ values of the conversion to phenytoin in human plasma at pH 7.4 and 37 °C were 23 s and 6.8 min, respectively. After oral administration of the prodrugs to dogs, plasma levels of phenytoin were several times higher than after administration of the drug itself.

A promising approach of broad potential is that of *O*-Mannich prodrugs of tertiary amines. Indeed, a number of poorly water soluble medicinal amines were quaternarized to *N*-phosphonooxymethyl derivatives of the general formula $R^1R^2R^3N^+CH_2OPO_3H_2$.[91] These derivatives showed a considerably increased water solubility over the parent drug, which was regenerated in near-quantitative amounts by alkaline phosphatase-catalyzed dephosphorylation followed by spontaneous breakdown of the *N*-hydroxymethyl intermediate. Shelf life and in vivo studies were also very encouraging.

There exist also Mannich bases with the NCH_2N or NCH_2O fragment enclosed in a five-membered ring. These are exemplified by oxazolidines as potential prodrugs of β-amino alcohols such as β-blockers and ephedrines, and imidazolidin-4-ones as potential prodrugs of peptides.[4]

5.44.4.2.3.4   Imines

The literature contains a number of examples of imines (i.e., Schiff bases) of medicinal interest, including prodrugs and drug metabolites. These compounds may undergo metabolic hydrolysis as described below:

$$R^1R^2C=NR^3 + H_2O \rightleftharpoons R^1R^2C=O + H_2NR^3$$

Such a reaction is reversible, since the condensation of a carbonyl compound and a primary amine yields the imine under dehydrating conditions. In fact, acidic conditions usually tend to favor hydrolysis, whereas neutral or slightly alkaline conditions will deprotonate the amine, and allow it to behave as a nucleophile toward the carbonyl.[4]

Azomethine prodrugs of (*R*)-α-methylhistamine (**52**, **Figure 18**) offer an apt illustration on the interest of this approach. (*R*)-α-Methylhistamine (**53**) is a potent and selective histamine H_3 receptor agonist, but its medicinal use is severely limited by its insufficient peroral absorption, poor brain penetration, and rapid metabolism in humans, especially by histamine *N*-methyltransferase. To circumvent these problems, potential prodrugs were prepared by reacting the drug with variously ring-substituted 2-hydroxybenzophenones.[92] Most compounds proved readily hydrolyzable in aqueous solution, with conversion yields at room temperature ranging from approximately 0.2% to 80% at pH 1, and from 1% to 34% at pH 4. Monohalogenation (i.e., R^1 or $R^2=X$) favored hydrolysis, and dihalogenation

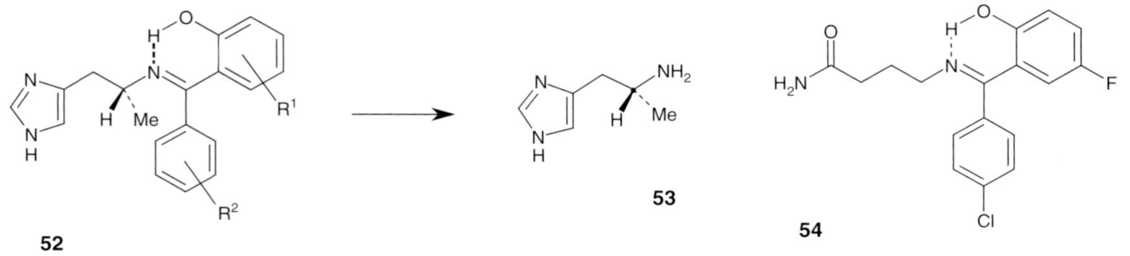

Figure 18   Imine prodrugs (**52**) of (*R*)-α-methylhistamine (**53**), and the structure of the GABA-ergic agent progabide (**54**).

(i.e., R^1 and $R^2 = X$) increased it further. The majority of the compounds appeared to hydrolyze slightly faster in acidic than in neutral solutions, whereas for others the rates were approximately equal or somewhat faster in neutral solutions.

Following oral administration to mice ($24 \mu mol\, kg^{-1}$), the plasma 'area-under-the-curve' concentration (AUC) of (R)-α-methylhistamine varied from not detectable (**52**, $R^1 = 5$-OMe, $R^2 = H$, **Figure 18**) to $1200\, ng\, h\, mL^{-1}$ (**52**, $R^1 = 4$-F, $R^2 = H$). For several of the compounds (e.g., $R^1 = H$, 4-F or 4-Cl; $R^2 = H$, 4-F or 4-Cl), the plasma AUC was around $300\, ng\, h\, mL^{-1}$. Only compounds having the highest plasma AUC produced measurable brain AUC, which varied from about 1% ($R^1 = 4$-F, $R^2 = H$) to 22% ($R^1 = R^2 = 4$-Cl) of the plasma AUC. These pharmacokinetic studies have been complemented by in vivo pharmacological investigations demonstrating anti-inflammatory and antinociceptive properties.[93] Taken together, the results clearly document the potential of this type of prodrugs to improve the pharmacokinetic profile of some primary amines.

The above prodrug candidates show an analogy with the well-known GABA-ergic agent progabide (**54**, **Figure 18**), a brain-penetrating analog or prodrug of the neurotransmitter GABA. Metabolic studies in the rat revealed two hydrolytic reactions. The major one was hydrolysis of the terminal amide group to form the corresponding carboxylic acid, whereas a less important reaction was hydrolysis of the imine link to liberate GABA and gabamide (i.e., the amide of GABA). The detection of GABA and gabamide in the brain of rats dosed with progabide implies slow in situ generation of these active metabolites.[4]

5.44.4.2.3.5   The case of triazenes

N,N-Disubstituted triazenes (**55**, **Figure 19**) have been developed as chemically stable triazene prodrugs capable of enzymatic hydrolysis under physiological conditions to liberate a cytotoxic *N*-methyltriazene antitumor agent (**56**). The latter then spontaneously releases the methyldiazonium cation (**57**), which acts by methylating DNA and RNA, most significantly the O-6 position of guanosine residues.

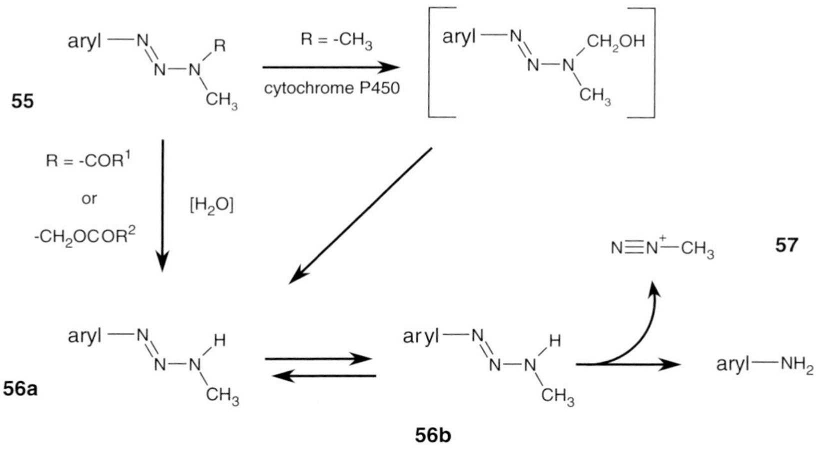

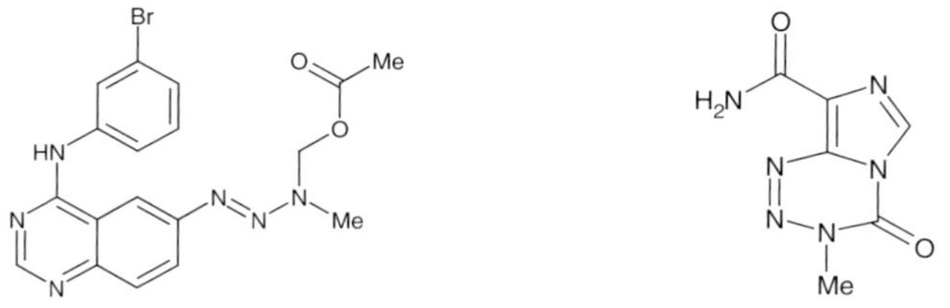

Figure 19 Mechanisms of activation of *N*-methyl, *N*-acyl, and *N*-acyloxymethyl derivatives (**55**) of *N*-methyltriazenes (**56**), the active antitumor agents which act by releasing the methyldiazonium cation **57**. In some of the cases discussed in the text (e.g., the double prodrug **58**), the aromatic amine released is also active. Compound **59** is the well-known antitumor drug temolozomide.

N,N-Dimethyltriazenes (**55**, R = Me, **Figure 19**) can undergo cytochrome P450-catalyzed *N*-dealkylation to *N*-methyltriazenes, but most investigations have focused on *N*-acyl-*N*-methyltriazenes and *N*-acyloxymethyl-*N*-methyltriazenes (**55**, R = –COCH₃ and –CH₂OCOCH₃), respectively.[94–96] Such prodrugs were found to hydrolyze in isotonic phosphate buffer and in human plasma. Thus, a β-alanyl derivative (**55**, R = –CH₂CH₂NH₂, aryl = 4-cyanophenyl) was more stable in phosphate buffer ($t_{1/2}$ = 180 min) than in plasma ($t_{1/2}$ = 53 min).[94] More recent studies report the promising profile of *N*-acyltriazene derivatives of pharmacologically active compounds, in other words, double prodrugs releasing both the methyldiazonium cation and another antitumor agent. This strategy is exemplified by *N*-acyltriazene analogs of the antiresistance drug *O*-benzylguanine,[95] and by compound **58**, which liberates 4-(3'-bromophenylamino)-6-aminoquinazoline, an inhibitor of the epidermal growth factor receptor tyrosine kinase.[96] It thus appears that *N*-acylation or *N*-acyloxymethylation of the triazene terminus is an effective and simple mean of reducing the chemical reactivity of *N*-methyltriazenes while retaining a rapid rate of activation. A look at the structure of the very successful antitumor drug temolozomide (**59**) reveals that it is a cyclized *N*-acyltriazene.

5.44.4.3   Bioprecursors and Bioreductive Agents

As defined above, bioprecursors are prodrugs containing a functional group whose biotransformation generates the active agent without release of any promoiety.[1,6,7] As for carrier-linked prodrugs, the bioactivation of bioprecursors can involve a reaction of oxidation, reduction, or hydrolysis/hydration. But while hydrolysis is by far the most frequent mechanism of activation of the carrier-linked prodrugs, a majority of known bioprecursors are activated by reduction. A further difference involves a very few cases of bioprecursors activated by conjugation (e.g., minoxidil sulfate).[97]

5.44.4.3.1   Bioprecursors activated by hydration

Bioprecursors activated by hydrolysis without loss of a promoiety imply a lactone ring opening. Here, the term 'hydration' appears preferable to 'hydrolysis,' given that the molecule is not split into two.[4] Some among the successful statins (i.e., HMG-CoA reductase inhibitors) exemplify this case.

Lactone ring hydration unveils the active hydroxy acid, a reaction accompanied by a marked decrease in lipophilicity. An argument of this type would suggest the value of lactone prodrugs in improving bioavailability and/or distribution, but other factors must also be considered, which only complicate any evaluation of the relative pharmacokinetic merits of a lactone prodrug compared with a hydroxy acid drug. The reversibility of lactone hydrolysis has been demonstrated at gastric pH and temperature (pH 2.0, 37 °C); for the prodrug lovastatin (**60**, **Figure 20**), reversible lactone hydrolysis to its active hydroxy acid occurred with an approximate $t_{1/2}$ of 1 h and an equilibrium constant close to one.[98] Similar results were obtained for some closely related compounds. In contrast, this reversible hydrolysis was much slower under the near-neutral pH conditions of the intestine.

Enzymatic hydration also occurs, in particular by serum paraoxonase. Thus, lovastatin was hydrolyzed at 37 °C in undiluted plasma at rates that were highly species-dependent but seemingly always faster than nonenzymatic hydrolysis (about 0.4% per minute in human plasma, about 0.8% for dogs and monkeys, and about 10% and 50% rats and mice, respectively).[99] In addition to oral absorption and enzymatic hydrolysis, hepatic extraction (passive or active) is also a major factor influencing the therapeutic activity of statins. This organ selectivity, together with the efficiency of enzymatic hydrolysis, does make some lactones useful prodrugs of HMG-CoA reductase inhibitors. However, additional factors may also influence the therapeutic response, in particular the extent and rate of metabolic reactions that compete with or follow hydrolysis, such as cytochrome P450-catalyzed oxidations, β-oxidation, and taurine conjugation.

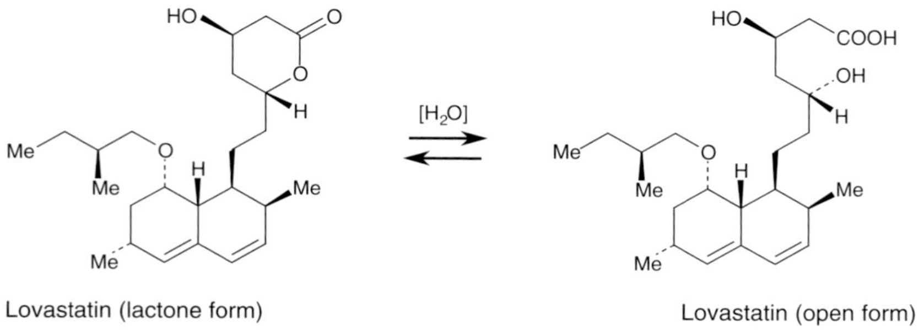

Lovastatin (lactone form)   Lovastatin (open form)

60

Figure 20   The reversible hydrolytic ring opening of the lactone prodrug lovastatin (**60**) to its active hydroxy acid.

As a result of these various factors, some HMG-CoA reductase inhibitors are used in the lactone prodrug form (e.g., lovastatin, simvastatin, and dalvastatin), while others are used as the active hydroxy acids (e.g., pravastatin and fluvastatine).[4,100]

5.44.4.3.2 Bioprecursors activated by metabolic oxidation

The number of oxidation-activated bioprecursors is limited, and no general rule can be proposed to help in their design. Yet some examples are of high interest, if only for the variety of situations they illustrate. Below, we examine one failure, one success, and one promising case.

A particularly informative, even if all but forgotten, example is provided by ethynylbiphenyl derivatives (61, Figure 21), investigated as potential bioprecursors of anti-inflammatory biphenylacetic acids. Indeed, it has been known for some decades that ethynylbiphenyl derivatives undergo fast and practically quantitative acetylenic oxidation to the corresponding arylacetic acid, but the detailed mechanism of the metabolic sequence was not fully understood. A few ethynylbiphenyls were thus investigated as potential bioprecursors until severe toxicity problems interrupted their development.[101] Since then, it has been clarified that cytochrome P450-catalyzed oxygenation of the arylethynyl group generates an oxirene that isomerizes to a highly unstable ketene (62); the latter in turn reacts with nucleophiles to form adducts, and with water to form the target biphenylacetic acids (63).

A therapeutically highly useful example is provided by the biguanide drug proguanil (64, Figure 21), a major antimalarial agent.[102] Although proguanil might have some specific activities of its own, it is in fact the fortuitous bioprecursor of cycloguanil (66). The latter is not useful as a drug, since it shows rather unfavorable pharmacokinetics in contrast to the well-absorbed and well-tolerated proguanil. It is now known that proguanil is a substrate of cytochrome P450 2C19. As shown in Figure 21, α-hydroxylation of the isopropyl substituent yields a carbinolamine intermediate, which N-dealkylates to the inactive primary amine 65 (minor route) or cyclizes to cycloguanil by an intramolecular nucleophilic substitution–elimination. Recent studies have taken benefit of this metabolic reaction to design, prepare, and evaluate analogs of proguanil, with promising profiles.[103]

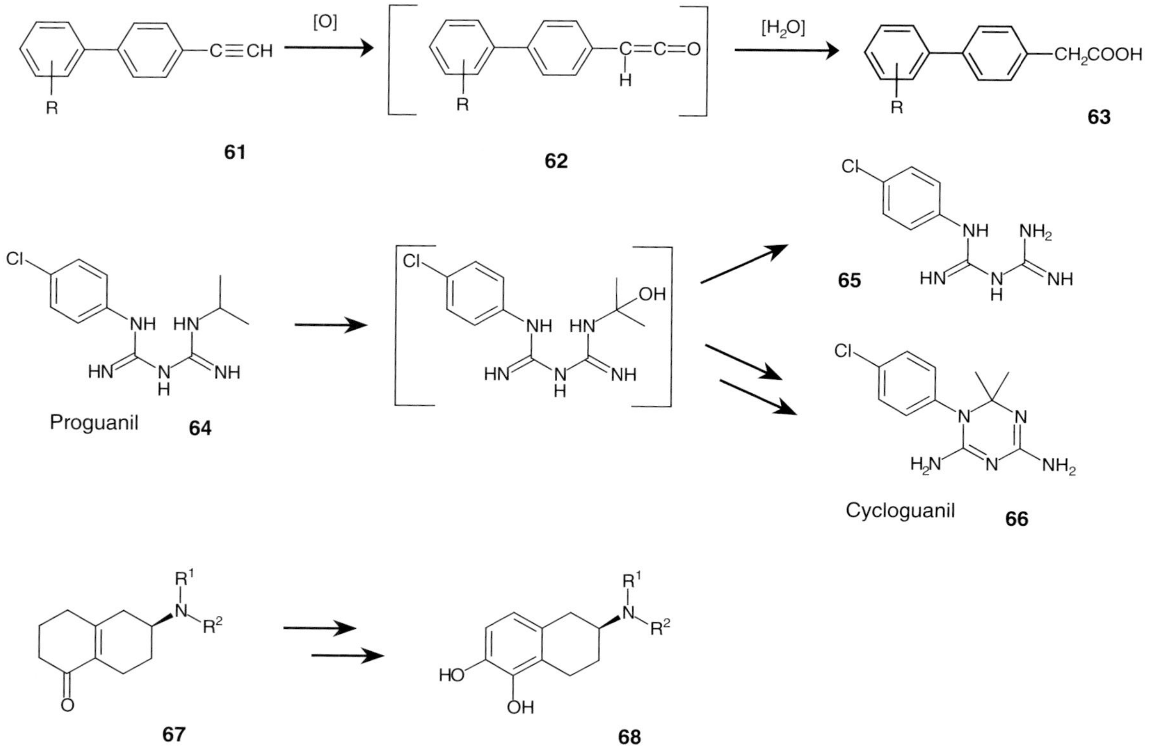

Figure 21 Examples of bioprecursors activated by metabolic oxidation. Ethynylbiphenyls (61) are failed bioprecursors of anti-inflammatory biphenylacetic acids (63), due to the intermediate formation of highly toxic ketenes (62). Proguanil (64) is a fortuitous and well-known bioprecursor of the antimalarial agent cycloguanil (66). Hexahydronaphthalen-1-one derivatives (67) are promising bioprecursors of dopaminergic catecholamines (68).

Our last example in this type of bioprecursor is 6-(*N*,*N*-disubstituted)-3,4,5,6,7,8-hexahydro-2*H*-naphthalen-1-one derivatives (**67, Figure 21**).[104] These enones were found to undergo metabolic oxidation and aromatization to the corresponding catechols (**68**). While this metabolic sequence does not appear to have been fully elucidated in terms of intermediates and enzymes, it seems that it begins with 2-hydroxylation followed by aromatization. What has been demonstrated, however, are the in vivo dopaminergic effects of some of these compounds (e.g., with $R^1 = R^2 = n$Pr; $R^1 = $Me, $R^2 = n$Pr; $R^1 = $Et, $R^2 = n$Pr) in a rat model of Parkinson's disease.

5.44.4.3.3 Bioprecursors activated by metabolic reduction (bioreductive agents)

Most (but not all) therapeutic applications of bioreductive agents are in the chemical targeting of antitumor agents. Such compounds are also called tumor-activated prodrugs (TAPs), and they belong to a few well-defined chemical classes such as aromatic *N*-oxides, aliphatic *N*-oxides, quinones, nitroaromatics, and metal complexes.[44] Some of these are exemplified below.

The case of *N*-oxides is eloquently illustrated by tirapazamine (**24**, see **Figure 9**), and will not be discussed further, except to mention that the utility of such compounds depends on a number of factors such as substrate affinity to one- versus two-electron reductases, redox potential, and cellular permeability.[44,105]

Antitumor nitroaromatic prodrugs are generally used in antibody-directed enzyme prodrug therapy (ADEPT) and gene-directed enzyme prodrug therapy (GDEPT). Also, interesting investigations are also in progress in different therapeutic areas using amidoximes (*N*-hydroxyamidines) as prodrugs of active amidines.[106]

An important class of TAPs is represented by quinones, particularly indolequinones and naphthoquinones, whose prototypal natural compound is the antitumor antibiotic mitomycin C (**69, Figure 22**). Such compounds undergo bioreductive activation to reduced metabolites, probably hydroquinones, but the mechanism of cytotoxicity of the latter is not fully understood. Oxygen-based radical species generated by the auto-oxidation of these reduced metabolites are prime candidates for the cytotoxicity of quinones.[107]

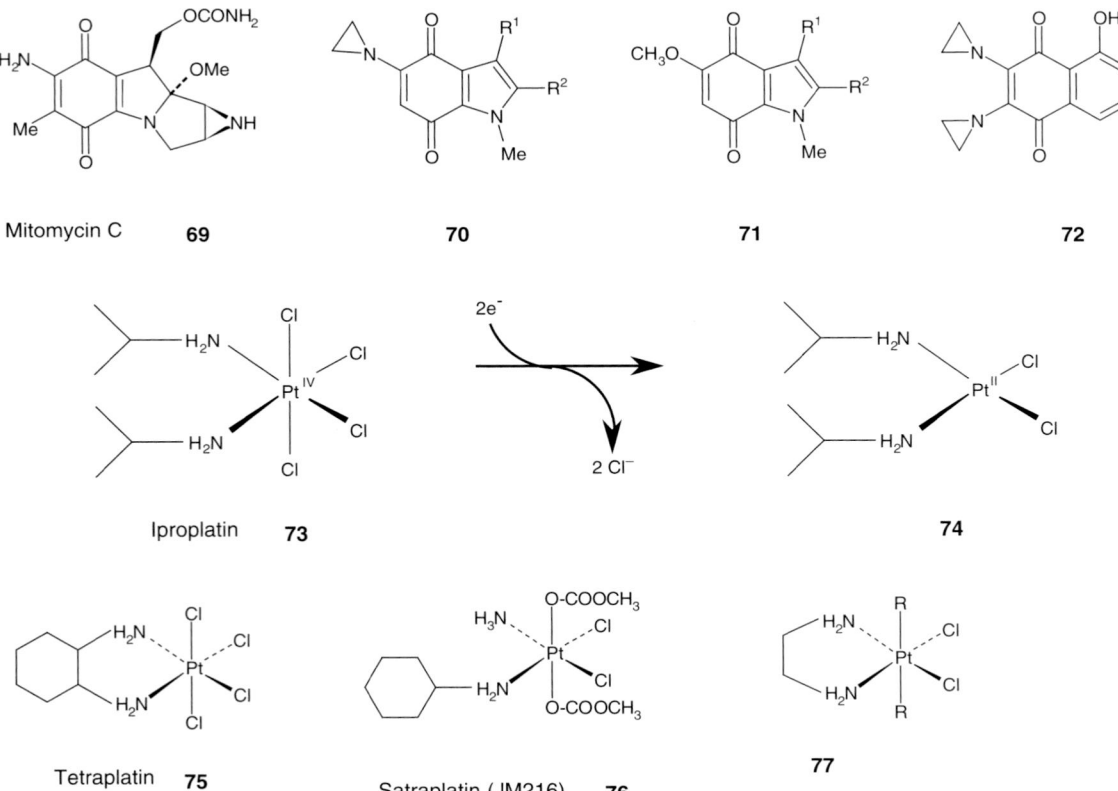

Figure 22 Structure of some antitumor bioprecursors activated by reduction. Compounds **69–72** are quinones, and compounds **73** and **75–77** are platinum(IV) complexes. The conversion of **73** to **74** illustrates the reductive bioactivation of platinum(IV) to platinum(II) complexes.

Many studies have focused on the substrate selectivity of potential antitumor quinones in being reduced by DT-diaphorase (NAD(P)H:quinone oxidoreductase, NQO1), an enzyme over-expressed in some tumor types. And, indeed, sufficiently fast reduction by purified NQO1 and in NQO1-rich cells is a criterion of screening. However, bioreduction under hypoxic conditions also appears as a favorable feature. Good substrate selectivity toward NQO1 is exemplified by two compounds of structure **70** in **Figure 22** ($R^1 = -CH_2OH$, $R^2 = H$; $R^1 = H$, $R^2 = -CH = CHCH_2OH$). Good capacity to be reduced in hypoxic cells was shown by two compounds of structure **71** ($R^1 = -CH_2OCONH_2$, $R^2 = i$Pr; $R^1 = -CH_2OCONH_2$, $R^2 = c$Pr).[108] Good in vivo pharmacokinetics such as a relatively long half-life and high AUC values are obviously additional criteria. A promising compound meeting all these criteria was reported as the naphthoquinone **72**.[109]

The last class of TAPs to be presented here are the platinum(IV) complexes.[110] Compared with the traditional platinum(II) complexes, these have a higher lipophilicity and a decreased reactivity, allowing some of them to be administered orally. Efficacy against some platinum(II)-resistant cell lines has also been reported. In fact, several investigations suggest that anticancer activity of platinum(IV) complexes requires bioreduction to the platinum(II) state (see the conversion of **73** to **74** in **Figure 22**).[111,112] Platinum(IV) complexes have two axial ligands in addition to the four equatorial ones characteristic of platinum(II) complexes. The axial ligands determine to a large degree the rate at which platinum(IV) complexes are reduced, with chlorine ligands allowing the fastest reduction, and hydroxy groups a slow reduction.

Iproplatin (**73**, **Figure 22**), tetraplatin (**75**), and JM216 (**76**) were highly promising candidates in the preclinical phases, but they failed in clinical trials, due to lower activity compared with cisplatin, neurotoxicity, and variable uptake, respectively. Yet active research continues in this field, with, for example, excellent in vitro results being reported for comparatively simple platinum(IV)–ethylenediamine complexes (**77**, R = Cl, OH).[112]

5.44.4.4 Principles of Directed Enzyme-Prodrug Therapies (DEPT)

Sections 5.44.4.2 and 5.44.4.3 above contain examples of prodrugs or bioprecursors targeted toward specific tissues and/or enzymes, and whose activation occur selectively around or in the target tissues. Note that prodrugs containing a polymer promoiety also show great promise in this context (see 5.45 Drug–Polymer Conjugates).[113]

Below, we briefly examine prodrug strategies targeting specific enzymes (directed enzyme-prodrug therapies, DEPT). Such strategies fall into a few categories, and by far their most important applications are in antitumor treatment. The first case is the targeting of endogenous enzymes over-expressed by the target tissues. A second case is seen when exogenous enzymes are imported to the target tissues by antibodies (ADEPT). A third case is seen when exogenous enzymes are produced from genes transferred to the tumor cells (GDEPT).

5.44.4.4.1 Targeting of over-expressed enzymes

Enzymes over-expressed in specific tissues allow their targeting by ad hoc designed prodrugs. This strategy offers great promises in tumor therapy and is illustrated here a glucuronide prodrug of doxorubicin, a well-known and highly active cytotoxic agent (**Figure 23**). This prodrug is made of the active agent (the 'warhead'), a linker, and glucuronic acid.

To avoid rapid renal and biliary excretion, the methyl ester (DOX-mGA3) was prepared and tested. Its hydrolysis (reaction [1] in **Figure 23**) occurred with a $t_{1/2}$ of 0.5 min and 2.5 h in mouse and human plasma, respectively,[114] strongly suggesting the involvement of carboxylesterases. The glucuronide prodrug so generated (DOX-GA3) is a good substrate of human β-glucuronidase (reaction [2]), liberating CO_2 and the drug linker molecule. The later is unstable, and will undergo a 1,6-elimination process (reaction [3]). This process releases the anthracyclin drug, CO_2, and iminoquinone methide, which spontaneously hydrates to 4-aminobenzyl alcohol. Incubation of an ovarian cancer cell line with DOX-mGA3 in combination with an excess of human β-glucuronidase resulted in similar growth inhibition to that of doxorubicin. Intravenous administration of DOX-mGA3 to mice resulted in an AUC of DOX-GA3 in tumor and most normal tissues that was 2.5- to 3-fold higher than after the same dose of DOX-GA3 itself. In tumor tissues, this was accompanied by a 2.7-fold increase in the AUC of doxorubicin compared with an administration of DOX-GA3.[114]

5.44.4.4.2 Antibody-directed enzyme prodrug therapies

In the straightforward form of antigen-targeted therapy, a drug is coupled to an antibody raised against tumor cells. Activation occurs by cleavage of the drug–antibody conjugate after its localization on target cells.[59]

ADEPT is different and more complex. Here, an exogenous enzyme is coupled to a monoclonal antibody (mAb) targeted to tumor cells. This enzyme–mAb conjugate is administered, and allowed sufficient time to localize on tumor cells and clear from the circulation. In a second step, a prodrug is administered, which, being a selective substrate of the exogenous enzyme, will be selectively activated at the tumor site. A schematic representation of this strategy is

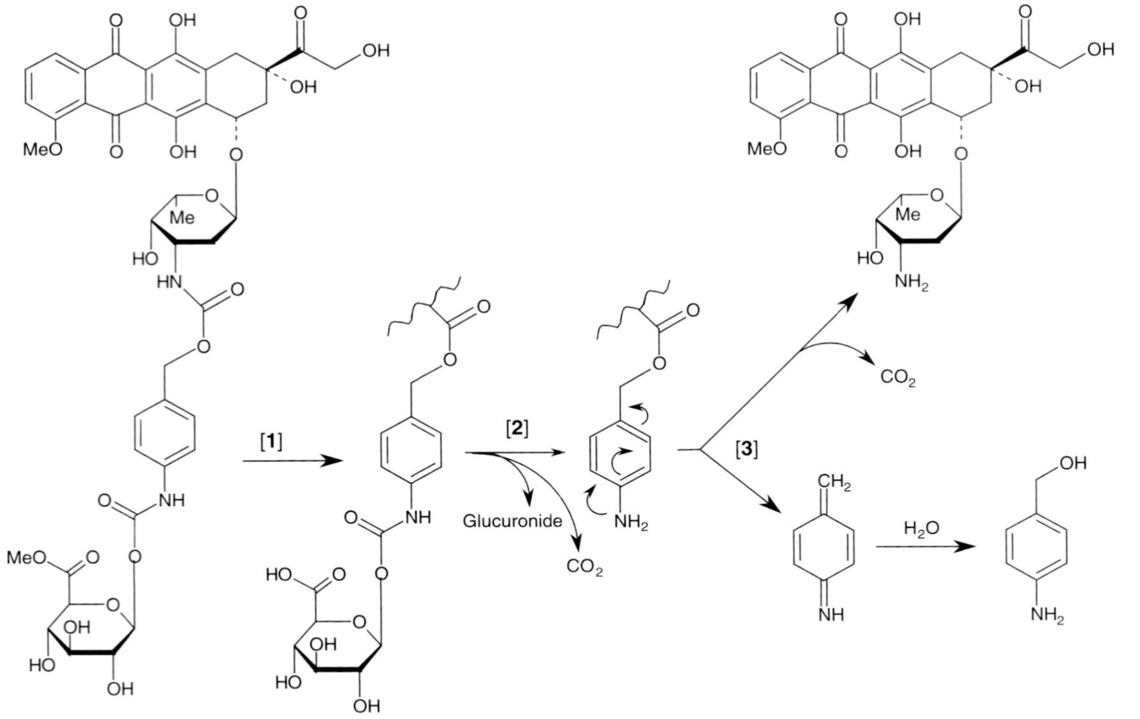

Figure 23 The activation pathway of DOX-mGA3, a multiple prodrug of the antitumor agent doxorubicin made of the drug, a linker, and glucuronic acid methyl ester as the pro-promoiety. Following systemic hydrolysis of the methyl ester group (reaction [1]), cleavage of the promoiety (reaction [2]) is catalyzed by human extracellular β-glucuronidase released in necrotic tumor areas. Spontaneous 1,6-elimination (reaction [3]) then liberates the drug.[114]

shown in **Figure 24**, with the enzyme–mAb conjugate binding selectively to tumor cells and activating the prodrug near their surface.[4,44,58,115,116]

Two relevant examples of ADEPT using carboxypeptidases are presented below. Other peptidases investigated in ADEPT projects include aminopeptidases, human lysosomal β-glucuronidase, and bacterial enzymes such as β-lactamases, penicillin amidases, and nitroreductases.[4]

Carboxypeptidase G2 (EC 3.4.17.11, CPG2, glutamate carboxypeptidase) is a bacterial enzyme unknown in mammalian cells. It has been used with promising success to target cytotoxic alkylating agents to tumor cells.[117,118] In a series of studies, a CPG2–monoclonal antibody conjugate was targeted to a human choriocarcinoma cell line and a human colorectal cell line. The compounds investigated were prodrugs of benzoic acid mustards designed as selective substrates of CPG2, namely 4-(bis(2-chloroethyl)amino)benzoyl-L-glutamic acid (**78**, **Figure 25**) and two analogs. And, indeed, the two tumor cell lines labeled with the CPG2–mAb conjugate effectively activated prodrug **78** to the anticancer agent 4-(bis(2-chloroethyl)amino)]benzoic acid (**79**, **Figure 25**), eliciting a strong cytotoxic response. Promising results were obtained in athymic mice with transplanted choriocarcinoma or colorectal xenografts. In a subsequent clinical study, an antibody-directed enzyme prodrug therapy using prodrug **78** met conditions for effective antitumor therapy, and gave evidence of tumor response in colorectal cancer.[117]

Using an exogenous enzyme such as CPG2 is certainly an attractive idea, but it presents the danger of eliciting an immune response in the patient. This is one of the reasons why another approach has been explored, namely the use of artificial mutants of human carboxypeptidases hCPA1 (EC 3.4.17.1) and hCPA2 (EC 3.4.17.15).[119] The drug used was methotrexate (**80**, R = H, **Figure 25**), an important agent in the treatment of solid malignancies, but one whose lack of cell selectivity is responsible for very serious toxicity to bone marrow and the gastrointestinal tract.

Molecular modeling studies followed by experimental verification had shown that bulky methotrexate prodrugs were good substrates of mutated hCPA1 and hCPA2, having an enlarged substrate-binding pocket obtained by site-directed mutagenesis of Thr268 or Ala268, respectively. Based on these promising results, the cytotoxic efficiency of hCPA1-T286G was tested in an in vitro ADEPT model. Here, hCPA1-T286G was chemically conjugated to ING-1 (an antibody that binds to the tumor antigen Ep-Cam) or to Campath-1H (an antibody that binds to the T and B cell

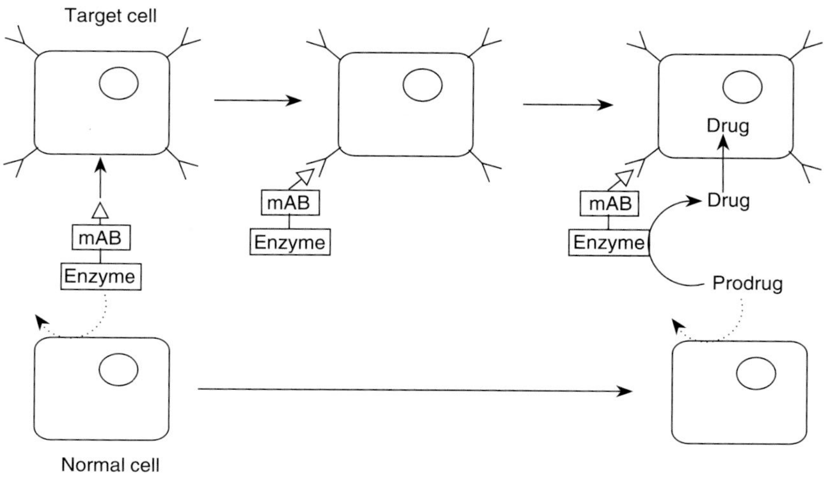

Figure 24 Schematic representation of a selective delivery obtained by ADEPT. An exogenous enzyme is coupled to the antibody mAb, targeted for tumor cells. In a second step, a prodrug is administered, which, being a selective substrate of the exogenous enzyme, will be selectively activated at the tumor site. (Reproduced from Testa, B.; Mayer, J. M. *Hydrolysis in Drug and Prodrug Metabolism – Chemistry, Biochemistry and Enzymology*; Wiley-Verlag Helvetica Chimica Acta: Zurich, Switzerland, 2003, with permission from Verlag Helvetica Chimica Acta.)

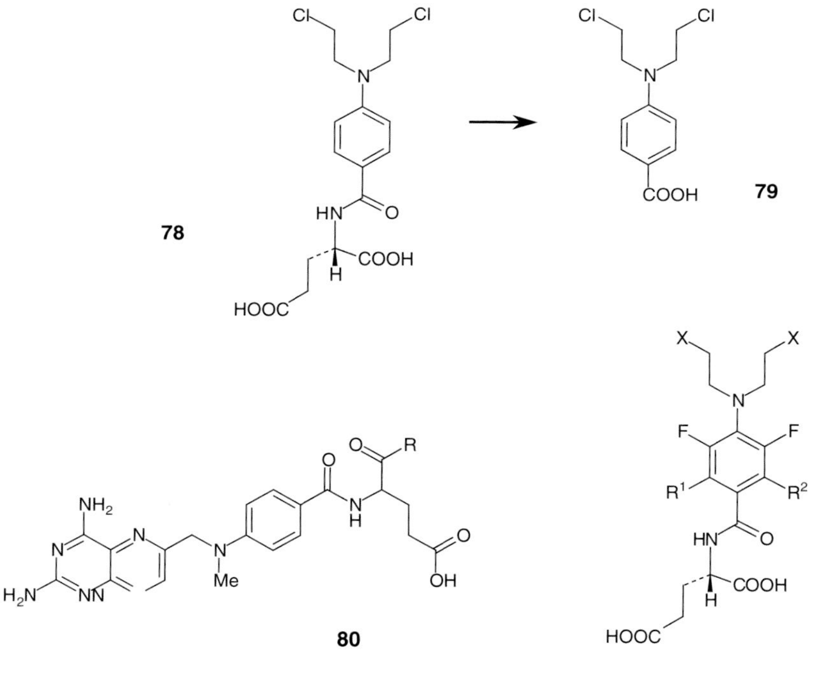

Figure 25 Examples of prodrugs used in ADEPT and GDEPT and discussed in the text.

antigen CDw52). These conjugates were then incubated with HT-29 human colon adenocarcinoma cells (which express Ep-Cam but not Campath-1H), followed by incubation of the cells with some of the in vivo stable methotrexate prodrugs. As expected, the ING-1:hCPA1-T286G conjugate produced excellent activation of the methotrexate prodrugs to kill HT-29 cells as efficiently as methotrexate itself. In contrast and again as expected, the enzyme–Campath-1H conjugate did not bind to the HT-29 cells (which as stated do not express Campath-1H) and was without effect on the prodrugs.

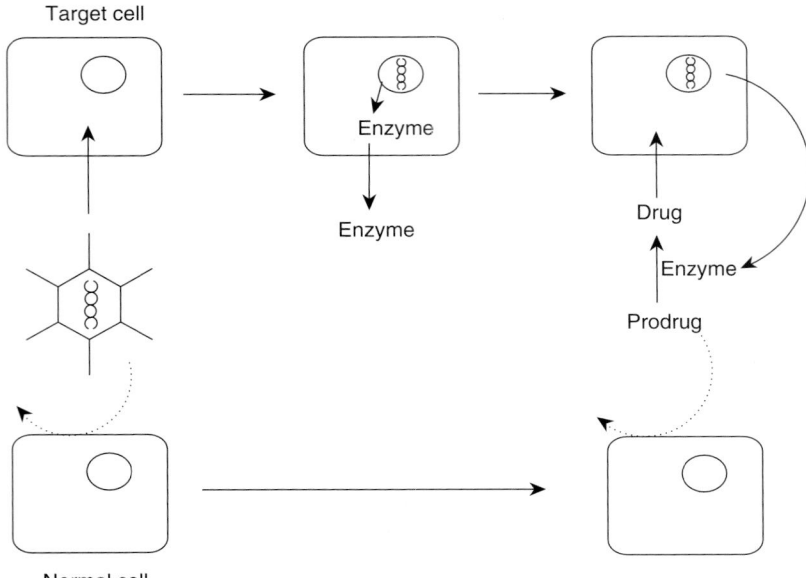

Target cell

Enzyme

Enzyme

Drug

Enzyme

Prodrug

Normal cell

Figure 26 Schematic representation of a selective delivery obtained by GDEPT. The gene encoding an exogenous enzyme is transferred to tumor cells, where it is to be expressed. In a second step, a prodrug is administered, which is selectively activated at the tumor site by the exogenous enzyme expressed by the tumor cells. (Reproduced from Testa, B.; Mayer, J. M. *Hydrolysis in Drug and Prodrug Metabolism – Chemistry, Biochemistry and Enzymology*; Wiley-Verlag Helvetica Chimica Acta: Zurich, Switzerland, 2003, with permission from Verlag Helvetica Chimica Acta.)

5.44.4.4.3   Gene-directed enzyme prodrug therapies

In GDEPT, the gene encoding an exogenous enzyme is transferred to tumor cells, where it is to be expressed. Two gene vectors are available, namely viral vectors (virus-directed enzyme prodrug therapy) and non-viral vectors composed of chemical gene delivery agents.[59,116,120] In a second step, a prodrug is administered that is selectively activated by the exogenous enzyme expressed by the tumor cells, as schematized in **Figure 26**.

As an example, this strategy was also used to activate methotrexate-Phe (**80**, R = Phe, **Figure 25**) and other methotrexate-α-peptides in the vicinity of tumor cells.[121] Carboxypeptidase A is normally synthesized as a zymogen that is inactive without proteolytic removal of its propeptide end by trypsin. To adapt this system to GDEPT, a mutant form of the enzyme (CPA$_{ST3}$) was engineered. This mutant did not require trypsin-dependent zymogen cleavage but was activated by ubiquitously expressed intracellular propeptidases. All evidence indicated that mature CPA$_{ST3}$ was structurally and functionally similar to the trypsin-activated, wild-type enzyme. Furthermore, tumors cells expressing CPA$_{ST3}$ were sensitized to the MTX prodrugs in a dose- and time-dependent manner.

To limit diffusion of CPA$_{ST3}$, a cell surface-localized form was generated by constructing a fusion protein between CPA$_{ST3}$ and the phosphatidylinositol linkage domain from decay accelerating factor (DAF). After retroviral transduction, both CPA$_{ST3}$ and CPA$_{ST3-DAF}$ exhibited a potent bystander effect, even when <10% of the cells were transduced, because extracellular production of MTX sensitized both transduced and nontransduced cells.

5.44.4.4.4   Antibody-directed enzyme prodrug therapies and gene-directed enzyme prodrug therapies: concluding remarks

In summary, ADEPT and GDEPT greatly expand the chemical space open to prodrug designers by allowing the use of promoieties and linkers entirely resistant to human enzymes. The activation of potential prodrugs by a promising exogenous enzyme can indeed be optimized computationally and experimentally. In a second stage, the enzyme or its gene is imported into the target tissue. A recent work with novel benzoic acid mustards is mentioned here as a concluding example.[122] A series of polyfluorinated analogs (**81**; X = Cl, Br or I; R¹ and R² = F or H; **Figure 25**) were prepared and screened. The di- and trifluorinated prodrugs were cytotoxic to human breast carcinoma cells (MDA MB361) engineered to express the bacterial CPG2, but were inactive against control cells that did not express the enzyme. Two difluorinated prodrugs (**81**; X = Br or I; R¹ = R² = H) were effective substrates for the enzyme, and showed excellent therapeutic activity in CPG2-expressing MDA MB361 xenografts, either curing or greatly extending the life of the animals.

The examples given above may not be the ones ultimately cleared for cancer therapy. They were selected here simply to illustrate the principles of the ADEPT and GDEPT strategies relevant to our context.

5.44.5   Challenges in Prodrug Research and Development

A number of challenges await medicinal chemists and biochemists carrying out prodrug research. These include the additional work involved in synthesis, physicochemical profiling, metabolic and pharmacokinetic profiling, and toxicological assessment.[4,6,8] Such challenges, indeed, may range from fair to prohibitive and can occur at all stages of the R&D process:

- Careful prodrug design is required to minimize the number of proposed candidates and maximize the explored space of physicochemical and pharmacokinetic properties. The ability to predict target properties (e.g., solubility, extent of absorption, and rate of activation) is a major need in rational prodrug design. At present, prodrug designers can rely on some local models or rules to make semiquantitative or even quantitative predictions. This situation should improve in the coming years for the prediction of physicochemical properties, absorption, and distribution, as discussed throughout this volume. In contrast, quantitative predictions of rates of biotransformation remain an elusive goal.
- One or several additional synthetic step(s) are needed for each prodrug candidate being prepared. This implies additional work and effort from chemists, and increased production costs, which may not be considered worthwhile.
- As far as molecular properties are concerned, a first issue is the physicochemical profile of the prodrug candidates and its adequacy with the goals of the project. Some physicochemical properties can be calculated or estimated at the design stage (see above), but experimental verification cannot be omitted. Except for the extra work involved, this is a comparatively straightforward issue.
- A truly critical difficulty is the pharmacokinetic behavior (absorption, distribution, etc.) of prodrug candidates and its adequacy with the goals of the project, first in vitro and ultimately in vivo. High-throughput methods are necessary to rapidly assess the in vitro pharmacokinetic profile of many prodrugs candidates, and to verify or falsify the predictions of the prodrug designers. As with drug candidates, the problem is extrapolation to humans. The danger is real, indeed, that prodrug candidates selected in in vitro pharmacokinetic screening programs may prove disappointing in vivo.
- The truly critical difficulty is the metabolic behavior and, particularly, the rate of activation of prodrug candidates. As discussed elsewhere in this volume (see 5.05 Principles of Drug Metabolism 1: Redox Reactions; 5.06 Principles of Drug Metabolism 2: Hydrolysis and Conjugation Reactions; 5.07 Principles of Drug Metabolism 3: Enzymes and Tissues), the huge diversity of drug-metabolizing enzymes and the large interspecies variations that exist make rational optimization of the rate of activation in humans an impossible task. Human liver microsomes have become a common tool in metabolic profiling, but even results so obtained may give a misleading preview of in vivo metabolism. Chemically activated prodrugs may offer the most viable alternative in a few cases.
- Another problem is toxicity relative to the underivatized active agent. By influencing the distribution and tissular concentrations of the active agent they deliver, prodrugs may elicit toxic effects not displayed by the active agent itself. Furthermore, the carrier moiety may generate toxic fragments (e.g., formaldehyde). Additional and careful in vitro (in the presence of activating enzymes) and in vivo toxicological investigations are therefore unavoidable and costly steps in prodrug development, whatever the lack of toxicity of the active agent.

The above problems appear to be the major sources of difficulty in prodrug R&D, not to mention possible complications in registration. No wonder, therefore, that so many medicinal chemists are critical of prodrugs. However, a lucid view cannot ignore the sunny side, in this case the mere existence of a number of successful prodrugs.[2] Several examples have been singled out in this chapter. They demonstrate that prodrug design may indeed achieve satisfy pharmaceutical, pharmacokinetic, and/or pharmacodynamic objectives that would be impossible to achieve otherwise.

5.44.6   Conclusion: What Therapeutic Benefit can Prodrugs Afford?

The gain in therapeutic benefit provided by prodrugs relative to the active agent is a question that knows no general answer. Depending on both the drug and its prodrug, the therapeutic gain may be negligible, modest, marked, or even significant. Nevertheless, a trend is apparent from innumerable data in the literature and when comparing marketed drugs and candidates in R&D.[2,12] In the case of marketed drugs endowed with useful qualities but displaying some unwanted property that a prodrug form should ameliorate (post hoc design), the therapeutic gain is usually modest yet real, but it will be negligible when the drug defect is tolerable or barely improved by transformation to a prodrug. In contrast, it will be marked or significant when much improved targeting is achieved, as exemplified by capecitabine.

In the case of difficult candidates showing excellent target properties but suffering from some severe and intractable physicochemical and/or pharmacokinetic drawback (e.g., high hydrophilicity restricting bioavailability), ad hoc design may lead to a marked or even significant benefit. Here, indeed, a prodrug form may prove necessary, and its design will be integrated into the iterative process of lead optimization. This is a case aptly illustrated by oseltamivir. In other words, well-designed or even fortuitous (e.g., clopidogrel) prodrugs allow the achievement of medicinal objectives that remain out of reach of the active drug.

In conclusion, a prodrug approach is most fruitful when optimization of a traditional lead candidate fails because the structural conditions for activity (i.e., the pharmacophore) are incompatible with the preset pharmaceutical, pharmacokinetic, or pharmacodynamic properties. In such cases, the gap between activity and other drug-like properties may be of such a nature that only a prodrug strategy can bridge it.[1]

References

1. Testa, B. *Biochem. Pharmacol.* **2004**, *68*, 2097–2106.
2. Ettmayer, P.; Amidon, G.; Clement, B.; Testa, B. *J. Med. Chem.* **2004**, *47*, 2393–2404.
3. Mackman, R. L.; Cihlar, T. *Annu. Rep. Med. Chem.* **2004**, *39*, 305–321.
4. Testa, B.; Mayer, J. M. *Hydrolysis in Drug and Prodrug Metabolism – Chemistry, Biochemistry and Enzymology*; Wiley-Verlag Helvetica Chimica Acta: Zurich, Switzerland, 2003.
5. Beaumont, K.; Webster, R.; Gardner, I.; Dack, K. *Curr. Drug Metab.* **2003**, *4*, 461–485.
6. Testa, B.; Soine, W. Principles of Drug Metabolism. In *Burger's Medicinal Chemistry and Drug Discovery*, 6th ed.; Abraham, D. J., Ed.; Wiley-Interscience: Hoboken, NJ, 2003; Vol. 2, pp 431–498.
7. Wermuth, C. G. Designing Prodrugs and Bioprecursors. In *The Practice of Medicinal Chemistry*; 2nd ed.; Wermuth, C. G., Ed.; Elsevier: Amsterdam, the Netherlands, 2003, pp 561–585.
8. Testa, B.; Mayer, J. M. Concepts in Prodrug Design to Overcome Pharmacokinetic Problem. In *Pharmacokinetic Optimization in Drug Research: Biological, Physicochemical and Computational Strategies*; Testa, B., van de Waterbeemd, H., Folkers, G., Guy, R., Eds.; Wiley-Verlag Helvetica Chimica Acta: Zurich, Switzerland, 2001, pp 85–95.
9. Wang, W.; Jiang, J.; Ballard, C. E.; Wang, B. *Curr. Pharm. Des.* **1999**, *5*, 265–287.
10. Lee, H. J.; You, Z.; Ko, D. H.; McLean, H. M. *Curr. Opin. Drug Disc. Dev.* **1998**, *1*, 235–244.
11. Bradley, D. A. *Adv. Drug Deliv. Rev.* **1996**, *19*, 171–202.
12. Testa, B.; Caldwell, J. *Med. Res. Rev.* **1996**, *16*, 233–241.
13. Waller, D. G.; George, C. F. *Br. J. Clin. Pharmacol.* **1989**, *28*, 497–507.
14. Stella, V. J.; Charman, W. N. A.; Naringrekar, V. H. *Drugs* **1985**, *29*, 455–473.
15. Harper, N. J. *Progr. Drug Res.* **1962**, *4*, 221–294.
16. Albert, A. *Nature* **1958**, *182*, 421–423.
17. Fura, A.; Shu, Y. Z.; Zhu, M.; Hanson, R. L.; Roongta, V.; Humphreys, W. G. *J. Med. Chem.* **2004**, *47*, 4339–4351.
18. Tréfouël, J.; Tréfouël, J.; Nitti, F.; Bovet, D. *C. R. Séanc. Soc. Biol.* **1935**, *120*, 756–762.
19. Bodor, N. *Adv. Drug Res.* **1984**, *13*, 255–331.
20. Bodor, N.; Buchwald, P. *Med. Res. Rev.* **2000**, *20*, 58–101.
21. Graffner-Nordberg, M.; Sjödin, K.; Tunek, A.; Hallberg, A. *Chem. Pharm. Bull.* **1998**, *46*, 591–601.
22. Durant, N. N.; Katz, R. L. *Br. J. Anaesth* **1982**, *54*, 195–207.
23. Hammer, R. H.; Amin, K.; Gunes, Z. E.; Brouillette, G.; Bodor, N. *Drug Des. Deliv.* **1988**, *2*, 207–219.
24. Bodor, N.; El-Koussi, A. A.; Kano, M.; Khalifa, M. M. *J. Med. Chem.* **1988**, *31*, 1651–1656.
25. Yang, H. S.; Wu, W. M.; Bodor, N. *Pharm. Res.* **1995**, *12*, 329–336.
26. Quon, C. Y.; Mai, K.; Patil, G.; Stampfli, H. F. *Drug Metab. Dispos.* **1988**, *16*, 425–428.
27. Achari, R.; Drissel, D.; Matier, W. L.; Hulse, J. D. *J. Clin. Pharmacol.* **1986**, *26*, 44–47.
28. Benfield, P.; Sorkin, E. M. *Drugs* **1987**, *33*, 392–412.
29. Loftsson, T.; Thorsteinsson, T.; Masson, M. *J. Pharm. Pharmacol.* **2005**, *57*, 721–727.
30. Pochopin, N. L.; Charman, W. N.; Stella, V. J. *Int. J. Pharmaceut.* **1995**, *121*, 157–167.
31. Mäntylä, A.; Garnier, T.; Rautio, J.; Nevalainen, T.; Vepsälainen, J.; Koskinen, A.; Croft, S. L.; Järvinen, T. *J. Med. Chem.* **2004**, *47*, 188–195.
32. Hayashi, Y.; Skwarczynski, M.; Hamada, Y.; Sohma, Y.; Kimura, T.; Kiso, Y. *J. Med. Chem.* **2004**, *47*, 3782–3784.
33. Skwarczynski, M.; Sohma, Y.; Noguchi, M.; Kimura, M.; Hayashi, Y.; Hamada, Y.; Kimura, T.; Kiso, Y. *J. Med. Chem.* **2005**, *48*, 2655–2666.
34. Amidon, G. L.; Lennernäs, H.; Shah, V. P.; Crison, J. R. *Pharm. Res.* **1995**, *12*, 413–420.
35. Bhadra, P. K.; Morris, G. A.; Barber, J. *J. Med. Chem.* **2005**, *48*, 3878–3884.
36. Abdel-Magid, A. F.; Maryanoff, C. A.; Mehrman, S. J. *Curr. Opin. Drug Disc. Dev.* **2001**, *4*, 776–791.
37. Sweeny, D. J.; Lynch, G.; Bidgood, A. M.; Lew, W.; Wang, K. Y.; Cundy, K. C. *Drug Metab. Dispos.* **2000**, *28*, 737–741.
38. Lew, W.; Chen, X.; Kim, C. U. *Curr. Med. Chem.* **2000**, *7*, 663–672.
39. Sloan, K. B.; Waso, S. *Med. Res. Rev.* **2003**, *23*, 763–793.
40. Tunek, A.; Levin, E.; Svensson, L. A. *Biochem. Pharmacol.* **1988**, *37*, 3867–3876.
41. Svensson, L. A. *Drug News Perspect.* **1991**, *4*, 544–549.
42. Persson, G.; Pahlm, O.; Gnosspelius, Y. *Curr. Therap. Res.* **1995**, *56*, 457–465.
43. Svensson, L. A.; Tunek, A. *Drug Metab. Rev.* **1988**, *19*, 165–194.
44. Denny, W. A. *Eur. J. Med. Chem.* **2001**, *36*, 577–595.

45. Huang, P. S.; Oliff, A. *Curr. Opin. Genet. Dev.* **2001**, *11*, 104–110.
46. Desmoulin, F.; Gilard, V.; Malet-Martino, M.; Martino, R. *Drug Metab. Dispos.* **2002**, *30*, 1221–1229.
47. Hwang, J. J.; Marshall, J. L. *Exp. Opin. Pharmacother.* **2002**, *3*, 733–743.
48. Tsukamoto, Y.; Kato, Y.; Ura, M.; Horii, I.; Ishitsuka, H.; Kusuhara, K.; Sugiyama, Y. *Pharm. Res.* **2001**, *18*, 1190–1202.
49. Clarke, T. A.; Waskell, L. A. *Drug Metab. Dispos.* **2003**, *31*, 53–59.
50. Pereillo, J. M.; Maftouh, M.; Andrieu, A.; Uzabiaga, M. F.; Fedeli, O.; Savi, P.; Pascal, M.; Herbert, J. M.; Maffrand, J. P.; Picard, C. *Drug Metab. Dispos.* **2002**, *30*, 1288–1295.
51. Richter, T.; Mürdter, T. E.; Heinkele, G.; Pleiss, J.; Tatzel, S.; Schwab, M.; Eichelbaum, M.; Zanger, U. M. *J. Pharmacol. Exp. Ther.* **2004**, *308*, 189–197.
52. Denny, W. A. Synthetic DNA-Targeted Chemotherapeutic Agents and Related Tumor-Activated Prodrugs. In *Burger's Medicinal Chemistry and Drug Discovery*, 6th ed.; Abraham, D. J. Ed.; Wiley-Interscience: Hoboken, NJ, 2003; Vol. 5, pp 51–105.
53. Riley, R. J.; Workman, P. *Biochem. Pharmacol.* **1992**, *43*, 167–174.
54. Delahoussaye, Y. M.; Evans, J. W.; Brown, J. M. *Biochem. Pharmacol.* **2001**, *62*, 1201–1209.
55. Pruijn, F. B.; Sturman, J. R.; Liyanaga, H. D. S.; Hicks, K. O.; Hay, M. P. Wilson, W. R. *J. Med. Chem.* **2005**, *48*, 1079–1087.
56. Veronese, F. M.; Morpurgo, M. *Farmaco* **1999**, *54*, 497–516.
57. Duncan, R. *Anti-Cancer Drugs* **1992**, *3*, 175–210.
58. Meyer, D. L.; Senter, P. D. *Annu. Rep. Med. Chem.* **2003**, *38*, 229–237.
59. Dubowchik, G. M.; Walker, M. A. *Pharmacol. Ther.* **1999**, *83*, 67–123.
60. Lawton, G.; Paciorek, P. M.; Waterfall, J. F. *Adv. Drug Res.* **1992**, *23*, 161–220.
61. Waldmeier, F.; Kaiser, G.; Ackermann, R.; Faigle, J. W.; Wagner, J.; Barner, A.; Lasseter, K. C. *Xenobiotica* **1991**, *21*, 251–261.
62. Bundgaard, H.; Nielsen, N. M. *Int. J. Pharmaceut.* **1988**, *43*, 101–110.
63. Bullingham, R. E. S.; Nicholls, A. J.; Kamm, B. R. *Clin. Pharmacokin.* **1998**, *34*, 429–455.
64. Testa, B.; Mayer, J. M. *Drug Metab. Rev.* **1998**, *30*, 787–807.
65. Raynaud, C.; Laneelle, M. A.; Senaratne, R. H.; Draper, P.; Laneelle, G.; Daffe, M. *Microbiology* **1999**, *145*, 1359–1367.
66. Sinha, V. R.; Kumria, R. *Pharm. Res.* **2001**, *18*, 557–564.
67. Jung, Y. J.; Lee, J. S.; Kim, Y. M. *J. Pharm. Sci.* **2001**, *90*, 1767–1775.
68. Niemi, R.; Vepsäläinen, J.; Taipale, H.; Järvinen, T. *J. Med. Chem.* **1999**, *42*, 5053–5058.
69. Niemi, R.; Turhanen, P.; Vepsäläinen, J.; Taipale, H.; Järvinen, T. *Eur. J. Pharm. Sci.* **2000**, *11*, 173–180.
70. Yuan, L. C.; Dahl, T. C.; Oliyai, R. *Pharm. Res.* **2001**, *18*, 234–237.
71. Schlienger, N.; Peyrottes, S.; Kassem, T.; Imbach, J. L.; Gosselin, G.; Périgaud, C. *J. Med. Chem.* **2000**, *43*, 4570–4574.
72. Shafiee, M.; Deferme, S.; Villard, A. L.; Egron, D.; Gosselin, G.; Imbach, J. L.; Lioux, T.; Pompon, A.; Varray, S.; Aubertin, A. M. et al. *J. Pharm. Sci.* **2001**, *90*, 448–463.
73. Larsen, C.; Kurtzhals, P.; Johansen, M. *Int. J. Pharmaceut.* **1988**, *41*, 121–129.
74. Mahfouz, N. M.; Hassan, M. A. *J. Pharm. Pharmacol.* **2001**, *53*, 841–848.
75. Bundgaard, H.; Larsen, C.; Thorbek, P. *Int. J. Pharmaceut.* **1984**, *18*, 67–77.
76. Rouquayrol, M.; Gaucher, B.; Roche, D.; Greiner, J.; Vierling, P. *Pharm. Res.* **2002**, *19*, 1704–1712.
77. Song, X.; Vig, B. S.; Lorenzi, P. L.; Drach, J. C.; Townsend, L. B.; Amidon, G. L. *J. Med. Chem.* **2005**, *48*, 1274–1277.
78. Pillai, O.; Hamad, M. O.; Crooks, P. A.; Stinchcomb, A. L. *Pharm. Res.* **2004**, *21*, 1146–1152; *Pharm. Res.* **1998**, *15*, 1154–1159.
79. Kim, D. K.; Lee, N.; Kim, Y. W.; Chang, K.; Kim, J. S.; Im, G. J.; Choi, W. S.; Jung, I.; Kim, T. S.; Hwang, Y. Y. et al. *J. Med. Chem.* **1998**, *41*, 3435–3441.
80. Xie, M.; Yang, D.; Wu, M.; Xue, B.; Yan, B. *Drug Metab. Dispos.* **2003**, *31*, 21–27.
81. Nolen, H. W., III.; Fedorak, R. N.; Friend, D. R. *J. Pharm. Sci.* **1995**, *84*, 677–681.
82. Juntunen, J.; Vepsäläinen, J.; Niemi, R.; Laine, K.; Järvinen, T. *J. Med. Chem.* **2003**, *46*, 5083–5086.
83. Witterland, A. H. I.; Koks, C. H. W.; Beijnen, J. H. *Pharm. World Sci.* **1996**, *18*, 163–170.
84. Heimbach, T.; Oh, D. M.; Li, L. Y.; Rodriguez-Hornedo, N.; Garcia, G.; Fleisher, D. *Int. J. Pharm.* **2003**, *26*, 81–92.
85. Herzog, R.; Leuschner, J. *Arzneim.-Forsch. (Drug Res.)* **1995**, *45*, 300–303.
86. Weiss, W. J.; Mikels, S. M.; Petersen, P. J.; Jacobus, N. V.; Bitha, P.; Lin, Y. I.; Testa, R. T. *Antimicrob. Agents Chemother.* **1999**, *43*, 460–464.
87. Ezra, A.; Hoffman, A.; Breuer, E.; Alferiev, I. S.; Mönkkönen, J.; El Hanany-Rozen, N.; Weiss, G.; Stepensky, D.; Gati, I.; Cohen, H. et al. *J. Med. Chem.* **2000**, *43*, 3641–3652.
88. Tanino, T.; Ogiso, T.; Iwaki, M.; Tanabe, G.; Muraoka, O. *Int. J. Pharmaceut.* **1998**, *163*, 91–102.
89. Shahrokh, Z.; Lee, E.; Olivero, A. G.; Matamoros, R. A.; Robarge, K. D.; Lee, A.; Weise, K. J.; Blackburn, B. K.; Powell, M. F. *Pharm. Res.* **1998**, *15*, 434–441.
90. Pitman, I. H. *Med. Res. Rev.* **1981**, *1*, 189–214.
91. Krise, J. P.; Zygmunt, J.; Georg, G. I.; Stella, V. J. *J. Med. Chem.* **1999**, *42*, 3094–3100.
92. Krause, M.; Rouleau, A.; Stark, H.; Luger, P.; Lipp, R.; Garbarg, M.; Schwartz, J. C.; Schunack, W. *J. Med. Chem.* **1995**, *38*, 4070–4079.
93. Rouleau, A.; Stark, H.; Schunack, W.; Schwartz, J. C. *J. Pharmacol. Exp. Therap.* **2000**, *295*, 219–225.
94. Carvalho, E.; Iley, J.; Perry, M. J.; Rosa, E. *Pharm. Res.* **1998**, *15*, 931–935.
95. Wanner, M. J.; Koch, M.; Koomen, G. J. *J. Med. Chem.* **2004**, *47*, 6875–6883.
96. Banerjee, R.; Rachid, Z.; McNamee, J.; Jean-Claude, B. J. *J. Med. Chem.* **2003**, *46*, 5546–5551.
97. Meisheri, K. D.; Johnson, G. A.; Puddington, L. *Biochem. Pharmacol.* **1993**, *45*, 271–279.
98. Kaufman, M. J. *Int. J. Pharmaceut.* **1990**, *66*, 97–106.
99. Duggan, D. E.; Chen, I. W.; Bayne, W. F.; Halpin, R. A.; Duncan, C. A.; Schwartz, M. S.; Stubbs, R. J.; Vickers, S. *Drug Metab. Dispos.* **1989**, *17*, 166–173.
100. Hamelin, B. A.; Turgeon, J. *Trends Pharmacol. Sci.* **1998**, *19*, 26–37.
101. Sullivan, H. R.; Roffey, P.; McMahon, R. E. *Drug Metab. Dispos.* **1979**, *7*, 76–80.
102. Casteel, D. A. Antimalarial Agents. In *Burger's Medicinal Chemistry and Drug Discovery*, 6th ed.; Abraham, D. J., Ed.; Wiley-Interscience: Hoboken, NJ, 2003; Vol. 5, pp 919–1031.
103. Shearer, T. W.; Kozar, M. P.; O'Neil, M. T.; Smith, P. L.; Schiehser, G. A.; Jacobus, D. P.; Diaz, D. S.; Yang, Y. S.; Milhous, W. K.; Skillman, D. R. *J. Med. Chem.* **2005**, *48*, 2805–2813.

104. Venhuis, B. J.; Dijkstra, D.; Wustrow, D. J.; Meltzer, L. T.; Wise, L. D.; Johnson, S. J.; Heffner, T. G.; Wikström, H. V. *J. Med. Chem.* **2003**, *46*, 584–590.
105. Ptiyadarsini, K. I.; Dennis, M. F.; Naylor, M. A.; Stratford, M. R. L.; Wardman, P. *J. Am. Chem. Soc.* **1996**, *118*, 5648–5654.
106. Clement, B. *Drug Metab. Rev.* **2002**, *34*, 565–579.
107. Bailey, S. M.; Lewis, A. D.; Knox, R. J.; Patterson, L. H.; Fisher, G. R.; Workman, P. *Biochem. Pharmacol.* **1998**, *56*, 613–621.
108. Phillips, R. M.; Naylor, M. A.; Jaffar, M.; Doughty, S. W.; Everett, S. A.; Breen, A. G.; Choudry, G. A.; Stratford, I. J. *J. Med. Chem.* **1999**, *42*, 4071–4080.
109. Phillips, R. M.; Jaffar, M.; Maitland, D. J.; Loadman, P. M.; Shnyder, S. D.; Steans, G.; Cooper, P. A.; Race, A.; Patterson, A. V.; Stratford, I. J. *Biochem. Pharmacol.* **2004**, *68*, 2107–2116.
110. Ho, Y. P.; Au-Yeung, S. C. F.; To, K. K. W. *Med. Res. Rev.* **2003**, *23*, 633–655.
111. Hall, M. D.; Hambley, T. W. *Coord. Chem. Rev.* **2002**, *232*, 49–67.
112. Hall, M. D.; Martin, C.; Ferguson, D. J. P.; Phillips, R. M.; Hambley, T. W.; Callaghan, R. *Biochem. Pharmacol.* **2004**, *67*, 17–30.
113. Cassidy, J. *Drug News Perspect.* **2000**, *13*, 477–480.
114. de Graaf, M.; Nevalainen, T. J.; Scheeren, H. W.; Pinedo, H. M.; Haisma, H. J.; Boven, E. *Biochem. Pharmacol.* **2004**, *68*, 2273–2281.
115. HariKrishna, D.; Raghu Ram Rao, A.; Krishna, D. R. *Drug News Persp.* **2003**, *16*, 309–318.
116. Niculescu-Duvaz, I.; Friedlos, F.; Niculescu-Duvaz, D.; Davies, L.; Springer, C. J. *Anti-Canc. Des.* **1999**, *14*, 517–538.
117. Napier, M. P.; Sharma, S. K.; Springer, C. J.; Bagshawe, K. D.; Green, A. J.; Martin, J.; Stribbling, S. M.; Cushen, N.; O'Malley, D.; Begent, R. H. *Clin. Canc. Res.* **2000**, *6*, 765–772.
118. Springer, C. J.; Antoniw, P.; Bagshawe, K. D.; Searle, F.; Bisset, G. M. F.; Jarman, M. *J. Med. Chem.* **1990**, *33*, 677–681.
119. Smith, G. K.; Banks, S.; Blumenkopf, T. A.; Cory, M.; Humphreys, J.; Laethem, R. M.; Miller, J.; Moxham, C. P.; Mullin, R.; Ray, P. H. et al. *J. Biol. Chem.* **1997**, *272*, 15804–15816.
120. Springer, C. J.; Niculescu-Duvaz, I. *J. Clin. Invest.* **2000**, *105*, 1161–1167.
121. Hamstra, D. A.; Pagé, M.; Maybaum, J.; Rehemtulla, A. *Canc. Res.* **2000**, *60*, 657–665.
122. Davies, L. C.; Friedlos, F.; Hedley, D.; Martin, J.; Ogilvie, L. M.; Scanlon, I. J.; Springer, C. J. *J. Med. Chem.* **2005**, *48*, 5321–5328.

Biography

Bernard Testa studied pharmacy because he was unable to choose between medicine and chemistry. Because he was incapable of working in a community pharmacy, he undertook a PhD thesis on the physicochemistry of drug–macromolecule interactions. Because he felt himself to be not gifted enough for the pharmaceutical industry, he applied for a postdoctoral research position at Chelsea College, University of London, where he worked for 2 years under the supervision of Prof Arnold H Beckett. And because these were easy times, he was called upon to become assistant professor at the University of Lausanne, Switzerland, becoming full professor and Head of Medicinal Chemistry in 1978. Since then, he has tried to repay his debts by fulfilling a number of local and international commitments, e.g., Dean of the Faculty of Sciences (1984–86), Director of the Geneva-Lausanne School of Pharmacy (1994–96 and 1999–2001), and President of the University Senate (1998–2000). He has written four books and edited 29 others, and (co)-authored 450 research and review articles in the fields of drug design and drug metabolism. During the years 1994–98, he was the Editor-Europe of Pharmaceutical Research, the flagship journal of the American Association of Pharmaceutical Scientists (AAPS), and he is now the co-editor of the new journal *Chemistry and Biodiversity*. He is also a member of the Editorial Board of several leading journals (e.g., *Biochemical Pharmacology*, *Chirality*, *Drug Metabolism Reviews*, *Helvetica Chimica Acta*, *Journal of Pharmacy and Pharmacology*, *Medicinal Research Reviews*, *Pharmaceutical Research*, and *Xenobiotica*). He holds Honorary Doctorates from the Universities of Milan, Montpellier, and Parma, and was the 2002 recipient of the Nauta Award on Pharmacochemistry given by the European Federation for

Medicinal Chemistry. He was elected a Fellow of the AAPS, and is a member of a number of scientific societies such as the French Academy of Pharmacy, the Royal Academy of Medicine of Belgium, and the American Chemical Society. His recently granted Emeritus status has freed him from administrative duties and gives him more time for writing, editing and collaborating in research projects. His hobbies, interests, and passions include jogging, science fiction, epistemology, teaching, and scientific exploration.

Comprehensive Medicinal Chemistry II
ISBN (set): 0-08-044513-6

ISBN (Volume 5) 0-08-044518-7; pp. 1009–1041

5.45 Drug–Polymer Conjugates

F M Veronese and G Pasut, University of Padua, Padua, Italy

5.45.1 Introduction

Proteins and peptides have always been seen as potent and specific therapeutic agents, either to rescue the activity of native biomolecules or to treat important diseases like cancer. Insulin and growth hormone are the most well-known examples of the former group,[1–4] while in the latter group some microbial enzymes used as anticancer drugs, such as asparaginase or methioninase,[5,6] have been reported to deplete essential nutrients for cancer cells. Despite the interesting opportunity that these therapeutic agents offer, unfortunately they have intrinsic limitations ascribed to low stability in vivo, a short period of residence in the body, and immunogenicity, the last being common for nonhuman proteins. In the case of low-molecular-weight drugs, a short in vivo half-life (due to rapid clearance by the kidneys or enzyme degradation) is often observed and sometimes there are physicochemical drawbacks, like low solubility or instability.

Several drug delivery systems have been developed in the last few years to improve the pharmacokinetic and pharmacodynamic profiles of many drugs[7] with high or low molecular weight and with different chemical structure. These approaches can be based on tailor-made formulation of the drug, such as liposomal preparation or controlled release formulations, or on a covalent modification of the drug molecule itself, such as polymer conjugation. When modified in this way, the drug can often achieve a prolonged body residence, an increased resistance to degradation, and decreased side effects, and there is generally improved patient compliance. In this chapter we will focus on the area of polymer conjugation to drugs, a technique that is growing fast and that has already resulted in several products available in the marketplace[8] (**Table 1**).

Table 1 Several polymer conjugates grouped according to the molecular weights of the conjugated drug

Conjugates	Indication	Year to market or status	Company
High-molecular-weight drugs			
SMANCS (Zinostatin, Stimalamer)[74]	Hepatocellular carcinoma	1993	Yamanouchi Pharmaceutical
PEG-adenosine deaminase (Adagen)[83]	SCID syndrome	1990	Enzon
PEG-asparaginase (Oncaspar)[84]	Acute lymphoblastic leukaemia	1994	Enzon
Linear PEG-interferon α2b (PEG-Intron)[81,151]	Hepatitis C, clinical evaluation on cancer, multiple sclerosis and HIV/AIDS	2000	Schering Plough/Enzon
Branched PEG-interferon α2a (Pegasys)[36]	Hepatitis C	2002	Roche/Nektar
PEG-growth hormone receptor antagonist (Pegvisomant, Somavert)[93]	Acromegaly	2002	Pfizer (Pharmacia)
PEG-G-CSF (Pegfilgrastim, Neulasta)[94]	Prevention of neutropenia associated with cancer chemotherapy	2002	Amgen
Branched PEG-anti-VEGF aptamer (Pegaptanib, Macugen)[95]	Age-related macular degeneration	2004	EyeTech
PEG-anti-TNF Fab (CDP870; Certolizumab pegol, Cimzia)[152,153]	Rheumatoid arthritis and Crohn's disease	Phase III	UCB (formerly Celltech)
A PEGylated diFab antibody. Targets VEGFR-2 (CDP791)	Solid tumors	Phase II	UCB-ImClone System
Low-molecular-weight drugs			
HPMA copolymer-doxorubicin (PK1; FCE28068)[43]	Cancer, in particular lung, breast cancers	Phase II	Pfizer (CRC/Pharmacia)
HPMA copolymer-doxorubicin-galactosamine (PK2; FCE28069)[45]	hepatocellular carcinoma	Phase I/II	Pfizer (CRC/Pharmacia)
HPMA copolymer-camptothecin (MAG-CPT; PNU166148)[49]	Clinical evaluation on several solid cancers	Phase I	Pfizer (Pharmacia)
HPMA copolymer-paclitaxel (PNU166945)[50]	Clinical evaluation on several solid cancers	Phase I	Pfizer (Pharmacia)
HPMA copolymer-platinate (AP5280)[51]	Clinical evaluation on several solid cancers	Phase II	Access Pharmaceutical
HPMA copolymer-platinate (AP5346)	Clinical evaluation on several solid cancers	Phase I/II	Access Pharmaceutical
Polyglutamate-paclitaxel (XYOTAX; CT-2103)[55–58]	Cancer, in particular lung, ovarian and esophageal cancers	Phase II/III	Cell Therapeutics
Polyglutamate-camptothecin (CT-2106)[61]	Clinical evaluation on colorectal, lung, ovarian cancer	Phase I/II	Cell Therapeutics
PEG-camptothecin (PROTHECAN)[97]	Clinical evaluation on several solid cancers	Phase II	Enzon
PEG-paclitaxel[101]	Clinical evaluation on several solid cancers	Phase I	Enzon

Adapted from Duncan, R. *Nat. Rev. Drug Disc.* **2003**, 2, 347–360.

Table 2 Most common advantages achieved using drug–polymer conjugation

Prolonged half-life and less frequent administrations
Increased water solubility (especially for insoluble small drugs)
Protection from proteolysis and chemical degradation
Reduction of immunogenicity and antigenicity
Reduced toxicity
Targeting to tumor tissue by 'EPR effect'
Modification of biodistribution
Drug release under specific conditions using proper spaces between the drug and the polymer

Reproduced with permission from Pasut, G.; Guiotto, A.; Veronese, F. M. *Exp. Op. Ther. Patents* **2004**, *14*, 859–894.

Many classes of therapeutic agents are amenable to polymer conjugation; however, almost all of the conjugated drugs so far available on the market are proteins, since these were the first to be studied with this technology. A number of different diseases can be treated with conjugated proteins, hence polymer conjugation is not limited to a small number of therapeutic areas. Several compounds produced by conjugation of polymers with nonpeptide drugs with low molecular weight are currently under clinical evaluation and are expected to be on the market in the near future. These are mainly antitumor drugs, as such drugs have the most side effects and limitations that might, at least in part, be solved by polymer conjugation. Recently, this technology has been applied in the field of gene delivery and antisense nucleotides. Polymer conjugation conveys several advantages to drugs (**Table 2**) by deeply changing their in vitro and in vivo properties. By conjugating drugs with hydrophilic polymers, it is possible to achieve: (1) an increased solubility of the conjugated drug, even for those with very low solubility (e.g., taxol); (2) an enhanced drug bioavailability, which is the result of reduced kidney clearance due to the increased hydrodynamic volume of conjugates; (3) protection from degrading enzymes; (4) prevention or reduction of several drawbacks such as aggregation, immunogenicity, and antigenicity; and (5) specific targeting in organs, tissues, or cells, or exploitation of the 'enhanced permeability and retention (EPR) effect' of solid tumor tissues. One limitation of this technology involves patient compliance; conjugates cannot be orally administered because their size prevents absorption through this route, and they have to be parenterally administered.

Research in the field of low-molecular-weight drug–polymer conjugation moved up a gear in the 1950s and 1960s when Peptamin was conjugated to the blood plasma expander polyvinylpyrrolidone.[9] Before then, the research in this area mainly involved the chemistry of conjugation. In the 1960s the first clinical trials of the polymeric anticancer agent divinylethermaleic anhydride/acid copolymer (DIVEMA) were conducted; unfortunately, a severe toxicity was demonstrated, which prevented its use.[10] Since then the concept of a polymer–drug conjugate has been better rationalized[11] and many interesting derivatives have been developed to enhance the activity of various drugs–peptides and proteins in particular, but also molecules such as oligonucleotides or chelating agents for diagnostic purposes.

The first studies on protein–polymer conjugation appeared in the 1960s and 1970s when dextran was investigated as a coupling polymer.[12] However, the real breakthrough came when other synthetic polymers were employed, and poly(ethylene glycol) (PEG) emerged as the best candidate for protein modification.[13] Its superiority is reflected in the fact that there are a number of interesting products already on the market containing PEG, and there are many more in advanced clinical trials (**Table 1**). Other synthetic polymers investigated for polymer conjugation include poly(styrene-co-maleic acid/anhydride) (SMA) and poly(N-(2-hydroxypropyl) methacrylamide) copolymers (HPMA), and polyglutamic acid (PGA).

The first application of PEG as a bioconjugation polymer was proposed in the late 1970s by Professor F. Davis and A. Abuchowski at Rutgers University.[31,32] Following his pioneering study a large number of drugs with different structure (proteins, peptides, low-molecular-weight drugs, polynucleotides) have been PEGylated, thus creating a new class of drugs,[13] some of which rapidly became blockbuster products.

The first polymers studied for drug delivery had a linear structure and were composed of one monomer, but polymer chemistry was soon expanded yielding a selection of interesting new structures such as dendrimers,[14–16] dendronized polymers,[17] graft polymers,[18,19] block copolymers,[20] branched polymers,[21] multivalent polymers,[22] stars,[23] and hybrid glycol[24] and peptide derivatives (**Figure 1**).[25]

5.45.2 Polymeric Conjugates

5.45.2.1 Conjugates with Low-Molecular-Weight Drugs

The term 'low-molecular-weight drugs' is used here to describe not only the nonpeptide drugs but also small peptides and oligonucleotides.

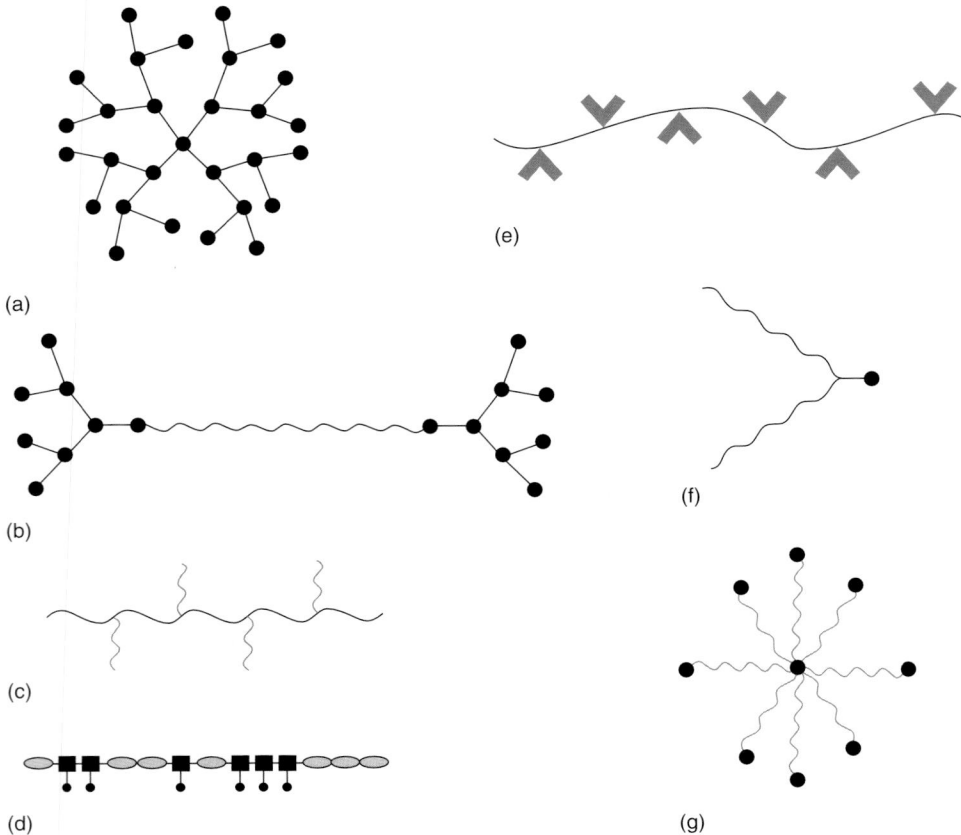

Figure 1 Different polymer structures: (a) dendrimers, (b) dendronized polymers, (c) graft polymers, (d) block copolymers, (e) multivalent polymers, (f) branched polymers, and (g) stars.

An important milestone in the development of these conjugates was the rationalization of the polymeric carrier proposed by Ringsdorf in 1975.[11] According to his model (**Figure 2**), the drug is linked to a polymeric backbone (biodegradable or not), directly or through a spacer; a targeting residue and eventually a solubilizing group may also be present. The spacer between the polymer and drug may control the release rate or target the release to certain cellular compartments. This may be achieved by the use of spacers cleavable under special conditions, such as acidic medium or lysosomal enzymes. The increased hydrodynamic volume of the conjugate generally extends the blood residence time since the kidney clearance rate is reduced and less frequent administrations are therefore needed. Moreover, hydrophobic drugs can be easily solubilized in water by coupling with hydrophilic polymers, which helps the formulation. The high hydrodynamic volume of the macromolecular carriers offers a passive targeting to solid tumor by the known 'enhanced permeability and retention (EPR) effect'[26] due to the anatomical and physiological modifications in such tissues, in particular the increased vascular density (the result of highly active angiogenesis), the presence of vessels with both wide fenestrations and lack of smooth muscle layer, and finally a decreased lymphatic drainage. In solid tumor there is also extensive production of vascular mediators, which again leads to increased vascular permeability. The EPR effect can therefore lead to concentrations of a macromolecular carrier in tumor tissue 5–10-fold higher than those in blood plasma, a situation that is difficult to reach with unmodified low-molecular-weight drugs. In addition, the polymeric prodrug can enter cells only through endocytosis,[27,28] a process that is increased in tumor cells, further enhancing drug specificity.[29]

In the last few years active targeting has been receiving great attention for the improvements that may come from a selective therapy toward selected cells, tissues, or organs.[30] An important requirement for drug–polymer conjugates is the absence or limited presence of free drug as impurity (1–2% with respect to the total amount of drug), which otherwise might lead to an overestimation of the conjugate's potency in a biological evaluation.

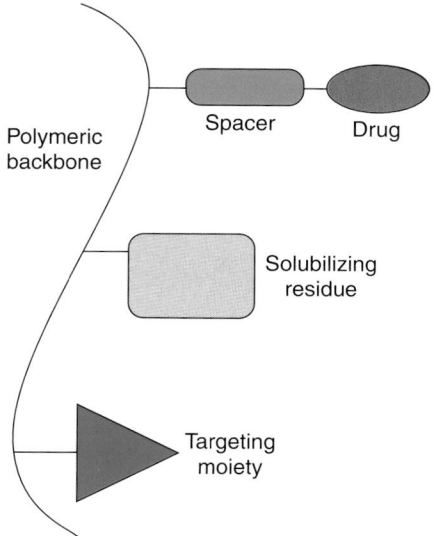

Figure 2 Ringsdorf's small drug–polymer model.

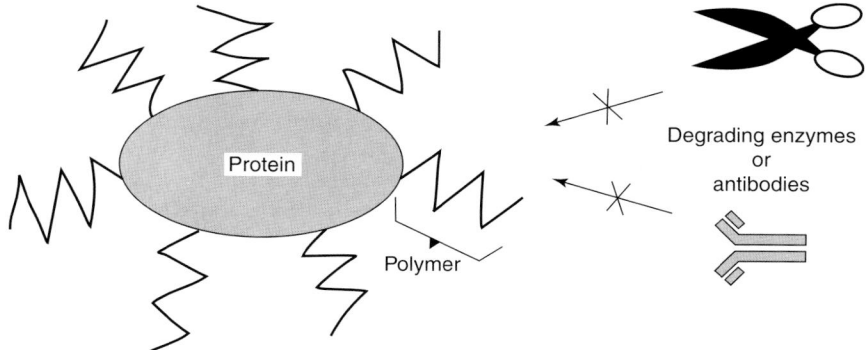

Figure 3 Protein surface protection offered by conjugated polymer chains.

5.45.2.2   Polymer–Protein Conjugates

Although successful results have been obtained in the clinic using streptokinase or neocarcinostatin conjugates, coupled to oxidized dextran and poly(styrene-co-maleic acid/anhydride), respectively, the research on protein–polymer conjugation really took off following two reports on the modification of bovine serum albumin and bovine liver catalase with PEG.[31,32]

Polymer–protein conjugates have all the benefits described above for low-molecular-weight drug conjugates, but some additional advantages can be achieved, namely the protection of sites of proteolysis and the reduction or prevention of immunogenicity and antigenicity (**Figure 3**).[33–35] These are the result of the shielding effect of polymer chains on the protein surface, which prevents the approach of other proteins or enzymes by steric hindrance. In general this also influences the activity of conjugated protein, which is lower compared to the native protein in in vitro tests, but is more than compensated for in vivo thanks to the great improvement of pharmacokinetic parameters.[36,37]

5.45.3   Polymers for Bioconjugation

Many polymers have been investigated as candidates for drug delivery,[38] for example, in the modification of a protein or as a carrier for low-molecular-weight drugs, with linear or branched structures of natural or synthetic origin. These include:

- synthetic polymers: PEG, HPMA copolymers, poly(ethyleneimine) (PEI), poly(acroloylmorpholine) (PacM), poly-(vinylpyrrolidone) (PVP), polyamidoamines, divinylethermaleic anhydride/acid copolymer (DIVEMA), poly-(styrene-co-maleic acid/anhydride) (SMA), polyvinylalcohol (PVA);

- natural polymers: dextran, pullulan, mannan, dextrin, chitosans, hyaluronic acid, proteins; and
- pseudosynthetic polymers: PGA, poly(L-lysine), poly(malic acid), poly(aspartamides), poly((N-hydroxyethyl)-L-glutamine) (PHEG).

A short description is given below of most of these polymers because of their limited application, while PEG is described in more detail.

5.45.3.1   Vinyl Polymers

These polymers are prepared by radical polymerization of the respective vinyl monomer or by copolymerization of two or more different monomers, to modulate the properties of the final product. Copolymerization in fact allows a tailor-made polymer that is better at responding to the specific requirements of drug delivery to be designed. This kind of carrier can reach high levels of drug loading since, in theory, it is possible to prepare a polymer in which each monomer can carry a drug molecule. Vinyl polymers are not biodegradable and therefore must be limited in size to allow body clearance, mainly by kidney filtration, thus avoiding accumulation in the body. Usually, their molecular weight needs to be below the limit for renal filtration (40–50 kDa). Among all synthesized vinyl polymers, HPMA emerged as the best candidate and led to a number of important conjugates.[39–41] The most studied HPMA conjugate is its derivative with the antitumor agent doxorubicin (FCE28068). The drug was linked to the polymeric backbone through a peptidyl spacer (Gly-Phe-Leu-Gly) specifically designed to be cleaved by lysosomal cathepsin B,[42] yielding a conjugate known as PK1 (**Figure 4a**).[43] PK1 was optimized to have certain features such as: (1) a molecular weight of 30 000 Da to ensure body clearance and, at the same time, to allow tumor targeting[44]; (2) a spacer sequence that was selected to be stable in plasma, but promptly released in lysosomes; and (3) an amount of bound drug (∼8.5 wt%) was required to reach therapeutic concentration in tumor cells.

From the PK1 studies the PK2 conjugate was produced (FC28069, **Figure 4b**), which, like the pure drug, possesses an N-acylated galactosamine residue as the targeting group.[45] This molecule conveys targeting to the liver thanks to its interaction with hepatocyte asialoglycoprotein receptors. PK2 has a molecular weight of 25 000 Da, a doxorubicin content of ∼7.5 wt%, and a galactosamine content of ∼1.5–2 wt%. PK1 reached phase II clinical trials for the treatment of breast, colon, and small-cell lung cancer, and it was the first HPMA conjugate to be administered to patients.[46,47] PK2 was the first targeted polymer to enter clinical trials, and it was used for the treatment of primary and secondary liver cancer.[48] PK1 and PK2 showed a two- to five-fold reduction in anthracycline toxicity, and despite the high cumulative doses of doxorubicin administered, no cardiotoxicity was observed.

HPMA copolymer was studied as a potential carrier for two other anticancer drugs apart from anthracyclines, namely, camptothecin (MAG-CPT[49]; **Figure 4c**) and paclitaxel (PNU166945[50]; **Figure 4d**). Both these drugs suffer from low solubility in water; polymer conjugation provides a solution to this problem while allowing the EPR effect for solid tumor targeting to be exploited. Camptothecin is linked to HPMA copolymer through a H-Gly-NH-(CH$_2$)$_6$-NH-Gly-OH spacer that forms an ester linkage with the C-20 hydroxyl group of the drug. The conjugate has a molecular weight ranging from 20 000 to 30 000 Da depending on the drug loading (5–10 wt%). Paclitaxel, on the other hand, is conjugated at the level of the C-2 hydroxyl group (drug loading ∼5 wt%) by an ester bond with the spacer Gly-Phe-Leu-Gly. The drug solubility in water is increased from 0.0001 mg mL^{-1} of the free drug to 2 mg mL^{-1} of the conjugate. Both HPMA conjugates entered phase I clinical trials, but the results indicated the need for optimization of the derivatives' design, in particular regarding the stability of the linkage between the polymer and the drugs (ester) and the drug loading. In this case, the products showed the toxicity of the respective free drugs due to the rapid hydrolysis of the ester bond. Another intensively studied and promising conjugate is the HPMA–platinate conjugate (AP5280, Access Pharmaceuticals, **Figure 4e**), where the metal is chelated by a malonate molecule, which is linked to a 25 000 Da HPMA chain by a Gly-Phe-Leu-Gly spacer. The loading of platinum is about 7 wt%. To be effective platinum has to be released from the conjugate and the malonate derivatives showed the best rate of hydrolysis. The conjugate has now entered phase II clinical trials.[51,52]

Many studies were carried out to investigate the potential toxicity of the HPMA copolymer alone or as a conjugate. For example, it was found that PK1 could be administered in clinical trials up to cumulative doses of >20 g^{-2} without problems of immunogenicity or toxicity.[46,53,54]

5.45.3.2   Poly(Amino Acids) and Analogs

PGA, poly(L-lysine), poly(aspartamides), poly(N-hydroxyethyl-L-glutamine) (PHEG), are easily synthesized and possess a peptidyl structure. They are biodegradable when obtained from the natural L-amino acids but are not if

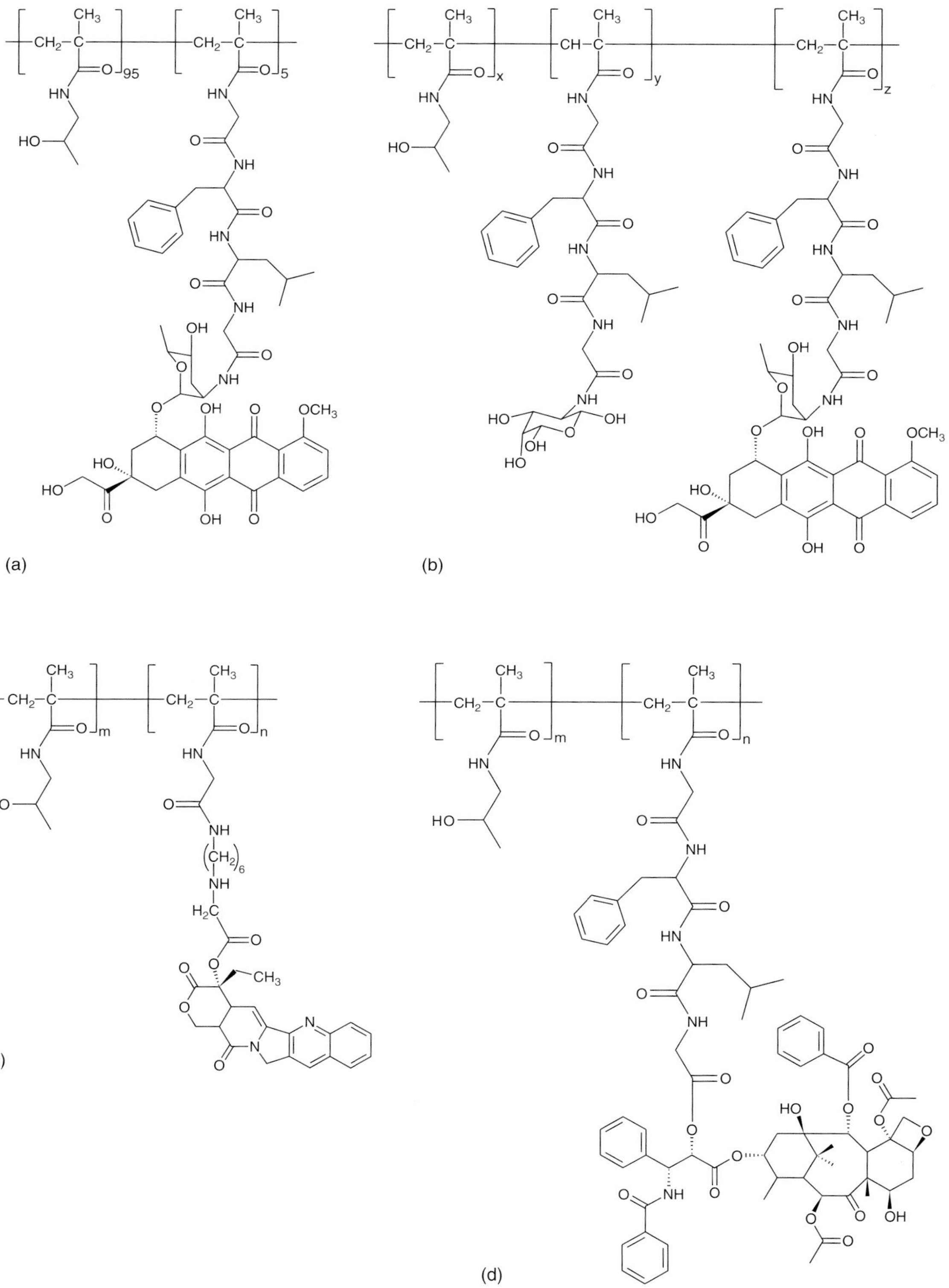

Figure 4 Several polymer conjugates of low-molecular-weight drugs: (a) HPMA copolymer-doxorubicin (PK1; FCE28068), (b) HPMA copolymer-doxorubicin-galactosamine (PK2; FCE28069), (c) HPMA copolymer-camptothecin (MAG-CPT; PNU166148), (d) HPMA copolymer-paclitaxel (PNU166945), (e) HPMA copolymer-platinate (AP5280), (f) polyglutamate-paclitaxel (XYOTAX; CT-2103), and (g) PEG-camptothecin (PROTHECAN).

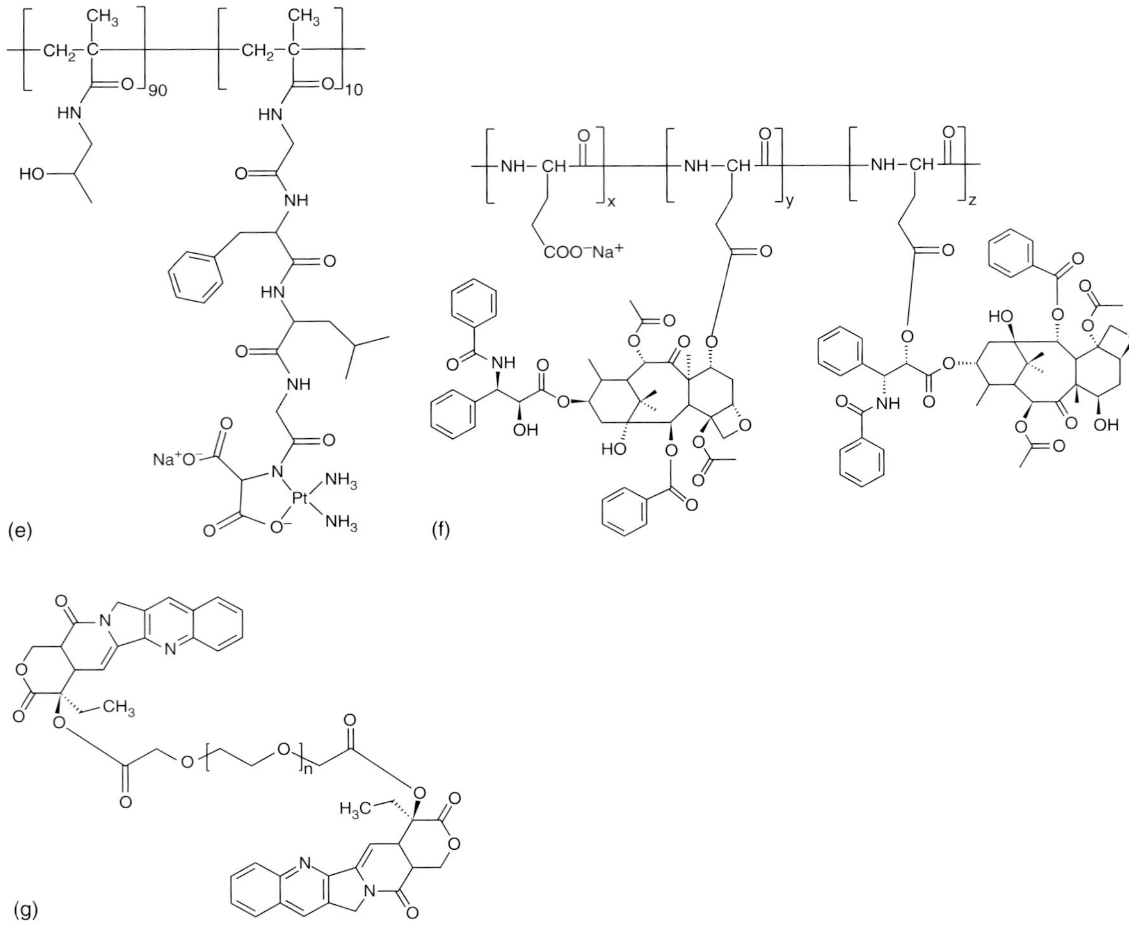

(e)

(f)

(g)

Figure 4 Continued

prepared from the D-enantiomers. The drug loading of these polymers, similarly to HPMA, is high since any monomer possesses a side reactive group (amine, carboxyl, or hydroxyl) that can be derivatized for drug coupling.

Of this class of polymers, the conjugate PGA-paclitaxel (CT-2103, XYOTAX,[55–58] **Figure 4f**) from Cell Therapeutics is at the most advanced clinical stage (phase II/III). In this case, a PGA with ~40 000 Da molecular weight was linked to paclitaxel through an ester bond reaching a loading of ~37 wt%. An interesting property of this polymer is its biodegradability by cathepsin B to form glutamyl-paclitaxel.[59] In early trials it was administered every 3 weeks in the treatment of mesothelioma, renal cell carcinoma, nonsmall cell lung cancer, and a paclitaxel-resistant ovarian cancer disease. A significant number of patients with partial response or stabilized disease was observed. Phase II confirmed its efficacy in patients with recurrent ovarian cancer, but in this case the conjugate displayed a more frequent neurotoxicity than predicted from phase I trials, probably related to the fact that the patients were heavily pretreated with platinates. Further investigations are ongoing.[60] PGA was also conjugated to camptothecin through an ester bond leading to the product CT-2106 (Cell Therapeutics) containing 33–35 wt% of drug and a molecular weight of 50 000 Da. The conjugate has entered a phase I trial.[61]

5.45.3.3   Polysaccharides

Several polysaccharides have been used in pharmaceutical applications. In terms of pharmacokinetics of the carriers themselves, molecular weight, electric charge, chemical modifications, and degree of polydispersity and/or branching largely determine their fate in vivo. Generally, large molecular weight (>40 kDa) polysaccharides have low clearance and relatively long plasma half-life, resulting in reticuloendothelial or tumor tissue accumulation. As with other macromolecular carriers the tumor accumulation is due to the EPR effect, but through the additional linking to the polymer of targeting molecules it is possible to reach specific cells. The most important application as polymeric carrier has been for the preparation of macromolecular prodrugs that are inactive as such, but after release of the free drug at the site of interest they perform their action.[62] Polysaccharides have also been used in the preparation of protein conjugates that, while still maintaining the activity of the starting proteins, increase the duration of effect and decrease the immunogenicity.[62]

The most commonly used polymer of this class was dextran[38] (mainly 1,6 poly α-D-glucose with some 1,4 branching links), which was first approved as a plasma expander. Its conjugate with doxorubicin (AD-70; mol.wt. dextran ~70 000 Da) entered phase I clinical trials, but displayed a toxicity attributed to uptake of dextran by the liver reticuloendothelial cells.[63] Dextran was also used as a carrier of proteins, in particular to improve the pharmacokinetics. The most important results were obtained with streptokinase (Streptodekase),[64] but superoxide dismutase was also studied.[65] For the coupling the polymer was oxidized by periodate yielding aldehyde groups that in turn were reacted with protein amino groups, mainly from lysines. This method, however, was abandoned due to the fact that the multiplicity of binding groups in the polymer might yield to undesirable cross-linking in the reaction with the protein, beyond the high heterogeneity of the products.

5.45.3.4 Proteins

Among all proteins, albumin and transferrin are the most studied as macromolecular carriers. Although the tissue distribution of these proteins is influenced by their functional role in the body, a series of investigations have shown that the anatomical and physiological characteristics of tumors can promote, by the EPR effect, the uptake in these tissues of drugs conjugated to serum proteins. Antitumor and antiviral agents have been conjugated to serum proteins and early clinical trials have been performed with transferrin conjugates of cisplatin and doxorubicin,[66] with adenine arabinoside monophosphate conjugated to lactosaminated albumin,[67] and with methotrexate bound to albumin.[68] Clinical trials with an albumin-doxorubicin prodrug and a long-acting opioid component are under way.[69] A limitation for the use of proteins is their structural complexity with several groups having different functionalities (amino, carboxylic, hydroxyl, and thiol groups). This may be an advantage where they are linked to simple drugs, since high loading is possible, but it also presents a great problem when conjugating to peptides or proteins, since the presence of many reactive sites may lead to cross-linking and heterogeneous products. The use of albumin as a carrier of therapeutic proteins was exploited 20 years ago with good results being achieved in the case of superoxide dismutase.[70]

However, in the case of albumin, it is possible to overcome the problem of the reactive group multiplicity by exploiting the lone thiol group,[71] which can react with specific thiol-reactive molecules. Thanks to the great potential of genetic engineering it is now possible to obtain the desired protein or peptide fused with albumin,[72] avoiding the risks of cross-linking.

5.45.3.5 Poly(Styrene-co-Maleic Acid/Anhydride)

The protein neocarcinostatin (NCS), which exhibits cytotoxicity against mammalian cells and Gram-positive bacteria at concentrations as low as $0.01\,\mu g\,mL^{-1}$, was conjugated to SMA: two small polymer chains (1.6 kDa) were linked to the amino group of Lys-20 and Ala-1. The conjugate, named SMANCS, showed a half-life 10–20 times higher than the native protein and by the EPR effect the accumulation in tumor was 30-fold that of muscle.[73] Thanks to its increased hydrophobicity, SMANCS can be solubilized in lipid media such as Lipiodol and administered intra-arterially for treatment of tumors. The derivative was marketed in Japan in 1993 (Zinostatin Stimalamer) for the treatment of hepatocellular carcinoma.[74] SMA conjugation, however, did not become a general method for protein coupling because, like dextran and HPMA, it possesses many sites of attachment along the polymer backbone, which can give rise to complex mixtures of products.

5.45.3.6 Poly(Ethylene Glycol)

As a modifying polymer PEG has the benefit of three decades of studies littered with important milestones, which has already resulted in seven products on the market (**Table 1**). This polymer is chosen for: (1) the lack of immunogenicity, antigenicity, and toxicity; (2) its high solubility in water and in many organic solvents; (3) the high hydration and flexibility of the chain, at the basis of the protein rejection properties; and (4) its approval by the FDA for human use. PEG is synthesized by ring opening polymerization of ethylene oxide. The reaction, which is initiated by methanol or water, gives polymers with one or two end chain hydroxyl groups termed monomethoxy-PEG (mPEG-OH) or diol PEG (HO-PEG-OH), respectively. The polymerization process produces a family of PEG molecules with a wide range of molecular weights. In this form PEG is not suitable for use in drug conjugation, although it is largely employed in pharmaceutical technology as excipient. As with any polymer, the biological properties of PEG, such as kidney excretion or protein surface coverage, are based on its size, so it is necessary to use samples that are as homogeneous as possible. Fractionation techniques (e.g., gel filtration chromatography) need to achieve narrow polydispersivity and must also exclude the presence of the diol form (formed from traces of water in the polymerization reaction) as an impurity in monomethoxy PEG batches. A wide series of activated PEGs have been developed to address different chemical groups in the drugs. Furthermore, PEGs with various shapes (branched, multifunctional, or heterobifunctional) are now available. Monofunctional polymers (mPEG-OH), linear or branched, are most suitable for protein modification since the diol or multifunctional polymers can give rise to cross-linking. Those with multiple reactive groups on the chain are useful to obtain an increased drug to polymer ratio. Increased loading is particularly needed in therapeutic agents with low biological activity, which would otherwise require the administration of a large amount of conjugate with consequent high viscosity of the solution.

The optimization of both the polymerization procedure and the purification process is now allowing the preparation of PEGs with low polydispersivity, M_w/M_n ranging from 1.01 for PEG with molelcular weights below 5 kDa to 1.1 for PEG as big as 50 kDa.

PEG has unique solvation properties that are due to the coordination of 2–3 water molecules per ethylene oxide unit.[75] This, together with the great flexibility of the molecule backbone, is responsible for the biocompatibility, nonimmunogenicity, and nonantigenicity of the polymer;[31] it also possesses an apparent molecular weight 5–10 times higher than that of a globular protein of comparable mass, as verified by gel permeation chromatography.[76] Owing to this large hydrodynamic volume, a single PEG molecule covers a large area of the surface of the conjugated protein, preventing the approach of proteolytic enzymes and antibodies.[77–79]

In vivo PEG undergoes limited chemical degradation and the body clearance depends upon its molecular weight: below 20 kDa it is easily secreted into the urine, while higher molecular weight PEGs are eliminated more slowly and the clearance through liver becomes predominant. The threshold for kidney filtration is about 40–60 kDa (a hydrodynamic radius of approximately 45 Å[80]), which represents the albumin excretion limit. Over this limit the polymer remains in circulation and mainly accumulates in liver. Alcohol dehydrogenase can degrade low-molecular-weight PEGs, and chain cleavage can be catalyzed by cytochrome P450 microsomial enzymes.[81] Some branched PEGs may also undergo a molecular weight reduction when the hydrolysis and loss of one polymer chain is catalyzed by anchimeric assistance.[82] However, regarding safety concerns when choosing the PEG's molecular weight in the design of a new drug–polymer conjugate 'it is commonly accepted to use PEGs below 40 kDa' to avoid accumulation in the body. This is the molecular weight employed in two successful products, Pegasys and Macugen (**Table 1**) belonging to the PEG–protein and PEG–small drug conjugate classes, respectively. Finally, several years of PEG use as an excipient in foods, cosmetics, and pharmaceuticals, without toxic effects, is clear proof of its safety.[76]

The first generation of PEG conjugates was based on low-molecular-weight polymers (≤ 12 kDa), most commonly in their linear monomethoxylated form (mPEG). Batches of this polymer often contained a significant percentage of PEG diol, an impurity that, once activated, could act as a potential cross-linking agent. Furthermore, the chemistry employed in mPEG synthesis often resulted in side reaction products or led to weak and reversible linkages. The drugs PEG-adenosine deaminase (Adagen)[83] and PEG-asparaginase (Oncaspar)[84] belong to this generation of PEG conjugates.

The second generation of PEGs is characterized by improved polymer purity (reduction of both polydispersivity and diol content for high-molecular-weight PEGs) and by a wide selection of activated PEGs that allow selectivity of reaction toward different protein sites. The availability of several new PEGs is now widening the opportunities in PEGylation technology. Second-generation PEGs currently on the market include:

- PEG-propionaldehyde, also in the form of the more stable acetal: the reaction with the amino group leads to a Shiff's base that, once reduced by sodium borohydride, yields a stable secondary amine that maintains the same neat charge of the parent drug.
- Several PEG-succinimidyl derivatives: highly reactive toward amino groups. The reaction rate of these derivatives may significantly change depending upon the length and the composition of the alkyl chain between PEG and the succinimidyl moiety.[85]
- 'Y' shaped branched PEG[86] (see **Figure 1f**): provides a higher surface shielding effect and is more effective in protecting the conjugated protein from proteolytic enzymes and antibodies (**Figure 5**). Moreover, proteins modified with this PEG retain higher activity than the same enzyme modified by linear PEGs. This effect is probably due to the branched polymer that prevents the PEG from entering the enzyme active site cleft or other sites involved in biological activity (**Figure 6**).

- PEGs reactive toward thiol groups: PEG-maleimide (MAL-PEG), PEG-vinylsulfone (VS-PEG), PEG-iodoacetamide (IA-PEG), and PEG-orthopyridyl-disulfide (OPSS-PEG) have been specifically developed for this conjugation, but only the last one strictly reacts with the thiol groups, avoiding any amino modification that might occur with the other three (**Figure 7**).
- Heterobifunctional PEGs[87,88]: these derivatives have different reactive groups at the two polymer ends that allow two different molecules to be linked separately to the same PEG chain. It is therefore possible to obtain conjugates that carry both a drug and a targeting molecule. Among the proposed and commercially available heterobifunctional PEGs, the most commonly used are H_2N-PEG-COOH, HO-PEG-COOH, and H_2N-PEG-OH.
- Multiarm or 'dendronized' PEGs (**Figure 1b** and **g**): the former are prepared by linking a linear PEG to a multimeric compound, whereas the latter are linear PEGs with a dentritic structure at one or both chain ends.[89,90] The aim of both derivatives is to increase the drug/polymer molar ratio, overcoming problems of high viscosity that may occur with solutions of monofunctional drug–polymer conjugates. This is particularly true for drugs that require a high amount of product for the therapeutic treatments.

Most of these PEGs, also in the activated form, are now commercially available.

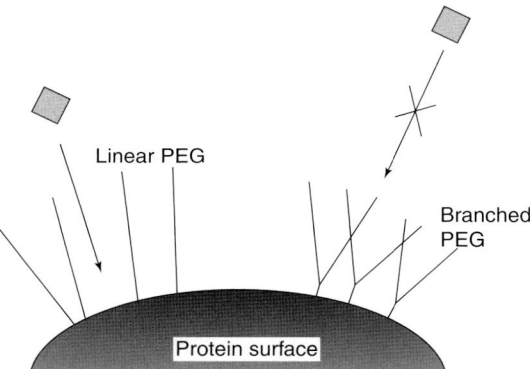

Figure 5 Shielding effect of different PEG structures – linear and branched. The higher steric hindrance of branched PEG compared with linear PEG explains its enhanced capacity to reject the approach of other proteins (e.g., degrading enzymes) or cells. (Reproduced with permission from Pasut, G.; Guiotto, A.; Veronese, F. M. *Exp. Op. Ther. Patents* **2004**, *14*, 859–894.)

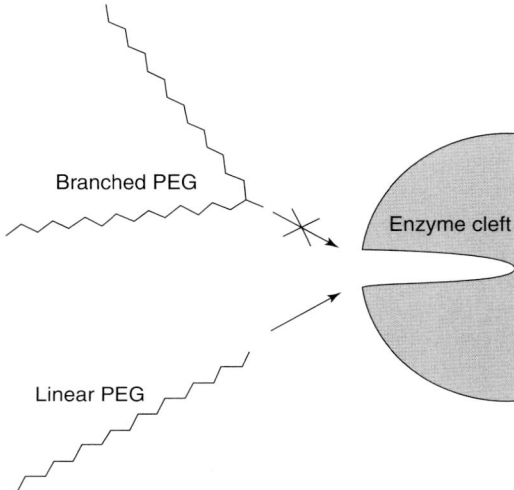

Figure 6 Effect of PEG hindrance on the enzyme active site access. The high steric hindrance of branched PEG, which makes it difficult for the active site cleft to be accessed, may explain the lower inactivation of enzymes as compared to linear PEG of the same size. (Reproduced with permission from Pasut, G.; Guiotto, A.; Veronese, F. M. *Exp. Op. Ther. Patents* **2004**, *14*, 859–894.)

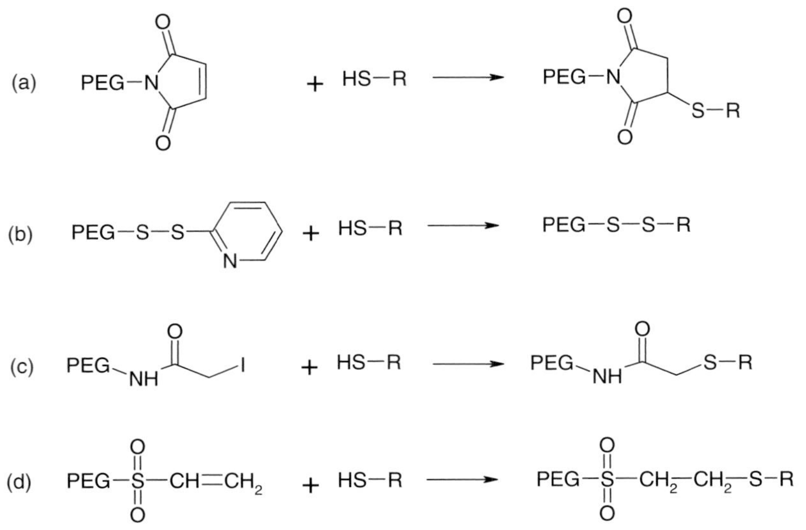

Figure 7   Examples of activated PEG molecules reactive toward thiol groups: (a) PEG maleimide, (b) PEG orthopyridyl-disulfide, (c) PEG iodoacetamide, and (d) PEG vinylsulfone.

Several protein bioconjugates derived from this second generation of PEGs have reached the market, such as linear PEG-interferon α-2b (PEG-Intron),[91] branched PEG-interferon α-2a (Pegasys),[36,92] PEG-growth hormone receptor antagonist (Pegvisomant, Somavert),[93] and PEG-granulocyte colony stimulating factor (G-CSF) (pegfilgrastim, Neulasta).[94] More recently also a PEG conjugate of a 24-mer oligonucleotide, a branched PEG-anti-VEGF aptamer (pegaptanib sodium injection, Macugen)[95] is also marketed, while many other peptide, protein, and oligonucleotides products are under clinical trials and hopefully will be available in the near future.

In PEGylation of low-molecular-weight drugs, PEG faces the limit of low drug loading since in its standard form the polymer can be derivatized only at the two chain extremes. However, several derivatives have been prepared and the most advanced is the PEG-camptothecin conjugate (PROTHECAN or Pegamotecan, Enzon), obtained by linking two drug molecules (using C-20 hydroxyl group) at the termini of a 40 000 Da polymer chain (**Figure 4g**). The drug loading corresponds to 1.7 wt%, which is rather low if compared with other multivalent polymers (i.e., PGA, HPMA copolymers). The product passed phase I trials[96]; the maximum tolerated dose is 100 mg m^{-2} (camptothecin equivalent) and the recommended dose for phase II studies is 7000 mg m^{-2} administered for 1 h i.v. every 3 weeks. Phase II studies are ongoing and preliminary data shows Pegamotecan to be a promising treatment for adenocarcinoma of the stomach and gastroesophageal (GE) junction. The conjugate appears to be well tolerated, with a low incidence of toxicities. Grade 2 cystitis, a known toxicity of camptothecin, was observed in 2 out of 15 patients.[97]

PEG-camptothecin conjugates with tri- or tetrapeptide linkers, to exploit the drug release inside the cell only, have been shown to be highly stable in plasma but rapidly hydrolyze in the presence of lysosomal enzyme cathepsin B.[98,99] In vitro and in vivo results are encouraging and indicate that the polymer bound camptothecin derivative SN-392 (10-amino-7-ethyl camptothecin) is less toxic than the free drug and the camptothecin analog CPT-11, but is still highly effective against Meth A fibrosarcoma cell lines. From pharmacokinetic studies it was immediately evident that the maximal plasma concentration (C_{max}) after administration of 10-amino-7-ethyl camptothecin is eightfold higher than that after PEG-SN392 injection, although AUC data are comparable. The mean residence time (MRT, a dose-independent measure of drug elimination expressing the average time a drug molecule remains in the body after intravenous injection) after PEG-peptide-camptothecin injection is threefold higher than that after injection of both forms of 10-amino-7-ethyl camptothecin. The data indicate that the conjugate acts as a circulating reservoir of the antitumor agents, since the conjugate itself it is not active until it enters the cells and the drug is released, by lysosomal enzymes, to exploit its action. This method of release, often employed for antitumor drugs, is discussed further in Section 5.45.5.

An interesting study reports the preparation of PEG-doxorubicin conjugates where several parameters, such as polymer molecular weight (5 000–20 000 Da), the structure of PEG (linear or branched), and the sequence of a peptide linker between drug and carrier (i.e., H-GFLG-OH, H-GLFG-OH, H-GLG-OH, H-GGRR-OH, or H-RGLG-OH), were taken into consideration.[100] Surprisingly, the conjugate with the greatest antitumor activity in rats had the lowest molecular weight PEG (mPEG$_{5000}$-GFLF-doxorubicin). Apparently, this contradicts the need to have conjugates with molecular weights above 20–30 kDa for a blood residence time long enough to allow accumulation in tumors by the EPR effect. It was demonstrated by light scattering that this low-molecular-weight PEG-doxorubicin conjugate forms micelles in solution, and consequently its apparent molecular weight rises to 120 000 Da, which ensures an effective EPR effect and explains the antitumor activity.

Enzon researchers also studied the preparation and optimization of PEG-paclitaxel conjugates using several PEG molecular weights.[101] In this case, the authors reported the importance of a molecular weight ≥ 30 kDa in order to prevent rapid elimination of the conjugates by the kidneys. A PEG derivative of this drug entered phase I clinical trials and preliminary data are encouraging; however, neutropenia, as with the free drug, is a predominant hematological toxicity.[102]

5.45.4 Considerations for PEGylation

To obtain conjugates with improved pharmacological properties several parameters of the polymer must be taken into consideration. In PEGylation the mass of the PEG chain and the binding site on the protein or drug are important parameters to be considered, as well as the chemistry of linkage and the use of a proper spacer. In the modification of proteins it is better to use few high-molecular-weight chains than a higher number of low-molecular-weight ones. A multipoint attachment of PEG or the use of too large a PEG is expected to reduce or prevent a protein's bioactivity by interfering with receptor binding; however, a low total mass of bound PEG cannot ensure protection from degradation and results in poor pharmacokinetic profiles. The effect of both number and mass of linked PEG chains on recognition and pharmacokinetic parameters are well documented in the literature.[103,104] The site of PEGylation is also critical since when this resides in, or close to, the protein recognition area or active site of enzymes, the polymer affects the biological activity. Furthermore, the modification of certain residues may cause conformational changes ending in denaturation or exposure of buried amino acids, which may lead to aggregation or facilitate proteolytic degradation.

5.45.5 Controlled Release of Conjugated Proteins or Drugs

In general, with nonprotein drugs a stable bond between the polymer and drug is not convenient since many molecules exploit their activity only in the free form. A chemical bond that releases the native molecule is therefore welcome, especially if the release can be triggered under specific controlled conditions. The polymeric prodrug of antitumor agents or substances that possess systemic toxicity should be stable in the bloodstream, but release the drug from the carrier in the desired tissue or cell. The only way to obtain such controlled release is by the use of a proper spacer or linkage between drug and polymer that can be hydrolytically or enzymatically cleaved. Examples of how to achieve a prompt or retarded release of conjugated drugs include:

- Linkers that respond to pH changes.[105] Following cell internalization by endocytosis the conjugates are exposed to the acidic pH of endosomes and lysosomes; furthermore, the pH of tumor tissue is slightly more acidic than that of healthy tissues.[106] This situation can be exploited using acid-labile spacers, such as N-cis-aconityl acid, which was the first approach used to reversibly link daunorubicin to aminoethyl polyacrylamide or poly(D-lysine).[107] In this case a prompt drug release was evident at pH 4 or lower, while in blood or at pH 6 the linker was stable. A hydrazon linkage can also take advantage of the fact that a low pH is needed to free the drug; this linker was exploited in several conjugates to release adriamycin[108] or streptomycin[109] from the carrier.
- Exploitation of lysosomal enzymes. In this case the macromolecular carriers enter the cell by endocytosis and the newly formed endosome fuses with the lysosome, exposing the polymer–drug to a series of degrading enzymes. A spacer (usually an oligopeptide) that is specifically designed to be cleaved by the lysosomal enzymes allows a lysosomotropic drug delivery. This strategy also takes advantage of the fact that lysosomes are overexpressed in tumor cells where cathepsins B or D and other metalloproteinases play a role in tumor growth. Examples of such oligopeptide linkers include those studied by Duncan and Kopecek for optimization of PK1; among these H-Gly-Phe-Leu-Gly-OH and H-Gly-Leu-Phe-Gly-OH appear to be the most effective.[42,110]
- Drug release by anchimeric-assisted hydrolysis. This sophisticated strategy uses linkers that are designed to form a double prodrug system. The drug-linker is first released from the polymer by hydrolysis (first prodrug), which triggers the linker (second prodrug) that finally releases the free and active drug. Examples of these double drug delivery systems include the 1,6-elimination reaction or trimethyl lock lactonization (**Figure 8**).[111,112]

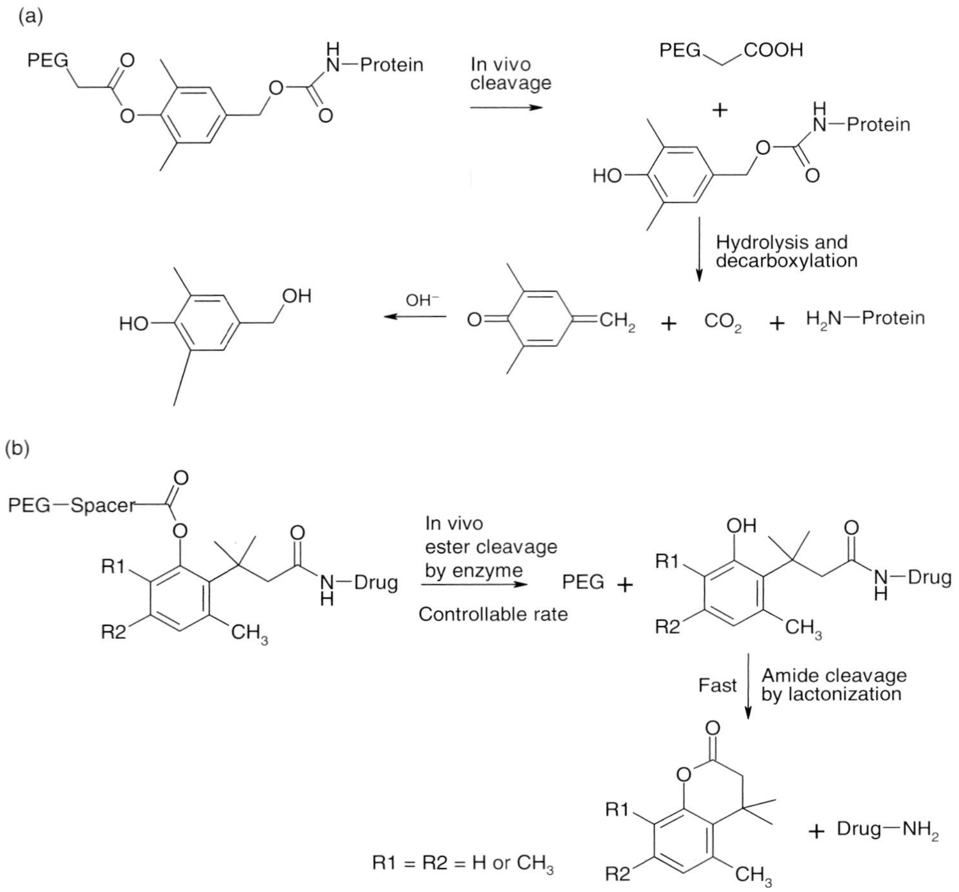

(a)

(b)

R1 = R2 = H or CH₃

Figure 8   Controlled release of active molecules from PEG based on the (a) 1,6-elimination system and (b) trimethyl lock lactonization system.

5.45.6   PEGylation Chemistry for Proteins

To appreciate the potential of PEGylation in drug delivery, it is important to understand the chemistry involved in the linking. A brief description is given below.

5.45.6.1   Amino Group PEGylation

PEGylation at the level of protein amino groups may be carried out with PEGs having different reactive groups at the end of the chain and often, although the coupling reaction is based on the same chemistry (for instance acylation), the obtained products are different. The difference may reside in the number of PEG chains linked per protein molecule, in the amino acids involved, and in the chemical bond between PEG and drug. The most common methods for random PEGylation are reported here, while procedures for site-direct modification will be discussed later.

Products available on the market so far mainly come from random PEGylation (Adagen, Oncaspar, PEG-Intron, and Pegasys) since Food and Drug Administration (FDA) authorities approve these conjugate mixtures upon demonstration of their reproducibility. The activated PEG for amino linking can be chosen from a range of commercially available polymers, the most common being PEG succinimidyl succinate (SS-PEG), PEG succinimidyl carbonate (SC-PEG), PEG p-nitrophenyl carbonate (pNPC-PEG), PEG benzotriazolyl carbonate (BTC-PEG), PEG trichlorophenyl carbonate (TCP-PEG), PEG carbonyldiimidazole (CDI-PEG), PEG tresylate, PEG dichlorotriazine, PEG aldehyde (AL-PEG), and a branched form of PEG (PEG₂-COOH) (**Figure 9**).

The difference between these PEGs lies in the kinetic rate of amino coupling and in the resulting link between polymer and drug. The derivatives with slower reactivity, such as the carbonate PEGs (pNPC-PEG, CDI-PEG, and

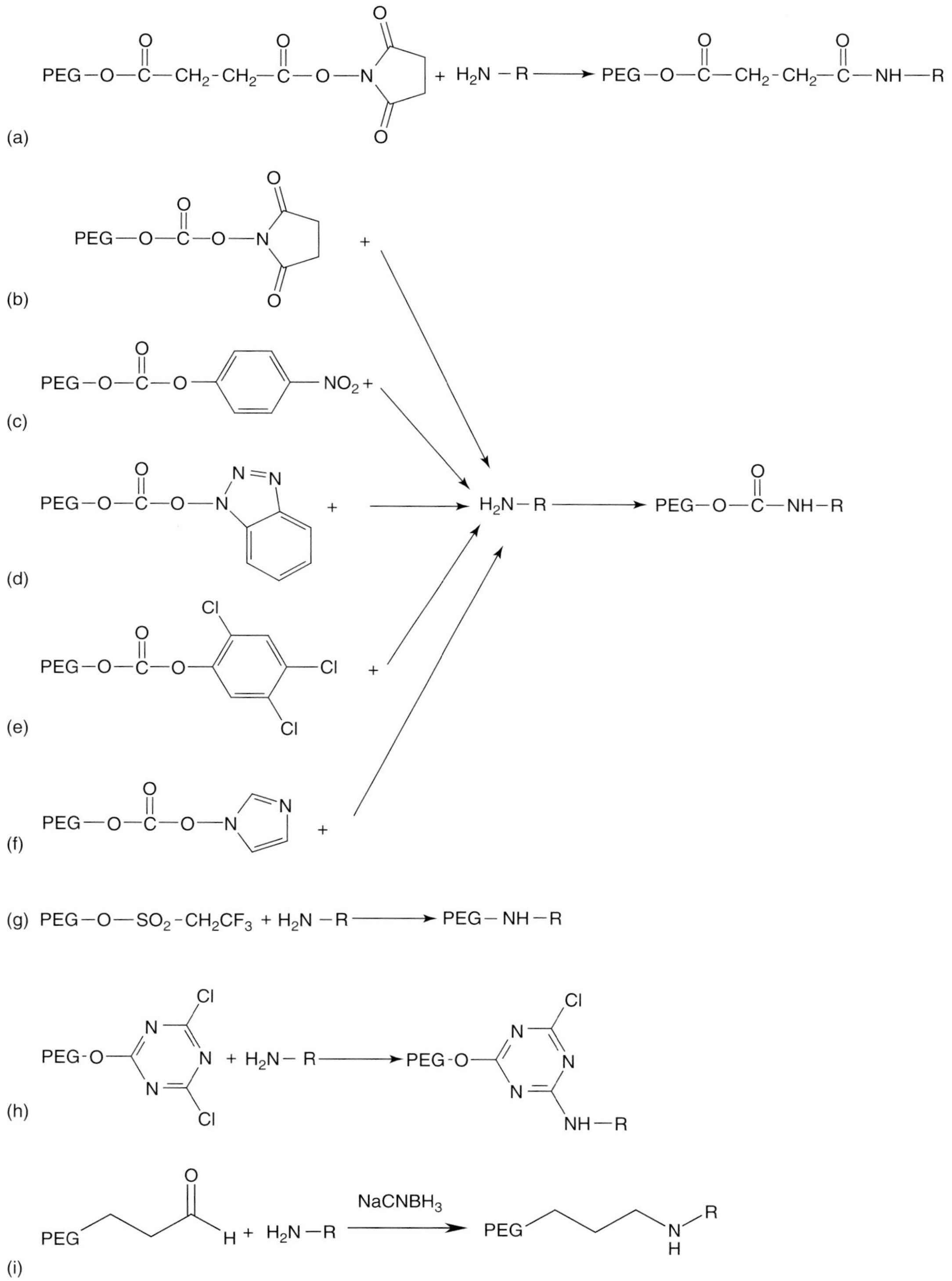

Figure 9 Examples of activated PEG molecules reactive toward amino groups: (a) PEG succinimidyl succinate, (b) PEG succinimidyl carbonate, (c) PEG p-nitrophenyl carbonate, (d) PEG benzotriazol carbonate, (e) PEG trichlorophenyl carbonate, (f) PEG carbonylimidazole, (g) PEG tresylate, and (h) PEG dichlorotriazine, and (i) PEG aldehyde (AL-PEG). (Adapted from Pasut, G.; Guiotto, A.; Veronese, F. M. *Exp. Op. Ther. Patents* **2004**, *14*, 859–894.)

TCP-PEG) or the aldehyde PEGs, allow a certain degree of selective conjugation within the amino groups present in a protein, according to their nucleophilicity or accessibility.[35] An important difference in reactivity is usually observed between the ε-amino and the α-amino group in proteins due to their pK_a: 9.3–9.5 for the ε-amino residue of lysine and 7.6–8 for the α-amino group. This was exploited for the α-amino modification reached by a conjugation at pH 5.5–6.0, as is the case for G-CSF with PEG-aldehyde.[94] The ε-amino groups of lysine, which possess high nucleophilicity at high pH, are instead the preferred site of conjugation at pH 8.5–9.

It is noteworthy that the conjugation performed using PEG dichlorotriazine, PEG tresylate, and PEG aldehyde (the latter after sodium cyanoborohydride reduction) maintains the same total charge on the native protein surface, since these derivatives react through an alkylation reaction, yielding a secondary amine. In contrast, PEGylation conducted with acylating PEGs (i.e., SS-PEG, SC-PEG, pNPC-PEG, CDI-PEG, TCP-PEG, and PEG$_2$-COOH) gives weakly acidic amide or carbamate linkages with loss of the positive charge.

The PEG derivatives described above may sometimes give side reactions involving the hydroxyl groups of serine, threonine, and tyrosine and the secondary amino group of histidine. These linkages, however, are generally hydrolytically unstable yielding the starting residue. The reaction conditions or particular conformational disposition may enhance the percentage of these unusual PEGylations; for example, α-interferon was found to be conjugated by SC-PEG or BTC-PEG also at His34 under slightly acidic conditions[113] (the pK_a value of histidine is between those of the α- and ε-amino groups). PEG was found to be linked to the hydroxyl groups of serine in the decapeptide antide or those of tyrosine in epidermal growth factor (EGF).[114,115]

Two examples of random amino PEGylation are reported here.

5.45.6.1.1 Random PEGylation of asparaginase

Asparaginase was one of the first PEGylated enzymes; it catalyzes the hydrolysis of asparagine to aspartic acid and ammonia. The resulting depletion in asparagine is fatal to leukemic lymphoblasts and certain other tumor cells, which by lacking or having very low levels of asparagine synthetase are unable to synthesize asparagine de novo and rely on asparagine supplied in the serum for survival. The free enzyme, however, is cleared too quickly from the body and it is immunogenic, which limits its therapeutic use. When asparaginase from *Escherichia coli* was modified with a PEG of 5000 Da the conjugate was shown to cause tumor regression in transplanted mice and to possess less immunogenicity than the native *Escherichia coli* form.[116–118] PEGylation also improves the enzyme chemical stability and the resistance to plasma proteases.[119] PEG-asparaginase first entered clinical trials in 1984 and since then it has been administered to thousands of patients with acute lymphoblastic leukemia (ALL).[120] The PEG-conjugated enzyme can be safely administered to most patients, even in cases where there is an allergic reaction to *E. coli* or *Erwinia* asparaginases. The longer serum half-life of PEG-asparaginase allows a longer interval between administered doses. PEG-asparaginase was developed by Enzon and was approved by the FDA in 1994 for the treatment of patients with ALL who are hypersensitive to native forms of the enzyme. It is now available commercially from Enzon as Oncaspar. The mean serum half-life of PEG-asparaginase is about 15 days as opposed to the 24 h of the nonmodified *E. coli* enzyme and the 10 h of the *Erwinia* form. The rate of total clearance of PEG-asparaginase was found to be 17-fold lower than that of the unmodified enzyme, whereas the volume of distribution was similar for the two preparations. L-Asparagine levels were undetectable immediately following the 1-h infusion of peg-asparaginase and remained low during the 14-day interval between doses. Interestingly, a recent pharmacoeconomic study[121] demonstrated that despite the higher cost of PEG-asparaginase versus the unmodified enzymes, the overall expense of treatment is comparable to that of unmodified enzymes.

Since FDA approval in 1994, drug monitoring has been performed by several phase IV clinical studies and detailed recent reviews are available in the literature.[84,122] Recent studies have been carried out on the rational basis that immunological, pharmacokinetic, and pharmacodynamic factors have a considerable impact on the efficacy of asparaginase therapy. Therefore, investigations are now aimed at defining the optimum dose and dosing schedule of the different asparaginase preparations that are used in the clinic[123] or correlating antibody levels with pharmacological response.[124]

5.45.6.1.2 Random PEGylation of interferon α-2a

Interferon α-2a (IFNα-2a) was modified with an *N*-hydroxysuccinimide activated PEG having a branched structure. A high-molecular-weight PEG (PEG$_2$, 40 kDa) was chosen on the basis of several preliminary studies disclosing the fact that: (1) the protein surface protection achieved with a single, long and hindered chain PEG is higher than that obtained with several small PEG chains linked at different sites[13]; (2) branched PEGs have lower distribution volumes than linear PEGs of identical molecular weight, and the delivery to organs such as liver and spleen is faster[125]; (3) proteins modified with branched PEG possess greater stability toward enzymes and pH degradation.[86]

Table 3 Pharmacokinetic properties of interferon α-2a and its PEGylated form in rats[92]

Protein	Half-life (h)	Plasma residence time (h)
Interferon α-2a	2.1	1.0
PEG2 (40 kDa)-interferon α-2a	15.0	20.0

Adapted from Reddy, K. R., Modi, M. W.; Pedder, S. *Adv. Drug Delivery Rev.* **2002**, *54*, 547–570.

The 40 kDa branched succinimidyl PEG (PEG$_2$-NHS) was linked to interferon α-2a using a 3:1 PEG:protein molar ratio in 50 mM sodium borate buffer pH 9 yielding a few protein isomers.[34] PEGylation under these conditions led to a mixture containing 45–50% monosubstituted protein, of 5–10% polysubstituted protein (essentially dimer), and 40–50% unmodified interferon. Identification of the major positional isomer within the mono-PEGylated fraction was carried out by a combination of high-performance cation exchange chromatography, peptide mapping, amino acid sequencing, and mass spectroscopy analysis. It was demonstrated that this branched PEG, thanks to its more hindered structure, was attached mainly to one of the following lysines: Lys-31, Lys-121, Lys-131, or Lys-134.[36] Even though the in vitro antiviral activity of PEG2-IFNα-2a was greatly reduced (only 7% was maintained with respect to the native protein) the in vivo activity, measured as ability to reduce the size of various human tumors, was higher than that of free IFNα-2a. The positive result could be related to the extended blood residence time of the conjugated form as shown in **Table 3**. These studies resulted in the release of the interferon conjugate Pegasys, which has a long period of residence in the blood and is effective in eradicating hepatic and extrahepatic hepatitis C virus (HCV) infection.[92]

5.45.6.2 Thiol PEGylation

The presence of a free cysteine residue represents an optimal opportunity to achieve site direct modification because it rarely occurs in proteins. PEG derivatives having specific reactivity toward the thiol group, i.e., MAL-PEG, OPSS-PEG, IA-PEG, and VS-PEG (**Figure 7**), are commercially available and allow thiol coupling with a good yield; however, differences among these polymers in terms of protein–polymer linkage and reaction conditions. Even if the thiol reaction rate of IA-, MAL-, or VS-PEGs is very rapid, some degree of amino coupling may also take place, especially if the reaction is carried out at pH values higher than 8. On the other hand, the reaction with OPSS-PEG is very specific for thiol groups, but the obtained conjugates may be cleaved in the presence of reducing agents as simple thiols or glutathione (present in vivo). PEGylation at the level of cysteine allows easier purification of the reaction mixture since the presence of only one or a few derivatizable sites (free cysteines) avoids the formation of a large number of positional isomers or products with different degrees of substitution, this being a common problem for amino PEGylation. The potential of thiol PEGylation may be further exploited by genetic engineering, which allows the introduction of an additional cysteine residue in a protein sequence or the switching of a nonessential amino acid to cysteine.

The use of this strategy with human growth factor was extensively studied. To overcome the common problems of random amino PEGylation, cysteine muteins were synthesized by recombinant DNA technology. The cysteine addition at the C-terminus of hGH leads to a fully active mutein that allows a site-specific PEGylation using the thiol-reactive PEG-maleimide (PEG-MAL, 8 kDa). It was necessary to treat the rhGH mutein with 1,4-dithio-DL-threitol (DTT) before the coupling step, to maintain the C-terminal cysteine in the reduced form and to prevent the formation of scrambled disulfide bridges. After removal of DTT excess by gel filtration, the conjugation leads to a mono-PEGylated derivative with a yield of over 80%.[126]

Hemoglobin (Hb) is another important example of thiol conjugation; in fact, PEGylation of this protein prevents the vasoactivity as a consequence of its extravasation.[76] After several unsuccessful attempts through random PEGylation, a site-specific modification was performed at Cys-93 (of the β-chain) with maleimidophenyl PEG (MAL-Phe-PEG; 5, 10, and 20 kDa), leading to PEGylated Hb carrying two polymer chains per Hb tetramer.[127] This product was found to be more efficient than polymerized Hb, the Hb-octamer, or Hb-dodecamer.

5.45.6.3 Carboxy PEGylation

PEGylation at the level of protein carboxylic groups needs their activation for the reaction with an amino PEG. This procedure, however, is not without its limitations since undesired intra- or intermolecular cross-links may occur by reaction with the amino groups of the protein itself.

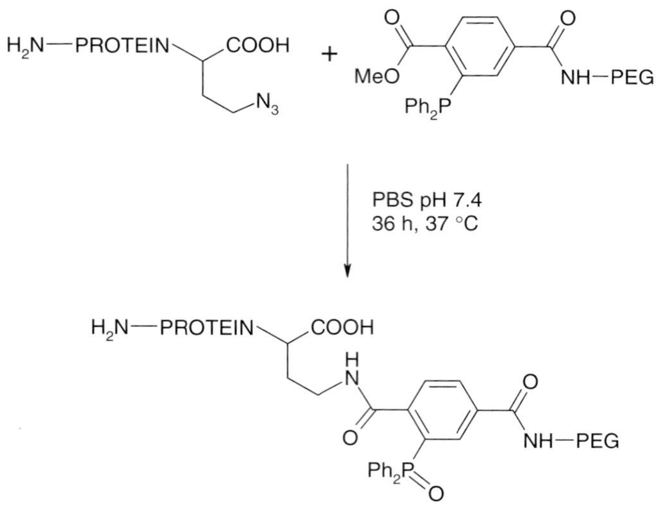

Figure 10 Staudinger ligation leading to a C-terminal mono-PEGylated protein by reaction of a mutated protein, containing a C-terminal azido-methionine, with an engineered PEG derivative, methyl-PEG-triarylphosphine.

To circumvent this problem it is possible to use PEG-hydrazide (PEG-CO-NH-NH₂) instead of the usual PEG-amino and to carry out the coupling at pH 4–5 thanks to the low pK_a of PEG-hydrazide. In this case, the protein's COOH groups, activated by water-soluble carbodiimide, do not react with the protein amino groups, which at low pHs are protonated, but with the amino group of PEG-hydrazide only.[128]

An alternative method for specific C-terminal PEGylation is based on the Staudinger ligation.[129] The protocol, using a truncated thrombomodulin mutant,[130] begins with the expression by *E. coli* of a mutated protein containing a C-terminal azido-methionine. This reacts specifically with an engineered PEG derivative, methyl-PEG-triarylphosphine, leading to a C-terminal mono-PEGylated protein (**Figure 10**). This method, however, involves the preparation of a gene encoding a protein with a C-terminal linker ending with methionine. Expression in *E. coli* is induced when the transformed bacteria are suspended in a medium where methionine is replaced by the azido-functionalized analog. Unfortunately, this method is applicable only to the rare case of proteins lacking methionine in the sequence; otherwise a methionine in the middle of the protein sequence will stop the protein transduction because the azido-analog does not permit the linking of the following amino acid.

5.45.7   Strategies in Protein PEGylation

To better exploit the potential of PEGylation several strategies have been developed with the purpose of: (1) obtaining homogeneous products; (2) forming PEGylated conjugates with high retention of activity; and (3) performing PEGylation under gentle conditions that are compatible with easily degradable proteins.

Site-selective conjugation is always preferred as it allows easy purification and characterization of products and, most importantly, better retention of biological activity. As described above, selective PEGylation may be achieved taking advantage of a free cysteine, but this is possible only when this amino acid is present in the native protein or is introduced by genetic engineering. Alternatively, it is possible to take advantage of the lower pK_a value of the α-amino group at the N-terminus with respect to the pK_a of ε-amine of lysines; in fact, performing the reaction under neutral or mildly acidic conditions, prevents the PEGylation at the level of lysine, but leaves the N-terminal amino group reactive.[131] The most successful example of this strategy is the alkylation of r-metHuG-CSF with PEG-aldehyde, proposed by Kinstler. The reaction was carried out at pH 5.5 in the presence of sodium cyanoborohydride to reduce the Shiff's base initially formed.[94,132] The conjugate obtained with a molecule of a 20 kDa PEG showed an improved pharmacokinetic profile due to the reduced excretion by the kidneys. The PEG-G-CSF conjugate Pegfilgastrim has been on the market since 2002.

Site-specific PEGylation can also be achieved by exploiting the different accessibility of protein amino groups, as reported for a truncated form of growth hormone-releasing hormone (hGRF₁₋₂₉). It was demonstrated that by using an appropriate solvent it is possible to alter the accessibility and reactivity of the three available amino groups. Nuclear magnetic resonance (NMR) and circular dichroism analysis indicated that the percentage of α-helix in hGRF₁₋₂₉, which is only 20% in water, rises to 90% in structure-promoting solvents such as methanol/water or 2,2,2-trifluoroethanol (TFE), thus facilitating a region-selective modification. When PEGylation was performed in TFE the monoPEGylated conjugate at the level of Lys-12 reached 80% for all PEGylated isomers;[133] however, the same reaction conducted in DMSO yielded an almost equimolar mixture of mono-PEGylated Lys-12 and Lys-21 isomers.[134]

Alternatively, specific PEGylation may be performed by blocking some of the reactive groups with a reversible protecting group as reported for insulin. This protein is formed by two polypeptide chains, A and B, and its three amino groups (Gly-A1, Phe-B1, and Lys-B29) are all candidates for PEGylation. Hinds proposed a site-directed PEGylation procedure involving the preliminary preparation of N-BOC (*tert*-butyl carbamate)-protected insulin.[135] As an example, in order to synthesize $N^{\alpha B1}$-PEG-insulin the intermediate $N^{\alpha A1}$, $N^{\varepsilon B29}$-BOC-protected insulin was prepared prior to conjugation with PEG-SPA at the level of free $N^{\alpha B1}$. The final conjugate was obtained upon BOC removal with TFA treatment, forming the $N^{\alpha B1}$-PEG$_{2000}$-insulin conjugate with 83% of the native insulin activity.

In general, in the PEGylation of an enzyme a requirement for high retention of the activity is that the PEG chains do not modify or obstruct the active site. Many strategies have been developed to achieve this goal: (1) the use of branched PEGs that, thanks to their higher hindrance with respect to linear polymers, have reduced accessibility to the active site (**Figure 6**); (2) to perform PEGylation in the presence of a substrate or an inhibitor that blocks polymer access to the active site; and (3) to conduct the modification after the enzyme is captured on an insoluble resin by substrates or inhibitors linked on it. In the last case, the obtained conjugate is eluted from the resin by changing the pH or adding denaturants, thus leading to a derivative that does not have linked PEG chains at the level of active site and its closer surroundings (**Figure 11**).[136]

A problem that may occur during protein PEGylation is the production of a low yield, especially when the modification is directed toward a buried or less accessible amino acid. This inconvenience is enhanced when the reaction is performed with high-molecular-weight PEGs due to the high steric hindrance. In the case of interferon beta (IFN-β), conjugation at cysteine 17 could only be achieved with a low-molecular-weight OPSS-PEG oligomer, but not with a high-molecular-weight polymer.[137] Modification with high-molecular-weight PEGs could be successfully attempted via a two-step procedure: in the first reaction, the protein is modified with a short-chain heterobifunctional PEG oligomer, while in the second, the obtained conjugate is linked to a higher molecular weight PEG, possessing specific reactivity toward the terminal end of the first oligomer (**Figure 12**). The heterobifunctional PEG oligomer had

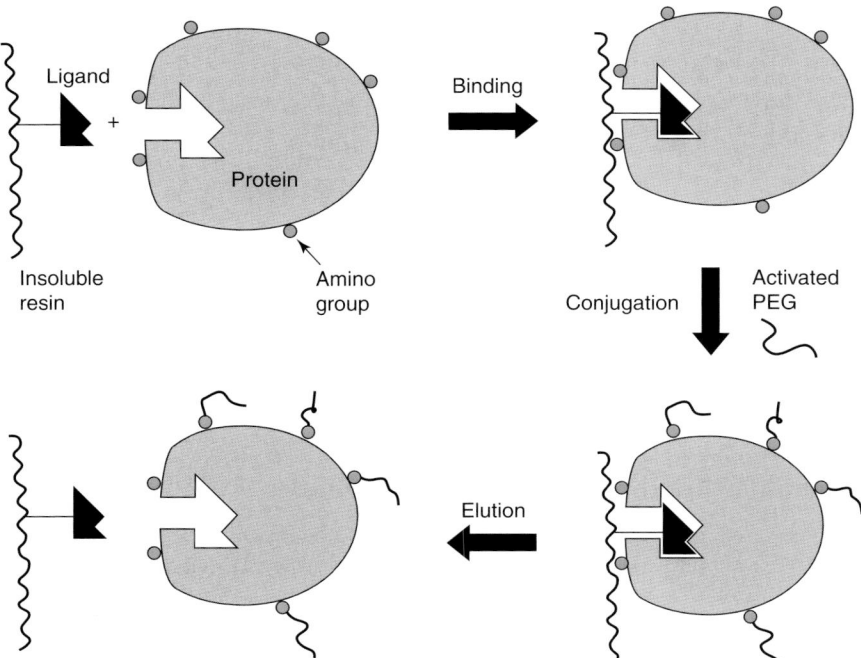

Figure 11 PEGylation strategy for the protection of an enzyme active site from polymer conjugation: firstly the enzyme is loaded into an affinity resin functionalized with the appropriate ligand. The enzyme's active site binds the ligand, thus protecting the active site itself and the area close to it from PEG modification. After reaction under heterogeneous conditions the modified enzyme is eluted from the column.

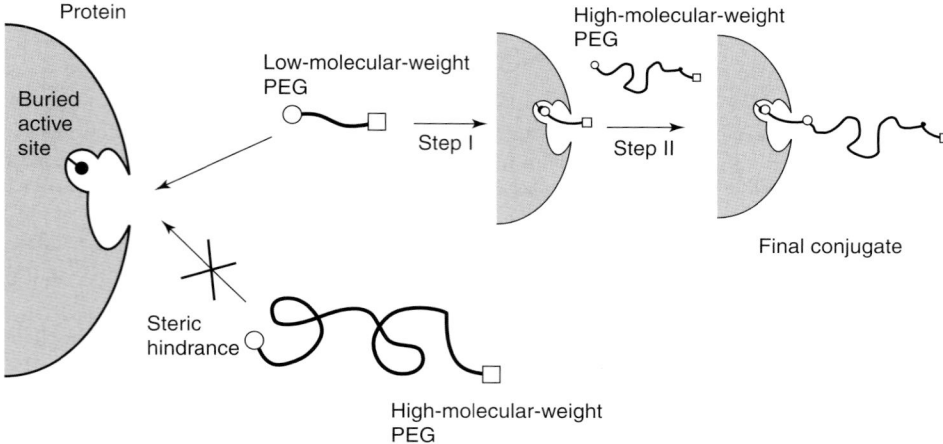

Figure 12   Two-step tagging PEGylation strategy for a buried SH group in a protein. Smaller PEG molecules are more reactive than high-molecular-weight PEGs toward the buried cysteine. (Reproduced with permission from Pasut, G.; Guiotto, A.; Veronese, F. M. *Exp. Op. Ther. Patents* **2004**, *14*, 859–894.)

R = lysine of protein, polymer, etc.

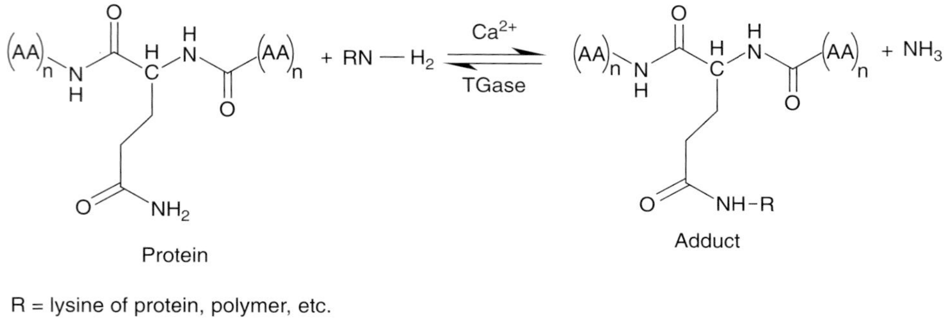

Figure 13   Reaction catalyzed by Tgase between a glutamine residue in a protein and an alkyl amine. (Reproduced with permission from Pasut, G.; Guiotto, A.; Veronese, F. M. *Exp. Op. Ther. Patents* **2004**, *14*, 859–894.)

a thiol reactive group at one end of the chain and a hydrazine group at the other (OPSS-PEG-Hz, 2 kDa). As mentioned above, hydrazine is still reactive at low pHs when a protein's amino groups are usually protonated and not reactive, thus preventing unwanted amino PEGylation of the protein. The INF-SS-PEG-Hz conjugate could therefore be selectively modified with PEG-aldehyde (30 kDa) by reductive alkylation. The overall yield was higher than 80%.

A recent study demonstrated that specific PEGylation at the lone, but not accessible, thiol groups of G-CSF could be achieved upon its exposure to partially denaturant conditions. After modification with OPSS-PEG, the native conformation of G-CSF was recovered by removal of the denaturant.[138]

5.45.8   Enzymatic Approach for Protein PEGylation

Recently, as an alternative solution to obtaining homogeneous PEG-protein conjugates under mild reaction conditions, some researchers have been investigating the possibility of exploiting the specificity of enzymes to catalyze PEGylation, which might preserve the protein activity better than with chemical methods. Since the first report, proposed by H. Sato involving the enzyme transglutaminase (Tgase),[139] other researchers have developed interesting approaches with different enzymes. From the data obtained so far it is predicted that these procedures will lead to further results of great interest and applicability.

Sato studied two methods for interleukin-2 (IL2) PEGylation using transglutaminases (Tgase) from guinea-pig liver (G-Tgase) or from *Streptoverticillium* sp. strain s-8112 (M-Tgase). Both enzymes catalyze the transfer of an amino group from a donor (for example PEG-NH$_2$) to a glutamine residue present in a protein (**Figure 13**). The two enzymes differ

Table 4 Comparision of IL2 conjugate activities between random PEGylation and site-directed PEGylation by Tgase[139]

Proteins	% Activity[a]
rhIL2	100
PEG10-rhIL2 (random PEGylation)	74
(PEG10)$_2$-rhIL2 (random PEGylation)	36
rTG1-IL2 (chimeric protein for enzymatic PEGylation)	72
PEG10-rTG1-IL2 (enzymatic PEGylation)	69
(PEG10)$_2$-rTG1-IL2 (enzymatic PEGylation)	72

The rTG1 abbreviation represents the N-terminal fused Pro-Lys-Pro-Gln-Gln-Phe-Met sequence introduced as TGase subtrate.
Adapted from Sato, H. *Adv. Drug Delivery Rev.* **2002**, *54*, 487–504.
[a] The amount of activity was expressed as the per cent residual bioactivity as compared to the rhIL2.

in the required amino acid sequence neighboring the glutamine of the substrate. Several tailor-made linear PEGs, differing in molecular weight and type of alkylamine present at the polymer end, have been synthesized, the best reactivity being shown by polymers terminating with a -$(CH_2)_6$-NH_2 group. IL2 contains six glutamines, but none of them is a suitable substrate for the more specific G-Tgase. The problem was overcome by preparing chimeric IL2 proteins that had at the N-terminus a peptide sequence that is a good G-Tgase substrate, like the Pro-Lys-Pro-Gln-Gln-Phe-Met (derived from Substance P[140]) or the Ala-Gln-Gln-Ile-Val-Met (derived from fibronectin[141]). In the former case, a mono-PEGylated conjugate was obtained while a mixture of mono- and di-PEGylated forms resulted in the latter case. The enzymatic coupling was carried out under mild conditions: 0.1 M Tris–HCl buffer, pH 7.5 at 25 °C for 12 h in the presence of $CaCl_2$ 10 mM.[136] The derivatives maintained almost the same activity of the native protein, whereas the classical chemical conjugation with mPEG-NHS yielded products with decreased activity (**Table 4**). Using the less specific M-Tgase, mPEG$_{12000}$-$(CH_2)_6$-NH_2 could be directly incorporated into rhIL2 at the level of Gln-74.[142] Compared to others site-specific chemical PEGylation, such as cysteine coupling or N-terminus modification at acidic pHs, the enzymatic method produces less undesired products, i.e., protein–protein dimers (due to cysteine oxidation) or εNH_2 lysine PEGylation.

A two-step enzymatic PEGylation, called GlycoPEGylation, was developed by Neose Technologies. In this case, *E. coli*-expressed proteins were glycosylated at the level of specific serine and threonine with *N*-acetylgalactosamine (GalNAc) by in vitro treatment with the recombinant O-GalNAc-transferase. The obtained glycosylated proteins were subsequently PEGylated using the O-GalNAc residue as the acceptor site of the cytidine monophosphate derivative of a sialic acid-PEG, a reaction selectively catalyzed by a sialyltransferase.[143] The great advantage of this technology is the possibility of PEGylating proteins produced in *E. coli* in order to mimic the mammalian ones, since the PEG chains replace the native sugar moiety at the precise site of glycosylation, forming conjugates that retain the correct structure for receptor recognition and an extended plasma half-life. This method, however, is still awaiting large-scale application.

5.45.9 PEGylated Protein Purification and Characterization

Theoretically, in a conjugation reaction conducted with an excess of PEG, one could expect all of the reactive groups of the protein to be modified, thus yielding mainly a single product. However, in order to avoid loss of biological activity, a smaller amount of PEG is generally employed even if this gives rise to a mixture of positional isomers.[144] Often the mixture can be fractionated by ionic exchange chromatography[145] due to the suitable differences in isoelectric point of each isomer. Reverse-phase chromatography was found to be less efficient in this respect, and gel filtration separates only species with different masses.

Once the isomers are separated it is necessary to identify the PEGylation site in the primary sequence. The usual approach involves enzymatic digestion of the conjugate, purification of the peptides, and their identification by mass spectroscopy or amino acid analysis. A good example is reported in the characterization of PEGylated interferon α-2a,[146] where the comparison of the peptide fingerprint of the conjugated protein with that of the native protein allowed the region of PEGylation to be identified on the basis of the disappeared peptide signal. Besides the lengthy procedure, the conjugated polymer may interfere by steric hindrance with the proteolytic enzymes, resulting in an incomplete cleavage that complicates the interpretation of the peptide finger printing. An alternative procedure exploits the use of

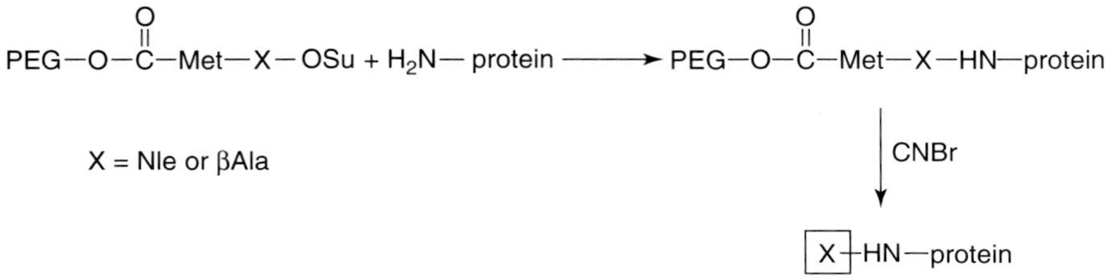

Figure 14 Use of PEG-Met-Nle-OSu or PEG-Met-βAla-OSu to introduce a reporter amino acid at the PEGylation site: PEG conjugation and release of PEG by CNBr. Nor-leucine (Nle) or β-alanine (βAla) can be identified on the protein by enzymatic digestion of the protein and mass spectrometry analysis of the obtained peptides.

PEG-Met-Nle-COOH or PEG-Met-βAla-COOH in the conjugation step. These possess a chemically labile bond in the peptide spacer at the level of methionine, which can be cleaved by treatment with CNBr (**Figure 14**) leaving only nor-leucine or β-alanine tags on the protein. The amino acid tagged peptides can be easily identified by standard sequence methods or by mass spectrometry analysis in the enzymatically digested mixture.[147]

5.45.10   Conclusion

In this chapter only a few drug conjugates out of the many described in the literature or patents have been discussed. The intention was to give examples of the potential of the method and how different problems were identified and solved. Polymer conjugation is often employed not only to improve the pharmacokinetic profiles of drugs, but also to give new properties to the modified drug, such as a higher solubility and stability, specific targeting, and for proteins removal or reduction of immunogenicity.

Many different polymers have been used in this field, both from natural or synthetic sources. Multifunctional polymers are preferred for low-molecular-weight drugs, since higher loading can be achieved, while monofunctional polymers are the essential choice for proteins to avoid unwanted cross-linking.

Among all polymers, PEG emerged as the most successful, and it has already led to several marketed products (**Table 1**). In fact, PEGylation, which was initially used to improve the therapeutic value of peptide and protein drugs, has now been extended to nonprotein compounds and, more recently, oligonucleotides. PEG has several useful properties that are finding application in other therapeutic approaches, such as the surface modification of liposomes (stealth liposome[148]), micro- and nanoparticles,[149] and, last but not least, cells,[150] making them biocompatible when implanted in vivo.

At the basis of this technology lies an in-depth understanding of the chemistry of polymers, which despite the number of advances made over the years is still open to new advancements.

References

1. Brange, J.; Skelbaek-Pedersen, B.; LangKjaer, L.; Damgaard, U.; Ege, H.; Havelund, S.; Heding, L. G.; Jorgensen, K. H.; Lykkeberg, J.; Markussen, J. et al. *Galenics of Insulin: The Physicochemical and Pharmaceutical Aspects of Insulin and Insulin Preparations*; Springer-Verlag: Berlin, Germany, 1987.
2. Barton, N. W.; Brady, R. O.; Dambrosia, J. M.; Di Bisceglie, A. M.; Doppelt, S. H.; Hill, S. C.; Mankin, H. J.; Murray, G. J.; Parker, R. I.; Argoff, C. E. et al. *N. Engl. J. Med.* **1991**, *324*, 1464–1470.
3. Eng, C. M.; Guffon, N.; Wilcox, W. R.; Germain, D. P.; Lee, P.; Waldek, S. et al. *N. Engl. J. Med.* **2001**, *345*, 9–16.
4. Lebrethon, M. C.; Grossman, A. B.; Afshar, F.; Plowman, P. N.; Besser, G. M.; Savage, M. O. *J. Clin. Endocrinol. Metab.* **2000**, *85*, 3262–3265.
5. Agrawal, N. R.; Bukowski, R. M.; Rybicki, L. A.; Kurtzberg, J.; Cohen, L. J.; Hussein, M. A. *Cancer* **2003**, *98*, 94–99.
6. Tan, Y.; Zavala, J., Sr.,; Xu, M.; Zavala, J., Jr.,; Hoffman, R. M. *Anticancer Res.* **1996**, *16*, 3937–3942.
7. Reddy, K. R. *Ann. Pharmacother.* **2000**, *34*, 915–923.
8. Duncan, R. *Nat. Rev. Drug Disc.* **2003**, *2*, 347–360.
9. Jatzkewitz, H. *Z. Naturforsch.* **1955**, *10b*, 27–31.
10. Breslow, D. S. *Pure Appl. Chem.* **1976**, *46*, 103–114.
11. Ringsdorf, H. J. *Polym. Ci. Symp.* **1975**, *51*, 135–153.
12. Molteni, L. Dextrans as Drug Carriers. In *Drug Carriers in Biology and Medicine*; Gregoriadis, G., Ed.; Academic Press: London, 1979, pp 107–125.
13. Bailon, P.; Berthold, W. *Pharm. Sci. Technol. Today* **1998**, *1*, 352–356.
14. Tomalia, D. A.; Baker, H.; Dewald, J.; Hall, M.; Kallos, G.; Martin, S.; Roeck, J.; Ryder, J.; Smith, P. *Polym. J.* **1985**, *17*, 117–132.
15. Frechet, J. M. J. *J. Mater. Sci. Pure Appl. Hem.* **1996**, *33*, 1399–1425.
16. Frechet, J. M. J.; Tomalia, D. A. *Dendrimers and Other Dendritic Polymers*; Wiley: Chichester, UK, 2001.
17. Malefant, P. R. L.; Frechet, J. M. J. In *Dendrimers and Other Dendritic Polymers*; Frechet, J. M. J., Tomalia, D. A., Eds.; Wiley: Chichester, UK, 2001, pp 171–196.
18. Ferruti, P.; Knobloch, S.; Ranucci, E.; Duncan, R.; Gianasi, E. *Macromol. Chem. Phys.* **1998**, *199*, 2565–2575.

19. Dautzenberg, H.; Zintchenko, A.; Konak, C.; Reschel, T.; Subr, V.; Ulbrich, K. *Langmuir* **2001**, *17*, 3096–3102.
20. Pechar, M.; Ulbrich, K.; Subr, V. *Bioconj. Chem.* **2000**, *11*, 131–139.
21. Stiriba, S. E.; Krautz, H.; Frey, H. *J. Am. Chem. Soc.* **2002**, *124*, 9698–9699.
22. Mammen, M.; Choi, S. K.; Whitesides, G. M. *Angew. Chem. Int. Ed. Engl.* **1998**, *37*, 2754–2794.
23. Mirhra, M. K.; Kobayashi, S. *Star and Hyperbranched Polymers*; Marcel Dekker: Basel, 1999.
24. Roy, R. *Top. Curr. Chem.* **1997**, *187*, 241–274.
25. Chaves, F.; Calvo, J. C.; Carvajal, C.; Rivera, Z.; Ramirez, L.; Pinto, M.; Trujillo, M.; Guzman, F.; Patarroyo, M. E. *J. Pept. Res.* **2001**, *58*, 307–316.
26. Matsumura, Y.; Maeda, H. *Cancer Res.* **1986**, *25*, 865–869.
27. Russell-Jones, G. J. *Adv. Drug Delivery Rev.* **1996**, *20*, 83–97.
28. Takakura, Y.; Mahoto, R. I.; Hashida, M. *Adv. Drug Delivery Rev.* **1998**, *34*, 93–108.
29. Okamoto, C. T. *Adv. Drug Delivery Rev.* **1998**, *29*, 215–228.
30. Allen, T. M. *Nat. Rev. Cancer* **2002**, *2*, 750–763.
31. Abuchowski, A.; Van Es, T.; Palczuk, N. C.; Davis, F. F. *J. Biol. Chem.* **1977**, *252*, 3578–3581.
32. Abuchowski, A.; McCoy, R.; Palczuk, N. C.; Van Es, T.; Davis, F. F. *J. Biol. Chem.* **1997**, *252*, 3582–3586.
33. Kopeček, J.; Kopeckova, P.; Minko, T.; Lu, Z. *Eur. J. Pharm. Biopharm.* **2000**, *50*, 61–81.
34. Hoste, K.; De Winne, K.; Schacht, E. *Int. J. Pharm.* **2004**, *277*, 119–131.
35. Veronese, F. M.; Morpurgo, M. *Farmaco* **1999**, *54*, 497–516.
36. Bailon, P.; Palleroni, A.; Schaffer, C. A.; Spence, C. L.; Fung, W. J.; Porter, J. E.; Ehrlich, G. K.; Pan, W.; Xu, Z. X.; Modi, M. W. et al. *Bioconj. Chem.* **2001**, *12*, 195–202.
37. Delgado, C.; Francis, G. E.; Fisher, D. *Crit. Rev. Ther. Drug Carrier Syst.* **1992**, *9*, 249–304.
38. Brocchini, S.; Duncan, R. In *Encyclopaedia of Controlled Drug Delivery*; Mathiowitz, E., Ed.; Wiley: New York, 1999, pp 786–816.
39. Kopecek, J.; Bazilova, H. *Eur. Polym. J.* **1973**, *9*, 7–14.
40. Lloyd, J. B.; Duncan, R.; Pratten, M. K. *Br. Polym. J.* **1983**, *15*, 158–159.
41. Duncan, R.; Kopecek, J.; Lloyd, J. B. Development of *N*-(-2-Hydroxypropyl) Methacrylamide Copolymers as Carriers of Therapeutic Agents. In *Polymers in Medicine: Biomedical and Pharmacological Applications*; Chielline, E., Giusti, P., Eds.; Plenum Press: New York, 1983, pp 97–113.
42. Duncan, R.; Cable, R.; Lloyd, H. C.; Rejmanova, P.; Kopecek, J. *Makromol. Chem.* **1983**, *184*, 1997–2008.
43. Duncan, R. Polymer Therapeutics. In *Business Briefing Pharmatech.*; Cooper, E., Ed.; World Markets Research Center, 2001, pp 178–184; ISBN1-903150-45-0.
44. Seymour, L. W.; Duncan, R.; Strohalm, J.; Kopecek, J. *J. Biomed. Mater. Res.* **1987**, *21*, 1341–1358.
45. Duncan, R.; Kopecek, J.; Rejmanova, P.; Lloyd, J. B. *Biochem. Biophys. Acta* **1983**, *755*, 518–521.
46. Duncan, R.; Coatsworth, J. K.; Burtles, S. *Hum. Exp. Toxicol.* **1998**, *17*, 93–104.
47. Vasey, P. A.; Kaye, S. B.; Morrison, R.; Twelves, C.; Wilson, P.; Duncan, R.; Thomson, A. H.; Murray, L. S.; Hilditch, T. E.; Murray, T. et al. *Clin. Cancer Res.* **1999**, *5*, 83–94.
48. Seymour, L. W.; Ferry, D. R.; Anderson, D.; Hesslewood, S.; Julyan, P. J.; Poyner, R.; Doran, J.; Young, A. M.; Burtles, S.; Kerr, D. J. Cancer Research Campaign Phase I/II Trials Committee. *J. Clin. Oncol.* **2002**, *20*, 1668–1676.
49. Schoemaker, N. E.; van Kesteren, C.; Rosing, H.; Jansen, S.; Swart, M.; Lieverst, J.; Fraier, D.; Breda, M.; Pellizoni, C.; Spinelli, M. et al. *Br. J. Cancer* **2002**, *87*, 608–614.
50. Meerum Terwogt, J. M.; ten Bokkel Huinink, W. W.; Schellens, J. H. M.; Schot, M.; Mandjes, I. A. M.; Zurlo, M. G.; Rocchetti, M.; Rosing, H.; Koopman, F. J.; Beijnen, J. H. et al. *Anticancer Drugs* **2001**, *12*, 315–323.
51. Gianasi, E.; Buckley, R. G.; Latigo, J.; Wasil, M.; Duncan, R. *J. Drug Targeting* **2002**, *10*, 549–556.
52. Rademaker-Lakhai, J. M.; Terret, C.; Howell, S. B.; Baud, C. M.; De Boer, R. F.; Pluim, D.; Beijnen, J. H.; Schellens, J. H.; Droz, J. P. *Clin. Cancer Res.* **2004**, *10*, 3386–3395.
53. Rihova, B. *Makromol. Chem.* **1983**, *9*, 13–24.
54. Rihova, B.; Bilej, M.; Vetvicka, V.; Ulbrich, K.; Strohalm, J.; Kopecek, J.; Duncan, R. *Biomaterials* **1989**, *10*, 335–342.
55. Langer, C. J. *Clin. Lung Cancer* **2004**, *6*, S85–S88.
56. Todd, R.; Sludden, J.; Boddy, A. V.; Griffin, M. J.; Robson, L.; Cassidy, J.; Bissett, D.; Bernareggi, A.; Verrill, M. W.; Calvert, A. H. ASCO, 2001, No. 439.
57. Sabbatini, P.; Brown, J.; Aghajanian, C.; Hensley, M. L.; Pezzulli, S.; O'Flaherty, C.; Lovegren, M.; Funt, S.; Warner, M.; Mitchell, P. et al. *Proc. Am. Soc. Clin. Oncol.* **2002**, *871*.
58. Kudelka, A. P.; Verschraegen, C. F.; Loyer, E.; Wallace, S.; Gershenson, D. M.; Han, J.; Ho, L.; Garzone, P. D.; Warner, M.; Bolton, M. G.; et al. *Proc. Am. Soc. Clin. Oncol.* **2002**, *2146*.
59. Singer, J. W.; Baker, B.; De Vries, P.; Kumar, A.; Shaffer, S.; Vawter, E.; Boltom, R.; Garzone, P. et al. *Adv. Exp. Med. Biol.* **2003**, *519*, 81–99.
60. Sabbatini, P.; Aghajanian, C.; Dizon, D.; Anderson, S.; Dupont, J.; Brown, J. V.; Peters, W. A.; Jacobs, A.; Mehdi, A.; Rivkin, S. et al. *J. Clin. Oncol.* **2004**, *22*, 4523–4531.
61. De Vries, P.; Bhatt, R.; Stone, I.; Klein, P.; Singer, J. *Proceedings of the International Conference on AACR-NCI-EORTC*, 2001, p. 21, No. 100.
62. Mehvar, R. *Curr. Pharm. Biotechnol.* **2003**, *4*, 283–302.
63. Danhauser-Riedl, S.; Hausmann, E.; Schick, H. D.; Bender, R.; Dietzfelbinger, H.; Rastetter, J.; Hanauske, A. R. *Invest. New Drugs* **1993**, *11*, 187–195.
64. Torchilin, V. P.; Voronkov, J. I.; Mazoev, A. V. *Ter. Ark. (Ther. Arch. Russ.)* **1982**, *54*, 21–25.
65. Fujita, T.; Nishikawa, M.; Tamaki, C.; Takakura, Y.; Hashida, M.; Sezaki, H. *J. Pharmacol. Exp. Ther.* **1992**, *263*, 971–978.
66. Faulk, W. P.; Taylor, C. G.; Yeh, C. J.; McIntyre, J. A. *Mol. Biother.* **1990**, *2*, 57–60.
67. Ponzetto, A.; Fiume, L.; Forzani, B.; Song, S. Y.; Busi, C.; Mattioli, A.; Spinelli, C.; Marinelli, M.; Smedile, A.; Chiaberge, E. et al. *Hepatology* **1991**, *14*, 16–24.
68. Wosikowski, K.; Biedermann, E.; Rattel, B.; Breiter, N.; Jank, P.; Loser, R.; Jansen, G.; Peters, G. J. *Clin. Cancer Res.* **2003**, *9*, 1917–1926.
69. Kratz, F. *Exp. Op. Ther. Patents* **2002**, *12*, 433–439.
70. Takeda, Y.; Hashimoto, H.; Kosaka, F.; Hirakawa, M.; Inoue, M. *Am. J. Physiol.* **1993**, *264*, H1708–H1715.
71. Kratz, F.; Warnecke, A.; Scheuermann, K.; Stockmar, C.; Schwab, J.; Lazar, P.; Druckes, P.; Esser, N.; Drevs, J.; Rognan, D. et al. *J. Med. Chem.* **2002**, *45*, 5523–5533.
72. Wang, W.; Ou, Y.; Shi, Y. *Pharm. Res.* **2004**, *21*, 2105–2111.
73. Maeda, H. *Adv. Drug Delivery Rev.* **1991**, *6*, 181–202.

74. Maeda, H.; Konno, T. In *Neocarzinostatin: The Past, Present, and Future of an Anticancer Drug*; Maeda, H., Edo, K., Ishida, N., Eds.; Springer: Berlin, Germany, 1997, pp 227–267.

75. Harris, J. M.; Chess, R. B. *Nat. Rev. Drug Disc.* **2003**, *2*, 214–221.

76. Manjula, B. N.; Tsai, A.; Upadhya, R.; Perumalsamy, K.; Smith, P. K.; Malavalli, A.; Vandegriff, K.; Winslow, R. M.; Intaglietta, M.; Prabhakaran, M. et al. *Bioconj. Chem.* **2003**, *14*, 464–472.

77. Working, P. K.; Newman, S. S.; Johnson, J.; Cornacoff, J. B. Safety of Poly(ethylene glycol) Derivatives. In *Poly(ethylene glycol) Chemistry and Biological Applications*; Harris, J. M., Zalipsky, S., Eds.; ACS Books: Washington, DC, 1997, pp 45–54.

78. Zaplisky, S; Harris, J. M. In *Chemistry and Biological Applications of Polyethylene Glycol*. American Chemical Society Symposium Series 680. ACS: San Francisco, 1997, pp 1–15.

79. Pasut, G.; Guiotto, A.; Veronese, F. M. *Exp. Op. Ther. Patents* **2004**, *14*, 859–894.

80. Petrak, K.; Goddard, P. *Adv. Drug Delivery Rev.* **1989**, *3*, 191–214.

81. Friman, S.; Egestad, B.; Sjovatt, J.; Svanvik, J. *J. Hepatol.* **1993**, *17*, 48–55.

82. Guiotto, A.; Canevari, M.; Pozzobon, M.; Moro, S.; Orsolini, P.; Veronese, F. M. *Bioorg. Med. Chem.* **2004**, *12*, 5031–5037.

83. Levy, Y.; Hershfield, M. S.; Fernandez-Mejia, C.; Polmar, S. H.; Scrudiery, D.; Berger, M.; Sorensen, R. U. *J. Pediatr.* **1988**, *113*, 312–317.

84. Graham, L. M. *Adv. Drug Delivery Rev.* **2003**, *10*, 1293–1302.

85. Harris, J. M.; Guo, L.; Fang, Z. H.; Morpurgo, M. In *PEG-Protein Tethering for Pharmaceutical Applications*. Proceedings of the Seventh International Symposium on Recent Advances in Drug Delivery, Salt Lake City, USA, 1995.

86. Monfardini, C.; Schiavon, O.; Caliceti, P.; Morpurgo, M.; Harris, J. M.; Veronese, F. M. *Bioconj. Chem.* **1995**, *6*, 62–69.

87. Akiyama, Y.; Otsuka, H.; Nagasaki, Y.; Kato, M.; Kataoka, K. *Bioconj. Chem.* **2000**, *11*, 947–950.

88. Zhang, S.; Du, J.; Sun, R.; Li, X.; Yang, D.; Zhang, S.; Xiong, C.; Peng, Y. *Reactive Funct. Polymer* **2003**, *56*, 17–25.

89. Choe, Y. H.; Conover, C. D.; Wu, D.; Royzen, M.; Gervacio, Y.; Borowski, V.; Mehlig, M.; Greenwald, R. B. *J. Control. Release* **2002**, *79*, 55–70.

90. Schiavon, O.; Pasut, G.; Moro, S.; Orsolini, P.; Guiotto, A.; Veronese, F. M. *Eur. J. Med. Chem.* **2004**, *39*, 123–133.

91. Wang, Y. S.; Youngster, S.; Grace, M.; Bausch, J.; Bordens, R.; Wyss, D. F. *Adv. Drug Delivery Rev.* **2002**, *54*, 547–570.

92. Reddy, K. R.; Modi, M. W.; Pedder, S. *Adv. Drug Delivery Rev.* **2002**, *54*, 571–586.

93. Trainer, P. J.; Drake, W. M.; Katznelson, L.; Freda, P. U.; Herman-Bonert, V.; Van der Lely, A. J.; Dimaraki, E. V.; Stewart, P. M.; Friend, K. E.; Vance, M. L. et al. *N. Engl. J. Med.* **2000**, *342*, 1171–1177.

94. Kinstler, O.; Moulinex, G.; Treheit, M.; Ladd, D.; Gegg, C. *Adv. Drug Delivery Rev.* **2002**, *54*, 477–485.

95. The EyeTech Study Group. *Retina* **2002**, *22*, 143–152.

96. Rowinsky, E. K.; Rizzo, J.; Ochoa, L.; Takimoto, C. H.; Forouzesh, B.; Schwartz, G.; Hammond, L. A.; Patnaik, A.; Kwiatek, J.; Goetz, A. et al. *J. Clin. Oncol.* **2003**, *21*, 148–157.

97. Scott, L. C.; Evans, T.; Yao, J. C.; Benson, A. I.; Mulcahy, M.; Thomas, A. M.; Decatris, M.; Falk, S.; Rudoltz, M.; Ajani, J. A. et al. *Journal of Clinical Oncology ASCO Annual Meeting Proceedings (Post-Meeting Edition)*, **2004**, *22* (July 15 Supplement), 4030.

98. Guiotto, A.; Canevari, M.; Orsolini, P.; Lavanchy, O.; Deuschel, C.; Kaneda, N.; Kurita, A.; Matsuzaki, T.; Yaegashi, T.; Sawada, S.; Veronese, F. M. *J. Med. Chem.* **2004**, *47*, 1280–1289.

99. Veronese, F. M.; Guiotto, A.; Sumiya, H. (DEBIO R. P.). Amino-substituted camptothecin polymer derivatives and use of the same for the manufacture of a medicament. World Patent 03,031,467, April, 17, 2003.

100. Veronese, F. M.; Schiavon, O.; Pasut, G.; Mendichi, R.; Andersson, L.; Tsirk, A.; Ford, J.; Wu, G.; Kneller, S.; Davies, J.; Duncan, R. *Bioconj. Chem.* **2005**, *16*, 775–784.

101. Greenwald, R. B.; Gilbert, C. W.; Pendri, A.; Conover, C. D.; Xia, J.; Martinez, A. *J. Med. Chem.* **1996**, *39*, 424–431.

102. Beeram, M.; Rowinsky, E. K.; Hammond, L.A.; Patnaik, A.; Schwartz, G. H.; de Bono, J. S.; Forero, L.; Forouzesh, B.; Berg, K. E.; Rubin E. H., et al. *A Phase I and Pharmacokinetic (PK) Study of PEG-Paclitaxel in Patients with Advanced Solid Tumors*; ASCO Annual Meeting; American Society of Clinical Oncology: Orlando, FL. 2002, No. 405.

103. Esposito, P.; Barbero, L.; Caccia, P.; Caliceti, P.; D'Antonio, M.; Piquet, G.; Veronese, F. M. *Adv. Drug Delivery Rev.* **2003**, *55*, 1279–1291.

104. Yamaoka, T.; Tabata, Y.; Ikata, Y. *J. Pharm. Sci.* **1994**, *83*, 601–606.

105. Kratz, F.; Beyer, U.; Schutte, M. T. *Crit. Rev. Ther. Drug Carrier Syst.* **1999**, *16*, 245–288.

106. Thistlethwaite, A. J.; Lepper, D. B.; Moylan, D. J.; Nerlinger, R. E. *Int. J. Radiat. Oncol.* **1985**, *11*, 1647–1652.

107. Shen, W. C.; Ryser, H. J. P. *Biochem. Biophys. Res. Commun.* **1981**, *102*, 1048–1054.

108. Kaneko, T.; Willner, D.; Monokovic, I.; Knipe, J. O.; Braslawsky, G. R.; Greenfield, R. S.; Dolotrai, M. V. *Bioconj. Chem.* **1991**, *2*, 133–141.

109. Coessen, V.; Schacht, E.; Domurando, D. *J. Control. Release* **1996**, *38*, 141–150.

110. Rejmanova, P; Pohl, J.; Baudys, M.; Kostka, V.; Kopecek, J. *Makromol. Chem.* **1984**, *184*, 2009–2020.

111. Greenwald, R. B.; Yang, K.; Zhao, H.; Conover, C. D.; Lee, S.; Filpula, D. *Bioconj. Chem.* **2003**, *14*, 395–403.

112. Greenwald, R. B.; Choe, Y. H.; Conover, C. D.; Shum, K.; Wu, D.; Royzen, M. *J. Med. Chem.* **2000**, *43*, 475–487.

113. Lee, S.; Mcnemar, C. (ENZON INC.). Substantially Pure Histidine-Linked Protein Polymer Conjugates. U.S. Patent 5,985,263, November 16, 1999.

114. El Tayar, N.; Zhao, X.; Bentley M. D. (Applied Research System). PEG-LHRH Analog Conjugates. World Patent 9,955,376, April 11, 1999.

115. Orsatti, L.; Veronese, F. M. *J. Bioact. Comp. Polymer.* **1999**, *14*, 429–436.

116. Park, Y. K.; Abuchowski, A.; Davis, S.; Davis, F. *Anticancer Res.* **1981**, *1*, 373–376.

117. Abuchowski, A.; Kazo, G. M.; Verhoest, C. R., Jr.; Van Es, T.; Kafkewitz, D.; Nucci, M. L.; Viau, A. T.; Davis, F. F. *Cancer Biochem. Biophys.* **1984**, *7*, 175–186.

118. Ho, D. H.; Brown, N. S.; Yen, A.; Holmes, R.; Keating, M.; Abuchowski, A.; Newman, R. A.; Krakoff, I. H. *Drug Metab. Dispos.* **1986**, *14*, 349–352.

119. Soares, A. L.; Guimaraes, G. M.; Polakiewicz, B.; de Moraes Pitombo, R. N.; Abrahao-Neto, J. *Int. J. Pharm.* **2002**, *237*, 163–170.

120. Kurtzberg, J. *Cancer Medicine*, 5th ed.; Gansler, T., Ed.; Decker Inc: Hamilton; Canada, 2000.

121. Kurre, H. A.; Ettinger, A. G.; Veenstra, D. L.; Gaynon, P. S.; Franklin, J.; Sencer, S. F.; Reaman, G. H.; Lange, B. J.; Holcenberg, J. S. *J. Pediatr. Hematol. Oncol.* **2002**, *24*, 175–181.

122. Davis, F. F. *Adv. Exp. Med. Biol.* **2003**, *519*, 51–58.

123. Hawkins, D. S.; Park, J. R.; Thomson, B. G.; Felgenhauer, J. L.; Holcenberg, J. S.; Panosyan, E. H.; Avramis, V. I. *Clin. Cancer Res.* **2004**, *10*, 5335–5341.

124. Panosyan, E. H.; Seibel, N. L.; Martin-Aragon, S.; Gaynon, P. S.; Avramis, I. A.; Sather, H.; Franklin, J.; Nachman, J.; Ettinger, L. J.; La, M. et al. *J. Pediatr. Hematol. Oncol.* **2004**, *26*, 217–226.

125. Pepinsky, R. B.; Le Page, D. J.; Gill, A.; Chakraborty, A.; Vaidyanathan, S.; Green, M.; Baker, D. P.; Whalley, E.; Hochman, P. S.; Martin, P. et al. *J. Pharmacol. Exp. Ther.* **2001**, *297*, 1059–1066.
126. Cox, G. N., III (Bolder Biotechnology Inc.). Derivatives of Growth Hormone and Related Proteins. World Patent 9,903,887, January 28, 1999.
127. Acharya, A. S.; Manjula B. N.; Smith P. K. (Eistein Coll Med.). Hemoglobin Crosslinkers. U.S. Patent 5,585,484, December 17, 1996.
128. Zalipsky, S.; Menon-Rudolph, S. Hydrazine Derivatives of Poly(ethylene glycol) and Their Conjugates. In *Poly(ethylene glycol) Chemistry: Biotechnical and Biomedical Applications*, 2nd ed.; Harris, J. M., Ed.; Plenum Press: New York, 1998, p 319.
129. Saxon, E.; Bertozzi, C. R. *Science* **2000**, *287*, 2007–2010.
130. Cazalis, C. S.; Haller, C. A.; Sease-Cargo, L.; Chaikof, E. L. *Bioconj. Chem.* **2004**, *15*, 1005–1009.
131. Wong, S. S. Reactive Groups of Proteins and Their Modifying Agents. In *Chemistry of Protein Conjugation and Cross-linking*; CRC Press: Boston, MA, 1991, p 13.
132. Kinstler, O. B.; Brems, D. N.; Lauren, S. L.; Paige, A. G.; Hamburger, J. B.; Treuheit, M. J. *Pharm. Res.* **1995**, *13*, 996–1002.
133. Piquet, G.; Barbero, L.; Traversa, S.; Gatti, M. (Ares Trading SA). Regioselective Liquid Phase pegylation. World Patent 0,228,437, April 11, 2002.
134. Piquet, G.; Gatti, M.; Barbero, L.; Traversa, S.; Caccia, P.; Esposito, P. *J. Chromatogr. A* **2002**, *A944*, 141–148.
135. Hinds, K. D.; Kim, S. W. *Adv. Drug Delivery Rev.* **2002**, *54*, 505–530.
136. Caliceti, P.; Schiavon, O.; Sartore, L.; Monfardini, C.; Veronese, F. M. *J. Bioact. Biocomp. Polym.* **1993**, *8*, 41–50.
137. El Tayar, N.; Roberts, M. J.; Harris, M.; Sawlivich, W. (Applied Research System). POLYOL-IFN-Beta Conjugates. World Patent 9,955,377, November 4, 1999.
138. Berna, M.; Spagnolo, L.; Veronese, F. M. *Site-Specific PEGylation of G-CSF by Reversible Denaturation*. 32nd International Symposium on Controlled Release of Bioactive Materials, 18–22 June, 2005, Miami, FL, No. 415.
139. Sato, H. *Adv. Drug Delivery Rev.* **2002**, *54*, 487–504.
140. Gorman, J. J.; Folk, J. E. *J. Biol. Chem.* **1980**, *255*, 1175–1180.
141. Mcdonagh, R. P.; McDonagh, J.; Petersen, T. E.; Thøgersen, H. C.; Skorstengaard, K.; Sottrup-Jensen, L. et al. *FEBS Lett.* **1981**, *127*, 174–178.
142. Sato, H.; Hayashi, E.; Yamada, N.; Yatagai, M.; Takahara, Y. *Bioconj. Chem.* **2001**, *12*, 701–710.
143. Defrees, S.; Zopf, D. A.; Bayer, R.; Bowe, C.; Hakes, D.; Chen, X. (Neose Technologies Inc.) Glycopegylation Methods and Proteins/Peptides Produced by the Methods. U.S. Patent 2004,132,640, July 8, 2004.
144. Hooftman, G.; Herman, S.; Schacht, E. *J. Bioact. Comp. Polymer* **1996**, *11*, 135–159.
145. Delgado, C.; Francis, G. E.; Malik, F.; Fisher, D.; Parkes, V. *Pharm. Sci.* **1997**, *3*, 59–66.
146. Monkarsh, S. P.; Ma, Y.; Aglione, A.; Bailon, P.; Ciolek, D.; DeBarbieri, B.; Graves, M. C.; Hollfelder, K.; Michel, H.; Palleroni, A. et al. *Anal. Biochem.* **1997**, *247*, 434–440.
147. Veronese, F. M.; Sacca, B.; Polverido de Laureto, P.; Sergi, M.; Caliceti, P.; Schiavon, O.; Orsolini, P. *Bioconj. Chem.* **2001**, *1*, 62–70.
148. Gabizon, A.; Martin, F. *Drugs* **1997**, *54*, 15–21.
149. Otsuka, H.; Nagasaki, Y.; Kataoka, K. *Adv. Drug Deliv. Rev.* **2003**, *55*, 403–419.
150. Scott, M. D.; Chen, A. M. *Transfus. Clinique Biologique* **2004**, *11*, 40–46.
151. Bukowski, R.; Ernstoff, M. S.; Gore, M. E.; Nemunaitis, J. J.; Amato, R.; Gupta, S. K.; Tendler, C. L. *J. Clin. Oncol.* **2002**, *20*, 3841–3849.
152. Chapman, A. P.; Antoniw, P.; Spitali, M.; West, S.; Stephens, S.; King, D. J. *Nat. Biotechnol.* **1999**, *17*, 780–783.
153. Schreiber, S.; Rutgeerts, P.; Fedorak, R. N.; Khaliq-Kareemi, M.; Kamm, M. A.; Bovin, M.; Bernstein, C. N.; Staun, M.; Thomsen, O. O.; Innes, A. CDP870 Crohn's Disease Study Group. *Gastreoenterology* **2005**, *129*, 807–818.

Biographies

Francesco M Veronese has a degree in Pharmacy from the University of Padova, Italy. In 1962–1970 he was a researcher for the Italian National Council of Researches, 1971–1972 an Associated Research Professor at the Department of Biochemistry, University of California-Los Angeles, CA, 1973–1981 an Assistant Professor of 'Pharmaceutical Industry,' University of Ferrara, and of 'Applied Biochemistry,' University of Padova, Italy, and from 1982 Full Professor of 'Applied Pharmaceutical Chemistry,' School of Pharmacy, University of Padova. His past research activities have included: specific chemical modification of peptides and proteins for structure–activity relationships; induction, purification, sequence, enzymatic, and structural properties of glutamate dehydrogenases; structural and enzyme activity investigation of thermostable enzymes; drug–protein interaction: covalent and noncovalent linkage between furocoumarins and proteins; covalent modification by amphiphylic polymers of polypeptides and proteins to improve their therapeutic potential and their use in biocatalysis; entrapment of bioactive substances by nonbiodegradable matrices (acrylate polymers obtained by γ-ray induced polymerization) or polyvinyl alcohol, and by biodegradable polymers (polyphosphazenes) for controlled release; and polymeric prodrugs with targeting properties. Currently, Prof Veronese is carrying out research in the following areas: covalent binding of polymers to low-molecular-weight drugs, peptides, and proteins for improved delivery and increased therapeutic index; modification of enzymes with amphiphylic polymers for biocatalysis in organic solvents; and noncovalent entrapment of drugs by insoluble nonbiodegradable or biodegradable polymers for slow and controlled release of bioactive compounds. He was chairman of the PhD program in Pharmaceutical Sciences at the University of Padova 1983–2003; Director of the 'Seminar of Pharmaceutical Sciences' 1982–2000; co-organizer of the 'Advanced School of Pharmaceutical Chemistry' 1980–1992; co-organizer of the 'National Seminars in Pharmaceutical Sciences' 1980–1992; and Past President of the 'Controlled Release Society,' Italian Chapter, since 2004. Author of over 190 peer research papers, 15 reviews, 11 book chapters, and 19 US or European patents.

Gianfranco Pasut has a degree in Pharmaceutical Chemistry from the University of Padova, Italy. He spent some time as an exchange student at the University of Pennsylvania, PA before becoming a researcher at the School of Pharmacy, University of Padova, in 2006. Dr Pasut's research activities have included: studies on doxorubicin conjugated to PEG and macromolecular carriers; modification of cis platinum and Ara-C with PEG; synthesis and characterization of dendronized PEGs as a high-loading carrier for low-molecular-weight drugs and development of a new labeling method with technetium of PEGylated chelating agent for diagnostic application.

Currently, Dr Pasut is carrying out research in the following areas: modification of proteins with PEG; preparation of specific targeted macromolecular molecules for diagnostic and therapeutic purposes; and synthesis of PEG drugs releasing nitric oxide.

Comprehensive Medicinal Chemistry II
ISBN (set): 0-08-044513-6

ISBN (Volume 5) 0-08-044518-7; pp. 1043–1068

List of Abbreviations

11β-HSD	11β-Hydroxysteroid dehydrogenase
2/4/A1	Conditionally immortalized cell line derived from fetal rat intestine
2D-NMR	Two-dimensional nuclear magnetic resonance
3α-HSD	3α-Hydroxysteroid dehydrogenase
3D	Three-dimensional
3-HPPA	3-Hydroxyphenylpropionic acid
5-ASA	5-Aminosalicylic acid
5-FU	5-Fluorouracil
5HT	5-Hydroxytryptamine
5MC	5-Methylcytosine
AAE	Average absolute error
AAMU	5-Acetylamino-6-amino-3-methyl uracil
ABC	ATP-binding cassette
ABCB1	Gene symbol of human multidrug resistant P-glycoprotein
ABCC1	Gene symbol of human multidrug resistant associated protein MRP1
ABCC2	Gene symbol of human multidrug resistant associated protein MRP2
ABCG2	Gene symbol of human breast cancer resistance protein
ACAT	Coenzyme A-cholesterol-*O*-acyltransferase
ACE	Angiotensin-converting enzyme
ACN	Acetonitrile
ACS	American Chemical Society
ACTH	Adrenocorticotropic hormone
ADEPT	Antibody-directed enzyme prodrug therapy
ADH	Alcohol dehydrogenase
ADME	Absorption, distribution, metabolism, and excretion
ADMET	Absorption, distribution, metabolism, excretion, and toxicity
ADR	Adverse drug reaction
AE	Average error
AGP	α_1-Acid glycoprotein
AHLs	Acyl homoserine lactones
AHR	Arylhydrocarbon receptor
AIDS	Acquired immune deficiency syndrome
AKR	Aldo-keto reductase
ALDH	Aldehyde dehydrogenase
ALL	Acute lymphoblastic leukemia
AL-PEG	PEG aldehyde
ALS	Adaptive least squares
AMP	Adenosine monophosphate

AMP-PNP	Adenosine 5'-(β,γ-imino) triphosphate (nonhydrolyzable ATP analog)
AMT	Absorptive-mediated transcytosis
ANDA	Abbreviated new drug application
ANN	Artificial neural networks
ANNE	Artificial neural networks ensemble
ANOVA	Analysis of variance
AOT	Dioctyl sulfosuccinate
AOX	Aldehyde oxidase
AP	Absorption potential
AP-1	Activating protein-1
API	Active pharmaceutical ingredient
aq	Aqueous (environment)
ARC	Accurate radioisotope counting
ARDS	Adult respiratory distress syndrome
ARNT	Arylhydrocarbon receptor nuclear translocator
ASA	Apolar surface area
ASBT	Apical sodium-dependent bile acid transporter
ATP	Adenosine triphosphate
ATR/FTIR	Attenuated total reflectance/Fourier transform infrared spectroscopy
AZT	Zidovudine
β-FNA	Beta-funaltrexamine
BBM	Brush-border membrane
BCRP1 (ABCG2)	Breast cancer resistance protein (ABCG2)
BCS	Biopharmaceutics classification system
bFGF	Basic fibroblast growth factor
BMC	Biopartitioning micellar chromatography
BMP	Bone morphogenetic protein
BNPP	Bis-nitrophenyl phosphate
BPH	Benign prostatic hypertrophy or hyperplasia
BSA	Bovine serum albumin
BTC-PEG	PEG benzotriazolyl carbonate
BtuCD	ATP dependent vitamin B_{12} uptake transporter
BUI	Brain uptake index
C6	Rat glioma cell line
Caco-2	Adenocarcinoma cell line derived from human colon (used for estimation of human absorption)
CADD	Computer-assisted drug design
CAT	Compartment absorption transit
CBR1	Cytosolic carbonyl reductase
CDI-PEG	PEG carbonyldiimidazole
CDK	Cyclin-dependent kinase (CDK-7, CDK-9, etc.)
CDNB	1-Chloro-2,4-dinitrobenzene

CDR	Convective diffusion with simultaneous chemical reaction
CE	Capillary electrophoresis
CEDD	Centers of excellence in drug discovery
cEND	Immortalized mouse brain endothelial cell line
CFC	Chlorofluorocarbon
CGH	Comparative genomic hybridization
CHAPs	Cyclic hydroxamic acid-containing compounds
CHD	Coronary heart disease
CHO	Chinese hamster ovary cell line
CI	Clearance index
CLND	Chemical luminescence nitrogen detector
ClogP	Calculated log P according to the method by Leo and Hansch (sometimes generic for calculated partition coefficient)
CMC	Critical micelle concentration
CMR	Calculated molar refractivity
CMV	Cytomegalovirus
CNS	Central nervous system
CNS −	Drugs that are not taken up by the brain or have no central effect
CNS +	Drugs that are taken up into the brain and have a central effect
CNS + /−	Central nervous system availability
CoA	Coenzyme A
COMET	Consortium for metabonomic toxicology
CoMFA	Comparative molecular field analysis
CoMMA	Comparative molecular moment analysis
CoMSA	Comparative molecular surface analysis
CoMSIA	Comparative molecular similarity indices analysis
COMT	Catechol O-methyltransferase
COSY	Correlation spectroscopy
COX	Cyclooxygenase
COX-1	Cyclooxygenase-1
CP	Carboxypeptidase
CPA	N^6-Cyclopentyladenosine
cPCA	Consensus principal component analysis
CPK	Corey, Pauling and Koltun
CPR	Cytochrome P450 reductase
CPSA	Charged partial surface area
CR	Controlled release
CREB	cAMP response element binding protein
CRF	Corticotropin-releasing factor
CRO	Contract research organization
CSF	Cerebrospinal fluid
CT-2103	PGA-paclitaxel conjugate

CV	Cyclic voltammetry
CYP	Cytochrome P450
CYP2D6	Cytochrome P450 2D6
CYP3A4	Cytochrome P450 3A4
CZE	Capillary zone electrophoresis
DA	Discriminant analysis
DAD	Diode assay detector
Daevent	Dermally absorbed dose per event
DAF	Decay accelerating factor
DAT	Dopamine transporter
DDI	Drug–drug interaction
DF	Decision forest
DHFR	Dihydrofolate reductase
DIDR	Disk intrinsic dissolution rate
DIVEMA	Divinylethermaleic anhydride/acid copolymer
DMPK	Drug metabolism and pharmacokinetics
DMSO	Dimethyl sulfoxide
DNA	Deoxyribonucleic acid
DOX	Doxorubicin
DPI	Dry powder inhaler
DR	Dissolution rate
DSC	Differential scanning calorimetry
E. coli	*Escherichia coli*
ECF	Extracellular fluid
ECG	Surface electrocardiogram
ECV304	Cell line with endothelial-epithelial phenotype
ECVAM	European Centre for Validation of Alternative Methods
EEG	Electro encephalogram
EGF	Epidermal growth factor
EH	Epoxide hydrolase
EI	Electron impact
EIA	Enzyme immunometric assay
EMEA	European Medicines Agency (European Agency for the Evaluation of Medicinal Products)
EMS	Ethylmethane sulfonate
ENU	N-Ethyl-N-nitrosourea
EPHX1	Microsomal epoxide hydrolase
EPHX2	Soluble epoxide hydrolase
EPR	Enhanced permeability and retention
ESI	Electrospray ionization
EST	Expressed sequence tag
E-State	Electrotopological state
EVA	Eigen-values analysis

FAB	Fast atom bombardment
FAD	Flavin adenine dinucleotide
FDA	Food and Drug Administration (USA)
FFA	Free fatty acid
FIPCO	Fully integrated pharmaceutical company
FITC	Fluorescein isothiocyanate
FKBP12	FK506 binding protein 12
FL	Fuzzy logic
FMN	Flavin mononucleotide
FMO	Flavin-containing monooxygenase
FRET	Fluorescence resonance energy transfer
FSCPX	8-Cyclopentyl-N-[3-(4-fluorosulfonyl) (benzoyl)-oxy)-propyl]-1-N-propyl-xanthine
FTE	Full-time employee
GABA	γ-Amino-butyric acid
GC	Gas chromatography
G-CSF	Granulocyte colony stimulating factor
GDEPT	Gene-directed enzyme prodrug therapy
GDNF	Glial cell-derived neurotrophic factor
GFAcT	Genome functionalization through arrayed cDNA transduction
GFP	Green fluorescent protein
GFR	Glomerular filtration rate
GI	Gastrointestinal
GLP	Good laboratory practice
GM1	Ganglioside GM1
GMDs	Gel microdroplets
GMP	Good manufacturing practice
GP	Genetic programming
GPCR	G protein-coupled receptor
GRE	Glucocorticoid receptor element
GRNN	General regression neural network
GSH	Glutathione (reduced)
GSK	Glycogen synthase kinase (e.g., GSK-3)
GSSG	Glutathione (oxidized)
GST	Glutathione S-transferase
Hb	Hemoglobin
HBA	Number of hydrogen bond acceptors
hCP	Human carboxypeptidase
HCPSA	High-charged polar surface area
HCV	Hepatitis C virus
HDAC	Histone deacetylase (HDAC1, HDAC4, etc.)
HDACI	Histone deacetylase inhibitor
HDL	High-density lipoproteins

HDM	Hexadecane membrane (artificial membrane)
HEK293	Human embryonic kidney cell line
hERG	Human ether-a-go-go related gene
HFA	Hydrofluoroalkane
HGH	Human growth factor
HIA	Human intestinal absorption
HIF	Hypoxia inducible factor
HINT	Hydrophobic interaction field
HIP	Hydrophobic ion pairing
HIV	Human immunodeficiency virus
HLB	Hydrophilic-lipophilic balance
HLM	Human liver microsomes
HMBA	Hexamethylene bisacetamide
HMEC	Human mammary epithelial cell
HOMO	Highest occupied molecular orbital
HPAC	High-performance affinity chromatography
hPEPT1	Human oligo-peptide carrier for di- and tripeptides
HPGL	Number of hepatocytes per gram of liver
HPLC	High-performance liquid chromatography
HPMA	N-(2-Hydroxypropyl)methacrylamide copolymer
HQSAR	Hologram quantitative structure–activity relationship
HRMAS	High-resolution magic angle spinning
HSA	Human serum albumin
HSP	Heat shock protein (hsp90, hsp100, etc.)
HT	High throughput
HT29	Mucus-producing adenocarcinoma cell line (used for estimation of human absorption)
HTS	High-throughput screening
IC	Inhibitor concentration
IAM	Immobilized artificial membrane
IA-PEG	PEG-iodoacetamide
IBD	inflammatory bowel disease
IBS	Irritable bowel syndrome
ICH	International Conference on Harmonization
IDR	Intrinsic dissolution rate
IFN	Interferon
IGF	Insulin-like growth factor
I_{kr}	Rapidly activating delayed rectifier potassium current
I_{ks}	Slowly activating delayed rectifier potassium current
IL2	Interleukin-2
ILC	Immobilized liposome chromatography
in silico	Computational
IND	Investigative new drug application

IPR	Intellectual property rights
IQ	2-Amino-3-methylimidazo[4,5-*f*]quinoline
IR	Infrared
IRES	Internal ribosome entry site
ISEF	Intersystem extrapolation factor
ISF	Interstitial fluid (also called extracellular fluid)
IT	Information technology
IUBMB	International Union of Biochemistry and Molecular Biology
IUPAC	International Union of Pure and Applied Chemistry
i.v.	Intravenous(ly)
IVIVC	In vitro/in vivo correlation
JAK	Janus kinase
kNN	k-Nearest neighbor analysis
LC	Liquid chromatography
LC/MS	Combined liquid chromatography/mass spectrometry
LC/MS/MS	Combined liquid chromatography/tandem mass spectrometry
LCM	Laser capture microdissection
LDA	Linear discriminant analysis
LDL	Low-density lipoproteins
LEKC	Liposome electrokinetic chromatography
LFER	Linear free-energy relationship
LHRH	Luteinizing hormone-releasing hormone
LIMS	Laboratory information management system
LLC-PK1	Cell line derived from pig kidney epithelia
LmrA	*Lactococcus lactis* multidrug resistance ABC transporter
LOF	Lack-of-fit statistic
LQTS	Long QT syndrome
LSCM	Laser scanning confocal microscopy
LT	Low throughput
LUMO	Lowest unoccupied molecular orbital
M	Response modifier
mAb	Monoclonal antibody
MAD	Maximum absorbable dose
MAE	Mean absolute error
MAL-PEG	PEG-maleimide
MAO	Monoamine oxidase
MAP	Mitogen activated protein
MAS	Magic angle spinning
MC	3-Methylcholanthrene
MCA	Medicines Control Agency
MCG	Membrane-coating granule
MCT	Monocarboxylic acid cotransporter

MD	Molecular dynamics
MDCK	Madin–Darby canine kidney cell line (used to estimate human oral absorption)
mDDI	Metabolic drug–drug interaction
MDI	Metered dose inhaler
MDN	Metered dose nebulizer
MDR	Multidrug resistance
mEH	Microsomal epoxide hydrolase
MEP	Molecular electrostatic potential
MHBP	Molecular hydrogen-bonding potential
MHC	Major histocompatibility complex
MI	Membrane-interaction
MIF	Molecular interaction field
miRNAs	microRNAs
MLP	Molecular lipophilicity potential
MLR	Multiple linear regression
MMAD	Mass median aerodynamic diameter
MMPs	Matrix metalloproteinases (e.g., MMP-9)
MO	Molecular orbital(s)
MOA	Mechanism of action
mp	Melting point
MPDP$^+$	1-Methyl-4-phenyl-2,3-dihydropyridinium
mPEG	Monomethoxy-poly(ethylene glycol)
MPP$^+$	1-Methyl-4-phenyl pyridinium salt
MPTP	1-Methyl-4-phenyl-1,2,3,6-tetrahydropyridine
MQSM	Molecular quantum similarity measures
MR	Molar refractivity
MRI	Magnetic resonance imaging
MRM	Multiple reaction monitoring (mode in MS/MS)
MRP	Multidrug resistance associated protein (ABCC)
MRP1	Multidrug resistance related protein-1 (ABCC1)
MRT	Mean residence time
MS	Mass spectrometry
MS/MS	Tandem mass spectrometry
MsbA	Lipid flippase multidrug resistance ABC transporter
MT	Methyltransferase
MTD	Minimum topological difference
MTS	Methanethiosulfonate
MTSF	Molecular tree structured fingerprints
MTT	Cell vitality assay using 3-(4,5-dimethylthiazol-2-yl)-2,5-diphenyltetrazolium bromide
Mu	Amount of drug excreted in urine
MudPIT	Multidimensional protein identification technology
MVP	Major vault protein

MW	Molecular weight
Na^+,K^+-ATPase	Sodium-potassium ATPase
NADPH	Nicotinamide adenine dinucleotide phosphate (reduced)
NAS	National Academy of Sciences
NAT	*N*-Acetyltransferase
NBD	Nucleotide-binding domain
NBE	New biological entity
NC	Nomenclature Committee
NCA	Normal coronary artery
NCBI	National Center for Biotechnology Information
NCE	New chemical entity
NCI	National Cancer Institute
NCS	Neocarcinostatin
NDA	New drug application
NEFA	Non-esterified fatty acid
NFκB	Nuclear factor kappa B
NHS	*N*-Hydroxysuccinimide
NIH	National Institutes of Health (also not-invented-here)
NMDA	*N*-Methyl-D-aspartate
NMR	Nuclear magnetic resonance (spectroscopy)
NN	Neural network
NNK	4-(Methylnitrosamino)-1-(3-pyridyl)-butan-1-one
NO	Nitric oxide
NORD	National Organization of Rare Disorders
NPSA	Nonpolar surface area
NQO	NAD(P)H quinone oxidoreductase (aka DT-diaphorase)
NQOR	NAD(P)H quinone oxidoreductase (DT-diaphorase)
NRPs	Nonribosomal peptides
NSAID	Nonsteroidal anti-inflammatory drug
Oa-ToF	Orthogonal acceleration time-of-flight
Oatp/OATP	Organic anion transporting polypeptide (animal/human, resp.)
OCT	Organic cation transporter family
ODA	Orphan Drug Act
OECD	Organization of Economic Co-operation and Development
o-NPOE	*Ortho*-nitrophenyloctyl ether
OP	Organophosphates
OPSS-PEG	PEG-orthopyridyl-disulfide
ORMUCS	Ordered multicategorical classification method using the simplex technique
OSC	Orthogonal signal correction
OTC	Over-the-counter
p.o.	*Per os* = oral(ly)
P450	Cytochrome P450

PA	Phosphatidic acid
PacM	Poly(acroloylmorpholine)
PAH	Polycyclic aromatic hydrocarbon
PAMPA	Parallel artificial membrane permeability assay
PAPS	3′-Phosphoadenosine 5′-phosphosulfate
PATQSAR	Population analysis by topology-based QSAR
PB	Phenobarbital
PBL	Porcine brain lipid
PBPK	Physiologically-based pharmacokinetics
PCA	Principal component analysis
PCM	Polarized continuum method
PcV	Phenoxymethylpenicillin
PD	Pharmacodynamic(s)
PDA	Photo-diode array
PDB	Protein Data Bank
PDMS	Polydimethylsiloxane
PE	Phosphatidylethanolamine
PECAM-1	Platelet endothelial adhesion molecule-1
PEG	Polyethyleneglycol
PEI	Poly(ethyleneimine)
PEPT1	(Oligo)peptide transporter 1
PET	Positron emission tomography
PG	Phosphatidylglycerol
PGA	Polyglutamic acid
P-gp	P-glycoprotein (MDR1, ABCB1)
PGS	Prostaglandin synthase
PHAH	Polyhalogenated aromatic hydrocarbon
PHEG	Poly((N-hydroxyethyl)-L-glutamine)
PhIP	2-Amino-1-methyl-6-phenylimidazo[4,5-b]pyridine
PI	Phosphatidylinositol
PI3K	Phosphoinositide-3-kinase
PK	Pharmacokinetic(s)
PK/PD	Pharmacokinetic–pharmacodynamic
PK1	HPMA-(Gly-Phe-Leu-Gly-doxorubicin)$_n$ conjugate
PK2	(N-acylated galactosamine)$_m$-HPMA-(Gly-Phe-Leu-Gly-doxorubicin)$_n$ conjugate
PKC	Protein kinase C
PKS	Polyketide synthase enzyme
PMT	Photo multiplier tube
PNN	Probabilistic neural network
pNPC-PEG	PEG p-nitrophenyl carbonate
PON	Paraoxonase
POPC	Palmitoyloleilphosphatidylcholine

PPAR-γ	Peroxisome proliferator-activated receptor gamma
PSA$_d$	Dynamic polar surface area
PSI	Protein structure initiative
PTSA	Partitioned total polar surface area
PVA	Polyvinylalcohol
PVP	Poly(vinylpyrrolidone)
QC	Quality control
QM	Quantum mechanical
QSAR	Quantitative structure–activity relationship
QSPeR	Quantitative structure–permeability relationship
QSPkR	Quantitative structure–pharmacokinetic relationship
QSPR	Quantitative structure–property relationship
QTL	Quantitative trait loci
R&D	Research and development
RAF	Relative activity factor
RBF	Radial basis function
RFP	Request for proposal
RH	Relative humidity
rhCYP	Recombinant human cytochrome P450
rhGH	Recombinant human growth hormone
RIA	Radioimmunoassay
RISC	RNA-induced silencing complex
RLIP-76 (RALBP1)	Ral-interacting protein, non-ABC transporter
r-metHuG-CSF	Recombinant-methionine human G-CSF
RMSE	Root mean square error
RMT	Receptor-mediated transcytosis
RNA	Ribonucleic acid
RNAi	RNA interference
ROI	Return on investment
ROS	Reactive oxygen species
RP	Recursive partitioning (also called decision tree)
RSF	Relative substrate activity factor
RT-PCR	Reverse-transcriptase polymerase chain reaction
S/N	Signal to noise ratio
S9	Supernant from homogenation and 9000g sedimentation
SAM	S-Adenosyl-L-methionine
Saos-2	Osteosarcoma cell line
SAR	Structure–activity relationship
SARS	Severe acute respiratory syndrome
SAS	Solvent accessible surface
SASA	Solvent accessible surface area
SAT	Site acceptance test

SATE	*S*-Acyl-2-thioethyl
SBDD	Structure-based drug design
SC	Stratum corneum
SC-PEG	PEG succinimidyl carbonate
SD	Standard deviation (see also s)
SDK	Substrate disappearance kinetics
SDR	Short-chain dehydrogenase/reductase
SERT	Serotonin transporter
SES	Solvent excluded surface
SF	Shake-flask
SGF	Simulated gastric fluid (USP)
SIF	Simulated intestinal fluid (USP)
siRNA	Small interfering RNA
SITT	Small intestinal transit time
SIWV	Small intestinal water volume
SLS	Sodium lauryl sulfate
SMA	Poly(styrene-co-maleic acid/anhydride)
SMILES	Simplified molecular input line entry system
SMM	Small molecule microarray
SNP	Single nucleotide polymorphism
SOM	Self-organizing map
SPA	Scintillation proximity assay
SPE	Solid phase extraction
SPEC	Solid phase extraction chromatography
SPECT	Single photon emission computed tomography
SRIF	Somatotropin-release inhibiting factor
SRP	Signal recognition particle
SRS	Substrate recognition site
SRS-A	Slow reacting substance of anaphylaxis
SSLM	Solid-supported lipid membrane
SS-PEG	PEG succinimidyl succinate
SSRI	Selective serotonin reuptake inhibitor
STAT	Signal-transducing activators of transcription
STI	Signal transduction inhibitor
SULT	Sulfotransferase
SVM	Support vector machine
TAP	Tumor-activated prodrug
TAR	Transactivation response
TCDD	2,3,7,8-Tetrachlorodibenzo-*p*-dioxin ('dioxin')
TC-NER	Transcription-coupled nucleotide excision repair
TCP-PEG	PEG trichlorophenyl carbonate
TDI	Time-dependent inhibition

TEER	Transendothelial electrical resistance
Tgase	Transglutaminase
TGFβ	Transforming growth factor beta
Thy-1	Thymus cell antigen-1 (CD90) expressed on pericytes/fibroblasts
TIAs	Tubulin interactive agents
TIC	Total ion current
TK	Tyrosine kinase
TMA	Trimethylamine
TMD	Transmembrane domain
TNF-α	Tumor necrosis factor alpha
TOCSY	Total correlation spectroscopy
ToF	Time-of-flight
TOPS-MODE	Topological substructural molecular design
TPA	Tissue plasminogen activator
TPGS	Tocopheryl polyethylene glycol succinate
TPSA	Topological polar surface area
TRD	Transcriptional repression domain
TRF	Time-resolved fluorescence
TVD	Triple vessel disease
UDP	Uridine diphosphate
UDPGA	UDP-glucuronic acid = uridine-5′-diphospho-α-D-glucuronic acid
UGT	UDP-glucuronosyltransferase = UDP-glucuronyltransferase
UPLC	Ultra performance liquid chromatography
URS	User requirement specification
US EPA or EPA (US)	United States Environmental Protection Agency
USP	United States Pharmacopeia
UTR	Untranslated region
UV	Ultraviolet
UWL	Unstirred water layer
V. cholerae	*Vibrio cholerae*
VEGFR	Vascular endothelial growth factor receptor
vHTS	Virtual high-throughput screening
VLDL	Very low-density lipoproteins
VS-PEG	PEG-vinylsulfone
WDI	World Drug Index
WHI	Women's Health Initiative
XED	Extended electron distribution
z	Charge number
ZO-1	Zonula occludens protein-1

Software and Vendors

Computer Programs		*URL Vendor*
ACD	Advanced Chemistry Development	www.acdlabs.com
ALOGPS	Artificial neural network program to predict lipophilicicy (log P) and aqueous solubility (log S) of chemicals	
CART	Recursive classification regression tree	www.salford-systems.com
CHARMM	Chemistry at Harvard molecular mechanics	www.charmm.org
CLOGP	Calculation of octanol/water partition coefficients for neutral species	www.biobyte.com
DAYLIGHT	Toolkit for database applications	www.daylight.com
DERMWIN	Dermal permeability coefficient program	www.syrres.com
DISCO	Pharmacophore perception module in SYBYL software system	
DRAGON	Set of over 1800 molecular descriptors	www.talete.mi.it
GASP	Genetic algorithm similarity program	
GOLPE	Generating optimal linear PLS estimation	www.miasrl.com
GRIND	GRID-Independent (descriptors derived from the ALMOND program)	www.moldiscovery.com
HYBOT	Hydrogen bond thermodynamics	http://software.timtec.net/hybot-plus.htm
MMFF94	Merck Molecular Force Field 94	
SIMCA	Soft-independent modelling of class analogy	www.umetrics.com
SPARC	Computer program to calculate pK_a values	
WHIM	Weighted holistic invariant molecular descriptor	

List of Symbols

%ppb	Percentage of bound drug in plasma
%PSA	Percentage polar surface area
$\%T$	Percentage transported through a membrane
%uptake	Per cent uptake efficiency of the trout gill
[...]	Square brackets used to indicate concentration of, e.g. $[Na^+]$ is 'the concentration of Na^+'
[A]	Concentration of agonist
[AR]	Concentration of agonist–receptor complex
$[I]_{av}$	Average systemic plasma concentration of inhibitor after repeated oral administration
$[I]_{in}$	Maximum hepatic input concentration of inhibitor
$[I]_u$	Inhibitor concentration, unbound
$[R_0]$	Total receptor concentration
0D,1D, 2D, 3D, 4D	Zero-, One-, two-, three-, four-dimensional
$^6\chi$	Sixth-order path molecular connectivity
$\underline{A}$	Generic symbol for the total concentration of the substance A; the sum of the reactant concentration and the concentrations of all the associated species containing the A reactant; usually expressed in units of mol cm^{-3} in kinetic equations
a	Constant reflecting the slope of a parabolic stimulus–effect relationship
$A\%$	Percentage oral absorption
A_D	Cross-sectional area of a molecule
$A \log P$	Calculated log P according to an atomic contribution method
A_m	Membrane area
A_s	Molecular cross-sectional area
AUC	Area under the curve of a plasma concentration vs time profile
AUC_i	AUC for substrate in presence of inhibitor
B	Hydrogen-bonding basicity (Abraham equation)
b	Stimulus at maximal effect of a parabolic stimulus–effect relationship
BBB	blood–brain barrier
C_0	Initial total concentration of drug in plasma
C_a	Binding capacity
C_A	Concentration in the water phases of the acceptor
C_a and C_d	Free energy factor of proton acceptors and proton donors, respectively (HYBOT, Raevsky)
C_b	Protein-bound drug concentration
C_D	Concentration in the water phases of the donor
C_e	Drug concentration in hypothetical effect compartment
C_{fa}	Fetal artery concentration
C_{fv}	Fetal vein concentration
C_{in}	Concentration of inhibitor
C_j	jth associated species concentration in kinetic equations

CL	Total in vivo plasma clearance
CL_b	In vivo blood clearance
CL_H	Hepatic clearance
$CL_{H,int}$	Hepatic intrinsic clearance
$CL_{H,int,u}$	Unbound hepatic intrinsic clearance
$CL_{H,u}$	Unbound hepatic clearance
CL_{int}	Intrinsic (metabolic, hepatic) clearance
$CL_{int,u}$	Unbound hepatic intrinsic clearance
$CL_{int,us}$	Intrinsic tubular secretion rate
CL_{max}	Maximum intrinsic clearance; when an enzyme is fully activated (used for enzymes which show positive cooperativity in vitro)
ClogP	Calculated log P according to the method by Leo and Hansch (sometimes generic for calculated partition coefficient)
CL_p	In vivo plasma clearance
CL_R	Renal clearance
C_{m0}	Concentration at position 0 in the membrane
C_{ma}	Maternal artery concentration
C_{max}	Maximum (peak) plasma concentration
$C_{max,ss}$	C_{max} at steady state
C_{mh}	Concentration at position h in the membrane
C_{min}	Minimum (peak) plasma concentration
$C_{min,ss}$	C_{min} at steady state
C_{on}	Concentration of surface activity onset
C_T	Concentration in tissue
C_u	Free (unbound) drug concentration
ΔE	Molecular energy difference
ΔG	(Change in) Gibbs free energy
$\Delta G(x)$	Free energy difference between the water phase and position x in the membrane
ΔH	(Change in) enthalpy
$\Delta \log P$	Difference log P (octanol/water) $-$ log P (alkane/water)
$\Delta \log P_{oct-dce}$	Difference between the logarithms of the octanol/water and 1,2-dichloroethane/water partition coefficients
ΔM	Amount of diffusant transported through the membrane
δ	Binding affinity (CYP)
δ_m	Solubility parameter for the membrane
δ_v	Solubility parameter for the vehicle
$D(x)$	Local diffusion coefficient at x position
$D_{barrier}$	Diffusion coefficient in the barrier domain
D_j	Diffusivity coefficient of reactant or associated species
D_m	Diffusion coefficient in the membrane
D_{oct}	Octanol/buffer distribution coefficient
Dose	Dose
$D_{T/B}$	Tissue/blood distribution coefficient

$D_{T/P}$	Tissue/plasma distribution coefficient
$D_{vo/w}$	Vegetable oil/water distribution coefficient
e	Proportionality constant denoting the power of a drug to produce a response
ε	Intrinsic efficacy
ϵ	Dielectric constant
E_0	Baseline drug effect
EC_{25}	Effective concentration needed to observe 25% of the biological response
EC_{50}	Drug concentration at half maximal drug effect
$E_{inter(vdW)}$	Energy of the Van der Waals' penetrant–membrane interaction
E_m	Maximum system response
E_{max}	Maximum effect of hyperbolic concentration–effect relationship
$E_{SS}(tor)$	Torsional energy at the total intermolecular system minimum energy for the solute
E_{top}	Maximum effect of parabolic stimulus–effect relationship
E_V	Free energy of vaporization
f	Molar fraction
$F\%$	Percentage bioavailable
f_a	Fraction of drug absorbed from the gut into the hepatic portal vein after oral administration
F_G	Gut availability (fraction of drug which escapes elimination by the gut)
f_g	Fraction of drug metabolized by the gut wall
f_h	Fraction of drug metabolized by the liver
$f_{i(7.4)}$	Fraction of ionized drug at pH 7.4
F_{inh}	Pulmonary bioavailability
f_j	Fraction of drug at ionization state j
$f_{m(CYP)}$	Fraction metabolized by a CYP isozyme
f_P	Fractional content of extracellular fluid in plasma
f_r	Fraction of drug reabsorbed
F_{rel}	Relative bioavailability comparing two different pharmaceutical
f_{rotb}	Fraction of rotatable bonds
f_T	Fractional content of extracellular fluid in tissues
$f_{u,b}$	Fraction of drug unbound (free) in blood
$f_{u,mic}$	Fraction of drug unbound (free) in microsomes
$f_{u,p}$	Fraction of drug unbound (free) in plasma
$f_{u,t}$	Fraction of drug unbound (free) in tissue
f_z	Fraction of each charged species
g	Gaseous environment
γ	Catalytic efficacy (CYP)
γ^0	Surface tension of the bare interface
h_j	Thickness of the aqueous boundary layer of reactant or associated species j
IC_{50}	Concentration of inhibitor required to give 50% inhibition
J	Flux across a membrane or dissolution of a solid
J_{max}	Maximal flux

J_{PDMS}	Maximum steady-state flux through PDMS
J_{SS}	Steady-state flux
K	Tissue/medium partition coefficient
k	Boltzmann constant
k_{1e}	Rate constant for distribution into the hypothetical effect compartment
$\kappa 3$	Third-order kappa index (topological descriptor)
k_a	Absorption rate constant
K_A	In vivo dissociation equilibrium constant
K_{aw}	Air/water partition coefficient
$K_{barrier/water}$	Partition coefficient of a solute from water to the barrier domain
K_{BB}	The ratio of concentration in brain over blood at equilibrium
K_{blood}	The ratio of concentration in blood divided by concentration in air, at equilibrium
K_{brain}	The ratio of concentration in brain divided by concentration in air, at equilibrium
k_{deg}	Endogenous degradation rate constant
k_{degrad}	Degradation rate constant
k_{dep}	Substrate depletion rate constant
K_E	Value midpoint location of transducer function
$k_{eff,GI}$	Effective rate of absorption
k_{el}	Elimination rate constant
k_{eO}	Rate constant for distribution out of the hypothetical effect compartment
k_f	Constant of passive transfer
k_{gill}	Rate of absorption across the gills in the guppy
K_i	Inhibition equilibrium constant
K_I	Inhibition equilibrium constant for mechanism-based inhibition
k_{IAM}	Capacity factor of IAM column chromatography
K_{ic}	Competitive inhibition equilibrium constant
k_{ILC}	Capacity factor of immobilized liposome chromatography
K_{in}	Unidirectional influx coefficient
k_{in}	Zero-order rate constant for production of a physiological entity
k_{inact}	Maximal inactivation rate constant
K_{iu}	Uncompetitive inhibition equilibrium constant
k_{loss}	First-order rate constant for release of precursor in central compartment
K_m	Michaelis–Menten constant – substrate concentration at which rate of reaction is half V_{max}
k_m	Rate constant for formation and dissipation of a response modifier
K_{memb}	Membrane partition coefficient
K_{mp}	Milk to plasma ratio
K_n	Aggregation constant, where n is the degree of aggregation
k_{out}	First-order rate constant for loss of a physiological entity
k_p	First-order permeability rate constant
$K_{p(epi)}$	Human epidermis permeability coefficient
$K_{p(sil)}$	Silicone permeability coefficient

$K_{\text{p-corneal}}$	Corneal permeability coefficient
K_s	Binding constant
λ	Rate constant of enzyme inactivation
λ_1	Apparent elimination constant
LBF	Liver blood flow; also Q_H
$\log BB$	log[brain:plasma ratio]
$\log D$	Logarithm of the distribution coefficient at given pH, typically in n-octanol/water and at pH 7.4
$\log D_{7.4}$	Logarithm of the distribution coefficient at given pH 7.4, typically in n-octanol/water
$\log D_{\text{oct}}$	Logarithm of the distribution coefficient at given pH, in n-octanol/water
$\log k_s$	Logarithm of the chromatography capacity factor
$\log k_{\text{sdc}}$	Logarithm of the chromatography capacity factor (capillary electrophoresis)
$\log P$	Logarithm of the partition coefficient for the neutral species, typically in n-octanol/water
$\log P_{\text{hept}}$	Logarithm of the partition coefficient for the neutral species in n-heptane/water
$\log P_{\text{ion}}$	Log P of the ionized species
$\log P_{\text{neutral}}$	Logarithm of the partition coefficient for the neutral species, typically in n-octanol/water
$\log P_{\text{oct}}$	Logarithm of the partition coefficient for the neutral species in n-octanol/water
μ	Dipole moment
M	Amount of drug
mp	Melting point
n	Number of datapoints (compounds or observations) used in the regression
N_A	Avogadro's number
O_{lumen}	Oversaturation number in the lumen
π^*	Dipolarity/polarizability
π_{memb}	Surface pressure of a lipid bilayer
ϕ	Membrane-to-water distribution coefficient
P	Partition coefficient between water and lipid or organic solvent (typically n-octanol) for a molecule in its unionized form (see also log P)
P_{chl}	Chloroform/buffer partition coefficient
$P(x)$	Local membrane permeability at position x
$p[H]$	$-\log_{10}[H^+]$, the degree of acidity of a solution, expressed on an concentration scale
P_0^{PAMPA}	Intrinsic PAMPA permeability of uncharged species
P_{app}	Apparent permeability coefficient
P_{cyc}	Water to cyclohexane partition coefficient
P_e	Permeability constant or coefficient
$P_{\text{e,PAMPA}}$	PAMPA permeability
P_{e0}	Intrinsic permeability of uncharged species
$P_{\text{e0}}^{\text{BBB}}$	Intrinsic BBB permeability of uncharged species
P_{eff}	Effective intestinal membrane permeability
$P_{\text{eg/hept}}$	Heptane to ethylene glycol partition coefficient
pH	Negative logarithm of the hydrogen ion concentration reflecting acidity/basicity
pK_a	Negative logarithm of the ionization/dissociation constant ($-\log_{10} K_a$)

pKa''	Negative logarithm of the ionization constant determined by direct titration in KCl/water-saturated n-octanol
pK_a^0	In Hammett and Taft equations, the pK_a of the unsubstituted acid or base
PLS	Partial least squares (or projection to latent structures)
PLS-DA	Partial least squares–discriminant analysis
P_m	Membrane permeability
$P_{m,app}$	Apparent membrane permeability
P_{oct}	Partition coefficient between water and n-octanol for a molecule in its unionized form
P_{para}	Paracellular pathway permeability
PPB	Plasma protein binding (see also %ppb)
P_{UWL}	Unstirred water layer permeability
$P_{vo/w}$	Vegetable oil/water partition coefficient
q	Electric charge
q (q^2)	Cross-validated correlation coefficient (squared cross-validated correlation coefficient)
q^-	Lowest atomic charge in the molecule
q^2	Square of the cross-validated correlation coefficient for external validation data set
Q_f	Fetal flow rate
Q_H	Hepatic blood flow
Q_R	Renal blood flow
r	Molecular radius
R	Universal gas constant
RMSE	Root mean square error
r (r^2)	Correlation coefficient (squared correlation coefficient)
$R_{e/i}$	Ratio of extravascular to intravascular proteins
$R_{E/P}$	Erythrocyte/plasma ratio
r_{gyr}	Radius of gyration
r_{para}	Apparent pore radius of the paracellular pathway
σ	In Hammett and Taft equations, a constant that is characteristic of a particular substituent
s	Standard deviation/error of a regression equation
S_0	Intrinsic solubility, solubility of the uncharged species
S_{50}	Substrate concentration at half V_{max} (used for enzymes which show positive cooperativity in vitro)
S_{pH}	The concentration of a substance in solution at equilibrium with excess solid at a given pH
S_s	Solubility at the interface
$\Sigma MHBP_{do}$	Sum of the calculated molecular hydrogen bond donor potential
τ	Dosing interval
T	(Body) temperature
$t_{1/2}$	Half-life
$t_{1/2deg}$	Degradation half-life
t_l	Lag time
t_{max}	Time to reach maximal (peak) plasma concentration C_{max}

TR	Placental transfer ratio
T_{SP}	Set-point of body temperature
TUI	Testis uptake index
V	Volume in experimental setup such as plate well
v	Volume fraction of component in tissue or blood
$V(x)$	Axial velocity in the rotating disk apparatus, as a function of the distance (x) from the surface of the rotating disk and directed into the disk
V_{ac}	Volume of the acceptor wells
V_D	Volume of distribution
$V_{D,ss}$	Volume of distribution at steady state
$V_{D,u}$	Unbound volume of distribution
V_{do}	Volume of the donor wells
VdWS	Van der Waals' surface
V_e	Effective Van der Waals' volume
V_E	Extracellular volume
V_{ery}	Volume of erythrocytes
V_F	Sorption of solvents into an artificial membrane
VISEF	ISEF based on V_{max} values in HLM and rhCYP
V_P	Plasma volume
V_R	Volume in which the drug distributes minus the extracellular space
V_T	Volume of tissue compartment
wt%	Weight per cent
x	Thickness of a membrane

INDEX FOR VOLUME 5

Notes

Cross-reference terms in italics are general cross-references, or refer to subentry terms within the main entry (the main entry is not repeated to save space). Readers are also advised to refer to the end of each article for additional cross-references – not all of these cross-references have been included in the index cross-references.

The index is arranged in set-out style with a maximum of three levels of heading. Major discussion of a subject is indicated by bold page numbers. Page numbers suffixed by T and F refer to Tables and Figures respectively. vs. indicates a comparison.

This index is in letter-by-letter order, whereby hyphens and spaces within index headings are ignored in the alphabetization. Prefixes and terms in parentheses are excluded from the initial alphabetization.

Any method, model or other subject, associated with the name of the developer (e.g. name's model) does NOT imply that Elsevier, nor the indexers, have assumed the right to name models/methods after the authors of the papers in which they are described. This is merely a succinct phrase to refer to a model/method developed/described by the relevant author, so that the subentry could be alphabetized under the most pertinent name.